Miscellaneous Formulas

SPECIAL TRIGONOMETRIC LIMITS: $\quad \lim\limits_{h \to 0} \dfrac{\sin h}{h} = 1 \qquad \lim\limits_{h \to 0} \dfrac{\cos h - 1}{h} = 0$

APPROXIMATION INTEGRATION METHODS: Let $\Delta x = \dfrac{b-a}{n}$ and, for the kth subinterval, $x_k = a + k\,\Delta x$; also suppose f is continuous throughout the interval $[a, b]$.

Rectangular Rule: (right endpoints)
$$\int_a^b f(x)\,dx \approx [f(a + \Delta x) + f(a + 2\Delta x) + \cdots + f(a + n\Delta x)]\Delta x$$

Trapezoidal Rule:
$$\int_a^b f(x)\,dx \approx \frac{1}{2}[f(x_0) + 2f(x_1) + 2f(x_2) + \cdots + 2f(x_{n-1}) + f(x_n)]\Delta x$$

Simpson Rule:
$$\int_a^b f(x)\,dx \approx \frac{1}{3}[f(x_0) + 4f(x_1) + 2f(x_2) + \cdots + 4f(x_{n-1}) + f(x_n)]\Delta x$$

ARC LENGTH: Let f be a function whose derivative f' is continuous on the interval $[a, b]$. Then the **arc length**, s, of the graph of $y = f(x)$ between $x = a$ and $x = b$ is given by the integral

$$s = \int_a^b \sqrt{1 + [f'(x)]^2}\,dx$$

If C is a parametrically described curve $x = x(t)$ and $y = y(t)$, and is simple, then

$$s = \int_a^b \sqrt{\left(\frac{dx}{dt}\right)^2 + \left(\frac{dy}{dt}\right)^2}\,dt$$

SERIES Suppose there is an open interval I containing c throughout which the function f and all its derivatives exist.

Taylor Series of f
$$f(x) = f(c) + \frac{f'(c)}{1!}(x - c) + \frac{f''(c)}{2!}(x - c)^2 + \frac{f'''(c)}{3!}(x - c)^3 + \cdots$$

Maclaurin Series of f
$$f(x) = f(0) + \frac{f'(0)}{1!}x + \frac{f''(x)}{2!}x^2 + \frac{f'''(0)}{3!}x^3 + \cdots$$

FIRST-ORDER LINEAR D.E. The general solution of the first-order linear differential equation

$$\frac{dy}{dx} + P(x)y = Q(x)$$

is given by

$$y = \frac{1}{I(x)}\left[\int Q(x)I(x)\,dx + C\right]$$

where $I(x) = e^{\int P(x)\,dx}$ and C is an arbitrary constant.

VECTOR-VALUED FUNCTIONS

Del operator:
$$\nabla = \frac{\partial}{\partial x}\mathbf{i} + \frac{\partial}{\partial y}\mathbf{j} + \frac{\partial}{\partial z}\mathbf{k}$$

Gradient:
$$\nabla f = \frac{\partial f}{\partial x}\mathbf{i} + \frac{\partial f}{\partial y}\mathbf{j} + \frac{\partial f}{\partial z}\mathbf{k} = f_x\mathbf{i} + f_y\mathbf{j} + f_z\mathbf{k}$$

Laplacian:
$$\nabla^2 f = \frac{\partial^2 f}{\partial x^2} + \frac{\partial^2 f}{\partial y^2} + \frac{\partial^2 f}{\partial z^2} = f_{xx} + f_{yy} + f_{zz}$$

Derivatives of a vector field: Normal derivative: $\dfrac{\partial f}{\partial n} = \nabla f \cdot \mathbf{N}$

$$\text{div } \mathbf{F} = \frac{\partial u}{\partial x} + \frac{\partial v}{\partial y} + \frac{\partial w}{\partial z} \qquad \textit{This is a scalar derivative, } \nabla \cdot \mathbf{F}$$

$$\text{curl } \mathbf{F} = \begin{vmatrix} \mathbf{i} & \mathbf{j} & \mathbf{k} \\ \frac{\partial}{\partial x} & \frac{\partial}{\partial y} & \frac{\partial}{\partial z} \\ u & v & w \end{vmatrix} \qquad \textit{This is a vector derivative, } \nabla \times \mathbf{F}$$

LOG IN INSTRUCTIONS

Browser Minimum: Internet Explorer 8.0 +, Firefox 3.6 +, Safari 4.0 +, and Chrome 9.0 + JavaScript must also be enabled in your browser. Some online publications may require additional free browser plug-ins (e.g., Flash, Adobe Acrobat).

Go To: http://www.grtep.com

You will only have to set up your account the first time you access this page. After you set up your account using a valid e-mail address and create a password, you will be able to enter the Registered User portion of this page.

To Set Up a New Account:

1. In the Register section enter the **Access Code** provided below.

2. Click the **REGISTER** button. This will bring you to the Create Your Account page where you will fill out your account information including creating your username & password. ***Your username will be the e-mail address that you enter. Write down your Password to ensure your success in entering the publication after the initial setup.*

3. Upon entering the information to create your account, click **SUBMIT**

4. You will now see a button "**Log in to WebCOM**." Click on this button to enter the publication.

5. Before you can log into WebCOM for the first time, you must verify your email address. After registering your code, an email will be sent to you. You must go to your email, open the email that was sent to you and click on the link provided in that email.

6. After verifying your email address, log into the publication by entering your Username and Password under the Login portion of the page. Click on the **LOGIN** button.

7. If you encounter difficulties with account setup, *DO NOT* attempt to create a second account. Please contact: **websupport@kendallhunt.com** for account setup difficulties or other technical assistance.

YOUR ONLINE ACCESS CODE

XSZ9R-NB24B-3VHPK-VZR2K

Calculus
Sixth Edition

Karl J. Smith
Santa Rosa Junior College

Monty J. Strauss
Texas Tech University

Magdalena D. Toda
Texas Tech University

Kendall Hunt
publishing company

CONTENTS

Contents

PREFACE

"Mathematics is the instrument by which the engineer tunnels our mountains, bridges our rivers, constructs our aqueducts, erects our factories and makes them musical by the busy hum of spindles. Take away the results of the reasoning of mathematics, and there would go with it nearly all the material achievements which give convenience and glory to modern civilization."

Edward Brooks, *Mental Science and Culture,*
Philadelphia: Normal Publishing, 1891, p. 255.

FOR THE STUDENT

THINKING IN A NEW DIRECTION

You have enrolled in a calculus course, and you might be thinking, "I've finally made it ... All my life I've studied mathematics, and now I'm enrolled in 'the big one.' Certainly, calculus is the pinnacle of my studies in mathematics!"

But wait! So you are here, ... but do you know why? Can you answer the question, "What is calculus?" In the first section of this text, we begin by answering this question, and if you look at college catalogues, you will see that calculus is the *first* course in college mathematics, not the last one. This course is a prerequisite for almost all the other nonremedial mathematics courses.

OK, so now you are here in a calculus course and you are asked to pay a great deal of money for this book. "Why does this book cost so much, and how could it possibly be worth all that money?" The value of any purchase is relative to many aspects which differ from person to person, but ultimately, value must be measured by the extent to which it changes our lives. The intent and goal of this book is to change your life by giving you life skills which enhance your problem-solving ability. By the time you finish this course you will have a book, which at different times, you will have loved, hated, cursed, and respected. You will have a reference book that should last you a lifetime.

Calculus is a difficult subject, and there is no magic key to success; it will require hard work. This book should make your calculus journey easier because it builds your problem-solving skills and helps you form some good study habits. You will need to read the book, work the examples in the book using your own pencil and paper, and make a commitment to do your mathematics homework on a daily basis. ☠ Read that last sentence once again. It is the best hint you will see about building success in mathematics. ☠

We have written this book so that it will be easy for you to know what is important.

- Each chapter begins with a preview and ends with a proficiency examination (generally 30 problems) consisting of concept problems (to help you know what is important) and practice problems (to help you with a self-test).
- Important terms are presented in **boldface type**.
- Important ideas, definitions, and procedures are enclosed in screened boxes: Definitions , Properties , and Procedures
- *Common pitfalls, helpful hints and explanations are shown using this font.*
- ☠ WARNINGS are given to call your attention to common mistakes.

- Color is used in a functional way to help you "see" what to do next.
- INTERACTIVE figures allow you to "see" a figure in a dynamic way—essential since calculus discusses dynamic, not static, processes.

Success in this course is a joint effort by the student, the instructor, and the author. The student must be willing to attend class and devote time to the course *on a daily basis*. There is no substitute for working problems in mathematics.

FOR THE INSTRUCTOR

As the instructor of a calculus course, you must select materials which are relevant to your students at your school or university. Some aspects of your course are necessitated by your environment, and others are selected by you or a committee in your department. We have put together a calculus textbook which we believe will meet your needs, as well as those of your department. Most importantly, our book will meet the needs of your students in the twenty-first century.

The writing of a calculus textbook is a monumental task, one that grows out of a love of teaching and of helping students struggle until we see "the light" in their eyes as he or she has an "ah-hah" moment. The creation of this text has been a life-long journey for its authors, and to understand the nature of this book we travel back to 1930 to tell you of the books which have influenced us as we have developed the materials in this text.

At one time, the defining and almost universally used calculus textbook was Granville, Smith, and Longley's calculus book published in 1929.* This 516 page book contained no color, few illustrations, and even fewer problems. At that time, there were about a million students enrolled in the American university system, and education was designed to be for the privileged few.

In 1957, Sputnik created a revolution in American education. Rigor was added to the teaching of calculus. The most influential calculus textbook of this era was a book by Thomas[†] which was initially published in 1952 and became the standard text in the 1960s. This book added rigor and problems, but still offered little pedagogy. By the end of 1968, Thomas was in its fourth edition, and had grown to 818 pages, which matched the growth of college enrollments at that time to almost four million students. Textbooks were still relatively inexpensive, but were colorless and without learning aids or technology. By 1970, *Calculus* by Apostol[‡] was being used as a rigid and rigorous alternative to calculus,... correct, but lifeless. Many instructors of that time adopted a "sink-or-swim" attitude as there seemed to be an unlimited supply of students, each year growing larger and larger.

By the 1980s, college enrollments had passed the ten million mark and calculus needed to change to keep up with the times. In 1987 the next generation of calculus books was born. Stewart's *Calculus*[§] changed both the content and pedagogy of calculus. During this period of time, there was a movement to "reform" calculus. In 1986 and 1987, Steve Maurer and Ronald Douglas held two conferences, "Toward a Lean and Lively Calculus" and "Calculus for a New Century: A Pump not a Filter." This movement gave rise to what has become known as the Harvard Calculus approach and the publication of a book commonly known as the Hughes-Hallett *Calculus*.[¶] The reform calculus movement divided the mathematics community into two "camps" and

---

*W. A. Granville, P. F. Smith, and W. R. Longley, *Elements of the Differential and Integral Calculus*. Boston: Ginn and Company, 1929.

[†]G. B. Thomas, *Calculus and Analytic Geometry, Classic Edition* (Reprint of the 1952 edition). Reading: Addison Wesley, 1983.

[‡]Tom Apostle, *Calculus* (2 volumes): Hoboken, Wiley, 1967

[§]James Stewart, *Calculus*. Belmont: Brooks/Cole Publishing Company, a division of Wadsworth, Inc., 1987

[¶]Deborah Hughes-Hallett, Andrew M. Gleason, et. al., *Calculus*. Hoboken: John Wiley & Sons, 1994.

the teaching of calculus went through great changes, most of which will have lasting benefit to students.

With these books (and we have either learned or taught from all of them) and others we found those features which are essential and those which are superficial. Over the years, we have class tested our ideas and with this edition we have brought to you a book which is as fresh as this century, but preserves all the tried-and-true techniques which have been found to provide the necessary foundation for future work in mathematics. As college enrollments soar past the twenty million mark, education must now be designed for the masses. Regardless of the equipment offered at school, individual students have access to more and more technology (computers, iPads, smartphones, calculators, and Facebook, etc.). They use that technology not only to learn, but to communicate with each other in a cyber community. The calculus book for next generation needs to "work" in *today's* world and with *today's* changing technology, but at the same time, not compromise the student's ability to work complicated mathematical problems *without* relying on technology. These issues were our guiding principle as we wrote this edition of *Calculus*.

FEATURES OF THIS BOOK

Some of the distinguishing characteristics of the earlier editions are continued with this edition:

- It is possible to begin the course with either Chapter 1 or Chapter 2 (where the calculus topics begin).
- This edition offers an early presentation of transcendental functions: Logarithms, exponential functions, and trigonometric functions are heavily integrated into all chapters of the book (especially Chapters 1–5).
- We have taken the introduction of differential equations seriously. Students in many allied disciplines need to use differential equations early in their studies and consequently cannot wait for a postcalculus course. In this edition, we introduce differential equations in a natural and reasonable way. Slope fields are introduced as a geometric view of antidifferentiation in Section 5.1, and then are used to introduce a graphical solution to differential equations in Section 5.6. We consider separable differential equations in Chapter 5 and first-order linear differential equations in Chapter 7, and demonstrate the use of both, modeling a variety of applied situations. Exact and homogeneous differential equations appear in Chapter 14, along with an introduction to second-order linear equations. The "early and often" approach to differential equations is intended to illustrate their value in continuous modeling and to provide a solid foundation for further study.
- We continue to utilize the *humanness* of mathematics. History is not presented as additional material to learn. Rather we have placed history into *problems* that lead the reader from the development of a concept to actually participating in the discovery process. The problems are designated as 𝔥*istorical* 𝒬*uest* problems. The problems are not designed to be "add-on or challenge problems," but rather to become an integral part of the assignment. The level of difficulty of 𝒬*uest* problems ranges from easy to difficult.
- This edition correctly reflects the precalculus mathematics being taught at most colleges and universities. We assume knowledge of the trigonometric functions, and in this edition we introduce e^x and $\ln x$ in Chapter 2 after we have defined the notion of a limit. We also assume a knowledge of the conic sections and their graphs.
- **Think Tank Problems** Thinking about and doing mathematics is different than solving textbook problems. When thinking about and doing mathematics, a proposition may be true or false, whereas the typical textbook problem has a nice and concise answer. In the **Think Tank** problems our task is to prove the proposition true or to find a counterexample to disprove the proposition. We believe this form of problem to

be important in preparing the student for future work in not only advanced mathematics courses, but also for analytically oriented courses.

- EXPLORATION PROBLEMS   It has been said that mathematical discovery is directed toward two major goals—the formulation of proofs and the construction of counterexamples. Most calculus books focus only on the first goal (the body of proofs and true statements), but we feel that some attention should be paid to the formulation, exploring concepts that may prove to be true or for which a counterexample is appropriate. These exploration problems go beyond the category of counterexample problem to provide opportunities for innovative thinking.

- **Journal Problems**   In an effort to show that "mathematicians work problems too," we have reprinted problems from leading mathematics journals. We have chosen problems that are within reach of the intended audience of this book. If students need help or hints for these problems, they can search out the original presentation and solution in the cited journal. In addition, we have included problems from various **Putnam examinations**. These problems, which are more challenging, are offered in the supplementary problems at the end of various chapters and are provided to give insight into the type of problems that are asked in mathematical competitions. The Putnam Examination is a national annual examination given under the auspices of the Mathematical Association of America and is designed to recognize mathematically talented college and university students.

- Modeling continues as a major theme in this edition. Modeling is discussed in Section, and then appears in almost every section of the book. These applications are designated **Modeling Problems** or **MODELING EXPERIMENTS**. Some authors use the words "Modeling Problem" to refer to any applied problem. We make a distinction between *modeling problems* and *application problems* by defining a modeling problem as follows. A **modeling problem** is a problem that requires the reader make some assumptions about the real world in order to derive or come up with the necessary mathematical formula or mathematical information to answer the question. These problems also include real-world examples of modeling by citing the source of the book or journal that shows the modeling process.

Much of the difficulty students encounter learning the ideas of calculus can be attributed to ways students study and learn mathematics in high school, which often involves stressing rote memorization over insight and understanding. On the other hand, some reform texts are perceived as spending so much time with the development of insight and understanding that students are not given enough exposure to important computational and problem-solving skills in order to perform well in more advanced courses. This text aims at a middle ground by providing sound development, simulating problems, and well-developed pedagogy within a framework of a traditional tropic structure. "Think, then do" is a fair summary of our approach.

TEXT CONTENT

The content of this text adapts itself to either semester or quarter systems, and both differentiation and integration can be introduced in the first course. We begin calculus with a minimum of review. Cumulative reviews are offered at locations that fit the way calculus is taught at most colleges and universities. The first one includes Chapters 1-5, the second includes Chapters 6-8. Furthermore there is a cumulative review for single-variable calculus (Chapters 1-10) and multivariable calculus (Chapters 11-13). Chapter 14, provides an introduction to differential equations, which is often considered as a separate course.

Sequence of Topics   We resisted the temptation to label certain sections as optional, because that is a prerogative of individual instructors and schools. However, the following sections could be skipped without any difficulty: 4.7, 5.9, 6.5 (delay until Sec. 12.6), 6.6,

7.8, 11.8, and 12.8. To assist instructors with the pacing of the course, we have written the material so each section can be reasonably covered in one classroom day, but to do so requires the students to read the text in order to tie together the ideas that might be discussed in a classroom setting.

Problems After the correct topics and sequence of topics, the important aspect of any book is its problems. We believe that students *learn* mathematics by *doing* mathematics. Therefore, the problems and applications are perhaps the most important feature of any calculus book, and you will find that the problems in this book extend from routine practice to challenging. The problem sets are divided into Level 1 Problems (routine), Level 2 Problems (requiring independent thought), and Level 3 Problems (challenging).

In this book we also include past Putnam examination problems as well as problems found in current mathematical journals. You will find the scope and depth of the problems in this book to be extraordinary. Even though engineering and physics examples and problems play a prominent role, we include applications from a wide variety of fields, such as biology, economics, ecology, psychology, and sociology. The problems have been in the developmental stages for over ten years and virtually all have been class tested. In addition, the chapter summaries provide not only topical review, but also many supplementary exercises. Although the chapter reviews are typical of examinations, the supplementary problems are not presented as graded problems, but rather as a random list of problems loosely tied to the ideas of that chapter. In addition, there are cumulative reviews located at natural subdivision points in the text.

The problem sets are uniform in length (60 problems each), which facilitates the assigning of problems from day-to-day. (For example 3-60, multiples of 3 works well with the paired nature of the problems.)

Parametric forms Parametric representation first appears when graphing lines in parametric form in Section 1.3 and is then generalized to parametric curves (and lines in $\mathbb{R}^3$) in Section 9.5. Parametric forms are essential in the development of vector-valued functions in Chapter 10 and area of a surface defined parametrically in Chapter 12.

Proofs

Precise reasoning has been, and we believe will continue to be, the backbone of good mathematics. While never sacrificing good pedagogy and student understanding, we present important results as *theorems*. We do not pretend to prove every theorem in this book; in fact, we often only outline the steps one would take in proving a theorem. We refer the reader to Appendix B for longer proofs, or sometimes to an advanced calculus text. Why then do we include the heading "PROOF" after each theorem? It is because we want the student to *know* that for a result to be a theorem there *must* be a proof. We use the heading not necessarily to give a complete proof in the text, but to give some direction to where a proof can be found, or an indication of how it can be constructed.

INNOVATIVE PRESENTATION

Calculus (first edition) by Bradley/Smith was first published in 1995, was revised (in the third edition) by Strauss/Bradley/Smith in 2002, and now ten years later has a new writing team of Smith/Strauss/Toda. We offer you a book designed to begin a new generation of calculus textbooks. This edition was developed through the lens of hindsight so as to blend the best aspects of calculus reform along with the goals and methodology of traditional calculus. We've used foresight to present it in a format which is enhanced, but not dominated, by new technology.

Conceptual Understanding through Verbalization Besides developing some minimal skills in algebraic manipulation and problem solving, today's calculus text should require students to develop verbal skills in a mathematical setting. This is not just because real mathematics wields its words precisely and compactly, but because verbalization should help students think conceptually.

Mathematical Communication   We have included several opportunities for mathematical communication in terms that can be understood by nonprofessionals. We believe that students will benefit from individual writing and research in mathematics. Shorter problems encouraging written communication are included in the problem sets and are designated by the logo ■ What does this say? Another pedagogical feature is the **"What this says:"** boxes in which we rephrase mathematical ideas in everyday language. In the problem sets we encourage students to summarize procedures and processes or to describe a mathematical result in everyday terms. Concept problems are found throughout the book as well as at the end of each chapter, and these problems are included to prove that mathematics is more than "working problems and getting answers." Mathematics education *must* include the communication of mathematical ideas.

Cooperative Learning (Group Research Projects)
We believe that encouraging students to work on significant projects in small groups acts as a counterbalance to traditional lecture methods and can serve to foster both conceptual understanding and the development of technical skills. However, many instructors still believe that mathematics can be learned only through independent work. We feel that independent work is of primary importance, but that students must also learn to work with others in group projects. After all, an individual's work in the "real world" is often done as part of a group and almost always involves solving problems for which there is no answer "in the back of the book." In response to this need, we have included challenging exercises (the journal and Putnam problems) and a number of group research projects, which appear at the end of each chapter. These projects involve intriguing questions whose mathematical content is tied loosely to the previous chapter. These projects have been developed and class-tested by their individual authors, to whom we are greatly indebted. Note that the complexity of these projects increase as we progress through the book and the mathematical maturity of the student is developed. Some instructors read the GROUP RESEARCH PROJECTS and say "I think you need to give more information." After teaching calculus for an average of 30 years, and assigning group projects many times, we have found that the *less* we say and the *more* we leave for the student to question, ask, refine, and define limitation, the *better* the quality of their projects. For most calculus classes, just give a minimal assignment and stand back! Be prepared to answer *their* questions, but when you see the results you will be pleasantly surprised.

Integration of Technology

Technology Notes   There is a need to embrace the benefits technology brings to the learning of mathematics. Simply adding a lab course to the traditional calculus is possible, but this may lead to unacceptable work loads for all involved. Today it is reasonable to assume that each student has access to a computer and the World Wide Web. Throughout the book, many of the figures are designated as INTERACTIVE which means that the curves can be manipulated online to observe the characteristics involved. For example, Figure 29, Section 2.1 shows the epsilon-delta definition of a limit. This is a difficult concept to illustrate with a static figure, but each when you can see the figure in an interactive format, it becomes crystal clear.

Most processes in calculus are now available on the web, so we do not give you specific programs or technology since those processes are often outdated by the time a book is printed. Work with your classmates to identify appropriate online software.

Significant Digits   We have included a brief treatment in Appendix C. On occasion, we show the entire calculator or computer output of 12 digits for clarity even though such a display may exceed the requisite number of significant digits.

Greater Text Visualization   Related to, but not exclusively driven by, the use of technology is the greater use of graphs and other mathematical pictures throughout this text. Over 1,500 graphs appear—more than any other calculus text—and included are over 100 INTERACTIVE pieces of art which show you the **dynamic** process of calculus. This increased visualization is intended to help develop greater student intuition, and to allow the student to experiment, test hypotheses, and formulate conclusions. Also, since tough calculus problems are often tough geometry (and algebra) problems, this increased emphasis on graphs will help students' problem-solving skills. Additional graphs are related to the student problems, including answer art.

Supplementary materials   A *Student's Solution and Survival Manual* by one of the authors of this text, Karl J. Smith, offers a running commentary of hints and suggestions to help ensure the students' success in calculus. The solutions given in the manual complement all of the procedures and development of the textbook. This manual includes the solutions to the odd-numbered problems.

An *Instructor's Solution Manual* also written by Karl J. Smith, provides solutions to all problems in the book.

ACKNOWLEDGMENTS

The writing and publishing of a calculus book is a tremendous undertaking. We take this responsibility very seriously because a calculus book is instrumental in transmitting knowledge from one generation to the next. We would like to thank the many people who helped us in the preparation of this book. First, we want to thank the entire team at Kendall Hunt, who worked so hard from the inception of this project through completion: Chris O' Brien-Assistant Vice President, Ray Wood-Director of Author Relations, Stephanie Aichele-Managing Editor, Lara McCombie-Author Account Manager, Carrie Maro-Senior Production Editor, and Angela Willenbring-Senior Editor. Second, the composition team from Laserwords, who handled so expertly all the difficult mathematical typesetting and the art, including the interactive pieces: Abel Robertson-Composition Management, Suresh Muruganandam-Composition/Typesetting, Yasodha Govindarajan-Quality Control, Leanne Binette and Stacy Proteau-Project Management and Vembuganesh Swaminathan and Rajesh Britto-Interactive Art. So much of the method of work was new to us as well as to many of them, particularly exchanging files full of mathematics electronically, and we appreciate the patience and concern shown by all.

Robert Byerly and David Gilliam worked diligently on implementing many problems for Webwork, for which we are extremely grateful.

Of primary concern is the accuracy of the book. We had the assistance of many: Melanie and Brian Fulton, who read the entire manuscript and offered us many valuable suggestions; Nancy Angle and Ann Ostberg also read large portions of the manuscript and accompanying problems and answers. All of these were meticulous in checking the manuscript and made many valuable suggestions. Thanks also to the accuracy checkers of the previous editions, Henry Finer, Jerry Alexanderson, Mike Ecker, Ken Sydel, Diana Gerardi, Kurt Norlin, Terri Bittner, Nancy Marsh, and Mary Toscano.

We also thank Jerry Bradley, who was a coauthor of the first three editions of this text.

We received helpful input from our colleagues at Texas Tech, especially Linda Allen, Eugenio Aulisa, Roger Barnard, Harold Bennett, Robert Byerly, Lance Drager, David Gilliam, Victoria Howle, Lourdes Juan, Wayne Lewis, Kevin Long, Kent Pearce, Carl Seaquist, Dean Victory, and Brock Williams, as well as users at Rutgers University.

Also extremely helpful with this edition were:
Dustin Brewer, Minerva Cordero-Epperson, James Epperson, Ruth Gornet, and Michaela
Vancliff, University of Texas at Arlington
Betsy Bennett, St. Albans School in Washington, D. C.
Bem Cayco, San Jose State University
Kimberly Drews, George Washington University
Mark Farris, Midwestern State University
Jim Hagler, University of Denver
David Jackson, Saint Louis University
Hae-Soo Oh, University of North Carolina at Charlotte
Tim Serino, Lynn English High School in Lynn, Massachusetts

Reviewers of the previous editions:
Gregory Adams, Bucknell University
Gerald Alexanderson, Santa Clara University
David Arterburn, New Mexico Tech
Robert Bakula, Ohio State University
Neil Berger, University of Illinois at Chicago
Michael L. Berry, West Virginia Wesleyan College
Linda A. Bolte, Eastern Washington University
Brian Borchers, New Mexico Tech
Barbara H. Briggs, Tennessee Technical University
Robert Broschat, South Dakota State University
Robert D. Brown, University of Kansas
J. Caggiano, Arkansas State University
James T. Campbell, University of Memphis
Dan Chiddix, Ricks College
Philip Crooke, Vanderbilt University
Lin Dearing, Clemson University
Stan Dick, George Mason University
Tevin Dray, Oregon State University
Ken Dunn, Dalhousie University
Michael W. Ecker, Pennsylvania State University, Wilkes-Barre Campus
John H. Ellison, Grove City College
Mark Farris, Midwestern State University
Sally Fïeschbeck, Rochester Institute of Technology
William P. Francis, Michigan Technological University
Anda Gadidov, Gannon University
Stuart Goldenberg, California Polytechnic State University, San Luis Obispo Campus
Ruth Gornet, Texas Tech University and University of Texas at Arlington
Harvey Greenwald, California Polytechnic San Luis Obispo
Julia Hassett, Oakton Community College
Isom H. Herron, Rensselaer Polytechnic Institute
Michael G. Hilgers, University State University
Richard Hitt, University of South Alabama
Jason P. Huffman, Jacksonville State University
Joel W. Irish, University of Southern Maine
James E. Jamison, University of Memphis
Roger Jay, Tomball College
John H. Jenkins, Embry Riddle Aeronautical University
Clement T. Jeske, University of Wisconsin-Platteville
Kathy Kepner, Paducah Community College
Daniel King, Sarah Lawrence College
Lawrence Kratz, Idaho State University
Jeuel G. LaTorre, Clemson University
Don Leftwich, Oklahoma Christian University

Sam Lessing, Northeast Missouri University
Ira Wayne Lewis, Texas Tech University
Estela S. Llinas, University of Pittsburgh at Greensburg
Pauline Lowman, Western Kentucky University
Ching Lu, Southern Illinois University at Edwardsville
William E. Mastrocola, Colgate University
Philip W. McCartney, Northern Kentucky University
E. D. McCune, Stephen F. Austin State University
John C. Michels, Chemeketa Community College
Judith Ann Miller, Delta College
Ann Morlet, Cleveland State University
Maura B. Mast, University of Massachusetts-Boston
Mark Naber, Monroe Country Community College
Dena Jo Perkins, Oklahoma Christian University
Chris Peterson, Colorado State University
Pamela B. Pierce, College of Wooster
Siew-Ching Pye, California State Polytechnic University-Pomona
Judith Reeves, California Polytechnic State University-Pomona
Joe Rody, Arizona State University
Yongwu Rong, The George Washington University
Eric Rowley, Utah State University
Jim Roznowski, Delta College
Peter Salamon, San Diego State University
Connie Schrock, Emporia State University
Carl Seaquist, Texas Tech University
Tatiana Shubin, San Jose State University
Jo Smith, Ketering University
Anita E. Solow, DePauw University
Lowell Stultz, Kalamazoo Valley Community College
Dennis Wacker, Saint Louis University
Tingxiu Wang, Oakton Community College
W. Thurmon Whitley, University of New Haven
Teri Woodington, Colorado School of Mines
Cathleen M. Zucco-Tevcloff, Trinity College

Karl J. Smith
Monty J. Strauss
Magdalena D. Toda

CHAPTER 1

FUNCTIONS AND GRAPHS

*E*very mathematical book that is worth reading must be read "backwards and forwards," if I may use the expression. I would modify Lagrange's advice a little and say, "Go on, but often return to strengthen your faith." When you come on a hard and dreary passage, pass it over; come back to it after you have seen its importance or found its importance or found the need for it further on.

George Chrystal,
Algebra, Part 2 (Edinburgh, 1889)

PREVIEW

This chapter uses several topics from algebra and trigonometry that are essential for the study of calculus. We begin by asking the question, "What is Calculus?" We shall see that calculus is used to model many aspects of the world about us, so we spend a little time discussing what is meant by mathematical modeling.

The concept of function is an especially important prerequisite for calculus. The idea of a function can be illustrated as a cause-and-effect relationship. In the 1930s, a cartoonist named Rube Goldberg became well known for depicting complex devices that performed simple tasks in indirect, convoluted ways. Today, he is remembered by engineering students in constructing deliberately over-engineered machines that perform simple tasks. Students often view mathematics as a "Rube Goldberg" machine, but the goal of calculus is to make the concepts crystal clear and as simple as possible. In fact, you might notice that the word *function* begins with fun.

Even if you wish to skip most of this initial chapter, you should review the basic notions of functions and inverse functions in Sections 1.4 and 1.5.

Courtesy of the Craig Engineering
Club, Janesville, WI

Rube Goldberg Machine
Contest

PERSPECTIVE

Change is a fact of our daily lives. Although modern science requires the use of many different skills and procedures, calculus is the primary mathematical tool for dealing with change. Sir Isaac Newton, one of the discoverers of calculus, once remarked that to accomplish his results, he "stood on the shoulders of giants." Indeed, calculus was not born in a moment of divine inspiration but developed gradually, as a variety of apparently different ideas and methods merged into a coherent pattern. The purpose of this initial chapter is to lay the foundation for the development of calculus.

1.1 WHAT IS CALCULUS?

IN THIS SECTION: *The limit: Zeno's paradox, the derivative: the tangent problem, the integral: the area problem, mathematical modeling*
We informally introduce you to the three main topics of calculus; the concepts of limit, derivative, and integral. We also introduce the idea of mathematical modeling.

ELEMENTARY MATHEMATICS

1. Slope of a line

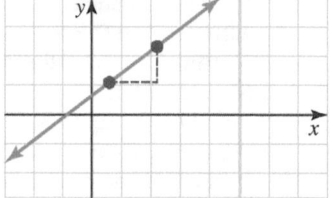

2. Tangent line to a circle

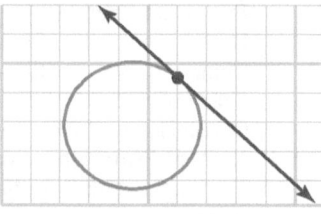

3. Area of a region bounded by line segments

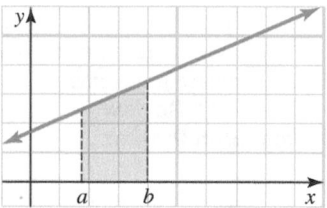

4. Average position and velocity

5. Average of a finite collection of numbers

Figure 1.1 Topics from elementary mathematics

If there is an event that marked the coming of age of mathematics in Western culture, it must surely be the essentially simultaneous development of the calculus by Newton and Leibniz in the 17th century. Before this remarkable synthesis, mathematics had often been viewed as merely a strange but harmless pursuit, indulged in by those with an excess of leisure time. After the calculus, mathematics became virtually the only acceptable language for describing the physical universe. This view of mathematics and its association with the scientific method has come to dominate the Western view of how the world ought to be explained.

What distinguishes calculus from algebra, geometry, and trigonometry is the transition from static or discrete applications (see Figure 1.1) to those that are dynamic or continuous (see Figure 1.2). For example, in elementary mathematics we consider the slope of a line, but in calculus we define the (nonconstant) slope of a nonlinear curve. In elementary mathematics we find average values of position and velocity, but in calculus we can find instantaneous values of changes of velocity and acceleration. In elementary mathematics we find the average of a finite collection of numbers, but in calculus we can find the average value of a function with infinitely many values over an interval.

Calculus is the mathematics of motion and change, which is why calculus is a prerequisite for many courses. Whenever we move from the static to the dynamic, we would consider using calculus.

The development of the calculus in the 17th century by Newton and Leibniz was the result of their attempt to answer some fundamental questions about the world and the way things work. These investigations led to two fundamental concepts of calculus, namely, the idea of a *derivative* and that of an *integral*. The breakthrough in the development of these concepts was the formulation of a mathematical tool called a *limit*, which we will consider in Chapter 2.

1. **Limit:** The limit is a mathematical tool for studying the *tendency* of a function as its variable *approaches* some value. Calculus is based on the concept of limit. We shall introduce the limit of a function informally in this section.
2. **Derivative:** The derivative is defined as a certain type of limit, and it is used initially to compute rates of change and slopes of tangent lines to curves. The study of derivatives is called *differential calculus*. Derivatives can be used in sketching graphs and in finding the extreme (largest and smallest) values of functions. We shall discuss derivatives in Chapter 3.
3. **Integral:** The integral is found by taking a special limit of a sum of terms, and the study of this process is called *integral calculus*. Area, volume, arc length, work, and hydrostatic force are a few of the many quantities that can be expressed as integrals. We shall discuss integrals in Chapter 5.

In this section, let us take an intuitive look at each of these three essential ideas of calculus.

The Limit: Zeno's Paradox

Zeno (ca. 500 B.C.) was a Greek philosopher known primarily for his famous paradoxes. One of those concerns a race between Achilles, a legendary Greek hero, and a tortoise. When the race begins, the (slower) tortoise is given a head start, as shown in Figure 1.3.

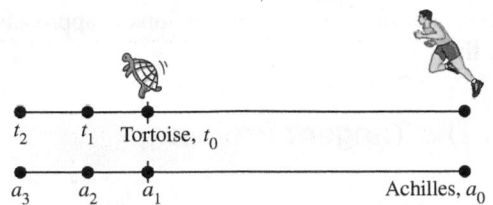

Figure 1.3 Achilles and the tortoise

Is it possible for Achilles to overtake the tortoise? Zeno pointed out that by the time Achilles reaches the tortoise's starting point, $a_1 = t_0$, the tortoise will have moved ahead to a new point t_1. When Achilles gets to this next point, a_2, the tortoise will be at a new point, t_2. The tortoise, even though much slower than Achilles, keeps moving forward. Although the distance between Achilles and the tortoise is getting smaller and smaller, the tortoise will apparently always be ahead.

Of course, common sense tells us that Achilles will overtake the slow tortoise, but where is the error in reasoning in the previous paragraph that always gives the tortoise the lead? The error is in the assumption that an infinite amount of time is required to cover a distance divided into an infinite number of segments. This discussion is getting at an essential idea in calculus—namely, the notion of a limit.

Consider the successive positions for both Achilles and the tortoise:

↓Starting position

Achilles: $a_0, a_1, a_2, a_3, \cdots$
Tortoise: $t_0, t_1, t_2, t_3, \cdots$

After the start, the positions for Achilles, as well as those for the tortoise, form sets of positions that are ordered with positive integers. Such ordered listings are called *sequences* (see Section 8.1).

For Achilles and the tortoise we have two sequences $\{a_0, a_1, a_2, a_3, \cdots, a_n, \cdots\}$ and $\{t_0, t_1, t_2, t_3, \cdots, t_n, \cdots\}$, where $a_n < t_n$ for all values of n. Both the sequence for Achilles' position and the sequence for the tortoise's position have limits, and it is precisely at that limit point that Achilles overtakes the tortoise. The idea of limit will be discussed in the next section and it is this limit idea that allows us to define the other two basic concepts of calculus: the derivative and the integral. Even if the solution to Zeno's paradox using limits seems unnatural at first, do not be discouraged. It took over 2000 years to refine the ideas of Zeno and provide conclusive answers to those questions about limits. The following example will provide an intuitive preview of a limit, which we consider in more detail in Section 8.1.

Example 1   Intuitively find a limit

The sequence $\frac{1}{2}, \frac{2}{3}, \frac{3}{4}, \frac{4}{5}, \cdots$ can be described by writing a *general term*: $\frac{n}{n+1}$, where $n = 1, 2, 3, 4, \cdots$. Can you guess the limit, L, of this sequence? We will say that L is the number that the sequence with general term $\frac{n}{n+1}$ tends toward as n becomes large without bound. We will define a notation to summarize this idea:

$$L = \lim_{n \to \infty} \frac{n}{n+1}$$

Solution   As you consider larger and larger values for n, you find a sequence of fractions:

$$\frac{1}{2}, \frac{2}{3}, \frac{3}{4}, \cdots, \frac{1,000}{1,001}, \frac{1,001}{1,002}, \cdots, \frac{9,999,999}{10,000,000}, \cdots$$

CALCULUS

1. Slope of a curve

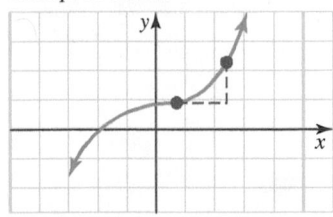

2. Tangent line to a general curve

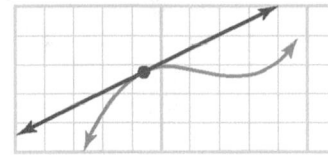

3. Area of a region bounded by curves

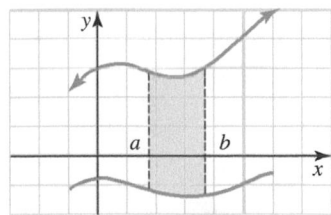

4. Instantaneous changes in position and velocity

5. Average of an infinite collection of numbers

Figure 1.2 Topics from calculus

It is reasonable to guess that the sequence of fractions is approaching the number 1. This number is called the **limit**. ■

The Derivative: The Tangent Problem

A tangent line (or, if the context is clear, simply say "tangent") to a circle at a given point P is a line that intersects the circle at P and only at P (see Figure 1.4**a**). This characterization does not apply for curves in general, as you can see by looking at Figure 1.4**b**. If we wish to have one tangent line, which line shown should be called the tangent to the curve?

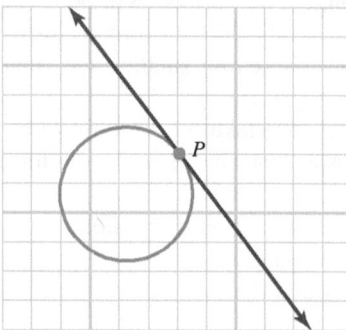

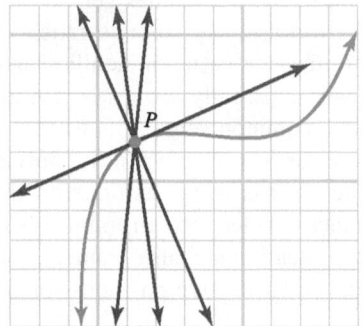

a. At each point P on a circle there is one line that intersects the circle exactly once

b. At a point P on a curve, there may be several lines that intersect that curve only once

Figure 1.4   Tangent line

To find a tangent line, begin by considering a line that passes through two points P and Q on the curve, as shown in Figure 1.5**a**. This line is called a **secant line**.

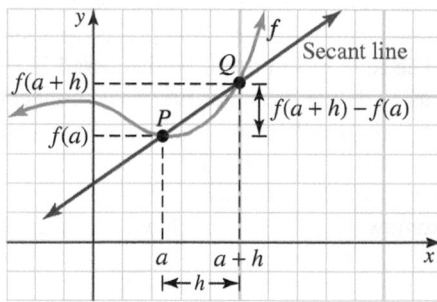

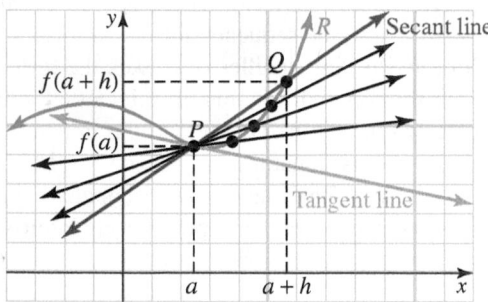

a. Locate points P and Q and draw secant line

b. Interactive Let point Q "slide" toward P and draw lines

Figure 1.5   Tangent line is the "limiting line" as Q approaches P

The coordinates of the two points P and Q are $P(a, f(a))$ and $Q(a + h, f(a + h))$. The slope of the secant line is

$$m = \frac{\text{RISE}}{\text{RUN}} = \frac{f(a + h) - f(a)}{h}$$

Now imagine that Q moves along the curve toward P, as shown in Figure 1.5**b**. You can see that the secant line approaches a limiting position as h approaches zero. We define this limiting position to be the **tangent line**. The slope of the tangent line is defined as the limit of the sequence of slopes of a set of secant lines. Once again, we can use limit notation to summarize this idea: We say that the slopes of the secant lines, as h becomes small, tend toward a number that we call the slope of the tangent line. We shall define the following notation to summarize this idea:

$$\lim_{h \to 0} \frac{f(a+h) - f(a)}{h}$$

Why would the slope of a tangent line be important? That is what we investigate in the first half of a calculus course, and it forms the definition of the derivative, which is the foundation for what is called **differential calculus** (see Chapter 3).

The Integral: The Area Problem

You probably know the formula for the area of a circle with radius r:

$$A = \pi r^2$$

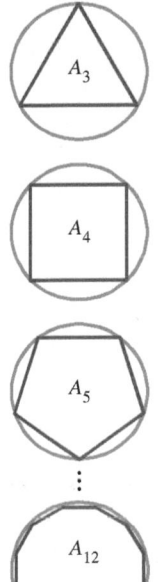

The Egyptians were the first to use this formula over 5,000 years ago, but the Greek Archimedes (ca. 300 B.C.) showed how to derive the formula for the area of a circle by using a limiting process. Consider the areas of inscribed polygons, as shown in Figure 1.6. Since we know the areas of these polygons, we can use them to estimate the area of a circle. Even though Archimedes did not use the following notation, here is the essence of what he did, using a method called "exhaustion":

Let A_3 be the area of the inscribed equilateral triangle;

A_4 be the area of the inscribed square;

A_5 be the area of the inscribed regular pentagon.

How can we find the area of this circle? As you can see from Figure 1.6, if we consider the area of A_3, then A_4, then A_5, $\cdots$, we should have a sequence of areas such that each successive area more closely approximates that of the circle. We write this idea as a limit statement:

$$A = \lim_{n \to \infty} A_n$$

Figure 1.6 Approximating the area of a circle

In this course we will use limits in yet a different way to find the areas of regions enclosed by curves. For example, consider the area shown in color in Figure 1.7.

We can approximate the area by using rectangles. If A_n is the area of the nth rectangle, then the total area can be approximated by finding the sum

$$A_1 + A_2 + A_3 + \cdots + A_{n-1} + A_n$$

This process is shown in Figure 1.8. Notice that in the Figure 1.8**a**, the area of the rectangles approximates the area under the curve, but there is some error. In Figure 1.8**b**, there is a little less error than in the first approximation. As more rectangles are included, the error decreases.

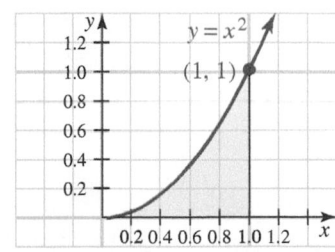

Figure 1.7 Area under a curve

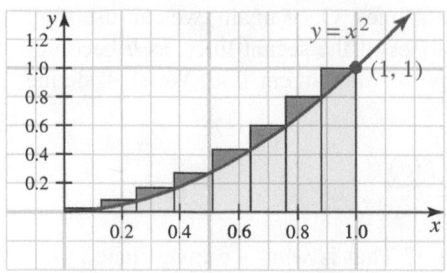
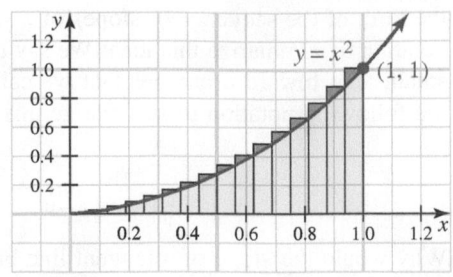

a. 8 approximating rectangles   **b. Interactive** 16 approximating rectangles

Figure 1.8 Approximating the area using rectangles

The area problem leads to a process called *integration*, and the study of integration forms what is called **integral calculus** (see Chapter 5). Similar reasoning allows us to calculate such quantities as volume, the length of a curve, the average value, or the amount of work required for a particular task.

Mathematical Modeling

A real life situation does not easily lend itself to mathematical analysis because the real world is far too complicated to be precisely and mathematically defined. It is therefore necessary to develop what is known as a **mathematical model**. The model is based on certain assumptions about the real world and is modified by experimentation and accumulation of data. It is then used to predict some future occurrence in the real world. A mathematical model is not static or unchanging, but is continually being revised and modified as additional relevant information becomes known.

Some mathematical models are quite accurate, particularly in the physical sciences. For example, the path of a projectile, the distance that an object falls in a vacuum, or the time of sunrise tomorrow have mathematical models that provide very accurate predictions about future occurrences. On the other hand, in the fields of social science, psychology, and management, models generally provide much less accurate predictions because they must deal with situations that are often random in character. It is therefore necessary to consider two types of models:

> **TYPES OF MODELS**   A **deterministic model** predicts the exact outcome of a situation because it is based on certain known laws. A **probabilistic model** deals with situations that are random in character and can predict the outcome within a certain stated or known degree of accuracy.

How do we construct a model? We need to observe a real world problem and make assumptions about the influencing factors. This is called *abstraction*.

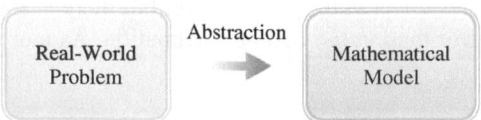

With the method of abstraction, certain assumptions about the real world are made, variables are defined, and appropriate mathematics is developed. The next step is to simplify the mathematics or derive related mathematical facts from the mathematical model.

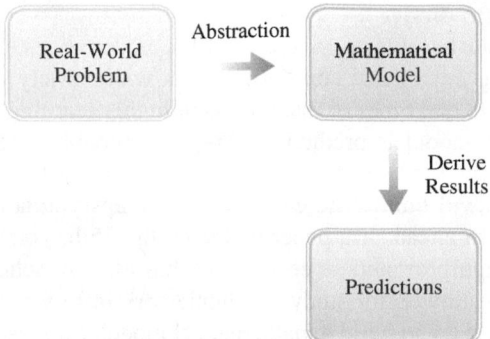

The results derived from the mathematical model should lead us to some predictions about the real world. The next step is to gather data from the situation being modeled, and then to compare those data with the predictions. If the two do not agree, then the gathered data are used to *modify* the assumptions used in the model.

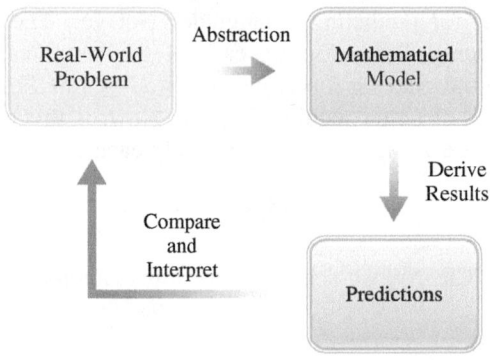

We begin with a rather artificial example of modeling (because at this time we have not developed any content in this course).

A Model For Studying In This Course

How do you plan for success in this course? Suppose we assume that the problem to solve is how you obtain a grade of C or better in this course. How do you expect to reach this goal? What are your past experiences with college-level courses in general and with mathematics courses in particular? Perhaps your model for studying in this course is quite simple, as shown in Figure 1.9.

Is the model shown in Figure 1.9 a good model? What do we mean by the word "good"? Is the model "true" in some absolute sense, or do we mean that it is "valid" in the sense that it has been checked successfully with a wide range of students and college-level courses?

A procedure that is often used in modeling says that if a question is difficult to answer, start by asking some easier questions. We might rephrase our question about this being a "good" model by asking some easier questions. You might ask yourself:

1. Has this procedure worked with success *for me* in previous college-level mathematics courses?
2. Is there a relationship between class attendance and final grade?
3. Is there a relationship between doing homework and final grade?
4. Will this course be typical of my past experiences in college-level mathematics courses?

How Global Climate Is Modeled

We find a good example of mathematical modeling by looking at the work being done with weather prediction. In theory, if the correct assumptions could be programmed into a computer, along with appropriate mathematical statements of the ways global climate conditions operate, we would have a model to predict the weather throughout the world. In the global climate model, a system of equations calculates time-dependent changes in wind as well as temperature and moisture changes in the atmosphere and on the land. The model may also predict alterations in the temperature of the ocean's surface. At the National Center for Atmospheric Research, a CRAY supercomputer is used to do this modeling.

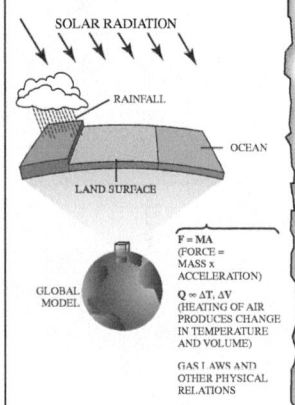

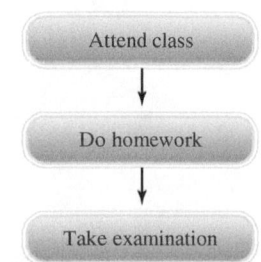

Figure 1.9 First model to obtain a C in this course

Models Should Be Predictive

A good model should be able to predict real-life occurrences. Is the model we build for student success in this course appropriate for a wide variety of students, instructors, methods of instruction, and types of institutions of higher learning? How can we go about making sure that our model is predictive? You are probably interested *only* in making sure that it is predictive for *you* at your school with your instructor.

Different people will build different models. Perhaps you study best under pressure and do not like to plan ahead. If a paper is due on the 15th, you will begin on the 14th. Perhaps you are a highly organized person who has already scheduled in advance how much time you have allowed for study and homework in this course.

Perhaps we should try to build a mathematical model. Suppose you count the number of sections you will study in this course. Let this number be n. Also, suppose you decide that your grade is determined by the amount of time, t, that you will spend in each section. You might begin by using the following model:

$$G = nt$$

Is *this* a good model for predicting your grade? Not yet. The variables are not well defined, nor are the units for measuring each of the variables. Let t be the time in minutes and let G be the grade as a percent, so $0 \leq G \leq 100$. Furthermore, suppose that obtaining a grade of C or better requires $G \geq 70$. Now, is this a good model? Not yet. For example, if there are 20 sections and you study each for 10 minutes, then

$$G = (20)(10) = 200$$

does not make sense. Clearly, we need to refine this mathematical model.

Criticize Your Model

Notice that building a model requires several steps and comparisons between the mathematics and the real world. Do not expect to come up with a working model on the first try. One of the most difficult aspects about teaching and learning how to model mathematics is to deal with the student's impatience. Do not expect to "get it right" on your first attempt. For this model, we might consider a scaling factor, k:

$$G = knt$$

Next, we could do some research and determine whether the time spent on each section by a large number of students and the grades obtained by those students are related in a linear way. Based on this research, we replace this formula by a more predictive formula, say

$$G = \frac{1}{3}(t - 30) + 50$$

Is this a good model? Not yet. We have not yet built into this model the way we prepare for examinations and other course-related requirements. Perhaps you are beginning to ask, "When will I finish this model-building process?" This is a reasonable question, but part of the impatience lies in wanting to "get an answer." The modeling process itself is never complete. We can simply build better and better models.

You can also see that we need to build some mathematical skills in order to develop some mathematical models. This is a topic of this book. In this section, we will see some simple models that are defined by a formula.

Models Defined By A Formula

The simplest type of model is one in which we are given a formula to model some phenomena.

Example 2 A physical model

A spring scale works on the principle that if a weight is hung from one end of a spring, the total length of the spring is a function of the amount of weight. Suppose that with no weight, a spring is 10 inches long. With a 10-lb weight the spring stretches to 14 inches, and with a 50-lb weight, the spring stretches to 30 inches. What would the stretch be for the maximum weight of 100 pounds? If the spring stretches to a length of 20 inches, what is the weight?

Solution   This is a physical model. There may be several factors that could affect the model we build. Does the temperature, humidity, or altitude affect the results of this experiment? Perhaps, but since none of these are mentioned in the problem, we will make some simplifying assumptions—namely, that the two variables, length of spring and weight, are the only variables to be considered. To build this model, we would need to make other observations, plot the data, and perhaps use some mathematics that will be developed later in this course. Suppose we assume this work has been completed and that the formula for this problem is

$$s = 0.4w + 10 \qquad 0 \le w \le 100$$

where s is the stretch of the spring in inches and w is the weight of the object in pounds.
   If $w = 100$, then

$$s = 0.4(100) + 10 = 50$$

The maximum stretch is 50 inches. If $s = 20$, then

$$20 = 0.4w + 10$$
$$10 = 0.4w$$
$$25 = w$$

The weight is 25 lbs.

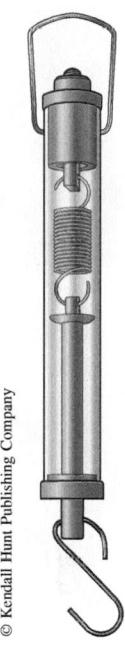

Example 3 Model from biology

It has been noticed that the rate at which certain crickets chirp depends on the temperature. Build a model to answer the following question: What is the temperature when 20 chirps are counted in 15 seconds? When 50 chirps are counted? If the temperature is 20°C, how many times will the cricket chirp in 1 minute?

Solution   To build a model, many observations must be made and much data gathered. We will assume that this work has been completed and that the formula for the temperature in degrees Celsius (C) is given as a function of the number of chirps (n) in 15 seconds:

$$C = 0.6n + 4$$

For 20 chirps, we let $n = 20$ and evaluate the formula:

$$C = 0.6(20) + 4 = 16$$

The temperature is 16°C. For 50 chirps, we let $n = 50$:

$$C = 0.6(50) + 4 = 34$$

The temperature is 34°C. For 20°C, we let $C = 20$:

$$20 = 0.6n + 4$$
$$16 = 0.6n \qquad \textit{Divide both sides by 0.6 and simplify.}$$
$$\frac{80}{3} = n$$

If $\frac{80}{3}$ chirps are heard in 15 seconds, then there will be

$$\frac{80}{3}(4) = 106\frac{2}{3}$$

chirps in 1 minute. Since we cannot count $\frac{2}{3}$ of a chirp, the answer is 106 chirps in 1 minute at 20°C. ■

Do you see why we do not round to 107 chirps for the last answer in Example 3? In general, *never round off in the middle of the problem*, but only when stating the final answer.

Cost Analysis—A Model From Business

In business, costs and prices can be analyzed over a long period or a short period. **Short-run analysis** has traditionally referred to a time period over which costs and prices remain constant. Over longer periods, economic factors such as inflation or supply and demand tend to influence costs and prices. The cost of every manufacturing process can be divided into fixed and variable costs. Certain costs, such as rent, taxes, insurance, and utilities, exist even if no product is actually manufactured. These are called **fixed costs** and usually remain constant over the short run. Other costs, such as materials, labor, and distribution, depend directly on the number of items actually produced. These are called **variable costs** and increase as more items are produced. The *total cost* can be given by the formula

$$C = ax + b$$

where ax represents the variable costs and b represents the fixed costs over the short-run manufacture of x items.

Example 4 Production based on costs

If the fixed costs for a certain item total $2,100 and the variable costs total $0.80 per item, what is the cost of producing 2,000 items? How many items can be produced for $10,000?

Solution In the formula $C = ax + b$, let $a = 0.8$ and $b = 2,100$ to obtain

$$C = 0.8x + 2,100$$

For $x = 2,000$:

$$C = 0.8(2,000) + 2,100$$
$$= 3,700$$

It costs $3,700 to produce 2,000 items. For $x = 5,000$:

$$C = 0.8(5,000) + 2,100$$
$$= 6,100$$

It costs $6,100 to produce 5,000 items. If the cost is $10,000, then

$$10,000 = 0.8x + 2,100$$
$$7,900 = 0.8x$$
$$9,875 = x$$

The company could produce 9,875 items for a cost of $10,000.

Example 5 Finding revenue

Suppose each item in Example 4 sells for $1.50. The **revenue**, R, is the amount collected and depends only on the price, p, and number of items sold, x, according to the formula

$$R = px$$

If $p = 1.50$, what is the revenue for 2,000 items? For 5,000 items? How many items must be sold for a revenue of $10,000?

Solution   We use the formula $R = 1.5x$. For $x = 2,000$:

$$R = 1.5(2,000) = 3,000$$

The revenue is $3,000 for 2,000 items. For $x = 5,000$:

$$R = 1.5(5,000) = 7,500$$

The revenue is $7,500 for 5,000 items. For $R = 10,000$:

$$10,000 = 1.5x$$
$$\frac{20,000}{3} = x$$

Since the variable x permits only positive integers, the company must sell 6,667 items to obtain $10,000 in revenue.

Modeling In This Textbook
It is not possible to build mathematical modeling into every task, due to various constraints. The best we can do is build a framework *in which you can practice building models*. We have tried to do this with the modeling applications that appear in each chapter.

PROBLEM SET 1.1

Level 1

1. ■ What does this say?* What are the three main topics of calculus?

2. ■ What does this say? What is a mathematical model? Why are mathematical models necessary or useful?

3. ■ What does this say? Based on your limited experience with mathematical models, why do we describe them as iterative?

4. ■ What does this say? Do a GOOGLE search of "Mathematical Model." Summarize your findings.

*Many problems in this book are labeled **What Does this Say?** Following this question will be a question for you to answer in your own words, or a statement for you to rephrase in your own words. These problems are intended to be similar to the "What This Says" boxes that appear throughout the book.

5. ■ **What does this say?** An analogy to Zeno's tortoise paradox can be made as follows.

> A woman standing in a room cannot walk to a wall. To do so, she would first have to go half the distance, then half the remaining distance, and then again half of what still remains. This process can always be continued and can never be ended.

Draw an appropriate figure for this problem and then present an argument using sequences to show that the woman will, indeed, reach the wall.

6. ■ **What does this say?** Zeno's paradoxes remind us of an argument that might lead to an absurd conclusion:

> Suppose I am playing baseball and decide to steal second base. To run from first to second base, I must first go half the distance, then half the remaining distance, and then again half of what remains. This process is continued so that I never reach second base. Therefore it is pointless to steal a base.

Draw an appropriate figure for this problem and then present a mathematical argument using sequences to show that the conclusion is absurd.

7. ■ **What does this say?** There is a news clipping from *Scientific American* on modeling global climatic change. *Without* doing research, discuss how this might be accomplished.

8. ■ **What does this say?** Global warming is a topic of concern and of political discussion. *Without* doing research, discuss how modeling might be used to determine what might be done to reduce global warming.

9. Consider the sequence 0.3, 0.33, 0.333, 0.3333, ⋯. What do you think is the appropriate limit of this sequence?

10. Consider the sequence 6, 6.6, 6.66, 6.666, ⋯. What do you think is the appropriate limit of this sequence?

11. Consider the sequence 0.9, 0.99, 0.999, 0.9999, ⋯. What do you think is the appropriate limit of this sequence?

12. Consider the sequence 3, 3.1, 3.14, 3.141, 3.1415, 3.14159, 3.141592, ⋯. What do you think is the appropriate limit of this sequence? Is there a unique answer to this question? Justify.

Copy the figures in Problems 13-18 *on your paper. Draw what you think is an appropriate tangent line for each curve at the point P by using the secant method.*

13.

14.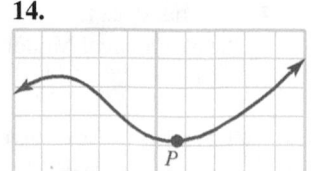

15. 16.

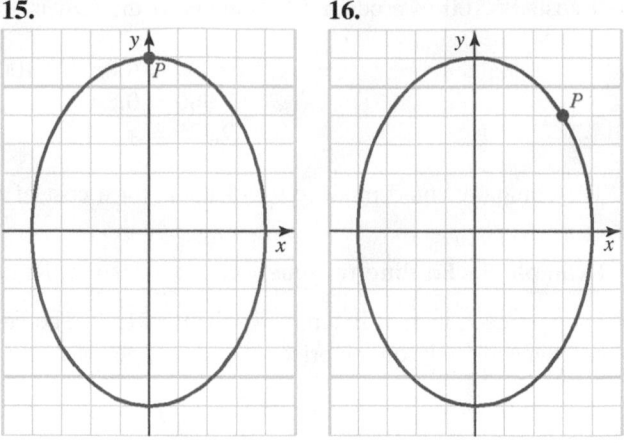

17. 18.

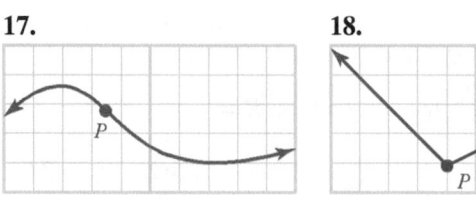

A Physical Model *Use the model given in Example 2 for Problems 19-24.*

19. What is the stretch on a spring with a weight of 15 pounds?

20. What is the stretch on a spring with a weight of 22.5 pounds?

21. What is the weight of an object that stretches a spring 35 inches?

22. What is the weight of an object that stretches a spring 20 inches?

23. A formula for the stretch of a certain spring, s in inches, for a weight w in pounds, is given by

$$s = 0.25w + 5$$

If an object stretches the spring 10 inches, what is the weight of the object?

24. A formula for the stretch of a certain spring, s in centimeters, for a weight w in kilograms (kg), is

$$w = 3s - 80$$

If an object weights 16 kg, what is the stretch on the spring?

A Biological Model *Use the model given in Example 3 for Problems 25-28.*

25. If 15 chirps are counted in 15 seconds, what is the Celsius temperature (to the nearest degree)?

26. If 12 chirps are counted in 15 seconds, what is the Celsius temperature (to the nearest degree)?

27. If the temperature is 40°C, how many cricket chirps would be heard in 1 minute?

28. If the temperature is 31°C, how many cricket chirps would be heard in 1 minute?

A Cost Analysis Model *Use the model given in Example 4 for Problems 29-34.*

29. A manufacturer has variable costs of $8.50/item and fixed costs of $3,600. What is the cost of producing 10 items?

30. A manufacturer has variable costs of $8.50/item and fixed costs of $3,600. What is the cost of producing 10,000 items?

31. A manufacturer has variable costs of $45/item and fixed costs of $405,000. What is the average cost of one item when only one item is produced?

32. A manufacturer has variable costs of $45/item and fixed costs of $405,000. What is the average cost per item when 100,000 items are produced?

33. If the item described in Problems 31-32 is to be sold for $25, what is the revenue for one item?

34. If the item described in Problems 31-32 is to be sold for $25, what is the revenue for 100,000 items?

Level 2

In Problems 35-40, **guess** *the requested limits.*

35. $\lim\limits_{n\to\infty} \dfrac{2n}{n+4}$

36. $\lim\limits_{n\to\infty} \dfrac{2n}{3n+1}$

37. $\lim\limits_{n\to\infty} \dfrac{n+1}{n+2}$

38. $\lim\limits_{n\to\infty} \dfrac{n+1}{2n}$

39. $\lim\limits_{n\to\infty} \dfrac{3n}{n^2+2}$

40. $\lim\limits_{n\to\infty} \dfrac{3n^2+1}{2n^2-1}$

Estimate the area in each figure shown in Problems 41-46.

41.

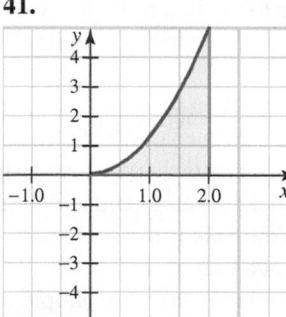

42.

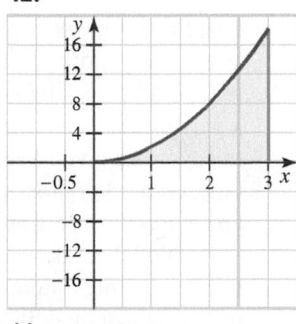

43.

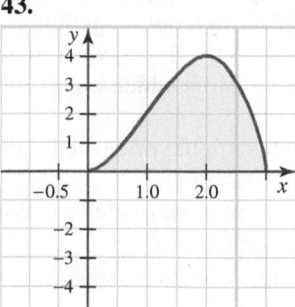

44.

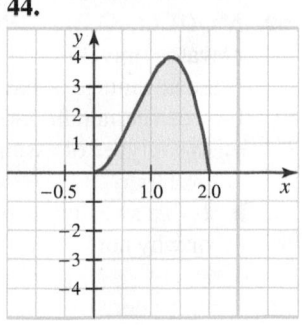

45.

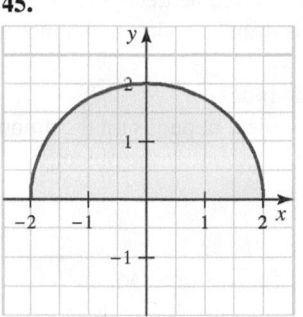

46.

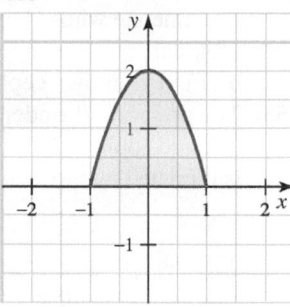

47. a. Calculate the sum of the areas of the rectangles shown in Figure 1.8**a**.
    b. Calculate the sum of the areas of the rectangles shown in Figure 1.8**b**.
    c. Make a guess about the shaded area under the curve.

Miscellaneous Models *Problems 48-54 deal with models taken from several different disciplines.*

48. The relationship between temperature measured in degrees Celsius and degrees Fahrenheit is

$$F = 1.8C + 32$$

    If the temperature is 40°C, what is the temperature measured in degrees Fahrenheit?

49. The relationship between temperature measured in degrees Celsius and degrees Fahrenheit is given in Problem 48. If the temperature is 70°F, what is the temperature (to the nearest degree) in degrees Celsius?

50. The cost of a multivitamin depends on the weight purchased according to the formula $C = 3w$, where C is the cost in dollars and w is the weight in grams. Give the cost for each of the following weights: $w = 2$; $w = 10$; $w = 30$.

51. In psychology, IQ is modeled by the formula

$$IQ = \frac{M}{C}(100)$$

    where M = mental age and C = chronological age.
    a. If a child has a chronological age of 15 and a mental age of 18, what is the child's IQ?
    b. If a 9-year-old has superior intelligence (IQ of 132), what is the child's mental age?

52. Anthropologists use the formula

$$c = \frac{w}{\ell}$$

    where w = width of the head and ℓ = length of the head to find the *cephalic index* of a person's head.

a. Find the cephalic index (to the nearest percent) for someone whose head is 4 inches wide and 6 inches long.

b. Find your own cephalic index.

53. The pressure, P, under an object depends on the amount of force applied, F, and the area under the object, A, and is modeled by the formula

$$P = \frac{F}{A}$$

If the blade on a knife is 2 in. long and is sharpened to a width of 0.001 in., find the pressure under the blade where a force of 4 lb is applied to the blade.

54. The force of gravity, F, is the force with which the earth, moon, or other massively large object attracts another object towards itself. In physics, it is shown that

$$F = mg$$

where m is the mass of the object (in kg) and g (on earth) is $g = 9.8$ m/s^2. What is the force of gravity for an object with a mass of 120 kg?

55. *Historical Quest* The notion of a mathematical model is not new. About 50 years ago, John Synge wrote an article for the Mathematical Education Notes section of the October 1961 issue of *The American Mathematical Monthly* (Vol. 68, p. 799). He described the modeling process as consisting of three stages:

1. A dive from the world of reality into the world of mathematics.

2. A swim in the world of mathematics.

3. A climb from the world of mathematics back to the world of reality, carrying a prediction in our teeth.

Relate these stages to the process of mathematical modeling described in this section.

56. *Historical Quest* Ptolemy (ca. 85 A.D.-165 A.D.) created a mathematical model based on a fixed earth, and imagined a physical realization of the universe as a set of nested spheres, in which he used the epicycles of his planetary model to compute the dimensions of the universe. He estimated the Sun was at an average distance of 1210 earth radii, while the radius of the sphere of the fixed stars was 20,000 times the radius of the earth. Over history, the movement of the earth and the stars has been modeled by a number of scientists. Discuss how modeling can be used to discuss the movement of planetary bodies.

57. MODELING EXPERIMENT Consider a collection of three bottles.

© 2013 Ronald Summers. Under license from Shutterstock, Inc.

a. For each bottle, sketch a graph predicting the water volume as it relates to the height of the bottle.

b. Use a graduated cylinder and fill each bottle 20 mL at a time. Measure the height of the water after each step. Plot these points on your graph.

c. Does it make sense to connect the data points for each separate graph? Why or why not?

d. Take another look at each bottle. Does your graphed prediction of volume/height data need any adjustments? Re-sketch your prediction, making any changes needed.

e. Is this an example of mathematical modeling? Why or why not?

58. MODELING EXPERIMENT Take out a piece of engineering paper ($8\frac{1}{2}$ in. by 11 in.). Cut the corners off so that you can fold up the sides to construct a box. The task is to construct a box of maximum volume. Bring your box to class and have an election to decide who has constructed the largest box. After the election, *calculate* the volumes of several of the best boxes. What is the relationship between the size of the square corners cut out of the paper to the box of largest volume? Is this an example of mathematical modeling? Why or why not?

59. MODELING EXPERIMENT You wish to construct a playground using a perimeter fence of 400 ft. To do this, you are given a strip of 400 attached tickets. Using tape, connect your strip of tickets end to end to define an enclosed fencing area. Your task is to use those tickets to model the perimeter of a playground that will provide the maximum play area. Is this an example of mathematical modeling? Why or why not?

60. MODELING EXPERIMENT In a building with elevators traveling to the building's floors, what (if any) is the advantage of having elevators which travel only between certain floors?

a. What types of assumptions might you make for this problem?

b. Is this an example of mathematical modeling? Why or why not?

1.2 PRELIMINARIES

> **IN THIS SECTION:** *Distance on a number line, absolute value, distance in the plane, trigonometry, solving trigonometric equations*
>
> We begin by reviewing one- and two-dimensional coordinate systems, absolute value, absolute value equations and inequalities, and trigonometric equations. These topics are needed for our work in calculus.

This section provides a quick review of some fundamental concepts and techniques from precalculus mathematics. If you have recently had a precalculus course, you may skip over this material.

In your previous mathematics courses, even those with prerequisites, the material you needed for that course was reviewed in that course. Calculus is a course that is different in that respect. Algebra, geometry, and trigonometry are important ingredients of calculus. Even though we will review many ideas from algebra, geometry, and trigonometry, we will not be able to develop every idea from these courses before we use them in calculus. For example, the law of cosines from trigonometry may be needed to solve a problem in a section that never mentions trigonometry in the exposition.

For this reason, it is important that you have ready access to algebra, geometry, and trigonometry textbooks. We suggest that you have these reference books close by as you read this book.

Distance on a Number Line

You are probably familiar with the set of **real numbers** and several of its subsets, including the counting or natural numbers, the integers, the rational numbers, and the irrational numbers. The real numbers can be visualized most easily by using a **one-dimensional coordinate system** called a **real number line**, as shown in Figure 1.10.

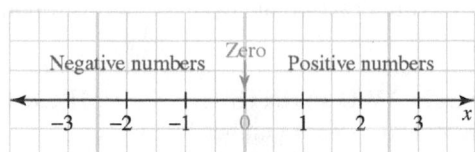

Figure 1.10   Real number line

Consider two numbers a and b on the real number line as shown in Figure 1.11. Notice that a number a is less than a number b if it is to the left of b on a real number line.

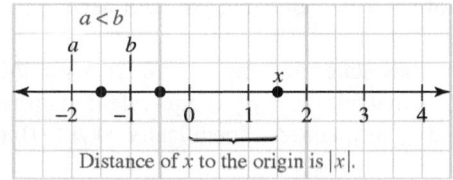

Figure 1.11   Geometric definition of *less than*

Similar definitions can be given for $a > b$, $a \le b$, and $a \ge b$.

The location of the number 0 is chosen arbitrarily, and a unit distance is picked (meters, feet, inches, ...). Numbers are ordered on the real number line according to the following order properties.

ORDER PROPERTIES   For all real numbers $a, b, c,$ and $d,$

Trichotomy law: Exactly one of the following is true:

$$a < b, \ a > b, \ \text{or} \ a = b$$

Transitive law of inequality: If $a < b$ and $b < c$, then $a < c$.

Additive law of inequality: If $a < c$ and $b < d$, then $a + b < c + d$.

Multiplicative law of inequality: If $a < b$, then

$$ac < bc \quad \text{if } c > 0 \qquad \text{and} \qquad ac > bc \quad \text{if } c < 0$$

Absolute Value

The distance of a point (a) from the origin is a if a is positive, $-a$ if a is negative, and, of course, is 0 if a is located at the origin. We summarize this with the following definition.

ABSOLUTE VALUE   The **absolute value** of a real number x denoted by $|x|$, is

$$|x| = \begin{cases} x, & \text{if } x \geq 0 \\ -x, & \text{if } x < 0 \end{cases}$$

☠ *$|a|$ is NOT the number a without its sign.* ☠

The number x is located $|x|$ units away from 0—to the right if $x > 0$ and to the left if $x < 0$.

Absolute value is used to describe the distance between points on a number line.

DISTANCE BETWEEN TWO POINTS ON A NUMBER LINE   The **distance** between the numbers x_1 and x_2 on a number line is

$$|x_2 - x_1|$$

☠ *Note that $|x_2 - x_1|$* *$= |x_1 - x_2|$.* ☠

For example, the distance between 2 and -3 is $|2 - (-3)| = 5$ units, and between -2 and -3 is $|-2 - (-3)| = 1$ unit.

Several properties of absolute value that you will need in this course are summarized in Table 1.1.

Property 7 is sometimes stated for any a or b as $|a| = |b|$ if and only if $a = \pm b$. Since $|b| = \pm b$, it follows that this property is equivalent to property 7. Also, properties 8 and 9 are true for $\leq$ and $\geq$ inequalities. Specifically, if $b \geq 0$, then

$$|a| \leq b \qquad \text{if and only if} \qquad -b \leq a \leq b$$

and

$$|a| \geq b \qquad \text{if and only if} \qquad a \geq b \text{ or } a \leq -b$$

A convenient notation for representing intervals on a number line is called **interval notation** and is summarized in the accompanying table. Note that a solid dot (•) at an

Table 1.1 Properties of Absolute Value

Let a and b be any real number	
Property	**Comment**
1. $\|a\| \geq 0$	**1.** Absolute value is nonnegative.
2. $\|-a\| = \|a\|$	**2.** The absolute value of a number and the absolute value of its opposite are equal.
3. $\|a\|^2 = a^2$	**3.** If an absolute value is squared, the absolute value signs can be dropped because both squares are nonnegative.
4. $\|ab\| = \|a\|\,\|b\|$	**4.** The absolute value of a product is the product of the absolute values.
5. $\left\|\dfrac{a}{b}\right\| = \dfrac{\|a\|}{\|b\|}, b \neq 0$	**5.** The absolute value of a quotient is the quotient of the absolute values.
6. $-\|a\| \leq a \leq \|a\|$	**6.** This property is true because $\|a\|$ is either a or $-a$.
7. Let $b \geq 0$; $\quad \|a\| = b$ if and only if $a = \pm b$	**7.** This property is useful in solving absolute value equations. See Example 2.
8. Let $b > 0$; $\quad \|a\| < b$ if and only if $-b < a < b$	**8/9.** These are the main properties used in solving absolute value inequalities. See Example 3.
9. Let $b > 0$; $\quad \|a\| > b$ if and only if $a > b$ or $a < -b$	
10. $\|a + b\| \leq \|a\| + \|b\|$	**10.** This property is called the **triangle inequality**. It is used in both theoretical and numerical computations involving inequalities.

☻ *"p if and only if q" is used to mean that both a statement and its converse are true. That is: If p, then q, and if q, then p. For example, property 8 has two parts: (i) If $\|a\| < b$, then $-b < a < b$; and (ii) If $-b < a < b$, then $\|a\| < b$.* ☻

endpoint of an interval indicates that it is included in the interval, whereas an open dot (∘) indicates that it is excluded. An interval is **bounded** if both of its endpoints are real numbers. A bounded interval is **open** if it includes neither of its endpoints, **half-open** if it includes only one endpoint, and **closed** if it includes both endpoints. The symbol "∞" (pronounced *infinity*) is used for intervals that are not limited in one direction or another. In particular, $(-\infty, \infty)$ denotes the entire real number line.

We summarize the various notations in Table 1.2.

Table 1.2 Interval notation

Name of Interval	Inequality notation	Interval notation	Graph
Closed interval	$a \leq x \leq b$	$[a, b]$	
	$a \leq x$	$[a, \infty)$	
	$x \leq b$	$(-\infty, b]$	
Open interval	$a < x < b$	(a, b)	
	$a < x$	(a, ∞)	
	$x < b$	$(-\infty, b)$	
Half-open interval	$a < x \leq b$	$(a, b]$	
	$a \leq x < b$	$[a, b)$	
Real number line	All real numbers, $\mathbb{R}$	$(-\infty, \infty)$	

We will use this interval notation to write the solutions of absolute value equations and absolute value inequality problems.

Absolute Value Equations

Absolute value property 7 allows us to solve absolute value equations easily. For this reason property 7 is called the **absolute value equation property**.

Example 1   Solving an equation with an absolute value on one side

Solve $|2x - 6| = x$.

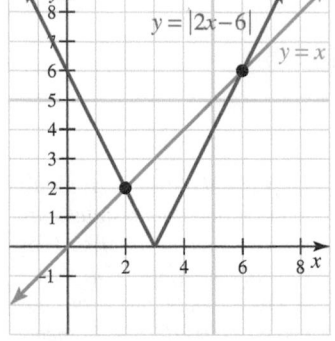

Figure 1.12 Interactive
Geometric solution

Algebraic Solution   If $2x - 6 \geq 0$, then $|2x - 6| = 2x - 6$ so that we solve

$$2x - 6 = x \text{ or } x = 6 \quad \textit{Property 7}$$

If $2x - 6 < 0$, then $|2x - 6| = -(2x - 6)$ so that we solve

$$-(2x - 6) = x$$
$$-3x = -6$$
$$x = 2$$

The solutions are $x = 6$ and $x = 2$.

Geometric Solution   Let $y_1 = |2x - 6|$ and $y_2 = x$ and graph each equation on the same coordinate system, as shown in Figure 1.12. By inspection, we see the graphs intersect at $x = 6$ and $x = 2$.

Example 2   Solving an equation with an absolute value on both sides

Solve $|x + 8| = |3x - 4|$.

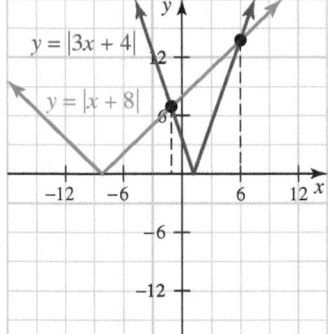

Figure 1.13 Interactive
Geometric solution

Algebraic Solution   We use absolute value property 7.

$$
\begin{array}{ll}
x + 8 = 3x - 4 & x + 8 = -(3x - 4) \\
-2x = -12 & 4x = -4 \\
x = 6 & x = -1
\end{array}
$$

Geometric Solution   Let $y_1 = |x + 8|$ and $y_2 = |3x - 4|$ and graph each equation on the same coordinate system, as shown in Figure 1.13. By inspection, we see the graphs intersect at $x = -1$ and $x = 6$.

The absolute value expression $|x - a|$ can be interpreted as the distance between x and a on a number line. An equation of the form

$$|x - a| = b$$

is satisfied by two values of x that are a given distance b from a when represented on a number line. For example, $|x - 5| = 3$ states that x is 3 units from 5 on a number line. Thus, x is either 2 or 8.

Geometric representation

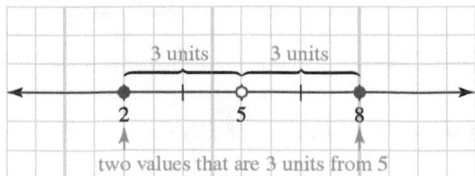

Algebraic representation

$$|x - 5| = 3$$
$$x - 5 = \pm 3$$
$$x = 5 \pm 3$$
$$x = 8 \text{ or } 2$$

Absolute Value Inequalities

Because $|x - 5| = 3$ states that the distance from x to 5 is 3 units, the inequality $|x - 5| < 3$ states that the distance from x to 5 is less than 3 units.

Unknown Distance

$$|\overset{\downarrow}{x} - \overset{\downarrow}{5}| < 3$$

$\uparrow$

Midpoint

On the other hand, $|x - 5| > 3$ states that x is any number greater than 3 units from 5.

Unknown Distance

$$|\overset{\downarrow}{x} - \overset{\downarrow}{5}| > 3$$

$\uparrow$

Midpoint

This number-line solution is a one-dimensional interpretation of an absolute value inequality. Note also that the first inequality gives an interval centered at the midpoint 5, whereas the second inequality gives two infinite intervals, each beginning 3 units from 5. For a two-dimensional interpretation, graph $y_1 = |x - 5|$ and $y_2 = 3$, and then look at the x-values for which $y_1 = y_2$, $y_1 < y_2$, or $y_1 > y_2$, as shown in Figure 1.14.

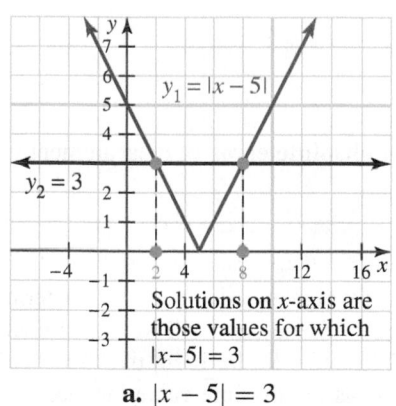
a. $|x - 5| = 3$

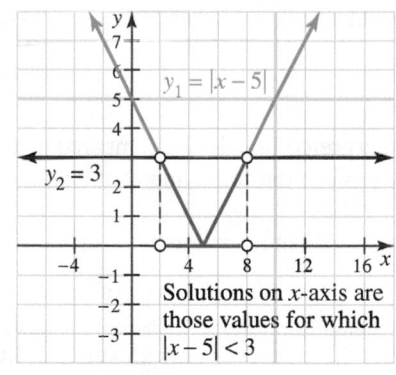
b. Interactive $|x - 5| < 3$

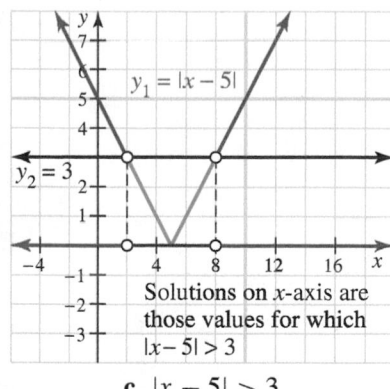
c. $|x - 5| > 3$

Figure 1.14 Two-dimensional graphs for absolute value inequalities

Example 3 Solving an absolute value inequality

Solve $|2x - 3| \le 4$.

Algebraic Solution

$$-4 \le \ \ 2x - 3 \ \ \le 4 \qquad \textit{Property 8}$$

$$-4 + 3 \le 2x - 3 + 3 \le 4 + 3$$

$$-1 \le \ \ \ \ 2x \ \ \ \ \le 7$$

$$-\frac{1}{2} \le \ \ \ \ x \ \ \ \ \le \frac{7}{2}$$

The solution is the interval $\left[-\frac{1}{2}, \frac{7}{2}\right]$.

Geometric Solution   Graph $y_1 = |2x - 3|$ and $y_2 = 4$, as shown in Figure 1.15.

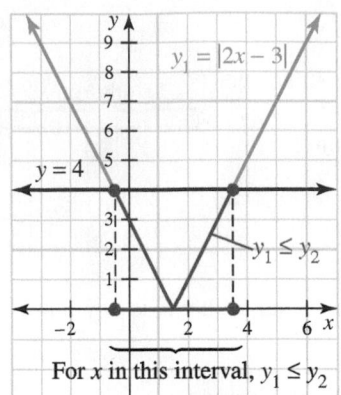

Figure 1.15 Interactive Solving an inequality by graphing

Because we are looking for $|2x - 3| \leq 4$, we note those x-values on the real number line for which the graph of y_1 is below the graph of y_2. We see that the interval is $[-0.5, 3.5]$ or $\left[-\frac{1}{2}, \frac{7}{2}\right]$.

When absolute value is applied to measurement, it is called **tolerance**. Tolerance is an allowable deviation from a standard. For example, a cement bag whose weight w lb is "90 lb plus or minus 2 lb" might be described as having a weight given by

$$|w - 90| \leq 2.$$

When considered as a tolerance, the expression

$$|x - a| \leq b$$

may be interpreted as x being compared to a with an **absolute error** of measurement of b units. Consider the following example.

Example 4   Absolute value as a tolerance

Suppose you purchase a 90-lb bag of cement. It will not weigh exactly 90 lb. The material must be measured, and the measurement is approximate. Some bags will weigh as much as 2 lb over 90 lb, and some will weigh as little as 2 lb under 90 lb. If so, the bag could weigh as much as 92 lb or as little as 88 lb. State this as an absolute value inequality.

Solution   Let $w = $ WEIGHT OF THE BAG (in pounds). Then

$$90 - 2 \leq \quad w \quad \leq 90 + 2$$
$$-2 \leq w - 90 \leq 2$$

Equivalently, $|w - 90| \leq 2$.

Distance in the Plane

Absolute value is used to find the distance between two points on a number line. To find the distance between two points in a coordinate plane, we use the *distance formula*, which is derived by using the Pythagorean theorem.

Theorem 1.1   Distance between two points in the plane

The distance d between the points $P_1(x_1, y_1)$ and $P_2(x_2, y_2)$ in the plane is given by

$$d = \sqrt{(\Delta x)^2 + (\Delta y)^2} = \sqrt{(x_2 - x_1)^2 + (y_2 - y_1)^2}$$

where Δx (read "delta x") is the **horizontal change** $x_2 - x_1$ (sometimes called **run**) and Δy (read "delta y") is the **vertical change** (sometimes called the **rise**).

Proof: Using the two points, form a right triangle by drawing lines through the given points parallel to the coordinate axes, as shown in Figure 1.16. The length of the horizontal side of the triangle is $|x_2 - x_1| = |\Delta x|$, and the length of the vertical side is $|y_2 - y_1| = |\Delta y|$. Then

$$d^2 = |\Delta x|^2 + |\Delta y|^2 \quad \textit{Pythagorean theorem}$$
$$d^2 = (\Delta x)^2 + (\Delta y)^2 \quad \textit{Absolute value property 3}$$
$$d = \sqrt{(\Delta x)^2 + (\Delta y)^2} \quad \textit{Solve for d}$$

◆

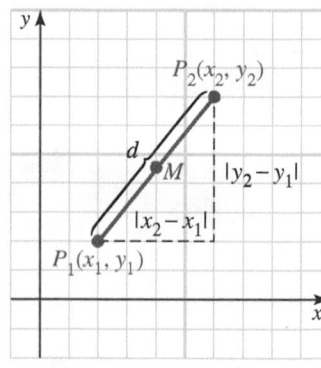

Figure 1.16   Form a triangle to derive the distance formula

Midpoint Formula

Related to the formula for the distance between two points is the formula for finding the midpoint of a line segment, as shown in Figure 1.16.

MIDPOINT FORMULA   The **midpoint**, M, of the segment with endpoints $P_1(x_1, y_2)$ and $P_2(x_2, y_2)$ has coordinates

$$M\left(\frac{x_1 + x_2}{2}, \frac{y_1 + y_2}{2} \right)$$

Notice that the coordinates of the midpoint of a segment are found by averaging the first and second components of the coordinates of the endpoints, respectively. You are asked to derive this formula in Problem 60.

Relationship Between an Equation and a Graph

Analytic geometry is that branch of geometry that ties together the geometric concept of position with an algebraic representation—namely, coordinates. For example, you remember from algebra that a line can be represented by an equation. Precisely what does this mean? Can we make such a statement that is true for any curve, not just for lines? We answer in the affirmative with the following definition.

GRAPH OF AN EQUATION   The **graph of an equation** in two variables x and y is the collection of all points $P(x, y)$ whose coordinates (x, y) satisfy the equation.

There are two frequently asked questions in analytic geometry.

- Given a graph (a geometrical representation), find the corresponding equation.
- Given an equation (an algebraic representation), find the corresponding graph.

In Example 5, we use the distance formula to derive the equation of a circle. This means that if x and y are numbers that satisfy the equation, then the point (x, y) must lie on the circle. Conversely, the coordinates of any point on the circle will satisfy the equation.

Example 5   Using the distance formula to derive an equation of a graph

Find the equation of a circle with center (h, k) and radius r.

Solution   Let (x, y) be any point on a circle. Recall that a circle is the set of all points in the plane a given distance from a given point. The given point (h, k) is the center and the given distance is the radius r.

$$r = \text{DISTANCE FROM } (h, k) \text{ TO } (x, y)$$
$$r = \sqrt{(x - h)^2 + (y - k)^2}$$
$$r^2 = (x - h)^2 + (y - k)^2$$

A **unit circle** is a circle of radius 1 and center at the origin, so its equation is $x^2 + y^2 = 1$.

Example 6   Finding the equation of a circle

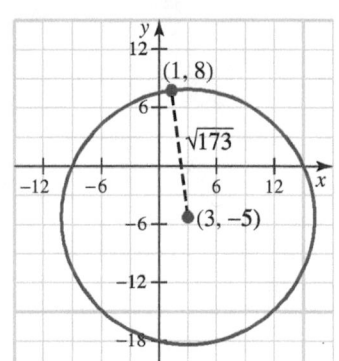

Figure 1.17 Graph of given circle

Find the equation of the circle with center $(3, -5)$ that passes through the point $(1, 8)$.

Solution   See Figure 1.17. The radius is the distance from the center to the given point:

$$r = \sqrt{(1 - 3)^2 + [8 - (-5)]^2}$$
$$= \sqrt{4 + 169}$$
$$= \sqrt{173}$$

Thus, the equation of the desired circle is

$$(x - 3)^2 + (y + 5)^2 = 173$$

Example 7   Graphing a circle given its equation

Sketch the graph of the circle whose equation is $4x^2 + 4y^2 - 4x + 8y - 5 = 0$.

Solution   We need to convert this equation into standard form. To do this we use a process called **completing the square**.

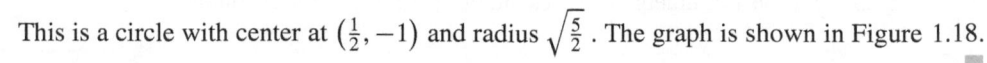

$$4x^2 + 4y^2 - 4x + 8y - 5 = 0 \qquad \textit{Given equation}$$

$$x^2 + y^2 - x + 2y = \frac{5}{4} \qquad \textit{Coefficients of squared terms should be 1.}$$

$$\left[x^2 - x + \left(-\frac{1}{2}\right)^2\right] + \left[y^2 + 2y + 1^2\right] = \frac{5}{4} + \frac{1}{4} + 1 \qquad \textit{Complete the squares by adding } \tfrac{1}{4} \textit{ and } 1 \textit{ to both sides.}$$

$$\left(x - \frac{1}{2}\right)^2 + (y + 1)^2 = \frac{10}{4} = \frac{5}{2} \qquad \textit{Factor and simplify.}$$

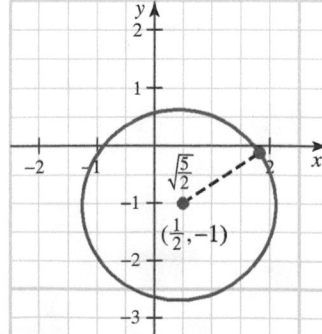

Figure 1.18 Sketch of circle

This is a circle with center at $\left(\frac{1}{2}, -1\right)$ and radius $\sqrt{\frac{5}{2}}$. The graph is shown in Figure 1.18.

Trigonometry

One of the prerequisites for this book is trigonometry. A summary of useful trigonometric formulas is found in Appendix E.

Angles are commonly measured in degrees and radians. A **degree** is defined to be $\frac{1}{360}$ revolution and a **radian** $\frac{1}{2\pi}$ revolution. Thus, to convert between degree and radian measure use the following formula:

$$\frac{\theta \text{ measured in degrees}}{360} = \frac{\theta \text{ measured in radians}}{2\pi}$$

In calculus, angles are usually measured in terms of radians rather than degrees since radian measure often leads to simpler formulas. You may assume that when we write expressions such as $\sin x$, $\cos x$, and $\tan x$, the angle x is in radians unless otherwise specified by a degree symbol.

Example 8   Converting degree measure to radian measure

Convert $255°$ to radian measure.

Solution

$$\frac{255}{360} = \frac{\theta}{2\pi}$$

$$\theta = \left(\frac{\pi}{180}\right)(255)$$

$$\approx 4.450589593$$

Example 9   Converting radian measure to degree measure

Express 1 radian in terms of degrees.

Solution

$$\frac{\theta}{360} = \frac{1}{2\pi}$$

$$\theta = \left(\frac{180}{\pi}\right)(1)$$

$$\approx 57.29577951°$$

Solving Trigonometric Equations

There will be many times in calculus when you will need to solve a trigonometric equation. As you may remember from trigonometry, solving a trigonometric equation is equivalent to evaluating an inverse trigonometric relation. Inverse functions will be introduced in Section 1.5; for now, we will solve trigonometric equations whose solutions involve the values in Table 1.3 (called a **table of exact values**).

Table 1.3 Exact Trigonometric Equations

Angle θ	0	$\frac{\pi}{6}$	$\frac{\pi}{4}$	$\frac{\pi}{3}$	$\frac{\pi}{2}$	π	$\frac{3\pi}{2}$
$\cos\theta$	1	$\frac{\sqrt{3}}{2}$	$\frac{\sqrt{2}}{2}$	$\frac{1}{2}$	0	-1	0
$\sin\theta$	0	$\frac{1}{2}$	$\frac{\sqrt{2}}{2}$	$\frac{\sqrt{3}}{2}$	1	0	-1
$\tan\theta$	0	$\frac{\sqrt{3}}{3}$	1	$\sqrt{3}$	undefined	0	undefined
$\sec\theta$	1	$\frac{2\sqrt{3}}{3}$	$\sqrt{2}$	2	undefined	-1	undefined
$\csc\theta$	undefined	2	$\sqrt{2}$	$\frac{2\sqrt{3}}{3}$	1	undefined	-1
$\cot\theta$	undefined	$\sqrt{3}$	1	$\frac{\sqrt{3}}{3}$	0	undefined	0

It is customary to use the values from Table 1.3 whenever possible. The approximate calculator values will be given only when necessary.

Example 10   Evaluating trigonometric functions

Evaluate $\cos \frac{\pi}{3}$; $\sin \frac{5\pi}{6}$; $\tan(\frac{-5\pi}{4})$; $\sec 1.2$; $\csc(-4.5)$; and $\cot 180°$.

Solution   If you use a calculator, make certain it is in the proper mode (radian or degree).

$$\cos \frac{\pi}{3} = \frac{1}{2} \qquad\qquad \textit{Exact value; quadrant I}$$

$$\sin \frac{5\pi}{6} = \frac{1}{2} \qquad\qquad \textit{Exact value; quadrant II}$$

$$\tan \left(\frac{-5\pi}{4} \right) = -1 \qquad\qquad \textit{Exact value; quadrant II}$$

$$\sec 1.2 \approx 2.759703601 \quad \textit{Approximate calculator value}$$

$$\csc (-4.5) \approx 1.022986384 \quad \textit{Approximate calculator value}$$

$$\cot 180° \text{ is undefined}$$

Example 11   Solving a trigonometric equation by factoring

Solve $2 \cos \theta \sin \theta = \sin \theta$ on $[0, 2\pi)$.

Solution

$$2 \cos \theta \sin \theta = \sin \theta$$
$$2 \cos \theta \sin \theta - \sin \theta = 0$$
$$\sin \theta (2 \cos \theta - 1) = 0$$

$$\sin \theta = 0 \qquad 2 \cos \theta - 1 = 0$$

Do not divide both sides by $\sin \theta$ because if $\theta = 0$ or π, then $\sin \theta = 0$ and you cannot divide by 0.

$$\theta = 0, \pi \qquad \cos \theta = \frac{1}{2}$$
$$\theta = \frac{\pi}{3}, \frac{5\pi}{3}$$

Example 12   Solving a trigonometric equation

Solve $\sin x + \sqrt{3} \cos x = 1$ on $[0, 2\pi)$.

Algebraic Solution

$$\sin x + \sqrt{3} \cos x = 1$$
$$\sqrt{3} \cos x = 1 - \sin x$$
$$3 \cos^2 x = 1 - 2 \sin x + \sin^2 x \qquad\qquad \textit{Square both sides.}$$
$$3(1 - \sin^2 x) = 1 - 2 \sin x + \sin^2 x$$
$$2 + 2 \sin x - 4 \sin^2 x = 0$$
$$2 \sin^2 x - \sin x - 1 = 0$$
$$(2 \sin x + 1)(\sin x - 1) = 0$$

$$2 \sin x + 1 = 0 \qquad\qquad\qquad \sin x - 1 = 0$$
$$\sin x = -\frac{1}{2} \qquad\qquad\qquad \sin x = 1$$
$$x = \frac{7\pi}{6}, \frac{11\pi}{6} \qquad\qquad\qquad x = \frac{\pi}{2}$$

However, since we squared both sides, we need to check for extraneous solutions by substituting into the *original equation*, since squaring sometimes introduces extraneous solutions (that is, solutions that do not satisfy the given equation). Checking, we see that $x = \frac{7\pi}{6}$ is extraneous, and the solution is $x = \frac{\pi}{2}, \frac{11\pi}{6}$.

Geometric Solution   It is often worthwhile to check by finding a geometric solution. We graph the left and right sides of the equation on the same axes: $y_1 = \sin x + \sqrt{3}\cos x$, $y_2 = 1$. The graphs are shown in Figure 1.19, and for a solution we look to the intersection points. Notice the closed dots on each curve for the endpoint $x = 0$ (which is included on each graph) and the open dots on each curve for the endpoint $x = 2\pi$ (which is not included on each graph). The intersection points are $x = \frac{\pi}{2}$ and $x = \frac{11\pi}{6}$.

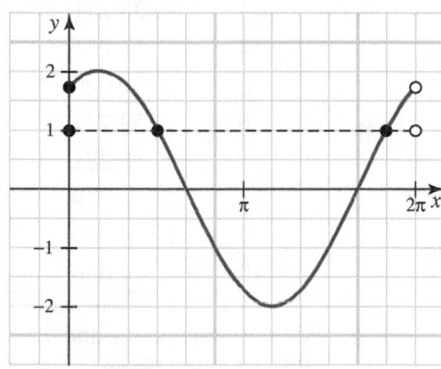

Figure 1.19   There are two intersection points

PROBLEM SET 1.2

Level 1

1. Fill in the missing parts in the table.

Inequality Notation	Interval Notation
$-3 \le x \le 4$	**a.**
b.	$[3, 5]$
c.	$[-2, 1)$
$2 < x \le 7$	**d.**

2. Fill in the missing parts in the table.

Inequality Notation	Interval Notation
a.	$(-\infty, -2)$
b.	$[\frac{\pi}{4}, \sqrt{2}]$
$x > -3$	**c.**
$-1 < x \le 5$	**d.**

3. Represent each of the following on a number line.
   a. $-1 \le x < 4$ **b.** $-1 \le x < 3$
   c. $(0, 2)$ **d.** $[-2, 1)$

4. Represent each of the following on a number line.
   a. $x \ge 2$ **b.** $x \le 4$
   c. $(-\infty, 2) \cup (2, \infty)$ **d.** $-2 < x \le 3$ or $x \ge 5$

In Problems 5-6, plot the given points P and Q on a Cartesian plane, find the distance between them, and find the coordinates of the midpoint of the line segment $\overline{PQ}$.

5. a. $P(2, 3)$, $Q(-2, 5)$ **b.** $P(-2, 3)$, $Q(4, 1)$
6. a. $P(-5, 3)$, $Q(-5, -7)$ **b.** $P(-4, 3)$, $Q(3, -4)$

Solve each equation in Problems 7-18. Assume that a, b, and c are known constants.

7. $x^2 - x = 0$ **8.** $2y^2 + y - 3 = 0$
9. $3x^2 - bx = c$ **10.** $|2x + 4| = 16$
11. $|3 - 2w| = 7$ **12.** $|5 - 3t| = 14$
13. $|2x + a| = -4$ **14.** $|1 - 5x| = -2$
15. $\sin x = -\frac{1}{2}$ on $[0, 2\pi)$
16. $(\sin x)(\cos x) = 0$ on $[0, 2\pi)$
17. $(2\cos x + \sqrt{2})(2\cos x - 1) = 0$ on $[0, 2\pi)$
18. $(3\tan x + \sqrt{3})(3\tan x - \sqrt{3}) = 0$ on $[0, 2\pi)$

Solve each inequality in Problems 19-28, and give your answer using interval notation.

19. $3x + 7 < 2$ **20.** $5(3 - x) > 3x - 1$
21. $-5 < 3x < 0$ **22.** $-3 < y - 5 \le 2$
23. $3 \le -y < 8$ **24.** $-5 \le 3 - 2x < 18$
25. $t^2 - 2t \le 3$ **26.** $s^2 + 3s - 4 > 0$
27. $|x - 8| \le 0.001$ **28.** $|x - 5| < 0.01$

In Problems 29-32, find an equation of the circle with given center C and radius r.

29. $C(-1, 2)$; $r = 3$ **30.** $C(3, 0)$; $r = 2$
31. $C(0, 1.5)$; $r = 0.25$ **32.** $C(-1, -5)$; $r = 4.1$

In Problems 33-36, find the centers and radii of the circles, and then graph.

33. $x^2 - 2x + y^2 + 2y + 1 = 0$
34. $4x^2 + 4y^2 + 4y - 15 = 0$
35. $x^2 + y^2 + 2x - 10y + 25 = 0$
36. $2x^2 + 2y^2 + 2x - 6y - 9 = 0$

Use the sum and difference formulas from trigonometry to find the exact values of the expressions in Problems 37-40. Check by finding a calculator approximation.

37. $\sin\left(-\frac{\pi}{12}\right)$ **38.** $\cos\frac{7\pi}{12}$
39. $\tan\frac{\pi}{12}$ **40.** $\sin 165°$

Level 2

41. ■ What does this say?* Describe a process for solving a quadratic equation.
42. ■ What does this say? Describe a process for solving absolute value equations.
43. ■ What does this say? Describe a process for solving absolute value inequalities.
44. ■ What does this say? Describe a process for solving trigonometric equations.

Specify the period for each graph in Problems 45-48. Also graph each curve.

45. a. $y = \sin x$ **b.** $y = \cos x$ **c.** $y = \tan x$
46. a. $y = 2\sin 2\pi x$ **b.** $y = 3\cos 3\pi x$
c. $y = 4\tan\left(\frac{\pi x}{5}\right)$

47. a. $y = \tan\left(2x - \frac{\pi}{2}\right)$ **b.** $y = 2\cos(3x + 2\pi) - 2$

48. a. $y = 4\sin\left(\frac{1}{2}x + 2\right) - 1$ **b.** $y = \tan\left(\frac{1}{2}x + \frac{\pi}{3}\right)$

49. Suppose a point P on a waterwheel with a 30-ft radius is d units from the water as shown in Figure 1.20.

Figure 1.20   Waterwheel

If it turns at 6 revolutions per minute, the height of the point P above the water is given by the following set of data points:

Time	Height	Time	Height
0	−1.000	11	4.729
1	4.729	12	19.729
2	19.729	13	38.271
3	38.271	14	53.271
4	53.271	15	59.000
5	59.000	16	53.271
6	53.271	17	38.271
7	38.271	18	19.729
8	19.729	19	4.729
9	4.729	20	−1.000
10	−1.000		

Plot the data points and draw a smooth curve passing through these points. Determine possible values of A, B, C, and D so that the graph of these data is approximated by the equation

$$y - A = B\cos\left[C(x - D)\right]$$

50. The sun and moon tide curves are shown here.[†] During a new moon, the sun and moon tidal bulges are centered at the same longitude, so their effects are added to produce maximum high tides and minimum low tides. This produces maximum tidal range (the distance between the low and high tides).

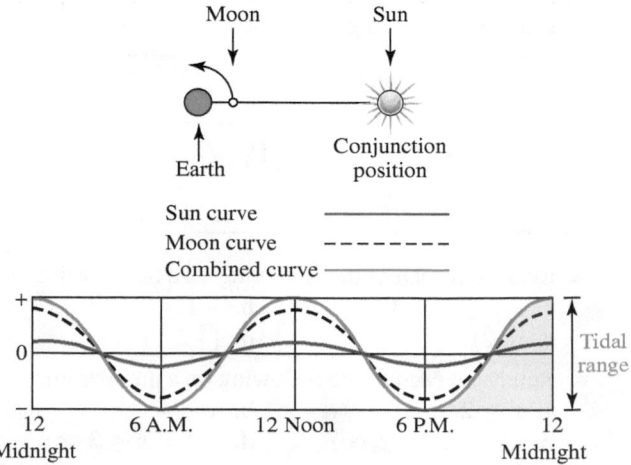

*Many problems in this book are labeled **What Does this Say**? Following this question will be a question for you to answer in your own words, or a statement for you to rephrase in your own words. These problems are intended to be similar to the "What This Says" boxes that appear throughout the book.

[†]From *Introductory Oceanography*, 5th ed., by H. V. Thurman, p. 253.

Write possible equations for the sun curve, the moon curve, and the combined curve. Assume the tidal range is 10 ft and the period is 12 hours.

51. **Journal problem:** The *Mathematics Student Journal* by Murray Klamkin.* Determine all the roots of the quartic equation $x^4 - 4x = 1$.

52. *Historical Quest* The division of one revolution into 360 equal parts (called degrees) is no doubt due to the sexagesimal (base 60) numeration system used by the Babylonians. Several explanations have been put forward to account for the choice of this number. (For example, see Howard Eves's **In Mathematical Circles**, Boston: Prindle, Weber & Schmidt, 1969.) One possible explanation is put forth by Otto Neugebauer, scholar and authority on early Babylonian mathematics and astronomy. In early Sumerian times, there existed a Babylonian mile, equal to about seven of our miles. Sometime in the first millennium B.C., when Babylonian astronomy reached the stage in which systematic records of the stars were kept, the Babylonian mile was adapted for measuring spans of time. Since a complete day was found to be equal to 12 time-miles, and one complete day is equivalent to one revolution of the sky, a complete circuit was divided into 12 equal parts. Then, for convenience, the Babylonian mile was subdivided into 30 equal parts. One complete circuit therefore has $(12)(30) = 360$ equal parts. Show that if a regular hexagon is inscribed in a circle, the length of each side is equal to the radius.

53. *Historical Quest* The numeration system we use (base 10) evolved over a long period of time. It is often called the **Hindu-Arabic** system because its origins can be traced back to the Hindus in Bactria (now Afghanistan). Later, in A.D. 700, India was invaded by the Arabs who used and modified the Hindu numeration system, and, in turn, introduced it to Western civilization. The Hindu Brahmagupta stated the rules for operations with positive and negative numbers in the 7th century A.D. There are some indications that the Chinese had some knowledge of negative numbers as early as 200 B.C. On the other hand, the Western mathematician Girolamo Cardano (1501-1576) was calling numbers such as (-1) absurd, as late as 1545. Write a paper on the history of the real number system.

Level 3

54. If $c \geq 0$, show that $|x| \leq c$ if and only if $-c \leq x \leq c$.
55. Show that $-|x| \leq x \leq |x|$ for any number x.
56. Prove that $|a| = |b|$ if and only if $a = b$ or $a = -b$.
57. Prove that if $|a| < b$ and $b > 0$, then $-b < a < b$.
58. Prove the triangle inequality:

$$|x + y| \leq |x| + |y|$$

59. Show that $||x| - |y|| \leq |x - y|$ for all x and y.
60. Derive the *midpoint formula*

$$M\left(\frac{x_1 + x_2}{2}, \frac{y_1 + y_2}{2}\right)$$

for the midpoint of a segment with endpoints $P(x_1, y_1)$ and $Q(x_2, y_2)$.

1.3 LINES IN THE PLANE; PARAMETRIC EQUATIONS

IN THIS SECTION: *Slope of a line, forms for the equation of a line, parallel and perpendicular lines*
A line is one of the simplest geometric concepts, yet one of the most useful from the standpoint of calculus. In this section, we review some of the basic facts about lines in a plane.

Slope of a Line

A distinguishing feature of a line is the fact that its *inclination* with respect to the horizontal is constant. It is common practice to specify inclination by means of a concept called *slope*. A carpenter might describe a roof line that rises 1 ft for every 3 ft of horizontal "run" as having a slope or pitch of 1 to 3.

Let Δx and Δy represent, as before, the amount of change in the variables x and y, respectively. Then a nonvertical line L that rises (or falls) Δy units (measured from

---

*Most mathematics journals have problem sections that solicit interesting problems and solutions for publication. From time to time, we will reprint a problem from a mathematics journal. If you have difficulty solving a journal problem, you may wish to use a library to find the problem and solution as printed in the journal. The title of the journal is included as part of the problem, and we will generally give you a reference as a footnote. This problem is found in the named journal in Volume 28 (1980), issue 3, p. 2.

bottom to top) for every Δx units of run (measured from left to right) is said to have a *slope* of $m = \Delta y / \Delta x$. (If Δy is negative, then the "rise" is actually a fall; and if Δx is negative, then the run is actually right to left.) In particular, if $P(x_1, y_1)$ and $Q(x_2, y_2)$ are two distinct points on L, then the changes in the variables x and y are given by $\Delta x = x_2 - x_1$ and $\Delta y = y_2 - y_1$, and the slope of L is

$$m = \frac{\Delta y}{\Delta x} \quad \text{for } \Delta x \neq 0 \quad \text{See Figure 1.21.}$$
$$= \frac{y_2 - y_1}{x_2 - x_1}$$

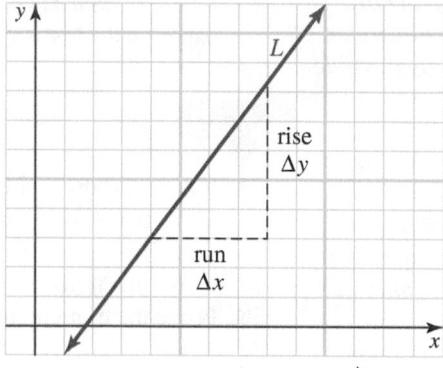

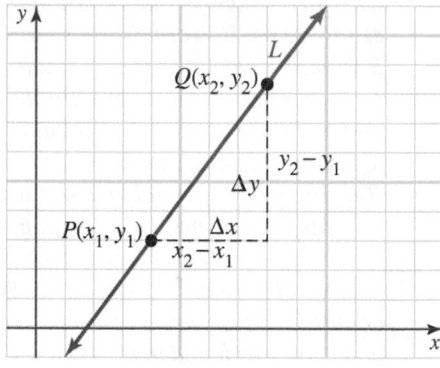

a. The slope of L is $m = \dfrac{\Delta y}{\Delta x}$

b. The slope is given by $m = \dfrac{y_2 - y_1}{x_2 - x_1}$

Figure 1.21   The slope of a line

SLOPE OF A LINE   A nonvertical line that contains the points $P(x_1, y_1)$ and $Q(x_2, y_2)$ has **slope**

$$m = \frac{\Delta y}{\Delta x} = \frac{y_2 - y_1}{x_2 - x_1}$$

We say that a line with slope m is *rising* (when viewed from left to right) if $m > 0$, *falling* if $m < 0$, and *horizontal* if $m = 0$.

There is a useful trigonometric formulation of slope. The *angle of inclination* of a line L is defined to be the nonnegative angle ϕ ($0 \leq \phi < \pi$) formed between L and the positive x-axis.

ANGLE OF INCLINATION   The **angle of inclination** of a line L is the angle ϕ ($0 \leq \phi < \pi$) between L and the positive x-axis. Then the *slope* of L with inclination ϕ is

$$m = \tan \phi$$

---

*Note that saying a line has no slope (vertical line) is not the same as saying it has zero slope (horizontal line). Sometimes we say a vertical line has **infinite slope**.*

We see that the line L is *rising* if $0 < \phi < \frac{\pi}{2}$ and is *falling* if $\frac{\pi}{2} < \phi < \pi$. The line is *horizontal* if $\phi = 0$ and is *vertical* if $\phi = \frac{\pi}{2}$. Notice that if $\phi = \frac{\pi}{2}$, $\tan \phi$ is not defined; therefore the slope, m, is undefined for a vertical line. A **vertical line** is said to have **no slope**.

To derive the trigonometric representation for slope we need to find the slope of a nonvertical line through $P(x_1, y_1)$ and $Q(x_2, y_2)$, where ϕ is the angle of inclination. From the definition of the tangent, we have

$$\tan \phi = \frac{y_2 - y_1}{x_2 - x_1}$$ *See Figure* 1.22.

$$= \frac{\Delta y}{\Delta x}$$

$$= m$$

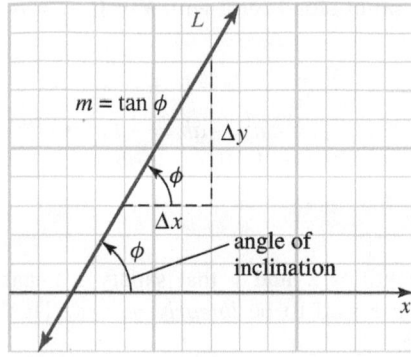

Figure 1.22 Trigonometric form of slope

Lines with various slopes are shown in Figure 1.23.

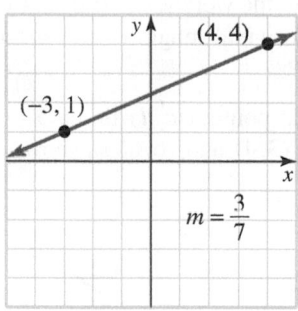

Positive slope; line rises

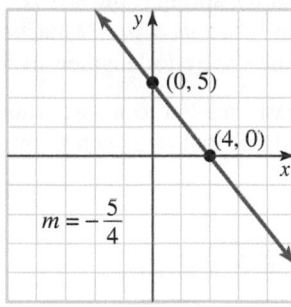

Negative slope; line falls

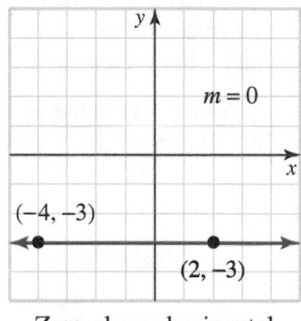

Zero slope; horizontal

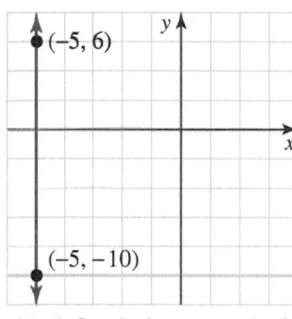

Undefined slope; vertical

Figure 1.23 Examples of lines with different slopes

Forms of the Equation of a Line

In algebra, you studied several forms of the equation of a line. The derivations of some of these are reviewed in the problems. Here is a summary of the forms most frequently used in calculus.

LINEAR EQUATIONS

Standard form:	$Ax + By + C = 0$	*A, B, C constant; A, B not both 0*
Slope-intercept form:	$y = mx + b$	*Slope m, y-intercept (0, b)*
Point-slope form:	$y - k = m(x - h)$	*Slope m through point (h, k)*
Horizontal line:	$y = k$	*Slope 0*
Vertical line:	$x = h$	*No slope*

Example 1 **Deriving the two-intercept form of the equation of a line**

Derive the equation of the line with intercepts $(a, 0)$ and $(0, b)$, $a \neq 0, b \neq 0$.

Solution The slope of the equation passing through the given points is

$$m = \frac{b - 0}{0 - a} = -\frac{b}{a}$$

Use the point-slope form with $h = 0$, $k = b$. (You can use either of the given points.)

$$y - b = -\frac{b}{a}(x - 0)$$
$$ay - ab = -bx$$
$$bx + ay = ab$$
$$\frac{x}{a} + \frac{y}{b} = 1 \qquad\qquad \textit{Divide both sides by ab.}$$

Two quantities x and y that satisfy a linear equation $Ax + By + C = 0$ (A and B not both 0) are said to be *linearly related*. This terminology is illustrated in the following example.

Example 2   Linearly related variables

When a weight is attached to a helical spring, it causes the spring to lengthen. According to Hooke's law, the length d of the spring is linearly related to the weight w.* If $d = 4$ cm when $w = 3$ g and $d = 6$ cm when $w = 6$ g, what is the original length of the spring, and what weight will cause the spring to lengthen to 5 cm?

Solution   Because d is linearly related to w, we know that points (w, d) lie on a line, and the given information tells us that two such points are $(3, 4)$ and $(6, 6)$ as shown in Figure 1.24.

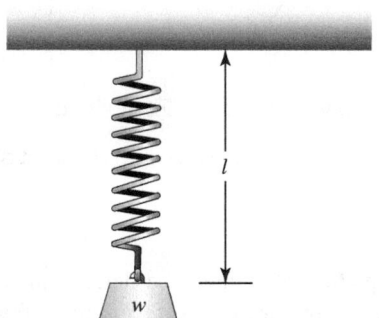
Interactive A weight at the end of a spring

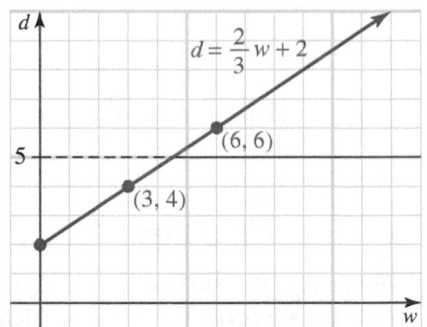
Graph of weight function

Figure 1.24   The length *d* of the spring is linearly related to the weight *w* of the attached object

We first find the slope of the line and then use the point-slope form to derive its equation.

$$m = \frac{6 - 4}{6 - 3} = \frac{2}{3}$$

Next, substitute into the point-slope form with $h = 3$ and $k = 4$:

$$d - 4 = \frac{2}{3}(w - 3)$$

$$d = \frac{2}{3}w + 2$$

The original length of the spring is found for $w = 0$:

$$d = \frac{2}{3}(0) + 2 = 2$$

---

*Hooke's law is useful for small displacements, but for larger displacements, it may not be a good model.

The original length was 2 cm. To find the weight that corresponds to $d = 5$, we solve the equation

$$5 = \frac{2}{3}w + 2$$
$$\frac{9}{2} = w$$

Therefore, the weight that corresponds to a length of 5 cm is 4.5 g.

Parametric Form

Sometimes it is convenient to define x and y in terms of another variable, say t. Let us suppose that the location of a point (x, y) is defined in terms of time, t (in seconds), as follows:

$$x = 1 + t \quad \text{and} \quad y = 2t$$

These equations must be interpreted by saying that although the first component has a 1-second "head start," the rate at which the second component changes (with respect to time) is twice as fast as the first component. We can tabulate the values of x and y by choosing values for t, where $t \geq 0$ (that is, time is considered to be positive).

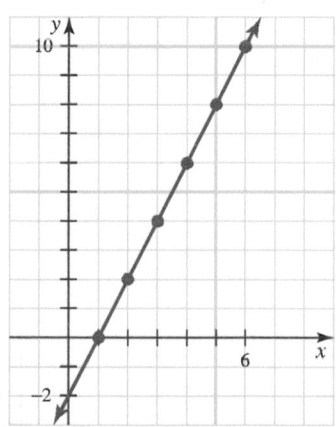

Figure 1.25 Graph of $x = 1 + t$, $y = 2t$

t	0	1	2	3	4	5
x	1	2	3	4	5	6
y	0	2	4	6	8	10

The variable t is called a **parameter**, and the equations $x = 1 + t$ and $y = 2t$ are called **parametric equations**. From the graph in Figure 1.25 we see that the data points seem to form a line. We can prove this by *eliminating the parameter*. We solve the first equation for t ($t = x - 1$) and substitute the result into the second equation

$$y = 2t = 2(x - 1) = 2x - 2$$

which is a linear equation.

PARAMETRIC FORMS OF THE EQUATION OF A LINE The graph of the parametric equations

$$x = x_1 + at \quad \text{and} \quad y = y_1 + bt$$

is a line passing through (x_1, y_1) with slope $m = \dfrac{b}{a}$ $(a \neq 0)$.

It is easy to derive this result by solving one of the equations for t and substituting the result into the other equation:

$$x - x_1 = at$$
$$\frac{x - x_1}{a} = t$$

Now, substitute into the second equation,

$$y = y_1 + bt$$
$$y = y_1 + b\left(\frac{x - x_1}{a}\right)$$
$$y - y_1 = \frac{b}{a}(x - x_1)$$

This is the equation of a line passing thorough (x_1, y_1) with slope $\frac{b}{a}$.

Parallel and Perpendicular Lines

It is often useful to know whether two given lines are either parallel or perpendicular. A vertical line can be parallel only to other vertical lines and perpendicular only to horizontal lines. Cases involving nonvertical lines may be handled by the criteria given in the following box.

SLOPE CRITERIA FOR PARALLEL AND PERPENDICULAR LINES   If L_1 and L_2 are nonvertical lines with slopes m_1 and m_2, then

$$L_1 \text{ and } L_2 \text{ are \textbf{parallel} if and only if } m_1 = m_2;$$

$$L_1 \text{ and } L_2 \text{ are \textbf{perpendicular} if and only if}$$

$$m_1 m_2 = -1 \text{ or } m_1 = -\frac{1}{m_2}.$$

■ **W**hat this says Nonvertical and nonhorizontal lines are parallel if and only if their slopes are equal. They are perpendicular if and only if their slopes are negative reciprocals of each other. See Problem 58.

The key ideas behind these two slope criteria are displayed in Figure 1.26.

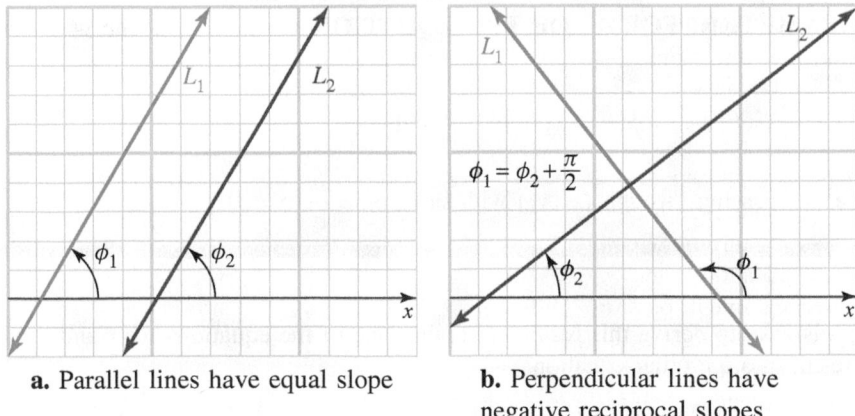

a. Parallel lines have equal slope

b. Perpendicular lines have negative reciprocal slopes

Figure 1.26   Parallel and perpendicular lines and their slopes

Example 3   Finding equations for parallel and perpendicular lines

Let L be the line $3x + 2y = 5$.

a. Find an equation of the line that is parallel to L and passes through $P(4, 7)$.
b. Find an equation of the line that is perpendicular to L and passes through $P(4, 7)$.

Solution   By rewriting the equation of L as $y = -\frac{3}{2}x + \frac{5}{2}$, we see that the slope of L is $m = -\frac{3}{2}$.

a. Any line that is parallel to L must also have slope $m_1 = -\frac{3}{2}$. The required line contains the point $P(4, 7)$. Use the point-slope form to find the equation and write your answer in standard form:

$$y - 7 = -\frac{3}{2}(x - 4)$$
$$2y - 14 = -3x + 12$$
$$3x + 2y - 26 = 0$$

b. Any line perpendicular to L must have slope $m_2 = \frac{2}{3}$ (negative reciprocal of the slope of L). Once again, the required line contains the point $P(4, 7)$, and we find

$$y - 7 = \frac{2}{3}(x - 4)$$
$$3y - 21 = 2x - 8$$
$$2x - 3y + 13 = 0$$

These lines are shown in Figure 1.27.

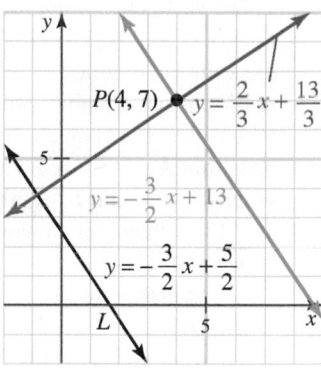

Figure 1.27   Graph of lines parallel and perpendicular to the given line

PROBLEM SET 1.3

Level 1

1. ■ **What does this say?** Outline a procedure for graphing a linear equation.

In Problems 2-15, find the equation in standard form for the line that satisfies the given requirements.

2. passing through $(1, 4)$ and $(3, 6)$
3. passing through $(-1, 7)$ and $(-2, 9)$
4. horizontal line through $(-2, -5)$
5. passing through the point $\left(1, \frac{1}{2}\right)$ with slope 0
6. slope 2 and y-intercept $(0, 5)$
7. vertical line through $(-2, -5)$
8. slope -3, x-intercept $(5, 0)$
9. x-intercept $(7, 0)$ and y-intercept $(0, -8)$
10. x-intercept $(4.5, 0)$ and y-intercept $(0, -5.4)$
11. passing through $(-1, 8)$ parallel to $3x + y = 7$
12. passing through $(4, 5)$ parallel to the line passing through $(2, 1)$ and $(5, 9)$
13. passing through $(3, -2)$ perpendicular to $4x - 3y + 2 = 0$
14. passing through $(-1, 6)$ perpendicular to the line through the origin with slope 0.5

15. perpendicular to the line whose equation is $x - 4y + 5 = 0$ where it intersects the line whose equation is $2x + 3y - 1 = 0$

In Problems 16-37, find, if possible, the slope, the y-intercept, and the x-intercept of the line whose equation is given. Sketch the graph of each equation.

16. $y = \frac{2}{3}x - 8$ **17.** $y = -\frac{5}{7}x + 3$
18. $y - 4 = 4.001(x - 2)$ **19.** $y - 9 = 6.001(x - 3)$
20. $5x + 3y - 15 = 0$ **21.** $3x + 5y + 15 = 0$
22. $2x - 3y - 2{,}550 = 0$ **23.** $6x - 10y - 3 = 0$
24. $\frac{x}{4} + \frac{y}{6} = 1$ **25.** $\frac{x}{2} - \frac{y}{3} = 1$
26. $y = 2x$ **27.** $x = 5y$
28. $y - 5 = 0$ **29.** $x + 3 = 0$
30. $x = 2 + 3t$ **31.** $x = 1 + 2t$
   $\quad y = 1 + t$ $\quad y = 2 + t$
32. $x = 5t$ **33.** $x = 3t$
   $\quad y = 2 + t$ $\quad y = 1 + 2t$
34. $x = 1 - 2t$ **35.** $x = 2 - 2t$
   $\quad y = 5 + 3t$ $\quad y = 1 + t$
36. $x = 1 + t$ **37.** $x = -2 + t$
   $\quad y = 3 - 2t$ $\quad y = 4 - 5t$

Level 2

38. Find an equation for a vertical line(s) L such that a region bounded by L, the x-axis, and the line $2y - 3x = 6$ has area 3.

39. Find an equation for a horizontal line(s) M such that the region bounded by M, the y-axis, and the line $2y - 3x = 6$ has an area of 3.

40. Find the equation of the perpendicular line passing through the midpoint of the line segment connecting $(-3, 7)$ and $(4, -1)$.

41. Three vertices of a parallelogram *include* $A(1, 3)$, $C(4, 11)$, and $B(3, -2)$. What is the fourth vertex? Is the solution unique?

42. A life insurance table indicates that a woman who is now A years old can expect to live E years longer. Suppose that A and E are linearly related and that $E = 50$ when $A = 24$ and $E = 20$ when $A = 60$.
  a. At what age may a woman expect to live 30 years longer?
  b. What is the life expectancy of a newborn female child?
  c. At what age is the life expectancy zero?

43. On the Fahrenheit temperature scale, water freezes at $32°$ and boils at $212°$; the corresponding temperatures on the Celsius scale are $0°$ and $100°$. Given that the Fahrenheit and Celsius temperatures are linearly related, first find numbers r and s so that $F = rC + s$, and then answer these questions.
  a. Mercury freezes at $-39°C$. What is the corresponding Fahrenheit temperature?
  b. For what value of C is $F = 0$?
  c. What temperature is the same in both scales?

44. The average SAT mathematics scores of incoming students at an eastern liberal arts college have been declining in recent years. In 2006, the average SAT score was 575; in 2011, it was 545. Assuming the average SAT score varies linearly with time, answer these questions.
  a. Express the average SAT score in terms of time measured from 2006.
  b. If the trend continues, what will the average SAT score of incoming students be in 2024?
  c. When will the average SAT score be 455?

45. A certain car rental agency charges $40 per day with 100 free miles plus 34¢ per mile after the first 100 miles. First express the cost of renting a car from this agency for one day in terms of the number of miles driven. Then draw the graph and use it to check your answers to these questions.
  a. How much does it cost to rent a car for a 1-day trip of 50 mi?
  b. How many miles were driven if the daily rental cost was $92.36?

46. A manufacturer buys $200,000 worth of machinery that depreciates linearly so that its trade-in value after 10 years will be $10,000. Express the value of the machinery as a function of its age and draw the graph. What is the value of the machinery after 4 years?

47. Show that if the point $P(x, y)$ is equidistant from $A(1, 3)$ and $B(-1, 2)$, its coordinates must satisfy the equation $4x + 2y - 5 = 0$. Sketch the graph of this equation.

48. Let $P_1(2, 6)$, $P_2(-1, 3)$, $P_3(0, -2)$, and $P_4(a, b)$ be vertices of a parallelogram.
  a. There are three possible choices for P_4. One is $A(3, 1)$. What are the others, which we will call B and C?
  b. The *centroid* of a triangle is the point where its three medians intersect. Find the centroid of $\triangle ABC$ and of $\triangle P_1 P_2 P_3$. Do you notice anything interesting?

49. MODELING EXPERIMENT Ethyl alcohol is metabolized by the human body at a constant rate (independent of concentration). Suppose the rate is 10 mL per hour.
  a. Express the time t (in hours) required to metabolize the effects of drinking ethyl alcohol in terms of the amount A of ethyl alcohol consumed.
  b. How much time is required to eliminate the effects of a liter of beer containing 3% ethyl alcohol?
  c. Discuss how the function in part **a** can be used to determine a reasonable "cutoff" value for the amount of ethyl alcohol A that each individual may be served at a party.

50. Since the beginning of the month, a local reservoir has been losing water at a constant rate (that is, the amount of water in the reservoir is a linear function of time). On the 12th of the month, the reservoir held 200 million gallons of water; on the 21st, it held only 164 million gallons. How much water was in the reservoir on the 8th of the month?

51. To encourage motorists to form car pools, the transit authority in a certain metropolitan area has been offering a special reduced rate at toll bridges for vehicles containing four or more persons. When the program began 30 days ago, 157 vehicles qualified for the reduced rate during the morning rush hour. Since then, the number of vehicles qualifying has increased at a constant rate (that is, the number is a linear function of time), and 247 vehicles qualified today. If the trend continues, how many vehicles will qualify during the morning rush hour 14 days from now?

52. MODELING EXPERIMENT The value of a certain rare book doubles every 10 years. The book was originally worth $3.
  a. How much is the book worth when it is 30 years old? When it is 40 years old?
  b. Is the relationship between the value of the book and its age linear? Explain.

53. *Historical Quest The region between the Tigris and Euphrates Rivers (present day Iraq) is rightly known as the Cradle of Civilization. During the so-called*

Babylonian period (roughly 2000-600 B.C.), important mathematical ideas began to germinate in the region, including positional notation for numeration. Unlike their Egyptian contemporaries who usually wrote on fragile papyrus, Babylonian mathematicians recorded their ideas on clay tablets. One of these tablets, in the Yale Collection, shows a system equivalent to*

$$\begin{cases} xy = 600 \\ (x+y)^2 - 150(x-y)^2 = 100 \end{cases}$$

Find a positive solution $(x > 0, y > 0)$ for this system correct to the nearest tenth.

54. *Historical Quest* *The Louvre Tablet from the Babylonian civilization is dated about 1500 B.C.*

The Louvre Tablet

It shows a system equivalent to

$$\begin{cases} xy = 1 \\ x + y = a \end{cases}$$

Solve this system for x and y in terms of a.

Level 3

55. Show that, in general, a line passing through $P(h,k)$ with slope m has the equation

$$y - k = m(x - h)$$

56. If $A(x_1, y_1)$ and $B(x_2, y_2)$ with $x_1 \neq x_2$ are two points on the graph of the line $y = mx + b$, show that

$$m = \frac{y_2 - y_1}{x_2 - x_1}$$

Use this fact to show that the graph of $y = mx + b$ is a line with slope m and then show that the line has y-intercept $(0,b)$.

57. Show that the distance s from the point (x_0, y_0) to the line $Ax + By + C = 0$ is given by the formula

$$s = \frac{|Ax_0 + By_0 + C|}{\sqrt{A^2 + B^2}}$$

58. Let L_1 and L_2 have slopes m_1 and m_2, respectively, and let ϕ be the angle between L_1 and L_2, as shown in Figure 1.28. Show that $\tan \phi = \frac{m_2 - m_1}{1 + m_1 m_2}$. Use this result to show that two lines have slopes that are negative reciprocals of each other if and only if they are perpendicular.

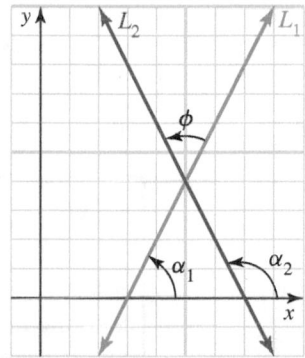

Figure 1.28 Angle ϕ between given lines

59. **Journal Problem** *(Ontario Secondary School Mathematics Bulletin[†])* Show that there is just one line in the family $y - 8 = m(x + 1)$ that is five units from the point $(2, 4)$.

60. The *centroid* of a triangle with vertices $A(x_1, y_1)$, $B(x_2, y_2)$, $C(x_3, y_3)$ is the point where its three medians intersect. It can be shown that the centroid $P(x_0, y_0)$ is located 2/3 the distance from each vertex to the midpoint of the opposite side. Use this fact to show that

$$x_0 = \frac{x_1 + x_2 + x_3}{3} \text{ and } y_0 = \frac{y_1 + y_2 + y_3}{3}$$

---

*For an interesting discussion of Mesopotamian mathematics, see *A History of Mathematics*, 2nd ed., by Carl B. Boyer, revised by Uta C. Merzbach, John Wiley and Sons, Inc., New York, 1968, pp. 26-47.

[†]Volume 18 (1982), issue 2, p. 7.

1.4 FUNCTIONS AND GRAPHS

IN THIS SECTION: *Definition of a function, functional notation, domain and range of a function, composition of functions, graph of a function, classification of functions*

The concept of *function* is the backbone of a calculus course. A thorough understanding of this section is essential for your work in calculus.

Scientists, economists, and other researchers study relationships between quantities. For example, an engineer may need to know how the illumination from a light source on an object is related to the distance between the object and the source; a biologist may wish to investigate how the population of a bacterial colony varies with time in the presence of a toxin; an economist may wish to determine the relationship between consumer demand for a certain commodity and its market price. The mathematical study of such relationships involves the concept of a *function*.

Definition of a Function

We begin with a definition.

> **FUNCTION** A **function** f is a rule that assigns to each element x of a set X a unique element y of a set Y. The element y is called the **image** of x under f and is denoted by $f(x)$ (read as "f of x"). The set X is called the **domain** of f, and the set of all images of elements of X is called the **range** of the function.

■ What this says To be called a function, the rule must have the property that it assigns to each legitimate input one and only one output. For instance, -2 is not a legitimate input for the function that computes the square root because we are working in the set of real numbers.

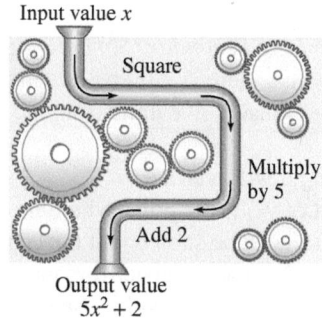

Input value x

Square

Multiply by 5

Add 2

Output value
$5x^2 + 2$

Figure 1.29 A function machine

A function whose *name* is f can be thought of as the set of ordered pairs (x, y) for which each member x of the domain is associated with exactly one member $y = f(x)$ of the range. The function can also be regarded as a rule that assigns a unique "output" in the set Y to each "input" from the set X, just as the function "machine" shown in Figure 1.29 generates an output of $5x^2 + 2$ for each input x.

A visual representation of a function is shown in Figure 1.30.

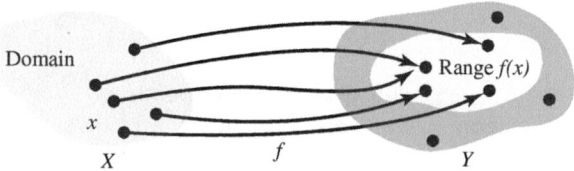

Domain

Range $f(x)$

x

X f Y

Figure 1.30 Function as a mapping

Note that it is quite possible for two different elements in the domain X to map into the same element in the range, and that it is possible for Y to include elements not in the range of f. If, however, the range of f does consist of all of Y, then f is said to map **X onto Y**. Furthermore, if each element in the range is the image of *one and only one* element in the domain, then f is said to be a **one-to-one function**. A function f is said to be **bounded** on $[a, b]$ if there exists a number B so that $|f(x)| \leq B$ for all x in $[a, b]$.

Most of our work will involve real-valued functions of a real variable, that is, functions whose domain and range are both sets of real numbers.*

Functional Notation

Functions can be represented in various ways, but usually they are specified by using a mathematical formula. It is traditional to let x denote the input and y the corresponding output, and to write an equation relating x and y. The letters x and y that appear in such an equation are called **variables**. Because the value of the variable y is determined by that of the variable x, we call y the **dependent variable** and x the **independent variable**.

In this book, when we define a function by an expression such as $f(x) = x^2 + 4x + 5$, we mean the function f is the set of all ordered pairs (x, y) satisfying the equation $y = x^2 + 4x + 5$. To **evaluate** a function f means to find the value of f for some particular value in the domain. For example, to evaluate f at $x = 2$ is to find $f(2)$.

Example 1   Using functional notation

Suppose $f(x) = 2x^2 - x$. Find $f(-1)$, $f(0)$, $f(2)$, $f(\pi)$, $f(x+h)$, and $\dfrac{f(x+h) - f(x)}{h}$, where x and h are real numbers and $h \neq 0$.

Solution   In this case, the defined function f tells us to subtract the independent variable x from twice its square. Thus, we have

$$f(-1) = 2(-1)^2 - (-1) = 3$$

$$f(0) = 2(0)^2 - (0) = 0$$

$$f(2) = 2(2)^2 - 2 = 6$$

$$f(\pi) = 2\pi^2 - \pi$$

To find $f(x+h)$ we begin by writing the formula for f in more neutral terms, say as

$$f(\square) = 2(\square)^2 - \square$$

Then we insert the expression $x + h$ inside each box, obtaining

$$f(\boxed{x+h}) = 2(\boxed{x+h})^2 - (\boxed{x+h})$$
$$= 2(x^2 + 2xh + h^2) - (x + h)$$
$$= 2x^2 + 4xh + 2h^2 - x - h$$

Finally, if $h \neq 0$,

$$\frac{f(x+h) - f(x)}{h} = \frac{[2x^2 + 4xh + 2h^2 - x - h] - [2x^2 - x]}{h}$$
$$= \frac{4xh + 2h^2 - h}{h}$$
$$= 4x + 2h - 1$$

Note: The expression $\dfrac{f(x+h) - f(x)}{h}$ is called a **difference quotient** and is used in Chapter 3 to compute the *derivative*. ∎

Certain functions are defined differently on different parts of their domain and are thus more naturally expressed in terms of more than one formula. We refer to such functions as **piecewise-defined functions**.

---

*Most functions that appear in this book belong to a very special class of **elementary functions**, defined by Joseph Liouville (1809-1882). You will study some nonelementary functions in complex analysis and in advanced calculus. Certain functions that appear in physics and higher mathematics are not elementary functions.

Example 2   Evaluating a piecewise-defined function

If $f(x) = \begin{cases} x \sin x & \text{if } x < 2 \\ 3x^2 + 1 & \text{if } x \geq 2 \end{cases}$, find $f(-0.5)$, $f\left(\frac{\pi}{2}\right)$, and $f(2)$.

Solution   To find $f(-0.5)$, we use the first line of the formula because $-0.5 < 2$:

$$f(-0.5) = -0.5 \sin(-0.5) = 0.5 \sin 0.5 \quad \textit{This is an exact value.}$$
$$\approx 0.2397 \quad \textit{This is the approximate value.}$$

To find $f\left(\frac{\pi}{2}\right)$, we use the first line of the formula because $\frac{\pi}{2} \approx 1.57 < 2$:

$$f\left(\frac{\pi}{2}\right) = \frac{\pi}{2} \sin \frac{\pi}{2} = \frac{\pi}{2} \quad \textit{This is the exact value.}$$

Finally, because $2 \geq 2$, we use the second line of the formula to find $f(2)$.

$$f(2) = 3(2)^2 + 1 = 13$$

Functional notation can be used in a wide variety of applied problems, as shown by Example 3 and again in the problem set.

Example 3   Applying functional notation

It is known that an object dropped from a height in a vacuum will fall a distance of s ft in t seconds according to the formula

$$s(t) = 16t^2, \quad t \geq 0$$

a. How far will the object fall in the first second? In the *next* 2 seconds?
b. How far will it fall during the time interval $t = 1$ to $t = 1 + h$ seconds?
c. What is the average rate of change of distance (in feet per second) during the time $t = 1$ sec to $t = 3$ sec?
d. What is the average rate of change of distance during the time $t = x$ seconds to $t = x + h$ seconds?

Solution

a. $s(1) = 16(1)^2 = 16$. In the first second the object will fall 16 ft. In the next two seconds the object will fall

$$s(1 + 2) - s(1) = s(3) - s(1)$$
$$= 16(3)^2 - 16(1)^2$$
$$= 128$$

The object will fall 128 ft in the next 2 sec.

b. $s(1 + h) - s(1) = 16(1 + h)^2 - 16(1)^2$

$$= 16 + 32h + 16h^2 - 16$$

$$= 32h + 16h^2$$

The object will fall $32h + 16h^2$ ft.

c. AVERAGE RATE $= \dfrac{\text{CHANGE IN DISTANCE}}{\text{CHANGE IN TIME}}$

$$= \frac{s(3) - s(1)}{3 - 1}$$

$$= \frac{128}{2}$$

$$= 64$$

The average rate of change is 64 ft/s.

d. $\dfrac{s(x+h)-s(x)}{(x+h)-x} = \dfrac{s(x+h)-s(x)}{h}$

$$= \dfrac{16(x+h)^2 - 16x^2}{h}$$

$$= \dfrac{16x^2 + 32xh + 16h^2 - 16x^2}{h}$$

$$= 32x + 16h$$

Does this look familiar?
See Example 1.

The average rate of change of distance is $32x + 16h$ ft/s.

Domain and Range of a Function

In this book, unless otherwise specified, the domain of a function is the set of real numbers for which the function is defined. We call this the **domain convention**.

This domain convention will be used throughout the text.

If a function f is **undefined** at x, it means that x is not in the domain of f. The most frequent exclusions from the domain are those values that cause division by 0 or negative values under a square root. In applications, the domain is often specified by the context. For example, if x is the number of people on an elevator, the context requires that negative numbers and nonintegers be excluded from the domain; therefore, x must be an integer such that $0 \le x \le c$, where c is the maximum capacity of the elevator.

Example 4 Domain of a function

Find the domain for the given functions.

a. $f(x) = 2x - 1$ **b.** $g(x) = 2x - 1, x \ne -3$

c. $h(x) = \dfrac{(2x-1)(x+3)}{x+3}$ **d.** $F(x) = \sqrt{x+2}$ **e.** $G(x) = \dfrac{4}{5 - \cos x}$

Solution

a. All real numbers: $(-\infty, \infty)$

b. All real numbers except -3. This can be written as $(-\infty, -3) \cup (-3, \infty)$.

c. Because the expression is meaningful for all $x \ne -3$, the domain is all real numbers except -3.

d. F has meaning if and only if $x + 2$ is nonnegative; therefore, the domain is $x \ge -2$, or $[-2, \infty)$.

e. G is defined whenever $5 - \cos x \ne 0$. This imposes no restriction on x since $|\cos x| \le 1$. Thus, the domain of G is the set of all real numbers; this can be written $(-\infty, \infty)$.

EQUALITY OF FUNCTIONS Two functions f and g are **equal** if and only if both

1. f and g have the same domain.
2. $f(x) = g(x)$ for all x in the domain.

Example 5 Equality of functions

Consider again the functions from Example 4.

$$f(x) = 2x - 1 \qquad g(x) = 2x - 1, x \ne -3 \qquad h(x) = \dfrac{(2x-1)(x+3)}{x+3}$$

Does $h(x)$ equal $f(x)$ or $g(x)$?

Solution   A common mistake is to "reduce" the function h to obtain the function f:

$$\text{\ding{56} WRONG \ding{56}} \quad h(x) = \frac{(2x-1)(x+3)}{x+3} = 2x - 1 = f(x)$$

This is wrong because the functions f and h have different domains. In particular, -3 is a valid input for f, but it is not a valid input for h. We say that the function h has a **hole** at $x = -3$. In other words, this WRONG calculation is valid only if $x \neq -3$, so the correct answer is

$$\text{RIGHT} \quad h(x) = \frac{(2x-1)(x+3)}{x+3} = 2x - 1, x \neq -3$$
$$\text{therefore, } h(x) = g(x) \quad \blacksquare$$

Composition of Functions

There are many situations in which a quantity is given as a function of one variable that, in turn, can be written as a function of a second variable. Suppose, for example, that your job is to ship X packages of a product via Federal Express to a variety of addresses. Let x be the number of packages to ship, and let $f(x)$ be the weight of the x objects and $g(w)$ be the cost of shipping a package of weight w. Then

> Weight is a function $f(x)$ of the number of objects x.
> Cost is a function $g[f(x)]$ of the weight.

So we have expressed cost as a function of the number of packages. This process of evaluating a function of a function is known as *functional composition*.

COMPOSITION OF FUNCTIONS   The **composite function** $f \circ g$ is defined by

$$(f \circ g)(x) = f[g(x)]$$

for each x in the domain of g for which $g(x)$ is in the domain of f.

■ **W**hat this says To visualize how functional composition works, think of $f \circ g$ in terms of an "assembly line" in which g and f are arranged in series, with output $g(x)$ becoming the input of f, as illustrated in Figure 1.31.

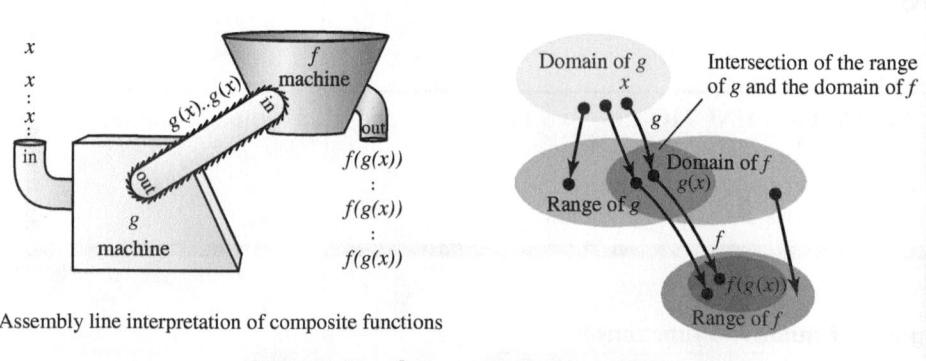

Assembly line interpretation of composite functions

Figure 1.31   Composition of functions

Example 6   Finding the composition of functions

If $f(x) = 3x + 5$ and $g(x) = \sqrt{x}$, find the composite functions $f \circ g$ and $g \circ f$.

Solution   The function $f \circ g$ is defined by $f[g(x)]$.

$$(f \circ g)(x) = f[g(x)] = f(\sqrt{x}) = 3\sqrt{x} + 5$$

The function $g \circ f$ is defined by $g[f(x)]$.

$$(g \circ f)(x) = g[f(x)] = g(3x + 5) = \sqrt{3x + 5}$$

That is, $f \circ g$ is **not**, in general, the same as $g \circ f$.

Example 6 illustrates that functional composition is not commutative.

Example 7   An application of composite functions

Air pollution is a problem for many metropolitan areas. Suppose a study conducted in a certain city indicates that when the population is p hundred thousand people, then the average daily level of carbon monoxide in the air is given by the formula

$$L(p) = 0.70\sqrt{p^2 + 3}$$

parts per million (ppm). A second study predicts that t years from now, the population will be $p(t) = 1 + 0.02t^3$ hundred thousand people. Assuming these formulas are correct, what level of air pollution should be expected in 4 years?

Solution   The level of pollution is $L(p) = 0.70\sqrt{p^2 + 3}$, where $p(t) = 1 + 0.02t^3$. Thus, the pollution level at time t is given by the composite function

$$(L \circ p)(t) = L(1 + 0.02t^3) = 0.70\sqrt{(1 + 0.02t^3)^2 + 3}$$

In particular, when $t = 4$, we have

$$(L \circ p)(4) = 0.70\sqrt{\left[1 + 0.02(4)^3\right]^2 + 3} \approx 2.00\,\text{ppm}$$

In calculus, it is frequently necessary to express a function as the composite of two simpler functions. Here is an example that illustrates how this can be done.

Example 8   Expressing a given function as a composite of two functions

Express each of the following functions as the composite of two functions u and g so that $f(x) = g[u(x)]$.

a. $f(x) = (x^2 + 5x + 1)^5$ **b.** $f(x) = \cos^3 x$ **c.** $f(x) = \sin x^3$ **d.** $f(x) = \sqrt{5x^2 - x}$

Solution   There are often many ways to express $f(x)$ as a composite $g[u(x)]$. Perhaps the most natural is to choose u to represent the "inner" portion of f and g as the "outer" portion. Such choices are indicated in the following table.

Given function $f(x) = g[u(x)]$	Inner function $u(x)$	Outer function $g(x)$
a. $f(x) = (x^2 + 5x + 1)^5$	$u(x) = x^2 + 5x + 1$	$g(u) = u^5$
b. $f(x) = \cos^3 x$	$u(x) = \cos x$	$g(u) = u^3$
c. $f(x) = \sin(x^3)$	$u(x) = x^3$	$g(u) = \sin u$
d. $f(x) = \sqrt{5x^2 - x}$	$u(x) = 5x^2 - x$	$g(u) = \sqrt{u}$

Graph of a Function

Graphs have visual impact. They also reveal information that may not be evident from verbal or algebraic descriptions. Two graphs depicting practical relationships are shown in Figure 1.32.

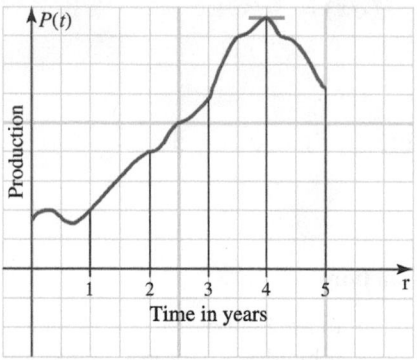

a. A production function

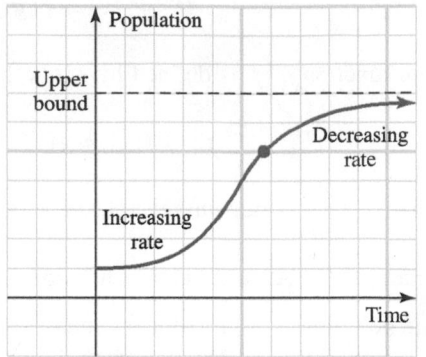

b. Bounded population growth

This graph describes the variation in total industrial production in a certain country over a five year time span. The fact that the graph has a peak suggests that production is greatest at the corresponding time

This graph represents the growth of a population when environmental factors impose an upper bound on the population. It indicates that the rate of population growth increases at first and then decreases as the size of the population gets closer and closer to the upper bound

Figure 1.32 Two graphs with practical interpretations

To represent a function $y = f(x)$ geometrically as a graph, it is traditional to use a Cartesian coordinate system on which units for the independent variable x are marked on the horizontal axis and units for the dependent variable y on the vertical axis.

> **GRAPH OF A FUNCTION** The **graph** of a function f consists of points whose coordinates (x, y) satisfy $y = f(x)$, for all x in the domain of f.

    In Chapter 4, we will discuss efficient techniques involving calculus that you can use to draw accurate graphs of functions. In algebra, you began sketching lines by plotting points, but you quickly discovered that this is not a very efficient way to draw more complicated graphs, especially without the aid of a graphing calculator or computer. Table 1.4 includes a few common graphs you have probably encountered in previous courses. We will assume that you are familiar with their general shape and know how to sketch each of them either by hand or with the assistance of your graphing calculator. There are, however, certain functions, such as $f(x) = \sin \frac{1}{x}$, that are difficult to graph even with a calculator or computer. That said, be assured that most functions that appear in this text or in practical applications can be graphed.

Table 1.4 Directory of Curves

Identity Function $y = x$	Standard Quadratic Function $y = x^2$	Standard Cubic Function $y = x^3$
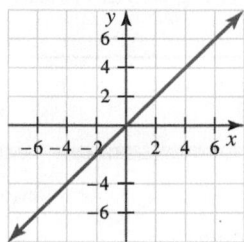	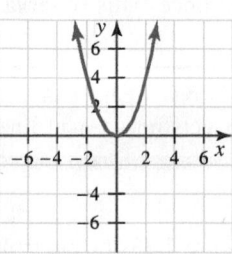	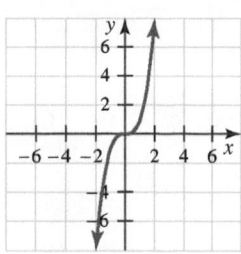
Absolute Value Function $y = \|x\| = \sqrt{x^2}$	Square Root Function $y = \sqrt{x}$	Cube Root Function $y = \sqrt[3]{x}$
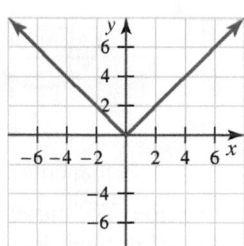	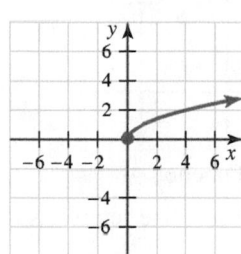	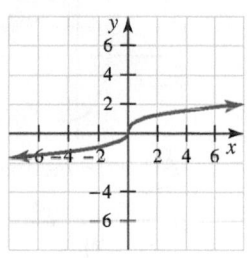
Standard Reciprocal $y = \frac{1}{x}$	Standard Reciprocal Squared $y = \frac{1}{x^2}$	Standard Square Root Reciprocal $y = \frac{1}{\sqrt{x}}$
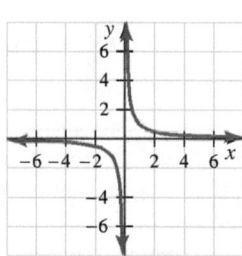	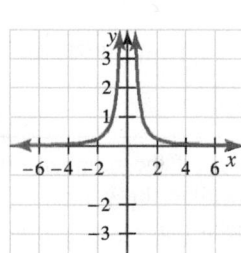	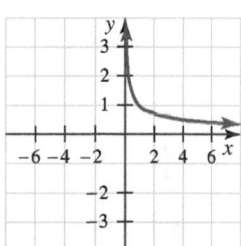
Cosine Function $y = \cos x$	Sine Function $y = \sin x$	Tangent Function $y = \tan x$
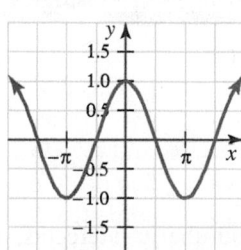	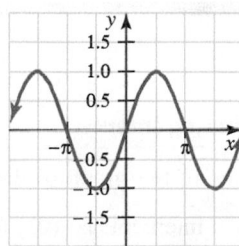	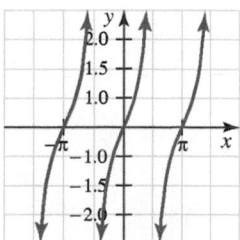
Secant Function $y = \sec x$	Cosecant Function $y = \csc x$	Cotangent Function $y = \cot x$
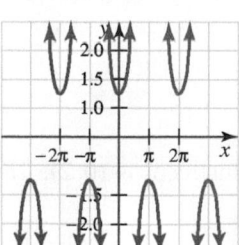	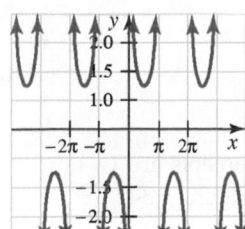	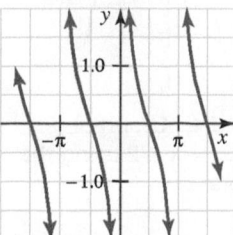

Vertical Line Test

By definition of a function, for a given x in the domain there is only one number y in the range. Geometrically, this means that any vertical line $x = a$ crosses the graph of a function at most once. This observation leads to the following useful criterion.

> **VERTICAL LINE TEST**   A curve in the plane is the graph of a function if and only if it intersects no vertical line more than once.

Look at Figure 1.33 for examples of the vertical line test.

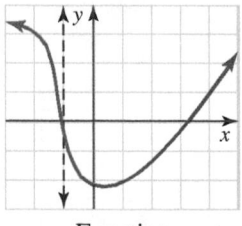   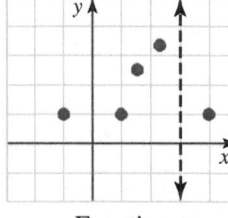   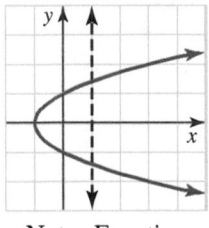   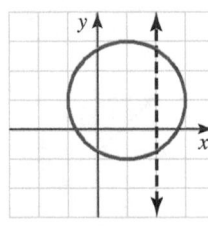

Function   Function   Not a Function   Not a Function

a. The graph of a function: No vertical line intersects the graph more than once

b. Not the graph of a function: The curve intersects at least one vertical line more than once

Figure 1.33   The vertical line test

Intercepts

The points where a graph intersects the coordinate axes are called *intercepts*. Here is a definition.

> **INTERCEPTS**   If the number zero is in the domain of f and $f(0) = b$, then the point $(0, b)$ is called the **y-intercept** of the graph of f. If a is a real number in the domain of f such that $f(a) = 0$, then $(a, 0)$ is an **x-intercept** of f.

■ **W**hat this says To find the x-intercepts, set y equal to 0 and solve for x. To find the y-intercept, set x equal to 0 and solve for y.

Example 9   Finding intercepts

Find all intercepts of the graph of the function $f(x) = -x^2 + x + 2$.

Solution   The y-intercept is $(0, f(0)) = (0, 2)$. To find the x-intercepts, solve the equation $f(x) = 0$. Factoring, we find that

$$f(x) = 0$$
$$-x^2 + x + 2 = 0$$
$$x^2 - x - 2 = 0$$
$$(x + 1)(x - 2) = 0$$
$$x = -1 \text{ or } x = 2$$

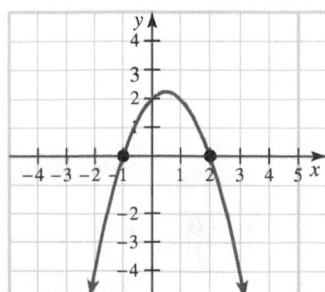

Figure 1.34 Graph of $f(x) = -x^2 + x + 2$ using the intercepts

Thus, the x-intercepts are $(-1, 0)$ and $(2, 0)$, and the y-intercept is $(0, 2)$. The graph of f is shown in Figure 1.34.   ■

Symmetry

There are two kinds of symmetry that help in graphing a function, as shown in Figure 1.35 and defined in the following box.

> **SYMMETRY** The graph of $y = f(x)$ is **symmetric with respect to the y-axis** if whenever $P(x, y)$ is a point on the graph, so is the point $(-x, y)$; that is, the mirror image of P about the y-axis. Thus, y-axis symmetry occurs if and only if $f(-x) = f(x)$ for all x in the domain of f. A function with this property is called an **even function**.
>
> The graph of $y = f(x)$ is **symmetric with respect to the origin** if whenever $P(x, y)$ is on the graph, so is $(-x, -y)$, the mirror image of P about the origin. Symmetry with respect to the origin occurs when $f(-x) = -f(x)$ for all x. A function that satisfies this condition is called an **odd function**.

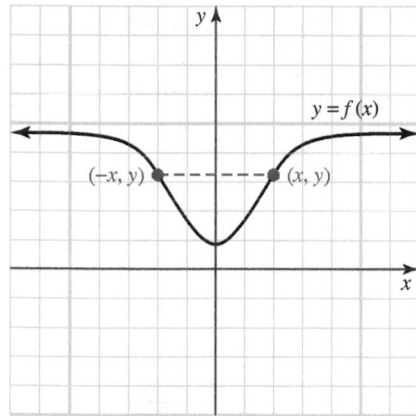

a. Graph of an even function f

$f(-x) = f(x)$; symmetry with respect to the y-axis

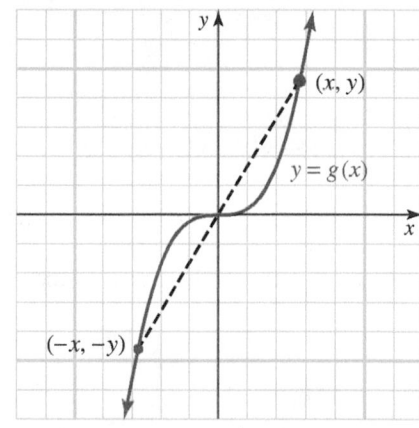

b. Graph of an odd function g

$g(-x) = -g(x)$; symmetry with respect to the origin

The graph of a nonzero function cannot be symmetric with respect to the x-axis. In other words, if the curve is symmetric with respect to the x-axis, such as the circle, then it will not be a function.

Figure 1.35 Graphs of even and odd functions

There are many functions that are neither odd nor even. You may wonder why we have said nothing about symmetry with respect to the x-axis, but such symmetry would require $f(x) = -f(x)$, which is precluded by the vertical line test (do you see why?).

Example 10 Even and odd functions

Classify the given functions as even, odd, or neither.

a. $f(x) = x^2$
b. $g(x) = x^3$
c. $h(x) = x^2 + 5x$

Graph each function, and comment on symmetry.

Solution

a. $f(x) = x^2$ is even because

$$f(-x) = (-x)^2 = x^2 = f(x)$$

Symmetric with respect to the y-axis.

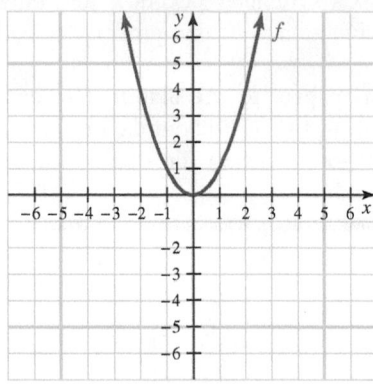

b. $g(x) = x^3$ is odd because

$$g(-x) = (-x)^3 = -x^3 = -g(x)$$

Symmetric with respect to the origin.

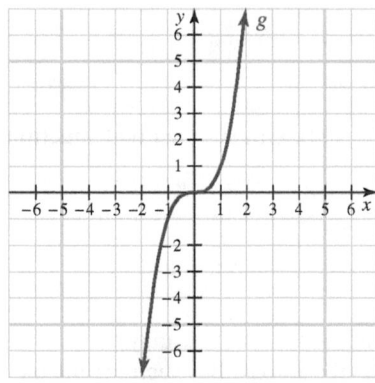

c. $h(x) = x^2 + 5x$ is neither even nor odd because

$$h(-x) = (-x)^2 + 5(-x) = x^2 - 5x$$

Note that $h(-x) \neq h(x)$ and $h(-x) \neq -h(x)$.

This curve is not symmetric with respect to either the y-axis or the origin.

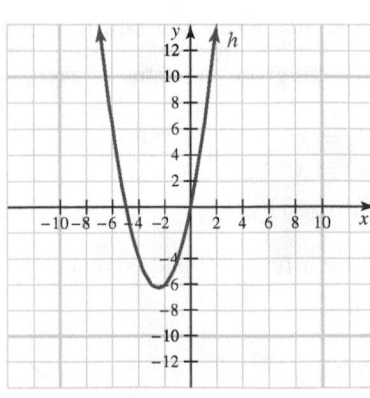

Classification of Functions

We will now describe some of the common types of functions used in this text.

POLYNOMIAL FUNCTION   A **polynomial function** is a function of the form

$$f(x) = a_n x^n + a_{n-1} x^{n-1} + \cdots + a_2 x^2 + a_1 x + a_0$$

where n is a nonnegative integer and $a_n, \cdots, a_2, a_1, a_0$ are constants. If $a_n \neq 0$, the integer n is called the **degree** of the polynomial. The constant a_n is called the **leading coefficient** and the constant a_0 is called the **constant term** of the polynomial function. In particular,

A **constant function** is zero degree: $f(x) = a$
A **linear function** is first degree: $f(x) = ax + b$
A **quadratic function** is second degree: $f(x) = ax^2 + bx + c$
A **cubic function** is third degree: $f(x) = ax^3 + bx^2 + cx + d$
A **quartic function** is fourth degree: $f(x) = ax^4 + bx^3 + cx^2 + dx + e$

Examples of polynomial functions

$$f(x) = 5$$
$$f(x) = 2x - \sqrt{2}$$
$$f(x) = 3x^2 + 5x - \tfrac{1}{2}$$
$$f(x) = \sqrt{2}x^3 - \pi x$$
$$f(x) = -\tfrac{1}{3}x^4 + \sqrt[3]{5}$$

The identity function, standard quadratic function, and standard cubic functions in Table 1.4 on page 43 are examples of polynomial functions.

A second important algebraic function is a *rational function*.

RATIONAL FUNCTION   A **rational function** is the quotient of two polynomial functions, $p(x)$ and $q(x)$:

$$f(x) = \frac{p(x)}{q(x)}, \quad q(x) \neq 0$$

Examples of rational functions

$$f(x) = x^{-1}$$
$$f(x) = \frac{x - 5}{x^2 + 2x - 3}$$
$$f(x) = x^{-3} + \sqrt{2}x$$

When we write $q(x) \neq 0$ we mean that all values c for which $q(c) = 0$ are excluded from the domain of q.

If r is any nonzero real number, the function $f(x) = x^r$ is called a **power function** with exponent r. You should be familiar with the following cases:

Integral powers ($r = n$, a positive integer): $f(x) = x^n = \underbrace{x \cdot x \cdot \ldots \cdot x}_{n \ factors}$

EXAMPLE : $f(x) = x^6$

Reciprocal powers (r is a negative integer): $f(x) = x^{-n} = \frac{1}{x^n}$ for $x \neq 0$.

EXAMPLE : $f(x) = x^{-4} = \dfrac{1}{x^4}$

Roots ($r = \frac{m}{n}$ is a positive rational number): $f(x) = x^{m/n} = \sqrt[n]{x^m} = \left(\sqrt[n]{x}\right)^m$ for $x \geq 0$ if n is even, $n \neq 0$ $\left(\frac{m}{n} \text{ is reduced}\right)$

EXAMPLE : $f(x) = x^{3/4}; \quad f(x) = \sqrt[5]{x^2}$

Power functions can also have irrational exponents (such as $\sqrt{2}$ or π), but such functions must be defined in a special way (see Section 2.4).

A function is called **algebraic** if it can be constructed using algebraic operations (such as adding, subtracting, multiplying, dividing, or taking roots) starting with polynomials. Any rational function is an algebraic function.

Functions that are not algebraic are called **transcendental**. The following functions are transcendental functions:

Trigonometric functions are the functions sine, cosine, tangent, secant, cosecant, and cotangent. You can review these functions by consulting a trigonometry or precalculus textbook or by looking in Appendix E.

Exponential functions are functions of the form $f(x) = b^x$, where b is a positive constant, $b \neq 1$. We will introduce these functions in Section 2.4.

Logarithmic functions are functions of the form $f(x) = \log_b x$, where b is a positive constant, $b \neq 1$. We will also study these functions in Section 2.4.

PROBLEM SET 1.4

Level 1

In Problems 1-10, find the domain of f and compute the indicated values or state that the corresponding x-value is not in the domain. Tell whether any of the indicated values are zeros of the function, that is, values of x that cause the functional value to be 0.

1. $f(x) = 2x + 3$; $f(-2), f(1), f(0)$

2. $f(x) = -x^2 + 2x + 3$; $f(0), f(1), f(-2)$

3. $f(x) = x + \dfrac{1}{x}$; $f(-1), f(1), f(2)$

4. $f(x) = \dfrac{(x+3)(x-2)}{x+3}$; $f(2), f(0), f(-3)$

5. $f(x) = (2x - 1)^{-3/2}$; $f(1), f\left(\frac{1}{2}\right), f(13)$

6. $f(x) = \sqrt{x^2 + 2x}$; $f(-1), f\left(\frac{1}{2}\right), f(1)$

7. $f(x) = \sin(1 - 2x)$; $f(-1), f\left(\frac{1}{2}\right), f(1)$

8. $f(x) = \sin x - \cos x$; $f(0), f\left(-\frac{\pi}{2}\right), f(\pi)$

9. $f(x) = \begin{cases} -2x + 4 & \text{if } x \le 1 \\ x + 1 & \text{if } x > 1 \end{cases}$; $f(3), f(1), f(0)$

10. $f(x) = \begin{cases} 3 & \text{if } x < -5 \\ x + 1 & \text{if } -5 \le x \le 5 \\ \sqrt{x} & \text{if } x > 5 \end{cases}$;

    $f(-6), f(-5), f(16)$

In Problems 11-18, evaluate the difference quotient $\dfrac{f(x+h) - f(x)}{h}$ for the given function.

11. $f(x) = 9x + 3$

12. $f(x) = 5 - 2x$

13. $f(x) = 5x^2$

14. $f(x) = 3x^2 + 2x$

15. $f(x) = |x|$ if $x < -1$ and $0 < h < 1$

16. $f(x) = |x|$ if $x > 1$ and $0 < h < 1$

17. $f(x) = \dfrac{1}{x}$

18. $f(x) = \dfrac{x+1}{x-1}$

State whether the functions f and g in Problems 19-22 are equal.

19. a. $f(x) = \dfrac{2x^2 + x}{x}$; $g(x) = 2x + 1$

    b. $f(x) = \dfrac{2x^2 + x}{x}$; $g(x) = 2x + 1, x \neq 0$

20. a. $f(x) = \dfrac{2x^2 - x - 6}{x - 2}$; $g(x) = 2x + 3, x \neq 2$

    b. $f(x) = \dfrac{2x^2 - x - 6}{x - 2}$; $g(x) = 2x + 3, x \neq 3$

21. a. $f(x) = \dfrac{3x^2 - 5x - 2}{x - 2}$; $g(x) = 3x + 1$

    b. $f(x) = \dfrac{3x^2 - 5x - 2}{x - 2}$; $g(x) = 3x + 1, x \neq 2$

22. a. $f(x) = \dfrac{(3x + 1)(x - 2)}{x - 2}$, $x \neq 6$;

    $g(x) = \dfrac{(3x + 1)(x - 6)}{x - 6}$, $x \neq 2$

    b. $f(x) = \dfrac{(3x + 1)(x - 2)}{x - 2}$, $x \neq 2$;

    $g(x) = \dfrac{(3x + 1)(x - 6)}{x - 6}$, $x \neq 6$

Classify the functions defined in Problems 23-26 as even, odd, or neither.

23. a. $f_1(x) = x^2 + 1$ b. $f_2(x) = \sqrt{x^2}$

24. a. $f_3(x) = \dfrac{1}{3x^3 - 4}$ b. $f_4(x) = x^3 + x$

25. a. $f_5(x) = |x|$ b. $f_6(x) = |x + x^3|$

26. a. $f_7(x) = \dfrac{1}{(x^3 + 3)^2}$ b. $f_8(x) = \dfrac{1}{(x^3 + x)^2}$

In Problems 27-30, find the composite functions $f \circ g$ and $g \circ f$.

27. $f(x) = x^2 + 1$ and $g(x) = 2x$

28. $f(x) = \sin x$ and $g(x) = 1 - x^2$

29. $f(x) = \sin x$ and $g(x) = 2x + 3$

30. $f(u) = \dfrac{u - 1}{u + 1}$ and $g(u) = \dfrac{u + 1}{1 - u}$

In Problems 31-38, express f as the composition of two functions u and g such that $f(x) = g(u)$, where u is a function of x.

31. $f(x) = (2x^2 - 1)^4$

32. $f(x) = |2x + 3|$

33. $f(x) = \sqrt{5x - 1}$

34. $f(x) = \tan^2 x$

35. $f(x) = \tan(x^2)$

36. $f(x) = \sin \sqrt{x}$

37. $f(x) = \sqrt{\sin x}$

38. $f(x) = \tan\left(\dfrac{2x}{1 - x}\right)$

39. If point A in Figure 1.36 has coordinates $(2, f(2))$, what are the coordinates of P and Q?

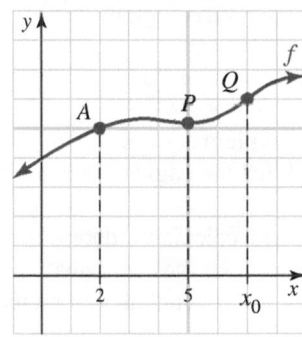
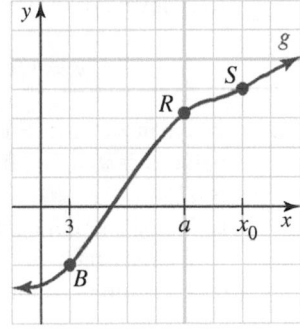

Problem 39 · · · · · · · · · · · · · · · Problem 40

Figure 1.36 Functions defined by a graph

40. If point B in Figure 1.36 has coordinates $(3, g(3))$, what are the coordinates of R and S?

Find the x-intercepts, if any, for the functions given in Problems 41-47.

41. $f(x) = 3x^2 - 5x - 2$

42. $f(x) = (x - 15)(2x + 25)(3x - 65)(4x + 1)$

43. $f(x) = (x^2 - 10)(x^2 - 12)(x^2 - 20)$

44. $f(x) = 5x^3 - 3x^2 + 2x$

45. $f(x) = x^4 - 41x^2 + 400$

46. $f(x) = \dfrac{x^2 - 1}{x^2 + 2}$

47. $f(x) = \dfrac{x(x^2 - 3)}{x^2 + 5}$

Level 2

48. Suppose the total cost (in dollars) of manufacturing q units of a certain commodity is given by

$$C(q) = q^3 - 30q^2 + 400q + 500$$

a. Compute the cost of manufacturing 20 units.

b. Compute the cost of manufacturing the 20th unit.

49. An efficiency study of the morning shift at a certain factory indicates that an average worker who arrives on the job at 8:00 A.M. will have assembled

$$f(x) = -x^3 + 6x + 15x^2$$

iPads x hours later ($0 \leq x \leq 8$).

a. How many units will such a worker have assembled by 10:00 A.M.?

b. How many units will such a worker assemble between 9:00 A.M. and 10:00 A.M.?

50. In physics, a light source of luminous intensity K candles is said to have *illuminance* $I = K/s^2$ on a flat surface s ft away. Suppose a small, unshaded lamp of luminous intensity 30 candles is connected to a rope that allows it to be raised and lowered between the floor and the top of a 10-ft-high ceiling. Assume that the lamp is being raised and lowered in such a way that at time t (in min) it is $s = 6t - t^2$ ft above the floor.

a. Express the illuminance on the floor as a composite function of t for $0 < t < 6$.

b. What is the illuminance when $t = 1$? When $t = 4$?

51. Biologists have found that the speed of blood in an artery is a function of the distance of the blood from the artery's central axis. According to *Poiseuille's law*, the speed (cm³/sec) of blood that is r cm from the central axis of an artery is given by the function

$$S(r) = C(R^2 - r^2)$$

where C is a constant and R is the radius of the artery.* Suppose that for a certain artery, $C = 1.76 \times 10^5$ cm/sec and $R = 1.2 \times 10^{-2}$ cm. (See Figure 1.37).

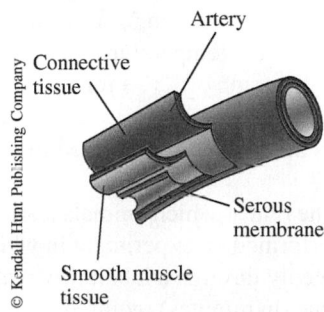

Figure 1.37 Artery

a. Compute the speed of the blood at the central axis of this artery.

b. Compute the speed of the blood midway between the artery's wall and central axis.

---

*The law and the unit *poise*, a unit of viscosity, are both named for the French physician Jean Louis Poiseuille (1799-1869).

52. At a certain factory, the total cost of manufacturing q units during the daily production run is

$$C(q) = q^2 + q + 900$$

dollars. On a typical workday, the number of units manufactured during the first t hours of a production run can be modeled by the function

$$q(t) = 25t$$

a. Express the total manufacturing cost as a function of t.

b. How much will have been spent on production by the end of the third hour?

c. When will the total manufacturing cost reach $11,000?

53. A ball is thrown directly upward from the edge of a cliff in such a way that t seconds later, it is

$$s = -16t^2 + 96t + 144$$

feet above the ground at the base of the cliff. Sketch the graph of this equation (making the t-axis the horizontal axis) and then answer these questions:

a. How high is the cliff?

b. When (to the nearest tenth of a second) does the ball hit the ground at the base of the cliff?

c. Estimate the time it takes for the ball to reach its maximum height. What is the maximum height?

54. *Charles' law* for gases states that if the pressure remains constant, then

$$V(T) = V_0 \left(1 + \frac{T}{273}\right)$$

where V is the volume (in.^{3}), V_0 is the initial volume (in.^{3}), and T is the temperature (in degrees Celsius).

a. Sketch the graph of $V(T)$ for $V_0 = 100$ and $T \geq -273$.

b. What is the temperature needed for the volume to double?

55. To study the rate at which animals learn, a psychology student performed an experiment in which a rat was sent repeatedly through a laboratory maze. Suppose that the time (in minutes) required for the rat to traverse the maze on the nth trial was approximately

$$f(n) = 3 + \frac{12}{n}$$

a. What is the domain of the function f?

b. For what values of n does $f(n)$ have meaning in the context of the psychology experiment?

c. How long did it take the rat to traverse the maze on the third trial?

d. On which trial did the rat first traverse the maze in 4 minutes or less?

e. According to the function f, what will happen to the time required for the rat to traverse the maze as the number of trials increases? Will the rat ever be able to traverse the maze in less than 3 minutes?

56. The trajectory of a cannonball shot from the origin with initial velocity v and angle of inclination α (measured from level ground), is given by the equation

$$y = mx - 16v^{-2}(1 + m^2)x^2$$

where $m = \tan \alpha$. For this problem assume the initial velocity is $v = 200$ ft/s and that the angle of inclination is $42°$.

a. Using a graphing utility of a calculator, determine the maximum height reached by the cannonball.

b. Estimate the point where the cannonball will hit the ground.

c. Which curve in Table 1.4 shows the basic shape of this graph?

57. Draw the path of a cannonball (see Problem 56) for an angle of inclination of $47°$. Determine the angle α that will maximize the distance the cannonball will travel.

58. It is estimated that t years from now, the population of a certain suburban community will be

$$P(t) = 20 - \frac{6}{t+1}$$

thousand people.

a. What will the population of the community be nine years from now?

b. By how much will the population increase during the ninth year?

c. What will happen to the size of the population in the "long run"?

Level 3

59. Journal Problem (*The Mathematics Student Journal.**) Given that $f(11) = 11$ and

$$f(x + 3) = \frac{f(x) - 1}{f(x) + 1}$$

for all x, find $f(2018)$.

60. *Historical Quest*

One of the best known mathematical theorems is the **Pythagorean theorem**, named after the Greek philosopher Pythagoras. Very little is known about the life of Pythagoras, but we do know he was born on the island of Samos. He founded a secret brotherhood called the

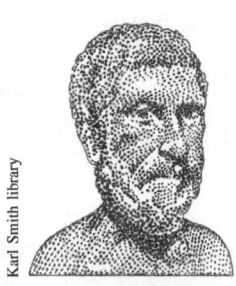

Karl Smith library

Pythagoras
(ca. 500 B.C.)

Pythagoreans that continued for at least 100 years after Pythagoras was murdered for political reasons. Even though the cult was called a "brotherhood," it did admit women. According to Lynn Osen in **Women in Mathematics** (MIT Press, Cambridge, MA, 1975), the order was carried on by his wife and daughters after his death. In fact, women were probably more welcome in the centers of learning in ancient Greece than in any other age from that time until now. State and prove the Pythagorean theorem.

1.5 INVERSE FUNCTIONS; INVERSE TRIGONOMETRIC FUNCTIONS

IN THIS SECTION: *Inverse functions, criteria for the existence of an inverse f^{-1}, graph of f^{-1}, inverse trigonometric functions, inverse trigonometric identities*

Intuitively, an inverse function f^{-1} "reverses" the effect of a given function f, so that both composite functions $f \circ f^{-1}$ and $f^{-1} \circ f$ have the effect of leaving x unchanged; that is

$$(f \circ f^{-1})(x) = x$$

and

$$(f^{-1} \circ f)(x) = x$$

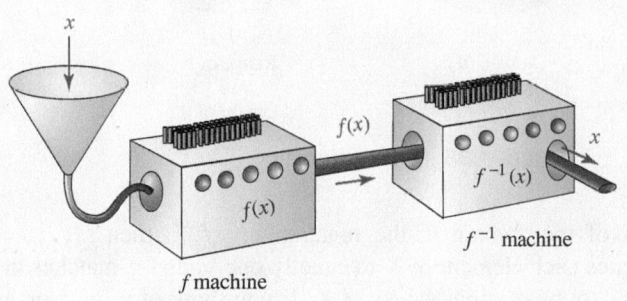

f machine

f⁻¹ machine

The inverse function

The function f^{-1} reverses the effect of the function f. This relationship can be illustrated by function "machines."

Inverse Functions

For a given function f, we write $y_0 = f(x_0)$ to indicate that f maps the number x_0 in its domain into the corresponding number y_0 in its range. **If f has an inverse f^{-1}**, it is the function that reverses the effect of f in the sense that

$$f^{-1}(y_0) = x_0$$

For example, if

$$f(x) = 2x - 3, \text{ then } f(0) = -3, \ f(1) = -1, \ f(2) = 1,$$

The symbol f^{-1} means the inverse of f and does not mean reciprocal, $1/f$.

and the inverse f^{-1} reverses f so that

$$f(0) = -3 \quad f^{-1}(-3) = 0 \quad \text{that is, } f^{-1}[f(0)] = 0$$

$$f(1) = -1 \quad f^{-1}(-1) = 1 \quad \text{that is, } f^{-1}[f(1)] = 1$$

$$f(2) = 1 \quad\ \ f^{-1}(1) = 2 \quad \text{that is, } f^{-1}[f(2)] = 2$$

In the case where the inverse of a function is itself a function, we have the following definition.

INVERSE FUNCTION   Let f be a function with domain D and range R. Then the function f^{-1} with domain R and range D is the **inverse of f** if

$$f^{-1}[f(x)] = x \qquad \text{for all } x \text{ in } D$$

and

$$f[f^{-1}(y)] = y \qquad \text{for all } y \text{ in } R$$

■ **W**hat this says Suppose we consider a function defined by a set of ordered pairs (x, y) where $y = f(x)$. The image of x is y, as shown in Figure 1.38.

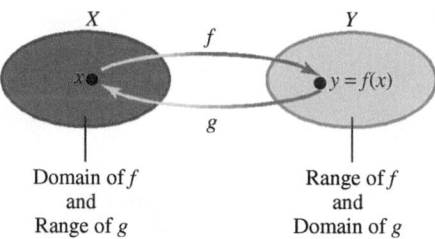

Figure 1.38   Inverse functions f and g

If y is a member of the domain of the function $g = f^{-1}$, then $f^{-1}(y) = x$. This means that f matches each element of X to exactly one y, and g matches those same elements of Y back to the original values of X. If you think of a function as a set of ordered pairs (x, y), the inverse of f is the set of ordered pairs with the components (y, x).

The language of this definition suggests that there is only one inverse function of f. Indeed, it can be shown (see Problem 59) that if f has an inverse, then the inverse is unique.

Example 1   Inverse of a given function defined as a set of ordered pairs

Let $f = \{(0, 3), (1, 5), (3, 9), (5, 13)\}$; find f^{-1}, if it exists.

Solution   The inverse simply reverses the ordered pairs:

$$f^{-1} = \{(3, 0), \ (5, 1), \ (9, 3), \ (13, 5)\}$$ ■

Example 2　Inverse of a given function defined by an equation

Let $f(x) = 2x - 3$; find f^{-1}, if it exists.

Solution　To find f^{-1}, let $y = f(x)$, then interchange the x and y variables, and finally solve for y.

$$\text{Given function: } y = 2x - 3 \quad \text{Then } x = 2y - 3$$
$$2y = x + 3$$
$$\text{Inverse:} \qquad y = \frac{1}{2}(x + 3)$$

Thus, we represent the inverse function as $f^{-1}(x) = \frac{1}{2}(x + 3)$. To verify that these functions are inverses of each other, we note that

$$f[f^{-1}(x)] = f\left[\frac{1}{2}(x+3)\right] = 2\left[\frac{1}{2}(x+3)\right] - 3 = x + 3 - 3 = x$$

and

$$f^{-1}[f(x)] = f^{-1}(2x - 3) = \frac{1}{2}[(2x - 3) + 3] = \frac{1}{2}(2x) = x$$

for all x.

Criteria for Existence of an Inverse f^{-1}

The inverse of a function may not exist. For example, both the functions

$$f = \{(0,0),\ (1,1),\ (-1,1),\ (2,4),\ (-2,4)\} \quad \text{and} \quad g(x) = x^2$$

do not have inverses because if we attempt to find the inverses, we obtain relations that are not functions. In the first case, we find

$$\text{Possible inverse of } f : \{(0,0),\ (1,1),\ (1,-1),\ (4,2),\ (4,-2)\}$$

This is not a function because not every member of the domain is associated with a single member in the range: $(1, 1)$ and $(1, -1)$, for example.

In the second case, if we interchange the x and y in the equation for the function g where $y = x^2$ and then solve for y, we find

$$x = y^2 \quad \text{or} \quad y = \pm\sqrt{x} \text{ for } x \geq 0$$

But this is not a function of x, because for any positive value of x, there are two corresponding values of y, namely, $\sqrt{x}$ and $-\sqrt{x}$.

A function f will have an inverse f^{-1} on the interval I when there is exactly one number in the domain associated with each number in the range. That is, f^{-1} exists if $f(x_1)$ and $f(x_2)$ are equal only when $x_1 = x_2$. A function with this property is said to be **one-to-one**. This is equivalent to the graphical criterion, called the *horizontal line test*, shown in Figure 1.39.

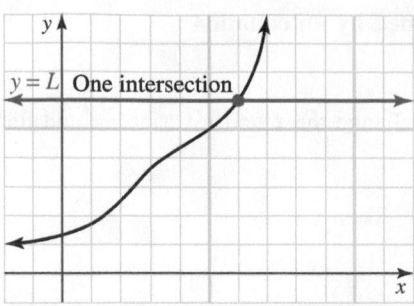 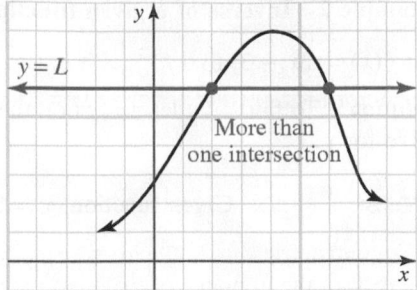

a. A function that has an inverse   **b.** A function that does not have an inverse

Figure 1.39   Horizontal line test

HORIZONTAL LINE TEST   A function f has an inverse if and only if no horizontal line intersects the graph of $y = f(x)$ at more than one point.

A function is said to be **strictly increasing** on an interval I if its graph is always rising on I, and **strictly decreasing** on I if the graph always falls on I. It is called **strictly monotonic** on I if it is either strictly increasing or strictly decreasing throughout that interval. A strictly monotonic function must be one-to-one and hence must have an inverse. For example, if f is strictly increasing on the interval I, we know that

$$x_1 > x_2 \qquad \text{implies} \qquad f(x_1) > f(x_2)$$

so there is no way to have $f(x_1) = f(x_2)$ unless $x_1 = x_2$. This observation is formalized in the following theorem.

Theorem 1.2   A strictly monotonic function has an inverse

Let f be a function that is strictly monotonic on an interval I. Then f^{-1} exists and is strictly monotonic on I (strictly increasing if f is strictly increasing and strictly decreasing if f is strictly decreasing).

Proof: We have already commented on why f^{-1} exists. To show that f^{-1} is strictly increasing whenever f is strictly increasing, let y_1 and y_2 be numbers in the range of f, with $y_2 > y_1$. We will show that $f^{-1}(y_2) > f^{-1}(y_1)$. Because y_1 and y_2 are in the range of f, there exist numbers x_1 and x_2 in the domain I such that $y_1 = f(x_1)$ and $y_2 = f(x_2)$. Because $y_2 > y_1$, it follows that $f(x_2) > f(x_1)$, and because f is strictly increasing, we must have $x_2 > x_1$. Thus, $f^{-1}(y_2) > f^{-1}(y_1)$, and f^{-1} is strictly increasing. Similarly, if f is strictly decreasing, then so is f^{-1}. (The details are left for the reader.)   ◆

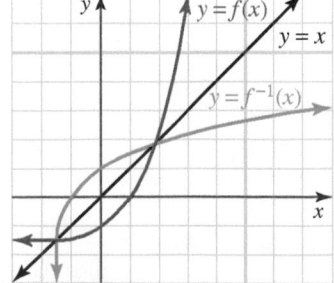

Figure 1.40   The graphs of f and f^{-1} are reflections in the line $y = x$

Graph of f^{-1}

The graphs of f and its inverse f^{-1} are closely related. In particular, if (a, b) is a point on the graph of f, then $b = f(a)$ and $a = f^{-1}(b)$, so (b, a) is on the graph of f^{-1}. It can be shown that (a, b) and (b, a) are reflections of one another in the line $y = x$. (See Figure 1.40). These observations yield the following procedure for sketching the graph of an inverse function.

OBTAINING THE GRAPH OF f^{-1} If f^{-1} exists, its graph may be obtained by reflecting the graph of f in the line $y = x$.

Inverse Trigonometric Functions

The trigonometric functions are not one-to-one, so their inverses do not exist. However, if we restrict the domains of the trigonometric functions, then the inverses exist on those domains.

Let us consider the sine function first. We know that the sine function is strictly increasing on the closed interval $\left[-\dfrac{\pi}{2}, \dfrac{\pi}{2}\right]$, and if we restrict $\sin x$ to this interval, it does have an inverse, as shown in Figure 1.41.

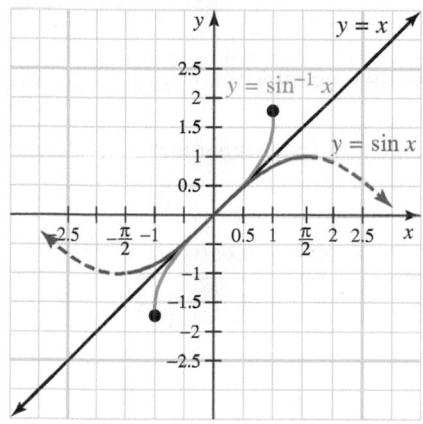

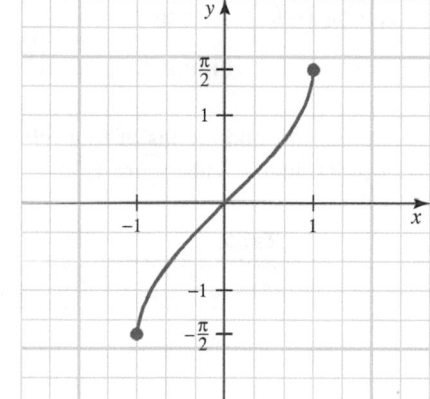

a. The graph of $y = \sin^{-1} x$ is obtained by reflecting the part of the sine on $\left[-\frac{\pi}{2}, \frac{\pi}{2}\right]$ about $y = x$

b. The graph of the inverse sine function, $y = \sin^{-1} x$

Figure 1.41 Graph of the inverse sine function

INVERSE SINE FUNCTION $y = \sin^{-1} x$ if and only if $x = \sin y$ and $-\frac{\pi}{2} \leq y \leq \frac{\pi}{2}$
The function $\sin^{-1} x$ is sometimes written $\arcsin x$.

The function $\sin^{-1} x$ is NOT the reciprocal of $\sin x$. To denote the reciprocal, write $(\sin x)^{-1}$. In other words, $\sin^{-1} x \neq \dfrac{1}{\sin x}$ whereas $(\sin x)^{-1} = \dfrac{1}{\sin x}$.

Inverses of the other five trigonometric functions may be constructed in a similar manner. For example, by restricting $\tan x$ to the open interval $\left(-\frac{\pi}{2}, \frac{\pi}{2}\right)$ where it is one-to-one, we can define the inverse tangent function as follows.

INVERSE TANGENT FUNCTION $y = \tan^{-1} x$ if and only if $x = \tan y$ and $-\frac{\pi}{2} < y < \frac{\pi}{2}$
The function $\tan^{-1} x$ is sometimes written $\arctan x$.

The graph of $y = \tan^{-1} x$ is shown in Figure 1.42.

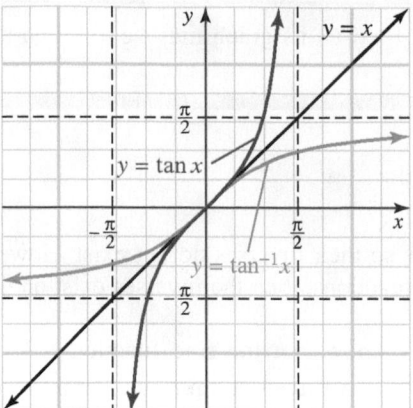
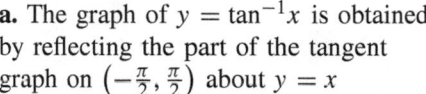
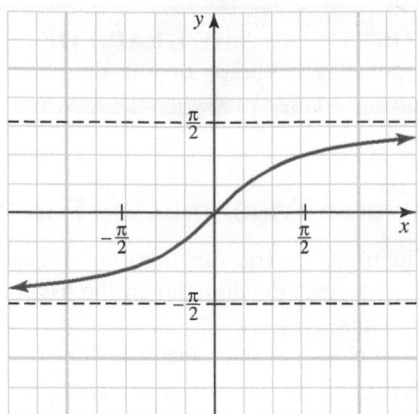

a. The graph of $y = \tan^{-1} x$ is obtained by reflecting the part of the tangent graph on $\left(-\frac{\pi}{2}, \frac{\pi}{2}\right)$ about $y = x$

b. The graph of the inverse tangent function, $y = \tan^{-1} x$

Figure 1.42   Graph of the inverse tangent function

Definitions and graphs of four other fundamental inverse trigonometric functions are given in Table 1.5 and Figure 1.43.

It is easier to remember the restrictions on the domain and range if you do so in terms of quadrants, as shown in Table 1.5

Table 1.5   Definition of inverse trigonometric functions

Inverse function	Domain	Range
$y = \sin^{-1} x$	$-1 \le x \le 1$	$-\dfrac{\pi}{2} \le y \le \dfrac{\pi}{2}$ *Quadrants I and IV*
$y = \cos^{-1} x$	$-1 \le x \le 1$	$0 \le y \le \pi$ *Quadrants I and II*
$y = \tan^{-1} x$	$-\infty < x < \infty$	$-\dfrac{\pi}{2} < y < \dfrac{\pi}{2}$ *Quadrants I and IV*
$y = \csc^{-1} x$	$x \ge 1$ or $x \le -1$	$-\dfrac{\pi}{2} \le y \le \dfrac{\pi}{2},\ y \ne 0$ *Quadrants I and IV*
$y = \sec^{-1} x$	$x \ge 1$ or $x \le -1$	$0 \le y \le \pi,\ y \ne \dfrac{\pi}{2} =$ *Quadrants I and II*
$y = \cot^{-1} x$	$-\infty < x < \infty$	$0 < y < \pi$ *Quadrants I and II*

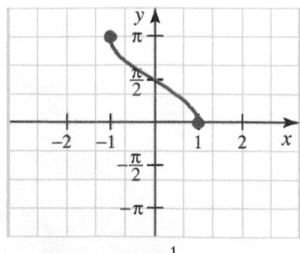
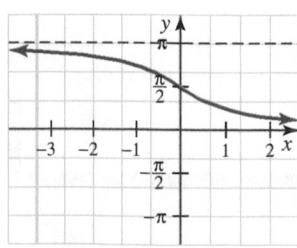
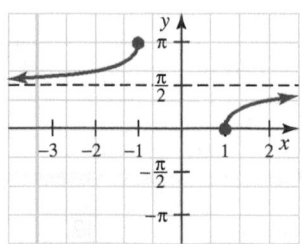
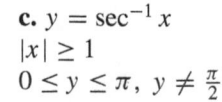
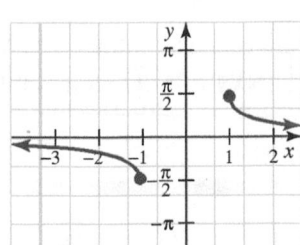

a. $y = \cos^{-1} x$
$-1 \le x \le 1$
$0 \le y \le \pi$

b. $y = \cot^{-1} x$
all x
$0 < y < \pi$

c. $y = \sec^{-1} x$
$|x| \ge 1$
$0 \le y \le \pi,\ y \ne \frac{\pi}{2}$

d. $y = \csc^{-1} x$
$|x| \ge 1$
$-\frac{\pi}{2} \le y \le \frac{\pi}{2}, y \ne 0$

Figure 1.43   Graphs of four inverse trigonometric functions

Example 3   Evaluating inverse trigonometric functions

Evaluate the given functions.

a. $\sin^{-1}\left(\frac{-\sqrt{2}}{2}\right)$ **b.** $\sin^{-1}0.21$ **c.** $\cos^{-1}0$ **d.** $\tan^{-1}\left(\frac{1}{\sqrt{3}}\right)$

Solution

a. $\sin^{-1}\left(\dfrac{-\sqrt{2}}{2}\right) = -\dfrac{\pi}{4}$ *Think: $x = \frac{-\sqrt{2}}{2}$ is negative, so y is in Quadrant IV, the reference angle is the angle whose sine is $\frac{\sqrt{2}}{2}$; it is $\frac{\pi}{4}$ so in Quadrant IV the angle is $-\frac{\pi}{4}$.*

b. $\sin^{-1}0.21 \approx 0.2115750$ *By calculator; be sure to use radian mode and inverse sine (not reciprocal).*

c. $\cos^{-1}0 = \dfrac{\pi}{2}$ *Memorized exact value*

d. $\tan^{-1}\left(\dfrac{1}{\sqrt{3}}\right) = \dfrac{\pi}{6}$ *Think: $x = \frac{1}{\sqrt{3}}$ is positive, so y is in Quadrant I; the reference angle is the same as the value of the inverse tangent in Quadrant I.*

Inverse Trigonometric Identities

The definition of inverse functions yields four formulas, which we call the inversion formulas for sine and tangent.

INVERSION FORMULAS

$$\sin(\sin^{-1}x) = x \quad \text{for } -1 \le x \le 1$$

$$\sin^{-1}(\sin y) = y \quad \text{for } -\frac{\pi}{2} \le y \le \frac{\pi}{2}$$

$$\tan(\tan^{-1}x) = x \quad \text{for all } x$$

$$\tan^{-1}(\tan y) = y \quad \text{for } -\frac{\pi}{2} < y < \frac{\pi}{2}$$

The inversion formulas for $\sin^{-1}x$ and $\tan^{-1}x$ are valid only on the specified domains.

Example 4   Inversion formula for x inside and outside domain

Evaluate the given functions.

a. $\sin(\sin^{-1}0.5)$ **b.** $\sin(\sin^{-1}2)$ **c.** $\sin^{-1}(\sin 0.5)$ **d.** $\sin^{-1}(\sin 2)$

Solution

a. $\sin(\sin^{-1}0.5) = 0.5$, because $-1 \le 0.5 \le 1$.
b. $\sin(\sin^{-1}2)$ does not exist, because 2 is not between -1 and 1
c. $\sin^{-1}(\sin 0.5) = 0.5$, because $-\frac{\pi}{2} \le 0.5 \le \frac{\pi}{2}$.
d. $\sin^{-1}(\sin 2) = 1.1415927$, by calculator. For exact values, notice that

$$\sin 2 = \sin(\pi - 2) \quad \text{and} \quad -\frac{\pi}{2} \le \pi - 2 \le \frac{\pi}{2}$$

so that we have $\sin^{-1}(\sin 2) = \sin^{-1}[\sin(\pi - 2)] = \pi - 2$.

Some trigonometric identities correspond to inverse trigonometric identities, but others do not. For example,

$$\sin(-x) = -\sin x \quad \text{and} \quad \cos(-x) = \cos x$$

It is true that

$$\sin^{-1}(-x) = -\sin^{-1}(x)$$

but in general,

$$\cos^{-1}(-x) \neq \cos^{-1}x$$

[For a counterexample, try $x = 1$: $\cos^{-1}(-1) = \pi$ and $\cos^{-1} 1 = 0$.]

Example 5 Proving inverse trigonometric identities

For $-1 \leq x \leq 1$, show that: **a.** $\sin^{-1}(-x) = -\sin^{-1}x$ **b.** $\cos(\sin^{-1} x) = \sqrt{1 - x^2}$

Solution

a. Let

$$
\begin{aligned}
y &= \sin^{-1}(-x) & \\
\sin y &= -x & \textit{Definition of inverse sine} \\
-\sin y &= x & \\
\sin(-y) &= x & \textit{Opposite angle identity} \\
-y &= \sin^{-1}x & \textit{Definition of inverse sine} \\
y &= -\sin^{-1}x &
\end{aligned}
$$

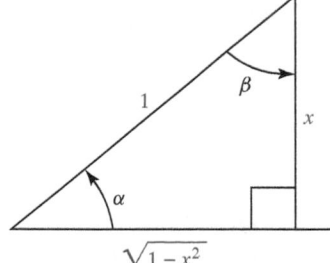

Figure 1.44   A reference triangle

b. Let $\alpha = \sin^{-1} x$ so $\sin \alpha = x$, where $-\frac{\pi}{2} \leq \alpha \leq \frac{\pi}{2}$. Here we consider the case where $0 \leq \alpha \leq \frac{\pi}{2}$, and the other case where $-\frac{\pi}{2} \leq \alpha < 0$ is handled similarly. Construct a right triangle with an acute angle α and hypotenuse 1, as shown in Figure 1.44. We call this triangle a **reference triangle**. The side opposite α is x (since $\sin \alpha = x$), and by the Pythagorean theorem, the adjacent side is $\sqrt{1 - x^2}$. Thus, we have

$$
\begin{aligned}
\cos(\sin^{-1} x) &= \cos \alpha \\
&= \frac{\pm\sqrt{1 - x^2}}{1} \\
&= \sqrt{1 - x^2} \quad \textit{Choose the positive value for the radical because} \\
& \qquad\qquad\quad \textit{α is in Quadrant I.}
\end{aligned}
$$

Reference triangles, such as the one shown in Figure 1.44, are extremely useful devices for obtaining inverse trigonometric identities. For instance, let α and β be angles of a right triangle with hypotenuse 1. If the side opposite α is x (so that $\sin \alpha = x$), then $\alpha = \sin^{-1}x$ and $\beta = \cos^{-1} x$

$$\sin^{-1}x + \cos^{-1}x = \alpha + \beta = \frac{\pi}{2} \qquad \text{for } 0 \leq x \leq 1$$

since the acute angles of a right triangle must sum to $\pi/2$. The same reasoning can also be used to show that

$$\tan^{-1}x + \cot^{-1}x = \frac{\pi}{2} \qquad \text{and} \qquad \sec^{-1}x + \csc^{-1}x = \frac{\pi}{2}$$

Identities involving inverse trigonometric functions have a variety of uses. For instance, most calculators have keys for evaluating $\sin^{-1}x$, $\cos^{-1}x$, and $\tan^{-1}x$, but what about the other three inverse trigonometric functions? The answer is given by the following theorem, which allows us to compute $\sec^{-1}x$, $\csc^{-1}x$, and $\cot^{-1}x$ using reciprocal identities involving the three inverse trigonometric functions the calculator does have.

Theorem 1.3   Reciprocal identities for inverse trigonometric functions

$$\sec^{-1}x = \cos^{-1}\frac{1}{x} \qquad \text{if } |x| \geq 1$$

$$\csc^{-1}x = \sin^{-1}\frac{1}{x} \qquad \text{if } |x| \geq 1$$

$$\cot^{-1}x = \frac{\pi}{2} - \tan^{-1}x$$

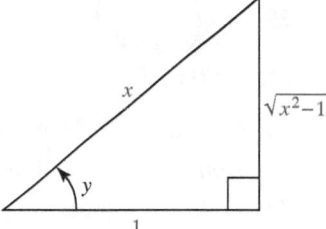

Figure 1.45   Reference triangle

Proof: We will prove that $\sec^{-1}x = \cos^{-1}\frac{1}{x}$ and leave the other two parts as an exercise. Let $y = \sec^{-1}x$. Then $\sec y = x$ and by using the reference triangle in Figure 1.45, we see that $\cos y = \frac{1}{x}$ so that

$$y = \sec^{-1}x = \cos^{-1}\frac{1}{x}$$

Note that the reference triangle is valid only if $|x| > 1$.   ◆

Example 6   Evaluating inverse reciprocal functions

Evaluate the given inverse functions using the inverse identities and a calculator.

a. $\sec^{-1}(-3)$ **b.** $\csc^{-1}7.5$ **c.** $\cot^{-1}2.4747$

Solution

a. $\sec^{-1}(-3) = \cos^{-1}\left(-\frac{1}{3}\right) \approx 1.9106332356$

b. $\csc^{-1}7.5 = \sin^{-1}\left(\frac{1}{7.5}\right) \approx 0.1337315894$

c. $\cot^{-1}2.4747 = \frac{\pi}{2} - \tan^{-1}2.4747 \approx 0.3840267299$ ◼

PROBLEM SET 1.5

Level 1

1. EXPLORATION PROBLEM   Discuss the restrictions on the domain and range in the definition of the inverse trigonometric functions.

2. EXPLORATION PROBLEM   Discuss the use of reference triangles with respect to the inverse trigonometric functions.

Determine which pairs of functions defined by the equations in Problems 3-10 are inverses of each other.

3. $f(x) = 5x + 3;\ g(x) = \dfrac{x-3}{5}$

4. $f(x) = \frac{2}{3}x + 2;\ g(x) = \frac{3}{2}x + 3$

5. $f(x) = \frac{4}{5}x + 4;\ g(x) = \frac{5}{4}x + 3$

6. $f(x) = \frac{1}{x}, x \neq 0;\ g(x) = \frac{1}{x},\ x \neq 0$

7. $f(x) = x^2,\ x < 0;\ g(x) = \sqrt{x},\ x > 0$

8. $f(x) = x^2,\ x \geq 0;\ g(x) = \sqrt{x},\ x \geq 0$

9. $f(x) = x^2,\ x > 0;\ g(x) = -\sqrt{x},\ x > 0$

10. $f(x) = x^2,\ x < 0;\ g(x) = -\sqrt{x},\ x > 0$

Find the inverse (if it exists) of each function given in Problems 11-20.

11. $f = \{(4,5),(6,3),(7,1),(2,4)\}$

12. $g = \{(3,9),(-3,9),(4,16),(-4,16)\}$

13. $f(x) = 2x + 3$ **14.** $g(x) = -3x + 2$

15. $f(x) = x^2 - 5,\ x \geq 0$ **16.** $g(x) = x^2 - 5,\ x < 0$

17. $F(x) = \sqrt{x} + 5$ **18.** $G(x) = 10 - \sqrt{x}$

19. $h(x) = \dfrac{2x-6}{3x+3}$ **20.** $h(x) = \dfrac{2x+1}{x}$

Give the exact values for functions in Problems 21-34.

21. a. $\cos^{-1}\frac{1}{2}$ **b.** $\cos^{-1}\left(-\frac{1}{2}\right)$

22. a. $\sin^{-1}\left(-\frac{1}{2}\right)$ **b.** $\sin^{-1}\frac{1}{2}$

23. a. $\tan^{-1}(-1)$ **b.** $\cot^{-1}(-\sqrt{3})$

24. a. $\cot^{-1}\sqrt{3}$ **b.** $\cot^{-1}(-1)$

25. a. $\sec^{-1}(-\sqrt{2})$ **b.** $\csc^{-1}(-\sqrt{2})$

26. a. $\sec^{-1}1$ **b.** $\sec^{-1}(-1)$

27. a. $\sin^{-1}\left(\dfrac{\sqrt{3}}{2}\right)$ **b.** $\sin^{-1}\left(-\dfrac{\sqrt{3}}{2}\right)$

28. a. $\sec^{-1}\left(\dfrac{2}{\sqrt{3}}\right)$ **b.** $\csc^{-1}\left(-\dfrac{2}{\sqrt{2}}\right)$

29. $\cos\left(\sin^{-1}\dfrac{1}{2}\right)$ **30.** $\sin\left(\cos^{-1}\dfrac{1}{\sqrt{2}}\right)$

31. $\cot\left(\tan^{-1}\frac{1}{3}\right)$ **32.** $\tan\left(\sin^{-1}\frac{1}{3}\right)$

33. $\cos\left(\sin^{-1}\frac{1}{5} + 2\cos^{-1}\frac{1}{5}\right)$

Hint: Use the addition law for $\cos(\alpha + \beta)$.

34. $\sin\left(\sin^{-1}\frac{1}{5} + \cos^{-1}\frac{1}{4}\right)$

Hint: Use the addition law for $\sin(\alpha + \beta)$.

35. Suppose that α is an acute angle where

$$\sin\alpha = \frac{s^2 - t^2}{s^2 + t^2} \qquad (s > t > 0)$$

Show that $\alpha = \tan^{-1}\left(\dfrac{s^2 - t^2}{2st}\right)$.

36. If $\sin\alpha + \cos\alpha = s$ and $\sin\alpha - \cos\alpha = t$, where α is an acute angle, show that

$$\alpha = \tan^{-1}\left(\frac{s + t}{s - t}\right)$$

Level 2

Sketch the graph of f in Problems 37-42 and then use the horizontal line test to determine whether f has an inverse. If f^{-1} exists, sketch its graph.

37. $f(x) = x^2$, for all x **38.** $f(x) = x^2$, $x \le 0$

39. $f(x) = \sqrt{1 - x^2}$, on $(-1, 1)$

40. $f(x) = x(x - 1)(x - 2)$, on $[1, 2]$

41. $f(x) = \cos x$, on $[0, \pi]$

42. $f(x) = \tan x$, on $\left(-\frac{\pi}{2}, \frac{\pi}{2}\right)$

Simplify each expression in Problems 43-48.

43. $\sin(2\sin^{-1} x)$ **44.** $\sin(2\tan^{-1} x)$

45. $\tan(\cos^{-1} x)$ **46.** $\tan(2\sin^{-1} x)$

47. $\sin(\sin^{-1} x + \cos^{-1} x)$

48. $\cos 2(\sin^{-1} x + \cos^{-1} x)$

49. Use reference triangles, if necessary, to justify

$$\cot^{-1}x = \frac{\pi}{2} - \tan^{-1}x \quad \text{for all } x$$

50. Use reference triangles, if necessary, to justify

$$\csc^{-1}x = \sin^{-1}\left(\frac{1}{x}\right) \quad \text{for all } |x| \ge 1$$

51. Use the identities in Theorem 1.3 to evaluate each function rounded to four decimal places.

a. $\cot^{-1} 0.67$ **b.** $\sec^{-1} 1.34$
c. $\csc^{-1} 2.59$ **d.** $\cot^{-1}(-1.54)$

52. Use the identities in Theorem 1.3 to evaluate each function rounded to four decimal places.

a. $\cot^{-1} 1.5$ **b.** $\cot^{-1}(-1.5)$
c. $\sec^{-1}(-1.7)$ **d.** $\csc^{-1}(-1.84)$

53. A painting 3 ft high is hung on a wall in such a way that its lower edge is 7 ft above the floor. An observer whose eyes are 5 ft above the floor stands x ft away from the wall. Express the angle θ subtended by the painting as a function of x.

54. To determine the height of a building (see Figure 1.46), select a point P and find the angle of elevation to be α. Then move out a distance of x units (on a level plane) to point Q and find that the angle of elevation is now β. Find the height h of the building, as a function of x.

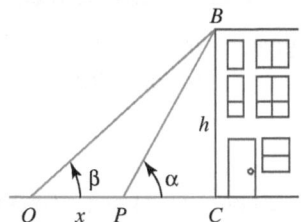

Figure 1.46 Determining height

Level 3

55. a. Find $\tan^{-1} 1 + \tan^{-1} 2 + \tan^{-1} 3$ using a calculator. Make a conjecture about the exact value. Prove your conjecture using reference triangles.

b. Prove the conjecture from part **a** using trigonometric identities. *Hint:* Find $\tan(\tan^{-1} 1 + \tan^{-1} 2 + \tan^{-1} 3)$.

56. Prove the conjecture from Problem 55a using right triangles as follows. You may use the figure shown in Figure 1.47. Assume that $\triangle ABD$, $\triangle ABC$, and $\triangle DEF$ are all right triangles with lengths of sides as shown.

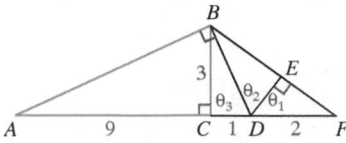

Figure 1.47 $\theta_1 + \theta_2 + \theta_3 = \pi$

57. We proved the identity

$$\csc^{-1}x = \sin^{-1}\left(\frac{1}{x}\right) \quad \text{for all } |x| \ge 1$$

By examining the graph of $y = \csc^{-1} x$, conjecture an identity of the form

$$\csc^{-1}(-x) = A + B \csc^{-1}x \text{ for } x > 1$$

Prove this identity, and then use it to evaluate $\csc^{-1}(-9.38)$.

58. **Think Tank Problem** In Theorem 1.3 we proved the identity

$$\sec^{-1}x = \cos^{-1}\left(\frac{1}{x}\right) \text{ for all } |x| \geq 1$$

The graph of $y = \sec^{-1} x$ suggests that

$$\sec^{-1}(-x) = \pi + \sec^{-1}x \text{ for } x > 1$$

Either prove this identity or find a counterexample.

59. Show that if f^{-1} exists, it is unique. *Hint*: If g_1 and g_2 both satisfy

$$g_1[f(x)] = x = f[g_1(x)]$$
$$g_2[f(x)] = x = f[g_2(x)]$$

then $g_1(x) = g_2(x)$.

60. Suppose that $\triangle ABC$ is *not* a right triangle, but has an obtuse angle γ. Draw $\overline{BD}$ perpendicular to $\overline{AC}$, forming right triangles $\triangle ABD$ and $\triangle BDC$ (with right angles at D). Show that

$$\frac{\sin \alpha}{a} = \frac{\sin \gamma}{c}$$

CHAPTER 1 REVIEW

*E*verything that the greatest minds of all time have accomplished toward the comprehension of forms by means of concepts is gathered into one great science, mathematics.

J. F. Herbart (1890)

Pestalazzi's Idee eines A, B, C der Anschauung, Werke [Kehrbach], (Langensalza, 1890), Bd. 1, p. 163

Proficiency Examination

Concept Problems

1. Characterize the following sets of numbers: natural numbers ($\mathbb{N}$), whole numbers ($\mathbb{W}$), integers ($\mathbb{Z}$), rational numbers ($\mathbb{Q}$), irrational numbers ($\mathbb{Q}'$), and real numbers ($\mathbb{R}$).
2. Define absolute value.
3. State the triangle inequality.
4. State the distance formula for points $P(x_1, y_1)$ and $Q(x_2, y_2)$.
5. Define slope in terms of angle of inclination.
6. List the following forms of the equation of a line:

    a. standard form   **b.** slope-intercept form   **c.** point-slope form   **d.** horizontal line

    e. vertical line
7. State the slope criteria for parallel and for perpendicular lines.
8. Define a function.
9. Define the composition of functions.
10. What is meant by the graph of a function?
11. Draw a quick sketch of an example of each of the following functions.

    a. identity function   **b.** standard quadratic function   **c.** standard cubic function

    d. absolute value function   **e.** cube root function   **f.** standard reciprocal function

    g. standard reciprocal squared function   **h.** cosine function

    i. sine function   **j.** tangent function   **k.** secant function   **l.** cosecant function

    m. cotangent function   **n.** inverse cosine function   **o.** inverse sine function

    p. inverse tangent function   **q.** inverse secant function   **r.** inverse cosecant function

    s. inverse cotangent function
12. What is a polynomial function?
13. What is a rational function?
14. **a.** Define an inverse function.

    b. What is the procedure for graphing the inverse of a given function?
15. What is the horizontal line test?
16. State the inversion formulas for sine and tangent.
17. State the reciprocal inverse trigonometric identities.

Practice Problems

18. Find an equation for the lines satisfying the given conditions:

    a. through $\left(-\frac{1}{2}, 5\right)$ with slope $m = -\frac{3}{4}$

    b. through $(-3, 5)$ and $(7, 2)$

    c. with x-intercept $(4, 0)$ and y-intercept $\left(0, -\frac{3}{7}\right)$

    d. through $\left(-\frac{1}{2}, 5\right)$ and parallel to the line $2x + 5y - 11 = 0$

    e. the perpendicular bisector of the line segment joining $P(-3, 7)$ and $Q(5, 1)$

Sketch the graph of each of the equations in Problems 19-26.

19. $3x + 2y - 12 = 0$ **20.** $y - 3 = |x + 1|$ **21.** $y - 3 = -2(x - 1)^2$

22. $y = x^2 - 4x - 10$ **23.** $y = 2\cos(x - 1)$ **24.** $y + 1 = \tan(2x + 3)$

25. $y = \sin^{-1}(2x)$ **26.** $y = \tan^{-1} x^2$

27. If $f(x) = \dfrac{1}{x + 1}$, what value(s) of x satisfy $f\left(\dfrac{1}{x + 1}\right) = f\left(\dfrac{2x + 1}{2x + 4}\right)$?

28. If $f(x) = \sqrt{\dfrac{x}{x - 1}}$ and $g(x) = \dfrac{\sqrt{x}}{\sqrt{x - 1}}$, does $f = g$? Why or why not?

29. If $f(x) = \sin x$ and $g(x) = \sqrt{1 - x^2}$, find the composite functions $f \circ g$ and $g \circ f$.

30. An open box with a square base is to be built for \$96. The sides of the box will cost \$3/ft^2 and the base will cost \$8/ft^2. Express the volume of the box as a function of the length of its base.

Supplementary Problems*

In Problems 1-4, *find, if possible, the slope, the y-intercept, and the x-intercept of the line whose equation is given. Sketch the graph of each equation.*

1. $y = \frac{4}{5}x - 5$ **2.** $3x + 4y - 12 = 0$ **3.** $\dfrac{x}{5} + \dfrac{y}{3} = 1$ **4.** $x = 1 + t,\ y = 3 - 2t$

Solve each inequality in Problems 5-8, *and give your answer using interval notation.*

5. $6(1 - x) < 2x + 1$ **6.** $-8 \le -2x < 0$ **7.** $x^2 - x \ge 6$ **8.** $|x - 4| < 0.0001$

In Problems 9-12, *find the perimeter and area of the given figure.*

9. the triangle with vertices $(-1, 3), (-1, 8)$, and $(11, 8)$

10. the triangle bounded by the lines $y = 5,\ 3y - 4x = 11$, and $12x + 5y = 25$

11. the trapezoid with vertices $A(-3, 0), B(5, 0), C(2, 8)$, and $D(0, 8)$

12. the trapezoid with vertices $A(0, 0), B(10, 0), C(12, 10)$, and $D(0, 10)$

In Problems 13-16, *find an equation for the indicated line or circle.*

13. the line through the two points where the circles $x^2 + y^2 - 5x + 7y = 3$ and $x^2 + y^2 + 4y = 0$ intersect

14. the vertical line through the point where the line $y = 2x - 7$ intersects the parabola $y = x^2 + 6x - 3$

15. the circle that is tangent to the x-axis and is centered at $(5, 4)$

16. the circle with center on the y-axis that passes through the origin and is tangent to the line $3x + 4y - 40 = 0$

In Problems 17-20, *find the domain of* f *and compute the indicated values or state that the corresponding x-value is not in the domain. Tell whether any of the indicated values are zeros of the function, that is, values of x that cause the functional value to be* 0.

17. $f(x) = x - \dfrac{3}{x};\ f(-1),\ f(1),\ f(3)$

18. $f(x) = \sqrt{1 - 2x};\ f(-1),\ f(0),\ f(-2)$

19. $f(x) = \cos x - \sin x;\ f(-\pi),\ f(\pi/2),\ f(\pi)$

20. $f(x) = \begin{cases} \sqrt{2x} & \text{if } x > 0 \\ 10 & \text{if } x = 0;\ f(-1),\ f(0),\ f(1) \\ x - 5 & \text{if } x < 0 \end{cases}$

Evaluate the difference quotient $\dfrac{f(x + h) - f(x)}{h}$ *for the functions given in Problems* 21-24.

21. $f(x) = 5x$ **22.** $f(x) = 3x^2$

23. $f(x) = \dfrac{1}{2x}$ **24.** $f(x) = |2x|$ if $x > \frac{1}{2}$ and $0 < h < 1$

---

*The supplementary problems are presented in a somewhat random order, not necessarily in order of difficulty.

Find the composite functions $f \circ g$ and $g \circ f$ for the functions given in Problems 25-28.

25. $f(x) = 3x, g(x) = x^2$

26. $f(x) = 1 + x, \ g(x) = \cos x$

27. $f(u) = u - 1, \ g(u) = u + 1$

28. $f(u) = \dfrac{1 - u}{1 + u}$ and $g(u) = \dfrac{2u + 1}{1 - u}$

Find the x-intercepts, if any, for the functions given in Problems 29-32.

29. $f(x) = 6x^2 + 13x - 10$

30. $f(x) = (x - 10)(x + 15)(3x - 69)(2x + 5)$

31. $f(x) = x^4 - 5x^2 + 4$

32. $f(x) = \dfrac{x(x^2 - 5)}{x^2 + 1}$

Find the inverse (if it exists) for each function given in Problems 33-36.

33. $f = \{(4, 3), (5, 1), (6, 1), (2, 5)\}$

34. $f(x) = 5x - 3$

35. $f(x) = x^2 + 1, \ x < 0$

36. $f(x) = \dfrac{x - 2}{x}$

Sketch the graph of f in Problems 37-40 and then use the horizontal line test to determine whether f has an inverse. If f^{-1} exists, sketch its graph as well.

37. $f(x) = -x^3, \ x \leq 0$

38. $f(x) = x(x - 2)(x + 1)$ on $[1,4]$

39. $f(x) = \cos\left(\frac{\pi}{2} - x\right)$ on $\left(0, \frac{\pi}{2}\right)$

40. $f(x) = \cot x$, on $(0, \pi)$

In Problems 41-44, find f^{-1} if it exists.

41. $f(x) = 2x^3 - 7$

42. $f(x) = \sqrt[7]{2x + 1}, x \geq -\frac{1}{2}$

43. $f(x) = \sqrt{\sin x}, 0 < x < \frac{\pi}{2}$

44. $f(x) = \dfrac{x + 5}{x - 7}, x \neq 7$

Simplify each expression in Problems 45-48.

45. $\cos(2 \cos^{-1} x)$

46. $\cos\left(2 \tan^{-1} x\right)$

47. $\cos\left(\sin^{-1} x + \cos^{-1} x\right)$

48. $\sin 2\left(\sin^{-1} x + \cos^{-1} x\right)$

Find constants A and B so that each of the statements in Problems 49-52 are true.

49. $A = \cos^{-1}\left(-\frac{\sqrt{3}}{2}\right)$

50. $B = \cos\left(\sin^{-1} \frac{1}{2}\sqrt{2}\right)$

51. $\sin^3 x = A \sin 3x + B \sin x$

52. $\tan\left(x + \dfrac{\pi}{3}\right) = \dfrac{A + \tan x}{1 + B \tan x}$

53. Use the vertical line test to determine whether the given curve is the graph of a function, and if it is the graph of a function, use the horizontal line test to determine whether it is one-to-one. Find the probable domain and range by looking at the graphs.

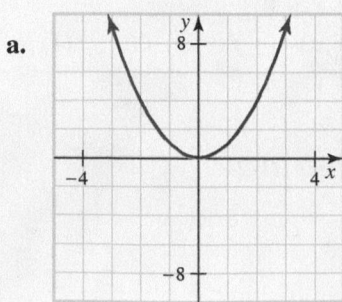

a.

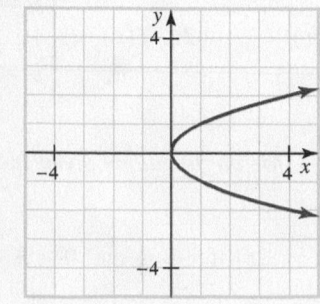

b.

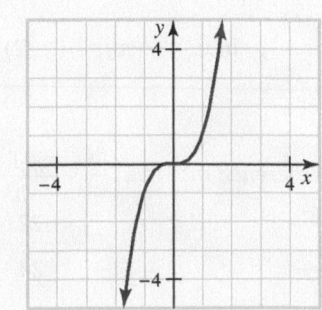

c.

d. 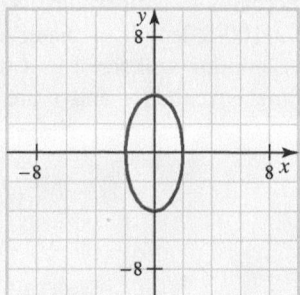   **e.**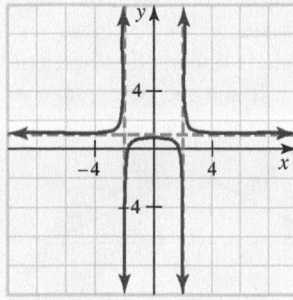

54. In a triangle, the perpendicular segment drawn from a given vertex to the opposite side is called the *altitude* on that side. Consider the triangle with vertices $(-2, 1), (5, 6)$ and $(3, -2)$. Find an equation for each line containing an altitude of this triangle.

55. Show that the three lines found in Problem 54 intersect at the same point. Find the coordinates of this point. Is this the same point where the three medians meet?

56. If an object is shot upward from the ground with an initial velocity of 256 ft/sec, its distance in feet above the ground at the end of t seconds is given by $s(t) = 256t - 16t^2$ (neglecting air resistance). What is the highest point for this projectile?

57. It is estimated that t years from now, the population of a certain suburban community will be

$$P(t) = \frac{11t + 12}{2t + 3}$$

thousand people. What is the current population of the community? What will the population be in 6 years? When will there be 5,000 people in the community?

58. Evaluate each of the given numbers (calculator approximations).

   a. $\tan^{-1} 2$ **b.** $\cot^{-1} 2$ **c.** $\sec^{-1}(-3.1)$

59. Evaluate each of the given numbers (exact values).

   a. $\sin\left(\cos^{-1}\frac{\sqrt{5}}{4}\right)$ **b.** $\sin(2\tan^{-1} 3)$ **c.** $\sin\left(\cos^{-1}\frac{13}{5} + \sin^{-1}\frac{5}{13}\right)$

60. Solve $\sqrt{x} = \cos^{-1} 0.317 + \sin^{-1} 0.317$.

61. Show that for any constant $a \neq 1$, the function $f(x) = \dfrac{x + a}{x - 1}$ is its own inverse.

62. Let $f(x) = \dfrac{ax + b}{cx + d}$, for constants a, b, c and d. Find $f^{-1}(x)$ in terms of $a, b, c,$ and d.

63. Find $f^{-1}(x)$ if $f(x) = \dfrac{x + 1}{x - 1}$. What is the domain of f^{-1}?

64. First show that $\tan^{-1}x + \tan^{-1}y = \tan^{-1}\left(\dfrac{x + y}{1 - xy}\right)$ for $xy \neq 1$ whenever $-\dfrac{\pi}{2} < \tan^{-1}\left(\dfrac{x + y}{1 - xy}\right) < \dfrac{\pi}{2}$. Then establish the following equations.

   a. $\tan^{-1}\frac{1}{2} + \tan^{-1}\frac{1}{3} = \frac{\pi}{4}$ **b.** $2\tan^{-1}\frac{1}{3} + \tan^{-1}\frac{1}{7} = \frac{\pi}{4}$ **c.** $4\tan^{-1}\frac{1}{5} - \tan^{-1}\frac{1}{239} = \frac{\pi}{4}$

   Note: The identity in part **c** will be used in Chapter 8 to estimate the value of π.

65. **Think Tank Problem** Each of the following equations may be either true or false. In each case, either show that the equation is generally true or find a counterexample.

   a. $\tan^{-1}x = \dfrac{\sin^{-1}x}{\cos^{-1}x}$ **b.** $\tan^{-1}x = \dfrac{1}{\tan x}$ **c.** $\cot^{-1}x = \frac{\pi}{2} - \tan^{-1}x$

   d. $\cos\left(\sin^{-1}x\right) = \sqrt{1 - x^2}$ **e.** $\sec^{-1}\left(\frac{1}{x}\right) = \cos^{-1}x$

66. Many materials, such as brick, steel, aluminum, and concrete, expand with increases in temperature. This is why spaces are placed between the cement slabs in sidewalks. Suppose you have a 100-ft length of material securely fastened at both ends, and assume that the buckle is linear. (It is not, but this assumption will serve as a worthwhile approximation.) If the height of the buckle is x ft and the percentage of swelling is y, then x and y are related as shown in Figure 1.48.

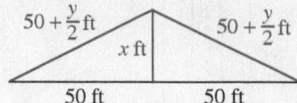

Figure 1.48 Buckling of a given material

Find the amount of buckling (to the nearest inch) for the following materials:
a. brick; $y = 0.03$ [This means $(0.03\%)(100\,\text{ft}) = 0.03$ ft, which is y in Figure 1.48.]

b. steel; $y = 0.06$ **c.** aluminum; $y = 0.12$ **d.** concrete; $y = 0.05$

67. A person is swimming at a depth of d units beneath the surface of the water. Because the light is refracted by the water, a viewer standing above the waterline sees the swimmer at an apparent depth of s units. (See Figure 1.49.)

Figure 1.49 Refracted light when viewing an underwater swimmer

In physics*, it is shown that if the person is viewed from an angle of incidence θ, then

$$s = \frac{3d \cos \theta}{\sqrt{7 + 9 \cos^2\theta}}$$

a. If $d = 5.0$ meters and $\theta = 37°$, what is the apparent depth of the swimmer?
b. If the actual depth is $d = 5.0$ meters, what angle of incidence yields an apparent depth of $s = 2.5$ meters?

68. A ball has been dropped from the top of a building. Its height (in feet) after t seconds is given by the function $h(t) = -16t^2 + 256$.
a. How high will the ball be after 2 sec?
b. How far will the ball travel during the third second?
c. How tall is the building?
d. When will the ball hit the ground?

69. Suppose the number of worker-hours required to distribute new telephone books to x percent of the households in a certain rural community is given by the function

$$f(x) = \frac{600x}{300 - x}$$

a. What is the domain of the function f?
b. For what values of x does $f(x)$ have a practical interpretation in this context?
c. How many worker-hours were required to distribute new telephone books to the first 50% of the households?
d. How many worker-hours were required to distribute new telephone books to the entire community?
e. What percentage of the households in the community had received new telephone books by the time 150 worker-hours had been expended?

70. Find the area of each of the following plane figures:
a. the circle with $P(0,0)$ and $Q(2,3)$ endpoints of a diameter
b. the trapezoid with vertices $A(0,0), B(4,0), C(1,3)$, and $D(2,3)$

---

*R. A. Serway, *Physics*, 3rd ed., Philadelphia: Saunders, 1992, p. 1007.

71. Find the volume and the surface area of each of the following solid figures:
   a. a sphere with radius 4
   b. a rectangular parallelepiped (box) with sides of length 2, 3, and 5
   c. a right circular cylinder (including top and bottom) with height 4 and radius 2
   d. an inverted cone with height 5 and top radius 3 (lateral surface only)

72. Consider the triangle with vertices $A(-1, 4), B(3, 2)$, and $C(3, -6)$. Determine the midpoints M_1 and M_2 of sides $\overline{AB}$ and $\overline{AC}$, respectively, and show that the line segment $\overline{M_1 M_2}$ is parallel to side $\overline{BC}$, with half its length.

73. Generalize the procedure of Problem 72 to show that the line segment joining the midpoints of any two sides of a given triangle is parallel to the third side and has half its length.

74. Let $ABCD$ be a quadrilateral in the plane, and let P, Q, R, and S be the midpoints of sides $\overline{AB}, \overline{BC}, \overline{CD}$, and $\overline{DA}$, as shown in Figure 1.50. Show that $PQRS$ is a parallelogram.

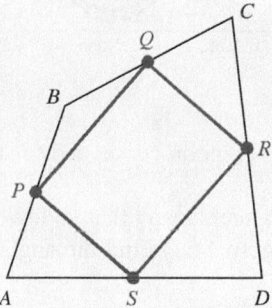

Figure 1.50 Problem 74

75. Find a constant c that guarantees that the graph of the equation $x^2 + xy + cy = 4$ will have a y-intercept of $(0, -5)$. What are the x-intercepts of the graph?

76. A bus charter company offers a travel club the following arrangements: If no more than 100 people go on a certain tour, the cost will be $500 per person, but the cost per person will be reduced by $4 for each person in excess of 100 who takes the tour.
   a. Express the total revenue R obtained by the charter company as a function of the number of people who go on the tour.
   b. Sketch the graph of R. Estimate the number of people that results in the greatest total revenue for the charter company.

77. The current I (in amperes) in a certain circuit (for some convenient unit of time) generates the following set of data points:

Time	Current	Time	Current
0	−60.00000	11	40.14784
1	−58.68886	12	48.54102
2	−54.81273	13	54.81273
3	−48.54102	14	58.68886
4	−40.14784	15	60.00000
5	−30.00000	16	58.68886
6	−18.54102	17	54.81273
7	−6.27171	18	48.54102
8	6.27171	19	40.14784
9	18.54102	20	30.00000
10	30.00000		

Plot the data points and draw a smooth curve passing through these points. Determine possible values of A, B, C, and D so that the graph of these data is approximated by the equation

$$y - A = B \sin[C(x - D)]$$

78. The sun/moon tide curves are shown below. During the third quarter (neap tide) the sun and moon tidal bulges are at right angles to each other. Thus, the sun tide reduces the effect of the moon tide.

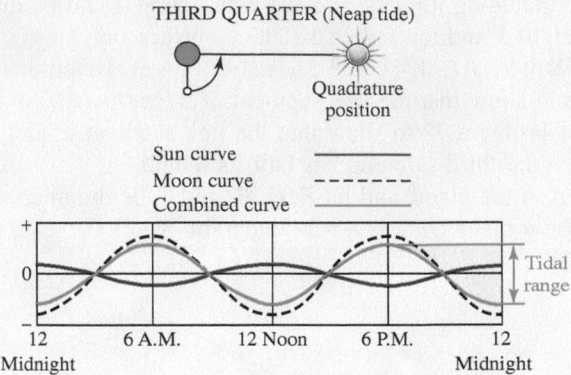

Write possible equations for the sun curve, moon curve, and for the combined curve. Assume the tidal range is 3 ft and the period is 12.

79. A mural 7 feet high is hung on a wall in such a way that its lower edge is 5 feet higher than the eye of an observer standing 12 feet from the wall. (See Figure 1.51) Find the angle θ subtended by the mural at the observer's eye.

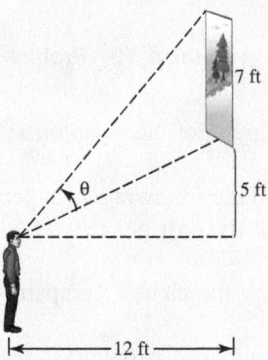

Figure 1.51 Problem 79

80. In Figure 1.52, ship A is at point P at noon and sails due east at 9 km/hr. Ship B arrives at point P at 1:00 P.M. and sails at 7 km/hr along a course that forms an angle of $60°$ with the course of ship A. Find a formula for the distance $s(t)$ separating the ships t hours after noon ($t \geq 1$). Approximately how far apart (to the nearest kilometer) are the ships at 4:00 P.M.?

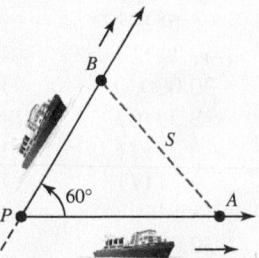

Figure 1.52 Problem 80

81. A manufacturer estimates that when the price for each unit is p dollars, the profit will be $N = -p^2 + 14p - 48$ thousand dollars. Sketch the graph of the profit function and answer these questions.

a. For what values of p is this a profitable operation? (That is, when is $N > 0$?)

b. What price results in maximum profit? What is the maximum profit?

82. Two jets bound for Los Angeles leave New York 30 minutes apart. The first travels 550 mph, and the second flies at 650 mph. How long will it take the second plane to pass the first?

83. To raise money, a service club has been collecting used bottles that it plans to deliver to a local glass company for recycling. Since the project began 8 days ago, the club has collected 2,400 pounds of glass, for which the glass company currently offers 15¢ per pound. However, since bottles are accumulating faster than they can be recycled, the company plans to reduce the price it pays by 1¢ per pound each day until the price reaches 0¢ fifteen days from now. Assuming that the club can continue to collect bottles at the same rate and that transportation costs make more than one trip to the glass company unfeasible, express the club's revenue from its recycling project as a function of the number of additional days the project runs. Draw the graph and estimate when the club should conclude the project and deliver the bottles in order to maximize its revenue.

84. An open box with a square base is to be built for $48. The sides of the box will cost $3/ft^2 and the base will cost $4/ft^2. Express the volume of the box as a function of the length of its base.

85. A closed box with a square base is to have a volume of 250 cubic feet. The material for the top and bottom of the box costs $2/ft^2, and the material for the sides costs $1/ft^2. Express the construction cost of the box as a function of the length of its base.

86. The famous author John Uptight must decide between two publishers who are vying for the rights to his new book, *Zen and the Art of Taxidermy*. Publisher A offers royalties of 1% of net proceeds on the first 30,000 copies sold and 3.5% on all copies in excess of that figure, and expects to net $2 on each copy sold. Publisher B will pay no royalties on the first 4,000 copies sold but will pay 2% on the net proceeds of all copies sold in excess of 4,000 copies, and expects to net $3 on each copy sold.

a. Express the revenue John should expect if he signs with Publisher A (as a function P_A of the number of books sold, x). Likewise, find the revenue function P_B associated with Publisher B.

b. Sketch the graphs of $P_A(x)$ and $P_B(x)$ on the same coordinate axes.

c. For what value(s) of x are the two offers equivalent?

d. With whom should he sign if he expects to sell 5,000 copies? 100,000 copies? 200,000 copies?

e. State a simple criterion for determining which publisher he should choose if he expects to sell N copies.

Use the Internet, a library, or references other than this textbook to research the information necessary to answer the questions in Problems 87-95.

87. *Historical Quest Sir Isaac Newton was one of the greatest mathematicians of all time. He was a genius of the highest order but was often absent-minded. One story about Newton is that, when he was a boy, he was sent to cut a hole in the bottom of the barn door for the cats to go in and out. He cut two holes—a large one for the cat and a small one for the kittens. Newton considered himself a theologian rather than a mathematician or a physicist. He spent years searching for clues about the end of the world and the geography of hell. One of Newton's quotations about himself is, "I seem to have been only like a boy playing on the seashore and diverting myself in now and then finding a smoother pebble or prettier shell than ordinary, whilst the great ocean of truth lay all undiscovered before me."* Write an essay about Isaac Newton and his discovery of calculus. This essay should be at least 500 words.

**Isaac Newton
(1642–1727)**

88. *Historical Quest At the age of 14, Gottfried Leibniz attempted to reform Aristotelian logic. He wanted to create a general method of reasoning by calculation. At the age of 20, he applied for his doctorate at the university in Leipzig and was refused (because, officials said, he was too young). He received his doctorate the next year at the University of Altdorf, where he made such a favorable impression that he was offered a professorship, which he declined, saying he had very different things in view. Leibniz went on to invent calculus, but not without a bitter controversy developing between Leibniz and Newton. Most historians agree that the bitterness over who invented calculus materially affected the history of mathematics. J. S. Mill characterized Leibniz by saying, "It would be difficult to name a man more remarkable for the greatness and universality of his intellectual powers than Leibniz."* Write a 500-word essay about Gottfried Leibniz and his discovery of calculus.

**Gottfried Leibniz
(1646–1716)**

89. *Historical Quest* Write a 500-word essay about the controversy surrounding the discovery of calculus by Newton and Leibniz.

90. *Historical Quest* The Greek mathematicians mentioned in this guest essay include Eudoxus, Euclid, and Archimedes. Write a short paper about contributions they made that might have been used by Newton.

91. *Historical Quest* What is the definition of elementary functions as given by Joseph Liouville?

92. *Historical Quest* Write a short paper about the contributions to the invention of calculus by de Roberval, Fermat, Cavalieri, Huygens, Wallis, and Barrow.

93. *Historical Quest* Suppose that calculus had not been invented by either Newton nor Leibniz. Would someone else have invented the calculus? If so, should we say that calculus was *discovered* rather than *invented*? Write a 500-word essay arguing that calculus was discovered rather than invented.

94. *Historical Quest Sophie Germain was one of the first women to publish original mathematics, but was not admitted to any first-rate university. Women, for the most part, were not taken seriously, so she was forced to write, at first, under the pseudonym Le Blanc. The situation is not too different from that portrayed by Barbra Streisand in the movie "Yentl." Even though Germain's most important research was in number theory, she was awarded the prize of the French Academy for a paper entitled "Memoir on the Vibrations of Elastic Plates."*

Sophie Germain (1776-1831)

As we progress through this book we will profile many mathematicians in the history of mathematics, and you will notice that most of them are white males. Why? Write a paper on the history of women mathematicians and their achievements. Your paper should include a list of many prominent women mathematicians and their primary contributions. It should also include a lengthy profile of at least one woman mathematician.

95. *Historical Quest The Navajo are a Native American people who, despite considerable interchange and assimilation with the surrounding dominant culture, maintain a world view that remains vital and distinctive. The Navajo believe in a dynamic universe. Rather than consisting of objects and situations, the universe is made up of processes. Central to our Western mode of thought is the idea that things are separable entities that can be subdivided into smaller discrete units. For us, things that change through time do so by going from one specific state to another specific state. While we believe time to be continuous, we often even break it into discrete units or freeze it and talk about an instant or point in time. Among the Navajo, where the focus is on process, change is ever present; interrelationship and motion are of primary significance. These incorporate and subsume space and time.*

There are, in every culture, groups or individuals who think more about some ideas than do others. For other cultures, we know about the ideas of some professional groups or some ideas of the culture at large. We know little, however, about the mathematical thoughts of individuals in those cultures who are specially inclined toward mathematical ideas. In Western culture, on the other hand, we focus on, and record much about, those special individuals while including little about everyone else. Realization of this difference should make us particularly wary of any comparisons across cultures. Even more important, it should encourage finding out more about the ideas of mathematically oriented innovators in other cultures and, simultaneously, encourage expanding the scope of Western history to recognize and include mathematical ideas held by different groups within our culture or by our culture as a whole.* Write a paper discussing this quotation.

96. **Putnam Examination Problem**[†] If f and g are real-valued functions of one real variable, show that there exist numbers x, y such that $0 \leq x \leq 1, 0 \leq y \leq 1$, and $|xy - f(x) - g(y)| \geq \frac{1}{4}$.

97. **Putnam Examination Problem** Consider a polynomial $f(x)$ with real coefficients having the property $f(g(x)) = g(f(x))$ for every polynomial $g(x)$ with real coefficients. What is $f(x)$?

---

*From Ethnomathematics by Marcia Ascher, pp. 128-129 and 188-189.

[†]The Putnam Examination is a national annual examination given under the auspices of the Mathematical Association of America and is designed to recognize mathematically talented college and university students. We include Putnam examination problems, which should be considered optional, to encourage students to consider taking the examination. Solutions to Putnam Examinations generally appear in *The American Mathematical Monthly*, approximately one year after the examination.

98. Putnam Examination Problem Figure 1.53 shows a rectangle of base b and height h inscribed in a circle of radius 1, surmounted by an isosceles triangle with the same area as the rectangle.

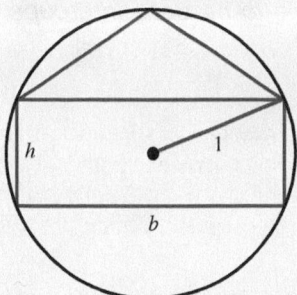

Figure 1.53 Putnam problem

Show that $h = \frac{2}{5}$.

99. **Book Report** "Ethnomathematics, as it is being addressed here, has the goal of broadening the history of mathematics to one that has a multicultural, global perspective." Read the book *Ethnomathematics* by Marcia Ascher (Pacific Grove: Brooks/Cole, 1991), and prepare a book report.

CHAPTER 1 GROUP RESEARCH PROJECT*

Working in small groups is typical of most work environments, and this book seeks to develop skills with group activities. We present a group project at the end of each chapter. These projects are to be done in groups of three or four students.

Rube Goldberg

Courtesy of the Craig Engineering
Club, Janesville, WI

A Rube Goldberg machine is a comically involved, complicated invention, laboriously contrived to perform a simple operation—*Webster's New World Dictionary*. Some examples of tasks from the past contents include putting toothpaste on a toothbrush (1987), raising a U.S. flag (2002) and inflating a balloon and popping it (2012).

© Kendall Hunt Publishing Company

The purpose of this first group project is two-fold. The first, and perhaps most important part, is to motivate you to find out about your classmates. The questions listed below will require that you get to know some of the students in your class. This project is to figure out how to best accomplish those goals.

Extended paper for further study. Part of your paper should include the following information about students in your class:

• Write down the number of students in your class.
• List the five students who live near to where you live.
• List five students who you would most like to know.
• Name some students in your class with whom you share some common interest.
  Name that interest.
• Select a group of three or four students for this project.
• Design a Rube Goldberg machine.

Extra credit. Construct your Rube Goldberg machine and prepare a classroom demonstration.

---

*The Rube Goldberg Machine Contest® is a registered trademark of Rube Goldberg, Inc. RGI is the licensee of the contest. Information about this contest can be found at http://www.rubegoldberg.com

CHAPTER 2

LIMITS AND CONTINUITY

*M*athematics is a vast adventure in ideas; its history reflects some of the noblest thoughts of countless generations.

Dirk J. Struik
A Concise History of Mathematics, Vol. 1,
New York: Dover, 1948, p. x

PREVIEW

As we discussed at the beginning of Chapter 1, calculus is the mathematics of motion and change, whereas algebra, geometry, and trigonometry are more static in nature. The development of calculus in the 17th century by Newton, Leibniz, and others grew out of attempts by these and earlier mathematicians to answer certain fundamental questions about dynamic real-world situations. These investigations led to two procedures, *differentiation* and *integration*, which can be formulated in terms of a concept called the *limit*. We will discuss the limit in this chapter and study its basic properties.

PERSPECTIVE

The limit is a mathematical tool for studying the tendency of a function as its independent variable approaches some value. One limit we mentioned in Chapter 1 was Zeno's paradox, which says:

Archilles fires an arrow, but it can never reach its target. Why? Because if the arrow takes T seconds to travel half the distance to the target, it will take T/2 seconds to finish half the rest, and so on for each half. No matter how close the arrow is to the target, it will take finite time to travel half the distance that remains, so the arrow will never reach its destination.

Common sense tells that the arrow will strike home in $2T$ seconds, so where is the error in reasoning? If we measure the total time of the arrow's flight by the "half, then half of that, and so on," approach of the paradox, it will take

$$T + \frac{T}{2} + \frac{T}{4} + \cdots$$

seconds for the arrow to strike the target; that is, an infinite sum of finite time intervals. The Greeks of Zeno's time assumed that such a sum would have to be infinite. However, the limit process can be used to show that this particular sum is not only finite, but equals $2T$, as common sense would suggest. We will study infinite sums in Chapter 8, but first it is necessary to introduce and explore finding the area of a circle using limits. In a very real sense, the concept of limit is the threshold to modern mathematics. You are about to cross that threshold, and beyond lies the fascinating world of calculus.

73

2.1 THE LIMIT OF A FUNCTION

IN THIS SECTION: *Intuitive notion of a limit, one-sided limits, limits that do not exist, formal definition of a limit.*

This section introduces you to the limit of a function, a concept that gives calculus its power and distinguishes it from other areas of mathematics, such as algebra.

The goal of this section is to define and explore what is meant by the *limit of a function*. We will begin with an informal discussion of the limit concept, emphasizing graphical and numerical computations. Then in Section 2.2, we will develop algebraic techniques for computing limits more systematically.

The development of the limit concept was a major breakthrough in the history of mathematics, and you should not be surprised or disappointed if certain aspects of this concept seem difficult to comprehend or apply. Be patient and this crucial concept will soon become part of your mathematical toolkit.

Intuitive Notion of a Limit

The limit of a function f is a tool for investigating the behavior of $f(x)$ as x gets closer and closer to a particular number c. That is, the limit concerns the *tendency* of $f(x)$ for x near c rather than the *value* of f at $x = c$. The following example illustrates how such a tendency can be used to compute velocity.

Example 1   Computing velocity as a limit

A freely falling body experiencing no air resistance falls $s(t) = 16t^2$ feet in t seconds. Express the body's velocity at time $t = 2$ as a limit.

Solution   What we want is the *instantaneous velocity* after 2 seconds, which can be thought of as the "speedometer reading" of the falling body at $t = 2$. To approximate the instantaneous velocity, we compute the average velocity of the body over smaller and smaller time intervals, either ending with or beginning with $t = 2$. For instance, over the time interval $1.9 \leq t \leq 2$, the average velocity, $\overline{v}(t)$, is

$$\overline{v}(t) = \frac{\text{DISTANCE TRAVELED}}{\text{ELAPSED TIME}} \qquad \textit{Where distance is in ft and time is in seconds.}$$

$$= \frac{s(2) - s(1.9)}{2 - 1.9}$$

$$= 62.4 \text{ ft/s}$$

Table 2.1   Average velocity

Time Interval	Interval length (sec)	Average velocity (ft/s)
$1.8 \leq t \leq 2$	0.2	60.8
$1.9 \leq t \leq 2$	0.1	62.4
$1.99 \leq t \leq 2$	0.01	63.84
$1.999 \leq t \leq 2$	0.001	63.98
$2 \leq t \leq 2.0001$	0.0001	64.0016
$2 \leq t \leq 2.001$	0.001	64.016
$2 \leq t \leq 2.01$	0.01	64.16

Similar computations of average velocity over short time intervals ending or beginning with time $t = 2$ are contained in Table 2.1.

Examining the rows of this table, we see that the average velocity seems to be approaching the value 64 as the time intervals become smaller and smaller. Thus, it is reasonable to expect the velocity at the instant when $t = 2$ to be 64 ft/s.

In symbols, the average velocity of the falling body over the time interval $2 \leq t \leq 2 + h$ is given by

$$\frac{s(2 + h) - s(2)}{(2 + h) - 2} = \frac{16(2 + h)^2 - 16(2)^2}{h}$$

We say that the average velocity has a limiting value of 64 as the length h of the time interval tends to zero (gets smaller and smaller), and we denote this tendency by writing

$$\lim_{h \to 0} \frac{16(2 + h)^2 - 16(2)^2}{h} = 64$$

Here is a general, though informal, definition of the limit of a function.

LIMIT OF A FUNCTION (Informal Definition)   The notation

$$\lim_{x \to c} f(x) = L$$

is read "the limit of $f(x)$ as x approaches c is L" and means that the functional values $f(x)$ can be made arbitrarily close to a unique number L by choosing x sufficiently close to c (but not equal to c).

■ **W**hat this says: If $f(x)$ becomes arbitrarily close to a single number L as x approaches c from either side, then we say that L is the limit of $f(x)$ as x approaches c. It means that the functional values of f "home in" on the number L as x gets closer and closer to c. The values of x can approach c from either side, but $x = c$ itself is excluded.

Notation: Sometimes we will find it convenient to represent the limit statement

$$\lim_{x \to c} f(x) = L$$

by writing $f(x) \to L$ as $x \to c$.

This informal definition of limit cannot be used in proofs until we give precise meaning to terms such as "arbitrarily close to L" and "sufficiently close to c." This will be done at the end of this section. For now, we will use this informal definition to explore a few basic features of the limiting process.

Example 2   Informal computation of a limit

Use a table to guess the value of

$$L = \lim_{x \to -2} \frac{x^2 + x - 2}{x + 2}$$

Solution   It would be nice if we could simply substitute $x = -2$ into the formula for

$$f(x) = \frac{x^2 + x - 2}{x + 2}$$

but note that $f(-2)$ is not defined. This is of no consequence since the limit process is concerned only when x *approaches* -2 and has nothing to do with the value of f at $x = -2$. We form a table of values of $f(x)$ for x near -2:

x	-2.3	-2.1	-2.05	-2.001	-2	-1.9997	-1.995
$f(x)$	-3.3	-3.1	-3.05	-3.001	undefined	-2.9997	-2.995

The numbers on the bottom row of this table suggest that $f(x) \to -3$ as $x \to -2$; that is,

$$L = \lim_{x \to -2} \frac{x^2 + x - 2}{x + 2} = -3$$

The graph of f is shown in Figure 2.1 Notice that the graph is a line with a "hole" at the point $(-2, -3)$.   ■

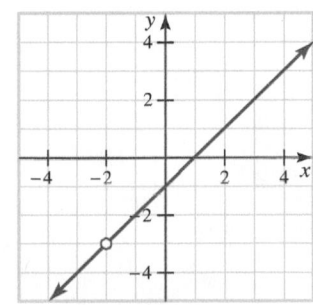

Figure 2.1   Graph of
$$f(x) = \frac{x^2 + x - 2}{x + 2}$$

Example 3 Finding limits of trigonometric functions

Evaluate $\lim\limits_{x\to 0} \sin x$ and $\lim\limits_{x\to 0} \cos x$.

Solution We can evaluate these limits by table.

x	1	0.5	0.1	0.01	−0.01	−0.1	−0.5
$\sin x$	0.84	0.48	0.0998	0.0099998	−0.0099998	−0.0998	−0.48
$\cos x$	0.54	0.88	0.9950	0.9999500	0.99995	0.9950	0.88

The pattern of numbers in the table suggests that

$$\lim_{x\to 0} \sin x = 0 = \sin 0 \qquad \text{and} \qquad \lim_{x\to 0} \cos x = 1 = \cos 0$$

Example 4 A computational dilemma

Use a table to guess the value of

$$L = \lim_{x\to 0} \frac{2\sqrt{x+1} - x - 2}{x^2}$$

Solution For x near 0:

x	−0.05	−0.1	−0.01	−0.001	0	0.001	0.005
$f(x)$	−0.2565	−0.2633	−0.2513	−0.2501	undefined	−0.2499	−0.2494

The number on the bottom line suggest that the limit is $L = -0.25$.

In Example 4, note that if you decide to try $x = 0.0000001$ just to make sure you have taken numbers close enough to 0, you may find that the calculator gives the value 0. Does this mean the limit is actually 0 instead of −0.25? No, the problem is that the numerator $2\sqrt{x+1} - x - 2$ is so close to 0 when $x = 0.0000001$ that the calculator gives a false value for f.

The point is this: When you use your calculator to estimate limits, always be aware that this method can introduce errors. In the next section, we will develop an algebraic approach for computing limits that is more reliable. The three basic approaches (examining appropriate data by using a calculator and/or a table, drawing a graph to find a limit, or using algebraic rules) each has its virtues and faults. As you study this section and the next, you need to learn all three approaches, along with their virtues and faults, and this knowledge will serve you well for the rest of your mathematical coursework.

One-Sided Limits

Sometimes we will be interested in the limiting behavior of a function from only one side; that is, the limit as x approaches a number c from the left or the analogous limit from the right.

ONE-SIDED LIMITS

Right-hand limit We write $\lim\limits_{x \to c^+} f(x) = L$ if we can make the number $f(x)$ as close to L as we please by choosing x sufficiently close to c on an interval (c, b) *immediately to the right of c.*

Left-hand limit We write $\lim\limits_{x \to c^-} f(x) = L$ if we can make the number $f(x)$ as close to L as we please by choosing x sufficiently close to c on the interval (a, c) *immediately to the left of c.*

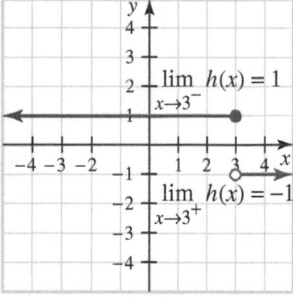

Figure 2.2 Graph of
$h(x) = \begin{cases} 1 & \text{if } x \le 3 \\ -1 & \text{if } x > 3 \end{cases}$ a graph
illustrating one-sided limits

For instance, note that the graph in Figure 2.2 is "broken" at $x = 3$. The limit as x approaches 3 from the left is 1, while the limit from the right at the same point is -1.

Observe that the "two-sided" limit, $\lim\limits_{x \to 3} h(x)$, defined earlier in this section, does not exist since $h(x)$ does not tend toward a single limiting value L as x approaches $c = 3$ from either side. It should be clear that, in general, a two-sided limit cannot exist if the corresponding pair of one-sided limits are different. Conversely, it can be shown that if the two one-sided limits of a given function f as $x \to c^-$ and $x \to c^+$ both exist and are equal, then the two-sided limit $\lim\limits_{x \to c} f(x)$ must also exist. These observations are so important that we state them in the form of a theorem.

Theorem 2.1 One-sided limit theorem

The two-sided limit $\lim\limits_{x \to c} f(x)$ exists if and only if the two one-sided limits $\lim\limits_{x \to c^-} f(x) = L$ and $\lim\limits_{x \to c^+} f(x) = L$ both exist and are equal. Furthermore, if

$$\lim\limits_{x \to c^-} f(x) = L = \lim\limits_{x \to c^+} f(x)$$

then $\lim\limits_{x \to c} f(x) = L$. ◆

One-sided and two-sided limits are illustrated in Figure 2.3.

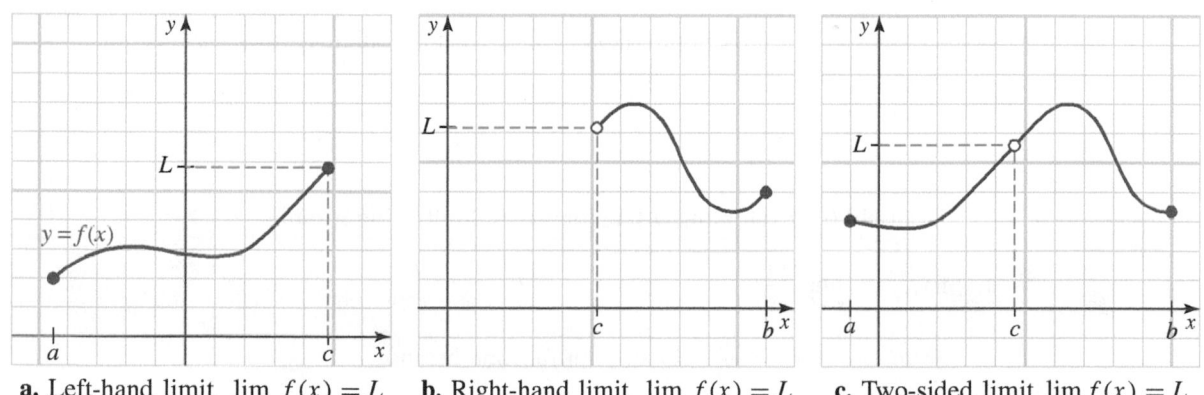

a. Left-hand limit $\lim\limits_{x \to c^-} f(x) = L$ **b.** Right-hand limit $\lim\limits_{x \to c^+} f(x) = L$ **c.** Two-sided limit $\lim\limits_{x \to c} f(x) = L$

Figure 2.3   We say that $\lim\limits_{x \to c} f(x) = L$ if and only if $\lim\limits_{x \to c} f(x) = L = \lim\limits_{x \to c^+} f(x)$

Example 5   Estimating limits by graphing

Given the functions defined by the graphs in Figure 2.4, find the requested limits by inspection, if they exist.

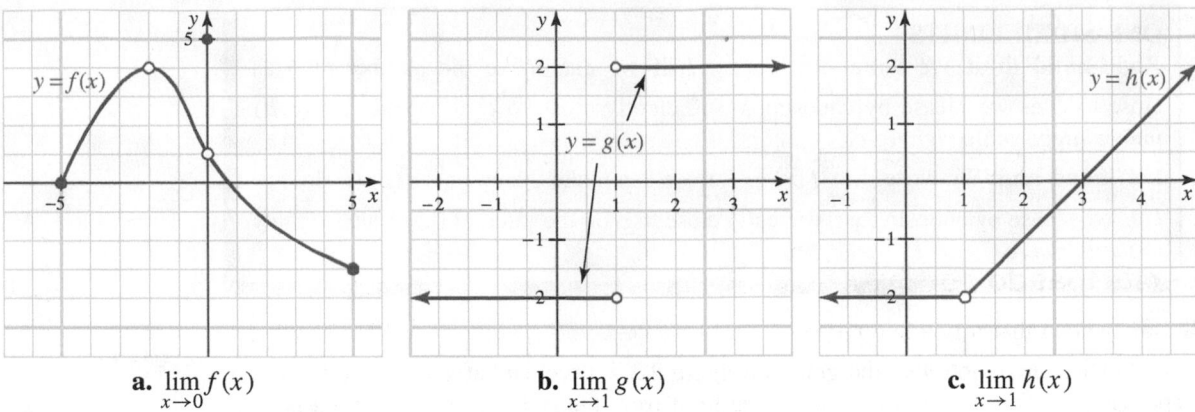

a. $\lim\limits_{x \to 0} f(x)$ **b.** $\lim\limits_{x \to 1} g(x)$ **c.** $\lim\limits_{x \to 1} h(x)$

Figure 2.4 Limits from graphs

Solution

a. Take a good look at the given graph; notice the open circles on the graph at $x = 0$ and $x = -2$ and also notice that $f(0) = 5$. To find $\lim\limits_{x \to 0} f(x)$ we need to look at both the left-hand and right-hand limits. Look at Figure 2.4**a** to find

$$\lim_{x \to 0^-} f(x) = 1 \qquad \text{and} \qquad \lim_{x \to 0^+} f(x) = 1$$

so $\lim\limits_{x \to 0} f(x)$ exists and $\lim\limits_{x \to 0} f(x) = 1$. Notice here that *the value of the limit as $x \to 0$ is not the same as the value of the function at $x = 0$.*

b. Look at Figure 2.4**b** to find

$$\lim_{x \to 1^-} g(x) = -2 \qquad \text{and} \qquad \lim_{x \to 1^+} g(x) = 2$$

so the *limit as $x \to 1$ does not exist.*

c. Look at Figure 2.4**c** to find

$$\lim_{x \to 1^-} h(x) = -2 \qquad \text{and} \qquad \lim_{x \to 1^+} h(x) = -2$$

so $\lim\limits_{x \to 1} h(x) = -2.$

Example 6   Evaluating a trigonometric limit using a table

Evaluate $\lim\limits_{x \to 0} \dfrac{\sin x}{x}$.

Solution $f(x) = \dfrac{\sin x}{x}$ is an even function because

$$f(-x) = \frac{\sin(-x)}{-x} = \frac{-\sin x}{-x} = \frac{\sin x}{x} = f(x)$$

This means that we need to find only the right-hand limit at 0 because the limiting behavior from the left will be the same as that from the right. Consider the following table.

x	$0.1 \rightarrow$	$0.05 \rightarrow$	$0.01 \rightarrow$	$0.001 \rightarrow$	0
$f(x)$	0.998334	0.999583	0.9999833	0.999999833	undefined

The table suggests that $\lim\limits_{x \to 0^+} \dfrac{\sin x}{x} = 1$ and hence that $\lim\limits_{x \to 0} \dfrac{\sin x}{x} = 1$. We will revisit this limit in Section 2.2. The graph of $f(x) = \dfrac{\sin x}{x}$ is shown in Figure 2.5. ■

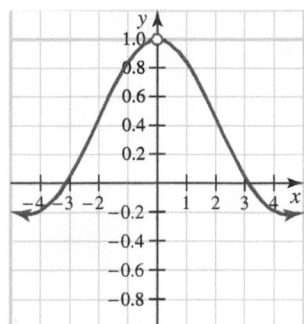

Figure 2.5 Interactive
Graph of $f(x) = \dfrac{\sin x}{x}$

Limits that do not Exist

It may happen that a function f does not have a (finite) limit as $x \to c$. When $\lim\limits_{x \to c} f(x)$ fails to exist, the functional values $f(x)$ are said to **diverge** as $x \to c$.

Example 7   A function that diverges

Evaluate $\lim\limits_{x \to 0} \dfrac{1}{x^2}$.

Solution   As $x \to 0$, the corresponding functional values of $f(x) = \frac{1}{x^2}$ grow arbitrarily large, as indicated in the following table.

x approaches 0 from the left; $x \to 0^-$ $\rightarrow |$	$\leftarrow$ x approaches 0 from the right; $x \to 0^+$

x	$-0.1 \rightarrow$	$-0.05 \rightarrow$	$-0.001 \rightarrow$	0	$\leftarrow 0.001$	$\leftarrow 0.005$	$\leftarrow 0.01$
$f(x) = \frac{1}{x^2}$	100	400	1×10^6	undefined	1×10^6	4×10^4	1×10^4

The graph of f is shown in Figure 2.6.

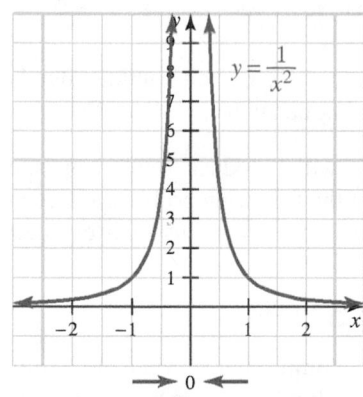

$$y = \frac{1}{x^2}$$

Figure 2.6 Interactive $\lim\limits_{x \to 0} \frac{1}{x^2}$ does not exist, and the graph illustrates that f rises without bound ■

Geometrically, the graph of $y = f(x)$ rises without bound as $x \to 0$. Thus, $\lim\limits_{x \to 0} \dfrac{1}{x^2}$ does not exist.

INFINITE LIMITS   A function f that increases or decreases without bound as x approaches c is said to **tend to infinity** (∞) at c. We indicate this behavior by writing

$$\lim_{x \to c} f(x) = \infty \qquad \text{if } f \text{ increases without bound}$$

and by

$$\lim_{x \to c} f(x) = -\infty \qquad \text{if } f \text{ decreases without bound.}$$

☠ *It is important to remember that ∞ is **not** a number, but is merely a symbol denoting unrestricted growth in the magnitude of the function.* ☠

Using this notation, we can rewrite the answer to Example 7 as

$$\lim_{x \to 0} \frac{1}{x^2} = \infty$$

Example 8   A function that diverges by oscillation

Evaluate $\lim_{x \to 0} \sin \dfrac{1}{x}$.

Solution   The values of $f(x) = \sin \dfrac{1}{x}$ oscillate infinitely often between 1 and -1 as x approaches 0. For example, $f(x) = 1$ for $x = \frac{2}{\pi}, \frac{2}{5\pi}, \frac{2}{9\pi}, \ldots$ and $f(x) = -1$ for $x = \frac{2}{3\pi}, \frac{2}{7\pi}, \frac{2}{11\pi}, \ldots$. The graph of $f(x)$ is shown in Figure 2.7.

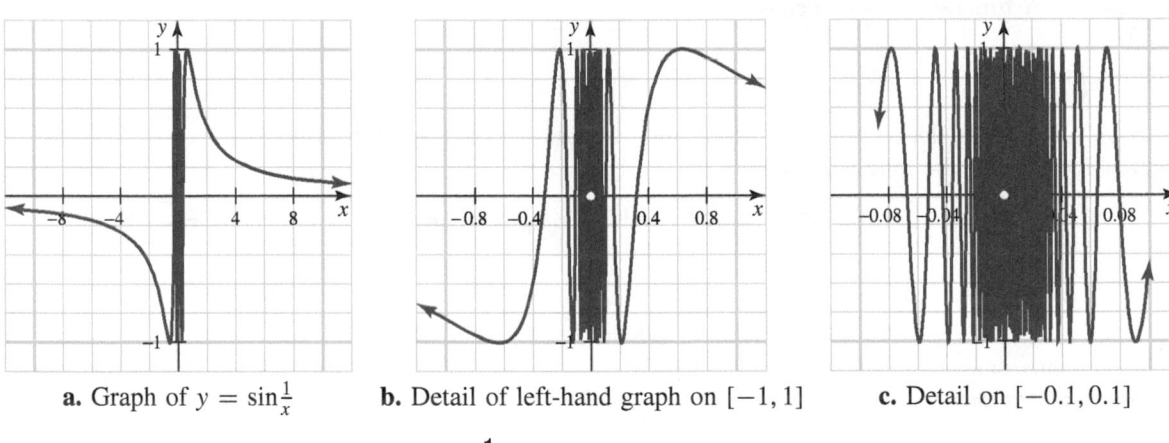

a. Graph of $y = \sin\frac{1}{x}$ **b.** Detail of left-hand graph on $[-1, 1]$ **c.** Detail on $[-0.1, 0.1]$

Figure 2.7 $\sin \dfrac{1}{x}$ diverges by oscillation as $x \to 0$

Because the values of $f(x)$ do not approach a unique number L as $x \to 0$, the limit does not exist. This kind of limiting behavior is called **divergence by oscillation**.   ■

In the next section, we will introduce some properties of limits that will help us evaluate limits efficiently. In the following problem set, remember that the emphasis is on an intuitive understanding of limits, including their evaluation by graphing and by table.

Formal Definition of a Limit

Our informal definition of the limit provides valuable intuition and allows you to develop a working knowledge of this fundamental concept. For theoretical work, however, the intuitive definition will not suffice, because it gives no precise, quantifiable meaning to the terms "arbitrarily close to L" and "sufficiently close to c." In the 19th century, leading mathematicians, including Augustin-Louis Cauchy (1789-1857) and Karl Weierstraß (1815-1897), sought to put calculus on a sound logical foundation by giving precise definitions for the foundational ideas of calculus. The following definition, derived from the work of Cauchy and Weierstraß, gives precision to the limit notion.

LIMIT OF A FUNCTION (Formal Definition)   The limit statement

$$\lim_{x \to c} f(x) = L$$

means that for each number $\epsilon > 0$, there is a number $\delta > 0$ such that

$$|f(x) - L| < \epsilon \qquad \text{whenever} \qquad 0 < |x - c| < \delta$$

We show this definition graphically in Figure 2.8.

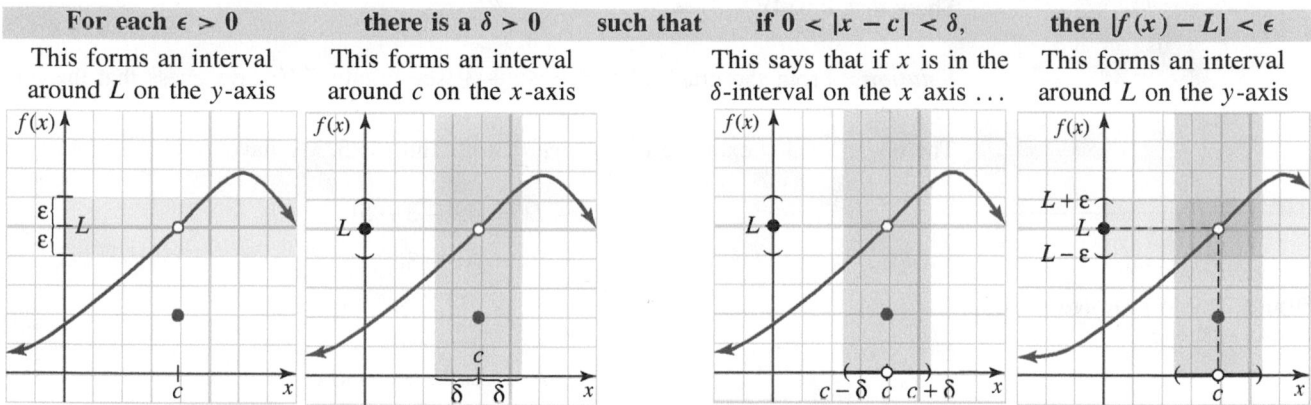

For each $\epsilon > 0$	there is a $\delta > 0$	such that	if $0 < |x - c| < \delta,$	then $|f(x) - L| < \epsilon$
This forms an interval around L on the y-axis	This forms an interval around c on the x-axis		This says that if x is in the δ-interval on the x axis ...	This forms an interval around L on the y-axis

Figure 2.8 Formal definition of limit: $\lim\limits_{x \to c} f(x) = L$

Because the Greek letters ϵ (epsilon) and δ (delta) are traditionally used in this context, the formal definition of limit is sometimes called the **epsilon-delta** definition of the limit. The goal of this section is to show how this formal definition embodies our intuitive understanding of the limit process and how it can be used rigorously to establish a variety of results.

Do not be discouraged if this material seems difficult — it is. Probably your best course of action is to read this section carefully and examine the details of a few examples closely. Then, using the examples as models, try some of the exercises. This material often takes several attempts, but if you persevere, you should come away with an appreciation of the epsilon-delta process — and with it, a better understanding of calculus.

Behind the formal language is a fairly straightforward idea. In particular, to establish a specific limit, say $\lim\limits_{x \to c} f(x) = L$, a number $\epsilon > 0$ is chosen first to establish a desired degree of proximity to L, and then a number $\delta > 0$ is found that determines how close x must be to c to ensure that $f(x)$ is within ϵ units of L.

The situation is summarized in Figure 2.9, which shows a function that satisfies the conditions of the definition.

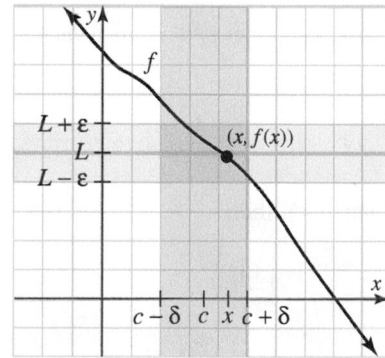

Figure 2.9 Interactive The epsilon-delta definition of limit

Notice that whenever x is within δ units of c (but not equal to c), the point $(x, f(x))$ on the graph of f must lie in the rectangle (shaded region) formed by the intersection of the horizontal band of width 2ϵ centered at L and the vertical band of width 2δ centered at c. The smaller the ϵ-interval around the proposed limit L, generally the smaller the δ-interval will need to be for $f(x)$ to lie in the ϵ-interval. If such a δ can be found no matter how small ϵ is, then $f(x)$ and L become arbitrarily close, so L must be the limit. The following examples illustrate epsilon-delta proofs, one in which the limit exists and one in which it does not.

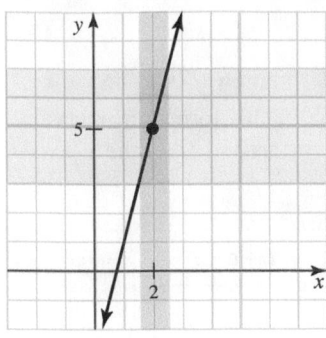

Figure 2.10 Interactive
$\lim_{x \to 2}(4x - 3) = 5$

Example 9   An epsilon-delta proof of a limit statement

Show that $\lim_{x \to 2}(4x - 3) = 5$.

Solution   From the graph of $f(x) = 4x - 3$ (see Figure 2.10), we guess that the limit as $x \to 2$ is 5.

The object of this example is to *prove* that the limit is 5. We have

$$|f(x) - L| = |4x - 3 - 5|$$
$$= |4x - 8|$$
$$= 4\,|x - 2|$$

This must be less than ϵ whenever $|x - 2| < \delta$.

For a given $\epsilon > 0$, choose $\delta = \frac{\epsilon}{4}$. Then

$$|f(x) - L| = 4\,|x - 2| < 4\delta = 4\left(\frac{\epsilon}{4}\right) = \epsilon$$

Example 10   An epsilon-delta proof that a limit does not exist

Show that $\lim_{x \to 0}\dfrac{1}{x}$ does not exist.

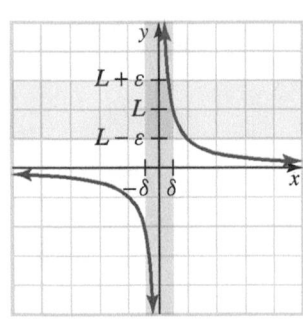

Figure 2.11 Interactive
$\lim_{x \to 0}\dfrac{1}{x}$

Solution   Let $f(x) = \dfrac{1}{x}$ and L be any number. By way of contradiction, assume that $\lim_{x \to 0}f(x) = L$. Look at the graph of f, as shown in Figure 2.11.

It would seem that no matter what value of ϵ is chosen, it would be impossible to find a corresponding δ. Consider the absolute value expression required by the definition of limit: If

$$|f(x) - L| < \epsilon, \ \text{ or, for this example,} \ \left|\frac{1}{x} - L\right| < \epsilon$$

then

$$-\epsilon < \frac{1}{x} - L < \epsilon \quad \textit{Property of absolute value (Table 1.1, p.17).}$$

and

$$L - \epsilon < \frac{1}{x} < L + \epsilon$$

If $\epsilon = 1$ (not a particularly small ϵ), then

$$\left|\frac{1}{x}\right| < |L| + 1$$

$$|x| > \frac{1}{|L| + 1}$$

In general, given any $\epsilon > 0$ then no matter how $\delta > 0$ is chosen, there will always be numbers x in the interval such that $\left|\frac{1}{x}\right| > L + \epsilon$. Since L was chosen arbitrarily, it follows that the limit does not exist.

Appendix B gives many of the important proofs in calculus, and if you look there, you will see many of them are given in ϵ-δ form. In the following problem set, remember that the emphasis is on an intuitive understanding of limits, including their evaluation by graphing and by table.

PROBLEM SET 2.1

Level 1

Given the functions defined by the graphs in Figure 2.12, ss *find the limits in Problems* 1-6.

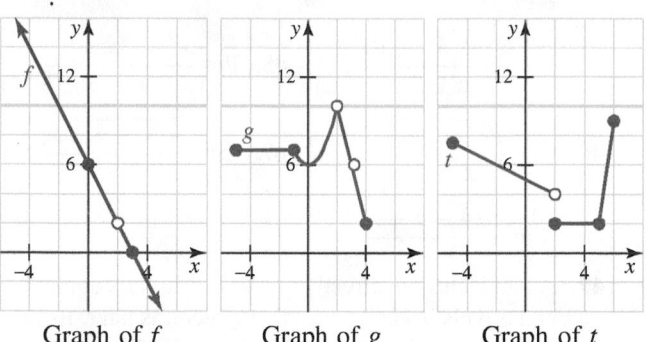

Graph of f Graph of g Graph of t

Figure 2.12 Graphs of three functions

1. **a.** $\lim_{x \to 3} f(x)$ **b.** $\lim_{x \to 2} f(x)$ **c.** $\lim_{x \to 0} f(x)$

2. **a.** $\lim_{x \to -3} g(x)$ **b.** $\lim_{x \to -1} g(x)$ **c.** $\lim_{x \to 4^-} g(x)$

3. **a.** $\lim_{x \to 4} t(x)$ **b.** $\lim_{x \to -4} t(x)$ **c.** $\lim_{x \to -5^+} t(x)$

4. **a.** $\lim_{x \to 2^-} f(x)$ **b.** $\lim_{x \to 2^+} f(x)$ **c.** $\lim_{x \to 2} f(x)$

5. **a.** $\lim_{x \to 3^-} g(x)$ **b.** $\lim_{x \to 3^+} g(x)$ **c.** $\lim_{x \to 3} g(x)$

6. **a.** $\lim_{x \to 2^-} t(x)$ **b.** $\lim_{x \to 2^+} t(x)$ **c.** $\lim_{x \to 2} t(x)$

Find the limits by filling in the appropriate values in the tables in Problems 7-10.

7. $\lim_{x \to 5^-} f(x)$, where $f(x) = 4x - 5$.

$$x \to 5^-$$

x	2	3	4	4.5	4.9	4.99
$f(x)$	3					

$$f(x) \to \ ?$$

8. $\lim_{x \to 2^-} g(x)$, where $g(x) = \dfrac{x^3 - 8}{x^2 + 2x + 4}$

$$x \to 2^-$$

x	1	1.5	1.9	1.99	1.999	1.9999
$g(x)$	-1					

$$g(x) \to \ ?$$

9. $\lim_{x \to 2} \dfrac{x^2 + 2x + 4}{x^3 - 8}$

$$x \to 2^- \qquad\qquad 2^+ \leftarrow x$$

x	1	1.9	1.99	1.999	2.001	2.1	2.5
$f(x)$							

$$f(x) \to \ ? \leftarrow f(x)$$

10. $\lim_{x \to 2} h(x)$, where $h(x) = \dfrac{3x^2 - 2x - 8}{x - 2}$

$$x \to 2^- \qquad\qquad 2^+ \leftarrow x$$

x	1	1.9	1.99	1.999	2.001	2.1	2.5	3
$h(x)$	7							

$$h(x) \to \ ? \leftarrow h(x)$$

11. Find $\lim_{x \to 0} \dfrac{\tan 2x}{\tan 3x}$ using the following procedure based on the fact that $f(x) = \tan x$ is an odd function.
If $f(x) = \dfrac{\tan 2x}{\tan 3x}$, then

$$f(-x) = \dfrac{\tan(-2x)}{\tan(-3x)} = \dfrac{-\tan 2x}{-\tan 3x} = f(x)$$

Thus, we simply need to check for $x \to 0^+$. Find the limit by completing the following table.

$$x \to 0^+$$

x	1	0.5	0.1	0.01	0.001
$f(x)$	15.33				

$$f(x) \to ?$$

Describe each illustration in Problems 12-17 *using a limit statement.*

12.

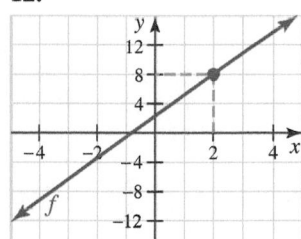

13.

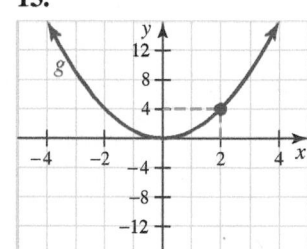

14.

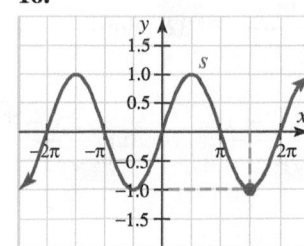

15.

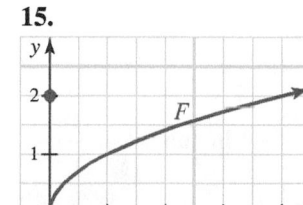

16.

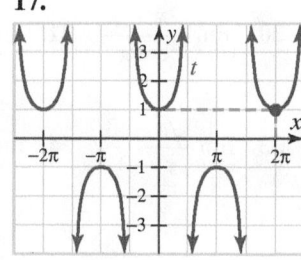

17.

Level 2

18. ■ **What does this say?*** Explain a process for finding a limit.

Evaluate the limits in Problems 19-48 to two decimal places by graphing or by using a table of values, and check your answer using a calculator. If the limit does not exist, explain why.

19. $\lim\limits_{x\to 0^+} x^4$

20. $\lim\limits_{x\to 0^+} \cos x$

21. $\lim\limits_{x\to 2^-} (x^2-4)$

22. $\lim\limits_{x\to 3^-} (x^2-4)$

23. $\lim\limits_{x\to 1^+} \dfrac{1}{x-3}$

24. $\lim\limits_{x\to -3^+} \dfrac{1}{x-3}$

25. $\lim\limits_{x\to 3} \dfrac{1}{x-3}$

26. $\lim\limits_{x\to \frac{\pi}{2}} \tan x$

27. **a.** $\lim\limits_{x\to 0} \dfrac{\cos x}{x}$
  b. $\lim\limits_{x\to \pi} \dfrac{\cos x}{x}$

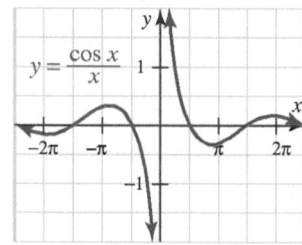

28. **a.** $\lim\limits_{x\to 0} \dfrac{1-\cos x}{x}$
  b. $\lim\limits_{x\to \pi} \dfrac{1-\cos x}{x}$

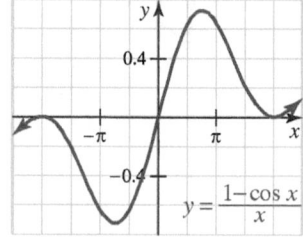

29. **a.** $\lim\limits_{x\to 0.4} |x|\sin\dfrac{1}{x}$
  b. $\lim\limits_{x\to 0} |x|\sin\dfrac{1}{x}$

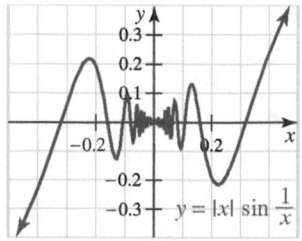

30. $\lim\limits_{x\to \frac{\pi}{2}} \dfrac{2x-\pi}{\cos x}$
31. $\lim\limits_{x\to 1} \dfrac{\sin\frac{\pi}{x}}{x-1}$
32. $\lim\limits_{x\to 9} \dfrac{\sqrt{x}-3}{x-9}$

33. $\lim\limits_{x\to 9} \dfrac{\sqrt{x}-3}{x-3}$
34. $\lim\limits_{x\to 1} \dfrac{\sqrt[3]{x}-1}{\sqrt{x}-1}$
35. $\lim\limits_{x\to 64} \dfrac{\sqrt[3]{x}-8}{\sqrt{x}-4}$

36. $\lim\limits_{x\to 4^+} \dfrac{\frac{1}{\sqrt{x}}-\frac{1}{2}}{x-4}$
37. $\lim\limits_{x\to 9^+} \dfrac{\frac{1}{\sqrt{x}}-\frac{1}{3}}{x-3}$
38. $\lim\limits_{x\to 1} \dfrac{1-\frac{1}{x}}{x-1}$

39. $\lim\limits_{x\to 0} \dfrac{1-\frac{1}{x+1}}{x}$
40. $\lim\limits_{x\to 0} \dfrac{\sin 2x}{x}$
41. $\lim\limits_{x\to 0} \dfrac{\sin 3x}{x}$

42. $\lim\limits_{x\to 0} \cos\dfrac{1}{x}$
43. $\lim\limits_{x\to 0} \tan\dfrac{1}{x}$

44. $\lim\limits_{x\to 3} \dfrac{x^2+3x-10}{x-2}$
45. $\lim\limits_{x\to 3} \dfrac{x^2+3x-10}{x-3}$

46. $\lim\limits_{x\to 2} \dfrac{\sqrt{x+2}-2}{x-2}$
47. $\lim\limits_{x\to 3^+} \dfrac{\sqrt{x-3}+x}{3-x}$

48. $\lim\limits_{x\to 4^-} \dfrac{\sqrt{4-x}+x}{4-x}$

49. A ball is thrown directly upward from the edge of a cliff and travels in such a way that t seconds later, its height (in feet) above the ground at the base of the cliff is
$$s(t) = -16t^2 + 40t + 24$$
a. Compute the limit
$$v(t) = \lim_{x\to t} \frac{s(x)-s(t)}{x-t}$$
to find the instantaneous velocity of the ball at time t.
b. What is the ball's initial velocity?
c. When does the ball hit the ground, and what is its impact velocity?
d. When does the ball have velocity 0? What physical interpretation should be given to this time?

50. Tom and Sue are driving along a straight, level road in a car whose speedometer needle is broken but that has a trip odometer that can measure the distance traveled from an arbitrary starting point in tenths of a mile. At 2:50 P.M., Tom says he would like to know how fast they are traveling at 3:00 P.M., so Sue takes down the odometer readings listed in the table, makes a few calculations, and announces the desired velocity. What is her result?

time t	2:50	2:55	2:59	3:00	3:01	3:03	3:06
odometer reading	33.9	38.2	41.5	42.4	43.2	44.9	47.4

In Problems 51-52, estimate the limits by plotting points or by using tables.

51. $\lim\limits_{x\to 13} \dfrac{x^3-9x^2-45x-91}{x-13}$

52. $\lim\limits_{x\to 13} \dfrac{x^3-9x^2-39x-86}{x-13}$

*Many problems in this book are labeled **What Does This Say?** Following the question will be a question for you to answer in your own words, or a statement for you to rephrase in your own words. These problems are intended to be similar to the "What This Says" boxes.

53. *Historical Quest*

By the second half of the 1700s, it was generally accepted that without logical underpinnings, calculus would be limited. Augustin-Louis Cauchy developed an acceptable theory of limits, and in doing so removed much doubt about the logical validity of calculus. Cauchy is described by the historian

Karl Smith library

Augustin-Louis Cauchy (1789-1857)

Howard Eves not only as a first-rate mathematician with tremendous mathematical productivity but also as a lawyer (he practiced law for 14 years), a mountain climber, and a painter (he worked in watercolors). Among other characteristics that distinguished him from his contemporaries, he advocated respect for the environment.

Cauchy wrote a treatise on integrals in 1814 that was considered a classic, and in 1816 his paper on wave propagation in liquids won a prize from the French Academy. It has been said that with his work the modern era of analysis began. In all, he wrote over 700 papers, which are, today, considered no less than brilliant.

Cauchy did not formulate the ϵ-δ definition of limit that we use today, but formulated instead a purely arithmetical definition. Consult some history of mathematics books to find a translation of Cauchy's definition, which appeared in his monumental treatise, *Cours d'Analyse de l'École Royale Polytechnique* (1821). As part of your research, find when and where the ϵ-δ definition of limit was first used.

Level 3

In Problems 54-59, use the formal definition of the limit to prove or disprove the given limit statement.

54. $\lim_{x \to 2}(x + 3) = 5$

55. $\lim_{t \to 0}(3t - 1) = 0$

56. $\lim_{x \to -2}(3x + 7) = 1$

57. $\lim_{x \to 1}(2x - 5) = -3$

58. $\lim_{x \to 2}(x^2 + 2) = 6$

59. $\lim_{x \to 2}\dfrac{1}{x} = \dfrac{1}{2}$

60. The tabular approach is a convenient device for discussing limits informally, but if it is not used very carefully, it can be misleading. For example, for $x \neq 0$, let $f(x) = \sin\dfrac{1}{x}$.

a. Construct a table showing the values of $f(x)$ for $x = \frac{-2}{\pi}, \frac{-2}{9\pi}, \frac{-2}{13\pi}, \frac{2}{19\pi}, \frac{2}{7\pi}, \frac{2}{3\pi}$. Based on this table, what would you say about $\lim_{x \to 0} f(x)$?

b. Construct a second table, this time showing the values of $f(x)$ for $x = \frac{-1}{2\pi}, \frac{-1}{11\pi}, \frac{-1}{20\pi}, \frac{1}{50\pi}, \frac{1}{30\pi}, \frac{1}{5\pi}$. Now what would you say about $\lim_{x \to 0} f(x)$?

c. Based on the results in parts **a** and **b**, what do you conclude about $\lim_{x \to 0} \sin\dfrac{1}{x}$?

2.2 ALGEBRAIC COMPUTATION OF LIMITS

IN THIS SECTION: *Computations with limits, using algebra to find limits, limits of piecewise-defined functions, two special trigonometric limits*

It is important to be able to find limits easily and efficiently, but the methods of graphing and table construction of the previous section are not always easy or efficient. In this section, we state some limit rules and then show how to use them along with algebra in order to evaluate limits.

Computations with Limits

In Section 2.1, we observed that finding limits by tables or by graphing is risky. In this section, we will explore a more exact way of computing limits that is based on the following properties.

BASIC PROPERTIES AND RULES FOR LIMITS For any real number c, suppose the functions f and g both have limits at $x = c$.

Constant rule $\displaystyle\lim_{x \to c} k = k$ for any constant k

Limit of x rule $\displaystyle\lim_{x \to c} x = c$

Multiple rule $\displaystyle\lim_{x \to c} \left[k f(x) \right] = k \lim_{x \to c} f(x)$ for any constant k
The limit of a constant times a function is the constant times the limit of the function.

Sum rule $\displaystyle\lim_{x \to c} \left[f(x) + g(x) \right] = \lim_{x \to c} f(x) + \lim_{x \to c} g(x)$
The limit of a sum is the sum of the limits.

Difference rule $\displaystyle\lim_{x \to c} \left[f(x) - g(x) \right] = \lim_{x \to c} f(x) - \lim_{x \to c} g(x)$
The limit of a difference is the difference of the limits.

Product rule $\displaystyle\lim_{x \to c} \left[f(x) g(x) \right] = \left[\lim_{x \to c} f(x) \right]\left[\lim_{x \to c} g(x) \right]$
The limit of a product is the product of the limits.

Quotient rule $\displaystyle\lim_{x \to c} \frac{f(x)}{g(x)} = \frac{\displaystyle\lim_{x \to c} f(x)}{\displaystyle\lim_{x \to c} g(x)}$ if $\displaystyle\lim_{x \to c} g(x) \neq 0$
The limit of a quotient is the quotient of the limits, as long as the limit of the denominator is not zero.

Power rule $\displaystyle\lim_{x \to c} \left[f(x) \right]^n = \left[\lim_{x \to c} f(x) \right]^n$ n is a rational number and the limit on the right exists.
The limit of a power is the power of the limit.

It is fairly easy graphically to justify the rules for the limit of a constant and the limit of x, as shown in Figure 2.13.

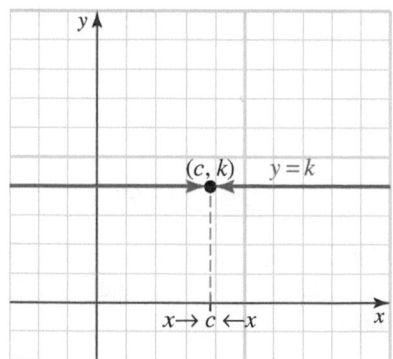

a. Limit of a constant: $\displaystyle\lim_{x \to c} k = k$

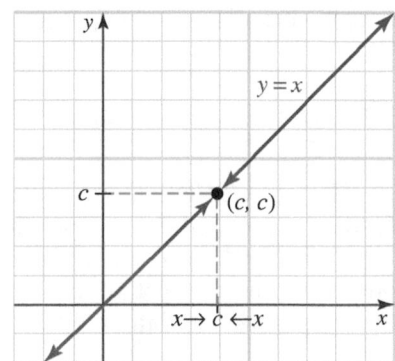

b. Limit of x: $\displaystyle\lim_{x \to c} x = c$

Figure 2.13 Two basic limits

All of these properties of limits (including the two basic limits shown in Figure 2.13) can be proved using the formal definition of limit.

Example 1   Finding the limit of a polynomial function

Evaluate $\lim\limits_{x \to 2} (2x^5 - 9x^3 + 3x^2 - 11)$.

Solution

$$\lim_{x \to 2} (2x^5 - 9x^3 + 3x^2 - 11) = \lim_{x \to 2} 2x^5 - \lim_{x \to 2} 9x^3 + \lim_{x \to 2} 3x^2 - \lim_{x \to 2} 11 \qquad \textit{Sum and difference rules}$$

$$= 2 \left[\lim_{x \to 2} x^5 \right] - 9 \left[\lim_{x \to 2} x^3 \right] + 3 \left[\lim_{x \to 2} x^2 \right] - 11 \quad \textit{Multiple and constant rules}$$

$$= 2 \left[\lim_{x \to 2} x \right]^5 - 9 \left[\lim_{x \to 2} x \right]^3 + 3 \left[\lim_{x \to 2} x \right]^2 - 11 \quad \textit{Power rule}$$

$$= 2(2)^5 - 9(2)^3 + 3(2)^2 - 11 \qquad \textit{Limit of } x \textit{ rule}$$

$$= -7$$

Comment: If you consider Example 1 carefully, it is easy to see that if f is any polynomial, then the limit at $x = c$ can be found by substituting $x = c$ into the formula for $f(x)$.

LIMIT OF A POLYNOMIAL FUNCTION   If P is a polynomial function, then

$$\lim_{x \to c} P(x) = P(c)$$

Example 2   Finding the limit of a rational function

Evaluate $\lim\limits_{z \to -1} \dfrac{z^3 - 3z + 7}{5z^2 + 9z + 6}$.

Solution

$$\lim_{z \to -1} \frac{z^3 - 3z + 7}{5z^2 + 9z + 6} = \frac{\lim\limits_{z \to -1} (z^3 - 3z + 7)}{\lim\limits_{z \to -1} (5z^2 + 9z + 6)} \qquad \textit{Quotient rule}$$

$$= \frac{(-1)^3 - 3(-1) + 7}{5(-1)^2 + 9(-1) + 6} \qquad \begin{array}{l}\textit{Both numerator and denominator}\\ \textit{are polynomials.}\end{array}$$

$$= \frac{9}{2}$$

You must be careful as to when you write the word "limit" and when you do not; pay particular attention to this when looking at the examples in this section.

Notice that if the denominator of the rational function is not zero, the limit can be found by substitution. If the limit of the denominator is zero, we find ourselves dividing by zero, an illegal operation.

LIMIT OF A RATIONAL FUNCTION   If R is a rational function defined by $R(x) = \dfrac{P(x)}{Q(x)}$, then $\lim\limits_{x \to c} R(x) = \dfrac{P(c)}{Q(c)}$ provided $\lim\limits_{x \to c} Q(x) \neq 0$.

Example 3   Finding the limit of a power (or root) function

Evaluate $\lim\limits_{x \to -2} \sqrt[3]{x^2 - 3x - 2}$.

Solution $\lim\limits_{x \to -2} \sqrt[3]{x^2 - 3x - 2} = \lim\limits_{x \to -2} (x^2 - 3x - 2)^{1/3}$

$$= \left[\lim\limits_{x \to -2} (x^2 - 3x - 2) \right]^{1/3} \quad \textit{Power rule}$$

$$= \left[(-2)^2 - 3(-2) - 2 \right]^{1/3}$$

$$= 8^{1/3}$$

$$= 2 \qquad \blacksquare$$

Once again, for values of the function for which $f(c)$ is defined, the limit can be found by substitution.

In the previous section we used a table to find that $\lim\limits_{x \to 0} \sin x = 0$ and $\lim\limits_{x \to 0} \cos x = 1$. In the following example we use this information, along with the properties of limits, to find other trigonometric limits.

Example 4   Finding trigonometric limits algebraically

Given that $\lim\limits_{x \to 0} \sin x = 0$ and $\lim\limits_{x \to 0} \cos x = 1$, evaluate:

a. $\lim\limits_{x \to 0} \sin^2 x$ **b.** $\lim\limits_{x \to 0} (1 - \cos x)$

Solution

a.
$$\lim\limits_{x \to 0} \sin^2 x = \left[\lim\limits_{x \to 0} \sin x \right]^2 \quad \textit{Power rule}$$

$$= 0^2 \qquad\qquad \textit{Given } \lim\limits_{x \to 0} \sin x = 0.$$

$$= 0$$

b.
$$\lim\limits_{x \to 0} (1 - \cos x) = \lim\limits_{x \to 0} 1 - \lim\limits_{x \to 0} \cos x \quad \textit{Difference rule}$$

$$= 1 - 1 \qquad\qquad \textit{Constant rule and given } \lim\limits_{x \to 0} \cos x = 1.$$

$$= 0 \qquad\qquad\qquad\qquad\qquad\qquad \blacksquare$$

The following theorem states that we can find limits of trigonometric functions by direct substitution, as long as the number that x is approaching is in the domain of the given function. Proofs of several parts are outlined in the problem set, and in fact an examination of the graphs of the functions indicates why they are true.

Theorem 2.2   Limits of trigonometric functions

If c is any number in the domain of a given function, then

$$\lim\limits_{x \to c} \cos x = \cos c \qquad \lim\limits_{x \to c} \sin x = \sin c \qquad \lim\limits_{x \to c} \tan x = \tan c$$

$$\lim\limits_{x \to c} \sec x = \sec c \qquad \lim\limits_{x \to c} \csc x = \csc c \qquad \lim\limits_{x \to c} \cot x = \cot c$$

Proof: We will show that $\lim\limits_{x \to c} \sin x = \sin c$. The other limit formulas may be proved in a similar fashion (see Problems 59-60). Let $h = x - c$. Then $x = h + c$, so $x \to c$ as $h \to 0$. Thus,

$$\lim\limits_{x \to c} \sin x = \lim\limits_{h \to 0} \sin(h + c)$$

Using the trigonometric identity

$$\sin(A + B) = \sin A \cos B + \cos A \sin B$$

and the limit formulas for sums and products, we find that

$$\lim_{x \to c} \sin x = \lim_{h \to 0} \sin(h + c)$$

$$= \lim_{h \to 0} [\sin h \cos c + \cos h \sin c]$$

$$= \lim_{h \to 0} \sin h \cdot \lim_{h \to 0} \cos c + \lim_{h \to 0} \cos h \cdot \lim_{h \to 0} \sin c$$

$$= 0 \cdot \cos c + 1 \cdot \sin c$$

$$= \sin c$$

Note that $\sin c$ and $\cos c$ do not change as $h \to 0$ because these are constants with respect to h. ◆

Example 5   Finding limits of trigonometric functions

Evaluate the following limits: **a.** $\lim_{x \to 1} (x^2 \cos \pi x)$ **b.** $\lim_{x \to 0} \dfrac{x}{\cos x}$

Solution

a. $\lim_{x \to 1} (x^2 \cos \pi x) = \left[\lim_{x \to 1} x \right]^2 \left[\lim_{x \to 1} \cos \pi x \right]$ **b.** $\lim_{x \to 0} \dfrac{x}{\cos x} = \dfrac{\lim_{x \to 0} x}{\lim_{x \to 0} \cos x}$

$$= 1^2 \cos \pi \qquad\qquad\qquad = \frac{0}{1}$$

$$= -1 \qquad\qquad\qquad\qquad = 0 \qquad ▇$$

Using Algebra to find Limits

Sometimes the limit of $f(x)$ as $x \to c$ *cannot* be evaluated by direct substitution. In such a case, we look for another function that agrees with f for all values of x *except at the troublesome value $x = c$*. We illustrate this procedure with some examples.

Example 6   Evaluating a limit using fractional reduction

Evaluate $\lim_{x \to 2} \dfrac{x^2 + x - 6}{x - 2}$.

Solution   If you try substitution on this limit, you will obtain

$$\lim_{x \to 2} \frac{x^2 + x - 6}{x - 2} = \frac{\lim_{x \to 2} (x^2 + x - 6)}{\lim_{x \to 2} (x - 2)}$$

$$= \frac{2^2 + 2 - 6}{2 - 2}$$

$$= \frac{0}{0}$$

The form $\frac{0}{0}$ is called an **indeterminate form** because the value of the limit cannot be determined without further analysis.

If the expression is a rational expression, the next step is to try to simplify the function by factoring and checking to see if the reduced form is a polynomial. For instance,

$$\lim_{x \to 2} \frac{x^2 + x - 6}{x - 2} = \lim_{x \to 2} \frac{(x + 3)(x - 2)}{x - 2} = \lim_{x \to 2} (x + 3)$$

This simplification is valid only if $x \neq 2$. Now complete the evaluation of the reduced function by direct substitution. This is not a problem, because the limit is concerned with values *as x approaches* 2, not the value where $x = 2$. In the preceding expression,

$$\lim_{x \to 2} \frac{x^2 + x - 6}{x - 2} = \lim_{x \to 2} (x + 3) = 5$$

Another algebraic technique for finding limits is to rationalize either the numerator or the denominator to obtain an algebraic form that is not indeterminate.

Example 7   Evaluating a limit by rationalizing

Evaluate $\lim\limits_{x \to 4} \dfrac{\sqrt{x} - 2}{x - 4}$.

Solution   Once again, notice that both the numerator and denominator of this rational expression are 0 when $x = 4$, so we cannot evaluate the limit by direct substitution. Instead, rationalize the numerator:

This method will work only if the resulting numerator allows the fraction to be simplified. Pay close attention to this example because it illustrates a procedure we will use over and over.

$$\lim_{x \to 4} \frac{\sqrt{x} - 2}{x - 4} = \lim_{x \to 4} \left[\frac{\sqrt{x} - 2}{x - 4} \cdot \frac{\sqrt{x} + 2}{\sqrt{x} + 2} \right] \quad \textit{Multiply by 1.}$$

$$= \lim_{x \to 4} \frac{x - 4}{(x - 4)(\sqrt{x} + 2)}$$

$$= \lim_{x \to 4} \frac{1}{\sqrt{x} + 2}$$

$$= \frac{1}{\sqrt{4} + 2}$$

$$= \frac{1}{4}$$

Limits of Piecewise-Defined Functions

In Section 1.4 we discussed *piecewise-defined functions*. To evaluate $\lim\limits_{x \to c} f(x)$ where the domain of f is divided into pieces, we first look to see whether c is a value separating two of the pieces. If so, we need to consider one-sided limits, as illustrated by the following examples.

Example 8   Limit of a piecewise-defined function

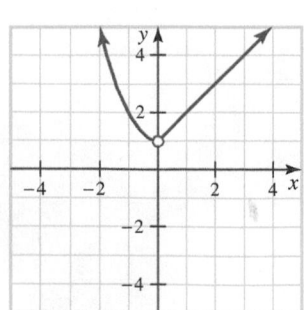

Figure 2.14 Graph of f

Find $\lim\limits_{x \to 0} f(x) = \begin{cases} x + 1 & \text{if } x > 0 \\ x^2 + 1 & \text{if } x < 0 \end{cases}$.

Solution   Notice that $f(0)$ is not defined, and that it is necessary to consider left- and right-hand limits because $f(x)$ is defined differently on each side of $x = 0$.

$$\lim_{x \to 0^-} f(x) = \lim_{x \to 0^-} (x^2 + 1) \quad f(x) = x^2 + 1 \quad \textit{to the left of } 0.$$

$$= 1$$

$$\lim_{x \to 0^+} f(x) = \lim_{x \to 0^+} (x + 1) \quad f(x) = x + 1 \quad \textit{to the right of } 0.$$

$$= 1$$

Because the left- and right-hand limits are equal, we conclude that $\lim_{x \to 0} f(x) = 1$. Looking at the graph in Figure 2.14, it is easy to see that the left- and right-hand limits are the same (even though the function is not defined at $x = 0$).

Example 9   Limit of a piecewise-defined function may not exist

Find $\lim_{x \to 0} g(x)$, where $g(x) = \begin{cases} x + 5 & \text{if } x > 0 \\ x & \text{if } x < 0 \end{cases}$.

Solution   Notice that $g(0)$ is not defined, and that it is necessary to consider left- and right-hand limits because $g(x)$ is defined differently on each side of $x = 0$. We have

$$\lim_{x \to 0^-} g(x) = \lim_{x \to 0^-} x \qquad g(x) = x \text{ to the left of } 0.$$

$$= 0$$

$$\lim_{x \to 0^+} g(x) = \lim_{x \to 0^+} (x + 5) \quad g(x) = x + 5 \text{ to the right of } 0.$$

$$= 5$$

Because the left- and right-hand limits are not the same, we conclude that $\lim_{x \to 0} g(x)$ does not exist. If we look at the graph, as shown in Figure 2.15, it is easy to see that the left- and right-hand limits are not the same. Compare this graph with the graph in Example 8.

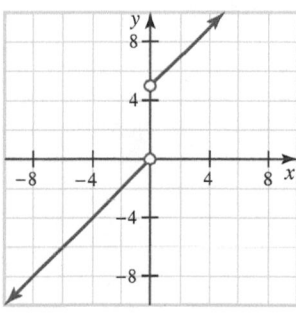

Figure 2.15 Graph of g

Two Special Trigonometric Limits

The following theorem contains two additional limits that play an important role in calculus. We evaluated the first of these limits by table and graphically in Example 6 of Section 2.1.

Theorem 2.3   Special limits involving sine and cosine

$$\lim_{h \to 0} \frac{\sin h}{h} = 1 \qquad \lim_{h \to 0} \frac{\cos h - 1}{h} = 0$$

Proof: A geometric argument justifying these rules is given at the end of this section ◆

For now, let us concentrate on illustrating how these rules can be used.

Example 10   Evaluation of trigonometric and inverse trigonometric limits

Find each of the following limits: **a.** $\lim_{x \to 0} \dfrac{\sin 3x}{5x}$ **b.** $\lim_{x \to 0} \dfrac{\sin^{-1} x}{x}$ **c.** $\lim_{h \to 0} \dfrac{\cos h - 1}{h^2}$

Solution

a. We prepare the limit for evaluation by Theorem 2.3 by writing

$$\frac{\sin 3x}{5x} = \frac{3}{5}\left(\frac{\sin 3x}{3x}\right) \qquad ☠ \sin 3x \neq 3 \sin x ☠$$

Since $3x \to 0$ as $x \to 0$, we can set $h = 3x$ in Theorem 2.3 to obtain

$$\lim_{x \to 0} \frac{\sin 3x}{5x} = \lim_{x \to 0} \frac{3}{3} \left(\frac{\sin 3x}{5x} \right) = \lim_{x \to 0} \frac{3}{5} \left(\frac{\sin 3x}{3x} \right) = \frac{3}{5} \lim_{x \to 0} \left(\frac{\sin 3x}{3x} \right)$$

$$= \frac{3}{5} \lim_{h \to 0} \frac{\sin h}{h} = \frac{3}{5}(1) = \frac{3}{5}$$

b. Let $u = \sin^{-1} x$, so $\sin u = x$. Thus $u \to 0$ as $x \to 0$, and

$$\lim_{x \to 0} \frac{\sin^{-1} x}{x} = \lim_{u \to 0} \frac{u}{\sin u} = 1$$

c. We begin by writing

$$\frac{\cos h - 1}{h^2} = \frac{(\cos h - 1)(\cos h + 1)}{h^2 (\cos h + 1)}$$

$$= \frac{\cos^2 h - 1}{h^2 (\cos h + 1)}$$

$$= \frac{-\sin^2 h}{h^2 (\cos h + 1)} \quad \textit{Since } \sin^2 h + \cos^2 h = 1.$$

Thus, we have

$$\lim_{h \to 0} \frac{\cos h - 1}{h^2} = \lim_{h \to 0} \frac{-\sin^2 h}{h^2 (\cos h + 1)}$$

$$= -\left[\lim_{h \to 0} \frac{\sin^2 h}{h^2} \right] \left[\lim_{h \to 0} \frac{1}{\cos h + 1} \right]$$

$$= -\left[\lim_{h \to 0} \frac{\sin h}{h} \right]^2 \left[\lim_{h \to 0} \frac{1}{\cos h + 1} \right]$$

$$= -(1)^2 \left(\frac{1}{1 + 1} \right)$$

$$= -\frac{1}{2}$$

We conclude this section with a justification of Theorem 2.3. Our argument is based on the following useful property that is proved in Appendix A.

SQUEEZE RULE* If $g(x) \le f(x) \le h(x)$ on an open interval containing c, and if

$$\lim_{x \to c} g(x) = \lim_{x \to c} h(x) = L, \text{ then } \lim_{x \to c} f(x) = L$$

What this says If a function can be squeezed between two functions whose limits at a particular point c have the same value L, then that function must also have limit L at $x = c$.

With this squeeze rule in our toolkit, we are now ready to justify Theorem 2.3.

---

*This is sometimes known as the **sandwich rule**.

Proof: Proof that $\lim\limits_{h \to 0} \dfrac{\sin h}{h} = 1$.

We will call this first part of Theorem 2.3 the **sine limit theorem**. Its proof requires some principles that are not entirely obvious. However, we can demonstrate its plausibility by considering Figure 2.16 in which AOC is a sector of a circle of radius 1 with angle h measured in radians. The line segments $\overline{AD}$ and $\overline{BC}$ are drawn perpendicular to segment $\overline{OC}$.

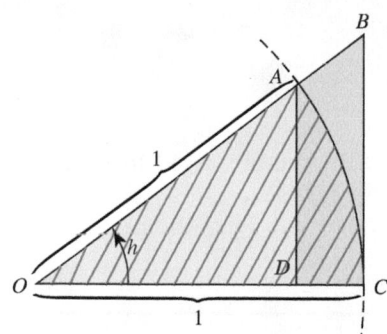

Assume $0 < h < \dfrac{\pi}{2}$; that is, h is in Quadrant I.

$$\left|\widehat{AC}\right| = h$$

$$\left|\overline{AD}\right| = \sin h$$

$$\left|\overline{BC}\right| = \tan h = \frac{\sin h}{\cos h}$$

$$\left|\overline{OD}\right| = \cos h$$

Figure 2.16 Trigonometric relationships in the proof that $\lim\limits_{h \to 0} \dfrac{\sin h}{h} = 1$

Now, compare the area of the sector AOC with those of $\triangle AOD$ and $\triangle BOC$. In particular, since the area of the circular sector of radius r and central angle θ is $\frac{1}{2} r^2 \theta$ (from geometry) sector AOC must have area

$$\frac{1}{2}(1)^2 h = \frac{1}{2} h$$

We also find that $\triangle AOD$ has area

$$\frac{1}{2} \left|\overline{OD}\right| \left|\overline{AD}\right| = \frac{1}{2} \cos h \, \sin h$$

and $\triangle BOC$ has area

$$\frac{1}{2} \left|\overline{BC}\right| \left|\overline{OC}\right| = \frac{1}{2} \left|\overline{BC}\right| (1) = \frac{1}{2} \frac{\sin h}{\cos h}$$

By comparing areas (see Figure 2.16), we have

AREA OF △ AOD		AREA OF SECTOR AOC		AREA OF △ BOC	
$\dfrac{1}{2} \cos h \, \sin h$	$\leq$	$\dfrac{1}{2} h$	$\leq$	$\dfrac{\sin h}{2 \cos h}$	
$\cos h$	$\leq$	$\dfrac{h}{\sin h}$	$\leq$	$\dfrac{1}{\cos h}$	*Divide by $\dfrac{1}{2} \sin h$.*
$\dfrac{1}{\cos h}$	$\geq$	$\dfrac{\sin h}{h}$	$\geq$	$\cos h$	*Take reciprocals (reverse order).*
$\cos h$	$\leq$	$\dfrac{\sin h}{h}$	$\leq$	$\dfrac{1}{\cos h}$	*Interchange left and right sides (reverse order again).*

The same inequalities hold in the interval $\left(-\frac{\pi}{2}, 0\right)$, which is Quadrant IV for the angle h. This can be shown by using the trigonometric identities $\cos(-h) = \cos h$ and $\sin(-h) = -\sin h$. Finally, we take the limit of all parts of the inequalities as $h \to 0$ to find

$$\lim_{h\to 0} \cos h \leq \lim_{h\to 0} \frac{\sin h}{h} \leq \lim_{h\to 0} \frac{1}{\cos h}$$

By Theorem 2.2, $\lim_{h\to 0} \cos h = \cos 0 = 1$. Thus,

$$1 \leq \lim_{h\to 0} \frac{\sin h}{h} \leq \frac{1}{1}$$

From the squeeze rule, we conclude that $\lim_{h\to 0} \frac{\sin h}{h} = 1$. You are asked to prove the second part of this theorem in Problem 57.

PROBLEM SET 2.2

Level 1

In Problems 1-30, evaluate each limit.

1. $\lim_{x\to -1} (x^2 + 3x - 7)$ **2.** $\lim_{t\to 0}(t^3 - 5t^2 + 4)$

3. $\lim_{x\to 3} (x + 5)(2x - 7)$

4. $\lim_{x\to 0} \frac{\sin^{-1} x}{x - 1}$

5. $\lim_{z\to 1} \frac{z^2 + z - 3}{z + 1}$

6. $\lim_{x\to 3} \frac{x^2 + 3x - 10}{3x^2 + 5x - 7}$

7. $\lim_{x\to \pi/3} \sec x$

8. $\lim_{x\to \pi/4} \frac{1 + \tan x}{\csc x + 1}$

9. $\lim_{x\to 1/3} \frac{x \sin \pi x}{1 + \cos \pi x}$

10. $\lim_{x\to 6} \frac{\tan(\pi/x)}{x - 1}$

11. $\lim_{u\to -2} \frac{4 - u^2}{2 + u}$

12. $\lim_{x\to 2} \frac{x^2 - 4x + 4}{x^2 - x - 2}$

13. $\lim_{x\to 1} \frac{\frac{1}{x} - 1}{x - 1}$

14. $\lim_{x\to 0} \frac{(x + 1)^2 - 1}{x}$

15. $\lim_{x\to 1} \frac{\sqrt{x} - 1}{x - 1}$

16. $\lim_{y\to 2} \frac{\sqrt{y + 2} - 2}{y - 2}$

17. $\lim_{x\to 0} \frac{\sqrt{x + 1} - 1}{x}$

18. $\lim_{x\to 0} \frac{\sqrt{x^2 + 4} - 2}{x}$

19. $\lim_{x\to 0^+} \frac{\sin x}{\sqrt{x}}$

20. $\lim_{x\to 0^+} \frac{1 - \cos \sqrt{x}}{x}$

21. $\lim_{x\to 0} \frac{\sin 2x}{x}$

22. $\lim_{x\to 0} \frac{\sin 4x}{9x}$

23. $\lim_{t\to 0} \frac{\tan 5t}{\tan 2t}$

24. $\lim_{x\to 0} \frac{\cot 3x}{\cot x}$

25. $\lim_{x\to 0} \frac{1 - \cos x}{\sin x}$

26. $\lim_{x\to 0} \frac{x^2 \cos 2x}{1 - \cos x}$

27. $\lim_{x\to 0} \frac{\sin^2 x}{2x}$

28. $\lim_{x\to 0} \frac{\sin^2 x}{x^2}$

29. $\lim_{x\to 0} \frac{\sec x - 1}{x \sec x}$

30. $\lim_{x\to \pi/4} \frac{1 - \tan x}{\sin x - \cos x}$

31. ■ What does this say? How do you find the limit of a polynomial function?

32. ■ What does this say? How do you find the limit of a rational function?

33. ■ What does this say? How do you find

$$\lim_{x\to 0} \frac{\sin ax}{x}$$

for $a \neq 0$?

Level 2

In Problems 34-41, compute the one-sided limit or use one-sided limits to find the given limit, if it exists.

34. $\lim_{x\to 2^-} (x^2 - 2x)$

35. $\lim_{x\to 1^+} \frac{\sqrt{x - 1} + x}{1 - 2x}$

36. $\lim_{x\to 2} |x - 2|$

37. $\lim_{x\to 3} |3 - x|$

38. $\lim_{x\to 0} \frac{|x|}{x}$

39. $\lim_{x\to -2} \frac{|x + 2|}{x + 2}$

40. $\lim\limits_{x \to 2} f(x)$, where $f(x) = \begin{cases} 3 - 2x & \text{if } x \le 2 \\ x^2 - 5 & \text{if } x > 2 \end{cases}$

41. $\lim\limits_{s \to 1} g(s)$, where $g(s) = \begin{cases} \frac{s^2 - s}{s - 1} & \text{if } s > 1 \\ \sqrt{1 - s} & \text{if } s \le 1 \end{cases}$

■ *What does this say? In Problems 42-49, explain why the given limit does not exist.*

42. $\lim\limits_{x \to 1} \dfrac{1}{x - 1}$

43. $\lim\limits_{x \to 2^+} \dfrac{1}{\sqrt{x - 2}}$

44. $\lim\limits_{t \to 2} \dfrac{t^2 - 4}{t^2 - 4t + 4}$

45. $\lim\limits_{x \to 3} \dfrac{x^2 + 4x + 3}{x - 3}$

46. $\lim\limits_{x \to \pi/2} \tan x$

47. $\lim\limits_{x \to 1} \csc \pi x$

48. $\lim\limits_{x \to 1} f(x)$, where $f(x) = \begin{cases} 2 & \text{if } x \ge 1 \\ -5 & \text{if } x < 1 \end{cases}$

49. $\lim\limits_{t \to -1} g(t)$, where $g(t) = \begin{cases} 2t + 1 & \text{if } t > -1 \\ 5t^2 & \text{if } t < -1 \end{cases}$

In Problems 50-55, either evaluate the limit or explain why it does not exist.

50. $\lim\limits_{x \to 1} \dfrac{\frac{1}{x} - 1}{\sqrt{x} - 1}$

51. $\lim\limits_{x \to 0} \left(\dfrac{1}{x} - \dfrac{1}{x^2} \right)$

52. $\lim\limits_{x \to 5} f(x)$, where $f(x) = \begin{cases} x + 3 & \text{if } x \ne 5 \\ 4 & \text{if } x = 5 \end{cases}$

53. $\lim\limits_{t \to 2} g(t)$, where

$$g(t) = \begin{cases} t^2 & \text{if } -1 \le t < 2 \\ 3t - 2 & \text{if } t \ge 2 \end{cases}$$

54. $\lim\limits_{x \to 2} f(x)$, where

$$f(x) = \begin{cases} 2(x + 1) & \text{if } x < 3 \\ 4 & \text{if } x = 3 \\ x^2 - 1 & \text{if } x > 3 \end{cases}$$

55. $\lim\limits_{x \to 3} f(x)$, where

$$f(x) = \begin{cases} 2(x + 1) & \text{if } x < 3 \\ 4 & \text{if } x = 3 \\ x^2 - 1 & \text{if } x > 3 \end{cases}$$

Level 3

56. INTERPRETATION PROBLEM Evaluate

$$\lim\limits_{x \to 0} \left[x^2 - \dfrac{\cos x}{1,000,000,000} \right]$$

Explain why a calculator solution may lead to an incorrect conclusion about the limit.

57. Prove the second part of Theorem 2.3:

$$\lim\limits_{h \to 0} \dfrac{\cos h - 1}{h} = 0$$

58. Let $f(x) = \dfrac{1}{x^2}$ with $x \ne 0$, and let L be any fixed positive integer. Show that

$$f(x) > 100L \quad \text{if} \quad |x| < \dfrac{1}{10\sqrt{L}}$$

What does this imply about $\lim\limits_{x \to 0} f(x)$?

59. Show that $\lim\limits_{x \to x_0} \cos x = \cos x_0$.

Hint: You will need to use the trigonometric identity

$$\cos(A + B) = \cos A \, \cos B - \sin A \, \sin B$$

60. Show that $\lim\limits_{x \to x_0} \tan x = \tan x_0$ whenever $\cos x_0 \ne 0$.

2.3 CONTINUITY

IN THIS SECTION: *Intuitive notion of continuity, definition of continuity, continuity theorems, continuity on an interval, the intermediate value theorem*

Informally, a continuous function is one whose graph has no "breaks" or "jumps." However, do not be misled by this simple geometric characterization. Continuity is an important and complex concept that plays a central role in the further development of calculus.

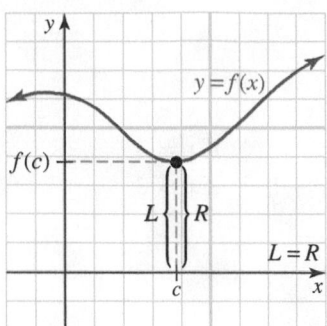

Figure 2.17 A function is continuous at a point $x = c$. Let $\lim\limits_{x \to c^-} f(x) = L$ and $\lim\limits_{x \to c^+} f(x) = R$, and notice that $L = R$

Intuitive Notion of Continuity

The idea of *continuity* may be thought of informally as the quality of having parts that are in immediate connection with one another, as shown in Figure 2.17. The idea evolved from the vague or intuitive notion of a curve "without breaks or jumps" to a rigorous definition first given toward the end of the 19th century (see *Historical Quest* Problem 55).

We begin with a discussion of *continuity at a point*. It may seem strange to talk about continuity *at a point,* but it should seem natural to talk about a curve being "discontinuous at a point." A few such discontinuities are illustrated in Table 2.2.

Table 2.2 Holes, Poles, Jumps, and Continuity

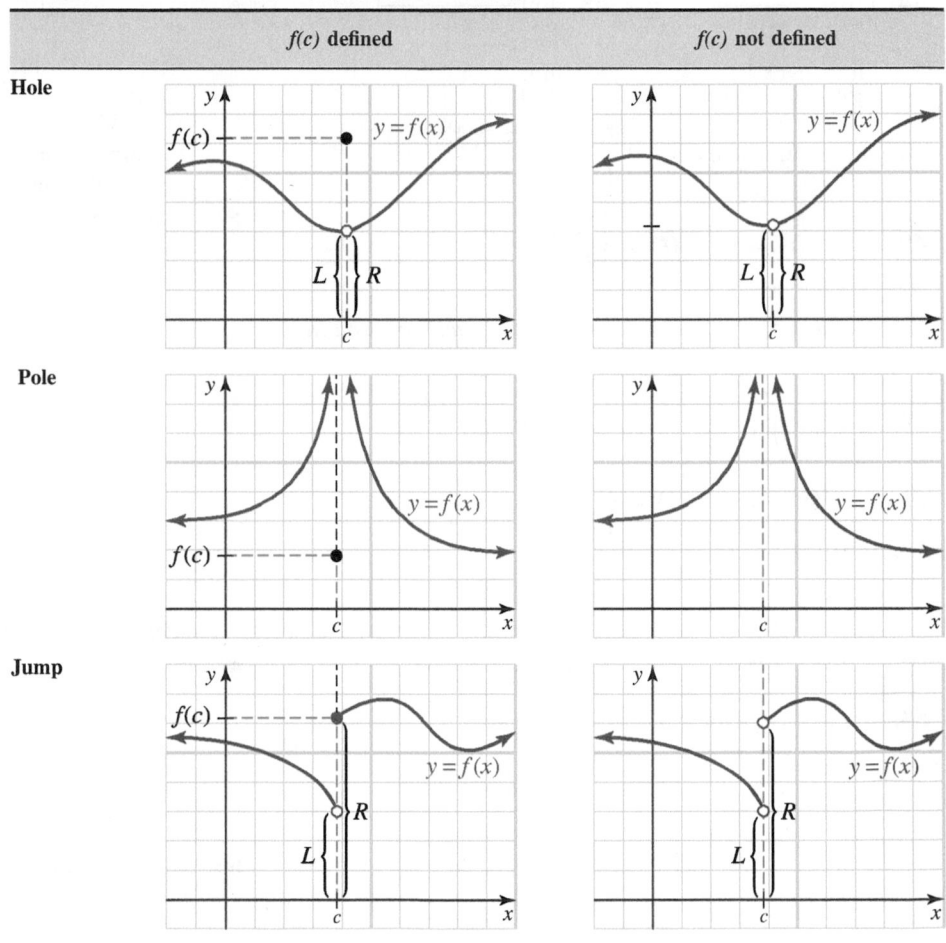

Definition of Continuity

Let us consider the conditions that must be satisfied for a function f to be continuous at a point c. First, $f(c)$ must be defined or we have a "hole" in the graph, as shown in Table 2.2. (An open dot indicates an excluded point.) Note there may be a "hole" even though $f(c)$ is defined. If one or both of the one-sided limits at c are infinite (∞ or $-\infty$), there will be a "pole" at $x = c$, as shown in the middle row of Table 2.2. Finally, if $\lim\limits_{x \to c} f(x)$ has one value as $x \to c^-$ and another as $x \to c^+$, then $\lim\limits_{x \to c} f(x)$ does not exist and there will be a "jump" in the graph of f, as shown in the last row of Table 2.2.

We now state a formal definition.

> **CONTINUITY OF A FUNCTION AT A POINT**   A function f is **continuous at a point** $x = c$ if the following three conditions are satisfied:
> 1. $f(c)$ is defined;
> 2. $\lim\limits_{x \to c} f(x)$ exists;
> 3. $\lim\limits_{x \to c} f(x) = f(c)$.
> A function that is not continuous at c is said to have a **discontinuity** at that point.

> ■ **W**hat this says Step 1 refers to the domain of the function and ignores what happens at points $x \neq c$, whereas step 2 refers to points close to c, but ignores the point $x = c$. The third condition looks at the entire picture: it says that if x is close to c, then $f(x)$ must be close to $f(c)$.

If f is continuous at $x = c$, the difference between $f(x)$ and $f(c)$ is small whenever x is close to c because $\lim\limits_{x \to c} f(x) = f(c)$. Geometrically, this means that the points $(x, f(x))$ on the graph of f converge to the point $(c, f(c))$ as $x \to c$, and this is what guarantees that the graph is unbroken at $(c, f(c))$ with no "gap" or "hole," as shown in Figure 2.18.

We should note that not every continuous function can be graphed (for example, see Example 2e later in this section), so you cannot always test continuity by graphing, even though graphing can help for many functions.

Example 1   Testing continuity

Test the continuity of each of the following functions at $x = 1$. If the given function is not continuous at $x = 1$, explain.

a. $f(x) = \dfrac{x^2 + 2x - 3}{x - 1}$

b. $g(x) = \dfrac{x^2 + 2x - 3}{x - 1}$ if $x \neq 1$ and $g(x) = 6$ if $x = 1$.

c. $h(x) = \dfrac{x^2 + 2x - 3}{x - 1}$ if $x \neq 1$ and $h(x) = 4$ if $x = 1$.

d. $F(x) = \dfrac{x + 3}{x - 1}$ if $x \neq 1$ and $F(x) = 4$ if $x = 1$.

e. $G(x) = 7x^3 + 3x^2 - 2$

f. $H(x) = 2 \sin x - \tan x$

Solution

a. The function f is not continuous at $x = 1$.
   Hole; $f(1)$ not defined.

b. 1. $g(1)$ is defined; $g(1) = 6$.

   2. $\lim\limits_{x \to 1} g(x) = \lim\limits_{x \to 1} \dfrac{x^2 + 2x - 3}{x - 1}$

   $\qquad = \lim\limits_{x \to 1} \dfrac{(x - 1)(x + 3)}{x - 1}$

   $\qquad = \lim\limits_{x \to 1} (x + 3)$

   $\qquad = 4$

   3. $\lim\limits_{x \to 1} g(x) \neq g(1)$, so g is not continuous at $x = 1$

   Hole; g(1) is defined, but the third condition is not satisfied.

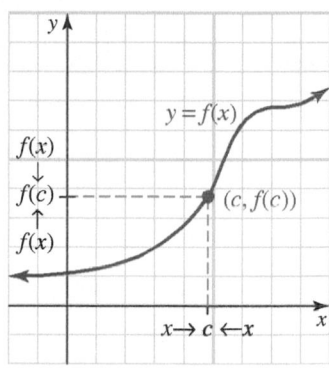

Figure 2.18   Geometric interpretation of continuity

Do not rely on graphing calculators to test for continuity. For example, the graph of $y = \dfrac{x - 2}{x - 2}$ looks like a constant function when done with a calculator; it does not show the discontinuity at $x = 2$.

c. Compare h with g of part **b**. We see that all three conditions of continuity are satisfied, so h is continuous at $x = 1$.

d. 1. $F(1)$ is defined; $F(1) = 4$.

2. $\lim\limits_{x \to 1} F(x) = \lim\limits_{x \to 1} \dfrac{x + 3}{x - 1}$; this limit does not exist.

Pole; the function F is not continuous at $x = 1$.

e. 1. $G(1)$ is defined; $G(1) = 8$.

2. $\lim\limits_{x \to 1} G(x) = \lim\limits_{x \to 1} (7x^3 + 3x^2 - 2)$

$$= 7 \left(\lim_{x \to 1} x \right)^3 + 3 \left(\lim_{x \to 1} x \right)^2 - \lim_{x \to 1} 2$$

$$= 7 + 3 - 2$$

$$= 8$$

3. $\lim\limits_{x \to 1} G(x) = G(1)$

Because the three conditions of continuity are satisfied, G is continuous at $x = 1$.

f. 1. $H(1)$ is defined; $H(1) = 2 \sin 1 - \tan 1$.

2. $\lim\limits_{x \to 1} H(x) = \lim\limits_{x \to 1} (2 \sin x - \tan x)$

$$= 2 \lim_{x \to 1} \sin x - \lim_{x \to 1} \tan x$$

$$= 2 \sin 1 - \tan 1$$

3. $\lim\limits_{x \to 1} H(x) = H(1)$

Because the three conditions of continuity are satisfied, H is continuous at $x = 1$.

Continuity Theorems

It is often difficult to determine whether a given function is continuous at a specified number. However, many common functions are continuous wherever they are defined.

Theorem 2.4   Continuity theorem

If f is a polynomial, a rational function, a power function, a trigonometric function, or an inverse trigonometric function, then f is continuous at any number $x = c$ for which $f(c)$ is defined.

Proof: The proof of the continuity theorem is based on the limit properties stated in the previous section. For instance, a polynomial is a function of the form

$$P(x) = a_n x^n + a_{n-1} x^{n-1} + \cdots + a_1 x + a_0$$

where $a_0, a_1, \ldots, a_n$ are constants. We know that $\lim\limits_{x \to c} a_0 = a_0$ and that $\lim\limits_{x \to c} x^m = c^m$ for $m = 1, 2, \cdots, n$. This is precisely the statement that the function $g(x) = ax^m$ is continuous at any number $x = c$, or simply continuous. Because P is a sum of functions of this form and a constant function, it follows from the limit properties that P is continuous.

The proofs of the other parts follow similarly.   ♦

The limit properties of the previous section can also be used to prove a second continuity theorem. This theorem tells us that continuous functions may be combined in various ways *without creating a discontinuity.*

Theorem 2.5   Properties of continuous functions

If f and g are functions that are continuous at $x = c$, then the following functions are also continuous at $x = c$.

Scalar multiple	sf	for any constant s (called a **scalar**)
Sum and difference	$f + g$ and $f - g$	
Product	fg	
Quotient	$\dfrac{f}{g}$	provided $g(c) \neq 0$
Composition	$f \circ g$	provided g is continuous at c and f is continuous at $g(c)$

Proof: The first four properties in this theorem follow directly from the basic limit rules given in Section 2.2. For instance, to prove the product property, note that since f and g are given to be continuous at $x = c$, we have

$$\lim_{x \to c} f(x) = f(c) \quad \text{and} \quad \lim_{x \to c} g(x) = g(c)$$

If $P(x) = (fg)(x) = f(x)g(x)$, then

$$\lim_{x \to c} P(x) = \lim_{x \to c} f(x)g(x)$$

$$= \left[\lim_{x \to c} f(x) \right] \left[\lim_{x \to c} g(x) \right]$$

$$= f(c)g(c)$$

$$= P(c)$$

so $P(x)$ is continuous at $x = c$, as required.   ◆

The continuous composition property is proved in a similar fashion, but requires the following limit rule, which we state without proof.

COMPOSITE LIMIT RULE   If $\lim\limits_{x \to c} g(x) = L$ and f is a function continuous at L, then

$$\lim_{x \to c} f\big[g(x)\big] = f(L)$$

That is,

$$\lim_{x \to c} f\big[g(x)\big] = f(L) = f\left[\lim_{x \to c} g(x)\right]$$

■ **W**hat this says The limit of a continuous function is the function of the limiting value. The continuous composition property says that a continuous function of a continuous function is continuous.

This property applies in the same way to other kinds of limits, in particular to one-sided limits. Now we can prove the continuous composition property of Theorem 2.5. Let $h(x) = (f \circ g)(x)$. Then we have

$$\lim_{x \to c} (f \circ g)(x) = \lim_{x \to c} f\big[g(x)\big] \quad \textit{Definition of composition}$$

$$= f\left[\lim_{x \to c} g(x)\right] \quad \textit{Composition limit rule}$$

$$= f\left[g(c)\right] \qquad g \text{ is continuous at } x = c$$

$$= (f \circ g)(c) \qquad \textit{Definition of composition}$$

We need to talk about a function being continuous on an interval. To do so, we must first know how to handle continuity at the endpoints of the interval, which leads to the following definition.

ONE-SIDED CONTINUITY   The function F is **continuous from the right at a** if and only if

$$\lim_{x \to a^+} f(x) = f(a)$$

and it is **continuous from the left at b** if and only if

$$\lim_{x \to b^-} f(x) = f(b)$$

Continuity on an Interval

The function f is said to be **continuous on the open interval (a, b)** if it is continuous at each number in this interval. Also note that the endpoints are not part of open intervals. If f is also continuous from the right at a, we say it is **continuous on the half-open interval $[a, b)$**. Similarly, f is **continuous on the half-open interval $(a, b]$** if it is continuous at each number between a and b and is continuous from the left at the endpoint b. Finally, f is **continuous on the closed interval $[a, b]$** if it is continuous at each number between a and b and is both continuous from the right at a and continuous from the left at b.

Example 2   Testing for continuity on an interval

Find the intervals on which each of the given functions is continuous.

a. $f_1(x) = \dfrac{x^2 - 1}{x^2 - 4}$ **b.** $f_2(x) = \left|x^2 - 4\right|$ **c.** $f_3(x) = \csc x$

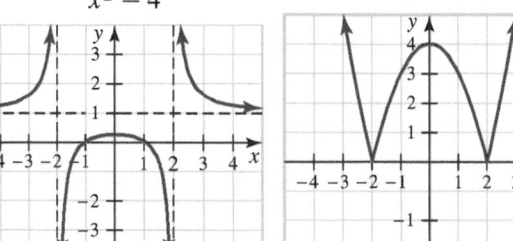
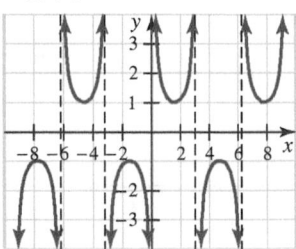

d. $f_4(x) = \sin \dfrac{1}{x}$ **e.** $f_5(x) = \begin{cases} x \sin \dfrac{1}{x} & \text{if } x \neq 0 \\ 0 & \text{if } x = 0 \end{cases}$

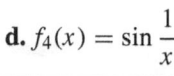

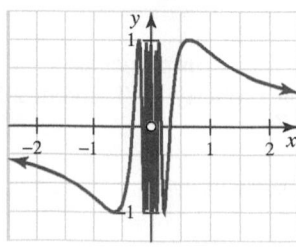
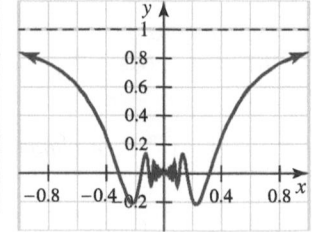

Solution

a. The function f_1 is not defined when $x^2 - 4 = 0$; that is, when $x = 2$ or $x = -2$. The function is continuous on $(-\infty, -2) \cup (-2, 2) \cup (2, \infty)$.
b. The function f_2 is continuous on $(-\infty, \infty)$.
c. The cosecant function is not defined at $x = n\pi$, n an integer. At all other points it is continuous.
d. Because $1/x$ is continuous except at $x = 0$ and the sine function is continuous everywhere, we need only check continuity at $x = 0$.

$$\lim_{x \to 0} \sin \frac{1}{x} \text{ does not exist}$$

Therefore, $f(x) = \sin \frac{1}{x}$ is continuous on $(-\infty, 0) \cup (0, \infty)$.
e. From the double inequality

$$-1 \le \sin \frac{1}{x} \le 1$$

it follows that

$$-|x| \le x \sin \frac{1}{x} \le |x|, x \ne 0$$

The plausibility of this inequality can be seen by drawing $y = |x|$ and $y = -|x|$ along with $y = f_5(x)$ on the same axis, as shown in Figure 2.19.
    We can now use the squeeze rule. Because $\lim_{x \to 0} |x| = 0$ and $\lim_{x \to 0} (-|x|) = 0$, it follows that $\lim_{x \to 0} x \sin \frac{1}{x} = 0$. Because $f_5(0) = 0$ we see that f is continuous at $x = 0$ and therefore is continuous on $(-\infty, \infty)$. ∎

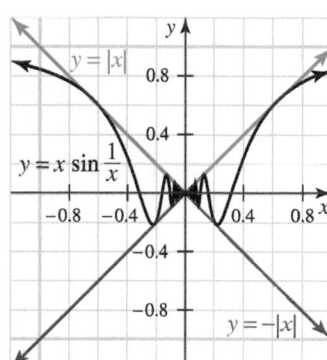

Figure 2.19 The graphs of $y = f_5(x), y = |x|$, and $y = -|x|$

    Usually there are only a few points in the domain of a given function f where a discontinuity can occur. We use the term **suspicious point** for a number c where either:

(1) The defining rule for f changes.
(2) Substitution of $x = c$ causes division by 0 in the function.

    For Example 2, the suspicious points can be listed:

a. $\dfrac{x^2 - 1}{x^2 - 4}$ has suspicious points for division by zero when $x = 2$ and $x = -2$.
b. $|x^2 - 4| = x^2 - 4$ when $x^2 - 4 \ge 0$ and $|x^2 - 4| = 4 - x^2$ when $x^2 - 4 < 0$. This means the definition of the function changes when $x^2 - 4 = 0$, namely, when $x = 2$ and $x = -2$.
c. There are no suspicious points; we know the function cannot be continuous at places where the function is not defined.
d. $\sin \frac{1}{x}$ has a suspicious point when $x = 0$ (division by 0).
e. $x \sin \frac{1}{x}$ has a suspicious point when $x = 0$ (definition changes).

Example 3   Checking continuity at suspicious points

Let $f(x) = \begin{cases} 3 - x & \text{if } -5 \le x < 2 \\ x - 2 & \text{if } 2 \le x < 5 \end{cases}$ and $g(x) = \begin{cases} 2 - x & \text{if } -5 \le x < 2 \\ x - 2 & \text{if } 2 \le x < 5 \end{cases}$

Find the intervals on which f and g are continuous.

Solution   We find that the domain for both functions is $[-5, 5)$; the continuity theorem tells us both functions are continuous everywhere on that interval except possibly at the suspicious points. Consider the graphs of f and g as shown in Figure 2.20.

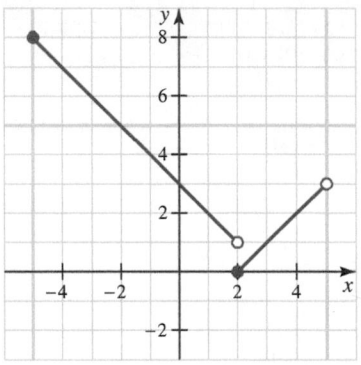 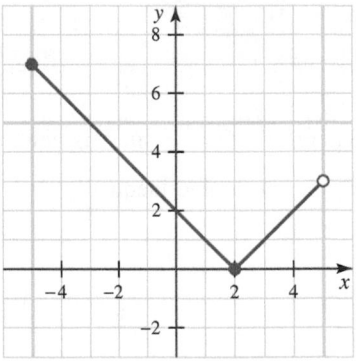

Figure 2.20   Graphs of f and g

Examining f, we see

$$\lim_{x \to 2^-} f(x) = \lim_{x \to 2^-} (3 - x) = 1 \qquad \text{and} \qquad \lim_{x \to 2^+} f(x) = \lim_{x \to 2^+} (x - 2) = 0$$

so $\lim_{x \to 2} f(x)$ does not exist and f is discontinuous at $x = 2$. Thus, f is continuous for $-5 \le x < 2$ and for $2 < x < 5$.

For g, the domain is also $-5 \le x < 5$, and again, the only suspicious point is $x = 2$. We have $g(2) = 0$ and

$$\lim_{x \to 2^-} g(x) = \lim_{x \to 2^-} (2 - x) = 0 \qquad \text{and} \qquad \lim_{x \to 2^+} g(x) = \lim_{x \to 2^+} (x - 2) = 0$$

Therefore, $\lim_{x \to 2} g(x) = 0 = g(2)$, and g is continuous at $x = 2$. Hence, g is continuous throughout the interval $[-5, 5)$. ■

Intermediate Value Theorem

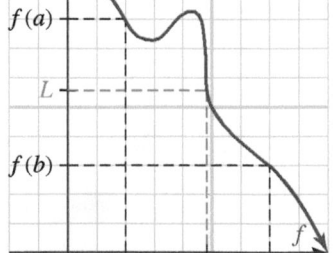

Figure 2.21   If L lies between $f(a)$ and $f(b)$, then $f(c) = L$ for some c between a and b

Intuitively, if f is continuous throughout an entire interval, its graph on that interval may be drawn "without the pencil leaving the paper." That is, if $f(x)$ varies continuously from $f(a)$ to $f(b)$ as x increases from a to b, then it must hit every number L between $f(a)$ and $f(b)$, as shown in Figure 2.21.

To illustrate the property shown in Figure 2.21, suppose f is a function defined as the weight of a person at age x. If we assume that weight varies continuously with time, a person who weighs 50 pounds at age 6 and 120 pounds at age 15 must weigh 100 pounds at some time between ages 6 and 15.

This feature of continuous functions is known as the *intermediate value property*. A formal statement of this property is contained in the following theorem.

Theorem 2.6   The intermediate value theorem

If f is a continuous function on the closed interval $[a, b]$ and L is some number strictly between $f(a)$ and $f(b)$, then there exists at least one number c on the open interval (a, b) such that $f(c) = L$.

> ■ **W**hat this says If f is a continuous function (with emphasis on the word *continuous*) on some *closed* interval $[a, b]$, then $f(x)$ must take on all values between $f(a)$ and $f(b)$.

Proof: This theorem is intuitively obvious, but it is not at all easy to prove. A proof may be found in most advanced calculus textbooks.   ◆

The intermediate value theorem can be used to estimate roots of the equation $f(x) = 0$. Suppose $f(a) > 0$ and $f(b) < 0$, so the graph of f is above the x-axis at $x = a$ and below the x-axis at $x = b$. Then, if f is continuous on the closed interval $[a, b]$, there must be a point $x = c$ between a and b where the graph crosses the x-axis, that is, where $f(c) = 0$. The same conclusion would be drawn if $f(a) < 0$ and $f(b) > 0$. The key is that $f(x)$ changes sign between $x = a$ and $x = b$. This is shown in Figure 2.22 and is summarized in the following theorem.

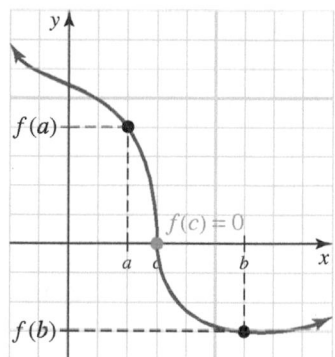

Theorem 2.7   Root location theorem

If f is continuous on the closed interval $[a, b]$ and if $f(a)$ and $f(b)$ have opposite algebraic signs (one positive and the other negative), then $f(c) = 0$ for at least one number c on the open interval (a, b).

Proof: This follows directly from the intermediate value theorem (see Figure 2.22). The details of this proof are left as a problem.   ◆

Figure 2.22 Because $f(a) > 0$ and $f(b) < 0$, $f(c) = 0$ for some c between a and b

Example 4   Using the root location theorem

Show that $\cos x = x^3 - x$ has at least one solution on the interval $\left[\frac{\pi}{4}, \frac{\pi}{2}\right]$.

Solution   We can use a graphing utility to estimate the point of intersection, as shown in Figure 2.23.

We will now use the root location theorem to confirm the intersection point shown in Figure 2.23. By calculator, we estimate $c \approx 1.16$.

Notice that the function $f(x) = \cos x - x^3 + x$ is continuous on $\left[\frac{\pi}{4}, \frac{\pi}{2}\right]$. We find that

$$f\left(\frac{\pi}{4}\right) = \cos\frac{\pi}{4} - \left(\frac{\pi}{4}\right)^3 + \frac{\pi}{4} \approx 1.008$$

$$f\left(\frac{\pi}{2}\right) = \cos\frac{\pi}{2} - \left(\frac{\pi}{2}\right)^3 + \frac{\pi}{2} \approx -2.305$$

Therefore, by the root location theorem there is at least one number c on $\left(\frac{\pi}{4}, \frac{\pi}{2}\right)$ for which $f(c) = 0$, and it follows that $\cos c = c^3 - c$.   ■

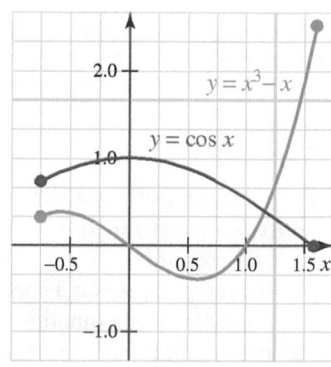

Figure 2.23 Estimate the point of intersection to be $x \approx 1.16$

Since the definition of continuity involves the limit, formal proofs of continuity theorems can be given in terms of the ϵ-δ definition of limit. Here is an example.

Example 5   Formal proof of the sum rule for continuity

Show that if f and g are continuous at $x = x_0$, then $f + g$ is also continuous at $x = x_0$.

Solution   The continuity of f and g at $x = x_0$ says that

$$\lim_{x \to x_0} f(x) = f(x_0) \qquad \text{and} \qquad \lim_{x \to x_0} g(x) = g(x_0)$$

This means that if $\epsilon > 0$ is given, there exist numbers δ_1 and δ_2 such that

$$|f(x) - f(x_0)| < \frac{\epsilon}{2} \text{ whenever } 0 < |x - x_0| < \delta_1$$

and

$$|g(x) - g(x_0)| < \frac{\epsilon}{2} \text{ whenever } 0 < |x - x_0| < \delta_2$$

Let δ be the smaller of δ_1 and δ_2; that is, $\delta = \min(\delta_1, \delta_2)$. Then

$$0 < |x - x_0| < \delta_1 \qquad \text{and} \qquad 0 < |x - x_0| < \delta_2$$

must both be true whenever $0 < |x - x_0| < \delta$ so that (by the triangle inequality)

$$\left| f(x) + g(x) - [f(x_0) + g(x_0)] \right| = \left| [f(x) - f(x_0)] + [g(x) - g(x_0)] \right|$$
$$\leq |f(x) - f(x_0)| + |g(x) - g(x_0)|$$
$$< \frac{\epsilon}{2} + \frac{\epsilon}{2} = \epsilon$$

Thus, $|f(x + g)(x) - (f + g)(x_0)| < \epsilon$ whenever $0 < |x - x_0| < \delta$, and it follows from the definition of limit that

$$\lim_{x \to x_0} [f(x) + g(x)] = f(x_0) + g(x_0)$$

In other words, $f + g$ is continuous at $x = x_0$, as claimed. ∎

PROBLEM SET 2.3

Level 1

Which of the functions described in Problems 1-6 represent continuous functions? State the domain, if possible, for each example.

1. the temperature on a specific day at a given location considered as a function of time
2. the humidity on a specific day at a given location considered as a function of time
3. the closing price of IBM stock on a specified day considered as a function of time
4. the number of unemployed people in the United States during the month of January 2012 considered as a function of time
5. the charges for a taxi ride across town considered as a function of mileage
6. the charges to mail a package as a function of its weight

Identify all suspicious points and determine all points of discontinuity in Problems 7-20.

7. $f(x) = x^3 - 7x + 3$

8. $f(x) = \dfrac{3x + 5}{2x - 1}$

9. $f(x) = \dfrac{3x}{x^2 - x}$

10. $f(t) = 3 - (5 + 2t)^3$

11. $h(x) = \sqrt{x} + \dfrac{3}{x}$

12. $f(u) = \sqrt[3]{u^2 - 1}$

13. $f(t) = \dfrac{1}{t} - \dfrac{3}{t + 1}$

14. $g(x) = \dfrac{x^2 - 1}{x^2 + x - 2}$

15. $f(x) = \begin{cases} x^2 - 2 & \text{if } x > 1 \\ 2x - 3 & \text{if } x \le 1 \end{cases}$

16. $g(t) = \begin{cases} 3t + 2 & \text{if } t \le 1 \\ 5 & \text{if } 1 < t \le 3 \\ 3t^2 - 1 & \text{if } t > 3 \end{cases}$

17. $f(x) = 3\tan x - 5\sin x \cos x$

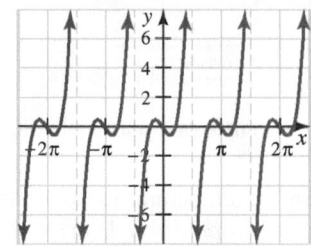

18. $g(x) = \dfrac{\cot x}{\sin x - \cos x}$

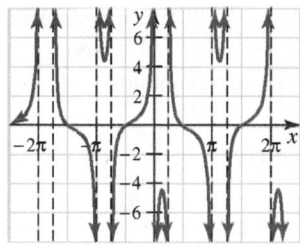

19. $h(x) = \csc x \cot x$

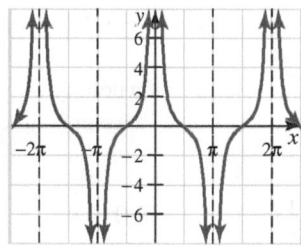

20. $f(x) \begin{cases} \dfrac{\sec x}{\sin x \tan x} & x \ne \dfrac{2n+1}{2}\pi \\ & n \text{ is an integer} \\ 1 & x = \dfrac{2n+1}{2}\pi \end{cases}$

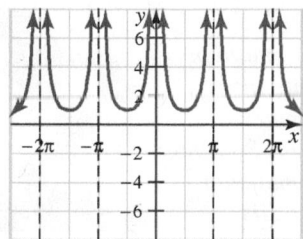

In Problems 21-26, the given function is defined for all $x > 0$ except at $x = 2$. In each case, find the value that should be assigned to $f(2)$, if any, to guarantee that f will be continuous at 2.

21. $f(x) = \dfrac{x^2 - x - 2}{x - 2}$ **22.** $f(x) = \sqrt{\dfrac{x^2 - 4}{x - 2}}$

23. $f(x) = \dfrac{\sin(\pi x)}{x - 2}$ **24.** $f(x) = \dfrac{\cos \frac{\pi}{x}}{x - 2}$

25. $f(x) = \begin{cases} 2x + 5 & \text{if } x > 2 \\ 15 - x^2 & \text{if } x < 2 \end{cases}$

26. $f(x) = \dfrac{\frac{1}{x} - 1}{x - 2}$

In Problems 27-32, determine whether or not the given function is continuous on the prescribed interval.

27. a. $f(x) = \dfrac{1}{x}$ on $[1, 2]$ **b.** $f(x) = \dfrac{1}{x}$ on $[0, 1]$

28. a. $g(x) = \dfrac{1}{x - 3}$ on $[4, 5]$ **b.** $g(x) = \dfrac{1}{x - 3}$ on $[0, 5]$

29. $f(x) = \begin{cases} x^2 & \text{if } 0 \le x < 2 \\ 3x + 1 & \text{if } 2 \le x < 5 \end{cases}$

30. $g(t) = \begin{cases} 15 - t^2 & \text{if } -3 < t \le 0 \\ 2t & 0 < t \le 1 \end{cases}$

31. $f(x) = x \sin x$ on $(0, \pi)$

32. $f(x) = \dfrac{\cos x - 1}{x}$ on $[-\pi, \pi]$

Level 2

In Problems 33-38, show that the given equation has at least one solution on the indicated interval.

33. $\sqrt[3]{x} = x^2 + 2x - 1$ on $(0, 1)$

34. $\sqrt[3]{x - 8} + 9x^{2/3} = 29$ on $(0, 8)$

35. $\dfrac{1}{x + 1} = x^2 - x - 1$ on $(1, 2)$

36. $\tan x = 2x^2 - 1$ on $\left(-\frac{\pi}{4}, 0\right)$

37. $\cos x - \sin x = x$ on $\left(0, \frac{\pi}{2}\right)$

38. $\cos x = x^2 - 1$ on $(0, \pi)$

In Problems 39-44, find constants a and b so that the given function will be continuous for all x.

39. $f(x) = \begin{cases} \dfrac{ax - 4}{x - 2} & \text{if } x \ne 2 \\ b & x = 2 \end{cases}$

40. $g(x) = \begin{cases} b & \text{if } x \leq 1 \\ \dfrac{\sqrt{x} - a}{x - 1} & x > 1 \end{cases}$

41. $F(x) = \begin{cases} x^2 + bx + 1 & \text{if } x < 5 \\ 8 & \text{if } x = 5 \\ ax + 3 & \text{if } x > 5 \end{cases}$

42. $G(x) = \begin{cases} ax + b & \text{if } x < 0 \\ \sqrt{3} & \text{if } x = 0 \\ 2\sin(a\,\cos^{-1} x) & \text{if } 0 < x < 1,\, 0 < a < 1 \end{cases}$

43. $M(x) = \begin{cases} \dfrac{\sin ax}{x} & \text{if } x < 0 \\ 5 & \text{if } x = 0 \\ x + b & \text{if } x > 0 \end{cases}$

44. $N(x) = \begin{cases} \dfrac{\tan ax}{\tan bx} & -\dfrac{\pi}{2} < ax < 0, -\dfrac{\pi}{2} < bx < 0 \\ 4 & \text{if } x = 0 \\ ax + b & \text{if } x > 0 \end{cases}$

45. Find constants a and b such that $f(2) + 3 = f(0)$ and f is continuous at $x = 1$.

$$f(x) = \begin{cases} ax + b & \text{if } x > 1 \\ 3 & \text{if } x = 1 \\ x^2 - 4x + b + 3 & \text{if } x \leq 1 \end{cases}$$

46. Let $f(x) = \begin{cases} x^2 & \text{if } x > 2 \\ x + 1 & \text{if } x \leq 2 \end{cases}$ Show that f is continuous from the left at $x = 2$, but not from the right.

47. Let $f(x) = \begin{cases} x & \text{if } x \neq 0 \\ 2 & \text{if } x = 0 \end{cases}$ and $g(x) = \begin{cases} 3x & \text{if } x \neq 0 \\ -2 & \text{if } x = 0 \end{cases}$

Show that $f + g$ is continuous at $x = 0$ even though f and g are both discontinuous there.

48. Think Tank Problem Find functions f and g such that f is discontinuous at $x = 1$, but fg is continuous there.

49. Think Tank Problem Give an example of a function defined for all real numbers which is continuous only at one point.

50. Prove the quotient rule: if f and g are both continuous at $x = c$ and $g(x) \neq 0$, then f/g must also be continuous at $x = c$.

51. INTERPRETATION PROBLEM Use the intermediate value theorem to explain why the hands of a clock coincide at least once between 1 P.M. and 1:15 P.M.

52. INTERPRETATION PROBLEM The graph shown in Figure 2.24 models how the growth rate of a bacterial colony changes with temperature.*

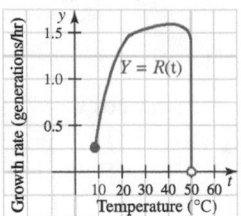

Figure 2.24 Growth rate of a bacterial colony

What happens when the temperature reaches 45°C? Does it make sense to compute $\lim\limits_{t \to 50} R(t)$? Write a paragraph describing how temperature affects the growth rate of a species.

53. A fish swims upstream at a constant speed v relative to the water, which in turn flows at a constant speed v_w ($v_w < v$) relative to the ground. The energy expended by the fish in traveling to a point upstream is given by

$$E(v) = \frac{Cv^k}{v - v_w}$$

where $C > 0$ is a physical constant and $k > 0$ is a number that depends on the type of fish.†

a. Compute $\lim\limits_{v \to v_w} E(v)$. Interpret your result in words.

b. What happens to $E(v)$ as $v \to \infty$? Interpret your result in words.

54. The population (in thousands) of a colony of bacteria t minutes after the introduction of a toxin is given by the function

$$P(t) = \begin{cases} t^2 + 1 & \text{if } 0 \leq t < 5 \\ -8t + 66 & \text{if } t \geq 5 \end{cases}$$

a. When does the colony die out?

b. Show that at some time between $t = 2$ and $t = 7$, the population is 9,000.

55. 𝔥*istorical Quest*

The first modern formulation of the notion of continuity appeared in a pamphlet published by Bernard Bolzano, a Bohemian mathematician and Catholic priest. His mathematical work was, for the most part, overlooked

Bernard Bolzano (1781-1848)

Karl Smith library

*Michael D. La Grega, Philip L. Buckingham, and Jeffery C. Evans, *Hazardous Waste Management,* New York: McGraw-Hill, 1994, pp. 565-566.

†E. Batschelet, *Introduction to Mathematics for Life Scientists,* 2nd ed., New York: Springer-Verlag, 1976, p. 280.

by his contemporaries. In explaining the concept of continuity, Bolzano said that one must understand the phrase, "A function f (x) (that) varies according to the law of continuity for all values of x which lie inside or outside certain limits, is nothing other than this: If x is any such value, the difference

$$f(x + \omega) - f(x)$$

can be made smaller than any given quantity, if one makes ω as small as one wishes."

*In his book, **Cours d'Analyse, Cauchy** (see ʆistorical Quest, Problem 53, Section 2.1) introduces the concept of continuity for a function defined on an interval in essentially the same way as Bolzano. In this book, Cauchy points out that the continuity of many functions is easily verified. As an example, he argues that sin x is continuous on every interval because "the numerical value of sin $\frac{1}{2}\alpha$, and consequently that of the difference*

$$\sin(x + \alpha) - \sin x = 2 \sin\left(\frac{1}{2}\alpha\right) \cos\left(x + \frac{1}{2}\alpha\right)$$

decreases indefinitely with that of α."*

Show that sin x is continuous, and also show that the given expression decreases as claimed by Cauchy.

We conclude this ʆistorical Quest by noting that the formal ε-δ definition of limits and continuity that we use today was first done by Karl Weierstraß (1815-1897), a German secondary school teacher who did his research at night. The historian David Burton describes Weierstraß as "the world's greatest analyst during the last third of the 19th century—the father of modern analysis."

56. Let f be a continuous function and suppose that f(a) and f(b) have opposite signs. The root location

theorem tells us that at least one root of f(x) = 0 lies between x = a and x = b.

a. Let c = (a + b)/2 be the midpoint of the interval [a, b]. Explain how the root location theorem can be used to determine whether the root lies in [a, c] or in [c, b]. Does anything special have to be said about the case where the interval [a, b] contains more than one root?

b. Based on your observation in part **a**, describe a procedure for approximating a root of f(x) = 0 more and more accurately. This is called the **bisection method** for root location.

c. Apply the bisection method to locate at least one root of

$$x^3 + x - 1 = 0$$

on [0, 1]. Check your answer using your calculator.

57. Apply the bisection method (see Problem 56) to locate at least one root of each of the given equations, and then check your answer using your calculator.

a. $\sin x + \cos x = x$ on $\left[0, \frac{\pi}{2}\right]$

b. $\cos(x + 1) = \sin(x - 1)$ on $[0, 1]$

Level 3

58. Prove the root location theorem, assuming the intermediate value theorem.

59. Show that f is continuous at c if and only if it is both continuous from the right and continuous from the left at c.

60. Let $u(x) = x$ and $f(x) = \begin{cases} 0 & \text{if } x \neq 0 \\ 1 & \text{if } x = 0 \end{cases}$

Show that $\lim_{x \to 0} f[u(x)] \neq f\left[\lim_{x \to 0} u(x)\right]$

2.4 EXPONENTIAL AND LOGARITHMIC FUNCTIONS

IN THIS SECTION: *Exponential functions, logarithmic functions, natural base e, natural logarithms, continuous compounding of interest*

In this section, we introduce the exponential function, an example of a transcendental function, as well as the irrational number *e*. We also discuss some graphical properties of exponential functions and an interesting application dealing with continuous compounding of interest.

**A. L. Cauchy, *Cours d'Analyse de l'Ecole Royale Polytechnique,Œuvres*, Ser. 2, Vol. 3. Paris: Gauthier-Villars, 1897, p. 44.

In Chapter 1, we introduced polynomial and rational functions, power functions, trigonometric functions, and inverse trigonometric functions. In this section, we will complete our list of the elementary functions typically used in calculus by using limits to introduce exponential functions and their inverses, the logarithmic functions.

Exponential Functions

We begin with a definition.

EXPONENTIAL FUNCTION   The function f is an **exponential function** if

$$f(x) = b^x$$

where b is a positive constant other than 1 and x is a real number.

Recall that if n is a natural number, then

$$b^n = \underbrace{b \cdot b \cdot b \cdots b}_{n \ factors}$$

If $b \neq 0$, then $b^0 = 1$, $b^{-n} = \dfrac{1}{b^n}$, and furthermore if $b > 0$, $b^{1/n} = \sqrt[n]{b}$. Also, if m and n are any integers, and m/n is a reduced fraction, then

$$b^{m/n} = \left(b^{1/n}\right)^m = (b^m)^{1/n}$$

This definition tells us what b^x means for rational values of x. However, we now wish to enlarge the domain of x to include all real numbers. To get a feeling for what is involved in this problem, let us examine the special case where $b = 2$. In Figure 2.25, we have plotted several points with coordinates $(r, 2^r)$, where r is a rational number.

To enlarge the domain of x to include all real numbers, we need the help of the following theorem.

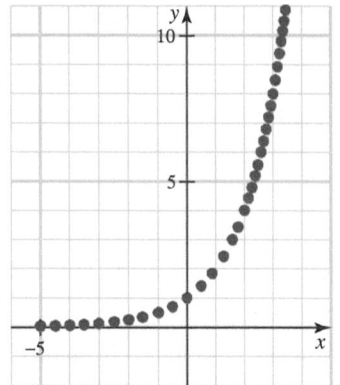

Figure 2.25   Graph of $y = 2^x$ for rational exponents

Theorem 2.8   Bracketing theorem for exponents

Suppose b is a real number greater than 1. Then for any real number x, there is a unique real number b^x. Moreover, if p and q are any two rational numbers such that $p < x < q$, then

$$b^p < b^x < b^q$$

In particular, the function b^x is continuous for all positive b.

Proof: A formal proof of this theorem is beyond the scope of this course, but we can outline the needed steps. The proof depends on a property of real numbers called the *completeness property* of the real numbers. What the completeness property says is that each real number can be approximated to any prescribed degree of accuracy by a rational number. This is equivalent to saying that if x is a real number, then for every positive integer n there exists a rational number r_n such that $|x - r_n| < \dfrac{1}{n}$. ♦

The bracketing theorem gives meaning to expressions such as $2^{\sqrt{3}}$, since

$$1.732 < \sqrt{3} < 1.733$$

implies

$$2^{1.732} < 2^{\sqrt{3}} < 2^{1.733}$$

Example 1   Graph of an exponential function

Graph $f(x) = 2^x$.

Solution   Begin by plotting points using rational x-values (see Figure 2.25). We see that this graph has well-defined shape, but it is riddled with "holes" that correspond to those points whose x-coordinates are irrational numbers. We complete the graph by connecting the points with a smooth curve as shown in Figure 2.26.

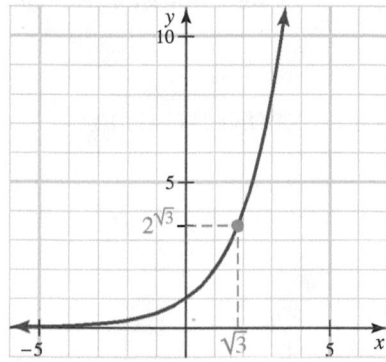

Figure 2.26   Graph of $f(x) = 2^x$

Let us take a moment to use the bracketing theorem for exponents to relate the completed graph to the definition of 2^x for irrational x-values. Consider the point $(x, 2^x)$ for $x = \sqrt{3}$. Because $\sqrt{3}$ is irrational, the decimal representation is nonterminating and nonrepeating. In other words, $\sqrt{3}$ is the limit of a sequence of rational numbers; specifically,

$$1, 1.7, 1.73, 1.732, 1.7320, 1.73205, 1.732050, 1.7320508, \cdots$$

Then, $2^{\sqrt{3}}$ is the limit of the sequence of numbers

$$2^1, 2^{1.7}, 2^{1.73}, 2^{1.732}, 2^{1.7320}, 2^{1.73205}, 2^{1.732050}, 2^{1.7320508}, \cdots$$

Graphically, this means that the rational numbers $1, 1.7, 1.73, \cdots$ tend toward $\sqrt{3}$, the points $(1, 2), (1.7, 2^{1.7}), (1.73, 2^{1.73}), \cdots$ tend toward the "hole" in the graph of $y = 2^x$ that corresponds to $x = \sqrt{3}$.

The shape of the graph of $y = b^x$ for any $b > 1$ is essentially the same as that of $y = 2^x$. The graph of $y = b^x$ for a typical base $b > 1$ is shown in Figure 2.27**a** and is rising (when viewed from left to right). For a base b where $0 < b < 1$ the graph is falling, as shown in Figure 2.27**b**.

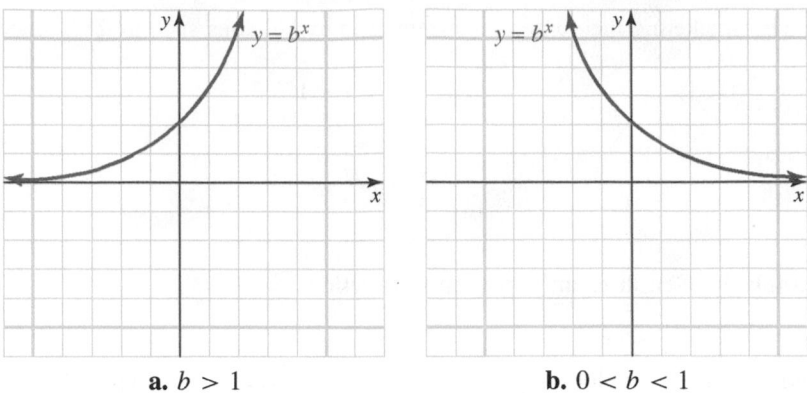

a. $b > 1$ **b.** $0 < b < 1$

Figure 2.27   Graph of $y = b^x$

     We summarize the basic properties of exponential functions with the following theorem. Many of these properties for bases with rational exponents should be familiar from previous courses.

BASIC PROPERTIES OF EXPONENTIAL FUNCTIONS   Let x and y be real numbers, a and b positive numbers.

Equality rule   If $b \neq 1$, then $b^x = b^y$ if and only if $x = y$.

Inequality rules If $x > y$ and $b > 1$, then $b^x > b^y$.

                  If $x > y$ and $0 < b < 1$, then $b^x < b^y$.

Product rule $b^x b^y = b^{x+y}$

Quotient rule $\dfrac{b^x}{b^y} = b^{x-y}$

Power rules $(b^x)^y = b^{xy}$

              $(ab)^x = a^x b^x$

              $\left(\dfrac{a}{b}\right)^x = \dfrac{a^x}{b^x}$

Graphical   The function $y = b^x$ is continuous for all x.

              The graph is always above the x-axis ($b^x > 0$).

              The graph is rising for $b > 1$.

              The graph is falling for $0 < b < 1$.

     Several parts of this theorem are used in the following example.

Example 2   Exponential equations

Solve each of the following exponential equations.

a. $2^{x^2+3} = 16$ **b.** $2^x 3^{x+1} = 108$ **c.** $\left(\sqrt{2}\right)^{x^2} = \dfrac{8^x}{4}$

Solution

a. $2^{x^2+3} = 16$

             $2^{x^2+3} = 2^4$ *Write 16 as 2^4 so that the equality rule can be used.*

           $x^2 + 3 = 4$ *Equality rule*

          $x^2 - 1 = 0$

    $(x - 1)(x + 1) = 0$

               $x = \pm 1$

b. $2^x 3^{x+1} = 108$

$2^x 3^x 3 = 3 \cdot 36$ *Product rule*

$(2 \cdot 3)^x = 6^2$ *Divide both sides by 3 and then use the power rule.*

$6^x = 6^2$

$x = 2$ *Equality rule*

c. $\left(\sqrt{2}\right)^{x^2} = \dfrac{8^x}{4}$

$\left(2^{1/2}\right)^{x^2} = \dfrac{\left(2^3\right)^x}{2^2}$

$2^{x^2/2} = 2^{3x-2}$ *Properties of exponents*

$\dfrac{x^2}{2} = 3x - 2$ *Equality rule*

$x^2 - 6x + 4 = 0$

$x = \dfrac{6 \pm \sqrt{36 - 4(1)(4)}}{2}$ *Quadratic equation*

$= 3 \pm \sqrt{5}$ *Simplify*

Logarithmic Functions

Since the exponential function $y = b^x$ for $b > 0$, $b \neq 1$ is monotonic (its graph is either always rising for all x or always falling), it must have an inverse that is itself monotonic. This inverse function is called the **logarithm of x to the base b**. Here is a definition, with some notation.

LOGARITHMIC FUNCTION   If $b > 0$ and $b \neq 1$, **the logarithm of x to the base b** is the function $y = \log_b x$ that satisfies $b^y = x$; that is

$$y = \log_b x \qquad \text{means} \qquad b^y = x$$

■ **W**hat this says: It is useful to think of a logarithm as an exponent. That is, consider the following sequence of interpretations:

$y = \log_b x$

y is the logarithm to the base b of x.

y is the **exponent to the base b** that gives x.

$b^y = x$

Notice that $y = \log_b x$ is defined only for $x > 0$ because $b^y > 0$ for all y. We have sketched the graph of $y = \log_b x$ for $b > 1$ in Figure 2.28 by reflecting the graph of $y = b^x$ in the line $y = x$.

Because $y = b^x$ is a continuous, increasing function that satisfies $b^x > 0$ for all x, $\log_b x$ must also be continuous and increasing, and its graph must lie entirely to the right of the y-axis.

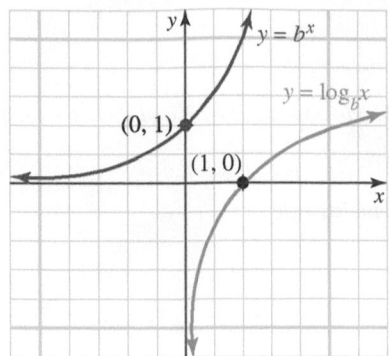

Figure 2.28 The graph of $y = \log_b x$ for $b > 1$

Also, because $b^0 = 1$ and $b^1 = b$, we have

$$\log_b 1 = 0 \quad \text{and} \quad \log_b b = 1$$

so $(1, 0)$ and $(b, 1)$ lie on the logarithmic curve.

We now list several general properties of logarithms. Notice the similarity between these properties and the basic properties of exponential functions.

Remember that $\log_b x$ is defined only for $x > 0$. Any undefined values of $\log_b x$ are excluded.

BASIC PROPERTIES OF LOGARITHMIC FUNCTIONS Assume $b > 0$ and $b \neq 1$.

Equality rule	$\log_b x = \log_b y$ if and only if $x = y$
Inequality rules	If $x > y$ and $b > 1$, then $\log_b x > \log_b y$.
	If $x > y$ and $0 < b < 1$, then $\log_b x < \log_b y$.
Product rule	$\log_b(xy) = \log_b x + \log_b y$
Quotient rule	$\log_b\left(\dfrac{x}{y}\right) = \log_b x - \log_b y$
Power rule	$\log_b x^p = p \log_b x$ for any real number p
Inversion rules	$b^{\log_b x} = x$
	$\log_b b^x = x$
Special values	$\log_b b = 1$
	$\log_b 1 = 0$
Graphical	$\log_b x$ is continuous for all $x > 0$
	The graph is to the right of the y-axis
	The graph is always rising if $b > 1$
	The graph is always falling if $b < 1$

Each of these properties can be derived by using the definition of the logarithm in conjunction with a suitable property of exponentials. (See Problems 55-56).

Example 3 Evaluating a logarithmic expression

Evaluate $\log_2 \frac{1}{8} + \log_2 128$.

Solution Using the basic properties of logarithmic functions, we find that

$$\log_2 \frac{1}{8} + \log_2 128 = \log_2 2^{-3} + \log_2 2^7$$
$$= \log_2\left[2^{-3}2^7\right]$$
$$= \log_2 2^4$$

$$= 4\log_2 2$$
$$= 4(1)$$
$$= 4$$

What we have shown here is longer than what you would do in your work, which would probably be shortened as follows:

$$\log_2 \frac{1}{8} + \log_2 128 = \log_2 16 \quad \textit{Because } \frac{1}{8}(128) = 16.$$

$$= 4 \qquad\qquad \textit{Because } 4 \textit{ is the exponent on } 2 \textit{ that yields } 16.$$

Example 4   Logarithmic equation

Solve the equation $\log_3(2x+1) - 2\log_3(x-3) = 2$.

Solution   Use the basic properties of logarithms and remember that both $2x+1 > 0$ and $x - 3 > 0$.

$$\log_3(2x+1) - 2\log_3(x-3) = 2$$
$$\log_3(2x+1) - \log_3(x-3)^2 = 2$$
$$\log_3 \frac{2x+1}{(x-3)^2} = 2$$
$$3^2 = \frac{2x+1}{(x-3)^2}$$
$$9(x-3)^2 = 2x+1$$
$$9x^2 - 56x + 80 = 0$$
$$(x-4)(9x-20) = 0$$
$$x = 4, \frac{20}{9}$$

Notice that $x - 3 < 0$ if $x = \frac{20}{9}$. The given logarithmic equation has only $x = 4$ as a solution because we cannot take the logarithm of a negative.

Natural Base e

In elementary algebra you probably studied logarithms to base 10 or possibly to base 2, but in calculus, it turns out to be convenient to use the number e whose decimal expansion is

$$e \approx 2.71828182845 \cdots$$

This number, called the **natural exponential base**, and can be defined by a limit:

$$e \text{ is the limiting value of } \left(1 + \frac{1}{x}\right)^x \text{ as } x \text{ becomes large}$$

In Chapter 4, we will write this as

$$e = \lim_{x \to \infty} \left(1 + \frac{1}{x}\right)^x$$

At first glance you may think this limit has the value 1. After all, $1 + \frac{1}{x}$ clearly approaches 1 as x increases without bound, and $1^x = 1$ for all x. But the limit process does not work that way, as indicated by the tabular computation in the margin.

You must be careful when using software to reach conclusions in calculus. For example, on most software or calculators, if you input

$$\left(1 + \frac{1}{10^{20}}\right)^{10^{20}}$$

the output is 1 (which, of course, is incorrect). We know (from the definition of e) that $\lim_{n \to \infty} (1 + 1/n)^n = e > 1$.

n	(1+1/n)^n
1	2
2	2.25
3	2.37037037
4	2.44140625
5	2.48832
6	2.521626372
7	2.546499697
8	2.565784514
9	2.581174792
10	2.59374246
20	2.653297705
30	2.674318776
40	2.685063838
50	2.691588029
60	2.695970139
70	2.699116371
80	2.701484941
90	2.703332461
100	2.704813829
200	2.711517123
300	2.713765158
400	2.714891744
500	2.715568521
600	2.716020049
700	2.716342738
800	2.716584847
900	2.716773208
1000	2.716923932
2000	2.717602569
3000	2.71782892
4000	2.717942121
5000	2.71801005
6000	2.71805534
7000	2.718087691
8000	2.718111955
9000	2.718130828
10000	2.718145927

The letter e was chosen to represent this number in honor of the great Swiss mathematician, Leonhard Euler (1707-1783), so it is sometimes known as **Euler's number**. Euler discovered many of its special properties and investigated applications in which e plays a vital role.

The $f(x) = e^x$ is called the **natural exponential function**, and obeys all of the basic rules of exponential functions with base $b > 1$. Because of many of the applications we consider in calculus, we have complicated exponents such as

$$e^{3x^2 - 2\sin x + 8}$$

We sometimes state this form by writing

$$\exp\left(3x^2 - 2\sin x + 8\right)$$

EXP NOTATION   If $f(x)$ is an expression in x, the notation

$$\exp\left(f(x)\right)$$

means $e^{f(x)}$.

Natural Logarithms

There are two frequently used bases for logarithms. If the base 10 is used, then the logarithm is called a *common logarithm*, and if the base e is used, the logarithm is called a *natural logarithm*. We use a special notation for each of these logarithms.

COMMON LOGARITHM   The **common logarithm**, $\log_{10} x$, is denoted by **log x**.
NATURAL LOGARITHM   The **natural logarithm**, $\log_e x$, is denoted by **ln x**.
(ln x is pronounced "ell n x" or "lawn x")

The following theorem summarizes the key properties of natural exponential and logarithmic functions.

Theorem 2.9   Basic properties of the natural logarithm

a. $\ln 1 = 0$ **b.** $\ln e = 1$ **c.** $e^{\ln x} = x$ for all $x > 0$ **d.** $\ln e^y = y$ for all y
e. $b^x = e^{x \ln b}$ for any $b > 0$, $b \neq 1$.

Proof: Parts **a** and **b** follow immediately from the definitions of $\ln x$ and e^x. Parts **c** and **d** are just the inversion rules for base e. We show the proof for part **e**.

$$e^{x \ln b} = e^{\ln b^x} \quad \text{\textit{Power rule for logarithmic functions.}}$$

$$= b^x \quad \text{\textit{Property c of this theorem.}}$$

On calculators, there are usually two logarithm keys, $\boxed{\text{LOG}}$ and $\boxed{\text{LN}}$, to correspond with common logarithms and natural logarithms, respectively. For other logarithms, you will need the useful conversion formula contained in the following theorem.

Theorem 2.10   Change-of-base theorem

$$\log_b x = \frac{\ln x}{\ln b} \qquad \text{for any } b > 0, b \neq 1$$

Proof: Let $y = \log_b x$. Then

$$b^y = x \qquad \text{\textit{Definition of logarithm}}$$

$$\ln b^y = \ln x \qquad \text{\textit{Equality rule of logarithms}}$$

$$y \ln b = \ln x \qquad \text{\textit{Power rule of logarithms}}$$

$$y = \frac{\ln x}{\ln b} \qquad \text{\textit{Divide both sides by} } \ln b.$$

If you are using a software package such as Mathematica, Derive, or Maple, sometimes referred to as CAS (Computer Algebra System), be careful about the notation. Some versions do not distinguish between $\log x$ (common logarithm) and $\ln x$ (natural logarithm). All logarithms on these versions are assumed to be natural logarithms, so to evaluate a common logarithm requires using the change-of-base theorem.

Example 5   Solving an exponential equation using the change-of-base theorem

Solve $6^x = 200$.

Solution $6^x = 200$

$$x = \log_6 200 \qquad \text{\textit{Definition of exponent}}$$

$$= \frac{\ln 200}{\ln 6} \qquad \text{\textit{Change-of-base theorem. This is the exact solution.}}$$

$$\approx 2.957 \qquad \text{\textit{Approximate calculator solution}}$$

Checking with a calculator shows that $6^{2.957} \approx 199.983$.

Example 6   Finding a velocity

An object moves along a straight line in such a way that after t seconds, its velocity (in ft/s) is given by

$$v(t) = 10 \log_5 t + 3 \log_2 t$$

How long (to the nearest hundredth of a second) will it take for the velocity to reach 20 ft/s?

Solution   We want to solve the equation

$$10 \log_5 t + 3 \log_2 t = 20$$

$$10 \frac{\ln t}{\ln 5} + 3 \frac{\ln t}{\ln 2} = 20 \qquad \text{\textit{Change of base formula}}$$

$$\left(\frac{10}{\ln 5} + \frac{3}{\ln 2} \right) \ln t = 20 \qquad \text{\textit{Factor}}$$

$$\ln t = 20 \left(\frac{10}{\ln 5} + \frac{3}{\ln 2} \right)^{-1}$$

$$\ln t \approx 1.89727$$

$$t \approx e^{1.89727} \approx 6.6677$$

It takes about 6.67 seconds for the velocity to reach 20 ft/s.

Example 7   Changing an exponential from base b to base e

Change 10^{2x} to an exponential using the natural base.

Solution   We wish to find N for

$$10^{2x} = e^N$$

$$\ln 10^{2x} = \ln e^N \quad \textit{ln of both sides}$$

$$2x \ln 10 = N \ln e \quad \textit{Power rule of logarithms}$$

$$2x \ln 10 = N \quad \quad \textit{ln } e = 1$$

Thus, $10^{2x} = e^{2x \ln 10}$.

Example 8   Solving an exponential equation with base e

Solve $\frac{1}{2} = e^{-0.000425t}$ to the nearest unit.

Solution
$$\frac{1}{2} = e^{-0.000425t}$$

$$-0.000425t = \ln 0.5 \quad \quad \textit{Definition of exponent}$$

$$t = \frac{\ln 0.5}{-0.000425} \quad \textit{This is the exact answer.}$$

$$\approx 1{,}630.9345$$

To the nearest unit, $t = 1{,}631$.

Example 9   Exponential growth

A biological colony grows in such a way that at time t (in minutes), the population is

$$P(t) = P_0 e^{kt}$$

where P_0 is the initial population and k is a positive constant. Suppose the colony begins with 5,000 individuals and contains a population of 7,000 after 20 min. Find k and determine the population (rounded to the nearest hundred individuals) after 30 min.

Solution   Because $P_0 = 5{,}000$, the population after t minutes will be

$$P(t) = 5{,}000 e^{kt}$$

In particular, because the population is 7,000 after 20 min,

$$P(20) = 5{,}000e^{k(20)}$$

$$7{,}000 = 5{,}000e^{20k}$$

$$\frac{7}{5} = e^{20k}$$

$$20k = \ln\left(\frac{7}{5}\right)$$

$$k = \frac{1}{20}\ln\left(\frac{7}{5}\right)$$

Finally, to determine the population after 30 min, substitute this value for k to find

$$P(30) = 5{,}000e^{30k}$$

$$\approx 8{,}282.5117$$

The expected population is approximately 8,300, to the nearest hundred individuals. ■

Continuous Compounding of Interest

One reason e is called the "natural" exponential base is that many natural growth phenomena can be described in terms of e^x. As an illustration of this fact, we close this section by showing how e^x can be used to describe the accounting procedure called *continuous compounding of interest*.

Suppose a sum of money is invested and the interest is compounded once during a particular period. If P is the initial investment (the **present value** or **principal**) and i is the **interest rate** during the period, then the **future value** (in dollars), A, after the interest is added will be

$$A = \underset{\underset{principal}{\uparrow}}{P} + \underset{\underset{interest}{\uparrow}}{Pi} = P(1 + i)$$

Thus, to compute the balance at the end of an interest period, we multiply the balance at the beginning of the period by the expression $1 + i$.

Interest is usually compounded more than once a year. The interest that is added to the account during one period will itself earn interest during subsequent periods. If the annual interest rate is r and the interest is compounded n times per year, then the year is divided into n equal interest periods and the interest rate during each such period is $i = \frac{r}{n}$. To compute the balance at the end of any period, you multiply the balance at the beginning of that period by the expression $1 + \frac{r}{n}$, so the balance at the end of the first period is

$$A_1 = P\left(1 + \frac{r}{n}\right)$$

At the end of the second period, the balance is

$$A_2 = A_1\left(1 + \frac{r}{n}\right) = \left[P\left(1 + \frac{r}{n}\right)\right]\left(1 + \frac{r}{n}\right) = P\left(1 + \frac{r}{n}\right)^2$$

At the end of the third period, the balance is

$$A_3 = A_2\left(1 + \frac{r}{n}\right) = \left[P\left(1 + \frac{r}{n}\right)^2\right]\left(1 + \frac{r}{n}\right) = P\left(1 + \frac{r}{n}\right)^3$$

and so on. At the end of t years, the interest has been compounded nt times, and the future value is

$$A(t) = P\left(1 + \frac{r}{n}\right)^{nt}$$

The first two graphs in Figure 2.29 show how an amount of money in an account over a one-year period of time grows, first with quarterly compounding and then with monthly compounding. Notice that these are "step" graphs, with jumps occurring at the end of each compounding period.

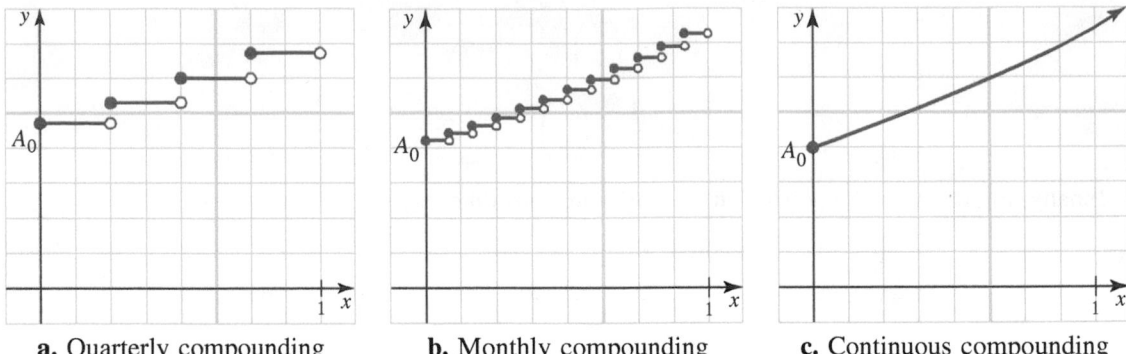

a. Quarterly compounding   **b.** Monthly compounding   **c.** Continuous compounding

Figure 2.29   The growth of an account over a one-year period with different compounding frequencies

With continuous compounding, we compound interest not quarterly, or monthly, or daily, or even every second, but *instantaneously*, so that the future amount of money A in the account grows continuously, as shown in Figure 2.29c. In other words, we compute A as the limiting value of

$$P\left(1 + \frac{r}{n}\right)^{nt}$$

as the number of compounding periods n grows without bound. We denote by $A(t)$ the future value after t years. As with the definition of e, we denote this by the limit notation $n \to \infty$:

$$A(t) = \lim_{n\to\infty} P\left(1 + \frac{r}{n}\right)^{nt} \qquad \textit{Let } k = \frac{n}{r}, \textit{ so that } krt = nt \textit{ and}$$

$$\frac{r}{n} = \frac{1}{k}; \textit{ also } k \to \infty \textit{ as } n \to \infty.$$

$$= \lim_{k\to\infty} P\left(1 + \frac{1}{k}\right)^{krt}$$

$$= \lim_{k\to\infty} P\left[\left(1 + \frac{1}{k}\right)^{k}\right]^{rt}$$

$$= P\left[\lim_{k\to\infty}\left(1 + \frac{1}{k}\right)^{k}\right]^{rt} \qquad \textit{Scalar rule for limits.}$$

$$= Pe^{rt} \qquad \textit{Definition of } e$$

These observations are now summarized.

FUTURE VALUE   If P dollars are compounded n times per year at an annual rate r, then the future value after t years is given by

$$A(t) = P\left(1 + \frac{r}{n}\right)^{nt}$$

and if the compounding is continuous, the future value is

$$A(t) = Pe^{rt}$$

Example 10 Continuous compounding of interest

If $12,000 is invested for 5 years at 4%, find the future value at the end of 5 years if interest is compounded: **a.** monthly   **b.** continuously   **c.** If the interest is compounded continuously, how long will it take for the money to double?

Solution $P = \$12,000$; $t = 5$; and $r = 4\% = 0.04$ are given.

a. $n = 12$; $A = P\left(1 + \frac{r}{n}\right)^{nt} = \$12,000\left(1 + \frac{0.04}{12}\right)^{12(5)} \approx \$14,651.96$

b. $A = \$12,000e^{0.04(5)} \approx \$14,656.83$

c. The account doubles when $A(t) = 2P = \$24,000$; that is, when

$$12,000e^{0.04t} = 24,000$$

$$e^{0.04t} = 2$$

$$0.04t = \ln 2 \qquad \textit{Definition of logarithm}$$

$$t = \frac{\ln 2}{0.04}$$

$$\approx 17.3287$$

This is about 17 years and 4 months.

PROBLEM SET 2.4

Level 1

Sketch the graph of the functions in Problems 1-4.

1. $y = 3^x$

2. $y = 4^{-x}$

3. $y = -e^{-x}$

4. $y = -e^x$

Evaluate the expressions given in Problems 5-12. Find the exact value and do not use a calculator.

5. $\log_2 4 + \log_3 \frac{1}{9}$

6. $2^{\log_2 3 - \log_2 5}$

7. $5\log_3 9 - 2\log_2 16$

8. $\left(\log_2 \frac{1}{8}\right)\left(\log_3 27\right)$

9. $e^{5\ln 2}$

10. $\log_3 3^4 - \ln e^{0.5}$

11. $\ln\left(\log 10^e\right)$

12. $\exp\left(\log_{e^2} 25\right)$

Solve the logarithmic and exponential equations to calculator accuracy in Problems 13-18.

13. $\log x = 5.1$

14. $e^{-3x} = 0.5$

15. $\ln x^2 = 9$

16. $7^{-x} = 15$

17. $e^{2x} = \ln(4 + e)$

18. $\frac{1}{2}\log_3 x = \log_2 8$

Solve the logarithmic and exponential equations in Problems 19-30.

19. $\log_x 16 = 2$

20. $\log_2\left(x^{\log_2 x}\right) = 4$

21. $3^{x^2-x} = 9$

22. $4^{x^2+x} = 16$

23. $2^x 5^{x+2} = 25,000$

24. $3^x 4^{x+1/2} = 3,456$

25. $\left(\sqrt[3]{2}\right)^{x+10} = 2^{x^2}$

26. $\left(\sqrt[3]{5}\right)^{x+2} = 5^{x^2}$

27. $e^{2x+3} = 1$

28. $\dfrac{e^{x^2}}{e^{x+6}} = 1$

29. $\log_3 x + \log_3(2x+1) = 1$

30. $\ln\left(\dfrac{x^2}{1-x}\right) = \ln x + \ln\left(\dfrac{2x}{1+x}\right)$

In Problems 31-34, evaluate the given limits.

31. a. $\lim\limits_{x \to 0} x^2 e^{-x}$

b. $\lim\limits_{x \to 1} x^2 e^{-x}$

32. a. $\lim\limits_{x \to 0} e^{-x^3}$

b. $\lim\limits_{x \to 1} e^{-x^3}$

33. a. $\lim\limits_{x \to 0^+} (1+x)^{1/x}$

b. $\lim\limits_{x \to 1} (1+x)^{1/x}$

34. a. $\lim\limits_{x \to 0^-} e^{1/x}$

b. $\lim\limits_{x \to 0^+} \ln x$

Level 2

35. If $\log_b 1,296 = 4$, what is $\left(\frac{3}{2}b\right)^{3/2}$?

36. If $\log_{\sqrt{b}} 106 = 2$, what is $\sqrt{b-15}$?

37. Solve $\log_2 x + \log_5(2x + 1) = \ln x$ correct to the nearest tenth.

38. Solve $\log_x 2 = \log_3 x$ correct to the nearest tenth.

Data points with a curve fit to those points are shown in Problems 39-42. Decide whether the data are better modeled by an exponential or a logarithmic function.

39.   **40.**

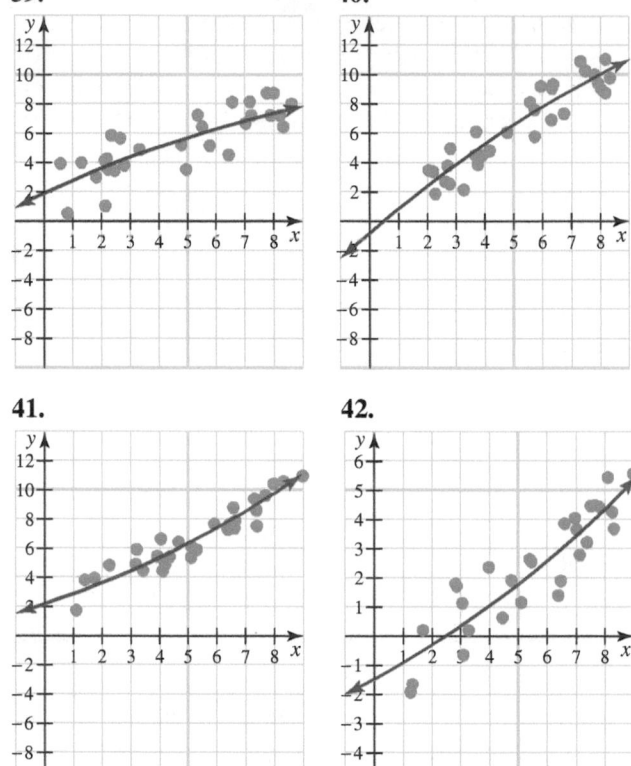

41.   **42.**

43. According to the *Bouguer-Lambert law*, a beam of light that strikes the surface of a body of water with intensity I_0 will have intensity I at a depth of x meters, where

$$I = I_0 e^{kx}$$

The constant k, called the *absorption coefficient*, depends on such things as the wavelength of the beam of light and the purity of the water. Suppose a given beam of light is only 5% as intense at a depth of 2 meters as it is at the surface. Find k and determine at what depth (to the nearest meter) the intensity is 1% of the intensity at the surface. (This explains why plant life exists only in the top 10 m of a lake or sea.)

44. A certain bank pays 6% interest compounded continuously. How long will it take for $835 to double?

45. a. If an amount of money is compounded continuously at an annual rate r, how long will it take for that amount of money to double?

  b. Explain why bankers sometimes call this the rule of 72, or more correctly, the rule of 69.

46. First National Bank pays 2.4% interest compounded monthly, and World Savings pays 2% interest compounded continuously. Which bank offers a better deal?

47. A person invests $8,500 in a bank that compounds interest continuously. If the investment doubles in 10 years, what is the (annual) rate of interest (correct to the nearest tenth percent)?

48. In 1626, Peter Minuit traded trinkets worth $24 for land on Manhattan Island. Assume that in 2014 the same land was worth $8 trillion. Find the annual rate of interest compounded continuously at which the $24 would have had to be invested during this time to yield the same amount.

49. Biologists estimate that the population of a bacterial colony is

$$P(t) = P_0 2^{kt}$$

at time t (in minutes). Suppose the population is found to be 1,000 after 20 minutes and that it doubles every hour.

  a. Find P_0 and k.

  b. When (to the nearest minute) will the population be 5,000?

50. The *Richter scale* measures the intensity of earthquakes. Specifically, if E is the energy (watt/cm^3) released by a quake, then it is said to have *magnitude M*, where

$$M = \frac{\log E - 11.4}{1.5}$$

  a. Express E in terms of M.

  b. How much more energy is released in an $M = 8.5$ earthquake (such as the devastating Alaska quake of 1964) than in an average quake of magnitude $M = 6.5$ (the Eureka earthquake of 2010)?

51. A manufacturer of car batteries estimates that p percent of the batteries will work for at least t months, where

$$p(t) = 100e^{-0.03t}$$

  a. What percent of the batteries can be expected to last at least 40 months?

  b. What percent can be expected to *fail* before 50 months?

  c. What percent can be expected to *fail* between the 40th and 50th months?

52. If a million dollars is invested at 1.5% compounded daily, how much money will there be in 7 years?

  a. Use a 365-day year; this is known as *exact interest*.

  b. Use a 360-day year; this is known as *ordinary interest*.

53. If P dollars are borrowed for a total of N months compounded monthly at an annual interest rate of r, then the monthly payment is found by the formula

$$m = \frac{P\left(\frac{r}{12}\right)}{1 - \left(1 + \frac{r}{12}\right)^{-N}}$$

a. Use this formula to determine the monthly car payment for a car costing \$17,487 with a down payment of \$7,487. The car is financed for 4 years at 12%.

b. A home loan is made for \$210,000 at 8% interest for 30 years. What is the monthly payment with a 20% down payment?

54. *Historical Quest*

John Napier was a Scottish landowner who was the Isaac Asimov of his day, having envisioned the tank, the machine gun, and the submarine. He also predicted that the end of the world would occur between 1688 and 1700. He is best known today as the inventor of logarithms,

John Napier (1550-1617)

Karl Smith library

which until the advent of the calculator were used extensively with complicated calculations. Napier's logarithms are not identical to the logarithms we use today. Napier chose to use $1 - 10^{-7}$ *as his given number, and then he multiplied by* 10^7. *That is, if*

$$N = 10^7 \left(1 - \frac{1}{10^7}\right)^L$$

then L is Napier's logarithm of the number N; that is $L = nog\ N$ *means*

$$N = 10^7 \left(1 - 10^{-7}\right)^L$$

One difference between Napier's logarithms (which for clarity we will call a nog) and modern logarithms is apparent when stating the product, quotient, and power rules for logarithms.

a. Show that if $L_1 = nog\ N_1$ and $L_2 = nog\ N_2$, then

$$nog\ N_1 + nog\ N_2 = nog\left(\frac{N_1 N_2}{10^7}\right)$$

b. State and prove similar results for a quotient rule and a power rule for Napier logarithms.

c. What is nog 10^7?

d. If nog $n = 10^7$, then $n \approx e^r$ for some rational number r. What is r?

e. Fortunately, Napier's 1614 paper on logarithms was read by a true mathematician, Henry Briggs (1561-1630), and together they decided that base 10 made a lot more sense. In the year Napier died, Briggs published a table of common logarithms (base 10) that at the time was a major accomplishment. In this paper he used the words "mantissa" and "characteristic." What is the meaning that Briggs gave to these words?

f. Who was the first person to publish a table of natural logarithms (base e)?

Level 3

55. For $b > 0$ and all positive integers m and n, show that:

a. $b^m b^n = b^{m+n}$ **b.** $\dfrac{b^m}{b^n} = b^{m-n}$

56. For $b > 0$ and all positive integers m and n, show that:

a. $(b^m)^n = b^{mn}$ **b.** $\left(\sqrt[n]{b}\right)^m = \sqrt[n]{b^m}$

57. Prove:

a. $\log_b x + \log_b y = \log_b (xy)$

b. $\log_b x - \log_b y = \log_b \left(\dfrac{x}{y}\right)$

58. Prove $\log_b x^p = p \log_b x$.

59. Let b be any positive number other than 1. Show that

$$x^x = b^{x \log_b x}$$

60. Let a and b be any positive numbers other than 1. Show that:

a. $\log_a x = \dfrac{\log_b x}{\log_b a}$ **b.** $\left(\log_a b\right)\left(\log_b a\right) = 1$

CHAPTER 2 REVIEW

C*alculus has its limits.*

Shelf Life T-Shirts

Proficiency Examination

Concept Problems

1. State the informal definition of a limit of a function. Discuss this informal definition.
2. State the formal definition of a limit.
3. State the following basic rules for limits:

   a. limit of a constant   **b.** multiple rule   **c.** sum rule

   d. difference rule   **e.** product rule   **f.** quotient rule

   g. power rule   **h.** limit of a polynomial function   **i.** limit of a rational function

   j. limit of transcendental functions
4. State the squeeze rule.
5. What are the values of the given limits?

   a. $\lim\limits_{x \to 0} \dfrac{\sin x}{x}$ **b.** $\lim\limits_{x \to 0} \dfrac{\cos x - 1}{x}$
6. Define the continuity of a function at a point and discuss.
7. State the continuity theorem.
8. State the composition limit rule.
9. State the intermediate value theorem.
10. State the root location theorem.
11. State the completeness property.
12. **a.** What is an exponential function?

    b. How is such a function related to a logarithmic function?
13. Define e.
14. **a.** What is a logarithmic function?   **b.** What is a common logarithm?   **c.** What is a natural logarithm?
15. Draw a quick sketch of:

    a. An exponential function b^x for $b > 1$.   **b.** An exponential function b^x for $0 < b < 1$.
16. Draw a quick sketch of:

    a. A logarithmic function $\log_b x$ for $b > 1$.   **b.** A logarithmic function $\log_b x$ for $0 < b < 1$.
17. State each of the following exponential properties:

    a. Equality rule   **b.** Inequality rules   **c.** Product rule   **d.** Quotient rule   **e.** Power rules
18. Complete the statement of each of the following natural logarithm properties:

    a. $\ln 1 = \underline{\quad}$ **b.** $\ln e = \underline{\quad}$ **c.** $e^{\ln x} = \underline{\quad}$ **d.** $\ln e^y = \underline{\quad}$

    e. $b^x = \underline{\quad}$ (write as a power of e)

Practice Problems

Evaluate the limits in Problems 19-24.

19. $\lim\limits_{x \to 3} \dfrac{x^2 - 4x + 9}{x^2 + x - 8}$ 20. $\lim\limits_{x \to 2} \dfrac{x^2 - 5x + 6}{x^2 - 4}$ 21. $\lim\limits_{x \to 4} \dfrac{\sqrt{x} - 2}{x - 4}$

22. $\lim\limits_{x \to 0} \dfrac{\sin 9x}{\sin 5x}$ 23. $\lim\limits_{x \to 0} \dfrac{1 - \cos x}{2 \tan x}$ 24. $\lim\limits_{x \to (1/2)^-} \dfrac{|2x - 1|}{2x - 1}$

25. Decide whether each of the curves is exponential or logarithmic.

   a. 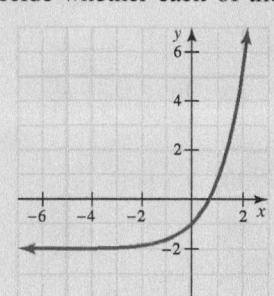   **b.** 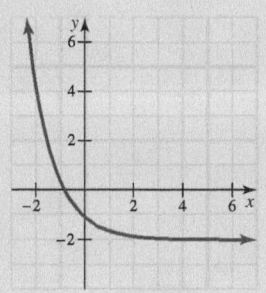   **c.** 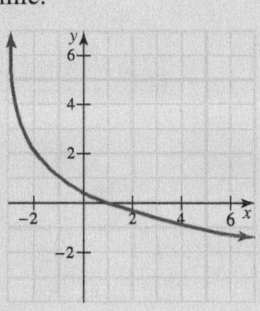   **d.**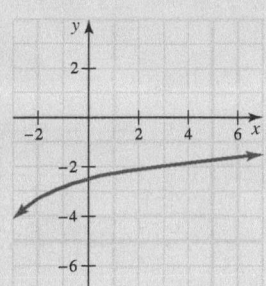

Determine whether the functions given in Problems 26-27 are continuous on the interval $[-5, 5]$.

26. $f(t) = \dfrac{1}{t} - \dfrac{3}{t+1}$ **27.** $g(x) = \dfrac{x^2 - 1}{x^2 + x - 2}$

28. How quickly will $2,000 grow to $5,000 when invested at an annual rate of 8% if interest is compounded:

   a. quarterly?   **b.** monthly?   **c.** continuously?

29. Find constants A and B such that f is continuous for all x:

$$f(x) = \begin{cases} Ax + 3 & \text{if } x < 1 \\ 2 & \text{if } x = 1 \\ x^2 + B & \text{if } x > 1 \end{cases}$$

30. Show that the equation

$$x + \sin x = \frac{1}{\sqrt{x + 3}}$$

has at least one solution on the interval $(0, \pi)$.

Supplementary Problems*

Evaluate each of the given numbers in problems 1-5 (exact answer without using a calculator).

1. $e^{\ln \pi}$ **2.** $\ln e^{\sqrt{2}}$ **3.** $32^{2/5} + 9^{3/2}$ **4.** $\dfrac{\left(e^{1.3}\right)^2 e^{-0.3}}{e^{1.3} + \sqrt{e^{-1.4}}}$

5. $(128)^{1/2}(81)^{3/4}$ **6.** $\ln 32 + \ln 8$

Evaluate each of the given numbers in Problems 7-11 (calculator answer rounded to the nearest hundredth).

7. $\ln 4.5$ **8.** $e^{2.8}$ **9.** $5,000 \left(1 + \frac{0.135}{12}\right)^{12(5)}$ **10.** $850,000 e^{-0.04(10)}$

11. $\log_2 7$ **12.** $\log_7 2$

Solve each equation in Problems 13-22 for x.

13. $4^{x-1} = 8$ **14.** $2^{x^2+4x} = \frac{1}{16}$ **15.** $\log_2 2^{x^2} = 4$

16. $\log_4 \sqrt{x(x-15)} = 1$ **17.** $\log_2 x + \log_2(x-15) = 4$ **18.** $3^{2x-1} = 6^x 3^{1-x}$

19. $\ln(x-1) + \ln(x+1) = 2 \ln \sqrt{12}$ **20.** $\log(x-1) + \log(x+1) = 2 \log \sqrt{12}$

21. $2 \ln x - \frac{1}{2} \ln 9 = \ln 3(x-2)$ **22.** $2 \log x - \frac{1}{2} \log 9 = \log 3(x-2)$

---

*The supplementary problems are presented in a somewhat random order, not necessarily in order of difficulty.

In Problems 23-46, evaluate the given limit.

23. $\lim\limits_{x\to 2} \dfrac{3x^2 - 7x + 2}{x - 2}$

24. $\lim\limits_{x\to -3} \dfrac{4x^2 + 11x - 3}{x^2 - x - 12}$

25. $\lim\limits_{x\to 1} \left(\dfrac{x^2 - 3x + 2}{x^2 + x - 2}\right)^2$

26. $\lim\limits_{x\to 1^+} \sqrt{\dfrac{x^2 - x}{x - 1}}$

27. $\lim\limits_{x\to 3} \sqrt{\dfrac{x^2 - 2x - 3}{x - 3}}$

28. $\lim\limits_{x\to e} x \ln x^2$

29. $\lim\limits_{x\to 1} \dfrac{x^3 - 1}{x^2 - 1}$

30. $\lim\limits_{x\to 4} |4 - x|$

31. $\lim\limits_{x\to 0^-} \dfrac{|x|}{x}$

32. $\lim\limits_{x\to 1} \dfrac{x^2 - 3x + 2}{x^2 - 1}$

33. $\lim\limits_{x\to e} \dfrac{\ln\sqrt{x}}{x}$

34. $\lim\limits_{x\to 1/\pi} \dfrac{1 + \cos\frac{1}{x}}{\pi x - 1}$

35. $\lim\limits_{x\to 0^+} (1 + x)^{4/x}$

36. $\lim\limits_{x\to 0^+} (1 + 2x)^{1/x}$

37. $\lim\limits_{x\to 1} \dfrac{x^5 - 1}{x - 1}$

38. $\lim\limits_{x\to e^2} \dfrac{(\ln x)^3 - 8}{\ln x - 2}$

39. $\lim\limits_{x\to 0} \dfrac{\sin 3x}{\sin 2x}$

40. $\lim\limits_{t\to 0} \dfrac{\tan^{-1} t}{\sin^{-1} t}$

41. $\lim\limits_{x\to 0} \dfrac{\sin(\cos x)}{\sec x}$

42. $\lim\limits_{x\to 0} \tan 2x \cot x$

43. $\lim\limits_{x\to 0} \dfrac{1 - \sin x}{\cos^2 x}$

44. $\lim\limits_{x\to 0} \dfrac{1 - 2\cos x}{\sqrt{3} - 2\sin x}$

45. $\lim\limits_{x\to 0} \dfrac{e^{3x} - 1}{e^x - 1}$

46. $\lim\limits_{x\to 3} \dfrac{\frac{1}{x} - \frac{1}{3}}{x - 3}$

Evaluate $\lim\limits_{\Delta x\to 0} \dfrac{f(x + \Delta x) - f(x)}{\Delta x}$ *for the functions f in Problems 47-59, wherever this limit exists.*

47. $f(x) = 7$

48. $f(x) = 3x + 5$

49. $f(x) = \sqrt{2x}$

50. $f(x) = x(x + 1)$

51. $f(x) = \dfrac{4}{x}$

52. $f(x) = \sin x$

53. $f(x) = \cos x$

54. $f(x) = e^x$

55. $f(x) = \ln x$

56. $f(x) = \dfrac{5}{x^2}$

57. $f(x) = \tan x$

58. $f(x) = x^3$

59. $f(x) = \log_2 x$

Decide whether the functions in Problems 60-63 are continuous on the given intervals. If not, define or redefine the function at one point to make it continuous everywhere on the given interval, or else explain why that cannot be done.

60. $f(x) = \dfrac{x + 4}{x - 8}$ on $[-5, 5]$

61. $f(x) = \dfrac{x + 4}{x - 8}$ on $[0, 10]$

62. $f(x) = \dfrac{\sqrt{x} - 8}{x - 64}$ on $\mathbb{R}$

63. $f(x) = \dfrac{\sqrt{x} - 6}{x - 36}$ on $[-5, 5]$

Decide whether the functions in Problems 64-67 are continuous on the given intervals. Check all of the suspicious points.

64. $r(x) = \begin{cases} 1 & \text{if } x \text{ is rational} \\ -1 & \text{if } x \text{ is irrational} \end{cases}$

65. $f(x) = |x - 2|$ on $[-5, 5]$

66. a. $f(x) = \dfrac{x^2 - x - 6}{x + 2}$ on $[0, 5]$

b. $f(x) = \dfrac{x^2 - x - 6}{x + 2}$ on $[-5, 5]$

c. $f(x) = \begin{cases} \frac{x^2 - x - 6}{x + 2} & \text{on } [-5, 5], x \neq -2] \\ -4 & \text{for } x = -2 \end{cases}$

d. $f(x) = \begin{cases} \frac{x^2 - x - 6}{x + 2} & \text{on } [-5, 5], x \neq -2] \\ -5 & \text{for } x = -2 \end{cases}$

67. **a.** $g(x) = \dfrac{x^2 - 3x - 10}{x + 2}$ on $[0, 5]$ **b.** $g(x) = \dfrac{x^2 - 3x - 10}{x + 2}$ on $[-5, 5]$

   c. $g(x) = \begin{cases} \dfrac{x^2 - 3x - 10}{x + 2} & \text{on } -5 \le x < -2 \\ -7 & \text{for } x = -2 \\ x - 5 & \text{for } -2 < x \le 5 \end{cases}$

*The **greatest integer function** $f(x) = [\![x]\!]$ is the largest integer that is less than or equal to x. This definition is used in Problems 68-70.*

68. **a.** Graph $f(x) = [\![x]\!]$ on $[-3, 6]$.
   b. Find $\lim_{x \to 3} f(x)$.
   c. For what values of a does $\lim_{x \to a} f(x)$ exist?

69. Repeat Problem 68 for $f(x) = [\![\frac{x}{2}]\!]$.

70. Let $f(x) = [\![x^2 + 1]\!]^{[\![x+1]\!]}$; find $\lim_{x \to 1} f(x)$.

71. Find constants A and B such that f is continuous for all x:

$$f(x) = \begin{cases} Ax + 3 & \text{if } x < 1 \\ 5 & \text{if } x = 1 \\ x^2 + B & \text{if } x > 1 \end{cases}$$

72. Find numbers a and b so that

$$\lim_{x \to 0} \frac{\sqrt{ax + b} - 1}{x} = 1$$

73. Find a number c so that

$$\lim_{x \to 3} \frac{x^3 + cx^2 + 5x + 12}{x^2 - 7x + 12}$$

exists. Then find the corresponding limit.

74. **Think Tank Problem** In the definition of the exponential function $f(x) = b^x$, we require that b be a positive constant. What happens if $b < 0$, for example, if $b = -2$? For what values of x is f defined? Describe the graph of f in this case.

75. Let $E(x) = 2^{x^2 - 2x}$.
   a. Use a calculator to sketch $E(x)$.
   b. Where does the graph cross the y-axis? What happens to the graph as $x \to \infty$ and as $x \to -\infty$?
   c. Use a calculator utility to find the value of x that minimizes $E(x)$. What is the minimum value?

76. A cool drink is removed from a refrigerator and is placed in a room where the temperature is $70°$F. According to a result in physics known as *Newton's law of cooling*, the temperature of the drink in t minutes will be

$$F(t) = 70 - Ae^{-kt}$$

where A and k are positive constants. Suppose the temperature of the drink was $35°$F when it left the refrigerator, and 30 minutes later, it was $50°$F [that is, $F(0) = 35$ and $F(30) = 50$].
   a. Find A and e^{-30k}.
   b. What will the temperature of the drink be after one hour (to the nearest degree)?
   c. What would you expect to happen to the temperature as $t \to \infty$?

77. A manufacturer of light bulbs estimates that the fraction $F(t)$ of bulbs that remain burning after t weeks is given by

$$F(t) = e^{-kt}$$

where k is a positive constant. Suppose twice as many bulbs are burning after 5 weeks as after 9 weeks.
   a. Find k and determine the fraction of bulbs still burning after 7 weeks.
   b. What fraction of the bulbs burn out before 10 weeks?
   c. What fraction of the bulbs can be expected to burn out between the 4th and 5th weeks?

78. How much should you invest now at an annual interest rate of 6.25% so that your balance 10 years from now will be $2,000 if interest is compounded

　　a. quarterly?　　　**b.** continuously?

79. A *decibel* (named for Alexander Graham Bell) is the smallest increase of the loudness of a sound that is detectable by the human ear. In physics, it is shown that when two sounds of intensity I_1 and I_2 (watts/cm^3) occur, the difference in loudness is D decibels, where

$$D = 10 \log\left(\frac{I_1}{I_2}\right)$$

When sound is rated in relation to the threshold of human hearing ($I_0 = 10^{-16}$) the level of normal conversation is 50 decibels, whereas that of a rock concert is 110 decibels. Show that a rock concert is 60 times as loud as normal conversation but a million times as intense.

80. If a function f is not continuous at $x = c$, but can be made continuous at c by being given a new value at the point, it is said to have a *removable discontinuity* at $x = c$. Which of the following functions has a removable discontinuity at $x = c$?

　　a. $f(x) = \dfrac{2x^2 + x - 15}{x + 3}$ at $c = -3$　　**b.** $f(x) = \dfrac{x - 2}{|x - 2|}$ at $c = 2$　　**c.** $f(x) = \dfrac{2 - \sqrt{x}}{4 - x}$ at $c = 4$

81. If $f(x) = x^3 - x^2 + x + 1$, show that there is a number c such that $f(c) = 0$.

82. Prove that $\sqrt{x + 3} = e^x$ has at least one real root, and then use a graphing utility to find the root correct to the nearest tenth.

83. Show that the tangent line to the circle $x^2 + y^2 = r^2$ at the point (x_0, y_0) has the equation $y_0 y + x_0 x = r^2$. *Hint*: Recall that the tangent line at a point P on a circle with center C is the line that passes through P and is perpendicular to the line segment $\overline{CP}$.

84. **Think Tank Problem** It is not necessarily true that $\lim_{x \to c} f(x)$ and $\lim_{x \to c} g(x)$ exist whenever $\lim_{x \to c} \big[f(x) \cdot g(x)\big]$ exists. Find functions f and g such that $\lim_{x \to c} \big[f(x) \cdot g(x)\big]$ exists and $\lim_{x \to c} g(x) = 0$, but $\lim_{x \to c} f(x)$ does not exist.

85. **Think Tank Problem** It is not necessarily true that $\lim_{x \to c} f(x)$ and $\lim_{x \to c} g(x)$ exist whenever $\lim_{x \to c} \dfrac{f(x)}{g(x)}$ exists. Find functions f and g such that $\lim_{x \to c} \dfrac{f(x)}{g(x)}$ exists but neither $\lim_{x \to 0} f(x)$ nor $\lim_{x \to 0} g(x)$ exists.

86. When analyzing experimental data involving two variables, a useful procedure is to pass a smooth curve through a number of plotted data points and then perform computations as if the curve were the graph of an equation relating the variables. Suppose the following data are gathered as a result of a physiological experiment in which skin tissue is subjected to external heat for t seconds, and a measurement is made of the change in temperature ΔT required to cause a change of $2.5°C$ at a depth of 0.5 mm in the skin:

t (sec)	1	2	3	4	10	20	25	30
ΔT (°C)	12.5	7.5	5.8	5.0	4.5	3.5	2.9	2.8

Plot the data points in a coordinate plane and then answer these questions.

　　a. What temperature difference ΔT would be expected if the exposure time is 2.2 sec?

　　b. Approximately what exposure time t corresponds to a temperature difference $\Delta T = 8.0°C$?

87. **Journal Problem** The theorem

$$\lim_{h \to 0} \frac{\sin h}{h} = 1$$

stated in this chapter has been the subject of much discussion in mathematical journals such as:

W. B. Gearhart and H. S. Schultz, "The Function sinx/x," *College Mathematics Journal* (1990): 90-99.

L. Gillman, "π and the Limit of (sin α)/α," *American Mathematical Monthly* (1991): 345-348.

F. Richman, "A Circular Argument," *College Mathematics Journal* (1993): 160-162.

D. A. Rose, "The Differentiability of sin$x$," *College Mathematics Journal* (1991):139-142

P. Ungar, "Reviews," *American Mathematical Monthly* (1986): 221-230.

Some of these articles argue that the demonstration shown in the text is circular, since we use the fact that the area of a circle is πr^2. How do we know the area of a circle? The answer, of course, is that we learned it in elementary school, but that does not constitute a proof. On the other hand, some of these articles argue that the reasoning is not necessarily circular. Read one or more of these journal articles and write a report, making certain you understand the concept of circular reasoning.

88. **Think Tank Problem** Give an example for which neither $\lim\limits_{x \to c} f(x)$ nor $\lim\limits_{x \to c} g(x)$ exists, but

$$\lim_{x \to c} \big[f(x) + g(x) \big]$$

does exist.

89. Use the sum rule to show that if

$$\lim_{x \to c} \big[f(x) + g(x) \big]$$

and $\lim\limits_{x \to c} f(x)$ both exist, then so does $\lim\limits_{x \to c} g(x)$.

90. **a.** Show that $\sin x < x$ if $0 < x < \frac{\pi}{2}$. Refer to Figure 2.30.
   b. Show that $|\sin x| < |x|$ if $0 < |x| < \frac{\pi}{2}$.
   c. Use the definition of continuity to show that $\sin x$ is continuous at $x = 0$.
   d. Use the formula
   $$\sin(\alpha + \beta) = \sin \alpha \cos \beta + \cos \alpha \sin \beta$$
   to show that $\sin x$ is continuous for all real x.

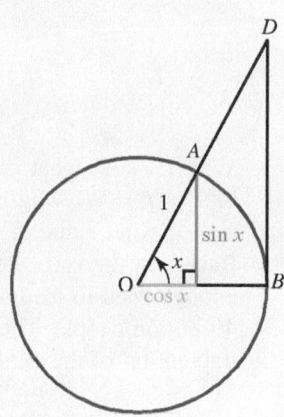

Figure 2.30 Figure for Problems 90 and 91

91. Follow the procedure outlined in Problem 90 to show that $\cos x$ is continuous for all x.
   Hint: After showing that
   $$\lim_{x \to 0} \cos x = 1$$
   you may need the identity
   $$\cos 2x = 1 - 2 \sin^2 x$$
   to show that $\cos x$ is continuous for $x \neq 0$.

92. A regular polygon of n sides is inscribed in a circle of radius R, as shown in Figure 2.31.

   a. Show that the perimeter of the polygon is given by
   $$P(\theta) = \frac{4\pi R}{\theta} \sin \frac{\theta}{2}$$
   where $\theta = \frac{2\pi}{n}$ is the central angle subtended by one side of the polygon.
   b. Use the formula in part **a** together with the fact that
   $$\lim_{x \to 0} \frac{\sin x}{x} = 1$$
   to show that a circle of radius R has circumference $2\pi R$.

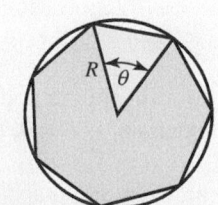

Figure 2.31 A regular polygon of seven sides with central angle θ (Problem 92)

93. **EXPLORATION PROBLEM** A cylindrical tank containing 50 L of water is drained into an empty rectangular trough that can hold 75 L. Explain why there must be a time when the height of water in the tank is the same as that in the trough.

94. The radius of the earth is roughly 4,000 mi, and an object located x miles from the center of the earth weighs w lb, where

$$w(x) = \begin{cases} Ax & \text{if } x \le 4{,}000 \\ \frac{B}{x^2} & \text{if } x > 4{,}000 \end{cases}$$

where A and B are positive constants. Assuming that w is continuous for all x, what must be true about A and B?

95. The wind chill temperature in degrees is a function of the air temperature T (in degrees Fahrenheit) and the wind speed v (in mi/h).* If we hold T constant and consider the wind chill as a function of v, we have

$$w(v) = \begin{cases} T & \text{if } 0 \le v \le 4 \\ 91.4 + (91.4 - T)(0.0203v - 0.304\sqrt{v} - 0.474) & \text{if } 4 < v < 45 \\ 1.6T - 55 & \text{if } v \ge 45 \end{cases}$$

a. If $T = 30$, what is the wind chill for $v = 20$? What is it for $v = 50$?
b. For $T = 30$, what wind speed corresponds to a wind chill temperature of $0°$
c. For what value of T is the wind chill function continuous at $v = 4$? At $v = 45$?

96. Based on the estimate that there are 10 billion acres of arable land on the earth and that each acre can produce enough food to feed 4 people, some demographers believe that the earth can support a population of no more than 40 billion people. The population of the earth reached approximately 5 billion in 1986 and 6 billion in 1999. If the population of the earth were growing according to the formula

$$P(t) = P_0 e^{rt}$$

where t is the time after the population is P_0 and r is the growth rate, when would the population reach the theoretical limit of 40 billion?

97. A function f is said to satisfy a *Lipschitz condition* (named for the 19th century mathematician, Rudolf Lipschitz, 1832-1903) on a given interval if there is a positive constant M such that

$$|f(x) - f(y)| < M\,|x - y|$$

for all x and y in the interval (with $x \ne y$). Suppose f satisfies a Lipschitz condition on an interval and let c be a fixed number chosen arbitrarily from the interval. Use the formal definition of limit to prove that f is continuous at c.

98. A population model employed at one time by the U.S. Census Bureau uses the formula

$$P(t) = \frac{202.31}{1 + e^{3.98 - 0.314t}}$$

to estimate the population of the United States (in millions) for every tenth year from the base year of 1790. For example, if $t = 0$, then the year is 1790, and if $t = 20$, the year is 1990.

a. Draw the graph using this population model and predict the population in the year 2010.
b. Consult an almanac or some other source to find the actual population figures for the years from 1790 to present.
c. What happens to the population P "in the long run," that is, as t increases without bound?

99. Putnam Examination Problem Evaluate

$$\lim_{x \to \infty} \left[\frac{1}{x} \cdot \frac{a^x - 1}{a - 1} \right]^{1/x} \quad \text{where } a > 0,\, a \ne 1$$

---

*From William Bosch and L. G. Cobb, "Windchill," *UMAP Module No. 658*, (1984), pp. 244-247.

CHAPTER 2 GROUP RESEARCH PROJECT*

Working in small groups is typical of most work environments, and this book seeks to develop skills with group activities. We present a group project at the end of each chapter. These projects are to be done in groups of three or four students.

Area of a circle

Everyone knows that the area of a circle of radius r is πr^2. Or at least, everyone thinks they know that result is correct because everyone has been told so by lots of teachers. But, how do *they* know?

It was Archimedes who derived the area of the circle, around 250 B.C. Here is a method similar to the one he used (see Figure 2.32).

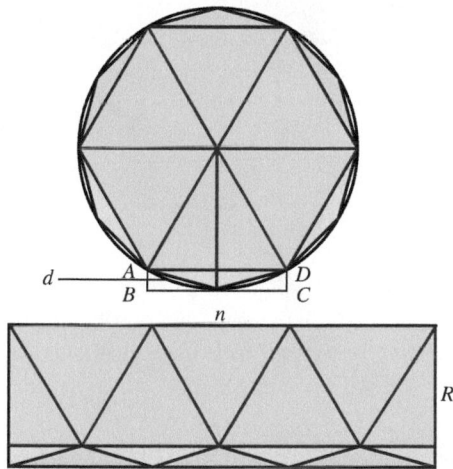

Figure 2.32 Archimedes method for finding the area of a circle

Let n be a positive integer and cut the circle into n equal sectors. In each sector, there is an isosceles triangle formed by the center of the circle and the points where the edges of the sector cut the circumference. (See Problem 92, Supplementary problems, page 69.) Archimedes found the area of each triangle, and then approximated the area of the circle by the sum of the areas of the triangles. Finally, he let the number of triangles (n) increase without bound to find the area of the circle.

Extended paper for further study. Repeat the procedure of Archimedes and prove that the formula for the area of a circle is correct.

---

*Adapted from *Student Research Projects in Calculus*, Cohen et al, © 1991 by The Mathematical Association of America. This is Project 17.

CHAPTER 3

DIFFERENTIATION

*C*ontrary to common belief, the calculus is not the highest of the so-called "higher mathematics." It is, in fact, only the beginning.

Morris Kline,
Mathematics in Western Culture, 1953

PREVIEW

In this chapter, we develop the main ideas of differential calculus. We begin by defining the *derivative*, which is the central concept of differential calculus. Then we develop a list of rules and formulas for finding the derivative of a variety of expressions involving polynomials, rational and root functions, and trigonometric, logarithmic, and inverse trigonometric functions. Along the way, we will see how the derivative can be used to find slopes of tangent lines and rates of change.

PERSPECTIVE

A *calculus* is a body of calculation or reasoning associated with a certain concept. For *differential calculus*, that concept is the *derivative*, one of the fundamental ideas in all mathematics and, arguably, a cornerstone of modern scientific thought. The basic ideas of what we now call calculus had been fermenting in intellectual circles throughout much of the 17th century. The genius of Newton and Leibniz centered not so much on the discovery of those ideas as on their systematization.

In this chapter, we will consider various ways of efficiently computing derivatives. We will also see how the derivative can be used to find rates of change and to measure the direction (slope) at each point on a graph. Falling body problems in physics and marginal analysis from economics are examples of applied topics to be discussed, and we will also explore several topics involving basic concepts. It is fair to say that your success with differential calculus hinges on understanding this material.

CONTENTS

3.1 AN INTRODUCTION TO THE DERIVATIVE: TANGENTS

IN THIS SECTION: *Tangent lines, the derivative, relationship between the graphs of f and f', existence of derivatives, continuity and differentiability, derivative notation*

Ancient Greek mathematicians were able to construct tangents to circles and a few other simple curves, but before the 17th century, there was no general procedure for finding the equation of a tangent to the graph of a function at a given point. To solve this "tangent problem," Sir Isaac Newton (see 𝔥*istorical Quest* on page 69) used an approach originally developed by the French mathematician Pierre de Fermat (see 𝔥*istorical Quest* on page 143.) This approach leads to the definition of the *derivative*.

Tangent Lines

Since classical times, mathematicians have known that the tangent line at a given point on a circle has the property that it is the only line that intersects the circle exactly once, but the same principle cannot be applied to finding tangent lines to more general curves. Indeed, as shown in Figure 3.1, the line we may intuitively think of as being tangent to the graph of $y = f(x)$ at point P can intersect the curve at other points, and for that matter, there may be many lines that intersect the curve only at P, but are clearly not tangent lines.

We begin this section by computing the slope of a tangent line as a limit, using a procedure originally developed by the French mathematician, Pierre de Fermat (1601-1665). Fermat's use of the "dynamic" limit process rather than the "static" procedures of classical geometry and algebra was a breakthrough in thinking that eventually led to the development of differential calculus by Isaac Newton (1642-1727) and Gottfried Leibniz (1646-1716) in the latter half of the 17th century. Following in the footsteps of these giants, we, too, will use what we discover about tangent lines as a springboard for introducing the derivative and studying its basic properties.

Suppose we wish to find the slope of the tangent line to $y = f(x)$ at the point $P(x_0, f(x_0))$. The strategy is to approximate the tangent line by other lines whose slopes can be computed directly. In particular, consider the line joining the given point P to the neighboring point Q on the graph of f as shown in Figure 3.2. This line is called a **secant line**. Compare the secant lines shown in Figure 3.2.

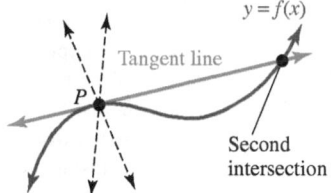

Figure 3.1   Tangent line at P

Do not confuse secant line (a line that intersects a curve in two or more points) with the secant function of trigonometry.

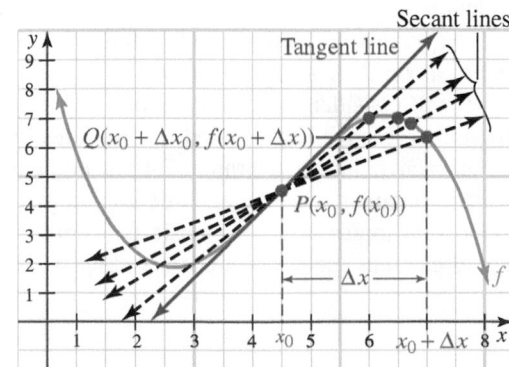

Figure 3.2 Interactive   The secant line $\overline{PQ}$

Notice that a secant line is a good approximation to the tangent line at point P as long as Q is close to P.

To compute the slope of a secant line, first label the coordinates of the neighboring point Q, as indicated in Figure 3.2. In particular, let Δx denote the change in the x-coordinate between the given point $P(x_0, f(x_0))$ and the neighboring point $Q(x_0 + \Delta x, f(x_0 + \Delta x))$.

The slope of this secant line, m_{sec}, is easy to calculate:

$$m_{sec} = \frac{\Delta y}{\Delta x} = \frac{f(x_0 + \Delta x) - f(x_0)}{\Delta x}$$

Δx *is a single symbol and does not mean delta times x. Do not forget that as $\Delta x \to 0$, Δx is getting close to 0, but is not equal to 0.*

To bring the secant line closer to the tangent line, let Q approach P *on the graph of f* by letting Δx approach 0. As this happens, the slope of the secant line should approach the slope of the tangent line at P. We denote the slope of the tangent line by m_{tan} to distinguish it from the slope of a secant line. These observations suggest the following definition.

SLOPE OF A LINE TANGENT TO A GRAPH AT A POINT   At the point $P(x_0, f(x_0))$, the **tangent line** to the graph of f has **slope** given by the formula

$$m_{tan} = \lim_{\Delta x \to 0} \frac{f(x_0 + \Delta x) - f(x_0)}{\Delta x}$$

provided this limit exists.

Example 1   Slope of a tangent line at a particular point

Find the slope of the tangent line to the graph of $f(x) = x^2$ at the point $P(-1, 1)$.

Solution   Figure 3.3 shows the tangent line to f at $x = -1$.

The slope of the tangent line is given by

$$m_{tan} = \lim_{\Delta x \to 0} \frac{f(-1 + \Delta x) - f(-1)}{\Delta x} \quad \textit{Because } f(x) = x^2; f(-1 + \Delta x) = (-1 + \Delta x)^2.$$

$$= \lim_{\Delta x \to 0} \frac{(-1 + \Delta x)^2 - (-1)^2}{\Delta x}$$

$$= \lim_{\Delta x \to 0} \frac{1 - 2\Delta x + (\Delta x)^2 - 1}{\Delta x}$$

$$= \lim_{\Delta x \to 0} \frac{-2\Delta x + (\Delta x)^2}{\Delta x}$$

$$= \lim_{\Delta x \to 0} \frac{(-2 + \Delta x)\Delta x}{\Delta x} \quad \textit{Factor out } \Delta x \textit{ and reduce.}$$

$$= \lim_{\Delta x \to 0} (-2 + \Delta x)$$

$$= -2$$

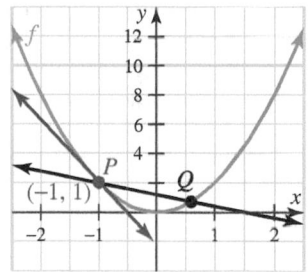

Figure 3.3 Tangent line to the graph of $y = x^2$ at $(-1, 1)$

In Example 1, we found the slope of the tangent line to the graph of $y = x^2$ at the point $(-1, 1)$. In Example 2, we perform the same calculation again, this time representing the given point algebraically as (x, x^2). This is the situation shown in Figure 3.4 for the slope of the tangent line to $y = x^2$ at *any* point (x, x^2).

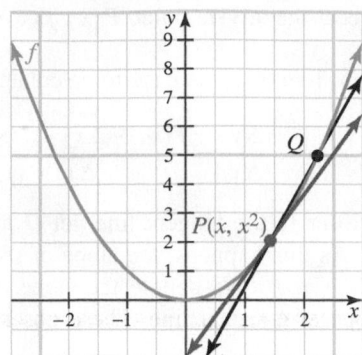

Figure 3.4   Tangent lines to the graph of $y = x^2$ at (x, x^2)

Example 2   Slope of a tangent line at an arbitrary point

Derive a formula for the slope of the tangent line to the graph of $f(x) = x^2$, and then use the formula to compute the slope at $(4, 16)$.

Solution   Figure 3.4 shows a tangent line at an arbitrary point $P(x, x^2)$ on the curve. From the definition of slope of the tangent line,

$$
\begin{aligned}
m_{\tan} &= \lim_{\Delta x \to 0} \frac{f(x + \Delta x) - f(x)}{\Delta x} \\[2mm]
&= \lim_{\Delta x \to 0} \frac{(x + \Delta x)^2 - x^2}{\Delta x} \qquad \text{\textit{Because } } f(x) = x^2, f(x + \Delta x) = (x + \Delta x)^2. \\[2mm]
&= \lim_{\Delta x \to 0} \frac{x^2 + 2x\,\Delta x + (\Delta x)^2 - x^2}{\Delta x} \\[2mm]
&= \lim_{\Delta x \to 0} \frac{2x\,\Delta x + (\Delta x)^2}{\Delta x} \\[2mm]
&= \lim_{\Delta x \to 0} \frac{(2x + \Delta x)\,\Delta x}{\Delta x} \qquad \text{\textit{Factor and reduce.}} \\[2mm]
&= \lim_{\Delta x \to 0} (2x + \Delta x) \\[2mm]
&= 2x
\end{aligned}
$$

At the point $(4, 16)$, $x = 4$, so $m_{\tan} = 2(4) = 8$.

The result of Example 2 gives a general formula for the slope of a line tangent to the graph of $f(x) = x^2$, namely, $m_{\tan} = 2x$. The answer from Example 1 can now be verified using this formula; if $x = -1$, then $m_{\tan} = 2(-1) = -2$.

The Derivative

The expression

$$
\frac{f(x + \Delta x) - f(x)}{\Delta x}
$$

which gives a formula for the slope of a secant line to the graph of a function f, is called the **difference quotient** of f. The limit of the difference quotient

$$
\lim_{\Delta x \to 0} \frac{f(x + \Delta x) - f(x)}{\Delta x}
$$

which gives a formula for the slope of the tangent line to the graph of f at the point $(x, f(x))$, is called the *derivative of* f and is frequently denoted by the symbol $f'(x)$

(read "eff prime of x"). To **differentiate** a function f at x means to find its derivative at the point $(x, f(x))$.

DERIVATIVE The **derivative** of f at x is given by

$$f'(x) = \lim_{\Delta x \to 0} \frac{f(x + \Delta x) - f(x)}{\Delta x}$$

provided this limit exists.

Alternatively, we can write

$$f'(x) = \lim_{t \to x} \frac{f(t) - f(x)}{t - x}$$

where $t = x + \Delta x$. Occasionally, this form is easier to use in computations.

The derivative is one of the fundamental concepts in calculus, and it is important to make some observations regarding this definition.

- If the limit for the difference quotient exists, then we say that the function f is **differentiable at x**.
- The value of a derivative depends only on the limit process and not on the symbols used in that process. In Example 2, we found that if $f(x) = x^2$, then $f'(x) = 2x$. This means that we also know

$$\text{If } g(t) = t^2, \text{ then } g'(t) = 2t.$$
$$\text{If } h(u) = u^2, \text{ then } h'(u) = 2u.$$
$$\vdots$$

- Notice that the derivative of a function is itself a function.

Finding the slope of a tangent line is just one of several applications of the derivative that we will discuss in this chapter. In Section 3.4, we will examine rectilinear motion and other rates of change, and in Section 3.8, marginal analysis from economics. In Chapter 4 we will examine more complex applications such as curve sketching and optimization.

Example 3 Derivative using the definition

Differentiate $f(t) = \sqrt{t}$.

Solution
$$\begin{aligned}
f'(t) &= \lim_{\Delta t \to 0} \frac{f(t + \Delta t) - f(t)}{\Delta t} \\[2mm]
&= \lim_{\Delta t \to 0} \frac{\sqrt{t + \Delta t} - \sqrt{t}}{\Delta t} \\[2mm]
&= \lim_{\Delta t \to 0} \frac{\sqrt{t + \Delta t} - \sqrt{t}}{\Delta t} \left(\frac{\sqrt{t + \Delta t} + \sqrt{t}}{\sqrt{t + \Delta t} + \sqrt{t}} \right) \quad \textit{Multiply by 1 to rationalize numerator.} \\[2mm]
&= \lim_{\Delta t \to 0} \frac{(t + \Delta t) - t}{\Delta t \left(\sqrt{t + \Delta t} + \sqrt{t} \right)}
\end{aligned}$$

☙ *Notice that $f(t) = \sqrt{t}$ is defined for all $t \geq 0$, whereas its derivative $f'(t) = \frac{1}{2\sqrt{t}}$ is defined for all $t > 0$. This shows that a function need not be differentiable throughout its entire domain.* ☙

$$= \lim_{\Delta t \to 0} \frac{1}{\sqrt{t + \Delta t} + \sqrt{t}} \qquad \textit{Reduce fraction.}$$

$$= \frac{1}{2\sqrt{t}} \qquad \textit{for } t > 0$$

Example 4   Estimating a derivative using a table

Estimate the derivative of $f(x) = \cos x$ at $x = \frac{\pi}{6}$ by evaluating the difference quotient

$$\frac{\Delta y}{\Delta x} = \frac{f(x + \Delta x) - f(x)}{\Delta x}$$

near the point $x = \frac{\pi}{6}$.

Solution $\dfrac{\Delta y}{\Delta x} = \dfrac{\cos\left(\frac{\pi}{6} + \Delta x\right) - \cos\frac{\pi}{6}}{\Delta x}$

Δx	$\frac{\pi}{6} + \Delta x$	Difference quotient $\frac{\Delta y}{\Delta x}$
1	1.523598776	−0.81885
0.5	1.023598776	−0.69146
0.125	0.648598776	−0.55276
0.0625	0.586098776	−0.52673
0.015625	0.539223776	−0.50675
0.0001220704	0.523720846	−0.50005
0.000007629	0.523606405	−0.50003
↓	↓	↓
0	$\frac{\pi}{6} \approx 0.5235987756$	$-\frac{1}{2}$

Choose a sequence of values for $\Delta x \to 0$: say $1, \frac{1}{2}, \frac{1}{4}, \frac{1}{8}, \cdots$ and use a calculator (or a computer) to estimate the difference quotient by table. We show selected elements from this sequence of calculations in the margin:

A similar table for negative values of Δx should also be considered. From the table, we would guess that $f'(\frac{\pi}{6}) = -\frac{1}{2}$.

> **Preview** An encouraging word is necessary after reading this first example of finding a derivative. By now you are aware of the fact that the derivative concept is one of the main ideas of calculus, and the previous example indicates that finding a derivative can be long and tedious. In the next section, we will begin to simplify the *process* of finding derivatives, so that you can quickly and efficiently find the derivative of many functions (without using this definition directly). For now, however, we focus on the derivative *concept* and *definition*.

Theorem 3.1   Equation of a line tangent to a curve at a point

If f is a differentiable function at x_0, the graph of $y = f(x)$ has a tangent line at the point $P(x_0, f(x_0))$ with slope $f'(x_0)$ and equation

$$y = f'(x_0)(x - x_0) + f(x_0)$$

Proof: To find the equation of the tangent line to the curve $y = f(x)$ at the point $P(x_0, y_0)$, we use the fact that the slope of the tangent line is the derivative $f'(x_0)$ and apply the point-slope formula for the equation of a line:

$$
\begin{aligned}
y - k &= m(x - h) & &\textit{Point-slope formula} \\
y - y_0 &= m_{\tan}(x - x_0) & &\textit{Given point } (x_0, y_0) \\
y - f(x_0) &= f'(x_0)(x - x_0) & &\textit{Substitute } y_0 = f(x_0) \textit{ and } m_{tan} = f'(x_0). \\
y &= f'(x_0)(x - x_0) + f(x_0) & &\textit{Add } f(x_0) \textit{ to both sides.}
\end{aligned}
$$

Example 5   Finding the equation of a tangent line

Find an equation for the tangent line to the graph of $f(x) = \frac{1}{x}$ at the point where $x = 2$.

Solution   The graph of the function $y = \frac{1}{x}$, the point where $x = 2$, and the tangent line at the point are shown in Figure 3.5.

First, find $f'(x)$:

$$f'(x) = \lim_{\Delta x \to 0} \frac{f(x + \Delta x) - f(x)}{\Delta x} \qquad \text{\textit{Definition of derivative}}$$

$$= \lim_{\Delta x \to 0} \frac{\frac{1}{x+\Delta x} - \frac{1}{x}}{\Delta x} \qquad \text{\textit{Substitute } } f(x) = \tfrac{1}{x} \text{ \textit{and} } f(x + \Delta x) = \frac{1}{x + \Delta x}.$$

$$= \lim_{\Delta x \to 0} \frac{x - (x + \Delta x)}{x \, \Delta x \, (x + \Delta x)} \qquad \text{\textit{Simplify the fraction.}}$$

$$= \lim_{\Delta x \to 0} \frac{-1}{x(x + \Delta x)}$$

$$= \frac{-1}{x^2}$$

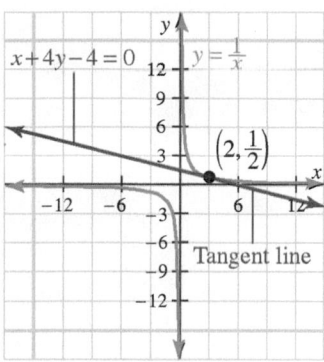

Figure 3.5 Tangent line to $y = \dfrac{1}{x}$ at $\left(2, \dfrac{1}{2}\right)$

Next, find the slope of the tangent line at $x = 2$: $m_{\text{tan}} = f'(2) = -\frac{1}{4}$. Since $f(2) = \frac{1}{2}$, the equation of the tangent line can now be found by using Theorem 3.1:

$$y = -\frac{1}{4}(x - 2) + \frac{1}{2}$$

or in standard form, $x + 4y - 4 = 0$.

The slope of f at x_0 is not the derivative f' but the value of the derivative at x_0. In Example 5 the function is defined by $f(x) = 1/x$, the derivative is $f'(x) = -1/x^2$, and the slope at $x = 2$ is the number $f'(2) = -\frac{1}{4}$.

Example 6 Finding a line that is perpendicular to a tangent line

Find the equation of the line that is perpendicular to the tangent line to $f(x) = \frac{1}{x}$ at $x = 2$ and intersects it at the point of tangency.

Solution In Example 5, we found that the slope of the tangent line is $f'(2) = -\frac{1}{4}$ and that the point of tangency is $(2, \frac{1}{2})$. In Section 1.3, we saw that two lines are perpendicular if and only if their slopes are negative reciprocals of each other. Thus, the perpendicular line we seek has slope 4 (the negative reciprocal of $m = -\frac{1}{4}$). The desired equation is

$$y - \frac{1}{2} = 4(x - 2) \qquad \text{\textit{Point-slope formula}}$$

In standard form, $4x - y - \frac{15}{2} = 0$; compare the coefficients of the variables in the tangent and perpendicular lines.

The perpendicular line we found in Example 6 has a name.

> **NORMAL LINE TO A GRAPH** The **normal line** to the graph of f at the point P is the line that is perpendicular to the tangent line to the graph at P.

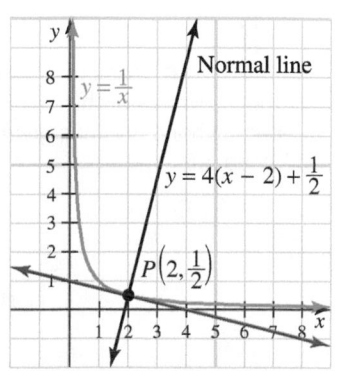

Figure 3.6 Graph of $y = \dfrac{1}{x}$ along with its normal

Figure 3.6 shows the graph of the function from Example 6, along with the graph of its normal line.

Relationship Between the Graphs of f and f'

It is important to take some time to study the relationship between the graph of a function and its derivative. Since slope is measured by f', it follows that at points x where $f'(x) > 0$, the tangent line must be tilted upward, and the graph is rising. Similarly, where $f'(x) < 0$, the tangent line is tilted downward, and the graph is falling. If $f'(x) = 0$, the tangent line is horizontal at x, so the graph "flattens." These observations are illustrated in Figure 3.7.

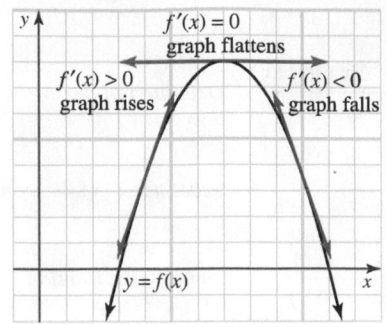

Figure 3.7 Interactive   How the sign of f' determines whether the tangent line slants up, slants down, or is horizontal

Example 7   Sketching the graph of f', given the graph of f.

The graph of a function f is shown in Figure 3.8.

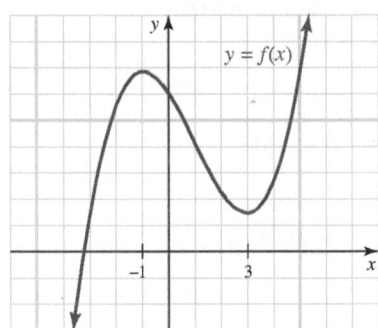

Figure 3.8 Interactive   Graph of a function f

Sketch a possible graph for the derivative f'.

Solution   Notice that the graph of f is rising for $x < -1$ and for $x > 3$; it is falling for $-1 < x < 3$; and that it has horizontal tangent lines at $x = -1$ and $x = 3$. Thus, the graph of f' is above the x-axis ($f'(x) > 0$) for $x < -1$ and $x > 3$, below the axis for $-1 < x < 3$ ($f'(x) < 0$), and crosses the axis at $x = -1$ and at $x = 3$. One possible graph with these features is shown in Figure 3.9.   ◼

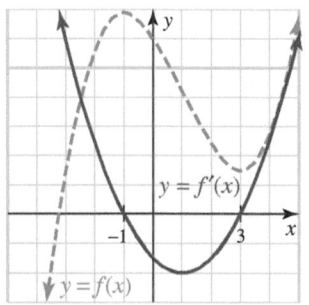

Figure 3.9   A possible graph for the derivative of the function whose graph is given in Figure 3.8

Existence of Derivatives

We observed that a function is differentiable only if *the limit* in the definition of derivative exists. At points where a function f is not differentiable, we say that *the derivative of f does not exist*. Three common ways for a derivative to fail to exist at a point $(c, f(c))$ in the domain of f are shown in Figure 3.10.

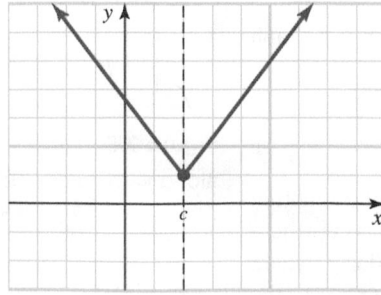

a. Corner point

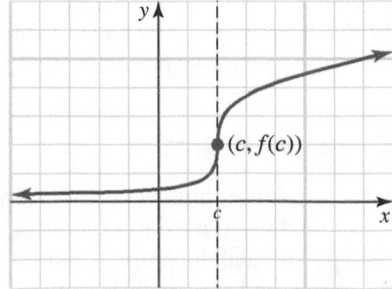

b. Vertical tangent

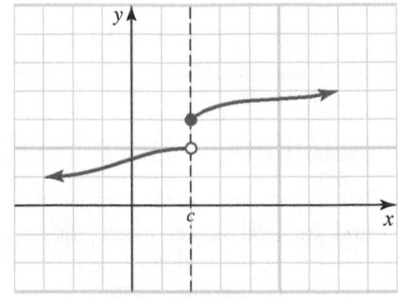

c. Point of discontinuity

Figure 3.10   Common examples where a derivative does not exist

Example 8   A function that does not have a derivative at a corner

Show that the absolute value function $f(x) = |x|$ is not differentiable at $x = 0$.

Solution   The graph of $f(x) = |x|$ is shown in Figure 3.11.

Note that because the slope "from the left" at $x = 0$ is -1 while the slope "from the right" is $+1$, the graph has a corner at the origin, which prevents a unique tangent line from being drawn there.

We can show this algebraically by using the definition of derivative:

$$f'(0) = \lim_{\Delta x \to 0} \frac{f(0 + \Delta x) - f(0)}{\Delta x}$$

$$= \lim_{\Delta x \to 0} \frac{f(\Delta x) - f(0)}{\Delta x}$$

$$= \lim_{\Delta x \to 0} \frac{|\Delta x|}{\Delta x}$$

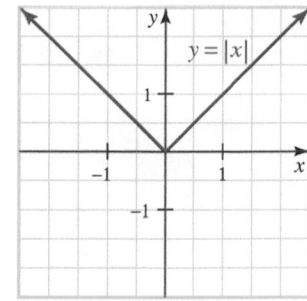

Figure 3.11 $f(x) = |x|$ is not differentiable at $x = 0$ because the slope from the left does not equal the slope from the right

We must now consider one-sided limits, because $|\Delta x| = \Delta x$ when $\Delta x > 0$, and $|\Delta x| = -\Delta x$ when $\Delta x < 0$.

$$\lim_{\Delta x \to 0^-} \frac{|\Delta x|}{\Delta x} = \lim_{\Delta x \to 0^-} \frac{-\Delta x}{\Delta x} = -1 \qquad \textit{Derivative from the left}$$

$$\lim_{\Delta x \to 0^+} \frac{|\Delta x|}{\Delta x} = \lim_{\Delta x \to 0^+} \frac{\Delta x}{\Delta x} = 1 \qquad \textit{Derivative from the right}$$

The left- and right-hand limits are not the same; therefore the limit does not exist. This means that the derivative does not exist at $x = 0$. ∎

The continuous function $f(x) = |x|$ in Example 8 failed to be differentiable at $x = 0$ because the one-sided limits of its difference quotient were unequal. A continuous function may also fail to be differentiable at $x = c$ if its difference quotient diverges to infinity. In this case, the function is said to have a *vertical tangent line* at $x = c$, as illustrated in Figure 3.12.

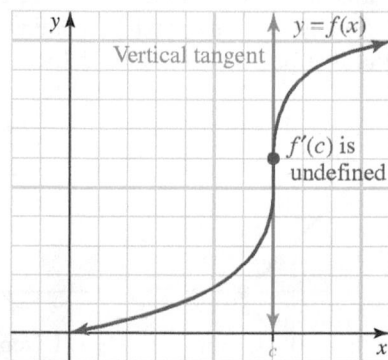

Figure 3.12   Vertical tangent line at $x = c$

We will have more to say about such functions when we discuss curve sketching with derivatives in Chapter 4.

Continuity and Differentiability

If the graph of a function has a tangent line at a point, we would expect to be able to draw the graph continuously (without the pencil leaving the paper). In other words, we expect the following theorem to be true.

Theorem 3.2   Differentiability implies continuity

If a function f is differentiable at c, then it is also continuous at c.

Proof: Recall that for f to be continuous at $x = c$:
1. $f(c)$ must be defined;   **2.** $\lim_{x \to c} f(x)$ exists; and   **3.** $\lim_{x \to c} f(x) = f(c)$.

Thus, continuity can be established by showing that

$$\lim_{\Delta x \to 0} f(c + \Delta x) = f(c)$$

or equivalently,

$$\lim_{\Delta x \to 0} [f(c + \Delta x) - f(c)] = 0$$

Because f is a differentiable function at $x = c$, $f'(c)$ exists and

$$\lim_{\Delta x \to 0} \frac{f(c + \Delta x) - f(c)}{\Delta x} = f'(c)$$

Therefore, by applying the product rule for limits, we find that

$$\lim_{\Delta x \to 0} [f(c + \Delta x) - f(c)] = \lim_{\Delta x \to 0} \left[\frac{f(c + \Delta x) - f(c)}{\Delta x} \cdot \Delta x \right]$$

$$= \left[\lim_{\Delta x \to 0} \frac{f(c + \Delta x) - f(c)}{\Delta x} \right] \left[\lim_{\Delta x \to 0} \Delta x \right]$$

$$= f'(x) \cdot 0$$

$$= 0$$

Thus, $\lim_{x \to c} f(x) = f(c)$, and we see that the conditions for continuity are satisfied.   ◆

☉ Be sure you understand what we have just shown with Example 8 and Theorem 3.2: If a function is differentiable at $x = c$, then it must be continuous at that point. The converse is not true: If a function is continuous at $x = c$, then it may or may not be differentiable at that point. Finally, if a function is discontinuous at $x = c$, then it cannot possibly have a derivative at that point. (See Figure 3.13c.) ☉

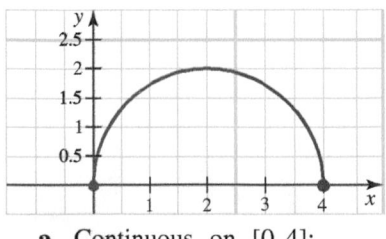

a. Continuous on $[0, 4]$; differentiable on $(0, 4)$

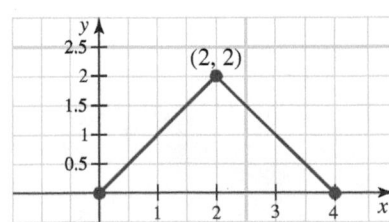

b. Continuous on $[0, 4]$; not differentiable at $x = 2$

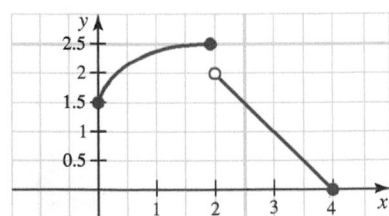

c. Discontinuous at $x = 2$; cannot be differentiable at $x = 2$

Figure 3.13   A function continuous at $x = 2$ may or may not be differentiable at $x = 2$. A function discontinuous at $x = 2$ cannot be differentiable at $x = 2$

Derivative Notation

In certain situations, it is convenient or suggestive to denote the derivative of $y = f(x)$ by $\frac{dy}{dx}$ instead of $f'(x)$. This notation is called the *Leibniz notation* because Leibniz was the first to use it.

For example, if $y = x^2$, the derivative is $y' = 2x$ or, using Leibniz notation, $\frac{dy}{dx} = 2x$. The symbol $\frac{dy}{dx}$ is read "the derivative of y with respect to x." When we wish to denote the value of the derivative at c in the Leibniz notation, we will write

$$\left. \frac{dy}{dx} \right|_{x=c}$$

For instance, we would evaluate $\frac{dy}{dx} = 4x^2$ at $x = 3$ by writing

$$\left. \frac{dy}{dx} \right|_{x=3} = 4x^2 \Big|_{x=3} = 4(3)^2 = 36$$

Another notation omits reference to y and f altogether, and we can write

$$\frac{d}{dx}(x^2) = 2x$$

which is read "the derivative of x^2 with respect to x is $2x$."

Despite its appearance, $\frac{dy}{dx}$ is a single symbol and is not a fraction. In Section 3.8, we introduce a concept called a differential that will provide independent meaning to symbols like dy and dx, but for now, these symbols have meaning only in connection with the Leibniz derivative symbol $\frac{dy}{dx}$.

Example 9   Derivative at a point with Leibniz notation

Find $\left. \dfrac{dy}{dx} \right|_{x=-1}$ if $y = x^3$.

Solution

$$\frac{dy}{dx} = \frac{d}{dx}(x^3)$$

$$= \lim_{\Delta x \to 0} \frac{(x + \Delta x)^3 - x^3}{\Delta x}$$

$$= \lim_{\Delta x \to 0} \frac{\left[x^3 + 3x^2 \Delta x + 3x(\Delta x)^2 + (\Delta x)^3 - x^3 \right]}{\Delta x}$$

$$= \lim_{\Delta x \to 0} \left[3x^2 + 3x(\Delta x) + (\Delta x)^2 \right]$$

$$= 3x^2$$

At $x = -1$,

$$\left. \frac{dy}{dx} \right|_{x=-1} = 3x^2 \Big|_{x=-1} = 3$$

If you are using technology to find derivatives, most formats with technology require that you input not only the function, but the variable in the expression and the value at which you wish to find the derivative. The usual format is nDeriv (*expression, variable, value*). The symbol "nDeriv" depends on the calculator or software, and sometimes is "nDer" (TI-85/86), "d" (TI-92), "diff" (Maple V), or "Dif" (Derive). Some software and calculators will find the exact value of a derivative, whereas some will do a numerical evaluation.

PROBLEM SET 3.1

Level 1

1. ■ What does this say? Describe the process of finding a derivative using the definition.
2. ■ What does this say? What is the definition of a derivative?

3. ■ What does this say? Discuss the truth or falsity of the following statements:
   a. If a function f is continuous on (a, b), then it is differentiable on (a, b).
   b. If a function f is differentiable on (a, b), then it is continuous on (a, b).

4. ■ What does this say? Discuss the relationship between the derivative of a function f at a point $x = x_0$ and the tangent line at that same point.

In each of Problems 5-10, the graph of a function f is given. Sketch a graph of f'.

5.

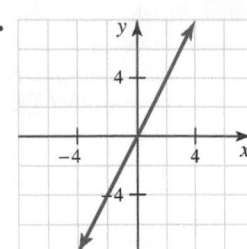

6.

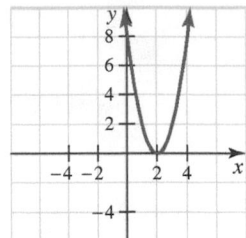

7.

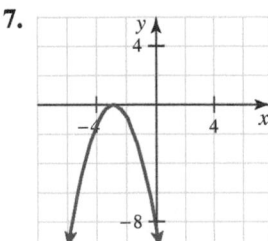

8.

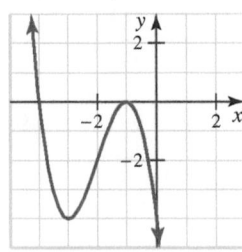

9.

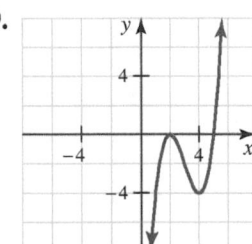

10.

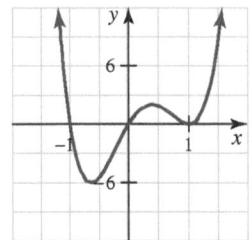

In each of Problems 11-16, a function f is given along with a number c in its domain.

a. *Find the difference quotient of f.*

b. *Find $f'(c)$ by computing the limit of the difference quotient.*

11. $f(x) = 3$ at $c = -5$ **12.** $f(x) = x$ at $c = 2$

13. $f(x) = 2x$ at $c = 1$ **14.** $f(x) = 2x^2$ at $c = 1$

15. $f(x) = 2 - x^2$ at $c = 0$ **16.** $f(x) = -x^2$ at $c = 2$

Use the definition to differentiate the functions given in Problems 17-24, and then describe the set of all numbers for which the function is differentiable.

17. $f(x) = 5$ **18.** $g(x) = 3x$

19. $F(x) = 3x - 7$ **20.** $G(x) = 4 - 5x$

21. $f(t) = 3t^2$ **22.** $g(t) = 4 - t^2$

23. $h(x) = \dfrac{1}{2x}$ **24.** $g(s) = \sqrt{5s}$

Find an equation for the tangent line to the graph of the function at the specified point in Problems 25-30.

25. $f(x) = 2x + 3$ at $(-1, 1)$ **26.** $g(x) = 3x^2$ at $(-2, 12)$

27. $f(s) = s^3$ at $s = -\frac{1}{2}$ **28.** $g(t) = 4 - t^2$ at $t = 0$

29. $h(x) = x - 1$ at $x = 0$ **30.** $g(x) = \sqrt{x - 5}$ at $x = 9$

Find an equation of the normal line to the graph of the function at the specified point in Problems 31-34.

31. $f(x) = 3x - 5$ at $(3, 4)$ **32.** $g(x) = 4 - 5x$ at $(0, 4)$

33. $f(x) = \dfrac{1}{x + 3}$ at $x = 2$ **34.** $f(x) = \sqrt{5x}$ at $x = 5$

Find $\left.\dfrac{dy}{dx}\right|_{x=c}$ for the functions and values of c given in Problems 35-38.

35. $y = 2x,\ c = -1$ **36.** $y = 4 - x,\ c = 2$

37. $y = 1 - x^2,\ c = 0$ **38.** $y = \dfrac{4}{x},\ c = 1$

Level 2

39. Suppose $f(x) = x^2$.
  a. Compute the slope of the secant line joining the points on the graph of f whose x-coordinates are -2 and -1.9.
  b. Use calculus to compute the slope of the line that is tangent to the graph when $x = -2$ and compare this slope with your answer in part **a**.

40. Suppose $f(x) = x^3$.
  a. Compute the slope of the secant line joining the points on the graph of f whose x-coordinates are 1 and 1.1.
  b. Use calculus to compute the slope of the line that is tangent to the graph when $x = 1$ and compare this slope to your answer from part **a**.

41. Sketch the graph of the function $y = x^2 - x$. Determine the value(s) of x for which the derivative is 0. What happens to the graph at the corresponding point(s)?

42. a. Find the derivative of $f(x) = x^2 - 3x$.
  b. Show that the parabola whose equation is $y = x^2 - 3x$ has one horizontal tangent line. Find the equation of this line.
  c. Find a point on the graph of f where the tangent line is parallel to the line $3x + y = 11$.
  d. Sketch the graph of the parabola whose equation is $y = x^2 - 3x$. Display the horizontal tangent line and the tangent line found in part **c**.

43. Show that the function $f(x) = |x - 2|$ is not differentiable at $x = 2$.

44. Is the function $f(x) = 2|x + 1|$ differentiable at $x = -1$? Why?

45. Let $f(x) = \begin{cases} -x^2 & \text{if } x < 0 \\ x^2 & \text{if } x \geq 0 \end{cases}$

Does $f'(0)$ exist? *Hint:* Find the difference quotient and take the limit as $\Delta x \to 0$ from the left and from the right.

46. Let $f(x) = \begin{cases} -2x & \text{if } x < 1 \\ \sqrt{x} - 3 & \text{if } x \geq 1 \end{cases}$

   a. Sketch the graph of f.

   b. Show that f is continuous but not differentiable at $x = 1$.

Estimate the derivative $f'(c)$ in Problems 47-52 by evaluating the difference quotient

$$\frac{\Delta y}{\Delta x} = \frac{f(c + \Delta x) - f(c)}{\Delta x}$$

at a succession of numbers near c.

47. $f(x) = (2x - 1)^2$ for $c = 1$

48. $f(x) = \dfrac{1}{x + 1}$ for $c = 2$

49. $f(x) = \sin x$ for $c = \frac{\pi}{3}$

50. $f(x) = \cos x$ for $c = \frac{\pi}{3}$

51. $f(x) = \sqrt{x}$ for $c = 4$

52. $f(x) = \sqrt[3]{x}$ for $c = 8$

53. Think Tank Problem

   a. Find the derivatives of the functions $y = x^2$ and $y = x^2 - 3$ and account geometrically for their similarity.

   b. Without further computation, find the derivative of $y = x^2 + 5$.

54. Think Tank Problem

   a. Find the derivative of $f(x) = x^2 + 3x$.

   b. Find the derivatives of the functions $g(x) = x^2$ and $h(x) = 3x$ separately. How are these derivatives related to the derivative in part **a**?

   c. In general, if $f(x) = g(x) + h(x)$, what would you guess is the relationship between the derivative of f and the derivatives of g and h?

55. Consider a graph of the function defined by $f(x) = x^{2/3}$.

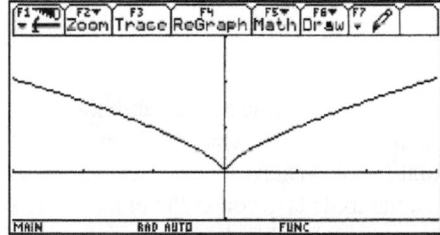

There is a tangent utility on many calculators that will draw tangent lines at a given point. Use this utility or use the preceding simulated graph to draw tangent lines as $x \to 0^-$. Next, draw tangent lines as $x \to 0^+$. Describe what happens, and use this description to support the conclusion that there is no tangent line at $x = 0$.

56. Consider the function

$$f(x) = (x - 2)^{2/3} + 2x^3$$

which gives trouble in seeking the tangent line at $(2, f(2))$. Attempt to compute $f'(2)$, either "by hand" or using a calculator. Describe what happens; in particular, do you see why the tangent line there is meaningless?

57. *Historical Quest*

Fermat is considered "one of the giants" of mathematics. He was a lawyer by profession, but he liked to do mathematics in his spare time. He wrote well over 3,000 mathematical papers and notes. Fermat developed a general procedure for finding tangent lines that is a precursor to the methods of Newton and Leibniz.

Pierre de Fermat (1601-1665)

We will explore this procedure by finding a tangent line to the curve

$$x^3 + y^3 - 2xy = 0$$

A graph is shown in Figure 3.14.

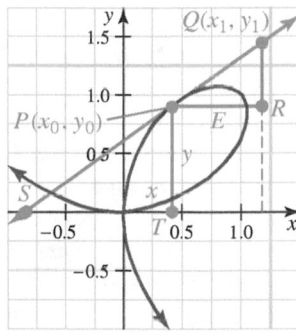

Figure 3.14 Fermat's method of subtangents

   a. Let $P(x_0, y_0)$ be a given point on the curve, and let S be the x-axis intercept of the tangent line at P. Let $Q(x_1, y_1)$ be another point on the tangent line and let T and R be located so $\triangle STP$ and $\triangle PRQ$ are right triangles. Finally, let $A = |\overline{ST}|$ and $E = |\overline{PR}|$. Express x_1 and y_1 in terms of x_0, y_0, A, and E.

   b. Fermat reasoned that if E were very small, then Q would "almost" be on the curve. Substitute the values for x_1 and y_1 you found in part **a** into the equation

$$x^3 + y^3 - 2xy = 0$$

That is, substitute the (x, y) values you found in part **a** into this equation.

c. With the answer from part **b**, you can follow the steps of Fermat by dividing both sides of the equation by E. Fermat reasoned that since E was close to 0, this step should be permitted. Now, after doing this, Fermat further reasoned that since E was close to 0, he could now set $E = 0$ and solve for A. Carry out these steps to write A in terms of x_0 and y_0.

d. Fermat then constructed the tangent line at P by joining P to S. Draw the tangent line for the given curve at the point $(0.5, 0.9)$ by plotting the point corresponding to the calculated value of A. Find an equation for the tangent line to the curve

$$x^3 + y^3 - 2xy = 0$$

at the point $P(0.5, 0.9)$.

58. *Historical Quest*

The groundwork for much of the mathematics we do today, and certainly a necessity for calculus, is the development of **analytic geometry** *by Descartes and Pierre de Fermat (see Problem 57). Descartes' ideas for analytic geometry were published in 1637 as one of three*

**René Descartes
(1596-1650)**

appendices to his **Discourse on the Method (of Reasoning Well and Seeking Truth in the Sciences).** *In that same year, Fermat sent an essay entitled* **"Introduction to Plane and Solid Loci"** *to Paris, and in this essay he laid the foundation for analytic geometry. Fermat's paper was more complete and systematic, but Descartes' was published first. Descartes is generally credited with the discovery of analytic geometry, and we speak today of the Cartesian coordinate system and Cartesian geometry to honor Descartes' discovery. Today we describe analytic geometry from two viewpoints: (1) given a curve, describe it by an equation (Descartes viewpoint); and (2) given an equation, describe it by a curve (Fermat's viewpoint).*

    In this Quest *we will describe Descartes' circle method for finding a tangent line to a given curve. This method uses algebra and geometry rather than limits. Descartes' method for finding the tangent line to the curve $y = f(x)$ at the point $P(x_0, f(x_0))$ involved first finding the point $Q(x_1, 0)$, which is the point of intersection of the normal line to the curve at P with the x-axis, as shown in Figure 3.15.*

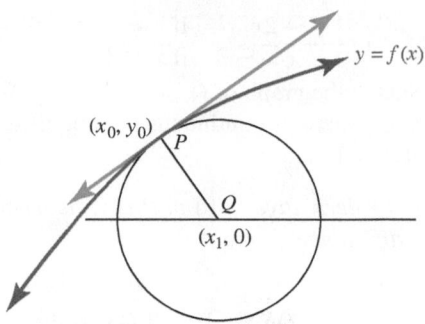

Figure 3.15 Descartes' circle method for finding tangent lines

Descartes then wrote the equation of the circle with center Q passing through P:

$$(x - x_1)^2 + y^2 = (x_0 - x_1)^2 + y_0^2$$

Descartes' next step was to use the equation of the given curve $y = f(x)$ (actually any equation involving two variables) to eliminate one of the variables (usually y) from the equation of the circle. Descartes reasoned that the circle will cut the given curve in two places **except** *when $\overline{PQ}$ is normal, in which case the two intersection points will coincide and the circle will be tangent to the curve at P. That is, Descartes imposed the condition that the resulting equation (after substitution) has only one root to solve for x_1. The point Q is thus found and the normal line was then known to Descartes. The tangent line can be taken as the perpendicular through P to the normal line. Carry out Descartes' circle method for the parabola $y^2 = 4x$ at the point $(4, 4)$.*

Level 3

59. Show that the tangent line to the parabola $y = Ax^2$ (for $A \neq 0$) at the point where $x = c$ will intersect the x-axis at the point $(\frac{c}{2}, 0)$. Where does it intersect the y-axis?

60. Suppose a parabola is given in the plane along with its axis of symmetry. Explain how you could construct the tangent line at a given point P on the parabola using only compass and straightedge methods.
Hint: You may assume that the parabola has an equation of the form $y = Ax^2$ in which the y-axis is the axis of symmetry and the vertex of the parabola is at the origin. Then use the result of Problem 59.

3.2 TECHNIQUES OF DIFFERENTIATION

IN THIS SECTION: *Derivative of a constant function, derivative of a power function, procedural rules for finding derivatives, higher-order derivatives*
In the last section, you learned how to find the derivative of a function f by computing the limit of a difference quotient. For even the simplest functions, this process is tedious and time-consuming. In this section, we describe some rules that simplify the process of differentiation.

If we had to compute the limit of a difference quotient every time we wished to find a derivative, differentiation would not be the valuable tool that it is. Fortunately, there is a better way, and in this section, we will derive several formulas that will enable us to compute derivatives indirectly, without evaluating any limits.

Derivative of a Constant Function

We begin by proving that the derivative of any constant function is zero. Notice that this is plausible because the graph of the constant function $f(x) = k$ is a horizontal line, and its slope is zero. Thus, for example, if $f(x) = 5$, then $f'(x) = 0$.

Theorem 3.3   Constant rule

A constant function $f(x) = k$ has derivative $f'(x) = 0$; in Leibniz notation,

$$\frac{d}{dx}(k) = 0$$

Proof: Note that if $f(x) = k$, then $f(x + \Delta x) = k$ for all Δx. Therefore, the difference quotient is

$$\frac{f(x + \Delta x) - f(x)}{\Delta x} = \frac{k - k}{\Delta x} = 0$$

and

$$f'(x) = \lim_{\Delta x \to 0} \frac{f(x + \Delta x) - f(x)}{\Delta x} = \lim_{\Delta x \to 0} 0 = 0$$

Remember $\Delta x \neq 0$ even though $\Delta x \to 0$.

as claimed.   ◆

Derivative of a Power Function

Recall that a **power function** is a function of the form $f(x) = x^n$, where n is a real number. For example, $f(x) = x^2, g(x) = x^{-3}, h(x) = x^{1/2}$ are all power functions. So are

$$F(x) = \frac{1}{x^2} = x^{-2} \quad \text{and} \quad G(x) = \sqrt[3]{x^2} = x^{2/3}$$

Here is a simple rule for finding the derivative of any power function.

Theorem 3.4   Power rule

For any real number n, the power function $f(x) = x^n$ has the derivative $f'(x) = nx^{n-1}$; in Leibniz notation,

$$\frac{d}{dx}(x^n) = nx^{n-1}$$

Proof: If the exponent n is a positive integer, we can prove the power rule by using the binomial theorem with the definition of derivative. Begin with the difference quotient:

$$\frac{f(x + \Delta x) - f(x)}{\Delta x} = \frac{(x + \Delta x)^n - x^n}{\Delta x}$$

$$= \frac{\left[x^n + nx^{n-1}\Delta x + \frac{n(n-1)}{2}x^{n-2}(\Delta x)^2 + \cdots + (\Delta x)^n\right] - x^n}{\Delta x}$$

$$= \frac{nx^{n-1}\Delta x + \frac{n(n-1)}{2}x^{n-2}(\Delta x)^2 + \cdots + (\Delta x)^n}{\Delta x}$$

$$= nx^{n-1} + \frac{n(n-1)}{2}x^{n-2}\Delta x + \cdots + (\Delta x)^{n-1}$$

Note that Δx is a factor of every term in this expression except the first. Hence, as $\Delta x \to 0$, we have

$$f'(x) = \lim_{\Delta x \to 0} \frac{f(x + \Delta x) - f(x)}{\Delta x}$$

$$= \lim_{\Delta x \to 0}\left[nx^{n-1} + \frac{n(n-1)}{2}x^{n-2}\Delta x + \cdots + (\Delta x)^{n-1}\right]$$

$$= nx^{n-1}$$

If $n = 0$, then $f(x) = x^0 = 1$, so $f'(x) = 0$. We will prove the power rule for negative integer exponents later in this section, and we will deal with the case in which the exponent is any real number in Section 3.6. Note, however, that we have already verified the power rule for $n = 3$ in Example 9 of Section 3.1 and for the rational exponent $\frac{1}{2}$ in Example 3 of Section 3.1, when we showed that the derivative of $f(t) = \sqrt{t} = t^{1/2}$ is

$$f'(t) = \frac{1}{2}t^{-1/2} = \frac{1}{2\sqrt{t}} \text{ for } t > 0$$

For the following examples, and the problems at the end of this section, you may assume that the power rule is valid when the exponent n is any real number. ♦

Example 1   Using the power rule to find a derivative

Differentiate each of the following functions.

a. $f(x) = x^8$ **b.** $g(x) = x^{3/2}$ **c.** $h(x) = \dfrac{\sqrt[3]{x}}{x^2}$

Solution

a. Applying the power rule with $n = 8$, we find that

$$\frac{d}{dx}(x^8) = 8x^{8-1} = 8x^7$$

b. Applying the power rule with $n = \frac{3}{2}$, we get

$$\frac{d}{dx}(x^{3/2}) = \frac{3}{2}x^{3/2-1} = \frac{3}{2}x^{1/2} = \frac{3}{2}\sqrt{x}$$

c. For this part you need to recognize that $h(x) = \dfrac{\sqrt[3]{x}}{x^2} = x^{-5/3}$

$$\frac{d}{dx}(x^{-5/3}) = -\frac{5}{3}x^{-5/3-1} = -\frac{5}{3}x^{-8/3}$$

Procedural Rules for Finding Derivatives

The next theorem expands the class of functions that we can differentiate easily by giving rules for differentiating certain combinations of functions, such as sums, differences, products, and quotients. We will see that the derivative of a sum (or difference) is the sum (or difference) of derivatives, but the derivative of a product (or quotient) does not have such a simple form. For example, to convince yourself that the derivative of a product is not the product of the separate derivatives, consider the power functions

$$f(x) = x \quad \text{and} \quad g(x) = x^2$$

and their product

$$p(x) = f(x)g(x) = x^3$$

Because $f'(x) = 1$ and $g'(x) = 2x$, the product of the derivatives is

$$f'(x)g'(x) = (1)(2x) = 2x$$

whereas the actual derivative of $p(x) = x^3$ is $p'(x) = 3x^2$. The product rule tells us how to find the derivative of a product.

Note the derivative of a product is not the product of derivatives. A similar warning applies to the derivative of a quotient. (See Theorem 3.5.)

Theorem 3.5 Basic rules for combining derivatives—Procedural forms

If f and g are differentiable functions at all x, and a, b, and c are any real numbers, then the functions cf, $f + g$, fg, and f/g (for $g(x) \neq 0$) are also differentiable, and their derivatives satisfy the following formulas:

Name of rule	Function notation	Leibniz notation
Constant multiple	$\left[cf(x)\right]' = cf'(x)$	$\dfrac{d}{dx}(cf) = c\dfrac{df}{dx}$
Sum rule	$\left[f(x) + g(x)\right]' = f'(x) + g'(x)$	$\dfrac{d}{dx}(f + g) = \dfrac{df}{dx} + \dfrac{dg}{dx}$
Difference rule	$\left[f(x) - g(x)\right]' = f'(x) - g'(x)$	$\dfrac{d}{dx}(f - g) = \dfrac{df}{dx} - \dfrac{dg}{dx}$
Linearity rule*	$\left[af(x) + bg(x)\right]' = af'(x) + bg'(x)$	$\dfrac{d}{dx}(af + bg) = a\dfrac{df}{dx} + b\dfrac{dg}{dx}$
Product rule	$\left[f(x)g(x)\right]' = f(x)g'(x) + f'(x)g(x)$	$\dfrac{d}{dx}(fg) = f\dfrac{dg}{dx} + g\dfrac{df}{dx}$
Quotient rule $(g(x) \neq 0)$	$\left[\dfrac{f(x)}{g(x)}\right]' = \dfrac{g(x)f'(x) - f(x)g'(x)}{\left[g(x)\right]^2}$	$\dfrac{d}{dx}\left(\dfrac{f}{g}\right) = \dfrac{g\frac{df}{dx} - f\frac{dg}{dx}}{g^2}$

Proof: We will prove the product rule in detail, leaving the other rules as problems.

Let $f(x)$ and $g(x)$ be differentiable functions of x and let $p(x) = f(x)g(x)$. We will add and subtract the term $f(x + \Delta x)g(x)$ to the numerator of the difference quotient for $p(x)$ to create difference quotients for $f(x)$ and $g(x)$. Thus,

$$
\begin{aligned}
p'(x) &= \lim_{\Delta x \to 0} \frac{p(x + \Delta x) - p(x)}{\Delta x} \\
&= \lim_{\Delta x \to 0} \frac{f(x + \Delta x)g(x + \Delta x) - f(x)g(x)}{\Delta x}
\end{aligned}
$$

Note that the order of terms in the quotient rule matters because $gf' - fg' \neq fg' - gf'$.

---

*The constant multiple, sum, and difference rules can be combined into a single rule, which is called the *linearity* rule.

$$= \lim_{\Delta x \to 0} \frac{f(x + \Delta x)g(x + \Delta x) - f(x + \Delta x)g(x) + f(x + \Delta x)g(x) - f(x)g(x)}{\Delta x}$$

$$= \lim_{\Delta x \to 0} \left\{ f(x + \Delta x) \left[\frac{g(x + \Delta x) - g(x)}{\Delta x} \right] + g(x) \left[\frac{f(x + \Delta x) - f(x)}{\Delta x} \right] \right\}$$

$$= \lim_{\Delta x \to 0} \left\{ f(x + \Delta x) \underbrace{\lim_{\Delta x \to 0} \left[\frac{g(x + \Delta x) - g(x)}{\Delta x} \right]}_{\text{This is the derivative of } g.} \right.$$

$$\left. + \lim_{\Delta x \to 0} g(x) \underbrace{\lim_{\Delta x \to 0} \left[\frac{f(x + \Delta x) - f(x)}{\Delta x} \right]}_{\text{This is the derivative of } f.} \right\}$$

$$= f(x)g'(x) + g(x)f'(x). \qquad \text{Note: } \lim_{\Delta x \to 0} f(x + \Delta x) = f(x) \text{ because } f \text{ is continuous. } \blacklozenge$$

Note that in each part of Theorem 3.5, we prove the differentiability of the appropriate functional combination at the same time we are establishing the differentiation formula.

Example 2 Using the basic rules to find derivatives

Differentiate each of the following functions.

a. $f(x) = 2x^2 - 5\sqrt{x}$ **b.** $p(x) = (3x^2 - 1)(7 + 2x^3)$

c. $q(x) = \dfrac{4x - 7}{3 - x^2}$ **d.** $g(x) = (4x + 3)^2$

e. $F(x) = \dfrac{2}{3x^2} - \dfrac{x}{3} + \dfrac{4}{5} + \dfrac{x + 1}{x}$

Solution

a. Apply the linearity rule (constant multiple, sum, and difference), and power rules:

$$f'(x) = 2(x^2)' - 5(x^{1/2})'$$
$$= 2(2x) - 5\left(\frac{1}{2}\right)(x^{-1/2})$$
$$= 4x - \frac{5}{2}x^{-1/2}$$

b. Apply the product rule; then apply the linearity and power rules:

$$p'(x) = (3x^2 - 1)(7 + 2x^3)' + (3x^2 - 1)'(7 + 2x^3)$$
$$= (3x^2 - 1)\left[0 + 2(3x^2)\right] + [3(2x) - 0](7 + 2x^3)$$
$$= (3x^2 - 1)(6x^2) + (6x)(7 + 2x^3)$$
$$= 6x(5x^3 - x + 7)$$

c. Apply the quotient rule, then the linearity and power rules:

$$q'(x) = \frac{(3 - x^2)(4x - 7)' - (4x - 7)(3 - x^2)'}{(3 - x^2)^2}$$
$$= \frac{(3 - x^2)(4 - 0) - (4x - 7)(0 - 2x)}{(3 - x^2)^2}$$
$$= \frac{12 - 4x^2 + 8x^2 - 14x}{(3 - x^2)^2}$$
$$= \frac{4x^2 - 14x + 12}{(3 - x^2)^2}$$

d. Write $g(x) = (4x + 3)(4x + 3)$ before applying the product rule:

$$g'(x) = (4x + 3)(4x + 3)' + (4x + 3)'(4x + 3)$$
$$= (4x + 3)(4) + (4)(4x + 3)$$
$$= 8(4x + 3)$$

Sometimes when the exponent is 2, it is easier to expand before differentiating:

$$g(x) = (4x + 3)^2 = 16x^2 + 24x + 9$$
$$g'(x) = 32x + 24$$

e. Write the function using negative exponents for reciprocal powers:

$$F(x) = \frac{2}{3}x^{-2} - \frac{1}{3}x + \frac{4}{5} + 1 + x^{-1}$$

Then apply the power rule term by term to obtain

$$F(x) = \frac{2}{3}(-2x^{-3}) - \frac{1}{3} + 0 + 0 + (-1)x^{-2}$$
$$= -\frac{4}{3}x^{-3} - \frac{1}{3} - x^{-2}$$

In applying the power rule term-by-term in Example 2e, we really used the following generalization of the linearity rule.

Corollary to Theorem 3.5   The extended linearity rule

If $f_1, f_2, \cdots, f_n$ are differentiable functions and $a_1, a_2, \cdots, a_n$ are constants, then

$$\frac{d}{dx}\left[a_1 f_1 + a_2 f_2 + \cdots + a_n f_n\right] = a_1 \frac{df_1}{dx} + a_2 \frac{df_2}{dx} + \cdots + a_n \frac{df_n}{dx}$$

Proof: The proof is a straightforward extension (using mathematical induction) of the proof of the linearity rule of Theorem 3.5.   ♦

Example 3 illustrates how the extended linearity rule can be used to differentiate a polynomial.

Example 3   Derivative of a polynomial function

Differentiate the polynomial function $p(x) = 2x^5 - 3x^2 + 8x - 5$.

Solution $p'(x) = \dfrac{d}{dx}\left[2x^5 - 3x^2 + 8x - 5\right]$

$$= 2\frac{d}{dx}(x^5) - 3\frac{d}{dx}(x^2) + 8\frac{d}{dx}(x) - \frac{d}{dx}(5) \quad \textit{Extended linearity rule}$$

$$= 2(5x^4) - 3(2x) + 8(1) - 0 \qquad\qquad \textit{Power and constant rules}$$

$$= 10x^4 - 6x + 8$$

Example 4   Derivative of a product of polynomials

Differentiate $p(x) = (x^3 - 4x + 7)(3x^5 - x^2 + 6x)$.

Solution   We could expand the product function $p(x)$ as a polynomial and proceed as in Example 3, but it is easier to use the product rule.

$$p'(x) = (x^3 - 4x + 7)(3x^5 - x^2 + 6x)' + (x^3 - 4x + 7)'(3x^5 - x^2 + 6x)$$
$$= (x^3 - 4x + 7)(15x^4 - 2x + 6) + (3x^2 - 4)(3x^5 - x^2 + 6x)$$

This form is an acceptable answer, but if you are using software or an algebraic calculator, more than likely you will obtain the expanded formula

$$p'(x) = 24x^7 - 72x^5 + 100x^4 + 24x^3 + 12x^2 - 62x + 42$$

Example 5   Equation of a tangent line

Find the standard form equation for the line tangent to the graph of

$$f(x) = \frac{3x^2 + 5}{2x^2 + x - 3}$$

at the point where $x = -1$.

Solution   Evaluating $f(x)$ at $x = -1$, we find that $f(-1) = -4$ (verify); therefore, the point of tangency is $(-1, -4)$. The slope of the tangent line at $(-1, -4)$ is $f'(-1)$. Find $f'(x)$ by applying the quotient rule:

$$f'(x) = \frac{(2x^2 + x - 3)(3x^2 + 5)' - (3x^2 + 5)(2x^2 + x - 3)'}{(2x^2 + x - 3)^2}$$

$$= \frac{(2x^2 + x - 3)(6x) - (3x^2 + 5)(4x + 1)}{(2x^2 + x - 3)^2}$$

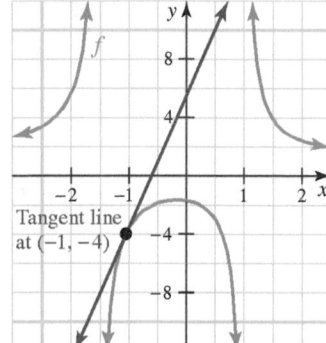

Thus, the slope of the tangent line is

$$f'(-1) = \frac{(2 - 1 - 3)(-6) - (3 + 5)(-4 + 1)}{(2 - 1 - 3)^2}$$

$$= \frac{(-2)(-6) - (8)(-3)}{(-2)^2}$$

$$= 9$$

From the formula (Theorem 3.1) $y = f'(x_0)(x - x_0) + f(x_0)$, we find that an equation for the tangent line at $(-1, -4)$ is

$$y = 9(x + 1) + (-4)$$

Figure 3.16 Graph of f and the tangent line at the point $(-1, -4)$

or, in standard form, $9x - y + 5 = 0$. The graphs of both f and its tangent line at $(-1, -4)$ are shown in Figure 3.16.

Example 6   Finding horizontal tangent lines

Let $y = (x - 2)(x^2 + 4x - 7)$. Find all points on this curve where the tangent line is horizontal.

Solution   The tangent line will be horizontal when $dy/dx = 0$, because the derivative dy/dx measures the slope and a horizontal line has slope 0 (see Figure 3.17).

Applying the product rule, we find

$$\frac{dy}{dx} = (x - 2)(x^2 + 4x - 7)' + (x - 2)'(x^2 + 4x - 7)$$

$$= (x - 2)(2x + 4) + (1)(x^2 + 4x - 7)$$

$$= 2x^2 - 8 + x^2 + 4x - 7$$

$$= 3x^2 + 4x - 15$$

$$= (3x - 5)(x + 3)$$

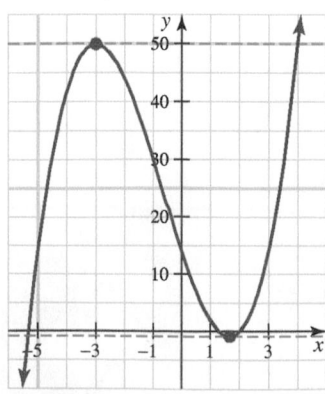

Figure 3.17 Graph of curve and the horizontal tangent lines

Thus, $\frac{dy}{dx} = 0$ when $x = \frac{5}{3}$ or $x = -3$. The corresponding points $\left(\frac{5}{3}, \frac{-22}{27}\right)$ and $(-3, 50)$ are the points on the curve at which the tangent line is horizontal.

In the following example, we use the quotient rule to extend the proof of the power rule to the case in which the exponent n is a negative integer.

Example 7 Proof of the power rule for negative exponents

Show that $\dfrac{d}{dx}(x^n) = nx^{n-1}$ if $n = -m$, where m is a positive integer.

Solution We have $f(x) = x^n = x^{-m} = 1/x^m$, so apply the quotient rule:

$$
\begin{aligned}
\frac{d}{dx}(x^n) &= \frac{d}{dx}\left(\frac{1}{x^m}\right) \\
&= \frac{x^m(1)' - (1)(x^m)'}{(x^m)^2} \\
&= \frac{x^m(0) - mx^{m-1}}{x^{2m}} \\
&= -mx^{(m-1)-2m} \\
&= -mx^{-m-1} \\
&= nx^{n-1} \qquad \text{(Substitute } -m = n)
\end{aligned}
$$

Higher-Order Derivatives

Occasionally, it is useful to differentiate the derivative of a function. In this context, we will refer to f' as the **first derivative** of f and to the derivative of f' as the **second derivative** of f. We could denote the second derivative by $(f')'$, but for simplicity we write f''. Other higher-order derivatives are defined and denoted similarly. Thus, the **third derivative** of f is the derivative of f'' and is denoted by f'''. In general, for $n > 3$, the **nth derivative** of f is denoted by $f^{(n)}$, for example $f^{(4)}$ or $f^{(5)}$. In Leibniz notation, higher-order derivatives for $y = f(x)$ are denoted as follows:

			Leibniz notation
First derivative:	y'	$f'(x)$	$\dfrac{dy}{dx}$ or $\dfrac{d}{dx}f(x)$
Second derivative	y''	$f''(x)$	$\dfrac{d}{dx}\left(\dfrac{dy}{dx}\right) = \dfrac{d^2y}{dx^2}$ or $\dfrac{d^2}{dx^2}f(x)$
Third derivative	y'''	$f'''(x)$	$\dfrac{d}{dx}\left(\dfrac{d^2y}{dx^2}\right) = \dfrac{d^3y}{dx^3}$ or $\dfrac{d^3}{dx^3}f(x)$
Fourth derivative	$y^{(4)}$	$f^{(4)}(x)$	$\dfrac{d^4y}{dx^4}$ or $\dfrac{d^4}{dx^4}f(x)$
$\vdots$	$\vdots$	$\vdots$	$\vdots$
nth derivative	$y^{(n)}$	$f^{(n)}(x)$	$\dfrac{d^ny}{dx^n}$ or $\dfrac{d^n}{dx^n}f(x)$

■ **W**hat this says Because the derivative of a function is a function, differentiation can be applied over and over, as long as the derivative itself is a differentiable function. That is, we can take derivatives of derivatives.

Notice also that for derivatives of higher orders than the third, the parentheses distinguish a derivative from a power. For example, $f^4 \neq f^{(4)}$.

You should note that all higher derivatives of a polynomial $p(x)$ will also be polynomials, and if p has degree n, then $p^{(k)}(x) = 0$ for $k \geq n + 1$, as illustrated in the following example.

Example 8 Higher derivatives for a polynomial function

Find all orders of derivatives of

$$p(x) = -2x^4 + 9x^3 - 5x^2 + 7$$

Solution $p'(x) = -8x^3 + 27x^2 - 10x$

$p''(x) = -24x^2 + 54x - 10$

$p'''(x) = -48x + 54$

$p^{(4)}(x) = -48$

$p^{(5)}(x) = 0, \cdots, p^{(k)}(x) = 0 \ (k \geq 5)$

PROBLEM SET 3.2

Level 1

To demonstrate the power of the theorems of this section, Problems 1-4 ask you to go back and rework some problems in Section 3.1, using the material of this section instead of the definition of derivative.

1. Find the derivatives of the functions given in Problems 11-16 of Problem Set 3.1.

2. Find the derivatives of the functions given in Problems 17-24 of Problem Set 3.1.

3. Find the derivatives of the functions given in Problems 25-29 of Problem Set 3.1.

4. Find the derivatives of the functions given in Problems 35-38 of Problem Set 3.1.

Differentiate the functions given in Problems 5-20. Assume that C is a constant.

5. a. $f(x) = 3x^4 - 9$ **b.** $g(x) = 3(9)^4 - x$

6. a. $f(x) = 5x^2 + x$ **b.** $g(x) = \pi^3$

7. a. $f(x) = x^3 + C$ **b.** $g(x) = C^2 + x$

8. a. $f(t) = 10t^{-1}$ **b.** $g(t) = \dfrac{7}{t}$

9. $r(t) = t^2 - \dfrac{1}{t^2} + \dfrac{5}{t^4}$ **10.** $f(x) = \pi^3 - 3\pi^2$

11. $f(x) = \dfrac{7}{x^2} + x^{2/3} + C$ **12.** $g(x) = \dfrac{1}{2\sqrt{x}} + \dfrac{x^2}{4} + C$

13. $f(x) = \dfrac{x^3 + x^2 + x - 7}{x^2}$ **14.** $g(x) = \dfrac{2x^5 - 3x^2 + 11}{x^3}$

15. $f(x) = (2x + 1)(1 - 4x^3)$

16. $g(x) = (x + 2)\left(2\sqrt{x} + x^2\right)$

17. $f(x) = \dfrac{3x + 5}{x + 9}$ **18.** $f(x) = \dfrac{x^2 + 3}{x^2 + 5}$

19. $g(x) = x^2(x + 2)^2$ **20.** $f(x) = x^2(2x + 1)^2$

In Problems 21-24, find f', f'', f''', and $f^{(4)}$.

21. $f(x) = x^5 - 5x^3 + x + 12$

22. $f(x) = \frac{1}{4}x^8 - \frac{1}{2}x^6 - x^2 + 2$

23. $f(x) = \dfrac{-2}{x^2}$

24. $f(x) = \dfrac{4}{\sqrt{x}}$

25. Find $\dfrac{d^2y}{dx^2}$, where $y = 3x^3 - 7x^2 + 2x - 3$.

26. Find $\dfrac{d^2y}{dx^2}$, where $y = (x^2 + 4)(1 - 3x^3)$.

In Problems 27-32, find the standard form equation for the tangent line to $y = f(x)$ at the specified point.

27. $f(x) = x^2 - 3x - 5$, where $x = -2$

28. $f(x) = x^5 - 3x^3 - 5x + 2$, where $x = 1$

29. $f(x) = (x^2 + 1)(1 - x^3)$, where $x = 1$

30. $f(x) = \dfrac{x + 1}{x - 1}$, where $x = 0$

31. $f(x) = \dfrac{x^2 + 5}{x + 5}$, where $x = 1$

32. $f(x) = 1 - \dfrac{1}{x} + \dfrac{2}{\sqrt{2}}$, where $x = 4$

Find the coordinates of each point on the graph of the given function where the tangent line is horizontal in Problems 33-38.

33. $f(x) = 2x^3 - 7x^2 + 8x - 3$

34. $g(x) = (3x - 5)(x - 8)$

35. $f(t) = \dfrac{1}{t^2} - \dfrac{1}{t^3}$ **36.** $f(x) = \sqrt{x}(x - 3)$

37. $h(u) = \dfrac{1}{\sqrt{u}}(u + 9)$ **38.** $h(x) = \dfrac{4x^2 + 12x + 9}{2x + 3}$

Level 2

39. a. Differentiate the function $f(x) = 2x^2 - 5x - 3$

   b. Factor the function in part **a** and differentiate by using the product rule. Show that the two answers are the same.

40. a. Use the quotient rule to differentiate $f(x) = \dfrac{2x - 3}{x^3}$.

   b. Rewrite the function in part **a** as $f(x) = x^{-3}(2x - 3)$ and differentiate by using the product rule.

   c. Rewrite the function in part **a** as
   $f(x) = 2x^{-2} - 3x^{-3}$ and differentiate.

   d. Show that the answers to parts **a, b,** and **c** are all the same.

41. Find equations for the tangent line to the curve with equation $y = x^4 - 2x + 1$ that is parallel to the line $2x - y - 3 = 0$.

42. Find equations for two tangent lines to the graph of $f(x) = \dfrac{3x + 5}{1 + x}$ that are perpendicular to the line $2x - y = 1$.

43. Let $f(x) = (x^3 - 2x^2)(x + 2)$.

   a. Find an equation for the tangent line to the graph of f at the point where $x = 1$. (See Figure 3.18.)

   b. Find an equation for the normal line to the graph of f at the point where $x = 0$.

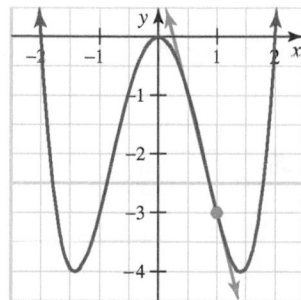

Figure 3.18 Graph of $f(x) = (x^3 - 2x^2)(x + 2)$ with tangent line at $x = 1$

44. Find an equation for a normal line to the graph of $f(x) = (x^3 - 2x^2)(x + 2)$ that is parallel to the line $x - 16y + 17 = 0$.

45. Find all points (x, y) on the graph of $y = 4x^2$ with the property that the tangent line at (x, y) passes through the point $(2, 0)$.

Determine which (if any) of the functions $y = f(x)$ given in Problems 46-49 satisfy the equation

$$y''' + y'' + y' = x + 1$$

46. $f(x) = x^2 + 2x - 3$

47. $f(x) = x^3 + x^2 + x$

48. $f(x) = \dfrac{1}{2}x^2 + 3$

49. $f(x) = 2x^2 + x$

50. *Historical Quest When working with rational expressions, we need to be careful about division by zero. One of the earliest recorded treatments of division by zero is attributed to the Hindu mathematician Āryabhata (476-550), in whose honor the first Indian satellite was named. He also gave rules for approximations of square roots and sums of arithmetic progressions as well as rules for basic algebraic manipulations. One example of his work is the following calculation for π: "Add four to one hundred, multiply by eight and add again sixty-two thousand; the result is the approximate value of the circumference of a circle whose diameter is twenty-thousand." Follow the steps of Āryabhata's approximation for π. After you have completed this demonstration, discuss the procedure and technology you used and contrast it with the tools that Āryabhata must have had available.*

Level 3

51. What is the relationship between the degree of a polynomial function P and the value of k for which $P^{(k)}(x)$ is first equal to 0?

52. Prove the constant multiple rule $(cf)' = cf'$.

53. Prove the sum rule $(f + g)' = f' + g'$.

54. Use the definition of the derivative to find the derivative of f^2, given that f is a differentiable function.

55. Prove the product rule by using the result of Problem 54 and the identity

$$fg = \frac{1}{2}\left[(f + g)^2 - f^2 - g^2\right]$$

56. Prove the quotient rule

$$\left(\frac{f}{g}\right)' = \frac{gf' - fg'}{g^2}$$

where $g(x) \neq 0$. *Hint*: First show that the difference quotient for f/g can be expressed as

$$\frac{\frac{f}{g}(x + \Delta x) - \frac{f}{g}(x)}{\Delta x}$$

$$= \frac{f(x + \Delta x)g(x) - f(x)g(x + \Delta x)}{(\Delta x)\,g(x + \Delta x)g(x)}$$

and then subtract and add the term $g(x)f(x)$ in the numerator.

57. Show that the reciprocal function

$$r(x) = \frac{1}{f(x)}$$

has the derivative

$$r'(x) = -\frac{f'(x)}{\left[f(x)\right]^2}$$

at each point x where f is differentiable and $f(x) \neq 0$.

58. If f, g, and h are differentiable functions, show that the product fgh is also differentiable and

$$(fgh)' = fgh' + fg'h + f'gh$$

59. Let f be a function that is differentiable at x.

   a. If $g(x) = \left[f(x) \right]^3$, show that

$$g'(x) = 3 \left[f(x) \right]^2 f'(x)$$

     Hint: Write $g(x) = \left[f(x) \right]^2$ and use the product rule.

   b. Show that $p(x) = \left[f(x) \right]^4$ has the derivative $p'(x) = 4 \left[f(x) \right]^3 f'(x)$.

60. Find constants A, B, and C so that $y = Ax^3 + Bx + C$ satisfies the equation

$$y''' + 2y'' - 3y' + y = x$$

3.3 DERIVATIVES OF TRIGONOMETRIC, EXPONENTIAL, AND LOGARITHMIC FUNCTIONS

IN THIS SECTION: *Derivatives of the sine and cosine functions, derivatives of the other trigonometric functions, derivatives of exponential and logarithmic functions*

In the last section, we developed general rules for differentiation and applied those rules to various algebraic functions involving power functions, polynomials, and rational functions. In this section, we develop differentiation formulas for the trigonometric, exponential, and logarithmic functions.

Derivatives of the Sine and Cosine Functions

In calculus we assume that the trigonometric functions are functions of real numbers or of angles measured in radians. We make this assumption because the trigonometric differentiation formulas rely on limit formulas that become complicated if degree measurement is used instead of radian measure.

We begin by finding the derivatives of the sine and cosine functions. This requires two limits established earlier in Theorem 2.3.

$$\lim_{h \to 0} \frac{\sin h}{h} = 1 \qquad \lim_{h \to 0} \frac{\cos h - 1}{h} = 0$$

Theorem 3.6   **Derivatives of the sine and cosine functions**

The functions $\sin x$ and $\cos x$ are differentiable for all x and

$$\frac{d}{dx} \sin x = \cos x \qquad \frac{d}{dx} \cos x = -\sin x$$

Proof: The proofs of these two formulas are similar. We will prove the first using the trigonometric identity

$$\sin(\alpha + \beta) = \sin \alpha \, \cos \beta + \cos \alpha \, \sin \beta$$

and leave the proof of the second formula as a problem. From the definition of the derivative

$$\frac{d}{dx} \sin x = \lim_{\Delta x \to 0} \frac{\sin(x + \Delta x) - \sin x}{\Delta x}$$

$$= \lim_{\Delta x \to 0} \frac{\sin x \, \cos \Delta x + \cos x \, \sin \Delta x - \sin x}{\Delta x}$$

$$= \lim_{\Delta x \to 0} \left[\frac{\sin x \, \cos \Delta x - \sin x}{\Delta x} + \frac{\cos x \, \sin \Delta x}{\Delta x} \right]$$

$$= \lim_{\Delta x \to 0} \left[\sin x \frac{\cos \Delta x - 1}{\Delta x} + \cos x \frac{\sin \Delta x}{\Delta x} \right]$$

$$= (\sin x) \lim_{\Delta x \to 0} \left(\frac{\cos \Delta x - 1}{\Delta x} \right) + (\cos x) \lim_{\Delta x \to 0} \left(\frac{\sin \Delta x}{\Delta x} \right)$$

$$= (\sin x)(0) + (\cos x)(1)$$

$$= \cos x$$

♦

Example 1   Derivative involving a trigonometric function

Differentiate $f(x) = 2x^4 + 3\cos x + \sin a$, for constant a.

Solution

$$f'(x) = \frac{d}{dx} \left(2x^4 + 3\cos x + \sin a \right)$$

$$= 2\frac{d}{dx} \left(x^4 \right) + 3\frac{d}{dx} (\cos x) + \frac{d}{dx} (\sin a) \quad \textit{Extended linearity rule}$$

$$= 2(4x^3) + 3(-\sin x) + 0 \qquad \textit{Power rule, derivative of cosine, derivative of a constant}$$

$$= 8x^3 - 3\sin x$$

Example 2   Derivative of a trigonometric function with product rule

Differentiate $f(x) = x^2 \sin x$.

Solution $f'(x) = \frac{d}{dx} \left(x^2 \sin x \right)$

$$= x^2 \frac{d}{dx} (\sin x) + (\sin x) \frac{d}{dx} \left(x^2 \right) \qquad \textit{Product rule}$$

$$= x^2 \cos x + 2x \sin x \qquad \textit{Power rule and derivative of sine}$$

Example 3   Derivative of a trigonometric function with quotient rule

Differentiate $h(t) = \dfrac{\sqrt{t}}{\cos t}$.

Solution   Write $\sqrt{t}$ as $t^{1/2}$. Then

$$h'(t) = \frac{d}{dt} \left[\frac{t^{1/2}}{\cos t} \right]$$

$$= \frac{\cos t \frac{d}{dt} \left(t^{1/2} \right) - t^{1/2} \frac{d}{dt} \cos t}{\cos^2 t} \qquad \textit{Quotient rule}$$

$$= \frac{\frac{1}{2} t^{-1/2} \cos t - t^{1/2} (-\sin t)}{\cos^2 t} \qquad \textit{Power rule and derivative of sine}$$

$$= \frac{\frac{1}{2} t^{-1/2} (\cos t + 2t \sin t)}{\cos^2 t} \qquad \textit{Common factor } \frac{1}{2} t^{-1/2}$$

$$= \frac{\cos t + 2t \sin t}{2\sqrt{t} \, \cos^2 t}$$

Differentiation of the Other Trigonometric Functions

You will need to be able to differentiate not only the sine and cosine functions, but also the other trigonometric functions. To find the derivatives of these functions you will need the following identities (which are called the **eight fundamental identities**):

$$\tan x = \frac{\sin x}{\cos x} \qquad\qquad \cot x = \frac{\cos x}{\sin x}$$

$$\sec x = \frac{1}{\cos x} \qquad \csc x = \frac{1}{\sin x} \qquad \cot x = \frac{1}{\tan x}$$

$$\cos^2 x + \sin^2 x = 1 \quad 1 + \tan^2 x = \sec^2 x \quad \cot^2 x + 1 = \csc^2 x$$

Theorem 3.7   Derivatives of the trigonometric functions

The six basic trigonometric functions $\sin x$, $\cos x$, $\tan x$, $\csc x$, $\sec x$, and $\cot x$ are all differentiable wherever they are defined, and

Trigonometric formulas use radian measure unless otherwise stated.

$$\frac{d}{dx}\sin x = \cos x \qquad \frac{d}{dx}\sec x = \sec x \tan x \qquad \frac{d}{dx}\tan x = \sec^2 x$$

$$\frac{d}{dx}\cos x = -\sin x \qquad \frac{d}{dx}\csc x = -\csc x \cot x \qquad \frac{d}{dx}\cot x = -\csc^2 x$$

Proof: The derivatives for sine and cosine were given in Theorem 3.6. All the other derivatives in this theorem are proved by using the appropriate quotient rules along with formulas for the derivatives of the sine and cosine. We will obtain the derivative of the tangent function and leave the rest as problems (see Problems 57-59).

$$\frac{d}{dx}\tan x = \frac{d}{dx}\left(\frac{\sin x}{\cos x}\right) \qquad\qquad\quad \textit{Trigonometric identity}$$

$$= \frac{\cos x \frac{d}{dx}(\sin x) - \sin x \frac{d}{dx}(\cos x)}{\cos^2 x} \qquad \textit{Quotient rule}$$

$$= \frac{\cos x (\cos x) - \sin x (-\sin x)}{\cos^2 x} \qquad \textit{Derivative of } \sin x \text{ and } \cos x$$

$$= \frac{\cos^2 x + \sin^2 x}{\cos^2 x}$$

$$= \frac{1}{\cos^2 x} \qquad\qquad\qquad\quad \textit{Trigonometric identity}$$

$$= \sec^2 x \qquad\qquad\qquad\quad \textit{Trigonometric identity} \qquad \blacklozenge$$

Example 4   Derivative of a trigonometric function with the product rule

Differentiate $f(\theta) = 3\theta \sec \theta$.

Solution $f'(\theta) = \dfrac{d}{d\theta}(3\theta \sec \theta)$

$$= 3\theta \frac{d}{d\theta}\sec\theta + \sec\theta \frac{d}{d\theta}(3\theta) \quad \textit{Product rule}$$

$$= 3\theta \sec\theta \tan\theta + 3\sec\theta$$

Example 5   Derivative of a product of trigonometric functions

Differentiate $f(x) = \sec x \tan x$.

Solution $f'(x) = \dfrac{d}{dx}(\sec x \tan x)$

$$= \sec x \frac{d}{dx}(\tan x) + \tan x \frac{d}{dx}(\sec x) \quad \textit{Product rule}$$

$$= \sec x \left(\sec^2 x\right) + \tan x \left(\sec x \tan x\right) \quad \textit{Derivative of } \tan x \textit{ and } \sec x$$
$$= \sec^3 x + \sec x \tan^2 x$$
$$= \sec^3 x + \sec x \left(\sec^2 x - 1\right)$$
$$= 2\sec^3 x - \sec x.$$

Example 6   Equation of a tangent line involving a trigonometric function

Find the equation of the tangent line to the curve $y = \cot x - 2 \csc x$ at the point where $x = \frac{2\pi}{3}$.

Solution   When $x = \frac{2\pi}{3}$, we have

$$\cot \frac{2\pi}{3} - 2 \csc \frac{2\pi}{3} = \frac{-\sqrt{3}}{3} - 2\left(\frac{2\sqrt{3}}{3}\right) = \frac{-5\sqrt{3}}{3}$$

so the point of tangency is $P\left(\frac{2\pi}{3}, \frac{-5\sqrt{3}}{3}\right)$. To find the slope of the tangent line at P, we first compute the derivative $\frac{dy}{dx}$:

$$\frac{dy}{dx} = \frac{d}{dx}\left(\cot x - 2 \csc x\right)$$
$$= \frac{d}{dx}\cot x - 2\frac{d}{dx}\csc x \qquad \textit{Linearity rule}$$
$$= -\csc^2 x - 2\left(-\csc x \cot x\right)$$
$$= 2\csc x \cot x - \csc^2 x$$

Then the slope of the tangent line is given by

$$\left.\frac{dy}{dx}\right|_{x=2\pi/3} = 2\csc \frac{2\pi}{3} \cot \frac{2\pi}{3} - \csc^2 \frac{2\pi}{3}$$
$$= 2\left(\frac{2}{\sqrt{3}}\right)\left(\frac{-1}{\sqrt{3}}\right) - \left(\frac{2}{\sqrt{3}}\right)^2$$
$$= -\frac{8}{3}$$

so the tangent line at $\left(\frac{2\pi}{3}, \frac{-5\sqrt{3}}{3}\right)$ with slope $-\frac{8}{3}$ is

$$y + \frac{5\sqrt{3}}{3} = -\frac{8}{3}\left(x - \frac{2\pi}{3}\right)$$
$$24x + 9y + 15\sqrt{3} - 16\pi = 0$$

Derivatives of Exponential and Logarithmic Functions

The next theorem, which is easy to prove and remember, is one of the most important results in all of differential calculus.

Theorem 3.8   Derivative rule for the natural exponential

The natural exponential function e^x is differentiable for all x, with derivative

$$\frac{d}{dx}\left(e^x\right) = e^x$$

Proof: We will proceed informally. Recall the definition of e:

$$\lim_{n \to \infty} \left(1 + \frac{1}{n}\right)^n = e$$

Let $n = \frac{1}{\Delta x}$, so that $\lim_{\Delta x \to 0} (1 + \Delta x)^{1/\Delta x} = e$. This means that for Δx very small, $e \approx (1 + \Delta x)^{1/\Delta x}$ or $e^{\Delta x} \approx 1 + \Delta x$ so that $e^{\Delta x} - 1 \approx \Delta x$. Thus, $\lim_{x \to 0} \dfrac{e^{\Delta x} - 1}{\Delta x} = 1$. Finally, using the limit in the definition of derivative for e^x, we obtain

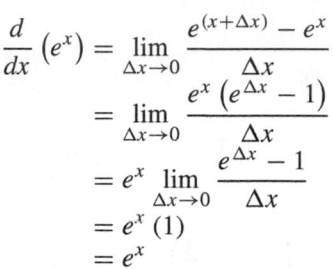

$$
\begin{aligned}
\frac{d}{dx}\left(e^x\right) &= \lim_{\Delta x \to 0} \frac{e^{(x + \Delta x)} - e^x}{\Delta x} \\
&= \lim_{\Delta x \to 0} \frac{e^x \left(e^{\Delta x} - 1\right)}{\Delta x} \\
&= e^x \lim_{\Delta x \to 0} \frac{e^{\Delta x} - 1}{\Delta x} \\
&= e^x \,(1) \\
&= e^x
\end{aligned}
$$

         ♦

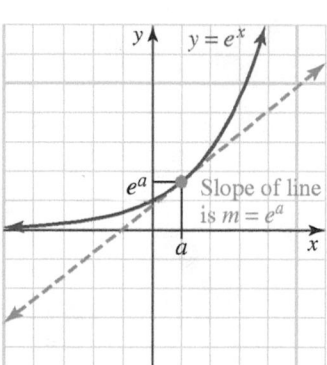

Figure 3.19 The slope of $y = e^x$ at each point (a, e^a) is $m = e^a$

Note: An easier verification of the derivative formula in Theorem 3.8 is given in Example 6 of Section 3.6 using methods developed in that section.

     The fact that $\frac{d}{dx}\left(e^x\right) = e^x$ means that the slope of the graph of $y = e^x$ at any point $x = a$ is $m = e^a$, the y-coordinate of the point, as shown in Figure 3.19. This is one of the features of the exponential function $y = e^x$ that makes it "natural."

Example 7   A derivative involving e^x

Differentiate $f(x) = x^2 e^x$. For what values of x does $f'(x) = 0$?

Solution   Using the product rule, we find

$$
\begin{aligned}
f'(x) &= \frac{d}{dx}\left(x^2 e^x\right) \\
&= x^2 \left[\frac{d}{dx} e^x\right] + \left[\frac{d}{dx} x^2\right] e^x \quad \text{\textit{Product rule}} \\
&= x^2 e^x + 2x e^x \quad\quad\quad\quad \text{\textit{Exponential and power rules}}
\end{aligned}
$$

To find where $f'(x) = 0$, we solve

$$
\begin{aligned}
x^2 e^x + 2x e^x &= 0 \\
x(x + 2)e^x &= 0 \\
x(x + 2) &= 0 \quad\quad \text{\textit{Since } } e^x \neq 0 \text{ \textit{for all} } x. \\
x &= 0, -2
\end{aligned}
$$

    ■

Example 8   A second derivative involving e^x

For $f(x) = e^x \sin x$, find $f'(x)$ and $f''(x)$.

Solution   Apply the product rule twice:

$$
\begin{aligned}
f'(x) &= e^x (\sin x)' + \left(e^x\right)' \sin x \\
&= e^x (\cos x) + e^x (\sin x) \\
&= e^x (\cos x + \sin x) \\
f''(x) &= e^x (\cos x + \sin x)' + \left(e^x\right)' (\cos x + \sin x) \\
&= e^x (-\sin x + \cos x) + e^x (\cos x + \sin x) \\
&= 2 e^x \cos x
\end{aligned}
$$

    ■

The following limit rule for the natural logarithm functions follows from the composition limit rule and is proved in advanced calculus.

LIMIT OF A LOG   If g is a continuous function and $b > 1$, then

$$\lim_{x \to \infty} \log_b g(x) = \log_b \left(\lim_{x \to \infty} g(x) \right)$$

■ **W**hat this says The limit of a log is the log of the limit.

Theorem 3.9   Derivative rule for the natural logarithm function

The natural logarithmic function $\ln x$ is differentiable for all $x > 0$, with derivative

$$\frac{d}{dx} (\ln x) = \frac{1}{x}$$

Proof: According to the definition of the derivative, we have

$$\frac{d}{dx} (\ln x) = \lim_{\Delta x \to 0} \frac{\ln (x + \Delta x) - \ln x}{\Delta x}$$

$$= \lim_{\Delta x \to 0} \frac{1}{\Delta x} \ln \left(\frac{x + \Delta x}{x} \right)$$

$$= \lim_{\Delta x \to 0} \frac{1}{\Delta x} \ln \left(1 + \frac{\Delta x}{x} \right) \qquad \text{\textit{Let }} h = \frac{x}{\Delta x}, \text{\textit{ so }} \Delta x = \frac{x}{h} \text{ \textit{and} } h \to \infty \text{ \textit{as} } \Delta x \to 0.$$

$$= \lim_{h \to \infty} \frac{h}{x} \ln \left(1 + \frac{1}{h} \right)$$

$$= \frac{1}{x} \left[\lim_{h \to \infty} h \ln \left(1 + \frac{1}{h} \right) \right]$$

$$= \frac{1}{x} \left[\lim_{h \to \infty} \ln \left(1 + \frac{1}{h} \right)^h \right] \qquad \textit{Power rule for logarithms}$$

$$= \frac{1}{x} \ln \left[\lim_{h \to \infty} \left(1 + \frac{1}{h} \right)^h \right] \qquad \textit{Since } g(x) = \left(1 + \frac{1}{h} \right)^h \textit{ is continuous for } h > 0,$$

$$\qquad\qquad\qquad\qquad\qquad\qquad\qquad \textit{the limit of the log is the log of the limit.}$$

$$= \frac{1}{x} \ln e \qquad\qquad\qquad \textit{Definition of } e$$

$$= \frac{1}{x} \qquad\qquad\qquad\qquad \ln e = 1 \qquad\qquad\qquad ◆$$

Example 9   Derivative of a quotient involving a natural logarithm

Differentiate $f(x) = \dfrac{\ln x}{\sin x}$.

Solution   We use the quotient rule.

$$f'(x) = \frac{(\sin x) \frac{d}{dx} (\ln x) - (\ln x) \frac{d}{dx} (\sin x)}{\sin^2 x}$$

$$= \frac{(\sin x) \left(\frac{1}{x} \right) - (\ln x)(\cos x)}{\sin^2 x}$$

$$= \frac{\sin x - x \ln x \cos x}{x \sin^2 x}$$

PROBLEM SET 3.3

Level 1

Differentiate the functions given in Problems 1-32.

1. $f(x) = \sin x + \cos x$

2. $f(x) = 2\sin x + \tan x$

3. $g(t) = t^2 + \cos t + \cos \dfrac{\pi}{4}$

4. $g(t) = 2\sec t + 3\tan t - \tan \dfrac{\pi}{3}$

5. $f(t) = \sin^2 t$ *Hint*: Use the product rule.

6. $g(x) = \cos^2 x$ *Hint*: Use the product rule.

7. $f(x) = \sqrt{x}\,\cos x + x\,\cot x$

8. $f(x) = 2x^3 \sin x - 3x\,\cos x$

9. $p(x) = x^2 \cos x$ **10.** $p(t) = (t^2 + 2)\sin t$

11. $q(x) = \dfrac{\sin x}{x}$ **12.** $r(x) = \dfrac{e^x}{\sin x}$

13. $h(t) = e^t \csc t$

14. $f(\theta) = \dfrac{\sec \theta}{2 - \cos \theta}$

15. $f(x) = x^2 \ln x$

16. $g(x) = \dfrac{\ln x}{x^2}$

17. $h(x) = e^x(\sin x - \cos x)$

18. $f(x) = \dfrac{\ln x}{x}$ **19.** $f(x) = \dfrac{\sin x}{e^x}$

20. $g(x) = \dfrac{x\,\cos x}{e^x}$ **21.** $f(x) = \dfrac{\tan x}{1 - 2x}$

22. $g(t) = \dfrac{1 + \sin t}{\sqrt{t}}$ **23.** $f(t) = \dfrac{2 + \sin t}{t + 2}$

24. $f(\theta) = \dfrac{\theta - 1}{2 + \cos \theta}$ **25.** $f(x) = \dfrac{\sin x}{1 - \cos x}$

26. $f(x) = \dfrac{x}{1 - \sin x}$ **27.** $f(x) = \dfrac{1 + \sin x}{2 - \cos x}$

28. $g(x) = \dfrac{\cos x}{1 + \cos x}$ **29.** $f(x) = \dfrac{\sin x + \cos x}{\sin x - \cos x}$

30. $f(x) = \dfrac{x^2 + \tan x}{3x + 2\tan x}$

31. $g(x) = \sec^2 x - \tan^2 x + \cos x$

32. $g(x) = \cos^2 x + \sin^2 x + \sin x$

Find the second derivative of each function given in Problems 33-44.

33. $f(\theta) = \sin \theta$ **34.** $f(\theta) = \cos \theta$

35. $f(\theta) = \tan \theta$ **36.** $f(\theta) = \cot \theta$

37. $f(\theta) = \sec \theta$ **38.** $f(\theta) = \csc \theta$

39. $f(x) = \sin x + \cos x$ **40.** $f(x) = x\sin x$

41. $f(x) = e^x \cos x$ **42.** $g(t) = t^3 e^t$

43. $h(t) = \sqrt{t}\,\ln t$ **44.** $f(t) = \dfrac{\ln t}{t}$

Level 2

Find an equation for the tangent line at the prescribed point for each function in Problems 45-52.

45. $f(\theta) = \tan \theta$ at $\left(\dfrac{\pi}{4}, 1\right)$

46. $f(\theta) = \sec \theta$ at $\left(\dfrac{\pi}{3}, 2\right)$

47. $f(x) = \sin x$, where $x = \dfrac{\pi}{6}$

48. $f(x) = \cos x$, where $x = \dfrac{\pi}{3}$

49. $y = x + \sin x$, where $x = 0$

50. $y = x\sec x$, where $x = 0$

51. $y = e^x \cos x$, where $x = 0$

52. $y = x\ln x$, where $x = 1$

53. For what values of A and B does

$$y = A\cos x + B\sin x$$

satisfy $y'' + 2y' + 3y = 2\sin x$?

54. For what values of A and B does

$$y = Ax\cos x + Bx\sin x$$

satisfy $y'' + y = -3\cos x$?

Level 3

55. Think Tank Problem Give an example of a differentiable function with a discontinuous derivative.

56. Complete the proof of Theorem 3.6 by showing that
$$\dfrac{d}{dx}\cos x = -\sin x.$$
Hint: You will need to use the identity
$$\cos(\alpha + \beta) = \cos \alpha \cos \beta - \sin \alpha \sin \beta.$$

Prove the requested parts of Theorem 3.7 in Problems 57-59.

57. $\dfrac{d}{dx}\cot x = -\csc^2 x$

58. $\dfrac{d}{dx}\sec x = \sec x \tan x$

59. $\dfrac{d}{dx}\csc x = -\csc x \cot x$

60. Use the limit of a difference quotient to prove that

$$\dfrac{d}{dx}\tan x = \sec^2 x$$

3.4 RATES OF CHANGE: MODELING RECTILINEAR MOTION

IN THIS SECTION: *Average and instantaneous rate of change, introduction to mathematical modeling, rectilinear motion (modeling in physics), falling body problems*
The derivative can be interpreted as a rate of change, which leads to a wide variety of applications. Viewed as rates of change, derivatives may represent such quantities as the speed of a moving object, the rate at which a population grows, a manufacturer's marginal cost, the rate of inflation, or the rate at which natural resources are being depleted.

In this section, we begin by showing how rates of change can be computed using differentiation. Then we discuss the general notion of mathematical modeling and conclude by demonstrating modeling principles in a variety of practical problems involving motion along a line.

Average and Instantaneous Rate of Change

The graph of a linear function $f(x) = ax + b$ is the line $y = ax + b$, whose slope $m = a$ can be thought of as the rate at which y is changing with respect to x (see Figure 3.20**a**). However, for another function g that is *not* linear, the average (and instantaneous) rate of change of $y = g(x)$ with respect to x varies from point to point, as shown in Figure 3.20**b**.

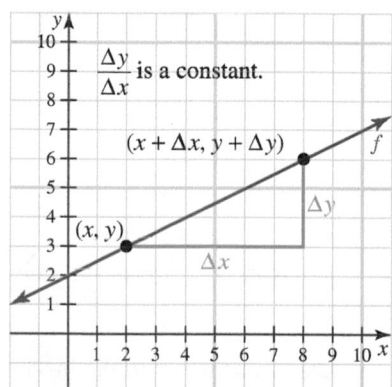

a. Linear function; rate of change $\Delta y / \Delta x$ is constant

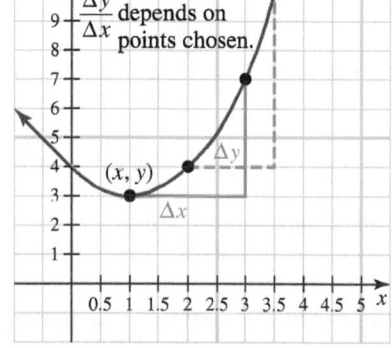

b. Nonlinear function; rate of change $\Delta y / \Delta x$ is variable

Figure 3.20 Rate of change is measured by the slope of a tangent line

Because the slope of the tangent line is given by the derivative of the function, the preceding geometric observations suggest that the rate of change of a function is measured by its derivative. This connection will be made more precise in the following discussion.

Suppose y is a function of x, say $y = f(x)$. Corresponding to a change from x to $x + \Delta x$, the variable y changes from $f(x)$ to $f(x + \Delta x)$. The change in y is $\Delta y = f(x + \Delta x) - f(x)$, and the **average rate of change** of y with respect to x is

$$\text{AVERAGE RATE OF CHANGE} = \frac{\text{CHANGE IN } y}{\text{CHANGE IN } x} = \frac{\Delta y}{\Delta x} = \frac{f(x + \Delta x) - f(x)}{\Delta x}$$

This formula for the change in y is important. Find Δy in Figure 3.21.

As the interval over which we are averaging becomes shorter (that is, as $\Delta x \to 0$), the average rate of change approaches what we would intuitively call the *instantaneous*

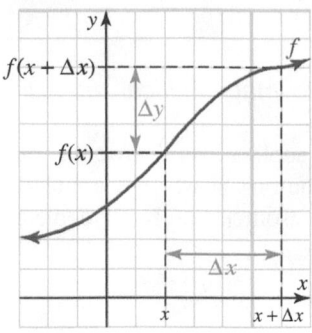

Figure 3.21   A change in Δy corresponding to a change Δx

rate of change of y with respect to x, and the difference quotient approaches the derivative $\dfrac{dy}{dx}$. Thus, we have

$$\text{INSTANTANEOUS RATE OF CHANGE} = \lim_{\Delta x \to 0} \frac{\Delta y}{\Delta x} = \lim_{\Delta x \to 0} \frac{f(x + \Delta x) - f(x)}{\Delta x} = f'(x)$$

To summarize:

INSTANTANEOUS RATE OF CHANGE   Suppose $f(x)$ is differentiable at $x = x_0$. Then the **instantaneous rate of change** of $y = f(x)$ with respect to x at x_0 is the value of the derivative of f at x_0. That is,

$$\text{INSTANTANEOUS RATE OF CHANGE} = f'(x_0) = \left.\frac{dy}{dx}\right|_{x=x_0}$$

Example 1   Instantaneous rate of change

Find the rate at which the function $y = x^2 \sin x$ is changing with respect to x when $x = \pi$.

Solution   For any x, the instantaneous rate of change is the derivative,

$$\frac{dy}{dx} = 2x \sin x + x^2 \cos x$$

Thus, the rate when $x = \pi$ is

$$\left.\frac{dy}{dx}\right|_{x=\pi} = 2\pi \sin \pi + \pi^2 \cos \pi = 2\pi (0) + \pi^2 (-1) = -\pi^2$$

The negative sign indicates that when $x = \pi$, the function is *decreasing* at the rate of $\pi^2 \approx 9.9$ units of y for each one-unit increase in x.

    Let us consider an example comparing the average rate of change with the instantaneous rate of change.

Example 2   Comparison between average rate and instantaneous rate of change

Let $f(x) = x^2 - 4x + 7$.

a. Find the average rate of change of f with respect to x between $x = 3$ and 5.
b. Find the instantaneous rate of change of f at $x = 3$.

Solution

a. The (average) rate of change from $x = 3$ to $x = 5$ is found by dividing the change in f by the change in x. The change in f from $x = 3$ to $x = 5$ is

$$f(5) - f(3) = \left[5^2 - 4(5) + 7\right] - \left[3^2 - 4(3) + 7\right] = 8$$

Thus, the average rate of change is

$$\frac{f(5) - f(3)}{5 - 3} = \frac{8}{2} = 4$$

The slope of the secant line is 4, as shown in black in Figure 3.22.

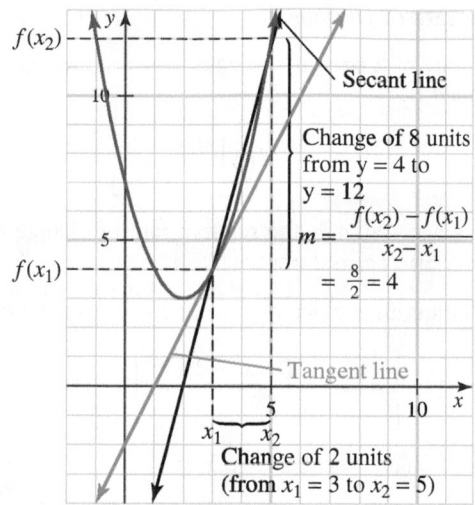

Figure 3.22 Comparison of rates of change
Average rate of change (black line) and instantaneous rate of change (blue line)

b. The derivative of the function is

$$f'(x) = 2x - 4$$

Thus, the instantaneous rate of change of f at $x = 3$ is

$$f'(3) = 2(3) - 4 = 2$$

The tangent line at $x = 3$ has slope 2, as shown in blue in Figure 3.22.

Often we are not as interested in the instantaneous rate of change of a quantity as its relative rate of change, defined as follows:

$$\text{RELATIVE RATE OF CHANGE} = \frac{\text{INSTANTANEOUS RATE OF CHANGE}}{\text{SIZE OF QUANTITY}}$$

For instance, if you are earning \$15,000/yr and receive a \$3,000 raise, you would probably be very pleased. However, that \$3,000 raise might be much less meaningful if you were earning \$100,000/yr. In both cases, the change is the quantity \$3,000, but in the first case the relative rate of change is $3,000/15,000 = 20\%$ whereas in the second, it is a much smaller $3,000/100,000 = 3\%$.

In terms of the derivative, the relative rate of change may be defined as follows:

RELATIVE RATE OF CHANGE The **relative rate of change** of $y = f(x)$ at $x = x_0$ is given by the ratio

$$\text{RELATIVE RATE OF CHANGE} = \frac{f'(x_0)}{f(x_0)}$$

Example 3   Relative rate of change

The aerobic rate is the rate of a person's oxygen consumption and is sometimes modeled by the function A defined by

$$A(x) = 110\left[\frac{\ln x - 2}{x}\right]$$

where x is the person's age. What is the relative rate of change of the aerobic rating of a 20-year old person? A 50-year old person?

Solution   Applying the quotient rule, we obtain the derivative of $A(x)$

$$A'(x) = 110\left[\frac{x\left(\frac{1}{x}\right) - (\ln x - 2)(1)}{x^2}\right] = 110\left[\frac{3 - \ln x}{x^2}\right]$$

Thus, the relative rate of change of A at age x is:

$$\text{RELATIVE RATE OF CHANGE} = \frac{A'(x)}{A(x)} = \frac{110\left[\frac{3-\ln x}{x^2}\right]}{110\left[\frac{\ln x - 2}{x}\right]} = \frac{3 - \ln x}{(\ln x - 2)x}$$

In particular, when $x = 20$, the relative rate of change is

$$\frac{A'(20)}{A(20)} = \frac{3 - \ln 20}{(\ln 20 - 2)20} \approx 0.0002143$$

In other words, at age 20, the person's aerobic rating is increasing at the rate of 0.02% per year. However, at age 50, the relative rate of change is

$$\frac{A'(50)}{A(50)} = \frac{3 - \ln 50}{(\ln 50 - 2)50} \approx -0.0095399$$

and the person's aerobic rating is *decreasing* (because of the negative sign) at the relative rate of 0.95% per year.   ◾

Introduction to Mathematical Modeling

A real-life situation is usually far too complicated to be precisely and mathematically defined. When confronted with a problem in the real world, therefore, it is usually necessary to develop a mathematical framework based on certain assumptions about the real world. This framework can then be used to find a solution to the real-world problem. The process of developing this body of mathematics is referred to as **mathematical modeling**. We introduced mathematical models in Chapter 1 but will do more with them now.

Some mathematical models are quite accurate, particularly many used in the physical sciences. For example, one of the models we will consider in calculus is a model for the path of a projectile. Other rather precise models predict such things as the time of sunrise and sunset, or the speed at which an object falls in a vacuum. Some mathematical models, however, are less accurate, especially those that involve examples from the life sciences and social sciences. Only recently has modeling in these disciplines become precise enough to be expressed in terms of calculus.

What, exactly, is a mathematical model? Sometimes, mathematical modeling can mean nothing more than a textbook word problem. But mathematical modeling can also mean choosing appropriate mathematics to solve a problem that has previously been unsolved. In this book, we use the term "mathematical modeling" to mean something between these two extremes. That is, it is a process we will apply to some real-life problem that does not have an obvious solution. It usually cannot be solved by applying a single formula.

Learning to use modeling techniques is an important part of your mathematical education. The first step in the modeling process involves *abstraction*, in which certain assumptions about a given practical problem are made in order to formulate a version of the problem in abstract terms. Next, the abstract version of the problem is *analyzed* using appropriate mathematical methods such as calculus, and finally, results derived from the abstract analysis are *interpreted* in terms of the original problem. The results obtained from the model often lead to predictions about the "real-world" situations related to the given problem. Data can be gathered to check the accuracy of these predictions and to modify the initial assumptions of the model to make it more precise. The process continues until a model is obtained that fits "reality" with acceptable accuracy. The steps in the modeling process are demonstrated in Figure 3.23, repeated from Chapter 1.

Figure 3.23 Mathematical modeling

Example 4 Identifying models

It is common to gather data when studying most phenomena. Classifying the data by comparing it with known models is an important mathematical skill. Identify each of the data sets as an example of a linear model, a quadratic model, a cubic model, a logarithmic model, an exponential model, or a sinusoidal model.

a.

x	y
−4	2
−3	1.28
−2	0.72
−1	0.32
0	0.08
1	0.00
2	0.08
3	0.32
4	0.72

b.

x	y
−4	−20.7
−3	−16.5
−2	−12.3
−1	−8.2
0	−3.8
1	0.4
2	4.7
3	8.9
4	13.12
5	17.3

c.

x	y
0.0	0.0
1.0	1.7
1.5	2.0
2.5	1.2
3.5	−0.7
4.5	−1.9
5.5	−1.4
6.0	−0.6
6.5	0.4

d.

x	y
−4	0.27
−3	0.30
−2	0.33
−1	0.36
0	0.40
1	0.44
2	0.49
3	0.52
4	0.60

e.

x	y
3	0.00
4	0.69
5	1.10
6	1.39
7	1.61
8	1.79
9	1.95
10	2.08
11	2.20
12	2.30

Solution Sketch each graph and observe the pattern (if any) of the graphs.

a.

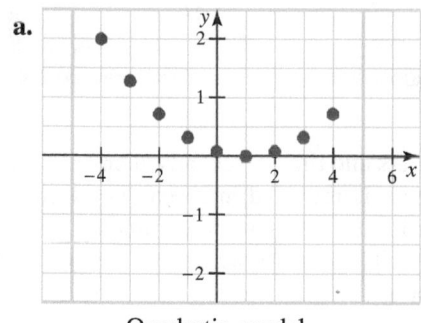

Quadratic model

b.

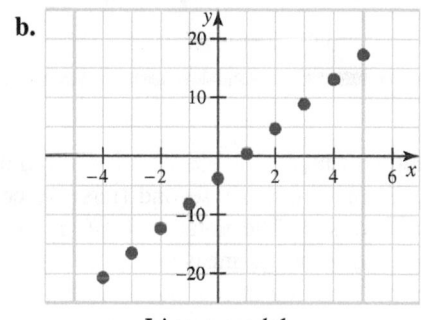

Linear model

c.

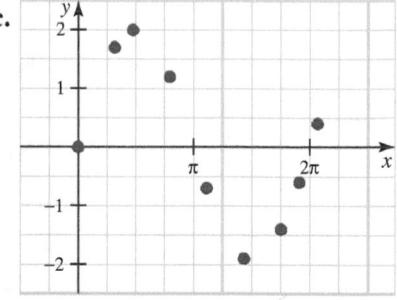

The data points seem to form a pattern, and we hypothesize that the data points are periodic, namely a sinusoidal model

d.

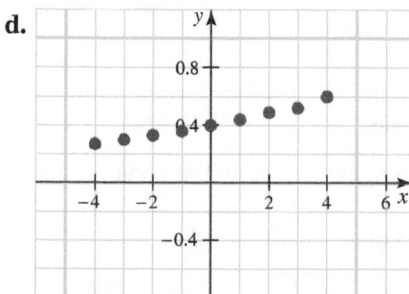

The data points look almost linear (but not quite), so we hypothesize that the data points may indicate an exponential model

e.

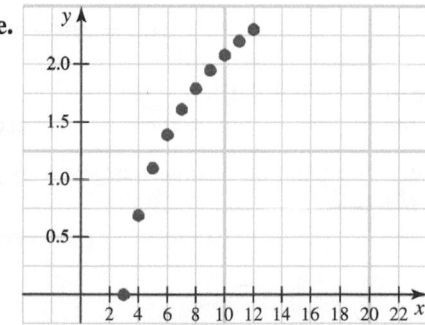

The data points seem to indicate a logarithmic model

Mathematical models will appear in both examples and exercises throughout this text. In the next subsection, we begin our work with modeling by studying rectilinear motion, an important topic from basic physics.   ■

Rectilinear Motion (Modeling in Physics)

Rectilinear motion is motion along a straight line. For example, the up and down motion of a yo-yo may be regarded as rectilinear, as may the motion of a rocket early in its flight.

When studying rectilinear motion, we may assume that the object is moving along a coordinate line. The *position* of the object on the line is a function of time t and is often expressed as $s(t)$. The rate of change of $s(t)$ with respect to time is the object's **velocity** $v(t)$, and the rate of change of the velocity with respect to t is its **acceleration** $a(t)$. The absolute value of the velocity is called **speed**. Thus,

☠ *Rectilinear motion involves position, velocity, and acceleration. Sometimes there is confusion between the words "speed" and "velocity." Because speed is the absolute value of the velocity, it indicates how fast an object is moving, whereas velocity indicates both speed and direction (relative to a given coordinate system).* ☠

$$\text{Velocity is } v(t) = \frac{ds}{dt} \quad \text{Speed is } |v(t)| = \left|\frac{ds}{dt}\right| \quad \text{Acceleration is } a(t) = \frac{dv}{dt} = \frac{d^2s}{dt^2}$$

If $v(t) > 0$, we say that the object is *advancing*, and if $v(t) < 0$, it is *retreating*. If $v(t) = 0$, the object is neither advancing nor retreating, and we say it is *stationary*. The object is *accelerating* when $a(t) > 0$ and is *decelerating* when $a(t) < 0$. The significance of the acceleration is that it gives the rate at which the velocity is changing. When you are riding in a car, you do not feel the velocity, but you do feel the acceleration. That is, you feel *changes* in the velocity. These ideas are summarized in the following box.

> **RECTILINEAR MOTION**   An object that moves along a straight line with position $s(t)$ has *velocity* $v(t) = \dfrac{ds}{dt}$ and *acceleration* $a(t) = \dfrac{dv}{dt} = \dfrac{d^2s}{dt^2}$ when these derivatives exist. The *speed* of the object is $|v(t)|$.

Notational comment: If distance is measured in meters and time in seconds, then velocity is measured in meters per second (m/s). Acceleration is recorded in meters per second per second (m/s/s). The notation m/s/s is awkward, so m/s^2 is more commonly used. Similarly, if distance is measured in feet, then velocity is measured in feet per second (ft/s) and acceleration in feet per second per second (ft/s^2).

Example 5   The position, velocity, and acceleration of a moving object

Assume that the position at time t of an object moving along a line is given by

$$s(t) = 3t^3 - 40.5t^2 + 162t$$

for t on $[0, 8]$. Find the initial position, velocity, and acceleration for the object and discuss the motion. Compute the total distance traveled.

Solution   The position at time t is given by the function s. The initial position occurs at time $t = 0$, so

$$s(0) = 0 \qquad \textit{The object starts at the origin.}$$

The velocity is determined by finding the derivative of the position function.

$$v(t) = s'(t)$$
$$= 9t^2 - 81t + 162$$
$$= 9(t - 3)(t - 6) \qquad \textit{The initial velocity is } v(0) = 162.$$

When $t = 3$ and when $t = 6$, the velocity v is 0, which means the *object is stationary* at those times. Furthermore,

$$v(t) > 0 \text{ on } [0, 3) \qquad \textit{Object is advancing.}$$
$$v(t) < 0 \text{ on } (3, 6) \qquad \textit{Object is retreating.}$$
$$v(t) > 0 \text{ on } (6, 8] \qquad \textit{Object is advancing.}$$

For the acceleration,

$$a(t) = s''(t)$$
$$= v'(t)$$
$$= 18t - 81$$
$$= 18(t - 4.5) \qquad \textit{The initial acceleration is } a(0) = -81.$$

We see that $\quad a(t) < 0$ on $[0, 4.5) \qquad$ *Velocity is decreasing; i.e., decelerating.*

$\qquad\qquad\quad a(t) > 0$ on $(4.5, 8) \qquad$ *Velocity is increasing; i.e., accelerating.*

The table in the margin gives certain values for $s, v,$ and a. We use these values to plot a few points, as shown in Figure 3.24. The actual path of the object is back and forth on the axis and the figure is for clarification only.

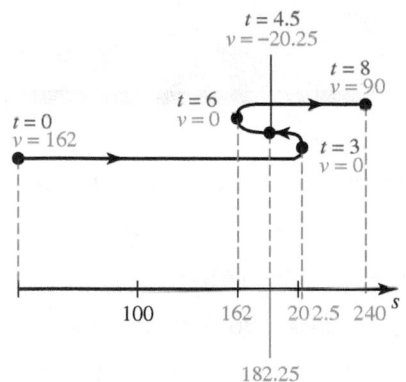
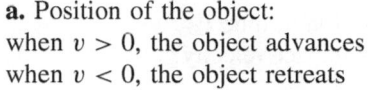

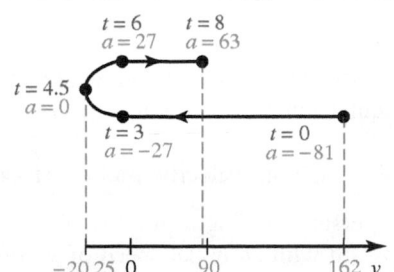

t	s	v	a
0	0	162	−81
1	124.5	90	−63
2	186	36	−45
3	202.5	0	−27
4	192	−18	−9
4.5	182.25	−20.25	0
5	172.5	−18	9
6	162	0	27
7	178.5	36	45
8	240	90	63

a. Position of the object:
when $v > 0$, the object advances
when $v < 0$, the object retreats

b. Velocity of the object:
when $a > 0$, the velocity increases
when $a < 0$, the velocity decreases

Figure 3.24 Interactive Analysis of rectilinear motion

Recall that the *speed* of the particle is the absolute value of its velocity. The speed decreases from 162 to 0 between $t = 0$ and $t = 3$ and increases from 0 to 20.25 as the velocity becomes negative between $t = 3$ and $t = 4.5$. Then for $4.5 < t < 6$, the particle slows down again, from 20.25 to 0, after which it speeds up. The total distance traveled is

$$|s(3) - s(0)| + |s(6) - s(3)| + |s(8) - s(6)| = 202.5 + 40.5 + 78$$

$$= 321$$

A common mistake is to think that a particle moving on a straight line speeds up when its acceleration is positive and slows down when the acceleration is negative, but this is not quite correct. Instead, the following is generally true:

- The speed *increases* (particle speeds up) when the velocity and acceleration have the same signs.
- The speed *decreases* (particle slows down) when the velocity and acceleration have opposite signs.

Falling Body Problems

As a second example of rectilinear motion, we will consider a *falling body problem.* In such a problem, it is assumed that an object is projected (that is, thrown, fired, dropped, etc.) vertically in such a way that the only acceleration acting on the object is the constant downward acceleration g due to gravity, which on the earth near sea level is approximately 32 ft/s^2 or 9.8 m/s^2.* Thus, we disregard air resistance and friction. At time t, the height of the projectile is given by the following formula (where s_0 and v_0 are the projectile's initial height and velocity, respectively):

FORMULA FOR THE HEIGHT OF A PROJECTILE

$\downarrow v_0$ *is the initial velocity*

$$h(t) = -\tfrac{1}{2}gt^2 + v_0 t + s_0 \leftarrow s_0 \text{ is the initial height.}$$

$\uparrow g$ *is the acceleration due to gravity*

You are asked to derive this formula for earth's gravity and a particular initial velocity and a particular initial position in Problem 59 in Section 5.6.

Example 6   Position, velocity, and acceleration of a falling object

Suppose a person standing at the top of the Tower of Pisa (176 ft high) throws a ball directly upward with an initial speed of 96 ft/s.

a. Find the ball's height, its velocity, and its acceleration at time t.
b. When does the ball hit the ground and what is its impact velocity?
c. How far does the ball travel during its flight?

---

*Although 32 ft/s^2 is the practical number we will use in this test, engineers in earthbound vehicular activity use 32.174 ft/s^2 as a better approximation for gravity.

Solution   First, draw a picture such as the one shown in Figure 3.25 to help you understand the problem.

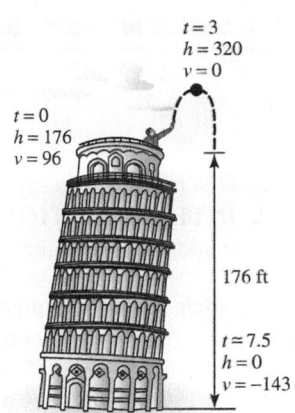

a. Substitute the known values into the formula for the height of an object:

$$\downarrow v_0 = 96 \text{ ft/s: Initial velocity}$$

$$h(t) = -\frac{1}{2}(32)t^2 + 96t + 176 \leftarrow h_0 = 176 \text{ ft is height of tower}$$

$$\uparrow g = 32 \text{ ft/s}^2 \text{: Constant downward gravitational acceleration}$$

$$= -16t^2 + 96t + 176 \text{ This is the position function. It gives the height of the ball.}$$

The velocity at time t is the derivative:

$$v(t) = \frac{dh}{dt} = -32t + 96$$

The acceleration is the derivative of the velocity function:

$$a(t) = \frac{dv}{dt} = \frac{d^2h}{dt^2} = -32$$

This means that the acceleration of the ball is always decreasing at the rate of 32 ft/s^2.

b. The ball hits the ground when $h(t) = 0$. Solve the equation

$$-16t^2 + 96t + 176 = 0$$

to find that this occurs when $t \approx -1.47$ and $t \approx 7.47$. Disregarding the negative value, we see that impact occurs at $t \approx 7.47$ s. The impact velocity is

$$v(7.47) \approx -143 \text{ ft/s}$$

The negative sign here means the ball is coming down at the moment of impact.

c. The ball travels upward for some time and then falls downward to the ground, as shown in Figure 3.25. We need to find the distance it travels upward plus the distance it falls to the ground. The turning point at the top (the highest point) occurs when the velocity is zero. Solve the equation

$$-32t + 96 = 0$$

to find that this occurs when $t = 3$. The ball starts at $h(0) = 176$ ft and rises to a maximum height when $t = 3$. Thus, the maximum height is

$$h(3) = -16(3)^2 + 96(3) + 176 = 320$$

and the **total distance** traveled is

$$\underbrace{(320 - 176)}_{\text{Upward distance}} + \underbrace{320}_{\text{Downward distance}} = 464$$

$$\uparrow \text{Initial height}$$

The *total distance* traveled is 464 ft.

PROBLEM SET 3.4

Level 1

1. ■ What does this say? What is a mathematical model?
2. **INTERPRETATION PROBLEM** Why are mathematical models necessary or useful?

For each function f given in Problems 3-14, find the instantaneous rate of change with respect to x at $x = x_0$.

3. $f(x) = x^2 - 3x + 5$ when $x_0 = 2$
4. $f(x) = 14 + x - x^2$ when $x_0 = 1$
5. $f(x) = \dfrac{-2}{x+1}$ for $x_0 = 1$
6. $f(x) = \dfrac{2x-1}{3x+5}$ when $x_0 = -1$
7. $f(x) = (x^2 + 2)(x + \sqrt{x})$ when $x_0 = 4$
8. $f(x) = x \cos x$ when $x_0 = \pi$
9. $f(x) = (x+1) \sin x$ when $x_0 = \frac{\pi}{2}$
10. $f(x) = x \ln \sqrt{x}$ when $x_0 = 1$
11. $f(x) = \dfrac{x^2}{e^x}$ when $x_0 = 0$
12. $f(x) = \sin x \cos x$ when $x_0 = \frac{\pi}{2}$
13. $f(x) = \left(x - \dfrac{2}{x}\right)^2$ when $x_0 = 1$
14. $f(x) = \sin^2 x$ when $x_0 = \frac{\pi}{4}$

The function s(t) in Problems 15-22 gives the position of an object moving along a line. In each case,

a. *Find the velocity at time t.*
b. *Find the acceleration at time t.*
c. *Describe the motion of the object; that is, tell when it is advancing and when it is retreating. Compute the total distance traveled by the object during the indicated time interval.*
d. *Tell when the object is accelerating and when it is decelerating.*

15. $s(t) = t^2 - 2t + 6$ on $[0, 2]$
16. $s(t) = 3t^2 + 2t - 5$ on $[0, 1]$
17. $s(t) = t^3 - 9t^2 + 15t + 25$ on $[0, 6]$
18. $s(t) = t^4 - 4t^3 + 8t$ on $[0, 4]$
19. $s(t) = \dfrac{2t+1}{t^2}$ for $1 \le t \le 3$
20. $s(t) = t^2 + t \ln t$ for $1 \le t \le e$
21. $s(t) = 3 \cos t$ for $0 \le t \le 2\pi$
22. $s(t) = 1 + \sec t$ for $0 \le t \le \frac{\pi}{4}$

Modeling Problems: *Identify each of the data sets in Problems 23-32 as an example of a linear model, a quadratic model, a cubic model, a logarithmic model, an exponential model, or a sinusoidal model.*

23.
x	y
-4	-140
-3	-96
-2	-60
-1	-32
0	-12
1	0
2	4
3	0
4	-12

24.
x	y
-4	-10.2
-3	-8.0
-2	-5.8
-1	-3.6
0	-1.4
1	0.8
2	3.0
3	5.2
4	7.4

25.
x	y
-4	-0.019531
-3	-0.078125
-2	-0.3125
-1	-1.25
0	-5
1	-20
2	-80
3	-320
4	-1,280

26.
x	y
-4	-0.0003
-3	-0.0009
-2	-0.0027
-1	-0.0082
0	-0.0247
1	-0.0741
2	-0.2222
3	-0.6667
4	-2.0000

27.
x	y
-4	238
-3	173
-2	118
-1	73
0	38
1	13
2	-2
3	-7
4	-2

28.
x	y
-4	4.84
-3	2.46
-2	-1.41
-1	-4.42
0	-4.76
1	-2.20
2	1.68
3	4.55
4	4.66

29.
x	y
-2.0	-28.0
-1.5	-8.5
-1.0	1.0
-0.5	3.5
0.0	2.0
0.5	-0.5
1.0	-1.0
1.5	3.5
2.0	16.0

30.
x	y
-4	-3.5
-3	-2.4
-2	-1.3
-1	-0.2
0	0.9
1	2.0
2	3.1
3	4.2
4	5.3

31.

x	y
2	0.0
3	−1.2
4	−1.9
5	−2.4
6	−2.8
7	−3.1
8	−3.4
9	−3.6
10	−3.8

32.

x	y
1	4.2
2	5.6
3	6.4
4	7.0
5	7.4
6	7.8
7	8.1
8	8.4
9	8.6

Level 2

33. It is estimated that x years from now, $0 \leq x \leq 10$, the average SAT score of the incoming students at a certain eastern liberal arts college will be

$$f(x) = -6x + 1{,}082$$

a. Derive an expression for the rate at which the average SAT score will be changing with respect to time x years from now.

b. What is the significance of the fact that the expression in part **a** is a negative constant?

34. A particle moving on the x-axis has position

$$x(t) = 2t^3 + 3t^2 - 36t + 40$$

after an elapsed time of t seconds.

a. Find the velocity of the particle at time t.

b. Find the acceleration at time t.

c. What is the total distance traveled by the particle during the first 3 seconds?

35. An object moving on the x-axis has position

$$x(t) = t^3 - 9t^2 + 24t + 20$$

after t seconds. What is the total distance traveled by the object during the first 8 seconds?

36. A car has position

$$s(t) = 50\,t\,\ln t + 80$$

ft after t seconds of motion. What is its acceleration (to the nearest hundredth ft/s^2) after 10 seconds?

37. An object moves along a straight line so that after t minutes, its position relative to its starting point (in meters) is

$$s(t) = 10\,t + \frac{t}{e^t}$$

a. At what speed (to the nearest thousandth m/min) is the object moving at the end of 4 min?

b. How far (to the nearest thousandth m) does the object actually travel during the fifth minute?

38. A bucket containing 5 gal of water has a leak. After t seconds, there are

$$Q(t) = 5\left(1 - \frac{t}{25}\right)^2$$

gallons of water in the bucket.

a. At what rate (to the nearest hundredth gal) is water leaking from the bucket after 2 seconds?

b. How long does it take for all the water to leak out of the bucket?

c. At what rate is the water leaking when the last drop leaks out?

Modeling Problems *In Problems 39-46, set up and solve an appropriate model to answer the given question. Be sure to state your assumptions.*

39. A person standing at the edge of a cliff throws a rock directly upward. It is observed that 2 seconds later the rock is at its maximum height (in ft) and that 5 seconds after that, it hits the ground at the base of the cliff.

a. What is the initial velocity of the rock?

b. How high is the cliff?

c. What is the velocity of the rock at time t?

d. With what velocity does the rock hit the ground?

40. A projectile is shot upward from the earth with an initial velocity of 320 ft/s.

a. What is its velocity after 5 seconds?

b. What is its acceleration after 3 seconds?

41. A rock is dropped from a height of 90 ft. One second later another rock is dropped from height H. What is H (to the nearest foot) if the two rocks hit the ground at the same time?

42. A ball is thrown vertically upward from the ground with an initial velocity of 160 ft/s.

a. When will the ball hit the ground?

b. With what velocity will the ball hit the ground?

c. When will the ball reach its maximum height?

43. An object is dropped (initial velocity $v_0 = 0$) from the top of a building and falls 3 seconds before hitting the pavement below. Determine the height of the building in ft.

44. An astronaut standing at the edge of a cliff on the moon throws a rock directly upward and observes that it passes her on the way down exactly 4 seconds later. Three seconds after that, the rock hits the ground at the base of the cliff. Use this information to determine the initial velocity v_0 and the height of the cliff. *Note*: $g = 5.5$ ft/s^2 on the moon.

45. Answer the question in Problem 44 assuming the astronaut is on Mars, where $g = 12$ ft/s^2.

46. A car is traveling at 88 ft/s (60 mi/h) when the driver applies the brakes to avoid hitting a child. After

t seconds, the car is $s(t) = 88t - 8t^2$ feet from the point where the brakes were first applied. How long does it take for the car to come to a stop, and how far does it travel before stopping?

47. It is estimated that t years from now, the circulation of a local newspaper can be modeled by the formula

$$C(t) = 100t^2 + 400t + 50t \ln t$$

a. Find an expression for the rate at which the circulation will be changing with respect to time t years from now.
b. At what rate will the circulation be changing with respect to time 5 years from now?
c. By how much will the circulation actually change during the sixth year?

48. An efficiency study of the morning shift at a certain factory indicates that an average worker who arrives on the job at 8:00 A.M. will have assembled

$$f(x) = -\frac{1}{3}x^3 + \frac{1}{2}x^2 + 50x$$

units x hours later.
a. Find a formula for the rate at which the worker will be assembling the units after x hours.
b. At what rate will the worker be assembling units at 9:00 A.M.?
c. How many units will the worker actually assemble between 9:00 A.M. and 10:00 A.M.?

49. An environmental study of a certain suburban community suggests that t years from now, the average level of carbon monoxide in the air can be modeled by the formula
$$q(t) = 0.05t^2 + 0.1t + 3.4$$

parts per million.
a. At what rate will the carbon monoxide level be changing with respect to time one year from now?
b. By how much will the carbon monoxide level change in the first year?
c. By how much will the carbon monoxide level change over the next (second) year?

50. According to *Newton's law of universal gravitation*, if an object of mass M is separated by a distance r from a second object of mass m, then the two objects are attracted to one another by a force that acts along the line joining them and has magnitude

$$F = \frac{GmM}{r^2}$$

where G is the gravitational constant. Show that the rate of change of F with respect to r is inversely proportional to r^3.

51. The population of a bacterial colony is approximately

$$P(t) = P_0 + 61t + 3t^2$$

thousand t hours after observation begins, where P_0 is the initial population. Find the rate at which the colony is growing after 5 hours.

52. The gross domestic product (GDP) of a certain country is
$$g(t) = t^2 + 5t + 106$$

billion dollars t years after 2010.
a. At what rate does the GDP change in 2013?
b. At what percentage rate does the GDP change in 2013?

53. It is projected that x months from now, the population of a certain town will be

$$P(x) = 2x + 4x^{3/2} + 5,000$$

a. At what rate will the population be changing with respect to time 9 months from now?
b. At what percentage rate will the population be changing with respect to time 9 months from now?

54. If y is a linear function of x, what will happen to the percentage rate of change of y with respect to x as x increases without bound?

55. According to Debye's formula in physical chemistry, the orientation polarization P of a gas satisfies

$$P = \frac{4}{3}\pi N \left(\frac{\mu^2}{3kT}\right)$$

where μ, k, and N are constants and T is the temperature of the gas. Find the rate of change of P with respect to T.

56. A disease is spreading in such a way that after t weeks, for $0 \le t \le 6$, it has affected

$$N(t) = 5 - t^2(t - 6)$$

hundred people. Health officials declare that this disease will reach epidemic proportions when the percentage rate of increase of $N(t)$ at the start of a particular week is at least 30% per week. The epidemic designation level is dropped when the percentage rate falls below this level.
a. Find the percentage rate of change of $N(t)$ at time t.
b. Between what weeks is the disease at the epidemic level?

57. An object attached to a helical spring is pulled down from its equilibrium position and then released, as shown in Figure 3.26. Suppose that t seconds later, its position (in centimeters, measured relative to the equilibrium position) is modeled by

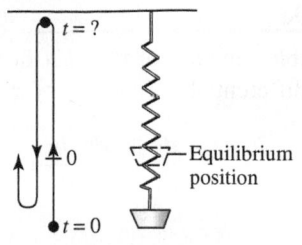

Figure 3.26 Helical spring

$$s(t) = 7 \cos t$$

a. Find the velocity and acceleration of the object at time t.

b. Find the length of time required for one complete oscillation. This is called the *period* of the motion.

c. What is the distance between the highest point reached by the object and the lowest point? Half of this distance is called the *amplitude* of the motion.

58. Two cars leave a town at the same time and travel at constant speeds along straight roads that meet at an angle of 60° in the town. If one car travels twice as fast as the other and the distance between them increases at the rate of 45 mi/h, how fast is the slower car traveling?

Level 3

59. Find the rate of change of the volume of a cube with respect to the length of one of its edges. How is this rate related to the surface area of the cube?

60. Show that the rate of change of the volume of a sphere with respect to its radius is equal to its surface area.

3.5 THE CHAIN RULE

IN THIS SECTION: *Introduction to the chain rule, extended derivative formulas, justification of the chain rule*
In many practical situations, a quantity is given as a function of one variable which, in turn, can be thought of as a function of a second variable. In such cases, we shall find that the rate of change of the quantity with respect to the second variable is equal to a product of rates.

Introduction to the Chain Rule

Suppose it is known that the carbon monoxide pollution in the air is changing at the rate of 0.02 ppm (parts per million) for each person in a town whose population is growing at the rate of 1,000 people per year. To find the rate at which the level of pollution is increasing with respect to time, we form the product

$$(0.02 \text{ ppm/person})(1,000 \text{ people/year}) = 20 \text{ ppm/year}$$

In this example, the level of pollution L is a function of the population P, which is itself a function of time t. Thus, L is a composite function of t, and

$$\begin{bmatrix} \text{RATE OF} \\ \text{CHANGE OF } L \\ \text{WITH RESPECT TO } t \end{bmatrix} = \begin{bmatrix} \text{RATE OF} \\ \text{CHANGE OF } L \\ \text{WITH RESPECT TO } P \end{bmatrix} \begin{bmatrix} \text{RATE OF} \\ \text{CHANGE OF } P \\ \text{WITH RESPECT TO } t \end{bmatrix}$$

Expressing each of these rates in terms of an appropriate derivative in Leibniz form, we obtain the following equation:

$$\frac{dL}{dt} = \frac{dL}{dP} \frac{dP}{dt}$$

These observations anticipate the following important theorem.

Recall our earlier warning in Section 3.1 against thinking of $\frac{dy}{dx}$ as a fraction. That said, there are certain times when this incorrect reasoning can be used as a mnemonic device. For instance, "canceling du," as indicated below, makes it easy to remember the chain rule:

$$\frac{dy}{dx} = \frac{dy}{du} \frac{du}{dx}$$

Theorem 3.10   Chain rule

If $y = f(u)$ is a differentiable function of u and u in turn is a differentiable function of x, then $y = f(u(x))$ is a differentiable function of x and its derivative is given by the product

$$\frac{dy}{dx} = \frac{dy}{du} \frac{du}{dx}$$

Proof: A rigorous proof involves a few details that make it inappropriate to include at this point in the text. A justification (partial proof) is given at the end of this section, and a full proof of the chain rule is included in Appendix B.   ◆

Example 1   The chain rule

Find $\dfrac{dy}{dx}$ if $y = u^3 - 3u^2 + 1$ and $u = x^2 + 2$.

Solution   Because $\dfrac{dy}{du} = 3u^2 - 6u$ and $\dfrac{du}{dx} = 2x$, it follows from the chain rule that

$$\frac{dy}{dx} = \frac{dy}{du} \frac{du}{dx}$$
$$= (3u^2 - 6u)(2x)$$

Notice that this derivative is expressed in terms of the variables x and u. To express dy/dx in terms of x alone, we substitute $u = x^2 + 2$ as follows:

$$\frac{dy}{dx} = \left[3(x^2 + 2)^2 - 6(x^2 + 2)\right](2x)$$
$$= 6x^3(x^2 + 2)$$

The chain rule is actually a rule for differentiating composite functions. In particular, if $y = f(u)$ and $u = u(x)$, then y is the composite function $y = (f \circ u)(x) = f[u(x)]$, and the chain rule can be rewritten as follows:

Theorem 3.10A   Chain rule (alternate form)

If u is a differentiable function at x and f is differentiable at $u(x)$, then the composite function $f \circ u$ is differentiable at x and

$$\frac{d}{dx} f[u(x)] = \frac{d}{du} f(u) \frac{du}{dx}$$

or

$$(f \circ u)'(x) = f'[u(x)] u'(x)$$ ◆

> ■ **W**hat this says   In Section 1.4 when we introduced composite functions, we talked about the "inner" and "outer" functions. With this terminology, the chain rule says that the derivative of the composite function $f[u(x)]$ is equal to the derivative of the outer function f evaluated at the inner function u times the derivative of the inner function u with respect to x.

Example 2   The chain rule applied to a power

Differentiate $y = (3x^4 - 7x + 5)^3$.

Solution   Here, the "inner" function is $u(x) = 3x^4 - 7x + 5$ and the "outer" function is u^3, so we have

$$y'(x) = (u^3)' \, u'(x) \quad \textit{The first differentiation is with respect to u and the second}$$
$$\textit{is with respect to x}$$
$$= (3u^2)(3x^4 - 7x + 5)'$$
$$= 3(3x^4 - 7x + 5)^2(12x^3 - 7)$$

With a lot of work, you could have found the derivative in Example 2 without using the chain rule either by expanding the polynomial or by using the product rule. The answer would be the same but would involve much more algebra. The chain rule often allows us to find derivatives quickly that would otherwise be very difficult to handle.

Example 3   Differentiation with quotient rule inside the chain rule

Differentiate $g(x) = \sqrt[4]{\dfrac{x}{1 - 3x}}$.

Solution   Write $g(x) = \left(\dfrac{x}{1 - 3x}\right)^{1/4} = u^{1/4}$, where $u = \dfrac{x}{1 - 3x}$ is the inner function and $u^{1/4}$ is the outer function.   Then

$$g'(x) = (u^{1/4})' \, u'(x) \quad \textit{The first differentiation is with respect to u and the second}$$
$$\textit{is with respect to x}$$
$$= \frac{1}{4}u^{-3/4}u'(x)$$
$$= \frac{1}{4}\left(\frac{x}{1 - 3x}\right)^{-3/4}\left(\frac{x}{1 - 3x}\right)'$$
$$= \frac{1}{4}\left(\frac{x}{1 - 3x}\right)^{-3/4}\left[\frac{(1 - 3x)(1) - x(-3)}{(1 - 3x)^2}\right] \quad \textit{Quotient rule}$$
$$= \frac{1}{4}\left(\frac{x}{1 - 3x}\right)^{-3/4}\left[\frac{1}{(1 - 3x)^2}\right]$$
$$= \frac{1}{4x^{3/4}(1 - 3x)^{5/4}}$$

Extended Derivative Formulas

The chain rule can be used to obtain generalized differentiation formulas for the standard functions, as displayed in the following box.

EXTENDED POWER RULE   If u is a differentiable function of x, then

$$\frac{d}{dx}u^n = nu^{n-1}\frac{du}{dx}$$

EXTENDED TRIGONOMETRIC RULES

$$\frac{d}{dx}\sin u = \cos u \frac{du}{dx} \qquad \frac{d}{dx}\sec u = \sec u \tan u \frac{du}{dx} \qquad \frac{d}{dx}\tan u = \sec^2 u \frac{du}{dx}$$

$$\frac{d}{dx}\cos u = -\sin u \frac{du}{dx} \qquad \frac{d}{dx}\csc u = -\csc u \cot u \frac{du}{dx} \qquad \frac{d}{dx}\cot u = -\csc^2 u \frac{du}{dx}$$

EXTENDED EXPONENTIAL RULE $\quad \dfrac{d}{dx}e^u = e^u \dfrac{du}{dx}$

EXTENDED LOGARITHMIC RULE $\quad \dfrac{d}{dx}\ln u = u^{-1}\dfrac{du}{dx}$

Example 4 Chain rule with a trigonometric function

Differentiate $f(x) = \sin\left(3x^2 + 5x - 7\right)$.

Solution Think of this as $f(u) = \sin u$, where $u = 3x^2 + 5x - 7$, and apply the chain rule:

$$f'(x) = \cos(3x^2 + 5x - 7) \cdot (3x^2 + 5x - 7)'$$

$$= (6x + 5)\cos(3x^2 + 5x - 7)$$

Example 5 Chain rule with other rules

Differentiate $g(x) = \cos x^2 + 5\left(\frac{3}{x} + 4\right)^6$.

Solution
$$\frac{dg}{dx} = \frac{d}{dx}\cos x^2 + 5\frac{d}{dx}\left(3x^{-1} + 4\right)^6$$

$$= -\sin x^2 \frac{d}{dx}(x^2) + 5\left[6(3x^{-1} + 4)^5 \frac{d}{dx}(3x^{-1} + 4)\right]$$

$$= -\sin x^2 (2x) + 30(3x^{-1} + 4)^5(-3x^{-2})$$

$$= -2x\sin x^2 - 90x^{-2}(3x^{-1} + 4)^5$$

Example 6 Extended power and cosine rules

Differentiate $y = \cos^4(3x + 1)^2$.

Solution

$$\frac{dy}{dx} = 4\cos^3(3x + 1)^2 \frac{d}{dx}\cos(3x + 1)^2 \qquad \textit{Extended power rule}$$

$$= 4\cos^3(3x + 1)^2 \cdot \left[-\sin(3x + 1)^2\right] \cdot \frac{d}{dx}(3x + 1)^2 \qquad \textit{Extended cosine rule}$$

$$= -4\cos^3(3x + 1)^2 \sin(3x + 1)^2 \cdot 2(3x + 1)(3) \qquad \textit{Extended power rule}$$

$$= -24(3x + 1)\cos^3(3x + 1)^2 \sin(3x + 1)^2$$

Example 7 Extended power rule inside quotient rule

Differentiate $p(x) = \dfrac{\tan 7x}{(1 - 4x)^5}$.

Solution
$$\frac{dp}{dx} = \frac{(1 - 4x)^5\left[\frac{d}{dx}\tan 7x\right] - \tan 7x\left[\frac{d}{dx}(1 - 4x)^5\right]}{\left[(1 - 4x)^5\right]^2}$$

$$= \frac{(1-4x)^5 \left[\sec^2 7x\right] \frac{d}{dx}(7x) - \tan 7x \left[5(1-4x)^4 \frac{d}{dx}(1-4x)\right]}{(1-4x)^{10}}$$

$$= \frac{(1-4x)^5 \left(\sec^2 7x\right)(7) - (\tan 7x)(5)(1-4x)^4(-4)}{(1-4x)^{10}}$$

$$= \frac{(1-4x)^4 \left[7(1-4x)\left(\sec^2 7x\right) + 20\left(\tan 7x\right)\right]}{(1-4x)^{10}}$$

$$= \frac{7(1-4x)\sec^2 7x + 20 \tan 7x}{(1-4x)^6}$$

Example 8 Extended exponential rule within product rule

Differentiate $e^{-3x} \sin x$.

Solution $\quad \dfrac{d}{dx}\left(e^{-3x} \sin x\right) = e^{-3x}\left[\dfrac{d}{dx}(\sin x)\right] + \left[\dfrac{d}{dx}(e^{-3x})\right] \sin x \qquad$ *Product rule*

$$= e^{-3x}(\cos x) + e^{-3x}(-3) \sin x$$

$$= e^{-3x}(\cos x - 3 \sin x)$$

Example 9 Finding a horizontal tangent line

Find the x-coordinate of each point on the graph of

$$f(x) = x^2(4x+5)^3$$

where the tangent line is horizontal.

Solution Because horizontal tangent lines have zero slope, we need to solve the equation $f'(x) = 0$. We find that

$$\begin{aligned}
f'(x) &= \left[x^2(4x+5)^3\right]' \\
&= \left[x^2\right]'(4x+5)^3 + x^2\left[(4x+5)^3\right]' && \textit{Product rule} \\
&= (2x)(4x+5)^3 + x^2\left[3(4x+5)^2(4)\right] && \textit{Extended power rule} \\
&= 2x(4x+5)^2\left[4x+5+6x\right] && \textit{Common factor} \\
&= 2x(4x+5)^2(10x+5) \\
&= 10x(4x+5)^2(2x+1)
\end{aligned}$$

From the factored form of the derivative, we see that $f'(x) = 0$ when $x = 0$, $x = -\frac{5}{4}$, and $x = -\frac{1}{2}$, so these are the x-coordinates of the points on the graph of f where horizontal tangent lines occur.

Example 10 A modeling problem using the chain rule

An environmental study of a certain suburban community suggests that the average daily level of carbon monoxide in the air may be modeled by the formula

$$C(p) = \sqrt{0.5p^2 + 17}$$

parts per million when the population is p thousand. It is estimated that t years from now, the population of the community will be

$$p(t) = 3.1 + 0.1t^2$$

thousand. At what rate will the carbon monoxide level be changing with respect to time 3 years from now?

Solution

$$\frac{dC}{dt} = \frac{dC}{dp}\frac{dp}{dt} = \left[\frac{1}{2}(0.5p^2 + 17)^{-1/2}(0.5)(2p)\right][0.2t]$$

When $t = 3$, $p(3) = 3.1 + 0.1(3)^2 = 4$, so

$$\frac{dC}{dt}\bigg|_{t=3} = \left[\frac{1}{2}(0.5 \cdot 4^2 + 17)^{-1/2}(4)\right][0.2(3)] = 0.24$$

The carbon monoxide level will be changing at the rate of 0.24 parts per million per year. It will be increasing because the sign of dC/dt is positive. ∎

Justification of the Chain Rule

To get a better feel for why the chain rule is true, suppose x is changed by a small amount Δx. This will cause u to change by an amount Δu, which, in turn, will cause y to change by an amount Δy. If Δu is *not zero*, we can write

$$\frac{\Delta y}{\Delta x} = \frac{\Delta y}{\Delta u}\frac{\Delta u}{\Delta x}$$

By letting $\Delta x \to 0$, we force Δu to approach zero as well, since

$$\Delta u = \left(\frac{\Delta u}{\Delta x}\right)\Delta x \qquad \text{so} \qquad \lim_{\Delta x \to 0} \Delta u = \left(\frac{du}{dx}\right)(0) = 0$$

It follows that

$$\lim_{\Delta x \to 0}\frac{\Delta y}{\Delta x} = \left(\lim_{\Delta u \to 0}\frac{\Delta y}{\Delta u}\right)\left(\lim_{\Delta x \to 0}\frac{\Delta u}{\Delta x}\right)$$

Historical Note

A calculus book written by the famous mathematician G. H. Hardy (1877-1947) contained essentially this "incorrect" proof rather than the one given in Appendix B. It is even more remarkable that the error was not noticed until the fourth edition.

or, equivalently,

$$\frac{dy}{dx} = \frac{dy}{du}\frac{du}{dx}$$

Unfortunately, there is a flaw in this "proof" of the chain rule. At the beginning we assumed that $\Delta u \neq 0$. However, it is theoretically possible for a small change in x to produce no change in u so that $\Delta u = 0$. This is the case that we consider in the proof in Appendix B.

PROBLEM SET 3.5

Level 1

1. ■ What does this say? What is the chain rule?

2. ■ What does this say? When do you need to use the chain rule?

In Problems 3-8, use the chain rule to compute the derivative dy/dx and write your answer in terms of x only.

3. $y = u^2 + 1$; $u = 3x - 2$

4. $y = 2u^2 - u + 5$; $u = 1 - x^2$

5. $y = \dfrac{2}{u^2}$; $u = x^2 - 9$

6. $y = \cos u$; $u = x^2 + 7$

7. $y = u \tan u$; $u = 3x + \dfrac{6}{x}$

8. $y = u^2$; $u = \ln x$

Differentiate each function in Problems 9-12 with respect to the given variable of the function.

9. **a.** $g(u) = u^3$ **b.** $u(x) = x^2 + 1$
     c. $f(x) = (x^2 + 1)^3$

10. **a.** $g(x) = u^5$ **b.** $u(x) = 3x - 1$
     c. $f(x) = (3x - 1)^5$

11. **a.** $g(u) = u^7$ **b.** $u(x) = 5 - 8x - 12x^2$
     c. $f(x) = (5 - 8x - 12x^2)^7$

12. a. $g(u) = u^{15}$ **b.** $u(x) = 3x^2 + 5x - 7$
    c. $f(x) = (3x^2 + 5x - 7)^{15}$

In Problems 13-34, find the derivative of the given function.

13. $f(x) = (5x - 2)^5$ **14.** $f(x) = (x^4 - 7x)^{15}$

15. $f(x) = (3x^2 - 2x + 1)^4$ **16.** $f(x) = (3 - x^2 - x^4)^{11}$

17. $s(\theta) = \sin(4\theta + 2)$ **18.** $c(\theta) = \cos(5 - 3\theta)$

19. $f(x) = e^{-x^2 + 3x}$ **20.** $y = e^{x^3 - \pi}$

21. $y = e^{\sec x}$ **22.** $g(x) = e^{\sin x}$

23. $f(t) = \exp(t^2 + t + 5)$ **24.** $g(t) = t^2 e^{-t} + (\ln t)^2$

25. $g(x) = x \sin 5x$ **26.** $h(x) = \dfrac{\tan ex}{x^2}$

27. $f(x) = \left(\dfrac{1}{1 - 2x}\right)^3$ **28.** $f(x) = \sqrt[3]{\dfrac{1}{2 - 3x}}$

29. $f(x) = \sqrt{\dfrac{x^2 + 3}{x^2 - 5}}$ **30.** $f(x) = \sqrt{\dfrac{2x^2 - 1}{3x^2 + 2}}$

31. $g(x) = \ln(3x^4 + 5x)$ **32.** $g(x) = \ln(\ln x)$

33. $f(x) = \ln(\sin x + \cos x)$ **34.** $T(x) = \ln(\sec x + \tan x)$

Find the x-coordinate of each point in Problems 35-40 where the graph of the given function has a horizontal tangent line.

35. $f(x) = x\sqrt{1 - 3x}$ **36.** $g(x) = x^2(2x + 3)^2$

37. $q(x) = \dfrac{(x - 1)^2}{(x + 2)^3}$ **38.** $f(x) = (2x^2 - 7)^3$

39. $T(x) = x^2 e^{1 - 3x}$ **40.** $V(x) = \dfrac{\ln \sqrt{x}}{x^2}$

Level 2

41. The graphs of $u = g(x)$ and $y = f(u)$ are shown in Figure 3.27.

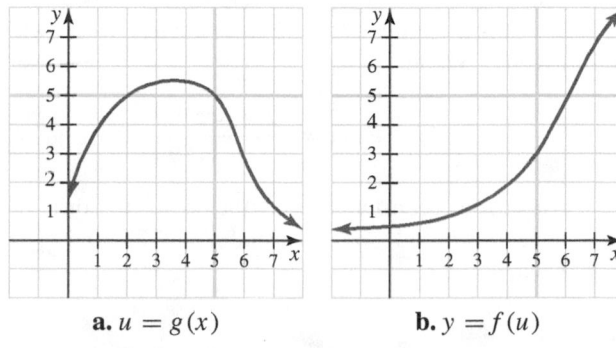

    a. $u = g(x)$ **b.** $y = f(u)$

Figure 3.27   Find the slope of $y = f[g(x)]$

   a. Find the approximate value of u at $x = 2$. What is the slope of the tangent line at that point?

   b. Find the approximate value of y at $x = 5$. What is the slope of the tangent line at that point?

   c. Find the slope of $y = f[g(x)]$ at $x = 2$.

42. The graphs of $y = f(x)$ and $y = g(x)$ are shown in Figure 3.28.

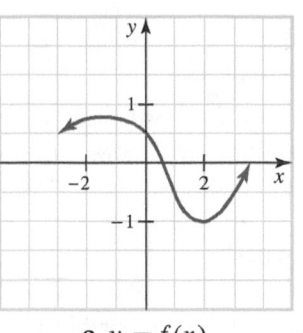 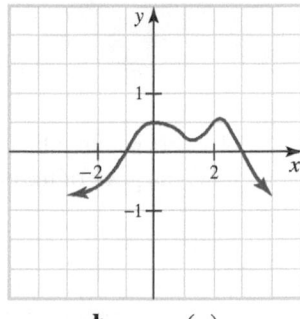

    a. $y = f(x)$ **b.** $y = g(x)$

Figure 3.28   Finding derivatives

Let $h(x) = f[g(x)]$. Estimate the following.

    a. $h'(-1)$ **b.** $h'(1)$ **c.** $h'(3)$

43. Repeat Problem 42 for $h(x) = g[f(x)]$.

44. If $g(x) = f[f(x)]$, use the table to find the value of $g'(2)$.

x	0.0	1.0	2.0	3.0	4.0	5.0
$f(x)$	18.5	9.4	4.0	2.6	8.3	14.0

45. If $h(x) = f[g(x)]$, use the table to find the value of $h'(1)$.

x	0.0	1.0	2.0	3.0	4.0	5.0
$f(x)$	6.9	4.3	3.1	2.8	2.2	2.0
$g(x)$	0.8	2.0	2.5	1.8	0.9	0.4

46. Assume that a spherical snowball melts in such a way that its radius decreases at a constant rate (that is, the radius is a linear function of time). Suppose it begins as a sphere with radius 10 cm and takes 2 hours to disappear.

    a. What is the rate of change of its volume after 1 hour? (Recall $V = \frac{4}{3}\pi r^3$.)

    b. At what rate is its surface area changing after 1 hour? (Recall $S = 4\pi r^2$.)

47. An importer of Rwandan coffee estimates that local consumers will buy approximately

$$D(p) = \frac{4{,}374}{p^2}$$

pounds of the coffee per week when the price is p dollars per pound. It is estimated that t weeks from now, the price of Rwandan coffee will be

$$p(t) = 0.02t^2 + 0.01t + 6$$

dollars per pound. At what rate will the weekly demand for the coffee be changing with respect to time 10 weeks from now? Will the demand be increasing or decreasing?

48. When electric blenders are sold for p dollars apiece, local consumers will buy

$$D(p) = \frac{8,000}{p}$$

blenders per month. It is estimated that t months from now, the price of the blenders will be

$$p(t) = 0.04t^{3/2} + 15$$

dollars. Compute the rate at which the monthly demand for the blenders will be changing with respect to time 25 months from now. Will the demand be increasing or decreasing?

When a point source of light of luminous intensity K (candles) shines directly on a point on a surface s meters away, the illuminance on the surface is given by the formula $I = Ks^{-2}$. Use this formula to answer the questions in Problems 49-50. (Note that 1 lux = 1 candle/m^2.)

49. Suppose a person carrying a 20-candlepower light walks toward a wall in such a way that at time t (seconds) the distance to the wall is $s(t) = 28 - t^2$ meters.
   a. How fast is the illuminance on the wall increasing when the person is 19 m from the wall?
   b. How far is the person from the wall when the illuminance is changing at the rate of 1 lux/s?

50. A lamp of luminous intensity 40 candles is 20 m above the floor of a room and is being lowered at the constant rate of 2 m/s. At what rate will the illuminance at the point on the floor directly under the lamp be increasing when the lamp is 15 m above the floor? Your answer should be in lux/s rounded to the nearest hundredth.

51. The average Fahrenheit temperature t hours after midnight in a certain city is modeled by the formula

$$T(t) = 58 + 17 \sin\left(\frac{\pi t}{10} - \frac{5}{6}\right)$$

At what rate is the temperature changing with respect to time at 2 A.M.? Is it getting hotter or colder at that time?

52. In physics, it is shown that when an object is viewed at depth d under water from an angle of incidence θ, then refraction causes the apparent depth of the object to be

$$s(\theta) = \frac{3d \cos \theta}{\sqrt{7 + 9 \cos^2 \theta}}$$

(See Figure 3.29). At what rate is the apparent depth changing with respect to θ when $d = 2$ m and $\theta = \frac{\pi}{6}$?

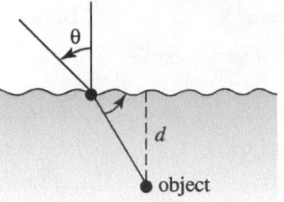

Figure 3.29 Light refraction

53. In the study of Frauenhofer diffraction in optics, a light beam of intensity I_0 from a source L passes through a narrow slit and is diffracted onto a screen (see Figure 3.30).

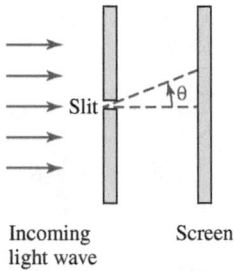

Incoming light wave   Screen

Figure 3.30 Optical diffraction

Experiments indicate that the intensity $I(\theta)$ of light on the screen depends on the diffraction angle θ shown in Figure 3.30 in such a way that

$$I(\beta) = I_0 \left(\frac{\sin \beta}{\beta}\right)^2$$

where $\beta = \pi a \sin(\theta/\lambda)$, λ is the wavelength of the light, and a is a constant related to the width of the screen. Use the chain rule to find $dI/d\theta$, the rate of change of intensity with respect to θ.

54. A worker pushes a heavy crate across a warehouse floor at a constant velocity. If the crate weighs W pounds and the worker applies a force F at an angle θ, as shown in Figure 3.31, then

$$F(\theta) = \frac{\mu W \sec \theta}{1 + \mu \tan \theta}$$

where μ is a constant (the coefficient of friction). For the case where $W = 100$ lb and $\mu = 0.25$, find $\frac{dF}{d\theta}$ when $\theta = 27°$.

Figure 3.31 Pushing a crate with variable force

55. Let $g(x) = f[u(x)]$, where $u(-3) = 5$, $u'(-3) = 2$, and $f(5) = 3$ and $f'(5) = -3$. Find an equation for the tangent line to the graph of g at the point where $x = -3$.

56. Let f be a function for which

$$f'(x) = \frac{1}{x^2 + 1}$$

a. If $g(x) = f(3x - 1)$, what is $g'(x)$?
b. If $h(x) = f\left(\frac{1}{x}\right)$, what is $h'(x)$?

57. a. If $F(x) = \ln|\cos x|$, show that $F'(x) = -\tan x$.
b. If $F(x) = \ln|\sec x + \tan x|$, show that $F'(x) = \sec x$.

58. If

$$\frac{df}{dx} = \frac{\sin x}{x}$$

and $u(x) = \cot x$, what is $\dfrac{df}{du}$?

59. Use the chain rule to find

$$\frac{d}{dx} f'\left[f(x)\right] \qquad \text{and} \qquad \frac{d}{dx} f\left[f'(x)\right]$$

assuming these derivatives exist.

60. Show that if a particle moves along a straight line with position $s(t)$ and velocity $v(t)$, then its acceleration satisfies

$$a(t) = v(t)\frac{dv}{ds}$$

Use this formula to find $\dfrac{dv}{ds}$ in the case where $s(t) = -2t^3 + 4t^2 + t - 3$.

3.6 IMPLICIT DIFFERENTIATION

IN THIS SECTION: *General procedure for implicit differentiation, derivative formulas for the inverse trigonometric functions, logarithmic differentiation*
We introduce implicitly defined functions and show how such functions can be differentiated when their derivatives exist, without first solving for y explicitly.

General Procedure for Implicit Differentiation

The equation $y = \sqrt{1 - x^2}$ **explicitly** defines $y = f(x) = \sqrt{1 - x^2}$ as a function of x for $-1 \le x \le 1$, but the same function can also be defined **implicitly** by the equation $x^2 + y^2 = 1$, as long as we restrict y by $0 \le y \le 1$ so that the vertical line test is satisfied. To find the derivative of the explicit form, we use the chain rule:

$$\begin{aligned}
\frac{d}{dx}\sqrt{1 - x^2} &= \frac{d}{dx}(1 - x^2)^{1/2} \\
&= \frac{1}{2}(1 - x^2)^{-1/2}(-2x) \\
&= \frac{-x}{\sqrt{1 - x^2}}
\end{aligned}$$

To obtain the derivative of the same function in its implicit form, we simply differentiate across the equation $x^2 + y^2 = 1$, remembering that y is a function of x:

$$\frac{d}{dx}(x^2 + y^2) = \frac{d}{dx}(1) \qquad \text{\textit{Derivative of both sides.}}$$

$$2x + 2y\frac{dy}{dx} = 0 \qquad \text{\textit{Chain rule for the derivative of } y}$$

$$\frac{dy}{dx} = -\frac{x}{y} \qquad \text{\textit{Solve for } } \frac{dy}{dx}.$$

$$= \frac{-x}{\sqrt{1 - x^2}} \qquad \text{\textit{Write as a function of } x, \text{ if desired.}}$$

The procedure we have just illustrated is called **implicit differentiation**. Our illustrative example was simple, but consider a differentiable function $y = f(x)$ defined by the equation

$$x^2 y^3 - 6 = 5y^3 + x$$

Implicit differentiation tells us that

$$x^2 \frac{d}{dx}(y^3) + y^3 \frac{d}{dx}(x^2) - \frac{d}{dx}(6) = 5\frac{d}{dx}(y^3) + \frac{d}{dx}(x)$$

$$x^2 \left(3y^2 \frac{dy}{dx}\right) + y^3(2x) - 0 = 5\left(3y^2 \frac{dy}{dx}\right) + 1$$

$$(3x^2 y^2 - 15y^2)\frac{dy}{dx} = 1 - 2xy^3$$

$$\frac{dy}{dx} = \frac{1 - 2xy^3}{3x^2 y^2 - 15y^2}$$

You may think that we are not finished since the derivative involves both x and y, but for many applications, that is enough as we may know both x and y simultaneously at a point. In this example, we could have found y as an explicit function of x, namely

$$y = \left(\frac{x + 6}{x^2 - 5}\right)^{1/3}$$

and then found dy/dx by the chain rule.

Let us now consider a different example, namely a differentiable function $y = f(x)$ defined by the equation

$$x^2 y + 2y^3 = 3x + 2y$$

Finding y as an explicit function of x is very difficult in this case, and it is not at all hard to imagine similar situations where solving for y in terms of x would be impossible or at least not worth the effort. We begin our work with implicit differentiation by using it to find the slope of a tangent line at a point on a circle.

Example 1   Slope of a tangent line using implicit differentiation

Find the slope of the tangent line to the circle $x^2 + y^2 = 10$ at the point $P(-1, 3)$.

Solution   We recognize the graph of $x^2 + y^2 = 10$ as not being the graph of a function. However, if we look at a small neighborhood around the point $(-1, 3)$, as shown in Figure 3.32, we see that this part of the graph *does pass the vertical line test* for functions. Thus, the required slope can be found by evaluating the derivative dy/dx at $(-1, 3)$. Instead of solving for y and finding the derivative, we *take the derivative of both sides of the equation*:

$$x^2 + y^2 = 10 \qquad \textit{Given equation}$$

$$\frac{d}{dx}(x^2 + y^2) = \frac{d}{dx}(10) \qquad \textit{Derivative of both sides}$$

$$2x + 2y\frac{dy}{dx} = 0 \qquad \textit{Do not forget that } y \textit{ is a function of } x.$$

$$\frac{dy}{dx} = -\frac{x}{y} \qquad \textit{Solve for } \frac{dy}{dx}.$$

The slope of the tangent line at $P(-1, 3)$ is

$$\left.\frac{dy}{dx}\right|_P = -\left.\frac{x}{y}\right|_{(-1,3)} = -\frac{-1}{3} = \frac{1}{3}$$

Here is a general description of the procedure for implicit differentiation.

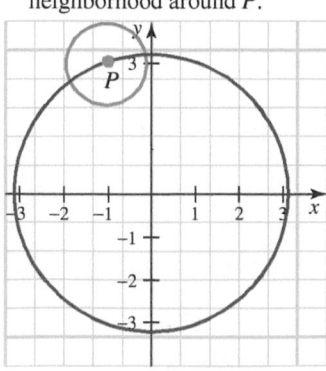

y is a function of x in this neighborhood around P.

Figure 3.32   Graph of $x^2 + y^2 = 10$ showing a neighborhood about $(-1, 3)$

PROCEDURE FOR IMPLICIT DIFFERENTIATION Suppose an equation defines y implicitly as a differentiable function of x.

To find $\dfrac{dy}{dx}$:

Step 1 Differentiate both sides of the equation with respect to x. Remember that y is really a function of x for part of the curve and use the chain rule when differentiating terms containing y.

Step 2 Solve the differentiated equation algebraically for $\dfrac{dy}{dx}$.

Example 2   Implicit differentiation

If $y = f(x)$ is a differentiable function of x such that

$$x^2 y + 2y^3 = 3x + 2y$$

find $\dfrac{dy}{dx}$.

Solution   The process is to differentiate both sides of the given equation with respect to x. To help remember that y is a function of x, replace y by the symbol $y(x)$:

$$x^2 y(x) + 2\left[y(x)\right]^3 = 3x + 2y(x)$$

Differentiate both sides of this equation term by term with respect to x:

$$\frac{d}{dx}\left\{x^2 y(x) + 2\left[y(x)\right]^3\right\} = \frac{d}{dx}\{3x + 2y(x)\}$$

$$\frac{d}{dx}\left[x^2 y(x)\right] + 2\frac{d}{dx}\left[y(x)\right]^3 = 3\frac{d}{dx}x + 2\frac{d}{dx}y(x)$$

$$\underbrace{x^2\frac{d}{dx}y(x) + y(x)\frac{d}{dx}x^2}_{\text{Product rule}} + \underbrace{2\left\{3[y(x)]^2\frac{d}{dx}y(x)\right\}}_{\text{Extended power rule}} = 3 + 2\frac{d}{dx}y(x)$$

$$x^2\frac{d}{dx}y(x) + 2xy(x) + 6\left[y(x)\right]^2\frac{d}{dx}y(x) = 3 + 2\frac{d}{dx}y(x)$$

$$x^2\frac{dy}{dx} + 2xy + 6y^2\frac{dy}{dx} = 3 + 2\frac{dy}{dx} \qquad y(x) \text{ is } y \text{ and } \frac{d}{dx}y(x) \text{ is } \frac{dy}{dx}.$$

$$x^2\frac{dy}{dx} + 6y^2\frac{dy}{dx} - 2\frac{dy}{dx} = 3 - 2xy \qquad \text{Solve for } \frac{dy}{dx}.$$

$$(x^2 + 6y^2 - 2)\frac{dy}{dx} = 3 - 2xy$$

$$\frac{dy}{dx} = \frac{3 - 2xy}{x^2 + 6y^2 - 2}$$

Notice that the formula for dy/dx contains both the independent variable x and the dependent variable y. This is usual when derivatives are computed implicitly.

WARNING: It is important to realize that implicit differentiation is a technique for finding dy/dx that is valid only if y is a differentiable function of x, and careless application of the technique can lead to errors. For example, there is clearly no real-valued function $y = f(x)$ that satisfies the equation $x^2 + y^2 = -1$, yet formal application of implicit differentiation yields the "derivative" $dy/dx = -x/y$. To be able to evaluate this "derivative," we must find some values of x and y for which $x^2 + y^2 = -1$. Because no such values exist, the derivative does not exist. ☠

In Example 2 it was suggested that you temporarily replace y by $y(x)$, so you would not forget to use the chain rule when first learning implicit differentiation. In the following example, we eliminate this unnecessary step and differentiate the given equation directly. Just keep in mind that y is really a function of x and remember to use the chain rule (or extended power rule) when it is appropriate.

Example 3   Implicit differentiation; simplified notation

Find $\dfrac{dy}{dx}$ if y is a differentiable function of x that satisfies

$$\sin(x^2 + y) = y^2(3x + 1)$$

Solution   There is no obvious way to solve the given equation explicitly for y. Differentiate implicitly to obtain

$$\frac{d}{dx}\left[\sin(x^2 + y)\right] = \frac{d}{dx}\left[y^2(3x + 1)\right]$$

$$\underbrace{\cos(x^2 + y)\frac{d}{dx}(x^2 + y)}_{\text{Chain rule}} = \underbrace{y^2\frac{d}{dx}(3x + 1) + (3x + 1)\frac{d}{dx}y^2}_{\text{Product rule}}$$

$$\cos(x^2 + y)\left(2x + \frac{dy}{dx}\right) = y^2(3) + (3x + 1)\left(2y\frac{dy}{dx}\right)$$

$$2x\cos(x^2 + y) + \cos(x^2 + y)\frac{dy}{dx} = 3y^2 + 2y(3x + 1)\frac{dy}{dx} \qquad \textit{Solve for } \frac{dy}{dx}.$$

$$\left[\cos(x^2 + y) - 2y(3x + 1)\right]\frac{dy}{dx} = 3y^2 - 2x\cos(x^2 + y)$$

$$\frac{dy}{dx} = \frac{3y^2 - 2x\cos(x^2 + y)}{\cos(x^2 + y) - 2y(3x + 1)}$$

Example 4   Slope of a tangent line using implicit differentiation

Find the slope of a line tangent to the circle $x^2 + y^2 = 5x + 4y$ at the point $P(5, 4)$.

Solution   The slope of a curve $y = f(x)$ is $\dfrac{dy}{dx}$, which we find implicitly.

$$x^2 + y^2 = 5x + 4y$$

$$\frac{d}{dx}(x^2 + y^2) = \frac{d}{dx}(5x + 4y)$$

$$2x + 2y\frac{dy}{dx} = 5 + 4\frac{dy}{dx}$$

$$2y\frac{dy}{dx} - 4\frac{dy}{dx} = 5 - 2x$$

$$(2y - 4)\frac{dy}{dx} = 5 - 2x$$

$$\frac{dy}{dx} = \frac{5 - 2x}{2y - 4}$$

At $(5,4)$, the slope of the tangent line is

$$\left.\frac{dy}{dx}\right|_{(5,4)} = \left.\frac{5-2x}{2y-4}\right|_{(5,4)} = \frac{5-2(5)}{2(4)-4} = \frac{-5}{4}$$

Note that the expression is undefined at $y = 2$; this makes sense when you see that the tangent line is vertical there. Look at the graph in Figure 3.33 and see whether you should exclude any other values.

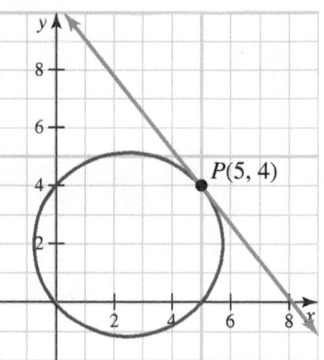

Figure 3.33 Graph of circle and tangent line

Example 5   Second derivative by implicit differentiation

Find $\dfrac{d^2 y}{dx^2}$ if $x^2 + y^2 = 10$.

Solution   In Example 1 we found (implicitly) that $\dfrac{dy}{dx} = -\dfrac{x}{y}$. Thus,

$$\underbrace{\frac{d^2 y}{dx^2} = \frac{d}{dx}\left(\frac{-x}{y}\right) = \frac{y\frac{d}{dx}(-x) - (-x)\frac{d}{dx}y}{y^2}}_{\textit{Quotient rule}} = \frac{-y + x\frac{dy}{dx}}{y^2}$$

Note that the expression for the second derivative contains the first derivative dy/dx. To simplify the answer, substitute the algebraic expression previously found for dy/dx:

$$\frac{d^2 y}{dx^2} = \frac{-y + x\frac{dy}{dx}}{y^2}$$

$$= \frac{-y + x\left(\frac{-x}{y}\right)}{y^2} \qquad \textit{Substitute } \frac{dy}{dx} = -\frac{x}{y}.$$

$$= \frac{-y^2 - x^2}{y^3}$$

$$= \frac{-(x^2 + y^2)}{y^3}$$

$$= \frac{-10}{y^3} \qquad \textit{Substitute } x^2 + y^2 = 10.$$

In Section 3.3, we found the derivative of $f(x) = e^x$ (Theorem 3.8). This derivative is more easily found using implicit differentiation, as shown in the following example.

Example 6   Derivative rule for the natural exponential

Using the derivative rule for the ln function show that $\dfrac{d}{dx}(e^x) = e^x$.

Solution   Let $v = e^x$, so that $x = \ln v$. Then

$$x = \ln v \qquad \textit{Given}$$

$$\frac{d}{dx}(x) = \frac{d}{dx}(\ln v) \qquad \textit{Derivative of both sides}$$

$$1 = \frac{1}{v}\frac{dv}{dx} \qquad \textit{Implicit differentiation}$$

$$v = \frac{dv}{dx}$$

$$e^x = \frac{d}{dx}(e^x) \qquad v = e^x$$

Implicit differentiation is a valuable theoretical tool. For example, in Section 3.2 we proved the power rule for the case where the exponent is an integer; implicit differentiation now allows us to extend the proof for all real exponents.

Example 7   Verification of power rule for real (rational and irrational) exponents

Show that $\dfrac{d}{dx}(x^r) = rx^{r-1}$ holds for all real numbers r if $x > 0$.

Solution   Let $y = x^r$.

$$y = x^r$$
$$y = e^{r \ln x} \qquad \textit{Property of logarithms}$$
$$r \ln x = \ln y \qquad \textit{Definition of logarithm}$$
$$r\left(\frac{1}{x}\right) = \frac{1}{y}\frac{dy}{dx} \qquad \textit{Implicit differentiation}$$
$$\frac{dy}{dx} = y\left(\frac{r}{x}\right) \qquad \textit{Solve for } \frac{dy}{dx}.$$
$$= x^r\left(\frac{r}{x}\right) \qquad \textit{Substitute } y = x^r.$$
$$= rx^{r-1} \qquad \textit{Simplify.}$$

Notice in Example 7 that $y = x^r = e^{r \ln x}$ is differentiable because it is defined as the composition of the differentiable functions $y = e^u$ and $u = r \ln x$.

Derivative Formulas for the Inverse Trigonometric Functions

Next, we will use implicit differentiation to obtain differentiation formulas for the six inverse trigonometric functions. Note that these derivatives are not inverse trigonometric functions or even trigonometric functions, but are instead rational functions or roots of rational functions. In fact, the usefulness of inverse trigonometric functions in many areas is related to the simplicity of their derivatives.

Theorem 3.11   Differentiation formulas for inverse trigonometric functions

If u is a differentiable function of x, then

☠ Note that the derivative of each inverse trigonometric function $y = \cos^{-1} x, y = \cot^{-1} x$, and $y = \csc^{-1} x$ is the opposite of the derivative of the corresponding inverse cofunction $y = \sin^{-1} x, y = \tan^{-1} x$, and $y = \sec^{-1} x.$ ☠

$$\frac{d}{dx}(\sin^{-1} u) = \frac{1}{\sqrt{1 - u^2}}\frac{du}{dx} \qquad \frac{d}{dx}(\sec^{-1} u) = \frac{1}{|u|\sqrt{u^2 - 1}}\frac{du}{dx}$$

$$\frac{d}{dx}(\cos^{-1} u) = \frac{-1}{\sqrt{1 - u^2}}\frac{du}{dx} \qquad \frac{d}{dx}(\csc^{-1} u) = \frac{-1}{|u|\sqrt{u^2 - 1}}\frac{du}{dx}$$

$$\frac{d}{dx}(\tan^{-1} u) = \frac{1}{1 + u^2}\frac{du}{dx} \qquad \frac{d}{dx}(\cot^{-1} u) = \frac{-1}{1 + u^2}\frac{du}{dx}$$

Proof: We will prove the first formula and leave the others as problems. Let $\alpha(x) = \sin^{-1} x$, so $x = \sin\alpha$. The inverse function theorem, proved in Advanced Calculus, says that any function that has a continuous derivative that is not zero in an interval has an inverse function that also has a continuous derivative on that interval. To find the derivative of the function $\alpha = \sin^{-1} x$ with values on $\left(-\frac{\pi}{2}, \frac{\pi}{2}\right)$ we proceed implicitly, and we will show that this derivative is defined on $(-1, 1)$.

$$\sin \alpha = x$$

$$\frac{d}{dx}(\sin \alpha) = \frac{d}{dx}(x)$$

$$\cos \alpha \frac{d\alpha}{dx} = 1$$

$$\frac{d\alpha}{dx} = \frac{1}{\cos \alpha}$$

$$= \frac{1}{\sqrt{1 - \sin^2 \alpha}} \qquad \cos \alpha \text{ is positive}$$

$$= \frac{1}{\sqrt{1 - x^2}} \qquad \textit{This exists on } (-1, 1); \textit{ see Figure } 3.34.$$

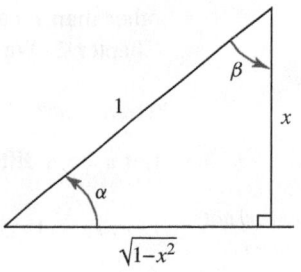

Figure 3.34 $\sin \alpha = x$ and $\cos \alpha = \sqrt{1 - x^2}$

If u is a differentiable function of x, then the chain rule gives

$$\frac{d}{dx}\left(\sin^{-1} u\right) = \frac{d}{du}\left(\sin^{-1} u\right)\frac{du}{dx} = \frac{1}{\sqrt{1 - u^2}}\frac{du}{dx} \qquad \blacklozenge$$

Example 8   Derivatives involving inverse trigonometric functions

Differentiate each of the following functions.

a. $f(x) = \tan^{-1} \sqrt{x}$ **b.** $g(t) = \sin^{-1}(1 - t)$ **c.** $h(x) = \sec^{-1} e^{2x}$

Solution

a. Let $u = \sqrt{x}$ in the formula for $\frac{d}{dx}(\tan^{-1} u)$.

$$f'(x) = \frac{d}{dx}(\tan^{-1}\sqrt{x})$$

$$= \frac{1}{1 + (\sqrt{x})^2}\frac{d}{dx}(\sqrt{x})$$

$$= \frac{1}{1 + x}\left(\frac{1}{2}\frac{1}{\sqrt{x}}\right)$$

$$= \frac{1}{2\sqrt{x}(1 + x)}$$

b. Let $u = 1 - t$.

$$g'(t) = \frac{d}{dt}\left[\sin^{-1}(1 - t)\right]$$

$$= \frac{1}{\sqrt{1 - (1 - t)^2}}\frac{d}{dt}(1 - t)$$

$$= \frac{-1}{\sqrt{1 - (1 - t)^2}}$$

$$= \frac{-1}{\sqrt{2t - t^2}}$$

c. Let $u = e^{2x}$.

$$h'(x) = \frac{d}{dx}\left[\sec^{-1} e^{2x}\right]$$

$$= \frac{1}{|e^{2x}|\sqrt{(e^{2x})^2 - 1}}\frac{d}{dx}(e^{2x})$$

$$= \frac{1}{e^{2x}\sqrt{e^{4x} - 1}}(2e^{2x})$$

$$= \frac{2}{\sqrt{e^{4x} - 1}}$$

The derivatives of b^u and $\log_b u$ for a differentiable function $u = u(x)$ and a base b other than e can be obtained using the chain rule and the change of base formulas from Chapter 2. We summarize the results in the following theorem.

Theorem 3.12   Derivatives of exponential and logarithmic functions

Let u be a differentiable function of x and b be a positive number (other than 1). Then

$$\frac{d}{dx}b^u = (\ln b)\, b^u \frac{du}{dx} \qquad \frac{d}{dx}\left(\log_b u\right) = \frac{1}{\ln b} \cdot \frac{1}{u}\frac{du}{dx}$$

$\frac{d}{dx}b^u \neq ub^{u-1}\frac{du}{dx}$ since b^u is a constant to a variable power, not a variable to a constant power.

$\blacksquare$ **W**hat this says The derivatives of b^x and $\log_b x$ are the same as the derivatives of e^x and $\ln x$, respectively, except for a factor of $\ln b$ that appears as a multiplier in the formula

$$\frac{d}{dx}(b^x) = (\ln b)\, b^x$$

and as a divisor in the formula

$$\frac{d}{dx}\left(\log_b x\right) = \frac{1}{(\ln b)\, x}$$

Proof: Because $b^u = e^{u \ln b}$, we can apply the chain rule as follows:

$$\frac{d}{dx}b^u = \frac{d}{dx}\left(e^{u \ln b}\right) = e^{u \ln b}\frac{d}{dx}\left(u \ln b\right) = e^{u \ln b}\left(\ln b\frac{du}{dx}\right) = (\ln b)\, b^u \frac{du}{dx}$$

To differentiate the logarithm, recall the change-of-base formula $\log_b u = \frac{\ln u}{\ln b}$ so that

$$\frac{d}{dx}\log_b u = \frac{d}{dx}\left(\frac{\ln u}{\ln b}\right) = \frac{1}{\ln b} \cdot \frac{1}{u}\frac{du}{dx} \qquad \blacklozenge$$

Example 9   Derivative of an exponential function with base $b \neq e$

Differentiate $f(x) = x\left(2^{1-x}\right)$.

Solution   Apply the product rule:

$$f'(x) = \frac{d}{dx}\left(x 2^{1-x}\right)$$
$$= x\frac{d}{dx}\left(2^{1-x}\right) + 2^{1-x}\frac{d}{dx}(x)$$

$$= x\,(\ln 2)\,(2^{1-x})(-1) + 2^{1-x}(1)$$
$$= 2^{1-x}(1 - x\,\ln 2)$$

The following theorem will prove useful in Chapter 5. $\blacksquare$

Theorem 3.13   Derivative of $\ln |u|$

If $f(x) = \ln |x|$, $x \neq 0$, then $f'(x) = \dfrac{1}{x}$. Also, if u is a differentiable function of x, then $\dfrac{d}{dx} \ln |u| = \dfrac{1}{u} \dfrac{du}{dx}$.

Proof: Using the definition of absolute value,

$$f(x) = \begin{cases} \ln x & \text{if } x > 0 \\ \ln(-x) & \text{if } x < 0 \end{cases}$$

so

$$f'(x) = \begin{cases} \dfrac{1}{x} & \text{if } x > 0 \\ \dfrac{1}{-x}(-1) = \dfrac{1}{x} & \text{if } x < 0 \end{cases}$$

Thus, $f'(x) = \dfrac{1}{x}$ for all $x \neq 0$.

The second part of the theorem (for u, a differentiable function of x) follows from the chain rule. ◆

Logarithmic Differentiation

Logarithmic differentiation is a procedure in which logarithms are used to trade the task of differentiating products and quotients for that of differentiating sums and differences. It is especially valuable as a means for handling complicated product or quotient functions and exponential functions where variables appear in both the base and the exponent.

Example 10   Logarithmic differentiation

Find the derivative of $y = \dfrac{e^{2x}(2x - 1)^6}{(x^3 + 5)^2(4 - 7x)}$ if $y > 0$.

Solution   The procedure called logarithmic differentiation requires that we first take the logarithm of both sides and then apply properties of logarithms before attempting to take the derivative.

$$y = \frac{e^{2x}(2x - 1)^6}{(x^3 + 5)^2(4 - 7x)}$$

$$\ln y = \ln \left[\frac{e^{2x}(2x - 1)^6}{(x^3 + 5)^2(4 - 7x)} \right]$$

$$= \ln e^{2x} + \ln (2x - 1)^6 - \ln(x^3 + 5)^2 - \ln(4 - 7x)$$

$$= 2x + 6 \ln(2x - 1) - 2 \ln(x^3 + 5) - \ln(4 - 7x)$$

Next, differentiate both sides with respect to x and then solve for $\dfrac{dy}{dx}$.

$$\frac{1}{y} \frac{dy}{dx} = 2 + 6 \left[\frac{1}{2x - 1}(2) \right] - 2 \left[\frac{1}{x^3 + 5}(3x^2) \right] - \left[\frac{1}{4 - 7x}(-7) \right]$$

$$\frac{dy}{dx} = y \left[2 + \frac{12}{2x - 1} - \frac{6x^2}{x^3 + 5} + \frac{7}{4 - 7x} \right]$$

This is the derivative in terms of x and y. If we want the derivative in terms of x alone, we can substitute the expression for y:

$$\frac{dy}{dx} = \frac{e^{2x}(2x - 1)^6}{(x^3 + 5)^2(4 - 7x)} \left[2 + \frac{12}{2x - 1} - \frac{6x^2}{x^3 + 5} + \frac{7}{4 - 7x} \right]$$

■

Example 11 Derivative with variables in both the base and the exponent

Find $\dfrac{dy}{dx}$, where $y = (x+1)^{2x}$.

Solution

$$\ln y = \ln\left[(x+1)^{2x}\right] = 2x\,\ln(x+1)$$

Differentiate both sides of this equation:

$$\frac{1}{y}\frac{dy}{dx} = 2x\left\{\frac{d}{dx}\,[\ln(x+1)]\right\} + \left[\frac{d}{dx}(2x)\right]\ln(x+1) \qquad \textit{Product rule}$$

$$= \frac{2x}{x+1} + 2\,\ln(x+1)$$

Finally, multiply both sides by $y = (x+1)^{2x}$:

$$\frac{dy}{dx} = \left[\frac{2x}{x+1} + 2\,\ln(x+1)\right](x+1)^{2x}$$

PROBLEM SET 3.6

Level 1

Find $\dfrac{dy}{dx}$ by implicit differentiation in Problems 1-10.

1. $x^2 + y^2 = 25$
2. $x^2 + y = x^3 + y^3$
3. $xy = 25$
4. $xy(2x + 3y) = 2$
5. $\dfrac{1}{y} + \dfrac{1}{x} = 1$
6. $\tan\dfrac{x}{y} = y$
7. $\cos xy = 1 - x^2$
8. $e^{xy} + 1 = x^2$
9. $\ln(xy) = e^{2x}$
10. $e^{xy} + \ln y^2 = x$

In Problems 11-14, find $\dfrac{dy}{dx}$ in two ways:

a. *By implicit differentiation of the equation*
b. *By differentiating an explicit formula for y*

11. $x^2 + y^3 = 12$
12. $xy + 2y = x^2$
13. $x + \dfrac{1}{y} = 5$
14. $xy - x = y + 2$

Find the derivative $\dfrac{dy}{dx}$ in Problems 15-24.

15. $y = \sin^{-1}(2x + 1)$
16. $y = \cos^{-1}(4x + 3)$
17. $y = \tan^{-1}\sqrt{x^2 + 1}$
18. $y = \cot^{-1}x^2$
19. $y = \sec^{-1}\left(e^{-x}\right)$
20. $y = \ln\left|\sin^{-1}x\right|$

21. $y = \tan^{-1}\left(\dfrac{1}{x}\right)$
22. $y = \cos^{-1}|\sin x|$
23. $\sin^{-1}y + y = 2xy$
24. $x\sin^{-1}y + y\tan^{-1}x = x$

In Problems 25-30, find an equation of the tangent line to the graph of each equation at the prescribed point.

25. $x^2 + y^2 = 13$ at $(-2, 3)$
26. $x^3 + y^3 = y + 21$ at $(3, -2)$
27. $\sin(x - y) = xy$ at $(0, \pi)$
28. $3^x + \log_2(xy) = 10$ at $(2, 1)$
29. $x\,\tan^{-1}y = x^2 + y$ at $(0, 0)$
30. $\sin^{-1}(xy) + \dfrac{\pi}{2} = \cos^{-1}y$ at $(1, 0)$

Find the slope of the tangent line to the graph at the points indicated in Problems 31-34.

31. bifolium
$$(x^2 + y^2)^2 = 4x^2y$$
at $(1, 1)$

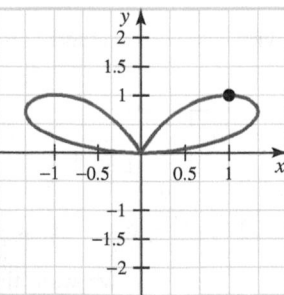

32. lemniscate of Bernoulli
$$(x^2 + y^2)^2 = \frac{25}{3}(x^2 - y^2)$$
at $(2, 1)$

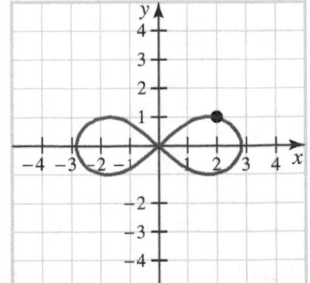

33. folium of Descartes

$$x^3 + y^3 - \frac{9}{2}xy = 0$$

at $(2, 1)$

34. cissoid of Diocles

$$y^2(6 - x) = x^3 \text{ at } (3, 3)$$

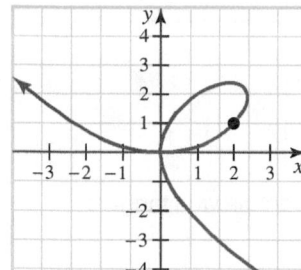
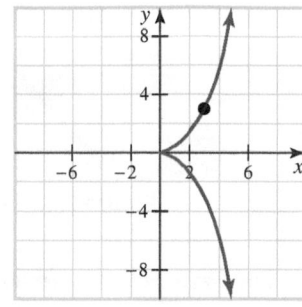

35. Find an equation of the normal line to the curve $x^2 + 2xy = y^3$ at $(1, -1)$.

36. Find an equation of the normal line to the curve $x^2\sqrt{y - 2} = y^2 - 3x - 5$ at $(1, 3)$.

Use implicit differentiation to find the second derivative y'' of the functions given in Problems 37-38.

37. $7x + 5y^2 = 1$

38. $x^2 + 2y^3 = 4$

Level 2

39. INTERPRETATION PROBLEM Compare and contrast the derivatives of the following functions.

a. $y = x^2$ **b.** $y = 2^x$ **c.** $y = e^x$ **d.** $y = x^e$

40. INTERPRETATION PROBLEM Compare and contrast the derivatives of the following functions:

a. $y = \log x$ **b.** $y = \ln x$

41. INTERPRETATION PROBLEM Discuss logarithmic differentiation.

42. Think Tank Problem

a. If $x^2 + y^2 = 6y - 10$ and $\dfrac{dy}{dx}$ exists, show that

$$\frac{dy}{dx} = \frac{x}{3 - y}.$$

b. Show that there are no real numbers x, y that satisfy the equation

$$x^2 + y^2 = 6y - 10$$

c. What can you conclude from the result found in part **a** in light of the observation in part **b**?

Use logarithmic differentiation in Problems 43-46 to find dy/dx. You may express your answer in terms of both x and y, and you do not need to simplify the resulting rational expressions.

43. $y = \sqrt[18]{(x^{10} + 1)^3(x^7 - 3)^8}$

44. $y = \dfrac{e^{3x^2}}{(x^3 + 1)^2(4x - 7)^{-2}}$

45. $y = x^x$

46. $y = x^{\ln\sqrt{x}}$

47. Let $\dfrac{u^2}{a^2} + \dfrac{v^2}{b^2} = 1$, where a and b are nonzero constants. Find:

a. $\dfrac{du}{dv}$ **b.** $\dfrac{dv}{du}$

48. Show that the tangent line at the point (a, b) on the curve whose equation is $2x^2 + 3xy + y^2 = -2$ is horizontal if $4a + 3b = 0$. Find two such points on the curve.

49. Find two points on the curve whose equation is $x^2 - 3xy + 2y^2 = -2$, where the tangent line is vertical.

50. Let g be a differentiable function of x that satisfies $g(x) < 0$ and $x^2 + g^2(x) = 10$ for all x.

a. Use implicit differentiation to show that

$$\frac{dg}{dx} = \frac{-x}{g(x)}$$

b. Show that $g(x) = -\sqrt{10 - x^2}$ satisfies the given requirements. Then use the chain rule to verify that

$$\frac{dg}{dx} = \frac{-x}{g(x)}$$

51. Find the equation of the tangent line and the normal line to the curve

$$x^3 + y^3 = 2Axy$$

at the point (A, A), where A is a nonzero constant.

52. Find all points on the lemniscate $(x^2 + y^2)^2 = 4(x^2 - y^2)$ where the tangent line is horizontal. (See Figure 3.35.)

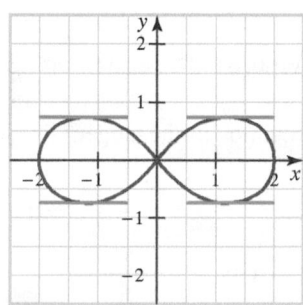

Figure 3.35 Lemniscate

53. Find all points on the cardioid $x^2 + y^2 = \sqrt{x^2 + y^2} + x$ where the tangent line is vertical. (See Figure 3.36.)

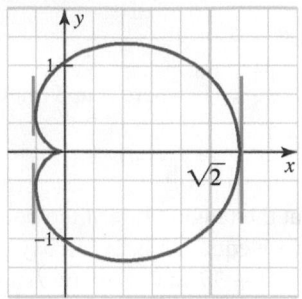

Figure 3.36   Cardioid

Level 3

54. Find two differentiable functions f that satisfy the equation

$$x - \left[f(x)\right]^2 = 9$$

Give the explicit form of each function, and sketch its graph.

55. Show that the tangent line to the ellipse

$$\frac{x^2}{a^2} + \frac{y^2}{b^2} = 1$$

at the point (x_0, y_0) is

$$\frac{x_0 x}{a^2} + \frac{y_0 y}{b^2} = 1$$

(See Figure 3.37.)

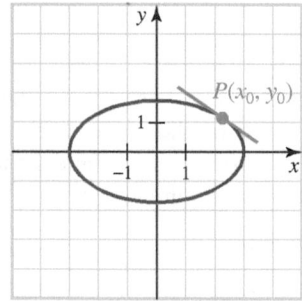

Figure 3.37   Ellipse with tangent

56. Find an equation of the tangent line to the hyperbola

$$\frac{x^2}{a^2} - \frac{y^2}{b^2} = 1$$

(See Figure 3.38.)

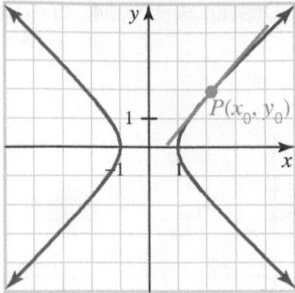

Figure 3.38   Hyperbola with tangent

*The **angle between curves** C_1 **and** C_2 at the point of intersection P is defined as the angle $0 \le \theta \le \dfrac{\pi}{2}$ between the tangent lines at P. Specifically, the angle between C_1 and C_2 is the angle between the tangent line to C_1 at P and the tangent line to C_2 at P as shown in Figure 3.39.*

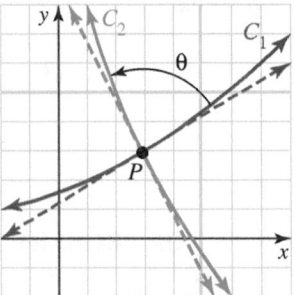

Figure 3.39   Angle between curves

Use this information for Problems 57-58.

57. If θ is the angle between curve C_1 and curve C_2 at P and the tangent lines to C_1 and C_2 at P have slopes m_1 and m_2, respectively, show that

$$\tan \theta = \frac{|m_2 - m_1|}{1 + m_1 m_2}$$

58. Find the angle between the circle $x^2 + y^2 = 1$ and the circle $x^2 + (y - 1)^2 = 1$ at each of the two points of intersection.

59. Use implicit differentiation to obtain the differentiation formulas for $y = \tan^{-1} u$ and $y = \sec^{-1} u$.

60. Use the identity $\sin^{-1} u + \cos^{-1} u = \frac{\pi}{2}$ to obtain the differentiation formula for $y = \cos^{-1} u$. Similar identities can be used to obtain differentiation formulas for $y = \cot^{-1} u$ and $y = \csc^{-1} u$.

3.7 RELATED RATES AND APPLICATIONS

IN THIS SECTION: In many practical applications of implicit differentiation, the derivatives represent rates of change (often with respect to time). In a typical situation, two or more quantities that vary with time may be related to each other, and implicit differentiation allows us to obtain corresponding relationships among their respective rates without first finding explicit formulas for the quantities as functions of time. Techniques for solving such problems are discussed in this section.

Certain practical problems involve a functional relationship $y = f(x)$ in which both x and y are themselves functions of another variable, such as time t. Implicit differentiation is then used to relate the rate of change dy/dt to the rate dx/dt. In this section, we will examine a variety of such **related rate problems**.

When working a related-rate problem, you must distinguish between the general situation and the specific situation. The *general situation* comprises properties that are true at *every* instant of time, whereas the *specific situation* refers to those properties that are guaranteed to be true only at the *particular* instant of time that the problem investigates. Here is an example.

Example 1 An application involving related rates

A spherical balloon is being filled with a gas in such a way that when the radius is 2 ft, the radius is increasing at the rate of 1/6 ft/min. How fast is the volume changing at this time?

Solution
The general situation: Let V denote the volume and r the radius, both of which are functions of time t (minutes). Since the container is a sphere, its volume is given by

$$V = \frac{4}{3}\pi r^3$$

Differentiating both sides implicitly with respect to time t yields

$$\frac{dV}{dt} = \frac{d}{dt}\left(\frac{4}{3}\pi r^3\right)$$

$$= 4\pi r^2 \frac{dr}{dt} \qquad \textit{Do not forget to use the chain rule because } r \textit{ is also a function of time.}$$

NOAA

The specific situation: Our goal is to find $\dfrac{dV}{dt}$ at the time when $r = 2$ ft and $\dfrac{dr}{dt} = \dfrac{1}{6}$.

$$\left.\frac{dV}{dt}\right|_{r=2} = 4\pi(2)^2\left(\frac{1}{6}\right) = \frac{8\pi}{3}$$

This means that the volume of the container is increasing at about 8.38 ft³/min when the radius is 2 ft. ■

Although each related rate problem has its own "personality," many can be handled by the following summary:

> **SOLVING RELATED RATE PROBLEMS**   The procedure for solving related rate problems follows these steps:
>
> **The General Situation**
>
> **Step 1** *Draw a figure, if appropriate, and assign variables to the quantities that vary.* Be careful not to label a quantity with a number unless it *never* changes in the problem.
>
> **Step 2** *Find a formula or equation that relates the variables.* Eliminate unnecessary variables; some of these "extra" variables may be constants, but others may be eliminated because of given relationships among the variables.
>
> **Step 3** *Differentiate the equations.* You will usually differentiate implicitly with respect to time.
>
> **The Specific Situation**
>
> **Step 4** *Substitute specific numerical values and solve algebraically for any required rate.* List the known quantities; list as unknown the quantity you wish to find. Substitute all values into the formula. The only remaining variable should be the unknown, which may be a variable or a rate. Solve for the unknown.

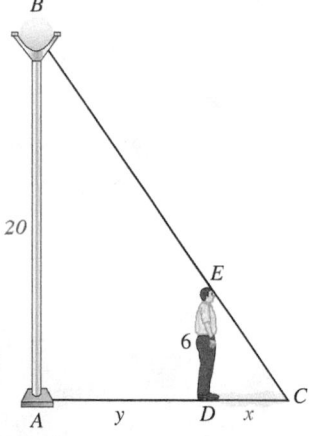

Figure 3.40   A person walking away from a street lamp

Example 2   Moving shadow problem

A person 6 ft tall is walking away from a street light 20 ft high at the rate of 7 ft/s. At what rate is the length of the person's shadow increasing?

Solution

The general situation:

Step 1. Let x denote the length (in feet) of the person's shadow, and y, the distance between the person and the street light, as shown in Figure 3.40.
Let t denote the time (in seconds).

Step 2. Since $\triangle ABC$ and $\triangle DEC$ are similar, we have

$$\frac{x+y}{20} = \frac{x}{6}$$

Step 3. Write this equation as $x + y = \frac{20}{6}x$, or $y = \frac{7}{3}x$, and differentiate both sides with respect to t.

$$\frac{dy}{dt} = \frac{7}{3}\frac{dx}{dt}$$

The specific situation:

Step 4. List the known quantities. We know that $dy/dt = 7$. Our goal is to find dx/dt. Substitute and then solve for the unknown value:

$$\frac{dy}{dt} = \frac{7}{3}\frac{dx}{dt}$$

$$7 = \frac{7}{3}\frac{dx}{dt} \qquad \textit{Substitute}$$

$$3 = \frac{dx}{dt} \qquad \textit{Multiply both sides by } \tfrac{3}{7}.$$

The length of the person's shadow is increasing at the rate of 3 ft/s.

Example 3   Leaning ladder problem

A bag is tied to the top of a 5-m ladder resting against a vertical wall. Suppose the ladder begins sliding down the wall in such a way that the foot of the ladder is moving away from the wall. How fast is the bag descending at the instant the foot of the ladder is 4 m from the wall and the foot is moving away at the rate of 2 m/s?

Solution

The general situation: Let x and y be the distances from the base of the wall to the foot and top of the ladder, respectively, as shown in Figure 3.41.

Notice that $\triangle TOB$ is a right triangle, so a relevant formula is the Pythagorean theorem:

$$x^2 + y^2 = 25$$

Differentiate both sides of this equation with respect to t:

$$2x\frac{dx}{dt} + 2y\frac{dy}{dt} = 0$$

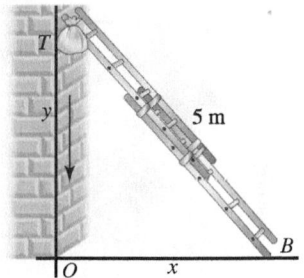

The specific situation: At the particular instant in question, $x = 4$ and $y = \sqrt{25 - 4^2} = 3$. We also know that $\dfrac{dx}{dt} = 2$, and the goal is to find $\dfrac{dy}{dt}$ at this instant. We have

Figure 3.41   Ladder problem

$$2(4)(2) + 2(3)\frac{dy}{dt} = 0$$
$$\frac{dy}{dt} = -\frac{8}{3}$$

This tells us that, at the instant in question, the bag is descending (since dy/dt is negative) at the rate of $\frac{8}{3} \approx 2.7$ m/sec. ∎

Example 4   Modeling a physical application involving related rates

When air expands *adiabatically* (that is, with no change in heat), the pressure P and the volume V satisfy the relationship

$$PV^{1.4} = C$$

where C is a constant. At a certain instant, the pressure is 20 lb/in.^{2} and the volume is 280 in.^{3}. If the volume is decreasing at the rate of 5 in.^{3}/s at this instant, what is the rate of change of the pressure?

Solution

The general situation: The required equation was given, so we begin by differentiating both sides with respect to t. Remember, because C is a constant, its derivative with respect to t is zero.

$$1.4PV^{0.4}\frac{dV}{dt} + V^{1.4}\frac{dP}{dt} = 0 \qquad \textit{Product rule}$$

The specific situation: At the instant in question, $P = 20$, $V = 280$, and $dV/dt = -5$ (negative because the volume is decreasing). The goal is to find dP/dt. First substitute to obtain

$$(1.4)(20)(280)^{0.4}(-5) + (280)^{1.4}\frac{dP}{dt} = 0$$

Now, solve for $\dfrac{dP}{dt}$:

$$\frac{dP}{dt} = \frac{5(1.4)(20)(280)^{0.4}}{(280)^{1.4}} = 0.5$$

Thus, at the instant in question, the pressure is increasing (because its derivative is positive) at the rate of 0.5 lb/in.^{2} per second. ■

Example 5 The water level in a cone-shaped tank

A tank filled with water is in the shape of an inverted cone 20 ft high with a circular base (on top) whose radius is 5 ft. Water is running out of the bottom of the tank at the constant rate of 2 ft^3/min. How fast is the water level falling when the water is 8 ft deep?

Solution

The general situation: Consider a conical tank with height 20 ft and circular base of radius 5 ft, as shown in Figure 3.42. Suppose that the water level is h ft and that the radius of the surface of the water is r.

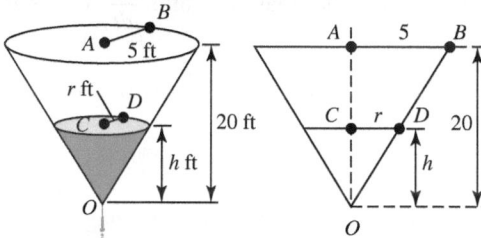

Figure 3.42 Dimensions of a conical tank

Let V denote the volume of water in the tank after t minutes. We know that

$$V = \frac{1}{3}\pi r^2 h$$

Once again, we use similar triangles (see Figure 3.42) to write $\frac{5}{20} = \frac{r}{h}$, or $r = \frac{h}{4}$. We substitute this into the formula to obtain

$$V = \frac{1}{3}\pi \left(\frac{h}{4}\right)^2 h = \frac{1}{48}\pi h^3$$

and then differentiate both sides of this equation with respect to t.

$$\frac{dV}{dt} = \frac{\pi}{16}h^2\frac{dh}{dt}$$

A common error in solving related-rate problems is to substitute numerical values too soon or, equivalently, to use relationships that apply only at a particular moment in time. This is the reason we have separated related-rate problems into two distinct parts. Be careful to work with general relationships among the variables and substitute specific numerical values only after you have found general rate relationships by differentiation.

The specific situation: Begin with the known quantities: We know that $dV/dt = -2$ (negative, because the volume is decreasing). The goal is to find dh/dt. At the particular instant in question, $h = 8$; we substitute to find

$$-2 = \frac{\pi}{16}(8)^2\frac{dh}{dt}$$
$$\frac{-1}{2\pi} = \frac{dh}{dt}$$

At the instant when the water is 8 ft deep, the water level is falling (since dh/dt is negative) at a rate of $\frac{1}{2\pi} \approx 0.16$ ft/min ≈ 2 in./min. ■

Example 6   Angle of elevation

Every day, a flight to Los Angeles flies directly over my home at a constant altitude of 4 mi. If I assume that the plane is flying at a constant speed of 400 mi/h, at what rate is the angle of elevation of my line of sight changing with respect to time when the horizontal distance between the approaching plane and my location is exactly 3 mi?

Solution

The general situation: Let x denote the horizontal distance between the plane and the observer, as shown in Figure 3.43. The height of the observer is insignificant when compared to the height of the plane.

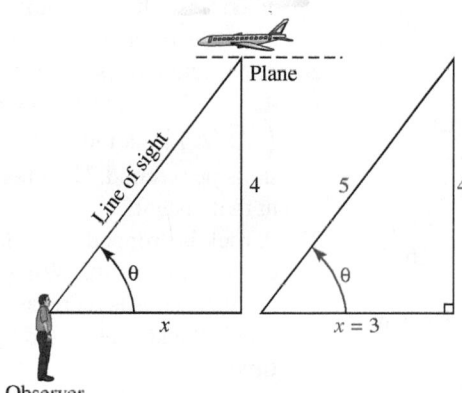

a. Observation of a plane   **b.** Related triangle

Figure 3.43   Angle of elevation problem

Then the angle of observation θ can be modeled by

$$\cot \theta = \frac{x}{4} \quad \text{or} \quad \theta = \cot^{-1}\left(\frac{x}{4}\right)$$

Differentiate both sides of this last equation with respect to t to obtain

$$\frac{d\theta}{dt} = \frac{-1}{1 + \left(\frac{x}{4}\right)^2}\left(\frac{1}{4}\right)\frac{dx}{dt} = \frac{-4}{16 + x^2}\frac{dx}{dt}$$

The specific situation: At the instant when $x = 3$, we are given that $\dfrac{dx}{dt} = -400$ (negative because the distance is decreasing). Thus, the angle of elevation is changing at the rate of

$$\frac{d\theta}{dt} = \frac{-4}{16 + 3^2}(-400) = 64 \text{ rad/hr}$$

The angle of elevation is changing at the rate of 64 radians per hour or, equivalently,

$$(64 \text{ rad/h})\left(\frac{360 \text{ deg}}{2\pi \text{ rad}}\right)\left(\frac{1 \text{ hr}}{3{,}600 \text{ sec}}\right) \approx 1.02 \text{ deg/s}$$

PROBLEM SET 3.7

Level 1

Find the indicated rates in Problems 1-22 using the given information. Assume that x and y are positive.

1. Find dy/dt where $2x + 3y - 5 = 0$ and $dx/dt = 3$ when $x = 5$.
2. Find dx/dt where $2x + 3y - 5 = 0$ and $dy/dt = 8$ when $x = 9$.
3. Find dx/dt when $y = 2x^2 + 3x + 4$ and $dy/dt = 46$ when $x = 5$.
4. Find dy/dt when $y = 2x^2 + 3x + 4$ and $dx/dt = 10$ when $x = 4$.
5. Find dx/dt where $5x^2 - y = 100$ and $dy/dt = 40$ when $x = 5$.
6. Find dy/dt where $5x^2 - y = 100$ and $dx/dt = 10$ when $x = 10$.
7. Find dx/dt where $4x^2 - y = 100$ and $dy/dt = -16$ when $x = 8$.
8. Find dy/dt where $4x^2 - y = 100$ and $dx/dt = 8$ when $x = 15$.
9. Find dy/dt where $y = \sqrt{x}$ and $dx/dt = 6$ when $x = 144$.
10. Find dx/dt where $y = \sqrt{x}$ and $dy/dt = -5$ when $x = 36$.
11. Find dx/dt where $y = 2\sqrt{x} - 9$ and $dy/dt = 5$ when $x = 9$.
12. Find dy/dt where $y = 2\sqrt{x} - 9$ and $dx/dt = 3$ when $x = 4$.
13. Find dx/dt where $y = 5\sqrt{x+9}$ and $dy/dt = 4$ when $x = 16$.
14. Find dy/dt where $y = 5\sqrt{x+9}$ and $dx/dt = 2$ when $x = 7$.
15. Find dy/dt where $xy = 10$ and $dx/dt = -2$ when $x = 5$.
16. Find dx/dt where $xy = 10$ and $dy/dt = 3$ when $x = 4$.
17. Find dy/dt where $x^3 + y^3 = 1,000$ and $dx/dt = 20$ when $x = \sqrt[3]{500}$.
18. Find dx/dt where $x^3 + y^3 = 1,000$ and $dy/dt = 10$ when $x = 8$.
19. Find dy/dt where $5xy^2 = 10$ and $dx/dt = -2$ when $x = 1$ and $y > 0$.
20. Find dx/dt where $5xy^2 = 10$ and $dy/dt = -6$ when $x = 4$ and $y > 0$.
21. Find dx/dt where $x^2 + xy - y^2 = 20$ and $dy/dt = 5$ when $x = 4$.
22. Find dy/dt where $x^2 + xy - y^2 = 20$ and $dx/dt = 1$ when $x = 6$.

In physics, Hooke's law says that when a spring is stretched x units beyond its natural length, the elastic force $F(x)$ exerted by the spring is $F(x) = -kx$, where k is a constant that depends on the spring. Assume $k = 12$ in Problems 23 and 24.

23. If a spring is stretched at the constant rate of $\frac{1}{4}$ in./s, how fast is the force $F(x)$ changing when $x = 2$ in.?
24. If a spring is stretched at the constant rate of $\frac{1}{4}$ in./s, how fast is the force $F(x)$ changing when $x = 3$ in.?
25. A particle moves along the parabolic path given by $y^2 = 4x$ in such a way that when it is at the point $(1, -2)$, its horizontal velocity (in the direction of the x-axis) is 3 ft/s. What is its vertical velocity (in the direction of the y-axis) at this instant?
26. A particle moves along the elliptical path given by $4x^2 + y^2 = 4$ in such a way that when it is at the point $\left(\sqrt{3}/2, 1 \right)$, its x-coordinate is increasing at the rate of 5 units per second. How fast is the y-coordinate changing at that instant?
27. A rock is dropped into a lake and an expanding circular ripple results. When the radius of the ripple is 8 in., the radius is increasing at a rate of 3 in./s. At what rate is the area enclosed by the ripple changing at this time?

28. A pebble dropped into a pond causes a circular ripple. Find the rate at which the radius of the ripple is changing at a time when the radius is one foot and the area enclosed by the ripple is increasing at the rate of 4 ft²/s.
29. An environmental study of a certain community indicates that there will be

$$Q(p) = p^2 + 3p + 1$$

units of a harmful pollutant in the air when the population is p thousand. The population is currently 30,000 and is increasing at a rate of 2,000 per year. At what rate is the level of air pollution increasing currently?
30. It is estimated that the annual advertising revenue received by a certain newspaper will be

$$R(x) = 0.5x^2 + 3x + 160$$

thousand dollars when its circulation is x thousand. The circulation of the paper is currently 10,000 and is increasing at a rate of 2,000 per year. At what rate will the annual advertising revenue be increasing with respect to time 2 years from now?

31. Hospital officials estimate that approximately

$$N(p) = p^2 + 5p + 900$$

people will seek treatment in the emergency room each year if the population of the community is p thousand. The population is currently 20,000 and is growing at the rate of 1,200 per year. At what rate is the number of people seeking emergency room treatment increasing?

32. Boyle's law states that when gas is compressed at constant temperature, the pressure P of a given sample satisfies the equation $PV = C$, where V is the volume of the sample and C is a constant. Suppose that at a certain time the volume is 30 in.^{3}, the pressure is 90 lb/in.^{2}, and the volume is increasing at the rate of 10 in.^{3}/s. How fast is the pressure changing at this instant? Is it increasing or decreasing?

 Level 2

33. **INTERPRETATION PROBLEM** What do we mean by a related rate problem?

34. **INTERPRETATION PROBLEM** Outline a procedure for solving related rate problems.

35. The volume of a spherical balloon is increasing at a constant rate of 3 in.^{3}/s. At what rate is the radius of the balloon increasing when the radius is 2 in.?

36. The surface area of a sphere is decreasing at the constant rate of 3π cm^2/s. At what rate is the volume of the sphere decreasing at the instant its radius is 2 cm?

37. A person 6 ft tall walks away from a street light at the rate of 5 ft/s. If the light is 18 ft above ground level, how fast is the person's shadow lengthening?

38. A ladder 13 ft long rests against a vertical wall and is sliding down the wall at the rate of 3 ft/s at the instant the foot of the ladder is 5 ft from the base of the wall. At this instant, how fast is the foot of the ladder moving away from the wall?

39. A car traveling north at 40 mi/h and a truck traveling east at 30 mi/h leave an intersection at the same time. At what rate will the distance between them be changing 3 hours later?

40. A person is standing at the end of a pier 12 ft above the water and is pulling in a rope attached to a rowboat at the waterline at the rate of 6 ft of rope per minute, as shown in Figure 3.44. How fast is the boat moving in the water when it is 16 ft from the pier?

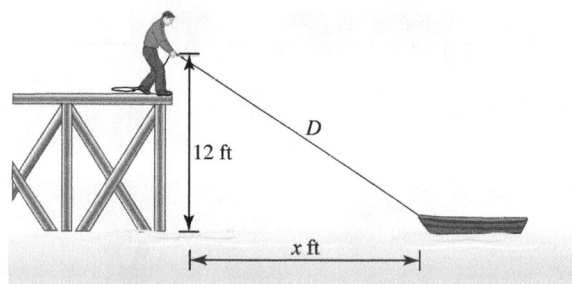

Figure 3.44 Problem 40

41. One end of a rope is fastened to a boat and the other end is wound around a windlass located on a dock at a point 4 m above the level of the boat. If the boat is drifting away from the dock at the rate of 2 m/min, how fast is the rope unwinding at the instant when the length of the rope is 5 m?

42. A ball is dropped from a height of 160 ft. A light is located at the same level, 10 ft away from the initial position of the ball. How fast is the ball's shadow moving along the ground one second after the ball is dropped?

43. A person 6 ft tall stands 10 ft from point P directly beneath a lantern hanging 30 ft above the ground, as shown in Figure 3.45.

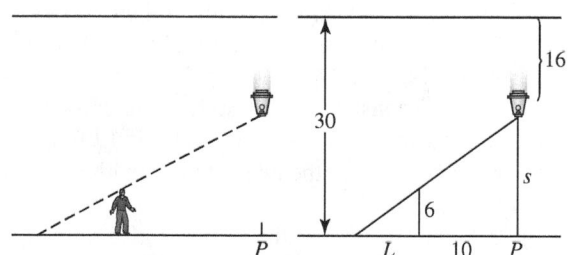

Figure 3.45 Problem 43

The lantern starts to fall, thus causing the person's shadow to lengthen. Given that the lantern falls $16 t^2$ ft in t seconds, how fast will the shadow be lengthening when $t = 1$?

44. A race official is watching a race car approach the finish line at the rate of 200 km/h. Suppose the official is sitting at the finish line, 20 m from the point where the car will cross, and let θ be the angle between the finish line and the official's line of sight to the car, as shown as Figure 3.46. At what rate is θ changing when the car crosses the finish line? Give your answer in terms of rad/s.

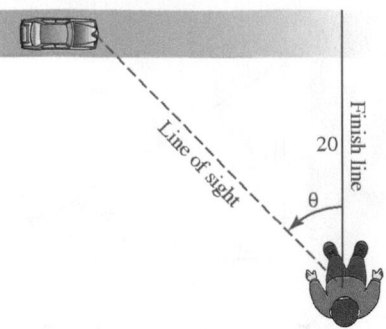

Figure 3.46   Problem 44

Modeling Problems: *In Problems* 45-48, *set up and solve an appropriate model to answer the given question. Be sure to state your assumptions.*

45. Consider a piece of ice in the shape of a sphere that is melting at the rate of 5 in.^{3}/min. Model the volume of ice by a function of the radius r. How fast is the radius changing at the instant when the radius is 4 in.? How fast is the surface area of the sphere changing at the same instant? What assumption are you making in this model about the shape of the ice?

46. A certain medical procedure requires that a balloon be inserted into the stomach and then inflated. Model the shape of the balloon by a sphere of radius r. If r is increasing at the rate of 0.3 cm/min, how fast is the volume changing when the radius is 4 cm?

47. Model a water tank by a cone 40 ft high with a circular base of radius 20 ft at the top. Water is flowing into the tank at a constant rate of 80 ft^3/min. How fast is the water level rising when the water is 12 ft deep? Give your answer to the nearest hundredth of a foot per minute.

48. In Problem 47, suppose that water is also flowing out the bottom of the tank. At what rate should the water be allowed to flow out so that the water level will be rising at a rate of only 0.05 ft/min when the water is 12 ft deep? Give your answer to the nearest tenth of a cubic foot per minute.

49. The air pressure $p(s)$ at a height of s meters above sea level is modeled by the formula $p(s) = e^{-0.000125s}$ atmospheres. An instrument box carrying a device for measuring pressure is dropped into the ocean from a plane and falls in such a way that after t seconds it is

$$s(t) = 3,000 - 49t - 245(e^{-t/5} - 1)$$

meters above the ocean's surface.

a. Find ds/dt and then use the chain rule to find dp/dt. How fast is the air pressure changing 2 seconds after the box begins to fall?

b. When does the box hit the water (rounded to the nearest second)? How fast is the air pressure changing at the time of impact?
What assumptions are you making in this model?

50. At noon on a certain day, a truck is 250 mi due east of a car. The truck is traveling west at a constant speed of 25 mi/h, while the car is traveling north at 50 mi/h.
a. At what rate is the distance between them changing at time t?
b. At what time is the distance between the car and the truck neither increasing nor decreasing?
c. What is the minimal distance between the car and the truck? *Hint*: This distance must occur at the time found in part **b**. Do you see why?

51. A weather balloon is rising vertically at the rate of 10 ft/s. An observer is standing on the ground 300 ft horizontally from the point where the balloon was released. At what rate is the distance between the observer and the balloon changing when the balloon is 400 ft high?

52. An observer watches a plane approach at a speed of 500 mi/h and an altitude of 3 mi. At what rate is the angle of elevation of the observer's line of sight changing with respect to time when the horizontal distance between the plane and the observer is 4 mi? Give your answer in radians per minute.

53. A person 6 ft tall is watching a street light 18 ft high while walking toward it at a speed of 5 ft/s, as shown in Figure 3.47. At what rate is the angle of elevation of the person's line of sight changing with respect to time when the person is 9 ft from the base of the light?

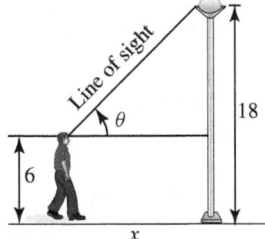

Figure 3.47   Problem 53

54. A revolving searchlight in a lighthouse 2 mi offshore is following a beachcomber along the shore, as shown in Figure 3.48.

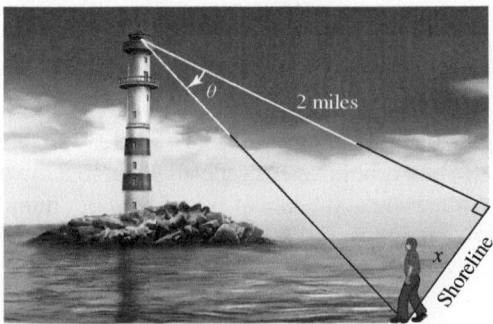

Figure 3.48   Problem 54

When the beachcomber is 1 mi from the point on the shore that is closest to the lighthouse, the searchlight is turning at the rate of 0.25 rev/h. How fast is the beach-comber walking at that moment? *Hint*: Note that 0.25 rev/h is the same as $\dfrac{\pi}{2}$ rad/h.

55. A water trough is 2 ft deep and 10 ft long. It has a trape-zoidal cross section with base lengths 2 ft and 5 ft, as shown in Figure 3.49.

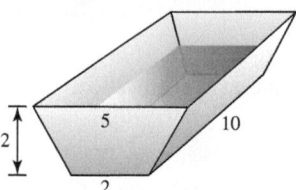

Figure 3.49   Water trough

a. Find a relationship between the volume of water in the trough at any given time and the depth of the water at that time.
b. If water enters the trough at the rate of 10 ft³/min, how fast is the water level rising (to the nearest $\frac{1}{2}$ in./min) when the water is 1 ft deep?

56. At noon, a ship sails due north from a point P at 8 knots (nautical miles per hour). Another ship, sailing at 12 knots, leaves the same point 1 h later on a course 60° east of north. How fast is the distance between the ships increasing at 2 P.M.? At 5 P.M.? *Hint*: Use the law of cosines.

57. A swimming pool is 60 ft long and 25 ft wide. Its depth varies uniformly from 3 ft at the shallow end to 15 ft at the deep end, as shown in Figure 3.50.

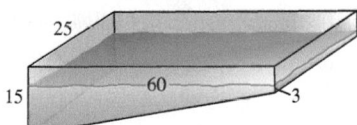

Figure 3.50   Swimming pool

Suppose the pool is being filled with water at the rate of 800 ft³/min. At what rate is the depth of water increasing at the deep end when it is 5 ft deep at that end?

58. Suppose a water bucket is modeled by the frustum of a cone with height 1 ft and upper and lower radii of 1 ft and 9 in., respectively, as shown in Figure 3.51.

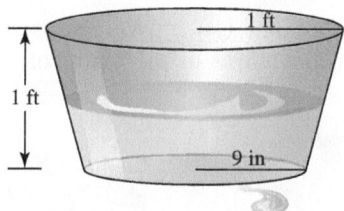

Figure 3.51   Water bucket

If water is leaking from the bottom of the bucket at the rate of 8 in.³/min, at what rate is the water level falling when the depth of water in the bucket is 6 in.?

Hint: The volume of the frustum of a cone with height h and base radii r and R is

$$V = \frac{\pi h}{3}\left(R^2 + rR + r^2\right)$$

59. A lighthouse is located 2 km directly across the sea from a point O on the shoreline, as shown in Figure 3.52.

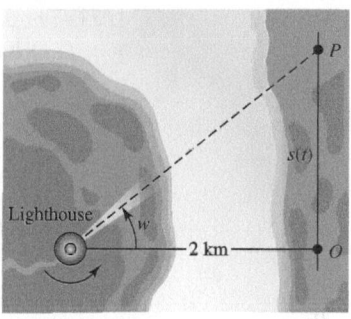

Figure 3.52   Problem 59

A beacon in the lighthouse makes 3 complete revolutions (that is, 6π radians) each minute, and during part of each revolution, the light sweeps across the face of a row of cliffs lining the shore.

a. Show that t minutes after it passes point O, the beam of light is at a point P located $s(t) = 2\tan(6\pi t)$ km from O.
b. How fast is the beam of light moving at the time it passes a point on the cliff that is 4 km from the lighthouse?

60. Modeling Problem   A car is traveling at the rate of 40 ft/s along a straight, level road that parallels the seashore. A rock with a family of seals is located 50 yd offshore.

a. Model the angle θ between the road and the driver's line of sight as a function of the distance x to the point P directly opposite the rocks.
b. As the distance x in Figure 3.53 approaches 0, what happens to $d\theta/dt$?

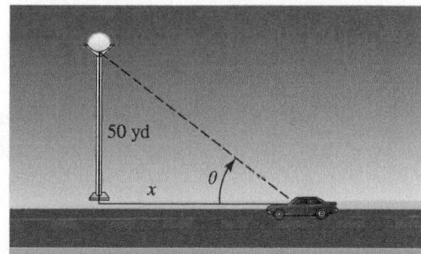

Figure 3.53   Problem 60

c. Suppose the car is traveling at v ft/s. Now what happens to $d\theta/dt$ as $x \to 0$? What effect does this have on a passenger looking at the seals if the car is traveling at a high rate of speed?

3.8 LINEAR APPROXIMATION AND DIFFERENTIALS

IN THIS SECTION: *Tangent line approximation, the differential, error propagation, marginal analysis in economics, the Newton-Raphson method for approximating roots*
We introduce a new symbol, dx, called the *differential*, which is used to estimate changes in functional values. A special case of the approximation procedure is the concept of *marginal analysis*, which plays an important role in business and economics.

Tangent Line Approximation

If $f(x)$ is differentiable at $x = a$, the tangent line at a point $P\,(a,\,f(a))$ on the graph of $y = f(x)$ has slope $m = f'(a)$ and equation

$$\frac{y - f(a)}{x - a} = f'(a) \qquad \text{or} \qquad y = f(a) + f'(a)(x - a)$$

In the immediate vicinity of P, the tangent line closely approximates the shape of the curve $y = f(x)$. For instance, if $f(x) = x^3 - 2x + 5$, the tangent line at $P(1, 4)$ has slope $f'(1) = 3(1)^2 - 2 = 1$ and equation

$$y = 4 + (1)(x - 1) = x + 3$$

The graph of $y = f(x)$, the tangent line at $P(1, 4)$, and two enlargements showing how the tangent line approximates the graph of f near P are shown in Figure 3.54.

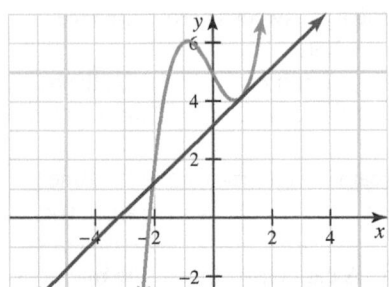 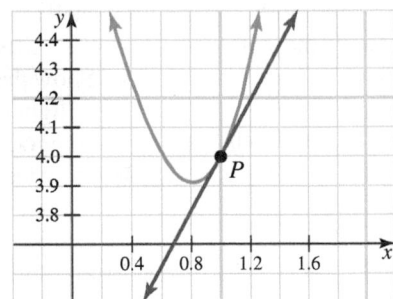 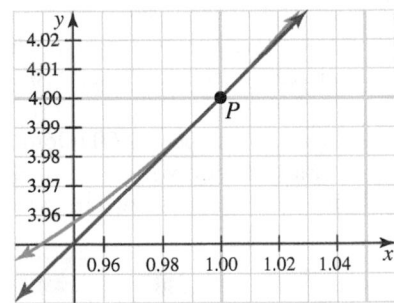

Figure 3.54   Tangent line approximation of $f(x) = x^3 - 2x + 5$ at $P(1, 4)$

Our observation about tangent lines suggests that if x_1 is near a, then $f(x_1)$ must be close to the point on the tangent line to $y = f(x)$ at $x = x_1$. That is,

$$f(x_1) \approx f(a) + f'(a)(x_1 - a)$$

We refer to this as a *linear approximation* of $f(x)$ at $x = a$, and the function

$$L(x) = f(a) + f'(a)(x - a)$$

is called a **linearization** of the function at a point $x = a$. We can use this line as an approximation of f as long as the line remains close to the graph of f, as shown in Figure 3.55.

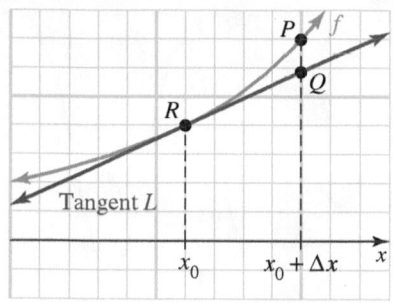

a. Tangent line to f at R

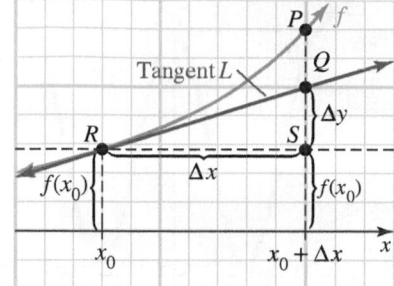

b. Tangent line approximation

Figure 3.55 Tangent line approximation

Recall that we use the notation Δx for horizontal change and Δy for vertical change. Here, we use this delta notation to represent the horizontal and vertical change from the point R to the point P. The linear approximation formula measures the vertical change from point S to point Q even though we really want the vertical change from the point S to the point P. If the distance Δx is small, then these two vertical distances should be approximately equal. This version of linear approximation is sometimes called the *incremental approximation formula*.

$$f(x_1) - f(a) \approx f'(a)(x_1 - a)$$

$$\Delta y \approx f'(a)\Delta x$$

Example 1 Incremental approximation

Show that if $f(x) = \sin x$, the function $\dfrac{\Delta f}{\Delta x}$ approximates the function $f'(x) = \cos x$ for small values of Δx.

Solution   The approximation formula

$$\Delta f = f(x_0 + \Delta x) - f(x_0) \approx f'(x_0)\Delta x$$

implies that

$$\frac{\Delta f}{\Delta x} = \frac{f(x_0 + \Delta x) - f(x_0)}{\Delta x} \approx f'(x_0)$$

Because $f(x) = \sin x$, $f'(x) = \cos x$, and

$$\frac{\sin(x + \Delta x) - \sin x}{\Delta x} \approx \cos x$$

Figure 3.56 shows the graphs for three different choices of Δx. Notice as Δx becomes smaller, it is more difficult to see the difference between f and g; in fact, for very small Δx, the graphs are virtually indistinguishable. ∎

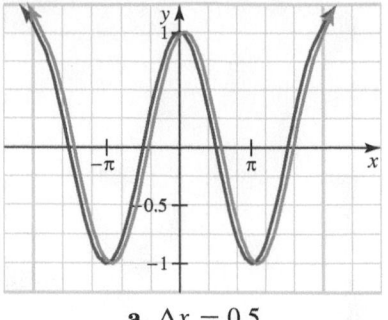

a. $\Delta x = 0.5$

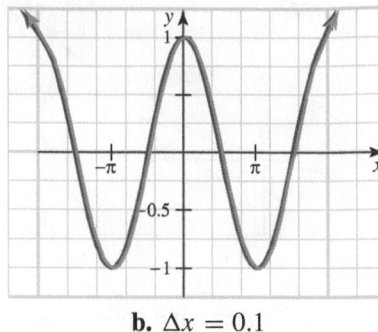

b. $\Delta x = 0.1$

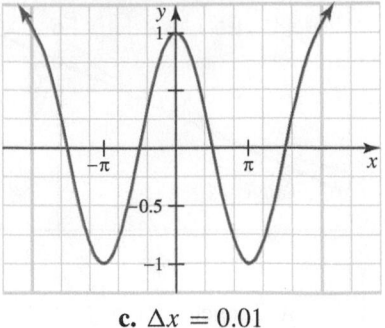

c. $\Delta x = 0.01$

Figure 3.56   Graphs of $f'(x) = \cos x$ and $g(x) = \dfrac{f(x + \Delta x) - f(x)}{\Delta x}$

Differential

We have already observed that writing the derivative of $f(x)$ in the Leibniz notation df/dx suggests that the derivative may be incorrectly regarded as a quotient of "df" by "dx." It is a tribute to the genius of Leibniz that this erroneous interpretation of his notation often turns out to make good sense.

To give dx and dy meaning as separate quantities, let x be fixed and define dx to be an independent variable equal to Δx, the change in x. That is, define dx, called the **differential of x**, to be an independent variable equal to Δx. Then, if f is differentiable at x, we define dy, called the **differential of y**, by the formula

$$dy = f'(x)\, dx \qquad \text{or, equivalently,} \qquad df = f'(x)\, dx$$

If we relate differentials to Figure 3.57, we see that $dx = \Delta x$ and that Δy is the rise of f that occurs for a change of Δx, whereas dy is the rise of a tangent line relative to the same change in x (Δy and dy are not the same thing). This is shown in Figure 3.57.

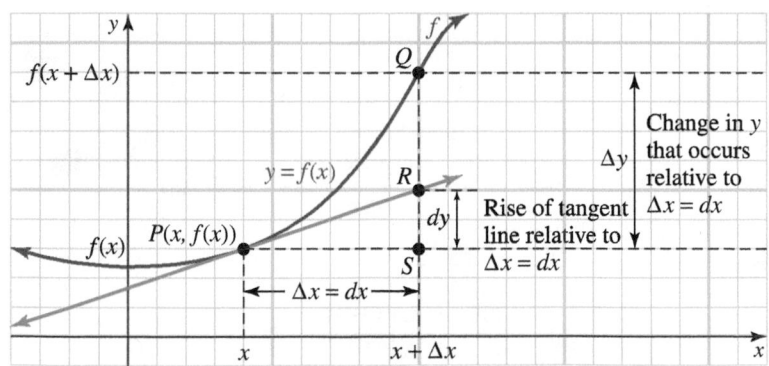

Figure 3.57   Geometrical definition of dx and dy

We now restate the standard rules and formulas for differentiation in terms of differentials. Remember that a and b are constants, while f and g are functions.

DIFFERENTIAL RULES

Linearity rule	$d(af + bg) = a\,df + b\,dg$
Product rule	$d(fg) = f\,dg + g\,df$
Quotient rule	$d\left(\dfrac{f}{g}\right) = \dfrac{g\,df - f\,dg}{g^2} \qquad (g \neq 0)$
Power rule	$d(x^n) = nx^{n-1}dx$

Trigonometric rules $\quad d(\sin x) = \cos x\,dx \qquad d(\cos x) = -\sin x\,dx$

$$d(\tan x) = \sec^2 x\,dx \qquad d(\cot x) = -\csc^2 x\,dx$$

$$d(\sec x) = \sec x \tan x\,dx \qquad d(\csc x) = -\csc x \cot x\,dx$$

Exponential rule $\quad d(e^x) = e^x dx$

Logarithmic rule $\quad d(\ln x) = \dfrac{1}{x}dx$

Inverse trigonometric rules

$$d\left(\sin^{-1} x\right) = \frac{dx}{\sqrt{1 - x^2}} \qquad d(\cos^{-1} x) = \frac{-dx}{\sqrt{1 - x^2}}$$

$$d(\tan^{-1} x) = \frac{dx}{1 + x^2} \qquad d(\cot^{-1} x) = \frac{-dx}{1 + x^2}$$

$$d(\sec^{-1} x) = \frac{dx}{|x|\sqrt{x^2 - 1}} \qquad d(\csc^{-1} x) = \frac{-dx}{|x|\sqrt{x^2 - 1}}$$

Note that each of the above differential formulas resembles the earlier corresponding derivative formula, and they can be remembered by formally "multiplying through" the corresponding derivative formula by "dx."

Example 2 Differential involving a product and a trigonometric function

Find $d(x^2 \sin x)$.

Solution   You can work with the differential, as follows:

$$\begin{aligned} d(x^2 \sin x) &= x^2 d(\sin x) + \sin x\,d(x^2) \\ &= x^2(\cos x\,dx) + \sin x\,(2x\,dx) \\ &= (x^2 \cos x + 2x \sin x)dx \end{aligned}$$

Or, you can work with the derivative, as shown:

$$\frac{d}{dx}(x^2 \sin x) = x^2 \cos x + 2x \sin x$$
$$d(x^2 \sin x) = (x^2 \cos x + 2x \sin x)\,dx$$

We can use differentials to approximate functional values as shown, in the following example.

Example 3 Comparing approximations of differential and calculator

Approximate $\dfrac{1}{3.98}$ using differentials, and compare with a calculator approximation.

Solution   Let $f(x) = \dfrac{1}{x} = x^{-1}$, so $f'(x) = -x^{-2}$.

For a very small quantity $|\Delta x|$, one can approximate:

$$f(x_0 + \Delta x) \approx f(x_0) + f'(x_0)\,dx$$

so we let $x_0 = 4$ and $\Delta x = dx = -0.02$ to find

$$\frac{1}{3.98} = \frac{1}{4 + (-0.02)}$$
$$\approx f(4) + f'(4)(-0.02)$$
$$= \frac{1}{4} + \frac{-1}{16}(-0.02)$$
$$= 0.25 + 0.00125$$
$$= 0.25125$$

By calculator, we find $\dfrac{1}{3.98} \approx 0.2512562814$

Error Propagation

In the next example, the approximation formula is used to study **propagation of error**, which is the term used to describe error that accumulates from other errors in an approximation. In particular, in the next example, the derivative is used to estimate the maximum error in a calculation that is based on figures obtained by imperfect measurement.

Example 4   Propagation of error in a volume measurement

You measure the side of a cube and find it to be 10 cm long. From this you conclude that the volume of the cube is $10^3 = 1,000$ cm^3. If your original measurement of the side is accurate to within 2% approximately how accurate is your calculation of the volume?

Solution   The volume of the cube is $V(x) = x^3$, where x is the length of a side. If you take the length of a side to be 10 when it is really $10 + \Delta x$, your error is Δx; and your corresponding error when computing the volume will be ΔV, given by

$$\Delta V = V(10 + \Delta x) - V(10) \approx V'(10)\Delta x$$

Remark that this is just a rough approximation, and the sign $\approx$ should never be confused with the equality sign $=$.

Now, $V'(x) = 3x^2$, so $V'(10) = 300$. Also, your measurement of the side can be off by as much as 2%—that is, by as much as $0.02(10) = 0.2$ cm in either direction. Substituting $\Delta x = \pm 0.2$ in the incremental approximation formula for ΔV, we get

$$\Delta V \approx 300(\pm 0.2) = \pm 60$$

Thus, the propagated error in computing the volume is approximately ± 60 cm^3. Hence the maximum error in your measurement of the side is $|\Delta x| = 0.2$ and the corresponding maximum error in your calculation for the volume is

$$|\Delta V| \approx V'(10)\,|\Delta x| = 300(0.2) = 60$$

This says that, at worst, your calculation of the volume as 1,000 cm^3 is off by 60 cm^3, or 6% of the calculated volume, when your maximum error in measuring the side is 2%.

ERROR PROPAGATION   If x represents the measured value of a variable and $x + \Delta x$ represents the exact value, then Δx is the **error in measurement**. The difference between $f(x + \Delta x)$ and $f(x)$ is called the **propagated error** at x and is defined by

$$\Delta f = f(x + \Delta x) - f(x)$$

The **relative error** is $\dfrac{\Delta f}{f} \approx \dfrac{df}{f}$.

The **percentage error** is $100\left(\dfrac{\Delta f}{f}\right)\%$

In Example 4, the approximate propagated error in measuring volume is ± 60, and the approximate relative error is $\dfrac{\Delta V}{V} \approx \dfrac{\pm 60}{10^3} = \pm 0.06$.

Example 5   Modeling relative error and percentage error

A certain container is modeled by a right circular cylinder whose height is twice the radius of the base. The radius is measured to be 17.3 cm, with a maximum measurement error of 0.02 cm. Estimate the corresponding propagated error, the relative error, and the percentage error when calculating the surface area S.

Solution   Figure 3.58 shows the container.

We have,

$$S = \underbrace{\overbrace{2\pi r}^{\text{Circumference}}\; \overbrace{(2r)}^{\text{Height}}}_{\text{Lateral side}} + \underbrace{\pi r^2}_{\text{Top}} + \underbrace{\pi r^2}_{\text{Bottom}} = 6\pi r^2$$

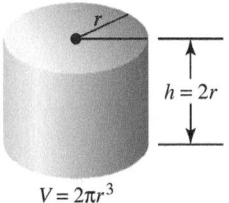

Figure 3.58   Right circular cylinder

where r is the radius of the cylinder's base. Then the approximate *propagated error* is

$$\Delta S \approx S'(r)\Delta r = 12\pi r\Delta r = 12\pi(17.3)(\pm 0.02) \approx \pm 13.0438927$$

Thus, the maximum error in the measurement of the surface area is about 13.04 cm². Is this a large or a small error? The *relative error* is found by computing the ratio

$$\frac{\Delta S}{S} \approx \frac{12\pi r\Delta r}{6\pi r^2} = 2r^{-1}\Delta r = 2(17.3)^{-1}(\pm 0.02) \approx \pm 0.0023121387$$

This tells us that the maximum error of approximately 13.04 is fairly small relative to the surface area S. The corresponding *percentage error* is found by

$$100\left(\frac{\Delta S}{S}\right)\% \approx 100\,(\pm 0.0023121387)\% = \pm 0.23121387\%$$

This means that the percentage error is about $\pm 0.23\%$.

Marginal Analysis in Economics

Marginal analysis is an area of economics concerned with estimating the effect on quantities such as cost, revenue, and profit when the level of production is changed by a unit amount. For example, if $C(x)$ is the cost of producing x units of a certain commodity, then the **marginal cost**, $MC(x)$, is the additional cost of producing one more unit and is given by the difference $MC(x) = C(x + 1) - C(x)$. Using the linear approximation formula with $\Delta x = 1$, we see that

$$MC(x) = C(x+1) - C(x) \approx C'(x)(1)$$

and for this reason, we will approximate marginal cost by the derivative $C'(x)$.

Similarly, if $R(x)$ is the revenue obtained from producing x units of a commodity, then the **marginal revenue**, $MR(x)$, is the additional revenue obtained from producing one more unit, and we approximate MR by the derivative $R'(x)$. To summarize:

MARGINAL COST   The marginal cost of producing x units of a commodity is computed by the derivative $C'(x)$. That is, $MC(x) \approx C'(x)$.

MARGINAL REVENUE   The marginal revenue of producing x units of a commodity is computed by the derivative $R'(x)$. That is, $MR(x) \approx R'(x)$.

Example 6   Modeling change in cost and revenue

A manufacturer models the total cost (in dollars) of a particular commodity by the function

$$C(x) = \frac{1}{8}x^2 + 3x + 98$$

and the price per item (in dollars) by

$$p(x) = \frac{1}{3}(75 - x)$$

where x is the number of items produced ($0 \leq x \leq 50$). The function p is called the **demand function** which is the market price per unit when x units are produced.

a. Find the marginal cost and the marginal revenue.
b. Use marginal cost to estimate the cost of producing the 9th unit. What is the actual cost of producing the 9th unit?
c. Use marginal revenue to estimate the revenue derived from producing the 9th unit. What is the actual revenue derived from producing the 9th unit?

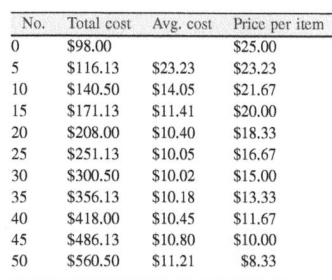

No.	Total cost	Avg. cost	Price per item
0	$98.00		$25.00
5	$116.13	$23.23	$23.23
10	$140.50	$14.05	$21.67
15	$171.13	$11.41	$20.00
20	$208.00	$10.40	$18.33
25	$251.13	$10.05	$16.67
30	$300.50	$10.02	$15.00
35	$356.13	$10.18	$13.33
40	$418.00	$10.45	$11.67
45	$486.13	$10.80	$10.00
50	$560.50	$11.21	$8.33

Solution   You can compare the values on $[0, 50]$ using a calculator in order to get a sense of the meaning of the problem.

a. The marginal cost is

$$C'(x) = \frac{1}{4}x + 3$$

To find the marginal revenue, we must first find the revenue function:

$$R(x) = xp(x) = x\left(\frac{1}{3}\right)(75 - x) = -\frac{1}{3}x^2 + 25x$$

Thus, the marginal revenue is

$$R'(x) = -\frac{2}{3}x + 25$$

b. The cost of producing the 9th unit is the change in cost as x increases from 8 to 9 and is estimated by

$$C'(8) = \frac{1}{4}(8) + 3 = 5$$

We estimate the cost of producing the 9th unit to be \$5. The actual cost is

$$\Delta C = C(9) - C(8)$$
$$= \left[\frac{1}{8}(9)^2 + 3(9) + 98\right] - \left[\frac{1}{8}(8)^2 + 3(8) + 98\right]$$
$$= 5.125 \qquad (\text{That is, \$5.13.})$$

c. The revenue (to the nearest cent) obtained from the sale of the 9th unit is approximated by the marginal revenue:

$$R'(8) = -\frac{2}{3}(8) + 25 = \frac{59}{3} \approx 19.67$$

(That is, \$19.67.) The actual revenue (to the nearest cent) obtained from the sale of the 9th unit is

$$\Delta R = R(9) - R(8)$$
$$= \frac{58}{3}$$
$$\approx 19.33 \qquad (\text{That is, \$19.33.})$$

Remember that the marginal revenue from the sale of the 9th item is not the revenue derived from selling 9 items. Rather, it is the additional revenue the company has earned by selling the 9th item—that is, the total revenue of 9 items minus the total revenue of 8 items.

The Newton-Raphson Method for Approximating Roots

The Newton-Raphson method is a different kind of tangent line approximation, one that uses tangent lines as a means for estimating roots of equations. The basic idea behind the procedure is illustrated in Figure 3.59.

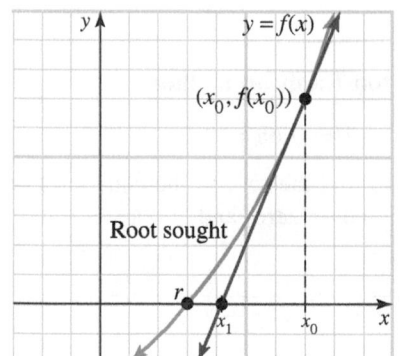

a. Interactive Estimating root r, of $y = f(x) = 0$

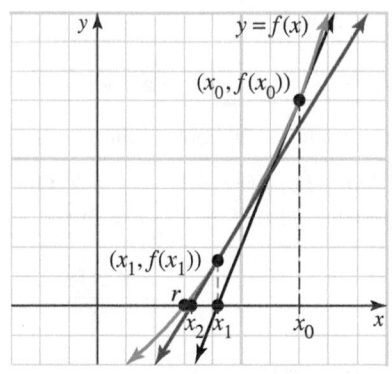

b. First second, and third estimates (x_0, x_1, and x_2 respectively)

Figure 3.59 The Newton-Raphson method

In this figure, r is a root of the equation $f(x) = 0$, x_0 is an approximation to r, and x_1 is a better approximation obtained by taking the x-intercept of the line that is tangent to the graph of f at $(x_0, f(x_0))$.

Theorem 3.14 The Newton-Raphson method

To approximate a root of the equation $f(x) = 0$, start with a preliminary estimate x_0 and generate a sequence $x_1, x_2, x_3, \cdots$ using the formula

$$x_{n+1} = x_n - \frac{f(x_n)}{f'(x_n)} \qquad f'(x_n) \neq 0$$

Either this sequence of approximations will approach a limit that is a root of the equation or else the sequence does not have a limit.

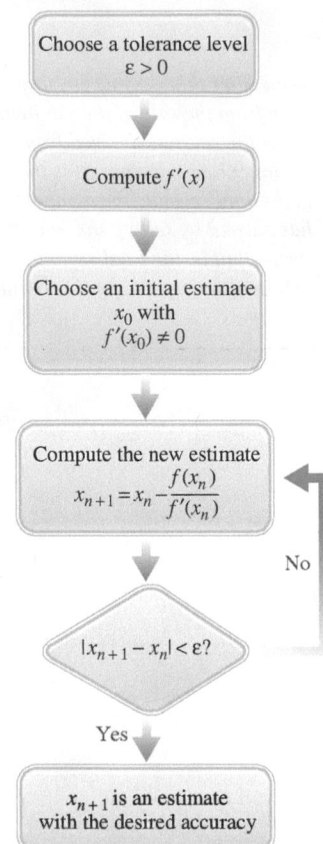

Figure 3.60 Flowchart for
the Newton-Raphson method

Proof: Rather than present a formal proof, we will present a geometric description of the procedure to help you understand what is happening. Let x_0 be an initial approximation such that $f'(x_0) \neq 0$. To find a formula for the improved approximation x_1, recall that the slope of the tangent line through $(x_0, f(x_0))$ is the derivative $f'(x_0)$. Therefore,

$$\underbrace{f'(x_0) = \frac{\Delta y}{\Delta x} = \frac{f(x_0) - 0}{x_0 - x_1}}_{\text{Slope of the tangent line through } (x_0, f(x_0))}$$

or, equivalently (by solving the equation for x_1),

$$x_1 = x_0 - \frac{f(x_0)}{f'(x_0)} \qquad f'(x_0) \neq 0$$

If this procedure is repeated using x_1 as the initial approximation, an even better approximation may often be obtained (see Figure 3.59b). This approximation, x_2, is related to x_1 as x_1 was related to x_0. That is,

$$x_2 = x_1 - \frac{f(x_1)}{f'(x_1)} \qquad f'(x_1) \neq 0$$

If this process produces a limit, it can be continued until the desired degree of accuracy is obtained. In general, the nth approximation x_n is related to the $(n-1)$st by the formula

$$x_n = x_{n-1} - \frac{f(x_{n-1})}{f'(x_{n-1})} \qquad f'(x_{n-1}) \neq 0 \qquad \blacklozenge$$

A step-by-step procedure for applying the Newton-Raphson method is shown in the flow chart in Figure 3.60.

Example 7   Estimating a root with the Newton-Raphson method

Approximate a real root of the equation $x^3 + x + 1 = 0$ on $[-2, 2]$.

Solution   Let $f(x) = x^3 + x + 1$. Our goal is to find the root of the equation $f(x) = 0$. The derivative of f is $f'(x) = 3x^2 + 1$, which is never zero, and so

$$x - \frac{f(x)}{f'(x)} = x - \frac{x^3 + x + 1}{3x^2 + 1} = \frac{2x^3 - 1}{3x^2 + 1}$$

Thus, for $n = 0, 1, 2, 3, \cdots$

$$x_{n+1} = x_n - \frac{f(x_n)}{f'(x_n)} = \frac{2x_n^3 - 1}{3x_n^2 + 1}$$

A convenient choice for the preliminary estimate is $x_0 = -1$. Then

$$x_1 = \frac{2x_0^3 - 1}{3x_0^2 + 1} = -0.75 \qquad \textit{You will need a calculator or a spreadsheet.}$$

$$x_2 = \frac{2x_1^3 - 1}{3x_1^2 + 1} \approx -0.6860465$$

$$x_3 = \frac{2x_3^3 - 1}{3x_3^2 + 1} \approx -0.6823396$$

So to two decimal places, the root seems to be approximately

$$x \approx -0.68$$

In general, we will stop finding new estimates when successive approximations x_n and x_{n+1} are within a desired tolerance of each other. Specifically, if we wish to have the solutions to be within ϵ ($\epsilon > 0$) of each other, we compute approximations until $|x_{n+1} - x_n| < \epsilon$ is satisfied. Using the Newton-Raphson method, we see that this condition is equivalent to

$$|x_{n+1} - x_n| = \left| \frac{-f(x_n)}{f'(x_n)} \right| < \epsilon \qquad f'(x_n) \neq 0$$

PROBLEM SET 3.8

Level 1

Find the differentials indicated in Problems 1-16.

1. $d(2x^3)$

2. $d(3 - 5x^2)$

3. $d(2\sqrt{x})$

4. $d(x^5 + \sqrt{x^2 + 5})$

5. $d(x \cos x)$

6. $d(x \sin 2x)$

7. $d\left(\dfrac{\tan 3x}{2x} \right)$

8. $d(xe^{-2x})$

9. $d(\ln |\sin x|)$

10. $d(x \tan^{-1} x)$

11. $d(e^x \ln x)$

12. $d\left(\dfrac{\sin^{-1} x}{e} \right)$

13. $d\left(\dfrac{x^2 \sec x}{x - 3} \right)$

14. $d(x\sqrt{x^2 - 1})$

15. $d\left(\dfrac{x - 5}{\sqrt{x + 4}} \right)$

16. $d\left(\dfrac{\ln \sqrt{x}}{x} \right)$

17. ■ What does this say? What is a differential?

18. ■ What does this say? Discuss error propagation, including relative error, and percentage error.

Use differentials to approximate the requested values in Problems 19-22 *and then determine the error as compared to the calculator value.*

19. $\sqrt{0.99}$

20. $\cos\left(\frac{\pi}{2} + 0.01 \right)$

21. $(3.01)^5 - 2(3.01)^3 + 3(3.01)^2 - 2$

22. $\sqrt[4]{4,100} + \sqrt[3]{4,100} + 3\sqrt{4,100}$

Level 2

23. You measure the radius of a circle to be 12 cm and use the formula $A = \pi r^2$ to calculate the area. If your measurement of the radius is accurate to within 3%, approximately how accurate (to the nearest percent) is your calculation of the area?

24. **EXPLORATION PROBLEM** Suppose a 12 oz can of Coke® has a height of 4.5 in. If your measurement of the radius has an accuracy to within 1%, how accurate is your measurement for volume? Check your answer by examining a Coke can.

25. You measure the radius of a sphere to be 6 in. Use the formula $V = \frac{4}{3}\pi r^3$ to calculate the volume. If your measurement of the radius is accurate to within 1%, approximately how accurate (to the nearest percent) is your calculation of the volume?

26. It is projected that t years from now the circulation of a local newspaper will be

$$C(t) = 100t^2 + 400t + 5{,}000$$

Estimate the amount by which the circulation will increase during the next 6 months.

27. An environmental study suggests that t years from now, the average level of carbon monoxide in the air will be

$$Q(t) = 0.05t^2 + 0.1t + 3.4$$

parts per million (ppm). By approximately how much will the carbon monoxide level change during the next 6 months?

28. A manufacturer's total cost (in dollars) is

$$C(q) = 0.1q^3 - 0.5q^2 + 500q + 200$$

when the level of production is q units. The current level of production is 4 units, and the manufacturer is planning to decrease this to 3.9 units. Estimate how the total cost will change as a result.

29. At a certain factory, the daily output is

$$Q(L) = 60{,}000L^{1/3}$$

units, where L denotes the size of the labor force measured in worker-hours. Currently 1,000 worker-hours of labor are used each day. Estimate the effect on output that will be produced if the labor force is cut to 940 worker-hours.

30. A soccer ball made of leather $\frac{1}{8}$ in. thick has an inner diameter of $8\frac{1}{2}$ in. Model the ball as a hollow sphere and estimate the volume of its leather shell.

31. A cubical box is to be constructed from three kinds of building materials. The material used in the four sides of the box costs 2¢ per in.^{2}, the material in the bottom costs 3¢ per in.^{2}, and the material used for the lid costs

4¢ per in.^{2}. Estimate the additional total cost of all the building materials if the length of a side is increased from 20 in. to 21 in.

32. In a healthy person of height x in., the average pulse rate in beats per minute is modeled by the formula

$$P(x) = \frac{596}{\sqrt{x}} \qquad 30 \le x \le 100$$

Estimate the change in pulse rate that corresponds to a height change from 59 to 60 in.

33. A drug is injected into a patient's bloodstream. The concentration of the drug in the bloodstream t hours after the drug is injected is modeled by the formula

$$C(t) = \frac{0.12t}{t^2 + t + 1}$$

milligrams per cubic centimeter. Estimate the change in concentration over the time period from 30 to 35 minutes after injection.

34. Modeling Problem   In a model developed by John Helms, the water evaporation $E(T)$ for a ponderosa pine is modeled by

$$E(t) = 4.6e^{17.3T/(T+237)}$$

where T (degrees Celsius) is the surrounding air temperature.* If the temperature is increased by 5% from 30°C, use differentials to estimate the corresponding percentage change in $E(T)$.

35. Modeling Problem   According to Poiseuille's law, the speed of blood flowing along the central axis of an artery of radius R is modeled by the formula $S(R) = cR^2$, where c is a constant.† What percentage error (rounded to the nearest percent) will you make in the calculation of $S(R)$ from this formula if you make a 1% error in the measurement of R?

36. Modeling Problem   One of the laws attributed to Poiseuille models the volume of a fluid flowing through a small tube in unit time under fixed pressure by the formula $V = kR^4$, where k is a positive constant and R is the radius of the tube. This formula is used in medicine to determine how wide a clogged artery must be opened to restore a healthy flow of blood. Suppose the radius of a certain artery is increased by 5%. Approximately what effect does this have on the volume of the blood flowing through the artery?‡

37. Modeling Problem   A certain cell is modeled as a sphere. If the formulas $S = 4\pi r^2$ and $V = \frac{4}{3}\pi r^3$ are used to compute the surface area and volume of the sphere, respectively, estimate the effect on S and V produced by a 1% increase in the radius r.

38. The period of a pendulum is given by the formula

$$T = 2\pi\sqrt{\frac{L}{g}}$$

where L is the length of the pendulum in feet, $g = 32$ ft/s^2 is the acceleration due to gravity, and T is time in seconds. If the pendulum has been heated enough to increase its length by 0.4%, what is the approximate percentage change in its period?

39. The *thermal expansion coefficient* of an object is defined to be

$$\sigma = \frac{L'(T)}{L(T)}$$

where $L(T)$ is the length of the object when the temperature is T. Suppose a 75-ft span of a bridge is built with steel with $\sigma = 1.4 \times 10^{-5}$ per degree Celsius. Approximately how much will the length change during a year when the temperature varies from -10°C in winter to 40°C in summer?

40. The radius R of a spherical ball is measured as 14 in.
  a. Use differentials to estimate the maximum propagated error in computing volume V if R is measured with a maximum error of $\frac{1}{8}$ inch.
  b. With what accuracy must the radius R be measured to guarantee an error of at most 2 in.^{3} in the calculated volume?

41. A thin horizontal beam of alpha particles strikes a thin vertical foil, and the scattered alpha particles will travel along a cone of vertex angle θ, as shown in Figure 3.61.

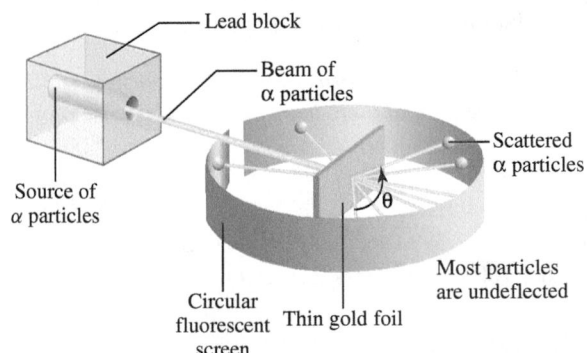

Figure 3.61   Paths of alpha particles

A vertical screen is placed at a fixed distance from the point of scattering. Physical theory predicts that the number N of alpha particles falling on a unit area of the screen is inversely proportional to $\sin^4\left(\frac{\theta}{2}\right)$. Suppose N

---

*John A. Helms, "Environmental Control of Net Photosynthesis in Naturally Grown Pinus Ponderosa Nets," *Ecology* (Winter 1972) p. 92.

†See "Introduction to Mathematics for Life Scientists," 2nd edition. New York: Springer-Verlag (1976); pp. 102-103.

‡Ibid.

is modeled by the formula

$$N = \frac{1}{\sin^4\left(\frac{\theta}{2}\right)}$$

Estimate the change in the number of alpha particles per unit area of the screen if θ changes from 1 to 1.1.

42. Suppose the total cost of manufacturing q units is

$$C(q) = 3q^2 + q + 500$$

dollars.
   a. Use marginal analysis to estimate the cost of manufacturing the 41st unit.
   b. Compute the actual cost of manufacturing the 41st unit.

43. A manufacturer's total cost is

$$C(q) = 0.1q^3 - 5q^2 + 500q + 200$$

dollars, where q is the number of units produced.
   a. Use marginal analysis to estimate the cost of manufacturing the 4th unit.
   b. Compute the actual cost of manufacturing the 4th unit.

44. Suppose the total cost of producing x units of a particular commodity is modeled by

$$C(x) = \frac{1}{7}x^2 + 4x + 100$$

and that each unit of the commodity can be sold for

$$p(x) = \frac{1}{4}(80 - x)$$

dollars.
   a. What is the marginal cost?
   b. What is the price when the marginal cost is 10?
   c. Estimate the cost of producing the 11th unit.
   d. Find the actual cost of producing the 11th unit.

45. At a certain factory, the daily output is modeled by the formula

$$Q(L) = 360L^{1/3}$$

units, where L is the size of the labor force measured in worker-hours. Currently, 1,000 worker-hours of labor are used each day. Use differentials to estimate the effect that one additional worker-hour will have on the daily output.

46. Approximate $\sqrt{2}$ to four decimal places by using the Newton-Raphson method.

47. Approximate $-\sqrt{2}$ to four decimal places by using the Newton-Raphson method.

48. Use the Newton-Raphson method to estimate a root of the equation

$$\cos x = x$$

You may start with $x_0 = 1$.

49. Use the Newton-Raphson method to estimate a root of the equation

$$e^{-x} = x$$

You may start with $x_0 = 1$.

50. Let $f(x) = -2x^4 + 3x^2 + \frac{11}{8}$
   a. Show that the equation $f(x) = 0$ has at least two solutions. *Hint*: Use the intermediate value theorem.
   b. Use $x_0 = 2$ in the Newton-Raphson method to find a root of the equation $f(x) = 0$.
   c. Show that the Newton-Raphson method fails if you choose $x_0 = \frac{1}{2}$ as the initial estimate. *Hint*: You should obtain $x_1 = -x_0, x_2 = x_0, \cdots$.

51. It can be shown that the volume of a spherical cap is given by

$$V = \frac{\pi}{3}H^2(3R - H)$$

where R is the radius of the sphere and H is the height of the cap, as shown in Figure 3.62.

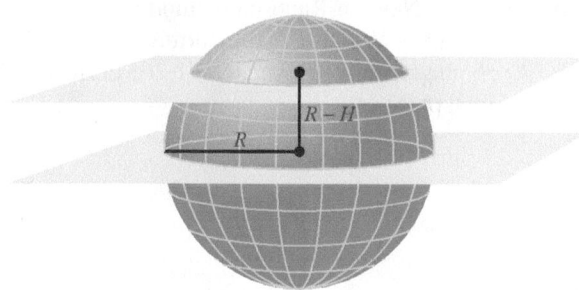

Figure 3.62 Problem 51

If $V = 8$ and $R = 2$, use the Newton-Raphson method to estimate the corresponding H.

52. *Historical Quest*

*The Greek geometer Archimedes is acknowledged to be one of the greatest mathematicians of all time. Ten treatises of Archimedes have survived the rigors of time (as well as traces of some lost works), and are masterpieces of mathematical exposition. In one of these works, **On the Sphere and Cylinder**, Archimedes asks*

Archimedes (287-212 B.C.)

where a sphere should be cut in order to divide it into two pieces whose volumes have a given ratio. Show that if a plane at distance x_c from the center of a sphere with $R = 1$ divides the sphere into two parts, one with volume five times that of the other, then

$$3x_c^3 - 9x_c^2 + 2 = 0$$

Use the Newton-Raphson method to estimate x_c.

53. *Historical Quest Among the peoples of the region between the Tigris and Euphrates rivers (at different times) during the period 2000-600 B.C. were Sumerians, Akkadians, Chaldeans, and Assyrians. Since the middle half of the 19th century, archeologists have found well over 50,000 clay tablets describing these great civilizations. Records show that they had highly developed religion, history, science (including alchemy, astronomy, botany, chemistry, math, and zoology). Mesopotamian culture had iterative formulas for computing algebraic quantities such as roots. In particular, they approximated $\sqrt{N}$ by repeatedly applying the formula*

$$x_{n+1} = \frac{1}{2}\left(x_n + \frac{N}{x_n}\right)$$

for $n = 0, 1, 2, 3, \cdots$.

a. Apply the Newton-Raphson method to $f(x) = x^2 - N$ to justify this formula.

b. Apply the formula to estimate $\sqrt{1{,}265}$ correct to five decimal places.

54. Suppose the plane described in Problem 52 is located so that it divides the sphere in the ratio $1 : 3$. Find an equation for x_c, and estimate the value of x_c by the Newton-Raphson method. (*Hint*: you may need the result of Problem 52.)

Level 3

55. Show that if h is sufficiently small, then
    a. $\sqrt{1+h}$ is approximately equal to $1 + \frac{h}{2}$.
    b. $\dfrac{1}{1+h}$ is approximately equal to $1 - h$.

56. If h is sufficiently small, find an approximate value for $\sqrt[n]{A^n + h}$ for constant A.

57. Tangent line approximations are useful only if Δx is small. Illustrate this fact by trying to approximate $\sqrt{97}$ by regarding 97 as being near 81 (instead of 100).

58. Let $f(x) = -x^4 + x^2 + A$ for constant A. What value of A should be chosen that guarantees that if $x_0 = \frac{1}{3}$ is chosen as the initial estimate, the Newton-Raphson method produces $x_1 = -x_0, x_2 = x_0, x_3 = -x_0, \cdots$?

59. Can you solve the equation $\sqrt{x} = 0$ using the Newton-Raphson method with the initial estimate $x_0 = 0.05$? Does it make any difference if we choose another initial estimate (other than $x_0 = 0$)?

60. **INTERPRETATION PROBLEM** Suppose that when we try to use the Newton-Raphson method to approximate a solution of $f(x) = 0$, we find that $f(x_n) = 0$ but $f'(x_n) \neq 0$ for some x_n. What does this imply about $x_{n+1}, x_{n+2}, \cdots$? Explain.

CHAPTER 3 REVIEW

P*upils should be taught to differentiate . . .*
before we attempt to teach them geometry and these other complicated things.

John Perry

Teaching of Mathematics, 1901

Proficiency Examination

Concept Problems

1. **a.** What is the slope of a tangent line?
   b. How does this compare to the slope of a secant line?
   c. What is a normal line to a graph?
2. **a.** Define the derivative of a function.
   b. List and explain some of the notations for derivative.
3. What is the relationship between continuity and differentiability?
4. State the following procedural rules for finding derivatives:

   a. constant multiple   **b.** sum rule   **c.** difference rule

   d. linearity rule   **e.** product rule   **f.** quotient rule

   g. State each of these rules again, this time in differential form.
5. State the following derivative rules:

   a. constant rule   **b.** power rule   **c.** trigonometric rules

   d. exponential rule   **e.** logarithmic rule **f.** inverse trigonometric rules

6. **a.** What is a higher-order derivative?
   b. List some of the different notations for higher-order derivatives.
7. **a.** What is meant by rate of change?
   b. Distinguish between average and instantaneous rate of change.
   c. What is relative rate of change?
8. How do you find the velocity and the acceleration for an object with position $s(t)$? What is speed?
9. State the chain rule.
10. Outline a procedure for logarithmic differentiation.
11. Outline a procedure for implicit differentiation.
12. Outline a procedure for solving related-rate problems.
13. **a.** What is meant by tangent line approximation?
    b. Define the terms propagated error, relative error, and percentage error.
14. Define the differential of x and the differential of y for a function $y = f(x)$. Draw a sketch showing $\Delta x, \Delta y, dx,$ and dy.
15. What is meant by marginal analysis?
16. What is the Newton-Raphson method?

Practice Problems

Find $\dfrac{dy}{dx}$ in Problems 17-25.

17. $y = x^3 + x\sqrt{x} + \cos 2x$ **18.** $y = \sqrt{\sin(3 - x^2)}$ **19.** $xy + y^3 = 10$

20. $y = x^2 e^{-\sqrt{x}}$ **21.** $y = \dfrac{\ln 2x}{\ln 3x}$ **22.** $y = \sin^{-1}(3x + 2)$

23. $y = \tan^{-1} 2x$ **24.** $y = \dfrac{\ln(x^2 - 1)}{\sqrt[3]{x}(1 - 3x)^3}$

25. $y = \sin^2\left(x^{10} + \sqrt{x}\right) + \cos^2\left(x^{10} + \sqrt{x}\right)$

26. Find $\dfrac{d^2y}{dx^2}$, the second derivative of $y = x^2(2x - 3)^3$.

27. Use the definition of the derivative to find $\dfrac{d}{dx}\left(x - 3x^2\right)$

28. Find the equation of the tangent line to the graph of

$$y = (x^2 + 3x - 2)(7 - 3x)$$

at the point where $x = 1$.

29. Let $f(x) = \sin^2\left(\dfrac{\pi x}{4}\right)$. Find equations of the tangent and normal lines to the graph of f at $x = 1$.

30. A rock tossed into a stream causes a circular ripple of water whose radius increases at a constant rate of 0.5 ft/s. How fast is the area contained inside of the ripple changing when the radius is 2 ft?

Supplementary Problems*

Find dy/dx in Problems 1-30.

1. $y = x^4 + 3x^2 - 7x + 5$ **2.** $y = x^5 + 3x^3 - 11$ **3.** $y = \sqrt{\dfrac{x^2 - 1}{x^2 - 5}}$

4. $y = \dfrac{\cos x}{x + \sin x}$ **5.** $2x^2 - xy + 2y = 5$ **6.** $y = (x^2 + 3x - 5)^7$

7. $y = \sqrt[3]{x}(x^3 + 1)^5$ **8.** $y = (x^2 + 3)^5(x^3 - 5)^8$ **9.** $y = \sqrt{\sin 5x}$

10. $\left(\sqrt{x} + \sqrt[3]{x}\right)^5$ **11.** $y = \exp(2x^2 + 5x - 3)$ **12.** $y = \ln(x^2 - 1)$

13. $y = x3^{2-x}$ **14.** $y = \log_3(x^2 - 1)$ **15.** $e^{xy} + 2 = \ln\dfrac{y}{x}$

16. $y = \sqrt{x}\sin^{-1}(3x + 2)$ **17.** $y = e^{\sin x}$ **18.** $y = 2^x \log_2 x$

19. $x2^y + y2^x = 3$ **20.** $\ln(x + y^2) = x^2 + 2y$ **21.** $y = e^{-x}\sqrt{\ln 2x}$

22. $y = \sin(\sin x)$ **23.** $y = \cos(\sin x)$ **24.** $x^{1/2} + y^{1/2} = x$

25. $4x^2 - 16y^2 = 64$ **26.** $\sin xy = y + x$ **27.** $y = xe^{1-2x}$

28. $y = \sin x^2 \cos x^2$ **29.** $y = \csc^2\sqrt{x}$ **30.** $y = x^3(2 - 4x)^2$

Find $\dfrac{d^2y}{dx^2}$ in Problems 31-36.

31. $y = x^5 - 5x^4 + 7x^3 + 17$ **32.** $y = 2x^3 - 2x - 5$ **33.** $x^2 + y^3 = 10$

34. $y = \dfrac{x - 5}{2x + 3} + (3x - 1)^2$ **35.** $x^2 + \sin y = 2$ **36.** $x^2 + \tan^{-1} y = 2$

In Problems 37-44, find an equation of the tangent line to the curve at the indicated point.

37. $y = x^4 - 7x^3 + x^2 - 3$ at $(0, -3)$ **38.** $y = (3x^2 + 5x - 7)^3$, where $x = 1$

39. $y = x\cos x$, where $x = \frac{\pi}{2}$ **40.** $y = (\cos x)^2$, where $x = \pi$

41. $xy^2 + x^2y = 2$ at $(1, 1)$ **42.** $y = x\ln(ex)$, where $x = 1$

43. $e^{xy} = x - y$ at $(1, 0)$ **44.** $y = (1 - x)^x$, where $x = 0$

45. a. Find the derivative of $f(x) = 4 - 2x^2$.
     b. The graph of f has one horizontal tangent line. What is its equation?
     c. At what point on the graph of f is the tangent line parallel to the line $8x + 3y = 4$?

---

*The supplementary probelms are presented in a somewhat random order, not necessarily in order of difficulty.

46. Let $f(x) = (x^3 - x^2 + 2x - 1)^4$. Find equations of the tangent and normal lines to the graph of f at $x = 1$.

47. Find $f''(x)$ if $f(x) = x^2 \sin x^2$.

48. Find $f^{(4)}$ if $f(x) = x^4 - \dfrac{1}{x^4}$.

49. Let $f(x) = \sqrt[3]{\dfrac{x^4 + 1}{x^4 - 2}}$. Find $f'(x)$ by implicitly differentiating $\left[f(x)\right]^3$.

50. Find y' if $x^3 y^3 + x - y = 1$. Leave your answer in terms of x and y.

51. Find y' and y'' if $x^2 + 4xy - y^2 = 8$. Your answer may involve x and y, but not y'.

52. Find the derivative of $f(x) = \begin{cases} x^2 + 5x + 4 & \text{for } x \leq 0 \\ 5x + 4 & \text{for } 0 < x < 6 \\ x^2 - 2 & \text{for } x \geq 6 \end{cases}$

53. **Think Tank Problem** Give an example of a function that is continuous on $(-\infty, \infty)$ but is not differentiable at $x = 5$.

54. If $f'(c) \neq 0$, what is the equation of the normal line to $y = f(x)$ at the point $P(c, f(c))$? What is the equation if $f'(c) = 0$?

55. Find the point(s) on the graph of $f(x) = -x^2$ such that the tangent line at that point passes through the point $(0, 9)$.

56. Compute the difference quotient for the function defined by

$$f(x) = \begin{cases} \frac{\sin x}{x} & \text{if } x \neq 0 \\ 1 & \text{if } x = 0 \end{cases}$$

Do you think $f(x)$ is differentiable at $x = 0$? If so, what is the equation of the tangent line at $x = 0$?

57. 𝔥istorical 𝔔uest *Pierre de Fermat (see Problem 57, Section 3.1) also obtained the first method for differentiating polynomials, but his real love was number theory. His most famous problem has come to be known as Fermat's last theorem. He wrote in the margin of a text: "To divide a cube into two cubes, a fourth power, or in general, any power whatever above the second, into powers of the same denominations, is impossible, and I have assuredly found an admirable proof of this, but the margin is too narrow to contain it."*

In 1993 an imaginative, but lengthy, proof was constructed by Andrew Wiles, and in 1997 Wiles was awarded the Wolfskehl Prize for proving this theorem. An engaging book, *Fermat's Enigma*, shows how a great mathematical puzzle of our age was solved: Lock yourself in a room and emerge seven years later. Write a paper on the history of Fermat's last theorem, including some discussion of the current status of its proof. [You might begin with "Fermat's Last Theorem, The Four Color Conjecture, and Bill Clinton for April Fools' Day," by Edward B. Burger and Frank Morgan, *American Mathematical Monthly*, March 1997, pp. 246-250. Next, see the NOVA film, *The Proof*, which was shown in 1997 on many PBS stations (check with your library or PBS Online). Finally, see "Fermat's Last Stand" by Simon Singh and Kenneth A. Ribet, *Scientific American*, November 1997, pp. 68-73.]

58. Book Report In Problem 57, we told the story of Fermat's last theorem. The book *Fermat's Enigma* by Simon Singh (New York: Walker & Company, 1997) tells the story of how Andrew Wiles solved the greatest mathematical puzzle of our age: He locked himself in a room and emerged seven years later. Read this book and prepare a book report.

59. The gross domestic product (GDP) of a certain country is growing at a constant rate. In 2009 the GDP was 125 billion dollars, and in 2011 it was 155 billion dollars. At what percentage rate will the GDP grow in 2014?

60. Van der Waal's equation states that a gas that occupies a volume V at temperature T (Kelvin) exerts pressure P, where

$$\left(P + \frac{A}{V^2}\right)(V - B) = kT$$

and A, B, k are physical constants. Find the rate of change of pressure with respect to volume, assuming fixed temperature.

61. It is estimated that t years from now, the population of a certain suburban community is modeled by the formula

$$p(t) = 20 - \frac{6}{t+1}$$

where $p(t)$ is in thousands of people. A separate environmental study indicates that the average daily level of carbon monoxide in the air will be

$$L(p) = 0.5\sqrt{p^2 + p + 58}$$

ppm when the population is p thousand. Find the rate at which the level of carbon monoxide will be changing with respect to time two years from now.

62. At a certain factory, the total cost of manufacturing q units during the daily production run is

$$C(q) = 0.2q^2 + q + 900$$

dollars. From experience, it has been determined that approximately

$$q(t) = t^2 + 100t$$

units are manufactured during the first t hours of a production run. Compute the rate at which the total manufacturing cost is changing with respect to time one hour after production begins.

63. To form a *simple pendulum*, a weight is attached to a rod that is then suspended by one end in such a way that it can swing freely in a vertical plane. Let θ be the angular displacement of the rod from the vertical, as shown in Figure 3.63.

It can be shown that as long as the maximum displacement θ_M is small, it is reasonable to assume that $\theta = \theta_M \sin kt$ at time t, where k is a constant that depends on the length of the rod and θ_M is a constant. Show that

$$\frac{d^2\theta}{dt^2} + k^2\theta = 0$$

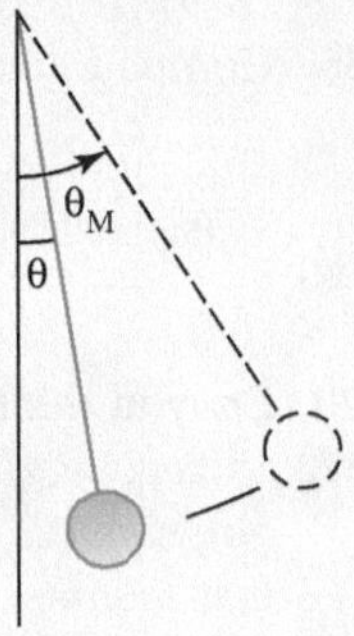

64. Using only the formula

$$\frac{d}{dx}\sin u = \cos u \frac{du}{dx}$$

and the identities $\cos x = \sin\left(\frac{\pi}{2} - x\right)$ and $\sin x = \cos\left(\frac{\pi}{2} - x\right)$, show that

$$\frac{d}{dx}\cos u = -\sin u \frac{du}{dx}$$

Figure 3.63 Problem 63

65. Let f be a function for which $f(2) = -3$ and $f'(x) = \sqrt{x^2 + 5}$. If

$$g(x) = x^2 f\left(\frac{x}{x - 1}\right)$$

what is $g'(2)$?

66. The tangent line to the curve $x^{2/3} + y^{2/3} = 8$ at the point $(8, 8)$ and the coordinate axes form a triangle, as shown in Figure 3.64. What is the area of this triangle?

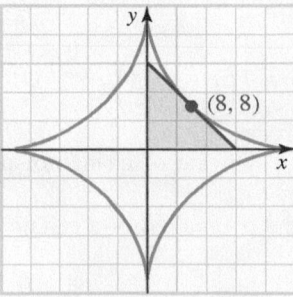

Figure 3.64 Area of triangle

67. **Modeling Problem** A worker stands 4 m from a hoist being raised at the rate of 2 m/s, as shown in Figure 3.65. Model the worker's angle of sight θ using an inverse trigonometric function, and then determine how fast θ is changing at the instant when the hoist is 1.5 m above eye level.

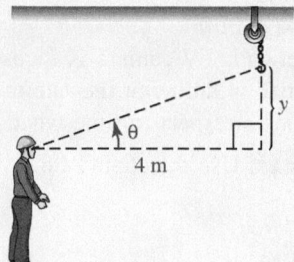

Figure 3.65   Modeling angle of elevation

68. Use implicit differentiation to find the second derivative y'', where y is a differentiable function of x that satisfies $ax^2 + by^2 = c$, $(a, b,$ and c are constants).

69. Show that the sum of the x-intercept and the y-intercept of any tangent line to the curve

$$\sqrt{x} + \sqrt{y} = C$$

is equal to C^2.

70. **Journal Problem** (*The Pi Mu Epsilon Journal* by Bruce W. King)*: When a professor asked the calculus class to find the derivative of y^2 with respect to x^2 for the function $y = x^2 - x$, one student found

$$\frac{dy}{dx} \cdot \frac{y}{x}$$

Was this answer correct? Suppose

$$y = x^3 - 3x^2 + \frac{7}{x}$$

What is the derivative of y^2 with respect to x^2 for this function?

71. Use the Newton-Raphson method to estimate $\sqrt[5]{23}$.

72. Use the Newton-Raphson method to estimate a root of the equation

$$x^6 - x^5 + x^3 = 3$$

73. Suppose that when x units of a certain commodity are produced, the total cost is $C(x)$ and the total revenue is $R(x)$. Let $P(x) = R(x) - C(x)$ denote the total profit, and let

$$A(x) = \frac{C(x)}{x}$$

be the average cost.
  a. Show that $P'(x) = 0$ when marginal revenue equals marginal cost.
  b. Show that $A'(x) = 0$ when average cost equals marginal cost.

74. **Advanced Placement Question** *The following problem is found on the 1982 BC AP Exam.* Let f be the function defined by

$$f(x) = \begin{cases} x^2 \sin \frac{1}{x} & \text{for } x \neq 0 \\ 0 & \text{for } x = 0 \end{cases}$$

a. Using the definition of derivative, prove that f is differentiable at $x = 0$.

b. Find $f'(x)$ for $x \neq 0$.

c. Show that f' is not continuous at $x = 0$.

75. Use differentials to approximate $(16.01)^{3/2} + 2\sqrt{16.01}$.

76. Use differentials to approximate $\cos \dfrac{101\pi}{600}$.

77. On New Year's Eve, a ground-level network TV camera is focusing on the descent of a lighted ball from the top of a building that is 600 ft away. The ball is falling at the rate of 20 ft/min. At what rate is the angle of elevation of the camera's line of sight changing with respect to time when the ball is 800 ft from the ground?

78. Let $f(x) = \sin 2x + \cos 3x$ and $g(x) = x^2$. Use the chain rule to find $\dfrac{d}{dx}(f \circ g)(x)$.

79. Let $f(x) = \begin{cases} x \sin \frac{1}{x} & \text{if } x \neq 0 \\ 0 & \text{if } x = 0 \end{cases}$.

Use the definition of the derivative to find $f'(0)$, if it exists.

80. Suppose the total cost of producing x units of a particular commodity is

$$C(x) = \frac{2}{5}x^2 + 3x + 10$$

and that each unit of the commodity can be sold for

$$p(x) = \frac{1}{5}(45 - x)$$

dollars.

a. What is the marginal cost?

b. What is the price when the marginal cost is 23?

c. Estimate the cost of producing the 11th unit.

d. Find the actual cost of producing the 11th unit.

81. A block of ice in the shape of a cube originally having volume 1,000 cm³ is melting in such a way that the length of each of its edges is decreasing at the rate of 1 cm/hr. At what rate is its surface area changing at the time its volume is 27 cm³? Assume that the block of ice maintains its cubical shape.

82. Show that the rate of change of the area of a circle with respect to its radius is equal to the circumference.

83. A charged particle is projected into a linear accelerator. The particle undergoes a constant acceleration that changes its velocity from 1,200 m/s to 6,000 m/s in 2×10^{-3} seconds. Find the acceleration of the particle.

84. A rocket is launched vertically from a point on level ground that is 3,000 feet from an observer with binoculars. If the rocket is rising vertically at the rate of 750 ft/s at the instant it is 4,000 ft above the ground, how fast must the observer change the angle of elevation of her line of sight in order to keep the rocket in sight at that instant?

85. Modeling Problem Assume that a certain artery in the body is modeled by a circular tube whose cross section has radius 1.2 mm. Fat deposits are observed to build up uniformly on the inside wall of the artery. Find the rate at which the cross-sectional area of the artery is decreasing relative to the thickness of the fat deposit at the instant when the deposit is 0.3 mm thick.

86. A processor who sells a certain raw material has analyzed the market and determined that the unit price should be modeled by the formula

$$p(x) = 60 - x^2$$

(thousand dollars) for x tons ($0 \leq x \leq 7$) produced. Estimate the change in the unit price that accompanies each change in sales:

a. from 2 tons to 2.05 tons

b. from 1 tons to 1.1 tons

c. from 3 tons to 2.95 tons

87. A company sends out a truck to deliver its products. To estimate costs, the manager models gas consumption by the formula

$$G(x) = \frac{1}{300}\left(\frac{1500}{x} + x\right)$$

gal/mi, under the assumption that the truck travels at a constant rate of x mi/h ($x \geq 5$). The driver is paid \$26 per hour to drive the truck 300 mi. Gasoline costs \$4 per gallon.
   a. Find an expression for the total cost $C(x)$ of the trip.
   b. Use differentials to estimate the additional cost if the truck is driven at 57 mi/h instead of 55 mi/h.

88. A viewer standing at ground level and 30 ft from a platform watches a balloon rise from that platform (at the same height as the viewer's eyes) at the constant rate of 3 ft/s. (See Figure 3.66.)

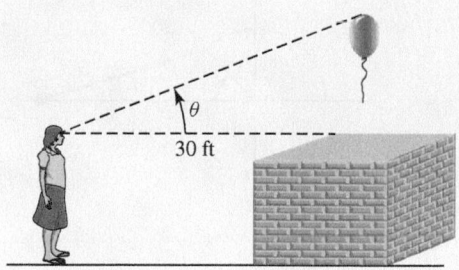

30 ft

Figure 3.66   Problem 88

   How fast is the angle of sight between the viewer and the object changing at the instant when $\theta = \frac{\pi}{4}$?

89. A lighthouse is 4,000 ft from a straight shore. Watching the beam on the shore from the point P on the shore that is closest to the lighthouse, an observer notes that the light is moving at the rate of 3 ft/s when it is 1,000 ft from P. How fast is the light revolving at this instant (in radians per second)?

90. A particle of mass m moves along the x-axis. The velocity $v = \dfrac{dx}{dt}$ and position $x = x(t)$ satisfy the equation

$$m(v^2 - v_0^2) = k(x_0^2 - x^2)$$

   where k, x_0, and v_0 are positive constants. The force F acting on the object is defined by $F = ma$, where a is the object's acceleration. Show that $F = -kx$.

91. The equation $\dfrac{d^2s}{dt^2} + ks = 0$ is called a **differential equation of simple harmonic motion**. Let A be any number. Show that the function $s(t) = A\sin 2t$ satisfies the equation

$$\frac{d^2s}{dt^2} + 4s = 0$$

92. Suppose $L(x)$ is a function with the property that $L'(x) = x^{-1}$. Use the chain rule to find the derivatives of the following functions.

   a. $f(x) = L\left(x^2\right)$ b. $f(x) = L\left(\dfrac{1}{x}\right)$ c. $f(x) = L\left(\dfrac{2}{3\sqrt{x}}\right)$ d. $f(x) = L\left(\dfrac{2x+1}{1-x}\right)$

93. Let f and g be differentiable functions such that $f\left[g(x)\right] = x$ and $g\left[f(x)\right] = x$ for all x (that is, f and g are inverses). Show that

$$\frac{dg}{dx} = \frac{1}{df/dx}$$

94. A baseball player is stealing second base, as shown in Figure 3.67.
    He runs at 30 ft/s and when he is 25 ft from second base, the catcher, while
    standing at home plate, throws the ball toward the base at a speed of 120 ft/s.
    At what rate is the distance between the ball and the player changing at the time
    the ball is thrown?

95. A 2 m connecting rod $\overline{OA}$ is rotating counterclockwise in a plane about O at
    the rate of 3 rev/s, as shown in Figure 3.68.

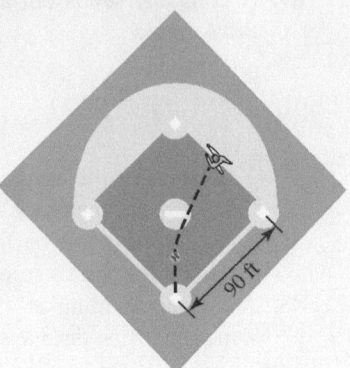

Figure 3.67   Problem 63

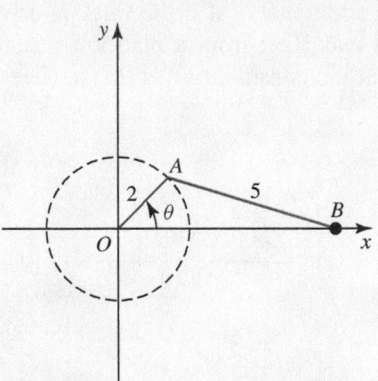

Figure 3.68   Connecting rod

The rod $\overline{AB}$ is attached to $\overline{OA}$ at A, and the end B slides along the x-axis. Suppose $\overline{AB}$ is 5 meters long. What are
the velocity and the acceleration of the motion of the point B along the x-axis?

96. The ideal gas law states that for an ideal gas, pressure P is given by the formula $P = kT/V$, where V is the
    volume, T is the temperature, and k is a constant. Suppose the temperature is kept fixed at 100°C, and the pressure
    decreases at the rate of 7 lb/in.^{2} per min. At what rate is the volume changing at the instant the pressure is 25 lb/in.^{2}
    and the volume is 30 in.^{3}?

97. **Putnam Examination Problem**   Suppose $f(x) = ax^2 + bx + c$, where $a, b,$ and c are real numbers and $|f(x)| \leq 1$
    for $|x| \leq 1$. Show that $\left|f'(x)\right| \leq 4$ for $|x| \leq 1$.

98. **Putnam Examination Problem**   A point P is taken on the curve $y = x^3$. The tangent line at P meets the curve
    again at Q. Prove that the slope of the curve at Q is four times the slope at P.

99. **Putnam Examination Problem**   A particle of unit mass moves on a straight line under the action of a force that is
    a function $f(v)$ of the velocity v of the particle, but the form of this function is not known. A motion is observed,
    and the distance x covered in time t is found to be related to t by the formula $x = at + bt^2 + ct^3$, where $a, b,$
    and c have numerical values determined by observation of the motion. Find the function $f(v)$ for the range of v
    covered by this experiment.

CHAPTER 3 GROUP RESEARCH PROJECT

Working in small groups is typical of most work environments, and this book seeks to develop skills with group activities. This project is to be done in groups of three or four students. Each group will submit one written report.

Chaos

This fractal image is an example of what is called mathematical chaos.

An exciting new topic in mathematics attempts to bring order to the universe by considering disorder. This topic is called **chaos theory**, and it shows how structures of incredible complexity and disorder really exhibit beauty and order.

Water flowing through a pipe offers one of the simplest physical models of chaos. Pressure is applied to the end of the pipe and the water flows in straight lines. More pressure increases the speed of the laminar flow until the pressure reaches a critical value, and a radically new situation evolves---turbulence. A simple laminar flow suddenly changes to a flow of beautiful complexity consisting of swirls within swirls. Before turbulence, the path of any particle was quite predictable. After a minute change in the pressure, turbulence occurs and predictability is lost. **Chaos** is concerned with systems in which minute changes suddenly transform predictability into unpredictability.*

For this research project begin with

$$f(x) = x^3 - x = x(x - 1)(x + 1)$$

Use the Newton-Raphson method to investigate what happens as you change the starting value, x_0, to solve $f(x) = 0$. What would you select as a starting value? Two important x-values in our study are

$$s_3 = \frac{1}{\sqrt{3}} \text{ and } s_5 = \frac{1}{\sqrt{5}}$$

One would *not* want to pick x_0 as either $\pm s_3$. Why not? Also, what happens if one picks $x_0 = s_5$? That is, what are $x_1, x_2, \cdots$?

Generate a good plot of $f(x)$ on $[-2, 2]$. Explain from the plots why you would *expect* that an initial value $x_0 > s_3$ would lead x_n to approach 1 as n becomes large.

Explain why if $|x_0| < s_5$, you would expect $x_n \to 0$. Numerically verify your assertions. Now to see the "chaos," use the following x_0 values and take 6 to 10 iterations, until convergence occurs, and report what happens: 0.448955; 0.447503; 0.447262; 0.447222; 0.447215; 0.447213. Write a paper telling what you see and why.

Extended paper for further study. Attempt to define chaos. Your paper should include, but not be limited to, answers to the questions on this page. Some references (to get you started) are listed:

Chaos: Making a New Science, James Gleick, Penguin Books, 1988.
Chaos and Fractals: New Frontiers of Science, H. O. Peitgen, H. Jurgens, and D. Saupe, Springer Verlag, 1992.
Newton's Method and Fractal Patterns, Phillip D. Straffin, Jr., UMAP Module 716, COMAP, 1991.

---

*Thanks to Jack Wadhams of Golden West College for this paragraph.

CHAPTER 4

ADDITIONAL APPLICATIONS OF THE DERIVATIVE

*E*verything that the greatest minds of all times have accomplished toward the comprehension of forms by means of concepts is gathered into one great science, mathematics.

J. F. Herbart (1890)

quoted in *Pestalozzi's Idee eines A, B, C der Anschauung, Werke [Kehrbach], (Langensalza, 1890), Bd. 1, p. 163*

PREVIEW

In Chapter 3, we used the derivative to find tangent lines and to compute rates of change. The primary goal of this chapter is to examine the use of calculus in curve sketching, optimization, and other applications. For example, homing pigeons and certain other birds are known to avoid flying over large bodies of water whenever possible. The reason for this behavior is not entirely known, but it is believed that the birds are minimizing the energy they expend while flying.*

© Kendall Hunt Publishing Company

In addition to these applied problems, we expand our knowledge of the derivative by studying the mean value theorem; and a very useful technique, called l'Hôpital's rule, helps us find certain limits.

PERSPECTIVE

Remember the three main ideas of calculus: limit, derivative, and integral. In this chapter we conclude our introduction of the second of these ideas, the derivative, and in the next chapter we introduce the integral.

Optimization (finding the maximum or minimum values) is one of the most important applications we will study in calculus, from maximizing profit and minimizing cost to maximizing the strength of a structure and minimizing the distance traveled. Optimization problems are considered in depth in Sections 4.6 and 4.7.

*See Problem 42, Section 4.7.

4.1 EXTREME VALUES OF A CONTINUOUS FUNCTION

IN THIS SECTION: *Extreme value theorem, relative extrema, absolute extrema, optimization*
It is often very important to know what the largest and smallest values of a function are and where these extreme values occur. In this section, we shall see how the derivative may be used to locate extreme values.

Extreme Value Theorem

One of the principal goals of calculus is to investigate the behavior of various functions. As part of this investigation, we will be laying the groundwork for solving a large class of problems that involve finding the maximum or minimum value of a function, if one exists. Such problems are called **optimization problems**. We begin by introducing some useful terminology.

> **ABSOLUTE MAXIMUM** Let f be a function defined on an interval I that contains the number c. Then $f(c)$ is the **absolute maximum** of f on D if
>
> $$f(c) \geq f(x) \text{ for all } x \text{ in } D$$
>
> **ABSOLUTE MINIMUM** $f(c)$ is the **absolute minimum** of f on D if
>
> $$f(c) \leq f(x) \text{ for all } x \text{ in } D.$$

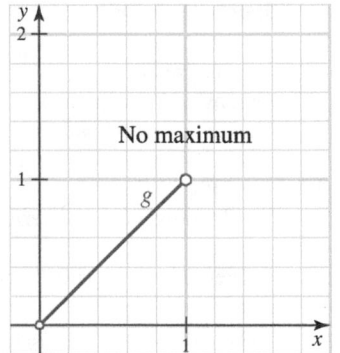

Figure 4.1 Continuous function without extrema

Sometimes we just use the terms *maximum* and/or *minimum* if the context is clear. Together, the absolute maximum and minimum of f on the interval I are called the **extreme values**, or the **absolute extrema**, of f on I. A function does not necessarily have extreme values on a given interval. For instance, the continuous function $g(x) = x$ has neither a maximum nor a minimum on the open interval $(0, 1)$, as shown in Figure 4.1.

The discontinuous function defined by

$$h(x) = \begin{cases} x^2 & \text{for } x \neq 0 \\ 1 & \text{for } x = 0 \end{cases}$$

has a maximum which is on the closed interval $[-1, 1]$, but no minimum, as shown in Figure 4.2.

Incidentally, this graph also illustrates the fact that a function may assume an absolute extremum at more than one point. In this case, the maximum occurs at the points $(-1, 1), (0, 1)$, and $(1, 1)$. If a function f is continuous and the interval I is closed and bounded, it can be shown that both an absolute maximum and an absolute minimum *must* occur. This result, called the **extreme value theorem**, plays an important role in our work.

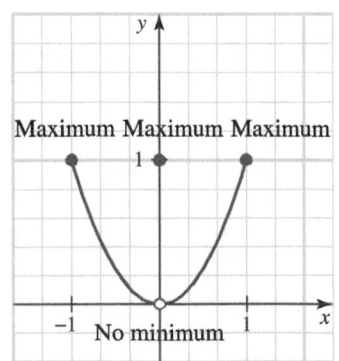

Figure 4.2 Discontinuous function with a maximum but no minimum

Theorem 4.1   Extreme value theorem

A function f has both an absolute maximum and an absolute minimum on any closed, bounded interval $[a, b]$ where it is continuous.

Proof: Even though this result may seem quite reasonable (see Figure 4.3), its proof requires concepts beyond the scope of this text and will be omitted.

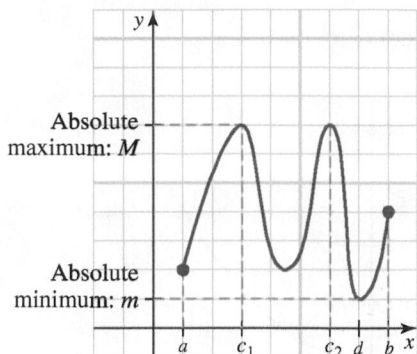

Absolute maximum: M

Absolute minimum: m

Absolute maximum at $x = c_1$ and $x = c_2$

Absolute minimum at $x = d$

Figure 4.3 Extreme value theorem ◆

If f is discontinuous or the interval is not both closed and bounded, you cannot conclude that f has both a largest and smallest value. You will be asked for appropriate counterexamples in the problem set. Sometimes there are extreme values even when the conditions of the theorem are not satisfied, but if the conditions hold, the extreme values are guaranteed.

Note that the maximum of a function occurs at the highest point on its graph and the minimum occurs at the lowest point. These properties are illustrated in Example 1.

Example 1   Extreme values of a continuous function

The graph of a function f is shown in Figure 4.4.

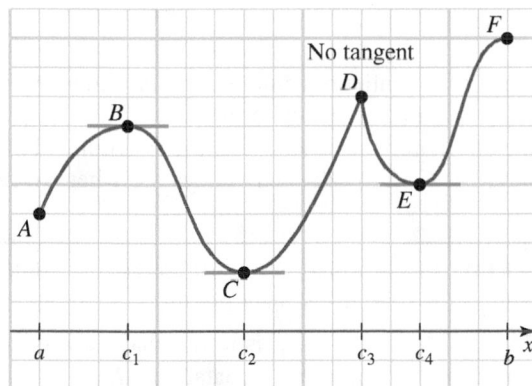

No tangent

Figure 4.4   A continuous function on a closed interval $[a, b]$

Locate the extreme values of f defined on the closed interval $[a, b]$.

Solution   The highest point on the graph occurs at the right endpoint F, and the lowest point occurs at C. Thus, the absolute maximum is $f(b)$, and the absolute minimum is $f(c_2)$.

In Example 1, the existence of a maximum and a minimum as required by the extreme value theorem may seem obvious, but there are times when it seems that the extreme value theorem fails. If this occurs, you need to see which of the conditions of the extreme value theorem are not satisfied. Consider Example 2.

Example 2   Conditions of the extreme value theorem

In each case, explain why the given function does not contradict the extreme value theorem:

a. $f(x) = \begin{cases} 2x & \text{if } 0 \le x < 1 \\ 1 & \text{if } 1 \le x \le 2 \end{cases}$ **b.** $g(x) = x^2$ on $0 < x \le 2$

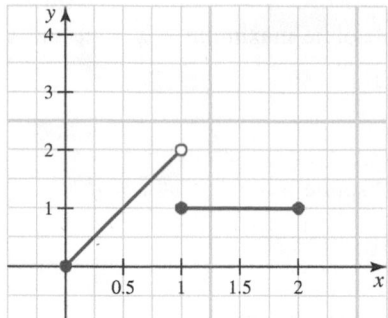

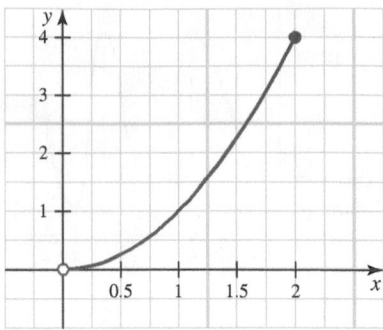

f does not have a maximum value. g does not have a minimum value.

Solution

a. The function f has no maximum. It takes on all values less than, but arbitrarily close to, 2. However, it never reaches the value 2. This function does not contradict the extreme value theorem because f is not continuous on $[0, 2]$.

b. Although the functional values of $g(x)$ become arbitrarily small as x approaches 0, $g(x)$ never reaches the value 0 so g has no minimum. The function g is continuous on the interval $(0, 2]$, but the extreme value theorem is not contradicted because the interval is not closed. ▪

Relative Extrema

Typically, the extrema of a continuous function occur either at endpoints of the interval or at points where the graph has a "peak" or a "valley" (points where the graph is higher or lower than all nearby points). For example, the function f in Figure 4.4 has "peaks" at B, D, and "valleys" at C, E. Peaks and valleys are what we call *relative extrema*.

Note that an extremum at an endpoint is, by definition, not a relative extremum, since there is not an open interval around the endpoint entirely contained in the domain of the function.

> **RELATIVE MAXIMUM**   A function f is said to have a **relative maximum** at the point c if $f(c) \ge f(x)$ for all x in an open interval containing c.
>
> **RELATIVE MINIMUM**   Likewise, f is said to have a **relative minimum** at d if $f(d) \le f(x)$ for all x in an open interval containing d.
>
> **RELATIVE EXTREMA**   Collectively, relative maxima and relative minima are called **relative extrema.**

Next we will formulate a procedure for finding relative extrema. By looking at Figure 4.4, we see that there are horizontal tangents at B, C, and E, while no tangent can be defined at the point D. This suggests that the relative extrema of f occur either where the derivative is zero (horizontal tangent) or where the derivative does not exist (no tangent). This notion leads us to the following definition.

*Note that if $f(c)$ is not defined, then c **cannot** be a critical number.*

> **CRITICAL NUMBER/CRITICAL POINT**   Suppose f is defined at c and either $f'(c) = 0$ or $f'(c)$ does not exist. Then the number c is called a **critical number** of f, and $P(c, f(c))$ on the graph of f is called a **critical point.**

Example 3 **Finding critical numbers**

Find the critical numbers for the given functions.

a. $f(x) = 4x^3 - 5x^2 - 8x + 20$ 　　**b.** $f(x) = \dfrac{e^x}{x-2}$ 　　**c.** $f(x) = 2\sqrt{x}(6-x)$

　　　　　　　　　　　　　　　　　 Note that $x \neq 2$. 　　　 *Note that* $x \geq 0$.

Solution

a. $f'(x) = 12x^2 - 10x - 8$ is defined for all values of x. Solve:

$$12x^2 - 10x - 8 = 0$$

$$2(3x - 4)(2x + 1) = 0$$

$$x = \frac{4}{3}, -\frac{1}{2} \quad \textit{These are the critical numbers.}$$

b. $f'(x) = \dfrac{(x-2)e^x - e^x(1)}{(x-2)^2}$

$\quad\;\; = \dfrac{e^x(x-3)}{(x-2)^2}$

The derivative is not defined at $x = 2$, but f is not defined at 2 either, so $x = 2$ is not a critical number. The actual critical numbers are found by solving $f'(x) = 0$:

$$\frac{e^x(x-3)}{(x-2)^2} = 0$$

$$e^x(x-3) = 0$$

$$x = 3 \quad \textit{This is the only critical number since } e^x > 0.$$

c. Write $f(x) = 12x^{1/2} - 2x^{3/2}$ so $f'(x) = 6x^{-1/2} - 3x^{1/2}$. The derivative is not defined at $x = 0$. We have

$$f(0) = 12(0)^{1/2} - 2(0)^{3/2} = 0$$

so we see that f is defined at $x = 0$ which means that $x = 0$ is a critical number. For other critical numbers, solve $f'(x) = 0$:

$$6x^{-1/2} - 3x^{1/2} = 0$$

$$3x^{-1/2}(2 - x) = 0$$

$$x = 2$$

The critical numbers are $x = 0, 2$.

Example 4 **Critical numbers and critical points**

Find the critical numbers and the critical points for the function

$$f(x) = (x-1)^2(x+2)$$

Solution Because the function f is a polynomial, we know that it is continuous and that its derivative exists for all x. Thus, we find the critical numbers by using the product rule and extended power rule to solve the equation $f'(x) = 0$:

$$f'(x) = (x-1)^2(1) + 2(x-1)(1)(x+2)$$

$$= (x-1)[(x-1) + 2(x+2)]$$

$$= 3(x-1)(x+1)$$

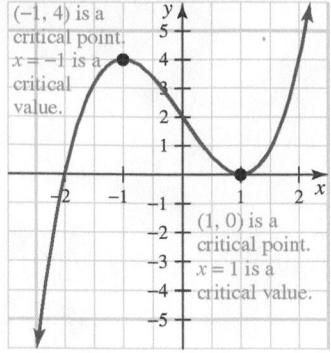

Figure 4.5 The graph of $f(x) = (x-1)^2(x+2)$

The critical numbers are $x = \pm 1$. To find the critical points, we need to find the y-coordinate for each critical number.

$$f(1) = (1-1)^2(1+2) = 0$$

$$f(-1) = (-1-1)^2(-1+2) = 4$$

Thus, the critical points are $(1, 0)$ and $(-1, 4)$. The graph of $f(x) = (x-1)^2(x+2)$ is shown in Figure 4.5. ∎

Note how the relative extrema occur at the critical points. Our observation that the relative extrema occur only at points on a graph where there is either a horizontal tangent line or no tangent at all is equivalent to the following result.

Theorem 4.2 Critical number theorem

If a continuous function f has a relative extremum at c, then c must be a critical number of f.

> ■ **W**hat this says If a point is a relative maximum or a relative minimum value for a function, then either the derivative is 0 or the derivative does not exist at that point. We do not claim the converse.

☘ *Theorem 4.2 tells us that a relative extremum of a continuous function f can occur **only** at a critical number, but it does not say that a relative extremum must occur at each critical number.* ☘

Proof: Since f has a relative extremum at c, $f(c)$ is defined. If $f'(c)$ does not exist, then c is a critical number by definition. We will show that if $f'(c)$ exists and a relative maximum occurs at c, then $f'(c) = 0$. Our approach will be to examine the difference quotient. (The case where $f'(c)$ exists and a relative minimum occurs at c is handled similarly in Problem 59.) Because a relative maximum occurs at c, we have $f(c) \geq f(x)$ for every number x in an open interval (a, b) containing c. Therefore, if Δx is small enough so $c + \Delta x$ is in (a, b), then

$$f(c) \geq f(c + \Delta x) \quad \text{\textit{Because a relative maximum occurs at } c}$$

$$f(c) - f(c + \Delta x) \geq 0$$

$$f(c + \Delta x) - f(c) \leq 0 \quad \text{\textit{Multiply both sides by } } -1, \text{\textit{ reversing the inequality.}}$$

For the next step we want to divide both sides by Δx (to write the left side as a difference quotient). However, as this is an inequality, we need to consider two possibilities:
Case I: Suppose $\Delta x > 0$ (the inequality does not reverse):

$$\frac{f(c + \Delta x) - f(c)}{\Delta x} \leq 0 \quad \text{\textit{Divide both sides by } } \Delta x.$$

$$\lim_{\Delta x \to 0^+} \frac{f(c + \Delta x) - f(c)}{\Delta x} \leq \lim_{\Delta x \to 0^+} 0 \quad \text{\textit{Now we take the limit of both sides as}} \atop \text{\textit{Δx approaches from the right (because}} \atop \text{\textit{Δx is positive).}}$$

$$f'(c) \leq 0$$

Case II: Next, suppose $\Delta x < 0$ (the inequality reverses). Then

$$\frac{f(c + \Delta x) - f(c)}{\Delta x} \geq 0 \quad \text{\textit{Divide both sides by } } \Delta x.$$

$$\lim_{\Delta x \to 0^-} \frac{f(c + \Delta x) - f(c)}{\Delta x} \geq \lim_{\Delta x \to 0^-} 0 \quad \text{\textit{Now we take the limit of both sides as}} \atop \text{\textit{Δx approaches from the left (because}} \atop \text{\textit{Δx is negative).}}$$

$$f'(c) \geq 0$$

Because we have shown that $f'(c) \leq 0$ and $f'(c) \geq 0$, it follows that $f'(c) = 0$. ◆

For example, if $f(x) = x^3$, then $f'(x) = 3x^2$ and $f'(0) = 0$, so 0 is a critical number. But there is no relative extremum at $c = 0$ on the graph of f because the graph is rising for $x < 0$ and also for $x > 0$, as shown in Figure 4.6**a**. It is also quite possible for a continuous function g to have no relative extremum at a point c where $g'(x)$ does not exist (see Figure 4.6**b**).

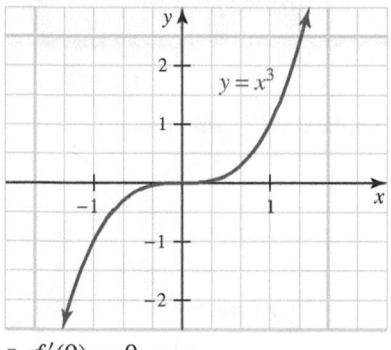

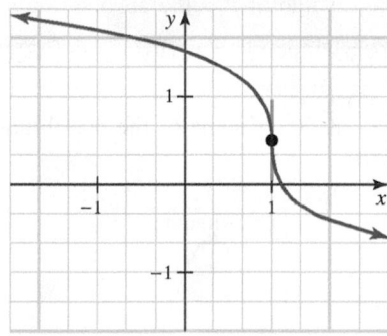

a. $f'(0) = 0$
No relative extremum occurs at $c = 0$

b. $g'(1)$ does not exist
No relative extremum occurs at $c = 1$

Figure 4.6 A relative extremum may not occur at each critical number

Example 5 Critical numbers where the derivative does not exist

Find the critical numbers for $f(x) = |x + 1|$ on $[-5, 5]$. The graph is shown in Figure 4.7.

Solution If $x > -1$, then $f(x) = x + 1$ and $f'(x) = 1$. However, if $x < -1$, then $f(x) = -x - 1$ and $f'(x) = -1$. We need to check what happens at $x = -1$.

$$f'(-1) = \lim_{\Delta x \to 0} \frac{f(-1 + \Delta x) - f(-1)}{\Delta x}$$

$$= \lim_{\Delta x \to 0} \frac{|\Delta x| - |0|}{\Delta x}$$

$$= \lim_{\Delta x \to 0} \frac{|\Delta x|}{\Delta x}$$

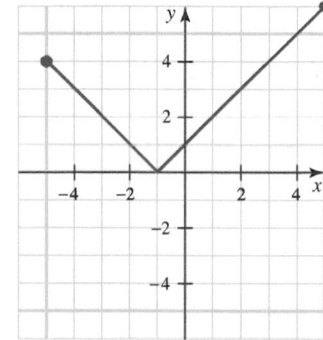

Figure 4.7 Graph of f

We consider the left and right limits:

$$\lim_{\Delta x \to 0^+} \frac{|\Delta x|}{\Delta x} = \lim_{\Delta x \to 0^+} \frac{\Delta x}{\Delta x} = 1 \quad \text{and} \quad \lim_{\Delta x \to 0^-} \frac{|\Delta x|}{\Delta x} = \lim_{\Delta x \to 0^-} \frac{-\Delta x}{\Delta x} = -1$$

Because these are not the same, this limit does not exist and thus the function $f(x) = |x + 1|$ is not differentiable at $x = -1$. Because $f(-1)$ is defined, it follows that -1 is the only critical number. ■

Absolute Extrema

Suppose we are looking for the absolute extrema of a continuous function f on the closed, bounded interval $[a, b]$. Theorem 4.1 tells us that these extrema exist and Theorem 4.2 enables us to narrow the list of "candidates" for points where extrema can occur from the entire interval $[a, b]$ to just the endpoints $x = a$, $x = b$, and the critical numbers between a and b. This suggests the following procedure.

> **PROCEDURE FOR FINDING ABSOLUTE EXTREMA** To find the absolute extrema of a continuous function f on $[a, b]$ follow these steps:
>
> **Step 1** Compute $f'(x)$ and find all critical numbers of f on $[a, b]$.
> **Step 2** Evaluate f at the endpoints a and b and at each critical number c.
> **Step 3** Compare the values in step 2.
>   The largest value of f is the absolute maximum of f on $[a, b]$. The smallest value of f is the absolute minimum of f on $[a, b]$.

Figure 4.8 shows some of the possibilities in the application of this procedure.

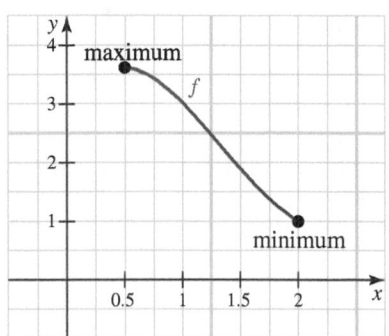

$f(x) = x^3 - 4x^2 + 3x + 3$ on $[0.5, 2]$
a. f is continuous with an extremum at each endpoint

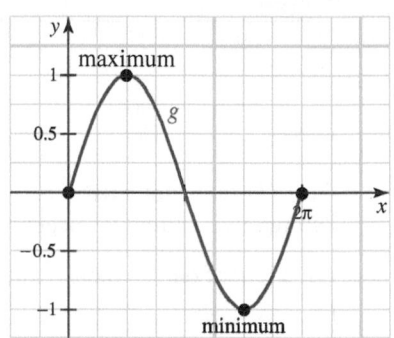

$g(x) = \sin x$ on $[0, 2\pi]$
b. g is continuous; neither extremum at an endpoint

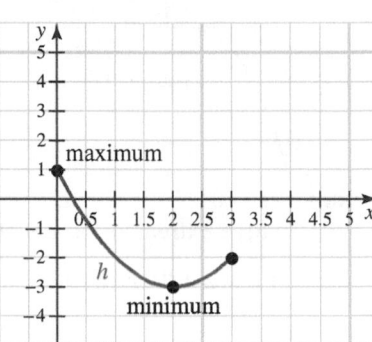

$h(x) = x^2 - 4x + 1$ on $[0, 3]$
c. h is continuous; one extremum at an endpoint

Figure 4.8 Absolute extrema

Example 6 Absolute extrema of a polynomial function

Find the absolute extrema of the function defined by the equation $f(x) = x^4 - 2x^2 + 3$ on the closed interval $[-1, 2]$.

Solution Because f is a polynomial function, it is continuous on the closed interval $[-1, 2]$. Theorem 4.1 tells us that there must be an absolute maximum and an absolute minimum on the interval.

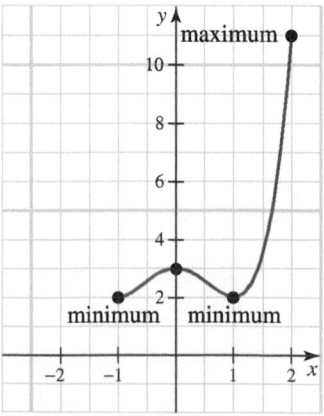

Figure 4.9 Graph of f

Step 1 $f'(x) = 4x^3 - 2(2x) + 0$

$\qquad = 4x(x^2 - 1)$

$\qquad = 4x(x - 1)(x + 1)$

The critical numbers are $x = 0, 1$, and -1.

Step 2 Values at endpoints: $f(-1) = 2$ $f(2) = 11$
          Critical numbers: $f(0) = 3$ $f(1) = 2$

Step 3 The absolute maximum of f occurs at $x = 2$ and is $f(2) = 11$; the absolute minimum of f occurs at $x = 1$ and $x = -1$ and is $f(1) = f(-1) = 2$. The graph of f is shown in Figure 4.9.

Example 7   Absolute extrema when the derivative does not exist

Find the absolute extrema of $g(x) = x^{2/3}(5 - 2x)$ on the interval $[-1, 2]$.

Solution

Step 1 To find the derivative, rewrite the given function as $g(x) = 5x^{2/3} - 2x^{5/3}$. Then

$$g'(x) = \frac{10}{3}x^{-1/3} - \frac{10}{3}x^{2/3} = \frac{10}{3}x^{-1/3}(1 - x)$$

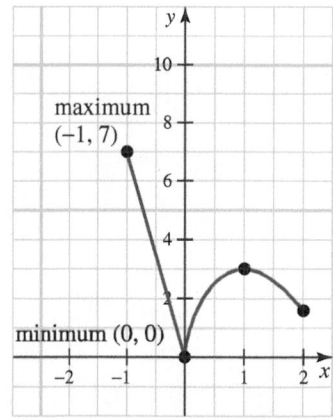

Critical numbers are found by solving $g'(x) = 0$ and by locating the places where the derivative does not exist. First, $g'(x) = 0$ when $x = 1$. Even though $g(0)$ exists, we note that $g'(x)$ does not exist at $x = 0$ (notice the division by zero when $x = 0$). Thus, the critical numbers are $x = 0$ and $x = 1$.

Step 2 Values at endpoints: $g(-1) = 7$ $g(2) = 2^{2/3} \approx 1.5874010520$
Critical numbers: $g(0) = 0$ $g(1) = 3$

Step 3 The absolute maximum of g occurs at $x = -1$ and is $g(-1) = 7$; the absolute minimum of g occurs at $x = 0$ and is $g(0) = 0$. The graph of g is shown in Figure 4.10. ■

Figure 4.10   Graph of g

Example 8   Absolute extrema for a trigonometric function

Find the absolute extrema of the continuous function

$$T(x) = \frac{1}{2}\left(\sin^2 x + \cos x\right) + 2\sin x - x$$

on the interval $\left[0, \frac{\pi}{2}\right]$.

Solution

Step 1 To find where $T'(x) = 0$, we begin by finding $T'(x)$:

$$T'(x) = \frac{1}{2}(2\sin x \cos x - \sin x) + 2\cos x - 1$$

$$= \frac{1}{2}\left(2\sin x \cos x - \sin x + 4\cos x - 2\right)$$

$$= \frac{1}{2}\left[2(\cos x)(\sin x + 2) - (\sin x + 2)\right]$$

$$= \frac{1}{2}\left[(\sin x + 2)(2\cos x - 1)\right]$$

Since the factor $(\sin x + 2)$ is never zero, it follows that $T'(x) = 0$ only when $2\cos x - 1 = 0$; that is, when $x = \frac{\pi}{3}$. This is the only critical number in $\left[0, \frac{\pi}{2}\right]$.

Step 2 Evaluate the function at the endpoints:

$$T(0) = \frac{1}{2}\left(\sin^2 0 + \cos 0\right) + 2\sin 0 - 0 \qquad T\left(\frac{\pi}{2}\right) = \frac{1}{2}\left(\sin^2 \frac{\pi}{2} + \cos \frac{\pi}{2}\right) + 2\sin \frac{\pi}{2} - \frac{\pi}{2}$$

$$= \frac{1}{2}(0 + 1) + 2(0) - 0 \qquad\qquad\qquad = \frac{1}{2}(1 + 0) + 2(1) - \frac{\pi}{2}$$

$$= 0.5 \qquad\qquad\qquad\qquad\qquad\qquad = \frac{5}{2} - \frac{\pi}{2}$$

$$\qquad\qquad\qquad\qquad\qquad\qquad\qquad\qquad \approx 0.9292036732$$

Evaluate the function at the critical number:

$$T\left(\frac{\pi}{3}\right) = \frac{1}{2}\left(\sin^2\frac{\pi}{3} + \cos\frac{\pi}{3}\right) + 2\sin\frac{\pi}{3} - \frac{\pi}{3}$$

$$= \frac{1}{2}\left(\frac{3}{4} + \frac{1}{2}\right) + 2\left(\frac{\sqrt{3}}{2}\right) - \frac{\pi}{3}$$

$$= \frac{5}{8} + \sqrt{3} - \frac{\pi}{3}$$

$$\approx 1.309853256$$

Step 3 The absolute maximum of T is approximately 1.31 at $x = \frac{\pi}{3}$ and the absolute minimum of T is 0.5 at $x = 0$. ◼

Optimization

In the next two examples, we examine applications involving optimization. Such problems are investigated in more depth in Sections 4.6 and 4.7.

Example 9 An applied maximum value problem

A box with a square base is constructed so that the length of one side of the base plus the height is 10 in. What is the largest possible volume of such a box?

Solution   We let b be the length of one side of the base and h be the height of the box. The volume, V, is

$$V = b^2 h$$

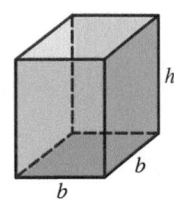

Because our methods apply only to functions of one variable, it may seem that we cannot deal with V as a function of two variables.

However, we know that $b + h = 10$ (such an equation is frequently called a **constraint**); therefore, $h = 10 - b$, and we can now write V as a function of b alone:

$$V(b) = b^2(10 - b)$$

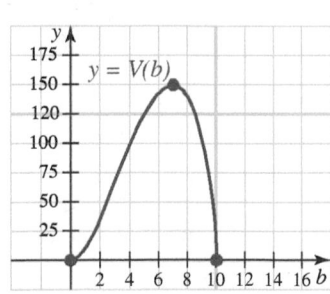

Figure 4.11 Volume of a box

The domain is not stated, but for physical reasons we must have $b \geq 0$ and $10 - b = h \geq 0$, so that $0 \leq b \leq 10$. Figure 4.11 shows the volume for various choices of b.

Now, find the value of b that produces the maximum volume. First, we find the critical numbers. Note that V is a polynomial function, so the derivative exists everywhere in the domain. Write $V(b) = 10b^2 - b^3$ and find $V'(b) = 20b - 3b^2$. Then

$$V'(b) = 0$$

$$20b - 3b^2 = 0$$

$$b(20 - 3b) = 0$$

$$b = 0, \frac{20}{3}$$

Alternatively, we could have solved the problem with V as a function of h: $V(h) = (10 - h)^2 h$, $0 \leq h \leq 10$, with the same result. We must write V as a function of one variable so that we can find its derivative.

Checking the endpoints and the critical numbers, we have

$$V(0) = 0$$

$$V(10) = 10^2(10 - 10) = 0$$

$$V\left(\frac{20}{3}\right) = \left(\frac{20}{3}\right)^2\left(10 - \frac{20}{3}\right) = \frac{4,000}{27}$$

Thus, the largest value for the volume V is $\frac{4,000}{27} \approx 148.1$ in.^{3}. It occurs when the square base has a side of length $\frac{20}{3}$ in. and the height is $h = 10 - \frac{20}{3} = \frac{10}{3}$ in.   ■

Example 10   A particle moves along the s-axis with position t

A particle moves along the s-axis with position t

$$s(t) = t^4 - 8t^3 + 18t^2 + 60t - 8$$

Find the largest and smallest values of its velocity for $1 \leq t \leq 5$.

Solution   The velocity is

$$v(t) = s'(t) = 4t^3 - 24t^2 + 36t + 60$$

(*Note*: v is a polynomial function, so it has a derivative for all t.) To find the largest value of $v(t)$, we compute the derivative of v:

$$v'(t) = 12t^2 - 48t + 36$$
$$= 12(t - 1)(t - 3)$$

Setting $v'(t) = 0$, we find that the critical numbers of $v(t)$ are $t = 1, 3$. Now we evaluate v at the critical numbers and endpoints.

$t = 1$ is both a critical number and an endpoint: $v(1) = 4(1)^3 - 24(1)^2 + 36(1) + 60 = 76$

$t = 3$ is a critical number: $v(3) = 4(3)^3 - 24(3)^2 + 36(3) + 60 = 60$

$t = 5$ is an endpoint: $v(5) = 4(5)^3 - 24(5)^2 + 36(5) + 60 = 140$

We see the largest value of the velocity is 140 at the endpoint when $t = 5$, and the smallest value is 60, when $t = 3$.   ■

PROBLEM SET 4.1

Level 1

In Problems 1-14, *find the critical numbers for each continuous function on the given closed, bounded interval, and then tell whether each yields a minimum, maximum, or neither.*

1. $f(x) = 5 + 10x - x^2$ on $[-3, 3]$
2. $f(x) = 10 + 6x - x^2$ on $[-4, 4]$
3. $f(x) = x^3 - 3x^2$ on $[-1, 3]$
4. $f(x) = x^3$ on $\left[-\frac{1}{2}, 1\right]$
5. $f(t) = t^4 - 8t^2$ on $[-3, 3]$
6. $f(t) = t^3 - 3t$ on $[-2, 2]$
7. $g(x) = x^5 - x^4$ on $[-1, 1]$
8. $g(t) = 3t^5 - 20t^3$ on $[-1, 2]$
9. $h(t) = te^{-t}$ on $[0, 2]$
10. $s(x) = \dfrac{\ln \sqrt{x}}{x}$ on $[1, 3]$
11. $f(x) = |x|$ on $[-1, 1]$
12. $f(x) = |x - 3|$ on $[-4, 4]$
13. $f(u) = \sin^2 u + \cos u$ on $[0, 2]$

14. $g(u) = \sin u - \cos u$ on $[0, \pi]$
15. ■ What does this say? Outline a procedure for finding the absolute extrema of a continuous function on a closed, bounded interval. Include in your outline a discussion of what is meant by critical numbers.
16. ■ What does this say? In Example 7, we found that the maximum value of $f(x) = x^{2/3}(5 - 2x)$ on $[-1, 2]$ is 7 and occurs at $x = -1$. The graph of this function on a leading brand of graphing calculator is shown:

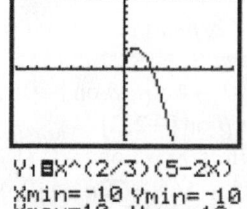

Y₁=X^(2/3)(5-2X)
Xmin=-10 Ymin=-10
Xmax=10 Ymax=10
Xscl=1   Yscl=1

This graph is not correct. Can you explain the discrepancy? It is not our intent to "make up" problems to use with a calculator, but *whenever* you use a calculator or computer to assist you with calculus, you must understand the nature of the functions with which you are working, and not rely only on the calculator or computer output.

In Problems 17-18, find the absolute maximum and absolute minimum (largest and smallest values, respectively) of each continuous function on the closed, bounded interval. If the function is not continuous on the interval, then the extreme value theorem does not apply; if this is the case, so state.

17. $f(u) = 1 - u^{2/3}$ on $[-1, 1]$

18. $f(t) = (50 + t)^{2/3}$ on $[-50, 14]$

19. $g(x) = 2x^3 - 3x^2 - 36x + 4$ on $[-4, 4]$

20. $g(x) = x^3 + 3x^2 - 24x - 4$ on $[-4, 4]$

21. $f(x) = \frac{8}{3}x^3 - 5x^2 + 8x - 5$ on $[-4, 4]$

22. $f(x) = \frac{1}{6}(x^3 - 6x^2 + 9x + 1)$ on $[0, 2]$

23. $h(x) = \tan x + \sec x$ on $[0, 2\pi]$

24. $s(t) = t \cos t - \sin t$ on $[0, 2\pi]$

25. $f(x) = e^{-x} \sin x$ on $[0, 2\pi]$

26. $g(x) = \cot^{-1}\left(\frac{x}{9}\right) - \cot^{-1}\left(\frac{x}{5}\right)$ on $[0, 10]$

27. $f(x) = \begin{cases} 9 - 4x & \text{if } x < 1 \\ -x^2 + 6x & \text{if } x \geq 1 \end{cases}$ on $[0, 4]$

28. $f(x) = \begin{cases} 8 - 3x & \text{if } x < 2 \\ -x^2 + 3x & \text{if } x \geq 2 \end{cases}$ on $[-1, 4]$

Find the required extremum in Problems 29-34, or explain why it does not exist.

29. the smallest value of $f(x) = \dfrac{1}{x(x+1)}$ on $[-0.5, 0)$

30. the smallest value of $g(x) = \dfrac{9}{x} + x - 3$ on $[1, 9]$

31. the smallest value of $g(x) = \dfrac{x^2 - 1}{x^2 + 1}$ on $[-1, 1]$

32. the largest value of $f(t) = \begin{cases} -t^2 - t + 2 & \text{if } t < 1 \\ 3 - t & \text{if } t \geq 1 \end{cases}$ on $[-2, 3]$

33. the smallest value of $f(x) = e^x + e^{-x} - x$ on $[0, 2]$

34. the largest value of $g(x) = \dfrac{\ln x}{\cos x}$ on $[2, 3]$ correct to the nearest tenth.

Level 2

In Problems 35-42, find the extrema (that is, the absolute maxima and minima).

35. $f(\theta) = \cos^3 \theta - 4 \cos^2 \theta$ on $[-0.1, \pi + 0.1]$

36. $g(\theta) = \theta \sin \theta$ on $[-2, 2]$

37. $f(x) = 20 \sin(378 \pi x)$ on $[-1, 1]$

38. $g(u) = 98u^3 - 4u^2 + 72u$ on $[0, 4]$

39. $f(w) = \sqrt{w}(w - 5)^{1/3}$ on $[0, 4]$

40. $h(x) = \sqrt[3]{x}\sqrt[3]{(x - 3)^2}$ on $[-1, 4]$

41. $h(x) = \cos^{-1} x \tan^{-1} x$ on $[0, 1]$ (answer correct to the nearest tenth)

42. $f(x) = e^{-x}(\cos x + \sin x)$ on $[0, 2\pi]$

In Problems 43-46, find functions that satisfy the stated conditions and each of the following side conditions, if possible:

a. *a minimum but no maximum*

b. *a maximum but no minimum*

c. *both a maximum and a minimum*

d. *neither a maximum nor a minimum.*

Note that in each problem, there may be a different solution function for each side condition.

43. Think Tank Problem For each of the four given conditions, find a function that is discontinuous and defined on an open interval.

44. Think Tank Problem For each of the four given conditions, find a function that is discontinuous and defined on a closed interval.

45. Think Tank Problem For each of the four given conditions, find a function that is continuous and defined on an open interval.

46. Think Tank Problem For each of the four given conditions, find a function that is continuous and defined on a closed interval.

47. Think Tank Problem Give a counterexample to show that the extreme value theorem does not necessarily apply if one disregards the condition that f be continuous.

48. Think Tank Problem Give a counterexample to show that the extreme value theorem does not necessarily apply if one disregards the condition that f be defined on a closed interval.

49. An object moves along the s-axis with position

$$s(t) = t^3 - 6t^2 - 15t + 11$$

Find the largest value of its velocity on $[0, 4]$.

50. An object moves along the s-axis with position

$$s(t) = t^4 - 2t^3 - 12t^2 + 60t - 10$$

Find the largest value of its velocity on $[0, 3]$.

51. Find two nonnegative numbers whose sum is 8 and the product of whose squares is as large as possible.

52. Find two nonnegative numbers such that the sum of one and twice the other is 12 if it is required that their product be as large as possible.

53. Under the condition that $3x + y = 80$, maximize xy^3 when $x \geq 0, y \geq 0$.

54. Under the condition that $3x + y = 126$ maximize xy when $x \geq 0$ and $y \geq 0$.

Level 3

55. Show that if a rectangle with fixed perimeter P is to enclose the greatest area, it must be a square.

56. Find all points on the circle

$$x^2 + y^2 = a^2 \qquad a \geq 0$$

such that the product of the x-coordinate and the y-coordinate is as large as possible.

57. a. Show that $\frac{1}{2}$ is the number that is greater than or equal to its own square by the greatest amount.

  b. Which nonnegative number is greater than or equal to its own cube by the greatest amount?

  c. Which nonnegative number is greater than or equal to its nth power ($n > 0$) by the greatest amount?

58. Given the constants $a_1, a_2, \ldots, a_n$, find the value of x that guarantees that the sum

$$S(x) = (a_1 - x)^2 + (a_2 - x)^2 + \cdots + (a_n - x)^2$$

will be as small as possible.

59. Without using Theorem 4.2, show that if $f'(c)$ exists and a relative minimum occurs at c, then $f'(c) = 0$.

60. Explain why the function

$$f(x) = \frac{8}{\sin x} + \frac{27}{\cos x}$$

must attain a minimum in the open interval $\left(0, \frac{\pi}{2}\right)$.

4.2 THE MEAN VALUE THEOREM

IN THIS SECTION: *Rolle's theorem, proof of the mean value theorem, the zero-derivative theorem*
The mean value theorem is an extremely important result in calculus, both in theory and for practical applications. The goal of this section is to prove the mean value theorem and to examine some of its consequences.

If a car travels smoothly down a straight, level road with average velocity 60 mi/h, we would expect the speedometer reading to be exactly 60 mi/h at least once during the trip. After all, if the car's velocity were always above 60 mi/h, the average velocity would also be above that level, and the same reasoning applies if the velocity were always below 60 mi/h.

More generally, if $s(t)$ is the car's position at time t during a trip over the time interval $a \leq t \leq b$, then there should be a time $t = c$ when the velocity $s'(c)$ equals the average velocity between times $t = a$ and $t = b$. That is, for some c with $a < c < b$,

$$\underbrace{\frac{s(b) - s(a)}{b - a}}_{Average\ velocity} = \overbrace{s'(c)}^{Instantaneous\ velocity}$$

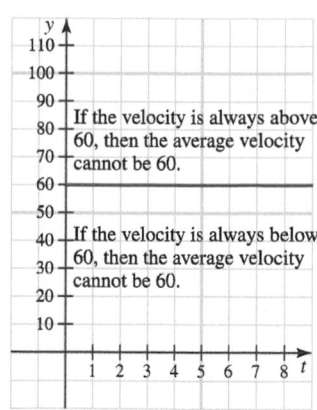

This example illustrates a result called the mean value theorem for derivatives (abbreviated MVT) that is fundamental in the study of calculus. Our proof of the MVT is based on the following special case, which is named for the French mathematician, Michel Rolle (1652-1719), who gave a version of the result in an algebra text published in 1690.*

Rolle's Theorem

The key to the proof of the mean value theorem is the following result, which is really just the MVT in the special case where $f(a) = f(b)$. In terms of our car example, Rolle's theorem says that if a moving car begins and ends at the same place, then somewhere during its journey, it must reverse direction, since $\dfrac{f(b) - f(a)}{b - a} = 0$ for $f(a) = f(b)$.

Theorem 4.3 Rolle's theorem

Suppose f is continuous on the closed interval $[a, b]$ and differentiable on the open interval (a, b). If $f(a) = f(b)$, then there exists at least one number c between a and b such that $f'(c) = 0$.

---

*Incidentally, Rolle was a number theorist at heart and thoroughly distrusted the methods of calculus, which he regarded as a "collection of ingenious fallacies."

Proof: To construct a formal proof, note that since f is continuous on the closed interval $[a, b]$, it follows from the extreme value theorem (Theorem 4.1) that f must have both a largest value and a smallest value on $[a, b]$. The case where f is constant on $[a, b]$ is easy since then $f'(x) = 0$ for all x in the interval (see Figure 4.12**a**).

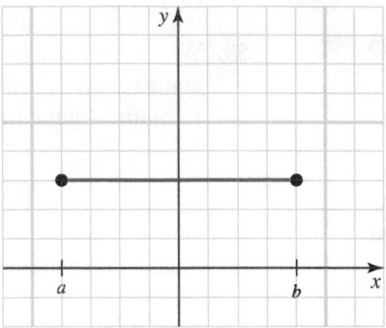

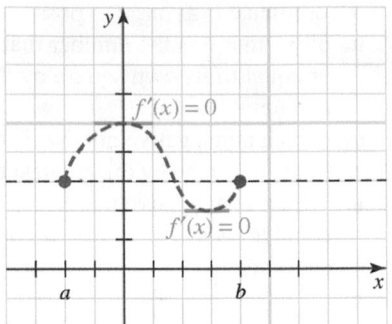

a. If f is constant on $[a, b]$ then $f'(x) = 0$ throughout (a, b)

b. If f is not constant on (a, b), the graph must have at least one horizontal tangent

Figure 4.12 A geometrical interpretation of Rolle's theorem

In the case where f is not constant throughout $[a, b]$, the largest and smallest values cannot be the same, and since $f(a) = f(b)$, we can conclude that at least one extreme value occurs at a number c that is not an endpoint; that is, $a < c < b$. Finally, since c is in the open interval (a, b) throughout which $f'(x)$ exists, it follows from the critical number theorem (Theorem 4.2) that $f'(c) = 0$. A graph illustrating this case is shown in Figure 4.12**b**.   ♦

Note: Rolle's theorem asserts only that at least one number c exists between a and b such that $f'(c) = 0$. As illustrated in Figure 4.12**b**, there may be more than one such number.

Proof of the Mean Value Theorem

Rolle's theorem can be interpreted as saying that there is at least one number c between a and b such that the tangent line to the graph of f at the point $(c, f(c))$ is parallel to the horizontal line through the points $(a, f(a))$ and $(b, f(b))$. It is reasonable to expect a similar result to hold if the endpoints of the graph are not necessarily at the same level of the graph; that is, if the graph in Figure 4.13**a** is tilted as shown in Figure 4.13**b**.

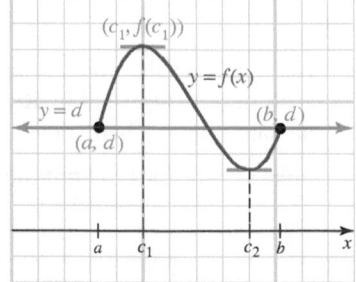

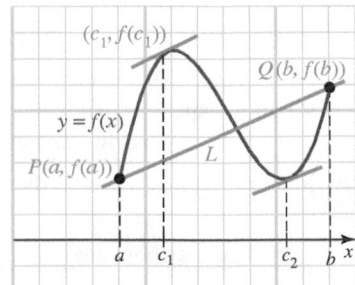

a. Rolle's theorem: The slope is 0 for some point c between a and b

b. Interactive Mean value theorem: The slope between $(a, f(a))$ and $(b, f(b))$ is the same as the slope at some point c between a and b

Figure 4.13 A geometrical comparison of Rolle's theorem with the mean value theorem

This "tilted" version can be interpreted as showing that the line segment joining points $(a, f(a))$ and $(b, f(b))$ on the graph of a differentiable function $y = f(x)$ has the same slope as the tangent line to the graph at some point c between a and b. This observation leads to the following statement of the mean value theorem.

Theorem 4.4   The mean value theorem for derivatives (MVT)

If f is continuous on the closed interval $[a, b]$ and differentiable on the open interval (a, b), then there exists in (a, b) at least one number c such that

$$\frac{f(b) - f(a)}{b - a} = f'(c)$$

Proof: Our strategy will be to apply Rolle's theorem to a function g related to f; namely the function

$$g(x) = \frac{f(b) - f(a)}{b - a}(x - a) + f(a) - f(x)$$

for $a \leq x \leq b$. (You may want to note that the function g in some sense measures the signed distance between the function f and the line segment.) Because f satisfies the hypotheses of the MVT, the function g is also continuous on the closed interval $[a, b]$ and differentiable on the open interval (a, b). In addition, we find that

$$g(a) = \frac{f(b) - f(a)}{b - a}(a - a) \quad g(b) = \frac{f(b) - f(a)}{b - a}(b - a)$$
$$\qquad\quad + f(a) - f(a) \qquad\qquad\quad + f(a) - f(b)$$
$$\quad = 0 \qquad\qquad\qquad\qquad = f(b) - f(a) + f(a) - f(b)$$
$$\qquad\qquad\qquad\qquad\qquad\qquad = 0$$

Thus, g satisfies the hypotheses of Rolle's theorem, so there exists at least one number c between a and b for which $g'(c) = 0$. Differentiating the function g, we find that

$$g'(x) = \frac{f(b) - f(a)}{b - a} - f'(x)$$
$$g'(c) = \frac{f(b) - f(a)}{b - a} - f'(c)$$
$$0 = \frac{f(b) - f(a)}{b - a} - f'(c)$$
$$f'(c) = \frac{f(b) - f(a)}{b - a} \qquad\qquad\qquad ◆$$

The MVT asserts the existence of a number, c, but does not tell us how to find it. The following example illustrates one method for finding such a number c.

Example 1   Finding the number c specified by the MVT

Show that the function $f(x) = x^3 + x^2$ satisfies the hypotheses of the MVT on the closed interval $[1, 2]$, and find a number c between 1 and 2 so that

$$f'(c) = \frac{f(2) - f(1)}{2 - 1}$$

Solution   Because f is a polynomial function, it is differentiable and also continuous on the entire interval $[1, 2]$. Thus, the hypotheses of the MVT are satisfied.

By differentiating f, we find that

$$f'(x) = 3x^2 + 2x$$

for all x. Therefore, we have $f'(c) = 3c^2 + 2c$, and the MVT equation

$$f'(c) = \frac{f(2) - f(1)}{2 - 1}$$

is satisfied when

$$3c^2 + 2c = \frac{f(2) - f(1)}{2 - 1} \quad \textit{MVT equation}$$

$$3c^2 + 2c = \frac{12 - 2}{1} \quad \textit{Simplify}$$

$$3c^2 + 2c = 10$$

$$3c^2 + 2c - 10 = 0$$

$$c = \frac{-1 \pm \sqrt{31}}{3} \quad \textit{Use the quadratic formula and simplify.}$$

$$\approx 1.522588121 \quad \textit{The negative value is not in the interval } (1, 2).$$

We see that this value satisfies the requirements of the MVT. ■

Example 2   Using the MVT to prove a trigonometric inequality

Show that $|\sin x_2 - \sin x_1| \le |x_2 - x_1|$ for all numbers x_1 and x_2 by applying the mean value theorem.

Solution   The inequality is true if $x_1 = x_2$. Suppose $x_1 \ne x_2$; then $f(x) = \sin x$ is differentiable and hence continuous for all x, with $f'(x) = \cos x$. By applying the MVT to f on the closed interval with endpoints x_1 and x_2, we see that

$$\frac{f(x_2) - f(x_1)}{x_2 - x_1} = f'(c)$$

for some c between x_1 and x_2. Since $f'(c) = \cos c$ and

$$f(x_2) - f(x_1) = \sin x_2 - \sin x_1,$$

it follows that

$$\frac{\sin x_2 - \sin x_1}{x_2 - x_1} = \cos c$$

Finally, we take the absolute value of the expression on each side, remembering that $|\cos c| \le 1$ for any number c:

$$\left| \frac{\sin x_2 - \sin x_1}{x_2 - x_1} \right| = |\cos c| \le 1$$

Thus, $|\sin x_2 - \sin x_1| \le |x_2 - x_1|$, as claimed. ■

The Zero-Derivative Theorem

The primary use of the MVT is as a tool for proving certain key theoretical results of calculus. For example, we know that the derivative of a constant is 0, and in Theorem 4.5, we prove a partial converse of this result by showing that a function whose derivative is 0 throughout the interval must be constant on that interval. This apparently simple result and its corollary (Theorem 4.6) turn out to be crucial to our development of integration in Chapter 5.

Theorem 4.5   Zero-derivative theorem

Suppose f is a continuous function on the closed interval $[a,b]$ and is differentiable on the open interval (a,b), with $f'(x) = 0$ for all x on (a,b). Then the function f is constant on $[a,b]$.

Proof: Let x_1 and x_2 be two distinct numbers $(x_1 \neq x_2)$ chosen arbitrarily from the closed interval $[a,b]$. The function f satisfies the requirements of the MVT on the interval with endpoints x_1 and x_2, which means that there exists a number c between x_1 and x_2 such that

$$\frac{f(x_2) - f(x_1)}{x_2 - x_1} = f'(c)$$

By hypothesis, $f'(x) = 0$ throughout the open interval (a,b), and because c lies within this interval, we have $f'(c) = 0$. Thus, by substitution, we have

$$\frac{f(x_2) - f(x_1)}{x_2 - x_1} = 0$$
$$f(x_2) = f(x_1)$$

Because x_1 and x_2 were chosen arbitrarily from $[a,b]$, we conclude that $f(x) = k$, a constant, for all x, as required.   ◆

Theorem 4.6   Constant difference theorem

Suppose the functions f and g are continuous on the closed interval $[a,b]$ and differentiable on the open interval (a,b). Then if $f'(x) = g'(x)$ for all x in (a,b), there exists a constant C such that

$$f(x) = g(x) + C$$

for all x on $[a,b]$.

> ■ **W**hat this says Two functions with equal derivatives on an open interval differ by a constant on that interval.

Proof: Let $h(x) = f(x) - g(x)$; then

$$h'(x) = f'(x) - g'(x)$$
$$= 0 \qquad \textit{Because } f'(x) = g'(x).$$

Thus, by Theorem 4.5, $h(x) = C$ for some constant C and all x on $[a,b]$, and because $h(x) = f(x) - g(x)$, it follows that

$$f(x) - g(x) = C$$
$$f(x) = g(x) + C \qquad\qquad\qquad ◆$$

PROBLEM SET 4.2

Level 1

1. ■ What does this say? What does Rolle's theorem say, and why is it important?
2. ■ What does this say? State the hypotheses used in the proof of the MVT. How are the hypotheses used in the proof? Can the conclusion of the MVT be true if any or all of the hypotheses are not satisfied?

In Problems 3-20, verify that the given function f satisfies the hypotheses of the MVT on the given interval $[a,b]$. Then find all numbers c between a and b for which

$$\frac{f(b) - f(a)}{b - a} = f'(c)$$

3. $f(x) = 2x^2 + 1$ on $[0, 2]$
4. $f(x) = -x^2 + 4$ on $[-1, 0]$
5. $f(x) = x^3 + x$ on $[1, 2]$
6. $f(x) = 2x^3 - x^2$ on $[0, 2]$
7. $f(x) = x^4 + 2$ on $[-1, 2]$
8. $f(x) = x^5 + 3$ on $[2, 4]$ 9. $f(x) = \sqrt{x}$ on $[1, 4]$
10. $f(x) = \dfrac{1}{\sqrt{x}}$ on $[1, 4]$ 11. $f(x) = \dfrac{1}{x+1}$ on $[0, 2]$
12. $f(x) = 1 + \dfrac{1}{x}$ on $[1, 4]$ 13. $f(x) = \cos x$ on $\left[0, \dfrac{\pi}{2}\right]$
14. $f(x) = \sin x + \cos x$ on $[0, 2\pi]$
15. $f(x) = e^x$ on $[0, 1]$
16. $f(x) = \frac{1}{2}\left(e^x + e^{-x}\right)$ on $[0, 1]$
17. $f(x) = \ln x$ on $\left[\frac{1}{2}, 2\right]$ 18. $f(x) = \dfrac{\ln \sqrt{x}}{x}$ on $[1, 3]$
19. $f(x) = \tan^{-1} x$ on $[0, 1]$
20. $f(x) = x \sin^{-1} x$ on $[0, 1]$

Decide whether Rolle's theorem can be applied to f on the interval indicated in Problems 21-30.

21. $f(x) = |x - 2|$ on $[0, 4]$ 22. $f(x) = \tan x$ on $[0, 2\pi]$

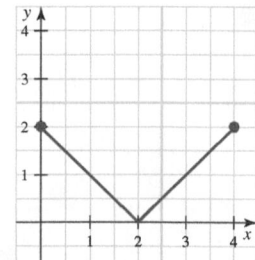 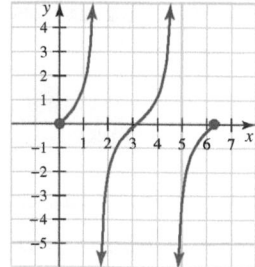

23. $f(x) = \sin x$ on $[0, 2\pi]$
24. $f(x) = |x| - 2$ on $[0, 4]$
25. $f(x) = \sqrt[3]{x} - 1$ on $[-8, 8]$
26. $f(x) = \dfrac{1}{x-2}$ on $[-1, 1]$
27. $f(x) = \dfrac{1}{x-2}$ on $[1, 2]$
28. $f(x) = 3x + \sec x$ on $[-\pi, \pi]$
29. $f(x) = \sin^2 x$ on $\left[-\frac{\pi}{2}, \frac{\pi}{2}\right]$

30. $f(x) = \sqrt{\ln x}$ on $[1, 2]$
31. Let $g(x) = 8x^3 - 6x + 8$. Find a function f with $f'(x) = g'(x)$ and $f(1) = 12$.
32. Let $g(x) = \sqrt{x^2 + 5}$. Find a function f with $f'(x) = g'(x)$ and $f(2) = 1$.
33. Show that $f(x) = \dfrac{x+4}{5-x}$ and $g(x) = \dfrac{-9}{x-5}$ differ by a constant. Are the conditions of the constant difference theorem satisfied? Does $f'(x) = g'(x)$?
34. Let $f(x) = (x - 2)^3$ and $g(x) = (x^2 + 12)(x - 6)$. Use f and g to demonstrate the constant difference theorem.
35. Let $f(x) = (x - 1)^3$ and $g(x) = (x^2 + 3)(x - 3)$. Use f and g to demonstrate the constant difference theorem.
36. Let f be defined as shown in Figure 4.14.

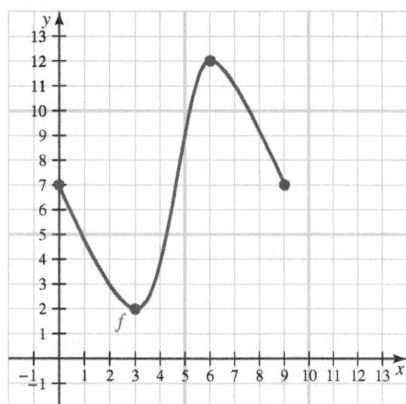

Figure 4.14   Function f on $[0, 9]$

Use the graph of f to estimate the values of c that satisfy the conclusion of Rolle's theorem on $[0, 9]$. What theorem would apply for the interval $[0, 5]$?

37. Let g be defined as shown in Figure 4.15.

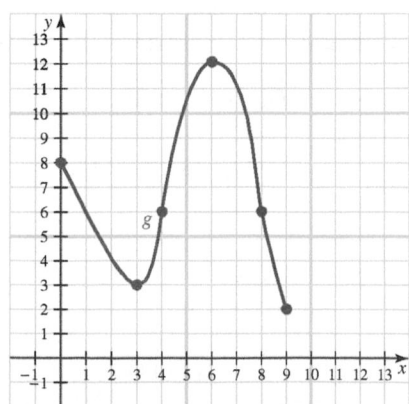

Figure 4.15   Function g on $[0, 9]$

Use the graph of g to estimate the values of c that satisfy the conclusion of the mean value theorem on $[0, 9]$. What theorem would apply for the interval $[4, 8]$?

38. Alternative form of the mean value theorem: If f is continuous on $[a, b]$ and differentiable on (a, b), then there exists a number c in (a, b) such that

$$f(b) = f(a) + (b - a)f'(c)$$

Prove this alternative form of the MVT.

39. Let x_1 and x_2 be any two numbers between $-\frac{\pi}{2}$ and $\frac{\pi}{2}$. Use the MVT to show that

$$|\tan x_2 - \tan x_1| \geq |x_2 - x_1|$$

40. If $f(x) = \dfrac{1}{x}$ on $[-1, 1]$, does the mean value theorem apply? Why why or not?

41. If $g(x) = |x|$ on $[-2, 2]$, does the mean value theorem apply? Why or why not?

42. Think Tank Problem Is it true that

$$|\cos x_2 - \cos x_1| \leq |x_2 - x_1|$$

for all x_1 and x_2? Either prove that the inequality is always valid or find a counterexample.

43. Think Tank Problem Consider

$$f(x) = \begin{cases} 1 & \text{if } x > 0 \\ -1 & \text{if } x < 0 \end{cases}$$

$f'(x) = 0$ for all x in the domain, but f is not a constant. Does this example contradict the zero-derivative theorem? Why or why not?

44. a. Let n be a positive integer. Show that there is a number c between 0 and x for which

$$\frac{(1 + x)^n - 1}{x} = n(1 + c)^{n-1}$$

b. Use part **a** to evaluate

$$\lim_{x \to 0} \frac{(1 + x)^n - 1}{x}$$

45. a. Show that there is a number w between 0 and x for which

$$\frac{\cos x - 1}{x} = -\sin w$$

b. Use part **a** to evaluate

$$\lim_{x \to 0} \frac{\cos x - 1}{x}$$

46. Use the MVT to evaluate

$$\lim_{x \to \pi^+} \frac{\cos x + 1}{x - \pi}$$

47. Let $f(x) = 1 + \dfrac{1}{x}$. If a and b are constants such that $a < 0$ and $b > 0$, show that there is no number w between a and b for which

$$f(b) - f(a) = f'(w)(b - a)$$

48. Show that for any $x > 4$, there is a number w between 4 and x such that

$$\frac{\sqrt{x} - 2}{x - 4} = \frac{1}{2\sqrt{w}}$$

Use this fact to show that if $x > 4$, then

$$\sqrt{x} < 1 + \frac{x}{4}$$

49. Show that if an object moves along a straight line in such a way that its velocity is the same at two different times (that is, for a differentiable function v, we are given $v(t_1) = v(t_2)$ for $t_1 \neq t_2$), then there is some intermediate time when the acceleration is zero.

50. Modeling Problem Two radar patrol cars are located at fixed positions 6 mi apart on a long, straight road where the speed limit is 55 mi/h. A sports car passes the first patrol car traveling at 53 mi/h, and then 5 min later, it passes the second patrol car going 48 mi/h. Analyze a model of this situation to show that at some time between the two clockings, the sports car exceeded the speed limit. *Hint*: Use the MVT.

© moodboard/Corbis

51. Modeling Problem Suppose two race cars begin at the same time and finish at the same time. Analyze a model to show that at some point in the race they had the same speed.

52. Use Rolle's theorem with

$$f(x) = (x - 1) \sin x$$

to show that the equation $\tan x = 1 - x$ has at least one solution for $0 < x < 1$.

53. Use the MVT to show that

$$\sqrt{1 + x} - 4 < \frac{1}{8}(x - 15)$$

if $x > 15$. *Hint*: Let $f(x) = \sqrt{1 + x}$.

54. Use the MVT to show that

$$\frac{1}{2x + 1} > \frac{1}{5} + \frac{2}{25}(2 - x)$$

if $0 \leq x \leq 2$.

Level 3

55. Let $f(x) = \tan x$. Note that $f(\pi) = f(0) = 0$. Show that there is no number w between 0 and π for which $f'(w) = 0$. Why does this fact not contradict the MVT?

56. Use Rolle's theorem or the MVT to show that there is no number a for which the equation

$$x^3 - 3x + a = 0$$

has *two* distinct solutions in the interval $[-1, 1]$.

57. If $a > 0$ is a constant, show that the equation

$$x^3 + ax - 1 = 0$$

has exactly one real solution. *Hint*: Let $f(x) = x^3 + ax - 1$ and use the intermediate value

theorem to show that there is at least one root. Then assume there are two roots, and use Rolle's theorem to obtain a contradiction.

58. For constants a and b, $a > 0$, and n a positive integer, use Rolle's theorem or the MVT to show that the polynomial

$$p(x) = x^{2n+1} + ax + b$$

can have at most one real root.

59. Show that if $f''(x) = 0$ for all x, then f is a linear function. (That is, $f(x) = Ax + B$ for constants $A \neq 0$ and B.)

60. Show that if $f'(x) = Ax + B$ for constants $A \neq 0$ and B, then $f(x)$ is a quadratic function. (That is, $f(x) = ax^2 + bx + c$ for constants a, b, and c, where $a \neq 0$.)

4.3 USING DERIVATIVES TO SKETCH THE GRAPH OF A FUNCTION

IN THIS SECTION: *Increasing and decreasing functions, the first-derivative test, concavity and inflection points, the second-derivative test, curve sketching using the first and second derivatives*
The sign of the derivative of a function can be used to determine whether the function is increasing or decreasing on a given interval. We shall use this information to develop a procedure called *the first and second derivative tests*.

Our next goal is to see how information about the derivative f' and the second derivative f'' can be used to determine the shape of the graph of f. We begin by showing how the sign of f' is related to whether the graph of f is rising or falling.

Increasing and Decreasing Functions

Suppose an ecologist has modeled the size of a population of a certain species as a function f of time t (months). If it turns out that the population is increasing until the end of the first year and is decreasing thereafter, it is reasonable to expect the population to be maximized at time $t = 12$ months and for the population curve to have a high point at $t = 12$, as shown in Figure 4.16.

If the graph of a function f, such as this population curve, is rising throughout the interval $0 < x < 12$, we say that f is *strictly increasing* on that interval. Similarly, the graph of the function in Figure 4.16 is *strictly decreasing* on the interval $12 < t < 20$. These terms may be defined more formally in the following box and illustrated in Figure 4.17.

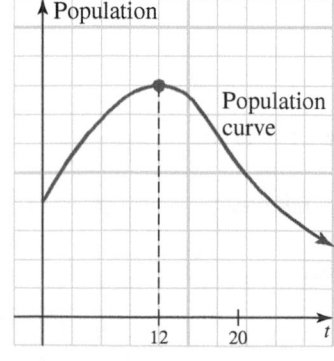

Figure 4.16 A population curve

STRICTLY INCREASING The function f is **strictly increasing** on an interval I if

$$f(x_1) < f(x_2)$$

whenever $x_1 < x_2$ for x_1 and x_2 on I.

STRICTLY DECREASING Likewise, f is **strictly decreasing** on I if

$$f(x_1) > f(x_2)$$

whenever $x_1 < x_2$ for x_1 and x_2 on I.

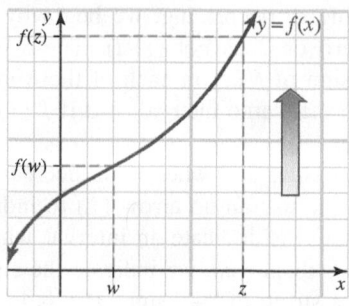

a. Strictly increasing function

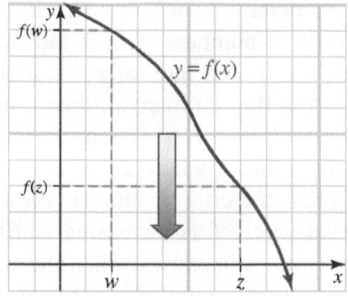

b. Strictly decreasing function

Figure 4.17 Increasing and decreasing functions

A function f is said to be (strictly) **monotonic** on an interval I if it is either strictly increasing on all of I or strictly decreasing on all of I. The monotonic behavior is closely related to the sign of the derivative $f'(x)$. In particular, if the graph of a function has tangent lines with positive slope on I, the graph will be inclined upward and f will be increasing on I (see Figure 4.18).

Note that we often do not use the word "strictly" every time we talk about a strictly monotonic function, or a strictly increasing function, or a strictly decreasing function.

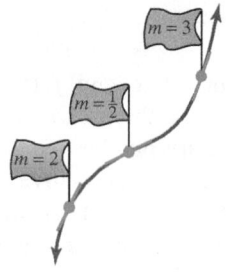

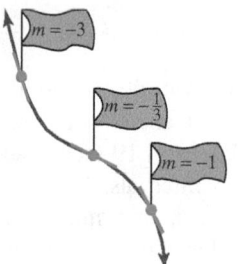

*A note on usage: We say that functions are **increasing** and that graphs are **rising**. For example if $f(x) = x^2$ we say that the function f is increasing for $x > 0$ and that the graph is rising for $x > 0$.*

Figure 4.18 The graph is rising where $f' > 0$ and falling where $f' < 0$. Notice that the small flags indicate the slope at various points on the graph.

Since the slope of the tangent at each point on the graph is measured by the derivative f', it is reasonable to expect f to be increasing on intervals where $f' > 0$. Similarly, it is reasonable to expect f to be decreasing on an interval when $f' < 0$. These observations are established formally in Theorem 4.7.

Theorem 4.7 Monotone function theorem

Let f be differentiable on the open interval (a, b).

If $f'(x) > 0$ on (a, b), then f is strictly increasing on (a, b).

If $f'(x) < 0$ on (a, b), then f is strictly decreasing on (a, b).

Proof: We will prove that f is strictly increasing on (a, b) if $f'(x) > 0$ throughout the interval. The strictly decreasing case is similar and is left as an exercise for the reader.

Suppose $f'(x) > 0$ throughout the interval (a, b), and let x_1 and x_2 be two numbers chosen arbitrarily from this interval, with $x_1 < x_2$. The MVT tells us that

$$\frac{f(x_2) - f(x_1)}{x_2 - x_1} = f'(c) \qquad \text{or} \qquad f(x_2) - f(x_1) = f'(c)(x_2 - x_1)$$

for some number c between x_1 and x_2. Because both $f'(c) > 0$ and $x_2 - x_1 > 0$, it follows that $f'(c)(x_2 - x_1) > 0$, and therefore

$$f(x_2) - f(x_1) > 0 \qquad \text{or} \qquad f(x_2) > f(x_1)$$

That is, if x_1 and x_2 are any two numbers in (a, b) such that $x_1 < x_2$, then $f(x_1) < f(x_2)$, which means that f is strictly increasing on (a, b). ◆

To determine where a function f is increasing or decreasing, we begin by finding the critical numbers (where the derivative is zero or does not exist). These numbers divide the x-axis into intervals, and we test the sign of $f'(x)$ in each of these intervals. If $f'(x) > 0$ in an interval, then f is increasing in that same interval, and if $f'(x) < 0$ in an interval, then f is decreasing in that same interval.

To indicate where a given function f is increasing and where it is decreasing, we will mark the critical numbers on a number line and use an up arrow ($\uparrow$) to indicate an interval where f is increasing and a down arrow ($\downarrow$) to indicate an interval where f is decreasing. Sometimes, when the full graph of f is displayed, we indicate the sign of f' on an interval bounded by critical numbers with a string of "+" signs if f is increasing on the interval or a string of "$-$" signs if f is decreasing there. This notation is illustrated in the following example.

Example 1   Finding intervals of increase and decrease

Determine where the function defined by $f(x) = x^3 - 3x^2 - 9x + 1$ is strictly increasing and where it is strictly decreasing.

Solution   First, we find the derivative:

$$f'(x) = 3x^2 - 6x - 9 = 3(x^2 - 2x - 3) = 3(x+1)(x-3)$$

Next, we determine the critical numbers: $f'(x)$ exists for all x and $f'(x) = 0$ at $x = -1$ and $x = 3$. These critical numbers divide the x-axis into three parts, as shown in Figure 4.19 (see dashed verticals), and we select any arbitrary number from each of these intervals. For example, we select $-2, 0,$ and 4, evaluate the derivative at these numbers, and mark each interval as increasing ($\uparrow$) or decreasing ($\downarrow$), according to whether the derivative is positive or negative, respectively. This is shown in Figure 4.19.

Pay attention to these number lines:

f shows when the graph is above or below the x-axis.

f' shows when the graph is rising or falling.

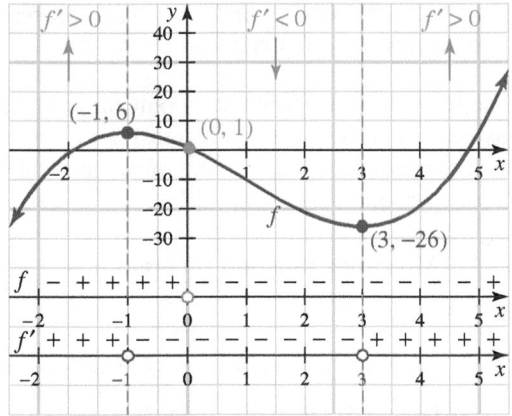

Figure 4.19   Graph of $f(x) = x^3 - 3x^2 - 9x + 1$

Example 2   Comparing the graphs of a function and its derivative

Graph $f(x) = x^3 - 3x^2 - 9x + 1$ and $f'(x) = 3x^2 - 6x - 9$ and compare.

a. When $f' > 0$, what can be said about the graph of f?
b. When the graph of f is falling, what can be said about the graph of f'?
c. Where do the critical numbers of f appear on the graph of f'?

Solution The graphs of f and f' are shown in Figure 4.20.

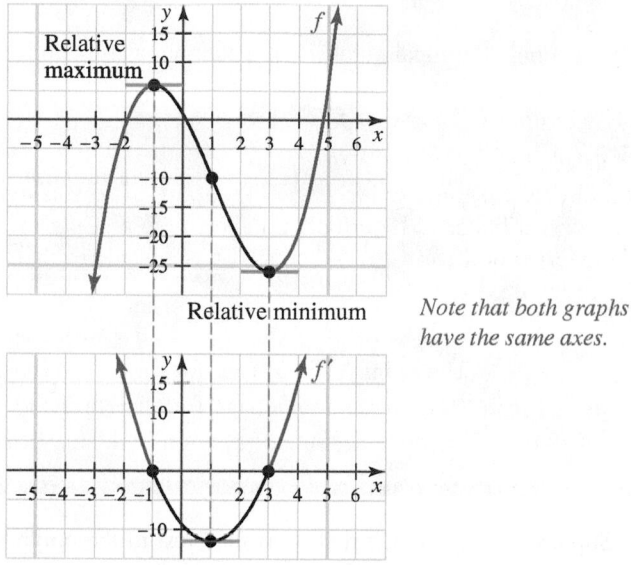

Note that both graphs have the same axes.

Figure 4.20 Interactive Graphs of f and f'

a. When $f'(x) > 0$, the graph of f is rising. This occurs when $x < -1$ and when $x > 3$.

b. When the graph of f is falling (for $-1 < x < 3$), we have $f'(x) < 0$, so the graph of f' is below the x-axis.

c. The critical numbers of f are where $f'(x) = 0$; that is, at $x = -1$ and $x = 3$, so they are the x-intercepts of the graph of f'.

The First-Derivative Test

Every relative extremum is a critical point. However, as we saw in Section 4.1, not every critical point of a continuous function is necessarily a relative extremum. If the derivative is positive to the immediate left of a critical number and negative to its immediate right, the graph changes from increasing to decreasing and the critical point must be a relative maximum, as shown in Figure 4.21**a**. If the derivative is negative to the immediate left of a critical number and positive to its immediate right, the graph changes from decreasing to increasing and the critical point is a relative minimum (Figure 4.21**b**). However, if the sign of the derivative is the same on both immediate sides of the critical number, then it is neither a relative maximum nor a relative minimum (Figure 4.21**c**). These observations are summarized in a procedure called the *first-derivative test for relative extrema*.

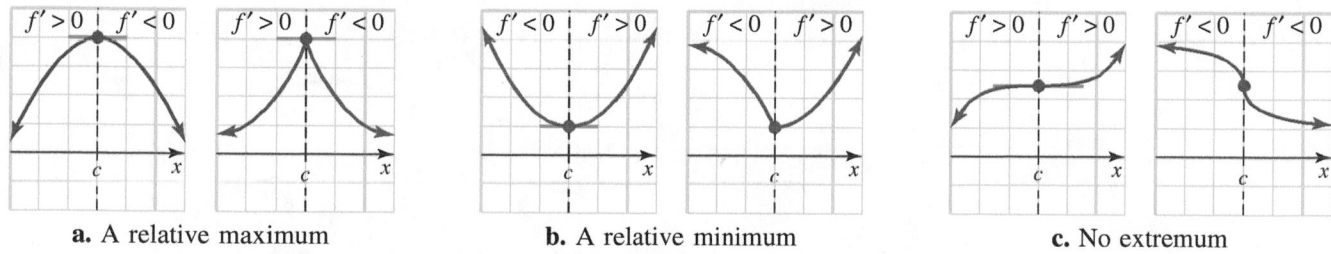

a. A relative maximum   **b.** A relative minimum   **c.** No extremum

Figure 4.21 Three patterns of behavior near a critical number

248

4.3 Using Derivatives to Sketch the Graph of a Function

FIRST-DERIVATIVE TEST   The **first-derivative test** for relative extrema provides a procedure for finding relative maxima and minima.

Step 1 Find all critical numbers of a continuous function f defined on an interval (a, b). That is, find all numbers c in (a, b) such that $f(c)$ is defined and either $f'(c) = 0$ or $f'(c)$ does not exist.

Step 2 Classify each critical point $(c, f(c))$ as follows:
The point $(c, f(c))$ is a **relative maximum** if

↗ ↘ **a.** $f'(x) > 0$ (graph rising) for all x in (a, c).
$f'(x) < 0$ (graph falling) for all x in (c, b).
The point $(c, f(c))$ is a **relative minimum** if

↘ ↗ **b.** $f'(x) < 0$ (graph falling) for all x in (a, c).
$f'(x) > 0$ (graph rising) for all x in (c, b).
The point $(c, f(c))$ is a **not an extremum** if

↗ ↗
or **c.** $f'(x)$ has the same sign for all x in the open
↘ ↘ intervals (a, c) and (c, b) on each side of c.

Suppose we apply this first-derivative test to the polynomial

$$f(x) = x^3 - 3x^2 - 9x + 1$$

In Example 2 we found that this function has the critical numbers -1 and 3 and that f is increasing when $x < -1$ and $x > 3$ and decreasing when $-1 < x < 3$ (see the arrow pattern above). The first-derivative test tells us there is a relative maximum at -1 (↗ ↘ pattern) and a relative minimum at 3 (↘ ↗ pattern).

Example 3   Relative extrema using the first-derivative test

Find all critical numbers of $g(t) = t - 2 \sin t$ for $0 \leq t \leq 2\pi$, and determine whether each corresponds to a relative maximum, a relative minimum, or neither. Sketch the graph of g.

Solution   Because $g'(t) = 1 - 2 \cos t$ exists for all t, the only critical numbers occur when $g'(t) = 0$; that is, when $\cos t = \frac{1}{2}$. Solving, we find that the critical numbers for $g(t)$ on the interval $[0, 2\pi]$ are $\frac{\pi}{3}$ and $\frac{5\pi}{3}$.

Next, we examine the sign of $g'(t)$. Thanks to the intermediate value theorem, because $g'(t)$ is continuous, it is enough to check the sign of $g'(t)$ at convenient numbers on each side of the critical numbers, as shown in Figure 4.22. Notice that the arrows show the increasing and decreasing pattern for g. According to the first-derivative test, there is a relative minimum at $\frac{\pi}{3}$ and a relative maximum at $\frac{5\pi}{3}$. The graph of g is shown in Figure 4.22.

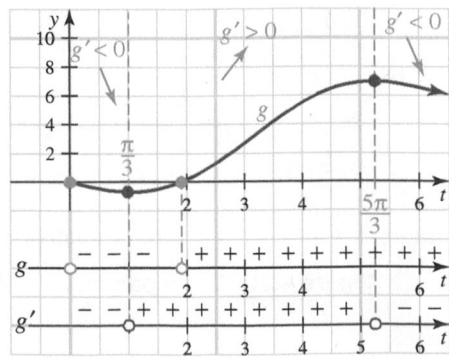

Figure 4.22   The first-derivative test for g

Concavity and Inflection Points

Knowing where a given graph is rising and falling gives only a partial picture of the graph. For example, suppose we wish to sketch the graph of $f(x) = x^3 + 3x + 1$. The derivative $f'(x) = 3x^2 + 3$ is positive for all x so the graph is always rising. But in what *way* is it rising? Each of the graphs in Figure 4.23 is a possible graph of f, but they are quite different from one another.

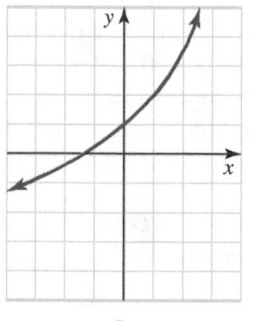   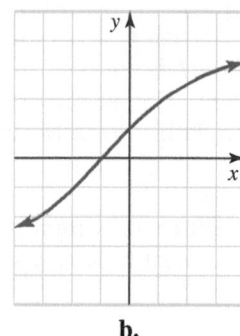   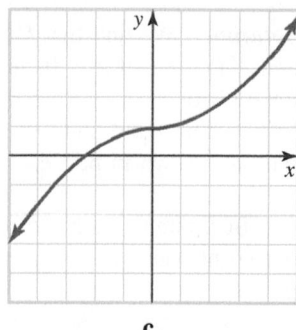

a.   **b.**   **c.**

Figure 4.23   Which curve is the graph of $f(x) = x^3 + 3x + 1$?

Figure 4.23 shows three possibilities for the graph of f. All three are always rising, but they differ in the way they "bend" as they rise. The "bending" of a curve is measured by its *concavity*. If a curve lies above its tangent lines on some interval (a, b), then we say the curve is *concave up*, and if the curve lies below its tangent lines, we say it is *concave down* on (a, b).

CONCAVITY   If the graph of a function f lies above all of its tangents on an interval I, then it is said to be **concave up** on I. If the graph of f lies below all of its tangents on I, it is said to be **concave down**.

We will examine concavity and then return in the following example to determine which of the three candidates in Figure 4.23 is indeed the graph of $f(x) = x^3 + 3x + 1$.

To be specific, a portion of a graph that is cupped upward is called *concave up*, and a portion that is cupped downward is *concave down*. Figure 4.24 shows a graph that is concave up between A and C and concave down between C and E.

At various points on the graph, the slope is indicated by "flags," and we observe that the slope increases from A to C and decreases from C to E. This is no accident! *The slope of a graph increases on an interval where the graph is concave up and decreases where the graph is concave down.*

Conversely, a graph will be concave up on any interval where the slope is increasing and concave down where the slope is decreasing. Because the slope is found by computing the derivative, it is reasonable to expect the graph of a given function f to be concave up where the derivative f' is strictly increasing. According to the monotone function theorem (Theorem 4.7), this occurs when $(f')' > 0$, which means that the graph of f is concave up where the *second derivative* f'' satisfies $f'' > 0$. Similarly, the graph is concave down where $f'' < 0$. We use this observation to characterize concavity.

Figure 4.24   Concave up between A and C. Concave down between C and E.

DERIVATIVE CHARACTERIZATION OF CONCAVITY   The graph of a function f is **concave up** on any open interval I where $f''(x) > 0$, and **concave down** where $f''(x) < 0$.

When discussing the concavity of a function f, we will display a diagram in which a "cup" symbol ($\cup$) above an interval indicates that f is concave up on the interval and a "cap" symbol ($\cap$) indicates that f is concave down there. This convention is illustrated in the following example.

Example 4 Concavity for a polynomial function

Find where the graph of $f(x) = x^3 + 3x + 1$ is concave up and where it is concave down.

Solution   We find that $f'(x) = 3x^2 + 3$ and $f''(x) = 6x$. Therefore, $f''(x) < 0$ if $x < 0$ and $f''(x) > 0$ if $x > 0$, so the graph of f is concave down ($\cap$) for $x < 0$ and concave up ($\cup$) for $x > 0$. Returning to Figure 4.23, we see the graph is concave down to the left of $x = 0$ and concave up to the right or $x = 0$. Hence, the correct graph of $f(x)$ is the one shown in part **c**.   ■

In Example 4, notice that the graph changes from concave down to concave up at the point where $x = 0$. It will be convenient to have a special name for such transition points.

☠ *An inflection point must be on the graph, meaning $f(c)$ must be defined if there is an inflection point at $x = c$.* ☠

> **INFLECTION POINT**   A point $P\,(c, f(c))$ on a curve is called an **inflection point** if the graph is concave up on one side of P and concave down on the other side.

Returning to Example 4, notice that the graph of $f(x) = x^3 + 3x + 1$ has exactly one inflection point, at $(0, 1)$, where the concavity changes from down to up.

Various kinds of graphical behavior are illustrated in Figure 4.25. Note that the graph is rising on the interval $[a, c_1]$, falling on $[c_1, c_2]$, rising on $[c_2, c_3]$, falling on $[c_3, c_4]$, rising on $[c_4, c_5]$, and falling on $[c_5, b]$.

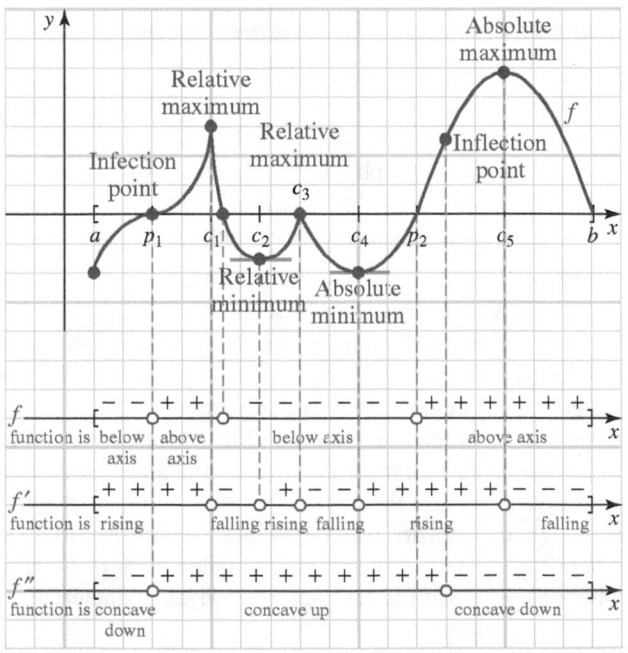

Figure 4.25   A graph of a function showing critical points and inflection points

The concavity is up on $[p_1, p_2]$, and down otherwise. In this figure, and elsewhere, when the graph of f is displayed, we will indicate the sign of f'' on an interval by a string of "+" signs if $f'' > 0$ on the interval, and a string of "−" signs if $f'' < 0$ there.

The graph has relative maxima at c_1, c_3, and c_5, and relative minima at c_2 and c_4. There are horizontal tangents ($f'(x) = 0$) at all of these points except at c_1 and c_3, where there are sharp points, called *corners*, ($f'(c_1)$ and $f'(c_3)$ do not exist). There is a horizontal tangent at p_1; that is, $f'(p_1) = 0$, but no relative extremum appears there. Instead, we have points of inflection at p_1 and p_2, because the concavity changes direction at each of these points.

In general, the concavity of the graph of f will change only at points where $f''(x) = 0$ or $f''(x)$ does not exist—that is, at critical numbers of the derivative $f'(x)$. We will call the number c a **second-order critical number if** $f''(c) = 0$ or $f''(c)$ does not exist, and in this context an "ordinary" critical number (where $f'(c) = 0$ or $f'(c)$ does not exist) will be referred to as a **first-order critical number**. If we do not specify otherwise, a critical number is always a first-order critical number. Inflection points correspond to second-order critical numbers and must actually be on the graph of f. Specifically, a number c such that $f''(c)$ is not defined and the concavity of f changes at c will correspond to an inflection point if and only if $f(c)$ is defined.

A continuous function f need not have an inflection point at every number c where $f''(c) = 0$. For instance, if $f(x) = x^4$, we have $f''(0) = 0$, but the graph of f is always concave up (see Figure 4.26).

Here is an example of one way inflection points may appear in applications.

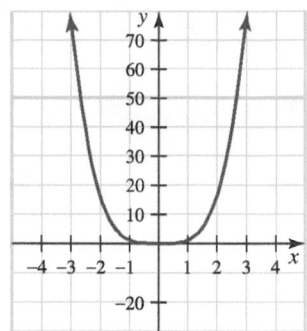

Figure 4.26   The graph of $f(x) = x^4$ has no inflection point at $(0,0)$ even though $f''(x) = 0$

Example 5   Peak worker efficiency

An efficiency study of the morning shift at a factory indicates that the number of units produced by an average worker t hours after 8:00 A.M. may be modeled by the formula $Q(t) = -t^3 + 9t^2 + 12t$. At what time in the morning is the worker performing most efficiently?

Solution   We assume that the morning shift runs from 8:00 A.M. until noon and that worker efficiency is maximized when the rate of production

$$R(t) = Q'(t) = -3t^2 + 18t + 12$$

is as large as possible for $0 \le t \le 4$. The derivative of R is

$$R'(t) = Q''(t) = -6t + 18$$

which is zero when $t = 3$; this is the critical number. Using the optimization criterion of Section 4.1, we know that the extrema of $R(t)$ on the closed interval $[0, 4]$ must occur at either the critical number 3; or at one (or both) of the endpoints (which are 0 and 4). We find that

$$R(0) = 12 \qquad R(3) = 39 \qquad R(4) = 36$$

so the rate of production $R(t)$ is greatest and the worker is performing most efficiently when $t = 3$, that is, at 11:00 A.M. The graphs of the production function Q and its derivative, the rate-of-production function R, are shown in Figure 4.27. Notice that the production curve is steepest and the rate of production is greatest when $t = 3$.

Note how the rate of production, as measured by the slope of the graph of the average worker's output, increases from 0 to the inflection point I and then decreases from I to E. Because the point I marks the point where the rate of production "peaks out," it is natural to refer to I as a point of **diminishing returns**. It is also an inflection point on the graph of Q. Knowing that this point occurs at 11:00 A.M., the manager of the factory might be able to increase the overall output of the labor force by scheduling a break near this time, since the production rate changes from increasing to decreasing at this time.

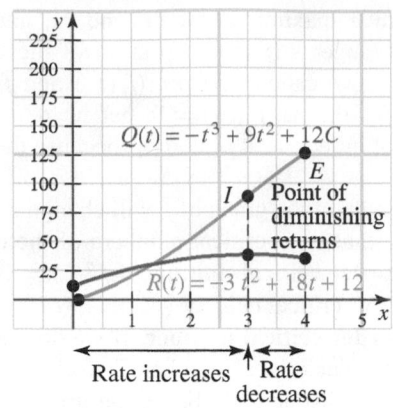

Figure 4.27   Graph showing point of diminishing returns

The Second-Derivative Test

It is often possible to classify a critical point P $(c, f(c))$ on the graph of f by examining the sign of $f''(c)$. Specifically, suppose $f'(c) = 0$ and $f''(c) > 0$. Then there is a horizontal tangent line at P and the graph of f is concave up in the neighborhood of P. This means that the graph of f is cupped upward from the horizontal tangent at P, and it is reasonable to expect P to be a relative minimum, as shown in Figure 4.28**a**. Similarly, we expect P to be a relative maximum if $f'(c) = 0$ and $f''(c) < 0$, because the graph is cupped down beneath the critical point P, as shown in Figure 4.28**b**.

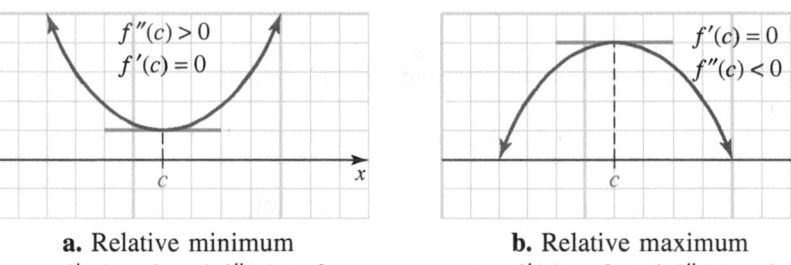

a. Relative minimum
$f'(c) = 0$ and $f''(c) > 0$

b. Relative maximum
$f'(c) = 0$ and $f''(c) < 0$

Figure 4.28   Second-derivative test for relative extrema

These observations lead to the *second-derivative test* for relative extrema.

SECOND-DERIVATIVE TEST   The **second-derivative test** for relative extrema provides a procedure for finding relative maxima and minima.

Step 1 Find all critical numbers of a continuous function f defined on an interval (a, b). That is, find all numbers c in (a, b) such that $f(c)$ is defined and either $f'(c) = 0$ or $f'(c)$ does not exist.

Step 2 Classify each critical point $(c, f(c))$ for which the second derivative exists on an open interval containing c as follows:
    a. If $f''(c) > 0$, there is a **relative minimum** at $x = c$.
    b. If $f''(c) < 0$, there is a **relative maximum** at $x = c$.
    c. If $f''(c) = 0$, then the second-derivative test fails. There may be a relative maximum, a relative minimum, or neither at $x = c$.

Example 6   Using the second-derivative test

Use the second-derivative test to determine whether each critical number of the function $f(x) = 3x^5 - 5x^3 + 2$ corresponds to a relative maximum, a relative minimum, or neither.

Solution   Once again, we begin by finding the first and second derivatives:

$$f'(x) = 15x^4 - 15x^2 = 15x^2(x - 1)(x + 1)$$
$$f''(x) = 60x^3 - 30x = 30x(2x^2 - 1)$$

Solving $f'(x) = 0$, we find that the critical numbers are $x = 0$, $x = 1$, and $x = -1$. To apply the second-derivative test, we evaluate $f''(x)$ at each critical number.

$f''(0) = 0$; test fails at $x = 0$

$f''(1) = 30 > 0$; positive, so the test tells us there is a relative minimum at $x = 1$.

$f'(-1) = -30 < 0$; negative, so the test tells us that there is a relative maximum at $x = -1$.

When the second-derivative test fails, as at $x = 0$ in Example 6, the critical point can often be classified using the first-derivative test. For instance, the first derivative in Example 6 is

$$f'(x) = 15x^2(x - 1)(x + 1)$$

We see $f'(x) = 0$ at $x = -1, 0$, and 1. We can plot these points on a number line, shown in Figure 4.29, and then evaluate $f'(x)$ at test numbers just to the left and just to the right of each critical point.

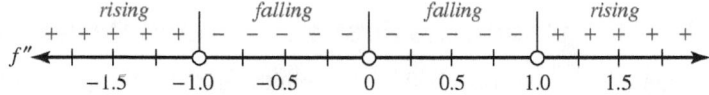

Figure 4.29   Sign graph of f'

We show a number line like this as part of the graph of many of our examples (as we did in Figure 4.25), but most often we show the derivative to the right and left of a critical number. For Example 6, the derivative is negative both to the immediate left and right of 0, which we note is the ↘↘ pattern telling us that neither kind of extremum occurs at $x = 0$.

Example 7   Finding inflection points

Find the inflection points for the function given in Example 6, namely $f(x) = 3x^5 - 5x^3 + 2$, and then graph the function.

Solution   We begin with the second derivative (from Example 6)

$$f''(x) = 30x(2x^2 - 1)$$

To find the inflection points, we look at the second-order critical numbers, namely where $f''(x) = 0$; that is at $x = 0$ and $x = \pm\sqrt{\frac{1}{2}} = \pm\frac{\sqrt{2}}{2}$. We can show these on a sign graph of the second derivative, as in Figure 4.30. Notice that the inflection points occur where the sign of the second derivative changes.

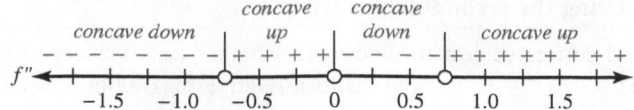

Figure 4.30   Sign graph of f''

The graph of f is shown in Figure 4.31, along with the sign graphs for both the first and second derivative. Spend some time studying this figure to see how the signs of the derivative indicate where the graph is rising and where it is falling, as well as the signs of the second derivative show the concavity. For completeness, the figure also shows the sign graph for the function itself: the graph is above the x-axis where f is positive, below the x-axis where f is negative, and crosses the x-axis where $f(x) = 0$.

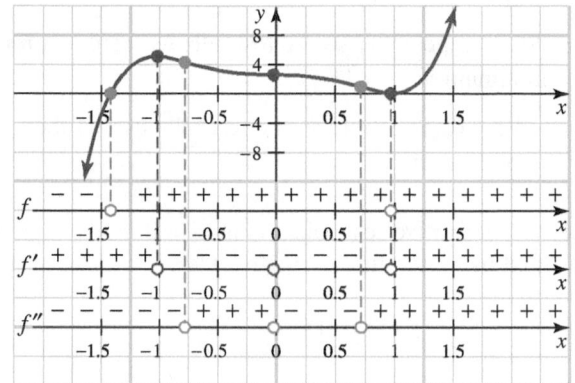

Figure 4.31 Interactive   Graph of $f(x) = 3x^5 - 5x^3 + 2$

Curve Sketching Using the First and Second Derivatives

Examples 6 and 7 demonstrate both the strength and the weakness of the second-derivative test. In particular, when it is relatively easy to find the second derivative (as with a polynomial) and if the zeros of this function are easy to find, then the second-derivative test provides a quick means for classifying the critical points. However, if it is difficult to compute $f''(c)$ or if $f''(c) = 0$, it may be easier, or even necessary, to apply the first-derivative test.

Example 8   Sketching the graph of a polynomial function

Determine where the function

$$f(x) = x^4 - 4x^3 + 10$$

is increasing, where it is decreasing, where its graph is concave up, and where its graph is concave down. Find the relative extrema and inflection points, and sketch the graph of f.

Solution

Functions	Critical numbers	Conclusions
$f(x) = x^4 - 4x^3 + 10$	$f'(x) = 0$	First derivative signs (increasing/decreasing):
$f'(x) = 4x^3 - 12x^2$	$4x^2(x-3) = 0$	
$\quad = 4x^2(x-3)$	$x = 0, 3$	Direction: (arrows)
		$4x^2(x-3)$: Negative 0 Negative 3 Positive x
$f''(x) = 12x^2 - 24x$	$f''(x) = 0$	Second derivative signs (concavity):
$\quad = 12x(x-2)$	$12x(x-2) = 0$	Shape: $\cup$ $\cap$ $\cup$
	$x = 0, 2$	$12x(x-2)$: Positive 0 Negative 2 Positive x

From the above table, we see there is a relative minimum at $x = 3$ and inflection points at $x = 0$ and $x = 2$ (because the second derivative changes sign at these points), and a horizontal tangent at $x = 0$ (because the first derivative is zero).

To find the y-values of the critical points and the inflection points, evaluate f at $x = 0, 2,$ and 3:

$$f(0) = (0)^4 - 4(0)^3 + 10 = 10 \quad \text{Point of inflection: } (0, 10)$$

$$f(2) = (2)^4 - 4(2)^3 + 10 = -6 \quad \text{Point of inflection: } (2, -6)$$

$$f(3) = (3)^4 - 4(3)^3 + 10 = -17 \quad \text{Relative minimum at } (3, -17);$$

$$\text{put a "cup" } (\cup) \text{ at this point.}$$

Construct a preliminary graph as shown in Figure 4.32**a**. Note that we include the above information. Complete the sketch by passing a smooth curve using the information in the preliminary sketch. The completed graph is shown in Figure 4.32**b**.

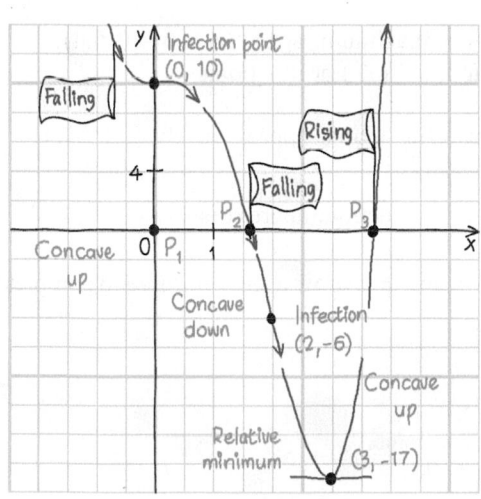

a. Preliminary sketch

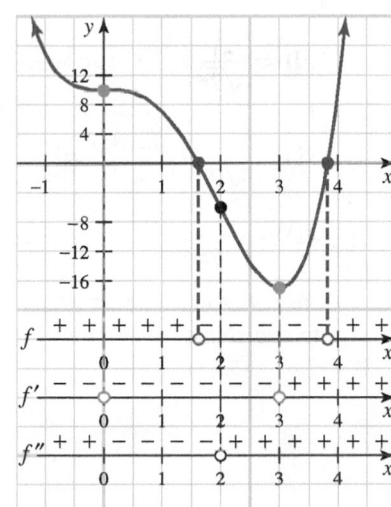

b. Completed graph (note sign graphs at bottom)

Figure 4.32   Graphing $f(x) = x^4 - 4x^3 + 10$

Example 9   Sketching the graph of an exponential function

Determine where the function

$$f(x) = \frac{1}{\sqrt{2\pi}} e^{-x^2/2}$$

is increasing, decreasing, concave up, and concave down. Find the relative extrema and inflection points and sketch the graph. This function plays an important role in statistics, where it is called the *standard normal density function*.

Solution

Functions	Critical numbers	Conclusions
$f(x) = \dfrac{1}{\sqrt{2\pi}} e^{-x^2/2}$		
$f'(x) = \dfrac{-x}{\sqrt{2\pi}} e^{-x^2/2}$	$f'(x) = 0$ $\dfrac{-x}{\sqrt{2\pi}} e^{-x^2/2} = 0$ $-xe^{-x^2/2} = 0$ $x = 0$	First derivative signs (increasing/ decreasing):
$f''(x) = \dfrac{x^2}{\sqrt{2\pi}} e^{-x^2/2} - \dfrac{1}{\sqrt{2\pi}} e^{-x^2/2}$ $= \dfrac{1}{\sqrt{2\pi}} (x^2 - 1) e^{-x^2/2}$	$f''(x) = 0$ $\dfrac{1}{\sqrt{2\pi}} (x^2 - 1) e^{-x^2/2} = 0$ $(x^2 - 1) e^{-x^2/2} = 0$ $(x - 1)(x + 1) = 0$ $x = \pm 1$	Second derivative signs (concavity):

From the above table, we see there is a relative maximum at $x = 0$ and inflection points at $x = -1$ and $x = 1$ (because the second derivative changes sign at these points). Also note that the graph has no x-intercepts, because $e^{-x^2/2}$ is always positive.

To find the y-values of the critical points and the inflection points, evaluate f at $x = 0, -1$, and 1:

$$f(0) = \frac{1}{\sqrt{2\pi}} e^{-(0)^2/2} = \frac{1}{\sqrt{2\pi}}$$ Relative maximum: approximately $(0, 0.4)$; put a "cap" ($\cap$) at this point

$$f(-1) = \frac{1}{\sqrt{2\pi}} e^{-(-1)^2/2} = \frac{1}{\sqrt{2e\pi}}$$ Point of inflection: approximately $(-1, 0.24)$

$$f(1) = \frac{1}{\sqrt{2\pi}} e^{-(1)^2/2} = \frac{1}{\sqrt{2e\pi}}$$ Point of inflection: approximately $(1, 0.24)$

Construct a preliminary graph (not shown), and then draw a smooth curve to complete the graph as shown in Figure 4.33.

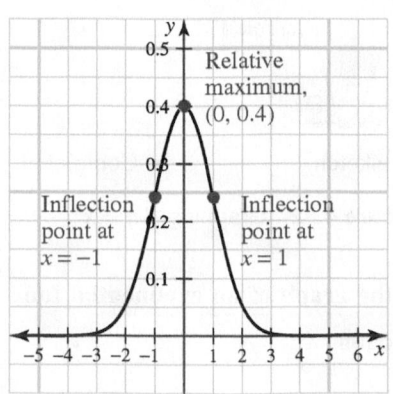

Figure 4.33 Graph of standard normal curve

Example 10   Sketching the graph of a trigonometric function

Sketch the graph of $T(x) = \sin x + \cos x$ on $[0, 2\pi]$.

Solution   You probably graphed this function in trigonometry by adding ordinates. However, with this example we wish to illustrate the power of calculus to draw the graph. Thus, we begin by finding the first and second derivatives on $[0, 2\pi]$.

Functions	Critical numbers	Conclusions
$T(x) = \sin x + \cos x$		
$T'(x) = \cos x - \sin x$	$T'(x) = 0$	First derivative signs (increasing/decreasing):
	$\cos x - \sin x = 0$	
	$\cos x = \sin x$	
	$x = \dfrac{\pi}{4}, \dfrac{5\pi}{4}$	
$T''(x) = -\sin x - \cos x$	$T''(x) = 0$	Second derivative signs (concavity):
	$-\sin x - \cos x = 0$	
	$\cos x = -\sin x$	
	$x = \dfrac{3\pi}{4}, \dfrac{7\pi}{4}$	

From the above table, we see there is a relative maximum at $x = \frac{\pi}{4}$, a relative minimum at $x = \frac{5\pi}{4}$, and inflection points at $x = \frac{3\pi}{4}$ and $x = \frac{7\pi}{4}$ (because the second derivative changes sign at these points).

To find the y-values of the critical points, the inflection points, and the endpoints, we evaluate T at $x = 0, \frac{\pi}{4}, \frac{3\pi}{4}, \frac{5\pi}{4}, \frac{7\pi}{4}$ and 2π:

$$T(0) = \sin(0) + \cos(0) = 1 \qquad \text{Left endpoint: } (0, 1)$$

$$T\left(\frac{\pi}{4}\right) = \sin\left(\frac{\pi}{4}\right) + \cos\left(\frac{\pi}{4}\right) \qquad \text{Relative maximum at } \left(\frac{\pi}{4}, \sqrt{2}\right);$$

$$= \frac{\sqrt{2}}{2} + \frac{\sqrt{2}}{2} = \sqrt{2} \qquad \text{put a "cap" } (\cap) \text{ at this point.}$$

$$T\left(\frac{3\pi}{4}\right) = \sin\left(\frac{3\pi}{4}\right) + \cos\left(\frac{3\pi}{4}\right) \qquad \text{Inflection point at } \left(\frac{3\pi}{4}, 0\right)$$

$$= \frac{\sqrt{2}}{2} - \frac{\sqrt{2}}{2} = 0$$

$$T\left(\frac{5\pi}{4}\right) = \sin\left(\frac{5\pi}{4}\right) + \cos\left(\frac{5\pi}{4}\right) \qquad \text{Relative minimum at } \left(\frac{5\pi}{4}, -\sqrt{2}\right);$$

$$= -\frac{\sqrt{2}}{2} - \frac{\sqrt{2}}{2} = -\sqrt{2} \qquad \text{put a "cup" } (\cup) \text{ at this point.}$$

$$T\left(\frac{7\pi}{4}\right) = \sin\left(\frac{7\pi}{4}\right) + \cos\left(\frac{7\pi}{4}\right) \qquad \text{Inflection point at } \left(\frac{7\pi}{4}, 0\right)$$

$$= -\frac{\sqrt{2}}{2} + \frac{\sqrt{2}}{2} = 0$$

$$T(2\pi) = \sin(2\pi) + \cos(2\pi) = 1 \qquad \text{Right endpoint: } (2\pi, 1)$$

We draw a preliminary curve and then pass a smooth curve through these key points to obtain the completed graph shown in Figure 4.34.

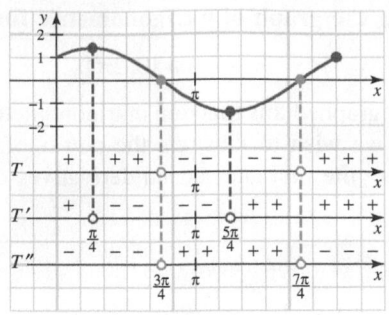

Figure 4.34   Graph of $T(x) = \sin x + \cos x$ on $[0, 2\pi]$

PROBLEM SET 4.3

Level 1

1. ■ What does this say?
   a. What is the first-derivative test?
   b. What is the second-derivative test?
   c. Which would you use first?
2. ■ What does this say? What is the relationship between concavity, points of inflection, and the second derivative?
3. **EXPLORATION PROBLEM**   What is the relationship between the graph of a function and the graph of its derivative?
4. **EXPLORATION PROBLEM**   A store posted a sign to entice shoppers that read: "Our prices are rising slower than anyplace else in town." Restate the sign using calculus terminology.

In Problems 5-6, identify which curve represents a function f and which curve represents its derivative f'.

5.
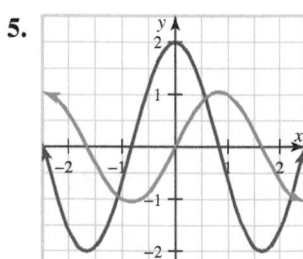

6.
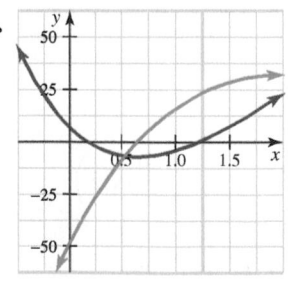

Draw a curve that represents the derivative of the function defined by the curves shown in Problems 7-10.

7.
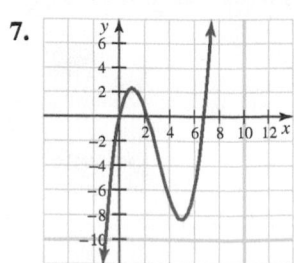

8.
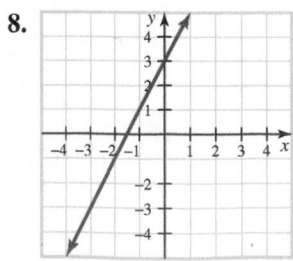

9.
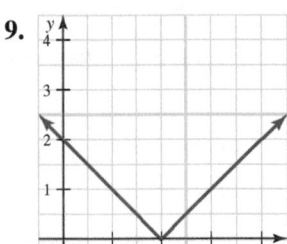

10.
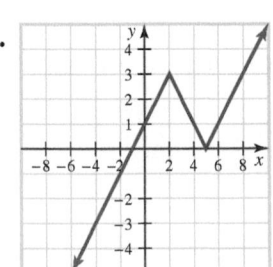

For the functions in Problems 11-17,

a. *Find all critical numbers.*
b. *Find where the function is increasing and decreasing.*
c. *Find the critical points and identify each as a relative maximum, relative minimum, or neither.*
d. *Find the second-order critical numbers and tell where the graph is concave up and where it is concave down.*
e. *Sketch the graph.*

11. $f(x) = \frac{1}{3}x^3 - 9x + 2$ 12. $f(x) = x^3 + 3x^2 + 1$
13. $f(x) = x^5 + 5x^4 - 550x^3 - 2{,}000x^2 + 60{,}000x$
    (use technology)
14. $f(x) = \dfrac{x-1}{x^2+3}$ 15. $f(x) = x + \dfrac{1}{x}$
16. $f(t) = (t+1)^2(t-5)$ 17. $f(x) = x \ln x$

In Problems 18-33, determine where the function is increasing and where it is decreasing, as well as where it is concave up and where it is concave down, and then use those intervals to help you sketch its graph.

18. $f(x) = (x-12)^4 - 2(x-12)^3$
19. $f(x) = 1 + 2x + 18x^{-1}$
20. $f(u) = 3u^4 - 2u^3 - 12u^2 + 18u - 5$
21. $g(u) = u^4 + 6u^3 - 24u^2 + 26$
22. $f(x) = \sqrt{x^2+1}$ 23. $g(t) = (t^3+t)^2$
24. $f(t) = (t^3+3t^2)^3$ 25. $f(x) = \dfrac{x}{x^2+1}$
26. $f(t) = t - \ln t$ 27. $f(t) = t^2 e^{-3t}$

28. $f(x) = e^x + e^{-x}$ **29.** $f(x) = (\ln x)^2$

30. $t(\theta) = \sin\theta - 2\cos\theta$ for $0 \le \theta \le 2\pi$

31. $t(\theta) = \theta + \cos 2\theta$ for $0 \le \theta \le \pi$

32. $f(x) = 2x - \sin^{-1} x$ for $-1 \le x \le 1$

33. $f(x) = x^3 + \sin x$ on $\left[-\frac{\pi}{2}, \frac{\pi}{2}\right]$

In Problems 34-37, use the first-derivative test to classify each of the given critical numbers as a relative minimum, a relative maximum, or neither.

34. $f(x) = (x^3 - 3x + 1)^7$ at $x = 1, x = -1$

35. $f(x) = \dfrac{e^{-x^2}}{3 - 2x}$ at $x = 1, x = \frac{1}{2}$

36. $f(x) = (x^2 - 4)^4(x^2 - 1)^3$ at $x = 1, x = 2$

37. $f(x) = \sqrt[3]{x^3 - 48x}$ at $x = 4$

In Problems 38-41, use the second-derivative test to classify each of the given critical numbers as a relative minimum, a relative maximum, or neither.

38. $f(x) = 2x^3 + 3x^2 - 12x + 11$ at $x = 1, x = -2$

39. $f(x) = \dfrac{x^2 - x + 5}{x + 4}$ at $x = -9, x = 1$

40. $f(x) = (x^2 - 3x + 1)e^{-x}$ at $x = 1, x = 4$

41. $f(x) = \sin x + \frac{1}{2}\cos 2x$ at $x = \frac{\pi}{6}, x = \frac{\pi}{2}$

Level 2

42. Think Tank Problem Sketch the graph of a function with the following properties:

$f'(x) > 0$ when $x < -1$
$f'(x) > 0$ when $x > 3$
$f'(x) < 0$ when $-1 < x < 3$
$f''(x) < 0$ when $x < 2$
$f'(x) > 0$ when $x > 2$

43. Think Tank Problem Sketch the graph of a function with the following properties:

$f'(x) > 0$ when $x < 2$ and when $2 < x < 5$
$f'(x) < 0$ when $x > 5$
$f'(2) = 0$
$f''(x) < 0$ when $x < 2$ and when $4 < x < 7$
$f''(x) > 0$ when $2 < x < 4$ and when $x > 7$

44. Think Tank Problem Sketch a graph of a function f that satisfies the following conditions:

$f'(x) > 0$ when $x < -5$ and when $x > 1$
$f'(x) < 0$ when $-5 < x < 1$
$f(-5) = 4$ and $f(1) = -1$

45. Think Tank Problem Sketch a graph of a function f that satisfies the following conditions:

$f'(x) < 0$ when $x < -1$
$f'(x) > 0$ when $-1 < x < 3$ and when $x > 3$
$f'(-1) = 0$ and $f'(3) = 0$

46. In physics, the formula

$$I(\theta) = I_0\left(\frac{\sin\theta}{\theta}\right)^2$$

where $I(0) = \lim\limits_{\theta \to 0} I(\theta)$ and I_0 is a constant, is used to model light intensity in the study of Fraunhofer diffraction.

a. Show that $I(0) = I_0$.

b. Sketch the graph for $[-3\pi, 3\pi]$. What are the critical points on this interval?

47. At a temperature of T (in degrees Celsius), the speed of sound in air is modeled by the formula

$$v = v_0\sqrt{1 + \frac{1}{273}T}$$

where v_0 is the speed at $0°$C. Sketch the graph of v for $T > 0$, and use calculus to check for critical points.

Modeling Problems: *In Problems 48-49, set up an appropriate model to answer the given question. Be sure to state your assumptions.*

48. At noon on a certain day, Frank sets out to assemble five stereo sets. His rate of assembly increases steadily throughout the afternoon until 4:00 P.M., at which time he has completed three sets. After that, he assembles sets at a slower and slower rate until he finally completes the fifth set at 8:00 P.M. Sketch a rough graph of a function that represents the number of sets Frank has completed after t hours of work.

49. An industrial psychologist conducts two efficiency studies at the Chilco appliance factory. The first study indicates that the average worker who arrives on the job at 8:00 A.M. will have assembled

$$-t^3 + 6t^2 + 13t$$

blenders in t hours (without a break), for $0 \le t \le 4$. The second study suggests that after a 15-minute coffee break, the average worker can assemble

$$-\frac{1}{3}t^3 + \frac{1}{2}t^2 + 25t$$

blenders in t hours after the break for $0 < t \le 4$. *Note:* The 15-minute break is not part of the work time.

a. Verify that if the coffee break occurs at 10:00 A.M., the average worker will assemble 42 blenders before the break and $49\frac{1}{3}$ blenders for the two hours after the break.

b. Suppose the coffee break is scheduled to begin x hours after 8:00 A.M. Find an expression for the total number of blenders $N(x)$ assembled by the average worker during the morning shift (8 A.M. to 12:15 P.M.).

c. At what time should the coffee break be scheduled so that the average worker will produce the maximum number of blenders during the morning shift?

50. Research indicates that the power P required by a bird to maintain flight is given by the formula

$$P = \frac{w^2}{2\rho S v} + \frac{1}{2}\rho a v^3$$

where v is the relative speed of the bird, w is its weight, ρ is the density of air, and S and A are constants associated with the bird's size and shape.* What speed will minimize the power? You may assume that v, w, ρ, S, and A are all positive.

51. The deflection of a hardwood beam of length ℓ is given by

$$D(x) = \frac{9}{4}x^4 - 7\ell x^3 + 5\ell^2 x^2$$

where x is the distance from one end of the beam. What value of x yields the maximum deflection?

> **Level 3**

52. 𝔥*istorical* 𝔔*uest*

Maria Agnesi
(1718-1799)

One of the most famous women in the history of mathematics is Maria Gaëtana Agnesi (pronounced än yā ̦zē). She was born in Milan, the first of 21 children. Her first publication was at age 9, when she wrote a Latin discourse defending higher education for women. Her most important work was a now-classic calculus textbook published in 1748. Maria Agnesi is primarily remembered for a curve defined a a positive constant by the equation

$$y = \frac{a^3}{x^2 + a^2}$$

The curve was named **versiera** *(from the Latin verb* **vertere** *(to turn)) by Agnesi, but John Colson, an Englishman who translated her work, confused the word* **versiera** *with the word* **avversiera**, *which means "wife of the devil" in Italian; the curve has ever since been called the "witch of Agnesi." This was particularly unfortunate because Colson wanted Agnesi's work to serve as a model for budding young mathematicians,*

especially young women. Graph this curve and find the critical numbers, extrema, and points of inflection.

53. Journal Problem *(Mathematics Magazine)*[†] Give an elementary proof that

$$f(x) = \frac{1}{\sin x} - \frac{1}{x}, \quad 0 < x \le \frac{\pi}{2}$$

is positive and increasing.

54. An important formula in physical chemistry is *van der Waals' equation,* which says that

$$\left(P + \frac{a}{V^2}\right)(v - b) = nRT$$

where P, V, and T are the pressure, volume, and temperature, respectively, of a gas, and a, b, n, and R are positive constants. The *critical temperature* T_C of the gas is the highest temperature at which the gaseous and liquid phases can exist as separate states.

  a. When $T = T_C$, the pressure P can be expressed as a function $P(V)$ of V alone. Show how this can be done, and then find $P'(V)$ and $P''(V)$.

  b. The *critical volume* V_C is the volume that satisfies $P'(V_C) = 0$ and $P''(V_C) = 0$. Find V_C, assuming $T = T_C$.

  c. Find T_C, the point where $P''(V) = 0$, using the V_C from part **b** to write it in terms of a, b, n, and R. Finally, find the *critical pressure* $P_C = P(V_C)$ in terms of a, b, n, and R.

  d. Sketch P as a function of V.

55. Use calculus to show that the graph of the quadratic function

$$y = Ax^2 + Bx + C$$

for constants A, B, and C, is concave up if $A > 0$ and concave down if $A < 0$.

56. Use calculus to prove that for constants a, b, and c, the vertex (relative extremum) of the quadratic function

$$y = ax^2 + bx + c$$

$(a \ne 0)$ occurs at $x = \dfrac{-b}{2a}$.

57. Find constants A, B, and C that guarantee that the function

$$f(x) = Ax^3 + Bx^2 + C$$

will have a relative extremum at $(2, 11)$ and an inflection point at $(1, 5)$. Sketch the graph of f.

---

*C. J. Pennycuick, "The Mechanics of Bird Migration," Ibis III, pp. 525-556.

†Volume 55 (1982), p. 300. "Elementary proof" in the question means that you should use only techniques from beginning calculus. For our purposes, you simply need to give a reasonable argument to justify the conclusion.

58. Find constants a, b, and c that guarantee that the graph of

$$f(x) = x^3 + ax^2 + bx + c$$

will have a relative maximum at $(-3, 18)$ and a relative minimum at $(1, -14)$.

59. Prove or disprove that if the graphs of the functions f and g are both concave up on an interval, then the graph of their sum $f + g$ is also concave up on that interval.

60. Find constants A, B, C, and D that guarantee that the graph of

$$f(x) = 3x^4 + Ax^3 + Bx^2 + Cx + D$$

will have horizontal tangents at $(2, -3)$ and $(0, 7)$. There is a third point that has a horizontal tangent. Find this point. Then, for all three points, determine whether each corresponds to a relative maximum, a relative minimum, or neither.

4.4 CURVE SKETCHING WITH ASYMPTOTES: LIMITS INVOLVING INFINITY

IN THIS SECTION: *Limits at infinity, infinite limits, graphs with asymptotes, vertical tangents and cusps, a general graphing strategy*
In order to complete our discussion of curve sketching, we need to introduce two concepts, asymptotes and limits involving infinity. We conclude this section with a summary which ties together a variety of curve-sketching skills.

Limits at Infinity

In applications, we are often concerned with "long run" behavior of a function. To indicate such behavior, we write

$$\lim_{x \to \infty} f(x) = L$$

to indicate that $f(x)$ approaches the number L as x increases without bound. Similarly, we write

$$\lim_{x \to -\infty} f(x) = M$$

to indicate that $f(x)$ approaches the number M as x decreases without bound. Here are the formal definitions of these **limits at infinity**.

LIMITS AT INFINITY The limit statement $\lim_{x \to \infty} f(x) = L$ means that for any number $\epsilon > 0$, there exists a number N_1 such that

$$|f(x) - L| < \epsilon \text{ whenever } x > N_1$$

for x in the domain of f. Similarly, $\lim_{x \to -\infty} f(x) = M$ means that for any $\epsilon > 0$, there exists a number N_2 such that

$$|f(x) - M| < \epsilon \text{ whenever } x < N_2$$

This definition can be illustrated graphically, as shown in Figure 4.35.

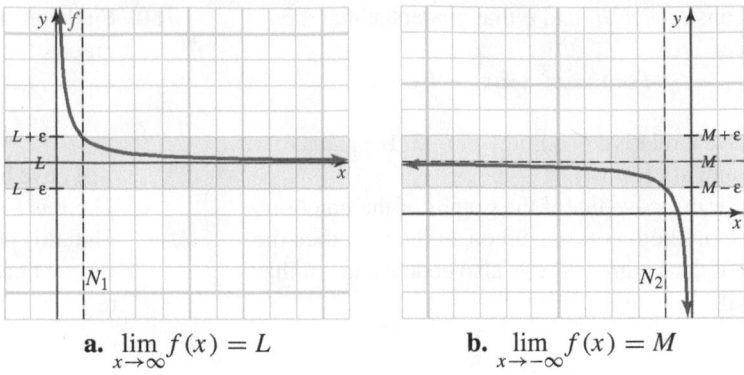

a. $\lim_{x \to \infty} f(x) = L$ **b.** $\lim_{x \to -\infty} f(x) = M$

Figure 4.35   Graphical representation of limits at infinity

With this formal definition, we can show that all the rules for limits established in Chapter 2 also apply to $\lim_{x \to \infty} f(x)$ and $\lim_{x \to -\infty} f(x)$.

LIMIT RULES (AT INFINITY)   If $\lim_{x \to \infty} f(x)$ and $\lim_{x \to \infty} g(x)$ exist, then for constants a and b:

Linearity rule $\lim_{x \to \infty} \left[af(x) + bg(x) \right] = a \lim_{x \to \infty} f(x) + b \lim_{x \to \infty} g(x)$

Product rule $\lim_{x \to \infty} \left[f(x)g(x) \right] = \left[\lim_{x \to \infty} f(x) \right] \left[\lim_{x \to \infty} g(x) \right]$

Quotient rule $\lim_{x \to \infty} \dfrac{f(x)}{g(x)} = \dfrac{\lim_{x \to \infty} f(x)}{\lim_{x \to \infty} g(x)}$ if $\lim_{x \to \infty} g(x) \neq 0$

Power rule $\lim_{x \to \infty} \left[f(x) \right]^n = \left[\lim_{x \to \infty} f(x) \right]^n$

Analogous results hold for $\lim_{x \to -\infty} f(x)$ and $\lim_{x \to -\infty} g(x)$, if they exist.

The following theorem will allow us to evaluate certain limits at infinity with ease.

Theorem 4.8   Special limits at infinity

If A is any real number and r is a positive rational number, then

$$\lim_{x \to \infty} \frac{A}{x^r} = 0$$

Furthermore, if r is such that x^r is defined for $x < 0$, then

$$\lim_{x \to -\infty} \frac{A}{x^r} = 0$$

Proof: We begin by proving that $\lim_{x \to \infty} \dfrac{1}{x} = 0$. For $\epsilon > 0$, let $N = \dfrac{1}{\epsilon}$. Then for $x > N$ we have

$$x > N = \frac{1}{\epsilon} \text{ so that } \frac{1}{x} < \epsilon$$

This means that $\left| \dfrac{1}{x} - 0 \right| < \epsilon$, so that, from the definition of limit we have $\lim_{x \to \infty} \dfrac{1}{x} = 0$. Now let r be a rational number, say $r = p/q$. Then

$$\lim_{x \to \infty} \frac{A}{x^r} = \lim_{x \to \infty} \frac{A}{x^{p/q}}$$

$$= A \lim_{x \to \infty} \left[\frac{1}{\sqrt[q]{x}} \right]^p$$

$$= A \left[\sqrt[q]{\lim_{x \to \infty} \frac{1}{x}} \right]^p$$

$$= A \left[\sqrt[q]{0} \right]^p$$

$$= A \cdot 0$$

$$= 0$$

The proof for the analogous limit as $x \to -\infty$ follows similarly.   ♦

When evaluating a limit of the form

$$\lim_{x \to \infty} \frac{p(x)}{q(x)} \quad \text{or} \quad \lim_{x \to -\infty} \frac{p(x)}{q(x)}$$

where $p(x)$ and $q(x)$ are polynomials, it is often useful to divide both $p(x)$ and $q(x)$ by the highest power of x that occurs in either. The limit can then be found by applying Theorem 4.8. This process is illustrated by the following examples.

Example 1   Evaluating a limit at infinity

Evaluate $\displaystyle \lim_{x \to \infty} \frac{3x^3 - 5x + 9}{5x^3 + 2x^2 - 7}$.

Solution   We may assume that $x \neq 0$, because we are interested only in very large values of x. Dividing both the numerator and denominator of the given expressions by x^3, the highest power of x appearing in the fraction, we find

$$\lim_{x \to \infty} \frac{3x^3 - 5x + 9}{5x^3 + 2x^2 - 7} = \lim_{x \to \infty} \frac{3x^3 - 5x + 9}{5x^3 + 2x^2 - 7} \cdot \frac{\frac{1}{x^3}}{\frac{1}{x^3}}$$

$$= \lim_{x \to \infty} \frac{3 - \frac{5}{x^2} + \frac{9}{x^3}}{5 + \frac{2}{x} - \frac{7}{x^3}}$$

$$= \frac{\lim_{x \to \infty} \left(3 - \frac{5}{x^2} + \frac{9}{x^3} \right)}{\lim_{x \to \infty} \left(5 + \frac{2}{x} - \frac{7}{x^3} \right)}$$

$$= \frac{3 - 0 + 0}{5 + 0 - 0}$$

$$= \frac{3}{5}$$

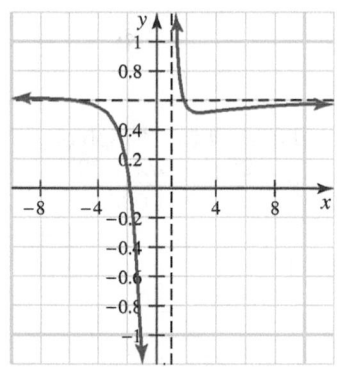

Figure 4.36 Graph of
$y = \dfrac{3x^3 - 5x + 9}{5x^3 + 2x^2 - 7}$

The graph of the given rational function is shown in Figure 4.36. Notice how the curve seems to approach $y = \frac{3}{5}$ as $x \to \infty$ and as $x \to -\infty$.   ■

Example 2 illustrates how Theorem 4.8 can be used along with the other limit properties to evaluate limits at infinity.

Example 2   Evaluating limits at infinity

Evaluate $\displaystyle \lim_{x \to \infty} \sqrt{\frac{3x - 5}{x - 2}}$ and $\displaystyle \lim_{x \to \infty} \left(\frac{3x - 5}{x - 2} \right)^3$.

Solution   Notice that for $x \neq 0$,

$$\frac{3x - 5}{x - 2} = \frac{3x - 5}{x - 2} \cdot \frac{\frac{1}{x}}{\frac{1}{x}} = \frac{3 - \frac{5}{x}}{1 - \frac{2}{x}}$$

We now find the limits using the quotient rule, the power rule, and Theorem 4.8:

$$\lim_{x \to \infty} \sqrt{\frac{3x-5}{x-2}} = \lim_{x \to \infty} \left(\frac{3x-5}{x-2}\right)^{1/2}$$

$$= \left(\lim_{x \to \infty} \frac{3x-5}{x-2}\right)^{1/2}$$

$$= \left(\frac{\lim_{x \to \infty}\left(3 - \frac{5}{x}\right)}{\lim_{x \to \infty}\left(1 - \frac{2}{x}\right)}\right)^{1/2}$$

$$= \left(\frac{3-0}{1-0}\right)^{1/2}$$

$$= \sqrt{3}$$

Similarly,

$$\lim_{x \to -\infty} \left(\frac{3x-5}{x-2}\right)^{3} = 3^3 = 27$$

Example 3   Evaluating a limit at negative infinity

Evaluate $\displaystyle\lim_{x \to -\infty} \frac{95x^3 + 57x + 30}{x^5 - 1,000}$.

Solution   Dividing the numerator and the denominator by the highest power, x^5, we find that

$$\lim_{x \to -\infty} \frac{95x^3 + 57x + 30}{x^5 - 1,000} = \lim_{x \to -\infty} \frac{95x^3 + 57x + 30}{x^5 - 1,000} \cdot \frac{\frac{1}{x^5}}{\frac{1}{x^5}}$$

$$= \lim_{x \to -\infty} \frac{\frac{95}{x^2} + \frac{57}{x^4} + \frac{30}{x^5}}{1 - \frac{1,000}{x^5}}$$

$$= \frac{0+0+0}{1-0}$$

$$= 0$$

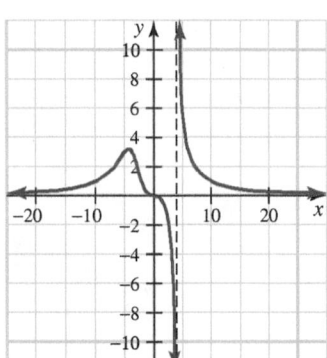

Figure 4.37   Graph of
$$y = \frac{95x^3 + 57x + 30}{x^5 - 1,000}$$

The graph of the rational function is shown in Figure 4.37.

Example 4   Evaluating a limit at infinity involving e^x

Find $\displaystyle\lim_{x \to \infty} e^{-x} \cos x$.

Solution   We cannot use the product rule for limits since $\lim_{x \to \infty} \cos x$ does not exist (it diverges by oscillation). Note, however, that the magnitude of

$$e^{-x} \cos x = \frac{\cos x}{e^x}$$

must become smaller and smaller as $x \to \infty$ since the numerator $\cos x$ is bounded between -1 and 1, while the denominator e^x grows relentlessly larger with x. Thus by the squeeze rule,

$$\lim_{x \to \infty} e^{-x} \cos x = 0$$

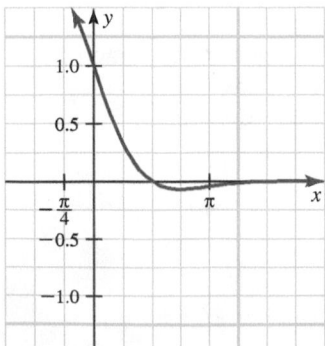

Figure 4.38   Graph of
$y = e^{-x} \cos x$

The graph of $y = e^{-x} \cos x$ is shown in Figure 4.38.

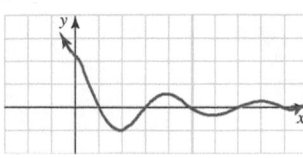

Try drawing the graph of $y = e^{-x} \cos x$ from Example 4 on your calculator, or other graphing software, and you will see that what you get does not properly illustrate the oscillatory behavior of this graph. This demonstrates the importance of NOT relying on technology for graphing. In fact, the accompanying graph (without scale) shows more clearly the behavior of the graph of the function $y = e^{-x} \cos x$ for large, positive values of x.

Infinite Limits

In mathematics, the symbol ∞ is not a number, but is used to describe either the process of unrestricted growth or the result of such growth. Thus, a limit statement such as

$$\lim_{x \to c} f(x) = \infty$$

means that the function f increases without bound as x approaches c from either side, while

$$\lim_{x \to c} g(x) = -\infty$$

means that g decreases without bound as x approaches c from either side. Such limits may be defined formally as follows.

INFINITE LIMITS   We write $\lim_{x \to c} f(x) = \infty$ if for any number $N > 0$ (no matter how large), it is possible to find a number $\delta > 0$ such that $f(x) > N$ whenever $0 < |x - c| < \delta$.

Similarly, $\lim_{x \to c} g(x) = -\infty$ if for any $N > 0$, it is possible to find a number $\delta > 0$ so that $g(x) < -N$ when $0 < |x - c| < \delta$.

■ **W**hat this says: Remember, ∞ *is **not** a number,* so an infinite limit does not exist in the sense that limits were defined in Chapter 2. However, there are several ways for a limit not to exist (for example, $\lim_{x \to \infty} \cos x$ fails to exist by oscillation), so saying that $\lim_{x \to c} f(x) = \infty$ or $\lim_{x \to c} g(x) = -\infty$ conveys more information than simply observing that the limit does not exist.

Example 5   Infinite limits

Find $\displaystyle \lim_{x \to 2^-} \frac{3x - 5}{x - 2}$ and $\displaystyle \lim_{x \to 2^+} \frac{3x - 5}{x - 2}$.

Solution   Notice that $\dfrac{1}{x - 2}$ increases without bound as x approaches 2 from the right and $\dfrac{1}{x - 2}$ decreases without bound as x approaches 2 from the left. That is,

$$\lim_{x \to 2^+} \frac{1}{x - 2} = \infty \qquad \text{and} \qquad \lim_{x \to 2^-} \frac{1}{x - 2} = -\infty$$

We also have $\lim_{x \to 2} (3x - 5) = 1$, so it follows that

$$\lim_{x \to 2^+} \frac{3x - 5}{x - 2} = \infty \qquad \text{and} \qquad \lim_{x \to 2^-} \frac{3x - 5}{x - 2} = -\infty \qquad ■$$

Graphs with Asymptotes

Figure 4.39 shows a graph that approaches the horizontal line $y = 2$ as $x \to \infty$ and $y = -1$ as $x \to -\infty$, and the vertical line $x = 3$ as x approaches 3 from either side.

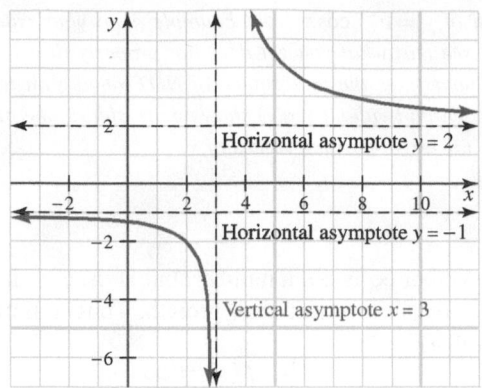

Figure 4.39   A typical graph with asymptotes

VERTICAL ASYMPTOTE   The line $x = c$ is a **vertical asymptote** of the graph of f if either of the one-sided limits

$$\lim_{x \to c^-} f(x) \qquad \text{or} \qquad \lim_{x \to c^+} f(x)$$

is infinite.

HORIZONTAL ASYMPTOTE   The line $y = L$ is a **horizontal asymptote** of the graph of f if

$$\lim_{x \to \infty} f(x) = L \qquad \text{or} \qquad \lim_{x \to -\infty} f(x) = L$$

Example 6   Graphing a rational function with asymptotes

Sketch the graph of $f(x) = \dfrac{3x - 5}{x - 2}$.

Solution

Vertical Asymptotes

First, make sure the rational function is written in simplified (reduced) form. Because vertical asymptotes for $f(x) = \dfrac{3x - 5}{x - 2}$ occur at values of c for which $\lim\limits_{x \to c^-} f(x)$ or $\lim\limits_{x \to c^+} f(x)$ is infinite, we look for values that cause the denominator to be zero (and the numerator not to be zero); that is, we solve $q(c) = 0$, where $q(x)$ is the denominator of $f(x)$, and then evaluate $\lim\limits_{x \to c^-} f(x)$ and $\lim\limits_{x \to c^+} f(x)$ to ascertain the behavior of the function at $x = c$. For this example, $x = 2$ is a value that causes division by zero, so we find

$$\lim_{x \to 2^+} \frac{3x - 5}{x - 2} = \infty \qquad \text{and} \qquad \lim_{x \to 2^-} \frac{3x - 5}{x - 2} = -\infty$$

(We found these limits in Example 5.) This means that $x = 2$ is a vertical asymptote and that the graph is moving downward as $x \to 2$ from the left and upward as $x \to 2$ from the right. This information is recorded on the preliminary graph shown in Figure 4.40**a** by a dashed vertical line with upward ($\uparrow$) and downward ($\downarrow$) arrows.

Horizontal Asymptotes

To find the horizontal asymptotes we compute

$$\lim_{x \to \infty} \frac{3x - 5}{x - 2} = \lim_{x \to \infty} \frac{3x - 5}{x - 2} \cdot \frac{\frac{1}{x}}{\frac{1}{x}} = \lim_{x \to \infty} \frac{3 - \frac{5}{x}}{1 - \frac{2}{x}} = \frac{3 - 0}{1 - 0} = 3$$

Similarly, we find $\lim\limits_{x \to -\infty} \dfrac{3x - 5}{x - 2} = 3$. This means that $y = 3$ is a horizontal asymptote. This information is recorded on the preliminary graph shown in Figure 4.40**a** by a dashed horizontal line with outbound arrows ($\leftarrow$, $\rightarrow$)

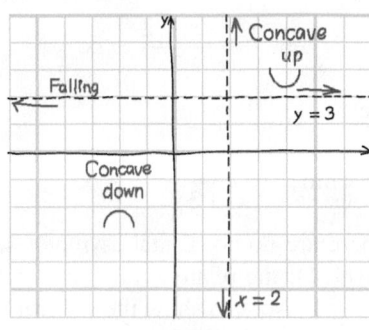

a. Preliminary sketch

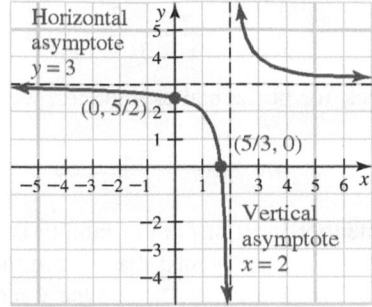

b. Completed sketch

Figure 4.40   Graph of $f(x) = \dfrac{3x - 5}{x - 2}$

The preliminary sketch gives us some valuable information about the graph, but it does not present the entire picture. Next, we use calculus to find where the function is increasing and decreasing (first derivative) and where it is concave up and concave down (second derivative):

$$f'(x) = \frac{-1}{(x - 2)^2} \quad \text{and} \quad f''(x) = \frac{2}{(x - 2)^3}$$

Neither derivative is ever zero, and both are undefined at $x = 2$. Checking the signs of the first and second derivatives, we find that it is concave up for $x > 2$ and concave down for $x < 2$, but does not have a point of inflection (function is not defined at $x = 2$). This information is added to the preliminary sketch shown in Figure 4.40**a**. The completed graph is shown in Figure 4.40**b**, which also shows the x- and y-intercepts at $\left(\frac{5}{3}, 0\right)$ and $\left(0, \frac{5}{2}\right)$.

Example 7   Sketching a curve with asymptotes

Discuss and sketch the graph of $f(x) = \dfrac{x^2 - x - 2}{x - 3}$.

Solution

Functions	Critical numbers	Conclusions
$f(x) = \dfrac{x^2 - x - 2}{x - 3}$		
$f'(x) = \dfrac{x^2 - 6x + 5}{(x - 3)^2}$	$f'(x) = 0$	First derivative signs (increasing/decreasing):
	$\dfrac{(x - 1)(x - 5)}{(x - 3)^2} = 0$	
	$x = 1, 5$	Positive 1 Negative 3 Negative 5 Positive x
$f''(x) = \dfrac{8}{(x - 3)^3}$	$f''(x) = 0$	Second derivative signs (concavity):
	$\dfrac{8}{(x - 3)^3} = 0$	
	no second-order critical numbers	Negative 3 Positive x

Note that there is a relative maximum at $x = 1$ ($\nearrow \searrow$) and a relative minimum at $x = 5$ ($\searrow \nearrow$). The concavity changes (from $\downarrow$ to $\uparrow$) at $x = 3$, but this does not correspond to an inflection point since $f(3)$ is not defined. We look for vertical asymptotes where $f(x)$ (in reduced form) is not defined; in this case, at $x = 3$. Testing with limits, we find that

$$\lim_{x \to 3^-} \frac{x^2 - x - 2}{x - 3} = -\infty \quad \text{and} \quad \lim_{x \to 3^+} \frac{x^2 - x - 2}{x - 3} = \infty$$

To check for horizontal asymptotes, we compute

$$\lim_{x \to \infty} \frac{x^2 - x - 2}{x - 3} = \infty \quad \text{and} \quad \lim_{x \to -\infty} \frac{x^2 - x - 2}{x - 3} = -\infty$$

Since neither limit of $f(x)$ at infinity is finite, there are no horizontal asymptotes.

We plot some points: the relative maximum $(1, 1)$; the relative minimum $(5, 9)$; the x-intercepts $(2, 0)$, $(-1, 0)$; and the y-intercept $\left(0, \frac{2}{3}\right)$. We also show the vertical asymptote $x = 3$, and using the intervals of increase and decrease and concavity we obtain the graph shown in Figure 4.41.

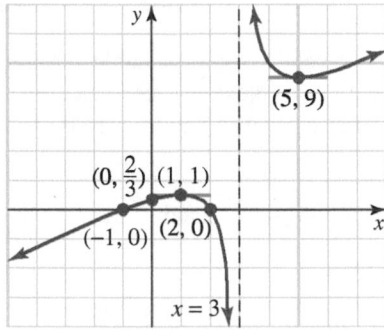

Figure 4.41   Graph of $f(x) = \dfrac{x^2 - x - 2}{x - 3}$

Vertical Tangents and Cusps

Suppose the function f is continuous at the point P where $x = c$ and that $f'(c)$ becomes infinite as x approaches c. Then the graph of f has a *vertical tangent* at P if the graph turns smoothly through P and a *cusp* at P if the graph changes direction abruptly there. These graphical features can be defined in terms of limits, as follows:

VERTICAL TANGENT Suppose the function f is continuous at the point $P\,(c, f(c))$.

Then, the graph of f has a **vertical tangent** at P if $\lim\limits_{x \to c^-} f'(x)$ and $\lim\limits_{x \to c^+} f'(x)$ are either both ∞ or both $-\infty$.

CUSP Furthermore, the graph has a **cusp** at P if $\lim\limits_{x \to c^-} f'(x)$ and $\lim\limits_{x \to c^+} f'(x)$ are both infinite with opposite signs (one ∞ and the other $-\infty$).

These possibilities are shown in Figure 4.42.

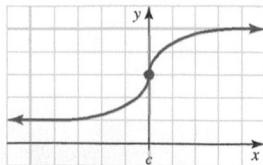

a. Vertical tangent
$$\lim_{x \to c^-} f'(x) = \infty$$
$$\lim_{x \to c^+} f'(x) = \infty$$

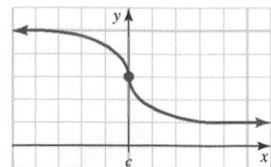

b. Vertical tangent
$$\lim_{x \to c^-} f'(x) = -\infty$$
$$\lim_{x \to c^+} f'(x) = -\infty$$

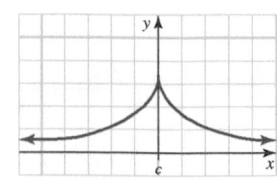

c. Cusp
$$\lim_{x \to c^-} f'(x) = \infty$$
$$\lim_{x \to c^+} f'(x) = -\infty$$

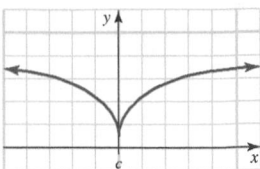

d. Cusp
$$\lim_{x \to c^-} f'(x) = -\infty$$
$$\lim_{x \to c^+} f'(x) = \infty$$

Figure 4.42   Vertical tangents and cusps

Example 8 Identifying vertical tangents and cusps

In each of the following cases, determine whether the graph of the given function has a vertical tangent or a cusp: **a.** $f(x) = x^{2/3}(2x + 5)$ **b.** $g(x) = x^{1/3}(x + 4)$

Solution

a. Writing $f(x) = 2x^{5/3} + 5x^{2/3}$, we find that

$$f'(x) = \frac{10}{3}x^{2/3} + \frac{10}{3}x^{-1/3} = \frac{10}{3}x^{-1/3}(x + 1)$$

so $f'(x)$ becomes infinite only at $x = 0$. Since

$$\lim_{x \to 0^-} f'(x) = \lim_{x \to 0^-} \frac{10}{3}x^{-1/3}(x + 1) = -\infty$$

and

$$\lim_{x \to 0^+} f'(x) = \lim_{x \to 0^+} \frac{10}{3}x^{-1/3}(x + 1) = \infty$$

it follows that there is a cusp on the graph of f at $(0, 0)$. The graph is shown in Figure 4.43**a**.

b. The derivative

$$g'(x) = \frac{4}{3}x^{-2/3}(x + 1)$$

becomes infinite only when $x = 0$. We find that

$$\lim_{x \to 0^-} g'(x) = \lim_{x \to 0^-} \frac{4}{3}x^{-2/3}(x + 1) = \infty$$

and

$$\lim_{x \to 0^+} g'(x) = \lim_{x \to 0^+} \frac{4}{3}x^{-2/3}(x + 1) = \infty$$

Since the limit approaches ∞ as x approaches zero from both the left and the right, we conclude that a vertical tangent occurs at the origin $(0, 0)$. The graph is shown in Figure 4.43**b**.

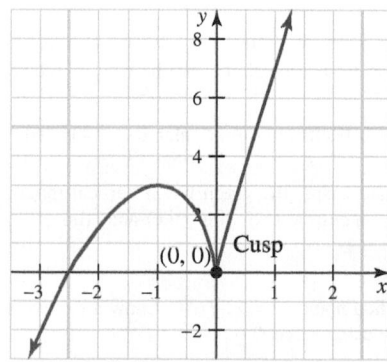

a. Graph of f showing cusp

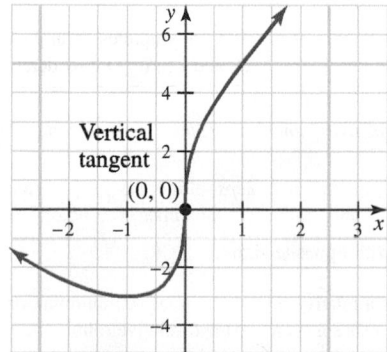

b. Graph of g showing vertical tangent

Figure 4.43 A graph with a cusp and another with a vertical tangent

A General Graphing Strategy

It is worthwhile to combine the techniques of curve sketching from calculus with those techniques studied in precalculus courses. You may be familiar with **extent** (finding the domain and the range of the function) and **symmetry** (with respect to the x-axis, y-axis, or origin). We now have all the tools we need to describe a general procedure for curve sketching, and this procedure is summarized in Table 4.1.

Table 4.1 Graphing Strategy for a Function Defined by $y = f(x)$

Step	Procedure
Simplify.	If possible, simplify algebraically. For example, if $f(x) = \frac{(x-2)(x+3)}{x-2}$, you must write this as $f(x) = x + 3$, $x \neq 2$ before beginning the procedures listed here.
Find derivatives and critical numbers.	Compute the first and second derivatives; set each equal to zero and solve. Find the first- and second-order critical numbers.
Determine intervals of increase and decrease.	Use the first-order critical numbers of f: $f'(x) > 0$, curve rising (indicate these regions by $\nearrow$). $f'(x) < 0$, curve falling (indicate these regions by $\searrow$)
Apply the second-derivative test.	Use the second-derivative test to find the relative maxima and minima; substitute the first-order critical numbers, c_1. $f''(c_1) > 0$, relative minimum $f''(c_1) < 0$, relative maximum $f''(c_1) = 0$, test fails; use first-derivative test.
Determine concavity and points of inflection.	Find the points of inflection. These points are located where the concavity changes from up to down or down to up, and are found by checking the intervals on either side of the second-order critical numbers.
Apply the first-derivative test. $\nearrow\ \searrow$ Maximum $\searrow\ \nearrow$ Minimum	Use the first-derivative test if the second-derivative fails or is too complicated. **1.** Let c_1 be a first-order critical number of a continuous function f. **2.**(a) $(c_1, f(c_1)$ is a relative maximum if $f'(x) > 0$ for all x in an open interval (a, c_1) to the left of c_1. $f'(x) < 0$ for all x in an open interval (c_1, b) to the right of c_1. (b) $(c_1, f(c_1))$ is a relative minimum if $f'(x) < 0$ for all x in an open interval (a, c_1) to the left of c_1. $f'(x) > 0$ for all x in an open interval (c_1, b) to the right of c_1. (c) $(c_1, f(c_1))$ is not an extremum if $f'(x)$ has the same sign in open intervals (a, c_1) and (c_1, b) on both sides of c_1.
Find asymptotes, vertical tangents, and cusps.	**1.** *Vertical asymptotes*—Vertical asymptotes, if any exist, will occur at values $x = c$ for which f is not defined. Use the limits $\lim\limits_{x \to c^-} f(x)$ and $\lim\limits_{x \to c^+} f(x)$ to determine the behaviors of the graphs near $x = c$. Show this behavior with arrows ($\uparrow \downarrow$) **2.** *Horizontal asymptotes*—Compute $\lim\limits_{x \to \infty} f(x)$ and $\lim\limits_{x \to -\infty} f(x)$. If either is finite, plot the associated horizontal asymptote. **3.** *Vertical tangents and cusps*—Find the vertical tangents and cusps; the graph has a vertical tangent at P if $\lim\limits_{x \to c^-} f'(x)$ and $\lim\limits_{x \to c^+} f'(x)$ are either both ∞ or $-\infty$. It has a cusp at P if $\lim\limits_{x \to c^-} f'(x)$ and $\lim\limits_{x \to c^+} f'(x)$ are both infinite with opposite signs.
Plot points.	**1.** If $f(c)$ is a relative maximum, relative minimum, or inflection point, plot $(c, f(c))$. Show a relative maximum point by using a "cap" ($\cap$) and relative minimum point by using a "cup" ($\cup$). **2.** x-intercepts: Set $y = 0$ and if $x = a$ is a solution, plot $(a, 0)$. **3.** y-intercepts: Set $x = 0$ and if $y = b$ is a solution, plot $(0, b)$; a function will have, at most, one y-intercept.
Sketch the curve.	Draw a curve using the above information.

PROBLEM SET 4.4

Level 1

1. ■ What does this say? Outline a method for curve sketching.
2. ■ What does this say? What are critical numbers? Discuss the importance of critical numbers in curve sketching.
3. ■ What does this say? Discuss the importance of concavity and points of inflection in curve sketching.
4. ■ What does this say? Discuss the importance of asymptotes in curve sketching.

Evaluate the limits in Problems 5-24.

5. $\displaystyle\lim_{x\to\infty} \frac{2{,}000}{x+1}$

6. $\displaystyle\lim_{x\to\infty} \frac{7{,}000}{\sqrt{x}+1}$

7. $\displaystyle\lim_{x\to\infty} \frac{3x+5}{x-2}$

8. $\displaystyle\lim_{x\to\infty} \frac{x+2}{3x-5}$

9. $\displaystyle\lim_{t\to\infty} \frac{9t^5+50t^2+800}{t^5-1{,}000}$

10. $\displaystyle\lim_{t\to-\infty} \frac{(2t+5)(t-3)}{(7t-2)(4t+1)}$

11. $\displaystyle\lim_{x\to\infty} \frac{x}{\sqrt{x^2+1{,}000}}$

12. $\displaystyle\lim_{x\to-\infty} \frac{3x}{\sqrt{4x^2+10}}$

13. $\displaystyle\lim_{x\to\infty} \frac{x^{5.916}+1}{x^{\sqrt{35}}}$

14. $\displaystyle\lim_{x\to\infty} \frac{x^{6.083}+1}{x^{\sqrt{37}}}$

15. $\displaystyle\lim_{x\to1^-} \frac{x-1}{|x^2-1|}$

16. $\displaystyle\lim_{x\to3^+} \frac{x^2-4x+3}{x^2-6x+9}$

17. $\displaystyle\lim_{x\to0^+} \frac{x^2-x+1}{x-\sin x}$

18. $\displaystyle\lim_{x\to\left(\frac{\pi}{4}\right)^+} \frac{\sec x}{\tan x-1}$

19. $\displaystyle\lim_{x\to\infty} \left(x\,\sin\frac{1}{x}\right)$

20. $\displaystyle\lim_{x\to0^+} \frac{x^2}{1-\cos x}$

21. $\displaystyle\lim_{x\to0^+} \frac{\ln\sqrt[3]{x}}{\sin x}$

22. $\displaystyle\lim_{x\to\infty} \frac{\ln\sqrt[3]{x}}{\sin x}$

23. $\displaystyle\lim_{x\to-\infty} e^x\sin x$

24. $\displaystyle\lim_{x\to\infty} \frac{\tan^{-1}x}{e^{0.1x}}$

Level 2

Find all vertical and horizontal asymptotes of the graph of the functions given in Problems 25-44. Find where each graph is rising and where it is falling, determine concavity, and locate all critical points and points of inflection. Finally, sketch the graph. Be sure to show any special features, such as cusps or vertical tangents.

25. $f(x) = \dfrac{3x+5}{7-x}$

26. $g(x) = \dfrac{15}{x+4}$

27. $f(x) = 4 + \dfrac{2x}{x-3}$

28. $g(x) = 1 - \dfrac{x}{4-x}$

29. $f(x) = \dfrac{x^3+1}{x^3-8}$

30. $f(x) = \dfrac{2x^2-5x+7}{x^2-9}$

31. $g(t) = (t^3+t)^2$

32. $g(t) = t^{-1/2} + \dfrac{1}{3}t^{3/2}$

33. $f(x) = (x^2-9)^2$

34. $g(x) = x(x^2-12)$

35. $f(x) = x^{1/3}(x-4)$

36. $f(u) = u^{2/3}(u-7)$

37. $f(x) = \tan^{-1}x^2$

38. $f(x) = \ln(4-x^2)$

39. $g(x) = \dfrac{8}{x-1} + \dfrac{27}{x+4}$

40. $f(x) = \dfrac{1}{x+1} + \dfrac{1}{x-1}$

41. $t(\theta) = \sin\theta - \cos\theta$ for $0 \le \theta \le 2\pi$

42. $f(x) = x - \sin 2x$ for $0 \le x \le \pi$

43. $f(x) = \sin^2 x - 2\sin x + 1$ for $0 \le x \le \pi$

44. $T(\theta) = \tan^{-1}\theta - \tan^{-1}\dfrac{\theta}{3}$

45. The *ideal speed* v for a banked curve on a highway is modeled by the equation

$$v^2 = gr\tan\theta$$

where g is the constant acceleration due to gravity, r is the radius of the curve, and θ is the angle of the bank. Assuming that r is constant, sketch the graph of v as a function of θ for $0 \le \theta \le \frac{\pi}{2}$.*

46. According to Einstein's special theory of relativity, the mass of a body is modeled by the expression

$$m = \frac{m_0}{\sqrt{1-\frac{v^2}{c^2}}}$$

where m_0 is the mass of the body at rest in relation to the observer, m is the mass of the body when it moves with speed v in relation to the observer, and c is the speed of light. Sketch the graph of m as a function of v. What happens as $v \to c^-$?

47. **Think Tank Problem** Sketch a graph of a function f with all the following properties: The graph has $y=1$ and $x=3$ as asymptotes; f is increasing for $x<3$ and $3<x<5$ and is decreasing elsewhere; the graph is concave up for $x<3$ and for $x>7$ and concave down for $3<x<7$; $f(0)=4=f(5)$ and $f(7)=2$.

48. **Think Tank Problem** Sketch a graph of a function g with all of the following properties:
(1) g is increasing for $x<-1$ and $-1<x<1$ and decreasing for $1<x<3$ and $x>3$;
(2) The graph has only one critical point $(1,-1)$, no inflection points;
(3) $\displaystyle\lim_{x\to-\infty} g(x) = -1;\ \lim_{x\to\infty} g(x) = 2;$
$\displaystyle\lim_{x\to-1^+} g(x) = \lim_{x\to3^-} g(x) = -\infty$

---

*In physics it is shown that if one travels around the curve at the ideal speed, no frictional force is required to prevent slipping. This greatly reduces wear on tires and contributes to safety.

49. In an experiment, a biologist introduces a toxin into a bacterial colony and then measures the effect on the population of the colony. Suppose that at time t (in minutes) the population is

$$P(t) = 5 + e^{-0.04t}(t + 1)$$

thousand. At what time will the population be the largest? What happens to the population in the long run (as $t \to \infty$)? Find where the graph of P has an inflection point, and interpret this point in terms of the population. Sketch the graph of P.

50. a. Show that, in general, the graph of the function

$$f(x) = \frac{ax^2 + bx + c}{rx^2 + sx + t}$$

will have $y = \dfrac{a}{r}$ as a horizontal asymptote and that when $br \neq as$, the graph will cross this asymptote at the point where

$$x = \frac{at - cr}{br - as}$$

b. Sketch the graph of each of the following functions:

$$g(x) = \frac{x^2 - 4x - 5}{2x^2 + x - 10}$$

$$h(x) = \frac{3x^2 - x - 7}{-12x^2 + 4x + 8}$$

51. Find constants a and b that guarantee that the graph of the function defined by

$$f(x) = \frac{ax + 5}{3 - bx}$$

will have a vertical asymptote at $x = 5$ and a horizontal asymptote at $y = -3$.

52. Journal Probelm (*Parabola**) Draw a careful sketch of the curve

$$y = \frac{x^2}{x^2 - 1}$$

indicating clearly any vertical or horizontal asymptotes, turning points, or points of inflection.

Level 3

Think Tank Problems *In Problems 53-55, either show that the statement is generally true or find a counterexample.*

53. If f is concave up and g is concave down on an interval I, then fg is neither concave up nor concave down on I.

54. If $f(x) > 0$ and $g(x) > 0$ for all x on I and if f and g are concave up on I, then fg is also concave up on I.

55. If $f(x) < 0$ and $f''(x) > 0$ for all x on I, then the function $g = f^2$ is concave up on I.

56. If $f(x) > 0$ and $f''(x) > 0$ for all x on I, then the function $g = f^2$ is concave up on I.

57. Think Tank Problem State what you think should be the formal definition of each of the following limit statements:

    a. $\lim\limits_{x \to c^+} f(x) = -\infty$ **b.** $\lim\limits_{x \to c^-} f(x) = \infty$

58. Consider the rational function

$$f(x) = \frac{a_n x^n + a_{n-1} x^{n-1} + \cdots + a_1 x + a_0}{b_m x^m + b_{m-1} x^{m-1} + \cdots + b_1 x + b_0}$$

    a. If $m > n$ and $b_m \neq 0$, show that the x-axis is the only horizontal asymptote of the graph of f.

    b. If $m = n$, show that the line $y = \dfrac{a_n}{b_m}$ is the only horizontal asymptote of the graph of f.

    c. If $m < n$, is it possible for the graph to have a horizontal asymptote? Explain.

59. Prove the following limit rule. If $\lim\limits_{x \to c} f(x) = \infty$ and $\lim\limits_{x \to c} g(x) = A \, (A > 0)$ then

$$\lim\limits_{x \to c} \left[f(x) g(x) \right] = \infty$$

Hint: Notice that because $\lim\limits_{x \to c} g(x) = A$ the function $g(x)$ is near A when x is near c. Therefore, because

$$\lim\limits_{x \to \infty} f(x) = \infty$$

the product $f(x)g(x)$ is large if x is near c. Formalize these observations for the proof.

60. Prove that if $\lim\limits_{x \to \infty} f(x)$ and $\lim\limits_{x \to \infty} g(x)$ both exist, so does $\lim\limits_{x \to \infty} \left[f(x) + g(x) \right]$ and

$$\lim\limits_{x \to \infty} \left[f(x) + g(x) \right] = \lim\limits_{x \to \infty} f(x) + \lim\limits_{x \to \infty} g(x)$$

Hint: The key is to show that if $|f(x) - L| < \frac{\epsilon}{2}$ for $x > N_1$ and $|g(x) - M| < \frac{\epsilon}{2}$ for $x > N_2$, then whenever $x > N$ for some number N,

$$\left| \left[f(x) + g(x) \right] - (L + M) \right| < \epsilon$$

You should also show that N relates to N_1 and N_2.

4.5 l'HÔPITAL'S RULE

IN THIS SECTION: *A rule to evaluate indeterminate forms, indeterminate forms 0/0 and ∞/∞, other indeterminate forms, special limits involving e^x and $\ln x$*
We can use limit theorems for sums, differences, products, and quotients (with nonzero division) provided certain meaningless expressions (such as 0/0) are not involved. We introduce a procedure, called l'Hôpital's rule, and illustrate its use in a variety of problems.

A Rule to Evaluate Indeterminate Forms

In curve sketching, optimization, and other applications, it is often necessary to evaluate a limit of the form $\lim_{x \to c} \dfrac{f(x)}{g(x)}$, where $\lim_{x \to c} f(x)$ and $\lim_{x \to c} g(x)$ are either both 0 or both ∞. Such limits are called 0/0 *indeterminate forms* and ∞/∞ *indeterminate forms* respectively, because their value cannot be determined without further analysis. There is a rule to evaluate such indeterminate forms, known as *l'Hôpital's rule*, which relates the evaluation to a computation of $\lim_{x \to c} \dfrac{f'(x)}{g'(x)}$, the limit of the ratio of the derivatives of f and g. Here is a precise statement of the rule.

Theorem 4.9   l'Hôpital's Rule

Let f and g be differentiable functions with $g'(x) \neq 0$ on an open interval containing c (except possibly at c itself). Suppose $\lim_{x \to c} \dfrac{f(x)}{g(x)}$ produces an indeterminate form $\dfrac{0}{0}$ or $\dfrac{\infty}{\infty}$ and that

$$\lim_{x \to c} \frac{f'(x)}{g'(x)} = L$$

where L is either a finite number, $+\infty$, or $-\infty$. Then

$$\lim_{x \to c} \frac{f(x)}{g(x)} = L$$

The theorem also applies to one-sided limits and to limits at infinity (where $x \to \infty$ and $x \to -\infty$).

> ■ **W**hat this says If the $\lim_{x \to c} \dfrac{f(x)}{g(x)}$ yields one of the indeterminate forms of the type $\dfrac{0}{0}$ or $\dfrac{\infty}{\infty}$, then you can evaluate the limit by replacing f with f' and g with g'.
> This *does not* mean you use the quotient rule for derivatives on $f(x)/g(x)$. Moreover, you can use l'Hôpital's rule on a reduced rational expression as long as you still get one of the forms
>
> $$\frac{0}{0} \quad \text{or} \quad \frac{\infty}{\infty} \quad \text{or} \quad \frac{\infty}{-\infty} \quad \text{or} \quad \frac{-\infty}{\infty} \quad \text{or} \quad \frac{-\infty}{-\infty}$$
>
> at $x = c$. Because l'Hôpital's rule can be applied repeatedly, you can keep applying it until you get something other than these forms.

Proof: The general proof is given in Appendix B. However, we can obtain a sense of why it is true. Suppose $f(x)$ and $g(x)$ are differentiable functions such that $f(c) = g(c) = 0$. Then, using the linearization formula for a differentiable function F:

$$F(x) \approx F(c) + F'(c)(x - c) \text{ from Section 3.8}$$

we can write

$$\frac{f(x)}{g(x)} \approx \frac{f(c)+f'(c)(x-c)}{g(c)+g'(c)(x-c)} = \frac{0+f'(c)(x-c)}{0+g'(c)(x-c)} = \frac{f'(c)}{g'(c)}$$

so

$$\lim_{x \to c} \frac{f(x)}{g(x)} = \frac{f'(c)}{g'(c)}$$ ◆

First *check the hypotheses (make sure it is of the form* $0/0$ *or* ∞/∞*), and* ***then*** *take* f'/g'*, not* $(f/g)'$*.*

l'Hôpital's rule is named after Guillaume François Antoine de l'Hôpital (1661-1704), the author of the first textbook on differential calculus, published in 1696. However, the result that bears l'Hôpital's name was actually due to Johann Bernoulli (see 𝔥*istorical* 𝔔*uest* in Problem 49).

Indeterminate Forms 0/0 and ∞/∞

We now consider a variety of problems involving indeterminate forms. We begin with a limit first computed in Chapter 2, but instead of using a geometric argument together with the squeeze rule of limits, we use l'Hôpital's rule.

Example 1   Using l'Hôpital's rule to compute a familiar trigonometric limit

Evaluate $\lim_{x \to 0} \dfrac{\sin x}{x}$.

Solution   Note that this is of indeterminate form because $\sin x$ and x both approach 0 as $x \to 0$. This means that l'Hôpital's rule applies:

$$\lim_{x \to 0} \frac{\sin x}{x} = \lim_{x \to 0} \frac{\cos x}{1} = 1$$ ∎

Example 2   l'Hôpital's rule with a 0/0 form

Evaluate $\lim_{x \to 2} \dfrac{x^7 - 128}{x^3 - 8}$.

Solution   For this example, $f(x) = x^7 - 128$ and $g(x) = x^3 - 8$, and the form is $0/0$ as $x \to 2$.

$$
\begin{aligned}
\lim_{x \to 2} \frac{x^7 - 128}{x^3 - 8} &= \lim_{x \to 2} \frac{7x^6}{3x^2} \qquad \textit{l'Hôpital's rule} \\[2mm]
&= \lim_{x \to 2} \frac{7x^4}{3} \qquad \textit{Simplify} \\[2mm]
&= \frac{7(2)^4}{3} \qquad \textit{Limit of a quotient} \\[2mm]
&= \frac{112}{3}
\end{aligned}
$$ ∎

Example 3   Limit is not an indeterminate form

Evaluate $\lim_{x \to 0} \dfrac{1 - \cos x}{\sec x}$.

Solution   You must always remember to check that you have an indeterminate form before applying l'Hôpital's rule. The limit is

$$\lim_{x \to 0} \frac{1 - \cos x}{\sec x} = \frac{\lim\limits_{x \to 0}(1 - \cos x)}{\lim\limits_{x \to 0} \sec x} = \frac{0}{1} = 0$$ ∎

☠ If you blindly apply l'Hôpital's rule in Example 3, you obtain the WRONG answer:

$$\lim_{x \to 0} \frac{1 - \cos x}{\sec x} = \lim_{x \to 0} \frac{\sin x}{\sec x \tan x} \quad \textit{This is NOT correct.}$$

$$= \lim_{x \to 0} \frac{\cos x}{\sec x}$$

$$= \frac{1}{1}$$

$$= 1$$

Example 4 l'Hôpital's rule applied more than once

Evaluate $\displaystyle\lim_{x \to 0} \frac{x - \sin x}{x^3}$.

Solution This is a 0/0 indeterminate form, and we find that

$$\lim_{x \to 0} \frac{x - \sin x}{x^3} = \lim_{x \to 0} \frac{1 - \cos x}{3x^2} \quad \textit{This is still the indeterminate form 0/0,}$$
$$\textit{so l'Hôpital's rule can be applied once again.}$$

$$= \lim_{x \to 0} \frac{-(-\sin x)}{6x}$$

$$= \frac{1}{6} \lim_{x \to 0} \frac{\sin x}{x}$$

$$= \frac{1}{6}(1)$$

$$= \frac{1}{6}$$

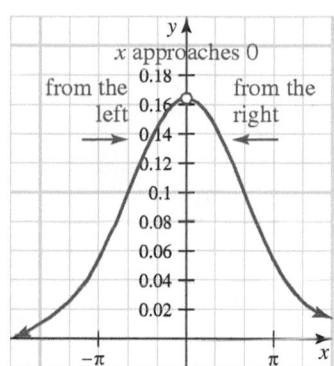

By looking at the graph of
$$\frac{x - \sin x}{x^3}$$
you can reinforce the result obtained using l'Hôpital's rule.

Example 5 l'Hôpital's rule with an ∞/∞ form

Evaluate $\displaystyle\lim_{x \to \infty} \frac{2x^2 + 3x + 1}{3x^2 + 5x - 2}$.

Solution Using the methods of Section 4.4, we could compute this limit by multiplying by $1 = (1/x^2)/(1/x^2)$. Instead, we note that this is of the form ∞/∞ and apply l'Hôpital's rule:

$$\lim_{x \to \infty} \frac{2x^2 + 3x + 1}{3x^2 + 5x - 2} = \lim_{x \to \infty} \frac{4x + 3}{6x + 5} \quad \textit{Apply l'Hôpital's rule again.}$$

$$= \lim_{x \to \infty} \frac{4}{6}$$

$$= \frac{2}{3}$$

It may happen that even when l'Hôpital's rule applies to a limit, it is not the best way to proceed, as illustrated by the following example.

Example 6 Using l'Hôpital's rule with other limit properties

Evaluate $\displaystyle\lim_{x \to 0} \frac{(1 - \cos x)\sin 4x}{x^3 \cos x}$.

Solution This limit has the form 0/0, but direct application of l'Hôpital's rule leads to a real mess (try it!). Instead, we compute the given limit by using the product rule for limits first, followed by two simple applications of l'Hôpital's rule. Specifically, using the product rule for limits (assuming the limits exist), we have

$$\lim_{x \to 0} \frac{(1 - \cos x) \sin 4x}{x^3 \cos x} = \left[\lim_{x \to 0} \frac{1 - \cos x}{x^2} \right] \left[\lim_{x \to 0} \frac{\sin 4x}{x} \right] \left[\lim_{x \to 0} \frac{1}{\cos x} \right]$$

$$= \left[\lim_{x \to 0} \frac{\sin x}{2x} \right] \left[\lim_{x \to 0} \frac{4 \cos 4x}{1} \right] \left[\lim_{x \to 0} \frac{1}{\cos x} \right]$$

$$= \left[\frac{1}{2} \right] [4] [1]$$

$$= 2$$

Example 7   Hypotheses of l'Hôpital's rule are not satisfied

Evaluate $\lim\limits_{x \to \infty} \dfrac{x + \sin x}{x - \cos x}$.

Solution   This limit has the indeterminate form ∞/∞. If you try to apply l'Hôpital's rule, you find

$$\lim_{x \to \infty} \frac{x + \sin x}{x - \cos x} = \lim_{x \to \infty} \frac{1 + \cos x}{1 + \sin x}$$

The limit on the right does not exist, because both $\sin x$ and $\cos x$ oscillate between -1 and 1 as $x \to \infty$. Recall that l'Hôpital's rule applies only if $\lim\limits_{x \to c} \dfrac{f'(x)}{g'(x)} = L$ or is $\pm\infty$. This does not mean that the limit of the original expression does not exist or that we cannot find it; it simply means that we cannot apply l'Hôpital's rule. To find this limit, factor out an x from the numerator and denominator and proceed as follows:

$$\lim_{x \to \infty} \frac{x + \sin x}{x - \cos x} = \lim_{x \to \infty} \frac{x \left(1 + \frac{\sin x}{x} \right)}{x \left(1 - \frac{\cos x}{x} \right)}$$

$$= \lim_{x \to \infty} \frac{1 + \frac{\sin x}{x}}{1 - \frac{\cos x}{x}}$$

$$= \frac{1 + 0}{1 - 0}$$

$$= 1$$

Example 8   Horizontal asymptotes with l'Hôpital's rule

Find all horizontal asymptotes of the graph of $y = xe^{-2x}$, and then graph the function.

Solution   To test for horizontal asymptotes, we compute

$$\lim_{x \to -\infty} xe^{-2x} \qquad \text{and} \qquad \lim_{x \to \infty} xe^{-2x}$$

For the first limit, we find

$$\lim_{x \to -\infty} xe^{-2x} = -\infty$$

and for the second,

$$\lim_{x \to \infty} xe^{-2x} = \lim_{x \to \infty} \frac{x}{e^{2x}} \qquad \text{Form } \infty/\infty$$

$$= \lim_{x \to \infty} \frac{1}{2e^{2x}} \qquad \textit{l'Hôpital's rule}$$

$$= 0$$

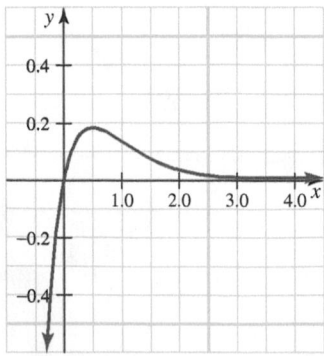

Figure 4.44   Graph of $y = xe^{-2x}$ showing the x-axis as a horizontal asymptote

Thus, $y = 0$ (the x-axis) is a horizontal asymptote. The graph of $y = xe^{-2x}$ is shown in Figure 4.44.

Other Indeterminate Forms

Other indeterminate forms such as 1^∞, 0^0, ∞^0, $\infty - \infty$, and $0 \cdot \infty$, can often be manipulated algebraically into one of the standard forms $0/0$ or ∞/∞, and then evaluated using l'Hôpital's rule. The following examples illustrate such procedures.

We begin by deriving a formula given without proof in Section 2.4.

Example 9   Limit of the form 1^∞

Show that $\lim\limits_{x \to \infty} \left(1 + \dfrac{1}{x} \right)^x = e$

Solution   Note that this limit is indeed of the indeterminate form 1^∞. Let $L = \lim\limits_{x \to \infty} \left(1 + \dfrac{1}{x} \right)^x$.

$$L = \lim_{x \to \infty} \left(1 + \frac{1}{x} \right)^x \qquad \textit{Given}$$

$$\ln L = \ln \left[\lim_{x \to \infty} \left(1 + \frac{1}{x} \right)^x \right] \qquad \textit{Take the logarithm of both sides.}$$

$$= \lim_{x \to \infty} \left[\ln \left(1 + \frac{1}{x} \right)^x \right] \qquad \textit{Limit of a log property}$$

$$= \lim_{x \to \infty} \left[x \ln \left(1 + \frac{1}{x} \right) \right] \qquad \textit{Property of logarithms}$$

$$= \lim_{x \to \infty} \frac{\ln \left(1 + \frac{1}{x} \right)}{\frac{1}{x}} \qquad \textit{This is now written in the form } 0/0.$$

$$= \lim_{x \to \infty} \frac{\frac{1}{1+\frac{1}{x}} \left(-\frac{1}{x^2} \right)}{-\frac{1}{x^2}} \qquad \textit{l'Hôpital's rule}$$

$$= \lim_{x \to \infty} \frac{1}{1 + \frac{1}{x}} \qquad \textit{Simplify}$$

$$= \frac{1}{1 + 0} \qquad \textit{Quotient rule}$$

$$= 1$$

$$L = e^1 \qquad \textit{Definition of logarithm}$$

Thus, $\lim\limits_{x \to \infty} \left(1 + \frac{1}{x} \right)^x = e$.

L'Hôpital's rule itself applies only to the indeterminate forms $0/0$ and ∞/∞.

☠ *Don't forget the last step, going from ln L to L.* ☠

Example 10   l'Hôpital's rule with the form $0 \cdot \infty$

Evaluate $L = \lim\limits_{x \to \left(\frac{\pi}{2} \right)^-} \left(x - \frac{\pi}{2} \right) \tan x$.

Solution   This limit has the form $0 \cdot \infty$, because

$$\lim_{x \to \left(\frac{\pi}{2} \right)^-} \left(x - \frac{\pi}{2} \right) = 0 \qquad \text{and} \qquad \lim_{x \to \left(\frac{\pi}{2} \right)^-} \tan x = \infty$$

Write $\tan x = \dfrac{1}{\cot x}$ to obtain

$$L = \lim_{x \to (\frac{\pi}{2})^-} \left(x - \frac{\pi}{2}\right)\tan x$$

$$= \lim_{x \to (\frac{\pi}{2})^-} \frac{x - \frac{\pi}{2}}{\cot x} \qquad \textit{Form } 0/0.$$

$$= \lim_{x \to (\frac{\pi}{2})^-} \frac{1}{-\csc^2 x} \qquad \textit{l'Hôpital's rule}$$

$$= \lim_{x \to (\frac{\pi}{2})^-} (-\sin^2 x)$$

$$= -1$$

Example 11 Limit of the form 0^0

Find $L = \lim\limits_{x \to 0^+} x^{\sin x}$.

Solution This is a 0^0 indeterminate form. From the graph shown in Figure 4.45, it looks as though the desired limit is 1.

We can verify this conjecture analytically. As in Example 9, we begin by using properties of logarithms.

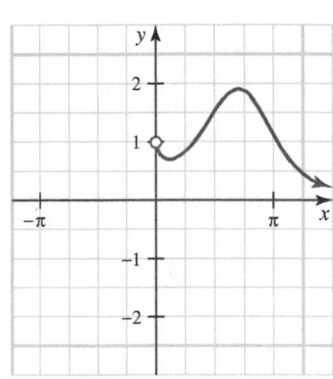

Figure 4.45 Graph of $y = x^{\sin x}$ suggests the limit as $x \to 0^+$

$$L = \lim_{x \to 0^+} x^{\sin x}$$

$$\ln L = \ln \lim_{x \to 0^+} x^{\sin x} \qquad \textit{Limit of a logarithm is the logarithm of a limit.}$$

$$= \lim_{x \to 0^+} (\sin x)\ln x \qquad \textit{Property of logarithms}$$

$$= \lim_{x \to 0^+} \frac{\ln x}{\csc x} \qquad \textit{This is } \infty/\infty \textit{ form.}$$

$$= \lim_{x \to 0^+} \frac{\frac{1}{x}}{-\csc x \cot x} \qquad \textit{l'Hôpital's rule}$$

$$= \lim_{x \to 0^+} \frac{-\sin^2 x}{x \cos x}$$

$$= \lim_{x \to 0^+} \left(\frac{\sin x}{x}\right)\left(\frac{-\sin x}{\cos x}\right)$$

$$= (1)(0)$$

$$= 0$$

$$L = e^0 = 1 \qquad \textit{Definition of logarithm}$$

Example 12 Limit of the form ∞^0

Find $L = \lim\limits_{x \to \infty} x^{1/x}$.

Solution This is a limit of the indeterminate form ∞^0.

$$L = \lim_{x \to \infty} x^{1/x}$$

$$\ln L = \ln \lim_{x \to \infty} x^{1/x}$$

$$= \lim_{x \to \infty} \ln x^{1/x} \qquad \textit{The limit of the log is the log of the limit.}$$

$$= \lim_{x \to \infty} \frac{1}{x} \ln x \qquad \textit{Property of logarithms}$$

$$= \lim_{x \to \infty} \frac{\ln x}{x} \qquad \textit{Form } \infty/\infty$$

$$= \lim_{x \to \infty} \frac{\frac{1}{x}}{1} \qquad \textit{l'Hôpital's rule}$$

$$= \frac{0}{1} \qquad \textit{Quotient rule}$$

$$= 0$$

$$L = e^0 = 1 \qquad \textit{Property of logarithms}$$

Example 13   l'Hôpital's rule with the form $\infty - \infty$

Evaluate $\displaystyle\lim_{x \to 0^+} \left(\frac{1}{x} - \frac{1}{\sin x} \right)$.

Solution   As it stands, this has the form $\infty - \infty$, because

$$\frac{1}{x} \to \infty \qquad \text{and} \qquad \frac{1}{\sin x} \to \infty$$

as $x \to 0$ from the right. However, using a little algebra, we find

$$\lim_{x \to 0^+} \left(\frac{1}{x} - \frac{1}{\sin x} \right) = \lim_{x \to 0^+} \frac{\sin x - x}{x \sin x}$$

This limit is now of the form $0/0$, so the hypotheses of l'Hôpital's rule are satisfied. Thus,

$$\lim_{x \to 0^+} \frac{\sin x - x}{x \sin x} = \lim_{x \to 0^+} \frac{\cos x - 1}{\sin x + x \cos x} \qquad \textit{Form } 0/0$$

$$= \lim_{x \to 0^+} \frac{-\sin x}{\cos x + x \, (-\sin x) + \cos x} \qquad \textit{l'Hôpital's rule}$$

$$= \frac{-(0)}{1 + (0)(0) + 1}$$

$$= 0$$

☠ Not all limits that appear indeterminate actually are indeterminate. The following table shows functions which appear to be indeterminate, but are, in fact, not indeterminate. Their forms and limits are also shown in the table.

Function	Form	Limit
$\displaystyle\lim_{x \to 0^+} (\sin x)^{1/x}$	0^∞	0
$\displaystyle\lim_{x \to 0^+} (\csc x - \ln x)$	$\infty - (-\infty)$	∞
$\displaystyle\lim_{x \to 0^+} \frac{\tan x}{\ln x}$	$\dfrac{0}{\infty}$	0

Other such "false indeterminate forms" include $\infty + \infty$, $\infty/0$, and $\infty \cdot \infty$, which are all actually infinite.

Special Limits Involving e^x and $\ln x$

We close this section with a theorem that summarizes the behavior of certain important special functions involving e^x and $\ln x$ near 0 and at $\pm\infty$.

Theorem 4.10   Limits involving natural logarithms and exponentials

If k and n are positive numbers, then

$$\lim_{x \to 0^+} \frac{\ln x}{x^n} = -\infty \qquad \lim_{x \to \infty} \frac{\ln x}{x^n} = 0$$

$$\lim_{x \to \infty} \frac{e^{kx}}{x^n} = \infty \qquad \lim_{x \to \infty} x^n e^{-kx} = 0$$

■ **W**hat this says The limit statements

$$\lim_{x \to \infty} \frac{e^{kx}}{x^n} = \infty \qquad \text{and} \qquad \lim_{x \to \infty} \frac{\ln x}{x^n} = 0$$

are especially important. They tell us that, in the long run, any exponential e^{kx} dominates any power x^n, for k and n positive, which, in turn, dominates any logarithm.

Proof: These can all be verified directly or by applying l'Hôpital's rule. For example,

$$\lim_{x \to \infty} \frac{\ln x}{x^n} = \lim_{x \to \infty} \frac{\frac{1}{x}}{nx^{n-1}} = \lim_{x \to \infty} \frac{1}{nx^n} = 0$$

The other parts are left for you to verify.   ◆

PROBLEM SET 4.5

Level 1

1. An incorrect use of l'Hôpital's rule is illustrated in the following limit computations. In each case, explain what is wrong and find the correct value of the limit.

   a. $\lim\limits_{x \to \pi} \dfrac{1 - \cos x}{x} = \lim\limits_{x \to \pi} \dfrac{\sin x}{1} = 0$

   b. $\lim\limits_{x \to \frac{\pi}{2}} \dfrac{\sin x}{x} = \lim\limits_{x \to \frac{\pi}{2}} \dfrac{\cos x}{1} = 0$

2. **EXPLORATION PROBLEM** Sometimes l'Hôpital's rule leads to faulty computations. For example, observe what happens when the rule is applied to

   $$\lim_{x \to \infty} \frac{x}{\sqrt{x^2 - 1}}$$

   Use any method you wish to evaluate this limit.

Find each of the limits in Problems 3-30.

3. $\lim\limits_{x \to 1} \dfrac{x^3 - 1}{x^2 - 1}$

4. $\lim\limits_{x \to 2} \dfrac{x^3 - 27}{x^2 - 9}$

5. $\lim\limits_{x \to 1} \dfrac{x^{10} - 1}{x - 1}$

6. $\lim\limits_{x \to -1} \dfrac{x^{10} - 1}{x + 1}$

7. $\lim\limits_{x \to 0^+} \dfrac{1 - \cos^2 x}{\sin^3 x}$

8. $\lim\limits_{x \to 0} \dfrac{1 - \cos^2 x}{3 \sin x}$

9. $\lim\limits_{x \to \pi} \dfrac{\cos \frac{x}{2}}{\pi - x}$

10. $\lim\limits_{x \to 0} \dfrac{\sin x}{\cos 2x}$

11. $\lim\limits_{x \to 0} \dfrac{1 - \cos x}{x^2}$

12. $\lim\limits_{x \to 0} \dfrac{x - \sin x}{x \tan x}$

13. $\lim\limits_{x \to 0} \dfrac{\tan 3x}{\sin 5x}$

14. $\lim\limits_{x \to \frac{\pi}{2}} \dfrac{3 \sec x}{2 + \tan x}$

15. $\lim\limits_{x \to 0} \dfrac{x + \sin^3 x}{x^2 + 2x}$

16. $\lim\limits_{x \to 0} \dfrac{x^2 + \sin x^2}{x^2 + x^3}$

17. $\lim\limits_{x \to 0} \dfrac{\sin 3x \sin 2x}{x \sin 4x}$

18. $\lim\limits_{x \to \pi/2} \dfrac{\sin 2x \cos x}{x \sin 4x}$

19. $\lim\limits_{x \to 0^+} x^2 \sin \dfrac{1}{x}$

20. $\lim\limits_{x \to \infty} x^{3/2} \sin \dfrac{1}{x}$

21. $\lim\limits_{x \to 0} \left(\cot x - \dfrac{1}{x} \right)$

22. $\lim\limits_{\theta \to 0} \dfrac{\theta - 1 + \cos^2 \theta}{\theta^2 + 5\theta}$

23. $\lim\limits_{x \to \infty} x^{-5} \ln x$

24. $\lim\limits_{x \to 0^+} x^{-5} \ln x$

25. $\lim\limits_{x \to 0^+} (\sin x) \ln x$

26. $\lim\limits_{x \to \left(\frac{\pi}{2}\right)^-} \sec 3x \cos 9x$

27. $\lim\limits_{x \to -\infty} \left(1 - \dfrac{3}{x} \right)^{2x}$

28. $\lim\limits_{x \to \infty} \left(1 + \dfrac{1}{2x} \right)^{3x}$

29. $\lim\limits_{x \to \infty} \dfrac{\ln(\ln x)}{x}$

30. $\lim\limits_{x \to \infty} (\ln x)^{1/x}$

Level 2

Find each of the limits in Problems 31-43.

31. $\lim\limits_{x\to(\frac{\pi}{2})^-}\left(\dfrac{1}{\pi-2x}+\tan x\right)$

32. $\lim\limits_{x\to\infty}\left(\sqrt{x^2-x}-x\right)$

33. $\lim\limits_{x\to\infty}\left[x-\ln(x^3-1)\right]$
Hint: $\ln e^x=x$.

34. $\lim\limits_{x\to0^+}\left(\dfrac{1}{x^2}-\ln\sqrt{x}\right)$

35. $\lim\limits_{x\to0}(e^x-1-x)^x$

36. $\lim\limits_{x\to0^+}(\ln x)(\cot x)$

37. $\lim\limits_{x\to0^+}(e^x-1)^{1/\ln x}$

38. $\lim\limits_{x\to\infty}\dfrac{x+\sin 3x}{x}$

39. $\lim\limits_{x\to\infty}\dfrac{x(\pi+\sin x)}{x^2+1}$

40. $\lim\limits_{x\to0^+}\left(\dfrac{2\cos x}{\sin 2x}-\dfrac{1}{x}\right)$

41. $\lim\limits_{x\to0}\dfrac{(2-x)(e^x-x-2)}{x^3}$

42. $\lim\limits_{x\to0}\dfrac{\tan^{-1}3x-3\tan^{-1}x}{x^3}$

43. $\lim\limits_{x\to\infty}x^5\left[\sin\dfrac{1}{x}-\dfrac{1}{x}+\dfrac{1}{6x^3}\right]$

In Problems 44-47, use l'Hôpital's rule to determine all horizontal asymptotes to the graph of the given function. You are NOT required to sketch the graph.

44. $f(x)=x^3e^{-0.01x}$

45. $f(x)=\dfrac{\ln x^5}{x^{0.02}}$

46. $f(x)=\left(\ln\sqrt{x}\right)^{2/x}$

47. $f(x)=\left(\dfrac{x+3}{x+2}\right)^{2x}$

48. A weight hanging by a spring is made to vibrate by applying a sinusoidal force, and the displacement at time t is given by

$$f(t)=\dfrac{C}{\beta^2-\alpha^2}(\sin\alpha t-\sin\beta t)$$

where C, α, and β are constants such that $\alpha\ne\beta$. What happens to the displacement as $\beta\to\alpha$? You may assume that α is fixed.

49. 𝔥istorical Quest The French mathematician Guillaume de l'Hôpital is best known today for the rule that bears his name, but the rule was discovered by l'Hôpital's teacher, Johann Bernoulli. Not only did l'Hôpital neglect to cite his sources in his book, but there is also evidence that he paid Bernoulli for his results and for keeping their arrangements for payment confidential. In a letter dated March 17, 1694, he asked Bernoulli "to communicate to me your discoveries ..."— with the request not to mention them to others—... it would not please me if they were made public." L'Hôpital's argument, which was originally given without using functional notation, can easily be reproduced:*

$$\dfrac{f(a+dx)}{g(a+dx)}=\dfrac{f(a)+f'(a)\,dx}{g(a)+g'(a)\,dx}$$
$$=\dfrac{f'(a)\,dx}{g'(a)\,dx}$$
$$=\dfrac{f'(a)}{g'(a)}$$

Supply reasons for this argument, and give necessary conditions for the functions f and g.

50. 𝔥istorical Quest The remarkable Bernoulli family of Switzerland produced at least eight noted mathematicians over three generations. Two brothers, Jacob (1654-1705) and Johann (1667-1748), were bitter rivals. These brothers were extremely influential advocates of the newly born calculus. Johann was the most prolific of the clan and was responsible for the discovery of l'Hôpital's rule (see Problem 49), Bernoulli numbers, Bernoulli polynomials, the lemniscate of Bernoulli, the Bernoulli equation, the Bernoulli theorem, and the Bernoulli distribution. He did a great deal of work with differential equations. Johann was jealous and cantankerous; he tossed a son (Daniel) out of the house for winning an award he had expected to win himself. Write a report on the Bernoulli family.

Find the limits in Problems 51-53 using the following methods.
a. graphically   **b.** analytically   **c.** numerically
Compare, contrast, and reconcile the three methods.

51. $\lim\limits_{x\to0^+}x^x$

52. $\lim\limits_{x\to0^+}(x^x)^x$

53. $\lim\limits_{x\to\infty}\left[x\sin^{-1}\left(\dfrac{1}{x}\right)\right]^{x^2}$

54. $\lim\limits_{x\to0^+}(e^x+x)^{1/x}$

Level 3

55. Find constants a and b so that

$$\lim\limits_{x\to0}\left(\dfrac{\sin 2x}{x^3}+\dfrac{a}{x^2}+b\right)=1$$

56. Find A so that $\lim\limits_{x\to\infty}\left(\dfrac{x+A}{x-2A}\right)^x=5$.

57. Find all values of B and C so that

$$\lim\limits_{x\to0}\dfrac{\sin Bx+Cx}{x^3}=36$$

*D. J. Struik, *A Source Book in Mathematics*, 1200-1800. Cambridge, MA: Harvard University Press, 1969, pp. 313-316.

58. For a certain value of D, the limit

$$\lim_{x \to \infty} \left(x^4 + 5x^3 + 3\right)^D - x$$

is finite and nonzero. Find D and then compute the limit.

59. For which values of constants E and F is it true that

$$\lim_{x \to 0} \left(x^{-3} \sin 7x + Ex^{-2} + F\right) = -2$$

60. For G and H positive constants, define

$$f(x) = \left(e^x + Gx\right)^{H/x}$$

a. Compute $L_1 = \lim_{x \to 0} f(x)$ and $L_2 = \lim_{x \to \infty} f(x)$.

b. What is the largest value of G for which the equation $L_1 = HL_2$ has a solution? What are L_1 and L_2 in this case?

4.6 OPTIMIZATION IN THE PHYSICAL SCIENCES AND ENGINEERING

IN THIS SECTION: *Optimization procedure, Fermat's principle of optics and Snell's Law*
In Chapter 3 we introduced the processes involved with model-building. Now that we have developed the necessary calculus skills, we shall use mathematical modeling and calculus to solve optimization problems.

Optimization Procedure

When light travels from one medium to another—say, from air into water—which path does it follow? In a mechanical system involving pulleys, which arrangement results in minimal potential energy? At what temperature does a given mass of water occupy the least volume (the answer is not 0°C)? These are typical *optimization problems* from the physical sciences and engineering, and will be examined in the examples and exercises of this section. The process of finding the maximum or minimum of a function is called **optimization**. Then, in the next section, we will explore optimization involving applications to business, economics, and the social and life sciences. We begin by outlining a general optimization procedure based on ideas originally suggested by George Pólya (1887-1985), one of the great problem-solvers of 20th century mathematics (see the *Historical Quest* in Problem 55 of Section 7.4).

OPTIMIZATION PROCEDURE To find the maximum or minimum value for an applied problem, follow these steps:

Step 1 Draw a figure (if appropriate) and label all quantities relevant to the problem.
Step 2 Focus on the quantity to be optimized. Name it. Find a formula for the quantity to be maximized or minimized.
Step 3 Use conditions in the problem to eliminate variables in order to express the quantity to be maximized or minimized in terms of a single variable.
Step 4 Find the practical domain for the variables in Step 3; that is, the interval of possible values determined from the physical restrictions in the problem.
Step 5 If possible, use the methods of calculus to obtain the required optimum value.

Example 1   Modeling Problem Maximizing a constrained area

You need to build a rectangular fence to enclose a play zone for children. What is the maximum area for this play zone if it is to fit into a right-triangular plot with sides measuring 4 m and 12 m?

Solution   A picture of the play zone is shown in Figure 4.46.

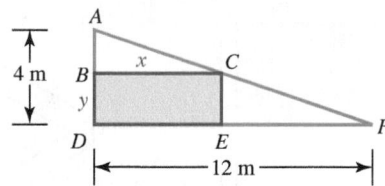

Figure 4.46   Children's play zone

Let x and y denote the length and width of the inscribed rectangle. The appropriate formula for the area is

$$A = (\text{LENGTH})(\text{WIDTH}) = xy$$

We wish to maximize A where $A = xy$, but first we must express A as a function of a single variable. To do this, note that, because $\triangle ABC$ is similar to $\triangle ADF$, the corresponding sides of these triangles are proportional,

$$\frac{4 - y}{4} = \frac{x}{12}$$

$$y = 4 - \frac{1}{3}x$$

We can now write A as a function of x alone:

$$A(x) = x\left(4 - \frac{1}{3}x\right) = 4x - \frac{1}{3}x^2$$

The domain for A is $0 \le x \le 12$. The critical numbers for A are values such that $A'(x) = 0$ (since $A'(x)$ exists for all x). Since

$$A'(x) = 4 - \frac{2}{3}x$$

the only critical number is $x = 6$. Evaluate $A(x)$ at the endpoints and the critical number:

$$A(6) = 4(6) - \frac{1}{3}(6)^2 = 12$$

$A(0) = 0; A(12) = 0$. The maximum area occurs when $x = 6$. This means that

$$y = 4 - \frac{1}{3}(6) = 2$$

Thus, the largest rectangular play zone that can be built in the triangular plot is a rectangle 6 m long and 2 m wide for an area of 12 m^2. ■

Example 2   Modeling Problem: Maximizing a volume

A carpenter wants to make an open-topped box out of a rectangular sheet of tin 24 in. wide and 45 in. long. The carpenter plans to cut congruent squares out of each corner of the sheet and then bend and solder the edges of the sheet upward to form the sides of the box, as shown in Figure 4.47.

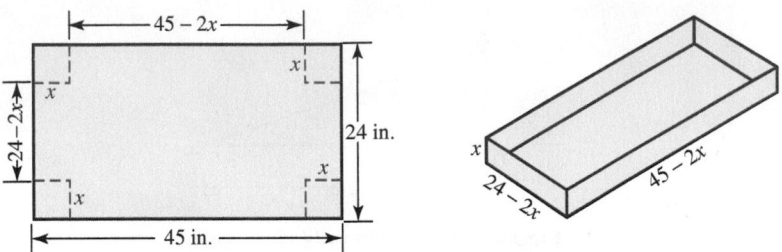

Figure 4.47   A box cut from a 24 in. by 45 in. piece of tin

For what dimensions does the box have the greatest possible volume?

Solution   If each square corner cut out has side x, the box will be x inches deep, $45 - 2x$ inches long, and $24 - 2x$ inches wide. The volume of the box shown in Figure 4.47 is

$$V(x) = x(45 - 2x)(24 - 2x)$$

$$= 4x^3 - 138x^2 + 1,080x$$

and this is the quantity to be maximized. To find the domain, we note that the dimensions must all be nonnegative; therefore, $x \geq 0$, $45 - 2x \geq 0$ (or $x \leq 22.5$), and $24 - 2x \geq 0$ (or $x \leq 12$). This implies that the domain is $[0, 12]$.

To find the critical numbers (the derivative is defined everywhere in the domain), we find values for which the derivative is 0:

$$V'(x) = 12x^2 - 276x + 1,080$$

$$= 12(x - 18)(x - 5)$$

The critical numbers are $x = 5$ and $x = 18$, but $x = 18$ is not in the domain, so the only relevant critical number is $x = 5$. Evaluating $V(x)$ at the critical number $x = 5$ and the endpoints $x = 0, x = 12$, we find

$$V(5) = 5(45 - 10)(24 - 10) = 2,450$$

$V(0) = 0$; $V(12) = 0$. Thus, the box with the largest volume is found when $x = 5$. Such a box has dimensions 5 in. $\times$ 14 in. $\times$ 35 in.   ∎

You will note in Example 2, we did not have to test whether the critical point was a maximum or a minimum, but knew that it was a maximum because the continuous function V is nonnegative on the interval $[0, 12]$ and is indeed zero at the endpoints. Since there is only one critical point, $x = 5$, that must yield the maximum. Similar reasoning can be used in many examples, allowing us to forgo having to test for a maximum or minimum with the first- or second- derivative test. We summarize this procedure with the following definition.

> **EVT CONVENTION**   If the hypotheses of the extreme value theorem apply to a particular problem, we can often avoid testing a candidate to see whether it is a maximum or a minimum. Suppose we have three candidates and two are obviously minima — the third *must be a maximum*. We will refer to this principle of not testing a candidate as the **EVT convention** because it is shorthand for using the extreme value theorem.

Example 3   Modeling Problem: Maximizing illumination

A lamp with adjustable height hangs directly above the center of a circular kitchen table that is 8 ft in diameter. Model the illumination I at the edge of the table to be directly proportional to the cosine of the angle θ and inversely proportional to the square of the distance d, where θ and d are as shown in Figure 4.48. How close to the table should the lamp be lowered to maximize the illumination at the edge of the table.

Solution   We note that

$$I(\theta) = \frac{k \cos \theta}{d^2}$$

where k is a (positive) constant of proportionality. Moreover, from the right triangle shown in Figure 4.48,

$$\sin \theta = \frac{4}{d} \quad \text{or} \quad d = \frac{4}{\sin \theta}$$

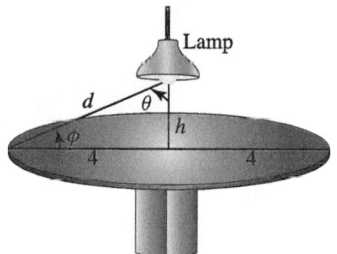

Figure 4.48   Illumination on a kitchen table

Hence,

$$I(\theta) = k \cos \theta \left[\frac{1}{\left(\frac{4}{\sin\theta}\right)^2} \right]$$

$$= \frac{k}{16} \cos \theta \sin^2 \theta$$

Only values of θ between 0 and $\frac{\pi}{2}$ are meaningful in the context of this problem. Hence, the goal is to find the absolute maximum of the function $I(\theta)$ on $\left[0, \frac{\pi}{2}\right]$.

Using the product rule and the chain rule to differentiate $I(\theta)$, we obtain

$$I'(\theta) = \frac{k}{16} \left[\cos \theta (2 \sin \theta \cos \theta) + \sin^2 \theta (-\sin \theta) \right]$$

$$= \frac{k}{16} (2 \cos^2 \theta \sin \theta - \sin^3 \theta)$$

$$= \frac{k}{16} \sin \theta (2 \cos^2 \theta - \sin^2 \theta)$$

This is zero when $\theta = 0$ (since $\sin 0 = 0$) and when

$$2 \cos^2 \theta - \sin^2 \theta = 0$$

$$\frac{\sin^2 \theta}{\cos^2 \theta} = 2$$

$$\tan^2 \theta = 2$$

$$\theta = \tan^{-1} \sqrt{2}$$

Evaluating $I(\theta)$ at the endpoints, $\theta = 0$ and $\theta = \frac{\pi}{2}$ and at the interior critical number $\tan^{-1}\sqrt{2} \approx 0.9553$, we find that

$$I(0) = \frac{k}{16}(\cos 0)(\sin^2 0) = 0$$

$$I\left(\frac{\pi}{2}\right) = \frac{k}{16}\left(\cos\frac{\pi}{2}\right)\left(\sin^2\frac{\pi}{2}\right) = 0$$

$$I(\tan^{-1}\sqrt{2}) \approx \frac{k}{16}(\cos 0.9553)\left(\sin^2 0.9553\right) \approx 0.0241k$$

Finally, to find the height h that maximizes the illumination I, observe from Figure 4.48 that

$$\tan\theta = \frac{4}{h}$$

Since $\tan\theta = \sqrt{2}$ when I is maximized, it follows that

$$h = \frac{4}{\tan\theta} = \frac{4}{\sqrt{2}} \approx 2.83$$

To maximize the illumination, the lamp should be placed about 2.83 ft (2 ft 10 in.) above the table. ■

Example 4   Modeling Problem: Minimizing time of travel

A dune buggy is on the desert at a point A located 40 km from a point B, which lies on a long, straight road, as shown in Figure 4.49. The driver can travel at 45 km/h on the desert and 75 km/h on the road. The driver will win a prize if she arrives at the finish line at point D, 50 km from B, in 84 min or less. What route should she travel to minimize the time of travel? Does she win the prize?

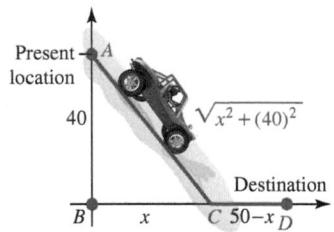

Figure 4.49 A path traveled by a dune buggy

Solution   Suppose the driver heads for a point C located x km down the road from B toward her destination, as shown in Figure 4.49. We want to minimize the time. We will need to remember the formula $d = rt$, or in terms of time, $t = \frac{d}{r}$.

$$\text{TIME} = \text{TIME FROM } A \text{ TO } C + \text{TIME FROM } C \text{ TO } D$$

$$= \frac{\text{DISTANCE FROM } A \text{ TO } C}{\text{RATE FROM } A \text{ TO } C} + \frac{\text{DISTANCE FROM } C \text{ TO } D}{\text{RATE FROM } C \text{ TO } D}$$

$$T(x) = \frac{\sqrt{x^2 + 1{,}600}}{45} + \frac{50 - x}{75}$$

The domain of T is $[0, 50]$. Next, find the derivative of time T with respect to x:

$$T'(x) = \frac{1}{45}\left[\frac{1}{2}(x^2 + 1{,}600)^{-1/2}(2x)\right] + \frac{1}{75}(-1)$$

$$= \frac{x}{45\sqrt{x^2 + 1{,}600}} - \frac{1}{75}$$

$$= \frac{5x - 3\sqrt{x^2 + 1{,}600}}{225\sqrt{x^2 + 1{,}600}}$$

The derivative exists for all x and is zero when

$$5x - 3\sqrt{x^2 + 1{,}600} = 0$$

Solving this equation, we find $x = 30$ (-30 is extraneous). Evaluating $T(x)$ here and at the endpoints, we find that

$$T(30) = \frac{\sqrt{30^2 + 1{,}600}}{45} + \frac{50 - 30}{75} \approx 1.3778 \text{ hr} \approx 83 \text{ min}$$

$$T(0) = \frac{\sqrt{0^2 + 1{,}600}}{45} + \frac{50 - 0}{75} \approx 1.5556 \text{ hr} \approx 93 \text{ min}$$

$$T(50) = \frac{\sqrt{50^2 + 1{,}600}}{45} + \frac{50 - 50}{75} \approx 1.42292 \text{ hr} \approx 85 \text{ min}$$

The driver can minimize the total driving time by heading for a point that is 30 km from the point B and then traveling on the road to point D. She wins the prize because this minimal route requires only 83 minutes.

Example 5 Modeling Problem: Optimizing a constrained area

A wire of length L is to be cut into two pieces, one of which will be bent to form a circle and the other to form a square. Determine how the wire should be cut to:

a. maximize the sum of the areas enclosed by the two pieces.
b. minimize the sum of the areas enclosed by the two pieces.

Solution   To understand the problem we draw a sketch as shown in Figure 4.50, and label the radius of the circle r and the side of the square s.

We find that the combined area is

$$\text{AREA} = \text{AREA OF CIRCLE} + \text{AREA OF SQUARE} = \pi r^2 + s^2$$

We need to write the radius r and the side s in terms of the length of wire, L.

$$L = \text{CIRCUMFERENCE OF CIRCLE} + \text{PERIMETER OF SQUARE}$$

$$= 2\pi r + 4s$$

Thus, $s = \frac{1}{4}(L - 2\pi r)$. (We could just as easily have solved for r.) By substitution,

$$A(r) = \pi r^2 + \left[\frac{1}{4}(L - 2\pi r)\right]^2$$

$$= \pi r^2 + \frac{1}{16}(L - 2\pi r)^2$$

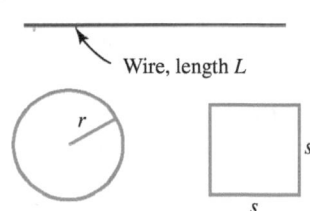

Wire, length L

Figure 4.50

To find the domain, we note that $r \geq 0$ and that $L - 2\pi r \geq 0$, so the domain is $0 \leq r \leq \frac{L}{2\pi}$. Note when $r = 0$, there is no circle, and when $r = \frac{L}{2\pi}$ there is no square. The derivative of $A(r)$ is

$$A'(r) = 2\pi r + \frac{1}{8}(L - 2\pi r)(-2\pi)$$

$$= 2\pi r - \frac{\pi}{4}(L - 2\pi r)$$

$$= \frac{\pi}{4}(8r - L + 2\pi r)$$

Solve $A'(r) = 0$ to find $r = \frac{L}{2(\pi + 4)}$.

Thus, the extreme values of the area function on $\left[0, \frac{L}{2\pi}\right]$ must occur either at the endpoints or at the critical number $\frac{L}{2\pi + 8}$.

Evaluating $A(r)$ at each of these numbers, we find

$$A(0) = \pi(0)^2 + \frac{1}{16}\left[L - 2\pi(0)\right]^2 = \frac{L^2}{16}$$

$$A\left(\frac{L}{2\pi}\right) = \pi\left(\frac{L}{2\pi}\right)^2 + \frac{1}{16}\left[L - 2\pi\left(\frac{L}{2\pi}\right)\right]^2 = \frac{L^2}{4\pi}$$

$$A\left(\frac{L}{2\pi + 8}\right) = \pi\left(\frac{L}{2\pi + 8}\right)^2 + \frac{1}{16}\left[L - 2\pi\left(\frac{L}{2\pi + 8}\right)\right]^2 = \frac{L^2}{4(\pi + 4)}$$

Comparing these values, we see that the smallest area occurs at $r = \frac{L}{2\pi + 8}$ and the largest area occurs at $r = \frac{L}{2\pi}$ To summarize:

1. To maximize the sum of the areas, do not cut the wire at all. Bend the wire to form a circle of radius $r = \frac{L}{2\pi}$.

2. To minimize the sum of the areas, cut the wire at the point located

$$2\pi r = \frac{2\pi L}{2\pi + 8} = \frac{\pi L}{\pi + 4}$$

units from one end, and form the circular part with resulting radius $r = \frac{L}{2\pi + 8}$.

If you use a graphing calculator or a computer, you can verify this result (for $L = 1$) by looking at the graph of the area function, as shown in Figure 4.51. ◼

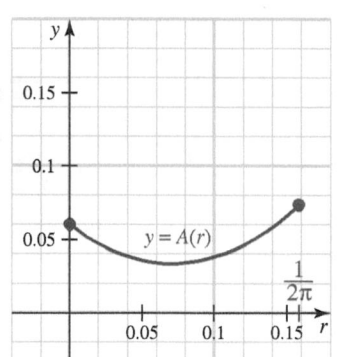

Figure 4.51 Graph of $y = A(r)$

Example 6   Modeling Problem: Optimizing an angle of observation

A painting is hung on a wall in such a way that its upper and lower edges are 10 ft and 7 ft above the floor, respectively. An observer whose eyes are 5 ft above the floor stands x feet from the wall, as shown in Figure 4.52.

How far away from the wall should the observer stand to maximize the angle subtended by the painting?

Solution   In Figure 4.52, θ is the angle whose vertex occurs at the observer's eyes at a point O located x feet from the wall. Note that α is the angle between the line $\overline{OA}$ drawn horizontally from the observer's eyes to the wall and the line $\overline{OB}$ from the eyes to the bottom edge of the painting. In $\triangle OAB$, the angle at O is α, with $\cot \alpha = \frac{x}{2}$. In

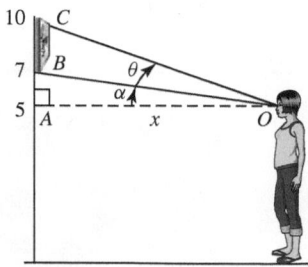

Figure 4.52 Viewing a painting

$\triangle OAC$, the angle at O is $(\alpha + \theta)$, and $\cot(\alpha + \theta) = \frac{x}{5}$. By using the definition of inverse cotangent we have

$$\theta = (\alpha + \theta) - \alpha = \cot^{-1}\left(\frac{x}{5}\right) - \cot^{-1}\left(\frac{x}{2}\right)$$

To maximize θ, we first compute the derivative

$$\frac{d\theta}{dx} = \frac{-1}{\left[1 + \left(\frac{x}{5}\right)^2\right]}\left(\frac{1}{5}\right) - \frac{-1}{\left[1 + \left(\frac{x}{2}\right)^2\right]}\left(\frac{1}{2}\right)$$

$$= \frac{-5}{25 + x^2} + \frac{2}{4 + x^2}$$

Solving the equation $\dfrac{d\theta}{dx} = 0$ yields

$$-5(4 + x^2) + 2(25 + x^2) = 0$$

$$-3x^2 + 30 = 0$$

$$x = \pm\sqrt{10}$$

Because distance must be nonnegative, we reject the negative value. We apply the first-derivative test to show that the positive critical number $\sqrt{10}$ corresponds to a relative maximum (verify). Thus, the angle θ is maximized when the observer stands $\sqrt{10}$ ft away from the wall. ▪

Fermat's Principle of Optics and Snell's Law

Light travels at different rates in different media; the more optically dense the medium, the slower the speed of transit. Consider the situation shown in Figure 4.53, in which a beam of light originates at a point A in one medium, then strikes the upper surface of a second, denser medium at a point P, and is refracted to a point B in the second medium.

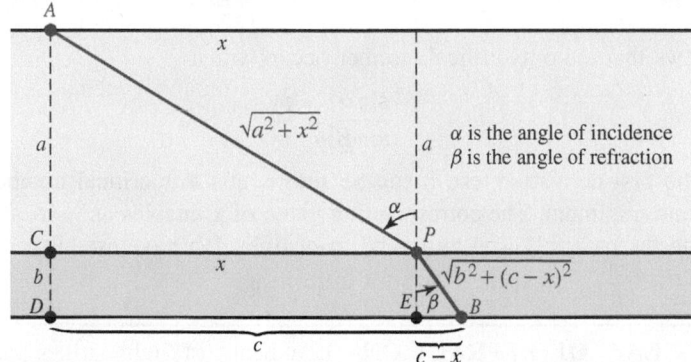

Figure 4.53   The path of a light beam through two media of different density

Suppose light travels with speed v_1 in the first medium and speed v_2 in the second. What can be said about the path followed by the beam of light?

Our method of investigating this question is based on the following optical property.

FERMAT'S PRINCIPLE OF OPTICS   Light travels between two points in such a way as to minimize the time of transit.

This problem is very similar to the dune buggy problem of Example 4. Light minimizes the time, and minimum time was the goal of that example. In Figure 4.53, let

$a = \left|\overline{AC}\right|$, $b = \left|\overline{CD}\right|$, and $c = \left|\overline{DB}\right|$, and let x denote the distance from C to P. Because A and B are fixed points, the path APB is determined by the location of P, which, in turn, is determined by x. Because v_1 is a given constant, the time required for the light to travel from A to P is given by

$$T_1 = \frac{\left|\overline{AP}\right|}{v_1} = \frac{\sqrt{a^2 + x^2}}{v_1}$$

and the time required for the light to go from P to B is

$$T_2 = \frac{\left|\overline{PB}\right|}{v_2} = \frac{\sqrt{b^2 + (c - x)^2}}{v_2}$$

for another given constant v_2. Therefore, the total time of transit is

$$T = T_1 + T_2 = \frac{\sqrt{a^2 + x^2}}{v_1} + \frac{\sqrt{b^2 + (c - x)^2}}{v_2}$$

where it is clear that $0 \le x \le c$. According to Fermat's principle, the path followed by the beam of light is the one that corresponds to the smallest possible value of T; that is, we want to minimize T as a function of x.

Toward this end, we begin by finding $\dfrac{dT}{dx}$.

$$\frac{dT}{dx} = \frac{x}{v_1\sqrt{a^2 + x^2}} - \frac{c - x}{v_2\sqrt{b^2 + (c - x)^2}}$$

Note from Figure 4.53 that if α is the angle of incidence of the beam of light and β is the angle of refraction, then (from the definition of sine)

$$\sin\alpha = \frac{x}{\sqrt{a^2 + x^2}} \qquad \text{and} \qquad \sin\beta = \frac{c - x}{\sqrt{b^2 + (c - x)^2}}$$

Therefore (by substitution), we see the derivative of T can be expressed as

$$\frac{dT}{dx} = \frac{\sin\alpha}{v_1} - \frac{\sin\beta}{v_2}$$

and it follows that the only critical number occurs when

$$\frac{\sin\alpha}{\sin\beta} = \frac{v_1}{v_2}$$

By using the first-derivative test, it can be shown that this critical number corresponds to an absolute minimum. The corresponding value of x enables us to locate P and hence to determine the path followed by the beam of light. We have established the following law of optics.

SNELL'S LAW OF REFRACTION   If a beam of light strikes the boundary between two media with angle of incidence α and is refracted through an angle β, then

$$\frac{\sin\alpha}{\sin\beta} = \frac{v_1}{v_2}$$

where v_1 and v_2 are the rates at which light travels through the first and second medium, respectively. The constant ratio

$$n = \frac{\sin\alpha}{\sin\beta}$$

is called the **relative index of refraction of the two media**.

PROBLEM SET 4.6

Level 1

For Problems 1-6, *decide whether it makes better sense to find a maximum or a minimum value for y. Find this value, and state the value of x for which this optimal value of y is obtained.*

1. $2x^2 + 2y + 16x + 8 = 0$
2. $3x^2 + 6x - 27y - 51 = 0$
3. $10x^2 + 100x + 20y + 90 = 0$
4. $10x^2 + 100x - 20y + 110 = 0$
5. $x^2 + 9y + 10x - 47 = 0$
6. $100x^2 - 120x - 25y + 41 = 0$
7. The sum of two numbers is 10. Find the largest possible product.
8. The difference of two numbers is 8. Find the smallest possible product.
9. The difference of two numbers is 6. Find the minimum value for the product of the numbers.
10. The sum of two numbers is 20. Find the smallest possible value for the sum of their squares.
11. ■ What does this say? Describe an optimization procedure.
12. ■ What does this say? What is Fermat's principle of optics?

Level 2

13. **EXPLORATION PROBLEM** Pull out a sheet of engineering or binder paper. Assume this paper is 22 cm by 28 cm. Cut squares from the corners and fold the sides up to form a container. Show that the maximum volume of such a container is about 1 liter.
14. **EXPLORATION PROBLEM** Draw a rectangle and find its perimeter. Now, experiment with that same perimeter to form other rectangles with different lengths and widths. Which of these rectangles has the largest area? Formulate a conjecture about this problem and then prove your conjecture.

The optimization Problems 15-20 *are presented in pairs. Each odd-numbered problem is followed by a generalization of that same problem.*

15. Find the rectangle of largest area that can be inscribed in a semicircle of radius 10 cm, assuming that one side of the rectangle lies on the diameter of the semicircle.

16. Find the rectangle of largest area that can be inscribed in a semicircle of radius R, assuming that one side of the rectangle lies on the diameter of the semicircle.
17. Find the dimensions of the right circular cylinder of largest volume that can be inscribed in a sphere of radius 20 in.
18. Find the dimensions of the right circular cylinder of largest volume that can be inscribed in a sphere of radius R.
19. Given a sphere of radius 40 ft, find the radius r and altitude $2h$ of the right circular cylinder with largest lateral surface area that can be inscribed in the sphere. *Hint:* The lateral surface area is $S = 4\pi rh$.
20. Given a sphere of radius R, find the radius r and altitude $2h$ of the right circular cylinder with largest lateral surface area that can be inscribed in the sphere. *Hint:* The lateral surface area is $S = 4\pi rh$.
21. Each edge of a square has length L, as shown in Figure 4.54.

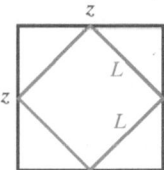

Figure 4.54 Square in a square

Determine the edge of the square of largest area that can be circumscribed about the given square.

22. Find the dimensions of the right circular cylinder of largest volume that can be inscribed in a right circular cone of radius R and altitude H. *Hint:* Begin with Figure 4.55.

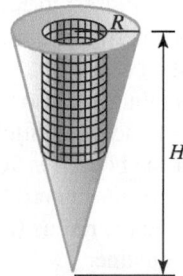

Figure 4.55 Cylinder in a cone

23. A farmer has 1,000 feet of fence and wishes to enclose a rectangular plot of land. The land has a barn on one edge and no fence is required on that side. What should the length of the side parallel to the barn be in order to include the largest possible area?

24. A farmer wishes to enclose a rectangular pasture with 360 ft of fence. Find the dimensions that give the maximum area in these situations:
    a. The fence is on all four sides of the pasture.
    b. The fence is on three sides of the pasture and the fourth side is bounded by a wall.

25. A fence must be built around a rectangular area of 1,600 ft². The fence along three sides is to be made of material that costs $120/ft. The material for the fourth side costs $40/ft. Find the dimensions (rounded to the nearest foot) of the rectangle that would be the least expensive to build.

26. **Journal Problem** (Parabola*) Farmer Jones has to build a fence to enclose a 1,200 m² rectangular area $ABCD$. Fencing costs $3 per meter, but Farmer Smith has agreed to pay half the cost of fencing $\overline{CD}$, which borders the property. Given x is the length of side $\overline{CD}$, what is the minimum amount (to the nearest cent) Jones has to pay?

27. A truck is 250 mi east of a sports car and is traveling west at a constant speed of 60 mi/h. Meanwhile, the sports car is going north at 80 mi/h. When will the truck and the car be closest to each other? What is the minimum distance between them? *Hint:* Minimize the square of the distance.

28. A cylindrical container with no top is to be constructed to hold a fixed volume V_0 of liquid. The cost of the material used for the bottom is 50¢/in.², and the cost of the material used for the curved lateral side is 30¢/in.² Use calculus to find the radius (in terms of V_0) of the least expensive container.

29. Show that of all rectangles with a given perimeter, the square has the largest area.

30. Show that of all rectangles with a given area, the square has the smallest perimeter.

31. The highest bridge in the world is the bridge over the Royal Gorge of the Arkansas River in Colorado.

It is 1,053 ft above the water. If a rock is projected vertically upward from this bridge with an initial velocity of 64 ft/s, the height h of the object above the river at time t is described by the function

$$h(t) = -16t^2 + 64t + 1,053$$

What is the maximum height possible for a rock projected vertically upward from the bridge with an initial velocity of 64 ft/s? After how many seconds does it reach that height?

32. In 1974, Evel Knievel attempted a skycycle ride across the Snake River.

Suppose the path of the skycycle is given by the equation

$$d(x) = -0.0005x^2 + 2.39x + 600$$

where $d(x)$ is the height in feet above the canyon floor for a horizontal distance of x feet from the launching ramp, as shown in Figure 4.56. What was Knievel's maximum height?

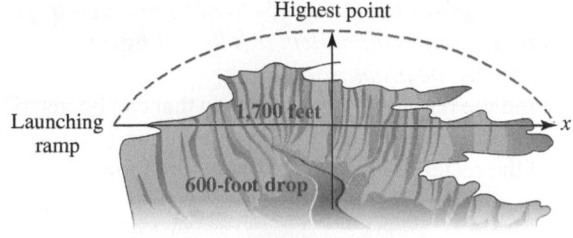

Figure 4.56 Evel Knievel's skycycle ride

33. An arch (see Figure 4.57) has equation

$$y - 18 = -\frac{2}{81}x^2$$

where both x and y are measured in feet.

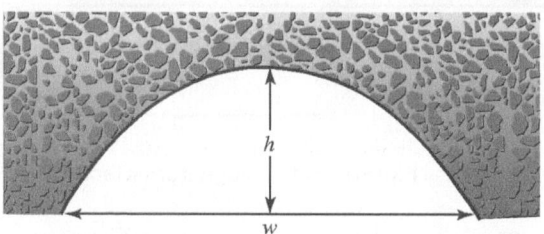

Figure 4.57 Stone archway

a. To relate the equation to the arch, you must super-impose a coordinate system over Figure 4.57. Where is the origin? How do x and y relate to h and w?

b. What is the maximum height, h?

c. What is the width, w, of the arch?

d. What is the height of the arch 9 ft from the center?

e. What is the height of the arch 18 ft from the center?

34. A closed box with square base is to be built to house an ant colony. The bottom of the box and all four sides are to be made of material costing $1/ft², and the top is to be constructed of glass costing $5/ft². What are the dimensions of the box of greatest volume that can be constructed for $72?

35. According to postal regulations, the girth plus the length of a parcel sent at the parcel post rate may not exceed 108 in. What is the largest possible volume of a rectangular parcel with two square sides that can be sent?

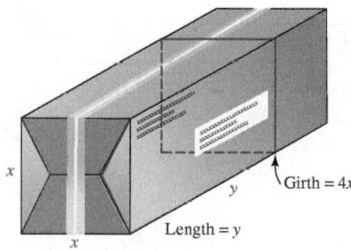

36. The bottom of an 8-ft-high mural painted on a vertical wall is 13 ft above the ground. The lens of a camera fixed to a tripod is 4 ft above the ground. How far from the wall should the camera be placed to photograph the mural with the largest possible angle?

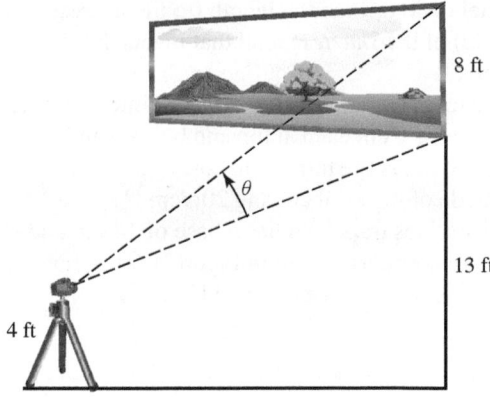

37. Søren Sovndal is at a point A on the north bank of a long, straight river 6 mi wide. Directly across from him on the south bank is a point B, and he wishes to reach a cabin C located s mi down the river from B, as shown in Figure 4.58.

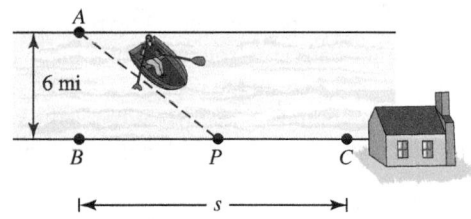

Figure 4.58 Søren going home

Given that Søren can row at 6 mi/h (including the effect of the current) and run at 10 mi/h, what is the minimum time (to the nearest minute) required for him to travel from A to C in each case?

a. $s = 4$ **b.** $s = 6$

38. Modeling Problem Two towns A and B are 12.0 mi apart and are located 5.0 and 3.0 mi, respectively, from a long, straight highway, as shown in Figure 4.59.

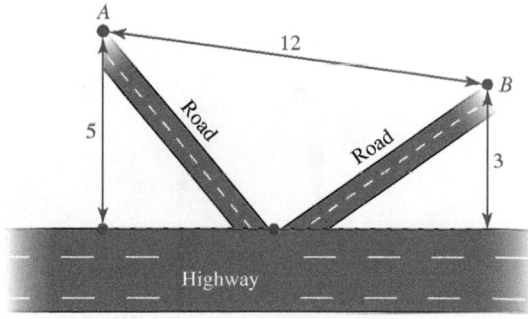

Figure 4.59 Building the shortest road

A construction company has a contract to build a road from A to the highway and then to B. Analyze a model to determine the length (to the nearest tenth of a mile) of the *shortest* road that meets these requirements.

39. A poster is to contain 108 cm² of printed matter, with margins of 6 cm each at top and bottom and 2 cm on the sides. What is the minimum cost of the poster if it is to be made of material costing 20¢/cm²?

40. An isosceles trapezoid has a base of 14 cm and slant sides of 6 cm, as shown in Figure 4.60. What is the largest area of such a trapezoid?

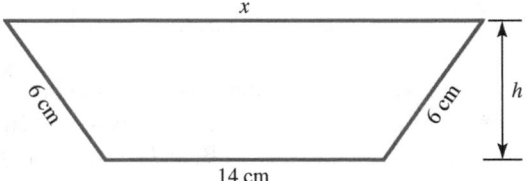

Figure 4.60   Area of a trapezoid

41. A storage bin is to be constructed by removing a sector with central angle θ from a circular piece of tin of radius 10 ft and folding the remainder of the tin to form a cone, as shown in Figure 4.61. What is the maximum volume of a storage bin formed in this fashion?

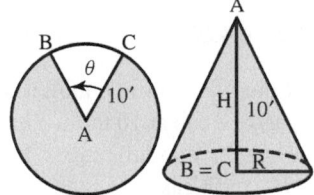

Figure 4.61   Fabricating a cone

42. A stained glass window in the form of an equilateral triangle is built on top of a rectangular window, as shown in Figure 4.62.

Figure 4.62   Maximize light

The rectangular part of the window is of clear glass and transmits twice as much light per square foot as the triangular part, which is made of stained glass. If the entire window has a perimeter of 20 ft, find the dimensions (to the nearest ft) of the window that will admit the most light.

43. Figure 4.63 shows a thin lens located p cm from an object AB and q cm from the image RS.

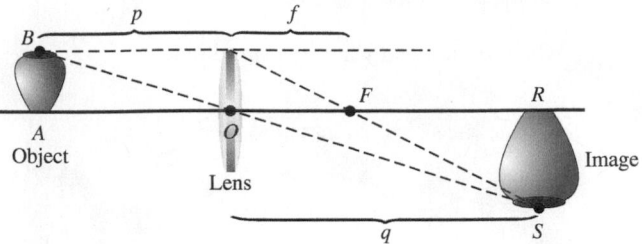

Figure 4.63   Image from a lens

The distance f from the center O of the lens to the point labeled F is called the **focal length** of the lens.

a. Using similar triangles, show that

$$\frac{1}{p} + \frac{1}{q} = \frac{1}{f}$$

b. Suppose a lens maker wishes to have $p + q = 24$. What is the largest value of f for which this condition can be satisfied?

44. One end of a cantilever beam of length L is built into a wall and the other end is supported by a single post. The deflection, or "sag," of the beam at a point located x units from the built-in end is modeled by the formula

$$D = k(2x^4 - 5Lx^3 + 3L^2x^2)$$

where k is a positive constant. Where does the maximum deflection occur on the beam?

45. When a mechanical system is at rest in an equilibrium position, its potential energy is minimized with respect to any small change in its position. Figure 4.64 shows a system involving a pulley, two small weights of mass m, and a larger weight of mass M.

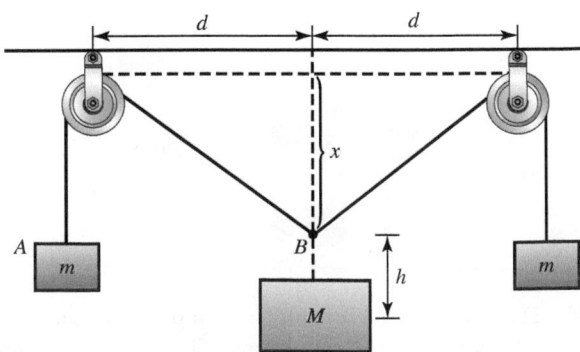

Figure 4.64   A pulley system

In physics, the total potential energy of the system is modeled by

$$E = -Mg(x + h) - 2mg\left(L - \sqrt{x^2 + d^2}\right)$$

where x, h, and d are the distances shown in Figure 4.64, L is the length of the cord from A around one pulley to B, and g is the constant acceleration due to gravity. All the other symbols represent constants. Use this information to find the critical number(s) for E.

46. A resistor of R ohms is connected across a battery of E volts whose internal resistance is r ohms. According to the principles of electricity, the formula

$$I = \frac{E}{R + r}$$

gives the current (amperes) in the circuit, while

$$P = I^2 R$$

gives the power (watts) in the external resistor. Assuming that E and r are constant, what value of R will result in maximum power in the external resistor?

47. **Modeling Problem** One model of a computer disk storage system uses the function

$$T(x) = N \left(k + \frac{c}{x} \right) p^{-x}$$

for the average time needed to send a file correctly by modem (including all retransmission of messages in which errors are detected), where x is the number of information bits, p is the (fixed) probability that any particular file will be received correctly, and N, k, and c are positive constants.*

  a. Find $T'(x)$.
  b. For what value of x is $T(x)$ minimized?
  c. Sketch the graph of T.

48. **Modeling Problem** In crystallography, a fundamental problem is the determination of the *packing fraction* of a crystal lattice, which is defined as the fraction of space occupied by the atoms in the lattice, assuming the atoms are hard spheres. When the lattice contains exactly two different kinds of atoms, the packing fraction is modeled by the formula

$$F(x) = \frac{K(1 + c^2 x^3)}{(1 + x)^3}$$

where $x = r/R$ is the ratio of the radii of the two kinds of atoms in the lattice and c and K are positive constants.†

  a. The function F has exactly one critical number. Find it and use the second-derivative test to determine whether it corresponds to a relative maximum or a relative minimum.

b. The numbers c and K and the domain of F depend on the cell structure in the lattice. For ordinary rock salt, it turns out that $c = 1$, $K = \frac{2\pi}{3}$, and the domain is $\left[\sqrt{2} - 1, 1 \right]$. Find the largest and smallest values of F on this domain.

c. Repeat part **b** for β-cristobalite, for which $c = \sqrt{2}$, $K = \sqrt{3}\pi/6$, and the domain is $[0, 1]$.

49. If air resistance is neglected, it can be shown that the stream of water emitted by a fire hose will have height

$$y = -16(1 + m^2) \left(\frac{x}{v} \right)^2 + mx$$

above a point located x ft from the nozzle, where m is the slope of the nozzle, and v is the velocity of the stream of water as it leaves the nozzle. (See Figure 4.65.) Assume v is constant.

  a. For fixed m, determine the distance x that results in maximum height.
  b. If m is allowed to vary but the firefighter must stand $x = x_0$ ft from the base of a burning building, what is the highest point on the building that the firefighter can reach with the water from her hose?

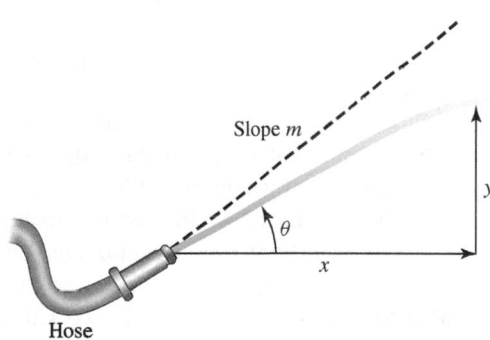

Figure 4.65 Stream of water

50. A telecommunications satellite is located

$$r(t) = \frac{4,831}{1 + 0.15 \cos 0.06t}$$

miles above the center of the earth t minutes after achieving orbit.

  a. Sketch the graph of $r(t)$.
  b. What are the lowest and highest points on the satellite's orbit? These are called the *perigee* and *apogee* positions, respectively.

---

*Paul J. Campbell, "Calculus Optimization in Information Technology," UMAP module 1991: *Tools for Teaching*. Lexington, MA: CUPM Inc., 1992; pp. 175-199.

†John C. Lewis and Peter P. Gillis, "Packing Factors in Diatomic Crystals," *American Journal of Physics*, Vol. 61, No. 5 (1993), pp. 434-438.

51. According to relativity theory, a particle's energy is related to its rest mass m_0, and its wavelength λ by the formula

$$E = \sqrt{\left(\frac{hc}{\lambda}\right)^2 + m_0^2 c^4}$$

where m_0 is the rest mass of the particle, c is the speed of light, and h is a constant (Planck's constant). This equation is used mostly in atomic, nuclear, and particle physics where the masses are not very large. Sketch the graph of $E(\lambda)$. What happens to E as $\lambda \to \infty$?

52. The theory of relativity models the Doppler shift s of an object traveling at a velocity v

$$s = \sqrt{\frac{c + v}{c - v}} - 1$$

where c is the speed of light and $v < c$.

a. Express v as a function of s and sketch the graph of $v(s)$.

b. Certain stellar objects called quasars appear to be moving away at velocities approaching c. The fastest one has a red shift of $s = 3.78$. What is its velocity in relation to c?

c. How fast is the velocity changing with respect to s when $s = 3.78$?

53. MODELING EXPERIMENT Use the fact that 12 oz ≈ 355 m$\ell = 355$ cm^3 to find the dimensions of the 12-oz Coke® can that can be constructed by using the least amount of metal. Compare these dimensions with a Coke from your refrigerator. What do you think accounts for the difference? An interesting article that discusses a similar question regarding tuna fish cans and the resulting responses is "What manufacturers Say about a Max/Min Application," by Robert F. Cunningham, Trenton State College, *The Mathematics Teacher*, March 1994, pp. 172-175.

54. MODELING EXPERIMENT Congruent triangles are cut out of a square piece of paper 20 cm on a side, leaving a star-like figure that can be folded to form a pyramid, as indicated in Figure 4.66.

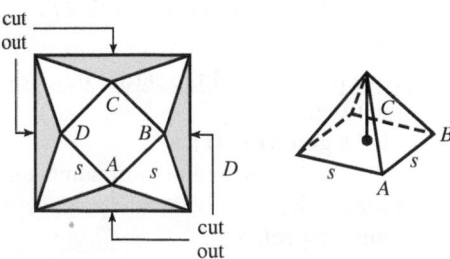

Figure 4.66   Constructing a pyramid

What is the largest volume of the pyramid that can be formed in this manner?

Level 3

55. The lower right-hand corner of a piece of paper is folded over to reach the leftmost edge, as shown in Figure 4.67.

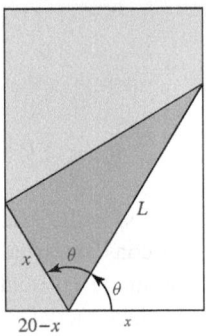

Figure 4.67   Paper folding problem

If the page is 20 cm wide and 30 cm long, what is the length L of the shortest possible crease? *Hint:* First express $\cos\theta$ and $\cos 2\theta$ in terms of x and then use the trigonometric identity

$$\cos 2\theta = 2\cos^2\theta - 1$$

to eliminate θ and express L in terms of x alone.

56. Find the length of the longest pipe that can be carried horizontally around a corner joining two corridors that are $2\sqrt{2}$ ft wide. *Hint:* Show that the length L can be written as

$$L(\theta) = \frac{2\sqrt{2}}{\sin\theta} + \frac{2\sqrt{2}}{\cos\theta}$$

and find the absolute minimum of $L(\theta)$ on an appropriate interval, as shown in Figure 4.68.

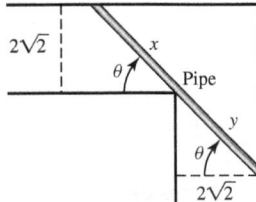

Figure 4.68   Corner problem

57. It is known that water expands and contracts according to its temperature. Physical experiments suggest that an amount of water that occupies 1 liter at 0°C will occupy

$$V(T) = 1 - 6.42 \times 10^{-5}T + 8.51 \times 10^{-6}T^2$$
$$- 6.79 \times 10^{-8}T^3$$

liters when the temperature is T°C. At what temperature is $V(T)$ minimized? How is this result related to the fact that ice forms only at the upper levels of a lake during winter?

58. **Modeling Problem** Light emanating from a source A is reflected by a mirror to a point B, as shown in Figure 4.69. Use Fermat's principle of optics to show that the angle of incidence α equals the angle of reflection β.

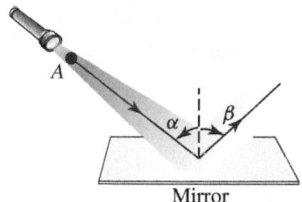

Figure 4.69 Reflection of light

59. **Modeling Problem** A universal joint is a coupling used in cars and other mechanical systems to join rotating shafts together at an angle. In a model designed to explore the mechanics of the universal joint mechanism, the angular velocity $\beta(t)$ of the output (driven) shaft is modeled by the formula

$$\beta(t) = \frac{\alpha \cos \gamma}{1 - \sin^2 \gamma \, \sin^2(\alpha t)}$$

where α (in rad/s) is the angular velocity of the input (driving) shaft, and γ is the angle between the two shafts, as shown in Figure 4.70.*

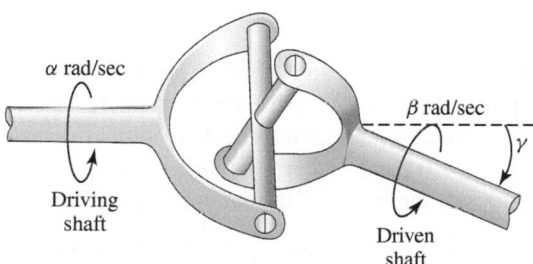

Figure 4.70 Universal joint

a. Sketch the graph of $\beta(t)$ for the case where α and γ are both positive constants.

b. What are the largest and smallest values of $\beta(t)$?

60. **Modeling Problem** Rainbows are formed when sunlight traveling through the air is both reflected and refracted by raindrops.†

Figure 4.71 shows a raindrop, which we assume to be a sphere for simplicity. An incoming beam of sunlight strikes the raindrop at point A with angle of incidence α, and some of the light is refracted through the angle β to point B. The process is then reversed, as indicated in the figure, and the light finally exits the drop at point C.

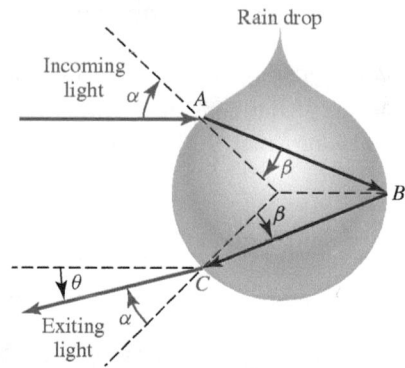

Figure 4.71 Light passing through a raindrop

a. At the interface points, A, B, and C, the light beam is deflected (from a straight line path) and loses intensity. For instance, at A, the incoming beam is deflected through a clockwise rotation of $(\alpha - \beta)$ radians. Show that the total deflection (at all three interfaces) is

$$D = \pi + 2\alpha - 4\beta$$

---

*Thomas O'Neil, "A Mathematical Model of a Universal Joint," *UMAP Modules 1982: Tools for Teaching*. Lexington, MA: Consortium for Mathematics and Its Applications, Inc., 1983, pp. 393-405.

†This problem is based on an article by Steve Janke, "Somewhere Within the Rainbow," *UMAP Modules 1992: Tools for Teaching*. Lexinton, MA: Consortium for Mathematics and Its Applications, Inc., 1993.

b. For raindrops falling through the air, Snell's law of refraction has the form $\sin \alpha = 1.33 \sin \beta$. Use this fact to express the derivative $D'(\alpha)$ in terms of α and β.

c. The intensity of rainbow light reaching the eye of an observer will be greatest when the total deflection $D(\alpha)$ is minimized. If the minimum deflection occurs when $\alpha = \alpha_0$, the corresponding angle of observation

$$\theta = \pi - D(\alpha_0)$$

is called the *rainbow angle* because the rainbow is brightest when viewed in that direction (see Figure 4.72).

　　Solve $D'(\alpha) = 0$ and show that the critical number you find minimizes $D(\alpha)$. Then find the rainbow angle θ. *Hint:* You may assume

$$0 \le \alpha - \beta \le \frac{\pi}{2}$$

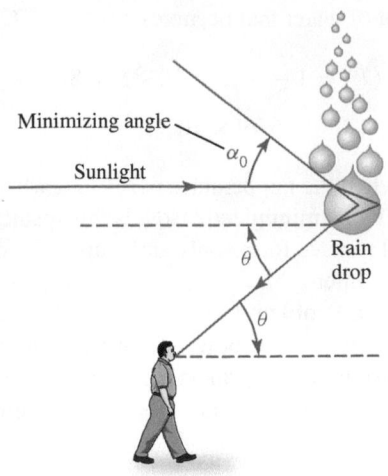

Figure 4.72 A rainbow is brightest when viewed at the rainbow angle

4.7 OPTIMIZATION IN BUSINESS, ECONOMICS, AND THE LIFE SCIENCES

IN THIS SECTION: *Economics:* maximizing profit and marginal analysis, **business management:** an inventory model and optimal holding time, **physiology:** concentration of a drug in the bloodstream and optimal angle for vascular branching

In the previous section, we saw how differentiation can be used to solve optimization problems in the physical sciences and engineering. The methods of calculus are also used in business and economics and are becoming increasingly prominent in the social and life sciences. We will examine a selection of such applications, and we begin with two modeling problems involving maximization of profit.

Economics

The most obvious application of optimization in economics and business it that of maximizing profit. We begin with an example.

Example 1 Maximizing profits

A manufacturer can produce a pair of earrings at a cost of $3. The earrings have been selling for $5 per pair and at this price, consumers have been buying 4,000 pairs per month. The manufacturer is planning to raise the price of the earrings and estimates that for each $1 increase in the price, 400 fewer pairs of earrings will be sold each month. At what price should the manufacturer sell the earrings to maximize profit?

Solution　Let x denote the number of $1 price increases, and let $P(x)$ represent the corresponding profit.

$$
\begin{aligned}
\text{PROFIT} &= \text{REVENUE} - \text{COST} \\
&= (\text{NUMBER SOLD})(\text{PRICE PER PAIR}) - (\text{NUMBER SOLD})(\text{COST PER PAIR}) \\
&= (\text{NUMBER SOLD})(\text{PRICE PER PAIR} - \text{COST PER PAIR})
\end{aligned}
$$

Recall that 4,000 pairs of earrings are sold each month when the price is $5 per pair and 400 fewer pairs will be sold each month for each added dollar in the price. Thus,

$$\text{NUMBER OF PAIRS SOLD} = 4{,}000 - 400 \, (\text{NUMBER OF } \$1 \text{ INCREASES})$$

$$= 4{,}000 - 400x$$

Knowing that the price per pair is $5 + x$, we can now write the profit as a function of x:

$$P(x) = (\text{NUMBER SOLD})(\text{PRICE PER PAIR} \; - \; \text{COST PER PAIR})$$

$$= (4{,}000 - 400x)[(5 + x) - 3]$$

$$= 400(10 - x)(2 + x)$$

To find the domain, we note that $x \geq 0$. And $400(10 - x)$, the number of pairs sold, should be nonnegative, so $x \leq 10$. Thus, the domain is $[0, 10]$.

The critical numbers are found when the derivative is 0 (P is a polynomial function, so there are no values for which the derivative is not defined):

$$P'(x) = 400(10 - x)(1) + 400(-1)(2 + x)$$

$$= 400(8 - 2x)$$

$$= 800(4 - x)$$

The critical number is $x = 4$ and the endpoints are $x = 0$ and $x = 10$. Checking for the maximum profit,

$$P(4) = 400(10 - 4)(2 + 4) = 14{,}400; \qquad P(0) = 8{,}000; \qquad P(10) = 0$$

The maximum possible profit is $14{,}400$, which will be generated if the earrings are sold for $5.00 + $4.00 = $9.00 per pair. The graph of the profit function is shown in Figure 4.73. ◼

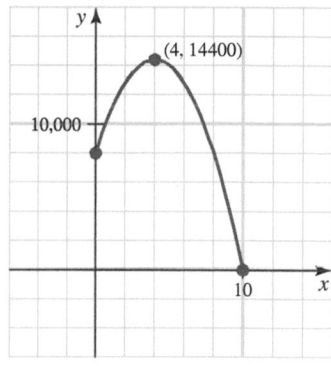

Figure 4.73 The profit function P

Sometimes the function to be optimized has practical meaning only when its independent variable is a positive integer. Such functions are said to have a discrete domain and are called **discrete functions.** Technically, the methods of calculus cannot be applied to discrete functions because the theorems we have developed apply only to continuous functions. However, useful information about a discrete function $f(n)$ can often be obtained by analyzing the related continuous function $f(x)$ obtained by replacing the discrete variable n by a real variable x. Here is an example that illustrates the general procedure.

Example 2   Maximizing a discrete revenue function

A travel company plans to sponsor a special tour to Africa. There will be accommodations for no more than 40 people, and the tour will be canceled if fewer than 10 people book reservations. Based on past experience, the manager determines that if n people book the tour, the profit (in dollars) may be modeled by the function

$$P(n) = -n^3 + 27.6n^2 + 970.2n - 4{,}235$$

For what size tour group is profit maximized?

Solution   The domain of $P(n)$ is the set of integers n between 10 and 40. To optimize this function, we apply the methods developed in this chapter to the related continuous function

$$P(x) = -x^3 + 27.6x^2 + 970.2x - 4{,}235$$

where x is a real variable defined on the interval $10 \leq x \leq 40$. To determine the critical number of $P(x)$, we solve

$$P'(x) = -3x^2 + 55.2x + 970.2 = 0$$

and obtain $x = 29.4$ and $x = -11$. Only $x = 29.4$ is in the domain, and by testing the sign of $P'(x)$ on each side of this critical number, we find that the continuous function $P(x)$ is increasing for $10 < x < 29.4$ and decreasing for $29.4 < x < 40$. Therefore, the original discrete function $P(n)$ is maximized either at $n = 29$ or at $n = 30$. Since

$$P(29) \approx 22{,}723.4 \qquad \text{and} \qquad P(30) \approx 22{,}711$$

we conclude that the maximum profit is \$22,723.40, and occurs when 29 people book the tour. The graphs of $P(n)$ and the related continuous function $P(x)$ are shown in Figure 4.74.

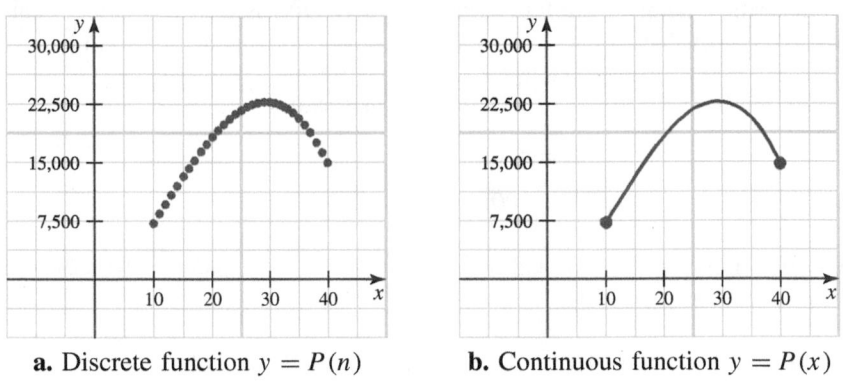

a. Discrete function $y = P(n)$ **b.** Continuous function $y = P(x)$

Figure 4.74 Graphs of profit function for different domains

In Section 3.8, we described **marginal analysis** as that branch of economics that is concerned with the way quantities such as price, cost, revenue, and profit vary with small changes in the level of production. Specifically, recall that if x is the number of units produced and brought to market, then the total cost of producing the units is denoted by $C(x)$. We call $p(x)$ the **demand** function and define it to be the price that consumers will pay for each unit of the commodity when x units are brought to market. Then $R(x) = xp(x)$ is the **total revenue** derived from the sale of the x units and $P(x) = R(x) - C(x)$ is the **total profit**. The relationship among revenue, cost, and profit is shown in Figure 4.75.

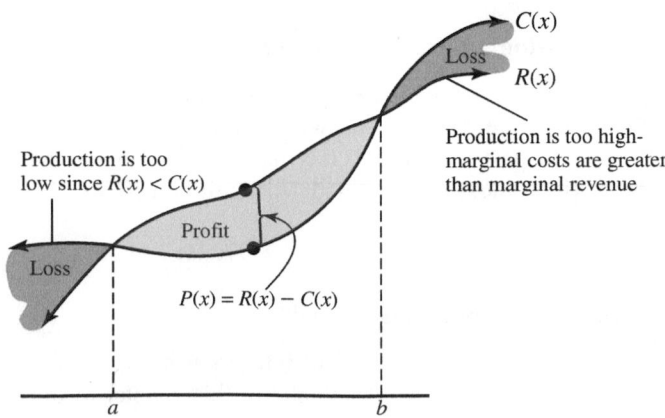

Figure 4.75 Cost, revenue, and profit functions

In Section 3.8, we worked with marginal quantities and their role as rates of change, namely, $C'(x)$, the *marginal cost*, and $R'(x)$, the *marginal revenue*.*

We will now consider these functions in optimization problems. For example, the manufacturer certainly would like to know what level of production results in maximum profit. To solve this problem, we want to maximize the profit function

$$P(x) = R(x) - C(x)$$

We differentiate with respect to x and find that

$$P'(x) = R'(x) - C'(x)$$

Thus, $P'(x) = 0$ when $R'(x) = C'(x)$, and by using economic arguments we can show that a maximum occurs at the corresponding critical point.

MAXIMUM PROFIT Profit is maximized when marginal revenue equals marginal cost.

Example 3   Using marginal analysis to maximize profit

A manufacturer estimates that when x units of a particular commodity are produced each month, the total cost (in dollars) will be

$$C(x) = \frac{1}{8}x^2 + 4x + 200$$

and all units can be sold at a price of $p(x) = 49 - x$ dollars per unit. Determine the price that corresponds to the maximum profit.

Solution   The marginal cost is $C'(x) = \frac{1}{4}x + 4$. The revenue is

$$R(x) = xp(x)$$
$$= x(49 - x)$$
$$= 49x - x^2$$

The marginal revenue is $R'(x) = 49 - 2x$. The profit is maximized when $R'(x) = C'(x)$:

$$R'(x) = C'(x)$$
$$49 - 2x = \frac{1}{4}x + 4$$
$$x = 20$$

Thus, the price that corresponds to the maximum profit is $p(20) = 49 - 20 = 29$ dollars/unit.   ∎

A second general principle of economics involves the following relationship between marginal cost and the **average cost** $A(x) = \dfrac{C(x)}{x}$.

---

*Recall (from Section 3.8) that marginal cost is the instantaneous rate of change of production cost C with respect to the number of units x_0 produced and the marginal revenue is the instantaneous rate of change of revenue with respect to the number of units x_0 produced.

> **MINIMUM AVERAGE COST**   Average cost is minimized at the level of production where the marginal cost equals the average cost.

To justify this second business principle, find the derivative of the average cost function:

$$A'(x) = \frac{xC'(x) - C(x)}{x^2}$$

$$= \frac{C'(x) - \frac{C(x)}{x}}{x}$$

$$= \frac{C'(x) - A(x)}{x}$$

Thus, $A'(x) = 0$ when $C'(x) = A(x)$. Once again, economic theory can be used to justify this result as a minimum.

Example 4   Using marginal analysis to minimize average cost

A manufacturer estimates that when x units of a particular commodity are produced each month, the total cost (in dollars) will be

$$C(x) = \frac{1}{8}x^2 + 4x + 200$$

and they can all be sold at a price of $p(x) = 49 - x$ dollars/unit. Determine the level of production at which the average cost is minimized.

Solution   The average cost is

$$A(x) = \frac{C(x)}{x} = \frac{1}{8}x + 4 + 200x^{-1}$$

and $A(x)$ is minimized when $C'(x) = A(x)$. Thus,

$$\frac{1}{4}x + 4 = \frac{1}{8}x + 4 + 200x^{-1}$$

$$x^2 = 1,600 \quad \textit{Multiply both sides by 8x and simplify.}$$

$$x = \pm 40$$

If we disregard the negative solution, it follows that the minimal average cost occurs when $x = 40$ units.

You might have noticed that we did not need the information $p(x) = 49 - x$ in arriving at the solution to this example. When doing real-world modeling, you must often make a choice about which parts of the available information are necessary to solve the problem.

Business Management

Sometimes mathematical methods can be used to assist managers in making certain business decisions. As an illustration, we shall show how calculus can be applied to a problem involving **inventory control**.

For each shipment of raw materials, a manufacturer must pay an ordering fee to cover handling and transportation. When the raw materials arrive, they must be stored until needed and storage costs result. If each shipment of raw materials is large, few shipments will be needed and ordering costs will be low, but storage costs will be high.

If each shipment is small, ordering costs will be high because many shipments will be needed, but storage cost will be low. Managers want to determine the shipment size that will minimize the total cost. Here is an example of how such a problem may be solved using calculus.

Example 5 Modeling Problem: Managing inventory to minimize cost

A retailer buys 6,000 calculator batteries a year from a distributor and is trying to decide how often to order the batteries. The ordering fee is $20 per shipment, the storage cost is $0.96 per battery per year, and each battery costs the retailer $0.25. Suppose that the batteries are sold at a constant rate throughout the year and that each shipment arrives uniformly throughout the year just as the preceding shipment has been used up. How many batteries should the retailer order each time to minimize the total cost?

Solution We begin by writing the cost function:

$$\text{TOTAL COST} = \text{STORAGE COST} + \text{ORDERING COST} + \text{COST OF BATTERIES}$$

We need to find an expression for each of these unknowns. Assume that the same number of batteries must be ordered each time an order is placed; denote this number by x so that $C(x)$ is the corresponding total cost. The average number of batteries in storage during the year is half of a given order (that is, $x/2$), and we assume that the total yearly storage cost is the same as if the $x/2$ batteries were kept in storage for the entire year. This situation is shown in Figure 4.76.

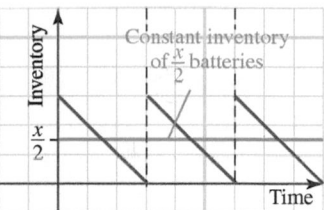

Figure 4.76 Inventory graph

$$\text{STORAGE COST} = \left(\begin{array}{c}\text{AVERAGE NUMBER}\\\text{IN STORAGE PER YR}\end{array}\right)\left(\begin{array}{c}\text{COST OF STORING 1}\\\text{BATTERY FOR 1 YR}\end{array}\right)$$

$$= \left(\frac{x}{2}\right)(0.96)$$

$$= 0.48x$$

To find the total ordering cost, we can multiply the ordering cost per shipment by the number of shipments. We also note that because 6,000 batteries are ordered during the year and because each shipment contains x batteries, the number of shipments is $6,000/x$.

$$\text{ORDERING COST} = \left(\begin{array}{c}\text{ORDERING COST}\\\text{PER SHIPMENT}\end{array}\right)\left(\begin{array}{c}\text{NUMBER OF}\\\text{SHIPMENTS}\end{array}\right)$$

$$= (20)\left(\frac{6,000}{x}\right)$$

$$= 120,000x^{-1}$$

Finally, we must also include the cost of purchasing batteries:

$$\text{COST OF BATTERIES} = \left(\begin{array}{c}\text{TOTAL NUMBER}\\\text{OF BATTERIES}\end{array}\right)\left(\begin{array}{c}\text{COST PER}\\\text{BATTERY}\end{array}\right)$$

$$= (6,000)(0.25)$$

$$= 1,500$$

Thus, we model the total cost by the function

$$C(x) = 0.48x + 120,000x^{-1} + 1,500$$

The goal is to minimize $C(x)$ on $(0, 6,000]$. To obtain the critical numbers, we find the derivative and then solve $C'(x) = 0$.

$$C'(x) = 0$$

$$0.48 - 120,000x^{-2} = 0$$

$$48x^2 = 12,000,000$$

$$x = \pm 500$$

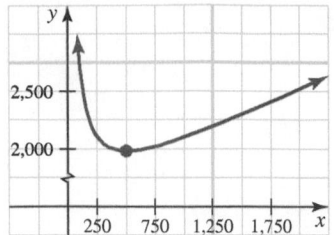

Figure 4.77 Total cost function

The root $x = -500$ does not lie in the interval $(0, 6{,}000]$. It is easy to check that C is decreasing on $(0, 500)$ and increasing on $(500, 6{,}000]$, as shown in Figure 4.77.

Thus, the absolute minimum of C on the interval $(0, 6{,}000]$ occurs when $x = 500$, and we conclude that to minimize cost, the manufacturer should order the batteries in lots of 500. ∎

Even with an asset that increases in value, there often comes a time when continuing to hold the asset is less advantageous than selling it and investing the proceeds of the sale. How should the investor decide when to sell? One way is to hold the asset until the time its present value at the current prevailing rate of interest is maximized. In other words, hold until today's dollar equivalent of the selling price is as large as possible and then sell. This is an example of an application known as **optimal holding time**.

Example 6   Modeling Problem: Optimal holding time

Suppose you own an asset whose market price t years from now is modeled by $V(t) = 10{,}000 e^{\sqrt{t}}$ dollars. If the prevailing rate of interest is 8% compounded continuously, when should the asset be sold?

Solution   The present value of the asset in t years is modeled by the function

$$P(t) = V(t)e^{-rt}$$

where r is the annual interest rate and t is the time in years. Thus,

$$P(t) = 10{,}000 e^{\sqrt{t}} e^{-0.08t} = 10{,}000 \exp(\sqrt{t} - 0.08t)$$

To maximize P, find $P'(t)$ and solve $P'(t) = 0$:

$$P'(t) = 10{,}000 \exp(\sqrt{t} - 0.08t)\left(\frac{1}{2} \cdot \frac{1}{\sqrt{t}} - 0.08\right)$$

$P'(t) = 0$ when $\dfrac{1}{2\sqrt{t}} - 0.08 = 0$ or when $t \approx 39.06$ years. Thus, the asset should be held for 39 years and then sold. ∎

Physiology

Calculus can be used to model a variety of situations in the biological and life sciences. We shall consider two examples from physiology. In the first example, we use a model for the **concentration of a drug in the bloodstream** of a patient to determine when the maximum concentration occurs.

Example 7   Modeling Problem: Maximum concentration of a drug

Let $C(t)$ denote the concentration in the blood at time t of a drug injected into the body intramuscularly. In a classic paper by E. Heinz, it was observed that the concentration may be modeled by

$$C(t) = \frac{k}{b - a}(e^{-at} - e^{-bt}) \qquad t \geq 0$$

where a, b (with $b > a$) and k are positive constants that depend on the drug.* At what time does the largest concentration occur? What happens to the concentration as $t \to \infty$?

---

*E. Heinz, "Probleme bei der Diffusion kleiner Substanzmengen innerhelb des menschlichen Körpers," *Biochem.*, Vol. 319 (1949), pp. 482-492.

Solution To locate the extrema, we solve $C'(t) = 0$.

$$C'(t) = \frac{d}{dt}\left[\frac{k}{b-a}(e^{-at} - e^{-bt})\right]$$

$$= \frac{k}{b-a}\left[(-a)e^{-at} - (-b)e^{-bt}\right]$$

$$= \frac{k}{b-a}\left(be^{-bt} - ae^{-at}\right)$$

We see that $C'(t) = 0$ when

$$be^{-bt} = ae^{-at}$$

$$\frac{b}{a} = e^{bt}e^{-at}$$

$$\frac{b}{a} = e^{bt-at}$$

$$bt - at = \ln\frac{b}{a} \qquad \textit{Definition of logarithm}$$

$$t = \frac{1}{b-a}\ln\frac{b}{a}$$

The second-derivative test can be used to show that the largest value of $C(t)$ occurs at this t (see Problem 57).

To see what happens to the concentration as $t \to \infty$, we compute the limit

$$\lim_{t\to\infty} C(t) = \lim_{t\to\infty}\frac{k}{b-a}\left[e^{-at} - e^{-bt}\right]$$

$$= \frac{k}{b-a}\left[\lim_{t\to\infty}\frac{1}{e^{at}} - \lim_{t\to\infty}\frac{1}{e^{bt}}\right]$$

$$= \frac{k}{b-a}[0 - 0]$$

$$= 0$$

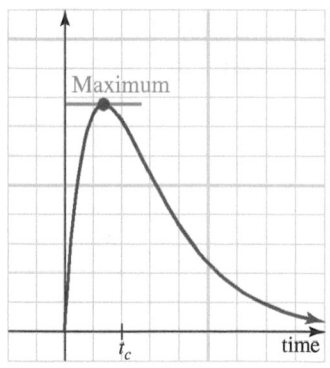

Figure 4.78 Graph of $C(t) = \dfrac{k}{b-a}(e^{-at} - e^{-bt})$

This tells us that the longer the drug is in the blood, the closer the concentration is to 0. The graph of C is shown in Figure 4.78.

Intuitively, we would expect the Heinz concentration function to begin at 0, increase to a maximum, and then gradually drop off to 0 in a finite amount of time. Figure 4.78 indicates that $C(t)$ does not have these characteristics, because it does not quite get back to 0 in finite time. This suggests that the Heinz model may apply most reliably to the period of time right after the drug has been injected. ∎

A final application in this section considers the **optimal angle for vascular branch-ing**. The blood vascular system operates in such a way that the circulation of blood—from the heart, through the organs of the body, and back to the heart—is accomplished with as little expenditure of energy as possible. Thus, it is reasonable to expect that when an artery branches, the angle between the "parent" artery and its "daughter" should mini-mize the total resistance to the flow of blood. Figure 4.79 shows a small artery of radius r branching from a larger artery of radius R.

Blood flows in the direction of the arrows from point A to the branch at B and then to points C and D. For simplicity, we assume that C and D are located in such a way that $\overline{CD}$ is perpendicular to the main line through A, B, and D. We wish to find the value of the branching angle θ that minimizes the total resistance to the flow of blood as it

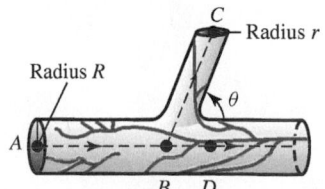

Figure 4.79 Vascular branching

moves from A to B and then to point C, which is located a fixed perpendicular distance h from the line through A and B.* (See Figure 4.79.)

POISEUILLE'S RESISTANCE TO FLOW LAW   The resistance to the flow of blood in an artery is directly proportional to the artery's length and inversely proportional to the fourth power of its radius.

According to Poiseuille's law, the resistance to flow, f_1, from A to B is

$$f_1 = \frac{ks_1}{R^4}$$

where R is the radius of the larger artery, and the resistance to flow, f_2, from B to C is

$$f_2 = \frac{ks_2}{r^4}$$

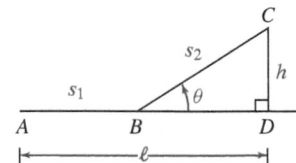

Figure 4.80   Minimizing the resistance to blood flow

where r is the radius of the smaller artery, k is a viscosity constant, $s_1 = \left| \overline{AB} \right|$, and $s_2 = \left| \overline{BC} \right|$. Thus, the total resistance to flow may be modeled by the sum

$$f = f_1 + f_2 = \frac{ks_1}{R^4} + \frac{ks_2}{r^4} = k \left(\frac{s_1}{R^4} + \frac{s_2}{r^4} \right)$$

The next task is to write f as a function of θ. To do this, reconsider Figure 4.79 by labeling s_1, s_2, h, and ℓ as shown in Figure 4.80.

We want to find equations for s_1 and s_2 in terms of h and θ; to this end we notice that

$$\sin \theta = \frac{h}{s_2} \qquad\qquad \tan \theta = \frac{h}{\ell - s_1}$$

$$s_2 = \frac{h}{\sin \theta} \qquad\qquad \ell - s_1 = \frac{h}{\tan \theta}$$

$$= h \csc \theta \qquad\qquad s_1 = \ell - \frac{h}{\tan \theta}$$

$$\qquad\qquad = \ell - h \cot \theta$$

We can now write f as a function of θ:

$$f(\theta) = k \left(\frac{s_1}{R^4} + \frac{s_2}{r^4} \right) = k \left(\frac{\ell - h \cot \theta}{R^4} + \frac{h \csc \theta}{r^4} \right)$$

To minimize f we need to find the critical numbers (remember that h and k are positive constants):

$$\frac{df}{d\theta} = 0$$

$$k \left[\frac{-h(-\csc^2 \theta)}{R^4} + \frac{h(-\csc \theta \cot \theta)}{r^4} \right] = 0$$

---

*The key to solving this problem is a result due to work of the 19th-century French physiologist, Jean Louis Poiseuille (1789-1869). Our discussion of vascular branching is adapted from *Introduction to Mathematics for Life Scientists,* 2nd edition, by Edward Batschelet (New York: Springer-Verlag, 1976, pp. 278-280). In this excellent little book, Batschelet develops a number of interesting applications of calculus, several of which appear in the problem set.

$$kh \csc\theta \left(\frac{\csc\theta}{R^4} - \frac{\cot\theta}{r^4} \right) = 0$$

$$\left(\frac{\csc\theta}{R^4} - \frac{\cot\theta}{r^4} \right) = 0$$

$$\frac{\csc\theta}{R^4} = \frac{\cot\theta}{r^4}$$

$$\frac{1}{R^4 \sin\theta} = \frac{\cos\theta}{r^4 \sin\theta}$$

$$\frac{r^4}{R^4} = \cos\theta$$

By finding the second derivative and noting that both r and R are both positive, we can show that any value θ_m that satisfies this equation yields a value $f(\theta_m)$ that is a minimum.

Thus, the required optimal angle is $\theta_m = \cos^{-1}\dfrac{r^4}{R^4}$.

PROBLEM SET 4.7

Level 1

In Problems 1-6, *we give the cost C of producing x units of a particular commodity and the selling price p when x units are produced. In each case, determine the level of production that maximizes profit.*

1. $C(x) = \frac{1}{8}x^2 + 5x + 98$ and $p(x) = \frac{1}{2}(75 - x)$
2. $C(x) = \frac{2}{3}x^2 + 3x + 10$ and $p(x) = \frac{1}{5}(45 - x)$
3. $C(x) = 50{,}000 + 50x$ and $p(x) = 200 - 0.04x$
4. $C(x) = 18{,}500 + 8.45x$ and $p(x) = 165 - 0.02x$
5. $C(x) = 2{,}000 + 3x^2$ and $p(x) = 200 - 2x$
6. $C(x) = \frac{1}{5}(x + 30)$ and $p(x) = \dfrac{70 - x}{x + 30}$

Just as marginal cost and marginal profit are the derivatives of the cost and profit functions, so are marginal average cost and marginal average profit the derivatives of the average cost and average profit functions, where average profit is defined similarly to average cost. To see whether you understand this, find the marginal average profit for the functions given in Problems 7-12.

7. $P(x) = x^3 - 8x^2 + 2x + 50$
8. $P(x) = x^3 - 10x^2 + 3x + 20$
9. $P(x) = 5x - 0.05x^2$
10. $P(x) = 4x - 0.06x^2$
11. $P(x) = x^3 - 50x^2 + 5x + 200$
12. $P(x) = x^3 - 40x^2 + 3x + 400$
13. Suppose the total cost of manufacturing x units of a certain commodity is

$$C(x) = 3x^2 + x + 48$$

dollars. Determine the minimum average cost.

14. Suppose the total cost of producing x units of a certain commodity is

$$C(x) = 2x^4 - 10x^3 - 18x^2 + x + 5$$

Determine the largest and smallest values of the marginal cost for $0 \le x \le 5$.

15. A business manager estimates that when p dollars are charged for every unit of a product, the sales will be $x = 380 - 20p$ units. At this level of production, the average cost is modeled by

$$A(x) = 5 + \frac{x}{50}$$

a. Find the total revenue and total cost functions, and express the profit as a function of x.
b. What price should the manufacturer charge to maximize profit? What is the maximum profit?

16. A toy manufacturer produces an inexpensive doll (Flopsy) and an expensive doll (Mopsy) in units of x hundreds and y hundreds, respectively. Suppose it is possible to produce the dolls in such a way that

$$y = \frac{82 - 10x}{10 - x}$$

for $0 \le x \le 8$ and that the company receives *twice* as much for selling a Mopsy doll as for selling a Flopsy doll. Find the level of production for both x and y for which the total revenue derived from selling these dolls is maximized. What vital assumption must be made about sales in this model?

17. A manufacturer can produce shoes at a cost of $50 a pair and estimates that if they are sold for x dollars a pair, consumers will buy approximately

$$s(x) = 1{,}000e^{-0.1x}$$

pairs of shoes per week. At what price should the manufacturer sell the shoes to maximize profit?

18. Suppose a manufacturer estimates that, when the market price of a certain product is p, the number of units sold will be

$$x = -6\ln\frac{p}{40}$$

It is also estimated that the cost of producing these x units will be

$$C(x) = 4xe^{-x/6} + 30$$

a. Find the average cost, the marginal cost, and the marginal revenue for this production process.
b. What level of production x corresponds to maximum profit?

19. It is projected that t years from now, the population of a certain country will be

$$P(t) = 50e^{0.02t}$$

million.
a. At what rate will the population be changing with respect to time 10 years from now?
b. At what percentage rate will the population be changing with respect to time t years from now?

20. A certain industrial machine depreciates in such a way that its value (in dollars) after t years is

$$Q(t) = 20,000e^{-0.4t}$$

a. At what rate is the value of the machine changing with respect to time after 5 years?
b. At what percentage rate is the value of the machine changing with respect to time after t years?

21. It is estimated that t years from now the population of a certain country will be

$$p(t) = \frac{160}{1 + 8e^{-0.01t}}$$

million. When will the population be growing most rapidly?

22. Some psychologists model a child's ability to memorize by a function of the form

$$g(t) = \begin{cases} t\ln t + 1 & \text{if } 0 < t \le 4 \\ 1 & \text{if } t = 0 \end{cases}$$

where t is time, measured in years. Determine when the largest and smallest values of g occur.

Level 2

23. **Modeling Problem** An electronics firm uses 18,000 cases of connectors each year. The cost of storing one case for a year is $4.50, and the ordering fee is $20 per shipment. Assume that the connectors are used at a constant rate throughout the year and that each shipment arrives just as the preceding shipment has been used

up. How many cases should the firm order each time to keep total cost to a minimum?

24. **Modeling Problem** The owner of the Pill Boxx drugstore expects to sell 600 bottles of hair spray each year. Each bottle costs $4, and the ordering fee is $30 per shipment. In addition, it costs 90¢ per year to store each bottle. Assuming that the hair spray sells at a uniform rate throughout the year and that each shipment arrives just as the last bottle from the previous shipment is sold, how frequently should shipments of hair spray be ordered to minimize the total cost?

25. Suppose the total revenue (in dollars) from the sale of x units of a certain commodity is

$$R(x) = -2x^2 + 68x - 128$$

a. At what level of sales is the average revenue per unit equal to the marginal revenue?
b. Verify that the average revenue is increasing if the level of sales is less than the level in part **a** and decreasing if the level of sales is greater than the level in part **a**.
c. On the same set of axes, graph the relevant portions of the average and marginal revenue functions.

26. Suppose the total cost (in dollars) of manufacturing x units of a certain commodity is

$$C(x) = 3x^2 + 5x + 75$$

a. At what level of production is the average cost per unit the smallest?
b. At what level of production is the average cost per unit equal to the marginal cost?
c. Graph the average cost and the marginal cost functions on the same set of axes, for $x > 0$.

27. Suppose you own a rare book whose value t years from now is modeled as $300e^{\sqrt{3t}}$. If the prevailing rate of interest remains constant at 8% compounded continuously, when will be the most advantageous time to sell?

28. A manufacturer finds that the demand function for a certain product is

$$x(p) = \frac{73}{\sqrt{p}}$$

Should the price p be raised or lowered to increase consumer expenditure? Explain your answer.

29. A store has been selling skateboards at the price of $40 per board, and at this price skaters have been buying 45 boards a month. The owner of the store wishes to raise the price and estimates that for each $1 increase in price, 3 fewer boards will be sold each month. If each board costs the store $29, at which price should the store sell the boards to maximize profit?

30. Suppose you own a parcel of land whose value t years from now is modeled as

$$P(t) = 200 \ln \sqrt{2t}$$

thousand dollars. If the prevailing rate of interest is 10% compounded continuously, when will be the most advantageous time to sell?

31. A tour agency is booking a tour and has 100 people signed up. The price of a ticket is $2,000 per person. The agency has booked a plane seating 150 people at a cost of $125,000. Additional costs to the agency are incidental fees of $500 per person. For each $10 that the price is lowered, a new person will sign up. How much should the price be lowered for all participants to maximize the profit to the tour agency?

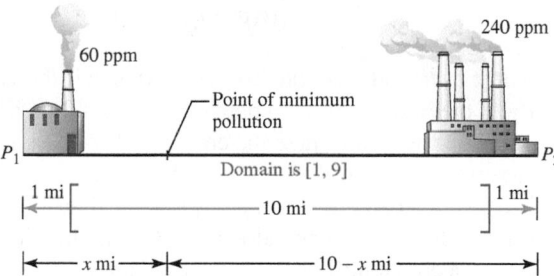

Point of minimum pollution
60 ppm
240 ppm
P_1
P_2
Domain is [1, 9]
1 mi
10 mi
1 mi
x mi
$10 - x$ mi

32. **Modeling Problem** As more and more industrial areas are constructed, there is a growing need for standards ensuring control of the pollutants released into the air. Suppose that the pollution at a particular location is based on the distance from the source of the pollution according to the principle that for distances greater than or equal to 1 mi, the concentration of particulate matter (in parts per million, ppm) decreases as the reciprocal of the distance from the source. This means that if you live 3 mi from a plant emitting 60 ppm, the pollution at your home is $\frac{60}{3} = 20$ ppm. On the other hand, if you live 10 mi from the plant, the pollution at your home is $\frac{60}{10} = 6$ ppm. Suppose that two plants 10 mi apart are releasing 60 and 240 ppm, respectively. At what point between the plants is the pollution a minimum? Where is it a maximum?

33. A Florida citrus grower estimates that if 60 orange trees are planted, the average yield per tree will be 400 oranges. The average yield will decrease by 4 oranges for each additional tree planted on the same acreage. How many trees should the grower plant to maximize the total yield?

34. A bookstore can obtain the best-seller *20,000 Leagues Under the Majors* from the publisher at a cost of $6 per book. The store has been offering the book at a price of $30 per copy and has been selling 200 copies per month at this price. The bookstore is planning to lower its price to stimulate sales and estimates that for each $2 reduction in the price, 20 more books will be sold per month. At what price should the bookstore sell the book to generate the greatest possible profit?

35. A viticulturist estimates that if 50 grapevines are planted per acre, each grapevine will produce 150 lb of grapes. Each additional grapevine planted per acre (up to 20) reduces the average yield per vine by 2 lb. How many grapevines should be planted to maximize the yield per acre?

36. Farmers can get $20 per bushel for their potatoes on July 1, and after that the price drops 20¢ per bushel per day. On July 1, a farmer has 80 bu of potatoes in the field and estimates that the crop is increasing at the rate of 1 bu per day. When should the farmer harvest the potatoes to maximize revenue?

37. **Modeling Problem** To raise money, a service club has been collecting used bottles, which it plans to deliver to a local glass company for recycling. Since the project began 80 days ago, the club has collected 24,000 pounds of glass, for which the glass company offers 1¢ per pound. However, because bottles are accumulating faster than they can be recycled, the company plans to reduce the price it will pay by 1¢ per day per 100 pounds of used glass.
    a. What is the most advantageous time for the club to conclude its project and deliver all the bottles?
    b. What assumptions must be made in part **a** to solve the problem with the given information?

38. A commuter train carries 600 passengers each day from a suburb to a city. It now costs $5 per person to ride the train. A study shows that 50 additional people will ride the train for each 25¢ reduction in fare. What fare should be charged to maximize total revenue?

39. Suppose the demand function for a certain commodity is linear, that is,

$$p(x) = \frac{b - x}{a}$$

for $0 \le x \le b$ where a and b are positive constants.
    a. Find the total revenue function explicitly and use its first derivative to determine its intervals of increase and decrease.
    b. Graph the relevant portions of the demand and revenue functions.

40. Suppose that the demand function for a certain commodity is expressed as

$$p(x) = \sqrt{\frac{120 - x}{0.1}}$$

$0 \le x \le 120$ where x is the number of items sold.
    a. Find the total revenue function explicitly and use its first derivative to determine the price at which revenue is maximized.
    b. Graph the relevant portions of the demand and revenue functions.

41. **Modeling Problem** In an experiment, a fish swims s meters upstream at a constant velocity v m/sec relative to the water, which itself has velocity v_1 relative to the

ground. The results of the experiment suggest that if the fish takes t seconds to reach its goal (that is, to swim s meters), the energy it expends may be modeled by

$$E = cv^k t$$

where $c > 0$ and $k > 2$ are physical constants. Assuming v_1 is known, what velocity v minimizes the energy?

42. **Modeling Problem** Homing pigeons will rarely fly over large bodies of water unless forced to do so, presumably because it requires more energy to maintain altitude in flight over the cool water. Suppose a pigeon is released from a boat floating on a lake 3 mi from a point A on the shore and 10 mi away from the pigeon's loft, as shown in Figure 4.81.*

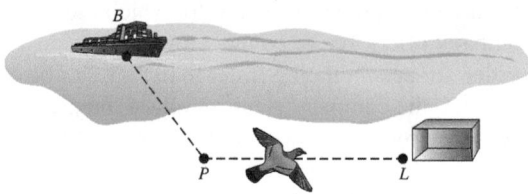

Figure 4.81　Flight path for a pigeon

Assuming the pigeon requires twice as much energy to fly over water as over land and it follows a path that minimizes total energy, find the angle θ of its heading as it leaves the boat. (This is the situation in the preview section of the chapter introduction.)

43. **Modeling Problem** The production of blood cells plays an important role in medical research involving leukemia and other so-called *dynamical diseases*. In 1977, a mathematical model was developed by A. Lasota that involved the cell production function

$$P(x) = Ax^s e^{-sx/r}$$

where A, s, and r are positive constants and x is the number of granulocytes (a type of white blood cell) present.[†]

a. Find the granulocyte level x that maximizes the production function P. How do you know it is a maximum?

b. If $s > 1$, show that the graph of P has two inflection points. Sketch the graph of P and give a brief interpretation of the inflection points.

c. Sketch the graph for the case where $0 < s < 1$. What is different about this case?

44. In a learning model, two responses (A and B) are possible for each of a series of observations. If there is a probability p of getting response A in any observation, the probability of getting a response A exactly n times in a series of M observations is

$$F(p) = p^n (1 - p)^{M-n}$$

The *maximum likelihood estimate* is the value of p that maximizes $F(p)$ on $[0, 1]$. For what value of p does this occur?

45. **Modeling Problem** The work of V. A. Tucker and K. Schmidt-Koenig models the energy expended in flight by a certain kind of bird by the function

$$E = \frac{1}{v} \left[a(v - b)^2 + c \right]$$

where a, b, and c are positive constants, v is the velocity of the bird, and the domain of v is $[16, 60]$.[‡] Which value of v will minimize the energy expenditure in the case where $a = 0.04$, $b = 36$, and $c = 9$?

46. **Modeling Problem** During a cough, the diameter of the trachea decreases. The velocity v of air in the trachea during a cough may be modeled by the formula

$$v = Ar^2 (r_0 - r)$$

where A is a constant, r is the radius of the trachea during the cough, and r_0 is the radius of the trachea in a relaxed state. Find the radius of the trachea when the velocity is greatest, and find the maximum velocity of air. (Notice that $0 \le r \le r_0$.)

47. **Modeling Problem** After hatching, the larva of the codling moth goes looking for food. The period between hatching and finding food is called the *searching period*. According to a model developed by P. L. Shaffer and H. J. Gold[¶] the length (in days) of the searching period is given by

$$S(T) = (-0.03T^2 + 1.67T - 13.67)^{-1}$$

where T (°C) is the air temperature ($20 \le T \le 30$), and the percentage of larvae that survive the searching period is

$$N(T) = -0.85T^2 + 45.45T - 547$$

a. Sketch the graph of N and find the largest and smallest survival percentages for the allowable range of temperatures, $20 \le T \le 30$.

---

*Edward Batschelet, *Introduction to Mathematics for Life Scientists*, 2nd ed. (New York: Springer-Verlag, 1979), pp. 276-277.

[†]See "A Blood Cell Population Model, Dynamical Diseases, and Chaos," by W. B. Gearhart and M. Martelli, UMAP Modules 1990: *Tools for Teaching*. Arlington, MA: Consortium for Mathematics and Its Applications (CUPM) Inc., 1991.

[‡]Tucker, V. A. and K. Schmidt-Koenig, "Flight of Birds in Relation to Energetics and Wind Directions," *The Auk* 88 (1971), pp. 97-107.

[¶]P. L. Shaffer and H. J. Gold, "A Simulation Model of Population Dynamics of the Codling Moth *Cydia Pomonella*," *Ecological Modeling*, Vol. 30, 1985, pp. 247-274.

b. Find $S'(T)$ and solve $S'(T) = 0$. Sketch the graph of S for $20 \leq T \leq 30$.

c. The percentage of codling moth eggs that hatch at a given temperature T is given by

$$H(T) = -0.53T^2 + 25T - 209$$

for $20 \leq T \leq 30$. Sketch the graph of H and determine the temperatures at which the largest and smallest hatching percentages occur.

48. Modeling Problem A plastics firm has received an order from the city recreation department to manufacture 8,000 special styrofoam kickboards for its summer swimming program. The firm owns 10 machines, each of which can produce 50 kickboards per hour. The cost of setting up the machines to produce the kickboards is $800 per machine. Once the machines have been set up, the operation is fully automated and can be overseen by a single production supervisor earning $35 per hour.

a. How many machines should be used to minimize the cost of production?

b. How much will the supervisor earn during the production run if the optimal number of machines is used?

Level 3

49. Modeling Problem Generalize Problem 48. Specifically, a manufacturing firm received an order for Q units of a certain commodity. Each of the firm's machines can produce n units per hour. The setup cost is S dollars per machine, and the operating cost is p dollars per hour.

a. Derive a formula for the number of machines that should be used to minimize the total cost of filling the order.

b. Show that when the total cost is minimal, the setup cost is equal to the cost of operating the machines.

50. According to a certain logistic model, the world's population (in billions) t years after 2000 is modeled by the function

$$P(t) = \frac{75}{1 + 12e^{-0.02t}}$$

a. If this model is correct, at what rate will the world's population be increasing with respect to time in the year 2020? At what percentage rate will it be increasing at this time?

b. Sketch the graph of P. What feature on the graph corresponds to the time when the population is growing most rapidly? What happens to $P(t)$ as $t \to \infty$ (that is, "in the long run")?

51. An epidemic spreads through a community in such a way that t weeks after its outbreak, the number of residents who have been infected is modeled by a function of the form

$$f(t) = \frac{A}{1 + Ce^{-kt}}$$

where A is the total number of susceptible residents. Show that the epidemic is spreading most rapidly when half the susceptible residents have been infected.

52. Modeling Problem (Continuation of Problem 47) Suppose you have 10,000 codling moth eggs. Find a function F for the number of moths that hatch and survive to have their first meal when the temperature is T. For what temperature T on $[20, 30]$ is the number of dining moths the greatest? How large is the optimum dining party and how long did it take to find its meal?

53. Modeling Problem A store owner expects to sell Q units of a certain commodity each year. It costs S dollars to order each new shipment of x units, and it costs t dollars to store each unit for a year. Assuming that the commodity is used at a constant rate throughout the year and that each shipment arrives just as the preceding shipment has been used up, show that the total cost of maintaining inventory is minimized when the ordering cost equals the storage cost.

54. In certain tissues, cells exist in the shape of circular cylinders. Suppose such a cylinder has radius r and height h. If the volume is fixed (say, at v_0), find the value of r that minimizes the total surface area

$$S = 2\pi rh + 2\pi r^2$$

of the cell.

55. An important quantity in economic analysis is *elasticity of demand*, defined by

$$E(x) = \frac{p}{x} \frac{dx}{dp}$$

where x is the number of units of a commodity demanded when the price is p dollars per unit.* Show that

$$\frac{dR}{dx} = \frac{R}{x}\left[1 + \frac{1}{E(x)}\right]$$

That is, marginal revenue is $\left[1 + \dfrac{1}{E(x)}\right]$ times average revenue.

---

*See the module with the intriguing title, "Price Elasticity of Demand: Gambling, Heroin, Marijuana, Prostitution, and Fish," by Yves Nievergelt, *UMAP Modules 1987: Tools for Teaching.* Arlington, MA: CUPM Inc., 1988, pp. 153-181.

56. Modeling Problem Sometimes investment managers determine the optimal holding time of an asset worth $V(t)$ dollars by finding when the relative rate of change $V'(t)/V(t)$ equals the prevailing rate of interest r. Show that the optimal time determined by this criterion is the same as that found by maximizing the present value of $V(t)$.

57. In this section we showed that the Heinz concentration function,

$$C(t) = \frac{k}{b-a}(e^{-at} - e^{-bt})$$

for $b > a$ has exactly one critical number, namely,

$$t_c = \frac{1}{b-a} \ln\left(\frac{b}{a}\right)$$

Find $C''(t)$ and use the second-derivative test to show that a relative maximum for the concentration $C(t)$ occurs at time $t = t_c$.

58. Modeling Problem There are alternatives to the inventory model analyzed in Example 5. Suppose a company must supply N units/month at a uniform rate. Assume the storage cost/unit is S_1 dollars/month and that the setup cost is S_2 dollars. Further assume that production is at a uniform rate of m units/month (with no units left over in inventory at the end of the month). Let x be the number of items produced in each run.

a. Explain why the total average cost per month may be modeled by

$$C(x) = \frac{S_1 x}{2}\left(1 - \frac{N}{m}\right) + \frac{S_2 N}{x}$$

b. Find an expression for the number of items that should be produced in each run in order to minimize the total average cost C.

Economists refer to the optimum found in this inventory model as the *economic production quantity* (EPQ), whereas the optimum found in the model analyzed in Example 5 is called the *economic order quantity* (EOQ).

59. Beehives are formed by packing together cells that may be modeled as regular hexagonal prisms open at one end, as shown in Figure 4.82.

© 2013 LilKar. Used under license from Shutterstock, Inc.

Beehive

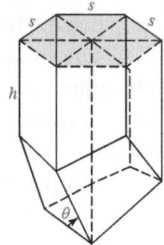

Figure 4.82 Beehive cell

It can be shown that a cell with hexagonal side of length s and prism height h has surface area

$$S(\theta) = 6sh + 1.5s^2(-\cot\theta + \sqrt{3}\csc\theta)$$

for $0 < \theta < \frac{\pi}{2}$. What is the angle θ (to the nearest degree) that minimizes the surface area of the cell (assuming that s and h are fixed)?

60. Let $f(\theta)$ be the resistance to blood flow obtained on page 306, namely,

$$f(\theta) = k\left(\frac{\ell - h\cot\theta}{R^4} + \frac{h\csc\theta}{r^4}\right)$$

We found that $\theta_m = \cos^{-1}\dfrac{r^4}{R^4}$ is a critical number of f. Find $f''(\theta)$ and use the second-derivative test to verify that a relative minimum occurs at $\theta = \theta_m$. Explain why this must be an *absolute minimum*.

CHAPTER 4 REVIEW

N*othing takes place in the world whose meaning is not that of some maximum or minimum.*

Leonhard Euler

Hilbert Space Problem Book

Proficiency Examination

Concept Problems

1. What is the difference between absolute and relative extrema of a function?
2. State the extreme value theorem.
3. What are the critical numbers of a function? What is the difference between critical numbers and critical points?
4. Outline a procedure for finding the absolute extrema of a continuous function on a closed interval $[a,b]$.
5. State both Rolle's theorem and the mean value theorem, and discuss the relationship between them.
6. What is the zero-derivative theorem?
7. State the first-derivative and second-derivative tests, and discuss when and in which order you would use them.
8. What is a vertical asymptote? A horizontal asymptote?
9. How would you identify a cusp? A vertical tangent?
10. State l'Hôpital's rule.
11. Define $\lim\limits_{x \to \infty} f(x) = L$ and $\lim\limits_{x \to c} f(x) = \infty$.
12. Outline a graphing strategy for a function defined by $y = f(x)$.
13. What do we mean by optimization? Outline an optimization procedure.
14. What is Fermat's principle of optics?
15. What is Snell's law of refraction?
16. When is the profit maximized in terms of marginal revenue and marginal cost? When is the average cost minimized?
17. What is Poiseuille's law of resistance to flow?

Practice Problems

Evaluate the limits in Problems 18-21.

18. $\lim\limits_{x \to \frac{\pi}{2}} \dfrac{\sin 2x}{\cos x}$
19. $\lim\limits_{x \to 1} \dfrac{1 - \sqrt{x}}{x - 1}$
20. $\lim\limits_{x \to \infty} \left(\dfrac{1}{x} - \dfrac{1}{\sqrt{x}} \right)$
21. $\lim\limits_{x \to \infty} \left(1 + \dfrac{2}{x} \right)^{3x}$

Sketch the graph of each function in Problems 22-27. *Include all key features, such as relative extrema, inflection points, asymptotes, intercepts, cusps, and vertical tangents.*

22. $f(x) = x^3 + 3x^2 - 9x + 2$
23. $f(x) = x^{1/3}(27 - x)$

24. $f(x) = \dfrac{x^2 - 1}{x^2 - 4}$
25. $f(x) = (x^2 - 3)e^{-x}$

26. $f(x) = x + \tan^{-1} x$
27. $f(x) = \sin^2 x - 2\cos x$ on $[0, 2\pi]$

28. Determine the largest and smallest values of

$$f(x) = x^4 - 2x^5 + 5$$

on the closed interval $[0, 1]$.

29. A box is to have a square base, an open top, and volume of 2 ft³. Find the dimensions of the box (to the nearest inch) that uses the least amount of material.

30. The personnel manager of a department store estimates that if N temporary salespersons are hired for the holiday season, the total net revenue derived (in hundreds of dollars) from their efforts may be modeled by the function

$$R(N) = -3N^4 + 50N^3 - 261N^2 + 540N$$

for $0 \leq N \leq 9$. How many salespersons should be hired to maximize total net revenue?

Supplementary Problems*

Sketch the graph of each function in Problems 1-20. Use as many as possible of the key features such as domain, relative extrema, inflection points, concavity, asymptotes, intercepts, cusps, and vertical tangents.

1. $f(x) = x^3 + 6x^2 + 9x - 1$ **2.** $f(x) = x^4 + 4x^3 + 4x^2 + 1$ **3.** $f(x) = 3x^4 - 4x^3 + 1$

4. $f(x) = 3x^4 - 4x^2 + 1$ **5.** $f(x) = 6x^5 - 15x^4 + 10x^3$ **6.** $f(x) = 3x^5 - 10x^3 + 15x + 1$

7. $f(x) = \dfrac{9 - x^2}{3 + x^2}$ **8.** $f(x) = \dfrac{x}{1 - x}$ **9.** $f(x) = \sin 2x - \sin x$ on $[-\pi, \pi]$

10. $f(x) = \sin x \sin 2x$ on $[-\pi, \pi]$ **11.** $f(x) = \dfrac{x^2 - 4}{x^2}$ **12.** $f(x) = \dfrac{x^2 + 2x - 3}{x^2 - 3x + 2}$

13. $f(x) = \dfrac{3x - 2}{(x + 1)^2(x - 2)}$ **14.** $f(x) = \dfrac{x^3 + 3}{x(x + 1)(x + 2)}$ **15.** $f(x) = x^2 \ln \sqrt{x}$

16. $f(x) = \sin^{-1} x + \cos^{-1} x$ **17.** $f(x) = x(e^{-2x} + e^{-x})$ **18.** $f(x) = \ln\left(\dfrac{x - 1}{x + 1}\right)$

19. $f(x) = \dfrac{5}{1 + e^{-x}}$ **20.** $f(x) = xe^{1/x}$

In Problems 21-24, the graph of the given function $f(x)$ for $x > 0$ is one of the six curves shown in Figure 4.83. In each case, match the function to a graph.

A. 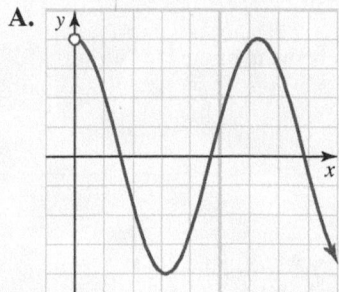   **B.** 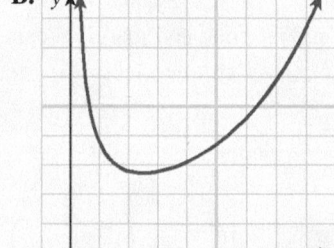   **C.**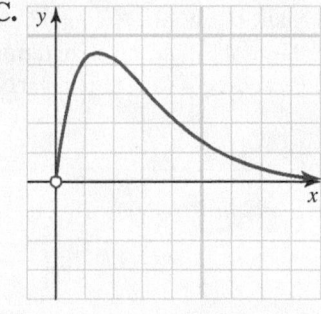

D. 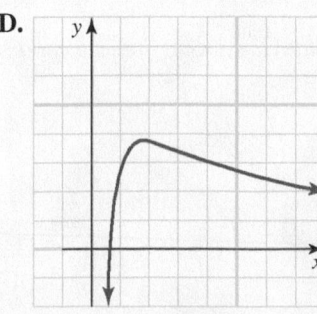   **E.** 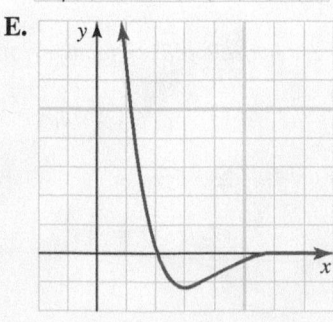   **F.**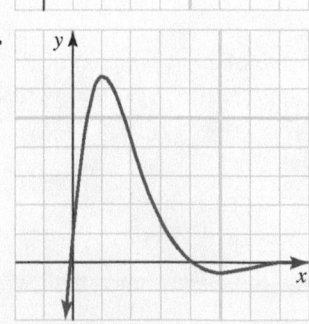

Figure 4.83   Problems 21-24

---

21. $f(x) = x2^{-x}$

22. $y = \dfrac{\ln \sqrt{x}}{x}$

23. $f(x) = \dfrac{e^x}{x}$

24. $y = e^{-x} \sin x$

Determine the absolute maximum and absolute minimum value of each function on the interval given in Problems 25-28.

25. $f(x) = x^4 - 8x^2 + 12$ on $[-1, 2]$

26. $f(x) = \sqrt{x}(x - 5)^{1/3}$ on $[0, 6]$

27. $f(x) = 2x - \sin^{-1} x$ on $[0, 1]$

28. $f(x) = e^{-x} \ln x$ on $\left[\frac{1}{2}, 2\right]$

Evaluate the limits in Problems 29-54.

29. $\displaystyle\lim_{x \to \infty} \dfrac{x \sin^2 x}{x^2 + 1}$

30. $\displaystyle\lim_{x \to \infty} \dfrac{2x^4 - 7}{6x^4 + 7}$

31. $\displaystyle\lim_{x \to \infty} \left(\sqrt{x^2 - x} - x \right)$

32. $\displaystyle\lim_{x \to \infty} \left[\sqrt{x(x + b)} - x \right]$

33. $\displaystyle\lim_{x \to 0} \dfrac{x \sin x}{x + \sin^3 x}$

34. $\displaystyle\lim_{x \to 0} \dfrac{x \sin x}{x^2 - \sin^3 x}$

35. $\displaystyle\lim_{x \to 0} \dfrac{x \sin^2 x}{x^2 - \sin^2 x}$

36. $\displaystyle\lim_{x \to 0} \dfrac{x - \sin x}{\tan^3 x}$

37. $\displaystyle\lim_{x \to 0} \dfrac{\sin^2 x}{\sin x^2}$

38. $\displaystyle\lim_{x \to 0} \left(\dfrac{1}{x^2} - \dfrac{1}{x^2 \sec x} \right)$

39. $\displaystyle\lim_{x \to (\frac{\pi}{2})^-} \dfrac{\sec^2 x}{\sec^2 3x}$

40. $\displaystyle\lim_{x \to (\frac{\pi}{2})^-} (1 - \sin x) \tan x$

41. $\displaystyle\lim_{x \to (\frac{\pi}{2})^-} (\sec x - \tan x)$

42. $\displaystyle\lim_{x \to 0^+} (\sin x)^{1/\ln \sqrt{x}}$

43. $\displaystyle\lim_{x \to 1} \dfrac{(x - 1)\sin(x - 1)}{1 - \cos(x - 1)}$

44. $\displaystyle\lim_{x \to 0} \left(\dfrac{1}{\sin 2x} - \dfrac{1}{2x} \right)$

45. $\displaystyle\lim_{x \to \infty} \left(\sqrt{x^2 + 4} - \sqrt{x^2 - 4} \right)$

46. $\displaystyle\lim_{x \to 0^+} (1 + x)^{4/x}$

47. $\displaystyle\lim_{x \to 1^-} \left(\dfrac{1}{1 - x} \right)^x$

48. $\displaystyle\lim_{x \to 0^+} x^{\tan x}$

49. $\displaystyle\lim_{x \to 0} \dfrac{5^x - 1}{x}$

50. $\displaystyle\lim_{x \to 0} \dfrac{\ln(x^2 + 1)}{x}$

51. $\displaystyle\lim_{x \to \infty} \left(\dfrac{1}{x} \right)^x$

52. $\displaystyle\lim_{x \to \infty} \left(4 - \dfrac{1}{x} \right)^x$

53. $\displaystyle\lim_{x \to \infty} \dfrac{e^x \cos x - 1}{x}$

54. $\displaystyle\lim_{x \to \infty} \left(\dfrac{x^3}{x^2 - x + 1} - \dfrac{x^3}{x^2 + x - 1} \right)$

55. Journal Problem (*Mathematics Teacher.**) Which of the graphs in Figure 4.84 is the derivative and which is the function?

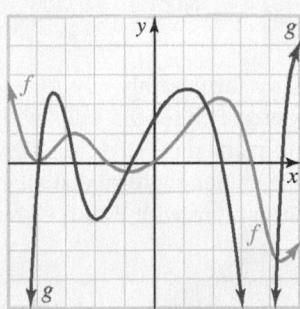

Figure 4.84 A function and its derivative

56. EXPLORATION PROBLEM Determine a, b, and c such that the graph of $f(x) = ax^3 + bx^2 + c$ has an inflection point and slope 1 at $(-1, 2)$.

57. EXPLORATION PROBLEM Consider the graph of $y = x^3 + bx^2 + cx + d$ for constants b, c, and d. What happens to the graph as b changes?

58. Think Tank Problem Explain why the graph of a quadratic polynomial cannot have a point of inflection. How many inflection points does the graph of a cubic polynomial have?

---

*December 1990, p. 718.

59. **Think Tank Problem** Sketch the graph of a function with the following properties:

$$f'(x) > 0 \text{ when } x < 1$$

$$f'(x) < 0 \text{ when } x > 1$$

$$f''(x) > 0 \text{ when } x < 1$$

$$f''(x) > 0 \text{ when } x > 1$$

What can you say about the derivative of f when $x = 1$?

60. Find the points on the hyperbola $x^2 - y^2 = 4$ that are closest to the point $(0, 1)$.

61. Find numbers A, B, C, and D that guarantee that the function

$$f(x) = Ax^3 + Bx^2 + Cx + D$$

will have a relative maximum at $(-1, 1)$ and a relative minimum at $(1, -1)$.

62. A Norman window consists of a rectangle with a semicircle surmounted on the top. What are the dimensions of the Norman window of largest area with a fixed perimeter of P_0 meters?

© Kendall Hunt Publishing Company

Norman window

63. An apartment complex has 200 units. When the monthly rent for each unit is $600, all units are occupied. Experience indicates that for each $20-per-month increase in rent, 5 units will become vacant. Each rented apartment costs the owners of the complex $80 per month to maintain. What monthly rent should be charged to maximize the owner's profit? What is maximum profit? How many units are rented when profit is a maximum?

64. A farmer wishes to enclose a rectangular pasture with 320 ft of fence. Find the dimensions that give the maximum area in these situations:

a. The fence is on all four sides of the pasture.

b. The fence is on three sides of the pasture and the fourth side is bounded by a wall.

65. A peach grower has determined that if 30 trees are planted per acre, each tree will average 200 lb of peaches per season. However, for each tree grown in addition to the 30 trees, the average yield for each of the trees in the grove drops by 5 lb per tree. How many peach trees should be planted on each acre in order to maximize the yield of peaches per acre? What is the maximum yield?

66. **Modeling Problem** To obtain the maximum price, a shipment of fruit should reach the market as early as possible after the fruit has been picked. If a grower picks the fruit immediately for shipment, 100 cases can be shipped at a profit of $10 per case. By waiting, the grower estimates that the crop will yield an additional 25 cases per week, but because the competitor's yield will also increase, the grower's profit will decrease by $1 per case per week. Use calculus to determine when the grower should ship the fruit to maximize profit. What will be the maximum profit?

67. **Modeling Problem** Oil from an offshore rig located 3 mi from the shore is to be pumped to a location on the edge of the shore that is east of the rig. The cost of constructing a pipe in the ocean from the rig to the shore is 1.5 times as expensive as the cost of construction on land. Set up and analyze a model to determine how the pipe should be laid to minimize cost.

68. The owner of a novelty store can obtain joy buzzers from the manufacturer for $1.40 apiece. It is estimated that 60 buzzers will be sold when the price is $3.20 per buzzer and that 10 more buzzers will be sold for every 10¢ decrease in price. What price should be charged to maximize profit?

69. **Journal Problem** (*AMATYC Review**) You are to build a rectangular enclosure with front side made of material costing $10 per foot and with the other three sides (back, left, right) of material costing $5 per foot. If the enclosure is to contain exactly 600 square feet and to be built at minimum total cost, how long should the side be and what is the total cost?

Answer: 20 *by* 30; $600.

Here is the real question: By coincidence, the answer matches the constrained area (both 600, aside from units, of course). *Question*: Characterize when this happens more generally, and thus describe the relationship that exists in such problems in which the resultant minimum cost matches the given area (all in appropriate units).

70. **Journal Problem** (*Quantum*[†]) Two students are pondering the following problem: A fish tank in the form of a right circular cylinder with radius R and height H. An ant is sitting on the border circle of one of its bases (point A in Figure 4.85.

It wants to crawl to the most distant point B at the border circle of the other base (symmetric to A with respect to the center of the tank). Find the shortest path for the ant.

a. What is the length of the shortest path?

b. *After* finding a solution, compare your solution with that of the first student as follows: "It's a simple problem!", the first student says confidently. "We just have to consider the planar development of the tank. Let's say, for the sake of definiteness, that the ant first crawls along the side surface and then across the upper base (of course, it's possible that the ant takes the symmetric route: first crawling across the lower base, then along the side surface; but the length of this route is the same)."

© Kendall Hunt Publishing Company

Figure 4.85 Cylindrical tank

c. Compare your solution with that of the second student, as follows: "Wait a second," the other student replies. "One can very comfortably develop the tank in a different way! Just throw away the lids and spread the side surface on the plane so that we get a rectangle. Then the shortest path will be the segment connecting points A and B, where the image of the ant's route on the tank will be part of a corkscrew line."

d. Either your solution matched one of the student's solutions or it did not. Which of the two (or possibly three) solutions provides the minimum? Before you answer, find out the conditions whereby the lengths of both of the student's routes are the equal. Critique the solutions given by the two students.

71. Westel Corporation manufactures telephones and has developed a new cellular phone. Production analysis shows that its price must not be less than $50; if x units are sold, then the price per unit is given by the formula $p(x) = 150 - x$. The total cost of producing x units is given by the formula $C(x) = 2{,}500 + 30x$. Find the maximum profit, and determine the price that should be charged to maximize the profit.

72. **Modeling Problem** A manufacturer receives an order for 5,000 items. There are 12 machines available, each of which can produce 25 items per hour. The cost of setting up a machine for a production run is $50. Once the machines are in operation, the procedure is fully automated and can be supervised by a single worker earning $20 per hour. Set up and analyze a model to determine the number of machines that should be used to minimize the total cost of filling the order. State any assumptions that must be made.

73. Show that the graph of a polynomial of degree n, with $n > 2$, has at most $n - 2$ inflection points.

---

*AMATYC Review, Spring 1995, pp. 67-68.

[†]September/October, 1997, Vol. 7, No. 4, pp. 50-53. We changed the problem to a fish tank instead of a tin can after we observed an ant at an aquarium.

74. Show that the graph of the function $f(x) = x^n$ with $n > 1$ has either one or no inflection points, depending on whether n is odd or even.

75. Each tangent line to the circle $x^2 + y^2 = 1$ at a point in the first quadrant will intersect the coordinate axes at points $(x_0, 0)$ and $(0, y_0)$. Determine the line for which $x_0 + y_0$ is a minimum.

76. An accelerated particle moving at speed close to the speed of light emits power P in the direction θ given by

$$P(\theta) = \frac{a \sin \theta}{(1 - b \cos \theta)^5}$$

for $0 < b < 1$ and $a > 0$. Find the value of θ ($0 \leq \theta \leq \pi$) for which P has the greatest value.

77. Suppose that f is a continuous function defined on the closed interval $[a, b]$ and that $f'(x) = c$ on the open interval (a, b) for some constant c. Use the MVT to show that

$$f(x) = c(x - a) + f(a)$$

for all x in the interval $[a, b]$.

78. Find the point of inflection of the curve

$$y = (x + 1) \tan^{-1} x$$

79. Find the critical numbers for the function

$$f(x) = \tan^{-1}\left(\frac{x}{a}\right) - \tan^{-1}\left(\frac{x}{b}\right) \qquad a > b$$

Classify each as corresponding to a relative maximum, a relative minimum, or neither.

80. Suppose that $f''(x)$ exists for all x and that $f''(x) + c^2 f(x) = 0$ for some number c with $f'(0) = 1$. Show that for any $x = x_0$, there is a number w between 0 and x_0 for which

$$f'(x_0) + c^2 f(w)x_0 = 1$$

81. Show that x^n grows more quickly than x^m for $n > m \geq 0$ for $x \to \pm\infty$, and thus the graph of a polynomial behaves like the graph of the term of highest degree for $x \to \pm\infty$.

Using the graphing and differentiation programs of your computer or calculator, graph $f(x)$, $f'(x)$, $f''(x)$ for each function in Problems 82-85. Print out a copy, if possible. On each graph, indicate where

a. $f'(x) > 0$ **b.** $f'(x) < 0$ **c.** $f''(x) > 0$ **d.** $f''(x) < 0$ **e.** $f'(x) = 0$
f. $f'(x)$ *does not exist*   **g.** $f''(x) = 0$

Describe how these inequalities qualitatively determine the shape of the graph of the functions over the given interval.

82. $f(x) = \sin 2x$ on $[-\pi, \pi]$ **83.** $f(x) = x^3 - x^2 - x + 1$ on $\left[-\frac{3}{2}, 2\right]$

84. $f(x) = x^4 - 2x^2$ on $[-2, 2]$ **85.** $f(x) = x^3 - x + \dfrac{1}{x} + 1$ on $[-2, 2]$

86. EXPLORATION PROBLEM Frank Kornerkutter has put off doing his math homework until the last minute, and he is now trying to evaluate

$$\lim_{x \to 0^+} \left(\frac{1}{x^2} - \frac{1}{x}\right)$$

At first he is stumped, but suddenly he has an idea: Because

$$\lim_{x \to 0^+} \frac{1}{x^2} = \infty \quad \text{and} \quad \lim_{x \to 0^+} \frac{1}{x} = \infty$$

it must surely be true that the limit in question has the value $\infty - \infty = 0$. Having thus "solved" his problem, he celebrates by taking a nap. Is he right, and if not, what is wrong with his argument?

87. Let

$$P(x) = a_n x^n + a_{n-1} x^{n-1} + \cdots + a_1 x + a_0$$

be polynomial with $a_n \neq 0$ and let

$$L = \lim_{x \to -\infty} P(x) \quad \text{and} \quad M = \lim_{x \to \infty} P(x)$$

Fill in the missing entries in the following table:

Sign of a_n	n	L	M
+	even	**a.**	∞
+	odd	$-\infty$	**b.**
−	even	**c.**	**d.**
−	odd	**e.**	**f.**

88. Consider a string 60 in. long that is formed into a rectangle. Using a graphing calculator or a graphing program, graph the area $A(x)$ enclosed by the string as a function of the length x of a given side ($0 < x < 30$). From the graph, deduce that the maximum area is enclosed when the rectangle is a square. What is the minimum area enclosed? Compare this problem to the exact solution. What conclusions do you draw about the desirability of analytical solutions and the role of the computer?

89. Modeling Problem The beach of a lake follows contours that are approximated by the curve $4x^2 + y^2 = 1$, and a nearby road lies along the curve $y = 1/x$ for $x > 0$. Using your graphing calculator or computer software, determine the closest approach of the road to the lake in the north-south direction. Take the positive y-axis as pointing north.

90. Modeling Problem Suppose you are a manager of a fleet of delivery trucks. Each truck is driven at a constant speed of x mi/h ($15 \leq x \leq 55$), and gas consumption (gal/mi) is modeled by the function

$$\frac{1}{250} \left(\frac{750}{x} + x \right)$$

Using a graphing and differentiation program (and/or root-solving program on your computer or calculator), answer the following questions:

a. If gas costs \$4.70/gal, estimate the steady speed that will minimize the cost of fuel for a 500-mi trip.

b. Estimate the steady speed that minimizes the cost if the driver is paid \$58 per hour and the price of gasoline remains constant at \$4.70/gal.

91. Using your graphing and differentiation programs, locate and identify all the relative extrema for the function defined by

$$f(x) = \frac{1}{5}x^5 - \frac{5}{4}x^4 + 2x^2$$

This may require judicious choices of the window settings.

92. *Historical Quest* * *The possibility of division by zero is a fact that causes special concern to mathematicians. One of the first recorded observations of division by zero comes from the 12th-century Hindu mathematician Bhaskaracharya (also known as Bhaskara)Bhaskara), who made the following observation: "The fraction whose denominator is zero, is termed an infinite quantity." Bhaskaracharya then went on to give a very beautiful conception of infinity that involved his view of God and creation. Bhaskaracharya gave a solution for the so-called **Pell equation***

$$x^2 = 1 + py^2 \quad (x, y \text{ both nonnegative})$$

which is related to the problem of cutting a given sphere so the volumes of the two parts have a specified ratio. Solve this equation for $p = \frac{2}{3}$ and write this Pell equation as $y = f(x)$. Find $\lim_{x \to \infty} f(x)$.

---

*From "Mathematics in India in the Middle Ages," by Chandra B. Sharma, Mathematical Spectrum, Volume 14(1), pp. 6-8, 1982.

93. *Historical Quest Leonhard Euler is one of the giants in the history of mathematics. His name is attached to almost every branch of mathematics. He was the most prolific writer on the subject of mathematics, and his mathematical textbooks were masterfully written. His writing was not at all slowed down by his total blindness for the last 17 years of his life. He possessed a phenomenal memory, had almost total recall, and could mentally calculate long and complicated problems. The basis for the historical development of calculus, as well as modern-day analysis, is the notion of a function. Euler's book* **Introductio in analysin infinitorum** *(1784) first used the function concept as the basic idea. It was the identification of functions, rather than curves, as the principal object of study, that permitted the advancement of mathematics in general, and calculus in particular. In Chapter 2, we noted that the number e is named in honor of Euler. Euler did not use the definition we use today, namely*

**Leonhard Euler
(1707-1783)**

$$e = \lim_{n \to \infty} \left(1 + \frac{1}{n}\right)^n$$

Instead, Euler used series in his work (which we will study in Chapter 8).

In this *Historical Quest* problem we will lay the groundwork for a *Historical Quest* in Chapter 8. Euler introduced $\log_a x$ (which he wrote as ℓx) as that exponent y such that $a^y = x$. This was done in 1748, which makes it the first appearance of a logarithm interpreted explicitly as an exponent. He does not define $a^0 = 1$, but instead writes

$$a^\epsilon = 1 + k\epsilon$$

for an infinitely small number ϵ. In other words,

$$k = \lim_{\epsilon \to 0} \frac{a^\epsilon - 1}{\epsilon}$$

Explain why $k = \ln a$.

When Euler was 13, he registered at the University of Basel and was introduced to another famous mathematician, Johann Bernoulli, who was an instructor there at the time (see Historical Quest, Section 4.5, Problems 49-50). If Bernoulli thought that a student was promising, he would provide, sometimes gratis, private instruction. Here is Euler's own account of this first encounter with Bernoulli **"I soon found an opportunity to gain introduction to the famous professor Johann Bernoulli, whose good pleasure it was to advance me further in the mathematical sciences. True, because of his business he flatly refused me private lessons, but he gave me much wiser advice, namely, to get some more difficult mathematical books and work through them with all industry, and wherever I should find some check or difficulties, he gave me free access to him every Saturday afternoon and was so kind as to elucidate all difficulties, which happened with such greatly desired advantage that whenever he had obviated one check for me, because of that ten others disappeared right away, which is certainly the way to make a happy advance in the mathematical sciences."**[*]

94. **Putnam examination problem** Given the parabola $y^2 = 2mx$, what is the length of the shortest chord that is normal to the curve at one end? *Hint:* If $\overline{AB}$ is normal to the parabola, where A and B are the points $(2mt^2, 2mt)$ and $(2ms^2, 2ms)$, show that the slope of the tangent at A is $\frac{1}{2t}$.

95. **Putnam examination problem** Prove that the polynomial

$$(a - x)^6 - 3a(a - x)^5 + \frac{5}{2}a^2(a - x)^4 - \frac{1}{2}a^4(a - x)^2$$

has only negative values for $0 < x < a$. *Hint:* Show that if $x = a(1 - y)$, the polynomial becomes $a^6 y^2 g(y)$, where

$$g(y) = y^4 - 3y^3 + \frac{5}{2}y^2 - \frac{1}{2}$$

and then prove that $g(y) < 0$ for $0 < y < 1$.

---

[*]From *Elements of Algebra* by Leonhard Euler. London: Longman, Orme, and Co. Reprinted by Springer-Verlag.

96. Putnam examination problem Find the maximum value of $f(x) = x^3 - 3x$ on the set of all real numbers x satisfying $x^4 + 36 \le 13x^2$.

97. Putnam examination problem Let T be an acute triangle. Inscribe a pair R and S of rectangles in T, as shown in Figure 4.86.

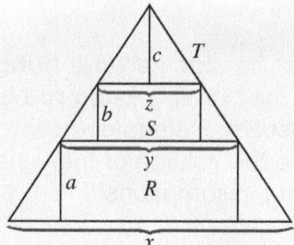

Figure 4.86 Problem 97

Let $A(x)$ denote the area of polygon X. Find the maximum value, or show that no maximum exists, of the ratio

$$\frac{A(R) + A(S)}{A(T)}$$

where T ranges over all triangles R, and S ranges over all rectangles, as shown in Figure 4.86.

98. Putnam examination problem Which is greater

$$\left(\sqrt{n}\right)^{\sqrt{n+1}} \qquad \text{or} \qquad \left(\sqrt{n+1}\right)^{\sqrt{n}}$$

where $n > 8$? *Hint:* Use the function $f(x) = \dfrac{\ln x}{x}$, and show that

$$x^y > y^x \qquad \text{when } e \le x < y$$

99. Putnam examination problem The graph of the equation $x^y = y^x$ for $x > 0$, $y > 0$ consists of a straight line and a curve. Find the coordinates of the point where the line and the curve intersect.

CHAPTER 4 GROUP RESEARCH PROJECT*

Working in small groups is typical of most work environments, and this book seeks to develop skills with group activities. We present a group project at the end of each chapter. These projects are to be done in groups of three or four students.

Wine Barrel Capacity

© 2013 Shcheko. Used under license from Shutterstock, Inc.

A wine barrel has a hole in the middle of its side called a **bung hole**. To determine the volume of wine in the barrel, a **bung rod** is inserted in the hole until it hits the lower seam. Determine how to calibrate such a rod so that it will measure the volume of the wine in the barrel. You should make the following assumptions:

1. The barrel is cylindrical.
2. The distance from the bung hole to the corner is λ.
3. The ratio of the height to the diameter of the barrel is t. This ratio should be chosen so that for a given λ value, the volume of the barrel is maximal.

Your paper is not limited to the following questions, but it should include these concerns: You should show that the volume of the cylindrical barrel is

$$V = 2\pi\lambda^3 t(4 + t^2)^{-3/2}$$

and you should find the approximate ideal value for t. Johannes Kepler was the first person to show mathematically why coopers were guided in their construction of wine barrels by one rule: *make the staves (the boards that make up the sides of the barrel) one and one-half times as long as the diameter.* (This is the approximate t-value.) You should provide dimensions for the barrel as well as for the bung rod.

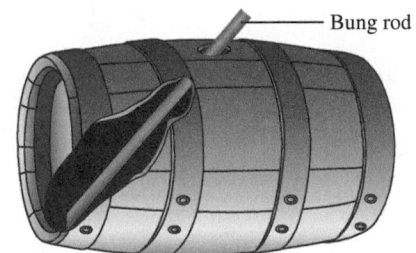

— Bung rod

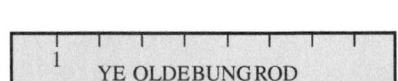

1
YE OLDE BUNG ROD

Extended paper for further study. Johannes Kepler (1571-1630) is usually remembered for his work in astronomy, in particular for his three laws of planetary motion. Tycho Brahe (1546-1601) was working for Rodolf II, Holy Roman Emperor in Prague in 1599, and he asked Kepler to work with him. The historian Burton describes this as a fortunate alliance. ''Tycho was a splendid observer but a poor mathematician, while Kepler was a splendid mathematician but a poor observer.'' Because Kepler was a Protestant during a time when most intellectuals were required to be Catholic, Kepler had trouble supporting himself, and consequently worked for many benefactors. While serving the Austrian emperor Matthew I, Kepler observed with admiration the ability of a young vintner to declare quickly and easily the capacities of a number of different wine casks. He describes how this can be done in his book *The New Stereometry of Wine Barrels, Mostly Austrian.* Write a paper telling of Kepler's work with coopers in deriving this formula.

---

*The idea for this group research project comes from research done at Iowa State University as part of a National Science Foundation grant. Our thanks to Elgin Johnston of Iowa State University.

CHAPTER 5

INTEGRATION

M*athematics is one of the oldest of the sciences; it is also one of the most active, for its strength is the vigor of perpetual youth.*

A. R. Forsythe,
Nature 84 (1910): 285

PREVIEW

The key concept in integral calculus is *integration,* a procedure that involves computing a special kind of limit of sums called the *definite integral.* We shall find that such limits can often be computed by reversing the process of differentiation; that is, given a function f, we find a function F such that $F' = f$. This is called *indefinite integration,* and the equation $F' = f$ is an example of a *differential equation.**

PERSPECTIVE

Finding integrals and solving differential equations are extremely important processes in calculus. We begin our study of these topics by defining definite and indefinite integration and showing how they are connected by a remarkable result called the *fundamental theorem of calculus.* Then we examine several techniques of integration and show how area, average value, and other quantities can be set up and analyzed by integration. Our study of differential equations begins in this chapter and will continue in appropriate sections throughout this text. We also establish a mean value theorem for integrals and develop numerical procedures for estimating the value of a definite integral. Finally, we close the chapter with an optional section that considers an alternate approach to the notion of logarithm. Instead of relying on the definitions from precalculus, in this section, we define the logarithm using the ideas from calculus to define a logarithm as an integral. This is a particularly elegant path, and the only reason why it is not used more widely is that it requires the delay of logarithms and the exponential number e until after the definition of an integral.

*Optional section.

CONTENTS

5.1 ANTIDIFFERENTIATION

IN THIS SECTION: *Reversing differentiation, antiderivative notation, antidifferentiation formulas, applications, area as an antiderivative*
Our goal in this section is to study a process called *antidifferentiation*, which reverses differentiation in much the same way that division reverses multiplication. We shall examine algebraic properties of the process and several useful applications.

Reversing Differentiation

A physicist who knows the acceleration of a particle may want to determine its velocity or its position at a particular time. An ecologist who knows the rate at which a certain pollutant is being absorbed by a particular species of fish might want to know the actual amount of pollutant in the fish's system at a given time. In each of these cases, a derivative f' is given and the problem is that of finding the corresponding function f. Toward this end, we make the following definition.

> **ANTIDERIVATIVE**   A function F is called an **antiderivative** of a given function f on an interval I if
> $$F'(x) = f(x)$$
> for all x in I.

Suppose we know $f(x) = 3x^2$. We want to find a function $F(x)$ such that $F'(x) = 3x^2$. It is not difficult to use the power rule in reverse to discover that $F(x) = x^3$ is such a function. However, that is not the only possibility:

Given	$F(x) = x^3$	$G(x) = x^3 - 5$	$H(x) = x^3 + \pi^2$
Find:	$F'(x) = 3x^2$	$G'(x) = 3x^2$	$H'(x) = 3x^2$

In fact, if F is an antiderivative of f, then so is $F + C$ for any constant C, because

$$[F(x) + C]' = F'(x) + 0 = f(x)$$

In the following theorem, we use the constant difference theorem of Section 4.2 to show that *any* antiderivative of f can be expressed in this form.

Theorem 5.1   Antiderivatives of a function differ by a constant

If F is an antiderivative of the continuous function f, then any other antiderivative, G, of f must have the form

$$G(x) = F(x) + C$$

> ■ **W**hat this says Two antiderivatives of the same function differ by a constant.

Proof: If F and G are both antiderivatives of f, then $F' = f$ and $G' = f$ and Theorem 4.6 (the constant difference theorem) tells us that

$$G(x) - F(x) = C \qquad \text{so} \qquad G(x) = F(x) + C \qquad \blacklozenge$$

Example 1 Finding antiderivatives

Find general antiderivatives for the given functions.

a. $f(x) = x^5$ **b.** $s(x) = \sin x$ **c.** $r(x) = \dfrac{1}{x}$

Solution

a. As we know very well, $\left(x^6\right)' = 6x^5$, so we see that a particular antiderivative of f is $F(x) = \dfrac{x^6}{6}$ to obtain $F'(x) = \dfrac{6x^5}{6} = x^5$. By Theorem 5.1, the most general antiderivative is $G(x) = \dfrac{x^6}{6} + C$.

b. If $S(x) = -\cos x$, then $S'(x) = \sin x$, so $G(x) = -\cos x + C$.

c. If $R(x) = \ln|x|$, then $R'(x) = \dfrac{1}{x}$, so $G(x) = \ln|x| + C$. ■

Recall that the slope of a function $y = f(x)$ at any point (x, y) on its graph is given by the derivative $f'(x)$. We can exploit this fact to obtain a "picture" of the graph of f. Reconsider Example 1c where $y' = 1/x$. There is an antiderivative $F(x)$ of $1/x$ such that the slope of F at each point $(x, F(x))$ is $1/x$ for each nonzero value of x. Let us draw a graph of these slopes following the procedure shown in Figure 5.1.

a. If $x = 1$, then the slope is $\frac{1}{1} = 1$. Draw short segments at $x = 1$, each with slope 1 for different y-values. If $x = -3$, then the slope is $-\frac{1}{3}$, so draw short segments, as before.

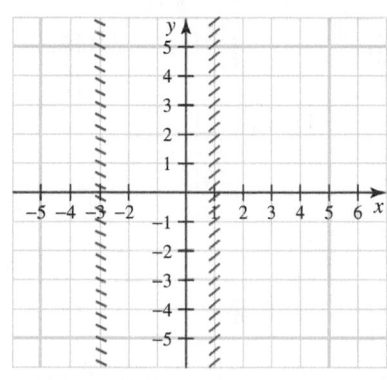

b. Continue to plot slope points for different values of x. The result is shown, and is known as a **slope field** for the the equation $y' = \frac{1}{x}$.

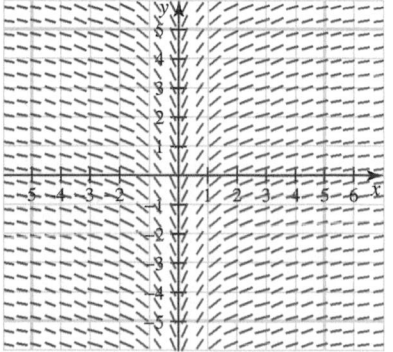

Figure 5.1 Slope field

Finally, notice the relationship between the slope field for $y' = \dfrac{1}{x}$ and its antiderivative $y = \ln|x| + C$ (found in Example 1c). If we choose particular values for C, say $C = 0$, $C = -\ln 2$, or $C = 2$ and draw these particular antiderivatives in Figure 5.2, we notice that these particular solutions are anticipated by the slope field drawn in Figure 5.1**b**. That is, the slope field shows the entire family of antiderivatives of the original equation.

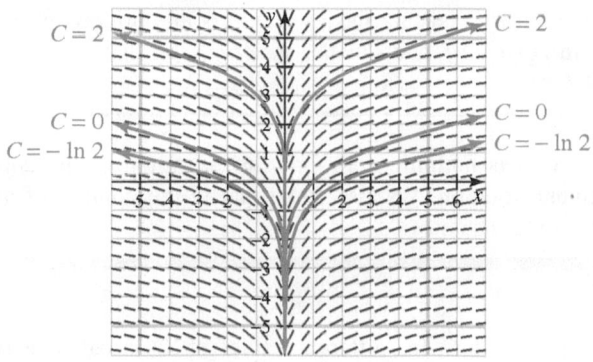

Figure 5.2 Sample antiderivatives

Slope fields will be discussed in more detail in Section 5.6, when we will use the term *direction fields* for what we are now calling *slope fields*. In general, slope fields, and antiderivatives obtained by using slope fields, are usually generated using technology in computers and calculators. Here is an example in which the antiderivatives cannot be obtained as elementary functions.

Example 2   Finding an antiderivative using a slope field

Consider the slope field for $y' = e^{x^2}$, which is shown in Figure 5.3.

Draw a possible graph of the antiderivative of e^{x^2} that passes through the point $(0, 0)$.

Solution   Each little segment represents the slope of e^{x^2} for a particular x value. For example, if $x = 0$, the slope is 1, if $x = 1$ the slope is e, and if $x = 2$, the slope is e^4 (quite steep). However, to draw the antiderivative, step back and take a large view of the graph, and "go with the flow." We sketch the apparent graph that passes through $(0, 0)$, as shown in Figure 5.4.

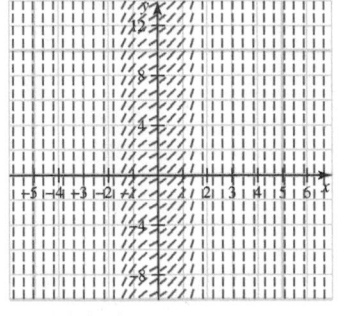

Figure 5.3 Interactive
Slope field for $y' = e^{x^2}$

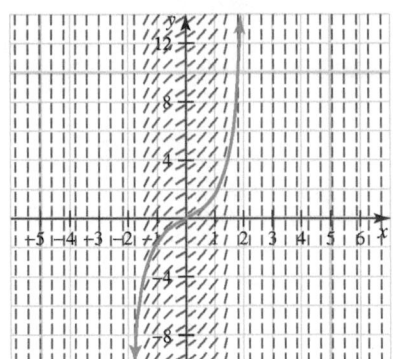

Figure 5.4 Interactive   Antiderivative of $y' = e^{x^2}$ passing through (0,0)

Antiderivative Notation

It is worthwhile to define a notation to indicate the operation of antidifferentiation.

> **INDEFINITE INTEGRAL**   The notation
>
> $$\int f(x)\, dx = F(x) + C$$
>
> where C is an arbitrary constant means that F is an antiderivative of f. It is called the **indefinite integral of** f and satisfies the condition that $F'(x) = f(x)$ for all x in the domain of f.

■ **W**hat this says This is nothing more than the definition of the antiderivative, along with a convenient notation. We also agree that in the context of antidifferentiation, C is an arbitrary constant.

It is important to remember that $\int f(x)\, dx$ represents a family of functions, not just a single function.

The graph of $F(x) + C$ for different values of C is called a **family of functions** (see Figure 5.5).

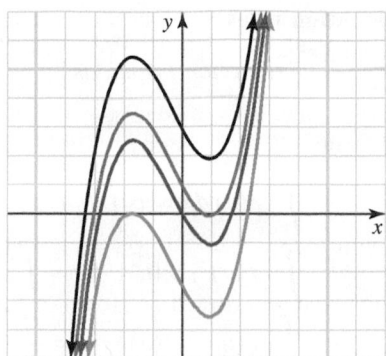

Figure 5.5 Several members of the family of curves $y = F(x) + C$

Because each member of the family $y = F(x) + C$ has the same derivative at x, the slope of the graph at x is the same. This means that the graph of all functions of the form $y = F(x) + C$ is a collection of parallel curves, as shown in Figure 5.5.

The process of finding indefinite integrals is called **indefinite integration**. Notice that this process amounts to finding an antiderivative of f and adding an arbitrary constant C, which is called the **constant of integration**.

Example 3   Antidifferentiation

Find each of the following indefinite integrals.

a. $\int 5x^3 dx$ **b.** $\int \sec^2 x\, dx$ **c.** $\int e^x dx$

Solution

a. Since $\dfrac{d}{dx}(x^4) = 4x^3$, it follows that $\dfrac{d}{dx}\left(\dfrac{5x^4}{4}\right) = 5x^3$. Thus,

$$\int 5x^3 dx = \frac{5x^4}{4} + C$$

b. Because $\dfrac{d}{dx}(\tan x) = \sec^2 x$, we have

$$\int \sec^2 x\, dx = \tan x + C$$

c. Since $\dfrac{d}{dx}(e^x) = e^x$, we have

$$\int e^x dx = e^x + C$$

Antidifferentiation Formulas

Example 3 leads us to state formulas for antidifferentiation. Theorem 5.2 summarizes several fundamental properties of indefinite integrals, each of which can be derived by reversing an appropriate differentiation formula. Assume that f and g are functions; u is a variable; a, b, c are given constants; and C is an arbitrary constant.

Theorem 5.2   Basic Integration rules

Differentiation Formulas   *Integration Formulas*

PROCEDURAL RULES

Constant multiple rule: $\dfrac{d}{du}(cf) = c\,\dfrac{df}{du}$ $\displaystyle\int cf(u)\,du = c\int f(u)\,du$

Sum rules: $\dfrac{d}{du}(f+g) = \dfrac{df}{du} + \dfrac{dg}{du}$ $\displaystyle\int [f(u)+g(u)]\,du = \int f(u)\,du + \int g(u)\,du$

Difference rules: $\dfrac{d}{du}(f-g) = \dfrac{df}{du} - \dfrac{dg}{du}$ $\displaystyle\int [f(u)-g(u)]\,du = \int f(u)\,du - \int g(u)\,du$

Linearity rules: $\dfrac{d}{du}(af+bg) = a\,\dfrac{df}{du} + b\,\dfrac{dg}{du}$ $\displaystyle\int [af(u)+bg(u)]\,du = a\int f(u)\,du + b\int g(u)\,du$

BASIC FORMULAS

Constant rules: $\dfrac{d}{dx}(c) = 0$ $\displaystyle\int 0\,du = 0 + C$

Exponential rules: $\dfrac{d}{du}(e^u) = e^u$ $\displaystyle\int e^u\,du = e^u + C$

Power rules: $\dfrac{d}{du}(u^n) = nu^{n-1}$ $\displaystyle\int u^n\,du = \begin{cases} \dfrac{u^{n+1}}{n+1} + C & n \neq -1 \\ \ln|u| + C & n = -1 \end{cases}$

Logarithmic rules: $\dfrac{d}{du}(\ln|u|) = \dfrac{1}{u}$ Included in the above rule

Trigonometric rules: $\dfrac{d}{dx}(\cos u) = -\sin u$ $\displaystyle\int \sin u\,du = -\cos u + C$

$\dfrac{d}{dx}(\sin u) = \cos u$ $\displaystyle\int \cos u\,du = \sin u + C$

$\dfrac{d}{dx}(\tan u) = \sec^2 u$ $\displaystyle\int \sec^2 u\,du = \tan u + C$

Inverse trigonometric rules: The other three inverse trigonometric rules are not needed since, for example, $\sin^{-1} u + \cos^{-1} u = \frac{\pi}{2}$.

$\displaystyle\int \dfrac{du}{\sqrt{1-u^2}} = \sin^{-1} u + C_1$
$= \dfrac{\pi}{2} - \cos^{-1} u + C_1$
$= -\cos^{-1} u + C$

$\dfrac{d}{dx}(\sec u) = \sec u \tan u$ $\displaystyle\int \sec u \tan u\,du = \sec u + C$

$\dfrac{d}{dx}(\csc u) = -\csc u \cot u$ $\displaystyle\int \csc u \cot u\,du = -\csc u + C$

$\dfrac{d}{dx}(\cot u) = -\csc^2 u$ $\displaystyle\int \csc^2 u\,du = -\cot u + C$

$\dfrac{d}{du}(\sin^{-1} u) = \dfrac{1}{\sqrt{1-u^2}}$ $\displaystyle\int \dfrac{du}{\sqrt{1-u^2}} = \sin^{-1} u + C$

$\dfrac{d}{du}(\tan^{-1} u) = \dfrac{1}{1+u^2}$ $\displaystyle\int \dfrac{du}{1+u^2} = \tan^{-1} u + C$

$\dfrac{d}{du}(\sec^{-1} u) = \dfrac{1}{|u|\sqrt{u^2-1}}$ $\displaystyle\int \dfrac{du}{|u|\sqrt{u^2-1}} = \sec^{-1} u + C$

Proof: Each of these parts can be derived by reversing the accompanying derivative formula. For example, to obtain the power rule, note that if n is any number other than -1, then

$$\dfrac{d}{dx}\left[\dfrac{1}{n+1}u^{n+1}\right] = \dfrac{1}{n+1}\left[(n+1)u^n\right] = u^n$$

so that $\frac{1}{n+1}u^{n+1}$ is an antiderivative of u^n and

$$\int u^n\,du = \dfrac{1}{n+1}u^{n+1} + C \text{ for } n \neq -1$$ ◆

Now we shall use these rules to compute a number of indefinite integrals.

Example 4   Indefinite integral of a polynomial function

Find $\int \left(x^5 - 3x^2 - 7\right) dx$.

Solution   The first two steps are usually done mentally:

$$\int \left(x^5 - 3x^2 - 7\right) dx = \int x^5 dx - \int 3x^2 dx - \int 7\, dx \quad \text{\textit{Sum and difference rules}}$$

$$= \int x^5 dx - 3\int x^2 dx - 7\int dx \quad \text{\textit{Constant multiple rule}}$$

$$= \frac{x^{5+1}}{5+1} - 3\frac{x^{2+1}}{2+1} - 7x + C \quad \text{\textit{Power rule}}$$

$$= \frac{1}{6}x^6 - x^3 - 7x + C$$

Example 5   Indefinite integral with a mixture of forms

Evaluate $\int \left(5\sqrt{x} + 4\sin x\right) dx$.

Solution

$$\int \left(5\sqrt{x} + 4\sin x\right) dx = 5\int x^{1/2} dx + 4\int \sin x\, dx \quad \text{\textit{Sum and constant rules}}$$

$$= 5\frac{x^{3/2}}{\frac{3}{2}} + 4(-\cos x) + C \quad \text{\textit{Power and trig rules}}$$

$$= \frac{10}{3}x^{3/2} - 4\cos x + C$$

Antiderivatives will be used extensively in integration in connection with a marvelous result called the fundamental theorem of calculus (Section 5.4).

Applications

In Chapter 3, we used differentiation to compute the slope at each point on the graph of a function. Example 6 shows how this procedure can be reversed.

Example 6   Find a function, given the slope

The graph of a certain function F has slope $4x^3 - 5$ at each point (x, y) and contains the point $(1, 2)$. Find the function F.

Solution   We will work this problem twice; first we find an analytic solution and the second time we approximate the solution using technology, which is often sufficient. *Analytic solution:* Because the slope of the tangent at each point (x, y) is given by $F'(x)$, we have

$$F'(x) = 4x^3 - 5$$

and it follows that

$$\int F'(x)\, dx = \int (4x^3 - 5)\, dx$$

$$F(x) = 4\left(\frac{x^4}{4}\right) - 5x + C$$

$$= x^4 - 5x + C$$

The family of curves is $y = x^4 - 5x + C$. To find the one that passes through $(1, 2)$, substitute:

$$2 = 1^4 - 5(1) + C$$
$$6 = C$$

The curve is $y = x^4 - 5x + 6$.

Technology solution: Begin by drawing the slope field, as shown in Figure 5.6**a**. We are interested in drawing the particular solution passing through $(1, 2)$. Remember to "go with the flow," as shown in Figure 5.6**b**.

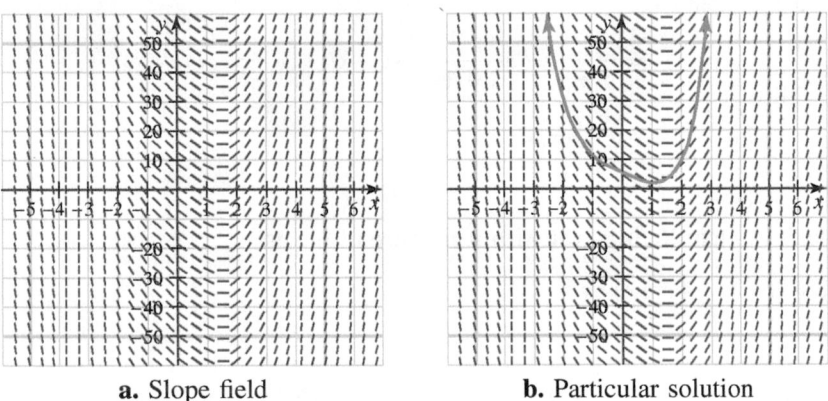

a. Slope field   **b.** Particular solution

Figure 5.6 Interactive   Slope field for $F'(x) = 4x^3 - 5$

If we compare the analytic solution and the graphical solution, we see that the graph of the equation in the analytic solution is the same as the one found by technology. ■

In Section 3.4, we observed that an object moving along a straight line with position $s(t)$ has velocity $v(t) = \dfrac{ds}{dt}$ and acceleration $a(t) = \dfrac{dv}{dt}$. Thus, we have

$$v(t) = \int a(t)\, dt \qquad \text{and} \qquad s(t) = \int v(t)\, dt$$

These formulas are used in Examples 7 and 8.

Example 7 Modeling Problem: The motion of a particle

A particle moves along a coordinate axis in such a way that its acceleration is modeled by $a(t) = 2t^{-2}$ for time $t > 0$. If the particle is at $s = 5$ when $t = 1$ and has velocity $v = -3$ at this time, where is it (to four decimal places) when $t = 4$?

Solution   Because $a(t) = v'(t)$, it follows that

$$v(t) = \int a(t)\, dt = \int 2t^{-2} dt = -2t^{-1} + C_1$$

and since $v(1) = -3$, we have

$$-3 = v(1) = \frac{-2}{1} + C_1 \qquad \text{so} \qquad C_1 = -3 + 2 = -1$$

We also know $v(t) = s'(t)$, so

$$s(t) = \int v(t)\, dt = \int \left(-2t^{-1} - 1\right) dt = -2\ln|t| - t + C_2$$

Since $s(1) = 5$, we have

$$5 = s(1) = -2\ln|1| - 1 + C_2 \qquad \text{or} \qquad C_2 = 6$$

Thus, $s(t) = -2\ln|t| - t + 6$ so that $s(4) \approx -0.7726$. The particle is at -0.7726 when $t = 4$.

Example 8   Stopping distance for an automobile

The brakes of a certain automobile produce a constant deceleration of 22 ft/s^2. If the car is traveling at 60 mi/h (88 ft/s) when the brakes are applied, how far will it travel before coming to a complete stop?

Solution   Let $a(t)$, $v(t)$, and $s(t)$ denote the acceleration, velocity, and position of the car t seconds after the brakes are applied. We shall assume that s is measured from the point where the brakes are applied, so that $s(0) = 0$.

$$\begin{aligned}
v(t) &= \int a(t)dt \\
&= \int (-22)dt \qquad \textit{Negative because the car is decelerating} \\
&= -22t + C_1 \qquad \textit{$v(0) = -22(0) + C_1 = 88$, so that $C_1 = 88$.} \\
&= -22t + 88
\end{aligned}$$

Similarly,

$$\begin{aligned}
s(t) &= \int v(t)dt \\
&= \int (-22t + 88)dt \\
&= -11t^2 + 88t + C_2 \qquad \textit{$s(0) = -11(0)^2 + 88(0) + C_2$, so that $C_2 = 0$.} \\
&= -11t^2 + 88t
\end{aligned}$$

Finally, the car comes to rest when its velocity is 0, so we need to solve $v(t) = 0$ for t:

$$\begin{aligned}
-22t + 88 &= 0 \\
t &= 4
\end{aligned}$$

This means that the car decelerates for 4 sec before coming to rest, and in that time it travels

$$s(4) = -11(4)^2 + 88(4) = 176 \text{ ft}$$

Indefinite integration also has applications in business and economics. Recall from Section 4.7, that the **demand function** for a particular commodity is the function $p(x)$, which gives the price p that consumers will pay for each unit of the commodity when x units are brought to market. Then the total revenue is $R(x) = xp(x)$, and the marginal revenue is $R'(x)$. The next example shows how the demand function can be determined from the marginal revenue.

Example 9   Finding the demand function given the marginal revenue

A manufacturer estimates that the marginal revenue of a certain commodity is $R'(x) = 240 + 0.1x$ when x units are produced. Find the demand function $p(x)$.

Solution

$$R(x) = \int R'(x)\,dx$$

$$= \int (240 + 0.1x)\,dx$$

$$= 240x + 0.1\left(\frac{1}{2}x^2\right) + C$$

$$= 240x + 0.05x^2 + C$$

Because $R(x) = xp(x)$, where $p(x)$ is the demand function, we must have $R(0) = 0$ so that

$$240(0) + 0.05(0)^2 + C = 0 \quad \text{or} \quad C = 0$$

Thus, $R(x) = 240x + 0.05x^2$. It follows that the demand function is

$$p(x) = \frac{R(x)}{x} = \frac{240x + 0.05x^2}{x} = 240 + 0.05x$$

Area as an Antiderivative

In the next section, we shall consider area as the limit of a sum, and we conclude this section by showing how area can be computed by antidifferentiation. The connection between area as a limit and area as an antiderivative is then made by a result called the fundamental theorem of calculus (see Section 5.4).

Theorem 5.3   Area as an antiderivative

If f is a continuous function such that $f(x) \geq 0$ for all x on the closed interval $[a, b]$, then the area bounded by the curve $y = f(x)$, the x-axis, and the vertical lines $x = a$, $x = t$, viewed as a function of t, is an antiderivative of $f(t)$ on $[a, b]$.

Proof: Define an **area function**, $A(t)$, as the area of the region bounded by the curve $y = f(x)$, the x-axis, and the vertical lines $x = a$, $x = t$ for $a \leq t \leq b$, as shown in Figure 5.7.

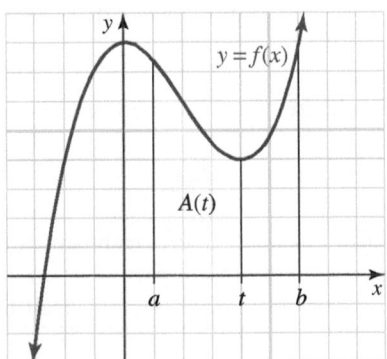

Figure 5.7   Area function $A(t)$

We need to show that $A(t)$ is an antiderivative of f on the interval $[a, b]$; that is, we need to show that $A'(t) = f(t)$.

Let $h > 0$ be small enough so that $t + h < b$ and consider the numerator of the difference quotient for $A(t)$, namely, the difference $A(t + h) - A(t)$. Geometrically, this difference is the area under the curve $y = f(x)$ between $x = t$ and $x = t + h$, as shown in Figure 5.8.

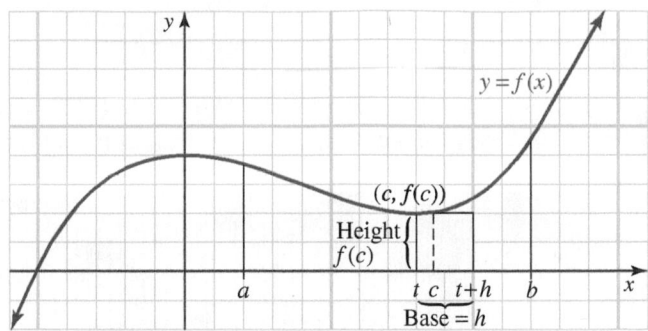

Figure 5.8 The area under the curve $y = f(x)$

If h is small enough, this area is approximately the same as the area of a rectangle with base h and height $f(c)$, where c is some point in the interval $[t, t+h]$, as shown in Figure 5.8. Thus, we have

$$\underbrace{A(t+h) - A(t)}_{\text{Area under the curve on } [t,\ t+h]} \approx \underbrace{hf(c)}_{\text{Area of rectangle}}$$

The difference quotient for $A(t)$ satisfies

$$\frac{A(t+h) - A(t)}{h} \approx f(c)$$

Finally, by taking the limit as $h \to 0^+$, we find the derivative of the area function $A(t)$ satisfies

$$\lim_{h \to 0^+} \frac{A(t+h) - A(t)}{h} = \lim_{h \to 0^+} \frac{hf(c)}{h}$$
$$A'(t) = f(t)$$

The limit on the left is the definition of derivative, and on the right we see that since f is continuous and c is in the interval $[t, t+h]$, c must approach t as $h \to 0^+$. A similar argument works as $h \to 0^-$. Thus, $A(t)$ is an antiderivative of $f(t)$. ♦

Example 10 Area as an antiderivative

Find the area under the parabola $y = x^2$ over x-axis on the interval $[0, 1]$. This area is shown in Figure 5.9.

Solution Let $A(t)$ be the area function for this example—namely, the area under $y = x^2$ over $y = 0$ on $[0, 1]$. Since f is continuous and $f(x) \geq 0$ on $[0, 1]$, Theorem 5.3 tells us that $A(t)$ is an antiderivative of $f(t) = t^2$ on $[0, 1]$. That is,

$$A(t) = \int t^2 dt = \frac{1}{3}t^3 + C$$

for all t in the interval $[0, 1]$. Clearly, $A(0) = 0$, so

$$A(0) = \frac{1}{3}(0)^3 + C \quad \text{or} \quad C = 0$$

and the area under the curve is

$$A(1) = \frac{1}{3}(1)^3 + 0 = \frac{1}{3}$$

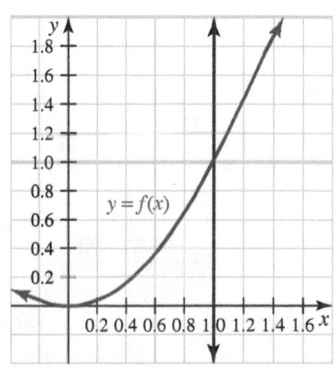

Figure 5.9 Area under $y = x^2$ over $[0, 1]$

PROBLEM SET 5.1

Level 1

Find the indefinite integral in Problems 1-30.

1. $\displaystyle\int 2\,dx$

2. $\displaystyle\int -4\,dx$

3. $\displaystyle\int (2x+3)\,dx$

4. $\displaystyle\int (4-5x)\,dx$

5. $\displaystyle\int (4t^3 + 3t^2)\,dt$

6. $\displaystyle\int (-8t^3 + 15t^5)\,dt$

7. $\displaystyle\int \frac{dx}{2x}$

8. $\displaystyle\int 14e^x\,dx$

9. $\displaystyle\int (6u^2 - 3\cos u)\,du$

10. $\displaystyle\int (5t^3 - \sqrt{t})\,dt$

11. $\displaystyle\int \sec^2\theta\,d\theta$

12. $\displaystyle\int \sec\theta\tan\theta\,d\theta$

13. $\displaystyle\int 2\sin\theta\,d\theta$

14. $\displaystyle\int \frac{\cos\theta}{3}\,d\theta$

15. $\displaystyle\int \frac{5}{\sqrt{1-y^2}}\,dy$

16. $\displaystyle\int \frac{dx}{10(1+x^2)}$

17. $\displaystyle\int x(x+\sqrt{x})\,dx$

18. $\displaystyle\int y(y^2-3y)\,dy$

19. $\displaystyle\int (u^{3/2} - u^{1/2} + u^{-10})\,du$

20. $\displaystyle\int (x^3 - 3x + \sqrt[4]{x} - 5)\,dx$

21. $\displaystyle\int \left(\frac{1}{t^2} - \frac{1}{t^3} + \frac{1}{t^4}\right)\,dt$

22. $\displaystyle\int \frac{1}{t}\left(\frac{2}{t^2} - \frac{3}{t^3}\right)\,dt$

23. $\displaystyle\int (2x^2 + 5)^2\,dx$

24. $\displaystyle\int (3 - 4x^3)^2\,dx$

25. $\displaystyle\int \left(\frac{x^2 + 3x - 1}{x^4}\right)\,dx$

26. $\displaystyle\int \frac{x^2 + \sqrt{x} + 1}{x^2}\,dx$

27. $\displaystyle\int \frac{x^2 + x - 2}{x^2}\,dx$

28. $\displaystyle\int \left(1 + \frac{1}{x}\right)\left(1 - \frac{4}{x^2}\right)\,dx$

29. $\displaystyle\int \frac{\sqrt{1-x^2} - 1}{\sqrt{1-x^2}}\,dx$

30. $\displaystyle\int \frac{x^2}{x^2+1}\,dx$

The slope $F'(x)$ at each point on a graph is given in Problems 31-38 along with one point (x_0, y_0) on the graph. Use this information to find F both graphically and analytically.

31. $F'(x) = x^2 + 3x$ with point $(0,0)$.

32. $F'(x) = (2x-1)^2$ with point $(1,3)$.

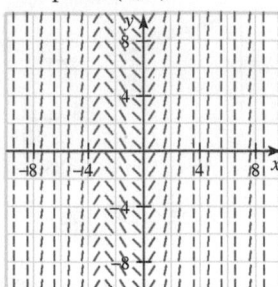 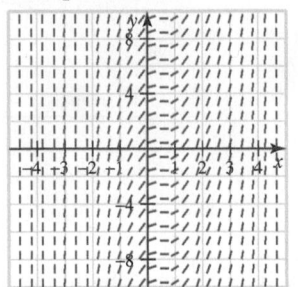

33. $F'(x) = \left(\sqrt{x} + 3\right)^2$ with point $(4, 36)$.

34. $F'(x) = 3 - 2\sin x$ with point $(0,0)$.

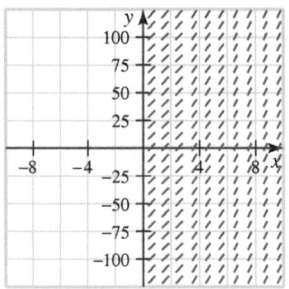 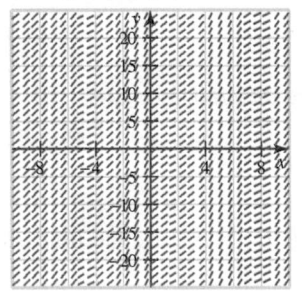

35. slope $\dfrac{x+1}{x^2}$ with point $(1, -2)$.

36. slope $\dfrac{2}{x\sqrt{x^2-1}}$ with point $(4, 1)$.

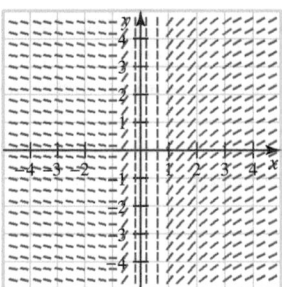 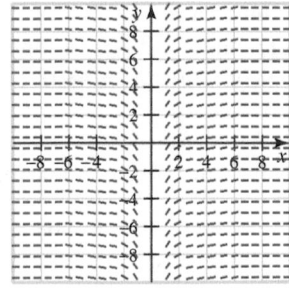

37. slope $x + e^x$ with point $(0, 2)$.

38. slope $\dfrac{x^2-1}{x^2+1}$ with point $(0,0)$.

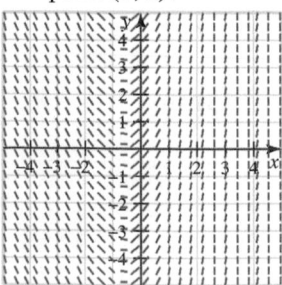 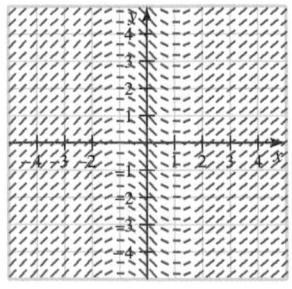

Level 2

39. $F(x) = \int \left(\dfrac{1}{\sqrt{x}} - 4 \right) dx$, find F so that $F(1) = 0$.
  a. Sketch the graphs of $y = F(x)$, $y = F(x) + 3$, and $y = F(x) - 1$.
  b. Find a constant C_0 so that the largest value of $G(x) = F(x) + C_0$ is 0.

40. A ball is thrown directly upward from ground level with an initial velocity of 96 ft/s. Assuming that the ball's only acceleration is that due to gravity (that is, $a(t) = -32$ ft/s^2), determine the maximum height reached by the ball and the time it takes to return to ground level.

41. The marginal cost of a certain commodity is $C'(x) = 6x^2 - 2x + 5$, where x is the level of production. If it costs \$500 to produce 1 unit, what is the total cost of producing 5 units?

42. The marginal revenue of a certain commodity is $R'(x) = -3x^2 + 4x + 32$, where x is the level of production (in thousands). Assume $R(0) = 0$.
  a. Find the demand function $p(x)$.
  b. Find the level of production that results in maximum revenue. What is the market price per unit at this level of production?

43. It is estimated that t months from now, the population of a certain town will be changing at the rate of $4 + 5t^{2/3}$ people per month. If the current population is 10,000, what will the population be 8 months from now?

44. A particle travels along the x-axis in such a way that its acceleration at time t is $a(t) = \sqrt{t} + t^2$. If it starts at the origin with an initial velocity of 2 (that is, $s(0) = 0$ and $v(0) = 2$), determine its position and velocity when $t = 4$.

45. An automobile starts from rest (that is, $v(0) = 0$) and travels with constant acceleration $a(t) = k$ in such a way that 6 sec after it begins to move, it has traveled 360 ft from its starting point. What is k?

46. The price of bacon is currently \$5.80/lb in Styxville. A consumer service has conducted a study predicting that t months from now, the price will be changing at the rate of $0.084 + 0.12\sqrt{t}$ dollars per month. How much will a pound of bacon cost 4 months from now?

47. An airplane has a constant acceleration while moving down the runway from rest. What is the acceleration of the plane (to the nearest tenth of a unit) at liftoff if the plane requires 900 ft of runway before lifting off at 88 ft/s (60 mi/h)?

48. After its brakes are applied, a certain sports car decelerates at a constant rate of 28 ft/s^2. Compute the stopping distance if the car is going 60 mi/h (88 ft/s) when the brakes are applied.

49. The brakes of a certain automobile produce a constant deceleration of k ft/s^2. The car is traveling at 60 mi/h (88 ft/s) when the driver is forced to hit the brakes, and it comes to rest at a point 121 ft from the point where the brakes were applied. What is k?

50. A particle moves along the x-axis in such a way that at time $t > 0$, its velocity (in ft/s) is

$$v(t) = t^{-1} + t$$

How far does it move between times $t = 1$ and $t = e^2$?

51. A manufacturer estimates that the marginal cost in a certain production process is

$$C'(x) = 0.1e^x + 21\sqrt{x}$$

when x units are produced. If the cost of producing 1 unit is \$100, what does it cost (to the nearest cent) to produce 4 units?

In Problems 52-57, find the area under the curve defined by the given equation, above the x-axis, and over the given interval.

52. $y = x^2$ over $[1, 4]$ **53.** $y = \sqrt{x}$ over $[1, 4]$

54. $y = e^x - x$ over $[0, 2]$ **55.** $y = \dfrac{x + 1}{x}$ over $[1, 2]$

56. $y = \cos x$ over $\left[0, \frac{\pi}{2}\right]$

57. $y = (1 - x^2)^{-1/2}$ over $\left[0, \frac{1}{2}\right]$

58. Find $\int \dfrac{dy}{2y\sqrt{y^2 - 1}}$ using the indicated methods.
  a. Use an inverse trigonometric differentiation rule.
  b. Use technology to find this integral.
  c. Reconcile your answers for parts **a** and **b**.

Level 3

59. If a, b, and c are constants, use the linearity rule twice to show that

$$\int \left[af(x) + bg(x) + ch(x) \right] dx$$

$$= a \int f(x)dx + b \int g(x)dx + c \int h(x)dx$$

60. Use the area as an antiderivative theorem (Theorem 5.3) to find the area under the line $y = mx + b$ over the interval $[c, d]$ where $m > 0$ and $mc + b > 0$. Check your result by using geometry to find the area of a trapezoid.

5.2 AREA AS THE LIMIT OF A SUM

IN THIS SECTION: *Area as the limit of a sum, the general approximation scheme, summation notation, area using summation formulas, area using special summation formulas*
In this section, we shall show that it is reasonable to *define* area as the limit of a sum. In the process, we shall introduce ideas that play a key role in our general development of integral calculus.

Area as the Limit of a Sum

Computing area has been a problem of both theoretical and practical interest since ancient times, but except for a few special cases, the problem is not easy. For example, you may know the formulas for computing the area of a rectangle, square, triangle, circle, and even a trapezoid. You have probably found the areas of regions that were more complicated but could be broken up into parts using these formulas. In Example 10 of the previous section we found the area between the parabola $y = x^2$ and the x-axis on the interval $[0, 1]$. (See Figure 5.10**a**.) We revisit this example to demonstrate a general procedure for computing area.

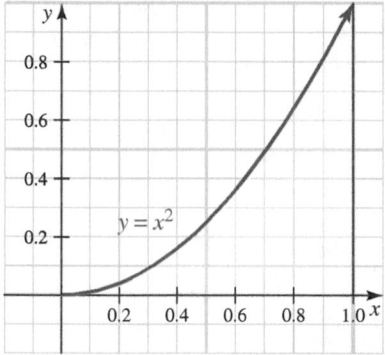
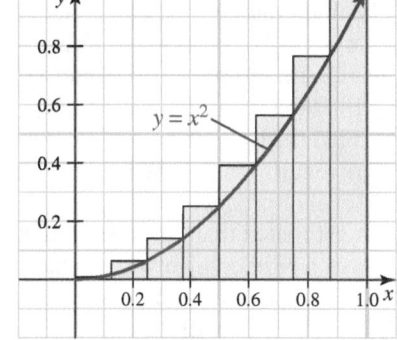

a. Compute area of shaded region   **b.** Required area is about the same as the shaded rectangles

Figure 5.10   Example of the area problem

Example 1   Estimating an area using rectangles and right endpoints

Estimate the area under the parabola $y = x^2$ and above the x-axis on the interval $[0, 1]$.

Solution   In the previous section we found this area using the area function. In this example, we will *estimate* the area by adding the areas of approximating rectangles constructed on subintervals of $[0, 1]$, as shown in Figure 5.10**b**. To simplify computations, we will require all approximating rectangles to have the same width and will take the height of each rectangle to be the y-coordinate of the parabola above the *right endpoint* of the subinterval on which it is based.*

For the first estimate, we divide the interval $[0, 1]$ into 5 subintervals, as shown in Figure 5.11**a**. Because the approximating rectangles all have the same width, the right endpoints are $x_1 = 0.2$, $x_2 = 0.4$, $x_3 = 0.6$, $x_4 = 0.8$, and $x_5 = 1$. This subdivision is called a *partition* of the interval.

The width of each subdivision is denoted by Δx and is found by dividing the length of the interval by the number of subintervals:

$$\Delta x = \frac{1 - 0}{5} = \frac{1}{5} = 0.2$$

---

*Actually, there is nothing special about right end-points, and we could just as easily have used any other point in the base subinterval—say, the left endpoint or the midpoint.

Let S_n be the total area of n rectangles. For the case where $n = 5$,

$$
\begin{aligned}
S_5 &= f(x_1)\Delta x + f(x_2)\Delta x + f(x_3)\Delta x + f(x_4)\Delta x + f(x_5)\Delta x \\
&= \left[f(x_1) + f(x_2) + f(x_3) + f(x_4) + f(x_5) \right] \Delta x \\
&= \left[f(0.2) + f(0.4) + f(0.6) + f(0.8) + f(1) \right] (0.2) \\
&= 0.44
\end{aligned}
$$

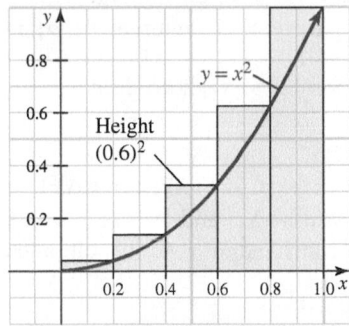 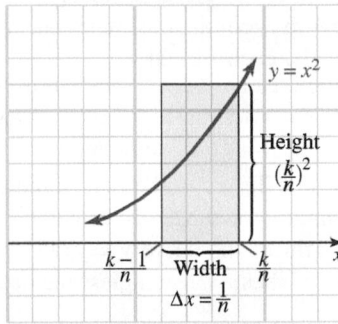

a. Partition into 5 subdivisions   **b.** Detail showing one rectangle

Figure 5.11   Partitioning of Figure 5.10

Even though $S_5 = 0.44$ serves as a reasonable approximation of the area, we see from Figure 5.11**a** that this approximation seems too large. Let us rework Example 1 using a general scheme rather than a specified number of rectangles. Partition the interval $[0, 1]$ into n equal parts, each with width

$$
\Delta x = \frac{1 - 0}{n} = \frac{1}{n}
$$

For $k = 1, 2, 3, \cdots, n$, the kth subinterval is $\left[\dfrac{k-1}{n}, \dfrac{k}{n} \right]$, and on this subinterval we then construct an approximating rectangle with width $\Delta x = \dfrac{1}{n}$ and height $\left(\dfrac{k}{n} \right)^2$, since $y = x^2$. The total area bounded by all n rectangles is

$$
S_n = \left[\left(\frac{1}{n} \right)^2 + \left(\frac{2}{n} \right)^2 + \cdots + \left(\frac{n}{n} \right)^2 \right] \left(\frac{1}{n} \right)
$$

Consider different choices for n, as shown in Figure 5.12.

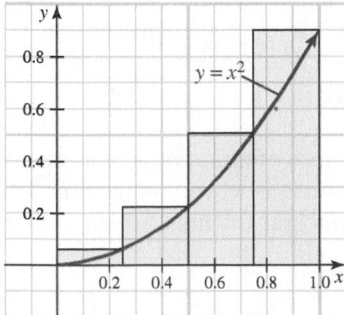 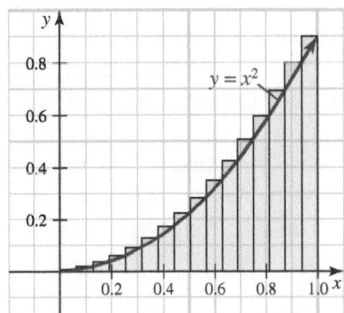 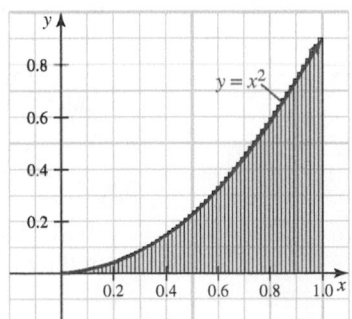

Figure 5.12 Interactive   The area estimate is improved by taking more rectangles

If we increase the number of subdivisions n, the width $\Delta x = \dfrac{1}{n}$ of each approximating rectangle will decrease, and we would expect the area estimates S_n to improve. Thus, it is reasonable to *define* the area A under the parabola to be the *limit of* S_n as $\Delta x \to 0$ or, equivalently, as $n \to \infty$. We can attempt to predict its value by seeing what happens to the sum as n grows large without bound. It is both tedious and difficult to evaluate such sums by hand, but fortunately we can use a calculator or computer to obtain some (rounded) values for S_n. For the area in Example 1, for example, we could use a computer to find $S_{100} = 0.338$, $S_{1,000} = 0.334$, and $S_{5,000} = 0.333$.

The General Approximation Scheme

We now compute the area under any curve $y = f(x)$ on an interval $[a, b]$, where f is a nonnegative continuous function. We first partition the interval $[a, b]$ into n equal subintervals, each of width

$$\Delta x = \frac{b-a}{n}$$

For $k = 1, 2, 3, \ldots, n$ the kth subinterval is $[a + (k-1)\Delta x, a + k\Delta x]$, and the kth approximating rectangle is constructed with width Δx and height $f(a + k\Delta x)$ equal to the height of the curve $y = f(x)$ above the right endpoint of the subinterval. Adding the areas of these n rectangles, we obtain

$$S_n = \overbrace{f(a + \Delta x)\Delta x}^{\text{Area of first rectangle}} + \overbrace{f(a + 2\Delta x)\Delta x}^{\text{Area of second rectangle}} + \cdots + \overbrace{f(a + n\Delta x)\Delta x}^{\text{Area of nth rectangle}}$$

as an estimate of the area under the curve. In advanced calculus, it is shown that the continuity of f guarantees the existence of $\lim_{\Delta x \to 0} S_n$, and we use this limit to define the required area, as shown in the following box.

AREA AS THE LIMIT OF A SUM   Suppose f is continuous and $f(x) \geq 0$ throughout the interval $[a, b]$. Then the **area** of the region under the curve $y = f(x)$ over this interval is

$$A = \lim_{\Delta x \to 0} \left[f(a + \Delta x) + f(a + 2\Delta x) + \cdots + f(a + n\Delta x) \right] \Delta x$$

where $\Delta x = \dfrac{b-a}{n}$.

The definition of area as a limit of a sum is consistent with the area concept introduced in plane geometry, and with the area function defined in Section 5.1. For example, it would not be difficult to use this formula to show that a rectangle has area $A = \ell w$ or that a triangle has area $A = \frac{1}{2}bh$. You will also note that we maintained everyday usage in saying that the *formula* for the area of a rectangle is $A = \ell w$ or the *area* of the region under the curve $y = f(x)$ is a limit, but what we are really doing is defining area as a limit.

The problem we now face is how to implement this definition of area as the limit of a sum. The immediate answer (discussed in this section) is to use summation formulas and technology. The long-range goal is to develop integral calculus, which is discussed in the next section.

Summation Notation

The expanded form of the sum for the definition of area makes it awkward to use. Therefore, we shall digress to introduce a more compact notation for sums, and that notation will motivate integral notation. Using this **summation notation**, we express the sum $a_1 + a_2 + \cdots + a_n$ as follows:

$$a_1 + a_2 + \cdots + a_n = \sum_{k=1}^{n} a_k$$

The summation notation is sometimes called the **sigma notation** because the upper-case Greek letter sigma (Σ) is used to denote the summation process. The **index** k is called the **index of summation** (or *running index*). The terminology used in connection with the summation notation is shown:

$$\overset{\text{Upper limit of summation}}{\underset{\underset{\text{Lower limit of summation}}{\uparrow\ \text{Index of summation}}}{\sum_{k=1}^{n}\ \overset{\text{General term}}{\overbrace{a_k}}}}$$

Note that in the summation process, the choice of summation index is immaterial. For example, the following sums are all exactly the same:

$$\sum_{k=3}^{7}k^2 = \sum_{j=3}^{7}j^2 = \sum_{i=3}^{7}i^2 = \sum_{\lambda=3}^{7}\lambda^2$$

In general, an index ($k, j, i,$ or λ) that represents a process in which it has no direct effect on the result is called a **dummy variable.**

Several useful properties of sums and sum formulas are listed in Theorem 5.4. We shall use the summation notation throughout the rest of this text, especially in this chapter, Chapter 6, and Chapter 8.

Theorem 5.4   Rules for summation

For any numbers c and d and positive integers m and n,

1. **Constant term rule**
$$\sum_{k=1}^{n} c = \underbrace{c + c + \cdots + c}_{n\ terms} = nc$$

2. **Sum rule**
$$\sum_{k=1}^{n}(a_k + b_k) = \sum_{k=1}^{n} a_k + \sum_{k=1}^{n} b_k$$

3. **Scalar multiple rule**
$$\sum_{k=1}^{n} ca_k = c\sum_{k=1}^{n} a_k = \left(\sum_{k=1}^{n} a_k\right)c$$

4. **Linearity rule**
$$\sum_{k=1}^{n}(ca_k + db_k) = c\sum_{k=1}^{n} a_k + d\sum_{k=1}^{n} b_k$$

5. **Subtotal rule**
If $1 < m < n$, then $\displaystyle\sum_{k=1}^{n} a_k = \sum_{k=1}^{m} a_k + \sum_{k=m+1}^{n} a_k$

6. **Dominance rule**
If $a_k \le b_k$ for $k = 1, 2, \cdots, n$, then $\displaystyle\sum_{k=1}^{n} a_k \le \sum_{k=1}^{n} b_k$

Proof: These properties can all be established by applying well-known algebraic rules. For example, to prove the linearity rule, use the associative, commutative, and distributive properties of real numbers.

$$\sum_{k=1}^{n}(ca_k + db_k) = (ca_1 + db_1) + (ca_2 + db_2) + \cdots + (ca_n + db_n)$$

$$= (ca_1 + ca_2 + \cdots + ca_n) + (db_1 + db_2 + \cdots + db_n)$$

$$= c(a_1 + a_2 + \cdots + a_n) + d(b_1 + b_2 + \cdots + b_n)$$

$$= c\sum_{k=1}^{n} a_k + d\sum_{k=1}^{n} b_k$$

See the Problems 57-60 for the proofs of the other parts. ◆

Area Using Summation Formulas

Using summation notation, we can streamline the symbolism in the formula for the area under the curve $y = f(x)$, $f(x) \geq 0$ on the interval $[a, b]$. In particular, note that the approximating sum S_n is

$$S_n = \left[f(a + \Delta x) + f(a + 2\Delta x) + \cdots + f(a + n\Delta x) \right] \Delta x$$

$$= \sum_{k=1}^{n} f(a + k\Delta x)\Delta x$$

where $\Delta x = \dfrac{b - a}{n}$. Thus, the formula for the definition of area is shown:

Area under $y = f(x)$ above the
x-axis between $x = a$ and $x = b$. n is the number of approximating rectangles

Width of each rectangle

$$A = \lim_{n \to \infty} S_n = \lim_{\Delta x \to 0} \sum_{k=1}^{n} f(a + k\Delta x) \, \Delta x$$

Height of the kth rectangle

From algebra we recall certain summation formulas (which can be proved using mathematical induction) that we will need in order to find areas using the limit definition.

SUMMATION FORMULAS

$$\sum_{k=1}^{n} 1 = n$$

$$\sum_{k=1}^{n} k = 1 + 2 + 3 + \cdots + n = \frac{n(n + 1)}{2}$$

$$\sum_{k=1}^{n} k^2 = 1^2 + 2^2 + 3^2 + \cdots + n^2 = \frac{n(n + 1)(2n + 1)}{6}$$

$$\sum_{k=1}^{n} k^3 = 1^3 + 2^3 + 3^3 + \cdots + n^3 = \frac{n^2(n + 1)^2}{4}$$

Example 2   Area using the definition and summation formulas

Use the summation definition of area to find the area under the parabola $y = x^2$ on the interval $[0, 1]$. You estimated this area in Example 1.

Solution   Partition the interval $[0, 1]$ into n subintervals with width $\Delta x = \frac{1-0}{n}$. The right endpoint of the kth subinterval is $a + k\Delta x = \frac{k}{n}$ and $f\left(\frac{k}{n}\right) = \frac{k^2}{n^2}$. Thus, from the definition of area we have

$$A = \lim_{\Delta x \to 0} \sum_{k=1}^{n} f(a + k\Delta x)\Delta x$$

$$= \lim_{n \to \infty} \sum_{k=1}^{n} \left(\frac{k^2}{n^2}\right)\left(\frac{1}{n}\right)$$

$$= \lim_{n \to \infty} \sum_{k=1}^{n} \frac{k^2}{n^3} \qquad \text{Note that } \frac{1}{n^3} \text{ is independent of the index of summation, } k.$$

$$= \lim_{n \to \infty} \frac{1}{n^3} \sum_{k=1}^{n} k^2 \qquad \text{Scalar multiple rule}$$

$$= \lim_{n \to \infty} \frac{1}{n^3} \left[\frac{n(n+1)(2n+1)}{6} \right] \qquad \text{Summation formula for squares}$$

$$= \lim_{n \to \infty} \frac{1}{6} \left[2 + \frac{3}{n} + \frac{1}{n^2} \right]$$

$$= \frac{1}{3}$$

This is the same as the answer found by antidifferentiation in Example 10 of Section 5.1.

Example 3   Tabular approach for finding area

Use a calculator or computer to estimate the area under the curve $y = \sin x$ on the interval $\left[0, \frac{\pi}{2}\right]$.

Solution   We see $a = 0$ and $b = \frac{\pi}{2}$, so $\Delta x = \frac{\frac{\pi}{2} - 0}{n} = \frac{\pi}{2n}$. The right endpoints of the subintervals are

$$a + \Delta x = 0 + \frac{\pi}{2n} = \frac{\pi}{2n}$$

$$a + 2\Delta x = \frac{2\pi}{2n} = \frac{\pi}{n}$$

$$a + 3\Delta x = \frac{3\pi}{2n}$$

$$\vdots$$

$$a + n\Delta n = \frac{n\pi}{2n} = \frac{\pi}{2} = b$$

Thus,

$$S_n = \sum_{k=1}^{n} f\left(0 + \frac{k\pi}{2n}\right)\left(\frac{\pi}{2n}\right) = \sum_{k=1}^{n} \left[\sin\left(\frac{k\pi}{2n}\right)\right]\left(\frac{\pi}{2n}\right)$$

Now we know from the definition that the actual area is

$$S = \lim_{n \to \infty} \sum_{k=1}^{n} \left[\sin\left(\frac{k\pi}{2n}\right)\right]\left(\frac{\pi}{2n}\right) = \frac{\pi}{2} \lim_{n \to \infty} \sum_{k=1}^{n} \frac{1}{n} \sin\left(\frac{k\pi}{2n}\right)$$

which we can estimate by computing S_n for successively large values of n, as summarized in the margin. Note that the table suggests that

$$\lim_{\Delta x \to 0} S_n = \lim_{n \to \infty} S_n = 1$$

Thus, we expect the actual area under the curve to be 1 square unit.

n	S_n
10	1.07648
20	1.03876
50	1.01563
100	1.00783
500	1.00157

PROBLEM SET 5.2

Level 1

Evaluate the sums in Problems 1-10 by using the summation formulas.

1. $\sum_{k=1}^{6} 1$

2. $\sum_{k=1}^{250} 2$

3. $\sum_{k=4}^{10} 3$

4. $\sum_{k=1}^{15} k$

5. $\sum_{k=1}^{10} (k+1)$

6. $\sum_{k=3}^{12} (k-1)$

7. $\sum_{k=1}^{5} k^3$

8. $\sum_{k=1}^{7} k^2$

9. $\sum_{k=1}^{100} (2k-3)$

10. $\sum_{k=1}^{100} (k-1)^2$

Use the properties of summation notation in Problems 11-18 to evaluate the given limits.

11. $\lim_{n\to\infty} \sum_{k=1}^{n} \dfrac{k}{n^2}$

12. $\lim_{n\to\infty} \sum_{k=1}^{n} \dfrac{k^2}{n^3}$

13. $\lim_{n\to\infty} \sum_{k=1}^{n} \dfrac{1}{n}$

14. $\lim_{n\to\infty} \sum_{k=1}^{n} \dfrac{5}{n}$

15. $\lim_{n\to\infty} \sum_{k=1}^{n} \dfrac{k}{n^3}$

16. $\lim_{n\to\infty} \sum_{k=1}^{n} \dfrac{k^3}{n^4}$

17. $\lim_{n\to\infty} \sum_{k=1}^{n} \left(1 + \dfrac{k}{n}\right)\left(\dfrac{2}{n}\right)$

18. $\lim_{n\to\infty} \sum_{k=1}^{n} \left(1 + \dfrac{2k}{n}\right)^2 \left(\dfrac{2}{n}\right)$

First sketch the region under the graph of $y = f(x)$ on the interval $[a, b]$ in Problems 19-28. Then approximate the area of each region by using right endpoints and the formula

$$S_n = \sum_{k=1}^{n} f(a + k\Delta x)\Delta x$$

for $\Delta x = \dfrac{b-a}{n}$ and the indicated values of n.

19. $f(x) = 4x + 1$ on $[0, 1]$ for

  a. $n = 4$ **b.** $n = 8$

20. $f(x) = 3 - 2x$ on $[0, 1]$ for

  a. $n = 3$ **b.** $n = 6$

21. $f(x) = x^2$ on $[1, 2]$ for

  a. $n = 4$ **b.** $n = 6$

22. $f(x) = 4 - x^2$ on $[1, 2]$ for

  a. $n = 4$ **b.** $n = 6$

23. $f(x) = \cos x$ on $\left[-\frac{\pi}{2}, 0\right]$ for $n = 4$

24. $f(x) = x + \sin x$ on $\left[0, \frac{\pi}{4}\right]$ for $n = 3$

25. $f(x) = \dfrac{1}{x^2}$ on $[1, 2]$ for $n = 4$

26. $f(x) = \dfrac{2}{x}$ on $[1, 2]$ for $n = 4$

27. $f(x) = \sqrt{x}$ on $[1, 4]$ for $n = 4$

28. $f(x) = \sqrt{1 + x^2}$ on $[0, 1]$ for $n = 4$

Level 2

Find the exact area under the given curve on the interval prescribed in Problems 29-34 by using area as the limit of a sum and the summation formulas.

29. $y = 4x^3 + 2x$ on $[0, 2]$

30. $y = 4x^3 + 2x$ on $[1, 2]$

31. $y = 6x^2 + 2x + 4$ on $[0, 3]$

32. $y = 6x^2 + 2x + 4$ on $[1, 3]$

33. $y = 3x^2 + 2x + 1$ on $[0, 1]$

34. $y = 4x^3 + 3x^2$ on $[0, 1]$

Think Tank Problems *Show that each statement about area in Problems 35-40 is generally true or provide a counterexample. It will probably help to sketch the indicated region for each problem.*

35. If $C > 0$ is a constant, the region under the line $y = C$ on the interval $[a, b]$ has area $A = C(b - a)$.

36. If $C > 0$ is a constant and $b > a \geq 0$, the region under the line $y = Cx$ on the interval $[a, b]$ has area $A = \frac{1}{2}C(b - a)$.

37. The region under the parabola $y = x^2$ on the interval $[a, b]$ has area less than $\frac{1}{2}(b^2 + a^2)(b - a)$.

38. The region under the curve $y = \sqrt{1 - x^2}$ on the interval $[-1, 1]$ has area $A = \frac{\pi}{2}$.

39. Let f be a function that satisfies $f(x) \geq 0$ for x in the interval $[a, b]$. Then the area under the curve $y = f^2(x)$ on the interval $[a, b]$ must always be greater than the area under $y = f(x)$ on the same interval.

40. Recall that a function f is said to be **even** if $f(-x) = f(x)$ for all x. If f is even and $f(x) \geq 0$ throughout the interval $[-a, a]$, then the area under the curve $y = f(x)$ on this interval is *twice* the area under $y = f(x)$ on $[0, a]$.

41. Show that the region under the curve $y = x^3$ on the interval $[0, 1]$ has area $\frac{1}{4}$ square units.

42. Use the definition of area to show that the area of a rectangle equals the product of its length ℓ and its width w.

43. Show that the triangle with vertices $(0, 0)$, $(0, h)$, and $(b, 0)$ has area $A = \frac{1}{2}bh$ using the area as the limit of a sum.

44. a. Compute the area under the parabola $y = 2x^2$ on the interval $[1, 2]$ as the limit of a sum.

b. Let $f(x) = 2x^2$ and note that $g(x) = \frac{2}{3}x^3$ defines a function that satisfies $g'(x) = f(x)$ on the interval $[1, 2]$. Verify that the area computed in part **a** satisfies $A = g(2) - g(1)$.

c. The function defined by

$$h(x) = \frac{2}{3}x^3 + C$$

for any constant C also satisfies $h'(x) = f(x)$. Is it true that the area in part **a** satisfies $A = h(2) - h(1)$?

Use the tabular approach to compute the area under the curve $y = f(x)$ on each interval given in Problems 45-51 as the limit of a sum of terms.

45. $f(x) = 4x$ on $[0, 1]$

46. $f(x) = x^2$ on $[0, 4]$

47. $f(x) = \cos x$ on $\left[-\frac{\pi}{2}, 0\right]$ (Compare with Problem 23.)

48. $f(x) = x + \sin x$ on $\left[0, \frac{\pi}{4}\right]$ (Compare with Problem 24.)

49. $f(x) = \ln(x^2 + 1)$ on $[0, 3]$

50. $f(x) = e^{-3x^2}$ on $[0, 1]$

51. $f(x) = \cos^{-1}(x + 1)$ on $[-1, 0]$

52. a. Use the tabular approach to compute the area under the curve $y = \sin x + \cos x$ on the interval $\left[0, \frac{\pi}{2}\right]$ as the limit of a sum.

b. Let $f(x) = \sin x + \cos x$ and note that $g(x) = -\cos x + \sin x$ satisfies $g'(x) = f(x)$ on the interval $\left[0, \frac{\pi}{2}\right]$. Verify that the area computed in part **a** satisfies $A = g\left(\frac{\pi}{2}\right) - g(0)$.

c. The function $h(x) = -\cos x + \sin x + C$ for constant C also satisfies $h'(x) = f(x)$. Is it true that the area in part **a** satisfies $A = h\left(\frac{\pi}{2}\right) - h(0)$?

Level 3

53. Derive the formula

$$\sum_{k=1}^{n} k = 1 + 2 + 3 + \cdots + n = \frac{n(n+1)}{2}$$

by completing these steps:

a. Use the basic rules for sums to show that

$$\sum_{k=1}^{n} k = \frac{1}{2}\sum_{k=1}^{n}\left[k^2 - (k-1)^2\right] + \frac{1}{2}\sum_{k=1}^{n} 1$$

$$= \frac{1}{2}\sum_{k=1}^{n}\left[k^2 - (k-1)^2\right] + \frac{1}{2}n$$

b. Show that

$$\sum_{k=1}^{n}\left[k^2 - (k-1)^2\right] = n^2$$

Hint: Expand the sum by writing out a few terms. Note the internal cancellation.

c. Combine parts **a** and **b** to show that

$$\sum_{k=1}^{n} k = \frac{n(n+1)}{2}$$

54. a. First find constants $a, b, c,$ and d such that

$$k^3 = a\left[k^4 - (k-1)^4\right] + bk^2 + ck + d$$

b. Modify the approach outlined in Problem 53 to establish the formula

$$\sum_{k=1}^{n} k^2 = \frac{n(n+1)(2n+1)}{6}$$

55. The purpose of this problem is to verify the results shown in Figure 5.12. Specifically, we shall find the area A under the parabola $y = x^2$ on the interval $[0, 1]$ using approximating rectangles with heights taken at the *left* endpoints. Verify that

$$\lim_{n \to \infty} \sum_{k=1}^{n} \left(\frac{k-1}{n}\right)^2 \left(\frac{1}{n}\right) = \frac{1}{3}$$

Compare this with the procedure outlined in Example 1. Note that when the interval $[0, 1]$ is subdivided into n equal parts, the kth subinterval is

$$\left[\frac{k-1}{n}, \frac{k}{n}\right]$$

56. Develop a formula for area based on approximating rectangles with heights taken at the *midpoints* of subintervals.

Use the properties of real numbers to establish the summation formulas in Theorem 5.4 in Problems 57-60.

57. constant term rule

58. sum rule

59. scalar multiple rule

60. subtotal rule

5.3 RIEMANN SUMS AND THE DEFINITE INTEGRAL

IN THIS SECTION: *Riemann sums, the definite integral, area as an integral, properties of the definite integral, distance as an integral*
In this section, we lay the groundwork for the third great idea of calculus (after *limit* and *derivative*), namely the idea of *integration*.

We shall soon discover that not just area, but other useful quantities such as distance, volume, mass, and work, can be first approximated by sums and then obtained exactly by taking a limit involving the approximating sums. The special kind of limit of a sum that appears in this context is called the *definite integral*, and the process of finding integrals is called *definite integration* or *Riemann integration* in honor of the German mathematician Georg Bernhard Riemann (1826-1866), who pioneered the modern approach to integration theory. We begin by introducing some special notation and terminology.

Riemann Sums

Recall from Section 5.1 that to find the area under the graph of the function $y = f(x)$ on the closed interval $[a, b]$ using sums where f is continuous and $f(x) \geq 0$, we proceed as follows:

1. Partition the interval into n subintervals of equal width

$$\Delta x = \frac{b - a}{n}$$

2. Evaluate f at the right endpoint $a + k\Delta x$ of the kth subinterval for $k = 1, 2, \cdots, n$.
3. Form the approximating sum of the areas of the n rectangles, which we denote by

$$S_n = \sum_{k=1}^{n} f(a + k\Delta x)\Delta x$$

4. Because we expect the estimates S_n to improve as Δx decreases, we *define* the area A under the curve, above the x-axis, and bounded by the lines $x = a$ and $x = b$, to be the limit of S_n as $\Delta x \to 0$. Thus, we write

$$A = \lim_{n \to \infty} \sum_{k=1}^{n} f(a + k\Delta x)\Delta x$$

if this limit exists. This means that A can be estimated to any desired degree of accuracy by approximating the sum S_n with Δx sufficiently small (or, equivalently, n sufficiently large).

This approach to the area problem contains the essentials of integration, but there is no compelling reason for the partition points to be evenly spaced or to insist on evaluating f at right endpoints. These conventions are for convenience of computation, but to accommodate easily as many applications as possible, it is useful to consider a more general type of approximating sum and to specify what is meant by the limit of such sums. The approximating sums that occur in integration problems are called **Riemann sums**, and the following definition contains a step-by-step description of how such sums are formed.

RIEMANN SUM   Suppose a bounded function f is given along with a closed interval $[a, b]$ on which f is defined. Then:

Step 1 Partition the interval $[a, b]$ into n subintervals by choosing points $\{x_0, x_1, \cdots, x_n\}$ arranged in such a way that $a = x_0 < x_1 < x_2 < \cdots < x_{n-1} < x_n = b$. Call this partition P. For $k = 1, 2, \cdots, n$, the kth subinterval width is $\Delta x_k = x_k - x_{k-1}$. The largest of these widths is called the **norm** of the partition P and is denoted by $\|P\|$; that is,

$$\|P\| = \max_{k=1,2,\cdots n}\{\Delta x_k\}$$

Step 2 Choose a number arbitrarily from each subinterval. For $k = 1, 2, \cdots, n$ the number x_k^* chosen from the kth subinterval $\left[x_{k-1}, x_k\right]$ is called the kth *subinterval representative* of the partition P.

Step 3 Form the sum

$$R_n = f(x_1^*)\Delta x_1 + f(x_2^*)\Delta x_2 + \cdots + f(x_n^*)\Delta x_n = \sum_{k=1}^{n} f(x_k^*)\Delta x_k$$

This is the **Riemann sum** associated with f, the given partition P, and the chosen subinterval representatives

$$x_1^*, x_2^*, \cdots, x_n^*$$

■ **W**hat this says: We will express quantities from geometry, physics, economics, and other applications in terms of a Riemann sum

$$\sum_{k=1}^{n} f(x_k^*)\Delta x_k$$

Riemann sums are generally used to model a quantity for a particular application. Note that the Riemann sum *does not* require that the function f be nonnegative, nor does it require that all the intervals must be the same length. In addition, x_k^* is *any* point in the kth subinterval and does not need to be something "nice" like the left or right endpoint, or the midpoint.

Example 1   Formation of the Riemann sum for a given function

Suppose the interval $[-2, 1]$ is partitioned into 6 subintervals with subdivision points $a = x_0 = -2$, $x_1 = -1.6$, $x_2 = -0.93$, $x_3 = -0.21$, $x_4 = 0.35$, $x_5 = 0.82$, $x_6 = 1 = b$. Find the norm of this partition P and the Riemann sum associated with the function $f(x) = 2x$, the given partition, and the subinterval representatives $x_1^* = -1.81$, $x_2^* = -1.12$, $x_3^* = -0.55$, $x_4^* = -0.17$, $x_5^* = 0.43$, $x_6^* = 0.94$.

Solution   Before we can find the norm of the partition or the required Riemann sum, we must compute the subinterval width Δx_k and evaluate f at each subinterval representative x_k^*. These values are shown in Figure 5.13 and the computations follow.

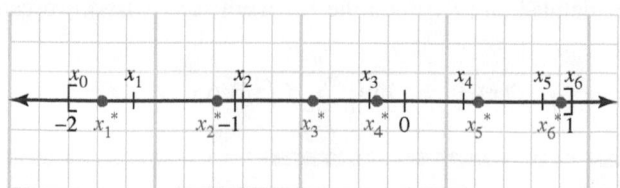

Figure 5.13   Riemann sums

k	Given $\overbrace{x_k - x_{k-1}}= \Delta x_k$	Given x_n^*	Given $\overbrace{f(x_n^*)} = 2x_n^*$
1	$-1.6 - (-2) = 0.40$	-1.81	$f(-1.81) = -3.62$
2	$-0.93 - (-1.6) = 0.67$	-1.12	$f(-1.12) = -2.24$
3	$-0.21 - (-0.93) = 0.72$	-0.55	$f(-0.55) = -1.10$
4	$0.35 - (-0.21) = 0.56$	-0.17	$f(-0.17) = -0.34$
5	$0.82 - 0.35 = 0.47$	0.43	$f(0.43) = 0.86$
6	$1.00 - 0.82 = 0.18$	0.94	$f(0.94) = 1.88$

From this table, we see that the largest subinterval width is $\Delta x_3 = 0.72$, so the partition has norm $\|P\| = 0.72$. Finally, by using the definition, we find the Riemann sum:

$$R_6 = (-3.62)(0.4) + (-2.24)(0.67) + (-1.10)(0.72) + (-0.34)(0.56)$$
$$+ (0.86)(0.47) + (1.88)(0.18)$$
$$= -3.1886$$

Notice from Example 1 that the Riemann sum does not necessarily represent an area. The sum found is negative (and areas must be nonnegative).

The Definite Integral

By comparing the formula for the Riemann sum with that of area in the previous section, we recognize that the sum S_n used to approximate area is actually a special kind of Riemann sum that has

$$\Delta x_k = \Delta x = \frac{b - a}{n} \qquad \text{and} \qquad x_k^* = a + k\Delta x$$

for $k = 1, 2, \cdots, n$. Because the subintervals in the partition P associated with S_n are equally spaced, it is called a **regular partition**. When we express the area under the curve $y = f(x)$ as $A = \lim_{\Delta x \to 0} S_n$, we are actually saying that A can be estimated to any desired accuracy by finding a Riemann sum of the form S_n with norm

$$\|P\| = \frac{b - a}{n}$$

sufficiently small. We use this interpretation as a model for the following definition.

DEFINITE INTEGRAL   If f is defined on the closed interval $[a, b]$, we say f is **integrable on $[a, b]$** if

$$I = \lim_{\|P\| \to 0} \sum_{k=1}^{n} f\left(x_k^*\right) \Delta x_k$$

exists. This limit is called the **definite integral** of f from a to b. The definite integral is denoted by

$$I = \int_a^b f(x)\, dx$$

 What this says   To say that f is *integrable* with definite integral I means the number I can be approximated to any prescribed degree of accuracy by *any* Riemann sum of f with norm sufficiently small. As long as the conditions of this definition are satisfied (that is, f is defined on $[a, b]$ and the Riemann sum exists), we can write

$$\int_a^b f(x)\, dx = \lim_{\|P\| \to 0} \sum_{k=1}^{n} f\left(x_k^*\right) \Delta x$$

Formally, I is the definite integral of f on $[a,b]$ if for each number $\epsilon > 0$, there exists a number $\delta > 0$ such that if

$$\sum_{k=1}^{n} f\left(x_k^*\right) \Delta x_k$$

is any Riemann sum of f whose norm satisfies $\|P\| < \delta$, then

$$\left| I - \sum_{k=1}^{n} f\left(x_k^*\right) \Delta x_k \right| < \epsilon$$

In advanced calculus, it is shown that when this limit exists, it is unique. Moreover, its value is independent of the particular way in which the partitions of $[a,b]$ and the subinterval representatives x_k^* are chosen.

🛈 Take a few minutes to make sure you understand the integral notation and terminology.

$$\int_a^b \overbrace{f(x)\,dx}^{\text{integrand}} \underbrace{dx}_{\text{variable of integration}}$$

upper limit of integration

lower limit of integration

The function f that is being integrated is called the **integrand**; the interval $[a,b]$ is the **interval of integration**; and the endpoints a and b are called, respectively, the **lower and upper limits of integration.**

In the special case where $a = b$, the interval of integration $[a,b]$ is really just a point, and the integral of any function on this "interval" is defined to be 0; that is,

$$\int_a^a f(x)\,dx = 0 \qquad \textit{Why does this make sense?}$$

Also, at times, we shall consider integrals in which the lower limit of integration is a larger number than the upper limit. To handle this case, we specify that the integral from b to a is the *opposite* of the integral from a to b:

$$\int_b^a f(x)\,dx = -\int_a^b f(x)\,dx$$

To summarize:

DEFINITE INTEGRAL AT A POINT / INTERCHANGING THE LIMITS

$$\int_a^a f(x)\,dx = 0 \qquad\qquad \int_b^a f(x)\,dx = -\int_a^b f(x)\,dx$$

At first, the definition of the definite integral may seem rather imposing. How are we to tell whether a given function f is integrable on an interval $[a,b]$? If f is integrable, how are we supposed to actually compute the definite integral? Answering these questions is not easy, but in advanced calculus, it is shown that f is integrable on a closed interval $[a,b]$ if it is continuous on the interval except at a finite number of points and if it is bounded on the interval (that is, there is a number $M > 0$ such that $|f(x)| < M$ for all x in the interval). We will state a special case of this result as a theorem.

Theorem 5.5   Integrability of a continuous function

If f is continuous on an interval $[a,b]$, then f is integrable on $[a,b]$.

Proof: The proof requires the methods of advanced calculus and is omitted here.   ◆

Our next example illustrates how to use the definition to find a definite integral.

Example 2   Evaluating a definite integral using the definition

Evaluate $\displaystyle\int_{-2}^{1} 4x\,dx$.

Solution   The integral exists because $f(x) = 4x$ is continuous on $[-2,1]$. Because the integral can be computed by any partition whose norm approaches 0 (that is, the integral is independent of the sequence of partitions *and* the subinterval representatives), we shall simplify matters by choosing a partition in which the points are evenly spaced. Specifically, we divide the interval $[-2,1]$ into n subintervals, each of width

$$\Delta x = \frac{1 - (-2)}{n} = \frac{3}{n}$$

For each k, we choose the kth subinterval representative to be the right endpoint of the kth subinterval; that is,

$$x_k^* = -2 + k\,\Delta x = -2 + k\left(\frac{3}{n}\right)$$

Finally, we form the Riemann sum

$$
\begin{aligned}
\int_{-2}^{1} 4x\,dx &= \lim_{\|P\|\to 0} \sum_{k=1}^{n} f\left(x_k^*\right)\Delta x \\
&= \lim_{n\to\infty} \sum_{k=1}^{n} 4\left(-2 + \frac{3k}{n}\right)\left(\frac{3}{n}\right) \qquad n\to\infty \ as \ \|P\|\to 0 \\
&= \lim_{n\to\infty} \frac{12}{n^2}\left(\sum_{k=1}^{n}(-2n) + \sum_{k=1}^{n}3k\right) \\
&= \lim_{n\to\infty} \frac{12}{n^2}\left\{(-2n)n + 3\left[\frac{n(n+1)}{2}\right]\right\} \quad \textit{Summation formula} \\
&= \lim_{n\to\infty} \frac{12}{n^2}\left(\frac{-4n^2 + 3n^2 + 3n}{2}\right) \\
&= \lim_{n\to\infty} \frac{-6n^2 + 18n}{n^2} \\
&= -6
\end{aligned}
$$

Area as an Integral

Because we have used the development of area in Section 5.2 as the model for our definition of the definite integral, it is no surprise to discover that the area under a curve can be expressed as a definite integral. However, integrals can be positive, zero, or negative (as in Example 2), and we certainly would not expect the area under a curve to be a negative number! The actual relationship between integrals and area under a curve is contained in the following observation, which follows from the definition of area as the limit of a sum, along with Theorem 5.5.

AREA AS AN INTEGRAL   Suppose f is continuous and $f(x) \geq 0$ on the closed interval $[a, b]$. Then the area under the curve $y = f(x)$ on $[a, b]$ is given by the definite integral of f on $[a, b]$. That is,

$$\text{AREA} = \int_a^b f(x)\,dx$$

We will find areas using a definite integral, but not every definite integral can be interpreted as an area.

Usually we find area by evaluating a definite integral, but sometimes area can be used to help us evaluate the integral. At this stage of our study, it is not easy to evaluate Riemann sums, so if you happen to recognize that the integral represents the area of some common geometric figure, you can use the known formula instead of the definite integral, as shown in Example 3.

Example 3   Evaluating an integral using an area formula

Evaluate $\displaystyle\int_{-3}^{3} \sqrt{9 - x^2}\,dx$.

Solution   Let $f(x) = \sqrt{9 - x^2}$. The curve $y = \sqrt{9 - x^2}$ is a semicircle centered at the origin of radius 3, as shown in Figure 5.14.

The given integral can be interpreted as the area under the semicircle on the interval $[-3, 3]$. From geometry, we know the area of the circle is $A = \pi r^2 = \pi(3)^2 = 9\pi$. Thus, the area of the semicircle is $\frac{9\pi}{2}$, and we conclude that

$$\int_{-3}^{3} \sqrt{9 - x^2}\,dx = \frac{9\pi}{2}$$

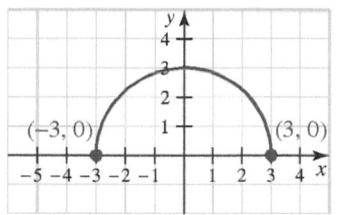

Figure 5.14   Graph of a semicircle

If f is continuous and $f(x) \leq 0$ on the closed interval $[a, b]$, then the area between the curve $y = f(x)$ and the x-axis on $[a, b]$ is the opposite of the definite integral of f on $[a, b]$, since $-f(x) \geq 0$. That is,

$$\text{AREA} = -\int_a^b f(x)\,dx$$

More generally, as illustrated in Figure 5.15, if a continuous function f is sometimes positive and sometimes negative, then

$$\int_a^b f(x)\,dx = A_1 - A_2$$

where A_1 is the sum of all areas of the regions above the x-axis and below the graph of f (that is, where $f(x) \geq 0$) and A_2 is the sum of all the areas of the regions below the x-axis and above the graph of f (that is, where $f(x) \leq 0$).

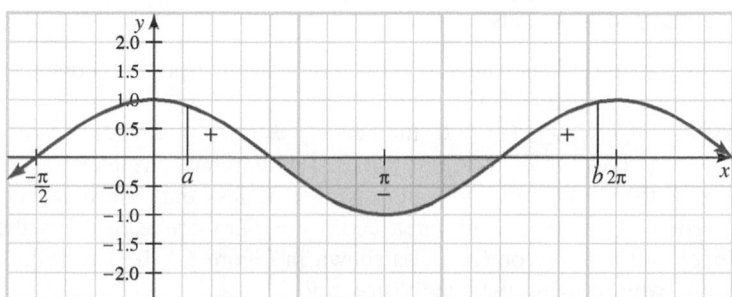

Figure 5.15 A_1 is the sum of the areas marked "+" and A_2 is the sum of the areas marked "−"

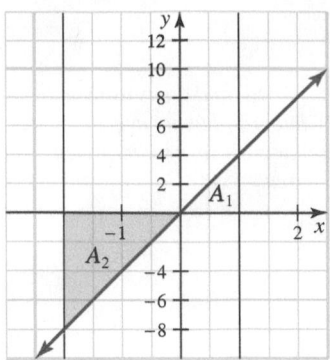

Figure 5.16 Area interpretation of Example 2

We just worked Example 2 (see Figure 5.16) using a Riemann sum.

It is easy to see from our knowledge of areas of triangles that the area of Triangle I (above the x-axis) is

$$A_1 = \frac{1}{2}bh = \frac{1}{2}(4)(1) = 2$$

and the area of Triangle II (below the x-axis) is

$$A_2 = \frac{1}{2}bh = \frac{1}{2}(8)(2) = 8$$

and we see that

$$A_1 - A_2 = 2 - 8 = -6$$

is the same as the value of the integral.

Properties of the Definite Integral

In computations involving integrals, it is often helpful to use the three general properties listed in the following theorem.

Theorem 5.6   General properties of the definite integral

Linearity rule
    If f and g are integrable on $[a, b]$, then so is $rf + sg$ for constants r and s

$$\int_a^b \left[rf(x) + sg(x)\right] dx = r \int_a^b f(x)\, dx + s \int_a^b g(x)\, dx$$

Dominance rule
    If f and g are integrable on $[a, b]$ and $f(x) \leq g(x)$ throughout this interval, then

$$\int_a^b f(x)\, dx \leq \int_a^b g(x)\, dx$$

Subdivision rule
    For any number c such that $a < c < b$,

$$\int_a^b f(x)\, dx = \int_a^c f(x)\, dx + \int_c^b f(x)\, dx$$

assuming all three integrals exist.

Proof: Each of these rules can be established by using a familiar property of sums or limits with the definition of the definite integral. For example, to derive the linearity rule, we note that any Riemann sum of the function $rf + sg$ can be expressed as

$$\sum_{k=1}^{n} \left[rf\left(x_k^*\right) + sg\left(x_k^*\right)\right] \Delta x_k = r \left[\sum_{k=1}^{n} f\left(x_k^*\right) \Delta x_k\right] + s \left[\sum_{k=1}^{n} g(x_k^*)\Delta x_k\right]$$

and the linearity rule then follows by taking the limit on each side of this equation as the norm of the partition tends to 0.   ◆

The dominance rule and the subdivision rule are interpreted geometrically for non-negative functions in Figure 5.17.

Notice that if $g(x) \geq f(x) \geq 0$, the curve $y = g(x)$ is always above (or touching) the curve $y = f(x)$, as shown in Figure 5.17**a**. The dominance rule expresses the fact that the area under the upper curve $y = g(x)$ cannot be less than the area under $y = f(x)$. The subdivision rule says that the area under $y = f(x)$ above $[a, b]$ is the sum of the area on $[a, c]$ and the area on $[c, b]$, as shown in Figure 5.17**b**. The following example illustrates one way of using the subdivision rule.

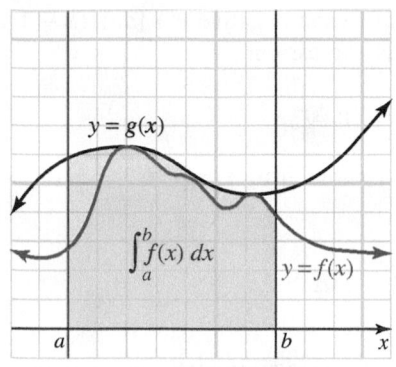

a. Dominance rule

If $g(x) \geq f(x)$, the area under $y = g(x)$ can be no less than the area under $y = f(x)$.

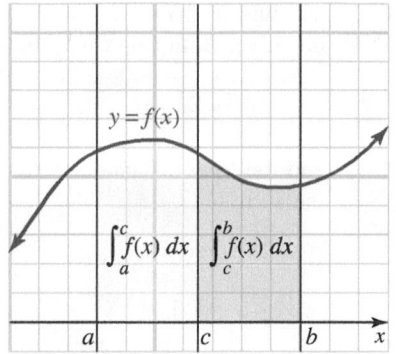

b. Subdivision rule

The area under $y = f(x)$ on $[a, b]$ equals the sum of the areas on $[a, c]$ and $[c, b]$.

Figure 5.17   The dominance and subdivision rules

Example 4   Subdivision rule

If $\displaystyle\int_{-2}^{1} f(x)\,dx = 3$ and $\displaystyle\int_{-2}^{7} f(x)\,dx = -5$, what is $\displaystyle\int_{1}^{7} f(x)\,dx$?

Solution   According to the subdivision rule, we have

$$\int_{-2}^{7} f(x)\,dx = \int_{-2}^{1} f(x)\,dx + \int_{1}^{7} f(x)\,dx$$

$$\int_{1}^{7} f(x)\,dx = \int_{-2}^{7} f(x)\,dx - \int_{-2}^{1} f(x)\,dx$$

$$= -5 - 3$$
$$= -8$$

Distance as an Integral

Many quantities other than area can be computed as the limit of a sum. For example, suppose an object moving along a line is known to have continuous velocity $v(t)$ for each time t between $t = a$ and $t = b$, and we wish to compute the total distance traveled by the object during this time period.

Let the interval $[a, b]$ be partitioned into n equal subintervals, each of length $\Delta t = \dfrac{b-a}{n}$ as shown in Figure 5.18.

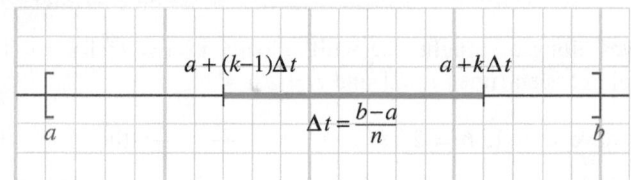

Figure 5.18   The distance traveled during the kth time subinterval

The kth subinterval is $[a + (k-1)\Delta t, a + k\Delta t]$, and if Δt is small enough, the velocity $v(t)$ will not change much over the subinterval so it is reasonable to approximate $v(t)$ by the constant velocity of $v[a + (k-1)\Delta t]$ throughout the entire subinterval.

The corresponding change in the object's position will be approximated by the product

$$v[a + (k-1)\Delta t]\,\Delta t$$

and will be positive if $v\left[a + (k - 1)\Delta t\right]$ is positive and negative if $v\left[a + (k - 1)\Delta t\right]$ is negative. Both cases may be summarized by the expression

$$\left|v\left[a + (k - 1)\Delta t\right]\right| \Delta t$$

and the total distance traveled by the object as t varies from $t = a$ to $t = b$ is given by the sum

$$S_n = \sum_{k=1}^{n} \left|v\left[a + (k - 1)\Delta t\right]\right| \Delta t$$

which we recognize as a Riemann sum. We can make the approximation more precise by taking more refined partitions (that is, shorter and shorter time intervals Δt). Therefore, it is reasonable to *define* the exact distance S traveled as the *limit* of the sum S_n as $\Delta t \to 0$ or, equivalently, as $n \to \infty$, so that

$$S = \lim_{n \to \infty} \sum_{k=1}^{n} \left|v\left[a + (k - 1)\Delta t\right]\right| \Delta t = \int_{a}^{b} |v(t)|\, dt$$

The reason that we use $|v|$ is that if we travel 300 miles east and then 300 miles west, for example, the *net* distance traveled is $300 - 300 = 0$ miles, but we are usually interested in the total distance traveled, $300 + 300 = 600$ miles.

DISTANCE   The **total distance traveled** by an object with continuous velocity $v(t)$ along a straight line from time $t = a$ to $t = b$ is

$$S = \int_{a}^{b} |v(t)|\, dt$$

Note: There is a difference between the *total distance* traveled by an object over a given time interval $[a, b]$, and the **net distance** or **displacement** of the object over the same interval, which is defined as the difference between the object's final and initial positions. It is easy to see that displacement is given by

$$D = \int_{a}^{b} v(t)\, dt$$

Example 5   Distance traveled by an object whose velocity is known

An object moves along a straight line with velocity $v(t) = t^2$ for $t > 0$. How far does the object travel between times $t = 1$ and $t = 2$?

Solution   We have $a = 1$, $b = 2$, and $\Delta t = \dfrac{2 - 1}{n} = \dfrac{1}{n}$; therefore, the required distance is

$$S = \int_{1}^{2} |v(t)|\, dt$$

$$= \lim_{n \to \infty} \sum_{k=1}^{n} \left|v\left[1 + (k - 1)\left(\frac{1}{n}\right)\right]\right| \left(\frac{1}{n}\right)$$

$$= \lim_{n \to \infty} \sum_{k=1}^{n} \left|v\left(\frac{n + k - 1}{n}\right)\right| \left(\frac{1}{n}\right)$$

$$= \lim_{n\to\infty} \sum_{k=1}^{n} \frac{(n+k-1)^2}{n^2}\left(\frac{1}{n}\right)$$

$$= \lim_{n\to\infty} \frac{1}{n^3} \sum_{k=1}^{n}\left[(n^2 - 2n + 1) + k^2 + 2(n-1)k\right]$$

$$= \lim_{n\to\infty} \frac{1}{n^3}\left[(n^2 - 2n + 1)\sum_{k=1}^{n}1 + \sum_{k=1}^{n}k^2 + 2(n-1)\sum_{k=1}^{n}k\right]$$

$$= \lim_{n\to\infty} \frac{1}{n^3}\left[(n^2 - 2n + 1)n + \frac{n(n+1)(2n+1)}{6} + 2(n-1)\frac{n(n+1)}{2}\right]$$

$$= \lim_{n\to\infty} \frac{14n^3 - 9n^2 + n}{6n^3}$$

$$= \frac{14}{6}$$

$$= \frac{7}{3}$$

Thus, we expect the object to travel $\frac{7}{3}$ units during the time interval $[1, 2]$. ◼

By considering the distance as an integral, we see again that zero or negative values of a definite integral can also be interpreted geometrically. When $v(t) > 0$ on a time interval $[a, b]$, then the total distance S traveled by the object between times $t = a$ and $t = b$ is the same as the area under the graph of $v(t)$ on $[a, b]$.

When $v(t) > 0$, the object moves forward (to the right), but when $v(t) < 0$, it reverses direction and moves backward (to the left). In the general case, where $v(t)$ changes sign on the time interval $[a, b]$, the integral

$$\int_a^b v(t)\,dt$$

measures the net distance or displacement of the object, taking into account both forward ($v > 0$) and backward ($v < 0$) motion. For instance, an object that moves right 2 units and left 3 on a given time interval has moved a total distance of 5 units, but its displacement is -1 because it ends up 1 unit to the left of its initial position.

For example, for the velocity function $v(t)$ graphed in Figure 5.19**a**, the net displacement is 0 because the area above the t-axis is the same as the area below, but in Figure 5.19**b**, there is more area below the t-axis, which means the net displacement is negative, and the object ends up "behind" (to the left of) its starting position.

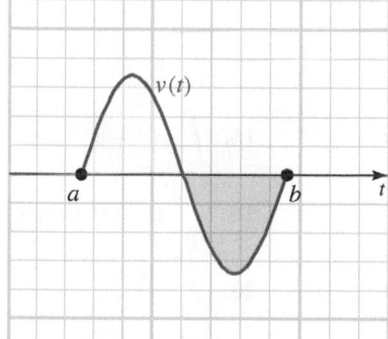

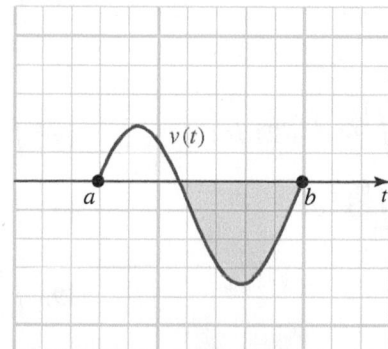

a. Net displacement of 0. There is as much area under the positive part of the velocity curve as there is above the negative part.

b. Negative net displacement. There is more area above the negative part than there is under the positive part.

Figure 5.19 Definite integral in terms of displacement

Preview Students often become discouraged at this point, thinking they will have to compute all definite integrals using the limit of a Riemann sum definition, so we offer an encouraging word. Recall the definition of the derivative in Chapter 3. It was difficult to find derivatives using the definition, but we soon proved some theorems to make it easier to find and evaluate derivatives. The same is true of integration. It is difficult to apply the definition of a definite integral, but you will soon discover that computing most definite integrals is no harder than finding an antiderivative.

As a preview of this result, consider an object moving along a straight line. We assume that its position is given by $s(t)$ and that its velocity $v(t) = s'(t)$ is positive ($v(t) > 0$) so that it is always moving forward. The total distance traveled by such an object between times $t = a$ and $t = b$ is clearly $S = s(b) - s(a)$, but earlier in this section, we showed that this distance is also given by the definite integral of $v(t)$ over the interval $[a, b]$. Thus we have

$$\text{TOTAL DISTANCE} = \int_a^b v(t)\, dt = s(b) - s(a)$$

where $s(t)$ is an antiderivative of $v(t)$.

This observation anticipates the *fundamental theorem of calculus*, which provides a vital link between differential and integral calculus. We will formally introduce the fundamental theorem in the next section. To illustrate, notice how our observation applies to Example 5, where $v(t) = t^2$ and the interval is $[1, 2]$. Since an antiderivative of $v(t)$ is $s(t) = \frac{1}{3}t^3$, we have

$$S = \int_1^2 t^2\, dt = s(2) - s(1) = \left[\frac{1}{3}(2)^3 - \frac{1}{3}(1)^3 \right] = \frac{7}{3}$$

which coincides with the result found numerically in Example 5.

PROBLEM SET 5.3

Level 1

In Problems 1-10, *estimate (using right endpoints) the given integral* $\int_a^b f(x)\, dx$ *by using a Riemann sum*

$$S_n = \sum_{k=1}^n f(a + k\Delta x)\Delta x \text{ for } n = 4.$$

Compare with Problems 11-20.

1. $\int_0^1 (2x + 1)\, dx$ 2. $\int_0^1 (4x^2 + 2)\, dx$

3. $\int_1^3 x^2\, dx$ 4. $\int_0^2 x^3\, dx$

5. $\int_0^1 (1 - 3x)\, dx$ 6. $\int_1^3 (x^2 - x^3)\, dx$

7. $\int_{-\frac{\pi}{2}}^0 \cos x\, dx$ 8. $\int_0^{\frac{\pi}{4}} (x + \sin x)\, dx$

9. $\int_0^1 e^x\, dx$ 10. $\int_1^2 \frac{dx}{x}$

In Problems 11-20, *estimate (using left endpoints) the given integral* $\int_a^b f(x)\, dx$ *by using a Riemann sum*

$$S_n = \sum_{k=1}^n f(a + (k-1)\Delta x)\Delta x \text{ for } n = 4.$$

Compare with Problems 1-10.

11. $\int_0^1 (2x + 1)\, dx$ 12. $\int_0^1 (4x^2 + 2)\, dx$

13. $\int_1^3 x^2\, dx$ 14. $\int_0^2 x^3\, dx$

15. $\int_0^1 (1 - 3x)\, dx$ 16. $\int_1^3 (x^2 - x^3)\, dx$

17. $\int_{-\frac{\pi}{2}}^0 \cos x\, dx$ 18. $\int_0^{\frac{\pi}{4}} (x + \sin x)\, dx$

19. $\int_0^1 e^x\, dx$ 20. $\int_1^2 \frac{dx}{x}$

In Problems 21-26, $v(t)$ is the velocity of an object moving along a straight line. Use the formula

$$S_n = \sum_{k=1}^n |v(a + k\Delta t)|\, \Delta t$$

where $\Delta t = \dfrac{b-a}{n}$ *to estimate (using right endpoints)
the total distance traveled by the object during the time
interval $[a,b]$. Let $n=4$. Compare with Problems
27-32.*

21. $v(t) = 3t + 1$ on $[1,4]$ **22.** $v(t) = 1 + 2t$ on $[1,2]$

23. $v(t) = \sin t$ on $[0,\pi]$ **24.** $v(t) = \cos t$ on $\left[0, \frac{\pi}{2}\right]$

25. $v(t) = e^{-t}$ on $[0,1]$ **26.** $v(t) = \dfrac{1}{t+1}$ on $[0,1]$

*In Problems 27-32, $v(t)$ is the velocity of an object moving
along a straight line. Use the formula*

$$S_n = \sum_{k=1}^{n} |v(a + (k-1)\Delta t)| \, \Delta t$$

*where $\Delta t = \dfrac{b-a}{n}$ to estimate (using left endpoints) the
total distance traveled by the object during the time interval $[a,b]$. Let $n=4$. Compare with Problems 21-26.*

27. $v(t) = 3t + 1$ on $[1,4]$ **28.** $v(t) = 1 + 2t$ on $[1,2]$

29. $v(t) = \sin t$ on $[0,\pi]$ **30.** $v(t) = \cos t$ on $\left[0, \frac{\pi}{2}\right]$

31. $v(t) = e^{-t}$ on $[0,1]$ **32.** $v(t) = \dfrac{1}{t+1}$ on $[0,1]$

*Evaluate each of the integrals in Problems 33-38 by using
the following information together with the linearity and
subdivision properties:*

$$\int_{-1}^{2} x^2\,dx = 3; \quad \int_{-1}^{0} x^2\,dx = \frac{1}{3}; \quad \int_{-1}^{2} x\,dx = \frac{3}{2}; \quad \int_{0}^{2} x\,dx = 2$$

33. $\displaystyle\int_{0}^{-1} x^2\,dx$ **34.** $\displaystyle\int_{-1}^{2} (x^2 + x)\,dx$

35. $\displaystyle\int_{-1}^{2} (2x^2 - 3x)\,dx$ **36.** $\displaystyle\int_{0}^{2} x^2\,dx$

37. $\displaystyle\int_{-1}^{0} x\,dx$ **38.** $\displaystyle\int_{-1}^{0} (3x^2 - 5x)\,dx$

*Use the dominance property of integrals to establish the
given inequality in Problems 39-42.*

39. $\displaystyle\int_{0}^{1} x^2\,dx \le \frac{1}{2}$ *Hint: Note that $x^2 \le x$ on $[0,1]$.*

40. $\displaystyle\int_{0}^{\pi} \sin x\,dx \le \pi$ *Hint: $\sin x \le 1$ for all x.*

41. $\displaystyle\int_{0}^{0.7} x^3\,dx \le \frac{1}{2}$ **42.** $\displaystyle\int_{0}^{\frac{\pi}{2}} \cos x\,dx \le \frac{\pi}{2}$

Level 2

43. Given $\displaystyle\int_{-2}^{4} \left[5f(x) + 2g(x)\right]dx = 7$ and

$$\int_{-2}^{4} \left[3f(x) + g(x)\right]dx = 4, \text{ find}$$

$$\int_{-2}^{4} f(x)\,dx \text{ and } \int_{-2}^{4} g(x)\,dx.$$

44. Suppose $\displaystyle\int_{0}^{2} f(x)\,dx = 3,$

$$\int_{0}^{2} g(x)\,dx = -1, \text{ and } \int_{0}^{2} h(x)\,dx = 3.$$

a. Evaluate

$$\int_{0}^{2} \left[2f(x) + 5g(x) - 7h(x)\right]dx$$

b. Find a constant s such that

$$\int_{0}^{2} \left[5f(x) + sg(x) - 6h(x)\right]dx = 0$$

45. Evaluate $\displaystyle\int_{-1}^{2} f(x)\,dx$ given that

$$\int_{-1}^{1} f(x)\,dx = 3, \int_{2}^{3} f(x)\,dx = -2$$

and $\displaystyle\int_{1}^{3} f(x)\,dx = 5.$

46. Let $f(x) = \begin{cases} 2 & \text{for } -1 \le x < 1 \\ 3 - x & \text{for } 1 \le x < 4 \\ 2x - 9 & \text{for } 4 \le x \le 5 \end{cases}$

Sketch the graph of f on the interval $[-1,5]$ and
show that f is continuous on this interval. Then use
Theorem 5.6 to evaluate

$$\int_{-1}^{5} f(x)\,dx$$

47. $f(x) = \begin{cases} 5 & \text{for } -3 \le x < -1 \\ 4 - x & \text{for } -1 \le x < 2 \\ 2x - 2 & \text{for } 2 \le x \le 5 \end{cases}$

Sketch the graph of f on the interval $[-3,5]$ and
show that f is continuous on this interval. Then use
Theorem 5.6 to evaluate

$$\int_{-3}^{5} f(x)\,dx$$

48. *Historical Quest*

Karl Smith library

During the 18th century, integrals were considered simply as antiderivatives. That is, there were no underpinnings for the concept of an integral until Cauchy formulated the definition of the integral in 1823 (see *Historical Quest* Problem 53, Problem Set 2.1). This formulation was later completed by Georg Friedrich

Georg Bernhard Friedrich Riemann (1826-1866)

*Riemann. In this section, we see that history honored Riemann by naming the process after him. In his personal life he was frail, bashful, and timid, but in his professional life he was one of the giants in mathematical history. Riemann used what are called topological methods in his theory of functions and in his work with surfaces and spaces. Riemann is remembered for his work in geometry (Riemann surfaces) and analysis. In his book **Space Through the Ages**, Cornelius Lanczos wrote, "Although Riemann's collected papers fill only one single volume of 538 pages, this volume weighs tons if measured intellectually. Every one of his many discoveries was destined to change the course of mathematical science."*

For this *Historical Quest*, investigate the Königsberg bridge problem, and its solution. *This famous problem was formulated by Leonhard Euler (1707-1783). The branch of mathematics known today as **topology** began with Euler's work on the bridge problem and other related questions and was extended in the 19th century by Riemann and others.*

Level 3

49. Generalize the subdivision property by showing that for $a \le c \le d \le b$,

$$\int_a^b f(x)\,dx = \int_a^c f(x)\,dx + \int_c^d f(x)\,dx + \int_d^b f(x)\,dx$$

whenever all these integrals exist.

50. If $Cx + D \ge 0$ for $a \le x \le b$, show that

$$\int_a^b (Cx + D)\,dx = (b - a)\left[\frac{C}{2}(b + a) + D\right]$$

Hint: Sketch the region under the line $y = Cx + D$, and express the integral as an area.

51. For $b > a > 0$, show that

$$\int_a^b x^2\,dx = \frac{1}{3}(b^3 - a^3)$$

In Problems 52-54, use the partition $P = \{-1, -0.2, 0.9, 1.3, 1.7, 2\}$ on the interval $[-1, 2]$.

52. Find the subinterval widths

$$\Delta x_k = x_k - x_{k-1}$$

for $k = 1, 2, \cdots, 5$. What is the norm of P?

53. Compute the Riemann sum on $[-1, 2]$ associated with $f(x) = 4 - 5x$; the partition P with $\Delta x_1 = 0.8$, $\Delta x_2 = 1.1$, $\Delta x_3 = 0.4$, $\Delta x_4 = 0.4$, and $\Delta x_5 = 0.3$; and the subinterval representatives $x_1^* = -0.5$, $x_2^* = 0.8, x_3^* = 1, x_4^* = 1.3$, $x_5^* = 1.8$. What is the norm of P?

54. Compute the Riemann sum on $[-1, 2]$ associated with $f(x) = x^3$; the partition P with $\Delta x_1 = 0.8$, $\Delta x_2 = 1.1$, $\Delta x_3 = 0.4$, $\Delta x_4 = 0.4$, and $\Delta x_5 = 0.3$; and subinterval representatives $x_1^* = -1$, $x_2^* = 0, x_3^* = 1$, $x_4^* = \frac{128}{81}$, $x_5^* = \frac{125}{64}$. What is the norm of P?

55. If the numbers a_k and b_k satisfy $a_k \le b_k$ for $k = 1, 2, \cdots, n$, then

$$\sum_{k=1}^{n} a_k \le \sum_{k=1}^{n} b_k$$

Use this dominance property of sums to establish the dominance property of integrals.

56. Think Tank Problem Either prove that the following result is generally true or find a counterexample: If f and g are continuous on $[a, b]$, then

$$\int_a^b f(x)g(x)\,dx = \left[\int_a^b f(x)\,dx\right]\left[\int_a^b g(x)\,dx\right]$$

57. Think Tank Problem Either prove that the following result is generally true or find a counterexample: The area between the x-axis and the continuous function f on the interval $[a, b]$ is zero if and only if

$$\int_a^b f(x)\,dx = 0$$

58. Recall that a function f is **odd** if $f(-x) = -f(x)$ for all x. If f is odd, show that

$$\int_{-a}^a f(x)\,dx = 0$$

for all a. This problem is analogous to Problem 40 of the previous section.

59. Prove the *bounding rule* for definite integrals: If f is integrable on the closed interval $[a, b]$ and $m \leq f(x) \leq M$ for constants m, M, and all x in the closed interval, then

$$m(b - a) \leq \int_a^b f(x)\,dx \leq M(b - a)$$

60. Use the definition of the definite integral to prove that

$$\int_a^b C\,dx = C(b - a)$$

5.4 THE FUNDAMENTAL THEOREMS OF CALCULUS

IN THIS SECTION: *The first fundamental theorem of calculus, the second fundamental theorem of calculus*
We have found that the value of a definite integral can be approximated numerically, and by using summation formulas we can compute some integrals algebraically. These methods are messy, but fortunately the fundamental theorems of calculus, presented in this section, give us a means of evaluating many integrals. We shall study techniques of integration in the next chapter, but one such technique, the *method of substitution*, is so fundamental that it is useful to have it now.

The First Fundamental Theorem of Calculus

In the previous section we observed that if $v(t)$ is the velocity of an object at time t as it moves along a straight line, then

$$\int_a^b v(t)\,dt = s(b) - s(a)$$

where $s(t)$ is the displacement of the object and satisfies $s'(t) = v(t)$. This result is an application of the following general theorem which was discovered by the English mathematician Isaac Barrow (1630-1677), Newton's mentor at Cambridge.

Theorem 5.7   The first fundamental theorem of calculus

If f is continuous on the interval $[a, b]$ and F is any function that satisfies $F'(x) = f(x)$ throughout this interval, then

$$\int_a^b f(x)\,dx = F(b) - F(a)$$

■ **W**hat this says The definite integral $\int_a^b f(x)\,dx$ can be computed by finding an antiderivative on the interval $[a, b]$ and evaluating it at the limits of integration a and b. It is a consequence of the first fundamental theorem of calculus that our two definitions of area between the graph of $y = f(x)$ for $y \geq 0$ and the x-axis from the first two sections of this chapter agree. Also notice that this theorem does not say *how* to find the antiderivative, nor does it say that an antiderivative F *exists*. The existence of an antiderivative is asserted by the second fundamental theorem of calculus, which is examined later in this section (page 360).

Proof: Let $P = \{x_0, x_1, x_2, \cdots, x_n\}$ be a regular partition of the interval, with subinterval widths $\Delta x = \frac{b-a}{n}$. Note that F satisfies the hypotheses of the mean value theorem (Theorem 2.4, Section 2.3) on each of the closed subintervals $[x_{k-1}, x_k]$. Thus, the MVT tells us that there is a point x_k^* in each open subinterval (x_{k-1}, x_k) for which

$$\frac{F(x_k) - F(x_{k-1})}{x_k - x_{k-1}} = F'(x_k^*)$$

$$F(x_k) - F(x_{k-1}) = F'(x_k^*)(x_k - x_{k-1})$$

Because $F'\left(x_k^*\right) = f(x_k^*)$ and $x_k - x_{k-1} = \Delta x = \dfrac{b-a}{n}$, we can write $F(x_k) - F(x_{k-1}) = f\left(x_k^*\right)\Delta x$, so that

$$F(x_1) - F(x_0) = f\left(x_1^*\right)\Delta x$$
$$F(x_2) - F(x_1) = f\left(x_2^*\right)\Delta x$$
$$\vdots$$
$$F(x_n) - F(x_{n-1}) = f\left(x_n^*\right)\Delta x$$

Thus, by adding both sides of all the equations, we obtain

$$\sum_{k=1}^{n} f\left(x_k^*\right)\Delta x = f\left(x_1^*\right)\Delta x + f\left(x_2^*\right)\Delta x + \cdots + f\left(x_n^*\right)\Delta x$$

$$= [F(x_1) - F(x_0)] + [F(x_2) - F(x_1)] + \cdots + [F(x_n) - F(x_{n-1})]$$
$$= F(x_n) - F(x_0)$$

Because $x_0 = a$ and $x_n = b$, we have

$$\sum_{k=1}^{n} f\left(x_k^*\right)\Delta x = F(b) - F(a)$$

Finally, we take the limit of the left side as $\|P\| \to 0$ (i.e., $n \to \infty$), and because $F(b) - F(a)$ is a constant, we have

$$\int_a^b f(x)\,dx = F(b) - F(a) \qquad \blacklozenge$$

To give you some insight into why this theorem is important enough to be named *the fundamental theorem of calculus*, we repeat Example 2 from Section 5.3 in the margin and then work the same example using the fundamental theorem on the right.

Example 1   Evaluating a definite integral

Solution from Section 5.3 using Riemann sums:

$$\int_{-2}^{1} 4x\,dx = \lim_{\|P\|\to 0}\sum_{k=1}^{n} f\left(x_k^*\right)\Delta x$$

$$= \lim_{n\to\infty}\sum_{k=1}^{n} 4\left(-2+\frac{3k}{n}\right)\left(\frac{3}{n}\right)$$

$$= \lim_{n\to\infty}\frac{12}{n^2}\left(\sum_{k=1}^{n}(-2n) + \sum_{k=1}^{n} 3k\right)$$

$$= \lim_{n\to\infty}\frac{12}{n^2}\left\{(-2n)n + 3\left[\frac{n(n+1)}{2}\right]\right\}$$

$$= \lim_{n\to\infty}\frac{12}{n^2}\left(\frac{-4n^2+3n^2+3n}{2}\right)$$

$$= \lim_{n\to\infty}\frac{-6n^2+18n}{n^2}$$

$$= -6$$

Evaluate $\int_{-2}^{1} 4x\,dx$ using the definition of the definite integral and also using the fundamental theorem.

Solution   Solution using the fundamental theorem of calculus: If $F(x) = 2x^2$, then $F'(x) = 4x$, so F is an antiderivative of f. Thus,

$$\int_{-2}^{1} 4x\,dx = F(1) - F(-2)$$
$$= 2(1)^2 - 2(-2)^2$$
$$= -6$$

Note that if we choose a *different* antiderivative, say $G(x) = 2x^2 + 3$, then

$$G(1) - G(-2) = -6$$

and this is the same for *any* antiderivative. ▪

The variable used in a definite integral is a *dummy variable* in the sense that it can be replaced by any other variable with no effect on the value of the integral. For instance, we have just found that

$$\int_{-2}^{1} 4x\,dx = -6$$

and without further computation, it follows that

$$\int_{-2}^{1} 4t\,dt = -6 \qquad \int_{-2}^{1} 4u\,du = -6 \qquad \int_{-2}^{1} 4N\,dN = -6$$

Henceforth, when evaluating an integral by the fundamental theorem, we shall denote the difference

$$F(b) - F(a) \qquad \text{by} \qquad F(x)\big|_a^b$$

Sometimes we also write

$$[F(x)]_a^b$$

where $F'(x) = f(x)$ on $[a, b]$. This notation is illustrated in Example 2.

Example 2   Area under a curve using the fundamental theorem of calculus

Find the area under the curve $y = \cos x$ on $\left[-\frac{\pi}{2}, \frac{\pi}{2}\right]$. See Figure 5.20.

Solution   Note that $f(x) = \cos x$ is continuous and $f(x) \geq 0$ on $\left[-\frac{\pi}{2}, \frac{\pi}{2}\right]$. Because the derivative of $\sin x$ is $\cos x$, it follows (from the definition of the antiderivative) that $\sin x$ is an antiderivative of $\cos x$. Thus, the required area is given by the integral

$$
\begin{aligned}
A &= \int_{-\frac{\pi}{2}}^{\frac{\pi}{2}} \cos x \, dx \\
&= \sin x \big|_{-\pi/2}^{\pi/2} \\
&= \sin \frac{\pi}{2} - \sin \left(-\frac{\pi}{2}\right) \\
&= 1 - (-1) \\
&= 2
\end{aligned}
$$

The region has area $A = 2$ square units.

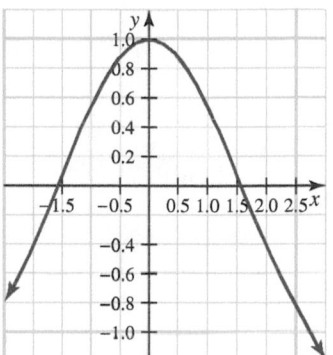

Figure 5.20 If you look at the graph of $y = f(x)$, you can see an area of 2 seems reasonable

☠ It is important to remember that, for fixed numerical values of a and b, the definite integral $\int_a^b f(x)\,dx$ is a number, whereas the indefinite integral $\int f(x)\,dx$ is a family of functions. ☠

The relationship between the indefinite and definite integral is given by

$$\int_a^b f(x)\,dx = \left[\int f(x)\,dx\right]_a^b$$

Example 3   Evaluating an integral using the fundamental theorem

Evaluate: **a.** $\displaystyle\int_{-3}^{5} (-10)\,dx$ **b.** $\displaystyle\int_{4}^{9}\left[\frac{1}{\sqrt{x}} + x\right]dx$ **c.** $\displaystyle\int_{-2}^{2} |x|\,dx$

Solution

a. $\displaystyle\int_{-3}^{5} (-10)\,dx = -10x\big|_{-3}^{5}$
$= -10(5 + 3)$
$= -80$

b. $\displaystyle\int_{4}^{9}\left[\frac{1}{\sqrt{x}} + x\right]dx = \int_{4}^{9}\left[x^{-1/2} + x\right]dx$
$= \left[\left(\frac{x^{1/2}}{\frac{1}{2}}\right) + \frac{x^2}{2}\right]_4^9$
$= \left(6 + \frac{81}{2}\right) - \left(4 + \frac{16}{2}\right)$
$= \frac{69}{2}$

c. $\displaystyle\int_{-2}^{2} |x|\, dx = \int_{-2}^{0} |x|\, dx + \int_{0}^{2} |x|\, dx$ *Subdivision rule* $\left\{\begin{array}{l}\textit{Note, we must use the}\\ \textit{subdivision rule because we}\\ \textit{do not know an antiderivative}\\ \textit{of } |x|, \textit{ where } x \textit{ is defined}\\ \textit{on an interval containing } 0.\end{array}\right.$

$$= \int_{-2}^{0} (-x)\, dx + \int_{0}^{2} x\, dx \quad \textit{Definition of absolute value}$$

$$= -\frac{x^2}{2}\Big|_{-2}^{0} + \frac{x^2}{2}\Big|_{0}^{2}$$

$$= -\left(0 - \frac{4}{2}\right) + \left(\frac{4}{2} + 0\right)$$

$$= 4$$

The Second Fundamental Theorem of Calculus

In certain circumstances, it is useful to consider an integral of the form

$$\int_{a}^{x} f(t)\, dt$$

where the upper limit of integration is a variable instead of a constant. As x varies, so does the value of the integral, and

$$F(x) = \int_{a}^{x} f(t)\, dt$$

is a function of the variable x. (Note that the *integrand* does not contain x.) For example,

$$F(x) = \int_{2}^{x} t^2 dt = \frac{1}{3}x^3 - \frac{1}{3}(2)^3 = \frac{1}{3}x^3 - \frac{8}{3}$$

The first fundamental theorem of calculus tells us that if f is continuous on $[a, b]$, then

$$\int_{a}^{b} f(x)\, dx = F(b) - F(a)$$

where F is an antiderivative of f on $[a, b]$. But in general, what guarantee do we have that such an antiderivative even exists? The answer is provided by the following theorem, which is often referred to as the second fundamental theorem of calculus.

Theorem 5.8   The second fundamental theorem of calculus

Let $f(t)$ be continuous on the interval $[a, b]$ and define the function G by the integral equation

$$G(x) = \int_{a}^{x} f(t)\, dt$$

for $a \leq x \leq b$. Then G is an antiderivative of f on $[a, b]$; that is,

$$G'(x) = \frac{d}{dx}\left[\int_{a}^{x} f(t)\, dt\right] = f(x)$$

on $[a, b]$.

Proof: Apply the definition of the derivative to G:

$$G'(x) = \lim_{\Delta x \to 0} \frac{G(x + \Delta x) - G(x)}{\Delta x} \qquad x \text{ is fixed, only } \Delta x \text{ changes}$$

$$= \lim_{\Delta x \to 0} \frac{1}{\Delta x} [G(x + \Delta x) - G(x)]$$

$$= \lim_{\Delta x \to 0} \frac{1}{\Delta x} \left[\int_a^{x+\Delta x} f(t)\, dt - \int_a^x f(t)\, dt \right]$$

$$= \lim_{\Delta x \to 0} \frac{1}{\Delta x} \left[\int_a^{x+\Delta x} f(t)\, dt + \int_x^a f(t)\, dt \right]$$

$$= \lim_{\Delta x \to 0} \frac{1}{\Delta x} \left[\int_x^{x+\Delta x} f(t)\, dt \right] \qquad \textit{Subdivision rule}$$

Since f is continuous on $[a, b]$, it is continuous on $[x, x + \Delta x]$ for any x in $\lfloor a, b]$. Let $m(x)$ and $M(x)$ be the smallest and largest values, respectively, for $f(t)$ on $[x, x + \Delta x]$. (*Note*: In general, m and M depend on x and Δx but are constants as far as t-integration is concerned.) Since $m(x) \le f(t) \le M(x)$, we have (by the dominance rule)

$$\int_x^{x+\Delta x} m(x)\, dt \le \int_x^{x+\Delta x} f(t)\, dt \le \int_x^{x+\Delta x} M(x)\, dt$$

$$m(x)\,[(x + \Delta x - x] \le \int_x^{x+\Delta x} f(t)\, dt \le M(x)\,[(x + \Delta x) - x]$$

$$m(x) \le \frac{1}{\Delta x} \int_x^{x+\Delta x} f(t)\, dt \le M(x)$$

Since

$$m(x) \le f(x) \le M(x)$$

on $[x, x + \Delta x]$, it follows that $m(x)$ and $M(x)$ are both "squeezed" toward $f(x)$ as $\Delta x \to 0$ and we have

$$G'(x) = \lim_{\Delta x \to 0} \frac{1}{\Delta x} \int_x^{x+\Delta x} f(t)\, dt = f(x) \qquad \blacklozenge$$

We also note that the function G is continuous, since it is differentiable. If F is *any* antiderivative of f on the interval $[a, b]$, then the antiderivative

$$G(x) = \int_a^x f(t)\, dt$$

found in Theorem 5.8 satisfies $G(x) = F(x) + C$ for some constant C and all x on the interval $[a, b]$. In particular, when $x = a$, we have

$$0 = \int_a^a f(t)\, dt = G(a) = F(a) + C$$

so that $C = -F(a)$. Finally, by letting $x = b$, we find that

$$\int_a^b f(t)\, dt = F(b) + C = F(b) + [-F(a)] = F(b) - F(a)$$

as claimed by the first fundamental theorem.

Example 4   Using the second fundamental theorem

Differentiate $F(x) = \displaystyle\int_7^x (2t - 3)\, dt$.

Solution   From the second fundamental theorem, we can obtain $F'(x)$ by simply replacing t with x in the integrand $f(t) = 2t - 3$. Thus,

$$F'(x) = \frac{d}{dx}\left[\int_7^x (2t - 3)\, dt\right] = 2x - 3$$

    The second fundamental theorem of calculus can also be applied to an integral function with a variable *lower* limit of integration. For example, to differentiate

$$G(z) = \int_z^5 \frac{\sin u}{u}\, du$$

reverse the order of integration and apply the second fundamental theorem of calculus as before:

$$G'(z) = \frac{d}{dx}\left[\int_z^5 \frac{\sin u}{u}\, du\right] = \frac{d}{dx}\left[-\int_5^z \frac{\sin u}{u}\, du\right] = -\frac{d}{dx}\left[\int_5^z \frac{\sin u}{u}\, du\right] = -\frac{\sin z}{z}$$

Example 5   Second fundamental theorem with the chain rule

Find the derivative of $F(x) = \displaystyle\int_0^{x^2} t^2\, dt$.

Solution $\displaystyle F'(x) = \frac{d}{dx}\left[\int_0^{x^2} t^2 dt\right]$

$\displaystyle \qquad\qquad = \frac{d}{du}\left[\int_0^{u} t^2 dt\right]\frac{du}{dx}$ *Chain rule; $u = x^2$*

$\displaystyle \qquad\qquad = u^2\frac{du}{dx}$ *Second fundamental theorem of calculus*

$\displaystyle \qquad\qquad = (x^2)^2\frac{d(x^2)}{dx}$ *Substitute $u = x^2$.*

$\displaystyle \qquad\qquad = x^4(2x)$

$\displaystyle \qquad\qquad = 2x^5$

PROBLEM SET 5.4

Level 1

In Problems 1-26, evaluate the definite integral.

1. $\displaystyle\int_{-10}^{10} 7\, dx$

2. $\displaystyle\int_{-5}^{7} (-3)\, dx$

3. $\displaystyle\int_{-3}^{5} (2x + a)\, dx$

4. $\displaystyle\int_{-2}^{2} (b - x)\, dx$

5. $\displaystyle\int_{-1}^{2} ax^3 dx$

6. $\displaystyle\int_{-1}^{1} (x^3 + bx^2)\, dx$

7. $\displaystyle\int_{1}^{2} \frac{c}{x^3}\, dx$

8. $\displaystyle\int_{-2}^{-1} \frac{p}{x^2}\, dx$

9. $\displaystyle\int_{0}^{9} \sqrt{x}\, dx$

10. $\displaystyle\int_{0}^{27} \sqrt[3]{x}\, dx$

11. $\displaystyle\int_{0}^{1} (5u^7 + \pi^2)\, du$

12. $\displaystyle\int_{0}^{1} (7x^8 + \sqrt{\pi})\, dx$

13. $\displaystyle\int_{1}^{2} (2x)^\pi\, dx$

14. $\displaystyle\int_{\ln 2}^{\ln 5} 5e^x\, dx$

15. $\displaystyle\int_{0}^{4} \sqrt{x}(x + 1)\, dx$

16. $\displaystyle\int_{0}^{1} \sqrt{t}(t - \sqrt{t})\, dt$

17. $\displaystyle\int_{1}^{2} \frac{x^3 + 1}{x^2}\, dx$

18. $\displaystyle\int_{1}^{4} \frac{x^2 + x - 1}{\sqrt{x}}\, dx$

19. $\displaystyle\int_{1}^{\sqrt{3}} \frac{6a}{1 + x^2}\, dx$

20. $\displaystyle\int_{0}^{0.5} \frac{b\, dx}{\sqrt{1 - x^2}}$

21. $\displaystyle\int_0^1 \frac{x^2 - 4}{x - 2}\, dx$ **22.** $\displaystyle\int_0^1 \frac{x^2 - 1}{x^2 + 1}\, dx$

23. $\displaystyle\int_{-1}^2 (x + |x|)\, dx$ **24.** $\displaystyle\int_0^2 (x - |x - 1|)\, dx$

25. $\displaystyle\int_{-2}^3 (\sin^2 x + \cos^2 x)\, dx$ **26.** $\displaystyle\int_0^{\frac{\pi}{4}} (\sec^2 x - \tan^2 x)\, dx$

In Problems 27-34, find the area of the region under the given curve over the prescribed interval.

27. $y = x^2 + 1$ on $[-1, 1]$

28. $y = \sqrt{t}$ on $[0, 1]$

29. $y = \sec^2 x$ on $\left[0, \frac{\pi}{4}\right]$

30. $y = \sin x + \cos x$ on $\left[0, \frac{\pi}{2}\right]$

31. $y = e^t - t$ on $[0, 1]$

32. $y = (x^2 + x + 1)\sqrt{x}$ on $[1, 4]$

33. $y = \dfrac{x^2 - 2x + 3}{x}$ on $[1, 2]$

34. $y = \dfrac{2}{1 + t^2}$ on $[0, 1]$

In Problems 35-40, find the derivative of the given function.

35. $F(x) = \displaystyle\int_0^x \frac{t^2 - 1}{\sqrt{t + 1}}\, dt$ **36.** $F(x) = \displaystyle\int_{-2}^x (t + 1)\sqrt[3]{t}\, dt$

37. $F(t) = \displaystyle\int_1^t \frac{\sin x}{x}\, dx$ **38.** $F(t) = \displaystyle\int_t^2 \frac{e^x}{x}\, dx$

39. $F(x) = \displaystyle\int_x^1 \frac{dt}{\sqrt{1 + 3t^2}}$ **40.** $F(x) = \displaystyle\int_{\frac{\pi}{3}}^x \sec^2 t \tan t\, dt$

Level 2

The formulas in Problems 41-46 are taken from a table of integrals. In each case, use differentiation to verify that the formula is correct.

41. $\displaystyle\int \cos^2(au)\, du = \frac{u}{2} + \frac{\sin 2au}{4a} + C$

42. $\displaystyle\int u \cos^2(au)\, du = \frac{u^2}{4} + \frac{u \sin 2au}{4a} + \frac{\cos 2au}{8a^2} + C$

43. $\displaystyle\int \frac{u\, du}{(a^2 - u^2)^{3/2}} = \frac{1}{\sqrt{a^2 - u^2}} + C$

44. $\displaystyle\int \frac{du}{u^2 - a^2} = \frac{1}{2a}\ln\left|\frac{u - a}{u + a}\right| + C$

45. $\displaystyle\int \frac{u\, du}{\sqrt{a^2 - u^2}} = -\sqrt{a^2 - u^2} + C$

46. $\displaystyle\int (\ln|u|)^2\, du = u\,(\ln|u|)^2 - 2u \ln|u| + 2u + C$

---

47. EXPLORATION PROBLEM What is the relationship between finding an area and evaluating an integral?

48. EXPLORATION PROBLEM If you use the first fundamental theorem for the following integral, you find

$$\int_{-1}^1 \frac{dx}{x^2} = \left[-\frac{1}{x}\right]_a^b = -1 - 1 = -2$$

But the function $y = \dfrac{1}{x^2}$ is never negative. What's wrong with this "evaluation"?

49. Journal Problem (FOCUS)* The area of the shaded region in Figure 5.21**a** is 8 times the area of the shaded region in Figure 5.21**b**. What is c in terms of a?

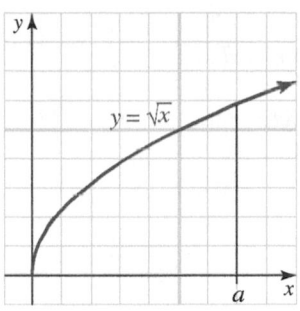
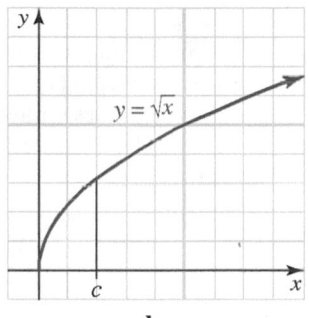

a.   b.

Figure 5.21 Comparing areas

50. a. If $F(x) = \displaystyle\int \left(\frac{1}{\sqrt{x}} - 4\right) dx$, find F so that $F(1) = 0$.
b. Sketch the graphs of $y = F(x)$, $y = F(x) + 3$, and $y = F(x) - 1$.
c. Find a constant K such that the largest value of $G(x) = F(x) + K$ is 0.

51. Evaluate $\displaystyle\int_0^2 f(x)\, dx$ where

$$f(x) = \begin{cases} x^3 & \text{if } 0 \le x < 1 \\ x^4 & \text{if } 1 \le x \le 2 \end{cases}$$

52. Evaluate $\displaystyle\int_0^\pi f(x)\, dx$ where

$$f(x) = \begin{cases} \cos x & \text{if } 0 \le x < \frac{\pi}{2} \\ x & \text{if } \frac{\pi}{2} \le x \le \pi \end{cases}$$

53. Let

$$g(x) = \int_0^x f(t)\, dt$$

where f is the function defined by the graph in Figure 5.22. Note that f crosses the x-axis at 3 points on $[0, 2]$; label these (from left to right), $x = a, x = b$, and $x = c$.

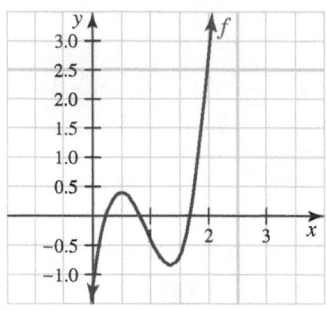

Figure 5.22   Graph of f

a. What can you say about $g(a)$?
b. Estimate $g'''(1)$.
c. Where does g have a maximum value on $[0, 2]$?
d. Sketch a rough graph of g.

54. Let

$$h(x) = \int_0^x f(t)\, dt$$

where f is the function defined by the graph in Figure 5.23.

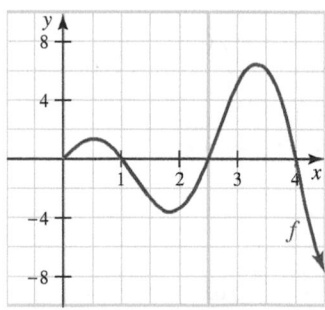

Figure 5.23   Graph of f

a. Where does h have a relative minimum on $[0, 5]$?
b. Where does h have a relative maximum value on $[0, 5]$?

c. If $h(1) = 1, h(2.5) = -2.5$, and $h(4) = 4$, sketch a rough graph of h.

55. Think Tank Problem The purpose of this problem is to provide a counterexample showing that the integral of a product (or quotient) is not necessarily equal to the product (quotient) of the respective integrals.

a. Show that $\displaystyle \int x\sqrt{x}\, dx \neq \left(\int x\, dx \right) \left(\int \sqrt{x}\, dx \right)$

b. Show that $\displaystyle \int \frac{\sqrt{x}}{x}\, dx \neq \frac{\int \sqrt{x}\, dx}{\int x\, dx}$

56. Suppose f is a function with the property that $f'(x) = f(x)$ for all x.

a. Show that $\displaystyle \int_a^b f(x)\, dx = f(b) - f(a)$

b. Show that $\displaystyle \int_a^b [f(x)]^2\, dx = \frac{1}{2} \left\{ [f(b)]^2 - [f(a)]^2 \right\}$

57. Find $\displaystyle \frac{d}{dx} \int_0^{x^4} (2t - 3)\, dt$ using the second fundamental theorem of calculus. Check your work by evaluating the integral directly and then taking the derivative of your answer.

58. Find $\displaystyle \frac{d}{dx} \int_0^{x^2} \sin t\, dt$ using the second fundamental theorem of calculus. Check your work by evaluating the integral directly and then taking the derivative of your answer.

59. Suppose $F(x)$ is the integral function

$$F(x) = \int_{u(x)}^{v(x)} f(t)\, dt$$

What is $F'(x)$? This result is called *Leibniz' rule*.

60. Suppose

$$F(x) = \int_{-x}^{x} e^t\, dt$$

Find $F'(x)$.

5.5 INTEGRATION BY SUBSTITUTION

IN THIS SECTION: *Substitution with indefinite integration, substitution with definite integration*
The fundamental theorem of calculus makes the process of integration much easier than using the definition, and we know that the process of integration reverses the process of differentiation. However, when differentiating we often need to use the chain rule which introduces an additional "integration factor." Now, we wish to reverse that process and need to account for that factor, which we do using the process discussed in this section.

Substitution with Indefinite Integration

Recall, that according to the chain rule, the derivative of $(x^2 + 3x + 5)^9$.

$$\frac{d}{dx}(x^2 + 3x + 5)^9 = 9(x^2 + 3x + 5)^8 (2x + 3)$$

Thus,

$$\int 9(x^2 + 3x + 5)^8(2x + 3)dx = (x^2 + 3x + 5)^9 + C$$

Note that the product is of the form $g(u)\dfrac{du}{dx}$ where, in this case, $g(u) = 9u^8$ and $u = x^2 + 3x + 5$. The following theorem generalizes this procedure by showing how products of the form $g(u)\dfrac{du}{dx}$ can be integrated by reversing the chain rule.

Theorem 5.9   Integration by substitution

Let f, g, and u be differentiable functions of x such that

$$f(x) = g(u)\frac{du}{dx}$$

Then

$$\int f(x)\, dx = \int g(u)\frac{du}{dx}\, dx = \int g(u)\, du = G(u) + C$$

where G is an antiderivative of g.

Proof: If G is an antiderivative of g, then $G'(u) = g(u)$ and, by the chain rule,

$$f(x) = \frac{d}{dx}[G(u)] = G'(u)\frac{du}{dx} = g(u)\frac{du}{dx}$$

Integrating both sides of this equation, we obtain

$$\int f(x)\, dx = \int \left[g(u)\frac{du}{dx} \right] dx = \int \left[\frac{d}{dx}G(u) \right] dx = G(u) + C$$

as required.   ◆

Example 1   Indefinite integral by substitution

Find $\displaystyle\int 9(x^2 + 3x + 5)^8(2x + 3)\, dx$.

Solution   Look at the integral and make the observations as shown in the boxes:

Let $u = x^2 + 3x + 5$.   ← If $u = x^2 + 3x + 5$, *then* $du = (2x + 3)\, dx$.

$$\int \underbrace{9(x^2 + 3x + 5)^8}_{\text{This is } g(u).}\ \underbrace{(2x + 3)\, dx}_{\text{This is } du.}$$

Now complete the integration of $g(u)$ and, when you are finished, back-substitute to express u in terms of x:

$$\int 9u^8\, du = 9\left(\frac{u^9}{9} \right) + C = (x^2 + 3x + 5)^9 + C$$

Example 2   Finding an integral by substitution

Find $\displaystyle\int (2x + 7)^5 dx$.

Solution

$$\int (2x + 7)^5 dx = \int u^5\left(\frac{1}{2}\, du \right)\quad \textit{Note: If } du = 2\, dx, \textit{ then } dx = \tfrac{1}{2}\, du.$$

Let $u = 2x + 7$, so $du = 2\, dx$.

$$= \frac{1}{2}\left[\frac{1}{6}u^6 \right] + C \qquad \textit{Power rule for integrals}$$

$$= \frac{1}{12}(2x + 7)^6 + C \qquad \textit{Replace } u = 2x + 7.$$

When choosing a substitution to evaluate $\int f(x)\,dx$, where $f(x)$ is a composite function $f(x) = f[u(x)]$, it is usually a good idea to pick the "inner" function $u = u(x)$. In Example 2, we chose $u = 2x + 7$, NOT the entire power $(2x + 7)^5$.

When making a substitution $u = u(x)$, remember that $du = u'(x)\,dx$, so you cannot simply replace dx by du in the transformed integral. For example, when you use a substitution $u = \sin x$, you have $du = \cos x\,dx$ and then replace the entire expression $\cos x\,dx$ by du in the transformed integral, as shown in the following example.

Example 3 Substitution with a trigonometric function

Find $\int (4 - 2\cos\theta)^3 \sin\theta\,d\theta$.

Solution Let $u = 4 - 2\cos\theta$, so $\dfrac{du}{d\theta} = -2(-\sin\theta) = 2\sin\theta$ and $d\theta = \dfrac{du}{2\sin\theta}$; substitute:

$$\int (4 - 2\cos\theta)^3 \sin\theta\,d\theta = \int u^3 \sin\theta\,\frac{du}{2\sin\theta}$$
$$= \frac{1}{2}\int u^3\,du$$
$$= \frac{1}{2}\left(\frac{u^4}{4}\right) + C$$
$$= \frac{1}{8}(4 - 2\cos\theta)^4 + C$$

Note: we will use a shorthand notation for substitution by placing a "box" under the integrand as follows:

$$\int (4 - 2\cos\theta)^3 \sin\theta\,d\theta = \int u^3 \sin\theta\,\frac{du}{2\sin\theta}$$
$$\boxed{u = 4 - 2\cos\theta;\ du = -2(-\sin\theta)\,d\theta}$$

This notation is used in the following example.

Example 4 Substitution in an exponential integral

Find $\int xe^{4-x^2}\,dx$.

Solution

$$\int xe^{4-x^2}\,dx = \int e^u\left(-\frac{1}{2}\,du\right)$$
$$\boxed{u = 4 - x^2;\ du = -2x\,dx}$$
$$= -\frac{1}{2}e^u + C$$
$$= -\frac{1}{2}e^{4-x^2} + C$$

☠ When using the method of substitution, all terms involving x in the original integral $\int f(x)\,dx$ must be converted into terms involving u in the transformed integral $\int g(u)\,du$. That is, after you have made your substitution and simplified, there should be no "leftover" x values in the integrand. ☠

Example 5 Substitution with leftover x-values

Find $\int x(4x - 5)^3\,dx$.

Solution $\displaystyle \int x(4x - 5)^3\,dx = \int xu^3\left(\frac{du}{4}\right) = \frac{1}{4}\int x\,u^3\,du$

$\boxed{u = 4x - 5;\ du = 4dx}$ *There is a leftover x-value.*

We are not ready to integrate until the leftover x-term has been eliminated. Because $u = 4x - 5$, we can solve for x:

$$x = \frac{u + 5}{4}$$

so

$$\begin{aligned}
\frac{1}{4}\int xu^3\,du &= \frac{1}{4}\int \left(\frac{u+5}{4}\right)u^3\,du \\
&= \frac{1}{16}\int (u^4 + 5u^3)\,du \\
&= \frac{1}{16}\left(\frac{u^5}{5} + 5\frac{u^4}{4}\right) + C \\
&= \frac{1}{80}(4x - 5)^5 + \frac{5}{64}(4x - 5)^4 + C
\end{aligned}$$

Note: Always introduce the constant of integration immediately after the last integration.

    The next example derives formulas that will be used many times throughout the text, often enough, in fact, that you should remember them or mark them for future reference.

Example 6   Integral formulas for the tangent and cotangent functions

Show that $\int \tan x\,dx = -\ln|\cos x| + C$ and $\int \cot x\,dx = \ln|\sin x| + C$

Solution $\displaystyle\int \tan x\,dx = \int \frac{\sin x}{\cos x}\,dx$ *Trigonometric identity* $\tan x = \frac{\sin x}{\cos x}$

$$= \int \left(\frac{1}{\cos x}\right)(\sin x\,dx)\qquad \boxed{u = \cos x;\ du = -\sin x\,dx}$$

$$= -\int \frac{du}{u}$$

$$= -\ln|u| + C$$

$$= -\ln|\cos x| + C$$

This can also be written as $\ln\left|(\cos x)^{-1}\right| + C = \ln|\sec x| + C$. Similarly,

$$\int \cot x\,dx = \int \frac{\cos x}{\sin x}\,dx = \ln|\sin x| + C$$

    Here is a problem in which the rate of change of a quantity is known and we use the method of substitution to find an expression for the quantity itself.

Example 7   Find the volume when the rate of flow is known

Water is flowing into a tank at the rate of $\sqrt{3t + 1}$ ft^3/min. If the tank is empty when $t = 0$, how much water does it contain 5 min later?

Solution   Because the rate at which the volume V is changing is dV/dt,

$$\frac{dV}{dt} = \sqrt{3t + 1}$$

$$V = \int \sqrt{3t + 1}\,dt$$

$$= \int \sqrt{u}\,\frac{du}{3}\qquad \boxed{u = 3t + 1;\ du = 3\,dt}$$

$$= \frac{1}{3} \int u^{1/2} du$$

$$= \frac{1}{3} \cdot \frac{2}{3} u^{3/2} + C$$

$$= \frac{2}{9} (3t + 1)^{3/2} + C$$

Since the initial volume is 0, we have

$$V(0) = \frac{2}{9} (3 \cdot 0 + 1)^{3/2} + C = 0, \text{ so } C = -\frac{2}{9}$$

$$V(t) = \frac{2}{9} (3t + 1)^{3/2} - \frac{2}{9} \text{ and } V(5) = \frac{2}{9} (16)^{3/2} - \frac{2}{9} = 14$$

The tank contains 14 ft³ of water after 5 minutes.

Substitution with Definite Integration

Example 7 could be considered as a definite integral:

$$\int_0^5 \sqrt{3t + 1}\, dt = \frac{2}{9} (3t + 1)^{3/2} \Big|_0^5 = \frac{2}{9} (16)^{3/2} - \frac{2}{9} (1)^{3/2} = 14$$

Notice that the definite integral eliminates the need for finding C. Furthermore, the following theorem eliminates the need for returning to the original variable (even though sometimes it is more convenient to do so).

Theorem 5.10   Substitution with the definite integral

If $f(u)$ is a continuous function of u and $u(x)$ is a differentiable function of x, then

$$\int_a^b f[u(x)]\, u'(x)\, dx = \int_{u(a)}^{u(b)} f(u)\, du$$

Proof: Let $F(u) = \int f(u)\, du$ be an antiderivative of $f(u)$. Then the first fundamental theorem of calculus tells us that

$$\int_a^b f[u(x)]\, u'(x)\, dx = F[u(x)]_{x=a}^{x=b} = F[u(b)] - F[u(a)]$$

But since $F(u)$ is an antiderivative of $f(u)$, it also follows that

$$\int_{u(a)}^{u(b)} f(u)\, du = F(u) \Big|_{u=u(a)}^{u=u(b)} = F[u(b)] - F[u(a)]$$

Thus,

$$\int_a^b f[u(x)]\, u'(x)\, dx = \int_{u(a)}^{u(b)} f(u)\, du \qquad \blacklozenge$$

We now have two valid methods for evaluating a definite integral of the form

$$\int_a^b f[u(x)]\, u'(x)\, dx$$

(1) Use substitution to evaluate the indefinite integral and then evaluate it between the limits of integration $x = a$ and $x = b$.

(2) Change the limits of integration to conform with the change of variable $u = u(x)$ and evaluate the transformed definite integral between the limits of integration $u = u(a)$ and $u = u(b)$.

These two methods are illustrated in Example 8.

Example 8   Substitution with the definite integral

Evaluate $\displaystyle\int_1^2 (4x - 5)^3 \, dx$.

Solution   *Method* (1): We first use the substitution with an indefinite integral.

$$\int (4x - 5)^3 dx = \int u^3 \left(\frac{1}{4} du \right) = \frac{1}{4} \left(\frac{u^4}{4} \right) + C = \frac{1}{16}(4x - 5)^4 + C$$

$$\boxed{u = 4x - 5; \; du = 4 \, dx}$$

Now evaluate the definite integral of $f(x)$ between $x = 1$ and $x = 2$ by using the fundamental theorem of calculus:

$$\int_1^2 (4x - 5)^3 dx = \frac{1}{16}(4x - 5)^4 \Big|_{x=1}^{x=2}$$

$$= \frac{1}{16}[4(2) - 5]^4 - \frac{1}{16}[4(1) - 5]^4$$

$$= \frac{1}{16}(81) - \frac{1}{16}(1)$$

$$= 5$$

Method (2): This is similar to method (1), except that as part of the integration process, we change the limits of integration when we are performing the substitution.

$$\boxed{u = 4x - 5; \; du = 4 \, dx}$$

If $x = 2$, then $u = 4(2) - 5 = 3$.

$$\int_1^2 (4x - 5)^3 dx = \int_{-1}^3 u^3 \frac{du}{4}$$

If $x = 1$, then $u = 4(1) - 5 = -1$.

☠ *You cannot change variables and keep the original limits of integration because the limits of integration are expressed in terms of the variable of integration.* ☠

$$= \frac{1}{4} \cdot \frac{u^4}{4} \Big|_{-1}^3$$

$$= \frac{1}{16}(81 - 1)$$

$$= 5 \qquad \blacksquare$$

PROBLEM SET 5.5

Level 1

Problems 1-8 present pairs of integration problems, one of which will require substitution and one of which will not. As you are working these problems, think about when substitution may be appropriate.

1. a. $\displaystyle\int_0^4 (2t + 4) \, dt$ **b.** $\displaystyle\int_0^4 (2t + 4)^{-1/2} dt$

2. a. $\displaystyle\int_0^{\frac{\pi}{2}} \sin\theta \, d\theta$ **b.** $\displaystyle\int_0^{\frac{\pi}{2}} \sin 2\theta \, d\theta$

3. a. $\displaystyle\int_0^{\pi} \cos t \, dt$ **b.** $\displaystyle\int_0^{\sqrt{\pi}} t \cos t^2 \, dt$

4. a. $\displaystyle\int_0^4 \sqrt{x}\,dx$ **b.** $\displaystyle\int_{-4}^0 \sqrt{-x}\,dx$

5. a. $\displaystyle\int_0^{16} \sqrt[4]{x}\,dx$ **b.** $\displaystyle\int_{-16}^0 \sqrt[4]{-x}\,dx$

6. a. $\displaystyle\int x(3x^2 - 5)\,dx$ **b.** $\displaystyle\int x(3x^2 - 5)^{50}\,dx$

7. a. $\displaystyle\int x^2\sqrt{2x^3}\,dx$ **b.** $\displaystyle\int x^2\sqrt{2x^3 - 5}\,dx$

8. a. $\displaystyle\int \frac{dx}{\sqrt{1-x^2}}$ **b.** $\displaystyle\int \frac{x\,dx}{\sqrt{1-x^2}}$

Use substitution to find the indefinite integrals in Problems 9-32.

9. $\displaystyle\int (2x+3)^4\,dx$ **10.** $\displaystyle\int \sqrt{3t-5}\,dt$

11. $\displaystyle\int (x^2 - \cos 3x)\,dx$ **12.** $\displaystyle\int \csc^2 5t\,dt$

13. $\displaystyle\int \sin(4-x)\,dx$

14. $\displaystyle\int \cot\left[\ln\left(x^2+1\right)\right]\frac{2x\,dx}{x^2+1}$

15. $\displaystyle\int \sqrt{t}(t^{3/2}+5)^3\,dt$ **16.** $\displaystyle\int \frac{(6x-9)\,dx}{(x^2-3x+5)^3}$

17. $\displaystyle\int x\sin(3+x^2)\,dx$ **18.** $\displaystyle\int \sin^3 t\cos t\,dt$

19. $\displaystyle\int \frac{x\,dx}{2x^2+3}$ **20.** $\displaystyle\int \frac{x^2\,dx}{x^3+1}$

21. $\displaystyle\int x\sqrt{2x^2+1}\,dx$ **22.** $\displaystyle\int \frac{4x\,dx}{2x+1}$

23. $\displaystyle\int \sqrt{x}\,e^{x\sqrt{x}}\,dx$ **24.** $\displaystyle\int \frac{e^{\sqrt[3]{x}}\,dx}{x^{2/3}}$

25. $\displaystyle\int x(x^2+4)^{1/2}\,dx$ **26.** $\displaystyle\int x^3(x^2+4)^{1/2}\,dx$

27. $\displaystyle\int \frac{\ln x}{x}\,dx$ **28.** $\displaystyle\int \frac{\ln(x+1)}{x+1}\,dx$

29. $\displaystyle\int \frac{dx}{\sqrt{x}(\sqrt{x}+7)}$ **30.** $\displaystyle\int \frac{dx}{x^{2/3}(\sqrt[3]{x}+1)}$

31. $\displaystyle\int \frac{e^t\,dt}{e^t+1}$ **32.** $\displaystyle\int \frac{e^{\sqrt{t}}\,dt}{\sqrt{t}(e^{\sqrt{t}}+1)}$

Evaluate the definite integrals given in Problems 33-42.

33. $\displaystyle\int_0^1 \frac{5x^2\,dx}{2x^3+1}$ **34.** $\displaystyle\int_1^4 \frac{e^{-\sqrt{x}}\,dx}{\sqrt{x}}$

35. $\displaystyle\int_{-\ln 2}^{\ln 2} \frac{1}{2}(e^x - e^{-x})\,dx$ **36.** $\displaystyle\int_0^2 (e^x - e^{-x})^2\,dx$

37. $\displaystyle\int_1^2 \frac{e^{1/x}\,dx}{x^2}$ **38.** $\displaystyle\int_0^2 x\sqrt{2x+1}\,dx$

39. $\displaystyle\int_0^{\frac{\pi}{6}} \tan 2x\,dx$ **40.** $\displaystyle\int_0^1 x^2(x^3+9)^{1/2}\,dx$

41. $\displaystyle\int_0^1 \frac{e^x\,dx}{1+e^x}$ **42.** $\displaystyle\int_0^1 \frac{e^{(x/5)}\,dx}{1+10e^{(x/5)}}$

43. *Historical Quest*

Karl Smith library

Lejeune Dirichlet
(1805-1859)

Johann Peter Gustav Lejeune Dirichlet was a professor of mathematics at the University of Berlin and is known for his role in formulating a rigorous foundation for calculus. He was not known as a good teacher. His nephew wrote that the mathematics instruction he received from Dirichlet was the most dreadful experience of his life. Howard Eves tells of the time Dirichlet was to deliver a lecture on definite integrals, but because of illness he posted the following note:

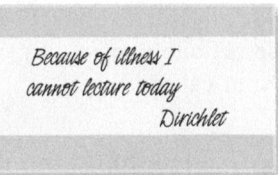

The students then doctored the note to read as follows:

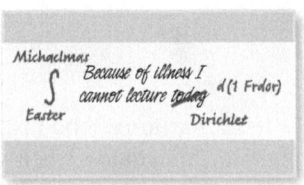

Michaelmas and Easter were school holidays, and 1 Frdor (Friedrichsd'or) was the customary honorarium for a semester's worth of lectures.

a. What is the answer when you integrate the student-doctored note?

b. The so-called *Dirichlet function* is often used for counterexamples in calculus. Look up the definition of this function. What special property does it have?

Level 2

Find the area of the region under the curves given in Problems 44-47.

44. $y = t\sqrt{t^2+9}$ on $[0,4]$

45. $y = \frac{1}{t^2}\sqrt{5 - \frac{1}{t}}$ on $[\frac{1}{5}, 1]$

46. $y = x(x-1)^{1/3}$ on $[2,9]$

47. $y = \frac{x}{\sqrt{x^2+1}}$ on $[1,3]$

48. a. In Section 5.3 (Problem 58), you were asked to prove that if f is continuous and *odd* [that is, $f(-x) = -f(x)$] on the interval $[-a, a]$, then

$$\int_{-a}^a f(x)\,dx = 0$$

b. In Section 5.2 (Problem 40), you were asked to find a counterexample or informally show that if f is continuous and *even* $[f(-x) = f(x)]$ on the interval $[-a, a]$, then

$$\int_{-a}^{a} f(x)\, dx = 2 \int_{0}^{a} f(x)\, dx = 2 \int_{-a}^{0} f(x)\, dx$$

For this problem, use the properties of integrals to prove each of these statements.

Use the results of Problem 48 to evaluate the integrals given in Problems 49-52.

49. $\displaystyle\int_{-\pi}^{\pi} \sin x\, dx$

50. $\displaystyle\int_{-\frac{\pi}{2}}^{\frac{\pi}{2}} \cos x\, dx$

51. $\displaystyle\int_{-3}^{3} x\sqrt{x^4 + 1}\, dx$

52. $\displaystyle\int_{-1}^{1} \frac{\sin x\, dx}{x^2 + 1}$

53. In each of the following cases, determine whether the given relationship is true or false.

a. $\displaystyle\int_{-175}^{175} \left(7x^{1,001} + 14x^{99}\right) dx = 0$

b. $\displaystyle\int_{0}^{\pi} \sin^2 x\, dx = \int_{0}^{\pi} \cos^2 x\, dx$

c. $\displaystyle\int_{-\frac{\pi}{2}}^{\frac{\pi}{2}} \cos x\, dx = \int_{-\pi}^{0} \sin x\, dx$

54. The slope at each point (x, y) on the graph of $y = F(x)$ is given by $x(x^2 - 1)^{1/3}$, and the graph passes through the point $(3, 1)$. Use this information to find F. Sketch the graph of F.

55. The slope at each point (x, y) on the graph of $y = F(x)$ is given by

$$\frac{2x}{1 - 3x^2}$$

What is $F(x)$ if the graph passes through $(0, 5)$?

56. A particle moves along the t-axis in such a way that at time t, its velocity is

$$v(t) = t^2(t^3 - 8)^{1/3}$$

a. At what time does the particle turn around?
b. If the particle starts at a position which we denote as 1, where does it turn around?

57. A rectangular storage tank has a square base 10 ft on a side. Water is flowing into the tank at the rate modeled by the function

$$R(t) = t(3t^2 + 1)^{-1/2} \text{ ft}^3/\text{s}$$

at time t seconds. If the tank is empty at time $t = 0$, how much water does it contain 4 sec later? What is the depth of the water (to the nearest quarter inch) at that time?

58. Journal Problem *(College Mathematics Journal*)*
Find

$$\int \left[(x^2 - 1)(x + 1)\right]^{-2/3} dx$$

59. Modeling Problem A group of environmentalists model the rate at which the ozone level is changing in a suburb of Los Angeles by the function

$$L'(t) = \frac{0.24 - 0.03t}{\sqrt{36 + 16t - t^2}}$$

parts per million per hour (ppm/h) t hours after 7:00 A.M.

© 2013 mikeledray. Used under license from Shutterstock, Inc.

a. Express the ozone level $L(t)$ as a function of t if L is 4 ppm at 7:00 A.M.
b. Use the graphing utility of your calculator to find the time between 7:00 A.M. and 7:00 P.M. when the highest level of ozone occurs. What is the highest level?
c. Use your graphing utility or another utility of your calculator to determine a second time during the day when the ozone level is the same as it is at 11:00 A.M.

Level 3

60. A *logistic* function is one of the form

$$Q(t) = \frac{B}{1 + Ae^{-rt}}$$

Find $\int Q(t)\, dt$.

5.6 INTRODUCTION TO DIFFERENTIAL EQUATIONS

IN THIS SECTION: *Introduction and terminology, direction fields, separable differential equations, modeling exponential growth and decay, orthogonal trajectories, modeling fluid flow through an orifice, modeling the motion of a projectile: escape velocity*

In this section, we introduce you to the basic terminology of *differential equations* and show how such equations can be used to model certain applications. Later in the text, we will investigate further, including more sophisticated issues regarding differential equations as we develop the necessary mathematical tools.

The study of differential equations is as old as calculus itself. Today, it would be virtually impossible to make a serious study of physics, astronomy, chemistry, or engineering without encountering physical models based on differential equations. In addition, differential equations are beginning to appear more frequently in the biological and social sciences, especially in economics. We begin by introducing some basic terminology and examining a few modeling procedures.

Introduction and Terminology

Any equation that contains a derivative or differential is called a **differential equation**. For example, the equations

$$\frac{dy}{dx} = 3x^3 + 5 \quad \frac{dP}{dt} = kP^2 \quad \left(\frac{dy}{dx}\right)^2 + 3\frac{dy}{dx} + 2y = xy \quad \frac{d^2x}{dt^2} + 2\frac{dx}{dt} + 5x = \sin t$$

are all differential equations.

Many practical situations, especially those involving rates, can be described mathematically by differential equations. For example, the assumption that population P grows at a rate proportional to its size can be expressed by the differential equation

$$\frac{dP}{dt} = kP$$

where t is time and k is the constant of proportionality.

A **solution** of a given differential equation is a function that satisfies the equation. A **general solution** is a characterization of all possible solutions of the equation. We say that the differential equation is **solved** when we find a general solution.

For example, $y = x^2$ is a solution of the differential equation

$$\frac{dy}{dx} = 2x$$

because

$$\frac{dy}{dx} = \frac{d}{dx}(y) = \frac{d}{dx}(x^2) = 2x$$

Moreover, because any solution of this equation must be an indefinite integral of $2x$, it follows that

$$y = \int 2x \, dx = x^2 + C$$

is the general solution of the differential equation.

Example 1   Verifying that a given function is a solution to a differential equation

If $4x - 3y^2 = 10$, and $y \neq 0$, verify that $\dfrac{dy}{dx} = \dfrac{2}{3y}$.

Solution

$$4x - 3y^2 = 10 \qquad \textit{Given equation}$$

$$\frac{d}{dx}(4x - 3y^2) = \frac{d}{dx}(10) \qquad \text{Take the derivative of both sides.}$$

$$4 - 6y\frac{dy}{dx} = 0 \qquad \textit{Don't forget the chain rule, since y is a function of x.}$$

$$\frac{dy}{dx} = \frac{4}{6y} \qquad \textit{Solve for } \frac{dy}{dx}, \, y \neq 0.$$

$$= \frac{2}{3y} \qquad \textit{Reduce}$$

Example 2   Finding future revenue

An oil well that yields 300 barrels of crude oil a day will run dry in 3 years. It is estimated that t days from now the price of the crude oil will be $p(t) = 100 + 0.3\sqrt{t}$ dollars per barrel. If the oil is sold as soon as it is extracted from the ground, what will be the total future revenue from the well?

Solution   Let $R(t)$ denote the total revenue up to time t. Then the rate of change of revenue is $\frac{dR}{dt}$, the number of dollars received per barrel is $p(t) = 100 + 0.3\sqrt{t}$, and the number of barrels sold per day is 300. Thus, we have

$$\begin{bmatrix} \text{RATE OF CHANGE} \\ \text{OF TOTAL} \\ \text{REVENUE} \end{bmatrix} = \begin{bmatrix} \text{NUMBER OF} \\ \text{DOLLARS} \\ \text{PER BARREL} \end{bmatrix} \begin{bmatrix} \text{NUMBER OF} \\ \text{BARRELS} \\ \text{SOLD PER DAY} \end{bmatrix}$$

$$\frac{dR}{dt} = (100 + 0.3\sqrt{t})(300)$$

$$= 30{,}000 + 90\sqrt{t}$$

This is actually a statement of the chain rule:

$$\frac{dR}{dt} = \frac{dR}{dB} \cdot \frac{dB}{dt}$$

where B denotes the number of barrels extracted and R denotes the revenue. It is often helpful to use the chain rule in this way when setting up differential equations.

     We solve this differential equation by integration:

$$R(t) = \int \frac{dR}{dt}\, dt$$

$$= \int (30{,}000 + 90\sqrt{t})\, dt$$

$$= 30{,}000t + 60t^{3/2} + C$$

Since $R(0) = 0$, it follows that $C = 0$. We also are given that the well will run dry in 3 years or 1,095 days, so that the total revenue obtained during the life of the well is

$$R(1{,}095) = 30{,}000(1{,}095) + 60(1{,}095)^{3/2}$$

$$\approx 35024064.52$$

The total future revenue is approximately \$35 million.

Direction Fields

We can use slope fields to help with a visualization of a differential equation. In Section 5.1 we looked at small segments with slope defined by the derivative of some function. The collection of all such line segments is called the *slope field* or **direction field** of the differential equation. Today, with the assistance of computers, we can sometimes draw the direction field in order to obtain a particular solution (or a family of solutions) as shown in the following example.

Example 3   Finding a solution, given the direction field

The direction field for the differential equation

$$\frac{dy}{dx} = y - x^2$$

is shown in Figure 5.24.

  a. Sketch a solution to the initial value problem passing through $(2, 1)$.
  b. Sketch a solution to the initial value problem passing through $(0, 1)$.

Solution

a. The initial value $y(2) = 1$ means the solution passes through $(2, 1)$, and is shown in Figure 5.25**a**.
b. Since $y(0) = 1$, the solution passes through $(0, 1)$, as shown in Figure 5.25**b**.

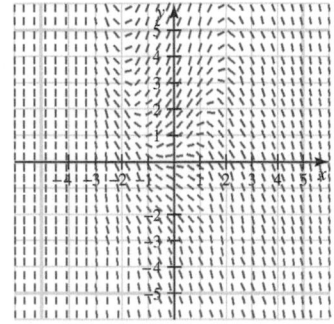

Figure 5.24 Interactive

Direction field for $\dfrac{dy}{dx} = y - x^2$

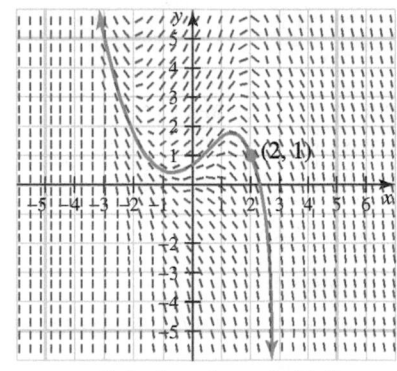

a. Solution through (2,1)

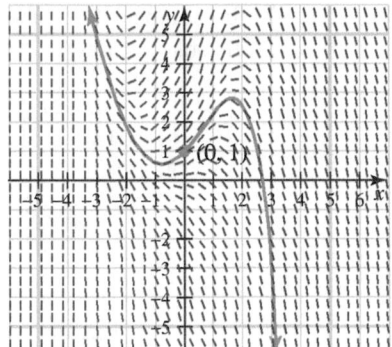

b. Solution through (0,1)

Figure 5.25 Interactive   Particular solutions for a given direction field

Separable Differential Equations

Solving a differential equation is often a complicated process. However, many important equations can be expressed in the form

$$\frac{dy}{dx} = \frac{g(x)}{f(y)}$$

assuming that the denominator does not vanish.

To solve such an equation, first separate the variables into the differential form

$$f(y)\,dy = g(x)\,dx$$

and then integrate both sides separately to obtain

$$\int f(y)\,dy = \int g(x)\,dx$$

A differential equation expressible in the form $dy/dx = g(x)/f(y)$ is said to be **separable**. This procedure is illustrated in Example 4.

Example 4   Solving a separable differential equation

Solve $\dfrac{dy}{dx} = \dfrac{x}{y}$.

Solution

$$\frac{dy}{dx} = \frac{x}{y}$$

$$y\,dy = x\,dx$$

$$\int y\,dy = \int x\,dx$$

$$\frac{1}{2}y^2 + C_1 = \frac{1}{2}x^2 + C_2$$

$$x^2 - y^2 = C \qquad \text{where } C = 2(C_1 - C_2)$$

Notice the treatment of constants in Example 4. Because all constants can be combined into a single constant, it is customary not to write $2(C_1 - C_2)$, but rather, whenever possible to simply replace all the arbitrary constants in the problem by a single arbitrary constant, C, immediately after the last integral is found, since an arbitrary constant minus an arbitrary constant is an arbitrary constant, and twice an arbitrary constant is an arbitrary constant.

The remainder of this section is devoted to selected applications involving separable differential equations.

Modeling Exponential Growth and Decay

A process is said to undergo **exponential change** if the relative rate of change of the process is modeled by a constant; in other words,

$$\frac{Q'(t)}{Q(t)} = k \qquad \text{or} \qquad \frac{dQ}{dt} = kQ(t)$$

If the constant k is positive, the exponential change is called **growth**, and if k is negative, it is called **decay**. Exponential growth occurs in certain populations, and exponential decay occurs in the disintegration of radioactive substances.

To solve the growth/decay equation, separate the variables and integrate both sides.

$$\frac{dQ}{dt} = kQ$$

$$\int \frac{dQ}{Q} = \int k\,dt$$

$$\ln |Q| = kt + C_1$$

$$e^{kt+C_1} = Q \qquad \textit{Definition of natural logarithm}$$

$$e^{kt}e^{C_1} = Q$$

Thus, $Q = Ce^{kt}$, where $C = e^{C_1}$. Finally, if we let Q_0 be the initial amount (that is, when $t = 0$), we see that

$$Q_0 = Ce^0 \qquad \text{or} \qquad C = Q_0 \qquad \text{so} \qquad Q(t) = Q_0 e^{kt}$$

> **GROWTH/DECAY EQUATION**   The **growth/decay equation** of a substance is
>
> $$Q(t) = Q_0 e^{kt}$$
>
> where $Q(t)$ is the amount of the substance present at time t, Q_0 is the initial amount of the substance, and k is a constant that depends on the substance. If $k > 0$, it is a *growth* equation; if $k < 0$, it is a *decay* equation.

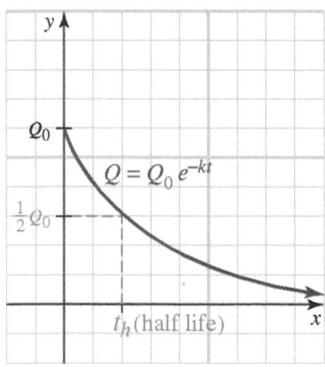

Figure 5.26   The decay curve for a radioactive substance

The graph of $Q = Q_0 e^{kt}$ for $k < 0$ is shown in Figure 5.26. In that figure we have also indicated the time t_h required for half of a given substance to disintegrate. The time t_h is called the **half-life** of the substance, and it provides a measure of the substance's rate of disintegration.

Example 5   Amount of a radioactive substance present

A particular radioactive substance has a half-life of 600 yr. Find k for this substance and determine how much of a 50-g sample will remain after 125 yr.

Solution   From the decay equation,

$Q(t) = Q_0 e^{kt}$	Growth/decay formula
$\dfrac{Q(t)}{Q_0} = e^{kt}$	Solve for exponential.
$\dfrac{1}{2} = e^{k(600)}$	Half-life is 600 years.
$600k = \ln \dfrac{1}{2}$	Definition of logarithm.
$k = \dfrac{\ln \frac{1}{2}}{600}$	This is the exact answer.
≈ -0.0011552453	Use a calculator to find an approximate answer.

Next, to see how much of a 50-g sample will remain after 125 yr, substitute $Q_0 = 50$, k (use the calculator output; do not round at this point in the solution), and $t = 125$ into the decay equation:

$$Q(125) \approx 50 e^{125k} \approx 43.27682805$$

There will be about 43 g present.

One of the more interesting applications of radioactive decay is a technique known as **carbon dating**, which is used by geologists, anthropologists, and archaeologists to estimate the age of various specimens.* The technique is based on the fact that all animal and vegetable systems (whether living or dead) contain both stable carbon ^{12}C and a radioactive isotope ^{14}C. Scientists assume that the ratio of ^{14}C to ^{12}C in the air has remained approximately constant throughout history. Living systems absorb carbon dioxide from the air, so the ratio of ^{14}C to ^{12}C in a living system is the same as that in the air itself. When a living system dies, the absorption of carbon dioxide ceases. The ^{12}C already in the system remains while the ^{14}C decays, and the ratio of ^{14}C to ^{12}C decreases exponentially. The half-life of ^{14}C is approximately 5,730 years. The ratio of ^{14}C to ^{12}C in a specimen t years after it was alive is approximately

$$R = R_0 e^{kt} \qquad \text{where } k = \frac{\ln \frac{1}{2}}{5,730}$$

and R_0 is the ratio of ^{14}C to ^{12}C in the atmosphere. By comparing $R(t)$ with R_0, scientists can estimate the age of the object. Here is an example.

Example 6   Modeling Problem: Carbon dating

An archaeologist has found a specimen in which the ratio of ^{14}C to ^{12}C is 20% the ratio found in the atmosphere. Approximately how old is the specimen?

Solution   The age of the specimen is the value of t for which $R(t) = 0.20 R_0$:

$$0.20 R_0 = R_0 e^{kt}$$
$$0.20 = e^{kt}$$
$$kt = \ln 0.20$$
$$t = \frac{1}{k} \ln 0.2 \qquad \text{where } k = \frac{\ln \frac{1}{2}}{5,730}$$
$$\approx 13304.64798$$

The specimen is approximately 13,000 yr old.

Orthogonal Trajectories

A curve that intersects each member of a given family of curves at right angles is called an **orthogonal trajectory** of that family, as shown in Figure 5.27. Orthogonal families arise in many applications. For example, in thermodynamics, the heat flow across a planar surface is orthogonal to the curves of constant temperature, called *isotherms*. In the theory of fluid flow, the flow lines are orthogonal trajectories of *velocity equipotential curves*. The basic procedure for finding orthogonal trajectories involves differential equations and is demonstrated in Example 7.

---

*Carbon dating is used primarily for estimating the age of relatively "recent" specimens. For example, it was used (along with other methods) to determine that the Dead Sea Scrolls were written and deposited in the Caves of Qumran approximately 2,000 years ago. For dating older specimens, it is better to use techniques based on radioactive substances with longer half-lives. In particular, potassium 40 is often used as a "clock" for events that occurred between 5 and 15 million years ago. Paleoanthropologists find this substance especially valuable because it often occurs in volcanic deposits and can be used to date specimens trapped in such deposits.

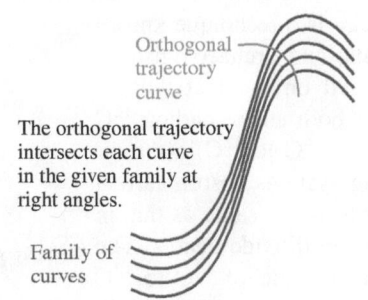

The orthogonal trajectory intersects each curve in the given family at right angles.

a. (above) Orthogonal trajectory
b. (right) Computer-generated family of orthogonal trajectories was a winner at the 2011 Joint Mathematics Meetings Mathematical Art Exhibition

Circles on Orthogonal Circles by Anne Burns

Figure 5.27   Orthogonal trajectories

Example 7   Finding orthogonal trajectories

Find the orthogonal trajectories of the family of curves of the form $xy = C$.

Solution   We are seeking a family of curves. Each curve in that family intersects each curve in the family $xy = C$ at right angles, as shown in Figure 5.28. Assume that a typical point on a given curve in the family $xy = C$ has coordinates (x, y) and that a typical point on the orthogonal trajectory curve has coordinates (X, Y).*

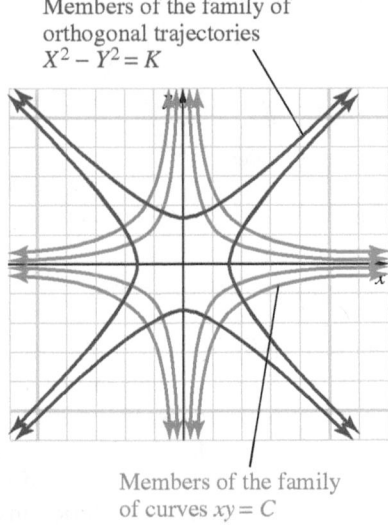

Members of the family of orthogonal trajectories
$X^2 - Y^2 = K$

Members of the family of curves $xy = C$

Figure 5.28   The family of curves $xy = C$ and their orthogonal trajectories

Let P be a point where a particular curve of the form $xy = C$ intersects the orthogonal trajectory curve. At P, we have $x = X$ and $y = Y$, and the slope dY/dX of the orthogonal trajectory is the same as the negative reciprocal of the slope dy/dx of the curve

---

*We use uppercase letters for one kind of curve and lowercase for the other to make it easier to tell which curve is being mentioned at each stage of the following discussion.

$xy = C$. Using implicit differentiation, we find

$$xy = C$$

$$x\frac{dy}{dx} + y = 0 \qquad \textit{Product rule}$$

$$\frac{dy}{dx} = \frac{-y}{x}$$

Thus, at the point of intersection P, the slope $\frac{dY}{dX}$ of the orthogonal trajectory is

$$\frac{dY}{dX} = -\frac{1}{\frac{dy}{dx}} = -\frac{1}{\frac{-y}{x}} = \frac{x}{y} = \frac{X}{Y}$$

According to this equation, the coordinates (X, Y) of the orthogonal trajectory curve satisfy the separable differential equation

$$\frac{dY}{dX} = \frac{X}{Y}$$

discussed in Example 4. Using the result of that example, we see that the orthogonal trajectories of the family $xy = C$ are the curves in the family

$$X^2 - Y^2 = K$$

where K is a constant. The given family of curves $xy = C$ and the family of orthogonal trajectory curves $X^2 - Y^2 = K$ are shown in Figure 5.28. ∎

Modeling Fluid Flow Through an Orifice

Consider a tank that is filled with a fluid being slowly drained through a small, sharp-edged hole in its base, as shown in Figure 5.29.

By using a principle of physics known as Torricelli's law,* we can show that the rate of discharge dV/dt of the volume V at time t is proportional to the square root of the depth h at that time. Specifically, if all dimensions are given in terms of feet, the drain hole has area A_0, and the height above the hole is h at time t (seconds), then

$$\frac{dV}{dt} = -4.8 A_0 \sqrt{h}$$

is the rate of flow of water in cubic feet per second. This formula is used in Example 7.

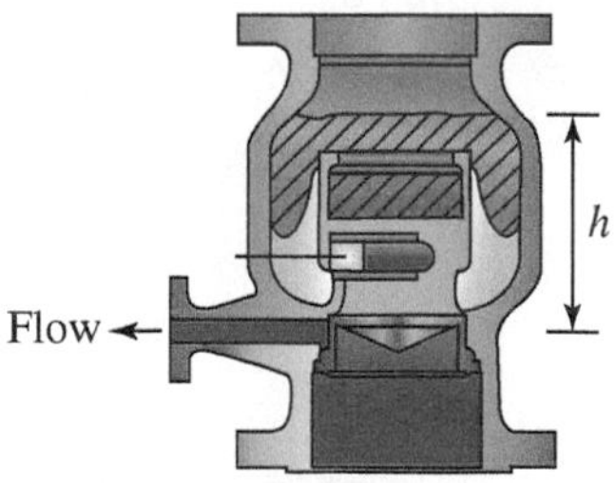

Figure 5.29 The flow of a fluid through a hole of area A_0

Example 8 Fluid flow through an orifice

A cylindrical tank (with a circular base) is filled with a liquid that is draining through a small circular hole in its base. If the tank is 9 ft high with radius 4 ft, and the drain hole has radius 1 in., how long does it take for the tank to empty?

Solution Because the drain hole is a circle of radius $\frac{1}{12}$ ft ($= 1$ in.), its area is $\pi r^2 = \pi \left(\frac{1}{12}\right)^2$, and the rate of flow is

$$\frac{dV}{dt} = -4.8 \left(\frac{\pi}{144}\right) \sqrt{h}$$

---

*Torricelli's law says that the stream of liquid through the orifice has velocity $\sqrt{2gh}$, where $g = 32$ ft/s^2 is the acceleration due to gravity, and h is the height of the liquid above the orifice. The factor 4.8 that appears in the rate of flow equation is required to compensate for the effect of friction.

Because the tank is cylindrical, the amount of fluid in the tank at any particular time will form a cylinder of radius 4 ft and height h. Also, because we are using $g = 32$ ft/s², it is implied that the time, t, is measured in seconds. The volume of such a liquid cylinder is

$$V = \pi r^2 h = \pi(4)^2 h = 16\pi h$$

and by differentiating both sides of this equation with respect to t, we obtain

$$\frac{dV}{dt} = 16\pi \frac{dh}{dt}$$

$$-4.8 \left(\frac{\pi}{144}\right) \sqrt{h} = 16\pi \frac{dh}{dt} \qquad \textit{Torricelli's law}$$

$$\frac{dh}{dt} = \frac{-4.8\sqrt{h}}{144(16)}$$

$$\frac{dh}{\sqrt{h}} = \frac{-4.8}{16(144)} dt \qquad \textit{Separate the variables.}$$

$$\int h^{-1/2} dh = \int \frac{-4.8}{16(144)} dt \qquad \textit{Integrate both sides.}$$

$$2h^{1/2} + C_1 = \frac{-4.8}{16(144)} t + C_2$$

$$\sqrt{h} = \frac{-2.4}{16(144)} t + C$$

To evaluate C, recall that the tank is full at time $t = 0$. Since $h = 9$ when $t = 0$, after replacing into the previous formula, we obtain

$$\sqrt{9} = C$$
$$3 = C$$

and the general formula is

$$\sqrt{h} = \frac{-2.4}{16(144)} t + 3$$

where time t is in seconds (because g is 32 ft/s²).

Now we can find the depth of the fluid at any given time or the time at which a prescribed depth occurs. In particular, the tank is empty at the time t_e when $h = 0$. By substituting $h = 0$ into this formula we find

$$\sqrt{0} = \frac{-2.4}{16(144)} t + 3$$

$$\frac{-3(16)(144)}{-2.4} = t$$

$$t = 2,880$$

Thus, 2,880 seconds or 48 min are required to drain the tank.   ■

Modeling the Motion of a Projectile: Escape Velocity

Consider a projectile that is launched with initial velocity v_0 from a planet's surface along a direct line through the center of the planet, as shown in Figure 5.30.

We shall find a general formula for the velocity of the projectile and the minimal value of v_0 required to guarantee that the projectile will escape the planet's gravitational attraction.

We assume that the only force acting on the projectile is that due to gravity, although in practice, factors such as air resistance must also be considered. With this assumption, Newton's law of gravitation can be used to show that when the projectile is at a distance s from the center of the planet, its acceleration is given by the formula

$$a = \frac{-gR^2}{s^2}$$

where R is the radius of the planet and g is the acceleration due to gravity at the planet's surface (see Problem 59).*

Our first goal is to express the velocity v of the projectile in terms of the height s. Because the projectile travels along a straight line, we know that

$$a = \frac{dv}{dt} \quad \text{and} \quad v = \frac{ds}{dt}$$

and by applying the chain rule, we see that

$$a = \frac{dv}{dt} = \frac{dv}{ds} \cdot \frac{ds}{dt} = \frac{dv}{ds}v$$

Therefore, by substitution for a we have

$$\frac{dv}{ds}v = \frac{-gR^2}{s^2}$$
$$v\,dv = -gR^2 s^{-2}\,ds$$
$$\int v\,dv = \int -gR^2 s^{-2}\,ds$$
$$\frac{1}{2}v^2 + C_1 = gR^2 s^{-1} + C_2$$
$$v^2 = 2gR^2 s^{-1} + C$$

To evaluate the constant C, recall that the projectile was fired from the planet's surface with initial velocity v_0. Thus, $v = v_0$ when $s = R$, and by substitution

$$v_0^2 = 2gR^2R^{-1} + C, \quad \text{which implies} \quad C = v_0^2 - 2gR$$

so

$$v^2 = 2gR^2 s^{-1} + v_0^2 - 2gR$$

Because the projectile is launched in a direction away from the center of the planet, we would expect it to keep moving in that direction until it stops. In other words, *the projectile will keep moving away from the planet until it reaches a point where $v = 0$.* Because $2gR^2 s^{-1} > 0$ for all $s > 0$, v^2 will always be positive if $v_0^2 - 2gR \geq 0$. On the other hand, if $v_0^2 - 2gR < 0$, then sooner or later v will become 0 and the projectile will eventually fall back to the surface of the planet.

Therefore, we conclude that the projectile will escape from the planet's gravitational attraction if $v_0^2 \geq 2gR$, that is, $v_0 \geq \sqrt{2gR}$. For this reason, the minimum speed for which this can occur, namely,

$$v_0 = \sqrt{2gR}$$

The first space shuttle mission was the *Columbia* launched on April 12, 1981 and the last one was the *Atlantis* launched on July 8, 2011.

NASA

Figure 5.30 A projectile launched from the surface of a planet

---

*According to Newton's law of gravitation, the force of gravity acting on a projectile of mass m has magnitude $F = mk/s^2$, where k is constant. If this is the only force acting on the projectile, then $F = ma$, and we have $ma = F = mk/s^2$.

is called the **escape velocity** of the planet. In particular, for the earth, $R = 3{,}956$ mi and $g = 32$ ft/s$^2 = 0.00606$ mi/s^2, and the escape velocity is

$$v_0 = \sqrt{2gR} \approx \sqrt{2(0.00606)(3{,}956)} \approx 6.924357$$

The escape velocity for the earth is about 6.92 mi/s.

PROBLEM SET 5.6

Level 1

Verify in Problems 1-8 that if y satisfies the prescribed relationship with x, then it will be a solution of the given differential equation.

1. If $x^2 + y^2 = 7$, then $\dfrac{dy}{dx} = -\dfrac{x}{y}$

2. If $5x^2 - 2y^2 = 3$, then $\dfrac{dy}{dx} = \dfrac{5x}{2y}$

3. If $xy = C$, then $\dfrac{dy}{dx} = \dfrac{-y}{x}$

4. If $x^2 - 3xy + y^2 = 5$, then
$(2x - 3y)\,dx + (2y - 3x)\,dy = 0$

5. If $y = \sin(Ax + B)$, then $\dfrac{d^2y}{dx^2} + A^2y = 0$

6. If $y = \dfrac{x^4}{20} - \dfrac{A}{x} + B$, then $x\dfrac{d^2y}{dx^2} + 2\dfrac{dy}{dx} = x^3$

7. If $y = 2e^{-x} + 3e^{2x}$, then $y'' - y' - 2y = 0$

8. If $y = Ae^x + Be^x \ln x$, then
$xy'' + (1 - 2x)y' - (1 - x)y = 0$

Find the particular solution of the first-order linear differential equation passing through the given point in Problems 9-14. A graphical solution within its direction field is shown.

9. $\dfrac{dy}{dx} = -\dfrac{x}{y}; \ (2, 2)$ **10.** $\dfrac{dy}{dx} - y = 10; \ (-2, -9)$

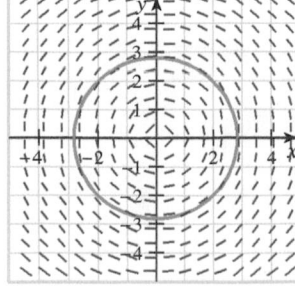 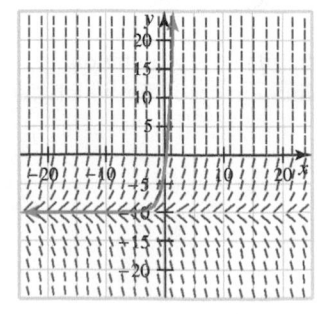

11. $\dfrac{dy}{dx} - y^2 = 1; \ (\pi, 1)$ **12.** $\dfrac{dy}{dx} = e^{x+y}; \ (0, 0)$

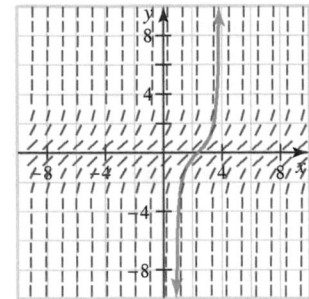 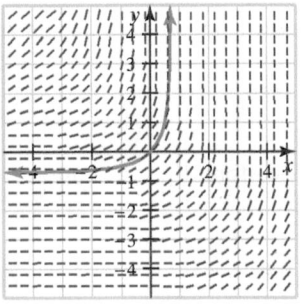

13. $\dfrac{dy}{dx} = \sqrt{\dfrac{x}{y}}; \ (4, 1)$ **14.** $\dfrac{dy}{dx} = y^2\sqrt{x}; \ \left(9, -\dfrac{1}{18}\right)$

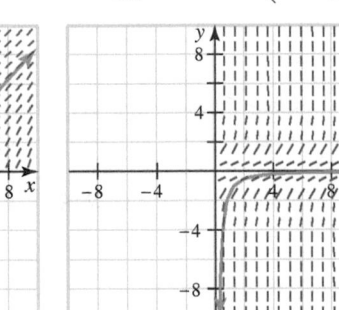

In each of Problems 15-20, sketch the particular solution passing through the given point for the differential equation whose direction field is given.

15. $(0, 1)$ **16.** $(0, 1)$

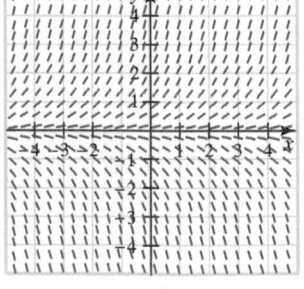 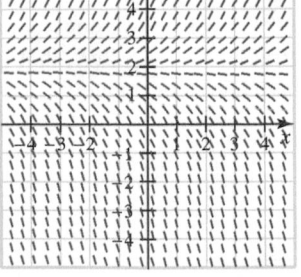

17. $(0, 0)$

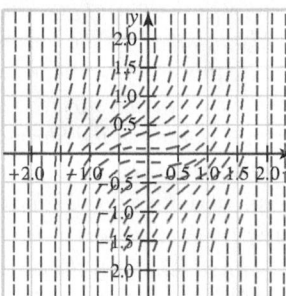

18. $(3, 3)$

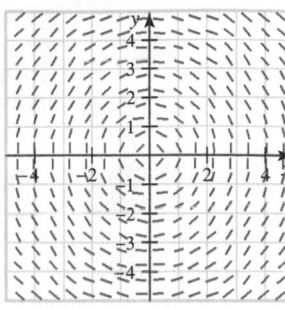

19. $(1, 0)$

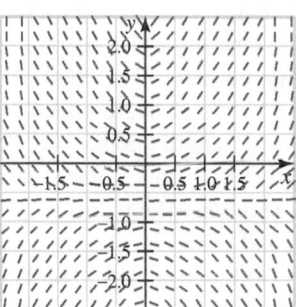

20. $(0, 0)$

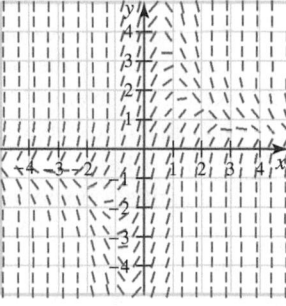

Find the general solution of the separable differential equations given in Problems 21-28. Note that for some problems, the general solution will be a relationship between x and y, rather than $y = f(x)$.

21. $\dfrac{dy}{dx} = 3xy$

22. $\dfrac{dy}{dx} = \sqrt{\dfrac{y}{x}}$

23. $\dfrac{dy}{dx} = \dfrac{x}{y}\sqrt{1 - x^2}$

24. $xy\, dx + \sqrt{xy}\, dy = 0$

25. $\dfrac{dy}{dx} = \dfrac{\sin x}{\cos y}$

26. $x^2\, dy + \sec y\, dx = 0$

27. $xy\dfrac{dy}{dx} = \dfrac{\ln x}{\sqrt{1 - y^2}}$

28. $\dfrac{dy}{dx} = e^{y-x}$

In Problems 29-32, find the general solution of the given differential equation by using either the product or the quotient rule.

29. $x\, dy + y\, dx = 0$

30. $\dfrac{x\, dy - y\, dx}{x^2} = 0$

31. $y\, dx = x\, dy, \ x > 0, \ y > 0$

32. $x^2 y\, dy + xy^2\, dx = 0$

Find the orthogonal trajectories of the family of curves given in Problems 33-40. In each case, sketch several members of the given family of curves and several members of the family of orthogonal trajectories on the same coordinate axes.

33. the lines $2x - 3y = C$

34. the lines $y = x + C$

35. the cubic curves $y = x^3 + C$

36. the curves $y = x^4 + C$

37. the curves $xy^2 = C$

38. the parabolas $y^2 = 4kx$

39. the circles $x^2 + y^2 = r^2$

40. the exponential curves $y = Ce^{-x}$

Modeling Problems *Write a differential equation to model the situation given in each of Problems 41-46. Be sure to define your unknown function first. Do not solve.*

41. The number of bacteria in a culture grows at a rate that is proportional to the number present.

42. A sample of radium decays at a rate that is proportional to the amount of radium present in the sample.

43. The rate at which the temperature of an object changes is proportional to the difference between its own temperature and the temperature of the surrounding medium.

44. When a person is asked to recall a set of facts, the rate at which the facts are recalled is proportional to the number of relevant facts in the person's memory that have not yet been recalled.
   Hint: Let Q denote the number of facts recalled and N the total number of relevant facts in the person's memory. Then dQ/dt is the rate of change of Q, and $(N - Q)$ is the number of relevant facts not recalled.

45. The rate at which an epidemic spreads through a community of P susceptible people is proportional to the product of the number of people who have caught the disease and the number who have not.

46. The rate at which people are implicated in a government scandal is proportional to the product of the number of people already implicated and the number of people involved who have not yet been implicated.

47. Think Tank Problem What do you think the orthogonal trajectories of the family of curves $x^2 - y^2 = C$ will be? Verify your conjecture.

48. **MODELING EXPERIMENT** A rectangular tank has a square base 2 ft on a side that is filled with water to a depth of 4 ft. It is being drained from the bottom of the tank through a sharp-edged square hole that is 2 in. on a side.

   a. Show that at time t, the depth h satisfies the differential equation

$$\frac{dh}{dt} = -\frac{1}{30}\sqrt{h}$$

   b. How long will it take to empty the tank?

   c. Construct this tank and then drain it out of the 4 in.^{2} hole. Is the time that it takes consistent with your answer to part **b**?

49. The following is a list of six families of curves. Sketch several members of each family and then determine which pairs are orthogonal trajectories of one another.

   a. the circles $x^2 + y^2 = A^2$
   b. the ellipses $2x^2 + y^2 = B^2$
   c. the ellipses $x^2 + 2y^2 = C^2$
   d. the lines $y = Cx$
   e. the parabolas $y^2 = Cx$
   f. the parabolas $y = Cx^2$

50. The Dead Sea Scrolls were written on parchment in about 100 B.C. What percentage of ^{14}C originally contained in the parchment remained when the scrolls were discovered in 1947?

51. Tests of an artifact discovered at the Debert site in Nova Scotia show that 28% of the original ^{14}C is still present. What is the probable age of the artifact?

52. *Historical Quest*

The Shroud of Turin is a rectangular linen cloth kept in the Chapel of the Holy Shroud in the cathedral of St. John the Baptist in Turin, Italy. It shows the image of a man whose wounds correspond with the biblical accounts of the crucifixion.

In 1389, Pierre d'Arcis, the Bishop of Troyes, wrote a memo to the Pope, accusing a colleague of passing off "a certain cloth, cunningly painted" as the burial shroud of Jesus Christ. Despite this early testimony of forgery, this so-called Shroud of Turin has survived as a famous relic. In 1988, a small sample of the Shroud of Turin was taken and scientists from Oxford University, the University of Arizona, and the Swiss Federal Institute of Technology were permitted to test it. It was found that the cloth contained 92.3% of the original ^{14}C. According to this information, how old is the Shroud?

53. A cylindrical tank of radius 3 ft is filled with water to a depth of 5 ft. Determine how long (to the nearest minute) it takes to drain the tank through a sharp-edged circular hole in the bottom with radius 2 in.

54. Rework Problem 53 for a tank with a sharp-edged drain hole that is square with side of length 1.5 in.

55. A toy rocket is launched from the surface of the earth with initial velocity $v_0 = 150$ ft/s. (The radius of the earth is roughly 3,956 mi, and $g = 32$ ft/s^2.)

   a. Determine the velocity (to the nearest ft/s) of the rocket when it is first 200 feet above the ground. (Remember, this is not the same as 200 ft from the center of the earth.)

   b. What is s when $v = 0$? Determine the maximum height above the ground that is attained by the rocket.

56. Determine the escape velocity of each of the following heavenly bodies:

   a. moon ($R = 1,080$ mi; $g = 5.5$ ft/s^2)
   b. Mars ($R = 2,050$ mi; $g = 12$ ft/s^2)
   c. Venus ($R = 3,800$ mi; $g = 28$ ft/s^2)

57. The shape of a tank is such that when it is filled to a depth of h feet, it contains $V = 9\pi h^3$ ft^3 of water. The tank is being drained through a sharp-edged circular hole of radius 1 in. If the tank is originally filled to a depth of 4 ft, how long does it take for the tank to empty?

58. A rectangular tank has a square base 4 ft on a side and is 10 ft high. Originally, it was filled with water to a depth of 6 feet but now is being drained from the bottom of the tank through a sharp-edged square hole 1 in. on a side.

   a. Find an equation involving the rate dh/dt.

   b. How long will it take to drain the tank?

Level 3

59. **Modeling Problem** A projectile is launched from the surface of a planet whose radius is R and where the constant acceleration due to gravity is g. According to Newton's law of gravitation, the force of gravity acting on a projectile of mass m has magnitude

$$F = \frac{mk}{s^2}$$

where k is a constant. If this is the only force acting on the projectile, then $F = ma$, where a is the acceleration of the projectile. Show that

$$a = \frac{-gR^2}{s^2}$$

60. Modeling Problem In physics, it is shown that the amount of heat Q (calories) that flows through an object by conduction will satisfy the differential equation

$$\frac{dQ}{dt} = -kA\frac{dT}{ds}$$

where t (seconds) is the time of flow, k is a physical constant (the *thermal conductivity* of the object), A is the surface area of the object measured at right angles to the direction of flow, and T is the temperature at a point s centimeters within the object, measured in the direction of the flow, as shown in Figure 5.31.

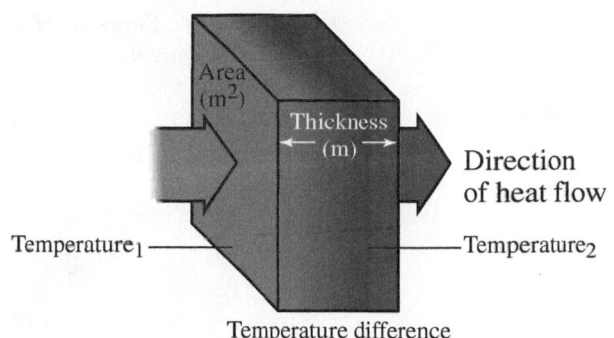

Figure 5.31 Heat conduction

Under certain conditions (equilibrium), the rate of heat flow dQ/dt will be constant. Assuming these conditions exist, find the number of calories that will flow each second across the face of a square pane of glass 2 cm thick and 50 cm on a side if the temperature on one side of the pane is 5° and on the other side is 60°. The thermal conductivity of glass is approximately $k = 0.0025$.

5.7 THE MEAN VALUE THEOREM FOR INTEGRALS; AVERAGE VALUE

IN THIS SECTION: *Mean value theorem for integrals, modeling average value of a function*
In this section, we establish a result called the mean value theorem for integrals which leads to the idea of the average value of a function.

Mean Value Theorem for Integrals

In Section 2.3, we established a very useful theoretical tool called the mean value theorem, which said that under reasonable conditions, there is at least one number c in the interval (a, b) such that

$$\frac{f(b) - f(a)}{b - a} = f'(c)$$

The mean value theorem for integrals is similar, and in the special case where $f(x) \geq 0$, it has a geometric interpretation that makes the theorem easy to understand. In particular, the theorem says that it is possible to find at least one number c on the interval (a, b) such that the area of the rectangle with height $f(c)$ and width $(b - a)$ has exactly the same area as the region under the curve $y = f(x)$ on $[a, b]$. This is illustrated in Figure 5.32.

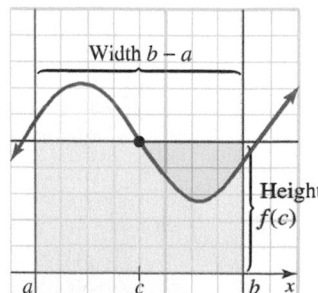

Figure 5.32 The shaded rectangle has the same area as the region under the curve $y = f(x)$ on $[a, b]$

Theorem 5.11   Mean value theorem for integrals

If f is continuous on the interval $[a, b]$, then there is at least one number c between a and b such that

$$\int_a^b f(x)\,dx = f(c)(b - a)$$

Proof: Suppose M and m are the absolute maximum and the absolute minimum of f, respectively, on $[a, b]$. This means that

$$m \leq \qquad f(x) \qquad \leq M \qquad\qquad \text{where } a \leq x \leq b$$

$$\int_a^b m \, dx \leq \int_a^b f(x) \, dx \leq \int_a^b M \, dx \qquad \text{Dominance rule}$$

$$m(b-a) \leq \int_a^b f(x) \, dx \leq M(b-a)$$

$$m \leq \frac{1}{b-a} \int_a^b f(x) \, dx \leq M$$

Because f is continuous on the closed interval $[a, b]$ and because the number

$$I = \frac{1}{b-a} \int_a^b f(x) \, dx$$

lies between m and M, the intermediate value theorem (Theorem 2.6 of Section 2.3) says that there exists a number c between a and b for which $f(c) = I$; that is,

$$\frac{1}{b-a} \int_a^b f(x) \, dx = f(c)$$

$$\int_a^b f(x) \, dx = f(c)(b-a) \qquad\qquad\qquad \blacklozenge$$

The mean value theorem for integrals does not specify how to determine c. It simply guarantees the existence of at least one number c in the interval. However, Example 1 shows how to find a value of c guaranteed by this theorem for a particular function and interval.

Example 1   Finding c using the mean value theorem for integrals

Find a value of c guaranteed by the mean value theorem for integrals for $f(x) = \sin x$ on $[0, \pi]$.

Solution

$$\int_0^\pi \sin x \, dx = -\cos x \big|_0^\pi = -\cos \pi + \cos 0 = -(-1) + 1 = 2$$

The region bounded by f and the x-axis on $[0, \pi]$ is shaded in Figure 5.33. The mean value theorem for integrals asserts the existence of a number c on $[0, \pi]$ such that $f(c)(b-a) = 2$. We can solve this equation to find this value:

$$f(c)(b-a) = 2$$
$$(\sin c)(\pi - 0) = 2$$
$$\sin c = \frac{2}{\pi}$$
$$c \approx 0.69 \text{ or } 2.45$$

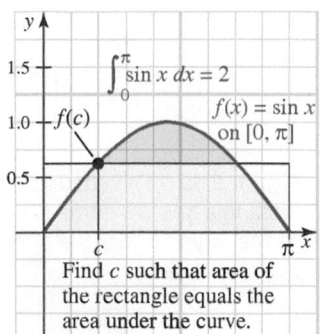

Find c such that area of the rectangle equals the area under the curve.

Figure 5.33   Graph of $f(x) = \sin x$. Find c on $[0, \pi]$

Because each choice of c is between 0 and π, we have found two values of c guaranteed by the mean value theorem for integrals.

Modeling Average Value of a Function

There are many practical situations in which one is interested in the *average value* of a continuous function on an interval, such as the average level of air pollution over a 24-hour period, the average speed of a truck during a 3-hour trip, or the average productivity of a worker during a production run.

You probably know that the average value of n numbers $x_1, x_2, \cdots, x_n$ is

$$\frac{x_1 + x_2 + \cdots + x_n}{n}$$

but what if there are infinitely many numbers? Specifically, what is the average value of $f(x)$ on the interval $a \leq x \leq b$? To see how the definition of finite average value can be used, imagine that the interval $[a, b]$ is divided into n equal subintervals, each of width

$$\Delta x = \frac{b - a}{n}$$

Then for $k = 1, 2, \cdots, n$, let x_k^* be a number chosen arbitrarily from the kth subinterval. Then the average value AV of f on $[a, b]$ is estimated by the sum

$$S_n = \frac{f(x_1^*) + f(x_2^*) + \cdots + f(x_n^*)}{n} = \frac{1}{n} \sum_{k=1}^{n} f(x_k^*)$$

Because $\Delta x = \dfrac{b - a}{n}$, we know that $\dfrac{1}{n} = \dfrac{1}{b - a} \Delta x$ and

$$S_n = \frac{1}{n} \sum_{k=1}^{n} f(x_k^*) = \left[\frac{1}{b - a} \Delta x \right] \sum_{k=1}^{n} f(x_k^*) = \frac{1}{b - a} \sum_{k=1}^{n} f(x_k^*) \Delta x$$

The sum on the right is a Riemann sum with norm $\|P\| = \dfrac{b - a}{n}$. It is reasonable to expect the estimating average S_n to approach the "true" average value AV of $f(x)$ on $[a, b]$ as $n \to \infty$. Thus, we model average value by

$$AV = \lim_{n \to \infty} S_n = \lim_{n \to \infty} \frac{1}{b - a} \sum_{k=1}^{n} f(x_k^*) \Delta x = \frac{1}{b - a} \int_a^b f(x)\, dx$$

We use this integral as the definition of average value.

AVERAGE VALUE If f is continuous on the interval $[a, b]$, the **average value** (AV) of f on this interval is given by the integral

$$AV = \frac{1}{b - a} \int_a^b f(x)\, dx$$

Example 2 Modeling average speed of traffic

Suppose a study suggests that between the hours of 1:00 P.M. and 4:00 P.M. on a normal weekday the speed of the traffic at a certain expressway exit is modeled by the formula

$$S(t) = 2t^3 - 21t^2 + 60t + 20$$

kilometers per hour, where t is the number of hours past noon. Compute the average speed of the traffic between the hours of 1:00 P.M. and 4:00 P.M.

Solution   Our goal is to find the average value of $S(t)$ on the interval $[1, 4]$. The average speed is

$$\frac{1}{4-1}\int_1^4 (2t^3 - 21t^2 + 60t + 20)\,dt = \frac{1}{3}\left[\frac{1}{2}t^4 - 7t^3 + 30t^2 + 20t\right]_1^4$$

$$= \frac{1}{3}[240 - 43.5]$$

$$= 65.5$$

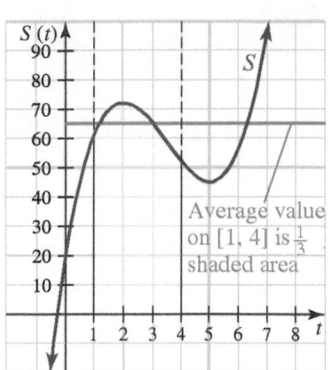

The function, as well as the average value, is shown in Figure 5.34.

According to the mean value theorem for integrals, we have

$$\frac{1}{b-a}\int_a^b f(x)\,dx = f(c)$$

which says that the *average value of a continuous function f on $[a, b]$ equals the value of f for at least one number c between a and b*. This is quite reasonable since the intermediate value theorem for continuous functions assures us that a continuous function f assumes every value between its minimum m and maximum M, and we would expect the average value to be between these two extremes. Example 3 illustrates one way of using these ideas.

Figure 5.34 Graph of S and the average value

Example 3   Modeling average temperature

Suppose that during a typical winter day in Minneapolis, the temperature (in degrees Celsius) x hours after midnight is modeled by the formula

$$T(x) = 2 - \frac{1}{7}(x - 13)^2$$

A graph of this formula is shown in Figure 5.35. Find the average temperature over the time period from 2:00 A.M. to 2:00 P.M., and find a time when the average temperature actually occurs.

Solution   We wish to find the average temperature T on the interval $[2, 14]$ (because 2 P.M. is 14 hours after midnight). The average value is

$$T = \frac{1}{14-2}\int_2^{14}\left[2 - \frac{1}{7}(x - 13)^2\right]dx$$

$$= \frac{1}{12}\left[2x - \frac{1}{7}\cdot\frac{1}{3}(x - 13)^3\right]_2^{14}$$

$$= \frac{1}{12}\left[\frac{587}{21} - \frac{1,415}{21}\right]$$

$$= -\frac{23}{7}$$

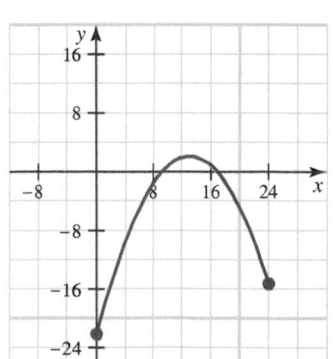

Figure 5.35 Graph of temperatures

Thus, the average temperature on the given time period is approximately 3.3°C below zero. To determine when this temperature actually occurs, solve the equation

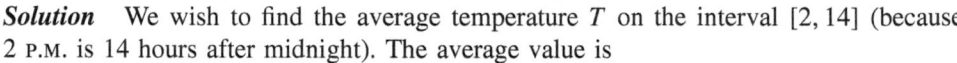

AVERAGE TEMPERATURE $=$ TEMPERATURE AT TIME x

$$-\frac{23}{7} = 2 - \frac{1}{7}(x - 13)^2$$

$$37 = (x - 13)^2$$

$$x = 13 \pm \sqrt{37}$$

$$\approx 19.082763 \text{ or } 6.9172375$$

The first value is to be rejected because it is not in the interval $[2, 14]$, so we find that the average temperature occurs 6.917 hr after midnight, or at approximately 6:55 A.M. ■

Example 4   Modeling average population

The logistic formula

$$P(t) = \frac{354.91}{1 + e^{4.4923 - 0.29052t}}$$

was developed by the United States Bureau of the Census to represent the population of the United States (in millions) during the period 1790-2010. Time t in the formula is the number of decades after 1790. Thus, $t = 0$ for 1790, $t = 22$ for 2010. Use this formula to compute the average population of the United States between 1790 and 2010. When did the average population actually occur?

Solution   The average population is given by the integral

$$A = \frac{1}{22 - 0} \int_0^{22} \frac{354.91 \, dt}{1 + e^{4.4923 - 0.29052t}} \qquad \text{Multiply by } \frac{e^{0.29052t}}{e^{0.29052t}}.$$

$$= \frac{1}{22} \int_0^{22} \frac{354.91 e^{0.29052t}}{e^{0.29052t} + e^{4.4923}} \, dt \qquad \boxed{\begin{array}{l} \text{Let } u = e^{0.29052t} + e^{4.4923} \\ du = 0.29052 \, e^{0.29052t} \, dt \end{array}}$$

$$= \frac{1}{22} \left(\frac{354.91}{0.29052} \right) \int_{t=0}^{t=22} \frac{du}{u}$$

$$= \frac{1}{22} \left(\frac{354.91}{0.29052} \right) \ln \left| e^{0.29052t} + e^{4.4923} \right| \Big|_0^{22}$$

$$= 112.5852959$$

To find when the average population of A million actually occurred, solve $P(t) = A$:

$$\frac{354.91}{1 + e^{4.4923 - 0.29052t}} = A \qquad \textit{Use the entire calculator value of}$$
$$\textit{A, and not a rounded value.}$$

$$e^{4.4923 - 0.29052t} = \frac{354.91}{A} - 1$$

$$4.4923 - 0.29052t = \ln \left(\frac{354.91}{A} - 1 \right)$$

$$t \approx 12.82435818$$

The average population occurred approximately 13 decades after 1790, around the year 1920. ▨

PROBLEM SET 5.7

Level 1

In Problems 1-14, *find c such that*

$$\int_a^b f(x) \, dx = f(c)(b - a)$$

as guaranteed by the mean value theorem for integrals. If you cannot find such a value, explain why the theorem does not apply.

1. $f(x) = 4x^3$ on $[1, 2]$
2. $f(x) = 5x^4$ on $[1, 2]$
3. $f(x) = x^2 + 4x + 1$ on $[0, 2]$

4. $f(x) = x^2 - 3x + 1$ on $[0, 4]$

5. $f(x) = 15x^{-2}$ on $[1, 5]$

6. $f(x) = 12x^{-3}$ on $[-3, 3]$

7. $f(x) = 2\csc^2 x$ on $\left[-\frac{\pi}{3}, \frac{\pi}{3}\right]$

8. $f(x) = \cos x$ on $\left[-\frac{\pi}{2}, \frac{\pi}{2}\right]$

9. $f(x) = e^{2x}$ on $\left[-\frac{1}{2}, \frac{1}{2}\right]$

10. $f(x) = e^{x-1}$ on $[-1, 1]$

11. $f(x) = \dfrac{x}{1+x}$ on $[0, 1]$

12. $f(x) = \dfrac{x+1}{1+x^2}$ on $[-1, 1]$

13. $f(x) = \cot x$ on $[-2, 2]$

14. $f(x) = \tan x$ on $[0, 2]$

Determine the area of the indicated region in Problems 15-20 and then draw a rectangle with base $(b - a)$ and height $f(c)$ for some c on $[a, b]$ so that the area of the rectangle is equal to the area of the given region.

15. $y = \dfrac{1}{2}x$ on $[0, 10]$ **16.** $y = x^2$ on $[0, 3]$

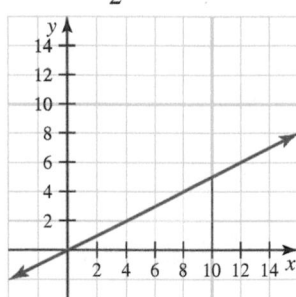 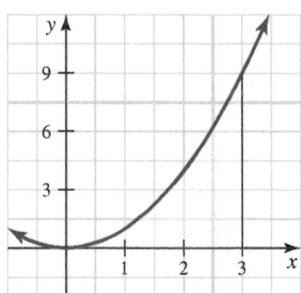

17. $y = x^2 + 2x + 3$ on $[0, 2]$ **18.** $y = \dfrac{1}{x^2}$ on $\left[\dfrac{1}{2}, 2\right]$

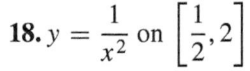

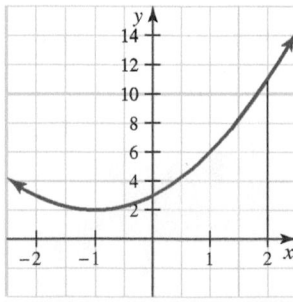 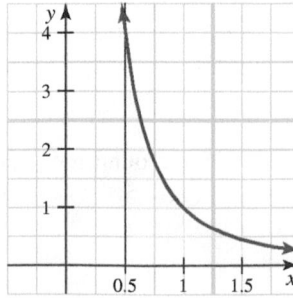

19. $y = \cos x$ on $[-1, 1.5]$ **20.** $y = x + \sin x$ on $[0, 10]$

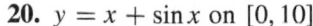

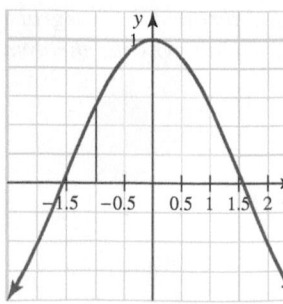 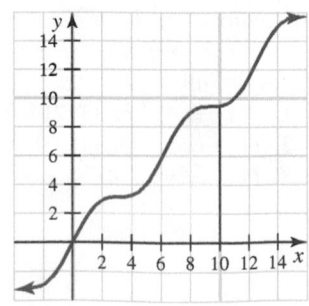

Find the average value of the function given in Problems 21-42 on the prescribed interval.

21. $f(x) = x^2 - x + 1$ on $[-1, 2]$

22. $f(x) = x^2 + x - 1$ on $[1, 2]$

23. $f(x) = x^3 - 3x^2$ on $[-2, 1]$

24. $f(x) = x^3 - 2x^2$ on $[1, 2]$

25. $f(x) = e^x - e^{-x}$ on $[-1, 1]$

26. $f(x) = e^x + e^{-x}$ on $[-1, 1]$

27. $f(x) = \dfrac{x}{2x+3}$ on $[0, 1]$

28. $f(x) = \dfrac{x}{3x-2}$ on $[-2, 0]$

29. $f(x) = \sin x$ on $\left[0, \frac{\pi}{4}\right]$

30. $f(x) = 2\sin x - \cos x$ on $\left[0, \frac{\pi}{2}\right]$

31. $f(x) = \sqrt{4 - x}$ on $[0, 4]$

32. $f(x) = \sqrt{9 - x}$ on $[0, 9]$

33. $f(x) = \sqrt[3]{1 - x}$ on $[-7, 0]$

34. $f(x) = \sqrt[3]{x - 1}$ on $[-1, 1]$

35. $f(x) = (2x - 3)^3$ on $[0, 1]$

36. $f(x) = (3x - 2)^2$ on $[-1, 1]$

37. $f(x) = x\sqrt{2x^2 + 7}$ on $[0, 1]$

38. $f(x) = x(x^2 + 1)^3$ on $[-2, 1]$

39. $f(x) = \dfrac{x}{\sqrt{x^2 + 1}}$ on $[0, 3]$

40. $f(x) = \dfrac{x}{\sqrt{x^2 + 1}}$ on $[-3, 0]$

41. $f(x) = \sqrt{9 - x^2}$ on $[-3, 3]$
   Hint: The integral can be evaluated as the area of part of a circle.

42. $f(x) = \sqrt{2x - x^2}$ on $[0, 2]$
   Hint: The integral can be evaluated as the area of part of a circle.

Level 2

43. A **supply function** is a function which defines a relationship between the supply price and quantity supplied. If the supply function is

$$p = S(x) = 10(e^{0.03x} - 1)$$

What is the average price (in dollars) of this function over the interval $[10, 50]$?

44. If the supply function (see Problem 43) is

$$p = S(x) = 5(e^{0.3x} - 1)$$

What is the average price (in dollars) over the interval $[1, 5]$?

45. Given the demand function

$$p(x) = 50e^{0.03x}$$

Find the average value of the price (in dollars) over the interval $[0, 100]$.

46. The number of rabbits in a limited geographical area (such as an island) is approximated by

$$P(t) = 500 + t - 0.25t^2$$

for the number t years on the interval $[0, 5]$. What is the average number of rabbits in the area over the 5-year time period?

47. The temperature (in degrees Fahrenheit) varies according to the formula

$$F(t) = 6.44t - 0.23t^2 + 30$$

Find the average daily temperature if t is the time of day (in hours) measured from midnight.

48. If an object is propelled upward from ground level with an initial velocity v_0, then its height at time t is given by

$$s = -\frac{1}{2}gt^2 + v_0 t$$

where g is the constant acceleration due to gravity. Show that between times t_0 and t_1, the average height of the object is $s = -\frac{1}{6}g\left[t_1^2 + t_1 t_0 + t_0^2\right] + \frac{1}{2}v_0(t_1 + t_0)$

49. What is the average velocity for the object described in Problem 48 during the same time period?

50. Records indicate that t hours past midnight, the temperature at the local airport was

$$f(t) = -0.1t^2 + t + 50$$

degrees Fahrenheit. What was the average temperature at the airport between 9:00 A.M. and noon?

51. What is the average temperature at the airport (Problem 50) between noon and 3:00 P.M.?

52. Suppose a study indicates that t years from now, the level of carbon dioxide in the air of a certain city will be

$$L(t) = te^{-0.01t^2}$$

parts per million (ppm) for $0 \leq t \leq 20$. What is the average level of carbon dioxide in the first three years?

53. At what time (or times) in the first three years does the average level of carbon dioxide (Problem 52) actually occur? Answer to the nearest month.

54. The number of bacteria (in thousands) present in a certain culture after t minutes is modeled by

$$Q(t) = \frac{2,000}{1 + 0.3e^{-0.276t}}$$

What was the average population during the *second* ten minutes ($10 \leq t \leq 20$)?

55. At what time during the period $10 \leq t \leq 20$ is the average population in Problem 54 actually attained?

Level 3

56. Let $f(t)$ be a function that is continuous and satisfies $f(t) \geq 0$ on the interval $[0, \frac{\pi}{2})$. Suppose it is known that for any number x between 0 and $\frac{\pi}{2}$, the region under the graph of f on $[0, x]$ has area $A(x) = \tan x$. Explain why

$$\int_0^x f(t)\, dt = \tan x$$

for $0 \leq x < \frac{\pi}{2}$.

57. Differentiate both sides of the equation in Problem 56 and deduce the identity of f.

58. Think Tank Problem Suppose that $f(t)$ is continuous for all t and that for any number x it is known that the average value of f on $[-1, x]$ is

$$A(x) = \sin x$$

Use this information to deduce the identity of f.

59. Modeling Problem In Example 7 of Section 4.7, we gave the Heinz function

$$f(t) = \frac{k}{b - a}(e^{-at} - e^{-bt})$$

where $t \geq 0$ represents time in hours and $f(t)$ is the concentration of a drug in a person's blood stream t hours after an intramuscular injection. The coefficients a and b ($b > a > 0$) are characteristics of the drug and the patient's metabolism and are called the *absorption* and *diffusion* rates, respectively. Show that for each fixed t, $f(t)$ can be thought of as the average value of a function of the form

$$g(\lambda) = (At^2 + Bt + C)e^{-\lambda t}$$

over the interval $a \leq \lambda \leq b$. Find A, B, and C.

60. Think Tank Problem Read Problem 59. Can you see any value in the interpretation of the Heinz concentration function as an average value? Explain.

5.8 NUMERICAL INTEGRATION: THE TRAPEZOIDAL RULE AND SIMPSON'S RULE

IN THIS SECTION: *Approximation by rectangles, trapezoidal rule, Simpson's rule, error estimation, summary of numerical integration techniques*

In this section, we examine approximation by rectangles, by trapezoids, and by parabolic arcs. It will help if you have access to a calculator or computer.

The fundamental theorem of calculus can be used to evaluate an integral whenever an appropriate antiderivative is known. However, certain functions, such as $f(x) = e^{x^2}$, have no simple antiderivatives. To find a definite integral of such a function, it is often necessary to use numerical approximation.

Approximation by Rectangles

If $f(x) \geq 0$ on the interval $[a, b]$, the definite integral $\int_a^b f(x)\,dx$ is equal to the area under the graph of f on $[a, b]$. As we saw in Section 5.2, one way to approximate this area is to use n rectangles, as shown in Figure 5.36.

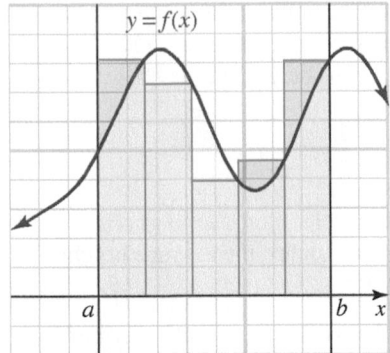

Figure 5.36 Interactive   Approximation by rectangles

In particular, divide the interval $[a, b]$ into n subintervals, each of width $\Delta x = \dfrac{b-a}{n}$, and let x_k^* denote the right endpoint of the kth subinterval. The base of the kth rectangle is the kth subinterval, and its height is $f(x_k^*)$. Hence, the area of the kth rectangle is $f(x_k^*)\Delta x$. The sum of the areas of all n rectangles is an approximation for the area under the curve and hence an approximation for the corresponding definite integral. Thus,

$$\int_a^b f(x)\,dx \approx f(x_1^*)\Delta x + f(x_2^*)\Delta x + \cdots + f(x_n^*)\Delta x$$

This approximation improves as the number of rectangles increases, and we can estimate the integral to any desired degree of accuracy by taking n large enough. However, because fairly large values of n are usually required to achieve reasonable accuracy, approximation by rectangles is rarely used in practice.

Trapezoidal Rule

The accuracy of the approximation generally improves significantly if trapezoids are used instead of rectangles. Figure 5.37 shows the area from Figure 5.36 approximated by n trapezoids instead of rectangles. Even from these rough illustrations you can see how much better the approximation is in this case.

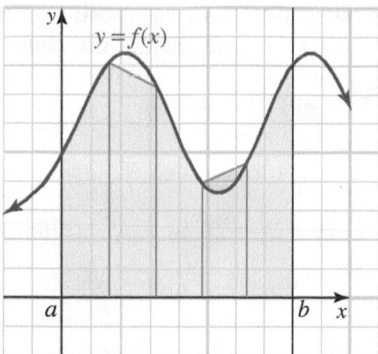

Figure 5.37 Interactive   Approximation by trapezoids

Suppose the interval $[a, b]$ is partitioned into n equal parts by the subdivision points $x_0, x_1, \cdots, x_n$, where $x_0 = a$ and $x_n = b$. The kth trapezoid is shown in greater detail in Figure 5.38.

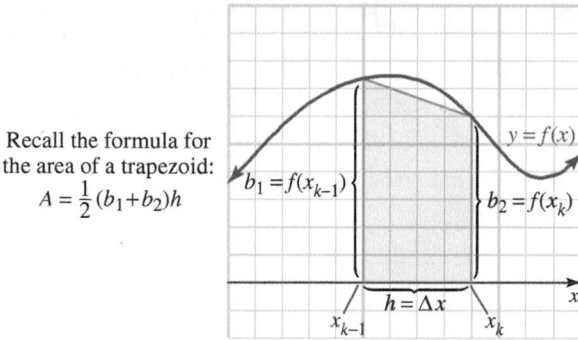

Figure 5.38   The kth trapezoid has area $\dfrac{1}{2}\left[f(x_{k-1}) + f(x_k)\right] \Delta x$

If we let T_n denote the sum of the areas of n trapezoids, we see that

$$T_n = \frac{1}{2}\left[f(x_0 + f(x_1)\right]\Delta x + \frac{1}{2}\left[f(x_1) + f(x_2)\right]\Delta x + \cdots + \frac{1}{2}\left[f(x_{n-1}) + f(x_n)\right]\Delta x$$

$$= \frac{1}{2}\left[f(x_0) + 2f(x_1) + \cdots + 2f(x_{n-1}) + f(x_n)\right]\Delta x$$

The sum T_n estimates the total area above the x-axis and under the curve $y = f(x)$ on the interval $[a, b]$ and hence also estimates the integral

$$\int_a^b f(x)\,dx$$

This approximation formula is known as the *trapezoidal rule* and applies as a means of approximating the integral, even if the function f is not positive.

TRAPEZOIDAL RULE   Let f be continuous on $[a, b]$. The **trapezoidal rule** is

$$\int_a^b f(x)\,dx \approx \frac{1}{2}\left[f(x_0) + 2f(x_1) + \cdots + 2f(x_{n-1}) + f(x_n)\right]\Delta x$$

where $\Delta x = \dfrac{b-a}{n}$ and, for the kth subinterval, $x_k = a + k\Delta x$. Moreover, the larger the value for n, the better the approximation.

Our first example uses the trapezoidal rule to estimate the value of an integral that we can compute exactly by using the fundamental theorem.

Example 1   Trapezoidal rule approximation

Use the trapezoidal rule with $n = 4$ to estimate $\int_{-1}^{2} x^2 \, dx$.

Solution   The interval is $[a, b] = [-1, 2]$, so $a = -1$ and $b = 2$. Then $\Delta x = \dfrac{2 - (-1)}{4} = \dfrac{3}{4} = 0.75$. Thus,

$$x_0 = a = -1 \qquad\qquad\qquad f(x_0) = f(-1) = (-1)^2 = 1$$

$$x_1 = a + 1 \cdot \Delta x = -1 + \frac{3}{4} = -\frac{1}{4} \qquad 2f(x_1) = 2\left(-\frac{1}{4}\right)^2 = \frac{1}{8} = 0.125$$

$$x_2 = a + 2 \cdot \Delta x = -1 + 2\left(\frac{3}{4}\right) = \frac{1}{2} \qquad 2f(x_2) = 2\left(\frac{1}{2}\right)^2 = \frac{1}{2} = 0.5$$

$$x_3 = a + 3 \cdot \Delta x = -1 + 3\left(\frac{3}{4}\right) = \frac{5}{4} \qquad 2f(x_3) = 2\left(\frac{5}{4}\right)^2 = \frac{25}{8} = 3.125$$

$$x_4 = a + 4 \cdot \Delta x = b = 2 \qquad\qquad f(x_4) = 2^2 = 4$$

$$T_4 = \frac{1}{2}[1 + 0.125 + 0.5 + 3.125 + 4](0.75) = 3.28125$$

The exact value of the integral in Example 1 is

$$\int_{-1}^{2} x^2 \, dx = \left.\frac{x^3}{3}\right|_{-1}^{2} = \frac{8}{3} - \frac{-1}{3} = 3$$

Therefore, the trapezoidal estimate T_4 involves an error, which we denote by E_4. We find that

$$E_4 = \int_{-1}^{2} x^2 \, dx - T_4 = 3 - 3.28125 = -0.28125$$

The negative sign indicates that the trapezoidal formula *overestimated* the true value of the integral in Example 1.

Simpson's Rule

Roughly speaking, the accuracy of a procedure for estimating the area under a curve depends on how well the upper boundary of each approximating area fits the shape of the given curve. Trapezoidal strips often result in a better approximation than rectangular strips, and it is reasonable to expect even greater accuracy to occur if the approximating regions have curved upper boundaries, as shown in Figure 5.39.

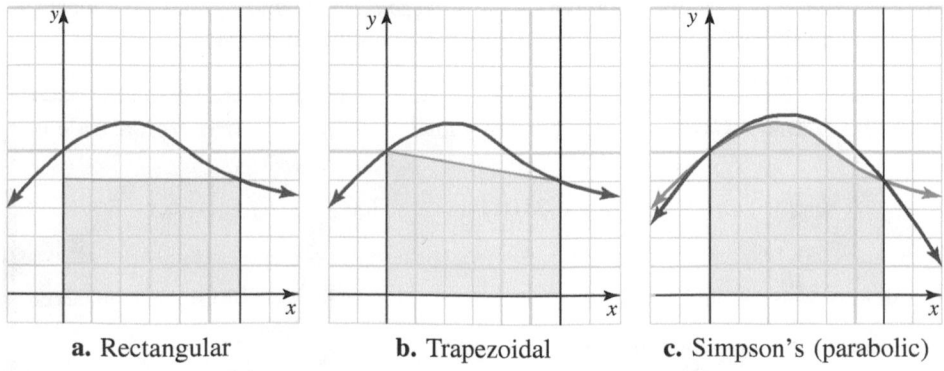

a. Rectangular   **b.** Trapezoidal   **c.** Simpson's (parabolic)

Figure 5.39 Interactive   A comparison of approximation methods

The name given to the procedure in which the approximating strip has a parabolic arc for its upper boundary is called *Simpson's rule*.*

As with the trapezoidal rule, we shall derive Simpson's rule as a means for approximating the area under the curve $y = f(x)$ on the interval $[a, b]$, where f is continuous and satisfies $f(x) \geq 0$. First, we partition the given interval into a number of equal subintervals, but this time, we require the number of subdivisions to be an *even* number (because this requirement will simplify the formula associated with the final result).

If $x_0, x_1, \cdots, x_n$ are the subdivision points in our partition (with $x_0 = a$ and $x_n = b$), we pass a parabolic arc through the points, three at a time (the points with x-coordinates x_0, x_1, x_2, then those with x_2, x_3, x_4, and so on). It can be shown (see Problem 60) that the region under the parabolic curve $y = f(x)$ on the interval $\left[x_{2k-2}, x_{2k}\right]$ has area

$$\frac{1}{3}\left[f(x_{2k-2}) + 4f(x_{2k-1}) + f(x_{2k})\right]\Delta x$$

where $\Delta x = \dfrac{b-a}{n}$. This procedure is illustrated in Figure 5.40.

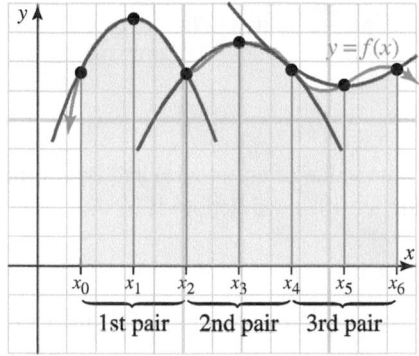

Figure 5.40 Approximation using parabolas

By adding the areas of the approximating parabolic strips and combining terms, we obtain the sum S_n of n parabolic regions:

$$S_n = \frac{1}{3}\left[f(x_0) + 4f(x_1) + f(x_2)\right]\Delta x + \frac{1}{3}\left[f(x_2) + 4f(x_3) + f(x_4)\right]\Delta x$$
$$+ \cdots + \frac{1}{3}\left[f(x_{n-2}) + 4f(x_{n-1}) + f(x_n)\right]\Delta x$$
$$= \frac{1}{3}\left[f(x_0) + 4f(x_1) + 2f(x_2) + 4f(x_3) + \cdots + 4f(x_{n-1}) + f(x_n)\right]\Delta x$$

These observations are summarized in the following box:

SIMPSON'S RULE   Let f be continuous on $[a, b]$. **Simpson's rule** is

$$\int_a^b f(x)\,dx \approx \frac{1}{3}\left[f(x_0) + 4f(x_1) + 2f(x_2) + 4f(x_3) + \cdots + 4f(x_{n-1}) + f(x_n)\right]\Delta x$$

where $\Delta x = \dfrac{b-a}{n}$, $x_k = a + k\,\Delta x$, k an integer and n an even integer. Moreover, the larger the value for n, the better the approximation.

---

*This rule is named for Thomas Simpson (1710-1761), an English mathematician who, curiously, neither discovered nor made any special use of the formula that bears his name.

Example 2   Approximation by Simpson's rule

Use Simpson's rule with $n = 10$ to approximate $\displaystyle\int_1^2 \frac{dx}{x}$.

Solution   We have $\Delta x = \dfrac{2-1}{10} = 0.1,$ and $x_0 = a = 1,\; x_1 = 1.1,\; x_2 = 1.2, \ldots,$ $x_9 = 1.9,\; x_{10} = b = 2.$

$$\int_1^2 \frac{1}{x}\, dx \approx S_{10}$$

$$= \frac{1}{3}\left(\frac{1}{1} + \frac{4}{1.1} + \frac{2}{1.2} + \frac{4}{1.3} + \frac{2}{1.4} + \frac{4}{1.5} + \frac{2}{1.6} + \frac{4}{1.7} + \frac{2}{1.8} + \frac{4}{1.9} + \frac{1}{2}\right)\left(\frac{1}{10}\right)$$

$$\approx 0.693\,15023069$$

This estimate compares well with the value found directly by applying the fundamental theorem of calculus:

$$\int_1^2 \frac{dx}{x} = \ln 2 - \ln 1 \approx 0.693\,14718056$$

Error Estimation

The difference between the value of an integral and its estimated value is called its **error**. Since this error is a function of n we denote it by E_n.

Theorem 5.12   Error in the trapezoidal rule and Simpson's rule

If f has a continuous second derivative on $[a, b]$, then the error E_n in approximating $\displaystyle\int_a^b f(x)\, dx$ by the trapezoidal rule satisfies the following:

Trapezoidal error: $|E_n| \le \dfrac{(b-a)^3}{12n^2}M$ where M is the maximum value of $\left|f''(x)\right|$ on $[a, b]$. Moreover, if f has a continuous fourth derivative on $[a, b]$, then, the error E_n (n even) in approximating $\displaystyle\int_a^b f(x)\, dx$ by Simpson's rule satisfies the following:

Simpson's error: $|E_n| \le \dfrac{(b-a)^5}{180n^4}K$ where K is the maximum value of $\left|f^{(4)}(x)\right|$ on $[a, b]$.

Proof: The proofs of these error estimates are beyond the scope of this book and can be found in many advanced calculus textbooks and in most numerical analysis books.   ♦

Example 3   Estimate of error when using Simpson's rule

Estimate the accuracy of the approximation of $\displaystyle\int_1^2 \frac{dx}{x}$ by Simpson's rule with $n = 10$ in Example 2.

Solution   If $f(x) = \dfrac{1}{x}$, we find that $f^{(4)}(x) = 24x^{-5}$. The maximum value of this function will occur at a critical number (there is none on $[1, 2]$) or at an endpoint. Thus, the largest value of $\left|f^{(4)}(x)\right|$ on $[1, 2]$ is $\left|f^{(4)}(1)\right| = 24$. Now, apply the error formula with $K = 24$, $a = 1$, $b = 2$, and $n = 10$ to obtain

$$|E_{10}| \leq \frac{K(b-a)^5}{180n^4} = \frac{24(2-1)^5}{180(10)^4} \approx 0.0000133$$

That is, the error in the approximation in Example 2 is guaranteed to be no greater than 0.0000133. ∎

With the aid of the error estimates, we can decide in advance how many subintervals to use to achieve a desired degree of accuracy.

Example 4 Number of subintervals to guarantee given accuracy

How many subintervals are required to guarantee that the estimate will be correct to four decimal places in the approximation of

$$\int_1^2 \frac{dx}{x}$$

on $[1, 2]$ using the trapezoidal rule?

Solution   To be correct to four decimal places means that $|E_n| < 0.00005$. Because $f(x) = x^{-1}$, we have $f''(x) = 2x^{-3}$. On $[1, 2]$ the largest value of $\left|f''(x)\right|$ is $\left|f''(1)\right| = 2$, so $M = 2$, $a = 1$, $b = 2$, and

$$|E_n| \leq \frac{2(2-1)^3}{12n^2} = \frac{1}{6n^2}$$

The goal is to find the smallest positive integer n for which

$$\frac{1}{6n^2} < 0.00005$$
$$10{,}000 < 3n^2 \qquad \qquad \textit{Multiply by the positive}$$
$$10{,}000 - 3n^2 < 0 \qquad \qquad \textit{number } 60{,}000n^2.$$
$$(100 - \sqrt{3}n)(100 + \sqrt{3}n) < 0$$
$$n < -\frac{100}{\sqrt{3}} \quad \text{or} \quad n > \frac{100}{\sqrt{3}} \approx 57.735$$

The smallest positive integer that satisfies this condition is $n = 58$; therefore, 58 subintervals are required to ensure the desired accuracy. ∎

If f is the linear function $f(x) = Ax + B$, then $f''(x) = 0$ and we can take $M = 0$ as the error estimate. In this case, the error in applying the trapezoidal rule satisfies $|E_n| \leq 0$. That is, the trapezoidal rule is exact for a linear function, which is what we would expect, because the region under a line on an interval is a trapezoid.

In discussing the accuracy of the trapezoidal rule as a means of estimating the value of the definite integral

$$I = \int_a^b f(x)\, dx$$

we have focused attention on the "error term," but this only measures the error that comes from estimating I by the trapezoidal or Simpson approximation sum. There are other kinds of error that must be considered in this or any other method of approximation. In particular, each time we cut off digits from a decimal, we incur what is known as a round-off error. For example, a round-off error occurs when we use 0.66666667 in place of $\frac{2}{3}$ or 3.1415927 for the number π. Round-off errors occur even in large computers and can accumulate to cause real problems. Specialized methods for dealing with these and other errors are studied in numerical analysis.

Summary of Numerical Integration Techniques

Calculators and computers are being used more and more to evaluate definite integrals. *Numerical integration* is the process of finding a numerical approximation for a definite integral *without actually carrying out the antidifferentiation process*. This means that you can easily write a computer or calculator program to evaluate definite integrals if you understand numerical integration. Numerical integration is also useful when it is difficult or impossible to find an antiderivative to carry out the integration process.

There are three common approximations to the definite integral, which for convenience, we designate by A_n, T_n, and P_n as defined in the following box.

APPROXIMATIONS FOR THE DEFINITE INTEGRAL Let a function f be continuous on $[a, b]$. Then the integral $\int_a^b f(x)dx$ can be approximated by

RECTANGLES: $A_n = \Delta x \left[f(x_0) + f(x_1) + \cdots + f(x_{n-1}) \right]$
or $\Delta x [f(x_1) + f(x_2) + \ldots + f(x_n)]$

TRAPEZOIDS: $T_n = \dfrac{\Delta x}{2} \left[f(x_0) + 2f(x_1) + \cdots + 2f(x_{n-1}) + f(x_n) \right]$

PARABOLAS: $P_n = \dfrac{\Delta x}{3} \left[f(x_0) + 4f(x_1) + 2f(x_2) + \cdots + 4f(x_{n-1}) + f(x_n) \right]$
(true for n an even integer).

where $\Delta x = \dfrac{b - a}{n}$ for some positive integer n.

The examples of numerical integration examined in this section are intended as illustrations and thus involve relatively simple computations, whereas in practice, such computations often involve hundreds of terms and can be quite tedious. Fortunately, these computations are extremely well suited to technology and the reader who is interested in pursuing computer methods in numerical integration should seek one of the many websites denoted to this topic.

PROBLEM SET 5.8

Level 1

1. ▪ What does this say? Describe the trapezoidal rule.
2. ▪ What does this say? Describe Simpson's rule.

3. a. A_1　　b. T_1　　c. S_2
4. a. A_2　　b. T_2　　c. S_4
5. a. A_3　　b. T_3　　c. S_6
6. a. A_4　　b. T_4　　c. S_8

Consider $\int_1^6 \sqrt{x + 3}\, dx$. Then,

$$\int_1^6 \sqrt{x + 3}\, dx = \frac{2}{3}(x + 3)^{3/2} \Big|_1^6$$

$$= \frac{2}{3}(27) - \frac{2}{3}(8)$$

$$= \frac{38}{3}$$

Approximate this integral by finding the requested values in Problems 3-6.

Consider $\displaystyle\int_2^4 \dfrac{x\, dx}{(1 + 2x)^2}$. Then,

$$\int_2^4 \frac{x\, dx}{(1 + 2x)^2} = \frac{1}{4} \left[\ln|1 + 2x| + \frac{1}{1 + 2x} \right]\Big|_2^4$$

$$= 0.25 \left(\ln 9 + \frac{1}{9} \right) - 0.25 \left(\ln 5 + \frac{1}{5} \right)$$

$$\approx 0.1247244440$$

Approximate this integral by finding the requested values in Problems 7-10.

7. a. A_1 **b.** T_1 **c.** S_2

8. a. A_2 **b.** T_2 **c.** S_4

9. a. A_3 **b.** T_3 **c.** S_6

10. a. A_4 **b.** T_4 **c.** S_8

Approximate the integrals in Problems 11-14 using the trapezoidal rule and Simpson's rule with the specified number of subintervals and then compare your answers with the exact value of the definite integral.

11. $\int_1^2 x^2\, dx$ with $n = 2$ **12.** $\int_1^2 x^2\, dx$ with $n = 4$

13. $\int_0^4 \sqrt{x}\, dx$ with $n = 4$ **14.** $\int_0^4 \sqrt{x}\, dx$ with $n = 6$

Approximate the integrals given in Problems 15-20 with the specified number of subintervals using:

     a. *the trapezoidal rule*   **b.** *Simpson's rule*

15. $\int_0^1 \dfrac{dx}{1 + x^2}$ with $n = 4$

16. $\int_{-1}^0 \sqrt{1 + x^2}\, dx$ with $n = 4$

17. $\int_2^4 \sqrt{1 + \sin x}\, dx$ with $n = 4$

18. $\int_0^2 x \cos x\, dx$ with $n = 6$

19. $\int_0^2 xe^{-x}\, dx$ with $n = 6$

20. $\int_1^2 \ln x\, dx$ with $n = 6$

Level 2

Estimate the value of the integrals in Problems 21-28 to within the prescribed accuracy.

21. $\int_0^1 \dfrac{dx}{x^2 + 1}$ with error less than 0.05. Use the trapezoidal rule.

22. $\int_{-1}^2 \sqrt{1 + x^2}\, dx$ with error less than 0.05. Use Simpson's rule.

23. $\int_0^1 \cos 2x\, dx$ accurate to three decimal places. Use Simpson's rule.

24. $\int_1^2 x^{-1}dx$ accurate to three decimal places. Use Simpson's rule.

25. $\int_0^2 x\sqrt{4 - x}\, dx$ with error less than 0.01. Use the trapezoidal rule.

26. $\int_0^\pi \theta \cos^2 \theta\, d\theta$ with error less than 0.01. Use Simpson's rule.

27. $\int_0^1 \tan^{-1} x\, dx$ with error less than 0.01. Use Simpson's rule.

28. $\int_0^\pi e^{-x} \sin x\, dx$ to three decimal places. Use Simpson's rule.

In Problems 29-34, determine how many subintervals are required to guarantee accuracy to within 0.00005 using:

     a. *the trapezoidal rule*   **b.** *Simpson's rule*

29. $\int_1^3 x^{-1}\, dx$ **30.** $\int_1^4 (x^3 + 2x^2 + 1)\, dx$

31. $\int_1^4 \dfrac{dx}{\sqrt{x}}$ **32.** $\int_0^2 \cos x\, dx$

33. $\int_0^1 e^{-2x}\, dx$ **34.** $\int_1^2 \ln \sqrt{x}\, dx$

35. A quarter-circle of radius 1 has the equation $y = \sqrt{1 - x^2}$ for $0 \le x \le 1$, which implies that

$$\int_0^1 \sqrt{1 - x^2}\, dx = \frac{\pi}{4}$$

Estimate π correct to one decimal place by applying the trapezoidal rule to this integral.

36. Estimate π correct to one decimal place by applying Simpson's rule to the integral in Problem 35.

37. Find the smallest value of n for which the trapezoidal rule estimates the value of the integral

$$\int_1^2 x^{-1}dx$$

with six-decimal-place accuracy.

38. The width of an irregularly shaped dam is measured at 5-m intervals, with the results indicated in Figure 5.41. Use the trapezoidal rule to estimate the area of the face of the dam.

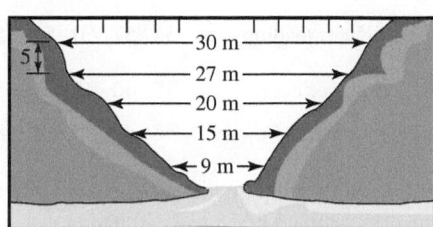

Figure 5.41   Area of the face of a dam

39. A landscape architect needs to estimate the number of cubic yards of fill necessary to level the area shown in Figure 5.42.

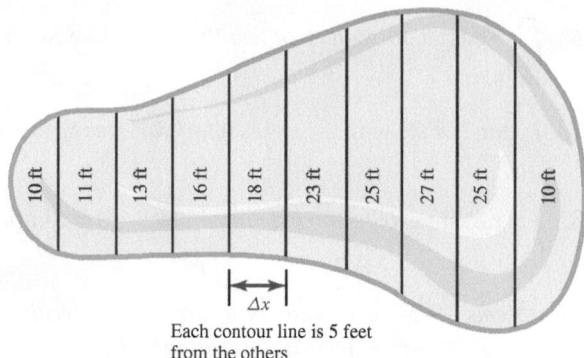

10 ft 11 ft 13 ft 16 ft 18 ft 23 ft 25 ft 27 ft 25 ft 10 ft

Δx

Each contour line is 5 feet
from the others

Figure 5.42   Volume of hole

x	$f(x)$
0	3.7
0.3	3.9
0.6	4.1
0.9	4.1
1.2	4.2
1.5	4.4
1.8	4.6
2.1	4.9
2.4	5.2
2.7	5.5
3	6

Use an approximation with $\Delta x = 5$ feet and y values equal to the distances measured in the figure to find the surface area. Assume that the depth averages 3 feet. (*Note:* 1 cubic yard equals 27 cubic feet.) Use a trapezoidal approximation.

40. An industrial plant spills pollutant into a lake. The pollutant spread out to form the pattern shown in Figure 5.43. All distances are in feet.

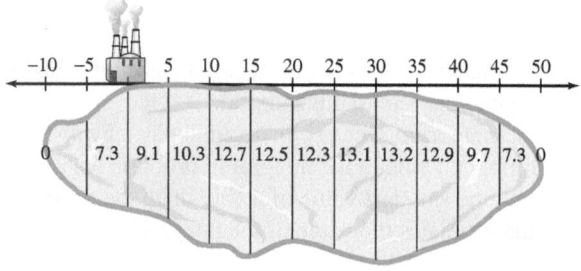

−10 −5   5 10 15 20 25 30 35 40 45 50

0   7.3 9.1 10.3 12.7 12.5 12.3 13.1 13.2 12.9 9.7 7.3 0

Figure 5.43   Pollutant spill

Use Simpson's rule to estimate the area of the spill.

41. Jack and Jill are traveling in a car with a broken odometer. In order to determine the distance they traveled between noon and 1:00 P.M., Jack (the passenger) takes a speedometer reading every 5 minutes.

Minutes (after noon)	0	5	10	15	20	25	30
Speedometer reading	54	57	50	51	55	60	49

Minutes (after noon)	35	40	45	50	55	60
Speedometer reading	53	47	39	42	48	53

Use the trapezoidal rule to estimate the total distance traveled by the couple from noon to 1:00 P.M.

42. Apply the trapezoidal rule to estimate

$$\int_0^3 f(x)\,dx$$

where the values for f are found on the following table:

43. Apply Simpson's rule to estimate

$$\int_0^5 f(x)\,dx$$

where the values for f are found on the following table:

x	$f(x)$
0	10
0.5	9.75
1	10
1.5	10.75
2	12
2.5	13.75
3	16
3.5	18.75
4	22
4.5	25.75
5	30

Think Tank Problems *In these problems, we explore the "order of convergence" of three numerical integration methods. A method is said to have* **order of convergence** n^k *if* $E_n \cdot n^k = C$, *where n is the number of intervals in the approximation, and k is a constant power. In other words, the error $E_n \to 0$ as $\frac{C}{n^k} \to 0$. In Problems 44-47, use the fact that*

$$I = \int_0^\pi \sin x\,dx = -\cos x\,|_0^\pi = 2$$

44. Use the trapezoidal rule to estimate I for $n = 10, 20, 40,$ and 80. Compute the error E_n in each case and compute $E_n \cdot n^k$ for $k = 1, 2, 3, 4$. Based on your results, you should be able to conclude that the order of convergence of the trapezoidal approximation is n^2.

45. Repeat Problem 44 using a rectangular approximation with right endpoints. What is the order of convergence for this method?

46. Repeat Problem 44 using a rectangular approximation with left endpoints. What is the order of convergence for this method?

47. Repeat Problem 44 using Simpson's rule. Based on your results, what is the order of convergence for Simpson's rule?

48. Let

$$I = \int_0^\pi (9x - x^3)\,dx$$

Estimate I using rectangles and using the trapezoidal rule for $n = 10, 20, 40,$ and 80.

49. Repeat Problem 48 using Simpson's rule. Something interesting happens. Explain.

50. Think Tank Problem Simpson struggles! In contrast to Problems 48-49, here is an example where Simpson's rule does not live up to expectations.

$$\int_0^2 \sqrt{4 - x^2}\,dx = \pi$$

For $n = 10, 20, 40,$ and 80, use Simpson's rule and the rectangular rule (select midpoints) for this integral and make a table of errors. Then try to explain Simpson's poor performance.

Hint: Look at the formula for the error.

51. 𝔥*istorical* 𝔔*uest*

The mathematician Seki Kōwa was born in Fujioka, Japan, the son of a samurai, but was adopted by a patriarch of the Seki family. Seki invented and used an early form of determinants for solving systems of equations, and he also invented a method for approximating areas that is

Takakazu Seki Kōwa (1642-1708)

very similar to the rectangular method introduced in this section. This method, known as the yenri (circle principle), found the area of a circle by dividing the circle into small rectangles, as shown in Figure 5.44.

The sample shown in Figure 5.44 was drawn by a student of Seki Kōwa. For this quest, draw a circle with radius 10 cm. Draw vertical chords through each centimeter on a diameter (you should have 18 rectangles). Measure the heights of the rectangles and approximate the area of the circle by adding the areas of the rectangles. Compare this with the formula for the area of this circle.

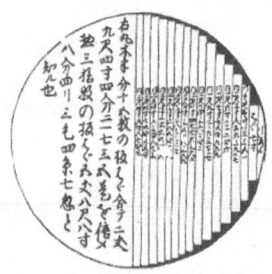

Figure 5.44 Early Asian calculus

52. 𝔥*istorical* 𝔔*uest In 1670, a predecessor of Seki Kōwa (see Problem 51), Kazuyuki Sawaguchi, wrote seven volumes that concluded with fifteen problems that he believed were unsolvable. In 1674, Seki Kōwa published solutions to all fifteen of Kazuyuki's unsolvable problems. One of the "unsolvable" problems was the following: Three circles are inscribed in a circle, each tangent to the other two and to the original circle. All three cover all but 120 square units of the circumscribing circle. The diameters of the two smaller inscribed circles are equal, and each is five units less than the diameter of the larger inscribed circle. Find the diameters of the three inscribed circles. The solution to this "impossible" problem is beyond the scope of this course (it involves solving a 6th degree equation with a horrendous amount of notation), but we can replace it with a simpler problem: Consider two circles inscribed in a circle, each tangent to the other and to the original circle. The diameter of the smaller circle is five units less than the diameter of the larger inscribed circle. The sum of the areas of the larger circle and twice the smaller circle cover all but 120 square units of the circumscribing circles. Find the diameters of the two inscribed circles.*

53. 𝔥*istorical* 𝔔*uest*

Isaac Newton (see 𝔥*istorical* 𝔔*uest* Problem 87 Chapter 1 supplementary problems) *invented a preliminary version of Simpson's rule. In 1779, Newton wrote an article to an addendum to* **Methodus Differentialis** *(1711) in which he gave the following example: If there are four ordinates at equal intervals, let A be the sum of the first and fourth, B the sum of the second and third, and R the interval between the first and fourth; then . . . the area between*

Roger Cotes (1682-1716)

the first and fourth ordinates is approximated by

$$\frac{1}{8}(A + 3B)R$$

*This is known today as the **Newton-Cotes three-eighths rule**, which can be expressed in the form*

$$\int_{x_0}^{x_3} f(x)\,dx \approx \frac{3}{8}(y_0 + 3y_1 + 3y_2 + y_3)\,\Delta x$$

Roger Cotes and James Stirling (1692-1770) both knew this formula, as well as what we called in this section Simpson's rule. In 1743, this rule was rediscovered by Thomas Simpson (1710-1761). Estimate the integral

$$\int_0^3 \tan^{-1} x\,dx$$

using the Newton-Cotes three-eighths rule and then compare with approximation using left endpoints (rectangles) and trapezoids with $n = 4$. Which of the three rules gives the most accurate estimate?

54. Show that if $p(x)$ is any polynomial of degree less than or equal to 3, then

$$\int_a^b p(x)\,dx = \frac{b-a}{6}\left[p(a) + 4p\left(\frac{a+b}{2}\right) + p(b)\right]$$

This result is often called the *prismoidal rule*.

55. Use the prismoidal rule (Problem 54) to evaluate

$$\int_{-1}^2 (x^3 - 3x + 4)\,dx$$

56. Use the prismoidal rule (Problem 54) to evaluate

$$\int_{-1}^3 (x^3 + 2x^2 - 7)\,dx$$

Level 3

57. Let $p(x)$ be a polynomial of degree at most 3. Find a number c between 0 and 1 such that

$$\int_{-1}^1 p(x)\,dx = p(c) + p(-c)$$

58. Let $p(x)$ be a polynomial of degree at most 3. Find a number c between $-\frac{1}{2}$ and $\frac{1}{2}$ such that

$$\int_{-\frac{1}{2}}^{\frac{1}{2}} p(x)\,dx = \frac{1}{3}\left[p(-c) + p(0) + p(c)\right]$$

59. Let $p(x) = Ax^3 + Bx^2 + Cx + D$ be a cubic polynomial. Show that Simpson's rule gives the exact value for

$$\int_a^b p(x)\,dx$$

60. The object of this exercise is to prove Simpson's rule for the special case involving three points.
    a. Let $P_1(-h, f(-h))$, $P_2(0, f(0))$, $P_3(h, f(h))$. Find the equation of the form

$$y = Ax^2 + Bx + C$$

   for the parabola through the points P_1, P_2, and P_3.
    b. If $y = p(x)$ is the quadratic function found in part **a**, show that

$$\int_{-h}^h p(x)\,dx = \frac{h}{3}\left[p(-h) + 4p(0) + p(h)\right]$$

    c. Let $Q_1(x_1, f(x_1))$, $Q_2(x_2, f(x_2))$, $Q_3(x_3, f(x_3))$ be points with $x_2 = x_1 + h$, and $x_3 = x_1 + 2h$. Explain why

$$\int_{x_1}^{x_3} p(x)\,dx = \frac{h}{3}\left[p(x_1) + 4p(x_2) + p(x_3)\right]$$

5.9 AN ALTERNATIVE APPROACH: THE LOGARITHM AS AN INTEGRAL[*]

IN THIS SECTION: *Natural logarithm as an integral, geometric interpretation, the natural exponential function*
There are two possible ways of teaching the exponential and logarithmic functions, and this section is included to give an alternate approach. In one method, the exponential function is defined first, as is done in high school. The logarithmic function is then defined as its inverse. In the alternative approach, the logarithm is defined first as an integral, followed by then defining the exponential as the inverse of a logarithm. Both of these methods are valid, and each has its advocates.

---

[*] Optional section.

Natural Logarithm as an Integral

You may have noticed in Section 2.4 that we did not prove the properties of exponential functions for all real number exponents. To treat exponentials and logarithms *rigorously,* we use the alternative approach provided in this section. Specifically, we use a definite integral to introduce the *natural logarithmic function* and then use this function to *define* the *natural exponential function.*

> **NATURAL LOGARITHM**   The **natural logarithm** is the function defined by
>
> $$\ln x = \int_1^x \frac{dt}{t} \qquad x > 0$$

At first glance, it appears there is nothing "natural" about this definition, but if this integral function has the properties of a logarithm, why should we not call it a logarithm? We begin with a theorem that shows that $\ln x$ does indeed have the properties we would expect of a logarithm.

Theorem 5.13   Properties of a logarithm as defined by an integral

Let $x > 0$ and $y > 0$ be positive numbers. Then

a. $\ln 1 = 0$ **b.** $\ln(xy) = \ln x + \ln y$ **c.** $\ln \dfrac{x}{y} = \ln x - \ln y$

d. $\ln x^p = p \ln x$ for all rational numbers p

Proof: We prove parts **a** and **b** here and leave the proofs of parts **c** and **d** for the Problem Set (Problems 20 and 59).

a. Let $x = 1$. Then $\displaystyle\int_1^1 \frac{dt}{t} = 0.$

b. For fixed positive numbers x and y, we use the additive property of integrals as follows:

$$
\begin{aligned}
\ln xy &= \int_1^{xy} \frac{dt}{t} &&\text{\textit{Definition of natural logarithm}}\\[2mm]
&= \int_1^x \frac{dt}{t} + \int_x^{xy} \frac{dt}{t} &&\text{\textit{Subdivision rule}}\\[2mm]
&= \int_1^x \frac{dt}{t} + \int_1^y \frac{x\,du}{ux} &&\text{\textit{Let $u = \frac{t}{x}$, so $t = ux$;}}\\
& &&\text{\textit{if $t = x$, then $u = 1$ and if $t = xy$, then $u = y$.}}\\[2mm]
&= \int_1^x \frac{dt}{t} + \int_1^y \frac{du}{u} &&\text{\textit{Simplify integrand}}\\[2mm]
&= \ln x + \ln y &&\text{\textit{Definition of natural logarithm}} \qquad\blacklozenge
\end{aligned}
$$

Geometric Interpretation

An advantage of defining the logarithm by the integral formula is that calculus can be used to study the properties of $\ln x$ from the beginning. For example, note that if $x > 1$, the integral

$$\ln x = \int_1^x \frac{dt}{t}$$

may be interpreted geometrically as the area under the graph of $y = \dfrac{1}{t}$ from $t = 1$ to $t = x$, as shown in Figure 5.45.

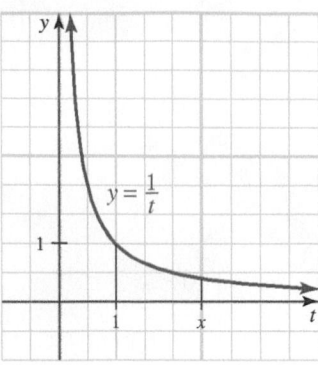

Figure 5.45 If $x > 1$,
$\ln x = \int\limits_{1}^{x} \dfrac{dt}{t}$ is the area under
$y = \dfrac{1}{t}$ on $[1, x]$

If $x > 1$, then $\ln x > 0$. On the other hand, if $0 < x < 1$, then

$$\ln x = \int_{1}^{x} \frac{dt}{t} = -\int_{x}^{1} \frac{dt}{t} < 0$$

so that

$$\ln x > 0 \text{ if } x > 1$$
$$\ln 1 = 0$$
$$\ln x < 0 \text{ if } 0 < x < 1$$

The definition $\ln x = \int\limits_{1}^{x} \dfrac{dt}{t}$ makes it easy to differentiate $\ln x$. Recall from Section 5.4 that according to the second fundamental theorem of calculus, if f is continuous on $[a, b]$, then

$$F(x) = \int_{a}^{x} f(t)\, dt$$

is a differentiable function of x with derivative $\dfrac{dF}{dx} = f(x)$ on any interval $[a, x]$. There-fore, because $\dfrac{1}{t}$ is continuous for all $t > 0$, it follows that $\ln x = \int\limits_{1}^{x} \dfrac{dt}{t}$ is differentiable for all $x > 0$ with derivative $\dfrac{d}{dx}(\ln x) = \dfrac{1}{x}$. By applying the chain rule, we also find that

$$\frac{d}{dx}(\ln u) = \frac{1}{u}\frac{du}{dx}$$

for any differentiable function u of x with $u > 0$.

To analyze the graph of $f(x) = \ln x$, we use the curve-sketching methods of Chapter 4:

- $\ln x$ is continuous for all $x > 0$ (because it is differentiable), so its graph is "unbroken."
- The graph of $\ln x$ is always *rising*, because the derivative

$$\frac{d}{dx}(\ln x) = \frac{1}{x}$$

is positive for $x > 0$. (Recall that the natural logarithm is defined only for $x > 0$.)
- The graph of $\ln x$ is *concave down,* because the second derivative

$$\frac{d^2}{dx^2}(\ln x) = \frac{d}{dx}\left(\frac{1}{x}\right) = \frac{-1}{x^2}$$

is negative for all $x > 0$.
- Note that $\ln 2 > 0$ because $\int\limits_{1}^{2} \dfrac{dt}{t} > 0$ and because $\ln 2^p = p \ln 2$, it follows that $\lim\limits_{p \to \infty} \ln 2^p = \infty$. But the graph of $f(x) = \ln x$ is always rising, and thus

$$\lim_{x \to \infty} \ln x = \infty$$

Similarly, it can be shown that

$$\lim_{x \to 0^+} \ln x = -\infty$$

- If b is any positive number, there is exactly one number a such that $\ln a = b$ (because the graph of $\ln x$ is always rising for $x > 0$). In particular, we define $x = e$ as the unique number that satisfies $\ln x = 1$.

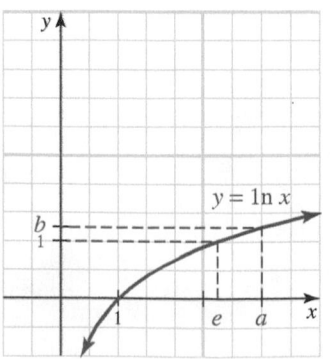

Figure 5.46 Graph of the natural logarithm function

These features are shown in Figure 5.46.

The Natural Exponential Function

Originally, we introduced the natural exponential function e^x and then defined the natural logarithm $\ln x$ as the inverse of e^x. In this alternative approach, we note that because the natural logarithm is an increasing function, it must be one-to-one. Therefore, it has an inverse function, which we denote by $E(x)$.

Because $\ln x$ and $E(x)$ are inverses, we have

$$E(x) = y \qquad \text{if and only if} \qquad \ln y = x$$

From the definition of inverse functions we have

$$E(\ln x) = x \qquad \text{and} \qquad \ln[E(x)] = x$$

We call these formulas the **inversion formulas.** Therefore,

$$
\begin{aligned}
E(0) &= E(\ln 1) & 0 &= \ln 1 & E(1) &= E(\ln e) & 1 &= \ln e \\
&= 1 & E(\ln x) &= x & &= e & E(\ln x) &= x
\end{aligned}
$$

To obtain the graph of $E(x)$, we reflect the graph of $y = \ln x$ in the line $y = x$. The graph is shown in Figure 5.47. Notice that $\ln x = 1$ has the unique solution $x = e$.

The following algebraic properties of the natural exponential function can be obtained by using the properties of the natural logarithm given in Theorem 5.13 along with the inversion formulas.

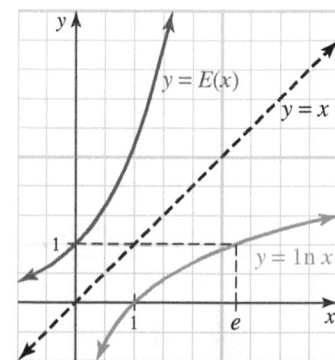

Theorem 5.14   Properties of the exponential as defined by $E(x)$

For any numbers x and y,

a. $E(0) = 1$ **b.** $E(x+y) = E(x)E(y)$ **c.** $E(x-y) = \dfrac{E(x)}{E(y)}$ **d.** $[E(x)]^p = E(px)$

Proof:

a. Proved above (inversion formula).
b. We use the fact that $\ln(AB) = \ln A + \ln B$ to show that

$$
\begin{aligned}
\ln[E(x)E(y)] &= \ln E(x) + \ln E(y) & &\textit{Property of logarithms} \\
&= x + y & &\textit{Inversion formula} \\
&= \ln E(x+y) & &\textit{Inversion formula}
\end{aligned}
$$

Because $\ln x$ is a one-to-one function, we conclude that

$$E(x)E(y) = E(x+y)$$

c. and **d.** The proofs are similar and are left as Problems 53 and 54.   ◆

Figure 5.47 Graph of $E(x)$ is the reflection of the graph of $\ln x$ in the line $y = x$

Next, we use implicit differentiation along with the differentiation formula

$$\frac{d}{dx}(\ln x) = \frac{1}{x}$$

to obtain a differentiation formula for $E(x)$.

Theorem 5.15   Derivative of $E(x)$

The function defined by $E(x)$ is differentiable and

$$\frac{dE}{dx} = E(x)$$

Proof: It can be shown (in advanced calculus) that the inverse of any differentiable function is also differentiable so long as the derivative is never zero. Since $\ln x$ is differentiable, it follows that its inverse $E(x)$ is also differentiable, and using implicit differentiation, we find that

$$\ln y = \ln[E(x)] \quad y = E(x)$$

$$\ln y = x \qquad\qquad \textit{Inversion formula}$$

$$\frac{1}{y}\frac{dy}{dx} = 1 \qquad\qquad \textit{Differentiate implicitly}$$

$$\frac{dy}{dx} = y$$

$$\frac{dE}{dx} = E(x) \qquad \frac{dy}{dx} = \frac{dE}{dx} \textit{ and } E(x) = y$$

$\blacklozenge$

Finally, we observe that if r is a rational number, then

$$\ln\left(e^r\right) = r\ln e = r$$

so that

$$E(r) = e^r$$

This means that $E(x) = e^x$ when x is a rational number, and we **define** $E(x)$ to be e^x for irrational x as well. In particular, we now have an alternative definition for e: $E(1) = e^1 = e$.

PROBLEM SET 5.9

Level 1

1. Write an equation that defines the exponential function with base $b > 0$.
2. Use your definition of Problem 1 to graph the exponential function for:

   a. $b > 1$ **b.** $b = 1$ **c.** $0 < b < 1$

3. ■ What does this say? Define e, approximate e, and then graph $f(x) = e^x$.
4. ■ What does this say? Lewis Carroll (1831-1898) wrote, "When I use a word," Humpty Dumpty said, "it means just what I choose it to mean—nothing more or less." Why do you think anyone would want to define the natural logarithm as an integral?

Simplify each expression in Problems 5-10.

5. $E(x+1)$ 6. $E(x^5)$
7. $E(\ln 5)$ 8. $E(\ln\sqrt{3})$
9. $E(\ln\sqrt{e})$ 10. $E(x+\ln x)$

Write each expression in problems 11-16 in terms of the exponential function $E(x)$ and $\ln x$.

11. 10^x 13. 8^x
12. 2^x 14. x^π 15. $x^{\sqrt{2}}$
                                          16. $2^{\cos x}$

Use the formula

$$\int_1^x \frac{1}{t}\,dt$$

$x > 0$, *to show the properties in Problems* 17-20.

17. $\ln 1 = 0$
18. $\frac{1}{2} < \ln 2 < \frac{3}{4}$
19. $-\frac{1}{2} > \ln\frac{1}{2} > -\frac{3}{4}$
20. $\ln\dfrac{x}{y} = \ln x - \ln y$ for all $x > 0, y > 0$.
21. Use the integral definition

$$\ln x = \int_1^x \frac{dt}{t}$$

to show that $\ln x \to -\infty$ as $x \to 0^+$.
Hint: What happens to $\ln(2^{-N})$ as N grows large without bound?

22. Use an area argument to show that $\ln 2 < 1$.
23. Show that $\ln 3 > 1$, and then explain why $2 < e < 3$.
24. Use a Riemann sum to approximate

$$\ln 3 = \int_1^3 \frac{1}{t}\,dt$$

25. Use a Riemann sum to approximate

$$\ln 5 = \int_1^5 \frac{1}{t}\,dt$$

26. Use Newton's method to approximate e by solving the equation $\ln x - 1 = 0$

Differentiate the functions in Problems 27-36.

27. $\dfrac{d}{dx}\left(\sqrt{x}\ln x\right)$

28. $\dfrac{d}{dx}\ln\sqrt{x}$

29. $\dfrac{d}{dx}(x^3 - 3^x)$

30. $\dfrac{d}{dx}3^{-1/x}$

31. $\dfrac{d}{dx}\ln(\sin x)$

32. $\dfrac{d}{dx}\sin(\ln x)$

33. $\dfrac{d}{dx}\log_2(x^2 - 4)$

34. $\dfrac{d}{dx}x^x$

35. $\dfrac{d}{dx}e^{\ln(\sin x + 1)}$

36. $\dfrac{d}{dx}\ln\left(e^{x^2 - 3\cos 2x}\right)$

Integrate the functions in Problems 37-44.

37. $\displaystyle\int_e^{e^2} \frac{2}{x}\,dx$

38. $\displaystyle\int_{-e^2}^{e} \frac{2}{x}\,dx$

39. $\displaystyle\int_0^2 \frac{dx}{2x + 3}$

40. $\displaystyle\int_0^3 \frac{dx}{3x + 2}$

41. $\displaystyle\int_1^2 10^x\,dx$

42. $\displaystyle\int \frac{\log x}{x}\,dx$

43. $\displaystyle\int \frac{e^x + \cos x}{e^x + \sin x}\,dx$

44. $\displaystyle\int \frac{\sec x}{\sqrt{\ln(\sec x + \tan x)}}\,dx$

Level 2

The product rule for logarithms states

$$\log_b(MN) = \log_b M + \log_b N$$

For this reason, we want the natural logarithm function $\ln x$ to satisfy the functional equation

$$f(xy) = f(x) + f(y)$$

Suppose $f(x)$ is a function that satisfies this equation throughout its domain S. Use this information in Problems 45-49.

45. Show that if $f(1)$ is defined, then $f(1) = 0$.

46. Show that if $f(-1)$ is defined, then $f(-1) = 0$.

47. Show that $f(-x) = f(x)$ for all x in D.

48. If $f'(x)$ is defined for each $x \neq 0$, show that

$$f'(x) = \frac{f'(1)}{x}$$

Then show that

$$f(x) = f'(1)\int_1^x \frac{dt}{t}$$

for $x > 0$.

49. Using the results of Problem 48, conclude that any solution of

$$f(xy) = f(x) + f(y)$$

that is not identically 0 and has a derivative for all $x \neq 0$ must be a multiple of

$$L(x) = \int_1^{|x|} \frac{dt}{t}$$

for $x \neq 0$.

50. Show that for each number A, there is only one number x for which $\ln x = A$.

Hint: If not, then $\ln x = \ln y = A$ for $x \neq y$. Why is this impossible?

51. $f(x) = e^{-3\sin^2(2x^2+1)}$. Find $f'(x)$.

52. Solve $\dfrac{d^2 y}{dx^2} = \sec^2 x$ with $y(0) = 0$ and $y'(0) = 1$.

53. Prove $E(x - y) = \dfrac{E(x)}{E(y)}$.

54. Prove $E(x)^p = E(px)$.

Level 3

55. Use Simpson's rule with $n = 8$ subintervals to estimate

$$\ln 3 = \int_1^3 \frac{dt}{t}$$

Compare your estimate with the value of $\ln 3$ obtained from your calculator.

56. Use Simpson's rule with $n = 8$ subintervals to estimate

$$\ln 5 = \int_1^5 \frac{dt}{t}$$

Compare your estimate with the value of $\ln 5$ obtained from your calculator.

57. Think Tank Problem

 a. Use the error estimate for Simpson's rule to determine the accuracy of the estimate in Problem 55. How many subintervals should be used in Simpson's rule to estimate $\ln 3$ with an error not greater than 0.00005?

 b. Calculate the actual error for your answers in Problem 55 (using $n = 8$). Does this contradict the results in part **a**; why or why not?

58. Use the error estimate for Simpson's rule to determine the accuracy of the estimate in Problem 56. How many subintervals should be used in Simpson's rule to estimate $\ln 5$ with an error not greater than 0.00005?

59. Show that $\ln x^p = p \ln x$ for $x > 0$ and all rational exponents p by completing the following steps.

a. Let $F(x) = \ln x^p$ and $G(x) = p \ln x$. Show that

$$F'(x) = G'(x)$$

for $x > 0$. Conclude that

$$F(x) = G(x) + C$$

b. Let $x = 1$ and conclude that $F(x) = G(x)$. That is,

$$\ln x^p = p \ln x$$

60. Use Rolle's theorem to show that

$$\ln M = \ln N$$

if and only if $M = N$.

Hint: Show that if $M \neq N$, Rolle's theorem implies that

$$\frac{d}{dx}(\ln x) = 0$$

for some number c between M and N. Why is this impossible?

CHAPTER 5 REVIEW

*T*he whole of mathematics consists in the organization of a series of aids to the imagination in the process of reasoning.

A. N. Whitehead

Universal Algebra (Cambridge, 1898), p. 12

Proficiency Examination

Concept Problems

1. What is an antiderivative?
2. **a.** State the integration rule for powers.   **b.** State the exponential rule for integration.
   c. State the integration rules we have learned for the trigonometric functions.
   d. State the integration rules for the inverse trigonometric functions.
3. **a.** What is an area function?   **b.** What do we mean by area as an antiderivative?
   c. What conditions are necessary for an integral to represent an area?
4. What is the formula for area as the limit of a sum?
5. What is a Riemann sum?   **6.** Define a definite integral.
7. Complete these statements summarizing the general properties of the definite integral.

   a. Definite integral at a point: $\displaystyle\int_a^a f(x)\,dx = ?$ **b.** Switching limits of a definite integral: $\displaystyle\int_b^a f(x)\,dx = ?$

8. How can distance traveled by an object be expressed as an integral?
9. **a.** State the first fundamental theorem of calculus.   **b.** State the second fundamental theorem of calculus.
10. Describe in your own words the process of integration by substitution.
11. **a.** What is a differential equation?   **b.** What is a separable differential equation?
    c. How do you solve such an equation?
12. **a.** What is the growth/decay equation?   **b.** Describe carbon dating.
13. What is an orthogonal trajectory?
14. State the mean value theorem for integrals.
15. What is the formula for the average value of a continuous function?
16. State the following approximation rules:

    a. rectangular approximation   **b.** trapezoidal rule   **c.** Simpson's rule

Practice Problems

17. Given that $\displaystyle\int_0^1 \left[f(x)\right]^4 dx = \frac{1}{5}$ and $\displaystyle\int_0^1 \left[f(x)\right]^2 dx = \frac{1}{3}$, find $\displaystyle\int_0^1 \left[f(x)\right]^2 \left\{2\left[f(x)\right]^2 - 3\right\} dx.$

18. Find $F'(x)$ if $\displaystyle F(x) = \int_3^x t^5 \sqrt{\cos(2t+1)}\, dt$

Find the integrals in Problems 19-24.

19. $\displaystyle\int \frac{dx}{1+4x^2}$ 20. $\displaystyle\int xe^{-x^2} dx$ 21. $\displaystyle\int_1^4 \left(\sqrt{x} + x^{-3/2}\right) dx$

22. $\displaystyle\int_0^1 (2x-6)\left(x^2 - 6x + 2\right)^2 dx$ 23. $\displaystyle\int_0^{\frac{\pi}{2}} \frac{\sin x\, dx}{(1+\cos x)^2}$ 24. $\displaystyle\int_{-2}^1 (2x+1)\sqrt{2x^2 + 2x + 5}\, dx$

25. Find the area under the curve $f(x) = 3x^2 + 2$ over $[-1, 3]$.

26. **Journal Problem** (*American Mathematical Monthly*[*]) Approximate the area under the curve $g(x) = e^{x^2}$ over $[1, 2]$.

27. **Journal Problem** (*Scientific American*[†]) In 1995, a team of archaeologists led by Michel Brunet of the University of Poitiers announced the identification in Chad of *Australopithecus* specimens believed to be about 3.5 Myr (million years) old. Using the exponential decay formula, explain why the archaeologists were reluctant to use carbon-14 to date their find.

28. Find the average value of $y = \cos 2x$ on the interval $\left[0, \frac{\pi}{2}\right]$.

29. A slope field for the differential equation $y' = x + y$ is given in Figure 5.48.

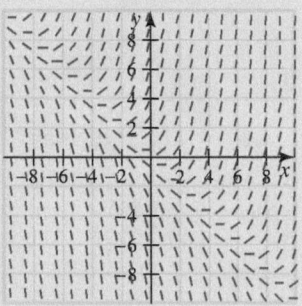

Figure 5.48   Direction field

Draw the particular solutions passing through the points requested in parts **a-d**.

a. $(0, -4)$ b. $(0, 2)$ c. $(4, 0)$ d. $(-4, 0)$

e. All of the curves you have drawn in parts **a-d** seem to have a common asymptote. From the slope field, write the equation for this asymptote.

30. a. Find the necessary n to estimate the value of $\int_0^{\frac{\pi}{2}} \cos x \, dx$ to within 0.0005 of its correct value using the trapezoidal rule?

   b. What is n if Simpson's rule is used?

Supplementary Problems[‡]

1. Given that $\int_{-1}^{0} f(x)\, dx = 3$, $\int_{0}^{1} f(x)\, dx = -1$, and $\int_{-1}^{1} g(x)\, dx = 7$, find $\int_{-1}^{1} \left[3g(x) + 2f(x)\right] dx$.

2. Use the definition of the definite integral to find $\int_{0}^{1} \left(3x^2 + 2x - 1\right) dx$.

3. Use the definition of the definite integral to find $\int_{0}^{1} \left(4x^3 + 6x^2 + 3\right) dx$.

Find the definite and indefinite integrals in Problems 4-35. If you do not have a technique for finding a closed (exact) answer, approximate the integral using numerical integration.

4. $\int_{0}^{1} (5x^4 - 8x^3 + 1)\, dx$

5. $\int_{-1}^{2} 30(5x - 2)^2\, dx$

6. $\int_{0}^{1} (x\sqrt{x} + 2)^2\, dx$

7. $\int_{1}^{2} \dfrac{x^2\, dx}{\sqrt{x^3 + 1}}$

8. $\int_{2}^{2} (x + \sin x)^3 dx$

9. $\int_{-1}^{0} \dfrac{dx}{\sqrt{1 - 2x}}$

---

[*]There is an interesting discussion of this integral in the May 1999 issue of *The American Mathematical Monthly* in an article entitled, "What Is a Closed-Form Number?" The function $\int e^{x^2}\, dx$ is an example of a function that is**not** an elementary function (that we mentioned in Section 1.4).

[†]"Early Hominid Fossils from Africa," by Meave Leakey and Alan Walker, *Scientific American*, June 1997, p. 79

[‡]The supplementary problems are presented in a somewhat random order, not necessarily in order of difficulty.

10. $\displaystyle\int_1^2 \frac{dx}{\sqrt{3x-1}}$ **11.** $\displaystyle\int_{-1}^0 \frac{dx}{\sqrt[3]{1-2x}}$ **12.** $\displaystyle\int \frac{1}{x^3}\,dx$

13. $\displaystyle\int \frac{5x^2-2x+1}{\sqrt{x}}\,dx$ **14.** $\displaystyle\int \frac{x+1}{2x}\,dx$ **15.** $\displaystyle\int \frac{3-x}{\sqrt{1-x^2}}\,dx$

16. $\displaystyle\int (e^{-x}+1)e^x\,dx$ **17.** $\displaystyle\int \frac{\sin x - \cos x}{\sin x + \cos x}\,dx$ **18.** $\displaystyle\int (x-1)^2\,dx$

19. $\displaystyle\int \sqrt{x}(x^2+\sqrt{x}+1)\,dx$ **20.** $\displaystyle\int \frac{x^2+1}{x^2}\,dx$ **21.** $\displaystyle\int (\sin^2 x + \cos^2 x)\,dx$

22. $\displaystyle\int x(x+4)\sqrt{x^3+6x^2+2}\,dx$ **23.** $\displaystyle\int x(2x^2+1)\sqrt{x^4+x^2}\,dx$ **24.** $\displaystyle\int \frac{dx}{\sqrt{x}(\sqrt{x}+1)^2}$

25. $\displaystyle\int x\sqrt{1-5x^2}\,dx$ **26.** $\displaystyle\int x^3 \cos\left(\tan^{-1}x\right)\,dx$ **27.** $\displaystyle\int \frac{\ln x}{x}\sqrt{1+\ln(2x)}\,dx$

28. $\displaystyle\int_{-10}^{10} \left[3+7x^{73}-100x^{101}\right]dx$ **29.** $\displaystyle\int_{-\frac{\pi}{4}}^{\frac{\pi}{4}} (\sin 4x + 2\cos 4x)\,dx$ **30.** $\displaystyle\int_1^2 x^{2a}\,dx,\; a \neq -\frac{1}{2}$

31. $\displaystyle\int_{e-2}^e \frac{dx}{x}$ **32.** $\displaystyle\int_0^1 (1-e^t)\,dt$ **33.** $\displaystyle\int_1^2 \frac{x^3+1}{x}\,dx$

34. $\displaystyle\int \sqrt{\sin x - \cos x}\,(\sin x + \cos x)\,dx$ **35.** $\displaystyle\int (\cos^{11}x \sin^9 x - \cos^9 x \sin^{11}x)\,dx$

In Problems 36-41, draw the indicated particular solution for the differential equations whose direction fields are given.

36. $y(3)=0$ **37.** $y(-2)=-2$ **38.** $y(0)=1$

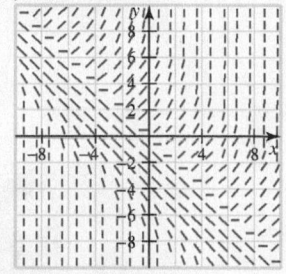 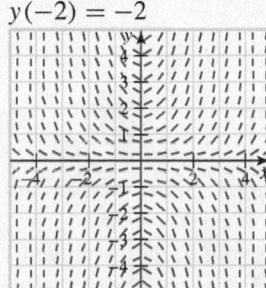 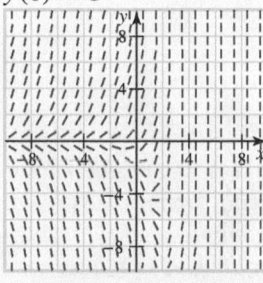

39. $y(1)=-1$ **40.** $y(2)=0$ **41.** $y(0)=-2$

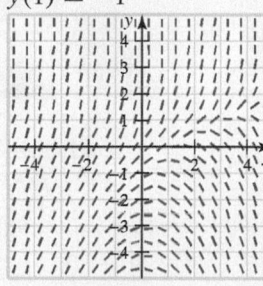 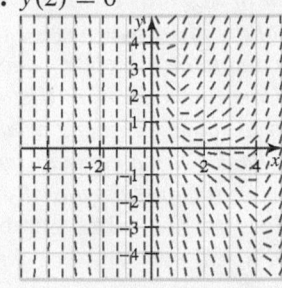 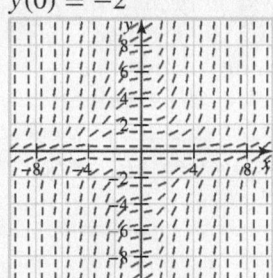

42. Find $F'(x)$, where $F(x) = \displaystyle\int_5^x t^2 \cos^4 t\,dt$.

43. Find the area above the x-axis and under the curve $f(x) = x^{-1}$ on $[1,4]$.

44. Find the area above the x-axis and under the curve $f(x) = 2 + x - x^2$ on $[-1,1]$.

45. Find the area above the x-axis and under the curve $f(x) = e^{4x}$ on $[0,2]$.

46. Find the area bounded by the curve $y = x\sqrt{x^2+5}$, the x-axis, and the vertical lines $x = -1$ and $x = 2$.

47. Find $f(t)$ if $f''(t) = \sin 4t - \cos 2t$ and $f\left(\frac{\pi}{2}\right) = f'\left(\frac{\pi}{2}\right) = 1$.

48. Find $f(x)$ if $f'''(x) = 2x^3 + x^2$, given that $f''(1) = 2$, $f'(1) = 1$, and $f(1) = 0$.

Solve the differential equations in Problems 49-58.

49. $\dfrac{dy}{dx} = (1-y)^2$ **50.** $\dfrac{dy}{dx} = \dfrac{\cos 4x}{y}$ **51.** $\dfrac{dy}{dx} = \left(\dfrac{\cos y}{\sin x}\right)^2$ **52.** $\dfrac{dy}{dx} = \dfrac{x}{y}$ **53.** $\dfrac{dy}{dx} = y(x^2+1)$

54. $\dfrac{dy}{dx} = \dfrac{x}{y}\sqrt{\dfrac{y^2+2}{x^2+1}}$ **55.** $\dfrac{dy}{dx} = \dfrac{\cos^2 y}{\cot x}$ **56.** $\dfrac{dy}{dx} = \sqrt{\dfrac{x}{y}}$ **57.** $\dfrac{dy}{dx} = (y-4)^2$ **58.** $\dfrac{dy}{dx} = \dfrac{y}{\sqrt{1-2x^2}}$

59. Find the average value of $f(x) = \dfrac{\sin x}{\cos^2 x}$ on the interval $\left[0, \frac{\pi}{4}\right]$.

60. Find the average value of $f(x) = \sin x$

    a. on $[0, \pi]$ **b.** on $[0, 2\pi]$

61. Use the trapezoidal rule with $n = 6$ to approximate

$$\int_0^\pi \sin x \, dx$$

Compare your result with the exact value of this integral.

62. Estimate $\displaystyle\int_0^1 \sqrt{1+x^3}\, dx$ using the trapezoidal rule with $n = 6$.

63. Estimate $\displaystyle\int_0^1 \dfrac{dx}{\sqrt{1+x^3}}$ using the trapezoidal rule with $n = 8$.

64. Estimate $\displaystyle\int_0^1 \sqrt{1+x^3}\, dx$ using Simpson's rule with $n = 6$.

65. Use the trapezoidal rule to approximate $\displaystyle\int_0^1 \dfrac{x^2\, dx}{1+x^2}$ with an error no greater than 0.005.

66. Use the trapezoidal rule to estimate to within 0.00005 the value of the integral

$$\int_1^2 \sqrt{x + \dfrac{1}{x}}\, dx$$

67. 𝔥istorical 𝔔uest

*Karl Gauss is considered to be one of the four greatest mathematicians of all time, along with Archimedes (𝔥istorical 𝔔uest Problem 52, Section 3.8), Newton (𝔥istorical 𝔔uest Supplementary Problem 87, Chapter 1), and Euler (𝔥istorical 𝔔uest Supplementary Problem 93, Chapter 4). Gauss graduated from college at the age of 15 and proved what was to become the fundamental theorem of algebra for his doctoral thesis at the age of 22. He published only a small portion of the ideas that seemed to storm his mind, because he believed that each published result had to be complete, concise, polished, and convincing. His motto was "Few, but ripe." Carl B. Boyer, in **A History of Mathematics**, describes Gauss as the last mathematician to know everything in his subject, and we could relate Gauss to nearly every topic of this book. Such a generalization is bound to be inexact, but it does emphasize the breadth of interest Gauss displayed. A **prime number**, p, is a counting number that has exactly two divisors. The **prime number theorem**, conjectured by Gauss in 1793, says that the number of primes $\pi(x)$ less than a real number x is a function that behaves like $x/\ln x$ as $x \to \infty$; that is,*

**Karl Gauss
(1777-1855)**

$$\lim_{x \to \infty} \dfrac{\pi(x)}{\frac{x}{\ln x}} = 1$$

Gauss estimated the prime distribution function $\pi(x)$ by the integral

$$G(x) = \int_2^x \dfrac{dt}{\ln t}$$

for $x \geq 2$ known as the integral logarithm. *History records many who constructed tables for $\pi(n)$, and Figure 5.49 shows values for $\pi(n)$ for various choices of n.*

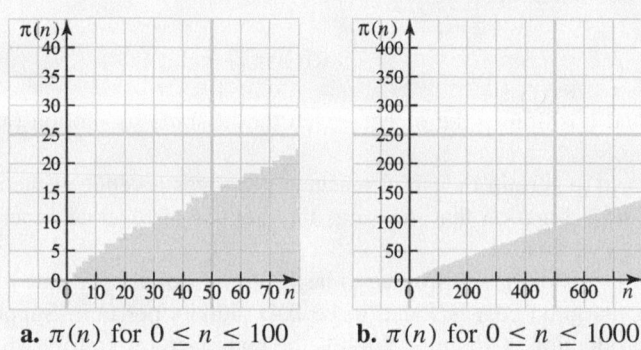

a. $\pi(n)$ for $0 \le n \le 100$ **b.** $\pi(n)$ for $0 \le n \le 1000$

Figure 5.49   Graphs of $\pi(n)$

Your 𝒬*uest* is to examine this estimate by using the trapezoidal rule to approximate $G(x)$ for $x = 100$, $1{,}000$, and $10{,}000$. The actual value of $\pi(x)$ for selected values of x are as follows:

n	π
10^2	25
10^3	168
10^4	1,229
10^5	9,592
10^6	78,498
10^{10}	455,052,511

68. 𝔥*istorical* 𝒬*uest*

There are many ways to estimate the prime distribution function $\pi(x)$ defined in Problem 67 other than by the integral function $G(x)$ used by Gauss. Adrien-Marie Legendre is best known for his work with elliptic integrals and mathematical physics. In 1794, he proved that π was an irrational number and formulated a conjecture that is equivalent to the prime number theorem. He made his estimate using the function

$$\pi(x) \approx L(x) = \frac{x}{\ln x - 1.08366}$$

For this 𝒬*uest*, use $L(x)$ to estimate $\pi(x)$ for $x = 100$, $1{,}000$, and $10{,}000$, and then compare with the results found using $G(x)$ in Problem 67.

Adrien-Marie Legendre (1752-1833)

69. Approximate the average value of the function defined by

$$f(x) = \frac{\cos x}{1 - \frac{x^2}{2}}$$

on $[0, 1]$. *Hint*: Use the trapezoidal rule with $n = 6$.

70. The brakes of a certain automobile produce a constant deceleration of k m/s^2. The car is traveling at 25 m/s when the driver is forced to hit the brakes. If it comes to rest at a point 50 m from the point where the brakes are applied, what is k?

71. A particle moves along the t-axis in such a way that $a(t) = -4s(t)$, where $s(t)$ and $a(t)$ are its position and acceleration, respectively, at time t. The particle starts from rest at $s = 5$.

a. Show that $v^2 + 4s^2 = 100$ where $v(t)$ is the velocity of the particle. *Hint*: First use the chain rule to show that $a(t) = v\left(\dfrac{dv}{ds}\right)$.

b. What is the velocity when the particle first reaches $s = 3$?

Note: At the time in question, the sign of v is determined by the direction the particle is moving.

72. An object experiences linear acceleration given by

$$a(t) = 2t + 1 \quad \text{ft/s}^2$$

Find the velocity and position of the object, given that it starts its motion (at $t = 0$) at $s = 4$ with initial velocity 2 ft/s.

73. When it is x years old, a certain industrial machine generates revenue at the rate of $R'(x) = 1{,}575 - 5x^2$ thousand dollars per year. Find a function that measures the amount of revenue generated in the first x years, and find the revenue for the first five years.

74. A manufacturer estimates marginal revenue to be $100x^{-1/3}$ dollars per unit when the level of production is x units. The corresponding marginal cost is found to be $0.4x$ dollars per unit. Suppose the manufacturer's profit is \$520 when the level of production is 16 units. What is the manufacturer's profit when the level of production is 25 units?

75. A tree has been transplanted and after t years is growing at a rate of

$$1 + \frac{1}{(t+1)^2}$$

feet per year. After 2 years it has reached a height of 5 ft. How tall was the tree when it was transplanted?

76. A manufacturer estimates that the marginal revenue of a certain commodity is

$$R'(x) = \sqrt{x}(x^{3/2} + 1)^{-1/2}$$

dollars per unit when x units are produced. Assuming no revenue is obtained when $x = 0$, how much revenue is obtained from producing $x = 4$ units?

77. Find a function whose tangent has slope $x\sqrt{x^2 + 5}$ for each value of x and whose graph passes through the point $(2, 10)$.

78. A particle moves along the x-axis in such a way that after t seconds its acceleration is $a(t) = 12(2t + 1)^{-3/2}$. If it starts at rest at $x = 3$, where will it be 4 seconds later?

79. An environmental study of a certain community suggests that t years from now, the level of carbon monoxide in the air will be changing at the rate of $0.1t + 0.2$ parts per million per year. If the current level of carbon monoxide in the air is 3.4 parts per million, what will the level be 3 years from now?

80. A woman, driving on a straight, level road at the constant speed v_0, is forced to apply her brakes to avoid hitting a deer. The car comes to a stop 3 seconds later, s_0 ft from the point where the brakes were applied. Continuing on her way, she increases her speed by 20 ft/s to make up time but is again forced to hit the brakes, and this time it takes her 5 seconds and s_1 feet to come to a full stop. Assuming that her brakes supplied a constant deceleration k ft/s^2 each time they were used, find k and determine v_0, s_0, and s_1.

81. A study indicates that x months from now the population of a certain town will be increasing at the rate of $10 + 2\sqrt{x}$ people per month. By how much will the population of the town increase over the next 9 months?

82. It is estimated that t days from now a farmer's crop will be increasing at the rate of $0.3t^2 + 0.6t + 1$ bushels per day. By how much will the value of the crop increase during the next 6 days if the market price remains fixed at \$20 per bushel?

83. Records indicate that t months after the beginning of the year, the price of turkey in local supermarkets was

$$P(t) = 0.06t^2 - 0.2t + 1.2$$

dollars per pound. What was the average price of turkey during the first 6 months of the year?

84. Modeling Problem   V. A. Tucker and K. Schmidt-Koenig have investigated the relationship between the velocity v (km/h) of a bird in flight and the energy $E(v)$ expended by the bird.* Their study showed that for a certain kind of parakeet, the rate of change of the energy expended with respect to velocity is modeled (for $v > 0$) by

$$\frac{dE}{dv} = \frac{0.074v^2 - 112.65}{v^2}$$

---

*Adapted from "Flight Speeds of Birds in Relation to Energies and Wind Directions," by V. A. Tucker and K. Schmidt-Koenig. *The Auk*, Vol. 88 (1971), pp. 97-107.

a. What is the most economical velocity for the parakeet? That is, find the velocity v_0 that minimizes the energy.

b. Suppose it is known that $E = E_0$ when $v = v_0$. Express E in terms of v_0 and E_0.

c. Express the average energy expended as the parakeet's velocity ranges from $v = \frac{1}{2}v_0$ to $v = v_0$ in terms of E_0.

85. **Modeling Problem** A toxin is introduced into a bacterial culture, and t hours later, the population $P(t)$ of the culture, in millions, is found to be changing at the rate

$$\frac{dP}{dt} = -(\ln 2)\, 2^{5-t}$$

If there were 1 million bacteria in the culture when the toxin was introduced, when will the culture die out?

86. **Modeling Problem The Snowplow Problem of R. P. Agnew*** One day it starts snowing at a steady rate sometime before noon. At noon, a snowplow starts to clear a straight, level section of road. If the plow clears 1 mile of road during the first hour but requires 2 hr to clear the second mile, at what time did it start snowing? Answer this question by completing the following steps:

a. Let t be the time (in hours) from noon. Let h be the depth of the snow at time t, and let s be the distance moved by the plow. If the plow has width w and clears snow at the constant rate p, explain why $wh\dfrac{ds}{dt} = p$.

b. Suppose it started snowing t_0 hours before noon. Let r denote the (constant) rate of snowfall. Explain why

$$h(t) = r(t + t_0)$$

By combining this equation with the differential equation in part **a**, note that

$$wr(t + t_0)\frac{ds}{dt} = p$$

Solve this differential equation (with appropriate conditions) to obtain t_0 and answer the question posed in the problem.

87. **a.** If f is continuous on $[a,b]$, show that

$$\left| \int_a^b f(x)\, dx \right| \leq \int_a^b |f(x)|\, dx$$

b. Show that $\left| \displaystyle\int_1^4 \frac{\sin x}{x}\, dx \right| \leq \frac{3}{2}$.

88. A company plans to hire additional advertising personnel. Suppose it is estimated that if x new people are hired, they will bring in additional revenue of $R(x) = \sqrt{2x}$ thousand dollars and that the cost of adding these x people will be $C(x) = \frac{1}{3}x$ thousand dollars. How many new people should be hired? How much total net revenue (that is, revenue minus cost) is gained by hiring these people?

89. The half-life of the radioactive isotope cobalt-60 is 5.25 years.

a. What percentage of a given sample of cobalt-60 remains after 5 years?

b. How long will it take for 90% of a given sample to disintegrate?

90. The rate at which salt dissolves in water is directly proportional to the amount that remains undissolved. If 8 lb of salt is placed in a container of water and 2 lb dissolves in 30 min, how long will it take for 1 lb to dissolve?

91. Scientists are observing a species of insect in a certain swamp region. The insect population is estimated to be 10 million and is expected to grow at the rate of 2% per year. Assuming that the growth is exponential and stays that way for a period of years, what will the insect population be in 10 yr? How long will it take to double?

92. Find an equation for the tangent line to the curve $y = F(x)$ at the point P where $x = 1$ and

$$F(x) = \int_1^{\sqrt{x}} \frac{2t+1}{t+2}\, dt$$

---

*There is an interesting discussion of this problem, along with a BASIC computer simulation, in the November 1995 (Vol. 26, No. 5) issue of *The College Mathematics Journal* in the article "The Meeting of the Plows: A Simulation" by Jerome Lewis (pp. 395-400).

93. The radius of planet X is one-fourth that of planet Y, and the acceleration due to gravity at the surface of X is eight-ninths that at the surface of Y. If the escape velocity of planet X is 6 ft/s, what is the escape velocity of planet Y?

94. A survey indicates that the population of a certain town is growing in such a way that the rate of growth at time t is proportional to the square root of the population P at the time. If the population was 4,000 ten years ago and is observed to be 9,000 now, how long will it take before 16,000 people live in the town?

95. The radioactive isotope gallium-67 (symbol ^{67}Ga) used in the diagnosis of malignant tumors has a half-life of 46.5 hours. If we start with 100 mg of ^{67}Ga, what percent is lost between the 30th and 35th hours? Is this the same as the percent lost over any other 5-hour period?

96. Modeling Problem A projectile is launched from the surface of a planet whose radius is R and where the acceleration due to gravity at the surface is g.

a. If the initial velocity v_0 of the projectile satisfies $v_0 < \sqrt{2gR}$, show that the maximum height above the surface of the planet reached by the projectile is

$$h = \frac{v_0^2 R}{2gR - v_0^2}$$

b. On a certain planet, it is known that $g = 25$ ft/s^2. A projectile is fired with an initial velocity of $v_0 = 2$ mi/s and attains a maximum height of 450 mi. What is the radius of the planet?

97. Solve the system of differential equations

$$\begin{cases} 2\dfrac{dx}{dt} + 5\dfrac{dy}{dt} = t \\ \dfrac{dx}{dt} + 3\dfrac{dy}{dt} = 7\cos t \end{cases}$$

Hint: Solve for $\dfrac{dx}{dt}$ and $\dfrac{dy}{dt}$ algebraically, and then integrate.

98. Putnam examination problem Where on the parabola $4ay = x^2$ ($a > 0$) should a chord be drawn so that it will be normal to the parabola and cut off a parabolic sector of minimum area? That is, find P so that the shaded area in Figure 5.50 is as small as possible.

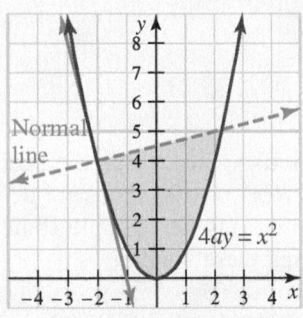

Figure 5.50 Minimum area problem

99. Putnam examination problem If $a_0, a_1, \cdots, a_n$ are real numbers that satisfy

$$\frac{a_0}{1} + \frac{a_1}{2} + \cdots + \frac{a_n}{n+1} = 0$$

show that the equation $a_0 + a_1 x + a_2 x^2 + \cdots + a_n x^n = 0$ has at least one real root.
Hint: Use the mean value theorem for integrals.

CHAPTER 5 GROUP RESEARCH PROJECT

Working in small groups is typical of most work environments, and this book seeks to develop skills with group activities. We present a group project at the end of each chapter. These projects are to be done in groups of three or four students.

Ups and Downs

You have been hired by Two Flags to help with the design of their new roller coaster. You are to plan the path design of ups and downs for a straight stretch with horizontal length of 185 ft. One possible design is shown in Figure 5.51.*

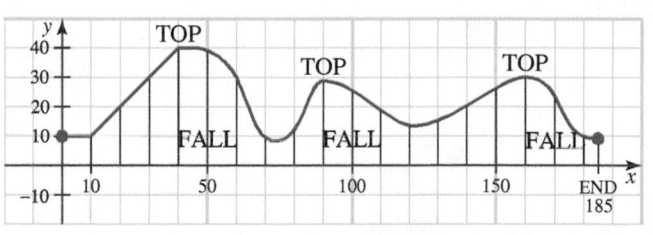

Figure 5.51   Roller Coaster path

Here is the information you will need:

- There must be a support every 10 ft.
- The descent can be no steeper than 80° at any point (angles refer to the angle that the path makes with a horizontal).
- The design must start with a 45° incline.
- The amount of material needed for support is the square of the height of the support. For example, a support that is 20 ft high requires $20^2 = 400$ ft of material.

We also define the **thrill** of the coaster as the sum of the angle of steepest descent in each fall in radians plus the number of tops.

You r paper should include, but is not limited to, the following concerns:

- Your paper should address where the path is increasing at an increasing rate, and decreasing at a decreasing rate.
- Your paper should address the safety criteria and show the graphs of the path, the slope, and the rate of change of the slope.

---

*This group project is one adapted from Diane Schwartz of Ithaca College, Ithaca, N.Y.

Cumulative Review Problems—Chapters 1-5

1. ■ What does this say? Define limit. Explain what this definition is saying using your own words.
2. ■ What does this say? Define derivative. Explain what this definition is saying using your own words.
3. ■ What does this say? Define a definite integral. Explain what this definition is saying using your own words.
4. ■ What does this say? Define a differential equation, and in your own words describe the procedure for solving a separable differential equation.

Evaluate the limits in Problems 5-13.

5. $\displaystyle \lim_{x \to 2} \frac{3x^2 - 5x - 2}{3x^2 - 7x + 2}$

6. $\displaystyle \lim_{x \to \infty} \frac{3x^2 + 7x + 2}{5x^2 - 3x + 3}$

7. $\displaystyle \lim_{x \to \infty} \left(\sqrt{x^2 + x} - x \right)$

8. $\displaystyle \lim_{x \to \frac{\pi}{2}} \frac{\cos^2 x}{\cos x^2}$

9. $\displaystyle \lim_{x \to 0} \frac{x \sin x}{x + \sin^2 x}$

10. $\displaystyle \lim_{x \to 0} \frac{\sin 3x}{x}$

11. $\displaystyle \lim_{x \to \infty} (1 + x)^{2/x}$

12. $\displaystyle \lim_{x \to 0} \frac{\tan^{-1} x - x}{x^3}$

13. $\displaystyle \lim_{x \to 0^+} x^{\sin x}$

Find the derivatives in Problems 14-22.

14. $y = 6x^3 - 4x + 2$

15. $y = (x^2 + 1)^3 (3x - 4)^2$

16. $y = \dfrac{x^2 - 4}{3x + 1}$

17. $x^2 + 3xy + y^2 = 9$

18. $y = \ln(5x^2 + 3x - 2)$

19. $y = (\sin x + \cos x)^3$

20. $y = \sqrt{\dfrac{x^3 - x}{1 - x^2}}$

21. $y = e^{-x} \log_5 3x$

22. $y = \sqrt{\cos \sqrt{x}}$

Find the integrals in Problems 23-28.

23. $\displaystyle \int_4^9 d\theta$

24. $\displaystyle \int_{-1}^{1} 50(2x - 5)^3 \, dx$

25. $\displaystyle \int_0^1 \frac{x \, dx}{\sqrt{9 + x^2}}$

26. $\displaystyle \int \csc 3\theta \cot 3\theta \, d\theta$

27. $\displaystyle \int \frac{e^x \, dx}{e^x + 2}$

28. $\displaystyle \int \frac{x^3 + 2x - 5}{x} \, dx$

29. Approximate $\displaystyle \int_0^4 \frac{dx}{\sqrt{1 + x^3}}$ using Simpson's rule with $n = 6$.

Find the equation of the tangent line in Problems 30-33.

30. $y = 2^x - \log_2 x$, where $x = 1$

31. $y = \dfrac{3x - 4}{3x^2 + x - 5}$ where $x = 1$

32. $x^{2/3} + y^{2/3} = 2$ at $(1, 1)$

33. $y = (x^3 - 3x^2 + 3)^3$, where $x = -1$

Solve the differential equations in Problems 34-35.

34. $\dfrac{dy}{dx} = x^2 y^2 \sqrt{4 - x^3}$

35. $(1 + x^2) \, dy = (x + 1)y \, dx$

Find the particular solution of the differential equations in Problems 36-37.

36. $\dfrac{dy}{dx} = 2(5 - y), y = 3$ when $x = 0$

37. $\dfrac{dy}{dx} = e^y \sin x, y = 5$ when $x = 0$

38. Use the chain rule to find $\dfrac{dy}{dt}$ when $y = x^3 - 7x$ and $x = t \sin t$.

39. Let $f(x) = 3x^2 + 1$ for all x. Use the chain rule to find $\dfrac{d}{dx} (f \circ f)(x)$.

40. Find f''' if $f(x) = x(x^2 + 1)^{7/2}$.

41. Find numbers a, b, and c that guarantee that the graph of the function

$$f(x) = ax^2 + bx + c$$

will have x-intercepts at $(0,0)$ and $(5,0)$ and a tangent line with slope 1, where $x = 2$.

42. Which of the following functions satisfy $y'' + y = 0$?

    a. $y_1 = 2 \sin x + 3 \cos x$ **b.** $y_2 = 4 \sin x - \pi \cos x$ **c.** $y_3 = x \sin x$ **d.** $y_4 = e^x \cos x$

43. Sketch the graph of $y = x^3 - 5x^2 + 2x + 8$.

44. Sketch the graph of $y = \dfrac{4 + x^2}{4 - x^2}$.

45. Find the largest value of $f(x) = \frac{1}{3}x^3 - 2x^2 + 3x - 10$ on $[0, 6]$.

46. Under the condition that $2x - 5y = 18$, minimize $x^2 y$ when $x \geq 0$ and $y \leq 0$.

47. Use differentials to estimate the change in the volume of a cone with a circular base if the height of the cone is increased from 10 cm to 10.01 cm while the radius of the base stays fixed at 2 cm.

48. Sketch the graph of a function with the following properties: There are relative extrema at $(-1, 7)$ and $(3, 2)$. There is an inflection point at $(1, 4)$. The graph is concave down only when $x < 1$. The x-intercept is $(-4, 0)$ and the y-intercept is $(0, 5)$.

49. The graph of a function f consists of a semicircle of radius 3 and two line segments as shown in Figure 5.52. Let F be the function defined by

$$F(x) = \int_0^x f(t)\, dt$$

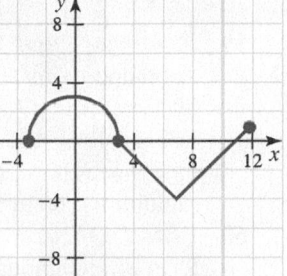

    a. Find $F(7)$.

    b. Find all values on the interval $(-3, 12)$ at which F has a relative maximum.

    c. Write an equation for the line tangent to the graph of F at $x = 7$.

    d. Find the x-coordinate of each point of inflection of the graph of F on the interval $(-3, 12)$.

Figure 5.52 Graph of f

50. An electric charge Q_0 is distributed uniformly over a ring of radius R. The electric field intensity at any point x along the axis of the ring is given by

$$E(x) = \frac{Q_0 x}{(x^2 + R^2)^{3/2}}$$

At what point on the axis is $E(x)$ maximized?

51. A rocket is launched vertically from a point on the ground 5,000 ft from a TV camera. If the rocket is rising vertically at the rate of 850 ft/s at the instant the rocket is 4,000 ft above the ground, how fast must the TV camera change the angle of elevation to keep the rocket in the picture?

52. A particle moves along the x-axis so that its velocity at any time $t \geq 0$ is given by $v(t) = 6t^2 - 2t - 4$. It is known that the particle is at position $x = 6$ for $t = 2$.

    a. Find the position of the particle at any time $t \geq 0$.

    b. For what value(s) of t, $0 \leq t \leq 3$, is the particle's instantaneous velocity the same as its average velocity over the interval $[0, 3]$.

    c. Find the total distance traveled by the particle from time $t = 0$ to $t = 3$.

53. A car and a truck leave an intersection at the same time. The car travels north at 60 mi/h and the truck travels east at 45 mi/h. How fast is the distance between them changing after 45 minutes?

54. A light is 4 miles from a straight shoreline. The light revolves at the rate of 2 rev/min. Find the speed of the spot of light along the shore when the light spot is 2 miles past the point on the shore closest to the source of light.

55. Let f be the function defined by $f(x) = 2\cos x$. Let $P(0,2)$ and $Q\left(\frac{\pi}{2}, 0\right)$ be the points where f crosses the y-axis and x-axis, respectively.

   a. Write an equation for the secant line passing through points P and Q.

   b. Write an equation for the tangent line of f at point Q.

   c. Find the x-coordinate of the point on the graph of f, between point P and Q, at which the tangent to the graph of f is parallel to $\overline{PQ}$. Cite a theorem or result that assures the existence of such a tangent.

56. Let f be the function defined by $f(x) = x^3 - 6x^2 + k$ for k an arbitrary constant.

   a. Find the relative maximum and minimum of f in terms of k.

   b. For which values of k does f have three distinct roots?

   c. Find the values of k for which the average value of f over $[-1, 2]$ is 2.

57. A scientist has discovered a radioactive substance that disintegrates in such a way that at time t, the rate of disintegration is proportional to the *square* of the amount present.

   a. If a 100-g sample of the substance dwindles to only 80 g in 1 day, how much will be left after 6 days?

   b. When will only 10 g be left?

58. Find an equation for the tangent line to the curve $y = F(x)$ at the point P where $x = 1$ and

$$F(x) = \int_1^{x^3} \frac{t^2 + 1}{t - 2}\, dt$$

59. *Historical Quest* "Lucy," the famous prehuman whose skeleton was discovered in East Africa, has been found to be approximately 3.8 million years old. About what percentage of ^{14}C would you expect to find if you tried to "date" Lucy by the usual carbon dating procedure? The answer you get to this question illustrates why it is reasonable to use carbon dating only on more recent artifacts, usually less than 50,000 years old (roughly the time since the last major ice age). Read an article on alternative dating procedures such as potassium-argon and rubidium-strontium dating. Write a paper comparing and contrasting such methods.

60. *Historical Quest* As we saw in a *Quest* Problem 51 in Section 5.8, the mathematician Seki Kōwa was doing a form of integration at about the same time that Newton and Leibniz were inventing the calculus. Write a paper on the history of calculus from the Eastern viewpoint.

CHAPTER 6

ADDITIONAL APPLICATIONS OF THE INTEGRAL

*T*he only way to learn mathematics is to do mathematics.

Paul Halmos
Hilbert Space Problem Book

PREVIEW

In Chapter 5, we found that area and average value can be expressed in terms of the definite integral. The goal of this chapter is to consider various other applications of integration such as computing volume, arc length, surface area, work, hydrostatic force, centroids of planar regions, and applications to business, economics, and life sciences.

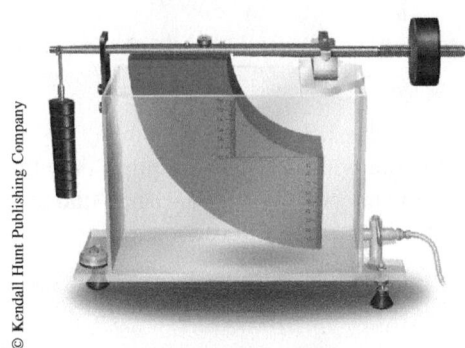

© Kendall Hunt Publishing Company

Apparatus for measuring hydrostatic pressure

PERSPECTIVE

What is the volume of the material that remains when a hole of radius $0.5R$ is drilled into a sphere of radius R? The "center" of a rectangle is the point where its diagonals intersect, but what is the center of a quarter circle? If a reservoir is filled to the top of a dam shaped like a parabola, what is the total force of the water on the face of the dam? If a worker is carrying a bag of sand up a ladder and the bag is leaking sand through a hole in such a way that all the sand is gone when the worker reaches the top, how much work is done by the worker? An investor asks a question seeking the future value of an income flow, or a physician asks how to measure the flow of blood through an artery. These questions are typical of the kind we shall answer in this chapter.

421

6.1 AREA BETWEEN TWO CURVES

IN THIS SECTION: *Area between curves, area using vertical strips, area using horizontal strips*
We have seen how the area between the curves defined by $y = f(x)$ and the x-axis can be computed by the integral $\int_a^b f(x)\,dx$ on an interval $a \leq x \leq b$ where $f(x) \geq 0$. In this section, we find that integration can be used to find the area of more general regions between curves.

Area Between Curves

In some practical problems, you may have to compute the area between two curves. Suppose f and g are functions such that $f(x) \geq g(x)$ on the interval $[a, b]$, as shown in Figure 6.1. We do not insist that both f and g be nonnegative functions, but we begin by showing that case in Figure 6.1.

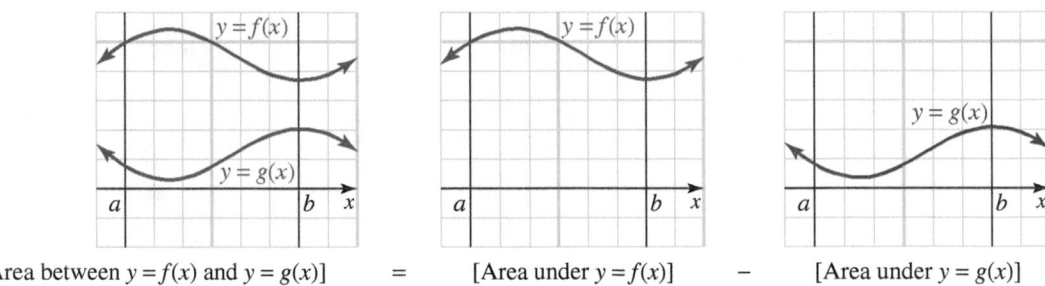

[Area between $y = f(x)$ and $y = g(x)$] = [Area under $y = f(x)$] − [Area under $y = g(x)$]

Figure 6.1 Area between two curves

To find the area of the region R between the curves defined by $y = f(x)$ and $y = g(x)$ from $x = a$ to $x = b$, we subtract the area between the lower curve $y = g(x)$ and the x-axis from the area between the upper curve $y = f(x)$ and the x-axis; that is,

$$\text{AREA OF } R = \int_a^b f(x)\,dx - \int_a^b g(x)\,dx = \int_a^b \left[f(x) - g(x) \right] dx$$

This formula seems obvious in the situation where both $f(x) \geq 0$ and $g(x) \geq 0$, as shown in Figure 6.1. However, the following derivation requires only that f and g be continuous and satisfy $f(x) \geq g(x)$ on the interval $[a, b]$. We wish to find the area between the curves $y = f(x)$ and $y = g(x)$ on this interval. Choose a partition

$$\{x_0 = a, x_1, x_2, \cdots, x_n = b\}$$

of the interval $[a, b]$ and a representative number x_k^* from each subinterval $\left[x_{k-1}, x_k \right]$. Next, for each index k, with $k = 1, 2, \cdots, n$, construct a rectangle of width

$$\Delta x_k = x_k - x_{k-1}$$

and height

$$f(x_k^*) - g(x_k^*)$$

equal to the vertical distance between the two curves at $x = x_k^*$. A typical approximating rectangle is shown in Figure 6.2. We refer to this approximating rectangle as a **vertical strip**.

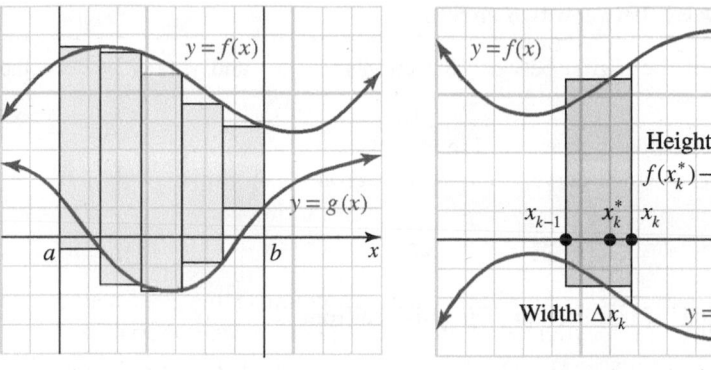

a. Approximating rectangles   **b.** A typical vertical strip

Figure 6.2   Using Riemann sums to find the area between two curves

The representative rectangle has area

$$\Delta A_k = \left[f(x_k^*) - g(x_k^*)\right] \Delta x_k$$

and the total area between the curves $y = f(x)$ and $y = g(x)$ can be estimated by the Riemann sum

$$A_n = \sum_{k=1}^{n} \left[f(x_k^*) - g(x_k^*)\right] \Delta x_k$$

It is reasonable to expect this estimate to improve if we increase the number of subdivision points in the partition P in such a way that the norm $\|P\|$ approaches 0. Thus, the region between the two curves has area

$$A = \lim_{\|P\| \to 0} \sum_{k=1}^{n} \left[f(x_k^*) - g(x_k^*)\right] \Delta x_k$$

which we recognize as the integral of the function $f(x) - g(x)$ over the interval $[a, b]$. These observations may be used to define the area between the curves.

AREA BETWEEN TWO CURVES   If f and g are continuous and satisfy $f(x) \geq g(x)$ on the closed interval $[a, b]$, then the **area between the two curves** $y = f(x)$ and $y = g(x)$ is given by

$$A = \int_a^b \left[f(x) - g(x)\right] dx$$

■ **W**hat this says To find the area between two curves on a given closed interval $[a, b]$ use the formula

$$A = \int_a^b [\text{TOP CURVE} - \text{BOTTOM CURVE}]\, dx$$

It is no longer necessary to require either curve to be above the x-axis. In fact, we will see later in this section that the curves might cross somewhere in the domain so that one curve is on top for part of the interval and the other curve is on top for the rest.

Example 1   Area between two curves

Find the area of the region between the curves $y = x^3$ and $y = x^2 - x$ on the interval $[0, 1]$.

Solution   The region is shown in Figure 6.3.
We need to know which curve is the *top curve* on $[0, 1]$. Solve

$$x^3 = x^2 - x \qquad \text{or} \qquad x(x^2 - x + 1) = 0$$

to find the points of intersection. The only real root is

$$x = 0$$

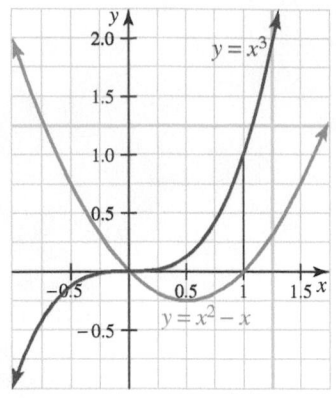

Figure 6.3   Area between curves

$(x^2 - x + 1 = 0$ has no real roots). Thus, by the intermediate value theorem, the same curve is on top throughout the interval $[0, 1]$. To see which curve is on top, take some representative value in $[0, 1]$, such as $x = 0.5$, and note that because $(0.5)^3 > 0.5^2 - 0.5$, the curve $y = x^3$ must be above $y = x^2 - x$. Thus, the required area is given by

$$A = \int_0^1 \left[\underset{\uparrow}{x^3} - \underbrace{(x^2 - x)} \right] dx = \left[\frac{1}{4}x^4 - \frac{1}{3}x^3 + \frac{1}{2}x^2 \right]_0^1 = \frac{5}{12}$$

Top curve   Bottom curve

In Section 5.2, we defined area for a continuous function f with the restriction that $f(x) \geq 0$. Example 1 shows us that when considering the area between two curves, we no longer need to be concerned with the nonnegative restriction but only with whether $f(x) \geq g(x)$. We now use this idea to find the area for a function whose graph is below the x-axis.

Example 2   Area with a function whose graph is below the x-axis

Find the area of the region bounded by the curve $y = e^{2x} - 3e^x + 2$ and the x-axis.

Solution   We find the points of intersection of the curve and the x-axis:

$$\begin{aligned} e^{2x} - 3e^x + 2 &= 0 & &\text{The curve intersects the } x\text{-axis where } y = 0. \\ (e^x - 1)(e^x - 2) &= 0 & &\text{Factor} \\ x &= 0, \ln 2 & &\text{Solve } e^x = 1 \text{ and } e^x = 2. \end{aligned}$$

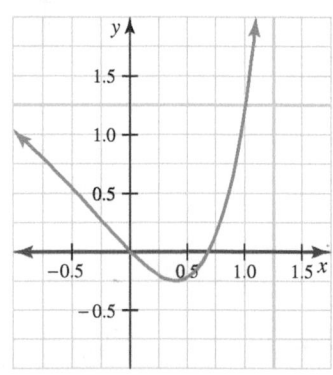

Figure 6.4   Graph of f with area shaded

The graph of $f(x) = e^{2x} - 3e^x + 2$ is shown in Figure 6.4. Notice the graph is below the x-axis.

We see that on the interval $[0, \ln 2]$, $f(x) \leq 0$, but we can find the area of the given region by considering the area between the curves defined by equations $y = 0$ and $y = e^{2x} - 3e^x + 2$.

$$A = \int_0^{\ln 2} [\ \underset{}{0} - \underbrace{(e^{2x} - 3e^x + 2)} \ ] \, dx$$

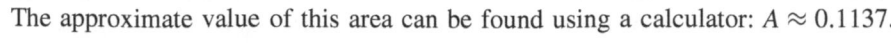

Top curve   Bottom curve

$$= \left[-\frac{e^{2x}}{2} + 3e^x - 2x \right]_0^{\ln 2}$$

$$= \frac{3}{2} - 2\ln 2$$

The approximate value of this area can be found using a calculator: $A \approx 0.1137$.

Area Using Vertical Strips

Although the only mathematically correct way to establish a formula involving integrals is to form Riemann sums and take the limit according to the definition of the definite integral, we can simulate this procedure with approximating strips. This simplification is especially useful for finding the area of a complicated region formed when two curves intersect one or more times, as shown in Figure 6.5.

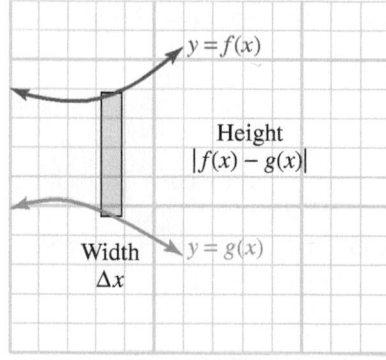

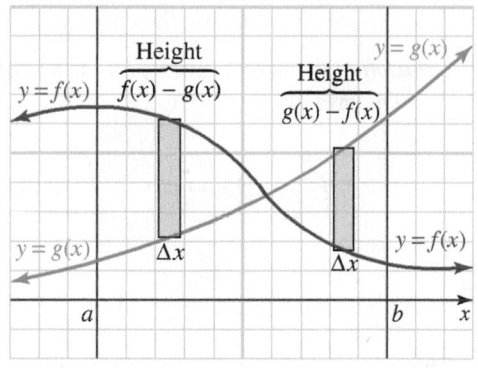

a. A typical vertical strip   **b.** Approximation by strips

Figure 6.5   Area by vertical strips

Note that the vertical strip has height $f(x) - g(x)$ if $y = f(x)$ is above $y = g(x)$, and height $g(x) - f(x)$ if $y = g(x)$ is above $y = f(x)$. In either case, the height can be represented by $|f(x) - g(x)|$, and the area of the vertical strip is

$$\Delta A = |f(x) - g(x)|\, \Delta x = |f(x) - g(x)|\, dx$$

Thus, we have a new integration formula for area; namely,

$$A = \int_a^b |f(x) - g(x)|\, dx$$

☠ *You cannot use the formula $A = \int [f - g]\, dx$ directly here because the hypothesis $f \geq g$ is not satisfied. In order to use $A = \int |f - g|\, dx$ over the entire interval, you must remember that $|f - g|$ might be $f - g$ over part of the interval and $g - f$ over another part (see Figure 6.5b). Make sure you check to see which curve is on top. Because the curves cross in Example 2, the interval must be subdivided accordingly.* ☠

Example 3   Area using vertical strips

Find the area of the region bounded by the line $y = 3x$ and the curve $y = x^3 + 2x^2$.

Solution   The region between the curve and the line is the shaded portion of Figure 6.6.

Part of the process of graphing these curves is to find which is the top curve and which is the bottom. To do this we need to find where the curves intersect:

$$x^3 + 2x^2 = 3x$$

$$x^3 + 2x^2 - 3x = 0$$

$$x(x + 3)(x - 1) = 0$$

The points of intersection occur at $x = -3, 0$, and 1. In the subinterval $[-3, 0]$, labeled A_1 in Figure 6.6, the curve $y = x^3 + 2x^2$ is on top (test a typical point in the subinterval, such as $x = -1$), and on $[0, 1]$, the region labeled A_2, the curve $y = 3x$ is on top. The representative vertical strips are shown in Figure 6.6, and the area between the curve and the line is given by the sum

$$A = \int_{-3}^0 \left[(x^3 + 2x^2) - (3x)\right]\, dx + \int_0^1 \left[(3x) - (x^3 + 2x^2)\right]\, dx$$

$$= \left[\frac{1}{4}x^4 + \frac{2}{3}x^3 - \frac{3}{2}x^2\right]_{-3}^0 + \left[\frac{3}{2}x^2 - \frac{1}{4}x^4 - \frac{2}{3}x^3\right]_0^1$$

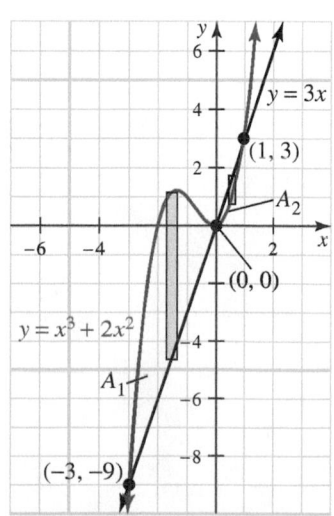

Figure 6.6 Interactive   Area between curves

$$= 0 - \left(\frac{81}{4} - \frac{54}{3} - \frac{27}{2}\right) + \left(\frac{3}{2} - \frac{1}{4} - \frac{2}{3}\right) - 0$$

$$= \frac{71}{6}$$

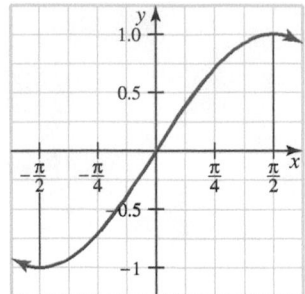

Figure 6.7 Area of a symmetric region

☠ *A common mistake is to work this problem by evaluating the integral*

$$\int_{-\frac{\pi}{2}}^{\frac{\pi}{2}} \sin x \, dx = (-\cos x)\Big|_{-\frac{\pi}{2}}^{\frac{\pi}{2}}$$

$$= -0 + 0$$
$$= 0$$

forgetting that the integral does not give the area unless the function is nonnegative. ☠

Example 4   Area using symmetry

Find the area bounded by the curve $y = \sin x$ and the x-axis between $x = -\frac{\pi}{2}$ and $x = \frac{\pi}{2}$.

Solution   We sketch the region as shown in Figure 6.7.
    The area we seek is found by

$$\int_{-\frac{\pi}{2}}^{0} (0 - \sin x) \, dx + \int_{0}^{\frac{\pi}{2}} (\sin x - 0) \, dx = \cos x\Big|_{-\frac{\pi}{2}}^{0} + (-\cos x)\Big|_{0}^{\frac{\pi}{2}}$$

$$= (1 - 0) + (0 + 1)$$

$$= 2$$

There is an easier procedure using symmetry of the curve $y = \sin x$ with respect to the origin. Symmetry uses the fact that the area bounded by the lines $x = -\frac{\pi}{2}$ and $x = 0$ is the same as the area bounded by the lines $x = 0$ and $x = \frac{\pi}{2}$. As a result, we often work this type of problem as shown here:

$$A = 2 \int_{0}^{\frac{\pi}{2}} (\sin x - 0) \, dx \qquad \textit{By symmetry}$$

$$= 2(-\cos x)\Big|_{0}^{\frac{\pi}{2}}$$

$$= 2(0 + 1)$$

$$= 2$$

Area Using Horizontal Strips

For many regions, it is easier to form horizontal strips rather than vertical strips. The procedure for horizontal strips is the same as the procedure for vertical strips. If we want to find the area between two curves of the form $x = F(y)$ and $x = G(y)$ on the interval $[c, d]$ on the y-axis, we form horizontal strips. In Figure 6.8, such a region and its typical horizontal approximating rectangles, or **horizontal strips**, are illustrated. Note that the width of the horizontal strip is denoted by Δy.

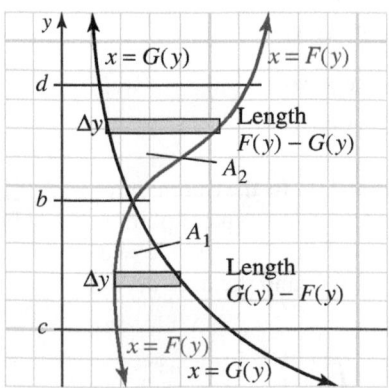

a. Approximation by horizontal strips of width Δy

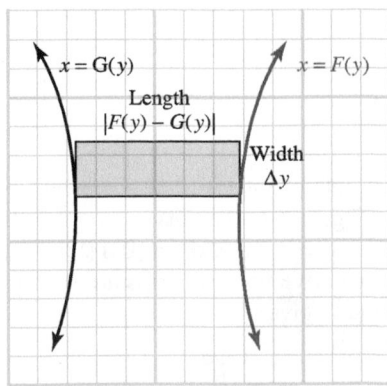

b. A typical horizontal strip

Figure 6.8   Area by horizontal strips

Regardless of which curve is "ahead" or "behind," the horizontal strip has length $|F(y) - G(y)|$ and area

$$\Delta A = |F(y) - G(y)|\, \Delta y$$

However, in practice, you must make sure to find the points of intersection of the curves and divide the integrals so that in each region one curve is the *leading curve* ("right curve") and the other is the *trailing curve* ("left curve"). Suppose the curves intersect where $y = b$ for b on the interval $[c, d]$, as shown in Figure 6.8. Then

$$A = \int_c^b \underbrace{[G(y) - F(y)]}_{G \; leading \; F}\, dy + \int_b^d \underbrace{[F(y) - G(y)]}_{F \; leading \; G}\, dy$$

Example 5 Area using horizontal strips

Find the area of the region, R, between the parabola $x = 4y - y^2$ and the line $x = 2y - 3$.

Solution Figure 6.9 shows the region between the parabola and the line, together with a typical horizontal strip.

To find where the line and the parabola intersect, solve

$$4y - y^2 = 2y - 3$$
$$y^2 - 2y - 3 = 0$$
$$(y + 1)(y - 3) = 0$$
$$y = -1, 3$$

Throughout the interval $[-1, 3]$ the parabola is to the right of the line (test a typical point between -1 and 3, such as $y = 0$). Thus, the horizontal strip has area

$$\Delta A = \underbrace{[(4y - y^2)}_{Right \; curve} - \underbrace{(2y - 3)]}_{Left \; curve}\, \Delta y$$

and the area between the parabola and the line is given by

$$A = \int_{-1}^{3} \left[(4y - y^2) - (2y - 3) \right] dy$$
$$= \int_{-1}^{3} (3 + 2y - y^2)\, dy$$
$$= \left[3y + y^2 - \frac{1}{3}y^3 \right]_{-1}^{3}$$
$$= (9 + 9 - 9) - \left(-3 + 1 + \frac{1}{3} \right)$$
$$= \frac{32}{3}$$

In Example 5, the area can also be found by using vertical strips, but the procedure is more complicated. Note in Figure 6.10 that on the interval $[-5, 3]$, a representative vertical strip would extend from the bottom half of the parabola $y^2 - 4y + x = 0$ to the line $y = \frac{1}{2}(x + 3)$, whereas on the interval $[3, 4]$, a typical vertical strip would extend from the bottom half of the parabola $y^2 - 4y + x = 0$ to the top half.

Thus, the area is given by the sum of two integrals, the second of which requires solving the equation $y^2 - 4y + x = 0$ for y using the quadratic formula. With some effort, it can be shown that the computation of area by vertical strips gives the same result as that found by horizontal strips in Example 5.

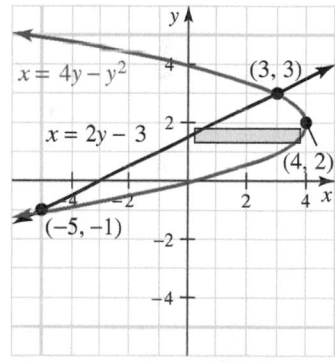

Figure 6.9 Interactive Area of R using horizontal strips

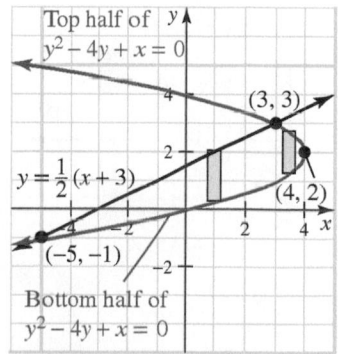

Figure 6.10 Area of R using vertical strips

PROBLEM SET 6.1

Level 1

Write integrals to express the area of the shaded regions in Problems 1-12. Both positive regions (pink) and negative regions (blue) should be included. Do not use absolute value symbols if possible.

1.

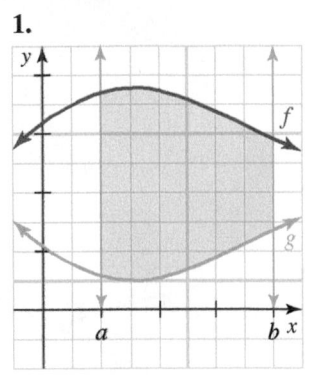

2.

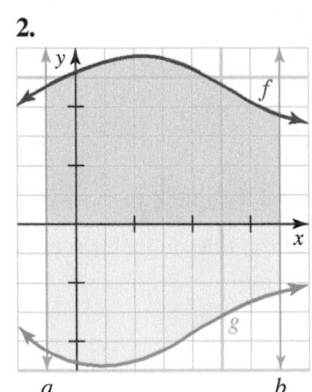

3.

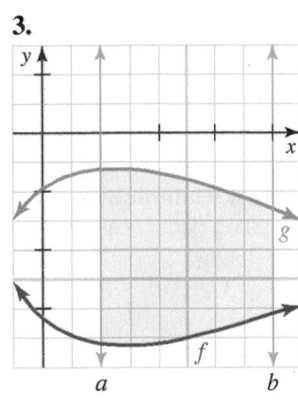

4.

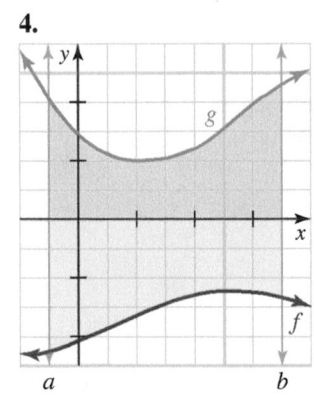

5.

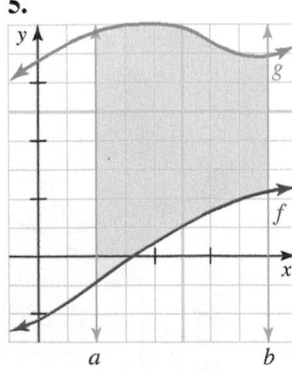

6.

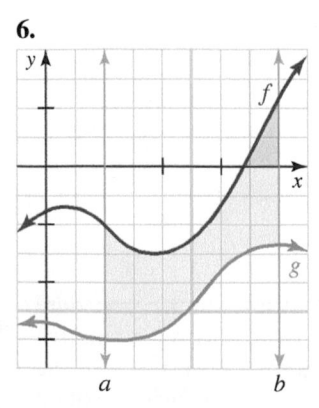

7.

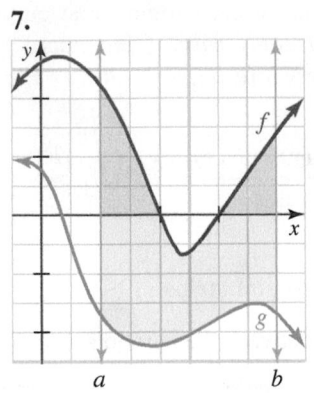

8.

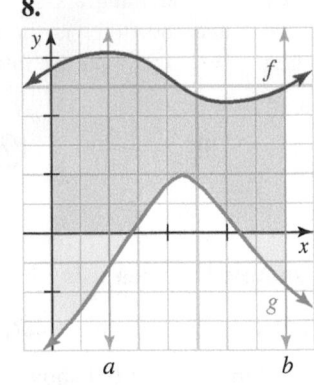

9.

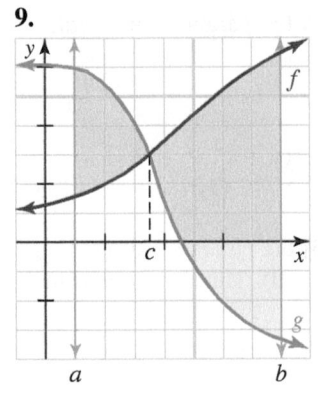

10.

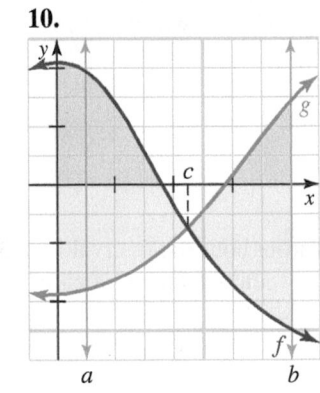

11.

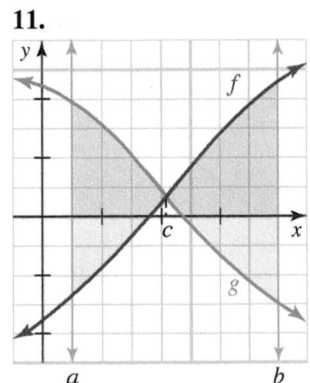

12.

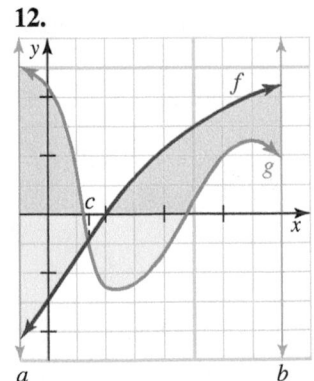

Sketch a representative vertical or horizontal strip and find the area of the given regions bounded by the specified curves in Problems 13-18.

13. $y = \dfrac{3}{2}x - \dfrac{3}{2}$,

$y = -x^2 + 6x - 5$

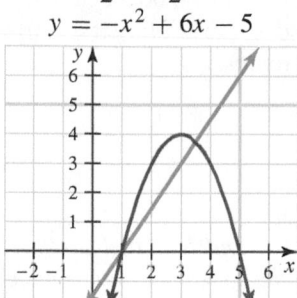

14. $y = x^2 - 8x$,

$y = 0$

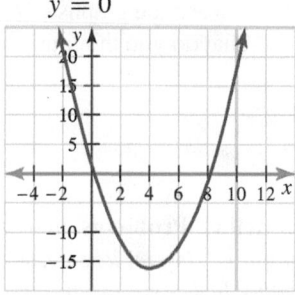

15. $y = \sin 2x$ on $[0, \pi]$

$y = 0$

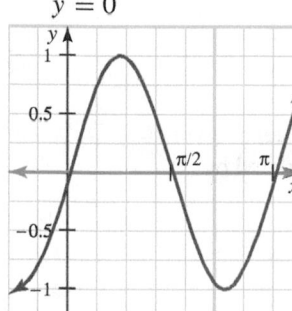

16. $y = (x-1)^3$,

$y = x - 1$

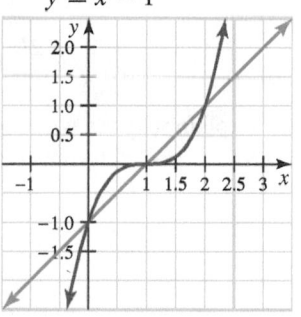

17. $x = y^2 - 5y$

$x = 0$

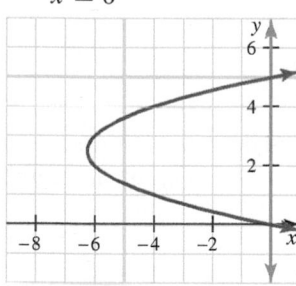

18. $x = y^2 - 6y$

$x = -y$

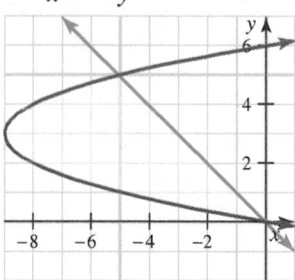

Sketch the region bounded between the given curves and then find the area of each region in Problems 19-43.

19. $y = x^2$, $y = x$, $x = -1$, $x = 1$
20. $y = 6x^2$, $y = 0$, $x = 1$, $x = 5$.
21. $y = 4x^3$, $y = 0$, $x = 1$, $x = 4$.
22. $y = x^3$, $y = x$, $x = -1$, $x = 1$
23. $y = x^{-1}$ and the x-axis on $[0.5, 1]$
24. $y = x^{-1}$ and the x-axis on $[0.1, 1]$
25. $y = x^2$, $y = x^3$
26. $y = x^2$, $y = \sqrt[3]{x}$
27. $y = x^2 - 1$, $x = -1$, $x = 2$, $y = 0$
28. $y = 4x^2 - 9$, $y = 0$, between $x = 0$ and $x = 3$
29. $y = x^4 - 3x^2$, $y = 6x^2$
30. $x = 8 - y^2$, $x = y^2$
31. $x = 2 - y^2$, $x = y$

32. $y = x^2 + 3x - 5$, $y = -x^2 + x + 7$
33. $y = 2x^3 + x^2 - x - 1$, $y = x^3 + 2x^2 + 5x - 1$
34. $y = \sin x$, $y = \cos x$, $x = 0$, $x = \frac{\pi}{4}$
35. $y = \sin x$, $y = \sin 2x$, $x = 0$, $x = \pi$
36. $y = |x|$, $y = x^2 - 6$
37. $y = |4x - 1|$, $y = x^2 - 5$, $x = 0$, $x = 4$
38. $y = e^x$, the x-axis on $[-1, 2]$
39. $y = e^{-x}$, the x-axis on $[-2, 1]$
40. x-axis, $y = x^3 - 2x^2 - x + 2$
41. y-axis, $x = y^3 - 3y^2 - 4y + 12$
42. $y = e^x$, $y = \frac{1}{2}e^x + \frac{1}{2}$, $x = -2$, $x = 2$
43. $y = \dfrac{1}{\sqrt{1 - x^2}}$, $y = \dfrac{2}{x + 1}$, y-axis

Level 2

44. ■ What does this say? When finding the area between two curves, discuss criteria for deciding between vertical and horizontal strips.
45. Find the number k (correct to two decimal places) so that the line $y = k$ bisects the area under the curve $y = \sin^{-1} x$ for $0 \le x \le 1$.
46. Find the area of the region that contains the origin and is bounded by the lines $2y = 11 - x$ and $y = 7x + 13$ and the parabola $y = x^2 - 5$.
47. Show that the region defined by the inequalities $x^2 + y^2 \le 8$, $x \ge y$, and $y \ge 0$ has area π.
48. Find the area of the region bounded by the curve $\sqrt{x} + \sqrt{y} = 1$ and the coordinate axes.
49. The Sorite Corporation has found that production costs are determined by the function

$$C(x) = \frac{100}{\sqrt{x + 25}} + 50$$

where $C(x)$ is the cost to produce the xth item. Use integration to find the total cost of producing 2,000 items.
50. The EPA (Environmental Protection Agency) estimates that pollutants are being dumped into the Russian River according to the formula

$$P(t) = \frac{1,000}{\sqrt[3]{(t + 5)^2}}$$

where t is the year and $P(t)$ is pollutants in tons. What is the total amount (to the nearest ton) of pollutants that will be dumped into the river in the next 25 years?
51. The marginal revenue and marginal cost (in dollars) per day are given in terms of the tth month by the formulas

$$R'(t) = 500e^{0.01t} \quad \text{and} \quad C'(t) = 50 - 0.1t$$

where $R(0) = C(0) = 0$. Find the total profit for the first year.

52. A chemical spill is known to kill fish according to the formula

$$N(t) = \frac{100,000}{\sqrt{t + 250}}$$

where N is the number of fish killed on the tth day since the spill. What is the approximate total number of fish killed in 30 days?

53. A culture is growing at a rate of

$$R'(t) = 200e^{0.05t} \qquad \text{for } 0 \le t \le 10$$

per hour. Find the area between the graph of this function and the t-axis. What do you think this area represents?

54. The rate of change of the demand for a product with respect to time (in weeks) is given by the function

$$D'(t) = 20 + 0.012t^2 \qquad \text{for } 0 \le t \le 25$$

Find the area between the graph of this function and the t-axis. What do you think this area represents?

55. Suppose the accumulated cost of a piece of equipment is $C(t)$ and the accumulated revenue is $R(t)$, where both of these are measured in thousands of dollars and t is the number of years after the piece of equipment was installed. If it is known that

$$C'(t) = 1 \qquad \text{and} \qquad R'(t) = 2e^{-0.1t}$$

find the area (to the nearest unit) in the bounded region between the graphs of C' and R' (do not forget $t \ge 0$). What do you think this area represents?

56. Suppose the accumulated cost of a piece of equipment is $C(t)$ and the accumulated revenue is $R(t)$, where both of these are measured in thousands of dollars and t

is the number of years after the piece of equipment was installed. It is known that

$$C'(t) = 18 \qquad \text{and} \qquad R'(t) = 21e^{-0.01t}$$

Find the area (to the nearest unit) in the bounded region between the graphs of C' and R' (do not forget $t \ge 0$). What do you think this area represents?

Level 3

Modeling Problems *Imagine a cylindrical fuel tank of length L lying on its side, where the ends are circular with radius b. Determine the amount of fuel in the tank for a given level by completing Problems 57-59.*

57. Explain why the volume of the tank may be modeled by

$$V = 2L \int_{-b}^{b} \sqrt{b^2 - y^2}\, dy$$

58. Explain why the volume of fuel at level $h\,(-b \le h \le b)$ may be modeled by

$$V(h) = 2L \int_{-b}^{h} \sqrt{b^2 - y^2}\, dy$$

59. Finally, for $b = 4$ and $L = 20$, numerically compute $V(h)$ for $h = -3, -2, \cdots, 4$. *Note:* $V(0)$ and $V(4)$ will serve as a check on your work.

60. The curve with the equation

$$(x^2 + y^2)^3 = x^2$$

has a loop for $0 \le x \le 1$. Find the area of the loop. *Hint:* Use the substitution $u = x^{1/3}$ in the area integral.

6.2 VOLUME

IN THIS SECTION: *Method of cross sections, method of disks and washers, method of cylindrical shells*
We have seen how to compute area by integration, and now we are ready to investigate several methods for computing volume.

The methods used in Section 6.1 to compute area by integration can be modified to compute the volume of a solid region. We shall develop methods for computing volume of several kinds of solids in this section, beginning with the case where the solid has a known cross section.

Method of Cross Sections

Let S be a solid and suppose that for $a \le x \le b$, the cross section of S that is perpendicular to the x-axis at x has area $A(x)$.

To find the volume of S, we first take a partition $x_0 = a, x_1, x_2, x_3, \cdots, x_n = b$ of the interval $[a, b]$ and choose a representative number x_k^* in each subinterval $[x_{k-1}, x_k]$. Think of cutting the solid with a knife at $x = x_k^*$ and removing a very thin slab whose face has area $A(x_k^*)$ and whose thickness is Δx_k, as shown in Figure 6.11.

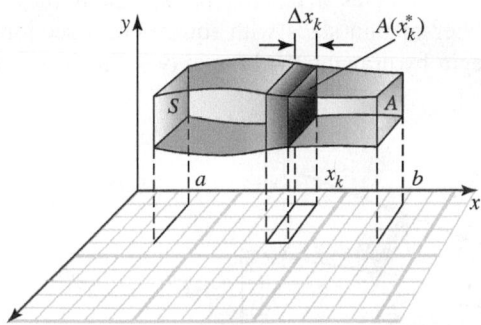

Figure 6.11   Volume of solids with known cross sections

Thus, we consider a slab with cross-sectional area $A(x_k^*)$ and width

$$\Delta x_k = x_k - x_{k-1}$$

This slab has volume

$$\Delta V_k = A(x_k^*)\Delta x_k$$

By adding up the volumes of all such slabs, we obtain an approximation to the volume of the solid S:

$$V_n = \sum_{k=1}^{n} A(x_k^*)\Delta x_k$$

The approximation improves as the number of partition points increases, and it is reasonable to *define* the volume V of the solid S as the limit of V_n as the norm of the partition $\|P\|$ tends to 0. That is,

$$V = \lim_{\|P\|\to 0} \sum_{k=1}^{n} A(x_k^*)\Delta x_k$$

which we recognize as the definite integral $\int_a^b A(x)\,dx$. To summarize:

VOLUME OF A SOLID WITH KNOWN CROSS-SECTIONAL AREA
A solid S with cross-sectional area $A(x)$ perpendicular to the x-axis at each point on the interval $[a, b]$ has **volume**

$$V = \int_a^b A(x)\,dx$$

Example 1 Volume of a solid using square cross sections

The base of a solid is the region in the xy-plane bounded by the y-axis and the lines $y = 1 - x$, $y = 2x + 5$, and $x = 3$. Each cross section perpendicular to the x-axis is a square. Find the volume of the solid.

Solution The solid resembles a tapered brick, and it may be constructed by "gluing" together a number of thin slabs with square cross sections, like the one shown in Figure 6.12**c**. We begin by drawing the base in two dimensions and then find the volume of the kth slice.

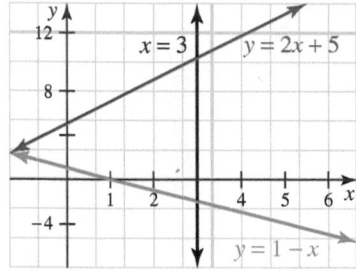

a. Two-dimensional graph of the base

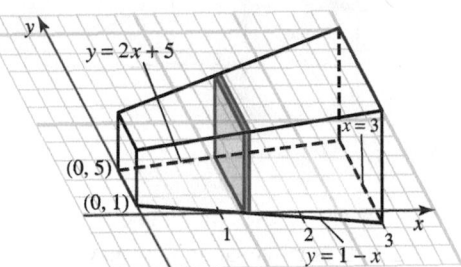

b. Three-dimensional solid

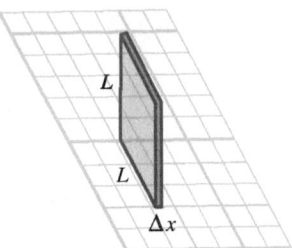

c. Cross-sectional representative element $\Delta V = L^2 \Delta x$

Figure 6.12 A solid with a square cross section

To model this construction mathematically, we subdivide the interval $[0, 3]$, form a vertical approximating rectangle on each resulting subinterval, and then construct a slab with square cross section on each approximating rectangle. If we choose the thickness of a typical slab to be Δx and the height and width to be L, the slab will have volume ΔV, where

$$\Delta V = L^2 \Delta x$$
$$= L^2(x) \Delta x$$
$$= [(2x + 5) - (1 - x)]^2 \, \Delta x$$
$$= (3x + 4)^2 \Delta x$$

The volume of the entire solid is obtained by integrating to "add up" all the volumes ΔV, and we find that the volume of the solid is

$$V = \int_0^3 (3x + 4)^2 dx$$
$$= \int_0^3 (9x^2 + 24x + 16) \, dx$$
$$= \left[3x^3 + 12x^2 + 16x \right]_0^3$$
$$= 237$$

Example 2   Volume of a regular pyramid with a square base

A regular pyramid has a square base of side L and its apex located H units above the center of its base. Derive the formula $V = \frac{1}{3}HL^2$.

Solution   The pyramid is shown in Figure 6.13.

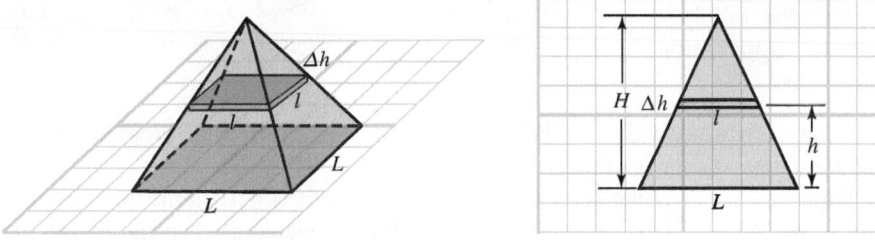

Figure 6.13   The volume of a pyramid

This pyramid can be constructed by stacking a number of thin square slabs. Suppose that a representative slab has side ℓ and thickness Δh, and that it is located h units above the base of the pyramid as shown in Figure 6.13. By creating a proportion from corresponding parts of similar triangles, we see that

$$\frac{\ell}{L} = \frac{H - h}{H} \qquad \text{so that} \qquad \ell = \left(1 - \frac{h}{H}\right)L$$

Therefore, the volume of the representative slab is

$$\Delta V = \ell^2 \Delta h$$
$$= \left(1 - \frac{h}{H}\right)^2 L^2 \Delta h$$

To compute the volume V of the entire pyramid, we integrate with respect to h from the base of the pyramid ($h = 0$) to the apex ($h = H$). Thus,

$$V = \int_0^H \left(1 - \frac{h}{H}\right)^2 L^2 dh$$
$$= L^2 \int_0^H \left(1 - \frac{2}{H}h + \frac{1}{H^2}h^2\right) dh$$
$$= L^2 \left[h - \frac{h^2}{H} + \frac{h^3}{3H^2}\right]_0^H$$
$$= L^2 \left(H - \frac{H^2}{H} + \frac{H^3}{3H^2}\right)$$
$$= \frac{1}{3}HL^2$$

Other volume formulas can be found in a similar fashion (see Problems 59 and 60).

Method of Disks and Washers

A **solid of revolution** is a solid figure S obtained by revolving a region D in the xy-plane about a line L (called the **axis of revolution**) that lies outside or on the boundary of D. Some familiar solids of revolution are shown in Figure 6.14.

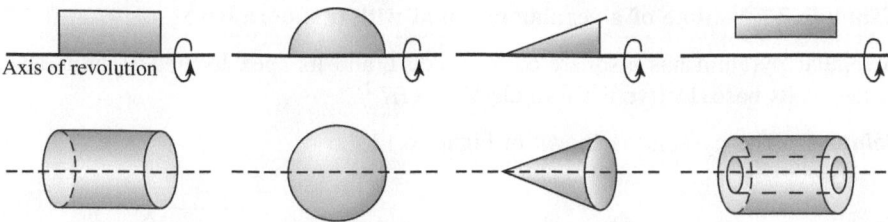

Figure 6.14   Solids of revolution

Note that such a solid S may be thought of as having circular cross sections in the direction perpendicular to L.

Suppose the function f is continuous and satisfies $f(x) \geq 0$ on the interval $[a,b]$, and suppose we wish to find the volume of the solid S generated when the region D under the curve $y = f(x)$ on $[a,b]$ is revolved about the x-axis. That is, *the axis of revolution is horizontal and it is a boundary of the region D*. Our strategy will be to form vertical strips and revolve them about the x-axis, generating what are called **disks** (that is, thin right circular cylinders) that approximate a portion of the solid of revolution S, as shown in Figure 6.15.

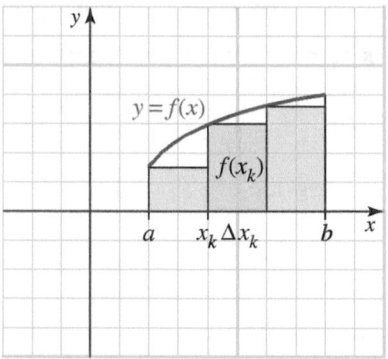
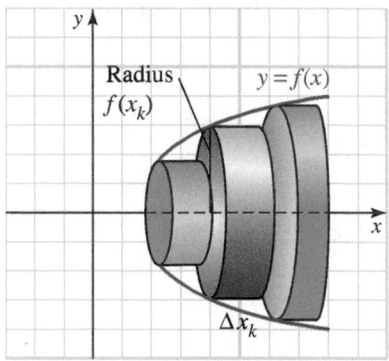

a. A representative vertical strip has height $f(x_k)$ and width Δx_k

b. The representative disk is formed by revolving the representative strip about the x-axis

Figure 6.15 Interactive   The disk method

Now we can compute the total volume of S by using integration to sum the volumes of all the approximating disks. Recall that the formula for the volume of a cylinder of height h and cross-sectional area A is Ah. Figure 6.15a shows a typical vertical strip with height $f(x_k)$ and width Δx_k. You might notice that the width of the strip is the same as the thickness of the disk. The solid of revolution can be thought of as having cross sections perpendicular to the x-axis that are circular disks of volume

$$\Delta V_k(x) = \underbrace{\pi [f(x_k)]^2}_{\text{Area of circular cross section}} \Delta x_k$$

The total volume may be found by integration:

$$V = \int_a^b \overbrace{A(x)}^{\text{Area of cross section}} \underbrace{dx}_{\text{Thickness}} = \int_a^b \pi [f(x)]^2 \, dx$$

This procedure may be summarized as follows:

THE DISK METHOD   The **disk method** is used to find a volume generated when a region D is revolved about an axis L that is *perpendicular* to a typical approximating strip in D. Suppose D is the region bounded by the curve $y = f(x)$, the x-axis, and the vertical lines $x = a$ and $x = b$. Then if D is revolved about the x-axis, it generates a solid with volume

$$V = \int_a^b \pi y^2 dx = \int_a^b \pi \left[f(x)\right]^2 dx$$

■ **W**hat this says The following diagram may help you remember the key ideas behind the disk method.

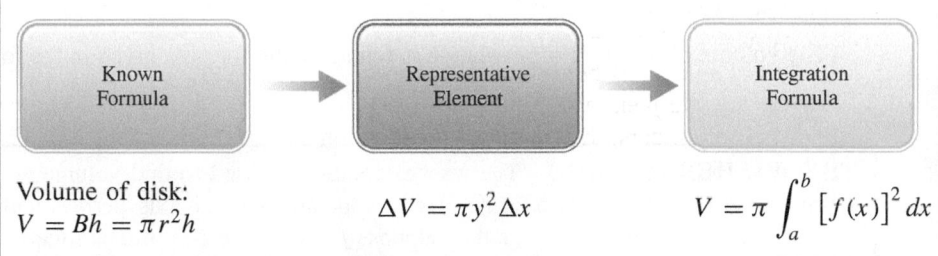

Volume of disk:
$V = Bh = \pi r^2 h$

$$\Delta V = \pi y^2 \Delta x$$

$$V = \pi \int_a^b \left[f(x)\right]^2 dx$$

Example 3   Volume by disks

Find the volume of the solid S formed by revolving the region D under the curve $y = x^2 + 1$ on the interval $[0, 2]$ about the x-axis.

Solution   The region is shown in Figure 6.16.

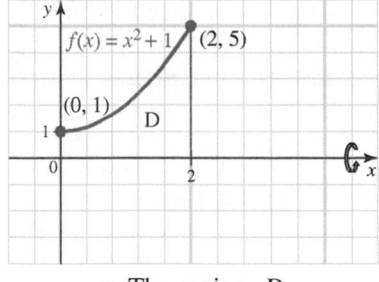

a. The region, D

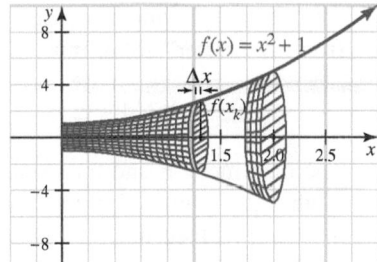

b. The solid of revolution, S

Figure 6.16 Interactive   Volume of a solid of revolution: The disk method

$$V = \pi \int_0^2 (x^2 + 1)^2 dx$$

$$= \pi \int_0^2 (x^4 + 2x^2 + 1)\, dx$$

$$= \pi \left[\frac{1}{5}x^5 + \frac{2}{3}x^3 + x \right]_0^2$$

$$= \frac{206}{15}\pi$$

The volume is about 43 cubic units.   ■

With a small modification of the disk method, we can find the volume of a solid figure generated by revolving about the x-axis the region between two curves $y = f(x)$ and $y = g(x)$ where $f(x) \geq g(x) \geq 0$ for $a \leq x \leq b$. When a typical vertical strip is revolved about the x-axis, a "washer" with cross-sectional area $\pi([f(x)]^2 - \lfloor g(x)]^2)$ is formed, as shown in Figure 6.17. This can be remembered by $\pi R^2 - \pi r^2$, where R is the outer radius and r is the inner radius.

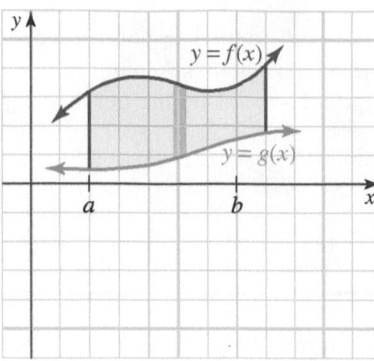
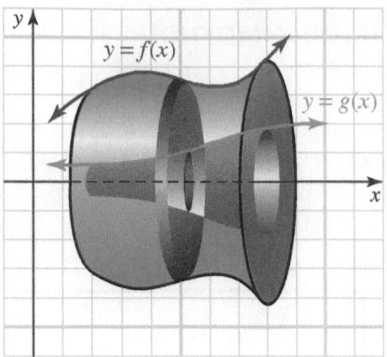

Figure 6.17 The washer method

The volume of the solid of revolution is found by the formula given in the following box.

> **THE WASHER METHOD** The **washer method** is used to find volume when a region between two curves is revolved about an external axis perpendicular to the approximating strip. In particular, suppose f and g are continuous functions on $[a,b]$ with $f(x) \geq g(x) \geq 0$. If R is the (outer) curve $y = f(x)$ and r is the (inner) curve $y = g(x)$, and the region bounded by $y = f(x)$, $y = g(x)$, $x = a$, and $x = b$ is revolved about the x-axis, then the volume thus formed is
>
> $$V = \int_a^b \pi \left([\underbrace{f(x)}]^2 - [\underbrace{g(x)}]^2 \right) dx$$
> $$\qquad\qquad\quad \text{Outer radius} \quad \text{Inner radius}$$

☠ $f^2 - g^2 \neq (f - g)^2$ ☠

■ What this says In practice, what this means is that the "inner" solid is "subtracted" from the "outer" solid, much like coring an apple.

The disk method and the washer method also apply when the axis of revolution is a line other than the x-axis. In Example 4, we consider what happens when a particular region D is revolved not only about the x-axis, but also about other axes parallel to a coordinate axis.

Example 4   Volume by washers

Let D be the solid region bounded by the parabola $y = x^2$ and the line $y = x$. Find the volume of the solid generated when D is revolved about the

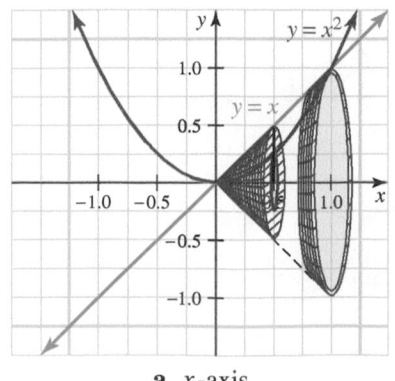
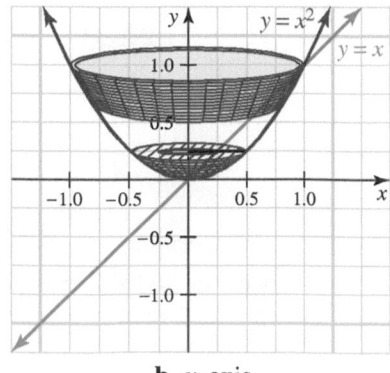
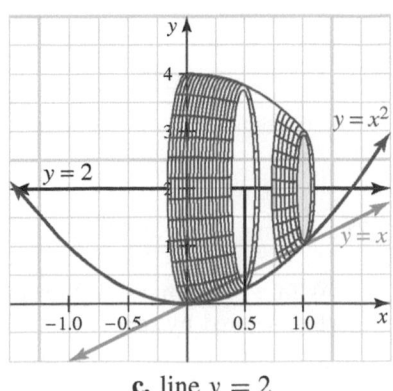

a. x-axis   **b.** y-axis   **c.** line $y = 2$

Figure 6.18 Interactive   Volume by washers

Solution The region D to be rotated is shown in Figure 6.19. Note that since $x = x^2$ at $x = 0$ and at $x = 1$, the line and parabola intersect at the origin and at $(1, 1)$.

a. The line is always above the parabola on the interval $[0, 1]$, so when we form a washer to approximate the volume of revolution, the outer radius is $R = x$ and the inner radius is $r = x^2$, as shown in Figure 6.19**a**. Thus, the required volume is

$$
\begin{aligned}
V &= \pi \int_0^1 \left[x^2 - (x^2)^2 \right] dx \\
&= \pi \int_0^1 (x^2 - x^4)\, dx \\
&= \pi \left[\frac{1}{3}x^3 - \frac{1}{5}x^5 \right]_0^1 \\
&= \frac{2\pi}{15}
\end{aligned}
$$

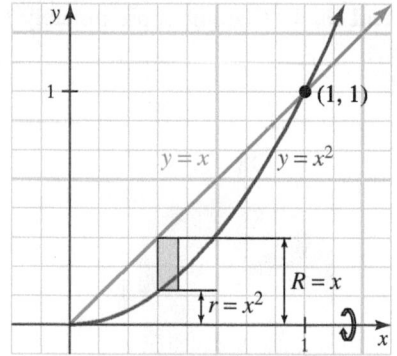

a. Vertical strip rotated about the x-axis

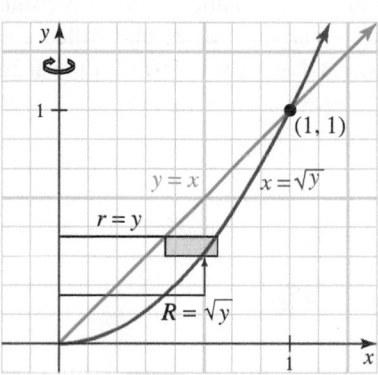

b. Horizontal strip rotated about the y-axis

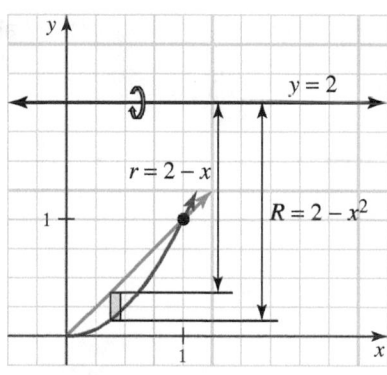

c. Vertical strip rotated about the line $y = 2$

Figure 6.19 Volume obtained by revolving a region about different lines

For the disk method or the washer method, make sure that the approximating strips are perpendicular to the axis of revolution.

b. Because we are revolving D about the y-axis, we use *horizontal* strips to approximate the solid of revolution, as shown in Figure 6.19**b**. Note that the parabola $x = \sqrt{y}$ is to the right of the line $x = y$ on the interval $[0, 1]$, so the approximating washer has outer radius $R = \sqrt{y}$ and inner radius $r = y$; thus,

$$
\begin{aligned}
V &= \pi \int_0^1 \left[(\sqrt{y})^2 - (y)^2 \right] dy \\
&= \pi \int_0^1 (y - y^2)\, dy \\
&= \pi \left[\frac{y^2}{2} - \frac{y^3}{3} \right]_0^1 \\
&= \frac{\pi}{6}
\end{aligned}
$$

c. Since the axis of rotation is the line $y = 2$, the outer radius of a typical approximating washer is $R = 2 - x^2$, and the inner radius is $r = 2 - x$, as shown in Figure 6.19**c**. Thus, the volume of revolution is

$$V = \pi \int_0^1 \left[(2 - x^2)^2 - (2 - x)^2 \right] dx$$

$$= \pi \int_0^1 (x^4 - 5x^2 + 4x)\, dx$$

$$= \pi \left[\frac{x^5}{5} - \frac{5x^3}{3} + \frac{4x^2}{2} \right]_0^1$$

$$= \frac{8\pi}{15}$$

The Method of Cylindrical Shells

Sometimes it is easier (or even necessary) to compute a volume by taking the approximating strip parallel to the axis of rotation instead of perpendicular to the axis as in the disk and washer methods. Figure 6.20**a** shows a region D under the curve $y = f(x) \geq 0$, on the interval $[a, b]$, together with a representative vertical strip. When this strip is revolved about the y-axis, it forms a solid called a **cylindrical shell** of height approximately $f(x)$ and thickness Δx, as shown in Figure 6.20**b**.

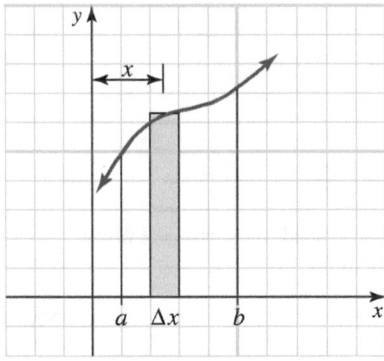

a. Vertical strip

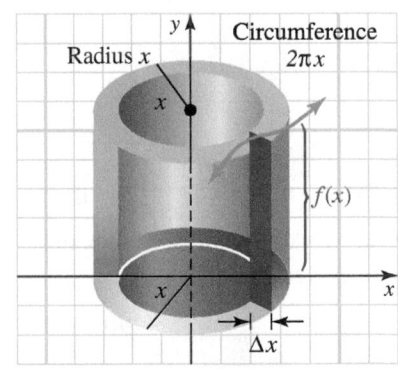

b. When the strip (part **a**) is revolved about the y-axis, a shell is generated

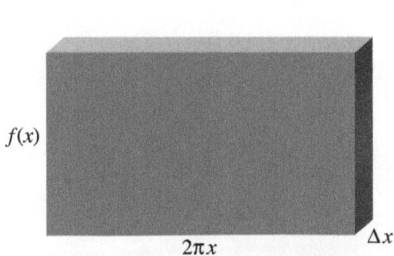

c. The "unwrapped" flattened shell has volume $\Delta V = 2\pi x f(x)\Delta x$

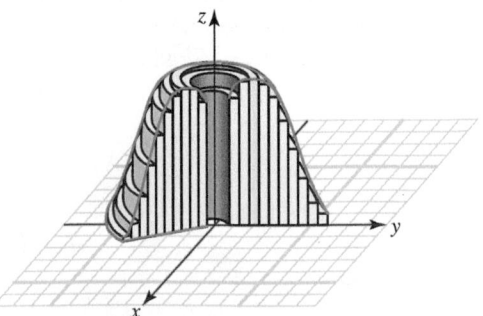

d. Approximating a solid of revolution using cylindrical shells

Figure 6.20   Method of cylindrical shells

Because the strip is x units from the axis of rotation and is assumed to be very thin, the cross section of the shell (perpendicular to the y-axis) will be a circle of radius x and circumference $2\pi x$. If we imagine the shell to be cut and flattened out, it is seen to be approximately a rectangular slab of volume

$$\Delta V = \underbrace{2\pi x f(x)}_{\text{Area of the rectangular slab}} \cdot \overbrace{\Delta x}^{\text{Thickness}}$$

(See Figure 6.20**c**.) Thus, the total volume of the solid is given by the integral

$$V = \int_a^b 2\pi x\, f(x)\, dx$$

The approximating shells are shown in Figure 6.20**d** and the formula is repeated in the following box.

METHOD OF CYLINDRICAL SHELLS The **shell method** is used to find a volume generated when a region D is revolved about an axis L *parallel* to a typical approximating strip in D.

In particular, if D is the region, as shown in Figure 6.21, bounded by the curve $y = f(x) \geq 0$, the x-axis, and the vertical lines $x = a$ and $x = b$ where $0 \leq a \leq b$, then the solid generated by revolving D about the y-axis has volume

$$V = \int_a^b 2\pi x f(x)\, dx$$

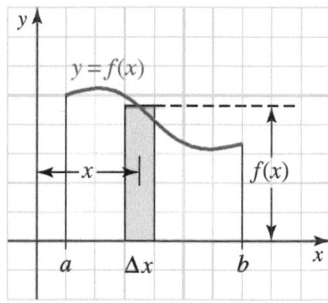

Figure 6.21 Shell method

Example 5 Volume using cylindrical shells

Find the volume of the solid formed by revolving the region bounded by the x-axis and the graphs of $y = x^3 + x^2 + 1$, $x = 1$, and $x = 3$ about the y-axis.

Solution The region to be rotated is shown in Figure 6.22, along with a typical vertical strip of height $f(x) = x^3 + x^2 + 1$ and width Δx.

The volume, by the method of shells, is

$$V = 2\pi \int_1^3 x(x^3 + x^2 + 1)\, dx$$

$$= 2\pi \int_1^3 (x^4 + x^3 + x)\, dx$$

$$= 2\pi \left[\frac{x^5}{5} + \frac{x^4}{4} + \frac{x^2}{2} \right]_1^3$$

$$= \frac{724}{5}\pi$$

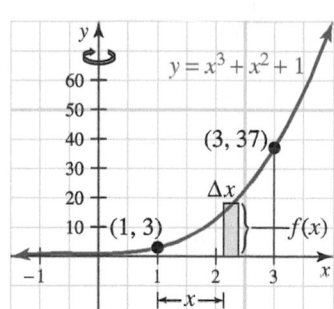

Figure 6.22 Shell method about y-axis

When the axis of rotation is a line $x = L$ parallel to the y-axis, the distance of a typical vertical strip from the axis will no longer be x. Here is an example that illustrates how to compute volume of revolution in such a case.

Example 6   Volume of revolution about a horizontal or vertical line

Find the volume of the solid formed by revolving the region D bounded by the curve $y = x^{-2}$ and the x-axis for $1 \le x \le 2$ about the line $x = -1$.

Solution   The region D is shown in Figure 6.23, along with a typical vertical strip of height $f(x) = x^{-2}$.

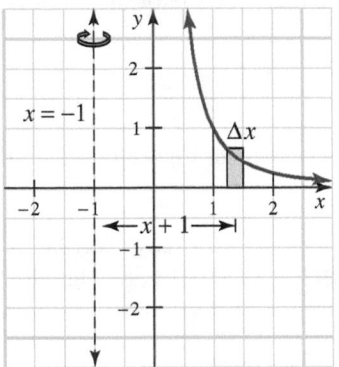

Figure 6.23   Shell method about $x = -1$

If we were revolving this strip about the y-axis, the distance from the axis to the strip would be $L = x$, but since the axis of rotation is $x = -1$, the distance is $L = x - (-1) = x + 1$.

$$V = 2\pi \int_1^2 (x + 1)\left(\frac{1}{x^2}\right) dx$$

$$= 2\pi \int_1^2 (x^{-1} + x^{-2}) \, dx$$

$$= 2\pi \left[\ln x - x^{-1}\right]_1^2$$

$$= 2\pi \ln 2 + \pi$$

Sometimes you have a choice between using shells or washers to compute a volume of revolution. The decision as to which method to use often comes down to which leads to the easier integration. Which method would you choose in the following example?

Example 7   Comparing the methods of shells and washers

A cylindrical hole of radius r is bored through the center of a solid sphere of radius R $(r < R)$, as shown in Figure 6.24.

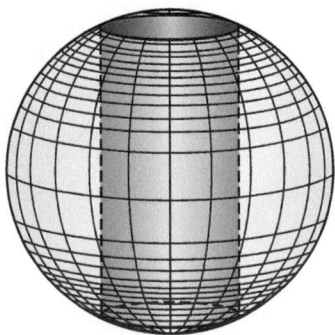

Figure 6.24   Drill a hole in a sphere

Find the volume of the solid that remains by using: **a.** cylindrical shells   **b.** washers

Solution

a. The required volume can be thought of as twice the volume generated when the shaded region of Figure 6.25a is revolved about the y-axis.

b. The required volume can be thought of as twice the volume generated when the shaded region of Figure 6.25b is revolved about the y-axis.

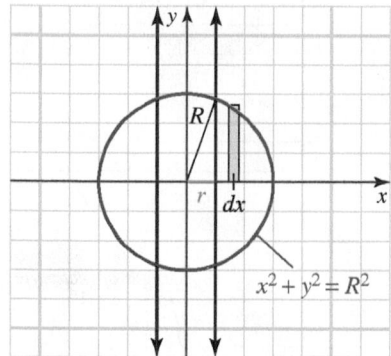

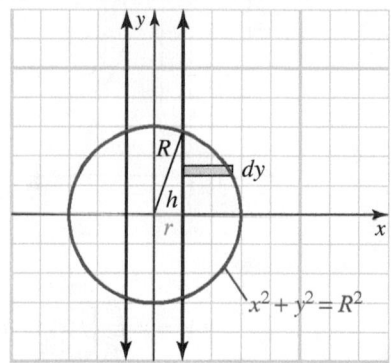

a. Strip *parallel* to the axis of rotation. Typical strip: height $y = \sqrt{R^2 - x^2}$ radius: x

b. Strip *perpendicular* to the axis of revolution. Typical washer: outer radius: $x = \sqrt{R^2 - y^2}$ inner radius: r
$h = \sqrt{R^2 - r^2}$

Figure 6.25   Volume of a solid

a. cylindrical shells $\displaystyle V = 2 \int_r^R 2\pi xy \, dx$

$$= 4\pi \int_r^R x\sqrt{R^2 - x^2} \, dx$$

$$= 4\pi \left[-\frac{1}{3}(R^2 - x^2)^{3/2} \right]_r^R$$

$$= \frac{4\pi}{3}(R^2 - r^2)^{3/2}$$

b. washers $\displaystyle V = 2 \int_0^h \pi \left[\left(\sqrt{R^2 - y^2} \right)^2 - r^2 \right] dy$

$$= 2\pi \int_0^{\sqrt{R^2 - r^2}} \left[R^2 - y^2 - r^2 \right] dy$$

$$= 2\pi \left[(R^2 - r^2) y - \frac{y^3}{3} \right]_0^{\sqrt{R^2 - r^2}}$$

$$= 2\pi \left[(R^2 - r^2)^{3/2} - \frac{1}{3} (R^2 - r^2)^{3/2} \right]$$

$$= \frac{4\pi}{3} (R^2 - r^2)^{3/2}$$

Note that in the special case $r = 0$, a sphere, we get the familiar $V = \frac{4\pi}{3} R^3$.

Table 6.1 compares and contrasts the disk, washer, and shell methods for computing volume.

Table 6.1 Volumes of revolution when the axis of revolution is either the *x*-axis or the *y*-axis

	Horizontal axis of revolution	Vertical axis of revolution

Disk method
A representative rectangle is **perpendicular** to the axis of revolution. The axis of revolution is a boundary of the region.

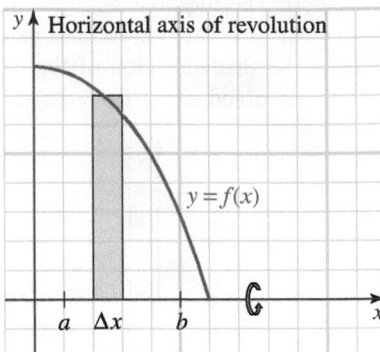

Length (height) of rectangle

$$V = \pi \int_a^b \overbrace{\left[f(x) \right]}^{2} \underbrace{dx}$$

Width of rectangle is Δx.

Length of rectangle

$$V = \pi \int_c^d \overbrace{\left[g(y) \right]}^{2} \underbrace{dy}$$

Width of rectangle is Δy.

Washer method
A representative rectangle is **perpendicular** to the axis of revolution. The axis of revolution is *not* part of the boundary of the region.

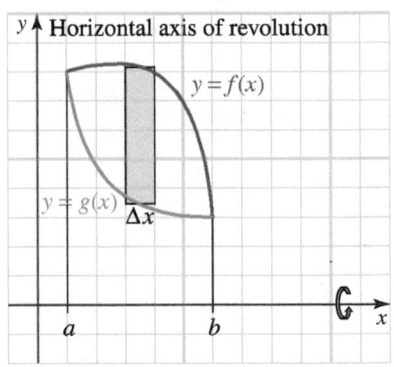 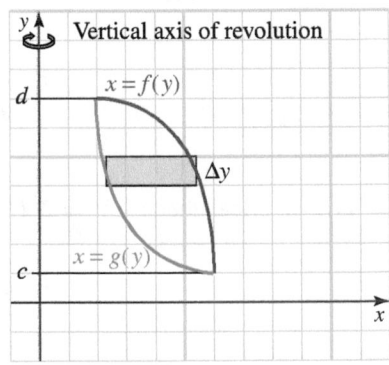

Top curve   Bottom curve

$$V = \pi \int_a^b \underbrace{\overbrace{\left[f(x) \right]}^{2} - \overbrace{\left[g(x) \right]}^{2}}_{\text{Length of rectangle}} \underbrace{dx}_{\text{Width}}$$

Right curve   Left curve

$$V = \pi \int_c^d \underbrace{\overbrace{\left[f(y) \right]}^{2} - \overbrace{\left[g(y) \right]}^{2}}_{\text{Length of rectangle}} \underbrace{dy}_{\text{Width}}$$

Shell method
A representative rectangle is **parallel** to the axis of revolution.

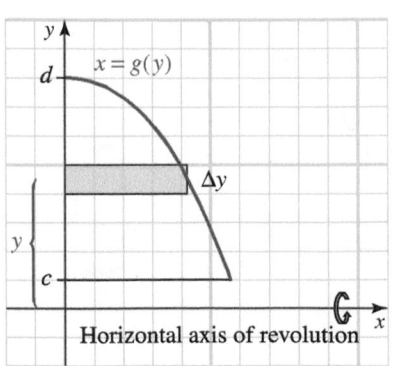 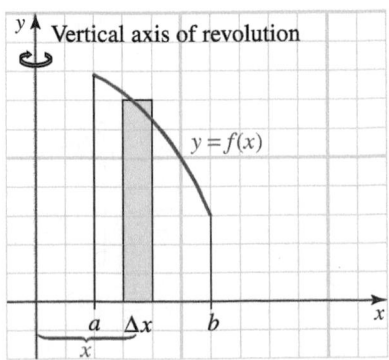

Length of rectangle

$$V = 2\pi \int_c^d \underset{\underset{\text{Distance to axis}}{\uparrow}}{y} \overbrace{\left[g(y) \right]} \underbrace{dy}_{\text{Width}}$$

Length (height)

$$V = 2\pi \int_a^b \underset{\underset{\text{Distance to axis}}{\uparrow}}{x} \overbrace{\left[f(x) \right]} \underbrace{dx}_{\text{Width}}$$

PROBLEM SET 6.2

*In Problems 1-4, sketch the given region and then find the volume of the solid whose base is the given region and which has the property that each cross section perpendicular to the x-axis is a **square**.*

1. the triangular region bounded by the coordinate axes and the line $y = 3 - x$
2. the region bounded by the x-axis and the semicircle $y = \sqrt{16 - x^2}$
3. the region bounded by the line $y = x + 1$ and the parabola $y = x^2 - 2x + 3$
4. the region bounded above by $y = \sqrt{\sin x}$ and below by the x-axis on the interval $[0, \pi]$

*In Problems 5-8, sketch the region and then find the volume of the solid whose base is the given region and which has the property that each cross section perpendicular to the x-axis is an **equilateral triangle**.*

5. the region bounded by the circle $x^2 + y^2 = 4$
6. the region bounded by the curves $y = x^3$ and $y = x^2$
7. the region bounded above by $y = \sqrt{\cos x}$ and below by the x-axis on the interval $\left[-\frac{\pi}{2}, \frac{\pi}{2}\right]$
8. the triangular region with vertices $A(1, 1)$, $B(3, 5)$, and $C(3, -2)$

In Problems 9-12, sketch the region and then find the volume of the solid where the base is the given region and which has the property that each cross section perpendicular to the x-axis is a semicircle.

9. the region bounded by the parabola $y = x^2$ and the line $2x + y - 3 = 0$
10. the region bounded above by $y = \cos x$, below by $y = \sin x$, and on the left by the y-axis
11. the region bounded above by the curve $y = \tan x$ and below by the x-axis, on the interval $\left[0, \frac{\pi}{4}\right]$
12. the region bounded by the x-axis and the curve $y = e^x$ between $x = 1$ and $x = 3$

In Problems 13-18, find the volume of the solid formed when the region described is revolved about the x-axis using washers or disks.

13. the region under the parabolic arc $y = \sqrt{x}$ on the interval $0 \le x \le 1$
14. the region under the curve $y = \sqrt[3]{x}$ on the interval $0 \le x \le 8$
15. the region bounded by the lines $y = x$, $y = 2x$, and $x = 1$
16. the region bounded by the lines $x = 0$, $x = 1$, $y = x + 1$, and $y = x + 2$

17. the region between $y = \sin x$ and $y = \cos x$ on $0 \le x \le \frac{\pi}{4}$
18. the region bounded by the curves $y = e^x$ and $y = e^{-x}$ on $[0, 2]$

In Problems 19-24, use shells to find the volume of the solid formed by revolving the given region about the y-axis.

19. the region bounded by the lines $y = 2x$, the y-axis, and $y = 1$
20. the region bounded by the parabolic arc $y = \sqrt{x}$, the y-axis, and the line $y = 1$
21. the region bounded by the parabola $y = 1 - x^2$, the y-axis, and the positive x-axis
22. the region bounded by the parabolas $y = x^2$, $y = 1 - x^2$, and the y-axis for $x \ge 0$
23. the region bounded by the curve $y = e^{-x^2}$, the y-axis, and the line $y = \frac{1}{2}$
24. the region inside the ellipse

$$2(x - 3)^2 + 3(y - 2)^2 = 6$$

about the y-axis

In Problems 25-32, set up but do not evaluate an integral using the shaded strips for the volume generated when the given region is revolved about:

    a. *the x-axis*   **b.** *the y-axis*
    c. *the line* $y = -1$ **d.** *the line* $x = -2$

25. $y = 4 - x$,   **26.** $y = 4 - x^2$,
    $0 \le x \le 4$ $0 \le x \le 2$

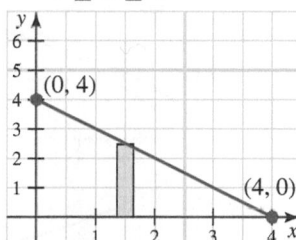 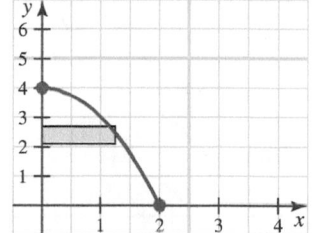

27. $y = \sqrt{4 - x^2}$,   **28.** $y = \sqrt{4 - x^2}$,
    $0 \le x \le 2$ $0 \le x \le 2$

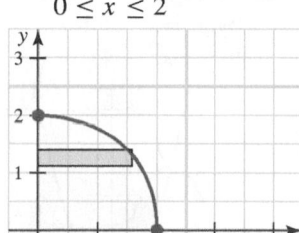 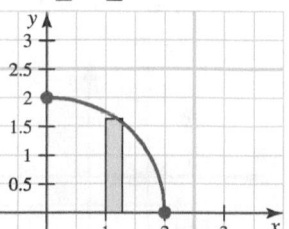

29. $y = e^{-x}$,
$0 \le x \le 1$

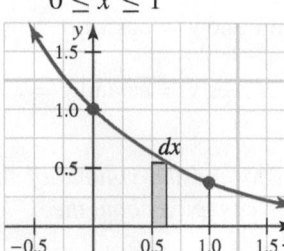

30. $y = \dfrac{x}{x+1}$,
$0 \le x \le 1$

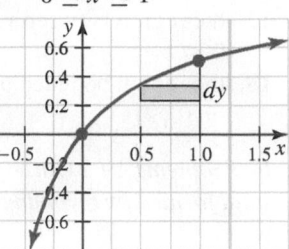

31. $y = \sin^{-1} x$,
$0 \le x \le 1$

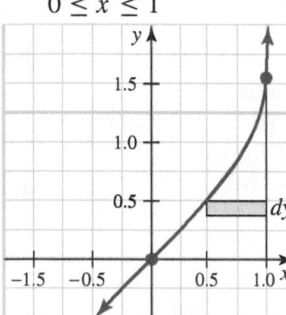

32. $y = \ln x$,
$1 \le x \le 2$

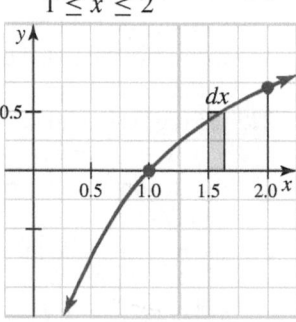

In Problems 33-38, *draw a representative strip and set up an integral for the volume of the solid formed by revolving the given region:*

  a. *about the x-axis*   **b.** *about the y-axis*

Set up the integral only; *DO NOT EVALUATE.*

33. the region bounded
by $x = y^2$ and $y = x^2$

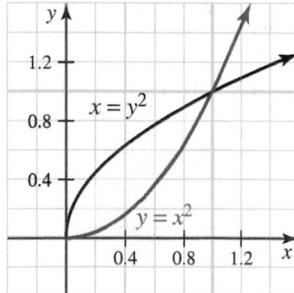

34. the region bounded
by $y = \dfrac{1}{3} x$ and $x = y^2$

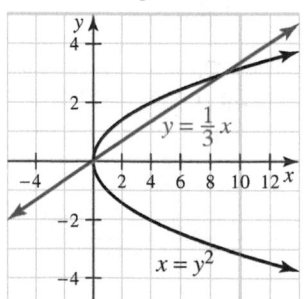

35. the region bounded
by $y = 1$, $x = 2$,
and $y = x^2 + 1$

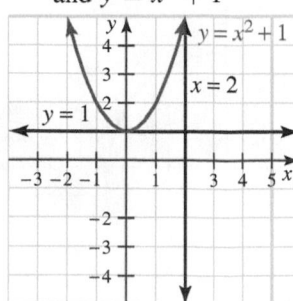

36. the region bounded
by $y = x^2$ and
$y = -x^2 - 4x$

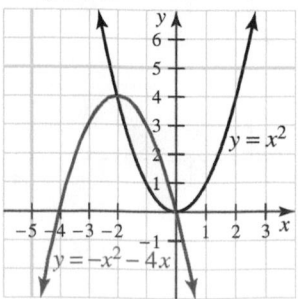

37. the region bounded
by $y = 0.1x^2$ and
$y = \ln x$

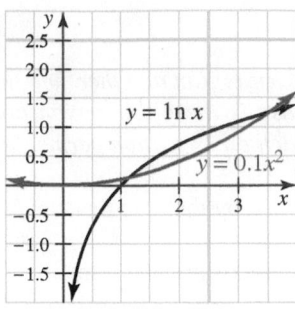

38. the region bounded
by $y = e^x - 1$,
$y = 2e^{-x}$, and
$x = 0$

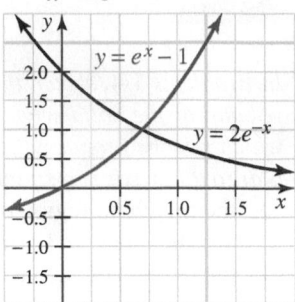

Level 2

In each of Problems 39-42, *find the volume by the*

a. *disk/washer method*
b. *shell method*

39. Revolve the region bounded by $y = x^2$ and $y = x^3$ about the x-axis.

40. Revolve the region bounded by $y = x^2$ and $y = x^3$ about the y-axis.

41. Revolve the region bounded by $y = x$, $y = 2x$, and $y = 1$ about the x-axis.

42. Revolve the region bounded by $y = x$, $y = 2x$, and $y = 1$ about the y-axis.

In Problems 43-46, *find the volume of the solid whose base is bounded by the circle* $x^2 + y^2 = 9$ *with the indicated cross sections taken perpendicular to the x-axis.*

43. squares

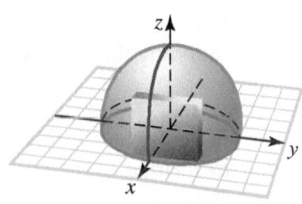

44. equilateral triangles

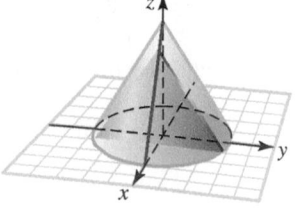

45. isosceles right triangles

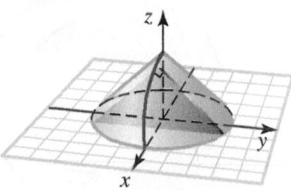

46. semicircles

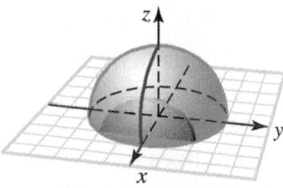

In Problems 47-48, find the volume V of the solid with the given information regarding its cross section.

47. The base of the solid is an equilateral triangle, each side of which has length 4. The cross sections perpendicular to a given altitude of the triangle are squares.

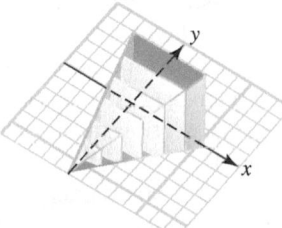

48. The base of the solid is an isosceles right triangle whose legs are each 4 units long. Each cross section perpendicular to a side is a semicircle.

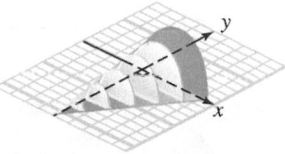

49. The great pyramid of Cheops is approximately 480 ft tall and 750 ft square at the base. Find the volume of this pyramid by using the cross section method.

Great pyramid of Cheops

50. Find the volume of the solid generated when the region $y = x^{-1/2}$ on the interval $[1, 4]$ is revolved about:
   a. the x-axis
   b. the y-axis
   c. the line $y = -2$

51. Let R be the region in the first quadrant bounded by the line $y = kx$ and the line $x = k$, $k > 0$. Find the volume when R is revolved about:
   a. the y-axis
   b. the line $x = 2k$

52. Derive the formula for the volume of a sphere of radius r by revolving the region bounded by the semicircle $y = \sqrt{r^2 - x^2}$ about the x-axis.

53. The portion of the ellipse

$$\frac{x^2}{9} + \frac{y^2}{4} = 1$$

with $x \geq 0$ is rotated about the y-axis to form a solid S. A hole of radius 1 is drilled through the center of S, along the y-axis. Find the volume of the part of S that remains.

54. Cross-sectional areas are measured at 1-foot intervals along the length of an irregularly shaped object, with the results listed in the following table (x is in ft and A is in ft^2):

x	0	1	2	3	4	5
A	1.12	1.09	1.05	1.03	0.99	1.01
x	6	7	8	9	10	
A	0.98	0.99	0.96	0.93	0.91	

Estimate the volume (correct to the nearest hundredth) by using the trapezoidal rule.

55. Cross-sectional areas are measured at 2 meter intervals along the length of an irregularly shaped object, with the results listed in the following table (x is in meters and A is in m^2):

x	0	2	4	6	8	10
A	1.12	1.09	1.05	1.03	0.99	1.01
x	12	14	16	18	20	
A	0.98	0.99	0.96	0.93	0.91	

Estimate the volume (correct to the nearest hundredth) by using Simpson's rule.

56. *Historical Quest*

Johannes Kepler is usually remembered for his work in astronomy (see Section 10.3.) Kepler also made interesting mathematical discoveries. In fact, he has been described as "number-intoxicated." He was looking for mathematical harmonies in the physical universe. He is quoted in

Karl Smith library

Johannes Kepler (1571-1630)

The World of Mathematics: *"Nothing holds me; I will indulge my sacred fury; I will triumph over mankind by the honest confession that I have stolen the golden vases of the Egyptians to build up a tabernacle for my God far away from the confines of Egypt."**

In this *Quest* we look at Kepler's derivation for the volume of a torus (Figure 6.26), generated by revolving a circle of radius a around a vertical axis at a distance b from its center ($b \geq a$).

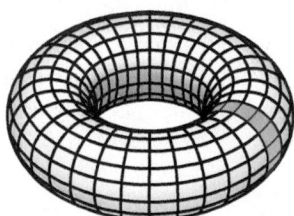

Figure 6.26 Torus

He found $V = 2\pi^2 a^2 b$. He derived this formula by dissecting a torus into infinitely many thin vertical circular slices by considering planes perpendicular to the axis of revolution, as shown in Figure 6.27.

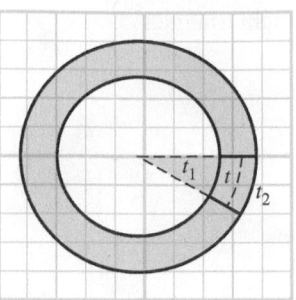

Figure 6.27 Torus cross section

Note that each slice is thinner (t_1) on the inside (nearer the axis) and thicker (t_2) on the outside. Kepler assumed that the volume of each slice is $\pi a^2 t$ where $t = \frac{1}{2}(t_1 + t_2)$. Because t is the average of its minimum and maximum thickness, it must be the thickness of the slice at its center. Use washers or shells to verify Kepler's formula for the volume of a torus.

Level 3

57. A hemisphere of radius r may be regarded as a solid whose base is the region bounded by the circle $x^2 + y^2 = r^2$ and with the property that each cross section perpendicular to the x-axis is a semicircle with a diameter in the base. Use this characterization and the method of cross sections to show that a sphere of radius r has volume $V = \frac{4}{3}\pi r^3$.

58. A student, Frank Kornercutter, conjectures that a tetrahedron is nothing more than a solid figure with an equilateral triangular base and cross sections perpendicular to that base that are also equilateral triangles. Is Frank correct this time? Either prove that he is or show that his conjecture must be false.

59. A "spherical cap" is formed by truncating a sphere of radius R at a point h units from the center, as shown in Figure 6.28. Find the volume of this cap.

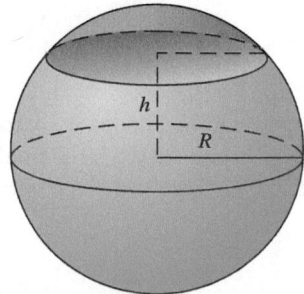

Figure 6.28 Volume of a spherical cap

---

*Vol. I, James R. Newman (New York Simon & Schuster, 1956), p. 220.

60. The *frustum of a cone* is the solid region bounded by a cone and two parallel planes, as shown in Figure 6.29.

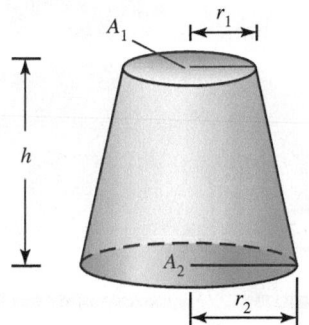

Figure 6.29 Volume of the frustum of a cone

Suppose the planes are h units apart and intersect the cone in plane regions of area A_1 and A_2, respectively. R_1 and R_2 respectively represent the radii of the circular bases.

Use integration to show that the frustum has volume

$$V = \frac{h}{3}\left(A_1 + \sqrt{A_1 A_2} + A_2\right)$$

or, equivalently,

$$V = \frac{\pi h}{3}\left(R_1^2 + R_1 R_2 + R_2^2\right)$$

6.3 POLAR FORMS AND AREA

IN THIS SECTION: *The polar coordinate system, polar graphs, summary of polar-form curves, intersection of polar-form curves, polar area*

Up to now, the only coordinate system we have used is a Cartesian coordinate system. In this section, we introduce the *polar coordinate system*, in which points are located by specifying their distance from a fixed point and their direction from a fixed line.

The Polar Coordinate System

In the **polar coordinate system**, points are plotted in relation to a fixed point O, called the origin or **pole** and a fixed ray emanating from the origin called the **polar axis**. We then associate with each point P in the plane an ordered pair of numbers $P(r, \theta)$, where r is the distance from O to P and θ is the angle measured from the polar axis to the ray OP, as shown in Figure 6.30. The number r is called the **radial coordinate** of P, and θ is the **polar angle**. The polar angle is regarded as positive if measured counterclockwise from the polar axis, and negative if measured clockwise. The origin O has radial coordinate 0, and it is convenient to say that O has polar coordinates $(0, \theta)$, for all angles θ.

If the point P has polar coordinates (r, θ), we say that the point Q obtained by reflecting P in the origin O has coordinates $(-r, \theta)$. Thus, if you think of a pencil lying along the directed line segment $\overline{OP}$, with its midpoint at O and tip at $P(r, \theta)$, then the eraser will be at $Q(-r, \theta)$, as illustrated in Figure 6.30. Note that $P(r, \theta)$ and $Q(-r, \theta)$ represent opposite points.

While each point in the plane is associated with exactly one pair of rectangular coordinates, a given point in polar coordinates has an infinite number of representations. For example, $\left(5, \frac{3\pi}{2}\right)$, $\left(-5, \frac{\pi}{2}\right)$, and $\left(5, -\frac{\pi}{2}\right)$ all represent the same point in polar coordinates. The non-uniqueness of representation in the polar coordinate system causes some difficulties, but this can be handled by exercising a little care. We shall point out situations in our examples where the multiplicity of polar representation must be taken into account.

There is a simple trigonometric relationship linking polar and rectangular coordinates. Specifically, if we assume that the origin of the rectangular coordinate system is the pole and the positive x-axis is the polar axis, then the rectangular coordinates (x, y) of each point are related to the polar coordinates (r, θ) of the same point by the formulas summarized in the following box.

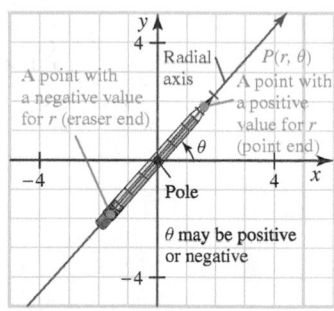

Figure 6.30 Polar-form points

CHANGING COORDINATES The procedure for changing from one coordinate system to another.

Step 1 To change *from polar to rectangular* form use the formulas:

$$x = r\cos\theta \qquad y = r\sin\theta$$

Step 2 To change from *rectangular to polar* form use the formulas

$$r = \sqrt{x^2 + y^2} \qquad \tan\theta = \frac{y}{x}$$

if $x \neq 0$.

Note that the preceding formula gives r as the distance from the pole to the point. This does **not** mean that ordered pairs in polar form need to have positive first components. For example $\left(1, \frac{3\pi}{2}\right)$ is the same point as $\left(-1, \frac{\pi}{2}\right)$.

You can remember these relationships by looking at Figure 6.31.

Also, note that $\tan\theta = \frac{y}{x}$ is the stated formula, but to find θ, we first find the reference angle $\overline{\theta}$ and then place it in the proper quadrant by noting the signs of x and y. Use $\overline{\theta} = \tan^{-1}\left|\frac{y}{x}\right|$, $x \neq 0$. If $x = 0$, $\overline{\theta} = \frac{\pi}{2}$.

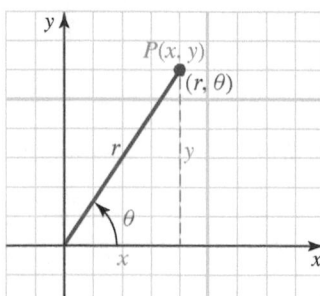

Figure 6.31 Relationship between rectangular and polar coordinate systems

Example 1   Converting from polar to rectangular coordinates

Change the polar coordinates $\left(-3, \frac{5\pi}{4}\right)$ to rectangular coordinates.

Solution $x = -3\cos\frac{5\pi}{4} = -3\left(-\frac{\sqrt{2}}{2}\right) = \frac{3\sqrt{2}}{2}$

$y = -3\sin\frac{5\pi}{4} = -3\left(-\frac{\sqrt{2}}{2}\right) = \frac{3\sqrt{2}}{2}$

The rectangular coordinates are $\left(\frac{3\sqrt{2}}{2}, \frac{3\sqrt{2}}{2}\right)$.

Example 2   Converting from rectangular to polar coordinates

Write polar-form coordinates for the point with rectangular coordinates $\left(\frac{5\sqrt{3}}{2}, -\frac{5}{2}\right)$.

Solution $r = \sqrt{\left(\frac{5\sqrt{3}}{2}\right)^2 + \left(-\frac{5}{2}\right)^2} = \sqrt{\frac{75}{4} + \frac{25}{4}} = 5$

Note that θ is in quadrant IV because x is positive and y is negative.

$\overline{\theta} = \tan^{-1}\left|\frac{-\frac{5}{2}}{\frac{5\sqrt{3}}{2}}\right| = \tan^{-1}\left(\frac{1}{\sqrt{3}}\right) = \frac{\pi}{6}$; thus, $\theta = \frac{11\pi}{6}$ (quadrant IV)

Polar form coordinates are $\left(5, \frac{11\pi}{6}\right)$. A representation with a negative first component is $\left(-5, \frac{5\pi}{6}\right)$.

Polar Graphs

The **graph** of an equation in polar coordinates is the set of all points P whose polar coordinates (r, θ) satisfy the given equation. Circles, lines through the origin, and rays emanating from the origin have particularly simple equations in polar coordinates. *

---

*Many books use what is called **polar graph paper**, but such paper is not really necessary. It also obscures the fact that polar curves and rectangular curves are both plotted as ordered pairs only with a different meaning attached to the ordered pairs. You can estimate the angles as necessary without polar graph paper.

Example 3   Graphing circles, lines, and rays

Graph: **a.** $r = 6$ **b.** $\theta = \dfrac{\pi}{6}$

Solution

a. The graph is the set of all points (r, θ) such that the first component is 6 for any angle θ. This is a circle with radius 6 centered at the origin, as shown in Figure 6.32**a**.

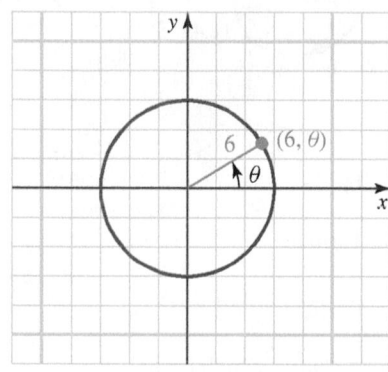
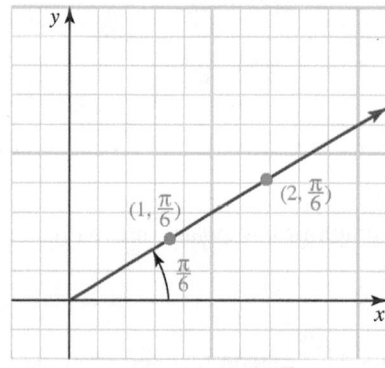

a. The circle $r = 6$ **b.** The line $\theta = \dfrac{\pi}{6}$

Figure 6.32   Two polar graphs

b. The graph is the set of all polar points (r, θ) with polar angle $\theta = \frac{\pi}{6}$. This is a line through the origin (pole) that makes an angle $\pi/6$ with the positive x-axis, as shown in Figure 6.32**b**. If we assume $r \geq 0$, then the graph is a ray (sometimes called a *half-line*).

Example 4   Graph by converting to rectangular coordinates

Graph $r = 4 \cos \theta$ by first converting the polar form to rectangular form.

Solution

$$r = 4 \cos \theta \quad \textit{Given}$$

$$r^2 = 4r \cos \theta \quad \textit{Multiply both sides by } r.$$

$$x^2 + y^2 = 4x \quad \textit{Change form (polar to rectangular).}$$

$$x^2 - 4x + y^2 = 0$$

$$(x - 2)^2 + y^2 = 2^2 \quad \textit{Complete the square; add } 2^2 \textit{ to both sides.}$$

This is a circle with center at $(2, 0)$ and radius 2, as shown in Figure 6.33.

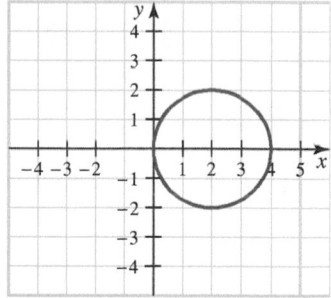

Figure 6.33   Graph of $r = 4 \cos \theta$

In our next example, we sketch a more complicated polar graph by plotting points in a polar coordinate plane. The graph is called a **cardioid** because of its heart-like shape.

Example 5   A cardioid by plotting points

Graph $r = 2(1 - \cos \theta)$.

Solution   Construct a table of values by choosing values for θ and approximating the corresponding values for r. Do not forget that even though we pick θ and find a value for r, we still represent the table values as ordered pairs of the form (r, θ).

θ	r
0	0
1	0.9193954
2	2.832294
3	3.979985
4	3.307287
5	1.432676
6	0.079659

The points are connected as shown in Figure 6.34.

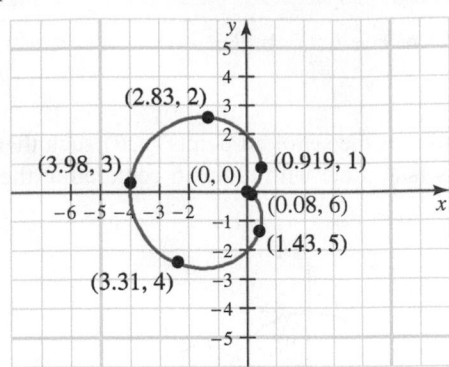

Figure 6.34   Graph of $r = 2(1 - \cos\theta)$

The graph of any polar equation of the general form

$$r = b \pm a\cos\theta \qquad \text{or} \qquad r = b \pm a\sin\theta$$

is called a **limaçon** (derived from the Latin word *limax*, which means "slug"; in French, it is the word for "*snail*"). The special case where $a = b$ is the *cardioid*. Figure 6.35 shows four different kinds of limaçons that can occur. Note how the appearance of the graph depends on the ratio b/a.

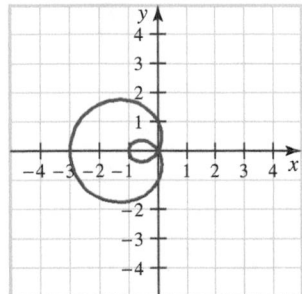

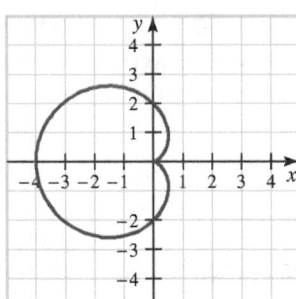

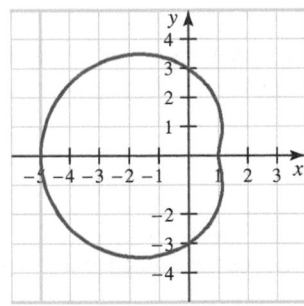

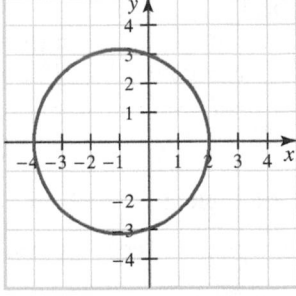

a. $\dfrac{b}{a} < 1$; inner loop

b. $\dfrac{b}{a} = 1$; cardioid

c. $1 < \dfrac{b}{a} < 2$; dimple

d. $\dfrac{b}{a} \geq 2$; convex

Case I: $(r = 1 - 2\cos\theta)$

Case II: $(r = 2 - 2\cos\theta)$

Case III: $(r = 3 - 2\cos\theta)$

Case IV: $(r = 3 - \cos\theta)$

Figure 6.35   Limaçons: $r = b \pm a\cos\theta$ or $r = b \pm a\sin\theta$

Summary of Polar-Form Curves

There are several polar form curves which you can graph by plotting points or by using a graphing calculator (be sure to convert to polar mode). Table 6.2 gives you the names of some of the special types of polar-form curves you will encounter. There are many others, some of which are represented in the problems.

Table 6.2 Directory of Polar-Form Curves

LIMAÇONS $r = b \pm a \cos \theta$ and $r = b \pm a \sin \theta$

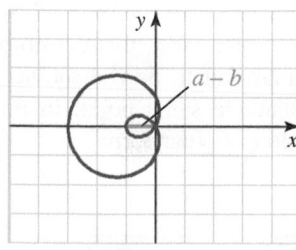

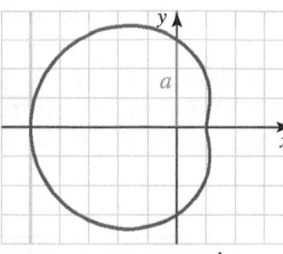

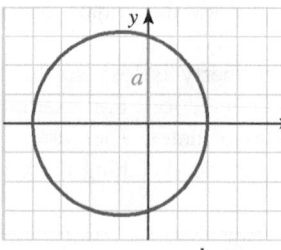

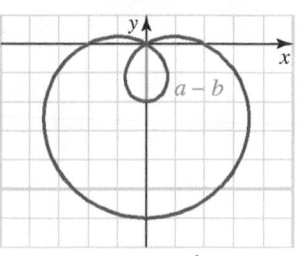

$r = b - a \cos \theta. \dfrac{b}{a} < 1$	$r = b - a \cos \theta, 1 < \dfrac{b}{a} < 2$	$r = b - a \cos \theta, \dfrac{b}{a} \geq 2$	$r = b - a \sin \theta, \dfrac{b}{a} < 1$
standard form, inner loop	standard form, dimple	standard form, convex	$\dfrac{\pi}{2}$ rotation; inner loop

---

CARDIOIDS $r = a(1 \pm \cos \theta)$ and $r = a(1 \pm \sin \theta)$ Limaçons in which $a = b$

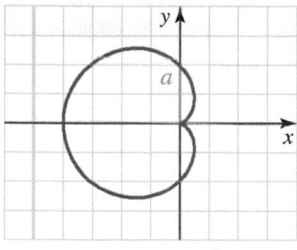

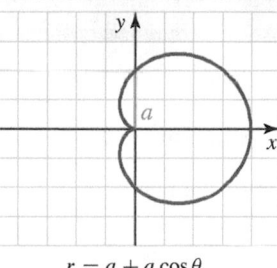

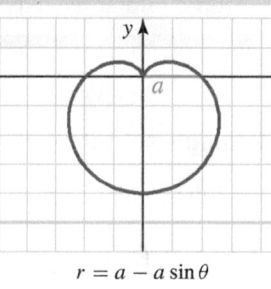

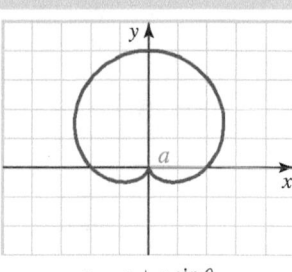

$r = a - a \cos \theta$	$r = a + a \cos \theta$	$r = a - a \sin \theta$	$r = a + a \sin \theta$
standard form	π rotation	$\dfrac{\pi}{2}$ rotation	$\dfrac{3\pi}{2}$ rotation

---

ROSE CURVES
$r = a \cos n\theta$ and $r = a \sin n\theta$

LEMNISCATES
$r^2 = a^2 \cos 2\theta$ and $r^2 = a^2 \sin 2\theta$

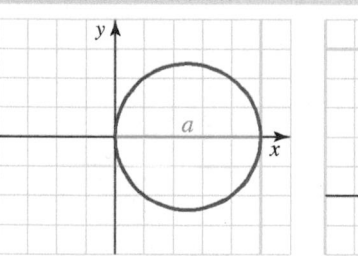

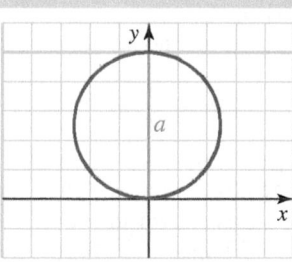

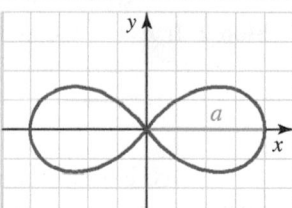

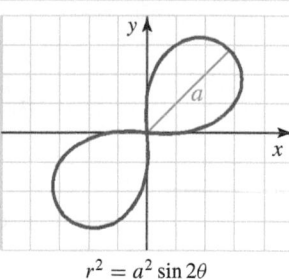

$r = a \cos \theta$; circle	$r = a \sin \theta$; circle	$r^2 = a^2 \cos 2\theta$	$r^2 = a^2 \sin 2\theta$
standard form; one petal	$\dfrac{\pi}{2}$ rotation; one petal	standard form	$\dfrac{\pi}{4}$ rotation

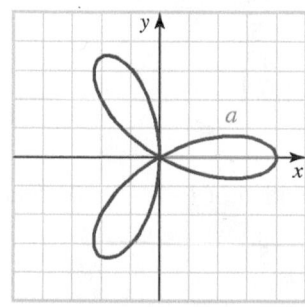

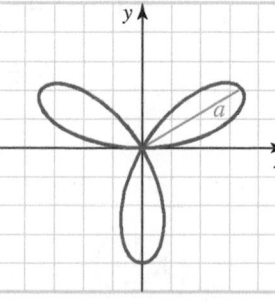

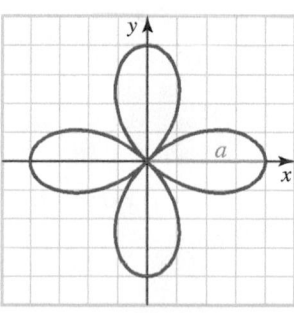

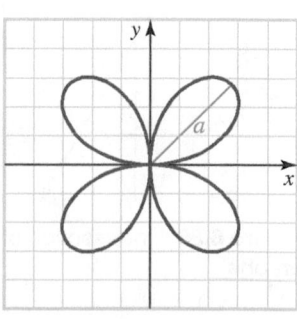

$r = a \cos 3\theta$	$r = a \sin 3\theta$	$r = a \cos 2\theta$	$r = a \sin 2\theta$
standard form; three petals	$\dfrac{\pi}{6}$ rotation; three petals	standard form; four petals	$\dfrac{\pi}{4}$ rotation; four petals

Intersection of Polar-Form Curves

To find the points of intersection of graphs in rectangular form, you need only find the simultaneous solution of the equations that define those graphs. It is not even necessary to draw the graphs, because there is a one-to-one correspondence between ordered pairs satisfying an equation and points on its graph. However, in polar form, this one-to-one property is lost, so that without drawing the graphs you may fail to find all points of intersection. For example, $(-1, \pi)$ and $(1, 0)$ both represent the same point, in polar coordinates. Therefore, our method for finding the intersection of polar-form curves will include sketching the graphs.

GRAPHICAL SOLUTION OF THE INTERSECTION OF POLAR FORM CURVES

Step 1 Find all simultaneous solutions of the given system of equations.
Step 2 Determine whether the pole $r = 0$ lies on the two graphs.
Step 3 Graph the curves to look for other points of intersection.

Example 6   Intersection of polar form curves

Find the points of intersection of the curves $r = \frac{3}{2} - \cos\theta$ and $\theta = \frac{2\pi}{3}$.

Step 1 Solve the system by substitution:

$$r = \frac{3}{2} - \cos\frac{2\pi}{3} = \frac{3}{2} - \left(-\frac{1}{2}\right) = 2$$

The solution is $\left(2, \frac{2\pi}{3}\right)$.

Step 2 If $r = 0$, the first equation has no solution because

$$0 = \frac{3}{2} - \cos\theta \qquad \text{or} \qquad \cos\theta = \frac{3}{2}$$

and a cosine cannot be larger than 1.

Step 3 Now look at the graphs, as shown in Figure 6.36.

We see that $\left(-1, \frac{2\pi}{3}\right)$ may also be a point of intersection. It satisfies the equation $\theta = \frac{2\pi}{3}$, but what about $r = \frac{3}{2} - \cos\theta$?

$$\text{Check } \left(-1, \frac{2\pi}{3}\right) \qquad -1 \overset{?}{=} \frac{3}{2} - \cos\frac{2\pi}{3}$$
$$= \frac{3}{2} - \left(-\frac{1}{2}\right)$$
$$= 2 \qquad \text{Not satisfied}$$

However, if you check an alternative representation of $\left(-1, \frac{2\pi}{3}\right)$, we find:

$$\text{Check } \left(1, \frac{5\pi}{3}\right) \qquad 1 \overset{?}{=} \frac{3}{2} - \cos\frac{5\pi}{3}$$
$$= \frac{3}{2} - \frac{1}{2}$$
$$= 1 \qquad \text{Satisfied}$$

Be sure to check for points of intersection that you may have missed.

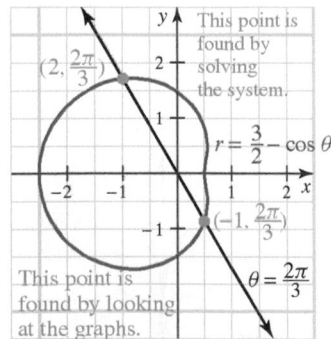

This point is found by solving the system.

$r = \frac{3}{2} - \cos\theta$

$\left(2, \frac{2\pi}{3}\right)$

$\left(-1, \frac{2\pi}{3}\right)$

This point is found by looking at the graphs.

$\theta = \frac{2\pi}{3}$

Figure 6.36 Intersection of graphs

Polar Area

To find the area of a region bounded by a polar graph, we use Riemann sums in much the same way as when we developed the integral formula for the area of a region described in rectangular form. However, instead of using rectangular areas as the basic units being summed, in polar form, we sum areas of *circular sectors*. A typical circular sector is shown in Figure 6.37. Recall the formula for the area of such a sector from trigonometry.

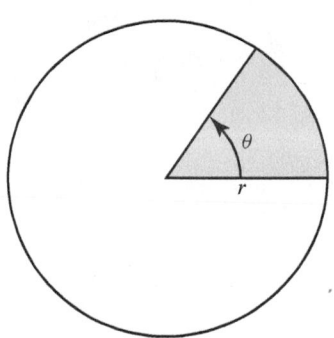

Figure 6.37 Sector of a circle

> **AREA OF A SECTOR**   The area of a circular sector of radius r is given by
>
> $$A = \frac{1}{2}r^2\theta$$
>
> where θ is the central angle of the sector measured in radians.

Using this formula for the area of a sector, we now state a theorem that gives us a formula for finding the area enclosed by a polar curve.

Theorem 6.1   Area in polar coordinates

Let $r = f(\theta)$ define a polar curve, where f is continuous and $f(\theta) \geq 0$ on the closed interval $\alpha \leq \theta \leq \beta$, where $0 \leq \beta - \alpha \leq 2\pi$. Then the region bounded by the curve $r = f(\theta)$ and the rays $\theta = \alpha$ and $\theta = \beta$ has area

$$A = \frac{1}{2}\int_\alpha^\beta r^2 d\theta = \frac{1}{2}\int_\alpha^\beta \left[f(\theta)\right]^2 d\theta$$

Proof: The region is shown in Figure 6.38. To find the area bounded by the graphs of the polar functions, partition the region between $\theta = \alpha$ and $\theta = \beta$ by a collection of rays, say $\theta_0, \theta_1, \cdots, \theta_n$. Let $\alpha = \theta_0$ and $\beta = \theta_n$.

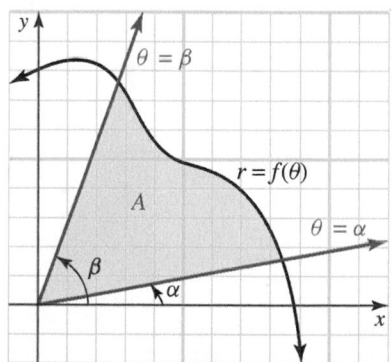

a. The region bounded by the polar curve $r = f(\theta)$ and the rays $\theta = a$ and $\theta = b$

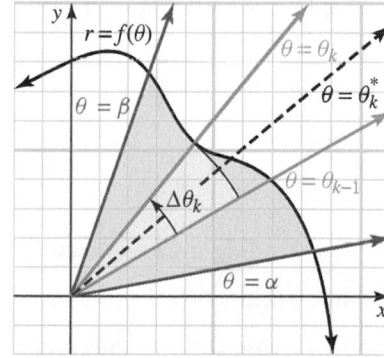

b. The area A can be estimated by adding the area of "small" (blue) circular sectors

Figure 6.38   Area in polar form

Pick any ray $\theta = \theta_k^*$, with $\theta_{k-1} \leq \theta_k^* \leq \theta_k$. Then the area ΔA_k of the circular sector is approximately the same as the area of the region bounded by the graph of f and the lines $\theta = \theta_{k-1}$ and $\theta = \theta_k$. Because this circular sector has radius $f(\theta_k^*)$ and central angle $\Delta\theta_k$, its area is

$$\Delta A_k \approx \frac{1}{2}(\text{radius})^2(\text{central angle}) = \frac{1}{2}\left[f(\theta_k^*)\right]^2 \Delta\theta_k$$

The sum $\sum_{k=1}^{n} \Delta A_k$ is an approximation to the total area A bounded by the polar curve, and by taking the limit as $n \to \infty$, we obtain

$$A = \lim_{n\to\infty} \sum_{k=1}^{n} \Delta A_k$$

$$= \lim_{n\to\infty} \frac{1}{2} \sum_{k=1}^{n} \left[f(\theta_k^*)\right]^2 \Delta\theta_k$$

$$= \frac{1}{2} \int_{\alpha}^{\beta} \left[f(\theta)\right]^2 d\theta \qquad \textit{Riemann sum} \qquad \blacklozenge$$

You will probably find that the most difficult part of the problem is deciding on the limits of integration. A sketch of the region will help with this.

Example 7   Finding area of part of a cardioid

Find the area of the top half $(0 \le \theta \le \pi)$ of the cardioid $r = 1 + \cos\theta$.

Solution   The cardioid is shown in Figure 6.39.

Note that the top half of the graph lies between the rays $\theta = 0$ and $\theta = \pi$. Hence the required area is given by:

$$A = \frac{1}{2} \int_0^{\pi} (1 + \cos\theta)^2 \, d\theta$$

$$= \frac{1}{2} \int_0^{\pi} (1 + 2\cos\theta + \cos^2\theta) \, d\theta$$

$$= \frac{1}{2} \int_0^{\pi} \left(1 + 2\cos\theta + \frac{1 + \cos 2\theta}{2}\right) d\theta \qquad \textit{Trigonometric half-angle identity}$$

$$= \frac{1}{2} \left[\theta + 2\sin\theta + \frac{\theta}{2} + \frac{\sin 2\theta}{4}\right]_0^{\pi}$$

$$= \frac{1}{2} \left[\pi + 2(0) + \frac{\pi}{2} + \frac{0}{4} - 0\right]$$

$$= \frac{3\pi}{4}$$

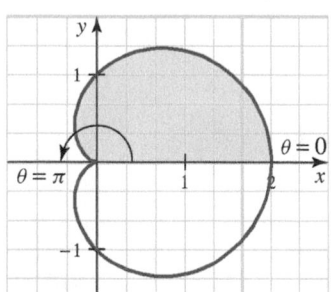

Figure 6.39   Area of top half of a cardioid

Example 8   Finding area enclosed by a four-leaved rose

Find the area enclosed by the four-leaved rose $r = \cos 2\theta$.

Solution   The rose curve is shown in Figure 6.40.

We will find the area of the top half of the right loop (shaded portion) to make sure $f(\theta) \ge 0$. We see that this corresponds to angles with measures from $\theta = 0$ to $\theta = \frac{\pi}{4}$. If you have a calculator, the easiest way to see this is to plot the graph and then use the *trace* feature to find the θ-values for the top half of the first leaf. By symmetry, the entire area enclosed by the four-leaved rose is 8 times the shaded region. Thus, the required area is given by

$$A = 8\left[\frac{1}{2} \int_0^{\frac{\pi}{4}} \cos^2 2\theta \, d\theta\right]$$

$$= 4 \int_0^{\frac{\pi}{4}} \frac{1 + \cos 4\theta}{2} \, d\theta \qquad \textit{Trigonometric half-angle identity}$$

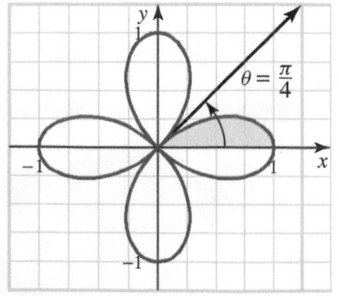

Figure 6.40   Area enclosed by a four-leaved rose

$$= 4 \left[\frac{\theta}{2} + \frac{\sin 4\theta}{8} \right]_0^{\frac{\pi}{4}}$$

$$= 4 \left[\frac{\pi}{8} + 0 - 0 \right]$$

$$= \frac{\pi}{2}$$

Example 9 Finding the area of a region between two polar curves

Find the area of the region common to the circles $r = a \cos\theta$ and $r = a \sin\theta$.

Solution The circles are shown in Figure 6.41.

Solve $a \cos\theta = a \sin\theta$ to find that the circles intersect at $\theta = \frac{\pi}{4}$ and at the pole. To set up the integrals properly, remember to think in terms of polar coordinates and not in terms of rectangular coordinates. Specifically, we must **scan radially**. This means that we need to find the area of intersection by adding the areas of the regions marked R_1 and R_2. Region R_1 is bounded by the circle $r = a \sin\theta$ and the rays $\theta = 0$, $\theta = \frac{\pi}{4}$. Region R_2 is bounded by the circle $r = a \cos\theta$ and $\theta = \frac{\pi}{4}$, $\theta = \frac{\pi}{2}$. Note the rays $\theta = 0$ and $\theta = \frac{\pi}{2}$ are not necessary as geometric boundaries for R_1 and R_2, but they are necessary to describe which part of each circle is being calculated.

$$A = \text{AREA OF } R_1 + \text{AREA OF } R_2$$

$$= \frac{1}{2} \int_0^{\frac{\pi}{4}} a^2 \sin^2\theta \, d\theta + \frac{1}{2} \int_{\frac{\pi}{4}}^{\frac{\pi}{2}} a^2 \cos^2\theta \, d\theta$$

$$= \frac{a^2}{2} \left[\frac{\theta}{2} - \frac{\sin 2\theta}{4} \right]_0^{\frac{\pi}{4}} + \frac{a^2}{2} \left[\frac{\theta}{2} + \frac{\sin 2\theta}{4} \right]_{\frac{\pi}{4}}^{\frac{\pi}{2}} \qquad \textit{Trigonometric half-angle identities}$$

$$= \frac{a^2}{2} \left[\frac{\pi}{8} - \frac{1}{4} - 0 \right] + \frac{a^2}{2} \left[\frac{\pi}{4} + 0 - \frac{\pi}{8} - \frac{1}{4} \right]$$

$$= \frac{a^2}{2} \left[\frac{\pi}{4} - \frac{1}{2} \right]$$

$$= \frac{1}{8} a^2 (\pi - 2)$$

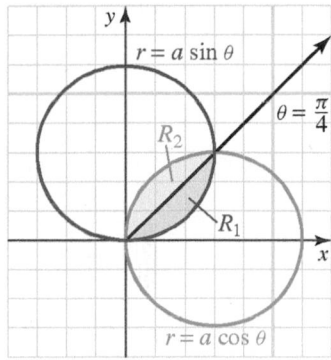

Figure 6.41 Area bounded by two polar curves

Example 10 Finding the area between a circle and a limaçon

Find the area between the circle $r = 5 \cos\theta$ and the limaçon $r = 2 + \cos\theta$. Round your answer to the nearest hundredth of a square unit.

Solution As usual, begin by drawing the graphs, as shown in Figure 6.42.

Note that both the limaçon and circle are symmetric with respect to the x-axis, so we can find the area in the first quadrant and multiply by 2.

Next, we need to find the points of intersection. We see that the curves do not intersect at the pole.

Now solve

$$5 \cos\theta = 2 + \cos\theta$$

$$\cos\theta = \frac{1}{2}$$

$$\theta = \frac{\pi}{3} \qquad \textit{Solution in the first quadrant}$$

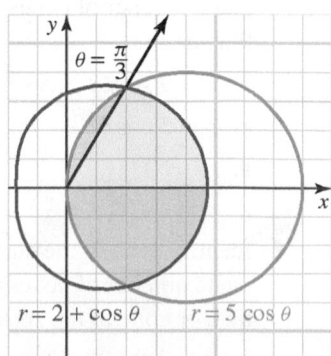

Figure 6.42 Area bounded by a circle and a limaçon

To find the area, divide the region into two parts, along the ray $\theta = \frac{\pi}{3}$. The right part (blue region above the x-axis) is bounded by the limaçon $r = 2 + \cos\theta$ and the rays $\theta = 0$ and $\theta = \frac{\pi}{3}$. The left part (shaded region shown in pink above the x-axis) is bounded by the circle $r = 5\cos\theta$ and the rays $\theta = \frac{\pi}{3}$ and $\theta = \frac{\pi}{2}$. Using this preliminary information, we can now set up and evaluate a sum of integrals for the required area

$$A = 2[\text{AREA OF RIGHT PART} + \text{AREA OF LEFT PART}]$$

$$= 2\left[\frac{1}{2}\int_0^{\frac{\pi}{3}}(2+\cos\theta)^2 d\theta + \frac{1}{2}\int_{\frac{\pi}{3}}^{\frac{\pi}{2}}(5\cos\theta)^2 d\theta\right]$$

$$= \int_0^{\frac{\pi}{3}}(4+4\cos\theta+\cos^2\theta)\,d\theta + \int_{\frac{\pi}{3}}^{\frac{\pi}{2}}25\cos^2\theta\,d\theta$$

$$= \left[4\theta + 4\sin\theta + \frac{\theta}{2} + \frac{\sin 2\theta}{4}\right]_0^{\frac{\pi}{3}} + 25\left[\frac{\theta}{2} + \frac{\sin 2\theta}{4}\right]_{\frac{\pi}{3}}^{\frac{\pi}{2}}$$

$$= \left[\frac{4\pi}{3} + \frac{4\sqrt{3}}{2} + \frac{\pi}{6} + \frac{\sqrt{3}}{8} - 0\right] + 25\left[\frac{\pi}{4} + 0 - \frac{\pi}{6} - \frac{\sqrt{3}}{8}\right]$$

$$= \frac{43\pi}{12} - \sqrt{3}$$

$$\approx 9.53$$

PROBLEM SET 6.3

1. ■ What does this say? Discuss a procedure for finding the intersection of polar-form curves.
2. ■ What does this say? Discuss a procedure for finding the area enclosed by a polar curve.
3. Identify each of the curves as a cardioid, rose curve (state number of petals), lemniscate, limaçon, circle, line, or none of the above.

   a. $r^2 = 9\cos 2\theta$ **b.** $r = 2\sin\frac{\pi}{6}$

   c. $r = 3\sin 3\theta$ **d.** $r = 3\theta$

   e. $r = 2 - 2\cos\theta$ **f.** $\theta = \frac{\pi}{6}$

   g. $r^2 = \sin 2\theta$ **h.** $r - 2 = 4\cos\theta$

4. Identify each of the curves as a cardioid, rose curve (state number of petals), lemniscate, limaçon, circle, line, or none of the above.

   a. $r = 2\sin 2\theta$ **b.** $r^2 = 2\cos 2\theta$

   c. $r = 5\cos 60°$ **d.** $r = 5\sin 8\theta$

   e. $r\theta = 3$ **f.** $r^2 = 9\cos(2\theta - \frac{\pi}{4})$

   g. $r = \sin 3\left(\theta + \frac{\pi}{6}\right)$ **h.** $\cos\theta = 1 - r$

5. Identify each of the curves as a cardioid, rose curve (state number of petals), lemniscate, limaçon, circle, line, or none of the above.

   a. $r = 2\cos 2\theta$ **b.** $r = 4\sin 30°$

   c. $r + 2 = 3\sin\theta$ **d.** $r + 3 = 3\sin\theta$

   e. $\theta = 4$ **f.** $\theta = \tan\frac{\pi}{4}$

   g. $r = 3\cos 5\theta$ **h.** $r\cos\theta = 2$

Graph the polar-form curves given in Problems 6-21.

6. $r = 3, 0 \le \theta \le \frac{\pi}{2}$ **7.** $\theta = -\frac{\pi}{2}, 0 \le r \le 3$

8. $r = 2\theta, \theta \ge 0$ **9.** $r = \theta + 1, 0 \le \theta \le \pi$

10. $r = 2\cos 2\theta$ **11.** $r = 5\sin 3\theta$

12. $r^2 = 16\cos 2\theta$ **13.** $r = 3\cos 3\left(\theta - \frac{\pi}{3}\right)$

14. $r = 5\cos 3\left(\theta - \frac{\pi}{4}\right)$ **15.** $r = \sin\left(2\theta + \frac{\pi}{3}\right)$

16. $r = 2 + \cos\theta$ **17.** $r^2 = 16\cos 2\left(\theta - \frac{\pi}{6}\right)$

18. $r = 1 + \sin\theta$ **19.** $r\cos\theta = 2$

20. $r = 1 + 3\cos\theta$ **21.** $r = -2\sin\theta$

Find the points of intersection of the curves given in Problems 22-29.

22. $\begin{cases} r = 4\cos\theta \\ r = 4\sin\theta \end{cases}$ **23.** $\begin{cases} r = 4\sin\theta \\ r = 2 \end{cases}$

24. $\begin{cases} r^2 = 9\cos 2\theta \\ r = 3 \end{cases}$ **25.** $\begin{cases} r = 3\theta \\ \theta = \frac{\pi}{3} \end{cases}$

26. $\begin{cases} r = 2(1 + \sin\theta) \\ r = 2(1 - \sin\theta) \end{cases}$ **27.** $\begin{cases} r^2 = \sin 2\theta \\ r = \sqrt{2}\sin\theta \end{cases}$

28. $\begin{cases} r = 2(1 - \cos\theta) \\ r = 4\sin\theta \end{cases}$ **29.** $\begin{cases} r = \dfrac{4}{1 - \cos\theta} \\ r = 2\cos\theta \end{cases}$

Find the area of each polar region enclosed by $f(\theta)$,
$\theta = a, \theta = b$, for $a \le \theta \le b$ in Problems 30-37.

30. $f(\theta) = \sin\theta, 0 \le \theta \le \frac{\pi}{6}$

31. $f(\theta) = \cos\theta, 0 \le \theta \le \frac{\pi}{6}$

32. $f(\theta) = \sec\theta, -\frac{\pi}{4} \le \theta \le \frac{\pi}{4}$

33. $f(\theta) = \sqrt{\sin\theta}, \frac{\pi}{6} \le \theta \le \frac{\pi}{2}$

34. $f(\theta) = e^{\theta/2}, 0 \le \theta \le 2\pi$

35. $f(\theta) = \sin\theta + \cos\theta, 0 \le \theta \le \frac{\pi}{4}$

36. $f(\theta) = \frac{\theta}{\pi}, 0 \le \theta \le 2\pi$

37. $f(\theta) = \frac{\theta^2}{\pi}, 0 \le \theta \le 2\pi$

38. Spirals are interesting mathematical curves.

Spiral staircase

There are three special types of spirals:

a. A **spiral of Archimedes** has the form $r = a\theta$.
   Graph $r = 2\theta$.

b. A **hyperbolic spiral** has the form $r\theta = a$.
   Graph $r\theta = 2$.

c. A **logarithmic spiral** has the form $r = a^{k\theta}$.
   Graph $r = 2^{\theta}$.

39. The **strophoid** is a curve of the form
$r = a\cos 2\theta \sec\theta$. Graph this curve for $a = 2$.

40. The **bifolium** has the form $r = a\sin\theta\cos^2\theta$. Graph this curve for $a = 1$.

41. The **folium of Descartes** has the form
$$r = \frac{3a\sin\theta\cos\theta}{\sin^3\theta + \cos^3\theta}.$$ Graph this curve for $a = 2$.

42. Find the area of one loop of the four-leaved rose $r = 2\sin 2\theta$.

43. Find the area enclosed by the three-leaved rose $r = a\sin 3\theta$.

44. Find the area of the region that is inside the circle $r = 4\cos\theta$ and outside the circle $r = 2$.

45. Find the area of the region that is inside the circle $r = a$ and outside the cardioid $r = a(1 - \cos\theta)$.

46. Find the area of the region that is inside the circle $r = \sin\theta$ and outside the cardioid $r = 1 - \cos\theta$.

47. Find the area of the region that is inside the circle $r = 6\cos\theta$ and outside the cardioid $r = 2(1 + \cos\theta)$.

48. Find the area of the portion of the lemniscate $r^2 = 8\cos 2\theta$ that lies in the region $r \ge 2$.

49. Find the area between the inner and outer loops of the limaçon $r = 2 - 4\sin\theta$.

50. Find the area to the right of the line $r\cos\theta = 1$ and inside the lemniscate $r^2 = 2\cos 2\theta$.

51. Find the area to the right of the line $r\cos\theta = 1$ and inside the lemniscate $r^2 = 2\sin 2\theta$.

52. Find the maximum value of the y-coordinate of points on the limaçon $r = 2 + 3\cos\theta$.

53. Find the maximum value of the x-coordinate of points on the cardioid $r = 3 + 3\sin\theta$.

54. a. Show that if the polar curve $r = f(\theta)$ is rotated about the pole through an angle α, the equation for the new curve is $r = f(\theta - \alpha)$.

b. Use a rotation to sketch the curve $r = 2\sec\left(\theta - \frac{\pi}{3}\right)$.

55. The **ovals of Cassini** have the polar form

$$r^4 + b^4 - 2b^2 r^2 \cos 2\theta = k^4.$$

Graph the curve, where $b = 2, k = 3$.

56. Sketch the graph of

$$r = \frac{\theta}{\cos\theta}$$

for $0 \le \theta \le \frac{\pi}{2}$. In particular, show that the graph has a vertical asymptote at $x = \frac{\pi}{2}$.

57. If f is a differentiable function of θ, then show that the tangent line to the polar curve $r = f(\theta)$ at the point $P(r, \theta)$ has slope

$$m = \frac{f(\theta)\cos\theta + f'(\theta)\sin\theta}{-f(\theta)\sin\theta + f'(\theta)\cos\theta}$$

whenever the denominator is not zero.

Hint: $\dfrac{dy}{dx} = \dfrac{dy/d\theta}{dx/d\theta}$.

58. Because the formula for the slope in polar-form is complicated (see Problem 57), it is more convenient to measure the inclination of the tangent line to a polar curve in terms of the angle α extended from the radial line to the tangent line, as shown in Figure 6.43.

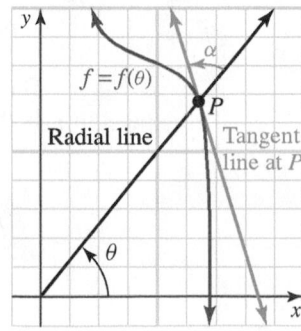

Figure 6.43 Problem 58

Let P be a point on the polar curve $r = f(\theta)$, and let α be the angle extending from the radial line to the tangent line, as shown in Figure 6.43. Assuming that $f'(\theta) \neq 0$, show that

$$\tan \alpha = \frac{f(\theta)}{f'(\theta)}$$

Hint: Use the formula
$$\tan \alpha = \tan(\phi - \theta) = \frac{\tan \phi - \tan \theta}{1 + \tan \phi \tan \theta}.$$

59. Use the formula in Problem 58 to find $\tan \alpha$ for each of the following:

    a. the circle $r = a \cos \theta$
    b. the cardioid $r = 2(1 - \cos \theta)$
    c. the logarithmic spiral $r = 2e^{3\theta}$

60. Journal Problem (*School Science and Mathematics**) Graph the polar curve

$$r = 4 + 2 \sin \frac{5\theta}{2}$$

and find the total area interior to this curve. Then find the area of the "star."

6.4 ARC LENGTH AND SURFACE AREA

IN THIS SECTION: *The arc length of a curve, the area of a surface of revolution, polar arc length and surface area*

If you were asked to measure the length of a given curve by hand, you might proceed by first fitting a piece of string to the curve and then measuring its length with a ruler. In this section, we see how such measurements can be carried out mathematically using integration.

The Arc Length of a Curve

If a function f has a derivative that is continuous on some interval, then f is said to be **continuously differentiable** on the interval. The portion of the graph of a continuously differentiable function f that lies between $x = a$ and $x = b$ is called the **arc** of the graph on the interval $[a, b]$. To find, and define, the length of this arc, let P be a partition of the interval $[a, b]$, with subdivision points $x_0, x_1, \cdots, x_n$, where $x_0 = a$ and $x_n = b$. Let P_k denote the point (x_k, y_k) on the graph, where $y_k = f(x_k)$. By joining the points $P_0, P_1, \cdots, P_n$, we obtain a polygonal path whose length approximates that of the arc. Figure 6.44 demonstrates the labeling for the case where $n = 6$.

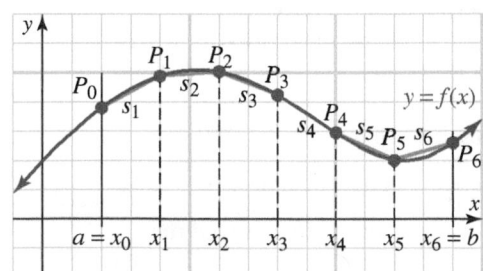

Figure 6.44 Interactive   A polygonal path approximating an arc of a curve

The length of the polygonal path connecting the points $P_0, P_1, \cdots, P_n$ on the graph of f is the sum

$$\sum_{k=1}^{n} \Delta s_k$$

where Δs_k is the length of the segment joining P_{k-1} to P_k. By applying the distance formula and rearranging terms with $\Delta x_k = x_k - x_{k-1}$ and $\Delta y_k = y_k - y_{k-1}$ we find that

$$\Delta s_k = \sqrt{(x_k - x_{k-1})^2 + (y_k - y_{k-1})^2}$$

$$= \sqrt{(\Delta x_k)^2 + (\Delta y_k)^2}$$

$$= \sqrt{\frac{(\Delta x_k)^2 + (\Delta y_k)^2}{(\Delta x_k)^2}} \Delta x_k$$

$$= \sqrt{1 + \left(\frac{\Delta y_k}{\Delta x_k}\right)^2} \Delta x_k$$

It is reasonable to expect a connection between the ratio $\Delta y_k / \Delta x_k$ and the derivative $dy/dx = f'(x)$. Indeed, using the MVT, it can be shown (see Problem 55) that

$$\Delta s_k = \sqrt{1 + \left[f'(x_k^*)\right]^2} \Delta x_k$$

for some number x_k^* between x_{k-1} and x_k. Therefore, the arc length of the graph of f on the interval $[a, b]$ may be estimated by the Riemann sum

$$\sum_{k=1}^{n} \sqrt{1 + \left[f'(x_k^*)\right]^2} \Delta x_k$$

This estimate may be improved by increasing the number of subdivision points in the partition P of the interval $[a, b]$ in such a way that the subinterval lengths tend to zero. Thus, it is reasonable to define the actual arc length to be the limit

$$\lim_{\|P\| \to 0} \sum_{k=1}^{n} \sqrt{1 + \left[f'(x_k^*)\right]^2} \Delta x_k$$

Notice that if f' is continuous on the interval $[a, b]$, then so is $\sqrt{1 + \left[f'(x)\right]^2}$. Thus, this limit exists and is the integral of $\sqrt{1 + \left[f'(x)\right]^2}$ with respect to x on the interval $[a, b]$. These observations lead us to the following definition.*

ARC LENGTH Let f be a function whose derivative f' is continuous on the interval $[a, b]$ and differentiable on (a, b). Then the **arc length**, s, of the graph of $y = f(x)$ between $x = a$ and $x = b$ is given by the integral

$$s = \int_a^b \sqrt{1 + \left[f'(x)\right]^2} \, dx$$

Similarly, for the graph of $x = g(y)$, where g' is continuous on the interval $[c, d]$, the arc length from $y = c$ to $y = d$ is

$$s = \int_c^d \sqrt{1 + \left[g'(y)\right]^2} \, dy$$

Here is an easy way to remember the arc length formula:

$$ds = \sqrt{(dx)^2 + (dy)^2}$$

$$= \sqrt{1 + \left(\frac{dy}{dx}\right)^2} \, dx$$

$$= \sqrt{\left(\frac{dx}{dy}\right)^2 + 1} \, dy.$$

---

*We have used integration to obtain a meaningful definition of arc length for a function $f(x)$ with a continuous derivative $f'(x)$. It is possible to define arc length for certain curves $y = f(x)$ when $f(x)$ does not have a continuous derivative, but we will not pursue this more general topic.

Example 1   Arc length of a curve

Find the arc length (rounded to two decimal places) of the curve $y = x^{3/2}$ on the interval $[0, 4]$.

Solution   The graph is shown in Figure 6.45.

Let $f(x) = x^{3/2}$; therefore, $f'(x) = \frac{3}{2}x^{1/2}$.

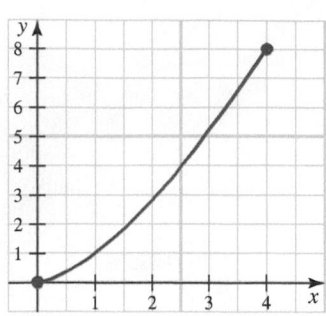

Figure 6.45 What is the length of this curve?

$$s = \int_0^4 \sqrt{1 + \left[\frac{3}{2}x^{1/2}\right]^2}\, dx$$

$$= \int_0^4 \sqrt{1 + \frac{9}{4}x}\, dx$$

$$= \left[\frac{4}{9} \cdot \frac{2}{3}\left(1 + \frac{9}{4}x\right)^{3/2}\right]_0^4$$

$$= \frac{8}{27}\left[10^{3/2} - 1^{3/2}\right]$$

$$\approx 9.0734$$

Example 2   Arc length of a curve $x = g(y)$

Find the arc length of the curve $x = \frac{1}{3}y^3 + \frac{1}{4}y^{-1}$ from $y = 1$ to $y = 3$.

Solution   Because $g(y) = \frac{1}{3}y^3 + \frac{1}{4}y^{-1}$, we have $g'(y) = y^2 - \frac{1}{4}y^{-2} = \frac{4y^4 - 1}{4y^2}$, which is continuous throughout the interval from $y = 1$ to $y = 3$. Therefore, the arc length is

$$s = \int_1^3 \sqrt{1 + \left[g'(y)\right]^2}\, dy$$

$$= \int_1^3 \sqrt{1 + \left(\frac{4y^4 - 1}{4y^2}\right)^2}\, dy$$

$$= \int_1^3 \sqrt{1 + \frac{16y^8 - 8y^4 + 1}{16y^4}}\, dy$$

$$= \int_1^3 \sqrt{\frac{\left(4y^4 + 1\right)^2}{\left(4y^2\right)^2}}\, dy$$

$$= \int_1^3 \frac{4y^4 + 1}{4y^2}\, dy$$

$$= \int_1^3 \left(y^2 + \frac{1}{4}y^{-2}\right) dy$$

$$= \left[\frac{1}{3}y^3 + \frac{1}{4} \cdot \frac{y^{-1}}{-1}\right]_1^3$$

$$= \frac{53}{6}$$

Example 3   Estimating arc length using numerical integration

Find the length of the curve defined by $y = \sin x$ on $[0, 2\pi]$.

Solution   The region is shown in Figure 6.46.

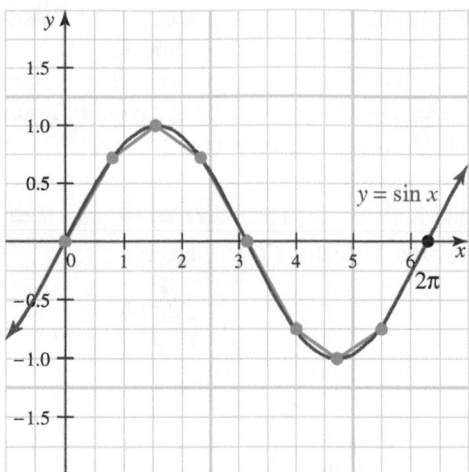

Figure 6.46 Interactive Finding the length of an arc (8 subdivisions shown)

Because $y' = \cos x$, we have, from the arc length formula,

$$s = \int_0^{2\pi} \sqrt{1 + \cos^2 x} \, dx$$

We do not have techniques to allow us to evaluate this integral, so we turn to numerical integration. If we consider four rectangles, we find, by various methods:

Method ($n = 4$)	Approximation
Rectangles:	
Left endpoints	7.584476
Right endpoints	7.584476
Midpoints	7.695299
Trapezoidal rule	7.584476
Simpson's rule	7.150712

In practice, you would choose just *one* of the methods whose approximate values are given. You might also note that, for this example, Simpson's rule performs worse than the trapezoidal, midpoint, or even the left- and right-endpoint methods. For more accurate results, you might wish to use a computer program. The actual arc length (rounded to six places) computed as an area is 7.640396. ∎

The Area of a Surface of Revolution

When the arc of a curve is revolved about a line L it generates a surface called a **surface of revolution**, as shown in Figure 6.47a. In particular, if the generating arc is a line segment, the frustum of a cone is generated, as shown in Figure 6.47b.

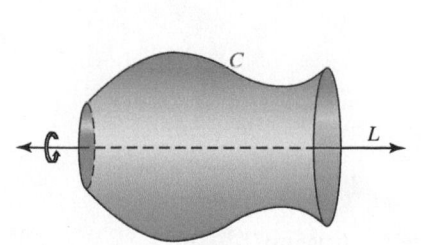

a. The surface generated by revolving the curve C about a line L

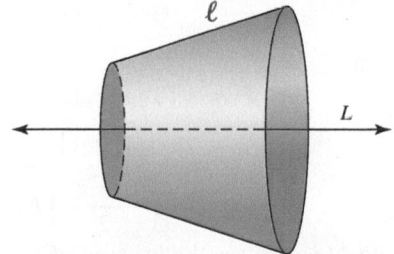

b. The line segment of length ℓ generates the frustum of a cone

Figure 6.47 Surfaces of revolution

We will obtain a formula for the lateral area of the frustum of a cone, which will then be used as part of an approximation scheme for the area of a general surface of revolution. Accordingly, first note that a cone with slant height s and base radius r has lateral surface area is $A = \pi rs$. To see this, cut the cone in Figure 6.48**a** along the dashed line and flatten it out as shown in Figure 6.48**b**. The angle θ is determined by noting that the circumference of the circular base of the cone is $2\pi r$, which is the same as the arc length $s\theta$ after the cone is flattened. Thus, $2\pi r = s\theta$, and $\theta = \dfrac{2\pi r}{s}$. The lateral surface area of the cone is the area of sector COD; namely,

$$A = \frac{1}{2}s^2\theta = \frac{1}{2}s^2\left(\frac{2\pi r}{s}\right) = \pi rs$$

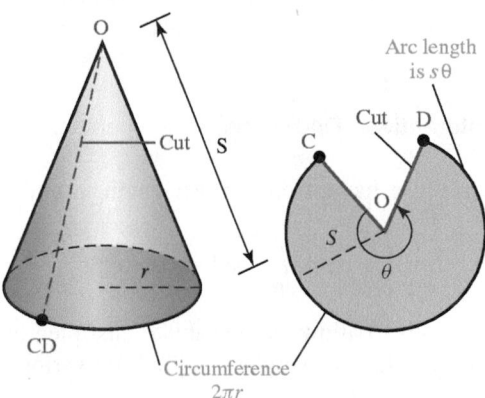

a. A cone with base radius r and slant height s is cut along the line from CD to O

b. The flattened cone is a sector with central angle θ in a circle of radius s

Figure 6.48   A cone with base radius r and slant height s has area $A = \pi rs$

Finally, the frustum of the cone shown in Figure 6.49 has slant height ℓ, top radius r_1, and bottom radius r_2, and can be formed by removing a small cone of base radius r_1 and slant height ℓ_1 from the larger cone of base radius r_2 and slant height $\ell_2 = \ell_1 + \ell$.

Thus, the lateral surface area F of the frustum can be found by subtracting the area of the smaller cone from that of the larger:

$$F = \pi r_2\ell_2 - \pi r_1\ell_1$$
$$= \pi r_2\left(\ell_1 + \ell\right) - \pi r_1\ell_1$$
$$= \pi\left[(r_2 - r_1)\ell_1 + r_2\ell\right]$$

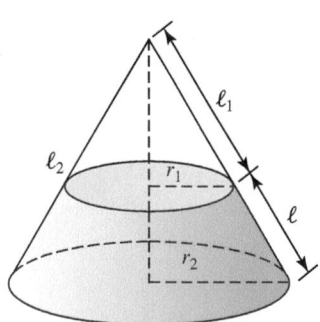

Figure 6.49   Frustum of a cone

Using similar triangles, we see that

$$\frac{\ell_1}{r_1} = \frac{\ell_1 + \ell}{r_2} \qquad \text{or} \qquad r_1\ell = (r_2 - r_1)\ell_1$$

so that

$$F = \pi\left[(r_2 - r_1)\ell_1 + r_2\ell\right]$$
$$= \pi\left[r_1\ell + r_2\ell\right] \qquad\qquad \textit{By substitution}$$
$$= 2\pi\left(\frac{r_1 + r_2}{2}\right)\ell$$

You might remember the formula $F = \pi(r_1 + r_2)\ell$.

where $\frac{1}{2}(r_1 + r_2)$ is the average of the radii of the frustum.

Next, assuming that the function $f(x)$ is continuous with $f(x) \geq 0$ on the interval $[a, b]$, we will derive a formula for the area of the surface generated by revolving the curve $y = f(x)$ about the x-axis. Rather than argue formally with Riemann sums, we will proceed heuristically. Figure 6.50**a** shows an approximating line segment with length Δs, whose endpoints are, respectively, distances r_1 and r_2 above the x-axis, and Figure 6.50**b** shows the surface of revolution.

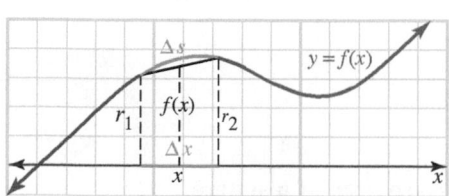

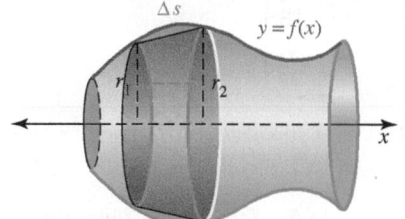

a. The approximating line segment of length Δs is at an average height $f(x)$ above the x-axis

b. Revolving the segment about the x-axis generates a frustum (gray) used to approximate surface area

Figure 6.50 Approximating the area of a surface

Let $f(x)$ be the average height of the segment above the x-axis, we generate the frustum of a cone, with surface area

$$\Delta S = 2\pi \left(\frac{r_1 + r_2}{2} \right) \Delta s = 2\pi f(x)\sqrt{1 + \left[f'(x) \right]^2}\, \Delta x$$

where Δx is the projection of the line segment on the x-axis. The area of the entire surface of revolution S can then be obtained by integrating this expression over the interval $[a, b]$. These considerations lead us to the following definition.

SURFACE AREA Suppose f' is continuous on the interval $[a, b]$. Then the surface generated by revolving about the x-axis the arc of the curve $y = f(x)$ on $[a, b]$ has **surface area**

$$S = 2\pi \int_a^b f(x)\sqrt{1 + \left[f'(x) \right]^2}\, dx$$

An easy-to-remember form for surface area is $S = 2\pi \int y\, ds$ since $ds = \sqrt{1 + \left[f'(x) \right]^2}\, dx$.

You may find it instructive to derive this formula by the more rigorous approach outlined in Problem 56. We close with two examples that illustrate the use of the surface area formula.

Example 4   Area of a surface of revolution

Find the area of the surface generated by revolving about the x-axis the arc of the curve $y = x^3$ on $[0, 1]$.

Solution   The graph is shown in Figure 6.51.
Because $f(x) = x^3$, we have $f'(x) = 3x^2$, which is certainly continuous on the interval $[0, 1]$.

$$S = 2\pi \int_0^1 x^3\sqrt{1 + \left(3x^2 \right)^2}\, dx$$

$$= 2\pi \int_0^1 x^3\sqrt{1 + 9x^4}\, dx$$

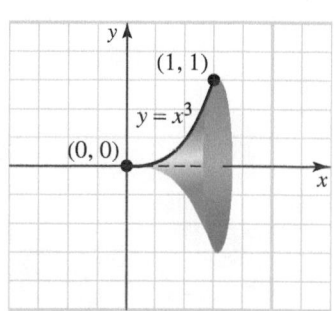

Figure 6.51 Graph of surface of revolution

$$= 2\pi \int_1^{10} u^{1/2} \left(\frac{du}{36}\right)$$

Let $u = 1 + 9x^4$; $du = 36x^3\, dx$
If $x = 0$, then $u = 1$; if $x = 1$, then $u = 10$.

$$= \frac{\pi}{18}\left[\frac{2}{3}u^{3/2}\right]_1^{10}$$

$$= \frac{\pi}{27}\left(10\sqrt{10} - 1\right)$$

This is a surface area of about 3.563 square units.

Example 5   Derive the formula for the surface area of a sphere

Find a formula for the surface area of a sphere of radius r.

Solution   We can generate the surface of the sphere by revolving the semicircle $y = \sqrt{r^2 - x^2}$ about the x-axis (see Figure 6.52)
We find that

$$y' = -x(r^2 - x^2)^{-1/2} = \frac{-x}{\sqrt{r^2 - x^2}}$$

Because the semicircle intersects the x-axis at $x = r$ and $x = -r$, the interval of integration is $[-r, r]$. Finally, by applying the formula for the surface area, we find

$$S = 2\pi \int_{-r}^{r} \sqrt{r^2 - x^2}\sqrt{1 + \left(\frac{-x}{\sqrt{r^2 - x^2}}\right)^2}\, dx$$

$$= 2\pi \int_{-r}^{r} \sqrt{(r^2 - x^2)\left(\frac{r^2 - x^2 + x^2}{r^2 - x^2}\right)}\, dx$$

$$= 2\pi \int_{-r}^{r} r\, dx$$

$$= 2\pi r(r + r)$$

$$= 4\pi r^2$$

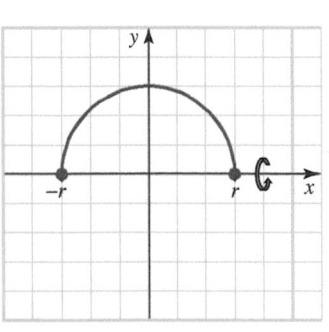

Figure 6.52   Graph of semicircle $y = \sqrt{r^2 - x^2}$

To generalize the formula for surface area to apply to any vertical or horizontal axis of revolution, suppose that an axis of revolution is $R(x)$ units from a typical element of arc on the graph of $y = f(x)$, as shown in Figure 6.53.

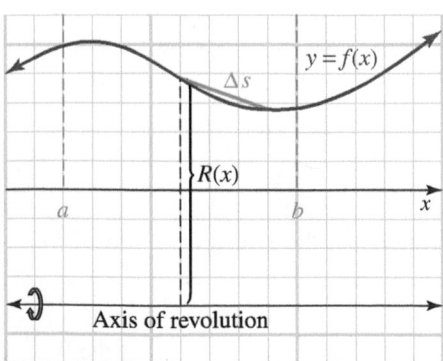
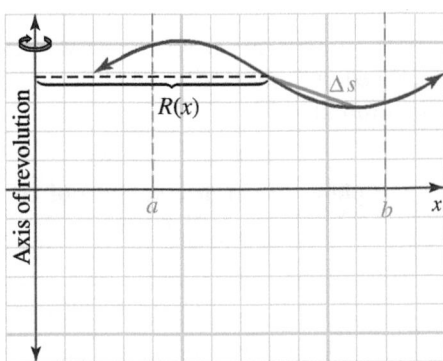

Figure 6.53   Surface of revolution about a general horizontal or vertical axis

Then $2\pi R(x)$ is the circumference of a circle of radius $R(x)$, and it can be shown that an element of surface area is

$$\Delta S = 2\pi R(x)\Delta s = 2\pi R(x)\sqrt{1 + \left[f'(x)\right]^2}\,\Delta x$$

Thus, the surface of revolution on the interval $[a,b]$ has area

$$S = 2\pi \int_a^b R(x)\sqrt{1 + [f'(x)]^2}\, dx$$

In particular, if the graph of $y = f(x)$ is revolved about the y-axis, an element of the arc is $R(x) = x$ units from the y-axis, and the resulting surface has area

$$S = 2\pi \int_a^b x\sqrt{1 + [f'(x)]^2}\, dx$$

Polar Arc Length and Surface Area

Sometimes we wish to find arc length and surface area in polar form. Figure 6.54 shows the portion of a polar curve subtended by a central angle $\Delta\theta$. If $\Delta\theta$ is small, the corresponding arc length Δs may be thought of as the hypotenuse of a right triangle with sides Δr and $r\Delta\theta$, the length of the circular arc AB with radius r.

We have

$$\Delta s = \sqrt{(r\Delta\theta)^2 + (\Delta r)^2} \qquad \textit{Pythagorean theorem}$$

$$= \sqrt{r^2 + \left(\frac{\Delta r}{\Delta\theta}\right)^2}\,\Delta\theta$$

By formalizing these observations, we are led to compute polar arc length by the following formula.

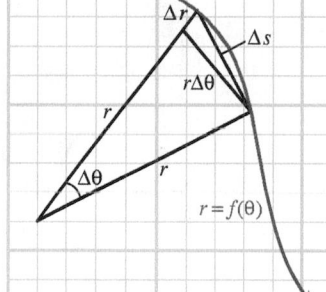

Figure 6.54 Polar arc length

ARC LENGTH IN POLAR COORDINATES The length of a polar curve $r = f(\theta)$ for $\alpha \le \theta \le \beta$ is given by the integral

$$S = \int_\alpha^\beta \sqrt{r^2 + \left(\frac{dr}{d\theta}\right)^2}\, d\theta$$

Example 6 Computing polar arc length

Find the length of the circle $r = 2\sin\theta$.

Solution This circle is shown in Figure 6.55, and since this is a unit circle, we know the circumference to be 2π.

We will verify this conclusion using the arc length formula. Note that $0 \le \theta \le \pi$.

$$s = \int_0^\pi \sqrt{r^2 + \left(\frac{dr}{d\theta}\right)^2}\, d\theta \qquad \textit{Arc length formula where } \alpha = 0 \text{ and } \beta = \pi$$

$$= \int_0^\pi \sqrt{(2\sin\theta)^2 + (2\cos\theta)^2}\, d\theta \qquad r = 2\sin\theta; \frac{dr}{d\theta} = 2\cos\theta$$

$$= \int_0^\pi 2\, d\theta \qquad \sin^2\theta + \cos^2\theta = 1$$

$$= 2\pi$$

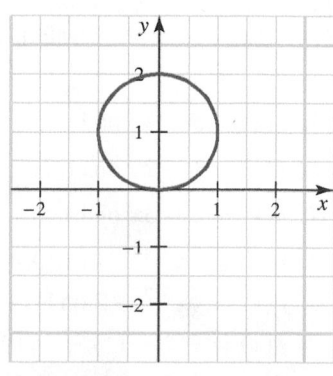

Figure 6.55 Circle $r = 2\sin\theta$

If a polar curve $r = f(\theta)$ between $\theta = \alpha$ and $\theta = \beta$ is revolved about the x-axis, it generates a surface of area

$$A = \int_{\alpha}^{\beta} 2\pi y \, ds$$

$$= 2\pi \int_{\alpha}^{\beta} (r \sin\theta) \sqrt{r^2 + \left(\frac{dr}{d\theta}\right)^2} \, d\theta$$

We conclude this section with an example showing the use of this formula.

Example 7   Computing surface area in polar coordinates

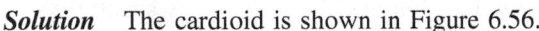

Find the area of the surface generated by revolving about the x-axis the top half of the cardioid $r = 1 + \cos\theta$.

Solution   The cardioid is shown in Figure 6.56.
Since $r = 1 + \cos\theta$ and $r' = -\sin\theta$, the surface area is given by

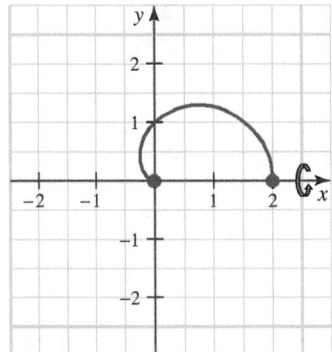

Figure 6.56   Top half of cardioid $r = 1 + \cos\theta$ $(0 \leq \theta \leq \pi)$

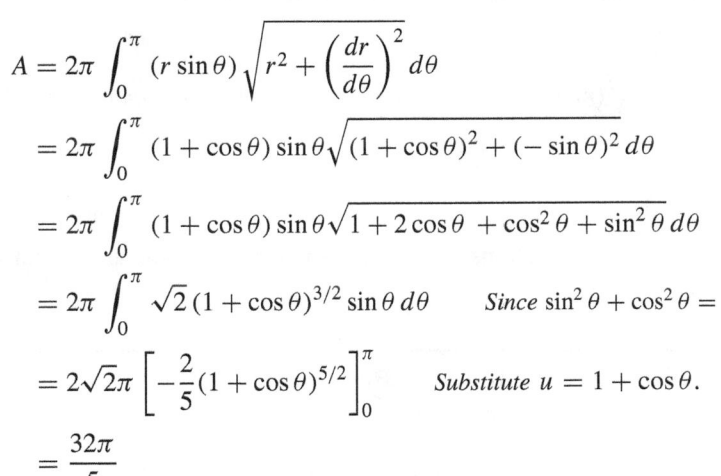

$$A = 2\pi \int_{0}^{\pi} (r \sin\theta) \sqrt{r^2 + \left(\frac{dr}{d\theta}\right)^2} \, d\theta$$

$$= 2\pi \int_{0}^{\pi} (1 + \cos\theta) \sin\theta \sqrt{(1 + \cos\theta)^2 + (-\sin\theta)^2} \, d\theta$$

$$= 2\pi \int_{0}^{\pi} (1 + \cos\theta) \sin\theta \sqrt{1 + 2\cos\theta + \cos^2\theta + \sin^2\theta} \, d\theta$$

$$= 2\pi \int_{0}^{\pi} \sqrt{2} (1 + \cos\theta)^{3/2} \sin\theta \, d\theta \qquad \text{Since } \sin^2\theta + \cos^2\theta = 1$$

$$= 2\sqrt{2}\pi \left[-\frac{2}{5}(1 + \cos\theta)^{5/2} \right]_{0}^{\pi} \qquad \text{Substitute } u = 1 + \cos\theta.$$

$$= \frac{32\pi}{5}$$

PROBLEM SET 6.4

Level 1

Find the length of the arc of the curve $y = f(x)$ on the intervals given in Problems 1-18.

1. $f(x) = 3x + 2$ on $[-1, 2]$

2. $f(x) = 5 - 4x$ on $[-2, 0]$

3. $f(x) = 1 - 2x$ on $[1, 3]$

4. $f(x) = 2 - 3x$ on $[0, 1]$

5. $f(x) = \frac{2}{3}x^{3/2} + 1$ on $[0, 4]$

6. $f(x) = \frac{2}{3}x^{3/2} + 2$ on $[1, 4]$

7. $f(x) = \frac{1}{3}(2 + x^2)^{3/2}$ on $[0, 3]$

8. $f(x) = \frac{1}{3}(2 + x^2)^{3/2}$ on $[1, 4]$

9. $f(x) = \frac{1}{12}x^5 + \frac{1}{5}x^{-3}$ on $[1, 2]$

10. $f(x) = \frac{1}{3}x^3 + \frac{1}{4}x^{-1}$ on $[1, 4]$

11. $f(x) = \frac{1}{8}x^2 - \ln x$ on $[1, 2]$

12. $f(x) = \frac{1}{4}x^4 + \frac{1}{8}x^{-2}$ on $[1, 2]$

13. $f(x) = \sqrt{e^{2x} - 1} - \sec^{-1}(e^x)$ on $[0, \ln 2]$

14. $f(x) = \sqrt{e^{2x} - 1} - \sec^{-1}(e^x)$ on $[0, \ln 3]$

15. $f(x) = x^3 + \frac{1}{12}x^{-1}$ on $[1, 2]$

16. $f(x) = \frac{1}{6}x^3 + \frac{1}{2}x^{-1}$ on $[1, 2]$

17. $f(x) = x^{3/2} - \frac{1}{3}x^{1/2}$ on $[1, 9]$

18. $f(x) = x^{3/2} - \frac{1}{3}x^{1/2}$ on $[4, 16]$

19. Find the length of the curve defined by $9x^2 = 4y^3$ between the points $(0, 0)$ and $\left(2\sqrt{3}, 3\right)$.

20. Find the length of the curve defined by $(y + 1)^2 = 4x^3$ between the points $(0, -1)$ and $(1, 1)$.

21. Find the length of the curve defined by

$$\int_{1}^{x} \sqrt{t^2 - 1} \, dt$$

on $[1, 2]$.

22. Find the length of the curve defined by

$$\int_{1}^{x} \sqrt{t^2 + 1}\, dt$$

on $[0, 1]$.

Find the surface area generated when the graph of each function given in Problems 23-28 on the prescribed interval is revolved about the x-axis.

23. $f(x) = 2x + 1$ on $[0, 2]$

24. $f(x) = 2x - 1$ on $[1, 2]$

25. $f(x) = \sqrt{x}$ on $[2, 6]$

26. $f(x) = \sqrt{1 - x^2}$ on $[0, 1/2]$

27. $f(x) = \frac{1}{3}x^3 + \frac{1}{4}x^{-1}$ on $[1, 2]$

28. $f(x) = \frac{1}{4}x^4 + \frac{1}{8}x^{-2}$ on $[1, 2]$

Find the length of the polar curves given in Problems 29-36.

29. $r = \cos\theta$

30. $r = \sin\theta + \cos\theta$

31. $r = e^{3\theta}, 0 \le \theta \le \frac{\pi}{2}$

32. $r = e^{1-\theta}, 0 \le \theta \le 1$

33. $r = \theta^2, 0 \le \theta \le 1$

34. $r = \theta^2, 1 \le \theta \le 2$

35. $r = \cos^2 \dfrac{\theta}{2}$

36. $r = \sin^2 \dfrac{\theta}{2}$

In Problems 37-40, find the surface area generated when the given polar curve is revolved about the x-axis.

37. $r = 5, 0 \le \theta \le \frac{\pi}{3}$

38. $r = 1 - \cos\theta, 0 \le \theta \le \pi$

39. $r = \csc\theta, \frac{\pi}{4} \le \theta \le \frac{\pi}{3}$

40. $r = \cos^2 \frac{\theta}{2}, 0 \le \theta \le \pi$

Level 2

In Problems 41 and 42, use numerical integration to find the arc length.

41. $f(x) = \sin x$ on $[0, \pi]$

42. $f(x) = \tan x$ on $[0, 1]$

43. Find the area of the surface generated when the arc of the curve

$$y = \frac{1}{3}x^3 + (4x)^{-1}$$

between $x = 1$ and $x = 3$ is revolved about

    a. the x-axis   **b.** the y-axis

44. Find the area of the surface generated when the arc of the curve

$$x = \frac{3}{5}y^{5/3} - \frac{3}{4}y^{1/3}$$

between $y = 0$ and $y = 1$ is revolved about

    a. the y-axis   **b.** the x-axis

    c. the line $y = -1$

Find the surface area generated when the graph of each function given in Problems 45-48 on the prescribed interval is revolved about the y-axis. Give your answers to the nearest hundredth.

45. $f(x) = \frac{1}{3}(12 - x)$ on $[0, 3]$

46. $f(x) = \frac{2}{3}x^{3/2}$ on $[0, 3]$

47. $f(x) = \frac{1}{3}\sqrt{x}(3 - x)$ on $[1, 3]$

48. $f(x) = 2\sqrt{4 - x}$ on $[1, 3]$

49. The graph of the equation $x^{2/3} + y^{2/3} = 1$ is an **astroid**.

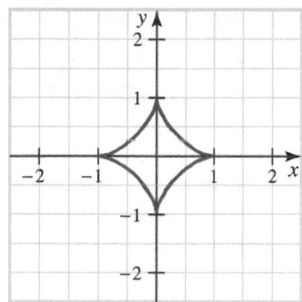

Find the length of this particular astroid by finding the length of the first quadrant portion,

$$y = (1 - x^{2/3})^{3/2} \text{ on } [0, 1]$$

By symmetry, the length of the entire curve is four times the length of the arc in the first quadrant.

50. Find the area of the surface generated by revolving the portion of the astroid with $y \ge 0$ (see Problem 49)

$$x^{2/3} + y^{2/3} = 1 \text{ on } [-1, 1]$$

about the x-axis.

51. The following table gives the slope $f'(x_k)$ for points x_k located at 0.3 unit intervals on a certain curve defined by $y = f(x)$. Use the trapezoidal rule to estimate the arc length of $y = f(x)$ over the interval $[0, 3]$.

x_k	$f'(x_k)$	x_k	$f'(x_k)$
0.0	3.7	1.8	4.6
0.3	3.9	2.1	4.9
0.6	4.1	2.4	5.2
0.9	4.1	2.7	5.5
1.2	4.2	3.0	6.0
1.5	4.4		

52. Use the table in Problem 51 to estimate the surface area generated when $y = f(x)$ is revolved about the y-axis over the interval $[0, 3]$.

53. Show that a cone of radius r and height h has lateral surface area

$$S = \pi r \sqrt{r^2 + h^2}$$

Hint: Revolve part of the line $hy = rx$ about the x-axis.

Level 3

54. Show that when the arc of the graph of $f(x)$ between $x = a$ and $x = b$ is revolved about the y-axis, the surface generated has area

$$S = 2\pi \int_a^b x \sqrt{1 + [f'(x)]^2}\, dx$$

55. If $f'(x)$ exists throughout the interval $[x_{k-1}, x_k]$, show that there exists a number x_k^* in this interval for which

$$\sqrt{(\Delta x_k)^2 + (\Delta y_k)^2} = \sqrt{1 + [f'(x_k^*)]^2}\,\Delta x$$

where $\Delta x_k = x_k - x_{k-1}$ and $\Delta y_k = f(x_k) - f(x_{k-1})$. *Hint*: Use the mean value theorem.

56. Verify the surface area formula by completing the following steps (see Figure 6.57).

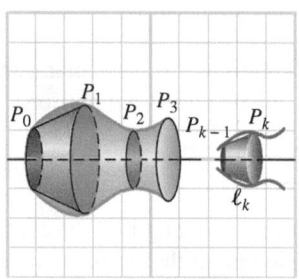

Figure 6.57 Surface area formula

a. Let $P = \{x_0, x_1, \cdots, x_n\}$ be a partition of the interval $[a, b]$, and for $k = 0, 1, \cdots, n$, let P_k denote the point (x_k, y_k) where $y_k = f(x_k)$. Show that when the line segment between P_{k-1} and P_k is revolved about the x-axis, it generates a frustum of a cone with surface area

$$\pi(y_{k-1} + y_k)\ell_k$$

where ℓ_k is the distance between P_{k-1} and P_k.
b. Notice that $\frac{1}{2}(y_{k-1} + y_k)$ is a number between y_{k-1} and y_k. Use the intermediate value theorem (Section 2.3) to show that

$$y_{k-1} + y_k = 2f(c_k)$$

for some number c_k between x_{k-1} and x_k.

c. Explain why the surface generated when the arc of the graph of f between $x = a$ and $x = b$ is revolved about the x-axis has area

$$\lim_{\|P\| \to 0} \sum_{k=1}^{n} 2\pi f(c_k) \sqrt{1 + [f'(x_k^*)]^2}\,\Delta x_k$$

d. Notice that as $\|P\| \to 0$, the numbers c_k and x_k^* are "squeezed" together. Use this observation to show that the surface area is given by the integral

$$S = 2\pi \int_a^b f(x) \sqrt{1 + [f'(x)]^2}\, dx$$

57. Let n be a positive integer and C, D be constants such that

$$n(n - 1) = \frac{1}{16CD}$$

Find the length of the curve

$$y = Cx^{2n} + Dx^{2(1-n)}$$

from $x = a$ to $x = b$.
58. The center of a circle of radius r is located at $(R, 0)$, where $R > r$. When the circle is revolved about the y-axis, it generates a *torus* (a doughnut-shaped solid). What is the surface area of the torus?
59. Suppose it is known that the arc length of the curve $y = f(t)$ on the interval $0 \le t \le x$ is

$$L(x) = \ln(\sec x + \tan x)$$

for every x on $0 \le x \le 1$. If the curve $y = f(x)$ passes through the origin, what is $f(x)$?
60. Find the length of

$$y = \frac{x^3}{24} + \frac{2}{x}$$

from $x = 2$ to $x = 3$. Find a general formula for the length of the curve

$$y = Cx^n + Dx^{2-n}$$

from $x = a$ to $x = b$ for

$$n(n - 2) = \frac{1}{4CD}$$

6.5 PHYSICAL APPLICATIONS: WORK, LIQUID FORCE, AND CENTROIDS

> **IN THIS SECTION:** *Work, modeling fluid pressure and force, modeling the centroid of a plane region, volume theorem of Pappus*
> Integration plays an important role in many different areas of physics. **Mechanics** is the branch of physics that deals with the effects of forces on objects. In this section, we show how integration can be used to compute work and the force exerted by a liquid.

Work

In physics, "force" is an influence that tends to cause motion in a body. When a constant force of magnitude F is applied to an object through a distance d, it performs **work** measured by the product $W = Fd$.

WORK DONE BY A CONSTANT FORCE If a body moves a distance d in the direction of an applied constant force F, the **work** W done is

$$W = Fd$$

If there is no movement, there is no work!

For example, the work done in lifting a 90 lb bag of concrete 3 ft is $W = Fd = (90 \text{ lb})(3 \text{ ft}) = 270$ ft-lb. Notice that this definition does not conform to everyday use of the word *work*. If you work at lifting the concrete all day, but you are not able to move the sack of concrete, then no work has been done. Table 6.3 shows some common units of work and force.

Table 6.3 Common units of work and force

Mass	Distance	Force	Work
kg	m	newton (N)	joule
g	cm	dyne (dyn)	erg
slug	ft	pound	ft-lb

To find the work done in moving an object against a variable force, calculus is required. Suppose F is a continuous function and $F(x)$ is a variable force that acts on an object moving along the x-axis from $x = a$ to $x = b$. We shall heuristically define the work done by this force. Partition the interval $[a, b]$ into n subintervals, and let Δx_k be the length of the kth subinterval I_k. If Δx_k is sufficiently small, we can expect the force to be essentially constant on I_k, equal, for instance, to $F(x_k^*)$, where x_k^* is a point chosen arbitrarily from the interval I_k. It is reasonable to expect that the work ΔW_k required to move the object on the interval I_k is approximately $\Delta W_k = F(x_k^*)\Delta x_k$, and the total work is estimated by the sum

$$\sum_{k=1}^{n} F(x_k^*)\Delta x_k$$

as indicated in Figure 6.58.

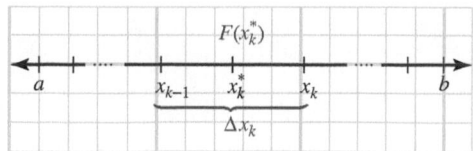

Figure 6.58 Work performed by a variable force

By taking the limit as the norm of the partition approaches 0, we find the total work may be modeled by

$$W = \lim_{\|P\|\to 0} \sum_{k=1}^{n} F(x_k^*)\Delta x_k = \int_a^b F(x)\, dx$$

> **WORK DONE BY A VARIABLE FORCE**   The work done by the variable force $F(x)$ in moving an object along the x-axis from $x = a$ to $x = b$ is given by
>
> $$W = \int_a^b F(x)\, dx$$

Example 1   Work done by a variable force

An object located x ft from a fixed starting position is moved along a straight road by a force of $F(x) = (3x^2 + 5)$ lb. What work is done by the force to move the object
a. through the first 4 ft?   **b.** from 1 ft to 4 ft?

Solution

a. $W = \int_0^4 (3x^2 + 5)\, dx = (x^3 + 5x)\big|_0^4 = 84$ ft-lb

b. $W = \int_1^4 (3x^2 + 5)\, dx = (x^3 + 5x)\big|_1^4 = 78$ ft-lb

Hooke's law, named for the English physicist Robert Hooke (1635-1703), states that *when a spring is pulled x units past its equilibrium (rest) position, there is a restoring force $F(x) = kx$ that pulls the spring back toward equilibrium.*[*] (See Figure 6.59.) The constant k in this formula is called the **spring constant**.

Hooke's law: A force $F(x) = kx$ acts to restore the spring to its equilibrium position.

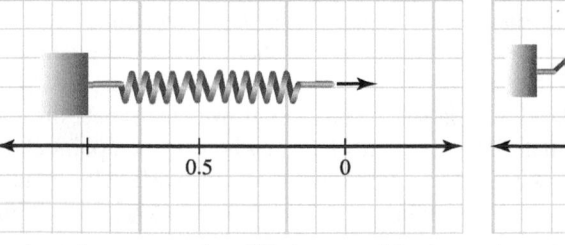
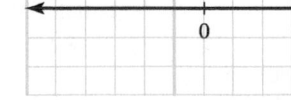

A spring at rest (equilibrium position)   A spring stretched x units past equilibrium

Figure 6.59   Hooke's law

---

[*]Actually, Hooke's law applies only in ideal circumstances, and a spring for which the law applies is sometimes called an ideal (or linear) spring.

Example 2 Modeling work using Hooke's formula

The natural length of a certain spring is 10 cm. If it requires 2 ergs of work to stretch the spring to a total length of 18 cm, how much work will be performed in stretching the spring to a total length of 20 cm? (See Figure 6.60.)

Solution Assume that the point of equilibrium is at 0 on a number line, and let x be the length the free end of the spring is extended past equilibrium. Because the stretching force of the spring is $F(x) = kx$, the work done in stretching the spring b cm beyond equilibrium is

$$W = \int_0^b F(x) = \int_0^b kx \; dx = \frac{1}{2}kb^2$$

We are given that $W = 2$ when $b = 8$ (since 18 cm is 8 cm beyond equilibrium), so

$$2 = \frac{1}{2}k(8)^2 \quad \text{implying} \quad k = \frac{1}{16}$$

Thus, the work done in stretching the spring b cm beyond equilibrium is

$$W = \frac{1}{2}\left(\frac{1}{16}\right)b^2 = \frac{1}{32}b^2 \text{ ergs}$$

In particular, when the total length of the spring is 20 cm, it is extended $b = 10$ cm, and the required work is

$$W = \frac{1}{32}(10)^2 = \frac{25}{8} = 3.125 \text{ ergs}$$

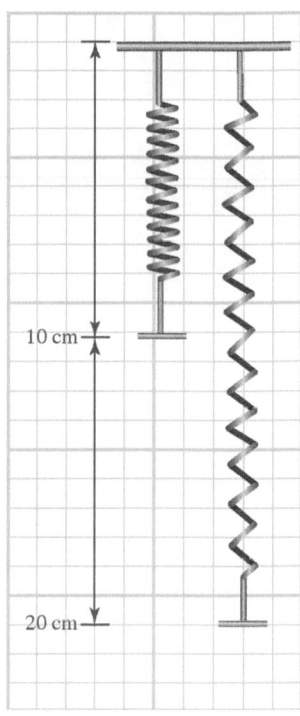

Figure 6.60 Stretching a spring

Example 3 Modeling the work performed in pumping water out of a tank

A tank in the shape of a right circular cone of height 12 ft and radius 3 ft is inserted into the ground with its vertex pointing down and its top at ground level, as shown in Figure 6.61**a**. If the tank is filled with water (weight density $\rho g = 62.4$ lb/ft^3)* to a depth of 6 ft, how much work is performed in pumping all the water in the tank to ground level? What changes if the water is pumped to a height of 3 ft above ground level?

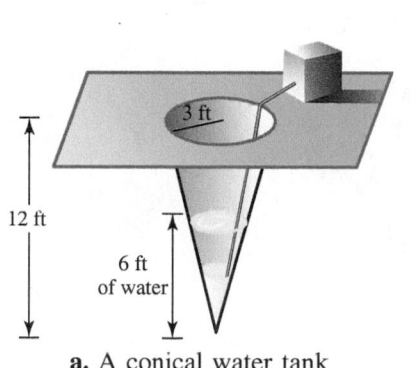

a. A conical water tank

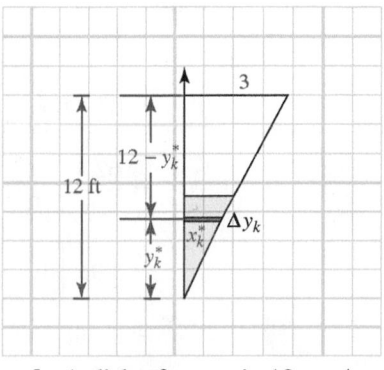

b. A disk of water is $12 - y_k^*$ units from the top

Figure 6.61 Work in pumping water

---

*In non-English-speaking countries, where the metric system is commonly used, density is generally considered to be mass density ρ. In English-speaking countries where the British and US system of measurement is commonly used, density is generally considered to be weight density $\delta = \rho g$, where g is the acceleration due to gravity.

Solution Set up a coordinate system with the origin at the vertex of the cone and the y-axis as the axis of symmetry (Figure 6.61**b**). Partition the interval $0 \leq y \leq 6$ (remember, the water only comes up 6 ft). Choose a representative point y_k^* in the kth subinterval, and construct a thin disklike slab of water y_k^* units above the vertex of the cone. Our plan is to think of the water in the tank as a collection of these "slabs" of water piled on top of each other. We shall find the work done to raise a typical water disk and then compute the total work by using integration to add the contributions of all such slabs.

Note that the force required to lift the slab is equal to the weight of the slab, which equals its volume multiplied by the weight per cubic foot of water. Let x_k^* be the radius of the kth slab. By similar triangles (Figure 6.61**b**), we have

$$\frac{x_k^*}{y_k^*} = \frac{3}{12} \quad \text{which implies} \quad x_k^* = \frac{y_k^*}{4}$$

Hence, the volume of the cylindrical slab is

$$\Delta V = \pi (x_k^*)^2 \Delta y_k = \pi \underbrace{\left(\frac{y_k^*}{4}\right)^2}_{Radius} \overbrace{\Delta y_k}^{Thickness}$$

so that the weight of the kth slab is modeled by

$$62.4\pi \left(\frac{y_k^*}{4}\right)^2 \Delta y_k \text{ lb}$$

From Figure 6.61**b**, we see that the slab of water must be raised $12 - y_k^*$ feet, which means that the work ΔW required to raise this kth slab is

$$\Delta W = \underbrace{62.4\pi \left(\frac{y_k^*}{4}\right)^2 \Delta y_k}_{Weight \ (force)} \underbrace{(12 - y_k^*)}_{Distance} = 62.4\pi \left(\frac{y_k^*}{4}\right)^2 (12 - y_k^*)\Delta y_k$$

Finally, to compute the total work, we add the work required to lift each thin slab and take the limit of the sum as the norm of the partition approaches 0.

$$W = \lim_{\|P\| \to 0} 62.4\pi \sum_{k=1}^{n} \left(\frac{y_k^*}{4}\right)^2 (12 - y_k^*)\Delta y_k$$

$$= 62.4\pi \int_0^6 \left(\frac{y}{4}\right)^2 (12 - y)dy$$

$$= \frac{62.4}{16}\pi \int_0^6 (12y^2 - y^3)dy$$

$$= 3.9\pi \left[4y^3 - \frac{1}{4}y^4\right]_0^6$$

$$= 2,106\pi$$

This is about 6,616 ft-lb.

If the water is pumped to a height of 3 ft above ground level, all that changes is the distance moved by the kth slab of water. It becomes $12 + 3 - y_k^* = 15 - y_k^*$, and the work is given by

$$W = 62.4\pi \int_0^6 \left(\frac{y}{4}\right)^2 (15 - y)\, dy \approx 9,263 \text{ ft-lb}$$

Modeling Fluid Pressure and Force

Anyone who has dived into water has probably noticed that the pressure (that is, the force per unit area) due to the water's weight increases with depth. Careful observations show that the water pressure at any given point is directly proportional to the depth at that point. The same principle applies to other fluids.

In physics, **Pascal's principle** (named for Blaise Pascal, 1623-1662, a French mathematician and scientist) states that fluid pressure is the same in all directions (see Figure 6.62).

This means that the pressure must be the same at all points on a surface submerged horizontally, and in this case, the fluid force on the surface is given by the formula in the following box.

FLUID FORCE If a surface of area A is submerged horizontally at a depth h in a fluid, the weight of the fluid exerts a force of

$$F = (\text{PRESSURE})(\text{AREA}) = \delta hA = \rho ghA$$

on the surface, where δ is the weight density, ρ is the mass density, and g is the acceleration due to gravity. This is called **fluid force** or **hydrostatic force**.

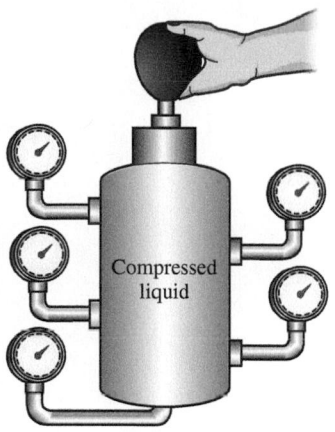

Figure 6.62 Pascal's principle

Table 6.4 shows the fluid density for some common fluids.

Table 6.4 Weight density, $\delta = \rho g$
(lb/ft^3)

Fluid	Density
water	62.4
seawater	64.0
gasoline	42.0
kerosene	51.2
SAE 20 oil	57.0
milk	64.5
mercury	849.0

It turns out that this fluid force does not depend on the shape of the container or its size. For instance, each of the containers in Figure 6.63 has the same pressure on its base because the fluid depth h and the base area A are the same in each case.

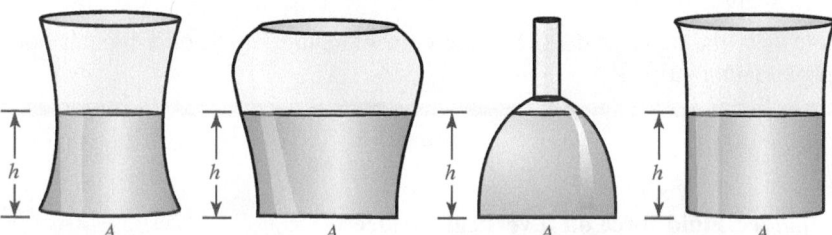

Figure 6.63 The fluid force $F = \delta h\,A$ does not depend on the shape or size of the container

When the surface is submerged vertically or at an angle, however, this simple formula does not apply because different parts of the surface are at different depths. We will use integration to compute the fluid force in such cases. Consider a plate that is submerged vertically in a fluid of density δ as shown in Figure 6.64**a**. We set up a coordinate system with the horizontal axis on the surface of the fluid and the positive h-axis (depth) pointing down. Thus, greater depths correspond to larger values of h (see Figure 6.64**b**). For simplicity, we assume the plate is oriented so its top and bottom are located a units and b units below the surface, respectively.

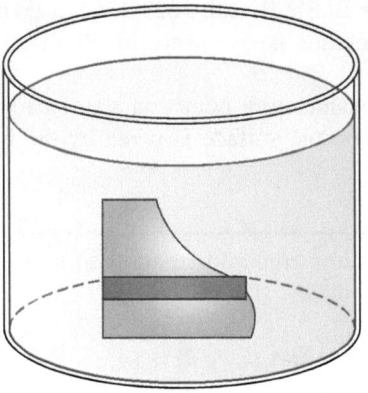
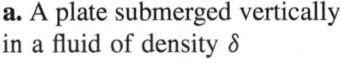
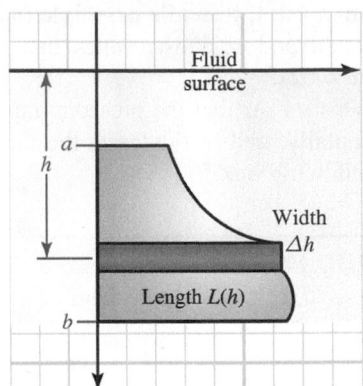

a. A plate submerged vertically in a fluid of density δ

b. A typical slab has area $A = L(h)\Delta h$

Figure 6.64   Hydrostatic force

Our strategy will be to think of the plate as a pile of subplates or slabs, each so thin that we may regard it as being at a constant depth below the surface. A typical slab (a horizontal strip) is shown in Figure 6.64**b**. Note that it has thickness Δh and length $L = L(h)$, so that its area is $\Delta A = L\Delta h$. Because the slab is at a constant depth h below the surface, the fluid force on its surface is

$$\Delta F = (\text{PRESSURE})(\text{AREA}) = \delta h\,\Delta A = \delta h L(h)\Delta h$$

and by integrating as h varies from a to b, we can model the total force on the plate.

FLUID (HYDROSTATIC) FORCE   Suppose a flat surface (a plate) is submerged vertically in a fluid of weight density $\delta = \rho g$ (lb/ft^3) and that the submerged portion of the plate extends from $h = a$ to $h = b$ on the vertical axis. Then the total force F exerted by the fluid is given by

$$F = \int_a^b \delta h L(h)\,dh$$

where h is the depth and $L(h)$ is the corresponding length of a typical horizontal approximating strip.

Example 4   Fluid force on a vertical surface

The cross sections of a certain trough are inverted isosceles triangles with height 6 ft and base 4 ft, as shown in Figure 6.65**a**. Suppose the trough contains water to a depth of 3 ft. Find the total fluid force on one end.

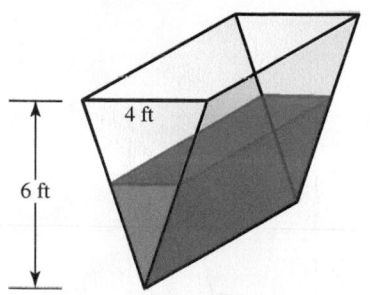

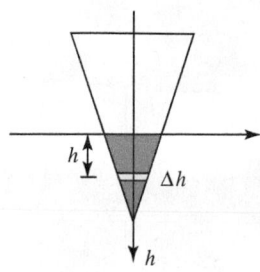

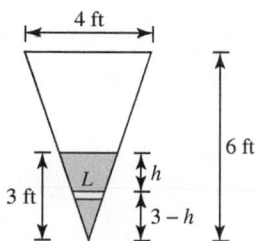

a. A trough with half-filled triangular cross sections

b. Side view of trough

c. Use similar triangles to find
$$\frac{L}{4} = \frac{3-h}{6}$$

Figure 6.65 Fluid force

Solution First, we set up a coordinate system in which the horizontal axis lies at the fluid surface and the positive vertical axis (the h-axis) points down (see Figure 6.65**b**). Next, we find expressions for the length and depth of a typical thin slab (horizontal strip) in terms of the variables. We assume the slab has thickness Δh and length L. Then by similar triangles (see Figure 6.65**c**) we have

$$\frac{L}{4} = \frac{3-h}{6} \qquad \text{so that} \qquad L = \frac{2}{3}(3-h)$$

We see that the approximating slab has area $\Delta A = L\Delta h = \frac{2}{3}(3-h)\Delta h$. Finally, we multiply the product of the fluid's weight density at a given depth (ρh) and the area of the approximating slab (ΔA) and integrate on the interval of depths occupied by the vertical plate.

$$F = \underbrace{\int_0^3 \frac{2}{3}\delta h(3-h)\,dh}_{\Delta F\,=\,force\ on\ a\ typical\ slab}$$

Note: (weight density)(depth)(area of strip)
$$\underset{\delta}{\downarrow} \quad \underset{h}{\downarrow} \quad \frac{2}{3}(3\underset{\downarrow}{-}h)\Delta h$$

$$= \frac{2}{3}(62.4)\int_0^3 (3h-h^2)\,dh \qquad \rho = 62.4\ lb/ft^3\ for\ water$$

$$= 41.6\left(\frac{3}{2}h^2 - \frac{1}{3}h^3\right)\Big|_0^3$$

$$= 187.2\ \text{lb}$$

Example 5 **Modeling the force on one face of a dam**

A reservoir is filled with water to the top of a dam. If the dam is in the shape of a parabola 40 ft high and 20 ft wide at the top, as shown in Figure 6.66**a**, what is the total fluid force on the face of the dam?

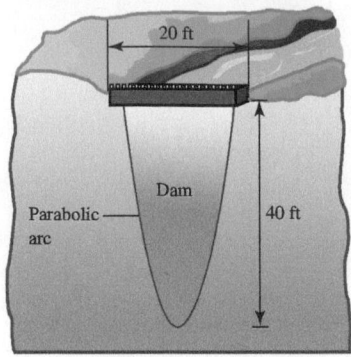

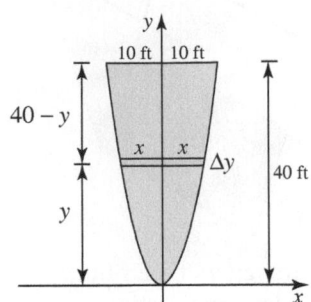

a. Cross section of a dam

b. A typical horizontal strip is
$40 - y$ ft below the water

Figure 6.66   Fluid force

Solution   Instead of putting the x-axis on the surface of the water (at the top of the dam), we place the origin at the vertex of the parabola and the positive y-axis along the parabola's axis of symmetry (Figure 6.66**b**). The advantage of this choice of axes is that the parabola can be represented by an equation of the form $y = cx^2$. Because we know that $y = 40$ when $x = 10$, it follows that

$$40 = c(10)^2 \quad \text{so that} \quad c = \frac{2}{5} \text{ and thus } y = \frac{2}{5}x^2$$

The typical horizontal strip on the face of the dam is located y feet above the x-axis, which means it is $h = 40 - y$ feet below the surface of the water. The strip is Δy feet wide and $L = 2x$ feet long. Write L as a function of y:

$$y = \frac{2}{5}x^2 \quad \text{so that} \quad x = \sqrt{\frac{5}{2}y}$$

and

$$L(y) = 2x = 2\left(\sqrt{\frac{5}{2}y}\right) = \sqrt{10y}$$

Therefore,

$$\Delta A = L\Delta y = \sqrt{10y}\,\Delta y$$

and the total force may be modeled by the integral

$$
\begin{array}{c}
\textit{Depth below water}: h = 40 - y \\
\downarrow \\
F = \int_0^{40} \underset{\uparrow}{\delta}\ h\ \underbrace{L(y)\,dy}_{\textit{Area of strip}} \\
\textit{Weight density of water} = 62.4
\end{array}
$$

$$= \int_0^{40} 62.4(40 - y)\sqrt{10y}\,dy$$

$$= 62.4\sqrt{10}\int_0^{40}(40y^{1/2} - y^{3/2})\,dy$$

$$= 62.4\sqrt{10}\left[\frac{80}{3}y^{3/2} - \frac{2}{5}y^{5/2}\right]_0^{40}$$

$$= 62.4\sqrt{10}\, y^{3/2} \left(\frac{80}{3} - \frac{2}{5}y \right) \Big|_0^{40}$$

$$= 532{,}480 \text{ lb}$$

Modeling the Centroid of a Plane Region

In mechanics, it is often important to determine the point where an irregularly shaped plate will balance. The **moment of a force** measures its tendency to produce rotation in an object and depends on the magnitude of the force and the point on the object where it is applied. Since the time of Archimedes (287-212 B.C.), it has been known that the balance point of an object occurs where all its moments cancel out (so there is no rotation).

The **mass** of an object is a measure of its **inertia**; that is, its propensity to maintain a state of rest or uniform motion. A thin plate whose material is distributed uniformly, so that its density ρ (mass per unit area) is constant, is called a **homogeneous lamina**. The balance point of such a lamina is called its **centroid** and may be thought of as its geometrical center. We shall see how centroids can be computed by integration in this section, and shall examine the topic even further in Section 12.6.

Consider a homogeneous lamina that covers a region R bounded by the curves $y = f(x)$ and $y = g(x)$ on the interval $[a, b]$, and consider a thin, vertical approximating strip within R, as shown in Figure 6.67.

Centroid

Theoretically, the lamina should balance on a point placed at its centroid.

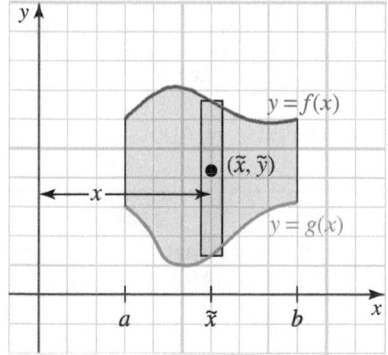

Figure 6.67 Homogeneous lamina with a vertical approximating strip

The mass of the strip is given by

$$\Delta m = \underset{\underset{Density}{\uparrow}}{\rho} \; \underbrace{[f(x) - g(x)]\Delta x}_{Area\ of\ strip}$$

and the total mass of the lamina may be found by integration:

$$m = \rho \int_a^b [f(x) - g(x)]\, dx$$

Note that $m = \text{AREA}$ if $\rho = 1$.

The centroid (geometrical center) of the approximating strip is $(\tilde{x}, \tilde{y})$, where $\tilde{x} = x$ and $\tilde{y} = \frac{1}{2}[f(x) + g(x)]$. The **moment of the approximating strip about the y-axis** is defined to be the product

$$\Delta M_y = \tilde{x} \cdot \underset{\underset{Mass\ of\ strip}{\uparrow}}{\underbrace{\Delta m}} = \tilde{x}\{\rho[f(x) - g(x)]\Delta x\}$$

Distance of the strip from the y-axis

This product provides a measure of the tendency of the strip to rotate about the y-axis. Similarly, the **moment of the strip about the x-axis** is defined to be the product

$$\Delta M_x = \tilde{y} \cdot \underbrace{\Delta m}_{mass\ of\ strip} = \frac{1}{2}[f(x) + g(x)]\{\rho[f(x) - g(x)]\Delta x\}$$

Distance of the strip from the y-axis   Average distance of the strip from the x-axis

$$= \frac{1}{2}\rho\{[f(x)]^2 - [g(x)]^2\}\Delta x$$

This product provides a measure of the tendency of the strip to rotate about the x-axis.

Integrating on $[a, b]$, we find the moments of the entire lamina R about the y-axis and x-axis may be modeled by

$$M_y = \rho \int_a^b x[f(x) - g(x)]\, dx \qquad \text{and} \qquad M_x = \frac{1}{2}\rho \int_a^b \{[f(x)]^2 - [g(x)]^2\}\, dx$$

If the entire mass m of the lamina R were located at the point $(\bar{x}, \bar{y})$, then its moments about the x-axis and y-axis would be $m\bar{y}$ and $m\bar{x}$, respectively. Thus, we have $m\bar{x} = M_y$ and $m\bar{y} = M_x$, so that

$$\bar{x} = \frac{M_y}{m} = \frac{\rho \int_a^b x[f(x) - g(x)]\, dx}{\rho \int_a^b [f(x) - g(x)]\, dx} \qquad \text{and}$$

$$\bar{y} = \frac{M_x}{m} = \frac{\frac{1}{2}\rho \int_a^b \{[f(x)]^2 - [g(x)]^2\}\, dx}{\rho \int_a^b [f(x) - g(x)]\, dx}$$

We can now summarize these results.

Let f and g be continuous and satisfy $f(x) \geq g(x)$ on the interval $[a, b]$, and consider a thin plate (lamina) of uniform density ρ that covers the region R between the graphs of $y = f(x)$ and $y = g(x)$ on the interval $[a, b]$. Then

MASS   The **mass** of R is: $m = \rho \int_a^b [f(x) - g(x)]\, dx$

CENTROID   The **centroid** of R is the point $(\bar{x}, \bar{y})$ such that

$$\bar{x} = \frac{M_y}{m} = \frac{\int_a^b x[f(x) - g(x)]\, dx}{\int_a^b [f(x) - g(x)]\, dx}$$

and

$$\bar{y} = \frac{M_x}{m} = \frac{\frac{1}{2}\int_a^b \{[f(x)]^2 - [g(x)]^2\}\, dx}{\int_a^b [f(x) - g(x)]\, dx}$$

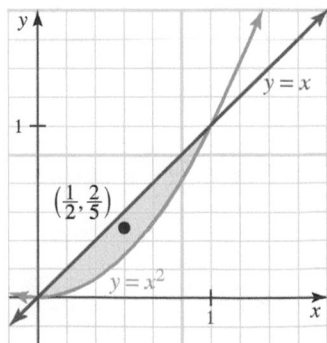

Figure 6.68 The centroid of a planar region

Example 6   Centroid of a thin plate

A homogeneous lamina R has constant density $\rho = 1$ and is bounded by the parabola $y = x^2$ and the line $y = x$. Find the mass and the centroid of R.

Solution   We see that the line and the parabola intersect at the origin and at the point $(1, 1)$, as shown in Figure 6.68.

Because $\rho = 1$, the mass is the same as the area of the region R. That is,

$$m = A = \int_0^1 (x - x^2)\, dx = \left(\frac{x^2}{2} - \frac{x^3}{3}\right)\Big|_0^1 = \frac{1}{6}$$

and we find that the region has moments M_y and M_x about the y-axis and x-axis, respectively, where

$$M_y = \int_0^1 x(x - x^2)\, dx = \left(\frac{x^3}{3} - \frac{x^4}{4}\right)\Big|_0^1 = \frac{1}{12}$$

and

$$\begin{aligned} M_x &= \frac{1}{2}\int_0^1 (x^2 - x^4)\, dx \\ &= \frac{1}{2}\left(\frac{x^3}{3} - \frac{x^5}{5}\right)\Big|_0^1 \\ &= \frac{1}{2}\left(\frac{1}{3} - \frac{1}{5}\right) \\ &= \frac{1}{15} \end{aligned}$$

Thus, the centroid of the region R has coordinates

$$\overline{x} = \frac{M_y}{m} = \frac{\frac{1}{12}}{\frac{1}{6}} = \frac{1}{2} \quad \text{and} \quad \overline{y} = \frac{M_x}{m} = \frac{\frac{1}{15}}{\frac{1}{6}} = \frac{2}{5}$$

Volume Theorem of Pappus

There is a remarkable result attributed to Pappus of Alexandria (ca. 300 A.D.), whom many regard as the last great Greek geometer, which gives a connection between centroids and volumes of revolution.

Theorem 6.2   Volume theorem of Pappus

The solid generated by revolving a region R about a line outside its boundary (but in the same plane) has volume $V = As$, where A is the area of R and s is the distance traveled by the centroid of R.

Proof: First, choose a coordinate system in which the y-axis coincides with the axis of revolution. Figure 6.69 shows a typical region R together with a vertical approximating rectangle. We shall assume that this rectangle has area ΔA and that it is located x units from the y-axis.

Notice that when the rectangle is revolved about the y-axis, it generates a shell of volume $\Delta V = 2\pi x \Delta A$. Thus, by partitioning the region R into a number of rectangles and taking the limit of the sum of the volumes of all the related approximating shells as the norm of the partition approaches 0, we find that

$$\begin{aligned} V &= \lim_{\|P\| \to 0} \sum_{k=1}^n 2\pi x_k \Delta A_k \\ &= \int 2\pi x\, dA \\ &= 2\pi \int x\, dA \end{aligned}$$

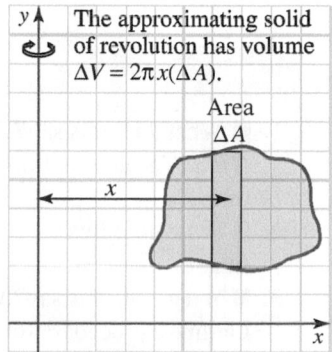

Figure 6.69 The volume theorem of Pappus

$$= 2\pi \bar{x} \int dA \qquad \text{Because } \bar{x} = \frac{\int x \, dA}{\int dA}$$

$$= 2\pi \bar{x} A \qquad \text{Because } \int dA = A$$

$$= As \qquad \text{Where } s = 2\pi\bar{x} \text{ is the distance traveled by the centroid.}$$

The circumference of a circle of radius $\bar{x}$.   ◆

We close this section with an example that illustrates the use of the volume theorem of Pappus.

Example 7   Volume of a torus using the volume theorem of Pappus

When a circle of radius r is revolved about a line in the plane of the circle located R units from its center $(R > r)$, the solid figure so generated is called a *torus*, as shown in Figure 6.70. Show that the torus has volume $V = 2\pi^2 r^2 R$.

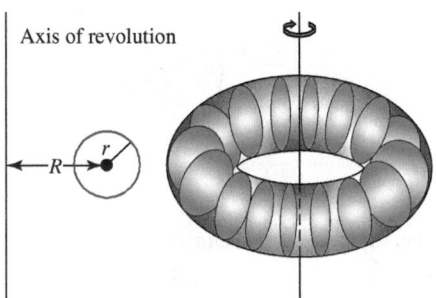

Axis of revolution

The centroid travels $2\pi R$ Units

Figure 6.70   The volume of a torus

Solution   The circle has area $A = \pi r^2$ and its center (which is its centroid) travels $2\pi R$ units. Comparing this example to the volume theorem of Pappus, we see $s = 2\pi R$. Therefore, the torus has volume

$$V = (2\pi R)A = 2\pi R(\pi r^2) = 2\pi^2 r^2 R$$

Incidentally, the volume found in Example 7 using the theorem of Pappus was worked in Problem 56 of Section 6.2 by using washers.

PROBLEM SET 6.5

Level 1

1. ■ What does this say? Discuss work and how to find it. (*Hint:* The answer is *not* to apply at the unemployment office!)
2. ■ What does this say? Discuss fluid force and how to find it.
3. ■ What does this say? What is meant by a centroid? Outline a procedure for finding both the mass and the centroid.
4. ■ What does this say? State and discuss the volume theorem of Pappus.
5. What is the work done in lifting an 850-lb billiard table 15 ft?

6. What is the work done in lifting a 50-lb bag of salt 5 ft?
7. A bucket weighing 75 lb when filled and 10 lb when empty is pulled up the side of a 100-ft building. How much more work is done in pulling up the full bucket than the empty bucket?
8. Suppose it takes 4 ergs of work to stretch a spring 10 cm beyond its natural length. How much more work is needed to stretch it 4 cm further?
9. A 5-lb force will stretch a spring 9 in. (= 0.75 ft) beyond its natural length. How much work is required to stretch it 1 ft beyond its natural length?
10. A spring whose natural length is 10 cm exerts a force of 30 newtons when stretched to a length of 15 cm. (*Note:* 15 cm − 10 cm = 5 cm = 0.05 m.) Find the spring constant, and then determine the work done in

stretching the spring 7 cm (= 0.07 m) beyond its natural length.

11. A 30-lb ball hangs at the bottom of a cable that is 50 ft long and weighs 20 lb. The entire length of cable hangs over a cliff. Find the work done to raise the cable and get the ball to the top of the cliff.

12. A 30-ft rope weighing 0.4 lb/ft hangs over the edge of a building 100 ft high. How much work is done in pulling the rope to the top of the building? Assume that the top of the rope is flush with the top of the building and the lower end of the rope is swinging freely.

13. An object moving along the x-axis is acted upon by a force $F(x) = |\sin x|$. Find the total work done by the force in moving the object from $x = 0$ cm to $x = 2\pi$ cm. *Note*: Distance is in cm, so force is in dynes.

14. An object moving along the x-axis is acted upon by a force $F(x) = x^4 + 2x^2$. Find the total work done by the force in moving the object from $x = 1$ cm to $x = 2$ cm. *Note*: Distance is in cm, so force is in dynes.

In Problems 15-20 the given figure is the vertical cross section of a tank containing the indicated fluid. Find the fluid force against the end of the tank. The weight densities are given in Table 6.4 on page 473.

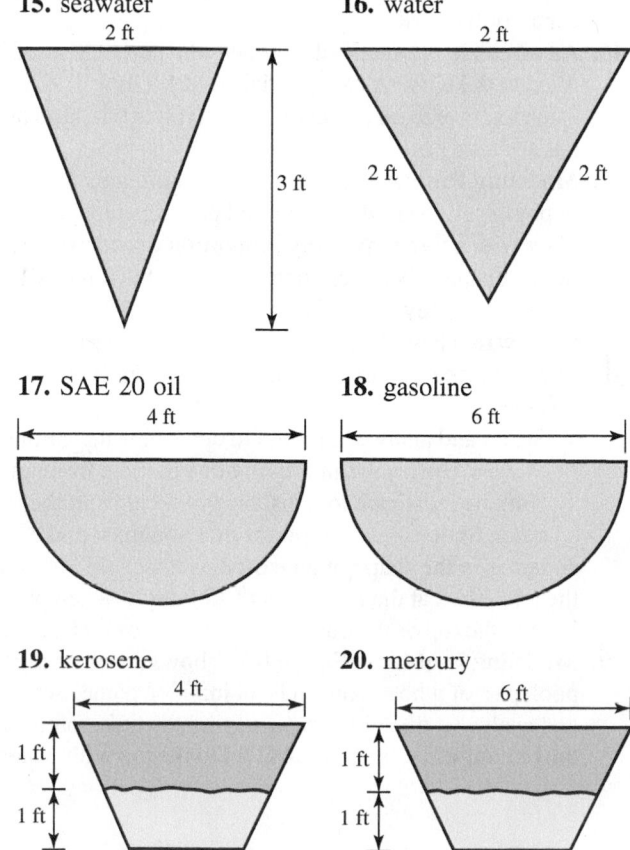

15. seawater

16. water

17. SAE 20 oil

18. gasoline

19. kerosene

20. mercury

In Problems 21-26, set up, but do not evaluate, an integral for the fluid force against the indicated vertical plate.

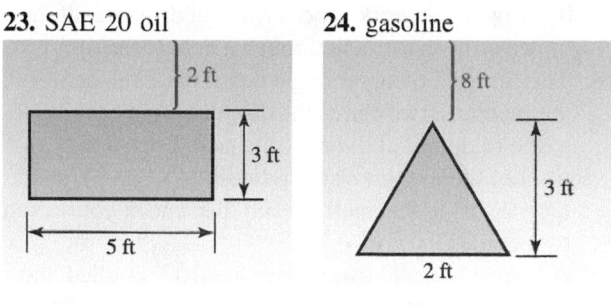

21. milk (half cookie)

22. water

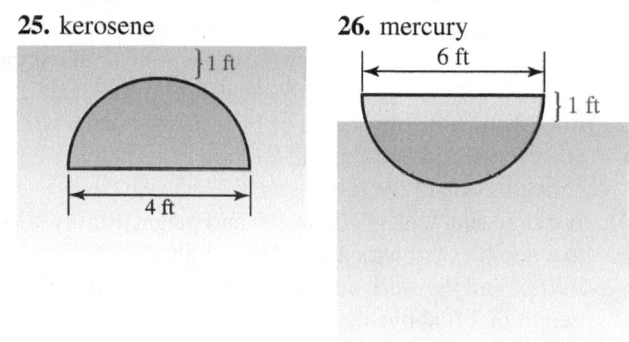

23. SAE 20 oil

24. gasoline

25. kerosene

26. mercury

Find the centroid ($\rho = 1$) of each planar region in Problems 27-32.

27. the region bounded by the parabola $y = x^2 - 9$ and the x-axis

28. the region in the first quadrant bounded by the curves $y = x^3$ and $y = \sqrt[3]{x}$

29. the region bounded by the curve $y = x^{-1}$, the x-axis, and the lines $x = 1$ and $x = 2$

30. the region bounded by the parabola $y = 4 - x^2$ and the line $y = x + 2$

31. the region bounded by $y = \dfrac{1}{\sqrt{x}}$ and the x-axis on $[1, 4]$

32. the region bounded by the curve $y = x^{-1}$ and the line $2x + 2y = 5$

Use the theorem of Pappus to compute the volume of the solids generated by revolving the regions given in Problems 33-36 about the prescribed axis.

33. the triangular region with vertices $(-3, 0)$, $(0, 5)$, and $(2, 0)$, about the line $y = -1$

34. the region bounded by the curve $y = \sqrt{x}$, the x-axis, and the line $x = 4$ about the line $x = -1$

35. the semicircular region $x = \sqrt{4 - y^2}$, about the line $x = -2$

36. the semicircular region $y = \sqrt{1 - x^2}$, about the line $y = -1$

Level 2

37. A tank in the shape of an inverted right circular cone of height 6 ft and top radius 3 ft is half full.
   a. How much work is performed in pumping all the water over the top edge of the tank?
   b. How much work is performed in draining all the water from the tank through a hole at the tip?

38. The centroid of any triangle lies at the intersection of the medians, two-thirds the distance from each vertex to the midpoint of the opposite side. Locate the centroid of the triangle with vertices $(0, 0)$, $(7, 3)$, and $(7, -2)$ using this method, and then check your result by calculus.

39. A hemispherical bowl with radius 10 ft is filled with water to a level of 6 ft. Find the work done to the nearest ft-lb to pump all the water to the top of the bowl.

40. A tank is formed by rotating the curve $y = x^2$ about the y-axis for $0 \le x \le 4$, where x is in feet. If the tank is filled with water to a depth of 12 feet, how much work is performed in pumping all the water to a height of 3 ft above the top of the tank?

41. A cylindrical tank of radius 3 ft and height 10 ft is filled to a depth of 2 ft with a liquid of weight density $\delta = 40$ lb/ft^3. Find the work done in pumping all the liquid to a height of 2 ft above the top of the tank.

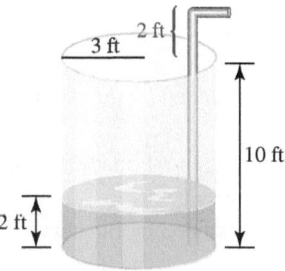

42. A holding tank has the shape of a rectangular parallelepiped 20 ft by 30 ft by 10 ft.
   a. How much work is done in pumping all the water to the top of the tank?

b. How much work is done in pumping all the water out of the tank to a height of 2 ft above the top of the tank?

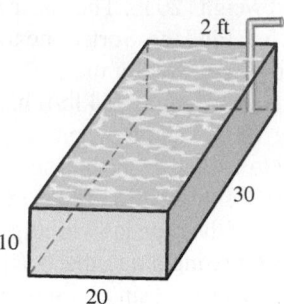

43. A swimming pool 20 ft by 15 ft by 10 ft deep is filled with water. A cube 1 ft on a side lies on the bottom of the pool. Find the total fluid force on the five exposed faces of the cube.

44. Suppose a log of radius 1 ft lies at the bottom of the pool in Problem 43. What is the total fluid force on one end of the log?

45. An object weighing 800 lb on the surface of the earth is propelled to a height of 200 mi above the earth. How much work is done against gravity? *Hint*: Assume the radius of the earth is 4,000 mi, and use Newton's law $F = -k/x^2$, for the force on an object x miles from the center of the earth.

46. An oil can in the shape of a rectangular parallelepiped is filled with kerosene (weight density 51.2 lb/ft^3). What is the total force on a side of the can if it is 6 in high and has a square base 4 in. on a side?

47. Modeling Problem According to *Coulomb's law* in physics, two similarly charged particles repel each other with a force inversely proportional to the square of the distance between them. Suppose the force is 12 dynes when they are 5 cm apart.
   a. How much work is done in moving one particle from a distance of 10 cm to a distance of 8 cm from the other?
   b. Set up and analyze a model to determine the amount of work performed in moving one particle from an "infinite" distance to a distance of 8 cm from the other. State any assumptions that you must make.

48. A dam is in the shape of an isosceles trapezoid 200 ft at the top, 100 ft at the bottom, and 75 ft high. When water is up to the top of the dam, what is the force on its face?

49. Modeling Problem Figure 6.71 shows a swimming pool, part of whose bottom is an inclined plane. Set up and analyze a model for determining the fluid force on the bottom when the pool is filled to the top with water.

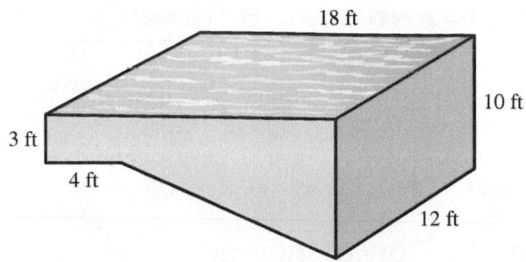

Figure 6.71 Swimming pool dimensions

50. Modeling Problem Figure 6.72 shows a dam whose face against the water is an inclined rectangle.

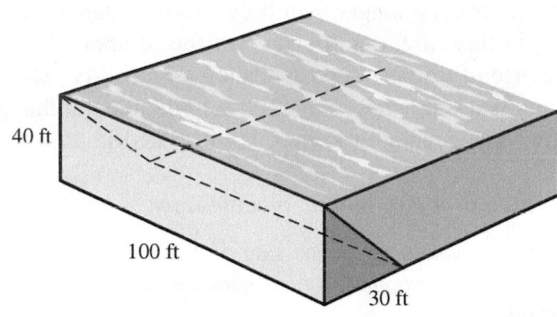

Figure 6.72 Dam with face at an inclined rectangle

a. Set up and analyze a model for determining the fluid force against the face of the dam when the water is level with the top.

b. What would the fluid force be if the face of the dam were a vertical rectangle (not inclined)?

*A region R with uniform density is said to have **moments of inertia** I_x about the x-axis and I_y about the y-axis, where*

$$I_x = \int y^2 \, dA \quad and \quad I_y = \int x^2 \, dA$$

The dA represents the area of an approximating strip. In Problems 51-52, find I_x and I_y for the given region.

51. the region bounded by the parabola $y = \sqrt{x}$, the x-axis, and the line $x = 4$

52. the region bounded by the curve $y = \sqrt[3]{1 - x^3}$ and the coordinate axes

53. Modeling Problem If a region has area A and moment of inertia I_y about the y-axis, then it is said to have a **radius of gyration**

$$\rho_y = \sqrt{\frac{I_y}{A}}$$

with respect to the y-axis. Set up and analyze a model to determine the radius of gyration for the triangular region bounded by the line $y = 4 - x$ and the positive coordinate axes.

54. An irregular plate is measured at 0.2-ft intervals from top to bottom, and the corresponding widths noted. The plate is then submerged vertically in water with its top 1 ft below the surface. Use Simpson's rule to estimate the total force on the face of the plate.

Depth, x_k	Width, L_k
1.0	0.0
1.2	3.3
1.4	3.6
1.6	3.7
1.8	3.7
2.0	3.6
2.2	3.4
2.4	2.9
2.6	0.0

Level 3

55. A right circular cone is generated when the region bounded by the line $y = x$ and the vertical lines $x = 0$ and $x = r$ is revolved about the x-axis. Use Pappus' theorem to show that the volume of this cone is $V = \frac{1}{3}\pi r^3$.

56. Show that the centroid of a rectangle is the point of intersection of its diagonals.

57. Find the volume of the solid figure generated when a square of side L is revolved about a line that is outside the square, parallel to two of its sides, and located s units from the closer side.

58. Let R be the triangular region bounded by the line $y = x$, the x-axis, and the vertical line $x = r$. When R is rotated about the x-axis, it generates a cone of volume $V = \frac{1}{3}\pi r^3$. Use Pappus' theorem to determine $\bar{y}$, the y-coordinate of the centroid of R. Then use similar reasoning to find $\bar{x}$.

59. Let R be a region in the plane, and suppose that R can be partitioned into two subregions R_1 and R_2. Let $(\bar{x}_1, \bar{y}_1)$ and $(\bar{x}_2, \bar{y}_2)$ be the centroids of R_1 and R_2, respectively, and let A_1 and A_2 be the areas of the subregions. Show that $(\bar{x}, \bar{y})$ is the centroid of R, where

$$\bar{x} = \frac{A_1 x_1 + A_2 x_2}{A_1 + A_2} \quad and \quad \bar{y} = \frac{A_1 y_1 + A_2 y_2}{A_1 + A_2}$$

60. Find the volume of the solid figure generated by revolving an equilateral triangle of side L about one of its sides.

6.6 APPLICATIONS TO BUSINESS, ECONOMICS, AND LIFE SCIENCES

IN THIS SECTION: *Future and present value of a flow of income; cumulative change: net earnings; consumer's and producer's surplus; survival and renewal, the flow of blood through an artery*
In Section 6.5, we saw how integration can be applied to the physical sciences. The goal of this section is to explore applications to a variety of other areas, especially business, economics, and the life sciences.

Future and Present Value of a Flow of Income

Recall from Chapter 2 that when P_0 dollars are invested at the continuous compounding annual rate r, then after t years, the account is worth $A = P_0 e^{rt}$ dollars. The amount A is called the *future value* of the initial value P_0 dollars, but what is the future value if the money were not deposited just once, in a lump sum P_0, but continuously over the entire time period? The difficulty is that at any given time, the "old" money deposited earlier in the time period has been accumulating interest longer than "new" money deposited more recently. To compute future value, in this case, our strategy will be to approximate the continuous income flow by a sequence of discrete deposits called an **annuity**. Then, the amount of the approximating annuity is a sum whose limit is a definite integral that gives the future value of the income flow. The general procedure is illustrated in Example 1.

Example 1   Computing future value of a continuous income flow

Money is transferred into an account continuously at the rate of 1,500 dollars/yr. If the account earns interest at the rate of 7% compounded continuously, what is the future value of the account 5 years from now?

Solution   To approximate the future value of the income flow, divide the 5-yr time period $0 \le t \le 5$ into n intervals, each of length $\Delta_n t = (5-0)/n$, and for $k = 1, 2, \cdots, n$, let t_{k-1} denote the beginning of the kth subinterval $[t_{k-1}, t_k]$. The amount of money deposited during the kth subinterval is $1{,}500 \Delta_n t$. Therefore, if as an approximation, we assume that all this money is deposited at time t_{k-1}, then since $5 - t_{k-1}$ years remain in the 5-yr investment period, it follows that this money would grow to $(1{,}500\Delta_n t)e^{0.07(5-t_{k-1})}$ dollars; that is,

$$\begin{bmatrix} \text{FUTURE VALUE OF MONEY DEPOSITED} \\ \text{DURING THE } k\text{TH SUBINTERAVAL} \end{bmatrix} = 1{,}500 e^{0.07(5-t_{k-1})} \Delta_n t$$

This approximation is illustrated in Figure 6.73.

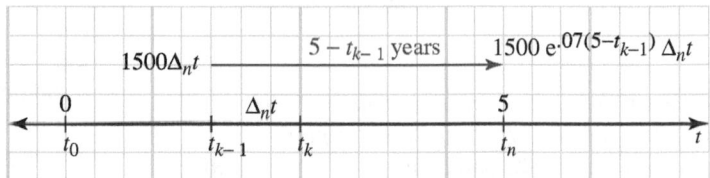

Figure 6.73   Approximate future value of money deposited during the kth subinterval

The future value of the entire income stream can be approximated by the sum of the future values of the money deposited during all n time subintervals. Hence, we have

$$[\text{FUTURE VALUE OF INCOME FLOW}] \approx \sum_{k=1}^{n} 1{,}500 e^{0.07(5-t_{k-1})} \Delta_n t$$

As n increases without bound, the length of each subinterval tends toward zero, and there is less and less error in our assumption that all the money in the kth subinterval is deposited at time $t = t_{k-1}$. Therefore, it is reasonable to compute the total future value of the income flow by taking the limit of the approximating sum as $n \to \infty$. We recognize this sum as a Riemann sum, and the limit as the definite integral. That is,

$$[\text{FUTURE VALUE OF INCOME FLOW}] = \lim_{n \to \infty} \sum_{k=1}^{n} 1{,}500 e^{0.07(5 - t_{k-1})} \Delta_n t$$

$$= \int_0^5 1{,}500 e^{0.07(5-t)} \, dt$$

$$= 1{,}500 e^{0.35} \int_0^5 e^{-0.07t} \, dt$$

$$= 1{,}500 e^{0.35} \left[\frac{e^{-0.07t}}{-0.07} \right]_0^5$$

$$= 8{,}980.02$$

Recall from Chapter 2, that the *present value*, *PV*, of A dollars invested at the annual interest rate r compounded continuously for a term T years is the amount of money $PV = Ae^{-rt}$ that must be deposited today at the prevailing interest rate to generate A dollars at the end of the term. Likewise, the present value of an income stream $f(t)$ generated continuously at a certain rate over a specified term is the amount of money that would have to be deposited at the beginning of the term at the prevailing interest rate to generate the income stream over the entire term. As with future value, the strategy is to approximate the continuous income stream by a sequence of discrete payments; that is, by an annuity. The present value of the approximating annuity is a sum, and by taking the limit of that sum, we obtain an integral

$$\int_0^T f(t) e^{-rt} \, dt$$

that represents the present value of the entire income flow. The details of this approximation scheme are outlined in the problem set. Here is a summary of the formulas to be used for computing future and present value of an income flow.

FUTURE VALUE AND PRESENT VALUE   Let $f(t)$ be the amount of money deposited at time t over the time period $[0, T]$ in an account that earns interest at the annual rate r compounded continuously. Then the **future value** of the income flow over the time period is given by

$$FV = \int_0^T f(t) e^{r(T-t)} \, dt$$

and the **present value** of the same income flow over the time period is

$$PV = \int_0^T f(t) e^{-rt} \, dt$$

Cumulative Change: Net Earnings

In certain applications, we are given the rate of change $Q'(x)$ of a quantity $Q(x)$ and are required to find the net change $Q(b) - Q(a)$ in $Q(x)$ as x varies from $x = a$ to $x = b$. But because $Q(x)$ is an antiderivative of $Q'(x)$, the fundamental theorem of calculus tells us that the net change is given by the definite integral

$$Q(b) - Q(a) = \int_a^b Q'(x)\,dx$$

Since the definite integral on the right "adds" all the instantaneous changes in $Q(x)$ over the time period $[a, b]$, the integral can also be thought of as **cumulative change**. For example, if $N(t)$ is the population of a certain species that is growing at the rate $N'(t)$ at time t, then the cumulative change in population over the time period $[t_1, t_2]$ is given by the integral

$$N(t_2) - N(t_1) = \int_{t_1}^{t_2} N'(t)\,dt$$

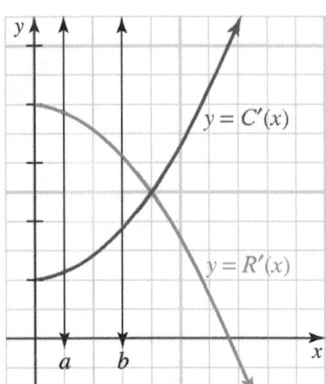

A business application involves the **net earnings** of an industrial process. Suppose $C(x)$ represents the total cost of producing x units and $R(x)$ represents the revenue generated by the sale of those units. Then $E(x) = R(x) - C(x)$ represents earnings, and if the marginal cost $C'(x)$ and marginal revenue $R'(x)$ are known, then the net earnings for the production range $a \le x \le b$ are given by the definite integral

$$E(b) - E(a) = \int_a^b E'(x)\,dx$$

$$= \int_a^b [R'(x) - C'(x)]\,dx$$

Figure 6.74 Net earnings as an area

Net earnings can be interpreted geometrically as an area between the marginal revenue and marginal cost curves as shown in Figure 6.74. These ideas are illustrated in Example 2.

Example 2   Computing net earnings

For a certain industrial process, the marginal cost and marginal revenue (in thousands of dollars) associated with producing x units are

$$C'(x) = 0.1x^2 + 4x + 10 \quad \text{and} \quad R'(x) = 70 - x$$

respectively. What are the net earnings of the process as x ranges from $x = 0$ to $x = x_m$, where x_m is the level of production at which marginal cost equals marginal revenue?

Solution   First, note that marginal cost equals marginal revenue when

$$0.1x^2 + 4x + 10 = 70 - x$$
$$0.1x^2 + 5x - 60 = 0$$
$$x^2 + 50x - 600 = 0$$
$$(x + 60)(x - 10) = 0$$
$$x = -60,\ 10$$

Reject the negative value, giving the solution $x_m = 10$, so that the net earnings as x ranges from $x = 0$ to $x = 10$ are given by

$$E(10) - E(0) = \int_0^{10} [R'(x) - C'(x)]\,dx$$

$$= \int_0^{10} [70 - x - (0.1x^2 + 4x + 10)]\,dx$$

$$= \int_0^{10} [-0.1x^2 - 5x + 60]\, dx$$

$$= \frac{950}{3}$$

The net earnings (rounded to the nearest dollar) are \$316,667, and are shown as the area between the marginal cost and marginal revenue curves in Figure 6.75. ∎

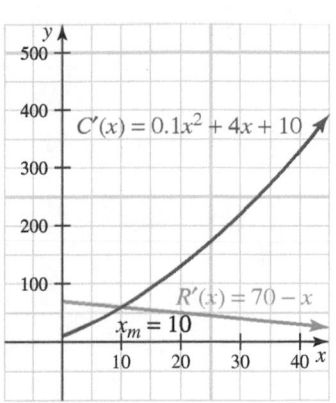

Figure 6.75 Net earnings for the marginal cost and marginal revenue

Consumer's and Producer's Surplus

In a competitive economy, the total amount that consumers actually spend on a commodity is usually less than the total amount they would have been willing to spend. The difference between the two amounts can be thought of as a savings realized by consumers and is known in economics as the **consumer's surplus**.

To get a better feel for the concept of consumer's surplus, consider an example of a couple who are willing to spend \$500 for their first flat-screen HDTV, \$300 for a second set, and only \$50 for a third set. If the market price is \$300, then the couple would buy only two sets and would spend a total of $2 \times \$300 = \$600$. This is less than the $\$500 + \$300 = \$800$ that the couple would have been willing to spend to get the two sets. The savings of $\$800 - \$600 = \$200$ is the couple's consumer's surplus. Consumer's surplus has a simple geometric interpretation, which is illustrated in Figure 6.76.

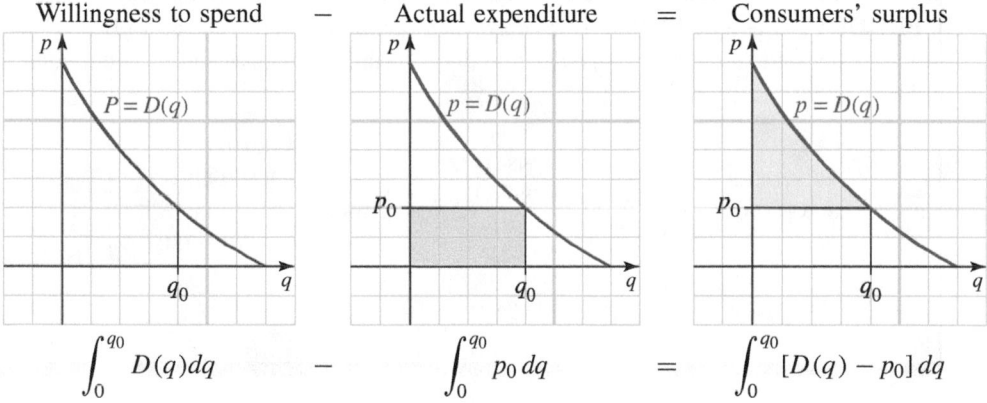

Figure 6.76 Geometric interpretation of consumer's surplus

Note that p and q denote the market price and the corresponding demand, respectively. The first figure shows the region under the demand curve $p = D(q)$ from $q = 0$ to $q = q_0$, and the area of this region represents the total amount that consumers are willing to spend to get q_0 units of the commodity. The rectangle in the middle has an area of $p_0 q_0$ and represents the actual consumer expenditure for q_0 units at p_0 dollars per unit. The difference between these two areas (figure at the right) represents the consumer's surplus. That is, consumer's surplus is the area of the region between the demand curve $p = D(q)$ and the horizontal line $p = p_0$ so that

$$\int_0^{q_0} [D(q) - p_0]\, dq = \int_0^{q_0} D(q)\, dq - \int_0^{q_0} p_0\, dq = \int_0^{q_0} D(q)\, dq - p_0 q_0$$

CONSUMER'S SURPLUS If q_0 units of a commodity are sold at a price of p_0 dollars per unit, and if $p = D(q)$ is the consumer's demand function for the commodity, then

$$
\begin{array}{c}
\text{CONSUMER'S} \\
\text{SURPLUS}
\end{array}
=
\begin{bmatrix}
\text{TOTAL AMOUNT} \\
\text{CONSUMERS ARE} \\
\text{WILLING TO SPEND} \\
\text{FOR } q_0 \text{ UNITS}
\end{bmatrix}
-
\begin{bmatrix}
\text{ACTUAL CONSUMER} \\
\text{EXPENDITURE FOR} \\
q_0 \text{ UNITS}
\end{bmatrix}
$$

$$
CS = \int_0^{q_0} D(q)\,dq - p_0 q_0
$$

Producer's surplus is the other side of the coin from consumer's surplus. In particular, a supply function $p = S(q)$ gives the price per unit that producers are willing to accept in order to supply q units to the marketplace. However, any producer who is willing to accept less than $p_0 = S(q_0)$ dollars for q_0 units gains from the fact that the price is actually p_0. Then **producer's surplus** is the difference between what producers would be willing to accept for supplying q units and the price they actually receive. Reasoning as we did with consumer's surplus, we obtain the following formula for producer's surplus.

PRODUCER'S SURPLUS If q_0 units of a commodity are sold at a price of p_0 dollars per unit, and if $p = S(q)$ is the producer's supply function for the commodity, then

$$
\begin{array}{c}
\text{PRODUCER'S} \\
\text{SURPLUS}
\end{array}
=
\begin{bmatrix}
\text{ACTUAL CONSUMER} \\
\text{EXPENDITURE} \\
\text{FOR } q_0 \text{ UNITS}
\end{bmatrix}
-
\begin{bmatrix}
\text{TOTAL AMOUNT} \\
\text{PRODUCER WILLING TO} \\
\text{ACCEPT WHEN } q_0 \\
\text{UNITS ARE SUPPLIED}
\end{bmatrix}
$$

$$
PS = p_0 q_0 - \int_0^{q_0} S(q)\,dq
$$

Example 3   Computing consumer's and producer's surplus

A tire manufacturer estimates that q (thousand) radial tires will be purchased (that is, demanded) by wholesalers when the price is

$$
p = D(q) = -0.1q^2 + 90
$$

dollars/tire, and the same number of tires will be supplied when the price is

$$
p = S(q) = 0.2q^2 + q + 50
$$

dollars/tire.

a. Find the equilibrium price (that is, where supply equals demand) and the quantity supplied and demanded at that price.

b. Determine the consumer's and producer's surplus at the equilibrium price.

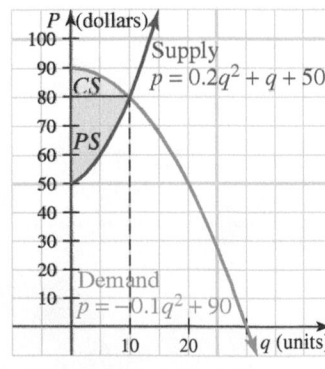

Figure 6.77 Consumer's and producer's surplus at equilibrium for the given supply and demand functions

Solution

a. The supply and demand curves are shown in Figure 6.77.

Supply equals demand when

$$0.2q^2 + q + 50 = -0.1q^2 + 90$$

$$0.3q^2 + q - 40 = 0$$

$$3q^2 + 10q - 400 = 0$$

$$(3q + 40)(q - 10) = 0$$

$$q = 10 \qquad \left(\text{reject } q = -\frac{40}{3}\right)$$

and $p = S(10) = D(10) = 80$. Thus, equilibrium occurs at a price of \$80/tire, when 10,000 tires are supplied and demanded.

b. Using $p_0 = 80$ and $q_0 = 10$, we find that the consumer's surplus is

$$\int_0^{10} (-0.1q^2 + 90)dq - (80)(10) = \left[-0.1\left(\frac{q^3}{3}\right) + 90q\right]_0^{10} - (80)(10)$$

$$= \frac{2{,}600}{3} - (80)(10)$$

$$= \frac{200}{3}$$

The producer's surplus is

$$(80)(10) - \int_0^{10} (0.2q^2 + q + 50)\, dq = (80)(10) - \left[0.2\left(\frac{q^3}{3}\right) + \left(\frac{q^2}{2}\right) + 50q\right]_0^{10}$$

$$= (80)(10) - \frac{1{,}850}{3}$$

$$= \frac{550}{3}$$

Since q is in thousands, the consumer's surplus is about \$66,667, and producer's surplus is about \$183,333. The consumer's surplus and the producer's surplus are the regions labeled *CS* and *PS*, respectively, in Figure 6.77. ▪

Survival and Renewal

There is an important category of problems in which a survival function gives the fraction of individuals in a group of the population that can be expected to remain in the group and a renewal function gives the rate at which new members can be expected to join the group. Such **survival and renewal** problems arise in many fields, including sociology, ecology, demography, and even finance, where the "population" is the number of dollars in an investment account and "survival and renewal" refers to results of an investment strategy. The following example illustrates the general procedure for modeling with survival and renewal functions.

Example 4   Modeling enrollment using survival and renewal

A new county mental health clinic has just opened. Statistics from similar facilities suggest that the fraction of patients who will still be receiving treatment at the clinic t months after their initial visit is given by the function $f(t) = e^{-t/20}$. The clinic initially accepts 300 people for treatment and plans to accept new patients at the rate of 10 per month. Approximately how many people will be receiving treatment at the clinic 15 months from now?

Solution   Since $f(15)$ is the fraction of patients whose treatment continues at least 15 months, it follows that of the current 300 patients, only $300 f(15)$ will still be receiving treatment 15 months from now.

To approximate the number of *new* patients who will be receiving treatment 15 months form now, divide the 15-month time interval $0 \leq t \leq 15$ into n equal subintervals of length Δt months and let t_{j-1} denote the beginning of the jth subinterval. Since new patients are accepted at the rate of 10/mo, the number of new patients accepted during the jth subinterval is $10 \Delta t$. Fifteen months from now, approximately $15 - t_{j-1}$ months will have elapsed since these $10 \Delta t$ new patients had their initial visits, and so approximately $(10 \Delta t) f(15 - t_{j-1})$ of them will still be receiving treatment at that time (see Figure 6.78).

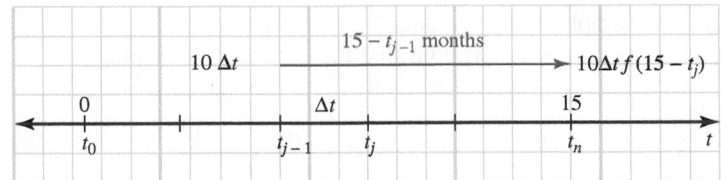

Figure 6.78   New members arriving during the jth subinterval

It follows that the total number of new patients still receiving treatment 15 months from now can be approximated by the sum

$$\sum_{j=1}^{n} 10 f(15 - t_{j-1}) \Delta t$$

Adding this to the number of current patients who will still be receiving treatment in 15 months, we obtain

$$P \approx 300 f(15) + \sum_{j=1}^{n} 10 f(15 - t_{j-1}) \Delta t$$

where P is the total number of patients who will be receiving treatment 15 months from now. As n increases without bound, the approximation improves and approaches the true value of P. That is,

$$P = 300 f(15) + \lim_{n \to \infty} \sum_{j=1}^{n} 10 f(15 - t_{j-1}) \Delta t$$

$$= 300 f(15) + \int_{0}^{15} 10 f(15 - t) \, dt$$

$$= 300 e^{-3/4} + 10 e^{-3/4} \int_{0}^{15} e^{t/20} dt \qquad \begin{aligned} & f(t) = e^{-t/20}; \ so f(15) = e^{-3/4} \\ & and f(15 - t) = e^{-(15-t)/20} = e^{-3/4} e^{t/20} \end{aligned}$$

$$= 300 e^{-3/4} + 200 e^{-3/4} e^{t/20} \Big|_{0}^{15}$$

$$= 300 e^{-3/4} + 200(1 - e^{-3/4})$$

$$\approx 247.24$$

That is, 15 months from now, the clinic will be treating approximately 247 patients. ∎

The Flow of Blood Through an Artery

In biology, Poiseuille's law of flow* asserts that the speed of blood in an artery is a function of the distance of the blood from the artery's central axis. That is, the speed (in cm/s) of blood that is r centimeters from the central axis of the artery may be modeled by

$$S(r) = k(R^2 - r^2)$$

where R is the radius of the artery and k is a constant. In the following example, you will see how to use this information to compute the volume rate (in cm³/s) at which blood flows through the artery.

Example 5   Modeling blood flow

Find an expression for the volume rate (in cm³/s) at which blood flows through an artery of radius R if the speed of the blood r cm from the central axis is $S(r) = k(R^2 - r^2)$, where k is a constant.

Solution   To find the approximate volume of blood that flows through a cross section of the artery each second, divide the interval $0 \le r \le R$ into n equal subintervals of width Δr cm and let r_j denote the beginning of the jth subinterval. These subintervals determine n concentric rings, as illustrated in Figure 6.79.

If Δr is small, the area of the jth ring is approximately the area of a rectangle whose length is the circumference of the (inner) boundary of the ring and whose width is Δr. That is,

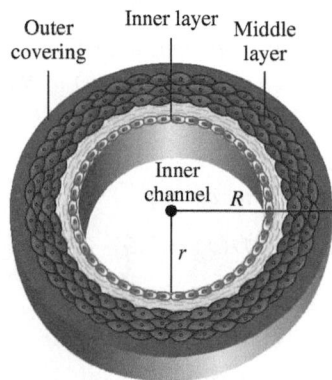

$$\text{AREA OF THE } j \text{TH RING} \approx 2\pi r_j \Delta r$$

If we multiply the area of the jth ring (cm²) by the speed (cm/s) of the blood flowing through this ring, we obtain the volume rate (cm³/s) at which blood flows through the jth ring. Since the speed of blood flowing through the jth ring is approximately $S(r_j)$ cm/s, it follows that

$$\begin{aligned}
\begin{matrix} \text{VOLUME RATE OF FLOW} \\ \text{THROUGH THE } j\text{TH RING} \end{matrix} &= (\text{AREA OF } j\text{TH RING})\begin{pmatrix} \text{SPEED OF BLOOD} \\ \text{THROUGH THE} \\ j\text{TH RING} \end{pmatrix} \\
&= (2\pi r_j \Delta r) S(r_j) \\
&= 2\pi r_j \Delta r)[k(R^2 - r_j^2)] \\
&= 2\pi k (R^2 r_j - r_j^3)\Delta r
\end{aligned}$$

Figure 6.79 **Interactive**
Subdividing a cross section of an artery into concentric rings

The volume rate of flow of blood through the entire cross section is the sum of n such terms, one for each of the n concentric rings. That is,

$$\begin{aligned}
\text{VOLUME RATE OF FLOW} &\approx \sum_{j=1}^{n} 2\pi k (R^2 r_j - r_j^3)\Delta r \\
&= \lim_{n \to \infty} \sum_{j=1}^{n} 2\pi k (R^2 r_j - r_j^3)\Delta r
\end{aligned}$$

As n increases without bound, this approximation approaches the true value of the volume rate of flow.

---

*The law of flow is named for the same French physician, Jean Louis Poiseuille (1799-1869), responsible for the resistance to flow law cited in Section 4.7.

$$= \int_0^R 2\pi k (R^2 r - r^3)\, dr$$

$$= 2\pi k \left(\frac{R^2}{2} r^2 - \frac{1}{4} r^4 \right) \Big|_0^R$$

$$= \frac{\pi k R^4}{2} \qquad \text{This is in cm}^3/\text{s.}$$

PROBLEM SET 6.6

Level 1

1. ■ What does this say? What is cumulative change for a quantity $Q(x)$?
2. ■ What does this say? What are the future and present values of an income flow?
3. ■ What does this say? What are consumer's and producer's surplus?
4. ■ What does this say? What do we mean by survival and renewal?

Find the consumer's surplus for the given demand function defined by $D(q)$ at the point that corresponds to the sales level q as given in Problems 5-8.

5. $D(q) = 150 - 6q$

   a. $q_0 = 5$ b. $q_0 = 12$

6. $D(q) = 3.5 - 0.5q$

   a. $q_0 = 1$ b. $q_0 = 1.5$

7. $D(q) = 2.5 - 1.5q$

   a. $q_0 = 1$ b. $q_0 = 0$

8. $D(q) = 100 - 8q$

   a. $q_0 = 4$ b. $q_0 = 10$

In Problems 9-12, $p = D(q)$ is the price (in dollars/unit) at which q units of a particular commodity will be demanded by the market (that is, all q units will be sold at this price), and q_0 is a specified level of production. In each case, find the price $p_0 = D(q_0)$ at which q_0 units will be demanded and compute the corresponding consumer's surplus. Sketch the demand curve $y = D(q)$ and shade the region whose area represents the consumer's surplus.

9. $D(q) = 150 - 2q - 3q^2$; $q_0 = 6$ units
10. $D(q) = 2(64 - q^2)$; $q_0 = 3$ units
11. $D(q) = 75e^{-0.04q}$; $q_0 = 3$ units
12. $D(q) = 40e^{-0.05q}$; $q_0 = 5$ units

In Problems 13-16, $p = S(q)$ is the price (in dollars/unit) at which q units of a particular commodity will be supplied to the market by producers, and q_0 is a specified

level of production. In each case, find the price $p_0 = S(q_0)$ at which q_0 units will be supplied and compute the corresponding producer's surplus. Sketch the supply curve $y = S(q)$ and shade the region whose area represents the producer's surplus.

13. $S(q) = 0.5q + 15$; $q_0 = 5$ units
14. $S(q) = 0.3q^2 + 30$; $q_0 = 4$ units
15. $S(q) = 17 + 11e^{0.01q}$; $q_0 = 7$ units
16. $S(q) = 10 + 15e^{0.03q}$; $q_0 = 3$ units

Level 2

In Problems 17-20, find the consumer's surplus at the point of market equilibrium [the level of production where supply $S(q)$ equals demand $D(q)$].

Demand function	Supply function
17. $D(q) = 27 - q^2$	$S(q) = \frac{1}{4}q^2 + \frac{1}{2}q + 5$
18. $D(q) = 14 - q^2$	$S(q) = 2q^2 + 2$
19. $D(q) = 25 - q^2$	$S(q) = 5q^2 + 1$
20. $D(q) = 32 - 2q^2$	$S(q) = \frac{1}{3}q^2 + 2q + 5$

21. It is estimated that t weeks from now, contributions in response to a fund-raising campaign will be coming in at the rate of

$$R'(t) = 5{,}000e^{-0.2t}$$

dollars/wk, while campaign expenses are expected to accumulate at the constant rate of \$676/wk.
   a. For how many weeks does the rate of revenue exceed the rate of cost?
   b. What net earnings will be generated by the campaign during the period of time determined in part **a**.
   c. Interpret the net earnings in part **b** as an area between two curves.

22. A manufacturer has determined that when q units of a particular commodity are produced, the price at which all the units can be sold is $p = D(q)$ dollars per unit, where D is the demand function given by

$$D(q) = \frac{300}{(0.1q + 1)^2}$$

a. How many units can the manufacturer expect to sell if the commodity is priced at $p_0 = \$12$/unit?

b. Find the consumer's surplus that corresponds to the level of production q_0 found in part **a.**

23. Money is transferred continuously into an account at the constant rate of $3,000/yr. The account earns interest at the annual rate of 8% compounded continuously.

a. What is the future value of the income flow in 3 years?

b. What is the present value of the income flow over a three-year term?

24. The management of a small firm expects to generate income at the rate of

$$f(t) = 2,000e^{0.03t}$$

dollars/yr. The prevailing rate of interest is 7% compounded continuously.

a. What is the future value for the firm over a five-year term?

b. What is the present value of the firm over the same five-year term?

25. Suppose an industrial machine that is x years old generates revenue

$$R'(x) = 6,025 - 10x^2$$

dollars and costs

$$C'(x) = 4,000 + 15x^2$$

dollars to operate and maintain.

a. For how many years is it profitable to use this machine? [Recall that profit is $P(x) = R(x) - C(x)$.]

b. What are the net earnings generated by the machine during its period of profitability? Interpret this amount as the area between two curves.

26. After t hours on the job, one factory worker is producing at the rate of

$$Q_1'(t) = 60 - 2(t - 1)^2$$

per hour, and a second is producing

$$Q_2'(t) = 50 - 5t$$

units per hour. If both arrive on the job at 8:00 A.M., how many more units (to the nearest unit) will the first worker have produced by noon than the second worker? Interpret your answer as the area between two curves.

27. Suppose that x years from now, one investment plan will be generating profit at the rate of

$$P_1'(x) = 100 + x^2$$

dollars per year, whereas a second plan will be generating profit at the rate of

$$P_2'(x) = 190 + 2x$$

dollars per year. Neither plan generates a profit in the beginning (when $x = 0$).

a. For how many years will the second plan be more profitable? That is, for how many years will P_2 be greater than P_1?

b. How much excess profit will you earn if you invest in the second plan instead of the first for the period of time in part **a**? Interpret the excess profit as the area between the two profit curves.

28. Parts for a piece of heavy machinery are sold in units of 1,000. The demand for the parts (in dollars) is given by $p(x) = 110 - x$. The total cost is given by $C(x) = x^3 - 25x^2 + 2x + 30$ (dollars).

a. For what value of x is the profit

$$P(x) = xp(x) - C(x)$$

maximized?

b. Find the consumer's surplus with respect to the price that corresponds to maximum profit.

29. Repeat Problem 28 with

$$p(x) = 124 - 2x$$

and

$$C(x) = 2x^3 - 59x^2 + 4x + 76$$

30. Suppose when q units of a commodity are produced, the demand is $p(q) = 45 - q^2$ dollars per unit, and the marginal cost is

$$\frac{dC}{dq} = 6 + \frac{1}{4}q^2$$

Assume there is no overhead (so $C(0) = 0$).

a. Find the total revenue and the marginal revenue.

b. Find the value of q (to the nearest unit) that maximizes profit.

c. Find the consumer's surplus at the value of q where profit is maximized. (Use the exact value of q.)

31. Repeat Problem 30 with

$$p(q) = \frac{1}{4}(10 - q)^2 \text{ and } \frac{dC}{dq} = \frac{3}{4}q^2 + 5$$

32. A company plans to hire additional personnel. Suppose it is estimated that as x new people are hired, it will cost $C(x) = 0.2x$ thousand dollars and that these x people will bring in $R(x) = \sqrt{3x}$ thousand dollars in additional revenue. How many new people would be hired to maximize the profit? How much net revenue would the company gain by hiring these people?

33. Suppose that the demand function for a certain commodity is

$$D(q) = 20 - 4q^2$$

and that the marginal cost is

$$C'(q) = 2q + 6$$

where q is the number of units produced. Find the consumer's surplus at the sales level q_0 where profit is maximized.

34. The resale value of a certain industrial machine decreases over a ten-year period at a rate that changes with time. When the machine is x years old, its value is decreasing at the rate of $220(x - 10)$ dollars/yr. By how much does the machine depreciate during the second year?

35. The promoters of a county fair estimate that t hours after the gates open at 9:00 A.M. visitors will be entering the fair at the rate of

$$-4(t + 2)^3 + 54(t + 2)^2$$

people/hr. How many people will enter the fair between 10:00 A.M. and noon?

36. At a certain factory, the marginal cost is $6(q - 5)^2$ dollars/unit when the level of production is q units. By how much will the total manufacturing cost increase if the level of production is raised from 10 to 13 units?

37. A certain oil well that yields 400 barrels of crude oil a month will run dry in two years. The price of crude oil is currently $108/barrel and is expected to rise at a constant rate of $0.03 per barrel per month. If the oil is sold as soon as it is extracted from the ground, what will be the total future revenue from the well?

38. It is estimated that t days from now a farmer's crop (in bushels) will be increasing at the rate of

$$0.3t^2 + 0.6t + 1$$

bu/day. By how much will the value of the crop increase during the next five days if the market price remains fixed at $3/bu?

39. It is estimated that the demand for a manufacturer's product is increasing exponentially at the rate of 2%/yr. If the current demand is 5,000 units/yr and if the price remains fixed at $400/unit, how much revenue will the manufacturer receive from the sale of the product over the next two years?

40. After t hours on the job, a factory worker can produce $100(2 + \sin t)$ units/hr. How many units does a worker who arrives on the job at 8:00 A.M. produce between 10:00 A.M. and noon?

41. Suppose that t years from now, one investment plan will be generating profit at the rate of

$$P_1'(t) = 100 + t^2$$

hundred dollars per year, while a second investment will be generating profit at the rate of $P_2'(t) = 220 + 2t$ hundred dollars per year.

a. For how many years does the rate of profitability of the second investment exceed that of the first?

b. Compute the net excess profit assuming that you invest in the second plan for the time period determined in part **a**.

c. Sketch the rate of profitability curves $y = P_1'(t)$ and $y = P_2'(t)$ and shade the region whose area represents the net excess profit computed in part **b**.

42. Answer the questions in Problem 41 for two investments with respective rates of profitability $P_1'(t) = 130 + t^2$ hundred dollars per year and $P_2'(t) = 306 + 5t$ hundred dollars per year.

43. Answer the questions in Problem 41 for two investments with respective rates of profitability

$$P_1'(t) = 60e^{0.12t}$$

thousand dollars per year and

$$P_2'(t) = 160e^{0.08t}$$

thousand dollars per year.

44. Answer the questions in Problem 41 for two investments with respective rates of profitability

$$P_1'(t) = 90e^{0.1t}$$

thousand dollars per year and

$$P_2'(t) = 140e^{0.07t}$$

thousand dollars per year.

45. Suppose that when it is t years old, a particular industrial machine generates revenue at the rate

$$R'(t) = 6{,}025 - 8t^2$$

dollars/yr and that operating and servicing costs accumulate at the rate of

$$C'(t) = 4{,}681 + 13t^2$$

dollars/yr.
  a. How many years pass before the machine ceases to be profitable?
  b. Compute the net earnings generated by the machine over the time period determined in part **a**.
  c. Sketch the revenue rate curve $y = R'(t)$ and the cost rate curve $y = C'(t)$ and shade the region whose area represents the net earnings computed in part **b**.

46. Answer the questions in Problem 45 for a machine that generates revenue at the rate

$$R'(t) = 7{,}250 - 18t^2$$

dollars/yr and for which costs accumulate at the rate

$$C'(t) = 3{,}620 + 12t^2$$

dollars/yr.

47. After t hours on the job, one factory worker is producing

$$Q_1'(t) = 60 - 2(t - 1)^2$$

units/hr, while a second worker is producing

$$Q_2'(t) = 50 - 5t$$

units/hr.
  a. If both arrive on the job at 8:00 A.M., how many more units will the first worker have produced by noon than the second worker?
  b. Interpret the answer in part **a** as the area between two curves.

48. A national consumers' association has compiled statistics suggesting that the fraction of its members who are still active t months after joining is given by the function $f(t) = e^{-0.2t}$. A new local chapter has 200 charter members and expects to attract new members at the rate of 10/mo. How many members can the chapter expect to have at the end of 8 months?

49. The population density r miles from the center of a certain city is

$$D(r) = 25{,}000e^{-0.05r^2}$$

people/mi^2. How many people live between 1 and 2 miles from the city center?

50. The population density r miles from the center of a certain city is

$$D(r) = 5{,}000e^{-0.1r^2}$$

people/mi^2. How many people live within 3 miles from the city center?

51. Calculate the volume rate (in cm^3/s) at which blood flows through an artery of radius 0.1 cm if the speed of the blood r cm from the central axis is $8(1 - 100r^2)$ cm/s.

52. Modeling Problem Blood flows through an artery of radius R. At a distance r cm from the central axis of the artery, the speed of the blood is given by

$$S(r) = k(R^2 - r^2)$$

Show that the average speed of the blood is one-half the maximum speed.

53. A study indicates that x months from now the population of a certain town will be increasing at the rate of $5 + 3x^{2/3}$ people/mo. By how much will the population of the town increase over the next 8 months?

54. An object is moving so that its speed after t minutes is $5 + 2t + 3t^2$ m/min. How far does the object travel during the second minute?

55. It is estimated that the demand for oil is increasing exponentially at the rate of 10%/yr. If the demand for oil is currently 30 billion barrels per year, how much oil will be consumed during the next 10 years?

Level 3

56. The administrators of a town estimate that the fraction of people who will still be residing in the town t years after they arrive is given by the function

$$f(t) = e^{-0.04t}$$

If the current population is 20,000 people and new townspeople arrive at the rate of 500/yr, what will the population be ten years from now? *Hint*: Divide the interval $0 \le t \le 10$ into n equal subintervals and form a Riemann sum.

57. The operators of a new computer dating service estimate that the fraction of people who will retain their membership in the service for at least t months is given by the function $f(t) = e^{-t/10}$. There are 8,000 charter members, and the operators expect to attract 200 new members/mo. How many members will the service have ten months from now? *Hint:* Divide the interval $0 \le t \le 10$ into n equal subintervals and form a Riemann sum.

58. Modeling Problem A certain nuclear power plant produces radioactive waste in the form of strontium-90 at the constant rate of 500 pounds/yr. The waste decays exponentially with a half-life of 28 years. How much of the radioactive waste from the nuclear plant will be present after 1400 years. *Hint*: Think of this as a survival and renewal problem and form a Riemann sum.

59. Modeling Problem A certain investment generates income continuously over a period of N years. After t years, the investment will be generating income at the rate of $f(t)$ dollars/yr. The prevailing annual rate of interest, compounded continuous, is r. Derive the formula

$$\int_0^N f(t)e^{-rt}\,dt$$

for the present value of the income flow by completing these steps.

a. Subdivide the time interval $[0, N]$ into n equal subintervals, the kth of which is $[t_{k-1}, t_k]$ for

$k = 1, 2, \cdots, n$. Assuming that the income for the entire kth subinterval is generated at a constant rate of $f(t_k)$, what is the present value of this income?

b. Approximate the present value of the total income flow by an appropriate sum.

c. Note that the sum in part **b** is a Riemann sum. Make the approximation precise by taking a limit to obtain the required integral formula for present value.

60. Modeling Problem A manufacturer receives N units of a certain raw material that are initially placed in storage and then withdrawn and used at a constant rate until the supply is exhausted one year later. Suppose storage costs remain fixed at p dollars per unit per year. Use definite integration to find an expression for the total storage costs the manufacturer will pay during the year. *Hint*: Let $Q(t)$ denote the number of units in storage after t years and find an expression for $Q(t)$. Then subdivide the interval $0 \leq t \leq 1$ into n equal subintervals and express the total storage cost as the limit of a sum.

CHAPTER 6 REVIEW

N*ature laughs at the difficulties of integration.*

Pierre-Simon de Laplace

Quoted in *The Armchair Science Reader, 1959*

Proficiency Examination

Concept Problems

1. Describe the process for finding the area between two curves.
2. Describe the process for finding volumes of solids with a known cross-sectional area.
3. Compare and contrast the methods of finding volumes by disks, washers, and shells.
4. What are the formulas for changing from polar to rectangular form? From polar to rectangular form?
5. Draw one example of each of the given polar-form curves.
   a. cardioid   **b.** limaçon   **c.** lemniscate   **d.** rose curve with three petals
6. Identify and then sketch each of the given curves.
   a. $r = 2\cos\theta$ **b.** $r = \cos\theta + 1$ **c.** $r = \theta, r \geq 0$ **d.** $r^2 = \cos 2\theta$
7. What is the procedure for finding the intersection of polar-form curves?
8. Outline a procedure for finding the area bounded by a polar graph.
9. What is the formula for arc length of a graph?
10. What is the formula for area of a surface of revolution?
11. How do you find the work done by a variable force?
12. How do you find the force exerted by a liquid on an object submerged vertically?
13. What are the formulas for the mass and centroid of a homogeneous lamina?
14. State the volume theorem of Pappus.
15. What formulas are used to compute future value and present value of an income flow?
16. How do you find the net change of a function f whose derivative f' is given?
17. Describe the process for finding consumer's and producer's surplus.
18. Describe a typical survival and renewal problem.
19. What formula is used for computing the flow of blood through an artery?

Practice Problems

20. **Journal Problem**   (*Mathematics Teacher**) *Use the integrals A-G to answer the given questions.*

    A. $\pi\int_a^b [f(x)]^2\, dx$ **B.** $\pi\int_c^d [f(y)]^2\, dy$ **C.** $\int_a^b A(x)\, dx$ **D.** $\int_c^d A(y)\, dy$

    E. $\pi\int_a^b [f(x) - g(x)]\, dx$ **F.** $\pi\int_a^b \{[f(x)]^2 - [g(x)]^2\}\, dx$ **G.** $\pi\int_c^d \{[f(y)]^2 - [g(y)]^2\}\, dy$

    a. Which formula does not seem to belong to this list?
    b. Which expressions represent volumes of solids of revolution?
    c. Which expressions represent volumes of solids containing holes?
    d. Which expressions represent volumes of a solid with cross sections that are perpendicular to an axis?
    e. Which expressions represent volumes of solids of revolution revolved about the x-axis?
    f. Which expressions represent volumes of solids of revolution revolved about the y-axis?

---

*Vol. 83, No. 9, December 1990, p. 695.

21. Find the area above the curve $f(x) = 3x^2 - 6$ and below the x-axis.
22. Find the area between the curves $y = x^2$ and $y^3 = x$.
23. The base of a solid in the xy-plane is the circular region given by $x^2 + y^2 = 4$. Every cross section perpendicular to the x-axis is a rectangle whose height is twice the length of the side that lies in the xy-plane. Find the volume of this solid.
24. Let R be a region bounded by the graphs of the equations $y = (x - 2)^2$ and $y = 4$.
   a. Sketch the graph of R and find its area.
   b. Find the volume of a solid of revolution of R about the x-axis.
   c. Find the volume of a solid of revolution of R about the y-axis.
25. Find the area of the intersection of the circles $r = 2a \cos \theta$ and $r = 2a \sin \theta$, where $a > 0$ is constant.
26. Find the arc length of the curve $y = 2 - x^{3/2}$ from $(0, 2)$ to $(1, 1)$. Give the exact value.
27. Find the surface area obtained by revolving the parabolic arc $y^2 = x$ from $(0, 0)$ to $(1, 1)$, about the x-axis. Give the exact value.
28. Find the centroid of the homogeneous region between $y = x^2 - x$ and $y = x - x^3$ on $[0, 1]$.
29. An observation porthole on a ship is circular with diameter of 2 feet and is located so its center is 4 feet below the waterline. Find the fluid force on the porthole, assuming that the boat is in seawater (weight density is 64 lb/ft^3). Set up the integral only, with the origin at the bottom of the porthole.
30. Money is transferred continuously into an account at the constant rate of $1,200/yr. The account earns interest at the annual rate of 8% compounded continuously. How much money will be in the account at the end of 5 years?

Supplementary Problems*

Name and sketch the curves whose equations are given in Problems 1-8.

1. $r = 2 \sin \theta - 3 \cos \theta$ **2.** $r = -2\theta, \theta \geq 0$ **3.** $r = -\csc \theta$ **4.** $r = \sin 3\theta$

5. $r = \cos(\theta - \frac{\pi}{3})$ **6.** $r^2 = 3 \cos 2\theta$ **7.** $r = \dfrac{4}{1 - \cos \theta}$ **8.** $r = \dfrac{5}{2 + 3 \sin \theta}$

Find the points of intersection of the polar-form curves in Problems 9-12.

9. $\begin{cases} r = 2 \cos \theta \\ r = 1 + \cos \theta \end{cases}$ **10.** $\begin{cases} r = 1 + \sin \theta \\ r = 1 + \cos \theta \end{cases}$

11. $\begin{cases} r \cos \theta + 2r \sin \theta = 4 \\ r = 2 \sec \theta \end{cases}$ **12.** $\begin{cases} r = 2 \sin 2\theta \\ r = 1 \text{ for } 0 \leq \theta \leq \frac{\pi}{2} \end{cases}$

Find the area of the regions bounded by the curves and lines in Problems 13-21.

13. $4y^2 = x$ and $2y = x - 2$ **14.** $y = x^3, y = x^4$
15. $x = y^{2/3}, x = y^2$ **16.** $y = \sqrt{3} \sin x, y = \cos x, x = 0,$ and $x = \frac{\pi}{2}$
17. the region bounded by the polar curve $r = 4 \cos 2\theta$
18. the region inside both the circles $r = 1$ and $r = 2 \sin \theta$
19. the region inside one loop of the lemniscate $r^2 = \cos 2\theta$
20. the region inside one petal of the rose curve $r = \sin 2\theta$
21. the region inside the circle $r = 2a \sin \theta$ and outside the circle $r = a$

Find the volume of a solid of revolution obtained by revolving the region R about the indicated axis in Problems 22-29.
22. R is bounded by $y = \sqrt{x}, x = 9, y = 0$; about the x-axis
23. R is bounded by $y = x^2, y = 9 - x^2,$ and $x = 0$; about the y-axis
24. R is bounded by $y = x^2$ and $y = x^4$; about the x-axis
25. R is bounded by $y = \sqrt{\cos x}, x = \frac{\pi}{4}, x = \frac{\pi}{3},$ and $y = 0$; about the x-axis.
26. R is bounded by $y = x, y = 2x, x = 1$; about the line $y = -1$.

---

*The supplementary probelms are presented in a somewhat random order, not necessarily in order of difficulty.

27. R is bounded by $y = x$, $y = 2x$, $x = 1$; around the line $y = 4$.

28. R is bounded by $y = x^2$ and $y = \sqrt{x}$; about the x-axis by using washers.

29. R is bounded by $y = x^2$ and $y = \sqrt{x}$; about the x-axis by using shells.

In Problems 30-31, find the volume of the solid formed when the region described is revolved about the x-axis using washers or disks.

30. the region under the curve $y = x^2 + x^3$ on the interval $0 \le x \le \pi$; approximate your answer to the nearest unit

31. the region under the curve $y = \sqrt{2 \sin x}$ on the interval $0 \le x \le \pi$

32. Find the centroid of a homogeneous lamina covering the region R that is bounded by the parabola $y = 2 - x^2$ and the line $y = x$.

33. Estimate the area of the surface obtained by revolving $y = \tan x$ about the x-axis on the interval $[0, 1]$.

34. The base of a solid is the region in the xy-plane that is bounded by the curve $y = 4 - x^2$ and the x-axis. Every cross section of the solid perpendicular to the y-axis is a rectangle whose height is twice the length of the side that lies in the xy plane. Find the volume of the solid.

35. The base of a solid is the region R in the xy plane bounded by the curve $y^2 = x$ and the line $y = x$. Find the volume of the solid if each cross section perpendicular to the x-axis is a semicircle with a diameter in the xy plane.

36. Use the method of cross sections to show that the volume of a regular tetrahedron of side a is $\frac{1}{12}\sqrt{2}a^3$.

37. A force of 100 lb is required to compress a spring from its natural length of 10 in. to a length of 8 in. Find the work required to compress it to a length of 7 in. from its natural length.

38. A bowl is formed by revolving the region bounded by the curve $y = 2x^2$ and the lines $x = 0$ and $y = 1$ about the y-axis (units are in ft). The bowl is filled with water to a height of 9 in.
  a. How much work is done in pumping the water to the top of the bowl?
  b. How much work is done in pumping the water to a point 0.5 ft above the top of the bowl?

39. Two solids S_1 and S_2 have as their base the region in the xy-plane bounded by the parabola $y = x^2$ and the line $y = 1$. For solid S_1, each cross section perpendicular to the x-axis is a square, and for S_2, each cross section perpendicular to the y-axis is a square. Which solid has the greater volume?

40. Use calculus to find the lateral surface area of a right circular cylinder with radius r and height h.

41. Use calculus to find the surface area of a right circular cone with (top) radius r and height h.

42. A 20-lb bucket in a well is attached to the end of a 60-ft chain that weighs 16 lb. If the bucket is filled with 50 lb of water, how much work is done in lifting the bucket and chain to the top of the well?

43. A swimming pool is 3 ft deep at one end and 8 ft deep at the other, as shown in Figure 6.80. The pool is 30 ft long and 25 ft wide with vertical sides. Set up and analyze a model to find the fluid force against one of the 30-ft sides.

44. The ends of a container filled with water have the shape of the region R bounded by the curve $y = x^4$ and the line $y = 16$. Find the fluid force (to two decimal places) exerted by the water on one end of the container.

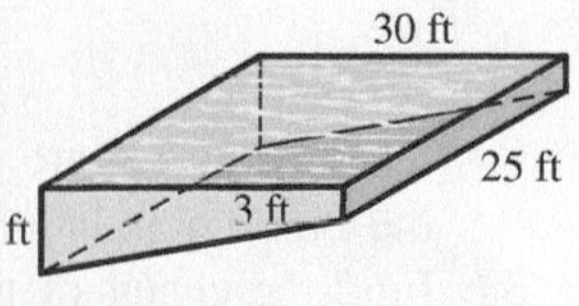

45. A conical tank whose vertex points down is 12 ft high with (top) diameter 6 ft. The tank is filled with a fluid of weight density $\delta = 22$ lb/ft^3. If the tank is filled to a depth of 6 ft, find the work required to pump all the fluid to a height of 2 ft above the top of the tank.

Figure 6.80 Cross section of a swimming pool

46. Each end of a container filled with water has the shape of an isosceles triangle, as shown in Figure 6.81. Find the force exerted by the water on an end of the container.

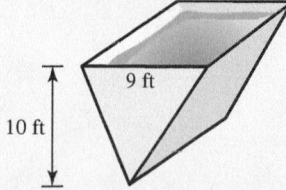

Figure 6.81 Water pressure in a tank

47. A container has the shape of the solid formed by rotating the region bounded by the curve $y = x^3$ and the lines $x = 0$ and $y = 1$ about the y-axis, where x and y are measured in feet. If the container is filled with water, how much work is required to pump all the water out of the container?

48. A gate in a water main has the shape of a circle with diameter 10 ft. If water stands 2 ft high in the main, what is the force exerted by the water on the gate?

Suppose a particle of mass m moves along a number line under the influence of a variable force F which always acts in a direction parallel to the line of motion. At a given time the particle is said to have **kinetic energy**

$$K = \frac{1}{2}mv^2$$

where v is the particle's velocity, and it has **potential energy**

$$P = -\int_{s_0}^{s} F\, dx$$

relative to the point s on the line of motion. Use these definitions in Problems 49-50.

49. A spring at rest has potential energy $P = 0$. If the spring constant is 30 lb/ft, how far must the spring be stretched so that its potential energy is 20 ft-lb?

50. A particle with mass m falls freely near the earth's surface. At time $t = 0$, the particle is s_0 units above the ground and is falling with velocity v_0.
   a. Find the potential energy P of the particle in terms of position, s, if the potential energy is 0 when $s = s_0$.
   b. Find the potential energy P and the kinetic energy K of the particle in terms of time t.

51. An object weighs $w = 1,000r^{-2}$ grams, where r is the object's distance from the center of the earth. What work is required to lift the object from the surface of the earth ($r = 6,400$ km) to a point 3,600 km above the surface of the earth?

52. Modeling Problem A certain species has population $P(t)$ at time t whose growth rate is affected by seasonal variations in the food supply. Suppose the population may be modeled by the differential equation

$$\frac{dP}{dt} = k(P - A)\cos t$$

where k and A are physical constants. Solve the differential equation and express $P(t)$ in terms of k, A, and P_0, the initial population.

53. Find the center of mass of a thin plate of constant density if the plate occupies the region bounded by the lines $2y = x, x + 3y = 5$, and the x-axis.

54. The portion of the ellipse (Figure 6.82)

$$\frac{x^2}{a^2} + \frac{y^2}{b^2} = 1$$

that lies in the first quadrant between $x = r$ and $x = a$, for $0 < r < a$, is revolved about the x-axis to form a solid S. Find the volume of S.

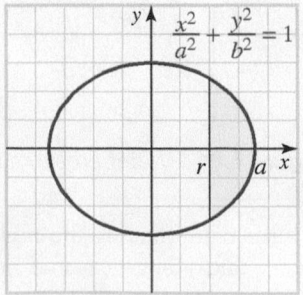

Figure 6.82 Ellipse

55. Find the volume of the football-shaped solid (called an *ellipsoid*) formed by revolving the ellipse

$$\frac{x^2}{a^2} + \frac{y^2}{b^2} = 1$$

about the x-axis. What is the volume if the ellipse is revolved about the y-axis?

56. When viewed from above, a swimming pool has the shape of the ellipse

$$\frac{x^2}{900} + \frac{y^2}{400} = 1$$

The cross sections of the pool perpendicular to the ground and parallel to the y-axis are squares. If the units are in feet, what is the volume of the pool?

57. The portion of the hyperbola (Figure 6.83)

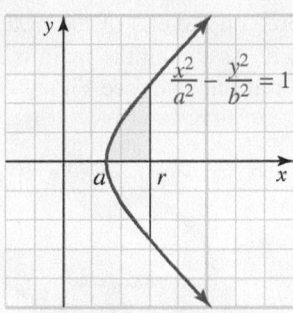

Figure 6.83 Hyperbola

$$\frac{x^2}{a^2} - \frac{y^2}{b^2} = 1$$

that lies in the first quadrant between $x = a$ and $x = r$, for $0 < a < r$, is revolved about the x-axis to form a solid S. Find the volume of S.

58. The base of the solid S is the hyperbola

$$\frac{x^2}{4} - \frac{y^2}{9} = 1$$

for $2 \leq x \leq 5$, and the cross sections perpendicular to the x-axis are squares. Find the volume of S.

59. What is the present value of an investment that will generate income continuously at a constant rate of $1,000/yr for 10 years if the prevailing annual interest rate remains fixed at 7% compounded continuously?

60. In a certain community the fraction of the homes placed on the market that remain unsold for at least t weeks is approximately $f(t) = e^{-0.2t}$. If 200 homes are currently on the market and if additional homes are placed on the market at the rate of 8 per week, approximately how many homes will be on the market 10 weeks from now?

61. The population density r miles from the center of a certain city is $D(r) = 5,000e^{-0.02r^2}$ people/mi^2. How many people live within 5 miles of the center of the city?

62. A certain oil well that yields 900 barrels of crude oil per month will run dry in 3 years. The price of crude oil is currently $110/barrel and is expected to rise at the constant rate of $1 per barrel per month. If the oil is sold as soon as it is extracted from the ground, what will be the total future revenue from the well?

63. Suppose that the consumer's demand function for a certain commodity is $D(q) = 50 - 3q - q^2$ dollars per unit.
  a. Find the number of units that will be bought if the market price is $32/unit.
  b. Compute the consumer's surplus when the market price is $32/unit.
  c. Graph the demand curve and interpret the consumer's surplus as an area in relation to this curve.

64. Let R be the part of the ellipse

$$\frac{x^2}{a^2} + \frac{y^2}{b^2} = 1$$

that lies in the first quadrant. Use the theorem of Pappus to find the coordinates of the centroid of R. You may use the facts that R has area $A = \frac{1}{4}\pi ab$, and the semiellipsoid it generates when it is revolved about the x-axis has volume $V = \frac{2}{3}\pi ab^2$.

65. Modeling Problem A container of a certain fluid is raised from the ground on a pulley. While resting on the ground, the container and fluid together weigh 200 lb. However, as the container is raised, fluid leaks out in such a way that at height x feet above the ground, the total weight of the container and fluid is $200 - 0.5x$ lb. Set up and analyze a model to find the work done in raising the container 100 ft.

66. A bag of sand is being lifted vertically upward at the rate of 31 ft/min. If the bag originally weighed 40 lb and leaks at the rate of 0.2 lb/s, how much work is done in raising it until it is empty?

67. The vertical end of a trough is an equilateral triangle, 3 ft on a side. Assuming the trough is filled with a fluid of weight density $\delta = 40$ lb/ft^3, find the force on the end of the trough.

68. The radius of a circular water main is 2 ft. Assuming the main is half full of water, find the total force exerted by the water on a gate that crosses the main at one end (see Figure 6.84).

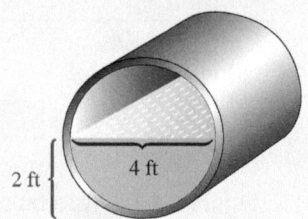

2 ft   4 ft

Figure 6.84 Circular water main

69. The force acting on an object moving along a straight line is known to be a linear function of its position. Find the force function F if it is known that 13 ft-lb of work is required to move the object 6 ft and 44 ft-lb of work is required to move it 12 ft.

70. A tank has the shape of a rectangular parallelepiped with length 10 ft, width 6 ft, and height 8 ft. The tank contains a liquid of weight density $\delta = 20$ lb/ft^3. Find the level of the liquid in the tank if it will require 2,400 ft-lb of work to move all the liquid in the tank to the top.

71. A tank is constructed by placing a right circular cylinder of height 10 ft and radius 2 ft on top of a hemisphere with the same radius (see Figure 6.85).

If the tank is filled with a fluid of weight density $\delta = 24$ lb/ft^3, how much work is done in bringing all the fluid to the top of the tank?

72. Modeling Problem A piston in a cylinder causes a gas either to expand or to contract. Assume the pressure P and the volume of gas V in the cylinder satisfy the adiabatic law (no exchange of heat)

$$P(V^{1.4}) = C$$

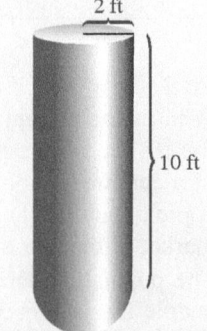

2 ft

10 ft

Figure 6.85 Cylindrical tank

where C is a constant.

a. If the gas goes from volume V_1 to volume V_2, show that the work done may be modeled by

$$W = \int_{V_1}^{V_2} CV^{-1.4}\,dV$$

b. How much work is done if 0.6 m^3 of steam at a pressure of 2,500 Newtons/m^2 expands 50%?

73. A thin sheet of tin is in the shape of a square, 16 cm on a side. A rectangular corner of area 156 cm^2 is cut from the square, and the centroid of the portion that remains is found to be 4.88 cm from one side of the original square, as shown in Figure 6.86. How far is it from the other three sides?

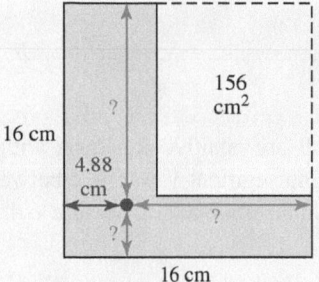

Figure 6.86 Centroid of a region

74. Let R be the region bounded by the semicircle $y = \sqrt{4-x^2}$ and the x-axis, and let S be the solid with base R and trapezoidal cross sections perpendicular to the x-axis. Assume that the two slant sides and the shorter parallel side are all half the length that lies in the base, as shown in Figure 6.87. Set up and evaluate an integral for the volume of S.

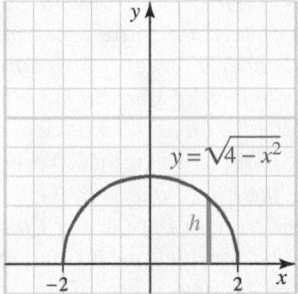 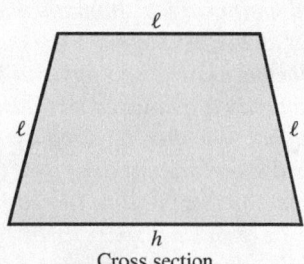

Figure 6.87 Volume of a region S

75. A child is building a sand structure at the beach.
    a. Find the work done if the weight density of the sand is $\delta = 140$ lb/ft^3 and the structure is a cone of height 1 ft and diameter 2 ft.
    b. What if the structure is a tower (cylinder) of height 1 ft and radius 4 in.
76. Let S be the solid generated by revolving about the x-axis the region bounded by the parabola $y = x^2$ and the line $y = x$. Find the number A so that the plane perpendicular to the x-axis at $x = A$ divides S in half. That is, the volume on one side of A equals the volume on the other side.
77. A plate in the shape of an isosceles triangle is submerged vertically in water, as shown in Figure 6.88.

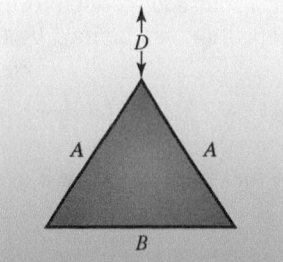

Figure 6.88 Hydrostatic force

a. Find the force on one side of the plate if the two equal sides have length 3, the third side has length 5, and the top vertex of the triangle is 4 units below the surface.

b. Find the fluid force for the general case where the triangle has equal sides of length A, the third side has length B, and the top vertex is D units below the surface.

78. Find the arc length of the curve

$$y = Ax^3 + \frac{B}{x}$$

on the interval $[a, b]$, where A and B are positive constants with $AB = \frac{1}{12}$.

79. Think Tank Problem When the line segment $y = x - 2$ between $x = 1$ and $x = 3$ is revolved about the x-axis, it generates part of a cone. The formula obtained in Section 6.4 says that the surface area of the cone is

$$\int_1^3 2\pi(x - 2)\sqrt{2}\, dx$$

but this integral is equal to 0. What is wrong? What is the actual surface area?

80. *Historical Quest* *Emmy Noether has been called the greatest woman mathematician in the history of mathematics. Once again, we see a woman overcoming great obstacles in achieving her education. At the time she received her Ph.D. at the University of Erlangen (in 1907), the Academic Senate declared that the admission of women students would "overthrow all academic order." Emmy Noether is famous for her work in physics and mathematics. In mathematics, she opened up entire areas for study in the theory of ideals, and in 1921 she published a paper on rings, which have since been called Noetherian rings (rings and ideals are objects of study in modern algebra). She was not only a renowned mathematician but also an excellent teacher. It has been reported that she often lectured by "working it out as she went." Nevertheless, her students flocked around her "as if following the Pied Piper." In the fall of 1934, pressure from the Nazi regime forced her to immigrate to Bryn Mawr College in Pennsylvania. In the years that followed, there were mass dismissals of "racially undesirable" professors, and many noted European Jewish mathematicians immigrated to American universities.* For this *Quest* compile a list of notable mathematicians who immigrated to the United States from 1933 to 1940.

Amalie ("Emmy") Noether
(1882–1935)

81. Modeling Problem An object below the surface of the earth is attracted by a gravitational force that is directly proportional to the square of the distance from the object to the center of the earth. Set up and analyze a model to find the work done in lifting an object weighing P lb from a depth of s feet (below the surface) to the surface. (Assume the earth's radius is 4,000 mi. and that P is constant between s and the surface.)

In Problems 82-83, consider a continuous curve $y = f(x)$ that passes through $(-1, 4)$, $(0, 3)$, $(1, 4)$, $(2, 6)$, $(3, 5)$, $(4, 6)$, $(5, 7)$, $(6, 6)$, $(7, 4)$, $(8, 3)$, *and* $(9, 4)$.

82. Use Simpson's rule to estimate the volume of the solid formed by revolving this curve about the x-axis for $[-1, 9]$.

83. Use the trapezoidal rule to estimate the volume of the solid formed by revolving this curve about the line $y = -1$ on $[-1, 9]$.

84. Modeling Problem In this computational problem we revisit a problem similar to the Wine Barrel Capacity group research project at the end of Chapter 4, but add a couple of complications. Imagine a cylindrical fuel tank lying on its side; its ends are elliptically shaped and are defined by the equation

$$\left(\frac{x}{a}\right)^2 + \left(\frac{y}{b}\right)^2 = 1$$

We are concerned with the amount of fuel in the tank for a given level.

a. Explain why the area of an end of the tank can be modeled by

$$A = 2a \int_{-b}^{b} \sqrt{1 - \left(\frac{y}{b}\right)^2} \, dy$$

b. By making the right substitution in the integral in part **a**, we can obtain an integral representing half the area inside the circle (with unit radius). Do this and show that the area inside the above ellipse is πab.

c. Explain why the volume of the fuel at level h $(-b \le h \le b)$ may be modeled by

$$V(h) = 2La \int_{-b}^{h} \sqrt{1 - \left(\frac{y}{b}\right)^2} \, dy$$

d. For $a = 5$, $b = 4$, and $L = 20$, numerically compute $V(h)$ for $h = -3, -2, \cdots, 3, 4$. *Note:* $V(0)$ and $V(4)$ will serve as a check on your work.

85. We think of the last part of Problem 84 as the "direct" problem: Given a value of h, compute $V(h)$ by a formula. Now we turn to the "inverse" problem: For a set value of V, say V_0, find h such that $V(h) = V_0$. Typically, the inverse problem is more difficult than the direct problem and requires some sort of iterative method to approximate the solution (for example, Newton-Raphson's method). For example, if $V = 500$ ft^3, we could define $f(h) = V(h) - 500$ and seek the relevant zero of f. Set up the necessary computer program to do this and find the h values, to two decimal place accuracy, corresponding to these values of V:

   a. 500   **b.** 800

   Hint: A key to this problem is thinking about $V'(h)$, where $V(h)$ is expressed in Problem 84c. Also, for starting values h_0, refer to your work in Problem 84**d**.

86. Find the centroid of a right triangle whose legs have lengths a and b. *Hint*: Use the volume theorem of Pappus.

87. The largest of the pyramids at Giza in Egypt (see Problem 49, p. 445) is roughly 480 ft high and has a square base 750 ft on a side. How much work was required to lift the stones to build this pyramid if we assume that the rock in the pyramid had an average weight density of 160 lb/ft^3?

88. Find the length of the curve $y = e^x$ on the interval $[0, 1]$.

89. Find the volume of the solid generated when the region under the curve

$$y = \frac{1}{\sqrt{9 - x^2}}$$

on the interval $[0, 2]$ is revolved about the y-axis.

90. Find the volume of the solid formed by revolving about the x-axis the region bounded by the curve

$$y = \frac{2}{\sqrt{3x - 2}}$$

the x-axis, and the lines $x = 1$ and $x = 2$.

91. Find the area between the curves $y = \tan^2 x$ and $y = \sec^2 x$ on the interval $\left[0, \frac{\pi}{4}\right]$.

92. Find the surface area of the solid generated by revolving the region bounded by the x-axis and the curve

$$y = e^x + \frac{1}{4}e^{-x}$$

on $[0, 1]$ about the x-axis.

93. Find the volume of the solid whose base is the region R bounded by the curve $y = e^x$ and the lines $y = 0$, $x = 0$, $x = 1$, if cross sections perpendicular to the x-axis are equilateral triangles.

94. Find the length of the curve

$$x = \frac{ay^2}{b^2} - \frac{b^2}{8a} \ln \frac{y}{b}$$

between $y = b$ and $y = 2b$.

95. Putnam Examination Problem Find the length of the curve $y^2 = x^3$ from the origin to the point where the tangent makes an angle of $45°$ with the x-axis.

96. Putnam Examination Problem A solid is bounded by two bases in the horizontal planes $z = \frac{h}{2}$ and $z = -\frac{h}{2}$, and by a surface with the property that every horizontal cross section has area given by an expression of the form

$$A = a_0 z^3 + a_1 z^2 + a_2 z + a_3$$

a. Show that the solid has volume $V = \frac{1}{6}h(B_1 + B_2 + 4M)$, where B_1 and B_2 are the area of the bases and M is the area of the middle horizontal cross section at $z = 0$.

b. Show that the formulas for the volume of a cone and a sphere can be obtained from the formula in part **a** for certain special choices of a_0, a_1, a_2, and a_3.

97. Putnam Examination Problem The horizontal line $y = c$ intersects the curve $y = 2x - 3x^3$ in the first quadrant as shown in Figure 6.89. Find c so that the areas of the two shaded regions are equal.

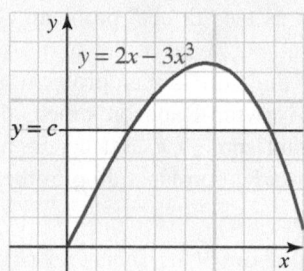

Figure 6.89 Putnam Problem 97

98. Putnam Examination Problem Find the volume of the region of points (x, y, z) such that $(x^2 + y^2 + z^2 + 8)^2 \le 36(x^2 + y^2)$. *Hint*: Let $r = \sqrt{x^2 + y^2}$ to generate a torus $(r - 3)^2 + z^2 \le 1$, the area being rotated is the disk $(x - 3)^2 + z^2 \le 1$ in the xz-plane. Use the volume theorem of Pappus.

99. Book Report Eli Maor, a native of Israel, has a long-standing interest in the relations between mathematics and the arts. Read the fascinating book *To Infinity and Beyond, A Cultural History of the Infinite* (Boston: Birkhäuser, 1987), and prepare a book report.

CHAPTER 6 GROUP RESEARCH PROJECT*

Working in small groups is typical of most work environments, and this book seeks to develop skills with group activities. We present a group project at the end of each chapter. These projects are to be done in groups of three or four students.

Houdini's Escape

Harry Houdini (born Ehrich Weiss) was a famous escape artist. In this project we relive a trick of his that challenged his mathematical prowess, as well as his skill and bravery. It may challenge these qualities in you as well.

Houdini had his feet shackled to the top of a concrete block which was placed on the bottom of a giant laboratory flask. The cross-sectional radius of the flask, measured in feet, was given as a function of height z from the ground by the formula $r(z) = 10z^{-1/2}$, with the bottom of the flask at $z = 1$ ft. The flask was then filled with water at a steady rate of 22π ft³/min. Houdini's job was to escape the shackles before he was drowned by the rising water in the flask.

Now Houdini knew it would take him exactly 10 minutes to escape the shackles. For dramatic impact, he wanted to time his escape so it was completed precisely at the moment the water level reached the top of his head. Houdini was exactly 6 ft tall. In the design of the apparatus, he was allowed to specify only one thing: the height of the concrete block he stood on.

Harry Houdini (1874–1926)

Extended paper for further study. Your paper is not limited to the following questions but should include these concerns: How high should the block be? (You can neglect Houdini's volume and the volume of the block.) How fast is the water level changing when the flask first starts to fill? How fast is the water level changing at the instant when the water reaches the top of his head? You might also help Houdini with any size flask by generalizing the derivation: Consider a flask with cross-sectional radius $r(z)$, an arbitrary function of z with a constant inflow rate of $\dfrac{dV}{dt} = A$. Can you find $\dfrac{dh}{dt}$ as a function of $h(t)$?

MAA Notes 17 (1991), "Priming the Calculus Pump: Innovations and Resources," by Marcus S. Cohen, Edward D. Gaughan, R. Arthur Knoebel, Douglas S. Kurtz, and David J. Pengelley.

CHAPTER 7

METHODS OF INTEGRATION

G*od does not care about our mathematical difficulties—*
he integrates empirically.

Albert Einsten

PREVIEW

In this chapter we increase the number of techniques and procedures for integrating a function. One of the most important techniques, substitution, is reviewed in the first section and is then expanded in several different contexts. Other important integration procedures include using tables, integration by parts, and partial fractions. In addition, improper integrals, hyperbolic functions, and inverse hyperbolic functions are discussed, along with first-order differential equations.

PERSPECTIVE

It is possible to differentiate most functions that arise in practice by applying a fairly short list of rules and formulas, but integration is a more complicated process. The purpose of this chapter is to increase your ability to integrate a variety of different functions. Learning to integrate is like learning to play a musical instrument: at first, it may seem impossibly complicated, but if you persevere, after a while music starts to happen. It should be noted that as more powerful technology becomes available, *techniques* become less important and *ideas* become more important.

CONTENTS

509

7.1 REVIEW OF SUBSTITUTION AND INTEGRATION BY TABLE

IN THIS SECTION: *Review of substitution, using tables of integrals*
In this section, we review our integration formulas as well as the method of integration by substitution. Today's technology has greatly enhanced the ability of professional mathematicians and scientists in techniques of integration, thereby minimizing the need for spending hours learning rarely used integration techniques. Along with this new technology, the use of integral tables has increased in importance. Most mathematical handbooks contain extensive integral tables, and we have included such an integration table in Appendix D.

For easy reference, the most important integration rules and formulas are listed inside the back cover. Notice that the first four integration formulas are the procedural rules, which allow us to break up integrals into simpler forms, whereas those on the remainder of the page form the building blocks of integration.

Review of Substitution

Remember that when doing substitution you must choose u, calculate du, and then substitute to make the form you are integrating look *exactly like* one of the known integration formulas. We will review substitution by looking at different situations and special substitutions that prove useful.

Example 1 Integration by substitution

Find $\displaystyle\int \frac{x^2\,dx}{\left(x^3-2\right)^5}$.

Solution Let u be the value in parentheses; that is, let $u = x^3 - 2$. Then $du = 3x^2\,dx$, so by substitution,

$$\int \frac{x^2\,dx}{\left(x^3-2\right)^5} = \int \frac{\frac{du}{3}}{u^5} \qquad\qquad \textit{Using substitution}$$

$$= \frac{1}{3}\int u^{-5}\,du$$

$$= \frac{1}{3}\frac{u^{-4}}{-4} + C$$

$$= -\frac{1}{12}\left(x^3-2\right)^{-4} + C$$

As with all indefinite integrals, you can check your work by finding the derivative of your answer to see if you obtain the integrand. For this example,

$$\frac{d}{dx}\left[-\frac{1}{12}\left(x^3-2\right)^{-4}+C\right] = -\frac{1}{12}\left[-4(x^3-2)^{-5}(3x^2)+0\right]$$

$$= \frac{x^2}{\left(x^3-2\right)^5}$$

Example 2 Fitting the form of a known integration formula by substitution

Find $\displaystyle\int \frac{t\,dt}{\sqrt{1-t^4}}$.

Solution We notice the similarity between this and the formula for inverse sine if we let $a = 1$ and $u = t^2$. Then $du = 2t\,dt$ and

$$\int \frac{t \, dt}{\sqrt{1-t^4}} = \int \frac{\frac{du}{2}}{\sqrt{1-u^2}}$$

> If $u = t^2$, then $du = 2t \, dt$ so $dt = \frac{du}{2}$.
> Also, $\sqrt{1-t^4} = \sqrt{1 - \left(t^2\right)^2} = \sqrt{1-u^2}$.

$$= \frac{1}{2} \int \frac{du}{\sqrt{1-u^2}}$$

$$= \frac{1}{2} \sin^{-1} u + C \quad \textit{Formula 22, Appendix D}$$

$$= \frac{1}{2} \sin^{-1} t^2 + C \quad \textit{Rewrite in terms of original variable t.}$$

The method of substitution (Section 5.5) is very important, because many of the techniques developed in this chapter will be used in conjunction with substitution. Examples 3 and 4 illustrate additional ways substitution can be used in integration problems.

Example 3   Multiplication by 1 to derive an integration formula

Find $\int \sec x \, dx$.

Solution   Multiply the integrand $\sec x$ by $\sec x + \tan x$ and divide by the same quantity:

$$\int \sec x \, dx = \int \frac{\sec x (\sec x + \tan x)}{\sec x + \tan x} dx \quad \textit{The advantage of this rearrangement is that the numerator is now the}$$

$$= \int \frac{\sec^2 x + \sec x \tan x}{\sec x + \tan x} dx \quad \textit{derivative of the denominator.}$$

> If $u = \tan x + \sec x$, then
> $du = (\sec^2 x + \sec x \tan x) dx$

$$= \int \frac{du}{u}$$

$$= \ln |u| + C$$

$$= \ln |\sec x + \tan x| + C$$

You may wonder why anyone would think to multiply and divide the integrand $\sec x$ in Example 3 by $\sec x + \tan x$. To say that we do it "because it works" is probably not a very satisfying answer. However, techniques like this are passed on from generation to generation, and it should be noted that multiplication by 1 is a common method in mathematics for changing the form of an expression to one that can be manipulated more easily.

Example 4   Substitution after an algebraic manipulation

Find $\int \frac{dx}{1+e^x}$.

Solution   The straightforward substitution $u = 1 + e^x$ does not work:

$$\int \frac{dx}{1+e^x} = \int \frac{\frac{du}{e^x}}{u} = \int \frac{du}{ue^x} \quad \textit{This is not an appropriate form because x has not been eliminated.}$$

Instead, rewrite the integrand as follows:

$$\int \frac{dx}{1+e^x} = \int \frac{1}{1+e^x} \cdot \frac{e^{-x}}{e^{-x}} dx \quad \textit{Multiply by 1.}$$

$$= \int \frac{e^{-x} dx}{e^{-x} + 1} \qquad \boxed{\textit{If } u = e^{-x} + 1, \textit{ then } du = -e^{-x} dx}$$

$$= \int \frac{-du}{u}$$

$$= -\ln |u| + C$$

$$= -\ln(e^{-x} + 1) + C \quad e^{-x} + 1 > 0 \textit{ for all } x, \textit{ so } \ln \left| e^{-x} + 1 \right| = \ln(e^{-x} + 1)$$

When the integrand involves terms with fractional exponents, it is usually a good idea to choose the substitution $x = u^n$, where n is the smallest positive integer that is divisible by all the denominators of the exponents (that is, the least common multiple of the denominators). For example, if the integrand involves terms such as $x^{1/4}$, $x^{2/3}$, and $x^{1/6}$, then the substitution $x = u^{12}$ is suggested, because 12 is the smallest integer divisible by the exponential denominators $4, 3$, and 6. The advantage of this policy is that it guarantees that each fractional power of x becomes an integral power of u. Thus,

$$x^{1/6} = (u^{12})^{1/6} = u^2, \qquad x^{1/4} = (u^{12})^{1/4} = u^3, \qquad x^{2/3} = (u^{12})^{2/3} = u^8$$

Example 5   Substitution with fractional exponents

Find $\displaystyle\int \frac{dx}{x^{1/3} + x^{1/2}}$.

Solution   Because 6 is the smallest integer divisible by the denominators 2 and 3, we set $u = x^{1/6}$, so that $u^6 = x$ and $6u^5 du = dx$. We now use substitution

$$\int \frac{dx}{x^{1/3} + x^{1/2}} = \int \frac{6u^5 \, du}{\left(u^6\right)^{1/3} + (u^6)^{1/2}} \qquad \textit{Let } x = u^6.$$

$$= \int \frac{6u^5 \, du}{u^2 + u^3}$$

$$= \int \frac{6u^5 \, du}{u^2(1 + u)}$$

Substitution does not guarantee an obvious integrable form. When the degree of the numerator is greater than or equal to the degree of the denominator, division is often helpful.

$$= \int \frac{6u^3 \, du}{1 + u} \qquad \textit{By long division, } \frac{6u^3}{1 + u} = 6u^2 - 6u + 6 + \frac{-6}{1 + u}.$$

$$= \int \left(6u^2 - 6u + 6 + \frac{-6}{1 + u}\right) du$$

$$= 2u^3 - 3u^2 + 6u - 6\ln|1 + u| + C \qquad \textit{Substitute } u = x^{1/6}.$$

$$= 2(x^{1/6})^3 - 3(x^{1/6})^2$$

$$\quad + 6(x^{1/6}) - 6\ln\left|1 + x^{1/6}\right| + C$$

$$= 2x^{1/2} - 3x^{1/3} + 6x^{1/6}$$

$$\quad - 6\ln(1 + x^{1/6}) + C \qquad \textit{Note: } 1 + x^{1/6} > 0. \qquad ■$$

Using Tables of Integrals

Example 5 (especially the part involving long division) seems particularly lengthy. When faced with the necessity of integrating a function such as

$$\int \frac{6u^3}{1 + u}$$

most people would turn to a computer, calculator, or a table of integrals. If you look, for example, at the short table of integrals (Appendix D), you will find (Formula 37)* that

$$\int \frac{u^3 \, du}{au + b} = \frac{(au + b)^3}{3a^4} - \frac{3b(au + b)^2}{2a^4} + \frac{3b^2(au + b)}{a^4} - \frac{b^3}{a^4}\ln|au + b| + C$$

---

*See the footnote at the beginning of the Integral Table that each entry in the table must add a constant of integration when using a formula from the table.

If we let $a = 1$ and $b = 1$, we find

$$\int \frac{6u^3\,du}{1+u} = 6\int \frac{u^3\,du}{u+1} = 2(u+1)^3 - 9(u+1)^2 + 18(u+1) - 6\ln|u+1| + C$$

When using integration tables, we note that the algebraic form does not always match the form we obtain by direct integration. You might wish to show algebraically that the form we have just found and the form

$$2u^3 - 3u^2 + 6u - 6\ln|1+u| + C$$

from Example 5 differ by a constant and hence are equivalent.

To use an integral table, first classify the integral by form. To facilitate substitution of forms, we use u as the variable of integration, and let a, b, c, m, and n represent constants. The forms listed in Appendix D are as follows:

Elementary forms (Formulas 1-29; these were developed in the text)
Linear and quadratic forms (Formulas 30-76)
  Forms involving $au + b$; $u^2 + a^2$; $u^2 - a^2$; $a^2 - u^2$; and $au^2 + bu + c$
Radical forms (Formulas 77-121)
  Forms involving $\sqrt{au + b}$; $\sqrt{u^2 + a^2}$; $\sqrt{u^2 - a^2}$; $\sqrt{a^2 - u^2}$
Trigonometric forms (Formulas 122-167)
  Forms involving $\cos au$; $\sin au$; both $\sin au$ and $\cos au$; $\tan au$; $\cot au$; $\sec au$; $\csc au$
Inverse trigonometric forms (Formula 168-182)
Exponential and logarithmic forms (Formulas 183-200)
  Forms involving e^{au}; $\ln|u|$

There is a common misconception that integration will be easy if a table is provided, but even with a table available there can be a considerable amount of work. After deciding which form applies, match the individual type with the problem at hand by making appropriate choices for the arbitrary constants. More than one form may apply, but the results derived by using different formulas will be the same (except for the constants) even though they may look quite different. We will not include the constants in the table listing, but you should remember to include them with your answers when using the table for integration.

Take a few moments to look at the integration table Appendix D. Notice that the table has two basic types of integration formulas. The first gives a formula that is the antiderivative, whereas the second, called a *reduction formula*, simply rewrites the integral in another form.

Example 6   Integration using a table of integrals

Find $\int x^2(3-x)^5\,dx$.

Solution   We can find this integral using substitution:

$$\int x^2(3-x)^5\,dx = \int (3-u)^2 u^5(-du) \qquad \boxed{\text{If } u = 3 - x, \text{ then } du = -dx}$$

$$= \int (-u^7 + 6u^6 - 9u^5)\,du$$

$$= -\frac{u^8}{8} + \frac{6u^7}{7} - \frac{9u^6}{6} + C$$

$$= -\frac{1}{8}(3-x)^8 + \frac{6}{7}(3-x)^7 - \frac{3}{2}(3-x)^6 + C$$

Even though this was not too difficult, it is a bit tedious, so we might think to find this integral by using a table of integrals. This is an integral involving an expression of the form $au + b$; we find that this is Formula 32 where $u = x, a = -1, b = 3$, and $n = 5$.

$$\int x^2 (3-x)^5 dx = \frac{(3-x)^{5+3}}{(5+3)(-1)^3} - \frac{2(3)(3-x)^{5+2}}{(5+2)(-1)^3} + \frac{3^2(3-x)^{5+1}}{(5+1)(-1)^3} + C$$

$$= -\frac{1}{8}(3-x)^8 + \frac{6}{7}(3-x)^7 - \frac{3}{2}(3-x)^6 + C$$

One of the difficult considerations when using tables or computer software to carry out integration is recognizing the variety of different forms for acceptable answers. Note that it is not easy to show that these forms are algebraically equivalent. If we use computer software for this evaluation, we find yet another algebraically equivalent form (remember you can check equivalence up to an arbitrary constant) by taking the derivative of both integrals and showing they are the same):

$$\int x^2 (3-x)^5 dx = -\frac{x^8}{8} + \frac{15x^7}{7} - 15x^6 + 54x^5 - \frac{405x^4}{4} + 81x^3$$

Once again, remember that you must add the $+C$ to the computer software form.   ▪

Example 7   Integration using a reduction formula from a table of integrals

Find $\int (\ln x)^4 \, dx$.

Solution   The integrand is in logarithmic form; from the table of integrals we see that Formula 198, Appendix D applies, where $u = x$ and $n = 4$. Take a close look at Formula 198, and note that it gives another integral as part of the result. This is called a **reduction formula** because it enables us to compute the given integral in terms of an integral of a similar type, only with a lower power in the integral.

$$\int (\ln x)^4 \, dx = x(\ln x)^4 - 4\int (\ln x)^{4-1}dx \qquad \textit{Formula 198}$$

$$= x(\ln x)^4 - 4\left[x(\ln x)^3 - 3\int (\ln x)^{3-1}dx \right] \qquad \textit{Formula 198 again}$$

$$= x(\ln x)^4 - 4x(\ln x)^3 + 12\int (\ln x)^2 \, dx$$

$$= x(\ln x)^4 - 4x(\ln x)^3 + 12\left[x(\ln x)^2 \right.$$

$$\left. - 2x\ln x + 2x \right] + C \qquad \textit{This is formula 197.}$$

$$= x(\ln x)^4 - 4x(\ln x)^3 + 12x(\ln x)^2$$

$$- 24x\ln x + 24x + C \qquad\qquad\qquad ▪$$

     Reduction formulas in a table of integrals, such as the one illustrated in Example 7, are usually obtained using substitution and integration by parts (which we will do in Section 7.2).

     Note from the previous example that we follow the convention of adding the constant C only after eliminating the last integral sign (rather than being technically correct and writing $C_1, C_2, \ldots$ for *each* integral). The reason we can do this is that $C_1 + C_2 + \cdots = C$, for arbitrary constants.

     It is often necessary to make substitutions before using one of the integration formulas, as shown in the following example.

Example 8   Using an integral table after substitution

Find $\displaystyle\int \frac{x\,dx}{\sqrt{8-5x^2}}$.

Solution   This is an integral of the form $\sqrt{a^2 - u^2}$, but it does not exactly match any of the formulas. Note, however, that except for the coefficient of 5, it is like Formula 111. Let $u = \sqrt{5}\,x$ (so $u^2 = 5x^2$); then $du = \sqrt{5}\,dx$:

$$\int \frac{x\,dx}{\sqrt{8-5x^2}} = \int \frac{\frac{u}{\sqrt{5}} \cdot \frac{du}{\sqrt{5}}}{\sqrt{8-u^2}}$$

> If $u = \sqrt{5}\,x$, so that $x = \frac{u}{\sqrt{5}}$
> $du = \sqrt{5}\,dx$, so that $dx = \frac{du}{\sqrt{5}}$

$$= \frac{1}{5}\int \frac{u\,du}{\sqrt{8-u^2}}$$

$$= \frac{1}{5}(-\sqrt{8-u^2}) + C \quad \textit{Formula } 111 \textit{ where } a^2 = 8.$$

As you can see from Example 8, using an integral table is not a trivial task. In fact, other methods of integration may be preferable. For Example 8, you can let $u = 8 - 5x^2$ and integrate by substitution:

$$\int \frac{x\,dx}{\sqrt{8-5x^2}} = \int \frac{x\left(\frac{du}{-10x}\right)}{\sqrt{u}}$$

> If $u = 8 - 5x^2$, then $du = -10x\,dx$

$$= -\frac{1}{10}\int u^{-1/2}\,du$$

$$= -\frac{1}{10}(2u^{1/2}) + C$$

$$= -\frac{1}{5}\sqrt{8-5x^2} + C$$

Of course, this answer is the same as the one we obtained in Example 8. The point of this calculation is to emphasize that you should try simple methods of integration before turning to the table of integrals.

Example 9   Integration by table (multiple forms with substitution)

Find $\displaystyle\int 5x^2\sqrt{3x^2+1}\,dx$.

Solution   This is similar to Formula 87, but you must take care of the 5 (constant multiple) and the 3 (by making a substitution).

$$\int 5x^2\sqrt{3x^2+1}\,dx = 5\int \left(\frac{u^2}{3}\right)\sqrt{u^2+1}\,\frac{du}{\sqrt{3}} \quad \boxed{\text{If } u = \sqrt{3}\,x, \text{ then } du = \sqrt{3}\,dx}$$

$$= \frac{5}{3\sqrt{3}}\int u^2\sqrt{u^2+1}\,du \quad \textit{Formula } 87 \textit{ with } a = 1$$

$$= \frac{5}{3\sqrt{3}}\left[\frac{u(u^2+1^2)^{3/2}}{4} - \frac{1^2 u\sqrt{u^2+1^2}}{8}\right.$$
$$\left. - \frac{1^4}{8}\ln\left(u+\sqrt{u^2+1^2}\right)\right] + C$$

$$= \frac{5}{24\sqrt{3}}\left[2\sqrt{3}x(3x^2+1)^{3/2} - \sqrt{3}\,x\sqrt{3x^2+1}\right.$$
$$\left. - \ln\left(\sqrt{3}\,x + \sqrt{3x^2+1}\right)\right] + C$$

$$= \frac{5}{24}\left[2x(3x^2+1)^{3/2} - x(3x^2+1)^{1/2}\right.$$
$$\left. - \frac{1}{\sqrt{3}}\ln\left(\sqrt{3}\,x + \sqrt{3x^2+1}\right)\right] + C$$

If you use a calculator or computer, you will probably obtain an alternate, but equivalent, form:

$$\int 5x^2\sqrt{3x^2+1}\,dx = \frac{5x\sqrt{3x^2+1}\,(6x^2+1)}{24} - \frac{5\sqrt{3}\,\ln\left(\sqrt{3x^2+1}+\sqrt{3}\,x\right)}{72}$$

You can verify that the two expressions are equivalent by differentiating both, as two functions whose derivatives are equal will differ by a constant (of integration, in this case). ■

PROBLEM SET 7.1

Level 1

Find each integral in Problems 1-12.

1. $\displaystyle\int \frac{2x+5}{\sqrt{x^2+5x}}\,dx$

2. $\displaystyle\int \frac{\ln x}{x}\,dx$

3. $\displaystyle\int \frac{dx}{x\ln x}$

4. $\displaystyle\int \cos x\, e^{\sin x}\,dx$

5. $\displaystyle\int \frac{x\,dx}{4+x^4}$

6. $\displaystyle\int \frac{t^2\,dt}{9+t^6}$

7. $\displaystyle\int (1+\cot x)^4 \csc^2 x\,dx$

8. $\displaystyle\int \frac{4x^3-4x}{x^4-2x^2+3}\,dx$

9. $\displaystyle\int \frac{x^3-x}{(x^4-2x^2+3)^2}\,dx$

10. $\displaystyle\int \frac{2x+4}{x^2+4x+3}\,dx$

11. $\displaystyle\int \frac{2x+1}{x^2+x+1}\,dx$

12. $\displaystyle\int \frac{2x-1}{(4x^2-4x)^2}\,dx$

Integrate the expressions in Problems 13-24 using the short table of integrals given in Appendix D.

13. $\displaystyle\int \frac{dx}{x^2\sqrt{x^2-a^2}}$

14. $\displaystyle\int \frac{dx}{x^2\sqrt{a^2-x^2}}$

15. $\displaystyle\int x\ln x\,dx$

16. $\displaystyle\int \ln x\,dx$

17. $\displaystyle\int xe^{ax}\,dx$

18. $\displaystyle\int \frac{dx}{a+be^{2x}}$

19. $\displaystyle\int \frac{x^2\,dx}{\sqrt{x^2+1}}$

20. $\displaystyle\int \frac{dx}{x^2\sqrt{x^2+16}}$

21. $\displaystyle\int \frac{x\,dx}{\sqrt{4x^2+1}}$

22. $\displaystyle\int \frac{dx}{x\sqrt{1-9x^2}}$

23. $\displaystyle\int e^{-4x}\sin 5x\,dx$

24. $\displaystyle\int x\sin^{-1}x\,dx$

Find the integrals in Problems 25-38. If you use the integral table, state the number of the formula used, and if you use substitution, show each step. If you use an alternative table of integrals, then cite the source as well as the formula number.

25. $\displaystyle\int (1+bx)^{-1}\,dx$

26. $\displaystyle\int \frac{x\,dx}{\sqrt{a^2-x^2}}$

27. $\displaystyle\int x(1+x)^3\,dx$

28. $\displaystyle\int x\sqrt{1+x}\,dx$

29. $\displaystyle\int xe^{4x}\,dx$

30. $\displaystyle\int x\ln 2x\,dx$

31. $\displaystyle\int \frac{dx}{1+e^{2x}}$

32. $\displaystyle\int \ln^3 x\,dx$

33. $\displaystyle\int \sec^3\left(\frac{x}{2}\right)\,dx$

34. $\displaystyle\int \frac{\sqrt{4x^2+1}}{x}\,dx$

35. $\displaystyle\int \sin^4 x\,dx$

36. $\displaystyle\int \frac{dx}{9x^2+6x+1}$

37. $\displaystyle\int \sqrt{9-x^2}\,dx$

38. $\displaystyle\int \frac{dx}{\sin^2 x\cos x}$

Level 2

39. Derive the **cosine squared formula** shown on the inside back cover

$$\int \cos^2 x\,dx = \frac{1}{2}x + \frac{1}{4}\sin 2x + C$$

Hint: Use the identity

$$\cos^2 x = \frac{1+\cos 2x}{2}$$

40. Derive the **sine squared formula** shown on the inside back cover

$$\int \sin^2 x\,dx = \frac{1}{2}x - \frac{1}{4}\sin 2x + C$$

Hint: Use the identity

$$\sin^2 x = \frac{1 - \cos 2x}{2}$$

Problems 41-44 use substitution to integrate certain powers of sine and cosine.

41. $\int \sin^3 x \cos^4 x \, dx$ *Hint*: Let $u = \cos x$.

42. $\int \sin^4 x \cos x \, dx$ *Hint*: Let $u = \sin x$.

43. $\int \sin^2 x \cos^2 x \, dx$

Hint: Use the identities shown in Problems 41 and 42.

44. Think Tank Problem Using Problems 41-43 formulate a procedure for integrals of the form

$$\int \sin^m x \cos^n x \, dx$$

Find each integral in Problems 45-48.

45. $\int \dfrac{dx}{x^{1/2} + x^{1/4}}$ **46.** $\int \dfrac{e^x \, dx}{1 + e^{x/2}}$

47. $\int \dfrac{18 \tan^2 t \sec^2 t}{(2 + \tan^3 t)^2} dt$ **48.** $\int \dfrac{4 \, dx}{x^{1/3} + 2x^{1/2}}$

49. Find the volume of the solid generated when the region under the curve

$$y = \frac{x^{3/2}}{\sqrt{x^2 + 9}}$$

between $x = 0$ and $x = 9$ is revolved about the x-axis.

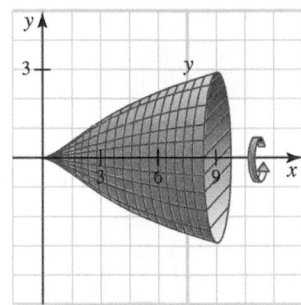

Interactive

50. Find the volume of the solid generated when the curve $y = x(1 - x^2)^{1/4}$ from $x = 0$ to $x = 1$ is revolved about the x-axis.

51. Find the volume of the solid generated when the region between the curve

$$y = \frac{1}{\sqrt{x}}(1 + \sqrt{x})^{1/3}$$

and the x-axis between $x = 1$ and $x = 4$ is revolved about the y-axis.

52. Find the volume of the solid generated when the region between the curve

$$x = \sqrt[4]{4 - y^2}$$

and the y-axis between $y = 1$ and $y = 2$ is revolved about the y-axis.

53. Let $y = f(x)$ be a function that satisfies the differential equation

$$xy' = \sqrt{(\ln x)^2 - x^2}$$

Find the arc length of $y = f(x)$ between $x = \frac{1}{4}$ and $x = \frac{1}{2}$.

54. Find the arc length of the curve $y = \ln(\cos x)$ on the interval $\left[0, \frac{\pi}{4}\right]$.

55. Find the area of the surface generated when the curve $y = x^2$ on the interval $[0, 1]$ is revolved about the x-axis.

56. Find the area of the surface generated when the curve $y = x^2$ on the interval $[0, 1]$ is revolved about the y-axis.

57. Show that

$$\int \csc x \, dx = -\ln |\csc x + \cot x| + C$$

Hint: Multiply the integrand by $\frac{\csc x + \cot x}{\csc x + \cot x}$.

58. Find

$$\int 2 \sin x \cos x \, dx$$

by using the indicated substitution.

 a. Let $u = \cos x$.
 b. Let $u = \sin x$.
 c. Write $2 \sin x \cos x = \sin 2x$ and carry out the integration.
 d. Show that the answers you obtained for parts **a-c** are the same.

Level 3

59. Derive the formula

$$\int_0^\pi x f(\sin x) dx = \frac{\pi}{2} \int_0^\pi f(\sin x) \, dx$$

60. Find the surface area of the torus generated when the circle

$$x^2 + (y - b)^2 = 1$$

$b > 1$ is revolved about the x-axis.

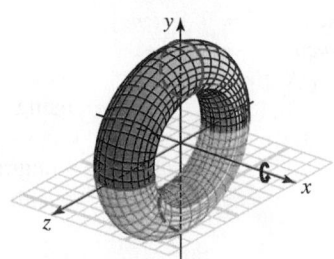

Interactive

7.2 INTEGRATION BY PARTS

IN THIS SECTION: *Integration by parts formula, repeated use of integration by parts, definite integration by parts*
Integration by parts is a procedure based on reversing the product rule for differentiation. We present this section not only as a technique of integration but also as a procedure that is necessary in a variety of useful applications.

Integration by Parts Formula

Recall the formula for differentiation of a product. If u and v are differentiable functions, then

$$d(uv) = u \, dv + v \, du$$

Integrate both sides of this equation to find the formula for integration by parts:

$$\int d(uv) = \int u \, dv + \int v \, du$$

$$uv = \int u \, dv + \int v \, du$$

Integration by parts is one of the most important techniques of integration, with many applications in engineering and physics problems. Due to a lack of proper understanding of the formula and integration process, students often face obstacles in applying this procedure correctly, so it is important to take your time, ask questions, and work through each example carefully.

If we rewrite this last equation, we obtain the formula summarized in the following box.

FORMULA FOR INTEGRATION BY PARTS

$$\int u \, dv = uv - \int v \, du$$

Example 1 Integration by parts

Find $\displaystyle\int xe^x \, dx$.

You may wonder why an arbitrary constant (call it K) was not included when performing the integration associated with $\int dv$ in integration by parts. The reason is that when applying integration by parts, we need just one function v whose differential is dv, so we take the simplest one–the one with $K = 0$. You may find it instructive to see that taking $v = e^x + K$ gives the same result.

Solution To use integration by parts, we must choose u and dv so that the new integral is easier to integrate than the original.

$$\int \underbrace{x}_{u} \underbrace{e^x \, dx}_{dv} = \underbrace{x}_{u} \underbrace{e^x}_{v} - \int \underbrace{e^x}_{v} \underbrace{dx}_{du}$$

$$\boxed{\begin{array}{ll} \text{Let } u = x, & dv = e^x \, dx \\ du = dx & v = \int e^x \, dx = e^x \end{array}}$$

$$= xe^x - e^x + C$$

You can check your work with integration by parts by differentiating, using software, or using the integration table found in Appendix D (Formula 184, with $a = 1$). ◾

Integration by parts is often difficult the first time you try to do it because there is no absolute choice for u and dv. In Example 1, you might have chosen

$$\boxed{\begin{array}{ll} \text{Let } u = e^x, & dv = x \, dx \\ du = e^x \, dx & v = \int x \, dx = \frac{x^2}{2} \end{array}}$$

Then

$$\int xe^x \, dx = \underbrace{e^x}_{u} \; \underbrace{\frac{x^2}{2}}_{v} - \int \underbrace{\frac{x^2}{2}}_{v} \; \underbrace{e^x \, dx}_{du}$$

Note, however, that this choice of u and dv leads to a more complicated form than the original. In general, when you are integrating by parts, if you make a choice for u and dv that leads to a more complicated form than when you started, consider going back and making another choice for u and dv.

Generally, you want to choose dv to be as difficult as possible (and still be something you can integrate), with the remainder being left for the u-factor.

Example 2   When the differentiable part is the entire integrand

Find $\displaystyle\int \ln x \, dx$ for $x > 0$.

Solution

Let $u = \ln x$,	$dv = dx$
$du = \frac{1}{x}dx$	$v = x$

$$\int \ln x \, dx = (\ln x)x - \int x\left(\frac{1}{x}dx\right)$$
$$= x \ln x - x + C$$

Check with Formula 196 (Appendix D), which we worked as Problem 16 of Problem Set 7.1. ▪

Repeated Use of Integration by Parts

Sometimes integration by parts must be applied several times to find a given integral.

Example 3   Repeated integration by parts

Find $\displaystyle\int x^2 e^{-x} dx$.

Solution

Let $u = x^2$,	$dv = e^{-x}dx$
$du = 2x \, dx$	$v = -e^{-x}$

$$\int x^2 e^{-x} dx = x^2(-e^{-x}) - \int (-e^{-x})(2x \, dx)$$

$$= -x^2 e^{-x} + 2\int xe^{-x}dx$$

Let $u = x$,	$dv = e^{-x}dx$
$du = dx$	$v = -e^{-x}$

$$= -x^2 e^{-x} + 2\left[x(-e^{-x}) - \int(-e^{-x})dx\right]$$

Do not forget the grouping symbols.

$$= -x^2 e^{-x} - 2xe^{-x} - 2e^{-x} + C$$
$$= -e^{-x}(x^2 + 2x + 2) + C$$

Check with Formula 185, where $a = -1$. ▪

In the following example, it is necessary to apply integration by parts more than once, but as you will see, when we do so a second time we return to the original integral. Note carefully how this situation can be handled algebraically.

Example 4   Repeated integration by parts with algebraic manipulation

Find $\displaystyle\int e^{2x} \sin x \ dx$.

Solution   For this problem you will see that it will be useful to call the original integral I. That is, we let

$$I = \int e^{2x} \sin x \ dx$$

$$\boxed{\begin{array}{ll} Let\ u = e^{2x}, & dv = \sin x \ dx \\ du = 2e^{2x}\ dx & v = -\cos x \end{array}}$$

$$= e^{2x}(-\cos x) - \int (-\cos x)(2e^{2x}\,dx)$$

$$= -e^{2x}\cos x + 2\int e^{2x}\cos x \ dx$$

$$\boxed{\begin{array}{ll} Let\ u = e^{2x}, & dv = \cos x \ dx \\ du = 2e^{2x}\ dx & v = \sin x \end{array}}$$

$$= -e^{2x}\cos x + 2\left[e^{2x}(\sin x) - \int \sin x \ (2e^{2x}\,dx) \right]$$

$$= -e^{2x}\cos x + 2e^{2x}\sin x - 4\int e^{2x}\sin x \ dx \qquad Notice \int e^{2x} \sin x \ dx = I.$$

$$I = -e^{2x}\cos x + 2e^{2x}\sin x - 4I \qquad\qquad Solve\ for\ I.$$

$$5I = -e^{2x}\cos x + 2e^{2x}\sin x + C$$

$$I = \frac{1}{5}e^{2x}(2\sin x - \cos x) + C$$

Thus, $\int e^{2x} \sin x \ dx = \frac{1}{5}e^{2x}(2\sin x - \cos x) + C$. Check with the integration table (Formula 192, where $a = 2$ and $b = 1$), or by taking the derivative. ■

Definite Integration by Parts

The integration by parts formula can be used for definite integrals, as summarized in the following box.

INTEGRATION BY PARTS FOR DEFINITE INTEGRALS

$$\int_a^b u\ dv = uv\Big|_a^b - \int_a^b v\ du$$

You should recognize that this formula for definite integrals is the same as the formula for indefinite integrals, where the first term after the equal sign has been evaluated at the appropriate limits of integration. This is illustrated by the following example.

Example 5   Integration by parts with a definite integral

Evaluate $\displaystyle\int_0^1 xe^{2x}\,dx$

Solution $\boxed{\begin{array}{ll} Let\ u = x, & dv = e^{2x}\ dx \\ du = dx & v = \frac{1}{2}e^{2x} \end{array}}$

$$\int_0^1 xe^{2x}\, dx = \frac{1}{2}xe^{2x}\Big|_0^1 - \frac{1}{2}\int_0^1 e^{2x}\, dx$$

↓ *It is sometimes easier to simplify the algebra and then do one*
evaluation here at the end of the integration part of the problem.

$$= \left[\frac{1}{2}xe^{2x} - \frac{1}{4}e^{2x}\right]_0^1$$

$$= \frac{1}{4}e^2 + \frac{1}{4}$$

Check in Appendix D (Formula 184, with $a = 2$).   ■

Example 6   Integration by parts with a definite integral followed by substitution

Evaluate $\displaystyle\int_0^1 \tan^{-1} x\, dx$.

Solution

$$\boxed{\begin{array}{ll} \text{Let } u = \tan^{-1} x, & dv = dx \\ du = \frac{dx}{1+x^2} & v = x \end{array}}$$

$$\int_0^1 \underbrace{\tan^{-1} x}_{u}\ \underbrace{dx}_{dv} = \underbrace{(\tan^{-1} x)}_{u}\ \underbrace{x}_{v}\Big|_0^1 - \int_0^1 \underbrace{\frac{x\, dx}{1+x^2}}_{v\, du}$$

$$= \left[x\tan^{-1} x - \frac{1}{2}\ln(1+x^2)\right]_0^1 \qquad \begin{array}{l} \textit{Use substitution where } w = 1+x^2 \textit{ and } dw = 2x\, dx. \\ \int \frac{x\, dx}{1+x^2} = \int \frac{dw}{2w} = \frac{1}{2}\ln w = \frac{1}{2}\ln(1+x^2) \end{array}$$

$$= \left[1\left(\tan^{-1} 1\right) - \frac{1}{2}\ln(1+1)\right] - \left[0 - \frac{1}{2}\ln 1\right]$$

$$= \frac{\pi}{4} - \frac{1}{2}\ln 2$$

Check in Appendix D (Formula 180, with $a = 1$).   ■

PROBLEM SET 7.2

Level 1

Find each integral in Problems 1-16 using integration by parts.

1. $\displaystyle\int xe^{-2x}\, dx$

2. $\displaystyle\int x\sin x\, dx$

3. $\displaystyle\int x\ln x\, dx$

4. $\displaystyle\int x\tan^{-1} x\, dx$

5. $\displaystyle\int \sin^{-1} x\, dx$

6. $\displaystyle\int x^2 \sin x\, dx$

7. $\displaystyle\int e^{-3x}\cos 4x\, dx$

8. $\displaystyle\int e^{2x}\sin 3x\, dx$

9. $\displaystyle\int x^2 \ln x\, dx$

10. $\displaystyle\int (x+\sin x)^2\, dx$

11. $\displaystyle\int \sin(\ln x)\, dx$

12. $\displaystyle\int x\sin x\cos x\, dx$

13. $\displaystyle\int \ln(x^2+1)\, dx$

14. $\displaystyle\int \sin\sqrt{x}\, dx$

15. $\displaystyle\int \frac{xe^{-x}}{(x-1)^2}\, dx$

16. $\displaystyle\int \frac{\ln(\sin x)\, dx}{\tan x}$

Find the exact value of the definite integrals in Problems 17-22 using integration by parts.

17. $\displaystyle\int_1^4 \sqrt{x}\ln x\, dx$

18. $\displaystyle\int_1^e x^3 \ln x\, dx$

19. $\displaystyle\int_1^e (\ln x)^2\, dx$ **20.** $\displaystyle\int_{1/3}^e 3(\ln 3x)^2\, dx$

21. $\displaystyle\int_0^\pi e^{2x}\cos 2x\, dx$ **22.** $\displaystyle\int_0^\pi x(\sin x + \cos x)\, dx$

Level 2

23. ■ *What does this say?* Describe the process known as *integration by parts*.

24. ■ *What does this say?* Contrast using integration by parts for definite and for indefinite integrals.

In Problems 25-28, first use an appropriate substitution and then integrate by parts to find the integral. Remember to give your answers in terms of x.

25. $\displaystyle\int [\sin 2x \ln(\cos x)]\, dx$

26. $\displaystyle\int \frac{\ln x \, \sin(\ln x)}{x}\, dx$

27. $\displaystyle\int [\sin x \ln(2 + \cos x)]\, dx$

28. $\displaystyle\int e^{2x}\sin e^x\, dx$

29. a. Find

$$\int \frac{x^3\, dx}{x^2 - 1}$$

using integration by parts.
   b. Find the integral in part **a** by first dividing the integrand.

30. a. Find

$$\int \frac{x^5\, dx}{x^3 - 1}$$

using integration by parts.
   b. Find the integral in part **a** by first dividing the integrand.

31. Find $\displaystyle\int \cos^2 x\, dx$.

32. Find $\displaystyle\int \frac{x\, dx}{\sqrt{x^2 + 1}}$

33. Use Problem 31 to find

$$\int x\cos^2 x\, dx$$

using integration by parts.

34. Use Problem 32 to find

$$\int \frac{x^3\, dx}{\sqrt{x^2 + 1}}$$

using integration by parts.

35. Find $\displaystyle\int x^n \ln x\, dx$, where n is any positive real number.

36. Show that

$$\int_0^1 x^m(1 - x)^n\, dx = \int_0^1 x^n(1 - x)^m\, dx$$

for positive integers m, n.

37. After t hours on the job, a factory worker can produce $100te^{-0.5t}$ units per hour. How many units does the worker produce during the first 3 hours?

38. After t seconds, an object is moving along a line with velocity of $te^{-t/2}$ meters per second. Express the position of the moving object as a function of time.

39. After t weeks, contributions in response to a local fund-raising campaign were coming in at the rate of $2,000te^{-0.2t}$ dollars per week. How much money was raised during the first 5 weeks?

40. Find the volume of the solid generated when the region under the curve $y = \sin x + \cos x$ on the interval $\left[0, \frac{\pi}{4}\right]$ is revolved about the y-axis.

41. Find the volume of the solid generated when the region under the curve $y = e^{-x}$ on the interval $[0, 2]$ is revolved about the y-axis.

42. Find the volume of the solid generated when the region under the curve $y = \ln x$ on the interval $[1, e]$ is revolved about the indicated axis:

   a. x-axis   **b.** y-axis

43. Find the centroid (with coordinates rounded to the nearest hundredth) of the region bounded by the curves $y = e^x, y = e^{-x}$, and the line $x = 1$.

44. Find the centroid (with coordinates rounded to the nearest hundredth) of the region in the first quadrant bounded by the curves $y = \sin x$ and $y = \cos x$ and the y-axis.

In Problems 45-46, solve the given separable differential equations.

45. $\displaystyle\frac{dy}{dx} = \sqrt{xy}\,\ln x$ **46.** $\displaystyle\frac{dy}{dx} = xe^{y-x}$

47. Find a function $y = f(x)$ whose graph passes through $(0, 1)$ and has the property that the normal line at each point (x, y) on the graph has slope $\dfrac{\sec x}{xy}$.

48. Find a function $y = f(x)$ whose graph passes through $(1, 1)$ and has the property that at each point (x, y) on the graph, the slope of the tangent line is $y \tan^{-1} x$.

49. Suppose it is known that $f(0) = 3$ and

$$\int_0^\pi \left[f(x) + f''(x)\right]\sin x\, dx = 0$$

What is $f(\pi)$?

50. In physics, it is known that loudness L of a sound is related to its intensity I by the equation

$$L = 10 \log \frac{I}{I_0}$$

decibels where $I_0 = 10^{-12}$ watt/m² is the threshold of audibility (the lowest intensity that can be heard). What is the average value of L as the intensity of a TV show ranges between I_0 and $I_1 = 3 \cdot 10^{-5}$ watt/m²?

51. Because a rocket burns fuel in flight, its mass decreases with time, and this in turn affects its velocity. It can be shown that the velocity $v(t)$ of the rocket at time t in its flight is given by

$$v(t) = -r \ln \frac{w - kt}{w} - gt$$

where w is the initial weight of the rocket (including its fuel) and r and k are, respectively, the expulsion speed and the rate of consumption of the fuel, which are assumed to be constant. As usual, $g = 32$ ft/s² is the acceleration due to gravity (assumed to be constant). Suppose $w = 30,000$ lb, $r = 8,000$ ft/s, and $k = 200$ lb/sec. What is the height of the rocket after 2 minutes (120 seconds)?

52. A photographer is taking a picture of a clever sign on the back of a truck. The sign is 5 ft high and its lower edge is 1 ft above the lens of the camera. At first the truck is 4 ft away from the photographer, but then it begins to move away. What is the average value of the angle θ (correct to two decimal places) subtended by the camera lens as the truck moves from 4 ft to 20 ft away from the photographer?

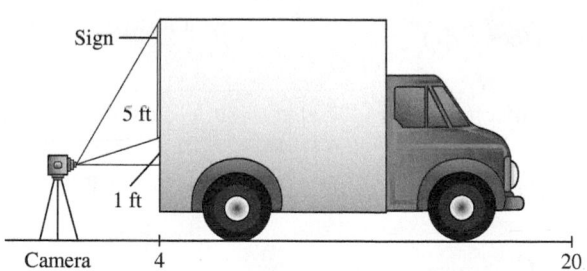

53. The displacement from equilibrium of a mass oscillating at the end of a spring hanging from the ceiling is given by

$$y = 2.3e^{-0.25t} \cos 5t$$

feet. What is the average displacement (rounded to the nearest hundredth) of the mass between times $t = 0$ and $t = \pi/5$ seconds?

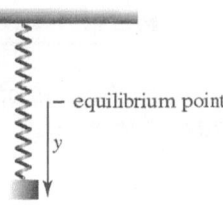

equilibrium point

54. If n moles of an ideal gas expand at constant temperature T, then its pressure p and volume V satisfy the equation $pV = nRT$, for constant R. It can be shown that the work done by the gas in expanding from volume V_1 to V is

$$W = nRT \ln \frac{V}{V_1}$$

What is the average work done as V increases from V_1 to $V_2 = 10V_1$?

55. Find $\displaystyle\int \frac{\sin x \, dx}{\sqrt{\cos 2x}}$.

56. Find $\displaystyle\int \frac{(2x - 1)}{x^2} e^{2x} \, dx$.

57. Derive the reduction formula

$$\int x^n e^x \, dx = x^n e^x - n \int x^{n-1} e^x \, dx$$

(This is Formula 186, with $a = 1$.)

58. Derive the reduction formula

$$\int (\ln x)^n \, dx = x(\ln x)^n - n \int (\ln x)^{n-1} dx$$

(This is Formula 198.)

Level 3

59. Wallis's formula If n is an even positive integer, use reduction formulas to show

$$\int_0^{\pi/2} \sin^n x \, dx = \int_0^{\pi/2} \cos^n x \, dx$$

$$= \left[\frac{1 \cdot 3 \cdot 5 \cdot \cdots \cdot (n - 1)}{2 \cdot 4 \cdot 6 \cdot \cdots \cdot n} \right] \frac{\pi}{2}$$

60. EXPLORATION PROBLEM State and prove a result similar to the one in Problem 59 for the case where n is an odd positive integer.

7.3 TRIGONOMETRIC METHODS

IN THIS SECTION: *Powers of sine and cosine, powers of secant and tangent, trigonometric substitutions, quadratic-form integrals*
If it has been awhile since you studied trigonometry, you might wish to review Appendix E.

Powers of Sine and Cosine

Problems 41-44 of Problem Set 7.1 anticipated the results of this section. We begin by considering a product of powers of sine and cosine, which we represent in the form

$$\int \sin^m x \cos^n x \, dx$$

There are essentially two cases that must be considered, depending on whether the powers m and n are both even or not. We will state the general strategy for handling each case and then illustrate the procedure with an example.

Case I: Either m or n is odd (or both)
General strategy: Suppose, for simplicity, that m is odd. Separate a factor of $\sin x$ from the rest of the integrand, so the remaining power of $\sin x$ is even; use the identity $\sin^2 x = 1 - \cos^2 x$ to express everything but the term $(\sin x \, dx)$ in terms of $\cos x$. Substitute $u = \cos x$, $du = -\sin x \, dx$ to convert the integral into a polynomial in u and integrate using the power rule. The case where n is odd is handled in analogous fashion, as shown in the following example by reversing the roles of $\sin x$ and $\cos x$.

Example 1   Power of cosine is odd

Find $\int \sin^4 x \cos^3 x \, dx$.

Solution   Since $n = 3$ is odd, peel off a factor of $\cos x$ and use $\cos^2 x = 1 - \sin^2 x$ to express the integral as a polynomial in $\sin x$.

$$\int \sin^4 x \cos^3 x \, dx = \int \sin^4 x \cos^2 x (\cos x \, dx)$$

$$= \int \sin^4 x (1 - \sin^2 x)(\cos x \, dx) \quad \boxed{\text{Let } u = \sin x, \text{ then } du = \cos x \, dx}$$

$$= \int u^4 (1 - u^2) du$$

$$= \frac{1}{5} u^5 - \frac{1}{7} u^7 + C$$

$$= \frac{1}{5} \sin^5 x - \frac{1}{7} \sin^7 x + C$$

Case II: Both m and n are even
General strategy: Convert this into Case I by using the identities

$$\sin^2 x = \frac{1}{2}(1 - \cos 2x) \quad \text{and} \quad \cos^2 x = \frac{1}{2}(1 + \cos 2x)$$

These are sometimes called *half-angle identities*.

Example 2   All powers are even

Find $\int \sin^2 x \cos^4 x \, dx$.

Solution

$$\int \sin^2 x \cos^4 x \; dx = \int \frac{1}{2}(1 - \cos 2x)\left(\frac{1}{4}\right)(1 + \cos 2x)^2 \; dx$$

$$= \frac{1}{8} \int (1 + \cos 2x - \cos^2 2x - \cos^3 2x) dx \quad \textit{Half-angle identities}$$

$$= \frac{1}{8} \int \left[1 + \cos 2x - \frac{1}{2}(1 + \cos 4x) - (1 - \sin^2 2x)(\cos 2x)\right] dx \quad \textit{Half-angle identities again}$$

$$= \frac{1}{8} \int \left[1 + \cos 2x - \frac{1}{2} - \frac{1}{2}\cos 4x - \cos 2x + (\sin^2 2x)\cos 2x\right] dx$$

$$= \frac{1}{8} \int \left(\frac{1}{2} - \frac{1}{2}\cos 4x\right) dx + \frac{1}{8} \int \sin^2 2x (\cos 2x) dx \quad \boxed{\begin{array}{l} \text{Let } u = \sin 2x, \\ \text{then } du = 2\cos 2x \; dx \end{array}}$$

$$= \frac{1}{16}x - \frac{1}{64}\sin 4x + \frac{1}{48}\sin^3 2x + C$$

TECHNOLOGY NOTE: One computer algebra system applied to this example yields

$$\int \sin^2 x \cos^4 x \; dx = -\frac{1}{6}\sin x \cos^5 x + \frac{1}{24}\sin x \cos^3 x + \frac{1}{16}\sin x \cos x + \frac{1}{16}x$$

(Recall, technology does not show the "$+C$" term.) Both answers are correct. They are related by trigonometric identities, and it can be shown that the two answers differ by a constant.

Powers of Secant and Tangent

The simplest integrals of this form are

$$\int \tan x \; dx = \ln |\sec x| + C \qquad \text{and} \qquad \int \sec x \; dx = \ln |\sec x + \tan x| + C$$

For the more general situation, which we write as

$$\int \tan^m x \sec^n x \; dx$$

there are three essentially different cases to consider.

Case I: n is even
General strategy: Peel off a factor of $\sec^2 x$ from the integrand and use the identity $\sec^2 x = \tan^2 + 1$ to express the integrand in powers of $\tan x$ except for ($\sec^2 x \; dx$); substitute $u = \tan x$, $du = \sec^2 x \; dx$, and integrate by using the power rule.

Example 3   Power of the secant is even

Find $\displaystyle\int \tan^2 x \sec^4 x \; dx$.

Solution

$$\int \tan^2 x \sec^4 x \, dx = \int \tan^2 x \sec^2 x (\sec^2 x \, dx)$$

$$= \int \tan^2 x (\tan^2 x + 1) \sec^2 x \, dx \quad \boxed{\text{Let } u = \tan x, \text{ then } du = \sec^2 x \, dx}$$

$$= \int u^2 (u^2 + 1) du$$

$$= \frac{1}{5} u^5 + \frac{1}{3} u^3 + C$$

$$= \frac{1}{5} \tan^5 x + \frac{1}{3} \tan^3 x + C$$

Case II: m is odd

General strategy: Peel off a factor of $\sec x \tan x$ from the integrand and use the identity $\tan^2 x = \sec^2 x - 1$ to express the integrand in powers of $\sec x$, except for $(\sec x \tan x \, dx)$; substitute $u = \sec x, du = \sec x \tan x \, dx$, and integrate using the power rule.

Example 4   Power of the tangent is odd

Find $\displaystyle\int \tan x \sec^6 x \, dx$.

Solution

$$\int \tan x \sec^6 x \, dx = \int \sec^5 x (\sec x \tan x \, dx) \quad \boxed{\text{Let } u = \sec x, \text{ then } du = \sec x \tan x \, dx}$$

$$= \int u^5 \, du$$

$$= \frac{1}{6} u^6 + C$$

$$= \frac{1}{6} \sec^6 x + C$$

Case III: m is even and n is odd

General strategy: Use the identity $\tan^2 x = \sec^2 x - 1$ to express the integrand in terms of powers of $\sec x$; then use the reduction Formula 161:

$$\int \sec^n au \, du = \frac{\sec^{n-2} au \tan au}{a(n-1)} + \frac{n-2}{n-1} \int \sec^{n-2} au \, du$$

Example 5   Power of the tangent is even and power of the secant is odd

Find $\displaystyle\int \tan^2 x \sec^3 x \, dx$.

Solution

$$\int \tan^2 x \sec^3 x \, dx = \int (\sec^2 x - 1) \sec^3 x \, dx$$

$$= \int \sec^5 x \, dx - \int \sec^3 x \, dx$$

$$= \left[\frac{\sec^3 x \tan x}{4} + \frac{3}{4} \int \sec^3 x \, dx \right] - \int \sec^3 x \, dx$$

$$= \frac{\sec^3 x \tan x}{4} - \frac{1}{4} \int \sec^3 x \, dx$$

$$= \frac{\sec^3 x \tan x}{4} - \frac{1}{4}\left[\frac{\sec x \tan x}{2} + \frac{1}{2}\int \sec x \ dx\right]$$

$$= \frac{\sec^3 x \tan x}{4} - \frac{\sec x \tan x}{8} - \frac{1}{8}\ln|\sec x + \tan x| + C$$

Trigonometric Substitutions

Trigonometric substitutions can also be useful. For instance, suppose an integrand contains the term $\sqrt{a^2 - u^2}$, where $a > 0$. Then by setting $u = a \sin\theta$ for an acute angle θ, and using the identity $\cos^2\theta = 1 - \sin^2\theta$, we obtain

$$\sqrt{a^2 - u^2} = \sqrt{a^2 - a^2\sin^2\theta} = a\sqrt{1 - \sin^2\theta} = a\cos\theta$$

Thus, the substitution $u = a\sin\theta$, $du = a\cos\theta \ d\theta$ eliminates the square root and may convert the given integral into one involving only sine and cosine. This substitution can best be remembered by setting up a reference triangle. This process is illustrated in the following example.

Example 6   Trigonometric substitution with form $\sqrt{a^2-u^2}$

Find $\displaystyle\int \sqrt{4 - x^2}\,dx$.

Solution   First, using a table of integration, we find (Formula 117; $a = 2$)

$$\int \sqrt{4 - x^2}\,dx = \frac{x\sqrt{4-x^2}}{2} + 2\sin^{-1}\left(\frac{x}{2}\right) + C$$

Our goal with this example is to show how we might obtain this formula with a trigonometric substitution. Refer to the triangle shown in Figure 7.1.

Let $x = 2\sin\theta$, so $dx = 2\cos\theta \ d\theta$. Then

$$\int \sqrt{4 - x^2}\,dx = \int \sqrt{4 - 4\sin^2\theta}\,(2\cos\theta \ d\theta)$$

$$= 4\int \cos^2\theta \ d\theta \qquad\qquad \textit{Since }\sqrt{1 - \sin^2\theta} = \cos\theta$$

$$= 4\int \frac{1 + \cos 2\theta}{2}\,d\theta \qquad\qquad \textit{Half-angle identity}$$

$$= 2\theta + \sin 2\theta + C$$

$$= 2\theta + 2\sin\theta\cos\theta + C$$

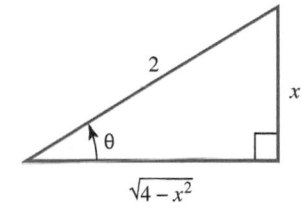

$$\cos\theta = \frac{\sqrt{4-x^2}}{2}; \quad \sin\theta = \frac{x}{2}$$

Figure 7.1   Reference triangle with form $\sqrt{a^2 - u^2}$

The final step is to convert this answer back to terms involving x. Using the reference triangle in Figure 7.1, we find that

$$\sin\theta = \frac{x}{2} \quad\text{and}\quad \cos\theta = \frac{1}{2}\sqrt{4 - x^2}$$

Thus, we have $\theta = \sin^{-1}\dfrac{x}{2}$

$$\int \sqrt{4 - x^2}\,dx = 2\sin^{-1}\left(\frac{x}{2}\right) + 2\left(\frac{x}{2}\right)\left(\frac{\sqrt{4-x^2}}{2}\right) + C$$

$$= 2\sin^{-1}\left(\frac{x}{2}\right) + \frac{x}{2}\sqrt{4 - x^2} + C$$

Similar methods can be used to convert integrals that involve terms of the form $\sqrt{a^2 + u^2}$ or $\sqrt{u^2 - a^2}$ into trigonometric integrals, as shown in Table 7.1. For this table, we require $0 \leq \theta < \frac{\pi}{2}$.

Table 7.1 Trigonometric substitution for integrands involving a radical form

If the integrand involves...	substitute	to obtain...
$\sqrt{a^2 - u^2}$	$u = a\sin\theta$	$\sqrt{a^2 - u^2} = a\cos\theta$
$\sqrt{a^2 + u^2}$	$u = a\tan\theta$	$\sqrt{a^2 + u^2} = a\sec\theta$
$\sqrt{u^2 - a^2}$	$u = a\sec\theta$	$\sqrt{u^2 - a^2} = a\tan\theta$

Example 7   Trigonometric substitution with form $\sqrt{a^2 + u^2}$

Find $\displaystyle\int x^2\sqrt{9 + x^2}\,dx$.

Solution   Let $x = 3\tan\theta$, $dx = 3\sec^2\theta\,d\theta$; then

$$\int x^2\sqrt{9 + x^2}\,dx = \int (3\tan\theta)^2\sqrt{9 + 9\tan^2\theta}\,(3\sec^2\theta\,d\theta)$$

$$= \int (9\tan^2\theta)(3\sec\theta)(3\sec^2\theta\,d\theta) \qquad 1 + \tan^2\theta = \sec^2\theta$$

$$= 81\int \tan^2\theta\,\sec^3\theta\,d\theta$$

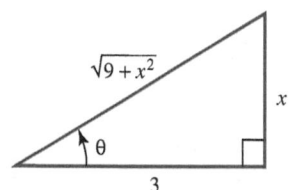

$\tan\theta = \dfrac{x}{3}; \quad \sec\theta = \dfrac{\sqrt{9 + x^2}}{3}$

Figure 7.2 Reference triangle with form $\sqrt{a^2 + u^2}$

This problem can now be completed by looking at Example 5. In order to express the antiderivative in terms of the original variable x, you can use the reference triangle in Figure 7.2.

Since $\tan\theta = \dfrac{x}{3}$, we have $\sec\theta = \dfrac{\sqrt{9 + x^2}}{3}$, and by substituting into the solution shown in Example 5, we find

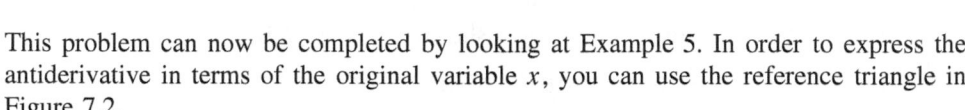

$$\int x^2\sqrt{9 + x^2}\,dx = 81\left[\frac{\sec^3\theta\,\tan\theta}{4} - \frac{\sec\theta\,\tan\theta}{8} - \frac{1}{8}\ln|\sec\theta + \tan\theta|\right] + C$$

$$= \frac{81}{4}\left(\frac{\sqrt{9 + x^2}}{3}\right)^3\left(\frac{x}{3}\right) - \frac{81}{8}\left(\frac{\sqrt{9 + x^2}}{3}\right)\left(\frac{x}{3}\right)$$

$$- \frac{81}{8}\ln\left|\frac{\sqrt{9 + x^2}}{3} + \frac{x}{3}\right| + C$$

$$= \frac{x}{4}(9 + x^2)^{3/2} - \frac{9x}{8}(9 + x^2)^{1/2} - \frac{81}{8}\ln\left|\frac{(9 + x^2)^{1/2}}{3} + \frac{x}{3}\right| + C$$

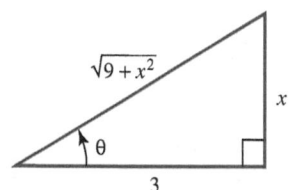

$\sec\theta = x; \quad \tan\theta = \sqrt{x^2 - 1}$

Figure 7.3 Reference triangle with form $\sqrt{u^2 - a^2}$

Example 8   Trigonometric substitution with form $\sqrt{u^2 - a^2}$

Find $\displaystyle\int x^3\sqrt{x^2 - 1}\,dx$.

Solution   Let $x = \sec\theta$, $dx = \sec\theta\tan\theta\,d\theta$; we use the reference triangle shown in Figure 7.3.

$$\int x^3\sqrt{x^2-1}\,dx = \int \sec^3\theta\sqrt{\sec^2\theta-1}\,(\sec\theta\tan\theta\,d\theta)$$

$$= \int \sec^4\theta\,\tan^2\theta\,d\theta \qquad\qquad \tan^2\theta = \sec^2\theta-1$$

$$= \int \sec^2\theta\,\tan^2\theta\,(\sec^2\theta\,d\theta) \qquad\quad \textit{Rearranging in view of substitution}$$

$$= \int (\tan^2\theta+1)\tan^2\theta\,(\sec^2\theta\,d\theta)$$

$$= \int (\tan^4\theta+\tan^2\theta)(\sec^2\theta\,d\theta)$$

$$= \int (u^4+u^2)\,du \qquad\qquad \textit{Let } u = \tan\theta.$$

$$= \frac{1}{5}u^5 + \frac{1}{3}u^3 + C$$

$$= \frac{1}{5}\tan^5\theta + \frac{1}{3}\tan^3\theta + C$$

$$= \frac{1}{5}(x^2-1)^{5/2} + \frac{1}{3}(x^2-1)^{3/2} + C$$

Quadratic-Form Integrals

An integral involving an expression of the form $Ax^2 + Bx + C$, with $A \neq 0$, $B \neq 0$, can often be evaluated by completing the square and making an appropriate substitution to convert it to one of the forms we have previously analyzed.

Example 9   Integration by completing the square

Find $\displaystyle\int \sqrt{16x - 2x^2 - 23}\,dx$.

Solution   Complete the square within the radical:

$$16x - 2x^2 - 23 = -2(x^2 - 8x) - 23$$
$$= -2(x^2 - 8x + 4^2) + 2\cdot 4^2 - 23$$
$$= -2(x-4)^2 + 9$$

Thus,

$$\int \sqrt{16x - 2x^2 - 23}\,dx = \int \sqrt{9 - 2(x-4)^2}\,dx$$

$$= \int \sqrt{9 - u^2}\left(\frac{du}{\sqrt{2}}\right) \qquad\qquad \textit{Where } u = \sqrt{2}\,(x-4)$$

$$= \int \sqrt{9 - (3\sin\theta)^2}\left(\frac{3}{\sqrt{2}}\cos\theta\,d\theta\right) \quad \textit{Where } u = 3\sin\theta$$

$$= \frac{3}{\sqrt{2}}\int 3\sqrt{1 - \sin^2\theta}\,\cos\theta\,d\theta$$

$$= \frac{9}{\sqrt{2}}\int \cos^2\theta\,d\theta$$

$$= \frac{9}{2\sqrt{2}} \left[\theta + \frac{\sin 2\theta}{2} \right] + C \qquad \textit{Half-angle identity}$$

$$= \frac{9}{2\sqrt{2}} \sin^{-1} \left[\frac{\sqrt{2}}{3}(x-4) \right] + \frac{x-4}{2} \sqrt{16x - 2x^2 - 23} + C$$

This last step requires back-substituting from θ to u and then from u to x. Details are left as an exercise. ∎

PROBLEM SET 7.3

Level 1

1. ■ What does this say? Explain how to integrate $\int \sin^m x \cos^n x \, dx$ when m and n are both even.
2. ■ What does this say? Explain how to integrate $\int \tan^m x \sec^n x \, dx$ when n is even.
3. ■ What does this say? Explain the process of using a trigonometric substitution on integrals of the form $\sqrt{a^2 + u^2}$.
4. ■ What does this say? Explain the process of using a trigonometric substitution on integrals of the form $\sqrt{a^2 - u^2}$. How is this different from handling an integral involving $\sqrt{u^2 - a^2}$?

Evaluate the integrals in Problems 5-50.

5. $\displaystyle\int \cos^3 x \, dx$

6. $\displaystyle\int \sin^5 x \, dx$

7. $\displaystyle\int \sin^2 x \cos^3 x \, dx$

8. $\displaystyle\int \sin^3 x \cos^3 x \, dx$

9. $\displaystyle\int \sqrt{\cos t} \sin t \, dt$

10. $\displaystyle\int \frac{\cos x \, dx}{1 + 3 \sin x}$

11. $\displaystyle\int e^{\cos x} \sin x \, dx$

12. $\displaystyle\int \cos^2(2t) \, dt$

13. $\displaystyle\int \sin^2 x \cos^2 x \, dx$

14. $\displaystyle\int \frac{\sin x \, dx}{\cos^5 x}$

15. $\displaystyle\int \tan 2\theta \, d\theta$

16. $\displaystyle\int \sec \frac{x}{2} \, dx$

17. $\displaystyle\int \tan^3 x \sec^4 x \, dx$

18. $\displaystyle\int \sec^5 x \tan x \, dx$

19. $\displaystyle\int (\tan^2 x + \sec^2 x) \, dx$

20. $\displaystyle\int (\sin x + \cos x)^2 \, dx$

21. $\displaystyle\int \tan^2 u \sec u \, du$

22. $\displaystyle\int \sec^4 x \, dx$

23. $\displaystyle\int \sqrt[3]{\tan x} \sec^2 x \, dx$

24. $\displaystyle\int e^x \sec e^x \, dx$

25. $\displaystyle\int x \sin x^2 \cos x^2 \, dx$

26. $\displaystyle\int x \sec^2 x \, dx$

27. $\displaystyle\int \tan^4 t \sec t \, dt$

28. $\displaystyle\int \csc 2\theta \, d\theta$

29. $\displaystyle\int \csc^3 x \cot x \, dx$

30. $\displaystyle\int \csc^2 x \cot^2 x \, dx$

31. $\displaystyle\int \csc^2 x \cos x \, dx$

32. $\displaystyle\int \tan x \csc^3 x \, dx$

33. $\displaystyle\int \sqrt{9 - 4t^2} \, dt$

34. $\displaystyle\int \frac{dx}{\sqrt{9 - x^2}}$

35. $\displaystyle\int \frac{x+1}{\sqrt{4 + x^2}} \, dx$

36. $\displaystyle\int \sqrt{9 + x^2} \, dx$

37. $\displaystyle\int \frac{dx}{\sqrt{x^2 - 7}}$

38. $\displaystyle\int \frac{dx}{5 + 2x^2}$

39. $\displaystyle\int \frac{dx}{\sqrt{5 - x^2}}$

40. $\displaystyle\int \frac{dx}{x\sqrt{7x^2 - 4}}$

41. $\displaystyle\int \frac{dx}{x^2\sqrt{4 - x^2}}$

42. $\displaystyle\int \frac{dx}{x\sqrt{x^2 + 9}}$

43. $\displaystyle\int \frac{\sqrt{x^2 - 4}}{x} \, dx$

44. $\displaystyle\int \frac{dx}{(x-1)^2 + 4}$

45. $\displaystyle\int \frac{dx}{9 - (x+1)^2}$

46. $\displaystyle\int \sqrt{2x - x^2} \, dx$

47. $\displaystyle\int \frac{dx}{\sqrt{x^2 - 2x + 6}}$

48. $\displaystyle\int \frac{dx}{\sqrt{x^2 + 8x + 3}}$

49. $\displaystyle\int \frac{\sin^3 u \, du}{\cos^5 u}$

50. $\displaystyle\int \frac{\sec^2 x \, dx}{\tan^2 x + \sec^2 x}$

Level 2

51. Find the average value of $f(x) = \sin^2 x$ over the interval $[0, \pi]$.
52. Find the centroid of the region (correct to two decimal places) bounded by the curve $y = \cos^2 x$, the x-axis, and the vertical lines $x = \frac{\pi}{4}$ and $x = \frac{\pi}{3}$.
53. Find the volume (correct to four decimal places) of the solid generated when the region bounded by the curve $y = \sin^2 x$ and the x-axis is revolved about the y-axis, $0 \le x \le \pi$.
54. A particle moves along the x-axis in such a way that the acceleration at time t is $a(t) = \sin^2 t$. What is the total distance traveled by the particle over the time interval $[0, \pi]$ if its initial velocity is $v(0) = 2$ units per second?

In Problems 55-58, use the following identities:

$$\sin A \cos B = \frac{1}{2}[\sin(A - B) + \sin(A + B)]$$

$$\sin A \sin B = \frac{1}{2}[\cos(A - B) - \cos(A + B)]$$

$$\cos A \cos B = \frac{1}{2}[\cos(A - B) + \cos(A + B)]$$

55. $\displaystyle\int \sin 3x \sin 5x \, dx$

56. $\displaystyle\int \cos \frac{x}{2} \sin 2x \, dx$

57. $\displaystyle\int \sin^2 3x \cos 4x \, dx$

58. $\displaystyle\int \cos 7x \cos(-3x) \sin 4x \, dx$

Level 3

59. Let f be a twice differentiable function that satisfies the initial value problem

$$f''(x) = -\frac{1}{2}\tan x \, f'(x) \qquad f'(0) = f(0) = 1$$

on the interval $\left[0, \frac{\pi}{2}\right]$. Find the arc length of the curve $y = f(x)$ over this interval.

60. Find $\displaystyle\int \frac{x \, dx}{9 - x^2 - \sqrt{9 - x^2}}$.

7.4 METHOD OF PARTIAL FRACTIONS

IN THIS SECTION: *Partial fraction decomposition, integrating rational functions, rational functions of sine and cosine*

Partial fraction decomposition has great value as a tool of integration. This process may be thought of as the "reverse" of adding fractional algebraic expressions, and it allows us to break up rational expressions into simpler terms.

Partial Fraction Decomposition

You are familiar with the algebraic procedure of adding a string of rational expressions to form a combined rational function with a common denominator. For example,

$$\frac{2}{x + 1} + \frac{-3}{x + 2} = \frac{2(x + 2) + (-3)(x + 1)}{(x + 1)(x + 2)} = \frac{-x + 1}{x^2 + 3x + 2}$$

In **partial fraction decomposition**, we do just the opposite: We start with the reduced fraction

$$\frac{-x + 1}{x^2 + 3x + 2}$$

and write it as the sum of fractions

$$\frac{2}{x + 1} + \frac{-3}{x + 2}$$

This procedure has great value for integration because the terms $\frac{2}{x+1}$ and $\frac{-3}{x+2}$ are easy to integrate. In particular,

$$\int \frac{-x + 1}{x^2 + 3x + 2}dx = \int \frac{2}{x + 1}dx + \int \frac{-3}{x + 2}dx$$

$$= 2\ln|x + 1| - 3\ln|x + 2| + C$$

In the following discussion, we shall consider rational functions

$$f(x) = \frac{P(x)}{D(x)}$$

that are reduced in the sense that $P(x)$ and $D(x)$ are polynomials in x which have no common factors and the degree of P is less than the degree of D. In algebra, it is shown that if $P(x)/D(x)$ is such an expression, then

$$\frac{P(x)}{D(x)} = F_1(x) + F_2(x) + \cdots + F_N(x)$$

where the $F_k(x)$ are expressions of the form

$$\frac{A}{(x-r)^n} \quad \text{or} \quad \frac{Ax+B}{(x^2+sx+t)^m}$$

If $\dfrac{P(x)}{D(x)}$ is not reduced, then we simply divide until a reduced form is obtained. For example $(x \neq 1)$,

$$\frac{(2x^3 + 7x^2 + 6x + 3)(x-1)}{(x-1)(x^2+3x+2)} = 2x + 1 + \frac{-x+1}{x^2+3x+2}$$

$$= 2x + 1 + \frac{2}{x+1} + \frac{-3}{x+2}$$

TECHNOLOGY NOTE: Software programs will decompose rational expressions quite handily using the direction "expand."

We will examine a nonreduced rational expression in Example 5.

We begin by focusing on the case where $D(x)$ can be expressed as a product of linear powers.

PARTIAL FRACTION DECOMPOSITION A Single Linear Power

Let $f(x) = \dfrac{P(x)}{(x-r)^n}$, where $P(x)$ is a polynomial of degree less than n and $P(r) \neq 0$. Then $f(x)$ can be **decomposed** into partial fractions in the following "cascading form"

$$\frac{A_1}{x-r} + \frac{A_2}{(x-r)^2} + \cdots + \frac{A_n}{(x-r)^n}$$

Example 1   Partial fraction decomposition with a single linear power

Decompose $\dfrac{x^2 - 6x + 3}{(x-2)^3}$ into a sum of partial fractions.

Solution

$$\frac{x^2 - 6x + 3}{(x-2)^3} = \frac{A_1}{x-2} + \frac{A_2}{(x-2)^2} + \frac{A_3}{(x-2)^3} \qquad \textit{Partial fraction decomposition.}$$

$$x^2 - 6x + 3 = A_1(x-2)^2 + A_2(x-2) + A_3 \qquad \textit{Multiply both sides by } (x-2)^3.$$

Let $x = 2$: $\quad (2)^2 - 6(2) + 3 = A_1(2-2)^2 + A_2(2-2) + A_3 \qquad \textit{This causes most terms to drop out.}$

$$-5 = A_3$$

We now substitute this value, $A_3 = -5$ and expand the right side to obtain

$$x^2 - 6x + 3 = A_1(x-2)^2 + A_2(x-2) + (-5)$$

$$= A_1 x^2 + (-4A_1 + A_2)x + (4A_1 - 2A_2 - 5)$$

This implies (by equating the coefficients of the similar terms) that

$$
\begin{array}{ll}
1 = A_1 & x^2 \text{ terms} \\
-6 = -4A_1 + A_2 & x \text{ terms} \\
3 = 4A_1 - 2A_2 - 5 & \text{constants}
\end{array}
$$

Because $A_1 = 1$, we find $A_2 = -2$, so the decomposition is

$$
\frac{x^2 - 6x + 3}{(x - 2)^3} = \frac{1}{x - 2} + \frac{-2}{(x - 2)^2} + \frac{-5}{(x - 2)^3}
$$

If there are two or more linear factors in the factorization of $D(x)$, there must be a separate cascade for each power. In particular, if $D(x)$ can be expressed as the product of n distinct linear factors, then

$$
\frac{P(x)}{(x - r_1)(x - r_2) \cdots (x - r_n)}
$$

is decomposed into separate terms

$$
\frac{A_1}{x - r_1} + \frac{A_2}{x - r_2} + \cdots + \frac{A_n}{x - r_n}
$$

This process is illustrated with the following example.

Example 2   Partial fraction decomposition with distinct linear factors

Decompose $\dfrac{8x - 1}{x^2 - x - 2}$.

Solution

$$
\frac{8x - 1}{x^2 - x - 2} = \frac{8x - 1}{(x - 2)(x + 1)}
$$
First, factor the denominator.

$$
= \frac{A_1}{x - 2} + \frac{A_2}{x + 1}
$$
Break up the fraction into parts, each with a linear denominator. The task is to find A_1 and A_2.

$$
= \frac{A_1(x + 1) + A_2(x - 2)}{(x - 2)(x + 1)}
$$
Obtain the least common denominator on the right.

$$
8x - 1 = A_1(x + 1) + A_2(x - 2)
$$
Multiply both sides by the least common denominator.

Note that the degree of the denominator is the same as the number of arbitrary constants, A_1 and A_2. This provides a quick intermediate check on the correct procedure.

Substitute, one at a time, the values that cause each of the factors in the least common denominator to be zero.

$$
8x - 1 = A_1(x + 1) + A_2(x - 2)
$$

$$
\text{Let } x = -1 \quad 8(-1) - 1 = A_1(-1 + 1) + A_2(-1 - 2)
$$

$$
-9 = -3A_2
$$

$$
3 = A_2
$$

$$
\text{Let } x = 2 \quad 8(2) - 1 = A_1(2 + 1) + A_2(2 - 2)
$$

$$
15 = 3A_1
$$

$$
5 = A_1
$$

Thus,

$$
\frac{8x - 1}{x^2 - x - 2} = \frac{5}{x - 2} + \frac{3}{x + 1}
$$

If there is a mixture of distinct and repeated linear factors, we combine the procedures illustrated in the preceding examples. For example,

$$\frac{5x^2 + 21x + 4}{(x+1)^3(x-3)} \text{ is decomposed as } \frac{A_1}{x+1} + \frac{A_2}{(x+1)^2} + \frac{A_3}{(x+1)^3} + \frac{A_4}{x-3}$$

Note that the degree of the denominator is 4, so we use four constants.

If the denominator $D(x)$ in the rational expression $P(x)/D(x)$ contains an irreducible quadratic power, the partial fraction decomposition contains a different kind of cascading sum.

PARTIAL FRACTION DECOMPOSITION A Single Quadratic Factor

Let $f(x) = \dfrac{P(x)}{(x^2 + sx + t)^m}$, where $P(x)$ is a polynomial of degree less than $2m$. Then $f(x)$ can be decomposed into partial fractions in the following "cascading form"

$$\frac{A_1 x + B_1}{x^2 + sx + t} + \frac{A_2 x + B_2}{(x^2 + sx + t)^2} + \cdots + \frac{A_m x + B_m}{(x^2 + sx + t)^m}$$

Because the degree of the denominator is $2m$, we have $2m$ arbitrary constants, namely $A_1, A_2, \cdots, A_m, B_1, B_2, \cdots, B_m$.

Note that the numerator in each term of a linear cascade is a constant A_k, while each numerator in a quadratic cascade is a linear term of the form $A_k x + B_k$.

Example 3   Partial fraction decomposition with a single quadratic power

Decompose $\dfrac{-3x^3 - x}{(x^2 + 1)^2}$.

Solution   The decomposition gives

$$\frac{-3x^3 - x}{(x^2+1)^2} = \frac{A_1 x + B_1}{x^2 + 1} + \frac{A_2 x + B_2}{(x^2 + 1)^2}$$

Multiply both sides of this equation by $(x^2 + 1)^2$, and simplify algebraically.

$$-3x^3 - x = (A_1 x + B_1)(x^2 + 1) + (A_2 x + B_2)$$
$$= A_1 x^3 + B_1 x^2 + (A_1 + A_2)x + (B_1 + B_2)$$

Next, equate the corresponding coefficients on each side of this equation and solve the resulting system of equations to find $A_1 = -3$, $A_2 = 2$, $B_1 = 0$, and $B_2 = 0$. This means that

$$\frac{-3x^3 - x}{(x^2+1)^2} = \frac{-3x}{x^2 + 1} + \frac{2x}{(x^2 + 1)^2}$$

Many of the examples we encounter will offer a mixture of linear and quadratic factors. For example,

$$\frac{x^2 + 4x - 23}{(x^2 + 4)(x + 3)^2} \text{ is decomposed as } \frac{A_1 x + B_1}{x^2 + 4} + \frac{A_2}{x + 3} + \frac{A_3}{(x + 3)^2}$$

The degree of the denominator is four, and there are four constants.

In algebra, the theory of equations tells us that any polynomial P with real coefficients can be expressed as a product of linear and irreducible quadratic powers, some of which may be repeated. This fact can be used to justify the following general procedure for obtaining the partial fraction decomposition of a rational function.

PARTIAL FRACTION DECOMPOSITION Let $f(x) = P(x)/D(x)$, where $P(x)$ and $D(x)$ are polynomials with no common factors and $D(x) \neq 0$.

Step 1 If the degree of P is greater than or equal to the degree of D, use long (or synthetic) division to express $\frac{P(x)}{D(x)}$ as the sum of a polynomial and a fraction $\frac{R(x)}{D(x)}$ in which the degree of the remainder polynomial $R(x)$ is less than the degree of the denominator polynomial $D(x)$.

Step 2 Factor the denominator $D(x)$ into the product of linear and irreducible quadratic powers.

Step 3 Express $\frac{P(x)}{D(x)}$ as a cascading sum of partial fractions of the form

$$\frac{A_i}{(x - r)^n} \quad \text{and} \quad \frac{A_k x + B_k}{(x^2 + sx + t)^m}$$

Integrating Rational Functions

We will now apply the procedure of partial fraction decomposition to integration.

Example 4 Integrating a rational function with a repeated linear factor

Find $\displaystyle\int \frac{x^2 - 6x + 3}{(x - 2)^3}\,dx$.

Solution From Example 1, we have

$$\int \frac{x^2 - 6x + 3}{(x - 2)^3}\,dx = \int \left[\frac{1}{x - 2} + \frac{-2}{(x - 2)^2} + \frac{-5}{(x - 2)^3} \right] dx$$

$$= \int (x - 2)^{-1}\,dx - 2 \int (x - 2)^{-2}\,dx - 5 \int (x - 2)^{-3}\,dx$$

$$= \ln|x - 2| + \frac{2}{x - 2} + \frac{5}{2(x - 2)^2} + C$$

Example 5 Integrating a rational expression with distinct linear factors

Find $\displaystyle\int \frac{x^4 + 2x^3 - 4x^2 + x - 3}{x^2 - x - 2}\,dx$.

Solution We have $P(x) = x^4 + 2x^3 - 4x^2 + x - 3$ and $D(x) = x^2 - x - 2$; because the degree of P is greater than the degree of D, we carry out the long division and write

$$\int \frac{x^4 + 2x^3 - 4x^2 + x - 3}{x^2 - x - 2}\,dx = \int \left(x^2 + 3x + 1 + \frac{8x - 1}{x^2 - x - 2} \right) dx$$

The polynomial part is easy to integrate. The rational expression was decomposed into partial fractions in Example 2.

$$\int \frac{x^4 + 2x^3 - 4x^2 + x - 3}{x^2 - x - 2}\,dx = \int \left(x^2 + 3x + 1 + \frac{8x - 1}{x^2 - x - 2} \right) dx$$

$$= \int \left(x^2 + 3x + 1 + \frac{5}{x - 2} + \frac{3}{x + 1} \right) dx$$

$$= \int x^2 \, dx + 3 \int x \, dx + \int dx + 5 \int (x-2)^{-1} dx + 3 \int (x+1)^{-1} dx$$

$$= \frac{x^3}{3} + \frac{3x^2}{2} + x + 5 \ln|x-2| + 3 \ln|x+1| + C \qquad\blacksquare$$

Example 6 Integrating a rational function with repeated quadratic factors

Find $\displaystyle\int \frac{-3x^3 - x}{(x^2+1)^2} dx$.

Solution From Example 3,

$$\int \frac{-3x^3 - x}{(x^2+1)^2} dx = \int \frac{2x}{(x^2+1)^2} dx + \int \frac{-3x}{x^2+1} dx \qquad \boxed{\text{Let } u = x^2 + 1, \text{ then } du = 2x \, dx}$$

$$= \int u^{-2} du - \frac{3}{2} \int u^{-1} du$$

$$= -u^{-1} - \frac{3}{2} \ln|u| + C$$

$$= \frac{-1}{x^2+1} - \frac{3}{2} \ln(x^2+1) + C \qquad\blacksquare$$

Example 7 Repeated linear factors

Find $\displaystyle\int \frac{5x^2 + 21x + 4}{(x+1)^2(x-3)} dx$.

Solution The partial fraction decomposition of the integrand is

$$\frac{5x^2 + 21x + 4}{(x+1)^2(x-3)} = \frac{A_1}{(x+1)^2} + \frac{A_2}{x+1} + \frac{A_3}{x-3}$$

Multiply both sides by $(x+1)^2(x-3)$:

$$5x^2 + 21x + 4 = A_1(x-3) + A_2(x+1)(x-3) + A_3(x+1)^2$$

As in Example 2, we substitute $x = -1$ and $x = 3$ on both sides of the equation to obtain $A_1 = 3$ and $A_3 = 7$, but A_2 cannot be obtained in this fashion. To find A_2, multiply out the polynomial on the right:

$$5x^2 + 21x + 4 = (A_2 + A_3)x^2 + (A_1 - 2A_2 + 2A_3)x + (-3A_1 - 3A_2 + A_3)$$

Equate the coefficients of x^2, x, and 1 (the constant term) on each side of the equation:

$$\begin{aligned} 5 &= A_2 + A_3 & & x^2 \text{ terms} \\ 21 &= A_1 - 2A_2 + 2A_3 & & x \text{ terms} \\ 4 &= -3A_1 - 3A_2 + A_3 & & \text{constants} \end{aligned}$$

Because we already know that $A_1 = 3$ and $A_3 = 7$, we use the equation $5 = A_2 + A_3$ to obtain $A_2 = -2$. We now turn to the integration:

$$\int \frac{5x^2 + 21x + 4}{(x+1)^2(x-3)} dx = \int \frac{3 \, dx}{(x+1)^2} + \int \frac{-2 \, dx}{x+1} + \int \frac{7 \, dx}{x-3}$$

$$= -3(x+1)^{-1} - 2 \ln|x+1| + 7 \ln|x-3| + C \qquad\blacksquare$$

Example 8 **Distinct linear and quadratic factors**

Find $\displaystyle\int \frac{x^2+4x-23}{(x^2+4)(x+3)}dx$.

Solution The partial fraction decomposition of the integrand has the form

$$\frac{x^2+4x-23}{(x^2+4)(x+3)} = \frac{A_1x+B_1}{x^2+4} + \frac{A_2}{x+3}$$

Multiply both sides by $(x^2+4)(x+3)$ and then combine the terms on the right:

$$x^2+4x-23 = (A_1x+B_1)(x+3) + A_2(x^2+4)$$
$$= (A_1+A_2)x^2 + (3A_1+B_1)x + (3B_1+4A_2)$$

Equate the coefficients to set up the following system of equations:

$$\begin{cases} A_1+A_2=1 & x^2 \text{ terms}\\ 3A_1+B_1=4 & x \text{ terms}\\ 3B_1+4A_2=-23 & \text{constants} \end{cases}$$

Solve this system (the details are not shown) to find $A_1=3$, $B_1=-5$, and $A_2=-2$. We now turn to the integration:

$$\int \frac{x^2+4x-23}{(x^2+4)(x+3)}dx = \int \frac{3x-5}{x^2+4}dx + \int \frac{-2\,dx}{x+3}$$
$$= 3\int \underbrace{\frac{x\,dx}{x^2+4}} - 5\int \frac{dx}{x^2+4} - 2\int \frac{dx}{x+3}$$

$$\boxed{\text{Let } u=x^2+4, \text{ then } du=2x\,dx}$$

$$= 3\left[\frac{1}{2}\ln(x^2+4)\right] - 5\left[\frac{1}{2}\tan^{-1}\frac{x}{2}\right] - 2\ln|x+3| + C$$

Rational Functions of Sine and Cosine

The German mathematician Karl Weierstraß (1815-1897) noticed that the substitution

$$u = \tan\frac{x}{2} \qquad -\pi < x < \pi$$

will convert any rational function of $\sin x$ and $\cos x$ into a rational function of x. To see why this is true, we use the double-angle identities for sine and cosine (see Figure 7.4):

$$\sin x = 2\sin\frac{x}{2}\cos\frac{x}{2} \qquad \text{and} \qquad \cos x = \cos^2\frac{x}{2} - \sin^2\frac{x}{2}$$
$$= 2\left(\frac{u}{\sqrt{1+u^2}}\right)\left(\frac{1}{\sqrt{1+u^2}}\right) \qquad\qquad = \left(\frac{1}{\sqrt{1+u^2}}\right)^2 - \left(\frac{u}{\sqrt{1+u^2}}\right)^2$$
$$= \frac{2u}{1+u^2} \qquad\qquad = \frac{1-u^2}{1+u^2}$$

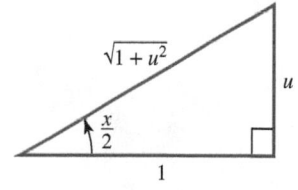

$$\tan\frac{x}{2}=u$$
$$\cos\frac{x}{2}=\frac{1}{\sqrt{1+u^2}}$$
$$\sin\frac{x}{2}=\frac{u}{\sqrt{1+u^2}}$$

Figure 7.4 Weierstrass substitution

Finally, because $u=\tan\frac{x}{2}$, we have $x=2\tan^{-1}u$, so $dx=\dfrac{2}{1+u^2}\,du$.

WEIRSTRASS SUBSTITUTION   For $-\pi < x < \pi$, let $u = \tan \frac{x}{2}$ so that

$$\sin x = \frac{2u}{1 + u^2}, \cos x = \frac{1 - u^2}{1 + u^2}, \text{ and } dx = \frac{2}{1 + u^2} \, du$$

This is called the **Weierstrass substitution**.

The following example illustrates how this substitution can be used along with partial fractions to integrate a rational trigonometric function.

Example 9   Integrating a rational trigonometric function

Find $\displaystyle\int \frac{dx}{3 \cos x - 4 \sin x}$.

Solution   Use the Weierstrass substitution — that is, let $u = \tan \frac{x}{2}$. Remember,

$$dx = \frac{2 \, du}{1 + u^2}; \cos x = \frac{1 - u^2}{1 + u^2}; \text{ and } \sin x = \frac{2u}{1 + u^2}$$

$$\int \frac{dx}{3 \cos x - 4 \sin x} = \int \frac{\frac{2 \, du}{1 + u^2}}{3 \left(\frac{1 - u^2}{1 + u^2} \right) - 4 \left(\frac{2u}{1 + u^2} \right)} \qquad \textit{Weierstrass substitution}$$

$$= \int \frac{-2 \, du}{3u^2 + 8u - 3} \qquad \textit{Simplify}$$

$$= \int \frac{-2 \, du}{(3u - 1)(u + 3)}$$

This integral can be handled by the method of partial fractions.

$$\frac{-2}{(3u - 1)(u + 3)} = \frac{A_1}{3u - 1} + \frac{A_2}{u + 3}$$

Solve this to find $A_1 = -\frac{3}{5}$ and $A_2 = \frac{1}{5}$. We continue with the integration.

$$\int \frac{dx}{3 \cos x - 4 \sin x} = \int \frac{-2 \, du}{(3u - 1)(u + 3)}$$

$$= \int \frac{-\frac{3}{5} \, du}{3u - 1} + \int \frac{\frac{1}{5} \, du}{u + 3}$$

$$= -\frac{3}{5} \cdot \frac{1}{3} \ln |3u - 1| + \frac{1}{5} \cdot \ln |u + 3| + C$$

$$= -\frac{1}{5} \ln \left| 3 \tan \frac{x}{2} - 1 \right| + \frac{1}{5} \ln \left| \tan \frac{x}{2} + 3 \right| + C$$

Problem 59 asks you to derive the formula

$$\int \sec x \, dx = \ln |\sec x + \tan x| + C$$

from scratch. You might recall that we derived this formula using an unusual algebraic step in Example 3 of Section 7.1. You can now derive it by using a Weierstrass substitution.

PROBLEM SET 7.4

Write each rational function given in Problems 1-14 as a sum of partial fractions.

1. $\dfrac{1}{x(x-3)}$ **2.** $\dfrac{3x-1}{x^2-1}$

3. $\dfrac{3x^2+2x-1}{x(x+1)}$ **4.** $\dfrac{2x^2+5x-1}{x(x^2-1)}$

5. $\dfrac{4}{2x^2+x}$ **6.** $\dfrac{x^2-x+3}{x^2(x-1)}$

7. $\dfrac{4x^3+4x^2+x-1}{x^2(x+1)^2}$ **8.** $\dfrac{x^2-5x-4}{(x^2+1)(x-3)}$

9. $\dfrac{x^3+3x^2+3x-4}{x^2(x+3)^2}$ **10.** $\dfrac{1}{x^3-1}$

11. $\dfrac{1}{1-x^4}$ **12.** $\dfrac{x^4-x^2+2}{x^2(x-1)}$

13. $\dfrac{x^2+x-1}{x(x+1)(2x-1)}$ **14.** $\dfrac{x^3-2x^2+x-5}{x(x^2-1)(3x+5)}$

Find the indicated integrals in Problems 15-30.

15. $\displaystyle\int \dfrac{2x^3+9x-1}{x^2(x^2-1)}dx$ **16.** $\displaystyle\int \dfrac{x^4-x^2+2}{x^2(x-1)}dx$

17. $\displaystyle\int \dfrac{x^2+1}{x^2+x-2}dx$ **18.** $\displaystyle\int \dfrac{dx}{x^3-8}$

19. $\displaystyle\int \dfrac{x^4+1}{x^4-1}dx$ **20.** $\displaystyle\int \dfrac{x^3+1}{x^3-1}dx$

21. $\displaystyle\int \dfrac{x\,dx}{(x+1)^2}$ **22.** $\displaystyle\int \dfrac{2x\,dx}{(x-2)^2}$

23. $\displaystyle\int \dfrac{dx}{x(x+1)(x-2)}$ **24.** $\displaystyle\int \dfrac{x+2}{x(x-1)^2}dx$

25. $\displaystyle\int \dfrac{x\,dx}{(x+1)(x+2)^2}$ **26.** $\displaystyle\int \dfrac{x+1}{x(x^2+2)}dx$

27. $\displaystyle\int \dfrac{5x+7}{x^2+2x-3}dx$ **28.** $\displaystyle\int \dfrac{5x\,dx}{x^2-6x+9}$

29. $\displaystyle\int \dfrac{3x^2-2x+4}{x^3-x^2+4x-4}dx$ **30.** $\displaystyle\int \dfrac{3x^2+4x+1}{x^3+2x^2+x-2}dx$

31. ■ *What does this say?* Describe the process of partial fraction decomposition.

32. ■ *What does this say?* What is the Weierstrass substitution and when would you use it?

Find the indicated integrals in Problems 33-50.

33. $\displaystyle\int \dfrac{e^x\,dx}{2e^{2x}-5e^x-3}$ **34.** $\displaystyle\int \dfrac{\cos x\,dx}{\sin^2 x-\sin x-2}$

35. $\displaystyle\int \dfrac{\sin x\,dx}{(1+\cos x)^2}$ **36.** $\displaystyle\int \dfrac{e^x\,dx}{e^{2x}-1}$

37. $\displaystyle\int \dfrac{\sec^2 x\,dx}{\tan x+4}$ **38.** $\displaystyle\int \dfrac{\tan x\,dx}{\sec^2 x+4}$

39. $\displaystyle\int \dfrac{dx}{x^{2/3}-x^{1/2}}$ **40.** $\displaystyle\int \dfrac{dx}{x^{1/4}-x}$

41. $\displaystyle\int \dfrac{dx}{3\cos x+4\sin x}$ **42.** $\displaystyle\int \dfrac{dx}{\sin x-\cos x}$

43. $\displaystyle\int \dfrac{\sin x-\cos x}{\sin x+\cos x}dx$ **44.** $\displaystyle\int \dfrac{dx}{5\sin x+4}$

45. $\displaystyle\int \dfrac{dx}{\sec x-\tan x}$ **46.** $\displaystyle\int \dfrac{dx}{4\cos x+5}$

47. $\displaystyle\int \dfrac{dx}{4\sin x-3\cos x-5}$ **48.** $\displaystyle\int \dfrac{dx}{3\sin x+4\cos x+5}$

49. $\displaystyle\int \dfrac{dx}{x(3-\ln x)(1-\ln x)}$ **50.** $\displaystyle\int \dfrac{dx}{2\csc x-\cot x+2}$

51. Find the area of the region bounded by the curve

$$y=\dfrac{1}{6-5x+x^2}$$

and the lines $x=\frac{4}{3}, x=\frac{7}{4}$, and $y=0$.

52. Find the area under the curve

$$y=\dfrac{1}{x^2+5x+4}$$

between $x=0$ and $x=3$.

53. Find the volume (to four decimal places) of the solid generated when the region under the curve

$$y=\dfrac{1}{\sqrt{x^2+4x+3}}$$

on the interval $[0,3]$ is revolved about

a. the x-axis   **b.** the y-axis

54. Find the volume (to four decimal places) of the solid generated when the curve

$$y=\dfrac{1}{x^2+5x+4}, \quad 0\le x\le 1$$

is revolved about

a. the y-axis
b. the x-axis
c. the line $x=-1$

55. *𝔥istorical 𝔔uest*

George Pólya was born in Hungary and attended universities in Budapest, Vienna, Göttingen, and Paris. He was a professor of mathematics at Stanford University. Pólya's research and winning personality earned him a place of honor not only among

Karl Smith library

George Pólya (1887–1985)

mathematicians, but among students and teachers as well. His discoveries spanned an impressive range of mathematics, real and complex analysis, probability, combinatorics, number theory, and geometry. Pólya's book, How to Solve It, *has been translated into 20 languages. His books have a clarity and elegance seldom seen in mathematics, making them a joy to read. For example, he explained why he became a mathematician by saying that essentially, he thought he was not good enough for physics but on the other hand, too good for philosophy, and mathematics is "in between". A story told by Pólya provides our next 𝔔uest.* He said that he had once intentionally put the problem

$$\int \frac{x\,dx}{x^2 - 9}$$

first on a test of techniques of integration, thinking that it would make his students confident at the start of the exam. If you substitute $u = x^2 - 9$, you can solve the problem very quickly and he assumed his students would do just that. Half of them did use it, but a quarter of them used the procedure of partial fractions, which is the correct method but it used up much of the time allowed, so they were not able to finish the exam. Another group of students used the trig substitution $x = 3\sin\theta$, which is also correct, but again, used up so much time that they weren't able to finish and did poorly on the exam. What Pólya found to be interesting was that the students who used the more time-consuming techniques demonstrated that they

knew more difficult mathematics than those who opted for the easy technique. "But they showed that 'it's not just what you know; it's how and when you use it.' It's nice when what you do is right, but it's much better when it's also *appropriate*."* Carry out all three methods of solution of the given integral Pólya described in this quotation.

56. EXPLORATION PROBLEM Consider the integral

$$\int \sqrt{1 - \cos x}\, dx, \quad 0 < x < \pi$$

a. If you have access to a CAS, report on your attempts to find this integral.

b. Use a Weierstrass substitution to find this integral.

c. Can you think of an even more clever way to perform the integration?

Level 3

57. Use partial fractions to derive the integration formula

$$\int \frac{dx}{x(ax + b)} = \frac{1}{b} \ln \left| \frac{x}{ax + b} \right| + C$$

58. Use partial fractions to derive the integration formula

$$\int \frac{dx}{a^2 - x^2} = \frac{1}{2a} \ln \left| \frac{a + x}{a - x} \right| + C$$

59. Derive the formula

$$\int \sec x\, dx = \ln|\sec x + \tan x| + C$$

using a Weierstrass substitution.

60. Derive the formula

$$\int \csc x\, dx = -\ln|\csc x + \cot x| + C$$

using a Weierstrass substitution.

7.5 SUMMARY OF INTEGRATION TECHNIQUES

IN THIS SECTION: *Integration strategy*

We conclude our study of integration techniques with a procedure summarizing integration techniques.

---

*"Pólya, Problem Solving, and Education," by Alan H. Schoenfeld, *Mathematics Magazine*, Vol. 60, No. 5, December 1987, p. 290.

INTEGRATION STRATEGY   In order to integrate a function, consider the following steps.

Step 1 Simplify.

Simplify the integrand, if possible, and use one of the procedural rules (see the inside back cover or the table of integrals).

Step 2 Use basic formulas.

Use the basic integration formulas (1-29 in the integration table, Appendix D). These are the fundamental building blocks for integration. Almost every integration will involve some basic formulas somewhere in the process, which means that you should memorize these formulas.

Step 3 Substitute.

Make any substitution that will transform the integral into one of the basic forms.

Step 4 Classify.

Classify the integrand according to form in order to use a table of integrals. You may need to use substitution to transform the integrand into a form contained in the integration table. Some special types of substitution are contained in the following list:

I. Integration by parts

    A. Forms $\int x^n e^{ax}\, dx$, $\int x^n \sin ax\, dx$, $\int x^n \cos ax\, dx$

        Let $u = x^n$.

    B. Forms $\int x^n \ln x\, dx$, $\int x^n \sin^{-1} ax\, dx$, $\int x^n \tan^{-1} ax\, dx$

        Let $dv = x^n dx$.

    C. Forms $\int e^{ax} \sin bx\, dx$, $\int e^{ax} \cos bx\, dx$

        Let $dv = e^{ax} dx$.

II. Trigonometric forms

    A. $\int \sin^m x \cos^n x\, dx$

        m odd:

        Peel off a factor of $(\sin x\, dx)$ and **let $u = \cos x$.**

        n odd:

        Peel off a factor of $(\cos x\, dx)$ and **let $u = \sin x$.**

        m and n both even:

        Use half-angle identities:

        $\cos^2 x = \frac{1}{2}(1 + \cos 2x)$ and $\sin^2 x = \frac{1}{2}(1 - \cos 2x)$

    B. $\int \tan^m x \sec^n x\, dx$

        n even:

        Peel off a factor of $(\sec^2 x\, dx)$ and **let $u = \tan x$.**

        m odd:

        Peel off a factor of $(\sec x \tan x)$ and **let $u = \sec x$.**

        m even, n odd:

        Write using powers of secant and **use a reduction formula** (Formula 161 in the integration tables).

    C. For a rational trigonometric integral, try the Weierstrass substitution. **Let $u = \tan \frac{x}{2}$,** so that

$$\sin x = \frac{2u}{1 + u^2}, \quad \cos x = \frac{1 - u^2}{1 + u^2}, \text{ and } dx = \frac{2}{1 + u^2}\, du$$

III. Radical forms; try a trigonometric substitution.

    A. Form $\sqrt{a^2 - u^2}$: **Let $u = a \sin \theta$.**

    B. Form $\sqrt{a^2 + u^2}$: **Let $u = a \tan \theta$.**

    C. Form $\sqrt{u^2 - a^2}$: **Let $u = a \sec \theta$.**

IV. Rational forms; try partial fraction decomposition.

Step 5 Try again.   Still stuck?
   1. Manipulate the integrand:
   Multiply by 1 (clever choice of numerator or denominator).
   Rationalize the numerator.
   Rationalize the denominator.
   2. Relate the problem to a previously worked problem.
   3. Look at another table of integrals or consult computer software that does integration.
   4. Some integrals do not have simple antiderivatives, so all these methods may fail. We will look at some of these forms later in the text.
   5. If dealing with a definite integral, an approximation may suffice. It may be appropriate to use a calculator, computer, or approximate integration.

Example 1   Deciding on an integration procedure

Indicate a procedure to set up each integral. It is not necessary to carry out the integration.

a. $\int \dfrac{\sin \sqrt{x}}{\sqrt{x}} dx$ **b.** $\int (1 + \tan^2 \theta) d\theta$ **c.** $\int \sin^3 x \cos^2 x \; dx$ **d.** $\int 4x^2 \cos 3x \; dx$

e. $\int \dfrac{x \; dx}{\sqrt{9 - x^2}}$ **f.** $\int \dfrac{x^2 \; dx}{\sqrt{9 - x^2}}$ **g.** $\int e^{3x} \sin 2x \; dx$ **h.** $\int \dfrac{\cos^4 x \; dx}{1 - \sin^2 x}$

Solution   Keep in mind that there may be several correct approaches, and the way you proceed from problem to solution is often a matter of personal preference. However, to give you some practice, we will show several hints and suggestions in the context of this example.

a. $\int \dfrac{\sin \sqrt{x}}{\sqrt{x}} \, dx$

The integrand is simplified and is not a fundamental type. Substitution will work well for this problem if you let $u = \sqrt{x}$. After you make this substitution you can integrate using a basic formula.

b. $\int (1 + \tan^2 \theta) d\theta$

The integrand can be simplified using a trigonometric identity $(1 + \tan^2 \theta = \sec^2 \theta)$, so that it can now be integrated using a basic formula.

c. $\int \sin^3 x \cos^2 x \; dx$

The integrand is simplified, it is not a basic formula, and there does not seem to be an easy substitution. Next, try to classify the integrand. We see that it involves powers of trigonometric functions of the type $\int \sin^m x \cos^n x \; dx$, where $m = 3$ (odd) and $n = 2$ (even), so we make the substitution $u = \cos x$.

d. $\int 4x^2 \cos 3x \; dx$

The integrand is simplified (you can bring the 4 out in front of the integral, if you wish), it is not a basic formula, and there is no easy substitution. You might try integration by parts (several steps) or you can check the table of integrals and find it to be Formula 124.

e. $\int \dfrac{x \; dx}{\sqrt{9 - x^2}}$

The integrand is simplified and is not a basic formula. However, we might try the substitution $u = 9 - x^2$. It looks like this will work because the degree of the numerator is one less than the degree of the denominator.

f. $\displaystyle\int \frac{x^2\,dx}{\sqrt{9-x^2}}$

The integrand is simplified, it is not a basic formula, and it looks as though an algebraic substitution will not work because of the degree of the numerator. Since the integrand involves $\sqrt{9-x^2}$, we use the trigonometric substitution $x = 3\sin\theta$, $dx = 3\cos\theta\,d\theta$. If the integral had been $\int \frac{x^2\,dx}{\sqrt{x^2-9}}$, you would have used the substitution $x = 3\sec\theta$, $dx = 3\tan\theta\sec\theta\,d\theta$.

g. $\displaystyle\int e^{3x}\sin 2x\,dx$

The integrand is simplified and it is not a basic formula. If we try to classify the integrand, we see that integration by parts will work with $u = \sin 2x$ and $dv = e^{3x}\,dx$. We also find that this form is Formula 192 in the integration table.

h. $\displaystyle\int \frac{\cos^4 x\,dx}{1-\sin^2 x}$

The integration can be simplified by writing $1-\sin^2 x = \cos^2 x$. After doing this substitution, you will obtain

$$\int \frac{\cos^4 x\,dx}{1-\sin^2 x} = \int \frac{\cos^4 x\,dx}{\cos^2 x} = \int \cos^2 x\,dx$$

This form can be integrated by using a half-angle identity or Formula 127 in the table of integrals. It can also be integrated by parts. ◼

PROBLEM SET 7.5

Level 1

Find each integral in Problems 1-34.

1. $\displaystyle\int \frac{2x-1}{(x-x^2)^3}\,dx$

2. $\displaystyle\int \frac{2x+3}{\sqrt{x^2+3x}}\,dx$

3. $\displaystyle\int (x\sec 2x^2)\,dx$

4. $\displaystyle\int (x^2\csc^2 2x^3)\,dx$

5. $\displaystyle\int (e^x\cot e^x)\,dx$

6. $\displaystyle\int \frac{\tan\sqrt{x}\,dx}{\sqrt{x}}$

7. $\displaystyle\int \frac{\tan(\ln x)\,dx}{x}$

8. $\displaystyle\int \sqrt{\cot x}\,\csc^2 x\,dx$

9. $\displaystyle\int \frac{e^{2t}\,dt}{1+e^{4t}}$

10. $\displaystyle\int \frac{\sin 2x\,dx}{1+\sin^4 x}$

11. $\displaystyle\int \frac{x^2+x+1}{x^2+9}\,dx$

12. $\displaystyle\int \frac{3x+2}{\sqrt{4-x^2}}\,dx$

13. $\displaystyle\int \frac{1+e^x}{1-e^x}\,dx$

14. $\displaystyle\int \frac{t^3\,dt}{2^8+t^8}$

15. $\displaystyle\int \frac{dx}{1+e^{2x}}$

16. $\displaystyle\int \frac{dx}{4-e^{-x}}$

17. $\displaystyle\int \frac{dx}{x^2+2x+2}$

18. $\displaystyle\int \frac{dx}{x^2+x+4}$

19. $\displaystyle\int e^{-x}\cos x\,dx$

20. $\displaystyle\int e^{2x}\sin 3x\,dx$

21. $\displaystyle\int \sin^3 x\,dx$

22. $\displaystyle\int \cos^5 x\,dx$

23. $\displaystyle\int \sin^3 x\cos^2 x\,dx$

24. $\displaystyle\int \sin^3 x\cos^3 x\,dx$

25. $\displaystyle\int \sin^2 x\cos^4 x\,dx$

26. $\displaystyle\int \sin^2 x\cos^5 x\,dx$

27. $\displaystyle\int \tan^5 x\sec^4 x\,dx$

28. $\displaystyle\int \tan^4 x\sec^4 x\,dx$

29. $\displaystyle\int \frac{\sqrt{1-x^2}}{x}\,dx$

30. $\displaystyle\int \frac{dx}{\sqrt{x^2-16}}$

31. $\displaystyle\int \frac{\cos x\,dx}{\sqrt{1+\sin^2 x}}$

32. $\displaystyle\int \frac{\sec^2 x\,dx}{\sqrt{\sec^2 x-2}}$

33. $\displaystyle\int \sin^5 x\,dx$

34. $\displaystyle\int \cos^6 x\,dx$

Find the exact value of the definite integrals in Problems 35-42.

35. $\displaystyle\int_0^2 \sqrt{4-x^2}\,dx$

36. $\displaystyle\int_0^1 \frac{dx}{\sqrt{9-x^2}}$

37. $\displaystyle\int_0^{\ln 2} e^t \sqrt{1+e^{2t}}\,dt$

38. $\displaystyle\int_0^1 \frac{dt}{4t^2+4t+5}$

39. $\displaystyle\int_1^2 \frac{dx}{x^4\sqrt{x^2+3}}$

40. $\displaystyle\int_0^2 \frac{x^3}{(3+x^2)^{3/2}}\,dx$

41. $\displaystyle\int_{-2}^{2\sqrt{3}} x^3\sqrt{x^2+4}\,dx$

42. $\displaystyle\int_0^{\sqrt{5}} x^2\sqrt{5-x^2}\,dx$

Level 2

Find each integral in Problems 43-50.

43. $\displaystyle\int \frac{e^x\,dx}{\sqrt{1+e^{2x}}}$

44. $\displaystyle\int \frac{(2x+1)dx}{\sqrt{4x^2+4x+2}}$

45. $\displaystyle\int \frac{x^2+4x+3}{x^3+x^2+x}\,dx$

46. $\displaystyle\int \frac{5x^2+3x-2}{x^3+2x^2}\,dx$

47. $\displaystyle\int \frac{3x+5}{x^2+2x+1}\,dx$

48. $\displaystyle\int \frac{3x^2+2x+1}{x^3+x^2+x}\,dx$

49. $\displaystyle\int \frac{5x^2+18x+34}{(x-7)(x+2)^2}\,dx$

50. $\displaystyle\int \frac{-3x^2+9x+21}{(x+2)^2(2x+1)}\,dx$

51. ■ What does this say? Outline a method of approaching integration problems. Comment on the strategy presented in this section.

52. ■ What does this say? Integrals of the general form

$$\int \cot^m x\, \csc^n x\, dx$$

are handled in much the same way as those of the form

$$\int \tan^m x\, \sec^n x\, dx$$

   a. What substitution would you use in the case where m is odd?
   b. What substitution would you use if n is even?
   c. What would you do if n is odd and m is even?

53. Find the arc length (correct to four decimal places) of the curve $y = x^2$ from $x = -1$ to $x = 1$.

54. Find the arc length (correct to four decimal places) of the curve $y = \ln x$ from $x = 2$ to $x = 3$.

55. Find the volume of the solid generated when the region under the curve $y = \cos x$ between $x = 0$ and $x = \frac{\pi}{2}$ is revolved about the x-axis.

56. What is the volume of the solid obtained when the region bounded by $y = \sqrt{x}\,e^{-x^2}$, $y = 0$, $x = 0$, and $x = 2$ is revolved about the x-axis?

Level 3

57. Generalize the result of Example 4, Section 7.2 by showing that

$$\int e^{ax} \sin bx\, dx = \frac{(a\sin bx - b\cos bx)e^{ax}}{a^2+b^2} + C$$

for constants a and b. (This is Formula 192.)

58. Let f'' be continuous on the closed interval $[a, b]$. Use integration by parts to show that

$$\int_a^b xf''(x)dx = bf'(b) - f(b) + f(a) - af'(a)$$

59. Derive the reduction formula

$$\int x^m (\ln x)^n dx$$
$$= \frac{x^{m+1}(\ln x)^n}{m+1} - \frac{n}{m+1}\int x^m (\ln x)^{n-1}dx$$

where m and n are positive integers. Use the formula to find

$$\int x^2 (\ln x)^3\, dx$$

60. Derive the reduction formula

$$\int \sin^n Ax\, dx$$
$$= \frac{-\sin^{n-1} Ax \cos Ax}{An} + \frac{n-1}{n}\int \sin^{n-2} Ax\, dx$$

Use this formula to evaluate

$$\int \sin^4 4x\, dx$$

7.6 FIRST-ORDER DIFFERENTIAL EQUATIONS

IN THIS SECTION: *First-order linear differential equations, applications of first-order equations*
In this section, we see how differential equations can be used to model several real-world applications.

First-Order Linear Differential Equations

We introduced separable differential equations in Section 5.6, but not all first-order differential equations have the separable form

$$\frac{dy}{dx} = \frac{f(x)}{g(y)}$$

The goal of this section is to examine a second class of differential equations, called **first-order linear**, that have the general form

$$\frac{dy}{dx} + P(x)y = Q(x)$$

For example, the differential equation

$$\frac{dy}{dx} - \frac{y}{x} = e^{-x}$$

is first-order linear with $P(x) = -\dfrac{1}{x}$ and $Q(x) = e^{-x}$, and

$$x^2\frac{dy}{dx} - (x^2 + 2)y = x^5$$

can be put into this form by dividing by x^2, the coefficient of dy/dx, to obtain

$$\frac{dy}{dx} - \left(\frac{x^2 + 2}{x^2}\right)y = \frac{x^5}{x^2}$$

$$\left[\text{Note that } P(x) = -\left(\frac{x^2 + 2}{x^2}\right) = -1 - 2x^{-2} \text{ and } Q(x) = \frac{x^5}{x^2} = x^3.\right]$$

First-order linear equations can be used to model a variety of important situations in the physical, social, and life sciences. We shall examine a few such models as well as additional models involving separable equations later in this section, but first we need to know how first-order linear differential equations may be solved.

Theorem 7.1   General solution of a first-order linear differential equation

The general solution of the first-order linear differential equation

$$\frac{dy}{dx} + P(x)y = Q(x)$$

is given by

$$y = \frac{1}{I(x)}\left[\int I(x)Q(x)dx + C\right]$$

where $I(x) = e^{\int P(x)dx}$, the exponent is any antiderivative of $P(x)$, and C is an arbitrary constant.

Coefficient of $\dfrac{dy}{dx}$ must be 1. The $\int P(x)dx$ that appears in the exponent of $I(x)$ is any antiderivative of $P(x)$.

Proof: First, note that

$$\frac{dI(x)}{dx} = e^{\int P(x)dx}\left[\frac{d}{dx}\int P(x)dx\right] = e^{\int P(x)dx}P(x) = I(x)P(x)$$

because $\dfrac{d}{dx}e^u = e^u\dfrac{du}{dx}$. Thus,

$$\frac{d}{dx}\big[I(x)y\big] = I(x)\frac{dy}{dx} + y\frac{d}{dx}I(x) = I(x)\frac{dy}{dx} + yI(x)P(x) \qquad \textit{Product rule}$$

We use this computation as the third step in the following.

$$\frac{dy}{dx} + P(x)y = Q(x) \qquad\qquad\qquad\qquad \textit{Given}$$

$$I(x)\left[\frac{dy}{dx} + P(x)y\right] = I(x)Q(x) \qquad\qquad \textit{Multiply both sides by } I(x).$$

$$I(x)\frac{dy}{dx} + I(x)P(x)y = I(x)Q(x)$$

$$\frac{d}{dx}\big[I(x)y\big] = I(x)Q(x) \qquad\qquad\qquad \textit{Substitute}$$

$$\int \frac{d}{dx}\big[I(x)y\big]\,dx = \int I(x)Q(x)\,dx \qquad\qquad \textit{Integrate both sides.}$$

$$I(x)y = \int I(x)Q(x)\,dx + C \qquad\qquad \textit{As usual, the constant of integration}$$
$$y = \frac{1}{I(x)}\left[\int I(x)Q(x)\,dx + C\right] \qquad\quad \textit{C is added after the last integration.}$$

$\blacklozenge$

Because multiplying both sides of the differential equation

$$\frac{dy}{dx} + P(x)y = Q(x)$$

by $I(x)$ makes the left side an exact derivative, the function $I(x)$ is called an **integrating factor** of the differential equation.

A first-order **initial value problem** involves a first-order equation and the value of y at a particular value $x = x_0$. Here is an example of such a problem.

Example 1 First-order linear initial value problem

Solve $\dfrac{dy}{dx} = e^{-x} - 2y, x \geq 0$, subject to the initial condition $y = 2$ when $x = 0$.

Solution A graphical solution (using technology) is found by looking at the direction field as shown in Figure 7.5.

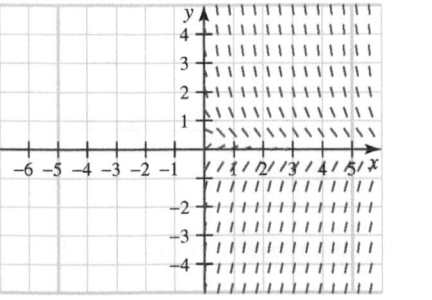

a. Direction field for
$$\frac{dy}{dx} = e^{-x} - 2y, x \geq 0$$

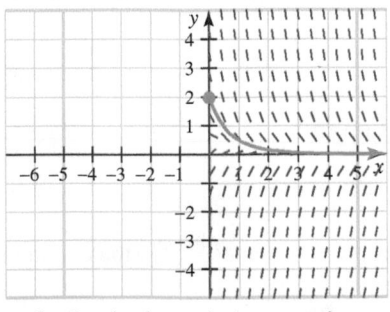

b. Particular solution passing through (0, 2)

Figure 7.5 Interactive   Graphical solution using a direction field

To find an analytic solution, we use Theorem 7.1. The differential equation can be expressed in first-order linear form by adding $2y$ to both sides:

$$\frac{dy}{dx} + 2y = e^{-x} \qquad \text{for } x \geq 0$$

We have $P(x) = 2$ and $Q(x) = e^{-x}$. Both $P(x)$ and $Q(x)$ are continuous in the domain $x \geq 0$. An integrating factor is given by

$$I(x) = e^{\int P(x)dx} = e^{\int 2\,dx} = e^{2x} \qquad \text{for } x \geq 0$$

We now use the first-order linear differential equation theorem, where $I(x) = e^{2x}$ and $Q(x) = e^{-x}$, to find y.

$$y = \frac{1}{e^{2x}} \left[\int e^{2x} e^{-x} dx + C \right]$$

$$= \frac{1}{e^{2x}} \left[e^x + C \right]$$

$$= e^{-x} + Ce^{-2x} \qquad \text{for } x \geq 0$$

To find C, note that because $y = 2$ when $x = 0$ it follows that $1 = C$. Thus,

$$y = e^{-x} + e^{-2x}, \; x \geq 0$$

You might wish to compare the analytic and graphical solutions by noting that the graph of the equation matches the solution shown in Figure 7.5**b**.

Integration by parts is often used in solving first-order linear differential equations. Here is an example.

Example 2   First-order linear differential equation

Find the general solution of the differential equation

$$x\frac{dy}{dx} + y = xe^{2x}, \; x > 0$$

Solution   As a first step, divide each term in the given equation by x so that we can use an integrating factor, as in Theorem 7.1.

$$\frac{dy}{dx} + \frac{y}{x} = e^{2x}$$

For this equation, $P(x) = \dfrac{1}{x}$ and $Q(x) = e^{2x}$. The integrating factor is

$$I(x) = e^{\int (1/x)dx} = e^{\ln x} = x$$

Thus, the general solution is

$$y = \frac{1}{x} \left[\int xe^{2x} dx + C \right] \qquad \textit{Integrate } \int xe^{2x} dx \textit{ by parts.}$$

$$= \frac{1}{x} \left[\frac{1}{2} xe^{2x} - \frac{1}{4} e^{2x} + C \right] \qquad \textit{See Example 5, Section 7.2.}$$

$$= \frac{1}{2} e^{2x} - \frac{e^{2x}}{4x} + \frac{C}{x}$$

Applications of First-Order Equations

Modeling Logistic Growth

When the population $Q(t)$ of a colony of living organisms (humans, bacteria, etc.) is small, it is reasonable to expect the relative rate of change of the population to be constant. In other words,

$$\frac{\frac{dQ}{dt}}{Q} = k \qquad \text{or} \qquad \frac{dQ}{dt} = kQ$$

where k is a constant (the **unrestricted growth rate**). This is called **exponential growth.** As long as the colony has plenty of food and living space, its population will obey this unrestricted growth rate formula and $Q(t)$ will grow exponentially.

However, in practice, there often comes a time when environmental factors begin to restrict the further expansion of the colony, and at this point, the growth ceases to be purely exponential in nature. To construct a population model that takes into account the effect of diminishing resources and crowding, we assume that the population of the species has a "cap" B, called the **carrying capacity** of the species. We further assume that the rate of change in population is jointly proportional to the current population $Q(t)$ and the difference $B - Q$, which can be thought of as the potential unborn population of the species. That is,

$$\frac{\frac{dQ}{dt}}{Q} = k(B - Q) \qquad \text{or} \qquad \frac{dQ}{dt} = kQ(B - Q)$$

This is called a **logistic equation**, and it arises not only in connection with population models, but also in a variety of other situations. The following example illustrates one way such an equation can arise.

Example 3   Logistic equation for the spread of an epidemic

The rate at which an epidemic spreads through a community is proportional to the product of the number of residents who have been infected and the number of susceptible residents. Express the number of residents who have been infected as a function of time.

Solution   Let $Q(t)$ denote the number of residents who have been infected by time t and B the total number of residents. Then the number of susceptible residents who have not been infected is $B - Q$, and the differential equation describing the spread of the epidemic is

$$\frac{dQ}{dt} = kQ(B - Q) \qquad \text{\textit{k is the constant of proportionality.}}$$

$$\frac{dQ}{Q(B - Q)} = k\, dt \qquad \text{\textit{Use separation of variables.}}$$

$$\int \frac{dQ}{Q(B - Q)} = \int k\, dt$$

$$\int \frac{1}{B}\left[\frac{1}{Q} + \frac{1}{B - Q}\right] dQ = \int k\, dt$$

Partial fraction decomposition:

$$\frac{1}{B}\left(\frac{1}{Q} + \frac{1}{B - Q}\right) = \frac{1}{B}\left[\frac{B - Q + Q}{Q(B - Q)}\right]$$

$$= \frac{1}{Q(B - Q)}$$

$$\frac{1}{B}\int \frac{dQ}{Q} + \frac{1}{B}\int \frac{dQ}{B - Q} = \int k\, dt$$

$$\frac{1}{B} \ln |Q| - \frac{1}{B} \ln |B - Q| = kt + C \qquad \text{\textit{Integrate each term.}}$$

$$\frac{1}{B} \ln \left| \frac{Q}{B - Q} \right| = kt + C \qquad \text{\textit{Division property of logarithms}}$$

$$\ln \left(\frac{Q}{B - Q} \right) = Bkt + BC \qquad \text{\textit{Multiply both sides by B}}$$
$$\text{\textit{(note that } B > Q > 0).}$$

$$\frac{Q}{B - Q} = e^{Bkt + BC} \qquad \text{\textit{Definition of logarithm}}$$

$$Q = (B - Q)e^{Bkt}e^{BC} \qquad \text{\textit{Because } } e^{BC} \text{ \textit{ is a constant, let } } A_1 = e^{BC}.$$

$$Q = A_1 B e^{Bkt} - A_1 Q e^{Bkt}$$

$$Q + A_1 Q e^{Bkt} = A_1 B e^{Bkt}$$

$$Q = \frac{A_1 B e^{Bkt}}{1 + A_1 e^{Bkt}}$$

To simplify the equation, we make another substitution; let $A = \dfrac{1}{A_1}$, so that (after several simplification steps) we have

$$Q = \frac{\frac{B e^{Bkt}}{A}}{1 + \frac{e^{Bkt}}{A}} = \frac{B}{1 + A e^{-Bkt}}$$

The graph of $Q(t)$ is shown in Figure 7.6.

Note that the curve has an inflection point where

$$Q(t) = \frac{B}{2}$$

since

$$\frac{d^2Q}{dt^2} = kQ \cdot (-1) + k(B - Q)$$

$$= k(B - 2Q)$$

$$= 0 \text{ when } Q = \frac{B}{2} \text{ and changes sign there.}$$

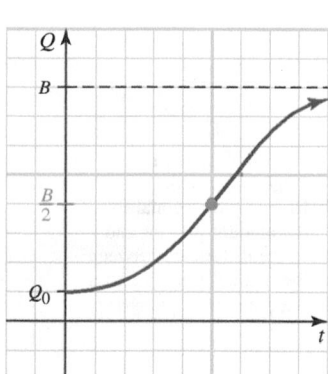

Figure 7.6 Interactive A logistic curve:
$$Q(t) = \frac{B}{1 + A e^{-Bkt}}$$

This corresponds to the fact that the epidemic is spreading most rapidly when half the susceptible residents have been infected (dQ/dt has a maximum there).

Note also in Example 3 that $y = Q(t)$ approaches the line $y = B$ asymptotically as $t \to \infty$. Thus, the number of infected people approaches the number of those susceptible in the long run.

A summary of growth models is given in Table 7.2.

Table 7.2   Summary of Growth Models

Model	Graph	Equation and solution	Sample applications

$Q(t)$ is the population at time t, Q_0 is the initial population, and k is a constant of proportionality. If $Q(t)$ has a limiting value, let it be denoted by B.

Uninhibited growth $(k > 0)$ Rate is proportional to the amount present.	J-curve 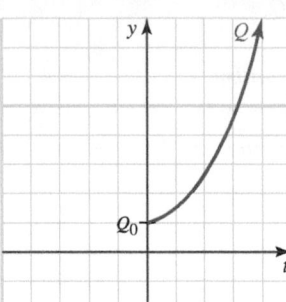	$\dfrac{dQ}{dt} = kQ$ ***Solution*** $Q = Q_0 e^{kt}$	Exponential growth Short-term population growth Interest compounded continuously Inflation Price-supply curves
Uninhibited decay $(k < 0)$ Rate is proportional to the amount present.	L-curve 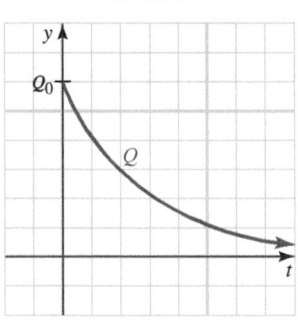	$\dfrac{dQ}{dt} = kQ$ ***Solution*** $Q = Q_0 e^{kt}$	Radioactive decay Depletion of natural resources Price-demand curves
Inhibited growth or **Logistic growth** Rate is proportional to the amount present and to the difference between the amount present and a fixed amount.	S-curve 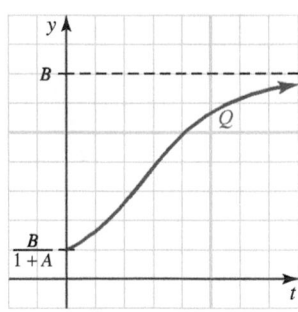	$\dfrac{dQ}{dt} = kQ(B - Q)$ ***Solution*** $Q = \dfrac{B}{1 + Ae^{-Bkt}}$ A is an arbitrary constant.	Long-term population growth Sales fads Growth of a business
Limited growth $(k > 0)$ Rate is proportional to the difference between the amount present and a fixed limit.	C-curve 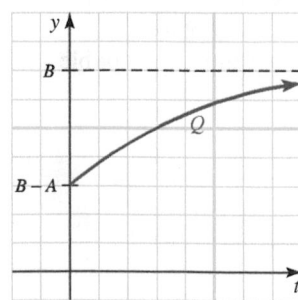	$\dfrac{dQ}{dt} = k(B - Q)$ ***Solution*** $Q = B - Ae^{-kt}$	Learning curve Diffusion of mass media Intravenous infusion of medication Newton's law of cooling Sales of new products

Compartmental Analysis: Modeling Dilution

Example 4   A dilution problem

A tank contains 20 lb of salt dissolved in 50 gal of water. Suppose 3 gal of brine containing 2 lb of dissolved salt per gallon runs into the tank every minute and that the mixture (kept uniform by stirring) runs out of the tank at the rate of 2 gal/min. Find the amount of salt in the tank at any time t. How much salt is in the tank at the end of one hour?

Solution   Let $S(t)$ denote the amount of salt in the tank at the end of t minutes. Because 3 gal of brine flows into the tank each minute and each gallon contains 2 lb of salt, it follows that $(3)(2) = 6$ lb of salt flows into the tank each minute (see Figure 7.7). This is the inflow rate.

For the outflow, note that at time t, there are $S(t)$ lb of salt and $50 + (3 - 2)t$ gallons of solution (because solution flows in at 3 gal/min and out at 2 gal/min). Thus, the concentration of salt in the solution at time t is

$$\frac{S(t)}{50 + t} \text{ lb/gal}$$

and the outflow rate of salt is

$$\left[\frac{S(t)}{50 + t} \text{ lb/gal} \right] \left[2 \text{ gal/min} \right] = \frac{2S(t)}{50 + t} \text{ lb/min}$$

Combining these observations, we see that the net rate of change of salt dS/dt is given by

$$\frac{dS}{dt} = \underbrace{6}_{\uparrow} - \underbrace{\frac{2S}{50 + t}}$$

$$\text{\textit{Inflow}} \quad \text{\textit{Outflow}}$$

$$\frac{dS}{dt} + \frac{2S}{50 + t} = 6$$

which we recognize as a first-order linear differential equation with

$$P(t) = \frac{2}{50 + t}$$

and $Q(t) = 6$. An integrating factor is

$$I(t) = e^{\int P(t)\,dt} = e^{\int (2\,dt)/(50+t)} = e^{2\ln|50+t|} = (50 + t)^2$$

and the general solution is

$$S(t) = \frac{1}{(50 + t)^2} \left[\int 6(50 + t)^2 dt + C \right]$$

$$= \frac{1}{(50 + t)^2} \left[2(50 + t)^3 + C \right]$$

$$= 2(50 + t) + \frac{C}{(50 + t)^2}$$

To evaluate C, we recall that there are 20 lb of salt initially in the solution. This means $S(0) = 20$, so that

$$S(0) = 2(50 + 0) + \frac{C}{(50 + 0)^2}$$

$$20 = 100 + \frac{C}{50^2}$$

$$-80(50^2) = C$$

Inflow rate of salt: 6 lb/min

Outflow rate of salt:

$$\frac{S(t)}{50 + t} \text{ lb/min}$$

Figure 7.7   Rate of flow equals inflow rate minus outflow rate

Thus, the solution to the given differential equation, subject to the initial condition $S(0) = 20$, is

$$S(t) = 2(50 + t) - \frac{80(50^2)}{(50 + t)^2}$$

Specifically, at the end of 1 hour (60 min), the tank contains

$$S(60) = 2(50 + 60) - \frac{80(50)^2}{(50 + 60)^2} \approx 203.4710744$$

The tank contains about 200 lb of salt.

Modeling *RL* Circuits

Another application of first-order linear differential equations involves the current in an *RL* electric circuit. An *RL* circuit is one with a constant resistance, R, and a constant inductance, L. Figure 7.8 shows an electric circuit with an electromotive force (**EMF**), a resistor, and an inductor connected in series.

The EMF source, which is usually a battery or generator, supplies voltage that causes a current to flow in the circuit. According to Kirchhoff's second law, if the circuit is closed at time $t = 0$, then the applied electromotive force is equal to the sum of the voltage drops in the rest of the circuit. It can be shown that this implies that the current $I(t)$ that flows in the circuit at time t must satisfy the first-order linear differential equation

$$L\frac{dI}{dt} + RI = E$$

where L (the inductance), and R (the resistance) are positive constants.* Write

$$\frac{dI}{dt} + \frac{R}{L}I = \frac{E}{L}$$

to see that $P(t) = \dfrac{R}{L}$ (a constant) and $Q(t) = \dfrac{E}{L}$ (a constant). Then

$$I(t) = e^{-\int(R/L)dt}\left[\int \frac{E}{L}e^{\int(R/L)dt}\,dt + C\right]$$

$$= e^{-(R/L)t}\left[\frac{E}{L}\int e^{(R/L)t}\,dt + C\right]$$

$$= e^{-(R/L)t}\left[\frac{E}{L}\cdot\frac{L}{R}e^{(R/L)t} + C\right]$$

$$= \frac{E}{R} + Ce^{-(R/L)t}$$

It is reasonable to assume that no current flows when $t = 0$. That is, $I = 0$ when $t = 0$, so we have $0 = E/R + C$ or $C = -E/R$. The solution of the initial value problem is

$$I = \frac{E}{R}\left(1 - e^{-Rt/L}\right)$$

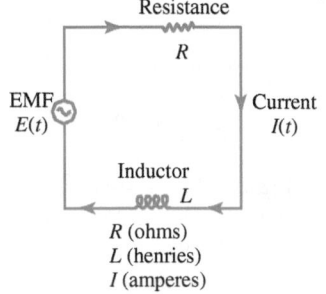

Resistance
R

EMF
$E(t)$

Current
$I(t)$

Inductor
L

R (ohms)
L (henries)
I (amperes)
V (volts)

Figure 7.8 An *RL* circuit diagram

---

*It is common practice in physics and applied mathematics to use I as the symbol for current. Of course, this has nothing to do with the concept of integrating factor introduced earlier in this section.

Notice that because $e^{-(Rt/L)} \to 0$ as $t \to \infty$, we have

$$\lim_{t \to \infty} I(t) = \lim_{t \to \infty} \frac{E}{R}\left(1 - e^{-Rt/L}\right)$$

$$= \frac{E}{R}(1 - 0)$$

$$= \frac{E}{R}$$

This means that, in the long run, the current I must approach $\dfrac{E}{R}$. The solution of the differential equation consists of two parts, which are given special names:

$\dfrac{E}{R}$ is the **steady-state current** $-\dfrac{E}{R}e^{-Rt/L}$ is the **transient current**

Figure 7.9 shows how the current $I(t)$ varies with time t.

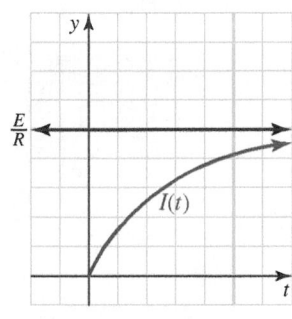

Figure 7.9 The current in an *RL* circuit with constant EMF

PROBLEM SET 7.6

Level 1

Solve the differential equations in Problems 1-16.

1. $\dfrac{dy}{dx} = x^2$

2. $\dfrac{dy}{dx} = 5$

3. $12\dfrac{dy}{dx} = \sqrt{5x+1}$

4. $\dfrac{3}{2}\dfrac{dy}{dx} = e^{4x-3}$

5. $\dfrac{dP}{dt} = 0.02P$

6. $\dfrac{dP}{dt} = 1.25P$

7. $\dfrac{dy}{dx} + \dfrac{3y}{x} = x$

8. $\dfrac{dy}{dx} - \dfrac{2y}{x} = \sqrt{x} + 1$

9. $x^4\dfrac{dy}{dx} + 2x^3 y = 5$

10. $x^2\dfrac{dy}{dx} + xy = 2$

11. $x\dfrac{dy}{dx} + 2y = xe^{x^3}$

12. $\dfrac{dy}{dx} + \left(\dfrac{2x+1}{x}\right)y = e^{-2x}$

13. $\dfrac{dy}{dx} + \dfrac{y}{x} = \tan^{-1} x$

14. $\dfrac{dy}{dx} + \dfrac{2y}{x} = \dfrac{\ln x}{x}$

15. $\dfrac{dy}{dx} + (\tan x)\,y = \sin x$ **16.** $\dfrac{dy}{dx} + (\sec x)\,y = \sin 2x$

Solve the differential equations in Problems 17-20 *for a particular solution.*

17. $\dfrac{dy}{dx} = x^2 y^{-2}$ for $y = 5$ when $x = 0$

18. $x^2\dfrac{dy}{dx} = y^{-2}$ for $y = 2$ when $x = 1$

19. $x\dfrac{dy}{dx} - y\sqrt{x} = 0$ for $y = 1$ when $x = 1$

20. $\dfrac{dy}{dx} = \dfrac{xy}{1+x^2}$ for $y = 2$ when $x = -1$

In Problems 21-26, *solve the initial value problems. A graphical solution is shown as a check for your work.*

21. $\dfrac{dy}{dx} + \dfrac{xy}{1+x} = x(1+x)$, $x > -1$ with $y = -1$ when $x = 0$

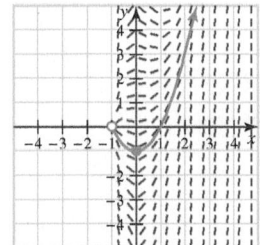

22. $\dfrac{dy}{dx} - \dfrac{xy}{1-x} = x(1+x)$, $x < 1$ with $y = 5$ when $x = 0$

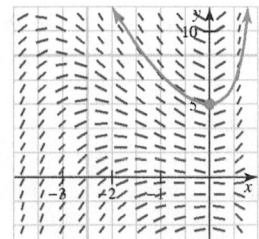

23. $\dfrac{dy}{dx} + \dfrac{2xy}{1+x^2} = \sin x$ with $y = 1$ when $x = 0$

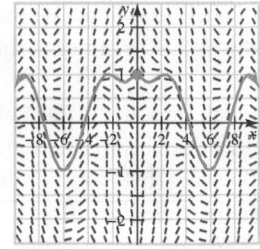

24. $\dfrac{dy}{dx} + \dfrac{2xy}{1+x^2} = \cos x$ with $y = 0$ when $x = 0$

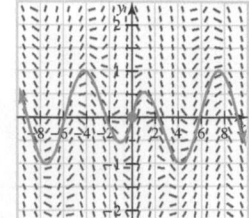

25. $x\dfrac{dy}{dx} - 2y = 2x^3$, $x > 0$ with $y = 0$ when $x = 3$

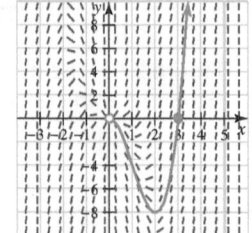

26. $\dfrac{dy}{dx} - \dfrac{y}{x} = x^2 e^{x^2}$, $x > 0$ with $y = 0$ when $x = 2$

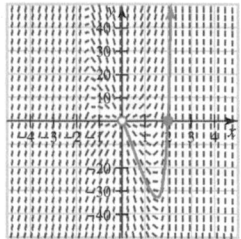

Level 2

In Problems 27-30, find the orthogonal trajectories of the given family of curves. Recall from Section 5.6 that a curve is an orthogonal trajectory of a given family if it intersects each member of that family at right angles.

27. the family of parabolas $y^2 = 4kx$
28. the family of hyperbolas $xy = c$
29. the family of circles $x^2 + y^2 = r^2$
30. the family of exponential curves $y = Ce^{-x}$

Modeling Problems: *In Problems 31-42, set up an appropriate model to answer the given question. Be sure to state your assumptions.*

31. In 2010, the gross domestic product (GDP) of the United States was $14.26 trillion. Suppose the growth rate from 2009 to 2010 was 1.80%. Predict the GDP in 2020.

32. In 2000, the gross domestic product (GDP) of the United States was $9.38 trillion. Suppose the growth rate from 2000 to 2010 was 3.10%. Predict the GDP in 2010. Do some research to find the GDP in 2010. Comment on your prediction. Cite your sources.

33. According to the U.S. Census Bureau, the number of U.S. marriages in 2005 was 2,230,000. If the marriage rate is 0.71%, how many marriages (to the nearest ten thousand) would you expect to be reported in 2010 if the marriage rate is constant?

34. According to the U.S. Census Bureau, the number of U.S. divorces was 957,200 in 2000. If the divorce rate is 0.35%, how many divorces (to the nearest hundred) would you expect to be reported in 2010 if the divorce rate is constant?

35. A tank contains 10 lb of salt dissolved in 30 gal of water. Suppose 2 gal of brine containing 1 lb of dissolved salt per gallon runs into the tank every minute and that the mixture (kept uniform by stirring) runs out at the same rate.
   a. Find the amount of salt in the tank at time t.
   b. How long does it take (to the nearest second) for the tank to contain 15 lb of salt?

36. In Problem 35, suppose the tank has a capacity of 100 gal and that the mixture flows out at the rate of 1 gal/min (instead of 2 gal/min).
   a. How long will it take for the tank to fill?
   b. How much salt will be in the tank when it is full?

37. The rate at which a drug is absorbed into the blood system is given by

$$\frac{db}{dt} = \alpha - \beta b$$

where $b(t)$ is the concentration of the drug in the bloodstream at time t. What does $b(t)$ approach in the long run (that is, as $t \to \infty$)? At what time is $b(t)$ equal to half this limiting value? Assume that $b(0) = 0$.

38. An RL circuit has a resistance of R ohms, inductance of L henries, and EMF of E volts, where R, L, and E are constant. Suppose no current flows in the circuit at time $t = 0$. If L is doubled and E and R are held constant, what effect does this have on the "long run" current in the circuit (that is, the current as $t \to \infty$)?

39. In 2000 there were 31.1 million Hispanics in the United States. If the growth is proportional to the population, how many Hispanics should have been recorded in 2010 if the 1990 Hispanic population was 15.5 million?

40. A population of animals on Catalina Island is limited by the amount of food available. Studies show there were 1,800 animals present in 2005 and 2,000 in 2011 and suggest that 5,000 animals can be supported by the conditions present on the island. Use a logistic model to predict the animal population in the year 2030.

41. In 1986 the Chernobyl nuclear disaster in the Soviet Union contaminated the atmosphere. The buildup of radioactive material in the atmosphere satisfies the differential equation

$$\frac{dM}{dt} = r\left(\frac{k}{r} - M\right)$$

$M = 0$ when $t = 0$ where M = mass of radioactive material in the atmosphere after time t (in years); k is the rate at which the radioactive material is introduced into the atmosphere; r is the annual decay rate of the radioactive material. Find the solution, $M(t)$, of this differential equation in terms of k and r.

42. **The Motion of a Body Falling through a Resisting Medium** A body of mass m is dropped from a great height and falls in a straight line. Assume that the only forces acting on the body are the earth's gravitational attraction mg and air resistance kv. (Recall $g = -32$ ft/s².)

   a. According to Newton's second law,

   $$m\frac{dv}{dt} = mg - kv$$

   Solve this equation, assuming the object has velocity $v_0 = 0$ at time $t = 0$.

   b. Find the distance $s(t)$ the body has fallen at time t. Assume $s = 0$ at time $t = 0$.

   c. If the body weighs $W = 100$ lb and $k = 0.35$, how long does it take for the body to reach the ground from a height of 10,000 ft?

Recall from Section 5.6 that a tank filled with water drains at the rate

$$\frac{dV}{dt} = -4.8A_0\sqrt{h}$$

where h is the height of water at time t (in seconds) and A_0 is the area (in ft²) of the drain hole. This formula, called Torricelli's law, is used in Problems 43 and 44.

43. A full tank of water has a drain with area 0.07 ft². If the tank has a constant cross-sectional area of $A = 5$ ft² and height of $h_0 = 4$ ft, how long does it take to empty?

44. A full tank of water of height 5 ft and constant cross-sectional area $A = 3$ ft² has two drains, both of area 0.02 ft². One drain is at the bottom and the other at height 2 ft. How long does it take for the tank to drain?

45. **The Euler beam model.** For a rigid beam with uniform loading, the deflection $y(x)$ is modeled by the differential equation $y^{(4)} = -k$, where $k > 0$ is a constant and x measures the distance along the beam from one of its ends. Assume that $y(0) = y(L) = 0$, where L is the length of the beam, and that $y''(0) = y''(L) = 0$.

   a. Solve the beam equation to find $y(x)$.

   b. Where does the maximum deflection occur? What is the maximum deflection?

   c. Suppose the beam is cantilevered, so that $y(0) = y(L) = 0$ and $y''(0) = y'(L) = 0$. Now where does the maximum deflection occur? Is the maximum deflection greater or less than the case considered in part **b**? Would you expect the graph of

the deflection $y = y(x)$ to be concave up or concave down on $[0, L]$? Prove your conjecture.

46. **Modeling Problem** A tank initially contains 5 lb of salt in 50 ft³ of solution. At time $t = 0$, brine begins to enter the tank at the rate of 2 ft³/h, and the mixed solution drains at the same rate. The brine coming into the tank has concentration

$$C(t) = 1 - e^{-0.2t}\,\text{lb/ft}^3$$

t hours after the dilution begins.

   a. Set up and solve a differential equation for the amount of salt $S(t)$ in the solution at time t.

   b. How much salt is eventually in the tank (as $t \to \infty$)?

   c. At what time is $S(t)$ maximized? What is the maximum amount of salt in the tank?

47. **Modeling Problem** Two 100-gallon tanks initially contain pure water. Brine containing 2 lb of salt per gallon enters the first tank at the rate of 1 gal/min, and the mixed solution drains into the second tank at the same rate. There, it is again thoroughly mixed and drains at the same rate, 1 gal/min.

   a. Set up and solve a differential equation of the amount of salt $S_1(t)$ in the first tank at time t (minutes).

   b. Set up and solve a second differential equation for the amount of salt $S_2(t)$ in the second tank at time t.

   c. Let $S(t) = S_1 - S_2$. Intuitively, $S(t) \geq 0$ for all t. At what time is the excess $S(t)$ maximized? What is the maximum excess?

48. **Modeling Problem** A chemical in a solution diffuses from a compartment with known concentration $C_1(t)$ across a membrane to a second compartment whose concentration $C_2(t)$ changes at a rate proportional to the difference $C_1 - C_2$. Set up and solve the differential equation for $C_2(t)$ in the following cases:

   a. $C_1(t) = 5e^{-2t}$; $k = 1.7$; $C_2(0) = 0$

   b. $k = 4$; $C_2(0) = 3$, and

$$C_1(t) = \begin{cases} 4 & \text{if } 0 \leq t \leq 3 \\ 5 & \text{if } t > 3 \end{cases}$$

49. The *Gompertz equation* for a population $P(t)$ is

$$\frac{dP}{dt} = kP(B - \ln P)$$

where k and B are positive constants. If the initial population is $P(0) = P_0$, and the ultimate population is

$$\lim_{t \to \infty} P(t) = P_\infty$$

find P(t).

50. **Modeling Problem** Uranium-234 (half-life 2.48×10^5 yr) decays to thorium-230 (half-life 80,000 yr).

a. If $U(t)$ and $T(t)$ are the amounts of uranium and thorium at time t, then

$$\frac{dU}{dt} = -k_1 U, \frac{dT}{dt} = -k_2 T + k_1 U$$

Solve this system of differential equations to obtain $U(t)$ and $T(t)$.

b. If we start with 100 g of pure U-234, how much Th-230 will there be after $t = 5{,}000$ yr?

51. Solve the differential equation

$$\frac{dy}{dx} = \frac{1+y}{xy + e^y(1+y)}$$

by regarding y as the independent variable (i.e., reverse the roles of x and y).

52. Certain biological processes occur periodically over the 24 hours of a day. A patient's metabolic excretion rate is modified by

$$R(t) = \frac{1}{4} - \frac{1}{8}\cos\frac{\pi}{12}(t-6)$$

and the rate of intake is $I(t)$ for $0 \le t \le 24$. The patient's body contains 200 g of the substance when $t = 0$. The amount of substance $Q(t)$ in a patient's body at time t satisfies $Q(t)$ is equal to the amount of intake minus the amount of input.

a. If $I(t) = 0$ (the patient intakes only water), the amount of substance $Q(t)$ in the patient's body at time t satisfies

$$\frac{dQ}{dt} = 0 - R(t)$$

Solve this equation for $Q(t)$.

b. Suppose the patient intakes (in gal/hr) the substance at a constant rate for part of the day. Specifically,

$$I(t) = \begin{cases} 0.4 & \text{for } 10 \le t \le 20 \\ 0 & \text{otherwise} \end{cases}$$

Set up and solve a differential equation for the amount of substance $Q(t)$ in the patient's body at time t. When is $Q(t)$ maximized?

53. An RL circuit has a resistance of $R = 10$ ohms and an inductance of $L = 5$ henries. Find the current $I(t)$ in the circuit at time t if $I(0) = 0$ and the electromotive force (EMF) is

a. $E = 15$ volts   **b.** $E = 5e^{-2t}\sin t$

54. An RL circuit has an inductance of $L = 3$ henries and a resistance $R = 6$ ohms in series with an EMF of $E = 50\sin 30t$. Assume $I(0) = 0$.

a. What is the current $I(t)$ at time t?

b. What is the transient current? The steady-state current?

55. Modeling Problem A lake has a volume of 6 billion ft^3, and its initial pollutant content is 0.22%. A river whose waters contain only 0.06% pollutants flows into the lake at the rate of 350 million ft^3/day, and another river flows out of the lake also carrying 350 million ft^3/day. Assume that the water in the two rivers and the lake is always well mixed. How long does it take for the pollutant content to be reduced to 0.15%?

Level 3

56. A **Bernoulli equation** is a differential equation of the form

$$y' + P(x)y = Q(x)y^n$$

where n is a real number, $n \ne 0, n \ne 1$.

a. Show that the change of variable $u = y^{1-n}$ transforms such an equation into one of the form

$$u'(x) + (1-n)P(x)u(x) = (1-n)Q(x)$$

This transformed equation is a first-order linear differential equation in u and can be solved by the methods of this section, yielding a solution to the given Bernoulli equation.

b. Use the change of variable suggested to solve the Bernoulli equation

$$y' + \frac{y}{x} = 2y^2$$

57. Consider a curve with the property that when horizontal and vertical lines are drawn from each point $P(x,y)$ on the curve to the coordinate axes, the area A_1 under the curve is *twice* the area A_2 above the curve, as shown in Figure 7.10

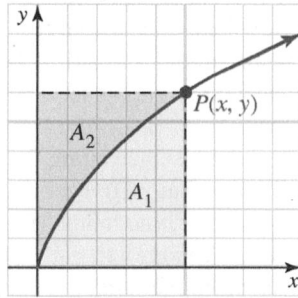

Figure 7.10   Problem 57

Show that x and y satisfy the differential equation

$$\frac{dy}{dx} = \frac{y}{2x}$$

Solve the equation and characterize the family of all curves that satisfy the given geometric condition.

58. *Historical Quest*
*Daniel Bernoulli was a
member of the famous
Bernoulli family (see
Historical Quest Problem 50,
Section 4.5). Between 1725
and 1749 he won ten prizes for
his work in astronomy, gravity,
tides, magnetism, ocean
currents, and the behavior of
ships at sea. While modeling the effects of a smallpox
epidemic, he obtained the differential equation*

**Daniel Bernoulli
(1700-1782)**

$$\frac{dS}{dt} = -pS + \left(\frac{S}{N}\right)\frac{dN}{dt} + \frac{pS^2}{mN}$$

*In this equation $S(t)$ is the number of people at age t
that are susceptible to smallpox; $N(t)$ is the number
of people at age t who survive; p is the probability of
a susceptible person getting the disease, and $1/m$ is
the proportion of those who die from the disease. Let
$y = S/N$, then solve the resulting equation for $y(t)$ and
then write N as a function of S. (Hint:* You will need
the result of Problem 56.)

59. An object of mass m is projected upward from ground
level with initial velocity v_0 against air resistance proportional to its velocity $v(t)$. Thus,

$$m\frac{dv}{dt} = -mg - kv$$

a. Solve this equation for $v(t)$, and then find $s(t)$ the
height of the object above the ground at time t.
b. When does the object reach its maximum height?
What is the maximum height?
c. Suppose $k = \frac{3}{4}$. For an object weighing 20 lb with
initial velocity $v_0 = 150$ ft/s, what is the maximum
height to the nearest foot? When to the nearest second does the object hit the ground?
d. Suppose the same object as in part **c** is launched
with the same initial velocity but in a vacuum, where
there is no air resistance ($k = 0$). Would you expect
the object to hit the ground sooner or later than the
object in part **c**? Prove your conjecture.

60. A body of mass m falls from a height of s_0 ft against air
resistance modeled to be proportional to the *square* of
its velocity v, so that

$$m\frac{dv}{dt} = mg - kv^2$$

a. Solve this differential equation to find $v(t)$. Then
find the height $s(t)$ of the object at time t (in seconds).
b. If the object weighs 1,000 lb, $s_0 = 100$ ft, and
$k = 0.01$, how long to the nearest hundredth of a
second does it take for the object to hit the ground?
Interpret this result.

7.7 IMPROPER INTEGRALS

IN THIS SECTION: *Improper integrals with infinite limits of integration, improper integrals with unbounded
integrands*
We have defined the definite integral $\int_a^b f(x)\,dx$ on a closed bounded interval $[a,b]$, where the integrand $f(x)$ is
bounded. In this section, we extend the concept of integral to the case where the interval of integration is infinite and
also to the case where f is unbounded at a finite number of points on the interval of integration. Collectively, these
are called **improper integrals.**

Improper Integrals with Infinite Limits of Integration

In physics, economics, probability and statistics, and other applied areas, it is useful to
have a concept of an integral that is defined on the entire real line or on half-lines of
the form $x \geq a$ or $x \leq a$. If $f(x) \geq 0$, the integral of f on the interval $x \geq a$ can be
thought of as the area under the curve $y = f(x)$ on this unbounded interval, as shown in
Figure 7.11.

A reasonable strategy for finding this area is first to use a definite integral to compute
the area from $x = a$ to some finite number $x = N$ and then to let N approach infinity in
the resulting expression. Here is a definition.

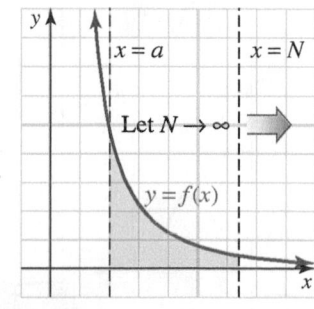

Figure 7.11 Area of an
unbounded region

IMPROPER INTEGRAL TYPE I Let a be a fixed number and assume $\int_a^N f(x)\,dx$ exists for all $N \geq a$. Then if $\lim_{N \to \infty} \int_a^N f(x)\,dx$ exists, we define the **improper integral** $\int_a^\infty f(x)\,dx$ by

$$\int_a^\infty f(x)\,dx = \lim_{N \to \infty} \int_a^N f(x)\,dx$$

The improper integral is said to **converge** if this limit is a finite number and to **diverge** otherwise.

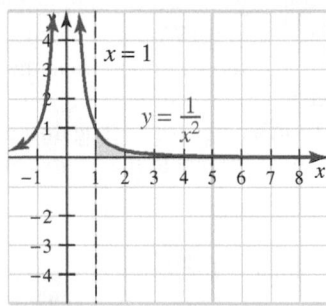

Figure 7.12 Graph of $y = \dfrac{1}{x^2}$

Example 1 Convergent improper integral

Evaluate $\displaystyle\int_1^\infty \frac{dx}{x^2}$.

Solution   The region under the curve $y = 1/x^2$ for $x \geq 1$ is shown in Figure 7.12.

This region is unbounded, so it might seem reasonable to conclude that the area is also infinite. Begin by computing the integral from 1 to N:

$$\int_1^N \frac{dx}{x^2} = -\frac{1}{x}\Big|_1^N = -\frac{1}{N} + 1$$

Let us consider the situation more carefully, as shown in Figure 7.13.

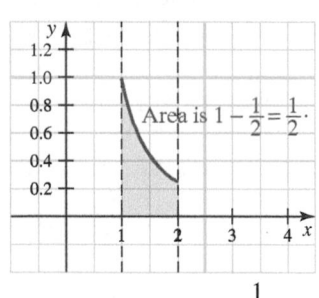

Let $N = 2$, $A = \dfrac{1}{2}$

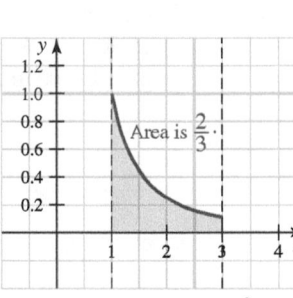

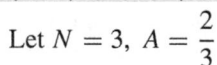

Let $N = 3$, $A = \dfrac{2}{3}$

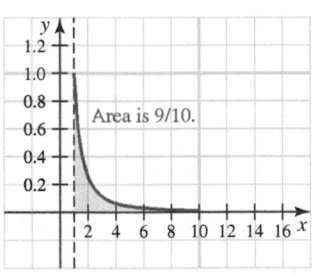

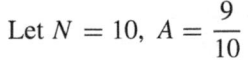

Let $N = 10$, $A = \dfrac{9}{10}$

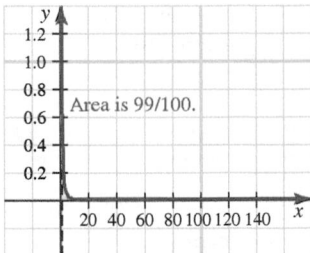

Let $N = 100$, $A = \dfrac{99}{100}$

Figure 7.13   Area enclosed by $y = \dfrac{1}{x^2}$, the x-axis, $x = 1$, and $x = N$

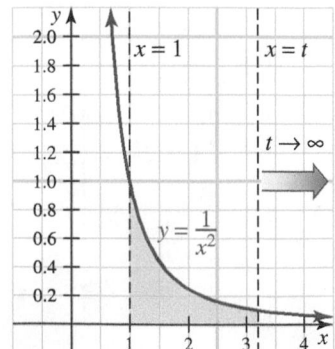

Figure 7.14   The region to the right of $x = 1$ bounded by the curve $y = \dfrac{1}{x^2}$ and the x-axis, has area 1

For $N = 2, 3, 10$, or 100, this integral has values $\frac{1}{2}, \frac{2}{3}, \frac{9}{10}$, and $\frac{99}{100}$, respectively, suggesting that the region under $y = \frac{1}{x^2}$ for $x \geq 1$ actually has finite area approaching 1 as N gets larger and larger.

To find this area analytically, we take the limit as $N \to \infty$, as shown in Figure 7.14. Thus, the improper integral converges and has the value 1. ∎

Example 2 A divergent improper integral

Evaluate $\displaystyle\int_1^\infty \frac{dx}{x}$.

Solution   The graph of $y = \dfrac{1}{x}$, shown in Figure 7.15, looks very much like the graph of $y = \dfrac{1}{x^2}$ in Figure 7.14.

We compute the integral from 1 to N and then let N go to infinity.

$$\int_1^\infty \frac{dx}{x} = \lim_{N \to \infty} \int_1^N \frac{dx}{x}$$

$$= \lim_{N \to \infty} \ln |x| \Big|_1^N$$

$$= \lim_{N \to \infty} \ln N$$

$$= \infty$$

The limit is not a finite number, which means the improper integral diverges. ■

We have shown that the improper integral $\displaystyle\int_1^\infty \frac{dx}{x^2}$ converges and the improper integral $\displaystyle\int_1^\infty \frac{dx}{x}$ diverges. In geometric terms, this says that the area to the right of $x = 1$ under the curve $y = \frac{1}{x^2}$ is finite, whereas the corresponding area under the curve $y = \frac{1}{x}$ is infinite. The reason for the difference is that as x increases, $\frac{1}{x^2}$ approaches zero more quickly than does $\frac{1}{x}$, as shown in Figure 7.16.

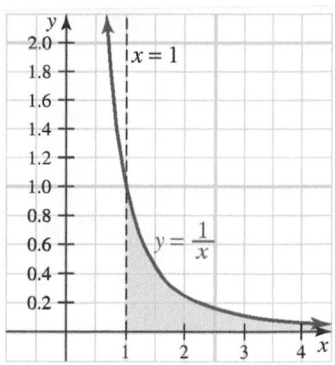

Figure 7.15 The area to the right of the line $x = 1$, bounded by the curve $y = \dfrac{1}{x}$ and the x-axis, is infinite

Example 3 Determine convergence of an improper integral

Show that the improper integral $\displaystyle\int_1^\infty \frac{dx}{x^p}$ converges only for constant $p > 1$.

Solution   For this example, remember that p is a constant and x is the variable. We have seen in Example 1 that the integral converges for $p = 2$. We also know from Example 2 that the integral diverges for $p = 1$, so we can assume $p \neq 1$ in the following computation:

$$\int_1^\infty \frac{dx}{x^p} = \lim_{N \to \infty} \int_1^N \frac{dx}{x^p}$$

$$= \lim_{N \to \infty} \left[\frac{x^{-p+1}}{-p+1} \right] \Bigg|_1^N$$

$$= \lim_{N \to \infty} \frac{1}{1-p} \left[N^{1-p} - 1 \right]$$

If $p > 1$, then $1 - p < 0$, so $N^{1-p} \to 0$ as $N \to \infty$, and

$$\int_1^\infty \frac{dx}{x^p} = \lim_{N \to \infty} \frac{1}{1-p} \left[N^{1-p} - 1 \right] = \frac{1}{1-p}[-1] = \frac{1}{p-1}$$

If $p < 1$, then $1 - p > 0$, so $N^{1-p} \to \infty$ as $N \to \infty$, and

$$\int_1^\infty \frac{dx}{x^p} = \lim_{N \to \infty} \frac{1}{1-p} \left[N^{1-p} - 1 \right] = \infty$$

Thus, the improper integral diverges for $p \leq 1$ and converges for $p > 1$. ■

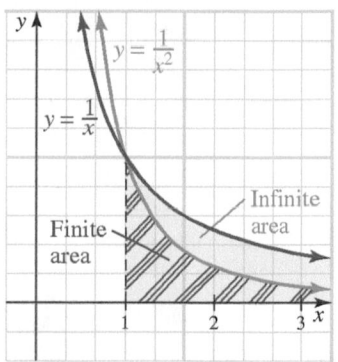

Figure 7.16 An unbounded region may have either a finite or an infinite area

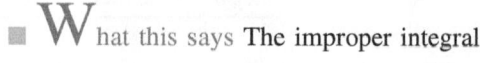

hat this says The improper integral

$$\int_1^\infty \frac{dx}{x^p} = \begin{cases} \dfrac{1}{p-1} & \text{if } p > 1 \\ \infty & \text{if } p \leq 1 \end{cases}$$

Example 4   Improper integral using l'Hôpital's rule and integration by parts

Evaluate $\displaystyle\int_0^\infty xe^{-2x}\,dx$.

Solution

$$\int_0^\infty xe^{-2x}\,dx = \lim_{N\to\infty}\int_0^N \underbrace{x}_{u}\,\underbrace{e^{-2x}\,dx}_{dv}$$

> By parts,
> Let $u = x$, $dv = e^{-2x}\,dx$
> $\quad du = dx\quad v = \frac{1}{-2}e^{-2x}$

$$= \lim_{N\to\infty}\left[\left(\underbrace{x}_{u}\,\underbrace{\frac{e^{-2x}}{-2}}_{v}\right)\Bigg|_0^N - \int_0^N \underbrace{\frac{1}{-2}e^{-2x}}_{v}\,\underbrace{dx}_{du}\right]$$

$$= \lim_{N\to\infty}\left[\frac{-xe^{-2x}}{2} - \frac{e^{-2x}}{4}\right]\Bigg|_0^N$$

$$= \lim_{N\to\infty}\left[-\frac{1}{2}Ne^{-2N} - \frac{1}{4}e^{-2N} + 0 + \frac{1}{4}\right] \qquad \lim_{N\to\infty}e^{-2N} = 0$$

$$= -\frac{1}{2}\lim_{N\to\infty}\left(\frac{N}{e^{2N}}\right) + \frac{1}{4} \qquad\qquad l'H\hat{o}pital's\ rule$$

$$= -\frac{1}{2}\lim_{N\to\infty}\left(\frac{1}{2e^{2N}}\right) + \frac{1}{4}$$

$$= \frac{1}{4}$$

Example 5   Gabriel's horn: a solid with a finite volume but infinite surface area

Gabriel's horn is the name given to the solid formed by revolving about the x-axis the unbounded region under the curve $y = \dfrac{1}{x}$ for $x \geq 1$. Show that this solid has finite volume but infinite surface area.

You can fill Gabriel's horn with a finite amount of paint, but it takes an infinite amount of paint to color its inside surface!

Solution   We will find the volume, V, by using disks, as shown in Figure 7.17.

$$V = \pi\int_1^\infty (x^{-1})^2\,dx = \pi\lim_{N\to\infty}\int_1^N x^{-2}\,dx = \pi\lim_{N\to\infty}\left(-\frac{1}{N}+1\right) = \pi$$

We also find the surface area, S:

$$S = 2\pi\int_1^\infty f(x)\sqrt{1+[f'(x)]^2}\,dx$$

$$= 2\pi\int_1^\infty \frac{1}{x}\sqrt{1+\frac{1}{x^4}}\,dx$$

$$= 2\pi\int_1^\infty \frac{1}{x}\sqrt{\frac{x^4+1}{x^4}}\,dx$$

> Let $u = x^2$, then $du = 2x\,dx$

$$= 2\pi\int_1^\infty \frac{\sqrt{u^2+1}}{2u^2}\,du$$

$$= \pi\lim_{N\to\infty}\int_1^N \frac{\sqrt{u^2+1}}{u^2}\,du$$

$$= \pi\lim_{N\to\infty}\left[\frac{-\sqrt{u^2+1}}{u} + \ln\left(u+\sqrt{u^2+1}\right)\right]_1^N \qquad \text{Formula 97}$$

$$= \infty$$

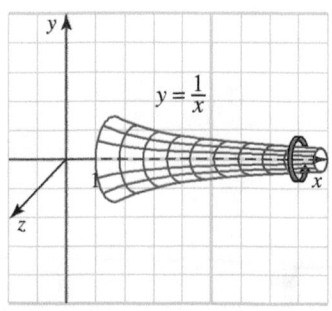

Figure 7.17 Interactive
Gabriel's horn

We can also define an improper integral on an interval that is unbounded to the left or on the entire real number line.

IMPROPER INTEGRAL TYPE I (Extended) Let b be a fixed number and assume $\int_t^b f(x)\,dx$ exists for all $t < b$. Then if $\lim\limits_{t \to -\infty} \int_t^b f(x)\,dx$ exists, we define the **improper integral**

$$\int_{-\infty}^b f(x)\,dx = \lim_{t \to -\infty} \int_t^b f(x)\,dx$$

The improper integral $\int_{-\infty}^b f(x)\,dx$ is said to **converge** if this limit is a finite number and to **diverge** otherwise. If both

$$\int_a^\infty f(x)\,dx \qquad \text{and} \qquad \int_{-\infty}^a f(x)\,dx$$

converge for some number a, the improper integral of $f(x)$ on the entire x-axis is defined by

$$\int_{-\infty}^\infty f(x)\,dx = \int_{-\infty}^a f(x)\,dx + \int_a^\infty f(x)\,dx$$

We say the integral on the left converges if both the integrals on the right converge, and it diverges otherwise.

Example 6   Improper integrals

Evaluate the following integrals.

a. $\displaystyle\int_1^\infty 2x^{-3}\,dx$ **b.** $\displaystyle\int_{-\infty}^{-3} \frac{dx}{x}$ **c.** $\displaystyle\int_{-\infty}^\infty xe^{-x^2}\,dx$

Solution

a. This is an improper integral at the upper limit.

$$\int_1^\infty 2x^{-3}\,dx = \lim_{b \to \infty} \int_1^b 2x^{-3}\,dx$$

$$= \lim_{b \to \infty} (-x^{-2})\Big|_1^b$$

$$= \lim_{b \to \infty} (-b^{-2} + 1)$$

$$= 1$$

b. This is an improper integral at the lower limit.

$$\int_{-\infty}^{-3} \frac{dx}{x} = \lim_{a \to -\infty} \int_a^{-3} x^{-1}\,dx$$

$$= \lim_{a \to -\infty} (\ln |x|)\Big|_a^{-3}$$

$$= \lim_{a \to -\infty} (\ln 3 - \ln |a|)$$

This limit does not exist, so we say that the integral diverges.

c. This is an improper integral at both limits, so we choose some point, say 0, between $-\infty$ and ∞.

Be sure that when an integral is improper for more than one reason (in this case because it starts at minus infinity and ends at (positive) infinity), you must break the integral into pieces. Each of these improper integrals must converge for the entire integral to converge.

$$\int_{-\infty}^{\infty} xe^{-x^2}\,dx = \int_{-\infty}^{0} xe^{-x^2}\,dx + \int_{0}^{\infty} xe^{-x^2}\,dx$$

$$= \lim_{a\to-\infty} \int_{a}^{0} xe^{-x^2}\,dx + \lim_{b\to\infty} \int_{0}^{b} xe^{-x^2}\,dx$$

$$= \lim_{a\to-\infty} \int_{x=a}^{x=0} \frac{-1}{2}e^{u}\,du + \lim_{b\to\infty} \int_{x=0}^{x=b} \frac{-1}{2}e^{u}\,du \qquad \boxed{\begin{array}{l}\text{Let } u = -x^2 \\ du = -2x\,dx\end{array}}$$

$$= \lim_{a\to-\infty} \left(-\frac{1}{2}e^{-x^2}\right)\Big|_{a}^{0} + \lim_{b\to\infty} \left(-\frac{1}{2}e^{-x^2}\right)\Big|_{0}^{b}$$

$$= \lim_{a\to-\infty} \left(-\frac{1}{2} + \frac{1}{2}e^{-a^2}\right) + \lim_{b\to\infty} \left(-\frac{1}{2}e^{-b^2} + \frac{1}{2}\right)$$

$$= -\frac{1}{2} + 0 - 0 + \frac{1}{2}$$

$$= 0$$

Example 7 Improper integral to negative infinity

a. Evaluate $\displaystyle\int_{-\infty}^{0} \frac{dx}{1+x^2}$.

b. How would you expect the value of $\displaystyle\int_{-\infty}^{\infty} \frac{dx}{1+x^2}$ to be related to the improper integral in **a**? Verify your conjecture by evaluating this integral.

Solution

a.
$$\int_{-\infty}^{0} \frac{dx}{1+x^2} = \lim_{t\to-\infty} \int_{t}^{0} \frac{dx}{1+x^2}$$

$$= \lim_{t\to-\infty} \tan^{-1} x \Big|_{t}^{0}$$

$$= \lim_{t\to-\infty} \left[\tan^{-1} 0 - \tan^{-1} t\right]$$

$$= 0 - \left(-\frac{\pi}{2}\right)$$

$$= \frac{\pi}{2}$$

b. Since $y(x) = y(-x)$, we see that $y = \dfrac{1}{1+x^2}$ is symmetric about the y-axis, and it is reasonable to conjecture that the integral in part **b** is twice the integral in part **a**. We have

$$\int_{-\infty}^{\infty} \frac{dx}{1+x^2} = \lim_{t\to-\infty} \int_{t}^{0} \frac{dx}{1+x^2} + \lim_{N\to\infty} \int_{0}^{N} \frac{dx}{1+x^2}$$

$$= \frac{\pi}{2} + \lim_{N\to\infty} (\tan^{-1} N - \tan^{-1} 0)$$

$$= \frac{\pi}{2} + \left(\frac{\pi}{2} - 0\right)$$

$$= \pi$$

We see this integral is indeed twice that in part **a**, as conjectured.

Example 8   An improper polar integral

Evaluate $\int_0^\infty \frac{1}{2} e^{-2\theta} d\theta$.

Solution
$$\int_0^\infty \frac{1}{2} e^{-2\theta} d\theta = \lim_{N \to \infty} \int_0^N \frac{1}{2} e^{-2\theta} d\theta$$

$$= \lim_{N \to \infty} \left(-\frac{1}{4} e^{-2\theta} \right) \Big|_0^N$$

$$= \lim_{N \to \infty} \left[-\frac{1}{4} e^{-2N} - \left(-\frac{1}{4} \right) \right]$$

$$= \frac{1}{4}$$

Improper Integrals with Unbounded Integrands

A function f is **unbounded** at c if it has arbitrarily large values near c. Geometrically, this occurs when the line $x = c$ is a vertical asymptote to the graph of f at c, as shown in Figure 7.18.

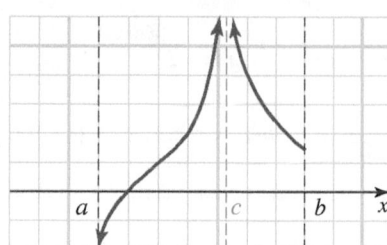 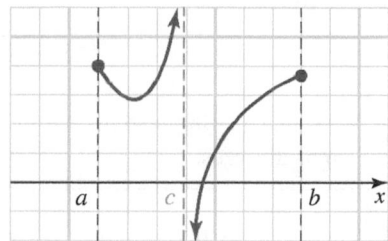

Figure 7.18   Two functions that are unbounded at c

If f is unbounded at c and $a \le c \le b$, the Riemann integral $\int_a^b f(x)\,dx$ is not even defined (only bounded functions are Riemann-integrable). However, it may still be possible to define $\int_a^b f(x)\,dx$ as an improper integral in certain cases.

Let us examine a specific problem. Suppose

$$f(x) = \frac{1}{\sqrt{x}} \qquad \text{for } 0 < x \le 1$$

Then f is unbounded at $x = 0$ and $\int_0^1 f(x)\,dx$ is not defined. However,

$$f(x) = \frac{1}{\sqrt{x}}$$

is continuous on every interval $[t, 1]$ for $t > 0$, as shown in Figure 7.19.

For any such interval $[t, 1]$, we have

$$\int_t^1 \frac{dx}{\sqrt{x}} = \int_t^1 x^{-1/2}\,dx = 2\sqrt{x}\,\Big|_t^1 = 2 - 2\sqrt{t}$$

If we let $t \to 0$ through positive values, we see that

$$\lim_{t \to 0^+} \int_t^1 \frac{dx}{\sqrt{x}} = \lim_{t \to 0^+} (2 - 2\sqrt{t}) = 2$$

This is called a **convergent improper integral** with value 2, and it seems reasonable to say

$$\int_0^1 \frac{dx}{\sqrt{x}} = \lim_{t \to 0^+} \int_t^1 \frac{dx}{\sqrt{x}} = 2$$

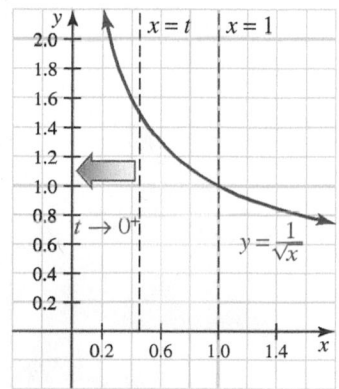

Figure 7.19 Interactive
Graph of $y = \dfrac{1}{\sqrt{x}}$

In this example, f is unbounded at the left endpoint of the interval of integration, but similar reasoning would apply if it were unbounded at the right endpoint or at an interior point. Here is a definition of this kind of improper integral.

IMPROPER INTEGRAL TYPE II   If f is unbounded at a and $\int_t^b f(x)\,dx$ exists for all t such that $a < t \leq b$, then

$$\int_a^b f(x)\,dx = \lim_{t \to a^+} \int_t^b f(x)\,dx$$

If the limit exists (as a finite number), we say that the improper integral **converges**; otherwise, the improper integral **diverges**. Similarly, if f is unbounded at b and $\int_a^t f(x)\,dx$ exists for all t such that $a \leq t < b$, then

$$\int_a^b f(x)\,dx = \lim_{t \to b^-} \int_a^t f(x)\,dx$$

If f is unbounded at c, where $a < c < b$, and the improper integrals $\int_a^c f(x)\,dx$ and $\int_c^b f(x)\,dx$ both converge, then

$$\int_a^b f(x)\,dx = \int_a^c f(x)\,dx + \int_c^b f(x)\,dx$$

We say that the integral on the left diverges if *either* (or both) of the integrals on the right diverges.

■ **W**hat this says: If a function is continuous on the open interval (a, b) but is unbounded at one of its endpoints of the interval, then replace that endpoint by t, evaluate the integral, and take the limit as t approaches that endpoint. On the other hand, if f is continuous on the interval $[a, b]$, except for some c in (a, b) where f has an infinite discontinuity, rewrite the integral as

$$\int_a^b f(x)\,dx = \int_a^c f(x)\,dx + \int_c^b f(x)\,dx$$

You will then need limits to evaluate each of the integrals on the right.

Example 9   Improper integral at a right endpoint

Evaluate $\displaystyle \int_0^1 \frac{dx}{(x-1)^{2/3}}$.

Solution $f(x) = (x-1)^{-2/3}$ is unbounded at the right endpoint of the interval of integration and is continuous on $[0, t]$ for any t with $0 \leq t < 1$. We find that

$$\int_0^1 \frac{dx}{(x-1)^{2/3}} = \lim_{t \to 1^-} \int_0^t \frac{dx}{(x-1)^{2/3}}$$

$$= \lim_{t \to 1^-} \left[3(x-1)^{1/3} \right] \Big|_0^t$$

$$= 3 \lim_{t \to 1^-} \left[(t-1)^{1/3} - (-1) \right]$$

$$= 3$$

That is, the improper integral converges and has the value 3.   ■

Example 10 Improper integral at a left endpoint

Evaluate $\displaystyle\int_{\pi/2}^{\pi} \sec x \ dx$.

Solution Because $\sec x$ is unbounded at the left endpoint $\frac{\pi}{2}$ of the interval of integration and is continuous on $[t, \pi]$ for any t with $\frac{\pi}{2} < t \le \pi$, we find that

$$\int_{\pi/2}^{\pi} \sec x \ dx = \lim_{t \to \left(\frac{\pi}{2}\right)^+} \int_{t}^{\pi} \sec x \ dx$$

$$= \lim_{t \to \left(\frac{\pi}{2}\right)^+} \ln|\sec x + \tan x| \Big|_{t}^{\pi}$$

$$= \lim_{t \to \left(\frac{\pi}{2}\right)^+} [\ln|-1 + 0| - \ln|\sec t + \tan t|]$$

$$= -\infty$$

Thus, the integral diverges.

Example 11 Improper integral at an interior point

Evaluate $\displaystyle\int_{0}^{3} (x - 2)^{-1} \ dx$.

Solution The integral is improper because the integrand is unbounded at $x = 2$. If the improper integral converges, we have

$$\int_{0}^{3} (x - 2)^{-1} dx = \int_{0}^{2} (x - 2)^{-1} dx + \int_{2}^{3} (x - 2)^{-1} dx$$

$$= \lim_{t \to 2^-} \int_{0}^{t} (x - 2)^{-1} dx + \lim_{t \to 2^+} \int_{t}^{3} (x - 2)^{-1} dx$$

If either of these limits fails to exist, then the original integral diverges.

$$= \lim_{t \to 2^-} \ln|x - 2| \Big|_{0}^{t} + \lim_{t \to 2^+} \int_{t}^{3} (x - 2)^{-1} dx$$

$$= \lim_{t \to 2^-} [\ln|t - 2| - \ln 2] + \lim_{t \to 2^+} \int_{t}^{3} (x - 2)^{-1} dx$$

$$= -\infty + \lim_{t \to 2^+} \int_{t}^{3} (x - 2)^{-1} dx$$

The original integral diverges.

☠ *A common mistake is to fail to notice the discontinuity at $x = 2$. It is WRONG to write*

$$\int_{0}^{3} (x - 2)^{-1} dx = \ln|x - 2| \big|_{0}^{3}$$

$$= \ln 1 - \ln 2 = -\ln 2$$

It is also WRONG to write
$\int_{0}^{3} (x - 2)^{-1} dx$
$= \int_{0}^{2} (x - 2)^{-1} dx + \int_{2}^{3} (x - 2)^{-1} dx$
$= \ln|x - 2| \big|_{0}^{2} + \ln|x - 2| \big|_{2}^{3}$
$= \ln 0 - \ln 2 + \ln 1 - \ln 0$
$= -\ln 2 + \infty - \infty$, *since*
$\infty - \infty$ *is indeterminate. Notice that either of these two mistakes leads to the conclusion that the integral converges, which, as you can see, is incorrect. You must also be cautious in using computer software with improper integrals, because it may not detect that the integral is improper.* ☠

Comparison Test for Convergence or Divergence

Sometimes it is difficult or even impossible to find the exact value of an improper integral, yet it is important to know whether or not it converges. If it converges we can often use numerical techniques to estimate its value, but first we must find out whether it converges. In those cases, the following result, stated for Improper Integrals of the First Type, can be useful. A similar theorem for Improper Integrals of the Second Type can be stated.

Theorem 7.2   Comparison test for improper integrals

Suppose f and g are continuous functions such that $f(x) \geq g(x) \geq 0$ for $x \geq a$. If $\int_a^\infty f(x)\,dx$ converges, then $\int_a^\infty g(x)\,dx$ converges to a value less than or equal to the integral of $f(x)$. If $\int_a^\infty g(x)\,dx$ diverges than $\int_a^\infty f(x)\,dx$ diverges.

Proof: We omit the proof of this but it is similar to the proof of the dominance rule (Theorem 5.6 in Section 5.3). See the illustration in the next example for an indication why it works.   ◆

Example 12   Comparison test

Show that $\displaystyle\int_1^\infty e^{-x^2}\,dx$ converges.

Solution   We can't evaluate this integral directly as the antiderivative of e^{-x^2} cannot be stated as an elementary function. However, $e^{-x^2} < e^{-x}$ for $x > 1$ (see Figure 7.20), so

$$\int_1^\infty e^{-x^2}\,dx < \int_1^\infty e^{-x}\,dx = \lim_{t \to \infty} \int_1^t e^{-x}\,dx = \lim_{t \to \infty}\left(-e^{-t} + e^{-1}\right) = e^{-1}$$

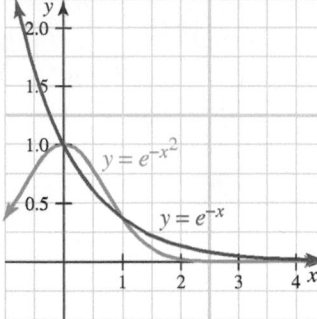

Figure 7.20 The graph of $y = e^{-x^2}$ is under the graph of $y = e^{-x}$ for $x > 1$

Notice from Figure 7.20 that the graph of $y = e^{-x^2}$ is under the graph of $y = e^{-x}$ for $x > 1$, so that Theorem 7.2 gives the result.   ■

PROBLEM SET 7.7

Level 1

1. ■ What does this say? What is an improper integral?
2. ■ What does this say? Discuss the different types of improper integrals.

In Problems 3-42, either show that the improper integral converges and find its value, or show that it diverges.

3. $\displaystyle\int_1^\infty \frac{dx}{x^3}$

4. $\displaystyle\int_1^\infty \frac{dx}{\sqrt[3]{x}}$

5. $\displaystyle\int_1^\infty \frac{dx}{x^{0.99}}$

6. $\displaystyle\int_1^\infty \frac{dx}{\sqrt{x}}$

7. $\displaystyle\int_1^\infty \frac{dx}{x^{1.01}}$

8. $\displaystyle\int_1^\infty x^{-2/3}\,dx$

9. $\displaystyle\int_3^\infty \frac{dx}{2x-1}$

10. $\displaystyle\int_3^\infty \frac{dx}{\sqrt[3]{2x-1}}$

11. $\displaystyle\int_3^\infty \frac{dx}{(2x-1)^2}$

12. $\displaystyle\int_0^\infty e^{-x}\,dx$

13. $\displaystyle\int_1^\infty \frac{x^2\,dx}{(x^3+2)^2}$

14. $\displaystyle\int_1^\infty \frac{x^2\,dx}{x^3+2}$

15. $\displaystyle\int_1^\infty \frac{x^2\,dx}{\sqrt{x^3+2}}$

16. $\displaystyle\int_0^\infty xe^{-x^2}\,dx$

17. $\displaystyle\int_1^\infty \frac{e^{-\sqrt{x}}}{\sqrt{x}}\,dx$

18. $\displaystyle\int_0^\infty xe^{-x}\,dx$

19. $\displaystyle\int_0^\infty 5xe^{10-x}\,dx$

20. $\displaystyle\int_1^\infty \frac{\ln x\,dx}{x}$

21. $\displaystyle\int_2^\infty \frac{dx}{x\ln x}$

22. $\displaystyle\int_2^\infty \frac{dx}{x\sqrt{\ln x}}$

23. $\displaystyle\int_{-\infty}^0 \frac{2x\,dx}{x^2+1}$

24. $\displaystyle\int_1^\infty \frac{x\,dx}{(1+x^2)^2}$

25. $\displaystyle\int_{-\infty}^0 \frac{dx}{\sqrt{2-x}}$

26. $\displaystyle\int_{-\infty}^4 \frac{dx}{(5-x)^2}$

27. $\displaystyle\int_{-\infty}^\infty xe^{-|x|}\,dx$

28. $\displaystyle\int_{-\infty}^\infty \frac{dx}{x^2+1}$

29. $\displaystyle\int_0^1 \frac{dx}{x^{1/5}}$

30. $\displaystyle\int_0^4 \frac{dx}{x\sqrt{x}}$

31. $\displaystyle\int_0^1 \frac{dx}{(1-x)^{1/2}}$

32. $\displaystyle\int_0^2 \frac{dx}{(1-x)^2}$

33. $\displaystyle\int_{-1}^1 \frac{e^x}{\sqrt[3]{1-e^x}}\,dx$

34. $\displaystyle\int_{-\infty}^\infty \frac{3x\,dx}{(3x^2+2)^3}$

35. $\displaystyle\int_0^1 \ln x\,dx$

36. $\displaystyle\int_1^\infty \ln x\,dx$

37. $\displaystyle\int_e^\infty \frac{dx}{x(\ln x)^2}$

38. $\displaystyle\int_0^1 \frac{x\,dx}{1-x^2}$

39. $\displaystyle\int_0^1 e^{-\frac{1}{2}\ln x}\,dx$

40. $\displaystyle\int_0^\infty \frac{dx}{e^x+e^{-x}}$

41. $\displaystyle\int_0^{\pi/3} \frac{\sec^2 x\,dx}{1-\tan x}$

42. $\displaystyle\int_0^{\pi/2} \frac{\sin x\,dx}{\sqrt[3]{1-2\cos x}}$

Level 2

43. Find the area of the unbounded region between the x-axis and the curve

$$y = \frac{2}{(x-4)^3}$$

for $x \geq 6$.

44. Find the area of the unbounded region between the x-axis and the curve

$$y = \frac{2}{(x-4)^3}$$

for $x \leq 2$.

45. The total amount of radioactive material present in the atmosphere at time T is modeled by

$$A = \int_0^T P e^{-rt} \, dt$$

where P is a constant and t is the number of years. Suppose a recent United Nations publication indicates that, at the present time, $r = 0.002$ and $P = 200$ millirads. Estimate the total future buildup of radioactive material in the atmosphere if these values remain constant.

46. Suppose that an oil well produces $P(t)$ thousand barrels of crude oil per month according to the formula

$$P(t) = 100e^{-0.02t} - 100e^{-0.1t}$$

where t is the number of months the well has been in production. What is the total amount of oil produced by the oil well?

47. Find all values of p for which

$$\int_2^\infty \frac{dx}{x(\ln x)^p}$$

converges, and find the value of the integral when it exists.

48. Find all values of p for which

$$\int_0^1 \frac{dx}{x^p}$$

converges, and find the value of the integral when it exists.

49. Find all values of p for which

$$\int_0^{1/2} \frac{dx}{x(\ln x)^p}$$

converges, and find the value of the integral when it exists.

50. Think Tank Problem Discuss the calculation

$$\int_{-1}^1 \frac{dx}{x^2} = \frac{-1}{x}\bigg|_{-1}^1$$

$$= -[1 - (-1)]$$

$$= -2$$

Is the calculation correct? Explain.

51. Journal Problem (*College Mathematics Journal**) Peter Lindstrom of North Lake College in Irving, Texas, had a student who handled an ∞/∞ form as follows:

$$\int_1^\infty (x-1)e^{-x}\,dx = \int_1^\infty \frac{x-1}{e^x}\,dx$$

$$= \int_1^\infty \frac{1}{e^x}\,dx \quad \text{l'Hôpital's rule}$$

$$= \frac{1}{e}$$

What is wrong, if anything, with this student's solution?

52. Find $\displaystyle\int_0^2 f(x)\,dx$, where

$$f(x) = \begin{cases} \dfrac{1}{\sqrt[4]{x^3}} & \text{for } 0 \leq x \leq 1 \\[3mm] \dfrac{1}{\sqrt[4]{(x-1)^3}} & \text{for } 1 < x < 2 \end{cases}$$

Level 3

*The **Laplace transform** of the function f is defined by the improper integral*

$$F(s) = \mathcal{L}\{f(t)\} = \int_0^\infty e^{-st} f(t)\,dt$$

where s is a constant. This notation is used in Problems 53-60.

53. Show that $\mathcal{L}\{af + bg\} = a\mathcal{L}\{f\} + b\mathcal{L}\{g\}$.

54. Show that for constant a (with $s - a > 0$):

a. $\mathcal{L}\{e^{at}\} = \dfrac{1}{s-a}$ **b.** $\mathcal{L}\{a\} = \dfrac{a}{s}$

c. $\mathcal{L}\{t\} = \dfrac{1}{s^2}$ **d.** $\mathcal{L}\{t^n\} = \dfrac{n!}{s^{n+1}}$

e. $\mathcal{L}\{\cos at\} = \dfrac{s}{s^2 + a^2}$ **f.** $\mathcal{L}\{\sin at\} = \dfrac{a}{s^2 + a^2}$

---

55. If $F(s) = \mathcal{L}\{f(t)\}$, show that

$$\mathcal{L}\{e^{at}f(t)\} = F(s - a)$$

56. Let $\mathcal{L}^{-1}\{F(s)\}$ denote the *inverse* Laplace transform of $F(s)$; that is, the function $f(t)$ such that $\mathcal{L}\{f(t)\} = F(s)$. Find the following inverse Laplace transformations (Problems 53 and 54 may help).

a. $\mathcal{L}^{-1}\left\{\dfrac{5}{s}\right\}$ **b.** $\mathcal{L}^{-1}\left\{\dfrac{s+2}{s^2+4}\right\}$

c. $\mathcal{L}^{-1}\left\{\dfrac{2s^2 - 3s + 3}{s^2(s-1)}\right\}$

   Hint: Use partial fraction decomposition.

d. $\mathcal{L}^{-1}\left\{\dfrac{3s^3 + 2s^2 - 3s - 17}{(s^2+4)(s^2-1)}\right\}$

57. a. If $F(s) = \mathcal{L}\{f(t)\}$, show that

$$\mathcal{L}\{tf(t)\} = -F'(s)$$

You may assume that

$$\frac{d}{ds}\int_0^\infty e^{-st}f(t)dt = \int_0^\infty \frac{d}{ds}\left[e^{-st}f(t)\right]dt$$

b. Use the result to find $\mathcal{L}\{t\cos 2t\}$.

58. Use the result in Problem 55, along with those of Problem 54, to find the following Laplace transforms and inverse transforms.

a. $\mathcal{L}\{t^3 e^{-2t}\}$ **b.** $\mathcal{L}\{e^{-3t}\cos 2t\}$

c. $\mathcal{L}^{-1}\left\{\dfrac{5}{(s-1)^2}\right\}$ **d.** $\mathcal{L}^{-1}\left\{\dfrac{4s}{s^2+4s+5}\right\}$

   Hint: Complete the square.

59. What is $\mathcal{L}\{t^2 f(t)\}$? What about $\mathcal{L}\{t^n f(t)\}$?

60. Let $\mathcal{L}\{f\} = F(s)$. Suppose that f is continuous on $[0, \infty)$ and that f' is piecewise continuous on $[0, k]$ for every positive number k. Further suppose that

$$\lim_{k\to\infty} e^{-sk}f(k) = 0$$

if $s > 0$.

a. Show that $\mathcal{L}\{f'\} = sF(s) - f(0)$.
   Hint: Use integration by parts.

b. Apply the result in part a to find $\mathcal{L}\{y\}$, where y satisfies the initial value problem

$$y' + 3y = \sin 2t$$

with $y(0) = 2$.

c. Solve the initial value problem in part **b** by using inverse Laplace transforms to obtain y.

7.8 HYPERBOLIC AND INVERSE HYPERBOLIC FUNCTIONS

IN THIS SECTION: *Hyperbolic functions, derivatives and integrals involving hyperbolic functions, inverse hyperbolic functions*
Hyperbolic and inverse hyperbolic functions are introduced and studied; integration formulas to be used throughout the rest of the text are derived.

Hyperbolic Functions

In physics, it is shown that a heavy, flexible cable (for example, a power line) that is suspended between two points at the same height assumes the shape of a curve called a **catenary** (see Figure 7.21), with an equation of the form

Figure 7.21 The hanging cable problem

$$y = \frac{a}{2}(e^{x/a} + e^{-x/a})$$

This is one of several important applications that involve sums of exponential functions. The goal of this section is to study such sums and their inverses.

In certain ways, the functions we shall study are analogous to the trigonometric functions, and they have essentially the same relationship to the hyperbola that the trigonometric functions have to the circle. For this reason, these functions are called **hyperbolic functions**. Three basic hyperbolic functions are the **hyperbolic sine** (denoted $\sinh x$ and pronounced "cinch"), the **hyperbolic cosine** ($\cosh x$; pronounced "kosh"), and the **hyperbolic tangent** ($\tanh x$; pronounced "tansh"). They are defined as follows.

HYPERBOLIC FUNCTIONS

$$\sinh x = \frac{e^x - e^{-x}}{2} \qquad \text{for all } x$$

$$\cosh x = \frac{e^x + e^{-x}}{2} \qquad \text{for all } x$$

$$\tanh x = \frac{\sinh x}{\cosh x} = \frac{e^x - e^{-x}}{e^x + e^{-x}} \qquad \text{for all } x$$

Graphs of the three basic hyperbolic functions are shown in Figure 7.22.

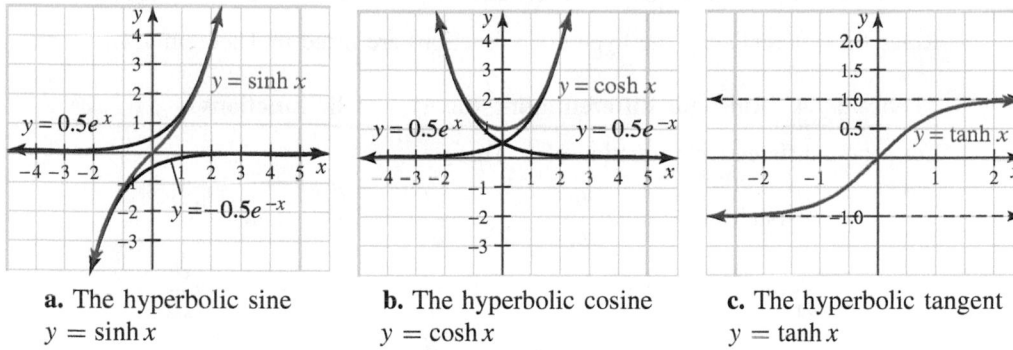

a. The hyperbolic sine $y = \sinh x$ **b.** The hyperbolic cosine $y = \cosh x$ **c.** The hyperbolic tangent $y = \tanh x$

Figure 7.22   Graphs of the three basic hyperbolic functions

The list of properties in the following theorem suggests that the basic hyperbolic functions are analogous to the trigonometric functions.

Theorem 7.3   Properties of the hyperbolic functions

$$\cosh^2 x - \sinh^2 x = 1$$

$$\sinh(-x) = -\sinh x \qquad\qquad \sinh x \text{ is odd}$$

$$\cosh(-x) = \cosh x \qquad\qquad \cosh x \text{ is even}$$

$$\tanh(-x) = -\tanh x \qquad\qquad \tanh x \text{ is odd}$$

$$\sinh(x + y) = \sinh x \cosh y + \cosh x \sinh y$$

$$\cosh(x + y) = \cosh x \cosh y + \sinh x \sinh y$$

Proof: We will verify the first identity and leave the others for the reader.

$$\cosh^2 x - \sinh^2 x = \left(\frac{e^x + e^{-x}}{2}\right)^2 - \left(\frac{e^x - e^{-x}}{2}\right)^2$$

$$= \frac{e^{2x} + 2 + e^{-2x}}{4} - \frac{e^{2x} - 2 + e^{-2x}}{4}$$

$$= 1$$

A major difference between the trigonometric and hyperbolic functions is that the trigonometric functions are periodic, but the hyperbolic functions are not. ◆

There are three additional hyperbolic functions: the **hyperbolic cotangent** (coth x), the **hyperbolic secant** (sech x), and the **hyperbolic cosecant** (csch x). These functions are defined as follows:

$$\coth x = \frac{1}{\tanh x} = \frac{e^x + e^{-x}}{e^x - e^{-x}} \qquad \operatorname{sech} x = \frac{1}{\cosh x} = \frac{2}{e^x + e^{-x}} \qquad \operatorname{csch} x = \frac{1}{\sinh x} = \frac{2}{e^x - e^{-x}}$$

TECHNOLOGY NOTE: Some of the software packages show these in simplified form as:

$$\coth x = \frac{e^{2x} + 1}{e^{2x} - 1} \qquad \operatorname{sech} x = \frac{2e^x}{e^{2x} + 1} \qquad \operatorname{csch} x = \frac{2e^x}{e^{2x} - 1}$$

Two identities involving these functions are

$$\operatorname{sech}^2 x = 1 - \tanh^2 x \qquad \text{and} \qquad \operatorname{csch}^2 x = \coth^2 x - 1$$

You will be asked to verify these identities in the problems.

Derivatives and Integrals Involving Hyperbolic Functions

Rules for differentiating the hyperbolic functions are listed in Theorem 7.4.

Theorem 7.4   Rules for differentiating the hyperbolic functions

Let u be a differentiable function of x. Then:

$$\frac{d}{dx}(\sinh u) = \cosh u \frac{du}{dx} \qquad\qquad \frac{d}{dx}(\cosh u) = \sinh u \frac{du}{dx}$$

$$\frac{d}{dx}(\tanh u) = \operatorname{sech}^2 u \frac{du}{dx} \qquad\qquad \frac{d}{dx}(\coth u) = -\operatorname{csch}^2 u \frac{du}{dx}$$

$$\frac{d}{dx}(\operatorname{sech} u) = -\operatorname{sech} u \tanh u \frac{du}{dx} \qquad \frac{d}{dx}(\operatorname{csch} u) = -\operatorname{csch} u \coth u \frac{du}{dx}$$

Proof: Each of these rules can be obtained by differentiating the exponential functions that make up the appropriate hyperbolic function. For example, to differentiate $\sinh x$, we use the definition of $\sinh x$:

$$\frac{d}{dx}(\sinh x) = \frac{d}{dx}\left(\frac{e^x - e^{-x}}{2}\right)$$

$$= \frac{1}{2}e^x - \frac{1}{2}e^{-x}(-1)$$

$$= \frac{1}{2}(e^x + e^{-x})$$

$$= \cosh x$$

The proofs for the other derivatives can be handled similarly.   ◆

Example 1   Derivatives involving hyperbolic functions

Find $\dfrac{dy}{dx}$ for each of the following functions:

a. $y = \cosh Ax$, A is a constant   **b.** $y = \tanh(x^2 + 1)$ **c.** $y = \ln(\sinh x)$

Solution

a. We have

$$\frac{d}{dx}(\cosh Ax) = \sinh(Ax)\frac{d}{dx}(Ax) = A \sinh Ax$$

b. We find that

$$\frac{d}{dx}\left[\tanh(x^2+1)\right] = \operatorname{sech}^2(x^2+1)\frac{d}{dx}(x^2+1) = 2x\operatorname{sech}^2(x^2+1)$$

c. Using the chain rule, with $u = \sinh x$, we obtain

$$\frac{d}{dx}\left[\ln(\sinh x)\right] = \frac{1}{\sinh x}\frac{d}{dx}(\sinh x) = \frac{1}{\sinh x}(\cosh x) = \coth x$$

Each differentiation formula for hyperbolic functions corresponds to an integration formula. These formulas are listed in the following theorem.

Theorem 7.5   Rules for integrating the hyperbolic functions

$$\int \sinh x\ dx = \cosh x + C \qquad \int \cosh x\ dx = \sinh x + C$$

$$\int \operatorname{sech}^2 x\ dx = \tanh x + C \qquad \int \operatorname{csch}^2 x\ dx = -\coth x + C$$

$$\int \operatorname{sech} x \tanh x\ dx = -\operatorname{sech} x + C \qquad \int \operatorname{csch} x \coth x\ dx = -\operatorname{csch} x + C$$

Proof: The proof of each of these formulas follows directly from the corresponding derivative formula.   ♦

Example 2   Integration involving hyperbolic forms

Find each of the following integrals:

a. $\displaystyle\int \cosh^3 x \sinh x\ dx$ **b.** $\displaystyle\int x \operatorname{sech}^2(x^2)\ dx$ **c.** $\displaystyle\int \tanh x\ dx$

Solution

a. $\displaystyle\int \cosh^3 x \sinh x\ dx = \int u^3\ du = \frac{u^4}{4} + C = \frac{1}{4}\cosh^4 x + C$

$\boxed{\text{Let } u = \cosh x, \text{ then } du = \sinh x\ dx}$

b.

$$\int x \operatorname{sech}^2(x^2)\ dx = \int \operatorname{sech}^2(x^2)(x\ dx) \quad \boxed{\text{Let } u = x^2, \text{ then } du = 2x\ dx}$$

$$= \int \operatorname{sech}^2 u\left(\frac{1}{2}\ du\right)$$

$$= \frac{1}{2}\tanh u + C$$

$$= \frac{1}{2}\tanh x^2 + C$$

c. $\displaystyle\int \tanh x\ dx = \int \frac{\sinh x\ dx}{\cosh x} = \int \frac{du}{u} = \ln|u| + C = \ln(\cosh x) + C$

$\boxed{\text{Let } u = \cosh x, \text{ then } du = \sinh x\ dx}$

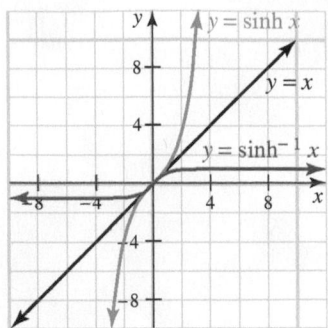

Figure 7.23 The graph of $y = \sinh^{-1} x$.
The graph of $y = \sinh^{-1} x$ is obtained by reflecting $y = \sinh x$ in the line $y = x$

Inverse Hyperbolic Functions

Inverse hyperbolic functions are also of interest primarily because they enable us to express certain integrals in simple terms. Because $\sinh x$ is continuous and strictly monotonic (increasing), it is one-to-one and has an inverse, which is defined by

$$y = \sinh^{-1} x \quad \text{if and only if} \quad x = \sinh y$$

for all x and y. This is called the **inverse hyperbolic sine** function, and its graph is obtained by reflecting the graph of $y = \sinh x$ in the line $y = x$, as shown in Figure 7.23. Other inverse hyperbolic functions are defined similarly (see Problem 58).

Because the hyperbolic functions are defined in terms of exponential functions, we may expect to be able to express the inverse hyperbolic functions in terms of logarithmic functions. We summarize these relationships in the following theorem.

Theorem 7.6   Logarithmic formulas for inverse hyperbolic functions

$$\sinh^{-1} x = \ln(x + \sqrt{x^2 + 1}), \text{ all } x \qquad \operatorname{csch}^{-1} x = \ln\left(\frac{1}{x} + \frac{\sqrt{1 + x^2}}{|x|}\right), \; x \neq 0$$

$$\cosh^{-1} x = \ln(x + \sqrt{x^2 - 1}), \; x \geq 1 \qquad \operatorname{sech}^{-1} x = \ln\left(\frac{1 + \sqrt{1 - x^2}}{x}\right), \; 0 < x \leq 1$$

$$\tanh^{-1} x = \frac{1}{2} \ln \frac{1 + x}{1 - x}, |x| < 1 \qquad \coth^{-1} x = \frac{1}{2} \ln \frac{x + 1}{x - 1}, |x| > 1$$

Proof: We will prove the first formula and leave the next two formulas for you to verify (see Problem 59). The other formulas are proved similarly.

Let $y = \sinh^{-1} x$; then its inverse is

$$x = \sinh y$$

$$x = \frac{1}{2}(e^y - e^{-y})$$

$$2x = e^y - \frac{1}{e^y}$$

$$e^{2y} - 2xe^y - 1 = 0 \qquad \textit{Quadratic formula with } a = 1, \; b = -2x, \; c = -1$$

$$e^y = \frac{2x \pm \sqrt{4x^2 + 4}}{2}$$

$$= x \pm \sqrt{x^2 + 1}$$

Because $e^y > 0$ for all y, the only solution is $e^y = x + \sqrt{x^2 + 1}$, and from the definition of logarithms,

$$y = \ln(x + \sqrt{x^2 + 1}) \qquad \blacklozenge$$

Differentiation and integration formulas involving inverse hyperbolic functions are listed in Theorem 7.7.

Theorem 7.7 **Differentiation and integration of the inverse hyperbolic functions**

$$\frac{d}{dx}(\sinh^{-1} u) = \frac{1}{\sqrt{1+u^2}}\frac{du}{dx} \qquad \int \frac{du}{\sqrt{1+u^2}} = \sinh^{-1} u + C$$

$$\frac{d}{dx}(\cosh^{-1} u) = \frac{1}{\sqrt{u^2-1}}\frac{du}{dx} \qquad \int \frac{du}{\sqrt{u^2-1}} = \cosh^{-1} u + C$$

$$\frac{d}{dx}(\tanh^{-1} u) = \frac{1}{1-u^2}\frac{du}{dx} \qquad \int \frac{du}{1-u^2} = \tanh^{-1} u + C$$

$$\frac{d}{dx}(\operatorname{csch}^{-1} u) = \frac{-1}{|u|\sqrt{1+u^2}}\frac{du}{dx} \qquad \int \frac{du}{u\sqrt{1+u^2}} = -\operatorname{csch}^{-1}|u| + C$$

$$\frac{d}{dx}(\operatorname{sech}^{-1} u) = \frac{-1}{u\sqrt{1-u^2}}\frac{du}{dx} \qquad \int \frac{du}{u\sqrt{1-u^2}} = -\operatorname{sech}^{-1}|u| + C$$

$$\frac{d}{dx}(\coth^{-1} u) = \frac{1}{1-u^2}\frac{du}{dx} \qquad \int \frac{du}{1-u^2} = \coth^{-1} u + C$$

Proof: The derivative of each inverse hyperbolic function can be found either by differentiating the appropriate logarithmic function or by using the definition in terms of hyperbolic functions. We will show how this is done for $y = \sinh^{-1} x$, and then will leave it for you to apply the chain rule for u, a differentiable function of x. By definition of inverse, we see

$$x = \sinh y \qquad \text{and} \qquad \frac{dx}{dy} = \cosh y$$

Thus,

$$\frac{d}{dx}(\sinh^{-1} x) = \frac{dy}{dx} = \frac{1}{\frac{dx}{dy}} = \frac{1}{\cosh y} = \frac{1}{\sqrt{1+\sinh^2 y}} = \frac{1}{\sqrt{1+x^2}}$$

since $\cosh^2 y - \sinh^2 y = 1$ and $\sinh y = x$. The integration formula that corresponds to this differentiation formula is

$$\int \frac{dx}{\sqrt{1+x^2}} = \sinh^{-1} x + C$$

The other differentiation and integration formulas follow similarly. ♦

Example 3 **Derivatives involving inverse hyperbolic functions**

Find $\dfrac{dy}{dx}$ for   **a.** $y = \sinh^{-1}(ax+b)$ **b.** $y = \cosh^{-1}(\sec x), 0 \le x < \frac{\pi}{2}$

Solution

a.
$$\frac{d}{dx}\left[\sinh^{-1}(ax+b)\right] = \frac{1}{\sqrt{1+(ax+b)^2}}\frac{d}{dx}(ax+b)$$
$$= \frac{a}{\sqrt{1+(ax+b)^2}}$$

b.
$$\frac{d}{dx}\left[\cosh^{-1}(\sec x)\right] = \frac{1}{\sqrt{\sec^2 x - 1}}\frac{d}{dx}(\sec x)$$
$$= \frac{\sec x \tan x}{\sqrt{\tan^2 x}} \qquad\qquad \tan x > 0 \text{ because } 0 \le x < \pi/2$$
$$= \sec x$$

The result of Example **3b** implies

$$\int \sec x \; dx = \cosh^{-1}(\sec x) + C$$

$$= \ln\left(\sec x + \sqrt{\sec^2 x - 1}\right) + C$$

$$= \ln(\sec x + \tan x) + C$$

Recall that the formula
$\int \sec x \; dx = \ln|\sec x + \tan x| + C$
was derived using algebraic
procedures in Section 7.1.

Notice that we do not need absolute values because $0 \le x < \frac{\pi}{2}$.

Example 4   Integral involving an inverse hyperbolic function

Evaluate $\displaystyle\int_0^1 \frac{dx}{\sqrt{1+x^2}}$.

Solution

$$\int_0^1 \frac{dx}{\sqrt{1+x^2}} = \left[\sinh^{-1} x\right]_0^1$$

$$= \ln(x + \sqrt{x^2 + 1})\Big|_0^1$$

$$= \ln(1 + \sqrt{2}) - \ln 1$$

$$= \ln(1 + \sqrt{2})$$

It may be a worthwhile exercise to evaluate this integral using a trigonometric substitution to obtain the same result. ∎

PROBLEM SET 7.8

Level 1

Evaluate (correct to four decimal places) the indicated hyperbolic or inverse hyperbolic functions in Problems 1-12.

1. $\sinh 2$ **2.** $\cosh 3$ **3.** $\tanh(-1)$

4. $\sinh^{-1} 0$ **5.** $\tanh^{-1} 0$ **6.** $\cosh^{-1} 1.5$

7. $\coth 1.2$ **8.** $\operatorname{sech} 1$ **9.** $\sinh(\ln 2)$

10. $\cosh(\ln 3)$ **11.** $\sec h^{-1} 0.2$ **12.** $\coth^{-1}(-3)$

Find $\dfrac{dy}{dx}$ in Problems 13-28.

13. $y = \sinh 3x$ **14.** $y = \sinh\sqrt{x}$

15. $y = \cosh(1 - 2x^2)$ **16.** $y = \cosh(2x^2 + 3x)$

17. $y = \sinh x^{-1}$ **18.** $y = \cosh^{-1} x^2$

19. $y = \sinh^{-1} x^3$ **20.** $y = x \tanh^{-1} 3x$

21. $y = \sinh^{-1}(\tan x)$ **22.** $y = \cosh^{-1}(\sec x)$

23. $y = \tanh^{-1}(\sin x)$ **24.** $y = \operatorname{sech}\left(\dfrac{1-x}{1+x}\right)$

25. $y = \dfrac{\sinh^{-1} x}{x}$ **26.** $y = \sinh^{-1} x - \sqrt{1+x^2}$

27. $y = x\cosh^{-1} x - \sqrt{x^2 - 1}$

28. $x\cosh y = y\sinh x + 5$

Compute the integrals in Problems 29-44.

29. $\displaystyle\int x\cosh(1 - x^2)dx$ **30.** $\displaystyle\int \dfrac{\operatorname{sech}^2(\ln x)dx}{x}$

31. $\displaystyle\int \dfrac{\sinh\frac{1}{x}dx}{x^2}$ **32.** $\displaystyle\int \coth x \; dx$

33. $\displaystyle\int \dfrac{dt}{\sqrt{9t^2 - 16}}$ **34.** $\displaystyle\int \dfrac{dx}{\sqrt{4x^2 + 16}}$

35. $\displaystyle\int \dfrac{\cos x \; dx}{\sqrt{1 + \sin^2 x}}$ **36.** $\displaystyle\int \dfrac{dt}{36 - 16t^2}$

37. $\displaystyle\int \dfrac{x^2 \; dx}{1 - x^6}$ **38.** $\displaystyle\int \dfrac{x \; dx}{\sqrt{1 + x^4}}$

39. $\displaystyle\int_2^3 \dfrac{dx}{1 - x^2}$ **40.** $\displaystyle\int_0^{\frac{1}{2}} \dfrac{dx}{1 - x^2}$

41. $\displaystyle\int_1^2 \dfrac{e^x \; dx}{\sqrt{e^{2x} - 1}}$ **42.** $\displaystyle\int_0^1 \dfrac{t^5 dt}{\sqrt{1 + t^{12}}}$

43. $\displaystyle\int_0^1 x\operatorname{sech}^2 x^2 \; dx$ **44.** $\displaystyle\int_0^{\ln 2} \sinh 3x \; dx$

45. Show that
 a. $\tanh(x + y) = \dfrac{\tanh x + \tanh y}{1 + \tanh x \tanh y}$
 b. $\sinh 2x = 2 \sinh x \cosh x$
 c. $\cosh 2x = \cosh^2 x + \sinh^2 x$

46. Show that
 a. $-1 < \tanh x < 1$ for all x
 b. $\lim\limits_{x \to \infty} \tanh x = 1$
 c. $\lim\limits_{x \to -\infty} \tanh x = -1$

47. Determine where the graph of $y = \tanh x$ is rising and falling and where it is concave up and concave down. Sketch the graph and compare with Figure 7.22**c**.

48. Sketch the graph of $y = \coth x$. Be sure to show key features such as intercepts, maxima and minima, and points of inflection.

49. First show that $\cosh x + \sinh x = e^x$, and then use this result to prove that

$$(\cosh x + \sinh x)^n = \cosh nx + \sinh nx$$

for positive integers n.

50. If $x = a \cosh t$ and $y = b \sinh t$ where a, b are positive constants and t is any number, show that

$$\frac{x^2}{a^2} - \frac{y^2}{b^2} = 1$$

51. a. Verify that $y = a \cosh cx + b \sinh cx$ satisfies the differential equation

$$y'' - c^2 y = 0$$

 b. Use part **a** to find a solution of the differential equation

$$y'' - 4y = 0$$

 subject to the initial conditions $y(0) = 1$, $y'(0) = 2$.

52. Find the volume V generated when the region bounded by the curves $y = \sinh x, y = \cosh x$, the y-axis, and the line $x = c$ ($c > 0$) is revolved about the x-axis. For what value of c does $V = 1$?

53. Find the length of the catenary

$$y = a \cosh \frac{x}{a}$$

between $x = -a$ and $x = a$.

54. Find the volume of the solid formed by revolving the region bounded by the curve $y = \tanh x$ on the interval $[0, 1]$ about the x-axis.

55. Find the surface area of the solid generated by revolving the curve $y = \cosh x$ on the interval $[-1, 1]$ about the x-axis.

56. Prove the following formulas
 a. $\operatorname{sech}^2 x + \tanh^2 x = 1$
 b. $\coth^2 x - \operatorname{csch}^2 x = 1$

57. Derive the differentiation formulas for $\cosh u$, $\tanh u$, and $\operatorname{sech} u$, where u is a differentiable function of x.

58. First give definitions for $\cosh^{-1} x$, $\tanh^{-1} x$, and $\operatorname{sech}^{-1} x$, and then derive the differentiation formulas for these functions by using these definitions.

59. Prove
 a. $\cosh^{-1} x = \ln(x + \sqrt{x^2 - 1})$, $x \geq 1$
 b. $\tanh^{-1} x = \dfrac{1}{2} \ln \dfrac{1 + x}{1 - x}$, $|x| < 1$

60. Show that

$$\int \frac{dx}{\sqrt{x^2 + a^2}} = \sinh^{-1} \frac{x}{a} + C$$

for constant $a > 0$. State and prove a similar formula for

$$\int \frac{dx}{\sqrt{x^2 - a^2}}$$

with $x > a > 0$.

CHAPTER 7 REVIEW

C*alculus is the most powerful weapon of thought yet devised by the wit of man.*

William. B. Smith

Infinitesimal Analysis, (1898)

Proficiency Examination

Concept Problems

1. Discuss the method of integration by substitution for both indefinite and definite integrals.
2. **a.** What is the formula for integration by parts?
   b. What is a reduction formula for integration?
3. **a.** When should you consider a trigonometric substitution?
   b. What are the substitutions to use when integrating rational trigonometric integrals?
4. When should you consider the method of partial fractions?
5. Outline a strategy for integration.
6. What is a first-order linear differential equation? Outline a procedure for solving such an equation.
7. What is an improper integral?
8. Define the hyperbolic sine, hyperbolic cosine, and hyperbolic tangent functions.
9. State the rules for differentiating and integrating the hyperbolic functions.
10. What are the logarithmic formulas for the inverse hyperbolic functions?
11. State the differentiation and integration formulas for the inverse hyperbolic functions.

Practice Problems

12. Evaluate each of the given numbers.

    a. $\tanh^{-1} 0.5$ **b.** $\sinh(\ln 3)$ **c.** $\coth^{-1} 2$

Find the integrals in Problems 13-18.

13. $\displaystyle\int \frac{2x+3}{\sqrt{x^2+1}} dx$ 14. $\displaystyle\int x \sin 2x \; dx$ 15. $\displaystyle\int \sinh(1-2x) dx$

16. $\displaystyle\int \frac{dx}{\sqrt{4-x^2}}$ 17. $\displaystyle\int \frac{x^2 \; dx}{(x^2+1)(x-1)}$ 18. $\displaystyle\int \frac{x^3 \; dx}{x^2-1}$

Evaluate the definite integrals in Problems 19-22.

19. $\displaystyle\int_1^2 x \ln x^3 \; dx$ 20. $\displaystyle\int_2^3 \frac{dx}{(x-1)^2(x+2)}$ 21. $\displaystyle\int_3^4 \frac{dx}{2x-x^2}$ 22. $\displaystyle\int_0^{\frac{\pi}{4}} \sec^3 x \tan x \; dx$

Determine whether each improper integral in Problems 23-26 *converges, and if it does, find its value.*

23. $\displaystyle\int_0^\infty xe^{-2x} dx$ 24. $\displaystyle\int_0^{\frac{\pi}{4}} \frac{\sec^2 x \; dx}{\sqrt{\tan x}}$ 25. $\displaystyle\int_0^1 \frac{2x+3}{x^2(x-2)} dx$ 26. $\displaystyle\int_0^\infty e^{-x} \sin x \; dx$

27. Find $\dfrac{dy}{dx}$ for $y = \sqrt{\tanh^{-1} 2x}$.

28. Find the volume of the solid generated when the region under the curve

$$y = \frac{1}{\sqrt{9-x^2}}$$

on the interval $[0, 2]$ is revolved about the y-axis.

29. Solve the first-order linear differential equation

$$\frac{dy}{dx} + \frac{xy}{x+1} = e^{-x}$$

subject to the condition $y = 1$ when $x = 0$.

30. A tank contains 200 gal of saturated brine with 2 lb of salt per gallon. A salt solution containing 1.3 lb of salt per gallon flows in at the rate of 5 gal/min, and the uniform mixture flows out at 3 gal/min. Find the amount of salt in the tank (to the nearest tenth of a lb) after 1 hour.

Supplementary Problems*

Find the derivative $\dfrac{dy}{dx}$ *in Problems* 1-5.

1. $y = \tanh^{-1}\dfrac{1}{x}$

2. $y = x\cosh^{-1}(3x + 1)$

3. $y = \dfrac{\sinh x}{e^x}$

4. $y = \sinh x \cosh x$

5. $y = x\sinh x + \sinh(e^x + e^{-x})$

Find the integrals in Problems 6-47.

6. $\displaystyle\int \cos^{-1} x \ dx$

7. $\displaystyle\int \frac{x^2 \ dx}{\sqrt{4 - x^2}}$

8. $\displaystyle\int \frac{x \ dx}{x^2 - 2x + 5}$

9. $\displaystyle\int \frac{3x - 2}{x^3 - 2x^2}dx$

10. $\displaystyle\int \frac{dx}{(x^2 + x + 1)^{3/2}}$

11. $\displaystyle\int \frac{dx}{\sqrt{x}(1 + \sqrt[4]{x})}$

12. $\displaystyle\int \frac{\sqrt{9x^2 - 1}\ dx}{x}$

13. $\displaystyle\int x^2 \tan^{-1} x \ dx$

14. $\displaystyle\int \frac{dx}{\sin x + \tan x}$

15. $\displaystyle\int e^x\sqrt{4 - e^{2x}}\ dx$

16. $\displaystyle\int \cos\frac{x}{2}\sin\frac{x}{3}dx$

17. $\displaystyle\int \frac{\sqrt{1 + \frac{1}{x^2}}}{x^5}dx$

18. $\displaystyle\int \frac{\sin x \ dx}{\cos^5 x}$

19. $\displaystyle\int \sqrt{1 + \sin x}\ dx$

20. $\displaystyle\int \cos x \ln(\sin x)dx$

21. $\displaystyle\int \sin(\ln x)dx$

22. $\displaystyle\int e^{2x}\operatorname{sech} e^{2x}dx$

23. $\displaystyle\int \frac{\sinh x \ dx}{2 + \cosh x}$

24. $\displaystyle\int \frac{\tanh^{-1} x \ dx}{1 - x^2}$

25. $\displaystyle\int x^2 \cot^{-1} x \ dx$

26. $\displaystyle\int x(1 + x)^{1/3}dx$

27. $\displaystyle\int \frac{x^2 + 2}{x^3 + 6x + 1}dx$

28. $\displaystyle\int \frac{\sin x - \cos x}{(\sin x + \cos x)^{1/4}}dx$

29. $\displaystyle\int \cos\sqrt{x + 2}\ dx$

30. $\displaystyle\int \sqrt{5 + 2\sin^2 x}\ \sin 2x \ dx$

31. $\displaystyle\int \frac{x^3 + 2x}{x^4 + 4x^2 + 3}dx$

32. $\displaystyle\int \frac{x \ dx}{\sqrt{5 - x^2}}$

33. $\displaystyle\int \frac{\sqrt{5 - x^2}\ dx}{x}$

34. $\displaystyle\int \frac{\sqrt{x^2 + x}\ dx}{x}$

35. $\displaystyle\int x^3(x^2 + 4)^{-1/2}dx$

36. $\displaystyle\int \sin^2\frac{x}{2}\cos^2\frac{x}{2}dx$

37. $\displaystyle\int \sec^3 x \tan x \ dx$

38. $\displaystyle\int \sec^5 x \tan^2 x \ dx$

39. $\displaystyle\int \frac{x^{1/3}dx}{x^{1/2} + x^{2/3}}$

40. $\displaystyle\int \frac{\sin x \ dx}{\sin x + \cos x}$

41. $\displaystyle\int \frac{\cos x + \sin x}{1 + \cos x - \sin x}dx$

42. $\displaystyle\int \frac{dx}{x^{1/4} + 1}$

43. $\displaystyle\int \frac{e^x \ dx}{e^{x/3} - e^{x/2}}$

44. $\displaystyle\int \tan^4 x \ dx$

45. $\displaystyle\int \sec^4 x \ dx$

46. $\displaystyle\int \frac{5x^2 - 4x + 9}{x^3 - x^2 + 4x - 4}dx$

47. $\displaystyle\int \frac{x \ dx}{(x + 1)(x + 2)(x + 3)}$

In Problems 48-53, *solve the given differential equations.*

48. $\dfrac{dy}{dx} = \sqrt{\dfrac{x^2 - 1}{y^2 + 1}}$

49. $\dfrac{dy}{dx} = \dfrac{1 - y}{1 + x}$

50. $\dfrac{dy}{dx} - \dfrac{y}{2x} = \dfrac{1}{x^{1/2} + 1}$

51. $xy' - yx^{1/2} = 0$

52. $y' + (\tan x)y = \sec^3 x$

53. $(\cos^3 x)y' = \sin^3 x \cos y$

---

*The supplementary probelms are presented in a somewhat random order, not necessarily in order of difficulty.

Determine whether the improper integral in Problems 54-59 converges, and if it does, find its value.

54. $\displaystyle\int_0^\infty 5e^{-2x}\,dx$

55. $\displaystyle\int_1^\infty e^{1-x}\,dx$

56. $\displaystyle\int_{-\infty}^\infty \frac{dx}{4+x^2}$

57. $\displaystyle\int_0^{\pi/2} \frac{\cos x\,dx}{\sqrt{\sin x}}$

58. $\displaystyle\int_1^\infty \frac{dx}{x^4+x^2}$

59. $\displaystyle\int_{-1}^1 \frac{dx}{(2x+1)^{1/3}}$

60. Think Tank Problem Each of the following equations may be either true or false. In each case, either show that the equation is generally true or find a number x for which it fails.

    a. $\tanh\left(\dfrac{1}{2}\ln x\right) = \dfrac{x-1}{x+1},\ x>0$
        b. $\sinh^{-1}(\tan x) = \tanh^{-1}(\sin x)$ for $-\frac{\pi}{2} < x < \frac{\pi}{2}$

61. Compute $\displaystyle\int_0^1 x\,f''(3x)\,dx$, given that $f'(0)$ is defined, $f(0)=1$, $f(3)=4$, and $f'(3)=-2$.

62. What is $\displaystyle\int_0^{\pi/2} \big[f(x)+f''(x)\big]\cos x\,dx$ if $f(0)$ is defined, $f(\frac{\pi}{2})=5$, and $f'(0)=-1$?

63. Newton's law of cooling says that the temperature T of a body satisfies

$$\frac{dT}{dt} = k(T - T_0)$$

where T_0 is the temperature of the surrounding medium. Solve this equation for T.

64. Modeling Problem The psychologist L. L. Thurstone investigated the way people learn using the differential equation

$$\frac{dS}{dt} = \frac{2k}{\sqrt{m}}\,[S(1-S)]^{3/2}$$

where $S(t)$ is the state of the learner at time t, and k and m are positive constants depending on the learner and the nature of the task.* Solve this equation.

65. If $s>0$, the integral function defined by

$$\Gamma(s) = \int_0^\infty e^{-t}t^{s-1}\,dt$$

is called the **gamma function**. It was introduced by Leonhard Euler in 1729 and has some useful properties.
    a. Show that $\Gamma(s)$ converges for all $s>0$.
    b. Show that $\Gamma(s+1) = s\Gamma(s)$.
    c. Show that $\Gamma(n+1) = n!$ for any positive integer n.

66. Find the centroid of a thin plate of constant density that occupies the region bounded by the graph of $y=\sin x + \cos x$, the coordinate axes, and the line $x=\frac{\pi}{4}$.

67. Repeat Problem 66 for a plate that occupies the region bounded by $y=\sec^2 x$, the coordinate axes, and the line $x=\frac{\pi}{3}$.

68. Find the volume obtained by revolving about the y-axis the region bounded by the curve $y=\sinh x$, the x-axis, and the line $x=1$.

69. Find the volume (to the nearest hundredth) obtained by revolving about the y-axis the region bounded by the curve $y=\cosh x$, the y-axis, and the line $y=2$.

70. The region bounded by the graph of $y=\sin x$ and the x-axis between $x=0$ and $x=\pi$ is revolved about the x-axis to generate a solid. Find the volume of the solid and its surface area.

71. Find the length of the arc of the curve $y=\frac{4}{5}x^{5/4}$ that lies between $x=0$ and $x=1$.

72. Find the volume of the solid obtained by revolving about the x-axis the region bounded by the curve $y=(9-x^2)^{1/4}$ and the x-axis.

73. Find the volume of the solid formed by revolving about the y-axis the region bounded by the curve

$$y = \frac{1}{1+x^4}$$

and the x-axis between $x=0$ and $x=4$.

---

* L. L. Thurstone, "The Learning Function," J. of General Psychology 3(1930), pp. 469-493.

74. Find the volume of the solid formed by revolving about the x-axis the region bounded by the curve

$$y = \frac{2}{\sqrt{3x-2}}$$

the x-axis, and the lines $x = 1$ and $x = 2$.

75. Find the surface area of the solid generated by revolving the region bounded by the curve

$$y = e^x + \frac{1}{4}e^{-x}$$

on $[0,1]$ about the x-axis.

76. Find the volume of the solid whose base is the region R bounded by the curve $y = e^x$ and the lines $y = 0$, $x = 0$, $x = 1$, if cross sections perpendicular to the x-axis are equilateral triangles.

77. Show that the area under $y = \dfrac{1}{x}$ on the interval $[1, a]$ equals the area under the same curve on $[k, ka]$ for any number $k > 0$.

78. Find the length of the curve $y = \ln(\csc x)$ on the interval $\left[\frac{\pi}{3}, \frac{\pi}{2}\right]$.

79. Find the length of the curve

$$x = \frac{ay^2}{b^2} - \frac{b^2}{8a}\ln\frac{y}{b}$$

between $y = b$ and $y = 2b$.

80. Derive the reduction formula

$$\int (a^2 - x^2)^n dx = \frac{x(a^2 - x^2)^n}{2n+1} + \frac{2a^2 n}{2n+1}\int (a^2 - x^2)^{n-1}dx$$

Use this formula to find $\displaystyle\int (9 - x^2)^{5/2}dx$.

81. Derive the reduction formula

$$\int \frac{\sin^n x\ dx}{\cos^m x} = \frac{1}{m-1}\frac{\sin^{n-1}x}{\cos^{m-1}x} - \frac{n-1}{m-1}\int \frac{\sin^{n-2}x\ dx}{\cos^{m-2}x}$$

82. Derive a reduction formula for $\displaystyle\int \frac{\cos^n x\ dx}{\sin^m x}$.

83. Derive a reduction formula for $\displaystyle\int x^n(x^2 + a^2)^{-1/2}dx$.

84. Derive the reduction formula

$$\int \frac{dx}{x^n\sqrt{ax+b}} = \frac{-\sqrt{ax+b}}{(n-1)bx^{n-1}} - \frac{(2n-3)a}{(2n-2)b}\int \frac{dx}{x^{n-1}\sqrt{ax+b}}$$

85. The improper integral

$$\int_1^\infty \left[\frac{2Ax^3}{x^4+1} - \frac{1}{x+1}\right]dx$$

converges for exactly one value of the constant A. Find A and then compute the value of the integral.

86. Evaluate $\displaystyle\int_0^\infty \frac{\sqrt{x}\ln x\ dx}{(x+1)(x^2+x+1)}$ if it exists. *Hint*: Let $u = \dfrac{1}{x}$.

87. Spectroscopic measurements based on the Doppler effect yield an observed mass m_0 for a certain type of binary star. The true mass m is then estimated by $m = \dfrac{m_0}{I}$, where

$$I = \int_0^{\pi/2} \sin^4 x\ dx$$

Determine the number I.

88. Two substances, A and B, are being converted into a single compound C. In the laboratory, it is shown that the time rate of change of the amount x of compound C is proportional to the product of the amount of unconverted substances A and B. Thus,

$$\frac{dx}{dt} = k(a - x)(b - x)$$

for initial concentrations a and b, of A and B, respectively.

   a. Solve this differential equation to express x in terms of time t. What happens to $x(t)$ as $t \to \infty$ if $b > a$? What if $a > b$?

   b. Suppose $a = b$. What is $x(t)$? What happens to $x(t)$ as $t \to \infty$ in this case?

89. Let $f(x) = \begin{cases} \dfrac{1}{x^2} & \text{for } x \geq 1 \\ 1 & \text{for } -1 < x < 1 \\ e^{x+1} & \text{for } x \leq -1 \end{cases}$

Sketch the graph of f and evaluate

$$\int_{-\infty}^{\infty} f(x)\,dx$$

90. Evaluate the improper polar integral

$$\int_0^{\infty} \theta e^{-\theta}\,d\theta$$

91. Find the total area between the spirals

$$r = e^{-2\theta} \text{ and } r = e^{-5\theta}$$

for $\theta \geq 0$.

92. **Modeling Problem** The residents of a certain community have voted to discontinue the fluoridation of their water supply. The local reservoir currently holds 200 million gallons of fluoridated water containing 1,600 lb of fluoride. The fluoridated water flows out of the reservoir at the rate of 4 million gallons per day and is replaced by unfluoridated water at the same rate. At all times, the remaining fluoride in the reservoir is evenly distributed. Set up and solve a differential equation for the amount $F(t)$ of fluoride in the reservoir at time t. When will 90% of the fluoride be replaced?

93. **Modeling Problem** The Nuclear Regulatory Commission puts atomic waste into sealed containers and dumps them into the ocean. It is important to dump the containers in water shallow enough to ensure they do not break when they hit bottom. Suppose $s(t)$ is the depth of the container at time t and let W and B denote the container's weight and the constant buoyancy force, respectively. Assume there is a "drag force" proportional to velocity v.

   a. Explain why the motion of the containers may be modeled by the differential equation

$$\frac{W}{g}\frac{dv}{dt} = W - B - kv$$

   where g is the constant acceleration due to gravity. Solve this equation to express $v(t)$ in terms of W, k, B, and g.

   b. Integrate the expression for $v(t)$ in part **a** to find $s(t)$.

   c. Suppose $W = 1,125$ newtons (about 250 lb), $B = 1,100$ newtons, and $k = 0.64$ kg/s. If the container breaks when the impact speed exceeds 10 m/s, what is the maximum depth (to the nearest meter) for safe dumping?

94. **The Evans price-adjustment model** assumes that if there is an excess demand D over supply S in any time period, the price p changes at a rate proportional to the excess, $D - S$; that is,

$$\frac{dp}{dt} = k(D - S)$$

Suppose for a certain commodity, demand is linear,

$$D(p) = c - pd$$

and supply is cyclical

$$S(t) = a\sin(bt)$$

Solve thc differential equation to express price $p(t)$ in terms of a, b, c, and d. What happens to $p(t)$ "in the long run" (as $t \to \infty$)? Use $p_0 = p(0)$.

95. Modeling Problem A country has 5 billion dollars of paper currency. Each day about 10 million dollars comes into banks and 12 million is paid out. The government decides to issue new currency, and whenever an "old" bill comes into the bank, it is replaced by a "new" bill. Set up and solve a differential equation to model the currency replacement. How long will it take for 95% of the currency in circulation to be "new" bills?

96. If n is a positive integer and a is a positive number, find

$$\int_0^\infty x^n e^{-ax}\, dx$$

97. Putnam Examination Problem Show that

$$\frac{22}{7} - \pi = \int_0^1 \frac{x^4(1-x)^4\, dx}{1+x^2}$$

98. Putnam Examination Problem Evaluate

$$\int_0^{\pi/2} \frac{dx}{1 + (\tan x)^{\sqrt{2}}}$$

99. Putnam Examination Problem Evaluate

$$\int_0^\infty t^{-1/2} \exp\left[-1985\left(t + \frac{1}{t}\right)\right] dt$$

You may assume that $\displaystyle\int_{-\infty}^\infty e^{-x^2}\, dx = \sqrt{\pi}$.

CHAPTER 7 GROUP RESEARCH PROJECT

Working in small groups is typical of most work environments, and this book seeks to develop skills with group activities. We present a group project at the end of each chapter. These projects are to be done in groups of three or four students.

Buoy Design

© 2013 Yory Frenklakh. Used under license from Shutterstock, Inc.

You have been hired as a special consultant by the U. S. Coast Guard to evaluate some proposed new designs for navigational aids (buoys).

The buoys are floating cans that need to be visible from some distance away, without rising too far out of the water. Each buoy has a circular cross section and will be fitted with a superstructure that carries equipment such as lights and batteries. To be acceptable, a fully equipped buoy must float with not less than 1.5 ft nor more than 3 ft of freeboard. (Freeboard is the distance from the water level to the top of the device.)
You should make the following assumptions:*

    a. A floating object will displace a volume of water whose weight equals the weight of the floating object. (This is Archimedes' principle.)

    b. Your devices will be floating in salt water, which weighs 64.0 lb/ft^3.

    c. The buoys will be constructed of $\frac{1}{2}$ in.-thick sheet metal that weighs 490 lb/ft^3.

    d. You can estimate an additional 20% in weight attributable to welds, bolts, and the like.

  e. Each buoy will be fitted with a superstructure and equipment weighing a total of 2,000 lb.

  f. Each design you are given to evaluate will be presented to you in the form of a curve $x = f(y)$, to be revolved about the y-axis.

Extended paper for further study. Write a report using the given assumptions to evaluate several designs for buoys, and formulate a conclusion about the design you recommend to the Coast Guard.

---

*This project is adapted from a computer project used at the U.S. Coast Guard Academy.

CHAPTER 8

INFINITE SERIES

*P*eople who don't count won't count.

Anatole France
W. F. Osgood (1907)

PREVIEW

Is it possible for the sum of infinitely many nonzero numbers to be finite? This concept, which may seem paradoxical at first, plays a central role in mathematics and has a variety of important applications. The goal of this chapter is to examine the theory and applications of infinite sums, which will be referred to as *infinite series*. Geometric series, introduced in Section 8.2, are among the simplest infinite series we will encounter and, in some ways, the most important. In Sections 8.3-8.6, we will develop *convergence tests*, which provide ways of determining quickly whether or not certain infinite series have a finite sum. Next, we will turn our attention to series in which the individual terms are functions instead of numbers. We will be especially interested in the properties of *power series*, which may be thought of as polynomials of infinite degree, although some of their properties are quite different from those of polynomials. We will find that many common functions, such as e^x, $\ln(x + 1)$, $\sin x$, $\cos x$, and $\tan^{-1} x$, can be represented as power series, and we will discuss some important theoretical and computational aspects of this kind of representation.

PERSPECTIVE

Series, or sums, arise in many different ways. For example, suppose it is known that a certain pollutant is released into the atmosphere at weekly intervals and is dissipated at the rate of 2% per week. If m grams of pollutant are released each week, then at the beginning of the first week, there will be $S_1 = m$ grams in the atmosphere, and at the beginning of the second, there will be $0.98m$ grams of "old" pollutant left plus m grams of "new" pollutant, to yield a total of $S_2 = m + 0.98m$ grams. Continuing, at the beginning of the nth week there will be $S_n = m + 0.98m + (0.98)^2 m + \cdots + (0.98)^{n-1} m$ grams. It is natural to wonder how much pollutant will accumulate in the "long run" (as $n \to \infty$). But just exactly what do we mean by such a sum, and if the total is a finite number, how can we compute its value? We seek answers to these questions in this chapter.

8.1 SEQUENCES AND THEIR LIMITS

IN THIS SECTION: *Sequences; limits of sequences; bounded, monotonic sequences*
Most phenomena we have considered occur continuously, but in practically every field of inquiry, there are situations that can be described by cataloging individual items in a numerical listing. In this section, we define a mathematical tool, called a *sequence*, in order to do this cataloging, and then we define what we mean by *the limit of a sequence*.

The making of a motion picture is a complex process, and editing all the film into a movie requires that all the frames of the action be labeled in chronological order. For example, R21-435 might signify the 435th frame of the 21st reel. A mathematician might refer to the movie editor's labeling procedure by saying the frames are arranged in a *sequence*.

16-mm movie film

Sequences

A sequence is a succession of numbers that are listed according to a given prescription or rule. Specifically, if n is a positive integer, the sequence whose nth term is the number a_n can be written as

$$a_1, a_2, \cdots, a_n, \cdots$$

or, more simply,

$$\{a_n\}$$

The number a_n is called the **general term** of the sequence. We will deal only with infinite sequences, so each term a_n has a **successor** a_{n+1} and for $n > 1$, a **predecessor** a_{n-1}. For example, by associating each positive integer n with its reciprocal $\frac{1}{n}$, we obtain the sequence denoted by $\{\frac{1}{n}\}$, which represents the succession of numbers $1, \frac{1}{2}, \frac{1}{3}, \cdots, \frac{1}{n}, \cdots$. The general term is denoted by $a_n = \frac{1}{n}$. The following examples illustrate the notation and terminology used in connection with sequences.

Example 1   Given the general term, find particular terms of a sequence

Find the 1st, 2nd, and 15th terms of the sequence $\{a_n\}$, where the general term is

$$a_n = \left(\frac{1}{2}\right)^{n-1}$$

Solution   If $n = 1$, then $a_1 = \left(\frac{1}{2}\right)^{1-1} = \left(\frac{1}{2}\right)^0 = 1$. Similarly,

$$a_2 = \left(\frac{1}{2}\right)^{2-1} = \frac{1}{2}$$

$$a_{15} = \left(\frac{1}{2}\right)^{15-1} = \left(\frac{1}{2}\right)^{14} = 2^{-14}$$

The reverse question, that of finding a general term given certain terms of a sequence, is a more difficult task, and even if we find a general term, we have no assurance that the general term is unique. For example, consider the sequence

$$2, 4, 6, 8, \cdots$$

This seems to have a general term $a_n = 2n$. However, the general term

$$a_n = (n-1)(n-2)(n-3)(n-4) + 2n$$

has the same first four terms, but $a_5 = 34$ (not 10, as we would expect from the sequence $2, 4, 6, 8$).

It is sometimes useful to start a sequence with a_0 instead of a_1; that is, to have a sequence of the form

$$a_0, a_1, a_2, \cdots.$$

So far, we have been discussing the concept of a sequence informally, without a definition. We have observed that a sequence $\{a_n\}$ associates the number a_n with the positive (or possibly, nonnegative) integer n. Hence, a sequence is really a special kind of function, one whose domain is a set of positive (or possibly, nonnegative) integers.

SEQUENCE A **sequence** $\{a_n\}$ is a function whose domain is a set of nonnegative integers and whose range is a subset of the real numbers. The functional values $a_1, a_2, a_3, \cdots$ are called the **terms** of the sequence, and a_n is called the **nth term**, or **general term**, of the sequence.

Limits of Sequences

It is often desirable to examine the behavior of a given sequence $\{a_n\}$ as n gets arbitrarily large. For example, consider the sequence

$$a_n = \frac{n}{n+1}$$

Even though we write a_n, remember this is a function, $f(n) = \frac{n}{n+1}$, where the domain is the set of positive integers.

Because $a_1 = \frac{1}{2}$, $a_2 = \frac{2}{3}$, $a_3 = \frac{3}{4}$, $\cdots$, we can plot the terms of this sequence on a number line, as shown in Figure 8.1**a**, or the sequence can be plotted in two dimensions, as shown in Figure 8.1**b**.

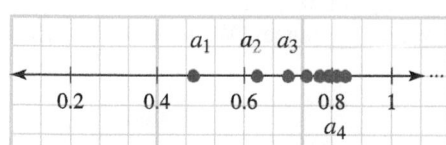

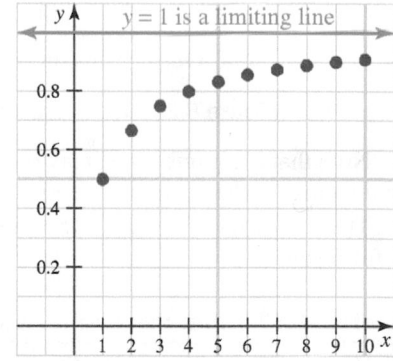

a. Graphing a sequence in one dimension   **b.** Graphing a sequence in two dimensions

Figure 8.1 Graphing the sequence $a_n = \dfrac{n}{n+1}$

By looking at either graph in Figure 8.1, we see that it appears the terms of the sequence are approaching 1. In general, if the terms of the sequence approach the number L as n increases without bound, we say that the sequence *converges to the limit L* and write

$$L = \lim_{n \to \infty} a_n$$

By looking at Figure 8.1, we would expect

$$\lim_{n \to \infty} a_n = \lim_{n \to \infty} \frac{n}{n+1} = 1$$

This limiting behavior is analogous to the continuous case (discussed in Section 4.4), and may be defined formally as follows.

CONVERGENT SEQUENCE   The sequence $\{a_n\}$ **converges** to the number L, and we write

$$L = \lim_{n \to \infty} a_n$$

if for every $\epsilon > 0$, there is an integer N such that

$$|a_n - L| < \epsilon \qquad \text{whenever} \qquad n > N$$

Otherwise, the sequence **diverges**.

■ **W**hat this says: The notation $L = \lim_{n \to \infty} a_n$ means that eventually the terms of the sequence $\{a_n\}$ can be made as close to L as may be desired by taking n sufficiently large.

A geometric interpretation of this definition is shown in Figure 8.2.

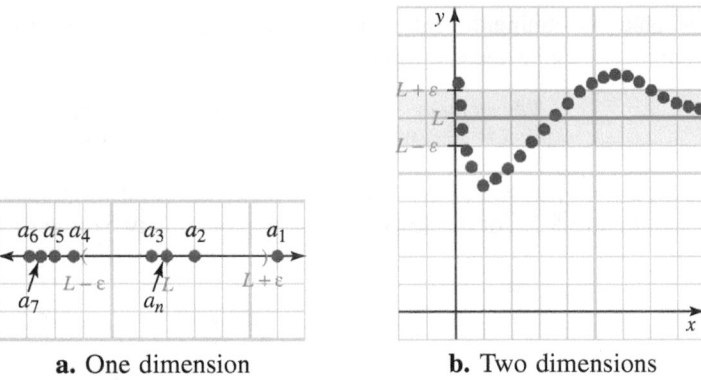

a. One dimension   **b.** Two dimensions

Figure 8.2   Geometric interpretation of a convergent sequence

Note that the numbers a_n may be practically anywhere at first (that is, for "small" n), but eventually, the a_n must "cluster" near the limiting value L.

The theorem on limits of functions carries over to sequences. We have the following useful result.

Theorem 8.1   Limit theorem for sequences

If $\lim_{n \to \infty} a_n = L$ and $\lim_{n \to \infty} b_n = M$, then

Linearity rule for sequences $\lim_{n \to \infty} (ra_n + sb_n) = rL + sM$

Product rule for sequences $\lim_{n \to \infty} (a_n b_n) = LM$

Quotient rule for sequences $\lim_{n \to \infty} \dfrac{a_n}{b_n} = \dfrac{L}{M}$ provided $M \neq 0$

Root rule for sequences $\lim_{n \to \infty} \sqrt[m]{a_n} = \sqrt[m]{L}$ provided $\sqrt[m]{a_n}$ is defined for all n and $\sqrt[m]{L}$ exists.

Proof: The proof of these rules is similar to the proof of the limit rules that were stated in Section 4.4.   ♦

Example 2   Convergent sequences

Find the limit of each of these convergent sequences:

a. $\left\{ \dfrac{100}{n} \right\}$ **b.** $\left\{ \dfrac{2n^2 + 5n - 7}{n^3} \right\}$ **c.** $\left\{ \dfrac{3n^4 + n - 1}{5n^4 + 2n^2 + 1} \right\}$

Solution

a. As n grows arbitrarily large, $\frac{100}{n}$ gets smaller and smaller. Thus,

$$\lim_{n \to \infty} \frac{100}{n} = 0$$

A graphical representation is shown in Figure 8.3.

b. We cannot use the quotient rule of Theorem 8.1 because neither the limit in the numerator nor the one in the denominator exists. However,

$$\frac{2n^2 + 5n - 7}{n^3} = \frac{2}{n} + \frac{5}{n^2} - \frac{7}{n^3}$$

and by using the linearity rule, we find that

$$\lim_{n \to \infty} \frac{2n^2 + 5n - 7}{n^3} = 2 \lim_{n \to \infty} \frac{1}{n} + 5 \lim_{n \to \infty} \frac{1}{n^2} - 7 \lim_{n \to \infty} \frac{1}{n^3}$$
$$= 2 \cdot 0 + 5 \cdot 0 - 7 \cdot 0$$
$$= 0$$

A graph is shown in Figure 8.4.

c. Divide the numerator and denominator by n^4, the highest power of n that occurs in the expression, to obtain

$$\lim_{n \to \infty} \frac{3n^4 + n - 1}{5n^4 + 2n^2 + 1} = \lim_{n \to \infty} \frac{3 + \frac{1}{n^3} - \frac{1}{n^4}}{5 + \frac{2}{n^2} + \frac{1}{n^4}} = \frac{3}{5}$$

Note that this procedure is very similar to the one we used in Section 4.4 when we evaluated limits of functions to infinity.

A graph of this sequence is shown in Figure 8.5.

Example 3   Divergent sequences

Show that the following sequences diverge:

a. $\{(-1)^n\}$ **b.** $\left\{ \dfrac{n^5 + n^3 + 2}{7n^4 + n^2 + 3} \right\}$

Solution

a. The sequence defined by $\{(-1)^n\}$ is $-1, 1, -1, 1, \cdots$ and this sequence diverges by oscillation because the nth term is always either 1 or -1. Thus a_n cannot approach one specific number L as n grows large.

b. $\lim_{n \to \infty} \dfrac{n^5 + n^3 + 2}{7n^4 + n^2 + 3} = \lim_{n \to \infty} \dfrac{1 + \frac{1}{n^2} + \frac{2}{n^5}}{\frac{7}{n} + \frac{1}{n^3} + \frac{3}{n^5}}$

The numerator tends toward 1 as $n \to \infty$, and the denominator approaches 0. Hence the quotient increases without bound, and the sequence must diverge.

If $\lim_{n \to \infty} a_n$ does not exist because the numbers a_n become arbitrarily large as $n \to \infty$, we write $\lim_{n \to \infty} a_n = \infty$. We summarize this more precisely in the following box.

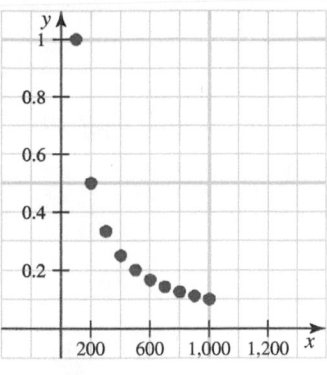

Figure 8.3 Graphical representation of $a_n = \dfrac{100}{n}$

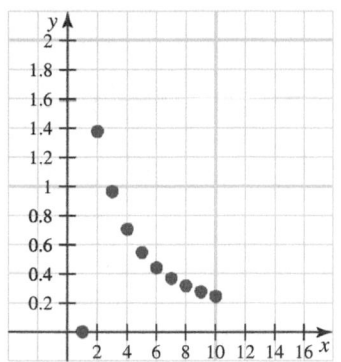

Figure 8.4 Graph of $a_n = \dfrac{2n^2 + 5n - 7}{n^3}$

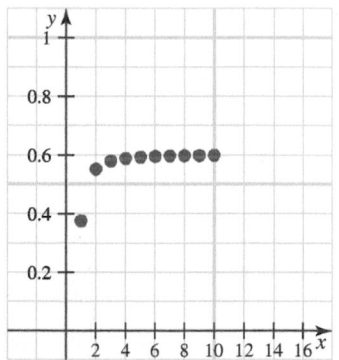

Figure 8.5 Graph of $\dfrac{3n^4 + n - 1}{5n^4 + 2n^2 + 1}$

LIMIT NOTATION

$\lim_{n \to \infty} a_n = \infty$ means that for any real number A, we have $a_n > A$ for all sufficiently large n.

$\lim_{n \to \infty} b_n = -\infty$ means that for any real number B, we have $b_n < B$ for all sufficiently large n.

Rewriting the answer to Example **3b** in this notation,

$$\lim_{n\to\infty} \frac{n^5 + n^3 + 2}{7n^4 + n^2 + 3} = \infty$$

Example 4   Determining the convergence or divergence of a sequence

Determine the convergence or divergence of the sequence $\left\{\sqrt{n^2 + 3n} - n\right\}$.

Solution   It would not be correct to apply the linearity property for sequences (because neither $\lim_{n\to\infty} \sqrt{n^2 + 3n}$ nor $\lim_{n\to\infty} n$ exists). It is also not correct to use this as a reason to say that the limit does not exist since infinity minus infinity is indeterminate. You might try some values of n (shown in Figure 8.6) to guess that there is some limit.

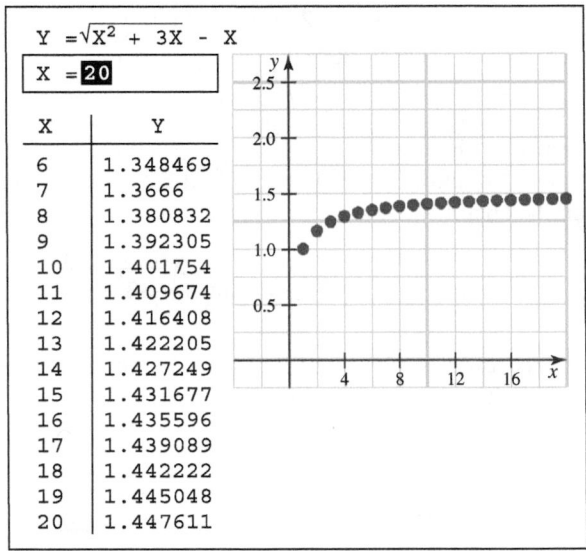

Figure 8.6 Interactive   Is this the graph of a convergent or divergent series?

To find the limit, however, we will rewrite the general term algebraically as follows:

$$\sqrt{n^2 + 3n} - n = \left(\sqrt{n^2 + 3n} - n\right)\frac{\sqrt{n^2 + 3n} + n}{\sqrt{n^2 + 3n} + n}$$

$$= \frac{n^2 + 3n - n^2}{\sqrt{n^2 + 3n} + n}$$

$$= \frac{n^2 + 3n - n^2}{\sqrt{n^2 + 3n} + n} \cdot \frac{\frac{1}{n}}{\frac{1}{n}}$$

$$= \frac{3}{\frac{\sqrt{n^2+3n}}{n} + 1}$$

$$= \frac{3}{\sqrt{1 + \frac{3}{n}} + 1}$$

Hence,

$$\lim_{n\to\infty}\left(\sqrt{n^2 + 3n} - n\right) = \lim_{n\to\infty}\frac{3}{\sqrt{1 + \frac{3}{n}} + 1} = \frac{3}{2}$$

Note the graph of the sequence in Example 4 in Figure 8.6. The graph of a sequence consists of a succession of isolated points. This can be compared with the graph of $y = \sqrt{x^2 + 3x} - x, x \geq 1$, which is a continuous curve (shown in Figure 8.7).

The only difference between $\lim_{n\to\infty} a_n = L$ and $\lim_{x\to\infty} f(x) = L$ is that n is required to be an integer. This is stated in the hypothesis of the following theorem.

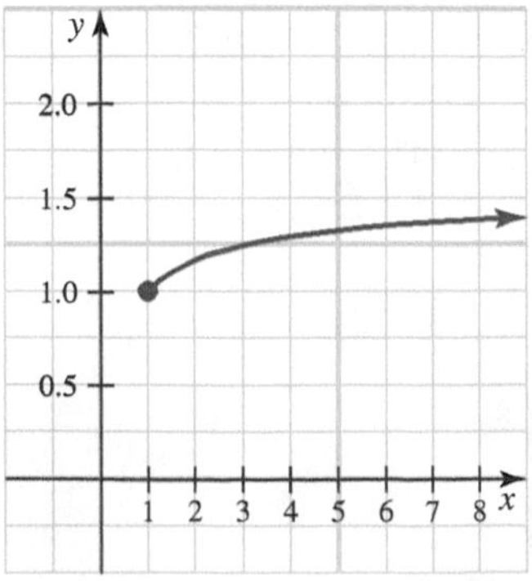

Figure 8.7   Graph of the function
$f(x) = \sqrt{x^2 + 3x} - x,\ x \geq 1$

Theorem 8.2   Limit of a sequence from the limit of a continuous function

Given the sequence $\{a_n\}$, let f be a continuous function such that $a_n = f(n)$ for $n = 1, 2, \cdots$. If $\lim\limits_{x \to \infty} f(x)$ exists and $\lim\limits_{x \to \infty} f(x) = L$, the sequence $\{a_n\}$ converges and $\lim\limits_{n \to \infty} a_n = L$.

Proof: Let $\epsilon > 0$ be given. Because $\lim\limits_{x \to \infty} f(x) = L$, there exists a number $N > 0$ such that

$$|f(x) - L| < \epsilon \text{ whenever } x > N$$

In particular, if $n > N$, it follows that $|f(n) - L| = |a_n - L| < \epsilon$.   ◆

Be sure you read this theorem correctly. In particular, note that it *does not* say that if $\lim\limits_{n \to \infty} a_n = L$, then $\lim\limits_{x \to \infty} f(x) = L$ (see Figure 8.8**b**).

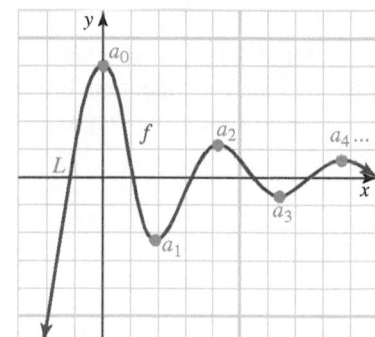

$\lim\limits_{n \to \infty} a_n = L$ but $\lim\limits_{x \to \infty} f(x) \neq L$

a. If $\lim\limits_{x \to \infty} f(x) = L$, then $\lim\limits_{n \to \infty} a_n = L$

b. ☠ If $\lim\limits_{n \to \infty} a_n = L$, then $\lim\limits_{x \to \infty} f(x)$ does NOT necessarily equal L

Figure 8.8   Comparison of the graphs of $\lim\limits_{x \to \infty} f(x)$ and $\lim\limits_{n \to \infty} a_n$ where $f(n) = a_n$ for $n = 1, 2, \cdots$

Example 5   Evaluating a limit using l'Hôpital's rule

Given that the sequence $\left\{ \dfrac{n^2}{1 - e^n} \right\}$ converges, evaluate $\lim\limits_{n \to \infty} \dfrac{n^2}{1 - e^n}$.

Solution   Let $L = \lim\limits_{x \to \infty} f(x)$, where $f(x) = \dfrac{x^2}{1 - e^x}$. Because $f(n) = a_n$ for $n = 1, 2, \cdots$, Theorem 8.2 tells us that $\lim\limits_{n \to \infty} \dfrac{n^2}{1 - e^n}$ is the same as $\lim\limits_{x \to \infty} f(x)$, provided this latter limit exists. Since the function f is continuous for all $x \neq 0$, we use l'Hôpital's rule.

$$\lim_{x \to \infty} \frac{x^2}{1 - e^x} = \lim_{x \to \infty} \frac{2x}{-e^x} \quad \text{l'Hôpital's rule}$$

$$= \lim_{x \to \infty} \frac{2}{-e^x} \quad \text{l'Hôpital's rule again}$$

$$= 0$$

Thus, by Theorem 8.2 , $\lim\limits_{n \to \infty} \dfrac{n^2}{1 - e^n} = \lim\limits_{x \to \infty} f(x) = L = 0$.   ▪

The squeeze rule (Section 2.2) can be reformulated in terms of sequences.

Theorem 8.3   Squeeze theorem for sequences

If $a_n \leq b_n \leq c_n$ for all $n > N$, and $\lim\limits_{n \to \infty} a_n = \lim\limits_{n \to \infty} c_n = L$, then

$$\lim_{n \to \infty} b_n = L$$

Proof: The proof of this theorem follows the same steps as the proof of the squeeze rule given in Appendix A.   ◆

Example 6   Using l'Hôpital's rule and the squeeze theorem

Show that the following sequences converge, and find their limits.

a. $\lim\limits_{n\to\infty} n^{1/n}$ **b.** $\lim\limits_{n\to\infty} \dfrac{n!}{n^n}$

Solution

a. Let $L = \lim\limits_{n\to\infty} n^{1/n}$; then we let $f(x) = x^{1/x}$. The function f is continuous for all $x > 0$, so we can apply l'Hôpital's rule.

$$\ln L = \lim_{n\to\infty} \frac{\ln n}{n} = \lim_{x\to\infty} \frac{\ln x}{x} = \lim_{x\to\infty} \frac{\frac{1}{x}}{1} = 0 \qquad \text{l'Hôpital's rule}$$

Thus, $L = e^0 = 1$.

b. We cannot use l'Hôpital's rule because $x!$ is not defined as an elementary function when x is not an integer. Instead, we use the squeeze theorem for sequences. Accordingly, note that if we let $a_n = 0$ and $c_n = \frac{1}{n}$, for all n, we have

$$\overbrace{0}^{a_n} \;\le\; \overbrace{\frac{n!}{n^n}}^{b_n} \;\le\; \overbrace{\frac{1}{n}}^{c_n}$$

The right inequality is true because

$$\begin{aligned}
\frac{n!}{n^n} &= \frac{n(n-1)(n-2)\cdots 1}{n\cdot n\cdot n\cdots n}\\[4pt]
&= \left(\frac{n}{n}\right)\left(\frac{n-1}{n}\right)\left(\frac{n-2}{n}\right)\cdots\left(\frac{1}{n}\right)\\[4pt]
&< (1)\left(\frac{n}{n}\right)\left(\frac{n}{n}\right)\cdots\left(\frac{n}{n}\right)\left(\frac{1}{n}\right)\\[4pt]
&= \frac{1}{n}
\end{aligned}$$

Because $\lim\limits_{n\to\infty} a_n = 0$ and $\lim\limits_{n\to\infty} c_n = 0$, it follows from the squeeze theorem that $\lim\limits_{n\to\infty} \dfrac{n!}{n^n} = 0$.   ■

Bounded, Monotonic Sequences

We introduce, along with a simple example, some terminology associated with sequences $\{a_n\}$.

Name	Condition	Example
strictly increasing	$a_1 < a_2 < \cdots < a_{k-1} < a_k < \cdots$	$1, 2, 3, 4, 5, \cdots$
increasing	$a_1 \le a_2 \le \cdots \le a_{k-1} \le a_k \le \cdots$	$1, 1, 2, 2, 3, 3, \cdots$
strictly decreasing	$a_1 > a_2 > \cdots > a_{k-1} > a_k > \cdots$	$1, \frac{1}{2}, \frac{1}{3}, \frac{1}{4}, \cdots$
decreasing	$a_1 \ge a_2 \ge \cdots \ge a_{k-1} \ge a_k \ge \cdots$	$1, 1, \frac{1}{2}, \frac{1}{2}, \frac{1}{3}, \frac{1}{3}, \cdots$
bounded above by M	$a_n \le M$ for $1, 2, 3, \cdots$	$M = 1$ for $1, \frac{1}{2}, \frac{1}{3}, \frac{1}{4}, \cdots$ Note $M = 2, M = 3$ are other choices.
bounded below by m	$m \le a_n$ for $1, 2, 3, \cdots$	$m = 1$ for $1, 2, 3, 4, \cdots$ Note $m = 0, m = -1$ are other choices.
bounded	It is bounded both above and below.	$1, \frac{1}{2}, \frac{1}{3}, \frac{1}{4}, \cdots$ $m = 0, M = 1$, so sequence is bounded.

We also say that a series is **monotonic** if it is increasing or decreasing or **strictly monotonic** if it is strictly increasing or strictly decreasing.

In general, it is difficult to tell whether a given sequence converges or diverges, but because of the following theorem, it is easy to make this determination if we know the sequence is monotonic.

Theorem 8.4   BMCT: The bounded, monotonic, convergence theorem

A monotonic sequence $\{a_n\}$ converges if it is bounded and diverges otherwise.

Proof: A formal proof of the BMCT is outlined in Problems 55-56. For the following informal argument, we will assume that $\{a_n\}$ is an increasing sequence. You might wish to see whether you can give a similar informal argument for the decreasing case. Because the terms of the sequence satisfy $a_1 \le a_2 \le \cdots \le a_{k-1} \le a_k \le \cdots$, we know that the sequence is bounded from below by a_1 and that the graph of the corresponding points (n, a_n) will be rising in the plane. Two possible cases are shown in Figure 8.9.

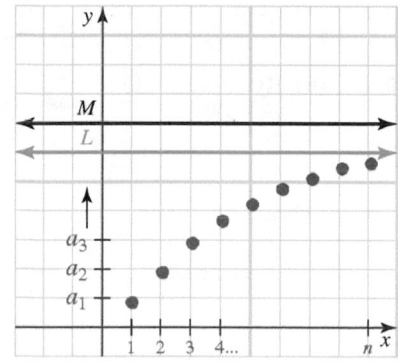 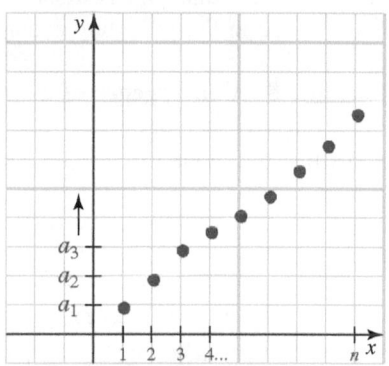

a. If $a_n < M$ for $n = 1, 2, \cdots$, the graph of the points (n, a_n) will approach a horizontal "barrier" line $y = L$

b. If $\{a_n\}$ is not bounded from above, the graph rises indefinitely

Figure 8.9   Graphical possibilities for the BMCT

Suppose the sequence $\{a_n\}$ is bounded from above by a number M, so that $a_1 \le a_n < M$ for $n = 1, 2, \cdots$. Then the graph of the points (n, a_n) must continually rise (because the sequence is monotonic) and yet it must stay below the line $y = M$. The only way this can happen is for the graph to approach a "barrier" line $y = L$ (where $L \le M$), and we have $\lim_{n \to \infty} a_n = L$, as shown in Figure 8.9**a**. However, if the sequence is not bounded from above, the graph will rise indefinitely (Figure 8.9**b**), and the terms in the sequence $\{a_n\}$ cannot approach any finite number L in the limit.   ◆

Example 7   Convergence using the BMCT

Show that the sequence $\left\{ \dfrac{1 \cdot 3 \cdot 5 \cdots (2n - 1)}{2 \cdot 4 \cdot 6 \cdots (2n)} \right\}$ converges.

Solution The first few terms of this sequence are

$$a_1 = \frac{1}{2} \qquad a_2 = \frac{1 \cdot 3}{2 \cdot 4} = \frac{3}{8} \qquad a_3 = \frac{1 \cdot 3 \cdot 5}{2 \cdot 4 \cdot 6} = \frac{5}{16}$$

Because $\frac{1}{2} > \frac{3}{8} > \frac{5}{16}$, it appears that the sequence is decreasing (that is, it is monotonic). We can prove this by showing that $a_{n+1} < a_n$ for all $n > 0$, or equivalently, $\dfrac{a_{n+1}}{a_n} < 1$. (Note that $a_n \ne 0$ for all n.)

$$\frac{a_{n+1}}{a_n} = \frac{\frac{1\cdot3\cdot5\cdots[2(n+1)-1]}{2\cdot4\cdot6\cdots[2(n+1)]}}{\frac{1\cdot3\cdot5\cdots(2n-1)}{2\cdot4\cdot6\cdots(2n)}}$$

$$= \frac{1\cdot3\cdot5\cdots[2(n+1)-1]}{2\cdot4\cdot6\cdots[2(n+1)]} \cdot \frac{2\cdot4\cdot6\cdots(2n)}{1\cdot3\cdot5\cdots(2n-1)}$$

$$= \frac{2n+1}{2n+2}$$

$$< 1$$

for any $n > 0$. Hence $a_{n+1} < a_n$ for all n, and $\{a_n\}$ is a decreasing sequence. Because $a_n > 0$ for all n, it follows that $\{a_n\}$ is bounded below by 0. Applying the BMCT, we see that $\{a_n\}$ converges, but the BMCT tells us nothing about the limit. ∎

The BMCT also applies to sequences whose terms are *eventually* monotonic. That is, it can be shown that the sequence $\{a_n\}$ converges if it is bounded and there exists an integer N such that $\{a_n\}$ is monotonic for all $n > N$. This modified form of the BMCT is illustrated in the following example.

Example 8   Convergence of a sequence that is eventually monotonic

Show that the sequence $\left\{\dfrac{\ln n}{\sqrt{n}}\right\}$ converges.

Solution   We will apply the BMCT. A few values are shown in Figure 8.10.

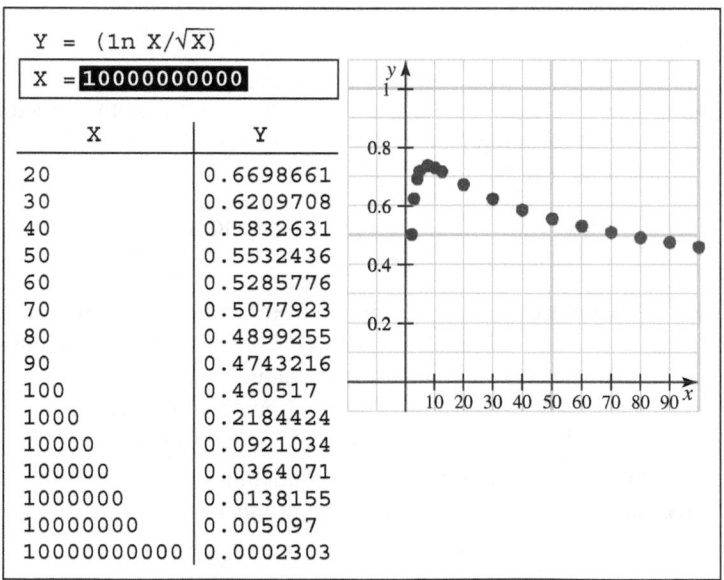

Figure 8.10 Interactive   Sequence $a_n = \dfrac{\ln n}{\sqrt{n}}$

The succession of numbers suggests that the sequence increases at first and then gradually begins to decrease. To verify this behavior, we let

$$f(x) = \frac{\ln x}{\sqrt{x}}$$

and find that

$$f'(x) = \frac{\sqrt{x}\left(\frac{1}{x}\right) - (\ln x)\left(\frac{1}{2}x^{-1/2}\right)}{x} = \frac{2\sqrt{x} - \sqrt{x}\ln x}{2x^2}$$

Find the critical values:

$$\frac{2\sqrt{x} - \sqrt{x}\ln x}{2x^2} = 0$$
$$2\sqrt{x} = \sqrt{x}\ln x$$
$$\ln x = 2$$
$$x = e^2$$

Thus, $x = e^2$ is the only critical value, and you can show that $f'(x) < 0$ for $x > e^2$ since $2 - \ln x < 0$ for $x > e^2$. This means that f is a decreasing function for $x > e^2$. Thus, the sequence $\left\{\dfrac{\ln n}{\sqrt{n}}\right\}$ must be decreasing for $n > 8$ (because e^2 is between 7 and 8).

We see that the sequence $\left\{\dfrac{\ln n}{\sqrt{n}}\right\}$ is bounded from below because

$$0 < \left\{\frac{\ln n}{\sqrt{n}}\right\} \qquad \text{for all } n \geq 2$$

Therefore, the given sequence is bounded from below and is eventually decreasing, so it must converge. ∎

The BMCT is an extremely valuable theoretical tool. For example, in Section 2.4, we defined the number e by the limit

$$\lim_{n\to\infty}\left(1 + \frac{1}{n}\right)^n = e$$

but to do so we assumed that this limit exists. We can now show this assumption is warranted, because it turns out that the sequence $\left\{\left(1 + \dfrac{1}{n}\right)^n\right\}$ is increasing and bounded from above by 3 (see Problems 57-59). Thus, the BMCT assures us that the sequence converges, and this in turn guarantees the existence of the limit. We end this section with a result that will be useful in our subsequent work. The proof makes use of the formal definition of BMCT.

Theorem 8.5   Convergence of a power sequence

If r is a fixed number such that $|r| < 1$, then $\lim\limits_{n\to\infty} r^n = 0$.

Proof: The case where $r = 0$ is trivial. We will prove the theorem for the case $0 < r < 1$ and leave the case $-1 < r < 0$ as an exercise (Problem 60).

$$0 \leq r^{n+1} = r^n r < r^n$$

The sequence $\{r^n\}$ is monotonic and bounded, so the limit exists. Either $\lim\limits_{n\to\infty} r^n = 0$ or $\lim\limits_{n\to\infty} r^n = L > 0$. Suppose

$$\lim_{n\to\infty} r^n = L$$
$$r\lim_{n\to\infty} r^n = rL \qquad \textit{Multiply both sides by } r.$$
$$\lim_{n\to\infty} rr^n = rL \qquad \textit{Properties of limits}$$
$$\lim_{n\to\infty} r^{n+1} = rL \qquad \textit{Properties of exponents}$$
$$< L \qquad \textit{Since } 0 < r < 1 \textit{ and } L > 0$$

But $\lim\limits_{n\to\infty} r^{n+1} = \lim\limits_{n\to\infty} r^n = L$ by the definition of limit. Hence $L = rL < L$ which is impossible, so $\lim\limits_{n\to\infty} r^n = 0$. ◆

PROBLEM SET 8.1

Level 1

1. ■ What does this say? What do we mean by the limit of a sequence?
2. ■ What does this say? What is meant by the bounded, monotonic, convergence theorem?

Write out the first five terms (beginning with $n = 1$) of the sequences given in Problems 3-12.

3. $\{1 + (-1)^n\}$

4. $\left\{\left(\frac{-1}{2}\right)^{n+2}\right\}$

5. $\left\{\frac{\cos 2n\pi}{n}\right\}$

6. $\left\{n \sin \frac{n\pi}{2}\right\}$

7. $\left\{\frac{3n + 1}{n + 2}\right\}$

8. $\left\{\frac{n^2 - n}{n^2 + n}\right\}$

9. $\{a_n\}$ where $a_1 = 256$ and $a_n = \sqrt{a_{n-1}}$ for $n \geq 2$

10. $\{a_n\}$ where $a_1 = 512$ and $a_n = \sqrt[3]{a_{n-1}}$ for $n \geq 2$

11. $\{a_n\}$ where $a_1 = 1$ and $a_n = (a_{n-1})^2 + a_{n-1} + 1$ for $n \geq 2$

12. $\{a_n\}$ where $a_1 = -1$ and $a_n = na_{n-1}$ for $n \geq 2$.

Compute the limit of the convergent sequences in Problems 13-36.

13. $\left\{\frac{5n}{n + 7}\right\}$

14. $\left\{\frac{5n + 8}{n}\right\}$

15. $\left\{\frac{4 - 7n}{8 + n}\right\}$

16. $\left\{\frac{2n + 1}{3n - 4}\right\}$

17. $\left\{\frac{100n + 7,000}{n^2 - n - 1}\right\}$

18. $\left\{\frac{8n^2 + 800n + 5,000}{2n^2 - 1,000n + 2}\right\}$

19. $\left\{\frac{n^3 - 6n^2 + 85}{2n^3 - 5n + 170}\right\}$

20. $\left\{\frac{8n^2 + 6n + 4,000}{n^3 + 1}\right\}$

21. $\left\{\frac{8n - 500\sqrt{n}}{2n + 800\sqrt{n}}\right\}$

22. $\left\{\frac{2n}{n + 7\sqrt{n}}\right\}$

23. $\left\{\frac{\ln n}{n^2}\right\}$

24. $\left\{\frac{3\sqrt{n}}{5\sqrt{n} + \sqrt[4]{n}}\right\}$

25. $\{n^{3/n}\}$

26. $\{2^{5/n}\}$

27. $\{(n + 4)^{1/n}\}$

28. $\left\{\left(1 + \frac{3}{n}\right)^n\right\}$

29. $\{(\ln n)^{1/n}\}$

30. $\{n^{1/(n+2)}\}$

31. $\left\{\sqrt{n^2 + n} - n\right\}$

32. $\left\{\int_0^\infty e^{-nx} dx\right\}$

33. $\{\sqrt[n]{n}\}$

34. $\left\{\sqrt{n + 5\sqrt{n}} - \sqrt{n}\right\}$

35. $\{(an + b)^{1/n}\}$ a and b positive constants

36. $\{\ln n - \ln(n + 1)\}$

Level 2

Show that each sequence given in Problems 37-42 converges either by showing it is increasing with an upper bound or decreasing with a lower bound.

37. $\left\{\ln\left(\frac{n + 1}{n}\right)\right\}$

38. $\left\{\frac{n}{2^n}\right\}$

39. $\left\{\frac{4n + 5}{n}\right\}$

40. $\left\{\frac{3n - 2}{n}\right\}$

41. $\{\sqrt[n]{n}\}$

42. $\left\{\frac{3n - 7}{2^n}\right\}$

Explain why each sequence in Problems 43-46 diverges.

43. $\{\cos n\pi\}$

44. $\{1 + (-1)^n\}$

45. $\{\sqrt{n}\}$

46. $\left\{\frac{n^3 - 7n + 5}{100n^2 + 219}\right\}$

47. **a.** Does the sequence $\{\frac{1}{n}(\sin n)\}$ converge, as $n \to \infty$? Give reasons.
    b. How about the sequence $\{n(\sin(\frac{1}{n}))\}$?
    c. Explain the relationship between the two sequences in this problem. *Hint:* Let $m = \frac{1}{n}$ in part **a.**

48. **Modeling Problem** A drug is administered into the body. At the end of each hour, the amount of drug present is half what it was at the end of the previous hour. What percent of the drug is present at the end of 4 hours? At the end of n hours?

49. Does the sequence $\left\{\sum_{k=1}^{n} \frac{n}{n^2 + k^2}\right\}$ converge? If so, to what? *Hint:* The sequence is related to an integral from 0 to 1, where k/n can be thought of as x_n.

50. *Historical Quest*

    *Leonardo de Pisa, also known as Fibonacci, was one of the best mathematicians of the Middle Ages. He played an important role in reviving ancient mathematics and introduced the Hindu-Arabic place-value decimal system to Europe. His book, **Liber Abaci**, published in 1202, introduced the Arabic numerals, as well as the famous **rabbit problem**, for which Fibonacci is best remembered today.*

Karl Smith library

Fibonacci (1170-1250)

To describe Fibonacci's rabbit problem, we consider a sequence whose nth term is defined by a *recursion formula* — that is, a formula in which the nth term is given in terms of previous terms in the sequence. Suppose rabbits breed in such a way that each pair of adult rabbits produces a pair of baby rabbits each month.

Number of Months	Number of Pairs	Pairs of Rabbits (the pairs shown in color are ready to reproduce in the next month)
Start	1	
1	1	
2	2	
3	3	
4	5	
5	8	
⋮	⋮	

Same pair
(rabbits
never die)

The first month after birth, the rabbits are adolescents and produce no offspring. However, beginning with the second month the rabbits are adults, and each pair produces a pair of offspring every month. The sequence of numbers describing the number of rabbits is called the *Fibonacci sequence*, and it has applications in many areas, including biology and botany.

In this *Quest* you are to examine some properties of the Fibonacci sequence. Let a_n denote the number of pairs of rabbits in the "colony" at the end of n months.

a. Explain why $a_1 = 1, a_2 = 1, a_3 = 2, a_4 = 3$, and, in general,

$$a_{n+1} = a_n + a_{n-1}$$

for $n = 2, 3, 4, \cdots$. The equation $a_{n+1} = a_n + a_{n-1}$ that specifies each term of the sequence $\{a_n\}$ for $n = 2, 3, \cdots$ is called a *recursion formula* because it defines later terms from those that precede them.

b. The *growth rate* of the colony during the $(n + 1)$th month is

$$r_n = \frac{a_{n+1}}{a_n}$$

Compute r_n for $n = 1, 2, 3, \cdots, 10$.

c. Assume that the growth rate sequence $\{r_n\}$ defined in part **b** converges. Find

$$L = \lim_{n \to \infty} r_n$$

Hint: Use the recursion formula in part **a**.

Level 3

51. The claim is made that

$$\lim_{n \to \infty} \frac{n}{n + 1} = 1$$

If $\epsilon = 0.01$ is chosen, find the smallest integer N so that

$$\left| \frac{n}{n + 1} - 1 \right| < 0.01$$

if $n > N$.

52. The claim is made that

$$\lim_{n \to \infty} \frac{2n + 1}{n - 3} = 2$$

If $\epsilon = 0.01$ is chosen, find the smallest integer N so that

$$\left| \frac{2n + 1}{n + 3} - 2 \right| < 0.01$$

if $n > N$.

53. The claim is made that

$$\lim_{n \to \infty} \frac{n^2 + 1}{n^3} = 0$$

If $\epsilon = 0.001$ is chosen, find the smallest integer N so that

$$\left| \frac{n^2 + 1}{n^3} \right| < 0.001$$

if $n > N$.

54. The claim is made that

$$\lim_{n \to \infty} e^{-n} = 0$$

If $\epsilon = 0.001$ is chosen, find the smallest integer N so that

$$\frac{1}{e^n} < 0.001$$

if $n > N$.

In Problems 55-56, we prove the BMCT. Suppose there exist a number M and a positive integer N so that

$$a_n \le a_{n+1} \le M$$

for all $n > N$.

55. It is a fundamental property of the real numbers that a sequence with an upper bound must have a *least* upper bound; that is, there must be a number A with the property that $a_n \le A$ for all n but if c is a number such that $c < A$, then $a_m > c$ for at least one m. Use this property to show that if $\epsilon > 0$ is given, there exists a positive integer N so that

$$A - \epsilon < a_n < A$$

for all $n > N$.

56. Show that $|a_n - A| < \epsilon$ for all $n > N$ and conclude that $\lim\limits_{n \to \infty} a_n = A$.

Problems 57-59 can be used to prove the convergence of the sequence

$$\left\{ \left(1 + \frac{1}{n} \right)^n \right\}$$

57. Suppose n is a positive integer. Use the binomial theorem to show that

$$\left(1 + \frac{1}{n} \right)^n = 1 + 1 + \frac{1}{2!} \left(1 - \frac{1}{n} \right)$$
$$+ \frac{1}{3!} \left(1 - \frac{1}{n} \right) \left(1 - \frac{2}{n} \right)$$
$$+ \cdots$$
$$+ \frac{1}{n!} \left(1 - \frac{1}{n} \right) \left(1 - \frac{2}{n} \right) \cdots \left(1 - \frac{n-1}{n} \right)$$

58. Use Problem 57 to show that

$$\left(1 + \frac{1}{n} \right)^n < 3$$

for all positive integers n.

59. Use Problem 57 to show that

$$\left\{ \left(1 + \frac{1}{n} \right)^n \right\}$$

is an increasing sequence. Then use the BMCT to show that this sequence converges.

60. Proof of the convergence power theorem In the text, we proved the case where $0 < r < 1$. Prove the remaining parts.
   a. $r = 0$
   b. $-1 < r < 0$

8.2 INTRODUCTION TO INFINITE SERIES; GEOMETRIC SERIES

IN THIS SECTION: *Definition of infinite series, general properties of infinite series, geometric series, applications of geometric series*

We define an infinite series (sum) as the limit of a special type of sequence. We then study a few basic properties of infinite series and examine *geometric series*, a particular kind of series with numerous applications.

Definition of Infinite Series

One way to add a list of numbers is to form subtotals until the end of the list is reached. Similarly, to give meaning to the infinite sum

$$S = a_1 + a_2 + a_3 + a_4 + \cdots$$

it is natural to examine the "partial sums"

$$a_1, \quad a_1 + a_2, \quad a_1 + a_2 + a_3, \quad a_1 + a_2 + a_3 + a_4, \quad \cdots$$

If it is possible to attach a numerical value to the infinite sum, we would expect the partial sums $S = a_1 + a_2 + a_3 + a_4 + \cdots + a_n$ to approach that value as n increases without bound. These ideas lead us to the following definition.

INFINITE SERIES An **infinite series** is an expression of the form

$$a_1 + a_2 + a_3 + \cdots = \sum_{k=1}^{\infty} a_k$$

and the **nth partial sum** of the series is

$$S_n = a_1 + a_2 + a_3 + a_4 + \cdots + a_n = \sum_{k=1}^{n} a_k$$

The *series* is said to **converge with sum S** if the *sequence* of partial sums $\{S_n\}$ converges to S. In this case, we write

$$\sum_{k=1}^{\infty} a_k = \lim_{n \to \infty} S_n = S$$

If the sequence $\{S_n\}$ does not converge, the series $\displaystyle\sum_{k=1}^{\infty} a_k$ **diverges** and has no sum.

■ **W**hat this says An infinite series converges if its sequence of partial sums converges and diverges otherwise. If it converges, its sum is defined to be the limit of the sequence of partial sums.

We will use the symbol $\displaystyle\sum_{k=1}^{\infty} a_k$ to denote the series $a_1 + a_2 + a_3 + \cdots$ regardless of whether this series converges or diverges. If the sequence of partial sums $\{S_n\}$ converges, then

$$\sum_{k=1}^{\infty} a_k = \lim_{n \to \infty} \left(\sum_{k=1}^{n} a_k \right)$$

and the symbol $\displaystyle\sum_{k=1}^{\infty} a_k$ is used to represent *both* the series and its sum.

We will also consider certain series in which the starting point is not 1; for example, the series

$$\frac{1}{3} + \frac{1}{4} + \frac{1}{5} + \cdots \text{ can be denoted by } \sum_{k=3}^{\infty} \frac{1}{k} \text{ or } \sum_{k=2}^{\infty} \frac{1}{k+1}$$

Example 1 Convergent series

Show that the series $\displaystyle\sum_{k=1}^{\infty} \frac{1}{2^k}$ converges.

Solution This series has the following partial sums.

$$S_1 = \frac{1}{2} \qquad\qquad\qquad \textit{Note:} \quad 1 - \frac{1}{2^1} = \frac{1}{2}$$

$$S_2 = \frac{1}{2} + \frac{1}{4} = \frac{3}{4} \qquad\qquad 1 - \frac{1}{2^2} = \frac{3}{4}$$

$$S_3 = \frac{1}{2} + \frac{1}{4} + \frac{1}{8} = \frac{7}{8} \qquad\qquad 1 - \frac{1}{2^3} = \frac{7}{8}$$

$$\vdots$$

$$S_n = \frac{1}{2} + \frac{1}{4} + \cdots + \frac{1}{2^n}$$

The sequence of partial sums is $\frac{1}{2}, \frac{3}{4}, \frac{7}{8}, \frac{15}{16}, \frac{31}{32}, \frac{63}{64}, \frac{127}{128}, \cdots$, and in general (by mathematical induction)

$$S_n = 1 - \frac{1}{2^n}$$

Because $\lim_{n \to \infty} \left(1 - \frac{1}{2^n}\right) = 1$, we conclude that the series converges and its sum is $S = 1$. ∎

Example 2 Divergent series

Show that the series $\sum_{k=1}^{\infty} (-1)^k$ diverges.

Solution   The series can be **expanded** (written out) as

$$\sum_{k=1}^{\infty} (-1)^k = -1 + 1 - 1 + 1 - 1 + \cdots\cdots$$

and we see that the nth partial sum is

$$S_n = \begin{cases} -1 & \text{if } n \text{ is odd} \\ 0 & \text{if } n \text{ is even} \end{cases}$$

Because the sequence $\{S_n\}$ has no limit, the given series must diverge. ∎

A series is called a **telescoping series** (or collapsing series) if there is internal cancellation in the partial sums, as illustrated by the following example.

Example 3 A telescoping series

Show that the series $\sum_{k=1}^{\infty} \frac{1}{k^2 + k}$ converges and find its sum.

Solution   Using partial fractions we find that

$$\frac{1}{k^2 + k} = \frac{1}{k(k+1)} = \frac{1}{k} + \frac{-1}{k+1}$$

Thus, the nth partial sum of the given series can be represented as follows:

$$\begin{aligned}
S_n &= \sum_{k=1}^{n} \frac{1}{k^2 + k} \\
&= \sum_{k=1}^{n} \left[\frac{1}{k} - \frac{1}{k+1}\right] \\
&= \left(1 - \frac{1}{2}\right) + \left(\frac{1}{2} - \frac{1}{3}\right) + \left(\frac{1}{3} - \frac{1}{4}\right) + \cdots + \left(\frac{1}{n} - \frac{1}{n+1}\right) \\
&= 1 + \left(-\frac{1}{2} + \frac{1}{2}\right) + \left(-\frac{1}{3} + \frac{1}{3}\right) + \cdots + \left(-\frac{1}{n} + \frac{1}{n}\right) - \frac{1}{n+1} \\
&= 1 - \frac{1}{n+1}
\end{aligned}$$

The limit of the sequence of partial sums is

$$\lim_{n \to \infty} S_n = \lim_{n \to \infty} \left[1 - \frac{1}{n+1}\right] = 1$$

so the series converges, with sum $S = 1$. ∎

General Properties of Infinite Series

Next, we will examine two general properties of infinite series. Here and elsewhere, *when the starting point of a series is not important, we may denote the series by writing*

$$\Sigma a_k \quad \text{instead of} \quad \sum_{k=1}^{\infty} a_k$$

Theorem 8.6 Linearity of infinite series

If Σa_k and Σb_k are convergent series, then so is $\Sigma(ca_k + db_k)$ for constants c, d, and

$$\Sigma(ca_k + db_k) = c\,\Sigma a_k + d\,\Sigma b_k$$

Proof: Compare this with the linearity property in Theorem 5.4 In Chapter 5, the limit properties are for finite sums, and this theorem is for infinite sums. The proof of this theorem is similar to the proof of Theorem 5.4, but in this case it follows from the linearity rule for sequences (Theorem 8.1). The details are left as a problem. ◆

Example 4 Linearity used to establish convergence

Show that the series $\displaystyle\sum_{k=1}^{\infty} \left[\frac{4}{k^2 + k} - \frac{6}{2^k} \right]$ converges, and find its sum.

Solution Because we know from Examples 1 and 3 that $\displaystyle\sum_{k=1}^{\infty} \frac{1}{k^2 + k}$ and $\displaystyle\sum_{k=1}^{\infty} \frac{1}{2^k}$ both converge with sum 1, the linearity property allows us to write the given series as

$$4\sum_{k=1}^{\infty} \frac{1}{k^2 + k} - 6\sum_{k=1}^{\infty} \frac{1}{2^k}$$

We can now conclude that the series converges and that the sum is

$$4\sum_{k=1}^{\infty} \frac{1}{k^2 + k} - 6\sum_{k=1}^{\infty} \frac{1}{2^k} = 4(1) - 6(1) = -2$$

The linearity property also provides useful information about a series of the form $\Sigma(ca_k + db_k)$ when exactly one of Σa_k or Σb_k diverges; that is, one converges and the other diverges.

Theorem 8.7 Divergence of the sum of a convergent and a divergent series

If either Σa_k or Σb_k diverges and the other converges, then the series $\Sigma(a_k + b_k)$ must diverge.

Proof: Suppose Σa_k diverges and Σb_k converges. Then if the series $\Sigma(a_k + b_k)$ also converges, the linearity property tells us that the series

$$\Sigma\left[(a_k + b_k) - b_k\right] = \Sigma a_k$$

must converge, contrary to hypotheses. It follows that the series $\Sigma(a_k + b_k)$ diverges. ◆

This theorem tells us, for example, that

$$\sum_{k=1}^{\infty} \left[\frac{1}{k^2 + k} + (-1)^k \right]$$

must diverge because even though $\displaystyle\sum_{k=1}^{\infty} \frac{1}{k^2 + k}$ converges (Example 3), the series $\displaystyle\sum_{k=1}^{\infty}(-1)^k$ diverges (Example 2).

Geometric Series

We are still very limited in the available techniques for determining whether a given series is convergent or divergent. In fact, the purpose of a great part of this chapter is to develop efficient techniques for making this determination. We begin this quest by considering an important special kind of series.

> **GEOMETRIC SERIES**   A **geometric series** is an infinite series in which the ratio of successive terms in the series is constant. If this constant ratio is r, then the series has the form
>
> $$\sum_{k=0}^{\infty} ar^k = a + ar + ar^2 + ar^3 + \cdots + ar^n + \cdots \quad a \neq 0$$

For example, $3 + \frac{3}{2} + \frac{3}{4} + \frac{3}{8} + \cdots$ is a geometric series because each term is one-half the preceding term. The ratio of a geometric series may be positive or negative. For example,

$$\sum_{k=0}^{\infty} \frac{2}{(-3)^k} = 2 - \frac{2}{3} + \frac{2}{9} - \frac{2}{27} + \cdots$$

is a geometric series with $r = -\frac{1}{3}$. The following theorem tells us how to determine whether a given geometric series converges or diverges and, if it does converge, what its sum must be.

Theorem 8.8   Geometric series theorem

The geometric series $\displaystyle\sum_{k=0}^{\infty} ar^k$ with $a \neq 0$ diverges if $|r| \geq 1$ and converges if $|r| < 1$ with sum

$$\sum_{k=0}^{\infty} ar^k = \frac{a}{1 - r}$$

Proof: Note that the nth partial sum of the geometric series is

$$S_n = a + ar + ar^2 + \cdots + ar^{n-1} \qquad \textit{Definition of a geometric series}$$

$$rS_n = ar + ar^2 + ar^3 + \cdots + ar^n \qquad \textit{Multiply both sides by } r.$$

$$rS_n - S_n = \left(ar + ar^2 + ar^3 + \cdots + ar^n\right) - \left(a + ar + ar^2 + \cdots + ar^{n-1}\right) \quad \textit{Subtract.}$$

$$(r - 1)S_n = ar^n - a \qquad \textit{Simplify.}$$

$$S_n = \frac{a(r^n - 1)}{r - 1} \qquad \textit{where} \quad r \neq 1$$

If $|r| > 1$, the sequence of partial sums $\{S_n\}$ has no limit since r^n is unbounded in this case, so the geometric series must diverge. However, if $|r| < 1$, Theorem 8.5 tells us that $r^n \to 0$ as $n \to \infty$, so we have

$$\sum_{k=0}^{\infty} ar^k = \lim_{n \to \infty} S_n = \lim_{n \to \infty} a\left(\frac{r^n - 1}{r - 1}\right) = \frac{a(0 - 1)}{r - 1} = \frac{a}{1 - r}$$

To complete the proof, it must be shown that the geometric series diverges when $|r| = 1$, and we leave this final step as a problem. ◆

Example 5   Testing for convergence in a geometric series

Determine whether each of the following geometric series converges or diverges. If the series converges, find its sum.

a. $\displaystyle\sum_{k=0}^{\infty} \frac{1}{7}\left(\frac{3}{2}\right)^k$ **b.** $\displaystyle\sum_{k=2}^{\infty} 3\left(-\frac{1}{5}\right)^k$

Solution

a. Because $r = \frac{3}{2}$ satisfies $|r| \geq 1$, the series diverges.
b. We have $r = -\frac{1}{5}$ so $|r| < 1$, and the geometric series converges. The first value of k is 2 (not 0), so the value a (the first value) is

$$a = 3\left(-\frac{1}{5}\right)^2 = \frac{3}{25}$$

and

$$\sum_{k=2}^{\infty} 3\left(-\frac{1}{5}\right)^k = \frac{a}{1 - r} = \frac{\frac{3}{25}}{1 - \left(-\frac{1}{5}\right)} = \frac{1}{10}$$ ∎

Applications of Geometric Series

Geometric series can be used in many different ways. Our next three examples illustrate several applications involving geometric series.

Recall that a rational number r is one that can be written as $r = p/q$ for integer p and nonzero integer q. It can be shown that any such number has a decimal representation in which a pattern of numbers repeats. For example,

$$\frac{5}{10} = 0.5 = 0.5\overline{0} \qquad \frac{5}{11} = 0.454545\cdots = 0.\overline{45}$$

where the bar indicates that the pattern repeats indefinitely. Example 6 shows how geometric series can be used to reverse this process by writing a given repeating decimal as a rational number.

Example 6   Expressing a repeating decimal as a rational number

Write $15.4\overline{23}$ as a rational number $\dfrac{p}{q}$.

Solution   The bar over the 23 indicates that this block of digits is to be repeated, that is, $15.4\overline{23} = 15.423232323\cdots$. The repeating part of the decimal can be written as a geometric series as follows.

$$15.4\overline{23} = 15 + \frac{4}{10} + \frac{23}{10^3} + \frac{23}{10^5} + \frac{23}{10^7} + \cdots$$

$$= 15 + \frac{4}{10} + \frac{23}{10^3}\left[1 + \frac{1}{10^2} + \frac{1}{10^4} + \cdots\right]$$

$$= 15 + \frac{4}{10} + \frac{23}{10^3}\left[\frac{1}{1 - \frac{1}{100}}\right] \qquad \textit{Sum of a geometric series with}$$
$$\textit{a = 1, r = 0.01.}$$

$$= 15 + \frac{4}{10} + \frac{23}{10^3}\left[\frac{100}{99}\right]$$

$$= 15 + \frac{4}{10} + \frac{23}{990}$$

$$= \frac{15{,}269}{990}$$

A tax rebate that returns a certain amount of money to taxpayers can result in spending that is many times this amount. This phenomenon is known in economics as the *multiplier effect*. It occurs because the portion of the rebate that is spent by one individual becomes income for one or more others, who, in turn, spend some of it again, creating income for yet other individuals to spend. If the fraction of income that is saved remains constant as this process continues indefinitely, the total amount spent as a result of the rebate is the sum of a geometric series.*

Example 7   Modeling Problem: Multiplier effect in economics

Suppose that nationwide approximately 90% of all income is spent and 10% is saved. How much total spending will be generated by a $40 billion tax rebate if savings habits do not change?

Solution   The amount (in billions) spent by original recipients of the rebate is 36. This becomes new income, of which 90%, or 0.9(36), is spent. This, in turn, generates additional spending of 0.9[0.9(36)], and so on. The total amount spent if this process continues indefinitely is

$$36 + 0.9(36) + 0.9^2(36) + 0.9^3(36) + \cdots = 36\big[1 + 0.9 + 0.9^2 + \cdots\big]$$

$$= 36 \sum_{k=0}^{\infty} 0.9^k \qquad \textit{Geometric series with}$$
$$\phantom{= 36 \sum_{k=0}^{\infty} 0.9^k} \qquad \textit{a = 1, r = 0.9}$$

$$= 36 \left(\frac{1}{1 - 0.9} \right)$$

$$= 360$$

The total spending generated by the $40 billion tax rebate is $360 billion.

Example 8   Modeling Problem: Accumulation of medication in a body

A patient is given an injection of 10 units of a certain drug every 24 hours. The drug is eliminated exponentially so that the proportion that remains in the patient's body after t days is $f(t) = e^{-t/5}$. If the treatment is continued indefinitely, approximately how many units of the drug will eventually be in the patient's body just prior to an injection?

Solution   Of the original dose of 10 units, only $10e^{-1/5}$ are left in the patient's body after the first day (just prior to the second injection). That is,

$$S_1 = 10e^{-1/5}$$

The medication in the patient's body after 2 days consists of what remains from the first two doses. Of the original dose, only $10e^{-2/5}$ units are left (because 2 days have elapsed), and of the second dose, $10e^{-1/5}$ units remain:

$$S_2 = 10e^{-1/5} + 10e^{-2/5}$$

Similarly, for n days,

$$S_n = 10e^{-1/5} + 10e^{-2/5} + \cdots + 10e^{-n/5} = \sum_{k=1}^{n} 10e^{-k/5}$$

---

*The fraction s of income that is saved is called the *marginal propensity to save* and $c = 1 - s$ is the *marginal propensity to consume*. In Example 7, we have $s = 0.1$ and $c = 0.9$.

The amount S of medication in the patient's body in the long run is the limit of S_n as $n \to \infty$. That is,

$$S = \lim_{n \to \infty} S_n$$
$$= \sum_{k=1}^{\infty} 10e^{-k/5} \quad \textit{Geometric series with } a = 10e^{-1/5} \textit{ and } r = e^{-1/5}.$$
$$= \frac{10e^{-1/5}}{1 - e^{-1/5}}$$
$$\approx 45.166566$$

We see that about 45 units remain in the patient's body.

STOP *As a final note, remember that a sequence is a succession of terms, whereas a series is a sum of such terms. Do not confuse the two concepts because they have very different properties. For example, a sequence of terms may converge, but the series of the same terms may diverge:*

$$\left\{ 1 + \frac{1}{2^n} \right\} \textit{ is the sequence } \frac{3}{2}, \frac{5}{4}, \frac{9}{8}, \frac{17}{16} \textit{ which converges to } 1.$$

$$\sum_{n=1}^{\infty} \left(1 + \frac{1}{2^n} \right) \textit{ is the series } \frac{3}{2} + \frac{5}{4} + \frac{9}{8} + \cdots, \textit{ which diverges.}$$

PROBLEM SET 8.2

Level 1

1. ■ What does this say? Explain the difference between sequences and series.
2. ■ What does this say? What is a geometric series?

Determine whether the geometric series given in Problems 3-22 converges or diverges, and find the sum of each convergent series.

3. $\displaystyle\sum_{k=0}^{\infty} \left(\frac{4}{5} \right)^k$

4. $\displaystyle\sum_{k=0}^{\infty} \left(-\frac{4}{5} \right)^k$

5. $\displaystyle\sum_{k=0}^{\infty} \frac{2}{3^k}$

6. $\displaystyle\sum_{k=0}^{\infty} \frac{2}{(-3)^k}$

7. $\displaystyle\sum_{k=1}^{\infty} \left(\frac{3}{2} \right)^k$

8. $\displaystyle\sum_{k=1}^{\infty} \frac{3}{2^k}$

9. $\displaystyle\sum_{k=0}^{\infty} \frac{(-2)^k}{5^{2k+1}}$

10. $\displaystyle\sum_{k=1}^{\infty} 5(0.9)^k$

11. $\displaystyle\sum_{k=1}^{\infty} e^{-0.2k}$

12. $\displaystyle\sum_{k=1}^{\infty} \frac{3^k}{4^{k+2}}$

13. $\displaystyle\sum_{k=2}^{\infty} \frac{(-2)^{k-1}}{3^{k+1}}$

14. $\displaystyle\sum_{k=2}^{\infty} (-1)^k \frac{2^{k+1}}{3^{k-3}}$

15. $\dfrac{1}{2} - \dfrac{1}{2^2} + \dfrac{1}{2^3} - \dfrac{1}{2^4} + \cdots$

16. $1 + \pi + \pi^2 + \pi^3 + \cdots$

17. $\dfrac{1}{4} + \left(\dfrac{1}{4} \right)^4 + \left(\dfrac{1}{4} \right)^7 + \left(\dfrac{1}{4} \right)^{10} + \cdots$

18. $\dfrac{2}{3} - \left(\dfrac{2}{3} \right)^3 + \left(\dfrac{2}{3} \right)^5 - \left(\dfrac{2}{3} \right)^7 + \cdots$

19. $2 + \sqrt{2} + 1 + \dfrac{1}{\sqrt{2}} + \cdots$

20. $3 - \sqrt{3} + 1 - \dfrac{1}{\sqrt{3}} + \cdots$

21. $\left(1 + \sqrt{2} \right) + 1 + \left(-1 + \sqrt{2} \right) + \left(3 - 2\sqrt{2} \right) + \cdots$

22. $\left(\sqrt{2} - 1 \right) + 1 + \left(\sqrt{2} + 1 \right) + (2\sqrt{2} + 3) + \cdots$

In Problems 23-30, each series telescopes. In each case, express the nth partial sum S_n in terms of n and determine whether the series converges or diverges by examining $\lim_{n \to \infty} S_n$. Notice that some of these problems may require partial fraction decomposition.

23. $\displaystyle\sum_{k=1}^{\infty} \left[\frac{1}{k^{0.1}} - \frac{1}{(k+1)^{0.1}} \right]$

24. $\displaystyle\sum_{k=1}^{\infty} \left[\frac{1}{2k+1} - \frac{1}{2k+3} \right]$

25. $\displaystyle\sum_{k=0}^{\infty} \frac{1}{(k+1)(k+2)}$ 26. $\displaystyle\sum_{k=2}^{\infty} \frac{1}{k(k+1)}$

27. $\displaystyle\sum_{k=1}^{\infty} \ln\left(1 + \frac{1}{k}\right)$ 28. $\displaystyle\sum_{k=1}^{\infty} \frac{1}{(2k-1)(2k+1)}$

29. $\displaystyle\sum_{k=1}^{\infty} \frac{2k+1}{k^2(k+1)^2}$ 30. $\displaystyle\sum_{k=1}^{\infty} \frac{\sqrt{k+1} - \sqrt{k}}{\sqrt{k^2 + k}}$

Express each repeating decimal given in Problems 31-34 as a common (reduced) fraction.

31. $0.\overline{01}$ 32. $2.23\overline{1}$ 33. $1.\overline{405}$ 34. $41.2\overline{010}$

Level 2

35. **a.** Find numbers A and B such that

$$\frac{k-1}{2^{k+1}} = \frac{Ak}{2^k} - \frac{B(k+1)}{2^{k+1}}$$

   b. Evaluate

$$\sum_{k=1}^{\infty} \frac{k-1}{2^{k+1}}$$

   as a telescoping series.

36. Evaluate

$$\sum_{k=1}^{\infty} \frac{2k-1}{3^{k+1}}$$

   Hint: Follow the procedure in Problem 35.

37. Evaluate

$$\sum_{n=1}^{\infty} \frac{\ln\left[\frac{n^{n+1}}{(n+1)^n}\right]}{n(n+1)}$$

   Hint: Use the properties of logarithms to show that the series telescopes.

38. Evaluate

$$\sum_{n=1}^{\infty} \frac{n}{(n+1)!}$$

   Hint: Express as a telescoping series.

39. Find $\displaystyle\sum_{k=0}^{\infty} (2a_k + 2^{-k})$ given that $\displaystyle\sum_{k=0}^{\infty} a_k = 3.57$.

40. Find $\displaystyle\sum_{k=0}^{\infty} \frac{b_k - 3^{-k}}{2}$ given that $\displaystyle\sum_{k=0}^{\infty} b_k = 0.54$.

41. Evaluate $\displaystyle\sum_{k=0}^{\infty} \left(\frac{1}{2^k} + \frac{1}{3^k}\right)^2$.

42. Evaluate $\displaystyle\sum_{k=0}^{\infty} \left[\left(\frac{2}{3}\right)^k + \left(\frac{3}{4}\right)^k\right]^2$.

Modeling Problems: *In Problems 43-53, set up an appropriate model to answer the given question.*

43. How much should you invest today at an annual interest rate of 1.5% compounded continuously so that, starting next year, you can make annual withdrawals of $2,000 forever?

44. A pendulum is released through an arc of length 20 cm from a vertical position and is allowed to swing free until it eventually comes to rest. Each subsequent swing of the bob of the pendulum is 90% as far as the preceding swing. How far will the bob travel before coming to rest?

45. A flywheel rotates at 500 rpm and slows in such a way that during each minute it revolves at two-thirds the rate of the preceding minute. How many total revolutions does the flywheel make before coming to rest?

46. A ball is dropped from a height of 10 ft. Each time the ball bounces, it rises 0.6 the distance it had previously fallen. What is the total distance traveled by the ball?

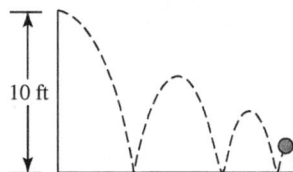

47. Suppose the ball in Problem 46 is dropped from a height of h feet and on each bounce rises 75% of the distance it had previously fallen. If it travels a total distance of 21 ft, what is h?

48. Suppose that nationwide approximately 92% of all income is spent and 8% is saved. How much total spending will be generated by a $50 billion tax cut if savings habits do not change?

49. Suppose that a piece of machinery costing $10,000 depreciates 20% of its current value each year. That is, the first year $10,000(0.20) = $2,000 is depreciated. The second year's depreciation is $8,000(0.20) = $1,600, because the value for the second year is $10,000 - $2,000 = $8,000. If the depreciation is calculated this way indefinitely, what is the total depreciation?

50. Winnie Winner wins $100 in a pie-baking contest run by the Hi-Do Pie Co. The company gives Winnie the $100. However, the tax collector wants 20% of the $100. Winnie pays the tax. But then she realizes that she didn't really win a $100 prize and tells her story to the Hi-Do Co. The friendly Hi-Do Co. gives Winnie the $20 she paid in taxes. Unfortunately, the tax collector now wants 20% of the $20. She pays the tax again and then goes back to the Hi-Do Co. with her story. Assume that this can go on indefinitely. How much money does the Hi-Do Co. have to give Winnie so that she will really win $100? How much does she pay in taxes?

51. A patient is given an injection of 20 units of a certain drug every 24 hr. The drug is eliminated exponentially, so the part that remains in the patient's body after t days is $f(t) = e^{-t/2}$. If the treatment is to continue indefinitely, approximately how many units of the drug will eventually be in the patient's body just before an injection?

52. A certain drug is injected into a body. It is known that the amount of drug in the body at the end of a given hour is $\frac{1}{4}$ the amount that was present at the end of the previous hour. One unit is injected initially, and to keep the concentration up, an additional unit is injected at the end of each subsequent hour. What is the highest possible amount of the drug that can ever be present in the body?

53. Each January 1, the administration of a certain private college adds six new members to its board of trustees. If the fraction of trustees who remain active for at least t years is $f(t) = e^{-0.2t}$, approximately how many active trustees will the college have in the long run? Assume that there were six members on the board on January 1 of the present year and that the trustees are counted on December 31.

54. 𝔥*istorical* 𝔔*uest*

A professor of mathematics at the University of Warsaw, Waclaw Sierpinski was the first person to give a lecture course devoted entirely to set theory. He became interested in set theory in 1907 when he came across a theorem that stated that points in the plane could be specified with a single coordinate.

Waclaw Sierpinski (1882-1969)

He wrote to a friend and asked how such a result could be possible, and he received a one word reply: "Cantor." Georg Cantor (1845-1918) wrote a series of eight papers which provided the basis for what we today call set theory. For this 𝔔*uest, we investigate an interesting geometrical figure named after Sierpinski, namely the Sierpinski gasket, shown in Figure 8.11.*

Figure 8.11 Interactive Sierpinski gasket

To form the gasket, begin by drawing an equilateral triangle, T, with sides of length 1. Remove the middle triangle formed by joining the midpoints of the sides of T, as shown in Figure 8.12**a**. This leaves three equal triangular regions, and the next step is to remove the middle triangle from each of these, as shown in Figure 8.12**b**. If you continue this process indefinitely, what remains is the Sierpinski gasket. Show that the Sierpinski gasket has area 0. *Hint*: What is the sum of the areas of the regions removed?

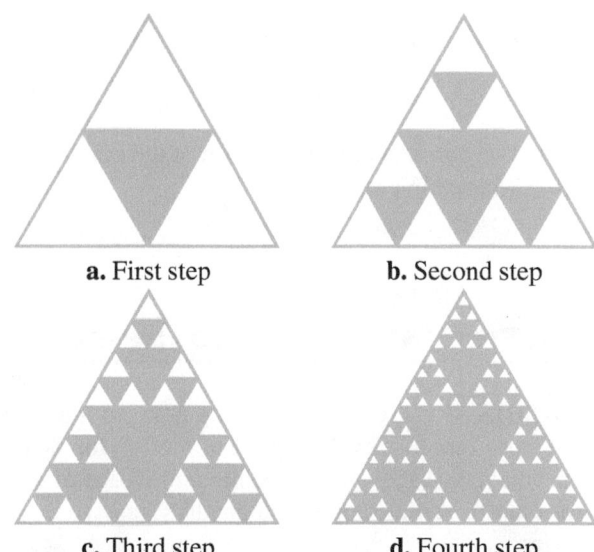

a. First step   **b.** Second step

c. Third step   **d.** Fourth step

Figure 8.12 Construction of a Sierpinski gasket

55. 𝔥*istorical* 𝔔*uest*

John von Neumann was one of the original six mathematics professors at the Institute for Advanced Study at Princeton; the other five were also world-class mathematicians or Nobel laureates: J. W. Alexander, Albert Einstein, Marston Morse, Oswald Veblen, and Hermann Weyl.

John von Neumann (1903-1957)

Von Neumann, a pioneer in computer science, received two Presidential Awards, the Medal for Merit and the Medal for Freedom. The fact that he had a brilliant mind is an understatement, as illustrated in the following anecdote, stated in the form of a quest problem. Two trains, each traveling at 10 mph approach each other on the same straight track. When the trains are 5 miles apart, a fly begins flying back and forth from train to train at the rate of 16 mi/h. Between the time the fly starts and the time the trains collide, how far does the fly fly?

a. Find the fly's total distance traveled by determining how long the fly flies.

b. *When von Neumann gave the correct answer almost immediately, the poser of the problem was convinced he had used the "trick" suggested in part* **a.** *"What trick?" asked von Neumann. "I summed the series."* Follow von Neumann's lead by summing an infinite series whose terms are the distances traveled by the fly as it flies between the trains.

Level 3

56. a. Let $\sum_{k=1}^{n} a_k = S_n$ be the nth partial sum of an infinite series. Show that

$$S_n - S_{n-1} = a_n$$

b. Use part **a** to find the nth term of the infinite series whose nth partial sum is

$$S_n = \frac{n}{2n + 3}$$

57. Show that if Σa_k converges and Σb_k diverges, then $\Sigma(a_k - b_k)$ diverges.
Hint: Note that $b_k = a_k - (a_k - b_k)$.

58. Show that $\sum_{k=0}^{\infty} ar^k$ diverges if $|r| \geq 1$.

59. Evaluate

$$\sum_{k=1}^{\infty} \frac{12^k}{(4^k - 3^k)(4^{k+1} - 3^{k+1})}$$

Hint: Find A and B so that

$$\frac{3^k A}{4^k - 3^k} + \frac{3^k B}{4^{k+1} - 3^{k+1}} = \frac{12^k}{(4^k - 3^k)(4^{k+1} - 3^{k+1})}$$

60. Square *ABCD* has sides of length 1. Square *EFGH* is formed by connecting the midpoints of the sides of the first square, as shown in Figure 8.13.

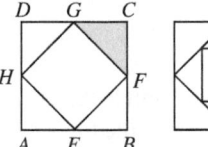 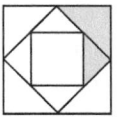 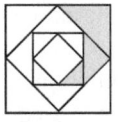 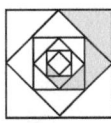

Figure 8.13   Find the area of the shaded regions

Assume that the pattern of shaded regions in the square is continued indefinitely. What is the total area of the shaded regions?

8.3 THE INTEGRAL TEST; *P*-SERIES

IN THIS SECTION: *Divergence test, series of nonnegative numbers; the integral test, p-series*
We now turn to developing some "tests" for determining the convergence of a given series. These tests have the advantage that they do not require knowledge of an explicit formula (such as the one we found for a geometric series) for the nth partial sum S_n and the disadvantage that they usually cannot be used to find the actual sum of a convergent series.

The convergence or divergence of an infinite series is determined by the behavior of its nth partial sum, S_n, as $n \to \infty$. In Section 8.2 we saw how algebraic methods can sometimes be used to find formulas for the nth partial sum of a series. Unfortunately, it is often difficult or even impossible to find a usable formula for the nth partial sum S_n of a series, and other techniques must be used to determine convergence or divergence.

Divergence Test

When investigating the infinite series Σa_k, it is easy to confuse the sequence of *general terms* $\{a_k\}$ with the sequence of *partial sums* $\{S_n\}$, where

$$S_n = \sum_{k=1}^{n} a_k$$

Because the sequence $\{a_k\}$ is usually more accessible than $\{S_n\}$, it would be convenient if the convergence of $\{S_n\}$ could be settled by examining

$$\lim_{k \to \infty} a_k$$

Even though we do not have one simple, definitive test for convergence, the following theorem tells us that if certain conditions are not met, the series must diverge.

Theorem 8.9   The divergence test

If $\lim_{k \to \infty} a_k \neq 0$, then the series Σa_k must diverge.

> ■ **W**hat this says The divergence test can only tell us that Σa_k diverges if $\lim_{k \to \infty} a_k \neq 0$, **but it *cannot* be used to show convergence of a series**. In Example 2 we show that the series $\Sigma \frac{1}{k}$ diverges even though $\lim_{k \to \infty} \frac{1}{k} = 0$.

Proof: Suppose the sequence of partial sums $\{S_n\}$ converges with sum L, so that $\lim_{n \to \infty} S_n = L$. Then we also have

$$\lim_{n \to \infty} S_{n-1} = L$$

Because $S_k - S_{k-1} = a_k$, it follows that

$$\lim_{k \to \infty} a_k = \lim_{k \to \infty} (S_k - S_{k-1}) = \lim_{k \to \infty} S_k - \lim_{k \to \infty} S_{k-1} = L - L = 0$$

If the series Σa_k converges, then $\lim_{k \to \infty} a_k = 0$. Thus, if $\lim_{k \to \infty} a_k \neq 0$, the series Σa_k must diverge. ◆

Theorem 8.9 gives a necessary but not a sufficient condition for convergence. For example, clouds are necessary for rain; clouds are not sufficient for rain, just as $\lim_{k \to \infty} a_k = 0$ is necessary for $\sum a_k$ to converge, but is not sufficient to make it converge. Note also that this is therefore called a DIVERGENCE test, not a CONVERGENCE test.

Example 1   Divergence test to show divergence

Show that the series $\displaystyle\sum_{k=0}^{\infty} \frac{k - 300}{4k + 750}$ diverges.

Solution   Taking the limit of the kth term as $k \to \infty$, we find

$$\lim_{k \to \infty} \frac{k - 300}{4k + 750} = \frac{1}{4}$$

Because this limit is not 0, the divergence test tells us that the series must diverge. ■

Series of Nonnegative Numbers; The Integral Test

Series whose terms are all nonnegative numbers play an important role in the general theory of infinite series and in applications. Our next goal is to develop methods for investigating the convergence of nonnegative-term series, and we begin by establishing the following general principle.

Theorem 8.10   Convergence criterion for series with nonnegative terms

A series Σa_k with $a_k \geq 0$ for all k converges if its sequence of partial sums is bounded from above and diverges otherwise.

Proof: Suppose Σa_k is a series of nonnegative terms, and let S_n denote its *n*th partial sum; that is

$$S_n = \sum_{k=1}^{n} a_k = a_1 + a_2 + \cdots + a_n$$

Because $S_{n+1} = S_n + a_{n+1}$ and because $a_k \geq 0$ for all k, it follows that

$$S_{n+1} \geq S_n$$

for all n, so that $\{S_n\}$ is an increasing sequence. According to the BMCT, the increasing sequence $\{S_n\}$ converges if it is bounded from above and diverges otherwise. Hence, because the series Σa_k represents the limit of the sequence $\{S_n\}$, the series Σa_k converges if the sequence $\{S_n\}$ is bounded from above and diverges if it is unbounded. ◆

This convergence criterion is often difficult to apply because it is not easy to determine whether the sequence of partial sums $\{S_n\}$ is bounded from above. Our next goal is to examine various "tests for convergence." These are procedures that allow us to determine indirectly whether a given series converges or diverges without actually having to compute the limit of the partial sums. We begin with a convergence test that relates the convergence of a series to that of an improper integral.

Theorem 8.11　The integral test

If $a_k = f(k)$ for $k = 1, 2, \cdots$, where f is a positive, continuous, and decreasing function of x for $x \geq 1$, then

$$\sum_{k=1}^{\infty} a_k \quad \text{and} \quad \int_1^{\infty} f(x)\,dx$$

either both converge or both diverge.

When applying the integral test, the function f need not be decreasing for all $x \geq 1$, only for all $x \geq N$ for some number N. That is, f must be decreasing "in the long run."

Proof: We will use a geometric argument to show that the sequence of partial sums $\{S_n\}$ of the series $\sum_{k=1}^{\infty} a_k$ is bounded from above if the improper integral $\int_1^{\infty} f(x)\,dx$ converges and that it is unbounded if the integral diverges. Accordingly, Figure 8.14**a** and Figure 8.14**b** show the graph of a continuous decreasing function f that satisfies $f(k) = a_k$ for $k = 1, 2, 3, \cdots$. Rectangles have been constructed at unit intervals in both figures, but in Figure 8.14**a**, kth rectangle has height $f(k+1) = a_{k+1}$, whereas in Figure 8.14**b**, the comparable rectangle has height $f(k) = a_k$.

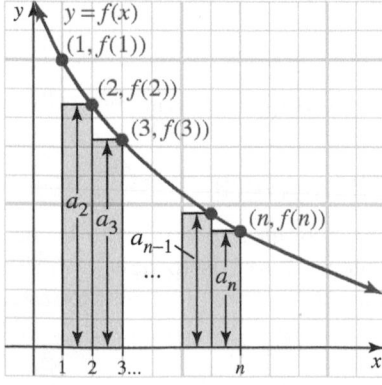

a. The kth rectangle has height $f(k+1) = a_{k+1}$

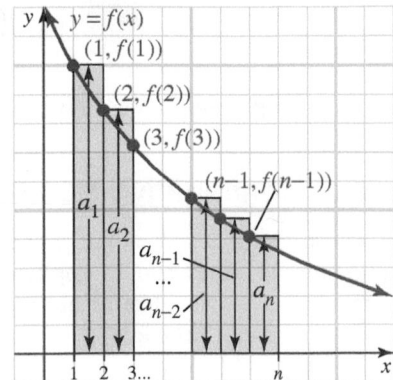

b. The kth rectangle has height $f(k) = a_k$

Figure 8.14 Interactive　Integral test

Notice that in Figure 8.14**a**, the rectangles are inscribed so that

AREA OF THE FIRST $n - 1$ RECTANGLES < AREA UNDER $y = f(x)$ OVER $[1, n]$

$$a_2 + a_3 + \cdots + a_n < \int_1^n f(x)\,dx$$

$$a_1 + a_2 + a_3 + \cdots + a_n < a_1 + \int_1^n f(x)\,dx \qquad \textit{Add } a_1 \textit{ to both sides.}$$

Similarly, in Figure 8.14**b**, the rectangles are circumscribed so that

AREA UNDER $y = f(x)$ OVER $[1, n+1]$ < AREA OF THE FIRST n RECTANGLES

$$\int_1^{n+1} f(x)\,dx < a_1 + a_2 + a_3 + \cdots + a_n$$

Let $S_n = a_1 + a_2 + a_3 + \cdots + a_n$ be the nth partial sum of the series so that

$$\int_1^{n+1} f(x)\,dx < S_n < a_1 + \int_1^n f(x)\,dx$$

Now, suppose the improper integral $\int_1^\infty f(x)\,dx$ converges and has the value I; that is,

$$\int_1^\infty f(x)\,dx = I$$

Then

$$S_n < a_1 + \int_1^n f(x)\,dx < a_1 + \int_1^\infty f(x)\,dx = a_1 + I$$

It follows that the sequence of partial sums is bounded from above (by $a_1 + I$), and the convergence criterion for nonnegative term series tells us that the series $\displaystyle\sum_{k=1}^\infty a_k$ must converge.

On the other hand, if the improper integral $\int_1^\infty f(x)\,dx$ diverges, then

$$\lim_{n \to \infty} \int_1^{n+1} f(x)\,dx = \infty$$

It follows that $\displaystyle\lim_{n \to \infty} S_n = \infty$ also because

$$\int_1^{n+1} f(x)\,dx < S_n$$

This means that the series must diverge. Thus, the series and the improper integral either both converge or both diverge, as claimed. ◆

A series that arises in connection with overtones produced by a vibrating string is

$$\sum_{k=1}^\infty \frac{1}{k} = 1 + \frac{1}{2} + \frac{1}{3} + \frac{1}{4} + \cdots$$

This is called the **harmonic series**, and in the next example, we use the integral test to show that this series diverges. Another proof that does not use the integral test can be found in the supplementary problems at the end of this chapter.

Example 2   Harmonic series diverges

Test the series $\sum\limits_{k=1}^{\infty} \frac{1}{k}$ for convergence.

Solution   Because $f(x) = \frac{1}{x}$ is positive, continuous, and decreasing for $x \geq 1$, the conditions of the integral test are satisfied.

$$\int_1^\infty \frac{1}{x}\,dx = \lim_{b \to \infty} \int_1^b \frac{1}{x}\,dx = \lim_{b \to \infty} [\ln b - \ln 1] = \infty$$

The integral diverges, so the harmonic series diverges.   ■

Example 3   Integral test for convergence

Test the series $\sum\limits_{k=1}^{\infty} \dfrac{k}{e^{k/5}}$ for convergence.

Solution   The function $f(x) = \dfrac{x}{e^{x/5}} = xe^{-x/5}$ is positive and continuous for all $x > 0$. We find that

$$f'(x) = x\left(-\frac{1}{5}e^{-x/5}\right) + e^{-x/5} = \left(1 - \frac{x}{5}\right)e^{-x/5}$$

The critical number is found when $f'(x) = 0$, so we solve

$$\left(1 - \frac{x}{5}\right)e^{-x/5} = 0$$
$$1 - \frac{x}{5} = 0 \qquad \textit{Since } e^{-x/5} \neq 0$$
$$x = 5$$

We see that $f'(x) < 0$ for $x > 5$, so it follows that f is decreasing for $x > 5$. Thus, the conditions for the integral test have been established, and the given series and the improper integral either both converge or both diverge. Computing the improper integral, we find

$$\int_5^\infty xe^{-x/5}\,dx = \lim_{b \to \infty} \int_5^b xe^{-x/5}\,dx$$

$$= \lim_{b \to \infty}\left[-5xe^{-x/5}\Big|_5^b - \int_5^b (-5e^{-x/5})\,dx\right] \qquad \boxed{\begin{array}{ll} \text{Let } u = x & dv = e^{-x/5}dx \\ du = dx & v = -5e^{-x/5} \end{array}}$$

$$= \lim_{b \to \infty}\left[-5xe^{-x/5} - 25e^{-x/5}\right]_5^b$$

$$= \lim_{b \to \infty}\left[-5be^{-b/5} - 25e^{-b/5} + 25e^{-1} + 25e^{-1}\right]$$

$$= -5 \lim_{b \to \infty} \frac{b+5}{e^{b/5}} + \lim_{b \to \infty} 50e^{-1}$$

$$= -5 \lim_{b \to \infty} \frac{1}{\left(\frac{1}{5}\right)e^{b/5}} + 50e^{-1} \qquad \textit{l'Hôpital's rule}$$

$$= 50e^{-1} \qquad\qquad\qquad\qquad\qquad\qquad \lim_{b \to \infty} \frac{1}{e^{b/5}} = 0$$

Note that the series converges because the integral converges to a finite number $50e^{-1}$, but this does not mean that the series

$$\sum_{k=1}^{\infty} \frac{k}{e^{k/5}}$$

converges to the same number $50e^{-1}$.

Thus, the improper integral converges, which in turn assures the convergence of the given series.   ■

p-Series

The harmonic series $\Sigma \frac{1}{k}$ is a special case of a more general series form called a p-series.

---

p-SERIES A series of the form

$$\sum_{k=1}^{\infty} \frac{1}{k^p} = \frac{1}{1^p} + \frac{1}{2^p} + \frac{1}{3^p} + \cdots$$

where p is a positive constant is called a **p-series**. The harmonic series (Example 2) is the case where $p = 1$.

The following theorem shows how the convergence of a p-series depends on the value of p.

Theorem 8.12 The p-series test

The p-series $\displaystyle\sum_{k=1}^{\infty} \frac{1}{k^p}$ converges if $p > 1$ and diverges if $p \leq 1$.

■ **W**hat this says It should be clear that all functions of the form $y = \dfrac{1}{x^p}$ with $p > 0$ are decreasing, but why should the series $\displaystyle\sum \frac{1}{k^p}$ converge only for $p > 1$? The answer lies in the *rate* of decrease for $y = \dfrac{1}{x^p}$. If $p > 1$, the curve $y = \dfrac{1}{x^p}$ falls fast enough to guarantee that the area under the curve for $p > 1$ is finite (see Figure 8.15a), while if $p \leq 1$, the curve $\dfrac{1}{x^p}$ falls more gradually, and the area under the curve is infinite (Figure 8.15b).

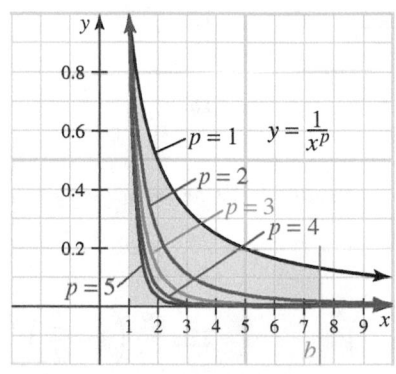

a. Areas are bounded

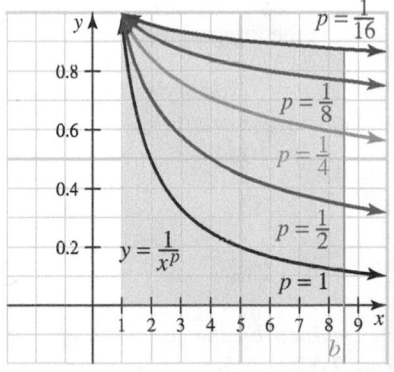

b. Areas are not bounded

Figure 8.15 Graphs for the p-series test

Proof: We leave it for the reader to verify that $f(x) = \dfrac{1}{x^p}$ is continuous, positive, and decreasing for $x \geq 1$ and $p > 0$. We know the harmonic series ($p = 1$) diverges. For $p > 0$, $p \neq 1$, we have

$$\int_1^{\infty} \frac{dx}{x^p} = \lim_{b \to \infty} \int_1^b x^{-p}\, dx = \lim_{b \to \infty} \frac{b^{1-p} - 1}{1 - p} = \begin{cases} \dfrac{1}{p-1} & \text{if } p > 1 \\ \infty & \text{if } 0 < p < 1 \end{cases}$$

That is, this improper integral converges if $p > 1$ and diverges if $0 < p < 1$. For $p = 0$, the series becomes

$$\sum_{k=1}^{\infty} \frac{1}{k^0} = \frac{1}{1} + \frac{1}{1} + \frac{1}{1} + \cdots$$

and if $p < 0$, we have $\lim\limits_{k \to \infty} \frac{1}{k^p} = \infty$, so the series diverges by the divergence test (Theorem 8.9). Thus, a p-series converges only when $p > 1$.   ◆

Example 4   *p*-series test

Test each of the following series for convergence.

a. $\displaystyle\sum_{k=1}^{\infty} \frac{1}{\sqrt{k^3}}$ **b.** $\displaystyle\sum_{k=1}^{\infty} \left(\frac{1}{e^k} - \frac{1}{\sqrt{k}} \right)$

Solution

a. Here $\sqrt{k^3} = k^{3/2}$, so $p = \dfrac{3}{2} > 1$, and the series converges.

b. $\displaystyle\sum_{k=1}^{\infty} \left(\frac{1}{e^k} - \frac{1}{\sqrt{k}} \right)$; We note that $\displaystyle\sum_{k=1}^{\infty} \frac{1}{e^k}$ converges, because it is geometric with $|r| = \dfrac{1}{e} < 1$ and $\displaystyle\sum_{k=1}^{\infty} \frac{1}{\sqrt{k}}$ diverges, because it is a p-series with $p = \dfrac{1}{2} < 1$. Because one series in the difference converges and the other diverges, the given series must diverge (Theorem 8.7). ▪

PROBLEM SET 8.3

Level 1

1. ■ What does this say? What is a p-series?

2. ■ What does this say? What is the integral test?

For the p-series given in Problems 3-6, specify the value of p and tell whether the series converges or diverges.

3. $\displaystyle\sum_{k=1}^{\infty} \frac{1}{k^3}$ 4. $\displaystyle\sum_{k=1}^{\infty} \frac{100}{\sqrt{k}}$

5. $\displaystyle\sum_{k=1}^{\infty} \frac{1}{\sqrt[3]{k}}$ 6. $\displaystyle\sum_{k=1}^{\infty} \frac{1}{2k\sqrt{k}}$

Show that the function f determined by the nth term of each series given in Problems 7-12 satisfies the hypotheses of the integral test. Then use the integral test to determine whether the series converges or diverges.

7. $\displaystyle\sum_{k=1}^{\infty} \frac{1}{(2+3k)^2}$ 8. $\displaystyle\sum_{k=1}^{\infty} (2+k)^{-3/2}$

9. $\displaystyle\sum_{k=2}^{\infty} \frac{\ln k}{k}$ 10. $\displaystyle\sum_{k=2}^{\infty} \frac{1}{k(\ln k)^2}$

11. $\displaystyle\sum_{k=1}^{\infty} \frac{(\tan^{-1} k)^2}{1+k^2}$ 12. $\displaystyle\sum_{k=1}^{\infty} ke^{-k^2}$

In Problems 13-18, either use the divergence test to show that the given series diverges or show that the divergence test does not apply.

13. $\displaystyle\sum_{k=1}^{\infty} \frac{k}{k+1}$ 14. $\displaystyle\sum_{k=1}^{\infty} \left(1 + \frac{1}{k} \right)^{-k}$

15. $\displaystyle\sum_{k=2}^{\infty} \frac{k}{\sqrt{k^2-1}}$ 16. $\displaystyle\sum_{k=1}^{\infty} e^{-1/k}$

17. $\displaystyle\sum_{k=0}^{\infty} \cos k\pi$ 18. $\displaystyle\sum_{k=1}^{\infty} ke^{-k}$

Test the series in Problems 19-38 for convergence.

19. $\displaystyle\sum_{k=1}^{\infty} \frac{\ln k}{k^2}$ 20. $\displaystyle\sum_{k=1}^{\infty} e^{-k}$

21. $\displaystyle\sum_{k=1}^{\infty} \left(2 + \frac{3}{k} \right)^k$ 22. $\displaystyle\sum_{k=1}^{\infty} \left(1 + \frac{2}{k} \right)^k$

23. $\displaystyle\sum_{k=1}^{\infty} \frac{1}{k^4}$ 24. $\displaystyle\sum_{k=1}^{\infty} \frac{5}{\sqrt{k}}$

25. $\displaystyle\sum_{k=1}^{\infty} k^{-3/4}$ 26. $\displaystyle\sum_{k=1}^{\infty} k^{-4/3}$

27. $\displaystyle\sum_{k=1}^{\infty} \frac{k - \sin k}{3k + 2\sin k}$ 28. $\displaystyle\sum_{k=1}^{\infty} \frac{1}{(0.25)^k}$

29. $\displaystyle\sum_{k=2}^{\infty} \frac{\ln \sqrt{k}}{\sqrt{k}}$ 30. $\displaystyle\sum_{k=2}^{\infty} \frac{1}{k\sqrt{\ln k}}$

31. $\displaystyle\sum_{k=1}^{\infty} \frac{k}{e^k}$ 32. $\displaystyle\sum_{k=1}^{\infty} \frac{k^2}{e^k}$

33. $\displaystyle\sum_{k=1}^{\infty} \frac{\tan^{-1} k}{1+k^2}$

34. $\displaystyle\sum_{k=1}^{\infty} \cot^{-1} k$

35. $\displaystyle\sum_{k=1}^{\infty} \tan^{-1} k$

36. $\displaystyle\sum_{k=1}^{\infty} \frac{1}{e^k + e^{-k}}$

37. $\displaystyle\sum_{k=1}^{\infty} k \sin \frac{1}{k}$

38. $\displaystyle\sum_{k=4}^{\infty} \ln\left(\frac{k}{k-3}\right)$

Determine which of the series in Problems 39-50 converge. Write "inconclusive" if there is not yet enough information to settle convergence.

39. $\displaystyle\sum_{k=1}^{\infty} \frac{k-1}{k+1}$

40. $\displaystyle\sum_{k=1}^{\infty} \frac{k^2+1}{k+5}$

41. $\displaystyle\sum_{k=1}^{\infty} \frac{k^2+1}{k^3}$

42. $\displaystyle\sum_{k=1}^{\infty} \frac{2k^4+3}{k^5}$

43. $\displaystyle\sum_{k=1}^{\infty} \frac{1}{k^2}$

44. $\displaystyle\sum_{k=1}^{\infty} \frac{1+k^2-k^5}{2+k^5}$

45. $\displaystyle\sum_{k=1}^{\infty} \frac{k^{\sqrt{3}}+1}{k^{2.7321}}$

46. $\displaystyle\sum_{k=1}^{\infty} \frac{k^{\sqrt{5}}+1}{k^{2.236}}$

47. $\displaystyle\sum_{k=1}^{\infty} \left[\frac{1}{k} + \frac{k+1}{k+2}\right]$

48. $\displaystyle\sum_{k=1}^{\infty} \left[\frac{1}{2k} + \left(\frac{3}{2}\right)^k\right]$

49. $\displaystyle\sum_{k=1}^{\infty} \left[\frac{1}{2^k} - \frac{1}{k}\right]$

50. $\displaystyle\sum_{k=1}^{\infty} \left[\frac{1}{2^k} + \frac{2k+3}{3k+4}\right]$

Level 2

Find all values of p for which each series given in Problems 51-56 converges.

51. $\displaystyle\sum_{k=2}^{\infty} \frac{k}{(k^2-1)^p}$

52. $\displaystyle\sum_{k=1}^{\infty} \frac{k^2}{(k^2+4)^p}$

53. $\displaystyle\sum_{k=3}^{\infty} \frac{1}{k^p \ln k}$

54. $\displaystyle\sum_{k=2}^{\infty} \frac{\ln k}{k^p}$

55. $\displaystyle\sum_{k=2}^{\infty} \frac{1}{k(\ln k)^p}$

56. $\displaystyle\sum_{k=3}^{\infty} \frac{1}{k \ln k\,[\ln(\ln k)]^p}$

Level 3

57. Think Tank Problem If Σa_k converges, Σb_k diverges, and $\lim_{k\to\infty} b_k = 0$, then $\Sigma a_k b_k$ must diverge. Either prove this statement or provide a counterexample to show it is not generally true.

58. Think Tank Problem If $\lim_{k\to\infty} a_k = 0$, $\lim_{k\to\infty} b_k = 0$, and Σa_k and Σb_k both diverge, then $\Sigma a_k b_k$ must also diverge. Either prove this statement or provide a counterexample to show it is not generally true.

59. Let $f(x)$ be a positive, continuous, decreasing function for $x \geq 1$, and let $a_k = f(k)$ for $k = 1, 2, \cdots$. If Σa_k converges, show that

$$\int_1^\infty f(x)\,dx \leq \sum_{k=1}^\infty a_k \leq a_1 + \int_1^\infty f(x)\,dx$$

60. Getting a feeling for a series by examining its graph is a good idea, but you should be careful because you cannot always obtain a graph. Decide whether

$$\sum \left(1 - \frac{100}{n}\right)^n$$

converges or diverges, and then see if you can verify your result graphically.

8.4 COMPARISON TESTS

IN THIS SECTION: *Direct comparison test, limit comparison tests*
Often a given series will have the same general appearance as another series whose convergence properties are known. In such case, it is convenient to use the properties of the known series to determine those of the given series. The goal of this section is to examine three comparison tests for making these determinations.

Direct Comparison Test

The first of the comparison tests is called the **direct comparison test**.

Theorem 8.13 Direct comparison test

Suppose $0 \leq a_k \leq c_k$ for all k.

$$\text{If } \sum_{k=1}^{\infty} c_k \text{ converges, then } \sum_{k=1}^{\infty} a_k \text{ also converges}$$

Let $0 \le d_k \le a_k$ for all k.

If $\displaystyle\sum_{k=1}^{\infty} d_k$ diverges, then $\displaystyle\sum_{k=1}^{\infty} a_k$ also diverges

---

■ **W**hat this says Let Σa_k, Σc_k and Σd_k be series with positive terms. The series Σa_k *converges* if it is "smaller" than (dominated by) a known convergent series Σc_k and *diverges* if it is "larger" than (dominates) a known divergent series Σd_k. That is, "smaller than convergent is convergent," and "bigger than divergent is divergent."

Proof: Suppose Σa_k is a given nonnegative term series that is **dominated by** a convergent series Σc_k with $0 \le a_k \le c_k$ for all k. Then, because the series Σc_k converges, its sequence of partial sums is bounded from above (say, by M), and we have

$$\sum_{k=1}^{n} a_k \le \sum_{k=1}^{n} c_k < M \qquad \text{for all } n$$

Thus, the sequence of partial sums of the dominated series Σa_k is also bounded from above by M, and it, too, must converge.

On the other hand, suppose the given series Σa_k **dominates** a divergent series Σd_k with $0 \le d_k \le a_k$. Then because the sequence of partial sums of Σd_k is unbounded, the same must be true for the partial sums of the dominating series Σa_k, and Σa_k must also diverge. ◆

Remark:

The above theorem is also true if the dominance only holds for all $k \ge N$ for some positive N because whether a series converges or diverges depends on only what happens for large k. That is, $\displaystyle\sum_{k=1}^{\infty} a_k$ converges if and only if $\displaystyle\sum_{k=N}^{\infty} a_k$ converges for any integer N.

Example 1   Convergence using the direct comparison test

Test the series $\displaystyle\sum_{k=1}^{\infty} \frac{1}{3^k + 1}$ for convergence.

Solution   For $k \ge 0$, we have $3^k + 1 > 3^k > 0$, and $0 < \dfrac{1}{3^k + 1} < \dfrac{1}{3^k}$. Thus, the given series is *dominated by* the convergent geometric series $\displaystyle\sum_{k=1}^{\infty} \frac{1}{3^k}$ (convergent because $r = \frac{1}{3}$). The direct comparison test tells us the given series must converge. ■

Example 2   Divergence using the direct comparison test

Test the series $\displaystyle\sum_{k=2}^{\infty} \frac{1}{\sqrt{k} - 1}$ for convergence.

Solution   For $k \ge 2$, we have

$$\frac{1}{\sqrt{k} - 1} > \frac{1}{\sqrt{k}} > 0$$

so the given series *dominates* the divergent p-series $\displaystyle\sum_{k=2}^{\infty} \frac{1}{k^{1/2}}$ (divergent because $p = \frac{1}{2} < 1$). The direct comparison test tells us the given series must diverge. ■

Example 3 Applying the direct comparison test

Test the series $\displaystyle\sum_{k=1}^{\infty} \frac{1}{k!}$ for convergence.

Solution We will compare the given series with a geometric series. Specifically, note that if $k \geq 2$, then $1/k! > 0$ and

$$k! = k(k-1)(k-2)\cdots 3\cdot 2\cdot 1 \geq \underbrace{2\cdot 2\cdot 2\cdots\cdots 2\cdot 2}_{k-1\ terms}\cdot 1 = 2^{k-1}$$

Therefore, we have $0 < \dfrac{1}{k!} \leq \dfrac{1}{2^{k-1}}$, and because the given series is *dominated by* the convergent geometric series $\displaystyle\sum_{k=1}^{\infty} \frac{1}{2^{k-1}}$ (with $r = \frac{1}{2}$), it must also converge. ■

Limit Comparison Tests

It is not always easy or even possible to make a suitable direct comparison between two similar series. For example, we would expect the series $\displaystyle\sum \frac{1}{2^k - 5}$ to converge because it is so much like the convergent geometric series $\displaystyle\sum \frac{1}{2^k}$. To compare the series, we must first note that $\dfrac{1}{2^k - 5}$ is negative for $k = 1$ and $k = 2$ and positive for $k \geq 3$. Thus, if $k \geq 3$,

$$0 \leq \frac{1}{2^k} < \frac{1}{2^k - 5}$$

That is, $\displaystyle\sum \frac{1}{2^k - 5}$ *dominates* the convergent series $\displaystyle\sum \frac{1}{2^k}$, which means the comparison test cannot be used to determine the convergence of $\displaystyle\sum \frac{1}{2^k - 5}$ by comparing with $\displaystyle\sum \frac{1}{2^k}$. In such situations, the following test is often useful.

Theorem 8.14 Limit comparison test

Suppose $a_k > 0$ and $b_k > 0$ for all sufficiently large k and that

$$\lim_{k\to\infty} \frac{a_k}{b_k} = L$$

where L is finite and positive ($0 < L < \infty$). Then Σa_k and Σb_k either both converge or both diverge.

> ■ **W**hat this says We have the following procedure for testing the convergence of Σa_k.
>
> *Step 1.* Find a series Σb_k whose convergence properties are known and whose general term b_k is "essentially the same" as a_k.
> *Step 2.* Verify that $\displaystyle\lim_{k\to\infty} \frac{a_k}{b_k}$ exists and is positive.
> *Step 3.* Determine whether Σb_k converges or diverges. Then the limit comparison test tells us that Σa_k does the same.

Proof: The limit comparison test is proved by using the direct comparison test. Details of the proof are given in Appendix B. ♦

Example 4   Convergence using the limit comparison test

Test the series $\sum\limits_{k=1}^{\infty} \dfrac{1}{2^k - 5}$ for convergence.

Solution   Because the given series has the same general appearance as the convergent geometric series $\Sigma 1/2^k$, compute the limit

$$\lim_{k \to \infty} \frac{\frac{1}{2^k - 5}}{\frac{1}{2^k}} = \lim_{k \to \infty} \frac{2^k}{2^k - 5} = 1$$

This limit is finite and positive, so the limit comparison test tells us that the given series will have the same convergence properties as $\Sigma 1/2^k$. Thus, the given series converges.

Example 5   Divergence using the limit comparison test

Test the series $\sum\limits_{k=1}^{\infty} \dfrac{3k + 2}{\sqrt{k}(3k - 5)}$ for convergence.

Solution   For large values of k, the general term of the given series

$$a_k = \frac{3k + 2}{\sqrt{k}(3k - 5)}$$

seems to be similar to

$$b_k = \frac{3k}{\sqrt{k}(3k)} = \frac{1}{\sqrt{k}}$$

To apply the limit comparison test, we first compute the limit

$$\lim_{k \to \infty} \frac{a_k}{b_k} = \lim_{k \to \infty} \frac{\frac{3k+2}{\sqrt{k}(3k-5)}}{\frac{1}{\sqrt{k}}} = \lim_{k \to \infty} \frac{3k + 2}{3k - 5} = 1$$

Because the limit is finite and positive, it follows that the given series behaves like the series $\Sigma 1/\sqrt{k}$, which is a divergent p-series ($p = \frac{1}{2}$). We conclude that the given series diverges.

Example 6   Applying the limit comparison test

Test the series $\sum\limits_{k=1}^{\infty} \dfrac{7k + 100}{e^{k/5} - 70}$ for convergence.

Solution   For large k,

$$a_k = \frac{7k + 100}{e^{k/5} - 70}$$

seems to behave like $b_k = \dfrac{k}{e^{k/5}}$ and indeed, we find that

$$\lim_{k \to \infty} \frac{\frac{7k+100}{e^{k/5}-70}}{\frac{k}{e^{k/5}}} = \lim_{k \to \infty} \frac{e^{k/5}(7k + 100)}{k(e^{k/5} - 70)} = 7$$

In Example 3, Section 8.3, we showed that the series $\Sigma \, k/e^{k/5}$ converges, and therefore the limit comparison test tells us that the given series also converges.

Occasionally, two series Σa_k and Σb_k appear to have similar convergence properties, but it turns out that

$$\lim_{k \to \infty} \frac{a_k}{b_k}$$

is either 0 or ∞, and the limit comparison test therefore does not apply. In such cases, it is often useful to have the following generalization of the limit comparison test.

Theorem 8.15   The zero-infinity limit comparison test

Suppose $a_k > 0$ and $b_k > 0$ for all sufficiently large k.

If $\displaystyle\lim_{k \to \infty} \frac{a_k}{b_k} = 0$ and Σb_k converges, then the series Σa_k converges.

If $\displaystyle\lim_{k \to \infty} \frac{a_k}{b_k} = \infty$ and Σb_k diverges, then the series Σa_k diverges.

Proof: The first part of the proof is outlined in Problem 59; the second part follows similarly.   ◆

Example 7   Convergence of the *q*-log series

Show that the series $\displaystyle\sum_{k=1}^{\infty} \frac{\ln k}{k^q}$ converges if $q > 1$ and diverges if $q \le 1$. We call this the *q*-**log** series.

Solution   We will carry out this proof in three parts:

Case I: $(q > 1)$
Let p be a number that satisfies $1 < p < q$, and let

$$a_k = \frac{\ln k}{k^q} \quad \text{and} \quad b_k = \frac{1}{k^p}$$

Then

$$
\begin{aligned}
\lim_{k \to \infty} \frac{a_k}{b_k} &= \lim_{k \to \infty} \frac{\frac{\ln k}{k^q}}{\frac{1}{k^p}} \\
&= \lim_{k \to \infty} \frac{\ln k}{k^{q-p}} \qquad \textit{This is the indeterminate form } \infty/\infty \textit{ because } q - p > 0. \\
&= \lim_{k \to \infty} \frac{\frac{1}{k}}{(q-p)k^{q-p-1}} \qquad \textit{l'Hôpital's rule} \\
&= \lim_{k \to \infty} \frac{1}{(q-p)k^{q-p}} \qquad \textit{Because } q - p > 0 \\
&= 0
\end{aligned}
$$

Because $\Sigma 1/k^p$ converges (p-series with $p > 1$), the series converges by the zero-infinity limit comparison test.

Case II: $(q < 1)$
Now, let p satisfy $q < p < 1$. Then with a_k and b_k defined as in case I, we have

$$\lim_{k \to \infty} \frac{a_k}{b_k} = \lim_{k \to \infty} \frac{\ln k}{k^{q-p}} = \lim_{k \to \infty} \left[(\ln k)k^{p-q} \right] = \infty$$

Because $\displaystyle\sum_{k=1}^{\infty} b_k = \sum_{k=1}^{\infty} \frac{1}{k^p}$ diverges (we know $p < 1$), it follows from the

zero-infinity comparison test that $\displaystyle\sum_{k=1}^{\infty} \frac{\ln k}{k^q}$ diverges when $q < 1$.

Case III: $(q = 1)$

Here, the series is $\displaystyle\sum_{k=1}^{\infty} \frac{\ln k}{k}$, which diverges by the integral test (where we let

$u = \ln x; \ du = \frac{1}{x}\,dx$):

$$\int_1^{\infty} \frac{\ln x}{x}\,dx = \lim_{b \to \infty} \frac{(\ln x)^2}{2}\Bigg|_1^b = \infty$$

PROBLEM SET 8.4

Level 1

The most common series used for comparison are given in Problems 1 and 2. Tell when each converges and when it diverges.

1. geometric series: $\displaystyle\sum_{k=0}^{\infty} r^k$ **2.** *p*-series: $\displaystyle\sum_{k=1}^{\infty} \frac{1}{k^p}$

Each series in Problems 3-12 can be compared to the geometric series or p-series given in Problems 1 and 2. State which, and then determine whether it converges or diverges.

3. $\displaystyle\sum_{k=1}^{\infty} \cos^k\left(\frac{\pi}{6}\right)$ **4.** $\displaystyle\sum_{k=0}^{\infty} (0.5)^k$

5. $\displaystyle\sum_{k=0}^{\infty} (1.5)^k$ **6.** $\displaystyle\sum_{k=0}^{\infty} 2^{k/2}$

7. $\displaystyle\sum_{k=1}^{\infty} \frac{1}{k}$ **8.** $\displaystyle\sum_{k=1}^{\infty} \frac{1}{k^{0.5}}$

9. $\displaystyle\sum_{k=1}^{\infty} \frac{1}{k^{3/2}}$ **10.** $\displaystyle\sum_{k=1}^{\infty} \sqrt{\frac{2}{k}}$

11. $\displaystyle\sum_{k=0}^{\infty} 1^k$ **12.** $\displaystyle\sum_{k=1}^{\infty} e^k$

Test the series in Problems 13-42 for convergence.

13. $\displaystyle\sum_{k=1}^{\infty} \frac{1}{k^2 + k}$ **14.** $\displaystyle\sum_{k=1}^{\infty} \frac{1}{k^2 + 3k + 2}$

15. $\displaystyle\sum_{k=1}^{\infty} \frac{1}{\sqrt{k}}$ **16.** $\displaystyle\sum_{k=1}^{\infty} \frac{1}{k\sqrt{k}}$

17. $\displaystyle\sum_{k=1}^{\infty} \frac{1}{\sqrt{2k+3}}$ **18.** $\displaystyle\sum_{k=1}^{\infty} \frac{1}{\sqrt{k(k+1)}}$

19. $\displaystyle\sum_{k=1}^{\infty} \frac{1}{\sqrt{k^3 + 2}}$ **20.** $\displaystyle\sum_{k=1}^{\infty} \frac{1}{\sqrt{k^2 + 1}}$

21. $\displaystyle\sum_{k=1}^{\infty} \frac{2k^2}{k^4 - 4}$ **22.** $\displaystyle\sum_{k=1}^{\infty} \frac{k+1}{k^2 + 1}$

23. $\displaystyle\sum_{k=1}^{\infty} \frac{(k+2)(k+3)}{k^{7/2}}$ **24.** $\displaystyle\sum_{k=1}^{\infty} \frac{(k+1)^3}{k^{9/2}}$

25. $\displaystyle\sum_{k=1}^{\infty} \frac{2k+3}{k^2 + 3k + 2}$ **26.** $\displaystyle\sum_{k=1}^{\infty} \frac{3k^2 + 2}{k^2 + 3k + 2}$

27. $\displaystyle\sum_{k=1}^{\infty} \frac{k}{(k+2)2^k}$ **28.** $\displaystyle\sum_{k=1}^{\infty} \frac{5}{4^k + 3}$

29. $\displaystyle\sum_{k=1}^{\infty} \frac{1}{k(k+2)}$ **30.** $\displaystyle\sum_{k=1}^{\infty} \frac{1}{(k+2)(k+3)}$

31. $\displaystyle\sum_{k=1}^{\infty} \frac{1}{\sqrt{k}\,2^k}$ **32.** $\displaystyle\sum_{k=1}^{\infty} \frac{1,000}{\sqrt{k}\,3^k}$

33. $\displaystyle\sum_{k=2}^{\infty} \frac{1}{\sqrt{k}\ln k}$ **34.** $\displaystyle\sum_{k=1}^{\infty} \frac{\ln k}{\sqrt{2k+3}}$

35. $\displaystyle\sum_{k=1}^{\infty} \frac{2k^3 + k + 1}{k^3 + k^2 + 1}$ **36.** $\displaystyle\sum_{k=1}^{\infty} \frac{6k^3 - k - 4}{k^3 - k^2 - 3}$

37. $\displaystyle\sum_{k=1}^{\infty} \frac{|\sin (k!)|}{k^2}$ **38.** $\displaystyle\sum_{k=1}^{\infty} \sin \frac{1}{k}$

39. $\displaystyle\sum_{k=1}^{\infty} \frac{6k^2 + 2k + 1}{k^{1.1}(4k^2 + k + 4)}$ **40.** $\displaystyle\sum_{k=1}^{\infty} \frac{6k^2 + 2k + 1}{k^{0.9}(4k^2 + k + 4)}$

41. $\displaystyle\sum_{k=1}^{\infty} \frac{\sqrt[6]{k}}{\sqrt[4]{k^3 + 2}\sqrt[8]{k}}$ **42.** $\displaystyle\sum_{k=1}^{\infty} \frac{\sqrt{k}}{\sqrt[3]{k^3 + 1}\sqrt[6]{k^5}}$

Level 2

Test the series given in Problems 43-50 for convergence.

43. $\displaystyle\sum_{k=1}^{\infty} \frac{1}{k^3 + 4}$

44. $\displaystyle\sum_{k=2}^{\infty} \frac{\ln k}{k - 1}$

45. $\displaystyle\sum_{k=1}^{\infty} \frac{\ln(k + 1)}{(k + 1)^3}$

46. $\displaystyle\sum_{k=1}^{\infty} \frac{\ln k}{k^2}$

47. $\displaystyle\sum_{k=2}^{\infty} \frac{1}{(k + 3)(\ln k)^{1.1}}$

48. $\displaystyle\sum_{k=2}^{\infty} \frac{1}{(k + 3)(\ln k)^{0.9}}$

49. $\displaystyle\sum_{k=1}^{\infty} k^{(1-k)/k}$

50. $\displaystyle\sum_{k=1}^{\infty} k^{(1+k)/k}$

51. Show that the series

$$\sum_{k=1}^{\infty} \frac{k^2}{(k + 3)!} = \frac{1}{4!} + \frac{4}{5!} + \frac{9}{6!} + \cdots$$

converges by using the limit comparison test.

52. Show that the series

$$1 + \frac{1}{1 \cdot 3} + \frac{1}{1 \cdot 3 \cdot 5} + \frac{1}{1 \cdot 3 \cdot 5 \cdot 7} + \cdots$$
$$+ \frac{2^k k!}{(2k + 1)!} + \cdots$$

converges. *Hint:* Compare with the convergent series $\Sigma 1/k!$.

53. Use a comparison test to show that

$$\sum_{k=2}^{\infty} \frac{1}{(\ln k)^{\ln k}}$$

converges. *Hint:* Use the fact that $\ln k > e^2$ for sufficiently large k.

Level 3

54. a. Let $\{a_n\}$ be a positive sequence such that

$$\lim_{k \to \infty} k^p a_k$$

exists. Use the limit comparison test to show that Σa_k converges if $p > 1$.
  b. Show that Σe^{-k^2} converges. *Hint:* Show that $\lim_{k \to \infty} k^2 e^{-k^2}$ exists and use part **a**.

55. Let Σa_k be a series of positive terms and let $\{b_k\}$ be a sequence of positive numbers that converge to a positive number. Show that Σa_k converges if and only if $\Sigma a_k b_k$ converges. *Hint*: Use the limit comparison test.

56. Suppose $0 \le a_k \le A$ for some number A and $b_k \ge k^2$. Show that $\Sigma a_k / b_k$ converges.

57. Suppose $a_k > 0$ for all k and that Σa_k converges. Show that $\Sigma 1/a_k$ diverges.

58. Think Tank Problem Suppose that $0 < a_k < 1$ for every k, and that Σa_k converges. Use a comparison test to show that Σa_k^2 converges. Find a counterexample where Σa_k^2 converges, but Σa_k diverges.

59. Show that if $a_k > 0, b_k > 0$, and

$$\lim_{k \to \infty} \frac{a_k}{b_k} = 0$$

then Σa_k converges whenever Σb_k converges. *Hint:* Show that there is an integer N such that $a_k < b_k$ if $k > N$. Then use a comparison test.

60. Show that

$$\sum_{k=2}^{\infty} \frac{1}{k(\ln k)^q}$$

converges if and only if $q > 1$.

8.5 THE RATIO TEST AND THE ROOT TEST

IN THIS SECTION: *Ratio test, root test*
In this section, we develop two additional tests, the *ratio test* and the *root test*.

Ratio Test

Intuitively, a series of positive terms Σa_k converges if and only if the sequence $\{a_k\}$ decreases *rapidly enough* toward 0. One way to measure the rate at which the sequence $\{a_k\}$ is decreasing is to examine the ratio a_{k+1}/a_k as k grows large. This approach leads to the following result.

Theorem 8.16 The ratio test

Given the series Σa_k with $a_k > 0$, suppose that

$$\lim_{k\to\infty} \frac{a_{k+1}}{a_k} = L$$

The **ratio test** states the following:

> If $L < 1$, then Σa_k converges.
> If $L > 1$, or if L is infinite, then Σa_k diverges.
> If $L = 1$, then the test is inconclusive.

Proof: In a sense, the ratio test is a limit comparison test in which Σa_k is compared to itself. We will use the direct comparison test to show that Σa_k converges if $L < 1$. Choose R such that $0 < L < R < 1$. Then there exists some $N > 0$ such that

$$\frac{a_{k+1}}{a_k} < R \text{ for all } k \geq N \quad \textit{The ratios } \frac{a_{k+1}}{a_k} \textit{ are eventually very close to } L.$$

Therefore,

$$a_{N+1} < a_N R$$

$$a_{N+2} < a_{N+1}R < a_N R^2$$

$$a_{N+3} < a_{N+2}R < a_{N+1}R^2 < a_N R^3$$

$$\vdots$$

The geometric series

$$\sum_{k=1}^{\infty} a_N R^k = a_N R + a_N R^2 + \cdots + a_N R^n + \cdots$$

converges because $0 < R < 1$, which means the "smaller" series

$$\sum_{k=1}^{\infty} a_{N+k} = a_{N+1} + a_{N+2} + \cdots + a_{N+n} + \cdots$$

also converges by the direct comparison test. Thus, Σa_k converges, because we can discard a finite number of terms ($k \leq N$).

The proof of the second part is similar, except now we choose R such that

$$\lim_{k\to\infty} \frac{a_{k+1}}{a_k} = L > R > 1$$

and show there exists some $M > 0$ so that $a_{M+k} > a_M R^k$.

To prove that the ratio test is inconclusive if $L = 1$, it is enough to note that the harmonic series

harmonic series diverges

$$\sum_{k=1}^{\infty} \frac{1}{k} \text{ diverges with } \lim_{k\to\infty} \frac{a_{k+1}}{a_k} = \lim_{k\to\infty} \frac{\frac{1}{k+1}}{\frac{1}{k}} = 1$$

while the p-series

p-series converges for p > 1

$$\sum_{k=1}^{\infty} \frac{1}{k^2} \text{ converges with } \lim_{k\to\infty} \frac{a_{k+1}}{a_k} = \lim_{k\to\infty} \frac{\frac{1}{(k+1)^2}}{\frac{1}{k^2}} = 1$$

◆

Example 1   Convergence using the ratio test

Test the series $\displaystyle\sum_{k=1}^{\infty}\frac{2^k}{k!}$ for convergence.

You will find the ratio test most useful with series involving factorials or exponentials.

Solution   Let $a_k = \dfrac{2^k}{k!}$ and note that

$$L = \lim_{k\to\infty} \frac{a_{k+1}}{a_k}$$

$$= \lim_{k\to\infty} \frac{\frac{2^{k+1}}{(k+1)!}}{\frac{2^k}{k!}}$$

$$= \lim_{k\to\infty} \frac{k!\,2^{k+1}}{(k+1)!\,2^k}$$

$$= \lim_{k\to\infty} \frac{2}{k+1}$$

$$= 0$$

Thus $L < 1$, and the ratio test tells us that the given series converges.

Example 2   Divergence using the ratio test

Test the series $\displaystyle\sum_{k=1}^{\infty}\frac{k^k}{k!}$ for convergence.

Solution   Let $a_k = \dfrac{k^k}{k!}$ and note that

$$L = \lim_{k\to\infty} \frac{a_{k+1}}{a_k}$$

$$= \lim_{k\to\infty} \frac{\frac{(k+1)^{k+1}}{(k+1)!}}{\frac{k^k}{k!}}$$

$$= \lim_{k\to\infty} \frac{k!\,(k+1)^{k+1}}{k^k(k+1)!}$$

$$= \lim_{k\to\infty} \frac{(k+1)^k}{k^k}$$

$$= \lim_{k\to\infty} \left(1 + \frac{1}{k}\right)^k$$

$$= e$$

$$> 1$$

Because $L > 1$, the given series diverges by the ratio test.

Window   **TECHNOLOGY NOTE**   — ▢ ✕

It is instructive to compare and contrast two different ways you can use technology to help you with problems such as Example 2. First, as you have seen, you can use a calculator or a computer to calculate the partial sums directly. This table is shown below at the left.

A second procedure is to use the ratio test for Example 2. In this case, we are interested in

$$\lim_{k \to \infty} \frac{a_{k+1}}{a_k} = \lim_{k \to \infty} \frac{\frac{(k+1)^{k+1}}{(k+1)!}}{\frac{k^k}{k!}} = \lim_{k \to \infty} \left(1 + \frac{1}{k}\right)^k$$

We can look at terms of the sequence $b_k = \left(1 + \frac{1}{k}\right)^k$ to see whether this limit is greater than, less than, or equal to 1. This is shown below on the right.

By table: By graph:

$$\sum_{k=1}^{\infty} \frac{k^k}{k!}$$ $$\lim_{k \to \infty} \left(1 + \frac{1}{k}\right)^k$$

N	F (1) + ... + F (N)
1	1
2	3
3	7.5
4	18.16667
5	44.20833
6	109.0083
7	272.4097
8	688.5113
9	1756.138
10	4511.87
11	11659.53
12	30273.46
13	78912.3
14	206375.3
15	541239.9

$Y = (1 + 1/X)^X$

$X = $ 1000000

X	Y
1	2
2	2.25
3	2.37037
4	2.44141
5	2.4883
6	2.521626
7	2.546500
8	2.565784
9	2.581175
10	2.593742
100	2.704814
1000	2.716924
10000	2.718146
100000	2.718268

Interactive The partial sums seem to diverge.   **Interactive** The ratio test gives a limit greater than 1.

Finally, you might test this using a calculator to verify that it seems to diverge. A word of caution: in some cases, plugging in certain large values for k on various types of calculators will give an erroneous answer (1) instead of a number that approaches Euler's constant e. This is due to the so-called "machine limit" and is a direct consequence of round-off error accumulations. A more detailed discussion about this topic can be found in Appendix C, section "Calculator Experiments."

Example 3   Ratio test fails

Test the series $\displaystyle\sum_{k=2}^{\infty} \frac{1}{2k - 3}$ for convergence.

Solution   Let $a_k = \dfrac{1}{2k - 3}$ and find

$$L = \lim_{k \to \infty} \frac{\frac{1}{2(k+1)-3}}{\frac{1}{2k-3}} = \lim_{k \to \infty} \frac{2k - 3}{2k - 1} = 1$$

The ratio test is inconclusive. We can use the integral test or the limit comparison test to determine convergence. Note that

$$\sum_{k=2}^{\infty} \frac{1}{2k-3} \text{ is similar to the known divergent series } \sum_{k=2}^{\infty} \frac{1}{2k}$$

so we conclude the given series is divergent. ∎

Example 4 Convergence of a series of power functions

Find all numbers $x > 0$ for which the series

$$\sum_{k=1}^{\infty} k^3 x^k = x + 2^3 x^2 + 3^3 x^3 + \cdots$$

converges. This is called a *power series*, and will be discussed in more detail in Sections 8.7 and 8.8.

Solution   Viewing x as fixed (that is, as a constant), we apply the ratio test:

$$L = \lim_{k\to\infty} \frac{(k+1)^3 x^{k+1}}{k^3 x^k} = \lim_{k\to\infty} \left(\frac{k+1}{k}\right)^3 x = x$$

Thus, the series converges if $L = x < 1$ and diverges if $x > 1$. When $x = 1$, the series becomes $\sum_{k=1}^{\infty} k^3$, which diverges by the divergence test. ∎

Root Test

Of all the tests we have developed, the divergence test is perhaps the easiest to apply, because it involves simply computing $\lim_{k\to\infty} a_k$ and observing whether that limit is zero. Unfortunately, most "interesting" series have $\lim_{k\to\infty} a_k = 0$, so the divergence test cannot be used to determine whether they converge or diverge. However, the following result shows that it may be possible to say more about the convergence of Σa_k by examining what happens to $\sqrt[k]{a_k}$ as $k \to \infty$. This test will prove particularly useful with a series involving a kth power.

Theorem 8.17   The root test

Given the series Σa_k with $a_k > 0$, suppose that

$$\lim_{k\to\infty} \sqrt[k]{a_k} = L$$

The **root test** states the following:

If $L < 1$, then Σa_k converges.
If $L > 1$, or if L is infinite, then Σa_k diverges.
If $L = 1$, then the test is inconclusive.

Proof: The proof of this theorem is similar to that of the ratio test and is outlined in Problem 59. ♦

Example 5   Convergence with the root test

Test the series $\displaystyle\sum_{k=2}^{\infty} \frac{1}{(\ln k)^k}$ for convergence.

Solution   Let $a_k = \dfrac{1}{(\ln k)^k}$ and note that

$$L = \lim_{k\to\infty} \sqrt[k]{a_k} = \lim_{k\to\infty} \sqrt[k]{(\ln k)^{-k}} = \lim_{k\to\infty} \frac{1}{\ln k} = 0$$

Because $L < 1$, the root test tells us that the given series converges.

Example 6   Divergence with the root test

Test the series $\displaystyle\sum_{k=1}^{\infty} \left(1 + \frac{1}{k}\right)^{k^2}$ for convergence.

Solution $\displaystyle L = \lim_{k\to\infty} \left[\left(1 + \frac{1}{k}\right)^{k^2}\right]^{1/k}$ *Root test*

$$= \lim_{k\to\infty} \left(1 + \frac{1}{k}\right)^k$$

$$= e$$

$$> 1$$

Since $L > 1$, the series diverges by the root test.

We have seen that the ratio test is especially useful for series Σa_k for which the general term a_k involves factorials or powers, and the root test applies naturally if a_k involves a power of k. However, both the ratio test and the root test fail with certain very simple series such as the p-series $\displaystyle\sum \frac{1}{k^p}$ since

$$\lim_{k\to\infty} \frac{\frac{1}{(k+1)^p}}{\frac{1}{k^p}} = \lim_{k\to\infty} \left(\frac{k}{k+1}\right)^p = 1^p = 1 \qquad \textit{Inconclusive ratio test}$$

$$\lim_{k\to\infty} \sqrt[k]{\frac{1}{k^p}} = \left[\lim_{k\to\infty} k^{1/k}\right]^{-p} = 1^{-p} = 1 \qquad \textit{Inclusive root test where l'Hôpital's}$$

$$\textit{rule is used to evaluate the limit.}$$

Example 7   Testing for convergence

Test the following series for convergence:

$$\sum_{k=0}^{\infty} \frac{k!}{1 \cdot 4 \cdot 7 \cdots (3k+1)} = 1 + \frac{1!}{1 \cdot 4} + \frac{2!}{1 \cdot 4 \cdot 7} + \frac{3!}{1 \cdot 4 \cdot 7 \cdot 10} + \cdots$$

Solution   Let $a_k = \dfrac{k!}{1 \cdot 4 \cdot 7 \cdots (3k+1)}$. Because a_k involves $k!$, we try the ratio test.

$$\frac{a_{k+1}}{a_k} = \frac{\frac{(k+1)!}{1 \cdot 4 \cdot 7 \cdot [3(k+1)+1]}}{\frac{k!}{1 \cdot 4 \cdot 7 \cdots (3k+1)}}$$

$$= \frac{(k+1)! \, [1 \cdot 4 \cdot 7 \cdots (3k+1)]}{k! \, [1 \cdot 4 \cdot 7 \cdots (3k+4)]}$$

$$= \frac{k+1}{3k+4}$$

Thus,

$$L = \lim_{k \to \infty} \frac{a_{k+1}}{a_k} = \lim_{k \to \infty} \frac{k+1}{3k+4} = \frac{1}{3}$$

Because $L < 1$, the given series converges by the ratio test.

Recall that $n!$ denotes the product of all positive integers up to n. So $(3k)!$ is not the same as $3k!$ and $(3k+1)!$ is not the same as $1 \cdot 4 \cdot 7 \cdots (3k+1)$.

Example 8   Testing for convergence

Test the series $\displaystyle\sum_{k=1}^{\infty} \frac{k!}{k^k}$ for convergence.

Solution   The general term $a_k = \dfrac{k!}{k^k}$ involves a kth power, which suggests using the root test, but the presence of the factorial $k!$ makes using the ratio test much more reasonable.

$$
\begin{aligned}
L &= \lim_{k \to \infty} \frac{\frac{(k+1)!}{(k+1)^{k+1}}}{\frac{k!}{k^k}} \\
&= \lim_{k \to \infty} \frac{(k+1)!\,k^k}{k!(k+1)^{k+1}} \\
&= \lim_{k \to \infty} \frac{(k+1)k^k}{(k+1)^{k+1}} \\
&= \lim_{k \to \infty} \left(\frac{k}{k+1}\right)^k \\
&= \lim_{k \to \infty} \frac{1}{\left(1 + \frac{1}{k}\right)^k} \\
&= \frac{1}{e} \\
&< 1
\end{aligned}
$$

Since $L < 1$, the series converges by the ratio test.

PROBLEM SET 8.5

Level 1

1. ■ What does this say? Describe the ratio test.
2. ■ What does this say? Describe the root test.

Use either the ratio test or the root test to determine the convergence of the series given in Problems 3-26.

3. $\displaystyle\sum_{k=1}^{\infty} \frac{1}{k!}$

4. $\displaystyle\sum_{k=1}^{\infty} \frac{k!}{2^k}$

5. $\displaystyle\sum_{k=1}^{\infty} \frac{k!}{2^{3k}}$

6. $\displaystyle\sum_{k=1}^{\infty} \frac{3^k}{k!}$

7. $\displaystyle\sum_{k=1}^{\infty} \frac{k}{2^k}$

8. $\displaystyle\sum_{k=1}^{\infty} \frac{2^k}{k^2}$

9. $\displaystyle\sum_{k=1}^{\infty} \frac{k^{100}}{e^k}$

10. $\displaystyle\sum_{k=1}^{\infty} ke^{-k}$

11. $\displaystyle\sum_{k=1}^{\infty} k\left(\frac{4}{3}\right)^k$

12. $\displaystyle\sum_{k=1}^{\infty} k\left(\frac{3}{4}\right)^k$

13. $\displaystyle\sum_{k=1}^{\infty} \left(\frac{2}{k}\right)^k$

14. $\displaystyle\sum_{k=1}^{\infty} \frac{k^{10}2^k}{k!}$

15. $\displaystyle\sum_{k=1}^{\infty} \frac{k^5}{10^k}$

16. $\displaystyle\sum_{k=1}^{\infty} \frac{3^k}{k^2}$

17. $\displaystyle\sum_{k=1}^{\infty} \left(\frac{k}{3k+1}\right)^k$

18. $\displaystyle\sum_{k=1}^{\infty} \frac{3k+1}{2^k}$

19. $\displaystyle\sum_{k=1}^{\infty} \frac{k!}{(k+2)^4}$

20. $\displaystyle\sum_{k=1}^{\infty} \frac{k^5+100}{k!}$

21. $\displaystyle\sum_{k=1}^{\infty} \frac{(k!)^2}{(2k)!}$

22. $\displaystyle\sum_{k=1}^{\infty} \frac{(k!)^2}{[(2k)!]^2}$

23. $\displaystyle\sum_{k=1}^{\infty} k^2 2^{-k}$

24. $\displaystyle\sum_{k=1}^{\infty} k^4 3^{-k}$

25. $\displaystyle\sum_{k=1}^{\infty} \left(\frac{k-2}{k}\right)^{k^2}$

26. $\displaystyle\sum_{k=1}^{\infty} \left(\frac{k}{2k+1}\right)^k$

Test the series in Problems 27-46 for convergence. Justify your answers (that is, state explicitly which test you are using).

27. $\displaystyle\sum_{k=1}^{\infty} \frac{1{,}000}{k}$

28. $\displaystyle\sum_{k=1}^{\infty} \frac{5{,}000}{k\sqrt{k}}$

29. $\displaystyle\sum_{k=1}^{\infty} \frac{5k+2}{k2^k}$

30. $\displaystyle\sum_{k=1}^{\infty} \frac{(k!)^2}{k^k}$

31. $\displaystyle\sum_{k=1}^{\infty} \frac{\sqrt{k!}}{2^k}$

32. $\displaystyle\sum_{k=1}^{\infty} \frac{3k+5}{k3^k}$

33. $\displaystyle\sum_{k=1}^{\infty} \frac{2^k k!}{k^k}$

34. $\displaystyle\sum_{k=1}^{\infty} \frac{2^{2k} k!}{k^k}$

35. $\displaystyle\sum_{k=1}^{\infty} \frac{\sqrt{k+1}}{k^{k+0.5}}$

36. $\displaystyle\sum_{k=1}^{\infty} \frac{1}{k^k}$

37. $\displaystyle\sum_{k=1}^{\infty} \frac{k!}{(k+1)!}$

38. $\displaystyle\sum_{k=1}^{\infty} \frac{2^{1{,}000k}}{k^{k/2}}$

39. $\displaystyle\sum_{k=1}^{\infty} \frac{k}{4k^3 - 5}$

40. $\displaystyle\sum_{k=1}^{\infty} \frac{k^2+1}{(k^2+2)k^2}$

41. $\displaystyle\sum_{k=1}^{\infty} \left(1+\frac{1}{k}\right)^{-k^2}$

42. $\displaystyle\sum_{k=1}^{\infty} \left(\frac{k+2}{k}\right)^{-k^2}$

43. $\displaystyle\sum_{k=1}^{\infty} \left|\frac{\cos k}{2^k}\right|$

44. $\displaystyle\sum_{k=1}^{\infty} \left|\frac{\sin k}{3^k}\right|$

45. $\displaystyle\sum_{k=2}^{\infty} \left(\frac{\ln k}{k}\right)^k$

46. $\displaystyle\sum_{k=2}^{\infty} \frac{1}{(\ln k)^k}$

Level 2

In Problems 47-54, assume $x \geq 0$ and find all x for which the given series converges.

47. $\displaystyle\sum_{k=1}^{\infty} k^2 x^k$

48. $\displaystyle\sum_{k=1}^{\infty} kx^k$

49. $\displaystyle\sum_{k=1}^{\infty} \frac{(x+0.5)^k}{k\sqrt{k}}$

50. $\displaystyle\sum_{k=1}^{\infty} \frac{(3x+0.4)^k}{k^2}$

51. $\displaystyle\sum_{k=1}^{\infty} \frac{x^k}{k!}$

52. $\displaystyle\sum_{k=1}^{\infty} \frac{x^{2k}}{k}$

53. $\displaystyle\sum_{k=1}^{\infty} (ax)^k, a > 0$

54. $\displaystyle\sum_{k=1}^{\infty} kx^{2k}$

55. Use the root test to show that $\Sigma k^p e^{-k}$ converges for any fixed positive real number p. What does this imply about the following improper integral?

$$\int_1^{\infty} x^p e^{-x}\, dx$$

56. Consider the series

$$\sum_{k=1}^{\infty} 2^{-k+(-1)^k}$$

What can you conclude about the convergence of this series if you use each of the following?
a. the ratio test
b. the root test

Level 3

57. Let $\{a_k\}$ be a sequence of positive numbers and suppose that

$$\lim_{k\to\infty} \frac{a_{k+1}}{a_k} = L$$

where $0 < L < 1$.
a. Show that $\displaystyle\lim_{k\to\infty} a_k = 0$.
  Hint: Note that Σa_k converges.
b. Show that

$$\lim_{k\to\infty} \frac{x^k}{k!} = 0$$

for any $x > 0$.

58. Consider the series

$$1 + \frac{1}{2} + \frac{1}{2} + \frac{1}{4} + \frac{1}{4} + \frac{1}{8} + \frac{1}{8} + \frac{1}{16} + \cdots$$

a. Show that the ratio test fails.
b. What is the result of applying the root test?

59. Prove the root test by completing the steps in the following outline:
a. Let

$$\lim_{n\to\infty} \sqrt[n]{a_n} = R$$

If $R < 1$, choose x so $R < x < 1$. Explain why there is an N such that

$$0 \leq \sqrt[n]{a_n} \leq x$$

for all $n \geq N$. Then use the direct comparison test to show that Σa_n converges.

b. If $R > 1$, explain why $a_n > 1$ for all but a finite number of n. Use the divergence test to show Σa_n diverges.

c. Find two series Σa_n and Σb_n with $R = 1$, so that Σa_n converges but Σb_n diverges.

60. EXPLORATION PROBLEM Recall from Problem 50, Section 8.1, that the Fibonacci sequence $\{a_n\}$ is

$$1, 1, 2, 3, 5, 8, 13, 21, \cdots$$

where $a_1 = 1$, $a_2 = 1$, and the rest of the terms are defined recursively by the formula

$$a_{n+1} = a_n + a_{n-1}$$

for $n = 2, 3, \cdots$. Determine whether the series of reciprocal Fibonacci numbers

$$\sum_{k=1}^{\infty} \frac{1}{a_k} = 1 + 1 + \frac{1}{2} + \frac{1}{3} + \frac{1}{5} + \cdots$$

converges or diverges.

8.6 ALTERNATING SERIES; ABSOLUTE AND CONDITIONAL CONVERGENCE

IN THIS SECTION: *Alternating series test, error estimates for alternating series, absolute and conditional convergence, summary of convergence tests, rearrangement of terms in an absolutely convergent series*
We now turn our attention from nonnegative term series to series with both positive and negative terms.

There are two classes of series for which the successive terms alternate in sign, and each of these series is appropriately called an **alternating series**:

$$\text{odd-indexed terms are negative: } \sum_{k=1}^{\infty}(-1)^k a_k = -a_1 + a_2 - a_3 + \cdots$$

$$\text{even-indexed terms are negative: } \sum_{k=1}^{\infty}(-1)^{k+1} a_k = a_1 - a_2 + a_3 - \cdots$$

where in both cases $a_k > 0$.

Alternating Series Test

In general, just knowing that $\lim_{k \to \infty} a_k = 0$ tells us very little about the convergence properties of the series Σa_k, but it turns out that an alternating series must converge if the absolute value of its terms decreases monotonically toward zero. This result, first established by Leibniz in the 17th century, may be stated as shown in the following theorem.

Theorem 8.18 Alternating series test

An alternating series

$$\sum_{k=1}^{\infty}(-1)^k a_k \quad \text{or} \quad \sum_{k=1}^{\infty}(-1)^{k+1} a_k$$

where $a_k > 0$ for all k converges if both of the following two conditions are satisfied:

1. $\lim_{k \to \infty} a_k = 0$

2. $\{a_k\}$ is a decreasing sequence; that is, $a_{k+1} \leq a_k$ for all k.

Proof: We will show that when an alternating series of the form

$$\sum_{k=1}^{\infty}(-1)^{k+1} a_k$$

The alternating series test only applies to series whose terms strictly alternate in sign, not to series with other patterns of $+$ and $-$ signs. See Example 6.

satisfies the two required properties, it converges. The steps for the other type of alternating series are similar. For this proof we are given that

$$\lim_{k \to \infty} a_k = 0 \quad \text{and} \quad a_{k+1} \le a_k \text{ for all } k$$

We need to prove that the sequence of partial sums $\{S_n\}$ converges, where

$$S_n = \sum_{k=1}^{n}(-1)^{k+1}a_k = a_1 - a_2 + a_3 - a_4 + \cdots + (-1)^{n+1}a_n$$

Our strategy will be to show first that the sequence of *even*-indexed partial sums $\{S_{2n}\}$ is *increasing* and converges to a certain limit L. Then we will show that the sequence of odd-indexed partial sums $\{S_{2n-1}\}$ also converges to L.

To understand why we would think to break up S_n into even- and odd-indexed partial sums, consider Figure 8.16. Start at the origin.

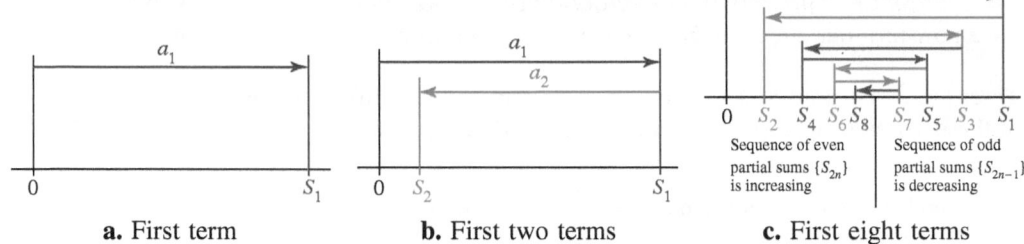

a. First term   **b.** First two terms   **c.** First eight terms

Figure 8.16   Alternating series test

Because $a_1 > 0$, S_1 will be somewhere to the right of 0 as shown in Figure 8.16**a**. We know that $a_2 \le a_1$ (decreasing sequence) so S_2 is to the left of S_1 but to the right of 0, as shown in Figure 8.16**b**. As you continue this process you can see that the partial sums oscillate back and forth. Because $a_k \to 0$, the successive steps are getting smaller and smaller, as shown in Figure 8.16**c**.

Note that the even partial sums of the alternating series satisfy

$$S_2 = a_1 - a_2$$
$$S_4 = (a_1 - a_2) + (a_3 - a_4)$$
$$\vdots$$
$$S_{2n} = (a_1 - a_2) + (a_3 - a_4) + \cdots + (a_{2n-1} - a_{2n})$$

Because $\{a_k\}$ is a decreasing sequence, each of the quantities in parentheses $(a_{2j-1} - a_{2j})$ is nonnegative, and it follows that $\{S_{2n}\}$ is an increasing sequence:

$$S_2 \le S_4 \le S_6 \le \cdots$$

Moreover, because

$$S_2 = a_1 - a_2 \le a_1$$
$$S_4 = a_1 - (a_2 - a_3) - a_4 \le a_1$$
$$\vdots$$
$$S_{2n} = a_1 - (a_2 - a_3) - (a_4 - a_5) - \cdots - (a_{2n-1} - a_{2n}) - a_{2n} \le a_1$$

the sequence of even partial sums $\{S_{2n}\}$ is bounded above by a_2, and because it is also an increasing sequence, the BMCT tells us that this sequence must converge, say, to L; that is,

$$\lim_{n \to \infty} S_{2n} = L$$

Next, consider $\{S_{2n-1}\}$, the sequence of odd partial sums. Because $S_{2n} - S_{2n-1} = a_{2n}$ and $\lim_{n \to \infty} a_n = 0$, it follows that $S_{2n-1} = S_{2n} - a_{2n}$ and

$$\lim_{n \to \infty} S_{2n-1} = \lim_{n \to \infty} S_{2n} - \lim_{n \to \infty} a_{2n} = L - 0 = L$$

Since the sequence of even partial sums and the sequence of odd partial sums both converge to L, it follows that $\{S_n\}$ converges to L, and the alternating series must converge (in fact, to L). ◆

NOTE: When you are testing the alternating series $\Sigma(-1)^{k+1}a_k$ for convergence, it is wise to begin by computing $\lim_{k \to \infty} a_k$. If $\lim_{k \to \infty} a_k \neq 0$, you know immediately that the series diverges (by the divergence test), and no further computation is necessary. However, if this limit is 0, you can show that the alternating series converges by verifying that $\{a_k\}$ is a decreasing sequence.

Example 1 Convergence using the alternating series test

Test the series $\displaystyle\sum_{k=1}^{\infty} \frac{(-1)^k}{k}$ for convergence. This series is called the **alternating harmonic series.**

Solution The series can be expressed as $\Sigma(-1)^k a_k$, where $a_k = \frac{1}{k}$. We have $\lim_{k \to \infty} \frac{1}{k} = 0$ and because

$$\frac{1}{k+1} < \frac{1}{k}$$

Remember that the a_k used in the alternating series test does NOT include the $(-1)^{k+1}$.

for all $k > 0$, the sequence $\{a_k\}$ is decreasing. Thus, the alternating series test tells us that the given series must converge. ∎

Example 2 Using l'Hôpital's rule with the alternating series test

Test the series $\displaystyle\sum_{k=1}^{\infty} \frac{(-1)^{k+1} \ln k}{k}$ for convergence.

Solution Express the series in the form $\Sigma(-1)^{k+1}a_k$, where $a_k = \dfrac{\ln k}{k}$ (note that $a_k > 0$ for $k > 1$). The graph of $y = (\ln x)/x$ is shown in Figure 8.17.

By applying l'Hôpital's rule, we find that

$$\lim_{x \to \infty} \frac{\ln x}{x} = \lim_{x \to \infty} \frac{\frac{1}{x}}{1} = 0$$

It remains to show that the sequence $\{a_k\}$ is decreasing. We can do this by setting $f(x) = (\ln x)/x$ and noting that the derivative

$$f'(x) = \frac{1 - \ln x}{x^2}$$

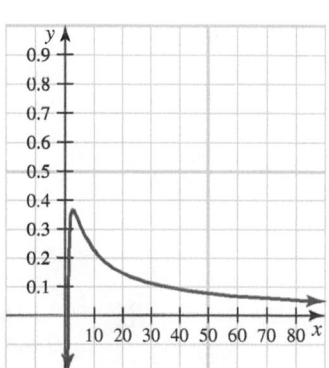

Figure 8.17 Graph of $y = \dfrac{\ln x}{x}$

satisfies $f'(x) < 0$ for all $x > e$. Thus, the sequence $\{a_k\}$ is (eventually) decreasing for all $k > 3 > e$. The conditions of the alternating series test are satisfied, and the given alternating series must converge. ∎

Example 3   Divergence of an alternating series

Test the series $\displaystyle\sum_{k=1}^{\infty} (-1)^{k+1} e^{1/k}$ for convergence.

Solution   Since

$$\lim_{k\to\infty} e^{1/k} = \exp\left(\lim_{k\to\infty} \frac{1}{k}\right) = e^0 = 1 \neq 0$$

the series diverges by the divergence test.   ■

ALTERNATING p-SERIES   The series $\displaystyle\sum_{k=2}^{\infty} \frac{(-1)^{k+1}}{k^p}$ is called the **alternating p-series.**

The following example shows that the alternating p-series converges for all $p > 0$.

Example 4   Alternating p-series

Prove that the alternating p-series $\displaystyle\sum_{k=1}^{\infty} \frac{(-1)^{k+1}}{k^p}$ converges for $p > 0$.

Solution   If $a^k = \dfrac{1}{k^p}$, then $\displaystyle\lim_{k\to\infty} a_k = 0$ for $p > 0$. To show that the sequence $\{a_k\}$ is decreasing, we note that

$$\frac{a_{k+1}}{a_k} = \frac{\frac{1}{(k+1)^p}}{\frac{1}{k^p}} = \frac{k^p}{(k+1)^p} < 1$$

so $a_k > a_{k+1}$. Thus, the alternating p-series converges for all $p > 0$.   ■

Error Estimates for Alternating Series

For any convergent series with sum L, the nth partial sum is an approximation to L that we expect to improve as n increases. In general, it is difficult to know how large an n to pick in order to ensure that the approximation will have a desired degree of accuracy. However, if the series in question satisfies the conditions of the alternating series test, the following theorem provides a convenient criterion for making this determination.

Theorem 8.19   The error estimate for an alternating series

Suppose an alternating series

$$\sum_{k=1}^{\infty} (-1)^k a_k \quad \text{or} \quad \sum_{k=1}^{\infty} (-1)^{k+1} a_k$$

satisfies the conditions of the alternating series test; namely, $\displaystyle\lim_{k\to\infty} a_k = 0$ and $\{a_k\}$ is a decreasing sequence $(a_{k+1} \leq a_k)$. If the series has sum S, then

$$|S - S_n| \leq a_{n+1}$$

where S_n is the nth partial sum of the series.

> ■ **W**hat this says If an alternating series satisfies the conditions of the alternating series test, you can approximate the sum of the series by using the nth partial sum $\{S_n\}$, and your error will have an absolute value no greater than the first term left off (namely, a_{n+1}).

Proof: We will prove the result for the second form of the alternating series and leave the first form for the reader. That is, let

$$S = \sum_{k=1}^{\infty} (-1)^{k+1} a_k \qquad \text{and} \qquad S_n = \sum_{k=1}^{n} (-1)^{k+1} a_k$$

Begin with $S - S_n$:

$$
\begin{aligned}
S - S_n &= \sum_{k=1}^{\infty} (-1)^{k+1} a_k - \sum_{k=1}^{n} (-1)^{k+1} a_k \\
&= \sum_{k=n+1}^{\infty} (-1)^{k+1} a_k \\
&= (-1)^{n+2} a_{n+1} + (-1)^{n+3} a_{n+3} + \cdots \\
&= (-1)^n (-1)^2 a_{n+1} + (-1)^n (-1)^3 a_{n+2} + (-1)^n (-1)^4 a_{n+3} + \cdots \\
&\quad\quad\quad\quad\quad\quad\quad\quad\vdots \\
&= (-1)^n \left[a_{n+1} - a_{n+2} + a_{n+3} - a_{n+4} + \cdots \right] \\
&= (-1)^n \left[a_{n+1} - (a_{n+2} - a_{n+3}) - (a_{n+4} - a_{n+5}) - \cdots \right]
\end{aligned}
$$

Because the sequence $\{a_n\}$ is decreasing, we have $a_k - a_{k+1} \geq 0$ for all k, and it follows that

$$|S - S_n| = |a_{n+1} - (a_{n+2} - a_{n+3}) - (a_{n+4} - a_{n+5}) - \cdots| \leq a_{n+1}$$

since every quantity in parentheses is positive.   ♦

Example 5   Error estimate for an alternating series

Consider the convergent alternating series $\displaystyle\sum_{k=1}^{\infty} \frac{(-1)^{k+1}}{k^4}$.

a. Estimate the sum of the series by taking the sum of the first four terms. How accurate is this estimate?

b. How many terms of the series are necessary to estimate its sum with three decimal-place accuracy? What is this estimate?

Solution

a. Let $a_k = \dfrac{1}{k^4}$ and let S denote the actual sum of the series. The error estimate tells us that

$$|S - S_4| \leq a_5$$

where S_4 is the sum of the first four terms of the series. Using a calculator, we find

$$S_4 = \frac{1}{1^4} - \frac{1}{2^4} + \frac{1}{3^4} - \frac{1}{4^4} \approx 0.9459394$$

and $a_5 = \dfrac{1}{5^4} = 0.0016$. This first term omitted is positive. Thus, if we estimate S by $S_4 \approx 0.9459$, we incur an error of at most 0.0016, which means that

$$|S - S_4| \leq 0.0016$$
$$0.9459394 \leq S \leq 0.9459394 + 0.0016$$
$$0.9459394 \leq S \leq 0.9475394$$

You may think that "three decimal place accuracy" means an error no greater than 0.001, *but to insure against round-off error, we really need* 0.0005.

b. Because we want to approximate S by a partial sum S_n with three-decimal-place accuracy, we can allow an error of no more than 0.0005. According to Theorem 8.19, we want to find an index n such that $a_{n+1} \leq 0.0005$; that is,

$$\frac{1}{(n+1)^4} \leq 0.0005$$
$$\frac{1}{0.0005} \leq (n+1)^4$$
$$\sqrt[4]{2{,}000} \leq n+1$$
$$6.687403 - 1 \leq n$$

Thus, $n \geq 5.687403$, so we need at least 6 terms to achieve the required accuracy. We find that

$$S_6 = \frac{1}{1^4} - \frac{1}{2^4} + \frac{1}{3^4} - \frac{1}{4^4} + \frac{1}{5^4} - \frac{1}{6^4} \approx 0.94677$$

The first term left off (namely $\dfrac{1}{7^4} \approx 0.00042$) is positive, so

$$0.94677 < S < 0.94677 + 0.00042$$
$$0.94677 < S < 0.94719$$

Both bounds round to 0.947, so $S \approx 0.947$, with three decimal-place accuracy.

Absolute and Conditional Convergence

The convergence tests we have developed cannot be applied to a series that has mixed terms or does not strictly alternate. In such cases, it is often useful to apply the following result.

Theorem 8.20   The absolute convergence test

A series of real numbers Σa_k must converge if the related absolute value series $\Sigma |a_k|$ converges.

Proof: Assume $\Sigma |a_k|$ converges and let $b_k = a_k + |a_k|$ for all k. Note that

$$b_k = \begin{cases} 2|a_k| & \text{if } a_k > 0 \\ 0 & \text{if } a_k \leq 0 \end{cases}$$

Thus, we have $0 \leq b_k \leq 2|a_k|$ for all k. Because the series $\Sigma |a_k|$ converges and both Σb_k and $\Sigma |a_k|$ are series of nonnegative terms, the direct comparison test tells us that the dominated series Σb_k also converges. Finally, because

$$a_k = b_k - |a_k|$$

and both Σb_k and $\Sigma |a_k|$ converge, it follows that Σa_k must also converge.   ◆

Example 6 Convergence using the absolute convergence test

Test the series

$$1 + \frac{1}{4} + \frac{1}{9} - \frac{1}{16} - \frac{1}{25} + \frac{1}{36} + \frac{1}{49} + \frac{1}{64} - \frac{1}{81} - \frac{1}{100} + \cdots$$

for convergence. This is the series in which the absolute value of the general term is $\frac{1}{k^2}$ and the pattern of the signs is $+++--+++---\cdots$.

Solution   This is not a series of positive terms, nor is it strictly alternating. However, we find that the corresponding series of absolute values is the convergent *p*-series

$$1 + \frac{1}{4} + \frac{1}{9} + \frac{1}{16} + \cdots = \sum_{k=1}^{\infty} \frac{1}{k^2}$$

Because the absolute value series converges, the absolute convergence test assures us that the given series also converges. ■

Example 7 Convergence of a trigonometric series

Test the series $\displaystyle\sum_{k=1}^{\infty} \frac{\sin k}{2^k}$ for convergence.

Solution   Because $\sin k$ takes on both positive and negative values, the series cannot be analyzed by methods that apply only to series of positive terms. Moreover, note that the series is not strictly alternating:

$$\sum_{k=1}^{\infty} \frac{\sin k}{2^k} = 0.421 + 0.227 + 0.018 - 0.047 - 0.030 - 0.004 + 0.005 + \cdots$$

The corresponding series of absolute values is

$$\sum_{k=1}^{\infty} \left| \frac{\sin k}{2^k} \right|$$

which is dominated by the convergent geometric series $\Sigma \frac{1}{2^k}$ because $|\sin k| \leq 1$ for all k; that is,

$$0 \leq \left| \frac{\sin k}{2^k} \right| \leq \frac{1}{2^k} \qquad \text{for all } k$$

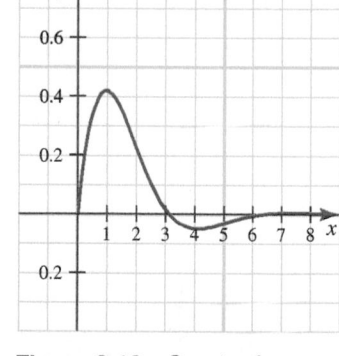

Figure 8.18   Graph of $y = \dfrac{\sin x}{2^x}$

Therefore, the absolute convergence test assures us that the given series converges. The graph of the related continuous function $y = (\sin x)/2^x$ is shown in Figure 8.18. ■

If Σa_k converges, then $\Sigma |a_k|$ may either converge or diverge. The two cases that can occur are given the following special names.

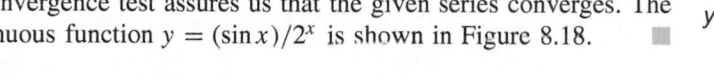

ABSOLUTELY AND CONDITIONALLY CONVERGENT   The series Σa_k is **absolutely convergent** (or converges absolutely) if the related series $\Sigma |a_k|$ converges. The series Σa_k is **conditionally convergent** (or converges conditionally) if it converges but $\Sigma |a_k|$ diverges.

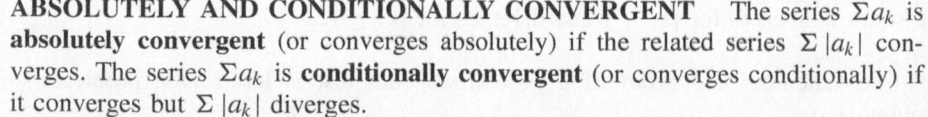

Note that a convergent series is either absolutely convergent or conditionally convergent, but it cannot be both.

We remark that the series in Examples 6 and 7 are both absolutely convergent, but the alternating harmonic series

$$\sum_{k=1}^{\infty} \frac{(-1)^{k+1}}{k}$$

is conditionally convergent because it converges (see Example 1), whereas the related series of absolute value terms $\Sigma 1/k$ (the harmonic series) diverges.

The ratio test and the root test for positive-term series can be generalized to apply to arbitrary series. The following is a statement and proof of the generalized ratio test.

Theorem 8.21   The generalized ratio test

For the series Σa_k, suppose $a_k \neq 0$ for $k \geq 1$ and that

$$\lim_{k \to \infty} \left| \frac{a_{k+1}}{a_k} \right| = L$$

where L is a real number or ∞. Then:

If $L < 1$, the series Σa_k converges absolutely and hence converges.

If $L > 1$ or if L is infinite, the series Σa_k diverges.

If $L = 1$, the test is inconclusive.

Proof: Assume $L < 1$; then the positive series $\Sigma |a_k|$ converges by the ratio test, and hence Σa_k converges absolutely. Assume $L > 1$; then the sequence $\{|a_k|\}$ is eventually increasing, which means that $\{a_k\}$ cannot converge to 0 as $k \to \infty$, so Σa_k diverges by the divergence test. Assume $L = 1$. See Problem 58 for this part of the proof.   ◆

Example 8   Convergence and divergence using the generalized ratio test

Find all values of x for which the series $\Sigma k x^k$ converges.

Solution   Let $a_k = k x^k$.

$$\begin{aligned}
L &= \lim_{k \to \infty} \left| \frac{a_{k+1}}{a_k} \right| \\
&= \lim_{k \to \infty} \left| \frac{(k+1)x^{k+1}}{k x^k} \right| \\
&= \lim_{k \to \infty} \left(\frac{k+1}{k} \right) \left| \frac{x^{k+1}}{x^k} \right| \\
&= \lim_{k \to \infty} \left(1 + \frac{1}{k} \right) |x| \\
&= |x|
\end{aligned}$$

Thus, by the generalized ratio test, since $L = |x|$ the series converges for $|x| < 1$ and diverges if $|x| > 1$. If $x = 1$, the series is $\Sigma k(1)^k$, which clearly diverges by the divergence test. Similarly, if $x = -1$, $\Sigma k(-1)^k$ diverges by the divergence test. Thus, the series converges for $|x| < 1$ and diverges for $|x| \geq 1$.   ■

The statement and proof of a generalized root test is left to the problems (see Problem 56).

Summary of Convergence Tests

This concludes our basic study of convergence tests. There are no firm rules for deciding how to test the convergence of a given series Σa_k, but we can offer a few guidelines (Table 8.1).

Table 8.1   Guidelines for determining convergence of the series Σa_k

Series with known convergence properties	
Geometric series, $\displaystyle\sum_{k=1}^{\infty} ar^k$	diverges if $\lvert r \rvert \geq 1$, and converges if $\lvert r \rvert < 1$ with sum $S = \frac{a}{1-r}$
p-series, $\displaystyle\sum_{k=1}^{\infty} \frac{1}{k^p}$	converges if $p > 1$ and diverges if $p \leq 1$
Harmonic series, $\displaystyle\sum_{k=1}^{\infty} \frac{1}{k}$	special case of the p-series where $p = 1$; it diverges
q-log series, $\displaystyle\sum_{k=1}^{\infty} \frac{\ln k}{k^q}$	converges if $q > 1$ and diverges if $q \leq 1$
Divergence test	Compute $\lim_{k \to \infty} a_k$; if this limit is anything but 0, the series diverges.
Tests for series Σa_k with $a_k > 0$	
Limit comparison test	Check to see whether Σa_k is similar in appearance to a series Σb_k. If $\lim_{k \to \infty} a_k/b_k = L$, where L is finite and positive, then Σa_k and Σb_k either both converge or both diverge.
Ratio test	If a_k involves $k!\, k^p$, or c^k, try the ratio test. Set $\displaystyle\lim_{k \to \infty} \frac{a_{k+1}}{a_k} = L$. Converges if $L < 1$, diverges if $L > 1$, inconclusive if $L = 1$.
Root test	If it is easy to find $\displaystyle\lim_{k \to \infty} \sqrt[k]{a_k} = L$, try the root test. Converges if $L < 1$, diverges if $L > 1$, is inconclusive if $L = 1$.
Integral test	If f is continuous, positive, and decreasing, and if $a_k = f(k)$ for all k, then $\displaystyle\sum_{k=1}^{\infty} a_k$ and the improper integral $\int_1^{\infty} f(x)\,dx$ either both converge or both diverge. Consider using this test when f is easy to integrate or if a_k involves a logarithm, a trigonometric function, or an inverse trigonometric function.
Direct comparison test	If $0 \leq a_k \leq c_k$ and $\displaystyle\sum_{k=1}^{\infty} c_k$ converges, then $\displaystyle\sum_{k=1}^{\infty} a_k$ converges. If $0 \leq d_k \leq a_k$ and $\displaystyle\sum_{k=1}^{\infty} d_k$ diverges, then $\displaystyle\sum_{k=1}^{\infty} a_k$ diverges.
Zero-infinity limit comparison test	If $\displaystyle\lim_{k \to \infty} \frac{a_k}{b_k} = 0$ and Σb_k converges, then the series Σa_k converges. If $\displaystyle\lim_{k \to \infty} \frac{a_k}{b_k} = \infty$ and Σb_k diverges, then the series Σa_k diverges.
Tests for series Σa_k that do not require $a_k > 0$	
Test for absolute convergence	Take the absolute value of each term and proceed to test for convergence of positive series, as summarized above. 1. If $\Sigma \lvert a_k \rvert$ converges, then Σa_k converges. 2. If $\Sigma \lvert a_k \rvert$ diverges, then the series is *not absolutely convergent*, but may still converge *conditionally*.
Alternating series test (conditional convergence)	The alternating series $\Sigma (-1)^k a_k$ or $\Sigma (-1)^{k+1} a_k$ with $a_k > 0$ converges if $a_{k+1} \leq a_k$ for $k = 1, 2, \cdots$ and $\displaystyle\lim_{k \to \infty} a_k = 0$.

Rearrangement of Terms in an Absolutely Convergent Series

You may be surprised to learn that if the terms in a conditionally convergent series are rearranged (that is, the order of the summands is changed), the new series may not converge or it may converge to a different sum from that of the original series!

For example, we know that the alternating harmonic series

$$\sum_{k=1}^{\infty} \frac{(-1)^{k+1}}{k}$$

converges conditionally, and it can be shown (see Problem 51) that

$$1 - \frac{1}{2} + \frac{1}{3} - \frac{1}{4} + \frac{1}{5} - \frac{1}{6} + \frac{1}{7} - \frac{1}{8} + \frac{1}{9} - \cdots = \ln 2$$

If we rearrange this series by placing two successive subtracted terms after each added term, we obtain

$$1 - \frac{1}{2} - \frac{1}{4} + \frac{1}{3} - \frac{1}{6} - \frac{1}{8} + \frac{1}{5} - \cdots = \left(1 - \frac{1}{2}\right) - \frac{1}{4} + \left(\frac{1}{3} - \frac{1}{6}\right) - \frac{1}{8} + \cdots$$
$$= \frac{1}{2} - \frac{1}{4} + \frac{1}{6} - \frac{1}{8} + \cdots$$
$$= \frac{1}{2}\left(1 - \frac{1}{2} + \frac{1}{3} - \frac{1}{4} + \cdots\right)$$
$$= \frac{1}{2}\ln 2$$

In general, it can be shown that if Σa_k is conditionally convergent, there is a rearrangement of the terms of Σa_k so that the sum is equal to *any* specified finite number. In Problem 52, for example, you are asked to rearrange the terms of the series

$$\sum_{k=1}^{\infty} \frac{(-1)^{k+1}}{k}$$

so that the sum is $\frac{3}{2}\ln 2$.

This information may be somewhat unsettling, because it is reasonable to expect a sum to be unaffected by the order in which the summands are taken. Absolutely convergent series behave more in the way we would expect. In fact, if the series Σa_k converges absolutely with sum S, then any rearrangement of the terms also converges absolutely to S. A detailed discussion of rearrangement of terms of a series requires the techniques of advanced calculus.

PROBLEM SET 8.6

Level 1

1. ■ What does this say? Discuss absolute versus conditional convergence.

2. ■ What does this say? Discuss the convergence of the alternating p-series.

Determine whether each series in Problems 3-30 converges absolutely, converges conditionally, or diverges.

3. $\displaystyle\sum_{k=1}^{\infty} \frac{(-1)^{k+1}k}{k^2+1}$

4. $\displaystyle\sum_{k=1}^{\infty} \frac{(-1)^{k+1}k^2}{k^3+1}$

5. $\displaystyle\sum_{k=1}^{\infty} \frac{(-1)^{k+1}k}{2k+1}$

6. $\displaystyle\sum_{k=1}^{\infty} \frac{(-1)^{k+1}k^2}{k^2+1}$

7. $\displaystyle\sum_{k=1}^{\infty} \frac{(-1)^{k+1}}{k^{3/2}}$

8. $\displaystyle\sum_{k=1}^{\infty} \frac{(-1)^{k+1}k}{2^k}$

9. $\displaystyle\sum_{k=1}^{\infty} (-1)^{k+1}\frac{k^2}{e^k}$

10. $\displaystyle\sum_{k=1}^{\infty} \frac{(-1)^k}{\sqrt{k}}$

11. $\displaystyle\sum_{k=1}^{\infty} (-1)^k\frac{(1+k^2)}{k^3}$

12. $\displaystyle\sum_{k=1}^{\infty} \frac{(-1)^{k+1}k!}{k^k}$

13. $\displaystyle\sum_{k=2}^{\infty} (-1)^k\frac{k!}{\ln k}$

14. $\displaystyle\sum_{k=1}^{\infty} \frac{(-1)^k(2k)!}{k^k}$

15. $\displaystyle\sum_{k=1}^{\infty} (-1)^{k+1}\frac{2^k}{k!}$

16. $\displaystyle\sum_{k=1}^{\infty} \frac{(-3)^k(k+1)}{k!}$

17. $\displaystyle\sum_{k=1}^{\infty} (-1)^{k+1} \frac{2^{2k+1}}{k!}$

18. $\displaystyle\sum_{k=2}^{\infty} \frac{(-1)^{k+1}}{\ln k}$

19. $\displaystyle\sum_{k=1}^{\infty} \frac{(-1)^{k+1}k}{(k+1)(k+2)}$

20. $\displaystyle\sum_{k=2}^{\infty} \frac{(-1)^{k+1}}{(\ln k)^4}$

21. $\displaystyle\sum_{k=2}^{\infty} \frac{(-1)^{k+1}}{\ln(\ln k)}$

22. $\displaystyle\sum_{k=2}^{\infty} \frac{(-1)^{k+1}}{k \ln k}$

23. $\displaystyle\sum_{k=1}^{\infty} \frac{(-1)^{k+1}\ln k}{k}$

24. $\displaystyle\sum_{k=2}^{\infty} \frac{(-1)^{k+1}k}{\ln k}$

25. $\displaystyle\sum_{k=1}^{\infty} (-1)^{k+1} \frac{\ln k}{k^2}$

26. $\displaystyle\sum_{k=1}^{\infty} \frac{(-1)^{k+1}k}{(k+2)^2}$

27. $\displaystyle\sum_{k=1}^{\infty} (-1)^{k+1} \left(\frac{k}{k+1}\right)^k$

28. $\displaystyle\sum_{k=2}^{\infty} (-1)^{k+1} \frac{\ln(\ln k)}{k \ln k}$

29. $\displaystyle\sum_{k=1}^{\infty} (-1)^{k+1} \left(\frac{1}{k}\right)^{1/k}$

30. $\displaystyle\sum_{k=1}^{\infty} (-1)^{k+1} \frac{k^5 5^{k+2}}{2^{3k}}$

Level 2

Given the series in Problems 31-36

a. *Estimate the sum of the series by taking the sum of the first four terms. How accurate is this estimate?*

b. *How many terms of the series are necessary to estimate its sum with three-place accuracy? What is this estimate?*

31. $\displaystyle\sum_{k=1}^{\infty} \frac{(-1)^{k+1}}{2^{2k-2}}$

32. $\displaystyle\sum_{k=1}^{\infty} \frac{(-1)^{k+1}}{k!}$

33. $\displaystyle\sum_{k=1}^{\infty} \frac{(-1)^k}{k^2}$

34. $\displaystyle\sum_{k=1}^{\infty} \left(\frac{-1}{3}\right)^{k+1}$

35. $\displaystyle\sum_{k=1}^{\infty} \frac{(-1)^{k+1}}{k^3}$

36. $\displaystyle\sum_{k=1}^{\infty} \left(\frac{-1}{5}\right)^k$

Use the generalized ratio test in Problems 37-42 to find all numbers x for which the given series converges.

37. $\displaystyle\sum_{k=1}^{\infty} \frac{x^k}{k}$

38. $\displaystyle\sum_{k=1}^{\infty} \frac{(2x)^k}{\sqrt{k}}$

39. $\displaystyle\sum_{k=1}^{\infty} \frac{2^k x^k}{k!}$

40. $\displaystyle\sum_{k=1}^{\infty} \frac{(k+2)x^k}{k^2(k+3)}$

41. $\displaystyle\sum_{k=1}^{\infty} (-1)^{k+1} \left(\frac{x}{k}\right)^k$

42. $\displaystyle\sum_{k=1}^{\infty} (-1)^k k^p x^k$,

for $p > 0$

43. Find an upper bound for the error if the convergent alternating series

$$\sum_{k=1}^{\infty} \frac{(-1)^{k+1}}{k}$$

is approximated by the partial sum

$$S_5 = 1 - \frac{1}{2} + \frac{1}{3} - \frac{1}{4} + \frac{1}{5}$$

44. Find an upper bound for the error if the convergent alternating series

$$\sum_{k=1}^{\infty} \frac{(-1)^{k+1}}{k^2}$$

is approximated by the partial sum

$$S_5 = 1 - \frac{1}{2^2} + \frac{1}{3^2} - \frac{1}{4^2} + \frac{1}{5^2}$$

45. Find an upper bound for the error if the convergent alternating series

$$\sum_{k=2}^{\infty} \frac{(-1)^k}{\ln k}$$

is approximated by the partial sum

$$S_7 = \frac{1}{\ln 2} - \frac{1}{\ln 3} + \frac{1}{\ln 4} - \frac{1}{\ln 5} + \frac{1}{\ln 6} - \frac{1}{\ln 7}$$

46. Find an upper bound for the error if the convergent alternating series

$$\sum_{k=1}^{\infty} \frac{(-1)^{k+1}k}{2^k}$$

is approximated by the partial sum

$$S_6 = \frac{1}{2} - \frac{2}{2^2} + \frac{3}{2^3} - \frac{4}{2^4} + \frac{5}{2^5} - \frac{6}{2^6}$$

47. For what numbers p does the convergent alternating series

$$\sum_{k=2}^{\infty} \frac{(-1)^{k+1}}{k(\ln k)^p}$$

converge? For what numbers p does it converge absolutely?

48. Test the series $\displaystyle\sum_{k=1}^{\infty} \frac{\sin \sqrt[k]{2}}{k^2}$ for convergence.

49. Use series methods to evaluate $\displaystyle\lim_{k\to\infty} \frac{x^k}{k!}$ where x is a real number.

Level 3

50. Show that the sequence $\{a_n\}$ converges, where

$$a_n = \frac{1}{1} + \frac{1}{2} + \frac{1}{3} + \cdots + \frac{1}{n} - \ln n$$

51. Show that $\displaystyle\sum_{k=1}^{\infty} \frac{(-1)^{k+1}}{k} = \ln 2$ by completing the following steps.

a. Let $S_m = \displaystyle\sum_{k=1}^{m} \frac{(-1)^{k+1}}{k}$ and let $H_m = \displaystyle\sum_{k=1}^{\infty} \frac{1}{k}$. Show that $S_{2m} = H_{2m} - H_m$.

b. It can be shown (see Problem 50)

$$\lim_{m\to\infty}\left(1 + \frac{1}{2} + \cdots + \frac{1}{m} - \ln m\right)$$

exists. The value of this limit is a number $\gamma \approx 0.57722$, called **Euler's constant**. Use this fact, along with the relationship in part **a**, to show that $\displaystyle\lim_{m\to\infty} S_n = \ln 2$.

Hint: Note that

$$S_{2m} = H_{2m} - H_m$$
$$= [H_{2m} - \ln(2m)] - [H_m - \ln m]$$
$$+ \ln(2m) - \ln m$$

52. Consider the following rearrangement of the conditionally convergent harmonic series $\displaystyle\sum_{k=1}^{\infty} \frac{(-1)^{k+1}}{k}$:

$$1 + \frac{1}{3} - \frac{1}{2} + \frac{1}{5} + \frac{1}{7} - \frac{1}{4} + \frac{1}{9} + \frac{1}{11} - \frac{1}{6} + \cdots$$

a. Let S_n and H_n denote the nth partial sum of the given rearranged series and the harmonic series $\displaystyle\sum_{k=1}^{\infty} \frac{1}{k}$, respectively. Show that if n is a multiple of 3, say $n = 3m$, then

$$S_{3m} = H_{4m} - \frac{1}{2}H_{2m} - \frac{1}{2}H_m$$

b. Show that $\displaystyle\lim_{n\to\infty} S_n = \frac{3}{2}\ln 2$.
Hint: You may find it helpful to use

$$\lim_{n\to\infty}\left(1 + \frac{1}{2} + \cdots + \frac{1}{n} - \ln n\right) = \gamma$$

given in Problem 51**b**.

53. Journal Problem (*School Science and Mathematics* by Michael Brozinsky)* Show that the series

$$1 + \frac{1}{2} + \frac{1}{3} - \frac{1}{4} - \frac{1}{5} - \frac{1}{6}$$
$$+ \frac{1}{7} + \frac{1}{8} + \frac{1}{9} - - - + + +$$

converges and find its sum.

54. Suppose $\{a_k\}$ is a sequence with the property that $|a_n| < A^n$ for some positive number A and all positive integers n. Show that the series $\Sigma a_k x^k$ converges absolutely for $|x| \le 1/A$.

55. Think Tank Problem What (if anything) is wrong with the following computation?

$$1 - \frac{1}{2} + \frac{1}{3} - \frac{1}{4} + \frac{1}{5} - \frac{1}{6} + \cdots$$
$$= 1 + \left(\frac{1}{2} - 1\right) + \frac{1}{3} + \left(\frac{1}{4} - \frac{1}{2}\right) + \frac{1}{5} + \left(\frac{1}{6} - \frac{1}{3}\right)$$
$$+ \cdots$$
$$= \left(1 + \frac{1}{2} + \frac{1}{3} + \frac{1}{4} + \cdots\right) - 1 - \frac{1}{2} - \frac{1}{3} - \frac{1}{4} - \cdots$$
$$= \left(1 + \frac{1}{2} + \frac{1}{3} + \cdots\right) - \left(1 + \frac{1}{2} + \frac{1}{3} + \cdots\right)$$
$$= 0$$

56. Prove the generalized root test, which may be stated as follows: Suppose that
If $L < 1$, the series Σa_k converges absolutely.

If $L > 1$ the series Σa_k diverges.

If $L = 1$, the test is inconclusive.

57. Show that if

$$L = \lim_{k\to\infty}\left|\frac{a_{k+1}}{a_k}\right| = 1$$

the series Σa_k can either converge or diverge.
Hint: Find a convergent series with $L = 1$ and a divergent series that satisfies the same condition.

58. Let $\Sigma(-1)^{k+1} a_k$ be an alternating series such that $\{a_k\}$ is decreasing and

$$\lim_{k\to 0} a_k = 0$$

Show that the sequence of odd partial sums $\{S_{2n-1}\}$ is decreasing.

59. Think Tank Problem Let Σa_k be a series that converges. Either prove that Σa_k^2 must also converge or find a counterexample.

60. Think Tank Problem Construct a divergent alternating series $\displaystyle\sum_{k=1}^{\infty}(-1)^{k+1} a_k$ such that $\displaystyle\lim_{k\to\infty} a_k = 0$, but $\{a_k\}$ is not decreasing.

8.7 POWER SERIES

IN THIS SECTION: *Convergence of a power series, term-by-term differentiation and integration of power series*
A *power series* is a series whose terms are power functions of the form $a_k(x-c)^k$. We shall examine the properties of such series and shall find that under reasonable conditions, they can be differentiated and integrated term-by-term.

An infinite series of the form

$$\sum_{k=0}^{\infty} a_k(x-c)^k = a_0 + a_1(x-c) + a_2(x-c)^2 + \cdots$$

is called a **power series** in $(x-c)$. The numbers $a_0, a_1, a_2, \cdots$ are the *coefficients* of the power series. If $c = 0$, the series has the form

$$\sum_{k=0}^{\infty} a_k x^k = a_0 + a_1 x + a_2 x^2 + a_3 x^3 + \cdots$$

A power series in x may be considered as an extension of a polynomial in x.

Convergence of a Power Series

For what numbers x does a given power series converge? The following theorem answers this question for the case where $c = 0$.

Theorem 8.22 Convergence of a power series

For a power series $\displaystyle\sum_{k=0}^{\infty} a_k x^k$, exactly one of the following is true:

1. The series converges for all x.
2. The series converges only for $x = 0$.
3. The series converges absolutely for all x in an open interval $(-R, R)$ and diverges for $|x| > R$. It may either converge or diverge at the endpoints of the interval, $x = R$ and $x = -R$.

Proof: The proof is found in most advanced calculus textbooks. ♦

The set of values for which a series converges is called the **convergence set** for the series. From this theorem we know it is always an interval, if one considers a point and the entire real line as intervals. The following three examples show each of these possibilities.

Example 1 When the convergence set is the entire x-axis

Show that the power series $\displaystyle\sum_{k=0}^{\infty} \frac{x^k}{k!}$ converges for all x.

Solution If $x = 0$, then the series is trivial and converges. For $x \neq 0$, we use the generalized ratio test to find

$$L = \lim_{k\to\infty} \left| \frac{\frac{x^{k+1}}{(k+1)!}}{\frac{x^k}{k!}} \right| = \lim_{k\to\infty} \left| \frac{x^{k+1}k!}{(k+1)!x^k} \right| = \lim_{k\to\infty} \frac{|x|}{k+1} = 0$$

Because $L = 0$ and thus $L < 1$, the series converges for all x. ∎

Example 2   Convergence only at the point $x = 0$

Show that the power series $\displaystyle\sum_{k=1}^{\infty} k!\, x^k$ converges only when $x = 0$.

Solution   We use the generalized ratio test to find

$$L = \lim_{k \to \infty} \left| \frac{(k+1)!\, x^{k+1}}{k!\, x^k} \right| = \lim_{k \to \infty} (k+1)\, |x|$$

For any x other than 0, we have $L = \infty$. Hence, the power series converges only when $x = 0$. ∎

Example 3   Convergence set is a bounded interval

Find the convergence set for the power series

$$\sum_{k=1}^{\infty} \frac{x^k}{\sqrt{k}}$$

Solution   Using the generalized ratio test, we find

$$L = \lim_{k \to \infty} \left| \frac{\frac{x^{k+1}}{\sqrt{k+1}}}{\frac{x^k}{\sqrt{k}}} \right| = \lim_{k \to \infty} \left| \frac{\sqrt{k}}{\sqrt{k+1}} \right| |x| = |x|$$

The power series converges absolutely if $|x| < 1$, and diverges if $|x| > 1$. We must also check the endpoints of the interval $|x| < 1$, namely $x = -1$ and $x = 1$:

$$x = -1: \quad \sum_{k=1}^{\infty} \frac{(-1)^k}{\sqrt{k}} \quad \text{converges (by the alternating series test)}$$

$$\frac{1}{\sqrt{k+1}} < \frac{1}{\sqrt{k}} \quad \text{and} \quad \lim_{k \to \infty} \frac{1}{\sqrt{k}} = 0$$

$$x = 1: \quad \sum_{k=1}^{\infty} \frac{(1)^k}{\sqrt{k}} \quad \text{diverges (p-series with $p = \frac{1}{2} < 1$).}$$

Thus, the given power series converges for $-1 \le x < 1$ and diverges otherwise. ∎

According to Theorem 8.22, the set of numbers for which the power series $\Sigma a_k x^k$ converges is an interval centered at $x = 0$. We call this the **interval of convergence** of the power series. If this interval has length $2R$, then R is called the **radius of convergence** of the series, as shown in Figure 8.19

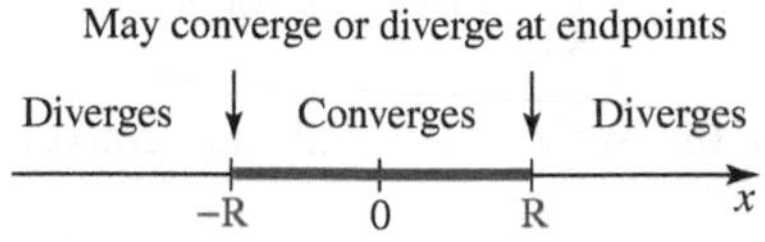

May converge or diverge at endpoints

Diverges   Converges   Diverges

$-R \qquad 0 \qquad R \qquad x$

Figure 8.19   The interval of convergence of a power series

If the series converges only for $x = 0$, the series has radius of convergence $R = 0$, and if it converges for all x, we say that $R = \infty$.

In Example 1, we found that $\Sigma x^k /k!$ converges for all x, so $R = \infty$ and the interval of convergence is the entire real line. In Example 2, we found that the power series $\Sigma k! x^k$ has radius of convergence 0. In Example 3, we found that $\Sigma x^k /\sqrt{k}$ converges on $[-1, 1)$, so $R = 1$.

The procedure for determining the convergence set of a power series is further illustrated in the following examples.

Example 4   Finding the interval of convergence for a power series

Find the interval of convergence for the power series $\displaystyle\sum_{k=1}^{\infty} \frac{2^k x^k}{k}$. What is the radius of convergence?

Solution   We apply the generalized ratio test to find

$$L = \lim_{k\to\infty} \left| \frac{\frac{2^{k+1}x^{k+1}}{k+1}}{\frac{2^k x^k}{k}} \right| = \lim_{k\to\infty} \left| \frac{2k}{k+1} \right| |x| = 2|x|$$

Thus, the series converges absolutely for $2|x| < 1$; that is for

$$-\frac{1}{2} < x < \frac{1}{2}$$

The radius of convergence is $R = \frac{1}{2}$. Checking the endpoints, we find:

Do not forget to check the endpoints.

$$x = \frac{1}{2}: \quad \sum_{k=1}^{\infty} \frac{2^k}{k}\left(\frac{1}{2}\right)^k = \sum_{k=1}^{\infty} \frac{1}{k} \quad \text{diverges (harmonic series)}$$

$$x = -\frac{1}{2}: \quad \sum_{k=1}^{\infty} \frac{2^k}{k}\left(-\frac{1}{2}\right)^k = \sum_{k=1}^{\infty} \frac{(-1)^k}{k} \quad \text{converges (alternating harmonic series)}$$

The interval of convergence is $-\frac{1}{2} \le x < \frac{1}{2}$. ∎

Example 5   Finding an interval of convergence

Find the radius of convergence and the interval of convergence for the power series $\displaystyle\sum_{k=1}^{\infty} \frac{2^k x^k}{k!}$.

Solution   Applying the generalized ratio test, we find

$$L = \lim_{k\to\infty} \left| \frac{\frac{2^{k+1}x^{k+1}}{(k+1)!}}{\frac{2^k x^k}{k!}} \right| = \lim_{k\to\infty} \left| \frac{2}{k+1} \right| |x| = 0$$

Thus, the power series converges absolutely for x, and the radius of convergence is infinite. ∎

Example 6   Radius of convergence using the root test

Find the radius of convergence of the power series $\displaystyle\sum_{k=1}^{\infty} \left(\frac{k+1}{k}\right)^{k^2} x^k$.

Solution   Using the root test to examine absolute convergence, we find that

$$L = \lim_{k \to \infty} \sqrt[k]{\left| \left(\frac{k+1}{k} \right)^{k^2} x^k \right|}$$

$$= \lim_{k \to \infty} \left| \left(\frac{k+1}{k} \right)^{k^2} x^k \right|^{1/k}$$

$$= \lim_{k \to \infty} \left(1 + \frac{1}{k} \right)^k |x|$$

$$= e\,|x|$$

Thus, the power series converges absolutely for $e\,|x| < 1$; that is, for $-e^{-1} < x < e^{-1}$. It follows that the radius of convergence is $R = e^{-1}$. We do not need to test the endpoints because we are not finding the interval of convergence. ∎

In some applications, we will encounter power series of the form

$$\sum_{k=0}^{\infty} a_k (x - c)^k = a_0 + a_1(x - c) + a_2(x - c)^2 + \cdots$$

in which each term is a constant times a power of $x - c$ where c is a constant. The intervals of convergence are intervals of the form $-R < x - c < R$, including possibly one or both of the endpoints $x = c - R$ and $x = c + R$. The procedure for determining the interval of convergence of such power series is illustrated in the following example.

Example 7   Power series in $(x - c)$

Find the interval of convergence of the power series $\displaystyle\sum_{k=0}^{\infty} \frac{(x+1)^k}{3^k}$.

Solution   Applying the generalized ratio rest, we obtain

$$L = \lim_{k \to \infty} \left| \frac{\frac{(x+1)^{k+1}}{3^{k+1}}}{\frac{(x+1)^k}{3^k}} \right| = \lim_{k \to \infty} \left| \frac{3^k}{3^{k+1}} \right| |x+1| = \lim_{k \to \infty} \frac{1}{3} |x+1|$$

Thus, the power series converges absolutely for

$$\frac{1}{3} |x+1| < 1$$
$$|x+1| < 3$$
$$-3 < x+1 < 3$$
$$-4 < \quad x \quad < 2$$

Checking the endpoints, we find

$$x = -4: \quad \sum_{k=0}^{\infty} \frac{(-4+1)^k}{3^k} = \sum_{k=0}^{\infty} (-1)^k \text{ diverges (by oscillation)}$$

$$x = 2: \quad \sum_{k=0}^{\infty} \frac{(2+1)^k}{3^k} = \sum_{k=0}^{\infty} 1^k \text{ diverges (by the divergence test)}$$

The interval of convergence is $(-4, 2)$. ∎

Term-by-Term Differentiation and Integration of Power Series

The power series $\sum_{k=0}^{\infty} a_k x^k$ can be thought of as a function defined on its interval of convergence. It is reasonable to ask if this function is differentiable and if it is integrable, and if so, what are its derivative and integral. This is useful because we can form new series using differentiation or integration of known series. If we regard a power series as an "infinite polynomial," we would expect to be able to differentiate and integrate it term by term. The following theorem shows that this procedure is legitimate on the interval of absolute convergence.

Theorem 8.23   Term-by-term differentiation and integration of a power series

A power series $\sum_{k=0}^{\infty} a_k x^k$ with radius of convergence $R > 0$ can be differentiated or integrated term by term on its interval of absolute convergence $-R < x < R$. More specifically, if $f(x) = \sum_{k=0}^{\infty} a_k x^k$ for $|x| < R$, then for $|x| < R$ we have

$$f'(x) = \sum_{k=1}^{\infty} k a_k x^{k-1} = a_1 + 2a_2 x + 3a_3 x^2 + 4a_4 x^3 + \cdots$$

and

$$\int f(x)\,dx = \int \left(\sum_{k=0}^{\infty} a_k x^k \right) dx = \sum_{k=0}^{\infty} \left(\int a_k x^k \, dx \right) = \sum_{k=0}^{\infty} \frac{a_k}{k+1} x^{k+1} + C$$

Whether or not the differentiated or integrated series converges at the endpoints of the interval of convergence needs to be checked in each case. From the theorem we only know what happens inside the interval of convergence.

■ **W**hat this says In many ways, a function defined by a power series behaves like a polynomial. It is differentiable, hence continuous, inside its interval of absolute convergence, and its derivative and integral can be determined by differentiating and integrating term by term, respectively.

Proof: A proof can be found in most advanced calculus texts. ♦

Example 8   Term-by-term differentiation of a power series

Let f be a function defined by the power series

$$f(x) = \sum_{k=0}^{\infty} \frac{x^k}{k!} \qquad \text{for all } x$$

Show that $f'(x) = f(x)$ for all x, and deduce that $f(x) = e^x$.

Solution   The given power series converges for all x (by the generalized ratio test—see Example 1), and Theorem 8.23 tells us that it is differentiable for all x. Differentiating term by term, we find

$$f'(x) = \frac{d}{dx}\left[1 + x + \frac{x^2}{2!} + \frac{x^3}{3!} + \frac{x^4}{4!} + \cdots \right]$$

$$= 0 + 1 + \frac{2x}{2!} + \frac{3x^2}{3!} + \frac{4x^3}{4!} + \cdots$$

$$= 1 + x + \frac{x^2}{2!} + \frac{x^3}{3!} + \cdots$$

$$= f(x)$$

In Chapter 5, we found that the differential equation $f'(x) = f(x)$ has the general solution $f(x) = Ce^x$. Substituting $x = 0$ into the power series for $f(x)$, we obtain

$$f(0) = 1 + 0 + \frac{0^2}{2!} + \frac{0^3}{3!} + \cdots = 1$$

and by solving the equation $1 = Ce^0$ for C, we find $C = 1$; therefore, $f(x) = e^x$. ■

A power series can be differentiated term by term—not just once, but infinitely often in its interval of absolute convergence. The key to this fact lies in showing that if f satisfies

$$f(x) = \sum_{k=0}^{\infty} a_k x^k$$

and R is the radius of convergence of the power series on the right, then the derivative series

$$f'(x) = \sum_{k=1}^{\infty} k a_k x^{k-1}$$

also has the radius of convergence R. (You are asked to show this in Problem 56). Therefore, Theorem 8.23 can be applied to the derivative series to obtain the second derivative

$$f''(x) = \sum_{k=2}^{\infty} k(k-1) a_k x^{k-2}$$

for $|x| < R$. Continuing in this fashion, we can apply Theorem 8.23 to $f'''(x)$, $f^{(4)}(x)$, and all other higher derivatives of f.

For example, we know that the geometric series

$$\sum_{k=0}^{\infty} x^k \qquad \text{converges absolutely to } f(x) = \frac{1}{1-x} \text{ for } |x| < 1$$

Thus, the term-by-term derivative

$$\frac{d}{dx}\left[\sum_{k=0}^{\infty} x^k \right] = \sum_{k=1}^{\infty} k x^{k-1} = 1 + 2x + 3x^2 + \cdots$$

converges to $f'(x) = \dfrac{1}{(1-x)^2}$ for $|x| < 1$, and the term-by-term *second derivative*

$$\frac{d^2}{dx^2}\left[\sum_{k=0}^{\infty} x^k \right] = \frac{d}{dx}\left[\sum_{k=1}^{\infty} k x^{k-1} \right] = \sum_{k=2}^{\infty} k(k-1) x^{k-2}$$

converges to $f''(x) = \dfrac{2}{(1-x)^3}$, and so on. These ideas are illustrated in our next example.

Example 9   Second derivative of a power series

Let f be the function defined by the power series

$$f(x) = \sum_{k=0}^{\infty} \frac{(-1)^k x^{2k}}{(2k)!}$$

Do not forget, $0! = 1$.

for all x. Show that $f''(x) = -f(x)$ for all x.

Solution   First, we use the ratio test to verify that the given power series converges absolutely for all x.

$$L = \lim_{k\to\infty} \left| \frac{\frac{(-1)^{k+1} x^{2(k+1)}}{[2(k+1)]!}}{\frac{(-1)^k x^{2k}}{(2k)!}} \right| = \lim_{k\to\infty} \frac{x^{2k+2}}{[2(k+1)]!} \cdot \frac{(2k)!}{x^{2k}} = \lim_{x\to\infty} \frac{x^2}{(2k+2)(2k+1)} = 0$$

Because $L < 1$, the series converges for all x. Next, differentiate the series term by term:

$$f'(x) = \frac{d}{dx}\left[1 - \frac{x^2}{2!} + \frac{x^4}{4!} - \frac{x^6}{6!} + \cdots \right]$$

$$= -\frac{2x}{2!} + \frac{4x^3}{4!} - \frac{6x^5}{6!} + \cdots$$

$$= -\frac{x}{1!} + \frac{x^3}{3!} - \frac{x^5}{5!} + \cdots$$

Finally, by differentiating term by term again, we obtain

$$f''(x) = \frac{d}{dx}\left[-\frac{x}{1!} + \frac{x^3}{3!} - \frac{x^5}{5!} + \cdots \right]$$

$$= -1 + \frac{3x^2}{3!} - \frac{5x^4}{5!} + \cdots$$

$$= -\left[1 - \frac{x^2}{2!} + \frac{x^4}{4!} + \cdots \right]$$

$$= -f(x)$$

Example 10   Term-by-term integration of a power series

By integrating an appropriate geometric series term by term, show that

$$\sum_{k=0}^{\infty} \frac{x^{k+1}}{k+1} = -\ln(1-x) \text{ for } -1 < x < 1$$

Solution   Integrating the geometric series $\displaystyle\sum_{k=0}^{\infty} u^k = \frac{1}{1-u}$ term-by-term in the interval $-1 < u < 1$, we obtain

$$\int_0^x \frac{1}{1-u}\,du = \int_0^x \left[\sum_{k=0}^{\infty} u^k \right] du$$

$$= \int_0^x \left[1 + u + u^2 + u^3 + \cdots \right] du$$

$$= x + \frac{x^2}{2} + \frac{x^3}{3} + \cdots$$

$$= \sum_{k=0}^{\infty} \frac{x^{k+1}}{k+1} \text{ for } -1 < x < 1$$

We also know

$$\int_0^x \frac{du}{1-u} = -\ln(1-x) \text{ for } -1 < x < 1$$

Thus,

$$-\ln(1-x) = \int_0^x \frac{du}{1-u} = \sum_{k=0}^{\infty} \frac{x^{k+1}}{k+1} \text{ for } -1 < x < 1$$

If we check endpoints, we find the series actually converges for $-1 < x < 1$.

Example 11   Term-by-term integration of a power series

Find a power series for $\tan^{-1} x$.

Solution   We will use the fact that $\tan^{-1} x = \int_0^x \frac{dt}{1+t^2}$.

Consider the integrand as a geometric series:

$$\frac{1}{1+t^2} = 1 - t^2 + t^4 - t^6 + \cdots = \sum_{k=0}^{\infty} (-1)^k t^{2k}$$

Thus,

$$
\begin{aligned}
\tan^{-1} x &= \int_0^x \frac{1}{1+t^2}\, dt \\
&= \int_0^x \left[1 - t^2 + t^4 - t^6 + \cdots \right] dt \\
&= \sum_{k=0}^{\infty} \int_0^x (-1)^k t^{2k}\, dt \\
&= \sum_{k=0}^{\infty} (-1)^k \left. \frac{t^{2k+1}}{2k+1} \right|_0^x \\
&= \sum_{k=0}^{\infty} \frac{(-1)^k x^{2k+1}}{2k+1} \text{ for } |x| < 1 \\
&= x - \frac{x^3}{3} + \frac{x^5}{5} - \frac{x^7}{7} + \cdots \text{ for } |x| < 1
\end{aligned}
$$

If we check the endpoints, we find the series converges for $|x| \le 1$.

This series for $\tan^{-1} x$ is called **Gregory's series** after the British mathematician James Gregory (1638-1675), who developed it in 1671 (see Figure 8.20).

Because it converges at $x = 1$ and $x = -1$, it can be used to represent $\tan^{-1} x$ on $[-1, 1]$. For example, at $x = 1$

$$\frac{\pi}{4} = \tan^{-1} 1 = 1 - \frac{1}{3} + \frac{1}{5} - \frac{1}{7} + \frac{1}{9} - \cdots$$

$$\pi = 4 \left[1 - \frac{1}{3} + \frac{1}{5} - \frac{1}{7} + \frac{1}{9} - \cdots \right]$$

This is the *Leibniz formula* for π. Unfortunately, as a means of estimating the actual value of π, it has a serious drawback—the convergence is very slow. Look at the partial sums for the 49th and 50th terms, and note that these are not very close to the limit (which is $\pi \approx 3.14159265359 \cdots$).

No. of terms	Gregory's Series term	partial sum
0	4	
1	-1.33333333333333	2.666666667
2	0.8	3.466666667
3	-0.57142857142857	2.895238095
4	0.44444444444444	3.33968254
5	-0.36363636363636	2.976046176
6	0.30769230769231	3.283738484
7	-0.26666666666667	3.017071817
8	0.23529411764706	3.252365935
9	-0.21052631578947	3.041839619
10	0.19047619047619	3.232315809
11	-0.17391304347826	3.058402766
12	0.16	3.218402766
13	-0.14814814814815	3.070254618
14	0.13793103448276	3.208185652
15	-0.12903225806452	3.079153394
16	0.12121212121212	3.200365515
17	-0.11428571428571	3.086079801
18	0.10810810810811	3.194187909
19	-0.1025641025641	3.091623807
20	0.09560975609756	3.189184782
21	-0.09302325581395	3.096161526
22	0.088888888888889	3.185050415
23	-0.08510638297872	3.099944032
24	0.081632653061224	3.181576685
25	-0.07843137254902	3.103145313
26	0.075471698113208	3.178617011
27	-0.07272727272727	3.105889738
28	0.070175438596491	3.176065177
29	-0.06779661016949	3.108268567
30	0.065573770491803	3.173842337
31	-0.06349206349206	3.110350274
32	0.061538461538462	3.171888735
33	-0.05970149253731	3.112187243
34	0.057971014492754	3.170158257
35	-0.05633802816901	3.113820229
36	0.054794520547945	3.16861475
37	-0.05333333333333	3.115281416
38	0.051948051948052	3.167229468
39	-0.05063291139241	3.116596557
40	0.0149382716049383	3.165979273
41	-0.04819277108434	3.117786502
42	0.047058823529412	3.164845325
43	-0.04597701149425	3.118868314
44	0.044943820224719	3.163812134
45	-0.04395604395604	3.11985609
46	0.043010752688172	3.162866843
47	-0.04210526315789	3.12076158
48	0.041237113402062	3.161998693
49	-0.04040404040404	3.121594653
50	0.03960396039604	3.161198613

Figure 8.20 Approximation of π using Gregory's series

In 1706, the mathematician John Machin (1680-1751) discovered another series that converges much faster. The identity

$$\tan(\alpha + \beta) = \frac{\tan \alpha + \tan \beta}{1 - \tan \alpha \tan \beta}$$

can be used to show that

$$\tan^{-1} 1 = 4 \tan^{-1} \frac{1}{5} - \tan^{-1} \frac{1}{239}$$

$$\frac{\pi}{4} = 4 \left[\frac{1}{5} - \frac{1}{3} \left(\frac{1}{5}\right)^3 + \frac{1}{5} \left(\frac{1}{5}\right)^5 - \cdots \right]$$

$$- \left[\frac{1}{239} - \frac{1}{3} \left(\frac{1}{239}\right)^3 + \frac{1}{5} \left(\frac{1}{239}\right)^5 - \cdots \right]$$

$$\pi = 16 \left[\frac{1}{5} - \frac{1}{3} \left(\frac{1}{5}\right)^3 + \frac{1}{5} \left(\frac{1}{5}\right)^5 - \cdots \right]$$

$$- 4 \left[\frac{1}{239} - \frac{1}{3} \left(\frac{1}{239}\right)^3 + \frac{1}{5} \left(\frac{1}{239}\right)^5 - \cdots \right]$$

No.of terms	Machin's Series term	partial sum
0	3.1832635983264	
1	-0.0426665690003	3.140597029
2	0.0010239999998974	3.141621029
3	-2.9257142857E-05	3.141591772
4	9.10222222222E-07	3.141592682
5	-2.9789090909E-08	3.141592653
6	1.00824615385E-09	3.141592654
7	-3.4952533333E-11	3.141592654

By comparing the table in Figure 8.21 showing the partial sums for this series with the one in Figure 8.20, we see that Machin's formula for π converges much more rapidly than the Leibniz formula. Indeed, the sum of the first seven terms $M_7 = 3.141592654$ already coincides with π for the first eight decimal places.

Figure 8.21 Approximation of π using Machin's series

PROBLEM SET 8.7

Level 1

Find the convergence set for the power series given in Problems 1-28.

1. $\sum_{k=1}^{\infty} \frac{kx^k}{k+1}$

2. $\sum_{k=1}^{\infty} \frac{k^2 x^k}{k+1}$

3. $\sum_{k=1}^{\infty} \frac{k(k+1)x^k}{k+2}$

4. $\sum_{k=1}^{\infty} \sqrt{k-1}\, x^k$

5. $\sum_{k=1}^{\infty} k^2 3^k (x-3)^k$

6. $\sum_{k=1}^{\infty} \frac{k^2 (x-2)^k}{3^k}$

7. $\sum_{k=0}^{\infty} \frac{3^k (x+3)^k}{4^k}$

8. $\sum_{k=0}^{\infty} \frac{4^k (x+1)^k}{3^k}$

9. $\sum_{k=1}^{\infty} \frac{k!(x-1)^k}{5^k}$

10. $\sum_{k=1}^{\infty} \frac{(x-15)^k}{\ln(k+1)}$

11. $\sum_{k=1}^{\infty} \frac{k^2}{2^k}(x-1)^k$

12. $\sum_{k=1}^{\infty} \frac{2^k (x-3)^k}{k(k+1)}$

13. $\sum_{k=1}^{\infty} \frac{k(3x-4)^k}{(k+1)^2}$

14. $\sum_{k=0}^{\infty} \frac{(2x+3)^k}{4^k}$

15. $\sum_{k=1}^{\infty} \frac{kx^k}{7^k}$

16. $\sum_{k=0}^{\infty} \frac{(2k)!x^k}{(3k)!}$

17. $\sum_{k=1}^{\infty} \frac{(k!)^2 x^k}{k^k}$

18. $\sum_{k=1}^{\infty} \frac{(-1)^k kx^k}{\ln(k+2)}$

19. $\sum_{k=2}^{\infty} \frac{(-1)^k x^k}{k(\ln k)^2}$

20. $\sum_{k=0}^{\infty} \frac{(3x)^k}{2^{k+1}}$

21. $\sum_{k=0}^{\infty} \frac{(2x)^{2k}}{k!}$

22. $\sum_{k=0}^{\infty} \frac{(x+2)^{2k}}{3^k}$

23. $\sum_{k=0}^{\infty} \frac{k!}{2^k}(3x)^{3k}$

24. $\sum_{k=1}^{\infty} \frac{(3x)^{2k}}{\sqrt{k}}$

25. $\sum_{k=0}^{\infty} \frac{2^k}{k!}(2x-1)^{2k}$

26. $\sum_{k=0}^{\infty} 2^k (3x)^{3k}$

27. $\sum_{k=1}^{\infty} \frac{x^k}{k\sqrt{k}}$

28. $\sum_{k=1}^{\infty} \frac{(\ln k)x^k}{k}$

Find the radius of convergence R in Problems 29-36.

29. $\displaystyle\sum_{k=1}^{\infty} k^2(x+1)^{2k+1}$ **30.** $\displaystyle\sum_{k=1}^{\infty} 2^{\sqrt{k}}(x-1)^k$

31. $\displaystyle\sum_{k=1}^{\infty} \frac{k!\,x^k}{k^k}$ **32.** $\displaystyle\sum_{k=1}^{\infty} \frac{k^k x^k}{k!}$

33. $\displaystyle\sum_{k=1}^{\infty} \frac{(k!)^2 x^k}{(2k)!}$ **34.** $\displaystyle\sum_{k=2}^{\infty} \frac{x^k}{(\ln k)^k}$

35. $\displaystyle\sum_{k=1}^{\infty} k(ax)^k$ for constant a.

36. $\displaystyle\sum_{k=1}^{\infty} (a^2 x)^k$ for constant a.

In Problems 37-42, find the derivative $f'(x)$ by differentiating term by term.

37. $\displaystyle f(x) = \sum_{k=0}^{\infty} \left(\frac{x}{2}\right)^k$ **38.** $\displaystyle f(x) = \sum_{k=1}^{\infty} \frac{x^k}{k}$

39. $\displaystyle f(x) = \sum_{k=0}^{\infty} (k+2)x^k$ **40.** $\displaystyle f(x) = \sum_{k=0}^{\infty} kx^k$

41. $\displaystyle f(x) = \sum_{k=0}^{\infty} x^k$ **42.** $\displaystyle f(x) = \sum_{k=0}^{\infty} \frac{(-1)^k}{k+1} x^k$

In Problems 43-48, find $\int_0^x f(u)\,du$ by integrating term by term.

43. $\displaystyle f(x) = \sum_{k=0}^{\infty} \left(\frac{x}{2}\right)^k$ **44.** $\displaystyle f(x) = \sum_{k=1}^{\infty} \frac{x^k}{k}$

45. $\displaystyle f(x) = \sum_{k=0}^{\infty} (k+2)x^k$ **46.** $\displaystyle f(x) = \sum_{k=0}^{\infty} kx^k$

47. $\displaystyle f(x) = \sum_{k=0}^{\infty} \frac{x^k}{k!}$ **48.** $\displaystyle f(x) = \sum_{k=0}^{\infty} (-1)^k x^k$

49. Use term-by-term differentiation of a geometric series to find a power series for

$$f(x) = \frac{1}{(1-x)^3}$$

For what values of x does this power series converge?

50. Find the radius of convergence for the power series

$$\sum_{k=0}^{\infty} \frac{(qk)!}{(k!)^q} x^k$$

where q is a positive integer.

51. Find the radius of convergence for

$$\sum_{k=1}^{\infty} \frac{(k+3)!\,x^k}{k!(k+4)!}$$

52. Find the radius of convergence for

$$\sum_{k=1}^{\infty} \frac{1\cdot 2\cdot 3\cdots k(-x)^{2k-1}}{1\cdot 3\cdot 5\cdots (2k-1)}$$

53. Let f be the function defined by the power series

$$f(x) = \sum_{k=0}^{\infty} \frac{(-1)^k x^{2k+1}}{(2k+1)!}$$

for all x. Show that $f''(x) = -f(x)$ for all x.

54. Let f be the function defined by the power series

$$f(x) = \sum_{k=0}^{\infty} \frac{x^{2k}}{(2k)!}$$

for all x. Show that $f''(x) = f(x)$.

55. Think Tank Problem Show that the series

$$S = \sum_{k=1}^{\infty} \frac{\sin(k!x)}{k^2}$$

converges for all x. Differentiate term by term to obtain the series

$$T = \sum_{k=1}^{\infty} \frac{k!\cos(k!x)}{k^2}$$

Show that this series diverges for all x. Why does this not violate Theorem 8.23?

56. Suppose $\{a_k\}$ is a sequence for which

$$\lim_{k\to\infty} \sqrt[k]{|a_k|} = \frac{1}{R}$$

Show that the power series

$$\sum_{k=1}^{\infty} k a_k x^{k-1}$$

has radius of convergence R.

57. Show that if $f(x) = \displaystyle\sum_{k=1}^{\infty} a_k x^k$ has radius of convergence $R > 0$, then the series

$$\sum_{k=1}^{\infty} a_k x^{kp}$$

where p is a positive integer, has radius of convergence $R^{1/p}$.

58. For what values of x does the series

$$\sum_{k=1}^{\infty} \frac{x}{k+x}$$

converge? *Note:* This is not a power series.

59. For what values of x does the series

$$\sum_{k=1}^{\infty} \frac{k}{x^k}$$

converge? *Note:* This is not a power series.

60. For what values of x does the series

$$\sum_{k=1}^{\infty} \frac{1}{kx^k}$$

converge? *Note:* This is not a power series.

8.8 TAYLOR AND MACLAURIN SERIES

IN THIS SECTION: *Taylor and Maclaurin polynomials, Taylor's theorem, Taylor and Maclaurin series, operations with Taylor and Maclaurin series*

In this section, we seek to represent a function as an infinite series in such a way that the value of the given function at $x = c$ and the value of the series differ by no more than some specified tolerance.

Taylor and Maclaurin Polynomials

Consider a function f that can be differentiated n times on some interval I. Our goal is to find a polynomial function that approximates f at a number c in its domain. For simplicity, we begin by considering an important special case where $c = 0$. For example, consider the function $f(x) = e^x$ at the point $x = 0$, as shown in Figure 8.22.

To approximate f by a polynomial function $M(x)$, we begin by making sure that both the polynomial function and f pass through the same point. That is, $M(0) = f(0)$. We say that the polynomial is **expanded about $c = 0$** or that it is **centered at 0**.

There are many polynomial functions that we could choose to approximate f at $x = 0$, and we proceed by making sure that both f and M have the same slope at $x = 0$. That is,

$$M'(0) = f'(0)$$

The graph in Figure 8.22 shows that we have $f(x) = e^x$, so $f'(x) = e^x$ and $f(0) = f'(0) = 1$.

To find M, we let

$$M_1(x) = a_0 + a_1 x \qquad M_1'(x) = a_1$$

and require $M_1(0) = 1$ and $M_1'(0) = 1$. Because $M_1(0) = a_0 = 1$ and $M_1'(0) = a_1 = 1$, we find that

$$M_1(x) = 1 + x$$

We see from Figure 8.22 that close to $x = c$ the approximation of f by M_1 is good, but as we move away from $(0, 1)$, M_1 no longer serves as a good approximation. To improve the approximation, we impose the requirement that the values of the second derivatives of M and f agree at $x = 0$. Using this criterion, we find

$$M_2(x) = a_0 + a_1 x + a_2 x^2$$

so that

$$M_2'(x) = a_1 + 2a_2 x \quad \text{and} \quad M_2''(x) = 2a_2$$

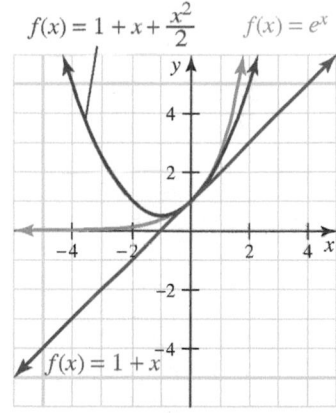

Figure 8.22 Interactive
Graph of $f(x) = e^x$ and approximating polynomials

Because $f''(x) = e^x$ and $f''(0) = 1$, we want

$$2a_2 = 1 \quad \text{or} \quad a_2 = \frac{1}{2}$$

Thus,

$$M_2(x) = 1 + x + \frac{1}{2}x^2$$

To improve the approximation even further, we can require that the values of the approximating polynomials $M_3, M_4, \cdots, M_n$ at $x = 0$ have derivatives that match those of f at $x = 0$. We find that

$$M_3(x) = 1 + x + \frac{1}{2}x^2 + \frac{1}{6}x^3$$

$$M_4(x) = 1 + x + \frac{1}{2}x^2 + \frac{1}{6}x^3 + \frac{1}{24}x^4$$

$$M_5(x) = 1 + x + \frac{1}{2}x^2 + \frac{1}{6}x^3 + \frac{1}{24}x^4 + \frac{1}{120}x^5$$

$$\vdots$$

$$M_n(x) = 1 + \frac{x}{1!} + \frac{x^2}{2!} + \frac{x^3}{3!} + \cdots + \frac{x^n}{n!}$$

This nth degree polynomial approximation of f at $x = 0$ is called a **nth degree Maclaurin polynomial** for f. If we repeat the steps for $x = c$ rather than for $x = 0$, we go through the same steps as above to get a polynomial $T_n(x)$ in powers of $(x - c)$ and we find

$$T_n(x) = e^c + \frac{(x - c)}{1!}e^c + \frac{(x - c)^2}{2!}e^c + \cdots + \frac{(x - c)^n}{n!}e^c$$

which is called the **nth-degree Taylor polynomial** of the function $f(x) = e^x$ at $x = c$.

Taylor's Theorem

Instead of stopping at the nth term, we will now approximate a function f by an infinite series. A function f is said to be represented by the power series

$$\sum_{k=0}^{\infty} a_k(x - c)^k$$

on an interval I if $f(x) = \sum_{k=0}^{\infty} a_k(x - c)^k$ for all x in I. Power series representation of functions are extremely useful, but before we can deal effectively with such representations, we must answer two questions.

1. **Existence:** Under what conditions does a given function have a power series representation?
2. **Uniqueness:** When f can be represented by a power series, is there only one such series, and, if so, what is it?

The uniqueness issue is addressed in the following theorem.

Theorem 8.24   The uniqueness theorem for power series representation

Suppose an infinitely differentiable function f is known to have the power series representation

$$f(x) = \sum_{k=0}^{\infty} a_k(x - c)^k$$

for $-R < x - c < R$. Then there is exactly one such representation, and the coefficients a_k must satisfy

$$a_k = \frac{f^{(k)}(c)}{k!} \text{ for } k = 0, 1, 2, \cdots$$

Proof: The uniqueness theorem may be established by differentiating the given power series term by term and evaluating successive derivatives at c. We start with

$$f(x) = \sum_{k=0}^{\infty} a_k(x - c)^k$$

and substitute $x = c$ to obtain

$$f(c) = a_0 + a_1(c - c) + a_2(c - c)^2 + \cdots = a_0$$

Next, differentiate the original series, term by term, to get

$$f'(x) = a_1 + 2a_2(x - c) + 3a_3(x - c)^2 + \cdots$$

and thus,

$$f'(c) = a_1 + 2a_2(c - c) + 3a_3(c - c)^2 + \cdots = a_1$$

Differentiating once again and substituting $x = c$, we find

$$f''(c) = 2a_2 \text{ so that } a_2 = \frac{f''(c)}{2}$$

In general, the kth derivative of f at $x = c$ is given by

$$f^{(k)}(c) = k!a_k \text{ so that } a_k = \frac{f^{(k)}(c)}{k!} \qquad \blacklozenge$$

The question of existence—that is, under what conditions a given function has a power series representation—is answered by considering what is called the **Taylor remainder function**:

$$R_n(x) = f(x) - T_n(x)$$

We see that f is represented by its Taylor series in I if and only if

$$\lim_{n \to \infty} R_n(x) = 0 \quad \text{for all } x \text{ in } I$$

This relationship among f, its Taylor polynomial $T_n(x)$, and the Taylor remainder function $R_n(x)$ is summarized in the following theorem.

Theorem 8.25   Taylor's theorem

If f and all its derivatives exist in an open interval I containing c, then for each x in I,

$$f(x) = f(c) + \frac{f'(c)}{1!} + \frac{f''(c)}{2!}(x - c)^2 + \cdots + \frac{f^{(n)}(c)}{n!}(x - c)^n + R_n(x)$$

where the remainder function $R_n(x)$ is given by

$$R_n(x) = \frac{f^{(n+1)}(z_n)}{(n + 1)!}(x - c)^{n+1}$$

for some z_n that depends on x and lies between c and x.

Proof: The proof of this theorem is given in Appendix B. $\qquad \blacklozenge$

The formula for $R_n(x)$ is called the *Lagrange form* of the Taylor remainder function, after the French/Italian mathematician Joseph Lagrange (1736-1813). There are others forms for the remainder, principally one developed by Cauchy (see *Historical Quest*, Problem 53 of Section 2.1), but we will consider only the Lagrange form in this text.

When applying Taylor's theorem, we do not expect to be able to find the exact value of z_n. However, we can often determine an upper bound for $|R_n(x)|$ in an open interval. We will illustrate this situation later in this section.

Taylor and Maclaurin Series

The uniqueness theorem tells us that if f has a power series representation at c, it must be the series

$$f(c) + \frac{f'(c)}{1!} + \frac{f''(c)}{2!}(x-c)^2 + \frac{f'''(c)}{3!}(x-c)^3 + \cdots$$

and Taylor's theorem says that $f(x)$ is represented by such a series precisely where the remainder term $R_n(x) \to 0$ as $n \to \infty$. It is convenient to have the following terminology.

The uniqueness theorem may be summarized by saying that the Taylor series of f at c is the only power series of the form $\Sigma a_k (x - c)^k$ that can possibly represent f on I. But it does not say that f is equal to the power series on I. All we really know is that if f has a power series representation at c, then that representation must be its Taylor series.

TAYLOR SERIES   Suppose there is an open interval I containing c throughout which the function f and all its derivatives exist. Then the power series

$$f(c) + \frac{f'(c)}{1!} + \frac{f''(c)}{2!}(x-c)^2 + \frac{f'''(c)}{3!}(x-c)^3 + \cdots$$

is called the **Taylor series of f at c.**

MACLAURIN SERIES   The special case where $c = 0$ is called the **Maclaurin series of f:**

$$f(0) + \frac{f'(0)}{1!}x + \frac{f''(0)}{2!}x^2 + \frac{f'''(0)}{3!}x^3 + \cdots$$

Example 1   Maclaurin series

Find the Maclaurin series for $f(x) = \cos x$.

Solution   First, note that f is infinitely differentiable at $x = 0$. We find

$$
\begin{aligned}
f(x) &= \cos x   &\qquad f(0) &= 1 \\
f'(x) &= -\sin x   &\qquad f'(0) &= 0 \\
f''(x) &= -\cos x   &\qquad f''(0) &= -1 \\
f'''(x) &= \sin x   &\qquad f'''(0) &= 0 \\
f^{(4)}(x) &= \cos x &\qquad f^{(4)}(0) &= 1 \\
&\;\;\vdots   &\qquad &\;\;\vdots
\end{aligned}
$$

Thus, by using the definition of a Maclaurin series we have

$$\cos x = 1 - \frac{x^2}{2!} + \frac{x^4}{4!} - \frac{x^6}{6!} + \cdots = \sum_{k=0}^{\infty} \frac{(-1)^k x^{2k}}{(2k)!}$$

You might ask about the relationship between a Maclaurin series and a Maclaurin polynomial. Figure 8.23 shows the function $f(x) = \cos x$ from Example 1 along with some successive Maclaurin polynomials. We use the notation $M_n(x)$ to represent the first $n + 1$ terms of the corresponding Maclaurin series. Note that the polynomials are quite close to the function near $x = 0$, but that they become very different as x moves away from the origin.

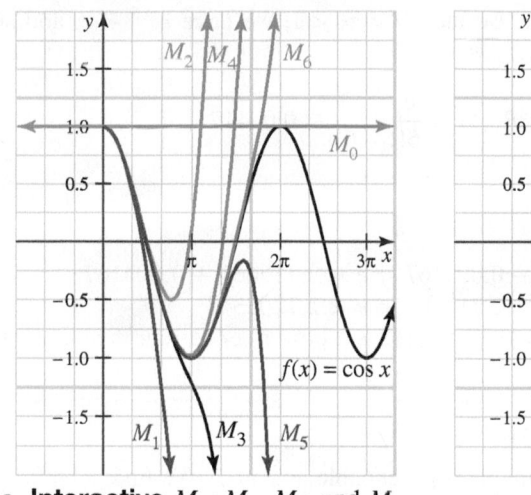

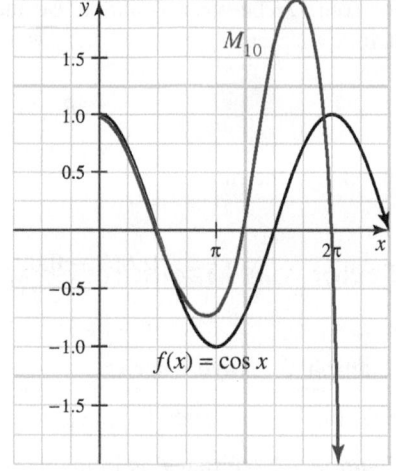

a. Interactive M_0, M_1, M_2, and M_3

b. M_{10}

Figure 8.23 Comparison of $\cos x$ and its Maclaurin polynomials

Example 2 Maximum error when using a Maclaurin polynomial approximation

Find the Maclaurin polynomial $M_5(x)$ for the function $f(x) = e^x$ and use this polynomial to approximate e. Use Taylor's theorem to determine the accuracy of this approximation.

Solution First find the Maclaurin series for $f(x) = e^x$.

$$
\begin{aligned}
f(x) &= e^x & f(0) &= 1 \\
f'(x) &= e^x & f'(0) &= 1 \\
f''(x) &= e^x & f''(0) &= 1 \\
&\;\vdots & &\;\vdots
\end{aligned}
$$

The Maclaurin series for e^x is

$$
e^x = 1 + \frac{1}{1!}x + \frac{1}{2!}x^2 + \frac{1}{3!}x^3 + \cdots = \sum_{k=0}^{\infty} \frac{x^k}{k!}
$$

Then find $M_5(x)$:

$$
M_5(x) = 1 + x + \frac{x^2}{2} + \frac{x^3}{6} + \frac{x^4}{24} + \frac{x^5}{120} \approx e^x
$$

A comparison of f and $M_5(x)$ is shown in Figure 8.24.

Notice from the graphs of these functions that the error seems to increase as x moves away from the origin. To determine the accuracy, we use Taylor's theorem

$$
\begin{aligned}
e &= 1 + 1 + \frac{1}{2} + \frac{1}{6} + \frac{1}{24} + \frac{1}{120} + R_5(1) \\
&= 2.71\overline{6} + R_5(1)
\end{aligned}
$$

The remainder term is given by

$$
R_5(1) = \frac{e^{z_5}}{(5+1)!}(1)^{5+1}
$$

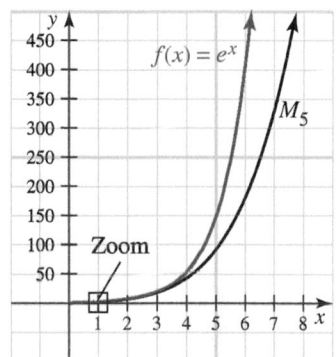

a. Comparison of f and M_5

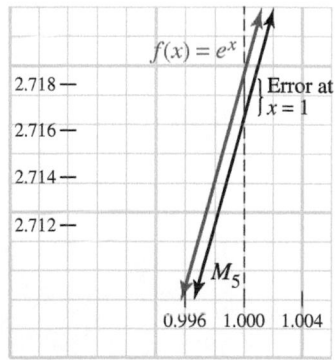

b. Zoom showing the error at $x = 1$

Figure 8.24 Comparison of $f(x) = e^x$ and $M_5(x)$

for some z_5 between 0 and 1. Because $0 < z_5 < 1$, we have $e^{z_5 5} < e$, and since $e < 3$, it follows that

$$R_5(1) < \frac{e}{6!} < \frac{3}{6!} \approx 0.00417$$

Thus, the number e satisfies

$$2.716667 - 0.004167 < e < 2.716667 + 0.004167$$
$$2.712500 < e < 2.720834$$

Example 3   Taylor series

Find the Taylor series for $f(x) = \ln x$ at $c = 1$.

Solution   Note that f is infinitely differentiable at $x = 1$. We find

$$
\begin{aligned}
f(x) &= \ln x & f(1) &= 0 \\
f'(x) &= \frac{1}{x} & f'(1) &= 1 \\
f''(x) &= \frac{-1}{x^2} & f''(1) &= -1 \\
f'''(x) &= \frac{2}{x^3} & f'''(1) &= 2 \\
f^{(4)}(x) &= \frac{-6}{x^4} & f^{(4)}(1) &= -6 \\
&\;\;\vdots & &\;\;\vdots \\
f^{(k)}(x) &= \frac{(-1)^{k+1}(k-1)!}{x^k} & f^{(k)}(1) &= (-1)^{k+1}(k-1)!
\end{aligned}
$$

Then, use the definition of a Taylor series to write

$$
\begin{aligned}
\ln x &= 0 + \frac{1}{1!}(x-1) - \frac{1}{2!}(x-1)^2 + \frac{2}{3!}(x-1)^3 - \frac{6}{4!}(x-1)^4 + \cdots \\
&= (x-1) - \frac{1}{2}(x-1)^2 + \frac{1}{3}(x-1)^3 - \frac{1}{4}(x-1)^4 + \cdots \\
&= \sum_{k=1}^{\infty} \frac{(-1)^{k+1}(x-1)^k}{k}
\end{aligned}
$$

Suppose f is a function that is infinitely differentiable at c. Now we have two mathematical quantities, f and its Taylor series. There are several possibilities:

1. The Taylor series of f may converge to f on the interval of absolute convergence, $-R < x - c < R$ (that is, $|x - c| < R$).
2. The Taylor series may converge only at $x = c$, in which case it certainly does not represent f on any interval containing c.
3. The Taylor series of f may have a positive radius of convergence (even $R = \infty$), but it may converge to a function g that does not equal f on the interval $|x - c| < R$.

Example 4   A function defined at points where its Taylor series does not converge

Show that the function $\ln x$ is defined at points for which its Taylor series at $c = 1$ does not converge.

Solution   We know that $\ln x$ is defined for all $x > 0$. From the previous example, we know the Taylor series for $\ln x$ at $c = 1$ is

$$\sum_{k=1}^{\infty} \frac{(-1)^{k-1}(x-1)^k}{k}$$

We find the interval of convergence for this series centered at $c = 1$ by computing

$$L = \lim_{k \to \infty} \left| \frac{\frac{(-1)^k(x-1)^{k+1}}{k+1}}{\frac{(-1)^{k-1}(x-1)^k}{k}} \right|$$

$$= \lim_{k \to \infty} \left| \frac{\frac{(-1)^k}{k+1}}{\frac{(-1)^{k-1}}{k}} \right| |x - 1|$$

$$= \lim_{k \to \infty} \left(\frac{k}{k+1} \right) |x - 1|$$

$$= |x - 1|$$

This means that the power series converges absolutely for

$$|x - 1| < 1$$
$$0 < x < 2$$

Test the endpoints:

$x = 0$: $\displaystyle\sum_{k=1}^{\infty} \frac{(-1)^{k-1}(0-1)^k}{k} = \sum_{k=1}^{\infty} \frac{-1}{k}$ diverges   *Opposite of harmonic series*

$x = 2$: $\displaystyle\sum_{k=1}^{\infty} \frac{(-1)^{k-1}(2-1)^k}{k} = \sum_{k=1}^{\infty} \frac{(-1)^{k-1}}{k}$ converges   *Alternating harmonic series*

Thus, the series converges only on $(0, 2]$, but the function $\ln x$ is defined for all $x > 0$. The function is compared with the sixth degree Taylor polynomial $T_6(x) = \displaystyle\sum_{k=1}^{6} \frac{(-1)^{k-1}}{k}(x - 1)^k$ approximation in Figure 8.25.

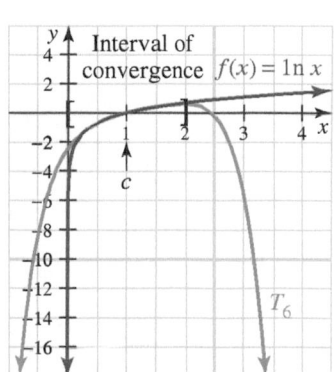

Figure 8.25 Interactive
Comparison of
$f(x) = \ln x$ and $T_6(x)$

The next example introduces a function whose Maclaurin series does not represent that function on *any* interval.

Example 5   A function represented by its Maclaurin series at a single point

Let f be the function defined by

$$f(x) = \begin{cases} e^{-1/x^2} & \text{if } x \neq 0 \\ 0 & \text{if } x = 0 \end{cases}$$

Show that f is represented by its Maclaurin series only at $x = 0$.

Solution   It can be shown that f is infinitely differentiable at 0 and that $f^{(k)}(0) = 0$ for all k. Therefore, f has the Maclaurin series

$$0 + \frac{0}{1!}x + \frac{0}{2!}x^2 + \frac{0}{3!}x^3 + \cdots$$

which converges to the zero function for all x and thus represents $f(x)$ only at $x = 0$. The graph of f and the Maclaurin series are shown in Figure 8.26.

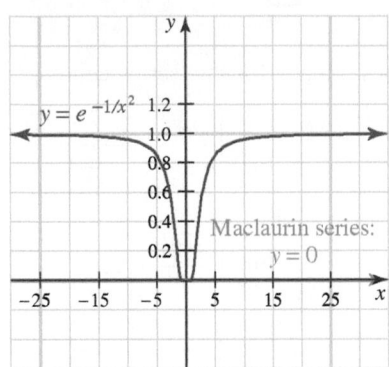

Figure 8.26 Interactive   Graph of f and its Maclaurin series at $y = 0$

Operations with Taylor and Maclaurin Series

According to the uniqueness theorem, the Taylor series for f at c is the only power series in $x - c$ that can satisfy $f(x) = \sum\limits_{k=0}^{\infty} a_k (x - c)^k$ for all x in some interval containing c. The coefficients in this series can always be found by substituting into the formula $a_k = \dfrac{f^{(k)}(c)}{k!}$, but occasionally they can also be found by algebraic manipulation.

In several of the examples that follow, we will use the geometric series equation

$$\frac{1}{1 - u} = \sum_{k=0}^{\infty} u^k \qquad \text{for } |u| < 1$$

to obtain Maclaurin series for rational functions. The general procedure is illustrated in Example 6.

Example 6   Maclaurin series using substitution in a geometric series

Find the Maclaurin series for $\dfrac{1}{1 + x^2}$.

Solution   We want a series of the form $\sum\limits_{k=0}^{\infty} a_k u^k$ that represents $\dfrac{1}{1 + x^2}$ on an interval containing 0. We could proceed directly using the definition of a Maclaurin series, but instead we will modify a known series. In fact, trying to proceed directly quickly becomes cumbersome for this series (try it and you will see). We know that if $|u| < 1$ we can write

$$\frac{1}{1 - u} = 1 + u + u^2 + \cdots$$

so by substitution ($u = -x^2$) we have

$$\frac{1}{1 + x^2} = \frac{1}{1 - (-x^2)} = 1 - x^2 + x^4 - x^6 + \cdots$$

provided $\left|-x^2\right| < 1$; that is, $-1 < x < 1$. Hence, by the uniqueness theorem, the desired representation is

$$\frac{1}{1+x^2} = \sum_{k=0}^{\infty}(-1)^k x^{2k} \qquad \text{for } -1 < x < 1$$

Example 7 Maclaurin series by modifying a geometric series

Find the Maclaurin series for $f(x) = \dfrac{5-2x}{3+2x}$ and determine the interval of convergence of this series.

Solution Our strategy is to rewrite $f(x)$ in the form $\dfrac{1}{1-u}$ of a geometric series:

$$\begin{aligned}
\frac{5-2x}{3+2x} &= -1 + \frac{8}{3+2x} & \textit{Long division}\\[2mm]
&= -1 + \frac{\frac{8}{3}}{1-\left(-\frac{2}{3}x\right)} & \textit{Write in the form } \frac{1}{1-u}.\\[2mm]
&= -1 + \sum_{k=0}^{\infty}\frac{8}{3}\left(-\frac{2}{3}x\right)^k\\[2mm]
&= -1 + \frac{8}{3} - \frac{16}{9}x + \frac{32}{27}x^2 - \frac{64}{81}x^3 + \cdots
\end{aligned}$$

The convergence set for $f(x)$ is the same as for the series $\sum\limits_{k=1}^{\infty}\left(-\frac{2}{3}x\right)^k$; namely,

$$\left|-\frac{2}{3}x\right| < 1$$
$$-\frac{3}{2} < x < \frac{3}{2}$$

What we see in these examples is that if one can find a Maclaurin series for a given function by some means, it is *the* Maclaurin series. A similar statement can be made about a Taylor series for a particular choice of c. This is because of the uniqueness theorem for power series representation above.

Example 8 Using a trigonometric identity with a known series

Find the Maclaurin series for **a.** $\cos x^2$ **b.** $\cos^2 x$

Solution

a. In Example 1 we found that

$$\cos u = 1 - \frac{u^2}{2!} + \frac{u^4}{4!} - \frac{u^6}{6!} + \cdots \qquad \text{for all } u$$

Therefore, by substituting $u = x^2$, we obtain

$$\begin{aligned}
\cos x^2 &= 1 - \frac{(x^2)^2}{2!} + \frac{(x^2)^4}{4!} - \frac{(x^2)^6}{6!} + \cdots & \textit{Let } u = x^2.\\[2mm]
&= 1 - \frac{x^4}{2!} + \frac{x^8}{4!} - \frac{x^{12}}{6!} + \cdots & \text{for all } x
\end{aligned}$$

b. For $\cos^2 x$ we could use the definition of a Maclaurin series, but instead we will use a double-angle trigonometric identity.

$$\cos^2 x = \frac{1}{2} + \frac{1}{2}\cos 2x$$

$$= \frac{1}{2} + \frac{1}{2}\left[1 - \frac{(2x)^2}{2!} + \frac{(2x)^4}{4!} - \frac{(2x)^6}{6!} + \cdots\right] \quad \textit{Let } u = 2x.$$

$$= \frac{1}{2} + \frac{1}{2} - \frac{2x^2}{2!} + \frac{2^3 x^4}{4!} - \frac{2^5 x^6}{6!} + \cdots$$

$$= 1 - x^2 + \frac{1}{3}x^4 - \frac{2}{45}x^6 + \cdots \quad \text{for all } x$$

Example 9 Maclaurin series using substitution

Find the Maclaurin series for $f(x) = \ln\left(\frac{1+x}{1-x}\right)$ and use this series to compute $\ln 2$ correct to five decimal places.

Solution For this function, we first use a property of logarithms:

$$f(x) = \ln\left(\frac{1+x}{1-x}\right) = \ln(1+x) - \ln(1-x)$$

From Example 3, we know

$$\ln x = (x-1) - \frac{1}{2}(x-1)^2 + \frac{1}{3}(x-1)^3 - \cdots$$

so that

$$\ln(1+x) = (1+x-1) - \frac{1}{2}(1+x-1)^2 + \frac{1}{3}(1+x-1)^3 - \cdots$$
$$= x - \frac{x^2}{2} + \frac{x^3}{3} - \frac{x^4}{4} + \cdots$$
$$\ln(1-x) = (1-x-1) - \frac{1}{2}(1-x-1)^2 + \frac{1}{3}(1-x-1)^3 - \cdots$$
$$= (-x) - \frac{(-x)^2}{2} + \frac{(-x)^3}{3} - \frac{(-x)^4}{4} + \cdots$$
$$= -x - \frac{x^2}{2} - \frac{x^3}{3} - \frac{x^4}{4} - \cdots$$

By subtracting series, we find

$$f(x) = \ln(1+x) - \ln(1-x)$$
$$= \left[x - \frac{x^2}{2} + \frac{x^3}{3} - \frac{x^4}{4} + \cdots\right] - \left[-x - \frac{x^2}{2} - \frac{x^3}{3} - \frac{x^4}{4} - \cdots\right]$$
$$= 2\left[x + \frac{x^3}{3} + \frac{x^5}{5} + \cdots\right]$$

Next, by solving the equation $\frac{1+x}{1-x} = 2$, we find $x = \frac{1}{3}$, so we are looking for $\ln\left(\frac{1+x}{1-x}\right)$ where $x = \frac{1}{3}$. It follows that

$$\ln 2 = 2\left[\frac{1}{3} + \frac{\left(\frac{1}{3}\right)^3}{3} + \frac{\left(\frac{1}{3}\right)^5}{5} + \cdots\right]$$
$$= 2\left[\frac{1}{3} + \frac{1}{3}\left(\frac{1}{3}\right)^3 + \frac{1}{5}\left(\frac{1}{3}\right)^5 + \cdots\right]$$

According to the Taylor theorem, we find

$$\ln 2 = \frac{2}{3} + \frac{2}{3}\left(\frac{1}{3}\right)^3 + \frac{2}{5}\left(\frac{1}{3}\right)^5 + \cdots + \frac{2}{2n+1}\left(\frac{1}{3}\right)^{2n+1} + R_n\left(\frac{1}{3}\right)$$

where $R_n\left(\frac{1}{3}\right)$ is the remainder term. To estimate the remainder, we note that $R_n\left(\frac{1}{3}\right)$ is the tail of the infinite series for $\ln 2$. Thus,

$$\left| R_n\left(\frac{1}{3}\right) \right| = \frac{2}{2n+3}\left(\frac{1}{3}\right)^{2n+3} + \frac{2}{2n+5}\left(\frac{1}{3}\right)^{2n+5} + \frac{2}{2n+7}\left(\frac{1}{3}\right)^{2n+7} + \cdots$$

$$< \frac{2}{2n+3}\left(\frac{1}{3}\right)^{2n+3} + \frac{2}{2n+3}\left(\frac{1}{9}\right)\left(\frac{1}{3}\right)^{2n+3} + \cdots$$

$$= \frac{2}{2n+3}\left(\frac{1}{3}\right)^{2n+3}\left[1 + \frac{1}{9} + \frac{1}{81} + \cdots\right]$$

Because the geometric series in brackets converges to $\dfrac{1}{1-\frac{1}{9}} = \dfrac{9}{8}$, we find that

$$\left| R_n\left(\frac{1}{3}\right) \right| < \frac{2}{2n+3}\left(\frac{9}{8}\right)\left(\frac{1}{3}\right)^{2n+3}$$

In particular, to achieve five-place accuracy, we must make sure the term on the right is less than 0.000005 (six places to account for round-off error). With a calculator we can see that if $n = 4$, we have

$$\frac{2}{2(4)+3}\left(\frac{9}{8}\right)\left(\frac{1}{3}\right)^{11} = 0.0000012$$

Thus, we approximate $\ln 2$ with $n = 4$:

$$\ln 2 \approx T_4\left(\frac{1}{3}\right) = \frac{2}{3} + \frac{2}{3}\left(\frac{1}{3}\right)^3 + \frac{2}{5}\left(\frac{1}{3}\right)^5 + \frac{2}{7}\left(\frac{1}{3}\right)^7 + \frac{2}{9}\left(\frac{1}{3}\right)^9 \approx 0.6931460$$

which is correct with an error of no more than 0.0000012; therefore,

$$0.6931460 - 0.0000012 < \ln 2 < 0.6931460 + 0.0000012$$
$$0.6931448 < \ln 2 < 0.6931472$$

Rounded to five-place accuracy, $\ln 2 \approx 0.69315$.

We can check with a calculator (to 10 decimal accuracy): $\ln 2 \approx 0.6931471806$. ∎

TECHNOLOGY NOTE: "Why bother with the accuracy part of Example 9?" you might ask. "I can just press $\ln 2$ on my calculator and obtain better accuracy anyway!" The answer is that the accuracy built into calculators and computers uses the same principles illustrated in Example 9. Calculators and computers have not changed the need for accuracy considerations but have simply pushed out the limits of these accuracy arguments. It is only for convenience that we ask for limits of accuracy that are less than what we can obtain with a calculator (or any available machine), but the method is just as valid if we ask for 100- or 1,000-place accuracy. Please see Appendix C, section "Significant Figures" for more details.

Some of the more common power series that we have derived are given in Table 8.2. Before we present this table, there is one last result we should consider. It is a generalization of the binomial theorem that was discovered by Isaac Newton while he was still a student at Cambridge University.

Theorem 8.26   Binomial series

If p is any real number and $-1 < x < 1$, then

$$(1+x)^p = 1 + px + \frac{p(p-1)}{2!}x^2 + \frac{p(p-1)(p-2)}{3!}x^3 + \cdots = \sum_{k=0}^{\infty} \binom{p}{k} x^k$$

where $\binom{p}{k} = \dfrac{p!}{k!(p-k)!}$ for k and p integers with $p \geq k \geq 0$ (no such shorthand exists for other values of p).

Proof: We begin by showing how the Maclaurin series is found. Let $f(x) = (1+x)^p$.

$$
\begin{aligned}
f(x) &= (1+x)^p &\qquad f(0) &= 1 \\
f'(x) &= p(1+x)^{p-1} &\qquad f'(0) &= p \\
f''(x) &= p(p-1)(1+x)^{p-2} &\qquad f''(0) &= p(p-1) \\
f'''(x) &= p(p-1)(p-2)(1+x)^{p-3} &\qquad f'''(0) &= p(p-1)(p-2) \\
&\;\;\vdots & &\;\;\vdots \\
f^{(n)}(x) &= p(p-1)\cdots(p-n+1)(1+x)^{p-n} &\qquad f^{(n)}(0) &= p(p-1)\cdots(p-n+1)
\end{aligned}
$$

We see this produces the Maclaurin series. We could use the ratio test to show that this series converges on the interval $(-1, 1)$, but to show that it converges to $(1+x)^p$ requires advanced calculus.   ◆

The binomial series converges on the open interval $-1 < x < 1$, but convergence at the endpoints $x = -1$ and $x = 1$ of this interval depends on the exponent p. In particular, it can be shown that if $-1 < p \leq 0$, the series converges at $x = 1$; if $p \geq 0$, the series converges at both $x = 1$ and $x = -1$. Moreover, if p is a nonnegative integer, the series terminates after only a finite number of terms (since $\binom{p}{k} = 0$ for $k > p$) and thus reduces to the ordinary binomial expansion, which, of course, converges for all x.

Table 8.2   Power series for elementary functions*

Name	Series	Interval of Convergence
Exponential series	$e^u = 1 + u + \frac{u^2}{2!} + \frac{u^3}{3!} + \frac{u^4}{4!} + \cdots + \frac{u^k}{k!} + \cdots$	$(-\infty, \infty)$
	$e^x = e^c + e^c(x-c) + \frac{e^c(x-c)^2}{2!} + \frac{e^c(x-c)^3}{3!} + \cdots + \frac{e^c(x-c)^k}{k!} + \cdots$	$(-\infty, \infty)$
Cosine series	$\cos u = 1 - \frac{u^2}{2!} + \frac{u^4}{4!} - \frac{u^6}{6!} + \cdots + \frac{(-1)^k u^{2k}}{(2k)!} + \cdots$	$(-\infty, \infty)$
	$\cos x = \cos c - (x-c)\sin c - \frac{(x-c)^2}{2!}\cos c + \frac{(x-c)^3}{3!}\sin c + \cdots$	$(-\infty, \infty)$
Sine series	$\sin u = u - \frac{u^3}{3!} + \frac{u^5}{5!} - \frac{u^7}{7!} + \cdots + \frac{(-1)^k u^{2k+1}}{(2k+1)!} + \cdots$	$(-\infty, \infty)$
	$\sin x = \sin c + (x-c)\cos c - \frac{(x-c)^2}{2!}\sin c - \frac{(x-c)^3}{3!}\cos c + \cdots$	$(-\infty, \infty)$
Geometric series	$\frac{1}{1-u} = 1 + u + u^2 + u^3 + \cdots + u^k + \cdots$	$(-1, 1)$
Reciprocal series	$\frac{1}{x} = 1 - (x-1) + (x-1)^2 - (x-1)^3 + (x-1)^4 - \cdots$	$(0, 2)$
Logarithmic series	$\ln x = (x-1) - \frac{(x-1)^2}{2} + \frac{(x-1)^3}{3} - \cdots + \frac{(-1)^{k-1}(x-1)^k}{k} + \cdots$	$(0, 2]$
	$\ln(1+u) = u - \frac{1}{2}u^2 + \frac{1}{3}u^3 - \frac{1}{4}u^4 + \cdots + \frac{(-1)^{k-1}u^k}{k} + \cdots$	$(-1, 1]$
	$\ln x = \ln c + \frac{x-c}{c} - \frac{(x-c)^2}{2c^2} + \frac{(x-c)^3}{3c^3} - \cdots + \frac{(-1)^{k-1}(x-c)^k}{kc^k} + \cdots$	$(0, 2c]$
Inverse tangent series	$\tan^{-1} u = u - \frac{u^3}{3} + \frac{u^5}{5} - \frac{u^7}{7} + \cdots + \frac{(-1)^k u^{2k+1}}{2k+1} + \cdots$	$[-1, 1]$
Inverse sine series	$\sin^{-1} u = u + \frac{u^3}{2\cdot3} + \frac{1\cdot3 u^5}{2\cdot4\cdot5} + \frac{1\cdot3\cdot5 u^7}{2\cdot4\cdot6\cdot7} + \cdots + \frac{1\cdot3\cdot5\cdots(2k-3)u^{2k-1}}{2\cdot4\cdot6\cdots(2k-2)(2k-1)} + \cdots$	$[-1, 1]$
Binomial series	$(1+u)^p = 1 + pu + \frac{p(p-1)}{2!}u^2 + \frac{p(p-1)(p-2)}{3!}u^3 + \cdots$	

---

*Taylor series at $x = c$ follows the Maclaurin series for each function.

The following example illustrates how the binomial series theorem can be used.

Example 10   Using the binomial series theorem to obtain a Maclaurin series expansion

Determine the Maclaurin series for $f(x) = \sqrt{9+x}$ and find its interval of convergence.

Solution   Write $f(x) = \sqrt{9+x} = (9+x)^{1/2} = 3\left(1+\frac{x}{9}\right)^{1/2}$.

Thus,

$$\sqrt{9+x} = 3\left(1+\frac{x}{9}\right)^{1/2}$$
$$= 3\left[1 + \frac{1}{2}\left(\frac{x}{9}\right) + \frac{\frac{1}{2}\left(\frac{1}{2}-1\right)}{2!}\left(\frac{x}{9}\right)^2 + \frac{\frac{1}{2}\left(\frac{1}{2}-1\right)\left(\frac{1}{2}-2\right)}{3!}\left(\frac{x}{9}\right)^3 + \cdots\right]$$
$$= 3\left[1 + \frac{1}{18}x - \frac{1}{648}x^2 + \frac{1}{11,664}x^3 - \cdots\right]$$
$$= 3 + \frac{1}{6}x - \frac{1}{216}x^2 + \frac{1}{3,888}x^3 - \cdots$$

Since the exponent $p = \frac{1}{2}$ satisfies $p > 0$, p not an integer, it follows from Theorem 8.26 that the series converges for $\left|\frac{x}{9}\right| \le 1$, that is for $|x| \le 9$. ∎

PROBLEM SET 8.8

Level 1

1. ▨ **What does this say?** Compare and contrast Maclaurin and Taylor series.
2. ▨ **What does this say?** Discuss the binomial series theorem.

Find the Maclaurin series for the functions given in Problems 3-26. Assume that a is any constant, and that all the derivatives of all orders exist at $x = 0$.

3. e^{2x}
4. e^{-x}
5. e^{x^2}
6. e^{ax}
7. $\sin x^2$
8. $\sin^2 x$
9. $\sin ax$
10. $\cos ax$
11. $\cos 2x^2$
12. $\cos x^3$
13. $x^2 \cos x$
14. $\sin \frac{x}{2}$
15. $x^2 + 2x + 1$
16. $x^3 - 2x^2 + x - 5$
17. xe^x
18. $e^{-x} + e^{2x}$
19. $e^x + \sin x$
20. $\sin x + \cos x$
21. $\dfrac{1}{1+4x}$
22. $\dfrac{1}{1-ax}$, $a \ne 0$
23. $\dfrac{1}{a+x}$, $a \ne 0$
24. $\dfrac{1}{a^2+x^2}$, $a \ne 0$
25. $\ln(3+x)$
26. $\log(1+x)$

Find the first four terms of the Taylor series of the functions in Problems 27-36 at the given value of c.

27. $f(x) = e^x$ at $c = 1$
28. $f(x) = \ln x$ at $c = 3$
29. $f(x) = \cos x$ at $c = \frac{\pi}{3}$
30. $f(x) = \sin x$ at $c = \frac{\pi}{4}$
31. $f(x) = \tan x$ at $c = 0$
32. $f(x) = \sqrt{x}$ at $c = 9$
33. $f(x) = \dfrac{1}{2-x}$ at $c = 5$
34. $f(x) = \dfrac{1}{4-x}$ at $c = -2$
35. $f(x) = \dfrac{3}{2x-1}$ at $c = 2$
36. $f(x) = \dfrac{5}{3x+2}$ at $c = 2$

Expand each function in Problems 37-42 as a binomial series. Give the interval of convergence of the series.

37. $f(x) = \sqrt{1+x}$
38. $f(x) = \dfrac{1}{\sqrt{1+x^2}}$
39. $f(x) = (1+x)^{2/3}$
40. $f(x) = (4+x)^{-1/3}$
41. $f(x) = \dfrac{x}{\sqrt{1-x^2}}$
42. $f(x) = \sqrt[4]{2-x}$

Level 2

43. Use the Maclaurin series for e^x and e^{-x} to find the Maclaurin series for $\sinh x = \dfrac{e^x - e^{-x}}{2}$.
44. Use the Maclaurin series for e^x and e^{-x} to find the Maclaurin series for $\cosh x = \dfrac{e^x + e^{-x}}{2}$.
45. Use the Maclaurin series expansion of e^x to estimate $\sqrt[3]{e}$ with three decimal place accuracy. How many terms are needed?

46. Use the Maclaurin series expansion of e^x to estimate $\sqrt{e}$ with three decimal place accuracy? How many terms are needed?

Find the Maclaurin series for the functions given in Problems 47-48. Hint: Begin by expressing each function as a sum of partial fractions.

47. $f(x) = \dfrac{1}{x^2 - 3x + 2}$

48. $f(x) = \dfrac{x^2}{(x+2)(x^2 - 1)}$

49. Use an appropriate identity to find the Maclaurin series for $f(x) = \sin x \cos x$.

50. a. It can be shown that

$$\cos 3x = 4\cos^3 x - 3\cos x$$

Use this identity to find the Maclaurin series for $\cos^3 x$.

b. Find the Maclaurin series for $\sin^3 x$.

51. Find the Maclaurin series for

$$\ln\left[\frac{1 + 2x}{1 - 3x + 2x^2}\right]$$

52. Find the Maclaurin series for

$$\ln\left[(1 + 2x)(1 + 3x)\right]$$

53. Find the Maclaurin series for
a. $f(x) = x + \sin x$ and then use this series to find

$$\lim_{x \to 0} \frac{x + \sin x}{x}$$

b. Find the Maclaurin series for $g(x) = e^x - 1$ and then use this series to find

$$\lim_{x \to 0} \frac{e^x - 1}{x}$$

In Problems 54-55, the Maclaurin series of a function appears with its remainder. Verify that the remainder satisfies the given inequality.

54. $e^x = \displaystyle\sum_{k=0}^{\infty} \frac{x^k}{k!}$; $|R_n(x)| \le \dfrac{e^p\,|x|^{n+1}}{(n+1)!}$ where p is some constant such that $0 < p < x$.

55. $\ln(x+1) = \displaystyle\sum_{k=0}^{\infty} \frac{(-1)^k x^{k+1}}{k+1}$ for $0 \le x < 1$;

$$|R_n(x)| \le \frac{|x|^{n+1}}{n+1}$$

56. 𝔥*istorical Quest*

The two mathematicians most closely associated with power series are Brook Taylor and Colin Maclaurin. The result stated in this section, called Taylor's theorem, was published by Taylor in 1712, but the result itself was discovered by the Scottish mathematician James Gregory (1638-1675) in 1670 and was obtained by Taylor in 1688, after Gregory's death. James Gregory apparently anticipated some of the key calculus discoveries, but died prematurely and never received proper recognition for his work. In 1694, Johann Bernoulli (see Quest *Problem 49, Section 4.5) published a series that was very similar to what we call Taylor's series. In fact, after Taylor published his work in 1714, Bernoulli accused him of plagiarism. In this* Quest *we seek to duplicate Bernoulli's work. He started by writing*

Karl Smith library

Brook Taylor
(1685-1731)

$$n\,dx = n\,dx + \left(x\,dn - x\frac{dn}{dx}\,dx\right)$$

$$= n\,dx + \left(x\,dn - x\frac{dn}{dx}\,dx\right)$$
$$- \left(\frac{x^2}{2!}\frac{d^2n}{dx^2}\,dx - \frac{x^2}{2!}\frac{d^2n}{dx^2}\,dx\right)$$

$$= n\,dx + \left(x\,dn - x\frac{dn}{dx}\,dx\right)$$
$$- \left(\frac{x^2}{2!}\frac{d^2n}{dx^2}\,dx - \frac{x^2}{2!}\frac{d^2n}{dx^2}\,dx\right)$$
$$+ \left(\frac{x^3}{3!}\frac{d^3n}{dx^3}\,dx - \frac{x^3}{3!}\frac{d^3n}{dx^3}\,dx\right) - \cdots$$

$$= (n\,dx + x\,dn)$$
$$- \left(x\frac{dn}{dx} + \frac{x^2}{2!}\frac{d^2n}{dx^2}\right)dx$$
$$+ \left(\frac{x^2}{2!}\frac{d^2n}{dx^2} + \frac{x^3}{3!}\frac{d^3n}{dx^3}\right)dx - \cdots$$

$$= d(nx) - d\left(\frac{x^2}{2!}\frac{dn}{dx}\right) + d\left(\frac{x^3}{3!}\frac{d^2n}{dx^2}\right) - \cdots$$

Integrate termwise to obtain what is known as Bernoulli's series. Next, see if you can show what angered Bernoulli by deriving Taylor's series from Bernoulli's series. *Incidentally, Bernoulli's assertion was unfounded. The first known explicit statement of the general Taylor series was in* **De Quadrature** *(1691) by Isaac Newton.*

57. ℌ*istorical* 𝔔*uest*

Colin Maclaurin
(1698-1746)

Karl Smith library

Colin Maclaurin was a mathematical prodigy. He entered the University of Glasgow at the age of 11. At 15 he took his master's degree and gave a remarkable public defense of his thesis on gravitational attraction. At 19 he was elected to the chair of mathematics at the Marischal College in Aberdeen and at 21 published his first important work, **Geometrica organica***. At 27 he became deputy, or assistant, to the professor of mathematics at the University of Edinburgh. There was some difficulty in obtaining a salary to cover his assistantship, and Newton offered to bear the cost personally so that the university could secure the services of such an outstanding young man. In time Maclaurin succeeded Newton and at the age of 44, published the first systematic exposition of Newton's work. In his work, Maclaurin uses what were called at the time fluxions, rather than derivatives. Here is what he concluded:* "If the first fluxion of the ordinate, with its fluxions of several subsequent orders vanish, the ordinate is a minimum or maximum, when the number of all those fluxions that vanish is 1, 3, 5, or any odd number. The ordinate is a *minimum*, when the fluxion next to those that vanish is positive; but a *maximum* when the fluxion is negative .... But if the number of all the fluxions of the first and successive orders that vanish be an even number, the ordinate is then neither a *maximum nor minimum*."[*] What is Maclaurin saying in this quotation?

Level 3

58. The functions $J_0(x)$ and $J_1(x)$ are defined as the following power series:

$$J_0(x) = \sum_{k=0}^{\infty} \frac{(-1)^k x^{2k}}{(k!)^2 2^{2k}}$$

and

$$J_1(x) = \sum_{k=0}^{\infty} \frac{(-1)^k x^{2k+1}}{k!(k+1)! 2^{2k+1}}$$

a. Show that $J_0(x)$ and $J_1(x)$ both converge for all x.
b. Show that $J_0'(x) = -J_1(x)$. These functions are called **Bessel functions of the first kind**. Bessel functions were first used in analyzing Kepler's laws of planetary motion. These functions play an important role in physics and engineering.

59. Show that J_0, the Bessel function of the first kind defined in Problem 58, satisfies the differential equation

$$x^2 J_0''(x) + x J_o'(x) + x^2 J_0(x) = 0$$

60. Journal Problem (*The Pi Mu Epsilon Journal* by Robert C. Gebhardt)[†] For what x does

$$\sum_{n=0}^{\infty} \frac{x^n}{(2n)!}$$

converge, and what is the sum?

CHAPTER 8 REVIEW

*T*here *is nothing now which has ever given me any thought or care in algebra except divergent series, which I cannot follow the French in rejecting.*

Augustus De Morgan

(New York: 1882-1889), p. 240

Proficiency Examination

Concept Problems

1. What is a sequence?
2. What is meant by the limit of a sequence?
3. Define the convergence and divergence of a sequence.
4. Explain each of the following terms:

   a. bounded sequence   **b.** monotonic sequence   **c.** strictly monotonic sequence

5. State the BMCT.
6. What is an infinite series?
7. Compare or contrast the convergence and divergence of sequences and series.
8. What is a telescoping series?
9. Describe the harmonic series. Does it converge or diverge?
10. Define a geometric series and give its sum.
11. State each indicated test.

    a. divergence test   **b.** integral test   **c.** *p*-series test

    d. direct comparison test   **e.** limit comparison test   **f.** zero-infinity limit comparison test

    g. ratio test   **h.** root test   **i.** alternating series test

12. How do you make an error estimate for an alternating series?
13. State the absolute convergence test.
14. What is meant by absolute and conditional convergence?
15. What is the generalized ratio test?
16. **a.** What is a power series?
    b. How do you find the convergence set of a power series?
17. What are the radius of convergence and interval of convergence for a power series?
18. **a.** What is a Taylor polynomial?   **b.** State Taylor's theorem.
19. What are a Taylor series and a Maclaurin series?
20. State the binomial series theorem.

Practice Problems

21. Find the limit of the sequence $\left\{ \left(1 + \frac{1}{n}\right)^n \right\}$.

22. **a.** Find the limit of the sequence $\left\{ \dfrac{e^n}{n!} \right\}$.   **b.** Test the series $\displaystyle\sum_{k=1}^{\infty} \dfrac{e^k}{k!}$ for convergence.

    c. Discuss the similarities and/or differences of parts **a** and **b**.

In Problems 23-26, test the given series for convergence.

23. $\displaystyle\sum_{k=2}^{\infty} \frac{1}{k \ln k}$ 24. $\displaystyle\sum_{k=1}^{\infty} \frac{\pi^k k!}{k^k}$ 25. $\displaystyle\sum_{k=2}^{\infty} \frac{1}{(\ln k)^{1/k}}$ 26. $\displaystyle\sum_{k=1}^{\infty} \frac{3k^2 - k + 1}{(1 - 2k)k}$

27. $1 - \frac{1}{4} + \frac{1}{9} - \frac{1}{16} + \frac{1}{25} - \frac{1}{36} + \frac{1}{49} - \frac{1}{64} + \cdots + \frac{(-1)^{n+1}}{n^2} + \cdots$

28. Find the convergence set for the series

$$1 - 2u + 3u^2 - 4u^3 + \cdots$$

29. Find the Maclaurin series for $f(x) = \sin 2x$.

30. Find the Taylor series for $f(x) = \dfrac{1}{x-3}$ at $c = \dfrac{1}{2}$.

Supplementary Problems*

Determine whether each sequence in Problems 1-15 converges or diverges. If it converges, find its limit.

1. $\left\{\dfrac{(-2)^n}{n^2+1}\right\}$ **2.** $\left\{\dfrac{(\ln n)^2}{\sqrt{n}}\right\}$ **3.** $\left\{\left(1 - \dfrac{2}{n}\right)^n\right\}$ **4.** $\left\{\dfrac{e^{0.1n}}{n^5 - 3n + 1}\right\}$

5. $\left\{\dfrac{n+(-1)^n}{n}\right\}$ **6.** $\left\{\dfrac{3^n}{3^n+2^n}\right\}$ **7.** $\left\{\dfrac{5n^4 - n^2 - 700}{3n^4 - 10n^2 + 1}\right\}$ **8.** $\left\{\left(1 + \dfrac{e}{n}\right)^{2n}\right\}$

9. $\left\{\sqrt{n+1} - \sqrt{n}\right\}$ **10.** $\left\{\sqrt{n^4 + 2n^2} - n^2\right\}$ **11.** $\left\{\dfrac{\ln n}{n}\right\}$ **12.** $\{1 + (-1)^n\}$

13. $\left\{\left(1 + \dfrac{4}{n}\right)^n\right\}$ **14.** $\left\{5^{2/n}\right\}$ **15.** $\left\{\dfrac{n^{3/4}\sin n^2}{n+4}\right\}$

16. Give an alternate proof that the harmonic series diverges by showing that $S_{2^n} > 1 + \dfrac{n}{2}$ for $n > 1$.

17. Find the sum $\displaystyle\sum_{k=-123,456,788}^{123,456,789} \dfrac{k}{370,370,367}$

Find the sum of the given convergent geometric or telescoping series in Problems 18-25.

18. $\displaystyle\sum_{k=1}^{\infty} 4\left(\dfrac{2}{3}\right)^k$ **19.** $\displaystyle\sum_{k=1}^{\infty} \left(\dfrac{e}{3}\right)^k$ **20.** $\displaystyle\sum_{k=2}^{\infty} \dfrac{1}{k^2-1}$ **21.** $\displaystyle\sum_{k=1}^{\infty} \dfrac{1}{4k^2-1}$

22. $\displaystyle\sum_{k=0}^{\infty} \left[\left(\dfrac{-3}{8}\right)^k + \left(\dfrac{3}{4}\right)^{2k}\right]$ **23.** $\displaystyle\sum_{k=0}^{\infty} \dfrac{e^k + 3^{k-1}}{6^{k+1}}$ **24.** $\displaystyle\sum_{k=1}^{\infty} (-1)^{k+1}\left(\dfrac{1}{k} + \dfrac{1}{k+1}\right)$

25. $\displaystyle\sum_{k=1}^{\infty} \dfrac{k}{(k+1)(k+2)(k+3)}$

Test the series in Problems 26-45 for convergence.

26. $\displaystyle\sum_{k=1}^{\infty} \dfrac{5^k k!}{k^k}$ **27.** $\displaystyle\sum_{k=1}^{\infty} \dfrac{1}{\sqrt{k^2+4}}$ **28.** $\displaystyle\sum_{k=0}^{\infty} \dfrac{k!}{2^k}$ **29.** $\displaystyle\sum_{k=1}^{\infty} \dfrac{1}{2k-1}$

30. $\displaystyle\sum_{k=0}^{\infty} ke^{-k}$ **31.** $\displaystyle\sum_{k=1}^{\infty} \dfrac{k^3}{k!}$ **32.** $\displaystyle\sum_{k=1}^{\infty} \dfrac{7^k}{k^2}$ **33.** $\displaystyle\sum_{k=1}^{\infty} \dfrac{k}{(2k-1)!}$

34. $\displaystyle\sum_{k=0}^{\infty} \dfrac{k^2}{(k^3+1)^2}$ **35.** $\displaystyle\sum_{k=0}^{\infty} \dfrac{k^2}{3^k}$ **36.** $\displaystyle\sum_{k=2}^{\infty} \dfrac{1}{k(\ln k)^{1.1}}$ **37.** $\displaystyle\sum_{k=2}^{\infty} \dfrac{1}{k(\ln k)^2}$

38. $\displaystyle\sum_{k=0}^{\infty} \left(\sqrt{k^3+1} - \sqrt{k^3}\right)$ **39.** $\displaystyle\sum_{k=1}^{\infty} \dfrac{1}{\sqrt{k(k+1)}}$ **40.** $\displaystyle\sum_{k=1}^{\infty} \dfrac{k-1}{k2^k}$ **41.** $\displaystyle\sum_{k=1}^{\infty} \dfrac{k!}{k^2(k+1)^2}$

42. $\displaystyle\sum_{k=0}^{\infty} \dfrac{1}{1+\sqrt{k}}$ **43.** $\displaystyle\sum_{k=1}^{\infty} \dfrac{k^2 3^k}{k!}$ **44.** $\displaystyle\sum_{k=1}^{\infty} \dfrac{k^2}{\sqrt{k^3+2}}$ **45.** $\displaystyle\sum_{k=1}^{\infty} \dfrac{k}{k^2+1}$

---

*The supplementary probelms are presented in a somewhat random order, not necessarily in order of difficulty.

Determine whether each series given in Problems 46-55 converges conditionally, converges absolutely, or diverges.

46. $\dfrac{1}{1\cdot 2} - \dfrac{1}{3\cdot 2} + \dfrac{1}{5\cdot 2} - \dfrac{1}{7\cdot 2} + \cdots$

47. $\dfrac{1}{1\cdot 2} - \dfrac{1}{2\cdot 3} + \dfrac{1}{3\cdot 4} - \dfrac{1}{4\cdot 5} + \cdots$

48. $\dfrac{3}{2} - \dfrac{4}{3} + \dfrac{5}{4} - \dfrac{6}{5} + \cdots$

49. $1 - \dfrac{1}{3} + \dfrac{1}{9} - \dfrac{1}{27} + \dfrac{1}{81} - \cdots$

50. $\dfrac{1}{5} - \dfrac{1}{7} + \dfrac{1}{9} - \dfrac{1}{11} + \cdots$

51. $-1 + \dfrac{1}{\sqrt{2}} - \dfrac{1}{\sqrt[3]{3}} + \dfrac{1}{\sqrt[4]{4}} - \cdots$

52. $\displaystyle\sum_{k=1}^{\infty} \dfrac{(-1)^k}{k\,[\ln(k+1)]^2}$

53. $\displaystyle\sum_{k=1}^{\infty} (-1)^k \left(\dfrac{3k+85}{4k+1}\right)^k$

54. $\displaystyle\sum_{k=1}^{\infty} (-1)^k \tan^{-1}\left(\dfrac{1}{2k+1}\right)$

55. $\displaystyle\sum_{k=1}^{\infty} \dfrac{(-1)^{k(k+1)/2}}{2^k}$

Find the interval of convergence for each power series in Problems 56-67.

56. $\displaystyle\sum_{k=1}^{\infty} \dfrac{kx^k}{3^k}$

57. $\displaystyle\sum_{k=1}^{\infty} k(x-1)^k$

58. $\displaystyle\sum_{k=1}^{\infty} \dfrac{x^k}{k(k+1)}$

59. $\displaystyle\sum_{k=1}^{\infty} \dfrac{\ln k\,(x^2)^k}{\sqrt{k}}$

60. $\displaystyle\sum_{k=1}^{\infty} \dfrac{k^2(x+1)^{2k}}{2^k}$

61. $\displaystyle\sum_{k=1}^{\infty} \dfrac{(-1)^k x^k}{k^k}$

62. $\displaystyle\sum_{k=1}^{\infty} \dfrac{(-1)^k(2x-1)^k}{k^2}$

63. $\displaystyle\sum_{k=1}^{\infty} \dfrac{(x+2)^k}{k\ln(k+1)}$

64. $\displaystyle\sum_{k=1}^{\infty} \dfrac{k(x-3)^k}{(k+3)!}$

65. $\displaystyle\sum_{k=1}^{\infty} \dfrac{(3k)!}{k!}x^k$

66. $\displaystyle\sum_{k=1}^{\infty} \dfrac{(-1)^k x^k}{9k^2-1}$

67. $x - \dfrac{x^3}{3!} + \dfrac{x^5}{5!} - \dfrac{x^7}{7!} + \cdots + \dfrac{(-1)^{k+1}x^{2k-1}}{(2k-1)!} + \cdots$

Find the Maclaurin series for each function given in Problems 68-73.

68. $f(x) = x^2 e^{-3x}$

69. $f(x) = x^3 \sin x$

70. $f(x) = 3^x$

71. $f(x) = x^2 + \tan^{-1} x$

72. $f(x) = \dfrac{5+7x}{1+2x-3x^2}$

73. $f(x) = \dfrac{11x-1}{2+x-3x^2}$

74. Compute the sum of the series

$$1 - \dfrac{1}{2} + \dfrac{1}{2!}\left(\dfrac{1}{2}\right)^2 - \dfrac{1}{3!}\left(\dfrac{1}{2}\right)^2 + \cdots$$

correct to three decimal places.

75. Find a bound on the error incurred by approximating the sum of the series

$$1 - \dfrac{1}{2!} + \dfrac{2}{3!} - \dfrac{3}{4!} + \cdots$$

by the sum of the first eight terms.

76. Use geometric series to write $12.342\overline{132}$ as a rational number.

77. Find the first three nonzero terms of the Maclaurin series for $\tan x$ and then use term-by-term integration to find the first three nonzero terms of the Maclaurin series for $\ln|\cos x|$.

78. *Historical Quest* *The Greek philosopher Zeno of Elea (495-435* B.C.*) presented a number of paradoxes that caused a great deal of concern to the mathematicians of his period. These paradoxes were important in the development of the notion of infinitesimals. Perhaps the most famous is the racecourse paradox, which can be stated as follows: A runner can never reach the end of a race: As he runs the track, he must first run $\frac{1}{2}$ the length of the track, then $\frac{1}{2}$ the remaining distance, so that at this point he has run $\frac{1}{2} + \frac{1}{4} = \frac{3}{4}$ of the length of the track, and $\frac{1}{4}$ remains to be run. After running half this distance, he finds he still has $\frac{1}{8}$ of the track to run, and so on, indefinitely.*

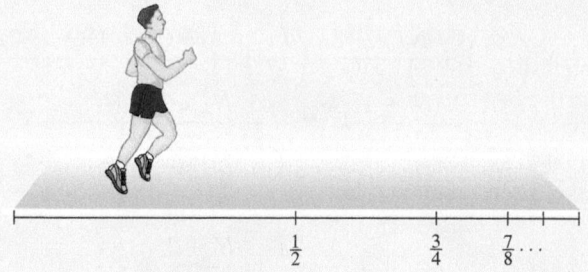

Suppose the runner runs at a constant pace and that it takes him T minutes to run the first half of the track. Set up an infinite series that gives the *total* time required to run the track, and verify that the total time is $2T$, as intuition would lead us to expect.

79. *Historical Quest*

Ramanujan was that rarest of mathematicians, an instinctive genius with virtually no formal training. Like his Hindu predecessor, Bhaskara (see the Historical Quest in Problem 92 of the supplementary problems of Chapter 4), Ramanujan had an uncanny instinct for numerical "truth" and conceived of his results in much the same way that a sculptor "sees" a statue in a raw block of stone. In his short life, he initiated new ways of thinking about number theory and made conjectures that are the subject of mathematical inquiry to this day.

A story related by his friend and mentor, the eminent British number theorist G. H. Hardy (1877-1947), serves to illustrate the resonance between Ramanujan's mind and the concept of number. Ramanujan was ill and Hardy came to visit him in the hospital. At a loss for how to begin the conversation, Hardy idly remarked that he had arrived in a taxi with the "dull" number 1729. Ramanujan immediately grew excited and exclaimed, "No, Hardy, it is a very interesting number, for it is the smallest integer that can be expressed as the sum of cubes in two different ways."

Srinivasa Ramanujan (1887-1920)

$$1729 = 1^3 + 12^3 = 9^3 + 10^3$$

In a letter from Ramanujan to G. H. Hardy at Cambridge University, Ramanujan stated*

$$1 - 5\left(\frac{1}{2}\right)^3 + 9\left(\frac{1\cdot 3}{2\cdot 4}\right)^3 - 13\left(\frac{1\cdot 3\cdot 5}{2\cdot 4\cdot 6}\right)^3 + \cdots = \frac{2}{\pi}$$

Hardy spent a great deal of time wondering how this sum could be equal to $2/\pi$. There is an interesting discussion of this identity, as well as biographical information on Ramanujan, in the article "Ramanujan the Phenomenon," by S. G. Gindikin, Quantum, pp. 4-9, April 1998.

a. Find a_k so this sum can be expressed as $1 + \sum_{k=1}^{\infty} a_k$.

b. There is no elementary way to prove this identity. Use technology to check that this formula is valid.

80. *Historical Quest Leonhard Euler (1707-1783) was introduced in Chapter 4 Supplementary Problem 93, where we saw that Euler writes $a^\epsilon - 1 + k\epsilon$ for an "infinitely small" number ϵ. In this Quest we will see why we define*

$$\lim_{n\to\infty}\left(1 + \frac{1}{n}\right)^n = e$$

by asking you to reconstruct the steps of Euler's work. Let x be a (finite) number. Euler introduced the "infinitely large" number $N = x/\epsilon$. Explain each step.

$$a^x = a^{N\epsilon}$$
$$= \left(a^\epsilon\right)^N$$
$$= \left(1 + \frac{kx}{N}\right)^N$$

$$= 1 + N\left(\frac{kx}{N}\right) + \frac{N(N-1)}{2!}\left(\frac{kx}{N}\right)^2 + \frac{N(N-1)(N-2)}{3!}\left(\frac{kx}{N}\right)^3 + \cdots$$

$$= 1 + kx + \frac{1}{2!}\frac{N(N-1)}{N^2}k^2x^2 + \frac{1}{3!}\frac{N(N-1)(N-2)}{N^3}k^3x^3 + \cdots$$

Because N is infinitely large, Euler assumed that

$$1 = \frac{N-1}{N} = \frac{N-2}{N} = \cdots$$

Substitute these values into the given derivation and also substitute e for a to obtain

$$e = 1 + \frac{1}{1!} + \frac{1}{2!} + \frac{1}{3!} + \cdots$$

Euler calculated e to 24 places:

$$e \approx 2.71828182845904523536028$$

Show that Euler's derivation is equivalent to $\lim\limits_{n\to\infty}\left(1+\frac{1}{n}\right)^n$.

81. Prove that if $\Sigma a_k = A$ and $\Sigma b_k = B$, then $\Sigma(a_k + b_k) = A + B$.

82. Think Tank Problem Either prove or disprove (by finding a counterexample) each of the following statements:
a. If Σa_k and Σb_k are both divergent series, then $\Sigma(a_k + b_k)$ is also divergent.
b. If Σa_k and Σb_k are both convergent series, then $\Sigma(a_k + b_k)$ is also convergent.

83. a. Show that $\displaystyle\sum_{k=1}^{n}(a_k - a_{k+2}) = (a_1 + a_2) - (a_{n+1} + a_{n+2})$

b. Use part **a** to show that if $\lim\limits_{k\to\infty} a_k = A$, then $\displaystyle\sum_{k=1}^{\infty}(a_k - a_{k+2}) = a_1 + a_2 - 2A$

c. Use part **b** to evaluate $\displaystyle\sum_{k=1}^{\infty}\left[k^{1/k} - (k+2)^{1/(k+2)}\right]$

d. Evaluate $\displaystyle\sum_{k=2}^{\infty}\frac{1}{k^2-1}$.

84. Find the Maclaurin series for $f(x) = \dfrac{2\tan x}{1 + \tan^2 x}$

85. Show that

$$\cos x = \cos c - (x-c)\sin c - \frac{(x-c)^2}{2!}\cos c + \frac{(x-c)^3}{3!}\sin c + \cdots$$

86. Show that

$$e^x = e^c + e^c(x-c) + \frac{e^c(x-c)^2}{2!} + \frac{e^c(x-c)^3}{3!} + \cdots + \frac{e^c(x-c)^k}{k!} + \cdots$$

87. Express the integral $\int \sqrt{x^3 + 1}\, dx$ as an infinite series by writing the first four nonzero terms.
Hint: Use the binomial theorem and then integrate term by term.

88. Estimate the value of the integral

$$\int_0^{0.4} \frac{dx}{\sqrt{1 + x^3}}$$

with five decimal-place accuracy.

89. Show that if $a > 0$, $a \neq 1$,

$$a^x = \sum_{k=0}^{\infty} \frac{(\ln a)^k}{k!} x^k \qquad \text{for all } x$$

90. Show that for $0 < x < 2c$,

$$\frac{1}{x} = \frac{1}{c} - \frac{1}{c^2}(x - c) + \frac{1}{c^3}(x - c)^2 - \frac{1}{c^4}(x - c)^3 + \cdots$$

91. Use series to find $\sin 0.2$ correct to five decimal places.

92. Use series to find $\ln 1.05$ correct to five decimal places.

93. Use a Maclaurin series to find a cubic polynomial that approximates e^{2x} for $|x| < 0.001$. Find an upper bound for the error in this approximation.

94. For what values of x can $\sin x$ be replaced by $x - \dfrac{x^3}{6}$ if the allowable error is 0.005?

95. For what values of x can $\cos x$ be replaced by $1 - \dfrac{x^2}{2}$ if the allowable error is 0.00005?

96. Modeling Problem When we studied carbon dating in Section 5.6, it was pointed out that radiocarbon methods apply only apply to objects that are not too "old" (roughly 40,000 years). To date older objects, radioactive isotopes other than ^{14}C are often employed. Regardless of the radioactive substance used, it can be shown that the ratio of stable isotope to radioactive isotope at any given time may be modeled by

$$\frac{S(t) - S(0)}{R(t)} + 1 = e^{(\ln 2)t/\lambda}$$

where $R(t)$ is the number of atoms of radioactive material at time t, $S(t)$ is the number of atoms of the stable product of radioactive decay, $S(0)$ is the number of atoms of stable product initially present (at $t = 0$), and λ is the half-life of the radioactive isotope.*

a. Use the first two terms of the Maclaurin series for e^x to approximate $e^{(\ln 2)t/\lambda}$, and then solve the modeling equation for t.

b. Suppose a piece of mica is analyzed, and it is found that 5% of the atoms in the rock are radioactive rubidium-87 and 0.04% of the atoms are strontium-87. Assuming that the strontium was produced by decay of the rubidium-87 originally in the rock, what is the approximate age of the sample? Use the result in part **a**, with $\lambda = 4.86 \times 10^9$ years as the half-life of rubidium-87.

97. Putnam Examination Problem Express

$$\sum_{k=1}^{\infty} \frac{6^k}{(3^{k+1} - 2^{k+1})(3^k - 2^k)}$$

as a rational number.

98. Putnam Examination Problem Find the sum of the convergent alternating series

$$\sum_{n=0}^{\infty} \frac{(-1)^{n+1}}{3n - 2}$$

99. Putnam Examination Problem For positive real x, let

$$B_n(x) = 1^x + 2^x + 3^x + \cdots + n^x$$

Test for convergence of the series

$$\sum_{k=2}^{\infty} \frac{B_k(\log_k 2)}{(k \log_2 k)^2}$$

*This model is discussed in detail in the article, "How Old Is the Earth?" Paul J. Campbell, *UMAP Modules 1992: Tools for Teaching*, Lexington, MA: Consortium for Mathematics and Its Applications, Inc. 1993, pp. 105-137.

CHAPTER 8 GROUP RESEARCH PROJECT*

Working in small groups is typical of most work environments, and this book seeks to develop skills with group activities. We present a group project at the end of each chapter. These projects are to be done in groups of three or four students.

Elastic Tightrope Project

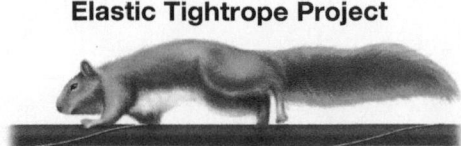

Suppose a squirrel starts at one end of a 100-m tightrope and runs toward the other end at a speed of 1 m/s. At the end of every minute, the tightrope stretches uniformly and instantaneously, increasing its length by 100 meters each time. Thus, after the first minute, the critter has traveled 60 meters and the length of the tightrope is 200 m. After the second minute, the squirrel has traveled another 60 meters and the tightrope has become 300 m long, and so on.

Does the squirrel ever reach the other end? If so, how long does it take?

Extended paper for further study. Your paper is not limited to the following questions but should include these concerns. Suppose it is you standing on the elastic rope b meters from the one end and c meters from the other end. Then suppose the entire rope stretches uniformly, increasing its length by d meters. Only the tightrope stretches; units of length and time remain constant. How far are you from each end? You might also try to generalize the tightrope problem.

---

*Adapted from a group project by David Pengelley, New Mexico State University, Las Cruces.

Cumulative Review Problems—Chapters 6-8

1. ■ What does this say? Describe a procedure for using integration to
   a. compute work done by a variable force   **b.** fluid (or hydrostatic) force
2. ■ What does this say? Describe integration by parts.
3. ■ What does this say? Describe the method of partial fractions.
4. ■ What does this say? Describe a geometric series. For what values does such a series converge? What is the sum of a convergent geometric series?
5. ■ What does this say? Describe a procedure for determining the convergence set of the power series $\Sigma a_k x^k$.
6. ■ What does this say? Outline a procedure for determining the convergence of Σa_k for $a_k > 0$.

Evaluate each of the integrals in Problems 7-20.

7. $\displaystyle\int \ln \sqrt{x}\, dx$

8. $\displaystyle\int \sin^2 x \cos^3 x\, dx$

9. $\displaystyle\int \frac{\tan^{-1} x}{1 + x^2}\, dx$

10. $\displaystyle\int \sec^{3/2} x \tan x\, dx$

11. $\displaystyle\int \sqrt{4 - x^2}\, dx$

12. $\displaystyle\int \frac{1 - x^2}{1 + x^2}\, dx$

13. $\displaystyle\int \frac{(3 + 2 \sin t)}{\cos t}\, dt$

14. $\displaystyle\int \frac{e^{1 - \sqrt{x}}\, dx}{\sqrt{x}}$

15. $\displaystyle\int \frac{dx}{x^2 + x + 1}$

16. $\displaystyle\int x^3 \sin x^2\, dx$

17. $\displaystyle\int \cos^{-1}(-x)\, dx$

18. $\displaystyle\int \frac{(2x + 1)dx}{\sqrt{4x - x^2 - 2}}$

19. $\displaystyle\int \frac{e^{-x} - e^x}{e^{2x} + e^{-2x} + 2}\, dx$

20. $\displaystyle\int \frac{dx}{(x + \frac{1}{2})\sqrt{4x^2 + 4x}}$

In Problems 21-24, either evaluate the given improper integral or show that it diverges.

21. $\displaystyle\int_0^\infty x e^{-2x}\, dx$

22. $\displaystyle\int_0^{\pi/2} \sec x\, dx$

23. $\displaystyle\int_0^\infty x^2 e^{-x}\, dx$

24. $\displaystyle\int_0^\infty x^3 e^{-x^2}\, dx$

25. Let $f(x) = \dfrac{x^2 + 17x - 8}{(2x + 3)(x - 1)^2}$
    a. Express $f(x)$ as a partial fraction sum.   **b.** Find $\int f(x)\, dx$.
    c. Either evaluate the improper integral $\displaystyle\int_2^\infty f(x)\, dx$ or show it diverges.

26. Evaluate $\displaystyle\lim_{x \to \infty} \frac{\sinh^{-1} x}{\cosh^{-1} x}$

In Problems 27-32, test the given series for convergence.

27. $\displaystyle\sum_{k=0}^\infty \frac{k^2 3^{k+1}}{4^k}$

28. $\displaystyle\sum_{k=1}^\infty \frac{k + 3k^{3/2}}{5k^2 - k + 3}$

29. $\displaystyle\sum_{k=1}^\infty \frac{(-1)^{k+1} k}{k + 1}$

30. $\displaystyle\sum_{k=1}^\infty \left(\frac{2k + 5}{3k - 2}\right)^k$

31. $\displaystyle\sum_{k=1}^\infty k e^{-k^2}$

32. $\displaystyle\sum_{k=1}^\infty \frac{k^2}{k!}$

In Problems 33 and 34, find the sum of the given convergent series.

33. $\displaystyle\sum_{k=1}^\infty 5\left(\frac{-2}{3}\right)^{2k-1}$

34. $\displaystyle\sum_{k=1}^\infty \frac{2}{(k + 1)(k + 3)}$

In Problems 35 and 36, find the convergence set of the given power series.

35. $\displaystyle\sum_{k=1}^\infty \frac{x^k}{k^3}$

36. $\displaystyle\sum_{k=1}^\infty \frac{(-x)^k}{k + 1}$

In Problems 37 *and* 38, *solve each of the first-order linear differential equations.*

37. $y' - \dfrac{y}{x} = \dfrac{x}{1+x^2}$

38. $y' + (\tan x)y = \tan x$, where $y = 3$ when $x = 0$

Find the Maclaurin series for the functions in Problems 39-41.

39. $f(x) = \sin x^2$

40. $f(x) = \sqrt{e^x}$

41. $f(x) = \dfrac{3x - 2}{5 + 2x}$

42. A triangular fin of a manta ray submerged in seawater ($\rho g = 64.0$ lb/ft^3) to a depth of 12 ft, as shown in Figure 8.27 (that is, the top of the fin is 12 ft below the surface).

© 2013 daulon. Used under license from Shutterstock, Inc.

Figure 8.28 Submerged manta ray

Find the fluid force on the triangular fin if we assume the sides of the fin are 5 ft and the base is 6 ft.

43. A tank has the shape of a circular cone with height 10 ft and top radius 4 ft. If the tank is filled with water ($\rho g = 62.4$ lb/ft^3) to a depth of 6 ft, how much work is done (rounded to the nearest ft-lb) in pumping all the water in the tank to a height 2 ft above the top?

44. Find the arc length of the curve $y = \frac{3}{2}x^{2/3}$ on the interval $\left[\frac{1}{8}, 8\right]$.

45. Let R be the region bounded by the x-axis, the curve $y = \dfrac{1}{\sqrt{x+1}}$, $x = 3$, and $x = 8$. Find the volume of the solid generated when R is revolved about:

  a. the x-axis   **b.** the y-axis   **c.** the line $y = -1$

46. Find both the exact and approximate volume of the solid generated when the region under the curve

$$y^2 = x^2 \left(\frac{4 - x}{4 + x}\right)$$

on the interval $[0, 4]$ is revolved about the x-axis.

47. Find the area of the region bounded by the graphs of

$$y = \frac{2x}{\sqrt{x^2 + 9}}$$

and $y = 0$ from $x = 0$ to $x = 4$.

48. Evaluate

$$\int \sqrt{1 + \cos x}\, dx$$

Hint: Use the identity

$$\cos x = 2\cos^2 \frac{x}{2} - 1$$

49. Show that

$$\sum_{k=0}^{\infty} \frac{1}{(a+k)(a+k+1)} = \frac{1}{a}$$

if $a > 0$.

50. Evaluate $\displaystyle\sum_{k=0}^{\infty} \frac{1}{(2+k)^2 + 2 + k}$.

51. If $\displaystyle\sum_{k=0}^{\infty} a_k^2 = \sum_{k=0}^{\infty} b_k^2 = 4$ and $\displaystyle\sum_{k=0}^{\infty} a_k b_k = 3$. what is $\displaystyle\sum_{k=0}^{\infty} (a_k - b_k)^2$?

52. Think Tank Problem Let Σa_k^2 be a series that converges. Either prove that Σa_k must diverge or find a counterexample.

53. Think Tank Problem Let Σa_k^2 be a series that converges. Either prove that Σa_k must also converge or find a counterexample.

54. Find the average value (to the nearest hundredth) of the function $f(x) = x \sin^3 x^2$ between $x = 0$ and $x = 1$.

55. An object moves along the x-axis in such a way that its velocity at time t is $v(t) = \sin t + \sin^2 t \cos^3 t$. Find the distance moved by the object between times $t = 0$ and $t = \frac{\pi}{3}$.

56. A ball is dropped from a height of 20 ft. Each time it bounces, it rises to $\frac{3}{4}$ of its previous height. What is the total distance traveled by the ball?

57. What is the volume of the solid obtained when the region bounded by $y = x\sqrt{9 - x^2}$ and $y = 0$ is revolved about the x-axis?

58. Find the length of the curve $y = \ln(\sec x)$ from $x = 0$ to $x = \frac{\pi}{4}$.

59. Find the centroid (to the nearest hundredth) of the region bounded by the curve $y = x^2 e^{-x}$ and the x-axis, between $x = 0$ and $x = 1$.

60. A tank contains 15 lb of salt dissolved in 70 gallons of water. Suppose four gallons of brine containing 3 lb/gal of salt run into the tank every minute and the mixture (kept uniform by stirring) runs out of the tank at the rate of 2 gal/min.

    a. Set up and solve a differential equation for the amount of salt $S(t)$ in the tank at time t.

    b. Suppose the capacity of the tank is 100 gal. How much salt is in the tank at the time it just begins to overflow?

CHAPTER 9

VECTORS IN THE PLANE AND IN SPACE

*M*athematics is the gate and key of the sciences.

Roger Bacon
Opus Majus, Part 4 (1897)

PREVIEW

In this chapter, we focus on various algebraic and geometric aspects of vector representations, and then in the next chapter, we see how vectors can be combined with calculus to study motion in space and other applications.

A Lamborghini Vector M12

PERSPECTIVE

A *scalar* quantity is one that can be described in terms of magnitude alone, whereas a *vector* quantity requires both magnitude and direction. Force, velocity, and acceleration are common vector quantities, and an important goal of this chapter is to develop methods for dealing with such quantities.

This chapter forms the building blocks for an important part of calculus introduced in the next chapter, *vector calculus*. This chapter and the next will complete what is known as single-variable calculus.

9.1 VECTORS IN $\mathbb{R}^2$

IN THIS SECTION: *Introduction to vectors, vector operations, standard representation of vectors in the plane*

Many applications of mathematics involve quantities that have both magnitude and direction, such as force, velocity, acceleration, and momentum. Vectors are an important tool in mathematics, and in this section we introduce you to the terminology and notation for vector representation.

Introduction to Vectors

A **vector** is a quantity (such as velocity or force) that has both magnitude and direction. We sometimes represent a vector as a directed line segment, an "arrow" with **initial point** P and **terminal point** Q. The direction of the vector is that of the arrow, and its magnitude is represented by the arrow's length, as shown in Figure 9.1**a**. We shall indicate such a vector by writing **PQ** in boldface print, but in your work you may write an arrow over the designated points: $\overrightarrow{PQ}$. The order of letters you write down is important: **PQ** means that the vector is from P to Q, but **QP** means that the vector is from Q to P. The first letter is always the initial point and the second letter is the terminal point. We shall denote the magnitude (length) of a vector by $\|\mathbf{PQ}\|$. Two vectors are regarded as **equal** (or **equivalent**) if they have the same magnitude and the same direction, even if they do not coincide (see Figure 9.1**b**). Thus, the initial point of a given vector may be moved to any convenient location by translating the vector parallel to its original position.

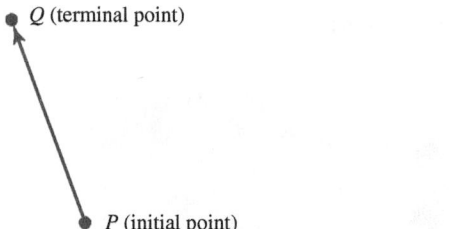

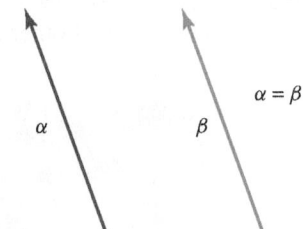

a. The vector **PQ** has magnitude $\|\mathbf{PQ}\|$

b. Two vectors are equal if they have the same magnitude and direction

Figure 9.1 Vectors in a Plane

A vector with magnitude 0 is called a **zero** (or **null**) **vector** and is denoted by **0**. The **0** vector has no specific direction, and we shall adopt the convention of assigning it any direction that may be convenient in a particular problem.

Vector Operations

Vectors can be multiplied by scalars and added to one another according to the rules developed by physics. If **v** is a vector other than **0**, then any vector **w** that is parallel to **v** is called a **scalar multiple** of **v** and satisfies $\mathbf{w} = s\mathbf{v}$ for some nonzero number s. A **scalar** quantity is one that has only magnitude and, in the context of vectors, is used to describe a real number. The scalar multiple $s\mathbf{v}$ has length $|s|$ times that of **v**; it points in the same direction as **v** if $s > 0$ and the opposite direction if $s < 0$ (see Figure 9.2). Notice that for any distinct points P and Q $\mathbf{PQ} = -\mathbf{QP}$. For the zero vector **0**, we define $s\mathbf{0} = \mathbf{0}$ for any scalar s.

Physical experiments indicate that force and velocity vectors can be added (or **resolved**) according to a **triangular rule** displayed in Figure 9.3**a**, and we use this rule as our definition of vector addition. In particular, to add the vector **v** to the vector **u**, we place the end (initial point) of **v** at the tip (terminal point) of **u** and define the sum,

Figure 9.2 Interactive
Some multiples of the vector **u**

also called the **resultant u + v**, to be the vector that extends from the initial point of **u** to the terminal point of **v**.

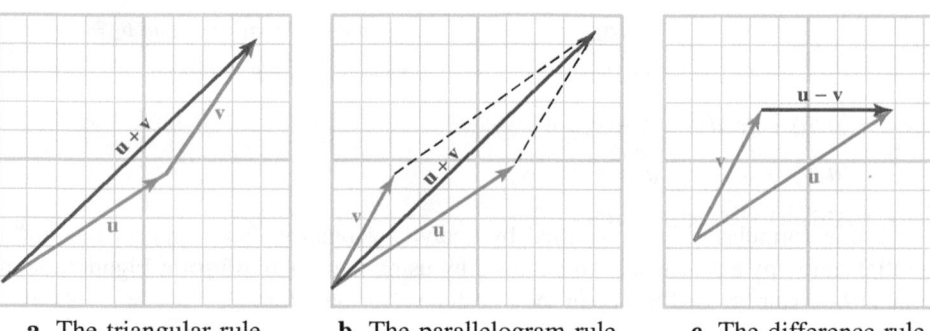

a. The triangular rule   **b.** The parallelogram rule   **c.** The difference rule

Figure 9.3 Adding and substracting vectors

Equivalently, the **parallelogram rule** states that **u + v** is the diagonal of the parallelogram formed with sides **u** and **v**, as shown in Figure 9.3**b**. **The difference u − v** is just the vector **w** that satisfies **v + w = u**, and it may be found by placing the initial points of **u** and **v** together and extending a vector from the terminal point of **v** to the terminal point of **u** (see Figure 9.3**c**). Note that **u + v = v + u**. In other words, addition is commutative.

The vector **v − u** may be found similarly and has the same magnitude, but points in the opposite direction to **u − v**.

Many issues involving vectors can be simplified by introducing a rectangular coordinate system. Suppose the vector **v** is positioned in a rectangular coordinate plane with its initial point at the origin $(0,0)$ and its terminal point at (v_1, v_2), as shown in Figure 9.4. Then v_1 and v_2 are called the **standard components** of **v**, and we write $\mathbf{v} = \langle v_1, v_2 \rangle$. The "angle" brackets "$\langle\ \rangle$" are used to distinguish the **component form** $\langle v_1, v_2 \rangle$ of **v** from the point (v_1, v_2) in $\mathbb{R}^2$.

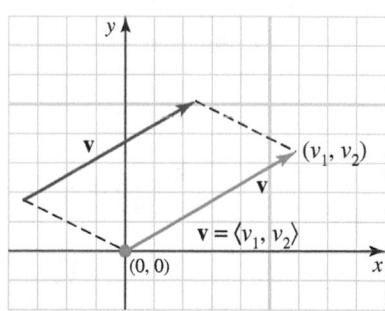

Figure 9.4 The standard component form of a vector v

Using the geometric relationships shown in Figure 9.5, the vector **PQ** with initial point $P(a,b)$ and terminal point $Q(c,d)$ equals the vector **OR**, where $O(0,0)$ is the origin and R is the point with coordinates $(c-a, d-b)$.

Thus, we can make the following general statement.

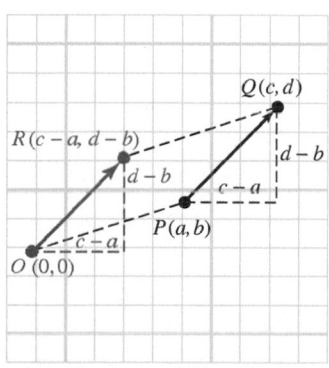

Figure 9.5 Representation of PQ

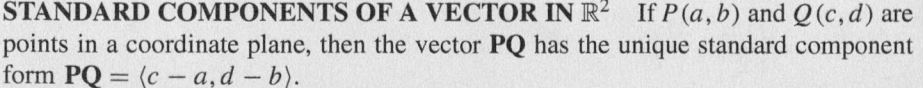

STANDARD COMPONENTS OF A VECTOR IN $\mathbb{R}^2$ If $P(a,b)$ and $Q(c,d)$ are points in a coordinate plane, then the vector **PQ** has the unique standard component form $\mathbf{PQ} = \langle c-a, d-b \rangle$.

Vector operations are easily represented when vectors are given in component form. In particular, we have

$$\langle a_1, b_1 \rangle = \langle a_2, b_2 \rangle \qquad \textit{If and only if } a_1 = a_2 \textit{ and } b_1 = b_2$$

$$k \langle a, b \rangle = \langle ka, kb \rangle \qquad \textit{For constant } k$$

$$\langle a, b \rangle + \langle c, d \rangle = \langle a + c, b + d \rangle$$

$$\langle a, b \rangle - \langle c, d \rangle = \langle a - c, b - d \rangle$$

These formulas may be verified by analytic geometry. For instance, the rule for multiplication by a scalar may be obtained by using the relationships in Figure 9.6**a**, and Figure 9.6**b** can be used to obtain the rule for vector addition.

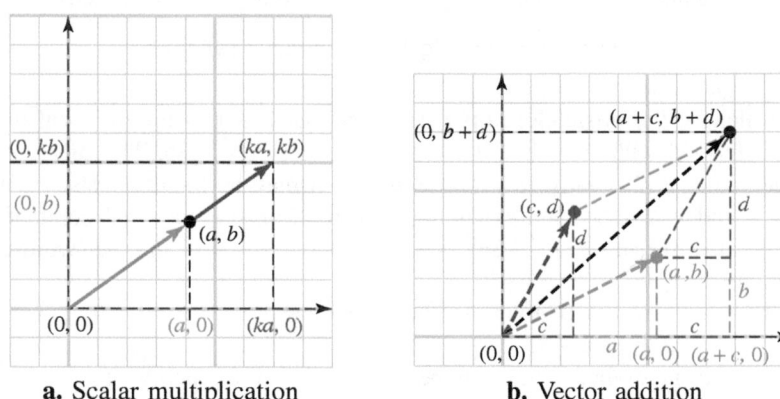

a. Scalar multiplication   **b.** Vector addition

Figure 9.6   Vector operations

Example 1   Vector operations

For the vectors $\mathbf{u} = \langle 2, -3 \rangle$ and $\mathbf{v} = \langle -1, 7 \rangle$, find:

a. $\mathbf{u} + \mathbf{v}$ **b.** $\frac{3}{4}\mathbf{u}$ **c.** $3\mathbf{u} - \frac{1}{2}\mathbf{v}$

Solution

a. $\mathbf{u} + \mathbf{v} = \langle 2, -3 \rangle + \langle -1, 7 \rangle$
$$= \langle 2 + (-1), -3 + 7 \rangle$$
$$= \langle 1, 4 \rangle$$

b. $\dfrac{3}{4}\mathbf{u} = \dfrac{3}{4}\langle 2, -3 \rangle$
$$= \left\langle \frac{3}{4}(2), \frac{3}{4}(-3) \right\rangle$$
$$= \left\langle \frac{3}{2}, \frac{-9}{4} \right\rangle$$

c. $3\mathbf{u} - \dfrac{1}{2}\mathbf{v} = 3\langle 2, -3 \rangle - \dfrac{1}{2}\langle -1, 7 \rangle$
$$= \langle 6, -9 \rangle + \left\langle \frac{1}{2}, -\frac{7}{2} \right\rangle \qquad \textit{Scalar multiplication}$$
$$= \left\langle 6 + \frac{1}{2}, -9 - \frac{7}{2} \right\rangle \qquad \textit{Add vectors.}$$
$$= \left\langle \frac{13}{2}, -\frac{25}{2} \right\rangle \qquad \textit{Simplify.}$$

In general, an expression of the form $a\mathbf{u} + b\mathbf{v}$ is called a **linear combination** of the vectors $\mathbf{u}$ and $\mathbf{v}$. Note that if $\mathbf{u} = \langle u_1, u_2 \rangle$ and $\mathbf{v} = \langle v_1, v_2 \rangle$, then

$$a\mathbf{u} + b\mathbf{v} = a\langle u_1, u_2 \rangle + b\langle v_1, v_2 \rangle = \langle au_1 + bv_1, au_2 + bv_2 \rangle$$

Vector addition and multiplication of a vector by a scalar behave very much like ordinary addition and multiplication. The following theorem lists several useful properties of these operations.

Theorem 9.1 Properties of vector operations

For any vectors $\mathbf{u}$, $\mathbf{v}$, and $\mathbf{w}$ in the plane and scalars s and t.

Commutativity of vector addition	$\mathbf{u} + \mathbf{v} = \mathbf{v} + \mathbf{u}$
Associativity of vector addition	$(\mathbf{u} + \mathbf{v}) + \mathbf{w} = \mathbf{u} + (\mathbf{v} + \mathbf{w})$
Associativity of scalar multiplication	$(st)\mathbf{u} = s(t\mathbf{u})$
Identity for addition	$\mathbf{u} + \mathbf{0} = \mathbf{u}$
Inverse property for addition	$\mathbf{u} + (-\mathbf{u}) = \mathbf{0}$
Vector distributivity	$(s + t)\mathbf{u} = s\mathbf{u} + t\mathbf{u}$
Scalar distributivity	$s(\mathbf{u} + \mathbf{v}) = s\mathbf{u} + s\mathbf{v}$

Proof: Each vector property can be established by using a corresponding property of real numbers. For example, to prove associativity of vector addition, let $\mathbf{u} = \langle u_1, u_2 \rangle$, $\mathbf{v} = \langle v_1, v_2 \rangle$, and $\mathbf{w} = \langle w_1, w_2 \rangle$. Then

$$
\begin{aligned}
(\mathbf{u} + \mathbf{v}) + \mathbf{w} &= (\langle u_1, u_2 \rangle + \langle v_1, v_2 \rangle) + \langle w_1, w_2 \rangle \\
&= \langle u_1 + v_1, u_2 + v_2 \rangle + \langle w_1, w_2 \rangle \\
&= \langle (u_1 + v_1) + w_1, (u_2 + v_2) + w_2 \rangle \\
&= \langle u_1 + (v_1 + w_1), u_2 + (v_2 + w_2) \rangle \quad \text{\textit{Associativity of addition for the real numbers}} \\
&= \langle u_1, u_2 \rangle + (\langle v_1, v_2 \rangle + \langle w_1, w_2 \rangle) \\
&= \mathbf{u} + (\mathbf{v} + \mathbf{w})
\end{aligned}
$$

You are asked to prove the other six properties in the problem set. ◆

Example 2 Vector proof of a geometric property

Show that the line segment joining the midpoints of two sides of a triangle is parallel to the third side and has half its length.

Solution Consider $\triangle ABC$ and let P and Q be the midpoints of sides $\overline{AC}$ and $\overline{BC}$, respectively, as shown in Figure 9.7.

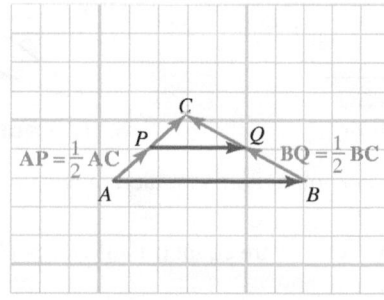

Figure 9.7 Vector proof of a geometric property

Given that $\mathbf{AP} = \frac{1}{2}\mathbf{AC}$ and $\mathbf{BQ} = \frac{1}{2}\mathbf{BC}$ we must prove that $\mathbf{PQ}$ is parallel to $\mathbf{AB}$ and $\|\mathbf{PQ}\| = \frac{1}{2}\|\mathbf{AB}\|$, which means that we must establish the vector equation $\mathbf{PQ} = \frac{1}{2}\mathbf{AB}$. Toward this end, we begin by noting that $\mathbf{AB}$ can be expressed as the following vector sum:

$$
\begin{aligned}
\mathbf{AB} &= \mathbf{AP} + \mathbf{PQ} + \mathbf{QB} \\
&= \frac{1}{2}\mathbf{AC} + \mathbf{PQ} - \mathbf{BQ} && AP = \tfrac{1}{2}AC \text{ and } QB = -BQ \\
&= \frac{1}{2}(\mathbf{AB} + \mathbf{BC}) + \mathbf{PQ} - \frac{1}{2}\mathbf{BC} && AC = (AB + BC) \text{ and } BQ = \tfrac{1}{2}BC \\
&= \frac{1}{2}\mathbf{AB} + \frac{1}{2}\mathbf{BC} + \mathbf{PQ} - \frac{1}{2}\mathbf{BC} \\
&= \frac{1}{2}\mathbf{AB} + \mathbf{PQ}
\end{aligned}
$$

$$\frac{1}{2}\mathbf{AB} = \mathbf{PQ} \qquad \textit{Subtract } \tfrac{1}{2}AB \textit{ from both sides.}$$

Therefore, the theorem is proved.

When a vector $\mathbf{u}$ is represented in component form $\mathbf{u} = \langle u_1, u_2 \rangle$, its length is given by the formula

$$\|\mathbf{u}\| = \sqrt{u_1^2 + u_2^2}$$

This is a simple application of the Pythagorean theorem as shown in Figure 9.8**a**. Another important relationship involving the length of vectors is the *triangle inequality*

$$\|\mathbf{u} + \mathbf{v}\| \leq \|\mathbf{u}\| + \|\mathbf{v}\|$$

for any vectors $\mathbf{u}$ and $\mathbf{v}$. Equality will occur precisely when $\mathbf{u}$ and $\mathbf{v}$ are multiples of one another (that is, when $\mathbf{u}$ and $\mathbf{v}$ have the same direction). To establish the inequality, we observe that $\mathbf{u}$ and $\mathbf{v}$ are two sides of a triangle in the plane, the third side has length $\|\mathbf{u} + \mathbf{v}\|$ and is "shorter" than the sum $\|\mathbf{u}\| + \|\mathbf{v}\|$ of the lengths of the other two sides, as shown in Figure 9.8**b**.

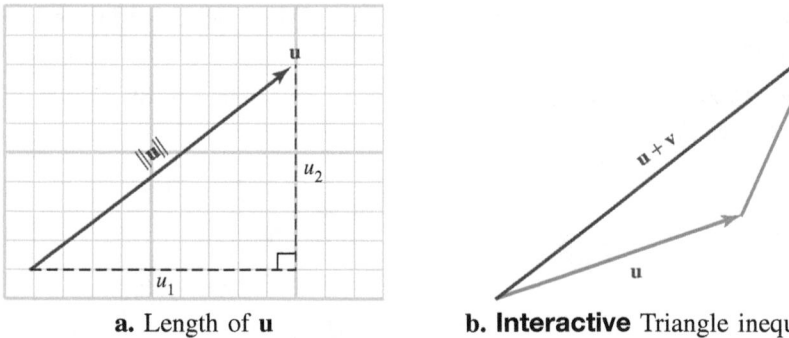

a. Length of $\mathbf{u}$ **b. Interactive** Triangle inequality

Figure 9.8   Geometric representation of two vector properties

In component form, two vectors are equal if their components are equal, as illustrated by the following example.

Example 3   Equal vectors in component form

If $\mathbf{u} = \langle 8, -2 \rangle$ and $\mathbf{v} = \langle -3, 5 \rangle$, find s and t so that $s\mathbf{u} + t\mathbf{v} = \mathbf{w}$, where $\mathbf{w} = \langle 2, 8 \rangle$.

Solution $s\mathbf{u} + t\mathbf{v} = s\langle 8, -2 \rangle + t\langle -3, 5 \rangle$

$$= \langle 8s, -2s \rangle + \langle -3t, 5t \rangle$$

$$= \langle 8s - 3t, -2s + 5t \rangle$$

Thus, if $s\mathbf{u} + t\mathbf{v} = \mathbf{w}$, we see

$$\begin{cases} 8s - 3t = 2 \\ -2s + 5t = 8 \end{cases}$$

Solving this system we find $(s, t) = (1, 2)$.

Example 4   Using vectors in a velocity problem

A special boat, the *Earthrace*, draws attention wherever it goes (see Figure 9.9).

A river 4 mi wide flows south with a current of 5 mi/h, and for an exhibition, the *Earthrace* must travel directly across the river from east to west past a viewing stand in 20 min. What is the desired heading for the boat?

Solution   Begin by drawing a diagram, as shown in Figure 9.10.

Let $\mathbf{B}$ be the velocity vector of the boat in the direction of the angle θ. If the river's current has velocity $\mathbf{C}$, the given information tells us that $\|\mathbf{C}\| = 5$ mi/h and that $\mathbf{C}$ points directly south. Moreover, because the boat is to cross the river from east to west in 20 min (that is, $\frac{1}{3}$ h), its *effective velocity* after compensating for the current is a vector $\mathbf{V}$ that points west. We will calculate $\|\mathbf{V}\|$ to find the effective velocity and then use this information to find the magnitude and direction of $\mathbf{B}$.

$$\|\mathbf{V}\| = \frac{\text{WIDTH OF THE RIVER}}{\text{TIME OF CROSSING}} = \frac{4 \text{ mi}}{\frac{1}{3} \text{ hr}} = 12 \text{ mi/h}$$

The effective velocity $\mathbf{V}$ is the resultant of $\mathbf{B}$ and $\mathbf{C}$; that is, $\mathbf{V} = \mathbf{B} + \mathbf{C}$. Because $\mathbf{V}$ and $\mathbf{C}$ act in perpendicular directions, we can determine $\mathbf{B}$ by referring to the right triangle with sides $\|\mathbf{V}\| = 12$ and $\|\mathbf{C}\| = 5$ with hypotenuse $\|\mathbf{B}\|$. We find that

$$\|\mathbf{B}\| = \sqrt{\|\mathbf{V}\|^2 + \|\mathbf{C}\|^2} = \sqrt{12^2 + 5^2} = 13$$

The direction of the velocity vector $\mathbf{V}$ is given by the angle θ in Figure 9.10, and we find that

$$\tan \theta = \frac{5}{12} \quad \text{so that} \quad \theta = \tan^{-1}\left(\frac{5}{12}\right) \approx 0.3948$$

Thus, the boat should travel at 13 mi/h in a of direction approximately 0.3948. In navigation, it is common to specify direction in degrees rather than radians; this is $22.6°$ north of west and to measure the direction as an angle from either north or south. In this case it is $90° - 22.6° = 67.4°$, or N67.4°W (that is, $67.4°$ west of north).

A **unit vector** is a vector with length 1, and a **direction vector** for a given nonzero vector $\mathbf{v}$ is a unit vector $\mathbf{u}$ that points in the same direction as $\mathbf{v}$. Such a vector can be found by simply "dividing" $\mathbf{v}$ by its length $\|\mathbf{v}\|$ $\left(\text{or multiplying by the scalar } \frac{1}{\|\mathbf{v}\|}\right)$; that is, $\mathbf{u} = \dfrac{\mathbf{v}}{\|\mathbf{v}\|}$.

Figure 9.9   The *Earthrace*, a high speed boat

© 2013 John Orsbun. Used under license from Shutterstock, Inc.

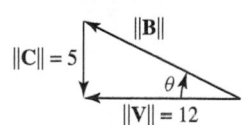

Figure 9.10   A velocity problem

Example 5   Finding a direction vector

Find a direction vector for the vector $\mathbf{v} = \langle 2, -3 \rangle$.

Solution   The vector $\mathbf{v}$ has length (magnitude) $\|\mathbf{v}\| = \sqrt{2^2 + (-3)^2} = \sqrt{13}$. Thus, the required direction vector is the unit vector

$$\mathbf{u} = \frac{\mathbf{v}}{\|\mathbf{v}\|} = \frac{\langle 2, -3 \rangle}{\sqrt{13}} = \frac{\sqrt{13}}{13}\langle 2, -3 \rangle = \left\langle \frac{2\sqrt{13}}{13}, \frac{-3\sqrt{13}}{13} \right\rangle$$

Standard Representation of Vectors in the Plane

The unit vectors $\mathbf{i} = \langle 1, 0 \rangle$ and $\mathbf{j} = \langle 0, 1 \rangle$ point in the directions of the positive x- and y-axes, respectively, and are called **standard basis vectors**. Any vector $\mathbf{v} = \langle v_1, v_2 \rangle$ in the plane can be expressed as a linear combination of the vectors $\mathbf{i}$ and $\mathbf{j}$, because

$$\mathbf{v} = \langle v_1, v_2 \rangle = v_1 \langle 1, 0 \rangle + v_2 \langle 0, 1 \rangle = v_1 \mathbf{i} + v_2 \mathbf{j}$$

This is called the **standard representation** of the vector $\mathbf{v}$, and it can be shown that the representation is unique in the sense that if $\mathbf{v} = a\mathbf{i} + b\mathbf{j}$, then $a = v_1$ and $b = v_2$ as shown in Figure 9.11**b**. In this context, the scalars v_1 and v_2 are called the **horizontal** and **vertical components** of $\mathbf{v}$, respectively.

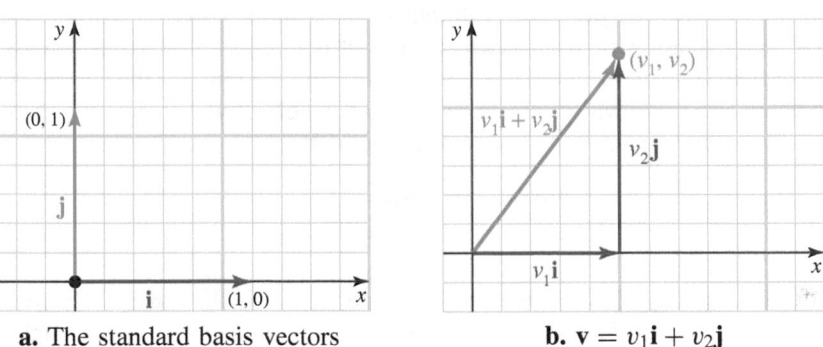

a. The standard basis vectors   **b.** $\mathbf{v} = v_1 \mathbf{i} + v_2 \mathbf{j}$

Figure 9.11   Standard representation of vectors in the plane

Example 6   Finding the standard representation of a vector

If $\mathbf{u} = 3\mathbf{i} + 2\mathbf{j}$, $\mathbf{v} = -2\mathbf{i} + 5\mathbf{j}$, and $\mathbf{w} = \mathbf{i} - 4\mathbf{j}$, what is the standard representation of the vector $2\mathbf{u} + 5\mathbf{v} - \mathbf{w}$?

Solution   Using properties of vectors, we find that

$$\begin{aligned}
2\mathbf{u} + 5\mathbf{v} - \mathbf{w} &= 2(3\mathbf{i} + 2\mathbf{j}) + 5(-2\mathbf{i} + 5\mathbf{j}) - (\mathbf{i} - 4\mathbf{j}) \\
&= [2(3) + 5(-2) - 1]\mathbf{i} + [2(2) + 5(5) - (-4)]\mathbf{j} \\
&= -5\mathbf{i} + 33\mathbf{j}
\end{aligned}$$

Example 7   A vector connecting two points

Find the standard representation of the vector **PQ**, for the points $P(3, -4)$ and $Q(-2, 6)$.

Solution   The component form of **PQ** is

$$\mathbf{PQ} = \langle (-2) - 3, 6 - (-4) \rangle = \langle -5, 10 \rangle$$

This means that **PQ** has the standard representation $\mathbf{PQ} = -5\mathbf{i} + 10\mathbf{j}$.

Example 8   Computing a resultant force

Two forces $\mathbf{F}_1$ and $\mathbf{F}_2$ act on the same body. Suppose it is known that $\mathbf{F}_1$ and $\mathbf{F}_2$ act on the same body. It is known that $\mathbf{F}_1$ has magnitude 3 newtons and acts in the direction of $-\mathbf{i}$, whereas $\mathbf{F}_2$ has magnitude 2 newtons and acts in the direction of the unit vector

$$\mathbf{u} = \frac{3}{5}\mathbf{i} - \frac{4}{5}\mathbf{j}$$

What additional force $\mathbf{F}_3$ must be applied in order to keep the body at rest? (See Figure 9.12.)

Solution   We will calculate $\mathbf{F}_1$ and $\mathbf{F}_2$ and we want to find $\mathbf{F}_3 = a\mathbf{i} + b\mathbf{j}$ so that $\mathbf{F}_1 + \mathbf{F}_2 + \mathbf{F}_3 = 0$. According to the given information, we have

$$\mathbf{F}_1 = 3(-\mathbf{i}) = -3\mathbf{i} \quad \text{and} \quad \mathbf{F}_2 = 2\left(\frac{3}{5}\mathbf{i} - \frac{4}{5}\mathbf{j}\right) = \frac{6}{5}\mathbf{i} - \frac{8}{5}\mathbf{j}$$

and we want to find $\mathbf{F}_3 = a\mathbf{i} + b\mathbf{j}$ so that $\mathbf{F}_1 + \mathbf{F}_2 + \mathbf{F}_3 = \mathbf{0}$. Substituting into this vector we obtain

$$(-3\mathbf{i}) + \left(\frac{6}{5}\mathbf{i} - \frac{8}{5}\mathbf{j}\right) + (a\mathbf{i} + b\mathbf{j}) = 0\mathbf{i} + 0\mathbf{j}$$

By combining terms on the left, we find that

$$\left(-3 + \frac{6}{5} + a\right)\mathbf{i} + \left(-\frac{8}{5} + b\right)\mathbf{j} = 0\mathbf{i} + 0\mathbf{j}$$

Because the standard representation is unique, we must have

$$-3 + \frac{6}{5} + a = 0 \quad \text{and} \quad -\frac{8}{5} + b = 0$$
$$a = \frac{9}{5} \qquad\qquad b = \frac{8}{5}$$

The required force is $\mathbf{F}_3 = \frac{9}{5}\mathbf{i} + \frac{8}{5}\mathbf{j}$. This is a force of magnitude

$$\|\mathbf{F}_3\| = \sqrt{\left(\frac{9}{5}\right)^2 + \left(\frac{8}{5}\right)^2} = \frac{1}{5}\sqrt{145} \text{ newtons}$$

which acts in the direction of the unit vector

$$\mathbf{v} = \frac{\mathbf{F}_3}{\|\mathbf{F}_3\|} = \frac{5}{\sqrt{145}}\left(\frac{9}{5}\mathbf{i} + \frac{8}{5}\mathbf{j}\right) = \frac{9}{\sqrt{145}}\mathbf{i} + \frac{8}{\sqrt{145}}\mathbf{j}$$

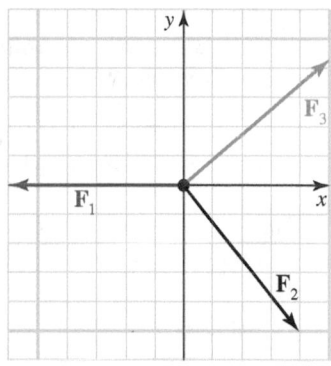

Figure 9.12 Keeping a body at rest

PROBLEM SET 9.1

Level 1

Sketch each vector given in Problems 1-4 assuming that its initial point is at the origin.

1. $3\mathbf{i} - 4\mathbf{j}$

2. $-2\mathbf{i} - 3\mathbf{j}$

3. $-\frac{1}{2}\mathbf{i} + \frac{5}{2}\mathbf{j}$

4. $-2(-\mathbf{i} + 2\mathbf{j})$

The initial point P and terminal point Q of a vector are given in Problems 5-8. Sketch each vector and then write it in component form and find $\|\mathbf{PQ}\|$.

5. $P(3, -1), Q(7, 2)$

6. $P(5, -2), Q(5, 8)$

7. $P(3, 4), Q(-2, 4)$

8. $P\left(\frac{1}{2}, 6\right), Q(-3, -2)$

Find the standard representation and length of each vector $\mathbf{PQ}$ in Problems 9-12.

9. $P(-1, -2), Q(1, -2)$

10. $P(5, 7), Q(6, 8)$

11. $P(-4, -3), Q(0, -1)$

12. $P(3, -5), Q(2, 8)$

Find a unit vector that points in the direction of each of the vectors given in Problems 13-16.

13. $\mathbf{i} + \mathbf{j}$

14. $\frac{1}{2}\mathbf{i} + \frac{1}{4}\mathbf{j}$

15. $3\mathbf{i} - 4\mathbf{j}$

16. $-4\mathbf{i} + 7\mathbf{j}$

Let $\mathbf{u} = \langle -3, 4 \rangle$ and $\mathbf{v} = \langle 1, -1 \rangle$. Find s and t so that $s\mathbf{u} + t\mathbf{v} = \mathbf{w}$ for the given vector in Problems 17-20.

17. $\mathbf{w} = \langle 6, 0 \rangle$

18. $\mathbf{w} = \langle 0, -3 \rangle$

19. $\mathbf{w} = \langle -2, 1 \rangle$

20. $\mathbf{w} = \langle 8, 11 \rangle$

Suppose $\mathbf{u} = 3\mathbf{i} - 4\mathbf{j}$, $\mathbf{v} = 4\mathbf{i} - 3\mathbf{j}$, and $\mathbf{w} = \mathbf{i} + \mathbf{j}$. Express each of the expressions in Problems 21-24 in standard form.

21. $2\mathbf{u} + 3\mathbf{v} - \mathbf{w}$

22. $\frac{1}{2}(\mathbf{u} + \mathbf{v}) - \frac{1}{4}\mathbf{w}$

23. $\|\mathbf{v}\|\mathbf{u} + \|\mathbf{u}\|\mathbf{v}$

24. $\|\mathbf{u}\|\|\mathbf{v}\|\mathbf{w}$

Find all real numbers x and y that satisfy the vector equations given in Problems 25-28.

25. $(x - y - 1)\mathbf{i} + (2x + 3y - 12)\mathbf{j} = \mathbf{0}$

26. $x\mathbf{i} - 4y^2\mathbf{j} = (5 - 3y)\mathbf{i} + (10 - 7x)\mathbf{j}$

27. $(x^2 + y^2)\mathbf{i} + y\mathbf{j} = 20\mathbf{i} + (x + 2)\mathbf{j}$

28. $(y - 1)\mathbf{i} + y\mathbf{j} = (\log x)\mathbf{i} + [\log 2 + \log(x + 4)]\mathbf{j}$

In Problems 29-32, find a unit vector $\mathbf{u}$ with the given characteristics.

29. $\mathbf{u}$ makes an angle of $30°$ with the positive x-axis.

30. $\mathbf{u}$ has the same direction as the vector $2\mathbf{i} - 3\mathbf{j}$.

31. $\mathbf{u}$ has the direction opposite that of $-4\mathbf{i} + \mathbf{j}$.

32. $\mathbf{u}$ has the direction of the vector from $P(-1, 5)$ to $Q(7, -3)$.

In Problems 33-34, let $\mathbf{u} = 4\mathbf{i} - \mathbf{j}$, $\mathbf{v} = \mathbf{i} + 2\mathbf{j}$, and $\mathbf{w} = -3\mathbf{i} + 4\mathbf{j}$.

33. Find a unit vector in the same direction as $\mathbf{u} + \mathbf{v}$.

34. Find a vector of length 3 with the same direction as $\mathbf{u} - 2\mathbf{v} + 2\mathbf{w}$.

35. Find the terminal point of the vector $5\mathbf{i} + 7\mathbf{j}$ if the initial point is $(-2, 3)$.

36. Find the initial point of the vector $-\mathbf{i} + 2\mathbf{j}$ if the terminal point is $(-1, -2)$.

37. a. Find the vector from the point P to the midpoint of the line segment joining the points $P(-3, -8)$ and $Q(9, -2)$.

  b. What is the vector from P to the point which is located $\frac{5}{6}$ of the distance from P to Q?

38. If $\|\mathbf{v}\| = 3$ and $-3 \leq r \leq 1$, what are the possible values of $r\|\mathbf{v}\|$?

39. Show that $\mathbf{v} = (\cos\theta)\mathbf{i} + (\sin\theta)\mathbf{j}$ is a unit vector for any angle θ.

Level 2

40. If $\mathbf{u}$ and $\mathbf{v}$ are nonzero vectors and

$$|r| = \frac{\|\mathbf{u}\|}{\|\mathbf{v}\|}$$

what is $\|r\mathbf{v}\|$?

41. If $\mathbf{u}$ and $\mathbf{v}$ are nonzero vectors with $\|\mathbf{u}\| = \|\mathbf{v}\|$, does it follow that $\mathbf{u} = \mathbf{v}$? Explain.

42. If $\mathbf{u} = 2\mathbf{i} - 3\mathbf{j}$ and $\mathbf{v} = x\mathbf{i} + y\mathbf{j}$, describe the set of points in the plane whose coordinates (x, y) satisfy $\|\mathbf{v} - \mathbf{u}\| \leq 2$.

43. Let $\mathbf{u_0} = x_0\mathbf{i} + y_0\mathbf{j}$ for constants x_0 and y_0, and let $\mathbf{u} = x\mathbf{i} + y\mathbf{j}$. Describe the set of all points in the plane whose coordinates satisfy:

  a. $\|\mathbf{u} - \mathbf{u_0}\| = 1$

  b. $\|\mathbf{u} - \mathbf{u_0}\| \leq r$

44. Let $\mathbf{u} = 3\mathbf{i} - \mathbf{j}$ and $\mathbf{v} = -6\mathbf{i} + 2\mathbf{j}$. Show that there are no numbers a, b for which $a\mathbf{u} + b\mathbf{v} = 2\mathbf{i} + 5\mathbf{j}$.

45. Suppose $\mathbf{u}$ and $\mathbf{v}$ are a pair of nonzero, nonparallel vectors. Find numbers a, b, c such that

$$a\mathbf{u} + b(\mathbf{u} - \mathbf{v}) + c(\mathbf{u} + \mathbf{v}) = \mathbf{0}$$

46. Let $\mathbf{u} = \langle 2, 1 \rangle$ and $\mathbf{v} = \langle -3, 4 \rangle$.

  a. Sketch the vector $c\mathbf{u} + (1 - c))\mathbf{v}$ for the cases where $c = 0$, $c = \frac{1}{4}$, $c = \frac{1}{2}$, $c = \frac{3}{4}$, and $c = 1$.

  b. In general, if the initial point of $c\mathbf{u} + (1 - c)\mathbf{v}$ for $0 \leq c \leq 1$ is at the origin, where is its terminal point (x, y)?

47. Two forces $\mathbf{F}_1 = 3\mathbf{i} + 4\mathbf{j}$ and $\mathbf{F}_2 = 3\mathbf{i} - 7\mathbf{j}$ act on an object. What additional force should be applied to keep the body at rest?

48. Three forces $\mathbf{F}_1 = \mathbf{i} - 2\mathbf{j}$, $\mathbf{F}_2 = 3\mathbf{i} - 7\mathbf{j}$, and $\mathbf{F}_3 = \mathbf{i} + \mathbf{j}$ act on an object. What additional force $\mathbf{F}_4$ should be applied to keep the body at rest?

49. A river 2.1 mi wide flows south with a current of 3.1 mi/h. What speed and heading should a motorboat assume to travel directly across the river from east to west in 30 min?

50. Four forces act on an object: $\mathbf{F}_1$ has magnitude 10 lb and acts at an angle of $\frac{\pi}{6}$ measured counterclockwise from the positive x-axis; $\mathbf{F}_2$ has magnitude 8 lb and acts in the direction of the vector $\mathbf{j}$; $\mathbf{F}_3$ has magnitude 5 lb and acts at an angle of $4\pi/3$ measured counterclockwise from the positive x-axis. What must the fourth force $\mathbf{F}_4$ be to keep the object at rest?

Level 3

51. Use vector methods to show that the diagonals of a parallelogram bisect each other.

52. In a triangle, let $\mathbf{u}, \mathbf{v},$ and $\mathbf{w}$ be the vectors from each vertex to the midpoint of the opposite side. Use vector methods to show that $\mathbf{u} + \mathbf{v} + \mathbf{w} = \mathbf{0}$.

53. Prove that the medians of a triangle intersect at a single point by completing the following argument (see Figure 9.13).

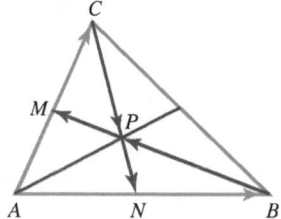

Figure 9.13   Median intersection

a. Let M and N be the midpoints of sides $\overline{AC}$ and $\overline{AB}$, respectively. Show that

$$\mathbf{CN} = \tfrac{1}{2}\mathbf{AB} - \mathbf{AC} \text{ and}$$

$$\mathbf{BM} = \tfrac{1}{2}\mathbf{AC} - \mathbf{AB}$$

b. Let P be the point where medians $\overline{BM}$ and $\overline{CN}$ intersect. Show that there are constants r and s such that

$$\mathbf{CP} = r\left(\tfrac{1}{2}\mathbf{AB} - \mathbf{AC}\right) \text{ and}$$

$$\mathbf{BP} = s\left(\tfrac{1}{2}\mathbf{AC} - \mathbf{AB}\right)$$

Note that $\mathbf{CP} + \mathbf{PB} = \mathbf{CB}$. Use this relationship to prove that $r = s = \frac{2}{3}$. Explain why this shows

that any pair of medians meet at a point located $\frac{2}{3}$ the distance from each vertex to the midpoint of the opposite side. Why does this show that *all three* medians meet at a single point?

c. The *centroid* of a triangle is the point where the medians meet. Show that if a triangle has vertices $A(x_1, y_1)$, $B(x_2, y_2)$, $C(x_3, y_3)$, then the centroid has coordinates

$$\left(\frac{x_1 + x_2 + x_3}{3}, \frac{y_1 + y_2 + y_3}{3} \right)$$

54. Show that the vector with initial point $P(c, d)$ and terminal point $Q(a, b)$ has the component form $\mathbf{PQ} = \langle a - c, b - d \rangle$. See Figure 9.5.

55. Two nonzero vectors $\mathbf{u}$ and $\mathbf{v}$ are said to be *linearly independent* in the plane if they are not parallel.

a. If $\mathbf{u}$ and $\mathbf{v}$ have this property and $a\mathbf{u} = b\mathbf{v}$ for constants a, b, show that $a = b = 0$.

b. Show that the standard representation of a vector is unique. That is, if the vector $\mathbf{u}$ has the representation $\mathbf{u} = a_1\mathbf{i} + b_1\mathbf{j}$ and $\mathbf{u} = a_2\mathbf{i} + b_2\mathbf{j}$ is another such representation, then $a_1 = a_2$ and $b_1 = b_2$.

56. a. Prove the commutativity property for vectors.

    b. Prove the associative property for scalar multiplication.

57. a. Prove the identity property for vectors.

    b. Prove the inverse property of addition for vectors.

58. a. Prove vector distributivity.

    b. Prove scalar distributivity.

59. Let $A, B, C,$ and D be any four points in the plane. If M and N are midpoints of $\overline{AC}$ and $\overline{BD}$, show that

$$\mathbf{MN} = \frac{1}{4}(\mathbf{AB} + \mathbf{AD} + \mathbf{CB} + \mathbf{CD})$$

60. Let A, B, C and D be vertices of a quadrilateral, and let $M, N, P,$ and Q, be the midpoints of the sides $\overline{AB}, \overline{BC}, \overline{CD},$ and $\overline{AD}$, respectively, as shown in Figure 9.14. Use vector methods to show that $MNPQ$ is a parallelogram.

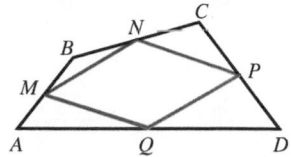

Figure 9.14   Quadrilateral *ABCD*

9.2 COORDINATES AND VECTORS IN $\mathbb{R}^3$

IN THIS SECTION: *Three-dimensional coordinate system, graphs in $\mathbb{R}^3$, vectors in $\mathbb{R}^3$*
Because we live in a three-dimensional world, it is essential that we be able to visualize surfaces in three dimensions. To that end, we consider ordered triples and their representation in three-dimensional space, as well as develop some knowledge of simple three-dimensional surfaces, namely, the quadric surfaces. In the next section, we will consider vectors in three dimensions to describe motion in space efficiently.

Three-Dimensional Coordinate System

We have already considered ordered pairs in $\mathbb{R}^2$, and because we exist in a three-dimensional world, it is also important to consider a three-dimensional system. We call this *three-space* and denote it by $\mathbb{R}^3$. We introduce a coordinate system to three-space by choosing three mutually perpendicular axes to serve as a frame of reference. The orientation of our reference system will be *right-handed* in the sense that if you stand at the origin with your right arm along the positive x-axis and your left arm along the positive y-axis, as shown in Figure 9.15, your head will then point in the direction of the positive z-axis.

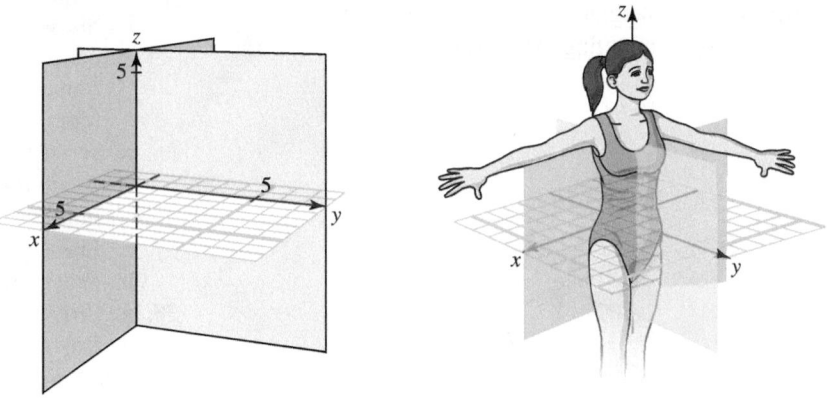

Figure 9.15   A "right-handed" rectangular coordinate system for $\mathbb{R}^3$

To orient yourself to a three-dimensional coordinate system, think of the x-axis and y-axis as lying in the plane of the floor and the z-axis as a line perpendicular to the floor. All the graphs that we have drawn in the first eight chapters of this book would now be drawn on the floor. If you orient yourself in a room as shown in Figure 9.16, you may notice some important planes. Assume that the room is 25 ft × 30 ft × 8 ft and fix the origin at a front corner (where the board hangs).

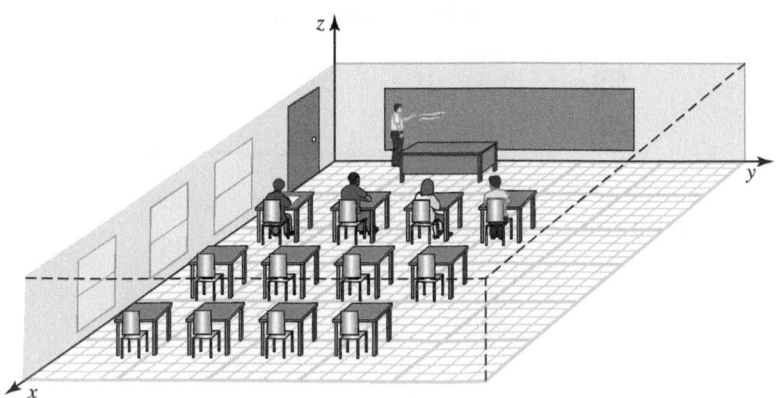

Figure 9.16   A typical classroom

Floor: **xy-plane**; equation is $z = 0$.
Ceiling: Plane parallel to the xy-plane; equation is $z = 8$.
Front wall: **yz-plane**; equation is $x = 0$.
Back wall: Plane parallel to the yz-plane; equation is $x = 30$.
Left wall: **xz-plane**; equation is $y = 0$.
Right wall: Plane parallel to the xz-plane; equation is $y = 25$.

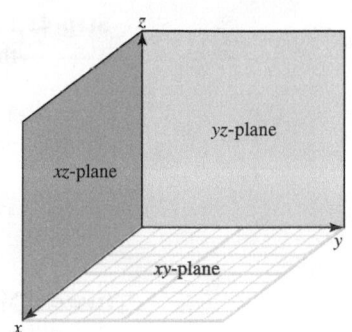

The xy-, xz-, and yz-planes are called the **coordinate planes**. We will obtain equations for planes in space after we discuss vectors in $\mathbb{R}^3$. However, in beginning to visualize objects in $\mathbb{R}^3$, we do not want to ignore planes, because they are so common (for example, the walls, ceiling, and floor in Figure 9.16). We will show in Section 9.5 that the graph of $Ax + By + Cz = D$ is a **plane** if A, B, C, and D are real numbers (not all zero). Points in $\mathbb{R}^3$ are located by their position in relation to the three coordinate planes and are given appropriate coordinates. Specifically, the point P is assigned coordinates (a, b, c) to indicate that it is a, b, and c units, respectively, from the yz-, xz-, and xy-planes.

Example 1   Drawing lesson   Points in three dimensions

Graph the following ordered triples without using technology:

a. $(3, 4, -5)$ **b.** $(10, 20, 10)$ **c.** $(-12, 6, 12)$ **d.** $(-12, -18, 6)$ **e.** $(20, -10, 18)$

Solution

a. Step 1: Sketch the x- and y-axes, adding tick marks. Outline the xy-plane.

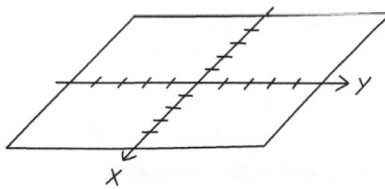

Step 2: Sketch the z-axis, adding tick marks. Use dashed segments for hidden parts.

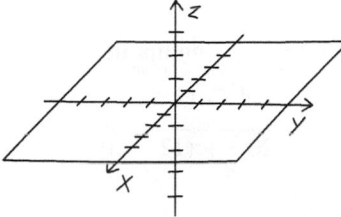

Step 3: Plot x-distance and y-distance on the xy-plane. Darken segments from each along grid lines. Colored pencil or highlighter may help you visualize the figure.

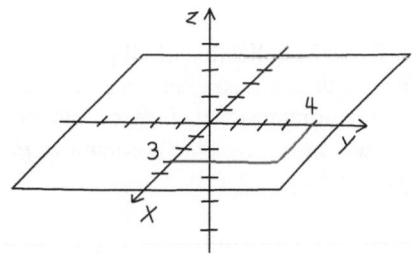

Step 4: Plot z-distance, using the unit size from the z-axis. Lightly sketch a grid on the xy-plane, using tick marks as guides.

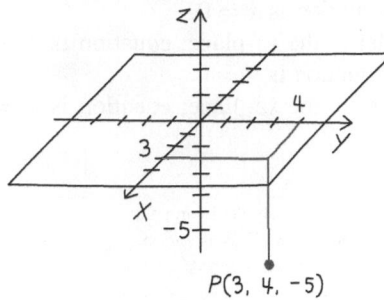

$P(3, 4, -5)$

The other points are plotted similarly as shown.

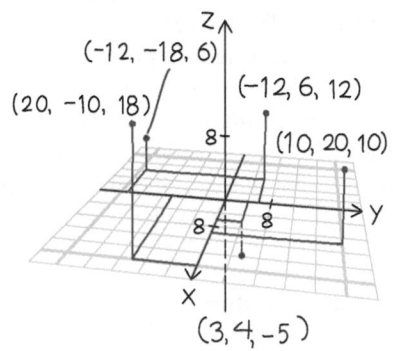

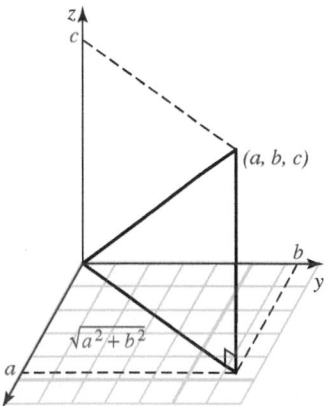

Figure 9.17 Distance from the origin (0,0,0) to the point (a, b, c)

In $\mathbb{R}^2$, the distance from the origin $(0,0)$ to the point (a, b) is $d = \sqrt{a^2 + b^2}$, and in $\mathbb{R}^3$, the distance from $(0, 0, 0)$ to (a, b, c) is $d = \sqrt{a^2 + b^2 + c^2}$, as shown in Figure 9.17. We can now summarize.

DISTANCE FORMULA   The distance $|P_1P_2|$ between $P_1(x_1, y_1, z_1)$ and $P_2(x_2, y_2, z_2)$ is

$$|P_1P_2| = \sqrt{(x_2 - x_1)^2 + (y_2 - y_1)^2 + (z_2 - z_1)^2}$$

Example 2   Finding the distance between points in $\mathbb{R}^3$

Find the distance between $(10, 20, 10)$ and $(-12, 6, 12)$.

Solution $d = \sqrt{(-12 - 10)^2 + (6 - 20)^2 + (12 - 10)^2}$

$\qquad\qquad = \sqrt{684}$

$\qquad\qquad = 6\sqrt{19}$

Graphs in $\mathbb{R}^3$

The **graph of an equation** in $\mathbb{R}^3$ is the collection of all points (x, y, z) whose coordinates satisfy a given equation. This graph is called a **surface**. You are not expected to spend a great deal of time graphing three-dimensional surfaces, but the drawing lessons in this section should help. You may also have access to a computer program to help you look at graphs in three dimensions.

Planes

We shall obtain equations for planes in space after we discuss vectors in space. As we said above, we need to be able to draw planes. The following drawing lesson will help you sketch planes in three dimensions. The task described begins with the xy-plane and point $P(3, 4, -5)$ from Example 1. Then you are asked to draw (on your paper) the planes $x = 2$ and $y = 0$.

Drawing Lesson: Drawing Vertical and Horizontal Planes

Begin with the xy-plane and the z-axis from Example 1, and then draw the line $x = 2$ on the xy-plane. Now, through each endpoint of the line segment you have drawn, draw a segment parallel to the z-axis. Then connect the endpoints. When you are finished, your drawing should look like this:

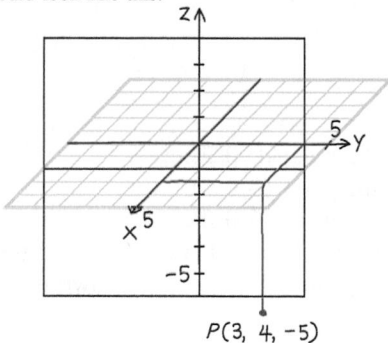

$P(3, 4, -5)$

Shade the plane $x = 2$ where it is not hidden by the xy-plane. Erase hidden parts of both planes, and use your eraser to dash hidden parts of the axis. When you are finished, your drawing should look like this:

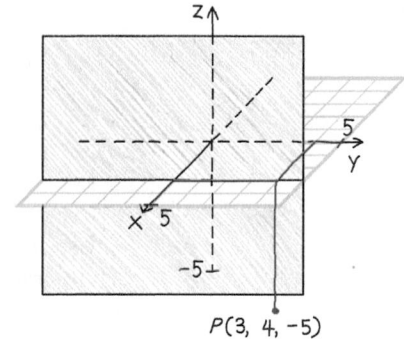

$P(3, 4, -5)$

Follow the same procedure to draw and shade the plane $y = 0$. Draw the intersection of the two planes. Here is what it should look like:

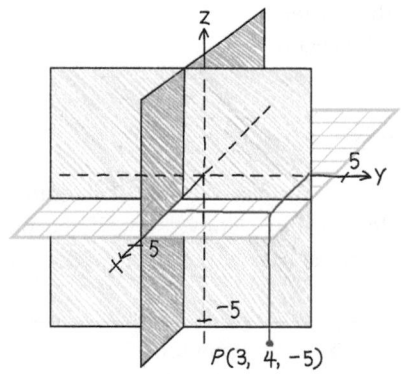

$P(3, 4, -5)$

Use colored pencils or highlighters to distinguish individual planes. Here is an example of a finished drawing:

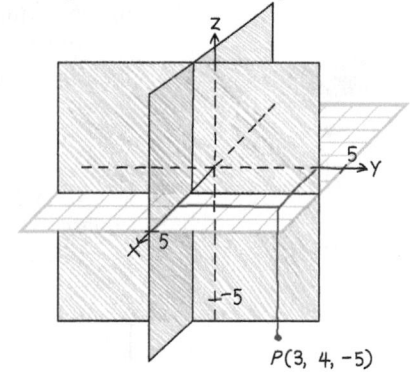

$P(3, 4, -5)$

The following example will also help you sketch planes.

Example 3   Graphing planes

Graph the planes defined by the given equations.

a. $x = 4$ **b.** $y + z = 5$ **c.** $x + 3y + 2z = 6$

Solution   To graph a plane, find some ordered triples satisfying the equation. The best ones to use are often those that fall on a coordinate axis (the intercepts).

a. When two variables are missing, then the plane is parallel to one of the coordinate planes, as shown in Figure 9.18**a**.

b. When one of the variables is missing from an equation of a plane, then that plane is parallel to the axis corresponding the missing variable; in this case it is parallel to the x-axis. Draw the line $y + z = 5$ on the yz-plane, and then complete the plane, as shown in Figure 9.18**b**.

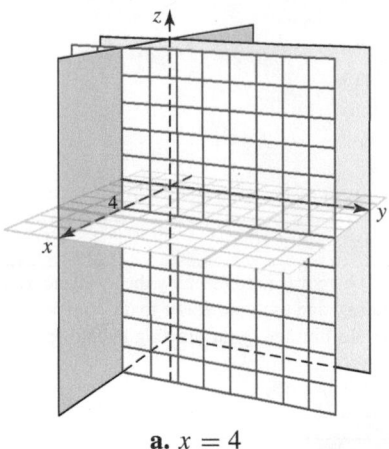

a. $x = 4$

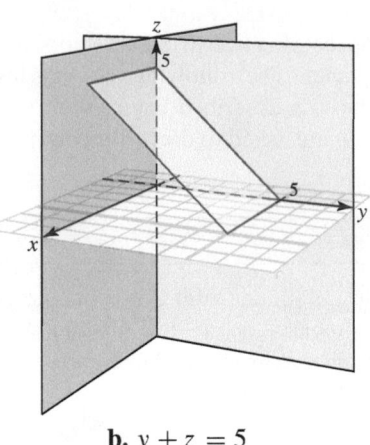

b. $y + z = 5$

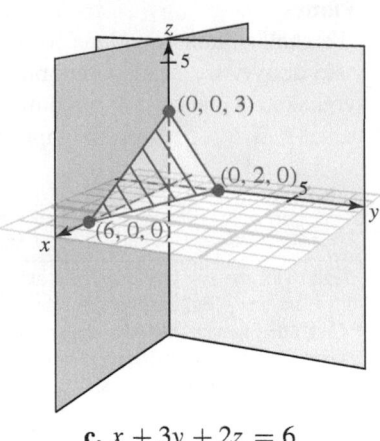

c. $x + 3y + 2z = 6$

Figure 9.18   Graphs of planes

c. Let $x = 0$ and $y = 0$; then $z = 3$; plot the point $(0, 0, 3)$.
Let $x = 0$ and $z = 0$; then $y = 2$; plot the point $(0, 2, 0)$.
Let $y = 0$ and $z = 0$; then $x = 6$; plot the point $(6, 0, 0)$.
Use these points to draw the intersection lines (called **trace lines**) of the plane you
are graphing with each of the coordinate planes. The result is shown in Figure 9.18**c**.

Spheres

A **sphere** (see Figure 9.19) is defined as the collection of all points located a fixed
distance (the **radius**) from a fixed point (the **center**).

In particular, if $P(x, y, z)$ is a point on the sphere with radius r and center $C(a, b, c)$,
then the distance from C to P is r. Thus,

$$r = \sqrt{(x - a)^2 + (y - b)^2 + (z - c)^2}$$

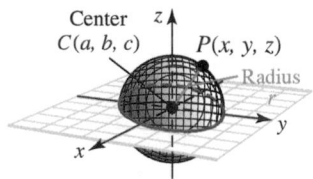

Figure 9.19   Graph of a
sphere with center
(a, b, c) and radius r

If you square both sides of this equation, you can see that it is equivalent to the
equation of a sphere displayed in the following box. Conversely, if the point (x, y, z)
satisfies an equation of this form, it must lie on a sphere with center (a, b, c) and radius r.

EQUATION OF A SPHERE   The graph of the equation

$$(x - a)^2 + (y - b)^2 + (z - c)^2 = r^2$$

is a sphere with center (a, b, c) and radius r, and any sphere has an equation of
this form. This is called the **standard form of the equation of a sphere** (*or simply
standard-form sphere*).

Example 4   Center and radius of a sphere from a given equation

Show that the graph of the equation $x^2 + y^2 + z^2 + 4x - 6y - 3 = 0$ is a sphere, and
find its center and radius.

Solution   By completing the square in both variables x and y, we have

$$(x^2 + 4x) + (y^2 - 6y) + z^2 = 3$$
$$(x^2 + 4x + 2^2) + (y^2 - 6y + (-3)^2) + z^2 = 3 + 4 + 9$$
$$(x + 2)^2 + (y - 3)^2 + z^2 = 16$$

Comparing this equation with the standard form, we see that it is the equation of a sphere with center $(-2, 3, 0)$ and radius 4. The graph is shown in Figure 9.20. Note that you cannot see the bottom half of the sphere because your view is obstructed by the xy-plane.

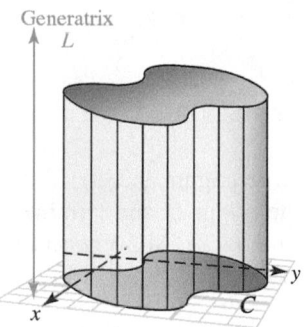

Figure 9.20 Sphere with center $(-2, 3, 0)$ and radius 4

Cylinders

A **cross-section** of a surface in space is a curve obtained by intersecting the surface with a plane. If parallel planes intersect a given surface in congruent cross-sectional curves, the surface is called a **cylinder**. We define a cylinder with **principal cross sections C and generating line L** to be the surface obtained by moving lines parallel to L along the boundary of the curve C, as shown in Figure 9.21. In this context, the curve C is called a **directrix** of the cylinder, and L is the **generatrix**.

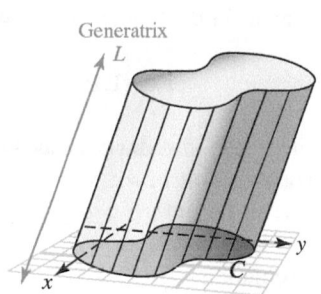

The generatrix is perpendicular to the plane containing C

The line L does not need to be perpendicular to the plane containing the curve C

Figure 9.21 Cylinders

We shall deal primarily with cylinders in which the directrix is a conic section and the generatrix L is one of the coordinate axes. Such a cylinder is often named for the type of conic section in its principal cross-sections and is described by an equation involving only two of the variables x, y, z. In this case, the generating line L is parallel to the coordinate axis of the missing variable. Thus,

$x^2 + y^2 = 5$ is a **circular cylinder** with L parallel to the z-axis.
$y^2 - z^2 = 9$ is a **hyperbolic cylinder** with L parallel to the x-axis.
$x^2 + 2z^2 = 25$ is an **elliptic cylinder** with L parallel to the y-axis.

The graphs of these cylinders are shown in Figure 9.22.

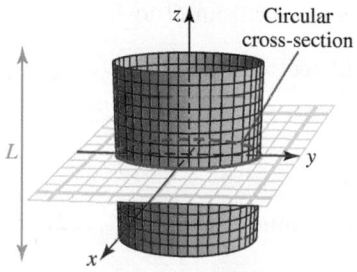

a. Circular cylinder

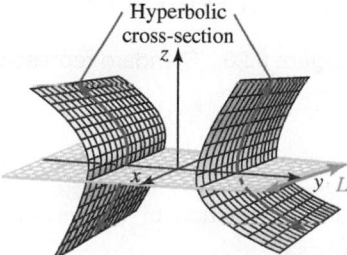
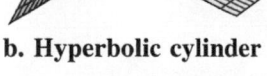

b. Hyperbolic cylinder

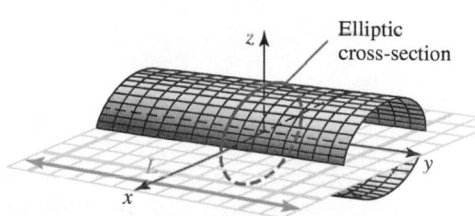

c. Elliptic cylinder

Figure 9.22 Cylinders

Vectors in $\mathbb{R}^3$

A vector in $\mathbb{R}^3$ is a directed line segment (an "arrow") in space. The vector $\mathbf{P_1P_2}$ with initial point $P_1(x_1, y_1, z_1)$ and terminal point $P_2(x_2, y_2, z_2)$ has the component form analogous to the $\mathbb{R}^2$ case

$$\mathbf{P_1P_2} = \langle x_2 - x_1, y_2 - y_1, z_2 - z_1 \rangle$$

Vector addition and multiplication of a vector by a scalar are defined for vectors in $\mathbb{R}^3$ in the same way as these operations were defined for vectors in $\mathbb{R}^2$. In addition, the properties of vector algebra listed in Theorem 9.1 apply to vectors in $\mathbb{R}^3$ as well as to those in $\mathbb{R}^2$.

For example, we observed that each vector in $\mathbb{R}^2$ can be expressed as a unique linear combination of the standard basis vectors $\mathbf{i}$ and $\mathbf{j}$. This representation can be extended to vectors in $\mathbb{R}^3$ by adding a vector $\mathbf{k}$ defined to be the unit vector in the direction of the positive z-axis. In component form, we have in $\mathbb{R}^3$,

$$\mathbf{i} = \langle 1, 0, 0 \rangle \qquad \mathbf{j} = \langle 0, 1, 0 \rangle \qquad \mathbf{k} = \langle 0, 0, 1 \rangle$$

We call these **the standard basis vectors in $\mathbb{R}^3$** (shown in Figure 9.23a). **The standard representation** of the vector with initial point at the origin O and terminal point $Q(a_1, a_2, a_3)$ is $\mathbf{OQ} = a_1\mathbf{i} + a_2\mathbf{j} + a_3\mathbf{k}$, as shown in Figure 9.23b. More generally, the vector $\mathbf{P_1P_2}$ with initial point $P_1(x_1, y_1, z_1)$ and terminal point $P_2(x_2, y_2, z_2)$ has the standard representation

$$\mathbf{P_1P_2} = (x_2 - x_1)\mathbf{i} + (y_2 - y_1)\mathbf{j} + (z_2 - z_1)\mathbf{k}$$

with magnitude

$$\|\mathbf{P_1P_2}\| = \sqrt{(x_2 - x_1)^2 + (y_2 - y_1)^2 + (z_2 - z_1)^2}$$

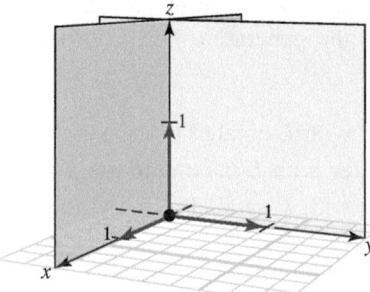 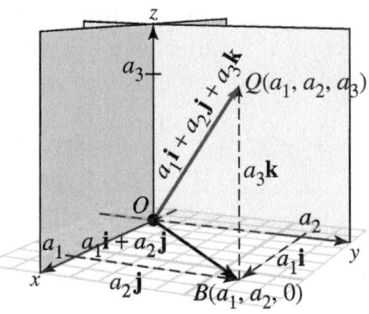

a. The standard basis vectors in $\mathbb{R}^3$ **b.** The vector from O to Q

Figure 9.23 Standard representation of vectors in $\mathbb{R}^3$

Example 5   Standard representation of a vector in $\mathbb{R}^3$

Find the standard representation of the vector $\mathbf{PQ}$ with initial point $P(-1, 2, 2)$ and terminal point $Q(3, -2, 4)$.

Solution   We have

$$\mathbf{PQ} = [3 - (-1)]\mathbf{i} + [-2 - 2]\mathbf{j} + [4 - 2]\mathbf{k} = 4\mathbf{i} - 4\mathbf{j} + 2\mathbf{k}$$

Example 6   Magnitude of a vector and distance between points in $\mathbb{R}^3$

Find the magnitude of the vector $\mathbf{v} = 2\mathbf{i} - 3\mathbf{j} + 5\mathbf{k}$ and the distance between the points $A(1, -1, -4)$ and $B(-2, 3, 8)$.

Solution $\|\mathbf{v}\| = \sqrt{2^2 + (-3)^2 + 5^2} = \sqrt{38}$ and

$$\|\overline{AB}\| = \sqrt{(-2 - 1)^2 + [3 - (-1)]^2 + [8 - (-4)^2]}$$
$$= 13$$

As in $\mathbb{R}^2$, if $\mathbf{v}$ is a given nonzero vector in $\mathbb{R}^3$, then the unit vector $\mathbf{u}$ that points in the same direction as $\mathbf{v}$ is

$$\mathbf{u} = \frac{\mathbf{v}}{\|\mathbf{v}\|}$$

Example 7   Finding a direction vector

Find the unit vector that points in the direction of the vector $\mathbf{PQ}$ from $P(-1, 2, 5)$ to $Q(0, -3, 7)$.

Solution $\mathbf{PQ} = [0 - (-1)]\,\mathbf{i} + [-3 - 2]\mathbf{j} + [7 - 5]\,\mathbf{k} = \mathbf{i} - 5\mathbf{j} + 2\mathbf{k}$

$$\|\mathbf{PQ}\| = \sqrt{1^2 + (-5)^2 + 2^2} = \sqrt{30}$$

Thus,

$$\mathbf{u} = \frac{\mathbf{PQ}}{\|\mathbf{PQ}\|}$$
$$= \frac{\mathbf{i} - 5\mathbf{j} + 2\mathbf{k}}{\sqrt{30}}$$
$$= \frac{1}{\sqrt{30}}\mathbf{i} - \frac{5}{\sqrt{30}}\mathbf{j} + \frac{2}{\sqrt{30}}\,\mathbf{k}$$

As in $\mathbb{R}^2$, two vectors in $\mathbb{R}^3$ are **parallel** if they are scalar multiples of one another. Similarly, nonzero vectors $\mathbf{u}$ and $\mathbf{v}$ are parallel if and only if $\mathbf{u} = s\mathbf{v}$ for some nonzero scalar s.

Example 8   Parallel vectors

A vector $\mathbf{PQ}$ has initial point $P(1, 0, -3)$ and length 3. Find Q so that $\mathbf{PQ}$ is parallel to $\mathbf{v} = 2\mathbf{i} - 3\mathbf{j} + 6\mathbf{k}$.

Solution   Let Q have coordinates (a_1, a_2, a_3). Then

$$\mathbf{PQ} = [a_1 - 1]\mathbf{i} + [a_2 - 0]\mathbf{j} + [a_3 - (-3)]\mathbf{k}$$
$$= (a_1 - 1)\mathbf{i} + a_2\mathbf{j} + (a_3 + 3)\mathbf{k}$$

Because $\mathbf{PQ}$ is parallel to $\mathbf{v}$, we have $\mathbf{PQ} = s\mathbf{v}$ for some scalar s, that is,

$$(a_1 - 1)\mathbf{i} + a_2\mathbf{j} + (a_3 + 3)\mathbf{k} = s(2\mathbf{i} - 3\mathbf{j} + 6\mathbf{k})$$

Since the standard representation is unique, this implies that

$$a_1 - 1 = 2s \qquad a_2 = -3s \quad a_3 + 3 = 6s$$
$$a_1 = 2s + 1 \qquad\qquad\qquad a_3 = 6s - 3$$

Because **PQ** has length 3, we have

$$
\begin{aligned}
3 &= \sqrt{(a_1 - 1)^2 + a_2^2 + (a_3 + 3)^2} \\
&= \sqrt{[(2s + 1) - 1]^2 + (-3s)^2 + [(6s - 3) + 3]^2} \\
&= \sqrt{4s^2 + 9s^2 + 36s^2} \\
&= \sqrt{49s^2} \\
&= 7\,|s|
\end{aligned}
$$

Thus, $s = \pm \frac{3}{7}$ and for the two cases, we have

$$s = \frac{3}{7}: \quad a_1 = 2\left(\frac{3}{7}\right) + 1 = \frac{13}{7} \quad a_2 = -3\left(\frac{3}{7}\right) = -\frac{9}{7} \quad a_3 = 6\left(\frac{3}{7}\right) - 3 = -\frac{3}{7}$$

$$s = -\frac{3}{7}: \quad a_1 = 2\left(-\frac{3}{7}\right) + 1 = \frac{1}{7} \quad a_2 = -3\left(-\frac{3}{7}\right) = \frac{9}{7} \quad a_3 = 6\left(-\frac{3}{7}\right) - 3 = -\frac{39}{7}$$

There are two points that satisfy the conditions for the required terminal point Q:
$\left(\frac{13}{7}, -\frac{9}{7}, -\frac{3}{7}\right)$ and $\left(\frac{1}{7}, \frac{9}{7}, -\frac{39}{7}\right)$.   ■

PROBLEM SET 9.2

Level 1

For the given vectors in Problems 1-4, find

a. u + v   **b. u − v**   **c.** $\frac{5}{2}$**u**   **d. 2u + 3v**

1. $\mathbf{u} = \langle 4, -3, 1 \rangle, \mathbf{v} = \langle -2, 5, 3 \rangle$
2. $\mathbf{u} = \langle 2, -1, 0 \rangle, \mathbf{v} = \langle 5, -3, 4 \rangle$
3. $\mathbf{u} = \langle 1, -2, 5 \rangle, \mathbf{v} = \langle 0, -1, 3 \rangle$
4. $\mathbf{u} = \langle 0, 3, -1 \rangle, \mathbf{v} = \langle 4, -2, 0 \rangle$

In Problems 5-8 plot the points P and Q in $\mathbb{R}^3$, and find the distance $\|\mathbf{PQ}\|$.

5. $P(3, -4, 5), Q(1, 5, -3)$
6. $P(3, 0, 0), Q(-2, 5, 7)$
7. $P(-3, -5, 8), Q(3, 6, -7)$
8. $P(0, 5, -3), Q(2, -1, 0)$

In Problems 9-12 find the standard form equation of the sphere with the given center C and radius r.

9. $C(0, 0, 0), r = 1$ 10. $C(-3, 5, 7), r = 2$
11. $C(0, 4, -5), r = 3$ 12. $C(-2, 3, -1), r = \sqrt{5}$

Find the center and radius of each sphere whose equations are given in Problems 13-16.

13. $x^2 + y^2 + z^2 - 2y + 2z - 2 = 0$
14. $x^2 + y^2 + z^2 + 4x - 2z - 8 - 0$
15. $x^2 + y^2 + z^2 - 6x + 2y - 2z + 10 = 0$
16. $x^2 + y^2 + z^2 - 2x - 4y + 8z + 17 = 0$

Follow the steps shown in the drawing lesson to plot the planes in Problems 17-20.

17. Sketch the planes $x = 0, y = 0,$ and $z = 0$.
18. Sketch the planes $x = 4, y = 0,$ and $z = 0$.
19. Sketch the planes $x = 0, y = 2,$ and $z = 0$.
20. Sketch the planes $x = 0, y = 3,$ and $z = 0$.

*Find the standard representation of the vector **PQ**, and then find $\|\mathbf{PQ}\|$ in Problems 21-24.*

21. $P(1, -1, 3), Q(-1, 1, 4)$
22. $P(0, 2, 3), Q(2, 3, 0)$
23. $P(1, 1, 1), Q(-3, -3, -3)$
24. $P(3, 0, -4), Q(0, -4, 3)$

In Problems 25-28, perform the indicated operations.

$\mathbf{u} = 2\mathbf{i} - \mathbf{j} + 3\mathbf{k}, \ \mathbf{v} = \mathbf{i} + \mathbf{j} - 5\mathbf{k}, \ \mathbf{w} = 5\mathbf{i} + 7\mathbf{k}$

25. $\mathbf{u} + \mathbf{v} - 2\mathbf{w}$ 26. $2\mathbf{u} - \mathbf{v} + 3\mathbf{w}$
27. $4\mathbf{u} + \mathbf{w}$ 28. $3\mathbf{u} - \mathbf{v} - 5\mathbf{w}$

*In Problems 29-32, find a unit vector that points in the same direction as the given vector **v**.*

29. $\mathbf{v} = \langle 3, -2, 1 \rangle$ 30. $\mathbf{v} = \langle -1, 1, \sqrt{2} \rangle$
31. $\mathbf{v} = \langle -5, 3, 4 \rangle$ 32. $\mathbf{v} = \langle 1, \sqrt{2}, 7 \rangle$

Sketch the cylindrical surfaces given in Problems 33-36.

33. $x^2 + y^2 = 4$ 34. $z = 9 - x^2$
35. $y = 3z^2$ 36. $z = \sin x$

37. Find an equation for a sphere, given that the endpoints of a diameter of the sphere are $(1, 2, -3)$ and $(-2, 3, 3)$.

38. Find an equation for a sphere with center $C(2, -1, 3)$ and radius $r = 4$.

Evaluate the expressions given in Problems 39-42.

39. $\|\mathbf{i} - \mathbf{j} + \mathbf{k}\|$

40. $\|\mathbf{i} + \mathbf{j} + \mathbf{k}\|$

41. $\|2(\mathbf{i} - \mathbf{j} + \mathbf{k}) - 3(2\mathbf{i} + \mathbf{j} - \mathbf{k})\|^2$

42. $\|2\mathbf{i} + \mathbf{j} - 3\mathbf{k}\|^2$

Let $\mathbf{v} = \mathbf{i} - 2\mathbf{j} + 2\mathbf{k}$ and $\mathbf{w} = 2\mathbf{i} + 4\mathbf{j} - \mathbf{k}$; find the vector or scalar requested in Problems 43-46.

43. $\|\mathbf{v}\|\mathbf{w}$

44. $2\|\mathbf{v}\| - 3\|\mathbf{w}\|$

45. $\|\mathbf{v} - \mathbf{w}\|(\mathbf{v} + \mathbf{w})$

46. $\|2\mathbf{v} - 3\mathbf{w}\|$

Determine whether each vector in Problems 47-50 is parallel to $\mathbf{u} = 2\mathbf{i} - 3\mathbf{j} + 5\mathbf{k}$.

47. $\mathbf{v} = \langle -2, 6, -10 \rangle$

48. $\mathbf{v} = \langle 4, -6, 10 \rangle$

49. $\mathbf{v} = \langle -1, \frac{3}{2}, -\frac{5}{2} \rangle$

50. $\mathbf{v} = \langle 1, -\frac{3}{2}, 2 \rangle$

The vertices $A, B,$ and C of a triangle in $\mathbb{R}^3$ are given in Problems 51-54. Find the lengths of the sides of the triangle and determine whether it is a right triangle, an isosceles triangle, both, or neither.

51. $A(1, 1, 1), B(3, 3, 2), C(3, -3, 5)$

52. $A(3, -1, 0), B(7, 1, 4), C(1, 3, 4)$

53. $A(2, 4, 3), B(-3, 2, -4), C(-6, 8, -10)$

54. $A(1, 2, 3), B(-3, 2, 4), C(1, -4, 3)$

In Problems 55 and 56, determine whether the given points are collinear (that is, lie on the same line). Note: for three points $A, B,$ and C to be collinear, $\mathbf{AC}$ must be a multiple of $\mathbf{AB}$.

55. $(2, 3, 2), (-1, 4, 0), (-4, 5, -2)$

56. $(3, 0, 3), (2, -3, 5), (4, 3, 1)$

57. A 150-lb professional television camera is mounted on a three-legged tripod. Suppose a coordinate system is chosen with the camera at point $P(0, 0, 6)$ and the feet of the tripod at $A(0, -2, 0)$, $B(-\sqrt{3}, 1, 0)$, and $C(\sqrt{3}, 1, 0)$, as shown in Figure 9.24. Find the force exerted on each of the three legs.

© 2013 Arogant. Used under license from Shutterstock, Inc.

Figure 9.24   T.V. camera

58. Let $\mathbf{u} = \langle -1, 1, 2 \rangle$, $\mathbf{v} = \langle 0, 2, -3 \rangle$, and $\mathbf{w} = \langle 5, -1, 0 \rangle$; find a vector $\mathbf{q}$ so that $2\mathbf{u} - \mathbf{v} + 5\mathbf{q} = 3\mathbf{w}$.

59. Find the point P that lies $\frac{2}{3}$ of the distance from the point $A(-1, 3, 9)$ to the midpoint of the line segment joining points $B(-2, 3, 7)$ and $C(4, 1, -3)$.

60. Let $P(3, 2, -1)$, $Q(-2, 1, c)$, and $R(c, 1, 0)$ be points in $\mathbb{R}^3$. For what values of c (if any) is PQR a right triangle?

9.3 THE DOT PRODUCT

IN THIS SECTION: *Definition of dot product, angle between two vectors, direction cosines, projections, work as a dot product*

In this section, we introduce a product of vectors that yields a scalar and use it to find the angle between three-dimensional vectors.

Definition of Dot Product

In the first two sections of this chapter, you have learned how to add vectors and multiply a vector by a scalar. We now turn our attention to the product of two vectors. The *dot (scalar) product* and the *cross (vector) product* are two important vector operations. We shall examine the cross product in the next section. The dot product is also known as a scalar product because it is a product of vectors that gives a scalar (that is, real number) as a result. Sometimes the dot product is called the **inner product**.

> **DOT PRODUCT** The **dot product** of vectors $\mathbf{v} = a_1\mathbf{i} + a_2\mathbf{j} + a_3\mathbf{k}$ and $\mathbf{w} = b_1\mathbf{i} + b_2\mathbf{j} + b_3\mathbf{k}$ is the scalar denoted by $\mathbf{v} \cdot \mathbf{w}$ and given by
>
> $$\mathbf{v} \cdot \mathbf{w} = a_1 b_1 + a_2 b_2 + a_3 b_3$$

The dot product of two vectors $\mathbf{v} = a_1\mathbf{i} + a_2\mathbf{j}$ and $\mathbf{w} = b_1\mathbf{i} + b_2\mathbf{j}$ in a plane is given by a similar formula with $a_3 = b_3 = 0$: $\mathbf{v} \cdot \mathbf{w} = a_1 b_1 + a_2 b_2$.

Example 1   Dot product

Find the dot product of $\mathbf{v} = -3\mathbf{i} + 2\mathbf{j} + \mathbf{k}$ and $\mathbf{w} = 4\mathbf{i} - \mathbf{j} + 2\mathbf{k}$.

Solution $\mathbf{v} \cdot \mathbf{w} = -3(4) + 2(-1) + 1(2) = -12$

Example 2   Dot product in component form

If $\mathbf{v} = \langle 4, -1, 3 \rangle$ and $\mathbf{w} = \langle -1, -2, 5 \rangle$ find the dot product, $\mathbf{v} \cdot \mathbf{w}$.

Solution $\mathbf{v} \cdot \mathbf{w} = 4(-1) + (-1)(-2) + 3(5) = 13$

Before we can apply the dot product to geometric and physical problems, we need to know how it behaves algebraically. A number of important general properties of the dot product are listed in the following theorem.

Theorem 9.2   Properties of dot product

If $\mathbf{u}$, $\mathbf{v}$, and $\mathbf{w}$ are vectors in $\mathbb{R}^2$ or $\mathbb{R}^3$ and c is a scalar, then:

Magnitude of a vector	$\mathbf{v} \cdot \mathbf{v} = \|\mathbf{v}\|^2$
Zero product	$\mathbf{0} \cdot \mathbf{v} = 0$
Commutativity	$\mathbf{v} \cdot \mathbf{w} = \mathbf{w} \cdot \mathbf{v}$
Scalar multiple	$c(\mathbf{v} \cdot \mathbf{w}) = (c\mathbf{v}) \cdot \mathbf{w} = \mathbf{v} \cdot (c\mathbf{w})$
Distributivity	$\mathbf{u} \cdot (\mathbf{v} + \mathbf{w}) = \mathbf{u} \cdot \mathbf{v} + \mathbf{u} \cdot \mathbf{w}$

Proof: Let $\mathbf{u} = a_1\mathbf{i} + a_2\mathbf{j} + a_3\mathbf{k}$, $\mathbf{v} = b_1\mathbf{i} + b_2\mathbf{j} + b_3\mathbf{k}$, and $\mathbf{w} = c_1\mathbf{i} + c_2\mathbf{j} + c_3\mathbf{k}$. We begin with the magnitude of a vector property.

$$\begin{aligned}
\|\mathbf{v}\|^2 &= \left(\sqrt{b_1^2 + b_2^2 + b_3^2} \right)^2 \\
&= b_1^2 + b_2^2 + b_3^2 \\
&= \mathbf{v} \cdot \mathbf{v}
\end{aligned}$$

Zero product, commutativity, and **scalar multiple** properties can be established in a similar fashion. For the distributive property, we have:

$$\begin{aligned}
\mathbf{u} \cdot (\mathbf{v} + \mathbf{w}) &= (a_1\mathbf{i} + a_2\mathbf{j} + a_3\mathbf{k}) \cdot [(b_1 + c_1)\mathbf{i} + (b_2 + c_2)\mathbf{j} + (b_3 + c_3)\mathbf{k}] \\
&= a_1(b_1 + c_1) + a_2(b_2 + c_2) + a_3(b_3 + c_3) \\
&= a_1 b_1 + a_1 c_1 + a_2 b_2 + a_2 c_2 + a_3 b_3 + a_3 c_3
\end{aligned}$$

Also,

$$\mathbf{u} \cdot \mathbf{v} + \mathbf{u} \cdot \mathbf{w} = (a_1 b_1 + a_2 b_2 + a_3 b_3) + (a_1 c_1 + a_2 c_2 + a_3 c_3)$$
$$= a_1 b_1 + a_1 c_1 + a_2 b_2 + a_2 c_2 + a_3 b_3 + a_3 c_3$$

Since these are the same, we see

$$\mathbf{u} \cdot (\mathbf{v} + \mathbf{w}) = \mathbf{u} \cdot \mathbf{v} + \mathbf{u} \cdot \mathbf{w}$$ ◆

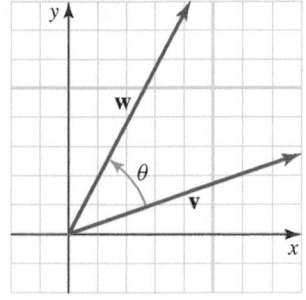

Angle Between Two Vectors

The angle between two nonzero vectors $\mathbf{v}$ and $\mathbf{w}$ is defined to be the angle θ with $0 \le \theta \le \pi$ that is formed when the vectors are in standard position (initial points at the origin), as shown in Figure 9.25.

The angle between two vectors plays an important role in certain applications and may be computed by using the following formula involving the dot product.

Figure 9.25 The angle between two vectors

Theorem 9.3   Angle between two vectors

If θ is the angle between the nonzero vectors $\mathbf{v}$ and $\mathbf{w}$, then

$$\cos \theta = \frac{\mathbf{v} \cdot \mathbf{w}}{\|\mathbf{v}\| \ \|\mathbf{w}\|}$$

Proof: We can derive this formula as follows. Suppose $\mathbf{v} = a_1 \mathbf{i} + a_2 \mathbf{j} + a_3 \mathbf{k}$ and $\mathbf{w} = b_1 \mathbf{i} + b_2 \mathbf{j} + b_3 \mathbf{k}$, and consider $\triangle AOB$ with vertices at the origin O and the points $A(a_1, a_2, a_3)$ and $B(b_1, b_2, b_3)$, as shown in Figure 9.26.

Note that sides $\overline{OA}$ and $\overline{OB}$ have lengths $\|\mathbf{v}\| = \sqrt{a_1^2 + a_2^2 + a_3^2}$ and $\|\mathbf{w}\| = \sqrt{b_1^2 + b_2^2 + b_3^2}$, respectively, and that side $\overline{AB}$ has length $\|\mathbf{v} - \mathbf{w}\| = \sqrt{(a_1 - b_1)^2 + (a_2 - b_2)^2 + (a_3 - b_3)^2}$. Next, we use the law of cosines.

$$a^2 = b^2 + c^2 - 2bc \cos \theta \qquad \textit{Law of cosines}$$

$$\|\mathbf{v} - \mathbf{w}\|^2 = \|\mathbf{v}\|^2 + \|\mathbf{w}\|^2 - 2\|\mathbf{v}\| \|\mathbf{w}\| \cos \theta \qquad \textit{See Figure 9.26.}$$

$$\cos \theta = \frac{\|\mathbf{v}\|^2 + \|\mathbf{w}\|^2 - \|\mathbf{v} - \mathbf{w}\|^2}{2\|\mathbf{v}\| \|\mathbf{w}\|} \qquad \textit{Solve for } \cos \theta.$$

$$= \frac{a_1^2 + a_2^2 + a_3^2 + b_1^2 + b_2^2 + b_3^2 - [(a_1 - b_1)^2 + (a_2 - b_2)^2 + (a_3 - b_3)^2]}{2\|\mathbf{v}\| \|\mathbf{w}\|}$$

$$= \frac{2a_1 b_1 + 2a_2 b_2 + 2a_3 b_3}{2\|\mathbf{v}\| \|\mathbf{w}\|}$$

$$= \frac{\mathbf{v} \cdot \mathbf{w}}{\|\mathbf{v}\| \|\mathbf{w}\|}$$ ◆

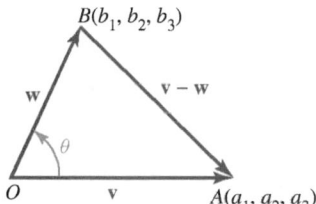

Figure 9.26 Angle between two vectors

Example 3   Angle between two given vectors

Let $\triangle ABC$ be the triangle with vertices $A(1, 1, 8), B(4, -3, -4)$, and $C(-3, 1, 5)$. Find the angle formed at A.

Solution   Draw $\triangle ABC$ and label the angle formed at A as α as shown in Figure 9.27. The angle α is the angle between vectors $\mathbf{AB}$ and $\mathbf{AC}$, where

$$\mathbf{AB} = (4 - 1)\mathbf{i} + (-3 - 1)\mathbf{j} + (-4 - 8)\mathbf{k} \qquad \mathbf{AC} = (-3 - 1)\mathbf{i} + (1 - 1)\mathbf{j} + (5 - 8)\mathbf{k}$$
$$= 3\mathbf{i} - 4\mathbf{j} - 12\mathbf{k} \qquad\qquad\qquad = -4\mathbf{i} - 3\mathbf{k}$$

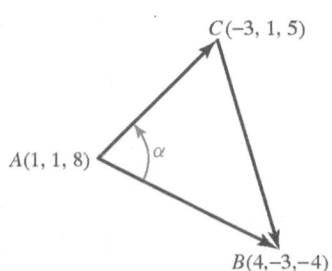

Figure 9.27 Angle in a triangle

Thus,

$$
\begin{aligned}
\cos \alpha &= \frac{\mathbf{AB} \cdot \mathbf{AC}}{\|\mathbf{AB}\| \, \|\mathbf{AC}\|} \\
&= \frac{3(-4) + (-4)(0) + (-12)(-3)}{\sqrt{3^2 + (-4)^2 + (-12)^2} \, \sqrt{(-4)^2 + (-3)^2}} \\
&= \frac{24}{\sqrt{169} \, \sqrt{25}} \\
&= \frac{24}{65}
\end{aligned}
$$

and the required angle is $\alpha = \cos^{-1}\left(\frac{24}{65}\right) \approx 1.19$ (or about 68°). ■

The formula for the angle between two vectors is often used in conjunction with the dot product. If we multiply both sides of the formula by $\|\mathbf{v}\| \, \|\mathbf{w}\|$, we obtain the following alternate form for the dot product formula.

GEOMETRICAL FORMULA FOR THE DOT PRODUCT

$$\mathbf{v} \cdot \mathbf{w} = \|\mathbf{v}\| \, \|\mathbf{w}\| \cos \theta$$

where θ is the angle ($0 \le \theta \le \pi$) between the vectors $\mathbf{v}$ and $\mathbf{w}$.

Two vectors are said to be **perpendicular**, or **orthogonal**, if the angle between them is $\theta = \frac{\pi}{2}$. The following theorem provides a useful condition for orthogonality.

Theorem 9.4   The orthogonal vector theorem

Nonzero vectors $\mathbf{v}$ and $\mathbf{w}$ are orthogonal if and only if

$$\mathbf{v} \cdot \mathbf{w} = 0$$

■ **W**hat this says The orthogonal vector theorem allows you to show that two vectors are orthogonal by finding the dot product of the vectors and showing this dot product is 0. On the other hand, if you know the vectors are orthogonal, then it follows that the dot product of the vectors is 0. In particular, $\mathbf{i} \cdot \mathbf{j} = \mathbf{j} \cdot \mathbf{k} = \mathbf{k} \cdot \mathbf{i} = 0$.

Proof: If the vectors are orthogonal, then the angle between them is $\frac{\pi}{2}$; therefore,

$$\mathbf{v} \cdot \mathbf{w} = \|\mathbf{v}\| \, \|\mathbf{w}\| \cos \frac{\pi}{2} = 0$$

Conversely, if $\mathbf{v} \cdot \mathbf{w} = 0$ and $\mathbf{v}$ and $\mathbf{w}$ are nonzero vectors, then $\cos \theta = 0$, so that $\theta = \frac{\pi}{2}$ and the vectors are orthogonal. ◆

Example 4   Orthogonal vectors

Determine which (if any) of the following vectors are orthogonal:

$$\mathbf{u} = 3\mathbf{i} + 7\mathbf{j} - 2\mathbf{k} \qquad \mathbf{v} = 5\mathbf{i} - 3\mathbf{j} - 3\mathbf{k} \qquad \mathbf{w} = \mathbf{j} - \mathbf{k}$$

Solution $\mathbf{u} \cdot \mathbf{v} = 3(5) + 7(-3) + (-2)(-3) = 0$; orthogonal vectors.
$\mathbf{u} \cdot \mathbf{w} = 3(0) + 7(1) + (-2)(-1) = 9$; not orthogonal vectors.
$\mathbf{v} \cdot \mathbf{w} = 5(0) + (-3)(1) + (-3)(-1) = 0$; orthogonal vectors.   ∎

Direction Cosines

The direction of a vector in $\mathbb{R}^2$ positioned with its initial point at the origin may be specified by determining its angle of inclination; that is, the angle ϕ is the smallest positive angle the vector makes with the positive x-axis, as shown in Figure 9.28a. In $\mathbb{R}^3$, it is common practice to measure the direction of a nonzero vector $\mathbf{v}$ in terms of angles α, β, and γ between $\mathbf{v}$ and the positive coordinate axes as shown in Figure 9.28b. These angles are called the **direction angles** of $\mathbf{v}$, and their cosines are the **direction cosines** of $\mathbf{v}$.

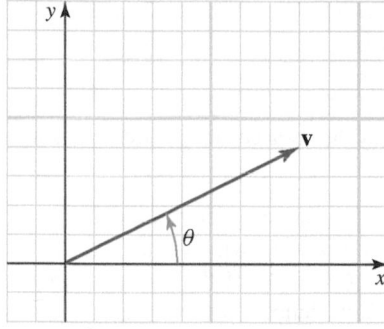

a. In $\mathbb{R}^2$, the direction of $\mathbf{v}$ is given by the angle of inclination θ

b. Interactive In $\mathbb{R}^3$, the direction of $\mathbf{v}$ is given by the angles α, β, and γ

Figure 9.28   The direction angles of a vector $\mathbf{v}$

The direction cosines of $\mathbf{v} = \langle v_1, v_2, v_3 \rangle$ can be computed by taking the dot product of $\mathbf{v}$ with the three standard basis vectors, $\mathbf{i} = \langle 1, 0, 0 \rangle$, $\mathbf{j} = \langle 0, 1, 0 \rangle$, and $\mathbf{k} = \langle 0, 0, 1 \rangle$. For example, to find $\cos \alpha$, note that

$$v_1 = \mathbf{v} \cdot \mathbf{i} = \|\mathbf{v}\| \, \|\mathbf{i}\| \cos \alpha, \text{ so that } \cos \alpha = \frac{v_1}{\|\mathbf{v}\|}$$

and, similarly,

$$\cos \beta = \frac{v_2}{\|\mathbf{v}\|} \qquad \text{and} \qquad \cos \gamma = \frac{v_3}{\|\mathbf{v}\|}$$

It can be shown (Problem 59) that

$$\cos^2 \alpha + \cos^2 \beta + \cos^2 \gamma = 1$$

This identity can be used to check whether you have computed the direction cosines correctly.

Example 5   Direction cosines and direction angles

Find the direction cosines and direction angles (to the nearest degree) of the vector $\mathbf{v} = -2\mathbf{i} + 3\mathbf{j} + 5\mathbf{k}$, and verify the formula

$$\cos^2 \alpha + \cos^2 \beta + \cos^2 \gamma = 1$$

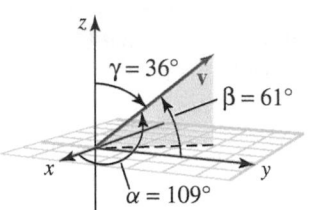

Figure 9.29 The vector **v** with direction angles α, β, and γ

Solution   We find $\|\mathbf{v}\| = \sqrt{(-2)^2 + 3^2 + 5^2} = \sqrt{38}$. Therefore,

$$\cos \alpha = \frac{v_1}{\|\mathbf{v}\|} = \frac{-2}{\sqrt{38}} \approx -0.324428 \quad \alpha \approx \cos^{-1}(-0.324428) \approx 109°$$

$$\cos \beta = \frac{v_2}{\|\mathbf{v}\|} = \frac{3}{\sqrt{38}} \approx 0.4866642 \quad \beta \approx \cos^{-1}(0.4866642) \approx 61°$$

$$\cos \gamma = \frac{v_3}{\|\mathbf{v}\|} = \frac{5}{\sqrt{38}} \approx 0.8111071 \quad \gamma \approx \cos^{-1}(0.8111071) \approx 36°$$

and $\cos^2 \alpha + \cos^2 \beta + \cos^2 \gamma = \left(\frac{-2}{\sqrt{38}}\right)^2 + \left(\frac{3}{\sqrt{38}}\right)^2 + \left(\frac{5}{\sqrt{38}}\right)^2 = 1$. These direction angles are shown in Figure 9.29.   ■

Projections

Let **v** and **w** be two vectors in $\mathbb{R}^2$ drawn so that they have a common initial point, as shown in Figure 9.30.*

If we drop a perpendicular from the head of **v** to the line determined by **w**, we determine a vector called the **vector projection of v onto w**, which we have labeled **u** in Figure 9.30. The **scalar projection of v onto w** (also called the **component of v along w**) is the length of the vector projection, so it is denoted by $\|\mathbf{u}\|$. Let θ be the acute angle between **v** and **w**. Then, by definition of cosine we have

$$\|\mathbf{u}\| = \|\mathbf{v}\| \cos \theta$$

$$= \|\mathbf{v}\| \left(\frac{\mathbf{v} \cdot \mathbf{w}}{\|\mathbf{v}\| \, \|\mathbf{w}\|} \right)$$

$$= \frac{\mathbf{v} \cdot \mathbf{w}}{\|\mathbf{w}\|} \qquad \text{Cosine is positive because } \theta \text{ is acute.}$$

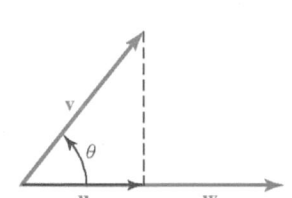

Figure 9.30 Projection of **v** onto **w**

☠ *Note that* $\left(\frac{v \cdot w}{w \cdot w}\right) w$ *is not the same as* $\left(\frac{v}{w}\right) w$*; you cannot "cancel" the vector w. Remember,* $v \cdot w$ *and* $w \cdot w$ *are real numbers whereas* v/w *is not defined.* ☠

If θ is obtuse, its cosine is negative and $\|\mathbf{u}\| = -\|\mathbf{v}\| \cos \theta$. To find a formula for the vector projection, we note that it is the scalar component of **v** in the direction of **w**. If θ is acute, then the vector projection has length $\|\mathbf{v}\| \cos \theta$ and has direction $\mathbf{w}/\|\mathbf{w}\|$ (the unit vector of **w**). If the angle is obtuse, the vector projection has length $-\|\mathbf{v}\| \cos \theta$ and has direction $-\mathbf{w}/\|\mathbf{w}\|$. In either case,

$$\mathbf{u} = (\|\mathbf{v}\| \cos \theta) \frac{\mathbf{w}}{\|\mathbf{w}\|}$$

$$= \frac{\mathbf{v} \cdot \mathbf{w}}{\|\mathbf{w}\|} \left(\frac{\mathbf{w}}{\|\mathbf{w}\|} \right)$$

$$= \frac{\mathbf{v} \cdot \mathbf{w}}{\|\mathbf{w}\|^2} \mathbf{w}$$

$$= \left(\frac{\mathbf{v} \cdot \mathbf{w}}{\mathbf{w} \cdot \mathbf{w}} \right) \mathbf{w}$$

---

*Even though Figure 9.30 is drawn in $\mathbb{R}^2$, the projection formula applies to $\mathbb{R}^3$ as well.

> **PROJECTIONS** If **v** and **w** are nonzero vectors, then the
> **vector projection** of **v** in the direction of **w** (a vector), denoted by $\text{proj}_{\mathbf{w}}\mathbf{v}$ is
>
> $$\text{proj}_{\mathbf{w}}\mathbf{v} = \left(\frac{\mathbf{v} \cdot \mathbf{w}}{\mathbf{w} \cdot \mathbf{w}}\right)\mathbf{w}$$
>
> **scalar projection** of **v** onto **w** (a number), denoted by $\text{comp}_{\mathbf{w}}\mathbf{v}$ is
>
> $$\text{comp}_{\mathbf{w}}\mathbf{v} = \frac{\mathbf{v} \cdot \mathbf{w}}{\|\mathbf{w}\|}$$

Note $\text{comp}_{\mathbf{w}}v > 0$ when $0 \le \theta < \frac{\pi}{2}$ and $\text{comp}_{\mathbf{w}}v < 0$ when $\frac{\pi}{2} < \theta \le \pi$.

In $\mathbb{R}^3$, the **i**, **j**, and **k** components of any vector v are the scalar projections of **v** onto the appropriate basis vector, as shown in Figure 9.31.

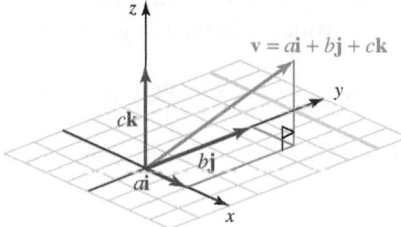

Figure 9.31 Basis vectors are vector projections

Example 6   Vector and scalar projections

Find the scalar and vector projections of $\mathbf{v} = 2\mathbf{i} + 3\mathbf{j} + 5\mathbf{k}$ onto $\mathbf{w} = 2\mathbf{i} - 2\mathbf{j} - \mathbf{k}$.

Solution   We first find the vector projection:

$$
\begin{aligned}
\text{proj}_{\mathbf{w}}\mathbf{v} &= \left(\frac{\mathbf{v} \cdot \mathbf{w}}{\mathbf{w} \cdot \mathbf{w}}\right)\mathbf{w} \\
&= \left(\frac{2(2) + 3(-2) + 5(-1)}{2^2 + (-2)^2 + (-1)^2}\right)(2\mathbf{i} - 2\mathbf{j} - \mathbf{k}) \\
&= -\frac{7}{9}(2\mathbf{i} - 2\mathbf{j} - \mathbf{k}) \\
&= -\frac{14}{9}\mathbf{i} + \frac{14}{9}\mathbf{j} + \frac{7}{9}\mathbf{k}
\end{aligned}
$$

To find the scalar projection, we can find the length of the vector projection, or we can use the scalar projection formula (which is usually easier than finding the length directly):

$$
\begin{aligned}
\text{comp}_{\mathbf{w}}\mathbf{v} &= \frac{\mathbf{v} \cdot \mathbf{w}}{\|\mathbf{w}\|} \\
&= \frac{2(2) + 3(-2) + 5(-1)}{\sqrt{2^2 + (-2)^2 + (-1)^2}} \\
&= \frac{-7}{3}
\end{aligned}
$$

Work as a Dot Product

One important application of the dot product and projections occurs in physics when calculating the amount of work done by a constant force. When a constant force of magnitude F is applied to an object through a distance d, the **work** performed is defined to be the product of force and distance. This means that the work, W, done by a constant force directed along the line of the motion of the object to be $W = Fd$ (Figure 9.32**a**). We now consider a force acting in some other direction (Figure 9.32**b**).

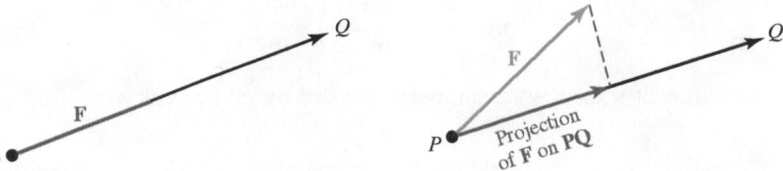

a. If the force **F** acts along the line of motion, then $W = \|\mathbf{F}\| \, \|\mathbf{PQ}\|$

b. If **F** acts at a nonezero angle with the line of motion, then $W = \|\text{proj. of } \mathbf{F} \text{ onto } \mathbf{PQ}\| \, \|\mathbf{PQ}\|$

Figure 9.32 Work W as a dot product

Physics experiments indicate that when a force F moves an object along the line from point P to point Q, the work performed is given by the product

$$W = [\text{SCALAR COMPONENT OF } \mathbf{F} \text{ ALONG } \mathbf{PQ}] \, [\text{DISTANCE MOVED BY THE OBJECT}]$$

The distance moved by the object is the length of the *displacement vector* **PQ**, and we find that

$$W = \overbrace{\frac{\mathbf{F} \cdot \mathbf{PQ}}{\|\mathbf{PQ}\|}}^{\text{Scalar component of } \mathbf{F} \text{ along } \mathbf{PQ}} \underbrace{\|\mathbf{PQ}\|}_{\text{Distance moved by the object}} = \mathbf{F} \cdot \mathbf{PQ}$$

WORK AS A DOT PRODUCT   An object that moves along a line with displacement **PQ** against a constant force **F** performs

$$W = \mathbf{F} \cdot \mathbf{PQ}$$

units of work.

Example 7   Work performed by a constant force

Suppose that the wind is blowing with a force **F** of magnitude 500 lb in the direction of N30°E over a boat's sail. How much work does the wind perform in moving the boat in a northerly direction a distance of 100 ft? Give your answer in foot-pounds.

Solution   We see that $\|\mathbf{F}\| = 500$ lb and is in the direction of N30°E, as shown in Figure 9.33.

The displacement direction is $\mathbf{PQ} = 100\mathbf{j}$, so $\|\mathbf{PQ}\| = 100$ ft. Thus,

$$\mathbf{F} = 500 \cos 60°\mathbf{i} + 500 \sin 60°\mathbf{j}$$
$$= 250\mathbf{i} + 250\sqrt{3}\mathbf{j}$$

Thus, the work performed is

$$W = \mathbf{F} \cdot \mathbf{PQ} = 100(250\sqrt{3}) = 25{,}000\sqrt{3}$$

Thus, the work is approximately 43,300 ft-lb.

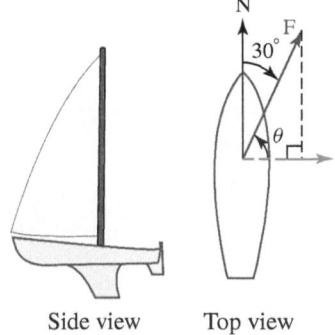

Side view   Top view

Figure 9.33 Work performed

In the U.S. system of measurements, work is typically expressed in *foot-pounds, inch-pounds,* or *foot-tons.* In the International System (SI), work is expressed in newton-meters (called *joules*), and in the centimeter-gram-second (CGS) system, the basic unit of work is the dyne-centimeter (called an *erg*). These units are shown in Table 9.1.

Table 9.1 Common Units of Work and Force

Mass	Distance	Unit of force	Work
kg	m	newton (N)	joule
g	cm	dyne (dyn)	erg
slug	ft	pound	ft-lb

PROBLEM SET 9.3

Level 1

1. ■ What does this say? Discuss how to find a dot product, and describe an application of the dot product.
2. ■ What does this say? What does it mean for vectors to be orthogonal? What is meant by the vector and scalar projections of a vector **v** onto another vector **w**?

Find the dot product **v·w** *in Problems 3-6.*

3. $\mathbf{v} = \langle 3, -2, 4 \rangle; \mathbf{w} = \langle 2, -1, -6 \rangle$
4. $\mathbf{v} = \langle 2, -6, 0 \rangle; \mathbf{w} = \langle 0, -3, 7 \rangle$
5. $\mathbf{v} = 2\mathbf{i} + 3\mathbf{j} - \mathbf{k}; \ \mathbf{w} = -3\mathbf{i} + 5\mathbf{j} + 4\mathbf{k}$
6. $\mathbf{v} = 3\mathbf{i} - \mathbf{j}; \ \mathbf{w} = 2\mathbf{i} + 5\mathbf{j}$

State whether the given pairs of vectors in Problems 7-10 are orthogonal.

7. $\mathbf{v} = \mathbf{i}; \mathbf{w} = \mathbf{k}$
8. $\mathbf{v} = \mathbf{j}; \mathbf{w} = -\mathbf{k}$
9. $\mathbf{v} = 3\mathbf{i} - 2\mathbf{j}; \mathbf{w} = 6\mathbf{i} + 9\mathbf{j}$
10. $\mathbf{v} = 4\mathbf{i} - 5\mathbf{j} + \mathbf{k}; \mathbf{w} = 8\mathbf{i} + 10\mathbf{j} - 2\mathbf{k}$

Let $\mathbf{v} = 3\mathbf{i} - 2\mathbf{j} + \mathbf{k}$ *and* $\mathbf{w} = \mathbf{i} + \mathbf{j} - \mathbf{k}$. *Evaluate the expressions in Problems 11-14.*

11. $(\mathbf{v}+\mathbf{w}) \cdot (\mathbf{v}-\mathbf{w})$
12. $(\mathbf{v}\cdot\mathbf{w})\mathbf{w}$
13. $(\|\mathbf{v}\|\mathbf{w}) \cdot (\|\mathbf{w}\| \mathbf{v})$
14. $\dfrac{2\mathbf{v} + 3\mathbf{w}}{\|3\mathbf{v} + 2\mathbf{w}\|}$

Let $\mathbf{v} = \mathbf{i} - 2\mathbf{j} + 2\mathbf{k}$ *and* $\mathbf{w} = 2\mathbf{i} + 4\mathbf{j} - \mathbf{k}$; *and find the vector or scalar requested in Problems 15-18.*

15. $2\mathbf{v}-3\mathbf{w}$
16. $\|\mathbf{v}\|\mathbf{w}$
17. $\|2\mathbf{v} - 3\mathbf{w}\|$
18. $\|\mathbf{v} + \mathbf{w}\|(\mathbf{v}+\mathbf{w})$

Evaluate the expressions given in Problems 19-22.

19. $\|\mathbf{i} + \mathbf{j} + \mathbf{k}\|$
20. $\|\mathbf{i} - \mathbf{j} + \mathbf{k}\|$
21. $\|2\mathbf{i} + \mathbf{j} - 3\mathbf{k}\|^2$
22. $\|2(\mathbf{i} - \mathbf{j} + \mathbf{k}) - 3(2\mathbf{i} + \mathbf{j} - \mathbf{k})\|^2$

Find the angle between the vectors given in Problems 23-26. Round to the nearest degree.

23. $\mathbf{v} = \mathbf{i} + \mathbf{j} + \mathbf{k}; \mathbf{w} = \mathbf{i} - \mathbf{j} + \mathbf{k}$
24. $\mathbf{v} = 2\mathbf{i} + \mathbf{k}; \mathbf{w} = \mathbf{j} - 3\mathbf{k}$
25. $\mathbf{v} = 2\mathbf{j} + \mathbf{k}; \mathbf{w} = \mathbf{i} - 2\mathbf{k}$
26. $\mathbf{v} = 4\mathbf{i} - \mathbf{j} + \mathbf{k}; \mathbf{w} = 2\mathbf{i} + 3\mathbf{j} + 5\mathbf{k}$

Find the scalar and vector projections of **v** *onto* **w** *in Problems 27-30.*

27. $\mathbf{v} = \mathbf{i} + \mathbf{j} + \mathbf{k}; \mathbf{w} = 2\mathbf{k}$
28. $\mathbf{v} = 2\mathbf{i} - 3\mathbf{j}; \mathbf{w} = 2\mathbf{j} - 3\mathbf{k}$
29. $\mathbf{v} = \mathbf{i} + 2\mathbf{k}; \mathbf{w} = -3\mathbf{j}$
30. $\mathbf{v} = \mathbf{i} + \mathbf{j} - 2\mathbf{k}; \mathbf{w} = \mathbf{i} + \mathbf{j} + \mathbf{k}$
31. Find two distinct unit vectors orthogonal to both $\mathbf{v} = \mathbf{i} + \mathbf{j} - \mathbf{k}$ and $\mathbf{w} = -\mathbf{i} + \mathbf{j} + \mathbf{k}$.
32. Find two distinct unit vectors orthogonal to both $\mathbf{v} = 2\mathbf{i} + \mathbf{j} + 2\mathbf{k}$ and $\mathbf{w} = -\mathbf{i} + 2\mathbf{j} - \mathbf{k}$.
33. Find a unit vector that points in the direction opposite to $\mathbf{v} = 2\mathbf{i} + 3\mathbf{j} - 2\mathbf{k}$.
34. Find a vector that points in the same direction as $\mathbf{v} = \mathbf{i} + 2\mathbf{j} - \mathbf{k}$ and has length one-third.
35. Find $x, y,$ and z that solve

$$x(\mathbf{i} + \mathbf{j} + \mathbf{k}) + y(\mathbf{i} - \mathbf{j} + 2\mathbf{k}) + z(\mathbf{i} + \mathbf{k}) = 2\mathbf{i} + \mathbf{k}$$

36. Find $x, y,$ and z that solve

$$x(\mathbf{i} - \mathbf{k}) + y(\mathbf{j} + \mathbf{k}) + z(\mathbf{i} - \mathbf{j}) = 5\mathbf{i} - \mathbf{k}$$

*Some of the measurements and units in this table may not be familiar to you. For example, a *slug* is a unit of measurement in physics defined to be "the unit of mass that is accelerated at the rate of one foot per second per second when acted on by a force of one pound weight." If you have not studied physics, this may be meaningless to you right now, but remember that our focus here is not on the units of measurement, but on how to calculate *work* using scalar multiplication of vectors. However, if you are interested, a *dyne* is defined to be "the force required to accelerate a mass of one centimeter per second per second to a mass of one gram" and a *newton* is defined to be "the unit of force required to accelerate a mass of one kilogram one meter per second per second." Thus, one newton is equal to 100,000 dynes.

37. Find a number a that guarantees that the vectors $3\mathbf{i} - 2\mathbf{j} + \mathbf{k}$ and $2\mathbf{i} + a\mathbf{j} - 2a\mathbf{k}$ will be orthogonal.

38. Find x if the vectors $\mathbf{v} = 3\mathbf{i} - x\mathbf{j} + 2\mathbf{k}$ and $\mathbf{w} = x\mathbf{i} + \mathbf{j} - 2\mathbf{k}$ are to be orthogonal.

Find the direction cosines and the direction angles for the vectors given in Problems 39-42.

39. $\mathbf{v} = 2\mathbf{i} - 3\mathbf{j} - 5\mathbf{k}$ 40. $\mathbf{v} = 3\mathbf{i} - 2\mathbf{k}$

41. $\mathbf{v} = 5\mathbf{i} - 4\mathbf{j} + 3\mathbf{k}$ 42. $\mathbf{v} = \mathbf{j} - 5\mathbf{k}$

43. Let $\mathbf{v} = 2\mathbf{i} - 3\mathbf{j} + 6\mathbf{k}$ and $\mathbf{w} = 4\mathbf{i} + 3\mathbf{k}$.
    Find
    a. $\mathbf{v} \cdot \mathbf{w}$
    b. $\cos\theta$, where θ is the angle between $\mathbf{v}$ and $\mathbf{w}$
    c. a scalar s such that $\mathbf{v}$ is orthogonal to $\mathbf{v} - s\mathbf{w}$
    d. a scalar t such that $\mathbf{v} + t\mathbf{w}$ is orthogonal to $\mathbf{w}$

44. Let $\mathbf{v} = 4\mathbf{i} - \mathbf{j} + \mathbf{k}$ and $\mathbf{w} = 2\mathbf{i} + 3\mathbf{j} - \mathbf{k}$.
    Find
    a. $\mathbf{v} \cdot \mathbf{w}$
    b. $\cos\theta$, where θ is the angle between $\mathbf{v}$ and $\mathbf{w}$
    c. a scalar s such that $\mathbf{v}$ is orthogonal to $\mathbf{v} - s\mathbf{w}$
    d. a scalar t such that $t\mathbf{v} + \mathbf{w}$ is orthogonal to $\mathbf{w}$

45. Find the cosine of the angle between the vectors $\mathbf{v} = \mathbf{i} - \mathbf{j} + 2\mathbf{k}$ and $\mathbf{w} = 2\mathbf{i} + \mathbf{j} - \mathbf{k}$. Then find the vector projection of $\mathbf{v}$ onto $\mathbf{w}$.

46. Find the scalar projection of the force $\mathbf{F} = 4\mathbf{i} - 2\mathbf{j} + 3\mathbf{k}$ in the direction of the vector $\mathbf{v} = \mathbf{i} - \mathbf{j} + 2\mathbf{k}$.

47. Find the work done by the constant force $\mathbf{F} = 2\mathbf{i} + 3\mathbf{j} + \mathbf{k}$ when it moves a particle along the line from $P(1, 0, -1)$ to $Q(3, 1, 2)$.

48. Find the work performed when a force $\mathbf{F} = \frac{6}{7}\mathbf{i} - \frac{2}{7}\mathbf{j} + \frac{6}{7}\mathbf{k}$ is applied to an object moving along the line from $P(-3, -5, 4)$ to $Q(4, 9, 11)$.

Level 2

49. Fred and his friend Sam are pulling a heavy log along flat horizontal ground by ropes attached to the front of the log. The ropes are 8 ft long. Fred holds his rope 2 ft above the log and 1 ft to the side, and Sam holds his end 1 ft above the log and 1 ft to the opposite side, as shown in Figure 9.34.

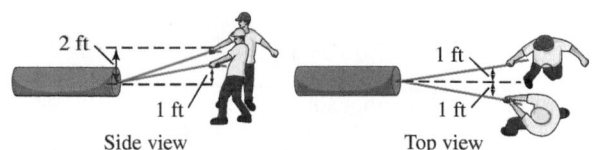

Figure 9.34   Problem 49

If Fred exerts a force of 30 lb and Sam exerts a force of 20 lb, what is the resultant force on the log?

50. Find the force required to keep a 5,000-lb van from rolling downhill if it is parked on a $10°$ slope (see Figure 9.35).

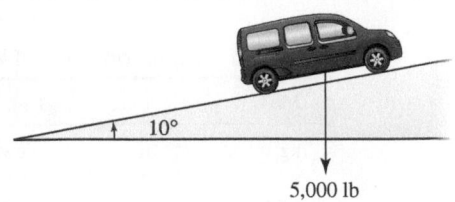

$10°$

$5,000$ lb

Figure 9.35   Problem 50

51. A trunk is dragged 20 ft across a floor, using a force of 50 lb, as shown in Figure 9.36.

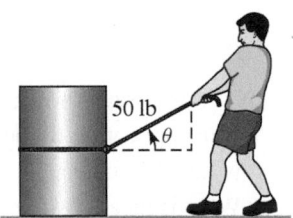

50 lb

θ

Figure 9.36   Problem 51

Find the work done if the direction of the force is inclined θ to the horizontal, where

a. $\theta = \frac{\pi}{3}$ b. $\theta = \frac{\pi}{4}$

52. Suppose that the wind is blowing with a 1,000-lb magnitude force $\mathbf{F}$ in the direction of N60°W behind a boat's sail. How much work does the wind perform in moving the boat in a northerly direction a distance of 50 ft? Give your answer in foot-pounds.

Orthogonality of functions *One of the big discoveries in modern mathematics was that it is very fruitful to extend some of the geometric notions of this chapter to functions—for example, taking the dot product of two functions or asking whether two functions are orthogonal, or attempting to project a function onto a second function (or onto a "plane" of functions). You are introduced to these important concepts in Problems 53 and 54.*

53. We say that two functions f and g are **orthogonal** on $[a, b]$ if

$$\int_a^b f(x)g(x)\,dx = 0$$

a. Show that the two functions x^2 and $x^3 - 5x$ are orthogonal on $[-b, b]$ for any positive b.

b. For distinct positive integers k and n, show that $\sin kx$ and $\sin nx$ are orthogonal on interval $[-\pi, \pi]$.

That is, the *family* of functions $\sin x, \sin 2x, \cdots$ are *mutually* orthogonal on $[-\pi, \pi]$. You may need the product-to-sum identity (trigonometric identity 28, Appendix E):

$$2 \sin \alpha \sin \beta = \cos(\alpha - \beta) - \cos(\alpha + \beta)$$

54. *Historical Quest*

Karl Smith library

Jean Fourier (1768-1830)

Jean Baptiste Joseph Fourier began his studies by training for the priesthood. In spite of this first calling, his interest in mathematics remained intense, and in 1789 he wrote, "Yesterday was my 21st birthday, and at that age Newton and Pascal had already acquired many claims to immortality."
He did not take his religious vows but continued his mathematical research while teaching. Fourier was renowned as an outstanding teacher, but it was not until around 1804-1807 that he did his important mathematical work on the theory of heat. During that time, he expanded on an idea of Daniel Bernoulli by showing how many functions, both continuous and discontinuous, can be represented by a trigonometric series of the general form

$$\frac{1}{2}a_0 + a_1 \cos x + a_2 \cos 2x + a_3 \cos 3x$$
$$+ \cdots + b_1 \sin x + b_2 \sin 2x + b_3 \sin 3x + \cdots$$

The study of such series, called **Fourier series**, is of fundamental importance in both theoretical and applied mathematics, and they are used for modeling phenomena in engineering and physics as well as in areas such as computer science and medical research.

The issue of whether a given function f can be represented by a Fourier series is a matter for a more advanced course. For this *Quest*, assume that $f(x)$ is such a function and that

$$f(x) = \frac{1}{2}a_0 + \sum_{k=1}^{\infty} a_k \cos kx + \sum_{k=1}^{\infty} b_k \sin kx$$

on the interval $[-\pi, \pi]$. In particular, if f is a continuous odd function (that is, $f(-x) = -f(x)$ on $[-\pi, \pi]$), it can be shown that $a_k = 0$ for $k = 0, 1, 2, \cdots$, and that

$$b_k = \frac{1}{\pi} \int_{-\pi}^{\pi} f(x) \sin kx \, dx = \frac{2}{\pi} \int_{0}^{\pi} f(x) \sin kx \, dx$$

for $k = 1, 2, \cdots$.

a. Using this result, show that the function $f(x) = x$ on $[-\pi, \pi]$ has the Fourier series representation

$$f(x) = 2 \sum_{k=1}^{\infty} \frac{(-1)^{k+1}}{k} \sin kx$$

b. Use the Fourier series in part **a** to evaluate the convergent alternating series

$$\sum_{k=1}^{\infty} \frac{(-1)^{k+1}}{2k - 1} = 1 - \frac{1}{3} + \frac{1}{5} - \frac{1}{7} + \frac{1}{9} - \cdots$$

Level 3

55. Show that the vector $\mathbf{B} = \|\mathbf{v}\| \mathbf{u} + \|\mathbf{u}\| \mathbf{v}$ bisects the angle θ between the two nonzero vectors $\mathbf{u}$ and $\mathbf{v}$.

56. Use vector methods to prove that an angle inscribed in a semicircle must be a right angle.

57. a. Show that
$$(\mathbf{v} + \mathbf{w}) \cdot (\mathbf{v} + \mathbf{w}) = \|\mathbf{v}\|^2 + \|\mathbf{w}\|^2 + 2(\mathbf{v} \cdot \mathbf{w})$$
b. Use part **a** to prove the **triangle inequality:**

$$\|\mathbf{v} + \mathbf{w}\| \leq \|\mathbf{v}\| + \|\mathbf{w}\|$$

Hint: Note that

$$\|\mathbf{v} + \mathbf{w}\|^2 = (\mathbf{v} + \mathbf{w}) \cdot (\mathbf{v} + \mathbf{w})$$

58. The **Cauchy-Schwarz** inequality in $\mathbb{R}^3$ states that for any vectors $\mathbf{v}$ and $\mathbf{w}$

$$|\mathbf{v} \cdot \mathbf{w}| \leq \|\mathbf{v}\| \|\mathbf{w}\|$$

a. Prove the Cauchy-Schwarz inequality. *Hint:* Use the formula for the angle between two vectors.
b. Show that equality $|\mathbf{v} \cdot \mathbf{w}| = \|\mathbf{v}\| \|\mathbf{w}\|$ occurs if and only if $\mathbf{v} = t\mathbf{w}$ for some scalar t.
c. Use the Cauchy-Schwarz inequality to prove the **triangle inequality:**

$$\|\mathbf{v} + \mathbf{w}\| \leq \|\mathbf{v}\| + \|\mathbf{w}\|$$

59. Show that $\cos^2 \alpha + \cos^2 \beta + \cos^2 \gamma = 1$, where α, β, and γ are the direction angles of a triangle $\triangle ABC$.

60. Find the angle (to the nearest degree) between the diagonal of a cube and a diagonal of one of its faces. Assume both diagonals have the same initial point.

9.4 THE CROSS PRODUCT

IN THIS SECTION: *Definition of the cross product, geometric interpretation of the cross product, properties of the cross product, the scalar triple product and volume, torque*

In this section, we introduce a second type of vector multiplication called *cross product*, which turns out to be a vector. We shall find that the cross product is a vector orthogonal to each of the given vectors. This geometric property will be particularly useful in the next section.

We have considered two types of products involving vectors. The first was *scalar multiplication*, which is the multiplication of a vector by a number (which is called a *scalar*). In the last section we defined a product of vectors that produces a number (i.e., scalar); this product is known as *scalar product* (because the answer is a scalar) or *dot product* since the notation we use for this product is a dot. We are now ready to consider a third product of vectors that produces a vector. It is called the *vector product* or *cross product* because we use the multiplication cross ($\times$) to denote this type of product.

Definition of the Cross Product*

The cross product is sometimes called **vector product** because the result is a vector. In other applications it is called the **outer product**. The definition requires a basis of **i**, **j**, and **k** and therefore is a definition that makes sense only in $\mathbb{R}^3$.

> **CROSS PRODUCT** If $\mathbf{v} = a_1\mathbf{i} + a_2\mathbf{j} + a_3\mathbf{k}$ and $\mathbf{w} = b_1\mathbf{i} + b_2\mathbf{j} + b_3\mathbf{k}$, the **cross product**, written $\mathbf{v} \times \mathbf{w}$, is the vector
>
> $$\mathbf{v} \times \mathbf{w} = (a_2 b_3 - a_3 b_2)\mathbf{i} + (a_3 b_1 - a_1 b_3)\mathbf{j} + (a_1 b_2 - a_2 b_1)\mathbf{k}$$

This definition seems very strange, indeed, until we show that these terms can be obtained by using a determinant

$$\mathbf{v} \times \mathbf{w} = \begin{vmatrix} \mathbf{i} & \mathbf{j} & \mathbf{k} \\ a_1 & a_2 & a_3 \\ b_1 & b_2 & b_3 \end{vmatrix}$$

We can verify this is the formula for cross product by expanding this determinant about the first row.

$$\mathbf{v} \times \mathbf{w} = \begin{vmatrix} \mathbf{i} & \mathbf{j} & \mathbf{k} \\ a_1 & a_2 & a_3 \\ b_1 & b_2 & b_3 \end{vmatrix} = \begin{vmatrix} a_2 & a_3 \\ b_2 & b_3 \end{vmatrix}\mathbf{i} - \begin{vmatrix} a_1 & a_3 \\ b_1 & b_3 \end{vmatrix}\mathbf{j} + \begin{vmatrix} a_1 & a_2 \\ b_1 & b_2 \end{vmatrix}\mathbf{k}$$

j is in row 1, column 2, so do not forget the negative sign here.

$$= (a_2 b_3 - a_3 b_2)\mathbf{i} - (a_1 b_3 - a_3 b_1)\mathbf{j} + (a_1 b_2 - a_2 b_1)\mathbf{k}$$
$$= (a_2 b_3 - a_3 b_2)\mathbf{i} + (a_3 b_1 - a_1 b_3)\mathbf{j} + (a_1 b_2 - a_2 b_1)\mathbf{k}$$

---

*Determinants (see Appendix F) are necessary for this section.

Example 1 Cross product

Find $\mathbf{v} \times \mathbf{w}$ where $\mathbf{v} = 2\mathbf{i} - \mathbf{j} + 3\mathbf{k}$ and $\mathbf{w} = 7\mathbf{j} - 4\mathbf{k}$.

Solution
$$\mathbf{v} \times \mathbf{w} = \begin{vmatrix} \mathbf{i} & \mathbf{j} & \mathbf{k} \\ 2 & -1 & 3 \\ 0 & 7 & -4 \end{vmatrix}$$

Do not forget minus here (j is negative position).

$$= [(-1)(-4) - 3(7)]\mathbf{i} \downarrow [2(-4) - 0(3)]\mathbf{j} + [2(7) - 0(-1)]\mathbf{k}$$

$$= -17\mathbf{i} + 8\mathbf{j} + 14\mathbf{k}$$

Properties of determinants can also be used to establish properties of the cross product. For instance, the following computation shows that the cross product is *not* commutative. This property is sometimes called **anticommutativity**.

$$\mathbf{v} \times \mathbf{w} = \begin{vmatrix} \mathbf{i} & \mathbf{j} & \mathbf{k} \\ a_1 & a_2 & a_3 \\ b_1 & b_2 & b_3 \end{vmatrix} = - \begin{vmatrix} \mathbf{i} & \mathbf{j} & \mathbf{k} \\ b_1 & b_2 & b_3 \\ a_1 & a_2 & a_3 \end{vmatrix} = -(\mathbf{w} \times \mathbf{v})$$

The cross product has interesting properties. We have already seen that $\mathbf{v} \times \mathbf{w} = -(\mathbf{w} \times \mathbf{v})$. This and other properties are listed in the following box.

Theorem 9.5   Properties of the cross product

If $\mathbf{u}$, $\mathbf{v}$, and $\mathbf{w}$ are vectors in $\mathbb{R}^3$ and s and t are scalars, then several properties can be derived:

Scalar distributivity	$(s\mathbf{v}) \times (t\mathbf{w}) = st(\mathbf{v} \times \mathbf{w})$
Vector distributivity*	$\mathbf{u} \times (\mathbf{v} + \mathbf{w}) = (\mathbf{u} \times \mathbf{v}) + (\mathbf{u} \times \mathbf{w})$
	$(\mathbf{u} + \mathbf{v}) \times \mathbf{w} = (\mathbf{u} \times \mathbf{w}) + (\mathbf{v} \times \mathbf{w})$
Anticommutativity	$\mathbf{v} \times \mathbf{w} = -(\mathbf{w} \times \mathbf{v})$
Product of a multiple	$\mathbf{v} \times \mathbf{v} = 0$; if $\mathbf{w} = s\mathbf{v}$, then $\mathbf{v} \times \mathbf{w} = 0$
Zero product	$\mathbf{v} \times \mathbf{0} = \mathbf{0} \times \mathbf{v} = \mathbf{0}$
Lagrange's identity	$\|\mathbf{v} \times \mathbf{w}\|^2 = \|\mathbf{v}\|^2 \|\mathbf{w}\|^2 - (\mathbf{v} \cdot \mathbf{w})^2$
***cab-bac* formula**	$\mathbf{a} \times (\mathbf{b} \times \mathbf{c}) = (\mathbf{c} \cdot \mathbf{a})\mathbf{b} - (\mathbf{b} \cdot \mathbf{a})\mathbf{c}$

Proof: Let

$$\mathbf{u} = a_1\mathbf{i} + a_2\mathbf{j} + a_3\mathbf{k}, \qquad \mathbf{v} = b_1\mathbf{i} + b_2\mathbf{j} + b_3\mathbf{k}, \qquad \text{and} \qquad \mathbf{w} = c_1\mathbf{i} + c_2\mathbf{j} + c_3\mathbf{k}$$

Scalar distributivity and *vector distributivity* are proved by using the definition of the cross product and the corresponding properties of real numbers. For *product of a multiple* we find:

$$\mathbf{v} \times s\mathbf{v} = \begin{vmatrix} \mathbf{i} & \mathbf{j} & \mathbf{k} \\ b_1 & b_2 & b_3 \\ sb_1 & sb_2 & sb_3 \end{vmatrix} = s \begin{vmatrix} \mathbf{i} & \mathbf{j} & \mathbf{k} \\ b_1 & b_2 & b_3 \\ b_1 & b_2 & b_3 \end{vmatrix} = \mathbf{0}$$

Property of determinants (two rows the same)

---

*Properly, it is the distributive property of vectors for cross product over addition, which, for convenience, we shorten to vector distributivity.

The *zero product* property is obvious. You are asked to prove *Lagrange's identity* in Problem 52 and the *cab-bac* formula in Problem 55. ♦

Geometric Interpretation of the Cross Product

We will show with the following property that the vector $(\mathbf{v} \times \mathbf{w})$ is orthogonal to both the vectors $\mathbf{v}$ and $\mathbf{w}$. The only way this can occur is in a three-dimensional setting. Any two distinct nonzero vectors in $\mathbb{R}^3$ that are not parallel (that is, not scalar multiples of one another) can be arranged to determine a plane. Then, according to the geometric property of the cross product the vector product *must* be orthogonal to this plane, as shown in Figure 9.37.

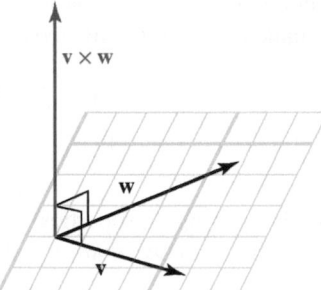

Figure 9.37   Vector product

☠ *This property is one of the most important properties when working in three dimensions. We will use this property extensively in Section 9.6.* ☠

Theorem 9.6   Geometric property of the cross product

If $\mathbf{v}$ and $\mathbf{w}$ are nonzero vectors in $\mathbb{R}^3$ that are not multiples of one another, then $\mathbf{v} \times \mathbf{w}$ is orthogonal to both $\mathbf{v}$ and $\mathbf{w}$.

Proof: We will show $(\mathbf{v} \times \mathbf{w})$ is orthogonal to $\mathbf{v}$ and leave the proof that $\mathbf{v} \times \mathbf{w}$ is orthogonal to $\mathbf{w}$ as an exercise. Let $\mathbf{v} = a_1\mathbf{i} + a_2\mathbf{j} + a_3\mathbf{k}$ and $\mathbf{w} = b_1\mathbf{i} + b_2\mathbf{j} + b_3\mathbf{k}$. Then,

$$\mathbf{v} \times \mathbf{w} = \begin{vmatrix} \mathbf{i} & \mathbf{j} & \mathbf{k} \\ a_1 & a_2 & a_3 \\ b_1 & b_2 & b_3 \end{vmatrix} = (a_2b_3 - a_3b_2)\mathbf{i} - (a_1b_3 - a_3b_1)\mathbf{j} + (a_1b_2 - a_2b_1)\mathbf{k}$$

To show this vector is orthogonal to $\mathbf{v}$, we find $\mathbf{v} \cdot (\mathbf{v} \times \mathbf{w})$:

$$\begin{aligned} \mathbf{v} \cdot (\mathbf{v} \times \mathbf{w}) &= a_1(a_2b_3 - a_3b_2) - a_2(a_1b_3 - a_3b_1) + a_3(a_1b_2 - a_2b_1) \\ &= a_1a_2b_3 - a_1a_3b_2 - a_1a_2b_3 + a_2a_3b_1 + a_1a_3b_2 - a_2a_3b_1 \\ &= 0 \end{aligned}$$

♦

Example 2   A vector orthogonal to two given vectors

Find a nonzero vector that is orthogonal to both $\mathbf{v} = -2\mathbf{i} + 3\mathbf{j} - 7\mathbf{k}$ and $\mathbf{w} = 5\mathbf{i} + 9\mathbf{k}$.

Solution   The cross product $\mathbf{v} \times \mathbf{w}$ is orthogonal to both $\mathbf{v}$ and $\mathbf{w}$.

$$\begin{aligned} \mathbf{v} \times \mathbf{w} &= \begin{vmatrix} \mathbf{i} & \mathbf{j} & \mathbf{k} \\ -2 & 3 & -7 \\ 5 & 0 & 9 \end{vmatrix} \\ &= (27 + 0)\mathbf{i} - (-18 + 35)\mathbf{j} + (0 - 15)\mathbf{k} \\ &= 27\mathbf{i} - 17\mathbf{j} - 15\mathbf{k} \end{aligned}$$

Because both $\mathbf{v} \times \mathbf{w}$ and $\mathbf{w} \times \mathbf{v}$ are orthogonal to the plane determined by $\mathbf{v}$ and $\mathbf{w}$, and because $(\mathbf{v} \times \mathbf{w}) = -(\mathbf{w} \times \mathbf{v})$, we see that one points up from the given plane and the other points down. In order to see which is which, we state the **right-hand rule**, which is described in Figure 9.38.

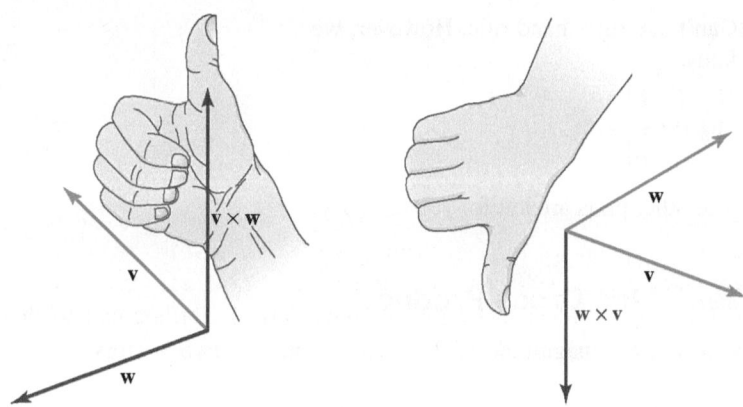

Figure 9.38 Right-hand rule

If you place the palm of your right hand along **v** in the positive direction and curl your fingers toward **w**, then your fingers are pointing in the direction of positive **v**, and your thumb points in the positive direction of **v** × **w**.

Example 3   Right-hand rule

Use the right hand rule to verify each of the following cross products.

$$\mathbf{i} \times \mathbf{j} = \mathbf{k} \quad \mathbf{j} \times \mathbf{i} = -\mathbf{k} \quad \mathbf{k} \times \mathbf{i} = \mathbf{j}$$

$$\mathbf{i} \times \mathbf{k} = -\mathbf{j} \quad \mathbf{j} \times \mathbf{k} = \mathbf{i} \quad \mathbf{k} \times \mathbf{j} = -\mathbf{i}$$

$$\mathbf{i} \times \mathbf{i} = 0 \quad \mathbf{j} \times \mathbf{j} = 0 \quad \mathbf{k} \times \mathbf{k} = 0$$

Solution

i × **j**   Place the palm of your right hand along **i** and curl your fingers toward **j**. The answer is **k**.

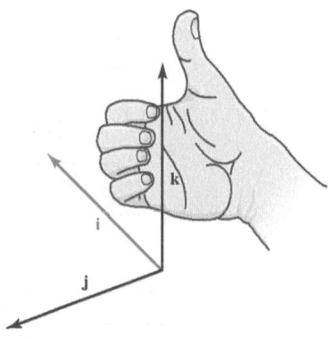

j × **i**   Place the palm of your right hand along **j** and curl your fingers toward **i**. The answer is −**k**.

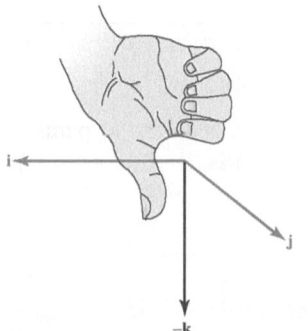

i × i   Can't use right-hand rule. However, we know

$$\begin{vmatrix} \mathbf{i} & \mathbf{j} & \mathbf{k} \\ 1 & 0 & 0 \\ 1 & 0 & 0 \end{vmatrix} = \mathbf{0}$$

The other parts are left for you to verify. ∎

Properties of the Cross Product

Next, we will find the magnitude of the cross product of two vectors.

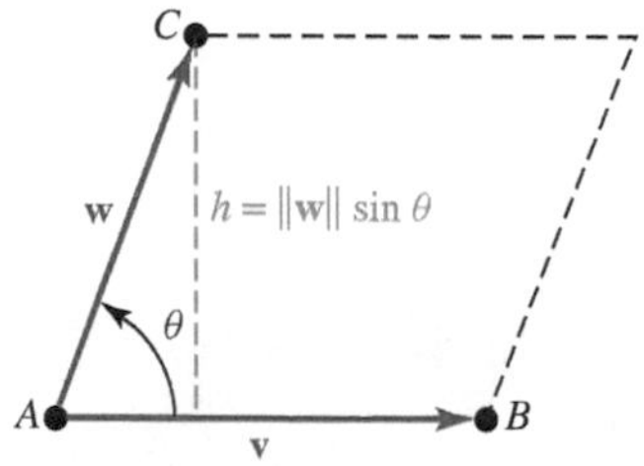

Figure 9.39 Area of a parallelogram

Theorem 9.7   Magnitude of a cross product

If $\mathbf{v}$ and $\mathbf{w}$ are nonzero vectors in $\mathbb{R}^3$ with θ the angle between $\mathbf{v}$ and $\mathbf{w}$ ($0 \le \theta \le \pi$), then

$$\|\mathbf{v} \times \mathbf{w}\| = \|\mathbf{v}\| \, \|\mathbf{w}\| \sin \theta$$

■ **W**hat this says The magnitude of the cross product of two vectors $\|\mathbf{v} \times \mathbf{w}\|$ is equal the area of the parallelogram having $\mathbf{v}$ and $\mathbf{w}$ as adjacent sides, as shown in Figure 9.39.

Proof: Let $\mathbf{v} = a_1\mathbf{i} + a_2\mathbf{j} + a_3\mathbf{k}$ and $\mathbf{w} = b_1\mathbf{i} + b_2\mathbf{j} + b_3\mathbf{k}$.

$$\|\mathbf{v}\| \, (\|\mathbf{w}\| \sin \theta) = \|\mathbf{v}\| \, \|\mathbf{w}\| \, \sqrt{1 - \cos^2\theta} \qquad \textit{From } \sin^2\theta + \cos^2\theta = 1.$$

$$= \|\mathbf{v}\| \, \|\mathbf{w}\| \sqrt{1 - \left[\frac{\mathbf{v} \cdot \mathbf{w}}{\|\mathbf{v}\| \, \|\mathbf{w}\|} \right]^2} \qquad \textit{Angle between two vectors: } \cos\theta = \frac{\mathbf{v} \cdot \mathbf{w}}{\|\mathbf{v}\| \|\mathbf{w}\|}$$

$$= \|\mathbf{v}\| \, \|\mathbf{w}\| \, \frac{\sqrt{\|\mathbf{v}\|^2 \, \|\mathbf{w}\|^2 - (\mathbf{v} \cdot \mathbf{w})^2}}{\|\mathbf{v}\| \, \|\mathbf{w}\|} \qquad \textit{Common denominator}$$

$$= \sqrt{\|\mathbf{v}\|^2 \, \|\mathbf{w}\|^2 - (\mathbf{v} \cdot \mathbf{w})^2}$$

$$= \sqrt{(a_1^2 + a_2^2 + a_3^2)(b_1^2 + b_2^2 + b_3^2) - (a_1 b_1 + a_2 b_2 + a_3 b_3)^2}$$

$$= \sqrt{(a_2 b_3 - a_3 b_2)^2 + (a_3 b_1 - a_1 b_3)^2 + (a_1 b_2 - a_2 b_1)^2}$$

$$= \|\mathbf{v} \times \mathbf{w}\|$$

◆

Consider the parallelogram shown in Figure 9.39. Then the area of this parallelogram is

$$\text{AREA} = (\text{BASE})(\text{HEIGHT})$$
$$= \|\mathbf{v}\| \, (\|\mathbf{w}\| \sin \theta)$$
$$= \|\mathbf{v} \times \mathbf{w}\|$$

Here is an example that uses this formula. You might also note that even though $\mathbf{v} \times \mathbf{w} \ne \mathbf{w} \times \mathbf{v}$, it is true that $\|\mathbf{v} \times \mathbf{w}\| = \|\mathbf{w} \times \mathbf{v}\|$.

Example 4 Area of a triangle

Find the area of the triangle with vertices $P(-2,4,5), Q(0,7,-4)$, and $R(-1,5,0)$.

Solution   Draw this triangle as shown in Figure 9.40.

Then $\triangle PQR$ has half the area of the parallelogram determined by the vectors **PQ** and **PR**; that is, the triangle has area

$$A = \frac{1}{2}\|\mathbf{PQ} \times \mathbf{PR}\|$$

First find

$$\mathbf{PQ} = (0+2)\mathbf{i} + (7-4)\mathbf{j} + (-4-5)\mathbf{k} = 2\mathbf{i} + 3\mathbf{j} - 9\mathbf{k}$$

$$\mathbf{PR} = (-1+2)\mathbf{i} + (5-4)\mathbf{j} + (0-5)\mathbf{k} = \mathbf{i} + \mathbf{j} - 5\mathbf{k}$$

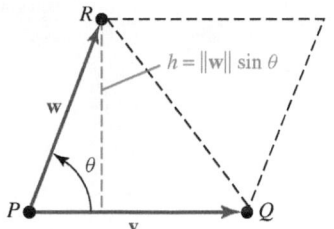

Figure 9.40   Area of a triangle

and compute the cross product:

$$\mathbf{PQ} \times \mathbf{PR} = \begin{vmatrix} \mathbf{i} & \mathbf{j} & \mathbf{k} \\ 2 & 3 & -9 \\ 1 & 1 & -5 \end{vmatrix}$$

$$= (-15+9)\mathbf{i} - (-10+9)\mathbf{j} + (2-3)\mathbf{k}$$

$$= -6\mathbf{i} + \mathbf{j} - \mathbf{k}$$

Thus, the triangle has area

$$A = \frac{1}{2}\|\mathbf{PQ} \times \mathbf{PR}\|$$

$$= \frac{1}{2}\sqrt{(-6)^2 + 1^2 + (-1)^2}$$

$$= \frac{1}{2}\sqrt{38}$$

The Scalar Triple Product and Volume

The cross product can also be used to compute the volume of a parallelepiped in $\mathbb{R}^3$. Consider the parallelepiped determined by three nonzero vectors **u**, **v**, and **w** that do not all lie in the same plane, as shown in Figure 9.41.

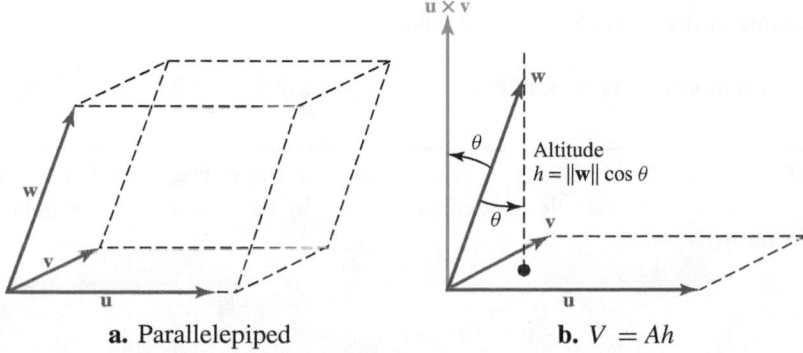

a. Parallelepiped   **b.** $V = Ah$

Figure 9.41   Computing the volume of a parallelepiped

It is known from solid geometry that this parallelogram has volume $V = Ah$, where A is the area of the face determined by $\mathbf{u}$ and $\mathbf{v}$ and h is the altitude from the tip of $\mathbf{w}$ to this face.

The face determined by $\mathbf{u}$ and $\mathbf{v}$ is a parallelogram with area $A = \|\mathbf{u} \times \mathbf{v}\|$, and we know that the cross-product vector $\mathbf{u} \times \mathbf{v}$ is perpendicular to both $\mathbf{u}$ and $\mathbf{v}$ and hence to the face determined by $\mathbf{u}$ and $\mathbf{v}$. From the alternative form of the dot product, we have

$$(\mathbf{u} \times \mathbf{v}) \cdot \mathbf{w} = \|\mathbf{u} \times \mathbf{v}\|\, \|\mathbf{w}\| \cos\,\theta$$

where θ is the angle between $\mathbf{u} \times \mathbf{v}$ and $\mathbf{w}$. Thus, the parallelepiped has altitude

$$h = |\|\mathbf{w}\| \cos\,\theta| = \left| \frac{(\mathbf{u} \times \mathbf{v}) \cdot \mathbf{w}}{\|\mathbf{u} \times \mathbf{v}\|} \right|$$

and the volume is given as

$$V = Ah = \|\mathbf{u} \times \mathbf{v}\|\, \left| \frac{(\mathbf{u} \times \mathbf{v}) \cdot \mathbf{w}}{\|\mathbf{u} \times \mathbf{v}\|} \right| = |(\mathbf{u} \times \mathbf{v}) \cdot \mathbf{w}|$$

Remark: Note the absolute value in the above formula. Volume can never be negative. The combined operation $(\mathbf{u} \times \mathbf{v}) \cdot \mathbf{w}$ is called the **scalar triple product** of $\mathbf{u}$, $\mathbf{v}$, and $\mathbf{w}$.

Example 5 Volume of a parallelepiped

Find the volume of the parallelepiped determined by the vectors $\mathbf{u} = \mathbf{i} - 2\mathbf{j} + 3\mathbf{k}$, $\mathbf{v} = -4\mathbf{i} + 7\mathbf{j} - 11\mathbf{k}$, and $\mathbf{w} = 5\mathbf{i} + 9\mathbf{j} - \mathbf{k}$.

Solution We first find the cross product.

$$\mathbf{u} \times \mathbf{v} = \begin{vmatrix} \mathbf{i} & \mathbf{j} & \mathbf{k} \\ 1 & -2 & 3 \\ -4 & 7 & -11 \end{vmatrix}$$
$$= (22 - 21)\mathbf{i} - (-11 + 12)\mathbf{j} + (7 - 8)\mathbf{k}$$
$$= \mathbf{i} - \mathbf{j} - \mathbf{k}$$

Thus,

$$V = |(\mathbf{u} \times \mathbf{v}) \cdot \mathbf{w}|$$
$$= |(\mathbf{i} - \mathbf{j} - \mathbf{k}) \cdot (5\mathbf{i} + 9\mathbf{j} - \mathbf{k})|$$
$$= |5 - 9 + 1|$$
$$= 3$$

The volume of the parallelepiped is 3 cubic units.

Computationally, there is an easier way to work Example 5.

SCALAR TRIPLE PRODUCT If $\mathbf{u} = a_1\mathbf{i} + a_2\mathbf{j} + a_3\mathbf{k}$, $\mathbf{v} = b_1\mathbf{i} + b_2\mathbf{j} + b_3\mathbf{k}$, and $\mathbf{w} = c_1\mathbf{i} + c_2\mathbf{j} + c_3\mathbf{k}$, then the volume determined by these vectors has volume using the scalar triple product

$$V = |(\mathbf{u} \times \mathbf{v}) \cdot \mathbf{w}| = \left\| \begin{vmatrix} a_1 & a_2 & a_3 \\ b_1 & b_2 & b_3 \\ c_1 & c_2 & c_3 \end{vmatrix} \right\|$$

Theorem 9.8   Determinant form for a scalar triple product

If $\mathbf{u} = a_1\mathbf{i} + a_2\mathbf{j} + a_3\mathbf{k}, \mathbf{v} = b_1\mathbf{i} + b_2\mathbf{j} + b_3\mathbf{k},$ and $\mathbf{w} = c_1\mathbf{i} + c_2\mathbf{j} + c_3\mathbf{k}$, then the scalar product can be found by evaluating the determinant

$$(\mathbf{u} \times \mathbf{v}) \cdot \mathbf{w} = \begin{vmatrix} a_1 & a_2 & a_3 \\ b_1 & b_2 & b_3 \\ c_1 & c_2 & c_3 \end{vmatrix} \begin{matrix} \leftarrow \textit{components of } u \\ \leftarrow \textit{components of } v \\ \leftarrow \textit{components of } w \end{matrix}$$

Proof: The proof follows by expanding the determinant (see Problem 49).   ♦

We can use this formula to rework Example 5.

$$(\mathbf{u} \times \mathbf{v}) \cdot \mathbf{w} = \begin{vmatrix} 1 & -2 & 3 \\ -4 & 7 & -11 \\ 5 & 9 & -1 \end{vmatrix} = -3$$

Thus, the volume is $|-3| = 3$, as computed directly in Example 5.
We summarize the area and volume vector formulas in the following box.

AREAS AND VOLUMES   Let $\mathbf{u}, \mathbf{v},$ and $\mathbf{w}$ be nonzero vectors that do not all lie in the same plane. Then,

Area of a parallelogram: $A = \|\mathbf{u} \times \mathbf{v}\|$

Area of a triangle: $A = \dfrac{1}{2} \|\mathbf{u} \times \mathbf{v}\|$

Volume of a parallelepiped: $V = |(\mathbf{u} \times \mathbf{v}) \cdot \mathbf{w}|$

In the problem set, we have included several exercises involving the scalar triple product and the vector triple product $\mathbf{u} \times \mathbf{v} \times \mathbf{w}$. In case you wonder why we neglect the product $(\mathbf{u} \cdot \mathbf{v}) \times \mathbf{w}$, notice that such a product makes no sense because $\mathbf{u} \cdot \mathbf{v}$ is a *scalar* and the cross product is an operation involving only vectors. Thus, the product

$$\mathbf{u} \cdot \mathbf{v} \times \mathbf{w} \text{ must mean } \mathbf{u} \cdot (\mathbf{v} \times \mathbf{w})$$

Torque

A useful physical application of the cross product involves **torque**. Suppose the force $\mathbf{F}$ is applied to the point Q. Then the torque of $\mathbf{F}$ around P is defined as the cross product of the "arm" vector $\mathbf{PQ}$ with the force $\mathbf{F}$, as shown in Figure 9.42.

The torque, $\mathbf{T}$, of $\mathbf{F}$ at Q about P is

$$\mathbf{T} = \mathbf{PQ} \times \mathbf{F}$$

The magnitude of the torque, $\|\mathbf{T}\|$, provides a measure of the tendency of the vector arm $\mathbf{PQ}$ to rotate counterclockwise about an axis perpendicular to the plane determined by $\mathbf{PQ}$ and $\mathbf{F}$.

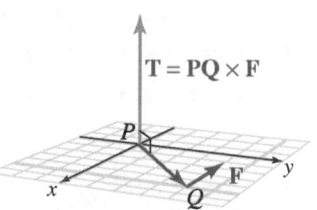

Figure 9.42   Torque as a cross product

Example 6   Torque on the hinge of a door

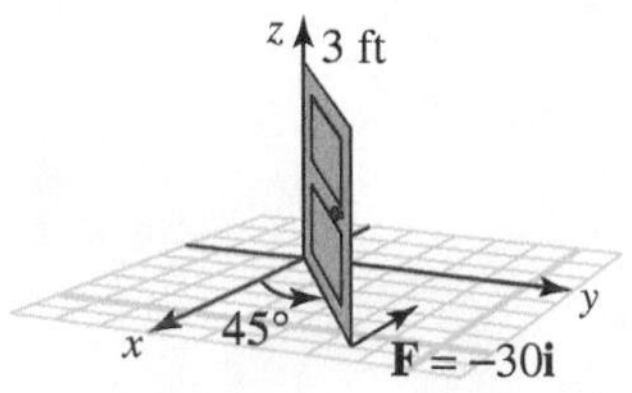

Figure 9.43 shows a half-open door that is 3 ft wide. A horizontal force of 30 lb is applied at the edge of the door. Find the torque of the force about the hinge on the door.

Solution   We represent the force by $\mathbf{F} = -30\mathbf{i}$. Because the door is half open, it makes an angle of $\frac{\pi}{4}$ with the horizontal, and we can represent the door (i.e., the "arm" **PQ**) by the vector

Figure 9.43   Swinging door

$$\mathbf{PQ} = 3\left(\cos \frac{\pi}{4}\mathbf{i} + \sin \frac{\pi}{4}\mathbf{j}\right) = 3\left(\frac{\sqrt{2}}{2}\mathbf{i} + \frac{\sqrt{2}}{2}\mathbf{j}\right) = \frac{3\sqrt{2}}{2}\mathbf{i} + \frac{3\sqrt{2}}{2}\mathbf{j}$$

The torque can now be found:

$$\mathbf{T} = \mathbf{PQ} \times \mathbf{F} = \begin{vmatrix} \mathbf{i} & \mathbf{j} & \mathbf{k} \\ \frac{3\sqrt{2}}{2} & \frac{3\sqrt{2}}{2} & 0 \\ -30 & 0 & 0 \end{vmatrix} = 45\sqrt{2}\,\mathbf{k}$$

The magnitude of the torque ($45\sqrt{2}$ ft-lb) is a measure of the tendency of the door to rotate about its hinges. ∎

PROBLEM SET 9.4

Level 1

Find $\mathbf{v} \times \mathbf{w}$ *for the vectors given in Problems* 1-10.
1. $\mathbf{v} = \mathbf{i}$; $\mathbf{w} = \mathbf{j}$
2. $\mathbf{v} = \mathbf{k}$; $\mathbf{w} = \mathbf{k}$
3. $\mathbf{v} = 3\mathbf{i} + 2\mathbf{k}$; $\mathbf{w} = 2\mathbf{i} + \mathbf{j}$
4. $\mathbf{v} = \mathbf{i} - 3\mathbf{j}$; $\mathbf{w} = \mathbf{i} + 5\mathbf{k}$
5. $\mathbf{v} = 3\mathbf{i} - 2\mathbf{j} + 4\mathbf{k}$; $\mathbf{w} = \mathbf{i} + 4\mathbf{j} - 7\mathbf{k}$
6. $\mathbf{v} = 5\mathbf{i} - \mathbf{j} + 2\mathbf{k}$; $\mathbf{w} = 2\mathbf{i} + \mathbf{j} - 3\mathbf{k}$
7. $\mathbf{v} = 3\mathbf{i} - \mathbf{j} + 2\mathbf{k}$; $\mathbf{w} = 2\mathbf{i} + 3\mathbf{j} - 4\mathbf{k}$
8. $\mathbf{v} = -\mathbf{j} + 4\mathbf{k}$; $\mathbf{w} = 5\mathbf{i} + 6\mathbf{k}$
9. $\mathbf{v} = \mathbf{i} - 6\mathbf{j} + 10\mathbf{k}$; $\mathbf{w} = -\mathbf{i} + 5\mathbf{j} - 6\mathbf{k}$
10. $\mathbf{v} = \cos\theta\,\mathbf{i} + \sin\theta\,\mathbf{j}$; $\mathbf{w} = -\sin\theta\,\mathbf{i} + \cos\theta\,\mathbf{j}$ for any angle θ

Use the cross product to find $\sin\theta$ *where* θ *is the angle between* $\mathbf{v}$ *and* $\mathbf{w}$ *in Problems* 11-14.

11. $\mathbf{v} = \mathbf{i} + \mathbf{k}$; $\mathbf{w} = \mathbf{i} + \mathbf{j}$ 12. $\mathbf{v} = \mathbf{i} + \mathbf{j}$; $\mathbf{w} = \mathbf{i} + \mathbf{j} + \mathbf{k}$
13. $\mathbf{v} = \mathbf{j} + \mathbf{k}$; $\mathbf{w} = \mathbf{i} + \mathbf{k}$ 14. $\mathbf{v} = \mathbf{i} + \mathbf{j}$; $\mathbf{w} = \mathbf{j} + \mathbf{k}$

Find a unit vector that is orthogonal to both $\mathbf{v}$ *and* $\mathbf{w}$ *in Problems* 15-18.
15. $\mathbf{v} = 2\mathbf{i} + \mathbf{k}$; $\mathbf{w} = \mathbf{i} - \mathbf{j} - \mathbf{k}$
16. $\mathbf{v} = \mathbf{j} - 3\mathbf{k}$; $\mathbf{w} = -\mathbf{i} + \mathbf{j} + \mathbf{k}$
17. $\mathbf{v} = \mathbf{i} + \mathbf{j} + \mathbf{k}$; $\mathbf{w} = 3\mathbf{i} + 12\mathbf{j} - 4\mathbf{k}$
18. $\mathbf{v} = 2\mathbf{i} - 2\mathbf{j} + \mathbf{k}$; $\mathbf{w} = 4\mathbf{i} + 2\mathbf{j} - 3\mathbf{k}$

Find the area of the parallelogram determined by the vectors in Problems 19-22.
19. $\mathbf{v} = 3\mathbf{i} + 4\mathbf{j}$ and $\mathbf{w} = \mathbf{i} + \mathbf{j} - \mathbf{k}$
20. $\mathbf{v} = 2\mathbf{i} - \mathbf{j} + 2\mathbf{k}$ and $\mathbf{w} = 4\mathbf{i} - 3\mathbf{j}$
21. $\mathbf{v} = 4\mathbf{i} - \mathbf{j} + \mathbf{k}$ and $\mathbf{w} = 2\mathbf{i} + 3\mathbf{j} - \mathbf{k}$
22. $\mathbf{v} = 2\mathbf{i} + 3\mathbf{k}$ and $\mathbf{w} = 2\mathbf{j} - 3\mathbf{k}$

Find the area of $\triangle PQR$ *in Problems* 23-26.
23. $P(0, 1, 1)$, $Q(1, 1, 0)$, $R(1, 0, 1)$
24. $P(1, 0, 0)$, $Q(2, 1, -1)$, $R(0, 1, -2)$
25. $P(1, 2, 3)$, $Q(2, 3, 1)$, $R(3, 1, 2)$
26. $P(-1, -1, -1)$, $Q(1, -1, -1)$, $R(-1, 1, -1)$

Determine whether each product in Problems 27-30 is a scalar or a vector or does not exist. Explain your reasoning.

27. a. $\mathbf{u} \times (\mathbf{v} \cdot \mathbf{w})$ b. $\mathbf{u} \cdot (\mathbf{v} \times \mathbf{w})$
28. a. $\mathbf{u} \times (\mathbf{v} \times \mathbf{w})$ b. $\mathbf{u} \cdot (\mathbf{v} \cdot \mathbf{w})$
29. a. $(\mathbf{u} \times \mathbf{v}) \cdot (\mathbf{u} \times \mathbf{w})$ b. $(\mathbf{u} \times \mathbf{v}) \times (\mathbf{u} \times \mathbf{w})$
30. a. $(\mathbf{u} \times \mathbf{v}) \cdot \mathbf{w}$ b. $|(\mathbf{u} \times \mathbf{v}) \cdot \mathbf{w}|$

In Problems 31-34, find the volume of the parallelepiped determined by vectors $\mathbf{u}, \mathbf{v},$ *and* $\mathbf{w}$.
31. $\mathbf{u} = \mathbf{j} + \mathbf{k}$; $\mathbf{v} = 2\mathbf{i} + \mathbf{j} + 2\mathbf{k}$; $\mathbf{w} = 5\mathbf{i}$
32. $\mathbf{u} = \mathbf{i} + \mathbf{j}$; $\mathbf{v} = \mathbf{j} + 2\mathbf{k}$; $\mathbf{w} = 3\mathbf{k}$
33. $\mathbf{u} = 2\mathbf{i} + \mathbf{j} - \mathbf{k}$; $\mathbf{v} = 3\mathbf{i} + \mathbf{k}$; $\mathbf{w} = \mathbf{j} + \mathbf{k}$
34. $\mathbf{u} = \mathbf{i} + \mathbf{j} + \mathbf{k}$; $\mathbf{v} = \mathbf{i} - \mathbf{j} - \mathbf{k}$; $\mathbf{w} = 2\mathbf{i} + 3\mathbf{k}$

Level 2

Explain your reasoning.

35. ■ What does this say? Contrast dot and cross products of vectors, including a discussion of some of their properties.

36. ■ What does this say? What is the right-hand rule?

37. Find a number s that guarantees that the vectors $\mathbf{i}$, $\mathbf{i} + \mathbf{j} + \mathbf{k}$, and $\mathbf{i} + 2\mathbf{j} + s\mathbf{k}$ will all be coplanar (in the same plane).

38. Find a number t that guarantees the vectors $\mathbf{i} + \mathbf{j}$, $2\mathbf{i} - \mathbf{j} + \mathbf{k}$, and $\mathbf{i} + \mathbf{j} + t\mathbf{k}$ will all be coplanar (in the same plane).

39. Let $\mathbf{u} = \mathbf{i} + \mathbf{j}$, $\mathbf{v} = 2\mathbf{i} - \mathbf{j} + \mathbf{k}$, and $\mathbf{w} = 3\mathbf{i}$. Compute $(\mathbf{u} \times \mathbf{v}) \times \mathbf{w}$ and $\mathbf{u} \times (\mathbf{v} \times \mathbf{w})$. What does this say about the associativity of cross product?

40. Show that $(a\mathbf{u}) \times (b\mathbf{v}) = ab(\mathbf{u} \times \mathbf{v})$ for any scalars a and b.

41. For a given vector $\mathbf{v}$ in $\mathbb{R}^3$, find all vectors $\mathbf{w}$ such that $\mathbf{v} \times \mathbf{w} = \mathbf{w}$.

42. What can be said about nonzero vectors $\mathbf{v}$ and $\mathbf{w}$ if $\mathbf{v} \cdot \mathbf{w} = 0$? What can be said if $\mathbf{v} \times \mathbf{w} = \mathbf{0}$?

43. One end of a 2-ft lever pivots about the origin in the yz-plane, as shown in Figure 9.44.

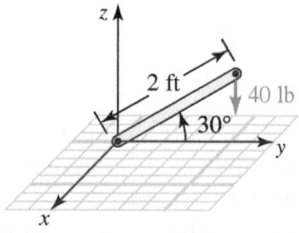

Figure 9.44 Finding the torque

If a vertical force of 40 lb is applied at the end of the lever, what is the torque of the lever about the pivot point (the origin) when the lever makes an angle of $30°$ with the xy-plane?

44. A 3-lb weight hangs from a rope at the end of Q of a 5-ft stick PQ that is held at an angle of $60°$ to the horizontal, as shown in Figure 9.45. What is the torque about the point P due to the weight?

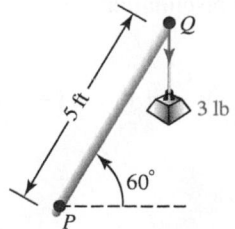

Figure 9.45 Finding the torque

45. Find the angle between the vector $2\mathbf{i} - \mathbf{j} + \mathbf{k}$ and the plane determined by the points $P(1, -2, 3)$, $Q(-1, 2, 3)$, and $R(1, 2, -3)$.

46. a. Show that the vectors $\mathbf{u}$, $\mathbf{v}$, and $\mathbf{w}$ are coplanar (all in the same plane) if

$$\mathbf{u} \cdot (\mathbf{v} \times \mathbf{w}) = 0 \text{ or } (\mathbf{u} \times \mathbf{v}) \cdot \mathbf{w} = 0$$

b. Are the vectors $\mathbf{u} = \mathbf{i} + 3\mathbf{j} + \mathbf{k}$, $\mathbf{v} = 2\mathbf{i} - \mathbf{j} - \mathbf{k}$, and $\mathbf{w} = 7\mathbf{j} + 3\mathbf{k}$ coplanar?

47. Show that the triangle with vertices (x_1, y_1), (x_2, y_2), (x_3, y_3) has area $A = \frac{1}{2}|D|$, where

$$D = \begin{vmatrix} x_1 & y_1 & 1 \\ x_2 & y_2 & 1 \\ x_3 & y_3 & 1 \end{vmatrix}$$

48. Using the properties of determinants, show that

$$\mathbf{u} \cdot (\mathbf{v} \times \mathbf{w}) = (\mathbf{u} \times \mathbf{v}) \cdot \mathbf{w}$$

for any vectors $\mathbf{u}, \mathbf{v}$, and $\mathbf{w}$.

Level 3

49. Prove the determinant formula for evaluating a scalar triple product.

50. Let A, B, C, and D be four points that do not lie in the same plane; see Figure 9.46, for example.

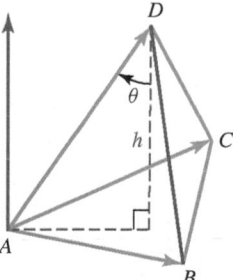

Figure 9.46 Problem 50

It can be shown that the volume of the tetrahedron with vertices A, B, C and D satisfies

$$[\text{VOLUME OF TETRAHEDRON } ABCD] = \frac{1}{3}\begin{pmatrix} \text{AREA OF} \\ \triangle ABC \end{pmatrix}\begin{pmatrix} \text{ALTITUDE FROM} \\ D \text{ TO } \triangle ABC \end{pmatrix}$$

Show that the volume is given by

$$V = \frac{1}{6}|(\mathbf{AB} \times \mathbf{AC}) \cdot \mathbf{AD}|$$

51. Show that if $\mathbf{u}, \mathbf{v},$ and $\mathbf{w}$ are vectors in $\mathbb{R}^3$ with $\mathbf{u} + \mathbf{v} + \mathbf{w} = \mathbf{0}$, then

$$\mathbf{u} \times \mathbf{v} = \mathbf{v} \times \mathbf{w} = \mathbf{w} \times \mathbf{u}$$

52. Prove Lagrange's identity

$$\|\mathbf{v} \times \mathbf{w}\|^2 = \|\mathbf{v}\|^2 \|\mathbf{w}\|^2 - (\mathbf{v} \cdot \mathbf{w})^2$$

53. Show that

$$\tan\theta = \frac{\|\mathbf{v} \times \mathbf{w}\|}{\mathbf{v} \cdot \mathbf{w}}$$

where θ $(0 \le \theta < \frac{\pi}{2})$ is the angle between $\mathbf{v}$ and $\mathbf{w}$.

54. Let $\mathbf{u}, \mathbf{v},$ and $\mathbf{w}$ be nonzero vectors in $\mathbb{R}^3$ that do not all lie in the same plane. Show that

$$|\mathbf{u} \cdot (\mathbf{v} \times \mathbf{w})| = |\mathbf{v} \cdot (\mathbf{u} \times \mathbf{w})|$$

What other scalar triple products involving $\mathbf{u}, \mathbf{v},$ and $\mathbf{w}$ have the same absolute values?

55. Let $\mathbf{a}, \mathbf{b},$ and $\mathbf{c}$ be vectors in space. Prove the *cab-bac* formula

$$\mathbf{a} \times (\mathbf{b} \times \mathbf{c}) = (\mathbf{c} \cdot \mathbf{a})\mathbf{b} - (\mathbf{b} \cdot \mathbf{a})\mathbf{c}$$

Establish the validity of the equations in Problems 56-59 for arbitrary vectors $\mathbf{u}, \mathbf{v}, \mathbf{w},$ and $\mathbf{z}$ in $\mathbb{R}^3$. You may use the result of Problem 55.

56. $(\mathbf{u} \times \mathbf{v}) \times (\mathbf{w} \times \mathbf{z}) = (\mathbf{u} \cdot \mathbf{w} \times \mathbf{z})\mathbf{v} - (\mathbf{v} \cdot \mathbf{w} \times \mathbf{z})\mathbf{u}$
57. $\mathbf{u} \times (\mathbf{v} \times \mathbf{w}) + \mathbf{v} \times (\mathbf{w} \times \mathbf{u}) + \mathbf{w} \times (\mathbf{u} \times \mathbf{v}) = \mathbf{0}$
58. $\mathbf{u} \times \mathbf{v} = (\mathbf{u} \cdot \mathbf{v} \times \mathbf{i})\mathbf{i} + (\mathbf{u} \cdot \mathbf{v} \times \mathbf{j})\mathbf{j} + (\mathbf{u} \cdot \mathbf{v} \times \mathbf{k})\mathbf{k}$
59. $\mathbf{u} \times [\mathbf{u} \times (\mathbf{u} \times \mathbf{v})] \cdot \mathbf{w} = -\|\mathbf{u}\|^2 \mathbf{u} \cdot \mathbf{v} \times \mathbf{w}$
60. Show that if vectors $\mathbf{OP}, \mathbf{OQ}, \mathbf{OR},$ and $\mathbf{OS}$ lie in the same plane, then

$$(\mathbf{OP} \times \mathbf{OQ}) \times (\mathbf{OR} \times \mathbf{OS}) = \mathbf{0}$$

9.5 LINES IN $\mathbb{R}^3$

IN THIS SECTION: *Equations of lines in $\mathbb{R}^3$, parametric equations, parameterizing a curve*
As we move from $\mathbb{R}^2$ to $\mathbb{R}^3$, we look for patterns. As we have seen, the number of components of points corresponds to the dimension. In $\mathbb{R}$, we locate a point with one component, in $\mathbb{R}^2$ we use ordered pairs, and in $\mathbb{R}^3$ we use triples. Circles in $\mathbb{R}^2$ are two points in $\mathbb{R}$ and spheres in $\mathbb{R}^3$. However, lines in $\mathbb{R}^2$ generalize to planes in $\mathbb{R}^3$, so we need to pay special attention to developing lines in $\mathbb{R}^3$.

Equations of Lines in $\mathbb{R}^3$

In precalculus, and in Section 1.3, we graphed lines in $\mathbb{R}^2$ using a variety of forms, and one of those forms was the parametric form of a line passing through (x_1, y_1) with slope $m = \frac{b}{a}, a \ne 0$, namely $x = x_1 + at$ and $y = y_1 + bt$. In order to graph a line in $\mathbb{R}^3$, we need to write the slope $m = \frac{b}{a}$ in a form which involves changes in three components instead of two. That is, instead of talking about the slope $m = \frac{b}{a}$, we talk about the direction $[a, b]$ where a is the change in x and b is the change in y, and we note that this line has the same direction as a vector $\mathbf{v} = a\mathbf{i} + b\mathbf{j}$. Using this representation, we now generalize to $\mathbb{R}^3$.

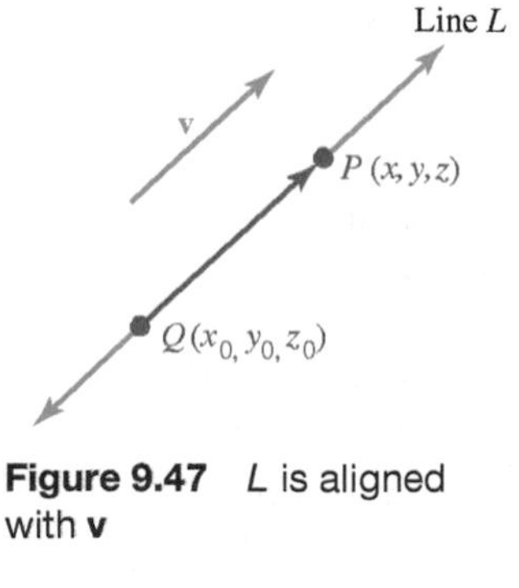

Suppose L is a line in space that contains $Q(x_0, y_0, z_0)$ whose location is determined by the vector $\mathbf{v} = A\mathbf{i} + B\mathbf{j} + C\mathbf{k}$. We say that L is **aligned with v**, as shown in Figure 9.47.

We also say that the line has **direction numbers** $A, B,$ and C and denote these direction numbers by $[A, B, C]$. If $P(x, y, z)$ is any point on L, then the vector $\mathbf{QP}$ is parallel to $\mathbf{v}$ and must satisfy the vector equation $\mathbf{QP} = t\mathbf{v}$ for some number t. If we introduce coordinates and use the standard representation, we can rewrite this vector equation as

$$(x - x_0)\mathbf{i} + (y - y_0)\mathbf{j} + (z - z_0)\mathbf{k} = t[A\mathbf{i} + B\mathbf{j} + C\mathbf{k}]$$

Figure 9.47 *L* is aligned with **v**

By equating components on both sides of this equation, we find that the coordinates of P must satisfy the linear system

$$x - x_0 = tA \qquad y - y_0 = tB \qquad z - z_0 = tC$$

where t is a real number.

PARAMETRIC FORM OF A LINE IN $\mathbb{R}^3$ If L is a line that contains the point (x_0, y_0, z_0) and is aligned with the vector $\mathbf{v} = A\mathbf{i} + B\mathbf{j} + C\mathbf{k}$, then the point (x, y, z) is on L if and only if its coordinates satisfy

$$x - x_0 = tA \qquad y - y_0 = tB \qquad z - z_0 = tC$$

for some number t.

⚲ The parameter t will be a different value for each point (x, y, z).⚲

Turning things around, if we are given the equation of a line with direction numbers $[A, B, C]$, then $\mathbf{v} = A\mathbf{i} + B\mathbf{j} + C\mathbf{k}$ is the **vector aligned with** L.

Example 1   Parametric equations of a line in space

Find the parametric equations for the line that contains the point $(3, 1, 4)$ and is aligned with the vector $\mathbf{v} = -\mathbf{i} + \mathbf{j} - 2\mathbf{k}$. Find where this line passes through the coordinate planes and sketch the line.

Solution   The direction numbers are $[-1, 1, -2]$ and $x_0 = 3$, $y_0 = 1$, $z_0 = 4$, so the line has the parametric form

$$
\begin{array}{ccc}
x - 3 = -t & y - 1 = t & z - 4 = -2t \\
x = 3 - t & y = 1 + t & z = 4 - 2t
\end{array}
$$

This line will intersect the xy-plane when $z = 0$; solve

$$0 = 4 - 2t \qquad \text{implies} \qquad t = 2$$

If $t = 2$, then $x = 3 - 2 = 1$ and $y = 1 + 2 = 3$. This is the point $(1, 3, 0)$. Similarly, the line intersects the xz-plane at $(4, 0, 6)$ and the yz-plane at $(0, 4, -2)$. Plot these points and draw the line, as shown in Figure 9.48.

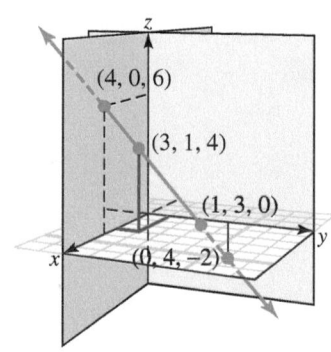

Figure 9.48   Graph of a line in space

Example 1 shows the graph of a line in $\mathbb{R}^3$, but it does not show how you will draw a line on your own paper. The following drawing lesson shows steps you can use to graph the line

$$x = 2 + 2t \qquad y = 10t \qquad z = 3 - 3t$$

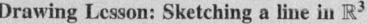

Drawing Lesson: Sketching a line in $\mathbb{R}^3$

Draw the three coordinate axes.

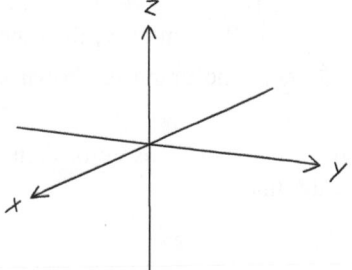

Next, draw the three coordinate planes.

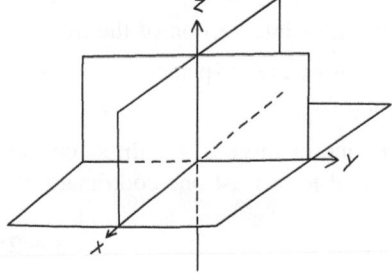

We focus, instead, on the points where the graph passes through each of the coordinate planes.

We see, by inspection, that the line passes through (2,0,3); plot this point.

Use colored pencils or highlighters to distinguish individual planes. Here is an example of a finished drawing:

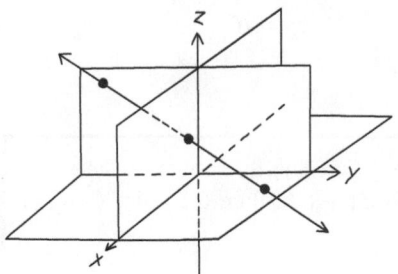

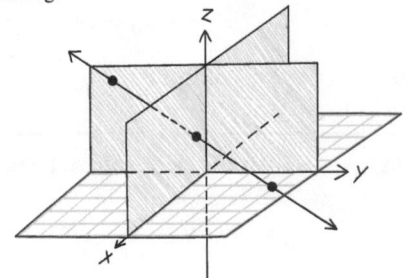

If $x = 0$, then $t = -1$, so $y = -10$ and $z = 6$; plot the point $(0, -10, 6)$ in the yz-plane. Finally, if $z = 0$, $t = 1$, so $x = 4, y = 10$; plot $(4, 10, 0)$ in the xy-plane.

In the special case where none of the direction numbers $A, B,$ or C is 0, we can solve each of the parametric-form equations for t to obtain the following **symmetric equations** for a line.

SYMMETRIC FORM OF A LINE IN $\mathbb{R}^3$ If L is a line that contains the point (x_0, y_0, z_0) and is aligned with the vector $\mathbf{v} = A\mathbf{i} + B\mathbf{j} + C\mathbf{k}$, ($A$, B, and C nonzero numbers) then the point (x, y, z) is L if and only if its coordinates satisfy

$$\frac{x - x_0}{A} = \frac{y - y_0}{B} = \frac{z - z_0}{C}$$

Example 2   Symmetric form of the equation of a line in space

Find symmetric equations for the line L through the points $P(-1, 3, 7)$ and $Q(4, 2, -1)$. Find the points of intersection with the coordinate planes and sketch the line.

Solution   The required line passes through P and is aligned with the vector

$$\mathbf{PQ} = [4 - (-1)]\mathbf{i} + [2 - 3]\mathbf{j} + [-1 - 7]\mathbf{k} = 5\mathbf{i} - \mathbf{j} - 8\mathbf{k}$$

Thus, the direction numbers of the line are $[5, -1, -8]$, and we can choose either P or Q as (x_0, y_0, z_0). Choosing P, we obtain:

$$\frac{x + 1}{5} = \frac{y - 3}{-1} = \frac{z - 7}{-8}$$

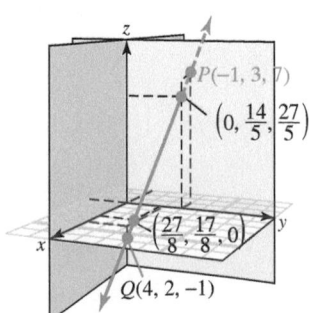

Figure 9.49 Interactive
Graph of
$$\frac{x + 1}{5} = \frac{y - 3}{-1} = \frac{z - 7}{-8}$$

Next, we find points of intersection with the coordinate planes:

xy-plane: $z = 0$, so $\dfrac{x + 1}{5} = \dfrac{7}{8}$ implies $x = \dfrac{27}{8}$ and $\dfrac{y - 3}{-1} = \dfrac{7}{8}$ implies $y = \dfrac{17}{8}$.

The point of intersection of the line with the xy-plane is $(\frac{27}{8}, \frac{17}{8}, 0)$. Similarly, the other intersections are xz-plane: $(14, 0, -17)$; yz-plane: $(0, \frac{14}{5}, \frac{27}{5})$. The graph is shown in Figure 9.49. ∎

If one or more of the direction numbers $[A, B, C]$ of a line L in $\mathbb{R}^3$ is zero, then L is parallel to at least one coordinate plane. For example, the line

$$x = 3 + 2t, \ y = 5, \ z = 4 - t$$

has direction numbers $[2, 0, -1]$ and lies in the plane $y = 5$, parallel to the xz-plane. The symmetric form for this line is

$$\frac{x-3}{2} = \frac{z-4}{-1}; \quad y = 5$$

On the other hand, consider a line with symmetric form

$$\frac{y+1}{4} = \frac{z-3}{-2}; \quad x = 7$$

lies in the plane $x = 7$, parallel to the yz-plane and has direction numbers $[0, 4, -2]$.

You might remember that two lines in $\mathbb{R}^2$ must intersect if their slopes are different (because they cannot be parallel), but two lines in space may have different direction numbers and still not intersect. In this case, the lines are said to be **skew**. The reason why the situation in $\mathbb{R}^3$ is different from that in $\mathbb{R}^2$ is that even though lines with different direction numbers cannot be parallel, there is still enough "room" in space for the lines to lie in parallel planes and be aligned with vectors that are not parallel. This situation is shown in Figure 9.50.

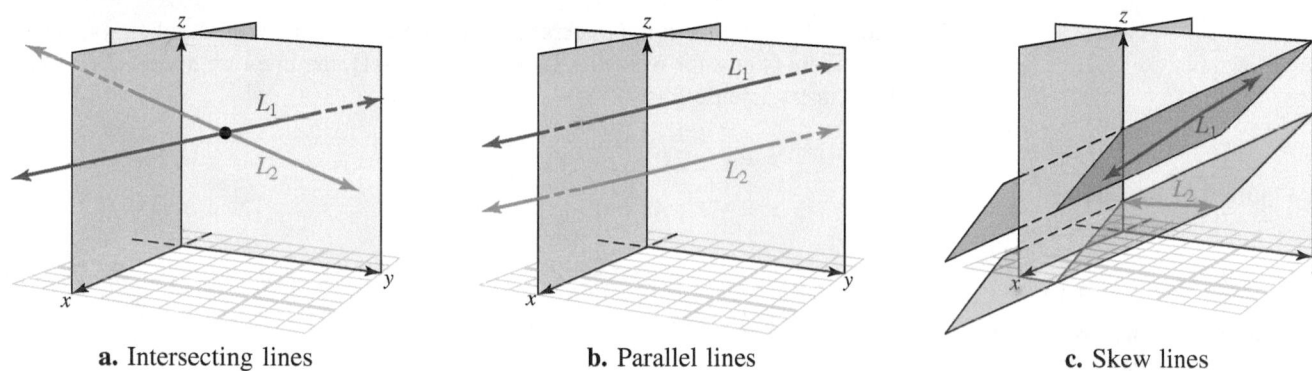

a. Intersecting lines   **b.** Parallel lines   **c.** Skew lines

Figure 9.50   Lines in space may intersect, be parallel, or skew

Example 3   Skew lines in space

Determine whether the following pair of lines intersect, are parallel, or are skew.

$$L_1: \frac{x-1}{2} = \frac{y+1}{1} = \frac{z-2}{4} \quad \text{and} \quad L_2: \frac{x+2}{4} = \frac{y}{-3} = \frac{z+1}{1}$$

Solution   Note that L_1 has direction numbers $[2, 1, 4]$ (that is, L_1 is aligned with $2\mathbf{i} + \mathbf{j} + 4\mathbf{k}$) and L_2 has direction numbers $[4, -3, 1]$. If we solve $\langle 2, 1, 4 \rangle = t \langle 4, -3, 1 \rangle$ for t, we find no possible solution for any value of t. This implies that the lines are not parallel. (See Figure 9.51.)

Next, we determine whether the lines intersect or are skew. Note that $S(1, -1, 2)$ lies on L_1 and $T(-2, 0, -1)$ lies on L_2. The lines intersect if and only if there is a point P that lies on both lines. To determine this, we write the equations of the lines in parametric form. We use a different parameter for each line so that points on one line do not depend on the value of the parameter of the other line:

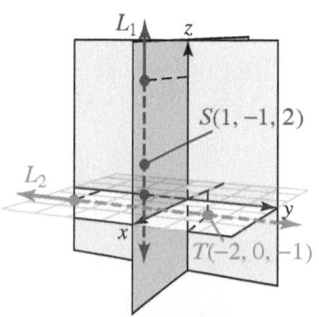

Figure 9.51 Interactive
Graphs of L_1 and L_2

L_1: $x = 1 + 2s$ $y = -1 + s$ $z = 2 + 4s$

L_2: $x = -2 + 4t$ $y = -3t$ $z = -1 + t$

Thus: $x = 1 + 2s = -2 + 4t$ $y = -1 + s = -3t$ $z = 2 + 4s = -1 + t$

or: $2s - 4t = -3$ $s + 3t = 1$ $4s - t = -3$

This is equivalent to the system of linear equations

$$\begin{cases} 2s - 4t = -3 \\ s + 3t = 1 \\ 4s - t = -3 \end{cases}$$

Any solution of this system must correspond to a point of intersection of L_1 and L_2, and if no solution exists, then L_1 and L_2 are skew. Because this is a system of three equations with two unknowns, we first solve the first two equations simultaneously to find $s = -\frac{1}{2}, t = \frac{1}{2}$. Because $s = -\frac{1}{2}$ and $t = \frac{1}{2}$ do not satisfy the third equation, it follows that L_1 and L_2 do not intersect, so they must be skew. ∎

Example 4 Intersecting lines

Show that the lines

$$L_1\colon \frac{x-1}{2} = \frac{y+1}{1} = \frac{z-2}{4} \quad \text{and} \quad L_2\colon \frac{x+2}{4} = \frac{y}{-3} = \frac{z-\frac{1}{2}}{-1}$$

intersect and find the point of intersection. (See Figure 9.52.)

Solution L_1 has direction numbers $[2, 1, 4]$ and L_2 has direction numbers $[4, -3, -1]$. Because there is no t for which $[2, 1, 4] = t[4, -3, -1]$, the lines are not parallel. Express the lines in parametric form:

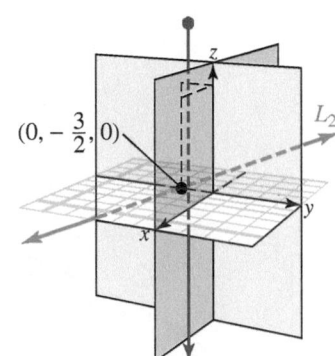

$(0, -\frac{3}{2}, 0)$

L_2

Figure 9.52 Interactive
Graph of L_1 and L_2

L_1:	$x = 1 + 2s$	$y = -1 + s$	$z = 2 + 4s$
L_2:	$x = -2 + 4t$	$y = -3t$	$z = \frac{1}{2} - t$
Thus:	$x = 1 + 2s = -2 + 4t$	$y = -1 + s = -3t$	$z = 2 + 4s = \frac{1}{2} - t$
or:	$2s - 4t = -3$	$s + 3t = 1$	$4s + t = -\frac{3}{2}$

Notice that we used two parameters (s and t) for the lines in Example 4. This is because it is quite possible for two lines to intersect at "different times" (that is, different values of the parameter). In Example 4, we have $s = -\frac{1}{2}$ and $t = \frac{1}{2}$ at the point of intersection, rather than $s = t$.

Solving the first two equations simultaneously, we find $s = -\frac{1}{2}, t = \frac{1}{2}$. This solution satisfies the third equation, namely,

$$4\left(-\frac{1}{2}\right) + \frac{1}{2} = -\frac{3}{2}$$

To find the coordinates of the point of intersection, substitute $s = -\frac{1}{2}$ into the parametric-form equations for L_1 (or substitute $t = \frac{1}{2}$ into L_2) to obtain

$$x_0 = 1 + 2\left(-\frac{1}{2}\right) = 0$$

$$y_0 = -1 + \left(-\frac{1}{2}\right) = -\frac{3}{2}$$

$$z_0 = 2 + 4\left(-\frac{1}{2}\right) = 0$$

Thus, the lines intersect at $P\left(0, -\frac{3}{2}, 0\right)$. ∎

Parametric Equations

In this book we have used parameters to graph lines in $\mathbb{R}^2$ (Section 1.3), and in $\mathbb{R}^3$ in this section. However, parametric representation can be used to graph a wide variety of curves and surfaces in both two and three dimensions. We begin with a definition.

PARAMETRIC REPRESENTATION   Let f_1, f_2, and f_3 be continuous functions of t on an interval I; then the equations

$$x = f_1(t), \quad y = f_2(t), \quad z = f_3(t)$$

are called **parametric equations** with **parameter** t. As t varies over the **parametric set** I, the points

$$(x, y, z) = (f_1(t), f_2(t), f_3(t))$$

trace out a **parametric curve** in $\mathbb{R}^3$. If $z = f_3(t) = 0$, then the curve is in the xy-plane and we say the parametric curve is in $\mathbb{R}^2$.

Note: The letter "t" used for the parameter does not necessarily denote time, although in many applications, time is a suitable parameter. Indeed, any letter or symbol may be used to denote a parameter. If $z = 0$ for all values of t, then the curve is called a **plane curve**, and if $z \neq 0$ for at least one value of t, then the curve is called a **space curve**. We will investigate space curves at length in the next chapter.

Example 5   Sketching the path of a parametric curve

Sketch the path of the curve $x = t^2 - 9$, $y = \frac{1}{3}t$ for $-3 \leq t \leq 2$.

Solution   Values of x and y corresponding to various choices of the parameter t are shown in the following table:

t	x	y	
-3	0	-1	(Starting or **initial** point)
-2	-5	$-\frac{2}{3}$	
-1	-8	$-\frac{1}{3}$	
0	-9	0	
1	-8	$\frac{1}{3}$	
2	-5	$\frac{2}{3}$	(Ending or **terminal** point)

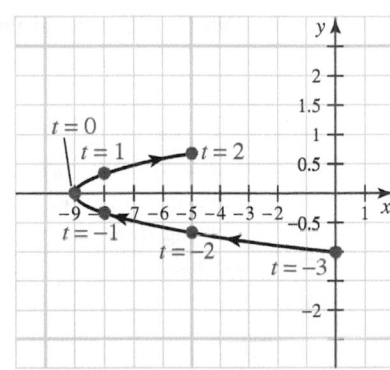

Figure 9.53   **Interactive** Graph of spiral with restricted domain

The graph is shown in Figure 9.53. Note how the arrows show the orientation as t increases from -3 to 2.

If you are using a computer or a graphing calculator, plotting points can be an effective way of sketching a parametric curve. Sometimes, however, we wish to eliminate the parameter to obtain a Cartesian equation. For instance, in Example 5, we have $y = \frac{1}{3}t$, so $t = 3y$, and by substituting into the equation $x = t^2 - 9$, we obtain

$$x = (3y)^2 - 9 = 9y^2 - 9$$

which is the Cartesian equation for a parabola that opens to the right. Because of the domain of the parameter t, we see that the parametric curve in Figure 9.53 is a subset of the set of points that satisfy the equation

$$x = 9y^2 - 9$$

Parameterizations are not unique. For example, the curve with parametric equations

$$x = 9(9s^2 - 1), \ y = 3s \ \text{for} \ -\frac{1}{3} \le s \le \frac{2}{9}$$

is the same as the curve in Figure 9.53.

Example 6 Sketching the path by eliminating the parameter

Describe the path $x = \sin \pi t$, $y = \cos 2\pi t$ for $0 \le t \le 0.5$.

Solution Using a double angle identity, we find

$$\cos 2\pi t = 1 - 2\sin^2 \pi t$$

so that

$$y = 1 - 2x^2$$

We recognize this as a Cartesian equation for a parabola. Because $y' = -4x$, we can find the critical number $x = 0$, which locates the vertex of the parabola at $(0, 1)$. The parabola is the curve shown in color as the dashed curve in Figure 9.54.

Because t is restricted to the interval $0 \le t \le 0.5$, the parametric representation involves only part of the right side of the parabola $y = 1 - 2x^2$. The curve is oriented from the point $(0, 1)$, where $t = 0$, to the point $(1, -1)$, where $t = 0.5$, and is the portion of the parabola shown in black in Figure 9.54. ∎

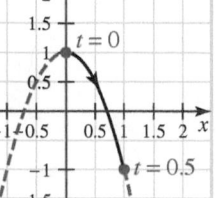

Figure 9.54 Interactive
Graph of parabola with restricted domain

When it is difficult to eliminate the parameter from a given parametric representation, we can sometimes get a good picture of the parametric curve by plotting points.

Example 7 Describing a spiraling path

Discuss the path of the curve described by the parametric equations

$$x = e^{-t} \cos t, \quad y = e^{-t} \sin t \qquad \text{for} \ t \ge 0$$

Solution We have no convenient way of eliminating the parameter so we write out a table of values (x, y) that correspond to various values of t. The curve is obtained by plotting these points in a Cartesian plane and passing a smooth curve through the plotted points, as shown in Figure 9.55.

t	x	y
0	1	0
$\frac{\pi}{4}$	0.32	0.32
1	0.20	0.31
$\frac{\pi}{2}$	0	0.21
2	-0.06	0.12
π	-0.04	0
$\frac{3\pi}{2}$	0	-0.01
2π	0.00	0

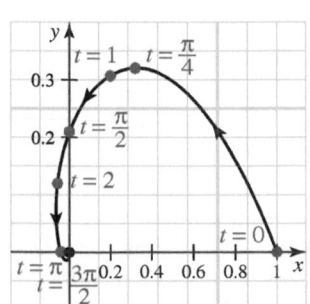

Figure 9.55 **Interactive** Graph of spiral with restricted domain

Note that for each value of t, the distance from $P(x, y)$ on the curve to the origin is

$$\sqrt{x^2 + y^2} = \sqrt{(e^{-t}\cos t)^2 + (e^{-t}\sin t)^2} = \sqrt{e^{-2t}(1)} = e^{-t}$$

Because e^{-t} decreases as t increases, it follows that P gets closer and closer to the origin as t increases. However, because $\cos t$ and $\sin t$ vary between -1 and $+1$, the approach is not direct but takes place along a spiral. ∎

Parameterizing a Curve

So far, our examples have dealt with sketching a parametric curve given the parametric equations. In general, this process may be tedious, but generally can be done easily using technology. However, the reverse process, finding a suitable set of parametric equations for a given curve, is an art for which there is no simple procedure. Indeed, a given curve can have many different parameterizations and there are curves for which no simple parameterization be given. The following two examples illustrate various methods for parameterizating a given curve.

Example 8   Parameterizing two curves

In each of the following cases, parameterize the given curve:

a. $y = 9x^2$ **b.** $r = 5\cos^3\theta$ in polar coordinates

Solution

a. The usual parameterization for a parabola is to let the parameter t be the variable that is squared: $x = t$, $y = 9t^2$. However, another parameterization is to let $t = 3x$ so that $x = \frac{1}{3}t$ and $y = t^2$.

b. In polar coordinates we have $x = r\cos\theta$, $y = r\sin\theta$, so we can parameterize x and y in terms of the parameter θ:

$$
\begin{aligned}
x &= r\cos\theta & y &= r\sin\theta \\
&= \left(5\cos^3\theta\right)\cos\theta & &= \left(5\cos^3\theta\right)\sin\theta \\
&= 5\cos^4\theta & &= 5\cos^3\theta\sin\theta
\end{aligned}
$$
∎

Example 9   Modeling Problem: Finding parametric equations for a trochoid

A bicycle wheel has radius a and a reflector is attached at a point P on a spoke of the bicycle wheel at a fixed distance d from the center. Find parametric equations for the curve described by P as the wheel rolls along a straight line without slipping. Such a curve is called a **trochoid**, and is shown in Figure 9.56.

Solution   Assume that the wheel rolls along the x-axis and that the center C of the wheel begins at $(0, a)$ on the y-axis. Further assume that P also starts on the y-axis, d units below C. Figure 9.57 shows the initial position of the wheel and its position after turning through an angle θ (radians).

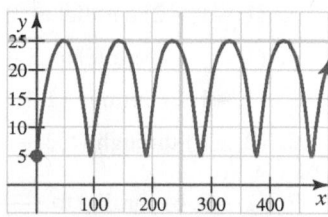

The path of a reflector placed 5 in. from the tire on the wheel of a bike with a 30-in wheel. This graph is an example of a trochoid.

Figure 9.56   Graph of a trochoid

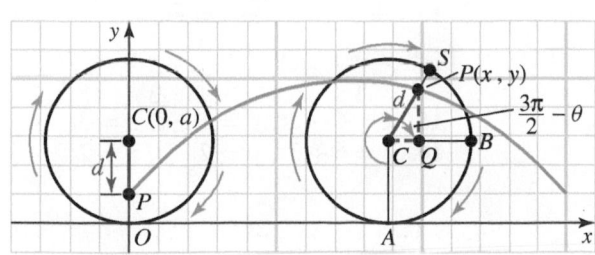

Figure 9.57   The path of a reflector on a bicycle

We begin by labeling some points: The point A is on the x-axis directly beneath C, whereas B is the point where the horizontal line through C meets the rim of the wheel. Finally, Q is the point on BC directly beneath P, and S is the point where the line through C and P intersects the rim. Let P have coordinates (x, y).

We need to find representations (in terms of a, d, and θ) for x and y.

$$x = |\overline{OA}| + |\overline{CQ}|$$
$$= a\theta + |\overline{CQ}|$$

Because the wheel rolls along the x-axis without slipping, $|\overline{OA}|$ is the same as the arc length from A to S, so $|\overline{OA}| = a\theta$.

$$y = |\overline{AC}| + |\overline{QP}|$$
$$= a + |\overline{QP}|$$

To complete our evaluation of x and y, we need to compute $|\overline{CQ}|$ and $|\overline{QP}|$. These are sides of $\triangle PCQ$. Note that $\angle PCQ = \frac{3\pi}{2} - \theta$; therefore, by the definition, of cosine and sine, we have

$$\cos\left(\frac{3\pi}{2} - \theta\right) = \frac{|\overline{CQ}|}{d} \qquad \text{so} \qquad |\overline{CQ}| = d\cos\left(\frac{3\pi}{2} - \theta\right) = -d\sin\theta$$

Similarly, $|\overline{QP}| = d\sin\left(\frac{3\pi}{2} - \theta\right) = -d\cos\theta$.

We can now substitute these values for $|\overline{CQ}|$ and $|\overline{QP}|$ into the equations we derived for x and y.

$$x = a\theta + |\overline{CQ}| = a\theta - d\sin\theta$$
$$y = a + |\overline{QP}| = a - d\cos\theta$$

The special case where P is on the rim of the wheel in Example 9 (when $d = a$) is a curve called a **cycloid**. There are several problems involving these and similar curves in the problem set.

PROBLEM SET 9.5

Level 1

1. ■ What does this say? What is a parameter?
2. ■ What does this say? Contrast the parametric and symmetric forms of the equation of a line.

Find the parametric and symmetric equations for the line(s) passing through the given points with the properties described in Problems 3-10.

3. $(1, -1, -2)$ parallel to $3\mathbf{i} - 2\mathbf{j} + 5\mathbf{k}$

4. $(1, 0, -1)$ parallel to $3\mathbf{i} + 4\mathbf{j}$

5. $(1, -1, 2)$ through $(2, 1, 3)$

6. $(2, 2, 3)$ through $(1, 3, -1)$

7. $(1, -3, 6)$ parallel to $\dfrac{x-5}{1} = \dfrac{y+2}{-3} = \dfrac{z}{-5}$

8. $(1, -1, 2)$ parallel to $\dfrac{x+3}{4} = \dfrac{y-2}{5} = \dfrac{z+5}{1}$

9. $(0, 4, -3)$ parallel to $\dfrac{x-1}{22} = \dfrac{y+2}{-6} = \dfrac{z-1}{10}$

10. $(1, 0, -4)$ parallel to $x = -2 + 3t$, $y = 4 + t$, $z = 2 + 2t$

11. Find the parametric form of the equation of the line passing through $(3, -1, 0)$ parallel to both the xy- and xz-planes.

12. Find the parametric form of the equation of the line passing through $(3, -1, 0)$ parallel to both the xy- and yz-planes.

Find the points of intersection of each line in Problems 13-16 with each of the coordinate planes.

13. $\dfrac{x-4}{4} = \dfrac{y+3}{3} = \dfrac{z+2}{1}$

14. $\dfrac{x+1}{1} = \dfrac{y+2}{2} = \dfrac{z-6}{3}$

15. $x = 6 - 2t$, $y = 1 + t$, $z = 3t$

16. $x = 6 + 3t$, $y = 2 - t$, $z = 2t$

In Problems 17-22, tell whether the two lines intersect, are parallel, are skew, or coincide. If they intersect, give the point of intersection.

17. $\dfrac{x-4}{2} = \dfrac{y-6}{-3} = \dfrac{z+2}{5}$

$\dfrac{x}{4} = \dfrac{y+2}{-6} = \dfrac{z-3}{10}$

18. $x = 4 - 2t,\ y = 6t,\ z = 7 - 4t$

$x = 5 + t,\ y = 1 - 3t,\ z = -3 + 2t$

19. $x = 3 + 3t,\ y = 1 - 4t,\ z = -4 - 7t$

$x = 2 + 3t,\ y = 5 - 4t,\ z = 3 - 7t$

20. $x = 2 - 4t,\ y = 1 + t,\ z = \frac{1}{2} + 5t$

$x = 3t,\ y = -2 - t,\ z = 4 - 2t$

21. $\dfrac{x-3}{2} = \dfrac{y-1}{-1} = \dfrac{z-4}{1}$

$\dfrac{x+2}{3} = \dfrac{y-3}{-1} = \dfrac{z-2}{1}$

22. $\dfrac{x+1}{2} = \dfrac{y-3}{-1} = \dfrac{z-2}{1}$

$\dfrac{x+1}{2} = \dfrac{y+1}{3} = \dfrac{z-3}{-4}$

Find an explicit relationship between x and y in Problems 23-36 by eliminating the parameter. In each case, sketch the path described by the parametric equations over the prescribed interval.

23. $x = t,\ y = t - 1,\ 0 \le t \le 3$

24. $x = -t,\ y = 3 - 2t,\ 0 \le t \le 1$

25. $x = 60t,\ y = 80t - 16t^2,\ 0 \le t \le 3$

26. $x = 30t,\ y = 60t - 9t^2,\ -1 \le t \le 2$

27. $x = t^3,\ y = t^2,\ t \ge 0$

28. $x = t^4,\ y = t^2,\ -1 \le t \le \sqrt{2}$

29. $x = 3\cos\theta,\ y = 3\sin\theta,\ 0 \le \theta \le 2\pi$

30. $x = 2\sin\theta,\ y = 2\cos\theta,\ 0 \le \theta \le 2\pi$

31. $x = 1 + \sin t,\ y = -2 + \cos t,\ 0 \le t \le 2\pi$

32. $x = 1 + \sin^2 t,\ y = -2 + \cos t,\ 0 \le t \le \pi$

33. $x = 4\tan 2t,\ y = 3\sec 2t,\ 0 \le t \le \pi$

34. $x = 4\sec 2t,\ y = 2\tan 2t,\ 0 \le t \le \pi$

35. $x = t^3,\ y = 3\ln t,\ t > 0$

36. $x = e^t,\ y = e^{-t},\ (-\infty, \infty)$

37. Find two unit vectors parallel to the line

$$\frac{x-3}{4} = \frac{y-1}{2} = \frac{z+1}{1}$$

38. Find two unit vectors parallel to the line

$$\frac{x-1}{2} = \frac{y+2}{4} = \frac{z+5}{1}$$

Level 2

Find parametric equations for each of the curves in Problems 39-44.

39. A circle of radius 3, centered at the origin, oriented counterclockwise.

40. A circle of radius 2 centered at the origin, oriented clockwise.

41. The ellipse

$$\frac{x^2}{9} + \frac{y^2}{4} = 1$$

oriented counterclockwise.

42. The parabola $y^2 = 4x + 9$, oriented from $(4, -5)$ to $(0, 3)$.

43. The hyperbola $\dfrac{x^2}{16} - \dfrac{y^2}{9} = 1$.

44. The ellipse $\dfrac{(x-2)^2}{3} + \dfrac{(y+3)^2}{5} = 1$.

45. Describe the path of the curve described by $x = \sin\pi t$, $y = \cos 2\pi t$ for $0 \le t \le 1$.

46. Describe the path of the curve described by $x = \cos\pi t$, $y = \cos 2\pi t$ for $0 \le t \le 1$.

47. Let $x = 4a\sin t$, $y = b\cos^2 t$. Express y as a function of x.

48. Let $x = a\cos t$, $y = 2b\sin^2 t$. Express y as a function of x.

49. Show that the vector $\mathbf{v} = 3\mathbf{i} - 4\mathbf{j} + \mathbf{k}$ is orthogonal to the line that passes through the points $P(0, 0, 1)$ and $Q(2, 1, -1)$.

50. Show that the vector $\mathbf{v} = 7\mathbf{i} + 4\mathbf{j} + 3\mathbf{k}$ is orthogonal to the line passing through the points $P(-2, 2, 7)$ and $Q(3, -3, 2)$.

51. Find constants a and b so that the following lines coincide:

$$L_1 : \frac{x-a}{2} = \frac{y-1}{4} = \frac{z+2}{1}$$

and

$$L_2 : \frac{x-2}{-4} = \frac{y-b}{-8} = \frac{z+1}{-2}$$

52. Find constants a and b so that the following lines coincide:

$$L_1 : \frac{x-1}{1} = \frac{y+a}{2} = \frac{z+2}{4}$$

and

$$L_2 : \frac{x-b}{-2} = \frac{y-1}{-4} = \frac{z+1}{-8}$$

53. Consider the lines

$$L_1: x = -1 + 2t, \, y = 3 - t, \, z = 2 + 2t$$
$$L_2: x = -2 - t, \, y = 5 + 2t, \, z = -2t$$

Show that L_1 and L_2 intersect and find the point of intersection and the acute angle θ (to the nearest degree) between them.

54. Consider the lines

$$L_1: \frac{x+2}{3} = \frac{y-1}{2} = \frac{z}{-1}$$
$$L_2: \frac{x+1}{-2} = \frac{y}{3} = \frac{z-2}{1}$$

Show that L_1 and L_2 are skew.

55. Find an equation for the line L_1 that contains the point $P(-1, 3, 1)$ and meets the line L_2 orthogonally, where

$$L_2: x = 2 - t, \, y = 1 - 2t, \, z = 5 + t$$

56. Find an equation for the line L_1 that contains the point $P(2, 3, -1)$ and meets the line L_2 orthogonally, where

$$L_2: \frac{x-2}{2} = \frac{y+1}{-2} = \frac{z}{1}$$

Level 3

57. What can be said about the lines

$$\frac{x - x_0}{a_1} = \frac{y - y_0}{b_1} = \frac{z - z_0}{c_1}$$

and

$$\frac{x - x_0}{a_2} = \frac{y - y_0}{b_2} = \frac{z - z_0}{c_2}$$

in the case where $a_1 a_2 + b_1 b_2 + c_1 c_2 = 0$?

58. Modeling Problem A circle of radius R rolls without slipping on the *outside* of a fixed circle of radius a. Assume the fixed circle is centered at the origin and that the moving point begins at $(a, 0)$. Use the angle t measured from the positive x-axis to the ray from the origin to the center of the rolling circle. Show that this *epicycloid* may be modeled by the parametric equations

$$x = (a + R)\cos t - R \cos\left(\frac{a+R}{R}\right) t$$
$$y = (a + R)\sin t - R \sin\left(\frac{a+R}{R}\right) t$$

59. Modeling Problem A circle of radius R rolls without slipping on the *inside* of a fixed circle of radius a. Find parametric equations to model the curve traced out by a point P on the circumference of the rolling circle of radius R. Let t be the angle measured from the positive x-axis to the ray that passes through the center of the rolling circle, and assume that the point P begins on the x-axis (that is, P has coordinates $(a, 0)$ when $t = 0$). Show that this *hypocycloid* has parametric equations

$$x = (a - R)\cos t + R \cos\left(\frac{a-R}{R}\right) t$$
$$y = (a - R)\sin t - R \sin\left(\frac{a-R}{R}\right) t$$

60. Journal Problem (*The MATYC Journal**) A rectangle is "rolled" along a straight line by rotations of $90°$ about each vertex in turn. One vertex traces out a series of arches where each arch is the union of three circular arcs. Show that the area of the region bounded by the straight line and one arch equals the area of the rectangle plus twice the area of the circumscribed circle.

9.6 PLANES IN $\mathbb{R}^3$

IN THIS SECTION: *Forms for the equation of a plane in $\mathbb{R}^3$, vector methods for measuring distances in $\mathbb{R}^3$*
Planes in three dimensions provide one of the foundational tools for working in $\mathbb{R}^3$.

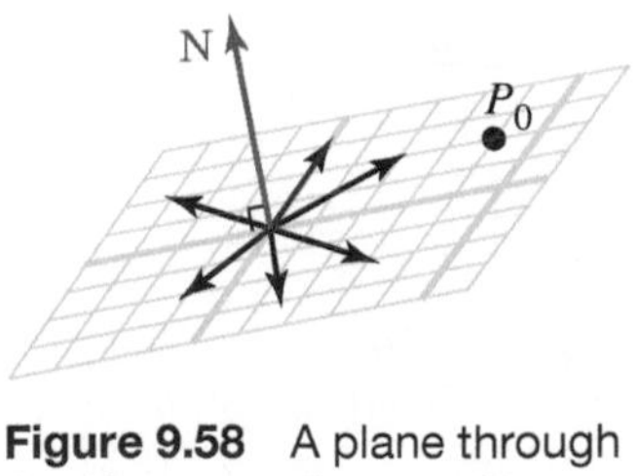

Forms for the Equation of a Plane in $\mathbb{R}^3$

Figure 9.58 A plane through P with a normal vector **N**

Planes in space can also be characterized by vector methods. In particular, any plane is completely determined once we know one of its points and its orientation; that is, the "direction it faces." A common way to specify the direction of a plane is by means of a vector **N** that is orthogonal to every vector in the plane, as shown in Figure 9.58. Such a vector is called a **normal** to the plane.

Example 1   Obtain the equation for a plane

Find an equation for the plane that contains the point $Q(3, -7, 2)$ and is normal to the vector $\mathbf{N} = 2\mathbf{i} + \mathbf{j} - 3\mathbf{k}$.

Solution   The normal vector $\mathbf{N}$ is orthogonal to every vector in the plane. In particular, if $P(x, y, z)$ is any point in the plane, then $\mathbf{N}$ must be orthogonal to the vector

$$\mathbf{QP} = (x - 3)\mathbf{i} + (y + 7)\mathbf{j} + (z - 2)\mathbf{k}$$

Because the dot (or scalar) product of two orthogonal vectors is 0, we have

$$\begin{aligned} \mathbf{N} \cdot \mathbf{QP} = 2(x - 3) + (1)(y + 7) + (-3)(z - 2) &= 0 \\ 2x - 6 + y + 7 - 3z + 6 &= 0 \\ 2x + y - 3z + 7 &= 0 \end{aligned}$$

Therefore, $2x + y - 3z + 7 = 0$ is the equation of the plane.   ■

By generalizing the approach illustrated in Example 1, we can show that the plane that contains the point (x_0, y_0, z_0) and has normal vector $\mathbf{N} = A\mathbf{i} + B\mathbf{j} + C\mathbf{k}$ must have the Cartesian equation

$$A(x - x_0) + B(y - y_0) + C(z - z_0) = 0$$

This is called the **point-normal form** of the equation of a plane. By rearranging terms, we can rewrite this equation in the form $Ax + By + Cz + D = 0$. This is called the **standard form** of the equation of a plane. The numbers $[A, B, C]$ called **attitude numbers** of the plane.

Notice from Figure 9.59 that *attitude numbers of a plane are the same as direction numbers of a normal line.* This means that you can find normal vectors to a plane *by inspecting the equation of the plane.*

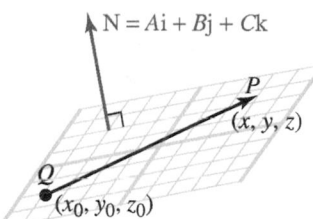

Figure 9.59   The graph of a plane with attitude numbers $[A, B, C]$

Example 2   Relationship between normal vectors and planes

Find normal vectors to the planes

a. $5x + 7y - 3z = 0$ **b.** $x - 5y + \sqrt{2}z = 6$ **c.** $3x - 7z = 10$

Solution

a. A normal to the plane $5x + 7y - 3z = 0$ is $\mathbf{N} = 5\mathbf{i} + 7\mathbf{j} - 3\mathbf{k}$.
b. For $x - 5y + \sqrt{2}z = 6$ the normal is $\mathbf{N} = \mathbf{i} - 5\mathbf{j} + \sqrt{2}\mathbf{k}$.
c. For $3x - 7z = 10$ it is $\mathbf{N} = 3\mathbf{i} - 7\mathbf{k}$.   ■

EQUATIONS OF A PLANE   A plane with normal $\mathbf{N} = A\mathbf{i} + B\mathbf{j} + C\mathbf{k}$ that contains the point (x_0, y_0, z_0) has the following equations:
Point-normal form: $A(x - x_0) + B(y - y_0) + C(z - z_0) = 0$
Standard form: $Ax + By + Cz + D = 0$ for some constants $A, B, C,$ and D

Example 3   Equation of a line orthogonal to a given plane

Find an equation of the line that passes through the point $Q(2, -1, 3)$ and is orthogonal to the plane $3x - 7y + 5z + 55 = 0$. Where does the line intersect the plane? (See Figure 9.60.)

Solution   By inspection of the equation of the plane, we see that $\mathbf{N} = 3\mathbf{i} - 7\mathbf{j} + 5\mathbf{k}$ is a normal vector to the plane. Because the required line is also orthogonal to the plane,

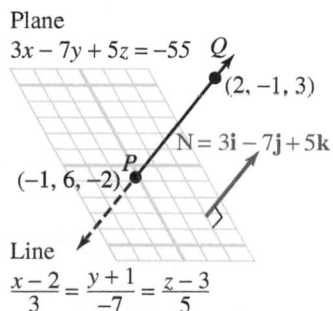

Figure 9.60   Graph of a plane

it must be parallel to **N**. Thus, the line contains the point $Q(2, -1, 3)$ and has direction numbers $[3, -7, 5]$, so that its equation is

$$\frac{x-2}{3} = \frac{y+1}{-7} = \frac{z-3}{5}$$

To find the point where this line intersects the plane, we rewrite it in parametric form:

$$x = 2 + 3t, y = -1 - 7t, \text{ and } z = 3 + 5t$$

Now, substitute into the equation of the plane:

$$
\begin{aligned}
3(2 + 3t) - 7(-1 - 7t) + 5(3 + 5t) &= -55 \\
6 + 9t + 7 + 49t + 15 + 25t &= -55 \\
83t &= -83 \\
t &= -1
\end{aligned}
$$

Thus, the point of intersection is found by substituting $t = -1$ for x, y, and z:

$$
\begin{aligned}
x &= 2 + 3(-1) = -1 \\
y &= -1 - 7(-1) = 6 \\
z &= 3 + 5(-1) = -2
\end{aligned}
$$

The point of intersection is $(-1, 6, -2)$.

Example 4 Equation of a plane containing three given points

Find the standard form equation of a plane containing $P(-1, 2, 1)$, $Q(0, -3, 2)$, and $R(1, 1, -4)$. (See Figure 9.61.)

Solution Because a normal **N** to the required plane is orthogonal to the vectors **PR** and **PQ**, we find **N** by computing the cross product $\mathbf{N} = \mathbf{PR} \times \mathbf{PQ}$.

$$
\begin{aligned}
\mathbf{PR} &= (1 + 1)\mathbf{i} + (1 - 2)\mathbf{j} + (-4 - 1)\mathbf{k} = 2\mathbf{i} - \mathbf{j} - 5\mathbf{k} \\
\mathbf{PQ} &= (0 + 1)\mathbf{i} + (-3 - 2)\mathbf{j} + (2 - 1)\mathbf{k} = \mathbf{i} - 5\mathbf{j} + \mathbf{k}
\end{aligned}
$$

$$
\mathbf{N} = \mathbf{PR} \times \mathbf{PQ} = \begin{vmatrix} \mathbf{i} & \mathbf{j} & \mathbf{k} \\ 2 & -1 & -5 \\ 1 & -5 & 1 \end{vmatrix}
$$

$$= (-1 - 25)\mathbf{i} - (2 + 5)\mathbf{j} + (-10 + 1)\mathbf{k}$$

$$= -26\mathbf{i} - 7\mathbf{j} - 9\mathbf{k}$$

We can now find the equation of the plane using this normal vector and any point in the plane. We will use the point P:

Attitude numbers of the plane: from $N = -26i - 7j - 9k$

$$-26(x + 1) - 7(y - 2) - 9(z - 1) = 0$$

Point on the plane; we are using $P(-1,2,1)$.

Thus, the equation of the plane is

$$
\begin{aligned}
-26x - 26 - 7y + 14 - 9z + 9 &= 0 \\
26x + 7y + 9z + 3 &= 0
\end{aligned}
$$

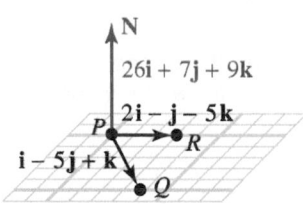

Figure 9.61 Plane passing through three points

Drawing Lesson: Sketching a plane in one octant of $\mathbb{R}^3$

Sketch the graph of
$4(x - 1) + 6y + 2(z - 4) = 0$
Begin by drawing the three coordinate axes.

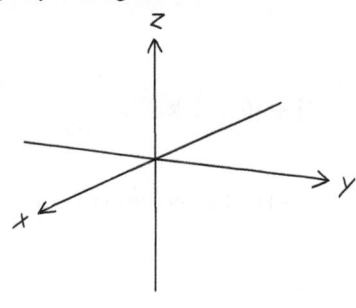

Next, draw the three coordinate planes, shaded for easy viewing.

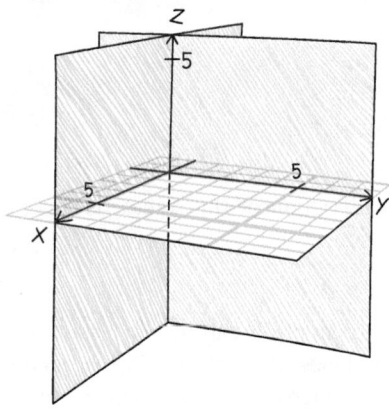

Plot the points where the plane passes through each of the coordinate axes.

If $y = z = 0$, then
$$4(x - 1) + 6(0) + 2(0 - 4) = 0$$
$$4x - 4 - 8 = 0$$
$$x = 3; \text{ Plot } (3, 0, 0)$$

If $x = y = 0$, then
$$4(0 - 1) + 6(0) + 2(z - 4) = 0$$
$$-4 + 0 + 2z - 8 = 0$$
$$z = 6; \text{ Plot } (0, 0, 6)$$

If $x = z = 0$, then
$$4(0 - 1) + 6y + 2(0 - 4) = 0$$
$$-4 + 6y - 8 = 0$$
$$y = 2; \text{ Plot } (0, 2, 0).$$

Connect the three points, (3,0,0), (0,0,6), and (0,2,0).

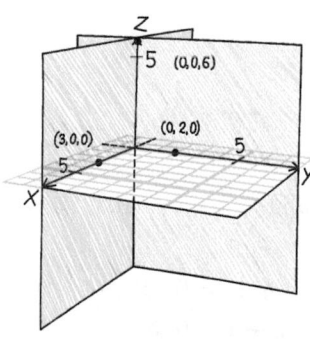

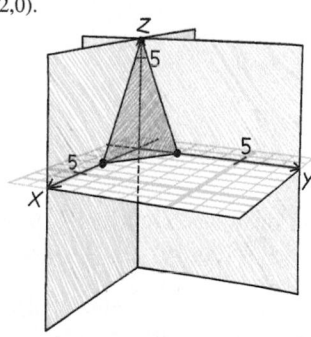

Shade the plane in order to add depth to the figure.

Example 5   Line parallel to the intersection of two planes

Find the equation of a line passing through $(-1, 2, 3)$ that is parallel to the line of intersection of the planes $3x - 2y + z = 4$ and $x + 2y + 3z = 5$. (See Figure 9.62.)

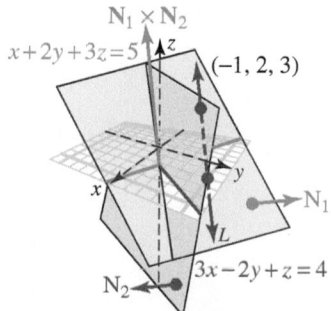

Figure 9.62 Given planes

Solution   By inspection, we see that the normals to the given planes are $N_1 = 3i - 2j + k$ and $N_2 = i + 2j + 3k$. The desired line is perpendicular to both of these normals, so the aligned vector is found by by computing with the cross computing with the cross product:

$$N_1 \times N_2 = \begin{vmatrix} i & j & k \\ 3 & -2 & 1 \\ 1 & 2 & 3 \end{vmatrix}$$

$$= (-6 - 2)i - (9 - 1)j + (6 + 2)k$$

$$= -8i - 8j + 8k$$

The direction of this vector is $\langle -8, -8, 8 \rangle = -8\langle 1, 1, -1 \rangle$. The equation of the desired line is

$$\frac{x + 1}{1} = \frac{y - 2}{1} = \frac{z - 3}{-1}$$

Example 5 can also be used to find the equation of the line of intersection of the two planes. Instead of using the given point $(-1, 2, 3)$, you will first need to find a point in the intersection and then proceed, using the steps of Example 5. We conclude this discussion by finding the equation of a plane containing two given (nonparallel) lines.

Example 6   Equation of a plane containing two intersecting lines

Find the standard-form equation of the plane determined by the intersecting lines

$$\frac{x - 2}{3} = \frac{y + 5}{-2} = \frac{z + 1}{4} \quad \text{and} \quad \frac{x + 1}{2} = \frac{y}{-1} = \frac{z - 16}{5}$$

Solution   Proceeding as in Example 4 of Section 9.5, we find that the lines intersect at $(-19, 9, -29)$.

The aligned vectors for these two lines are $v_1 = 3i - 2j + 4k$ and $v_2 = 2i - j + 5k$ and the normal to the desired plane is orthogonal to both v_1 and v_2. (See Figure 9.63.)

Thus, we take the normal to be the cross product:

$$N = v_1 \times v_2$$

$$= \begin{vmatrix} i & j & k \\ 3 & -2 & 4 \\ 2 & -1 & 5 \end{vmatrix}$$

$$= (-10 + 4)i - (15 - 8)j + (-3 + 4)k$$

$$= -6i - 7j + k$$

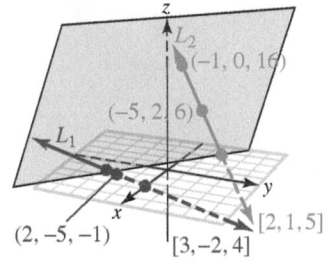

Figure 9.63 Plane determined by lines

The point of intersection $P(-19, 9, -29)$ is certainly in the plane, so we can use $(2, -5, -1)$, $(-1, 0, 16)$, or $(-19, 9, -29)$ to obtain

$$-6(x - 2) - 7(y + 5) + 1(z + 1) = 0$$

$$-6x + 12 - 7y - 35 + z + 1 = 0$$

$$-6x - 7y + z - 22 = 0$$

$$6x + 7y - z + 22 = 0$$

Vector Methods for Measuring Distances in $\mathbb{R}^3$

Vector methods can also be used for measuring distances in $\mathbb{R}^3$, between points, lines, and planes. We begin by deriving the following formula for the distance from a point to a plane.

Theorem 9.9 Distance from a point to a plane in $\mathbb{R}^3$

The distance from the point $P(x_0, y_0, z_0)$ to the plane $Ax + By + Cz + D = 0$ is given by

$$d = \frac{|\mathbf{QP} \cdot \mathbf{N}|}{\|\mathbf{N}\|} = \frac{|Ax_0 + By_0 + Cz_0 + D|}{\sqrt{A^2 + B^2 + C^2}}$$

where Q is any point in the given plane and $\mathbf{N}$ is a normal to the given plane.

Proof: The required distance is found by projecting the vector $\mathbf{QP}$ onto the normal $\mathbf{N}$. Thus, the distance from the point to the plane is given by

$$d = \|\mathbf{QP}\| \, |\cos\theta| = \frac{\|\mathbf{QP}\| \, \|\mathbf{N}\| \, |\cos\theta|}{\|\mathbf{N}\|} = \frac{|\mathbf{QP} \cdot \mathbf{N}|}{\|\mathbf{N}\|}$$

where θ is the angle between $\mathbf{QP}$ and $\mathbf{N}$ (see Figure 9.64.)

Suppose $Q(x_1, y_1, z_1)$ is any particular point in the plane, so that

$$\mathbf{QP} = (x_0 - x_1)\mathbf{i} + (y_0 - y_1)\mathbf{j} + (z_0 - z_1)\mathbf{k}$$

The dot product of $\mathbf{QP}$ with $\mathbf{N} = A\mathbf{i} + B\mathbf{j} + C\mathbf{k}$ is given by

$$\begin{aligned}
\mathbf{QP} \cdot \mathbf{N} &= (x_0 - x_1)A + (y_0 - y_1)B + (z_0 - z_1)C \\
&= (Ax_0 + By_0 + Cz_0) - (Ax_1 + By_1 + Cz_1) \\
&= Ax_0 + By_0 + Cz_0 - (-D) \qquad \textit{Because } Ax_1 + By_1 + Cz_1 + D = 0.
\end{aligned}$$

Because the normal vector has length $\|\mathbf{N}\| = \sqrt{A^2 + B^2 + C^2}$, we can substitute into the formula for d to obtain

$$d = \frac{|\mathbf{QP} \cdot \mathbf{N}|}{\|\mathbf{N}\|} = \frac{|Ax_0 + By_0 + Cz_0 + D|}{\sqrt{A^2 + B^2 + C^2}} \qquad \blacklozenge$$

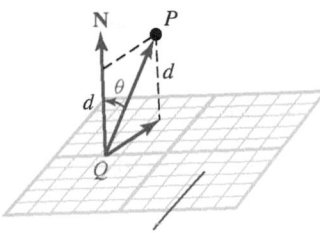

Plane $Ax + By + Cz + D = 0$

Figure 9.64 The distance from a point to a plane in $\mathbb{R}^3$

Example 7 Finding an equation for a sphere tangent to a given plane

Find an equation for the sphere with center $C(-3, 1, 5)$ that is tangent to the plane $6x - 2y + 3z = 9$ is shown in Figure 9.65.

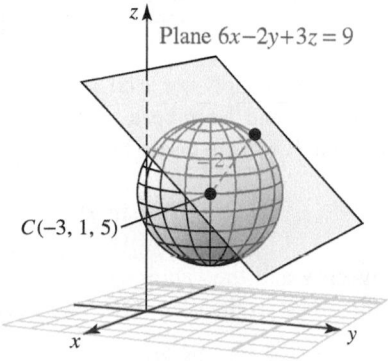

Plane $6x-2y+3z = 9$

$C(-3, 1, 5)$

Figure 9.65 Sphere with a tangent plane

Solution The radius r of the sphere is the distance from the center C to the given plane, as shown in Figure 9.65; namely,

$$r = \left| \frac{6(-3) + (-2)(1) + 3(5) - 9}{\sqrt{6^2 + (-2)^2 + 3^2}} \right| = \left| \frac{-14}{7} \right| = 2$$

Therefore, an equation of the sphere is

$$(x+3)^2 + (y-1)^2 + (z-5)^2 = 2^2$$

Vector methods can also be used to derive a formula for the distance between two skew lines L_1 and L_2 in $\mathbb{R}^3$.

Example 8 Computing the distance between two skew lines

Find the distance between the two skew lines

$$L_1: x = 1+2s, \; y = -1+s, \; z = 2+4s$$
$$L_2: x = -2+4t, \; y = -3t, \; z = -1+t$$

Solution Recall from Section 9.5 that two skew lines do not intersect and are not parallel, but do lie in two parallel planes. Thus, the distance d between the lines L_1 and L_2 is the same as the distance between the parallel planes p_1 and p_2 that contain them (see Figure 9.66). Our strategy for finding the distance d is to find the distance from a point Q on p_1 to the plane p_2.

First, find a point Q on p_1; set $s = 0$ in the parametric equations for L_1. Then $x = 1, y = -1$, and $z = 2$, so $Q(1, -1, 2)$ is such a point. Next, vectors $\mathbf{v}_1 = \langle 2, 1, 4 \rangle$ and $\mathbf{v}_2 = \langle 4, -3, 1 \rangle$ are parallel to L_1 and L_2, respectively, and the cross product $\mathbf{N} = \mathbf{v}_1 \times \mathbf{v}_2$ is orthogonal to both p_1 and p_2. We find that

$$\mathbf{N} = \begin{vmatrix} \mathbf{i} & \mathbf{j} & \mathbf{k} \\ 2 & 1 & 4 \\ 4 & -3 & 1 \end{vmatrix} = 13\mathbf{i} + 14\mathbf{j} - 10\mathbf{k}$$

Setting $t = 0$, we see that the point $P(-2, 0, -1)$ is on p_2, so an equation for p_2 is

$$13(x+2) + 14(y-0) - 10(z+1) = 0$$
$$13x + 14y - 10z + 16 = 0$$

The distance d between the lines L_1 and L_2 is the same as the distance from the point $Q(1, -1, 2)$ to the plane p_2. Thus

$$d = \frac{|13(1) + 14(-1) - 10(2) + 16|}{\sqrt{13^2 + 14^2 + (-10)^2}} = \frac{5}{\sqrt{465}} \approx 0.2319$$

In the final example of this section, we find the distance from a point P to a line L in $\mathbb{R}^3$. Let Q be a point on L and let $\mathbf{v}$ be a vector parallel to L. Then, as shown in Figure 9.67, the distance from P to L is given by

$$d = \|\mathbf{QP}\| \, |\sin \theta|$$

where θ is the angle between $\mathbf{v}$ and the vector $\mathbf{QP}$.

This reminds us of the cross product $\mathbf{v} \times \mathbf{QP}$, and because

$$\|\mathbf{v} \times \mathbf{QP}\| = \|\mathbf{v}\| \, \|\mathbf{QP}\| \sin \theta$$

we have

$$d = \|\mathbf{QP}\| \sin \theta = \frac{\|\mathbf{v} \times \mathbf{QP}\|}{\|\mathbf{v}\|}$$

These observations are summarized in the following theorem.

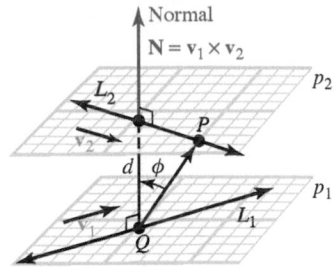
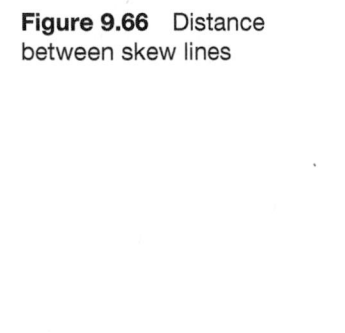

Figure 9.66 Distance between skew lines

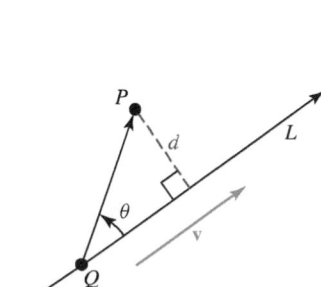

Figure 9.67 The distance from a point to a line in $\mathbb{R}^3$

Theorem 9.10   Distance from a point to a line

The distance from point P to the line L is given by the formula

$$d = \frac{\|\mathbf{v} \times \mathbf{QP}\|}{\|\mathbf{v}\|}$$

where $\mathbf{v}$ is a vector parallel to L and Q is any point on L.

Proof: A sketch of the proof precedes the statement of the theorem.   ◆

Example 9   Distance from a point to a line

Find the distance from the point $P(3, -8, 1)$ to the line

$$\frac{x-3}{3} = \frac{y+7}{-1} = \frac{z+2}{5}$$

Solution We need to find a point Q on the line. We see that $Q(3, -7, -2)$ is on the line and that $\mathbf{QP} = \langle 0, -1, 3 \rangle$. A vector parallel to L is $\mathbf{v} = \langle 3, -1, 5 \rangle$, so that

$$\mathbf{v} \times \mathbf{QP} = \begin{vmatrix} \mathbf{i} & \mathbf{j} & \mathbf{k} \\ 3 & -1 & 5 \\ 0 & -1 & 3 \end{vmatrix} = 2\mathbf{i} - 9\mathbf{j} - 3\mathbf{k}$$

Thus,

$$d = \frac{\|\mathbf{v} \times \mathbf{QP}\|}{\|\mathbf{v}\|} = \frac{\sqrt{(2)^2 + (-9)^2 + (-3)^2}}{\sqrt{3^2 + (-1)^2 + 5^2}} = \frac{\sqrt{94}}{\sqrt{35}} \approx 1.64$$

PROBLEM SET 9.6

Level 1

1. ■ What does this say? Describe the relationship between normal vectors and planes.
2. ■ What does this say? Describe a procedure for finding the indicated distances
   a. from a point to a plane.
   b. from a point to a line in $\mathbb{R}^3$.

Write each equation for a plane given in Problems 3-6 in standard form.

3. $-2(x + 1) + 4(y - 3) - 8z = 0$
4. $4(x + 1) - 2(y + 1) + 6(z - 2) = 0$
5. $5(x - 2) - 3(y + 2) + 4(z + 3) = 0$
6. $-3(x - 4) + 2(y + 1) - 2(z + 1) = 0$

Sketch each plane in Problems 7-10.

7. $4(x - 5) + 3(y - 4) + 2(z - 7) = 0$
8. $-3(x + 4) + 5(y - 2) + 4(z - 1) = 0$
9. $3(x - 4) + 2(y - 7) + 2(z + 4) = 0$
10. $-4(x - 5) + 2(y + 3) + 8(z - 4) = 0$

Find an equation for the plane that contains the point P and has the normal vector $\mathbf{N}$ given in Problems 11-16.

11. $P(0, -7, 1); \mathbf{N} = -\mathbf{i} + \mathbf{k}$
12. $P(0, -3, 0); \mathbf{N} = -2\mathbf{j} + 3\mathbf{k}$
13. $P(1, 1, -1); \mathbf{N} = -\mathbf{i} - 2\mathbf{j} + 3\mathbf{k}$
14. $P(0, 0, 0); \mathbf{N} = \mathbf{k}$
15. $P(0, 0, 0); \mathbf{N} = \mathbf{i}$
16. $P(-1, 3, 5); \mathbf{N} = 2\mathbf{i} + 4\mathbf{j} - 3\mathbf{k}$
17. Find two unit vectors perpendicular to the plane
    $$5x - 3y + 2z = 15$$
18. Find two unit vectors perpendicular to the plane
    $$2x + 4y - 3z = 4$$

Find the distance between the point and the plane given in Problems 19-22. Assume $a \neq 0$.

19. $P(0, 0, 0); 2x - 3y + 5z = 10$
20. $P(1, 0, -1); x + y - z = 1$
21. $P(a, 2a, 3a); \; 3x - 2y + z = -\dfrac{1}{a}$
22. $P(a, -a, 2a); 2ax - y + az = 4a$

Find the distance from the point $(-1, 2, 1)$ *to each plane given in Problems 23-26.*

23. the plane through the points $(-1, 1, 1)$, $(4, 3, 7)$ and $(3, -1, 0)$

24. the plane through the points $(0, 0, 0)$, $(1, 2, 4)$, and $(-2, -1, 1)$

25. the plane through the point $(1, 0, 1)$ with normal vector $2\mathbf{i} - \mathbf{j} + 2\mathbf{k}$

26. the plane through the point $(-3, 5, 1)$ with normal vector $3\mathbf{i} + \mathbf{j} + 5\mathbf{k}$

Find the distance from the point P to the line L in Problems 27-32.

27. $P(9, -3)$; $3x - 4y + 8 = 0$

28. $P(4, 5)$; $3x - 4y + 8 = 0$

29. $P(4, -3)$; $12x + 5y - 2 = 0$

30. $P(1, 0, 1)$; $\dfrac{x}{3} = \dfrac{y - 1}{2} = \dfrac{z}{1}$

31. $P(1, 0, -1)$; $\dfrac{x - 2}{3} = \dfrac{y + 1}{1} = \dfrac{z - 1}{2}$

32. $P(1, -2, 2)$; $\dfrac{x}{1} = \dfrac{2y}{1} = \dfrac{z}{-1}$

Find the distance between the lines given in Problems 33-36.

33. $x = 2 - t$, $y = 5 + 2t$, $z = 3t$ and $x = 2t$, $y = -1 - t$, $z = 1 + 2t$

34. $\dfrac{x + 1}{3} = \dfrac{y - 2}{-2} = \dfrac{z - 1}{1}$ and $\dfrac{x - 2}{5} = \dfrac{y + 1}{1} = \dfrac{z}{3}$

35. $x = -1 + t$, $y = -2t$, $z = 3$ and the line passing through $(0, -1, 2)$ and $(1, -2, 3)$

36. $\dfrac{x + 1}{1} = \dfrac{y - 3}{2} = \dfrac{z + 2}{3}$ and the line passing through $(1, 3, -2)$ and $(0, 1, -1)$

Level 2

37. a. Show that the line

$$\frac{x - 1}{3} = \frac{y}{-2} = \frac{z + 1}{1}$$

is parallel to the plane $x + 2y + z = 1$.

b. Find the distance from the line to the plane in part **a**.

38. Three of the four vertices of a parallelogram $QRTS$ in $\mathbb{R}^3$ are $Q(-1, 3, 5)$, $R(6, -3, 2)$, and $S(2, 4, -3)$. (Note that the fourth vertex of the parallelogram may be located in one of several locations.) What is the area of the parallelogram?

39. Find an equation for the set of all points $P(x, y, z)$ such that the distance from P to the point $P_0(-1, 2, 4)$ is the same as the distance from P to the plane $2x - 5y + 3z = 7$. (Do not expand binomials.)

40. Find an equation for the set of all points $P(x, y, z)$ such that the distance from P to the line

$$\frac{x - 1}{4} = \frac{y + 1}{-1} = \frac{z}{3}$$

is 5. (Do not expand.)

41. In Figure 9.68, $\mathbf{N}$ is normal to the plane p and L is a line that intersects p.

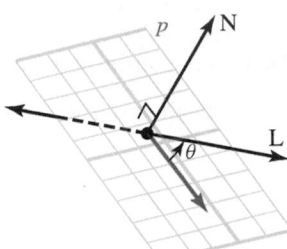

Figure 9.68 Problem 41

Assume $\mathbf{N} = a\mathbf{i} + b\mathbf{j} + c\mathbf{k}$ and that L is given by $x = x_0 + At$, $y = y_0 + Bt$, $z = z_0 + Ct$

a. Let $\phi = 90° - \theta$, where θ is the angle between L and the plane p. Find $\cos \phi$.

b. Find the angle between the plane $x + y + z = 10$ and the line

$$\frac{x - 1}{2} = \frac{y + 3}{3} = \frac{z - 2}{-1}$$

42. Find the equation of the sphere with center $C(-2, 3, 7)$ that is tangent to the plane $2x + 3y - 6z = 5$.

43. Find a vector that is parallel to the line of intersection of the planes $2x + 3y = 0$ and $3x - y + z = 1$.

44. Find an equation for the line of intersection of the planes $2x - y + z = 8$ and $x + y - z = 5$.

45. Find the direction cosines of a vector determined by the line of intersection of the planes $x + y + z = 3$ and $2x + 3y - z = 4$.

46. Find an equation for the plane that passes through $P(1, -1, 2)$ and is normal to $\mathbf{PQ}$ where Q is $Q(2, 1, 3)$.

47. Find an equation for the line that passes through the point $(1, -5, 3)$ and is orthogonal to the plane $2x - 3y + z = 1$.

48. Find two unit vectors that are parallel to the line of intersection of the planes $x + y = 1$ and $x - 2z = 3$.

49. Find two unit vectors that are parallel to the line of intersection of the planes $x + y + z = 3$ and $x - y + z = 1$.

50. Find an equation for the plane that contains the point $(2, 1, -1)$ and is orthogonal to the line

$$\frac{x - 3}{3} = \frac{y + 1}{5} = \frac{z}{2}$$

51. Find a plane that passes through the point $(1, 2, -1)$ and is parallel to the plane $2x - y + 3z = 1$.

52. Show that the line

$$\frac{x-1}{2} = \frac{y+1}{3} = \frac{z-2}{4}$$

is parallel to the plane $x - 2y + z = 6$.

53. Find the point where the line

$$\frac{x-1}{2} = \frac{y+1}{-1} = \frac{z}{3}$$

intersects the plane $3x + 2y - z = 5$.

54. The *angle between two planes* is defined to be the acute angle between their normal vectors. Find the angle between the planes $2x + y - 4z = 3$ and $x - y + z = 2$, rounded to the nearest degree.

55. Find an equation for the line that passes through the point $P(2, 3, 1)$ and is parallel to the line of intersection of the planes $x + 2y - 3z = 4$ and $x - 2y + z = 0$.

Level 3

56. The planes $Ax + By + Cz = D_1$ and $Ax + By + Cz = D_2$ are parallel.
   a. Is the distance between the planes $|D_1 - D_2|$? If not, what is the correct formula?
   b. Find the distance between the parallel planes $x + y + 2z = 2$ and $x + y + 2z = 4$.

57. Show that the angle between the planes $A_1x + B_1y + C_1z + D_1 = 0$ and $A_2x + B_2y + C_2z + D_2 = 0$ is $\pi/2$ if and only if $A_1A_2 + B_1B_2 + C_1C_2 = 0$.

58. Show that a plane with x-intercept a, y-intercept b, and z-intercept c has the equation

$$\frac{x}{a} + \frac{y}{b} + \frac{z}{c} = 1$$

assuming a, b, and c are all nonzero.

59. Suppose planes p_1 and p_2 intersect. If $\mathbf{v}_1$ and $\mathbf{w}_1$ are nonzero vectors on p_1 and $\mathbf{v}_2$ and $\mathbf{w}_2$ are nonzero vectors on plane p_2, then show that

$$(\mathbf{v}_1 \times \mathbf{w}_1) \times (\mathbf{v}_2 \times \mathbf{w}_2)$$

is parallel to the line of intersection of the planes.

60. Let L_1 and L_2 be skew lines that contain the points P_1 and P_2, respectively, and are parallel to the vectors $\mathbf{v}_1$ and $\mathbf{v}_2$. Show that the distance between the lines is given by

$$d = \left| \frac{(\mathbf{v}_1 \times \mathbf{v}_2) \cdot \mathbf{P}_1\mathbf{P}_2}{\|\mathbf{v}_1 \times \mathbf{v}_2\|} \right|$$

9.7 QUADRIC SURFACES

> **IN THIS SECTION:** *Catalog of quadric surfaces; methods for graphing quadric surfaces*
> In this section we develop some of the most common three-dimensional surfaces, which are the three-dimensional counterparts for the conic sections, namely those two-dimensional curves whose equations are of the form
>
> $$Ax^2 + Bxy + Cy^2 + Dx + Ey + F = 0$$
>
> Conic sections are graphs in $\mathbb{R}^2$ that can be represented by a second degree equation in two variables, and quadric surfaces are graphs in $\mathbb{R}^3$ which can be represented by a second-degree equation in three variables.

Catalog of Quadric Surfaces

Spheres and elliptic, parabolic, and hyperbolic cylinders are examples of **quadric surfaces**. In general, such a surface is the graph of an equation of the form

$$Ax^2 + By^2 + Cz^2 + Dxy + Exz + Fyz + Gx + Hy + Iz + J = 0$$

where $A, B, C, D, E, F, G, H, I$, and J are constants. By using rotations and translations, it can be shown that any such equation can be expressed in one of the following standard forms:

$$z = Mx^2 + Ny^2 \qquad \text{or} \qquad Px^2 + Qy^2 + Rz^2 = S$$

where M, N, P, Q, R, and S are constants. We will concentrate on examining graphs of the standard forms.

The quadric surfaces are shown in Table 9.2.

Table 9.2 Quadric Surfaces

Surface	Description	Surface	Description
Elliptic cone 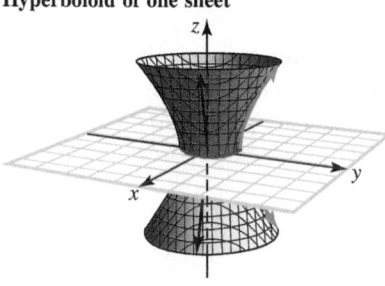	The trace in the xy-plane is a point; in planes parallel to the xy-plane, it is an ellipse. Traces in the xz- and yz-planes are intersecting lines; in planes parallel to these, they are hyperbolas: $$z^2 = \frac{x^2}{a^2} + \frac{y^2}{b^2}$$ The axis of this cone is the z-axis.	**Circular cone** 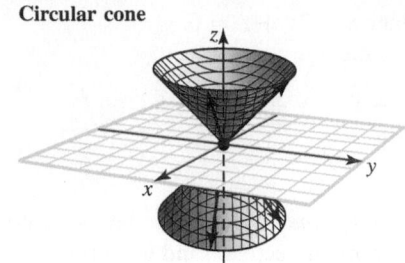	The circular cone is a special kind of elliptic cone for which $a = b = r$. $$z^2 = \frac{x^2}{r^2} + \frac{y^2}{r^2}$$ The axis of this cone is the z-axis.
Hyperboloid of one sheet 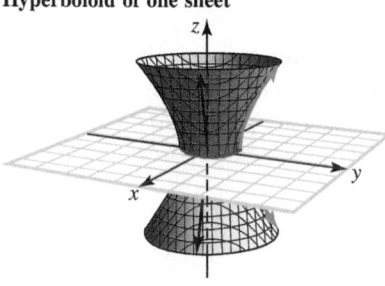	The trace in the xy-plane is an ellipse; in planes parallel to the xz- and yz-planes the traces are hyperbolas: $$\frac{x^2}{a^2} + \frac{y^2}{b^2} - \frac{z^2}{c^2} = 1$$ One negative on equation; the axis of this one is the z-axis.	**Hyperboloid of two sheets** 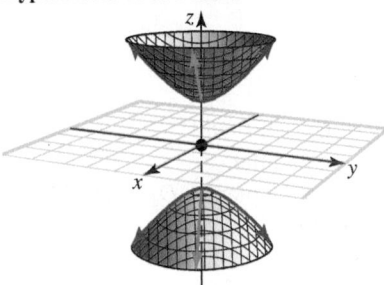	There is no trace in the xy-plane. In planes parallel to the xy-plane which intersect the surface, the traces are ellipses. Traces in the xz- and yz-planes are hyperbolas. $$\frac{x^2}{a^2} + \frac{y^2}{b^2} - \frac{z^2}{c^2} = -1$$ Two negatives; the axis of this one is the z-axis.
Ellipsoid 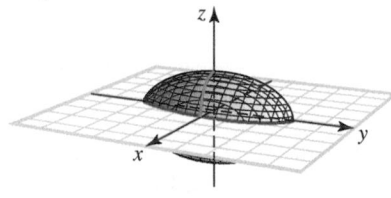	The traces in the coordinate planes are ellipses. $$\frac{x^2}{a^2} + \frac{y^2}{b^2} + \frac{z^2}{c^2} = 1$$	**Sphere** 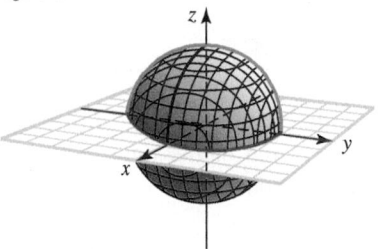	The sphere is a special kind of ellipsoid for which $a = b = c = r$ $x^2 + y^2 + z^2 = r^2$
Elliptic paraboloid 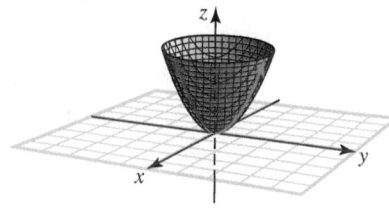	The trace in the xy-plane is a point; in planes parallel to the xy-plane, it is an ellipse. Traces in the xz- and yz-planes are parabolas $$z = \frac{x^2}{a^2} + \frac{y^2}{b^2}$$ Axis is z-axis (z first degree). If $a = b = r$, then the cross-section is a circle, and it is called a *circular paraboloid*.	**Hyperbolic paraboloid** 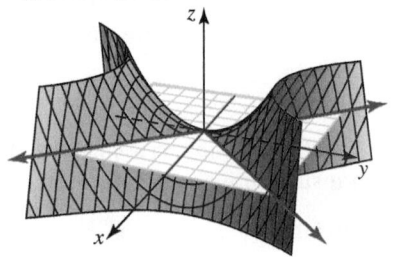	The trace in the xy-plane is two intersecting lines; in the plane parallel to the xy-plane, the traces are hyperbolas. Traces in the xz-plane and yz-planes are parabolas. $$z = \frac{y^2}{b^2} - \frac{x^2}{a^2}$$ This is also called a **saddle**.

Example 1   Identifying a quadric surface

Describe the given surfaces.

a. $7x^2 + 4y^2 - 28z^2 = 0$ **b.** $3x^2 + 5y^2 - 15z = 0$

Solution

a. We rewrite the equation in one of the standard forms given in Table 9.2:

$$z^2 = \frac{x^2}{4} + \frac{y^2}{7}$$

We see this is an elliptic cone.

b. Rewrite as

$$z = \frac{x^2}{5} + \frac{y^2}{3}$$

We see that the surface is an elliptic paraboloid.

Methods for Graphing Quadric Surfaces

The intersection of a surface with a plane is called the **trace** of the surface in that plane. The trace of a curve is found by setting one of the variables equal to a constant and then graphing the resulting curve. If $x = k$ (k a constant), the resulting curve is drawn in the plane $x = k$, which is parallel to the yz-plane; if $y = k$ the resulting curve is drawn in the plane $y = k$ which is parallel to the xz-plane; and if $z = k$, the curve is drawn in the plane $z = k$, parallel to the xy-plane.

Example 2   Using traces to sketch the graph of an ellipsoid

Sketch the graph of the quadric surface given by

$$\frac{x^2}{4} + \frac{y^2}{5} + \frac{z^2}{9} = 1$$

Solution   Consider the trace of the given surface in the plane $z = 0$. We substitute $z = 0$ into the equation to obtain

$$\frac{x^2}{4} + \frac{y^2}{5} = 1$$

which we recognize as an ellipse. More generally, the trace in the plane $z = k$ is, by substitution,

$$\frac{x^2}{4} + \frac{y^2}{5} + \frac{k^2}{9} = 1 \qquad \text{or} \qquad \frac{x^2}{4} + \frac{y^2}{5} = 1 - \frac{k^2}{9}$$

which is also an ellipse for $k^2 \leq 9$, since the right hand side is a constant. Similarly, the trace of the surface in a plane of the form $y = k$ is

$$\frac{x^2}{4} + \frac{k^2}{5} + \frac{z^2}{9} = 1 \qquad \text{or} \qquad \frac{x^2}{4} + \frac{z^2}{9} = 1 - \frac{k^2}{5}$$

which is an ellipse for $k^2 \leq 5$, as is the trace in the plane $x = k$ (for $k^2 \leq 4$):

$$\frac{k^2}{4} + \frac{y^2}{5} + \frac{z^2}{9} = 1 \qquad \text{or} \qquad \frac{y^2}{5} + \frac{z^2}{9} = 1 - \frac{k^2}{4}$$

Sketch this ellipse in $\mathbb{R}^3$ as shown in Figure 9.69. Since the traces in all three principal directions are ellipses, we call this surface an **ellipsoid**.

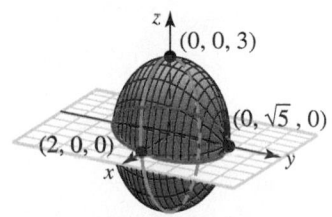

Figure 9.69   Graph of an ellipsoid

Example 3   Using traces to sketch the graph of a hyperboloid

Sketch the graph of the surface $9x^2 + 36y^2 - 4z^2 + 36 = 0$.

Solution   We begin by setting $z = k$, to obtain

$$9x^2 + 36y^2 = 4k^2 - 36$$

which is impossible for $k^2 < 9$, and represents an ellipse for $k^2 \geq 9$. For $x = k$, we have

$$4z^2 - 36y^2 = 36 + 9k^2$$

and for $y = k$

$$4z^2 - 9x^2 = 36 + 36k^2$$

both of which represent hyperbolas for all k.

    To summarize, the surface has hyperbolic traces for planes parallel to the yz-plane and the xz-plane and elliptical traces for planes parallel to the xy-plane, except for a "gap" corresponding to the planes of the form $z = k$ with $|k| < 3$. This surface is a hyperboloid of two sheets. The graph is shown in Figure 9.70. ∎

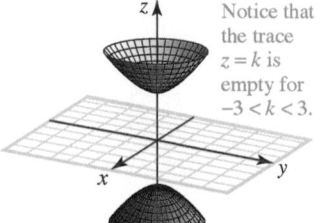

Notice that the trace $z = k$ is empty for $-3 < k < 3$.

Figure 9.70 Graph of a hyperboloid of two sheets

Example 4   Identifying and sketching a quadric surface

Identify and sketch the surface with equation $9x^2 - 16y^2 + 144z = 0$.

Solution   Look at Table 9.2 and note that the equation is second degree in x and y but first degree in z. This means it is an elliptic paraboloid or a hyperbolic paraboloid. Solve the equation for z:

$$9x^2 - 16y^2 + 144z = 0$$
$$144z = 16y^2 - 9x^2$$
$$z = \frac{y^2}{9} - \frac{x^2}{16}$$

We recognize this as a hyperbolic paraboloid. Next, we take cross sections of $9x^2 - 16y^2 + 144z = 0$.

Cross Section	Chosen Value	Equation	Description
xy-plane	$z = 0$	$\dfrac{y^2}{9} - \dfrac{x^2}{16} = 0$	Two intersecting lines
Parallel to the xy-plane	$z = 4$	$\dfrac{y^2}{36} - \dfrac{x^2}{64} = 1$	Hyperbola
xz-plane	$y = 0$	$z = -\dfrac{x^2}{16}$	Parabola opens downward
Parallel to the xz-plane	$y = 10$	$z - \dfrac{100}{9} = -\dfrac{x^2}{16}$	Parabola opens downward
yz-plane	$x = 0$	$z = \dfrac{y^2}{9}$	Parabola opens upward
Parallel to the yz-plane	$x = 5$	$z + \dfrac{25}{16} = \dfrac{y^2}{9}$	Parabola opens upward

These traces are shown in Figure 9.71.

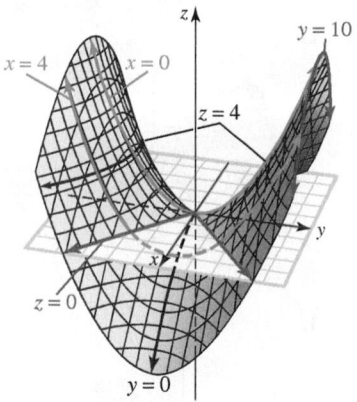

Figure 9.71   Graph of quadric surface

You can use Figure 9.71 to visualize the surface in Example 4, but that does not help you to draw a quadratic surface on your own paper. The following drawing lesson may help you do to this. For example, suppose you seek to sketch $z = \dfrac{1}{1 + x^2 + y^2}$.

Drawing Lesson: Sketching a surface in $\mathbb{R}^3$

Draw the xy-plane in three dimensions adding the z-axis.

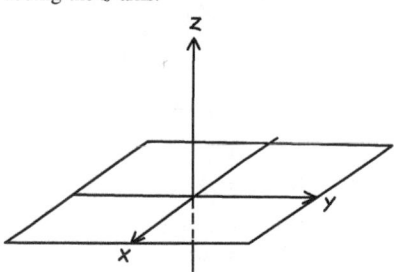

Draw a trace in one of the coordinate planes (in this case, the plane is $x = 0$). If necessary, adjust the z-scale to show the trace more clearly.

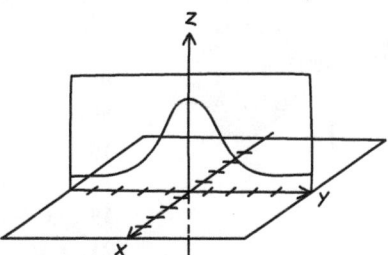

Draw a trace in another coordinate plane (in this case, the plane $y = 0$).

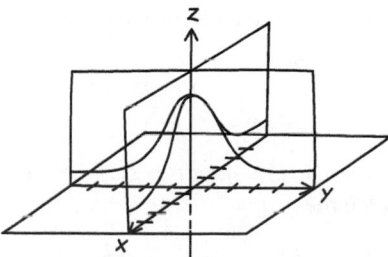

Draw several additional trace curves to reveal the contours of the surface.

Erase all hidden lines. Use highlighters to color the surface and the xy-plane. Plot $(4,10,0)$ in the xy-plane.

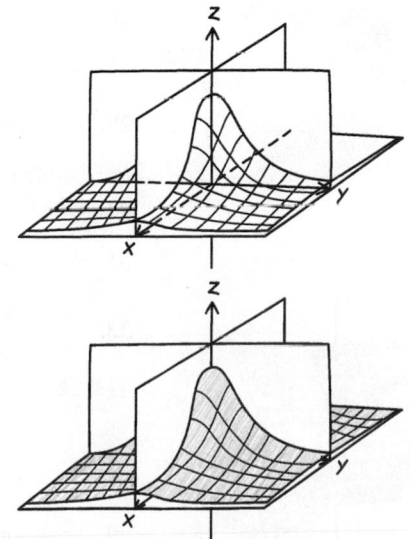

PROBLEM SET 9.7

Level 1

1. ■ What does this say? What is a quadric surface?
2. ■ What does this say? What is the trace of a surface?

In Problems 3-14, match the equation with its graph (A-L).

3. $x^2 = z^2 + y^2$

4. $z^2 = \dfrac{x^2}{4} + \dfrac{y^2}{9}$

5. $x^2 + y^2 + z^2 = 9$

6. $y = x^2 + z^2$

7. $9z = x^2 + y^2$

8. $z = \dfrac{x^2}{16} - \dfrac{y^2}{4}$

9. $x = \dfrac{y^2}{25} + \dfrac{z^2}{16}$

10. $y = \dfrac{z^2}{4} - \dfrac{x^2}{9}$

11. $y^2 + z^2 - x^2 = -1$

12. $\dfrac{x^2}{9} + \dfrac{y^2}{16} - \dfrac{z^2}{4} = 1$

13. $\dfrac{x^2}{2} + \dfrac{y^2}{4} + \dfrac{z^2}{9} = 1$

14. $\dfrac{x^2}{4} - \dfrac{y^2}{9} + \dfrac{z^2}{9} = -1$

A.

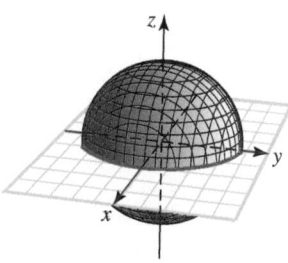

B.

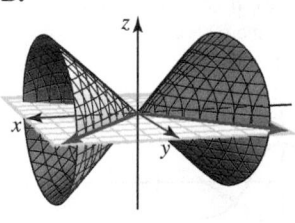

C.

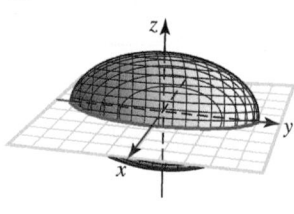

D.

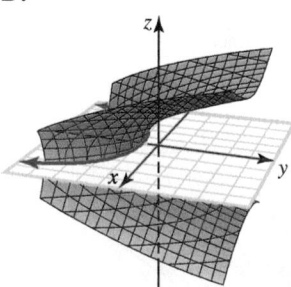

E.

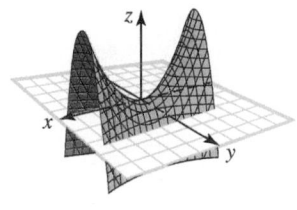

F.

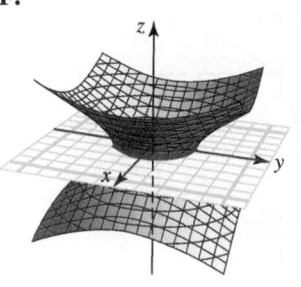

G.

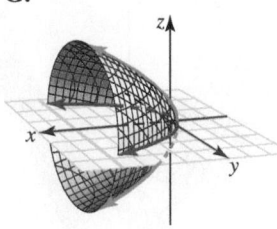

H.

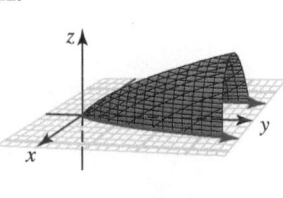

I.

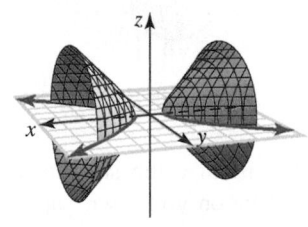

J.

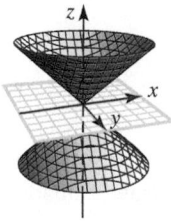

K.

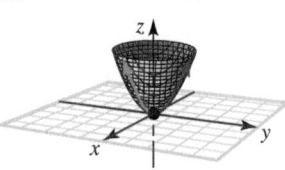

L.

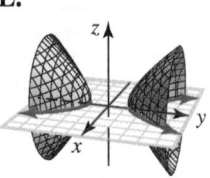

In Problems 15-26, sketch the graph of each equation in $\mathbb{R}^3$.

15. $2x + y + 3z = 6$

16. $x = 4$

17. $x + 2y + 5z = 10$

18. $x + y + z = 1$

19. $3x + 2y + z = 6$

20. $y = x^2$

21. $z = x - 1$

22. $x^2 + y^2 = z$

23. $z = e^y$

24. $y = \ln x$

25. $2z + 5y^2 = 1$

26. $z = \sin y$

Sketch the graph of each surface given in Problems 27-30,

27. $y = \dfrac{1}{1 + x^2 + z^2}$

28. $x = \dfrac{1}{1 + y^2 + z^2}$

29. $x^2 + y^2 + z^2 = 1$

30. $y = x^2 + y^2$

In Problems 31-40, identify the quadric surface and describe the traces. Sketch the graph.

31. $9x^2 + 4y^2 + z^2 = 1$

32. $\dfrac{x^2}{4} + \dfrac{y^2}{9} - z^2 = 1$

33. $\dfrac{x^2}{4} + \dfrac{y^2}{9} - z^2 = -1$

34. $\dfrac{x^2}{9} - y^2 - z^2 = 1$

35. $z^2 = x^2 + \dfrac{y^2}{4}$

36. $z = x^2 + \dfrac{y^2}{4}$

37. $z = \dfrac{x^2}{9} - \dfrac{y^2}{16}$

38. $x^2 - 2y^2 = 9z$

39. $x^2 + 2y^2 = 9z^2$

40. $z^2 = 1 + \dfrac{x^2}{9} + \dfrac{y^2}{4}$

The equations given in Problems 41-44 represent quadric surfaces whose axis of symmetry is either the x-axis or the y-axis. Identify and sketch the surface.

41. $y - x^2 - 9z^2 = 0$

42. $x = \dfrac{y^2}{4} - \dfrac{z^2}{9}$

43. $x^2 = y^2 + 3z^2$

44. $y^2 - x^2 + z^2 = 1$

Level 2

Describe each of the translated quadric surfaces in Problems 45-50.

45. $\dfrac{(x-1)^2}{4} + (y+2)^2 + \dfrac{(z-3)^2}{9} = 1$

46. $z = (2x+1)^2 + \dfrac{(y-1)^2}{9} + 3$

47. $7x^2 + y^2 + 3z^2 - 9x + 4y + 7z = 1$

48. $z^2 = x^2 - y^2 + 2x + y - z + 5$

49. $y = x^2 - x + z^2 + z + 5$

50. $\dfrac{(x-1)^2}{2} - \dfrac{(y+1)^2}{4} - \dfrac{z^2}{9} = 1$

Describe the curve of intersection of the quadric surfaces in Problems 51 and 52.

51. $\dfrac{x^2}{9} + \dfrac{y^2}{4} + \dfrac{z^2}{5} = 1$ and $z^2 = \dfrac{x^2}{3} + \dfrac{y^2}{2}$

52. $z = \dfrac{y^2}{9} - \dfrac{x^2}{16}$ and $\dfrac{x^2}{16} + \dfrac{y^2}{2} - \dfrac{z^2}{3} = 1$

53. Use the method of cross-sections (Section 6.2) to show that the ellipsoid

$$\frac{x^2}{a^2} + \frac{y^2}{b^2} + \frac{z^2}{c^2} = 1$$

has volume $V = \frac{4}{3}\pi abc$. *Hint:* You may assume that the ellipse

$$\frac{x^2}{A^2} + \frac{y^2}{B^2} = 1$$

has area πAB.

54. Use the method of cross-sections (Section 6.2) to find the volume bounded by the plane $z = 5$ and the paraboloid $9z = x^2 + y^2$.

55. What is the intersection of the surfaces $z = x^2 + y^2$ and $z = 1 - y^2$?

56. What is the intersection of the surfaces $z = x^2 + y^2$ and $z = y$?

Level 3

57. Rotational forces cause the earth to flatten at the poles so its shape is really an ellipsoid rather than a true sphere. Assume that the equatorial radius of the earth is 6,378.2 km and that its polar radius is 6,356.5 km.
   a. Assuming that the earth is an ellipsoid with the general equation

$$\frac{x^2}{a^2} + \frac{y^2}{a^2} + \frac{z^2}{b^2} = 1$$

   find a and b.
   b. Use the formula in Problem 53 to find the volume of the earth.

58. The line

$$x = 1 - t, \; y = 2 + t, \; z = 3t$$

intersects the hyperboloid

$$\frac{z^2}{9} - \frac{x^2}{1} + \frac{y^2}{4} = 1$$

in two points P and Q. Find the distance between P and Q.

59. Find an equation for the surface S comprised of all points (x, y, z) the sum of whose distances from the points $A(1, 0, 0)$ and $B(-1, 0, 0)$ is 3. Is S a quadric surface?

60. Find an equation for the surface S comprised of all points (x, y, z) whose distance from the origin is the same as their distance from the plane $x + y + z = 5$. Is S a quadric surface?

CHAPTER 9 REVIEW

*T*ime is said to have only **one dimension**, and space to have **three dimensions**...
The mathematical **quaternion** partakes of **both** of these elements; in technical language it may be said to be "time plus space," or "space plus time"; and in the sense it has, or at least involves a reference to, **four dimensions**.

W. R. Hamilton
Graves' *Life of Hamilton* (New York, 1882-1889)
Vol. 4, p. 635

Even though quaternions are not particularly important today, they preceded the modern concept of a vector. Our use of **i**, **j**, and **k** derives from Hamilton's quaternions.

Proficiency Examination

Concept Problems

1. **a.** What is a vector?
   b. What is a scalar?
   c. Give both an algebraic and a geometric interpretation of multiplication of a vector by a scalar.
   d. What is the parallelogram rule for vector sums?
2. State each of the following properties of vector operations:
   a. commutativity of vector addition
   c. identity for vector addition
   e. magnitude of a vector for dot product
   g. multiple of a dot product
   i. cross product of a zero vector
   k. distributivity for cross product over addition

   b. associativity of vector addition
   d. inverse property for vector addition
   f. commutativity for dot product
   h. distributivity for dot product over addition
   j. anticommutativity for cross product
3. **a.** Define cross product.
   b. Give a geometric interpretation of cross product.
   c. What is the formula for the magnitude of a cross product?
   d. If **u** and **v** are parallel vectors, what can be said about **u** × **v**?
4. **a.** How do you find the length of a vector?
   b. What is a unit vector?
   c. How do you find a unit vector **u** in the direction of a given vector **v**?
   d. State the triangle inequality.
   e. What are the standard basis vectors in $\mathbb{R}^3$?
5. **a.** What is the standard-form equation of a sphere?   **b.** What is a cylinder?
6. **a.** What is Lagrange's identity?   **b.** What is the vector formula for work?
7. **a.** Define dot product.   **b.** What are the direction cosines of a vector in $\mathbb{R}^3$?
8. **a.** What is the formula for the angle between two vectors?
   b. What is meant by orthogonal vectors?
   c. What is the algebraic condition for orthogonality?
9. **a.** What is a vector projection?
   b. Give a formula for finding the vector projection of **v** onto **w**.
   c. What is a scalar projection?
   d. Give a formula for finding the scalar projection of **v** onto **w**.
10. What is the right-hand rule for a coordinate system?
11. **a.** What is the determinant form for the scalar triple product?
    b. How do you find the volume of a parallelepiped?

12. **a.** What is the parametric form of a line in $\mathbb{R}^3$?
    b. What is the symmetric form of a line in $\mathbb{R}^3$?
13. **a.** What is the normal vector of a plane in $\mathbb{R}^3$?
    b. What is the point-normal form of a plane in $\mathbb{R}^3$?
    c. What is the standard form of a plane in $\mathbb{R}^3$?
14. **a.** What is the distance between two points in $\mathbb{R}^3$?
    b. What is the formula for the distance from a point to a plane?
    c. What is the formula for the distance from a point to a line?
15. What is a quadric surface?

Practice Problems

16. Given $\mathbf{v} = 2\mathbf{i} - 3\mathbf{j} + \mathbf{k}$, $\mathbf{w} = 3\mathbf{i} - 2\mathbf{j}$. Find each of the following vectors or scalars.
    a. $2\mathbf{v} + 3\mathbf{w}$ **b.** $\|\mathbf{v}\|^2 - \|\mathbf{w}\|^2$ **c.** $\mathbf{v} \cdot \mathbf{w}$ **d.** $\mathbf{v} \times \mathbf{w}$
    e. vector projection of $\mathbf{v}$ onto $\mathbf{w}$ **f.** scalar projection of $\mathbf{w}$ onto $\mathbf{v}$
17. Given $\mathbf{u} = 2\mathbf{i} - 3\mathbf{j} + \mathbf{k}$, $\mathbf{v} = \mathbf{i} + \mathbf{j} - 2\mathbf{k}$, and $\mathbf{w} = 3\mathbf{i} + 5\mathbf{k}$. In each of the following cases, either perform the indicated computation or explain why it is not defined.
    a. $(\mathbf{u} \times \mathbf{v}) \cdot \mathbf{w}$ **b.** $(\mathbf{u} \cdot \mathbf{v}) \times \mathbf{w}$ **c.** $(\mathbf{u} \times \mathbf{v}) \times \mathbf{w}$ **d.** $(\mathbf{u} \cdot \mathbf{v}) \cdot \mathbf{w}$

Find the equations for the lines and planes in Problems 18-21.
18. the line through the points $P(-1, 4, -3)$ and $Q(0, -2, 1)$
19. the plane that contains the point $P(1, 1, 3)$ and is normal to the vector $\mathbf{v} = 2\mathbf{i} + 3\mathbf{k}$
20. the line of intersection of the planes $2x + 3y + z = 2$ and $y - 3z = 5$
21. the plane that contains the points $P(0, 2, -1)$, $Q(1, -3, 5)$ and $R(3, 0, -2)$
22. Find the direction cosines and the direction angles of the vector $\mathbf{u} = -2\mathbf{i} + 3\mathbf{j} + \mathbf{k}$. Round to the nearest degree.
23. In each case, determine whether the lines intersect, are parallel, or are skew. If they intersect, find the point of intersection.
    a. $x = 2t - 3$, $y = 4 - t$, $z = 2t$; and $\dfrac{x+2}{3} = \dfrac{y-3}{5}$; $z = 3$

    b. $\dfrac{x-7}{5} = \dfrac{y-6}{4} = \dfrac{z-8}{5}$; and $\dfrac{x-8}{6} = \dfrac{y-6}{4} = \dfrac{z-9}{6}$
24. Let $\mathbf{u} = 2\mathbf{i} + \mathbf{j}$, $\mathbf{v} = \mathbf{i} - \mathbf{j} - \mathbf{k}$, and $\mathbf{w} = 3\mathbf{i} + \mathbf{k}$. Find the volume of the parallelepiped determined by these vectors.
25. For the vectors in the previous problem, find a positive number A that guarantees that the tetrahedron determined by $A\mathbf{u}$, $A\mathbf{v}$, and $\mathbf{w}$ has volume that is twice the volume of the tetrahedron determined by $\mathbf{u}, \mathbf{v},$ and $\mathbf{w}$. *Hint:* See Problem 50, Section 9.4.
26. Find the following distances:
    a. from $P(-1, 1, 4)$ to $2x + 5y - z = 3$ **b.** from $P(4, 5, 0)$ to $\dfrac{x-2}{3} = \dfrac{y}{5} = \dfrac{z+1}{-1}$
    c. between the skew lines $x = t, y = 2t, z = 3t - 1$ and $x = 1 - t, y = t + 2, z = t$
27. An airplane flies at 200 mi/h parallel to the ground at an altitude of 10,000 ft. If the plane flies due south and the wind is blowing toward the northeast at 50 mi/h, what is the ground speed of the plane (that is, effective speed)?
28. Sketch the graph of $x = 2^t$, $y = 2^{t+1}$ for $t > 0$ by
    a. using the parametric form   **b.** eliminating the parameter
29. **a.** Graph the line $\dfrac{x}{4} = -\dfrac{y}{9}$
         i. using rectangular form   **ii.** by parameterizing
    b. Graph the line $\dfrac{x}{4} = -\dfrac{y}{9} = \dfrac{z}{1}$
         i. using rectangular form   **ii.** by parameterizing
30. Identify the graph of each given equation.
    a. $x^2 = z^2 + y^2$ **b.** $x^2 = z + y^2$ **c.** $\dfrac{x^2}{2} - \dfrac{y^2}{4} - \dfrac{z^2}{9} = 1$
    d. $\dfrac{x^2}{2} + \dfrac{y^2}{4} - \dfrac{z^2}{9} = 1$ **e.** $\dfrac{x}{2} - \dfrac{y}{4} - \dfrac{z}{9} = 1$ **f.** $\dfrac{x}{2} = \dfrac{y}{4} = \dfrac{z}{9}$
    g. $x^2 + y^2 + z^2 = 9$ **h.** $y = x^2 + z^2$ **i.** $y^2 + z^2 - x^2 = -1$

Supplementary Problems*

In Problems 1-4, a triangle in $\mathbb{R}^3$ has vertices $A(0, 2, -1)$, $B(1, 1, 3)$, and $C(1, 0, -4)$.
1. Find the perimeter of the triangle.
2. Find the area of the triangle.
3. Find the three vertex angles of the triangle. (Round to the nearest degree.)
4. Find a number p such that the points A, B, C and $D(p, p, 0)$ form a tetrahedron of volume $V = 100$ cubic units. *Hint:* See Problem 50, Section 9.4.

Find equations, in both parametric and symmetric forms, of the lines described in Problems 5-10. Find two additional points on each line.
5. passing through $A(4, -3, 2)$ and perpendicular to $2x - y + z = 4$
6. passing through $A(1, -2, 3), B(4, -1, 2)$
7. passing through $P(1, 4, 0)$ with direction numbers $[2, 0, 1]$
8. passing through $P(2, 1, -3)$ orthogonal to the plane $5x - 2y + z = 1$
9. passing through $P(3, 4, -1)$ and parallel to the line of intersection of the planes $x + 2y + 2z + 5 = 0$ and $2x + y - 3z - 6 = 0$
10. passing through the point $P(0, 1, -1)$ and is parallel to the line of intersection of the planes $2x + y - 2z = 5$ and $3x - 6y - 2z = 7$

Find an equation for the plane satisfying the conditions given in Problems 11-18.
11. the xy-plane
12. parallel to the xz-plane passing through $(4, 3, 7)$
13. passing through $(1, -3, 4)$ with attitude numbers $[3, 4, -1]$
14. passing through $(-1, 4, 5)$ and orthogonal to a line with direction numbers $[4, 4, -3]$
15. passing through $(4, -3, 2)$ and parallel to the plane $5x - 2y + 3z - 10 = 0$
16. containing the lines $\dfrac{x - 3}{4} = \dfrac{z - 1}{2}; y = -2$ and $\dfrac{x - 3}{3} = \dfrac{y + 2}{1} = \dfrac{z - 1}{-2}$
17. passing through $P(4, 1, 3), Q(-4, 2, 1)$ and $R(1, 0, 2)$
18. passing through $(4, -1, 2)$ and parallel to the lines $\dfrac{x + 2}{3} = \dfrac{y - 2}{-1} = \dfrac{z + 1}{2}$ and $\dfrac{x - 2}{1} = \dfrac{y - 3}{2} = \dfrac{z - 4}{3}$

Given $\mathbf{v} = 4\mathbf{i} + 2\mathbf{j} + \mathbf{k}$, $\mathbf{w} = 2\mathbf{i} + \mathbf{j} - 5\mathbf{k}$, find the vectors or scalars in Problems 19-25.

19. $\|\mathbf{v}\|$ 　　　　20. $\mathbf{v} - \mathbf{w}$ 　　　　21. $2\mathbf{v} + 3\mathbf{w}$ 　　　　22. $5\mathbf{v} - 3\mathbf{w}$ 　　　　23. $\|2\mathbf{v} - \mathbf{w}\|$

24. vector projection of $\mathbf{v}$ onto $\mathbf{w}$ 　　　　25. scalar projection of $\mathbf{w}$ onto $\mathbf{v}$

Graph the curves, planes, or surfaces in $\mathbb{R}^3$ defined by the equations in Problems 26-33.
26. $x + 3y + 2z = 6$ 　　　　27. $x + y = 3$
28. $x = 4$ 　　　　29. $x = 3 - t, y = 4 + 2t, z = 6 - 3t$
30. $3(x - 2) - 2y + 3z = 0$ 　　　　31. $z = \dfrac{1}{1 + x^2 + y^2}$
32. $x = 2\cos t, y = 2\sin t, 0 \le t < 2\pi$ 　　　　33. $x = t, y = t^3 + t^2, 0 \le t \le 2$

Find the distance between the point and the line or the point and the plane given in Problems 34-37.
34. $P(1, 1, -1); x - y + 2z = 4$ 　　　　35. $P(2, 1, -2); 3x - 4y + z = -1$
36. $P(-4, 3, 2)$ to the plane containing $A(0, 0, 0), B(1, 4, -1)$, and $C(2, -1, 3)$
37. $P(3, -2, 1); \dfrac{x - 3}{2} = \dfrac{y + 1}{1} = \dfrac{z - 2}{-1}$

---

*The supplementary probelms are presented in a somewhat random order, not necessarily in order of difficulty.

38. Suppose that the wind is blowing with a 1,000-lb magnitude force **F** in the direction of N60°E over a boat's sail. How much work does the wind perform in moving the boat in an easterly direction a distance of 50 ft? Give your answer in foot-pounds.

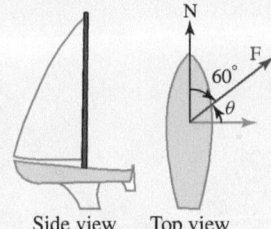

Side view   Top view

39. Given $\mathbf{u} = \mathbf{i} - \mathbf{j} + \mathbf{k}$, $\mathbf{v} = 3\mathbf{i} - 2\mathbf{j} + 5\mathbf{k}$, and $\mathbf{w} = \mathbf{i} + \mathbf{j} - \mathbf{k}$, find $(\mathbf{u} - \mathbf{v}) \cdot \mathbf{w}$ and $(2\mathbf{u} + \mathbf{v}) \times (\mathbf{u} - \mathbf{w})$.
40. If **u** and **v** are orthogonal unit vectors, show that $(\mathbf{u} \times \mathbf{v}) \times \mathbf{u} = \mathbf{v}$. What is $(\mathbf{u} \times \mathbf{v}) \times \mathbf{v}$?
41. Show that three planes whose normals **u**, **v**, **w** satisfy $\mathbf{u} \times \mathbf{v} \cdot \mathbf{w} \neq 0$ intersect in exactly one point.
42. Show that $\mathbf{u} \times (\mathbf{v} \times \mathbf{w}) = (\mathbf{u} \times \mathbf{v}) \times \mathbf{w}$ if and only if $\mathbf{v} \times (\mathbf{w} \times \mathbf{u}) = \mathbf{0}$.
43. Find the direction cosines for $\mathbf{v} = (2\mathbf{i} + \mathbf{j}) \times (\mathbf{i} + \mathbf{j} - 3\mathbf{k})$.
44. Find the direction angles of the vector $\mathbf{u} = -2\mathbf{i} + 3\mathbf{j} + \mathbf{k}$.
45. Find the (acute) angle, rounded to the nearest degree, between the intersecting lines

$$\frac{x - 1}{3} = \frac{y - 3}{-1} = \frac{z + 5}{2} \quad \text{and} \quad \frac{x - 1}{2} = \frac{y - 3}{-1} = \frac{z + 5}{-2}$$

46. Find the area of the parallelogram determined by $3\mathbf{i} - 4\mathbf{j}$ and $-\mathbf{i} - \mathbf{j} + \mathbf{k}$.
47. Find the center and the radius of the sphere

$$4x^2 + 4y^2 + 4z^2 + 12y - 4z + 1 = 0$$

48. Find the points of intersection of the line $x = 6 + 3t, y = 10 - 2t, z = 5t$ with each of the coordinate planes.
49. Find the point of intersection of the planes $3x - y + 4z = 15, 2x + y - 3z - 1 = 0$, and $x + 3y + 5z = 2$.
50. Find the equation of the plane determined by the intersecting lines

$$\frac{x + 3}{3} = \frac{y}{-2} = \frac{z - 7}{6} \quad \text{and} \quad \frac{x + 6}{1} + \frac{y + 5}{-3} = \frac{z - 1}{2}$$

51. Find an equation for the set of all points that are equidistant from the planes $3x - 4y + 12z = 6$ and $4x + 3z = 7$.
52. Vertices B and C of $\triangle ABC$ lie along the line

$$\frac{x + 2}{2} = \frac{y - 1}{1} = \frac{z}{4}$$

Find the area of the triangle given that A has coordinates $(1, -1, 2)$ and line segment $\overline{BC}$ has length 5.
53. Find two unit vectors that are parallel to the line

$$\frac{x}{6} = \frac{y}{2} = \frac{z - 1}{6}$$

54. A boy mowing the lawn wants to impress his dad by calculating the amount of work he is doing (see Figure 9.72).

Component of **F** in direction of motion

F (Constant force)

Figure 9.72   Work mowing the lawn

The *displacement*, *d*, the distance the lawn mower is moved, is 2,450 ft and the angle in which the boy is pressing down on the handle is 40° with a force of 48.0 lb. What is the amount of work done by the boy?

55. A dad pulls his child in a wagon (see Figure 9.73) for 150 ft by exerting a constant force of 25 lb along the handle. How much work is done?

Figure 9.73 Work pulling a wagon

56. A girl pulls a sled 50 ft on level ground with a rope inclined at an angle of 30° with the horizontal (the ground). If she applies 3 lb of tension to the rope, how much work is performed on the sled? Give the exact answer.

57. Find the work done by the constant force $\mathbf{F} = 5\mathbf{i} + 4\mathbf{j} + \mathbf{k}$ in moving a particle along the line from $P(2, 1, -1)$ to $Q(4, 1, 2)$.

58. How much work does it take to move a container 25 m along a horizontal loading platform onto a truck using a constant force of 100 newtons at an angle of $\frac{\pi}{6}$ from the horizontal?

59. Find a formula for the surface area of the tetrahedron determined by vectors $\mathbf{u}$, $\mathbf{v}$, and $\mathbf{w}$. Assume the vectors do not all lie in the same plane.

60. Find the parametric and symmetric equations for the line passing through the point $(-1, 1, 6)$ perpendicular to $3x + y - 2z = 5$.

61. Use vectors to show that the sum of the squares of the lengths of the sides of a parallelogram equals the sum of the squares of the lengths of the diagonals.

62. Suppose $\mathbf{v}$ and $\mathbf{w}$ are nonzero vectors. Show that $\|\mathbf{v}\| \mathbf{w} + \|\mathbf{w}\| \mathbf{v}$ and $\|\mathbf{v}\| \mathbf{w} - \|\mathbf{w}\| \mathbf{v}$ are orthogonal vectors.

63. Let $\mathbf{v} = \cos\theta\mathbf{i} + \sin\theta\mathbf{j}$ and $\mathbf{w} = \cos\phi\mathbf{i} + \sin\phi\mathbf{j}$. Find $\mathbf{v} \times \mathbf{w}$. Interpret this cross product geometrically, and use it to derive a well-known trigonometric identity.

64. Find the three vertex angles (rounded to the nearest degree) of the triangle whose vertices are $A(1, -2, 3)$, $B(-1, 2, -3)$, and $C(2, 1, -3)$.

65. Find a relationship between the numbers a_1, b_1, and c_1 so that the angle between the vectors $\mathbf{v} = a_1\mathbf{i} + b_1\mathbf{j} + c_1\mathbf{k}$ and $\mathbf{w}_1 = \mathbf{i} - 2\mathbf{j}$ is the same as the angle between $\mathbf{v}$ and $\mathbf{w}_2 = 2\mathbf{i} + \mathbf{k}$.

66. Find an equation of the plane that passes through $(a, 0, 0)$, $(0, a, 0)$, and $(0, 0, a)$.

67. Find the area of the triangle with vertices $A(0, -1, 2)$, $B(1, 2, -1)$, and $C(3, -1, 2)$.

68. Find an equation for the surface comprised of all points equidistant from $D(0, 0, 6)$ and the xy-plane. Is this a quadric surface?

69. Find the area of the triangle determined by the vectors $\mathbf{v} = \mathbf{i} - \mathbf{j} + \mathbf{k}$ and $\mathbf{w} = 2\mathbf{i} + \mathbf{j} - 2\mathbf{k}$.

70. Find an equation for the plane that passes through the origin and is parallel to the vectors $\mathbf{v} = \mathbf{i} - 2\mathbf{j} - 3\mathbf{k}$ and $\mathbf{w} = -\mathbf{i} + \mathbf{j} + 2\mathbf{k}$.

71. Find an equation for the plane that passes through the origin and whose normal vector is parallel to the line of intersection of the planes $2x - y + z = 4$ and $x + 3y - z = 2$.

72. Find a number A such that the angle between the planes $2Ax + 3y + z = 1$ and $x - Ay + 3z = 5$ is $\pi/2$.

73. The lines L_1 and L_2 are parallel to the vectors $\mathbf{v}_1 = \mathbf{i} - \mathbf{j}$ and $\mathbf{v}_2 = \mathbf{i} - \mathbf{j} + 2\mathbf{k}$, respectively. Find an equation for the line L that passes through the point $(-1, 2, 0)$ and is orthogonal to both L_1 and L_2.

74. A parallelepiped is determined by the vectors $\mathbf{u} = \mathbf{i} - \mathbf{j} + \mathbf{k}$, $\mathbf{v} = \mathbf{i} + 2\mathbf{j} - \mathbf{k}$, and $\mathbf{w} = 2\mathbf{i} + \mathbf{j} + \mathbf{k}$. Find the altitude from the tip of $\mathbf{w}$ to the side determined by $\mathbf{u}$ and $\mathbf{v}$.

75. In the next chapter, we will show that $\mathbf{T} = \mathbf{i} + 2x\mathbf{j}$ is a vector in the direction of the tangent line at each point $P(x, x^2)$ on the parabola $y = x^2$. Find a unit vector normal to the parabola at the point $(3, 9)$.

76. For nonzero vectors $\mathbf{u}$, $\mathbf{v}$, and $\mathbf{w}$, show that the vector

$$(\mathbf{u} \times \mathbf{v}) \times (\mathbf{u} \times \mathbf{w})$$

is parallel to $\mathbf{u}$.

77. Let $\mathbf{v} = a\mathbf{i} + b\mathbf{j} + c\mathbf{k}$ and $\mathbf{w} = A\mathbf{i} + B\mathbf{j} + C\mathbf{k}$, where $a, b, c, A, B,$ and C are constants. Describe the set of vectors $\mathbf{v} + t\mathbf{w}$, where t is any scalar.

78. A 40-lb child sits on a seesaw, 3 ft from the fulcrum, as shown in Figure 9.74. What torque is exerted when the child is 2 ft above the horizontal?

Figure 9.74   Seesaw torque

79. Let $\mathbf{u}_0 = a\mathbf{i} + b\mathbf{j} + c\mathbf{k}$ and let $\mathbf{u} = x\mathbf{i} + y\mathbf{j} + z\mathbf{k}$. Describe the set of points in $\mathbb{R}^3$ defined by

$$\|\mathbf{u}_0 - \mathbf{u}\| < r$$

where $r > 0$ and $a, b, c,$ are constants.

80. The vectors $\mathbf{u}, \mathbf{v},$ and $\mathbf{w}$ are said to be *linearly independent* in $\mathbb{R}^3$ if the only solution to the equation $a\mathbf{u} + b\mathbf{v} + c\mathbf{w} = \mathbf{0}$ is $a = b = c = 0$. Otherwise, the are *linearly dependent*. The simplest example of a set of linearly independent vectors is $\mathbf{i}, \mathbf{j},$ and $\mathbf{k}$. Determine whether the vectors $\mathbf{u} = -\mathbf{i} + 2\mathbf{k}$, $\mathbf{v} = 2\mathbf{i} - \mathbf{j} + 3\mathbf{k}$, $\mathbf{w} = \mathbf{i} + 3\mathbf{j} - 2\mathbf{k}$ are linearly independent or dependent.

81. Figure 9.75 shows a parallelogram $ABCD$. In Problem 82, we show that if M is the midpoint of side $\overline{AB}$, then the line $\overline{CM}$ intersects diagonal $\overline{BD}$ at a point P located one-third of the distance from B to D. In order to complete this, we must first show the following:

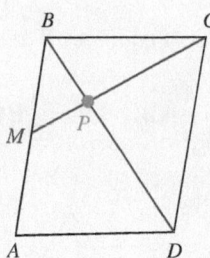

Figure 9.75   Parallelogram $ABCD$

Let a and b be scalars such that $\mathbf{MP} = a\mathbf{MC}$ and $\mathbf{BP} = b\mathbf{BD}$. Show that

$$\frac{1}{2}\mathbf{AB} + b(\mathbf{AD} - \mathbf{AB}) = a\left(\mathbf{AD} - \frac{1}{2}\mathbf{AB}\right)$$

82. Use the fact that $\mathbf{AB}$ and $\mathbf{AD}$ in Figure 9.75 are linearly independent along with the results of Problem 81 to show that

$$\frac{1}{2} - b - \frac{1}{2}a = 0 \text{ and } a - b = 0$$

Solve this system of equations to show that P has the required location.

83. Show that $\mathbf{u} = a_1\mathbf{i} + a_2\mathbf{j} + a_3\mathbf{k}$, $\mathbf{v} = b_1\mathbf{i} + b_2\mathbf{j} + b_3\mathbf{k}$ and $\mathbf{w} = c_1\mathbf{i} + c_2\mathbf{j} + c_3\mathbf{k}$ are linearly dependent (see Problem 80), if and only if

$$\begin{vmatrix} a_1 & a_2 & a_3 \\ b_1 & b_2 & b_3 \\ c_1 & c_2 & c_3 \end{vmatrix} = 0$$

84. Show that

$$\begin{vmatrix} \mathbf{u}_1 \cdot \mathbf{v}_1 & \mathbf{u}_1 \cdot \mathbf{v}_2 \\ \mathbf{u}_2 \cdot \mathbf{v}_1 & \mathbf{u}_2 \cdot \mathbf{v}_2 \end{vmatrix} = (\mathbf{u}_1 \times \mathbf{u}_2) \cdot (\mathbf{v}_1 \times \mathbf{v}_2)$$

85. Let $A(-2, 3, 7), B(1, 5, -3)$, and $C(2, 8, -1)$ be the vertices of a triangle in $\mathbb{R}^3$. What are the coordinates of the point M where the medians of the triangle meet (the centroid)?

86. The medians of a triangle meet at a point (the centroid) located two-thirds of the distance from each vertex to the midpoint of the opposite side. Generalize this result by showing that the four lines that join each vertex of a tetrahedron to the centroid of the opposite face meet at a point located three-fourths of the distance from the vertex to the centroid.

87. A triangle in $\mathbb{R}^3$ is determined by the vectors $\mathbf{v}$ and $\mathbf{w}$ as shown in Figure 9.76.

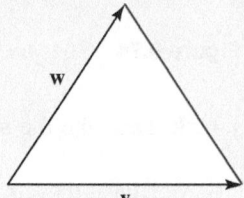

Figure 9.76 Area of a triangle

Show that the triangle has area

$$A = \frac{1}{2}\sqrt{\|\mathbf{v}\|^2 \|\mathbf{w}\|^2 - (\mathbf{v} \cdot \mathbf{w})^2}$$

88. Think Tank Problem

   a. If $\mathbf{u} \times \mathbf{w} = \mathbf{v} \times \mathbf{w}$, does it follow that $\mathbf{u} = \mathbf{v}$?   **b.** If $\mathbf{u} \cdot \mathbf{w} = \mathbf{v} \cdot \mathbf{w}$, does it follow that $\mathbf{u} = \mathbf{v}$?

Think Tank Problems *In Figure 9.77, $\triangle ABC$ is equilateral and the points M, N, and O are located so that*

$$\mathbf{AM} = \frac{1}{3}\mathbf{AB} \quad \mathbf{BN} = \frac{1}{3}\mathbf{BC} \quad \mathbf{CO} = \frac{1}{3}\mathbf{CA}$$

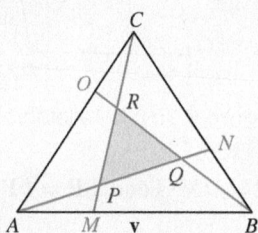

Figure 9.77 Problems 89-91

It can be shown that $\triangle PQR$ is also equilateral, and

$$\|PM\| = \|QN\| = \|RO\| \quad \text{and} \quad \|AP\| = \|BQ\| = \|CR\|$$

Use this information in Problems 89-91.

89. Show that $\mathbf{PQ} = \frac{3}{7}\mathbf{AN}$, then show that $\|\mathbf{AN}\|^2 = \frac{7}{9}\|\mathbf{AB}\|^2$. *Hint:* Use the law of cosines.

90. Show that $\triangle PQR$ has area $\frac{1}{7}$ that of $\triangle ABC$.

91. Do you think the results from Problems 89 and 90 would hold if $\triangle ABC$ were not equilateral? Investigate your conjecture.

Think Tank Problems **The Gram-Schmidt orthogonalization process** *In Problems* 92-93, *let* **u, v,** *and* **w** *be nonzero vectors in* $\mathbb{R}^3$ *that do not lie on the same plane. Define vectors* $\boldsymbol{\alpha}$ *and* $\boldsymbol{\beta}$ *as follows:*

$$\boldsymbol{\alpha} = \mathbf{v} - \left[\frac{\mathbf{v}\cdot\mathbf{u}}{\|\mathbf{u}\|^2}\right]\mathbf{u} \quad \text{and} \quad \boldsymbol{\beta} = \mathbf{w} - \left[\frac{\mathbf{w}\cdot\mathbf{u}}{\|\mathbf{u}\|^2}\right]\mathbf{u} - \left[\frac{\mathbf{w}\cdot\boldsymbol{\alpha}}{\|\boldsymbol{\alpha}\|^2}\right]\boldsymbol{\alpha}$$

92. Show that **u**, $\boldsymbol{\alpha}$, $\boldsymbol{\beta}$ are mutually orthogonal (any pair is orthogonal).
93. If $\boldsymbol{\gamma}$ is any vector in $\mathbb{R}^3$, show that

$$\boldsymbol{\gamma} = \left[\frac{\boldsymbol{\gamma}\cdot\mathbf{u}}{\|\mathbf{u}\|^2}\right]\mathbf{u} + \left[\frac{\boldsymbol{\gamma}\cdot\boldsymbol{\alpha}}{\|\boldsymbol{\alpha}\|^2}\right]\boldsymbol{\alpha} + \left[\frac{\boldsymbol{\gamma}\cdot\boldsymbol{\beta}}{\|\boldsymbol{\beta}\|^2}\right]\boldsymbol{\beta}$$

94. Suppose **u, v,** and **w** are nonzero vectors in $\mathbb{R}^3$ with $\mathbf{u}\times\mathbf{v}=\mathbf{w}$ and $\mathbf{u}\cdot\mathbf{v}=0$. Show that $\mathbf{v}=s(\mathbf{w}\times\mathbf{u})$ and $\mathbf{u}=t(\mathbf{v}\times\mathbf{w})$ for scalars s and t.

95. **Think Tank Problem** In Figure 9.78, $ABCD$ is a rectangle, with M the midpoint of side $\overline{CD}$ and $\mathbf{AR}_k = \dfrac{1}{2k+1}\mathbf{AB}$. If P is the intersection of $\overline{AM}$ and $\overline{CR_k}$, and R_{k+1} is the foot of the perpendicular drawn from P to $\overline{AB}$, show that

$$\mathbf{AR}_{k+1} = \frac{1}{2k+3}\mathbf{AB}$$

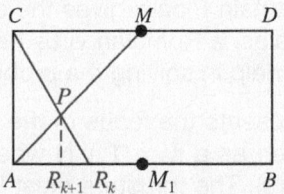

Figure 9.78 Problem 95

96. **Think Tank Problem** It is easy to subdivide a line segment line $\overline{AB}$ into halves, or fourths, or eighths, etc. However, dividing it into thirds or fifths, $\cdots$ is more difficult. Use the result obtained in Problem 95 to describe a procedure for subdividing $\overline{AB}$ into an odd number of equal parts.

97. **Putnam Examination Problem** Find the equations of two straight lines, each of which cuts all four of the following lines:

$$L_1: x = 1, y = 0 \qquad L_2: y = 1, z = 0 \qquad L_3: z = 1, x = 0 \qquad L_4: x = y = -6z$$

98. **Putnam Examination Problem** Find the equation of the smallest sphere that is tangent to both the lines

$$L_1: x = t + 1, \, y = 2t + 4, \, z = -3t + 5 \qquad L_2: x = 4t - 12, y = t + 8, z = t + 17$$

99. **Putnam Examination Problem** The hands of an accurate clock have lengths 3 cm and 4 cm. Find the distance between the tips of the hands when the distance is increasing most rapidly.

CHAPTER 9 GROUP RESEARCH PROJECT*

Working in small groups is typical of most work environments, and this book seeks to develop skills with group activities. We present a group project at the end of each chapter. These projects are to be done in groups of three or four students.

Star Trek Quest

© Mark E. Gibson/CORBIS

The starship *Enterprise* has been captured by the evil Romulans and is being held in orbit by a Romulan tractor beam. The orbit is elliptical with the planet Romulus at one focus of the ellipse. Repeated efforts to escape have been futile and have almost exhausted the fuel supplies. Morale is low and food reserves are dwindling.

In searching the ship's log, Lieutenant Commander Data discovers that the Enterprise had been captured long ago by a Romulan tractor beam and had escaped. The key to that escape was to fire the ship's thrusters at exactly the right position in the orbit. Captain Picard gives the command to feed the required information into the computer to find that position. But, alas, a Romulan virus has rendered the computer useless for this task. Everyone turns to you and asks for your help in solving the problem.

Here is what Data discovered. If F represents the focus of the ellipse and P is the position of the ship on the ellipse, then the vector **FP** can be written as a sum $\mathbf{T} + \mathbf{N}$ where $\mathbf{T}$ is tangent to the ellipse and $\mathbf{N}$ is normal to the ellipse (not necessarily unit vectors). The thrusters must be fired when the ratio $\|\mathbf{T}\|/\|\mathbf{N}\|$ is equal to the eccentricity of the ellipse.

Extended paper for further study. Your mission is to save the starship from the evil Romulans. Write a paper describing your plan.

---

*Adapted from *MAA Notes*: 17 (1991) "Priming the Calculus Pump: Innovations and Resources," Marcus S. Cohen, Edward D. Gaughan, R. Arthur Knoebel, Douglas S. Kurtz, and David J. Pengelley.

CHAPTER 10

VECTOR-VALUED FUNCTIONS

*W*here there is matter, there is geometry.

Johannes Kepler *(1571 – 1630)*

Quoted in Conversation with the Sidereal Messenger (an open letter to Galileo Galilei)

PREVIEW

The marriage of calculus and vector methods forms what is called *vector calculus*. The key to using vector calculus is the concept of a *vector-valued function*. In this chapter we introduce such functions and examine some of their properties. We will see that vector-valued functions behave much like the *scalar-valued functions* studied earlier in this text.

PERSPECTIVE

A car travels down a curved road at a constant speed of 55 mi/h. What additional information do we need about the car to determine whether it will stay on the road or skid off as it rounds a particular curve?

Can we modify the road (say, by banking) so an average-sized car can travel at moderate speeds without skidding? How does a highway department decide what warning sign to install on a particular curve? A soldier fires a howitzer whose muzzle speed and angle of elevation are known. If the shell overshoots its target by 40 yd, how should the angle of elevation be changed to ensure a hit on the next shot? If a satellite is in orbit above the earth, how fast must it travel to remain stationary above a particular point on the equator? These and other similar questions can be answered using vector calculus.

751

10.1 INTRODUCTION TO VECTOR FUNCTIONS

IN THIS SECTION: *Vector-valued functions, operations with vector-valued functions, limits and continuity of vector functions*
Vector-valued functions are defined and are used to study curves in the plane and in space. We will also study limits and continuity of vector-valued functions.

Vector-Valued Functions

In Section 9.5, we described a *plane curve* using parametric equations

$$x = f_1(t) \text{ and } y = f_2(t)$$

where f_1 and f_2 are functions of t on some interval. We extend this definition to three dimensions. A **curve** in $\mathbb{R}^3$ is the set of all ordered triples $(f_1(t), f_2(t), f_3(t))$ satisfying the parametric equations $x = f_1(t)$, $y = f_2(t)$, $z = f_3(t)$, where f_1, f_2, and f_3 are functions of t on some domain D, which lies in $\mathbb{R}$.

The concept of a vector-valued function is fundamental to the ideas we plan to explore. Here is a definition.

VECTOR-VALUED FUNCTION A **vector-valued function** (or, simply, a **vector function**) $\mathbf{F}$ of a real variable with *domain D* assigns to each number t in the set D a unique vector $\mathbf{F}(t)$. The set of all vectors $\mathbf{v}$ of the form $\mathbf{v} = \mathbf{F}(t)$ for t in D is the *range* of $\mathbf{F}$. That is,

$$\mathbf{F}(t) = f_1(t)\mathbf{i} + f_2(t)\mathbf{j} \qquad Plane\,(\mathbb{R}^2)$$
$$\mathbf{F}(t) = f_1(t)\mathbf{i} + f_2(t)\mathbf{j} + f_3(t)\mathbf{k} \quad Three\text{-}Space\,(\mathbb{R}^3)$$

where f_1, f_2, and f_3 are real-valued (**scalar-valued**) functions of the real number t defined on the domain set D. In this context, f_1, f_2, and f_3 are called the **components** of $\mathbf{F}$. A vector function may also be denoted by $\mathbf{F}(t) = \langle f_1(t), f_2(t) \rangle$ or $\mathbf{F}(t) = \langle f_1(t), f_2(t), f_3(t) \rangle$.

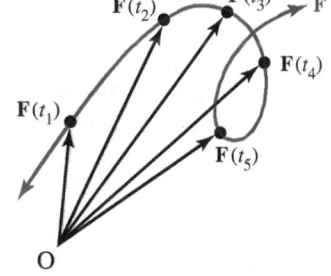

Figure 10.1 The graph of the vector function $\mathbf{F}(t)$ is traced out by the terminal point of $\mathbf{F}(t)$ as t varies over the domain D

Let $\mathbf{F}$ be a vector function, and suppose the initial point of the vector $\mathbf{F}(t)$ is at the origin. The graph of $\mathbf{F}$ is the curve traced out by the terminal point of the vector $\mathbf{F}(t)$ as t varies over the domain set D, as shown in Figure 10.1. In this context, $\mathbf{F}(t)$ is called the **position vector** at t for the point $P(f_1(t), f_2(t), f_3(t))$ on C.

Example 1 Graph of a vector function

Sketch the graph of the vector function

$$\mathbf{F}(t) = (3 - t)\mathbf{i} + (2t)\mathbf{j} + (3t - 4)\mathbf{k}$$

for all t.

Solution The graph is the collection of all points (x, y, z) with

$$x = 3 - t \qquad y = 2t \qquad z = -4 + 3t$$

for all t. We recognize these as the parametric equations for the line in $\mathbb{R}^3$ that contains the point $P_0(3, 0, -4)$ and is parallel to the vector

$$\mathbf{V} = -\mathbf{i} + 2\mathbf{j} + 3\mathbf{k}$$

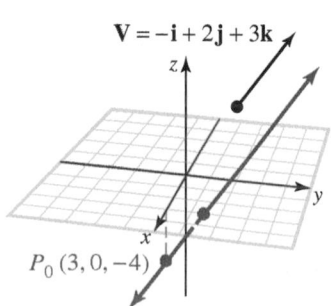

Figure 10.2 Graph of **F**

as shown in Figure 10.2.

Example 2   Graph of a circular helix

Sketch the graph of the vector function

$$\mathbf{F}(t) = (2\sin t)\mathbf{i} - (2\cos t)\mathbf{j} + (3t)\mathbf{k}$$

Solution   The graph of **F** is the collection of all points (x, y, z) in $\mathbb{R}^3$ whose coordinates satisfy

$$x = 2\sin t \qquad y = -2\cos t \qquad z = 3t \qquad \text{for all } t$$

The first two components satisfy

$$x^2 + y^2 = (2\sin t)^2 + (2\cos t)^2 = 4(\sin^2 t + \cos^2 t) = 4$$

which means that the graph lies on the surface of the circular cylinder with radius 2 as shown in Figure 10.3.

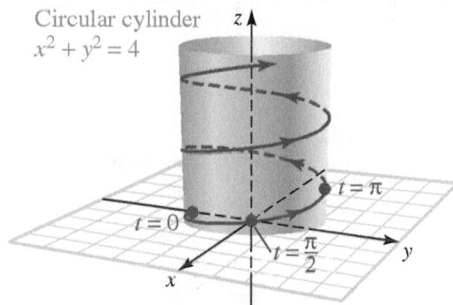

Figure 10.3   Graph of a helix

Next notice the axis of symmetry for the graph in Figure 10.3 is the z-axis. We also know that as t increases, the z-coordinate of the point $P(x, y, z)$ on the graph of **F** increases according to the formula $z = 3t$, which means that the point (x, y, z) on the graph rises in a spiral on the surface of the cylinder $x^2 + y^2 = 4$. The point on the graph of **F** that corresponds to $t = 0$ is $(0, -2, 0)$, and the points that correspond to $t = \frac{\pi}{2}$ and $t = \pi$ are $(2, 0, \frac{3\pi}{2})$ and $(0, 2, 3\pi)$, respectively. Thus, the graph spirals upward counterclockwise (as viewed from above). The graph is known as a **right circular helix**.

A well-known example of a helix is the DNA (deoxyribonucleic acid) molecule, which has a structure consisting of two intertwined helixes, as shown in Figure 10.4. Some other computer-generated helixes are also shown.

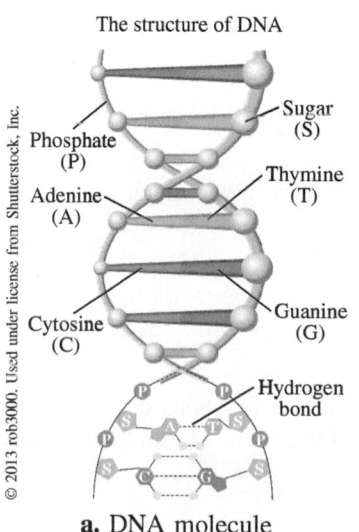

a. DNA molecule

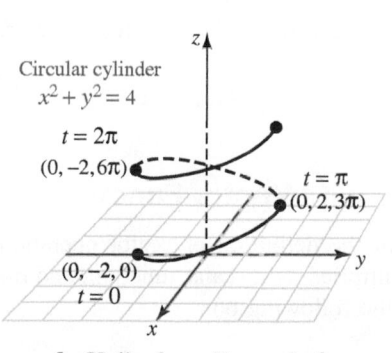

b. Helix from Example 2

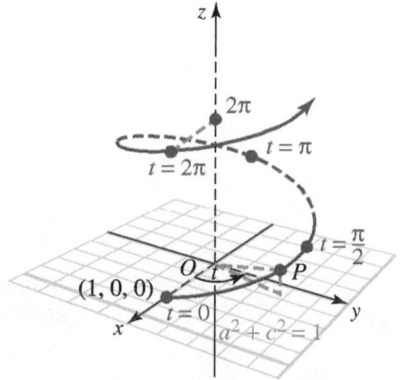

c. Interactive Computer generated helix
$\mathbf{F}(t) = (\cos t)\mathbf{i} + (\sin t)\mathbf{j} + 0.2t\mathbf{k}$

Figure 10.4   Examples of helixes

Examples 1 and 2 illustrate how the graph of a vector function

$$\mathbf{F}(t) = f_1(t)\mathbf{i} + f_2(t)\mathbf{j} + f_3(t)\mathbf{k}$$

can be obtained by examining the parametric equations

$$x = f_1(t) \qquad y = f_2(t) \qquad z = f_3(t)$$

In Example 3, we turn things around and find a vector function whose graph is a given curve.

Example 3   Find a vector function

Find a vector function $\mathbf{F}$ whose graph is the curve of intersection of the hemisphere

$$z = \sqrt{4 - x^2 - y^2} \text{ and the parabolic cylinder } y = x^2$$

as shown in Figure 10.5.

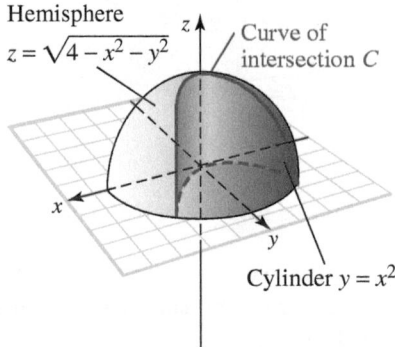

Figure 10.5   The curve of intersection of the hemisphere and the cylinder

Solution   Finding a parametric representation is sometimes called **parameterizing the curve**. There are several ways this can be done, but one choice is to let $x = t$. Then $y = t^2$ (from the equation of the parabolic cylinder), and by substituting into the equation for the hemisphere, we find

$$z = \sqrt{4 - x^2 - y^2}$$
$$= \sqrt{4 - (t)^2 - (t^2)^2}$$
$$= \sqrt{4 - t^2 - t^4}$$

We can now state a formula for a vector-valued function of the given graph:

$$\mathbf{F}(t) = t\mathbf{i} + t^2\mathbf{j} + \sqrt{4 - t^2 - t^4}\mathbf{k}$$

Operations with Vector Functions

It follows from the definition of vector operations that vector functions can be added, subtracted, multiplied by a scalar function, and multiplied together. We summarize these operations in the following box.

VECTOR FUNCTION OPERATIONS Let $\mathbf{F}$ and $\mathbf{G}$ be vector functions of the real variable t, and let $f(t)$ be a scalar function. Then, $\mathbf{F} + \mathbf{G}$, $\mathbf{F} - \mathbf{G}$, $f\mathbf{F}$, and $\mathbf{F} \times \mathbf{G}$ are vector functions, and $\mathbf{F} \cdot \mathbf{G}$ is a scalar function. These operations are defined as follows:

Vector functions:

$$(\mathbf{F} + \mathbf{G})(t) = \mathbf{F}(t) + \mathbf{G}(t) \quad (\mathbf{F} - \mathbf{G})(t) = \mathbf{F}(t) - \mathbf{G}(t)$$
$$(f\mathbf{F})(t) = f(t)\mathbf{F}(t) \quad\quad (\mathbf{F} \times \mathbf{G})(t) = \mathbf{F}(t) \times \mathbf{G}(t)$$

Scalar function: $\quad (\mathbf{F} \cdot \mathbf{G})(t) = \mathbf{F}(t) \cdot \mathbf{G}(t)$

These operations are defined on the intersection of the domain of the vector and scalar functions that occur in their definitions.

Remember, $(\mathbf{F} \times \mathbf{G})(t)$, is a vector and $(\mathbf{F} \cdot \mathbf{G})(t)$ is a scalar.

Example 4   Vector function operations

Let $\mathbf{F}(t) = t^2\mathbf{i} + t\mathbf{j} - (\sin t)\mathbf{k}$ and $\mathbf{G}(t) = t\mathbf{i} + \frac{1}{t}\mathbf{j} + 5\mathbf{k}$. Find

a. $(\mathbf{F} + \mathbf{G})(t)$ **b.** $(e^t\mathbf{F})(t)$ **c.** $(\mathbf{F} \times \mathbf{G})(t)$ **d.** $(\mathbf{F} \cdot \mathbf{G})(t)$

Solution
a. $(\mathbf{F} + \mathbf{G})(t) = [t^2\mathbf{i} + t\mathbf{j} - (\sin t)\mathbf{k}] + \left[t\mathbf{i} + \frac{1}{t}\mathbf{j} + 5\mathbf{k} \right]$

$$= (t^2 + t)\mathbf{i} + \left(t + \frac{1}{t} \right)\mathbf{j} + (5 - \sin t)\mathbf{k}$$

b. $(e^t\mathbf{F})(t) = e^t\mathbf{F}(t) = e^t t^2\mathbf{i} + e^t t\mathbf{j} - (e^t \sin t)\mathbf{k}$
c. $(\mathbf{F} \times \mathbf{G})(t) = \mathbf{F}(t) \times \mathbf{G}(t)$

$$= [t^2\mathbf{i} + t\mathbf{j} - (\sin t)\mathbf{k}] \times \left[t\mathbf{i} + \frac{1}{t}\mathbf{j} + 5\mathbf{k} \right]$$

$$= \begin{vmatrix} \mathbf{i} & \mathbf{j} & \mathbf{k} \\ t^2 & t & -\sin t \\ t & \frac{1}{t} & 5 \end{vmatrix}$$

$$= \left[5t + \frac{\sin t}{t} \right]\mathbf{i} - [5t^2 + t \sin t]\mathbf{j} + [t - t^2]\mathbf{k}$$

d. $(\mathbf{F} \cdot \mathbf{G})(t) = \mathbf{F}(t) \cdot \mathbf{G}(t)$

$$= [t^2\mathbf{i} + t\mathbf{j} - (\sin t)\mathbf{k}] \cdot \left[t\mathbf{i} + \frac{1}{t}\mathbf{j} + 5\mathbf{k} \right]$$

$$= t^3 + 1 - 5\sin t$$

Limits and Continuity of Vector Functions

We begin with a definition.

LIMIT OF A VECTOR FUNCTION Suppose the components f_1, f_2, f_3 of the vector function

$$\mathbf{F}(t) = f_1(t)\mathbf{i} + f_2(t)\mathbf{j} + f_3(t)\mathbf{k}$$

all have finite limits as $t \to t_0$, where t_0 is any number or ∞ or $-\infty$. Then the **limit** of $\mathbf{F}(t)$ as $t \to t_0$ is the vector

$$\lim_{t \to t_0} \mathbf{F}(t) = \left[\lim_{t \to t_0} f_1(t) \right]\mathbf{i} + \left[\lim_{t \to t_0} f_2(t) \right]\mathbf{j} + \left[\lim_{t \to t_0} f_3(t) \right]\mathbf{k}$$

Example 5   **Limit of a vector function**

Find $\lim\limits_{t \to 2} \mathbf{F}(t)$, where $\mathbf{F}(t) = (t^2 - 3)\mathbf{i} + e^t\mathbf{j} + (\sin \pi t)\mathbf{k}$.

Solution

$$\lim_{t \to 2} \mathbf{F}(t) = \left[\lim_{t \to 2}(t^2 - 3)\right]\mathbf{i} + \left[\lim_{t \to 2}(e^t)\right]\mathbf{j} + \left[\lim_{t \to 2}(\sin \pi t)\right]\mathbf{k}$$

$$= 1\mathbf{i} + e^2\mathbf{j} + (\sin 2\pi)\mathbf{k}$$

$$= \mathbf{i} + e^2\mathbf{j}$$

Vector limits behave like scalar limits. The following theorem contains some useful general properties of such limits.

Theorem 10.1   Rules for Vector Limits

If the vector functions $\mathbf{F}$ and $\mathbf{G}$ are functions of a real variable t and $h(t)$ is a scalar function such that all three functions have finite limits as $t \to t_0$, then

Limit of a sum	$\lim\limits_{t \to t_0}[\mathbf{F}(t) + \mathbf{G}(t)] = \lim\limits_{t \to t_0} \mathbf{F}(t) + \lim\limits_{t \to t_0} \mathbf{G}(t)$
Limit of a difference	$\lim\limits_{t \to t_0}[\mathbf{F}(t) - \mathbf{G}(t)] = \lim\limits_{t \to t_0} \mathbf{F}(t) - \lim\limits_{t \to t_0} \mathbf{G}(t)$
Limit of a scalar multiple	$\lim\limits_{t \to t_0}[h(t)\mathbf{F}(t)] = \left[\lim\limits_{t \to t_0} h(t)\right]\left[\lim\limits_{t \to t_0} \mathbf{F}(t)\right]$
Limit of a dot product	$\lim\limits_{t \to t_0}[\mathbf{F}(t) \cdot \mathbf{G}(t)] = \left[\lim\limits_{t \to t_0} \mathbf{F}(t)\right] \cdot \left[\lim\limits_{t \to t_0} \mathbf{G}(t)\right]$
Limit of a cross product	$\lim\limits_{t \to t_0}[\mathbf{F}(t) \times \mathbf{G}(t)] = \left[\lim\limits_{t \to t_0} \mathbf{F}(t)\right] \times \left[\lim\limits_{t \to t_0} \mathbf{G}(t)\right]$

These limit formulas are also valid as $t \to \infty$ or as $t \to -\infty$, assuming all expressions have finite limits.

Proof: We will establish the formula for the limit of a dot product and leave the rest of the proof as an exercise. Let

$$\mathbf{F}(t) = f_1(t)\mathbf{i} + f_2(t)\mathbf{j} + f_3(t)\mathbf{k} \quad \text{and} \quad \mathbf{G}(t) = g_1(t)\mathbf{i} + g_2(t)\mathbf{j} + g_3(t)\mathbf{k}$$

Apply the limit of a vector function along with the sum rule and product rule for scalar limits to write

$$\lim_{t \to t_0}[\mathbf{F}(t) \cdot \mathbf{G}(t)] = \lim_{t \to t_0}[f_1(t)g_1(t) + f_2(t)g_2(t) + f_3(t)g_3(t)]$$

$$= \left[\lim_{t \to t_0} f_1(t)\right]\left[\lim_{t \to t_0} g_1(t)\right] + \left[\lim_{t \to t_0} f_2(t)\right]\left[\lim_{t \to t_0} g_2(t)\right]$$

$$+ \left[\lim_{t \to t_0} f_3(t)\right]\left[\lim_{t \to t_0} g_3(t)\right]$$

$$= \left(\left[\lim_{t \to t_0} f_1(t)\right]\mathbf{i} + \left[\lim_{t \to t_0} f_2(t)\right]\mathbf{j} + \left[\lim_{t \to t_0} f_3(t)\right]\mathbf{k}\right) \cdot$$

$$\left(\left[\lim_{t \to t_0} g_1(t)\right]\mathbf{i} + \left[\lim_{t \to t_0} g_2(t)\right]\mathbf{j} + \left[\lim_{t \to t_0} g_3(t)\right]\mathbf{k}\right)$$

$$= \left[\lim_{t \to t_0} \mathbf{F}(t)\right] \cdot \left[\lim_{t \to t_0} \mathbf{G}(t)\right]$$

Example 6 Limit of a cross product of vector functions

Show that $\lim\limits_{t \to 1}[\mathbf{F}(t) \times \mathbf{G}(t)] = \left[\lim\limits_{t \to 1} \mathbf{F}(t)\right] \times \left[\lim\limits_{t \to 1} \mathbf{G}(t)\right]$ for the vector functions

$\mathbf{F}(t) = t\mathbf{i} + (1 - t)\mathbf{j} + t^2\mathbf{k}$ and $\mathbf{G}(t) = e^t\mathbf{i} - (3 + e^t)\mathbf{k}$

Solution

$$\lim_{t \to 1}[\mathbf{F}(t) \times \mathbf{G}(t)] = \lim_{t \to 1} \begin{vmatrix} \mathbf{i} & \mathbf{j} & \mathbf{k} \\ t & 1-t & t^2 \\ e^t & 0 & -(3+e^t) \end{vmatrix}$$

$$= \lim_{t \to 1} \left\{ [(1-t)(-3-e^t) - 0]\mathbf{i} - [-t(3+e^t) - t^2e^t]\mathbf{j} + [0 - e^t(1-t)]\mathbf{k} \right\}$$

$$= \lim_{t \to 1} \left\{ [(1-t)(-3-e^t) - 0]\mathbf{i} - \lim_{t \to 1}[-t(3+e^t) - t^2e^t]\mathbf{j} + \lim_{t \to 1}[0 - e^t(1-t)]\mathbf{k} \right\}$$

$$= (0 - 0)\mathbf{i} + (e + e + 3)\mathbf{j} + (0 - 0)\mathbf{k}$$

$$= (2e + 3)\mathbf{j}$$

Now we find the cross product of the limits.

$$\lim_{t \to 1} \mathbf{F}(t) = \left[\lim_{t \to 1} t\right]\mathbf{i} + \left[\lim_{t \to 1}(1-t)\right]\mathbf{j} + \left[\lim_{t \to 1} t^2\right]\mathbf{k} = \mathbf{i} + \mathbf{k}$$

$$\lim_{t \to 1} \mathbf{G}(t) = \left[\lim_{t \to 1} e^t\right]\mathbf{i} + \left[\lim_{t \to 1}(-3 - e^t)\right]\mathbf{k} = e\mathbf{i} + (-3 - e)\mathbf{k}$$

so that

$$\left[\lim_{t \to 1} \mathbf{F}(t)\right] \times \left[\lim_{t \to 1} \mathbf{G}(t)\right] = \begin{vmatrix} \mathbf{i} & \mathbf{j} & \mathbf{k} \\ 1 & 0 & 1 \\ e & 0 & -3 - e \end{vmatrix}$$

$$= (0 - 0)\mathbf{i} - (-3 - e - e)\mathbf{j} + (0 - 0)\mathbf{k}$$

$$= (3 + 2e)\mathbf{j}$$

Thus,

$$\lim_{t \to 1}[\mathbf{F}(t) \times \mathbf{G}(t)] = \left[\lim_{t \to 1} \mathbf{F}(t)\right] \times \left[\lim_{t \to 1} \mathbf{G}(t)\right] \qquad \blacksquare$$

CONTINUITY OF A VECTOR FUNCTION A vector function $\mathbf{F}(t)$ is said to be **continuous** at t_0 if t_0 is in the domain of $\mathbf{F}$ and $\lim\limits_{t \to t_0} \mathbf{F}(t) = \mathbf{F}(t_0)$.

■ **W**hat this says: This is the same as requiring each component of $\mathbf{F}(t)$ to be continuous at t_0. That is,

$$\mathbf{F}(t) = f_1(t)\mathbf{i} + f_2(t)\mathbf{j} + f_3(t)\mathbf{k}$$

is continuous at t_0 when t_0 is in the domain of the component functions $f_1(t)$, $f_2(t)$, and $f_3(t)$ and

$$\lim_{t \to t_0} f_1(t) = f_1(t_0); \quad \lim_{t \to t_0} f_2(t) = f_2(t_0); \quad \lim_{t \to t_0} f_3(t) = f_3(t_0)$$

The rules for vector limits listed in Theorem 10.1 can be used to derive general properties of vector function continuity.

Example 7   Continuity of a vector function

For what values of t is $\mathbf{F}(t) = (\sin t)\mathbf{i} + (1-t)^{-1}\mathbf{j} + (\ln t)\mathbf{k}$ continuous?

Solution   The vector function $\mathbf{F}$ is continuous where its component functions

$$f_1(t) = \sin t \qquad f_2(t) = (1-t)^{-1} \qquad f_3(t) = \ln t$$

are continuous. The function f_1 is continuous for all t; f_2 is continuous whenever $1-t \neq 0$ (that is, when $t \neq 1$); f_3 is continuous for $t > 0$. Thus, $\mathbf{F}$ is continuous when t is a positive number other than one; that is, $t > 0$, $t \neq 1$.   ■

PROBLEM SET 10.1

Level 1

1. ■ What does this say? Discuss the concept of a vector-valued function.
2. ■ What does this say? Discuss the concept of the limit of a vector-valued function.

Find the domain of the vector functions given in Problems 3-10.

3. $\mathbf{F}(t) = 2t\mathbf{i} - 3t\mathbf{j} + \dfrac{1}{t}\mathbf{k}$

4. $\mathbf{F}(t) = (1-t)\mathbf{i} + \sqrt{t}\mathbf{j} - \dfrac{1}{t-2}\mathbf{k}$

5. $\mathbf{F}(t) = (\sin t)\mathbf{i} + (\cos t)\mathbf{j} + (\tan t)\mathbf{k}$

6. $\mathbf{F}(t) = (\cos t)\mathbf{i} - (\cot t)\mathbf{j} + (\csc t)\mathbf{k}$

7. $h(t)\mathbf{F}(t)$ where $h(t) = \sin t$ and
$$\mathbf{F}(t) = \dfrac{1}{\cos t}\mathbf{i} + \dfrac{1}{\sin t}\mathbf{j} + \dfrac{1}{\tan t}\mathbf{k}$$

8. $\mathbf{F}(t) + \mathbf{G}(t)$, where $\mathbf{F}(t) = 3t\mathbf{j} + t^{-1}\mathbf{k}$ and $\mathbf{G}(t) = 5t\mathbf{i} + \sqrt{10-t}\,\mathbf{j}$

9. $\mathbf{F}(t) - \mathbf{G}(t)$, where
$\mathbf{F}(t) = \ln t\,\mathbf{i} + 3t\mathbf{j} - t^2\mathbf{k}$ and
$\mathbf{G}(t) = \mathbf{i} + 5t\mathbf{j} - t^2\mathbf{k}$

10. $\mathbf{F}(t) \times \mathbf{G}(t)$ where $\mathbf{F}(t) = t^2\mathbf{i} - t\mathbf{j} + 2t\mathbf{k}$ and $\mathbf{G}(t) = \dfrac{1}{t+2}\mathbf{i} + (t+4)\mathbf{j} - \sqrt{-t}\mathbf{k}$

Describe the graph of the vector functions given in Problems 11-20 or sketch a graph in $\mathbb{R}^3$. A graph in $\mathbb{R}^2$ may help with your description.

11. $\mathbf{F}(t) = 2t\mathbf{i} + t^2\mathbf{j}$

12. $\mathbf{G}(t) = (1-t)\mathbf{i} + \dfrac{1}{t}\mathbf{j}$

13. $\mathbf{G}(t) = (\sin t)\mathbf{i} - (\cos t)\mathbf{j}$

14. $\mathbf{F}(t) = (2\cos t)\mathbf{i} + (\sin t)\mathbf{j}$

15. $\mathbf{F}(t) = t\mathbf{i} - 4\mathbf{k}$

16. $\mathbf{G}(t) = e^t\mathbf{j} + t\mathbf{k}$

17. $\mathbf{F}(t) = (\cos t)\mathbf{i} + (\sin t)\mathbf{j} + t\mathbf{k}$

18. $\mathbf{F}(t) = e^t\mathbf{i} + e^t\mathbf{j} + e^{-t}\mathbf{k}$

19. $\mathbf{G}(t) = (1-t)\mathbf{i} + t^2\mathbf{j} + t\mathbf{k}$

20. $\mathbf{G}(t) = (2\sin t)\mathbf{i} + (2\cos t)\mathbf{j} + 3\mathbf{k}$

Perform the operations indicated in Problems 21-32 with

$$\mathbf{F}(t) = 2t\mathbf{i} - 5\mathbf{j} + t^2\mathbf{k}$$

$$\mathbf{G}(t) = (1-t)\mathbf{i} + \dfrac{1}{t}\mathbf{k}$$

$$\mathbf{H}(t) = (\sin t)\mathbf{i} + e^t\mathbf{j}$$

21. $2\mathbf{F}(t) - 3\mathbf{G}(t)$
22. $t^2\mathbf{F}(t) - 3\mathbf{H}(t)$
23. $\mathbf{F}(t) \cdot \mathbf{G}(t)$
24. $\mathbf{F}(t) \cdot \mathbf{H}(t)$
25. $\mathbf{G}(t) \cdot \mathbf{H}(t)$
26. $\mathbf{F}(t) \times \mathbf{G}(t)$
27. $\mathbf{F}(t) \times \mathbf{H}(t)$
28. $\mathbf{G}(t) \times \mathbf{H}(t)$
29. $2e^t\mathbf{F}(t) + t\mathbf{G}(t) + 10\mathbf{H}(t)$
30. $\mathbf{F}(t) \cdot [\mathbf{H}(t) \times \mathbf{G}(t)]$
31. $\mathbf{G}(t) \cdot [\mathbf{H}(t) \times \mathbf{F}(t)]$
32. $\mathbf{H}(t) \cdot [\mathbf{G}(t) \times \mathbf{F}(t)]$

Level 2

33. Show that the curve given by
$\mathbf{R}(t) = (2\sin t)\mathbf{i} + (2\sin t)\mathbf{j} + (\sqrt{8}\cos t)\mathbf{k}$
lies on a sphere centered at the origin.

34. Sketch the graph of the curve given by
$\mathbf{R}(t) = (2\cos t)\mathbf{i} + (\sin t)\mathbf{j} - 2t\,\mathbf{k}$.

Find a vector function $\mathbf{F}$ whose graph is the curve given in Problems 35-40.

35. $y = x^2; z = 2$

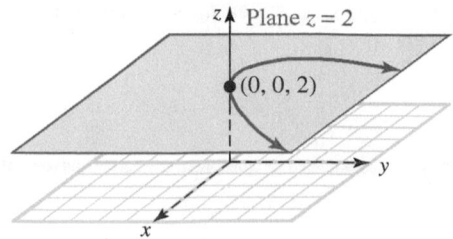

36. $x^2 + y^2 = 4; z = -1$

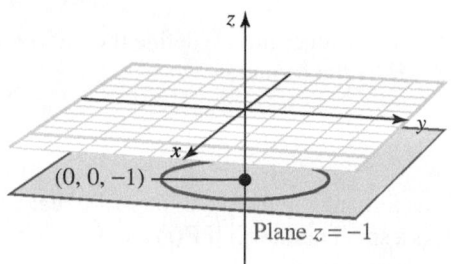

37. $x = 2t, y = 1 - t, z = \sin t$

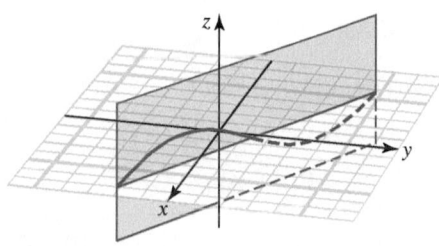

38. $\dfrac{x-2}{3} = \dfrac{y-1}{2} = \dfrac{z}{4}$

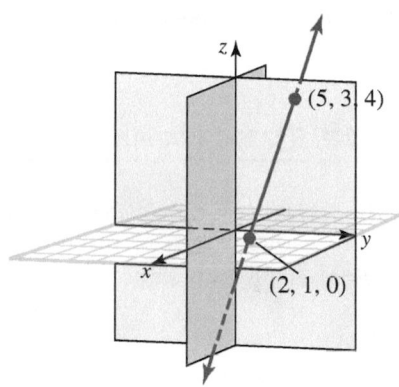

39. The curve of intersection of the hemisphere $z = \sqrt{9 - x^2 - y^2}$ and the parabolic cylinder $x = y^2$

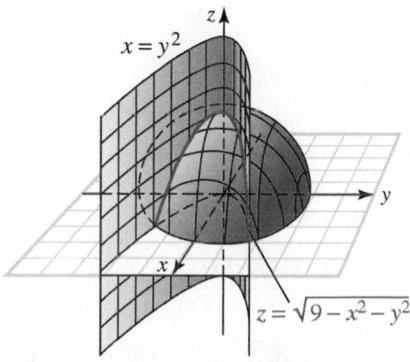

40. The line of intersection of the planes $2x + y + 3z = 6$ and $x - y - z = 1$

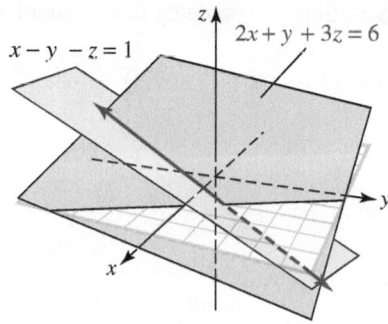

Find each limit indicated in Problems 41-48.

41. $\lim\limits_{t \to 1} \left[2t\mathbf{i} - 3\mathbf{j} + e^t \mathbf{k} \right]$

42. $\lim\limits_{t \to 1} \left[3t\mathbf{i} + e^{2t}\mathbf{j} + (\sin \pi t)\mathbf{k} \right]$

43. $\lim\limits_{t \to 0} \left[\dfrac{(\sin t)\mathbf{i} - t\mathbf{k}}{t^2 + t - 1} \right]$

44. $\lim\limits_{t \to 1} \left[\dfrac{t^3 - 1}{t - 1}\mathbf{i} + \dfrac{t^2 - 3t + 2}{t^2 + t - 2}\mathbf{j} + (t^2 + 1)e^{t-1}\mathbf{k} \right]$

45. $\lim\limits_{t \to 0} \left[\dfrac{te^t}{1 - e^t}\mathbf{i} + \dfrac{e^{t-1}}{\cos t}\mathbf{j} \right]$

46. $\lim\limits_{t \to 0} \left[\dfrac{\sin t}{t}\mathbf{i} + \dfrac{1 - \cos t}{t}\mathbf{j} + e^{1-t}\mathbf{k} \right]$

47. $\lim\limits_{t \to 0^+} \left[\dfrac{\sin 3t}{\sin 2t}\mathbf{i} + \dfrac{\ln(\sin t)}{\ln(\tan t)}\mathbf{j} + (t \ln t)\mathbf{k} \right]$

48. $\lim\limits_{t \to 2} \left[(2\mathbf{i} - t\mathbf{j} + e^t \mathbf{k}) \times (t^2\mathbf{i} + 4 \sin t\mathbf{j}) \right]$

Determine all values of t for which the vector function given in Problems 49-54 is continuous.

49. $\mathbf{F}(t) = t\mathbf{i} + 3\mathbf{j} - (1 - t)\mathbf{k}$

50. $\mathbf{G}(t) = t\mathbf{i} - \dfrac{1}{t}\mathbf{k}$

51. $\mathbf{G}(t) = \dfrac{\mathbf{i} + 2\mathbf{j}}{t^2 + t}$

52. $\mathbf{F}(t) = (e^t \sin t)\mathbf{i} + (e^t \cos t)\mathbf{k}$

53. $\mathbf{F}(t) = e^t \left[t\mathbf{i} + \tfrac{1}{t}\mathbf{j} + 3\mathbf{k} \right]$

54. $\mathbf{G}(t) = \dfrac{\mathbf{u}}{\|\mathbf{u}\|}$, where $\mathbf{u} = t\mathbf{i} + \sqrt{t}\mathbf{j}$

55. The graph of

$$\mathbf{R}(t) = t\mathbf{i} + \left(\frac{1-t}{t} \right)\mathbf{j} + \left(\frac{1-t^2}{t} \right)\mathbf{k}$$

lies in a plane. What is the equation of this plane?

56. How many revolutions are made by the circular helix

$$\mathbf{R}(t) = (2\sin t)\mathbf{i} + (2\cos t)\mathbf{j} + \frac{5}{8}t\mathbf{k}$$

in a vertical distance of 8 units?

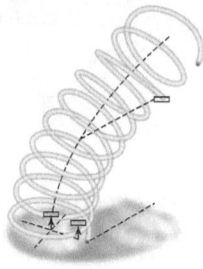

57. Given the vector functions

$\mathbf{F}(t) = t\mathbf{i} + t^2\mathbf{j} + t^3\mathbf{k}$ and

$\mathbf{G}(t) = \frac{1}{t}\mathbf{i} - e^t\mathbf{j}$, directly verify each of the following limit formulas (that is, without using Theorem 10.1).

a. $\lim\limits_{t\to 0} e^t\mathbf{F}(t) = \left[\lim\limits_{t\to 0} e^t\right]\left[\lim\limits_{t\to 0} \mathbf{F}(t)\right]$

b. $\lim\limits_{t\to 1} \mathbf{F}(t)\cdot\mathbf{G}(t) = [\lim\limits_{t\to 1}\mathbf{F}(t)]\cdot\left[\lim\limits_{t\to 1}\mathbf{G}(t)\right]$

c. $\lim\limits_{t\to 1}[\mathbf{F}(t)\times\mathbf{G}(t)] = [\lim\limits_{t\to 1}\mathbf{F}(t)]\times[\lim\limits_{t\to 1}\mathbf{G}(t)]$

Level 3

58. If $\mathbf{H}(t)$ is a vector function, we define the **difference operator** $\Delta\mathbf{H}$ by the formula

$$\Delta\mathbf{H} = \mathbf{H}(t + \Delta t) - \mathbf{H}(t)$$

where Δt is a change in the parameter t. (*Note*: Usually, $|\Delta t|$ is a small number.) If $\mathbf{F}(t)$ and $\mathbf{G}(t)$ are vector functions, show that

$$\Delta(\mathbf{F}\times\mathbf{G})(t) = \mathbf{F}(t+\Delta t)\times\Delta\mathbf{G}(t) + \Delta\mathbf{F}(t)\times\mathbf{G}(t)$$

Hint: Observe that

$$\Delta(\mathbf{F}\times\mathbf{G})(t) = \mathbf{F}(t+\Delta t)\times\mathbf{G}(t+\Delta t)$$
$$- \mathbf{F}(t+\Delta t)\times\mathbf{G}(t)$$
$$+ \mathbf{F}(t+\Delta t)\times\mathbf{G}(t) - \mathbf{F}(t)\times\mathbf{G}(t)$$

59. Prove the limit of a sum rule of Theorem 10.1. (The difference rule is proved similarly.)

60. a. Prove the limit of a scalar multiple rule of Theorem 10.1.

b. Prove the limit of a cross product rule of Theorem 10.1.

10.2 DIFFERENTIATION AND INTEGRATION OF VECTOR FUNCTIONS

IN THIS SECTION: *Vector derivatives, tangent vectors, properties of vector derivatives, modeling the motion of an object in $\mathbb{R}^3$, vector integrals*
Our next objective is to introduce the derivative and integral of a vector function, along with some of their basic properties and applications.

Vector Derivatives

In Chapter 3, we defined the derivative of the function f to be the limit as $\Delta x \to 0$ of the difference quotient $\Delta f/\Delta x$. In the context of this chapter, this definition would be called the derivative of the *scalar* function f. The **difference quotient of a vector function F** is the vector expression

$$\frac{\Delta\mathbf{F}}{\Delta t} = \frac{\mathbf{F}(t+\Delta t) - \mathbf{F}(t)}{\Delta t}$$

and we define the derivative of $\mathbf{F}$ as follows.

DERIVATIVE OF A VECTOR FUNCTION   The **derivative** of the vector function $\mathbf{F}$ is the vector function $\mathbf{F}'$ determined by the limit

$$\mathbf{F}'(t) = \lim\limits_{\Delta t\to 0}\frac{\Delta\mathbf{F}}{\Delta t} = \lim\limits_{\Delta t\to 0}\frac{\mathbf{F}(t+\Delta t) - \mathbf{F}(t)}{\Delta t}$$

wherever this limit exists. In the Leibniz notation, the derivative of $\mathbf{F}$ is denoted by $\dfrac{d\mathbf{F}}{dt}$. We say that the vector function $\mathbf{F}$ is **differentiable** at $t = t_0$ if $\mathbf{F}'(t)$ is defined at t_0.

The following theorem establishes a convenient method for computing the derivative of a vector function.

Theorem 10.2 Derivative of a vector function

The vector function $\mathbf{F}(t) = f_1(t)\mathbf{i} + f_2(t)\mathbf{j} + f_3(t)\mathbf{k}$ is differentiable whenever the component functions f_1, f_2, and f_3 are each differentiable, and in this case

$$\mathbf{F}'(t) = f_1'(t)\mathbf{i} + f_2'(t)\mathbf{j} + f_3'(t)\mathbf{k}$$

Proof: We use the definition of the derivative, along with rules of vector limits (Theorem 10.1), and the fact that the scalar derivatives $f_1'(t)$, $f_2'(t)$, and $f_3'(t)$ all exist.

$$\mathbf{F}'(t) = \lim_{\Delta t \to 0} \frac{\mathbf{F}(t + \Delta t) - \mathbf{F}(t)}{\Delta t}$$

$$= \lim_{\Delta t \to 0} \frac{[f_1(t + \Delta t)\mathbf{i} + f_2(t + \Delta t)\mathbf{j} + f_3(t + \Delta t)\mathbf{k}] - [f_1(t)\mathbf{i} + f_2(t)\mathbf{j} + f_3(t)\mathbf{k}]}{\Delta t}$$

$$= \left[\lim_{\Delta t \to 0} \frac{f_1(t + \Delta t) - f_1(t)}{\Delta t}\right]\mathbf{i} + \left[\lim_{\Delta t \to 0} \frac{f_2(t + \Delta t) - f_2(t)}{\Delta t}\right]\mathbf{j} + \left[\lim_{\Delta t \to 0} \frac{f_3(t + \Delta t) - f_3(t)}{\Delta t}\right]\mathbf{k}$$

$$= f_1'(t)\mathbf{i} + f_2'(t)\mathbf{j} + f_3'(t)\mathbf{k} \qquad \blacklozenge$$

Example 1 Differentiability of a vector function

For what values of t is $\mathbf{G}(t) = |t|\mathbf{i} + (\cos t)\mathbf{j} + (t - 5)\mathbf{k}$ differentiable?

Solution The component functions $\cos t$ and $t - 5$ are differentiable for all t, but $|t|$ is not differentiable at $t = 0$. Thus, the vector function $\mathbf{G}$ is differentiable for all $t \neq 0$. ■

Example 2 Derivative of a vector function

Find the derivative of the vector function

$$\mathbf{F}(t) = e^t\mathbf{i} + (\sin t)\mathbf{j} + (t^3 + 5t)\mathbf{k}$$

Solution Differentiating each component separately, we find that

$$\mathbf{F}'(t) = (e^t)'\mathbf{i} + (\sin t)'\mathbf{j} + (t^3 + 5t)'\mathbf{k} = e^t\mathbf{i} + (\cos t)\mathbf{j} + (3t^2 + 5)\mathbf{k} \qquad ■$$

Tangent Vectors

Recall that the scalar derivative $f'(x_0)$ gives the slope of the tangent line to the graph of f at the point where $x = x_0$ and thus provides a measure of the graph's direction at that point. Our first goal is to extend this interpretation by showing how the vector derivative can be used to find tangent vectors to curves in space. Let t be a number in the domain of the vector function $\mathbf{F}(t)$, and let P_0 be the point on the graph of $\mathbf{F}$ that corresponds to t_0, as shown in Figure 10.6.

Then for any positive number Δt, the difference quotient

$$\frac{\Delta \mathbf{F}}{\Delta t} = \frac{\mathbf{F}(t_0 + \Delta t) - \mathbf{F}(t_0)}{\Delta t}$$

is a vector that points in the same direction as the secant vector

$$\mathbf{P_0Q} = \mathbf{F}(t_0 + \Delta t) - \mathbf{F}(t_0)$$

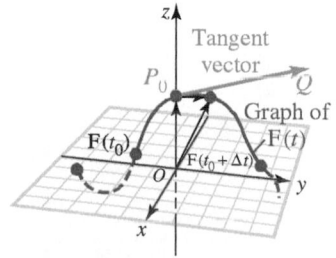

Figure 10.6 The difference quotient is a multiple of the secant line vector

where Q is the point on the graph of $\mathbf{F}$ that corresponds to $t = t_0 + \Delta t$ (see Figure 10.6). Suppose the difference quotient $\Delta\mathbf{F}/\Delta t$ has a limit as $\Delta t \to 0$ and that

$$\lim_{\Delta t \to 0} \frac{\Delta\mathbf{F}}{\Delta t} \neq \mathbf{0}$$

Then, as $\Delta t \to 0$, the direction of the secant vector $\mathbf{P_0Q}$, and hence that of the difference quotient $\Delta\mathbf{F}/\Delta t$, will approach the direction of the tangent vector at P_0, as shown in Figure 10.7.

Thus, we expect the tangent vector at P_0 to be the limit vector

$$\lim_{\Delta t \to 0} \frac{\Delta\mathbf{F}}{\Delta t}$$

which we recognize as the vector derivative $\mathbf{F}'(t_0)$. These observations lead to the following interpretation

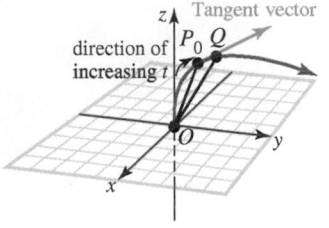

Figure 10.7 As $\Delta t \to 0$, the vector $\mathbf{P_0Q}$ and hence the difference quotient $\Delta\mathbf{F}/\Delta t$ approaches the tangent vector at P_0

TANGENT VECTOR　Suppose $\mathbf{F}(t)$ is differentiable at t_0 and that $\mathbf{F}'(t_0) \neq \mathbf{0}$. Then $\mathbf{F}'(t_0)$ is a **tangent vector** to the graph of $\mathbf{F}(t)$ at the point where $t = t_0$ and points in the direction determined by increasing t.

Example 3　Finding a tangent vector

Find a tangent vector at the point P_0 where $t = 0.2$ on the graph of the vector function

$$\mathbf{F}(t) = e^{2t}\mathbf{i} + (t^2 - t)\mathbf{j} + (\ln t)\mathbf{k}$$

What is an equation for the tangent line at P_0?

Solution　The derivative of $\mathbf{F}(t)$ is

$$\mathbf{F}'(t) = 2e^{2t}\mathbf{i} + (2t - 1)\mathbf{j} + t^{-1}\mathbf{k}$$

so a tangent vector at the point where $t = 0.2$ is

$$\mathbf{F}'(0.2) = 2e^{0.4}\mathbf{i} - 0.6\mathbf{j} + 5\mathbf{k}$$

The tangent line to the graph of $\mathbf{F}(t)$ at P_0 is the line that passes through and is parallel to the vector $\mathbf{F}'(0.2)$. Since

$$\mathbf{F}(0.2) = e^{0.4}\mathbf{i} - 0.16\mathbf{j} + (\ln 0.2)\mathbf{k}$$

the point of tangency is $(e^{0.4}, -0.16, \ln 0.2)$. Thus, the tangent line has parametric equations

$$x = e^{0.4} + 2e^{0.4}t, \qquad y = -0.16 - 0.6t, \qquad z = \ln 0.2 + 5t$$

If $\mathbf{F}'(t_0) \neq \mathbf{0}$ and we also require the derivative $\mathbf{F}'$ be continuous at t_0, the tangent vector at each point of the graph of $\mathbf{F}$ near P_0 will be close to the tangent vector at P_0. In this case, the graph is said to be *smooth* at P_0.

SMOOTH CURVE   The graph of the vector function defined by $\mathbf{F}(t)$ is **smooth** on any interval of t where $\mathbf{F}'$ is continuous and $\mathbf{F}'(t) \neq \mathbf{0}$. The graph is **piecewise smooth** on an interval that can be subdivided into a finite number of subintervals on which $\mathbf{F}$ is smooth.

Example 4   Determining whether a curve is smooth

Determine whether the graph of the vector function

$$\mathbf{F}(t) = \langle t^2 + 1, \cos t, e^t + e^{-t} \rangle$$

is smooth for all t.

Solution   The derivative

$$\mathbf{F}'(t) = \langle 2t, -\sin t, e^t - e^{-t} \rangle$$

is continuous for all t, but $\mathbf{F}'(0) = \langle 0, 0, 1 - 1 \rangle = \mathbf{0}$. Thus, the graph is not smooth for all t, but it is smooth on any interval not containing $t = 0$.

A graph will not be smooth on any interval containing a point where there is an abrupt change in direction. For example, the graph in Figure 10.8 is not smooth on any interval containing the point corresponding to the sharp corner. In Example 4, such a point occurs when $t = 0$, namely $P_0(1, 1, 2)$.

We say this curve is *piecewise smooth* if it consists of a finite number of smooth pieces.

Properties of Vector Derivatives

Higher-order derivatives of a vector function $\mathbf{F}$ are obtained by successively differentiating the components of $\mathbf{F}(t) = f_1(t)\mathbf{i} + f_2(t)\mathbf{j} + f_3(t)\mathbf{k}$. For instance, the **second derivative** of $\mathbf{F}$ is the vector function

$$\mathbf{F}''(t) = [\mathbf{F}'(t)]' = f_1''(t)\mathbf{i} + f_2''(t)\mathbf{j} + f_3''(t)\mathbf{k}$$

whereas the **third derivative** $\mathbf{F}'''(t)$ is the derivative of $\mathbf{F}''(t)$, and so on. In the Leibniz notation, the second vector derivative of $\mathbf{F}$ with respect to t is denoted by $\dfrac{d^2\mathbf{F}}{dt^2}$, and the third derivative, by $\dfrac{d^3\mathbf{F}}{dt^3}$.

Example 5   Higher-order derivatives of a vector function

Find the second and third derivatives of the vector function

$$\mathbf{F}(t) = e^{2t}\mathbf{i} + (1 - t^2)\mathbf{j} + (\cos 2t)\mathbf{k}$$

Solution

$$\mathbf{F}'(t) = 2e^{2t}\mathbf{i} + (-2t)\mathbf{j} + (-2\sin 2t)\mathbf{k}$$
$$\mathbf{F}''(t) = 4e^{2t}\mathbf{i} - 2\mathbf{j} - (4\cos 2t)\mathbf{k}$$
$$\mathbf{F}'''(t) = 8e^{2t}\mathbf{i} + (8\sin 2t)\mathbf{k}$$

Several rules for computing derivatives of vector functions are listed in the following theorem.

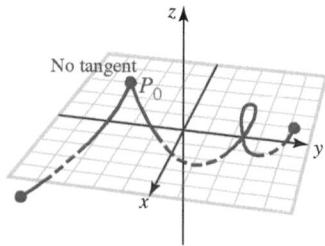

Figure 10.8   A curve that is not smooth

Theorem 10.3   Rules for differentiating vector functions

If the vector functions $\mathbf{F}$ and $\mathbf{G}$ and the scalar function h are differentiable at t, then $a\mathbf{F}+b\mathbf{G}$, $h\mathbf{F}$, $\mathbf{F}\cdot\mathbf{G}$, and $\mathbf{F}\times\mathbf{G}$ are also differentiable at t, and:

Linearity rule $(a\mathbf{F}+b\mathbf{G})'(t) = a\mathbf{F}'(t) + b\mathbf{G}'(t)$ for constants a,b

Scalar multiple rule $(h\mathbf{F})'(t) = h'(t)\mathbf{F}(t) + h(t)\mathbf{F}'(t)$

Dot product rule $(\mathbf{F}\cdot\mathbf{G})'(t) = (\mathbf{F}'\cdot\mathbf{G})(t) + (\mathbf{F}\cdot\mathbf{G}')(t)$

Cross product rule* $(\mathbf{F}\times\mathbf{G})'(t) = (\mathbf{F}'\times\mathbf{G})(t) + (\mathbf{F}\times\mathbf{G}')(t)$

Chain rule $[\mathbf{F}(h(t))]' = h'(t)\mathbf{F}'(h(t))$

Proof: We will prove the linearity rule and leave the rest of the proof as exercises. Note that if $\mathbf{F}$ and $\mathbf{G}$ are vector functions differentiable at t, and a,b are constants, then the linear combination $a\mathbf{F}+b\mathbf{G}$ has the difference quotient

$$\frac{\Delta(a\mathbf{F}+b\mathbf{G})}{\Delta t} = \frac{a\Delta\mathbf{F}}{\Delta t} + \frac{b\Delta\mathbf{G}}{\Delta t}$$

We can now find the derivative:

$$(a\mathbf{F}+b\mathbf{G})'(t) = \lim_{\Delta t\to 0}\left[\frac{\Delta(a\mathbf{F}+b\mathbf{G})}{\Delta t}\right]$$

$$= \lim_{\Delta t\to 0}\left[\frac{a\Delta\mathbf{F}}{\Delta t} + \frac{b\Delta\mathbf{G})}{\Delta t}\right]$$

$$= a\lim_{\Delta t\to 0}\frac{\Delta\mathbf{F}}{\Delta t} + b\lim_{\Delta t\to 0}\frac{\Delta\mathbf{G}}{\Delta t}$$

$$= a\mathbf{F}'(t) + b\mathbf{G}'(t) \qquad\qquad \blacklozenge$$

Example 6   Derivative of a cross product

Let $\mathbf{F}(t) = \mathbf{i}+t\mathbf{j}+t^2\mathbf{k}$ and $\mathbf{G}(t) = t\mathbf{i}+e^t\mathbf{j}+3\mathbf{k}$. Verify that

$$(\mathbf{F}\times\mathbf{G})'(t) = (\mathbf{F}'\times\mathbf{G})(t) + (\mathbf{F}\times\mathbf{G}')(t)$$

Solution   First find the derivative of the cross product:

$$(\mathbf{F}\times\mathbf{G})(t) = \begin{vmatrix} \mathbf{i} & \mathbf{j} & \mathbf{k} \\ 1 & t & t^2 \\ t & e^t & 3 \end{vmatrix} = (3t - t^2 e^t)\mathbf{i} - (3 - t^3)\mathbf{j} + (e^t - t^2)\mathbf{k}$$

so that $(\mathbf{F}\times\mathbf{G})'(t) = (3 - 2te^t - t^2 e^t)\mathbf{i} + (3t^2)\mathbf{j} + (e^t - 2t)\mathbf{k}$.

Next, find $(\mathbf{F}'\times\mathbf{G})(t) + (\mathbf{F}\times\mathbf{G}')(t)$ by first finding the derivatives of $\mathbf{F}$ and $\mathbf{G}$ and then the appropriate cross products.

$$\mathbf{F}'(t) = \mathbf{j} + 2t\mathbf{k} \qquad \text{and} \qquad \mathbf{G}'(t) = \mathbf{i} + e^t\mathbf{j}$$

$$(\mathbf{F}'\times\mathbf{G})(t) = \begin{vmatrix} \mathbf{i} & \mathbf{j} & \mathbf{k} \\ 0 & 1 & 2t \\ t & e^t & 3 \end{vmatrix} = (3 - 2te^t)\mathbf{i} - (-2t^2)\mathbf{j} + (-t)\mathbf{k}$$

$$(\mathbf{F}\times\mathbf{G}')(t) = \begin{vmatrix} \mathbf{i} & \mathbf{j} & \mathbf{k} \\ 1 & t & t^2 \\ 1 & e^t & 0 \end{vmatrix} = (-t^2 e^t)\mathbf{i} - (-t^2)\mathbf{j} + (e^t - t)\mathbf{k}$$

---

*The order of the factors is important in the cross product rule, because the cross product of vectors is not commutative.

Finally, add these vector functions:

$$\mathbf{F}' \times \mathbf{G}(t) + (\mathbf{F} \times \mathbf{G}')(t) = [(3 - 2te^t) - t^2 e^t]\mathbf{i} + [2t^2 + t^2]\mathbf{j} + [-t + e^t - t]\mathbf{k}$$
$$= (3 - 2te^t - t^2 e^t)\mathbf{i} + (3t^2)\mathbf{j} + (e^t - 2t)\mathbf{k}$$

Thus, $(\mathbf{F} \times \mathbf{G})' = (\mathbf{F}' \times \mathbf{G}) + (\mathbf{F} \times \mathbf{G}')$.

Example 7   Derivatives of vector function expressions

Let $\mathbf{F}(t) = \mathbf{i} + e^t \mathbf{j} + t^2 \mathbf{k}$ and $\mathbf{G}(t) = 3t^2 \mathbf{i} + e^{-t}\mathbf{j} - 2t\mathbf{k}$. Find

a. $\dfrac{d}{dt}[2\mathbf{F}(t) + t^3\mathbf{G}(t)]$ **b.** $\dfrac{d}{dt}[\mathbf{F}(t) \cdot \mathbf{G}(t)]$

Solution   Begin by finding $\dfrac{d\mathbf{F}}{dt}$ and $\dfrac{d\mathbf{G}}{dt}$:

$$\frac{d\mathbf{F}}{dt} = e^t \mathbf{j} + 2t\mathbf{k} \qquad \text{and} \qquad \frac{d\mathbf{G}}{dt} = 6t\mathbf{i} - e^{-t}\mathbf{j} - 2\mathbf{k}$$

a. $\dfrac{d}{dt}[2\mathbf{F}(t) + t^3\mathbf{G}(t)] = 2\dfrac{d\mathbf{F}}{dt} + \left[t^3 \dfrac{d\mathbf{G}}{dt} + \dfrac{d}{dt}(t^3)\mathbf{G}(t) \right]$

$$= 2(e^t \mathbf{j} + 2t\mathbf{k}) + t^3(6t\mathbf{i} - e^{-t}\mathbf{j} - 2\mathbf{k}) + 3t^2(3t^2\mathbf{i} + e^{-t}\mathbf{j} - 2t\mathbf{k})$$
$$= (6t^4 + 9t^4)\mathbf{i} + (2e^t - t^3 e^{-t} + 3t^2 e^{-t})\mathbf{j} + (4t - 2t^3 - 6t^3)\mathbf{k}$$
$$= 15t^4\mathbf{i} + [2e^t + e^{-t}t^2(3 - t)]\mathbf{j} + (4t - 8t^3)\mathbf{k}$$

b. $\dfrac{d}{dt}[\mathbf{F}(t) \cdot \mathbf{G}(t)] = \dfrac{d\mathbf{F}}{dt} \cdot \mathbf{G}(t) + \mathbf{F}(t) \cdot \dfrac{d\mathbf{G}}{dt}$

$$= (e^t \mathbf{j} + 2t\mathbf{k}) \cdot (3t^2\mathbf{i} + e^{-t}\mathbf{j} - 2t\mathbf{k}) + (\mathbf{i} + e^t\mathbf{j} + t^2\mathbf{k}) \cdot (6t\mathbf{i} - e^{-t}\mathbf{j} - 2\mathbf{k})$$
$$= [0(3t^2) + e^t(e^{-t}) + 2t(-2t)] + [1(6t) + e^t(-e^{-t}) + t^2(-2)]$$
$$= (1 - 4t^2) + (6t - 1 - 2t^2)$$
$$= -6t^2 + 6t$$

We will use Theorem 10.3 in both applied and theoretical problems. In the following theorem, we establish an important geometric property of vector functions.

Theorem 10.4   Orthogonality of a function of constant length and its derivative

If the nonzero vector function $\mathbf{F}(t)$ is differentiable and has constant length, then $\mathbf{F}(t)$ is orthogonal to the derivative vector $\mathbf{F}'(t)$.

Proof: We are given that $\|\mathbf{F}(t)\| = r$ for some constant r and all t. To prove that $\mathbf{F}$ and $\mathbf{F}'$ are orthogonal, we will show that $\mathbf{F}(t) \cdot \mathbf{F}'(t) = 0$ for all t. First, note that

$$r^2 = \|\mathbf{F}(t)\|^2 = \mathbf{F}(t) \cdot \mathbf{F}(t)$$

for all t. We take the derivative of both sides (remember that the derivative of r^2 is zero because it is a constant).

$$[r^2]' = [\mathbf{F}(t) \cdot \mathbf{F}(t)]'$$
$$0 = \mathbf{F}'(t) \cdot \mathbf{F}(t) + \mathbf{F}(t) \cdot \mathbf{F}'(t)$$
$$0 = 2\mathbf{F}(t) \cdot \mathbf{F}'(t)$$
$$0 = \mathbf{F}(t) \cdot \mathbf{F}'(t)$$

Thus, $\mathbf{F}$ and $\mathbf{F}'$ are orthogonal.

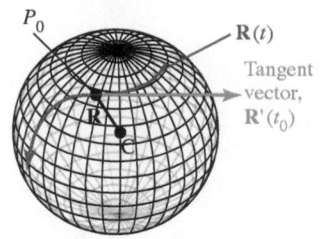

Figure 10.9 The tangent $\mathbf{R}'(t_0)$ at P_0 is orthogonal to the radius vector $\mathbf{R}(t_0)$ from the center of the sphere to P_0

In plane geometry, it is shown that a tangent line to a circle is always perpendicular to the radial line from the center of a circle to the point of tangency. Theorem 10.4 can be used to extend this property to spheres in $\mathbb{R}^3$.

Suppose $\mathbf{R}(t) = \langle f(t), g(t), h(t) \rangle$ is a vector function whose graph lies entirely on the sphere $x^2 + y^2 + z^2 = r^2$ and let P_0 be the point on the graph that corresponds to $t = t_0$, as shown in Figure 10.9. Then $\mathbf{R}(t_0)$ is the vector from the center C to P_0, and $\mathbf{R}'(t_0)$ is a tangent vector to the graph of $\mathbf{R}(t)$ at P_0. Thus $\mathbf{R}'(t_0)$ is also tangent to the sphere at P_0. Note that $\mathbf{R}(t)$ has constant length,

$$\|\mathbf{R}(t)\| = \sqrt{x^2(t) + y^2(t) + z^2(t)} = r \qquad \text{for all } t$$

It follows from Theorem 10.4 that $\mathbf{R}$ is orthogonal to $\mathbf{R}'$. That is, the tangent vector $\mathbf{R}'(t_0)$ at P_0 is orthogonal to the vector $\mathbf{R}(t_0)$ from the center of the sphere to the point of tangency.

Modeling the motion of an object in $\mathbb{R}^3$

Recall from Chapter 3 that the derivative of an object's position with respect to time is the velocity, and the derivative of the velocity is the acceleration. We frequently know the acceleration and can use integration to find the velocity and the position. We will now express these concepts in terms of vector functions.

VECTOR MOTION   An object that moves in such a way that its position at time t is given by the vector function $\mathbf{R}(t)$ is said to have

Position vector, $\mathbf{R}(t)$ and

Velocity, $\mathbf{V} = \dfrac{d\mathbf{R}}{dt}$

At any time t,

the **speed** is $\|\mathbf{V}\|$, the magnitude of the velocity,

the **direction of motion** is the unit vector $\dfrac{\mathbf{V}}{\|\mathbf{V}\|}$, and

the **acceleration vector** is the derivative of the velocity:

$$\mathbf{A} = \frac{d\mathbf{V}}{dt} = \frac{d^2\mathbf{R}}{dt^2}$$

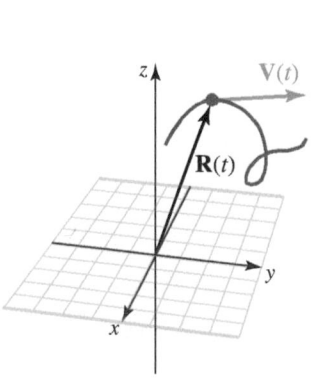

Figure 10.10 The velocity vector is tangent to the trajectory of the moving object

The graph of the position vector $\mathbf{R}(t)$ is called the **trajectory** of the moving object. According to the results obtained earlier in this section, the velocity $\mathbf{V}(t) = \mathbf{R}'(t)$ is a tangent vector to the trajectory at any point where $\mathbf{V}(t)$ exists and $\mathbf{V}(t) \neq 0$ (see Figure 10.10).

The *direction of motion is given* by the unit vector $\mathbf{V}/\|\mathbf{V}\|$, so the velocity satisfies

$$\mathbf{V} = \|\mathbf{V}\| \left(\frac{\mathbf{V}}{\|\mathbf{V}\|} \right) = (\text{SPEED})(\text{DIRECTION})$$

Appropriately, whenever $\mathbf{V}(t) = \mathbf{0}$, the object is **stationary**.

In practice, the position vector is often represented in the form

$$\mathbf{R}(t) = f_1(t)\mathbf{i} + f_2(t)\mathbf{j} + f_3(t)\mathbf{k}$$

and in this case, the velocity and the acceleration vectors are given by

$$\mathbf{V}(t) = \mathbf{R}'(t) = f_1'(t)\mathbf{i} + f_2'(t)\mathbf{j} + f_3'(t)\mathbf{k}$$

and

$$\mathbf{A}(t) = \mathbf{V}'(t) = \mathbf{R}''(t) = f_1''(t)\mathbf{i} + f_2''(t)\mathbf{j} + f_3''(t)\mathbf{k}$$

Example 8   Speed and direction of a particle

A particle's position at time t is determined by the vector

$$\mathbf{R}(t) = (\cos t)\mathbf{i} + (\sin t)\mathbf{j} + t^3\mathbf{k}$$

Analyze the particle's motion. In particular, find the particle's velocity, speed, acceleration, and direction of motion at time $t = 2$.

Solution

$$\mathbf{V} = \frac{d\mathbf{R}}{dt} = (-\sin t)\mathbf{i} + (\cos t)\mathbf{j} + 3t^2\mathbf{k}$$

$$\mathbf{A} = \frac{d\mathbf{V}}{dt} = (-\cos t)\mathbf{i} - (\sin t)\mathbf{j} + 6t\mathbf{k}$$

The velocity at time $t = 2$ is $\mathbf{V}(2)$:

$$(-\sin 2)\mathbf{i} + (\cos 2)\mathbf{j} + 3(2)^2\mathbf{k} \approx -0.91\mathbf{i} - 0.42\mathbf{j} + 12\mathbf{k}$$

The acceleration at $t = 2$ is $\mathbf{A}(2)$:

$$(-\cos 2)\mathbf{i} - (\sin 2)\mathbf{j} + 6(2)\mathbf{k} \approx 0.42\mathbf{i} - 0.91\mathbf{j} + 12\mathbf{k}$$

The speed is $\|\mathbf{V}\| = \sqrt{(-\sin t)^2 + (\cos t)^2 + (3t^2)^2} = \sqrt{1 + 9t^4}$
    At time $t = 2$, the speed is

$$\|\mathbf{V}(2)\| = \sqrt{1 + 144} \approx 12.04$$

The direction of motion at time $t = 2$ is $\dfrac{\mathbf{V}}{\|\mathbf{V}\|} = \dfrac{1}{\sqrt{145}}[(-\sin t)\mathbf{i} + (\cos t)\mathbf{j} + 3t^2\mathbf{k}]$

At time $t = 2$, the direction is $\mathbf{V}(2)/\|\mathbf{V}(2)\|$:

$$\frac{1}{\sqrt{145}}[(-\sin 2)\mathbf{i} + (\cos 2)\mathbf{j} + 12\mathbf{k}] \approx -0.08\mathbf{i} - 0.03\mathbf{j} + 1.00\mathbf{k}$$

The geometry of this example is shown in Figure 10.11.

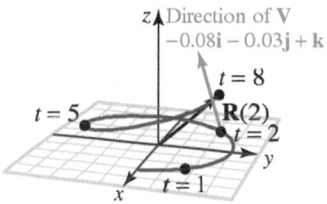

Figure 10.11   Graph of **R** and **V** at $t = 2$

Vector Integrals

Like vector limits and derivatives, vector integration is performed in a componentwise fashion.

VECTOR INTEGRALS   Let $\mathbf{F}(t) = f_1(t)\mathbf{i} + f_2(t)\mathbf{j} + f_3(t)\mathbf{k}$, where f_1, f_2, and f_3 are continuous on the closed interval $a \leq t \leq b$. Then the **indefinite integral** of $\mathbf{F}(t)$ is the vector function

$$\int \mathbf{F}(t)\, dt = \left[\int f_1(t)\, dt\right]\mathbf{i} + \left[\int f_2(t)\, dt\right]\mathbf{j} + \left[\int f_3(t)\, dt\right]\mathbf{k} + \mathbf{C}$$

where $\mathbf{C} = C_1\mathbf{i} + C_2\mathbf{j} + C_3\mathbf{k}$ is an arbitrary constant vector. The **definite integral** of $\mathbf{F}(t)$ on $a \leq t \leq b$ is the vector

$$\int_a^b \mathbf{F}(t)\, dt = \left[\int_a^b f_1(t)\, dt\right]\mathbf{i} + \left[\int_a^b f_2(t)\, dt\right]\mathbf{j} + \left[\int_a^b f_3(t)\, dt\right]\mathbf{k}$$

Example 9   Integral of a vector function

Find $\int_0^\pi [t\mathbf{i} + 3\mathbf{j} - (\sin t)\mathbf{k}]\, dt$.

Solution

$$\int_0^\pi [t\mathbf{i} + 3\mathbf{j} - (\sin t)\mathbf{k}]\, dt = \left[\int_0^\pi t\, dt\right]\mathbf{i} + \left[\int_0^\pi 3\, dt\right]\mathbf{j} - \left[\int_0^\pi \sin t\, dt\right]\mathbf{k}$$

$$= \left(\frac{t^2}{2}\right)\Big|_0^\pi \mathbf{i} + (3t)|_0^\pi\, \mathbf{j} - (-\cos t)|_0^\pi\, \mathbf{k}$$

$$= \frac{1}{2}\pi^2\mathbf{i} + 3\pi\mathbf{j} + (\cos\pi - \cos 0)\mathbf{k}$$

$$= \frac{1}{2}\pi^2\mathbf{i} + 3\pi\mathbf{j} - 2\mathbf{k}$$

Example 10   Position of a particle given its velocity

The velocity of a particle moving in space is

$$\mathbf{V}(t) = e^t\mathbf{i} + t^2\mathbf{j} + (\cos 2t)\mathbf{k}$$

Find the particle's position as a function of t if the position at time $t = 0$ is $\mathbf{R}(0) = 2\mathbf{i} + \mathbf{j} - \mathbf{k}$.

Solution   We need to solve the initial value problem that consists of

The differential equation: $\mathbf{V}(t) = \dfrac{d\mathbf{R}}{dt} = e^t\mathbf{i} + t^2\mathbf{j} + (\cos 2t)\mathbf{k}$

The initial condition: $\mathbf{R}(0) = 2\mathbf{i} + \mathbf{j} - \mathbf{k}$

We integrate both sides of the differential equation with respect to t:

$$\int d\mathbf{R} = \left[\int e^t\, dt\right]\mathbf{i} + \left[\int t^2\, dt\right]\mathbf{j} + \left[\int \cos 2t\, dt\right]\mathbf{k}$$

$$\mathbf{R}(t) = (e^t + C_1)\mathbf{i} + \left(\frac{1}{3}t^3 + C_2\right)\mathbf{j} + \left(\frac{1}{2}\sin 2t + C_3\right)\mathbf{k}$$

$$= e^t\mathbf{i} + \frac{1}{3}t^3\mathbf{j} + \left(\frac{1}{2}\sin 2t\right)\mathbf{k} + \underbrace{C_1\mathbf{i} + C_2\mathbf{j} + C_3\mathbf{k}}_{\mathbf{C}}$$

Now use the initial condition to find $\mathbf{C}$:

$$\underbrace{2\mathbf{i} + \mathbf{j} - \mathbf{k}}_{\mathbf{R}(0)} = e^0\mathbf{i} + \frac{1}{3}(0)^3\mathbf{j} + \left(\frac{1}{2}\sin 0\right)\mathbf{k} + \mathbf{C} = \mathbf{i} + \mathbf{C}$$

We solve three separate equations:

$$2\mathbf{i} = \mathbf{i} + C_1\mathbf{i} \qquad \mathbf{j} = C_2\mathbf{j} \qquad -\mathbf{k} = C_3\mathbf{k}$$

$$C_1 = 1 \qquad\qquad C_2 = 1 \qquad\qquad C_3 = -1$$

so

$$\mathbf{i} + \mathbf{j} - \mathbf{k} = \mathbf{C}$$

Thus, the particle's position at any time t is

$$\mathbf{R}(t) = e^t\mathbf{i} + \frac{1}{3}t^3\mathbf{j} + \left(\frac{1}{2}\sin 2t\right)\mathbf{k} + \mathbf{i} + \mathbf{j} - \mathbf{k}$$

$$= (e^t + 1)\mathbf{i} + \left(\frac{1}{3}t^3 + 1\right)\mathbf{j} + \left(\frac{1}{2}\sin 2t - 1\right)\mathbf{k}$$

PROBLEM SET 10.2

1. ■ What does this say? Describe the concept of a smooth curve.

© 2013 PeterPhoto123. Used under license from Shutterstock, Inc.

2. ■ What does this say? Compare or contrast the notions of speed and velocity of a moving object.

Find the vector derivative $\mathbf{F}'$ in Problems 3-6.

3. $\mathbf{F}(t) = t\mathbf{i} + t^2\mathbf{j} + (t + t^3)\mathbf{k}$

4. $\mathbf{F}(s) = (s\mathbf{i} + s^2\mathbf{j} + s^2\mathbf{k}) + (2s^2\mathbf{i} - s\mathbf{j} + 3\mathbf{k})$

5. $\mathbf{F}(s) = (\ln s)[s\mathbf{i} + 5\mathbf{j} - e^s\mathbf{k}]$

6. $\mathbf{F}(\theta) = (\cos\theta)[\mathbf{i} + (\tan\theta)\mathbf{j} + 3\mathbf{k}]$

Find $\mathbf{F}'$ and $\mathbf{F}''$ for the vector functions given in Problems 7-10.

7. $\mathbf{F}(t) = t^2\mathbf{i} + t^{-1}\mathbf{j} + e^{2t}\mathbf{k}$

8. $\mathbf{F}(s) = (1 - 2s^2)\mathbf{i} + (s\cos s)\mathbf{j} - s\mathbf{k}$

9. $\mathbf{F}(s) = (\sin s)\mathbf{i} + (\cos s)\mathbf{j} + s^2\mathbf{k}$

10. $\mathbf{F}(\theta) = (\sin^2\theta)\mathbf{i} + (\cos 2\theta)\mathbf{j} + \theta^2\mathbf{k}$

Differentiate the scalar functions in Problems 11-14.

11. $f(x) = [x\mathbf{i} + (x + 1)\mathbf{j}] \cdot [(2x)\mathbf{i} - (3x^2)\mathbf{j}]$

12. $f(x) = \langle \cos x, x, -x \rangle \cdot \langle \sec x, -x^2, 2x \rangle$

13. $g(x) = \|\langle \sin x, -2x, \cos x \rangle\|$

14. $f(x) = \|[x\mathbf{i} + x^2\mathbf{j} - 2\mathbf{k}] + [(1 - x)\mathbf{i} - e^x\mathbf{j}]\|$

In Problems 15-20, $\mathbf{R}$ is the position vector for a particle in space at time t. Find the particle's velocity and acceleration vectors and then find the speed and direction of motion for the given value of t.

15. $\mathbf{R}(t) = t\mathbf{i} + t^2\mathbf{j} + 2t\mathbf{k}$ at $t = 1$

16. $\mathbf{R}(t) = (1 - 2t)\mathbf{i} - t^2\mathbf{j} + e^t\mathbf{k}$ at $t = 0$

17. $\mathbf{R}(t) = (\cos t)\mathbf{i} + (\sin t)\mathbf{j} + 3t\mathbf{k}$ at $t = \frac{\pi}{4}$

18. $\mathbf{R}(t) = (2\cos t)\mathbf{i} + t^2\mathbf{j} + (2\sin t)\mathbf{k}$ at $t = \frac{\pi}{2}$

19. $\mathbf{R}(t) = e^t\mathbf{i} + e^{-t}\mathbf{j} + e^{2t}\mathbf{k}$ at $t = \ln 2$

20. $\mathbf{R}(t) = (\ln t)\mathbf{i} + \frac{1}{2}t^3\mathbf{j} - t\mathbf{k}$ at $t = 1$

Find a tangent vector to the graph of the given vector function $\mathbf{F}$ at the points indicated in Problems 21-24.

21. $\mathbf{F}(t) = t^2\mathbf{i} + 2t\mathbf{j} + (t^3 + t^2)\mathbf{k}$; $t = 0$, $t = 1$, $t = -1$

22. $\mathbf{F}(t) = \dfrac{t\mathbf{i} + t^2\mathbf{j} + t^3\mathbf{k}}{1 + 2t}$; $t = 0, t = 2$

23. $\mathbf{F}(t) = t^2\mathbf{i} + (\cos t)\mathbf{j} + (t^2\cos t)\mathbf{k}$; $t = 0$, $t = \frac{\pi}{2}$

24. $\mathbf{F}(t) = (e^t\sin\pi t)\mathbf{i} + (e^t\cos\pi t)\mathbf{j} + (\sin\pi t + \cos\pi t)\mathbf{k}$; $t = 0$, $t = 1$, $t = 2$

In Problems 25 and 26, find parametric equations for the tangent line to the graph of the given vector function $\mathbf{F}(t)$ at the point P_0 corresponding to t_0.

25. $\mathbf{F}(t) = \langle t^{-3}, t^{-2}, t^{-1} \rangle$; $t_0 = -1$

26. $\mathbf{F}(t) = \langle \sin t, \cos t, 3t \rangle$; $t_0 = 0$

Find the indefinite vector integrals in Problems 27-32.

27. $\displaystyle\int \langle t, -e^{3t}, 3 \rangle \, dt$

28. $\displaystyle\int \langle \cos t, \sin t, -2t \rangle \, dt$

29. $\displaystyle\int \langle \ln t, -t, 3 \rangle \, dt$

30. $\displaystyle\int e^{-t} \langle 3, t, \sin t \rangle \, dt$

31. $\displaystyle\int \langle t\ln t, -\sin(1 - t), t \rangle \, dt$

32. $\displaystyle\int \langle \sinh t, -3, \cosh t \rangle \, dt$

Find the position vector $\mathbf{R}(t)$, given the velocity $\mathbf{V}(t)$, and the initial position $\mathbf{R}(0)$ in Problems 33-36.

33. $\mathbf{V}(t) = t^2\mathbf{i} - e^{2t}\mathbf{j} + \sqrt{t}\,\mathbf{k}$; $\mathbf{R}(0) = \mathbf{i} + 4\mathbf{j} - \mathbf{k}$

34. $\mathbf{V}(t) = t\mathbf{i} - \sqrt[3]{t}\mathbf{j} + e^t\mathbf{k}$; $\mathbf{R}(0) = \mathbf{i} - 2\mathbf{j} + \mathbf{k}$

35. $\mathbf{V}(t) = 2\sqrt{t}\,\mathbf{i} + (\cos t)\mathbf{j}$; $\mathbf{R}(0) = \mathbf{i} + \mathbf{j}$

36. $\mathbf{V}(t) = -3t\mathbf{i} + (\sin^2 t)\mathbf{j} + (\cos^2 t)\mathbf{k}$; $\mathbf{R}(0) = \mathbf{j}$

Find the position vector $\mathbf{R}(t)$ and velocity vector $\mathbf{V}(t)$, given the acceleration $\mathbf{A}(t)$ and initial position and velocity vectors $\mathbf{R}(0)$ and $\mathbf{V}(0)$, respectively, in Problems 37-38.

37. $\mathbf{A}(t) = (\cos t)\mathbf{i} - (t\sin t)\,\mathbf{k}$; $\mathbf{R}(0) = \mathbf{i} - 2\mathbf{j} + \mathbf{k}$; $\mathbf{V}(0) = 2\mathbf{i} + 3\mathbf{k}$

38. $\mathbf{A}(t) = t^2\mathbf{i} - 2\sqrt{t}\,\mathbf{j} + e^{3t}\mathbf{k}$; $\mathbf{R}(0) = 2\mathbf{i} + \mathbf{j} - \mathbf{k}$; $\mathbf{V}(0) = \mathbf{i} - \mathbf{j} - 2\mathbf{k}$

39. The velocity of a particle moving in space is $\mathbf{V}(t) = e^t\mathbf{i} + t^2\mathbf{j}$. Find the particle's position as a function of t if $\mathbf{R}(0) = \mathbf{i} - \mathbf{j}$.

40. The velocity of a moving particle is $\mathbf{V}(t) = (t\cos t)\mathbf{i} + e^{2t}\mathbf{k}$. Find the particle's position as a function of t if $\mathbf{R}(0) = \mathbf{i} + 2\mathbf{k}$.

Level 2

Let $\mathbf{v} = 2\mathbf{i} - \mathbf{j} + 5\mathbf{k}$ *and* $\mathbf{w} = \mathbf{i} + 2\mathbf{j} - 3\mathbf{k}$. *Find the derivatives in Problems 41-44.*

41. $\dfrac{d^2}{dt^2}(\mathbf{v} \cdot t^4\mathbf{w})$

42. $\dfrac{d}{dt}(\mathbf{v} + t\mathbf{w})$

43. $\dfrac{d}{dt}(t\mathbf{v} \times t^2\mathbf{w})$

44. $\dfrac{d^2}{dt^2}(t\|\mathbf{v}\| + t^2\|\mathbf{w}\|)$

In Problems 45-46, verify the indicated equations for the vector functions $\mathbf{F}(t) = (3 + t^2)\mathbf{i} - (\cos 3t)\mathbf{j} + t^{-1}\mathbf{k}$ *and* $\mathbf{G}(t) = \sin(2 - t)\mathbf{i} - e^{2t}\mathbf{k}$

45. $(\mathbf{F} \cdot \mathbf{G})'(t) = (\mathbf{F}' \cdot \mathbf{G})(t) + (\mathbf{F} \cdot \mathbf{G}')(t)$

46. $(3\mathbf{F} - 2\mathbf{G})'(t) = 3\mathbf{F}'(t) - 2\mathbf{G}'(t)$

In Problems 47-48, find a value of a that satisfies the given equation.

47. $\displaystyle\int_0^{2a} [(\cos t)\mathbf{i} + (\sin t)\mathbf{j} + (\sin t \cos t)\mathbf{k}\, dt$
$= \mathbf{i} + \mathbf{j} + \dfrac{1}{2}\mathbf{k}$

48. $\displaystyle\int_0^a \left[\left(t\sqrt{1 + t^2}\right)\mathbf{i} + \left(\dfrac{1}{1 + t^2}\right)\mathbf{j}\right] dt$
$= \dfrac{1}{3}(2\sqrt{2} - 1)\mathbf{i} + \dfrac{\pi}{4}\mathbf{j}$

In Problems 49-50, show that $\mathbf{F}(t)$ *and* $\mathbf{F}''(t)$ *are parallel for all t if k is a constant.*

49. $\mathbf{F}(t) = (\cos kt)\mathbf{i} + (\sin kt)\mathbf{j}$

50. $\mathbf{F}(t) = e^{kt}\mathbf{i} + e^{-kt}\mathbf{j}$

Determine whether the given vector function $\mathbf{F}(t)$ *in Problems 51-52 is smooth over the interval I.*

51. $\mathbf{F}(t) = \langle e^t - t, t^{3/2}, \cos t\rangle$ over $[-1, 2]$

52. $\mathbf{F}(t) = \langle t^2 - t, \cos t, \sin t^2\rangle$ over $[-\pi, \pi]$

Level 3

53. **Think Tank Problem** Is it true that whenever vector functions $\mathbf{F}(t)$ and $\mathbf{G}(t)$ are smooth over the interval $[a, b]$, then $\mathbf{F}(t) + \mathbf{G}(t)$ is also smooth over the same interval? Either prove that this statement is generally true, or find a counterexample.

54. Let $\mathbf{F}(t) = u(t)\mathbf{i} + v(t)\mathbf{j} + w(t)\mathbf{k}$, where u, v, and w are differentiable scalar functions of t. Show that

$$\mathbf{F}(t) \cdot \mathbf{F}'(t) = \|\mathbf{F}(t)\| (\|\mathbf{F}(t)\|)'$$

55. Let $\mathbf{F}(t)$ be a differentiable vector function and let $h(t)$ be a differentiable scalar function of t. Show that

$$[h(t)\mathbf{F}(t)]' = h(t)\mathbf{F}'(t) + h'(t)\mathbf{F}(t)$$

56. If $\mathbf{F}$ and $\mathbf{G}$ are differentiable vector functions of t, prove

$$(\mathbf{F} \cdot \mathbf{G})'(t) = (\mathbf{F} \cdot \mathbf{G}')(t) + (\mathbf{F}' \cdot \mathbf{G})(t)$$

57. If $\mathbf{F}$ and $\mathbf{G}$ are differentiable vector functions of t, prove

$$(\mathbf{F} \times \mathbf{G})'(t) = (\mathbf{F}' \times \mathbf{G})(t) + (\mathbf{F} \times \mathbf{G}')(t)$$

58. If $\mathbf{F}$ is a differentiable vector function such that $\mathbf{F}(t) \neq \mathbf{0}$, show that

$$\frac{d}{dt}\left(\frac{\mathbf{F}(t)}{\|\mathbf{F}(t)\|}\right) = \frac{\mathbf{F}'(t)}{\|\mathbf{F}(t)\|} - \frac{[\mathbf{F}(t) \cdot \mathbf{F}'(t)]\mathbf{F}(t)}{\|\mathbf{F}(t)\|^3}$$

59. Show that $\dfrac{d}{dt}\|\mathbf{R}(t)\| = \dfrac{1}{\|\mathbf{R}\|}\mathbf{R} \cdot \dfrac{d\mathbf{R}}{dt}$

60. If $\mathbf{G} = \mathbf{F} \cdot (\mathbf{F}' \times \mathbf{F}'')$, what is $\mathbf{G}'$?

10.3 MODELING BALLISTICS AND PLANETARY MOTION

IN THIS SECTION: *Modeling the motion of a projectile in a vacuum, Kepler's second law*
Historically, the study of projectiles and planetary motion were two of the earliest applications of calculus. In this section, we explore these applications using the methods of vector calculus.

Modeling the Motion of a Projectile in a Vacuum

In general, it can be quite difficult to analyze the motion of a projectile, but the problem becomes manageable if we assume that the acceleration due to gravity is constant and that the projectile travels in a vacuum, so we will assume throughout that we are working in a vacuum. A model of the motion in the real world (with air resistance, for example) based on these assumptions is fairly realistic as long as the projectile is reasonably heavy, travels at relatively low speed, and stays close to the surface of the earth.

Human cannonball daredevil Jennifer Smith is fired into the air and her trajectory is shown in Figure 10.12. Her father, Dave "Cannonball" Smith, held the Guinness World's Record for the longest human cannonball flight—traveling more than 70 mi/h for a distance of more than 185 ft.

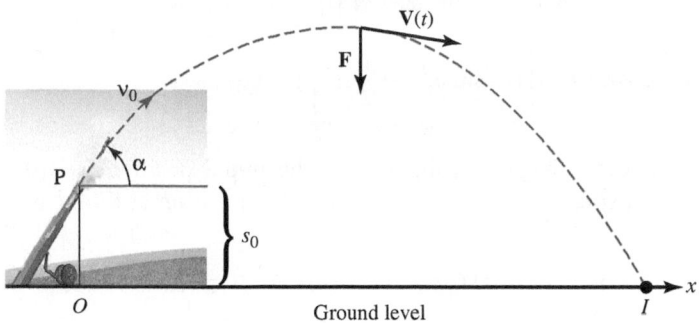

Figure 10.12   The path followed by a projectile illustrates Jennifer Smith's feat

Suppose that Smith, the projectile, is fired from a point P at the end of the cannon and travels in a vertical plane coordinatized so that P is directly above the origin O and the impact point I is on the x-axis at ground level. We let s_0 be the height of P above O, v_0 be the initial speed of the projectile (the **muzzle speed**), and α be the angle of elevation (the angle at which the projectile is launched).

Because the projectile is to travel in a vacuum, we will assume that the only force acting on the projectile at any given time is the force due to gravity. This force $\mathbf{F}$ is directed downward. The magnitude of this force is the projectile's weight. If m is the mass of the projectile and g is the "free-fall" *acceleration due to gravity* (g is approximately 32 ft/s^2 or 9.8 m/s^2), then the magnitude of $\mathbf{F}$ is mg. Also, according to **Newton's second law of motion**, the sum of all the forces acting on the projectile must equal $m\mathbf{A}(t)$, where $\mathbf{A}(t)$ is the acceleration of the projectile at time t. Because $\mathbf{F} = -mg\mathbf{j}$ is the only force acting on the projectile, it follows that

$$\mathbf{F} = -mg\mathbf{j}$$

$$m\mathbf{A}(t) = -mg\mathbf{j} \quad \textit{Since } \mathbf{F} = m\mathbf{A}(t)$$

$$\mathbf{A}(t) = -g\mathbf{j} \quad \text{for all } t \geq 0$$

The velocity $\mathbf{V}(t)$ of the projectile can now be obtained by integrating $\mathbf{A}(t)$. Specifically, we have

$$\mathbf{V}(t) = \int \mathbf{A}(t)\,dt = \int (-g\mathbf{j})\,dt = -gt\mathbf{j} + \mathbf{C}_1$$

where $\mathbf{C}_1 = \mathbf{V}(0)$ is the velocity when $t = 0$ (that is, the *initial velocity*). Because the projectile is fired with initial speed $v_0 = \|\mathbf{V}(0)\|$ at an angle of elevation α, the initial velocity must be

$$\mathbf{C}_1 = \mathbf{V}(0) = (v_0 \cos \alpha)\mathbf{i} + (v_0 \sin \alpha)\mathbf{j}$$

(as shown in Figure 10.13).

By substitution we find

$$\mathbf{V}(t) = -gt\mathbf{j} + (v_0 \cos \alpha)\mathbf{i} + (v_0 \sin \alpha)\mathbf{j}$$

$$= (v_0 \cos \alpha)\mathbf{i} + (v_0 \sin \alpha - gt)\mathbf{j}$$

Figure 10.13 Components of the initial velocity vector for a javelin

Next, by integrating $\mathbf{V}$ with respect to t, we find that the projectile has displacement

$$\mathbf{R}(t) = \int \mathbf{V}(t)\, dt$$

$$= \int [(v_0 \cos \alpha)\mathbf{i} + (v_0 \sin \alpha - gt)\mathbf{j}]\, dt$$

$$= (v_0 \cos \alpha)t\mathbf{i} + \left[(v_0 \sin \alpha)t - \frac{1}{2}gt^2\right]\mathbf{j} + \mathbf{C}_2 \quad \text{\textit{where} } \mathbf{C}_2 = \mathbf{R}(0) \text{ \textit{is the position}} \\ \text{\textit{at time} } t = 0$$

Because the projectile begins its flight at P, the initial position is $\mathbf{R}(0) = s_0\mathbf{j}$, and by substituting this expression for $\mathbf{C}_2$, we find that the position at time t is

$$\mathbf{R}(t) = [(v_0 \cos \alpha)t]\mathbf{i} + \left[(v_0 \sin \alpha)t - \frac{1}{2}gt^2 + s_0\right]\mathbf{j}$$

Specifically, $\mathbf{R}(t)$ gives the position of the projectile at any time t after it is fired and before it hits the ground. If we let $x(t)$ and $y(t)$ denote the horizontal and vertical components of $\mathbf{R}(t)$, respectively, we can make the following statement.

> **MOTION OF A PROJECTILE IN A VACUUM**   Consider a projectile that travels in a vacuum in a coordinate plane, with the x-axis along level ground. If the projectile is fired from a height of s_0 with initial speed v_0 and angle of elevation α, then at time t ($t \geq 0$) it will be at the point $(x(t), y(t))$, where
>
> $$x(t) = (v_0 \cos \alpha)t \qquad \text{and} \qquad y(t) = -\frac{1}{2}gt^2 + (v_0 \sin \alpha)t + s_0$$

The parametric equations for the motion of a projectile in a vacuum provide useful general information about a projectile's motion. For instance, note that if $\alpha \neq 90°$, we can eliminate the parameter t by solving the first equation, obtaining

$$t = \frac{x}{v_0 \cos \alpha}$$

By substituting into the second equation, we find

$$y = -\frac{1}{2}g\left[\frac{x}{v_0 \cos \alpha}\right]^2 + (v_0 \sin \alpha)\left[\frac{x}{v_0 \cos \alpha}\right] + s_0$$

$$= \left[\frac{-g}{2(v_0 \cos \alpha)^2}\right]x^2 + (\tan \alpha)x + s_0$$

This is the Cartesian equation for the trajectory of the projectile, and because the equation has the general form

$$y = ax^2 + bx + c, \text{ with } a < 0$$

the trajectory must be part of a downward-opening parabola. That is, *if a projectile is not fired vertically (so that $\tan \alpha$ is defined), its trajectory will be part of a downward-opening parabolic arc*, as shown in Figure 10.14.

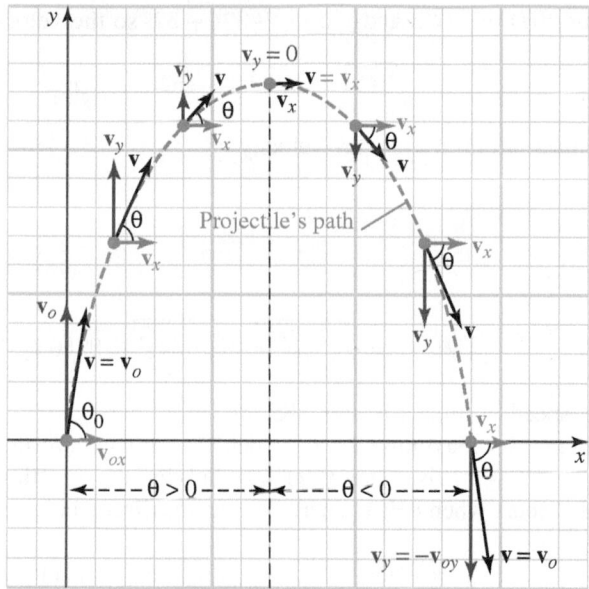

Figure 10.14 Interactive The motion of a projectile. The trajectory is part of a parabolic arc

The *time of flight* T_f of a projectile is the elapsed time between launch and impact, and the *range* is the total horizontal distance R_f traveled by the projectile during its flight. Because $y = 0$ at impact, it follows that T_f must satisfy the quadratic equation

$$-\frac{1}{2}gT_f^2 + (v_0 \sin\ \alpha)T_f + s_0 = 0$$

Then, because the flight begins at $x = 0$ when $t = 0$ and ends at $x = R_f$ when $t = T_f$, the range R_f can be computed by evaluating $x(t)$ at $t = T_f$. That is,

$$R_f = (v_0 \cos\ \alpha)T_f$$

Example 1 Modeling the path of a projectile

A boy standing at the edge of a cliff throws a ball upward at a 30° angle of elevation with an initial speed of 64 ft/s. Suppose that when the ball leaves the boy's hand, it is 48 ft above the ground at the base of the cliff.

a. What are the time of flight of the ball and its range?
b. What are the velocity of the ball and its speed at impact?
c. What is the highest point reached by the ball during its flight?

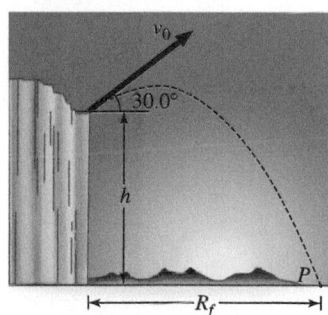

Solution We know that $g = 32$ ft/s^2, and we are given $s_0 = 48$ ft, $v_0 = 64$ ft/s, and $\alpha = 30°$. Also, $\mathbf{R}_f = 166$ ft so the ball's trajectory has parametric equations

$$x(t) = (v_0 \cos\ \alpha)t \qquad y(t) = -\frac{1}{2}gt^2 + (v_0 \sin\alpha)t + s_0$$

$$= (64 \cos 30°)t \qquad = -\frac{1}{2}(32)t^2 + (64 \sin 30°)t + 48$$

$$= 32\sqrt{3}t \qquad\qquad = -16t^2 + 32t + 48$$

a. The ball hits the ground when $y = 0$, and by solving the equation $-16t^2 + 32t + 48 = 0$ for $t \geq 0$, we find that $t = 3$ (we discard the negative solution $t = -1$), so the time of flight is $T_f = 3$ seconds. The range is

$$x(3) = 32\sqrt{3}(3) \approx 166.27688$$

That is, the ball hits the ground about 166 ft from the base of the cliff.

b. We find that $x'(t) = 32\sqrt{3}$ and $y'(t) = -32t + 32$, so the velocity at time t is

$$\mathbf{V}(t) = 32\sqrt{3}\,\mathbf{i} + (-32t + 32)\mathbf{j}$$

Thus, at impact (when $t = 3$), the velocity is

$$\mathbf{V}(3) = 32\sqrt{3}\,\mathbf{i} - 64\mathbf{j}$$

and its speed is

$$\|\mathbf{V}(3)\| = \sqrt{(32\sqrt{3})^2 + (-64)^2} \approx 84.664042$$

That is, the speed at impact is about 85 ft/s.

c. The ball attains its maximum height when the upward (vertical) component of its velocity $\mathbf{V}(t)$ is 0—that is, when $y'(t) = 0$. Solving the equation $-32t + 32 = 0$, we find that this occurs when $t = 1$. Therefore, the maximum height attained by the ball is

$$
\begin{array}{ll}
y_m = y(1) & x_m = x(1) \\[4pt]
\quad = -16(1)^2 + 32(1) + 48 & \quad = 32\sqrt{3}(1) \\[4pt]
\quad = 64 & \quad = 55.425626
\end{array}
$$

The highest point reached by the ball has coordinates (rounded to the nearest unit) of $(55, 64)$. ∎

In general, if the projectile is fired from ground level (so $s_0 = 0$), the time of flight T_f satisfies

$$-\frac{1}{2}gT_f^2 + (v_0 \sin \alpha)T_f = 0$$

$$T_f = \frac{2}{g}v_0 \sin \alpha$$

Thus, the range R_f is

$$R_f = x(T_f) = (v_0 \cos\alpha)\left(\frac{2v_0 \sin \alpha}{g}\right) = \frac{v_0^2}{g}(2\sin\alpha\cos\alpha) = \frac{v_0^2}{g}\sin 2\alpha$$

Notice that for a given initial speed v_0, the range assumes its largest value when $\sin 2\alpha = 1$; that is, when $\alpha = 45° = \frac{\pi}{4}$. To summarize

---

TIME OF FLIGHT/RANGE OF FLIGHT A projectile fired from *ground level* has **time of flight** T_f and **range** R_f given by the equations

$$T_f = \frac{2}{g}v_0 \sin \alpha \qquad \text{and} \qquad R_f = \frac{v_0^2}{g}\sin 2\alpha$$

The maximal range is $R_m = \dfrac{v_0^2}{g}$, and it occurs when $\alpha = \frac{\pi}{4}$.

Example 2   Flight time and range for the motion of a projectile

A projectile is fired from ground level at an angle of $40°$ with muzzle speed 110 ft/s. Find the time of flight and the range.

Solution With $g = 32 \text{ ft/s}^2$, we see that the flight time T_f and range R are

$$T_f = \frac{2}{g} v_0 \sin\ \alpha \qquad\qquad R = \frac{v_0^2}{g} \sin 2\alpha$$

$$= \frac{2}{32}(110)(\sin 40°) \qquad\qquad = \frac{(110)^2}{32}(\sin 80°)$$

$$\approx 4.4191648 \qquad\qquad\qquad \approx 372.38043$$

That is, the projectile travels about 372 ft horizontally before it hits the ground, and the flight takes a little more than 4 sec. ∎

Kepler's Second Law

In the 17th century, the German astronomer Johannes Kepler (1571-1630) formulated three useful laws for describing planetary motion.

KEPLER'S LAWS

FIRST LAW
The planets move about the sun in elliptical orbits, with the sun at one focus.
SECOND LAW
The radius vector joining a planet to the sun sweeps out equal areas in equal intervals of time.
THIRD LAW
The square of the time of one complete revolution of a planet about its orbit is proportional to the cube of the length of the semimajor axis of its orbit.

Kepler's approach was empirical, but his three laws were proved eighty years later by Isaac Newton, in the *Principia Mathematica* (1687). We will prove the second law using vector methods. The other two laws can be established similarly.

We begin by introducing some useful notation involving polar coordinates. Let $\mathbf{u}_r$ and $\mathbf{u}_\theta$ denote unit vectors along the radial axis and orthogonal to that axis, respectively, as shown in Figure 10.15.

Then, in terms of the unit Cartesian vectors $\mathbf{i}$ and $\mathbf{j}$, we have

$$\mathbf{u}_r = (\cos\ \theta)\mathbf{i} + (\sin\ \theta)\mathbf{j} \qquad \text{and} \qquad \mathbf{u}_\theta = (-\sin\ \theta)\mathbf{i} + (\cos\ \theta)\mathbf{j}$$

The derivatives $\dfrac{d\mathbf{u}_r}{d\theta}$ and $\dfrac{d\mathbf{u}_\theta}{d\theta}$ satisfy

$$\frac{d\mathbf{u}_r}{d\theta} = (-\sin\theta)\mathbf{i} + (\cos\theta)\mathbf{j} = \mathbf{u}_\theta \qquad \text{and} \qquad \frac{d\mathbf{u}_\theta}{d\theta} = (-\cos\theta)\mathbf{i} + (-\sin\theta)\mathbf{j} = -\mathbf{u}_r$$

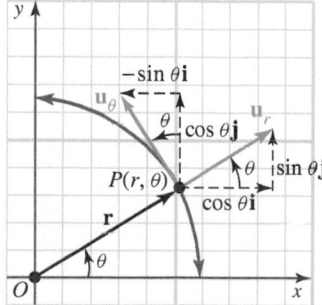

Figure 10.15 Description of the motion of a particle along a curve

Now, suppose the sun S is at the origin (pole) of a polar coordinate system, and consider the motion of a body B (planet, comet, artificial satellite) about S. The radial vector $\mathbf{R} = \mathbf{SB}$ can be expressed as

$$\mathbf{R} = r\mathbf{u}_r = (r\cos\theta)\mathbf{i} + (r\sin\theta)\mathbf{j}$$

where $r = \|\mathbf{R}\|$, and the velocity $\mathbf{V}$ satisfies

$$
\begin{aligned}
\mathbf{V} &= \frac{dr}{dt}\mathbf{u}_r + r\frac{d\mathbf{u}_r}{dt} \qquad \textit{Derivative of a scalar multiple} \\
&= \frac{dr}{dt}\mathbf{u}_r + r\frac{d\mathbf{u}_r}{d\theta}\frac{d\theta}{dt} \quad \textit{Chain rule} \\
&= \frac{dr}{dt}\mathbf{u}_r + r\frac{d\theta}{dt}\mathbf{u}_\theta
\end{aligned}
$$

A similar formula for acceleration can be derived (see Problem 59). We summarize these formulas in the following box.

POLAR FORMULAS FOR VELOCITY AND ACCELERATION If the position of a radial vector is $\mathbf{R} = r\mathbf{u}_r = (r\cos\theta)\mathbf{i} + (r\sin\theta)\mathbf{j}$ then the velocity is $\mathbf{V}(t) = \dfrac{dr}{dt}\mathbf{u}_r + r\dfrac{d\theta}{dt}\mathbf{u}_\theta$ and the acceleration is

$$
\begin{aligned}
\mathbf{A}(t) &= \frac{d\mathbf{V}(t)}{dt} \\
&= \frac{d^2\mathbf{R}}{dt^2} \\
&= \left[\frac{d^2r}{dt^2} - r\left(\frac{d\theta}{dt}\right)^2\right]\mathbf{u}_r + \left[r\frac{d^2\theta}{dt^2} + 2\frac{dr}{dt}\frac{d\theta}{dt}\right]\mathbf{u}_\theta
\end{aligned}
$$

Example 3 Velocity of a moving body

The position vector of a moving body is $\mathbf{R}(t) = 2t\mathbf{i} - t^2\mathbf{j}$ for $t \geq 0$. Express $\mathbf{R}$ and the velocity vector $\mathbf{V}(t)$ in terms of $\mathbf{u}_r$ and $\mathbf{u}_\theta$.

Solution   We note that on the trajectory, $x = 2t$, $y = -t^2$, so $\mathbf{R}$ has length

$$r = \|\mathbf{R}(t)\| = \sqrt{(2t)^2 + (-t^2)^2} = \sqrt{4t^2 + t^4} = t\sqrt{t^2 + 4}$$

and $\mathbf{R} = r\mathbf{u}_r = t\sqrt{t^2 + 4}\,\mathbf{u}_r$. Because $\mathbf{V}(t) = \dfrac{dr}{dt}\mathbf{u}_r + r\dfrac{d\theta}{dt}\mathbf{u}_\theta$, we need $\dfrac{dr}{dt}$ and $\dfrac{d\theta}{dt}$. We find that

$$\frac{dr}{dt} = \sqrt{t^2 + 4} + t\left(\frac{1}{2}\right)(t^2 + 4)^{-1/2}(2t) = \frac{2t^2 + 4}{\sqrt{t^2 + 4}}$$

and because the polar angle satisfies $\theta = \tan^{-1}\left(\dfrac{y}{x}\right)$ for all points (x, y) on the curve (except where $x = 0$), we have

$$\theta = \tan^{-1}\left(\frac{-t^2}{2t}\right) = \tan^{-1}\left(-\frac{t}{2}\right)$$

$$\frac{d\theta}{dt} = \frac{1}{1 + \left(-\frac{t}{2}\right)^2}\left(-\frac{1}{2}\right) = \frac{-2}{t^2 + 4}$$

Thus,

$$\mathbf{V}(t) = \frac{2t^2+4}{\sqrt{t^2+4}}\mathbf{u}_r + t\sqrt{t^2+4}\left(\frac{-2}{t^2+4}\right)\mathbf{u}_\theta = \frac{(2t^2+4)\mathbf{u}_r - 2t\mathbf{u}_\theta}{\sqrt{t^2+4}}$$

We now have the tools we need to establish Kepler's second law.

Theorem 10.5 Kepler's second law

The radius vector from the sun to a planet in its orbit sweeps out equal areas in equal intervals of time.

Proof: The situation described by Kepler's second law is illustrated in Figure 10.16. We will assume that the only force acting on a planet is the gravitational attraction of the sun. According to the universal law of gravitation, the force of attraction is given by

$$\mathbf{F} = -G\frac{mM}{r^2}\mathbf{u}_r$$

where G is a physical constant, and m and M are the masses of the planet and the sun, respectively. Because this is the only force acting on the planet, Newton's second law of motion tells us that $\mathbf{F} = m\mathbf{A}$, where $\mathbf{A}$ is the acceleration of the planet in its orbit. By equating these two expressions for the force $\mathbf{F}$, we find

$$m\mathbf{A} = -G\frac{mM}{r^2}\mathbf{u}_r$$

$$\mathbf{A} = \frac{-GM}{r^2}\mathbf{u}_r$$

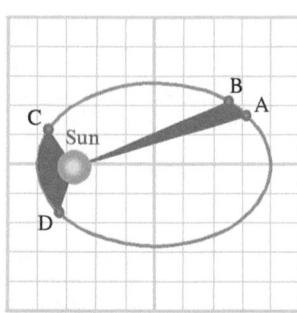

Figure 10.16 The radial line sweeps out equal areas in equal time

This says that *the acceleration of a planet in its orbit has only a radial component.* Note how this conclusion depends on our assumption that the only force on the planet is the sun's gravitational attraction, which acts along radial lines from the sun to the planet.

This means that the $\mathbf{u}_\theta$ component of the planet's acceleration is 0, and by examining the polar formula for acceleration, we see that this condition is equivalent to the differential equation

$$\left[r\frac{d^2\theta}{dt^2} + 2\frac{dr}{dt}\frac{d\theta}{dt}\right] = 0$$

If we set $w = \dfrac{d\theta}{dt}$, we obtain $\dfrac{d^2\theta}{dt^2} = \dfrac{dw}{dt}$, so that

$$r\frac{dw}{dt} + 2w\frac{dr}{dt} = 0 \qquad \textit{Given differential equation}$$

$$w^{-1}\frac{dw}{dt} = -2r^{-1}\frac{dr}{dt} \qquad \textit{Separation of variables}$$

$$\int w^{-1}dw = \int (-2r^{-1})dr \qquad \textit{Integrate both sides.}$$

$$\ln|w| + C_1 = -2\ln|r| + C_2$$

$$\ln w = \ln(Cr^{-2}) \qquad \textit{Since } w > 0 \textit{ and } r > 0$$

$$w = Cr^{-2}$$

$$\frac{d\theta}{dt} = Cr^{-2}$$

$$r^2\frac{d\theta}{dt} = C$$

Now, let $[t_1, t_2]$ and $[t_3, t_4]$ be two time intervals of equal length, so $t_2 - t_1 = t_4 - t_3$. According to Theorem 6.1, the area swept out in the time period $[t_1, t_2]$ (see Figure 10.17) is

$$S_1 = \int_{t_1}^{t_2} \frac{1}{2} r^2 d\theta = \int_{t_1}^{t_2} \frac{1}{2} \left[r^2 \frac{d\theta}{dt} \right] dt = \int_{t_1}^{t_2} \frac{1}{2} C \, dt \qquad \text{Since } r^2 \frac{d\theta}{dt} = C.$$

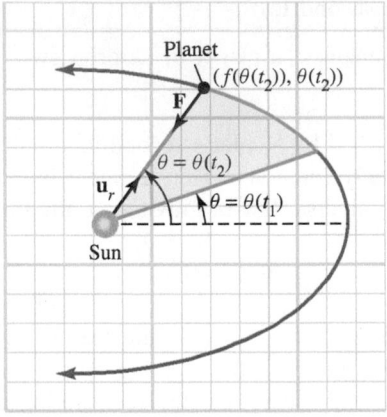

Figure 10.17 Interactive The radial line sweeps out area at the rate $\dfrac{d\mathbf{A}}{dt} = \dfrac{1}{2} r^2 \dfrac{d\theta}{dt}$

Thus, $S_1 = \frac{1}{2} C(t_2 - t_1)$. Similarly, the area swept out in time period $[t_3, t_4]$ is $S_2 = \frac{1}{2} C(t_4 - t_3)$, and we have

$$S_1 = \frac{1}{2} C(t_2 - t_1) = \frac{1}{2} C(t_4 - t_3) = S_2$$

so equal areas are swept out in equal time, as claimed by Kepler.   ♦

An object is said to move in a **central force field** if it is subject to a single force that is always directed toward a particular point. Our derivation of Kepler's second law is based on the assumption that planetary motion occurs in a central force field. In particular, the formula

$$\mathbf{A} = \frac{-GM}{r^2} \mathbf{u}_r$$

shows that a planet's acceleration has only a radial component, and this turns out to be true for an object moving in a plane in any central force field.

Much of what we know about the motion of the planets is derived from Kepler's laws. Some of these data are listed in Table 10.1.

Table 10.1 Data on Planetary Orbits

Planet	Moon Orbit (in millions of miles)	Eccentricity	Period (in days)	Comparative Mass (Earth = 1.00)
Mercury	36.0	0.2056	88.0	0.06
Venus	67.3	0.0068	224.7	0.81
Earth	**93.0**	**0.0167**	**365.256**	**1.00**
Mars	141.7	0.0934	687.0	0.1074
Jupiter	484.0	0.0484	4,322.589	317.83
Saturn	887.0	0.0543	10,759.22	95.16
Uranus	1,787.0	0.0460	30,685.4	14.5
Neptune	2,797.0	0.0082	60,189.0	17.2

PROBLEM SET 10.3

Level 1

In Problems 1-10, *find the time of flight T_f (to the nearest tenth second) and the range R_f (to the nearest unit) of a projectile fired (in a vacuum) from ground level at the given angle α with the indicated initial speed v_0. Assume that $g = 32$ ft/s^2 or $g = 9.8$ m/s^2.*

1. $\alpha = 35°, v_0 = 128$ ft/s   **2.** $\alpha = 45°, v_0 = 80$ ft/s

3. $\alpha = 48.5°, v_0 = 850$ m/s **4.** $\alpha = 43.5°, v_0 = 185$ m/s

5. $\alpha = 23.74°,$ **6.** $\alpha = 31.04°,$
 $v_0 = 23.3$ m/s $v_0 = 38.14$ m/s

7. $\alpha = 14.11°,$ **8.** $\alpha = 24.23°,$
 $v_0 = 100$ ft/s $v_0 = 200$ ft/s

9. $\alpha = 49.31°,$ **10.** $\alpha = 78.09°, v_0 = 88$ ft/s
 $v_0 = 450$ ft/s

In Problems 11-18, *an object moves along the given curve in the plane (described in either polar or parametric form). Find its velocity and acceleration in terms of the unit polar vectors $\mathbf{u}_r$ and $\mathbf{u}_\theta$.*

11. $x = 2t, y = t$ **12.** $x = 3t, y = 4t$

13. $x = \cos t, y = \sin t$ **14.** $x = \sin t, y = \cos t$

15. $r = \sin\theta, \theta = 2t$ **16.** $r = e^{-\theta}, \theta = 1 - t$

17. $r = 5(1 + \cos\theta), \theta = 2t + 1$

18. $r = \dfrac{1}{1 - \cos\theta}, \theta = t$

Level 2

In Problems 19-24, *consider a particle moving on a circular path of radius a described by the equation*

$$\mathbf{R}(t) = (a \cos\omega t)\mathbf{i} + (a \sin\omega t)\mathbf{j}$$

where $\omega = \dfrac{d\theta}{dt}$ is the constant angular velocity.

19. Find the velocity vector.
20. Show that the velocity vector is orthogonal to $\mathbf{R}(t)$.
21. Find the speed of the particle.
22. Find the acceleration vector.
23. Show that the direction of the acceleration vector is always toward the center of the circle.
24. Find the magnitude of the acceleration vector.
25. A shell fired from ground level at an angle of 45° hits the ground 2,000 m away. What is the muzzle speed of the shell?

26. A gun at ground level has muzzle speed of 300 ft/s. What angle of elevation (to the nearest tenth of a degree) should be used to hit an object 1,500 ft away?

27. At what angle (to the nearest tenth of a degree) should a projectile be fired from ground level if its muzzle speed is 167.1 ft/s and the desired range is 600 ft?

28. A shell is fired at ground level with a muzzle speed of 280 ft/s and at an elevation of 45° from ground level.
 a. Find the maximum height attained by the shell.
 b. Find the time of flight and the range of the shell.
 c. Find the velocity and speed of the shell at impact.

29. Buster Posey, 2010 rookie player of the year, hits a baseball at a 30° angle with a speed of 144 ft/s. If the ball is 4 ft above ground level when it is hit, what is the maximum height reached by the ball? How far will it travel from home plate before it lands? If it just barely clears a 5-foot wall in the outfield (see Figure 10.18), how far (to the nearest foot) is the wall from home plate?

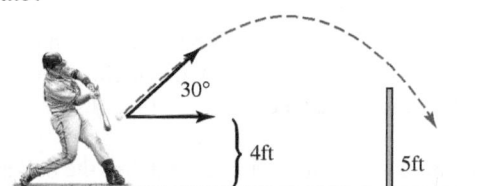

30°   4ft   5ft

Figure 10.18   Home run distance

30. A baseball hit at a 24° angle from 3 ft above the ground just goes over the 9-ft fence 400 ft from home plate. About how fast was the ball traveling, and how long did it take the ball to reach the wall?

31. A nozzle discharges a stream of water with an initial velocity $v_0 = 50$ ft/s into the end of a horizontal pipe of inside diameter $d = 5$ ft (see Figure 10.19). What is the maximum horizontal distance that the stream can reach?

Figure 10.19   Water Stream

32. If a shotputter throws a shot from a height of 5 ft with an angle of 46° and initial speed of 25 ft/s, what is the horizontal distance of the throw?

33. From the laws of motion of a projectile in a vacuum launched at an angle, deduce the laws of motion for an object thrown upwards vertically.

34. Basketball player Shaquille O'Neal could hit a shot while standing 20 ft from the basket. If "The Shaq" shot from a height of 7 ft and the ball reaches a maximum height of 12 ft before passing through the 10-ft high basket, what was the initial speed of the ball?

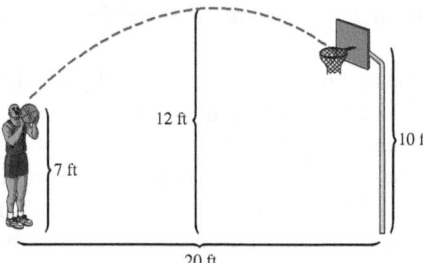

35. A golf ball is hit from the tee to a green with an initial speed of 125 ft/s at an angle of elevation of 45°. How long will it take for the ball to hit the green should be Denver Broncos!?

36. Peyton Manning, a quarterback for the Indianapolis Colts, throws a pass at a 45° angle from a height of 6.5 ft with a speed of 50 ft/s (see Figure 10.20). A receiver races straight down field on a fly pattern at a constant speed of 32 ft/s and catches the ball at a height of 6 ft. If Manning and the receiver both line up on the 50 yard line at the start of the play, how far does Manning fade back (retreat) from the line of scrimmage before he releases the ball?

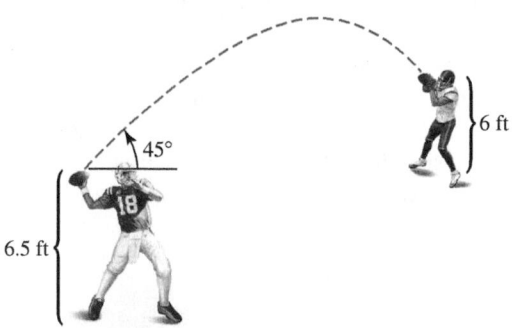

Figure 10.20   Downfield pass

37. Modeling Problem In 1974, Evel Knievel attempted a skycycle ride across the Snake River (see Figure 10.21), and at the time there was a great deal of hype about "will he make it?" If the angle of the launching ramp was 45° and if the horizontal distance the sky-cycle needed to travel was 1,700 ft, at what speed did Evel have to leave the ramp to make it across the Snake River Canyon?

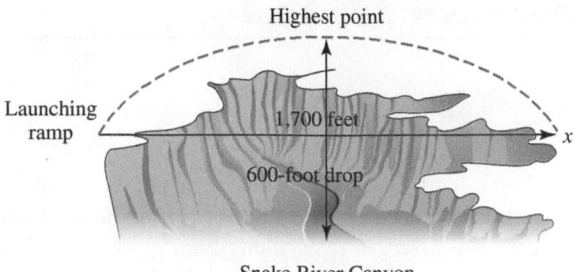

Figure 10.21   Evel Knievel's jump

Assume that the opposite edges of the canyon are at the same height and neglect wind resistance.

38. Modeling Problem A 3-oz paddleball attached to a string is swung in a circular path with a 1-ft radius, as shown in Figure 10.22. If the string will break under a force of 2 lb, find the maximum speed the ball can attain without breaking the string. (*Hint*: 3 oz is 3/16 lb.)

Figure 10.22   Paddleball path

39. A calculus text (other than the one you are now reading) is hurled downward from the top of a 120-ft-high building at an angle of 30° from the horizontal, as shown in Figure 10.23. Assume that the initial speed of the book is 8 ft/s. How far from the base of the building will the text land?

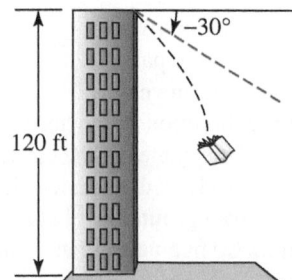

Figure 10.23   A moving calculus book

40. A child, running along level ground at the top of a 30-ft-high vertical cliff at a speed of 15 ft/s, throws a rock over the cliff into the sea below, as shown in Figure 10.24. Suppose the child's arm is 3 ft above the ground and her arm speed is 25 ft/s. If the rock is released 10 ft from the edge of the cliff at an angle of 30°, how long does it take for the rock to hit the water? How far from the base of the cliff does it hit?

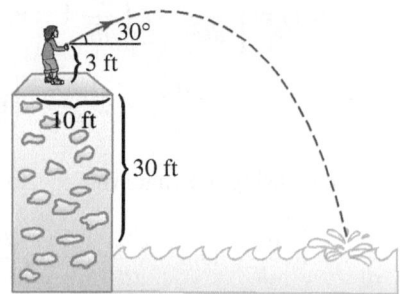

Figure 10.24 Tossing an object off a cliff

41. A particle moves along the polar path (r, θ), where $r(t) = 4 + 3\cos t$, $\theta(t) = t^3$. Find $\mathbf{V}(t)$ and $\mathbf{A}(t)$ in terms of $\mathbf{u}_r$ and $\mathbf{u}_\theta$.

42. A particle moves along the polar path (r, θ), where $r(t) = 3 + 2\sin t$, $\theta(t) = t^3$. Find $\mathbf{V}(t)$ and $\mathbf{A}(t)$ in terms of $\mathbf{u}_r$ and $\mathbf{u}_\theta$.

43. A gun is fired with muzzle speed $v_0 = 550$ ft/s at an angle of $\alpha = 22.0°$. The shell overshoots the target by 50 ft. At what angle should a second shot be fired with the same muzzle speed to hit the target?

44. Find two angles of elevation α_1 and α_2 (to the nearest degree) so that a shell fired at ground level with muzzle speed of 650 ft/s will hit a target 6,000 ft away (also at ground level).

45. A shell is fired from ground level with muzzle speed of 750 ft/s at an angle of 25°. An enemy gun 20,000 ft away fires a shot 2 seconds later and the shells collide 50 ft above the ground at the same speed. What are the muzzle speed v_0 and angle of elevation α of the second gun?

46. A gun is fired with muzzle speed $v_0 = 700$ ft/s at an angle of $\alpha = 25°$. It overshoots the target by 60 ft. The target moves away from the gun at a constant speed of 10 ft/s. If the gunner takes 30 seconds to reload, at what angle should a second shot be fired with the same muzzle speed to hit the target?

47. 𝔥istorical 𝒬uest

    In a previous 𝔥istorical 𝒬uest (Problem 56, Section 6.2), we outlined a procedure by Johannes Kepler for finding the volume of a torus. This 𝒬uest asks you to use Kepler's reasoning to find the area of an ellipse.

Karl Smith library

Johannes Kepler (1571-1630)

a. Kepler divided a circle of radius r and circumference C into a large number of very thin sectors. By regarding each sector as a thin isosceles triangle of altitude r and base equal to the arc of the sector, he heuristically arrived at the formula $A = \frac{1}{2}rC$ for the area of a circle. Show how this was done.

b. Use Kepler's reasoning to obtain a volume V of a sphere of radius r and surface area S as

$$V = \frac{1}{3}rS$$

c. Use Kepler's reasoning to show that an ellipse of semimajor axis a and semiminor axis b has area $A = \pi ab$.

48. Prove Kepler's first law. You may search the literature or the Internet for help with this proof.

49. Prove Kepler's third law. You may search the literature or the Internet for help with this proof.

50. Where is a planet in its orbit when its speed is the greatest? Support your response.

51. Two hypothetical planets are moving about the sun in elliptical orbits having equal major axes. The minor axis of one, however, is half that of the other. How do the periods of the two planets compare? Support your response.

A cannon is placed at the top of a downward-sloping hill as shown in Figure 10.25. In Problems 52-55, assume that a shell is fired with muzzle speed v_0 at an angle α with respect to the horizontal. Furthermore, assume the hill is sloped at an angle β with the horizontal.

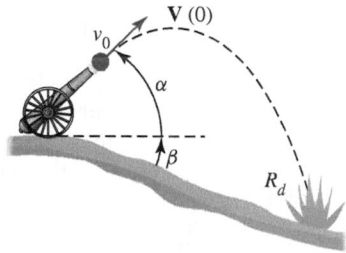

Figure 10.25 Path of a projectile fired from a sloping surface

52. If the hill slopes downward at a 20° angle, and a shell is fired with muzzle speed of $v_0 = 100$ m/s at an angle of 40° relative to the horizontal, how far down the slope does the shell land?

53. What is the impact speed of the shell in Problem 52.

54. Derive a formula for the time of flight T_f of the shell and the downhill range R_d (measured along the slope).

55. What should α be in order to maximize the downhill range R_d? *Hint*: The maximum R_d occurs when the

initial velocity vector bisects the angle between the vertical and the slope of the hill. Your answer will be in terms of β.

Level 3

Suppose a shell is fired with muzzle speed v_0 and an angle α from a height s_0 above ground in Problems 56-58.

56. Show that the range R_f must satisfy

$$g(\sec^2 \alpha)R_f^2 - 2v_0^2(\tan \alpha)R_f - 2v_0^2 s_0 = 0$$

57. Show that the maximum range R_m occurs at angle α_m, where

$$R_m \tan \alpha_m = \frac{v_0^2}{g}$$

58. Show that $R_{max} = \dfrac{v_0}{g}\sqrt{v_0^2 + 2gs_0}$ and

$$\alpha_m = \tan^{-1}\left(\frac{v_0}{\sqrt{v_0^2 + 2gs_0}}\right).$$

59. Use the formula

$$\mathbf{V(t)} = \frac{dr}{dt}\mathbf{u_r} + r\frac{d\theta}{dt}\mathbf{u}_\theta$$

to derive the formula for polar acceleration:

$$\mathbf{A}(t) = \left[\frac{d^2 r}{dt^2} - r\left(\frac{d\theta}{dt}\right)^2\right]\mathbf{u}_r$$

$$+ \left[r\frac{d^2\theta}{dt^2} + 2\frac{dr}{dt}\frac{d\theta}{dt}\right]\mathbf{u}_\theta$$

Hint: Begin by using the chain rule to show

$$\frac{d\mathbf{u}_r}{dt} = \mathbf{u}_\theta\frac{d\theta}{dt} \quad \text{and} \quad \frac{d\mathbf{u}_\theta}{dt} = -\mathbf{u}_r\frac{d\theta}{dt}$$

60. Use the results of Problem 59 to show that the force $\mathbf{F}(t)$ acting on an object moving with acceleration $\mathbf{A}(t)$ acts only in a radial direction (parallel to $\mathbf{u}_r$) if and only if

$$mr^2\frac{d\theta}{dt} = C$$

where C is a constant. Under what conditions will $\mathbf{F}(t)$ act only in a direction orthogonal to $\mathbf{u}_r$?

10.4 UNIT TANGENT AND PRINCIPAL UNIT NORMAL VECTORS; CURVATURE

IN THIS SECTION: *Unit tangent and principal unit normal vectors, arc length as a parameter, curvature*
It is useful to describe the motion of an object in terms of unit vectors that are tangent and normal at each point on the trajectory, and in this section, we learn how they can be computed.

Unit Tangent and Principal Unit Normal Vectors

In Section 10.2, we observed that if $\mathbf{R}(t)$ is a vector function whose graph is a smooth curve C, then the nonzero derivative $\mathbf{R}'(t)$ is tangent to C at the point P that corresponds to t. Since the curve C is smooth, we have $\mathbf{R}'(t) \neq \mathbf{0}$ and a *unit tangent vector* $\mathbf{T}$ to C can be obtained by dividing $\mathbf{R}'(t)$ by its length, that is,

$$\mathbf{T}(t) = \frac{\mathbf{R}'(t)}{\|\mathbf{R}'(t)\|}$$

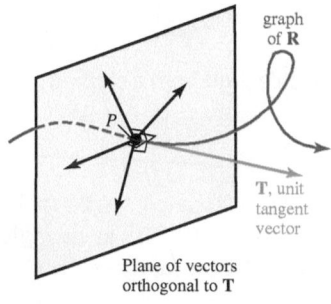

Figure 10.26 Unit tangent vector **T** and vectors orthogonal to **T** on a given curve

Figure 10.26 shows a smooth curve C and the unit tangent $\mathbf{T}$ at a particular point P. According to Theorem 10.4 (orthogonality of a function of constant length and its derivative), since $\mathbf{T}(t)$ has constant magnitude (namely $\|\mathbf{T}(t)\| = 1$), it follows that its derivative $\mathbf{T}'(t)$ is orthogonal to $\mathbf{T}(t)$ and hence is normal to the curve C at P. As indicated in Figure 10.26, there are infinitely many vectors that are orthogonal to $\mathbf{T}$ and thus deserve to be called normal vectors. To distinguish $\mathbf{T}'(t)$ from the others, we normalize it by dividing by its length and refer to it as the *principal unit normal* $\mathbf{N}(t)$, so

$$\mathbf{N}(t) = \frac{\mathbf{T}'(t)}{\|\mathbf{T}'(t)\|}$$

To summarize,

UNIT TANGENT VECTOR AND PRINCIPAL UNIT NORMAL   If $\mathbf{R}(t)$ is a vector function that defines a smooth graph, then at each point a **unit tangent** is

$$\mathbf{T}(t) = \frac{\mathbf{R}'(t)}{\|\mathbf{R}'(t)\|}$$

and the **principal unit normal** vector is

$$\mathbf{N}(t) = \frac{\mathbf{T}'(t)}{\|\mathbf{T}'(t)\|}$$

Note these are pairs of unit vectors that change direction depending on where you are on the curve.

Example 1   Unit tangent and principal unit normal

Find the unit tangent vector $\mathbf{T}(t)$ and the principal unit normal vector $\mathbf{N}(t)$ at each point on the graph of the vector function $\mathbf{R}(t) = \langle 3\sin t, 4t, 3\cos t \rangle$.

Solution   We have $\mathbf{R}'(t) = \langle 3\cos t,\ 4, -3\sin t \rangle$, so

$$\begin{aligned}
\|\mathbf{R}'(t)\| &= \sqrt{(3\cos t)^2 + 4^2 + (-3\sin t)^2} \\
&= \sqrt{9\cos^2 t + 16 + 9\sin^2 t} \\
&= \sqrt{9 + 16} \qquad \textit{Since } \sin^2 t + \cos^2 t = 1 \\
&= 5
\end{aligned}$$

Thus, the unit tangent vector is

$$\begin{aligned}
\mathbf{T}(t) &= \frac{\mathbf{R}'(t)}{\|\mathbf{R}'(t)\|} \\
&= \frac{1}{5}\langle 3\cos t, 4, -3\sin t \rangle \\
&= \left\langle \frac{3}{5}\cos t, \frac{4}{5}, -\frac{3}{5}\sin t \right\rangle
\end{aligned}$$

To find the principal unit normal $\mathbf{N}$, we first find $\mathbf{T}'(t)$ and then its magnitude:

$$\mathbf{T}'(t) = \left\langle -\frac{3}{5}\sin t, 0, -\frac{3}{5}\cos t \right\rangle$$

$$\begin{aligned}
\|\mathbf{T}'(t)\| &= \sqrt{\left(-\frac{3}{5}\sin t\right)^2 + 0 + \left(-\frac{3}{5}\cos t\right)^2} \\
&= \sqrt{\frac{9}{25}(\sin^2 t + \cos^2 t)} \\
&= \frac{3}{5}
\end{aligned}$$

So, the principal unit normal is

$$\mathbf{N}(t) = \frac{\mathbf{T}'(t)}{\|\mathbf{T}'(t)\|}$$

$$= \frac{\langle -\frac{3}{5}\sin t, 0, -\frac{3}{5}\cos t\rangle}{\frac{3}{5}}$$

$$= \langle -\sin t, 0, -\cos t\rangle$$

Suppose an object moves along the graph C of a vector function $\mathbf{R}(t)$ in the direction of increasing t. Then at each point P on the trajectory C, the unit tangent $\mathbf{T}(t)$ points in the direction of motion, while the principal unit normal $\mathbf{N}(t)$ points in the direction the object is turning, as shown in Figure 10.27**a**. For a trajectory in a plane, this means that $\mathbf{N}(t)$ points toward the concave side of the trajectory (Figure 10.27**b**).

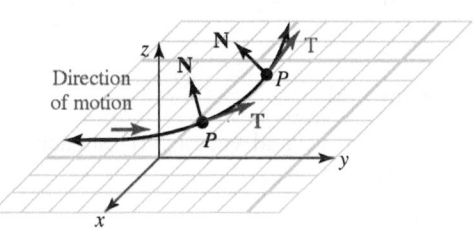

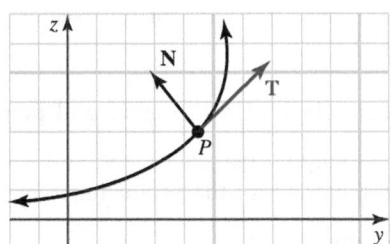

a. T points in the direction of motion while **N** points in the direction the moving object is turning

b. In the plane, **N** points toward the concave side of the trajectory

Figure 10.27 Unit tangent **T** and principal unit normal **N** at each point on the trajectory of a moving object

Arc Length as a Parameter

Recall from Section 6.4 that the arc length of the graph of $y = F(x)$ on the interval $[a, b]$ is given by the integral

$$s = \int_a^b \sqrt{1 + \left(\frac{dy}{dx}\right)^2}\, dx$$

If the graph is described by parametric equations $x = x(t)$ and $y = y(t)$, then by the chain rule $\dfrac{dy}{dt} = \dfrac{dy}{dx}\dfrac{dx}{dt}$, so we have

$$\frac{dy}{dx} = \frac{\frac{dy}{dt}}{\frac{dx}{dt}} \qquad \text{for } \frac{dx}{dt} \neq 0$$

Thus, if the graph is traced out exactly once over the parameter interval $t_1 \le t \le t_2$, its arc length is given by

$$s = \int_a^b \sqrt{1 + \left(\frac{\frac{dy}{dt}}{\frac{dx}{dt}}\right)^2}\, dx$$

$$= \int_{t_1}^{t_2} \sqrt{\frac{\left(\frac{dx}{dt}\right)^2 + \left(\frac{dy}{dt}\right)^2}{\left(\frac{dx}{dt}\right)^2}}\left(\frac{dx}{dt}\right) dt$$

$$= \int_{t_1}^{t_2} \sqrt{\left(\frac{dx}{dt}\right)^2 + \left(\frac{dy}{dt}\right)^2} \, dt$$

Similarly, if $x = x(t), y = y(t), z = z(t)$ are parametric equations for a smooth curve C in $\mathbb{R}^3$ that is traced out exactly once over the parameter interval $t_1 \le t \le t_2$, then the arc length of C is given by the integral

$$s = \int_{t_1}^{t_2} \sqrt{\left(\frac{dx}{dt}\right)^2 + \left(\frac{dy}{dt}\right)^2 + \left(\frac{dz}{dt}\right)^2} \, dt$$

Often, however, we want a formula for arc length that does not depend on t or any other particular parameter used in its computation. It can be shown that the following definition of the arc length function is, indeed, independent of parameterization.

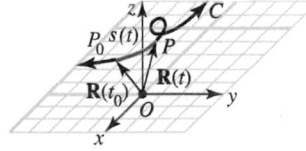

Figure 10.28 $s(t)$ is the length of the curve C from the point $P(t_0)$ to $P(t)$ if $t > t_0$

ARC LENGTH FUNCTION   Let C be a piecewise-smooth curve that is the graph of the vector described parametrically by $\mathbf{R}(t) = x(t)\mathbf{i} + y(t)\mathbf{j} + z(t)\mathbf{k}$, and let $P_0 = P(t_0)$ be a particular point on C (called the **base point**). Then the length of C from the base point P_0 to the variable point $P(t)$ is given by the **arc length function** $s(t)$ defined by

$$s(t) = \int_{t_0}^{t} \sqrt{\left(\frac{dx}{du}\right)^2 + \left(\frac{dy}{du}\right)^2 + \left(\frac{dz}{du}\right)^2} \, du$$

■ **W**hat this says: The arc length function $s(t)$ measures the distance along the curve C from $P(t_0)$ to $P(t)$ if $t > t_0$ and the opposite of that distance if $t < t_0$ (see Figure 10.28). Since the length along C is a geometric property, it does not depend on the fact that we have used t as a parameter for describing $\mathbf{R}(t)$.

This definition allows us to use arc length $s(t)$ as a parameter for describing a curve. While time is a natural parameter for studying motion in space, it is usually more fruitful to use the arc length parameter for describing a curve's geometric features.

Example 2   Finding the arc length of a curve defined by parametric equations

Find the arc length of the curve defined by

$$\mathbf{R}(t) = \langle 12t, 5\cos t, 3 - 5\sin t \rangle$$

from $t = 0$ to $t = 2$.

Solution   We find that

$$\frac{dx}{dt} = 12, \frac{dy}{dt} = -5\sin t, \frac{dz}{dt} = -5\cos t$$

Thus,

$$s = \int_{t_1}^{t_2} \sqrt{\left(\frac{dx}{dt}\right)^2 + \left(\frac{dy}{dt}\right)^2 + \left(\frac{dz}{dt}\right)^2} \, dt$$

$$= \int_0^2 \sqrt{12^2 + (-5\sin t)^2 + (-5\cos t)^2} \, dt$$

$$= \int_0^2 13 \, dt$$

$$= 26$$

The formula for arc length can be used to obtain the following useful formula for the speed of an object moving in space.

Theorem 10.6   Speed as the derivative of arc length
___

Suppose an object moves along a smooth curve C that is the graph of the position function $\mathbf{R}(t) = \langle x(t), y(t), z(t) \rangle$, where $\mathbf{R}'(t)$ is continuous on the interval $[t_1, t_2]$. Then the object has speed

$$\frac{ds}{dt} = \|\mathbf{V}(t)\| = \|\mathbf{R}'(t)\| \qquad \text{for } t_1 \leq t \leq t_2$$

Proof: Apply the second fundamental theorem of calculus to the arc length formula (where we use u as the variable of integration to distinguish it from t):

$$s(t) = \int_{t_0}^{t} \sqrt{\left(\frac{dx}{du}\right)^2 + \left(\frac{dy}{du}\right)^2 + \left(\frac{dz}{du}\right)^2} \, du$$

$$\frac{ds}{dt}(t) = \sqrt{\left(\frac{dx}{dt}\right)^2 + \left(\frac{dy}{dt}\right)^2 + \left(\frac{dz}{dt}\right)^2}$$

$$= \|\mathbf{R}'(t)\| \qquad \qquad \blacklozenge$$

Example 3   Speed and distance traveled by an object moving in space
___

The position vector of a moving object is $\mathbf{R}(t) = \langle e^t, \sqrt{2}\,t + 3, e^{-t} \rangle$. Find the speed of the object at time t and compute the distance the object travels between times $t = 0$ and $t = 1$.

Solution   By differentiating $\mathbf{R}(t)$ with respect to t, we find the velocity:

$$\mathbf{V}(t) = \mathbf{R}'(t) = \langle e^t, \sqrt{2}, -e^{-t} \rangle$$

Therefore, the speed at time t is

$$\frac{ds}{dt} = \|\mathbf{R}'(t)\|$$

$$= \sqrt{(e^t)^2 + \left(\sqrt{2}\right)^2 + (-e^{-t})^2}$$

$$= \sqrt{e^{2t} + 2 + e^{-2t}}$$

$$= \sqrt{(e^t + e^{-t})^2}$$

$$= e^t + e^{-t}$$

The distance traveled by the object between times $t = 0$ and $t = 1$ is the arc length and is given by

$$s = \int_0^1 (e^t + e^{-t})\, dt = [e^t - e^{-t}]_0^1 = e - e^{-1} - 1 + 1 \approx 2.3504024$$

To change a given parameterization from a given form, say $\mathbf{R}(t)$, to a form involving the parameter s, we first find $s(t)$ using the arc length formula (as in the proof of Theorem 10.6), and then solve for t as a function of s. The graph of $\mathbf{R}(t)$ can then be parameterized in terms of s by substituting $t = t(s)$. The general procedure is illustrated in the following example.

Example 4   Using arc length to parameterize a curve

Express the helix $\mathbf{R}(t) = \langle \sin t, \cos t, 2t \rangle$ in terms of arc length measured from the point $P_0(0, 1, 0)$ in the direction of increasing t.

Solution   The point P_0 corresponds to $t = 0$, so arc length $s(t)$ is given by the integral

$$s(t) = \int_0^t \sqrt{\left(\frac{dx}{du}\right)^2 + \left(\frac{dy}{du}\right)^2 + \left(\frac{dz}{du}\right)^2}\, du$$

$$= \int_0^t \sqrt{(\cos u)^2 + (-\sin u)^2 + 2^2}\, du$$

$$= \int_0^t \sqrt{5}\, du$$

$$= \sqrt{5}\, t$$

Solving $s = \sqrt{5}t$ for t, we obtain $t = \dfrac{1}{\sqrt{5}} s$, and the required reparameterization is

$$x = \sin\left(\frac{s}{\sqrt{5}}\right),\ y = \cos\left(\frac{s}{\sqrt{5}}\right),\ z = \frac{2s}{\sqrt{5}}$$

Expressing a given position vector $\mathbf{R}$ in terms of the arc length parameter s often leads to complicated computation. However, once we have such a parameterization, the corresponding unit tangent vector $\mathbf{T}$ and principal unit normal vector $\mathbf{N}$ can be expressed in the relatively simple forms shown in the following theorem.

Theorem 10.7   Formulas for T and N in terms of arc length s

If $\mathbf{R}(t)$ has a piecewise-smooth graph and is represented as $\mathbf{R}(s)$ in terms of the arc length parameters, then the unit tangent vector $\mathbf{T}$ and the principal unit normal vector $\mathbf{N}$ satisfy

$$\mathbf{T} = \frac{d\mathbf{R}}{ds} \qquad \text{and} \qquad \mathbf{N} = \frac{1}{\kappa}\frac{d\mathbf{T}}{ds}$$

where $\kappa = \left\|\dfrac{d\mathbf{T}}{ds}\right\|$ is a scalar function of s.

Proof: By the chain rule, we have

$$\frac{d\mathbf{R}}{dt} = \frac{d\mathbf{R}}{ds}\frac{ds}{dt}$$

and since $\dfrac{ds}{dt} = \|\mathbf{R}'(t)\|$, it follows that

$$\frac{d\mathbf{R}}{ds} = \frac{\frac{d\mathbf{R}}{dt}}{\frac{ds}{dt}}$$

$$= \frac{\mathbf{R}'(t)}{\|\mathbf{R}'(t)\|}$$

$$= \mathbf{T}$$

Next, to obtain the formula for the principal unit normal vector

$$\mathbf{N} = \frac{\mathbf{T}'(t)}{\|\mathbf{T}'(t)\|}$$

in terms of the arc length parameter s, we first note that

$$\mathbf{T}'(t) = \frac{d\mathbf{T}}{dt}$$

$$= \frac{d\mathbf{T}}{ds}\frac{ds}{dt} \qquad \text{Chain rule}$$

Because $\dfrac{ds}{dt} > 0$, it follows that the vector derivatives $\dfrac{d\mathbf{T}}{dt}$ and $\dfrac{d\mathbf{T}}{ds}$ point in the same direction, and $\mathbf{N}$ can be computed by finding $\dfrac{d\mathbf{T}}{ds}$ and dividing by its length; that is,

$$\mathbf{N} = \frac{\frac{d\mathbf{T}}{ds}}{\left\|\frac{d\mathbf{T}}{ds}\right\|}$$

$$= \frac{1}{\kappa}\frac{d\mathbf{T}}{ds}$$

as claimed.   ◆

Curvature

Let C be a smooth curve that is the graph of a vector function $\mathbf{R}(s)$ parameterized in terms of arc length s. Since the unit tangent $\mathbf{T}(s)$ is a unit vector, only its direction changes with s and the rate of change of that direction may be measured by the magnitude of the derivative $\dfrac{d\mathbf{T}}{ds}$. As shown in Figure 10.29a, if the curve is a line, then the direction remains constant so that $\dfrac{d\mathbf{T}}{ds} = \mathbf{0}$. But if the curve is bent sharply, the direction of $\mathbf{T}$ changes rapidly so $\dfrac{d\mathbf{T}}{ds}$ is relatively large in magnitude (Figure 10.29b). The magnitude of $\dfrac{d\mathbf{T}}{ds}$ is called the *curvature* of C.

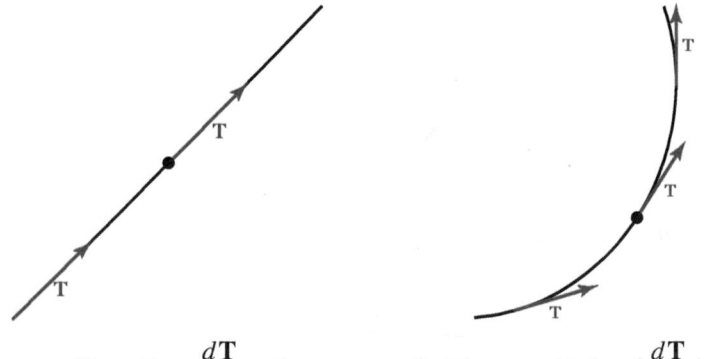

a. For a line, $\dfrac{d\mathbf{T}}{ds} = 0$ **b.** The magnitude of $\dfrac{d\mathbf{T}}{ds}$ is large where the curve is sharply bent

Figure 10.29   The curvature of a curve C is related to the magnitude of $\dfrac{d\mathbf{T}}{ds}$

CURVATURE Suppose the smooth curve C is the graph of the vector function $\mathbf{R}(s)$, parameterized in terms of the arc length s. Then the **curvature** of C is the function

$$\kappa(s) = \left\| \frac{d\mathbf{T}}{ds} \right\|$$

where $\mathbf{T}(s)$ is the unit tangent vector.

■ W hat this says: The curvature κ measures the rate at which the curve bends away from the tangent line at a particular point, as shown in Figure 10.30.

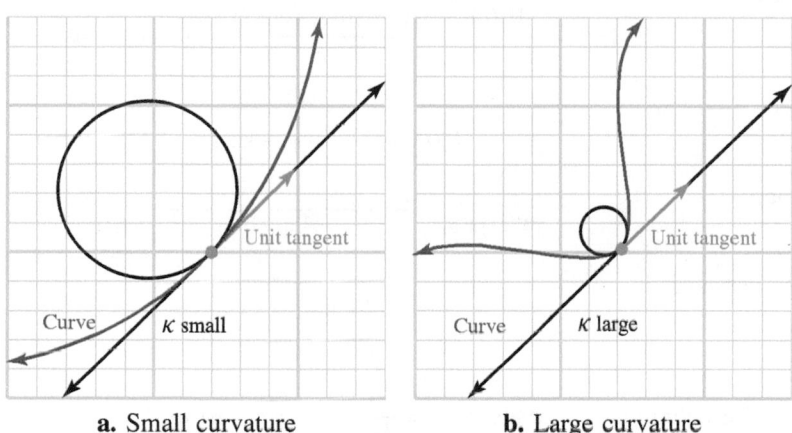

a. Small curvature   **b.** Large curvature

Figure 10.30 Curves with small and large curvature

First, notice that κ in Theorem 10.7 is the curvature of $\mathbf{R}(s)$. Since a vector function $\mathbf{R}$ is often represented in terms of a parameter other than arc length s, it may not be especially convenient to compute the curvature using the formula $\kappa = \left\| \dfrac{d\mathbf{T}}{ds} \right\|$. For instance, if the parameter is t, then according to the chain rule, we have

$$\frac{d\mathbf{T}}{dt} = \frac{d\mathbf{T}}{ds}\frac{ds}{dt}$$

so that

$$\left\| \frac{d\mathbf{T}}{ds} \right\| = \frac{\left\| \frac{d\mathbf{T}}{dt} \right\|}{\frac{ds}{dt}}$$

(Remember, s increases in the same direction as t, so $\dfrac{ds}{dt} > 0$). Since $\dfrac{ds}{dt} = \left\| \mathbf{R}'(t) \right\|$, it follows that

$$\kappa = \frac{\left\| \mathbf{T}'(t) \right\|}{\left\| \mathbf{R}'(t) \right\|}$$

which we will refer to as the **two-derivatives form** of the curvature. This form is often used in practice to compute κ because it involves only the magnitudes of $\mathbf{R}'(t)$ and $\mathbf{T}'(t)$, both of which are usually easier to compute than $\left\| \dfrac{d\mathbf{T}}{ds} \right\|$. In Example 5, we illustrate the use of the two-derivatives form by computing the curvature of a circle.

Example 5 Finding the curvature of a circle

Show that the circle with radius r

$$\mathbf{R}(t) = \langle r \cos t, r \sin t \rangle$$

has curvature equal to the reciprocal of its radius; that is, $\kappa = \dfrac{1}{r}$.

Solution Since $\mathbf{R}'(t) = \langle -r \sin t, r \cos t \rangle$, we have

$$\|\mathbf{R}'(t)\| = \sqrt{(-r \sin t)^2 + (r \cos t)^2} = r$$

Then

$$\mathbf{T}(t) = \frac{\mathbf{R}'(t)}{\|\mathbf{R}'(t)\|} = \frac{\langle -r \sin t, r \cos t \rangle}{r} = \langle -\sin t, \cos t \rangle$$

and

$$\mathbf{T}'(t) = \langle -\cos t, -\sin t \rangle$$

Since

$$\|\mathbf{T}'(t)\| = \sqrt{(-\cos t)^2 + (-\sin t)^2} = 1$$

The curvature is

$$\kappa = \frac{\|\mathbf{T}'(t)\|}{\|\mathbf{R}'(t)\|} = \frac{1}{r}$$

The definition of curvature, $\kappa = \dfrac{\|\mathbf{T}'(t)\|}{\|\mathbf{R}'(t)\|}$, can lead to difficult computations when $\|\mathbf{R}'(t)\|$ is not a constant. The formula for curvature derived in the following theorem is often easier to apply.

Theorem 10.8 The cross product derivative formula for curvature

Suppose the smooth curve C is the graph of the vector function $\mathbf{R}(t)$. Then the curvature is given by

$$\kappa = \frac{\|\mathbf{R}' \times \mathbf{R}''\|}{\|\mathbf{R}'\|^3}$$

Proof: Since $\|\mathbf{R}'\| = \dfrac{ds}{dt}$, we have

$$\mathbf{R}' = \|\mathbf{R}'\| \mathbf{T} \qquad \mathbf{R}'' = \frac{d}{dt}\left(\frac{ds}{dt}\right)\mathbf{T} + \frac{ds}{dt}\frac{d\mathbf{T}}{dt}$$

$$= \left(\frac{ds}{dt}\right)\mathbf{T} \qquad = \frac{d^2 s}{dt^2}\mathbf{T} + \frac{ds}{dt}\mathbf{T}'$$

Therefore, we have

$$\mathbf{R}' \times \mathbf{R}'' = \left(\frac{ds}{dt}\mathbf{T}\right) \times \left(\frac{d^2 s}{dt^2}\mathbf{T} + \frac{ds}{dt}\mathbf{T}'\right)$$

$$= \left(\frac{ds}{dt}\right)\left(\frac{d^2 s}{dt^2}\right)(\mathbf{T} \times \mathbf{T}) + \left(\frac{ds}{dt}\right)^2 (\mathbf{T} \times \mathbf{T}')$$

$$= \left(\frac{ds}{dt}\right)^2 (\mathbf{T} \times \mathbf{T}') \qquad \text{Since } \mathbf{T} \times \mathbf{T} = \mathbf{0}$$

Since $\|\mathbf{T}\| = 1$, it follows from the orthogonality of a function of constant length and its derivative (Theorem 10.4) that $\mathbf{T} \cdot \mathbf{T}' = 0$, so the angle between $\mathbf{T}$ and $\mathbf{T}'$ is $\theta = \frac{\pi}{2}$. Thus,

$$\|\mathbf{R}' \times \mathbf{R}''\| = \left(\frac{ds}{dt}\right)^2 \|\mathbf{T} \times \mathbf{T}'\|$$

$$= \left(\frac{ds}{dt}\right)^2 \|\mathbf{T}\| \, \|\mathbf{T}'\| \sin \theta$$

$$= \left(\frac{ds}{dt}\right)^2 \|\mathbf{T}\| \, \|\mathbf{T}'\| \qquad \textit{Since } \sin \tfrac{\pi}{2} = 1$$

$$= \left(\frac{ds}{dt}\right)^2 \|\mathbf{T}'\| \qquad \textit{Since } \|\mathbf{T}\| = 1$$

$$= \|\mathbf{R}'\|^2 \|\mathbf{T}'\|$$

The curvature is given by

$$\kappa = \frac{\|\mathbf{T}'\|}{\|\mathbf{R}'\|}$$

$$= \frac{\dfrac{\|\mathbf{R}' \times \mathbf{R}''\|}{\|\mathbf{R}'\|^2}}{\|\mathbf{R}'\|}$$

$$= \frac{\|\mathbf{R}' \times \mathbf{R}''\|}{\|\mathbf{R}'\|^3} \qquad \blacklozenge$$

Example 6   Curvature of a helix

Given that a and b are both nonnegative, express the curvature κ of

$$\mathbf{R}(t) = (a \cos t)\mathbf{i} + (a \sin t)\mathbf{j} + bt\mathbf{k}$$

in terms of a and b. The curve defined by $\mathbf{R}$ is shown in Figure 10.31.

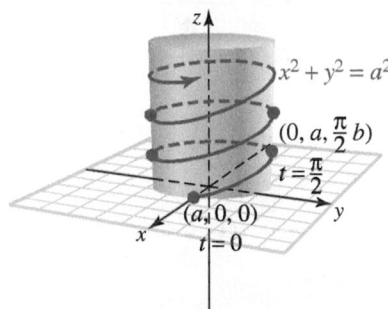

Figure 10.31   Graph of helix

Solution $\mathbf{R}'(t) = (-a \sin t)\mathbf{i} + (a \cos t)\mathbf{j} + b\mathbf{k}$

$\mathbf{R}''(t) = (-a \cos t)\mathbf{i} + (-a \sin t)\mathbf{j}$

$$\mathbf{R}' \times \mathbf{R}'' = \begin{vmatrix} \mathbf{i} & \mathbf{j} & \mathbf{k} \\ -a \sin t & a \cos t & b \\ -a \cos t & -a \sin t & 0 \end{vmatrix} = (ab \sin t)\mathbf{i} - (ab \cos t)\mathbf{j} + a^2\mathbf{k}$$

We now need to find the magnitude of this vector, as well as that of $\mathbf{R}'$:

$$\|\mathbf{R}' \times \mathbf{R}''\| = \sqrt{(ab \sin t)^2 + (ab \cos t)^2 + a^4}$$

$$= \sqrt{a^2 b^2 (\sin^2 t + \cos^2 t) + a^4}$$

$$= \sqrt{a^2 b^2 + a^4}$$

$$= a\sqrt{a^2 + b^2} \qquad a > 0 \ is \ given$$

and

$$\|\mathbf{R}'\| = \sqrt{a^2 \sin^2 t + a^2 \cos^2 t + b^2} = \sqrt{a^2 + b^2}$$

Then we have

$$\kappa = \frac{\|\mathbf{R}' \times \mathbf{R}''\|}{\|\mathbf{R}'\|^3} = \frac{a\sqrt{a^2 + b^2}}{(\sqrt{a^2 + b^2})^3} = \frac{a\sqrt{a^2 + b^2}}{(a^2 + b^2)\sqrt{a^2 + b^2}} = \frac{a}{a^2 + b^2}$$

In the special case where C is a curve in the plane with the equation $y = f(x)$, the curvature has the form given in the following theorem.

Theorem 10.9   Curvature of a planar curve using $y = f(x)$

The graph C of the function $y = f(x)$ has curvature

$$\kappa = \frac{|f''(x)|}{(1 + [f'(x)]^2)^{3/2}}$$

wherever $f(x)$ and its derivatives $f'(x)$ and $f''(x)$ all exist.

Proof: Using x as a parameter, we can regard the curve C as the graph of the vector function $\mathbf{R}(x) = x\mathbf{i} + f(x)\mathbf{j}$. Then, $\mathbf{R}' = \mathbf{i} + f'(x)\mathbf{j}$ and $\mathbf{R}''(x) = f''(x)\mathbf{j}$, so

$$\mathbf{R}' \times \mathbf{R}'' = \begin{vmatrix} \mathbf{i} & \mathbf{j} & \mathbf{k} \\ 1 & f'(x) & 0 \\ 0 & f''(x) & 0 \end{vmatrix} = f''(x)\mathbf{k}$$

Therefore, since $\|\mathbf{R}'\| = \sqrt{1 + [f'(x)]^2}$, the curvature of C is given by

$$\kappa = \frac{\|\mathbf{R}' \times \mathbf{R}''\|}{\|\mathbf{R}'\|^3}$$

$$= \frac{|f''(x)|}{\left(\sqrt{1 + [f'(x)]^2}\right)^3}$$

$$= \frac{|f''(x)|}{(1 + [f'(x)]^2)^{3/2}}$$

◆

Example 7   Maximizing the curvature of a planar curve

Find the curvature of $y = x^{-1}$ for $x > 0$. For what value of x is the curvature the largest?

Solution   For $f(x) = x^{-1}$, we have $f'(x) = -x^{-2}$ and $f''(x) = 2x^{-3}$, so the curvature is

$$\kappa = \frac{|f''(x)|}{(1 + [f'(x)]^2)^{3/2}}$$

$$= \frac{2x^{-3}}{(1 + [-x^{-2}]^2)^{3/2}}$$

$$= \frac{2x^3}{(x^4 + 1)^{3/2}}$$

To find the largest value of κ, we find the derivative of $\kappa(x)$ and solve the equation $\kappa'(x) = 0$:

$$\kappa'(x) = \frac{6x^2(1 - x^4)}{(x^4 + 1)^{5/2}}$$

Now $\kappa'(x) = 0$ when

$$6x^2(1 - x^4) = 0$$

$$6x^2(1 - x)(1 + x)(1 + x^2) = 0$$

$$x = 0, 1, -1 \quad \textit{The 0 and } -1 \textit{ are not in the domain.}$$

Since $\kappa'(x) > 0$ for $x < 1$ and $\kappa'(x) < 0$ for $x > 1$, it follows that the largest value of the curvature must occur at $x = 1$.

Suppose a curve C in the plane has curvature $\kappa \neq 0$ at a point P. Then the circle of radius $\rho = 1/\kappa$ that is centered on the concave side of the curve and shares a common tangent with C at P is called the **osculating** (kissing) **circle** at P. The radius ρ of the osculating circle is called the **radius of curvature** and the center of the circle is called the **center of curvature**. The osculating circle and the curve C are tangent and have the same curvature at P, and in this sense, the osculating circle is the circle that best approximates the curve C at P.

Figure 10.32 shows the curve $y = x^{-1}$ for $x > 0$ whose curvature we found in Example 7, along with the osculating circle at the point $P(1, 1)$.

Note that the curvature at P is

$$\kappa(1) = \frac{2(1)^3}{(1^4 + 1)^{3/2}} = \frac{1}{\sqrt{2}}$$

so the radius of curvature is

$$\rho(1) = \frac{1}{\kappa(1)} = \sqrt{2}$$

In Problem 60, you are asked to show that the osculating circle at P has the equation

$$(x - 2)^2 + (y - 2)^2 = 2$$

We conclude this section by providing a summary of various formulas for computing curvature (Table 10.2).

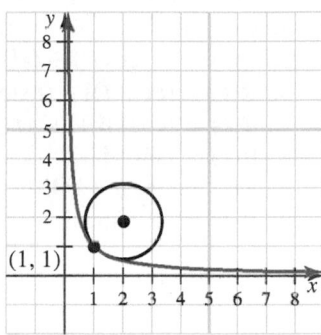

Figure 10.32 Osculating circle at (1,1)

Table 10.2 Curvature Formulas

Type	Given information	Formula	Where derived
Arc length parameter	$\mathbf{R}(s)$	$\left\| \dfrac{d\mathbf{T}}{ds} \right\|$	Definition
Two-derivatives form	$\mathbf{R}(t)$	$\dfrac{\|\mathbf{T}'(t)\|}{\|\mathbf{R}'(t)\|}$	Page 789
Cross-derivative form	$\mathbf{R}(t)$	$\dfrac{\|\mathbf{R}' \times \mathbf{R}''\|}{\|\mathbf{R}'\|^3}$	Theorem 10.8
Functional form	$y = f(x)$	$\dfrac{\|f''(x)\|}{(1 + [f'(x)]^2)^{3/2}}$	Theorem 10.9
Parametric form	$x = x(t), \ y = y(t)$	$\dfrac{\|x'y'' - y'x''\|}{[(x')^2 + (y')^2]^{3/2}}$	Problem 46
Polar form	$r = f(\theta)$	$\dfrac{\|r^2 + 2r'^2 - rr''\|}{(r^2 + r'^2)^{3/2}}$	Problem 48

PROBLEM SET 10.4

1. ■ What does this say? Describe what is meant by using arc length as a parameter.
2. ■ What does this say? Discuss the various curvature formulas given in Table 10.2.

In Problems 3-10, find the unit tangent vector **T**(t) *and the principal unit normal vector* **N**(t) *for the curve given by* **R**(t).

3. $\mathbf{R}(t) = t^2\mathbf{i} + t^3\mathbf{j}, t \neq 0$
4. $\mathbf{R}(t) = t^2\mathbf{i} + \sqrt{t}\,\mathbf{j}, t > 0$
5. $\mathbf{R}(t) = (e^t \cos t)\mathbf{i} + (e^t \sin t)\mathbf{j}$
6. $\mathbf{R}(t) = (t \cos t)\mathbf{i} + (t \sin t)\mathbf{j}$
7. $\mathbf{R}(t) = (\cos t)\mathbf{i} + (\sin t)\mathbf{j} + t\mathbf{k}$
8. $\mathbf{R}(t) = (\sin t)\mathbf{i} - (\cos t)\mathbf{j} + t\mathbf{k}$
9. $\mathbf{R}(t) = (\ln t)\mathbf{i} + t^2\mathbf{k}$
10. $\mathbf{R}(t) = (e^{-t}\sin t)\mathbf{i} + e^{-t}\mathbf{j} + (e^{-t}\cos t)\mathbf{k}$

In Problems 11-16, find the length of the given curve over the given interval.

11. $\mathbf{R}(t) = 2t\mathbf{i} + t\mathbf{j}$, over $[0, 4]$
12. $\mathbf{R}(t) = t\mathbf{i} + 3t\mathbf{j}$, over $[0, 4]$
13. $\mathbf{R}(t) = (\cos^3 t)\mathbf{i} + (\cos^2 t)\mathbf{k}$, over $\left[0, \frac{\pi}{2}\right]$
14. $\mathbf{R}(t) = t\mathbf{i} + 2t\mathbf{j} + 3t\mathbf{k}$, over $[0, 2]$
15. $\mathbf{R}(t) = (4 \cos t)\mathbf{i} + (4 \sin t)\mathbf{j} + 5t\mathbf{k}$, over $[0, \pi]$
16. $\mathbf{R}(t) = 3t\mathbf{i} + (3 \cos t)\mathbf{j} + (3 \sin t)\mathbf{k}$, over $\left[0, \frac{\pi}{2}\right]$

Find the curvature of the plane curves at the points indicated in Problems 17-22.

17. $y = 4x - 2$ at $x = 2$
18. $y = x - \frac{1}{9}x^2$ at $x = 3$
19. $y = x + \frac{1}{x}$ at $x = 1$
20. $y = \sin x$ at $x = \frac{\pi}{2}$
21. $y = \ln x$ at $x = 1$
22. $y = e^x$ at $x = 0$

Express the vector function **R**(t) *in Problems 23-26 in terms of the arc length parameter s measured from the point where* $t = 0$ *in the direction of increasing t.*

23. $\mathbf{R}(t) = \langle e^{-t}, 1 - e^{-t}\rangle$
24. $\mathbf{R}(t) = \langle \sin t, \cos t\rangle$
25. $\mathbf{R}(t) = \langle 2 - 3t, 1 + t, -4t\rangle$
26. $\mathbf{R}(t) = \langle 3 \cos t + 3t \sin t, 2t^2, 3 \sin t - 3t \cos t\rangle$
27. Let **u** and **v** be constant, nonzero vectors. Show that the line given by $\mathbf{R}(t) = \mathbf{u} + t\mathbf{v}$ has curvature 0 at each point.
28. Using the appropriate formula in table 10.2, compute the curvature of the curve

$$\mathbf{R}(t) = \langle 3 \sin t, 3 \cos t, 4t\rangle$$

29. Find the unit tangent **T** for

$$\mathbf{R}(t) = [\ln(\sin t)]\mathbf{i} + [\ln(\cos t)]\mathbf{j}$$

at the point where $t = \frac{\pi}{3}$.

30. Find the unit tangent **T** for

$$\mathbf{R}(t) = (\cosh t)\mathbf{i} + (\sinh t)\mathbf{j}$$

at the point where $t = 0$.

31. For the curve given by

$$\mathbf{R}(t) = (\sin t)\mathbf{i} + (\cos t)\mathbf{j} + t\mathbf{k}$$

a. Find a unit tangent vector **T** at the point on the curve where $t = \pi$.
b. Find the curvature when $t = \pi$.
c. Find the length of the curve from $t = 0$ to $t = \pi$.

32. A curve C in the plane is given parametrically by $x = 32t, y = 16t^2 - 4$.
a. Sketch the graph of the curve.
b. Find the unit tangent vector when $t = 3$.
c. Find the radius of curvature of the point P on C where $t = 3$.

33. Find the point (or points) where the ellipse $9x^2 + 4y^2 = 36$ has maximum curvature.

34. Let C be the curve given by

$$\mathbf{R}(t) = (t - \sin t)\mathbf{i} + (1 - \cos t)\mathbf{j} + \left(4 \sin \frac{1}{2}t\right)\mathbf{k}$$

a. Find the unit tangent vector **T**(t) to C.
b. Find $\dfrac{d\mathbf{T}}{ds}$ and the curvature $\kappa(t)$.

35. Find the radius of curvature at each relative extremum of the graph of $y = x^6 - 3x^2$.
36. Find the maximum curvature of the curve $y = e^{2x}$.
37. Find the curvature of the cycloid defined by $x = t - \sin t, y = 1 - \cos t$. Sketch the curve for $0 \leq t \leq 2\pi$ and sketch the osculating circle at the point where $t = \frac{\pi}{2}$.

There are several different formulas for finding curvature, as summarized in Table 10.2. In Problems 38-45, find the curvature of each of the given curves using the indicated formula.

38. $\mathbf{R}(t) = (t - \cos t)\mathbf{i} + (\sin t)\mathbf{j} + 3\mathbf{k}$; cross-derivative form
39. $\mathbf{R}(t) = t\mathbf{i} + t^2\mathbf{j} + t^3\mathbf{k}$; cross-derivative form
40. $y = x^3$; functional form
41. $y = x^2$; functional form
42. $y = \sin x$; functional form
43. $y = x^{-2}$; functional form
44. the cardioid $r = 1 + \cos\theta, 0 \leq \theta \leq 2\pi$; polar form
45. the spiral $r = e^\theta$; polar form

Let C be a smooth curve in $\mathbb{R}^2$ described by the parametric equations $x = x(t)$ and $y = y(t)$. In Problems 46-51, use the parametric form for curvature formula

$$\frac{|x'y'' - y'x''|}{[(x')^2 + (y')^2]^{3/2}}$$

46. Prove the curvature formula for the parametric form.
47. Find the curvature of the curve described by
$x = 1 - \cos t$, $y = 2 + \sin t$.
48. Let f be a twice differentiable function in $\mathbb{R}^2$ described by the polar function $r = f(\theta)$. Show that the curvature is
$$\frac{|r^2 + 2r'^2 - rr''|}{(r^2 + r'^2)^{3/2}}$$

Hint: let $x = f(\theta)\cos\theta$, $y = f(\theta)\sin\theta$ in the parametric form.
49. Use the formula in Problem 48 to find the curvature of the cardioid with the equation $r = 1 + \cos\theta$.
50. Use the formula in Problem 48 to find the curvature of the spiral with the equation $r = \theta$.
51. Show that the formula for polar curvature in Problem 48 reduces to $\kappa = \dfrac{2}{|r'|}$ at the pole (where $r = 0$).

52. A *pestus houseflyus* is observed to zip around a room in such a way that at time t its position with respect to the nose of an observer is given by the vector function

$$\mathbf{R}(t) = t\mathbf{u} + t^2\mathbf{v} + 2\left(\frac{2}{3}t\right)^{3/2}(\mathbf{u} \times \mathbf{v})$$

where $\mathbf{u}$ and $\mathbf{v}$ are unit vectors separated by an angle of $60°$. Compute the fly's speed and find how long it takes to move a distance of 20 units along its path (starting from the nose), as shown in Figure 10.33.

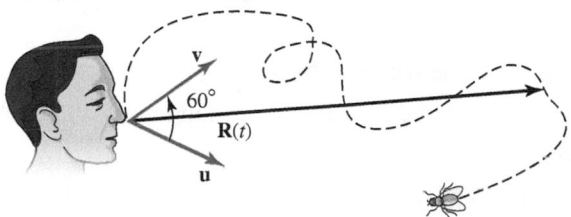

Figure 10.33 Path of a *pestus houseflyus*

53. The *Cornu spiral* is a parametric curve described by the equations

$$x = \int_0^t \cos\left(\frac{\pi u^2}{2}\right)du, \quad y = \int_0^t \sin\left(\frac{\pi u^2}{2}\right)du$$

It is used in optics to study amplitudes of light waves, and in the design of highway exit ramps. It also has interesting mathematical properties.
a. Show that the arc length of the spiral measured from the origin $(0, 0)$ satisfies $s = t$.

b. Show that the curvature of the spiral satisfies $\kappa(s) = \pi|s|$. Thus, for $s \geq 0$, the curvature increases at a constant rate.
c. What happens to the curvature, κ, as $s \to \infty$? That is, evaluate $\lim_{x \to \infty}\kappa(s)$.
54. Let $P(a, b)$ be a point on the graph C of the vector function $\mathbf{R}(t)$.
a. Describe a general procedure for finding an equation for the osculating circle to C at P.
b. Find an equation for the osculating circle at the point $P(32, 12)$ on the curve C defined by $x = 32t$, $y = 16t^2 - 4$.

Level 3

55. Prove that if a curve is parameterized by arc length, the velocity is perpendicular to acceleration.
56. If $\mathbf{T}$ and $\mathbf{N}$ are the unit tangent and normal vectors, respectively, on the trajectory of a moving body, then the cross product vector $\mathbf{B} = \mathbf{T} \times \mathbf{N}$ is called the unit **binormal** of the trajectory. Three planes determined by $\mathbf{T}, \mathbf{N}$, and $\mathbf{B}$ are shown in Figure 10.34.

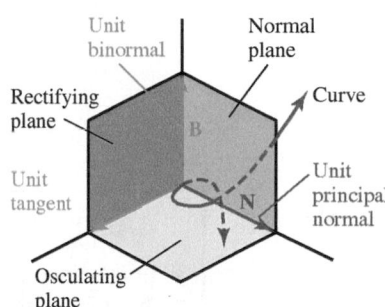

Figure 10.34 Three planes determined by $\mathbf{T}, \mathbf{N}$, and $\mathbf{B}$

a. Show that $\mathbf{T}$ is orthogonal to $\dfrac{d\mathbf{B}}{ds}$.
Hint: Differentiate $\mathbf{B} \cdot \mathbf{T}$.
b. Show that $\mathbf{B}$ is orthogonal to $\dfrac{d\mathbf{B}}{ds}$.
Hint: Differentiate $\mathbf{B} \cdot \mathbf{B}$.
c. Show that $\dfrac{d\mathbf{B}}{ds} = -\tau\mathbf{N}$ for some constant τ.
Note: τ is called the **torsion** of the trajectory.
57. Prove the Frenet-Serret formulas:

$$\frac{d\mathbf{T}}{ds} = \kappa\mathbf{N}$$
$$\frac{d\mathbf{N}}{ds} = -\kappa\mathbf{T} + \tau\mathbf{B}$$
$$\frac{d\mathbf{B}}{ds} = -\tau\mathbf{N}$$

where κ is the curvature and $\tau = \tau(s)$ is a scalar function called the torsion, which provides a measure of the amount of twisting at each point on the trajectory.

58. Show that the torsion may be computed by the formula

$$\tau = \frac{[\mathbf{R}'(t) \times \mathbf{R}''(t)] \cdot \mathbf{R}'''(t)}{\|\mathbf{R}'(t) \times \mathbf{R}''(t)\|^2}$$

59. A highway has an exit ramp that begins at the origin and follows the curve $y = \frac{1}{32}x^{5/2}$ to the point $(4, 1)$.

Then it follows the shape of the osculating circle at $(4, 1)$ until the point where $y = 3$. What is the total length of the exit ramp?

60. Show that the osculating circle at the point $(1, 1)$ on the curve $y = x^{-1}$ has the equation

$$(x - 2)^2 + (y - 2)^2 = 2$$

10.5 TANGENTIAL AND NORMAL COMPONENTS OF ACCELERATION

IN THIS SECTION: *Components of acceleration, applications*
We see how the velocity and acceleration vectors of a moving body can be decomposed into tangential and normal components.

Components of Acceleration

When a body is caused to accelerate or brakes are applied, it is of interest to know how much of the acceleration acts in the direction of the body's motion, as indicated by the unit tangent vector $\mathbf{T}$. This question is answered by the following theorem.

Theorem 10.10   Tangential and normal components of acceleration

An object moving along a smooth curve (with $\mathbf{T}' \neq \mathbf{0}$) has velocity $\mathbf{V}$ and acceleration $\mathbf{A}$, where

$$\mathbf{V} = \left(\frac{ds}{dt}\right)\mathbf{T} \quad \text{and} \quad \mathbf{A} = \left(\frac{d^2 s}{dt^2}\right)\mathbf{T} + \kappa\left(\frac{ds}{dt}\right)^2 \mathbf{N}$$

and s is the arc length along the trajectory.

> **W**hat this says At each point on the trajectory of a moving object, the velocity $\mathbf{V}$ points in the direction of the unit tangent $\mathbf{T}$, but the acceleration $\mathbf{A}$ may have both a tangential and normal component. The trajectory may twist and turn, but the acceleration is always in the plane determined by $\mathbf{T}$ and the unit normal $\mathbf{N}$.

Proof: The formula for $\mathbf{V}$ follows from the fact that the velocity vector has magnitude $\|\mathbf{V}\| = \dfrac{ds}{dt}$ from Theorem 10.6 and points in the direction of the unit tangent. To establish the formula for $\mathbf{A}$, first note that

$$\frac{d\mathbf{T}}{ds} = \left\|\frac{d\mathbf{T}}{ds}\right\|\mathbf{N} = \kappa\mathbf{N}$$

since the derivative of $\mathbf{T}(s)$ points in a direction orthogonal to $\mathbf{T}$. Using this formula, we find

$$\mathbf{A} = \frac{d\mathbf{V}}{dt} \qquad\qquad \textit{Definition of } \mathbf{A}$$

$$= \frac{d}{dt}\left[\mathbf{T}\frac{ds}{dt}\right] \qquad \mathbf{V} = \left(\frac{ds}{dt}\right)\mathbf{T}$$

$$= \frac{d^2 s}{dt^2}\mathbf{T} + \frac{ds}{dt}\frac{d\mathbf{T}}{dt} \qquad \textit{Product rule}$$

$$= \frac{d^2s}{dt^2}\mathbf{T} + \frac{ds}{dt}\left[\frac{d\mathbf{T}}{ds}\frac{ds}{dt}\right] \qquad \textit{Chain rule}$$

$$= \frac{d^2s}{dt^2}\mathbf{T} + \left(\frac{ds}{dt}\right)^2 \frac{d\mathbf{T}}{ds}$$

$$= \frac{d^2s}{dt^2}\mathbf{T} + \left(\frac{ds}{dt}\right)^2 (\kappa\mathbf{N}) \qquad\qquad \kappa\mathbf{N} = \frac{d\mathbf{T}}{ds} \qquad\qquad \blacklozenge$$

The two components of acceleration have special names.

TANGENTIAL COMPONENT/NORMAL COMPONENT The acceleration $\mathbf{A}$ of a moving object can be written as

$$\mathbf{A} = A_T\mathbf{T} + A_N\mathbf{N}$$

where

$A_T = \dfrac{d^2s}{dt^2}$ is the **tangential component of acceleration**

and

$A_N = \kappa\left(\dfrac{ds}{dt}\right)^2$ is the **normal component of acceleration**

The tangential and normal components are shown in Figure 10.35.

The formulas $A_T = \dfrac{d^2s}{dt^2}$ and $A_N = \kappa\left(\dfrac{ds}{dt}\right)^2$ provide useful information about the way a moving object travels along its trajectory, but they are often not the most convenient computational forms. The following theorem provides alternative formulas for A_T and A_N that use only the derivatives $\mathbf{R}'$ and $\mathbf{R}''$ of the position vector $\mathbf{R}(t)$.

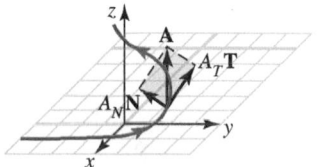

Figure 10.35 Components of acceleration
$$\mathbf{A} = A_T\mathbf{T} + A_N\mathbf{N}$$

Theorem 10.11   Formulas for the components of acceleration

Let $\mathbf{R}(t)$ be the position vector for an object moving along a smooth curve C. Then the tangential and normal components of the object's acceleration are given by

$$A_T = \frac{\mathbf{R}' \cdot \mathbf{R}''}{\|\mathbf{R}'\|} \qquad \text{and} \qquad A_N = \frac{\|\mathbf{R}' \times \mathbf{R}''\|}{\|\mathbf{R}'\|}$$

Proof: Let θ be the angle between $\mathbf{R}'$ and $\mathbf{R}''$. Then, since the acceleration is given by $\mathbf{A} = \mathbf{R}''$, we have

$$A_T = \|\mathbf{A}\|\cos\theta = \frac{\|\mathbf{R}'\|\,\|\mathbf{A}\|\cos\theta}{\|\mathbf{R}'\|} = \frac{\mathbf{R}' \cdot \mathbf{R}''}{\|\mathbf{R}'\|}$$

and

$$A_N = \|\mathbf{A}\|\sin\theta = \frac{\|\mathbf{R}'\|\,\|\mathbf{A}\|\sin\theta}{\|\mathbf{R}'\|} = \frac{\|\mathbf{R}' \times \mathbf{R}''\|}{\|\mathbf{R}'\|} \qquad\qquad \blacklozenge$$

Example 1   Finding tangential and normal components of acceleration

Find the tangential and normal components of the acceleration of an object that moves with displacement $\mathbf{R}(t) = \langle t^3, t^2, t \rangle$.

Solution   Since $\mathbf{R}'(t) = \mathbf{V}(t) = \langle 3t^2, 2t, 1 \rangle$ and $\mathbf{R}''(t) = \mathbf{A}(t) = \langle 6t, 2, 0 \rangle$, we have

$$\mathbf{R}' \cdot \mathbf{R}'' = \mathbf{V} \cdot \mathbf{A} = (3t^2)(6t) + (2t)(2) = 18t^3 + 4t$$

and

$$\mathbf{R}' \times \mathbf{R}'' = \mathbf{V} \times \mathbf{A} = \begin{vmatrix} \mathbf{i} & \mathbf{j} & \mathbf{k} \\ 3t^2 & 2t & 1 \\ 6t & 2 & 0 \end{vmatrix} = -2\mathbf{i} + 6t\mathbf{j} - 6t^2\mathbf{k}$$

Therefore, the components of acceleration are

$$A_T = \frac{\mathbf{R}' \cdot \mathbf{R}''}{\|\mathbf{R}'\|} \qquad\qquad A_N = \frac{\|\mathbf{R}' \times \mathbf{R}''\|}{\|\mathbf{R}'\|}$$

$$= \frac{\mathbf{V} \cdot \mathbf{A}}{\|\mathbf{V}\|} \qquad\qquad = \frac{\|\mathbf{V} \times \mathbf{A}\|}{\|\mathbf{V}\|}$$

$$= \frac{18t^3 + 4t}{\sqrt{(3t^2)^2 + (2t)^2 + 1}} \qquad\qquad = \frac{\sqrt{(-2)^2 + (6t)^2 + (-6t^2)^2}}{\sqrt{9t^4 + 4t^2 + 1}}$$

$$= \frac{18t^3 + 4t}{\sqrt{9t^4 + 4t^2 + 1}} \qquad\qquad = 2\sqrt{\frac{9t^4 + 9t^2 + 1}{9t^4 + 4t^2 + 1}}$$

Note in Figure 10.35 that $\|\mathbf{A}\|$ is the hypotenuse of a right triangle whose other two sides have lengths $\|A_T\mathbf{T}\| = |A_T|$ and $\|A_N\mathbf{N}\| = A_N$. Therefore, by the Pythagorean theorem, we have

$$\|\mathbf{A}\|^2 = A_T^2 + A_N^2$$

$$A_N = \sqrt{\|\mathbf{A}\|^2 - A_T^2}$$

This formula is used to compute A_N in the following example.

Example 2   Finding components of acceleration on a helix

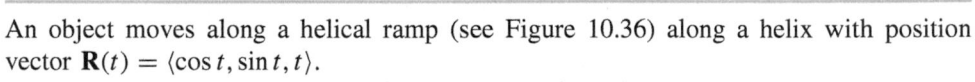

An object moves along a helical ramp (see Figure 10.36) along a helix with position vector $\mathbf{R}(t) = \langle \cos t, \sin t, t \rangle$.

Find the tangential and normal components of acceleration.

Solution

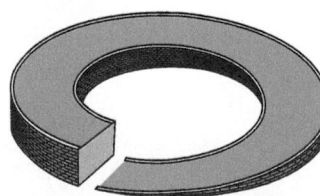

Figure 10.36   Helical ramp

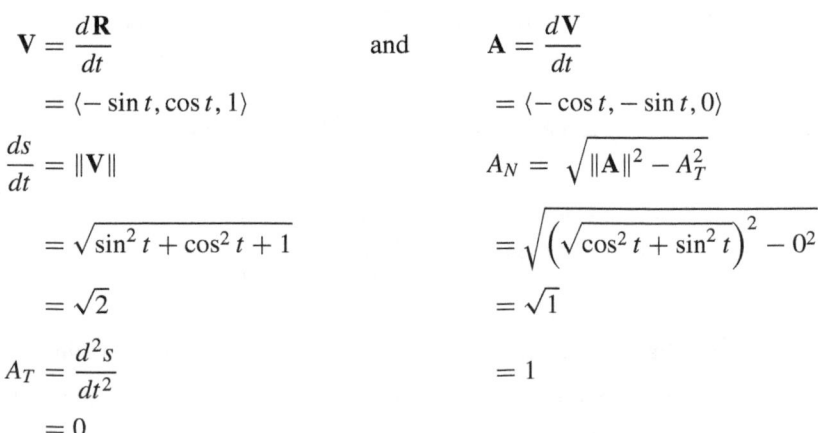

$$\mathbf{V} = \frac{d\mathbf{R}}{dt} \qquad\qquad \text{and} \qquad\qquad \mathbf{A} = \frac{d\mathbf{V}}{dt}$$

$$= \langle -\sin t, \cos t, 1 \rangle \qquad\qquad\qquad = \langle -\cos t, -\sin t, 0 \rangle$$

$$\frac{ds}{dt} = \|\mathbf{V}\| \qquad\qquad\qquad\qquad A_N = \sqrt{\|\mathbf{A}\|^2 - A_T^2}$$

$$= \sqrt{\sin^2 t + \cos^2 t + 1} \qquad\qquad = \sqrt{\left(\sqrt{\cos^2 t + \sin^2 t}\right)^2 - 0^2}$$

$$= \sqrt{2} \qquad\qquad\qquad\qquad\qquad = \sqrt{1}$$

$$A_T = \frac{d^2 s}{dt^2} \qquad\qquad\qquad\qquad\qquad = 1$$

$$= 0$$

The tangential and normal components of acceleration are 0 and 1, respectively. This means that the acceleration satisfies

$$\mathbf{A} = A_T\mathbf{T} + A_N\mathbf{N} = (0)\mathbf{T} + (1)\mathbf{N} = \mathbf{N}$$

That is, the acceleration vector is the principal unit normal $\mathbf{N}$, and the acceleration is always normal to the trajectory of the helix.

Applications

Now that we know how to compute the tangential and normal components of acceleration, A_T and A_N, we will examine some applications. First, according to Newton's second law of motion, the total force acting on a moving object of mass m satisfies $\mathbf{F} = m\mathbf{A}$, where $\mathbf{A}$ is the acceleration of the object. Because $\mathbf{A} = A_T\mathbf{T} + A_N\mathbf{N}$, we have

$$\mathbf{F} = m\mathbf{A} = (mA_T\mathbf{T}) + (mA_N)\mathbf{N} = F_T\mathbf{T} + F_N\mathbf{N}$$

where

$$F_T = m\frac{d^2s}{dt^2} \qquad \text{and} \qquad F_N = m\kappa\left(\frac{ds}{dt}\right)^2$$

For instance, experience leads us to expect a car to skid if it makes a sharp turn at moderate speed or even a gradual turn at high speed. Mathematically, a "sharp turn" occurs when the radius of curvature $\rho = 1/\kappa$ is small (that is, when κ is large), and "high speed" means that ds/dt is large. In either case,

$$F_N = m\kappa\left(\frac{ds}{dt}\right)^2$$

will be relatively large, and the car will stay on the road (its trajectory) only if (assuming no banking) there is a correspondingly large frictional force between the tires of the car and the surface of the road (see Figure 10.37).

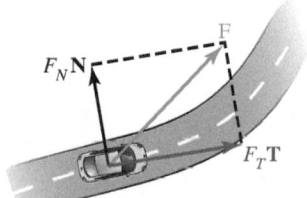

Figure 10.37 Tendency to skid

When forces, masses, and accelerations are involved, it is customary to express the mass m in slugs of an object whose weight W has been given in pounds.* From the definition of weight,

$$m = \frac{W}{g}$$

where g is the acceleration of gravity ($g \approx 32$ ft/s^2).

Example 3   Modeling Application: Tendency of a vehicle to skid

A car weighing 2,700 lb makes a turn on a flat road while traveling at 56 ft/s (about 38 mi/h). If the radius of the turn is 70 ft, how much frictional force is required to keep the car from skidding?

Solution   The force tending to push the car off the road is $F_N = m\kappa\left(\frac{ds}{dt}\right)^2$, where m is the mass of the car and κ is the curvature of the road. This is the force that must be compensated for by friction if the car is not to skid. We know that $\dfrac{ds}{dt} = 56$ ft/s,

---

*A slug is a unit of measurement defined as the unit of mass that receives an acceleration of 1 ft/s^2 when a force of 1 lb is applied to it. That is,

$$1\text{ slug} = \frac{1\text{ lb}}{1\text{ ft/s}^2} = \frac{\text{lb s}^2}{\text{ft}}$$

and because the car weighs $W = 2{,}700$ lb, its mass is $m = \dfrac{W}{g} = \dfrac{2{,}700}{32} \approx 84.38$ slugs. Because the turn radius is 70 ft, we have $\kappa = \frac{1}{70}$, so that

$$F_N = \left(\frac{2{,}700}{32} \text{ slugs} \right) \left(\frac{1}{70 \text{ ft}} \right) \left(56 \, \frac{\text{ft}}{\text{s}} \right)^2 = 3{,}780 \, \frac{\text{lb} \cdot \text{s}^2}{\text{ft}} \, \frac{1}{\text{ft}} \, \frac{\text{ft}^2}{\text{s}^2} = 3{,}780 \, \text{lb} \qquad \blacksquare$$

There are certain important applications in which an object moves along its trajectory with constant speed ds/dt, and when this occurs, the acceleration $\mathbf{A}$ can have only a normal component, because $d^2 s/dt^2 = 0$.

Theorem 10.12   Acceleration of an object with constant speed

The acceleration of an object moving with constant speed is always orthogonal to the direction of motion.

Proof: Notice that we really do not need to know anything about components of acceleration to prove this result. Saying that the object has constant speed means that $\|\mathbf{R}'(t)\|$ is constant, and by Theorem 10.4, we conclude that $\mathbf{R}'(t)$ is orthogonal to its derivative $\mathbf{R}''(t) = \mathbf{A}(t)$. But $\mathbf{R}'(t)$ points in the direction of the object's motion along its trajectory, which means that the acceleration $\mathbf{A}$ is orthogonal to the direction of motion. $\qquad \blacklozenge$

As an application of this result, note that when an object moves with constant speed v_0 along a circular path of radius R (so that $\kappa = 1/R$), its acceleration is directed toward the center of the path and has magnitude

$$A_N = \kappa \left(\frac{ds}{dt} \right)^2 = \frac{1}{R} v_0^2$$

(See Figure 10.38.)

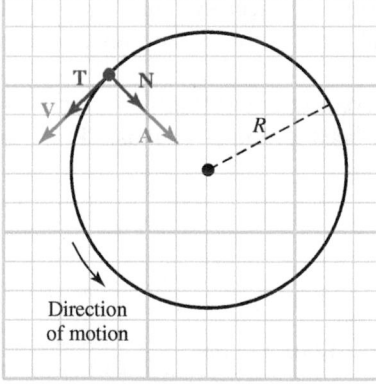

If the constant speed is v_0 and the path has radius R, the acceleration is

$$\mathbf{A} = \frac{v_0^2}{R} \mathbf{N}$$

Figure 10.38   An object moving with constant speed on a circular path

Example 4   Modeling Application: Period of a satellite

An artificial satellite travels at constant speed in a stable circular orbit 20,000 km above the earth's surface. How long does it take for the satellite to make one complete circuit of the earth?

Solution   Let m denote the satellite's mass and v denote its speed. We will assume that the earth is a sphere of radius 6,440 km, so the curvature of the path is $\kappa = 1/R$, where

$$R = \underbrace{6{,}440 \text{ km}}_{\textit{radius of the earth}} + \overbrace{20{,}000 \text{ km}}^{\textit{height}} = 26{,}440 \text{ km}$$

is the distance of the satellite from the center of the earth. The satellite maintains a stable orbit when the magnitude of the force $F_c = mv^2/R$ produced by the rate of change of tangential velocity, called *centripetal* (center directed) *acceleration,* equals the magnitude of the force F_g due to gravity, but is opposite in direction. (See Figure 10.39.)

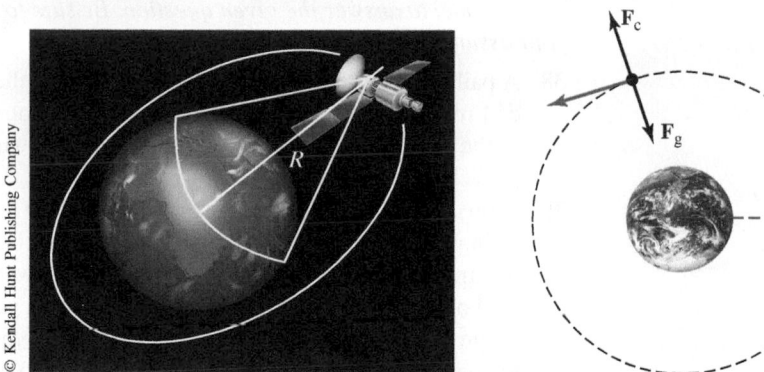

Figure 10.39 A satellite in a stable orbit; the magnitude of the centripetal force **F**$_c$ equals the magnitude of the force **F**$_g$ due to gravity, but points in the opposite direction

According to Newton's law of universal gravitation, $\|\mathbf{F_g}\| = GmM/R^2$, where M is the mass of the earth and G is the *gravitational constant*. Thus, for stability we must have

$$\frac{mv^2}{R} = \frac{GmM}{R^2}$$

so $v = \sqrt{GM/R}$. Experiments indicate that $GM = 398{,}600$ km^3/s^2, and by substituting $R = 26{,}440$, we obtain

$$v = \sqrt{\frac{GM}{R}} = \sqrt{\frac{398{,}600}{26{,}440}} \approx 3.88273653$$

For this example, we see that the speed of the satellite is approximately 3.883 km/s.

Finally, suppose T is the time required for the satellite to make one complete circuit of the earth (called the **period** of the satellite). In each period, the satellite travels a distance equal to the circumference of a circle of radius $R = 26{,}440$ km, and because it travels at v km/s, we must have $vT = 2\pi R$, so that the time (in seconds) is

$$T = \frac{2\pi R}{v} \approx \frac{2\pi (26{,}440 \text{ km})}{3.88273653} \approx 42786.16852$$

or approximately 713 minutes (11 h 53 min).

PROBLEM SET 10.5

Level 1

In Problems 1-10, $\mathbf{R}(t)$ *is the position vector of a moving object. Find the tangential and normal components of the object's acceleration.*

1. $\mathbf{R}(t) = t\mathbf{i} + t^2\mathbf{j}$
2. $\mathbf{R}(t) = t\mathbf{i} + e^t\mathbf{j}$
3. $\mathbf{R}(t) = \langle t\sin t, t\cos t\rangle$
4. $\mathbf{R}(t) = \langle 3\cos t, 2\sin t\rangle$
5. $\mathbf{R}(t) = \langle t, t^2, t\rangle$
6. $\mathbf{R}(t) = \langle t, -t, 2t\rangle$

7. $\mathbf{R}(t) = \langle 0, 2\sin t, 2\cos t\rangle$
8. $\mathbf{R}(t) = \langle 4\cos t, 0, \sin t\rangle$
9. $\mathbf{R}(t) = (\sin t)\mathbf{i} + (\cos t)\mathbf{j} + (\sin t)\mathbf{k}$
10. $\mathbf{R}(t) = \left(\frac{5}{13}\cos t\right)\mathbf{i} + \frac{12}{13}(1 - \cos t)\mathbf{j} + (\sin t)\mathbf{k}$

In Problems 11-18, the velocity $\mathbf{V}$ *and acceleration* $\mathbf{A}$ *of a moving object are given at a certain instant. Find* A_T *and* A_N *at this instant.*

11. $\mathbf{V}_0 = \langle 1, -3 \rangle; \mathbf{A}_0 = \langle 2, 5 \rangle$
12. $\mathbf{V}_0 = \langle -2, 3 \rangle; \mathbf{A}_0 = \langle 1, 4 \rangle$
13. $\mathbf{V}_0 = \mathbf{i} + 5\mathbf{k}; \mathbf{A}_0 = 3\mathbf{i} - \mathbf{k}$
14. $\mathbf{V}_0 = -\mathbf{i} + 7\mathbf{j}; \mathbf{A}_0 = 4\mathbf{i} + \mathbf{j}$
15. $\mathbf{V}_0 = 2\mathbf{i} + 3\mathbf{j} - \mathbf{k}; \mathbf{A}_0 = -\mathbf{i} - 5\mathbf{j} + 2\mathbf{k}$
16. $\mathbf{V}_0 = \mathbf{i} - \mathbf{j} - 2\mathbf{k}; \mathbf{A}_0 = \mathbf{i} - 2\mathbf{j} + \mathbf{k}$
17. $\mathbf{V}_0 = \langle 1, -2, 3 \rangle; \mathbf{A}_0 = \langle 0, 1, -3 \rangle$
18. $\mathbf{V}_0 = \langle 5, -1, 2 \rangle; \mathbf{A}_0 = \langle 1, 0, -7 \rangle$

In Problems 19-26, the speed $\|\mathbf{V}\|$ *of a moving object is given. Find* A_T, *the tangential component of acceleration, at the indicated time.*

19. $\|\mathbf{V}\| = \sqrt{5t^2 + 3}; t = 1$
20. $\|\mathbf{V}\| = \sqrt{3t^2 + 1}; t = 2$
21. $\|\mathbf{V}\| = \sqrt{2t^2 + t + 3}; t = 1$
22. $\|\mathbf{V}\| = \sqrt{t^2 + t + 1}; t = 3$
23. $\|\mathbf{V}\| = \sqrt{\sin^2 t + \cos 2t}; t = 0$
24. $\|\mathbf{V}\| = \sqrt{2 \sin^2 t + \cos 2t}; t = 0$
25. $\|\mathbf{V}\| = \sqrt{e^t + t^2}; t = 0$
26. $\|\mathbf{V}\| = \sqrt{e^{-t} + t^4}; t = 0$

Level 2

Use the formulas in Theorem 10.1 to find A_T *and* A_N *for the given position vector* $\mathbf{R}(t)$ *in Problems 27-32.*

27. $\mathbf{R}(t) = t\mathbf{i} + 2t\mathbf{j} + t^2\mathbf{k}$
28. $\mathbf{R}(t) = t^3\mathbf{i} + t^2\mathbf{j} + t\mathbf{k}$
29. $\mathbf{R}(t) = (-2 \sin t)\mathbf{i} + (2 \cos t)\mathbf{j} + \mathbf{k}$
30. $\mathbf{R}(t) = (\cos t)\mathbf{i} + (\sin t)\mathbf{j} + \mathbf{k}$
31. $\mathbf{R}(t) = (e^t \cos t)\mathbf{i} + (e^t \sin t)\mathbf{j} + e^t\mathbf{k}$
32. $\mathbf{R}(t) = e^t\mathbf{i} + e^{-t}\mathbf{j} + \mathbf{k}$
33. Find the maximum and minimum speeds of a particle whose position vector is

$$\mathbf{R}(t) = (4 \sin 2t)\mathbf{i} - (3 \cos 2t)\mathbf{j}$$

34. Where on the trajectory of

$$\mathbf{R}(t) = (2t^2 - 5t)\mathbf{i} + (5t + 2)\mathbf{j} + 4t^2\mathbf{k}$$

is the speed minimized? What is the minimum speed?

35. The position of an object at time t is (x, y), where $x = 1 + \cos 2t$, $y = \sin 2t$. Find the velocity, the acceleration, and the tangential and normal components of acceleration of the object at time t.

36. An object moves with constant angular velocity ω around the circle $x^2 + y^2 = r^2$ in the xy-plane. Find a parameterization for the circle and compute the tangential and normal components of acceleration for the moving object.

37. Find the tangential and normal components of the acceleration of an object that moves along the parabolic path $y = 4x^2$ at the instant the speed is $ds/dt = 20$.

Modeling Problems *In Problems 38-41, set up an appropriate model to answer the given question. Be sure to state your assumptions.*

38. A pail attached to a rope 1 yd long is swung at the rate of 1 rev/s. Find the tangential and normal components of the pail's acceleration. Assume the rope is swung in a level plane.

39. A boy holds onto a pail of water weighing 2 lb and swings it in a vertical circle with a radius of 3 ft. If the pail travels at ω rev/s, what is the force of the water on the bottom of the pail at the highest and lowest points of the swing? What is the *smallest* value of ω required to keep the water from spilling from the pail? Assume the pail is held by a handle so that its bottom is straight up when it is at its highest point.

40. A car weighing 2,700 lb (1.35 tons) moves along the elliptic path $900x^2 + 400y^2 = 1$, where x and y are measured in miles. If the car travels at the constant speed of 45 mi/h, how much frictional force is required to keep it from skidding as it turns the "corner" at $(\frac{1}{30}, 0)$? What about the corner $(0, \frac{1}{20})$?

41. What is the smallest radius that should be used for a circular highway exchange as shown in Figure 10.40, if the normal component of the acceleration of a car traveling at 45 mi/h is not to exceed 2.4 ft/s²? Note that you may use the inner radius or the outer radius, or even the median one.

Figure 10.40 Highway exchange

Modeling Problems *Curved sections of road such as expressway exit ramps are often banked to protect against skidding. Consider a road with a circular curve of radius 150 feet, as shown in Figure 10.40. Assume the magnitude of the force of static friction* $\mathbf{F}_s$ *(the force resisting the tendency to skid) is proportional to the car's weight; that is* $\|\mathbf{F}_s\| = \mu W$, *where* μ *is a constant called the* coefficient *of static friction. Use this information in Problems 42-47.*

42. Suppose the road is not banked and $\mu = 0.47$. What is the largest speed v_m that a 3,500-lb car can travel around the curve without skidding?

43. The car in Problem 42 is unloaded, so it now weighs only 2,000 lb. Will the largest speed v_m for which the car can travel around the curve without skidding change compared to the previous problem?

44. Use the information in Problem 42, but suppose the curve is banked at 17°. What is the largest speed v_m for which the car can travel around the curve without skidding?

45. Use the information in Problem 43, but suppose the curve is banked at 17°. Would the result of this problem differ from that of the previous problem? That is, would anything change if the mass is different?

46. Suppose the highway engineers want the maximum safe speed to be 50 mi/h. At what angle should the road be banked if you assume $\mu = 0.47$?

47. Suppose the highway engineers want the maximum safe speed to be 55 mi/h. At what angle should the road be banked if you assume $\mu = 0.47$?

48. **Modeling Problem** A Ferris wheel (see Figure 10.41) with radius 15 ft rotates in a vertical plane at ω rpm.

Figure 10.41 Ferris wheel

What is the maximum value of ω for a Ferris wheel carrying a person of weight W? (That is, the largest ω so the passenger is not be "thrown off.")

49. **Modeling Problem** An amusement park ride consists of a large (25-ft radius), flat, horizontal wheel.

Figure 10.42 Carnival fun ride

Customers board the wheel while it is stationary and try to stay on as long as possible as it begins to rotate. The purpose of this problem is to discover how fast the wheel can rotate without losing its passengers. Suppose the wheel rotates at ω revolutions per minute and that a volunteer weighing W lb sits 15 feet from the center of the wheel, as shown in Figure 10.42. Find $\mathbf{F}_N = m\mathbf{A}_N$,

where $m = W/g$ is the volunteer's mass. This is the force tending to push the volunteer off the wheel.

50. **Modeling Problem** If the frictional force of the volunteer in Problem 49 to stay on the wheel is $0.12W$, find the largest value of ω that will allow the volunteer to stay in place. Wait, we did not tell you the volunteer's weight. Does it matter?

51. **Modeling Problem** A curve in a railroad track has the shape of the parabola $x = y^2/120$. If a train is loaded so that its scalar normal component of acceleration cannot exceed 30 units/s², what is its maximum possible speed as it rounds the curve at $(0, 0)$?

Level 3

52. Let $\mathbf{T}$ be the unit tangent vector, $\mathbf{N}$ the principal unit normal, and $\mathbf{B}$ the unit binormal vector (see Problem 56, Section 10.4) to a given curve C. Show that

$$\frac{d\mathbf{B}}{ds} = \mathbf{T} \times \frac{d\mathbf{N}}{ds}$$

53. Suppose an object moves in the plane along the curve $y = f(x)$. Use the formulas obtained in Theorem 10.11 to show that

$$A_T = \frac{f'(x)f''(x)}{\sqrt{1 + [f'(x)]^2}}; \quad A_N = \frac{|f''(x)|}{\sqrt{1 + [f'(x)]^2}}$$

54. If the graph of the function f has an inflection point at $x = a$ and $f''(a)$ exists, show that the graph of f has curvature 0 at $(a, f(a))$.

55. A projectile is fired from ground level with an angle of elevation α and muzzle speed v_0. Find formulas for the tangential and normal components of the projectile's acceleration at time t. What are A_T and A_N at the time the projectile is at its maximum height?

56. An object connected to a string of length r is spun counterclockwise in a circular path in a horizontal plane. Let ω be the constant angular velocity of the object.
a. Show that the position vector of the object is

$$\mathbf{R}(t) = (r \cos \omega t)\mathbf{i} + (r \sin \omega t)\mathbf{j}$$

b. Find the normal component of the object's acceleration.

c. If the angular velocity ω is doubled, what is the effect on the normal component of acceleration?

d. What is the effect on **A** if ω is unchanged but the length of the string is doubled?

e. What happens to A_N if ω is doubled and r is halved?

57. An artificial satellite travels at constant speed in a stable circular orbit R km above the surface of the Earth (as in Example 4).

a. Show that the satellite's period is given by the formula

$$T = 2\pi \frac{(R + R_e)^{3/2}}{\sqrt{GM}}$$

where R_e is the radius of the earth, $R_e = 6,440$ kilometers, approximately, while $GM = 398,600.44 \text{ km}^3/\text{s}^2$.

b. A satellite is said to be in **geosynchronous orbit** if it completes one orbit every sidereal day (23 hours, 56 minutes). How high above the Earth should the satellite be to achieve such an orbit?

c. What is the speed of a satellite in geosynchronous orbit?

58. Modeling Problem For Mars the following facts are known (all in relation to the Earth):

diameter	0.533
length of day	1.029
mass	0.1074
gravity	0.3776

Use these facts to determine how high above the surface of Mars a satellite must be to achieve Mars synchronous orbit (see Problem 57)? How fast would such a satellite be traveling?

59. Perform an Internet search on the role of the normal component of acceleration in aeronautics and write a 500-word essay on the topic.

60. Perform an Internet search on the role of the tangential component of acceleration in aeronautics and write a 500-word essay on the topic.

CHAPTER 10 REVIEW

*T*he science of mathematics has grown to such vast proportion that probably no living mathematician can claim to have achieved its mastery as a whole.

Quote attributed to Alfred North Whitehead, 15 February 1861 – 30 December 1947

Proficiency Examination

Concept Problems

1. What is a vector-valued function?
2. What are the components of a vector function?
3. Describe what is meant by the graph of a vector function.
4. Define the limit of a vector function.
5. Define the derivative of a vector function.
6. Define the integral of a vector function.
7. What is a smooth curve?
8. State the following rules for differentiating vector functions:

   a. Linearity rule
   b. Scalar multiple rule
   c. Dot product rule

   d. Cross product rule
   e. Chain rule

9. State the theorem about the orthogonality of a vector function of constant length and its derivative.
10. What are the position, velocity, and acceleration vectors?
11. What is the speed of a particle moving on a curve C?
12. What are the formulas for the motion of a projectile in a vacuum?
13. What are the formulas for time of flight and range of a projectile?
14. State Kepler's laws.
15. What are the unit polar vectors, $\mathbf{u}_r$ and $\mathbf{u}_\theta$?
16. What are the unit tangent and normal vectors?
17. What is speed in terms of arc length?
18. Give a formula for the arc length of a curve in $\mathbb{R}^3$.
19. What is the curvature of a graph?
20. What is a formula for curvature in terms of the velocity and acceleration vectors?
21. What are the formulas for unit tangent and principal unit normal in terms of the arc length parameter?
22. What is the radius of curvature?
23. What are the tangential and normal components of acceleration?

Practice Problems

24. Sketch the graph of $\mathbf{R}(t) = (3\cos t)\mathbf{i} + (3\sin t)\mathbf{j} + t\mathbf{k}$, and find the length of this curve from $t = 0$ to $t = 2\pi$.
25. If $\mathbf{F}(t) = \dfrac{t}{1+t}\mathbf{i} + \dfrac{\sin t}{t}\mathbf{j} + (\cos t)\mathbf{k}$, find $\mathbf{F}'(t)$ and $\mathbf{F}''(t)$.
26. Evaluate $\displaystyle\int_1^2 \langle 3t, 0, 3\rangle \times \langle 0, \ln t, -t^2\rangle\, dt$.
27. Find a vector function $\mathbf{F}$ such that $\mathbf{F}''(t) = \langle e^t, -t^2, 3\rangle$ and $\mathbf{F}(0) = \langle 1, -2, 0\rangle$, $\mathbf{F}'(0) = \langle 0, 0, 3\rangle$.
28. Find the velocity $\mathbf{V}$, the speed $\dfrac{ds}{dt}$, and the acceleration $\mathbf{A}$ for the body with position vector $\mathbf{R}(t) = t\mathbf{i} + 2t\mathbf{j} + te^t\mathbf{k}$.
29. Find $\mathbf{T}$, $\mathbf{N}$, A_T and A_N (the tangential and normal components of acceleration), and κ (curvature) for an object with position vector $\mathbf{R}(t) = t^2\mathbf{i} + 3t\mathbf{j} - 3t\mathbf{k}$.
30. A projectile is fired from ground level with initial velocity 50 ft/s at an angle of elevation of $\alpha = 30°$.
    a. What is the maximum height reached by the projectile?
    b. What are the time of flight and the range?

Supplementary Problems*

Find the vector limits in Problems 1-6.

1. $\lim\limits_{t \to 0} \left[t^2 \mathbf{i} - 3te^t \mathbf{j} + \dfrac{\sin 2t}{t} \mathbf{k} \right]$

2. $\lim\limits_{t \to 0} \left[\dfrac{\mathbf{i} + t\mathbf{j} - e^{-t}\mathbf{k}}{1 - t} \right]$

3. $\lim\limits_{t \to 0} \langle t, 0, 5 \rangle \cdot \langle \sin t, 3t, -(1 - t) \rangle$

4. $\lim\limits_{t \to \pi} \langle 1 + t, -3, 0 \rangle \times \langle 0, t^2, \cos t \rangle$

5. $\lim\limits_{t \to 0^+} \left[\left(1 + \dfrac{1}{t}\right)^t \mathbf{i} - \left(\dfrac{\sin t}{t}\right)\mathbf{j} - t\mathbf{k} \right]$

6. $\lim\limits_{t \to \infty} \left[\left(\dfrac{1 - \cos t}{t}\right)\mathbf{i} + 4\mathbf{j} + \left(1 + \dfrac{3}{t}\right)^t \mathbf{k} \right]$

Find $\mathbf{F}'(t)$ and $\mathbf{F}''(t)$ for the vector functions in Problems 7-14.

7. $\mathbf{F}(t) = te^t\mathbf{i} + t^2\mathbf{j}$

8. $\mathbf{F}(t) = (t \ln 2t)\mathbf{i} + t^{3/2}\mathbf{k}$

9. $\mathbf{F}(t) = \langle t^2, t^{3/2}, t^{-3} \rangle$

10. $\mathbf{F}(t) = \langle -t^{-3}, t^{-1}, 4t^3 \rangle$

11. $\mathbf{F}(t) = \langle 2t^{-1}, -2t, te^{-t} \rangle$

12. $\mathbf{F}(t) = \langle t, -(1 - t), 0 \rangle \times \langle 0, t^2, e^{-t} \rangle$

13. $\mathbf{F}(t) = (t^2 + e^t)\mathbf{i} + (te^{-t})\mathbf{j} + (e^{t+1})\mathbf{k}$

14. $\mathbf{F}(t) = (1 - t)^{-1}\mathbf{i} + (\sin 2t)\mathbf{j} + (\cos^2 t)\mathbf{k}$

Sketch the graph of the vector functions in Problems 15-18.

15. $\mathbf{F}(t) = te^t\mathbf{i} + t^2\mathbf{j}$

16. $\mathbf{F}(t) = t^2\mathbf{i} - 3t\mathbf{j}$

17. $\mathbf{F}(t) = (1 - \cos t)\mathbf{i} + (\sin t)\mathbf{j}$

18. $\mathbf{F}(t) = (\cos t)\mathbf{j} + (\sin t)\mathbf{k}$

Describe the graph of the vector functions given in Problems 19-26 or sketch a graph in $\mathbb{R}^3$. A graph in $\mathbb{R}^2$ may help with your description.

19. $\mathbf{F}(t) = \langle 2 \sin t, 2 \cos t, 5t \rangle$

20. $\mathbf{F}(t) = \langle 3 \cos t, 3 \sin t, t \rangle$

21. $\mathbf{F}(t) = t\mathbf{i} + 3\mathbf{j} - 5t\mathbf{k}$

22. $\mathbf{F}(t) = 2t^2\mathbf{i} + (1 - t)\mathbf{j} + 3\mathbf{k}$

23. $\mathbf{F}(t) = t\mathbf{i} + (t^2 + 1)\mathbf{j} + t^2\mathbf{k}$

24. $\mathbf{F}(t) = \left(\dfrac{\sqrt{2}}{2} \sin t\right)\mathbf{i} + \left(\dfrac{\sqrt{2}}{2} \sin t\right)\mathbf{j} + (\cos^2 t)\mathbf{k}$

25. $\mathbf{F}(t) = -t\mathbf{i} + t^2\mathbf{j} + 2t\mathbf{k}$

26. $\mathbf{F}(t) = 2t\mathbf{i} + (1 - t^2)\mathbf{j} + t\mathbf{k}$

27. Which of the following curves lies on the surface $z = x^2 + y^2$?

    a. $x = \sin t, y = \cos t, z = t$

    b. $x = e^{-t} \cos t, y = e^{-t} \sin t, z = e^{-2t}$

    c. $x = 1 + \dfrac{1}{t}, y = 1 - \dfrac{1}{t}, z = 2 + \dfrac{2}{t^2}$

28. Find parametric equations for the curve of intersection of the elliptical cylinder $x^2 + 3y^2 = 1$ and the parabolic cylinder $z = 2x^2 - 1$.

Find the indicated derivative in terms of $\mathbf{F}$ and $\mathbf{F}'$ in Problems 29-34.

29. $\dfrac{d}{dt}[\mathbf{F}(t) \cdot \mathbf{F}(t)]$

30. $\dfrac{d}{dt}[\|\mathbf{F}(t)\| \mathbf{F}(t)]$

31. $\dfrac{d}{dt}\left[\dfrac{\mathbf{F}(t)}{\|\mathbf{F}(t)\|}\right]$

32. $\dfrac{d}{dt}\|\mathbf{F}(t)\|$

33. $\dfrac{d}{dt}[\mathbf{F}(t) \times \mathbf{F}(t)]$

34. $\dfrac{d}{dt}\mathbf{F}(e^t)$

Evaluate the definite and indefinite vector integrals in Problems 35-40.

35. $\displaystyle\int_{-1}^{1} (e^{-t}\mathbf{i} + t^3\mathbf{j} + 3\mathbf{k})\,dt$

36. $\displaystyle\int_{1}^{2} [(1 - t)\mathbf{i} - t^{-1}\mathbf{j} + e^t\mathbf{k}]\,dt$

37. $\displaystyle\int [te^t\mathbf{i} - (\sin 2t)\mathbf{j} + t^2\mathbf{k}]\,dt$

38. $\displaystyle\int e^{2t}[2\mathbf{i} - t\mathbf{j} + (\sin t)\mathbf{k}]\,dt$

39. $\displaystyle\int t[e^t\mathbf{i} + (\ln t)\mathbf{j} + 3\mathbf{k}]\,dt$

40. $\displaystyle\int [e^t\mathbf{i} + 2\mathbf{j} - t\mathbf{k}] \cdot [e^{-t}\mathbf{i} - t\mathbf{j}]\,dt$

---

*The supplementary problems are presented in a somewhat random order, not necessarily in order of difficulty.

In Problems 41-44, **R** *is the position vector of a moving body. Find the velocity* **V**, *the speed ds /dt, and the acceleration* **A**.

41. $\mathbf{R}(t) = t\mathbf{i} + (3 - t)\mathbf{j} + 2\mathbf{k}$

42. $\mathbf{R}(t) = (\sin 2t)\mathbf{i} + 2\mathbf{j} - (\cos 2t)\mathbf{k}$

43. $\mathbf{R}(t) = \langle t \sin t, te^{-t}, -(1 - t) \rangle$

44. $\mathbf{R}(t) = \langle \ln t, e^t, - \tan t \rangle$

Find **T** *and* **N** *for the vector functions given in Problems* 45-48.

45. $\mathbf{R}(t) = t\mathbf{i} - t^2\mathbf{j}$

46. $\mathbf{R}(t) = (3 \cos t)\mathbf{i} - (3 \sin t)\mathbf{j}$

47. $\mathbf{R}(t) = \langle 4 \cos t, -3t, 4 \sin t \rangle$

48. $\mathbf{R}(t) = \langle e^t \sin t, e^t, e^t \cos t \rangle$

In Problems 49-52, find the tangential and normal components of acceleration, and the curvature of a moving object with position vector **R**(t).

49. $\mathbf{R}(t) = t^2\mathbf{i} + 2t\mathbf{j} + e^t\mathbf{k}$

50. $\mathbf{R}(t) = t^2\mathbf{i} - 2t\mathbf{j} + (t^2 - t)\mathbf{k}$

51. $\mathbf{R}(t) = (4 \sin t)\mathbf{i} + (4 \cos t)\mathbf{j} + 4t\mathbf{k}$

52. $\mathbf{R}(t) = (a \sin 3t)\mathbf{i} + (a + a \cos 3t)\mathbf{j} + (3a \sin t)\mathbf{k}$, for constant $a \neq 0$

A polar curve C is given by $r = f(\theta)$. *If* $f''(\theta)$ *exists (see Problem 48, Section 10.4), then the curvature can be found by the formula*

$$\kappa = \frac{|r^2 + 2r'^2 - rr''|}{(r^2 + r'^2)^{3/2}}$$

Find the curvature at the given point on each of the polar curves given in Problems 53-58.

53. $r = 4 \cos \theta$, where $\theta = \frac{\pi}{3}$

54. $r = \theta^2$, where $\theta = 2$

55. $r = e^{-\theta}$, where $\theta = 1$

56. $r = 1 + \cos \theta$, where $\theta = \frac{\pi}{2}$

57. $r = 4 \cos 3\theta$, where $\theta = \frac{\pi}{6}$

58. $r = 1 - 2 \sin \theta$, where $\theta = \frac{\pi}{4}$

In Problems 59-62, find $\mathbf{F}'(x)$.

59. $\mathbf{F}(x) = \displaystyle\int_1^x [t\mathbf{i} - t^{-1}\mathbf{j} + e^{-t}\mathbf{k}]\, dt$

60. $\mathbf{F}(x) = \displaystyle\int_1^{3x} [t^2\mathbf{i} + e^t\mathbf{j} + \sqrt{t}\mathbf{k}]\, dt$

61. $\mathbf{F}(x) = \displaystyle\int_1^x [(\sin t)\mathbf{i} - (\cos 2t)\mathbf{j} + e^{-t}\mathbf{k}]\, dt$

62. $\mathbf{F}(x) = \displaystyle\int_1^{2x} [t^2\mathbf{i} + (\sec e^{-t})\mathbf{j} - (\tan e^{2t})\mathbf{k}]\, dt$

In Problems 63-66, solve the initial value problems for **F** *as a vector function of t.*

63. differential equation: $\mathbf{F}'(t) = t\mathbf{i} + t\mathbf{j} - t\mathbf{k}$; initial condition: $\mathbf{F}(0) = \mathbf{i} + 2\mathbf{j} - 3\mathbf{k}$

64. differential equation: $\mathbf{F}'(t) = \frac{3}{2}\sqrt{t + 1}\,\mathbf{i} + (t + 1)^{-1}\mathbf{j} + e^t\mathbf{k}$; initial condition: $\mathbf{F}(0) = \mathbf{j} + 2\mathbf{k}$

65. differential equation: $\mathbf{F}'(t) = (\sin 2t)\mathbf{i} + (e^t \cos t)\mathbf{j} - \left(\dfrac{3}{t + 1}\right)\mathbf{k}$; initial condition: $\mathbf{F}(0) = \mathbf{i} - 3\mathbf{k}$

66. differential equation: $\dfrac{d^2\mathbf{F}}{dt^2} = -32\mathbf{j}$; initial conditions: $\mathbf{F} = 50\mathbf{j}$ and $\dfrac{d\mathbf{F}}{dt} = 5\mathbf{i} + 5\mathbf{k}$ at $t = 0$

67. For what values of t is the following vector function continuous?

$$\mathbf{F}(t) = (2t - 1)\mathbf{i} + \left(\frac{t^2 - 1}{t - 1}\right)\mathbf{j} + 4\mathbf{k}$$

68. Find the length of the graph of the vector function

$$\mathbf{R}(t) = \left(\frac{t^2 - 2}{2}\right)\mathbf{i} + \frac{(2t + 1)^{3/2}}{3}\mathbf{j}$$

from $t = 0$ to $t = 6$.

69. Find a vector function **F** such that $\mathbf{F}(0) = \mathbf{F}'(0) = \mathbf{i}$, $\mathbf{F}''(0) = 2\mathbf{j} + \mathbf{k}$, and

$$\mathbf{F}'''(t) = (\cos t)\mathbf{i} + (\sin t)\mathbf{j} + \frac{t}{\pi}\mathbf{k}$$

70. Find parametric equations for the tangent line to the graph of

$$\mathbf{R}(t) = te^{-t}\mathbf{i} + t^2\mathbf{j} + te^{-2t}\mathbf{k}$$

at the highest point on the graph (that is, where z has the largest value).

71. Let $\mathbf{R}(t) = t\mathbf{i} + t\mathbf{j} + t^3\mathbf{k}$.
   a. Find parametric equations for the tangent line to the graph of $\mathbf{R}$ at the point where $t = 1$.
   b. The tangent line in part a intersects the graph of $\mathbf{R}(t)$ in a second point. Find the coordinates of this point.

72. An object moves in space with acceleration $\mathbf{A}(t) = \langle -t, 2, 2 - t \rangle$. When $t = 0$, it is known that the object is at the point $(1, 0, 0)$ and that it has velocity $\mathbf{V}(0) = \langle 2, -4, 0 \rangle$.
   a. Find the velocity $\mathbf{V}(t)$ and the position vector $\mathbf{R}(t)$.
   b. What are the speed and location of the object when $t = 1$?
   c. When is the object stationary and what is its position at that time?

73. The acceleration of a moving particle is $\mathbf{A}(t) = 24t^2\mathbf{i} + 4\mathbf{j}$. Find the particle's position as a function of t if $\mathbf{R}(0) = \mathbf{i} + 2\mathbf{j}$ and $\mathbf{V}(0) = \mathbf{0}$.

74. Find the radius of curvature of the curve given by $y = 1 + \sin x$ at the points where x is
   a. $\frac{\pi}{6}$
   b. $\frac{\pi}{4}$
   c. $\frac{3\pi}{2}$

75. The position vector for a curve is given in terms of arc length s by $\mathbf{R}(s) = \langle a \cos \frac{s}{a}, a \sin \frac{s}{a}, 2s \rangle$ for $0 \le s \le 2\pi a$, $a \ne 0$. Find the unit tangent vector $\mathbf{T}(s)$ and the principal unit normal $\mathbf{N}(s)$.

76. **Modeling Problem:** A stunt pilot flying horizontally at an altitude of 4,000 ft with a speed of 180 mi/h drops a weighted marker, attempting to hit a target on the ground below, as shown in Figure 10.43. How far away from the target (measured horizontally) should the pilot be when she releases the marker? You may neglect air resistance.

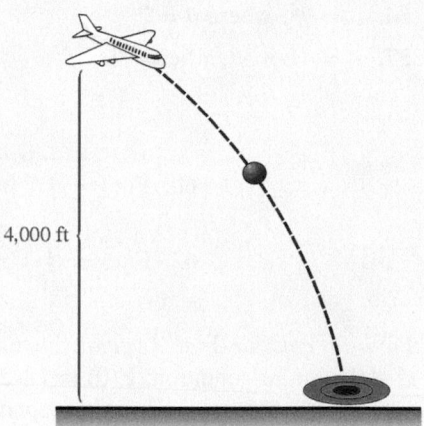

4,000 ft

Figure 10.43   Stunt pilot target

77. **Modeling Problem** A car weighing 3,000 lb travels at a constant speed of 60 mi/h on a flat road and then makes a circular turn on an interchange (see Figure 10.44). If the radius of the turn is 160 ft, what frictional force is needed to keep the car from skidding?

Figure 10.44   Highway interchange

78. Find the point or points on the curve $y = e^{ax}$ $(a > 0)$ where the curvature is maximized.

79. The position of an object moving in space is given by

$$\mathbf{R}(t) = (e^{-t} \cos t)\mathbf{i} + (e^{-t} \sin t)\mathbf{j} + e^{-t}\mathbf{k}$$

    a. Find the velocity, speed, and acceleration of the object at arbitrary time t.

    b. Determine the curvature of the trajectory at time t.

80. Verify that

$$\frac{d}{dt}[\mathbf{F} \cdot (\mathbf{G} \times \mathbf{H})] = \frac{d\mathbf{F}}{dt} \cdot (\mathbf{G} \times \mathbf{H}) + \mathbf{F} \cdot \left(\frac{d\mathbf{G}}{dt} \times \mathbf{H}\right) + \mathbf{F} \cdot \left(\mathbf{G} \times \frac{d\mathbf{H}}{dt}\right)$$

81. Verify that

$$\frac{d}{dt}[\mathbf{F} \times (\mathbf{G} \times \mathbf{H})]' = [(\mathbf{H} \cdot \mathbf{F})\mathbf{G}]' - [(\mathbf{G} \cdot \mathbf{F})\mathbf{H}]'$$

82. Find the radius of curvature of the ellipse given by $\mathbf{R}(t) = \langle a \cos t, b \sin t \rangle$, where $a > 0$, $b > 0$, $a \neq b$, and $0 \leq t \leq 2\pi$ at the points where $t = 0$ and $\pi/2$.

83. A particle moves along a path given in parametric form, where $r(t) = 1 + \cos at$ and $\theta(t) = e^{-at}$ (for positive constant a). Find the velocity and acceleration of the particle in terms of the unit polar vectors $\mathbf{u}_r$ and $\mathbf{u}_\theta$.

84. The position vector of an object in space is

$$\mathbf{R}(t) = (a \cos \omega t)\mathbf{i} + (a \sin \omega t)\mathbf{j} + \omega^2 t\mathbf{k}$$

Find $\omega \neq 0$ so that the sum of the object's tangential and normal components of acceleration equal half its speed.

85. Sketch the graph of the vector function

$$\mathbf{R}(t) = \left(\frac{3t}{1 + t^3}\right)\mathbf{i} + \left(\frac{3t^2}{1 + t^3}\right)\mathbf{j}$$

then find parametric equations for the tangent line at the point where $t = 2$. This curve is called the *folium of Descartes*.

86. Modeling Problem A fireman stands 5.5 m from the front of a burning building 15 m high. His fire hose discharges water at a speed of v_0 m/s from a height of 1.2 m at an angle of 62°, as shown in Figure 10.45.

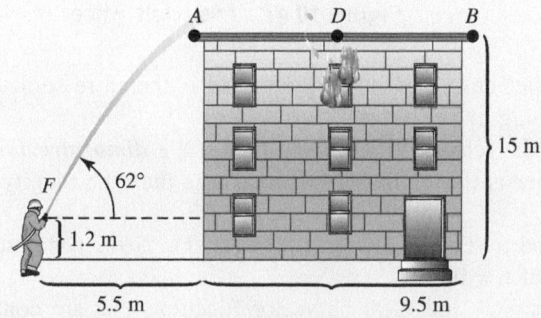

Figure 10.45 Firefighter's problems

    a. What is the smallest value of v_0 that will allow water from the hose to reach a window 11 m above the ground?

    b. Suppose $v_0 = 50$ m/s. At what angle should the hose be held so the water can be directed onto a fire on the roof of the building (at a height of 15 m)?

87. A DNA molecule has the shape of a double helix (see Figure 10.46). The radius of each helix is about 10^{-8} μm. Each helix rises about 3×10^{-8} μm during each complete turn and there are about 3×10^8 complete turns. Estimate the length of each helix.

88. Show that the tangential component of acceleration of a moving object is 0 if the object has constant speed. Is the converse statement also true? That is, if $A_T = 0$, can we conclude that the speed is constant?

89. The path of a particle P is an Archimedean spiral. The motion of the particle is described by the polar coordinates $r = 10t$ and $\theta = 2\pi t$, where r is expressed in inches and t is in seconds. Determine the velocity of the particle (in terms of $\mathbf{u}_r$ and $\mathbf{u}_\theta$) when
   a. $t = 0$ sec   **b.** $t = 0.25$ sec

90. *Historical Quest The three laws of Kepler described in this section forever changed the way we view the universe, but it was not Kepler who correctly proved these laws. Isaac Newton (Chapter 1, Supplementary Problem 49) proved these laws from the inverse-square law of gravitation. In Section 10.3 we proved Kepler's second law. Write a 500-word essay on the life and mathematics of Johannes Kepler, and include, as part of your essay, where Kepler made his mistake in his proof of his second law.*

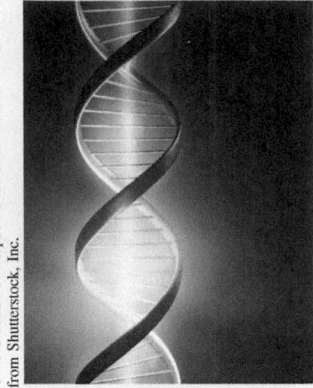

Figure 10.46   Double helix DNA molecule

91. *Historical Quest*
    Write a 500-word essay on the life and mathematics of Galileo Galilei.

92. *Historical Quest In 1638, Galileo published his ideas about dynamics in his book **Discorsie dimostrazioni matematiche intorno a due nuove scienze**. In this book, he considers the following problem:* Suppose the larger circle of Figure 10.47 has made one revolution in rolling along the straight line from A to B, so that $|\overline{AB}|$ is equal to the circumference of the large circle. Then the small circle, fixed to the large one, has also made one revolution, so that $|\overline{CD}|$ is equal to the circumference of the small circle. It follows that the *two circles have equal circumferences.*

Galileo Galilei
(1564-1642)

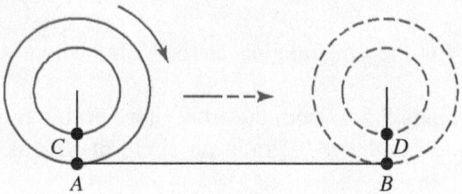

Figure 10.47   Aristotle's wheel

This paradox had been earlier described by Aristotle and is therefore sometimes referred to as *Aristotle's wheel*. Can you explain what is going on?

93. *Historical Quest* Explain the remark in Galileo's *Discorsi e dimonstrazioni matematiche intorno a due nuove scienze* of 1638 that "neither is the number of squares less than the totality of all numbers, nor the latter greater than the former."

94. **Think Tank Problem** Find a vector function $\mathbf{F}(t) = \langle f_1(t), f_2(t), f_3(t) \rangle$ such that $\|\mathbf{F}(t)\|$ is continuous at $t = 0$ but $\mathbf{F}(t)$ is not continuous at $t = 0$.

95. **Think Tank Problem** Let $\mathbf{F}(t)$ and $\mathbf{G}(t)$ be vector functions that are continuous at t_0, and let $h(t)$ be a scalar function continuous at t_0. In each of the following cases, either show that the given function is continuous at t_0 or provide a counterexample to show that it is not continuous.
   a. $3\mathbf{F}(t) + 5\mathbf{G}(t)$ **b.** $\mathbf{F}(t) \cdot \mathbf{G}(t)$ **c.** $(h(t))^{-1}\mathbf{F}(t)$ **d.** $\mathbf{F}(t) \times \mathbf{G}(t)$

96. **Think Tank Problem** Let f be a function that is twice differentiable. Either prove that the largest curvature of the curve $y = f(x)$ occurs at a relative extremum of f or find a counterexample.

97. Putnam Examination Problem A shell strikes an airplane flying at a height h above the ground. It is known that the shell was fired from a gun on the ground with muzzle speed v_0, but the position of the gun and its angle of elevation are both unknown. Deduce that the gun is situated within a circle whose center lies directly below the airplane and whose radius is

$$\frac{v_0}{g}\sqrt{v_0^2 - 2gh}$$

Neglect resistance of the atmosphere.

98. Putnam Examination Problem A coast artillery gun can fire at any angle of elevation between $0°$ and $90°$ in a fixed vertical plane. If air resistance is neglected and the muzzle speed is constant $(v = v_0)$, determine the set H of points in the plane that can be hit. Consider only those points above the horizontal.

99. Putnam Examination Problem A particle moves on a circle with center O, starting from rest at a point P and coming to rest again at a point Q, without coming to rest at any intermediate point. Prove that the acceleration vector of the particle does not vanish at any point between P and Q, and that, at some point R between P and Q, the acceleration vector points in along the radius $\overline{RO}$.

CHAPTER 10 GROUP RESEARCH PROJECT*

Working in small groups is typical of most work environments, and this book seeks to develop skills with group activities. We present a group project at the end of each chapter. These projects are to be done in groups of three or four students.

Saint Louis Arch Problem

© 2013 Rudy Balasko. Used under license from Shutterstock, Inc.

A mathematical problem appeared in the *St. Louis Post-Dispatch* on July 4, 1982. The dimensions of the St. Louis Arch (as given in this article) are:

height (outside surface)	630 ft
height (inside surface)	615 ft
width (outside surface)	630 ft
width (inside surface)	530 ft

Legend tells us that *The Spirit of St. Louis* airplane was flown through the arch 200 ft above the base. The problem in the newspaper asked readers to find the distance from the wingtip to the inside edge of the arch.

Here are some hints and suggestions to be included in your paper.

- Assume the airplane has a wingspan of 46.00 ft.
- Assume the inside edge of the Arch is a catenary curve.
- Some solvers (incorrectly) assumed that the curve of the inner edge is a parabola. If you make this assumption, show that the answer is 194.69 ft.
- Many of the newspaper readers took the curve of the inner edge of the arch to be an inverted catenary. Show that the answer found using this equation is 208.81 ft.
- Show that the curve of the arch's inner edge is not a simple inverted catenary.
- Show that the answer you find is unique; that is, show that there is only one solution.
- Wikipedia's article on the catenary says that Galileo mistakenly confused a catenary and a parabola. Address this issue.

---

*This project is adapted from *The St. Louis Arch Problem* by William V. Thayer, *UMAP Module* 638, COMAP, Inc., © 1984., pp. 447-464.

Cumulative Review—Chapters 1-10 (Single variable calculus)

Calculus is sometimes divided into *single variable calculus* and *multivariable calculus*. The end of this chapter marks the boundary between these major divisions of calculus, and as such it is a good time to take a look back at what we have done. There are three main ideas in single variable calculus. The first is the notion of *limit*, which was introduced in Chapter 2. In Chapter 3, we introduced the idea of a *derivative*, the second major topic of calculus. In Chapter 5, we introduced one of the most important and revolutionary ideas in the history of mathematics—namely, that of a Riemann sum, and *integration*.

Definition of limit   The limit statement

$$\lim_{x \to c} f(x) = L$$

means that for each $\epsilon > 0$, there corresponds a number $\delta > 0$ with the property that

$$|f(x) - L| < \epsilon \qquad \text{whenever} \qquad 0 < |x - c| < \delta$$

The notion of a limit is fundamental for many of the ideas of calculus, primarily in the definitions of both the derivative and integral.

Definition of derivative   The derivative of f at x is given by

$$f'(x) = \lim_{\Delta x \to 0} \frac{f(x + \Delta x) - f(x)}{\Delta x}$$

provided this limit exists. If this limit exists the function f is said to be differentiable.

There are many applications of the derivative, including related rates, curve sketching, and optimization.

Definition of integral   If f is defined on the closed interval $[a, b]$, we say f is integrable on $[a, b]$ if

$$I = \lim_{\|P\| \to 0} \sum_{k=1}^{n} f(x_k^*) \Delta x_k$$

exists, where $\|P\| = \max\{\Delta x_k\}$, $k = 1, 2, \cdots n$. This limit is called the definite integral of f from a to b. The definite integral is denoted by

$$I = \int_a^b f(x)\, dx$$

Table 10.3 summarizes the relationship between modeling and the Riemann sum.

Table 10.3 Modeling Formulas Involving 813 Integration

Section	Situation	Riemann Sum Model	Outcome
5.3	**Introduction** Model a quantity in terms of one or more continuous functions on a closed interval $[a, b]$.	1. Partition interval 2. Choose a number x_k^* arbitrarily 3. Form a sum of the type $$R_n = \sum_{k=1}^{n} f(x_k^*)\Delta x_k$$	The approximations improve as the norm of the partitions goes to 0. The Riemann sums approach a limiting integral. Use this integral to define and calculate what we originally wanted to measure.
5.1	**Area** Find the area under a curve defined by $y = f(x)$, where $f(x) \geq 0$ above the x-axis and bounded by the lines $x = a$ and $x = b$.	Find the sum of n rectangles, $$\sum_{k=1}^{n} \underbrace{f(x_k^*)\Delta x_k}_{\text{Area of } k\text{th rectangle}}$$	$\text{Area} = \int_a^b f(x)\,dx$
5.3	**Displacement and distance traveled** Find the displacement and distance traveled for an object that is moving along a line having continuous velocity $v(t)$ for each t between $t = a$ and $t = b$.	Find the sum of n distances $$\sum_{k=1}^{n} \underbrace{\lvert v(a + (k-1)\Delta t \rvert \Delta t}_{\substack{\text{Distance traveled over} \\ k\text{th time interval.}}}$$	$\text{Distance} = \displaystyle\int_a^b \lvert v(t) \rvert\, dt$ $\text{Displacement} = \displaystyle\int_a^b v(t)\,dt$ $= s(b) - s(a)$
5.7	**Average value** Find the average value of a continuous function on the interval $[a, b]$.	Find the average of a weighted sum of n values. Let $\Delta x = \dfrac{b-a}{n}$, then $$\frac{1}{b-a}\sum_{k=1}^{n} f(x_k^*)\Delta x_k$$	$\text{Average value} = \dfrac{1}{b-a}\displaystyle\int_a^b f(x)\,dx$
6.1	**Area between curves** Find the area between the curves $y = f(x)$ and $y = g(x)$.	$$\sum_{k=1}^{n} \underbrace{\lvert f(x_k^*) - g(x_k^*) \rvert \Delta x_k}_{\text{Area between curves}}$$	$\text{Area} = \displaystyle\int_a^b \lvert f(x) - g(x) \rvert\, dx$
6.2	**Volume by cross section** Find the volume of a solid whose cross section perpendicular to the x-axis has area $A(x)$.	$$\sum_{k=1}^{n} A(x_k^*)\Delta x_k$$	$\text{Volume} = \displaystyle\int_a^b A(x)\,dx$
6.2	**Volume of revolution** Find the volume of the region bounded by $y = f(x)$ and $y = g(x)$ revolved about the specified axis.		
	Disks: x-axis	$$\sum_{k=1}^{n} \pi [f(x_k^*)]^2 \Delta x_k$$	$\text{Volume} = \pi \displaystyle\int_a^b [f(x)]^2\, dx$
	Washers: x-axis	$$\sum_{k=1}^{n} \pi \left([f(x_k^*)]^2 - [g(x_k^*)]^2\right) \Delta x_k$$	$\text{Volume} = \pi \displaystyle\int_a^b \left([f(x)]^2 - [g(x)]^2\right)\, dx$
	Shells: y-axis	$$\sum_{k=1}^{n} 2\pi x_k^* f(x_k^*)\Delta x_k$$	$\text{Volume} = 2\pi \displaystyle\int_a^b x f(x)\, dx$
6.4	**Arc length** Find the length of a curve $y = f(x)$ on $[a, b]$.	$$\sum_{k=1}^{n} \sqrt{1 + [f'(x_k^*)]^2}\,\Delta x_k$$	$\text{Volume} = \pi \displaystyle\int_a^b \sqrt{1 + [f'(x)]^2}\, dx$
6.4	**Area of a surface of revolution** Find the area of a surface formed by rotating the curve $y = f(x)$ about the x-axis over $[a, b]$.	$$\sum_{k=1}^{n} 2\pi f(x_k^*)\Delta x_k$$	$\text{Area} = 2\pi \displaystyle\int_a^b f(x)\sqrt{1 + [f'(x)]^2}\, dx$
6.5	**Mass** Find the mass of a thin plate with constant density ρ, bounded by $y = f(x)$ and $y = g(x)$.	$$\sum_{k=1}^{n} \rho [f(x_k^*) - g(x_k^*)]\Delta x_k$$	$m = \rho \displaystyle\int_a^b [f(x) - g(x)]\, dx$

Continued.

Table 10.3 (Continued)

Section	Situation	Riemann Sum Model	Outcome
6.5	**First moment** Find the moment of a thin plate with constant density ρ bounded by $y = f(x)$ and $y = g(x)$ about the y-axis.	First moment about the y-axis $$\sum_{k=1}^{n} \rho x_k^* [f(x_k^*) - g(x_k^*)] \Delta x_k$$ First moment about the x-axis $$\sum_{k=1}^{n} \frac{\rho}{2} \left\{ [f(x_k^*)]^2 - [g(x_k^*)]^2 \right\} \Delta x_k$$	$$M_y = \rho \int_a^b x[f(x) - g(x)]\, dx$$ $$M_x = \frac{\rho}{2} \int_a^b \left\{ [f(x)]^2 - [g(x)]^2 \right\} dx$$
	Centroid Find the centroid of a homogeneous lamina. (ρ constant.)		$$(\bar{x}, \bar{y}) = \left(\frac{M_y}{m}, \frac{M_x}{m} \right)$$
6.5	**Work** Find the work done when an object moves along the x-axis from a to b when a variable force $F(x)$ is applied.	$$\sum_{k=1}^{n} F(x_k^*) \Delta x_k$$	$$\text{Work} = \int_a^b F(x)\, dx$$
6.5	**Hydrostatic force** Find the total force of a fluid against one side of a vertical plate, where h is the depth.	$$\sum_{k=1}^{n} \delta(h_k^*) L(h_k^*) \Delta h_k$$	$$F = \int_a^b \delta h L(h)\, dh$$

In the last two chapters we considered vectors in $\mathbb{R}^2$ and $\mathbb{R}^3$, as well as vector-valued functions. We summarize these ideas in Table 10.4.

Table 10.4 Summary of Velocity, Acceleration, and Curvature

Topic	Formula
Position vector	$\mathbb{R}^2$ (plane) $\quad \mathbf{R}(t) = x(t)\mathbf{i} + y(t)\mathbf{j}$ $\mathbb{R}^3$ (space) $\quad \mathbf{R}(t) = x(t)\mathbf{i} + y(t)\mathbf{j} + z(t)\mathbf{k}$ The graph is called the **trajectory** of the object's motion.
Velocity vector	$\mathbf{V}(t) = \dfrac{d\mathbf{R}}{dt} = \mathbf{R}'(t)$ $\|\mathbf{V}\| = \dfrac{ds}{dt}$ is called **speed.**
Unit vector	$\dfrac{\mathbf{V}}{\|\mathbf{V}\|}$ is the unit vector in the direction of $\mathbf{V}$ or *direction of motion.*
Unit tangent vector	$\mathbf{T} = \dfrac{\mathbf{V}}{\|\mathbf{V}\|} = \dfrac{\mathbf{R}'(t)}{\|\mathbf{R}'(t)\|} = \dfrac{d\mathbf{R}}{ds}$ is the **unit tangent** vector in the direction of motion.
Principal unit normal vector	$\mathbf{N} = \dfrac{\mathbf{T}'(t)}{\|\mathbf{T}'(t)\|} = \dfrac{1}{\kappa} \dfrac{d\mathbf{T}}{ds}$ is the **principal unit normal vector**.
Acceleration vector	$\mathbf{A}(t) = \dfrac{d^2\mathbf{R}}{dt^2} = \dfrac{d\mathbf{V}}{dt} = A_T \mathbf{T} + A_N \mathbf{N}$ where $A_T = \dfrac{d^2 s}{dt^2} = \dfrac{\mathbf{R}' \cdot \mathbf{R}''}{\|\mathbf{R}'\|}$ This is the **tangential component.** $A_N = \kappa \left(\dfrac{ds}{dt} \right)^2 = \dfrac{\|\mathbf{R}' \times \mathbf{R}''\|}{\|\mathbf{R}'\|} = \sqrt{\|\mathbf{A}\|^2 - A_T^2}$ This is the **normal component.**
Curvature	$\kappa = \left\| \dfrac{d\mathbf{T}}{ds} \right\| = \dfrac{\|\mathbf{R}' \times \mathbf{R}''\|}{\|\mathbf{R}'\|^3}$; if $y = f(x)$ then $\kappa = \dfrac{|f''(x)|}{\{1 + [f'(x)]^2\}^{3/2}}$
Torsion	$\tau = \left\| \dfrac{d\mathbf{B}}{ds} \right\|$

Cumulative Review Problems—Chapters 1-10

1. ■ What does this say? Outline a procedure for integrating a given function.
2. ■ What does this say? Outline a procedure for deciding whether a given series converges or diverges.
3. ■ What does this say? What is a vector? What is a vector function? In your own words, discuss what is meant by vector calculus.

Evaluate the limits in Problems 4-9, or explain why the limit does not exist.

4. $\displaystyle\lim_{x\to 1}\frac{7x^3 - 5x - 2}{3x^2 + 5x - 8}$

5. $\displaystyle\lim_{x\to\infty}\frac{(2x^2 - 3x + 1)(1 - 5x)}{5x^3 + 4x - 9}$

6. $\displaystyle\lim_{x\to 0}\frac{\sin 3x}{\tan 2x}$

7. $\displaystyle\lim_{x\to\infty}\left(1 - \frac{2}{x}\right)^{3x}$

8. $\displaystyle\lim_{x\to 0^+}\left(\frac{2x + 3}{x} - \csc x\right)$

9. $\displaystyle\lim_{x\to 0}\frac{\cos x - 1}{e^x - x - 1}$

Find the derivatives in Problems 10-15.

10. $y = 3x^4 - 7\sqrt{2x} + 4$

11. $y = \sin^3 x + 2\tan x$

12. $y = \ln\sqrt[3]{x} + xe^{-2x}$

13. $y = \sin^{-1} x + 2\tan^{-1}\dfrac{1}{x}$

14. $y = \dfrac{e^{-2x}(1 - x^2)}{\sqrt[3]{x^2 - x^3 + 3x^4}}$

15. $xy^3 + x^2 e^{-y} = 4$

Find the integrals in Problems 16-21.

16. $\displaystyle\int x\ln\sqrt[3]{x}\,dx$

17. $\displaystyle\int \sin^2 x\cos^3 x\,dx$

18. $\displaystyle\int x\sqrt{16 - x}\,dx$

19. $\displaystyle\int \frac{dx}{1 + \cos x}$

20. $\displaystyle\int \frac{dx}{x^2(x^2 + 5)}$

21. $\displaystyle\int \frac{dx}{\sqrt{2x - x^2}}$

22. A student graphs the function $f(x) = 2x^3 + 51x^2 - 360x$ using a calculator, and the graph is shown in Figure 10.48. Is this graph sufficient to get a idea of what the function looks like? Why or why not?

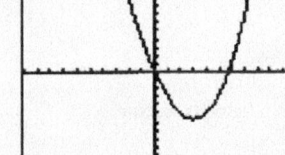

23. Let f be a continuously differentiable function with $f(0) = 0$ and $-2 \le f'(x) \le 4$ for $-3 \le x \le 3$.
    a. Can $f(2)$ be negative? Give a reason for your answer.
    b. Can $f(-2)$ be negative? Give a reason for your answer.
    c. Must f have a critical point between -3 and 3? Give a reason for your answer.

Figure 10.48 Calculator graph

24. Find the area bounded by the curve $y = x\sqrt{x^2 + 8}$ and the x-axis on $[-2, 1]$.
25. Find the area bounded by the curves $y = \sin x$ and $y = \cos x$ and the lines $x = 0$ and $x = 1$.
26. Find the area of the region inside the circle $r = 4$ and to the right of the line $r = 2\sec\theta$.
27. Find the volume of the solid formed by revolving about the x-axis the region bounded by the curve

$$y = \frac{2}{\sqrt{3x - 2}}$$

between $x = 1$ and $x = 2$.

28. Find the equation of the line through $(-1, 2, 5)$ and perpendicular to the plane $2x - 3y + z = 11$ in symmetric form.
29. Find the equation of the plane satisfying the given conditions.
    a. passing through $(5, 1, 2)$, $(3, 1, -2)$, and $(3, 2, 5)$
    b. passing through $(-2, -1, 4)$ and perpendicular to the line $\dfrac{x}{2} = \dfrac{y + 1}{5} = \dfrac{3 - z}{2}$

Test the series in Problems 30-35 for convergence.

30. $\displaystyle\sum_{k=0}^{\infty} \frac{k^3}{k^4+2}$

31. $\displaystyle\sum_{k=1}^{\infty} \frac{1}{k4^k}$

32. $\displaystyle\sum_{k=0}^{\infty} \frac{3k^2-7k+2}{(2k-1)(k+3)}$

33. $\displaystyle\sum_{k=1}^{\infty} \frac{k!}{2^k k}$

34. $\displaystyle\sum_{k=0}^{\infty} \frac{(-1)^{k+1}k}{k^2+k-1}$

35. $1 + \frac{1}{8} - \frac{1}{27} - \frac{1}{64} + \frac{1}{125} + \frac{1}{216} - - + + \cdots$

36. In each case, find the sum of the convergent series

a. $\displaystyle\sum_{k=1}^{\infty} \frac{2^{k-1}}{5^{k+3}}$

b. $\displaystyle\sum_{k=1}^{\infty} \frac{1}{(3k-1)(3k+2)}$

37. In each case, determine whether the given improper integral converges, and if it does, find its value.

a. $\displaystyle\int_{1}^{\infty} x^2 e^{-x}\, dx$

b. $\displaystyle\int_{0}^{2} \frac{dx}{\sqrt{4-x^2}}$

38. Find $\mathbf{v} \cdot \mathbf{w}$ and $\mathbf{v} \times \mathbf{w}$ for the vectors $\mathbf{v} = 3\mathbf{i} - 2\mathbf{j} + 5\mathbf{k}$ and $\mathbf{w} = \mathbf{i} - 3\mathbf{j} - \mathbf{k}$.

39. Find $\mathbf{F}'$ and $\mathbf{F}''$ for $\mathbf{F}(t) = 2t\mathbf{i} + e^{-3t}\mathbf{j} + t^4\mathbf{k}$.

40. Find $\displaystyle\int [e^t\mathbf{i} - \mathbf{j} - t\mathbf{k}] \cdot [e^{-t}\mathbf{i} + t\mathbf{j} - \mathbf{k}]dt$.

41. Find $\mathbf{T}$ and $\mathbf{N}$ for $\mathbf{R}(t) = 2(\sin 2t)\mathbf{i} + (2 + 2\cos 2t)\mathbf{j} + 6t\mathbf{k}$.

42. Show that the alternating series

$$S = \sum_{k=1}^{\infty} \frac{(-1)^{k+1}}{\sqrt{k}}$$

converges and determine what N must be to guarantee that the partial sum

$$S_N = \sum_{k=1}^{N} \frac{(-1)^{k+1}}{\sqrt{k}}$$

approximates the entire sum S with four-decimal-place accuracy.

43. a. Find the Maclaurin series representation for $f(x) = x^2 e^{-x^2}$.

b. Use the representation in **a** to approximate

$$\int_{0}^{1} x^2 e^{-x^2}\, dx$$

with three-decimal-place accuracy.

44. Let $P(x) = 7 - 3(x-4) + 5(x-4)^2 - 2(x-4)^3 + 6(x-4)^4$ be the fourth-degree Taylor polynomial for the function f about 4.

a. Find $f(4)$ and $f'''(4)$.

b. Use the third-degree Taylor polynomial for f' about $x = 4$ to approximate $f'(4.2)$.

c. Write a fifth-degree Taylor polynomial for $F(x) = \int_{4}^{x} f(t)dt$ about 4.

45. Solve the first-order linear differential equation

$$\frac{dy}{dx} + 2y = x^2$$

subject to the condition $y = 2$ when $x = 0$.

46. During the time period from $t = 0$ to $t = 6$ seconds, a particle moves along the path given by
    $x(t) = 3\cos(\pi t)$, $y(t) = 5\sin(\pi t)$.
    a. What is the position of the particle when $t = 1.5$?
    b. Graph the path of the particle from $t = 0$ to $t = 6$. Show direction.
    c. How many times does the particle pass through the point found in part a?
    d. Find the velocity vector for the particle at any time t.
    e. What is the distance traveled (to the nearest hundredth) by the particle from $t = 1.25$ to $t = 1.75$.

47. Find an equation in terms of x and y for the line tangent to the curve given by

$$x = t^2 - 2t - 1, \; y = t^4 - 4t^2 + 2$$

    at the point where $t = 1$.

48. A particle moves in space with position vector

$$\mathbf{R}(t) = (\sin t)\mathbf{i} - (\cos\ t)\mathbf{j} + \mathbf{k}$$

    a. Find the velocity and acceleration vectors for the particle and find its speed.
    b. Find the Cartesian equation for the particle's trajectory.
    c. Find the curvature κ and the tangential and normal components of acceleration for the particle's motion.

49. A mason lifts a 25-lb bucket of mortar from ground level to the top of a 50-ft building using a rope that weighs 0.25 lb/ft. Find the work done in lifting the bucket.

50. **Modeling Problem** At what speed must a satellite travel to maintain a circular orbit 1,000 mi above the surface of the earth? Assume the earth is a sphere of radius 4,000 mi and that GM in Newton's law of universal gravitation is approximately 9.56×10^4 mi^3/s^2.

51. Find equations for the tangent and normal lines to the graph of $y = \left(2x + \dfrac{1}{x}\right)^3$ at the point $(1, 27)$.

52. Suppose f is a differentiable function whose derivative satisfies

$$f'(x) = 2x^2 + 3$$

    Find $\dfrac{d}{dx}f(x^3 - 1)$.

53. Suppose f is a differentiable function such that $f'(x) = x^2 + x$. Find $\dfrac{d}{dx}f(x^2 + x)$.

54. Solve the differential equation $\dfrac{dy}{dx} = y^2 \sin 3x$.

55. A spherical balloon is being filled with air in such a way that its radius is increasing at a constant rate of 2 cm/s. At what rate is the volume of the balloon increasing at the instant when its surface has area 4π cm^2?

56. Let f be the function defined by $f(x) = \sqrt{x - 2}$.
    a. Sketch f and shade the region R enclosed by the graph of f, the x-axis, and the vertical line $x = 5$.
    b. Find the area of R.

57. A pebble is thrown from a hot air balloon directly downward with velocity 5 ft/s, and t seconds later is falling with acceleration $a(t) = 32 - 0.08v$ where $v(t)$ is the object's velocity at time t.
    a. Find an expression for v in terms of t.
    b. If the terminal velocity is defined as $\lim\limits_{t\to\infty} v(t)$, find the terminal velocity of the object (to the nearest ft/s).

58. A certain artifact is tested by carbon dating and found to contain 73% of its original carbon-14. As a cross-check, it is also dated using radium, and is found to contain 32% of the original amount. Assuming the dating procedures are accurate, what is the half-life of radium?

59. *Historical Quest* The derivative is one of the great ideas of calculus. Write an essay of at least 500 words about some application of the derivative that is not discussed in this text.

60. *Historical Quest* The concept of the integral is one of the great ideas of calculus. Write an essay of at least 500 words about the relationship of integration and differentiation as it relates to the history of calculus.

CHAPTER 11

PARTIAL DIFFERENTIATION

T*he book of nature is written in the language of mathematics.**

Christoph Clavis,
16th Century

PREVIEW

This chapter extends the methods of single-variable differential calculus to functions of two or more independent variables. You will learn how to take derivatives of such functions and to interpret those derivatives as slopes and rates of change. You will also learn a more general version of the chain rule and extended procedures for optimizing a function. The vector methods developed in Chapters 9 and 10 play an important role in this chapter, and indeed, we will find that the closest analogue of the single-variable derivative is a certain vector function called the *gradient*. In physics, the gradient is the rate at which a variable quantity, such as temperature or pressure, changes in value. In this chapter, we will define the *gradient of a function*, which is a vector whose components along the axes are related to the rate at which the function changes in the direction of the given component.

PERSPECTIVE

In many practical situations, the value of one quantity depends on the values of two or more others. For example, the amount of water in a reservoir depends on the amount of rainfall and on the amount of water consumed by local residents and thus may be regarded as a function of two independent variables. The current in an electrical circuit is a function of four variables: the electromotive force, the capacitance, the resistance, and the inductance. We will analyze a variety of models using techniques and tools that will be developed in this chapter.

*This quotation is usually attributed to Galileo, but according to the historian Rebekah Higgitt it was probably written by Christoph Clavis, one of Galileo's friends, supporters, and teachers.

11.1 FUNCTIONS OF SEVERAL VARIABLES

> **IN THIS SECTION:** *Basic concepts, level curves and surfaces, graphs of functions of two variables*
> The basic terminology of functions of several variables is introduced, along with the notion of level curves and surfaces. We also examine various methods for sketching graphs of functions of two variables.

Basic Concepts

Physical quantities often depend on two or more variables (see the Perspective at the beginning of this chapter). For example, we might be concerned with the temperature, T, at various points (x, y) on a metal plate. In this situation, T might be regarded as a function of the two location variables, x and y. Extending the notation for a function of a single variable, we can denote this relationship by $T(x, y)$.

> **FUNCTION OF TWO VARIABLES**   A **function of two variables** is a rule f that assigns to each ordered pair (x, y) in a set D a unique number $f(x, y)$. The set D is called the **domain** of the function, and the corresponding values of $f(x, y)$ constitute the **range** of f.

Functions of three or more variables can be defined in a similar fashion. For example, the temperature in our introductory example may vary not only on the plate, but also with time t, in which case it would be denoted by $T(x, y, t)$. Occasionally, we will examine functions of four or more variables, but for simplicity, we will focus most of our attention on functions of two or three variables.

When dealing with a function f of two variables, we may write $z = f(x, y)$ and refer to x and y as the **independent variables** and to z as the **dependent variable**. Often, the functional "rule" will be given as a formula, and unless otherwise stated, *we will assume that the domain is the largest set of points in the plane (or in $\mathbb{R}^3$) for which the functional formula is defined and real-valued.* These definitions and conventions are illustrated in Example 1 for a function of two variables.

Example 1   Domain, range, and evaluating a function of two variables

Let $f(x, y) = \sqrt{9 - x^2 - 4y^2}$.

a. Evaluate $f(2, 1)$ and $f(2t, t^2)$.   **b.** Describe the domain and range of f.

Solution

a.
$$f(2, 1) = \sqrt{9 - 2^2 - 4(1)^2} \qquad f(2t, t^2) = \sqrt{9 - (2t)^2 - 4\left(t^2\right)^2}$$
$$= \sqrt{9 - 4 - 4} \qquad\qquad\qquad = \sqrt{9 - 4t^2 - 4t^4}$$
$$= 1$$

b. The domain of f is the set of all ordered pairs (x, y) for which $\sqrt{9 - x^2 - 4y^2}$ is defined. We must have $9 - x^2 - 4y^2 \geq 0$, or, equivalently, $x^2 + 4y^2 \leq 9$, in order for the square root to be defined. Thus, the domain of f is the set of all points (x, y) inside or on the ellipse $x^2 + 4y^2 = 9$, as shown in Figure 11.1. The range of f is the set of all numbers $z = \sqrt{9 - x^2 - 4y^2}$ for (x, y) in the domain $x^2 + 4y^2 \leq 9$. Thus, the range is the interval $0 \leq z \leq 3$.

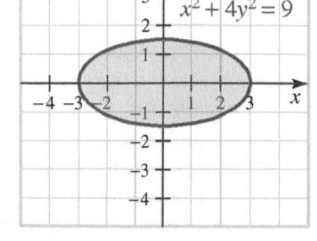

Figure 11.1   The domain for the given function

Functions of several variables can be combined in much the same way as functions of a single variable, as noted in the following definition.

OPERATIONS WITH FUNCTIONS OF TWO VARIABLES   If $f(x,y)$ and $g(x,y)$ are functions of two variables with domain D, then

Sum $(f+g)(x,y) = f(x,y) + g(x,y)$

Difference $(f-g)(x,y) = f(x,y) - g(x,y)$

Product $(fg)(x,y) = f(x,y)g(x,y)$

Quotient $\left(\dfrac{f}{g}\right)(x,y) = \dfrac{f(x,y)}{g(x,y)}$ $g(x,y) \neq 0$

A **polynomial function in x and y** is a sum of functions of the form

$$Cx^m y^n$$

with nonnegative integers m and n and C a constant; for example,

$$3x^5 y^3 - 7x^2 y + 2x - 3y + 11$$

is a polynomial in x and y. A **rational function** is a quotient of a polynomial function divided by a nonzero polynomial function. Similar notation and terminology apply to functions of three or more variables.

Level Curves and Surfaces

By analogy with the single-variable case, we define the **graph of the function $f(x,y)$** to be the collection of all 3-tuples—ordered triples—(x,y,z) such that (x,y) is in the domain of f and $z = f(x,y)$. The graph of $f(x,y)$ is a surface in $\mathbb{R}^3$ whose projection onto the xy-plane is the domain D.

It is usually not easy to sketch the graph of a function of two variables without the assistance of technology.* One way to proceed is illustrated in Figure 11.2. When we graphed quadric surfaces in Section 9.7 we used the *trace* of a graph in a plane, and we use that idea here.

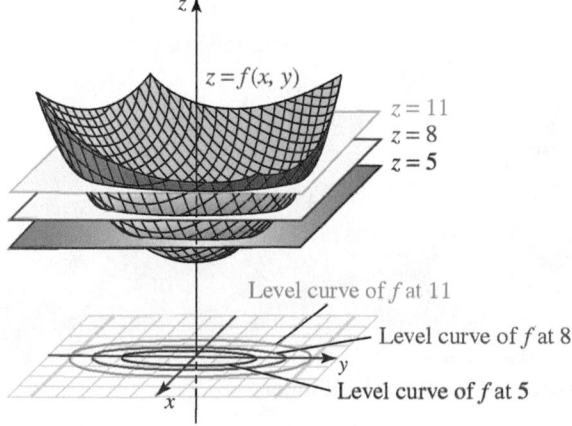

Figure 11.2   Graph of a function of two variables

Notice that when the plane $z = C$ intersects the surface $z = f(x,y)$, the result is the trace, the equation $f(x,y) = C$. The set of points (x,y) in the xy-plane that satisfy

---

*See the Computational Window at the end of this section.

$f(x, y) = C$ is called the **level curve** of f at C, and an entire family of level curves is generated as C varies over the range of f. We can think of a trace as a "slice" of the surface at a particular location and a level curve as its projection onto the xy-plane. By sketching members of this family on the xy-plane, we obtain a useful topographical map of the surface $z = f(x, y)$. Because level curves are used to show the shape of a surface (a mountain, for example, as shown in Figure 11.3), they are sometimes called **contour curves**.

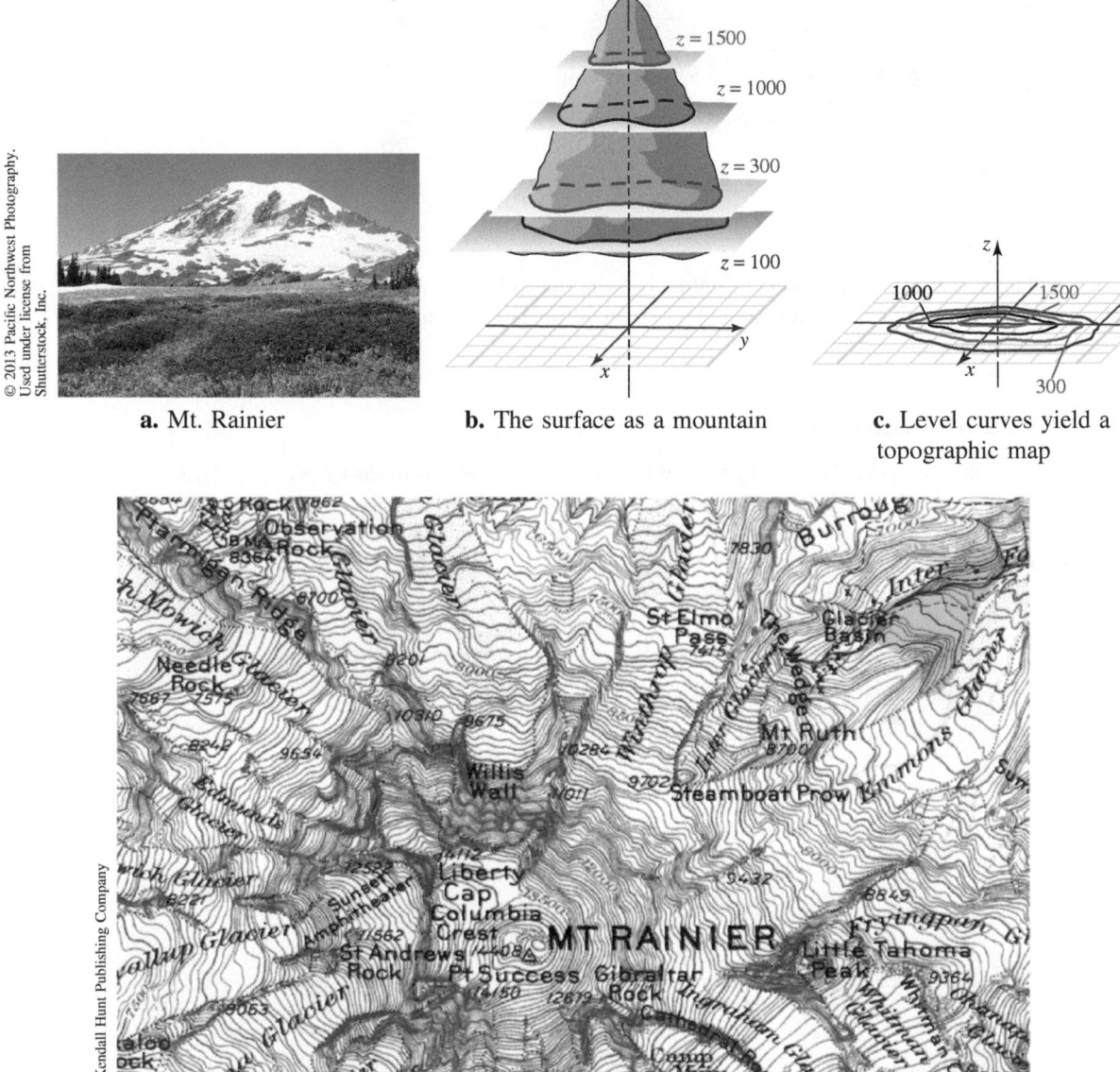

a. Mt. Rainier   **b.** The surface as a mountain   **c.** Level curves yield a topographic map

d. Topographic map of Mt. Rainier

Figure 11.3 Level curves of a surface

For example, imagine that the surface $z = f(x, y)$ is a "mountain," and that we wish to draw a two-dimensional "profile" of its shape. To draw such a profile, we indicate the paths of constant elevation by sketching the family of level curves in the plane and pinning a "flag" to each curve to show the elevation to which it corresponds, as shown in

Figure 11.3**c**. Notice that regions in the map where paths are crowded together correspond to the steeper portions of the mountain. An actual topographical map of Mount Rainier is shown in Figure 11.3**d**.

You probably have seen level curves on the weather report in the newspaper or television news shows, where level curves of equal temperature are called **isotherms** (see Figure 11.4). Other common uses of level curves involve curves of equal pressure (called)**isobars**) and curves of electric potential (called **equipotential lines**).

Figure 11.4 Isotherms

Example 2 Level curves

Sketch some level curves of the function $f(x,y) = 10 - x^2 - y^2$.

k	level curve $z = k$
1	$1 = 10 - x^2 - y^2$
	or $x^2 + y^2 = 9$
6	$6 = 10 - x^2 - y^2$
	or $x^2 + y^2 = 4$
9	$9 = 10 - x^2 - y^2$
	or $x^2 + y^2 = 1$

Solution A computer graph of $z = f(x,y)$ is the surface shown in Figure 11.5**a**. Figure 11.5**b** shows traces of the graph of f in the planes $z = 1$, $z = 6$, and $z = 9$, and the corresponding level curves are shown in Figure 11.5**c**. A table of values is shown in the margin.

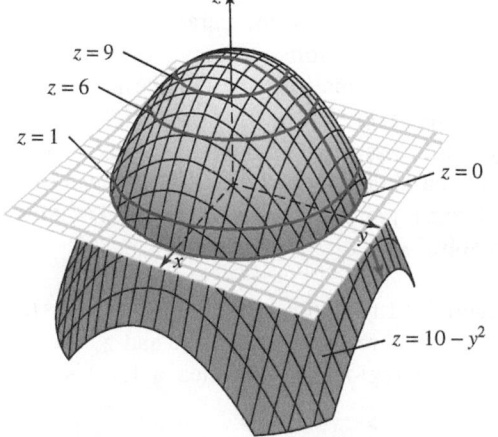

a. Interactive Computer generated graph of surface

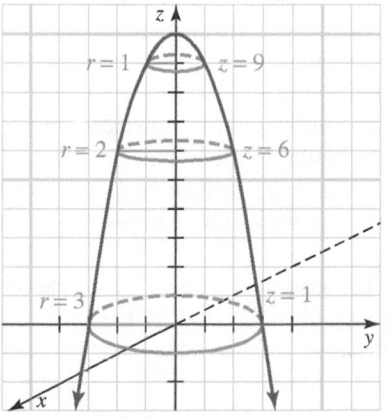

b. Simplified graph showing traces

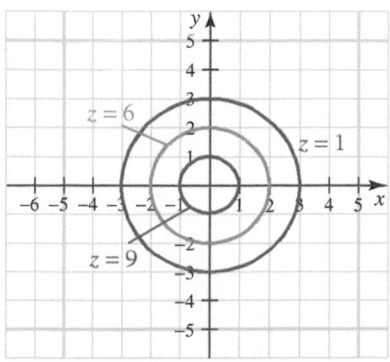

c. Level curves

Figure 11.5 Three ways of viewing the surface $z = 10 - x^2 - y^2$

Graphs of Functions of Two Variables

The level curves of a function $f(x, y)$ provide information about the cross sections of the surface $z = f(x, y)$ perpendicular to the z-axis. However, a more complete picture of the surface can often be obtained by examining cross sections in other directions as well. This procedure is used to graph a function in Example 3.

Example 3   Level curves

Use the level curves of the function $f(x, y) = x^2 + y^2$ to sketch the graph of f.

Solution   The level curve $x^2 + y^2 = 0$ (that is, $C = 0$) is the point $(0, 0)$, and for $C > 0$, the level curve $x^2 + y^2 = C$ is the circle with center $(0, 0)$ and radius $\sqrt{C}$ (Figure 11.6a). There are no points (x, y) that satisfy $x^2 + y^2 = C$ for $C < 0$.

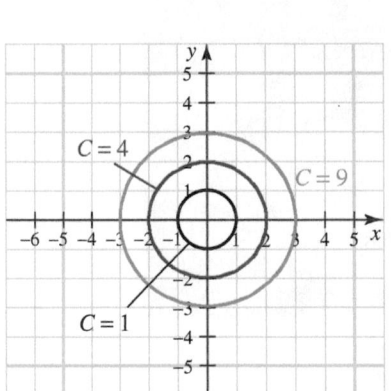
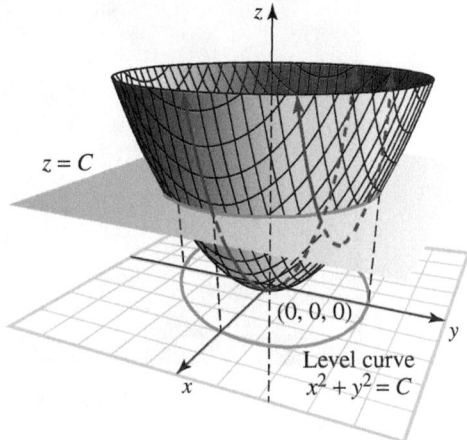

a. Interactive The level curves are circles   **b.** The cross sections perpendicular to the y-axis are parabolas (shown in white)

Figure 11.6   The graph of the function $f(x, y) = x^2 + y^2$

We can gain additional information about the appearance of the surface by examining cross sections perpendicular to the other two principal directions; that is, to the x-axis and the y-axis. Cross-sectional planes perpendicular to the x-axis have the form $x = A$ and intersect the surface $z = x^2 + y^2$ in parabolas of the form $z = A^2 + y^2$. For example, the cross section of the plane $x = 5$ intersects the surface in the parabola $z = 25 + y^2$. That is, it is the set of points $(5, y, 25 + y^2)$ as y-varies. Similarly, cross-sectional planes perpendicular to the y-axis have the form $y = B$ and intersect the surface in parabolas of the form $z = x^2 + B^2$ (shown in white in Figure 11.6b).

To summarize, the surface $z = x^2 + y^2$ has cross sections that are circles in planes perpendicular to the z-axis (Figure 11.6a) and parabolas in the other two principal directions. Since the surface is formed by revolving a parabola about its axis, it is called a **circular paraboloid** or a **paraboloid of revolution**. ∎

The concept of level curve can be generalized to apply to functions of more than two variables. In particular, if f is a function of three variables x, y, and z, then the solution set of the equation $f(x, y, z) = C$ is a region of $\mathbb{R}^3$ called a **level surface** of f at C.

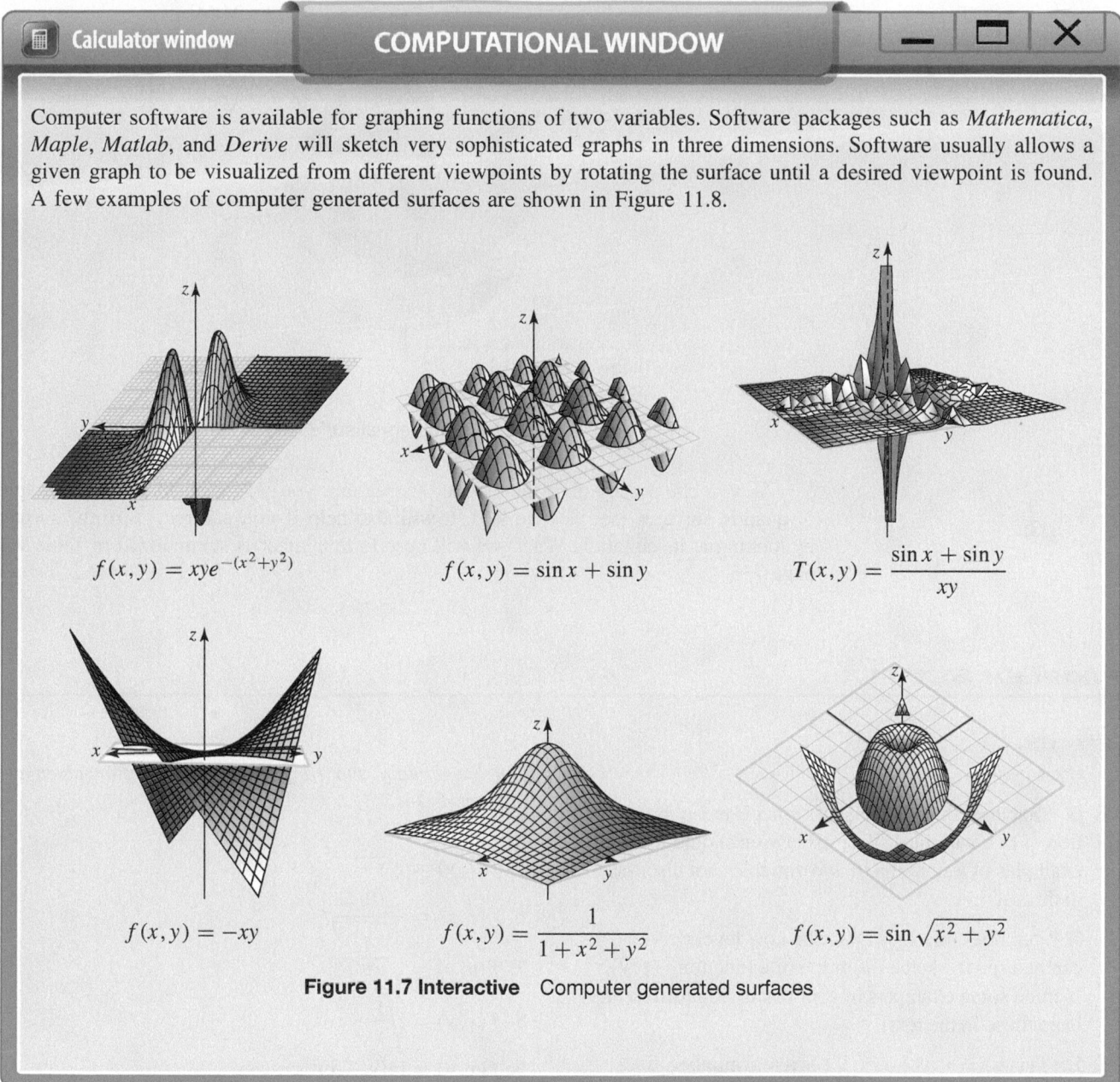

Figure 11.7 Interactive Computer generated surfaces

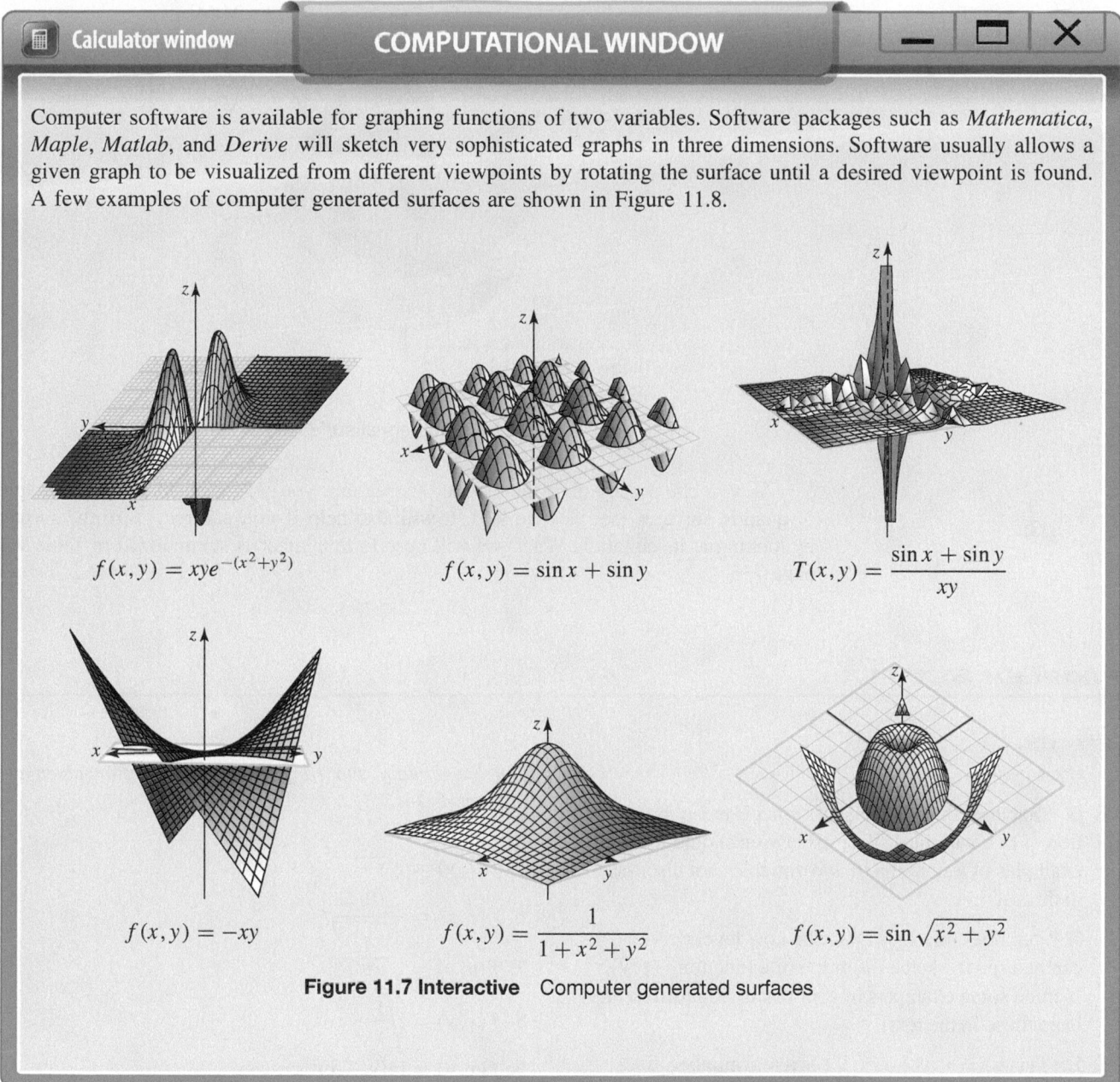Computer software is available for graphing functions of two variables. Software packages such as *Mathematica*, *Maple*, *Matlab*, and *Derive* will sketch very sophisticated graphs in three dimensions. Software usually allows a given graph to be visualized from different viewpoints by rotating the surface until a desired viewpoint is found. A few examples of computer generated surfaces are shown in Figure 11.8.

$f(x,y) = xye^{-(x^2+y^2)}$

$f(x,y) = \sin x + \sin y$

$T(x,y) = \dfrac{\sin x + \sin y}{xy}$

$f(x,y) = -xy$

$f(x,y) = \dfrac{1}{1 + x^2 + y^2}$

$f(x,y) = \sin \sqrt{x^2 + y^2}$

Calculator window   **COMPUTATIONAL WINDOW**

Example 4 Isothermal surface

Suppose a region of $\mathbb{R}^3$ is heated so that its temperature T at each point (x,y,z) is given by $T(x,y,z) = 100 - x^2 - y^2 - z^2$ degrees Celsius. Describe the isothermal surfaces for $T > 0$.

Solution   The isothermal surfaces are given by $T(x,y,z) = k$ for constant k; that is, $x^2 + y^2 + z^2 = 100 - k$. If $100 - k > 0$, the graph of $x^2 + y^2 + z^2 = 100 - k$ is a sphere of radius $\sqrt{100 - k}$ and center $(0,0,0)$. When $k = 100$, the graph is a single point (the origin), and $T(0,0,0) = 100°$C. As the temperature drops, the constant k gets smaller, and the radius $\sqrt{100 - k}$ of the sphere gets larger. Hence, the isothermal surfaces are spheres, and the larger the radius, the cooler the surface. This situation is illustrated in Figure 11.7.

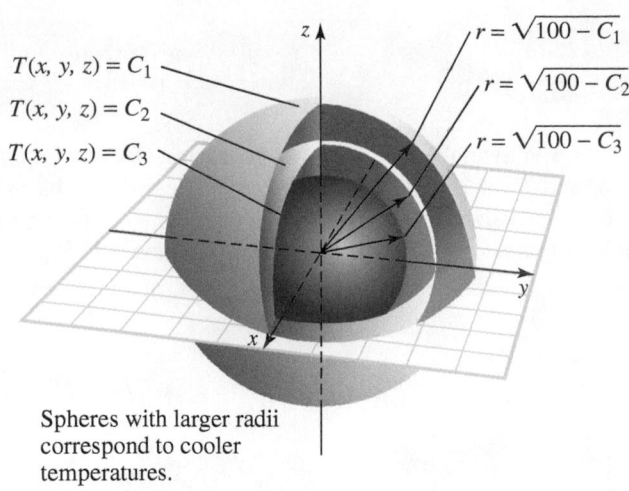

$$T(x, y, z) = C_1$$
$$T(x, y, z) = C_2$$
$$T(x, y, z) = C_3$$

$$r = \sqrt{100 - C_1}$$
$$r = \sqrt{100 - C_2}$$
$$r = \sqrt{100 - C_3}$$

Spheres with larger radii
correspond to cooler
temperatures.

Figure 11.8   Isothermal surfaces for T

As you can see by the examples in this section, you will need to recall the graphs of quadric surfaces (see Section 9.7). It will also help if you can recognize the surface by looking at its equation. What you will need to remember is summarized in Table 9.2, page 736.

PROBLEM SET 11.1

Level 1

1. ■ What does this say? Discuss what is meant by a function of two variables. Your discussion should include examples of a function of two variables not discussed in this section.

2. ■ What does this say? Describe how level curves can be used to sketch the graph of a function $f(x, y)$. Include some examples in your discussion (different from those in the text).

3. Let $f(x, y) = x^2 y + xy^2$. If t is a real number, find:

   a. $f(0, 0)$ **b.** $f(-1, 0)$ **c.** $f(0, -1)$

   d. $f(1, 1)$ **e.** $f(2, 4)$ **f.** $f(t, t)$

   g. $f(t, t^2)$ **h.** $f(1 - t, t)$

4. Let $f(x, y, z) = x^2 y e^{2x} + (x + y - z)^2$. Find

   a. $f(0, 0, 0)$ **b.** $f(1, -1, 1)$

   c. $f(-1, 1, -1)$ **d.** $\dfrac{d}{dx} f(x, x, x)$

   e. $\dfrac{d}{dy} f(1, y, 1)$ **f.** $\dfrac{d}{dz} f(1, 1, z^2)$

Find the domain and range for each function given in Problems 5-14.

5. $f(x, y) = \sqrt{x - y}$

6. $f(x, y) = \dfrac{1}{\sqrt{x - y}}$

7. $f(u, v) = \sqrt{uv}$

8. $f(x, y) = \sqrt{\dfrac{y}{x}}$

9. $f(x, y) = \ln(y - x)$

10. $f(x, y) = \sqrt{u \sin v}$

11. $f(x, y) = \sqrt{(x + 3)^2 + (y - 1)^2}$

12. $f(x, y) = e^{(x+1)/(y-2)}$

13. $f(x, y) = \dfrac{1}{\sqrt{x^2 - y^2}}$

14. $f(x, y) = \dfrac{1}{\sqrt{9 - x^2 - y^2}}$

Sketch some level curves $f(x, y) = C$ for $C \geq 0$ for each function given in Problems 15-20.

15. $f(x, y) = 2x - 3y$

16. $f(x, y) = x^2 - y^2$

17. $f(x, y) = x^3 - y$

18. $g(x, y) = x^2 - y$

19. $h(x, y) = x^2 + \dfrac{y^2}{4}$

20. $f(x, y) = \dfrac{x}{y}$

In Problems 21-26, sketch and describe the level surface $f(x, y, z) = C$ for the given value of C.

21. $f(x, y, z) = y^2 + z^2$ for $C = 1$

22. $f(x, y, z) = x^2 + z^2$ for $C = 1$

23. $f(x, y, z) = x + y - z$ for $C = 1$

24. $f(x, y, z) = x + y - z$ for $C = 0$

25. $f(x, y, z) = (x + 1)^2 + (y - 2)^2 + (z - 3)^2$ for $C = 4$

26. $f(x, y, z) = 2x^2 - 2y^2 - z$ for $C = 1$

In Problems 27-34, describe the traces of the given quadric surface in each coordinate plane, then sketch the surface.

27. $9x^2 + 4y^2 + z^2 = 1$

28. $\dfrac{x^2}{4} + y^2 + \dfrac{z^2}{9} = 1$

29. $\dfrac{x^2}{4} + \dfrac{y^2}{9} - z^2 = 1$

30. $\dfrac{x^2}{9} - y^2 - z^2 = 1$

31. $z = x^2 + \dfrac{y^2}{4}$

32. $z = \dfrac{x^2}{9} - \dfrac{y^2}{16}$

33. $x^2 + 2y^2 = 9z^2$

34. $z^2 = 1 + \dfrac{x^2}{9} + \dfrac{y^2}{4}$

Match each family of level curves given in Problems 35-40 with one of the surfaces labeled A-F.

35. $f(x, y) = x^2 - y^2$

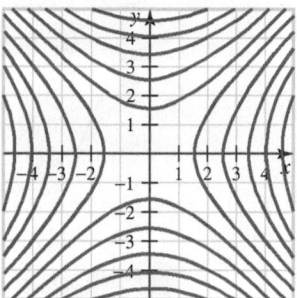

36. $f(x, y) = e^{1 - x^2 + y^2}$

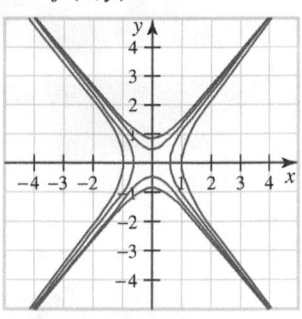
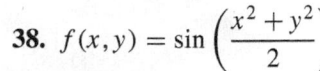

37. $f(x, y) = \dfrac{1}{x^2 + y^2}$

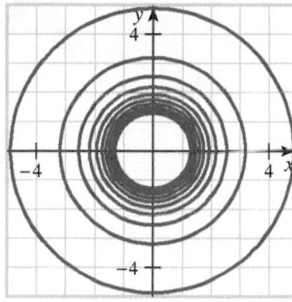

38. $f(x, y) = \sin\left(\dfrac{x^2 + y^2}{2}\right)$

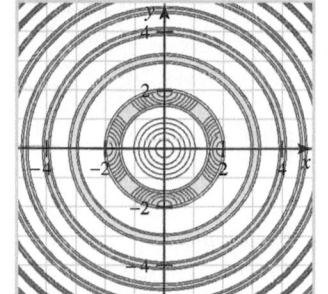
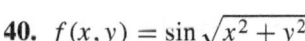

39. $f(x, y) = \dfrac{\cos xy}{x^2 + y^2}$

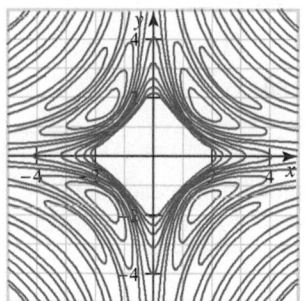

40. $f(x, y) = \sin\sqrt{x^2 + y^2}$

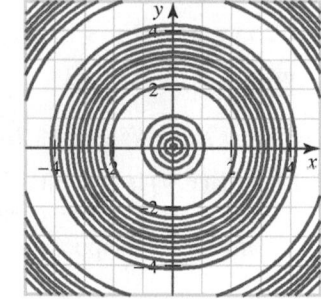

A.

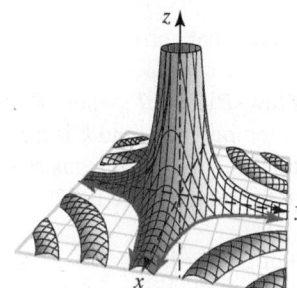

B.

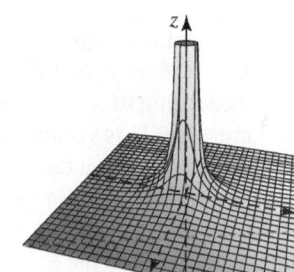

C.
D.

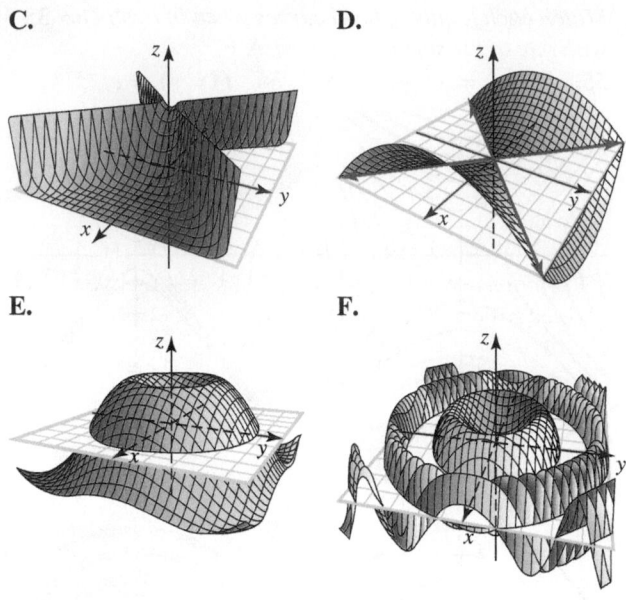

E.
F.

where d_o is the distance of an object from a thin, spherical lens, d_i is the distance of its image on the other side of the lens, and L is the *focal length* of the lens (see Figure 11.9). Write L as a function of d_o and d_i and describe some level curves of the function (these are curves of constant focal length).

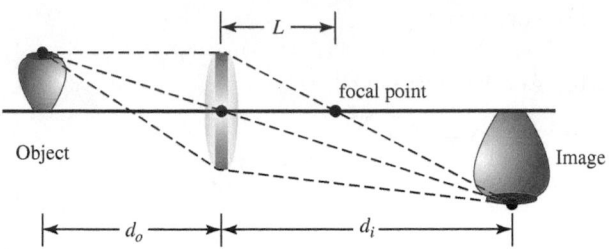

Figure 11.9 Image of an object through a lens

Level 2

Sketch the graph of each function given in Problems 41-52.

41. $f(x,y) = -4$ **42.** $f(x,y) = x$

43. $f(x,y) = y^2 + 1$ **44.** $f(x,y) = x^3 - 1$

45. $f(x,y) = 2x - 3y$ **46.** $f(x,y) = x^2 - y$

47. $f(x,y) = 2x^2 + y^2$ **48.** $f(x,y) = x^2 - y^2$

49. $f(x,y) = \dfrac{x}{y}$ **50.** $f(x,y) = \sqrt{x+y}$

51. $f(x,y) = x^2 + y^2 + 2$ **52.** $f(x,y) = \sqrt{1 - x^2 - y^2}$

53. If $E(x,y)$ is the voltage (potential) at each point (x,y) in the plane, the level curves of E are called *equipotential curves*. Suppose

$$E(x,y) = \frac{7}{\sqrt{3 + x^2 + 2y^2}}$$

Sketch the equipotential curves that correspond to $E = 1, E = 2$, and $E = 3$.

54. According to the ideal gas law, $PV = kT$, where P is pressure, V is volume, T is temperature, and k is a constant. Suppose a tank contains 3,500 in.^{3} of a gas at a pressure of 24 lb/in.^{2} when the temperature is $270°$K (degrees Kelvin).
  a. Determine the constant of proportionality k.
  b. Express the temperature T as a function of P and V using the constant found in part **a** and sketch the isotherms for the temperature functions.

55. The *lens equation* in optics states that

$$\frac{1}{d_o} + \frac{1}{d_i} = \frac{1}{L}$$

56. The EZGRO agricultural company estimates that when $100x$ worker-hours of labor are employed on y acres of land, the number of bushels of wheat produced is $f(x,y) = Ax^a y^b$, where A, a, and b are nonnegative constants. Suppose the company decides to double the production factors x and y. Determine how this decision affects the production of wheat in each of these cases:

  a. $a + b > 1$ **b.** $a + b < 1$ **c.** $a + b = 1$

57. Suppose that when x machines and y worker-hours are used each day, a certain factory will produce $Q(x,y) = 10xy$ cellphones. Describe the relationship between the "inputs" x and y that result in an "output" of 1,000 phones each day. *Note:* You are finding a level curve of the production function Q.

58. A publishing house has found that in a certain city each of its salespeople will sell approximately

$$\frac{r^2}{2,000p} + \frac{s^2}{100} - s$$

units per month, at a price of p dollars/unit, where s denotes the total number of salespeople employed and r is the amount of money spent each month on local advertising. Express the total revenue R as a function of p, r, and s.

Level 3

59. Modeling Problem A manufacturer with exclusive rights to a sophisticated new industrial machine is planning to sell a limited number of the machines to both foreign and domestic firms. The price that the manufacturer can expect to receive for the machines will depend on the number of machines made available. It is estimated that if the manufacturer supplies x machines to the domestic market and y machines to the foreign market, the machines will sell for

$$60 - \frac{x}{5} + \frac{y}{20}$$

thousand dollars apiece at home and

$$50 - \frac{x}{10} + \frac{y}{20}$$

thousand dollars apiece abroad. Express the revenue R as a function of x and y.

60. **Modeling Problem** At a certain factory, the daily output is modeled by

$$Q = CK^r L^{1-r}$$

units, where K denotes capital investment, L is the size of the labor force, and C and r are constants, with $0 < r < 1$ (this is called a *Cobb-Douglas production function*). What happens to Q if K and L are both doubled? What if both are tripled?

11.2 LIMITS AND CONTINUITY

IN THIS SECTION: *Open and closed sets in $\mathbb{R}^2$ and $\mathbb{R}^3$, limit of a function of two variables, continuity, limits and continuity for functions of three variables*

As in the single-variable case, the concept of limit plays a fundamental role in the calculus of functions of several variables. Our focus is on functions of two variables, but the notions discussed also apply to functions of three or more variables.

Most commonly considered functions of a single variable have domains that can be described in terms of intervals. However, dealing with functions of two or more variables requires special terminology and notation that we will introduce in this section and then use to discuss limits and continuity for functions of two variables.

Open and Closed Sets in $\mathbb{R}^2$ and $\mathbb{R}^3$

An **open disk** centered at the point $C(a, b)$ in $\mathbb{R}^2$ is the set of all points $P(x, y)$ such that

$$\sqrt{(x - a)^2 + (y - b^2)} < r$$

for $r > 0$ (see Figure 11.10a). If the boundary of the disk is included (that is, if $\sqrt{(x - a)^2 + (y - b)^2} \leq r$), the disk is said to be **closed** (see Figure 11.10b). Open and closed disks are analogous to open and closed intervals on a coordinate line.

A point P_0 is said to be an **interior point** of a set S in $\mathbb{R}^2$ if some open disk centered at P_0 is contained entirely within S (as shown in Figure 11.11a). If S is the empty set, or if every point of S is an interior point, then S is called an **open set** (Figure 11.11b). A point P_0 is called a **boundary point** of S if every open disk centered at P_0 contains both points that belong to S and points that do not. The collection of all boundary points of S is called the **boundary** of S, and S is said to be **closed** if it contains its boundary (Figure 11.11c). The empty set and $\mathbb{R}^2$ are both open and closed.

Similarly, an **open ball** centered at $C(a, b, c)$ is the set of all points $P(x, y, z)$ such that $\sqrt{(x - a)^2 + (y - b)^2 + (z - c)^2} < r$, for $r > 0$. A point P_0 is an **interior point** of a set S in $\mathbb{R}^3$ if there exists an open ball centered at P_0 that is contained entirely within S, and the nonempty set S is an **open set** if all its points are interior points. A point P_0 is a **boundary point** of S if every open disk centered at P_0 contains both points that belong to S and points that do not, and S is **closed** if it contains all its boundary points. As with $\mathbb{R}^2$, the set $\mathbb{R}^3$ is both open and closed.

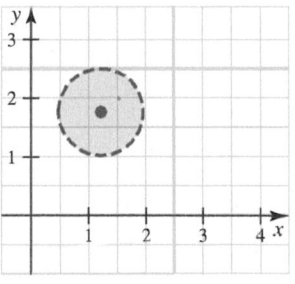

a. Open disk

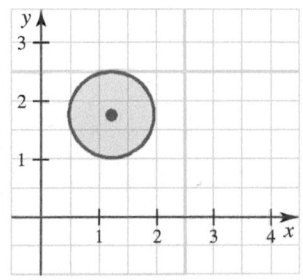

b. Closed disk

Figure 11.10 Disks in the xy-plane

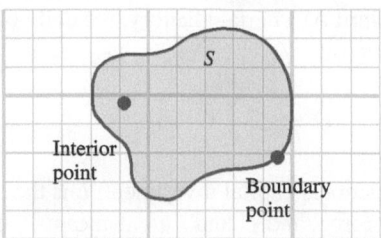

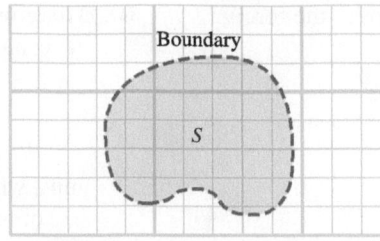

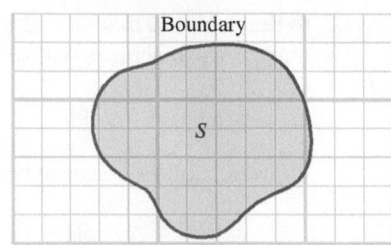

a. Interior points and boundary points of a set S

b. The set S is open if all its points are interior points

c. The set S is closed if it contains all its boundary points

Figure 11.11 Open and closed points in $\mathbb{R}^2$

Limit of a Function of Two Variables

In Chapter 2, we informally defined the limit statement $\lim_{x \to c} f(x) = L$ to mean that $f(x)$ can be made arbitrarily close to L by choosing x sufficiently close (but not equal) to c. For a function of two variables we write

$$\lim_{(x,y) \to (x_0,y_0)} f(x,y) = L$$

to mean that the functional values $f(x,y)$ can be made arbitrarily close to the number L by choosing the point (x,y) sufficiently close to the point (x_0, y_0). The following is a more formal definition of this limiting behavior.

LIMIT OF A FUNCTION OF TWO VARIABLES   The limit statement

$$\lim_{(x,y) \to (x_0,y_0)} f(x,y) = L$$

means that for each given number $\epsilon > 0$, there exists a number $\delta > 0$ so that whenever (x, y) is a point in the domain D of f such that

$$0 < \sqrt{(x - x_0)^2 + (y - y_0)^2} < \delta$$

then

$$|f(x,y) - L| < \epsilon$$

■ **W**hat this says: If $\lim_{(x,y) \to (x_0,y_0)} f(x,y) = L$, then given $\epsilon > 0$, the function value of $f(x,y)$ must lie in the interval $(L - \epsilon, L + \epsilon)$ whenever (x,y) is a point in the domain of f other than $P_0(x_0, y_0)$ that lies inside the disk of radius δ centered at P_0. This is illustrated in Figure 11.12.

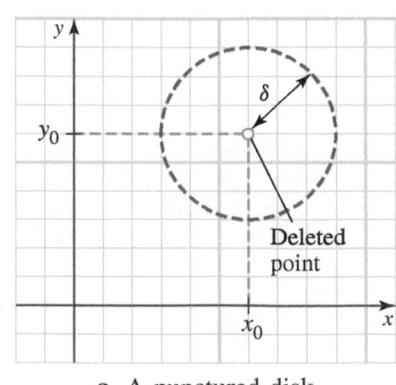

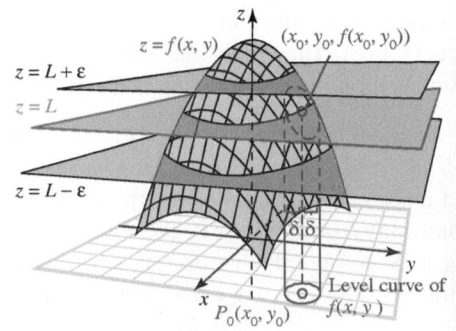

a. A punctured disk

b. $\displaystyle\lim_{(x,y) \to (x_0,y_0)} f(x,y) = L$

Figure 11.12 Limit of a function of two variables

When considering $\lim_{x \to c} f(x)$ we need to examine the approach of x to c from two directions (the left- and right-hand limits). However, for a function of two variables, we write $(x, y) \to (x_0, y_0)$ to mean that the point (x, y) is allowed to approach (x_0, y_0) along *any* curve in the domain of f that passes through (x_0, y_0). For example, no matter how you get to the 15th step of Denver's state capitol, you will still be at 5,280 ft when you get there. If the limit

$$\lim_{(x,y) \to (x_0,y_0)} f(x, y) = L$$

is not the same for *all* approaches, or **paths** within the domain of f, then *the limit does not exist*.

Example 1 Evaluating a limit of a function of two variables

$$\lim_{(x,y) \to (x_0,y_0)} \frac{x^2 + x - xy - y}{x - y}.$$

Solution First, note that for $x \neq y$,

$$f(x, y) = \frac{x^2 + x - xy - y}{x - y} = \frac{(x + 1)(x - y)}{x - y} = x + 1$$

Therefore,

$$\lim_{(x,y) \to (0,0)} \frac{x^2 + x - xy - y}{x - y} = \lim_{(x,y) \to (0,0)} (x + 1) = 1$$

Example 2 Showing a limit does not exist

If $f(x, y) = \dfrac{2xy}{x^2 + y^2}$, show

$$\lim_{(x,y) \to (0,0)} f(x, y)$$

does not exist by evaluating this limit along the x-axis, the y-axis, and along the line $y = x$.

Solution First note that the denominator is zero at $(0, 0)$ so $f(0, 0)$ is not defined. If we approach the origin along the x-axis (where $y = 0$), we find that

$$\lim_{x \to 0} \frac{2x(0)}{x^2 + 0} = \lim_{x \to 0} 0 = 0$$

so $f(x, y) \to 0$ as $(x, y) \to (0, 0)$ along $y = 0$ (and $x \neq 0$). We find a similar result if we approach the origin along the y-axis (where $x = 0$); see Figure 11.13. However, along the line $y = x$, the functional values (for $x \neq 0$) are

$$f(x, x) = \frac{2x^2}{x^2 + x^2} = 1$$

so that $\dfrac{2xy}{x^2 + y^2} \to 1$ as $(x, y) \to (0, 0)$ along $y = x$. Because $f(x, y)$ tends toward different numbers as $(x, y) \to (0, 0)$ along different curves, it follows that f has no limit at the origin.

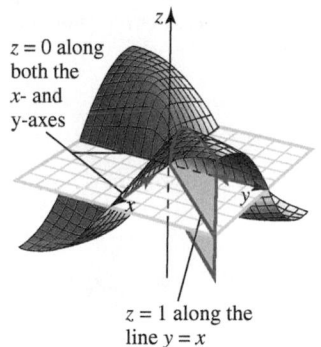

$z = 0$ along both the x- and y-axes

$z = 1$ along the line $y = x$

Figure 11.13 Graph of $z = \dfrac{2xy}{x^2 + y^2}$ and limits as $(x, y) \to (0, 0)$

Example 3   Showing a limit does not exist

Show that

$$\lim_{(x,y)\to(0,0)} \frac{x^2 y}{x^4 + y^2}$$

does not exist.

Solution   If we approach this problem as we did the one in Example 2, we would take limits as $(x, y) \to (0, 0)$ along the x-axis, the y-axis, and the line $y = x$. All these approach lines have the general form $y = mx$. We find:

$$\text{Along } y = mx: \quad \lim_{(x,y)\to(0,0)} \frac{x^2 y}{x^4 + y^2} = \lim_{(x,y)\to(0,0)} \frac{x^2(mx)}{x^4 + (mx)^2}$$

$$= \lim_{(x,y)\to(0,0)} \frac{mx}{x^2 + m^2}$$

$$= 0$$

However, if we approach $(0, 0)$ along the parabola $y = x^2$, we find:

It is often possible to show that a limit does not exist by the methods illustrated in Examples 2 and 3. However, it is impossible to try to prove that $\lim_{(x,y)\to(0,0)} f(x,y)$ exists by showing the limiting value of $f(x, y)$ is the same along every curve that passes through (x_0, y_0) because there are too many such curves.

$$\text{Along } y = x^2: \quad \lim_{(x,y)\to(0,0)} \frac{x^2 y}{x^4 + y^2} = \lim_{(x,y)\to(0,0)} \frac{x^2(x^2)}{x^4 + (x^2)^2}$$

$$= \lim_{(x,y)\to(0,0)} \frac{x^4}{2x^4}$$

$$= \frac{1}{2}$$

Therefore, since approaching $(0, 0)$ along the lines $y = mx$ gives a different limiting value from approaching along the parabola $y = x^2$, we conclude that the given limit does not exist.

We have just observed that it is difficult to show that a given limit exists. However, limits of functions of two variables that are known to exist can be manipulated in much the same way as limits involving functions of a single variable. Here is a list of the basic rules for manipulating such limits.

BASIC FORMULAS AND RULES OF LIMITS OF TWO VARIABLES

Suppose $\lim_{(x,y)\to(0,0)} f(x,y) = L$ and $\lim_{(x,y)\to(x_0,y_0)} g(x,y) = M$.

Then for any constant a,

Scalar multiple rule $\lim_{(x,y)\to(x_0,y_0)} [af](x,y) = aL$

Sum rule $\lim_{(x,y)\to(x_0,y_0)} [f + g](x,y) = L + M$

Product rule $\lim_{(x,y)\to(x_0,y_0)} [fg](x,y) = LM$

Quotient rule $\lim_{(x,y)\to(x_0,y_0)} \left[\dfrac{f}{g}\right](x,y) = \dfrac{L}{M}$ if $M \neq 0$

Example 4   Manipulating limits of functions of two variables

Assuming each limit exists, evaluate

a. $\displaystyle\lim_{(x,y)\to(4,3)} (x^2 + xy + y^2)$ **b.** $\displaystyle\lim_{(x,y)\to(1,2)} \frac{2xy}{x^2 + y^2}$

Solution

a. $\displaystyle\lim_{(x,y)\to(4,3)} (x^2 + xy + y^2) = (4)^2 + (4)(3) + (3)^2$

$$= 37$$

A graph is shown below.

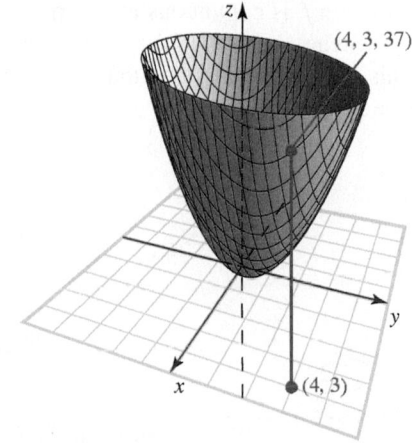

b. $\displaystyle\lim_{(x,y)\to(1,2)} \frac{2xy}{x^2 + y^2} = \frac{\displaystyle\lim_{(x,y)\to(1,2)} 2xy}{\displaystyle\lim_{(x,y)\to(1,2)} (x^2 + y^2)}$

$$= \frac{2(1)(2)}{1^2 + 2^2}$$

$$= \frac{4}{5}$$

A graph is shown below.

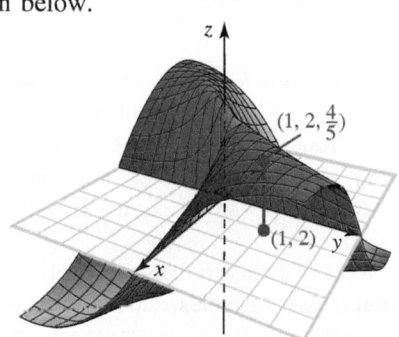

Continuity

Recall that a function of a single variable x is continuous at $x = c$ if

1. $f(c)$ is defined;   **2.** $\displaystyle\lim_{x\to c} f(x)$ exists;   **3.** $\displaystyle\lim_{x\to c} f(x) = f(c)$

Using the definition of the limit of a function of two variables, we can now define the continuity of a function of two variables analogously.

CONTINUITY AS A FUNCTION OF TWO VARIABLES   The function $f(x, y)$ is **continuous** at the point (x_0, y_0) if and only if

1. $f(x_0, y_0)$ is defined;
2. $\displaystyle\lim_{(x,y)\to(x_0,y_0)} f(x, y)$ exists;
3. $\displaystyle\lim_{(x,y)\to(x_0,y_0)} f(x, y) = f(x_0, y_0)$.

The function f is **continuous on a set S** if it is continuous at each point in S.

■ **W**hat this says: The function f is continuous at (x_0, y_0) if the functional value of $f(x, y)$ is close to $f(x_0, y_0)$ whenever (x, y) in the domain of f is sufficiently close to (x_0, y_0). Geometrically, this means that f is continuous if the surface $z = f(x, y)$ has no "holes" or "gaps" (see Figure 11.14).

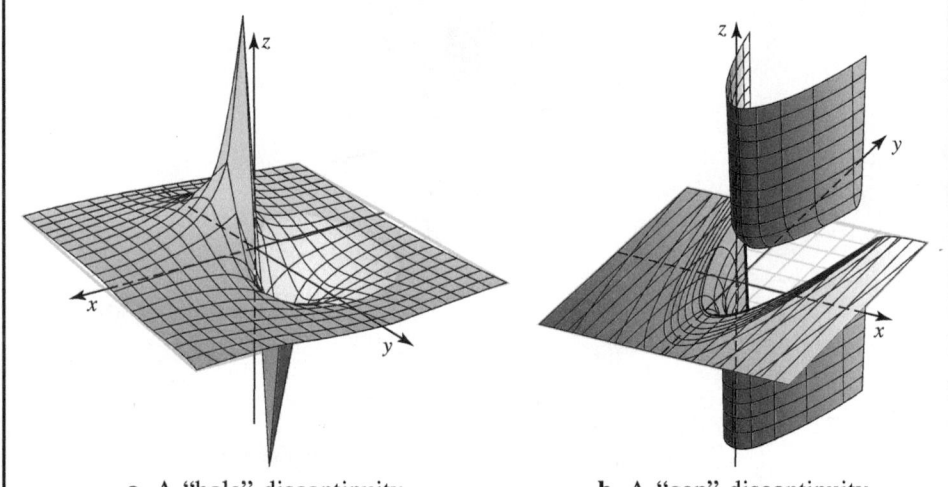

a. A "hole" discontinuity   **b.** A "gap" discontinuity

Figure 11.14   Two types of discontinuity

The basic properties of limits can be used to show that if f and g are both continuous on the set S, then so are the sum $f + g$, the multiple af, the product fg, the quotient f/g (whenever $g \neq 0$), and the root $\sqrt[n]{f}$ wherever it is defined. Also, if F is a function of two variables, that is $F(x_0, y_0)$ continuous at (x_0, y_0) and G is a function of one variable that is continuous at x_0 (or y_0), it can be shown that the composite function $G \circ F$ is continuous at (x_0, y_0).

Many common functions of two variables are continuous wherever they are defined. For instance, a polynomial in two variables, such as $x^3 y^2 + 3xy^3 - 7x + 2$, is continuous throughout the plane, and a rational function in two variables is continuous wherever the denominator polynomial is not zero. Unless otherwise stated, the functions of two or more variables considered in this text will be continuous wherever they are defined.

Example 5   Testing for continuity

Test the continuity of the functions:

a. $f(x, y) = \dfrac{x - y}{x^2 + y^2}$ **b.** $f(x, y) = \dfrac{1}{y - x^2}$

These are the functions whose graphs are shown in Figure 11.14.

Solution

a. The function f is a rational function of x and y (since $x - y$ and $x^2 + y^2$ are both polynomials), so it is discontinuous only where the denominator is 0; namely at $(0,0)$.

b. Again, this is a rational function and is discontinuous only where the denominator is 0; that is, where

$$y - x^2 = 0$$

Thus, the function is continuous at all points except those lying on the parabola $y = x^2$.

Example 6   Continuity using the limit definition

Show that f is continuous at $(0,0)$ where

$$f(x,y) = \begin{cases} y \sin \frac{1}{x} & x \neq 0 \\ 0 & x = 0 \end{cases}$$

The graph is shown in Figure 11.15.

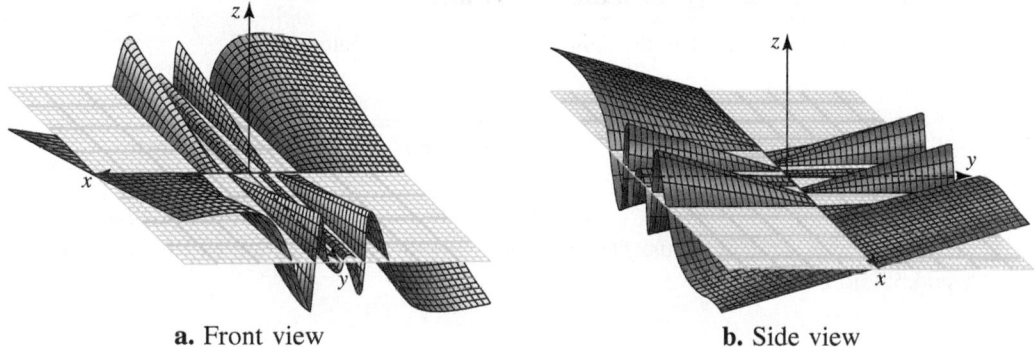

a. Front view   **b.** Side view

Figure 11.15   Graph of f

Solution   The function appears to be continuous. Since the function vanishes along the y-axis, we only need to analyze it at points (x,y) with $x \neq 0$. To prove continuity at $(0,0)$, we must show that for any $\epsilon > 0$, there exists a $\delta > 0$ such that

$$|f(x,y) - f(0,0)| = |f(x,y)| < \epsilon \text{ whenever } 0 < x^2 + y^2 < \delta^2, x \neq 0$$

(We use $x^2 + y^2$ and δ^2 here instead of $\sqrt{x^2 + y^2}$ and δ for convenience.) If $x = 0$, then $|f(x,y)| = 0$. If $x \neq 0$ note that $|f(x,y)| \leq |y|$ because $\left|\sin \frac{1}{x}\right| \leq 1$ for $x \neq 0$. Thus, in either case, $|f(x,y)| \leq |y|$. Also, if (x,y) lies in the disk $x^2 + y^2 < \delta^2$, then the points $(0,y)$ that satisfy $y^2 < \delta^2$ lie in the same disk (let $x = 0$ in $x^2 + y^2 < \delta^2$). In other words, points satisfying $|y| < \delta$ lie in the disk, and if we let $\delta = \epsilon$ it follows that

$$|f(x,y) - f(0,0)| \leq |y| < \delta = \epsilon \text{ whenever } 0 < \left|x^2 + y^2\right| < \delta^2.$$

Limits and Continuity for Functions of Three Variables

The concepts we have just introduced for functions of two variables in $\mathbb{R}^2$ extend naturally to functions of three variables in $\mathbb{R}^3$. In particular, the limit statement

$$\lim_{(x,y,z) \to (x_0,y_0,z_0)} f(x,y,z) = L$$

means that for each number $\epsilon > 0$, there exists a number $\delta > 0$ such that

$$|f(x,y,z) - L| < \epsilon$$

whenever (x,y,z) is a point in the domain of f such that

$$0 < \sqrt{(x-x_0)^2 + (y-y_0)^2 + (z-z_0)^2} < \delta$$

The function $f(x,y,z)$ is **continuous** at the point $P_0(x_0,y_0,z_0)$ if

1. $f(x_0,y_0,z_0)$ is defined;

2. $\displaystyle\lim_{(x,y,z)\to(x_0,y_0,z_0)} f(x,y,z)$ exists;

3. $\displaystyle\lim_{(x,y,z)\to(x_0,y_0,z_0)} f(x,y,z) = f(x_0,y_0,z_0)$

Most commonly considered functions of three variables are continuous wherever they are defined. We close with an example that illustrates how to determine the set of discontinuities of a function of three variables.

Example 7   Continuity for a function of three variables

For what points (x,y,z) is the following function continuous?

$$f(x,y,z) = \frac{3}{\sqrt{x^2 + y^2 - 2z}}$$

Solution   The function $f(x,y,z)$ is continuous except where it is not defined; that is, for $x^2 + y^2 - 2z \leq 0$. Thus, $f(x,y,z)$ is continuous at any point not inside or on the paraboloid $z = \frac{1}{2}(x^2 + y^2)$.

PROBLEM SET 11.2

Level 1

1. ■ **What does this say?** Describe the notion of a limit of function of two variables.

2. ■ **What does this say?** Describe the basic formulas and rules for limits of a function of two variables.

In Problems 3-20, find the given limit, assuming it exists.

3. $\displaystyle\lim_{(x,y)\to(-1,0)} (xy^2 + x^3y + 5)$

4. $\displaystyle\lim_{(x,y)\to(0,0)} (5x^2 - 2xy + y^2 + 3)$

5. $\displaystyle\lim_{(x,y)\to(1,3)} \frac{x+y}{x-y}$

6. $\displaystyle\lim_{(x,y)\to(3,4)} \frac{x-y}{\sqrt{x^2+y^2}}$

7. $\displaystyle\lim_{(x,y)\to(1,0)} e^{xy}$

8. $\displaystyle\lim_{(x,y)\to(1,0)} (x+y)e^{xy}$

9. $\displaystyle\lim_{(x,y)\to(0,1)} \left[e^{x^2+x} \ln(ey^2) \right]$

10. $\displaystyle\lim_{(x,y)\to(e,0)} \ln(x^2 + y^2)$

11. $\displaystyle\lim_{(x,y)\to(0,0)} \frac{x^2 - 2xy + y^2}{x - y}$

12. $\displaystyle\lim_{(x,y)\to(1,2)} \frac{(x^2-1)(y^2-4)}{(x-1)(y-2)}$

13. $\displaystyle\lim_{(x,y)\to(0,0)} \frac{e^x \tan^{-1} y}{y}$

14. $\displaystyle\lim_{(x,y)\to(0,0)} \frac{x^2 y^2}{x^2 + y^2}$

15. $\displaystyle\lim_{(x,y)\to(0,0)} \frac{\sin(x+y)}{x+y}$

16. $\displaystyle\lim_{(x,y)\to(0,0)} (\sin x + \cos y)$

17. $\displaystyle\lim_{(x,y)\to(5,5)} \frac{x^4 - y^4}{x^2 - y^2}$

18. $\displaystyle\lim_{(x,y)\to(a,a)} \frac{x^4 - y^4}{x^2 - y^2}$; a is a constant

19. $\displaystyle\lim_{(x,y)\to(2,1)} \frac{x^2 - 4y^2}{x - 2y}$

20. $\displaystyle\lim_{(x,y)\to(0,0)} \frac{x^2 + y^2}{\sqrt{x^2 + y^2 + 4} - 2}$

In Problems 21-30, evaluate the indicated limit, if it exists. If it does not exist, give reasons.

21. $\displaystyle\lim_{(x,y)\to(2,1)} (xy^2 + x^3 y)$

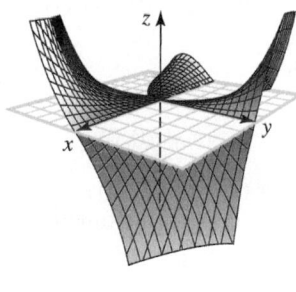

22. $\displaystyle\lim_{(x,y)\to(1,2)} (5x^2 - 2xy + y^2)$

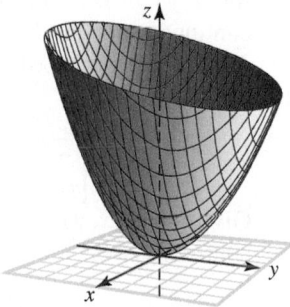

23. $\displaystyle\lim_{(x,y)\to(0,0)} \frac{x+y}{x-y}$

24. $\displaystyle\lim_{(x,y)\to(0,0)} \frac{x-y}{\sqrt{x^2 + y^2}}$

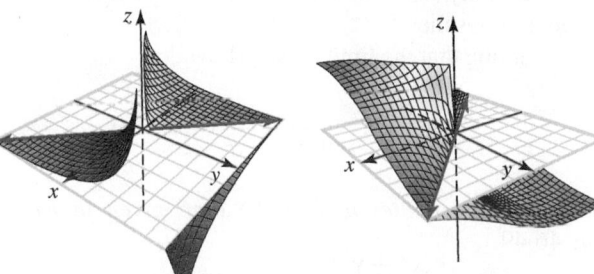

25. $\displaystyle\lim_{(x,y)\to(0,0)} e^{xy}$

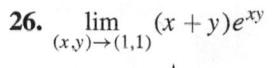

26. $\displaystyle\lim_{(x,y)\to(1,1)} (x+y)e^{xy}$

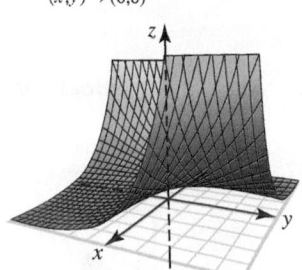

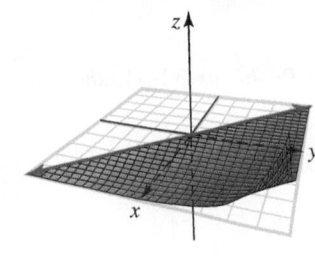

27. $\displaystyle\lim_{(x,y)\to(0,0)} (\sin x - \cos y)$

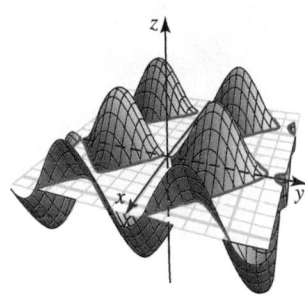

28. $\displaystyle\lim_{(x,y)\to(0,0)} \left[1 - \frac{\sin(x^2 + y^2)}{x^2 + y^2}\right]$

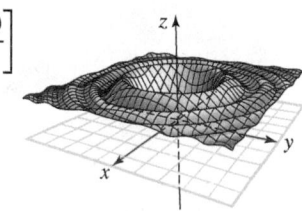

29. $\displaystyle\lim_{(x,y)\to(0,0)} \left[\frac{\sin(x^2 + y^2)}{(x^2 + y^2)}\right]$

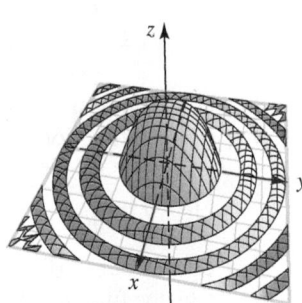

30. $\displaystyle\lim_{(x,y)\to(0,0)} \left[1 - \frac{\cos(x^2 + y^2)}{x^2 + y^2}\right]$

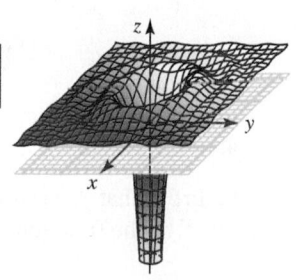

Level 2

In Problems 31-34, show that $\lim\limits_{(x,y)\to(0,0)} f(x,y)$ *does not exist.*

31. $f(x,y) = \dfrac{x^2y^2}{x^4+y^4}$

32. $f(x,y) = \dfrac{x^4y^4}{(x^2+y^4)^3}$

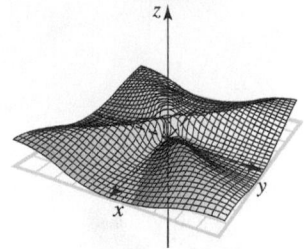 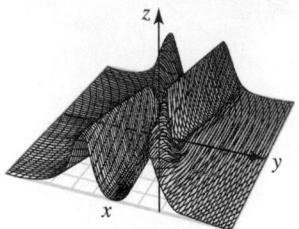

33. $f(x,y) = \dfrac{x-y^2}{x^2+y^2}$

34. $f(x,y) = \dfrac{x^2+y}{x^2+y^2}$

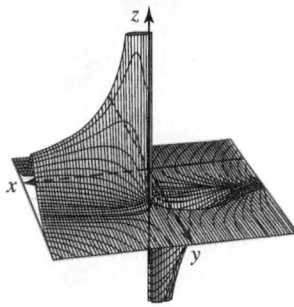 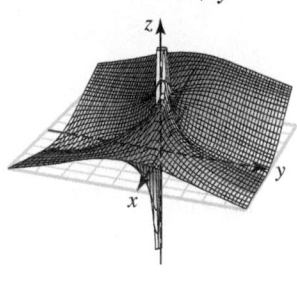

35. Let f be the function defined by
$$f(x,y) = \begin{cases} \dfrac{xy^3}{x^2+y^6} & \text{for } (x,y) \neq (0,0) \\ 0 & \text{for } (x,y) = (0,0) \end{cases}$$
Is f continuous at $(0,0)$? Explain.

36. Let f be the function defined by
$$f(x,y) = \begin{cases} \dfrac{xy^2}{x^2+y^4} & \text{for } (x,y) \neq (0,0) \\ 0 & \text{for } (x,y) = (0,0) \end{cases}$$
Is f continuous at $(0,0)$? Explain.

37. Let f be the function defined by
$$f(x,y) = \dfrac{x^2+2y^2}{x^2+y^2}$$
for $(x,y) \neq (0,0)$
a. Find $\lim\limits_{(x,y)\to(3,1)} f(x,y)$.
b. Prove that f has no limit at $(0,0)$.

38. Let f be the function defined by
$$f(x,y) = \dfrac{x^2-y^2}{x^2+y^2}$$
for $(x,y) \neq (0,0)$
a. Find $\lim\limits_{(x,y)\to(2,1)} f(x,y)$
b. Prove that f has no limit at $(0,0)$.

39. Given that the function
$$f(x,y) = \begin{cases} \dfrac{3x^3-3y^3}{x^2-y^2} & \text{for } x^2 \neq y^2 \\ A & \text{otherwise} \end{cases}$$
is continuous at the origin, what is A?

40. Given that the function
$$f(x,y) = \begin{cases} \dfrac{x^3+y^3}{x^2+y^2} & \text{for } (x,y) \neq (0,0) \\ B & \text{for } (x,y) = (0,0) \end{cases}$$
is continuous at the origin, what is B?

41. Given that the function
$$f(x,y) = \begin{cases} \dfrac{2x^2-x^2y^2+2y^2}{x^2+y^2} & \text{for } (x,y) \neq (0,0) \\ C & \text{for } (x,y) = (0,0) \end{cases}$$
is continuous at the origin, what is C?
Hint: Use polar coordinates.

42. Let
$$f(x,y) = \dfrac{\sin(x^2+y^2)}{x^2+y^2}$$
for $(x,y) \neq (0,0)$. For what value of $f(0,0)$ is $f(x,y)$ continuous at $(0,0)$?
Hint: Use polar coordinates.

43. Let
$$f(x,y) = \begin{cases} \dfrac{xy^3}{x^2+y^2} & \text{for } (x,y) \neq (0,0) \\ 0 & \text{for } (x,y) = (0,0) \end{cases}$$
Given that $f(x,y)$ has a limit at $(0,0)$, is the function f continuous there?

44. Let
$$f(x,y) = \begin{cases} \dfrac{x^2y^2}{x^2+y^2} & \text{for } (x,y) \neq (0,0) \\ 0 & \text{for } (x,y) = (0,0) \end{cases}$$
Given that $f(x,y)$ has a limit at $(0,0)$, is the function f continuous there?

45. Assuming that the limit exists, show that
$$\lim\limits_{(x,y,z)\to(0,0,0)} \dfrac{xyz}{x^2+y^2+z^2} = 0$$

Use polar coordinates to find the limit given in Problems 46-49.

46. $\lim\limits_{(x,y)\to(0,0)} \dfrac{\tan(x^2+y^2)}{x^2+y^2}$

47. $\lim\limits_{(x,y)\to(0,0)} (1+x^2+y^2)^{1/(x^2+y^2)}$

48. $\lim\limits_{(x,y)\to(0,0)} \dfrac{x^2y^2}{x^2+y^2}$

49. $\displaystyle\lim_{(x,y)\to(0,0)} x \ln \sqrt{x^2 + y^2}$

Think Tank Problems *In Problems 50-52, either show that the given statement is true or find a counterexample.*

50. If $\displaystyle\lim_{y\to 0} f(0,y) = 0$, then $\displaystyle\lim_{(x,y)\to(0,0)} f(x,y) = 0$

51. If $f(x,y)$ is continuous for all $x \neq 0$ and $y \neq 0$, and $f(0,0) = 0$, then $\displaystyle\lim_{(x,y)\to(0,0)} f(x,y) = 0$.

52. If $f(x)$ and $g(y)$ are continuous functions of x, and y, respectively, then

$$h(x,y) = f(x) + g(y)$$

is a continuous function of x and y.

Level 3

Use the ϵ-δ definition of limit to verify the limit statements given in Problems 53-56.

53. $\displaystyle\lim_{(x,y)\to(0,0)} (2x^2 + 3y^2) = 0$

54. $\displaystyle\lim_{(x,y)\to(0,0)} (x + y^2) = 0$

55. $\displaystyle\lim_{(x,y)\to(0,0)} \frac{x^2 - y^2}{x + y} = 0$

56. $\displaystyle\lim_{(x,y)\to(1,-1)} \frac{x^2 - y^2}{x + y} = 2$

57. Prove that if f is continuous and $f(a,b) > 0$, then there exists a δ-neighborhood about (a,b) such that $f(x,y) > 0$ for every point (x,y) in the neighborhood.

58. Prove the scalar multiple rule:

$$\lim_{(x,y)\to(x_0,y_0)} [af](x,y) = a \lim_{(x,y)\to(0,0)} f(x,y)$$

59. Prove the sum rule:

$$\lim_{(x,y)\to(x_0,y_0)} [f + g](x,y) = L + M$$

where $L = \displaystyle\lim_{(x,y)\to(x_0,y_0)} f(x,y)$ and $M = \displaystyle\lim_{(x,y)\to(x_0,y_0)} g(x,y)$.

60. A function of two variables $f(x,y)$ may be continuous in each separate variable at $x = x_0$ and $y = y_0$ without being itself continuous at (x_0, y_0). Let $f(x,y)$ be defined by

$$f(x,y) = \begin{cases} \dfrac{xy}{x^2 + y^2} & \text{for } (x,y) \neq (0,0) \\ 0 & \text{for } (0,0) \end{cases}$$

Let $g(x) = f(x,0)$ and $h(y) = f(0,y)$. Show that both $g(x)$ and $h(y)$ are continuous at 0 but that $f(x,y)$ is not continuous at $(0,0)$.

11.3 PARTIAL DERIVATIVES

IN THIS SECTION: *Partial differentiation, partial derivative as a slope, partial derivative as a rate, higher-order partial derivatives*

In many problems involving functions of several variables, the goal is to find the derivative of the function with respect to one of its variables when all the others are held constant. In this section, we consider this concept and we will see how it can be used to find slopes and rates of change.

Partial Differentiation

It is often important to know how a function of two variables changes with respect to one of the variables. For example, according to the ideal gas law, the pressure of a gas is related to its temperature and volume by the formula $P = \dfrac{kT}{V}$, where k is a constant. If the temperature is kept constant while the volume is allowed to vary, we might want to know the effect on the rate of change of pressure. Similarly, if the volume is kept constant while the temperature is allowed to vary, we might want to know the effect on the rate of change of pressure.

The process of differentiating a function of several variables with respect to one of its variables while keeping the other variable(s) fixed is called **partial differentiation**, and the resulting derivative is a **partial derivative** of the function.

Recall that the derivative of a function of a single variable f is defined to be the limit of a difference quotient, namely,

$$f'(x) = \lim_{\Delta x \to 0} \frac{f(x + \Delta x) - f(x)}{\Delta x}$$

Partial derivatives with respect to x or y are defined similarly.

PARTIAL DERIVATIVE OF A FUNCTION OF TWO VARIABLES
If $z = f(x,y)$, then the partial derivatives of f with respect to x and y are the functions f_x and f_y, respectively, defined by

$$f_x(x,y) = \lim_{\Delta x \to 0} \frac{f(x + \Delta x, y) - f(x,y)}{\Delta x}$$

and

$$f_y(x,y) = \lim_{\Delta y \to 0} \frac{f(x, y + \Delta y) - f(x,y)}{\Delta y}$$

provided the limits exist.

■ **W**hat this says: For the partial differentiation of a function of two variables, $z = f(x,y)$, we find the partial derivative with respect to x by regarding y as constant while differentiating the function with respect to x. Similarly, the partial derivative with respect to y is found by regarding x as constant while differentiating with respect to y.

Example 1   Partial derivatives

If $f(x,y) = x^3y + x^2y^2$, find: **a.** f_x **b.** f_y

Solution

a. For f_x, hold y constant and find the derivative with respect to x:

$$f_x(x,y) = 3x^2y + 2xy^2$$

b. For f_y, hold x constant and find the derivative with respect to y:

$$f_y(x,y) = x^3 + 2x^2y$$ ■

Several different symbols are used to denote partial derivatives, as indicated in the following box.

ALTERNATIVE NOTATION FOR PARTIAL DERIVATIVES   For $z = f(x,y)$, the partial derivatives f_x and f_y are denoted by

$$f_x(x,y) = \frac{\partial f}{\partial x} = \frac{\partial z}{\partial x} = \frac{\partial}{\partial x}f(x,y) = z_x = D_x(f)$$

and

$$f_y(x,y) = \frac{\partial f}{\partial y} = \frac{\partial z}{\partial y} = \frac{\partial}{\partial y}f(x,y) = z_y = D_y(f)$$

The values of the partial derivatives of $f(x,y)$ at the point (a,b) are denoted by

$$\left.\frac{\partial f}{\partial x}\right|_{(a,b)} = f_x(a,b) \quad \text{and} \quad \left.\frac{\partial f}{\partial y}\right|_{(a,b)} = f_y(a,b)$$

Example 2   Finding and evaluating a partial derivative

Let $z = x^2 \sin(3x + y^3)$.

a. Evaluate $\left. \dfrac{\partial z}{\partial x} \right|_{(\frac{\pi}{3}, 0)}$ **b.** Evaluate z_y at $(1, 1)$.

Solution

a. $\dfrac{\partial z}{\partial x} = 2x \sin(3x + y^3) + x^2 \cos(3x + y^3)(3)$ **b.** $z_y = x^2 \cos(3x + y^3)(3y^2)$

$\qquad\qquad = 2x \sin(3x + y^3) + 3x^2 \cos(3x + y^3)$ $\qquad\qquad = 3x^2 y^2 \cos(3x + y^3)$

$\left. \dfrac{\partial z}{\partial x} \right|_{(\frac{\pi}{3}, 0)} = 2\left(\dfrac{\pi}{3}\right) \sin\pi + 3\left(\dfrac{\pi}{3}\right)^2 \cos\pi$ $\qquad z_y(1, 1) = 3(1)^2(1)^2 \cos(3 + 1)$

$\qquad\qquad\quad = \dfrac{2\pi}{3}(0) + \dfrac{\pi^2}{3}(-1)$ $\qquad\qquad\qquad\quad = 3\cos 4$

$\qquad\qquad\quad = -\dfrac{\pi^2}{3}$

Example 3   Partial derivative of a function of three variables

Let $f(x, y, z) = x^2 + 2xy^2 + yz^3$; determine: **a.** f_x **b.** f_y **c.** f_z

Solution

a. For f_x, think of f as a function of x alone with y and z treated as constants:

$$f_x(x, y, z) = 2x + 2y^2$$

b. $f_y(x, y, z) = 4xy + z^3$ **c.** $f_z(x, y, z) = 3yz^2$

Example 4   Partial derivative of an implicitly defined function

Let z be defined implicitly as a function of x and y by the equation

$$x^2 z + yz^3 = x$$

Determine $\partial z / \partial x$ and $\partial z / \partial y$.

Solution   Differentiate implicitly with respect to x, treating y as a constant:

$$2xz + x^2 \frac{\partial z}{\partial x} + 3yz^2 \frac{\partial z}{\partial x} = 1$$

Then solve this equation for $\partial z / \partial x$:

$$\frac{\partial z}{\partial x} = \frac{1 - 2xz}{x^2 + 3yz^2}$$

Similarly, holding x constant and differentiating implicitly with respect to y, we find

$$x^2 \frac{\partial z}{\partial y} + z^3 + 3yz^2 \frac{\partial z}{\partial y} = 0$$

so that

$$\frac{\partial z}{\partial y} = \frac{-z^3}{x^2 + 3yz^2}$$

Partial Derivative as a Slope

A useful geometric interpretation of partial derivatives is indicated in Figure 11.16. In Figure 11.16**a**, the plane $y = y_0$ intersects the surface $z = f(x,y)$ in a curve C parallel to the xz-plane. That is, C is the trace of the surface in the plane $y = y_0$. An equation for this curve is $z = f(x, y_0)$, and because y_0 is fixed, the function depends only on x. Thus, we can compute the slope of the tangent line to C at the point $P(x_0, y_0, z_0)$ in the plane $y = y_0$ by differentiating $f(x, y_0)$ with respect to x and evaluating the derivative at $x = x_0$. That is, the slope is $f_x(x_0, y_0)$, the value of the partial derivative f_x at (x_0, y_0). The analogous interpretation for $f_y(x_0, y_0)$ is shown in Figure 11.16**b**.

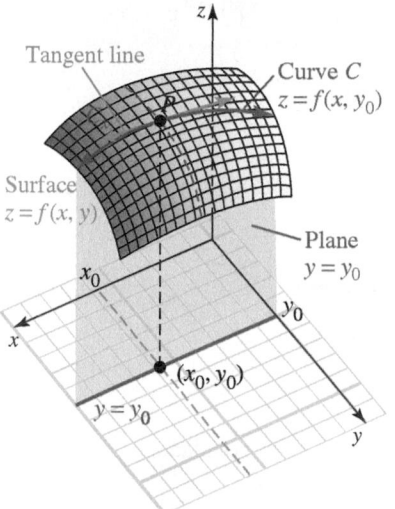

a. The tangent line in the plane $y = y_0$ to the curve C at the point P has slope $f_x(x_0, y_0)$

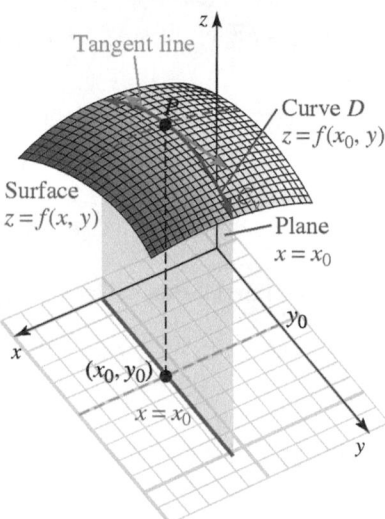

b. The tangent line in the plane $x = x_0$ to the curve D at the point P has slope $f_y(x_0, y_0)$

Figure 11.16   Slope interpretation of the partial derivative

---

PARTIAL DERIVATIVE AS THE SLOPE OF A TANGENT LINE   The line parallel to the xz-plane and tangent to the surface $z = f(x,y)$ at the point $P_0(x_0, y_0, z_0)$ has slope $f_x(x_0, y_0)$.

Likewise, the tangent line to the surface at P_0 that is parallel to the yz-plane has slope $f_y(x_0, y_0)$.

Example 5   Slope of a line parallel to the xz-plane

Find the slope of the line that is parallel to the xz-plane and tangent to the surface $z = x\sqrt{x + y}$ at the point $P(1, 3, 2)$.

Solution   If $f(x, y) = x\sqrt{x + y} = x(x + y)^{1/2}$, then the required slope is $f_x(1, 3)$.

$$f_x(x, y) = x\left(\frac{1}{2}\right)(x + y)^{-1/2}(1 + 0) + (1)(x + y)^{1/2}$$

$$= \frac{x}{2\sqrt{x + y}} + \sqrt{x + y}$$

$$f_x(1, 3) = \frac{1}{2\sqrt{1 + 3}} + \sqrt{1 + 3}$$

$$= \frac{9}{4}$$

Partial Derivative as a Rate

The derivative of a function of one variable can be interpreted as a rate of change, and the analogous interpretation of partial derivative may be described as follows.

> **PARTIAL DERIVATIVE AS A RATE**   As the point (x, y) moves from the fixed point $P_0(x_0, y_0)$, the function $f(x, y)$ changes at a rate given by $f_x(x_0, y_0)$ in the direction of the positive x-axis and by $f_y(x_0, y_0)$ in the direction of the positive y-axis.

Example 6   Partial derivatives as rates of change

In an electrical circuit with electromotive force (EMF) of E volts and resistance R ohms, the current is $I = E/R$ amperes. Find the partial derivatives $\partial I / \partial E$ and $\partial I / \partial R$ at the instant when $E - 120$ and $R = 15$ and interpret these derivatives as rates.

Solution   Since $I = ER^{-1}$, we have

$$\frac{\partial I}{\partial E} = R^{-1} \quad \text{and} \quad \frac{\partial I}{\partial R} = -ER^{-2}$$

and thus, when $E = 120$ and $R = 15$, we find that

$$\frac{\partial I}{\partial E} = 15^{-1} \approx 0.0667 \quad \text{and} \quad \frac{\partial I}{\partial R} = -(120)(15)^{-2} \approx -0.5333$$

This means that if the resistance is fixed at 15 ohms, the current is increasing (because the derivative is positive) with respect to voltage at the rate of 0.0667 ampere per volt when the EMF is 120 volts. Likewise, with the same fixed EMF, the current is decreasing (because the derivative is negative) with respect to resistance at the rate of 0.5333 ampere per ohm when the resistance is 15 ohms.   ∎

Higher-Order Partial Derivatives

The partial derivative of a function is a function, so it is possible to take the partial derivative of a partial derivative. This is very much like taking the second derivative of a function of one variable if we take two consecutive partial derivatives with respect to the same variable, and the resulting derivative is called the **second-order partial derivative** with respect to that variable. However, we can also take the partial derivative with respect to one variable and then take a second partial derivative with respect to a different variable, producing what is called a **mixed second-order partial derivative**. The higher-order partial derivatives for a function of two variables $f(x, y)$ are denoted as indicated in the following box:

> **HIGHER-ORDER PARTIAL DERIVATIVES**   Given $z = f(x, y)$.
> **Second-order partial derivatives**
>
> $$\frac{\partial^2 f}{\partial x^2} = \frac{\partial}{\partial x}\left(\frac{\partial f}{\partial x}\right) = (f_x)_x = f_{xx}$$
> $$\frac{\partial^2 f}{\partial y^2} = \frac{\partial}{\partial y}\left(\frac{\partial f}{\partial y}\right) = (f_y)_y = f_{yy}$$
>
> **Mixed second-order partial derivatives**
>
> $$\frac{\partial^2 f}{\partial x \partial y} = \frac{\partial}{\partial x}\left(\frac{\partial f}{\partial y}\right) = (f_y)_x = f_{yx}$$
> $$\frac{\partial^2 f}{\partial y \partial x} = \frac{\partial}{\partial y}\left(\frac{\partial f}{\partial x}\right) = (f_x)_y = f_{xy}$$

The notation f_{xy} means that we differentiate first with respect to x and then with respect to y, while $\frac{\partial^2 f}{\partial x \partial y}$ means just the opposite (differentiate with respect to y first and then with respect to x).

Example 7   Higher-order partial derivatives of a function of two variables

For $z = f(x, y) = 5x^2 - 2xy + 3y^3$, determine these higher-order partial derivatives.

a. $\dfrac{\partial^2 z}{\partial x\, \partial y}$ **b.** $\dfrac{\partial^2 f}{\partial y\, \partial x}$ **c.** $\dfrac{\partial^2 z}{\partial x^2}$ **d.** $f_{xy}(3, 2)$

Solution

a. First differentiate with respect to y; then differentiate with respect to x.

$$\frac{\partial z}{\partial y} = -2x + 9y^2$$

$$\frac{\partial^2 z}{\partial x\, \partial y} = \frac{\partial}{\partial x}\left(\frac{\partial z}{\partial y}\right) = \frac{\partial}{\partial x}(-2x + 9y^2) = -2$$

b. Differentiate first with respect to x and then with respect to y:

$$\frac{\partial f}{\partial x} = 10x - 2y$$

$$\frac{\partial^2 f}{\partial y\, \partial x} = \frac{\partial}{\partial y}\left(\frac{\partial f}{\partial x}\right) = \frac{\partial}{\partial y}(10x - 2y) = -2$$

c. Differentiate with respect to x twice:

$$\frac{\partial^2 z}{\partial x^2} = \frac{\partial}{\partial x}\left(\frac{\partial z}{\partial x}\right) = \frac{\partial}{\partial x}(10x - 2y) = 10$$

d. Evaluate the mixed partial found in part **b** at the point $(3, 2)$:

$$f_{xy}(3, 2) = -2$$

Notice from parts **a** and **b** of Example 7 that $\dfrac{\partial^2 z}{\partial x\, \partial y} = \dfrac{\partial^2 z}{\partial y\, \partial x}$. This equality of mixed partials does not hold for all functions, but for most functions we will encounter, it will be true. The following theorem provides sufficient conditions for this equality to occur.

Theorem 11.1   Equality of mixed partials

If the function $f(x, y)$ has mixed second-order partial derivatives f_{xy} and f_{yx} that are continuous in an open disk containing (x_0, y_0), then

$$f_{yx}(x_0, y_0) = f_{xy}(x_0, y_0)$$

Proof: This proof is omitted.   ♦

Example 8   Partial derivatives of functions of two variables

Determine f_{xy}, f_{yx}, f_{xx}, and f_{xxy}, where $f(x, y) = x^2 y e^y$.

Solution   We have the partial derivatives

$$f_x = 2xye^y \qquad f_y = x^2 e^y + x^2 y e^y$$

The mixed partial derivatives (which must be the same by the previous theorem) are

$$f_{xy} = (f_x)_y = 2xe^y + 2xye^y \qquad f_{yx} = (f_y)_x = 2xe^y + 2xye^y$$

Finally, we compute the second- and higher-order partial derivatives:

$$f_{xx} = (f_x)_x = 2ye^y \qquad \text{and} \qquad f_{xxy} = (f_{xx})_y = 2e^y + 2ye^y \qquad ▨$$

An equation involving partial derivatives is called a **partial differential equation**. An important partial differential equation is the diffusion or heat equation

$$\frac{\partial T}{\partial t} = c^2 \frac{\partial^2 T}{\partial x^2}$$

where $T(x, t)$ is the temperature in a thin rod at position x and time t. The constant c is called the **diffusivity** of the material in the rod. In the following example, we verify that a certain function satisfies this heat equation.

Example 9   Verifying that a function satisfies the heat equation

Verify that $T(x, t) = e^{-t} \cos \frac{x}{c}$ satisfies the heat equation, $\dfrac{\partial T}{\partial t} = c^2 \dfrac{\partial^2 T}{\partial x^2}$.

Solution $\quad \dfrac{\partial T}{\partial t} = -e^{-t} \cos \dfrac{x}{c}$
and

$$\begin{aligned}
\frac{\partial^2 T}{\partial x^2} &= \frac{\partial}{\partial x} \left(-\frac{1}{c} e^{-t} \sin \frac{x}{c} \right) \\
&= -\frac{1}{c^2} e^{-t} \cos \frac{x}{c}
\end{aligned}$$

Thus, T satisfies the heat equation $\dfrac{\partial T}{\partial t} = c^2 \dfrac{\partial^2 T}{\partial x^2}$. $\qquad ▨$

Analogous definitions can be made for functions of more than two variables. For example,

$$f_{zzz} = \frac{\partial^3 f}{\partial z^3} = \frac{\partial}{\partial z} \left[\frac{\partial}{\partial z} \left(\frac{\partial f}{\partial z} \right) \right] \qquad \text{or} \qquad f_{xyz} = \frac{\partial^3 f}{\partial z \, \partial y \, \partial x} = \frac{\partial}{\partial z} \left[\frac{\partial}{\partial y} \left(\frac{\partial f}{\partial x} \right) \right]$$

Example 10   Higher-order partial derivatives of a function of several variables

By direct calculation, show that $f_{xyz} = f_{yzx} = f_{zyx}$ for the function

$$f(x, y, z) = xyz + x^2 y^3 z^4$$

Solution   First, compute the partials:

$$\begin{aligned}
f_x(x, y, z) &= yz + 2xy^3 z^4 \\
f_y(x, y, z) &= xz + 3x^2 y^2 z^4 \\
f_z(x, y, z) &= xy + 4x^2 y^3 z^3
\end{aligned}$$

Next, determine the mixed partials:

$$\begin{aligned}
f_{xy}(x, y, z) &= (yz + 2xy^3 z^4)_y = z + 6xy^2 z^4 \\
f_{yz}(x, y, z) &= (xz + 3x^2 y^2 z^4)_z = x + 12x^2 y^2 z^3 \\
f_{zy}(x, y, z) &= (xy + 4x^2 y^3 z^3)_y = x + 12x^2 y^2 z^3
\end{aligned}$$

Finally, obtain the required higher mixed partials:

$$\begin{aligned}
f_{xyz}(x, y, z) &= (z + 6xy^2 z^4)_z = 1 + 24xy^2 z^3 \\
f_{yzx}(x, y, z) &= (x + 12x^2 y^2 z^3)_x = 1 + 24xy^2 z^3 \\
f_{zyx}(x, y, z) &= (x + 12x^2 y^2 z^3)_x = 1 + 24xy^2 z^3
\end{aligned}$$

▨

PROBLEM SET 11.3

Level 1

1. ■ What does this say? What is a partial derivative?
2. ■ What does this say? Describe two funda-mental interpretations of the partial derivatives $f_x(x,y)$ and $f_y(x,y)$

Determine f_x, f_y, f_{xx}, and f_{yx} in Problems 3-8.

3. $f(x,y) = x^3 + x^2y + xy^2 + y^3$

4. $f(x,y) = (x + xy + y)^3$ 5. $f(x,y) = \dfrac{x}{y}$

6. $f(x,y) = xe^{xy}$ 7. $f(x,y) = \ln(2x + 3y)$

8. $f(x,y) = \sin x^2 y$

Determine f_x and f_y in Problems 9-16.

9. **a.** $f(x,y) = (\sin x^2)\cos y$
   b. $f(x,y) = \sin(x^2 \cos y)$
10. **a.** $f(x,y) = (\sin \sqrt{x})\ln y^2$
    b. $f(x,y) = \sin(\sqrt{x} \ln y^2)$
11. $f(x,y) = \sqrt{3x^2 + y^4}$ 12. $f(x,y) = xy^2 \ln(x+y)$
13. $f(x,y) = x^2 e^{x+y}\cos y$ 14. $f(x,y) = xy^3 \tan^{-1} y$
15. $f(x,y) = \sin^{-1}(xy)$ 16. $f(x,y) = \cos^{-1}(xy)$

Determine f_x, f_y, and f_z in Problems 17-22.

17. $f(x,y,z) = xy^2 + yz^3 + xyz$

18. $f(x,y,z) = xye^z$

19. $f(x,y,z) = \dfrac{x + y^2}{z}$

20. $f(x,y,z) = \dfrac{xy^2 + yz}{xz}$

21. $f(x,y,z) = \ln(x + y^2 + z^3)$
22. $f(x,y,z) = \sin(xy + z)$

In Problems 23-28, determine $\dfrac{\partial z}{\partial x}$ and $\dfrac{\partial z}{\partial y}$ by differentiating implicitly.

23. $\dfrac{x^2}{9} - \dfrac{y^2}{4} + \dfrac{z^2}{2} = 1$

24. $3x^2 + 4y^2 + 2z^2 = 5$

25. $3x^2y + y^3z - z^2 = 1$

26. $x^3 - xy^2 + yz^2 - z^3 = 4$

27. $\sqrt{x} + y^2 + \sin xz = 2$

28. $\ln(xy + yz + xz) = 5 \ (x > 0, y > 0, z > 0)$

In Problems 29-32, compute the slope of the tangent line to the graph of f at the given point P_0 in the direction parallel to

a. *the xz-plane*   **b.** *the yz-plane*

29. $f(x,y) = xy^3 + x^3y; P_0(1, -1, -2)$

30. $f(x,y) = \dfrac{x^2 + y^2}{xy}; P_0(1, -1, -2)$

31. $f(x,y) = x^2 \sin(x+y); P_0(\frac{\pi}{2}, \frac{\pi}{2}, 0)$
32. $f(x,y) = x\ln(x + y^2); P_0(e, 0, e)$

Level 2

33. Determine f_x and f_y for

$$f(x,y) = \int_x^y (t^2 + 2t + 1)dt$$

34. Determine f_x and f_y for

$$f(x,y) = \int_{x^2}^{2y} (e^t + 3t)\, dt$$

*A function $f(x,y)$ is said to be **harmonic** on the open set S if f_{xx} and f_{yy} are continuous and*

$$f_{xx} + f_{yy} = 0$$

throughout S. Show that each function in Problems 35-38 is harmonic on the given set.

35. $f(x,y) = 3x^2y - y^3$; S is the entire plane.
36. $f(x,y) = \ln(x^2 + y^2)$; S is the plane with the point $(0,0)$ removed.
37. $f(x,y) = e^x \sin y$; S is the entire plane.
38. $f(x,y) = \sin x \cosh y$; S is the entire plane.
39. For $f(x,y) = \cos xy^2$, show $f_{xy} = f_{yz}$.
40. For $f(x,y) = (\sin^2 x)(\sin y)$, show

$$f_{xy} = f_{yx}$$

41. Find $f_{xzy} - f_{yzz}$, where

$$f(x,y,z) = x^2 + y^2 - 2xy \cos z$$

42. Two commodities Q_1 and Q_2 are said to be *substitute commodities* if an increase in the demand for either results in a decrease in the demand of the other. Let $D_1(p_1, p_2)$ and $D_2(p_1, p_2)$ be the demand functions for Q_1 and Q_2, respectively, where p_1 and p_2 are the respective unit prices for the commodities.
   a. Explain why $\dfrac{\partial D_1}{\partial p_1} < 0$ and $\dfrac{\partial D_2}{\partial p_2} < 0$.
   b. Are $\dfrac{\partial D_1}{\partial p_2}$ and $\dfrac{\partial D_2}{\partial p_1}$ positive or negative? Explain.
   c. Give examples of substitute commodities.

43. Two commodities Q_1 and Q_2 are said to be *complementary commodities* if a decrease in the demand for either results in a decrease in the demand of the other. Let $D_1(p_1, p_2)$ and $D_2(p_1, p_2)$ be the demand functions for Q_1 and Q_2, respectively, where p_1 and p_2 are the respective unit prices for the commodities.

a. Is it true that $\dfrac{\partial D_1}{\partial p_1} < 0$ and $\dfrac{\partial D_2}{\partial p_2} < 0$? Explain.

b. Determine whether $\dfrac{\partial D_1}{\partial p_2}$ and $\dfrac{\partial D_2}{\partial p_1}$ are positive or negative? Explain.

c. Give examples of complementary commodities.

44. Modeling Problem The flow (in cm^3/s) of blood from an artery into a small capillary can be modeled by

$$F(x, y, z) = \frac{c\pi x^2}{4}\sqrt{y - z}$$

for constant $c > 0$, where x is the diameter of the capillary, y is the pressure in the artery, and z is the pressure in the capillary. Compute the rate of change of the flow of blood with respect to

a. the diameter of the capillary.

b. the arterial pressure.

c. the capillary pressure.

45. Modeling Problem Biologists have studied the oxygen consumption of certain furry mammals. They have found that if the mammal's body temperature is T degrees Celsius, fur temperature is t degrees Celsius, and the mammal does not sweat, then its relative oxygen consumption can be modeled by

$$C(m, t, T) = \sigma(T - t)m^{-0.67}$$

(kg/h) where m is the mammal's mass (in kg) and σ is a physical constant. Compute the rate at which the oxygen consumption changes with respect to

a. the mass m

b. the body temperature T

c. the fur temperature t

46. Modeling Problem A gas that gathers on a surface in a condensed layer is said to be *adsorbed* on the surface, and the surface is called an *adsorbing surface*. The amount of gas adsorbed per unit area on an adsorbing surface can be modeled by

$$S(p, T, h) = ape^{h/(bT)}$$

where p is the gas pressure, T is the temperature of the gas, h is the heat of the adsorbed layer of gas, and a and b are physical constants. Compute the rate of change of S with respect to:

a. p **b.** h **c.** T

47. Modeling Problem At a certain factory, the output is given by the production function $Q = 120K^{2/3}L^{2/5}$, where K denotes the capital investment (in units of

$1,000) and L measures the size of the labor force (in worker-hours).

a. Determine the *marginal productivity of capital*, $\partial Q/\partial K$, and the *marginal productivity of labor*, $\partial Q/\partial L$.

b. Determine the signs of the second-order partial derivatives $\partial^2 Q/\partial L^2$ and $\partial^2 Q/\partial K^2$, and give an economic interpretation.

48. The ideal gas law says that $PV = kT$, where P is the pressure of a confined gas, V is the volume, T is the temperature, and k is a physical constant.

a. Calculate $\dfrac{\partial V}{\partial T}$. **b.** Calculate $\dfrac{\partial P}{\partial V}$.

c. Show that $\dfrac{\partial P}{\partial V} \cdot \dfrac{\partial V}{\partial T} \cdot \dfrac{\partial T}{\partial P} = -1$.

49. The temperature, T in degrees, at a point (x, y) on a given metal plate in the xy-plane is determined according to the formula

$$T(x, y) = x^3 + 2xy^2 + y$$

Compute the rate at which the temperature changes with distance if we start at $(2, 1)$ and move

a. parallel to the vector $\mathbf{j}$

b. parallel to the vector $\mathbf{i}$

50. In physics, the **wave equation** is

$$\frac{\partial^2 z}{\partial t^2} = c^2 \frac{\partial^2 z}{\partial x^2}$$

and the **heat equation** is

$$\frac{\partial z}{\partial t} = c^2 \frac{\partial^2 z}{\partial x^2}$$

In each of the following cases, determine whether z satisfies the wave equation, the heat equation, or neither.

a. $z = e^{-t}(\sin \frac{x}{c} + \cos \frac{x}{c})$

b. $z = \sin 3ct \sin 3x$

c. $z = \sin 5ct \cos 5x$

51. The *Cauchy-Riemann equations* are

$$\frac{\partial u}{\partial x} = \frac{\partial v}{\partial y} \quad \text{and} \quad \frac{\partial u}{\partial y} = -\frac{\partial v}{\partial x}$$

where $u = u(x, y)$ and $v = v(x, y)$. Which (if any) of the following functions satisfy the Cauchy-Riemann equations?

a. $u = e^{-x}\cos y, v = e^{-x}\sin y$

b. $u = x^2 + y^2, v = 2xy$

c. $u = \ln(x^2 + y^2), v = 2\tan^{-1}\left(\frac{y}{x}\right)$

52. When two resistors R_1 and R_2 are connected in parallel, their combined resistance R satisfies

$$\frac{1}{R} = \frac{1}{R_1} + \frac{1}{R_2}$$

What is $\dfrac{\partial^2 R}{\partial R_1^2}\dfrac{\partial^2 R}{\partial R_2^2}$?

53. Show that the production function $P(L,K) = L^\alpha K^\beta$, where α, β constants, satisfies

$$L\frac{\partial P}{\partial L} + K\frac{\partial P}{\partial K} = (\alpha + \beta)P$$

54. The kinetic energy of a body of mass m and velocity v is $K = \frac{1}{2}mv^2$. Show that

$$\frac{\partial K}{\partial m}\frac{\partial^2 K}{\partial v^2} = K$$

55. A study of ground penetration near a toxic waste dump models the amount of pollution P at a depth x (in feet) and time t (in years) by the function

$$P(x,t) = P_0 + P_1 e^{-kx}\sin(At - kx)$$

where A, k, P_0, and P_1 are constants. Show that $P(x,t)$ satisfies the diffusion equation

$$\frac{\partial P}{\partial t} = c^2\frac{\partial^2 P}{\partial x^2}$$

for a certain constant c.

Level 3

56. Think Tank Problem Let

$$f(x,y) = \begin{cases} xy\left(\dfrac{x^2 - y^2}{x^2 + y^2}\right) & \text{if } (x,y) \neq (0,0) \\ 0 & \text{if } (x,y) = (0,0) \end{cases}$$

Show that $f_x(0,y) = -y$ and $f_y(x,0) = x$, for all x and y. Then show that $f_{xy}(0,0) = -1$ and $f_{yx}(0,0) = 1$. Why does this not violate the equality of mixed partials theorem?

57. Show that $f_x(0,0) = 0$ but $f_y(0,0)$ does not exist, where

$$f(x,y) = \begin{cases} (x^2 + y)\sin\left(\dfrac{1}{x^2 + y^2}\right) & \text{if } (x,y) \neq (0,0) \\ 0 & \text{if } (x,y) = (0,0) \end{cases}$$

58. Modeling Problem Suppose a substance is injected into a tube containing a liquid solvent and that the

tube is placed so that its axis is parallel to the x-axis as shown in Figure 11.17.

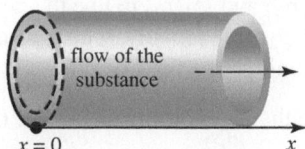

flow of the substance

$x = 0$

x

Figure 11.17 Problem 58

Assume that the concentration of the substance varies only in the x-direction, and let $C(x,t)$ denote the concentration at position x and time t. Because the number of molecules in this substance is very large, it is reasonable to assume that C is a continuous function whose partial derivatives exist. One model for the flow yields the **diffusion equation in one dimension**, namely,

$$\frac{\partial C}{\partial t} = \delta\frac{\partial^2 C}{\partial x^2}$$

where δ is the **diffusion constant**.
  a. What must δ be for a function of the form

$$C(x,t) = e^{ax+bt}$$

(a and b are constants) to satisfy the diffusion equation?
  b. Verify that

$$C(x,t) = t^{-1/2}e^{-x^2/(4\delta t)}$$

satisfies the diffusion equation.

59. The area of a triangle is $A = \frac{1}{2}ab\sin\gamma$, where γ is the angle between sides of length a and b.
  a. Determine $\dfrac{\partial A}{\partial a}, \dfrac{\partial A}{\partial b}$, and $\dfrac{\partial A}{\partial\gamma}$.
  b. Suppose a is given as a function of b, A, and γ. What is $\dfrac{\partial a}{\partial\gamma}$?

60. Journal Problem (*Crux*, problem by John A. Winterink.)* Prove the validity of the following simple method for finding the center of a conic: For the central conic,

$$\phi(x,y) = ax^2 + 2hxy + by^2 + 2gx + 2fy + c = 0$$

where $ab - h^2 \neq 0$, show that the center is the intersection of the lines $\partial\phi/\partial x = 0$ and $\partial\phi/\partial y = 0$.

11.4 TANGENT PLANES, APPROXIMATIONS, AND DIFFERENTIABILITY

IN THIS SECTION: *Tangent planes, incremental approximations, the total differential, differentiability*
We now find the equation of a plane tangent to a surface at a given point on the surface and then use this tangent plane as an approximation for the value of a corresponding nearby point on the surface. This leads us to introduce the increment of a function of two variables and to show how this concept can be used to define differentiability.

Tangent Planes

Suppose S is a surface with the equation $z = f(x, y)$, where f has continuous first partial derivatives f_x and f_y. Let $P_0(x_0, y_0, z_0)$ be a point on S, and let C_1 be the curve of intersection of S with the plane $x = x_0$ and let C_2 be the intersection of S with the plane $y = y_0$, as shown in Figure 11.18**a**. The tangent lines T_1 and T_2 to C_1 and C_2, respectively, at P_0 determine a unique plane, and we will find that this plane actually contains the tangent to *every* smooth curve on S that passes through P_0. We call this plane the **tangent plane** to S at P_0 (Figure 11.18**b**).

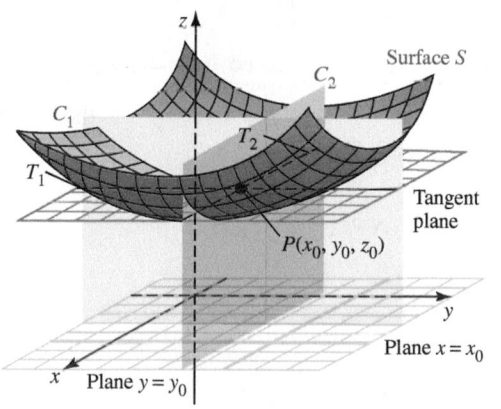

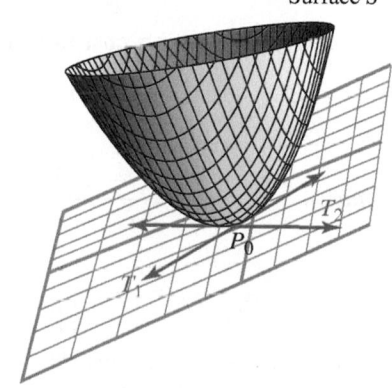

a. C_1 and C_2 are the curves of intersection of the surface S with the planes $x = x_0$ and $y = y_0$

b. The tangent plane to S at P_0 contains the tangent lines T_1 and T_2 to C_1 and C_2, respectively

Figure 11.18   Tangent plane to the surface S at the point P_0

To find an equation for the tangent plane at P_0, consider its normal $\mathbf{N} = A\mathbf{i} + B\mathbf{j} + C\mathbf{k}$. The plane can be expressed as

$$A(x - x_0) + B(y - y_0) + C(z - z_0) = 0$$

If $C \neq 0$, divide both sides by C and let $a = -A/C$ and $b = -B/C$ to obtain

$$z - z_0 = a(x - x_0) + b(y - y_0)$$

The intersection of this plane and the plane $x = x_0$ is the tangent line T_1, which we know has slope $f_y(x_0, y_0)$ from the geometric interpretation of partial derivatives. Setting $x = x_0$ in the equation for the tangent plane, we find that T_1 has the point-slope form

$$z - z_0 = b(y - y_0)$$

so we must have $b = f_y(x_0, y_0) = \left.\dfrac{\partial z}{\partial y}\right|_{(x_0, y_0)}$. Similarly, setting $y = y_0$, we obtain $z - z_0 = a(x - x_0)$ which represents the tangent line T_2, with slope $a = f_x(x_0, y_0)$. To summarize:

EQUATION OF THE TANGENT PLANE Suppose S is a surface with the equation $z = f(x, y)$ and let $P_0(x_0, y_0, z_0)$ be a point on S at which a tangent plane exists. Then an **equation for the tangent plane** to S at P_0 is

$$z - z_0 = f_x(x_0, y_0)(x - x_0) + f_y(x_0, y_0)(y - y_0)$$

If the equation is written in the form

$$Ax + By + Cz + D = 0$$

we say the equation of the plane is in **standard form**.

Example 1   Equation of a tangent plane for a surface defined by $z = f(x,y)$

Find an equation for the tangent plane to the surface $z = \tan^{-1}\frac{y}{x}$ at the point $P_0(1, \sqrt{3}, \frac{\pi}{3})$.

Solution $f_x(x, y) = \dfrac{-yx^{-2}}{1 + \left(\frac{y}{x}\right)^2} = \dfrac{-y}{x^2 + y^2}$ $f_x(1, \sqrt{3}) = \dfrac{-\sqrt{3}}{1 + 3} = \dfrac{-\sqrt{3}}{4}$

$f_y(x, y) = \dfrac{x^{-1}}{1 + \left(\frac{y}{x}\right)^2} = \dfrac{x}{x^2 + y^2}$ $f_y(1, \sqrt{3}) = \dfrac{1}{1 + 3} = \dfrac{1}{4}$

The equation of the tangent plane is

$$z - \frac{\pi}{3} = \left(\frac{-\sqrt{3}}{4}\right)(x - 1) + \frac{1}{4}(y - \sqrt{3})$$

or, in standard form,

$$3\sqrt{3}x - 3y + 12z - 4\pi = 0$$

Incremental Approximations

In Chapter 3, we observed that the tangent line to the curve $y = f(x)$ at the point $P(x_0, y_0)$ is the line that best fits the shape of the curve in the immediate vicinity of P. That is, if f is differentiable at $x = x_0$ and the increment Δx is sufficiently small, then

$$\Delta y = f(x_0 + \Delta x) - f(x_0) \approx f'(x_0)\Delta x$$

or equivalently,

$$f(x_0 + \Delta x) \approx f(x_0) + f'(x_0)\Delta x$$

Similarly, the tangent plane at $P(x_0, y_0, z_0)$ is the plane that best fits the shape of the surface $z = f(x, y)$ near P, and the analogous **incremental** (or **linear**) approximation formula is as follows.

> **INCREMENTAL APPROXIMATION OF A FUNCTION OF TWO VARIABLES** If $f(x, y)$ and its partial derivatives f_x and f_y are defined in an open region R containing the point $P(x_0, y_0)$ and f_x and f_y are continuous at P, then
>
> $$\Delta f = f(x_0 + \Delta x, y_0 + \Delta y) - f(x_0, y_0)$$
> $$\approx f_x(x_0, y_0)\Delta x + f_y(x_0, y_0)\Delta y$$
>
> so that
>
> $$f(x_0 + \Delta x, y_0 + \Delta y) \approx f(x_0, y_0) + f_x(x_0, y_0)\Delta x + f_y(x_0, y_0)\Delta y$$

A graphical interpretation of this incremental approximation formula is shown in Figure 11.19.

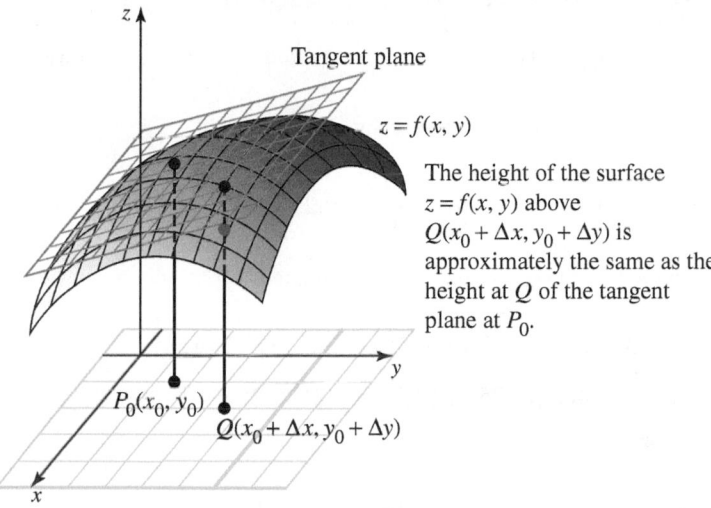

Figure 11.19 Incremental approximation to a function of two variables

The tangent plane to the surface $z = f(x, y)$ has the equation

$$z - z_0 = f_x(x_0, y_0)(x - x_0) + f_y(x_0, y_0)(y - y_0)$$

or

$$z - f(x_0, y_0) = f_x(x_0, y_0)\Delta x + f_y(x_0, y_0)\Delta y$$

As long as we are near (x_0, y_0), the height of the tangent plane is approximately the same as the height of the surface. Thus, if $|\Delta x|$ and $|\Delta y|$ are small, the point $(x_0 + \Delta x, y_0 + \Delta y)$ will be near (x_0, y_0) and we have

$$\underbrace{f(x_0 + \Delta x, y_0 + \Delta y)}_{\textit{Height of } z = f(x, y) \textit{ above } Q(x_0 + \Delta x, y_0 + \Delta y)} \approx \underbrace{f(x_0, y_0) + f_x(x_0, y_0)\Delta x + f_y(x_0, y_0)\Delta y}_{\textit{Height of the tangent plane above } Q(x_0 + \Delta x, y_0 + \Delta y)}$$

Increments of a function of three variables $f(x, y, z)$ can be defined in a similar fashion. Suppose f has continuous partial derivatives f_x, f_y, f_z in a ball centered at the point (x_0, y_0, z_0). Then if the numbers Δx, Δy, Δz are all sufficiently small, we have

$$\Delta f = f(x_0 + \Delta x, y_0 + \Delta y, z_0 + \Delta z) - f(x_0, y_0, z_0)$$
$$\approx f_x(x_0, y_0, z_0)\Delta x + f_y(x_0, y_0, z_0)\Delta y + f_z(x_0, y_0, z_0)\Delta z$$

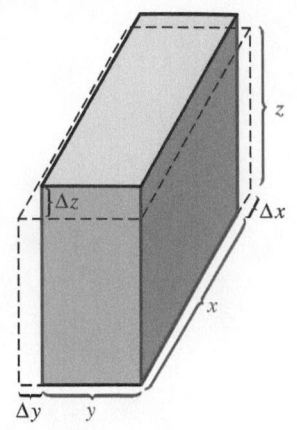

Figure 11.20 Construction of a box

Example 2   Using increments to estimate the change of a function

An open box has length 3 ft, width 1 ft, and height 2 ft, and is constructed from material that costs $2/ft^2 for the sides and $3/ft^2 for the bottom (see Figure 11.20).
Compute the cost of constructing the box, and then use increments to estimate the change in cost if the length and width are each increased by 3 in. and the height is decreased by 4 in.

Solution   An open (no top) box with length x, width y, and height z has surface area

$$S = \underbrace{xy}_{Bottom} + \underbrace{2xz + 2yz}_{Four\ side\ faces}$$

Because the sides cost $2/ft^2 and the bottom $3/ft^2, the total cost is

$$C(x, y, z) = 3xy + 2(2xz + 2yz)$$

The partial derivatives of C are

$$C_x = 3y + 4z \qquad C_y = 3x + 4z \qquad C_z = 4x + 4y$$

and the dimensions of the box change by

$$\Delta x = \frac{3}{12} = 0.25 \text{ ft} \qquad \Delta y = \frac{3}{12} = 0.25 \text{ ft} \qquad \Delta z = \frac{-4}{12} \approx -0.33 \text{ ft}$$

Thus, the change in the total cost is approximated by

$$\Delta C \approx C_x(3, 1, 2)\Delta x + C_y(3, 1, 2)\Delta y + C_z(3, 1, 2)\Delta z$$
$$= [3(1) + 4(2)](0.25) + [3(3) + 4(2)](0.25) + [4(3) + 4(1)]\left(-\frac{4}{12}\right)$$
$$\approx 1.67$$

That is, the cost increases by approximately $1.67.

Example 3   Maximum percentage error using differentials

The radius and height of a right circular cone are measured with errors of at most 3% and 2% respectively. Use incremental approximations to estimate the maximum percentage error in computing the volume of the cone using these measurements and the formula $V = \frac{1}{3}\pi R^2 H$.

Solution   We are given that

$$\left|\frac{\Delta R}{R}\right| \leq 0.03 \qquad \text{and} \qquad \left|\frac{\Delta H}{H}\right| \leq 0.02$$

The partial derivatives of V are

$$V_R = \frac{2}{3}\pi RH \qquad \text{and} \qquad V_H = \frac{1}{3}\pi R^2$$

so the change in V is approximated by

$$\Delta V \approx \left(\frac{2}{3}\pi RH\right)\Delta R + \left(\frac{1}{3}\pi R^2\right)\Delta H$$

Dividing by the volume $V = \frac{1}{3}\pi R^2 H$, we obtain

$$\frac{\Delta V}{V} \approx \frac{\frac{2}{3}\pi RH\,\Delta R + \frac{1}{3}\pi R^2\,\Delta H}{\frac{1}{3}\pi R^2 H} = 2\left(\frac{\Delta R}{R}\right) + \left(\frac{\Delta H}{H}\right)$$

so that

$$\left|\frac{\Delta V}{V}\right| \le 2\left|\frac{\Delta R}{R}\right| + \left|\frac{\Delta H}{H}\right| = 2(0.03) + (0.02) = 0.08$$

Thus, the maximum percentage error in computing the volume V is approximately 8%.

The Total Differential

For a function of one variable, $y = f(x)$, we defined the differential dy to be $dy = f'(x)\,dx$. For the two-variable case, we make the following analogous definition.

> **TOTAL DIFFERENTIAL** The **total differential of the function of two variables** $f(x,y)$ is
>
> $$df = \frac{\partial f}{\partial x}dx + \frac{\partial f}{\partial y}dy = f_x(x,y)\,dx + f_y(x,y)\,dy$$
>
> where x and y are independent variables. Similarly, for a function of three variables $w = f(x,y,z)$ the **total differential of the function of three variables** is
>
> $$df = \frac{\partial f}{\partial x}dx + \frac{\partial f}{\partial y}dy + \frac{\partial f}{\partial z}dz$$

Example 4 Total differential

Determine the total differential of the given functions:
a. $f(x,y) = x^2 \ln(3y^2 - 2x)$ **b.** $f(x,y,z) = 2x^3 + 5y^4 - 6z$

Solution

a. $df = \dfrac{\partial f}{\partial x}dx + \dfrac{\partial f}{\partial y}dy$

$$= \left[2x\ln(3y^2 - 2x) + x^2\frac{-2}{3y^2 - 2x}\right]dx + \left[x^2\frac{6y}{3y^2 - 2x}\right]dy$$

$$= \left[2x\ln(3y^2 - 2x) - \frac{2x^2}{3y^2 - 2x}\right]dx + \frac{6x^2 y}{3y^2 - 2x}dy$$

b. $df = \dfrac{\partial f}{\partial x}dx + \dfrac{\partial f}{\partial y}dy + \dfrac{\partial f}{\partial z}dz$

$$= 6x^2\,dx + 20y^3\,dy - 6\,dz$$

Example 5 Application of the total differential

At a certain factory, the daily output is $Q = 60K^{1/2}L^{1/3}$ units, where K denotes the capital investment (in units of \$1,000) and L the size of the labor force (in worker-hours). The current capital investment is \$900,000, and 1,000 worker-hours of labor are used each day. Estimate the change in output that will result if capital investment is increased by $1,000 and labor is decreased by 2 worker-hours.

Solution   The change in output is represented by quantity ΔQ. We have $K = 900$, $L = 1,000$, $\Delta K = 1$, and $\Delta L = -2$. The total differential of $Q(x, y)$ is

$$dQ = \frac{\partial Q}{\partial K} dK + \frac{\partial Q}{\partial L} dL$$

$$= 60 \left(\frac{1}{2} \right) K^{-1/2} L^{1/3} dK + 60 \left(\frac{1}{3} \right) K^{1/2} L^{-2/3} dL$$

$$= 30 K^{-1/2} L^{1/3} dK + 20 K^{1/2} L^{-2/3} dL$$

Likewise, we have:

$$\Delta Q \approx \frac{\partial Q}{\partial K} \Delta K + \frac{\partial Q}{\partial L} \Delta L$$

$$\Delta Q \approx 30(900)^{-1/2}(1,000)^{1/3}(1) + 20(900)^{1/2}(1,000)^{-2/3}(-2)$$
$$= -2$$

Thus, the output decreases by approximately 2 units when the capital investment is increased by $1,000 and labor is decreased by 2 worker-hours.

Example 6   Maximum percentage error in an electrical circuit

When two resistances R_1 and R_2 are connected in parallel, the total resistance R satisfies

$$\frac{1}{R} = \frac{1}{R_1} + \frac{1}{R_2}$$

If R_1 is measured as 300 ohms with a maximum error of 2% and R_2 is measured as 500 ohms with a maximum error of 3%, use incremental approximations to estimate the maximum percentage error in R.

Solution   We are given that

$$\left| \frac{\Delta R_1}{R_1} \right| \leq 0.02 \quad \text{and} \quad \left| \frac{\Delta R_2}{R_2} \right| \leq 0.03$$

and we wish to find the maximum value of $\left| \frac{\Delta R}{R} \right|$. Because $R = \frac{R_1 R_2}{R_1 + R_2}$ (solve the given equation for R), we have

$$\frac{\partial R}{\partial R_1} = \frac{R_2^2}{(R_1 + R_2)^2} \quad \text{and} \quad \frac{\partial R}{\partial R_2} = \frac{R_1^2}{(R_1 + R_2)^2} \qquad \textit{Quotient rule}$$

the total differential of R is

$$\Delta R \approx \frac{\partial R}{\partial R_1} \Delta R_1 + \frac{\partial R}{\partial R_2} \Delta R_2$$

$$= \frac{R_2^2}{(R_1 + R_2)^2} \Delta R_1 + \frac{R_1^2}{(R_1 + R_2)^2} \Delta R_2$$

We now find $\frac{\Delta R}{R}$ by dividing both sides by R; however, since $\frac{1}{R} = \frac{R_1 + R_2}{R_1 R_2}$, it follows that

$$\Delta R \cdot \frac{1}{R} \approx \left[\frac{R_2^2}{(R_1 + R_2)^2} \Delta R_1 + \frac{R_1^2}{(R_1 + R_2)^2} \Delta R_2 \right] \cdot \frac{R_1 + R_2}{R_1 R_2}$$

$$\frac{\Delta R}{R} \approx \frac{R_2}{R_1 + R_2} \cdot \frac{\Delta R_1}{R_1} + \frac{R_1}{R_1 + R_2} \cdot \frac{\Delta R_2}{R_2}$$

Finally, apply the triangle inequality (Table 1.1, p. 17) to this relationship:

$$\left|\frac{\Delta R}{R}\right| \leq \left|\frac{R_2}{R_1 + R_2}\right| \left|\frac{\Delta R_1}{R_1}\right| + \left|\frac{R_1}{R_1 + R_2}\right| \left|\frac{\Delta R_2}{R_2}\right|$$

$$\leq \frac{500}{300 + 500}(0.02) + \frac{300}{300 + 500}(0.03)$$

$$= 0.02375$$

The maximum percentage is approximately 2.4%.

Differentiability

Recall from Chapter 3 that if $f(x)$ is differentiable at x_0, its *increment* is

$$\Delta f = f(x_0 + \Delta x) - f(x_0) = f'(x_0)\Delta x + \epsilon \Delta x$$

where $\epsilon \to 0$ as $\Delta x \to 0$ (see Figure 11.21).

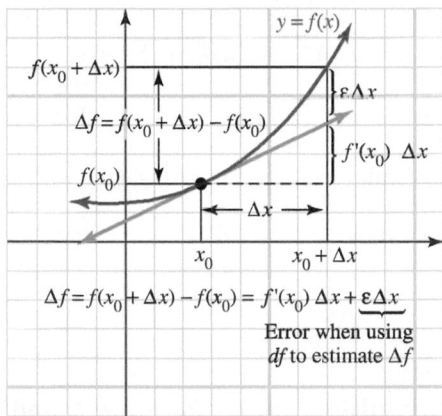

Figure 11.21   Increment of a function f

For a function of two variables, the increment of x is an independent variable denoted by Δx, the increment of y is an independent variable denoted by Δy, and the increment of f at (x_0, y_0) is defined as

$$\Delta f = f(x_0 + \Delta x, y_0 + \Delta y) - f(x_0, y_0)$$

We use this increment representation to define differentiability as follows.

DIFFERENTIABILITY   The function $f(x, y)$ is **differentiable** at (x_0, y_0) if the increment of f can be expressed as

$$\Delta f = f_x(x_0, y_0)\Delta x + f_y(x_0, y_0)\Delta y + \epsilon_1 \Delta x + \epsilon_2 \Delta y$$

where $\epsilon_1 \to 0$ and $\epsilon_2 \to 0$ as both $\Delta x \to 0$ and $\Delta y \to 0$. In addition, $f(x, y)$ is said to be **differentiable in the region R** of the plane if f is differentiable at each point in R.

In Section 3.1 (Theorem 3.2) we showed that a function of one variable is continuous wherever it is differentiable. The following theorem establishes the same result for a function of two variables.

Theorem 11.2   Differentiability implies continuity

If $f(x,y)$ is differentiable at (x_0, y_0), it is also continuous there.

Proof: We wish to show that $f(x,y) \to f(x_0, y_0)$ as $(x,y) \to (x_0, y_0)$ or, equivalently, that

$$\lim_{(x,y) \to (x_0, y_0)} [f(x,y) - f(x_0, y_0)] = 0$$

If we set $\Delta x = x - x_0$ and $\Delta y = y - y_0$ and let Δf denote the increment of f at (x_0, y_0), we have (by substitution)

$$f(x,y) - f(x_0, y_0) = f(x_0 + \Delta x, y_0 + \Delta y) - f(x_0, y_0) = \Delta f$$

Then, because $(\Delta x, \Delta y) \to (0,0)$ as $(x,y) \to (x_0, y_0)$, we wish to prove that

$$\lim_{(\Delta x, \Delta y) \to (0,0)} \Delta f = 0$$

Be careful about how you use the word differentiable. In the single variable case, a function is differentiable at a point if its derivative exists there. However, the word is used differently for a function of two variables. In particular, the existence of the partial derivatives f_x and f_y does not guarantee that the function is differentiable, as illustrated in the following example.

Since f is differentiable at (x_0, y_0), we have

$$\Delta f = f_x(x_0, y_0)\Delta x + f_y(x_0, y_0)\Delta y + \epsilon_1 \Delta x + \epsilon_2 \Delta y$$

where $\epsilon_1 \to 0$ and $\epsilon_2 \to 0$ as $(\Delta x, \Delta y) \to (0,0)$. It follows that

$$\begin{aligned}
\lim_{(\Delta x, \Delta y) \to (0,0)} \Delta f &= \lim_{(\Delta x, \Delta y) \to (0,0)} [f_x(x_0, y_0)\Delta x + f_y(x_0, y_0)\Delta y + \epsilon_1 \Delta x + \epsilon_2 \Delta y] \\
&= [f_x(x_0, y_0)] \cdot 0 + [f_y(x_0, y_0)] \cdot 0 + 0 + 0 \\
&= 0
\end{aligned}$$

as required.   ◆

Example 7   A nondifferentiable function for which f_x and f_y exist

Let

$$f(x,y) = \begin{cases} 1 & \text{if } x > 0 \text{ and } y > 0 \\ 0 & \text{otherwise} \end{cases}$$

That is, the function f has the value 1 when (x,y) is in the first quadrant and is 0 elsewhere. Show that the partial derivatives f_x and f_y exist at the origin, but f is not differentiable there.

Solution   Since $f(0,0) = 0$, we have

$$f_x(0,0) = \lim_{\Delta x \to 0} \frac{f(0 + \Delta x, 0) - f(0,0)}{\Delta x} = 0$$

and similarly, $f_y(0,0) = 0$. Thus, the partial derivatives both exist at the origin.

If $f(x,y)$ were differentiable at the origin, it would have to be continuous there (Theorem 11.2). Thus, we can show f is *not* differentiable by showing that it is *not* continuous at $(0,0)$. Toward this end, note that $\lim_{(x,y) \to (0,0)} f(x,y)$ is 1 along the line $y = x$ in the first quadrant but is 0 if the approach is along the x-axis. This means that the limit does not exist. Thus, f is not continuous at $(0,0)$ and consequently is also not differentiable there.   ▪

Although the existence of partial derivatives at $P(x_0, y_0)$ is not enough to guarantee that $f(x,y)$ is differentiable at P, we do have the following sufficient condition for differentiability.

Theorem 11.3 **Sufficient condition for differentiability**

If f is a function of x and y, and f, f_x, and f_y are continuous in a disk D centered at (x_0, y_0), then f is differentiable at (x_0, y_0).

Proof: The proof is found in advanced calculus texts.   ♦

Note that the function in Example 7 does not contradict Theorem 11.3 because there is no disk centered at (0,0) on which f is continuous.

Example 8   **Establishing differentiability**

Show that $f(x, y) = x^2 y + xy^3$ is differentiable for all (x, y).

Solution   Compute the partial derivatives

$$f_x(x, y) = \frac{\partial}{\partial x}(x^2 y + xy^3) = 2xy + y^3$$

$$f_y(x, y) = \frac{\partial}{\partial y}(x^2 y + xy^3) = x^2 + 3xy^2$$

Because f, f_x, and f_y are all polynomials in x and y, they are continuous throughout the plane. Therefore, the sufficient condition for differentiability theorem assures us that f must be differentiable for all x and y.   ▧

Finally, let us remark that although continuity of the first-order partial derivatives in some neighborhood of the point (x_0, y_0) guarantees that f is differentiable at (x_0, y_0), it is possible for a function to be differentiable at (x_0, y_0) when its partial derivatives are not continuous there. Examples that illustrate this fact can be seen in advanced calculus.

PROBLEM SET 11.4

Level 1

1. ■ What does this say? Discuss the notion of differentiability of a function of two variables.
2. ■ What does this say? Even though it is important to read and understand three-dimensional figures in the book, we have also given drawing lessons to help you draw three-dimensional figures. On your paper, draw a surface S and a plane tangent to the surface at a point P.

In Problems 3-8, determine the standard-form equations for the tangent plane to the given surface at the prescribed point P_0.

3. $z = \sqrt{x^2 + y^2}$ at $P_0(3, 1, \sqrt{10})$

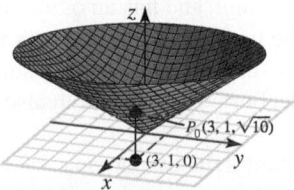

Interactive

4. $z = 10 - x^2 - y^2$ at $P_0(2, 2, 2)$

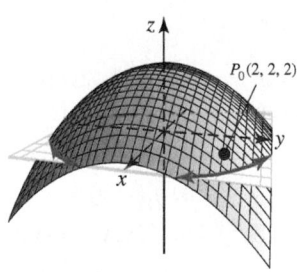

Interactive

5. $f(x, y) = x^2 + y^2 + \sin xy$ at $P_0 = (0, 2, 4)$

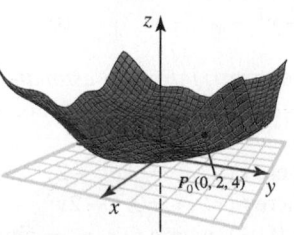

Interactive

6. $f(x,y) = e^{-x} \sin y$ at $P_0(0, \frac{\pi}{2}, 1)$

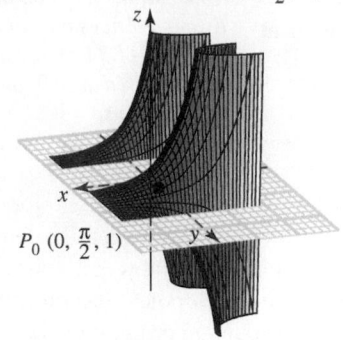

$P_0\ (0, \frac{\pi}{2}, 1)$

Interactive

7. $z = \tan^{-1} \frac{y}{x}$ at $P_0(2, 2, \frac{\pi}{4})$

8. $z = \ln \left| x + y^2 \right|$ at $P_0(-3, -2, 0)$

Determine the total differential of the functions given in Problems 9-20.

9. $f(x,y) = 5x^2 y^3$

10. $f(x,y) = 8x^3 y^2 - x^4 y^5$

11. $f(x,y) = \sin xy$

12. $f(x,y) = \cos x^2 y$

13. $f(x,y) = \dfrac{y}{x}$

14. $f(x,y) = \dfrac{x^2}{y}$

15. $f(x,y) = ye^x$

16. $f(x,y) = e^{x^2 + y}$

17. $f(x,y,z) = 3x^3 - 2y^4 + 5z$

18. $f(x,y,z) = \sin x + \sin y + \cos z$

19. $f(x,y,z) = z^2 \sin(2x - 3y)$

20. $f(x,y,z) = 3y^2 z \cos x$

Show that the functions in Problems 21-26 are differentiable for all (x,y).

21. $f(x,y) = xy^3 + 3xy^2$ **22.** $f(x,y) = x^2 + 4x - y^2$

23. $f(x,y) = e^{2x + y^2}$ **24.** $f(x,y) = e^{x^2 + y^2}$

25. $f(x,y) = \cos(2x - 3y^2)$ **26.** $f(x,y) = \sin(x^2 + 3y)$

Find $f_x, f_y,$ and f_λ for the functions in Problems 27-30.

27. $f(x,y,\lambda) = x + 2xy + \lambda(xy - 10)$

28. $f(x,y,\lambda) = 2x + 2y + \lambda xy$

29. $f(x,y,\lambda) = x^2 + y^2 - \lambda(3x + 2y - 6)$

30. $f(x,y,\lambda) = x^2 - y^2 - \lambda(5x - 3y + 10)$

Use an incremental approximation to estimate the functions at the values given in Problems 31-36. Check by using a calculator.

31. $f(1.01, 2.03)$, where $f(x,y) = 3x^4 + 2y^4$

32. $f(0.98, 1.03)$, where $f(x,y) = x^5 - 2y^3$

33. $f(\frac{\pi}{2} + 0.01, \frac{\pi}{2} - 0.01)$, where $f(x,y) = \sin(x + y)$

34. $f\left(\sqrt{\frac{\pi}{2}} + 0.01, \sqrt{\frac{\pi}{2}} - 0.01\right)$, where $f(x,y) = \sin(xy)$

35. $f(1.01, 0.98)$, where $f(x,y) = e^{xy}$

36. $f(1.01, 0.98)$, where $f(x,y) = e^{x^2 y^2}$

Level 2

37. Find an equation for each horizontal tangent plane to the surface

$$z = 5 - x^2 - y^2 + 4y$$

38. Find an equation for each horizontal tangent plane to the surface

$$z = 4(x - 1)^2 + 3(y + 1)^2$$

39. Show that if x and y are sufficiently close to zero and f is differentiable at $(0,0)$, then

$$f(x,y) \approx f(0,0) + x f_x(0,0) + y f_y(0,0)$$

40. Use the approximation formula in Problem 39 to show that

$$\frac{1}{1 + x - y} \approx 1 - x + y$$

for small x and y.

41. If x and y are sufficiently close to zero, what is the approximate value of the expression

$$\frac{1}{(x + 1)^2 + (y + 1)^2}$$

Hint: Use Problem 39.

42. When two resistors with resistances P and Q ohms are connected in parallel, the combined resistance is R, where

$$\frac{1}{R} = \frac{1}{P} + \frac{1}{Q}$$

If P and Q are measured at 6 and 10 ohms, respectively, with errors no greater than 1%, what is the maximum percentage error in the computation of R?

43. A closed box is found to have length 2 ft, width 4 ft, and height 3 ft, where the measurement of each dimension is made with a maximum possible error of ± 0.02 ft. The top of the box is made from material that costs \$2/ft^2; and the material for the sides and bottom costs only \$1.50/ft^2. Use increments to approximate the maximum error in the computation of the cost of the box.

44. A cylindrical tank is 4 ft high and has an outer diameter of 2 ft. The walls of the tank are 0.2 in. thick. Approximate the volume of the interior of the tank assuming the tank has a top and a bottom that are both also 0.2 in. thick.

45. a. The Higrade Company sells two brands, X and Y, of a commercial soap, in thousand-pound units. If x units of brand X and y units of brand Y are sold, the unit price for brand X is

$$p(x) = 4,000 - 500x$$

and that of brand Y is

$$q(y) = 3,000 - 450y$$

Find an expression for the total revenue R in terms of p and q.

b. Suppose brand X sells for \$500 per unit and brand Y sells for \$750 per unit. Estimate the change in total revenue if the unit prices are increased by \$20 for brand X and \$18 for brand Y.

46. The output at a certain factory is

$$Q = 150K^{2/3}L^{1/3}$$

where K is the capital investment in units of \$1,000 and L is the size of the labor force, measured in worker-hours. The current capital investment is \$500,000 and 1,500 worker-hours of labor are used.

a. Estimate the change in output that results when capital investment is increased by \$700 and labor is increased by 6 worker-hours.

b. What if capital investment is increased by \$500 and labor is decreased by 4 worker-hours?

47. According to Poiseuille's law, the resistance to the flow of blood offered by a cylindrical blood vessel of radius r and length x is

$$R(r, x) = \frac{cx}{r^4}$$

for a constant $c > 0$. A certain blood vessel in the body is 8 cm long and has a radius of 2 mm. Estimate the percentage change in R when x is increased by 3% and r is decreased by 2%.

48. For 1 mole of an ideal gas, the volume V, pressure P, and absolute temperature T are related by the equation $PV = kT$, where k is a certain fixed constant that depends on the gas. Suppose we know that if $T = 400$ (absolute) and $P = 3,000$ lb/ft^2, then $V = 14$ ft^3. Approximate the change in pressure if the temperature and volume are increased to 403 and 14.1 ft^3, respectively.

49. Modeling Problem If x gram-moles of sulfuric acid are mixed with y gram-moles of water, the heat liberated is modeled by

$$F(x, y) = \frac{1.786xy}{1.798x + y} \text{ cal}$$

Approximately how much additional heat is generated if a mixture of 5 gram-moles of acid and 4 gram-moles of water is increased to a mixture of 5.1 gram-moles of acid and 4.04 gram-moles of water?

50. Modeling Problem A business analyst models the sales of a new product by the function

$$Q(x, y) = 20x^{3/2}y$$

where x thousand dollars are spent on development and y thousand dollars on promotion. Current plans call for the expenditure of \$36,000 on development and \$25,000 on promotion. Use the total differential of Q to estimate the change in sales that will result if the amount spent on development is increased by \$500 and the amount spent on promotion is decreased by \$500.

51. Using x hours of skilled labor and y hours of unskilled labor, a manufacturer can produce $f(x, y) = 10xy^{1/2}$ units. Currently, the manufacturer has used 30 hr of skilled labor and 36 hr of unskilled labor and is planning to use 1 additional hour of skilled labor. Use calculus to estimate the corresponding change that the manufacturer should make in the level of unskilled labor so that the total output will remain the same.

52. Modeling Problem A grocer's weekly profit from the sale of two brands of juice is modeled by

$$P(x, y) = (x - 30)(70 - 5x + 4y)$$
$$+ (y - 40)(80 + 6x - 7y)$$

dollars, where x cents is the price per can of the first brand and y cents is the price per can of the second. Currently the first brand sells for 50¢ per can and the second for 52¢ per can. Use the total differential to estimate the change in the weekly profit that will result if the grocer raises the price of the first brand by 1¢ per can and lowers the price of the second brand by 2¢ per can.

53. Modeling Problem It is known that the period T of a simple pendulum with small oscillations is modeled by

$$T = 2\pi\sqrt{\frac{L}{g}}$$

where L is the length of the pendulum and g is the acceleration due to gravity. For a certain pendulum, it is known that $L = 4.03$ ft. It is also known that $g = 32.2$ ft/s^2. What is the approximate error in calculating T by using $L = 4$ and $g = 32$?

54. A juice can is 12 cm tall and has a radius of 3 cm. A manufacturer is planning to reduce the height of the can by 0.2 cm and the radius by 0.3 cm. Use a total differential to estimate the percentage decrease in volume that occurs when the new cans are introduced. (Round to the nearest percent.)

55. If the weight of an object that does not float in water is x pounds in the air and its weight in water is y pounds, then the specific gravity of the object is

$$S = \frac{x}{x - y}$$

For a certain object, x and y are measured to be 1.2 lb and 0.5 lb, respectively. It is known that the measuring instrument will not register less than the true weights, but it could register more than the true weights by as much as 0.01 lb. Use incremental approximations to estimate the maximum possible error in the computation of the specific gravity.

56. A football has the shape of the ellipsoid

$$\frac{x^2}{9} + \frac{y^2}{36} + \frac{z^2}{9} = 1$$

where the dimensions are in inches, and is made of leather 1/8 inch thick. Use differentials to estimate the volume of the leather shell. *Hint*: The ellipsoid

$$\frac{x^2}{a^2} + \frac{y^2}{b^2} + \frac{z^2}{c^2} = 1$$

has volume $V = \frac{4}{3}\pi abc$.

Level 3

57. Show that the following function is not differentiable at $(0,0)$:

$$f(x,y) = \begin{cases} \dfrac{xy}{x^2 + y^2} & \text{if } (x,y) \neq (0,0) \\ 0 & \text{if } (x,y) = (0,0) \end{cases}$$

58. Compute the total differentials

$$d\left(\frac{x}{x-y}\right) \quad \text{and} \quad d\left(\frac{y}{x-y}\right)$$

Why are these differentials equal?

59. Let A be the area of a triangle with sides a and b separated by an angle θ, as shown in Figure 11.22.

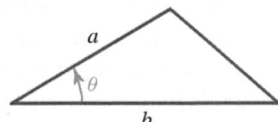

Figure 11.22 Problem 59

Suppose $\theta = \frac{\pi}{6}$, and a is increased by 4% while b is decreased by 3%. Use differentials to estimate the percentage change in A.

60. In Problem 59, suppose that θ also changes by no more than 2%. What is the maximum percentage change in A?

11.5 CHAIN RULES

IN THIS SECTION: *Chain rule for one parameter, extensions of the chain rule*
Functional composition is as important in several-variable calculus as in the single-variable case, and now because more than one variable are available, it is possible to form a "function of a function" in a variety of ways. We will examine several derivatives and partial derivatives in each case.

Chain Rule for One Parameter

We begin with a differentiable function of two variables $f(x,y)$. If $x = x(t)$ and $y = y(t)$ are, in turn, functions of a single parameter t, then $z = f(x(t), y(t))$ is a composite function of a parameter t. In this case, the chain rule for finding the derivative with respect to one parameter can now be stated.

Theorem 11.4 The chain rule for one independent parameter

Let $f(x,y)$ be a differentiable function of x and y, and let $x = x(t)$ and $y = y(t)$ be differentiable functions of t. Then $z = f(x,y)$ is a differentiable function of t, and

$$\frac{dz}{dt} = \frac{\partial z}{\partial x}\frac{dx}{dt} + \frac{\partial z}{\partial y}\frac{dy}{dt}$$

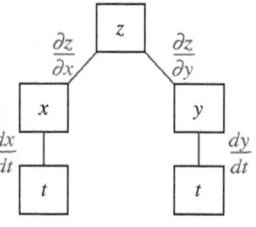

■ $\mathbf{W}$hat this says The tree diagram shown in the margin is a device for remembering the chain rule. The diagram begins at the top with the dependent variable z and cascades downward in two branches, first to the independent variables x and y, and then to the parameter t on which they each depend. Each branch segment is labeled with a derivative, and the chain rule is obtained by first multiplying the derivatives on the two segments of each branch and then adding to obtain

$$\frac{dz}{dt} = \underbrace{\frac{\partial z}{\partial x}\frac{dx}{dt}}_{left\ branch} + \underbrace{\frac{\partial z}{\partial y}\frac{dy}{dt}}_{right\ branch}$$

Proof: Recall that because $z = f(x,y)$ is differentiable, we can write the increment Δz in the following form:

$$\Delta z = \frac{\partial z}{\partial x}\Delta x + \frac{\partial z}{\partial y}\Delta y + \epsilon_1 \Delta x + \epsilon_2 \Delta y$$

where $\epsilon_1 \to 0$ and $\epsilon_2 \to 0$ as both $\Delta x \to 0$ and $\Delta y \to 0$. Dividing by $\Delta t \neq 0$, we obtain

$$\frac{\Delta z}{\Delta t} = \frac{\partial z}{\partial x}\frac{\Delta x}{\Delta t} + \frac{\partial z}{\partial y}\frac{\Delta y}{\Delta t} + \epsilon_1 \frac{\Delta x}{\Delta t} + \epsilon_2 \frac{\Delta y}{\Delta t}$$

Because x and y are functions of t, we can write their increments as

$$\Delta x = x(t + \Delta t) - x(t) \qquad \text{and} \qquad \Delta y = y(t + \Delta t) - y(t)$$

We know that x and y both vary continuously with t (remember, they are differentiable), and it follows that $\Delta x \to 0$ and $\Delta y \to 0$ as $\Delta t \to 0$, so that $\epsilon_1 \to 0$ and $\epsilon_2 \to 0$ as $\Delta t \to 0$. Therefore, we have

$$\begin{aligned}\frac{dz}{dt} &= \lim_{\Delta t \to 0} \frac{\Delta z}{\Delta t} \\ &= \lim_{\Delta t \to 0} \left[\frac{\partial z}{\partial x}\frac{\Delta x}{\Delta t} + \frac{\partial z}{\partial y}\frac{\Delta y}{\Delta t} + \epsilon_1 \frac{\Delta x}{\Delta t} + \epsilon_2 \frac{\Delta y}{\Delta t} \right] \\ &= \frac{\partial z}{\partial x}\frac{dx}{dt} + \frac{\partial z}{\partial y}\frac{dy}{dt} + 0\frac{dx}{dt} + 0\frac{dy}{dt} \\ &= \frac{\partial z}{\partial x}\frac{dx}{dt} + \frac{\partial z}{\partial y}\frac{dy}{dt} \end{aligned}$$

◆

Example 1 Verifying the chain rule explicitly

Let $z = x^2 + y^2$, where $x = \dfrac{1}{t}$ and $y = t^2$. Compute $\dfrac{dz}{dt}$ in two ways:

a. by first expressing z explicitly in terms of t
b. by using the chain rule

Solution

a. By substituting $x = \dfrac{1}{t}$ and $y = t^2$, we find that (for $t \neq 0$)

$$z = x^2 + y^2 = \left(\frac{1}{t}\right)^2 + (t^2)^2 = t^{-2} + t^4$$

Thus, $\dfrac{dz}{dt} = -2t^{-3} + 4t^3$.

b. Because $z = x^2 + y^2$ and $x = t^{-1}$, $y = t^2$,

$$\frac{\partial z}{\partial x} = 2x; \qquad \frac{\partial z}{\partial y} = 2y; \qquad \frac{dx}{dt} = -t^{-2}; \qquad \frac{dy}{dt} = 2t$$

Use the chain rule for one independent parameter:

$$\begin{aligned}
\frac{dz}{dt} &= \frac{\partial z}{\partial x}\frac{dx}{dt} + \frac{\partial z}{\partial y}\frac{dy}{dt} \\
&= (2x)(-t^{-2}) + 2y(2t) && \text{\textit{Chain rule}} \\
&= 2(t^{-1})(-t^{-2}) + 2(t^2)(2t) && \text{\textit{Substitute}} \\
&= -2t^{-3} + 4t^3
\end{aligned}$$

Example 2 Chain rule for one independent parameter

Let $z = \sqrt{x^2 + 2xy}$, where $x = \cos\theta$ and $y = \sin\theta$. Find $\dfrac{dz}{d\theta}$ in terms of x, y, and θ.

Solution $\dfrac{\partial z}{\partial x} = \dfrac{1}{2}(x^2 + 2xy)^{-1/2}(2x + 2y)$ and $\dfrac{\partial z}{\partial y} = \dfrac{1}{2}(x^2 + 2xy)^{-1/2}(2x)$

Also, $\dfrac{dx}{d\theta} = -\sin\theta$ and $\dfrac{dy}{d\theta} = \cos\theta$. Use the chain rule for one independent parameter to find

$$\begin{aligned}
\frac{dz}{d\theta} &= \frac{\partial z}{\partial x}\frac{dx}{d\theta} + \frac{\partial z}{\partial y}\frac{dy}{d\theta} \\
&= \frac{1}{2}(x^2 + 2xy)^{-1/2}(2x + 2y)(-\sin\theta) + \frac{1}{2}(x^2 + 2xy)^{-1/2}(2x)(\cos\theta) \\
&= (x^2 + 2xy)^{-1/2}(x\cos\theta - x\sin\theta - y\sin\theta)
\end{aligned}$$

Example 3 Related rate application using the chain rule

A right circular cylinder (see Figure 11.23) is changing in such a way that its radius r is increasing at the rate of 3 in./min and its height h is decreasing at the rate of 5 in./min. At what rate is the volume of the cylinder changing when the radius is 10 in. and the height is 8 in.?

Solution The volume of the cylinder is $V = \pi r^2 h$, and we are given $\dfrac{dr}{dt} = 3$ and $\dfrac{dh}{dt} = -5$. We find that

$$\frac{\partial V}{\partial r} = \pi(2r)h \qquad \text{and} \qquad \frac{\partial V}{\partial h} = \pi r^2(1)$$

By the chain rule for one parameter,

$$\frac{dV}{dt} = \frac{\partial V}{\partial r}\frac{dr}{dt} + \frac{\partial V}{\partial h}\frac{dh}{dt} = 2\pi rh\frac{dr}{dt} + \pi r^2\frac{dh}{dt}$$

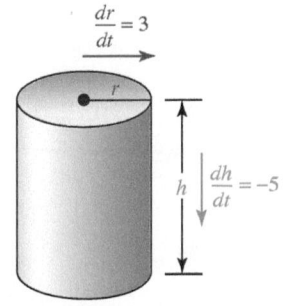

Figure 11.23 Right circular cylinder

Thus, at the instant when $r = 10$ and $h = 8$, we have

$$\frac{dV}{dt} = 2\pi(10)(8)(3) + \pi(10)^2(-5) = -20\pi$$

The volume is decreasing at the rate of about 62.8 in.^{3}/min.

The chain rule can be used to prove a useful alternative to the procedure for implicit differentiation introduced in Section 3.6. If $F(x,y) = 0$ defines y implicitly as a differentiable function of x, we can regard x as a parameter and apply the chain rule to obtain

$$0 = \frac{\partial F}{\partial x}\frac{dx}{dx} + \frac{\partial F}{\partial y}\frac{dy}{dx} = \frac{\partial F}{\partial x} + \frac{\partial F}{\partial y}\frac{dy}{dx}$$

so

$$\frac{dy}{dx} = \frac{-\frac{\partial F}{\partial x}}{\frac{\partial F}{\partial y}} = -\frac{F_x}{F_y}$$

This formula provides a useful alternative to the procedure for implicit differentiation, so we state it formally as a theorem.

Theorem 11.5 Implicit function theorem

Let F be defined on a disk containing (a,b) as an interior point, such that $F(a,b) = 0$, and assume that F_x and F_y are both continuous on this disk, with $F_y(a,b) \neq 0$. Then there **exist** an **interval** I on the real line containing a as an interior point and a **unique function** $y = y(x)$ defined on the interval I, such that $y(a) = b$ and $F(x, y(x)) = 0$, for every value x on the interval I. Furthermore, the derivative of y is given by

$$\frac{dy}{dx} = -\frac{F_x}{F_y}$$

Proof: The proof is left for advanced calculus. Note that the formula for $\frac{dy}{dx}$ is a direct consequence of the chain rule. ◆

This alternative procedure is illustrated in the following example.

Example 4 Implicit differentiation using partial derivatives

If y is a differentiable function of x such that

$$\sin(x + y) + \cos(x - y) = y$$

find $\dfrac{dy}{dx}$.

Solution   Let $F(x,y) = \sin(x + y) + \cos(x - y) - y$, so that $F(x,y) = 0$. Then

$$F_x = \cos(x + y) - \sin(x - y)$$
$$F_y = \cos(x + y) - \sin(x - y)(-1) - 1$$

so

$$\frac{dy}{dx} = -\frac{F_x}{F_y} = \frac{-[\cos(x + y) - \sin(x - y)]}{\cos(x + y) + \sin(x - y) - 1}$$

When z is defined implicitly in terms of x and y by an equation $F(x, y, z) = 0$, the chain rule can be used to find $\dfrac{\partial z}{\partial x}$ and $\dfrac{\partial z}{\partial y}$ in terms of F_x, F_y, and F_z. The procedure is outlined in Problem 57.

Example 5 Second derivative of a function of two variables

Let $z = f(x, y)$, where $x = at$ and $y = bt$ for constants a and b. Assuming all necessary differentiability, find d^2z/dt^2 in terms of the partial derivatives of z.

Solution By using the chain rule, we have

$$\frac{dz}{dt} = \frac{\partial z}{\partial x}\frac{dx}{dt} + \frac{\partial z}{\partial y}\frac{dy}{dt}$$

and

$$\frac{d^2z}{dt^2} = \frac{d}{dt}\left(\frac{dz}{dt}\right) = \frac{d}{dt}\left(\frac{\partial z}{\partial x}\frac{dx}{dt} + \frac{\partial z}{\partial y}\frac{dy}{dt}\right)$$

$$= \left[\frac{d}{dt}\left(\frac{\partial z}{\partial x}\right)\right]\frac{dx}{dt} + \frac{\partial z}{\partial x}\left[\frac{d}{dt}\left(\frac{dx}{dt}\right)\right] + \left[\frac{d}{dt}\left(\frac{\partial z}{\partial y}\right)\right]\frac{dy}{dt} + \frac{\partial z}{\partial y}\left[\frac{d}{dt}\left(\frac{dy}{dt}\right)\right]$$

$$= \left[\frac{\partial z}{\partial x}\frac{d^2x}{dt^2} + \frac{dx}{dt}\left(\frac{\partial^2 z}{\partial x^2}\frac{dx}{dt} + \frac{\partial^2 z}{\partial y \partial x}\frac{dy}{dt}\right)\right] + \left[\frac{\partial z}{\partial y}\frac{d^2y}{dt^2} + \frac{dy}{dt}\left(\frac{\partial^2 z}{\partial x \partial y}\frac{dx}{dt} + \frac{\partial^2 z}{\partial y^2}\frac{dy}{dt}\right)\right]$$

Substituting $\dfrac{dx}{dt} = a$ and $\dfrac{dy}{dt} = b$, we obtain

$$\frac{d^2z}{dt^2} = \left[\frac{\partial z}{\partial x}(0) + a\left(\frac{\partial^2 z}{\partial x^2}a + \frac{\partial^2 z}{\partial x \partial y}b\right)\right] + \left[\frac{\partial z}{\partial y}(0) + b\left(\frac{\partial^2 z}{\partial y \partial x}a + \frac{\partial^2 z}{\partial y^2}b\right)\right]$$

$$= a^2\frac{\partial^2 z}{\partial x^2} + 2ab\frac{\partial^2 z}{\partial x \partial y} + b^2\frac{\partial^2 z}{\partial y^2} \qquad \textit{Note:} \; \frac{\partial^2 z}{\partial x \partial y} = \frac{\partial^2 z}{\partial y \partial x}.$$

Extensions of the Chain Rule

Next, we will consider the kind of composite function that occurs when x and y are both functions of *two* parameters. Specifically, let $z = F(x, y)$, where $x = x(u, v)$ and $y = y(u, v)$ are both functions of two independent parameters u and v. Then $z = F[x(u, v), y(u, v)]$ is a composite function of u and v, and with suitable assumptions regarding differentiability, we can find the partial derivatives $\partial z/\partial u$ and $\partial z/\partial v$ by applying the chain rule obtained in the following theorem.

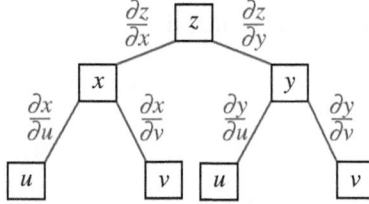

Theorem 11.6 The chain rule for two independent parameters

Suppose $z = f(x, y)$ is differentiable at (x, y) and that the partial derivatives of $x = x(u, v)$ and $y = y(u, v)$ exist at (u, v). Then the composite function $z = f[x(u, v), y(u, v)]$ is differentiable at (u, v) with

$$\frac{\partial z}{\partial u} = \frac{\partial z}{\partial x}\frac{\partial x}{\partial u} + \frac{\partial z}{\partial y}\frac{\partial y}{\partial u} \qquad \text{and} \qquad \frac{\partial z}{\partial v} = \frac{\partial z}{\partial x}\frac{\partial x}{\partial v} + \frac{\partial z}{\partial y}\frac{\partial y}{\partial v}$$

Proof: This version of the chain rule follows immediately from the chain rule for one independent parameter. For example, if v is fixed, the composite function

$z = f[x(u, v), y(u, v)]$ depends on u alone, and we have the situation described in the chain rule of one independent variable. We apply this chain rule with a partial derivative (because x and y are functions of more than one variable):

$$\frac{\partial z}{\partial u} = \frac{\partial z}{\partial x}\frac{\partial x}{\partial u} + \frac{\partial z}{\partial y}\frac{\partial y}{\partial u}$$

The formula for $\dfrac{\partial z}{\partial v}$ can be established in a similar fashion. ◆

Example 6   Chain rule for two independent parameters

Let $z = 4x - y^2$, where $x = uv^2$ and $y = u^3v$. Find $\dfrac{\partial z}{\partial u}$ and $\dfrac{\partial z}{\partial v}$.

Solution   First find the partial derivatives:

$$\frac{\partial z}{\partial x} = \frac{\partial}{\partial x}(4x - y^2) = 4 \qquad\qquad \frac{\partial z}{\partial y} = \frac{\partial}{\partial y}(4x - y^2) = -2y$$

and

$$\frac{\partial x}{\partial u} = \frac{\partial}{\partial u}(uv^2) = v^2 \qquad\qquad \frac{\partial y}{\partial u} = \frac{\partial}{\partial u}(u^3v) = 3u^2v$$

$$\frac{\partial x}{\partial v} = \frac{\partial}{\partial v}(uv^2) = 2uv \qquad\qquad \frac{\partial y}{\partial v} = \frac{\partial}{\partial v}(u^3v) = u^3$$

Therefore, the chain rule for two independent parameters gives

$$
\begin{aligned}
\frac{\partial z}{\partial u} &= \frac{\partial z}{\partial x}\frac{\partial x}{\partial u} + \frac{\partial z}{\partial y}\frac{\partial y}{\partial u} \\
&= (4)(v^2) + (-2y)(3u^2v) \\
&= 4v^2 - 2(u^3v)(3u^2v) \\
&= 4v^2 - 6u^5v^2
\end{aligned}
\qquad \text{and} \qquad
\begin{aligned}
\frac{\partial z}{\partial v} &= \frac{\partial z}{\partial x}\frac{\partial x}{\partial v} + \frac{\partial z}{\partial y}\frac{\partial y}{\partial v} \\
&= (4)(2uv) + (-2y)(u^3) \\
&= 8uv - 2(u^3v)u^3 \\
&= 8uv - 2u^6v
\end{aligned}
$$

Example 7   Implicit differentiation using the chain rule

If f is differentiable and $z = u + f(u^2v^2)$, show that $u\dfrac{\partial z}{\partial u} - v\dfrac{\partial z}{\partial v} = u$.

Solution   Let $w = u^2v^2$, so $z = u + f(w)$. Then, according to the chain rule,

$$
\begin{aligned}
\frac{\partial z}{\partial u} &= 1 + \frac{df}{dw}\frac{\partial w}{\partial u} \\
&= 1 + f'(w)(2uv^2)
\end{aligned}
\qquad \text{and} \qquad
\begin{aligned}
\frac{\partial z}{\partial v} &= \frac{df}{dw}\frac{\partial w}{\partial v} \\
&= f'(w)(2u^2v)
\end{aligned}
$$

so that

$$
\begin{aligned}
u\frac{\partial z}{\partial u} - v\frac{\partial z}{\partial y} &= u\left[1 + f'(w)(2uv^2)\right] - v\left[f'(w)(2u^2v)\right] \\
&= u + f'(w)\left[u(2uv^2) - v(2u^2v)\right] \\
&= u
\end{aligned}
$$

The chain rules can be extended to functions of three or more variables. For instance, if $w = f(x, y, z)$ is a differentiable function of three variables and $x = x(t)$, $y = y(t)$, $z = z(t)$ are each differentiable functions of t, then w is a differentiable composite function of t and

$$\frac{dw}{dt} = \frac{\partial w}{\partial x}\frac{dx}{dt} + \frac{\partial w}{\partial y}\frac{dy}{dt} + \frac{\partial w}{\partial z}\frac{dz}{dt}$$

In general, if $w = f(x_1, x_2, \cdots, x_n)$ is a differentiable function of the n variables $x_1, x_2, \cdots, x_n$, which in turn are differentiable functions of m parameters $t_1, t_2, \cdots, t_m$, then

$$\frac{\partial w}{\partial t_1} = \frac{\partial w}{\partial x_1}\frac{\partial x_1}{\partial t_1} + \frac{\partial w}{\partial x_2}\frac{\partial x_2}{\partial t_1} + \cdots + \frac{\partial w}{\partial x_n}\frac{\partial x_n}{\partial t_1}$$

$$\vdots$$

$$\frac{\partial w}{\partial t_m} = \frac{\partial w}{\partial x_1}\frac{\partial x_1}{\partial t_m} + \frac{\partial w}{\partial x_2}\frac{\partial x_2}{\partial t_m} + \cdots + \frac{\partial w}{\partial x_n}\frac{\partial x_n}{\partial t_m}$$

Example 8   Chain rule for a function of three variables with three parameters

Find $\dfrac{\partial w}{\partial s}$ if $w = 4x + y^2 + z^3$, where $x = e^{rs^2}$, $y = \ln\dfrac{r+s}{t}$, and $z = rst^2$.

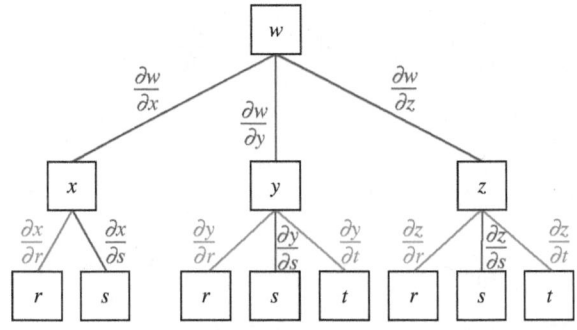

Solution

$$\frac{\partial w}{\partial s} = \frac{\partial w}{\partial x}\frac{\partial x}{\partial s} + \frac{\partial w}{\partial y}\frac{\partial y}{\partial s} + \frac{\partial w}{\partial z}\frac{\partial z}{\partial s}$$

$$= \left[\frac{\partial}{\partial x}(4x + y^2 + z^3)\right]\left[\frac{\partial}{\partial s}(e^{rs^2})\right] + \left[\frac{\partial}{\partial y}(4x + y^2 + z^3)\right]\left[\frac{\partial}{\partial s}\left(\ln\frac{r+s}{t}\right)\right]$$

$$+ \left[\frac{\partial}{\partial z}(4x + y^2 + z^3)\right]\left[\frac{\partial}{\partial s}(rst^2)\right]$$

$$= 4\left[e^{rs^2}(2rs)\right] + 2y\left(\frac{1}{\frac{r+s}{t}}\right)\left(\frac{1}{t}\right) + 3z^2(rt^2)$$

$$= 8rse^{rs^2} + \frac{2y}{r+s} + 3rt^2z^2$$

In terms of r, s, and t, the partial derivative is

$$\frac{\partial w}{\partial s} = 8rse^{rs^2} + \frac{2}{r+s}\ln\frac{r+s}{t} + 3r^3s^2t^6$$

PROBLEM SET 11.5

1. ■ What does this say? Discuss the various chain rules and the need for such chain rules.

2. ■ What does this say? Discuss the usefulness of the schematic (tree) representation for the chain rules.

3. ■ What does this say? Write out a chain rule for a function of two variables and three independent parameters.

In Problems 4-7, the function $z = f(x, y)$ depends on x and y, which in turn are each functions of t. In each case, find dz/dt in two different ways:

   a. *Express z explicitly in terms of t.*

   b. *Use the chain rule for one parameter.*

4. $f(x,y) = 2xy + y^2$, where $x = -3t^2$ and $y = 1 + t^3$

5. $f(x,y) = (4 + y^2)x$, where $x = e^{2t}$ and $y = e^{3t}$

6. $f(x,y) = (1 + x^2 + y^2)^{1/2}$, where $x = \cos 5t$ and $y = \sin 5t$

7. $f(x,y) = xy^2$, where $x = \cos 3t$ and $y = \tan 3t$

In Problems 8-11, the function $F(x,y)$ depends on x and y. Let $x = x(u,v)$ and $y = y(u,v)$ be given functions of u and v. Let $z = F[x(u,v), y(u,v)]$ and find the partial derivatives $\partial z/\partial u$ and $\partial z/\partial v$ in these two ways:

    a. *Express z explicitly in terms of u and v*

    b. *Apply the chain rule for two independent parameters.*

8. $F(x,y) = x + y^2$, where $x = u + v$ and $y = u - v$

9. $F(x,y) = x^2 + y^2$, where $x = u \sin v$ and $y = u - 2v$

10. $F(x,y) = e^{xy}$, where $x = u - v$ and $y = u + v$

11. $F(x,y) = \ln xy$, where $x = e^{uv^2}$ and $y = e^{uv}$

Write out the chain rule for the functions given in Problems 12-15.

12. $z = f(x,y)$, where $x = x(s,t)$, $y = y(s,t)$

13. $w = f(x,y,z)$, where $x = x(s,t)$, $y = y(s,t)$, $z = z(s,t)$

14. $t = f(u,v)$, where $u = u(x,y,z,w)$, $v = v(x,y,z,w)$

15. $w = f(x,y,z)$, where $x = x(s,t,u)$, $y = y(s,t,u)$, $z = z(s,t,u)$

Find the indicated derivatives or partial derivatives in Problems 16-21. Leave your answers in mixed form (x,y,z,t).

16. Find $\dfrac{dw}{dt}$, where $w = \ln(x + 2y - z^2)$ and $x = 2t - 1$, $y = \dfrac{1}{t}$, $z = \sqrt{t}$.

17. Find $\dfrac{dw}{dt}$, where $w = \sin xyz$ and $x = 1 - 3t$, $y = e^{1-t}$, $z = 4t$.

18. Find $\dfrac{dw}{dt}$, where $w = ze^{xy^2}$ and $x = \sin t$, $y = \cos t$, $z = \tan 2t$.

19. Find $\dfrac{dw}{dt}$, where $w = e^{x^3 + yz}$ and $x = \dfrac{2}{t}$, $y = \ln(2t - 3)$, $z = t^2$.

20. Find $\dfrac{\partial w}{\partial r}$, where $w = e^{2x - y + 3z^2}$, and $x = r + s - t$, $y = 2r - 3s$, $z = \cos rst$.

21. Find $\dfrac{\partial w}{\partial r}$ and $\dfrac{\partial w}{\partial t}$, where $w = \dfrac{x + y}{2 - z}$ and $x = 2rs$, $y = \sin rt$, $z = st^2$.

In Problems 22-27, assume the given equations define y as a differentiable function of x and find dy/dx using the procedure illustrated in Example 4.

22. $x^2 y + \sqrt{xy} = 4$

23. $(x^2 - y)^{3/2} + x^2 y = 2$

24. $x^2 y + \ln(2x + y) = 5$

25. $x \cos y + y \tan^{-1} x = x$

26. $xe^{xy} + ye^{-xy} = 3$

27. $\tan^{-1}\left(\dfrac{x}{y}\right) = \tan^{-1}\left(\dfrac{y}{x}\right)$

Level 2

Find the following higher-order partial derivatives in Problems 28-33.

    a. $\dfrac{\partial^2 z}{\partial x \partial y}$ **b.** $\dfrac{\partial^2 z}{\partial x^2}$ **c.** $\dfrac{\partial^2 z}{\partial y^2}$

28. $x^3 + y^2 + z^2 = 5$ **29.** $xyz = 2$

30. $\ln(x + y) = y^2 + z$ **31.** $x^{-1} + y^{-1} + z^{-1} = 3$

32. $x \cos y = y + z$ **33.** $z^2 + \sin x = \tan y$

34. Let $f(x,y)$ be a differentiable function of x and y, and let $x = r \cos \theta$, $y = r \sin \theta$ for $r > 0$ and $0 < \theta < 2\pi$.

    a. If $z = f[x(r,\theta), y(r,\theta)]$, find $\dfrac{\partial z}{\partial r}$ and $\dfrac{\partial z}{\partial \theta}$.

    b. Show that

$$\left(\frac{\partial z}{\partial r}\right)^2 + \frac{1}{r^2}\left(\frac{\partial z}{\partial \theta}\right)^2 = \left(\frac{\partial z}{\partial x}\right)^2 + \left(\frac{\partial z}{\partial y}\right)^2$$

35. Let $z = f(x,y)$, where $x = au$ and $y = bv$, with a, b constants. Express $\partial^2 z/\partial u^2$ and $\partial^2 z/\partial v^2$ in terms of the partial derivatives of z with respect to x and y. Assume the existence and continuity of all necessary first and second partial derivatives.

36. The earth may be viewed as an ellipsoid with equation

$$\frac{x^2}{a^2} + \frac{y^2}{b^2} + \frac{z^2}{c^2} = 1$$

for appropriate constants a, b, and c.

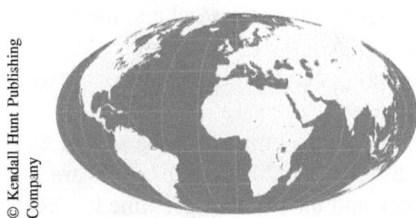

Let (x,y,z) lie on this ellipsoid. Without solving for z explicitly in terms of x and y, compute the higher-order partial derivatives

$$\frac{\partial^2 z}{\partial x^2} \quad \text{and} \quad \frac{\partial^2 z}{\partial x \partial y}$$

37. The dimensions of a rectangular box are linear functions of time, $\ell(t)$, $w(t)$, and $h(t)$. If the length and width are increasing at 2 in./sec and the height is decreasing at 3 in./sec, find the rates at which the volume V and the surface area S are changing with respect to time. If $\ell(0) = 10$, $w(0) = 8$, and $h(0) = 20$, is V increasing or decreasing when $t = 5$ sec? What about S when $t = 5$?

38. Van der Waal's equation in physical chemistry states that a gas occupying volume V at temperature T (Kelvin) exerts pressure P, where

$$\left(P + \frac{A}{V^2}\right)(V - B) = kT$$

for physical constants A, B, and k. Compute the following rates:

a. the rate of change of volume with respect to temperature

b. the rate of change of pressure with respect to volume

39. The concentration of a drug in the blood of a patient t hours after the drug is injected into the body intramuscularly is modeled by the Heinz function

$$C = \frac{1}{b-a}(e^{-at} - e^{-bt}) \qquad b > a > 0$$

where a and b are parameters that depend on the patient's metabolism and the particular kind of drug being used.

a. Compute the rates $\dfrac{\partial C}{\partial a}$, $\dfrac{\partial C}{\partial b}$, and $\dfrac{\partial C}{\partial t}$.

b. Explore the assumption that $a = (\ln b)/t$, b constant for $t > (\ln b)/b$. In particular, what is dC/dt?

40. A paint store carries two brands of latex paint. An analysis of sales figures indicates that the demand Q for the first brand is modeled by

$$Q(x, y) = 210 - 12x^2 + 18y$$

gallons/month, where x, y are the prices of the first and second brands, respectively. A separate study indicates that t months from now, the first brand will cost $x = 4 + 0.18t$ dollars/gal and the second brand will cost $y = 5 + 0.3\sqrt{t}$ dollars/gal. At what rate will the demand Q be changing with respect to time 9 months from now?

41. To model the demand for the sale of bicycles, it is assumed that if 24-speed bicycles are sold for x dollars apiece and the price of gasoline is y cents per gallon, then

$$Q(x, y) = 240 - 21\sqrt{x} + 4(0.2y + 12)^{3/2}$$

bicycles will be sold each month. For this model it is further assumed that t months from now, bicycles will be selling for $x = 120 + 6t$ dollars apiece, and the price of gasoline will be $y = 380 + 10\sqrt{4t}$ cents/gal. At what rate will the monthly demand for the bicycles be changing with respect to time 4 months from now?

42. At a certain factory, the amount of air pollution generated each day is modeled by the function

$$Q(E, T) = 127E^{2/3}T^{1/2}$$

where E is the number of employees and T ($^\circ$C) is the average temperature during the workday. Currently, there are 142 employees and the average temperature is $18\,^\circ$C. If the average daily temperature is falling at the rate of $0.23\,^\circ$/day and the number of employees is increasing at the rate of 3/month, what is the corresponding effect on the rate of pollution? Express your answer in units/day. For this model, assume there are 22 workdays/month.

43. The combined resistance R produced by three variable resistances R_1, R_2, and R_3 connected in parallel is modeled by the formula

$$\frac{1}{R} = \frac{1}{R_1} + \frac{1}{R_2} + \frac{1}{R_3}$$

Suppose at a certain instant, $R_1 = 100$ ohms, $R_2 = 200$ ohms, $R_3 = 300$ ohms. R_1 and R_3 are decreasing at the rate of 1.5 ohms/s while R_2 is increasing at the rate of 2 ohms/s. How fast is R changing with respect to time at this instant? Is it increasing or decreasing?

In Problems 44-48, assume that all functions have whatever derivatives or partial derivatives are necessary for the problem to be meaningful.

44. If $z = f(uv^2)$, show that

$$2u\frac{\partial z}{\partial u} - v\frac{\partial z}{\partial v} = 0$$

45. If $z = f(u - v, v - u)$, show that

$$\frac{\partial z}{\partial u} + \frac{\partial z}{\partial v} = 0$$

46. If $z = u + f(uv)$, show that

$$u\frac{\partial z}{\partial u} - v\frac{\partial z}{\partial v} = u$$

47. If $w = f\left(\dfrac{r-s}{s}\right)$, show that

$$r\frac{\partial w}{\partial r} + s\frac{\partial w}{\partial s} = 0$$

48. If $z = xy + f(x^2 + y^2)$, show that

$$y\frac{\partial z}{\partial x} - x\frac{\partial z}{\partial y} = y^2 - x^2$$

49. Let $w = f(t)$ be a differentiable function of t, where $t = (x^2 + y^2 + z^2)^{1/2}$. Show that

$$\left(\frac{dw}{dt}\right)^2 = \left(\frac{\partial w}{\partial x}\right)^2 + \left(\frac{\partial w}{\partial y}\right)^2 + \left(\frac{\partial w}{\partial z}\right)^2$$

50. Suppose f is a twice differentiable function of one variable, and let $z = f(x^2 + y^2)$. Find

a. $\dfrac{\partial^2 z}{\partial x^2}$　　**b.** $\dfrac{\partial^2 z}{\partial y^2}$　　**c.** $\dfrac{\partial^2 z}{\partial x \partial y}$

51. Find $\dfrac{d^2 z}{d\theta^2}$, where z is a twice differentiable function of one variable θ and can be written $z = f(\cos\theta, \sin\theta)$. *Hint:* Let $x = \cos\theta$ and $y = \sin\theta$. Leave your answer in terms of x and y.

52. Let f and g be twice differentiable functions of one variable, and let

$$u(x, t) = f(x + ct) + g(x - ct)$$

for a constant c. Show that

$$\frac{\partial^2 u}{\partial t^2} = c^2 \frac{\partial^2 u}{\partial x^2}$$

Hint: Let $r = x + ct$; $s = x - ct$.

53. Suppose $z = f(x, y)$ has continuous second-order partial derivatives. If $x = e^r \cos\theta$ and $y = e^r \sin\theta$, show that

$$\frac{\partial^2 z}{\partial x^2} + \frac{\partial^2 z}{\partial y^2} = e^{-2r} \left[\frac{\partial^2 z}{\partial r^2} + \frac{\partial^2 z}{\partial \theta^2} \right]$$

54. If $f(u, v, w)$ is differentiable and $u = x - y$, $v = y - z$, and $w = z - x$, what is

$$\frac{\partial f}{\partial x} + \frac{\partial f}{\partial y} + \frac{\partial f}{\partial z}?$$

Level 3

55. The *Cauchy-Riemann equations* are

$$\frac{\partial u}{\partial x} = \frac{\partial v}{\partial y} \quad \text{and} \quad \frac{\partial u}{\partial y} = -\frac{\partial v}{\partial x}$$

(See Problem 51, Section 11.3). Show that if x and y are expressed in terms of polar coordinates, the Cauchy-Riemann equations become

$$\frac{\partial u}{\partial r} = \frac{1}{r}\frac{\partial v}{\partial \theta} \quad \text{and} \quad \frac{\partial v}{\partial r} = -\frac{1}{r}\frac{\partial u}{\partial \theta}$$

56. Let $T(x, y)$ be the temperature at each point (x, y) in a portion of the plane that contains the ellipse $x = 2\cos t$, $y = \sin t$ for $0 \le t \le 2\pi$. Suppose

$$\frac{\partial T}{\partial x} = y \quad \text{and} \quad \frac{\partial T}{\partial y} = x$$

a. Find $\dfrac{dT}{dt}$ and $\dfrac{d^2 T}{dt^2}$ by using the chain rule.

b. Locate the maximum and minimum temperatures on the ellipse.

57. Let $F(x, y, z)$ be a function of three variables with continuous partial derivatives F_x, F_y, F_z in a certain region where $F(x, y, z) = C$ for some constant C. Use the chain rule for two parameters and the fact that x and y are independent variables to show that (for $F_z \ne 0$)

$$\frac{\partial z}{\partial x} = -\frac{F_x}{F_z} \quad \text{and} \quad \frac{\partial z}{\partial y} = -\frac{F_y}{F_z}$$

Use these formulas to find $\dfrac{\partial z}{\partial x}$ and $\dfrac{\partial z}{\partial y}$ if z is defined implicitly by the equation $x^2 + 2xyz + y^3 + e^z = 4$.

58. Suppose the system

$$\begin{cases} xu + yv - uv = 0 \\ yu - xv + uv = 0 \end{cases}$$

can be solved for u and v in terms of x and y, so that $u = u(x, y)$ and $v = v(x, y)$. Use implicit differentiation to find the partial derivatives $\dfrac{\partial u}{\partial x}$ and $\dfrac{\partial v}{\partial x}$.

59. A function $f(x, y)$ is said to be *homogeneous of degree* n if $f(tx, ty) = t^n f(x, y)$ for all $t > 0$.

a. Show that $f(x, y) = x^2 y + 2y^3$ is homogeneous and find its degree.

b. If $f(x, y)$ is homogeneous of degree n, show that

$$x\frac{\partial f}{\partial x} + y\frac{\partial f}{\partial y} = nf$$

60. Suppose that F and G are functions of three variables and that it is possible to solve the equations $F(x, y, z) = 0$ and $G(x, y, z) = 0$ for y and z in terms of x, so that $y = y(x)$ and $z = z(x)$. Use the chain rule to express dy/dx and dz/dx in terms of the partial derivatives of F and G. Assume these partials are continuous and that

$$\frac{\partial F}{\partial y}\frac{\partial G}{\partial z} \ne \frac{\partial F}{\partial z}\frac{\partial G}{\partial y}$$

11.6 DIRECTIONAL DERIVATIVES AND THE GRADIENT

IN THIS SECTION: *The directional derivative, the gradient, maximal property of the gradient, functions of three variables, normal property of the gradient, tangent planes and normal lines*
In this section, we see how to find a derivative in various specified directions, called *directional derivatives*. After we find the directional derivative, we will express it in terms of a vector function called a *gradient*. We then conclude by looking at some applications of the gradient.

Suppose $z = T(x, y)$ gives the temperature at each point (x, y) in a region R of the plane, and let $P_0(x_0, y_0)$ be a particular point in R. Then we know that the partial derivative $T_x(x_0, y_0)$ gives the rate at which the temperature changes for a move from P_0 in the x-direction, while the rate of temperature change in the y-direction is given by $T_y(x_0, y_0)$. Suppose we want to find the direction of greatest temperature change, which may be in a direction not parallel to either coordinate axis. To answer this question, we will introduce the concept of *directional derivative* in this section and examine its properties.

The Directional Derivative

In Chapter 3 we defined the *slope of a curve* at a point to be the ratio of the change in the dependent variable to the change in the independent variable at the given point. To determine the slope of the tangent line at a point $P_0(x_0, y_0)$ on a surface defined by $z = f(x, y)$, we need to specify the *direction* in which we wish to measure. We do this by using vectors. In Section 11.3 we found the slope parallel to the xz-plane to be the partial derivative $f_x(x_0, y_0)$. We could have specified this direction in terms of the unit vector **i** (x-direction), while $f_y(x, y)$ could have been specified in terms of the unit vector **j**. Finally, to measure the slope of the tangent line in an *arbitrary* direction, we use a unit vector $\mathbf{u} = u_1\mathbf{i} + u_2\mathbf{j}$ in that direction.

To find the desired slope, we look at the intersection of the surface with the vertical plane passing through the point P_0 parallel to the vector **u**, as shown in Figure 11.24.

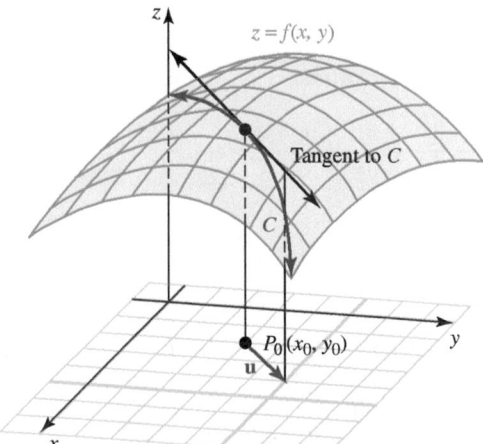

Figure 11.24   The directional derivative

This vertical plane intersects the surface to form a curve C, and we define the slope of the surface at P_0 in the direction of **u** to be the slope of the tangent line to the curve C defined by **u** at that point. We summarize this idea of slope *in a particular direction* with the following definition.

DIRECTIONAL DERIVATIVE Let f be a function of two variables, and let $\mathbf{u} = u_1\mathbf{i} + u_2\mathbf{j}$ be a *unit vector*. The **directional derivative of** f **at** $P_0(x_0, y_0)$ **in the direction of u** is given by

$$D_{\mathbf{u}}f(x_0, y_0) = \lim_{h \to 0} \frac{f(x_0 + hu_1, y_0 + hu_2) - f(x_0, y_0)}{h}$$

provided the limit exists.

Remember, $\mathbf{u}$ must be a unit vector.

At a particular point $P_0(x_0, y_0)$, there are infinitely many directional derivatives for the graph of $z = f(x, y)$, one for each direction radiating from P_0. Two of these are the partial derivatives $f_x(x_0, y_0)$ and $f_y(x_0, y_0)$. To see this, note that if $\mathbf{u} = \mathbf{i}$ (so $u_1 = 1$ and $u_2 = 0$), then

$$D_{\mathbf{i}}f(x_0, y_0) = \lim_{h \to 0} \frac{f(x_0 + h, y_0) - f(x_0, y_0)}{h} = f_x(x_0, y_0)$$

and if $\mathbf{u} = \mathbf{j}$ (so $u_1 = 0$ and $u_2 = 1$),

$$D_{\mathbf{j}}f(x_0, y_0) = \lim_{h \to 0} \frac{f(x_0, y_0 + h) - f(x_0, y_0)}{h} = f_y(x_0, y_0)$$

The definition of the directional derivative is similar to the definition of the derivative of a function of a single variable. Just as with a single variable, it is difficult to apply the definition directly. Fortunately, the following theorem allows us to find directional derivatives more efficiently than by using the definition.

Theorem 11.7 Directional derivatives using partial derivatives

Let $f(x, y)$ be a function that is differentiable at $P_0(x_0, y_0)$. Then f has a directional derivative in the direction of the unit vector $\mathbf{u} = u_1\mathbf{i} + u_2\mathbf{j}$ given by

$$D_{\mathbf{u}}f(x_0, y_0) = f_x(x_0, y_0)u_1 + f_y(x_0, y_0)u_2$$

Proof: We define a function F of a single variable h by $F(h) = f(x_0 + hu_1, y_0 + hu_2)$, so that

$$\begin{aligned} D_{\mathbf{u}}f(x_0, y_0) &= \lim_{h \to 0} \frac{f(x_0 + hu_1, y_0 + hu_2) - f(x_0, y_0)}{h} \\ &= \lim_{h \to 0} \frac{F(h) - F(0)}{h} \\ &= F'(0) \end{aligned}$$

Apply the chain rule with $x = x_0 + hu_1$ and $y = y_0 + hu_2$:

$$F'(h) = \frac{dF}{dh} = \frac{\partial f}{\partial x}\frac{dx}{dh} + \frac{\partial f}{\partial y}\frac{dy}{dh} = f_x(x, y)u_1 + f_y(x, y)u_2$$

When $h = 0$, we have $x = x_0$ and $y = y_0$, so that

$$D_{\mathbf{u}}f(x_0, y_0) = F'(0) = \frac{\partial f}{\partial x}u_1 + \frac{\partial f}{\partial y}u_2 = f_x(x_0, y_0)u_1 + f_y(x_0, y_0)u_2 \qquad \blacklozenge$$

Example 1   Finding a directional derivative using partial derivatives

Find the directional derivative of $f(x, y) = 3 - 2x^2 + y^3$ at the point $P(1, 2)$ in the direction of the unit vector $\mathbf{u} = \frac{1}{2}\mathbf{i} - \frac{\sqrt{3}}{2}\mathbf{j}$.

Solution   First, find the partial derivatives $f_x(x, y) = -4x$ and $f_y(x, y) = 3y^2$. Then since $u_1 = \frac{1}{2}$ and $u_2 = -\frac{\sqrt{3}}{2}$, we have

$$D_{\mathbf{u}}f(1, 2) = f_x(1, 2)\left(\frac{1}{2}\right) + f_y(1, 2)\left(-\frac{\sqrt{3}}{2}\right)$$

$$= -4(1)\left(\frac{1}{2}\right) + 3(2)^2\left(-\frac{\sqrt{3}}{2}\right)$$

$$= -2 - 6\sqrt{3}$$

Notice that the directional derivative is a number. This number can be interpreted as the slope of a tangent line to $z = f(x, y)$ or as a rate of change of the function $z = f(x, y)$. For Example 1 we see this number $-2 - 6\sqrt{3} \approx -12.4$. We use this example to illustrate the directional derivative as a slope and as a rate.

Directional derivative as a slope
Geometrically, we look at the surface defined in Example 1, namely $z = 3 - 2x^2 + y^3$. The intersection of this surface with a plane aligned with the vector $\mathbf{u} = \frac{1}{2}\mathbf{i} - \frac{\sqrt{3}}{2}\mathbf{j}$ is a curve C, and the directional derivative is the slope of the tangent line to C at the point on the surface above $P(1, 2)$. This directional derivative is shown in Figure 11.25**d**.

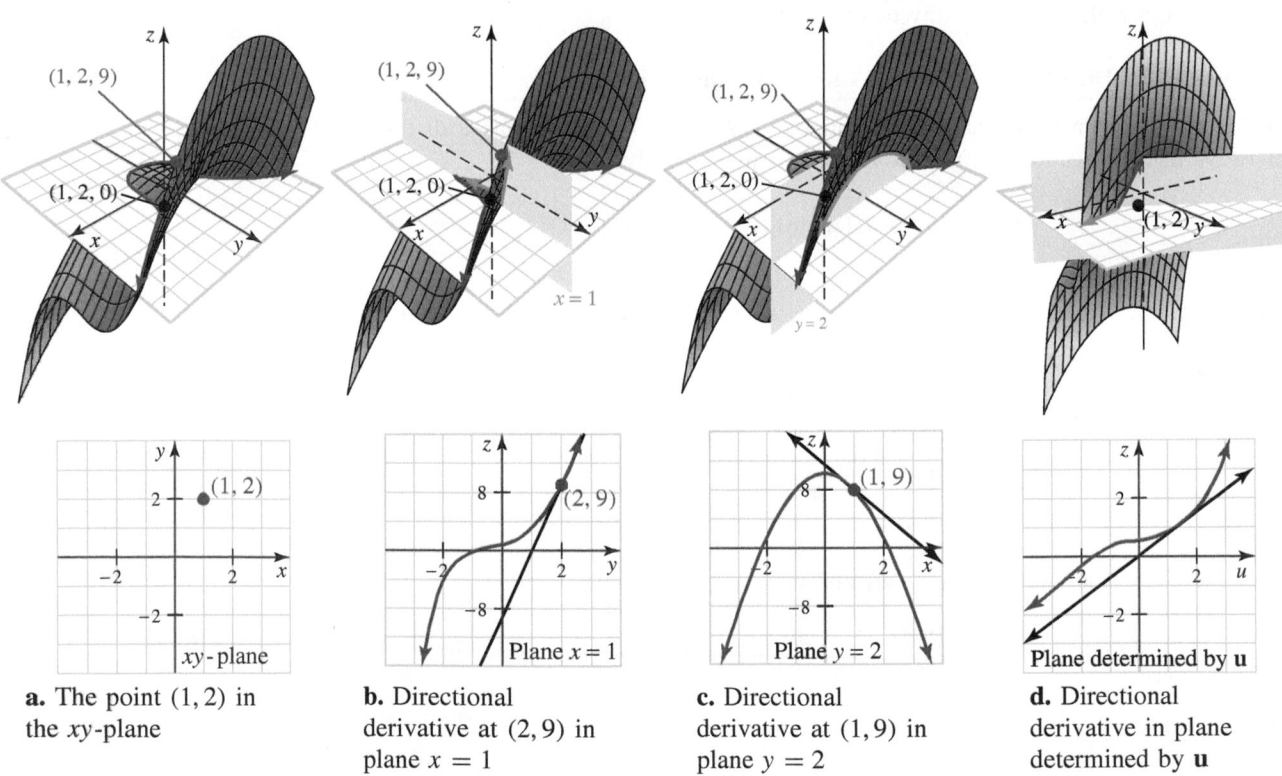

a. The point $(1, 2)$ in the xy-plane

b. Directional derivative at $(2, 9)$ in plane $x = 1$

c. Directional derivative at $(1, 9)$ in plane $y = 2$

d. Directional derivative in plane determined by $\mathbf{u}$

Figure 11.25   Graph of $z = 3 - 2x^2 + y^3$ and some derivatives

Directional derivative as a rate

The directional derivative $D_{\mathbf{u}} f(x_0, y_0)$ can also be interpreted as the rate at which the function $z = f(x, y)$ changes as a point moves from $P_0(x_0, y_0)$ in the direction of the unit vector $\mathbf{u}$. Thus, in Example 1, we say that the function $f(x, y) = 3 - 2x^2 + y^3$ changes at the rate of $-2 - 6\sqrt{3}$ as we move from $P(1, 2)$ in the direction of the unit vector $\mathbf{u} = \frac{1}{2}\mathbf{i} - \frac{\sqrt{3}}{2}\mathbf{j}$.

The Gradient

The directional derivative $D_{\mathbf{u}} f(x, y)$ can be expressed concisely in terms of a vector function called the *gradient*, which has many important uses in mathematics. The gradient of a function of two variables may be defined as follows.

GRADIENT   Let f be a differentiable function at (x, y) and let $f(x, y)$ have partial derivatives $f_x(x, y)$ and $f_y(x, y)$. Then the **gradient** of f, denoted by ∇f (pronounced "del eff"), is a vector given by

$$\nabla f(x, y) = f_x(x, y)\mathbf{i} + f_y(x, y)\mathbf{j}$$

The value of the gradient at the point $P_0(x_0, y_0)$ is denoted by

$$\nabla f_0 = f_x(x_0, y_0)\mathbf{i} + f_y(x_0, y_0)\mathbf{j}$$

Think of the symbol ∇ as an "operator" on a function that produces a vector. Another notation for ∇f is **grad** $f(x, y)$.

Example 2   Finding the gradient of a given function

Find $\nabla f(x, y)$ for $f(x, y) = x^2 y + y^3$.

Solution   Begin with the partial derivatives:

$$f_x(x, y) = \frac{\partial}{\partial x}(x^2 y + y^3) = 2xy \quad \text{and} \quad f_y(x, y) = \frac{\partial}{\partial y}(x^2 y + y^3) = x^2 + 3y^2$$

Then

$$\nabla f(x, y) = 2xy\mathbf{i} + (x^2 + 3y^2)\mathbf{j}$$

The following theorem shows how the directional derivative can be expressed in terms of the gradient.

Theorem 11.8   The gradient formula for the directional derivative

If f is a differentiable function of x and y, then the directional derivative of f at the point $P_0(x_0, y_0)$ in the direction of the unit vector $\mathbf{u}$ is

$$D_{\mathbf{u}} f(x_0, y_0) = \nabla f_0 \cdot \mathbf{u}$$

Proof: Because $\nabla f_0 = f_x(x_0, y_0)\mathbf{i} + f_y(x_0, y_0)\mathbf{j}$ and $\mathbf{u} = u_1\mathbf{i} + u_2\mathbf{j}$, we have

$$D_{\mathbf{u}} f(x_0, y_0) = \nabla f_0 \cdot \mathbf{u} = f_x(x_0, y_0)u_1 + f_y(x_0, y_0)u_2 \qquad \blacklozenge$$

Example 3   Using the gradient formula to compute a directional derivative

Find the directional derivative of $f(x, y) = \ln(x^2 + y^3)$ at $P_0(1, -3)$ in the direction of $\mathbf{v} = 2\mathbf{i} - 3\mathbf{j}$.

Solution $\qquad f_x(x, y) = \dfrac{2x}{x^2 + y^3},$ and $\qquad f_y(x, y) = \dfrac{3y^2}{x^2 + y^3},$

$$\text{so } f_x(1, -3) = -\frac{2}{26} \qquad\qquad \text{so } f_y(1, -3) = -\frac{27}{26}$$

$$\nabla f_0 = \nabla f(1, -3) = -\frac{2}{26}\mathbf{i} - \frac{27}{26}\mathbf{j}$$

A unit vector in the direction of $\mathbf{v}$ is

$$\mathbf{u} = \frac{\mathbf{v}}{\|\mathbf{v}\|} = \frac{2\mathbf{i} - 3\mathbf{j}}{\sqrt{2^2 + (-3)^2}} = \frac{1}{\sqrt{13}}(2\mathbf{i} - 3\mathbf{j})$$

Thus,

$$D_{\mathbf{u}}(x, y) = \nabla f \cdot \mathbf{u}$$

$$= \left(-\frac{2}{26}\right)\left(\frac{2}{\sqrt{13}}\right) + \left(-\frac{27}{26}\right)\left(-\frac{3}{\sqrt{13}}\right)$$

$$= \frac{77\sqrt{13}}{338}$$

Although a differentiable function of one variable $f(x)$ has exactly one derivative $f'(x)$, a differentiable function of two variables $F(x, y)$ has two partial derivatives and an infinite number of directional derivatives. Is there any single mathematical concept for functions of several variables that is the analogue of the derivative of a function of a single variable? The properties listed in the following theorem suggest that the gradient plays this role.

Theorem 11.9   Basic properties of the gradient

Let f and g be differentiable functions. Then

$\qquad$ **Constant rule** $\quad \nabla c = \mathbf{0}$ $\quad$ for any constant $\mathbf{c}$

$\qquad$ **Linearity rule** $\quad \nabla(af + bg) = a\nabla f + b\nabla g$ $\qquad$ for constants a and b

$\qquad$ **Product rule** $\quad \nabla(fg) = f\nabla g + g\nabla f$

$\qquad$ **Quotient rule** $\quad \nabla\left(\dfrac{f}{g}\right) = \dfrac{g\nabla f - f\nabla g}{g^2}$ $\quad g \neq 0$

$\qquad$ **Power rule** $\quad \nabla(f^n) = nf^{n-1}\nabla f$

PROOF OF **linearity rule**

$$\nabla(af + bg) = (af + bg)_x\mathbf{i} + (af + bg)_y\mathbf{j}$$
$$= (af_x + bg_x)\mathbf{i} + (af_y + bg_y)\mathbf{j}$$
$$= af_x\mathbf{i} + bg_x\mathbf{i} + af_y\mathbf{j} + bg_y\mathbf{j}$$
$$= a(f_x\mathbf{i} + f_y\mathbf{j}) + b(g_x\mathbf{i} + g_y\mathbf{j})$$
$$= a\nabla f + b\nabla g$$

PROOF OF **power rule**
$$\nabla\left(f^n\right) = [f^n]_x\mathbf{i} + [f^n]_y\mathbf{j}$$
$$= nf^{n-1}f_x\mathbf{i} + nf^{n-1}f_y\mathbf{j}$$
$$= nf^{n-1}[f_x\mathbf{i} + f_y\mathbf{j}]$$
$$= nf^{n-1}\nabla f$$

The other rules are left for the problem set (Problem 54).

Maximal Property of the Gradient

In applications, it is often useful to compute the greatest rate of increase (or decrease) of a given function at a specified point. The direction in which this occurs is called the direction of **steepest ascent** (or **steepest descent**). For example, suppose the function $z = f(x, y)$ gives the altitude of a skier coming down a slope, and we want to state a theorem that will give the skier the *compass direction* of the path of steepest descent (see Figure 11.26).

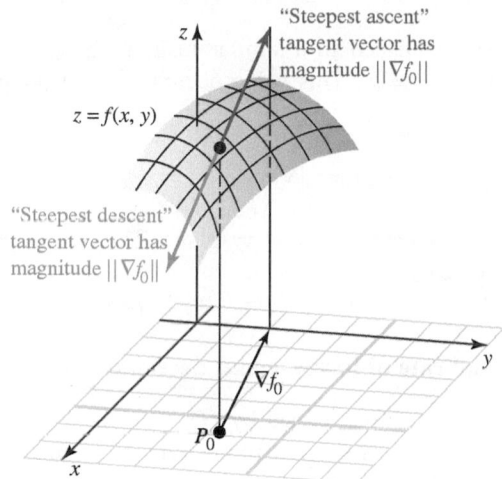

Figure 11.26 Curve of steepest ascent or descent

We emphasize the words "compass direction" because the gradient gives direction in the xy-plane and does not itself point up or down the mountain. The following theorem shows how the direction of maximum change is determined by the gradient.

Theorem 11.10 Maximal direction property of the gradient

Suppose f is differentiable at the point P_0 and that the gradient of f at P_0 satisfies $\nabla f_0 \neq \mathbf{0}$. Then

a. The largest value of the directional derivative $D_{\mathbf{u}}f$ at P_0 is $\|\nabla f_0\|$ and occurs when the unit vector $\mathbf{u}$ points in the direction of ∇f_0.

b. The smallest value of $D_{\mathbf{u}}f$ at P_0 is $- \|\nabla f_0\|$ and occurs when $\mathbf{u}$ points in the direction of $-\nabla f_0$.

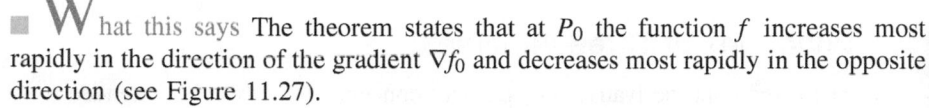

■ **W**hat this says The theorem states that at P_0 the function f increases most rapidly in the direction of the gradient ∇f_0 and decreases most rapidly in the opposite direction (see Figure 11.27).

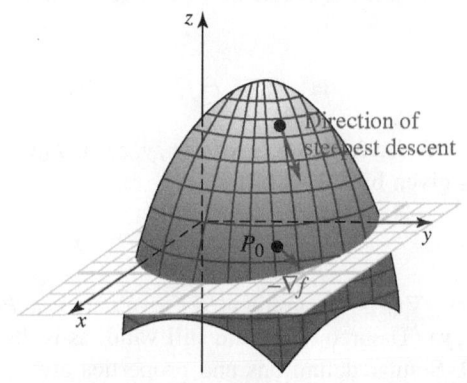

Figure 11.27 The direction of steepest descent

Proof: If **u** is any unit vector, then

$$D_{\mathbf{u}}f = \nabla f_0 \cdot \mathbf{u} = \|\nabla f_0\| \left(\|\mathbf{u}\| \cos\theta\right) = \|\nabla f_0\| \cos\theta$$

where θ is the angle between ∇f_0 and **u**. But $\cos\theta$ assumes its largest value 1 at $\theta = 0$; that is, when **u** points in the direction ∇f_0. Thus, the largest possible value of $D_{\mathbf{u}}f$ is

$$D_{\mathbf{u}}f = \|\nabla f_0\| \,(1) = \|\nabla f_0\|$$

Statement **b** may be established in a similar fashion by noting that $\cos\theta$ assumes its smallest value -1 when $\theta = \pi$. This value occurs when **u** points toward $-\nabla f_0$, and in this direction

$$D_{\mathbf{u}}f = \|\nabla f_0\| \,(-1) = -\|\nabla f_0\|$$

The gradient of f is the vector ∇f in the xy-plane. This vector points in the direction of steepest ascent at a given point. ♦

Example 4 Maximal rate of increase and decrease

In what direction is the function defined by $f(x,y) = xe^{2y-x}$ increasing most rapidly at the point $P_0(2,1)$, and what is the maximum rate of increase? In what direction is f decreasing most rapidly?

Solution   We begin by finding the gradient of f:

$$
\begin{aligned}
\nabla f &= f_x\mathbf{i} + f_y\mathbf{j} \\
&= [e^{2y-x} + xe^{2y-x}(-1)]\mathbf{i} + [xe^{2y-x}(2)]\mathbf{j} \\
&= e^{2y-x}[(1-x)\mathbf{i} + 2x\mathbf{j}]
\end{aligned}
$$

At $(2,1)$,
$$
\begin{aligned}
\nabla f_0 &= e^{2(1)-2}[(1-2)\mathbf{i} + 2(2)\mathbf{j}] \\
&= -\mathbf{i} + 4\mathbf{j}
\end{aligned}
$$

The most rapid rate of increase is

$$\|\nabla f_0\| = \sqrt{(-1)^2 + (4)^2} = \sqrt{17}$$

and it occurs in the direction of $-\mathbf{i} + 4\mathbf{j}$. The most rapid rate of decrease occurs in the direction of $-\nabla f_0 = \mathbf{i} - 4\mathbf{j}$. ■

Functions of Three Variables

The directional derivative and gradient concepts can easily be extended to functions of three or more variables. For a function of three variables, $f(x,y,z)$, the gradient ∇f is defined by

$$\nabla f = f_x\mathbf{i} + f_y\mathbf{j} + f_z\mathbf{k}$$

and the directional derivative $D_{\mathbf{u}}f$ of $f(x,y,z)$ at $P_0(x_0,y_0,z_0)$ in the direction of the unit vector **u** is given by

$$D_{\mathbf{u}}f = \nabla f_0 \cdot \mathbf{u}$$

where, as before, ∇f_0 is the gradient ∇f evaluated at P_0. The basic properties of the gradient of $f(x,y)$ (Theorem 11.9) are still valid, as is the maximal direction property of Theorem 11.10. Similar definitions and properties are valid for functions of more than three variables.

Example 5   Directional derivative of a function of three variables

Let $f(x, y, z) = xy \sin(xz)$. Find ∇f_0 at the point $P_0(1, -2, \pi)$ and then compute the directional derivative (rounded to the nearest hundredth) of f at P_0 in the direction of the vector

$$\mathbf{v} = -2\mathbf{i} + 3\mathbf{j} - 5\mathbf{k}$$

Solution   Begin with the partial derivatives:

$$f_x = y \sin(xz) + xy\,(z\cos(zx)) \qquad\qquad f_y = x\sin(xz) \qquad\qquad f_z = xy\,(x\cos(xz))$$

$$f_x(1, -2, \pi) = -2\sin\pi - 2\pi\cos\pi \qquad f_y(1, -2, \pi) = 1\sin\pi \qquad f_z(1, -2, \pi) = (1)(-2)(1)\cos\pi$$

$$= 2\pi \qquad\qquad\qquad\qquad = 0 \qquad\qquad\qquad\qquad = 2$$

Thus, the gradient of f at P_0 is

$$\nabla f_0 = 2\pi\mathbf{i} + 2\mathbf{k}$$

To find $D_{\mathbf{u}}f$ we need $\mathbf{u}$, the unit vector in the direction of $\mathbf{v}$:

$$\mathbf{u} = \frac{\mathbf{v}}{\|\mathbf{v}\|} = \frac{-2\mathbf{i} + 3\mathbf{j} - 5\mathbf{k}}{\sqrt{(-2)^2 + (3)^2 + (-5)^2}} = \frac{1}{\sqrt{38}}(-2\mathbf{i} + 3\mathbf{j} - 5\mathbf{k})$$

Finally,

$$D_{\mathbf{u}}f(1, -2, \pi) = \nabla f_0 \cdot \mathbf{u} = \tfrac{1}{\sqrt{38}}(-4\pi - 10) \approx -3.66 \qquad\blacksquare$$

Normal Property of the Gradient

Suppose S is a level surface of the function defined by $f(x, y, z)$; that is, $f(x, y, z) = K$ for some constant K. Then if $P_0(x_0, y_0, z_0)$ is a point on S, the following theorem shows that the gradient ∇f_0 at P_0 is a vector that is **normal** (that is, orthogonal) to the tangent plane surface at P_0 (see Figure 11.28).

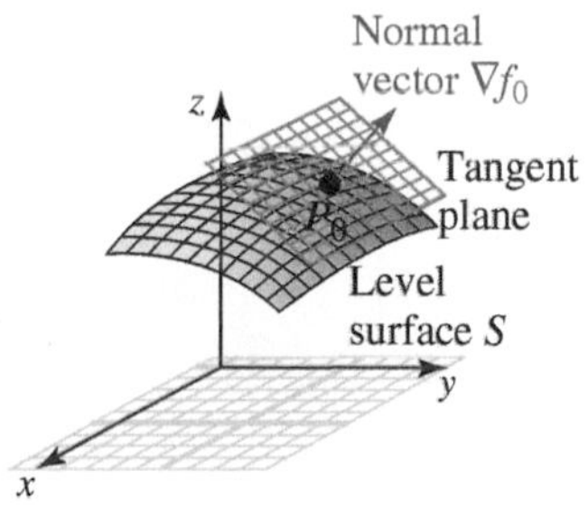

Figure 11.28 The normal property of the gradient

Theorem 11.11   The normal property of the gradient

Suppose the function f is differentiable at the point P_0 and that the gradient at P_0 satisfies $\nabla f_0 \neq \mathbf{0}$. Then ∇f_0 is orthogonal to the level surface of f through P_0.

> $\blacksquare$ **W**hat this says   The gradient ∇f_0 at each point P_0 on the surface $f(x, y, z) = K$ is orthogonal at P_0 to the tangent vector $\mathbf{T} = \dfrac{d\mathbf{R}}{dt}$ of each curve $C: \mathbf{R} = \mathbf{R}(t)$ on the surface that passes through P_0. Thus, all these tangent vectors lie in a single plane through P_0 with normal vector $\mathbf{N} = \nabla f_0$. This plane is the *tangent plane* to the surface at P_0.

Proof: Let C be any smooth curve on the level surface $f(x, y, z) = K$ that passes through $P_0(x_0, y_0, z_0)$, and describe the curve C by the vector function $\mathbf{R}(t) = x(t)\mathbf{i} + y(t)\mathbf{j} + z(t)\mathbf{k}$ for all t in some interval I. We will show that the gradient ∇f_0 is orthogonal to the tangent vector $d\mathbf{R}/dt$ at P_0.

Because C lies on the level surface, any point $P\,(x(t), y(t), z(t))$ on C must satisfy $f[x(t), y(t), z(t)] = K$, and by applying the chain rule, we obtain

$$\frac{d}{dt}[f(x(t), y(t), z(t))] = f_x(x, y, z)\frac{dx}{dt} + f_y(x, y, z)\frac{dy}{dt} + f_z(x, y, z)\frac{dz}{dt}$$

Suppose $t = t_0$ at P_0.

Then $\dfrac{d}{dt}[f(x(t), y(t), z(t))]\Big|_{t=t_0}$

$$= f_x(x(t_0), y(t_0), z(t_0))\frac{dx}{dt}(t_0) + f_y(x(t_0), y(t_0), z(t_0))\frac{dy}{dt}(t_0)$$

$$+ f_z(x(t_0), y(t_0), z(t_0))\frac{dz}{dt}(t_0)$$

$$= \nabla f_0 \cdot \frac{d\mathbf{R}}{dt}(t_0)$$

since $\dfrac{d\mathbf{R}}{dt} = \dfrac{dx}{dt}\mathbf{i} + \dfrac{dy}{dt}\mathbf{j} + \dfrac{dz}{dt}\mathbf{k}$. We also know that $f(x(t), y(t), z(t)) = K$ for all t in I (because the curve C lies on the level surface $f(x, y, z) = K$). Thus, we have

$$\frac{d}{dt}[f(x(t), y(t), z(t))] = \frac{d}{dt}(K) = 0$$

and it follows that $\nabla f_0 \cdot \dfrac{d\mathbf{R}}{dt}(t_0) = 0$. We are given that $\nabla f_0 \neq \mathbf{0}$, and $d\mathbf{R}/dt \neq \mathbf{0}$ because the curve C is smooth. Therefore, ∇f_0 is orthogonal to $d\mathbf{R}/dt$ at the point P_0, as required. ♦

Example 6   Finding a vector that is normal to a level surface

Find a vector that is normal to the level surface $x^2 + 2xy - yz + 3z^2 = 7$ at the point $P_0(1, 1, -1)$.

Solution   Since the gradient vector at P_0 is perpendicular to the level surface, we have

$$\nabla f = f_x\mathbf{i} + f_y\mathbf{j} + f_z\mathbf{k} = (2x + 2y)\mathbf{i} + (2x - z)\mathbf{j} + (6z - y)\mathbf{k}$$

At the point $(1, 1, -1)$, $\nabla f_0 = 4\mathbf{i} + 3\mathbf{j} - 7\mathbf{k}$ is the required normal. ∎

Here is an example in which f involves only two variables, so $f(x, y) = K$ is a level curve in the plane instead of a level surface in space.

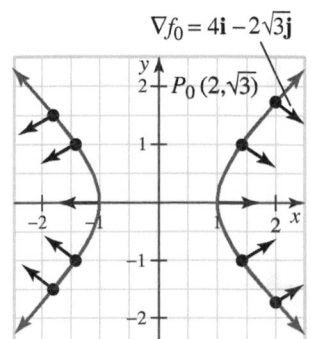

$\nabla f_0 = 4\mathbf{i} - 2\sqrt{3}\mathbf{j}$

$P_0\,(2, \sqrt{3})$

Figure 11.29 The level curve $x^2 - y^2 = 1$

Example 7   Finding a vector normal to a level curve

Sketch the level curve corresponding to $C = 1$ for the function $f(x, y) = x^2 - y^2$ and find a normal vector at the point $P_0(2, \sqrt{3})$.

Solution   The level curve for $C = 1$ is a hyperbola given by $x^2 - y^2 = 1$, as shown in Figure 11.29.

The gradient vector is perpendicular to the level curve, so we have

$$\nabla f = f_x i + f_y\mathbf{j} = 2x\mathbf{i} - 2y\mathbf{j}$$

At the point $(2, \sqrt{3})$, $\nabla f_0 = 4\mathbf{i} - 2\sqrt{3}\mathbf{j}$ is the required normal. This normal vector and a few others are shown in Figure 11.29. ∎

Example 8 Heat-flow application

The set of points (x, y) with $0 \le x \le 5$ and $0 \le y \le 5$ is a square in the first quadrant of the xy-plane. Suppose this square is heated in such a way that $T(x, y) = x^2 + y^2$ is the temperature at the point $P(x, y)$. In what direction will heat flow from the point $P_0(3, 4)$?

Solution The flow of heat in the region is given by a vector function $\mathbf{H}(x, y)$, whose value at each point (x, y) depends on x and y. From physics it is known that $\mathbf{H}(x, y)$ will be perpendicular to the isothermal curves $T(x, y) = C$ for C constant. The gradient ∇T and all its multiples point in such a direction. Therefore, we can express the heat flow as $\mathbf{H} = -k \nabla T$, where k is a positive constant (called the **thermal conductivity**) and the negative sign is introduced to account for the fact that heat flows "downhill" (that is, in the direction of decreasing temperature).

Because $T(3, 4) = 25$, the point $P_0(3, 4)$ lies on the isotherm $T(x, y) = 25$, which is part of the circle $x^2 + y^2 = 25$, as shown in Figure 11.30.
We know that the heat flow $\mathbf{H}_0$ at P_0 will satisfy $\mathbf{H}_0 = -k \nabla T_0$, where ∇T_0 is the gradient at P_0. Because $\nabla T = 2x\mathbf{i} + 2y\mathbf{j}$, we see that $\nabla T_0 = 6\mathbf{i} + 8\mathbf{j}$. Thus, the heat flow at P_0 satisfies

$$\mathbf{H}_0 = -k \nabla T_0 = -k(6\mathbf{i} + 8\mathbf{j})$$

Because the thermal conductivity k is positive, we can say that heat flows from P_0 in the direction of the unit vector $\mathbf{u}$ given by

$$\mathbf{u} = \frac{-(6\mathbf{i} + 8\mathbf{j})}{\sqrt{(-6)^2 + (-8)^2}} = -\frac{3}{5}\mathbf{i} - \frac{4}{5}\mathbf{j}$$

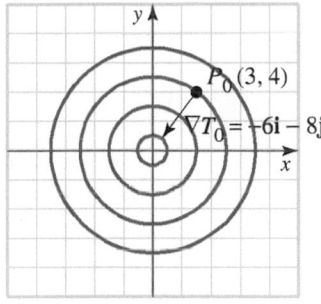

Figure 11.30 Isotherms

Tangent Planes and Normal Lines

Tangent planes and normal lines to a surface are the natural extensions to $\mathbb{R}^3$ of the tangent and normal lines we examined in $\mathbb{R}^2$. Suppose S is a surface and $\mathbf{N}$ is a vector normal to S at the point P_0. We would intuitively expect the normal line and the tangent plane to S at P_0 to be, respectively, the line through P_0 with the direction of $\mathbf{N}$ and the plane through P_0 with normal $\mathbf{N}$ (see Figure 11.31).
These observations lead us to the following definition.

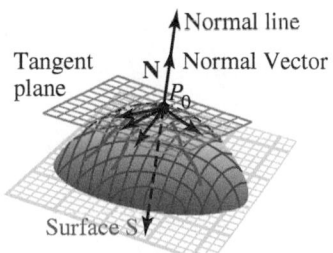

Figure 11.31 Tangent plane and normal line

> **NORMAL LINE AND TANGENT LINE** Suppose the surface S has a nonzero normal vector $\mathbf{N}$ at the point P_0. Then the line through P_0 parallel to $\mathbf{N}$ is called the **normal line** to S at P_0, and the plane through P_0 with normal vector $\mathbf{N}$ is the **tangent plane** to S at P_0.

We would expect a surface S with the representation $z = f(x, y)$ to have a nonvertical tangent plane at each point where $\nabla f \ne \mathbf{0}$. In particular, if S has an equation of the form $F(x, y, z) = C$, where C is a constant and F is a function differentiable at P_0, the normal property of a gradient tells us that the gradient ∇F_0 at P_0 is normal to S (if $\nabla F_0 \ne \mathbf{0}$) and that S must therefore have a tangent plane at P_0.

Example 9 Finding the tangent plane and normal line to a given surface

Find equations for the tangent plane and the normal line at the point $P_0(1, -1, 2)$ on the surface S given by $x^2 y + y^2 z + z^2 x = 5$.

Solution We need to rewrite this problem so that the normal property of the gradient theorem applies. Let $F(x, y, z) = x^2 y + y^2 z + z^2 x$, and consider S to be the level surface

$F(x, y, z) = 5$. The gradient ∇F is normal to S at P_0. We find that

$$\nabla F(x, y, z) = (2xy + z^2)\mathbf{i} + (x^2 + 2yz)\mathbf{j} + (y^2 + 2xz)\mathbf{k}$$

so the normal vector at P_0 is

$$\mathbf{N} = \nabla F_0 = \nabla F(1, -1, 2) = 2\mathbf{i} - 3\mathbf{j} + 5\mathbf{k}$$

Hence, the required tangent plane is

$$2(x - 1) - 3(y + 1) + 5(z - 2) = 0$$
$$2x - 3y + 5z = 15$$

The normal line to the surface at P_0 is

$$x = 1 + 2t, \quad y = -1 - 3t, \quad z = 2 + 5t$$

By generalizing the procedure illustrated in the preceding example, we are led to the following formulas for the tangent plane and normal line. (Also see Figure 11.32.)

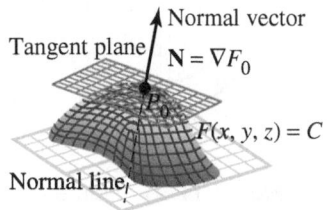

Figure 11.32 The tangent plane and normal line to a surface

FORMULAS FOR THE TANGENT PLANE AND NORMAL LINES TO A SURFACE Suppose S is a surface with the equation $F(x, y, z) = C$ and let $P_0(x_0, y_0, z_0)$ be a point on S where F is differentiable with $\nabla F_0 \neq \mathbf{0}$. Then the **equation of the tangent plane** to S at P_0 is

$$F_x(x_0, y_0, z_0)(x - x_0) + F_y(x_0, y_0, z_0)(y - y_0) + F_z(x_0, y_0, z_0)(z - z_0) = 0$$

and the **normal line** to S at P_0 has parametric equations

$$x = x_0 + F_x(x_0, y_0, z_0)t$$
$$y = y_0 + F_y(x_0, y_0, z_0)t$$
$$z = z_0 + F_z(x_0, y_0, z_0)t$$

Note that in the special case where $z = f(x, y)$, we have $F(x, y, z) = f(x, y) - z = 0$. Then $F_x = f_x$, $F_y = f_y$, and $F_z = -1$ and the equation of the tangent plane becomes

$$f_x(x_0, y_0, z_0)(x - x_0) + f_y(x_0, y_0, z_0)(y - y_0) - (z - z_0) = 0$$

which is equivalent to the tangent plane formula given in Section 11.4.

Example 10   Equations of the tangent plane and the normal line

Find the equations for the tangent plane and the normal line to the cone $z^2 = x^2 + y^2$ at the point where $x = 3$, $y = 4$, and $z > 0$.

Solution   If $P_0(x_0, y_0, z_0)$ is the point of tangency and $x_0 = 3$, $y_0 = 4$, and $z_0 > 0$, then

$$z_0 = \sqrt{x_0^2 + y_0^2} = \sqrt{9 + 16} = 5$$

If we consider $F(x, y, z) = x^2 + y^2 - z^2$, then the cone can be regarded as the level surface $F(x, y, z) = 0$. The partial derivatives of F are

$$F_x = 2x; \qquad F_y = 2y; \qquad F_z = -2z$$

so at $P_0(3,4,5)$,

$$F_x(3,4,5) = 6 \; ; \; F_y(3,4,5) = 8 \; ; \; F_z(3,4,5) = -10$$

Thus the tangent plane has the equation

$$6(x-3) + 8(y-4) - 10(z-5) = 0$$

or $3x + 4y - 5z = 0$, and the normal line is given parametrically by the equations

$$x = 3 + 6t, \qquad y = 4 + 8t, \qquad z = 5 - 10t \qquad \blacksquare$$

PROBLEM SET 11.6

Level 1

Find the gradient of the functions given in Problems 1-10.

1. $f(x,y) = x^2 - 2xy$
2. $f(x,y) = 3x + 4y^2$
3. $f(x,y) = \dfrac{y}{x} + \dfrac{x}{y}$
4. $f(x,y) = \ln(x^2 + y^2)$
5. $f(x,y) = xe^{3-y}$
6. $f(x,y) = e^{x+y}$
7. $f(x,y) = \sin(x + 2y)$
8. $f(x,y,z) = xyz^2$
9. $f(x,y,z) = xe^{y+3z}$
10. $f(x,y,z) = \dfrac{xy - 1}{z + x}$

Compute the directional derivative of the functions given in Problems 11-16 at the point P_0 in the direction of the given vector **v**.

Function	Point P_0	Vector **v**
11. $f(x,y) = x^2 + xy$	$(1,-2)$	$\mathbf{i} + \mathbf{j}$
12. $f(x,y) = \dfrac{e^{-x}}{y}$	$(2,-1)$	$-\mathbf{i} + \mathbf{j}$
13. $f(x,y) = \ln(x^2 + 3y)$	$(1,1)$	$\mathbf{i} + \mathbf{j}$
14. $f(x,y) = \ln(3x + y^2)$	$(0,1)$	$\mathbf{i} - \mathbf{j}$
15. $f(x,y) = \sec(xy - y^3)$	$(2,0)$	$-\mathbf{i} - 3\mathbf{j}$
16. $f(x,y) = \sin xy$	$(\sqrt{\pi}, \sqrt{\pi})$	$3\pi\mathbf{i} - \pi\mathbf{j}$

Find a unit vector that is normal to each surface given in Problems 17-24 at the prescribed point, and the standard form of the equation of the tangent plane at that point.

17. $x^2 + y^2 + z^2 = 3$ at $(1, -1, 1)$

18. $x^4 + y^4 + z^4 = 3$ at $(1, -1, -1)$

19. $\cos z = \sin(x + y)$ at $\left(\frac{\pi}{2}, \frac{\pi}{2}, \frac{\pi}{2}\right)$

20. $\sin(x + y) + \tan(y + z) = 1$ at $\left(\frac{\pi}{4}, \frac{\pi}{4}, -\frac{\pi}{4}\right)$

21. $\ln\left(\dfrac{x}{y - z}\right) = 0$ at $(2, 5, 3)$

22. $\ln\left(\dfrac{x - y}{y + z}\right) = x - z$ at $(1, 0, 1)$

23. $ze^{x+2y} = 3$ at $(2, -1, 3)$

24. $ze^{x^2 - y^2} = 3$ at $(1, 1, 3)$

Find the direction from P_0 in which the given function f increases most rapidly and compute the magnitude of the greatest rate of increase in Problems 25-32.

25. $f(x,y) = 3x + 2y - 1$; $P_0(1, -1)$
26. $f(x,y) = 1 - x^2 - y^2$; $P_0(1, 2)$
27. $f(x,y) = x^3 + y^3$; $P_0(3, -3)$
28. $f(x,y) = ax + by + c$; $P_0(a, b)$
29. $f(x,y) = \ln\sqrt{x^2 + y^2}$; $P_0(1, 2)$
30. $f(x,y) = \sin xy$; $P_0\left(\frac{\sqrt{\pi}}{3}, \frac{\sqrt{\pi}}{2}\right)$
31. $f(x,y,z) = (x + y)^2 + (y + z)^2 + (x + z)^2$; $P_0(2, -1, 2)$
32. $f(x,y,z) = z\ln\left(\dfrac{y}{x}\right)$; $P_0(1, e, -1)$

In Problems 33-36, find a unit vector that is normal to the given graph at the point $P_0(x_0, y_0)$ on the graph. Assume that a, b, and c are constants.

33. the line $ax + by = c$

34. the circle $x^2 + y^2 = a^2$

35. the ellipse $\dfrac{x^2}{a^2} + \dfrac{y^2}{b^2} = 1$

36. the hyperbola $\dfrac{x^2}{a^2} - \dfrac{y^2}{b^2} = 1$

37. Find the directional derivative of

$$f(x,y) = x^2 + y^2$$

at the point $P_0(1, 1)$ in the direction of the unit vector **u** shown in Figure 11.33.

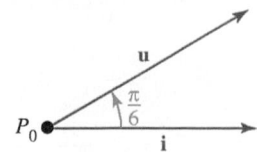

Figure 11.33 Problem 37

38. Find the directional derivative of

$$f(x,y) = x^2 + xy + y^2$$

at $P_0(1, -1)$ in the direction toward the origin.

39. Find the directional derivative of

$$f(x,y) = e^{x^2 y^2}$$

at $P_0(1, -1)$ in the direction toward $Q(2, 3)$).

40. Let

$$f(x,y,z) = 2x^2 - y^2 + 3z^2 - 8x - 4y + 201$$

and let P_0 be the point $\left(2, -\frac{3}{2}, \frac{1}{2}\right)$.

a. Find ∇f_0.

b. Find $\cos \theta$, where θ is the angle between ∇f_0 and the vector toward the origin from P_0.

Level 2

41. Let $f(x,y,z) = xyz$, and let $\mathbf{u}$ be a unit vector perpendicular to both $\mathbf{v} = \mathbf{i} - 2\mathbf{j} + 3\mathbf{k}$ and $\mathbf{w} = 2\mathbf{i} + \mathbf{j} - \mathbf{k}$. Find the directional derivative of f at $P_0(1, -1, 2)$ in the direction of $\mathbf{u}$.

42. Let $f(x,y,z) = ye^{x+z} + ze^{y-x}$. At the point $P(2, 2, -2)$, find the unit vector pointing in the direction of most rapid increase of f.

43. Modeling Problem Suppose a box in space given by $0 \le x \le 2, 0 \le y \le 2, 0 \le z \le 2$ is temperature controlled so that the temperature at a point $P(x,y,z)$ in the box is modeled by $T(x,y,z) = xy + yz + xz$. A heat-seeking missile is located at $P_0(1, 1, 1)$. In what direction will the missile move for the temperature to increase as quickly as possible? What is the maximum rate of change of the temperature at the point P_0?

44. Modeling Problem A metal plate covering the rectangular region $0 < x \le 6, 0 < y \le 5$ is charged electrically in such a way that the potential at each point (x,y) is inversely proportional to the square of its distance from the origin. If an object is at the point $(3, 4)$, in which direction should it move to increase the potential most rapidly?

45. A skier is speeding down a mountain path. If the surface of the mountain is modeled by $z = 1 - 3x^2 - \frac{5}{2}y^2$ (where x, y, and z are in miles) and the skier begins at the point $P_0\left(\frac{1}{4}, -\frac{1}{2}, \frac{3}{16}\right)$, in what direction should the skier head to descend the mountainside most rapidly?

46. Let f have continuous partial derivatives, and assume the maximal directional derivative of f at $(0, 0)$ is equal to 100 and is attained in the direction toward $(3, -4)$. Find the gradient ∇f at $(0, 0)$.

47. Suppose at the point $P_0(-1, 2)$, a certain function $f(x,y)$ has directional derivative 8 in the direction of $\mathbf{v}_1 = 3\mathbf{i} - 4\mathbf{j}$ and 1 in the direction of $\mathbf{v}_2 = 12\mathbf{i} + 5\mathbf{j}$. What is the directional derivative of f at P_0 in the direction of $\mathbf{v} = 3\mathbf{i} - 5\mathbf{j}$?

48. The directional derivative of $f(x,y,z)$ at the point P_0 is greatest in the direction of $\mathbf{v} = \mathbf{i} + \mathbf{j} - \mathbf{k}$ and has value $5\sqrt{3}$ in this direction. What is the directional derivative of f at P_0 in the direction of $\mathbf{w} = \mathbf{i} + \mathbf{j}$?

49. Let f have continuous partial derivatives and suppose the maximal directional derivative of f at $P_0(1, 2)$ has magnitude 50 and is attained in the direction from P_0 toward $Q(3, -4)$. Use this information to find $\nabla f(1, 2)$.

50. Let $T(x,y) = 1 - x^2 - 2y^2$ be the temperature at each point $P(x,y)$ in the plane. A heat-loving bug is placed in the plane at the point $P_0(-1, 1)$. Find the path that the bug should take to stay as warm as possible. *Hint:* Assume that at each point on the bug's path, the tangent line will point in the direction in which T increases most rapidly.

51. a. Show that the ellipsoid

$$\frac{x^2}{a^2} + \frac{y^2}{b^2} + \frac{z^2}{c^2} = 1$$

has a tangent plane at $P_0(x_0, y_0, z_0)$ with the equation

$$\frac{x_0 x}{a^2} + \frac{y_0 y}{b^2} + \frac{z_0 z}{c^2} = 1$$

b. Find the equation for the tangent plane to the hyperboloid of one sheet

$$\frac{x^2}{a^2} + \frac{y^2}{b^2} - \frac{z^2}{c^2} = 1$$

at the point $P_0(x_0, y_0, z_0)$.

c. Find the equation for the tangent plane to the elliptic paraboloid

$$\frac{z}{c} = \frac{x^2}{a^2} + \frac{y^2}{b^2}$$

at the point $P_0(x_0, y_0, z_0)$.

Level 3

52. Modeling Problem A particle P_1 with mass m_1 is located at the origin, and a particle P_2 with mass 1 unit is located at the point (x, y, z). According to Newton's law of universal gravitation, the force P_1 exerts on P_2 is modeled by

$$\mathbf{F} = \frac{-Gm_1(x\mathbf{i} + y\mathbf{j} + z\mathbf{k})}{r^3}$$

where r is the distance between P_1 and P_2, and G is the gravitational constant.

a. Starting from the fact that $r^2 = x^2 + y^2 + z^2$, show that

$$\frac{\partial}{\partial x}\left(\frac{1}{r}\right) = \frac{-x}{r^3}, \frac{\partial}{\partial y}\left(\frac{1}{r}\right) = \frac{-y}{r^3}, \frac{\partial}{\partial z}\left(\frac{1}{r}\right) = \frac{-z}{r^3}$$

b. The function $V = -Gm_1/r$ is called the **potential energy** function for the system. Show that $\mathbf{F} = -\nabla V$.

53. Recall that an ellipse is the set of all points $P(x, y)$ such that the sum of the distances from P to two fixed points (the *foci*) is constant. Let $P(x, y)$ be a point on the ellipse, and let r_1 and r_2 denote the respective distances from P to the two foci, F_1 and F_2 as shown in Figure 11.34.

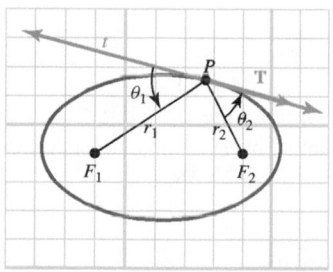

Figure 11.34 Unit tangent to an ellipse

a. Show that $\mathbf{T} \cdot \nabla(r_1 + r_2) = 0$, where $\mathbf{T}$ is the unit tangent to the ellipse at P.

b. Use part **a** to show that the tangent line to the ellipse at P makes equal angles with the lines joining P to the foci (that is, $\theta_1 = \theta_2$ in Figure 11.34).

54. Verify each of the following properties for functions of two variables.

a. $\nabla(c) = \mathbf{0}$ for constant c

b. $\nabla(f + g) = \nabla f + \nabla g$

c. $\nabla(f/g) = \dfrac{g\nabla f - f\nabla g}{g^2}, g \neq 0$

d. $\nabla(fg) = f\nabla g + g\nabla f$

55. Find a general formula for the directional derivative $D_{\mathbf{u}}f$ of the function $f(x, y)$ at the point $P(x_0, y_0)$ in the direction of the unit vector $\mathbf{u} = (\cos\theta)\mathbf{i} + (\sin\theta)\mathbf{j}$. Apply your formula to obtain the directional derivative of $f(x, y) = xy^2e^{x-2y}$ at $P_0(-1, 3)$ in the direction of the unit vector

$$\mathbf{u} = \left(\cos\frac{\pi}{6}\right)\mathbf{i} + \left(\sin\frac{\pi}{6}\right)\mathbf{j}$$

56. Suppose that $\mathbf{u}$ and $\mathbf{v}$ are unit vectors and that f has continuous partial derivatives. Show that

$$D_{\mathbf{u}+\mathbf{v}}f = \frac{1}{\|\mathbf{u} + \mathbf{v}\|}(D_{\mathbf{u}}f + D_{\mathbf{v}}f)$$

57. Think Tank Problem If f, f_x, and f_y are continuous and $\nabla f(x, y) = \mathbf{0}$ inside the disk $x^2 + y^2 < 1$, then either show that $f(x, y)$ is a constant function inside the disk or find a counterexample.

58. Think Tank Problem If f is differentiable at $P_0(x_0, y_0)$ and $D_{\mathbf{u}}f(x_0, y_0) = 0$ for unit vectors $\mathbf{u}_1$ and $\mathbf{u}_2$, where $\mathbf{u}_1 \times \mathbf{u}_2 \neq \mathbf{0}$, then show that $D_{\mathbf{u}}f(x_0, y_0) = 0$ for every unit vector $\mathbf{u}$, or find a counterexample.

59. Let $\mathbf{R} = x\mathbf{i} + y\mathbf{j} + z\mathbf{k}$, and let

$$r = \|\mathbf{R}\| = \sqrt{x^2 + y^2 + z^2}$$

a. Show that ∇r is a unit vector in the direction of $\mathbf{R}$.

b. Show that $\nabla(r^n) = nr^{n-2}\mathbf{R}$, for any positive integer n.

60. Suppose the surfaces $F(x, y, z) = 0$ and $G(x, y, z) = 0$ both pass through the point $P_0(x_0, y_0, z_0)$ and that the gradients ∇F_0 and ∇G_0 both exist. Show that the two surfaces are tangent at P_0 if and only if $\nabla F_0 \times \nabla G_0 = \mathbf{0}$.

11.7 EXTREMA OF FUNCTIONS OF TWO VARIABLES

IN THIS SECTION: *Relative extrema, second partials test, absolute extrema of continuous functions, least squares approximation of data*
In the real world, we encounter a variety of problems involving optimization. In Chapter 4, we considered extrema of functions of a single variable, and in this section we extend our work to functions of two variables.

There are many practical situations in which it is necessary or useful to know the largest and smallest values of a function of two variables. For example, if $T(x, y)$ is the temperature at a point (x, y) in a plate, where are the hottest and coldest points in the plate and what are these extreme temperatures? A hazardous waste dump is bounded by the curve $F(x, y) = 0$. What are the largest and smallest distances to the boundary from a given interior point P_0? We begin our study of extrema with some terminology.

> **ABSOLUTE EXTREMA**   The function $f(x, y)$ is said to have an **absolute maximum** at (x_0, y_0) if $f(x_0, y_0) \geq f(x, y)$ for all (x, y) in the domain D of f. Similarly, f has an **absolute minimum** at (x_0, y_0) if $f(x_0, y_0) \leq f(x, y)$ for all (x, y) in D. Collectively, absolute maxima and minima are called **absolute extrema**.

In Chapter 4, we located absolute extrema of a function of one variable by first finding *relative extrema*, those values of $f(x)$ that are larger or smaller than those at all nearby points. The relative extrema of a function of two variables may be defined as follows.

> **RELATIVE EXTREMA**   Let f be a function defined on a region containing (x_0, y_0). Then
>
> $$f(x_0, y_0) \text{ is a \textbf{relative maximum} if } f(x, y) \leq f(x_0, y_0)$$
>
> for all (x, y) in an open disk containing (x_0, y_0).
>
> $$f(x_0, y_0) \text{ is a \textbf{relative minimum} if } f(x, y) \geq f(x_0, y_0)$$
>
> for all (x, y) in an open disk containing (x_0, y_0).
>
> Collectively, relative maxima and minima are called **relative extrema**.

Relative Extrema

In Chapter 4, we observed that relative extrema of the function f correspond to "peaks and valleys" on the graph of f, and the same observation can be made about relative extrema in the two-variable case, as seen in Figure 11.35.

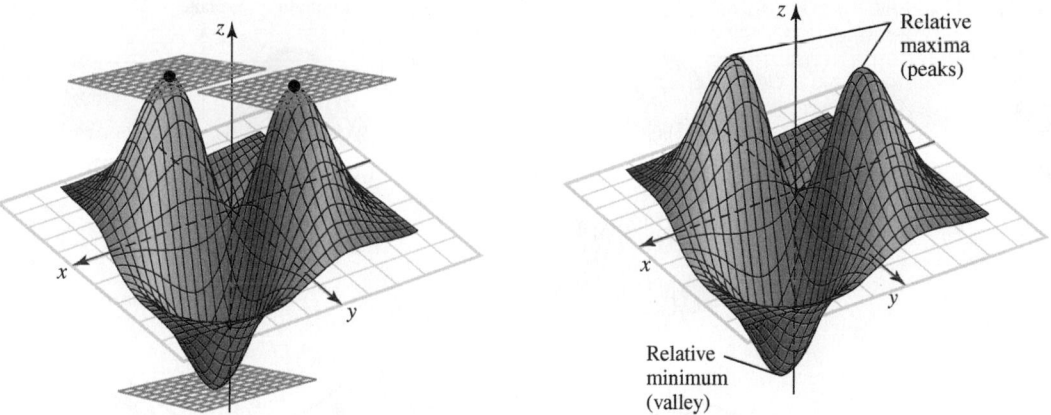

Figure 11.35 Relative extrema correspond to peaks and valleys

For a function f of one variable, we found that the relative extrema occur where $f'(x) = 0$ or $f'(x)$ does not exist. The following theorem shows that the relative extrema of a function of two variables can be located similarly.

Theorem 11.12 Partial derivative criteria for relative extrema

If f has a relative extremum (maximum or minimum) at $P_0(x_0, y_0)$ and partial derivatives f_x and f_y both exist at (x_0, y_0), then

$$f_x(x_0, y_0) = f_y(x_0, y_0) = 0$$

Proof: Let $F(x) = f(x, y_0)$. Then $F(x)$ must have a relative extremum at $x = x_0$, so $F'(x_0) = 0$, which means that $f_x(x_0, y_0) = 0$. Similarly, $G(y) = f(x_0, y)$ has a relative extremum at $y = y_0$, so $G'(y_0) = 0$ and $f_y(x_0, y_0) = 0$. Thus, we must have *both* $f_x(x_0, y_0) = 0$ and $f_y(x_0, y_0) = 0$, as claimed. ◆

In single variable calculus, we referred to a number x_0 where $f'(x_0)$ does not exist or $f'(x_0) = 0$ as a *critical number*. This terminology is extended to functions of two variables as follows.

> **STOP** *There is a horizontal tangent plane at each extreme point where the first partial derivatives exist. However, this **does not** say that whenever a horizontal tangent plane occurs at a point P, there must be an extremum there. All that can be said is that such a point P is a possible location for a relative extremum.*

CRITICAL POINTS   A **critical point** of a function f defined on an open set D is a point (x_0, y_0) in D where either one of the following is true:

a. $f_x(x_0, y_0) = f_y(x_0, y_0) = 0$.
b. At least one of $f_x(x_0, y_0)$ or $f_y(x_0, y_0)$ does not exist

---

Example 1 Distinguishing critical points

Discuss the nature of the critical point $(0, 0)$ for the quadric surfaces
a. $z = x^2 + y^2$ **b.** $z + x^2 + y^2 = 1$ **c.** $z = y^2 - x^2$

Solution   The graphs of these quadric surfaces are shown in Figure 11.36.

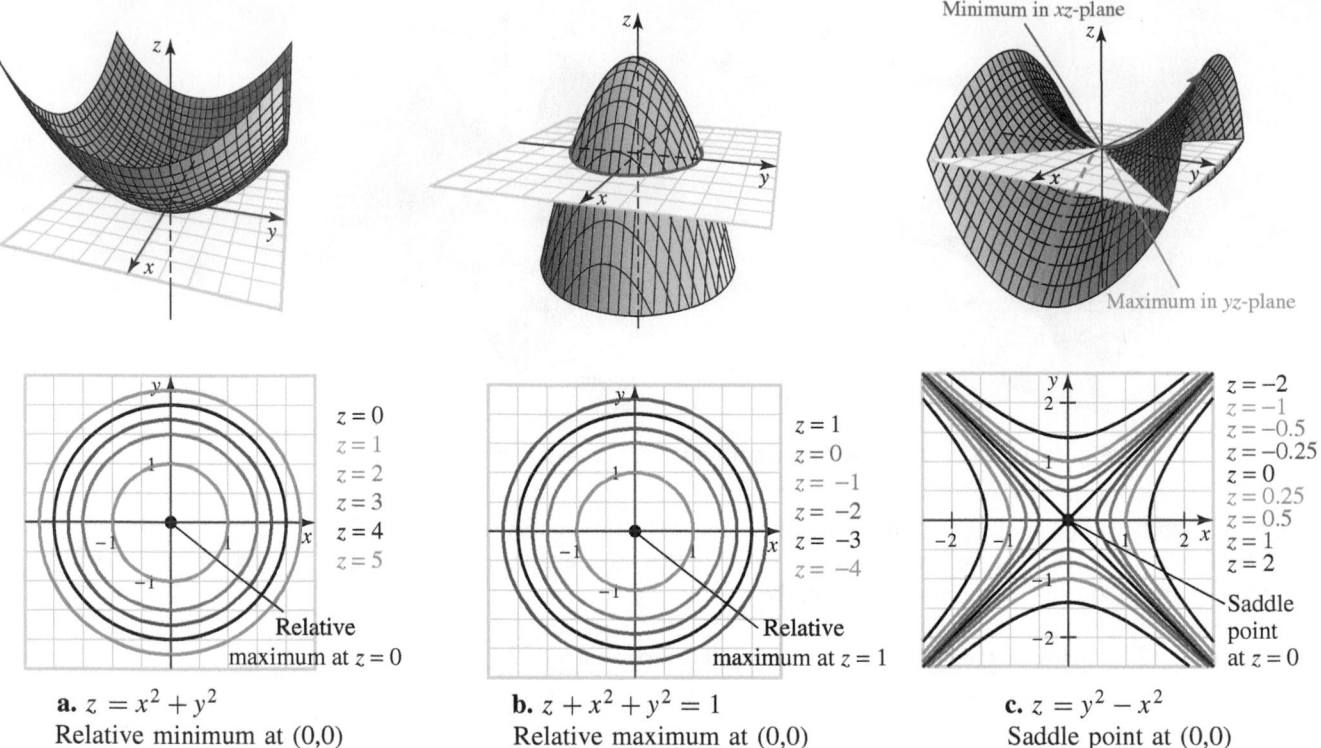

a. $z = x^2 + y^2$
Relative minimum at $(0,0)$

b. $z + x^2 + y^2 = 1$
Relative maximum at $(0,0)$

c. $z = y^2 - x^2$
Saddle point at $(0,0)$

Figure 11.36 Interactive   Classification of critical points

Let $f(x,y) = x^2 + y^2$, $g(x,y) = 1 - x^2 - y^2$, and $h(x,y) = y^2 - x^2$.
We find the critical points:

a. $f_x(x,y) = 2x$, $f_y(x,y) = 2y$; critical point $(0,0)$. The function f has a relative minimum at $(0,0)$ because x^2 and y^2 are both nonnegative, yielding $x^2 + y^2 > 0$ for all
nonzero x and y.

b. $g_x(x,y) = -2x$, $g_y(x,y) = -2y$; critical point $(0,0)$. Since $z = 1 - x^2 - y^2$, it follows that $z \leq 1$ with a relative maximum occurring where x^2 and y^2 are both 0; that
is, at $(0,0)$.

c. $h_x(x,y) = -2x$, $h_y(x,y) = 2y$; critical point $(0,0)$. The function h has neither a relative maximum nor a relative minimum at $(0,0)$. When $z = 0$, h is a minimum on
the y-axis (where $x = 0$) and a maximum on the x-axis (where $y = 0$).   ∎

A critical point $P_0(x_0, y_0)$ is called a **saddle point** of $f(x,y)$ if every open disk
centered at P_0 contains points in the domain of f that satisfy $f(x,y) > f(x_0, y_0)$ as well
as points in the domain of f that satisfy $f(x,y) < f(x_0, y_0)$. An example of a saddle point
is $(0,0)$ on the hyperbolic paraboloid $z = y^2 - x^2$, as shown in Figure 1.36c.

Second Partials Test

The previous example points to the need for some sort of a test to determine the nature
of a critical point. In Chapter 4, we developed the second derivative test for functions
of one variable as a means for determining whether a particular critical number c of
f corresponds to a relative maximum or minimum. If $f'(x) = 0$, then according to this
test, a relative maximum occurs at $x = c$ if $f''(c) < 0$ and a relative minimum occurs if
$f''(c) > 0$. If $f''(c) = 0$, the test is inconclusive. The analogous result for the two-variable
case may be stated as follows.

Theorem 11.13   Second partials test

Let $f(x,y)$ have a critical point at $P_0(x_0, y_0)$ and assume that f has continuous second-order partial derivatives in a disk centered at (x_0, y_0). The *discriminant* of f is the expression

$$D = f_{xx}f_{yy} - f_{xy}^2$$

Then

A **relative maximum** occurs at P_0 if $D(x_0, y_0) > 0$ and $f_{xx}(x_0, y_0) < 0$ (or equivalently $D(x_0, y_0) > 0$ and $f_{yy}(x_0, y_0) < 0$).

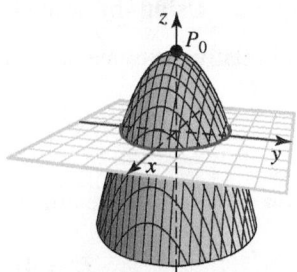

A **relative minimum** occurs at P_0 if $D(x_0, y_0) > 0$ and $f_{xx}(x_0, y_0) > 0$ (or $f_{yy}(x_0, y_0) > 0$).

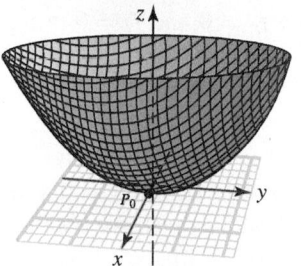

Summary: For a critical point (a, b)

$D(a,b)$	$f_{xx}(a,b)$	Type
$+$	$-$	Rel. max.
$+$	$+$	Rel. min.
$-$	NA	Saddle point
0	NA	Inconclusive

A **saddle point** occurs at P_0 if $D(x_0, y_0) < 0$.

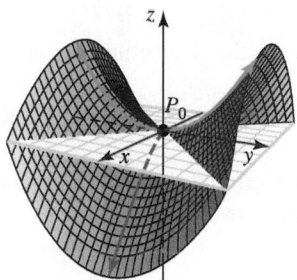

If $D(x_0, y_0) = 0$, then the test is **inconclusive**. Nothing can be said about the nature of the surface at (x_0, y_0) without further analysis.

Proof: The second partials test can be proved by using an extension of the Taylor series expansion (Section 8.8) that applies to functions of two variables $f(x,y)$. Details can be found in most advanced calculus texts (also see Problem 58). ◆

The discriminant $D = f_{xx}f_{yy} - f_{xy}^2$ may be easier to remember in the equivalent determinant form

$$D = \begin{vmatrix} f_{xx} & f_{xy} \\ f_{xy} & f_{yy} \end{vmatrix}$$

Note that if $D > 0$ at the critical point $P_0(x_0, y_0)$, then f_{xx} and f_{yy} must have the same sign. This is the reason that when $D > 0$, either $f_{xx} > 0$ or $f_{yy} > 0$ is enough to guarantee that a relative minimum occurs at P_0 (or a relative maximum if $f_{xx} < 0$ or $f_{yy} < 0$).

Geometrically, if $D > 0$ and $f_{xx} > 0$ and $f_{yy} > 0$ at P_0, then the surface $z = f(x,y)$ curves upward in all directions from the point $Q(x_0, y_0, z_0)$, so there is a relative minimum at Q. Likewise, if $D > 0$ and $f_{xx} < 0$ and $f_{yy} < 0$ at P_0, then the surface curves downward in all directions from P_0, which must therefore be a relative maximum. However, if $D < 0$ at P_0, the surface curves up from Q in some directions, and down in others, so Q must be a saddle point.

Example 2   Using the second partials test to classify critical points.

Find all relative extrema and saddle points of the function

$$f(x,y) = 2x^2 + 2xy + y^2 - 2x - 2y + 5$$

Solution   First, find the critical points:

$$f_x = 4x + 2y - 2 \qquad f_y = 2x + 2y - 2$$

Setting $f_x = 0$ and $f_y = 0$, we obtain the system of equations

$$\begin{cases} 4x + 2y - 2 = 0 \\ 2x + 2y - 2 = 0 \end{cases}$$

and solve to obtain $x = 0, y = 1$. Thus, $(0, 1)$ is the only critical point. To apply the second partials test, we obtain

$$f_{xx} = 4 \qquad f_{yy} = 2 \qquad f_{xy} = 2$$

and form the discriminant

$$D = f_{xx}f_{yy} - f_{xy}^2 = (4)(2) - 2^2 = 4$$

For the critical point $(0, 1)$ we have $D = 4 > 0$ and $f_{xx} = 4 > 0$, so there is a relative minimum at $(0, 1)$.   ▨

Example 3   Second partials test with a relative minimum and a saddle point

Find all critical points on the graph of $f(x,y) = 8x^3 - 24xy + y^3$, and use the second partials test to classify each point as a relative extremum or a saddle point.

Solution $f_x(x,y) = 24x^2 - 24y$, $f_y(x,y) = -24x + 3y^2$. To find the critical points, solve

$$\begin{cases} 24x^2 - 24y = 0 \\ -24x + 3y^2 = 0 \end{cases}$$

From the first equation, $y = x^2$; substitute this into the second equation to find

$$-24x + 3(x^2)^2 = 0$$
$$x(x^3 - 8) = 0$$
$$x(x - 2)(x^2 + 2x + 4) = 0$$
$$x = 0, 2 \qquad \textit{The solutions of } x^2 + 2x + 4 = 0 \textit{ are not real.}$$

If $x = 0$, then $y = 0$, and if $x = 2$, then $y = 4$, so the critical points are $(0,0)$ and $(2,4)$. To obtain D, we first find $f_{xx}(x,y) = 48x$, $f_{xy}(x,y) = -24$, and $f_{yy}(x,y) = 6y$ to find $D(x,y) = (48x)(6y) - (-24)^2$ and then compute:

$$D = \begin{vmatrix} f_{xx} & f_{xy} \\ f_{xy} & f_{yy} \end{vmatrix} = \begin{vmatrix} 48x & -24 \\ -24 & 6y \end{vmatrix} = 288xy - 576$$

(a,b)	$D(a,b)$	$f_{xx}(a,b)$	Type
$(0,0)$	$-$	NA	Saddle point
$(2,4)$	$+$	$+$	Rel. min.

At $(0,0)$, $D = -576 < 0$, so there is a saddle point at $(0,0)$.

At $(2,4)$, $D = 288(2)(4) - 576 = 1{,}728 > 0$ and $f_{xx}(2,4) = 96 > 0$, so there is a relative minimum at $(2,4)$.

To view the situation graphically, we calculate the coordinates of the saddle point $(0,0,0)$, and the relative minimum $(2,4,-64)$, as shown in Figure 11.37.

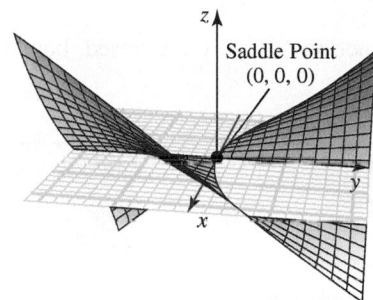

a. View near the origin (showing the saddle point)

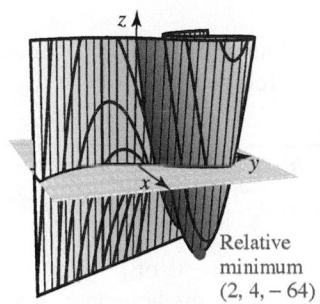

b. View away from the origin (showing the relative minimum point)

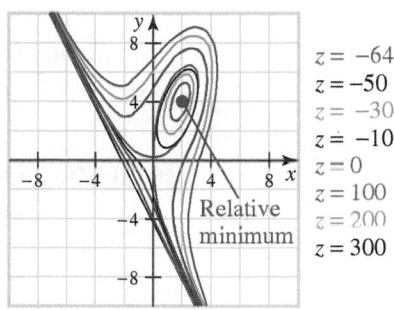

c. Level curves

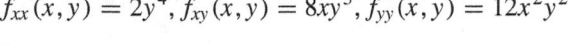

Figure 11.37 Interactive   Graph of $f(x,y) = 8x^3 - 24xy + y^3$

Example 4   Extrema when the second partials test fails

Find all relative extrema and saddle points on the graph of

$$f(x,y) = x^2 y^4$$

The graph is shown in Figure 11.38.

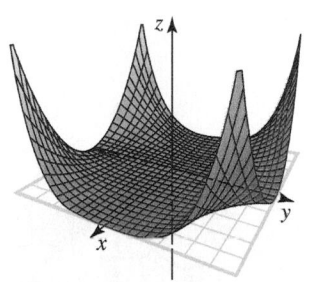

Figure 11.38 Interactive
Graph of $f(x,y) = x^2 y^4$

Solution   Since $f_x(x,y) = 2xy^4$, $f_y(x,y) = 4x^2 y^3$, we see that the critical points occur only when $x = 0$ or $y = 0$; that is, every point on the x-axis or y-axis is a critical point. Because

$$f_{xx}(x,y) = 2y^4, \quad f_{xy}(x,y) = 8xy^3, \quad f_{yy}(x,y) = 12x^2 y^2$$

the discriminant is

$$D = \begin{vmatrix} f_{xx} & f_{xy} \\ f_{xy} & f_{yy} \end{vmatrix} = \begin{vmatrix} 2y^4 & 8xy^3 \\ 8xy^3 & 12x^2 y^2 \end{vmatrix} = 24x^2 y^6 - 64x^2 y^6 = -40x^2 y^6$$

Since $D = 0$ for any critical point $(x_0, 0)$ or $(0, y_0)$, the second partials test fails. However, $f(x,y) = 0$ for every critical point (because either $x = 0$ or $y = 0$, or both), and because $f(x,y) = x^2 y^4 > 0$ when $x \neq 0$ and $y \neq 0$, it follows that each critical point must be a relative minimum.

Absolute Extrema of Continuous Functions

The extreme value theorem (Theorem 4.1) says that a function of a single variable f must attain both an absolute maximum and an absolute minimum on any closed, bounded interval $[a, b]$ on which it is continuous. In $\mathbb{R}^2$, a nonempty set S is *closed* if it contains its boundary (see the introduction to Section 11.2) and is *bounded* if it is contained in a disk. A similar statement can be made in $\mathbb{R}^3$. The extreme value theorem can be extended to functions of two variables in the following form.

Theorem 11.14   Extreme value theorem for a function of two variables

---

A function of two variables $f(x, y)$ attains both an absolute maximum and an absolute minimum on any closed, bounded set S where it is continuous.

Proof: The proof is found in most advanced calculus texts.   ◆

To find the absolute extrema of a continuous function $f(x, y)$ on a closed, bounded set S, we proceed as follows:

ABSOLUTE EXTREMA   To find the absolute extrema of a continuous function f that is continuous on a closed, bounded set S follow this procedure:

Step 1 Find all critical points of f in S.
Step 2 Find all points on the boundary of S where absolute extrema can occur (boundary points, critical points, endpoints, etc.).
Step 3 Compute the value of $f(x_0, y_0)$ for each of the points (x_0, y_0) found in steps 1 and 2.
Step 4 The absolute maximum of f on S is the largest of the values computed in step 3, and the absolute minimum is the smallest of the computed values.

Example 5   Finding absolute extrema

Find the absolute extrema of the function $f(x, y) = e^{x^2 - y^2}$ over the disk $x^2 + y^2 \leq 1$. The graph is shown in Figure 11.39.

Solution

Step 1: $f_x(x, y) = 2xe^{x^2 - y^2}$ and $f_y(x, y) = -2ye^{x^2 - y^2}$. These partial derivatives are defined for all (x, y). Because $f_x(x, y) = f_y(x, y) = 0$ only when $x = 0$ and $y = 0$, it follows that $(0, 0)$ is the only critical point of f and it is inside the disk.

Step 2: Examine the values of f on the boundary curve $x^2 + y^2 = 1$. Because $y^2 = 1 - x^2$ on the boundary of the disk, we find that

$$f(x, y) = e^{x^2 - (1 - x^2)} = e^{2x^2 - 1}$$

We need to find the largest and smallest values of $F(x) = e^{2x^2 - 1}$ for $-1 \leq x \leq 1$. Since

$$F'(x) = 4xe^{2x^2 - 1}$$

we see that $F'(x) = 0$ only when $x = 0$ (since $e^{2x^2 - 1}$ is always positive). At $x = 0$, we have $y^2 = 1 - 0^2$, so $y = \pm 1$; thus $(0, 1)$ and $(0, -1)$ are boundary

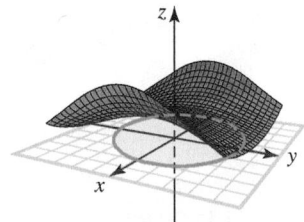

Figure 11.39   Graph of f over a disk

critical points. At the endpoints of the interval $-1 \leq x \leq 1$, the corresponding points are $(1, 0)$ and $(-1, 0)$.

Step 3: Compute the value of f for the points found in steps 1 and 2:

Points to check	Compute $f(x_0, y_0) = e^{x_0^2 - y_0^2}$
$(0, 0)$	$f(0, 0) = e^0 = 1$
$(0, 1)$	$f(0, 1) = e^{-1};$ minimum
$(0, -1)$	$f(0, -1) = e^{-1};$ minimum
$(1, 0)$	$f(1, 0) = e;$ maximum
$(-1, 0)$	$f(-1, 0) = e;$ maximum

Step 4: As indicated in the preceding table, the absolute maximum value of f on the given disk is e, which occurs at $(1, 0)$ and $(-1, 0)$, and the absolute minimum value is e^{-1}, which occurs at $(0, 1)$ and $(0, -1)$.   ◼

In general, it can be difficult to show that a relative extremum is actually an absolute extremum. In practice, however, it is often possible to make the determination using physical or geometric considerations.

Example 6   Minimum distance from a point to a plane

Find the point on the plane $x + 2y + z = 5$ that is closest to the point $P(0, 3, 4)$.

Solution   If $Q(x, y, z)$ is a point on the plane $x + 2y + z = 5$, then $z = 5 - x - 2y$ and the distance from P to Q is

$$d = \sqrt{(x - 0)^2 + (y - 3)^2 + (z - 4)^2}$$
$$= \sqrt{x^2 + (y - 3)^2 + (5 - x - 2y - 4)^2}$$

Instead of minimizing d, we minimize

$$f(x, y) = d^2 = x^2 + (y - 3)^2 + (1 - x - 2y)^2$$

since the minimum of d will occur at the same points where d^2 is also minimized.

To minimize $f(x, y)$, we first determine the critical points of f by solving the system

$$\begin{cases} f_x = 2x - 2(1 - x - 2y) = 4x + 4y - 2 = 0 \\ f_y = 2(y - 3) - 4(1 - x - 2y) = 4x + 10y - 10 = 0 \end{cases}$$

We obtain $x = -\frac{5}{6}$, $y = \frac{4}{3}$, and since

$$f_{xx} = 4, \ f_{yy} = 10, \ f_{xy} = 4$$

we find that

$$D = f_{xx} f_{yy} - f_{xy}^2 = 4(10) - 4^2 > 0 \quad \text{and} \quad f_{xx} = 4 > 0$$

so a relative minimum occurs at $\left(-\frac{5}{6}, \frac{4}{3}\right)$. Intuitively, we see that this relative minimum must also be an absolute minimum because there must be exactly one point on the plane that is closest to the given point. The corresponding z-value is $z = 5 - \left(-\frac{5}{6}\right) - 2\left(\frac{4}{3}\right) = \frac{19}{6}$. Thus, the closest point on the plane is $Q\left(-\frac{5}{6}, \frac{4}{3}, \frac{19}{6}\right)$, and the minimum distance is

$$d = \sqrt{\left(\frac{5}{6}\right)^2 + \left(\frac{4}{3} - 3\right)^2 + \left(\frac{19}{6} - 4\right)^2} = \sqrt{\frac{25}{6}} = \frac{5}{\sqrt{6}}$$

Check: You might want to check your work by using the formula for the distance from a point to a plane in $\mathbb{R}^3$ (Theorem 9.9):

$$d = \left| \frac{Ax_0 + By_0 + Cz_0 - D}{\sqrt{A^2 + B^2 + C^2}} \right| = \left| \frac{0 + 2(3) + 4 - 5}{\sqrt{1^2 + 2^2 + 1^2}} \right| = \frac{5}{\sqrt{6}}$$

Least Squares Approximation of Data

In the following example, calculus is applied to justify a formula used in statistics and in many applications in the social and physical sciences.

Example 7   Least squares approximation of data

Suppose data consisting of n points $P_1, \cdots, P_n$ are known, and we wish to find a function $y = f(x)$ that fits the data reasonably well. In particular, suppose we wish to find a line $y = mx + b$ that "best fits" the data in the sense that the sum of the squares of the vertical distances from each data point to the line is minimized.

Solution   We wish to find values of m and b that minimize the sum of the squares of the differences between the y-values and the line $y = mx + b$. The line that we seek is called the **regression line**. Suppose that the point P_k has components (x_k, y_k). Now at this point the value on the regression line is $y = mx_k + b$ and the value of the data point is y_k. The "error" caused by using the point on the regression line rather than the actual data point can be measured by the difference

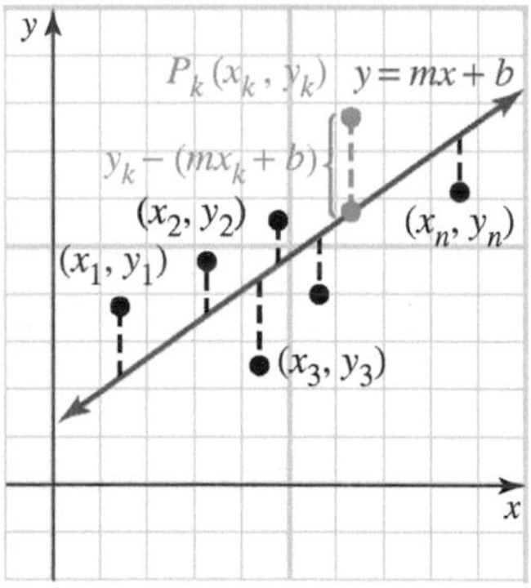

$$y_k - (mx_k + b)$$

The data points may be above the regression line for some values of k and below the regression line for other values of k. We see that we need to minimize the function that represents the sum of the *squares* of all these differences:

$$F(m, b) = \sum_{k=1}^{n} [y_k - (mx_k + b)]^2$$

Figure 11.40   Least squares approximation

The situation is illustrated in Figure 11.40.

Because F is a function of two variables, we use the second partials test.

$$F_m(m, b) = \sum_{k=1}^{n} 2[y_k - (mx_k + b)](-x_k) \qquad\qquad F_b(m, b) = \sum_{k=1}^{n} 2[y_k - (mx_k + b)](-1)$$

$$= 2m \sum_{k=1}^{n} x_k^2 + 2b \sum_{k=1}^{n} x_k - 2 \sum_{k=1}^{n} x_k y_k \qquad\qquad = 2m \sum_{k=1}^{n} x_k + 2b \sum_{k=1}^{n} 1 - 2 \sum_{k=1}^{n} y_k$$

$$= 2m \sum_{k=1}^{n} x_k + 2bn - 2 \sum_{k=1}^{n} y_k$$

Set each of these partial derivatives equal to 0 to find the critical values (see Problem 59).

$$m = \frac{n \sum_{k=1}^{n} x_k y_k - \left(\sum_{k=1}^{n} x_k \right) \left(\sum_{k=1}^{n} y_k \right)}{n \sum_{k=1}^{n} x_k^2 - \left(\sum_{k=1}^{n} x_k \right)^2} \quad \text{and} \quad b = \frac{\sum_{k=1}^{n} x_k^2 \sum_{k=1}^{n} y_k - \left(\sum_{k=1}^{n} x_k \right) \left(\sum_{k=1}^{n} x_k y_k \right)}{n \sum_{k=1}^{n} x_k^2 - \left(\sum_{k=1}^{n} x_k \right)^2}$$

We leave it to you to complete the second partials test to verify that these values of m and b yield a relative minimum.

Most applications of the **least squares formula** stated in Example 7 involve using a calculator or computer software. If you use a calculator, check the owner's manual for specifics.

PROBLEM SET 11.7

Level 1

1. ■ What does this say? Describe a procedure for classifying relative extrema.
2. ■ What does this say? Describe a procedure for determining absolute extrema on a closed, bounded set S.

Find the critical points in Problems 3-22, and classify each point as a relative maximum, a relative minimum, or a saddle point.

3. $f(x, y) = \dfrac{9x}{x^2 + y^2 + 1}$

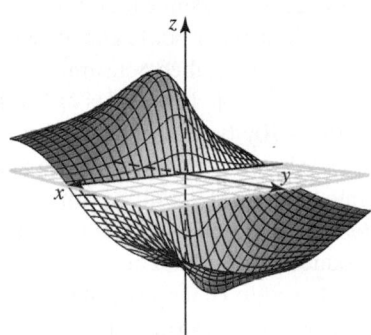

4. $f(x, y) = 2x^2 - 4xy + y^3 + 2$

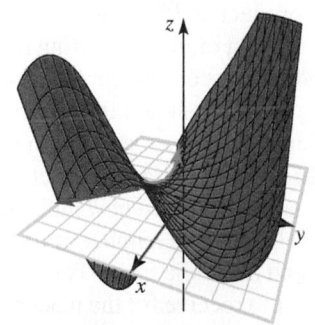

5. $f(x, y) = (x - 2)^2 + (y - 3)^4$

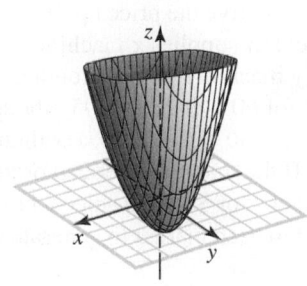

6. $f(x, y) = e^{-x} \sin y$

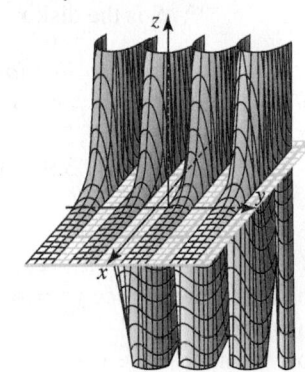

7. $f(x, y) = (1 + x^2 + y^2)e^{1-x^2-y^2}$

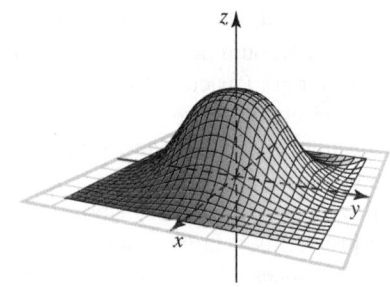

8. $f(x, y) = 3x^2 + 12x + 8y^3 - 12y^2 + 7$
9. $f(x, y) = x^2 + xy + y^2$
10. $f(x, y) = xy - x + y$
11. $f(x, y) = -x^3 + 9x - 4y^2$
12. $f(x, y) = e^{-(x^2+y^2)}$
13. $f(x, y) = (x^2 + 2y^2)e^{1-x^2-y^2}$
14. $f(x, y) = e^{xy}$
15. $f(x, y) = x^{-1} + y^{-1} + 2xy$
16. $f(x, y) = (x - 4)\ln(xy)$
17. $f(x, y) = x^3 + y^3 + 3x^2 - 18y^2 + 81y + 5$
18. $f(x, y) = 2x^3 + y^3 + 3x^2 - 3y - 12x - 4$
19. $f(x, y) = x^2 + y^2 - 6xy + 9x + 5y + 2$
20. $f(x, y) = x^2 + y^2 + \dfrac{32}{xy}$

21. $f(x, y) = x^2 + y^3 + \dfrac{768}{x + y}$

22. $f(x, y) = 3xy^2 - 2x^2y + 36xy$

Level 2

Find the absolute extrema of f on the closed bounded set S in the plane as described in Problems 23-28.

23. $f(x,y) = x^2 + xy + y^2$; S is the disk $x^2 + y^2 \leq 1$.
24. $f(x,y) = 2x^2 - y^2$; S is the disk $x^2 + y^2 \leq 1$.
25. $f(x,y) = xy - 2x - 5y$; S is the triangular region with vertices $(0,0)$, $(7,0)$, and $(7,7)$.
26. $f(x,y) = x^2 + 3y^2 - 4x + 2y - 3$; S is the square region with vertices $(0,0)$, $(3,0)$, $(3,-3)$, and $(0,-3)$.
27. $f(x,y) = 2\sin x + 5\cos y$; S is the rectangular region with vertices $(0,0)$, $(2,0)$, $(2,5)$, and $(0,5)$.
28. $f(x,y) = e^{x^2 + 2x + y^2}$; S is the disk $x^2 + 2x + y^2 \leq 0$.

Find the least squares regression line for the data points given in Problems 29-32.

29. $(-2,-3), (-1,-1), (0,1), (1,3), (3,5)$
30. $(0,1), (1,1.6), (2.2,3), (3.1,3.9), (4,5)$
31. $(3,5.72), (4,5.31), (6.2,5.12),$
    $(7.52,5.32), (8,03,5.67)$
32. $(-4,2), (-3,1), (0,0), (1,-3), (2,-1), (3,-2)$
33. Find all points on the surface $y^2 = 4 + xz$ that are closest to the origin.
34. Find all points in the plane $x + 2y + 3z = 4$ in the first octant where $f(x,y,z) = x^2yz^3$ has a maximum value.
35. A rectangular box with no top is to have a fixed volume. What should its dimensions be if we want to use the least amount of material in its construction?
36. A wire of length L is cut into three pieces that are bent to form a circle, a square, and an equilateral triangle. How should the cuts be made to minimize the sum of the total area?
37. Find three positive numbers whose sum is 54 and whose product is as large as possible.
38. A dairy produces whole milk and skim milk in quantities x and y pints, respectively. Suppose the price (in cents) of whole milk is $p(x) = 100 - x$ and that of skim milk is $q(y) = 100 - y$, and also assume that $C(x,y) = x^2 + xy + y^2$ is the joint-cost function of the commodities. Maximize the profit

$$P(x,y) = px + qy - C(x,y)$$

39. Let R be the triangular region in the xy-plane with vertices $(-1,-2)$, $(-1,2)$, and $(3,2)$. A plate in the shape of R is heated so that the temperature at (x,y) is

$$T(x,y) = 2x^2 - xy + y^2 - 2y + 1$$

(in degrees Celsius). At what point in R or on its boundary is T maximized? Where is T minimized? What are the extreme temperatures?

40. A particle of mass m in a rectangular box with dimensions x, y, z has ground state energy

$$E(x,y,z) = \frac{k^2}{8m}\left(\frac{1}{x^2} + \frac{1}{y^2} + \frac{1}{z^2}\right)$$

where k is a physical constant. If the volume of the box is fixed (say $V_0 = xyz$), find the values of x, y, and z that minimize the ground state energy.

41. A manufacturer produces two different kinds of smart phones, A and B, in quantities x and y (units of 1,000), respectively. If the revenue function (in dollars) is

$$R(x,y) = -x^2 - 2y^2 + 2xy + 8x + 5y$$

Find the quantities of A and B that should be produced to maximize revenue.

42. Suppose we wish to construct a closed rectangular box with volume 32 ft³. Three different materials will be used in the construction. The material for the sides costs $1 per square foot, the material for the bottom costs $3 per square foot, and the material for the top costs $5 per square foot. What are the dimensions of the least expensive such box?

43. **Modeling Problem** A store carries two competing brands of bottled water, one from California and the other from upstate New York. To model this situation, assume the owner of the store can obtain both at a cost of $2/bottle. Also assume that if the California water is sold for x dollars per bottle and the New York water for y dollars per bottle, then consumers will buy approximately $40 - 50x + 40y$ bottles of California water and $20 + 60x - 70y$ bottles of the New York water each day. How should the owner price the bottled water to generate the largest possible profit?

44. **Modeling Problem** A telephone company is planning to introduce two types of communications systems that it hopes to sell to its largest commercial customers. To create a model to determine the maximum profit, it is assumed that if the first product is priced at x hundred dollars each and the second type at y hundred dollars each, approximately $40 - 8x + 5y$ consumers will buy the first product and $50 + 9x - 7y$ will buy the second type. If the cost of manufacturing the first type is $1,000 each and the cost of manufacturing the second type is $3,000 each, how should the telephone company price the systems to generate maximum profit?

45. **Modeling Problem** A manufacturer with exclusive rights to a sophisticated new industrial machine is planning to sell a limited number of the machines to both foreign and domestic firms. The price the manufacturer can expect to receive for the machines will depend on the number of machines made available. For example, if only a few of the machines are placed on the market, competitive bidding among prospective purchasers will tend to drive the price up. It is estimated that if the manufacturer supplies x machines to the domestic market and y machines to the foreign market, the machines will sell for $60 - 0.2x + 0.05y$ thousand dollars apiece at home and $50 - 0.1y + 0.05x$ thousand dollars apiece abroad. If the manufacturer can produce the machines at a total cost of $10,000 apiece, how many should be supplied to each market to generate the largest possible profit?

46. A college admissions officer, Dr. Westfall, has compiled the following data relating students' high-school and college GPAs:

HS GPA	2.0	2.5	3.0	3.0	3.5	3.5	4.0	4.0
College GPA	1.5	2.0	2.5	3.5	2.5	3.0	3.0	3.5

Plot the data points on a graph and find the equation of the regression line for these data. Then use the regression line to predict the college GPA of a student whose high school GPA is 3.75.

47. It is known that if an ideal spring is displaced a distance y from its natural length by a force (weight) x, then $y = kx$, where k is the so-called spring constant. To compute this constant for a particular spring, a scientist obtains the following data:

x (lb)	0	5.2	7.3	8.4	10.12	12.37
y (in.)	0	11.32	15.56	17.44	21.96	26.17

Based on these data, what is the "best" choice for k?

48. EXPLORATION PROBLEM The following table gives the approximate U.S. census figures (in millions):

Year	1920	1930	1940	1950	1960
Population	106.0	123.2	132.1	151.3	179.3

Year	1970	1980	1990	2000	2010
Population	203.3	226.5	248.7	281.4	308.7

a. Find the least squares regression line for the given data and use this line to "predict" the population in 2017.

b. Use the least squares linear approximation to estimate the population at the present time. Check your answer by looking up the population using the Internet. Comment on the accuracy (or inaccuracy) of your prediction.

49. EXPLORATION PROBLEM The following table gives the Dow Jones Industrial Average (DJIA) Stock Index opening prices every five years along with the per capita consumption of wine (in gallons) for those years.

Year	1965	1970	1975	1980	1985
DJIA	874	800	619	839	1,212
Consumption	0.98	1.31	1.71	2.11	2.43

Year	1990	1995	2000	2005	2010
DJIA	2,753	3,834	11,502	10,784	10,431
Consumption	2.05	1.79	2.05	2.14	2.24

a. Plot these data on a graph, with the DJIA Index on the x-axis and consumption on the y-axis.

b. Find the equation of the least squares line.

c. Determine whether the consumption figures predicted by the least squares line in part **b** are approximately correct by using the most recent figures available.

50. Linearizing nonlinear data In this problem, we turn to data that do *not* tend to change linearly. Often we can "linearize" the data by taking the logarithm or exponential of the data and then doing a linear fit as described in this section.

a. Suppose we have, or suspect, a relationship $y = kx^m$. Show that by taking the natural logarithm of this equation we obtain a linear relationship: $Y = K + mX$. Explain the new variables and constant K.

b. Below are data relating the periods of revolution t (in days) of the six inner planets and their semimajor axis a (in 10^6 km). Kepler conjectured the relationship $t = ka^m$, which is very accurate for the correct k and a. You are to "transform" the data as in part **a** (thus obtaining $T_i = \ln t_1, \cdots$); and do a linear fit to the new data, thus finding k and m.

t-data	87.97	224.7	365.26	686.98	4,332.59	10,759.2
a-data	58	108	149	228	778	1,426

51. Following are data pertaining to a recent Olympic weight-lifting competition. The x-data are the "class data" giving eight weight classes (in kg) from featherweight to heavyweight-2. The w-data are the combined weights lifted by the winners in each class. Theoretically, we would expect a relationship $w = kx^m$, where $m = \frac{2}{3}$. (Can you see why?)

a. Linearize the data as in Problem 50 and use the least squares approximation to find k and m.

x-data	56	60	68	75	83	90	100	110
w-data	293	343	340	375	378	413	425	455

b. Comment on the 60-kg entry. Do you see why this participant (N. Suleymanoglu of Turkey) was referred to as the strongest man in the world?

Level 3

52. This problem is designed to show, by example, that if $D = 0$ at a critical point, then almost anything can happen.

a. Show that $f(x, y) = x^4 - y^4$ has a saddle point at $(0, 0)$.

b. Show that $g(x, y) = x^2 y^2$ has a relative minimum at $(0, 0)$.

c. Show that $h(x, y) = x^3 + y^3$ has no extremum or saddle point at $(0, 0)$.

53. Think Tank Problem If f is a continuous function of one variable with two relative maxima on a given interval, there must be a relative minimum between the maxima. By considering the function*

$$f(x, y) = 4x^2 e^y - 2x^4 - e^{4y}$$

---

*Based on the problem, "Two Mountains Without a Valley," by Ira Rosenholtz, *Mathematics Magazine*, 1987, Vol. 60, No. 1, p. 48.

show that it is possible for a continuous function of two variables to have only two relative maxima and no relative minima.

54. **Think Tank Problem** If a continuous function of one variable has only one critical number on an interval, then a relative extremum must also be an absolute extremum. Show that this result does not extend to functions of two variables by considering the function*

$$f(x,y) = 3xe^y - x^3 - e^{3y}$$

In particular, show that it has exactly one critical point, which corresponds to a relative maximum. Is this also an absolute maximum? Explain.

55. Tom, Dick, and Mary are participating in a cross-country relay race. Tom will trudge as fast as he can through thick woods to the edge of a river, then Dick will take over and row to the opposite shore. Finally, Mary will take the baton and run along the river road to the finish line. The course is shown in Figure 11.41.

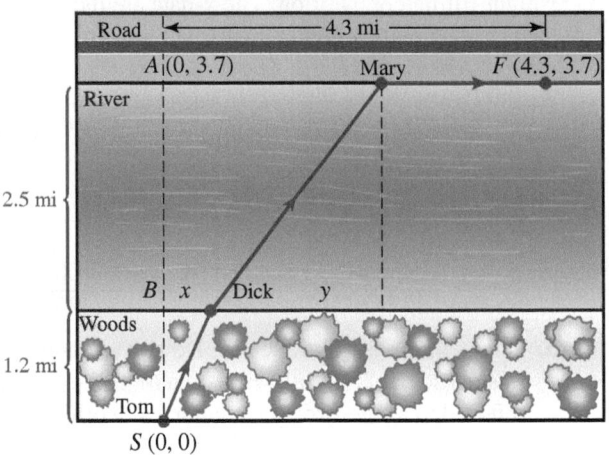

Figure 11.41   Course for the relay

Teams must start at point S and finish at point F, but they may position one member anywhere along the shore of the river and another anywhere along the river road. Suppose Tom can trudge at 2 mi/h, Dick can row at 4 mi/h, and Mary can run at 6 mi/h. Where should

Dick and Mary wait to receive the baton in order for the team to finish the course as quickly as possible?

56. **EXPLORATION PROBLEM**   Consider the function

$$f(x,y) = (y - x^2)(y - 2x^2)$$

Discuss the behavior of this function at $(0,0)$.

57. **EXPLORATION PROBLEM**   Sometimes the critical points of a function can be classified by looking at the level curves. In each case shown in Figure 11.42, determine the nature of the critical point(s) of $z = f(x,y)$ at $(0,0)$.

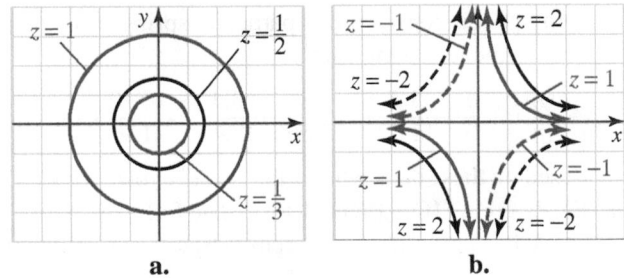

Figure 11.42   Problem 57

58. Prove the second partials test. *Hint*: Compute the second directional derivative of f in the direction of the unit vector $\mathbf{u} = h\mathbf{i} + k\mathbf{j}$ and complete the square.

59. Verify the formulas for m and b associated with the least squares approximation.

60. This problem involves a generalization of the least-squares procedure, in which a "least squares plane" is found to produce the best fit for a given set of data. A researcher knows that the quantity z is related to x and y by a formula of the form $z = k_1 x + k_2 y$, where k_1 and k_2 are physical constants. To determine these constants, she conducts a series of experiments, the results of which are tabulated as follows:

x	1.20	0.86	1.03	1.65	−0.95	−1.07
y	0.43	1.92	1.52	−1.03	1.22	−0.06
z	3.21	5.73	2.22	0.92	−1.11	−0.97

Modify the method of least squares to find a "best approximation" for k_1 and k_2.

---

*Based on material in the article, "The Only Critical Point in Town Test" by Ira Rosenholtz and Lowell Smythe, *Mathematics Magazine*, 1985, Vol. 58, No. 3, pp. 149-150.

11.8 LAGRANGE MULTIPLIERS

IN THIS SECTION: *Method of Lagrange multipliers, constrained optimization problems, Lagrange multipliers with two parameters, a geometric interpretation of Lagrange's theorem*
In the real world, we encounter a variety of problems involving optimization. In Chapter 4, we considered extrema of functions of a single variable, and in this section we extend our work to functions of two variables.

Method of Lagrange Multipliers

In many applied problems, a function of two variables is to be optimized subject to a restriction or **constraint** on the variables. For example, consider a container heated in such a way that the temperature at the point (x, y, z) in the container is given by the function $T(x, y, z)$. Suppose that the surface $z = f(x, y)$ lies in the container, and that we wish to find the point on $z = f(x, y)$ where the temperature is the greatest. In other words, *What is the maximum value of T subject to the constraint $z = f(x, y)$, and where does this maximum value occur?*

Theorem 11.15 Lagrange's theorem

Assume that f and g have continuous first partial derivatives and that f has an extremum at $P_0(x_0, y_0)$ when restricted to the smooth constraint curve $g(x, y) = c$. If $\nabla g(x_0, y_0) \neq \mathbf{0}$, there is a number λ such that

$$\nabla f(x_0, y_0) = \lambda \nabla g(x_0, y_0)$$

Proof: Denote the constraint curve $g(x, y) = c$ by C, and note that C is smooth. We represent this curve by the vector function

$$\mathbf{R}(t) = x(t)\mathbf{i} + y(t)\mathbf{j}$$

for all t in an open interval I, including t_0 corresponding to P_0, where $x'(t)$ and $y'(t)$ exist and are continuous. Let $F(t) = f(x(t), y(t))$ for all t in I, and apply the chain rule to obtain

$$F'(t) = f_x(x(t), y(t))\frac{dx}{dt} + f_y(x(t), y(t))\frac{dy}{dt} = \nabla f(x(t), y(t)) \cdot \mathbf{R}'(t)$$

Because $f(x, y)$ has an extremum at P_0, we know that $F(t)$ has an extremum at t_0, the value of t that corresponds to P_0 (that is, P_0 is the point on C where $t = t_0$). Therefore, we have $F'(t_0) = 0$ and

$$F'(t_0) = \nabla f(x(t_0), y(t_0)) \cdot \mathbf{R}'(t_0) = 0$$

If $\nabla f(x(t_0), y(t_0)) = 0$, then $\lambda = 0$, and the condition $\nabla f = \lambda \nabla g$ is satisfied trivially. If $\nabla f(x(t_0), y(t_0)) \neq 0$, then $\nabla f(x(t_0), y(t_0))$ is orthogonal to $\mathbf{R}'(t_0)$. Because $\mathbf{R}'(t_0)$ is tangent to the constraint curve C, it follows that $\nabla f(x_0, y_0)$ is normal to C. But $\nabla g(x_0, y_0)$ is also normal to C (because C is a level curve of g), and we conclude that ∇f and ∇g must be *parallel* at P_0. Thus, there is a scalar λ such that

$$\nabla f(x_0, y_0) = \lambda \nabla g(x_0, y_0)$$

as required.   ♦

Constrained Optimization Problems

The general procedure for the method of Lagrange multipliers may be described as follows.

METHOD OF LAGRANGE MULTIPLIERS   Suppose f and g satisfy the hypotheses of Lagrange's theorem, and that $f(x, y)$ has an extremum subject to the constraint $g(x, y) = c$. Then to find the extreme value, proceed as follows:

Step 1 Simultaneously solve the following three equations for x, y, and λ:

$$f_x(x, y) = \lambda g_x(x, y), \quad f_y(x, y) = \lambda g_y(x, y), \quad g(x, y) = c$$

Step 2 Evaluate f at all points found in step 1 and all points on the boundary of the constraint. The extremum we seek must be among these values.

Example 1   Optimization with Lagrange multipliers

Given that the largest and smallest values of $f(x, y) = 1 - x^2 - y^2$ subject to the constraint $x + y = 1$ with $x \geq 0$, $y \geq 0$ exist, use the method of Lagrange multipliers to find these extrema.

Solution   Because the constraint is $x + y = 1$, let $g(x, y) = x + y$

$$f_x(x, y) = -2x \qquad f_y(x, y) = -2y \qquad g_x(x, y) = 1 \qquad g_y(x, y) = 1$$

Form the system

$$\begin{cases} -2x = \lambda(1) & \leftarrow f_x(x, y) = \lambda g_x(x, y) \\ -2y = \lambda(1) & \leftarrow f_y(x, y) = \lambda g_x(x, y) \\ x + y = 1 & \leftarrow g(x, y) = 1 \end{cases}$$

The only solution is $x = \frac{1}{2}, y = \frac{1}{2}$.

$$f\left(\frac{1}{2}, \frac{1}{2}\right) = 1 - \left(\frac{1}{2}\right)^2 - \left(\frac{1}{2}\right)^2 = \frac{1}{2}$$

The endpoints of the line segment

$$x + y = 1 \text{ for } x \geq 0, \, y \geq 0$$

are at $(1, 0)$ and $(0, 1)$, and we find that

$$f(1, 0) = 1 - 1^2 - 0^2 = 0$$
$$f(0, 1) = 1 - 0^2 - 1^2 = 0$$

Therefore, the maximum value is $\frac{1}{2}$ at $\left(\frac{1}{2}, \frac{1}{2}\right)$, and the minimum value is 0 at $(1, 0)$ and $(0, 1)$. See Figure 11.43 to visualize the maximum value. ∎

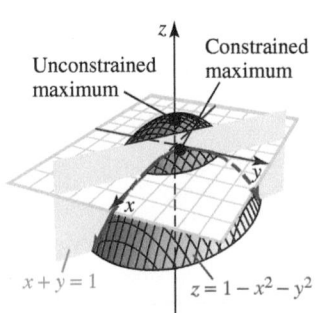

Figure 11.43 The maximum value is the high point of the curve of intersection of the surface and the plane

The method of Lagrange multipliers extends naturally to functions of three or more variables. If a function $f(x, y, z)$ has an extreme value subject to a constraint $g(x, y, z) = c$, then the extremum occurs at a point (x_0, y_0, z_0) such that $g(x_0, y_0, z_0) = c$ and $\nabla f(x_0, y_0, z_0) = \lambda \nabla g(x_0, y_0, z_0)$ for some number λ. Here is an example.

Example 2   Hottest and coldest points on a plate

A container in $\mathbb{R}^3$ has the shape of the cube given by $0 \le x \le 1, 0 \le y \le 1, 0 \le z \le 1$. A plate is placed in the container in such a way that it occupies that portion of the plane $x + y + z = 1$ that lies in the cubical container. If the container is heated so that the temperature at each point (x, y, z) is given by

$$T(x, y, z) = 4 - 2x^2 - y^2 - z^2$$

in hundreds of degrees Celsius, what are the hottest and coldest points on the plate? You may assume these extreme temperatures exist.

Solution   The cube and plate are shown in Figure 11.44.

We will use Lagrange multipliers to find all critical points in the interior of the plate, and then we will examine the plate's boundary. To apply the method of Lagrange multipliers, we must solve $\nabla T = \lambda \nabla g$, where $g(x, y, z) = x + y + z$. We obtain the partial derivatives.

$$T_x = -4x \qquad T_y = -2y \qquad T_z = -2z \qquad g_x = g_y = g_z = 1$$

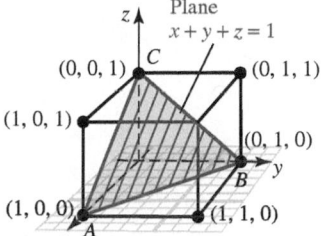

Figure 11.44   Cube and plate

We must solve the system

$$\begin{cases} -4x = \lambda & \leftarrow T_x = \lambda g_x \\ -2y = \lambda & \leftarrow T_y = \lambda g_y \\ -2z = \lambda & \leftarrow T_z = \lambda g_z \\ x + y + z = 1 & \leftarrow g(x, y, z) = 1 \end{cases}$$

The solution of this system is $\left(\frac{1}{5}, \frac{2}{5}, \frac{2}{5}\right)$. The boundary of the plate is a triangle with vertices $A(1, 0, 0)$, $B(0, 1, 0)$, and $C(0, 0, 1)$. The temperature along the edges of this triangle may be found as follows:

$$T_1(x) = 4 - 2x^2 - (0)^2 - (1 - x)^2 = 3 \quad -3x^2 + 2x \quad 0 \le x \le 1 \qquad \text{Edge } AC: \quad x + z = 1, y = 0$$

$$T_2(x) = 4 - 2x^2 - (1 - x)^2 - (0)^2 = 3 \quad -3x^2 + 2x \quad 0 \le x \le 1 \qquad \text{Edge } AB: \quad x + y = 1, z = 0$$

$$T_3(y) = 4 - 2(0)^2 - y^2 - (1 - y)^2 = 3 \quad +2y - 2y^2 \quad 0 \le y \le 1 \qquad \text{Edge } BC: \quad y + z = 1, x = 0$$

Edge AC: Differentiating, $T_1'(x) = T_2'(x) = -6x + 2$, which equals 0 when $x = \frac{1}{3}$. If $x = \frac{1}{3}$, then $z = \frac{2}{3}$ (because $x + z = 1$, $y = 0$ on edge AC), so we have the critical point $\left(\frac{1}{3}, 0, \frac{2}{3}\right)$.

Edge AB: Because $T_2 = T_1$, we see $x = \frac{1}{3}$. If $x = \frac{1}{3}$, then $y = \frac{2}{3}$ (because $x + y = 1$, $z = 0$ on edge AB), so we have another critical point $\left(\frac{1}{3}, \frac{2}{3}, 0\right)$.

Edge BC: Differentiating, $T_3'(y) = 2 - 4y$, which equals 0 when $y = \frac{1}{2}$. Because $y + z = 1$ and $x = 0$, we have the critical point $\left(0, \frac{1}{2}, \frac{1}{2}\right)$.

Endpoints of the edges: $(1, 0, 0)$, $(0, 1, 0)$, and $(0, 0, 1)$.

The last step is to evaluate T at the critical points and the endpoints:

$$T\left(\frac{1}{5}, \frac{2}{5}, \frac{2}{5}\right) = 3\frac{3}{5};$$

$$T\left(\frac{1}{3}, 0, \frac{2}{3}\right) = 3\frac{1}{3}; \quad T\left(\frac{1}{3}, \frac{2}{3}, 0\right) = 3\frac{1}{3}; \quad T\left(0, \frac{1}{2}, \frac{1}{2}\right) = 3\frac{1}{2}$$

$$T(1, 0, 0) = 2; \qquad T(0, 1, 0) = 3; \qquad T(0, 0, 1) = 3$$

Comparing these values (remember that the temperature is in hundreds of degrees Celsius), we see that the highest temperature is $360°C$ at $\left(\frac{1}{5}, \frac{2}{5}, \frac{2}{5}\right)$ and the lowest temperature is $200°C$ at $(1, 0, 0)$. ∎

Notice that the multiplier is used only as an intermediary device for finding the critical points and plays no role in the final determination of the constrained extrema. However, the value of λ is more important in certain problems, thanks to the interpretation given in the following theorem.

Theorem 11.16   Rate of change of the extreme value

Suppose E is an extreme value (maximum or minimum) of f subject to the constraint $g(x, y) = c$. Then the Lagrange multiplier λ is the rate of change of E with respect to c; that is, $\lambda = dE/dc$.

Proof: Note that at the extreme value (x, y) we have

$$f_x = \lambda g_x, \qquad f_y = \lambda g_y, \qquad \text{and} \qquad g(x, y) = c$$

The coordinates of the optimal ordered pair (x, y) depend on c (because different constraint levels will generally lead to different optimal combinations of x and y). Thus,

$$E = E(x, y) \qquad \text{where } x \text{ and } y \text{ are functions of } c$$

By the chain rule for partial derivatives,

$$
\begin{aligned}
\frac{dE}{dc} &= \frac{\partial E}{\partial x}\frac{dx}{dc} + \frac{\partial E}{\partial y}\frac{dy}{dc} \\
&= f_x \frac{dx}{dc} + f_y \frac{dy}{dc} & \text{\textit{Because } } E = f(x, y) \\
&= \lambda g_x \frac{dx}{dc} + \lambda g_y \frac{dy}{dc} & \text{\textit{Because } } f_x = \lambda g_x \text{ and } f_y = \lambda g_y \\
&= \lambda \left(g_x \frac{dx}{dc} + g_y \frac{dy}{dc} \right) \\
&= \lambda \frac{dg}{dc} & \text{\textit{Chain rule}} \\
&= \lambda & \text{\textit{Because } } g(x, y) = c
\end{aligned}
$$
◆

This theorem can be interpreted as saying that the multiplier estimates the change in the extreme value E that results when the constraint c is increased by 1 unit. This interpretation is illustrated in the following example.

Example 3   Maximum output for a Cobb-Douglas production function

If x thousand dollars is spent on labor, and y thousand dollars is spent on equipment, it is estimated that the output of a certain factory will be

$$Q(x, y) = 50x^{2/5}y^{3/5}$$

units. If \$150,000 is available, how should this capital be allocated between labor and equipment to generate the largest possible output? How does the maximum output change if the money available for labor and equipment is increased by \$1,000? In economics, an output function of the general form $Q(x, y) = cx^{\alpha}y^{1-\alpha}$ is known as a **Cobb-Douglas production function**.

Solution   Because x and y are given in units of \$1,000, the constraint equation is $x + y = 150$. If we set $g(x, y) = x + y$, we wish to maximize Q subject to $g(x, y) = 150$. To apply the method of Lagrange multipliers, we first find

$$Q_x = 20x^{-3/5}y^{3/5} \qquad Q_y = 30x^{2/5}y^{-2/5} \qquad g_x = 1 \qquad g_y = 1$$

Cobb-Douglas functions are widely used to represent the relationship of an output to inputs

Next, solve the system

$$\begin{cases} 20x^{-3/5}y^{3/5} = \lambda(1) \\ 30x^{2/5}y^{-2/5} = \lambda(1) \\ \qquad\qquad x+y = 150 \end{cases}$$

From the first two equations we have

$$20x^{-3/5}y^{3/5} = 30x^{2/5}y^{-2/5}$$
$$20y = 30x$$
$$y = 1.5x$$

Substitute $y = 1.5x$ into the equation $x + y = 150$ to find $x = 60$. This leads to the solution $y = 90$, so that the maximum output is

$$Q(60,90) = 50(60)^{2/5}(90)^{3/5} \approx 3{,}826.273502 \text{ units}$$

We also find that

$$\lambda = 20(60)^{-3/5}(90)^{3/5} \approx 25.50849001$$

Thus, the maximum output is about 3,826 units and occurs when \$60,000 is allocated to labor and \$90,000 to equipment. We also note that an increase of \$1,000 (1 unit) in the available funds will increase the maximum output by approximately $\lambda \approx 25.51$ units (from 3,826.27 to 3,851.78 units). Note that we do not need to check the endpoints $x = 0$ and $y = 0$, as they lead to a minimum, not a maximum. ■

Lagrange Multipliers with Two Parameters

The method of Lagrange multipliers can also be applied in situations with more than one constraint equation. Suppose we wish to locate an extremum of a function defined by $f(x,y,z)$ subject to two constraints, $g(x,y,z) = c_1$ and $h(x,y,z) = c_2$, where g and h are also differentiable and ∇g and ∇h are not parallel. By generalizing Lagrange's theorem, it can be shown that if (x_0,y_0,z_0) is the desired extremum, then there are numbers λ and μ such that $g(x_0,y_0,z_0) = c_1$, $h(x_0,y_0,z_0) = c_2$, and

$$\nabla f(x_0,y_0,z_0) = \lambda \nabla g(x_0,y_0,z_0) + \mu \nabla h(x_0,y_0,z_0)$$

As in the case of one constraint, we proceed by first solving this system of equations simultaneously to find λ, μ, x_0, y_0, z_0 and then evaluating $f(x,y,z)$ at each solution and comparing to find the required extremum. This approach is illustrated in our final example of this section.

Example 4   Optimization with two constraints

Find the point on the intersection of the plane $x + 2y + z = 10$ and the paraboloid $z = x^2 + y^2$ that is closest to the origin (see Figure 11.45). You may assume that such a point exists.

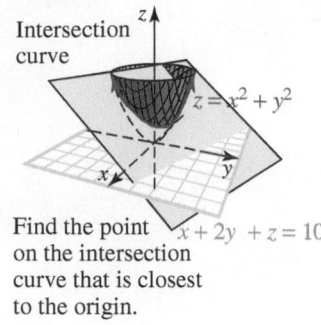

Intersection curve

$z = x^2 + y^2$

Find the point $x + 2y + z = 10$ on the intersection curve that is closest to the origin.

Figure 11.45 Graphical representation of Example 4

Solution   The distance from a point (x, y, z) to the origin is $s = \sqrt{x^2 + y^2 + z^2}$, but instead of minimizing this quantity, it is easier to minimize its square. That is, we will minimize

$$f(x, y, z) = x^2 + y^2 + z^2$$

subject to the joint constraints

$$g(x, y, z) = x + 2y + z = 10 \quad \text{and} \quad h(x, y, z) = x^2 + y^2 - z = 0$$

Compute the partial derivatives of f, g, and h:

$$\begin{array}{ccc} f_x = 2x & f_y = 2y & f_z = 2z \\ g_x = 1 & g_y = 2 & g_z = 1 \\ h_x = 2x & h_y = 2y & h_z = -1 \end{array}$$

To apply the method of Lagrange multipliers, we use the formula

$$\nabla f(x_0, y_0, z_0) = \lambda \nabla g(x_0, y_0, z_0) + \mu \nabla h(x_0, y_0, z_0)$$

which leads to the following system of equations:

$$\begin{cases} 2x = \lambda(1) + \mu(2x) \\ 2y = \lambda(2) + \mu(2y) \\ 2z = \lambda(1) + \mu(-1) \\ x + 2y + z = 10 \\ z = x^2 + y^2 \end{cases}$$

This is not a linear system, so solving it requires ingenuity. Multiply the first equation by 2 and subtract the second equation to obtain

$$4x - 2y = (4x - 2y)\mu$$
$$(4x - 2y) - (4x - 2y)\mu = 0$$
$$(4x - 2y)(1 - \mu) = 0$$
$$4x - 2y = 0 \qquad \text{or} \qquad 1 - \mu = 0$$

CASE I: **If $4x - 2y = 0$**, then $y = 2x$. Substitute this into the two constraint equations:

$$\begin{array}{cc} x + 2y + z = 10 & x^2 + y^2 - z = 0 \\ x + 2(2x) + z = 10 & x^2 + (2x)^2 - z = 0 \\ z = 10 - 5x & z = 5x^2 \end{array}$$

By substitution we have $5x^2 = 10 - 5x$, which has solutions $x = 1$ and $x = -2$. This implies

$$\begin{array}{cc} x = 1 & x = -2 \\ y = 2x = 2(1) = 2 & y = 2x = 2(-2) = -4 \\ z = 5x^2 = 5(1)^2 = 5 & z = 5x^2 = 5(-2)^2 = 20 \end{array}$$

Thus, the points $(1, 2, 5)$ and $(-2, -4, 20)$ are candidates for the minimal distance.

CASE II: **If $1 - \mu = 0$**, then $\mu = 1$, and we look at the system of equations involving x, y, z, λ, and μ.

$$\begin{cases} 2x = \lambda(1) + \mu(2x) \\ 2y = \lambda(2) + \mu(2y) \\ 2z = \lambda(1) + \mu(-1) \\ x + 2y + z = 10 \\ z = x^2 + y^2 \end{cases}$$

The top equation becomes $2x = \lambda + 2x$, so that $\lambda = 0$. We now find z from the third equation:

$$2z = -1 \quad \text{or} \quad z = -\frac{1}{2}$$

Next, turn to the constraint equations:

$$
\begin{array}{ll}
x + 2y + z = 10 & x^2 + y^2 - z = 0 \\[4pt]
x + 2y - \dfrac{1}{2} = 10 & x^2 + y^2 + \dfrac{1}{2} = 0 \\[4pt]
x + 2y = 10 + \dfrac{1}{2} & x^2 + y^2 = -\dfrac{1}{2}
\end{array}
$$

There is no solution because $x^2 + y^2$ cannot equal a negative number. We check the candidates for the minimal distance:

$$
\begin{aligned}
f(x,y,z) &= x^2 + y^2 + z^2 \text{ so that} \\
f(1,2,5) &= 1^2 + 2^2 + 5^2 = 30 \\
f(-2,-4,20) &= (-2)^2 + (-4)^2 + 20^2 = 420
\end{aligned}
$$

Because $f(x,y,z)$ represents the square of the distance, the minimal distance is $\sqrt{30}$ and the point on the intersection of the two surfaces nearest to the origin is $(1,2,5)$. Note that we do not need to check the boundaries, as there are no boundaries on the plane and the parabola. ∎

A Geometrical Interpretation of Lagrange's Theorem

Lagrange's theorem can be interpreted geometrically. Suppose the constraint curve $g(x,y) = c$ and the level curves $f(x,y) = k$ are drawn in the xy-plane, as shown in Figure 11.46.

To maximize $f(x,y)$ subject to the constraint $g(x,y) = c$, we must find the "highest" (rightmost, actually) level curve of f that intersects the constraint curve. As the sketch in Figure 11.46 suggests, this critical intersection occurs at a point where the constraint curve is tangent to a level curve—that is, where the slope of the constraint curve $g(x,y) = c$ is equal to the slope of a level curve $f(x,y) = k$. According to the implicit function theorem (Theorem 11.5, p. 863),

Slope of constraint curve $g(x,y) = c$ is $\dfrac{-g_x}{g_y}$.

Slope of each level curve is $\dfrac{-f_x}{f_y}$.

The condition that the slopes are equal can be expressed by

$$\frac{-f_x}{f_y} = \frac{-g_x}{g_y} \qquad \text{or, equivalently} \qquad \frac{f_x}{g_x} = \frac{f_y}{g_y}$$

Let λ equal this common ratio,

$$\lambda = \frac{f_x}{g_x} \qquad \text{and} \qquad \lambda = \frac{f_y}{g_y}$$

so that

$$f_x = \lambda g_x \qquad \text{and} \qquad f_y = \lambda g_y$$

and

$$\nabla f = f_x \mathbf{i} + f_y \mathbf{j} = \lambda(g_x \mathbf{i} + g_y \mathbf{j}) = \lambda \nabla g$$

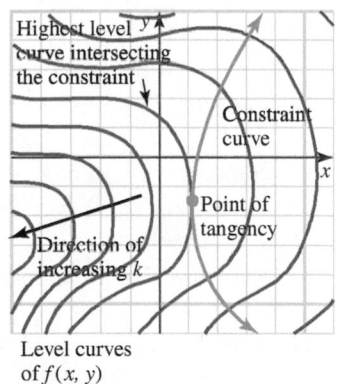

Figure 11.46 Increasing level curves and the constraint curve

Because the point in question must lie on the constraint curve, we also have $g(x,y) = c$. If these equations are satisfied at a certain point (a,b), then f will reach its constrained *maximum* at (a,b) if the highest level curve that intersects the constraint curve does so at this point. On the other hand, if the *lowest* level curve that intersects the constraint curve does so at (a,b), then f achieves its constrained *minimum* at this point.

PROBLEM SET 11.8

For the problems in this set, you may assume that the requested extreme value(s) exist.

Level 1

Use the method of Lagrange multipliers to find the required constrained extrema in Problems 1-18.

1. Maximize $f(x,y) = xy$ subject to $2x + 2y = 5$.
2. Maximize $f(x,y) = xy$ subject to $x + y = 20$.
3. Minimize $f(x,y) = x^2 + y^2$ subject to $x + y = 24$.
4. Minimize $f(x,y) = x^2 + y^2$ subject to $x + y = 2$.
5. Minimize $f(x,y) = x^2 + y^2$ subject to $x + y = 9$.
6. Maximize $f(x,y) = 16 - x^2 - y^2$ subject to $x + 2y = 6$.
7. Minimize $f(x,y) = x^2 + y^2$ subject to $xy = 1$.
8. Maximize $f(x,y) = x^2 + y^2$ subject to $x + y - 1 = 0$.
9. Minimize $f(x,y) = x^2 + y^2 - xy - 4$ subject to $x + y = 6$
10. Minimize $f(x,y) = x^2 - xy + 2y^2$ subject to $2x + y = 22$.
11. Minimize $f(x,y) = x^2 - y^2$ subject to $x^2 + y^2 = 4$.
12. Maximize $f(x,y) = x^2 - 2y - y^2$ subject to $x^2 + y^2 = 1$.
13. Maximize $f(x,y) = \cos x + \cos y$ subject to $y = x + \frac{\pi}{4}$.
14. Maximize $f(x,y) = e^{xy}$ subject to $x^2 + y^2 = 3$.
15. Maximize $f(x,y) = \ln(xy^2)$ subject to $2x^2 + 3y^2 = 8$ for $x > 0, y > 0$
16. Maximize $f(x,y,z) = xyz$ subject to $3x + 2y + z = 6$.
17. Minimize $f(x,y,z) = x^2 + y^2 + z^2$ subject to $x - 2y + 3z = 4$.
18. Minimize $f(x,y,z) = x^2 + y^2 + z^2$ subject to $4x^2 + 2y^2 + z^2 = 4$.

Level 2

19. Find the smallest value of $f(x,y,z) = 2x^2 + 4y^2 + z^2$ subject to $4x - 8y + 2z = 10$. What, if anything, can be said about the largest value of f subject to this constraint?
20. Find the largest value of $f(x,y,z) = x^2y^2z^2$ on the sphere $x^2 + y^2 + z^2 = R^2$.

21. Find the maximum and minimum values of $f(x,y,z) = x - y + z$ on the sphere $x^2 + y^2 + z^2 = 100$.
22. Find the maximum and minimum values of $f(x,y,z) = 4x - 2y - 3z$ on the sphere $x^2 + y^2 + z^2 = 100$.
23. Use Lagrange multipliers to find the distance from the origin to the plane $Ax + By + Cz = D$ where at least one of A, B, C is nonzero.
24. Find the maximum and minimum distance from the origin to the ellipse $5x^2 - 6xy + 5y^2 = 4$.
25. Find the point on the plane $2x + y + z = 1$ that is nearest to the origin.
26. Find the largest product of positive numbers x, y, and z such that their sum is 24.
27. Write the number 12 as the sum of three positive numbers x, y, z in such a way that the product xy^2z is a maximum.
28. A rectangular box with no top is to be constructed from 96 ft^2 of material. What should be the dimensions of the box if it is to enclose maximum volume?

29. The temperature T at point (x, y, z) in a region of space is given by the formula $T = 100 - xy - xz - yz$. Find the lowest temperature on the plane $x + y + z = 10$.
30. A farmer wishes to fence off a rectangular pasture along the bank of a river. The area of the pasture is to be 3,200 yd^2, and no fencing is needed along the river bank. Find the dimensions of the pasture that will require the least amount of fencing.
31. There are 320 yd of fencing available to enclose a rectangular field. How should the fencing be used so that the enclosed area is as large as possible?
32. **EXPLORATION PROBLEM** Use the fact that 12 fl oz is approximately 6.89π in.^{3} to find the dimensions of the 12-oz Coke® can that can be constructed using the least amount of metal. Compare your answer with an actual can of Coke. Explain what might cause the discrepancy.

33. EXPLORATION PROBLEM A cylindrical can is to hold 4π in.^{3} of orange juice. The cost per square inch of constructing the metal top and bottom is twice the cost per square inch of constructing the cardboard side. What are the dimensions of the least expensive can?

34. Find the volume of the largest rectangular parallelepiped that can be inscribed in the ellipsoid

$$x^2 + \frac{y^2}{4} + \frac{z^2}{9} = 1$$

35. A manufacturer has $8,000 to spend on the development and promotion of a new product. It is estimated that if x thousand dollars is spent on development and y thousand is spent on promotion, sales will be approximately

$$f(x, y) = 50x^{1/2}y^{3/2}$$

units. How much money should the manufacturer allocate to development and how much to promotion to maximize sales?

36. Modeling Problem If x thousand dollars is spent on labor and y thousand dollars is spent on equipment, the output at a certain factory may by modeled by

$$Q(x, y) = 60x^{1/3}y^{2/3}$$

units. Assume $120,000 is available.

a. How should money be allocated between labor and equipment to generate the largest possible output?

b. Use the Lagrange multiplier λ to estimate the change in the maximum output of the factory that would result if the money available for labor and equipment is increased by $1,000.

37. Modeling Problem An architect decides to model the usable living space in a building by the volume of space that can be used comfortably by a person 6 feet tall—that is, by the largest 6-foot-high rectangular box that can be inscribed in the building. Find the dimensions of an A-frame building y ft long with equilateral triangular ends x ft on a side that maximizes usable living space if the exterior surface area of the building cannot exceed 500 ft^2.

38. Find the radius of the largest cylinder of height 6 in. that can be inscribed in an inverted cone of height H, radius R, and lateral surface area 250 in.^{2}.

39. In Problem 40 of Problem Set 11.7, you were asked to minimize the ground state energy

$$E(x, y, z) = \frac{k^2}{8m}\left(\frac{1}{x^2} + \frac{1}{y^2} + \frac{1}{z^2}\right)$$

subject to the volume constraint $V = xyz = C$. Solve the problem using Lagrange multipliers.

40. Modeling Problem A university extension agricultural service concludes that, on a particular farm, the yield of wheat per acre is a function of water and fertilizer. Let x be the number of acre-feet of water applied, and y the number of pounds of fertilizer applied during the growing season. The agricultural service then concluded that the yield B (measured in bushels), can be modeled by the formula

$$B(x, y) = 500 + x^2 + 2y^2$$

Suppose that water costs $20 per acre-foot, fertilizer costs $12 per pound, and the farmer will invest $236 per acre for water and fertilizer. How much water and fertilizer should the farmer buy to maximize the yield?

41. How would the farmer of Problem 40 maximize the yield if the amount spent is $100 instead of $236?

42. Present post office regulations specify that a box (that is, a package in the form of a rectangular parallelepiped) can be mailed parcel post only if the sum of its length and girth does not exceed 108 inches, as shown in Figure 11.47.

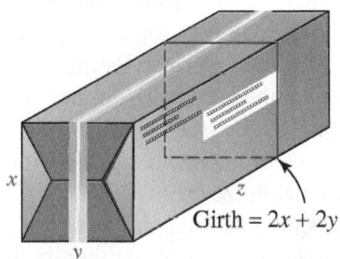

Girth $= 2x + 2y$

Figure 11.47 What is the maximum volume that can be mailed?

Find the maximum volume of such a package. (Compare your solution here with the one you might have given to Problem 35, Section 4.6, page 293.)

43. Heron's formula says that the area of a triangle with sides a, b, c is

$$A = \sqrt{s(s - a)(s - b)(s - c)}$$

where $s = \frac{1}{2}(a + b + c)$ is the semiperimeter of the triangle. Use this result and the method of Lagrange multipliers to show that of all triangles with a given fixed perimeter P, the equilateral triangle has the largest area.

44. If x, y, z are the angles of a triangle, what is the maximum value of the product

$$P(x, y, z) = \sin x \sin y \sin z?$$

What about

$$Q(x, y, z) = \cos x \cos y \cos z?$$

Use the method of Lagrange multipliers in Problems 45-48 to find the required extrema for the two given constraints.

45. Find the minimum of $f(x,y,z) = x^2 + y^2 + z^2$ subject to $x + y = 4$ and $y + z = 6$.

46. Find the maximum of $f(x,y,z) = xyz$ subject to $x^2 + y^2 = 3$ and $y = 2z$.

47. Maximize $f(x,y,z) = xy + xz$ subject to $2x + 3z = 5$ and $xy = 4$.

48. Minimize $f(x,y,z) = 2x^2 + 3y^2 + 4z^2$ subject to $x + y + z = 4$ and $x - 2y + 5z = 3$.

49. Modeling Problem A manufacturer is planning to sell a new product at the price of $150 per unit and estimates that if x thousand dollars is spent on development and y thousand dollars on promotion, then approximately

$$\frac{320y}{y+2} + \frac{160x}{x+4}$$

units of the product will be sold. The cost of manufacturing the product is $50 per unit.
  a. If the manufacturer has a total of $8,000 to spend on the development and promotion, how should this money be allocated to generate the largest possible profit?
  b. Suppose the manufacturer decides to spend $8,100 instead of $8,000 on the development and promotion of the new product. Estimate how this change will affect the maximum possible profit.
  c. If unlimited funds are available, how much should the manufacturer spend on development and promotion to maximize profit?
  d. What is the Lagrange multiplier in part **c**? Your answer should suggest another method for solving the problem in part **c**. Solve the problem using this alternative approach.

50. Modeling Problem A jewelry box with a square base has an interior partition and is required to have volume 800 cm³ (see Figure 11.48).

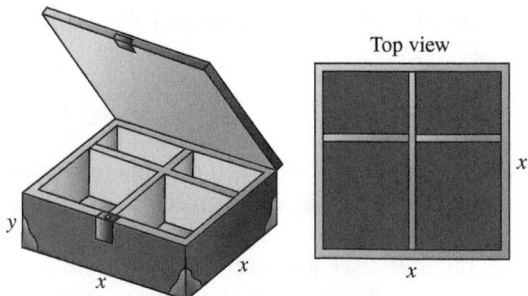

Top view

Figure 11.48 Constructing a jewelry box

  a. The material in the top costs twice as much as that in the sides and the bottom, which in turn, costs twice as much as the material in the partitions. Find the dimensions of the box that minimize the total cost of construction. *Does it matter that you have not been told where the partitions are located?*
  b. Suppose the volume constraint changes from 800 cm³ to 801 cm³. Estimate the appropriate effect on the minimal cost.

51. Three alleles, A, B, and O, determine the four blood types A, B, O, and AB. The *Hardy-Weinberg law* says that the proportions of individuals in a population who carry two different alleles is

$$P = 2pq + 2pr + 2rq$$

where p, q, and r are the proportions of blood types A, B, and O in the population. Given that $p + q + r = 1$, what is the largest value of P?

Level 3

52. A farmer wants to build a metal silo in the shape of a right circular cylinder with a right circular cone on the top (the bottom of the silo will be a concrete slab). What is the least amount of metal that can be used if the silo is to have a fixed volume V_0?

53. Find the volume of the largest rectangular parallelepiped (box) that can be inscribed in the ellipsoid

$$\frac{x^2}{a^2} + \frac{y^2}{b^2} + \frac{z^2}{c^2} = 1$$

(See Problem 34.)

54. EXPLORATION PROBLEM The method of Lagrange multipliers gives a constrained extremum only if one exists. If the method is applied to optimizing $f(x,y) = x + y$ subject to $xy = 1$, the method yields two candidates for an extremum. Is one a maximum and the other a minimum? Explain.

In Problems 55-58, let $Q = f(x,y)$ be a production function in which x and y represent units of labor and capital, respectively. If p and q represent unit costs of labor and capital, respectively, then $C(x,y) = px + qy$ represents the total cost of production.

55. Use Lagrange multipliers to show that subject to a fixed production level c, the total cost is smallest when

$$\frac{f_x}{p} = \frac{f_y}{q} \quad \text{and} \quad Q = f(x, y) = c$$

(provided $\nabla f \neq \mathbf{0}$, and $p \neq 0$, $q \neq 0$). This is often referred to as the **minimum cost problem**, and its solution is called the **least cost combination of inputs**.

56. Show that the inputs x, y that maximize the production level $Q = f(x, y)$ subject to a fixed cost k satisfy

$$\frac{f_x}{p} = \frac{f_y}{q} \quad \text{with } px + qy = k$$

(assume $p \neq 0$, $q \neq 0$). This is called a **fixed-budget problem**.

57. A Cobb-Douglas production function is an output function of the form

$$Q(x, y) = cx^\alpha y^\beta$$

with $\alpha + \beta = 1$.

a. Show that such a function is maximized with respect to the fixed cost $px + qy = k$ when $x = \alpha k / p$ and $y = \beta k / q$.

b. Where does the maximum occur if we drop the condition $\alpha + \beta = 1$? How does the maximum output change if k is increased by 1 unit?

58. Show that the cost function

$$C(x, y) = px + qy$$

is minimized subject to the fixed production level $Ax^\alpha y^\beta = k$, with $\alpha + \beta = 1$ when

$$x = \frac{k}{A}\left(\frac{\alpha q}{\beta p}\right)^\beta \qquad y = \frac{k}{A}\left(\frac{\beta p}{\alpha q}\right)^\alpha$$

59. a. The **geometric mean** of three positive numbers x, y, z is $G = (xyz)^{1/3}$ and the **arithmetic mean** is $A = \frac{1}{3}(x + y + z)$. Use the method of Lagrange multipliers to show that

$$G(x, y, z) \leq A(x, y, z)$$

for all x, y, z.

b. Generalize the result in part **a** to n variables, $x_1, x_2, \cdots, x_n$.

60. *Historical Quest*

Joseph Lagrange (1736-1813)

Joseph Lagrange is generally acknowledged as one of the two greatest mathematicians of the 18th century, along with Leonhard Euler (see Historical Quest 93 of the Supplementary Problems of Chapter 4). There is a distinct difference in style between Lagrange and Euler. Lagrange has been characterized as the first true analyst in the sense that he attempted to write concisely and with rigor. On the other hand, Euler wrote using intuition and with an abundance of detail. Lagrange was described by Napoleon Bonaparte as "the lofty pyramid of the mathematical sciences" and followed Euler as the court mathematician for Frederick the Great. He was the first to use the notation $f'(x)$ and $f''(x)$ for derivatives. In this section we were introduced to the method of Lagrange multipliers, which provide a procedure for constrained optimization. This method was contained in a paper on mechanics that Lagrange wrote when he was only 19 years old.*

For this *Quest*, we consider Lagrange's work with solving *algebraic* equations. You are familiar with the quadratic formula, which provides a general solution for any second-degree equation $ax^2 + bx + c = 0$ $a \neq 0$. Lagrange made an exhaustive study of the general solution for the first four degrees. Here is what he did. Suppose you are given a general algebraic expression involving letters $a, b, c, \cdots$; how many *different* expressions can be derived from the given one if the letters are interchanged in all possible ways? For example, from $ab + cd$ we obtain $ad + cb$ by interchanging b and d.

This problem suggests another closely related problem, also part of Lagrange's approach. Lagrange solved general algebraic equations of degrees 2, 3, and 4. It was proved later (not by Lagrange, but by Galois and Abel), that no general solution for equations greater than 4 can be found. Do some research and find the general solution for equations of degrees 1, 2, 3, and 4.

---

*From *Men of Mathematics* by E. T. Bell, Simon & Schuster, New York, 1937, p. 165.

CHAPTER 11 REVIEW

*T*he *pseudomath is a person who handles mathematics as a monkey handles the razor. The creature tried to shave himself as he had seen his master do; but, not having any notion of the angle at which the razor was to be held, he cut his own throat. He never tried it a second time, poor animal! But the pseudomath keeps on in his work, proclaims himself clean shaved, and all the rest of the world hairy.*

A. De Morgan *Budget of Paradoxes* (London, 1872), p. 473

Proficiency Examination

Concept Problems

1. **a.** What is a function of two variables?
   b. What are the domain and range of a function of two variables?
2. What do we mean by the limit of a function of two variables?
3. State the following properties of a limit of functions of two variables.
   a. scalar rule   **b.** sum rule   **c.** product rule   **d.** quotient rule
4. Define the continuity of a function defined by $f(x,y)$ at a point (x_0, y_0) in its domain and continuity on a set S.
5. If $z = f(x,y)$,
   a. define the first partial derivatives of f with respect to x and y.
   b. represent the second partial derivatives.
   c. what are the increments of x, y, and z?
6. What is the slope of a tangent line to the surface defined by $z = f(x,y)$ that is parallel to the xy-plane at a point P_0 on f? In what direction?
7. What does it mean for a function of two variables to be differentiable at (x_0, y_0)?
8. **a.** State the incremental approximation of $f(x,y)$.
   b. Define the total differential of $z = f(x,y)$.
9. **a.** State the chain rule for a function of one parameter.
   b. State the chain rule for a function of two independent parameters.
10. Define the directional derivative of a function defined by $z = f(x,y)$.
11. **a.** Define the gradient $\nabla f(x,y)$.
    b. Express the directional derivative in terms of the gradient.
    c. State the normal property of the gradient.
12. State the following basic properties of the gradient.
    a. constant rule   **b.** linearity rule   **c.** product rule   **d.** quotient rule   **e.** power rule
13. State the optimal direction property of the gradient (that is, the steepest ascent and steepest descent).
14. Define the normal line and tangent plane to a surface S at a point P_0.
15. **a.** Define the absolute extrema of a function of two variables.
    b. Define the relative extrema of a function of two variables.
    c. What is a critical point of a function of two variables?
16. State the second partials test.
17. State the extreme value theorem for a function of two variables.
18. What is the least squares approximation of data, and what is a regression line?
19. State Lagrange's theorem.
20. State the procedure for the method of Lagrange multipliers.

Practice Problems

21. If $f(x,y) = \sin^{-1} xy$, verify that $f_{xy} = f_{yx}$.

22. Let $w = x^2 y + y^2 z$, where $x = t \sin t$, $y = t \cos t$, and $z = 2t$. Use the chain rule to find $\dfrac{dw}{dt}$, where $t = \pi$.

23. Let $f(x,y,z) = xy + yz + xz$, and let P_0 denote the point $(1, 2, -1)$.
  a. Find the gradient of f at P_0.
  b. Find the directional derivative of f in the direction from P_0 toward the point $Q(-1, 1, -1)$.
  c. Find the direction from P_0 in which the directional derivative has its largest value. What is the magnitude of the largest directional derivative at P_0?

24. Show that the function defined by

$$f(x,y) = \begin{cases} \dfrac{x^2 y}{x^3 + y^3} & \text{if } (x,y) \neq (0,0) \\ 0 & \text{if } (x,y) = (0,0) \end{cases}$$

is not continuous at $(0,0)$.

25. If $f(x,y) = \ln \left(\dfrac{y}{x}\right)$, find f_x, f_y, f_{xx}, f_{yy}, and f_{xy}.

26. Show that if $f(x,y,z) = x^2 y + y^2 z + z^2 x$, then

$$\frac{\partial f}{\partial x} + \frac{\partial f}{\partial y} + \frac{\partial f}{\partial z} = (x + y + z)^2$$

27. Let $f(x,y) = (x^2 + y^2)^2$. Find the directional derivative of f at $(2, -2)$ in the direction that makes an angle of $\frac{2\pi}{3}$ with the positive x-axis.

28. Find all critical points of $f(x,y) = 12xy - 2x^2 - y^4$ and classify them using the second partials test.

29. Use the method of Lagrange multipliers to find the maximum and minimum values of the function $f(x,y) = x^2 + 2y^2 + 2x + 3$ subject to the constraint $x^2 + y^2 = 4$. You may assume these extreme values exist.

30. Find the largest and smallest values of the function

$$f(x,y) = x^2 - 4y^2 + 3x + 6y$$

on the region defined by $-2 \leq x \leq 2$, $0 \leq y \leq 1$. You may assume these extreme values exist.

Supplementary Problems*

Describe the domain of each function given in Problems 1-4.

1. $f(x,y) = \sqrt{16 - x^2 - y^2}$ **2.** $f(x,y) = \dfrac{x^2 - y^2}{x - y}$ **3.** $f(x,y) = \sin^{-1} x + \cos^{-1} y$

4. $f(x,y) = e^{x+y} \tan^{-1}\left(\dfrac{y}{x}\right)$

Find the partial derivatives f_x and f_y for the functions defined in Problems 5-10.

5. $f(x,y) = \dfrac{x^2 - y^2}{x + y}$ **6.** $f(x,y) = x^3 e^{3y/(2x)}$ **7.** $f(x,y) = x^2 y + \sin \dfrac{y}{x}$

8. $f(x,y) = \ln\left(\dfrac{xy}{x + 2y}\right)$ **9.** $f(x,y) = 2x^3 y + 3xy^2 + \dfrac{y}{x}$ **10.** $f(x,y) = xy e^{xy}$

For each function given in Problems 11-16, describe the level curve or level surface $f = c$ for the given values of the constant c.

11. $f(x,y) = x^2 - y$; $c = 2$, $c = -2$ **12.** $f(x,y) = 6x + 2y$; $c = 0$, $c = 1$, $c = 2$

---

*The supplementary problems are presented in a somewhat random order, not necessarily in order of difficulty.

13. $z = f(x, y) = \begin{cases} \sqrt{x^2 + y^2} & \text{if } x \geq 0 \\ -|y| & \text{if } x < 0 \end{cases}$

    $c = 0,\ c = 1,\ c = -1$

14. $f(x, y, z) = x^2 + y^2 + z^2;$

    $c = 16,\ c = 0,\ c = -25$

15. $f(x, y, z) = x^2 + \dfrac{y^2}{2} + \dfrac{z^2}{2};\ c = 1,\ c = 2$

16. $f(x, y, z) = \dfrac{x^2}{2} - \dfrac{y^2}{2} + z^2;\ c = 1,\ c = 2$

Evaluate the limits in Problems 17 and 18, assuming they exist.

17. $\displaystyle\lim_{(x,y)\to(0,0)} \dfrac{x + ye^{-x}}{1 + x^2}$

18. $\displaystyle\lim_{(x,y)\to(1,1)} \dfrac{xy}{x^2 + y^2}$

Show that each limit in Problems 19 and 20 does not exist.

19. $\displaystyle\lim_{(x,y)\to(0,0)} \dfrac{x^3 y^2}{x^6 + y^4}$

20. $\displaystyle\lim_{(x,y)\to(0,0)} \dfrac{x^3 - y^3}{x^3 + y^3}$

Find the derivatives in Problems 21-24 using the chain rule. You may leave your answers in terms of $x, y, t, u,$ and v.

21. Find $\dfrac{dz}{dt}$, where $z = xy + y^2$, and $x = e^t t^{-1}$, $y = \tan t$.

22. Find $\dfrac{dz}{dt}$, where $z = -xy + y^3$, and $x = -3t^2$, $y = 1 + t^3$.

23. Find $\dfrac{\partial z}{\partial u}$ and $\dfrac{\partial z}{\partial v}$, where $z = x \tan \dfrac{x}{y}$, and $x = uv$, $y = \dfrac{u}{v}$.

24. Find $\dfrac{\partial z}{\partial u}$ and $\dfrac{\partial z}{\partial v}$, where $z = x^2 - y^2$, and $x = u + 2v$, $y = u - 2v$.

Use implicit differentiation to find $\dfrac{\partial z}{\partial x}$ and $\dfrac{\partial z}{\partial y}$ in Problems 25-28.

25. $x + 2y - 3z = \ln z$ 26. $x^2 + 6y^2 + 2z^2 = 5$ 27. $e^x + e^y + e^z = 3$ 28. $x^3 + 2xz - yz^2 - z^3 = 1$

In Problems 29-34, find f_{xx} and f_{yx}.

29. $f(x, y) = \displaystyle\int_x^y \sin(\cos t)\, dt$

30. $f(x, y) = \tan^{-1} xy$

31. $f(x, y) = \sin^{-1} xy$

32. $f(x, y) = x^2 + y^3 - 2xy^2$

33. $f(x, y) = e^{x^2 + y^2}$

34. $f(x, y) = x \ln y$

Find equations for the tangent plane and normal line to the surfaces given in Problems 35-38 at the prescribed point.

35. $z = x^2 - y^2$ at $P_0 = (1, 1, 0)$

36. $x^2 y^3 z = 8$ at $P_0(2, -1, -2)$

37. $x^3 + 2xy^2 - 7x^3 + 3y + 1 = 0$ at $P_0(1, 1, 1)$

38. $z = \dfrac{-4}{2 + x^2 + y^2}$ at $P_0(1, 1, -1)$

Find all critical points of $f(x, y)$ in Problems 39-44 and classify each as a relative maximum, a relative minimum, or a saddle point.

39. $f(x, y) = x^2 - 6x + 2y^2 + 4y - 2$

40. $f(x, y) = x^3 + y^3 - 6xy$

41. $f(x, y) = (x - 1)(y - 1)(x + y - 1)$

42. $f(x, y) = x^2 + y^3 + 6xy - 7x - 6y$

43. $f(x, y) = x^3 + y^3 + 3x^2 - 18y^2 + 81y + 5$

44. $f(x, y) = \sin(x + y) + \sin x + \sin y$ for $0 < x < \pi$, $0 < y < \pi$ (See Figure 11.49.)

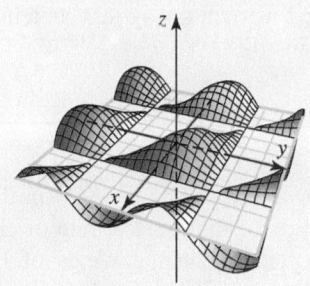

Figure 11.49   Graph for Problem 44

In Problems 45-48, *find the largest and smallest values of the function f on the specified closed, bounded set S.*

45. $f(x,y) = xy - 2y$; S is the rectangular region $0 \le x \le 3$, $-1 \le y \le 1$

46. $f(x,y) = x^2 + 2y^2 - x - 2y$; S is the triangular region with vertices $(0,0)$, $(2,0)$, $(2,2)$

47. $f(x,y) = x^2 + y^2 - 3y$; S is the disk $x^2 + y^2 \le 4$

48. $f(x,y) = 6x - x^2 + 2xy - y^4$; S is the square $0 \le x \le 3$, $0 \le y \le 3$.

49. Use the chain rule to find $\dfrac{dz}{dt}$ if $z = x^2 - 3xy^2$; $x = 2t$, $y = t^2$.

50. Use the chain rule to find $\dfrac{dz}{dt}$ if $z = x \ln y$; $x = 2t$, $y = e^t$.

51. Let $z = ue^{u^2 - v^2}$, where $u = 2x^2 + 3y^2$ and $v = 3x^2 - 2y^2$. Use the chain rule to find $\dfrac{\partial z}{\partial x}$ and $\dfrac{\partial z}{\partial y}$.

52. Use implicit differentiation to find $\dfrac{\partial z}{\partial x}$ and $\dfrac{\partial z}{\partial y}$, where x, y, and z are related by the equation
$x^3 + 2xz - yz^2 - z^3 = 1$.

53. Find the slope of the level curve of $x^2 + y^2 = 2$, where $x = 1$, $y = 1$.

54. Find the slope of the level curve of $xe^y = 2$, where $x = 2$.

55. Find the equations for the tangent plane and normal line to the surface $z = \sin x + e^{xy} + 2y$ at the point $P_0(0,1,3)$.

56. The electric potential at each point (x,y) in the disk $x^2 + y^2 < 4$ is $V = 2(4 - x^2 - y^2)^{-1/2}$ volts. Draw the equipotential curves $V = c$ for $c = \sqrt{2}$, $2/\sqrt{3}$, and 8.

57. Let $f(x,y,z) = x^3 y + y^3 z + z^3 x$. Find a function $g(x,y,z)$ such that
$$\frac{\partial f}{\partial x} + \frac{\partial f}{\partial y} + \frac{\partial f}{\partial z} = x^3 + y^3 + z^3 + 3g(x,y,z).$$

58. Let $u = \sin \dfrac{x}{y} + \ln \dfrac{y}{x}$. Show that $y\dfrac{\partial u}{\partial y} + x\dfrac{\partial u}{\partial x} = 0$.

59. Let $w = \ln(1 + x^2 + y^2) - 2\tan^{-1} y$, where $x = \ln(1 + t^2)$ and $y = e^t$. Use the chain rule to find $\dfrac{dw}{dt}$.

60. Let $f(x,y) = \tan^{-1}\dfrac{y}{x}$. Find the directional derivative of f at $(1,2)$ in the direction that makes an angle of $\dfrac{\pi}{3}$ with the positive x-axis.

61. Let $f(x,y) = y^x$. Find the directional derivative of f at $P_0(3,2)$ in the direction toward the point $Q(1,1)$.

62. According to postal regulations, the largest cylindrical can that can be sent has a girth $(2\pi r)$ plus length ℓ of 108 inches. What is the largest volume cylindrical can that can be mailed?

63. Let $f(x,y,z) = z(x - y)^5 + xy^2z^3$.
   a. Find the directional derivative of f at $(2, 1 - 1)$ in the direction of the outward normal to the sphere $x^2 + y^2 + z^2 = 6$.
   b. In what direction is the directional derivative at $(2,1,-1)$ largest?

64. Find positive numbers x and y for which xyz is a maximum, given that $x + y + z = 1$. Assume that the extreme value exists.

65. Maximize $f(x,y,z) = x^2 yz$ given that x, y, and z are all positive numbers and $x + y + z = 12$. Assume that the extreme value exists.

66. If $z = f(x^2 - y^2)$, evaluate $y\dfrac{\partial z}{\partial x} + x\dfrac{\partial z}{\partial y}$.

67. Find the shortest distance from the origin to the surface $z^2 = 3 + xy$. Assume the extreme value exists.

68. **Modeling Problem** A plate is heated in such a way that its temperature at a point (x, y) measured in centimeters on the plate is given in degrees Celsius by

$$T(x,y) = \frac{64}{x^2 + y^2 + 4}$$

a. Find the rate of change in temperature at the point $(3,4)$ in the direction $2\mathbf{i} + \mathbf{j}$.

b. Find the direction and the magnitude of the greatest rate of change of the temperature at the point $(3,4)$.

69. **Modeling Problem** The beautiful patterns on the wings of butterflies have long been a subject of curiosity and scientific study. Mathematical models used to study these patterns often focus on determining the level of morphogen (a chemical that affects change). In a model dealing with eyespot patterns, a quantity of morphogen is released from an eyespot and the morphogen concentration t days later is modeled by

$$S(r,t) = \frac{1}{\sqrt{4\pi t}} e^{-(\gamma k t + r^2/(4t))} \qquad t > 0$$

where r measures the radius of the region on the wing affected by the morphogen, and k and γ are positive constants.[*]

a. Find t_m so that $\partial S/\partial t = 0$ at t_m. Show that the function $S_m(t)$ formed from $S(r,t)$ by fixing r has a relative maximum at t_m. Is this the same as saying that the function of two variables $S(r,t)$ has a relative maximum?

b. Let $M(r)$ denote the maximum found in part **a**; that is, $M(r) = S(r, t_m)$. Find an expression for M in terms of $z = (1 + 4\gamma k r^2)^{1/2}$.

c. Show that $\dfrac{dM}{dr} < 0$ and interpret this result.

70. **Modeling Problem** Certain malignant tumors that do not respond to conventional methods of treatment (surgery, chemotherapy, etc.) may be treated by *hyperthermia*, a process involving the application of extreme heat using microwave transmission (see Figure 11.50). For one particular kind of microwave application used in such therapy, the temperature at each point located r units from the central axis of the tumor and h units inside it is modeled by the formula

$$T(r,h) = K e^{-pr^2} [e^{-qh} - e^{-sh}]$$

where K, p, q, and s are positive constants that depend on the properties of the patient's blood and the heating application.[†]

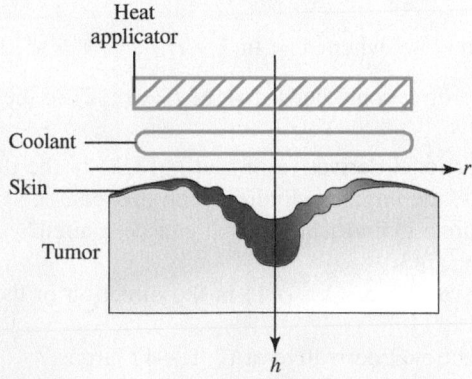

Figure 11.50 Hyperthermia treatment

[*]J. D. Murray, *Mathematical Biology*, 2nd Edition, Springer-Verlag, New York, 1993, p. 464.

[†]"Heat Therapy for Tumors," by Leah Edelstein-Keshet, *UMAP Modules 1991: Tools for Teaching*, Consortium for Mathematics and Its Applications, Inc., Lexington MA, 1992, pp. 73-101.

a. At what depth inside the tumor does the maximum temperature occur? Express your answers in terms of K, p, q, and s.

b. The article on which this problem is based discusses the physiology of hyperthermia in addition to raising several other interesting mathematical issues. Read this article and discuss assumptions made in the model.

71. **Modeling Problem** The marketing manager for a certain company has compiled the following data relating monthly advertising expenditure and monthly sales (in units of $1,000).

Advertising	3	4	7	9	10
Sales	78	86	138	145	156

a. Plot the data on a graph and find the least squares line.

b. Use the least squares line to predict monthly sales if the monthly advertising expenditure is $5,000.

72. Let $z = f(x, y)$, where $x = t + \cos t$ and $y = e^t$.

a. Suppose $f_x(1, 1) = 4$ and $f_y(1, 1) = -3$. Find $\dfrac{dz}{dt}$ when $t = 0$.

b. Suppose $f_x(0, 2) = -1$ and $f_y(0, 2) = 3$. Find $\dfrac{\partial z}{\partial r}$ and $\dfrac{\partial z}{\partial \theta}$ at the point where $r = 2$, $\theta = \frac{\pi}{2}$, and $x = r \cos \theta$, $y = r \sin \theta$.

73. Suppose f has continuous partial derivatives in some region D in the plane, and suppose $f(x, y) = 0$ for all (x, y) in D. If $(1, 2)$ is in D and $f_x(1, 2) = 4$ and $f_y(1, 2) = 6$, find dy/dx when $x = 1$ and $y = 2$.

74. Suppose $\nabla f(x, y, z)$ is parallel to the vector $x\mathbf{i} + y\mathbf{j} + z\mathbf{k}$ for all (x, y, z). Show that $f(0, 0, a) = f(0, 0, -a)$ for any a.

75. Find two unit vectors that are normal to the surface given by $z = f(x, y)$ at the point $(0, 1)$, where $f(x, y) = \sin x + e^{xy} + 2y$.

76. Let $f(x, y) = 3(x - 2)^2 - 5(y + 1)^2$. Find all points on the graph of f where the tangent plane is parallel to the plane $2x + 2y - z = 0$.

77. Let z be defined implicitly as a function of x and y by the equation $\cos(x + y) + \cos(x + z) = 1$. Find $\dfrac{\partial^2 z}{\partial y \partial x}$ in terms of x, y, and z.

78. Suppose F and F' are continuous functions of t and that $F'(t) = C$. Define f by $f(x, y) = F(x^2 + y^2)$. Show that the direction of $\nabla f(a, b)$ is the same as the direction of the line joining (a, b) to $(0, 0)$.

79. Let $f(x, y) = 12x^{-1} + 18y^{-1} + xy$, where $x > 0, y > 0$. How do you know that f must necessarily have a minimum in the region $x > 0, y > 0$? Find the minimum.

80. Let $f(x, y) = 3x^4 - 4x^2y + y^2$. Show that f has a minimum at $(0, 0)$ on every line $y = mx$ that passes through the origin. Then show that f has no relative minimum at $(0, 0)$.

In Problems 81-82, you may assume the required extremum exists.

81. Find the minimum of $x^2 + y^2 + z^2$ subject to the constraint $ax + by + cz = 1$ (with $a \neq 0$, $b \neq 0$, $c \neq 0$).

82. Suppose $0 < a < 1$ and $x \geq 0$, $y \geq 0$. Find the maximum of $x^a y^{1-a}$ subject to the constraint $ax + (1 - a)y = 1$.

83. For the production function given by $Q(x, y) = x^a y^b$, where $a > 0$ and $b > 0$, show that

$$x \frac{\partial Q}{\partial x} + y \frac{\partial Q}{\partial y} = (a + b)Q$$

In particular, if $b = 1 - a$ with $0 < a < 1$,

$$x \frac{\partial Q}{\partial x} + y \frac{\partial Q}{\partial y} = Q$$

84. Liquid flows through a tube with length L centimeters and internal radius r centimeters. The total volume V of fluid that flows each second is related to the pressure P and the viscosity a of the fluid by the formula $V = \dfrac{\pi P r^4}{8aL}$. What is the maximum error that can occur in using this formula to compute the viscosity a, if errors of $\pm 1\%$ can be made in measuring r and L, $\pm 2\%$ in measuring V, and $\pm 3\%$ in measuring P?

85. Suppose the functions f and g have continuous partial derivatives and satisfy

$$\frac{\partial f}{\partial x} = \frac{\partial g}{\partial y} \quad \text{and} \quad \frac{\partial f}{\partial y} = -\frac{\partial g}{\partial x}$$

These are called the *Cauchy-Riemann equations*.
a. Show that level curves of f and g intersect at right angles provided $\nabla f \neq 0$ and $\nabla g \neq 0$.
b. Assuming that the second partials of f and g are continuous, show that f and g satisfy Laplace's equations

$$f_{xx} + f_{yy} = 0 \quad \text{and} \quad g_{xx} + g_{yy} = 0$$

86. Show that if $z = f(r, \theta)$, where r and θ are defined as functions of x and y by the equations $x = r \cos \theta$, $y = r \sin \theta$, then the equation $\dfrac{\partial^2 z}{\partial x^2} + \dfrac{\partial^2 z}{\partial y^2} = 0$ becomes

$$\frac{\partial^2 z}{\partial r^2} + \frac{1}{r^2} \frac{\partial^2 z}{\partial \theta^2} + \frac{1}{r} \frac{\partial z}{\partial r} = 0$$

This is Laplace's equation in polar coordinates.

87. A circular sector of radius r and central angle θ has area $A = \frac{1}{2} r^2 \theta$. Find θ and r for the sector of fixed area A_0 for which the perimeter of the sector is minimized. Use the method of Lagrange multipliers and assume that the minimum exists.

88. A right circular cone is measured and is found to have base radius $r = 40$ cm and altitude $h = 20$ cm. If it is known that each measurement is accurate to within 2% what is the maximum percentage error in the measurement of the volume?

89. An elastic cylindrical container is filled with air so that the radius of the base is 2.02 cm and the height is 6.04 cm. If the container is deflated so that the radius of the base reduces to 2 cm and the height to 6 cm, approximately how much air has been removed? (Ignore the thickness of the container.)

90. Suppose f is a differentiable function of two variables with f_x and f_y also differentiable, and assume that f_{xx}, f_{yy} and f_{xy} are continuous. The **second directional derivative** of f at the point (x, y) in the direction of the unit vector $\mathbf{u} = a\mathbf{i} + b\mathbf{j}$ is defined by $D_{\mathbf{u}}^2 f(x, y) = D_{\mathbf{u}}[D_{\mathbf{u}} f(x, y)]$. Show that

$$D_{\mathbf{u}}^2 f(x, y) = a^2 f_{xx}(x, y) + 2ab f_{xy}(x, y) + b^2 f_{yy}(x, y)$$

91. A capsule is a cylinder of radius r and length ℓ, capped on each end by a hemisphere. Assume that the capsule dissolves in the stomach at a rate proportional to the ratio $R = S/V$, where V is the volume and S is the surface area of the capsule. Show that

$$\frac{\partial R}{\partial r} < 0 \quad \text{and} \quad \frac{\partial R}{\partial \ell} < 0$$

92. It can be shown that if $z = f(x, y)$ has the least surface area of all surfaces with a given boundary, then

$$(1 + z_y^2) z_{xx} - 2 z_x z_y z_{xy} + (1 + z_x^2) z_{yy} = 0$$

A surface satisfying such an equation is called a **minimal surface** (or, more precisely, a **minimal graph**).
a. Find constants A, B so that

$$z = \ln \left(\frac{A \cos y}{B \cos x} \right)$$

is a minimal surface.

b. Is it possible to find C and D so that

$$z = C \ln(\sin x) + D \ln(\sin y)$$

is a minimal surface?*

93. Based on the definition in Problem 92, prove that the plane

$$z = Ax + By + C$$

represents a minimal surface, where A, B, and C are real numbers.

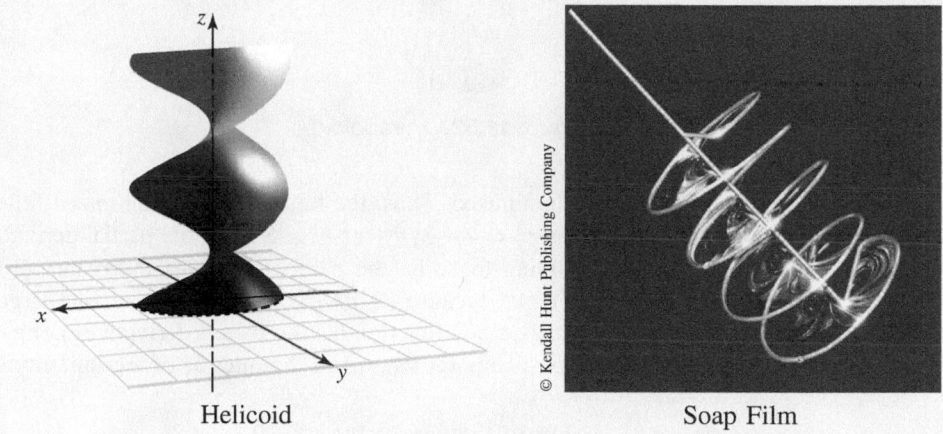

Helicoid   Soap Film

© Kendall Hunt Publishing Company

Figure 11.51   Helicoid

94. In problem 92, the notion of a minimal graph $z = f(x, y)$ was introduced. Consider the position vector

$$\mathbf{R}(u, v) = (u \cos v)\mathbf{i} + (u \sin v)\mathbf{j} + bv\mathbf{k},$$

which defines a parameterized helicoid (see Figure 11.51).

a. Can this surface be expressed as a graph of the form $z = f(x, y)$? If the answer is positive, find the surface representation as a graph. *Hint*: For non-zero values x, the ratio of the first two coordinates of the vector $\mathbf{R}$ eliminates the parameter u.

b. Based on the equation $z = f(x, y)$ determined in part **a**, prove that the helicoid is a minimal surface by verifying that

$$(1 + z_y^2)z_{xx} - 2z_x z_y z_{xy} + (1 + z_x^2)z_{yy} = 0$$

95. Think Tank Problem Find the minimum distance from the origin to the paraboloid $z = 4 - x^2 - 4y^2$. The graph is shown in Figure 11.52.

The distance from $P(x, y, z)$ to the origin is

$$d = \sqrt{x^2 + y^2 + z^2}$$

*For an interesting discussion, see "The Geometry of Soap Films and Soap Bubbles," by Frederick J. Almgren, Jr., and Jean E. Taylor, *Scientific American*, July 1976, pp. 82-93.

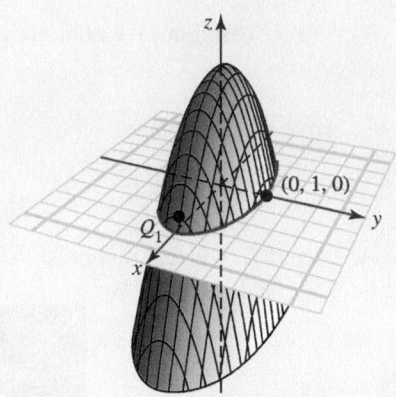

Figure 11.52 Paraboloid

This distance will be minimized when d^2 is minimized. Thus, the function to be minimized (after replacing x^2 by $4 - 4y^2 - z$) is $D(x, y) = 4 - 4y^2 - z + y^2 + z^2 = 4 - 3y^2 + z^2 - z$. Setting the partial derivatives D_y and D_z to 0 gives the critical point $y = 0$, $z = 0.5$. Solving for x on the paraboloid gives the points $Q_1(\sqrt{3.5}, 0, 0.5)$ and $Q_2(-\sqrt{3.5}, 0, 0.5)$. These points are NOT minimal because $(0, 1, 0)$ is closer. Explain what is going on here.*

96. **Putnam Examination Problem** Let f be a real-valued function having partial derivatives defined for $x^2 + y^2 < 1$ that satisfies $|f(x, y)| \leq 1$. Show that there exists a point (x_0, y_0) in the interior of the unit circle such that

$$[f_x(x_0, y_0)]^2 + [f_y(x_0, y_0)]^2 \leq 16$$

97. **Putnam Examination Problem** Find the smallest volume bounded by the coordinate planes and a tangent plane to the ellipsoid

$$\frac{x^2}{a^2} + \frac{y^2}{b^2} + \frac{z^2}{c^2} = 1$$

98. **Putnam Examination Problem** Find the shortest distance between the plane $Ax + By + Cz + 1 = 0$ and the ellipsoid

$$\frac{x^2}{a^2} + \frac{y^2}{b^2} + \frac{z^2}{c^2} \leq 1$$

99. **Book Report** "Science is that body of knowledge that describes, defines, and where possible, explains the universe... we think of the history of science as a history of men. .... [History] is the story of thousands of people who contributed to the knowledge and theories that constituted the science of their eras and made the 'great leaps' possible. Many of these people were women." begins a history of women in science entitled *Hypatia's Heritage* by Margaret Alic (Boston: Beacon Press, 1986). Read this book and write a book report.

---

*Our thanks to Herbert R. Bailey, who presented this problem in *The College Mathematics Journal* ("'Hidden' Boundaries in Constrained Max-Min Problems," May 1991, Vol. 22, p. 227).

CHAPTER 11 GROUP RESEARCH PROJECT*

Working in small groups is typical of most work environments, and this book seeks to develop skills with group activities. At the end of each chapter, we present a group project. These problems are to be done in groups of three or four students.

Desertification

© 2013 apdesign. Used under license from Shutterstock, Inc.

A friend of yours named Maria is studying the causes of the continuing expansion of deserts (a process known as *desertification*). She is working on a biological index of vegetation disturbance, which she has defined. By seeing how this index and other factors change through time, she hopes to discover the role played in desertification. She is studying a huge tract of land bounded by a rectangle. This piece of land surrounds a major city but does not include the city. She needs to find an economical way to calculate for this piece of land the important vegetation disturbance index $J(x, y)$.

Maria has embarked upon an ingenious approach of combining the results of photographic and radar images taken during flights over the area to calculate the index J. She is assuming that J is a smooth function. Although the flight data do not directly reveal the values of the function J, they give the rate at which the values of J change as the flights sweep over the landscape surrounding the city. Maria's staff has conducted numerous flights, and from the data she believes she has been able to find actual formulas for the rates at which J changes in the east-west and north-south directions. She has given these functions the names M and N. Thus, $M(x, y)$ is the rate at which J changes as one sweeps in the positive x-direction and $N(x, y)$ is the corresponding rate in the y-direction. Maria shows you these formulas:

$$M(x, y) = 3.4e^{x(y-7.8)^2} \text{ and } N(x, y) = 22\sin(75 - 2xy)$$

Extended paper for further study. Convince Maria that these two formulas cannot possibly be correct. Do this by showing her that there is a condition that the two functions M and N must satisfy if they are to be the east-west and north-south rates of change of the function J and that her formulas for M and N do not meet this condition. However, show Maria that if she can find formulas for M and N that satisfy the condition that you showed her, it is possible to find a formula for the function J from the formulas for M and N.

---

*Adapted from Marcus S. Cohen, Edward D. Gaughan, R. Arthur Knoebel, Douglas S. Kurtz, and David J. Pengelley, "Priming the Calculus Pump: Innovations and Resources," MAA Notes 17(1991).

CHAPTER 12

MULTIPLE INTEGRATION

"Mathematics... furnishes the peculiar study that gives us... the command of nature."

W.T. Harris
Psychological Foundations of Education (New York: 1898), p. 325

PREVIEW

The *single integral* $\int_a^b f(x)\,dx$ introduced in Chapter 5 has many uses, as we have seen. In this chapter, we will generalize the single integral to define *multiple* integrals in which the integrand is a function of several variables. We will find that multiple integration is used in much the same way as single integration, by "adding" small quantities to define and compute area, volume, surface area, moments, centroids, and probability.

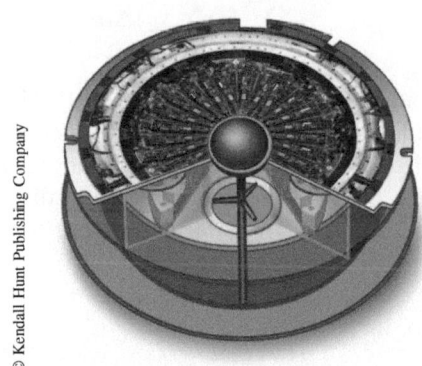

© Kendall Hunt Publishing Company

Multiple integral field spectrograph of the VLT

PERSPECTIVE

What is the volume of a doughnut (torus)? Given the joint probability function for the amount of time a typical shopper spends shopping at a particular store and the time spent in the checkout line, how likely is it that a shopper will spend no more than 30 minutes altogether in the store? If the temperature in a solid body is given at each point (x, y, z) and time t, what is the average temperature of the body over a particular time period? Where should a security watch tower be placed in a parking lot to ensure the most comprehensive visual coverage? We will answer these and other similar questions in this chapter using multiple integration.

CONTENTS

919

12.1 DOUBLE INTEGRATION OVER RECTANGULAR REGIONS

IN THIS SECTION: *Definition of the double integral, properties of double integrals, volume interpretation, iterated integration, an informal argument for Fubini's theorem*
We introduce the concept of a *double integral*—that is, an integral with respect to two independent variables.

Definition of the Double Integral

Recall that in Chapter 5, we defined the definite integral of a single variable $\int_a^b f(x)\,dx$ as a limit involving Riemann sums

$$\sum_{k=1}^{n} f(x_k^*)\Delta x_k, \text{ where } a = x_0 < x_1 < x_2 \cdots < x_n = b$$

are points in a partition of the interval $[a, b]$ and x_k^* is a representative point in the subinterval $[x_{k-1}, x_k]$. We now apply the same ideas to define a definite integral of two variables $\iint_R f(x, y)\,dA$, over the rectangle R: $a \leq x \leq b$, $c \leq y \leq d$.

The definition requires the ideas and notation described in the following three steps:

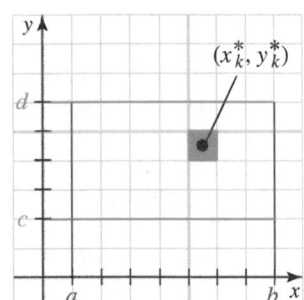

Figure 12.1 A partition of the rectangle R into mn cells showing the kth cell representative

Step 1: Partition the interval $a \leq x \leq b$ into m subintervals and the interval $c \leq y \leq d$ into n subintervals. Using these subdivisions, partition the rectangle R into
$N = mn$ **cells** (subrectangles), as shown in Figure 12.1. Call this partition P.

Step 2: Choose a representative point (x_k^*, y_k^*) from each cell in the partition of the rectangle. Form the sum

$$\sum_{k=1}^{N} f(x_k^*, y_k^*)\Delta A_k$$

where ΔA_k is the area of the kth representative cell. This is called the **Riemann sum** of $f(x, y)$ with respect to the partition P and cell representatives (x_k^*, y_k^*).

Step 3: To measure the size of the rectangles in the partition P, we define the **norm** $\|P\|$ of the partition to be the length of the longest diagonal of any rectangle in the partition. We **refine** the partition by subdividing the cells in such a way that the norm decreases.

When this process is applied to the Riemann sum and the norm decreases indefinitely to zero, we write

$$\lim_{\|P\| \to 0} \sum_{k=1}^{N} f(x_k^*, y_k^*)\Delta A_k$$

If this limit exits, its value is called the *double integral* of f over the rectangle R.

DOUBLE INTEGRAL If f is defined on a closed, bounded rectangular region R in the xy-plane, then the **double integral of f over R** is defined by

$$\iint_R f(x, y)\,dA = \lim_{\|P\| \to 0} \sum_{k=1}^{N} f(x_k^*, y_k^*)\Delta A_k$$

provided this limit exists, in which case f is said to be **integrable** over R.

More formally, the limit statement

$$I = \lim_{\|P\| \to 0} \sum_{k=1}^{N} f(x_k^*, y_k^*) \Delta A_k$$

means that for any $\epsilon > 0$, there exists a $\delta > 0$ such that

$$\left| I - \sum_{k=1}^{N} f(x_k^*, y_k^*) \Delta A_k \right| < \epsilon$$

whenever $\sum_{k=1}^{N} f(x_k^*, y_k^*) \Delta A_k$ is a Riemann sum whose norm satisfies $\|P\| < \delta$. In this process, the number of cells N depends on the partition P, and as $\|P\| \to 0$, it follows that $N \to \infty$.

It can be shown that if $f(x, y)$ is continuous on a rectangle R, then it must be integrable on R, although it is also true that certain discontinuous functions are integrable as well. Moreover, it can also be shown that if the limit that defines the definite integral exists, then it is unique in the sense that the same limiting value results no matter how the partitions and subinterval representatives are chosen. We will extend the definition of the definite integral to nonrectangular regions in Section 12.2. However, issues involving the existence and uniqueness of the definite integral are generally dealt with in a course in advanced calculus.

Properties of Double Integrals

Double integrals have many of the same properties as single integrals. Three of these properties are contained in the following theorem.

Theorem 12.1   Properties of double integrals

Assume that all the given integrals exist on a rectangular region R.
Linearity rule: For constants a and b,

$$\iint\limits_{R} [af(x, y) + bg(x, y)]dA = a \iint\limits_{R} f(x, y)\, dA + b \iint\limits_{R} g(x, y)\, dA$$

Dominance rule: If $f(x, y) \geq g(x, y)$ throughout a rectangular region R, then

$$\iint\limits_{R} f(x, y)\, dA \geq \iint\limits_{R} g(x, y)\, dA$$

Subdivision rule:

If the rectangular region of integration R is subdivided into two subrectangles R_1 and R_2, then

$$\iint\limits_{R} f(x, y)\, dA = \iint\limits_{R_1} f(x, y)\, dA + \iint\limits_{R_2} f(x, y)\, dA$$

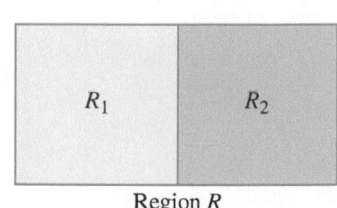

Region R

Proof: This proof is omitted.   ◆

Volume Interpretation

If $g(x) \geq 0$ on the interval $[a, b]$, the single integral $\int_a^b g(x)\, dx$ can be interpreted as the area under the curve $y = g(x)$ over $[a, b]$. The double integral $\iint\limits_R f(x, y)\, dA$ has a similar interpretation in terms of volume. To see this, note that if $f(x, y) \geq 0$ on the rectangular region R and we partition R, then the product $f(x_k^*,\, y_k^*)\Delta A_k$ is the volume of a parallelepiped (a box) with height $f(x_k^*,\, y_k^*)$ and base area ΔA_k, as shown in Figure 12.2.

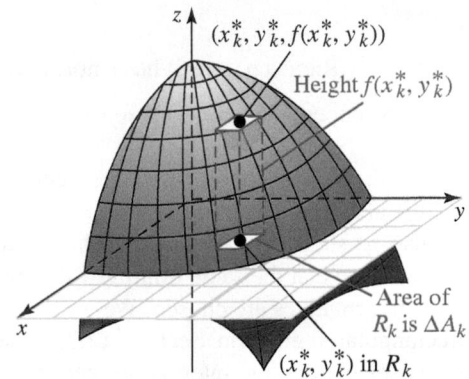

Figure 12.2 The approximating parallelepiped has volume $\Delta V_k = f(x_k^*,\, y_k^*)\Delta A_k$

Thus, the Riemann sum

$$\sum_{k=1}^{N} f(x_k^*,\, y_k^*)\Delta A_k$$

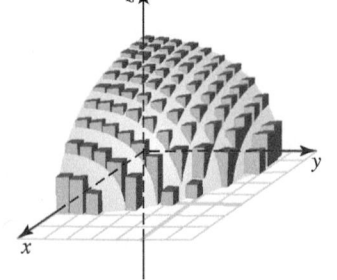

Figure 12.3 Volume approximated by rectangular parallelepipeds

provides an estimate of the total volume under the surface $z = f(x, y)$ over R, and if f is continuous, we expect the approximation to improve by using more refined partitions (that is, more rectangles with smaller norm). Thus, it is natural to *define* the total volume under the surface as the limit of Riemann sums as the norm tends to 0. That is, the volume under $z = f(x, y)$ over the domain R is given by

$$V = \lim_{\|P\| \to 0} \sum_{k=1}^{N} f(x_k^*,\, y_k^*)\Delta A_k = \iint\limits_R f(x, y)\, dA$$

The approximation by the Riemann sum is illustrated in Figure 12.3.

Example 1   Evaluating a double integral by relating it to a volume

Evaluate $\iint\limits_R (2 - y)\, dA$, where R is the rectangle in the xy-plane with vertices $(0, 0)$, $(3, 0)$, $(3, 2)$, and $(0, 2)$.

Solution   Because $z = 2 - y$ satisfies $z \geq 0$ for all points in R, the value of the double integral is the same as the volume of the solid bounded above by the plane $z = 2 - y$ and below by the rectangle R. The solid is shown in Figure 12.4. Looking at it end-wise, it has a triangular cross section of area B and has length 3. We use the formula $V = Bh$.

Because the base is a triangle of side 2 and altitude 2, we have

$$V = Bh = \left[\frac{1}{2}(2)(2)\right](3) = 6$$

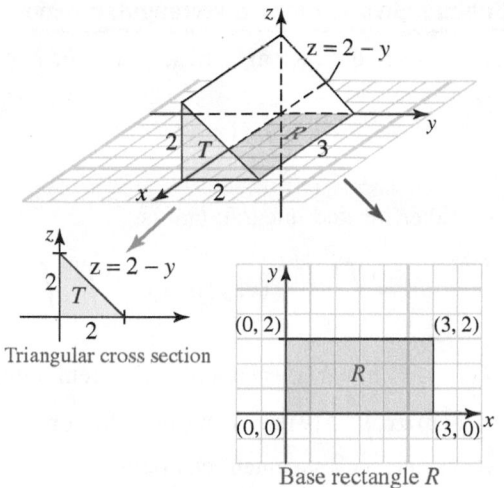

Figure 12.4 Evaluation of $\iint\limits_{R} (2-y)\,dA$ as volume

Therefore, the value of the integral is also 6; that is,

$$\iint\limits_{R} (2-y)\,dA = 6$$

Iterated Integration

As with single integrals, it is often not practical to evaluate a double integral even over a simple rectangular region by using the definition. Instead, we will compute double integrals by a process called **iterated integration** which is like partial differentiation in reverse.

Suppose $f(x, y)$ is continuous over the rectangle R: $a \leq x \leq b$, $c \leq y \leq d$. Then we write $\int_{c}^{d} f(x, y)\,dy$ to denote the integral obtained by integrating $f(x, y)$ with respect to y over the interval $[c, d]$ with x held constant. The integral obtained by this partial integration is a function of x alone, $G(x) = \int_{c}^{d} f(x, y)\,dy$, which we integrate over the interval $[a, b]$ to obtain the **iterated integral**

$$\int_{a}^{b} G(x)\,dx = \int_{a}^{b} \left[\int_{c}^{d} f(x, y)\,dy \right] dx \qquad \begin{array}{l}\textit{Integrate with respect to y first,}\\ \textit{keeping x constant, and then}\\ \textit{integrate with respect to x.}\end{array}$$

Similarly, if we integrate $f(x, y)$ first with respect to x over $[a, b]$ holding y constant, and then with respect to y over $[c, d]$, we obtain the iterated integral

$$\int_{a}^{b} G(x)\,dx = \int_{c}^{d} \left[\int_{a}^{b} f(x, y)\,dx \right] dy \qquad \begin{array}{l}\textit{Integrate with respect to x}\\ \textit{first, keeping y constant, and then}\\ \textit{integrate with respect to y.}\end{array}$$

The following theorem, which shows how the double integral

$$\iint\limits_{R} f(x, y)\,dA$$

can be evaluated in terms of iterated integrals, was proved in a more general form by the Italian mathematician Guido Fubini (1879-1943) in 1907.

Theorem 12.2 Fubini's theorem over a rectangular region

If $f(x, y)$ is continuous over the rectangle $R: a \leq x \leq b, \ c \leq y \leq d$, then the double integral

$$\iint\limits_{R} f(x, y) \, dA$$

may be evaluated by either iterated integral; that is,

$$\iint\limits_{R} f(x, y) \, dA = \int_{c}^{d} \int_{a}^{b} f(x, y) \, dx \, dy = \int_{a}^{b} \int_{c}^{d} f(x, y) \, dy \, dx$$

In the case that $f(x, y) = g(x)h(y)$, Fubini's Theorem allows the double integral to be written as $\int_{a}^{b} g(x) \, dx \int_{c}^{d} h(y) \, dy$. Note that this only works for multiplication. If $f(x, y) = g(x) + h(y)$ no such statement can be made.

■ **W**hat this says Instead of using the definition of a double integral of $f(x, y)$ over the rectangle $R: a \leq x \leq b, \ c \leq y \leq d$, evaluate *either* of the iterated integrals:

Limits of x (variable outside brackets)   *Limits of y (variable outside brackets)*

$$\int_{a}^{b} \left[\int_{c}^{d} f(x, y) \, dy \right] dx \qquad \text{or} \qquad \int_{c}^{d} \left[\int_{a}^{b} f(x, y) \, dx \right] dy$$

Limits of y (variable inside brackets)   *Limits of x (variable inside brackets)*

Note that Fubini's theorem applies only when a, b, c, and d are constants.

Proof: We will provide an informal, geometric argument at the end of this section. The formal proof may be found in most advanced calculus textbooks. Here, we will consider the case where $f(x, y) = g(x)h(y)$.

$$\iint\limits_{R} f(x, y) \, dA = \int_{a}^{b} \int_{c}^{d} f(x, y) \, dy \, dx = \int_{a}^{b} \left[\int_{c}^{d} g(x)h(y) \, dy \right] dx$$

In the inner integral, x is constant, so $g(x)$ is constant and we can write

$$\int_{a}^{b} \left[\int_{c}^{d} g(x)h(y) \, dy \right] dx = \int_{a}^{b} g(x) \left[\int_{c}^{d} h(y) \, dy \right] dx = \int_{a}^{b} g(x) \, dx \int_{c}^{d} h(y) \, dy$$

since $\int_{c}^{d} h(y) \, dy$ is constant.   ◆

Note. We have simplified the presentation of the Fubini-Tonelli Theorem, making it accessible to a first course of multivariable calculus. In advanced calculus, it is shown that it is sufficient but **not** necessary for the function $f(x, y)$ to be continuous over the rectangle $R: a \leq x \leq b, \ c \leq y \leq d$, for the iterated integration to be possible. We often encounter cases where the integrand is either piecewise continuous, or discontinuous exactly at the boundary, or at certain points on the boundary, in which cases Fubini's integral is still applicable if certain conditions on integrability are satisfied.

A more "advanced" version of Fubini's Theorem without introducing advanced notions, such as Riemann integrability or Lebesque integrability, can be stated as follows:

Remark: Fubini-Tonelli theorem over a rectangular region (adapted version)

Let us consider the rectangle R: $a \leq x \leq b$, $c \leq y \leq d$ and a function $f(x, y)$ which is bounded and continuous over the rectangle R except for a subset S of R with area zero, such that the integral

$$\iint_R f(x, y)\, dA$$

exists. Assume further that for each value x such that $a \leq x \leq b$, that subset S contains only finitely many points (possibly none) with first coordinate x. Then the interated integral also exists, and

$$\iint_R f(x, y)\, dA = \int_a^b \int_c^d f(x, y)\, dy\, dx$$

Example: R: $0 \leq x \leq 1$, $0 \leq y \leq 1$, and $f(x, y) = \sqrt{\dfrac{x}{y}}$. Although the integrand is discontinuous and unbounded at all the points whose coordinate $y = 0$, it can be shown that the double integral exists on the domain R and can be computed using interated integrals (note that one of the two simple integrals is improper!).

Another way to explain this phenomenon can be stated simply like this: *If a function of two variables over a rectangle is integrable one at a time with respect to either variable, then it is integrable over the entire rectangle, and the double integral can be computed "one variable at a time".* The proof of the relaxed version of Fubini's Theorem exceeds the level of the present text, and is usually presented in advanced calculus courses.

Let us see how Fubini's theorem can be used to evaluate double integrals. We begin by taking another look at Example 2.

Example 2 Evaluating a double integral by using Fubini's theorem

Use an iterated integral with y-integration first to compute $\iint_R (2 - y)\, dA$, where R is the rectangle with vertices $(0,0)$, $(3,0)$, $(3,2)$, and $(0,2)$.

Solution The region of integration is the rectangle $0 \leq x \leq 3$, $0 \leq y \leq 2$ (see Example 1 and Figure 12.4). Thus, by Fubini's theorem, the double integral can be evaluated by the iterated integral:

$$\iint_R (2 - y)\, dA = \int_0^3 \int_0^2 (2 - y)\, dy\, dx \qquad \textit{Integrate inner integral with respect to } y.$$

$$= \int_0^3 \left[2y - \frac{y^2}{2} \right]_0^2 dx$$

$$= \int_0^3 \left[4 - \frac{4}{2} - (0) \right] dx$$

$$= \int_0^3 2\, dx$$

$$= 2x \big|_0^3$$

$$= 6$$

which is the same as the result obtained geometrically in Example 1.

Example 3 Double integral using an iterated integral

Evaluate $\iint_R x^2 y^5 dA$, where R is the rectangle $1 \leq x \leq 2$, $0 \leq y \leq 1$, using an iterated integral with   **a.** y-integration first   **b.** x-integration first

Solution   The graph of the surface $z = x^2 y^5$ over the rectangle is shown in Figure 12.5a, and the rectangular region of integration in Figure 12.5b.

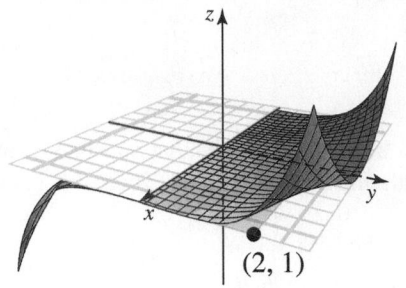

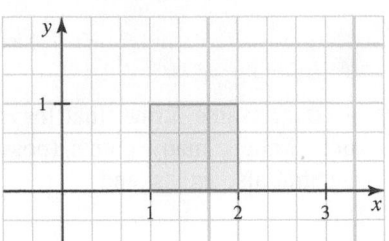

a. **Interactive** Graph of the surface over a rectangle   **b.** Sketch of rectangular region

Figure 12.5   Graph of a surface over a rectangle

a.
$$\iint_R x^2 y^5 \, dA = \int_1^2 \int_0^1 x^2 y^5 \, dy \, dx$$

$$= \int_1^2 \left[x^2 \frac{y^6}{6} \right]_0^1 dx$$

$$= \int_1^2 \left[x^2 \left(\frac{1}{6} - \frac{0}{6} \right) \right] dx$$

$$= \frac{x^3}{18} \Big|_1^2$$

$$= \frac{8}{18} - \frac{1}{18}$$

$$= \frac{7}{18}$$

b.
$$\iint_R x^2 y^5 \, dA = \int_0^1 \int_1^2 x^2 y^5 \, dx \, dy$$

$$= \int_0^1 y^5 \left[\frac{x^3}{3} \right]_1^2 dy$$

$$= \int_0^1 y^5 \left(\frac{8}{3} - \frac{1}{3} \right) dy$$

$$= \frac{7y^6}{18} \Big|_0^1$$

$$= \frac{7}{18} - \frac{0}{18}$$

$$= \frac{7}{18}$$

In many problems, the order of integration in a double integral is largely a matter of personal choice, but occasionally, choosing the order correctly makes the difference between a straightforward evaluation and one that is either difficult or impossible. Consider the following example.

Example 4   Choosing the order of integration for a double integral

Evaluate $\iint_R x \cos(xy) \, dA$ for R: $0 \leq x \leq \frac{\pi}{2}$, $0 \leq y \leq 1$.

Solution   Suppose we integrate with respect to x first:

$$\int_0^1 \left[\int_0^{\pi/2} x \cos(xy) \, dx \right] dy$$

The inner integral requires integration by parts. However, integrating with respect to y first is much simpler:

$$\int_0^{\pi/2} \left[\int_0^1 x \cos(xy) \, dy \right] dx = \int_0^{\pi/2} \left[\frac{x \sin(xy)}{x} \right]_0^1 dx$$

$$= \int_0^{\pi/2} (\sin x - \sin 0) \, dx$$

$$= -\cos x \big|_0^{\pi/2}$$

$$= 1$$

An Informal Argument for Fubini's Theorem

We can make Fubini's theorem plausible with a geometric argument in the case where $f(x, y) \geq 0$ on R. If $\iint\limits_R f(x, y)\, dA$ is defined on a rectangle $R: a \leq x \leq b,\ c \leq y \leq d$, it represents the volume of the solid D bounded above by the surface $z = f(x, y)$ and below by the rectangle R. If $A(y_k^*)$ is the cross-sectional area perpendicular to the y-axis at the point y_k^*, then $A(y_k^*)\Delta y_k$ represents the volume of a "slab" that approximates the volume of part of the solid D, as shown in Figure 12.6.

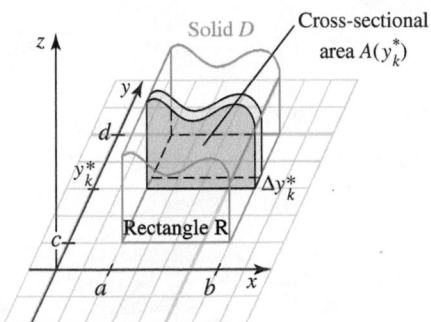

Figure 12.6 Cross-sectional volume parallel to the xz-plane

By using a limit to "add up" all such approximating volumes, we obtain the volume, V, of the entire solid D; that is,

$$\iint\limits_R f(x, y)\, dA = V = \lim_{\|P\|\to 0} \sum_{k=1}^{N} A(y_k^*)\Delta y_k$$

The limit on the right is just the integral of $A(y)$ over the interval $c \leq y \leq d$, where $A(y)$ is the area of a cross section with fixed y. In Chapter 5, we found that the area $A(y)$ can be computed by the integral

$$A(y) = \int_a^b f(x, y)\, dx \qquad \textit{Integration with respect to } x \quad (y \textit{ is fixed})$$

We can now make this substitution for $A(y)$ to obtain:

$$\iint\limits_R f(x, y)\, dA = V = \lim_{\|P\|\to 0} \sum_{k=1}^{N} A(y_k^*)\Delta y_k$$

$$= \int_c^d A(y)\, dy$$

$$= \int_c^d \left[\underbrace{\int_a^b f(x, y)\, dx}_{A(y)} \right] dy \qquad \textit{Substitution}$$

$$= \int_c^d \int_a^b f(x, y)\, dx\, dy$$

The fact that

$$\iint\limits_R f(x, y)\, dA = V = \int_a^b \int_c^d f(x, y)\, dy\, dx$$

can be justified in a similar fashion (you are asked to do this in Problem 58). Thus, we have

$$\int_c^d \left[\int_a^b f(x, y) \, dx \right] dy = \iint_R f(x, y) \, dA = V = \int_a^b \left[\int_c^d f(x, y) \, dy \right] dx$$

PROBLEM SET 12.1

Level 1

1. ■ What does this say? Discuss the definition of the double integral.
2. ■ What does this say? Explain how Fubini's theorem is used to evaluate double integrals.

In Problems 3-12, evaluate the iterated integrals.

3. $\displaystyle\int_0^2 \int_0^1 (x^2 + xy + y^2) \, dy \, dx$

4. $\displaystyle\int_1^2 \int_0^3 x^2 y^3 \, dx \, dy$

5. $\displaystyle\int_0^{\pi/2} \int_1^2 y \sin x \, dy \, dx$ 6. $\displaystyle\int_0^1 \int_0^\pi x \cos y \, dy \, dx$

7. $\displaystyle\int_1^{e^2} \int_1^2 \left[\frac{1}{x} + \frac{1}{y} \right] dy \, dx$ 8. $\displaystyle\int_1^3 \int_1^{e^3} \left[\frac{1}{x} \right] dx \, dy$

9. $\displaystyle\int_0^{\ln 5} \int_0^1 e^{2x+y} \, dy \, dx$ 10. $\displaystyle\int_0^{\ln 2} \int_0^1 e^{x+2y} \, dx \, dy$

11. $\displaystyle\int_3^4 \int_1^2 \frac{x}{x-y} \, dy \, dx$ 12. $\displaystyle\int_2^3 \int_{-1}^2 \frac{1}{(x+y)^2} \, dy \, dx$

Use an appropriate volume formula to evaluate the double integral given in Problems 13-18.

13. $\displaystyle\iint_R 4 \, dA;$ 14. $\displaystyle\iint_R 5 \, dA;$
    $R: 0 \le x \le 2, 0 \le y \le 4$ $R: 2 \le x \le 5, 1 \le y \le 3$

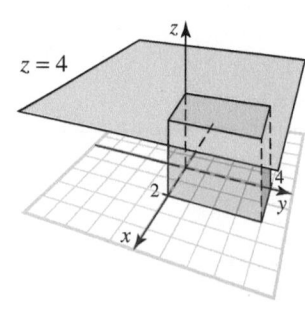   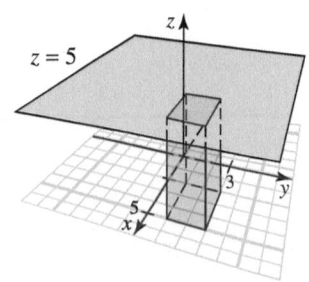

15. $\displaystyle\iint_R (4 - y) \, dA;$ 16. $\displaystyle\iint_R (4 - 2y) \, dA;$
    $R: 0 \le x \le 3, 0 \le y \le 4$ $R: 0 \le x \le 4, 0 \le y \le 2$

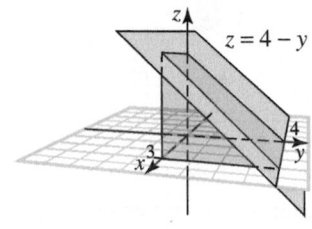   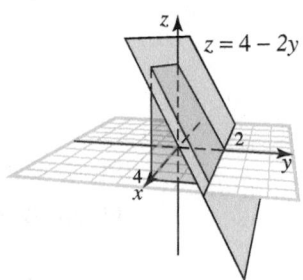

17. $\displaystyle\iint_R \frac{y}{2} \, dA;$ 18. $\displaystyle\iint_R \frac{y}{4} \, dA;$
    $R: 0 \le x \le 6, 0 \le y \le 4$ $R: 0 \le x \le 2, 0 \le y \le 8$

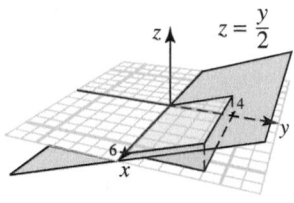   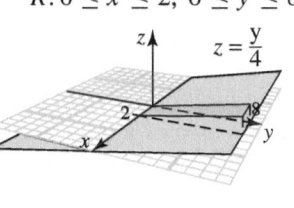

Use iterated integration to compute the double integrals in Problems 19-28 over the specified rectangle.

19. $\displaystyle\iint_R x^2 y \, dA; \; R: 1 \le x \le 2, 0 \le y \le 1$

20. $\displaystyle\iint_R (x^2 + 2xy + y^2) \, dA; \; R: 0 \le x \le 1, 0 \le y \le 2$

21. $\displaystyle\iint_R \left(\frac{x^2 + 1}{y^2 + 2} \right) dA; \; R: 0 \le x \le 1, 0 \le y \le 1$

22. $\displaystyle\iint_R \left(\frac{4 + x^2}{1 + y^2} \right) dA; \; R: 0 \le x \le 1, 0 \le y \le 1$

23. $\displaystyle\iint_R 2xe^y \, dA; \; R: -1 \le x \le 0, 0 \le y \le \ln 2$

24. $\displaystyle\iint_R x^2 e^{xy} \, dA; \; R: 0 \le x \le 1, 0 \le y \le 1$

25. $\displaystyle\iint\limits_{R} \frac{2xy\,dA}{x^2+1};\ R: 0\le x\le 1, 1\le y\le 3$

26. $\displaystyle\iint\limits_{R} y\sqrt{x-y^2}\,dA;\ R: 1\le x\le 5, 0\le y\le 1$

27. $\displaystyle\iint\limits_{R} \sin(x+y)\,dA;\ R: 0\le x\le \frac{\pi}{4}, 0\le y\le \frac{\pi}{2}$

28. $\displaystyle\iint\limits_{R} x\sin(xy)\,dA;\ R: 0\le x\le \pi, 0\le y\le 1$

Find the volume of the solid bounded below by the rectangle R in the xy-plane and above by the graph of $z = f(x, y)$ in Problems 29-42.

29. $f(x, y) = 2x + 3y; R: 0\le x\le 1, 0\le y\le 2$

30. $f(x, y) = 5x + 2y; R: 0\le x\le 1, 0\le y\le 2$

31. $f(x, y) = \dfrac{x}{y} + \dfrac{y}{x}; R: 1\le x\le 2, 1\le y\le 2$

32. $f(x, y) = \dfrac{1}{6xy}\ R: 1\le x\le 4, 1\le y\le 6$

33. $f(x, y) = x\ln(xy); R: 1\le x\le 2, 1\le y\le e$

34. $f(x, y) = \dfrac{\ln x}{y}; R: 1\le x\le e, 1\le y\le 2$

35. $f(x, y) = \sqrt{xy}; R: 0\le x\le 1, 0\le y\le 4$

36. $f(x, y) = \sqrt{\dfrac{x}{y}};\ R: 1\le x\le 2;\ 1\le y\le 4$

37. $f(x, y) = \dfrac{xy}{\sqrt{x^2+y^2+1}}; R: 0\le x\le 1, 0\le y\le 1$

38. $f(x, y) = \sqrt{x+y}; R: 0\le x\le 1, 0\le y\le 1$

39. $f(x, y) = (x+y)^5; R: 0\le x\le 1, 0\le y\le 1$

40. $f(x, y) = xe^{xy}; R: 0\le x\le 1, 0\le y\le \ln 3$

41. $f(x, y) = x\cos y + y\sin x; R: 0\le x\le \frac{\pi}{2}, 0\le y\le \frac{\pi}{2}$

42. $f(x, y) = x\sin y + y\cos x; R: 0\le x\le \frac{\pi}{2}, 0\le y\le \frac{\pi}{2}$

Level 2

43. Modeling Problem Suppose R is a rectangular region within the boundary of a Colorado great plain that contains 20,000 head of cattle per square mile. Model the total number of cattle, C, per square mile in the great plain as a double integral. Assume that x and y are measured in miles.

44. Modeling Problem Suppose R is a rectangular region within the boundary of a certain national forest that contains 600,000 trees per square mile. Model the total number of trees, T, per square mile in the forest as a double integral. Assume that x and y are measured in miles.

45. Modeling Problem Suppose mass is distributed on a rectangular region R in the xy-plane so that the density (mass per unit area) at the point (x, y) is $\rho(x, y)$. Model the total mass as a double integral.

46. Modeling Problem Let R be a rectangular region within the boundary of a certain city defined by R: $-2\le x\le 2, -1\le y\le 1$, where units are in miles and $(0,0)$ is the city center. Assume the population density is $12e^{-0.07r}$ thousand people per square mile, where $r = \sqrt{x^2+y^2}$. Model the total population of the region of the city as a double integral, but do not evaluate the integral.

In Problems 47-50, evaluate the integral. Note that one order of integration may be considerably easier than the other.

47. Compute $\iint\limits_{R} x\sqrt{1-x^2}e^{3y}\,dA$, where R is the rectangle $0\le x\le 1, 0\le y\le 2$.

48. Compute $\iint\limits_{R} \dfrac{\ln\sqrt{y}}{xy}\,dA$, where R is the rectangle $1\le x\le 4, 1\le y\le e$.

49. Compute (correct to the nearest hundredth) $\iint\limits_{R} \dfrac{xy}{x^2+y^2}\,dA$, where R is the rectangle $1\le x\le 3, 1\le y\le 2$.

50. Evaluate $\iint\limits_{R} xe^{xy}\,dA$, where R is the rectangle $0\le x\le 1, 1\le y\le 2$.

51. Think Tank Problem Explain why

$$\iint\limits_{R} (4 - x^2 - y^2)\,dA > 2$$

where R is the rectangular domain in the plane given by $0\le x\le 1, 0\le y\le 1$.

52. Think Tank Problem Explain why

$$\iint\limits_{R} (9 - x^2 - y^2)\,dA < 81$$

where R is the rectangular domain in the plane given by $0\le x\le 3, 0\le y\le 3$.

53. Use a grid with 16 cells to approximate the volume of the solid lying between the surface

$$f(x, y) = x^2 + y^2 + 1$$

and the square region R given by $0\le x\le 1$, $0\le y\le 1$. Use the left lower corner point in each square, to create the approximating sum.

54. Use a grid with 16 cells to approximate the volume of the solid lying between the surface

$$f(x, y) = 4 - x^2 - y^2$$

and the square region R given by $0 \le x \le 1$, $0 \le y \le 1$.

55. *Historical Quest*

Guido Fubini taught at the Institute for Advanced Study at Princeton. He was nicknamed the "Little Giant," because of his small body but large mind. Even though the conclusion of Fubini's theorem was known for a long time and successfully applied in various instances, it was not

Guido Fubini (1879-1943)

satisfactorily proved in a general setting until 1907. His most important work was in differential projective geometry. In 1938 he was forced to leave Italy because of the Fascist government, and he immigrated to the United States. For this *Quest*, write several paragraphs about the nature of differential projective geometry.

56. *Historical Quest*

William Rowan Hamilton has been called the most renowned Irish mathematician. He was a child prodigy who read Greek, Hebrew, and Latin by the time he was five, and by the age of ten he knew over a dozen languages. Many mathematical advances are credited to Hamilton.

William Rowan Hamilton (1805-1865)

For example, he developed vector methods in analytic geometry and calculus, as well as a system of algebraic quantities called **quaternions**, *which occupied his energies for the last 22 years of his life. Hamilton pursued the study of quaternions with an almost religious fervor, but by the early 20th century, the notation and terminology of vectors dominated. Much of the credit for the eventual emergence of vector methods goes not only to Hamilton, but also to the scientists James Clerk Maxwell (1831-1879), J. Willard Gibbs (1839-1903), and Oliver Heaviside (1850-1925).* For this *Quest*, write a paper on quaternions.

Level 3

57. Let f be a function with continuous second partial derivatives on a rectangular domain R with vertices $(x_1, y_1), (x_1, y_2), (x_2, y_2)$, and (x_2, y_1), where $x_1 < x_2$ and $y_1 < y_2$. Use the fundamental theorem of calculus to show that

$$\iint_R \frac{\partial^2 f}{\partial y\, \partial x}\, dA$$

$$= f(x_1, y_1) - f(x_2, y_1) + f(x_2, y_2) - f(x_1, y_2)$$

58. Let f be a continuous function defined on the rectangle $R: a \le x \le b, c \le y \le d$. Show that

$$\iint_R f(x, y)\, dA = \int_a^b \int_c^d f(x, y)\, dx\, dy$$

Hint: Modify the argument given in the text by taking cross-sectional areas perpendicular to the x-axis.

59. Think Tank Problem Show that the iterated integrals

$$\int_0^1 \int_0^1 \frac{y - x}{(x + y)^3}\, dy\, dx$$

and

$$\int_0^1 \int_0^1 \frac{y - x}{(x + y)^3}\, dx\, dy$$

have different values. Why does this not contradict Fubini's theorem?

60. Think Tank Problem You want to evaluate

$$\iint_D y \sin(xy) \sin^2(\pi y)\, dA$$

over the rectangle $D: 0 \le x \le \pi, 0 \le y \le \frac{1}{2}$. Which order of integration is easier?

12.2 DOUBLE INTEGRATION OVER NONRECTANGULAR REGIONS

IN THIS SECTION: *Double integrals over type I and type II regions, nonrectangular regions, more on area and volume, choosing the order of integration in a double integral*
In this section, we show how double integrals can be evaluated over regions that are not rectangles.

Let $f(x, y)$ be a function that is continuous on the region D which can be contained in a rectangle R. (See Figure 12.7.)

Define the function $F(x, y)$ on R as $f(x, y)$ if (x, y) is in D and 0 otherwise. That is,

$$F(x, y) = \begin{cases} f(x, y) & \text{for } (x, y) \text{ in } D \\ 0 & \text{for } (x, y) \text{ in } R, \text{ but not in } D \end{cases}$$

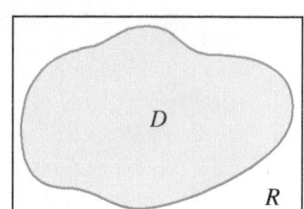

Then, if F is integrable over R, we say that f is *integrable* over D, and the *double integral* of f over D is defined as

Figure 12.7 The region D bounded by a rectangle R

$$\iint\limits_{D} f(x, y) \, dA = \iint\limits_{R} F(x, y) \, dA$$

The function $F(x, y)$ may have discontinuities on the boundary of D, but if $f(x, y)$ is continuous on D and the boundary of D is fairly "well behaved", then it can be shown that $\iint_{R} F(x, y) \, dA$ exists and hence that $\iint_{D} f(x, y) \, dA$ exists. This procedure is valid for the type I and type II regions we discuss next, although a general notion of what constitutes a "well-behaved" boundary is a topic for advanced calculus.

Double Integrals over Type I and Type II Regions

A type I, or vertically simple region, D, in the plane is a region that can be described by the inequalities

    type I D_1: $a \le x \le b$, $g_1(x) \le y \le g_2(x)$

where $g_1(x)$ and $g_2(x)$ are continuous functions of x on $[a.b]$.

Likewise, a type II, or horizontally simple region, D, is one that can be described by the inequalities

    type II D_2: $c \le y \le d$, $h_1(y) \le x \le h_2(y)$

where $h_1(y)$ and $h_2(y)$ are continuous functions of y on $[c, d]$. Vertically and horizontally simple regions are illustrated in the following box.

Type I Region (vertically simple) A **type I** region is the set of all points (x, y) such that for each fixed x between $x = a$ and $x = b$ the vertical line segment $g_1(x) \le y \le g_2(x)$ lies in the region.	**Type II Region (horizontally simple)** A **type II** region is the set of all points (x, y) such that for each fixed y between $y = c$ and $y = d$ the horizontal line segment $h_1(y) \le x \le h_2(y)$ lies in the region.
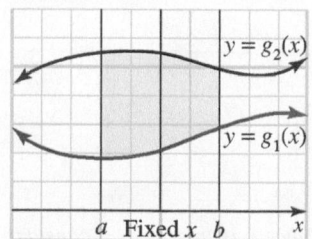	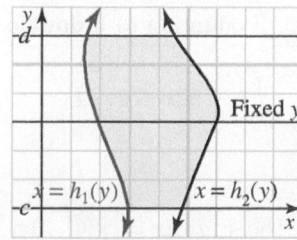

To evaluate a double integral $\int \int_{D_1} f(x, y)\, dA$ when D_1 is a type I region, first note that D_1 is contained in the rectangle R: $a \le x \le b$, $c \le y \le d$. Then, if

$$F(x, y) = \begin{cases} f(x, y) & \text{for } (x, y) \text{ in } D_1 \\ 0 & \text{for } (x, y) \text{ in } R, \text{ but not in } D_1 \end{cases}$$

we have

$$\int_c^d F(x, y)\, dy = \int_{g_1(x)}^{g_2(x)} f(x, y)\, dy$$

because for each fixed x in the interval $[a, b]$, $F(x, y) = 0$ if $y < g_1(x)$ and also if $y > g_2(x)$, and $F(x, y) = f(x, y)$ for $g_1(x) \le y \le g_2(x)$. Therefore,

$$\int \int_{D_1} f(x, y)\, dA = \iint_R F(x, y)\, dA$$

$$= \int_a^b \left[\int_c^d F(x, y)\, dy \right] dx$$

$$= \int_a^b \int_{g_1(x)}^{g_2(x)} f(x, y)\, dy\, dx$$

Similarly, if D_2 is a type II region, then

$$\int \int_{D_2} f(x, y)\, dA = \int_c^d \left[\int_a^b F(x, y)\, dx \right] dy$$

$$= \int_c^d \int_{h_1(y)}^{h_2(y)} f(x, y)\, dx\, dy$$

These observations are summarized in the following theorem.

Theorem 12.3   Fubini's theorem for nonrectangular regions

TYPE I (vertically simple): x fixed, y varies (form $dy\, dx$)	If D_1 is a type I region, then $\displaystyle \iint_{D_1} f(x, y)\, dA = \int_a^b \int_{g_1(x)}^{g_2(x)} f(x, y)\, dy\, dx$ whenever both integrals exist. Similarly, for a type II region, D_2,
TYPE II (horizontally simple): y fixed, x varies (form $dx\, dy$)	$\displaystyle \iint_{D_2} f(x, y)\, dA = \int_c^d \int_{h_1(y)}^{h_2(y)} f(x, y)\, dx\, dy$ whenever both integrals exist.

Proof: This proof is found in most advanced calculus textbooks.   ◆

Example 1   Evaluation of a double integral

Evaluate $\displaystyle \int_0^1 \int_{x^2}^{\sqrt{x}} 160xy^3\, dy\, dx$

Solution

$$\int_0^1 \int_{x^2}^{\sqrt{x}} 160xy^3\, dy\, dx = \int_0^1 \left[40xy^4 \Big|_{y=x^2}^{y=\sqrt{x}} \right] dx \quad \textit{Since } x \textit{ is treated as a constant}$$

$$= \int_0^1 \left[40x \left(\sqrt{x} \right)^4 - 40x(x^2)^4 \right] dx$$

$$= \int_0^1 \left[40x^3 - 40x^9 \right] dx$$

$$= \left[10x^4 - 4x^{10} \right]_0^1$$

$$= 6$$

When using Fubini's theorem for nonrectangular regions, it helps to sketch the region of integration D and to find equations for all boundary curves of D. Such a sketch often provides the information needed to determine whether D is a type I or type II region (or neither, or both) and to set up the limits of integration of an iterated integral.

Example 2   Double integral over a triangular region

Let T be the triangular region enclosed by the lines $y = 0$, $y = 2x$, and $x = 1$. Evaluate the double integral

$$\iint_T (x + y)\, dA$$

using an iterated integral with: **a.** y-integration first   **b.** x-integration first

Solution

a. To set up the limits of integration in the iterated integral, we draw the graph as shown in Figure 12.8 and note that for fixed x, the variable y varies from $y = 0$ (the x-axis) to the line $y = 2x$.

These are the limits of integration for the inner integral (with respect to y first). The outer limits of integration are the numerical limits of integration for x; that is, x varies between $x = 0$ and $x = 1$.

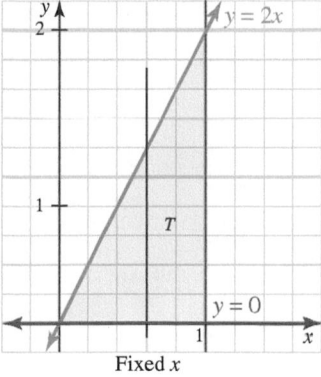

Figure 12.8   For each fixed $x\,(0 \le x \le 1)$, y varies from $y = 0$ to $y = 2x$

$$\iint_T (x + y)\, dA = \int_0^1 \int_0^{2x} (x + y)\, dy\, dx$$

$$= \int_0^1 \left[xy + \frac{1}{2}y^2 \right]_{y=0}^{y=2x} dx$$

$$= \int_0^1 \left[x(2x) + \frac{1}{2}(2x)^2 - \left(x(0) + \frac{1}{2}(0)^2 \right) \right] dx$$

$$= \int_0^1 4x^2\, dx$$

$$= \frac{4}{3}x^3 \Big|_{x=0}^{x=1}$$

$$= \frac{4}{3}$$

b. Reversing the order of integration, we see from Figure 12.9 that for each fixed y, the variable x varies (left to right) from the line $x = y/2$ to the vertical line $x = 1$. The outer limits of integration are for y as y varies from $y = 0$ to $y = 2$.

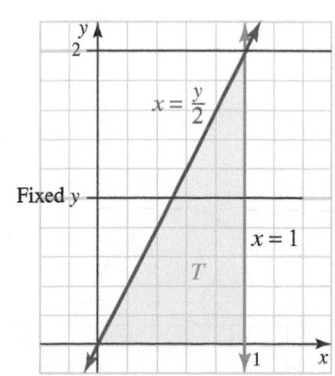

Figure 12.9   For each fixed $y\,(0 \le y \le 2)$, x varies from $y/2$ to 1

$$\iint_T (x + y)\, dA = \int_0^2 \int_{y/2}^1 (x + y)\, dx\, dy$$

$$= \int_0^2 \left[\frac{1}{2}x^2 + xy \right]_{x=y/2}^{x=1} dy$$

$$= \int_0^2 \left[\frac{1}{2} + y - \frac{y^2}{8} - \frac{y^2}{2} \right] dy$$

$$= \left[\frac{y}{2} + \frac{y^2}{2} - \frac{5y^3}{24} \right]_{y=0}^{y=2}$$

$$= \left[1 + 2 - \frac{5(8)}{24}\right] - [0]$$

$$= \frac{4}{3}$$

More on Area and Volume

Even though we can find the area between curves with single integrals, it is often easier to compute area using a double integral. If $f(x, y) \geq 0$ over a region D in the xy-plane, then $\iint_D f(x, y)\, dA$ gives the *volume of the solid* bounded above by the surface $z = f(x, y)$ and below by the region D. In the special case where $f(x, y) = 1$, we have $\iint_D 1\, dA = \text{AREA OF } D$.

> **DOUBLE INTEGRAL AS AREA AND VOLUME**   The **area** of the region D in the xy-plane is given by
>
> $$A = \iint_D dA$$
>
> If f is continuous and $f(x, y) \geq 0$ on the region D, the **volume** of the solid under the surface $z = f(x, y)$ above the region D is given by
>
> $$V = \iint_D f(x, y)\, dA$$

Example 3   Area of a region in the xy-plane using a double integral

Find the area of the region D between $y = \cos x$ and $y = \sin x$ over the interval $0 \leq x \leq \frac{\pi}{4}$ using   **a.** a single integral   **b.** a double integral

Solution

a. The graph is shown in Figure 12.10.

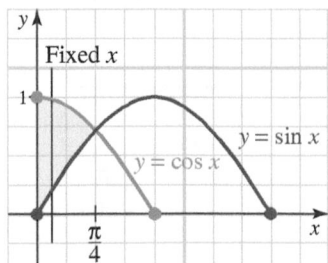

Figure 12.10   Find the area of shaded region

We find that

$$\int_0^{\pi/4} (\cos x - \sin x)\, dx = [\sin x + \cos x]_0^{\pi/4} = \sqrt{2} - 1$$

b. $A = \iint_D dA$

$$= \int_0^{\pi/4} \int_{\sin x}^{\cos x} 1\, dy\, dx$$

$$= \int_0^{\pi/4} y \Big|_{y=\sin x}^{y=\cos x}\, dx$$

$$= \sqrt{2} - 1$$

The area is $\sqrt{2} - 1 \approx 0.41$ square unit.

In comparing the single and double integral solutions for area in Example 3, you might ask, "Why bother with the double integral, because it reduces to the single integral case after one step?" The answer is that it is often easier to begin with the double integral

$$A = \iint_D dA$$

and then let the *evaluation* lead to the proper form.

Example 4 Volume using a double integral

Find the volume of the solid bounded above by the plane $z = y$ and below in the xy-plane by the part of the disk $x^2 + y^2 \le 1$ in the first quadrant.

Solution The projection in the xy-plane is shown in Figure 12.11.

We can regard D as either a type I or type II region, and because we worked with a type I (vertical) region in Example 3, we will use a type II (horizontal) region for this example. Accordingly, note that for each fixed number y between 0 and 1, x varies between $x = 0$ on the left and $x = \sqrt{1 - y^2}$ on the right. Thus,

$$
\begin{aligned}
V &= \iint_D f(x, y)\, dA \\
&= \int_0^1 \int_0^{\sqrt{1-y^2}} y\, dx\, dy \qquad f(x, y) = y \text{ is given; } dA = dx\, dy. \\
&= \int_0^1 yx \Big|_{x=0}^{x=\sqrt{1-y^2}} dy \\
&= \int_0^1 y\sqrt{1 - y^2}\, dy \qquad \text{Let } u = 1 - y^2 \text{ to integrate.} \\
&= \left[-\frac{1}{3}(1 - y^2)^{3/2} \right]\Big|_{y=0}^{y=1} \\
&= \frac{1}{3}
\end{aligned}
$$

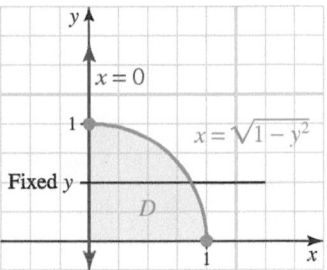

Figure 12.11 The quarter disk in Example 4

The volume is $\frac{1}{3}$ cubic unit. ◼

Choosing the Order of Integration in a Double Integral

Often a region D is both vertically and horizontally simple, and to evaluate an integral $\iint_D f(x, y)\, dA$ you have a choice between performing x-integration before y-integration, or vice-versa. In the following example, you are given one order of integration and are asked to reverse the order of integration.

Example 5 Reversing order of integration in a double integral

Reverse the order of integration in the iterated integral

$$\int_0^2 \int_1^{e^x} f(x, y)\, dy\, dx$$

Solution Draw the region D by looking at the limits of integration for both x and y in the double integral. For this example, we see that the y-integration is done first, so D is a type I region. The inner limits are

$$y = e^x \text{ (top curve)} \qquad \text{and} \qquad y = 1 \text{ (bottom curve)}$$

These are shown in Figure 12.12**a**. Next, draw the appropriate limits of integration for x:

$$x = 0 \text{ (left point)} \quad \text{and} \quad x = 2 \text{ (right point)}$$

The vertical lines $x = 0$ and $x = 2$ are also drawn in Figure 12.12**a**.

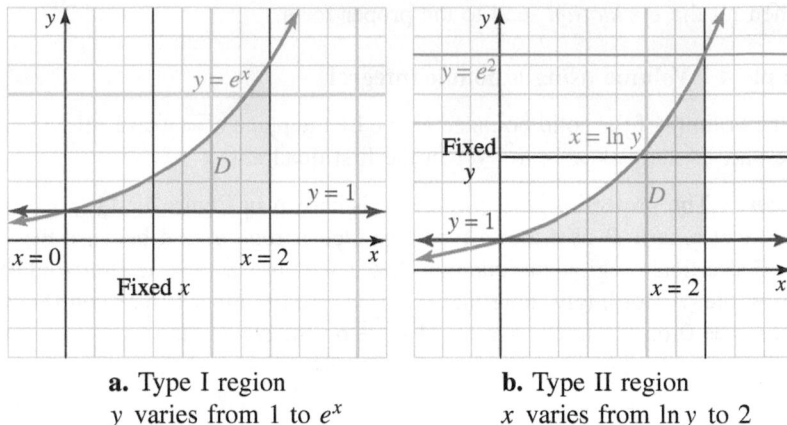

a. Type I region
y varies from 1 to e^x

b. Type II region
x varies from $\ln y$ to 2

Figure 12.12   The region of integration for Example 5

To reverse the order of integration, we need to regard D as a type II region (Figure 12.12**b**). Note that the region varies from $y = 1$ to $y = e^2$ (corresponding to where $y = e^x$ intersects $x = 0$ and $x = 2$, respectively). For each fixed y between 1 and e^2, the region extends from the curve $x = \ln y$ (that is, $y = e^x$) on the left to the line $x = 2$ on the right. Thus, reversing the order of integration, we find that the given integral becomes

$$\int_1^{e^2} \int_{\ln y}^2 f(x, y)\, dx\, dy$$

The two different ways of representing the integral in this example may be summarized as follows:

Type I: x fixed (vertically simple)	Type II: y fixed (horizontally simple)
y-integration first; *varies from* $y = 1$ *to* $y = e^x$ $\displaystyle\int_0^2 \int_1^{e^x} f(x, y)d\overset{\downarrow}{y}\, dx$ $\underset{\uparrow}{}$ *x varies from* 0 *to* 2	*x-integration first;* *varies from* $x = \ln y$ *to* $x = 2$ $\displaystyle\int_1^{e^2} \int_{\ln y}^2 f(x, y)d\overset{\downarrow}{x}\, dy$ $\underset{\uparrow}{}$ *y varies from* 1 *to* e^2

Example 6   Choosing the order of integration

The region D bounded by the parabola $y = x^2 - 2$ and the line $y = x$ is both vertically and horizontally simple. To find the area of D, would you prefer to use a type I or a type II description?

Solution   The parabola and the line intersect where

$$x^2 - 2 = x$$

$$x^2 - x - 2 = 0$$

$$(x - 2)(x + 1) = 0$$

$$x = 2, -1$$

Thus, they intersect at $(2, 2)$ and $(-1, -1)$. The graph of D is shown in Figure 12.13.

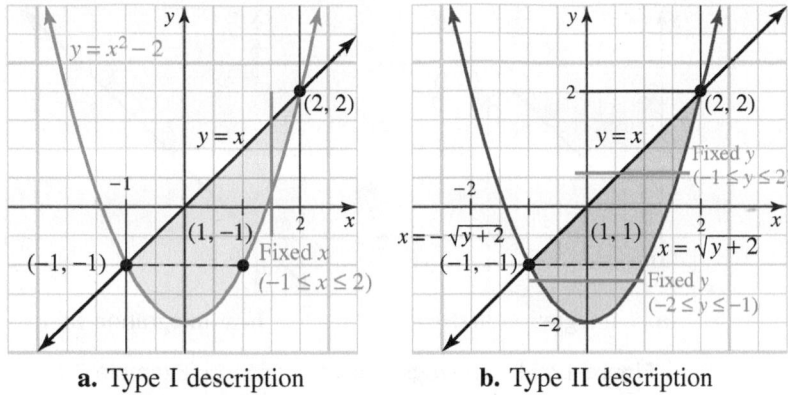

a. Type I description   **b.** Type II description

Figure 12.13   The region D in Example 6

As a type I region, D can be described as the set of all points (x, y) such that for each fixed x in $[-1, 2]$, y varies from $x^2 - 2$ to x (see Figure 12.13**a**). The area, A, of D is given by

$$A = \int_{-1}^{2} \int_{x^2-2}^{x} dy\, dx$$

$$= \int_{-1}^{2} [x - (x^2 - 2)] dx$$

$$= \left[\frac{x^2}{2} - \frac{x^3}{3} + 2x \right]_{-1}^{2}$$

$$= \frac{9}{2}$$

If, however, D, is regarded as a type II region, it is necessary to split up the description into two parts (see Figure 12.13**b**): The set of all (x, y) such that for each fixed y between -1 and 2, x varies from the line $x = y$ to the right branch of the parabola $x = \sqrt{y + 2}$, and the set of all (x, y) such that for each fixed y between -2 and -1, x varies from the left branch of the parabola $x = -\sqrt{y + 2}$ to the right branch $x = \sqrt{y + 2}$. Thus, the area of D is given by the sum of the integrals

$$A = \int_{-2}^{-1} \int_{-\sqrt{y+2}}^{\sqrt{y+2}} dx\, dy + \int_{-1}^{2} \int_{y}^{\sqrt{y+2}} dx\, dy$$

Clearly, it is preferable to use the type I representation, although both methods give the same result (you might wish to verify this fact).

As illustrated in Example 6, the shape of the region of integration D may determine which order of integration is more suitable for a given integral. However, our final example of this section shows that the integrand also plays a role in determining which order of integration is preferable.

Example 7   Evaluating a double integral by reversing the order

Evaluate $\displaystyle\int_{0}^{1} \int_{x}^{1} e^{y^2} dy\, dx$.

Solution   We cannot evaluate the integral in the given order (y-integration first) because the integrand e^{y^2} has no elementary antiderivative. We will evaluate the integral by reversing the order of integration. The region of integration is sketched in Figure 12.14**a**. Note that for any fixed x between 0 and 1, y varies from x to 1.

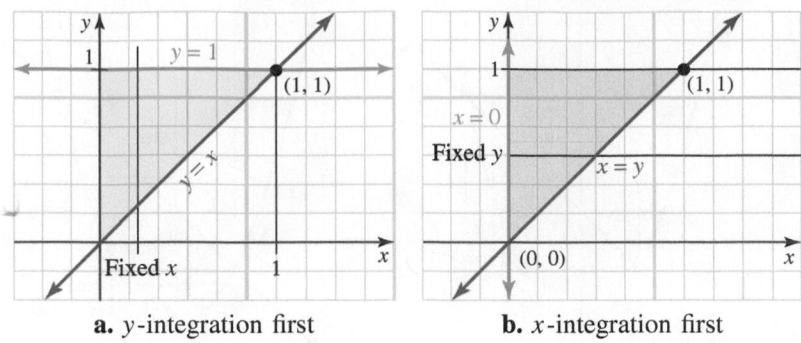

a. y-integration first   **b.** x-integration first

Figure 12.14   The region of integration for Example 7

To reverse the order of integration, observe that for each fixed y between 0 and 1, x varies from 0 to y, as shown in Figure 12.14**b**.

$$\int_0^1 \int_x^1 e^{y^2} \, dy \, dx = \int_0^1 \int_0^y e^{y^2} \, dx \, dy$$

$$= \int_0^1 xe^{y^2} \Big|_{x=0}^{x=y} \, dy$$

$$= \int_0^1 ye^{y^2} \, dy \qquad Let\ u = y^2.$$

$$= \left[\frac{1}{2}e^{y^2}\right]_0^1$$

$$= \frac{1}{2}(e^1 - e^0)$$

$$= \frac{1}{2}(e - 1)$$

PROBLEM SET 12.2

Level 1

1. ■ What does this say? Describe the process for finding volume using a double integral.
2. ■ What does this say? What is Fubini's theorem?

Sketch the region of integration in Problems 3-12, and compute the double integral (either in the order of integration given or with the order reversed).

3. $\displaystyle\int_0^4 \int_0^{4-x} xy \, dy \, dx$

4. $\displaystyle\int_0^4 \int_{x^2}^{4x} dy \, dx$

5. $\displaystyle\int_0^{2\sqrt{2}} \int_{y^2/4}^{\sqrt{12-y^2}} dx \, dy$

6. $\displaystyle\int_{-2}^1 \int_{y^2+4y}^{3y+2} dx \, dy$

7. $\displaystyle\int_0^1 \int_0^x (x^2 + 2y^2) \, dy \, dx$

8. $\displaystyle\int_{-1}^1 \int_{-1}^x (3x + 2y) \, dy \, dx$

9. $\displaystyle\int_0^2 \int_0^{\sin x} y \cos x \, dy \, dx$

10. $\displaystyle\int_0^{\pi/2} \int_0^{\sin x} e^y \cos x \, dy \, dx$

11. $\displaystyle\int_0^{\pi/3} \int_0^{y^2} \frac{1}{y} \sin \frac{x}{y} dx \, dy$

12. $\displaystyle\int_0^1 \int_0^y y^2 e^{xy} \, dx \, dy$

Evaluate the double integrals in Problems 13-16.

13. $\displaystyle\int_0^1 \int_0^{y^3} e^{x/y} \, dx \, dy$

14. $\displaystyle\int_{\pi/3}^{\pi} \int_0^{y^2} \frac{1}{y} \cos \frac{x}{y} dx \, dy$

15. $\displaystyle\int_0^1 \int_{\tan^{-1} y}^{\pi/4} \sec x \, dx \, dy$

16. $\displaystyle\int_0^1 \int_{\sin^{-1} y}^{\pi/2} \sqrt{1 + \cos^2 x} \cos x \, dx \, dy$

Evaluate the double integral given in Problems 17-24 for the specified region of integration D.

17. $\displaystyle\iint_D (x + y) \, dA$

   D is the triangle with vertices $(0, 0)$, $(0, 1)$, $(1, 1)$.

18. $\displaystyle\iint_D (2y - x) \, dA$

   D is the region bounded by $y = x^2$ and $y = 2x$.

19. $\displaystyle\iint_D y\,dA$

D is the region bounded by $y = \sqrt{x}$, $y = 2 - x$, and $y = 0$.

20. $\displaystyle\iint_D 4x\,dA$

D is the region bounded by $y = 4 - x^2$, $y = 3x$, and $x = 0$.

21. $\displaystyle\iint_D \frac{dA}{y^2 + 1}$

D is the triangle bounded by $x = 2y$, $y = -x$, and $y = 2$.

22. $\displaystyle\iint_D x\sqrt{y^2 - x^2}\,dx\,dy$

D is the triangle with vertices $(0,0)$, $(1,1)$, and $(0,1)$.

23. $\displaystyle\iint_D \frac{\sin x}{x}\,dx\,dy$

D is the triangle bounded by the lines $y = 0$, $x = \pi$, and $y = x$.

24. $\displaystyle\iint_D \frac{4 - 2x}{y^2}\,dy\,dx$

D is the triangle with vertices $(0,-2)$, $(3,-2)$, and $(0,4)$.

Sketch the region of integration in Problems 25-30, and then compute the integral in two ways:
   a. *with the given order of integration, and*
   b. *with the order of integration reversed.*

25. $\displaystyle\int_0^4 \int_0^{4-x} xy\,dy\,dx$ **26.** $\displaystyle\int_0^4 \int_{x^2}^{4x} dy\,dx$

27. $\displaystyle\int_0^1 \int_{-x^2}^{x^2} dy\,dx$ **28.** $\displaystyle\int_1^e \int_0^{\ln x} xy\,dy\,dx$

29. $\displaystyle\int_0^{2\sqrt{3}} \int_{y^2/6}^{\sqrt{16-y^2}} dx\,dy$ **30.** $\displaystyle\int_{-2}^1 \int_{y^2+4y}^{3y+2} dx\,dy$

Sketch the region of integration in problems 31-36, and write an equivalent integral with the order of integration reversed.

31. $\displaystyle\int_0^1 \int_0^{2y} f(x,y)\,dx\,dy$ **32.** $\displaystyle\int_0^1 \int_{x^2}^{\sqrt{x}} f(x,y)\,dy\,dx$

33. $\displaystyle\int_0^3 \int_{y/3}^{\sqrt{4-y}} f(x,y)\,dx\,dy$ **34.** $\displaystyle\int_0^1 \int_x^{2-x} f(x,y)\,dy\,dx$

35. $\displaystyle\int_0^7 \int_{x^2-6x}^x f(x,y)\,dy\,dx$ **36.** $\displaystyle\int_0^1 \int_{\tan^{-1}x}^{\pi/4} f(x,y)\,dy\,dx$

Level 2

Set up a double integral for the volume of the solid region described in Problems 37-43.

37. The tetrahedron that lies in the first octant and is bounded by the coordinate planes and the plane $z = 7 - 3x - 2y$

38. The solid bounded above by the paraboloid $z = 6 - 2x^2 - 3y^2$ and below by the plane $z = 0$

39. The solid that lies inside both the cylinder $x^2 + y^2 = 3$ and the sphere $x^2 + y^2 + z^2 = 7$

40. The solid that lies inside both the sphere $x^2 + y^2 + z^2 = 3$ and the paraboloid $2z = x^2 + y^2$

41. The solid bounded by the ellipsoid

$$\frac{x^2}{a^2} + \frac{y^2}{b^2} + \frac{z^2}{c^2} = 1$$

42. The solid bounded above by the plane $z = 2 - 3x - 5y$ and below by the region shown in Figure 12.15.

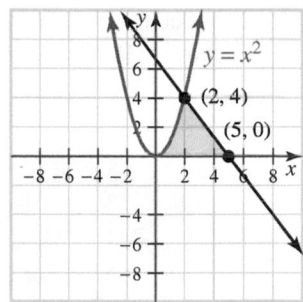

Figure 12.15   Region for Problem 42

43. The solid that remains when a square hole of side 2 is drilled through the center of a sphere of radius $\sqrt{2}$.

Express the area of the region D bounded by the curve given in Problems 44-45 as a double integral in two different ways (with different orders of integration). Evaluate one of these integrals to find the area.

44. Let D denote the region in the first quadrant of the xy-plane that is bounded by the curves $y = \dfrac{4}{x^2}$ and $y = 5 - x^2$.

45. Let D denote the region bounded by the ellipse $\dfrac{x^2}{a^2} + \dfrac{y^2}{b^2} = 1$.

46. Find the volume under the surface $z = x + y + 2$ and above the region D bounded by the curves $y = x^2$ and $y = 2$.

47. Find the volume under the plane $z = 4x$ and above the region D bounded by $y = x^2$, $y = 0$, and $x = 1$.

48. Evaluate

$$\int_0^1 \int_0^y (x^2 + y^2)\,dx\,dy + \int_1^2 \int_0^{2-y} (x^2 + y^2)\,dx\,dy$$

by reversing the order of integration.

49. Reverse the order of integration in

$$\int_1^2 \int_x^{x^3} f(x, y)\, dy\, dx + \int_2^8 \int_x^8 f(x, y)\, dy\, dx$$

50. Evaluate $\iint_D xy\, dA$, where D is the triangular region in the xy-plane with vertices $(0, 0)$, $(1, 0)$, and $(4, 1)$.

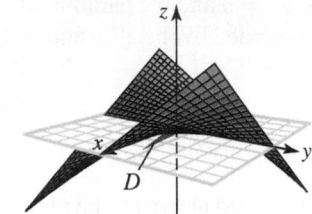

51. Compute $\iint_D (x^2 - xy - 1)\, dA$, where D is the triangular region bounded by the lines $x - 2y + 2 = 0$, $x + 3y - 3 = 0$, and $y = 0$.

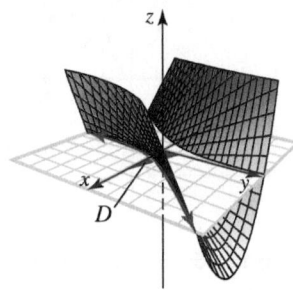

52. Find the volume under the plane $z = x + 2y + 4$ and above the region D bounded by the lines $y = 2x$, $y = 3 - x$, and $y = 0$.

53. Find the volume under the plane $3x + y - z = 0$ and above the elliptic region bounded by $4x^2 + 9y^2 \le 36$, with $x \ge 0$ and $y \ge 0$.

54. Find the volume under the surface $z = x^2 + y^2$ and above the square region bounded by $|x| \le 1$ and $|y| \le 1$.

Evaluate each integral in Problems 55-57 by relating it to the volume of a simple solid.

55. Evaluate

$$\iint_R \left(3 - \sqrt{x^2 + y^2}\right) dA$$

where R is the disk $x^2 + y^2 \le 9$ in the xy-plane.

56. Evaluate

$$\iint_R \left(1 - \sqrt{x^2 + z^2}\right) dA$$

where R is the disk $x^2 + z^2 \le 1$ in the xz-plane.

57. Evaluate

$$\iint_R \left(4 - \sqrt{y^2 + z^2}\right) dA$$

where R is the disk $y^2 + z^2 \le 16$ in the yz-plane.

Level 3

58. Show that if $f(x, y)$ is continuous on a region D and $m \le f(x, y) \le M$ for all points (x, y) in D, then

$$mA \le \iint_D f(x, y)\, dA \le MA$$

where A is the area of D.

59. Use the result of Problem 58 to estimate the value of the double integral

$$\iint_D e^{y \sin x}\, dA$$

where D is the triangle with vertices $(-1, 0)$, $(2, 0)$, and $(0, 1)$.

60. The region D shown in Figure 12.16 is the union of a type I and a type II region. Evaluate

$$\iint_D xy\, dA$$

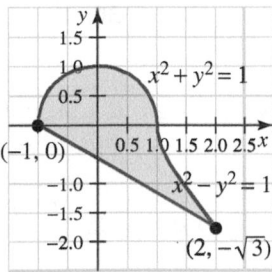

Figure 12.16 Union of type I and type II regions

12.3 DOUBLE INTEGRALS IN POLAR COORDINATES

IN THIS SECTION: *Change of variables to polar form, area and volume in polar form*
In general, changing variables in a double integral is more complicated than in a single integral. In this section, we focus our attention on using polar coordinates in a double integral.

Change of Variables to Polar Form

Polar coordinates are used in double integrals primarily when the integrand or the region of integration (or both) have relatively simple polar descriptions. As a preview of the ideas we plan to explore, consider Figure 12.17 and let us examine the double integral

$$\iint\limits_{R} (x^2 + y^2 + 1)\, dA$$

where R is the region (disk) in the xy-plane bounded by the circle $x^2 + y^2 = 4$.

Interpreting R as a vertically simple region (type I), we see that for each fixed x between -2 and 2, y varies from the lower boundary semicircle with equation $y = -\sqrt{4-x^2}$ to the upper semicircle $y = \sqrt{4-x^2}$, as shown in Figure 12.18a.

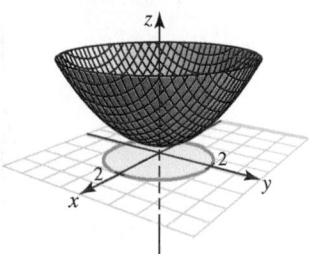

Figure 12.17 Graph of a surface and a region R

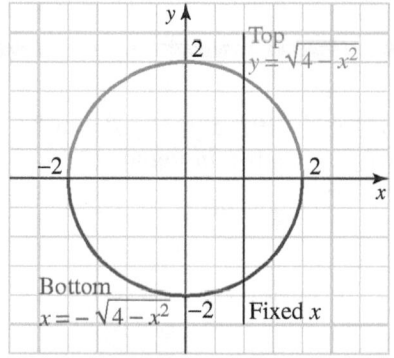

a. Type I description

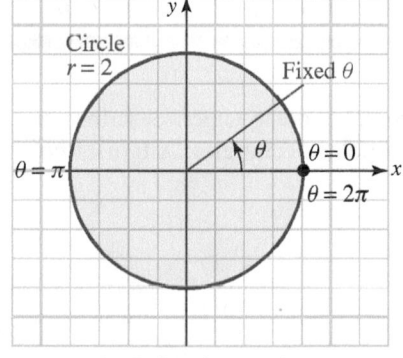

b. Polar description

Figure 12.18 Two interpretations of a region R

Using the type I description, we have

$$\iint\limits_{D} (x^2 + y^2 + 1)\, dA = \int_{-2}^{2} \int_{-\sqrt{4-x^2}}^{\sqrt{4-x^2}} (x^2 + y^2 + 1)\, dy\, dx$$

The iterated integral on the right is difficult to evaluate, but both the integrand and the domain of integration can be represented quite simply in terms of polar coordinates. Specifically, using the polar conversion formulas

$$x = r\cos\theta \qquad y = r\sin\theta \qquad r = \sqrt{x^2 + y^2} \qquad \tan\theta = \frac{y}{x}$$

we find that the integrand $f(x, y) = x^2 + y^2 + 1$ can be rewritten

$$f(r\cos\theta, r\sin\theta) = (r\cos\theta)^2 + (r\sin\theta)^2 + 1 = r^2 + 1$$

and the region of integration D is just the interior of the circle $r = 2$. Thus, D can be described as the set of all points (r, θ) so that for each fixed angle θ between 0 and 2π, r varies from the origin ($r = 0$) to the circle $r = 2$, as shown in Figure 12.18b.

But what is the differential of integration dA in polar coordinates? Can we simply substitute "$dr\, d\theta$" for dA and perform the integration with respect to r and θ? The answer is no, and the correct formula for expressing a given double integral in polar form is given in the following theorem.

Theorem 12.4   Double integral in polar coordinates

If f is continuous in the polar region D described by $0 \leq r_1(\theta) \leq r \leq r_2(\theta)$, $\alpha \leq \theta \leq \beta$ where $0 \leq \beta - \alpha \leq 2\pi$, then

$$\iint\limits_{D} f(r,\ \theta)\, dA = \int_{\alpha}^{\beta} \int_{r_1(\theta)}^{r_2(\theta)} f(r,\theta)\, r\, dr\, d\theta$$

■ **W**hat this says The procedure for changing from a Cartesian integral

$$\iint\limits_{R} f(x, y)\, dA$$

into a polar integral requires two steps. First, substitute $x = r\cos\theta$, $y = r\sin\theta$, and $dx\, dy = r\, dr\, d\theta$ into the Cartesian integral. Then convert the region of integration R to polar form D. Thus,

$$\iint\limits_{R} f(x, y)\, dA = \iint\limits_{D} f(r\cos\theta, r\sin\theta)\, r\, dr\, d\theta$$

Proof: A polar region described by $r_1(\theta) \leq r \leq r_2(\theta)$, $\alpha \leq \theta \leq \beta$ can be subdivided into **polar rectangles**. A typical polar rectangle is shown in Figure 12.19.

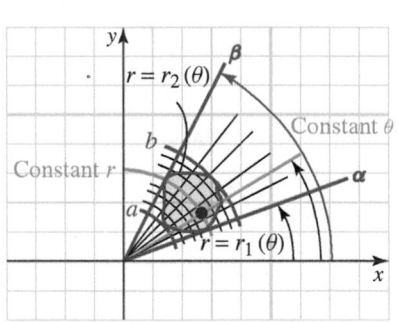

a. A partition of a region into polar rectangles

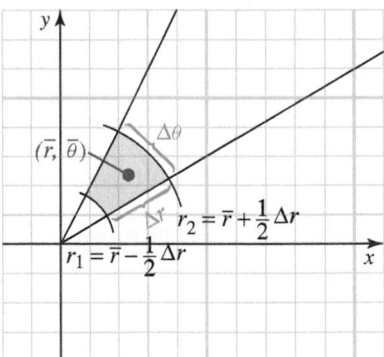

b. A typical polar rectangle in the partition

Figure 12.19   A polar rectangle

We begin by subdividing the region of integration into polar rectangles. Next we pick an arbitrary polar-form point (r_k^*, θ_k^*) in each polar rectangle in the partition and then take the limit of an appropriate Riemann sum

$$\sum_{k=1}^{N} f(r_k^*, \theta_k^*)\, \Delta A_k$$

where ΔA_k is the area of the kth polar rectangle.

To find the area of a typical polar rectangle, let (r_k^*, θ_k^*) be the center of the polar rectangle—that is, the point midway between the arcs and rays that form the polar rectangle, as shown in Figure 12.19**b**. If the circular arcs that bound the polar rectangle are Δr apart, then the arcs are given by

$$r_1 = r_k^* - \frac{1}{2}\Delta r_k \qquad \text{and} \qquad r_2 = r_k^* + \frac{1}{2}\Delta r_k$$

In Section 6.4 we used the fact that a circular section of radius r and central angle θ has area $\frac{1}{2}r^2\theta$. Thus, a typical polar rectangle has area

$$\Delta A_k = \left[\underbrace{\frac{1}{2}(r_k^* + \frac{1}{2}\Delta r_k)^2}_{\text{Radius of outside arc}} - \underbrace{\frac{1}{2}(r_k^* - \frac{1}{2}\Delta r_k)^2}_{\text{Radius of inside arc}} \right] \Delta\theta_k = r_k^* \Delta r_k \, \Delta\theta_k$$

Finally, we compute the given double integral in polar form by taking the limit

$$\iint\limits_{D} f(r,\theta)dA = \lim_{\|P\|\to 0} \sum_{k=1}^{N} f(r_k^*, \theta_k^*)\Delta A_k$$

$$= \lim_{\|P\|\to 0} \sum_{k=1}^{N} f(r_k^*, \theta_k^*) \, r_k^* \Delta r_k \, \Delta\theta_k$$

$$= \int_{\alpha}^{\beta} \int_{r_1(\theta)}^{r_2(\theta)} f(r,\theta) \, r \, dr \, d\theta \qquad \blacklozenge$$

Preview It can be shown that under reasonable conditions, the change of variable $x = x(u,v)$, $y = y(u,v)$ transforms the integral $\iint f(x,y) \, dA$ into $\iint f(u,v) |J(u,v)| \, du \, dv$, where

$$J(u,v) = \begin{vmatrix} \dfrac{\partial x}{\partial u} & \dfrac{\partial x}{\partial v} \\[2mm] \dfrac{\partial y}{\partial u} & \dfrac{\partial y}{\partial v} \end{vmatrix}$$

This determinant is known as the **Jacobian** of the transformation. In the case of polar coordinates, we have $x = r\cos\theta$ and $y = r\sin\theta$, so the Jacobian is

$$J(r,\theta) = \begin{vmatrix} \dfrac{\partial}{\partial r}(r\cos\theta) & \dfrac{\partial}{\partial \theta}(r\cos\theta) \\[2mm] \dfrac{\partial}{\partial r}(r\sin\theta) & \dfrac{\partial}{\partial \theta}(r\sin\theta) \end{vmatrix}$$

$$= \begin{vmatrix} \cos\theta & -r\sin\theta \\ \sin\theta & r\cos\theta \end{vmatrix}$$

$$= r\cos^2\theta + r\sin^2\theta$$

$$= r$$

This yields the result of Theorem 12.4:

$$\iint\limits_{R} f(x,y) \, dA = \iint\limits_{D} f(r,\theta) \, r \, dr \, d\theta = \int_{\alpha}^{\beta} \int_{r_1(\theta)}^{r_2(\theta)} f(r\cos\theta, r\sin\theta) \, r \, dr \, d\theta$$

We discuss this more completely in Section 12.8.

If you carefully compare the result in the preview box with the result of Theorem 12.4, you will see that they are not *exactly* the same. The region D in Theorem 12.4 already is in polar coordinates. The more common situation is that we are given f and a region R in rectangular coordinates that we need to change to polar coordinates.

Area and Volume in Polar Form

You will need to be familiar with the graphs of many polar-form curves. These can be found in Table 6.2 on page 451. We now present the example we promised in the introduction to this section.

Example 1   Double integral in polar form

Evaluate $\iint_R (x^2 + y^2 + 1)\,dA$, where D is the region inside the circle $x^2 + y^2 = 4$.

Solution   In this example, the region R is given in rectangular form. We will describe this as a polar region D. Earlier, we observed that in D, for each fixed angle θ between 0 and 2π, r varies from $r = 0$ (the pole) to $r = 2$ (the circle). Thus,

$$\iint_R \underbrace{(x^2 + y^2 + 1)}_{f(x,y)}dA = \int_0^{2\pi} \int_0^2 \underbrace{(r^2 + 1)}_{f(r\cos\theta,\, r\sin\theta)}\ \underbrace{r\,dr\,d\theta}_{dA}$$

$$= \int_0^{2\pi} \int_0^2 (r^3 + r)\,dr\,d\theta$$

$$= \int_0^{2\pi} \left[\frac{r^4}{4} + \frac{r^2}{2}\right]_0^2 d\theta$$

$$= \int_0^{2\pi} 6\,d\theta$$

$$= 6\theta \Big|_0^{2\pi}$$

$$= 12\pi$$

Example 2   Computing area in polar form using a double integral

Compute the area of the region D bounded above by the line $y = x$ and below by the circle $x^2 + y^2 - 2y = 0$.

Solution   The circle $x^2 + y^2 - 2y = 0$ and the line $y = x$ are shown in Figure 12.20. The polar form for the line $y = x$ is $\theta = \pi/4$ and for the circle is

$$x^2 + y^2 = 2y$$

$$r^2 = 2r\sin\theta$$

$$r = 2\sin\theta$$

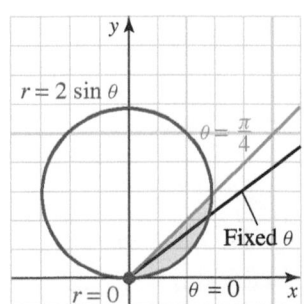

Figure 12.20   The region D

Thus, the region D is determined as r varies from 0 (the pole) to $2\sin\theta$ (the circle) for θ between 0 and $\pi/4$. The area is given by the integral

$$A = \iint_D dA$$

$$= \int_0^{\pi/4} \int_0^{2\sin\theta} r\,dr\,d\theta$$

$$= \int_0^{\pi/4} \left[\frac{1}{2}r^2\right]_{r=0}^{r=2\sin\theta} d\theta$$

$$= \int_0^{\pi/4} \frac{1}{2} \left(4 \sin^2 \theta \right) d\theta$$

$$= 2 \int_0^{\pi/4} \frac{1 - \cos 2\theta}{2} d\theta$$

$$= \left[\theta - \frac{\sin 2\theta}{2} \right] \Big|_0^{\pi/4}$$

$$= \frac{\pi - 2}{4}$$ ▪

Example 3   Volume in polar form

Use a polar double integral to show that a sphere of radius a has volume $\frac{4}{3}\pi a^3$.

Solution   We will compute the required volume by doubling the volume of the solid hemisphere $x^2 + y^2 + z^2 \leq a^2$, with $z \geq 0$. This hemisphere (see Figure 12.21) may be regarded as a solid bounded below by the circular disk $x^2 + y^2 \leq a^2$ and above by the spherical surface.

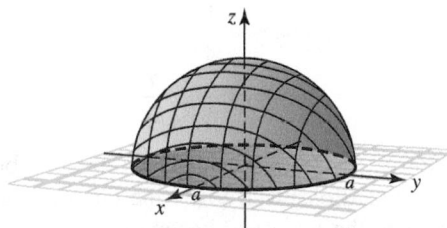

Figure 12.21   A solid hemisphere bounded above by the sphere and below by a disk

We need to change the equation of the hemisphere to polar form:

In rectangular form: $z = \sqrt{a^2 - x^2 - y^2}$

In polar form: $z = \sqrt{a^2 - r^2}$ *Because $r^2 = x^2 + y^2$*

Describing the disk D in polar terms, we see that for each fixed θ between 0 and 2π, r varies from the origin to the circle $x^2 + y^2 = a^2$, which has the polar equation $r = a$, as shown in Figure 12.22.

Thus, the volume is given by the integral

$$V = 2 \iint_D z \, dA$$ *Rectangular form*

$$= 2 \int_0^{2\pi} \int_0^a \sqrt{a^2 - r^2} \, r \, dr \, d\theta$$ $\boxed{\text{Let } u = a^2 - r^2; \, du = -2r \, dr}$

$$= 2 \int_0^{2\pi} \left[-\frac{1}{3}(a^2 - r^2)^{3/2} \right] \Big|_{r=0}^{r=a} d\theta$$

$$= -\frac{2}{3} \int_0^{2\pi} \left[(a^2 - a^2)^{3/2} - (a^2 - 0)^{3/2} \right] d\theta$$

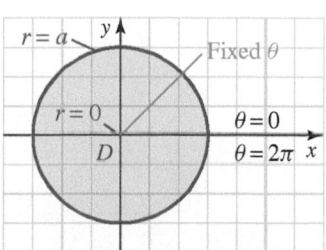

Figure 12.22   The disk D for θ between 0 and 2π, r varies from 0 to a

$$= \frac{2}{3} \int_0^{2\pi} a^3 d\theta$$

$$= \frac{2}{3} a^3 \theta \Big|_0^{2\pi}$$

$$= \frac{4}{3} \pi a^3$$

Example 4 Region of integration between two polar curves

Evaluate $\iint_D \frac{1}{x} \, dA$, where D is the region that lies inside the circle $r = 3 \cos \theta$ and outside the cardioid $r = 1 + \cos \theta$.

Solution Begin by sketching the given curves, as shown in Figure 12.23.
Next, find the points of intersection:

$$3 \cos \theta = 1 + \cos \theta$$

$$\cos \theta = \frac{1}{2}$$

$$\theta = -\frac{\pi}{3}, \frac{\pi}{3}$$

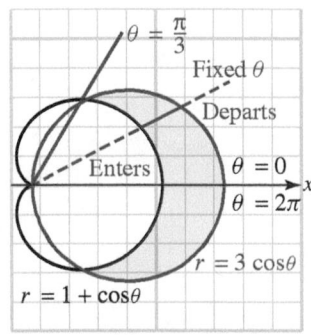

Figure 12.23 Interactive
Region D

Notice that D is the region such that, for each fixed angle θ between $-\pi/3$ and $\pi/3$, r varies from $1 + \cos \theta$ (the cardioid) to $3 \cos \theta$ (the circle). This gives us the limits of integration, and in polar form, the integrand becomes:

$$\underbrace{\frac{1}{x}}_{Rectangular\ form} = \underbrace{\frac{1}{r \cos \theta}}_{Polar\ form}$$

Thus,

$$\iint_D \frac{1}{x} \, dA = \iint_D \frac{1}{r \cos \theta} r \, dr \, d\theta$$

$$= \int_{-\pi/3}^{\pi/3} \int_{1+\cos\theta}^{3\cos\theta} \frac{1}{\cos\theta} dr \, d\theta$$

$$= 2 \int_0^{\pi/3} \left[\frac{r}{\cos\theta} \right] \Big|_{r=1+\cos\theta}^{r=3\cos\theta} d\theta \qquad \textit{By symmetry}$$

$$= 2 \int_0^{\pi/3} \frac{1}{\cos\theta} \left[3\cos\theta - (1 + \cos\theta) \right] d\theta$$

$$= 2 \int_0^{\pi/3} (2 - \sec\theta) \, d\theta$$

$$= 2 \left[2\theta - \ln|\sec\theta + \tan\theta| \right] \Big|_0^{\pi/3}$$

$$= \frac{4\pi}{3} - 2\ln(2 + \sqrt{3})$$

Example 5   Converting an integral to polar form

Evaluate $\int_0^2 \int_0^{\sqrt{2x-x^2}} y\sqrt{x^2+y^2}\,dy\,dx$ by converting to polar coordinates.

Solution   The region of integration D is the set of all points (x, y) such that for each point x in the interval $[0, 2]$, y varies from $y = 0$ to $y = \sqrt{2x - x^2}$. Note that

$$y = \sqrt{2x - x^2}$$
$$y^2 = 2x - x^2$$
$$x^2 - 2x + y^2 = 0$$
$$(x - 1)^2 + y^2 = 1$$

This is the equation of a circle of radius 1 centered at $(1, 0)$. Thus, $y = \sqrt{2x - x^2}$ is the semicircle shown in Figure 12.24.

The top boundary of this region is the semicircle

$$y = \sqrt{2x - x^2}$$
$$y^2 = 2x - x^2$$
$$x^2 + y^2 = 2x$$

which has the polar form

$$r^2 = 2r \cos \theta$$
$$r = 2 \cos \theta$$

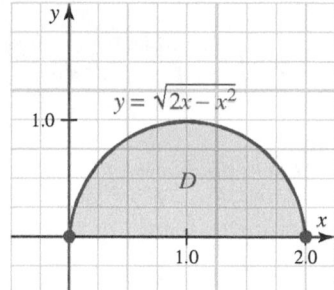

Figure 12.24   Region of integration for Example 5

In polar coordinates, D is the region in which r varies from 0 to $2 \cos \theta$ for θ between 0 and $\frac{\pi}{2}$. Thus, the integral is

$$\int_0^2 \int_0^{\sqrt{2x-x^2}} y\sqrt{x^2+y^2}\,dy\,dx = \iint\limits_{D} y\sqrt{x^2+y^2}\,dA$$

$$= \int_0^{\pi/2} \int_0^{2\cos\theta} (r \sin\theta) r (r\,dr\,d\theta)$$

$$= \int_0^{\pi/2} \sin\theta \left[\frac{r^4}{4} \right]_0^{2\cos\theta} d\theta$$

$$= \frac{1}{4} \int_0^{\pi/2} 16 \cos^4 \theta \sin\theta\,d\theta$$

$$= 4 \left[-\frac{1}{5} \cos^5 \theta \right]_0^{\pi/2}$$

$$= \frac{4}{5}$$

PROBLEM SET 12.3

Level 1

In Problems 1–8, sketch the region D and then evaluate the double integral $\iint\limits_{D} f(r,\theta)\,dr\,d\theta$.

1. $\displaystyle\int_{0}^{\pi/2}\int_{1}^{3} re^{-r^2}\,dr\,d\theta$ **2.** $\displaystyle\int_{0}^{\pi/2}\int_{1}^{2}\sqrt{4-r^2}\,r\,dr\,d\theta$

3. $\displaystyle\int_{0}^{\pi}\int_{0}^{4} r^2\sin^2\theta\,dr\,d\theta$ **4.** $\displaystyle\int_{0}^{\pi/2}\int_{1}^{3} r^2\cos^2\theta\,dr\,d\theta$

5. $\displaystyle\int_{0}^{2\pi}\int_{0}^{4} 2r^2\cos\theta\,dr\,d\theta$ **6.** $\displaystyle\int_{0}^{2\pi}\int_{0}^{1-\sin\theta}\cos\theta\,dr\,d\theta$

7. $\displaystyle\int_{0}^{\pi/2}\int_{0}^{2\sin\theta} dr\,d\theta$ **8.** $\displaystyle\int_{0}^{\pi}\int_{0}^{1+\sin\theta} dr\,d\theta$

Use a double integral to find the area of the shaded region in Problems 9–22.

9. $r = 4$

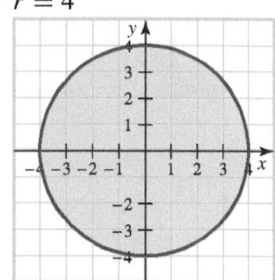

10. $r = 2\cos\theta$

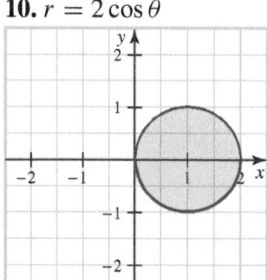

11. $r = 2(1-\cos\theta)$

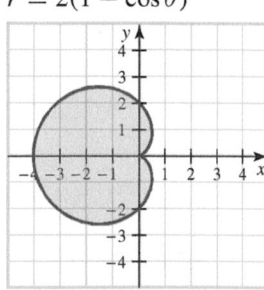

12. $r = 1 + \sin\theta$

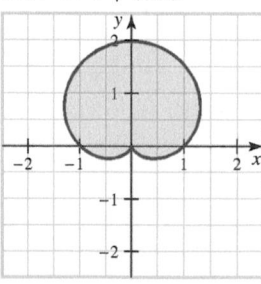

13. $r = 4\cos 3\theta$

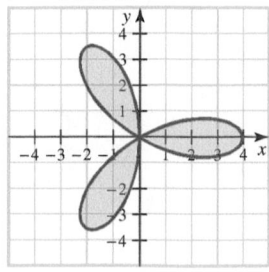

14. $r = 5\sin 2\theta$

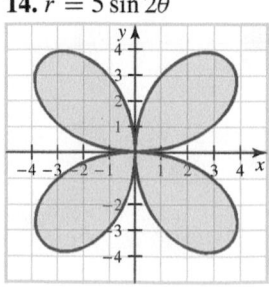

15. $r = \cos 2\theta$

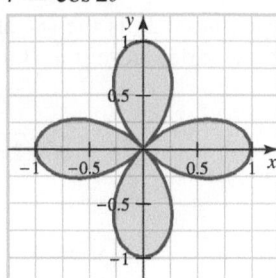

16. $r = 1$ and $r = 2\sin\theta$

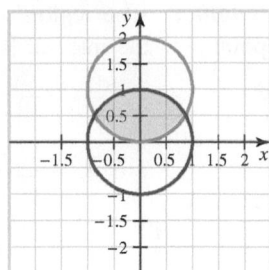

17. $r = 1$ and $r = 2\sin\theta$

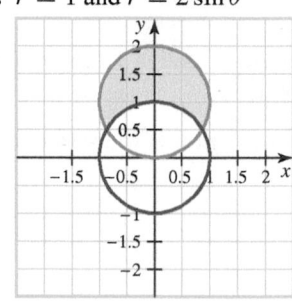

18. $r = 1$ and $r = 1 + \cos\theta$

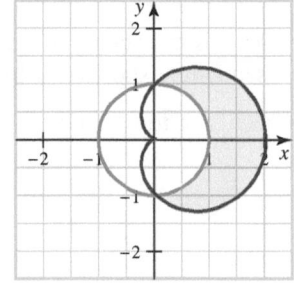

19. $r = 1$ and $r = 1 + \cos\theta$

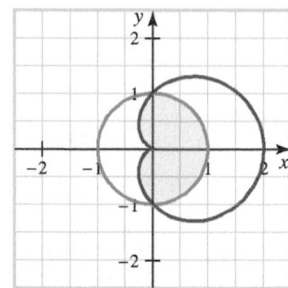

20. $r = 1$ and $r = 1 + \cos\theta$

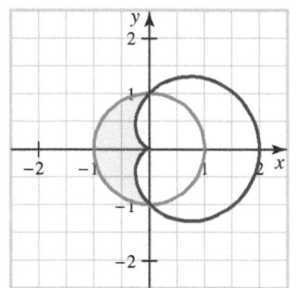

21. $r = 3\cos\theta$ and $r = 1 + \cos\theta$

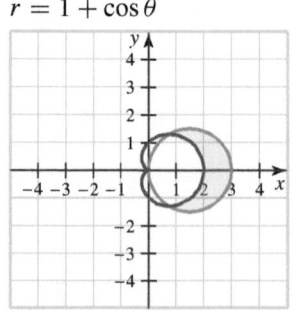

22. $r = 3$ and $r = 2 - \cos\theta$

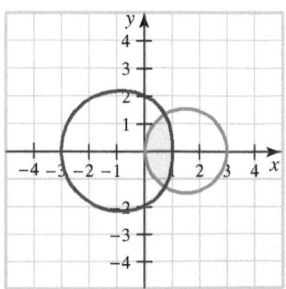

23. Use a double integral to find the area inside the inner loop of the limaçon $r = 1 + 2\cos\theta$.

24. Use a double integral to find the area bounded by the parabola

$$r = \frac{5}{1+\cos\theta}$$

and the lines $\theta = 0$, $\theta = \frac{\pi}{3}$, and $r = \frac{5}{2}\sec\theta$.

Use polar coordinates in Problems 25-30 to find $\iint_D f(x, y)\, dA$, where D is the disk defined by

$x^2 + y^2 \leq a^2$ *for constant $a > 0$.*

25. $f(x, y) = y^2$ **26.** $f(x, y) = x^2 + y^2$

27. $f(x, y) = \dfrac{1}{a^2 + x^2 + y^2}$ **28.** $f(x, y) = e^{-(x^2+y^2)}$

29. $f(x, y) = \dfrac{1}{a + \sqrt{x^2 + y^2}}$

30. $f(x, y) = \ln(a^2 + x^2 + y^2)$

Use polar coordinates in Problems 31-36 to evaluate the given double integral.

31. $\iint_D y\, dA,$

     where D is the disk $x^2 + y^2 \leq 4$

32. $\iint_D (x^2 + y^2)\, dA,$

     where D is the region in the first quadrant bounded by the x-axis, the line $y = x$, and the circle $x^2 + y^2 = 1$

33. $\iint_D e^{x^2+y^2}\, dA,$

     where D is the region inside the circle $x^2 + y^2 = 9$

34. $\iint_D \sqrt{x^2 + y^2}\, dA,$

     where D is the region inside the circle $(x - 1)^2 + y^2 = 1$ in the first quadrant

35. $\iint_D \ln(x^2 + y^2 + 2)\, dA,$

     where D is the region inside the circle $x^2 + y^2 = 4$ in the first quadrant

36. $\iint_D \sin(x^2 + y^2)\, dA,$

     where D is the region bounded by the circles $x^2 + y^2 = 1$ and $x^2 + y^2 = 4$, and the lines $y = 0$ and $x = \sqrt{3}\, y$

In Problems 37-44, evaluate the given integral by converting to polar coordinates.

37. $\displaystyle\int_0^3 \int_0^{\sqrt{9-x^2}} x\, dy\, dx$

38. $\displaystyle\int_0^2 \int_y^{\sqrt{8-y^2}} \dfrac{1}{\sqrt{1 + x^2 + y^2}}\, dx\, dy$

39. $\displaystyle\int_0^2 \int_0^{\sqrt{4-y^2}} e^{x^2+y^2}\, dx\, dy$

40. $\displaystyle\int_0^3 \int_0^{\sqrt{9-x^2}} \cos(x^2 + y^2)\, dy\, dx$

41. $\displaystyle\int_0^4 \int_0^{\sqrt{4y-y^2}} \dfrac{1}{\sqrt{x^2 + y^2}}\, dx\, dy$

42. $\displaystyle\int_0^1 \int_x^{\sqrt{x}} (x^2 + y^2)^{3/2}\, dy\, dx$

43. $\displaystyle\int_0^2 \int_0^{\sqrt{2x-x^2}} \dfrac{x - y}{x^2 + y^2}\, dy\, dx$

44. $\displaystyle\int_{-3}^3 \int_{-\sqrt{9-x^2}}^{\sqrt{9-x^2}} \ln(x^2 + y^2 + 9)\, dy\, dx$

Level 2

45. Find the volume of the solid bounded by the paraboloid $z = 4 - x^2 - y^2$ and the xy-plane.

46. Find the volume of the solid bounded above by the cone $z = 6\sqrt{x^2 + y^2}$ and below by the circular region $x^2 + y^2 \leq a^2$ in the xy-plane, where a is a positive constant.

47. Use polar coordinates to evaluate $\iint_D xy\, dA$, where D is the intersection of the circular disks $r \leq 4\cos\theta$ and $r \leq 4\sin\theta$. Sketch the region of integration.

48. Let D be the region formed by intersecting the regions (in the xy-plane) described by $y \leq x, y \geq 0$, and $x \leq 1$.

     a. Evaluate $\iint_D dA$ as an iterated integral in Cartesian coordinates.

     b. Evaluate $\iint_D dA$ in terms of polar coordinates.

49. Example 9 of Section 6.3 used a single integral to find the area of the region common to the circles $r = a\cos\theta$ and $r = a\sin\theta$. Rework this example using double integrals.

50. Example 10 of Section 6.3 used a single integral to find the area between the circle $r = 5\cos\theta$ and the limaçon $r = 2 + \cos\theta$. Use double integrals to find the same area.

In Problems 51-56, find the volume of the given solid region.

51. The solid that lies inside the sphere $x^2 + y^2 + z^2 = 25$ and outside the cylinder $x^2 + y^2 = 9$

52. The solid bounded above by the sphere $x^2 + y^2 + z^2 = 4$ and below by the paraboloid $3z = x^2 + y^2$

53. The solid region common to the sphere $x^2 + y^2 + z^2 = 4$ and the cylinder $x^2 + y^2 = 2x$

54. The solid region common to the cylinder $x^2 + y^2 = 2$ and the ellipsoid $3x^2 + 3y^2 + z^2 = 7$

55. The solid bounded above by the paraboloid $z = 1 - x^2 - y^2$, below by the plane $z = 0$, and on the sides by the cylinder $x^2 + y^2 = x$

56. The solid region bounded above by the cone $z = x^2 + y^2$, below by the plane $z = 0$, and on the sides by the cylinder $x^2 + y^2 = y$

57. Use polar coordinates to evaluate the double integral $\iint\limits_{D} x \, dA$ over the shaded region.

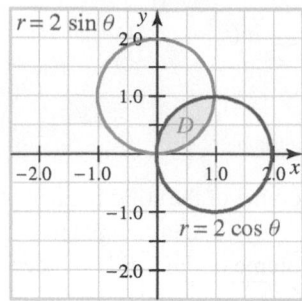

58. Use polar coordinates to evaluate the double integral $\iint\limits_{D} (x^2 + y^2) \, dA$ over the shaded region.

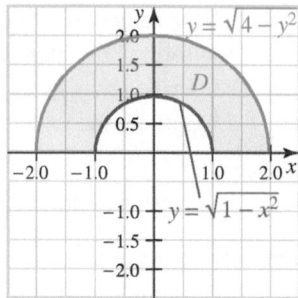

Level 3

59. EXPLORATION PROBLEM If we evaluate the integral $\iint\limits_{R} r^2 \, dr \, d\theta$ where R is the region in the xy-plane bounded by $r = 2\cos\theta$, we obtain

$$\int_0^\pi \int_0^{2\cos\theta} r^2 \, dr \, d\theta = \int_0^\pi \frac{r^3}{3}\Big|_0^{2\cos\theta} d\theta$$

$$= \frac{8}{3} \int_0^\pi \cos^3\theta \, d\theta$$

$$= \frac{8}{3}\left[\sin\theta - \frac{\sin^3\theta}{3}\right]_0^\pi$$

$$= 0$$

Alternatively, we can set up the integral as

$$\iint\limits_{R} r^2 \, dr \, d\theta$$

$$= \int_0^{\pi/2} \int_0^{2\cos\theta} r^2 dr \, d\theta + \int_{\pi/2}^\pi \int_0^{2\cos\theta} r^2(-dr)\, d\theta$$

$$= \frac{32}{9}$$

Both of these answers cannot be correct. Which procedure (if either) is correct and why?

60. 𝔥𝔦𝔰𝔱𝔬𝔯𝔦𝔠𝔞𝔩 𝔔𝔲𝔢𝔰𝔱

Newton and Leibniz have been credited with the discovery of calculus, but much of its development was due to the mathematicians Pierre-Simon Laplace, Lagrange (𝔥𝔦𝔰𝔱𝔬𝔯𝔦𝔠𝔞𝔩 𝔔𝔲𝔢𝔰𝔱 Problem 60, Section 11.8) and Gauss (𝔥𝔦𝔰𝔱𝔬𝔯𝔦𝔠𝔞𝔩 𝔔𝔲𝔢𝔰𝔱, Supplementary Problem 67, Chapter 5).

Pierre-Simon Laplace (1749-1827)

These three great mathematicians of calculus were contrasted by W. W. Rouse Ball:

> The great masters of modern analysis are Lagrange, Laplace, and Gauss, who were contemporaries. It is interesting to note the marked contrast in their styles. Lagrange is perfect both in form and matter, he is careful to explain his procedure, and though his arguments are general they are easy to follow. Laplace on the other hand explains nothing, is indifferent to style, and, if satisfied that his results are correct, is content to leave them either with no proof or with a faulty one. Gauss is as exact and elegant as Lagrange, but even more difficult to follow than Laplace, for he removes every trace of the analysis by which he reached his results, and strives to give a proof which, while rigorous, shall be as concise and synthetic as possible.*

Pierre-Simon Laplace has been called the Newton of France. He taught Napoleon Bonaparte, was appointed for a time as Minister of Interior, and was at times granted favors from his powerful friend. Today, Laplace is best known as the major contributor to probability, taking it from gambling to a true branch of mathematics. He was one of the earliest to evaluate the improper integral

$$I = \int_{-\infty}^{\infty} e^{-x^2} dx$$

which plays an important role in the theory of probability. Show that $I = \sqrt{\pi}$. Hint: Note that

$$\int_{-\infty}^{\infty} e^{-x^2} dx \cdot \int_{-\infty}^{\infty} e^{-y^2} dy = \int_{-\infty}^{\infty}\int_{-\infty}^{\infty} e^{-(x^2+y^2)} dx \, dy$$

where the integral on the right is over the entire xy-plane, described in polar terms by $\theta \le r < \infty$ and $0 \le \theta \le 2\pi$.

---

*A Short Account of the History of Mathematics, as quoted in Mathematical Circles Adieu by Howard Eves (Boston: Prindle, Weber & Schmidt, Inc., 1977).

12.4 SURFACE AREA

IN THIS SECTION: *Definition of surface area, surface area projections, area of a surface defined parametrically*

In Chapter 6, we found that if f has a continuous derivative on the closed, bounded interval $[a.b]$, then the length L of the portion of the graph of $y = f(x)$ between the points where $x = a$ and $x = b$ may be defined by

$$L = \int_a^b \sqrt{1 + [f'(x)]^2}\, dx$$

The goal of this section is to study an analogous formula for the surface area of the graph of a differentiable function of two variables.

Definition of Surface Area

Consider a surface defined by $z = f(x, y)$ defined over a region R of the xy-plane. Enclose the region R in a rectangle partitioned by a grid with lines parallel to the coordinate axes, as shown in Figure 12.25**a**. This creates a number of cells, and we let $R_1, R_2, \ldots, R_n$ denote those that lie entirely within R.

For $m = 1, 2, \cdots, n$, let $P_m(x_m^*, y_m^*)$ be any corner of the rectangle R_m, and let T_m be the tangent plane above P_m on the surface of $z = f(x, y)$. Finally, let ΔS_m denote the area of the "patch" of surface that lies directly above R_m. The rectangle R_m projects onto a parallelogram $ABDC$ in the tangent plane T_m, and if R_m is "small," we would expect the area of this parallelogram to approximate closely the element of surface area ΔS_m (see Figure 12.25**b**).

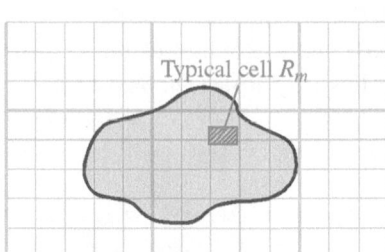

a. The region R is partitioned by the rectangular grid

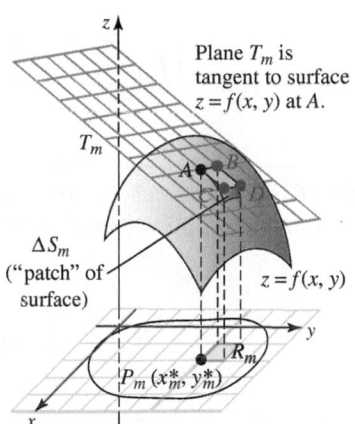

b. The surface area above R_m is approximated by the area of the parallelogram on the tangent plane

Figure 12.25 Surface area

If Δx_m and Δy_m are the lengths of the sides of the rectangle R_m, the approximating parallelogram will have sides determined by the vectors

$$\mathbf{AB} = \Delta x_m \mathbf{i} + [f_x(x_m^*, y_m^*)\Delta x_m]\mathbf{k} \text{ and } \mathbf{AC} = \Delta y_m \mathbf{j} + [f_y(x_m^*, y_m^*)\Delta y_m]\mathbf{k}$$

In Chapter 9, we showed that such a parallelogram with sides $\mathbf{AB}$ and $\mathbf{AC}$ has area

$$\|\mathbf{AB} \times \mathbf{AC}\|$$

This is shown in Figure 12.26.
If K_m is the area of the approximating parallelogram, we have

$$K_m = \|\mathbf{AB} \times \mathbf{AC}\|$$

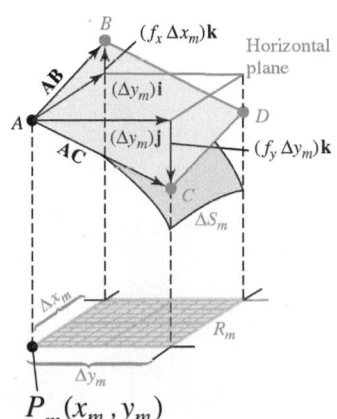

Figure 12.26 An element of surface area has area
$\|\mathbf{AB} \times \mathbf{AC}\|$

To compute K_m, we first find the cross product:

$$\mathbf{AB} \times \mathbf{AC} = \begin{vmatrix} \mathbf{i} & \mathbf{j} & \mathbf{k} \\ \Delta x_m & 0 & f_x(x_m^*, y_m^*)\Delta x_m \\ 0 & \Delta y_m & f_y(x_m^*, y_m^*)\Delta y_m \end{vmatrix}$$

$$= -f_x(x_m^*, y_m^*)\Delta x_m \Delta y_m \mathbf{i} - f_y(x_m^*, y_m^*)\Delta x_m \Delta y_m \mathbf{j} + \Delta x_m \Delta y_m \mathbf{k}$$

Then, we calculate the norm:

$$K_m = \|\mathbf{AB} \times \mathbf{AC}\|$$

$$= \sqrt{[f_x(x_m^*, y_m^*)]^2 \Delta x_m^2 \Delta y_m^2 + [f_y(x_m^*, y_m^*)]^2 \Delta x_m^2 \Delta y_m^2 + \Delta x_m^2 \Delta y_m^2}$$

$$= \sqrt{[f_x(x_m^*, y_m^*)]^2 + [f_y(x_m^*, y_m^*)]^2 + 1} \; \Delta x_m \Delta y_m$$

Finally, summing over the entire partition, we see that the surface area over R may be approximated by the sum

$$\Delta S_n = \sum_{m=1}^{n} \sqrt{[f_x(x_m^*, y_m^*)]^2 + [f_y(x_m^*, y_m^*)]^2 + 1} \; \Delta A_m$$

where $\Delta A_m = \Delta x_m \Delta y_m$. This is a Riemann sum, and by taking an appropriate limit (as the partitions become more and more refined), we find that surface area, S, satisfies

$$S = \lim_{n \to \infty} \sum_{m=1}^{n} \sqrt{[f_x(x_m^*, y_m^*)]^2 + [f_y(x_m^*, y_m^*)]^2 + 1} \; \Delta A_m$$

$$= \iint\limits_{R} \sqrt{[f_x(x, y)]^2 + [f_y(x, y)]^2 + 1} \; dA$$

SURFACE AREA AS A DOUBLE INTEGRAL   Assume that the function $f(x, y)$ has continuous partial derivatives f_x and f_y in a region R of the xy-plane. Then the portion of the surface $z = f(x, y)$ that lies over R has **surface area**

$$S = \iint\limits_{R} \sqrt{[f_x(x, y)]^2 + [f_y(x, y)]^2 + 1} \; dA$$

Note that $dS = \sqrt{[f_x(x, y)]^2 + [f_y(x, y)]^2 + 1} \; dA$ and this is called the **surface area element.**

The region R may be regarded as the projection of the surface $z = f(x, y)$ on the xy-plane. If there were a light source with rays perpendicular to the xy-plane, R would be the "shadow" of the surface on the plane. You will notice the shadows drawn in the figures shown in this section. It is also worthwhile to make the following comparisons:

Length on x-axis: **Arc length:**

$$\int_a^b dx \qquad\qquad \int_a^b ds = \int_a^b \sqrt{[f'(x)]^2 + 1} \; dx$$

Area in xy-plane: **Surface area:**

$$\iint\limits_{R} dA \qquad\qquad \iint\limits_{R} dS = \iint\limits_{R} \sqrt{[f_x(x, y)]^2 + [f_y(x, y)]^2 + 1} \; dA$$

Example 1 Surface area of a plane region

Find the surface area of the portion of the plane $x + y + z = 1$ that lies in the first octant (where $x \geq 0$, $y \geq 0$, $z \geq 0$).

Solution The plane $x + y + z = 1$ and the triangular region T that lies beneath it in the xy-plane are shown in Figure 12.27.

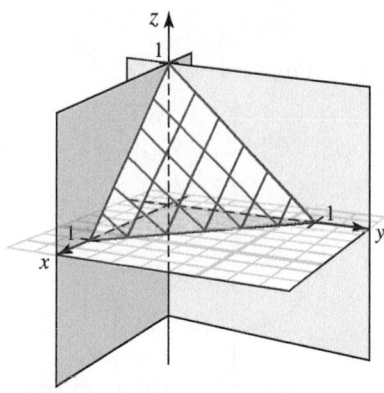

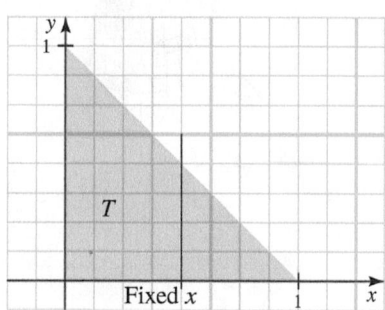

a. The surface of the plane $x + y + z = 1$ in the first octant

b. The surface projects onto a triangle in the xy-plane

Figure 12.27 Surface area

We begin by letting $f(x, y) = 1 - x - y$. Then $f_x(x, y) = -1$ and $f_y(x, y) = -1$, and

$$S = \iint\limits_{T} \sqrt{[f_x(x, y)]^2 + [f_y(x, y)]^2 + 1}\, dA$$

$$= \iint\limits_{T} \sqrt{(-1)^2 + (-1)^2 + 1}\, dA$$

$$= \int_0^1 \int_0^{1-x} \sqrt{3}\, dy\, dx \qquad \text{\textit{Look at Figure 12.27 to see the region}} \\ \text{\textit{T and the limits for both x and y.}}$$

$$= \sqrt{3} \int_0^1 y\big|_0^{1-x}\, dx$$

$$= \sqrt{3} \int_0^1 (1 - x)\, dx$$

$$= \sqrt{3} \left[x - \frac{x^2}{2} \right]_0^1$$

$$= \frac{\sqrt{3}}{2}$$

The surface area is $\frac{\sqrt{3}}{2} \approx 0.866$ square units.

Example 2 Surface area by changing to polar coordinates

Find the surface area (to the nearest hundredth square unit) of that part of the paraboloid $x^2 + y^2 + z = 5$ that lies above the plane $z = 1$.

Solution Let $f(x, y) = 5 - x^2 - y^2$. Then $f_x(x, y) = -2x$, $f_y(x, y) = -2y$, and

$$S = \iint\limits_{D} \sqrt{[f_x(x, y)]^2 + [f_y(x, y)]^2 + 1}\, dA = \iint\limits_{D} \sqrt{4x^2 + 4y^2 + 1}\, dA$$

To determine the limits for the region D note that the paraboloid intersects the plane $z = 1$ in the circle $x^2 + y^2 = 4$. Thus, the part of the paraboloid whose surface area we seek projects onto the disk $x^2 + y^2 \leq 4$ in the xy-plane, as shown in Figure 12.28a (note the shadow of the base in the xy-plane).

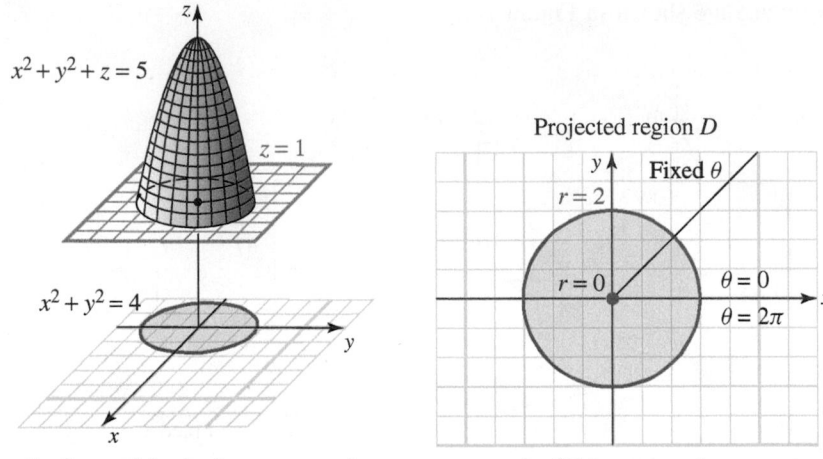

a. Surface with shadow on xy-plane   **b.** Disk on xy-plane

Figure 12.28   The surface $x^2 + y^2 + z = 5$ above $z = 1$ projects onto a disk

It is easier if we convert to polar coordinates by letting $x = r\cos\theta$, $y = r\sin\theta$. Because the region is bounded by $0 \leq r \leq 2$ and $0 \leq \theta \leq 2\pi$, we have

It is usually best to set up an integral first in rectangular coordinates and then switch to polar form, if appropriate.

$$S = \iint_D \sqrt{4x^2 + 4y^2 + 1} \; dA$$

$$= \int_0^{2\pi} \int_0^2 \sqrt{4r^2 + 1} \; r \, dr \, d\theta$$

> Let $u = 4r^2 + 1$; $du = 8r \, dr$
> If $r = 2$, then $u = 17$ and if $r = 0$, $u = 1$.

$$= \int_0^{2\pi} \int_1^{17} u^{1/2} \frac{du}{8} \, d\theta$$

$$= \int_0^{2\pi} \frac{1}{8} \cdot \frac{2}{3} u^{3/2} \Big|_1^{17} \, d\theta$$

$$= \frac{1}{12} \int_0^{2\pi} (17^{3/2} - 1) \, d\theta$$

$$= \frac{1}{12} (17^{3/2} - 1)(2\pi - 0)$$

$$= \frac{\pi}{6} (17^{3/2} - 1)$$

The surface area is approximately 36.18 square units.

Surface Area Projections

We have been projecting the given surface onto the xy-plane, but we can easily modify our procedure to handle other cases. For instance, if the surface $y = f(x, z)$ is projected onto the region Q in the xz-plane, the formula for surface area becomes

$$S = \iint_Q \sqrt{[f_x(x, z)]^2 + [f_z(x, z)]^2 + 1} \; dx \; dz$$

An analogous formula for projection onto the region T in the yz-plane would be

$$S = \iint\limits_{T} \sqrt{[f_y(y,z)]^2 + [f_z(y,z)]^2 + 1} \; dy \; dz$$

Example 3 Surface area of a cylinder

Find the lateral surface area of the cylinder $x^2 + y^2 = 4$, $0 \le z \le 3$.

Solution The cylinder is shown in Figure 12.29**a**.

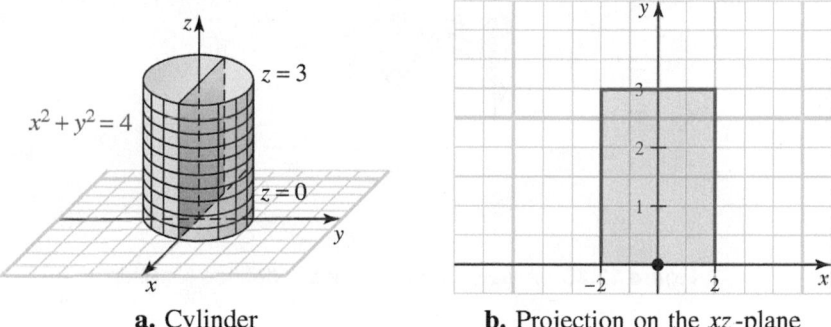

a. Cylinder   **b.** Projection on the xz-plane

Figure 12.29 Surface area of a cylinder

We can find the lateral surface area using the formula for a rectangle by imagining the cylinder "opened up." The length of one side is the height of the cylinder (3 units), and the length of the other side of the rectangle is the circumference of the bottom, namely, $2\pi(2) = 4\pi$. Thus, the lateral surface area is 12π.

We will illustrate our integral formula by using it to recompute this surface area. Specifically, we will compute the surface area of the right half of the cylinder and then double it to obtain the required area. If we project the half-cylinder onto the xy-plane, we will not be accounting for the height of the cylinder. Instead, we will project the surface onto the xz-plane to obtain the rectangle Q bounded by $x = 2$, $x = -2$, $z = 0$, and $z = 3$, as shown in Figure 12.29**b**. Because we are projecting onto the xz-plane, we solve for y to express the surface as $y = f(x, z)$. Since

$$y = f(x, z) = \sqrt{4 - x^2}$$

we have $f_x(x, z) = \dfrac{-x}{\sqrt{4 - x^2}}$ and $f_z(x, z) = 0$ so that

$$[f_x(x, z)]^2 + [f_z(x, z)]^2 + 1 = \frac{x^2}{4 - x^2} + 1 = \frac{4}{4 - x^2}$$

Finally, we note the limits of integration:

$$-2 \le x \le 2 \qquad \text{and } 0 \le z \le 3$$

The surface area can now be calculated:

$$S = \iint\limits_{Q} \sqrt{\frac{4}{4 - x^2}} \; dA$$

$$= \int_{-2}^{2} \int_{0}^{3} \frac{2}{\sqrt{4 - x^2}} dz \; dx$$

$$= 6 \sin^{-1} \frac{x}{2} \Big|_{-2}^{2}$$

$$= 6\pi$$

The lateral surface area of the whole cylinder is $2(6\pi) = 12\pi$.

Area of a Surface Defined Parametrically

Suppose a surface S (see Figure 12.30) is defined parametrically by the vector function

$$\mathbf{R}(u, v) = x(u, v)\mathbf{i} + y(u, v)\mathbf{j} + z(u, v)\mathbf{k}$$

for parameters u and v.

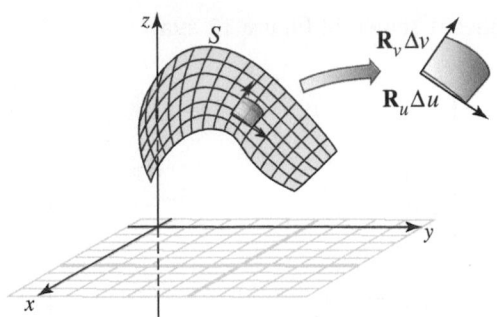

Figure 12.30   Area of a parametrically defined function

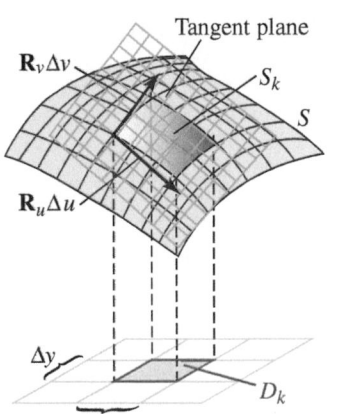

Figure 12.31   Area of a surface defined by a vector function

Let D be a region in the uv-plane on which x, y, and z, as well as their partial derivatives with respect to u and v, are continuous. The partial derivatives of $\mathbf{R}(u, v)$ are given by

$$\mathbf{R}_u = \frac{\partial \mathbf{R}}{\partial u} = \frac{\partial x}{\partial u}\mathbf{i} + \frac{\partial y}{\partial u}\mathbf{j} + \frac{\partial z}{\partial u}\mathbf{k} \qquad \mathbf{R}_v = \frac{\partial \mathbf{R}}{\partial v} = \frac{\partial x}{\partial v}\mathbf{i} + \frac{\partial y}{\partial v}\mathbf{j} + \frac{\partial z}{\partial v}\mathbf{k}$$

Suppose the region D is subdivided into cells, as shown in Figure 12.31.

Consider a typical rectangle in this partition, with dimensions Δx and Δy, where Δx and Δy are small. If we project this rectangle onto the surface $R(u, v)$, we obtain a **curvilinear parallelogram** with adjacent sides $\mathbf{R}_u(u, v)\Delta u$ and $\mathbf{R}_v(u, v)\Delta v$. The area of this rectangle is approximated by

$$\Delta S = \|\mathbf{R}_u(u, v)\Delta u \times \mathbf{R}_v(u, v)\Delta v\| = \|\mathbf{R}_u(u, v) \times \mathbf{R}_v(u, v)\| \ \Delta u \ \Delta v$$

By taking an appropriate limit, we find the surface area to be a double integral.

SURFACE AREA PARAMETRIC DEFINITION   Let S be a surface defined parametrically by

$$\mathbf{R}(u, v) = x(u, v)\mathbf{i} + y(u, v)\mathbf{j} + z(u, v)\mathbf{k}$$

on the region D in the uv-plane, and assume that S is smooth in the sense that $\mathbf{R}_u$ and $\mathbf{R}_v$ are continuous with $\mathbf{R}_u \times \mathbf{R}_v \neq 0$ on D. Then the **surface area**, S, is defined by

$$S = \iint\limits_{D} \|\mathbf{R}_u \times \mathbf{R}_v\| \ du \, dv$$

The quantity $\mathbf{R}_u \times \mathbf{R}_v$ is called the **fundamental cross product**.

Example 4   Area of a surface defined parametrically

Find the surface area (to the nearest square unit) of the surface given parametrically by $\mathbf{R}(u, v) = (u \sin v)\mathbf{i} + (u \cos v)\mathbf{j} + u^2\mathbf{k}$ for $0 \leq u \leq 3$, $0 \leq v \leq 2\pi$.

Solution   We find that $\mathbf{R}_u = \sin v\ \mathbf{i} + \cos v\mathbf{j} + 2u\mathbf{k}$

$$\mathbf{R}_v = u\cos v\mathbf{i} + (-u\sin v)\mathbf{j}$$

so the fundamental cross product is

$$\mathbf{R}_u \times \mathbf{R}_v = \begin{vmatrix} \mathbf{i} & \mathbf{j} & \mathbf{k} \\ \sin v & \cos v & 2u \\ u\cos v & -u\sin v & 0 \end{vmatrix} = (2u^2\sin v)\mathbf{i} + (2u^2\cos v)\mathbf{j} - u\mathbf{k}$$

We find that

$$\|\mathbf{R}_u \times \mathbf{R}_v\| = \sqrt{4u^4\sin^2 v + 4u^4\cos^2 v + u^2} = \sqrt{4u^4 + u^2} = u\sqrt{4u^2 + 1}$$

and we can now compute the surface area:

$$S = \int_0^{2\pi}\int_0^3 u\sqrt{4u^2 + 1}\ du\ dv$$

$$= \int_0^{2\pi}\left[\frac{1}{12}(4u^2 + 1)^{3/2}\right]_{u=0}^{u=3} dv$$

$$= \int_0^{2\pi}\left[\frac{1}{12}(37^{3/2} - 1)\right] dv$$

$$= \frac{37^{3/2} - 1}{12}[2\pi - 0]$$

$$\approx 117.3187007$$

The surface area is approximately 117 square units.   ◼

Notice that in the special case where the surface under consideration has the explicit representation $z = f(x, y)$, it can also be represented in the vector form

$$\mathbf{R}(x, y) = x\mathbf{i} + y\mathbf{j} + f(x, y)\mathbf{k}$$

where x and y are used as parameters ($x = u$ and $y = v$). With this vector representation, we find that

$$\mathbf{R}_x = \mathbf{i} + f_x\mathbf{k} \qquad \text{and} \qquad \mathbf{R}_y = \mathbf{j} + f_y\mathbf{k}$$

so the fundamental cross product is

$$\mathbf{R}_x \times \mathbf{R}_y = \begin{vmatrix} \mathbf{i} & \mathbf{j} & \mathbf{k} \\ 1 & 0 & f_x \\ 0 & 1 & f_y \end{vmatrix} = -f_x\mathbf{i} - f_y\mathbf{j} + \mathbf{k}$$

Therefore, in the case where $z = f(x, y)$, the surface area over the region D is given by

$$S = \iint_D \|\mathbf{R}_x \times \mathbf{R}_y\|\ dx\ dy$$

$$= \iint_D \sqrt{(-f_x)^2 + (-f_y)^2 + 1^2}\ dx\ dy$$

$$= \iint_D \sqrt{f_x^2 + f_y^2 + 1}\ dx\ dy$$

which is the formula obtained at the beginning of this section.

PROBLEM SET 12.4

Level 1

1. ■ What does this say? Describe the process for finding a surface area.
2. ■ What does this say? Describe what is meant by a surface area projection.
3. ■ What does this say? Compare and contrast the following formulas.
   a. length on the x-axis and arc length
   b. area in the xy-plane and surface area
4. ■ What does this say? Describe two methods for finding surface area.

Find the surface area of each surface given in Problems 5-26.

5. The portion of the plane $2x + y + 4z = 8$ that lies in the first octant
6. The portion of the plane $4x + y + z = 9$ that lies in the first octant
7. The portion of the plane $2x + y + z = 2$ that lies in the first octant
8. The portion of the plane $x + 2y - z = 4$ that lies in the first octant
9. The portion of the paraboloid $z = x^2 + y^2$ that lies inside the cylinder $x^2 + y^2 = 1$
10. The portion of the paraboloid $z = 3x^2 + 3y^2$ that lies inside the cylinder $x^2 + y^2 = 4$.
11. The portion of the plane $3x + 6y + 2z = 12$ that is above the triangular region in the plane with vertices $(0, 0, 0)$, $(1, 0, 0)$, and $(1, 1, 0)$
12. The portion of the plane $2x + 2y - z = 0$ that is above the square region in the plane with vertices $(0, 0, 0)$, $(1, 0, 0)$, $(0, 1, 0)$, $(1, 1, 0)$
13. The portion of the surface $x^2 + z = 9$ above the square region in the plane with vertices $(0, 0, 0)$, $(2, 0, 0)$, $(0, 2, 0)$, $(2, 2, 0)$
14. The portion of the surface $z = x^2$ that lies over the triangular region in the plane with vertices $(0, 0, 0)$, $(0, 1, 0)$, and $(1, 0, 0)$
15. The portion of the surface $z = x^2$ over the square region with vertices $(0, 0, 0)$, $(0, 4, 0)$, $(4, 0, 0)$, $(4, 4, 0)$
16. The portion of the surface $z = 2x + y^2$ over the square region with vertices $(0, 0, 0)$, $(3, 0, 0)$, $(0, 3, 0)$, $(3, 3, 0)$
17. The portion of the paraboloid $z = x^2 + y^2$ that lies below the plane $z = 1$
18. The portion of the sphere $x^2 + y^2 + z^2 = 25$ that lies above the plane $z = 3$
19. The part of the cylinder $x^2 + z^2 = 4$ that is in the first octant and is bounded by the plane $y = 2$

20. The portion of the plane $x + y + z = 4$ that lies inside the cylinder $x^2 + y^2 = 16$

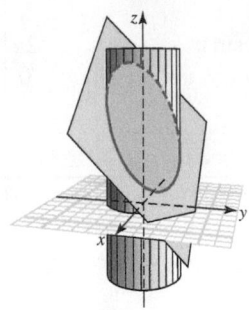

21. The portion of the sphere $x^2 + y^2 + z^2 = 8$ that is inside the cone $x^2 + y^2 - z^2 = 0$

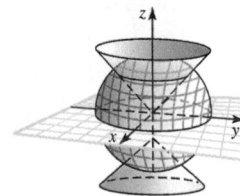

22. The portion of the sphere $x^2 + y^2 + z^2 = 4$ that lies inside the cylinder $x^2 + y^2 = 2y$

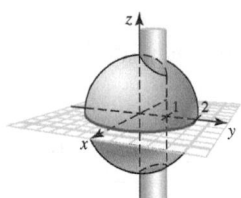

23. The portion of the surface $z = x^2 + y$ above the rectangle $0 \le x \le 2, 0 \le y \le 5$
24. The portion of the surface $z = x^2 - y^2$ that lies inside the cylinder $x^2 + y^2 = 9$
25. The portion of the paraboloid $z = 4 - x^2 - y^2$ that lies above the xy-plane
26. The portion of the paraboloid $z = x^2 + y^2$ that lies inside the sphere $x^2 + y^2 + z^2 = 2$

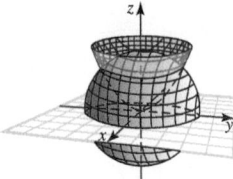

Level 2

27. On a given map, a city parking lot is shown to be a rectangle that is 300 ft by 400 ft. However, the parking lot slopes in the 400-ft direction. It rises uniformly 1 ft for every 5 ft of horizontal displacement. What is the actual surface area of the parking lot?

28. Consider a building whose floor is rectangular, of area 1,000 square feet, and whose roof consists of a single rectangular surface that is inclined at an angle of 45 degrees with respect to the floor. Assume that the roof projects perfectly over the floor. Compute the area of the roof and write it as a double integral.

29. In Example 3, we verified the formula for the surface area of a particular cylinder by using calculus. Similarly, we know the surface area of a cylinder of radius a and height h has surface area $2\pi ah$. Use calculus to verify this formula.

30. In Example 3, we verified the formula for the surface area of a cylinder by using calculus. Similarly, we know the surface area of a sphere of radius a is $4\pi a^2$. Use calculus to verify this formula.

31. Find the surface area of that portion of the sphere $x^2 + y^2 + z^2 = 9z$ that lies inside the paraboloid $x^2 + y^2 = 4z$.

32. Find the surface area of that portion of the sphere $x^2 + y^2 + z^2 = 4z$ that lies inside the paraboloid $z^2 + y^2 = z$.

33. Find a formula for the area of the conical surface $z = \sqrt{x^2 + y^2}$ between the planes $z = 0$ and $z = h$.

34. Find a formula for the area of the conical surface $y = \sqrt{x^2 + z^2}$ between the planes $y = 1$ and $y = k$.

35. Find the surface area of that portion of the cylinder $x^2 + z^2 = 4$ that is above the triangle with vertices $(0,0,0)$, $(1,1,0)$, and $(1,0,0)$.

36. Find the surface area of that portion of the hyperbolic paraboloid (saddle surface) $z = xy + 10$ that lies above the standard unit disk in the xy-plane, centered at $(0,0)$.

37. Find the surface area of the portion of the cylinder $x^2 + z^2 = 9$ that lies inside the cylinder $y^2 + z^2 = 9$.

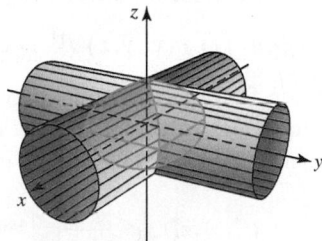

38. Find a formula for the surface area of the frustum of the cone $z = 4\sqrt{x^2 + y^2}$ between the planes $z = h_1$ and $z = h_2$, where $h_1 > h_2$.

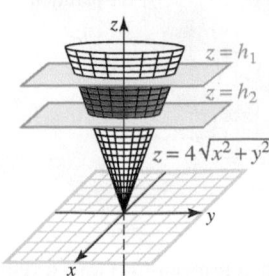

39. Find the surface area of the portion of the plane $Ax + By + Cz = D$ (A, B, C, and D all positive) that lies in the first octant.

40. Show that the surface area of the portion of the plane $Ax + By + Cz = D$ (A, B, C, and D all positive) which projects onto the standard unit disk in the xy-plane is a constant multiple of π.

41. Find the surface area of the portion of the plane $x + 2y + 3z = 12$ that "lies over" (that is, projects onto) the triangular region in the xy-plane with vertices $(0,0,0)$, $(0,a,0)$, and $(a,a,0)$.

42. Find the surface area of the portion of the plane $z = Ax + By + C$ that lies above a triangular domain T in the xy-plane, as a function of the area of the triangle.

43. Find the surface area of that portion of the plane

$$x + y + z = a$$

that lies between the concentric cylinders $x^2 + y^2 = \frac{a^2}{4}$ and $x^2 + y^2 = a^2$ ($a > 0$).

44. Find the surface area of the surface $z = x^2 + y^2$ that lies above the standard unit disk in the xy-plane.

*In Problems 45-50, **set up** (but do not evaluate) the double integral for the surface area of the given portion of surface.*

45. The surface given by $z = e^{-x} \sin y$ over the triangle with vertices $(0,0,0)$, $(0,1,1)$, $(0,1,0)$

46. The surface given by $x = z^3 - yz + y^3$ over the square $(0,0,0)$, $(0,0,2)$, $(0,2,0)$, $(0,2,2)$

47. The surface given by $z = \cos(x^2 + y^2)$ over the disk $x^2 + y^2 \le \frac{\pi}{2}$

48. The surface given by $z = e^{-x} \cos y$ over the disk $x^2 + y^2 \le 2$.

49. The surface given by $z = x^2 + 5xy + y^2$ over the region in the xy-plane bounded by the curve $xy = 5$ and the line $x + y = 6$

50. The surface given by $z = x^2 + 3xy + y^2$ over the region in the xy-plane bounded by $0 \le x \le 4$, $0 \le y \le x$

Compute the magnitude of the fundamental cross product for the surface defined parametrically in Problems 51-54.

51. $\mathbf{R}(u, v) = (2u \sin v)\mathbf{i} + (2u \cos v)\mathbf{j} + u^2\mathbf{k}$

52. $\mathbf{R}(u, v) = (4 \sin u \cos v)\mathbf{i} + (4 \sin u \sin v)\mathbf{j}$
$\qquad + (5 \cos u)\mathbf{k}$

53. $\mathbf{R}(u, v) = u\mathbf{i} + v^2\mathbf{j} + u^3\mathbf{k}$

54. $\mathbf{R}(u, v) = (2u \sin v)\mathbf{i} + (2u \cos v)\mathbf{j} + (u^2 \sin 2v)\mathbf{k}$

55. Find the area of the surface given parametrically by the equation

$$\mathbf{R}(u, v) = uv\mathbf{i} + (u - v)\mathbf{j} + (u + v)\mathbf{k}$$

for $u^2 + v^2 \le 1$
Hint: Use polar coordinates with $u = r \cos \theta$ and $v = r \sin \theta$.

56. A *spiral ramp* has the vector parametric equation

$$\mathbf{R}(u, v) = (u \cos v)\mathbf{i} + (u \sin v)\mathbf{j} + v\mathbf{k}$$

for $0 \le u \le 1, 0 \le v \le \pi$.

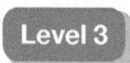

Find the surface area of this ramp.

Level 3

57. The surface given parametrically by

$$\mathbf{R}(u, v) = (u \sin v)\mathbf{i} + (u \cos v)\mathbf{j} + v\mathbf{k}$$

for $0 \le u \le a$ and $0 \le v \le b$ is called a *helicoid*.
a. Compute $\mathbf{R}_u \times \mathbf{R}_v$.
b. Find the surface area of the helicoid.
   Hint: express your answer in terms of a and b.

58. Find the surface area of the torus defined by
$$\mathbf{R}(u, v) = (a + b \cos v) \cos u\mathbf{i}$$
$$+(a + b \cos v) \sin u\mathbf{j}+b \sin v\mathbf{k}$$
for $0 < b < a, 0 \le u \le 2\pi, 0 \le v \le 2\pi$

59. Suppose a surface is given implicitly by $F(x, y, z) = 0$. If the surface can be projected onto a region D in the xy-plane, show that the surface area is given by

$$A = \iint\limits_{D} \frac{\sqrt{F_x^2 + F_y^2 + F_z^2}}{|F_z|} dA$$

where $F_z \ne 0$. Use this formula to find the surface area of a sphere of radius R.

60. Let S be the surface defined by $f(x, y, z) = C$, and let R be the projection of S on a plane. Show that the surface area of S can be computed by the integral

$$\iint\limits_{R} \frac{\|\nabla f\|}{|\nabla f \cdot \mathbf{u}|} dA$$

where $\mathbf{u}$ is a unit vector normal to the plane containing R and $\nabla f \cdot \mathbf{u} \ne 0$. This is a practical formula sometimes used in calculating the surface area.

12.5 TRIPLE INTEGRALS

IN THIS SECTION: *Definition of the triple integral, iterated integration, volume by triple integrals*
Triple integrals are defined and developed in essentially the same way as double integrals. We evaluate triple integrals by iterated integration and interpret them geometrically in terms of volume.

Definition of the Triple Integral

A double integral $\iint\limits_{D} f(x, y) \, dA$ is evaluated over a closed, bounded region in the plane, and in essentially the same way, a *triple integral* $\iiint\limits_{D} f(x, y, z) \, dV$ is evaluated over a closed, bounded solid region D in $\mathbb{R}^3$. Suppose $f(x, y, z)$ is defined on a closed region D, which in turn is contained in a "box" B in space. Partition B into a finite number of smaller boxes using planes parallel to the coordinate planes, as shown in Figure 12.32.

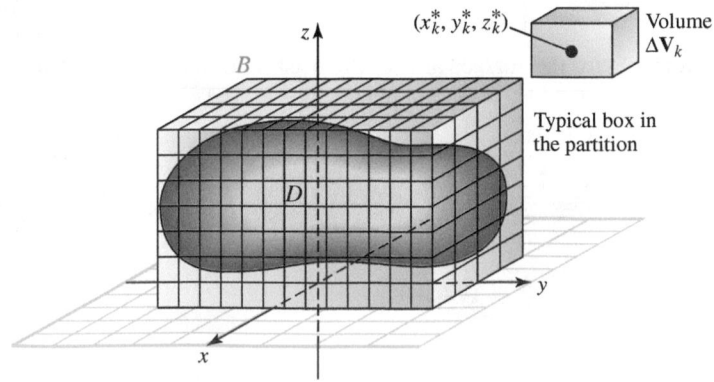

Figure 12.32   The box B contains D and is subdivided into smaller boxes

We exclude from consideration any boxes with points outside D. Let $\Delta V_1, \Delta V_2, \cdots, \Delta V_n$ denote the volumes of the boxes that remain, and define the norm $\|P\|$ of the partition to be the length of the longest diagonal of any box in the partition. Next, choose a representative point (x_k^*, y_k^*, z_k^*) from each box in the partition and form the Riemann sum

$$\sum_{k=1}^{n} f(x_k^*, y_k^*, z_k^*)\Delta V_k$$

If we repeat the process with more subdivisions, so that the norm approaches zero, we are led to the following definition.

TRIPLE INTEGRAL If f is a function defined over a closed bounded solid region D, then the **triple integral of f over D** is defined to be the limit

$$\iiint\limits_{D} f(x, y, z)\, dV = \lim_{\|P\| \to 0} \sum_{k=1}^{n} f(x_k^*, y_k^*, z_k^*)\Delta V_k$$

provided this limit exists.

A surface is said to be **piecewise smooth** if it is made up of a finite number of smooth surfaces. It can be shown that the triple integral exists if $f(x, y, z)$ is continuous on D and the surface of D is piecewise smooth, and even under more general conditions on f and D. It can also be shown that triple integrals have the following properties, which are analogous to those of double integrals listed in Theorem 12.1. In each case, assume the indicated integrals exist.

Linearity rule   For constants a and b

$$\iiint\limits_{D} [af(x, y, z) + bg(x, y, z)]\, dV = a \iiint\limits_{D} f(x, y, z)\, dV + b \iiint\limits_{D} g(x, y, z)\, dV$$

Dominance rule   If $f(x, y, z) \geq g(x, y, z)$ on D, then

$$\iiint\limits_{D} f(x, y, z)\, dV \geq \iiint\limits_{D} g(x, y, z)\, dV$$

Subdivision rule   If the solid region of integration D can be subdivided into two solid subregions D_1 and D_2 (see Figure 12.33, then)

$$\iiint\limits_{D} f(x, y, z)\, dV = \iiint\limits_{D_1} f(x, y, z)\, dV + \iiint\limits_{D_2} f(x, y, z)\, dV$$

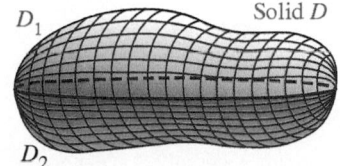

Figure 12.33 Dividing a solid D into two solid subregions

Iterated Integration

As with double integrals, we evaluate triple integrals by iterated integration. However, setting up the limits of integration in an iterated triple integral is often difficult, especially if the solid region of integration, D, is hard to visualize. The relatively simple case where B is a rectangular solid (box) may be handled by applying the following theorem.

Theorem 12.5   Fubini's theorem over a parallelepiped in space

If $f(x, y, z)$ is continuous over a rectangular box B: $a \leq x \leq b$, $c \leq y \leq d$, $r \leq z \leq s$, then the triple integral may be evaluated by the iterated integral

$$\iiint\limits_{B} f(x, y, z)\, dV = \int_{r}^{s} \int_{c}^{d} \int_{a}^{b} f(x, y, z)\, dx\, dy\, dz$$

The iterated integration can be performed in any order, with appropriate adjustments to the limits of integration:

$$dx\, dy\, dz \quad dx\, dz\, dy \quad dz\, dx\, dy$$
$$dy\, dx\, dz \quad dy\, dz\, dx \quad dz\, dy\, dx$$

Similar comments about Fubini's Theorem for double integrals can be made for Fubini's Theorem for triple integrals.

Proof: The proof, which is similar to the two-dimensional case, can be found in an advanced calculus course.   ◆

Remark:

As in the case for double integrals earlier in this chapter, if $f(x, y, z) = f_1(x)f_2(y)f_3(z)$ the integration can be written as

$$\iiint\limits_{B} f(x, y, z)\, dV = \int_{a}^{b} f_1(x)\, dx \int_{c}^{d} f_2(y)\, dy \int_{r}^{s} f_3(z)\, dz$$

See Problem 55. We will not use this shortcut in the examples to follow, so that we can show the more general cases, but leave it to the reader to verify that the shortcut works in such circumstances.

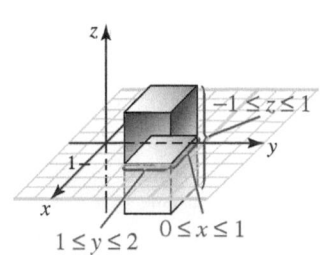

Figure 12.34   Region B

Example 1   Evaluating a triple integral using Fubini's theorem

Evaluate $\iiint\limits_{B} z^2 y e^x\, dV$, where B is the box given by $0 \leq x \leq 1$, $1 \leq y \leq 2$, $-1 \leq z \leq 1$

This box is shown in Figure 12.34.

Solution   We will evaluate the integral in the order $dx\, dy\, dz$.

$$\iiint\limits_{B} f(x, y, z)\, dV = \int_{-1}^{1} \int_{1}^{2} \underbrace{\int_{0}^{1} z^2 y e^x\, dx}\, dy\, dz$$

Treat y and z as constants.

$$= \int_{-1}^{1} \int_{1}^{2} z^2 y \left[e^x\right]\Big|_{x=0}^{x=1}\, dy\, dz$$

$$= \int_{-1}^{1} \underbrace{\int_{1}^{2} z^2 y \left[e - 1\right]\, dy}\, dz$$

Treat z as a constant

$$= (e - 1) \int_{-1}^{1} z^2 \left[\frac{y^2}{2}\right]\Big|_{y=1}^{y=2}\, dz$$

$$= (e - 1) \int_{-1}^{1} z^2 \left[\frac{2^2}{2} - \frac{1^2}{2}\right] dz$$

$$= \frac{3}{2}(e-1) \int_{-1}^{1} z^2 \, dz$$

$$= \frac{3}{2}(e-1) \frac{z^3}{3} \Big|_{z=-1}^{z=1}$$

$$= \frac{3}{2}(e-1) \left[\frac{1^3}{3} - \frac{(-1)^3}{3} \right]$$

$$= e - 1$$

As an exercise, verify that the same result is obtained by using any other order of integration—for example, $dz \, dy \, dx$.

Next, we will see how triple integrals can be evaluated over solid regions that are not rectangular boxes. We will assume the solid region of integration D is *z-simple* in the sense that it has a lower bounding surface $z = u(x, y)$ and an upper bounding surface $z = v(x, y)$, and that it projects onto a region A in the xy-plane that is of either type I or type II. Such a solid is shown in Figure 12.35.

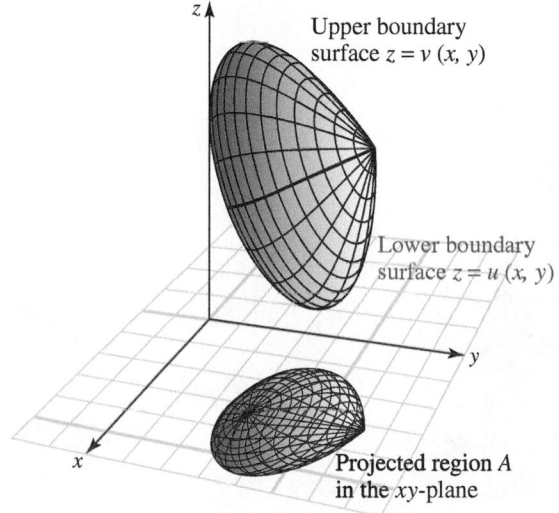

Upper boundary surface $z = v(x, y)$

Lower boundary surface $z = u(x, y)$

Projected region A in the xy-plane

Figure 12.35 A z-simple solid D

The region D can be described as the set of all points (x, y, z) such that $u(x, y) \leq z \leq v(x, y)$ for all (x, y) in A. Then the triple integral of $f(x, y, z)$ over the solid region D can be expressed as an iterated integral with inner limits of integration (with respect to z) $u(x, y)$ and $v(x, y)$. We summarize in the following theorem.

Theorem 12.6 Triple integral over a z-simple solid region

Suppose D is a solid region bounded below by the surface $z = u(x, y)$ and above by $z = v(x, y)$ that projects onto the region A in the xy-plane. If A is of either type I or type II, then the integral of the continuous function $f(x, y, z)$ over D is

$$\iiint_D f(x, y, z) \, dV = \int \int_A \left(\int_{u(x,y)}^{v(x,y)} f(x, y, z) \, dz \right) dA$$

Proof: Even though a proof is beyond the scope of this text, we note that if A is vertically simple (type I), then for each fixed x in an interval $[a, b]$, y varies from $g_1(x)$ to $g_2(x)$, and the triple integral of $f(x, y, z)$ over D can be expressed as

$$\iiint_D f(x, y, z)\, dV = \int_a^b \int_{g_1(x)}^{g_2(x)} \int_{u(x,y)}^{v(x,y)} f(x, y, z)\, dz\, dy\, dx$$

Likewise, if A is horizontally simple (type II), then for each fixed y in an interval $[c, d]$, x varies from $h_1(y)$ to $h_2(y)$, and

$$\iiint_D f(x, y, z)\, dV = \int_c^d \int_{h_1(y)}^{h_2(y)} \int_{u(x,y)}^{v(x,y)} f(x, y, z)\, dz\, dx\, dy$$

We illustrate this procedure in the following example.   ◆

Example 2   Evaluating a triple integral over a general region

Evaluate $\iiint_D x\, dV$, where D is the solid in the first octant bounded by the cylinder $x^2 + y^2 = 4$ and the plane $2y + z = 4$.

Solution   The solid is shown in Figure 12.36.

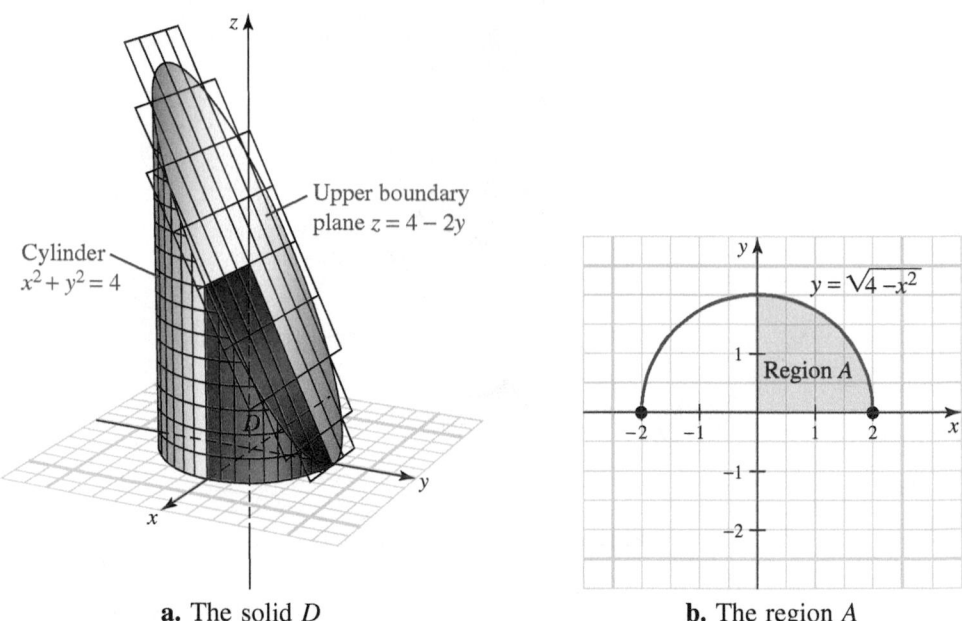

a. The solid D **b.** The region A

Figure 12.36   A solid D with its projection A in the xy-plane

The upper boundary surface of D is the plane $z = 4 - 2y$, and the lower boundary surface is the xy-plane, $z = 0$. The projection A of the solid on the xy-plane is the quarter disk $x^2 + y^2 \leq 4$ with $x \geq 0$, $y \geq 0$ (because D lies in the first octant). This projection may be described in type I form as the set of all (x, y) such that for each fixed x between 0 and 2, y varies from 0 to $\sqrt{4 - x^2}$. Thus, we have

$$\iiint_D x\, dV = \iint_A \int_0^{4-2y} x\, dz\, dA$$

$$= \int_0^2 \int_0^{\sqrt{4-x^2}} \int_0^{4-2y} x\, dz\, dy\, dx$$

$$= \int_0^2 \int_0^{\sqrt{4-x^2}} x[(4 - 2y) - 0]\, dy\, dx$$

$$= \int_0^2 \int_0^{\sqrt{4-x^2}} (4x - 2xy)\, dy\, dx$$

$$= \int_0^2 \left[4xy - xy^2\right]_{y=0}^{y=\sqrt{4-x^2}} dx$$

$$= \int_0^2 \left[4x\sqrt{4-x^2} - x(4-x^2)\right] dx$$

$$= \left[-\frac{4}{3}(4-x^2)^{3/2} - 2x^2 + \frac{1}{4}x^4\right]_0^2$$

$$= \left[0 - 8 + 4 + \frac{32}{3} + 0 - 0\right]$$

$$= \frac{20}{3}$$

Volume by Triple Integrals

Just as a double integral can be interpreted as the area of the region of integration, a triple integral may be interpreted as the **volume** of a solid. That is, if V is the volume of the solid region D, then

$$V = \iiint_D dV$$

Example 3   Volume of a tetrahedron

Find the volume of the tetrahedron T bounded by the plane $2x + y + 3z = 6$ and the coordinate planes $x = 0$, $y = 0$, and $z = 0$.

Solution   The tetrahedron T is shown in Figure 12.37**a**.

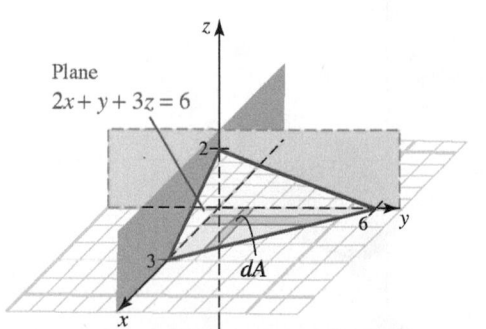

a. Tetrahedron in the first octant

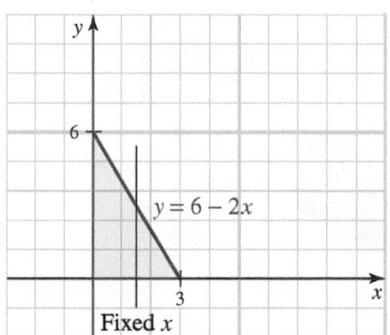

b. Projection onto the xy-plane

Figure 12.37   Volume of a tetrahedron

The upper surface of T is the plane $z = \frac{1}{3}(6 - 2x - y)$ and its lower surface is $z = 0$. Note that T projects onto a triangle, A, in the xy-plane, as shown in Figure 12.37**b**. Described in type I form, the triangle A is the set of all (x, y) such that for each fixed x between 0 and 3, y varies from 0 to $6 - 2x$. Thus,

$$V = \iiint_T dV$$

$$= \iint_A \int_0^{\frac{1}{3}(6-2x-y)} dz\, dA$$

$$= \int_0^3 \int_0^{6-2x} \int_0^{\frac{1}{3}(6-2x-y)} dz\, dy\, dx$$

$$= \int_0^3 \int_0^{6-2x} \left[\frac{1}{3}(6-2x-y) - 0\right] dy\, dx$$

$$= \int_0^3 \left[2y - \frac{2}{3}xy - \frac{1}{6}y^2\right]_{y=0}^{y=6-2x} dx$$

$$= \int_0^3 \left[2(6-2x) - \frac{2}{3}x(6-2x) - \frac{1}{6}(6-2x)^2 - 0\right] dx$$

$$= \int_0^3 \left[6 - 4x + \frac{2}{3}x^2\right] dx$$

$$= 6$$

The volume of the tetrahedron is 6 cubic units.

Sometimes it is easier to evaluate a triple integral by integrating first with respect to x or y instead of z. For instance, if the solid region of integration D is bounded by $x = x_1(y,z)$ and $x = x_2(y,z)$, and the boundary surfaces project onto a region A in the yz-plane, as shown in Figure 12.38a, then

$$\iiint_D f(x,y,z)\, dV = \iint_A \int_{x_1}^{x_2} f(x,y,z)\, dx\, dA$$

STOP *Spend some time with this figure.*

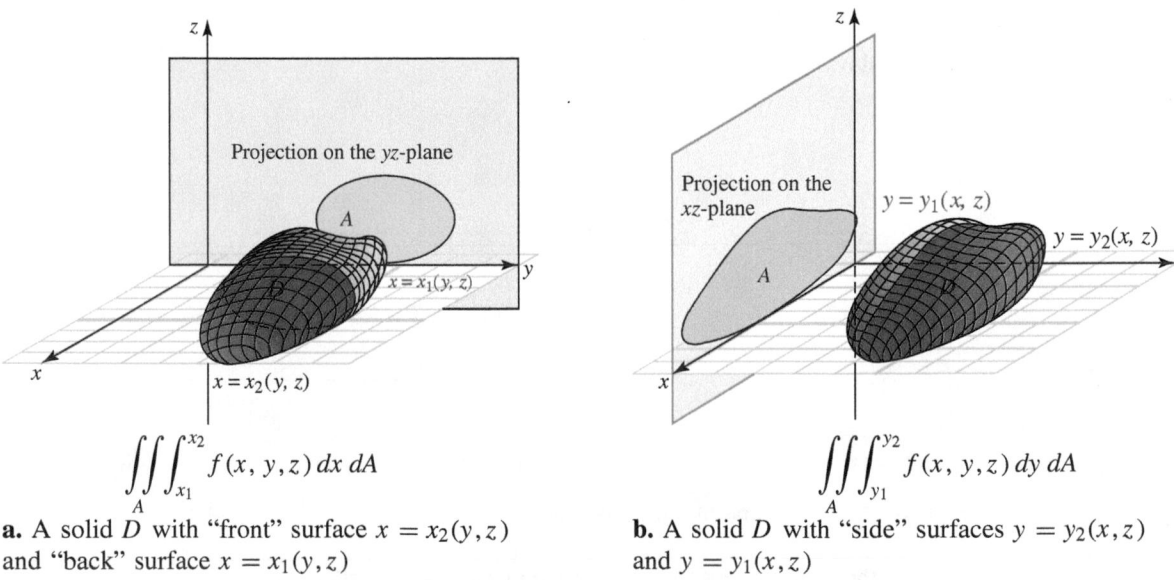

$$\iint_A \int_{x_1}^{x_2} f(x,y,z)\, dx\, dA$$

a. A solid D with "front" surface $x = x_2(y,z)$ and "back" surface $x = x_1(y,z)$

$$\iint_A \int_{y_1}^{y_2} f(x,y,z)\, dy\, dA$$

b. A solid D with "side" surfaces $y = y_2(x,z)$ and $y = y_1(x,z)$

Figure 12.38 Iterated integration with respect to x or y first

On the other hand, if the solid region of integration D is bounded by the surfaces $y = y_1(x,z)$ and $y = y_2(x,z)$, and the boundary surfaces project onto a region A in the xz-plane as shown in Figure 12.38, then

$$\iiint_D f(x,y,z)\, dV = \iint_A \int_{y_1}^{y_2} f(x,y,z)\, dy\, dA$$

As an illustration, we will now rework Example 3 by projecting the tetrahedron T onto the yz-plane.

Example 4   Volume of a tetrahedron by changing the order of integration

Find the volume of the tetrahedron T bounded by the coordinate planes and the plane $2x + y + 3z = 6$ in the first octant by projecting onto the yz-plane.

Solution   Note that T is bounded by the yz-plane and the plane $2x + y + 3z = 6$, which we express as $x = \frac{1}{2}(6 - y - 3z)$. (See Figure 12.39a.) The volume is given by

$$V = \iiint_T dV = \iint_B \int_0^{\frac{1}{2}(6-y-3z)} dx\, dA$$

where B is the projection in the yz-plane. This projection is the triangle bounded by the lines $z = 0$, $y = 0$, and $z = \frac{1}{3}(6 - y)$, as shown in Figure 12.39b.

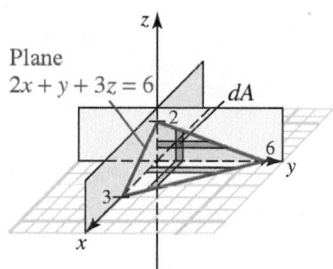

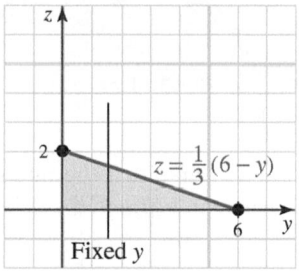

a. The tetrahedron bounded by the plane $2x + y + 3z = 6$ and the positive coordinate axes

b. The region projected onto the yz-plane is a triangle

Figure 12.39   Volume of a tetrahedron; alternative projection

Thus, for each fixed y between 0 and 6, z varies from 0 to $\frac{1}{3}(6 - y)$, and we have

$$V = \int_0^6 \int_0^{\frac{1}{3}(6-y)} \int_0^{\frac{1}{2}(6-y-3z)} dx\, dz\, dy$$

$$= \int_0^6 \int_0^{\frac{1}{3}(6-y)} \frac{1}{2}(6 - y - 3z)\, dz\, dy$$

$$= \int_0^6 \left[3z - \frac{1}{2}yz - \frac{3}{4}z^2 \right]_{z=0}^{z=\frac{1}{3}(6-y)} dy$$

$$= \int_0^6 \left[(6 - y) - \frac{1}{6}y(6 - y) - \frac{1}{12}(6 - y)^2 - 0 \right] dy$$

$$= 6$$

This is the same result we obtained in Example 3 by projecting onto the xy-plane.   ◼

Example 5   Setting up a triple integral to find a volume

Set up (but do not evaluate) a triple integral for the volume of the solid D that is bounded above by the sphere $x^2 + y^2 + z^2 = 4$ and below by the plane $y + z = 2$. The projection on the xy-plane is shown as a shadow in Figure 12.40.

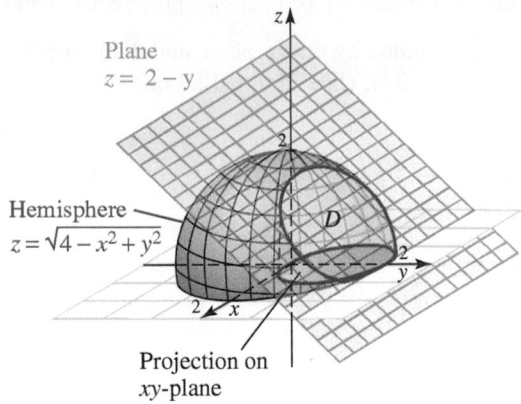

Figure 12.40   Region bounded above by a sphere and below by a plane

Solution   First, note that the intersection of the plane and the sphere occurs above the xy-plane (where $z \geq 0$), so the sphere can be represented by the equation

$$z = \sqrt{4 - x^2 - y^2}$$

(the upper hemisphere). To find the limits of integration for x and y we consider the projection of D onto the xy-plane. To this end, consider the intersection of the hemisphere and the plane $z = 2 - y$:

$$\sqrt{4 - x^2 - y^2} = 2 - y$$
$$4 - x^2 - y^2 = 4 - 4y + y^2$$
$$x^2 + 2y^2 - 4y = 0$$
$$x^2 + 2(y - 1)^2 = 2$$

Although this intersection occurs in $\mathbb{R}^3$, its equation does not contain z. Therefore, the equation serves as a projection on the xy-plane, where $z = 0$. This is an ellipse centered at $(0, 1)$, as shown in Figure 12.41.

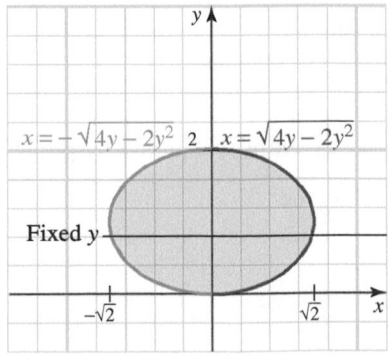

Figure 12.41   The projection of D onto the xy-plane

We consider this as a type II region, which means we will integrate first with respect to x, then with respect to y. Because $x^2 + 2(y - 1)^2 = 2$ we see that for fixed y between 0 and 2, x varies from $-\sqrt{2 - 2(y - 1)^2} = -\sqrt{4y - 2y^2}$ to $\sqrt{4y - 2y^2}$. However, using symmetry, we see the required volume V is twice the integral as x varies from 0 to $\sqrt{4y - 2y^2}$. This leads us to evaluate V by the following triple integral.

$$V = 2 \int_0^2 \int_0^{\sqrt{4y-2y^2}} \int_{2-y}^{\sqrt{4-x^2-y^2}} dz\, dx\, dy$$

Example 6 Choosing an order of integration to compute volume

Find the volume of the solid D bounded below by the paraboloid $z = x^2 + y^2$ and above by the plane $2x + z = 3$.

Solution The graph of the solid D is shown in Figure 12.42**a**. The projection of the solid onto the xy-plane is the graph of the equation

$$x^2 + y^2 = 3 - 2x \quad \text{\textit{Substitute the second equation into the first.}}$$

$$(x + 1)^2 + y^2 = 4 \quad \text{\textit{Complete the square for x.}}$$

This is a circle of radius 2 centered at $(-1, 0)$, as shown in Figure 12.42**b**.

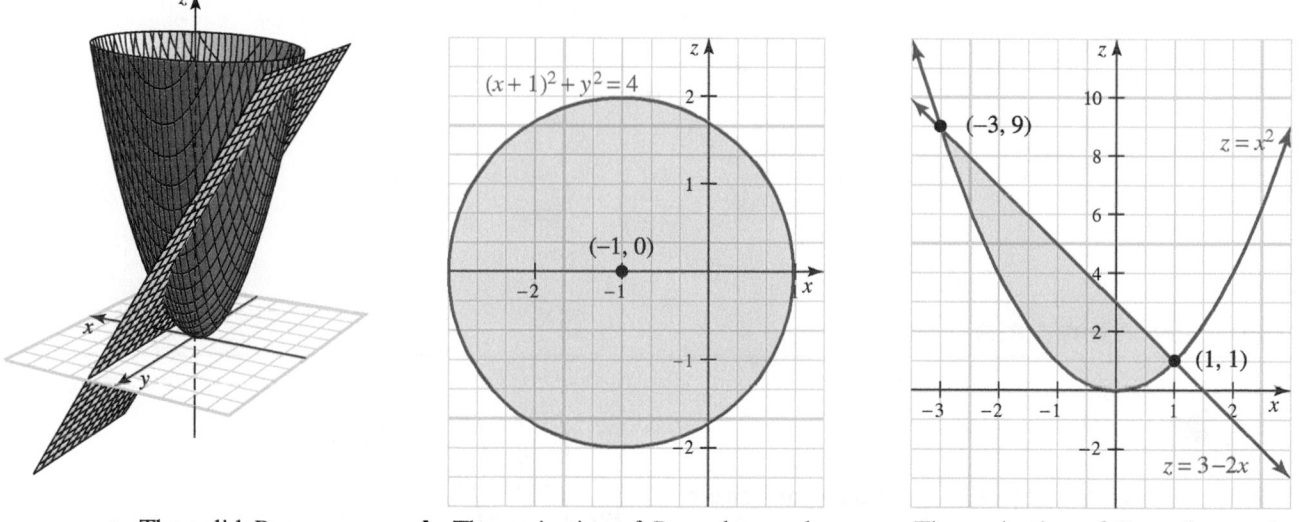

a. The solid D **b.** The projection of D on the xy-plane   **c.** The projection of D on the xz-plane

Figure 12.42   The solid D showing projections on two coordinate planes

Proceeding as in Example 5, we find that the volume is given by the integral

$$V = 2 \int_{-3}^1 \int_0^{\sqrt{3-2x-x^2}} \int_{x^2+y^2}^{3-2x} dz\, dy\, dx$$

which is not especially easy to integrate (try it!).

However, if we project the solid D onto the xz-plane, the projection A is the region bounded by the line $2x + z = 3$ and the parabola $z = x^2$ (take $y = 0$ in the equation $z = x^2 + y^2$), which intersect at the points $(1, 1)$ and $(-3, 9)$, as shown in Figure 12.42**c**. The solid D is symmetric with respect to the xz-plane, so we can integrate with respect to y from $y = 0$ to $y = \sqrt{z - x^2}$ and double the result. We can then describe A as the following type I region.

A: set of all (x, z) such that for each fixed x in the interval

$$-3 \leq x \leq 1, z \text{ varies from } z = x^2 \text{ to } z = 3 - 2x$$

The volume of D is given by the integral

$$V = 2 \int_{-3}^{1} \int_{x^2}^{3-2x} \int_{0}^{\sqrt{z-x^2}} dy\, dz\, dx$$

$$= 2 \int_{-3}^{1} \int_{x^2}^{3-2x} \sqrt{z - x^2}\, dz\, dx$$

$$= 2 \int_{-3}^{1} \frac{2}{3}(z - x^2)^{3/2} \Big|_{x^2}^{3-2x} dx$$

$$= \frac{4}{3} \int_{-3}^{1} [3 - 2x - x^2]^{3/2}\, dx$$

$$= \frac{4}{3} \int_{-3}^{1} [4 - (x + 1)^2]^{3/2}\, dx \qquad \textit{Complete the square.}$$
$$\boxed{\text{Let } x + 1 = 2\sin\theta;\ dx = 2\cos\theta\, d\theta}$$

$$= \frac{4}{3} \int_{-\pi/2}^{\pi/2} 8\cos^3\theta (2\cos\theta\, d\theta)$$

$$= \frac{128}{3} \int_{0}^{\pi/2} \cos^4\theta\, d\theta \qquad \textit{Symmetry}$$

$$= \frac{128}{3} \left(\frac{3\pi}{16}\right) \qquad \textit{Formula 130}$$

$$= 8\pi$$

PROBLEM SET 12.5

Level 1

1. ■ What does this say? State Fubini's theorem for a continuous function over a parallelepiped in $\mathbb{R}^3$.

2. ■ What does this say? Set up integrals, with appropriate limits of integration for the six possible orders of integration for $\iiint_D f(x, y, z)\, dV$, where D is the solid described by $D: y^2 \leq x \leq 4, 0 \leq y \leq 2,$ $0 \leq z \leq 4 - x$

Compute the iterated triple integrals in Problems 3-18.

3. $\displaystyle\int_{1}^{4} \int_{-2}^{3} \int_{2}^{5} dx\, dy\, dz$

4. $\displaystyle\int_{1}^{3} \int_{-2}^{1} \int_{4}^{5} dx\, dz\, dy$

5. $\displaystyle\int_{-3}^{3} \int_{0}^{1} \int_{-1}^{2} dy\, dx\, dz$

6. $\displaystyle\int_{-1}^{3} \int_{0}^{2} \int_{-2}^{2} dy\, dz\, dx$

7. $\displaystyle\int_{0}^{4} \int_{1}^{4} \int_{-2}^{3} dz\, dx\, dy$

8. $\displaystyle\int_{0}^{2} \int_{1}^{3} \int_{2}^{6} dz\, dy\, dx$

9. $\displaystyle\int_{1}^{2} \int_{0}^{1} \int_{-1}^{2} 8x^2yz^3 dx\, dy\, dz$

10. $\displaystyle\int_{4}^{7} \int_{-1}^{2} \int_{0}^{3} x^2y^2z^2 dx\, dy\, dz$

11. $\displaystyle\int_{0}^{2} \int_{0}^{x} \int_{0}^{x+y} xyz\, dz\, dy\, dx$

12. $\displaystyle\int_{0}^{1} \int_{\sqrt{x}}^{\sqrt{1+x}} \int_{0}^{xy} y^{-1}z\, dz\, dy\, dx$

13. $\displaystyle\int_{-1}^{2} \int_{0}^{\pi} \int_{1}^{4} yz\cos(xy)dz\, dx\, dy$

14. $\displaystyle\int_{0}^{\pi} \int_{0}^{1} \int_{0}^{1} x^2y\cos(xyz)dz\, dy\, dx$

15. $\displaystyle\int_{0}^{1} \int_{3}^{y} \int_{0}^{\ln y} e^{z+2x} dz\, dx\, dy$

16. $\displaystyle\int_{0}^{3} \int_{0}^{2z} \int_{0}^{\ln y} ye^{-x} dx\, dy\, dz$

17. $\displaystyle\int_{1}^{4} \int_{-1}^{2z} \int_{0}^{\sqrt{3x}} \frac{x - y}{x^2 + y^2} dy\, dx\, dz$

18. $\displaystyle\int_0^1 \int_{x-1}^{x^2} \int_{-x}^{y} (x+y)\,dz\,dy\,dx$

Evaluate the triple integrals in Problems 19-26.

19. $\displaystyle\iiint_D (x^2 y + y^2 z)\,dV$,

where D is the box $1 \le x \le 3, -1 \le y \le 1, 2 \le z \le 4$

20. $\displaystyle\iiint_D (xy + 2yz)\,dV$,

where D is the box $2 \le x \le 4, 1 \le y \le 3, -2 \le z \le 4$

21. $\displaystyle\iiint_D xyz\,dV$,

where D is the tetrahedron with vertices
$(0,0,0),\ (1,0,0),\ (0,1,0),$ and $(0,0,1)$

22. $\displaystyle\iiint_D x^2 y\,dV$,

where D is the tetrahedron with vertices
$(0,0,0),\ (3,0,0),\ (0,2,0),$ and $(0,0,1)$

23. $\displaystyle\iiint_D xyz\,dV$,

where D is the region given by $x^2 + y^2 + z^2 \le 1$,
$y \ge 0, z \ge 0$

24. $\displaystyle\iiint_D x\,dV$,

where D is bounded by the paraboloid $z = x^2 + y^2$ and
the plane $z = 1$

25. $\displaystyle\iiint_D e^z\,dV$,

where D is the region described by the inequalities
$0 \le x \le 1, 0 \le y \le x,$ and $0 \le z \le x+y$

26. $\displaystyle\iiint_D yz\,dV$,

where D is the solid in the first octant bounded by the
hemisphere $x = \sqrt{9 - y^2 - z^2}$ and the coordinate
planes

*Find the volume V of the solids bounded by the graphs of
the equations given in Problems 27-36 by using triple integration.*

27. $x + y + z = 1$ and the coordinate planes

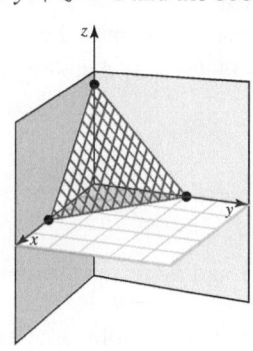

28. $z = 9 - x^2, z = 0, z = y$

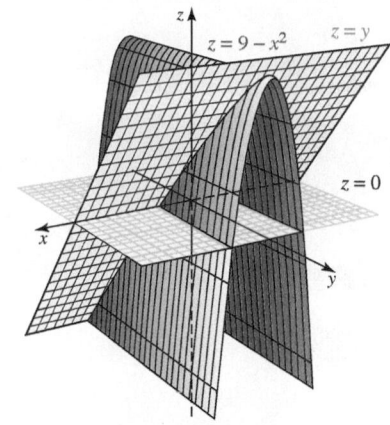

29. $(x - 1)^2 + (y - 2)^2 + (z - 3)^2 = 1$

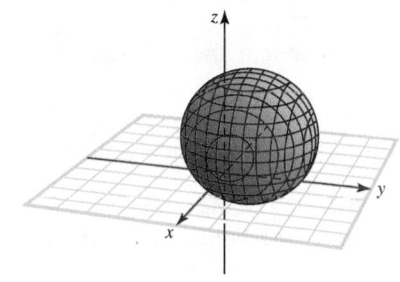

30. $z = 4 - 4x^2 - 4y^2, z = 0$

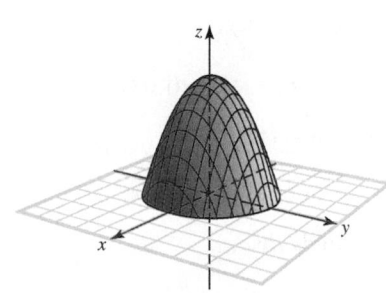

31. $x^2 + 3y^2 = z$ and the cylinder $y^2 + z = 4$

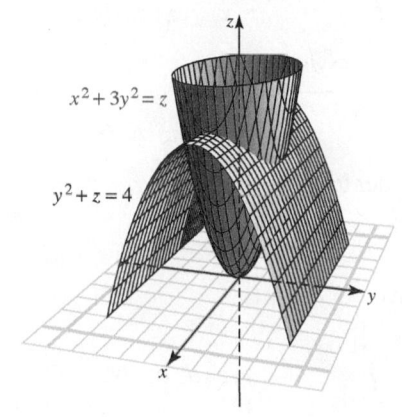

32. $x^2 + y^2 + z^3 = 9$, $z = 0$

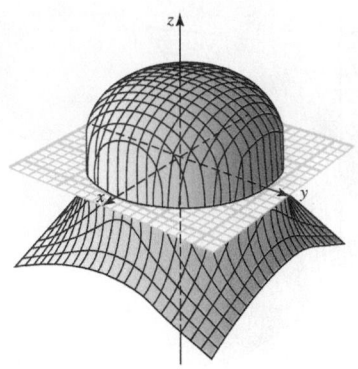

33. The solid bounded above by the paraboloid $z = 6 - x^2 - y^2$ and below by $z = 2x^2 + y^2$

34. The solid bounded by the sphere $x^2 + y^2 + z^2 = 2$ and the paraboloid $x^2 + y^2 = z$

35. The solid region common to the cylinders $x^2 + z^2 = 1$ and $x^2 + y^2 = 1$

36. The solid bounded by the cylinders $y = z^2$ and $y = 2 - z^2$ and the planes $x = 1$ and $x = -2$.

For each given iterated integral there are five other equivalent iterated integrals. Find the one with the requested order in Problems 37-44.

37. $\displaystyle\int_0^1 \int_0^{1-x} \int_0^{1-x-y} f(x,y,z)dz\,dy\,dx$,
change the order to $dz\,dx\,dy$.

38. $\displaystyle\int_0^1 \int_0^{1-x} \int_0^{1-x-y} f(x,y,z)dz\,dy\,dx$,
change the order to $dy\,dx\,dz$.

39. $\displaystyle\int_0^1 \int_0^{1-x} \int_0^{1-x-y} f(x,y,z)dz\,dy\,dx$,
change the order to $dx\,dy\,dz$.

40. $\displaystyle\int_1^2 \int_0^{z-1} \int_0^x f(x,y,z)\,dy\,dx\,dz$,
change the order to $dy\,dz\,dx$.

41. $\displaystyle\int_0^2 \int_0^{\sqrt{4-x^2}} \int_0^{\sqrt{4-x^2-y^2}} f(x,y,z)\,dz\,dy\,dx$
change the order to $dx\,dy\,dz$

42. $\displaystyle\int_0^2 \int_0^{\sqrt{4-x^2}} \int_0^{\sqrt{4-x^2}} f(x,y,z)\,dz\,dy\,dx$
change the order to $dy\,dx\,dz$.

43. $\displaystyle\int_0^1 \int_0^{1-y} \int_0^{y^3} f(x,y,z)\,dx\,dz\,dy$,
change to the order $dz\,dy\,dx$.

44. $\displaystyle\int_0^{1/2} \int_0^{1-4x^2} \int_0^{1-2x} f(x,y,z)\,dz\,dy\,dx$,
change to the order $dy\,dx\,dz$.

Level 2

45. Find the volume of the ellipsoid

$$\frac{x^2}{4} + \frac{y^2}{9} + \frac{z^2}{16} = 1$$

46. Find the volume of the region between the two elliptic paraboloids $z = \dfrac{x^2}{9} + y^2 - 4$ and $z = -\dfrac{x^2}{9} - y^2 + 4$.

47. Find the volume of the region bounded by the paraboloids $z = 16 - x^2 - 2y^2$ and $z = 3x^2 + 2y^2$.

48. A wedge is cut from a right-circular cylinder of radius R by a plane perpendicular to the axis of the cylinder and a second plane that meets the first on the axis at an angle of θ degrees, as shown in Figure 12.43.

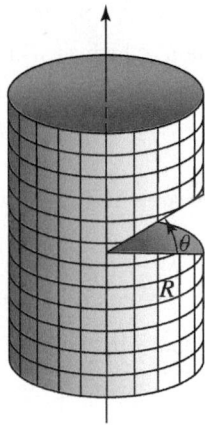

Figure 12.43   Cutting a wedge from a cylinder

Set up and evaluate a triple integral for the volume of the wedge.

49. Find the volume of the region that is bounded above by the elliptic paraboloid

$$z = \frac{x^2}{9} + y^2$$

on the sides by the cylinder

$$\frac{x^2}{9} + y^2 = 1$$

and below by the xy-plane.

50. Find the volume of the solid region in the first octant that is bounded by the planes $z = 8 + 2x + y$ and $y = 3 - 2x$.

Level 3

STOP Look at Problems 51 and 52 to see the source of two famous volume formulas.

51. Use triple integration to find the volume of a sphere of radius R.

© 2013 Katherine Welles. Used under license from Shutterstock, Inc.

52. Use triple integration to find the volume of a right pyramid with height H and a square base of side S.

53. Use triple integration to find the volume of the ellipsoid

$$\frac{x^2}{a^2} + \frac{y^2}{b^2} + \frac{z^2}{c^2} = 1$$

(assume $a > 0, b > 0, c > 0$).

54. Find the volume of the solid region common to the paraboloid $z = k(x^2 + y^2)$ and the sphere $x^2 + y^2 + z^2 = 2k^{-2}$, where $k > 0$.

55. Let B be the box defined by $a \le x \le b, c \le y \le d, r \le z \le s$. Show that

$$\iiint_B f(x)g(y)h(z)\,dV$$

$$= \left[\int_a^b f(x)\,dx\right]\left[\int_c^d g(y)\,dy\right]\left[\int_r^s h(z)\,dz\right]$$

if f, g, and h are continuous.

56. Change the order of integration to show that

$$\int_0^x \int_0^v f(u)\,du\,dv = \int_0^x (x - u)f(u)\,du$$

Also, show that

$$\int_0^x \int_0^v \int_0^u f(w)\,dw\,du\,dv = \frac{1}{2}\int_0^x (x - w)^2 f(w)\,dw$$

57. Evaluate the triple integral

$$\iiint_D \sin(\pi - z)^3\,dz\,dy\,dx$$

where D is the solid region bounded below by the xy-plane, above by the plane $x = z$, and laterally by the planes $x = y$ and $y = \pi$. *Hint:* See Problem 56.

58. One of the following integrals has the value 0. Which is it and why?

A. $\displaystyle\int_{-2}^2 \int_{-\sqrt{4-y^2}}^{\sqrt{4-y^2}} \int_{-\sqrt{4-x^2-y^2}}^{\sqrt{4-x^2-y^2}} (x + z^2)\,dz\,dx\,dy$

B. $\displaystyle\int_0^1 \int_x^{2-x^2} \int_{-3}^3 z^2 \sin(xz)\,dz\,dy\,dx$

Higher-dimensional multiple integrals can be defined and evaluated in essentially the same way as double integrals and triple integrals. Evaluate the given multiple integrals in Problems 59-60.

59. $\displaystyle\iiiint_H xyz^2 w^2\,dx\,dy\,dz\,dw,$

where H is the four-dimensional "hyperbox" defined by $0 \le x \le 1, 0 \le y \le 2, -1 \le z \le 1, 1 \le w \le 2$.

60. $\displaystyle\iiiint_H e^{x-2y+z+w}\,dw\,dz\,dy\,dx,$

where H is the four-dimensional region bounded by the hyperplane $x + y + z + w = 4$ and the coordinate spaces $x = 0, y = 0, z = 0$, and $w = 0$ in the first hyperoctant (where $x \ge 0, y \ge 0, z \ge 0, w \ge 0$)

12.6 MASS, MOMENTS, AND PROBABILITY DENSITY FUNCTIONS

IN THIS SECTION: *Mass and center of mass, moments of inertia, joint probability density functions*
We investigate applications of double and triple integrals to problems involving center of mass for both homogeneous and nonhomogeneous laminas. We also examine moments of inertia and joint probability density functions.

Mass and Center of Mass

A (planar) **lamina** is a flat plate that occupies a region R in the plane and is so thin it can be regarded as two dimensional. If m is the lamina's mass and A is the area of the region R, then $\rho = \dfrac{m}{A}$ is the density of the lamina. The lamina is **homogeneous** if its density $\rho(x, y)$ is constant over R and **nonhomogeneous** if $\rho(x, y)$ varies from point to point. We considered homogeneous lamina in Section 6.5, and in this section, we examine nonhomogeneous lamina.

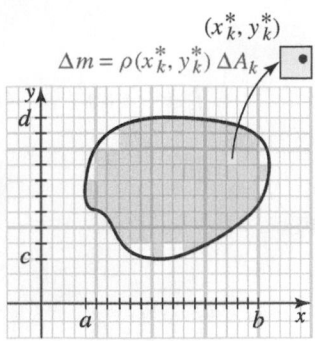

Figure 12.44 Partition of a lamina in the plane

Figure 12.44 shows a lamina covering a region R that has been partitioned into a number of rectangles.

Choose a representative point (x_k^*, y_k^*) in each rectangle of the partition, and note that the mass of the part of the lamina that lies in the kth subrectangle is approximated by

$$\Delta m_k \approx \rho(x_k^*, y_k^*)\Delta A_k$$

We then approximate the total mass m by the Riemann sum

$$m \approx \sum_{k=1}^{n} \rho(x_k^*, y_k^*)\Delta A_k$$

so that

$$m = \lim_{\|P\|\to 0} \sum_{k=1}^{n} \rho(x_k^*, y_k^*)\Delta A_k = \iint\limits_{R} \rho(x, y)\, dA$$

We use these observations as the basis for the following definition.

MASS OF A PLANAR LAMINA OF VARIABLE DENSITY   If ρ is a continuous density function on the lamina corresponding to a plane region R, then the mass m of the lamina is given by

$$m = \iint\limits_{R} \rho(x, y)\, dA$$

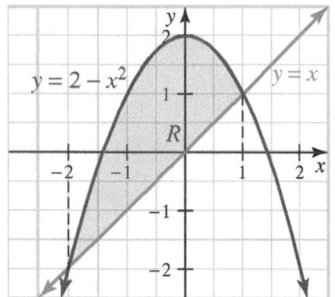

Figure 12.45 A lamina

Example 1   Mass of a planar lamina

Find the mass of the lamina of density $\rho(x, y) = x^2$ that occupies the region R bounded by the parabola $y = 2 - x^2$ and the line $y = x$.

Solution   Begin by drawing the parabola and the line, and by finding their points of intersection, as shown in Figure 12.45.

By substitution,

$$x = 2 - x^2$$
$$x^2 + x - 2 = 0$$
$$x = -2, 1$$

We see that the region R is the set of all (x, y) such that for each x between -2 and 1, y varies from x to $2 - x^2$. Thus, we have

$$m = \iint\limits_{R} x^2\, dA$$

$$= \int_{-2}^{1} \int_{x}^{2-x^2} x^2\, dy\, dx$$

$$= \int_{-2}^{1} x^2(2 - x^2 - x)\, dx$$

$$= \int_{-2}^{1} (2x^2 - x^4 - x^3) \, dx$$

$$= \left[\frac{2x^3}{3} - \frac{x^5}{5} - \frac{x^4}{4} \right]_{-2}^{1}$$

$$= \frac{63}{20}$$

The *moment* of an object about an axis measures the tendency of the object to rotate about that axis. It is defined as the product of the object's mass and the signed distance from the axis (see Figure 12.46).

Thus, by partitioning the region R as before (with mass), we see that the moments M_x and M_y of the lamina about the x-axis and y-axis, respectively, are approximated by the Riemann sums

$$M_x \approx \underbrace{\sum_{k=1}^{n} y_k^* \rho(x_k^*, y_k^*) \Delta A_k}_{\text{Distance to } x\text{-axis}} \quad \text{and} \quad M_y \approx \underbrace{\sum_{k=1}^{n} x_k^* \rho(x_k^*, y_k^*) \Delta A_k}_{\text{Distance to } y\text{-axis}}$$

By taking the limit as the norm of the partition tends to 0, we obtain

$$M_x = \iint\limits_{R} y\rho(x, y) \, dA \quad \text{and} \quad M_y = \iint\limits_{R} x\rho(x, y) \, dA$$

The *center of mass* of the lamina covering R is the point $(\overline{x}, \overline{y})$ where the mass m can be concentrated without affecting the moments M_x and M_y; that is,

$$m\overline{x} = M_y \qquad \text{and} \qquad m\overline{y} = M_x$$

The center of mass $(\overline{x}, \overline{y})$ may also be thought of as the point from which the lamina may be suspended without movement. For future reference, these observations are summarized in the following box.

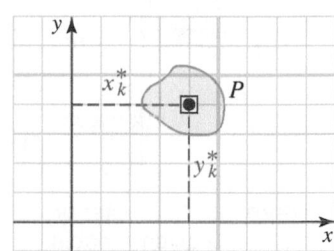

Figure 12.46 The distance from P to the x-axis is y_k^* and the distance to the y-axis is x_k^*

MOMENTS OF MASS

PLANAR LAMINA If $\rho(x, y)$ is a continuous density function on a lamina corresponding to a plane region R, then the **moments of mass** of a variable density function with respect to the x- and y-axes, respectively, are

$$M_x = \iint\limits_{R} y\,\rho(x, y) \, dA \quad \text{and} \quad M_y = \iint\limits_{R} x\,\rho(x, y) \, dA$$

M_x has a factor of y
M_y has a factor of x.

CENTER OF MASS Furthermore, if m is the mass of the lamina, the **center of mass** is $(\overline{x}, \overline{y})$, where

$$\overline{x} = \frac{M_y}{m} \quad \text{and} \quad \overline{y} = \frac{M_x}{m}$$

CENTROID If the density ρ is constant, the point $(\overline{x}, \overline{y})$ is called the **centroid** of the region.

Example 2 Finding a center of mass

Locate the center of mass of the lamina of density $\rho(x, y) = x^2$ that occupies the region R bounded by the parabola $y = 2 - x^2$ and the line $y = x$. This is the lamina defined in Example 1.

Solution   In Example 1, we found that for each fixed x between -2 and 1, y varies from x to $2 - x^2$ (see Figure 12.47). Thus, we have

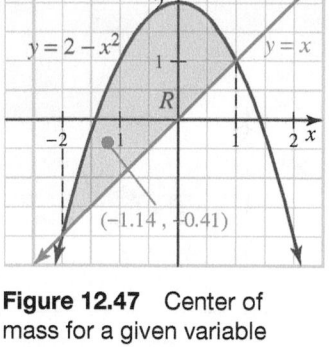

Figure 12.47 Center of mass for a given variable density lamina

$$M_x = \iint\limits_{R} y(x^2)\, dA \qquad\qquad M_y = \iint\limits_{R} x(x^2)\, dA$$

$$= \int_{-2}^{1} \int_{x}^{2-x^2} yx^2\, dy\, dx \qquad\qquad = \int_{-2}^{1} \int_{x}^{2-x^2} x^3\, dy\, dx$$

$$= \int_{-2}^{1} \left[\frac{1}{2}x^2 y^2\right]_{y=x}^{y=2-x^2} dx \qquad\qquad = \int_{-2}^{1} x^3[(2 - x^2) - x]\, dx$$

$$= \frac{1}{2}\int_{-2}^{1} x^2(x^4 - 5x^2 + 4)\, dx \qquad\qquad = \int_{-2}^{1} (2x^3 - x^5 - x^4)\, dx$$

$$= \frac{1}{2}\left[\frac{1}{7}x^7 - x^5 + \frac{4}{3}x^3\right]_{-2}^{1} \qquad\qquad = \left[\frac{2x^4}{4} - \frac{x^6}{6} - \frac{x^5}{5}\right]_{-2}^{1}$$

$$= -\frac{9}{7} \qquad\qquad\qquad\qquad = -\frac{18}{5}$$

From Example 1, $m = \frac{63}{20}$, so the center of mass is $(\bar{x}, \bar{y})$, where

$$\bar{x} = \frac{M_y}{m} = \frac{-\frac{18}{5}}{\frac{63}{20}} = -\frac{8}{7} \approx -1.14 \qquad \bar{y} = \frac{M_x}{m} = \frac{-\frac{9}{7}}{\frac{63}{20}} = -\frac{20}{49} \approx -0.41$$

The center of mass for Example 1 is shown as a blue dot in Figure 12.47.

In a completely analogous way, we can use the triple integral to find the mass and center of mass of a solid in $\mathbb{R}^3$ with density $\rho(x, y, z)$. The mass m, moments M_{yz}, M_{xz}, M_{xy} about the yz-, xz-, and xy-planes, respectively, and coordinates $\bar{x}, \bar{y}, \bar{z}$, of the center of mass (see Figure 12.48) are given by:

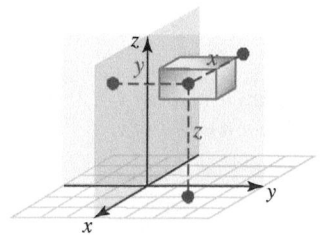

Figure 12.48 Note the distances to the coordinate axes

Mass
$$m = \iiint\limits_{R} \rho(x, y, z)\, dV$$

Moments
$$M_{yz} = \iiint\limits_{R} x\,\rho(x, y, z)\, dV$$
Distance to the yz-plane.

$$M_{xz} = \iiint\limits_{R} y\,\rho(x, y, z)\, dV$$
Distance to the xz-plane.

$$M_{xy} = \iiint\limits_{R} z\,\rho(x, y, z)\, dV$$
Distance to the xy-plane.

Center of mass
$$(\bar{x}, \bar{y}, \bar{z}) = \left(\frac{M_{yz}}{m}, \frac{M_{xz}}{m}, \frac{M_{xy}}{m}\right)$$

As before, if the density is constant, the center of mass is called the **centroid**. Example 3 illustrates how this point can be found by multiple integration.

Example 3 Centroid of a tetrahedron

A solid tetrahedron has vertices $(0,0,0)$, $(1,0,0)$, $(0,1,0)$, and $(0,0,1)$ and constant density $\rho = 6$. Find the centroid.

Solution The tetrahedron may be described as the region in the first octant that lies beneath the plane $x + y + z = 1$, as shown in Figure 12.49**a**.

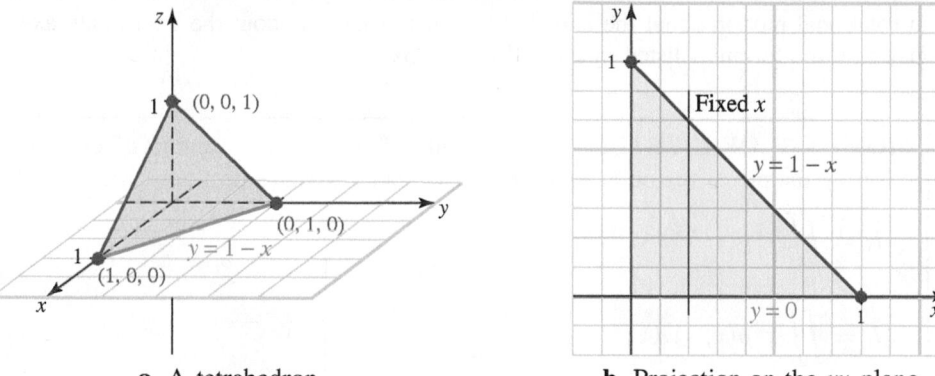

a. A tetrahedron   **b.** Projection on the xy-plane

Figure 12.49 The centroid of a tetrahedron

The boundary of the projection of the top face of the tetrahedron in the xy-plane is found by solving the equations $x + y + z = 1$ and $z = 0$ simultaneously. We find that the projection is the region bounded by the coordinate axes and the line $x + y = 1$, as shown in Figure 12.49**b**. This means that for each fixed x between 0 and 1, y varies from 0 to $1 - x$.

We find that

$$M_{yz} = \iiint\limits_R 6x \, dV = \int_0^1 \int_0^{1-x} \int_0^{1-x-y} 6x \, dz \, dy \, dx = \frac{1}{4}$$

$$M_{xz} = \iiint\limits_R 6y \, dV = \int_0^1 \int_0^{1-x} \int_0^{1-x-y} 6y \, dz \, dy \, dx = \frac{1}{4}$$

$$M_{xy} = \iiint\limits_R 6z \, dV = \int_0^1 \int_0^{1-x} \int_0^{1-x-y} 6z \, dz \, dy \, dx = \frac{1}{4}$$

$$\bar{x} = \frac{M_{yz}}{m} = \frac{\frac{1}{4}}{1} = 0.25$$

$$\bar{y} = \frac{M_{xz}}{m} = \frac{\frac{1}{4}}{1} = 0.25$$

$$\bar{z} = \frac{M_{xy}}{m} = \frac{\frac{1}{4}}{1} = 0.25$$

$$m = \iiint\limits_R \rho \, dV$$

$$= \int_0^1 \int_0^{1-x} \int_0^{1-x-y} 6 \, dz \, dy \, dx$$

$$= \int_0^1 \int_0^{1\ x} 6(1 - x - y) \, dy \, dx$$

$$= \int_0^1 \left[6y - 6xy - 3y^2 \right]_{y=0}^{y=1-x} dx$$

$$= \int_0^1 3(x - 1)^2 dx$$

$$= \left[(x - 1)^3 \right]_0^1$$

$$= 1$$

The centroid is $(0.25, 0.25, 0.25)$. ∎

Moments of Inertia

In general, a lamina of density $\rho(x, y)$ covering the region R in the first quadrant of the plane has (first) moment about a line L given by the integral

$$M_L = \iint\limits_R s \, dm$$

where $dm = \rho(x, y)\, dA$ and $s = s(x, y)$ is the distance from a typical point $P(x, y)$ in R to L. Similarly, the *second* moment, or *moment of inertia*, of R about L is defined by

$$I_L = \iint\limits_R s^2\, dm$$

In physics, the moment of inertia measures the tendency of the lamina to resist a change in rotational motion about the axis L. Moments of inertia about the coordinate axes are given by the formulas listed in the following box.

MOMENTS OF INERTIA   The **moments of inertia** of a lamina of density ρ covering the planar region R about the x-, y-, and z-axes, respectively, are given by

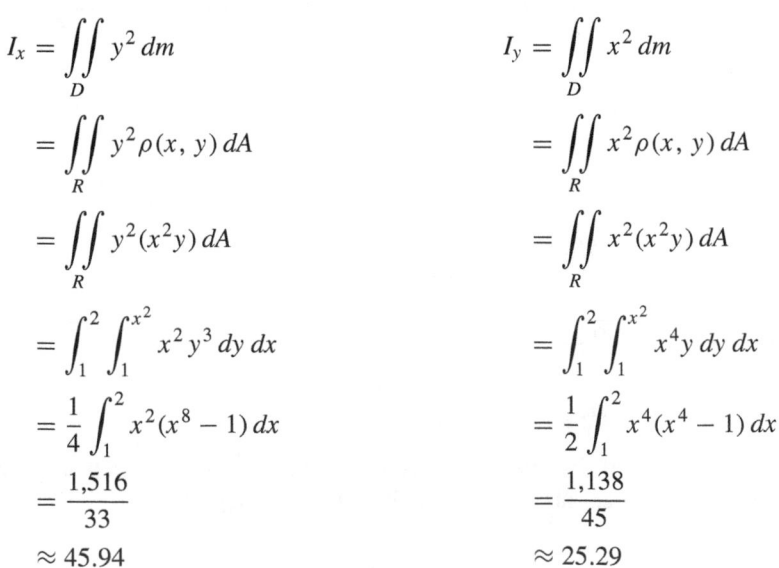

$$I_x = \iint\limits_R y^2 \rho(x, y)\, dA$$

$$I_y = \iint\limits_R x^2 \rho(x, y)\, dA$$

$$I_z = \iint\limits_R (x^2 + y^2)\rho(x, y)\, dA$$

$$= I_x + I_y$$

Example 4   Finding the moments of inertia

A lamina occupies the region R in the plane that is bounded by the parabola $y = x^2$ and the lines $x = 2$ and $y = 1$. The density of the lamina at each point (x, y) is $\rho(x, y) = x^2 y$. Find the moments of inertia (rounded to the nearest hundredth) of the lamina about the x-axis and the y-axis.

Solution   The graph of R is shown in Figure 12.50.

We see that for each fixed x between 1 and 2, y varies from 1 to x^2.

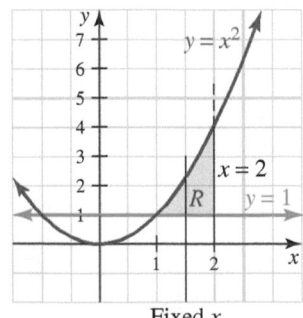

Figure 12.50 Moments of inertia of a lamina

$$I_x = \iint\limits_D y^2\, dm \qquad\qquad I_y = \iint\limits_D x^2\, dm$$

$$= \iint\limits_R y^2 \rho(x, y)\, dA \qquad\qquad = \iint\limits_R x^2 \rho(x, y)\, dA$$

$$= \iint\limits_R y^2 (x^2 y)\, dA \qquad\qquad = \iint\limits_R x^2 (x^2 y)\, dA$$

$$= \int_1^2 \int_1^{x^2} x^2 y^3\, dy\, dx \qquad\qquad = \int_1^2 \int_1^{x^2} x^4 y\, dy\, dx$$

$$= \frac{1}{4} \int_1^2 x^2 (x^8 - 1)\, dx \qquad\qquad = \frac{1}{2} \int_1^2 x^4 (x^4 - 1)\, dx$$

$$= \frac{1{,}516}{33} \qquad\qquad\qquad\qquad = \frac{1{,}138}{45}$$

$$\approx 45.94 \qquad\qquad\qquad\qquad\quad \approx 25.29$$

A simple generalization enables us to compute the moment of inertia of a solid figure about an arbitrary axis L. Specifically, suppose the solid occupies a region R (see Figure 12.51 and that the density at each point (x, y, z) in R is given by $\rho(x, y, z)$.

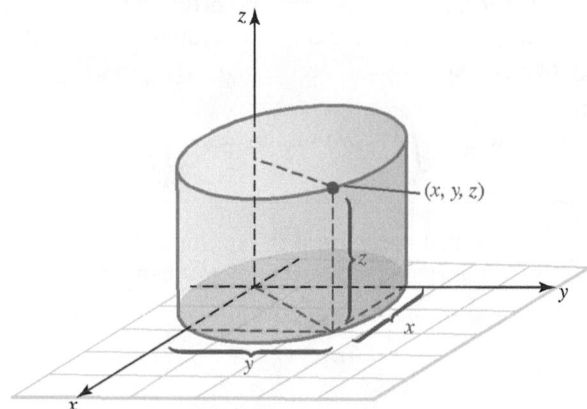

Figure 12.51   A typical solid R

Because the square of the distance of a typical cell in R from the x-axis is $y^2 + z^2$, the moment of inertia about the x-axis is

$$I_x = \iiint\limits_R \underbrace{(y^2 + z^2)}_{\substack{\\ \textit{Square of the distance to the x-axis}}} \underbrace{\rho(x, y, z) \, dV}_{\uparrow \quad \textit{Increment of mass}}$$

Similarly, the moments of inertia of the solid about the y-axis and the z-axis are, respectively,

$$I_y = \iiint\limits_R \underbrace{(x^2 + z^2)}_{\substack{\\ \textit{Square of the distance to the y-axis}}} \underbrace{\rho(x, y, z) \, dV}_{\uparrow \quad \textit{Increment of mass}} \qquad I_z = \iiint\limits_R \underbrace{(x^2 + y^2)}_{\substack{\\ \textit{Square of the distance to the z-axis}}} \underbrace{\rho(x, y, z) \, dV}_{\uparrow \quad \textit{Increment of mass}}$$

Example 5   Moment of inertia of a solid

Find the moment of inertia about the z-axis of the solid tetrahedron S with vertices $(0, 0, 0)$, $(0, 1, 0)$, $(1, 0, 0)$, $(0, 0, 1)$, and density $\rho(x, y, z) = x$.

Solution   In Example 3, we observed that the solid S can be described as the set of all (x, y, z) such that for each fixed x between 0 and 1, y lies between 0 and $1 - x$, and $0 \le z \le 1 - x - y$. Thus,

$$
\begin{aligned}
I_z &= \iiint\limits_S (x^2 + y^2)\rho(x, y, z) \, dV \\
&= \int_0^1 \int_0^{1-x} \int_0^{1-x-y} x(x^2 + y^2) \, dz \, dy \, dx \\
&= \int_0^1 \int_0^{1-x} x(x^2 + y^2)(1 - x - y) \, dy \, dx \\
&= \int_0^1 \left[\frac{x^3(1-x)^2}{2} + \frac{x(1-x)^4}{12} \right] dx \\
&= \frac{1}{90}
\end{aligned}
$$

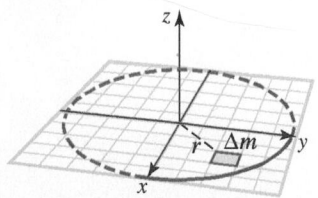

Figure 12.52 Kinetic energy of a rotating disk

Moments of inertia have a useful interpretation in physics. The **kinetic energy** of a body of mass m moving with velocity v along a straight line is defined in physics as $K = \frac{1}{2}mv^2$. Suppose a lamina covering a circular disk R centered at the origin (see Figure 12.52 is rotating around the z-axis with angular speed ω radians/second.

A cell of mass Δm located r units from the origin has linear velocity $v = r\omega$ and linear kinetic energy $K_{\text{lin}} = \frac{1}{2}(\Delta m)v^2$, and by integrating, we find that the entire disk R has kinetic energy of rotation

$$K_{\text{rot}} = \iint\limits_R \frac{1}{2}\omega^2 r^2\, dm = \frac{1}{2}\omega^2 \iint\limits_R r^2\, dm$$

Since $r^2 = x^2 + y^2$, we see that the integral in this formula is just the moment of inertia of R about the z-axis, so the rotational kinetic energy can be expressed as

$$K_{\text{rot}} = \frac{1}{2}I_z\omega^2$$

Comparing this formula to the kinetic energy formula $K_{\text{lin}} = \frac{1}{2}mv^2$, we see that the moment of inertia may be thought of as the rotational analogue of mass or, equivalently, as rotational inertia.

Joint Probability Density Functions

Suppose we consider the useful life of a randomly chosen automobile of a particular type, or the weight of an infant chosen at random from a certain hospital, or the level of CO_2 pollution in the air of a city randomly selected from a region in India. These quantities are called **continuous random variables** because their values range over an interval of real numbers. It is often important to know the probability that a random variable takes on a given set of values. For instance, if X represents the useful life of a Honda Accord, then the probability that a randomly selected Accord will last no more than 10 years may be denoted by $P(0 \leq X \leq 10)$.

Every continuous random variable X has a **probability density function** f with the property that the probability of X lying between the numbers a and b is given by the integral

$$P(a \leq X \leq b) = \int_a^b f(x)\, dx$$

Note the identification between the random variable X and the real number x. For the integration itself, one could use either notation.

In general, $f(x) \geq 0$ for all x, and since the value of x is always some real number, it follows that $P(-\infty < X < \infty) = 1$, so

$$\int_{-\infty}^{\infty} f(x)\, dx = 1$$

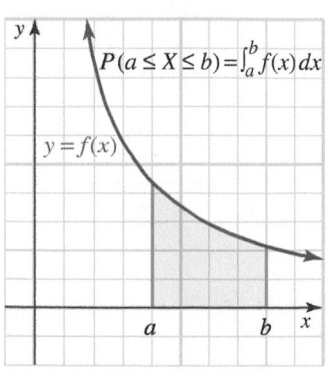

Figure 12.53 Probability as the area under the graph of the probability density function

In geometric terms, the probability $P(a \leq X \leq b)$ is the area under the graph of f over the interval $a \leq x \leq b$. (See Figure 12.53.)

If X and Y are both continuous random variables, then the **joint probability density function** for two random variables X and Y is a function of two variables $f(x, y)$ such that $f(x, y) \geq 0$ for all $f(x, y)$ and

$$P[(X, Y) \text{ in } D] = \iint\limits_D f(x, y)\, dA$$

where $P[(X, Y)$ in $D]$ denotes the probability that (X, Y) is in the region D. Note that

$$P[(X, Y) \text{ in the } XY\text{-plane}] = \int_{-\infty}^{\infty} \int_{-\infty}^{\infty} f(x, y)\, dx\, dy = 1$$

Geometrically, $P[(X, Y)$ in $D]$ may be thought of as the volume under the surface $Z = f(x, y)$ above the region D.

The techniques for constructing joint probability density functions from experimental data are outside the scope of this text and are discussed in many texts on probability and statistics. The use of a double integral to compute a probability with a given joint density function is illustrated in the next example.

Example 6 Compute a probability

Suppose X measures the time (in minutes) that a customer at a particular grocery store spends shopping and Y measures the time the customer spends in the checkout line. A study suggests that the joint probability function for X and Y may be modeled by

$$f(X, Y) = \begin{cases} \frac{1}{200} e^{-x/10} e^{-y/20} & \text{for } x \geq 0 \text{ and } y \geq 0 \\ 0 & \text{otherwise} \end{cases}$$

Find the probability (as a percent) that the customer's total time in the store will be no greater than 30 min.

Solution The goal is to find the probability that $X + Y \leq 30$. Stated geometrically, we wish to find the probability that a randomly selected point (x, y) lies in the region R in the first quadrant that is bounded by the coordinate axes and the line $x + y = 30$ (see Figure 12.54.)

This probability is given by the double integral

$$P[(X, Y) \text{ is in } R] = \iint\limits_{R} f(x, y)\, dA$$

$$= \int_{0}^{30} \int_{0}^{30-x} \frac{1}{200} e^{-x/10} e^{-y/20}\, dy\, dx$$

$$= \frac{1}{200} \int_{0}^{30} e^{-x/10} \left[\frac{e^{-y/20}}{-\frac{1}{20}} \right]_{0}^{30-x} dx$$

$$= \frac{-20}{200} \int_{0}^{30} e^{-x/10} \left[e^{-(1/20)(30-x)} - 1 \right] dx$$

$$= \frac{-1}{10} \left[\frac{e^{-3/2} e^{-x/20}}{-\frac{1}{20}} - \frac{e^{-x/10}}{-\frac{1}{10}} \right]_{0}^{30}$$

$$= e^{-3} - 2e^{-3/2} + 1$$

$$\approx 0.603526748$$

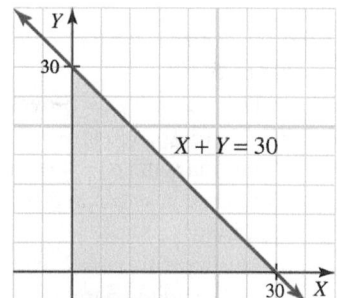

Figure 12.54 Triangle comprised of all points (X, Y) such that $X + Y \leq 30$, $X \geq 0$, $Y \geq 0$

Thus, it is about 60% likely that the shopper will spend no more than 30 minutes in the store.

A useful property of probability states that

$$P(E) + P(\overline{E}) = 1$$

where E and $\overline{E}$ together constitute the set of all possible outcomes (that is, they are *complementary* events). For Example 6, since the events $E = \{$the shopper will spend no more than 30 minutes in the store$\}$ and $\overline{E} = \{$the shopper will spend more than 30 minutes in the store$\}$ are complementary, we can find the probability that the shopper will spend more than 30 minutes in the store by using this complementary property:

$$1 - 60\% = 40\%$$

PROBLEM SET 12.6

Level 1

1. ■ What does this say? Discuss the procedure for finding the center of mass of a lamina.
2. ■ What does this say? Discuss a procedure for finding moments in three dimensions.
3. ■ What does this say? How is probability related to a joint probability function?
4. ■ What does this say? Discuss moments of inertia.

Find the centroid for the regions described in Problems 5-12.

5. A lamina with $\rho = 5$ over the rectangle with vertices $(0,0)$, $(3,0)$, $(3,4)$, $(0,4)$
6. A lamina with $\rho = 4$ over the region bounded by the curve $y = \sqrt{x}$ and the line $x = 4$ in the first quadrant
7. A lamina with $\rho = 2$ over the region between the line $y = 2x$ and the parabola $y = x^2$
8. A lamina with $\rho = 4$ over the region bounded by $y = \sin\frac{\pi}{2}x$, $x = 0$, $y = 0$, $x = \frac{1}{2}$. Give the centroid rounded to the nearest hundredth.
9. A thin homogeneous plate of density $\rho = 1$ with the shape of the region bounded above by the parabola $y = 2 - 3x^2$ and below by the line $3x + 2y = 1$.
10. The part of the spherical solid with density $\rho = 2$ described by $x^2 + y^2 + z^2 \leq 9$, $x \geq 0$, $y \geq 0, z \geq 0$
11. The solid tetrahedron of density $\rho = 4$ bounded by the plane $x + y + z = 4$ in the first octant
12. The solid bounded by the surface $z = \sin x$, $x = 0$, $x = \pi$, $y = 0$, $z = 0$, and $y + z = 1$, where the density is $\rho = 1$

Use double integration in Problems 13-18 to find the center of mass of a lamina covering the given region in the plane and having the specified density ρ.

13. $\rho(x,y) = x^2 + y^2$
    over $x^2 + y^2 \leq 9, y \geq 0$
14. $\rho(x, y) = k(x^2 + y^2)$
    over $x^2 + y^2 \leq a^2, y \geq 0$
15. $\rho(x, y) = 7x$ over the triangle with vertices $(0,0), (6,5)$, and $(12,0)$
16. $\rho(x, y) = 3x$ over the region bounded by $y = 0$, $y = x^2$, and $x = 6$
17. $\rho(x, y) = x^{-1}$ over the region bounded by $y = \ln x$, $y = 0, x = 2$
18. $\rho(x, y) = y$ over the region bounded by $y = e^{-x}$, $x = 0$, $x = 2$, $y = 0$
19. A lamina in the xy-plane has the shape of the semi-circular region $x^2 + y^2 \leq a^2$, $y \geq 0$. Find the center of mass if the density at any point in the lamina is:
    a. directly proportional to the distance of the point from the origin
    b. directly proportional to the polar angle
20. A lamina has the shape of a semicircular region $x^2 + y^2 \leq a^2, y \geq 0$. Find the center of mass of the lamina if the density at each point is directly proportional to the square of the distance from the point to the origin.
21. Find the centroid of a homogeneous lamina that covers the region bounded by the curve $y = \ln x$ and the lines $x = e^2$, and $y = 0$.
22. Find I_x, the moment of inertia about the x-axis, of the lamina that covers the region bounded by the graph of $y = 1 - x^2$ and the x-axis, if the density is $\rho(x, y) = x^2$.
23. Find I_z, the moment of inertia about the z-axis, of the lamina that covers the square in the plane with vertices $(-1, -1), (1, -1), (1, 1)$, and $(-1, 1)$, if the density is $\rho(x, y) = x^2y^2$.

Level 2

24. Find the center of mass of the cardioid $r = 1 + \sin\theta$ if the density at each point (r, θ) is $\rho(r, \theta) = r$.

25. Find the centroid (correct to the nearest hundredth) of the loop of the lemniscate $r^2 = 2\sin 2\theta$ that lies in the first quadrant.

26. Find the centroid (correct to the nearest hundredth) of the part of the large loop of the limaçon $r = 1 + 2\cos\theta$ that does not include the small loop.

27. Find the center of mass of the lamina that covers the triangular region with vertices $(0,0), (a,0), (a,b)$, if a and b are both positive and the density at $P(x, y)$ is directly proportional to the distance of P from the y-axis.

28. A rectangular lamina has vertices $(0,0), (a,0), (a,b), (0,b)$ and its density at any point (x, y) is the product $\rho(x, y) = xy$. Find the center of mass of the plate.

29. A homogeneous lamina covers the circular disk with boundary $x^2 + y^2 = ax$. Find the moment of inertia of this circular plate about the vertical line passing through the center of the lamina. Assume $\rho = 1$. *Hint:* Use polar coordinates.

30. Show that a homogeneous lamina of density ρ and mass m that covers the circular region $x^2 + y^2 = a^2$ will have moment of inertia $ma^2/4$ with respect to both the x- and y-axes. What is the moment of inertia of the lamina with respect to the z-axis?

31. Show that a homogeneous lamina of mass m that covers the ellipse

$$\frac{x^2}{a^2} + \frac{y^2}{b^2} \leq 1$$

has moment of inertia about the x-axis equal to

$$I_x = \frac{\pi}{4}ab^3\rho = \frac{1}{4}mb^2$$

where $m = \underbrace{\rho(\pi ab)}_{\text{Area of ellipse}}$

32. Find the center of mass of the tetrahedron in the first octant bounded by the plane

$$\frac{x}{a} + \frac{y}{b} + \frac{z}{c} = 1$$

where a, b, and c are all positive constants. Assume the density is $\rho = x$.

33. A solid has the shape of the homogeneous sphere $x^2 + y^2 + z^2 \leq a^2$. Find the centroid of the part of the solid in the first octant $(x \geq 0, y \geq 0, z \geq 0)$.

34. Suppose the joint probability density function for the random variables X and Y is

$$f(x, y) = \begin{cases} 2e^{-2x}e^{-y} & \text{if } x \geq 0, y \geq 0 \\ 0 & \text{otherwise} \end{cases}$$

Find the probability that $X + Y \leq 1$.

35. Suppose the joint probability density function for the random variables X and Y is

$$f(x, y) = \begin{cases} xe^{-x}e^{-y} & \text{if } x \geq 0, y \geq 0 \\ 0 & \text{otherwise} \end{cases}$$

Find the probability that $X + Y \leq 1$.

36. Modeling Problem Suppose X measures the length of time (in days) that a person stays in the hospital after abdominal surgery, and Y measures the length of time (in days) that a person stays in the hospital after orthopedic surgery. On Monday, the patient in bed 107A undergoes an emergency appendectomy (abdominal surgery), while the patient's roommate in bed 107B undergoes orthopedic surgery for the repair of torn knee cartilage. Suppose the joint probability density function for X and Y is

$$f(x, y) = \begin{cases} \frac{1}{6}e^{-x/2}e^{-y/3} & \text{if } x \geq 0, y \geq 0 \\ 0 & \text{otherwise} \end{cases}$$

Find the probability (to the nearest percent) that both patients will be discharged from the hospital within 3 days.

37. Modeling Problem Suppose X measures the time (in minutes) that a person stands in line at a certain bank and Y, the duration (in minutes) of a routine transaction at the teller's window. You arrive at the bank to deposit a check. If the joint probability density function for X and Y is modeled by

$$f(x, y) = \begin{cases} \frac{1}{8}e^{-x/2}e^{-y/4} & \text{if } x \geq 0, y \geq 0 \\ 0 & \text{otherwise} \end{cases}$$

Find the probability to the nearest percent that you will complete your business at the bank within 8 min.

38. Modeling Problem Suppose X measures the time (in minutes) that a person spends with an insurance agent choosing a life insurance policy and Y, the time (in minutes) that the agent spends doing the paperwork once the client has decided. If the joint probability density function for X and Y is

$$f(x, y) = \begin{cases} \frac{1}{300}e^{-x/30}e^{-y/10} & \text{if } x \geq 0, y \geq 0 \\ 0 & \text{otherwise} \end{cases}$$

Find the probability to the nearest percent that the entire transaction will occur in a half hour or less.

39. **Modeling Problem** Racing yachts, such as those in the America's Cup competition, benefit from sophisticated, computer-enhanced construction techniques.* For example, define the *center of pressure* on a boat's sail as the point $(\bar{x}, \bar{y})$ where all aerodynamic forces appear to act. Suppose a sail occupies the triangular region R in the plane, as illustrated in Figure 12.55, and that $\bar{x}$ and $\bar{y}$ are modeled by the formulas

$$\bar{x} = \frac{\iint\limits_{R} xy\, dA}{\int\int\limits_{R} y\, dA} \quad \bar{y} = \frac{\iint\limits_{R} y^2\, dA}{\iint\limits_{R} y\, dA}$$

Calculate the center of pressure on this sail.

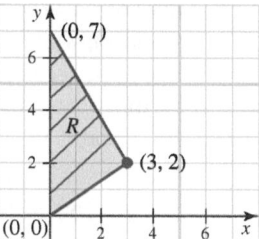

Figure 12.55 Triangular sail

The average value of the continuous function f over R is given by
One variable:

$$\frac{1}{\text{length of segment } R}\int\limits_{R} f(x)\, dx$$

Two variables:

$$\frac{1}{\text{area of region } R}\iint\limits_{R} f(x, y)\, dA$$

Three variables:

$$\frac{1}{\text{volume of solid } R}\iiint\limits_{R} f(x, y, z)\, dV$$

Use these definitions in Problems 40-43.

40. Find the average value of $f(x, y) = e^{x^3}$, where R is the region in the first quadrant bounded by $y = x^2$, $y = 0$, and $x = 1$.

41. Find the average value of $f(x, y) = e^x y^{-1/2}$ where R is the region in the first quadrant bounded by $y = x^2$, $x = 0$, and $y = 1$.

42. Find the average value of the function $f(x, y, z) = x + 2y + 3z$ over the solid region S bounded by the tetrahedron with vertices $(0, 0, 0)$, $(1, 0, 0)$, $(0, 1, 0)$ and $(0, 0, 1)$.

43. Find the average value of the function $f(x, y, z) = xyz$ over the solid sphere $x^2 + y^2 + z^2 \leq 1$.

44. **EXPLORATION PROBLEM** Perform a literature search in order to find a general formula for the moment of inertia of a continuous solid body rotating about a given axis. Who contributed to this theoretical result? Provide references.

*The **radius of gyration** for revolving a region R with mass m about an axis of rotation L, with moment of inertia I, is*

$$d = \sqrt{\frac{I}{m}}$$

Note that if the entire mass m of R is located at a distance d from the axis of rotation L, then R would have the same moment of inertia. Use this definition in Problems 45-47.

45. A homogeneous lamina has the shape of the right triangle in the xy-plane with vertices $(0, 0)$, $(a, 0)$, and $(0, b)$, $a > 0$, $b > 0$. Find the radius of gyration of the lamina about the z-axis.

46. Find the radius of gyration about the x-axis of the semicircular region $x^2 + y^2 \leq a$, $y \geq 0$ given that the density at (x, y) is directly proportional to the distance of the point from the x-axis.

47. Let R be the lamina bounded by the parabola $y = x^2$ and the lines $x = 2$ and $y = 1$, with density $\rho(x, y) = x^2 y$. What is the radius of gyration (to four decimal places) about the x-axis?

48. **EXPLORATION PROBLEM** A solid has the shape of the rectangular parallelepiped given by $-a \leq x \leq a, -b \leq y \leq b, -c \leq z \leq c$, and its density is $\rho(x, y, z) = x^2 y^2 z^2$.

 a. Guess the location of the center of mass and value of the moment of inertia about the z-axis.

 b. Check your response to part **a** by direct calculation.

49. **Modeling Problem** An industrial plant is located on a narrow river. Suppose C_0 units of pollutant are released into the river at time $t = 0$ and that the concentration of pollutant t hours later at a point x miles downstream from the plant is modeled by the diffusion function

$$C(x, t) = \frac{C_0}{\sqrt{k\pi t}}e^{-x^2/(4kt)}$$

where k is a physical constant.

---

*See *Scientific American* (August 1987).

a. At what time $t_m(x_0)$ does the maximum pollution occur at point $x = x_0$ miles from the plant? What is the maximum concentration $C_m(x_0)$?

b. Define the *danger zone* to be the portion of the riverbank such that $0 \le x \le x_m$ where x_m is the largest value of x such that $C_m(x_m) \ge 0.25C_0$. Find x_m.

c. *Set up* a double integral for the average concentration of pollutant over the set of all (x, t) such that $0 \le t \le t_m(x)$ for each fixed x between $x = 0$ and $x = x_m$.

d. How would you define the "dangerous period" for the pollution spill?

50. Modeling Problem The *stiffness* of a horizontal beam is modeled to be proportional to the moment of inertia of its cross section with respect to a horizontal line L through its centroid, as illustrated for three shapes shown in Figure 12.56.

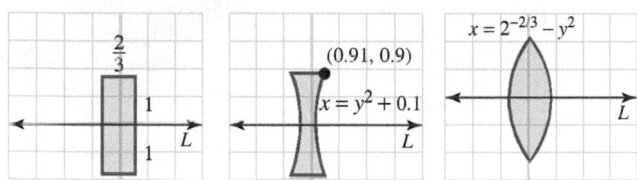

Figure 12.56 Cross sections of a horizontal beam with an area 4/3

Which of the illustrated beams is the stiffest? For this model, assume the constant of proportionality is the same for all three cases and $\rho = 1$.

EXPLORATION PROBLEMS *Often the knowledge of the center of mass allows us to greatly simplify a problem, but sometimes it can lead us to an incorrect answer. We explore this in Problems 51-53.*

51. Centers of mass in physics Suppose a volume of liquid (or granular solid) is located so that its center of mass is at the origin. The material is to be lifted to a height of h units above the center of mass. Starting with a small element of volume, ΔV, to be lifted to height h, derive the formula for the total work done:

$$\text{WORK} = \delta \iiint (h - z)\, dV$$

where δ is the weight (density) of the material.

52. Explain why the integral in Problem 51 simplifies to the formula

$$\text{WORK} = \delta V h = \text{FORCE} \times \text{DISTANCE}$$

where h is the distance between the center of mass and the level to which the material is lifted.

53. Picture a cylindrical tank of radius 6 ft and height 10 ft positioned so its center of mass is at the origin. Compute the work required to lift (that is, pump) the contents (of density δ) to a level of 15 ft above the center of mass. Compute this two ways: by the integral in Problem 51 (suggestion — do the $dx\, dy$ integral by inspection), and also by simply lifting the center of mass to the required height. These results should agree.

EXPLORATION PROBLEMS *Often the knowledge of the center of mass allows us to greatly simplify a problem, but sometimes it can lead us to an incorrect answer. In Problems 54-58, we recall that if two masses m and M are each concentrated at (or nearly at) a point and are separated by a distance p and G is the gravitational constant, then the attractive force is expressed by*

$$F = GmM / p^2$$

What physicists and others prefer to do in the case of real-life masses (which occupy some volume) is to use p, the distance between the two centers of mass. The question you are to explore here is whether, and when, this simplifying way of computing attracting forces is justified.

Apparatus for testing Newton's inverse square law.

54. Newton's inverse square law Consider a one point mass, m, that is located at $(0, 0, h)$. A second mass, M, corresponds to a volume V with constant density ρ. We are interested in the total *resultant* force that M exerts on m, and we assume this to be in the z-direction. Hence we will sum the vertical components of the elementary forces acting on m. Consider, in M, an infinitesimal element of volume dV (centered at (x, y, z)) and argue that the magnitude of the force exerted on m and its vertical component are:

$$dF_{\text{mag}} = \frac{Gm\rho \, dV}{p^2} \quad \text{and} \quad dF = \frac{Gm\rho \, dV(h-z)}{p^3}$$

where $p^2 = x^2 + y^2 + (h-z)^2$. Note that the term $(h-z)/p$ projects the force vector onto the z-axis, giving the vertical component. Do you see why?

55. Using the result in the previous problem, integrate the elemental forces in order to obtain the total force of attraction. Prove that you obtain the formula:

$$F = Gm\rho \iiint \frac{(h-z)\, dV}{p^3}$$

56. Consider a solid (with density $\rho = 5$) positioned at $-2 \le x \le 2, -4 \le y \le 4$, and $-1 \le z \le 1$. (Note the center of mass is at the origin.) Compute the attracting force between this mass and a point mass, m, at $(0, 0, 8)$ using the center of mass formula $F = GmM/p^2$.

57. For the masses in Problem 56, compute F using the integration formula in Problem 55; compare the result with that of Problem 56.

58. Repeat the calculation in Problem 57 with $h = 200$ and compare with the center of mass approximation. Comment on the role that the separating distance p plays.

Level 3

59. **Area theorem of Pappus** Prove the following area theorem of Pappus: Let C be a curve of length L in the plane. Then the surface obtained by rotating C about the axis L in the plane has area $2\pi Lh$, where h is the distance from the centroid of C to the axis of rotation.

60. A torus (doughnut) can be formed by rotating the circle $(x - b)^2 + y^2 = a^2$ for $b > a$ about the y-axis. Find the surface area of the torus by applyingy Pappus' area theorem (see Problem 59 and Figure 12.57). Compare your result with the area found parametrically in Problem 58, Section 12.4.

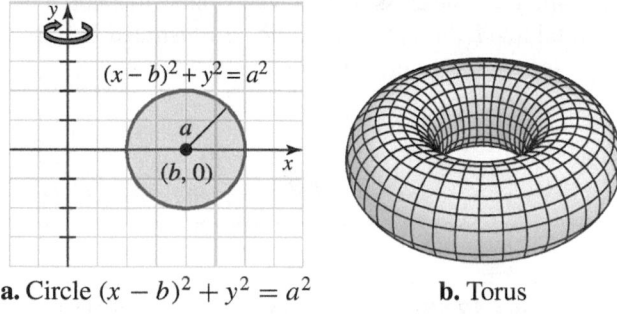

a. Circle $(x - b)^2 + y^2 = a^2$ **b.** Torus

Figure 12.57   Problem 60

12.7 CYLINDRICAL AND SPHERICAL COORDINATES

IN THIS SECTION: *Cylindrical coordinates, integration with cylindrical coordinates, spherical coordinates, integration with spherical coordinates*

We have seen that certain curves have simpler descriptions in polar coordinates than in rectangular coordinates. In this section, we introduce *cylindrical* and *spherical* coordinate systems, which are equally useful for describing certain surfaces and solids in space.

Cylindrical Coordinates

Cylindrical Coordinates are a generalization of polar coordinates to surfaces in $\mathbb{R}^3$. Recall that the point P with rectangular coordinates (x, y, z) is located z units above the point $Q(x, y, 0)$ in the xy-plane (below if $z < 0$). In cylindrical coordinates, we measure the point in the xy-plane in polar coordinates, with the same z-coordinate as in the Cartesian coordinate system. These relationships are shown in Figure 12.58.

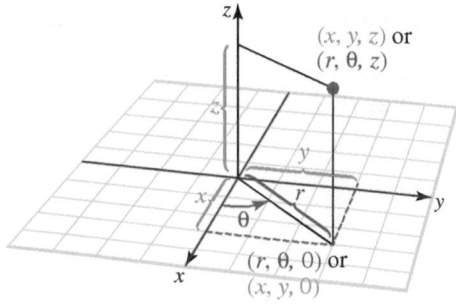

Figure 12.58   The cylindrical.coordinate system

Cylindrical coordinates are convenient for representing cylindrical surfaces and surfaces of revolution for which the z-axis is the axis of symmetry. Some examples are shown in Figure 12.59.

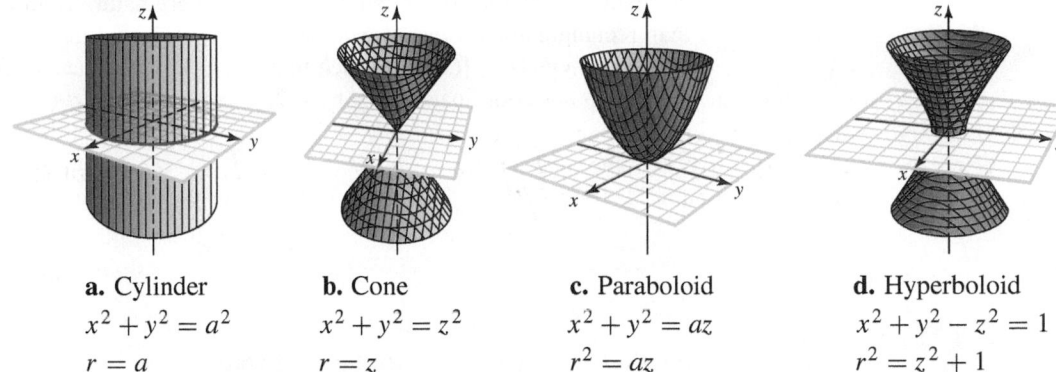

	a. Cylinder	**b.** Cone	**c.** Paraboloid	**d.** Hyperboloid
Rectangular equation:	$x^2 + y^2 = a^2$	$x^2 + y^2 = z^2$	$x^2 + y^2 = az$	$x^2 + y^2 - z^2 = 1$
Cylindrical equation:	$r = a$	$r = z$	$r^2 = az$	$r^2 = z^2 + 1$

Figure 12.59 Surfaces with convenient cylindrical coordinates

We have the following conversion formulas, which follow directly from the rectangular-polar conversions.

CONVERSION FORMULAS RECTANGULAR/CYLINDRICAL

Cylindrical to rectangular: **Rectangular to cylindrical:**

(r, θ, z) to (x, y, z) $x = r \cos \theta$ (x, y, z) to (r, θ, z) $r = \sqrt{x^2 + y^2}$

$\qquad\qquad\qquad\qquad y = r \sin \theta \qquad\qquad\qquad\qquad\qquad\qquad \tan \theta = \dfrac{y}{x}$

$\qquad\qquad\qquad\qquad z = z \qquad\qquad\qquad\qquad\qquad\qquad\qquad\quad z = z$

Example 1   **Rectangular-form equation converted to cylindrical-form equation**

Find an equation in cylindrical coordinates for the elliptic paraboloid $z = x^2 + 3y^2$.

Solution   We use the conversion formulas $x = r \cos \theta$ and $y = r \sin \theta$.

$$z = x^2 + 3y^2$$
$$= (r \cos \theta)^2 + 3 (r \sin \theta)^2$$
$$= r^2(\cos^2 \theta + 3 \sin^2 \theta)$$
$$= r^2 \left[(1 - \sin^2 \theta) + 3 \sin^2 \theta \right]$$
$$= r^2 (1 + 2 \sin^2 \theta)$$

Integration with cylindrical coordinates

A triple integral $\iiint_D f(x, y, z)\,dV$ can often be evaluated by transforming to cylindrical coordinates if the region of integration D is z-simple and the projection of D onto the xy-plane is a region A that can be described more naturally in terms of polar coordinates than rectangular coordinates. Suppose $f(x, y, z)$ is continuous over the region of integration D, where $D = \{(x, y, z)$ such that $u(x, y) \leq z \leq v(x, y)$ for all (x, y) in $A\}$. Then, since in polar coordinates, $x = r\cos\theta$, $y = r\sin\theta$, and $dA = r\,dr\,d\theta$, we have:

$$\iiint_D f(x, y, z)\,dV = \iint_A \left[\int_{u(x,y)}^{v(x,y)} f(x, y, z)\,dz \right] dA$$

$$= \iint_A \int_{u(x,y)}^{v(x,y)} f(r\cos\theta, r\sin\theta, z)\, r\,dz\,dr\,d\theta$$

(recall Theorem 12.4 and see Figure 12.60).
Thus, in cylindrical coordinates $dV = r\,dr\,d\theta\,dz$. Finally, the projected region is A, where $A = \{(r, \theta)$ such that $g_1(\theta) \leq r \leq g_2(\theta)$ for $\alpha \leq \theta \leq \beta\}$. Transforming the given integral to cylindrical coordinates yields the form shown in the following box.

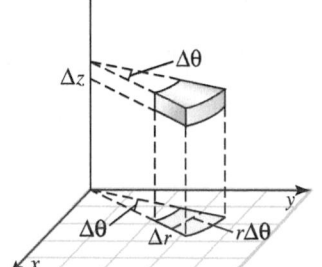

Figure 12.60 Volume element in cylindrical coordinates

> **TRIPLE INTEGRAL IN CYLINDRICAL COORDINATES** Let D be a solid with upper surface $z = v(r, \theta)$ and lower surface $z = u(r, \theta)$, and let A be the projection of the solid onto the xy-plane expressed in polar coordinates. Then, if $f(x, y, z)$ is continuous on D, the **triple integral of f over D** is
>
> $$\iiint_D f(x, y, z)\,dV = \int_\alpha^\beta \int_{g_1(\theta)}^{g_2(\theta)} \int_{u(r,\theta)}^{v(r,\theta)} f(r\cos\theta, r\sin\theta, z)\, r\,dz\,dr\,d\theta$$

Figure 12.61 illustrates how the region of integration D can be determined by using cylindrical coordinates.

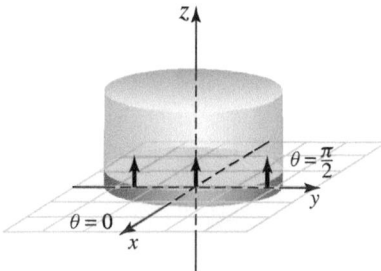

a. Integrate with respect to z

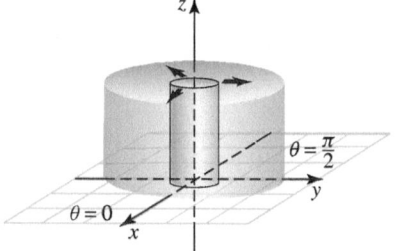

b. Integrate with respect to r

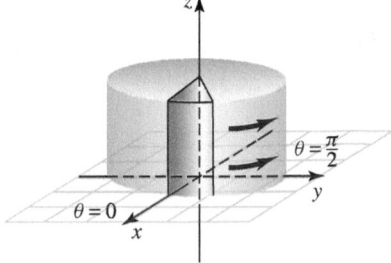

c. Integrate with respect to θ

Figure 12.61 Determining regions of integration in cylindrical coordinates

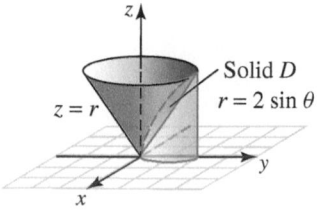

Figure 12.62 Solid in Example 2

Example 2 Finding volume in cylindrical coordinates

Find the volume of the solid in the first octant that is bounded by the cylinder $x^2 + y^2 = 2y$, the half-cone $z = \sqrt{x^2 + y^2}$, and the xy-plane.

Solution Let D be the region occupied by the solid as shown in Figure 12.62.

This surface is most easily described in cylindrical coordinates.

Cylinder:	Cone:
$x^2 + y^2 = 2y$	$z = \sqrt{x^2 + y^2}$
$r^2 = 2r\sin\theta$	$z = r$
$r = 2\sin\theta$	

Since the region D lies in the first octant, we have $0 \le \theta \le \frac{\pi}{2}$, so D may be described by

$$0 \le z \le r \qquad 0 \le r \le 2\sin\theta \qquad 0 \le \theta \le \frac{\pi}{2}$$

$$V = \iiint_D dV$$

$$= \iint_A \int_{u(r,\theta)}^{v(r,\theta)} r\, dz\, dr\, d\theta$$

$$= \int_0^{\pi/2} \int_0^{2\sin\theta} \int_0^r r\, dz\, dr\, d\theta$$

$$= \int_0^{\pi/2} \int_0^{2\sin\theta} r^2\, dr\, d\theta$$

$$= \int_0^{\pi/2} \left. \frac{r^3}{3} \right|_{r=0}^{r=2\sin\theta} d\theta$$

$$= \frac{8}{3} \int_0^{\pi/2} \sin^3\theta\, d\theta$$

$$= \frac{8}{3} \int_0^{\pi/2} (1 - \cos^2\theta)\sin\theta\, d\theta$$

$$= \frac{8}{3} \left[-\cos\theta + \frac{\cos^3\theta}{3} \right]_0^{\pi/2}$$

$$= \frac{16}{9}$$

Example 3 Centroid in cylindrical coordinates

A homogeneous solid D with constant density ρ is bounded below by the xy-plane, on the sides by the cylinder $x^2 + y^2 = a^2$ $(a > 0)$, and above by the paraboloid $z = x^2 + y^2$. Find the centroid of the solid.

Solution The solid D is shown in Figure 12.63.
Because the solid is bounded by a cylinder, we will carry out the integration in cylindrical coordinates.

Cylinder:	Paraboloid:
$x^2 + y^2 = a^2$	$z = x^2 + y^2$
$r^2 = a^2$	$z = r^2$
$r = a \quad (a > 0)$	

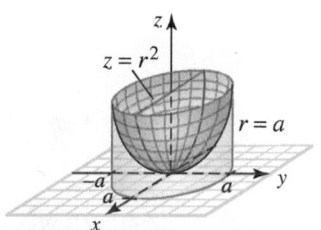

Figure 12.63 The solid D

Let $(\bar{x}, \bar{y}, \bar{z})$ denote the centroid. Using symmetry and the fact that the density function is constant, we have $\bar{x} = \bar{y} = 0$. Let m denote the mass of D. Since the projected region is $r = a$ for $0 \leq \theta \leq 2\pi$, we find that

$$\bar{z} = \frac{M_{xy}}{m} = \frac{\iiint\limits_{D} zr \; \rho \; dz \; dr \; d\theta}{\iiint\limits_{D} r \; \rho \; dz \; dr \; d\theta} = \frac{\int_0^{2\pi} \int_0^a \int_0^{r^2} zr \; dz \; dr \; d\theta}{\int_0^{2\pi} \int_0^a \int_0^{r^2} r \; dz \; dr \; d\theta} = \frac{\frac{\pi}{6}a^6}{\frac{\pi}{2}a^4} = \frac{a^2}{3}$$

The centroid is $\left(0, 0, \dfrac{a^2}{3} \right)$. ∎

Spherical Coordinates

In **spherical coordinates** we label a point P by a triple (ρ, θ, ϕ), where ρ, θ, and ϕ are numbers determined as follows (refer to Figure 12.64):

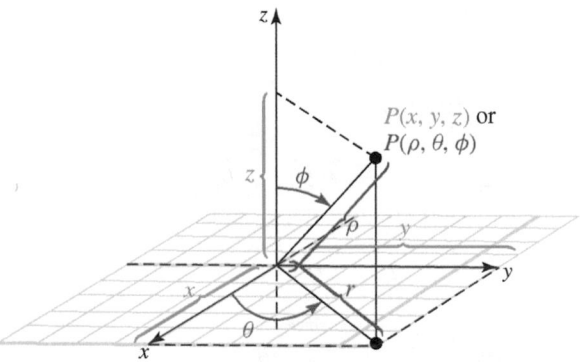

$\rho = $ the distance from the origin to the point P; we require $\rho \geq 0$.
$\theta = $ the polar angle (as in polar coordinates); we require $0 \leq \theta \leq 2\pi$.
$\phi = $ the angle measured down from the positive z-axis to the ray from the origin through P we require $0 \leq \phi \leq \pi$.

Figure 12.64   The spherical coordinate system

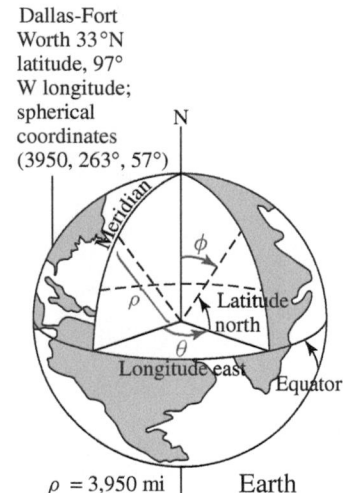

Dallas-Fort Worth 33°N latitude, 97° W longitude; spherical coordinates (3950, 263°, 57°)

$\rho = 3{,}950$ mi | Earth

Figure 12.65   Spherical coordinates on the earth's surface

You might recognize that spherical coordinates are related to the longitude and latitude coordinates used in navigation. To be more specific, consider a rectangular coordinate system with the origin at the center of the earth, with the positive z-axis passing through the north pole and the xz-plane passing through the prime meridian. Then, a particular location on the surface is denoted by (ρ, θ, ϕ), where ρ is the distance from the center of the earth, θ is the longitude, and $\frac{\pi}{2} - \phi$ is the latitude (since latitude is the angle *up* from the equator). For example, Dallas-Forth Worth has coordinates $(\rho, \theta, \phi) = (3{,}950, 263°, 57°)$, as shown in Figure 12.65.

Spherical coordinates are desirable when representing spheres, cones, or certain planes. Some examples are shown in Figure 12.66.

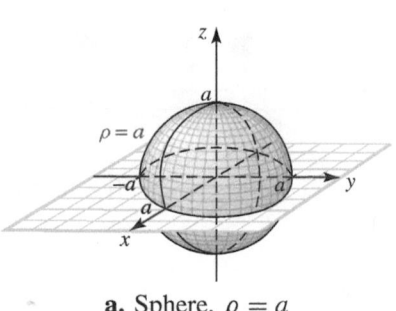

a. Sphere, $\rho = a$

$a > 0$

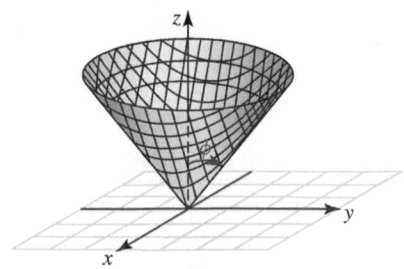

b. Half-cone, $\phi = a$

$0 < a < \dfrac{\pi}{2}$

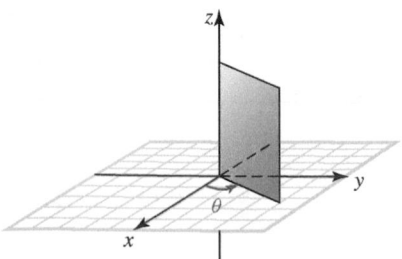

c. Vertical half-plane, $\theta = a$

$0 \leq \theta < 2\pi$

Figure 12.66   Surfaces with convenient spherical coordinates

We use the relationships in Figure 12.64 to obtain the remaining conversion formulas. All of these formulas can be derived by considering the relationships among the variables. We repeat Figure 12.64 for convenience.

CONVERSION FORMULAS SPHERICAL COORDINATES

Spherical to rectangular $x = \rho \sin\phi \cos\theta$

(ρ, θ, ϕ) **to** (x, y, z) $y = \rho \sin\phi \sin\theta$

$z = \rho \cos\phi$

Spherical to cylindrical $r = \rho \sin\phi$

(ρ, θ, ϕ) **to** (r, θ, z) $\theta = \theta$

$z = \rho \cos\phi$

Rectangular to spherical $\rho = \sqrt{x^2 + y^2 + z^2}$

(x, y, z) **to** (ρ, θ, ϕ) $\tan\theta = \dfrac{y}{x}$

$\phi = \cos^{-1}\left(\dfrac{z}{\sqrt{x^2 + y^2 + z^2}}\right)$

Cylindrical to spherical $\rho = \sqrt{r^2 + z^2}$

(r, θ, z) **to** (ρ, θ, ϕ) $\theta = \theta$

$\phi = \cos^{-1}\left(\dfrac{z}{\sqrt{r^2 + z^2}}\right)$

Example 4   Converting rectangular-form equations to spherical-form equations

Rewrite each of the given equations in spherical form.

a. the sphere $x^2 + y^2 + z^2 = a^2$ $(a > 0)$
b. the paraboloid $z = x^2 + y^2$

Solution

a. Because $\rho = \sqrt{x^2 + y^2 + z^2}$, we see $x^2 + y^2 + z^2 = \rho^2$, so we can write

$$\rho^2 = a^2$$

$$\rho = a \qquad \textit{Because } \rho \geq 0$$

b. $z = x^2 + y^2$

$$\rho \cos\phi = (\rho \sin\phi \cos\theta)^2 + (\rho \sin\phi \sin\theta)^2$$

$$= \rho^2 \sin^2\phi \, \cos^2\theta + \rho^2 \sin^2\phi \, \sin^2\theta$$

$$= \rho^2 \sin^2\phi (\cos^2\theta + \sin^2\theta)$$

$$\rho = \frac{\cos\phi}{\sin^2\phi}$$

$$= \cot\phi \csc\phi$$

Integration with Spherical Coordinates

For a solid D in spherical coordinates, the fundamental element of volume is a spherical "wedge" bounded in such a way that

$$\rho_1 \le \rho \le \rho_1 + \Delta\rho \qquad \phi_1 \le \phi \le \phi_1 + \Delta\phi, \qquad \theta_1 \le \theta \le \theta_1 + \Delta\theta$$

This "wedge" is shown in Figure 12.67.

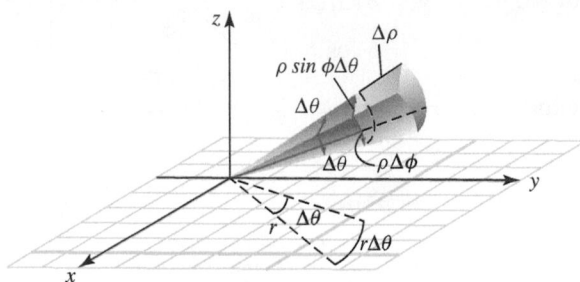

Figure 12.67   A spherical wedge

In Section 12.8 we show that the volume of the wedge is approximated by

$$dV = \rho^2 \sin\phi \, d\rho \, d\phi \, d\theta$$

Using this formula, we can form partitions and take a limit of a Riemann sum as the partitions are refined to obtain the integral form shown in the following box.

TRIPLE INTEGRAL IN SPHERICAL COORDINATES   If f is continuous on the bounded solid region D, then the **triple integral of f over D** is given by

$$\iiint\limits_{D} f(x, y, z) \, dV$$

$$= \iiint\limits_{\overline{D}} f(\rho \sin\phi \cos\theta, \rho \sin\phi \sin\theta, \rho \cos\phi) \, \rho^2 \sin\phi \, d\rho \, d\theta \, d\phi$$

where $\overline{D}$ is the region D expressed in spherical coordinates.

Example 5   Volume of a sphere

In geometry, it is shown that a sphere of radius R has volume $V = \frac{4}{3}\pi R^3$. Verify this formula using integration.

Solution   It seems clear that we should work in spherical coordinates with the origin of the coordinate system at the center of the sphere, because the equation of the sphere is $\rho = R$ for $0 \le \theta \le 2\pi$ and $0 \le \phi \le \pi$.

$$V = \iiint\limits_{D} dV$$

$$= \int_0^{2\pi} \int_0^{\pi} \int_0^{R} \underbrace{\rho^2 \sin\phi \, d\rho \, d\phi \, d\theta}_{dV}$$

$$= \int_0^{2\pi} \int_0^{\pi} \frac{\rho^3}{3} \sin \phi \Big|_{\rho=0}^{\rho=R} d\phi \, d\theta$$

$$= \frac{R^3}{3} \int_0^{2\pi} \int_0^{\pi} \sin \phi \, d\phi \, d\theta$$

$$= \frac{R^3}{3} \int_0^{2\pi} (-\cos \phi) \Big|_{\phi=0}^{\phi=\pi} d\theta$$

$$= \frac{R^3}{3} \int_0^{2\pi} 2 \, d\theta$$

$$= \frac{2R^3}{3} \theta \Big|_{\theta=0}^{\theta=2\pi}$$

$$= \frac{4}{3} \pi R^3$$

Example 6   Moment of inertia using spherical coordinates

A toy top of constant density ρ_0 is constructed from a portion of a solid hemisphere with a conical base, as shown in Figure 12.68.

The center of the spherical cap is at the point where the top spins, and the height of the conical base is equal to its radius. Find the moment of inertia of the top about its axis of symmetry.

Solution   We use a Cartesian coordinate system in which the z-axis is the axis of symmetry of the top. Suppose the cap is part of the hemisphere $z = \sqrt{R^2 - x^2 - y^2}$. Since the height of the conical base is equal to its radius, the cone makes an angle of $\pi/4$ radians (45°) with the z-axis. Let D denote the solid region occupied by the top. Then the moment of inertia I_z with respect to the z-axis is given by

$$I_z = \iiint_D (x^2 + y^2) dm = \iiint_D (x^2 + y^2) \rho_0 \, dV$$

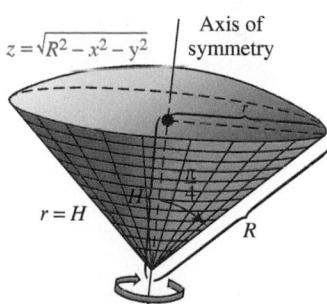

Figure 12.68 Moment of inertia for a spinning top

The shape of the top suggests that we convert to spherical coordinates, and we find that D can be described as $0 \le \theta \le 2\pi, 0 \le \phi \le \frac{\pi}{4}, 0 \le \rho \le R$. Also,

$$x^2 + y^2 = \rho^2 \sin^2 \phi \cos^2 \theta + \rho^2 \sin^2 \phi \, \sin^2 \theta = \rho^2 \sin^2 \phi$$

We now evaluate the integral using spherical coordinates:

$$I_z = \iiint_D (x^2 + y^2) \rho_0 \, dV$$

$$= \rho_0 \int_0^{\pi/4} \int_0^{2\pi} \int_0^R \underbrace{\rho^2 \sin^2 \phi}_{x^2+y^2} \underbrace{\rho^2 \sin \phi \, d\rho \, d\theta \, d\phi}_{dV}$$

$$= \rho_0 \int_0^{\pi/4} \int_0^{2\pi} \int_0^R \rho^4 \sin^3 \phi \, d\rho \, d\theta \, d\phi$$

$$= \rho_0 \int_0^{\pi/4} \int_0^{2\pi} \frac{\rho^5}{5} \sin^3\phi \Big|_0^R d\theta \, d\phi$$

$$= \frac{R^5 \rho_0}{5} \int_0^{\pi/4} (2\pi - 0) \sin^3 \phi \, d\phi$$

$$= \frac{2R^5 \rho_0 \pi}{5} \left[-\cos\phi + \frac{1}{3}\cos^3\phi \right] \Bigg|_0^{\pi/4}$$

$$= \frac{2R^5 \rho_0 \pi}{5} \left[-\frac{1}{2}\sqrt{2} + \frac{1}{12}\sqrt{2} + 1 - \frac{1}{3} \right]$$

$$= \frac{\pi \rho_0 R^5}{30} (8 - 5\sqrt{2})$$

It was clear in Example 5 that we should use the spherical coordinate system to find the volume of the interior of a sphere. However, sometimes it is not obvious which coordinate system to use. We address this question with the following example.

Example 7   Deciding which coordinate system to apply to an integral

Evaluate the integral

$$I = \int_{-1}^1 \int_{-\sqrt{1-x^2}}^{\sqrt{1-x^2}} \int_{x^2+y^2}^{\sqrt{2-x^2-y^2}} z \, dz \, dy \, dx$$

using the given rectangular form, or by transforming to either cylindrical or spherical coordinates.

Solution   The region of integration is D, where $D = \{(x, y, z)$ such that $x^2 + y^2 \le z \le \sqrt{2 - x^2 - y^2}$ for $-\sqrt{1-x^2} \le y \le \sqrt{1-x^2}$ and $-1 \le x \le 1\}$. Geometrically, D is the region bounded above by the sphere $x^2 + y^2 + z^2 = 2$ and below by the paraboloid $z = x^2 + y^2$. The bounding surfaces intersect where $z = x^2 + y^2$ and $x^2 + y^2 + z^2 = 2$ so

$$z + z^2 = 2 \quad \textit{Substitute the first into the second.}$$

$$(z - 1)(z + 2) = 0$$

$$z = 1 \quad \textit{Reject } z = -2 \textit{ since } z \ge 0.$$

It follows that the projected region in the xy-plane is the disk $x^2 + y^2 \le 1$ (see Figure 12.69).

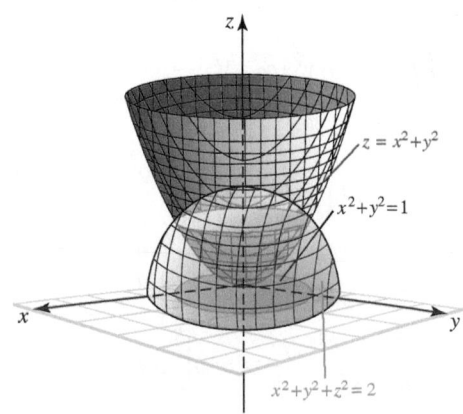

Figure 12.69   The solid D and the projected region in the xy-plane

It may seem that transforming to spherical coordinates may result in a less compli-cated integration, since the upper boundary of D has the simple form $\rho = \sqrt{2}$, but the lower boundary $z = x^2 + y^2$ has the form

$$z = x^2 + y^2 = r^2$$

$$\rho \cos\phi = (\rho\ \sin\ \phi)^2$$

$$\rho = \frac{\cos\phi}{\sin^2\phi} = \cot\phi\ \csc\phi$$

Moreover, for $0 \le \phi \le \frac{\pi}{4}$, we have $0 \le \rho \le \sqrt{2}$, but $0 \le \rho \le \cot\phi\,\csc\phi$ for $\frac{\pi}{4} \le \phi \le \frac{\pi}{2}$, which means we need two integrals to evaluate the integral in spherical coordinates:

$$I = \int_0^{2\pi} \int_0^{\pi/4} \int_0^{\sqrt{2}} (\rho\cos\phi)\rho^2 \sin\phi\, d\rho\, d\phi\, d\theta$$

$$+ \int_0^{2\pi} \int_{\pi/4}^{\pi/2} \int_0^{\cot\phi\,\csc\phi} (\rho\cos\phi)\rho^2 \sin\phi\, d\rho\, d\phi\, d\theta$$

However, if we use cylindrical coordinates, the integral has the relatively simple form

$$I = \int_0^{2\pi} \int_0^1 \int_{r^2}^{\sqrt{2-r^2}} zr\, dz\, dr\, d\theta$$

$$= \int_0^{2\pi} \int_0^1 r\left[\frac{1}{2}z^2\right]_{r^2}^{\sqrt{2-r^2}} dr\, d\theta$$

$$= \int_0^{2\pi} \int_0^1 \frac{1}{2}r\left[(2-r^2) - r^4\right] dr\, d\theta$$

$$= \frac{1}{2}\int_0^{2\pi} \left[r^2 - \frac{1}{4}r^4 - \frac{1}{6}r^6\right]\Big|_0^1 d\theta$$

$$= \frac{1}{2}\left(\frac{7}{12}\right)\int_0^{2\pi} d\theta$$

$$= \frac{7}{24}(2\pi)$$

$$= \frac{7\pi}{12}$$

PROBLEM SET 12.7

Level 1

1. ■ *What does this say?* Compare and contrast the rect-angular, cylindrical, and spherical coordinate systems.
2. ■ *What does this say?* Suppose you need to evaluate a particular triple integral. Discuss some criteria for choosing a coordinate system.

In Problems 3-6, convert from rectangular coordinates to
a. *cylindrical*   b. *spherical*

3. $(0, 4, \sqrt{3})$

4. $(\sqrt{2}, -2, \sqrt{3})$

5. $(1, 2, 3)$

6. (π, π, π)

In Problems 7-10 convert from cylindrical coordinates to
a. *rectangular*   b. *spherical*

7. $\left(3, \frac{2\pi}{3}, -3\right)$

8. $\left(4, \frac{\pi}{6}, -2\right)$

9. $\left(2, \frac{\pi}{4}, \pi\right)$

10. (π, π, π)

In Problems 11-14, convert from spherical coordinates to
a. *rectangular*   **b.** *cylindrical*

11. $\left(2, \frac{\pi}{6}, \frac{2\pi}{3}\right)$ **12.** $\left(1, \frac{\pi}{6}, 0\right)$

13. $(1, 2, 3)$ **14.** (π, π, π)

Convert each equation in Problems 15-18 to cylindrical coordinates and sketch its graph in $\mathbb{R}^3$.

15. $z = x^2 - y^2$ **16.** $x^2 - y^2 = 1$

17. $\dfrac{x^2}{4} - \dfrac{y^2}{9} + z^2 = 0$ **18.** $z = x^2 + y^2$

Convert each equation in Problems 19-22 to spherical coordinates and sketch its graph in $\mathbb{R}^3$.

19. $z^2 = x^2 + y^2$, $z \geq 0$ **20.** $2x^2 + 2y^2 + 2z^2 = 1$

21. $4z = x^2 + 3y^2$ **22.** $x^2 + y^2 - 4z^2 = 1$

Convert each equation in Problems 23-28 to rectangular coordinates and sketch its graph in $\mathbb{R}^3$.

23. $z = r^2 \sin 2\theta$ **24.** $r = \sin \theta$

25. $z = r^2 \cos 2\theta$ **26.** $\rho^2 \sin^2 \phi = 1$

27. $\rho^2 \sin \phi \cos \phi \cos \theta = 1$ **28.** $\rho = \sin \phi \cos \theta$

Evaluate each iterated integral in Problems 29-36.

29. $\displaystyle\int_0^\pi \int_0^2 \int_0^{\sqrt{4-r^2}} r \sin \theta \, dz \, dr \, d\theta$

30. $\displaystyle\int_0^{\pi/4} \int_0^1 \int_0^{\sqrt{r}} r^2 \sin \theta \, dz \, dr \, d\theta$

31. $\displaystyle\int_0^{\pi/2} \int_0^{2\pi} \int_0^2 \cos \phi \sin \phi \, d\rho \, d\theta \, d\phi$

32. $\displaystyle\int_0^{\pi/2} \int_0^{\pi/4} \int_0^{\cos \phi} \rho^2 \sin \phi \, d\rho \, d\theta \, d\phi$

33. $\displaystyle\int_0^{2\pi} \int_0^4 \int_0^1 zr \, dz \, dr \, d\theta$

34. $\displaystyle\int_{-\pi/4}^{\pi/3} \int_0^{\sin \theta} \int_0^{4 \cos \theta} r \, dz \, dr \, d\theta$

35. $\displaystyle\int_0^{\pi/2} \int_0^{\cos \theta} \int_0^{1-r^2} r \sin \theta \, dz \, dr \, d\theta$

36. $\displaystyle\int_0^{\pi/3} \int_0^{\cos \theta} \int_0^{\phi} \rho^2 \sin \theta \, d\rho \, d\phi \, d\theta$

Level 2

37. Let D be a homogeneous solid (with density 1) that has the shape of a right circular cylinder with height h and radius R. Use cylindrical coordinates to find the moment of inertia of S about its axis of symmetry.

38. Use cylindrical coordinates to compute the integral

$$\iiint_D xy \, dx \, dy \, dz$$

where D is the cylindrical solid $x^2 + y^2 \leq 1$ with $0 \leq z \leq 1$.

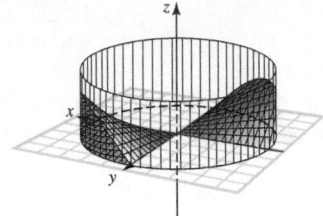

39. Use cylindrical coordinates to compute the integral

$$\iiint_D (x^4 + 2x^2y^2 + y^4) \, dx \, dy \, dz$$

where D is the cylindrical solid $x^2 + y^2 \leq a^2$ with $0 \leq z \leq \frac{1}{\pi}$

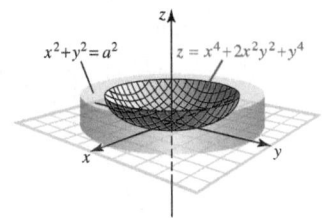

40. Use cylindrical coordinates to compute the integral

$$\iiint_D z(x^2 + y^2)^{-1/2} dx \, dy \, dz$$

where D is the solid bounded above by the plane $z = 2$ and below by the surface $2z = x^2 + y^2$.

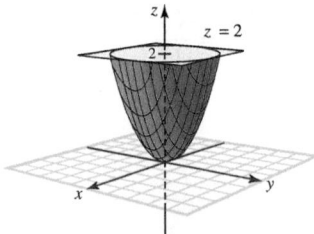

41. Find the centroid of the solid bounded by the surface $z = \sqrt{x^2 + y^2}$ and the plane $z = 9$.

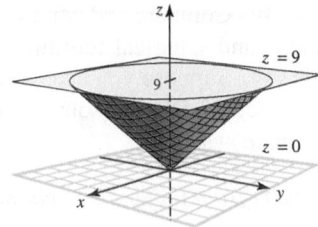

42. Find the centroid of the region bounded by the cone $z = \sqrt{x^2 + y^2}$ and the plane $z = 1$.

43. Suppose the density at each point in the hemisphere $z = \sqrt{9 - x^2 - y^2}$ is $\rho(x, y, z) = xy + z$. Set up integrals for the following quantities:
  a. the mass of the hemisphere
  b. the x-coordinate of the center of mass
  c. the moment of inertia about the z-axis

44. Find the moment of inertia about the z-axis of the portion of the homogeneous hemisphere $z = \sqrt{4 - x^2 - y^2}$ that lies between the cones $z = \sqrt{x^2 + y^2}$ and $2z = \sqrt{x^2 + y^2}$. Assume $\delta = 1$.

45. Find the mass of the torus $\rho = 2 \sin \phi$ if the density is ρ.

Recall (from Problems 40-43 of Section 12.6) that the average value of a function $f(x, y, z)$ over a solid region D is given by

$$\frac{1}{\text{volume of solid } D} \iiint_D f(x, y, z) dV$$

Use this definition in Problems 46-47.

46. Find the average value of the function $f(x, y, z) = x + y + z$ over the sphere $x^2 + y^2 + z^2 = 4$.

47. **EXPLORATION PROBLEM** What do you think the average values of θ and ϕ are over the solid sphere $\rho \le a$ for $a > 0$? Prove your conjecture.

48. Evaluate

$$\iiint_D \sqrt{x^2 + y^2 + z^2}\, dx\, dy\, dz$$

where D is defined by $x^2 + y^2 + z^2 \le 2$.

49. Evaluate

$$\iiint_D (x^2 + y^2 + z^2)\, dx\, dy\, dz$$

where D is defined by $x^2 + y^2 + z^2 \le 2$.

50. Evaluate

$$\iiint_D z^2\, dx\, dy\, dz$$

where D is the solid hemisphere $x^2 + y^2 + z^2 \le 1$, $z \ge 0$.

51. Evaluate

$$\iiint_D \frac{dx\, dy\, dz}{\sqrt{x^2 + y^2 + z^2}}$$

where D is the sphere $x^2 + y^2 + z^2 \le 3$.

Find the volume of the solid D given in Problems 52-55 by using integration in any convenient system of coordinates.

52. D is bounded by the paraboloid $z = 1 - 4(x^2 + y^2)$ and the xy-plane.

53. D is bounded above by the paraboloid $z = 4 - (x^2 + y^2)$, below by the plane $z = 0$, and laterally by the cylinder $x^2 + y^2 = 1$.

54. D is the intersection of the solid sphere $x^2 + y^2 + z^2 \le 9$ and the solid cylinder $x^2 + y^2 \le 1$.

55. D is the region bounded laterally by the cylinder $r = 2 \sin \theta$, below by the plane $z = 0$, and above by the paraboloid $z = 4 - r^2$.

Level 3

56. **EXPLORATION PROBLEM** In Problems 54-58 of Section 12.6, we noted that it is not always accurate to compute the attractive force between two masses by simply using the distance between their centers of mass. This approximation was applied first in astronomy, where the bodies are typically spherical in shape. As we will see, in this case the simple calculation is very appropriate. Consider a sphere of radius a with density δ centered at the origin and a point mass m located at $(x, y, z) = (0, 0, R)$. In this problem you are to set up the integral to compute the attractive force between the masses; and in the next problem you will do some computing.
  a. Show that the distance p between point $(x, y, z) = (0, 0, R)$ and a point in the sphere, (ρ, θ, ϕ), can be expressed $p^2 = R^2 + \rho^2 - 2R\rho \cos \phi$.
  Hint: Use the law of cosines.
  b. In the sphere, consider the infinitesimal element of volume dV centered at point (ρ, θ, ϕ). Argue that the elemental force exerted on m, and the corresponding vertical component, are, respectively:
  $$dF_{\text{mag}} = Gm(\delta dV)/p^2$$
  $$dF = Gm(\delta dV)(R - \rho \cos \phi)/p^3$$
  Note that the term $(R - \rho \cos \phi)/p$ projects the force vector onto the z-axis, giving the vertical component. Do you see why?
  c. Integrate in order to obtain the force acting between point mass m and the sphere of radius a:

  $$F = \iiint \frac{Gm\delta(R - \rho \cos \phi)dV}{p^3}$$

  $$= 2\pi Gm\delta \int_0^a \int_0^\pi p^2 \frac{(R - \rho \cos \phi) \sin \phi}{(R^2 + p^2 - 2R\rho \cos \phi)^{3/2}}\, d\phi\, d\rho$$

  Note: The integrand is independent of θ.

57. **EXPLORATION PROBLEM**
  a. For an unspecified sphere of radius a and $R > a$, compute the attractive force between the masses in the previous problem using the center of mass calculation.

b. Set a and R to numerical values and compute the attractive force by the integral in part **c** of the previous problem. Try two or three different values of R until you are satisfied that the center of mass "approximation" is exact in this case.

c. Compare this result with the corresponding calculations of Problems 54-58 of Section 12.6 and conjecture why the center of mass argument is sound in the case of the sphere and not for the rectangular region.

58. How much volume remains from a spherical ball of radius a when a cylindrical hole of radius b $(0 < b < a)$ is bored out of its center? (See Figure 12.70.)

59. Find the sum $I_x + I_y + I_z$ of the moments of inertia of the solid sphere

$$x^2 + y^2 + z^2 \leq 1$$

about the coordinate axes. Assume the density is 1.

60. **Journal Problem** (*UMAP* Journal* Tumors can be modeled by what are called *bumpy spheres* or *wrinkled spheres* (see Figure 12.71). Find the volume of the bumpy sphere

$$\rho = 1 + 0.2 \sin 4\theta \, \sin 3\phi$$

$(0 \leq \theta \leq 2\pi, 0 \leq \phi \leq \pi)$ as follows:
a. by direct evaluation (exact value)
b. by calculator or computer (approximate value)

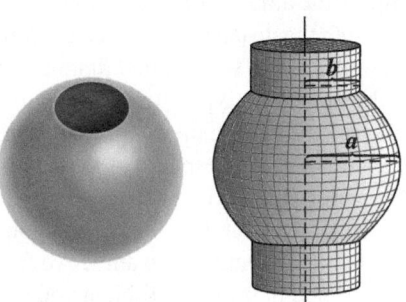

Figure 12.70   Boring a hole in a spherical ball

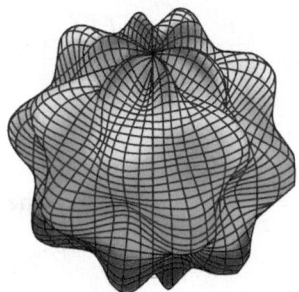

Figure 12.71   Bumpy sphere

12.8 JACOBIANS: CHANGE OF VARIABLES

IN THIS SECTION: *Change of variables in a double integral, change of variables in a triple integral*
We have seen how a change of variables to polar coordinates often simplifies the evaluation of a double integral, and likewise, how cylindrical and spherical coordinates can be used in evaluating triple integrals. These are not the only substitutions that may be used effectively, and the goal of this section is to examine some general ideas about changing variables in multiple integrals.

Change of Variables in a Double Integral

When the change of variable $x = g(u)$ is made in the single integral, we know

$$\int_a^b f(x)\, dx = \int_c^d f(g(u))\, g'(u)\, du$$

where the limits of integration c and d satisfy $a = g(c)$ and $b = g(d)$. By changing variables in a double integral $\iint_D f(x, y)\, dA$, we want to transform the integrand $f(x, y)$ and the region of integration D so that the modified integral is easier to evaluate than the original. In general, this process involves introducing a "mapping factor" analogous to the term $g'(u)$ in the single variable case. This factor is called a *Jacobian* in honor of the German mathematician Karl Gustav Jacobi (1804-1851; see ℌistorical 𝔔uest Problem 59), who made the first systematic study of change of variables in multiple integrals in the middle of the 19th century.

---

*Heat Therapy for Tumors" by Leah Edelstine-Keshet (*UMAP Journal*, Summer 1991).

Suppose we want to evaluate the double integral $\iint_D f(x, y)\, dy\, dx$ by converting it to an equivalent integral involving the variables u and v. This conversion is determined by a transformation (function) T that maps the uv-plane onto the xy-plane. If $T(u, v) = (x, y)$, then (x, y) is the *image* of (u, v) under T, and if no two points in the uv-plane map into the same point (x, y) in the xy-plane, then T is *one-to-one*. In this case, it may be possible to solve the equations $x = x(u, v)$ and $y = y(u, v)$ for u and v in terms of x and y to obtain the equations $u = u(x, y)$ and $v = v(x, y)$, which defines a **transformation** from the xy-plane back to the uv-plane, called the **inverse transformation** of T, and is denoted by T^{-1}. This terminology is illustrated in Figure 12.72.

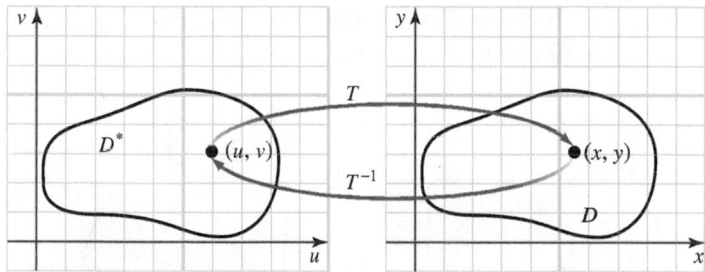

Figure 12.72 A one-to-one transformation T and its inverse T^{-1}

The basic result we will use for changing variables in a double integral is stated in the following theorem.

Theorem 12.7   Change of variables in a double integral

Let f be a continuous function on the interior of a region D in the xy-plane and bounded on that region, and let T be a one-to-one transformation except possibly on the boundary that maps the region D^* in the uv-plane onto D under the change of variables $x = g(u, v)$, $y = h(u, v)$, where g and h are continuously differentiable functions in D^*. Then

$$\iint_D f(x, y)\, dy\, dx = \iint_{D^*} f[g(u, v), h(h, v)]\ |J(u, v)|\ du\, dv$$

where

$$J(u, v) = \begin{vmatrix} \frac{\partial x}{\partial u} & \frac{\partial x}{\partial v} \\ \frac{\partial y}{\partial u} & \frac{\partial y}{\partial v} \end{vmatrix} = \frac{\partial x}{\partial u}\frac{\partial y}{\partial v} - \frac{\partial y}{\partial u}\frac{\partial x}{\partial v}$$

is nonzero and does not change sign on D^*. The mapping factor $J(u, v)$ is called the **Jacobian** and is also denoted by $\dfrac{\partial(x, y)}{\partial(u, v)}$.

☠ *Be sure you use the absolute value of the Jacobian $J(u, v)$. Thus it does not matter whether one uses $du\, dv$ or $dv\, du$.* ☠

Proof: A formal proof is a matter for advanced calculus, but a geometric argument for this theorem is presented in Appendix B.

This theorem represents a simplified version again, and the condition that f be a continuous function on the interior of a region D in the xy-plane and bounded on that region is sufficient, but not necessary. This theorem can be presented in advanced calculus with a more permissive hypothesis, which would allow integrability of the function $f(x, y)$ with respect to either variable. The conditions on the map $x = g(u, v)$, $y = h(u, v)$, however, are very important as stated. This must represent a one-to-one transformation through functions g and h whose partial derivatives exist and are continuous, and such that the Jacobian never vanishes inside the domain (except possibly on an appropriate set of area zero such that the theorem may still apply). ◆

Example 1   Finding a Jacobian

Find the Jacobian for the change of variables from rectangular to polar coordinates, namely, $x = r\cos\theta$ and $y = r\sin\theta$.

Solution The Jacobian of the change of variables is

$$\frac{\partial(x, y)}{\partial(r, \theta)} = \begin{vmatrix} \frac{\partial x}{\partial r} & \frac{\partial x}{\partial \theta} \\ \frac{\partial y}{\partial r} & \frac{\partial y}{\partial \theta} \end{vmatrix} = \begin{vmatrix} \cos\theta & -r\sin\theta \\ \sin\theta & r\cos\theta \end{vmatrix} = r\cos^2\theta + r\sin^2\theta = r$$

This result justifies the formula we have previously used:

$$\iint\limits_{D} f(x, y)\, dy\, dx = \iint\limits_{D^*} f(r\cos\theta, r\sin\theta)\, r\, dr\, d\theta$$

where D^* is the region in the (r, θ) plane that is mapped into the region D in the (x, y) plane by the polar transformation $x = r\cos\theta$, $y = r\sin\theta$, as illustrated in Figure 12.73.

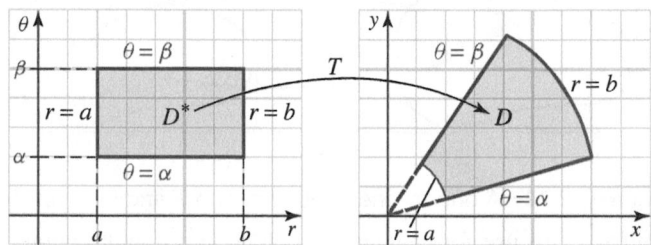

Figure 12.73 Transformation of a region D^* by $T: x = r\cos\theta$, $y = r\sin\theta$

Sometimes it is easier to express u and v in terms of x and y, and to compute the Jacobian $\frac{\partial(u, v)}{\partial(x, y)}$. The Jacobian $\frac{\partial(x, y)}{\partial(u, v)}$ needed for converting the integral $\iint\limits_{D} f(x, y)\, dy\, dx$ can then be computed by the formula

$$\frac{\partial(x, y)}{\partial(u, v)} \frac{\partial(u, v)}{\partial(x, y)} = 1$$

(see Problem 60). This procedure is illustrated by the following example.

Example 2 Finding the Jacobian when $u = u(x, y)$ and $v = v(x, y)$ are given

If $u = xy$ and $v = x^2 - y^2$, express the Jacobian $\frac{\partial(x, y)}{\partial(u, v)}$ in terms of u and v.

Solution It is not easy to express x and y in terms of u and v. However,

$$\frac{\partial(u, v)}{\partial(x, y)} = \begin{bmatrix} \frac{\partial u}{\partial x} & \frac{\partial u}{\partial y} \\ \frac{\partial v}{\partial x} & \frac{\partial v}{\partial y} \end{bmatrix} = \begin{bmatrix} y & x \\ 2x & -2y \end{bmatrix} = -2y^2 - 2x^2 = -2(x^2 + y^2)$$

since

$$(x^2 - y^2)^2 + 4x^2y^2 = (x^2 + y^2)^2$$

$$v^2 + 4u^2 = (x^2 + y^2)^2 \quad \textit{Substitute } v = x^2 - y^2 \textit{ and } u^2 = x^2y^2.$$

$$x^2 + y^2 = \sqrt{4u^2 + v^2} \quad \textit{Note: } x^2 + y^2 \textit{ is nonnegative.}$$

Therefore, since $\dfrac{\partial(x, y)}{\partial(u, v)} \dfrac{\partial(u, v)}{\partial(x, y)} = 1$, we have

$$\frac{\partial(x, y)}{\partial(u, v)} = \frac{1}{\frac{\partial(u,v)}{\partial(x,y)}} = \frac{1}{-2(x^2 + y^2)} = \frac{-1}{2\sqrt{4u^2 + v^2}}$$

Example 3 Calculating a double integral by changing variables

Compute $\iint\limits_{D} \left(\frac{x-y}{x+y}\right)^4 dy\, dx$, where D is the triangular region bounded by the line $x + y = 1$ and the coordinate axes.

Solution   This is a rather difficult computation if no substitution is made. The form of the integral suggests we make the substitution

$$u = x - y, \; v = x + y$$

Solving for x and y, we obtain

$$x = \frac{1}{2}(u + v), \; y = \frac{1}{2}(v - u)$$

and the Jacobian is

$$\frac{\partial(x, y)}{\partial(u, v)} = \begin{vmatrix} \frac{\partial}{\partial u}\left(\frac{u+v}{2}\right) & \frac{\partial}{\partial v}\left(\frac{u+v}{2}\right) \\ \frac{\partial}{\partial u}\left(\frac{v-u}{2}\right) & \frac{\partial}{\partial v}\left(\frac{v-u}{2}\right) \end{vmatrix} = \begin{vmatrix} \frac{1}{2} & \frac{1}{2} \\ -\frac{1}{2} & \frac{1}{2} \end{vmatrix} = \frac{1}{2}$$

To find the image D^* of D in the uv-plane, note that the boundary lines $x = 0$ and $y = 0$ for D map into the lines $u = -v$ and $u = v$, respectively, while $x + y = 1$ maps into $v = 1$. Therefore, the transformed region of integration D^* is the triangular region shown in Figure 12.47**b**, with vertices $(0, 0), (1, 1)$, and $(-1, 1)$.

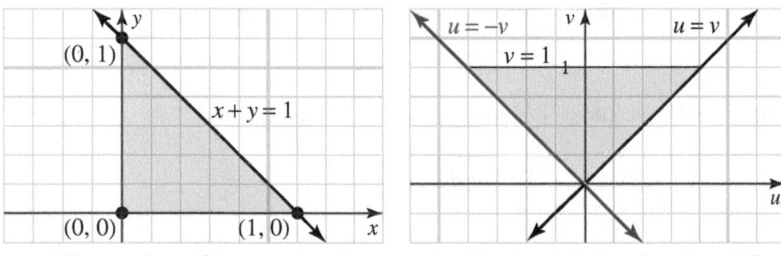

a. The region of integration D **b.** The transformed region D^*

Figure 12.74   Transformation of D to D^*

We now evaluate the given integral:

$$\int\limits_{D} \int \left(\frac{x-y}{x+y}\right)^4 dy\, dx = \int\limits_{D^*} \int \left(\frac{u}{v}\right)^4 \left|\frac{1}{2}\right| du\, dv = \frac{1}{2}\int_0^1 \int_{-v}^{v} u^4 v^{-4} du\, dv = \frac{1}{10} \; \blacksquare$$

In Example 3, the change of variables was chosen to simplify the integrand, but sometimes it is useful to introduce a change of variables that simplifies the region of integration.

Example 4 Change of variables to simplify a region

Find the area of the region E bounded by the ellipse $\dfrac{x^2}{a^2} + \dfrac{y^2}{b^2} = 1$

Solution   The area is given by the integral

$$A = \iint\limits_{E} dy\, dx$$

where E is the region shown in Figure 12.75**a**.

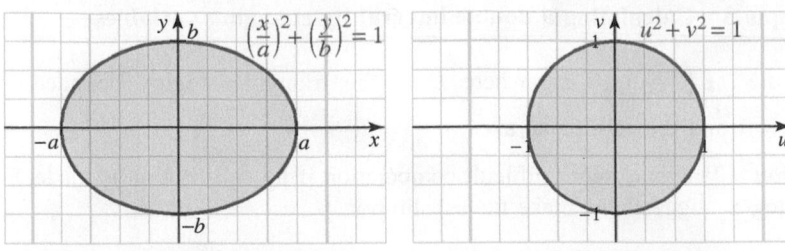

a. The elliptical region E **b.** The transformed region C is a circular disk

Figure 12.75   Transformation of the ellipse E to the circle C

Because E can be represented by

$$\left(\frac{x}{a}\right)^2 + \left(\frac{y}{b}\right)^2 \leq 1$$

we consider the substitution $u = \dfrac{x}{a}$ and $v = \dfrac{y}{b}$, which will map the elliptical region E onto the circular disk

$$C: \ u^2 + v^2 \leq 1$$

as shown in Figure 12.75**b**. Then $x = au$ and $y = bv$, and the Jacobian is

$$\frac{\partial(x, y)}{\partial(u, v)} = \begin{vmatrix} \frac{\partial}{\partial u}(au) & \frac{\partial}{\partial v}(au) \\ \frac{\partial}{\partial u}(bv) & \frac{\partial}{\partial v}(bv) \end{vmatrix} = \begin{vmatrix} a & 0 \\ 0 & b \end{vmatrix} = ab$$

Because $ab > 0$, we have $|ab| = ab$, and the area of E is given by

$$\iint\limits_{E} dy\,dx = \iint\limits_{C} ab\,du\,dv$$

$$= ab \iint\limits_{C} du\,dv$$

$$= ab \iint\limits_{C} 1\,dA$$

$$= ab\left[\pi(1)^2\right] \qquad \textit{Because C is a circle of radius } 1.$$

$$= \pi ab$$

Example 5   Using a change of variables to find a centroid

Find the centroid (correct to the nearest tenth) of the region D^* in the xy-plane that is bounded by the lines $y = \frac{1}{4}x$ and $y = \frac{5}{2}x$ and the hyperbolas $xy = 1$ and $xy = 5$.

Solution   If A is the area of D^*, the centroid is $(\overline{x}, \overline{y})$, where

$$\overline{x} = \frac{1}{A} \iint\limits_{D^*} x\,dA \qquad \text{and} \qquad \overline{y} = \frac{1}{A} \iint\limits_{D^*} y\,dA$$

We wish to find a transformation that simplifies the boundary curves of the region of integration, so we let $u = \dfrac{y}{x}$ and $v = xy$ and the boundaries of the transformed region D become:

$$u = \frac{1}{4} \qquad u = \frac{5}{2} \qquad v = 1 \qquad v = 5$$

$$(y = \frac{1}{4}x) \quad (y = \frac{5}{2}x) \quad (xy = 1) \quad (xy = 5)$$

The regions D^* and D are shown in Figure 12.76.

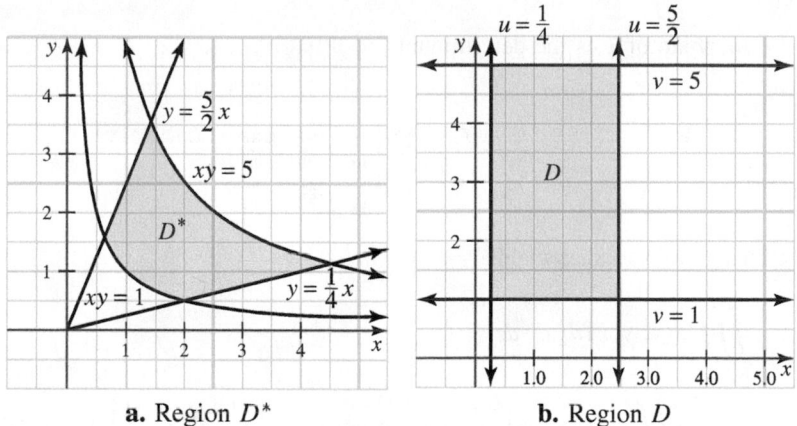

a. Region D^* **b.** Region D

Figure 12.76   Change of variables

Solving the equations $u = \dfrac{y}{x}$ and $v = xy$ for x and y, we obtain $x = \sqrt{\dfrac{v}{u}}$ and $y = \sqrt{uv}$, so the Jacobian is

$$\frac{\partial(x, y)}{\partial(u, v)} = \begin{bmatrix} \frac{\partial x}{\partial u} & \frac{\partial x}{\partial v} \\ \frac{\partial y}{\partial u} & \frac{\partial y}{\partial v} \end{bmatrix} = \begin{bmatrix} -\frac{1}{2}\frac{1}{u}\sqrt{\frac{v}{u}} & \frac{1}{2}\frac{1}{\sqrt{uv}} \\ \frac{1}{2}\sqrt{\frac{v}{u}} & \frac{1}{2}\sqrt{\frac{u}{v}} \end{bmatrix} = -\frac{1}{4u} - \frac{1}{4u} = -\frac{1}{2u}$$

The area of the region D^* is

$$A = \int\!\!\int_{D^*} dy \, dx$$

$$= \int_1^5 \int_{1/4}^{5/2} \left| \frac{-1}{2u} \right| du \, dv$$

$$= \frac{1}{2} \int_1^5 \ln u \Big|_{1/4}^{5/2} dv$$

$$= \frac{1}{2} \left[\ln \frac{5}{2} - \ln \frac{1}{4} \right] \int_1^5 dv$$

$$= \frac{1}{2} \ln \left(\frac{5/2}{1/4} \right) v \Big|_1^5$$

$$= 2 \ln 10$$

To find the centroid $(\overline{x}, \overline{y})$, we compute

$$\overline{x} = \frac{1}{A} \int\!\!\int_{D^*} x \, dA = \frac{1}{2 \ln 10} \int_1^5 \int_{1/4}^{5/2} \sqrt{\frac{v}{u}} \left| \frac{-1}{2u} \right| du \, dv \approx 2.0154$$

$$\overline{y} = \frac{1}{A} \int\!\!\int_{D^*} y \, dA = \frac{1}{2 \ln 10} \int_1^5 \int_{1/4}^{5/2} \sqrt{uv} \left| \frac{-1}{2u} \right| du \, dv \approx 1.5933$$

Thus, the centroid is at approximately $(2.0, 1.6)$ in the xy-plane.

Change of Variables in a Triple Integral

The change of variables formula for triple integrals is similar to the one given for double integrals. Let T be a change of variables that maps a region R^* in uvw-space onto a region R in xyz-space, where

$$T: \quad x = x(u, v, w) \qquad y = y(u, v, w) \qquad z = z(u, v, w)$$

Then the Jacobian of T is the determinant

$$\frac{\partial(x, y, z)}{\partial(u, v, w)} = \begin{vmatrix} \dfrac{\partial x}{\partial u} & \dfrac{\partial x}{\partial v} & \dfrac{\partial x}{\partial w} \\[2mm] \dfrac{\partial y}{\partial u} & \dfrac{\partial y}{\partial v} & \dfrac{\partial y}{\partial w} \\[2mm] \dfrac{\partial z}{\partial u} & \dfrac{\partial z}{\partial v} & \dfrac{\partial z}{\partial w} \end{vmatrix}$$

and the change of variables yields

$$\iiint\limits_{R} f(x, y, z)\, dx\, dy\, dz$$

$$= \iiint\limits_{R^*} f[x(u, v, w),\ y(u, v, w),\ z(u, v, w)] \left| \frac{\partial(x, y, z)}{\partial(u, v, w)} \right| du\, dv\, dw$$

Example 6   Formula for integrating with spherical coordinates

Obtain the formula for converting a triple integral in rectangular coordinates to one in spherical coordinates.

Solution   The conversion formulas from rectangular coordinates to spherical coordinates are:

$$x = \rho \sin \phi \cos \theta \qquad y = \rho \sin \phi \sin \theta \qquad z = \rho \cos \phi$$

The Jacobian of this transformation is

$$\frac{\partial(x, y, z)}{\partial(\rho, \theta, \phi)} = \begin{vmatrix} \dfrac{\partial}{\partial \rho}(\rho \sin \phi \cos \theta) & \dfrac{\partial}{\partial \theta}(\rho \sin \phi \cos \theta) & \dfrac{\partial}{\partial \phi}(\rho \sin \phi \cos \theta) \\[2mm] \dfrac{\partial}{\partial \rho}(\rho \sin \phi \sin \theta) & \dfrac{\partial}{\partial \theta}(\rho \sin \phi \sin \theta) & \dfrac{\partial}{\partial \phi}(\rho \sin \phi \sin \theta) \\[2mm] \dfrac{\partial}{\partial \rho}(\rho \cos \phi) & \dfrac{\partial}{\partial \theta}(\rho \cos \phi) & \dfrac{\partial}{\partial \phi}(\rho \cos \phi) \end{vmatrix}$$

$$= \begin{vmatrix} \sin \phi \cos \theta & -\rho \sin \phi \sin \theta & \rho \cos \phi \cos \theta \\[2mm] \sin \phi \sin \theta & \rho \sin \phi \cos \theta & \rho \cos \phi \sin \theta \\[2mm] \cos \phi & 0 & -\rho \sin \phi \end{vmatrix}$$

$$= -\rho^2 \sin \phi$$

Since $0 \le \phi \le \pi$, we have $\sin \phi \ge 0$, so

$$\left| \frac{\partial(x, y, z)}{\partial(\rho, \theta, \phi)} \right| = \left| -\rho^2 \sin \phi \right| = \rho^2 \sin \phi$$

and the desired formula is

$$\iiint\limits_{R} f(x, y, z)\, dz\, dx\, dy$$

$$= \iiint\limits_{R^*} f(\rho \sin \phi \cos \theta, \rho \sin \phi \sin \theta, \rho \cos \phi)\, \rho^2 \sin \phi \, d\rho \, d\theta \, d\phi$$

PROBLEM SET 12.8

Level 1

1. ■ What does this say? Discuss the process of changing variables in a triple integral.
2. ■ What does this say? Discuss the necessity of including the Jacobian when changing variables in a double integral.

Find the Jacobian $\dfrac{\partial(x, y)}{\partial(u, v)}$ *or* $\dfrac{\partial(x, y, z)}{\partial(u, v, w)}$ *in Problems 3-14.*

3. $x = u - v, y = u + v$

4. $x = u + 2v, y = 3u - 4v$

5. $x = u^2, y = u + v$

6. $x = u^2 v^2, y = v^2 - u^2$

7. $x = e^{u+v}, y = e^{u-v}$

8. $x = u \cos v, \; y = u \sin v$

9. $x = e^u \sin v, y = e^u \cos v$

10. $x = \dfrac{v}{u^2 + v^2}, y = \dfrac{u}{u^2 + v^2}$

11. $x = u + v - w, \; y = 2u - v + 3w,$
    $z = -u + 2v - w$

12. $x = 2u - w, \; y = u + 3v, z = v + 2w$

13. $x = u \cos v, \; y = u \sin v, z = we^{uv}$

14. $x = \dfrac{u}{v}, \; y = \dfrac{v}{w}, z = \dfrac{w}{u}$

Find the Jacobian $\dfrac{\partial(x, y)}{\partial(u, v)}$ *in Problems 15-22 either by solving the given equations for x and y or by first computing* $\dfrac{\partial(u, v)}{\partial(x, y)}$. *Express your answers either in terms of u and v, or x and y.*

15. $u = 2x - 3y, v = x + 4y$

16. $u = 3x + y, v = -x + 2y$

17. $u = ye^{-x}, v = e^x$

18. $u = ye^x, \; v = e^{-x}$

19. $u = \sqrt{x^2 + y^2}, \; v = \cot^{-1}\left(\dfrac{y}{x}\right)$

20. $u = xy, v = \dfrac{y}{x}$ for $x > 0, \; y > 0$

21. $u = x^2 - y^2, v = x^2 + y^2$

22. $u = \dfrac{x}{x^2 + y^2}, v = \dfrac{y}{x^2 + y^2}$

A region R is given in Problems 23-26. Sketch the corresponding region R* in the uv-plane using the given transformations.

23.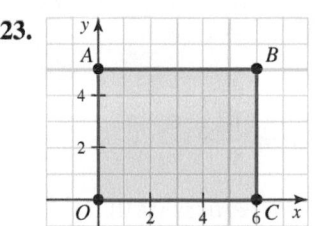

$u = x + y, \; v = x - y$

24.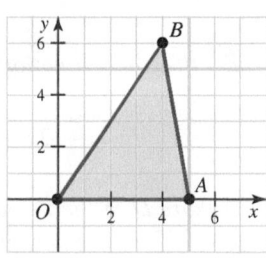

$u = 2x, v = x + y$

25.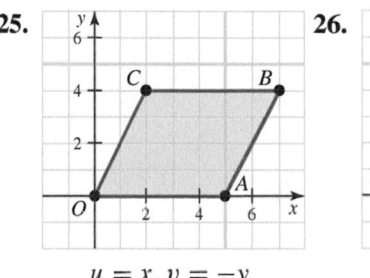

$u = x, v = -y$

26.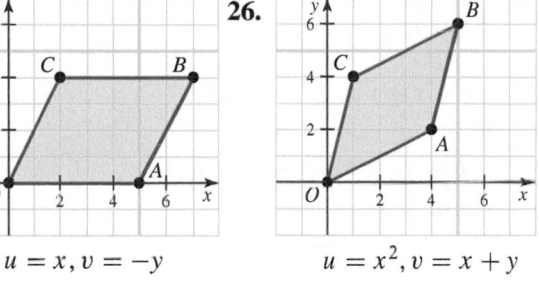

$u = x^2, v = x + y$

27. Suppose the uv-plane is mapped onto the xy-plane by the equations $x = u(1 - v), y = uv$. Express $dx\, dy$ in terms of $du\, dv$.

28. Suppose the uv-plane is mapped onto the xy-plane by the equations $x = u^2 - v^2, y = 2uv$. Express $dx\, dy$ in terms of $du\, dv$.

Use a suitable change of variables to find the area of the region specified in Problems 29 and 30.

29. The region R bounded by the hyperbolas $xy = 1$ and $xy = 4$ and the lines $y = x$ and $y = 4x$

30. The region R bounded by the parabolas $y = x^2$, $y = 4x^2$, $y = \sqrt{x}$, and $y = \frac{1}{2}\sqrt{x}$

Let D be the region in the xy-plane that is bounded by the coordinate axes and the line $x + y = 1$. Use the change of variables $u = x - y, v = x + y$ to compute the integrals given in Problems 31-34.

31. $\displaystyle\iint_D \left(\dfrac{x - y}{x + y}\right)^5 dy\, dx$

32. $\displaystyle\iint_D \left(\dfrac{x - y}{x + y}\right)^4 dy\, dx$

33. $\displaystyle\iint_D (x - y)^5 (x + y)^3 \, dy\, dx$

34. $\displaystyle\iint_D (x - y)e^{x^2+y^2} \, dy\, dx$

Under the transformation

$$u = \frac{1}{5}(2x + y) \quad v = \frac{1}{5}(x - 2y)$$

the square D in the xy-plane with vertices $(0,0)$, $(1,-2)$, $(3,-1)$, $(2,1)$ *is mapped onto a square in the uv-plane. Use this information to find the integrals in Problems 35-40.*

35. $\iint_D \left(\frac{2x+y}{x-2y+5}\right)^2 dy\,dx$

36. $\iint_D (2x+y)(x-2y)^2 dy\,dx$

37. $\iint_D (2x+y)^2(x-2y)\,dy\,dx$

38. $\iint_D \sqrt{(2x+y)(x-2y)}\,dy\,dx$

39. $\iint_D (2x+y)\tan^{-1}(x-2y)\,dy\,dx$

40. $\iint_D \cos(2x+y)\sin(x-2y)\,dy\,dx$

41. Evaluate

$$\iint_R e^{x+y}\,dA$$

where R is the triangle with vertices $(1,0), (0,1)$, and $(0,0)$.

42. Let R be the region in the xy-plane that is bounded by the parallelogram with vertices $(0,0), (1,1), (2,1)$ and $(1,0)$. Use the linear transformation $x = u+v$, $y = v$ to compute

$$\iint_R (2x - y)\,dy\,dx$$

43. Use a suitable linear transformation $u = ax+by$, $v = rx + sy$ to evaluate the integral

$$\iint_R \left(\frac{x+y}{2}\right)^2 e^{(y-x)/2}dy\,dx$$

where R is the region inside the square with vertices $(0,0), (1,1), (0,2)$, and $(-1,1)$.

44. Under the change of variables $x = s^2 - t^2$, $y = 2st$, the quarter circular region in the st-plane given by $s^2 + t^2 \leq 1, s \geq 0, t \geq 0$ is mapped onto a certain region D^* of the xy-plane. Evaluate

$$\iint_{D^*} \frac{dy\,dx}{\sqrt{x^2 + y^2}}$$

45. Use the change of variables $x = ar\cos\theta$, $y = br\sin\theta$ to evaluate

$$\iint_{D^*} \exp\left(-\frac{x^2}{a^2} - \frac{y^2}{b^2}\right)dy\,dx$$

where D^* is the region bounded by the quarter ellipse

$$\frac{x^2}{a^2} + \frac{y^2}{b^2} = 1$$

in the first quadrant.

46. Evaluate

$$\iint_{D^*} \frac{2x - 5y}{3x + y}dA$$

where D^* is the region bounded by the lines $2x - 5y = 1, 2x - 5y = 3, 3x + y = 1, 3x + y = 4$.

47. Evaluate

$$\iint_{D^*} e^{-(4x^2+5y^2)}dA$$

where D^* is the elliptical disk

$$\frac{x^2}{5} + \frac{y^2}{4} \leq 1$$

48. Evaluate

$$\iiint_{D^*} y^3(2x - y)\cos(2x - y)\,dy\,dx$$

where D^* is the region bounded by the parallelogram with vertices $(0,0), (2,0), (3,2)$, and $(1,2)$.

49. Find the centroid correct to the nearest hundredth of the region D^* bounded by the curves $y = x, y = 2x$, $xy^3 = 3$, and $xy^3 = 6$.

50. Evaluate

$$\iint_{D^*} (x^4 - y^4)e^{xy}\,dA$$

where D^* is the region bounded by the hyperbolas $xy = 1, xy = 2, x^2 - y^2 = 1$, and $x^2 - y^2 = 4$.

51. Use the change of variables $x = au, y = bv, z = cw$ to find the volume of the solid ellipsoid

$$\frac{x^2}{a^2} + \frac{y^2}{b^2} + \frac{z^2}{c^2} = 1$$

52. Use a change of variables to find the moment of inertia about the z-axis of the solid ellipsoid with $\rho = 1$

$$\frac{x^2}{a^2} + \frac{y^2}{b^2} + \frac{z^2}{c^2} = 1$$

53. Evaluate

$$\iint\limits_{R} \ln\left(\frac{x-y}{x+y}\right) dy\, dx$$

where R is the triangular region with vertices $(1, 0)$, $(4, -3)$, and $(4, 1)$.

54. A rotation of the xy-plane through the fixed angle θ is given by

$$x = u\cos\theta - v\sin\theta$$

$$y = u\sin\theta + v\cos\theta$$

(See Figure 12.77.)

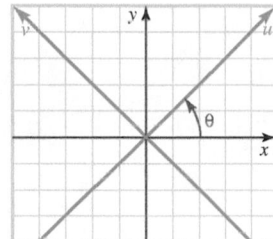

Figure 12.77 Rotational transformation

a. Compute the Jacobian $\dfrac{\partial(x, y)}{\partial(u, v)}$.

b. Let E denote the ellipse

$$x^2 + xy + y^2 = 3$$

Use a rotation of $\frac{\pi}{4}$ to obtain an integral that is equivalent to $\iint\limits_{E} y\, dy\, dx$. Evaluate the transformed integral.

55. Find the area of the rotated ellipse

$$5x^2 - 4xy + 2y^2 = 1$$

Note that $5x^2 - 4xy + 2y^2 = Au^2 + Bv^2$, where A and B are constants and $u = x + 2y$, $v = 2x - y$.

56. Find the Jacobian of the cylindrical coordinate transformations

$$x = r\cos\theta, \quad y = r\sin\theta, \quad z = z$$

57. EXPLORATION PROBLEM Let R be the region bounded by the parallelogram with vertices $(0, 0)$, $(2, 0)$, $(1, 1)$, and $(-1, 1)$. Show that under a suitable change of variables of the form $x = au + bv$, $y = cu + dv$, we have

$$\iint\limits_{R} f(x+y)\, dy\, dx = \int_{0}^{2} f(t)\, dt$$

where f is continuous on $[0, 2]$.

58. Find the volume of the solid shown in Figure 12.78 that is under the surface

$$z = \frac{xy}{1 + x^2 y^2}$$

over the region bounded by $xy = 1$, $xy = 5$, $x = 1$, and $x = 5$.

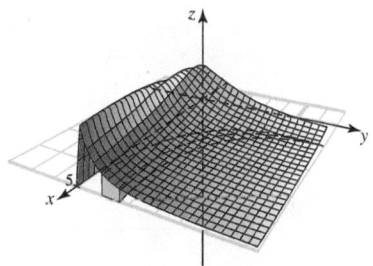

Figure 12.78 Volume of a solid

Level 3

59. *𝔥istorical 𝔔uest*

Karl G. Jacobi was a gifted teacher and one of Germany's most distinguished mathematicians during the first half of the 19th century. He made major contributions to the theory of elliptic functions, but his work with functional determinants is what secured his place in history. In 1841 he

Karl G. Jacobi (1804-1851)

published a long memoir called "De determinantibus functionalibus," devoted to what we today call the Jacobian and pointing out that this determinant is in many ways the multivariable analogue to the differential of a single variable. The memoir was published in what is usually known as **Crelle's Journal**, *one of the first journals devoted to serious mathematics. It was begun in 1826 by August Crelle (1780-1855) with the title Journal für die reine und angewandte Mathematik. For this 𝔔uest, set up an integral for the circumference of the ellipse*

$$\frac{x^2}{a^2} + \frac{y^2}{b^2} = 1$$

This is an example of an elliptic integral.

60. Let $T: x = x(u, v)$, $y = y(u, v)$ be a one-to-one transformation on a set D so that T^{-1} has the form $u = u(x, y)$, $v = v(x, y)$. Use the multiplicative property of determinants, along with the chain rule, to show that

$$\frac{\partial(x, y)}{\partial(u, v)} \frac{\partial(u, v)}{\partial(x, y)} = 1$$

You may assume that all the necessary partial derivatives exist.

CHAPTER 12 REVIEW

*T*he infinite! No other question has ever moved so profoundly the spirit of man; no other idea has so fruitfully stimulated his intellect; yet no other concept stands in greater need of clarification than that of the infinite.

David Hilbert

To Infinity and Beyond by Eli Maor (Birkhäuser, 1987), p. vii

Proficiency Examination

Concept Problems

1. Define a double integral.
2. State Fubini's theorem over a rectangular region.
3. What is a type I region? State Fubini's theorem for a type I region.
4. What is a type II region? State Fubini's theorem for a type II region.
5. State the formula for finding an area using a double integral.
6. State the formula for finding a volume using a double integral.
7. State the following properties of a double integral.

   a. linearity rule   **b.** dominance rule   **c.** subdivision rule
8. What is the formula for a double integral in polar coordinates?
9. State the double integral formula for surface area.
10. State the formula for the area of a surface defined parametrically.
11. State Fubini's theorem over a parallelepiped in space.
12. Give a triple integral formula for volume.
13. Explain how to use a double integral to find the mass of a planar lamina of variable density.
14. How do you find the moment of a lamina with respect to the x-axis?
15. What are the formulas for the centroid of a region?
16. Define moment of inertia.
17. What is a joint probability density function?
18. Complete the following table for the conversion of coordinates.

To: From:	(x, y, z) Rectangular	(r, θ, z) Cylindrical	(ρ, θ, ϕ) Spherical
Rectangular (x, y, z)	$x = x$ $y = y$ $z = z$	$r = ?$ $\tan\theta = ?$ $z = ?$	$\rho = ?$ $\tan\theta = ?$ $\phi = ?$
Cylindrical (r, θ, z)	$x = ?$ $y = ?$ $z = ?$	$r = r$ $\theta = \theta$ $z = z$	$\rho = ?$ $\theta = ?$ $\phi = ?$
Spherical (ρ, θ, ϕ)	$x = ?$ $y = ?$ $z = ?$	$r = ?$ $\theta = ?$ $z = ?$	$\rho = \rho$ $\theta = \theta$ $\phi = \phi$

19. **a.** State the formula for a triple integral in cylindrical coordinates.
    b. State the formula for a triple integral in spherical coordinates.
20. Define the Jacobian for a mapping $x = x(u, v)$, $y = y(u, v)$ from u, v variables to x, y variables.
21. State the formula for the change of variables in a double integral.

Practice Problems

22. Evaluate $\displaystyle\int_0^{\pi/3}\int_0^{\sin y} e^{-x}\cos y\,dx\,dy$.

23. Evaluate $\displaystyle\int_{-1}^{1}\int_0^z\int_y^{y-z}(x+y-z)\,dx\,dy\,dz$

24. Use a double integral to compute the area of the region R that is bounded by the x-axis and the parabola $y = 9 - x^2$.

25. Use polar coordinates to evaluate

$$\int_0^1\int_0^{\sqrt{1-x^2}}\cos(x^2+y^2)\,dy\,dx$$

26. Use a triple integral to find the volume of the tetrahedron that is bounded by the coordinate planes and the plane

$$\frac{x}{a}+\frac{y}{b}+\frac{z}{c}=1$$

with $a > 0$, $b > 0$, $c > 0$.

27. A certain appliance consisting of two independent electronic components will be usable as long as either one of its components is still operating. The appliance carries a warranty from the manufacturer guaranteeing replacement if the appliance becomes unusable within one year of the date of purchase. To model this situation, let the random variables X and Y measure the life span (in years) of the first and second components, respectively, and assume that the joint probability density function for X and Y is

$$f(x, y) = \begin{cases} \frac{1}{4}e^{-x/2}e^{-y/2} & \text{if } x \geq 0,\ y \geq 0 \\ 0 & \text{otherwise} \end{cases}$$

Suppose the quality assurance department selects one of these appliances at random. What is the probability that the appliance will fail during the warranty period?

28. A solid D is bounded above by the plane $z = 4$ and below by the surface $z = x^2 + y^2$. Its density $\rho(x, y, z)$ at each point P is equal to the distance from P to the z-axis. Find the total mass of D.

29. Set up and evaluate a triple integral for the volume of the solid region in the first octant that is bounded above by the plane $z = 4x$ and below by the paraboloid $z = x^2 + 2y^2$. See Figure 12.79.

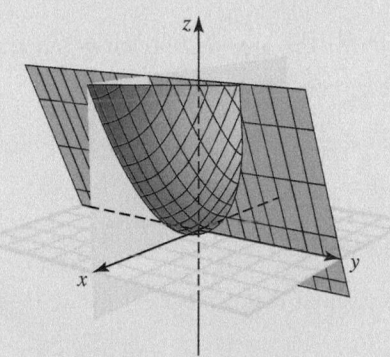

Figure 12.79 Volume of a solid

30. Use a linear change of variables to evaluate the double integral

$$\iint_R (x+y)e^{x-2y}\,dy\,dx$$

where R is the triangular region with vertices $(0,0),(2,0),(1,1)$.

Supplementary Problems*

In Problems 1-14, *sketch the region of integration, exchange the order, and evaluate the integral using either order of integration.*

1. $\displaystyle\int_0^1 \int_{-x^2}^{x^2} dy\, dx$

2. $\displaystyle\int_1^e \int_0^{\ln x} xy\, dy\, dx$

3. $\displaystyle\int_0^3 \int_1^{4-x} (x+y)\, dy\, dx$

4. $\displaystyle\int_0^1 \int_{x^3}^1 (x+y^2)\, dy\, dx$

5. $\displaystyle\int_0^1 \int_0^{3x} x^2 y^2\, dy\, dx$

6. $\displaystyle\int_0^4 \int_0^1 \sqrt{\frac{y}{x}}\, dy\, dx$

7. $\displaystyle\int_1^2 \int_0^y \frac{1}{x^2+y^2}\, dx\, dy$

8. $\displaystyle\int_0^{\pi/2} \int_0^{\pi/2} \cos(x+y)\, dx\, dy$

9. $\displaystyle\int_0^1 \int_0^1 x\sqrt{x^2+y}\, dx\, dy$

10. $\displaystyle\int_0^{\pi/2} \int_0^{\sqrt{\sin x}} xy\, dy\, dx$

11. $\displaystyle\int_0^3 \int_0^{9-x^2} \frac{xe^{y/3}}{9-y}\, dy\, dx$

12. $\displaystyle\int_0^1 \int_{\sqrt{y}}^1 \sqrt{1-x^3}\, dx\, dy$

13. $\displaystyle\int_0^1 \int_{x^2}^1 x^3 \sin y^3\, dy\, dx$

14. $\displaystyle\int_0^2 \int_0^{\sqrt{4-x^2}} \sqrt{4-x^2-y^2}\, dy\, dx$

Sketch the region of integration in problems 15-18, *and write an equivalent integral with the order of integration reversed.*

15. $\displaystyle\int_0^4 \int_{y/2}^{\sqrt{y}} f(x, y)\, dx\, dy$

16. $\displaystyle\int_1^{e^2} \int_{\ln x}^2 f(x, y)\, dy\, dx$

17. $\displaystyle\int_{-3}^2 \int_{x^2}^{6-x} f(x, y)\, dy\, dx$

18. $\displaystyle\int_2^4 \int_x^{16/x} f(x, y)\, dy\, dx$

Evaluate the integrals given in Problems 19-26.

19. $\displaystyle\int_0^1 \int_{\sqrt{x}}^1 e^{y^3}\, dy\, dx$

20. $\displaystyle\int_0^2 \int_x^2 \frac{y}{(x^2+y^2)^{3/2}}\, dy\, dx$

21. $\displaystyle\int_0^1 \int_1^4 \int_x^y z\, dz\, dy\, dx$

22. $\displaystyle\int_1^2 \int_0^1 \int_0^{\sqrt{1-x^2}} e^{\sqrt{x^2+y^2}}\, dy\, dx\, dz$

23. $\displaystyle\int_0^{\pi/4} \int_0^{2\pi} \int_0^{\theta} r^2 \sin\phi\, dr\, d\theta\, d\phi$

24. $\displaystyle\int_0^1 \int_0^x \int_0^y x^2 y\, dz\, dy\, dx$

25. $\displaystyle\int_1^2 \int_x^{x^2} \int_0^{\ln x} xe^z\, dz\, dy\, dx$

26. $\displaystyle\int_0^1 \int_{1-x}^{1+x} \int_0^{xy} xz\, dz\, dy\, dx$

Evaluate the integrals in Problems 27-46 *for the specified region of integration.*

27. $\displaystyle\iint_D x^3\sqrt{4-y^2}\, dA$, where D is the circular disk $x^2+y^2 \le 4$

28. $\displaystyle\iint_D x^2 y\, dA$, where D is the circular disk $x^2+y^2 \le 4$

29. $\displaystyle\iint_D (x^2+y^2+1)\, dA$, where D is the circular disk $x^2+y^2 \le 4$

30. $\displaystyle\iint_D e^{x^2+y^2}\, dA$, where D is the circular disk $x^2+y^2 \le 4$

31. $\displaystyle\iint_D (x^2+y^2)^n\, dA$, where D is the circular disk $x^2+y^2 \le 4, n \ge 0$

32. $\displaystyle\iint_D \frac{2y}{x}\, dA$, where D is the region above the line $x+y=1$ and inside the circle $x^2+y^2=1$

33. $\displaystyle\iint_D 2x\, dA$, where D is the region bounded by $x^2 y=1$, $y=x$, $x=2$, and $y=0$

34. $\displaystyle\iint_D 12x^2 e^{y^2}\, dA$, where D is the region in the first quadrant bounded by $y=x^3$ and $y=x$

35. $\displaystyle\iint_D x\, dy\, dx$, where D is the region between the parabola $y=x^2$ and the line $y=2x+3$.

---

*The supplementary problems are presented in a somewhat random order, not necessarily in order of difficulty.

36. $\displaystyle\iint_D \cos e^x \, dA$, where D is the region bounded by $y = e^x$, $y = -e^x$, $x = 0$, and $x = \ln 2$

37. $\displaystyle\int_{-\sqrt{3}}^{\sqrt{3}} \int_{-\sqrt{3-y^2}}^{\sqrt{3-y^2}} \int_{(x^2+y^2)^2}^{9} y^2 dz \, dx \, dy$ *Hint:* Use cylindrical coordinates.

38. $\iint_D x^3 \, dA$, where D is the shaded portion shown in Figure 12.80**a**

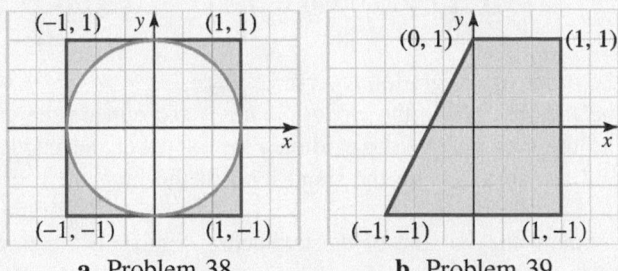

a. Problem 38   **b.** Problem 39

Figure 12.80   Defined regions

39. $\iint_D xy^2 \, dA$, where D is the shaded portion shown in Figure 12.80**b**

40. $\displaystyle\iint_D \exp\left(\frac{y-x}{y+x}\right) dy \, dx$, where D is the triangular region with vertices $(0,0), (2,0), (0,2)$

Hint: Make a suitable change of variables.

41. $\displaystyle\iint_R \frac{\partial^2 f}{\partial x \partial y} \, dx \, dy$, where R is the rectangle $a \le x \le b, c \le y \le d$ and $\dfrac{\partial^2 f}{\partial x \partial y}$ is continuous over R with
$f(a,c) = 4, f(a,d) = -3, f(b,c) = 1, f(b,d) = 5$

42. $\displaystyle\iiint_D \sqrt{x^2 + y^2 + z^2} \, dV$, where D is the portion of the solid sphere $x^2 + y^2 + z^2 \le 1$ that lies in the first octant

43. $\displaystyle\iiint_H z^2 \, dV$, where H is the solid hemisphere $x^2 + y^2 + z^2 \le 1$ with $z \ge 0$

44. $\displaystyle\iiint_H \frac{dV}{\sqrt{x^2 + y^2 + z^2}}$, where H is the solid hemisphere $x^2 + y^2 + z^2 \le 1$ with $z \ge 0$

45. $\displaystyle\iint_D \int \frac{dV}{x^2 + y^2 + z^2}$, where D is the solid region bounded below by the paraboloid $2z = x^2 + y^2$ and above by the sphere $x^2 + y^2 + z^2 = 8$

46. $\displaystyle\iint_D \int z(x^2 + y^2)^{-1/2} \, dV$, where D is the solid bounded by the surface $2z = x^2 + y^2$ and the plane $z = 2$

47. Set up (but do not evaluate) an integral for the mass of a lamina with density ρ that covers the region outside the circle $r = \sqrt{2}a$ and inside the lemniscate $r^2 = 4a^2 \sin 2\theta$ in the first quadrant.

48. Rewrite the triple integral $\displaystyle\int_0^1 \int_0^x \int_0^{\sqrt{xy}} f(x, y, z) \, dz \, dy \, dx$ as a triple integral in the order $dy \, dx \, dz$.

49. Convert the double integral

$$\int_0^{\pi/2} \int_0^{2a\cos\theta} r \sin 2\theta \, dr \, d\theta$$

to rectangular coordinates and evaluate.

50. Reverse the order of integration

$$\int_1^2 \int_{y^2}^{y^5} e^{x/y^2} dx \, dy$$

and evaluate using either order.

51. Express the integral

$$\int_0^1 \int_0^y f(x, y)\, dx\, dy + \int_1^4 \int_0^{(4-y)/3} f(x, y)\, dx\, dy$$

as a double integral with the order of integration reversed.

52. Express the integral

$$\int_0^1 \int_{-\sqrt{y}}^{\sqrt{y}} f(x, y)\, dx\, dy + \int_1^3 \int_{-1}^{2-y} f(x, y)\, dx\, dy$$

as a double integral with the order of integration reversed.

53. Use a double integral to find the area inside the circle $r = \cos\theta$ and outside the cardioid $r = 1 - \cos\theta$.

54. Use a double integral to find the area outside the cardioid $r = 1 + \cos\theta$ and inside the cardioid $r = 1 + \sin\theta$.

55. Use a double integral to find the area outside the small loop of the limaçon $r = 1 + 2\sin\theta$ and inside the large loop.

56. Use a double integral to compute the volume of the tetrahedral region in the first octant that is bounded by the coordinate planes and the plane $3x + y + 2z = 6$.

57. Use a double integral to find the volume of the solid that is bounded above by the paraboloid $z = x^2 + y^2$ and below by the square region $0 \le x \le 1$, $0 \le y \le 1$, $z = 0$.

58. Find the volume of the solid that is bounded above by the surface $z = xy$, below by the xy-plane, and on the sides by the circular disk $x^2 + y^2 \le a^2$ $(a > 0)$ that lies in the first quadrant.

59. Find the volume of the solid that is bounded above by the paraboloid $z = 4 - x^2 - y^2$ and below by the plane $z = 4 - 2x$.

60. Find the surface area of that portion of the cone $z = \sqrt{x^2 + y^2}$ that is contained in the cylinder $x^2 + y^2 = 1$.

61. Find the surface area of the portion of the paraboloid $z = x^2 + y^2$ that lies below the plane $z = 9$.

62. Find the surface area of the portion of the cylinder $x^2 + z^2 = 4$ that is bounded by the planes $x = 0$, $x = 1$, $y = 0$, and $y = 2$.

63. Find the volume of the region bounded by the paraboloid $y^2 + z^2 = 2x$ and the plane $x + y = 1$.

64. Find the volume of the region bounded above by the paraboloid $z = 4 - x^2 - y^2$ and below by the plane $z + 2y = 4$.

65. Find the volume of the region bounded by the ellipsoid $z^2 = 4 - 4r^2$ and the cylinder $r = \cos\theta$.

66. Find the volume of the region bounded above by the hemisphere $z = \sqrt{9 - x^2 - y^2}$, below by the xy-plane, and on the sides by the cylinder $x^2 + y^2 = 1$.

67. Show that the solid bounded below by the cone $z = \sqrt{x^2 + y^2}$ and above by the sphere $x^2 + y^2 + z^2 = 2az$ for $a > 0$ has volume $V = \pi a^3$.

68. Find the volume of the solid region bounded above by the sphere given in spherical coordinates by $\rho = a$ $(a > 0)$ and below by the cone $\phi = \phi_0$ where $0 < \phi_0 < \frac{\pi}{2}$.

69. A homogeneous solid S is bounded above by the sphere $R = a(a > 0)$ and below by the cone $\phi = \phi_0$ where $0 < \phi_0 < \frac{\pi}{2}$. Find the moment of inertia of S about the z-axis. Assume $\rho = 1$.

70. The region between the circles $x^2 + y^2 = 1$ and $x^2 + y^2 = 4$ with $0 \le z \le 5$ forms a "washer." Find the moment of inertia of the washer about the z-axis if the density ρ is given by:

    a. $\rho = k$ (k a constant)   **b.** $\rho(x, y) = x^2 + y^2$ **c.** $\rho(x, y) = x^2 y^2$

71. Find $\bar{x}$, the x-coordinate of the center of mass, of a triangular lamina with vertices $(0, 0)$, $(1, 0)$, and $(0, 1)$ if the density is $\rho(x, y) = x^2 + y^2$.

72. Find the mass of a cone of top radius R and height H if the density at each point P is proportional to the distance from P to the tip of the cone.

73. Find $\bar{z}$, the z-coordinate of the centroid of a homogeneous right circular cone with height H and base radius R. Assume the cone is positioned with its vertex at the origin and the z-axis is its axis of symmetry.

74. Find the Jacobian $\dfrac{\partial(u, v, w)}{\partial(x, y, z)}$ of the change of variables $u = 2x - 3y + z$, $v = 2y - z$, $w = 2z$.

75. Find the Jacobian $\dfrac{\partial(u, v, w)}{\partial(x, y, z)}$ of the change of variables $u = x^2 + y^2 + z^2$, $v = 2y^2 + z^2$, $w = 2z^2$.

76. Let $u = 2x - y$ and $v = x + 2y$. Find the image of the unit square given by $0 \le x \le 1$, $0 \le y \le 1$.

77. Find the mass of a lamina with density $\rho = r\theta$ that covers the region enclosed by the rose $r = \cos 3\theta$ for $0 \le \theta \le \frac{\pi}{6}$.

78. Suppose $u = \frac{1}{2}(x^2 + y^2)$ and $v = \frac{1}{2}(x^2 - y^2)$, with $x > 0$, $y > 0$.

a. Find the Jacobian $\dfrac{\partial(u, v)}{\partial(x, y)}$.

b. Solve for x and y in terms of u and v, and find the Jacobian $\dfrac{\partial(x, y)}{\partial(u, v)}$.

c. Verify that $\dfrac{\partial(u, v)}{\partial(x, y)} \dfrac{\partial(x, y)}{\partial(u, v)} = 1$.

79. Let D be the region given by $u \geq 0$, $v \geq 0$, $1 \leq u + v \leq 2$.

a. Show that under the transformation $u = x + y$, $v = x - y$, D is mapped into the region R such that $-x \leq y \leq x, \frac{1}{2} \leq x \leq 1$.

b. Find the Jacobian $\dfrac{\partial(u, v)}{\partial(x, y)}$, and compute $\iint\limits_{D} (u + v)\, du\, dv$.

80. Use a change of variables after a transformation of the form $x = Au$, $y = Bv$, $z = Cw$ to find the volume of the region bounded by the surface $\sqrt{x} + \sqrt{2y} + \sqrt{3z} = 1$ and the coordinate planes.

81. Find the centroid of a homogeneous lamina that covers the part of the plane $Ax + By + Cz = 1$, $A > 0$, $B > 0$, $C > 0$ that lies in the first quadrant.

82. A homogeneous plate has the shape of the region in the first quadrant of the xy-plane that is bounded by the circle $x^2 + y^2 = 1$ and the lines $y = x$ and $x = 0$. Sketch the region and find $\bar{x}$, the x-coordinate of the centroid.

83. Let R be a lamina covering the region in the xy-plane that is bounded by the parabola $y = 1 - x^2$ and the positive coordinate axes. Assume the lamina has density $\rho(x, y) = xy$.

a. Find the center of mass of the lamina.

b. Find the moment of inertia of the plate about the z-axis.

84. A solid has the shape of the cylinder $x^2 + y^2 \leq a$, with $0 \leq z \leq b$. Assume the solid has density $\rho(x, y, z) = x^2 + y^2$.

a. Find the mass of the solid.   **b.** Find the center of mass of the solid.

85. Use cylindrical coordinates to find the volume of the solid bounded by the circular cylinder $r = 2a \cos\theta$ $(a > 0)$, the cone $z = r$, and the xy-plane.

86. In a psychological experiment, x units of stimulus A and y units of stimulus B are applied to a subject, whose performance on a certain task is modeled by the function

$$f(x, y) = 10 + xy e^{1 - x^2 - y^2}$$

Suppose the stimuli are controlled in such a way that the subject is exposed to every possible combination (x, y) with $x \geq 0$, $y \geq 0$, and $x + y \leq 1$. What is the subject's average response to stimuli, rounded to the nearest hundredth?

87. The point (x, y, z) lies on an ellipsoid if

$$x = aR \sin\phi \cos\theta$$
$$y = bR \sin\phi \sin\theta$$
$$z = cR \cos\phi$$

for a constant R. Find an equation for this ellipsoid in rectangular coordinates.

88. Let u be everywhere continuous in the plane, and define the functions f and g by

$$f(x) = \int_a^x u(x, y)\, dy \qquad g(y) = \int_y^b u(x, y),\, dx$$

Show that

$$\int_a^b f(x)\, dx = \int_a^b g(y)\, dy$$

Hint: Show that both integrals equal $\iint\limits_{D} u(x, y)\, dA$ for a certain region D.

89. Find the volume of the solid region bounded by the surface

$$x^{2/3} + y^{2/3} + z^{2/3} = a^{2/3}$$

where a is a positive constant.

90. **Modeling Problem** The parking lot for a certain shopping mall has the shape shown in Figure 12.81.

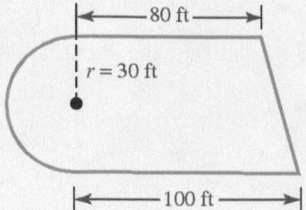

Figure 12.81 Shopping mall parking lot

Assuming a security observation tower can be located anywhere in the parking lot, where would you put it (answer to the nearest foot)? State all assumptions made in setting up and analyzing your model.

91. Find the centroid of the solid common to the cylindrical solids $x^2 + z^2 \le 1$ and $y^2 + z^2 \le 1$.

92. Show that if $z = f(r, \theta)$ is the equation of a surface S in polar coordinates, the surface area of S is given by

$$\iint\limits_R \sqrt{1 + \left(\frac{\partial z}{\partial r}\right)^2 + \frac{1}{r^2}\left(\frac{\partial z}{\partial \theta}\right)^2}\ r\,dr\,d\theta$$

where R is the projected region in the $r\theta$-plane.

93. Find the center of mass of the solid S that lies inside the sphere $x^2 + y^2 + z^2 = 1$ in the first octant $x \ge 0$, $y \ge 0$, $z \ge 0$ if the density is $\rho(x, y, z) = (x^2 + y^2 + z^2 + 1)^{-1}$.

94. Suppose we drill a square hole of side c through the center of a sphere of radius c, as shown in Figure 12.82. What is the volume of the solid that remains? Give your answer as a factor of c^3 rounded to the nearest hundredth.

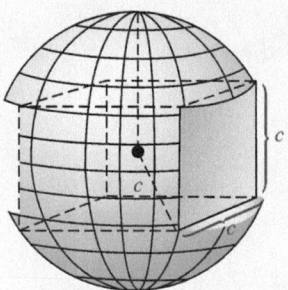

Figure 12.82 Sphere with a square hole

95. A cube of side 2 is surmounted by a hemisphere of radius 1, as shown in Figure 12.83.

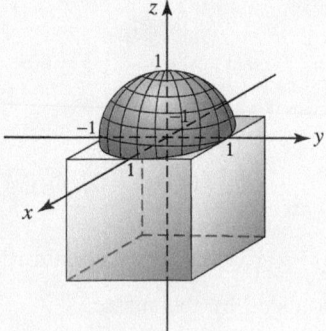

Figure 12.83 Center of mass of a complex solid

Suppose the origin is at the center of the hemispherical dome. If the entire solid is made of the same material with density $\rho = 1$, where is the centroid?

96. **Putnam Examination Problem** The function $K(x, y)$ is positive and continuous for $0 \leq x \leq 1$, $0 \leq y \leq 1$ and the functions $f(x)$ and $g(x)$ are positive and continuous for $0 \leq x \leq 1$. Suppose that for all $0 \leq x \leq 1$, we have

$$\int_0^1 f(y)K(x, y)\, dy = g(x) \qquad \text{and} \qquad \int_0^1 g(y)K(x, y)\, dy = f(x)$$

Show that $f(x) = g(x)$ for $0 \leq x \leq 1$.

97. **Putnam Examination Problem** A circle of radius a is revolved through $180°$ about a line in its plane, distant b from the center of the circle ($b > a$). For what value of the ratio b/a does the center of gravity of the solid thus generated lie on the surface of the solid?

98. **Putnam Examination Problem** For $f(x)$ a positive, monotone, decreasing function defined in $0 \leq x \leq 1$, prove that

$$\frac{\int_0^1 x[f(x)]^2\, dx}{\int_0^1 x f(x)\, dx} \leq \frac{\int_0^1 [f(x)]^2\, dx}{\int_0^1 f(x)\, dx}$$

99. **Putnam Examination Problem** Show that the integral equation

$$f(x, y) = 1 + \int_0^x \int_0^y f(u, v)\, du\, dv$$

has at most one continuous solution for $0 \leq x \leq 1$, $0 \leq y \leq 1$.

CHAPTER 12 GROUP RESEARCH PROJECT*

Working in small groups is typical of most work environments, and this book seeks to develop skills with group activities. We present a group project at the end of each chapter. These projects are to be done in groups of three or four students.

Space Capsule Design

Suppose you were part of a team of engineers designing the Apollo space capsule. The capsule is composed of two parts:

1. A cone with a height of 4 meters and a base of radius 3 meters
2. A reentry shield in the shape of a parabola revolved about the axis of the cone, which is attached to the cone along the edge of the base of the cone. Its vertex is a distance D below the base of the cone. Find values of the design parameters D and ρ so that the capsule will float with the vertex of the cone pointing up and with the waterline 2 m below the top of the cone, in order to keep the exit port $1/3$ m above water. You may make the following assumptions:

   a. The capsule has uniform density ρ.
   b. The center of mass of the capsule should be below the center of mass of the displaced water because this will give the capsule better stability in heavy seas.
   c. A body floats in a fluid at the level at which the weight of the displaced fluid equals the weight of the body (Archimedes' principle).

Extended paper for further study. Your paper is not limited to the following questions but should include these concerns: Show the project director that the task is impossible; that is, there are no values of D and ρ that satisfy the design specifications. However, you can solve this dilemma by incorporating a flotation collar in the shape of a torus. The collar will be made by taking hollow plastic tubing with a circular cross section of radius 1 m and wrapping it in a circular ring about the capsule, so that it fits snugly. The collar is designed to float just submerged with its top tangent to the surface of the water. Show that this flotation collar makes the capsule plus collar assembly satisfy the design specifications. Find the density ρ needed to make the capsule float at the 2 meter mark. Assume the weight of the tubing is negligible compared to the weight of the capsule, that the design parameter D is equal to 1 meter, and the density of the water is 1.

---

*Adapted from *MAA Notes* 17 (1991), "Priming the Calculus Pump: Innovations and Resources," Marcus S. Cohen, Edward D. Gaughan, R. Arthur Knoebel, Douglas S. Kurtz, and David J. Pengelley.

CHAPTER 13

VECTOR ANALYSIS

I*n mathematics, you never understand things; you just get used to them.*

John von Neumann

As quoted in *The Dancing Wu Li Masters: An Overview of the New Physics* (1984), p. 206.

PREVIEW

In this chapter, we combine what we have learned about differentiation, integration, and vectors to study the calculus of vector functions defined on a set of points in $\mathbb{R}^2$ or $\mathbb{R}^3$. We introduce *line integrals* and *surface integrals* to study such things as fluid flow and then obtain a result called *Green's theorem* that enables line integrals to be computed in terms of ordinary double integrals. This result is extended into $\mathbb{R}^3$ to obtain *Stokes' theorem* and the *divergence theorem*, which have extensive applications in areas such as fluid dynamics and electromagnetic theory.

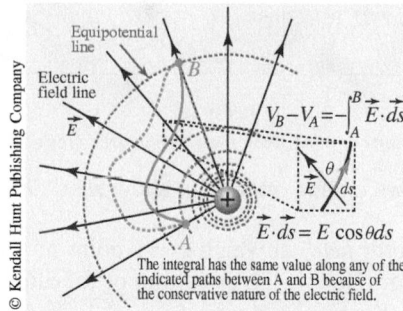

Voltage from an electric field using line integrals

PERSPECTIVE

How much work is done by a variable force acting along a given curve in space? How can the amount of heat flowing across a particular surface in unit time be measured, and is the measurement similar to measuring the flow of water or electricity? We will use line integrals and surface integrals to answer these and other questions from physics and engineering mathematics.

13.1 PROPERTIES OF A VECTOR FIELD: DIVERGENCE AND CURL

IN THIS SECTION: *Definition of a vector field, divergence, curl*
In order to model properties of electricity, magnetism, fluid dynamics, and other applications, we use the notion of a *vector field*. After introducing some basic terminology, we consider two concepts, called the *divergence* and the *curl*, which involve vector differentiation. We conclude by considering applications of these concepts in the study of fluid motions.

Definition of a Vector Field

The satellite photograph in Figure 13.1 shows wind measurements over the world's oceans. Wind direction is indicated by directed line segments, showing wind direction and wind speed.

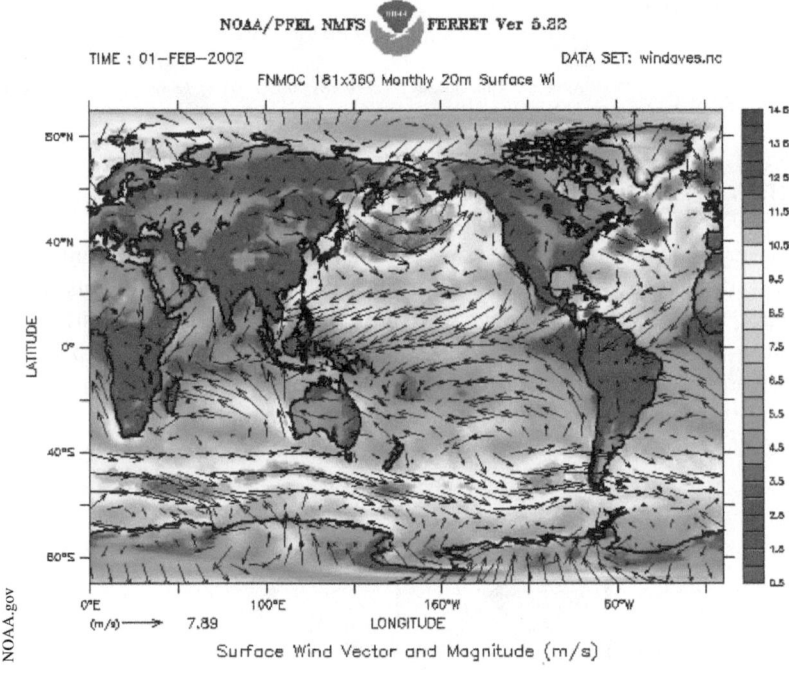

Figure 13.1   The arrows on this world map show wind direction and magnitude

This is an example of a *vector field*, in which every point in a given region of the plane or space is assigned a vector. Here is the definition of a vector field in $\mathbb{R}^3$.

> **VECTOR FIELD**   A **vector field** in $\mathbb{R}^3$ is a function $\mathbf{F}$ that assigns a vector to each point in its domain. A vector field with domain D in $\mathbb{R}^3$ has the form
>
> $$\mathbf{F}(x,y,z) = M(x,y,z)\mathbf{i} + N(x,y,z)\mathbf{j} + P(x,y,z)\mathbf{k}$$
>
> where the scalar functions M, N, and P are called the **components** of $\mathbf{F}$. A **continuous** vector field $\mathbf{F}$ is one whose components M, N, and P are continuous, and a **differentiable** vector field is one for which all partial derivatives of M, N, and P exist.

For example,

$$\mathbf{F} = 2x^2 y\mathbf{i} + e^{yz}\mathbf{j} + \left(\tan\frac{x}{2}\right)\mathbf{k}$$

is a vector field with **i**-component $2x^2 y$, **j**-component e^{yz}, and **k**-component $\tan\frac{x}{2}$.

A vector field in $\mathbb{R}^2$ can be thought of as a special case where there are no z-coordinates and no **k**-components. That is, a vector field in $\mathbb{R}^2$ has the form

$$\mathbf{F}(x,y) = M(x,y)\mathbf{i} + N(x,y)\mathbf{j}$$

To visualize a particular vector field $\mathbf{F}(x,y,z)$, it often helps to select a number of points in the domain of $\mathbf{F}$ and then draw an arrow emanating from each point $P(a,b,c)$ with the direction of $\mathbf{F}(a,b,c)$ and length representing the magnitude $\|\mathbf{F}(a,b,c)\|$. We will refer to such a representation as the **graph of F**. Here is an example involving the graph of a vector field in $\mathbb{R}^2$.

Example 1   Graph of a vector field

Sketch the graph of the vector field $\mathbf{F}(x,y) = y\mathbf{i} - x\mathbf{j}$.

Solution   We will evaluate $\mathbf{F}$ at various points. For example,

$$\mathbf{F}(3,4) = 4\mathbf{i} - 3\mathbf{j} \qquad \text{and} \qquad \mathbf{F}(-1,2) = 2\mathbf{i} - (-1)\mathbf{j} = 2\mathbf{i} + \mathbf{j}$$

We can generate as many such vector values of $\mathbf{F}$ as we wish. Several are shown in Figure 13.2.

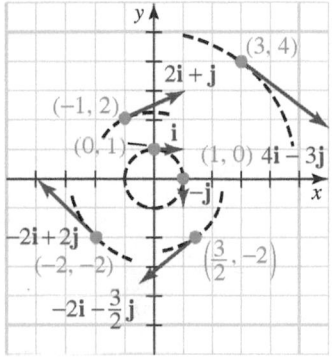

Figure 13.2   The graph of a vector field

Vector fields are difficult to draw by hand, so most of the vector fields you see in books are computer generated. It would be worthwhile to investigate drawing a vector field using a calculator or a computer.

One of the most important applications of vector fields is in fluid dynamics, which is a part of fluid mechanics which deals with fluid flows. A one-dimensional flow is a flow that has spatial variations in one direction only. A two-dimensional flow is a flow in which special variations occur in some planar surface (thus involving two directions). A three-dimensional flow has spatial variations everywhere in a three-dimensional space. The graph of a vector field often yields useful information about the properties of the field. A flow is said to be *irrotational* if the angular velocity at any point of the flow is zero, at any moment in time. Otherwise, the flow is said to be rotational.

The flow in Figure 13.3**a** is an irrotational one, whereas Figure 13.3**b** suggests a rotational flow. Sometimes, an irrotational flow is informally said to be *parallel*, and a rotational flow is said to be *circular*, but these terms are imprecise.

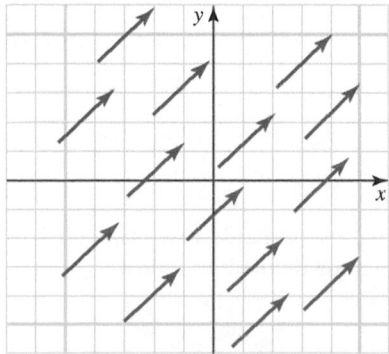

a. An irrotational fluid flow

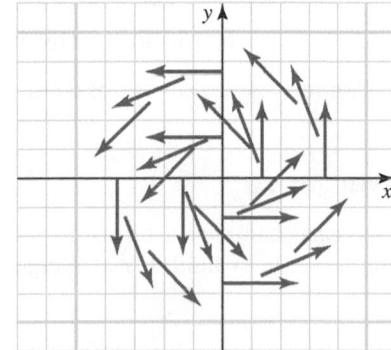

b. A rotational flow

Figure 13.3   Flow diagrams

In fluid dynamics, if time derivatives of a flow field vanish, then the flow is considered to be a *steady flow*. Otherwise, the fluid flow is said to be *unsteady*. Note that a flow field can be irrotational and at the same time unsteady! One should not confuse the trajectories of the flow particles with its other physical properties.

Gravitational, electrical, and magnetic vector fields play an important role in physical applications. We will discuss gravitational fields now, and electrical and magnetic fields later in this section. Accordingly, we begin with Newton's law of gravitation which says that a point mass (particle) m at the origin exerts on a unit point mass located at the point $P(x, y, z)$ a force $\mathbf{F}(x, y, z)$ given by

$$\mathbf{F}(x, y, z) = \frac{Gm}{x^2 + y^2 + z^2} \mathbf{u}(x, y, z)$$

where G is a constant (the universal gravitational constant) and $\mathbf{u}$ is the unit vector extending from the point P toward the origin. The vector field $\mathbf{F}(x, y, z)$ is called the **gravitational field** of the point mass m. Because

$$\mathbf{u}(x, y, z) = \frac{-1}{\sqrt{x^2 + y^2 + z^2}} (x\mathbf{i} + y\mathbf{j} + z\mathbf{k})$$

it follows that

$$\mathbf{F}(x, y, z) = \frac{-Gm}{(x^2 + y^2 + z^2)^{3/2}} (x\mathbf{i} + y\mathbf{j} + z\mathbf{k})$$

Note that the gravitational field $\mathbf{F}$ always points toward the origin and has the same magnitude for any point m located $r = \sqrt{x^2 + y^2 + z^2}$ units from the origin. Such a vector field is called a **central force field**. This force field is shown in Figure 13.4a. Other physical vector fields are shown in Figure 13.4b and Figure 13.4c.

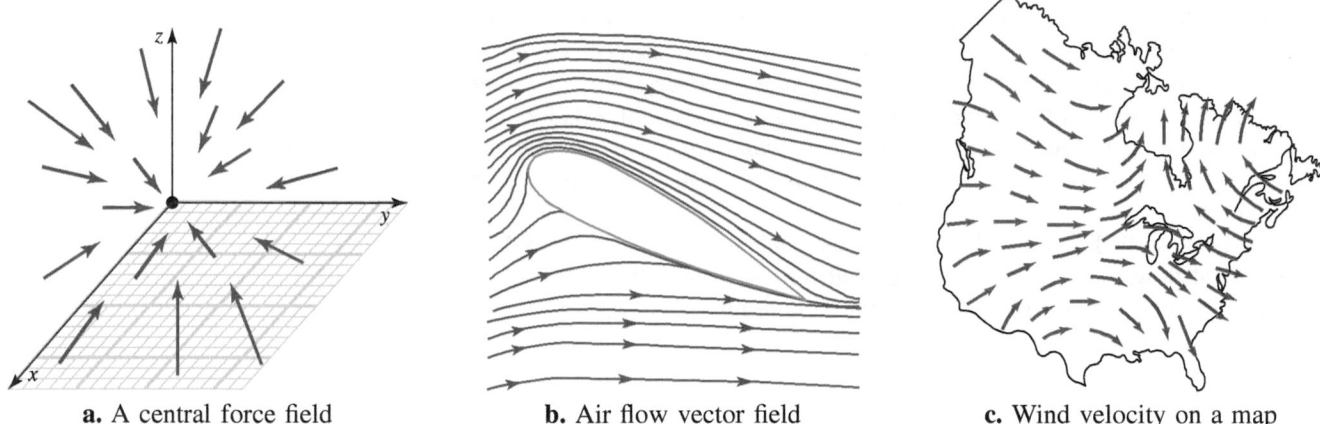

a. A central force field   b. Air flow vector field   c. Wind velocity on a map

Figure 13.4   Examples of physical vector fields

Divergence

Divergence and *curl* are two operations on vector fields that originated in connection with the study of fluid flow. Divergence may be defined as follows.

DIVERGENCE The **divergence** of a differentiable vector field

$$\mathbf{V}(x,y,z) = u(x,y,z)\mathbf{i} + v(x,y,z)\mathbf{j} + w(x,y,z)\mathbf{k}$$

is denoted by div $\mathbf{V}$ and is given by

$$\text{div } \mathbf{V} = \frac{\partial u}{\partial x}(x,y,z) + \frac{\partial v}{\partial y}(x,y,z) + \frac{\partial w}{\partial z}(x,y,z)$$

STOP *The divergence of a vector field is a scalar function.*

Example 2 Divergence of a vector field

Find the divergence of each of the following vector fields.

a. $\mathbf{F}(x,y) = x^2 y\mathbf{i} + xy^3\mathbf{j}$　　**b.** $\mathbf{G}(x,y,z) = x\mathbf{i} + y^3 z^2\mathbf{j} + xz^3\mathbf{k}$

Solution

a. div $\mathbf{F} = \dfrac{\partial}{\partial x}(x^2 y) + \dfrac{\partial}{\partial y}(xy^3) = 2xy + 3xy^2$

b. div $\mathbf{G} = \dfrac{\partial}{\partial x}(x) + \dfrac{\partial}{\partial y}(y^3 z^2) + \dfrac{\partial}{\partial z}(xz^3) = 1 + 3y^2 z^2 + 3xz^2$

Suppose the vector field

$$\mathbf{V}(x,y,z) = u(x,y,z)\mathbf{i} + v(x,y,z)\mathbf{j} + w(x,y,z)\mathbf{k}$$

represents the velocity of a fluid with density $\rho(x,y,z)$ at a point (x,y,z) in a certain region R in $\mathbb{R}^3$. Then the vector field $\rho\mathbf{V}$ is called the **flux density** and is denoted by $\mathbf{D}$. We can think of $\mathbf{D} = \rho\mathbf{V}$ as measuring the "mass flow" of the liquid.

Assuming there are no external processes acting on the fluid that would tend to create or destroy fluid, it can be shown that div $\mathbf{D}$ gives the negative of the rate of change of density with respect to time, that is,

$$\text{div } \mathbf{D} = -\frac{\partial \rho}{\partial t}$$

This is often referred to as the **continuity equation** of fluid dynamics. (A derivation is given in Section 13.7.) When div $\mathbf{D} = 0$, $\mathbf{D}$ is said to be **incompressible**. By virtue of the continuity equation, incompressible flow represents flow in which the material density is constant within an infinitesimal volume that moves with the velocity of the fluid. If div $\mathbf{D} > 0$ at a point (x_0, y_0, z_0), the point is called a **source**; if div $\mathbf{D} < 0$, the point is called a **sink** (see Figure 13.5). The terms *sink, source,* and *incompressible* apply to any vector field $\mathbf{F}$ and are not reserved only for fluid applications.

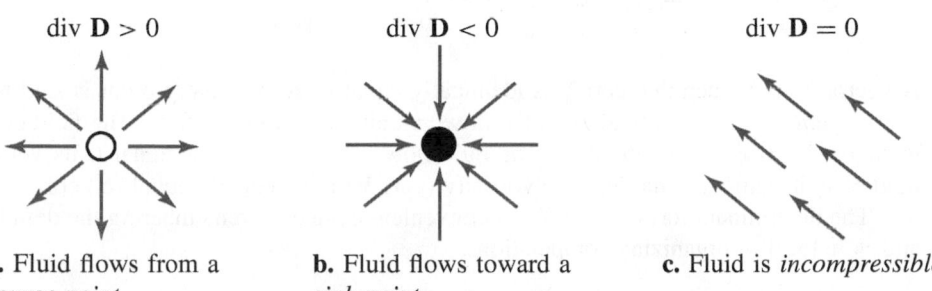

div $\mathbf{D} > 0$　　　div $\mathbf{D} < 0$　　　div $\mathbf{D} = 0$

a. Fluid flows from a source point　　**b.** Fluid flows toward a sink point　　**c.** Fluid is *incompressible*

Figure 13.5 Flow of a fluid across a plane region, **D**

A useful way to think of the divergence div $\mathbf{V}$ is in terms of the **del operator** defined by

$$\nabla = \frac{\partial}{\partial x}\mathbf{i} + \frac{\partial}{\partial y}\mathbf{j} + \frac{\partial}{\partial z}\mathbf{k}$$

Recall (from Section 11.6) that applying the del operator to the differentiable function $f(x, y, z)$ produces the **gradient field**

$$\nabla f = \frac{\partial f}{\partial x}\mathbf{i} + \frac{\partial f}{\partial y}\mathbf{j} + \frac{\partial f}{\partial z}\mathbf{k}$$

Similarly, by taking the dot product of the operator ∇ with the vector field $\mathbf{V} = u(x, y, z)\mathbf{i} + v(x, y, z)\mathbf{j} + w(x, y, z)\mathbf{k}$, we obtain the divergence

$$\begin{aligned}
\nabla \cdot \mathbf{V} &= \left(\frac{\partial}{\partial x}\mathbf{i} + \frac{\partial}{\partial y}\mathbf{j} + \frac{\partial}{\partial z}\mathbf{k}\right) \cdot (u\mathbf{i} + v\mathbf{j} + w\mathbf{k}) \\
&= \frac{\partial}{\partial x}(u) + \frac{\partial}{\partial y}(v) + \frac{\partial}{\partial z}(w) \\
&= \frac{\partial u}{\partial x} + \frac{\partial v}{\partial y} + \frac{\partial w}{\partial z} \\
&= \operatorname{div} \mathbf{V}
\end{aligned}$$

Curl

The del operator may also be used to describe another derivative operation for vector fields, called the *curl*.

CURL   The **curl** of a differentiable vector field

$$\mathbf{V}(x, y, z) = u(x, y, z)\mathbf{i} + v(x, y, z)\mathbf{j} + w(x, y, z)\mathbf{k}$$

is denoted by curl $\mathbf{V}$ and is defined by

$$\operatorname{curl} \mathbf{V} = \left(\frac{\partial w}{\partial y} - \frac{\partial v}{\partial z}\right)\mathbf{i} + \left(\frac{\partial u}{\partial z} - \frac{\partial w}{\partial x}\right)\mathbf{j} + \left(\frac{\partial v}{\partial x} - \frac{\partial u}{\partial y}\right)\mathbf{k}$$

STOP *The curl of a vector field is a vector function.*

Note that

$$\begin{aligned}
\operatorname{curl} \mathbf{V} &= \left(\frac{\partial w}{\partial y} - \frac{\partial v}{\partial z}\right)\mathbf{i} - \left(\frac{\partial w}{\partial x} - \frac{\partial u}{\partial z}\right)\mathbf{j} + \left(\frac{\partial v}{\partial x} - \frac{\partial u}{\partial y}\right)\mathbf{k} \\
&= \nabla \times \mathbf{V} \\
&= \begin{vmatrix} \mathbf{i} & \mathbf{j} & \mathbf{k} \\ \frac{\partial}{\partial x} & \frac{\partial}{\partial y} & \frac{\partial}{\partial z} \\ u & v & w \end{vmatrix}
\begin{matrix} \leftarrow \text{Standard basis vectors} \\ \leftarrow \nabla \\ \leftarrow \mathbf{V} \end{matrix}
\end{aligned}$$

A vector field $\mathbf{V}$ such that curl $\mathbf{V}$ is identically equal to zero at every point is said to be *irrotational*. In case of a fluid flow, the flow velocity $\mathbf{V}$ is a vector field. The field curl $\mathbf{V}$ is called the **vorticity** of the flow. The fluid flow is said to be irrotational if its velocity field $\mathbf{V}$ is irrotational, that is, if its vorticity (curl $\mathbf{V}$) is identically equal to zero.

The determinant form of curl $\mathbf{V}$ is a convenient device for remembering the definition and is helpful in organizing computations.

DEL OPERATOR FORMS FOR DIVERGENCE AND CURL

Consider a differentiable vector field

$$\mathbf{V}(x,y,z) = u(x,y,z)\mathbf{i} + v(x,y,z)\mathbf{j} + w(x,y,z)\mathbf{k}$$

The divergence and curl of $\mathbf{V}$ are given by

$$\text{div } \mathbf{V} = \nabla \cdot \mathbf{V} \qquad \text{and} \qquad \text{curl } \mathbf{V} = \nabla \times \mathbf{V}$$

Note that div *is a scalar and* curl *is a vector.*

Example 3 Curl of a vector field

Find the curl of each of the following vector fields:

$$\mathbf{F} = x^2 yz\,\mathbf{i} + xy^2 z\,\mathbf{j} + xyz^2\,\mathbf{k} \qquad \text{and} \qquad \mathbf{G} = (x\cos y)\mathbf{i} + xy^2\mathbf{j}$$

Solution

$$\text{curl } \mathbf{F} = \begin{vmatrix} \mathbf{i} & \mathbf{j} & \mathbf{k} \\ \frac{\partial}{\partial x} & \frac{\partial}{\partial y} & \frac{\partial}{\partial z} \\ x^2 yz & xy^2 z & xyz^2 \end{vmatrix}$$

$$= \left[\frac{\partial}{\partial y}(xyz^2) - \frac{\partial}{\partial z}(xy^2 z)\right]\mathbf{i} - \left[\frac{\partial}{\partial x}(xyz^2) - \frac{\partial}{\partial z}(x^2 yz)\right]\mathbf{j}$$

$$+ \left[\frac{\partial}{\partial x}(xy^2 z) - \frac{\partial}{\partial y}(x^2 yz)\right]\mathbf{k}$$

$$= (xz^2 - xy^2)\mathbf{i} + (x^2 y - yz^2)\mathbf{j} + (y^2 z - x^2 z)\mathbf{k}$$

$$\text{curl } \mathbf{G} = \begin{vmatrix} \mathbf{i} & \mathbf{j} & \mathbf{k} \\ \frac{\partial}{\partial x} & \frac{\partial}{\partial y} & \frac{\partial}{\partial z} \\ x\cos y & xy^2 & 0 \end{vmatrix}$$

$$= \left[0 - \frac{\partial}{\partial z}(xy^2)\right]\mathbf{i} - \left[0 - \frac{\partial}{\partial z}(x\cos y)\right]\mathbf{j} + \left[\frac{\partial}{\partial x}(xy^2) - \frac{\partial}{\partial y}(x\cos y)\right]\mathbf{k}$$

$$= (y^2 + x\sin y)\mathbf{k}$$

Example 4 A vector field with constant components has divergence and curl zero

Let $\mathbf{F} = a\mathbf{i} + b\mathbf{j} + c\mathbf{k}$. Show that div $\mathbf{F} = 0$ and curl $\mathbf{F} = \mathbf{0}$.

Solution Let $\mathbf{F} = a\mathbf{i} + b\mathbf{j} + c\mathbf{k}$ for constants a, b, and c. Then

$$\text{div } \mathbf{F} = \frac{\partial}{\partial x}(a) + \frac{\partial}{\partial y}(b) + \frac{\partial}{\partial z}(c) \qquad \text{curl } \mathbf{F} = \begin{vmatrix} \mathbf{i} & \mathbf{j} & \mathbf{k} \\ \frac{\partial}{\partial x} & \frac{\partial}{\partial y} & \frac{\partial}{\partial z} \\ a & b & c \end{vmatrix}$$

$$= 0$$

$$= 0\mathbf{i} - 0\mathbf{j} + 0\mathbf{k}$$

$$= \mathbf{0}$$

*Example 4 shows that the divergence and curl of a constant vector field are zero, but this does **not** mean that if div $\mathbf{F} = 0$ and curl $\mathbf{F} = 0$, then $\mathbf{F}$ must be a constant. For instance, the nonconstant vector field*

$$\mathbf{F}(x,y,z) = x\mathbf{i} + y\mathbf{j} - 2z\mathbf{k}$$

has both div $\mathbf{F} = 0$ and curl $\mathbf{F} = \mathbf{0}$.

Combinations of the gradient, divergence, and curl appear in a variety of applications. In particular, note that if f is a differentiable scalar function, its gradient ∇f is a vector field, and we can compute

$$\text{div } \nabla f = \left(\frac{\partial}{\partial x}\mathbf{i} + \frac{\partial}{\partial y}\mathbf{j} + \frac{\partial}{\partial z}\mathbf{k} \right) \cdot \left(\frac{\partial f}{\partial x}\mathbf{i} + \frac{\partial f}{\partial y}\mathbf{j} + \frac{\partial f}{\partial z}\mathbf{k} \right)$$

$$= \frac{\partial^2 f}{\partial x^2} + \frac{\partial^2 f}{\partial y^2} + \frac{\partial^2 f}{\partial z^2}$$

$$= \nabla \cdot \nabla f$$

In the following box, we introduce some special notation and terminology for this operation.

LAPLACIAN OPERATOR   Let $f(x,y,z)$ define a function with continuous first and second partial derivatives. Then the **Laplacian of f** is

$$\nabla^2 f = \nabla \cdot \nabla f$$

$$= \frac{\partial^2 f}{\partial x^2} + \frac{\partial^2 f}{\partial y^2} + \frac{\partial^2 f}{\partial z^2}$$

$$= f_{xx} + f_{yy} + f_{zz}$$

The equation $\nabla^2 f = 0$ is called **Laplace's equation**, and a function that satisfies such an equation in a region D is said to be **harmonic** in D.

The Laplacian is named for the French mathematician Pierre Laplace (1749-1827). See the *Historical Quest*, Section 12.3, Problem 60.

Note that if $f(x,y)$ is a function of two variables x and y, $\nabla^2 f = f_{xx}(x,y) + f_{yy}(x,y)$.

Example 5   Showing a function is harmonic

Show that $f(x,y) = e^x \cos y$ is harmonic.

Solution $f_x(x,y) = e^x \cos y$

$f_{xx}(x,y) = e^x \cos y$

$f_y(x,y) = -e^x \sin y$

$f_{yy} = -e^x \cos y$

The Laplacian of f is given by

$$\nabla^2 f(x,y) = f_{xx}(x,y) + f_{yy}(x,y)$$
$$= e^x \cos y - e^x \cos y$$
$$= 0$$

Thus, f is harmonic.

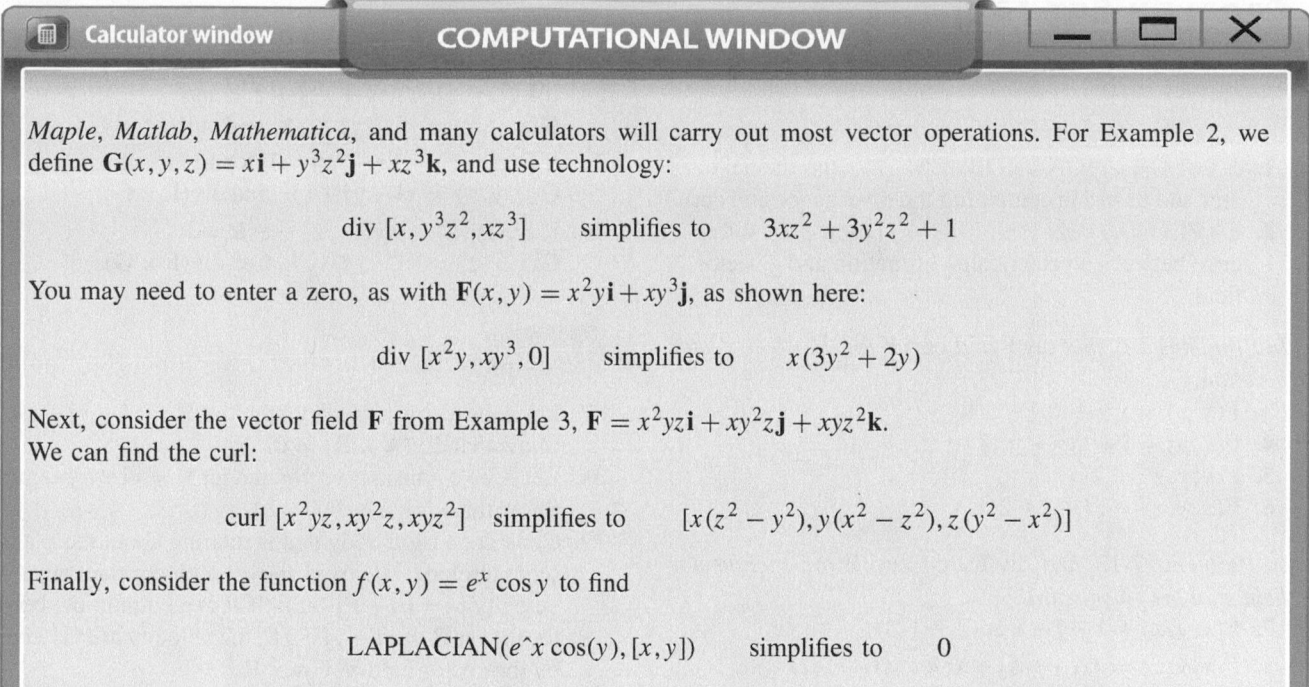

Maple, Matlab, Mathematica, and many calculators will carry out most vector operations. For Example 2, we define $\mathbf{G}(x, y, z) = x\mathbf{i} + y^3 z^2 \mathbf{j} + xz^3 \mathbf{k}$, and use technology:

$$\text{div } [x, y^3 z^2, xz^3] \qquad \text{simplifies to} \qquad 3xz^2 + 3y^2 z^2 + 1$$

You may need to enter a zero, as with $\mathbf{F}(x, y) = x^2 y\mathbf{i} + xy^3 \mathbf{j}$, as shown here:

$$\text{div } [x^2 y, xy^3, 0] \qquad \text{simplifies to} \qquad x(3y^2 + 2y)$$

Next, consider the vector field $\mathbf{F}$ from Example 3, $\mathbf{F} = x^2 yz\mathbf{i} + xy^2 z\mathbf{j} + xyz^2 \mathbf{k}$. We can find the curl:

$$\text{curl } [x^2 yz, xy^2 z, xyz^2] \quad \text{simplifies to} \quad [x(z^2 - y^2), y(x^2 - z^2), z(y^2 - x^2)]$$

Finally, consider the function $f(x, y) = e^x \cos y$ to find

$$\text{LAPLACIAN}(e\hat{\ }x \cos(y), [x, y]) \qquad \text{simplifies to} \qquad 0$$

In many ways, the study of electricity and magnetism is analogous to that of fluid dynamics, and the curl and divergence play an important role in this study. In electromagnetic theory, it is often convenient to regard interaction between electrical charges as forces somewhat like the gravitational force between masses and then to seek quantitative measure of these forces.

One of the great scientific achievements of the 19th century was the discovery of the laws of electromagnetism by the English scientist James Clerk Maxwell (see *Historical Quest*, Section 13.7, Problems 57-60). These laws have an elegant expression in terms of the divergence and curl. It is known empirically that the force acting on a charge due to an electromagnetic field depends on the position, velocity, and amount of the particular charge, and not on the number of other charges that may be present or how those other charges are moving. Suppose a charge is located at the point (x, y, z) at time t, and consider the electric intensity field $\mathbf{E}(x, y, z, t)$ and the magnetic intensity field $\mathbf{H}(x, y, z, t)$. Then the behavior of the resulting electromagnetic field is determined by

$$\text{div } \mathbf{E} = \frac{Q}{\epsilon} \qquad\qquad \text{div } (\mu\mathbf{H}) = 0$$

$$\text{curl } \mathbf{E} = -\frac{\partial}{\partial t}(\mu\mathbf{H}) \qquad c^2(\text{curl } \mathbf{B}) = \frac{\partial E}{\partial t} + \frac{\mathbf{J}}{\epsilon}$$

where Q is the *electric charge density* (charge per unit volume), $\mathbf{J}$ is the *electric current density* (rate at which the charge flows through a unit area per second), $\mathbf{B}$ is the *magnetic flux density*, c is the speed of light, and μ and ϵ are constants called the *permeability* and *permittivity*, respectively. Working with these equations and terms is beyond the scope of this course, but if you are interested there are many references you can consult. One of the best (despite being almost 50 years old) is the classic *Feynman Lectures in Physics* (Reading, Mass: Addison-Wesley, 1963), by Nobel laureate Richard Feynman, Robert Leighton, and Matthew Sands.

PROBLEM SET 13.1

1. **EXPLORATION PROBLEM** Discuss the del operator and its use in computing the divergence and curl.
2. **EXPLORATION PROBLEM** Discuss the difference between a vector valued function and a vector field.

In Problems 3-6, find div **F** *and* curl **F** *for the given vector function.*

3. $\mathbf{F}(x,y) = x^2\mathbf{i} + xy\mathbf{j} + z^3\mathbf{k}$
4. $\mathbf{F}(x,y) = \mathbf{i} + (x^2 + y^2)\mathbf{j}$
5. $\mathbf{F}(x,y,z) = 2y\mathbf{j}$
6. $\mathbf{F}(x,y,z) = z\mathbf{i} - \mathbf{j} + 2y\mathbf{k}$

In Problems 7-12, find div **F** *and* curl **F** *for each vector field* **F** *at the given point.*

7. $\mathbf{F}(x,y,z) = \mathbf{i} + \mathbf{j} + \mathbf{k}$ at $(2,-1,3)$
8. $\mathbf{F}(x,y,z) = xz\mathbf{i} + y^2z\mathbf{j} + xz\mathbf{k}$ at $(1,-1,2)$
9. $\mathbf{F}(x,y,z) = xyz\mathbf{i} + y\mathbf{j} + x\mathbf{k}$ at $(1,2,3)$
10. $\mathbf{F}(x,y,z) = (\cos y)\mathbf{i} + (\sin y)\mathbf{j} + \mathbf{k}$ at $(\frac{\pi}{4}, \pi, 0)$
11. $\mathbf{F}(x,y,z) = e^{-xy}\mathbf{i} + e^{xz}\mathbf{j} + e^{yz}\mathbf{k}$ at $(3,2,0)$
12. $\mathbf{F}(x,y,z) = (e^{-x}\sin y)\mathbf{i} + (e^{-x}\cos y)\mathbf{j} + \mathbf{k}$ at $(1,3,-2)$

Find div **F** *and* curl **F** *for each vector field* **F** *given in Problems 13-26.*

13. $\mathbf{F} = (\sin x)\mathbf{i} + (\cos y)\mathbf{j}$ 14. $\mathbf{F} = (-\cos x)\mathbf{i} + (\sin y)\mathbf{j}$
15. $\mathbf{F} = x\mathbf{i} - y\mathbf{j}$ 16. $\mathbf{F} = -x\mathbf{i} + y\mathbf{j}$
17. $\mathbf{F} = \dfrac{x}{\sqrt{x^2+y^2}}\mathbf{i} + \dfrac{y}{\sqrt{x^2+y^2}}\mathbf{j}$
18. $\mathbf{F} = x^2\mathbf{i} - y^2\mathbf{j}$ 19. $\mathbf{F} = x^2\mathbf{i} - z^2\mathbf{k}$
20. $\mathbf{F} = x^2\mathbf{i} + y^2\mathbf{j} + z^2\mathbf{k}$ 21. $\mathbf{F} = y\mathbf{i} + z\mathbf{j} + x\mathbf{k}$
22. $\mathbf{F} = xyz\mathbf{i} + x^2y^2z^2\mathbf{j} + y^2z^3\mathbf{k}$
23. $\mathbf{F} = (\ln z)\mathbf{i} + e^{xy}\mathbf{j} + \tan^{-1}\left(\frac{x}{z}\right)\mathbf{k}$
24. $\mathbf{F} = (z^2 e^{-x})\mathbf{i} + (y^3 \ln z)\mathbf{j} + (xe^{-y})\mathbf{k}$
25. $\mathbf{F} = \dfrac{x\mathbf{i} + y\mathbf{j} + z\mathbf{k}}{\sqrt{x^2+y^2+z^2}}$ 26. $\mathbf{F} = yz\mathbf{i} + xz\mathbf{j} + xy\mathbf{k}$

Determine whether each scalar function in Problems 27-30 is harmonic.

27. $u(x,y,z) = e^{-x}(\cos y - \sin y)$
28. $v(x,y,z) = (x^2 + y^2 + z^2)^{1/2}$
29. $w(x,y,z) = (x^2 + y^2 + z^2)^{-1/2}$
30. $r(x,y,z) = xyz$
31. Find div **F**, given that $\mathbf{F} = \nabla f$, where $f(x,y,z) = xy^3z^2$.
32. Find div **F**, given that $\mathbf{F} = \nabla f$, where $f(x,y,z) = x^2yz^3$.
33. If $\mathbf{F}(x,y,z) = 2\mathbf{i} + 2x\mathbf{j} + 3y\mathbf{k}$ and $\mathbf{G}(x,y,z) = x\mathbf{i} - y\mathbf{j} + z\mathbf{k}$, find curl($\mathbf{F} \times \mathbf{G}$).

34. If $\mathbf{F}(x,y,z) = xy\mathbf{i} + yz\mathbf{j} + z^2\mathbf{k}$ and $\mathbf{G}(x,y,z) = x\mathbf{i} + y\mathbf{j} - z\mathbf{k}$, find curl($\mathbf{F} \times \mathbf{G}$).
35. If $\mathbf{F}(x,y,z) = 2\mathbf{i} + 2x\mathbf{j} + 3y\mathbf{k}$ and $\mathbf{G}(x,y,z) = x\mathbf{i} - y\mathbf{j} + z\mathbf{k}$, find div($\mathbf{F} \times \mathbf{G}$).
36. If $\mathbf{F}(x,y,z) = xy\mathbf{i} + yz\mathbf{j} + z^2\mathbf{k}$ and $\mathbf{G}(x,y,z) = x\mathbf{i} + y\mathbf{j} - z\mathbf{k}$, find div($\mathbf{F} \times \mathbf{G}$).

37. Let **A** be a constant vector and let $\mathbf{R} = x\mathbf{i} + y\mathbf{j} + z\mathbf{k}$. Show that div($\mathbf{A} \times \mathbf{R}$) $= 0$.
38. Let **A** be a constant vector and let $\mathbf{R} = x\mathbf{i} + y\mathbf{j} + z\mathbf{k}$. Show that curl($\mathbf{A} \times \mathbf{R}$) $= 2\mathbf{A}$.
39. Consider a rigid body that is rotating about the z-axis (counterclockwise from above) with constant angular velocity $\boldsymbol{\omega} = a\mathbf{i} + b\mathbf{j} + c\mathbf{k}$. If P is a point in the body located at $\mathbf{R} = x\mathbf{i} + y\mathbf{j} + z\mathbf{k}$, the velocity at P is given by the vector field $\mathbf{V} = \boldsymbol{\omega} \times \mathbf{R}$.
    a. Express **V** in terms of the vectors **i**, **j**, and **k**.
    b. Find div **V** and curl **V**.
40. **EXPLORATION PROBLEM** If $\mathbf{F} = \langle f, g, h \rangle$ is an arbitrary vector field whose components are twice differentiable, what can be said about curl(curl **F**)?
41. Which (if any) of the following is the same as div($\mathbf{F} \times \mathbf{G}$) for all vector fields **F** and **G**?
    I. (div **F**)(div **G**)
    II. (curl **F**) $\cdot$ **G** $-$ **F** $\cdot$ (curl **G**)
    III. **F**(div **G**) $+$ (div **F**)**G**
    IV. (curl **F**) $\cdot$ **G** $+$ **F** $\cdot$ (curl **G**)
42. Show that the field

$$\mathbf{B} = y^2z\mathbf{i} + xz^3\mathbf{j} + y^2x^2\mathbf{k}$$

is incompressible.

43. If $\mathbf{F}(x,y) = u(x,y)\mathbf{i} + v(x,y)\mathbf{j}$, show that curl $\mathbf{F} = \mathbf{0}$ if and only if $\dfrac{\partial u}{\partial y} = \dfrac{\partial v}{\partial x}$.

In Problems 44-51, prove the given property for the vector fields **F** *and* **G**, *scalar c, and scalar functions f and g. Assume that all required partial derivatives exist and are continuous.*

44. div($c\mathbf{F}$) $= c$ div **F**
45. div($\mathbf{F} + \mathbf{G}$) $=$ div **F** $+$ div **G**
46. curl($\mathbf{F} + \mathbf{G}$) $=$ curl **F** $+$ curl **G**
47. curl($c\mathbf{F}$) $= c$ curl **F**
48. curl($f\,\mathbf{F}$) $= f$ curl **F** $+ (\nabla f \times \mathbf{F})$
49. div($f\mathbf{F}$) $= f$ div **F** $+ (\nabla f \cdot \mathbf{F})$
50. curl($\nabla f +$ curl **F**) $=$ curl(∇f) $+$ curl(curl **F**)
51. div($f\nabla g$) $= f$ div $\nabla g + \nabla f \cdot \nabla g$

52. The curl of the gradient of a function is always $\mathbf{0}$. That is, $\nabla \times (\nabla f) = \mathbf{0}$.

53. The divergence of the curl of a vector field is 0. That is, $\operatorname{div}(\operatorname{curl} \mathbf{F}) = \mathbf{0}$.

In Problems 54-58, $\mathbf{R} = \langle x, y, z \rangle$, and $r = \|\mathbf{R}\| = \sqrt{x^2 + y^2 + z^2}$. In each case, verify the given identity or answer the question.

54. $\operatorname{curl} \mathbf{R} = \mathbf{0}$; what is $\operatorname{div} \mathbf{R}$?

55. $\operatorname{div}\left(\frac{1}{r^3}\mathbf{R}\right) = 0$

56. $\operatorname{curl}\left(\frac{1}{r^3}\mathbf{R}\right) = \mathbf{0}$

57. $\operatorname{div}(r\mathbf{R}) = 4r$

58. $\operatorname{div}(\nabla r) = \frac{2}{r}$

59. EXPLORATION PROBLEM State and prove an identity for $\operatorname{div}(\nabla(fg))$, where f and g are differentiable scalar functions of x, y, and z.

60. Think Tank Problem Let $\mathbf{F} = \langle x^2 y, yz^2, zy^2 \rangle$. Either find a vector field $\mathbf{G}$ such that $\mathbf{F} = \operatorname{curl} \mathbf{G}$, or show that no such $\mathbf{G}$ exists.

13.2 LINE INTEGRALS

IN THIS SECTION: *Definition of a line integral; line integrals with respect to x, y, and z; line integrals of vector fields; applications of line integrals: mass and work*

In Section 5.3, we introduced a Riemann integral and then in Section 6.5 we used it to compute the work done when an object moves along a line segment against a given force. In this section, we ask the question, what if the object moves along a curve in space? To answer that question, we introduce the notion of a line integral.

A line integral is an integral whose integrand is evaluated at points along a curve in $\mathbb{R}^2$ or in $\mathbb{R}^3$. We will introduce line integrals in this section and show how they can be used for a variety of purposes in mathematics and physics.

Definition of a Line Integral

Let C be a smooth curve, with parametric equations $x = x(t)$, $y = y(t)$, $z = z(t)$ for $a \le t \le b$, that lies within the domain of a function $f(x, y, z)$. We say that C is **orientable** if it is possible to describe direction along the curve for increasing t.

To define a line integral, we begin by partitioning C into n subarcs, the kth of which has length Δs_k. Let (x_k^*, y_k^*, z_k^*) be a point chosen arbitrarily from the kth subarc (see Figure 13.6).

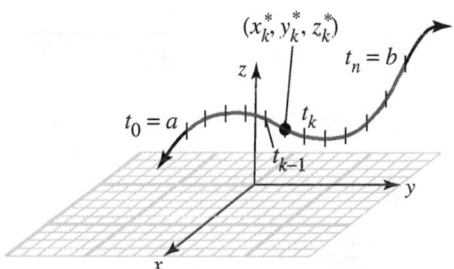

Figure 13.6 The curve C partitioned into subarcs

Form the Riemann sum

$$\sum_{k=1}^{n} f(x_k^*, y_k^*, z_k^*)\Delta s_k$$

and let $\|\Delta s\|$ denote the largest subarc length in the partition. Then, if the limit

$$\lim_{\|\Delta s\| \to 0} \sum_{k=1}^{n} f(x_k^*, y_k^*, z_k^*)\Delta s_k$$

exists, we call this limit the *line integral* of f over C and denote it by $\int_C f(x, y, z)\, ds$.

> **LINE INTEGRAL** If $f(x,y,z)$ is defined on the smooth curve C with parametric equations $x = x(t)$, $y = y(t)$, $z = z(t)$, then the **line integral** of f over C is given by
>
> $$\int_C f(x,y,z)\,ds = \lim_{\|\Delta s\| \to 0} \sum_{k=1}^{n} f(x_k^*, y_k^*, z_k^*)\Delta s_k$$
>
> provided that this limit exists. If C is a closed curve, we sometimes indicate the line integral of f around C by $\oint_C f\,ds$.

It can be shown that the limit that defines the line integral $\int_C f\,ds$ always exists if f is continuous at each point of C. Also, since the curve C is smooth, the component functions $x(t)$, $y(t)$, and $z(t)$ will all be continuously differentiable. Thus, we have

$$ds = \sqrt{[x'(t)]^2 + [y'(t)]^2 + [z'(t)]^2}\,dt$$

so the line integral can be written entirely in terms of t:

$$\int_C f(x,y,z)\,ds = \int_a^b f\,(x(t),y(t),z(t))\,\sqrt{[x'(t)]^2 + [y'(t)]^2 + [z'(t)]^2}\,dt$$

Finally, if $f(x,y)$ is a function of only two variables and C is a curve in the plane, then

$$\int_C f(x,y)\,ds = \int_a^b f\,(x(t),y(t))\,\sqrt{[x'(t)]^2 + [y'(t)]^2}\,dt$$

Example 1   Evaluating a line integral in three variables

Evaluate the line integral $\int_C x^2 z\,ds$, where C is the helix $x = \cos t, y = 2t, z = \sin t$, for $0 \le t \le \pi$.

Solution   Since $x'(t) = -\sin t$, $y'(t) = 2$, $z'(t) = \cos t$, the line integral is

$$\begin{aligned}
\int_C x^2 z\,ds &= \int_0^\pi [x(t)]^2 z(t)\sqrt{[x'(t)]^2 + [y'(t)]^2 + [z'(t)]^2}\,dt \\
&= \int_0^\pi (\cos t)^2 (\sin t)\sqrt{(-\sin t)^2 + (2)^2 + (\cos t)^2}\,dt \\
&= \int_0^\pi \sqrt{5}\cos^2 t \sin t\,dt \\
&= \left. \frac{-\sqrt{5}}{3}\cos^3 t \right|_0^\pi \\
&= \frac{2\sqrt{5}}{3}
\end{aligned}$$

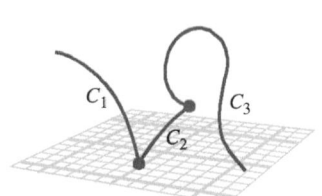

Figure 13.7 Piecewise smooth curve

The definition of a line integral can be extended to curves that are *piecewise smooth* in the sense that they are the union of a finite number of smooth curves with only endpoints in common, as shown in Figure 13.7.

In particular, if C is comprised of a number of smooth subarcs $C_1, C_2, \ldots, C_n$, then

$$\int_C f(x,y,z)\,ds = \int_{C_1} f(x,y,z)\,dx + \int_{C_2} f(x,y,z)\,ds + \cdots + \int_{C_n} f(x,y,z)\,ds$$

This definition of line integration is illustrated in the following example.

Example 2   Evaluating a line integral over a union of curves

Evaluate the line integral $\int_C xy\, ds$, where C consists of the line segment C_1 from $(-3, 3)$ to $(0, 0)$, followed by the portion of the curve C_2: $16y = x^4$ between $(0, 0)$ and $(2, 1)$.

Solution   The curve C is shown in Figure 13.8.

The segment C_1 is part of the line $y = -x$ and can be parameterized by the equations $x = t$, $y = -t$ over the interval $-3 \leq t \leq 0$. On this curve, we have

$$x'(t) = 1, y'(t) = -1, \text{ so } ds = \sqrt{(1)^2 + (-1)^2}\, dt = \sqrt{2}\, dt$$

and the line integral of $f(x, y) = xy$ over C_1 is

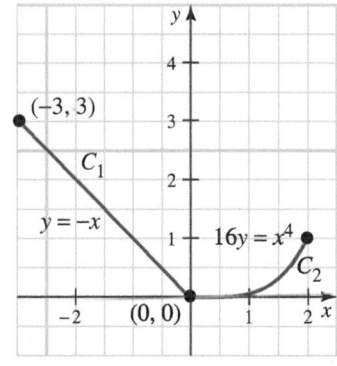

$$\int_{C_1} xy\, ds = \int_{-3}^{0} (t)(-t)\sqrt{2}\, dt$$

$$= \int_{-3}^{0} -\sqrt{2}\, t^2\, dt$$

$$= -9\sqrt{2}$$

Figure 13.8 The curve C in Example 2

The curve C_2, $16y = x^4$, can be parameterized by the equations $x = 2t$, $y = \frac{1}{16}x^4 = \frac{1}{16}(2t)^4 = t^4$ for $0 \leq t \leq 1$. We find that

$$x'(t) = 2, y'(t) = 4t^3, \text{ so } ds = \sqrt{(2)^2 + (4t^3)^2}\, dt$$

and the line integral over C_2 is

$$\int_{C_2} xy\, ds = \int_{0}^{1} (2t)(t^4)\sqrt{4 + 16t^6}\, dt$$

$$= \frac{1}{9}(1 + 4t^6)^{3/2}\Big|_{0}^{1}$$

$$= \frac{1}{9}(5\sqrt{5} - 1)$$

Thus, the line integral over C is given by the sum

$$\int_C xy\, ds = \int_{C_1} xy\, ds + \int_{C_2} xy\, ds = -9\sqrt{2} + \frac{1}{9}(5\sqrt{5} - 1)$$

Theorem 13.1   Properties of line integrals

Let f be a given scalar function defined with respect to s on a piecewise smooth, orientable curve C. Then, for any constant k,

Constant multiple rule: $\displaystyle\int_C kf\, ds = k\int_C f\, ds$

Sum rule: $\displaystyle\int_C (f_1 + f_2)\, ds = \int_C f_1\, ds + \int_C f_2\, ds$
where f_1 and f_2 are scalar functions defined with respect to s on C.

Opposite direction rule: $\displaystyle\int_{-C} f\, ds = -\int_C f\, ds$
where $-C$ denotes the curve C traversed in the opposite direction.

Subdivision rule:
$$\int_C f \, ds = \int_{C_1} f \, ds + \int_{C_2} f \, ds + \cdots + \int_{C_n} f \, ds$$

where C is the union of smooth orientable subarcs $C = C_1 \cup C_2 \cup \cdots \cup C_n$ with only endpoints in common.

Proof: The proof follows directly from the properties of limits and the definition of a line integral. ◆

Line Integrals with Respect to x, y, and z

If Δs is replaced by Δx in the discussion leading to the definition of the line integral $\int_C f(x,y,z) ds$, we obtain a definition for the line integral $\int_C f(x,y,z) dx$. Since $x = x(t)$ is differentiable, we have $dx = x'(t) \, dt$ and the line integral of f with respect to x can be evaluated as follows:

$$\int_C f(x,y,z) \, dx = \int_a^b f[x(t), y(t), z(t)] \, x'(t) \, dt$$

Similarly, if g and h are continuous on C, then

$$\int_C g \, dy = \int_a^b g[x(t), y(t), z(t)] \, y'(t) \, dt \qquad \int_C h \, dz = \int_a^b h[x(t), y(t), z(t)] \, z'(t) \, dt$$

By combining the line integrals with respect to the coordinate variables x, y, and z, we obtain a line integral of the form

$$\int_C [f(x,y,z) \, dx + g(x,y,z) \, dy + h(x,y,z) \, dz]$$

Example 3 Evaluating a line integral with respect to coordinate variables

Evaluate the line integral

$$\int_C [y \, dx - z \, dy + x \, dz]$$

where C is the curve with parametric equations $x = t^2$, $y = e^{-t}$, $z = e^t$ for $0 \le t \le 1$.

Solution Since $x'(t) = 2t$, $y'(t) = -e^{-t}$, and $z'(t) = e^t$, we have

$$\int_C [y \, dx - z \, dy + x \, dz] = \int_0^1 [e^{-t}(2t \, dt) - e^t(-e^{-t} dt) + t^2(e^t \, dt)]$$

$$= \int_0^1 [2t \, e^{-t} + 1 + t^2 e^t] \, dt$$

$$= [-2e^{-t}(t+1) + t + e^t(t^2 - 2t + 2)]\Big|_0^1$$

$$= [-4e^{-1} + 1 + e(1 - 2 + 2)] - [-2 + 0 + 2]$$

$$= e - 4e^{-1} + 1 \qquad \blacksquare$$

Line Integrals of Vector Fields

We will now discuss what it means to compute the *line integral of a vector field*.

LINE INTEGRAL OF A VECTOR FIELD

Let $\mathbf{F}(x, y, z) = u(x, y, z)\mathbf{i} + v(x, y, z)\mathbf{j} + w(x, y, z)\mathbf{k}$ be a vector field, and let C be a piecewise smooth orientable curve with parametric representation

$$\mathbf{R}(t) = x(t)\mathbf{i} + y(t)\mathbf{j} + z(t)\mathbf{k} \qquad \text{for } a \le t \le b$$

Using $d\mathbf{R} = dx\,\mathbf{i} + dy\,\mathbf{j} + dz\,\mathbf{k}$, we define the **line integral of F along C** by

$$
\begin{aligned}
\int_C \mathbf{F} \cdot d\mathbf{R} &= \int_C (u\,dx + v\,dy + w\,dz) \\
&= \int_C \mathbf{F}[\mathbf{R}(t)] \cdot \mathbf{R}'(t)\,dt \\
&= \int_a^b \left[u\left[x(t), y(t), z(t)\right] \frac{dx}{dt} \right. \\
&\qquad + v\left[x(t), y(t), z(t)\right] \frac{dy}{dt} \\
&\qquad \left. + w\left[x(t), y(t), z(t)\right] \frac{dz}{dt} \right] dt
\end{aligned}
$$

Remark: Note that since $d\mathbf{R} = \mathbf{R}'(t)\,dt$, we may write:

$$\int_C \mathbf{F} \cdot d\mathbf{R} = \int_a^b \mathbf{F}[\mathbf{R}(t)] \cdot \mathbf{R}'(t)\,dt = \int_a^b \mathbf{F}[\mathbf{R}(t)] \cdot \frac{\mathbf{R}'(t)}{\|\mathbf{R}'(t)\|} \|\mathbf{R}'(t)\|\,dt = \int_a^b \mathbf{F}[\mathbf{R}(t)] \cdot \mathbf{T} \|\mathbf{R}'(t)\|\,dt$$

Example 4   Evaluating a line integral of a vector field

Evaluate $\int_C \mathbf{F} \cdot d\mathbf{R}$, where $\mathbf{F} = (y^2 - z^2)\mathbf{i} + (2yz)\mathbf{j} - x^2\mathbf{k}$ and C is the curve defined parametrically by $x = t^2, y = 2t$, and $z = t$ for $0 \le t \le 1$.

Solution   Rewrite $\mathbf{F}$ using the parameter t:

$$\mathbf{F} = [(2t)^2 - (t)^2]\mathbf{i} + [2(2t)(t)]\mathbf{j} - [(t^2)^2]\mathbf{k} = 3t^2\mathbf{i} + 4t^2\mathbf{j} - t^4\mathbf{k}$$

Because $\mathbf{R}(t) = t^2\mathbf{i} + 2t\mathbf{j} + t\mathbf{k}$, we have $d\mathbf{R} = (2t\,dt)\mathbf{i} + (2\,dt)\mathbf{j} + dt\,\mathbf{k}$, so

$$
\begin{aligned}
\mathbf{F} \cdot d\mathbf{R} &= (3t^2)(2t\,dt) + (4t^2)(2\,dt) + (-t^4)(dt) \\
&= (6t^3 + 8t^2 - t^4)\,dt
\end{aligned}
$$

Thus,

$$\int_C \mathbf{F} \cdot d\mathbf{R} = \int_0^1 (6t^3 + 8t^2 - t^4)\,dt = \left[\frac{3}{2}t^4 + \frac{8}{3}t^3 - \frac{1}{5}t^5 \right]_0^1 = \frac{119}{30}$$

A line integral over a curve C does not depend on the parameterization used for C. This property of line integrals is illustrated in the following example.

Example 5   The value of a line integral is independent of parameterization

Let $\mathbf{F} = y\mathbf{i} + x\mathbf{j}$ and let C be the top half of the circle $x^2 + y^2 = 4$ traversed counterclockwise from $(2, 0)$ to $(-2, 0)$, as shown in Figure 13.9.
Evaluate the line integral $\int_C \mathbf{F} \cdot d\mathbf{R}$ for each of the following parameterizations of the curve C.

a. $x = 2\cos\theta, y = 2\sin\theta, 0 \le \theta \le \pi$ **b.** $x = -t, y = \sqrt{4 - t^2}, -2 \le t \le 2$

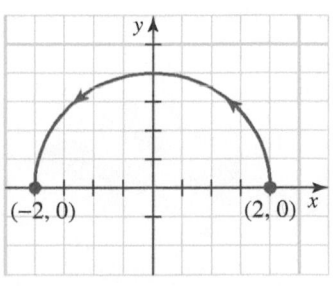

Figure 13.9 Graph of the semicircle C

Solution

a. With the parameterization $x = 2\cos\theta$, $y = 2\sin\theta$, we find that $x'(\theta) = -2\sin\theta$, and $y'(\theta) = 2\cos\theta$

$$\int_C \mathbf{F} \cdot d\mathbf{R} = \int_C [y\,dx + x\,dy]$$
$$= \int_0^\pi [(2\sin\theta)(-2\sin\theta) + (2\cos\theta)(2\cos\theta)]\,d\theta$$
$$= \int_0^\pi 4(\cos^2\theta - \sin^2\theta)\,d\theta$$
$$= \int_0^\pi 4\cos 2\theta\,d\theta$$
$$= 0$$

b. For the parameterization $x = -t$, $y = \sqrt{4 - t^2}$, we have $x'(t) = -1$, $y'(t) = \dfrac{-t}{\sqrt{4 - t^2}}$. We find that

$$\int_C [y\,dx + x\,dy] = \int_{-2}^2 \left[\sqrt{4 - t^2}(-1) + (-t)\left(\frac{-t}{\sqrt{4 - t^2}}\right)\right] dt$$
$$= \int_{-2}^2 \frac{-4 + 2t^2}{\sqrt{4 - t^2}}\,dt$$
$$= -t\sqrt{4 - t^2}\,\Big|_{-2}^2$$
$$= 0$$

This is the same result that was obtained using the parameterization in part **a.**   ◼

Example 6   Evaluating line integrals along different paths

Let $\mathbf{F} = xy^2\mathbf{i} + x^2 y\mathbf{j}$ and evaluate the line integral $\int_C \mathbf{F} \cdot d\mathbf{R}$ between the points $(0,0)$ and $(2,4)$ along the following paths:

a. the line segment connecting the points
b. the parabolic arc $y = x^2$ connecting the points

Solution   The two paths we are considering are shown in Figure 13.10.

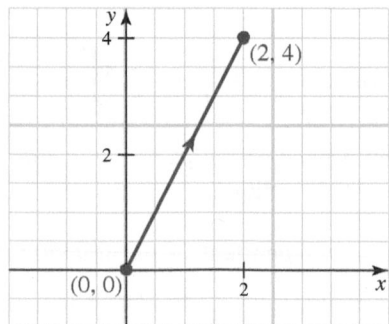

a. The line segment path

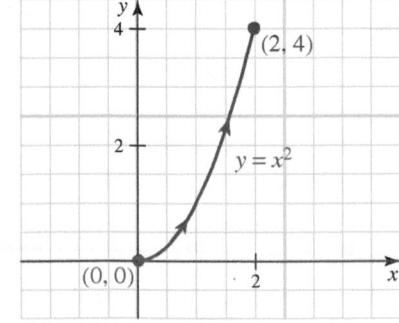

b. The parabolic path

Figure 13.10   A line interval along different paths

Note: The "line" integral should be thought of as a "curve" integral. The equation(s) of a curve are needed so that the integral can be expressed in terms of a single variable.

a. The line joining the given points has equation $y = 2x$, which may be parameterized by setting $x = t, y = 2t$ for $0 \le t \le 2$. Thus, $\mathbf{R}(t) = t\mathbf{i} + 2t\mathbf{j}$ so that $d\mathbf{R} = dt\,\mathbf{i} + 2\,dt\,\mathbf{j}$. In terms of t, we find $\mathbf{F} = 4t^3\mathbf{i} + 2t^3\mathbf{j}$ and $\mathbf{F} \cdot d\mathbf{R} = 4t^3\,dt + 4t^3\,dt = 8t^3\,dt$

$$\int_C \mathbf{F} \cdot d\mathbf{R} = \int_0^2 8t^3 \, dt = \left[2t^4\right]\Big|_0^2 = 32$$

b. The parabola $y = x^2$ can be parameterized by setting $x = t, y = t^2$ for $0 \le t \le 2$. Thus, $\mathbf{R}(t) = t\mathbf{i} + t^2\mathbf{j}$ so that $d\mathbf{R} = dt\,\mathbf{i} + 2t\,dt\,\mathbf{j}$. In terms of t,

$$\mathbf{F} = xy^2\mathbf{i} + x^2y\mathbf{j} = (t)(t^2)^2\mathbf{i} + (t)^2(t^2)\mathbf{j} = t^5\mathbf{i} + t^4\mathbf{j} \qquad \text{and}$$

$$\mathbf{F} \cdot d\mathbf{R} = t^5\,dt + 2t^5\,dt = 3t^5\,dt$$

We then have

$$\int_C \mathbf{F} \cdot d\mathbf{R} = \int_0^2 3t^5 \, dt = \left[\frac{1}{2}t^6\right]_0^2 = 32$$

In Example 6 we see that the value of the line integral is the same for both paths. Indeed, it can be shown that for the vector field $\mathbf{F} = xy^2\mathbf{i} + x^2y\,\mathbf{j}$, the line integral $\int_C \mathbf{F} \cdot d\mathbf{R}$ along *any* path C joining $(0,0)$ to $(2,4)$ has the same value. This is not true for every $\mathbf{F}$ (see Problem 60) but when it is true, the line integral is said to be **independent of path** or **path independent**. Path independence is an important feature of certain vector fields and will be discussed in detail in the next section.

Applications of Line Integrals: Mass and Work

Line integrals were developed in the 19th century, primarily to deal with problems in physics involving force, fluid flow, electricity, and magnetism. We will show how line integration can be used to compute the mass of a thin wire and the work performed on an object moving along a curve in a force field. Additional applications are examined in the problem set and in Sections 13.3 and 13.4.

Consider a thin wire with the shape of a curve C and let $\rho(x, y, z)$ be the density (mass per unit length) at each point $P(x, y, z)$ on the wire. Suppose the curve is described by parametric equations $x = x(t)$, $y = y(t)$, $z = z(t)$ for $a \le t \le b$, and subdivide the parameter interval $[a, b]$ into n parts. This induces a partition of the portion of the curve covered by the wire into n subarcs. Let Δs_k be the length of the kth subarc, and let $P_k(x_k^*, y_k^*, z_k^*)$ be a point chosen arbitrarily from this subarc. Then, the mass of the subarc is approximately

$$\Delta m_k = \rho(x_k^*, y_k^*, z_k^*)\Delta s_k$$

and the sum $\sum_{k=1}^{n} \Delta m_k$ approximates the total mass of the wire. We improve the approximation by taking more and more subdivision points in such a way that the maximum value of Δs_k (of all values Δs_k, $k = 1, 2, \cdots, n$) decreases to 0, and the actual mass of the wire is then given by the limiting value

$$m = \lim_{n \to \infty} \sum_{k=1}^{n} \rho(x_k^*, y_k^*, z_k^*)\Delta s_k = \int_C \rho(x, y, z)\,ds$$

The **center of mass** of the wire is then the point (x, y, z), where

$$\bar{x} = \frac{1}{m}\int_C x\rho(x, y, z)\,ds \qquad \bar{y} = \frac{1}{m}\int_C y\rho(x, y, z)\,ds \qquad \bar{z} = \frac{1}{m}\int_C z\rho(x, y, z)\,ds$$

Example 7 **Computing the mass of a thin wire using line integration**

A wire has the shape of the curve

$$x = \sqrt{2}\sin t \qquad y = \cos t \qquad z = \cos t \qquad \text{for } 0 \le t \le \pi$$

If the wire has density $\rho(x,y,z) = xyz$ at each point (x,y,z), what is its mass?

Solution   We note that $x'(t) = \sqrt{2}\cos t$, $y'(t) = z'(t) = -\sin t$. The mass is given by the line integral

$$
\begin{aligned}
m &= \int_C \rho(x,y,z)\,ds \\
&= \int_C xyz\sqrt{[x'(t)]^2 + [y'(t)]^2 + [z'(t)]^2}\,dt \\
&= \int_C \sqrt{2}\sin t\cos^2 t\sqrt{\left(\sqrt{2}\cos t\right)^2 + (-\sin t)^2 + (-\sin t)^2} \\
&= \int_0^\pi \sqrt{2}\sin t\cos^2 t\sqrt{2(\cos^2 t + \sin^2 t)}\,dt \\
&= 2\int_0^\pi \cos^2 t\sin t\,dt \\
&= \frac{4}{3}
\end{aligned}
$$

In Problem 51, you are asked to determine the center of mass of this wire. Note that the center of mass may not lie on the curve, or wire, itself.

One of the most important physical applications of line integration is in computing work. Recall from Section 9.3, that if an object moves along a line with displacement **D** in a constant force field **F**, the work done is $W = \mathbf{F}\cdot\mathbf{D}$. We now consider the case where **F** is a variable force field and the object moves along an orientable curve C. Assume that C is parameterized by $\mathbf{R}(t)$ and that the object moves in the direction of increasing t. Partition C with subdivision points $P_0, P_1, \ldots, P_n$, as shown in Figure 13.11.

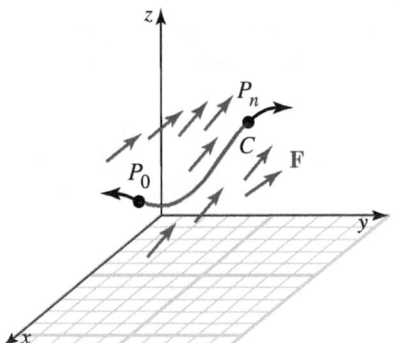

a. An object moves along a curve C in a force field **F**

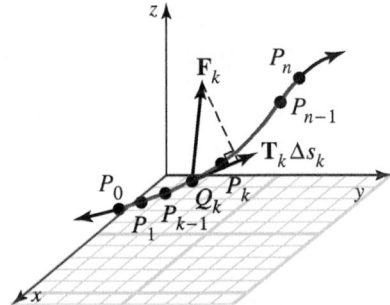

b. The work performed as the object moves along the kth subarc is $W_k \approx \mathbf{F}_k\cdot\mathbf{T}_k\Delta s_k$

Figure 13.11   Work performed as an object moves in a force field F along a curve C

For $k = 1, 2, \ldots, n$, let $Q_k(x_k^*, y_k^*, z_k^*)$ be a point chosen arbitrarily from the kth subarc C_k (the one with endpoints P_{k-1} and P_k), and let $\mathbf{F}_k = \mathbf{F}(x_k^*, y_k^*, z_k^*)$. If the length Δs_k of the subarc C_k is small, the force will be approximately constant and we assume it has the constant value $\mathbf{F}_k$ on the subarc. Also, the direction of motion will not change much over the subarc, so we can assume that the object will move a distance of Δs_k in the direction of the unit tangent $\mathbf{T}_k = \mathbf{T}(x_k^*, y_k^*, z_k^*)$, for a linear displacement of $\mathbf{T}_k\Delta s_k$. Therefore, we can approximate the work performed over the kth subarc by

$$W_k \approx \mathbf{F}_k\cdot\mathbf{T}_k\Delta s_k$$

By adding the contributions along all n subarcs, we obtain the sum

$$\sum_{k=1}^{n} \mathbf{F}_k \cdot \mathbf{T}_k \Delta s_k$$

This is an approximation to the total work performed as the object moves along C in the force field $\mathbf{F}$. As the length of the largest subarc $\|\Delta s\|$ tends to 0, this approximating sum approaches the value of the line integral $\int_C \mathbf{F} \cdot \mathbf{T}\, ds$; that is,

$$W = \lim_{\|\Delta s\| \to 0} \sum_{k=1}^{n} \mathbf{F}_k \cdot \mathbf{T}_k \Delta s_k = \int_C \mathbf{F} \cdot \mathbf{T}\, ds$$

These observations lead us to consider **work as a line integral**.

WORK AS A LINE INTEGRAL Let $\mathbf{F}$ be a continuous force field over a domain D. Then the work W performed as an object moves along a smooth curve C in D is given by the integral

$$W = \int_C \mathbf{F} \cdot \mathbf{T}\, ds$$

where $\mathbf{T}$ is the unit tangent at each point on C.

Recall (from Theorem 10.7 in Section 10.4) that $\mathbf{T} = \dfrac{d\mathbf{R}}{ds}$ where $\mathbf{R}$ is the position vector of the object moving on C. Thus, the work can also be given by the line integral

$$W = \int_C \mathbf{F} \cdot \frac{d\mathbf{R}}{ds}\, ds = \int_C \mathbf{F} \cdot d\mathbf{R}$$

This form of the line integral for work is used in the following example.

Example 8 Work as a line integral

An object moves in the force field $\mathbf{F} = y^2\mathbf{i} + 2(x+1)y\mathbf{j}$. How much work is performed as the object moves from the point $(2, 0)$ counterclockwise along the elliptical path $x^2 + 4y^2 = 4$ to $(0, 1)$, then back to $(2, 0)$ along the line segment joining the two points, as shown in Figure 13.12.

Solution   If C is the trajectory of the moving object, the work performed is

$$W = \oint_C \mathbf{F} \cdot d\mathbf{R}$$

Let C_1 be the top (elliptical) part of C, and let C_2 be the bottom (linear) part. The curve C_1 can be parameterized by the equations $x = 2\cos t, y = \sin t$ for $0 \le t \le \frac{\pi}{2}$ (since $x^2 + 4y^2 = 4$), so the position vector for C_1 is $\mathbf{R}_1 = \langle 2\cos t, \sin t \rangle$. Thus, we have $\mathbf{R}_1' = \langle -2\sin t, \cos t \rangle$ and the work performed as the object moves along C_1 is

$$
\begin{aligned}
W_1 &= \int_{C_1} \mathbf{F} \cdot d\mathbf{R} \\
&= \int_0^{\pi/2} \mathbf{F}[\mathbf{R}_1(t)] \cdot \mathbf{R}_1'(t)\, dt \\
&= \int_0^{\pi/2} [(\sin^2 t)(-2\sin t) + 2(2\cos t + 1)\sin t(\cos t)]\, dt
\end{aligned}
$$

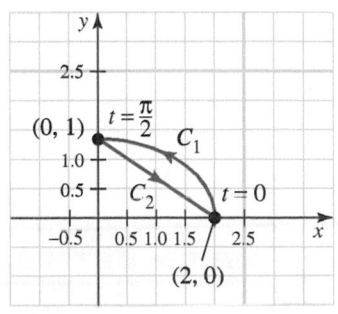

Figure 13.12   The curve C

$$= \int_0^{\pi/2} (-2\sin^3 t + 4\cos^2 t \sin t + 2 \sin t \cos t) \, dt$$

$$= \int_0^{\pi/2} (-2\sin^2 t + 4\cos^2 t + 2\cos t) \sin t \, dt$$

$$= \int_0^{\pi/2} (6\cos^2 t + 2\cos t - 2) \sin t \, dt \qquad \boxed{\begin{array}{l} \text{Let } u = \cos t; \ du = -\sin t \, dt \\ \text{If } t = 0, u = 1, \text{ and if } t = \frac{\pi}{2}, u = 0. \end{array}}$$

$$= -\int_1^0 (6u^2 + 2u - 2) \, du$$

$$= 1$$

For the line segment C_2, a parameterization is

$$x = 2t, \qquad y = 1 - t, \qquad 0 \le t \le 1$$

The position vector is $\mathbf{R}_2 = \langle 2t, 1 - t \rangle$, so $\mathbf{R}_2' = \langle 2, -1 \rangle$ and the work performed as the object moves along C_2 is

$$\begin{aligned} W_2 &= \int_{C_2} \mathbf{F} \cdot d\mathbf{R}_2 \\ &= \int_0^1 \mathbf{F}[\mathbf{R}_2(t)] \cdot \mathbf{R}_2'(t) \, dt \\ &= \int_0^1 [(1 - t)^2 (2) + 2(2t + 1)(1 - t)(-1)] \, dt \\ &= \int_0^1 [6t^2 - 6t] \, dt \\ &= -1 \end{aligned}$$

Thus, the total work performed as the object moves along the curve $C = C_1 \cup C_2$ is

$$W = \oint_C \mathbf{F} \cdot d\mathbf{R} = \int_{C_1} \mathbf{F} \cdot d\mathbf{R} + \int_{C_2} \mathbf{F} \cdot d\mathbf{R} = 1 + (-1) = 0$$

It can be shown that the work performed as an object moves around *any* closed path in the force field in Example 8 will always be 0. When this occurs, the force is said to be **conservative**. We will discuss conservative force fields in Sections 13.3 and 13.4.

PROBLEM SET 13.2

1. ■ What does this say? Explain the difference between $\int_C f \, ds$ and $\int_C f \, dx$.

2. ■ What does this say? Discuss the evaluation of the line integral $\int_C \mathbf{F} \cdot d\mathbf{R}$.

Evaluate each line integral given in Problems 3-8.

3. $\displaystyle\int_C (-y \, dx + x \, dy)$

C is the parabolic path $y = 4x^2$ from $(1, 4)$ to $(0, 0)$.

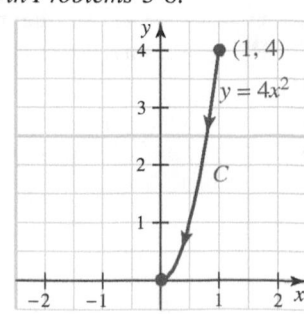

1037

4. $\displaystyle\int_C (-y\,dx + 3x\,dy)$

C is the parabolic

path $y^2 = x$ from

$(1, 1)$ to $(9, 3)$.

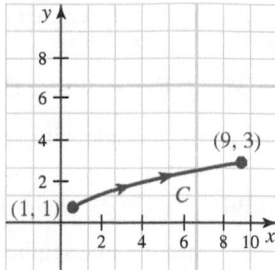

5. $\displaystyle\int_C (x\,dy - y\,dx)$

C is the line segment

$2x - 4y = 1$ as x

varies from 4 to 8.

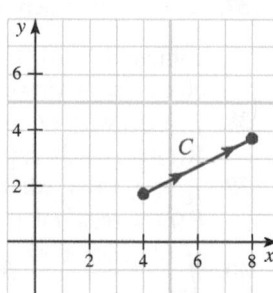

6. $\displaystyle\int_C [(y - x)dx + x^2 y\,dy]$

C is the curve

$y^2 = x^3$ from $(1, -1)$

to $(1, 1)$.

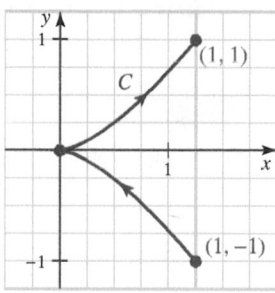

7. $\displaystyle\int_C [(x + y)^2 dx - (x - y)^2 dy]$

C is the curve $y = |2x|$

from $(-1, 2)$ to $(1, 2)$.

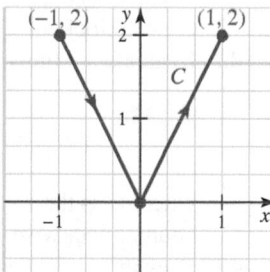

8. $\displaystyle\int_C [(y^2 - x^2)dx - x\,dy]$

C is the quarter-circle

$x^2 + y^2 = 4$ from $(0, 2)$

to $(2, 0)$.

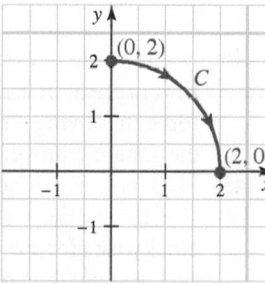

In Problems 9-14, evaluate the given line integral over the curve C with the prescribed parameterization.

9. $\displaystyle\int_C \frac{1}{3 + y}\,ds$

for $C: x = 2t^{3/2},\ y = 3t, 0 \le t \le 1$

10. $\displaystyle\int_C \frac{1}{x + 1}\,ds$

for $C: x = 2t, y = t, 0 \le t \le 1$

11. $\displaystyle\int_C (3x - 2y)\,ds$

for $C: x = \sin t,\ y = \cos t,\ 0 \le t \le \pi$

12. $\displaystyle\int_C \frac{y^2}{x^3}\,ds$

for $C: x = 2t, y = t^4, 0 \le t \le 1$

13. $\displaystyle\int_C \frac{x^2}{y^2}\,ds$

for $C: x = t,\ y = 3t,\ 0 \le t \le 1$

14. $\displaystyle\int_C (x^2 + y^2)\,ds$

for $C: x = e^{-t}\cos t,\ y = e^{-t}\sin t,\ 0 \le t \le \frac{\pi}{2}$

15. Evaluate $\int_C [(x^2 + y^2)dx + 2xy\,dy]$ for these choices of the curve C:

 a. C is the quarter circle $x^2 + y^2 = 1$ from $(1, 0)$ to $(0, 1)$.

 b. C is the segment of the line $y = 1 - x$ from $(1, 0)$ to $(0, 1)$.

16. Evaluate $\int_C [x^2 y\,dx + (x^2 - y^2)\,dy]$ for these choices of the curve C:

 a. C is the arc of the parabola $y = x^2$ from $(0, 0)$ to $(2, 4)$.

 b. C is the segment of the line $y = 2x$ for $0 \le x \le 2$.

17. Evaluate the integral in Problem 15 for the path C that consists of the horizontal line segment $(0, 0)$ to $(2, 0)$, followed by the vertical segment from $(2, 0)$ to $(2, 4)$.

18. Evaluate the integral in Problem 16 for the path C that consists of the horizontal line segment $(0, 0)$ to $(2, 0)$, followed by the vertical segment from $(2, 0)$ to $(2, 4)$.

19. Evaluate $\int_C (-xy^2 dx + x^2 dy)$ where C is the path shown in Figure 13.13.

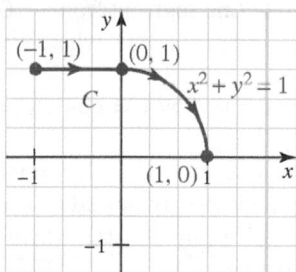

Figure 13.13 Path for C in Problem 19

20. Evaluate $\int_C (-y^2 dx + x^2 dy)$, where C is the path shown in Figure 13.14.

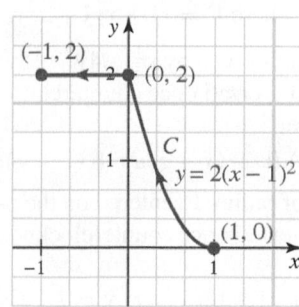

Figure 13.14 Path for C in Problem 20

21. Evaluate

$$\oint_C [(x^2 - y^2)dx + x\, dy]$$

where C is the circular path given by $x = 2\cos\theta$, $y = 2\sin\theta, 0 \le \theta \le 2\pi$.

22. Evaluate

$$\oint_C (x^2 y\, dx - xy\, dy)$$

where C is the path that begins at $(0,0)$, goes to $(1, 1)$ along the parabola $y = x^2$, and then returns to $(0,0)$ along the line $y = x$.

In Problems 23-26, evaluate $\int_C \mathbf{F} \cdot d\mathbf{R}$, where
$\mathbf{F} = (5x + y)\mathbf{i} + x\mathbf{j}$ *and C is the specified curve.*

23. C is the straight line segment from $(0,0)$ to $(2, 1)$.

24. C is the vertical line from $(0,0)$ to $(0, 1)$, followed by the horizontal line from $(0, 1)$ to $(2, 1)$.

25. C is the curve given by $\mathbf{R}(t) = 2t\mathbf{i} + t\mathbf{j}$ for $0 \le t \le 1$.

26. C is the curve given by $\mathbf{R}(t) = t^2\mathbf{i} - t\mathbf{j}$ for $0 \le t \le 1$.

Evaluate the line integrals in Problems 27-40.

27. $\int_C (y\, dx - x\, dy + dz)$, where C is the helical path given by

    a. $x = 3\sin t, y = 3\cos t, z = t$ for $0 \le t \le \frac{\pi}{2}$

    b. $x = a\sin t, y = a\cos t, z = t$ for constant a and $0 \le t \le \frac{\pi}{2}$

28. $\int_C (x\, dx + y\, dy + z\, dz)$, where C is the following path:

    a. the helix defined by $x = \cos t, y = \sin t, z = t$ for $0 \le t \le \frac{\pi}{2}$

    b. the straight line segment from $(1, 0, 0)$ to $(0, 1, \frac{\pi}{2})$

29. $\int_C (-y\, dx + x\, dy + xz\, dz)$, where C is the following path:

    a. the helix defined by $x = \cos t, y = \sin t, z = t$ for $0 \le t \le 2\pi$

    b. the unit circle $x^2 + y^2 = 1, z = 0$, traversed once counterclockwise as viewed from above

30. $\int_C (5xy\, dx + 10yz\, dy + z\, dz)$, where C is the following path:

    a. the parabolic arc $x = y^2$ from $(0, 0, 0)$ to $(1, 1, 0)$ followed by the line segment given by $x = 1, y = 1$, $0 \le z \le 1$

    b. the straight line segment from $(0, 0, 0)$ to $(1, 1, 1)$

31. $\oint_C \mathbf{F} \cdot d\mathbf{R}$, where $\mathbf{F} = x\mathbf{i} + xy\mathbf{j} + x^2 yz\mathbf{k}$ and C is the elliptical path given by $x^2 + 4y^2 - 8y + 3 = 0$ in the xy-plane, traversed once counterclockwise as viewed from above

32. $\oint_C [(y + z)dx + (x + z)dy + (x + y)dz]$, where C is the circle of radius 1 centered on the z-axis in the plane $z = 2$, traversed once counterclockwise as viewed from above.

33. $\int_C 2xy^2 z\, ds$, for $C: x = t, y = t^2, z = \frac{2}{3}t^3$ for $0 \le t \le 1$

34. $\int_C ye^{xz}\, ds$ where C is the line segment from $(0, 0, 0)$ to $(2, 1, 3)$

35. $\oint_C \mathbf{F} \cdot d\mathbf{R}$, where $\mathbf{F} = y^2\mathbf{i} + x^2\mathbf{j} - (x + z)\mathbf{k}$ and C is the triangle with vertices $(0, 0, 0)$, $(1, 0, 0)$, $(1, 1, 0)$, traversed once clockwise, as viewed from above.

36. $\oint_C \mathbf{F} \cdot \mathbf{T}\, ds$, where $\mathbf{F} = -3y\mathbf{i} + 3x\mathbf{j} + 3x\mathbf{k}$ and C is the straight line segment from $(0, 0, 1)$ to $(1, 1, 1)$

37. $\oint_C \mathbf{F} \cdot \mathbf{T}\, ds$, where $\mathbf{F} = -x\mathbf{i} + 2\mathbf{j}$ and C is the trapezoid with vertices $(0, 0)$, $(1, 0)$, $(2, 1)$, $(0, 1)$, traversed once clockwise as viewed from above

38. $\int_C y\, ds$, where C is the curve given by $\mathbf{R}(t) = t\mathbf{i} + 2t^3\mathbf{j}, 0 \le t \le 2$

39. $\int_C (x + y)\, ds$, where C is given by $\mathbf{R}(t) = (\cos^2 t)\mathbf{i} + (\sin^2 t)\mathbf{j}, -\frac{\pi}{4} \le t \le 0$

40. $\int_C \dfrac{x^2 + xy + y^2}{z^2}\, ds$, where C is the path given by $\mathbf{R(t)} = (\cos t)\mathbf{i} + (\sin t)\mathbf{j} - \mathbf{k}$ for $0 \le t \le 2\pi$

Level 2

41. Evaluate the line integral

$$\oint_C \frac{x\, dy - y\, dx}{x^2 + y^2}$$

where C is the unit circle $x^2 + y^2 = 1$ traversed once counterclockwise.

42. Evaluate the line integral

$$\oint_C \frac{dx + dy}{|x| + |y|}$$

where C is the square $|x| + |y| = 1$ traversed once counterclockwise.

43. How much work is done by a constant force $\mathbf{F} = a\mathbf{i} + \mathbf{j}$ when a particle moves along the line $y = ax$ from $x = a$ to $x = 0$?

44. A force field in the plane is given by $\mathbf{F} = (x^2 - y^2)\mathbf{i} + 2xy\mathbf{j}$. Find the total work done by this force in moving a point mass counterclockwise around the square with vertices $(0, 0)$, $(2, 0)$, $(2, 2)$, $(0, 2)$.

45. Find the work done by the force field $\mathbf{F} = (x^2 + y^2)\mathbf{i} + (x + y)\mathbf{j}$ as an object moves counterclockwise along the circle $x^2 + y^2 = 1$ from $(1, 0)$ to $(-1, 0)$, and then back to $(1, 0)$ along the x-axis.

46. A force acting on a point mass located at (x, y) is given by $\mathbf{F} = y\mathbf{i} + 2x\mathbf{j}$. Find the work done by this force as the point mass moves along a straight line from $(1, 0)$ to $(0, 1)$.

Find the work done by the force $\mathbf{F}(x, y, z)$ on an object moving along the curve C in Problems 47-50.

47. $\mathbf{F} = (y^2 - z^2)\mathbf{i} + 2yz\mathbf{j} - x^2\mathbf{k}$, and C is the path given by $x(t) = t, y(t) = t^2, z(t) = t^3$, for $0 \le t \le 1$.

48. $\mathbf{F} = 2xy\mathbf{i} + (x^2 + 2)\mathbf{j} + y\mathbf{k}$, and C is the line segment from $(1, 0, 2)$ to $(3, 4, 1)$.

49. $\mathbf{F} = x\mathbf{i} + y\mathbf{j} + (xz - y)\mathbf{k}$, and C is the line segment from $(0, 0, 0)$ to $(2, 1, 2)$.

50. $\mathbf{F} = x\mathbf{i} + y\mathbf{j} + (xz - y)\mathbf{k}$, and C is the path given by $\mathbf{R}(t) = t^2\mathbf{i} + 2t\mathbf{j} + 4t^3\mathbf{k}$ for $0 \le t \le 1$.

51. A wire has the shape of the curve $C: x = \sqrt{2}\sin t$, $y = \cos t, z = \cos t$ for $0 \le t \le \pi$, and the density at the point (x, y, z) on the curve is $\rho(x, y, z) = xyz$. The mass of this wire was computed in Example 7. Find the center of mass.

52. Find the center of mass of a wire in the shape of the helix $x = 3\sin t$, $y = 3\cos t$, $z = 2t$ for $0 \le t \le \pi$ and the following choices of density $\rho(x, y, z)$.
    a. $\rho(x, y, z) = z(x, y, z)$
    b. $\rho(x, y, z) = x$

53. Find the centroid of a thin wire in the shape of the curve $x = 2t, y = t^2, z = \frac{1}{3}t^3$ for $0 \le t \le 2$.

54. Find the centroid of the arch of the cycloid $C: x = t - \sin t, y = 1 - \cos t$ for $0 \le t \le 2\pi$. *Note*: The centroid may be thought of as the center of mass of a wire of constant density.

55. A 180-lb laborer carries a bag of sand weighing 40 lb up a circular helical staircase (Figure 13.15) on the outside of a tower 50 ft high and 20 ft in diameter. How much work is done as the laborer climbs to the top in exactly five revolutions?

Figure 13.15   A circular helical staircase

56. Repeat Problem 55, assuming that the bag leaks 1 lb of sand for every 10 ft of ascent. How much work is done during the laborer's climb to the top?

57. A 5,000-lb satellite orbits the earth in a circular orbit 5,000 mi from the center of the earth. How much work is done as the satellite moves through one complete revolution?

Level 3

58. Suppose a particle with charge Q and mass m moves with velocity $\mathbf{V}$ under the influence of an electric field $\mathbf{E}$ and a magnetic field $\mathbf{B}$. Then the total force on the particle is $\mathbf{F} = Q(\mathbf{E} + \mathbf{V} \times \mathbf{B})$, called the *Lorentz force*. Use Newton's second law of motion, $\mathbf{F} = m\mathbf{A}$, to show that

$$m\frac{d\mathbf{V}}{dt} \cdot \mathbf{V} = Q\mathbf{E} \cdot \frac{d\mathbf{R}}{dt}$$

and then evaluate the line integral $\int_C \mathbf{E} \cdot d\mathbf{R}$, where C is the trajectory of a particle traveling with constant speed.

59. **Think Tank Problem** If

$$\int_C f(x, y, z)\, ds = 0$$

is it true that $f(x, y, z) = 0$ on C? Either prove that it is, or find a counterexample.

60. **Think Tank Problem** This problem shows that not all line integrals are path independent. Let $\mathbf{F} = \langle y, -x \rangle$, and let C_1 and C_2 be the following two paths joining $(0, 0)$ to $(1, 1)$.
    $C_1: y = x$ for $0 \le x \le 1$ and
    $C_2: y = x^2$ for $0 \le x \le 1$.
    Show that

$$\int_{C_1} \mathbf{F} \cdot d\mathbf{R} \ne \int_{C_2} \mathbf{F} \cdot d\mathbf{R}$$

13.3 THE FUNDAMENTAL THEOREM AND PATH INDEPENDENCE

IN THIS SECTION: *Fundamental theorem for line integrals, conservative vector fields, independence of path*
In this section, we will study path independence and characterize the kind of vector field $\mathbf{F}$ for which it occurs.

In general, the value of the line integral $\int_C \mathbf{F} \cdot d\mathbf{R}$ depends on the path of integration C, but in certain cases, the integral will be the same for all paths in a given region D with the same initial point P and terminal point Q. In this case, we say the line integral is **independent of path** in D (see Figure 13.16).

Fundamental Theorem for Line Integrals

The fundamental theorem of calculus (Section 5.4) says that if the function f is continuous on $[a, b]$, then

$$\int_a^b f(x)\,dx = F(b) - F(a)$$

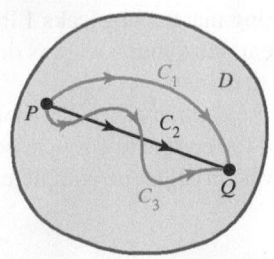

Figure 13.16 A line integral $\int_C \mathbf{F} \cdot d\mathbf{R}$ is independent of path in D if its value is the same for all curves joining any two points in D

where F is any antiderivative of f; that is, $F'(x) = f(x)$. For a function of two or three variables, the analogue of the derivative is the gradient, and the corresponding analogue of the fundamental theorem is the following theorem.

Theorem 13.2   Fundamental theorem for line integrals

Let C be a piecewise smooth curve that is parameterized by the vector function $\mathbf{R}(t)$ for $a \le t \le b$, and let $\mathbf{F}(t)$ be a vector field that is continuous on C. If f is a scalar function such that $\mathbf{F} = \nabla f$, then

$$\int_C \mathbf{F} \cdot d\mathbf{R} = f(Q) - f(P)$$

where $Q = \mathbf{R}(b)$ and $P = \mathbf{R}(a)$ are the endpoints of C.

Proof: We will prove this theorem for the case where $f(x,y,z)$ is a function of three variables, such that $\mathbf{F} = \nabla f(x,y,z)$. Suppose $\mathbf{R}(t) = \langle x(t), y(t), z(t) \rangle$ and let G be the composite function $G(t) = f[x(t), y(t), z(t)]$. Then, according to the chain rule

$$\frac{dG}{dt} = \frac{\partial f}{\partial x}\frac{dx}{dt} + \frac{\partial f}{\partial y}\frac{dy}{dt} + \frac{\partial f}{\partial z}\frac{dz}{dt}$$

and we have

$$
\begin{aligned}
\int_C \mathbf{F} \cdot d\mathbf{R} &= \int_C \nabla f \cdot d\mathbf{R} \\
&= \int_C \left[\frac{\partial f}{\partial x}dx + \frac{\partial f}{\partial y}dy + \frac{\partial f}{\partial z}dz \right] \\
&= \int_a^b \left[\frac{\partial f}{\partial x}\frac{dx}{dt} + \frac{\partial f}{\partial y}\frac{dy}{dt} + \frac{\partial f}{\partial z}\frac{dz}{dt} \right]dt \\
&= \int_a^b \frac{dG}{dt}dt && \textit{Substitution} \\
&= G(b) - G(a) && \textit{Fundamental theorem of calculus} \\
&= f[x(b),y(b),z(b)] - f[x(a),y(a),z(a)] && \textit{Substitution} \\
&= f[\mathbf{R}(b)] - f[\mathbf{R}(a)] && \textit{Substitution} \\
&= f(Q) - f(P)
\end{aligned}
$$

◆

Example 1   Using the fundamental theorem to evaluate a line integral

Evaluate the line integral $\int_C \mathbf{F} \cdot d\mathbf{R}$, where

$$\mathbf{F} = \nabla(e^x \sin y - xy - 2y)$$

and C is the path described by $\mathbf{R}(t) = \left[t^3 \sin \frac{\pi}{2}t \right]\mathbf{i} - \left[\frac{\pi}{2}\cos\left(\frac{\pi}{2}t + \frac{\pi}{2}\right) \right]\mathbf{j}$ for $0 \le t \le 1$.

Solution   First, note that the hypotheses of the fundamental theorem for line integrals are satisfied since

$$f(x,y) = e^x \sin y - xy - 2y$$

has continuous partial derivatives on the smooth curve C. At the endpoints of C, we find

left endpoint $(t = 0)$: $\mathbf{R}(0) = \langle 0,0 \rangle$ $f(0,0) = e^0 \sin 0 - 0 - 0 = 0$

right endpoint $(t = 1)$: $\mathbf{R}(1) = \left\langle 1, \dfrac{\pi}{2} \right\rangle$ $f\left(1, \dfrac{\pi}{2}\right) = e^1 \sin \dfrac{\pi}{2} - \dfrac{\pi}{2} - 2\left(\dfrac{\pi}{2}\right) = e - \dfrac{3\pi}{2}$

Thus, according to the fundamental theorem for line integrals, we have

$$
\begin{aligned}
\int_C \mathbf{F} \cdot d\mathbf{R} &= f(Q) - f(P) \\
&= f\left(1, \dfrac{\pi}{2}\right) - f(0,0) \\
&= \left(e - \dfrac{3\pi}{2}\right) - 0 \\
&= e - \dfrac{3\pi}{2}
\end{aligned}
$$

Conservative Vector Fields

A key requirement for evaluating the line integral $\int_C \mathbf{F} \cdot d\mathbf{R}$ by the fundamental theorem for line integrals is that $\mathbf{F}$ be the gradient of some scalar function f; that is, $\mathbf{F} = \nabla f$. Our next goal is to learn how to determine when a given vector field $\mathbf{F}$ can be expressed as a gradient field, and we begin by introducing some terminology.

> **CONSERVATIVE VECTOR FIELD** A vector field $\mathbf{F}$ is said to be **conservative** in a region D if $\mathbf{F} = \nabla f$ for some scalar function f in D. The function f is called a **scalar potential** of $\mathbf{F}$ in D. That is,
>
> $$\mathbf{F} = \nabla f \quad \text{for } (x,y) \text{ on } D$$
>
> *Conservative vector field*   *Scalar potential*

The term *conservative* comes from physics and is related to the law of conservation of energy (see Problem 59).

Example 2   Verifying that a vector field is conservative

Verify that the vector field $\mathbf{F} = 2xy\mathbf{i} + x^2\mathbf{j}$ is conservative, with scalar potential $f = x^2 y$.

Solution $\nabla f = 2xy\mathbf{i} + x^2\mathbf{j}$ and this is the same as $\mathbf{F}$, so $\mathbf{F}$ is conservative.

It is one thing to verify that a vector field is conservative as we did in Example 2, but it is more common to be given a vector field $\mathbf{F}$ and then be asked to determine whether it is conservative without knowing the answer in advance. We will establish a simple criterion for $\mathbf{F}$ to be conservative on any set D in the plane that has these properties:

(1) Any two points P and Q in D can be joined by a piecewise-smooth curve entirely within D.

(2) Every closed curve in D encloses only points that are also in D.

A region D with property (1) is **connected,** and a connected region with property (2) is **simply connected**. Roughly speaking, a simply connected region is one with no "holes," as shown in Figure 13.17**a**.

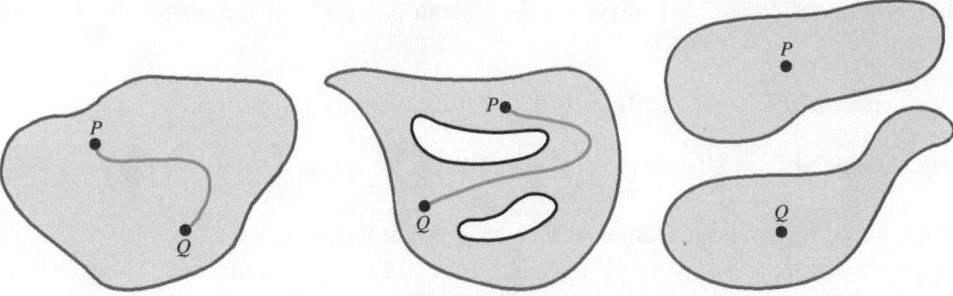

a. A region that is simply connected (no holes)   **b.** A connected region that is not simply connected   **c.** A region that is not connected

Figure 13.17   Types of regions

Theorem 13.3   Cross-partials test for a conservative vector field in the plane

Consider the vector field $\mathbf{F}(x,y) = u(x,y)\mathbf{i} + v(x,y)\mathbf{j}$, where u and v have continuous first partials in the open, simply connected region D in the plane. Then $\mathbf{F}(x,y)$ is conservative in D if and only if

$$\frac{\partial u}{\partial y} = \frac{\partial v}{\partial x} \qquad \text{throughout } D$$

Proof: We outline the proof in Problem Set 13.4 after our discussion of Green's theorem.   ◆

Example 3   Finding a scalar potential function

Show that the vector field $\mathbf{F} = (e^x \sin y - y)\mathbf{i} + (e^x \cos y - x - 2)\mathbf{j}$ is conservative and then find a scalar potential function f for $\mathbf{F}$.

Solution   First note that the component functions $u(x,y) = e^x \sin y - y$ and $v(x,y) = e^x \cos y - x - 2$ have continuous partial derivatives. Then

$$\frac{\partial u}{\partial y} = e^x \cos y - 1 \qquad \text{and} \qquad \frac{\partial v}{\partial x} = e^x \cos y - 1$$

Since $\dfrac{\partial u}{\partial y} = \dfrac{\partial v}{\partial x}$, it follows that $\mathbf{F}$ is conservative. To find a scalar potential function f such that $\nabla f = \mathbf{F}$, we note that f must satisfy $u(x,y) = f_x(x,y)$ and $v(x,y) = f_y(x,y)$.

$$f(x,y) = \int u(x,y)\,dx = \underbrace{\int (e^x \sin y - y)\,dx}$$

This is the "partial integral" in the sense that y is held constant while the integration is performed with respect to x alone.

$$= e^x \sin y - yx + k(y)$$

Note that $k(y)$ is a function of y alone—a "constant" as far as x-integration is concerned.

Because f must also satisfy $f_y(x,y) = v(x,y)$, we compute the partial derivative of this f with respect to y:

$$f_y(x,y) = \frac{\partial}{\partial y}[e^x \sin y - yx + k(y)] = e^x \cos y - x + \frac{dk}{dy}$$

Set this equal to $v = e^x \cos y - x - 2$ and solve for $\dfrac{dk}{dy}$:

$$e^x \cos y - x + \frac{dk}{dy} = e^x \cos y - x - 2$$
$$\frac{dk}{dy} = -2$$
$$k(y) = -2y + C$$

Thus, any function $f(x, y) = e^x \sin y - xy - 2y + C$. Any such function is a scalar potential of $\mathbf{F}$ and, for simplicity, we pick $C = 0$:

$$f(x, y) = e^x \sin y - xy - 2y$$

In Example 3, we began by using the fact that $u = f_x$. In general, the issue of whether to start with $u = f_x$ or $v = f_y$ is often determined by which equation leads to the simpler integration.

Example 4   Testing for a conservative vector field in the plane

Determine whether the vector field $\mathbf{F} = ye^{xy}\mathbf{i} + (xe^{xy} + x)\mathbf{j}$ is conservative; if it is, find a scalar potential.

Solution   We have $u(x, y) = ye^{xy}$ and $v(x, y) = xe^{xy} + x$.

$$\frac{\partial u}{\partial y} = xye^{xy} + e^{xy} \qquad \frac{\partial v}{\partial x} = xye^{xy} + e^{xy} + 1$$

so $\dfrac{\partial u}{\partial y} \neq \dfrac{\partial v}{\partial x}$, and $\mathbf{F}$ is not conservative.

In $\mathbb{R}^3$, the following generalization of the cross-partials test may be used as a criterion for determining whether a given vector field is conservative.

Theorem 13.4   The curl criterion for a conservative vector field in $\mathbb{R}^3$

Suppose the vector field $\mathbf{F}$ and curl $\mathbf{F}$ are continuous in the simply connected region D of $\mathbb{R}^3$. Then $\mathbf{F}$ is conservative in D if and only if curl $\mathbf{F} = \mathbf{0}$.

Proof: As in $\mathbb{R}^2$, a simply connected region in $\mathbb{R}^3$ can be described informally as one with no "holes." A proof using Stokes' theorem is given in Section 13.6. ◆

Note that a vector field $\mathbf{F} = \langle u(x, y), v(x, y) \rangle$ in $\mathbb{R}^2$ can be regarded as the vector field $G = \langle u(x, y, 0), v(x, y, 0), 0 \rangle$ in $\mathbb{R}^3$. Since

$$\text{curl } \mathbf{G} = \begin{vmatrix} \mathbf{i} & \mathbf{j} & \mathbf{k} \\ \frac{\partial}{\partial x} & \frac{\partial}{\partial y} & \frac{\partial}{\partial z} \\ u(x, y, 0) & v(x, y, 0) & 0 \end{vmatrix}$$
$$= 0\mathbf{i} + 0\mathbf{j} + \left(\frac{\partial v}{\partial x} - \frac{\partial u}{\partial y} \right)\mathbf{k}$$

We have curl $\mathbf{G} = \mathbf{0}$ if and only if $\dfrac{\partial v}{\partial x} = \dfrac{\partial u}{\partial y}$. Thus, Theorem 13.4 becomes the cross-partials test if $\mathbf{F}$ is in $\mathbb{R}^2$.

Example 5   Determining a scalar potential for a conservative vector field in $\mathbb{R}^3$

Show that the vector field

$$\mathbf{F} = \langle 20x^3z + 2y^2, 4xy, 5x^4 + 3z^2 \rangle$$

is conservative in $\mathbb{R}^3$ and find a scalar potential function for $\mathbf{F}$.

Solution   To show that **F** is conservative in $\mathbb{R}^3$, note that the components of **F** are continuous with continuous partial derivatives and that

$$\text{curl } \mathbf{F} = \begin{vmatrix} \mathbf{i} & \mathbf{j} & \mathbf{k} \\ \frac{\partial}{\partial x} & \frac{\partial}{\partial y} & \frac{\partial}{\partial z} \\ 20x^3z + 2y^2 & 4xy & 5x^4 + 3z^2 \end{vmatrix}$$

$$= (0 - 0)\mathbf{i} - (20x^3 - 20x^3)\mathbf{j} + (4y - 4y)\mathbf{k}$$

$$= \mathbf{0}$$

Therefore, according to Theorem 13.4, the vector field **F** is conservative.

To determine a scalar potential function for **F**, we proceed as in Example 3, except that now we have three component equations, each involving three variables:

$$\frac{\partial f}{\partial x} = 20x^3z + 2y^2 \qquad \frac{\partial f}{\partial y} = 4xy \qquad \frac{\partial f}{\partial z} = 5x^4 + 3z^2$$

Integrating the first equation with respect to x (holding y and z constant), we obtain

$$f(x, y, z) = 5x^4z + 2xy^2 + g(y, z)$$

where $g(y, z)$ is a constant with respect to x-integration. Taking the partial derivative of this expression with respect to y, and then comparing the result to the required equation $\frac{\partial f}{\partial y} = 4xy$, we see that

$$4xy + \frac{\partial g}{\partial y} = 4xy$$

Thus,

$$\frac{\partial g}{\partial y} = 0 \qquad \text{and} \qquad g(y, z) = h(z)$$

where $h(z)$ is a constant with respect to x- and y-integration, so

$$f(x, y, z) = 5x^4z + 2xy^2 + h(z)$$

Finally, we compute the partial derivative of f with respect to z from this equation, and compare it to the required equation $\frac{\partial f}{\partial z} = 5x^4 + 3z^2$:

$$5x^4 + h'(z) = 5x^4 + 3z^2$$

$$h'(z) = 3z^2$$

$$h(z) = z^3 + C$$

Therefore, any function of the form

$$F(x, y, z) = 5x^4z + 2xy^2 + z^3 + C$$

is a scalar potential for the vector field **F**.

Independence of Path

We now have the tools to discuss path independence for a line integral. Here is the definition we will use.

INDEPENDENCE OF PATH   The line integral $\int_C \mathbf{F} \cdot d\mathbf{R}$ is **independent of path** in a region D if for any two points P and Q in D the line integral along every piecewise smooth curve in D from P to Q has the same value.

The following theorem provides three equivalent ways of determining whether a given line integral is path independent.

Theorem 13.5   Equivalent conditions for path independence

If **F** is a continuous vector field on the open connected set D, then the following three conditions are either all true or all false:

i. **F** is conservative on D; that is, $\mathbf{F} = \nabla f$ for some function f defined on D.
ii. $\oint_C \mathbf{F} \cdot d\mathbf{R} = 0$ for every piecewise smooth closed curve C in D.
iii. $\int_C \mathbf{F} \cdot d\mathbf{R}$ is independent of path within D.

Proof: To prove this theorem, it is enough to prove that (i) implies (ii), that (ii) implies (iii), and that (iii) implies (i). Then, if any one is true, all are true.

(i) implies (ii) Assume that **F** is conservative, and let C be a closed curve in D, as shown in Figure 13.18a. Then any point P on C can serve as both the initial point and the terminal point of the curve, and according to the fundamental theorem for line integrals, we have

$$\oint_C \mathbf{F} \cdot d\mathbf{R} = f(P) - f(P) = 0$$

where f is a scalar potential for **F**.

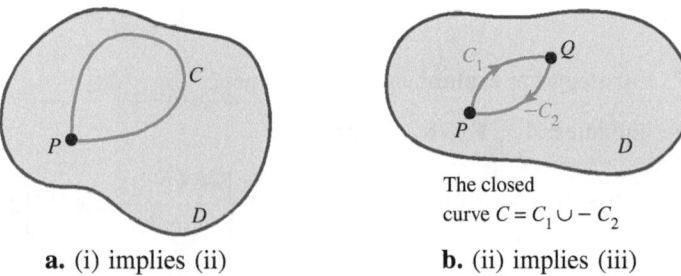

a. (i) implies (ii)   **b.** (ii) implies (iii)

The closed curve $C = C_1 \cup -C_2$

Figure 13.18   Equivalent conditions for path independence

(ii) implies (iii) Let C_1 and C_2 be two curves in D with the same initial point P and terminal point Q. Then the curve C formed by C_1 followed by $-C_2$ (the reverse of C_2) is a closed curve beginning and ending at P. According to condition **(ii)**, the line integral around this closed curve must be 0, so we have

$$0 = \underbrace{\oint_C \mathbf{F} \cdot d\mathbf{R}}_{\text{Condition (ii)}} = \underbrace{\int_{C_1} \mathbf{F} \cdot d\mathbf{R} + \int_{-C_2} \mathbf{F} \cdot d\mathbf{R}}_{\text{Additivity of line integrals}}$$

and it follows that (see Figure 13.18**b**)

$$\int_{C_1} \mathbf{F} \cdot d\mathbf{R} = -\int_{-C_2} \mathbf{F} \cdot d\mathbf{R} = \int_{C_2} \mathbf{F} \cdot d\mathbf{R}$$

Since C_1 and C_2 were chosen as any two curves in D with the same endpoints, it follows that $\oint_C \mathbf{F} \cdot d\mathbf{R}$ is independent of path in D.

(iii) implies (i) For this implication, we assume that $\oint_C \mathbf{F} \cdot d\mathbf{R}$ is independent of path in D and construct a scalar function f such that $\mathbf{F} = \nabla f$ in order to show that **F** must be conservative. Details of this implication are outlined in Problem 60. ♦

Example 6   Work along a closed path in a conservative force field

Show that no work is performed when an object moves along a closed path in a connected domain where the force field is conservative.

Solution   In such a force field $\mathbf{F}$, we have $\nabla f = \mathbf{F}$, where f is a scalar potential of $\mathbf{F}$, and because the path of motion is closed, it begins and ends at the same point P. Thus, the work is given by

$$W = \oint_C \mathbf{F} \cdot d\mathbf{R} = f(P) - f(P) = 0 \qquad \blacksquare$$

We now have several ways for evaluating a given line integral $\int_C \mathbf{F} \cdot d\mathbf{R}$. We can:

(1) Parameterize C and use the parameterization to convert the line integral into an "ordinary" integral in t over an interval $a \le t \le b$.

(2) Check to see whether $\mathbf{F}$ is conservative. If it is, find a scalar potential function f and then use the fundamental theorem for line integrals.

(3) If $\mathbf{F}$ is conservative, find a convenient path C_1 with the same endpoints as C and use the fact that

$$\int_{C_1} \mathbf{F} \cdot d\mathbf{R} = \int_C \mathbf{F} \cdot d\mathbf{R}$$

since the line integral is independent of path.

Here is an example that illustrates these options.

Example 7   Strategy for evaluating a line integral

Evaluate the line integral $\int_C \mathbf{F} \cdot d\mathbf{R}$, where

$$\mathbf{F} = [(2x - x^2 y)e^{-xy} + \tan^{-1} y]\mathbf{i} + \left[\frac{x}{y^2 + 1} - x^3 e^{-xy}\right]\mathbf{j}$$

for each of the following curves:

a. C_1: the ellipse $9x^2 + 4y^2 = 36$
b. C_2: the curve with parametric equations $x = t^2 \cos \pi t, y = e^{-t} \sin \pi t, 0 \le t \le 1$

Solution   First, we check to see whether $\mathbf{F}$ is conservative using the cross-partials test:

$$\frac{\partial}{\partial x}\left[\frac{x}{y^2 + 1} - x^3 e^{-xy}\right] = \frac{1}{y^2 + 1} + (x^3 y - 3x^2)e^{-xy}$$

$$= \frac{\partial}{\partial y}\left[(2x - x^2 y)e^{-xy} + \tan^{-1} y\right]$$

Thus, $\mathbf{F}$ is conservative. We could attempt to find a potential function for $\mathbf{F}$, but it isn't really necessary since $\mathbf{F}$ is conservative.

a. Since $\mathbf{F}$ is conservative, and the ellipse C_1 is a closed curve, we must have $\oint_C \mathbf{F} \cdot d\mathbf{R} = 0$.

b. The curve C_2 has initial point $P(0,0)$, where $t = 0$, and terminal point $Q(-1,0)$, where $t = 1$. Since $\mathbf{F}$ is conservative, the line integral $\int_C \mathbf{F} \cdot d\mathbf{R}$ is independent of path, so the given line integral has the same value as $\int_{C_3} \mathbf{F} \cdot d\mathbf{R}$, where C_3 is the line segment from P to Q. A parameterization for C_3 is $x = -t$, $y = 0$. Thus, if $\mathbf{R} = \langle -t, 0 \rangle$, we have $\mathbf{R}' = \langle -1, 0 \rangle$ and

$$
\begin{aligned}
\int_{C_2} \mathbf{F} \cdot d\mathbf{R} &= \int_{C_3} \mathbf{F} \cdot d\mathbf{R} \\
&= \int_0^1 \mathbf{F}[\mathbf{R}(t)] \cdot \mathbf{R}'(t)\, dt \\
&= \int_0^1 \left\{ \left[2(-t) - 0)e^0 + \tan^{-1} 0 \right](-1) + \left[\frac{-t}{0+1} - (-t)^3 e^0 \right](0) \right\} dt \\
&= \int_0^1 2t\, dt \\
&= 1
\end{aligned}
$$

In the next section, we will develop another criterion for path independence as part of our study of an important result known as Green's theorem.

PROBLEM SET 13.3

Level 1

1. ■ What does this say? Describe the fundamental theorem for line integrals.
2. ■ What does this say? Explain what is meant by independence of path. Describe various equivalent conditions for path independence.

Determine whether or not each vector field in Problems 3-8 is conservative, and if it is, find a scalar potential.

3. $(e^{2x} \sin y)\mathbf{i} + (e^{2x} \cos y)\mathbf{j}$
4. $y^2 \mathbf{i} + 2xy \mathbf{j}$
5. $2xy^3 \mathbf{i} + 3y^2 x^2 \mathbf{j}$
6. $(xe^{xy} \sin y)\mathbf{i} + (e^{xy} \cos xy + y)\mathbf{j}$
7. $(-y + e^x \sin y)\mathbf{i} + [(x+2)e^x \cos y]\mathbf{j}$
8. $(y - x^2)\mathbf{i} + (2x + y^2)\mathbf{j}$

Evaluate

$$
\int_C [(3x + 2y)\, dx + (2x + 3y)\, dy]
$$

for each of the paths given in Problems 9-12.

9.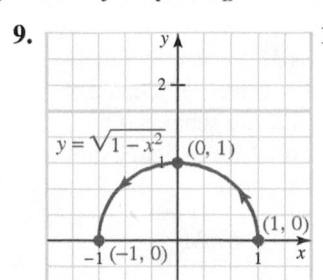
10.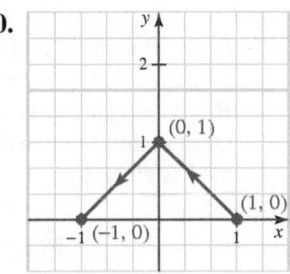

11.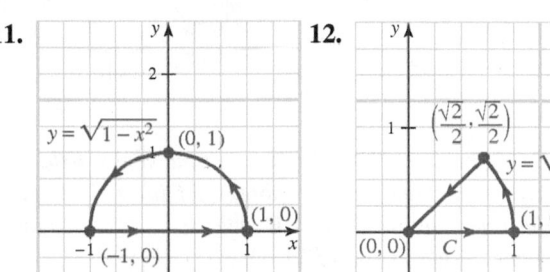
12.

Evaluate

$$
\int_C [2x^2 y\, dx + x^3 dy]
$$

for each of the paths given in Problems 13-16.

13.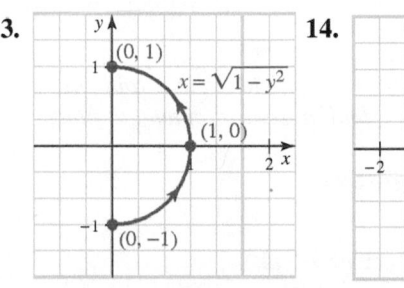
14.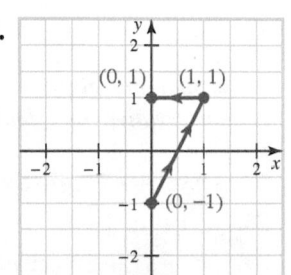

15.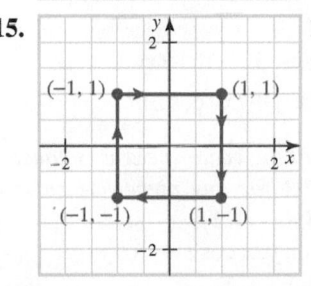
16.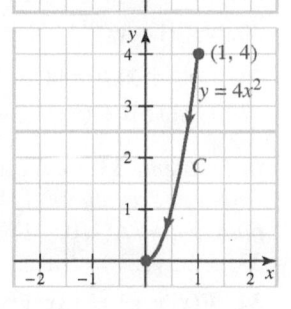

Evaluate

$$\int_C [2xy\,dx + x^2\,dy]$$

for each of the paths given in Problems 17-20.

17. 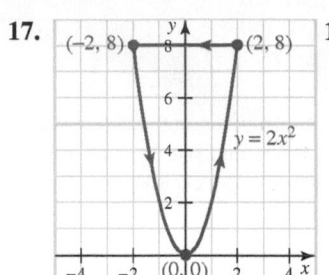 **18.**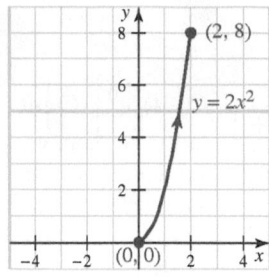

19. 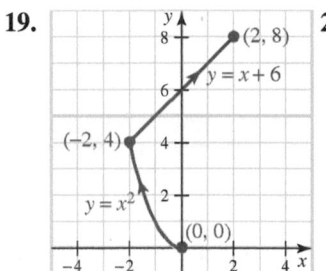 **20.**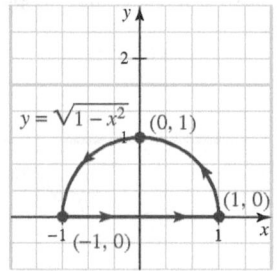

In each of Problems 21-26, show that the vector field **F** *is conservative and find a scalar potential f for* **F**. *Then evaluate the line integral* $\int_C \mathbf{F} \cdot d\mathbf{R}$, *where C is any smooth path connecting A(0, 0) to B(1, 1).*

21. $\mathbf{F}(x,y) = (2x - y)\mathbf{i} + (y^2 - x)\mathbf{j}$

22. $\mathbf{F}(x,y) = (x + 2y)\mathbf{i} + (2x + y)\mathbf{j}$

23. $\mathbf{F}(x,y) = 2xy\mathbf{i} + x^2\mathbf{j}$

24. $\mathbf{F}(x,y) = (y - x^2)\mathbf{i} + (x + y^2)\mathbf{j}$

25. $\mathbf{F}(x,y) = \dfrac{(y + 1)\mathbf{i} - x\mathbf{j}}{(y + 1)^2}$

26. $\mathbf{F}(x,y) = e^{-y}\mathbf{i} - xe^{-y}\mathbf{j}$

Show that the given vector field **F** *in Problems 27-32 is conservative and find a scalar potential function f for* **F**.

27. $e^{xy}yz\mathbf{i} + e^{xy}xz\mathbf{j} + e^{xy}\mathbf{k}$

28. $yz^2\mathbf{i} + xz^2\mathbf{j} + 2xyz\mathbf{k}$

29. $(x^2 + y^2 + z^2)(x\mathbf{i} + y\mathbf{j} + z\mathbf{k})$

30. $yz^{-1}\mathbf{i} + xz^{-1}\mathbf{j} - xyz^{-2}\mathbf{k}$

31. $(xy^2 + yz)\mathbf{i} + (x^2y + xz + 3y^2z)\mathbf{j} + (xy + y^3)\mathbf{k}$

32. $(y\sin z)\mathbf{i} + (x\sin z + 2y)\mathbf{j} + (xy\cos z)\mathbf{k}$

In Problems 33-36, show that the vector field **F** *is conservative and evaluate* $\int_C \mathbf{F} \cdot d\mathbf{R}$ *for any piecewise smooth path joining A(1, 0, -1) to B(0, -1, 1).*

33. $\mathbf{F}(x,y,z) = \langle \sin z, -z\sin y, x\cos z + \cos y \rangle$

34. $\mathbf{F}(x,y,z) = \langle 3x^2y^2z, 2x^3yz, x^3y^2 - e^{-z} \rangle$

35. $\mathbf{F}(x,y,z) = \left\langle \dfrac{y}{1+x^2} + \tan^{-1}z, \tan^{-1}x, \dfrac{x}{1+z^2} \right\rangle$

36. $\mathbf{F}(x,y,z) = \langle 2xz^3 - e^{-xy}y\sin z, -xe^{-xy}\sin z, 3x^2z^2 + e^{-xy}\cos z \rangle$

Level 2

Verify that each line integral in Problems 37-42 is independent of path and then find its value.

37. $\int_C [(3x^2 + 2x + y^2)\,dx + (2xy + y^3)\,dy]$, where C is any path from (0, 0) to (1, 1)

38. $\int_C [(xy\cos xy + \sin xy)\,dx + (x^2\cos xy)\,dy]$, where C is any path from $\left(0, \frac{\pi}{18}\right)$ to $\left(1, \frac{\pi}{6}\right)$

39. $\int_C [(y - x^2)\,dx + (x + y^2)\,dy]$, where C is any path from (-1, -1) to (0, 3)

40. $\int_C [(3x^2y + y^2)\,dx + (x^3 + 2xy)\,dy]$, where C is the path given parametrically by $\mathbf{R}(t) = t\mathbf{i} + (t^2 + t - 2)\mathbf{j}$ for $0 \le t \le 2$

41. $\int_C [\sin y\,dx + (3 + x\cos y)\,dy]$, where C is the path given parametrically by $\mathbf{R}(t) = 2\sin\left(\frac{\pi t}{2}\right)\cos(\pi t)\mathbf{i} + (\sin^{-1} t)\mathbf{j}$ for $0 \le t \le 1$.

42. $\int_C [e^x\cos y\,dx - e^x\sin y\,dy]$, where C is the path given parametrically by $\mathbf{R}(t) = (\cos t)\mathbf{i} + (\sin t)\mathbf{j}$ for $0 \le t \le \frac{\pi}{2}$.

Evaluate the line integrals given in Problems 43-46 using the fundamental theorem for line integrals.

43. $\int_C (y\mathbf{i} + x\mathbf{j}) \cdot d\mathbf{R}$, where C is any path from (0, 0) to (2, 4)

44. $\int_C (xy^2\mathbf{i} + x^2y\mathbf{j}) \cdot d\mathbf{R}$, where C is any path from (4, 1) to (0, 0)

45. $\int_C (2y\,dx + 2x\,dy)$, where C is the line segment from (0, 0) to (4, 4)

46. $\int_C (e^x\sin y\,dx + e^x\cos y\,dy)$, where C is any smooth curve from (0, 0) to $(0, 2\pi)$

47. Find a function g so $g(x)\mathbf{F}(x, y)$ is conservative, where

$$\mathbf{F}(x,y) = (x^4 + y^4)\mathbf{i} - (xy^3)\mathbf{j}$$

48. Find a function g so $g(x)\mathbf{F}(x, y)$ is conservative, where

$$\mathbf{F}(x,y) = (x^2 + y^2 + x)\mathbf{i} + xy\mathbf{j}$$

49. Let A be a particle of mass M located at the origin, and let B be a particle of mass m located at $\mathbf{R} = x\mathbf{i} + y\mathbf{j} + z\mathbf{k}$. If A and B are separated by a distance r, then the **gravitational force field** exerted on B by A is given by

$$\mathbf{F}(x,y,z) = -\frac{KmM}{r^3}\mathbf{R}$$

where K is the gravitational constant.

Classical force field

a. Show that $\mathbf{F}$ is conservative by finding a scalar potential for $\mathbf{F}$. The scalar potential function f is often called the **Newtonian potential**.

b. Compute the amount of work done by the force field $\mathbf{F}$ in moving an object from the point $P(a_1, b_1, c_1)$ to $Q(a_2, b_2, c_2)$.

50. a. Over what region in the xy-plane will the line integral

$$\int_C [(-yx^{-2} + x^{-1})\, dx + x^{-1}\, dy]$$

be independent of path?

b. Evaluate the line integral in part **a** if C is defined by

$$\mathbf{R}(t) = (\cos^3 t)\mathbf{i} + (\sin 3t)\mathbf{j}$$

for $0 \le t \le \frac{\pi}{3}$.

51. Let $\mathbf{F}(x, y) = \dfrac{-y\mathbf{i} + x\mathbf{j}}{x^2 + y^2}$

a. Compute the line integral $\int_{C_1} \mathbf{F} \cdot d\mathbf{R}$, where C_1 is the upper semicircle $y = \sqrt{1 - x^2}$ traversed counterclockwise. What is the value of $\int_{C_2} \mathbf{F} \cdot d\mathbf{R}$ if C_2 is the lower semicircle $y = -\sqrt{1 - x^2}$ also traversed counterclockwise?

b. Show that if $\mathbf{F} = M\mathbf{i} + N\mathbf{j}$, then

$$\frac{\partial M}{\partial y} = \frac{\partial N}{\partial x}$$

but $\mathbf{F}$ is not conservative on the unit disk $x^2 + y^2 \le 1$

EXPLORATION PROBLEMS *In Problems 52-58, you are to experiment with the notion of computing work along a path, with and without the benefit of independence of path. Suppose you are to power a boat of some kind from point $A(0, 0)$ to point $B(2, 1)$, and the primary consideration is the force of the wind, which generally opposes you. You are to investigate the effect of taking different paths from A to B.*

52. Suppose the wind force is $\mathbf{F} = \langle -a, -b \rangle$, for a and b positive. Compute the work involved along the straight

line path between A and B; then along a second path, of your choice. Does the path matter here? Why or why not?

53. Due to the effect of the harbor you are entering, suppose the wind force is $\langle -a, -ae^{-y} \rangle$. Again, compute the work along two paths as in Problem 52. Does the path matter here? Why or why not?

54. Repeat Problem 53 for $\langle -a, -ae^{-y+x/9} \rangle$.

55. An important type of problem in several fields of application is, "Can we find the optimal path to minimize the work?" You are to explore this issue in regard to the wind force in Problem 54,

$$\mathbf{F} = \langle -a, -ae^{-y+x/9} \rangle$$

You should have the work computed for two different paths from Problems 52-53. By looking at these numbers and carefully studying $\mathbf{F}$, you should see that we will be rewarded (or punished) by changing the path slightly from the straight-line path. What do your observations suggest regarding trying to minimize the work?

56. Attempt to find an (approximate) optimal path, starting with your observations in Problem 55 and common sense. For example, the path could be described by a parabola (or higher-degree polynomial) or by some trigonometric function. Find a "good" path for this purpose.

57. Explore the following question: "Is there a realistic optimal path from A to B?" Consider two conditions: first, there is no restriction on the path; second, suppose there is a shoreline at $y = 1$ so that one's path cannot exceed this limit. *Hint*: One approach might be to consider all parabolic paths of the form $y = x(b - ax)$, where a and b are nonnegative, and $y(2) = 1$. In this case, you can eliminate b and do a one-parameter study.

58. Assume the path in Problem 57 cannot go above $y = 1$. You may have discovered the optimal path. What is it? If you did the parabolic study suggested in Problem 57, there is an optimal path in this case. What is it?

Level 3

59. An object of mass m moves along a trajectory $\mathbf{R}(t)$ with velocity $\mathbf{V}(t)$ in a conservative force field $\mathbf{F}(t)$. Let $\mathbf{R}(t_0) = Q_0$ and $\mathbf{R}(t_1) = Q_1$ be the initial and terminal points on the trajectory.

a. Show that the work done on the object is $W = K(t_1) - K(t_0)$ where $K(t) = \frac{1}{2}m \, \|\mathbf{V}(t)\|^2$ is the object's *kinetic energy*.

b. Let f be a scalar potential for $\mathbf{F}$. Then $P(t) = -f(t)$ is the *potential energy* of the object. Prove the *law of conservation of energy*—namely,

$$P(t_0) + K(t_0) = P(t_1) + K(t_1)$$

60. Complete the proof of Theorem 13.5 by showing that if $\int_C \mathbf{F} \cdot d\mathbf{R}$ is independent of path in the open connected set D, then $\mathbf{F}$ is conservative. Do this by completing the following steps:

a. Let $P(a,b)$ be a fixed point in D, and let $Q(x,y)$ be any point in D. Define the function $f(x,y)$ by

$$f(x,y) = \int_P^Q \mathbf{F} \cdot d\mathbf{R}$$

Let $Q_1(x,y)$ be a point in D other than Q that lies on the same horizontal line as Q. Let C_1 be any curve in D from P to Q_1, and let C_2 be the horizontal line segment joining Q_1 to Q. Show that if

$$\mathbf{F} = M(x,y)\mathbf{i} + N(x,y)\mathbf{j}, \text{ then } \frac{\partial f}{\partial x} = M(x,y)$$

b. Using a similar argument (only with vertical line segments), show that

$$\frac{\partial f}{\partial y} = N(x,y)$$

Conclude that since $\mathbf{F} = \nabla f$, $\mathbf{F}$ must be conservative in D.

13.4 GREEN'S THEOREM

> **IN THIS SECTION:** *Green's theorem, area as a line integral, Green's theorem for multiply-connected regions, alternative forms of Green's theorem, normal derivatives*
> Green's theorem is the two variable form of the fundamental theorem of calculus, and it is presented in this section.

The fundamental theorem of calculus, $\int_a^b \frac{dF}{dx} \, dx = F(b) - F(a)$, can be described as saying that when the derivative $\frac{dF}{dx}$ is integrated over the closed interval $a \le x \le b$, the result is the same as that obtained by evaluating $F(x)$ at the "boundary points" a and b and forming the difference $F(b) - F(a)$. In Section 13.3, we obtained the analogous result

$$\int_C \nabla f \cdot d\mathbf{R} = f(Q) - f(P)$$

where $Q = \mathbf{R}(b)$ and $P = \mathbf{R}(a)$ are the endpoints of C. Our next goal is to obtain a different kind of analogue to the fundamental theorem of calculus called *Green's theorem* after the English mathematician George Green (*Historical Quest* in Problem 55).

Green's Theorem

Green's theorem relates a line integral around a closed curve to a double integral over the region contained by the curve. We begin with some terminology. A **Jordan curve**, named for the French mathematician Camille Jordan (1838-1922), is a closed curve C that does not intersect itself (see Figure 13.19). A simply connected region D in the plane has the property that it is connected and the interior of every Jordan curve C in D also lies in D, as shown in Figure 13.19.

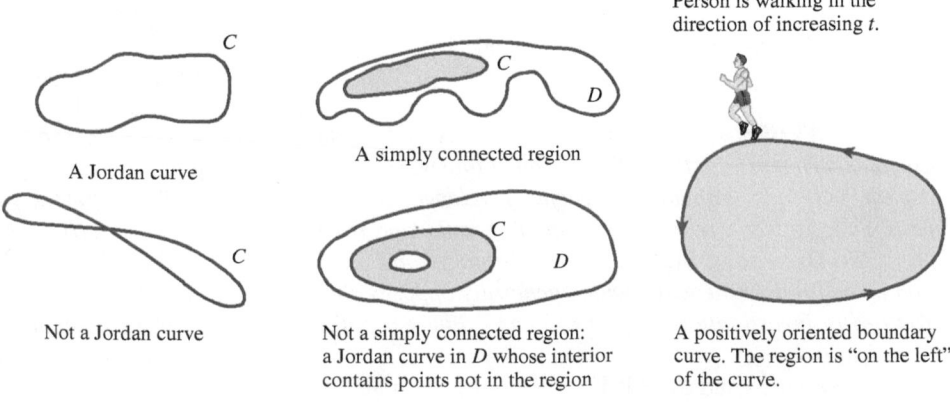

A Jordan curve

Not a Jordan curve

A simply connected region

Not a simply connected region: a Jordan curve in D whose interior contains points not in the region

Person is walking in the direction of increasing t.

A positively oriented boundary curve. The region is "on the left" of the curve.

Figure 13.19   Jordan curves and simply connected regions

Picture yourself as a point moving along a curve. If the region D stays on your *left* as you, the point, move along the curve C with increasing t, then C is said to be **positively oriented** (see the right illustration in Figure 13.19). Now we are ready to state Green's theorem.

Theorem 13.6 Green's theorem
───

Let D be a simply connected region that is bounded by the positively oriented piecewise smooth Jordan curve C. Then if the vector field $\mathbf{F}(x,y) = M(x,y)\mathbf{i} + N(x,y)\mathbf{j}$ is continuously differentiable on D, we have

$$\oint_C (M\,dx + N\,dy) = \iint_D \left(\frac{\partial N}{\partial x} - \frac{\partial M}{\partial y} \right) dA$$

THINK GREEN!

Graffiti on the wall of a high school playground in Tel Aviv, Isreal.

Courtesy of Regev Nathansohn and Eli Maor

■ **W**hat this says This theorem expresses an important relationship between a line integral around a simple closed curve (a curve is **simple** if there are no self-intersections and **closed** if there are no endpoints) in the plane and a double integral over the region bounded by the curve. It is one of the most important and elegant theorems in calculus. Take special note that D is required to be simply connected with a *positively oriented* boundary C. We use the notation $\oint$ to indicate such a line integral and, in fact, for an integral over a closed curve either $\int$ or $\oint$ is acceptable (see the definition of the line integral earlier in this chapter).

Proof: A **standard region** is one in which no vertical or horizontal line can intersect the boundary curve more than twice (see Figure 13.20).

We will prove Green's theorem for the special case where D is a standard region, and then we will indicate how to extend the proof to more general regions. Suppose D is a standard region with boundary curve C. We begin by showing that

$$\iint_D \frac{\partial M}{\partial y}\,dx\,dy = -\oint_C M\,dx$$

Because D is a standard region, as shown in Figure 13.20, the boundary curve C is composed of a lower portion C_L and an upper portion C_U, which are the graphs of functions $f_1(x)$ and $f_2(x)$, respectively, on a certain interval $a \le x \le b$. Then we can evaluate the double integral by iterated integration:

$$\iint_D \frac{\partial M}{\partial y}\,dx\,dy = \iint_D \frac{\partial M}{\partial y}\,dy\,dx$$

$$= \int_a^b \left[\int_{f_1(x)}^{f_2(x)} \frac{\partial M}{\partial y}\,dy \right] dx$$

$$= \int_a^b M[x, f_2(x)]\,dx - \int_a^b M[x, f_1(x)]\,dx$$

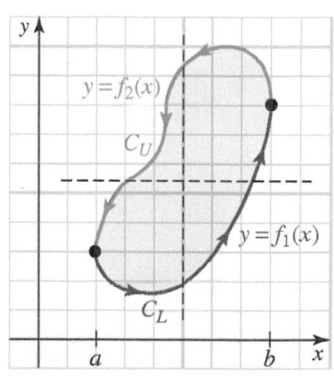

Figure 13.20 Standard region: no vertical or horizontal line intersects the boundary more than twice

$$= \int_{-C_U} M\,dx - \int_{C_L} M\,dx$$

$$= -\left[\int_{C_U} M\,dx + \int_{C_L} M\,dx\right]$$

$$= -\oint_C M\,dx$$

A similar argument shows that $\displaystyle\iint_D \frac{\partial N}{\partial x}\,dx\,dy = \oint_C N\,dy$.

Thus,

$$\iint_D \left(\frac{\partial N}{\partial x} - \frac{\partial M}{\partial y}\right)dA = \iint_D \frac{\partial N}{\partial x}\,dx\,dy - \iint_D \frac{\partial M}{\partial y}\,dx\,dy$$

$$= \oint_C N\,dy - \oint_C (-M)\,dx$$

$$= \oint_C (M\,dx + N\,dy)$$

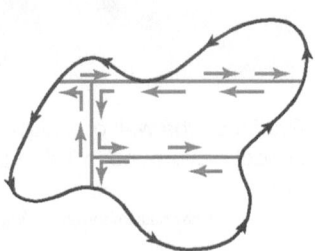

Figure 13.21 General case: The region is decomposed into a finite number of standard regions by cuts

This completes the proof for a standard region. If D is not a standard region, it can be decomposed into a number of standard subregions by using horizontal and vertical "cuts," as shown in Figure 13.21.

The proof for the standard region is then applied to each of these subregions, and the results are added. The line integrals along the cuts cancel in pairs, and after cancellation, the only remaining line integral is the one along the outer boundary C. Thus,

$$\int_C (M\,dx + N\,dy) = \iint_R \left(\frac{\partial N}{\partial x} - \frac{\partial M}{\partial y}\right)dA$$

The case where there is one cut is considered in Problem 56.   ♦

Example 1   Using Green's theorem

Show that Green's theorem is true for the line integral $\int_C (-y\,dx + x\,dy)$, where C is the closed path shown in Figure 13.22.

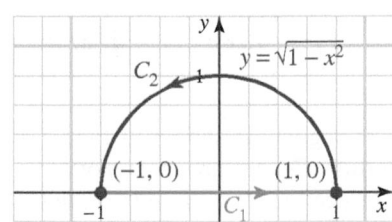

Figure 13.22 Path C

Solution   First, evaluate the line integral directly. The curve C consists of the line segment C_1 from $(-1, 0)$ to $(1, 0)$, followed by the semicircular arc C_2 from $(1, 0)$ back to $(-1, 0)$. We parameterize each of these:

$$C_1: \quad x = t, \quad y = 0, \quad -1 \le t \le 1$$
$$dx = dt \quad dy = 0$$

$$C_2: \quad x = \cos u, \qquad y = \sin u, \qquad 0 \le u \le \pi$$
$$dx = -\sin u \, du \quad dy = \cos u \, du$$

$$\oint_C (-y \, dx + x \, dy) = \int_{C_1} (-y \, dx + x \, dy) + \int_{C_2} (-y \, dx + x \, dy)$$

$$= \int_{-1}^{1} [-0 \, dt + t \cdot 0] + \int_0^{\pi} [-\sin u(-\sin u \, du) + \cos u(\cos u \, du)]$$

$$= \int_0^{\pi} (\sin^2 u + \cos^2 u) \, du$$

$$= \int_0^{\pi} 1 \, du$$

$$= \pi$$

Next, we use Green's theorem to evaluate this integral. Note that the boundary C is a Jordan curve and $M = -y$, $N = x$, so that $\mathbf{F}(x,y) = -y\mathbf{i} + x\mathbf{j}$ is continuously differentiable. We now apply Green's theorem:

$$\oint_C (-y \, dx + x \, dy) = \iint_D \left(\frac{\partial}{\partial x}(x) - \frac{\partial}{\partial y}(-y) \right) dA$$

$$= \iint_D 2 \, dA$$

$$= 2(\text{AREA OF SEMICIRCLE})$$

$$= 2 \left[\frac{1}{2} \pi (1)^2 \right]$$

$$= \pi$$

Example 2 Computing work with Green's theorem

A closed path C in the plane is defined by Figure 13.23.

Find the work done on an object moving along C in the force field

$$\mathbf{F}(x,y) = (x + xy^2)\mathbf{i} + 2(x^2 y - y^2 \sin y)\mathbf{j}$$

Solution The work done, W, is given by the line integral $\oint_C \mathbf{F} \cdot d\mathbf{R}$. Note that $\mathbf{F}$ is continuously differentiable on the region D enclosed by C, and since D is simply connected with a positively oriented boundary (namely, C), the hypotheses of Green's theorem are satisfied. We find that

$$W = \oint_C \mathbf{F} \cdot d\mathbf{R}$$

$$= \iint_D \left[\frac{\partial}{\partial x}(2x^2 y - 2y^2 \sin y) - \frac{\partial}{\partial y}(x + xy^2) \right] dA$$

$$= \iint_D (4xy - 2xy) \, dA$$

$$= 2 \int_0^1 \int_{x^2}^1 xy \, dy \, dx$$

$$= 2 \int_0^1 \frac{1}{2} xy^2 \Big|_{y=x^2}^{y=1} dx$$

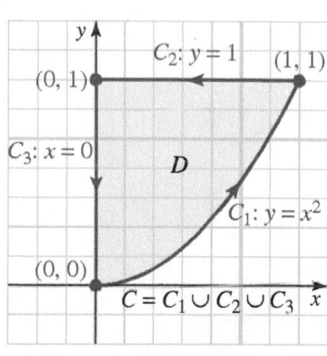

Figure 13.23 Closed path C

$$= \int_0^1 (x - x^5)\, dx$$

$$= \left[\frac{1}{2}x^2 - \frac{1}{6}x^6 \right]_0^1$$

$$= \frac{1}{3}$$

∎

Area as a Line Integral

A line integral can be used to compute an area of a region in the plane by applying the following theorem.

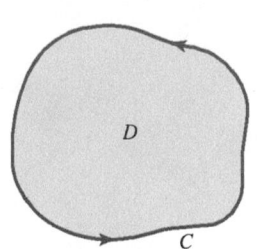

Figure 13.24 Area of region D

Theorem 13.7   Area as a line integral

Let D be a simply connected region in the plane with piecewise smooth, positively oriented closed boundary C, as shown in Figure 13.24.

Then the area A of region D is given by each of the following line integrals:

$$A = \oint_C x\, dy = -\oint_C y\, dx = \frac{1}{2}\oint_C [x\, dy - y\, dx]$$

> ■ What this says This gives us yet another technique for finding the area of a region, especially when its boundary is specified in parametric form. Do not forget the factor one-half after you finish the integration.

Proof: We will prove that

$$A = \frac{1}{2}\oint_C [x\, dy - y\, dx]$$

and leave the other two area formulas for the reader (see Problem 60).

Let $\mathbf{F}(x,y) = -y\mathbf{i} + x\mathbf{j}$. Then since $\mathbf{F}$ is continuously differentiable on D, Green's theorem applies. We have

$$\oint_C (-y\, dx + x\, dy) = \iint_D \left[\frac{\partial}{\partial x}(x) - \frac{\partial}{\partial y}(-y) \right] dA = \iint_D 2\, dA = 2A$$

so that

$$A = \frac{1}{2}\oint_C (-y\, dx + x\, dy). \qquad\qquad \blacklozenge$$

Example 3   Area enclosed by an ellipse

Show that the ellipse $\dfrac{x^2}{a^2} + \dfrac{y^2}{b^2} = 1$ has area πab.

Solution   The elliptical path E is given parametrically by $x = a\cos\theta, y = b\sin\theta$ for $0 \le \theta \le 2\pi$. We find $dx = -a\sin\theta\, d\theta, dy = b\cos\theta\, d\theta$. If A is the area of this ellipse, then

$$A = \frac{1}{2} \oint_C (-y\, dx + x\, dy)$$

$$= \frac{1}{2} \int_0^{2\pi} [-(b \sin\theta)(-a \sin\theta\, d\theta) + (a \cos\theta)(b \cos\theta\, d\theta)]$$

$$= \frac{1}{2} \int_0^{2\pi} ab(\sin^2\theta + \cos^2\theta)\, d\theta$$

$$= \frac{1}{2} \int_0^{2\pi} ab\, d\theta$$

$$= \frac{1}{2} ab(2\pi - 0)$$

$$= \pi ab$$

Green's Theorem for Multiply-Connected Regions

In the statement of Green's theorem, we require the region R inside the boundary curve C to be simply connected, but the theorem can be extended to multiply connected regions, that is, regions with one or more "holes." A region with a single hole is shown in Figure 13.25**a**. The boundary of this region consists of an "outer" curve C_1 and an "inner" curve C_2, oriented so that the region R is always on the left as we travel around the boundary, which means that C_1 is oriented counterclockwise and C_2 clockwise.

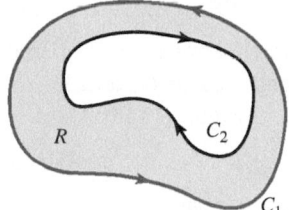
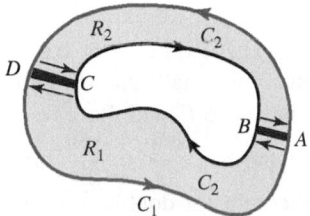

a. A doubly-connected region with oriented boundary curves

b. Two "cuts" are made through the hole

Figure 13.25 Multiply-connected region

We now make cuts AB and CD through the region to the hole, as indicated in Figure 13.25**b**. Let R_1 be the simply connected region contained by the closed curve C_3 that begins at A, extends along the cut to B, and then clockwise along the bottom of the curve C_2 to C, along the cut to D, and counterclockwise along the bottom of C_1 back to A. Similarly, let R_2 be the region contained by the curve C_4 that begins at D and extends to C along the cut, to B along the top of C_2, to A along the second cut, and back to D along the top part of C_1. Then, if the vector field $\mathbf{F} = M\mathbf{i} + N\mathbf{j}$ is continuously differentiable on R, we can apply Green's theorem to show

$$\iint_R \left(\frac{\partial N}{\partial x} - \frac{\partial M}{\partial y} \right) dA = \iint_{R_1} \left(\frac{\partial N}{\partial x} - \frac{\partial M}{\partial y} \right) dA + \iint_{R_2} \left(\frac{\partial N}{\partial x} - \frac{\partial M}{\partial y} \right) dA$$

$$= \oint_{C_3} (M\, dx + N\, dy) + \oint_{C_4} (M\, dx + N\, dy)$$

But the line integrals from A to B and C to D cancel those from B to A and D to C, leaving only the line integrals along the original boundary curves C_1 and C_2. Thus, we have

$$\iint_R \left(\frac{\partial N}{\partial x} - \frac{\partial M}{\partial y} \right) dA = \oint_{C_1} (M\, dx + N\, dy) + \oint_{C_2} (M\, dx + N\, dy)$$

> **GREEN'S THEOREM FOR DOUBLY CONNECTED REGIONS**   Let R be a doubly connected region (one hole) in the plane, with outer boundary C_1 oriented counterclockwise and boundary C_2 the hole oriented clockwise. If the boundary curves and $\mathbf{F}(x, y) = M(x, y)\mathbf{i} + N(x, y)\mathbf{j}$ satisfy the hypotheses of Green's theorem, then
>
> $$\iint\limits_{R} \left(\frac{\partial N}{\partial x} - \frac{\partial M}{\partial y} \right) dA = \oint_{C_1} (M\,dx + N\,dy) + \oint_{C_2} (M\,dx + N\,dy)$$

Example 4 illustrates one way this result can be used.

Example 4   **Green's theorem for a region containing a singular point**

Show that $\displaystyle\oint_{C} \frac{-y\,dx + x\,dy}{x^2 + y^2} = 2\pi$, where C is any piecewise smooth Jordan curve enclosing the origin $(0, 0)$.

Solution   Let $M(x, y) = \dfrac{-y}{x^2 + y^2}$ and $N(x, y) = \dfrac{x}{x^2 + y^2}$. Then

$$\frac{\partial N}{\partial x} = \frac{y^2 - x^2}{(x^2 + y^2)^2} = \frac{\partial M}{\partial y}$$

at any point (x, y) other than the origin. Next, let C_1 be a circle centered at the origin with radius r so small that the entire circle is contained in C, and let R be the region between the curve C and the circle C_1, as shown in Figure 13.26.

We know that $\dfrac{\partial N}{\partial x} = \dfrac{\partial M}{\partial y}$ throughout R (since R does not contain the origin) and Green's theorem for doubly connected regions tells us that

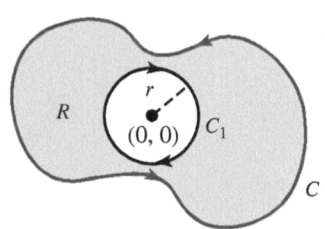

$$\oint_{C} \frac{-y\,dx + x\,dy}{x^2 + y^2} + \oint_{C_1} \frac{-y\,dx + x\,dy}{x^2 + y^2} = \iint\limits_{R} \left(\frac{\partial N}{\partial x} - \frac{\partial M}{\partial y} \right) dA = 0$$

Figure 13.26   The region R for doubly connected regions

so that

$$\oint_{C} \frac{-y\,dx + x\,dy}{x^2 + y^2} = -\oint_{C_1} \frac{-y\,dx + x\,dy}{x^2 + y^2} = \oint_{-C_1} \frac{-y\,dx + x\,dy}{x^2 + y^2}$$

where $-C_1$ is the circle C_1 traversed counterclockwise instead of clockwise. In other words, we can find the value of the given line integral about the curve C by finding the line integral about the circle $-C_1$. To do this, we parameterize $-C_1$ by $x = r\cos\theta$, $y = r\sin\theta$ for $0 \leq \theta \leq 2\pi$ and find that

$$\begin{aligned}
\oint_{-C_1} \frac{-y\,dx + x\,dy}{x^2 + y^2} &= \int_{0}^{2\pi} \frac{-r\sin\theta(-r\sin\theta\,d\theta) + r\cos\theta(r\cos\theta\,d\theta)}{r^2\cos^2\theta + r^2\sin^2\theta} \\
&= \int_{0}^{2\pi} \frac{r^2(\sin^2\theta + \cos^2\theta)}{r^2(\sin^2\theta + \cos^2\theta)}\,d\theta \\
&= \int_{0}^{2\pi} 1\,d\theta \\
&= 2\pi
\end{aligned}$$

Thus,

$$\oint_{C} \frac{-y\,dx + x\,dy}{x^2 + y^2} = \oint_{-C_1} \frac{-y\,dx + x\,dy}{x^2 + y^2} = 2\pi$$

Note that the line integral in Example 4 is not independent of path in R because there are closed paths (like C_1) along which the line integral is not 0.

Alternate Forms of Green's Theorem

Green's theorem can be expressed in two forms that generalize nicely to $\mathbb{R}^3$. For the first, note that the curl of the vector field $\mathbf{F}(x,y) = M(x,y)\mathbf{i} + N(x,y)\mathbf{j}$ is given by

$$\text{curl } \mathbf{F} = \begin{bmatrix} \mathbf{i} & \mathbf{j} & \mathbf{k} \\ \frac{\partial}{\partial x} & \frac{\partial}{\partial y} & \frac{\partial}{\partial z} \\ M(x,y) & N(x,y) & 0 \end{bmatrix}$$

$$= 0\mathbf{i} + 0\mathbf{j} + \left[\frac{\partial N}{\partial x} - \frac{\partial M}{\partial y} \right] \mathbf{k}$$

$$= \left[\frac{\partial N}{\partial x} - \frac{\partial M}{\partial y} \right] \mathbf{k}$$

so

$$\iint\limits_{D} \left(\frac{\partial N}{\partial x} - \frac{\partial M}{\partial y} \right) dA = \iint\limits_{D} (\text{curl } \mathbf{F} \cdot \mathbf{k}) \, dA$$

On the other hand, the line integral in Green's theorem can be expressed as

$$\oint_C \mathbf{F} \cdot d\mathbf{R} = \oint_C \mathbf{F} \cdot \frac{d\mathbf{R}}{ds} ds = \oint_C \mathbf{F} \cdot \mathbf{T} \, ds$$

By combining these results, we obtain

$$\oint_C \mathbf{F} \cdot d\mathbf{R} = \iint\limits_{D} \left(\frac{\partial N}{\partial x} - \frac{\partial M}{\partial y} \right) dA \qquad \textit{Green's theorem}$$

$$\oint_C \mathbf{F} \cdot \mathbf{T} \, ds = \iint\limits_{D} (\text{curl } \mathbf{F} \cdot \mathbf{k}) \, dA$$

When we extend this result to surfaces in $\mathbb{R}^3$ it will be called *Stokes' theorem*. We will examine Stokes' theorem in Section 13.6 and show how it can be interpreted in terms of the circulation of a fluid flow.

We have just seen how Green's theorem can be written in terms of $\mathbf{F} \cdot \mathbf{T}$, the tangential component of the vector field $\mathbf{F}$. The following example shows how Green's theorem can also be expressed in terms of $\mathbf{F} \cdot \mathbf{N}$, the normal component of $\mathbf{F}$. When this result is extended to $\mathbb{R}^3$ in Section 13.7, it will be called the *divergence theorem*.

Example 5   Alternate form for Green's theorem

Suppose $\mathbf{F}(x,y) = M(x,y)\mathbf{i} + N(x,y)\mathbf{j}$ in a domain D with a piecewise smooth boundary C. Show that

$$\oint_C \mathbf{F} \cdot \mathbf{N} \, ds = \iint\limits_{D} \text{div } \mathbf{F} \, dA$$

Solution   Parameterize C with the arc length parameter s, so that it is positively oriented with $\mathbf{R}(s) = x(s)\mathbf{i} + y(s)\mathbf{j}$ is the position vector on C. A unit tangent vector $\mathbf{T}$ to the curve C is $\mathbf{T} = \mathbf{R}'(s) = x'(s)\mathbf{i} + y'(s)\mathbf{j}$, which means that an outward normal vector is $\mathbf{N} = y'(s)\mathbf{i} - x'(s)\mathbf{j}$ (see Figure 13.27).

We now apply Green's theorem to find a representation using div $\mathbf{F}$.

$$\oint_C \mathbf{F} \cdot \mathbf{N} \, ds = \int_a^b (M\mathbf{i} + N\mathbf{j}) \cdot [y'(s)\mathbf{i} - x'(s)\mathbf{j}] \, ds$$

$$= \int_a^b \left(M \frac{dy}{ds} - N \frac{dx}{ds} \right) ds$$

$$= \oint_C (-N \, dx + M \, dy)$$

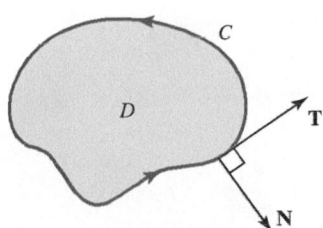

Figure 13.27 The outward unit normal and tangential vectors to a point on C

$$= \iint_D \left(\frac{\partial M}{\partial x} + \frac{\partial N}{\partial y} \right) dx\, dy \qquad \text{Green's theorem}$$

$$= \iint_D \text{div } \mathbf{F}\, dA$$

We list these alternative forms of Green's theorem for easy reference.

ALTERNATIVE FORMS OF GREEN'S THEOREM Let D be a simply connected region with a positively oriented boundary C. Then if the vector field $\mathbf{F} = M\mathbf{i} + N\mathbf{j}$ is continuously differentiable on D, we have

$$\oint_C \underbrace{\mathbf{F} \cdot d\mathbf{R}}_{\uparrow} = \oint_C (M\, dx + N\, dy)$$

tangential component of $\mathbf{F}$

$$= \iint_D \left(\frac{\partial N}{\partial x} - \frac{\partial M}{\partial y} \right) dA = \iint_D (\text{curl } \mathbf{F} \cdot \mathbf{k})\, dA$$

$$\oint_C \underbrace{\mathbf{F} \cdot \mathbf{N}}_{\uparrow}\, ds = \iint_D \left(\frac{\partial M}{\partial x} + \frac{\partial N}{\partial y} \right) dA = \iint_D \text{div } \mathbf{F}\, dA$$

normal component of $\mathbf{F}$

Normal Derivatives

In physics, some important applications of Green's theorem involve the *normal derivative* of a scalar function f, which is defined as the directional derivative of f in the direction of the outward normal vector $\mathbf{N}$ to some curve or surface.

NORMAL DERIVATIVE The **normal derivative** of f, denoted by $\partial f/\partial n$, is the directional derivative of f in the direction of the normal vector pointing to the exterior of the domain of f. In other words

$$\frac{\partial f}{\partial n} = \nabla f \cdot \mathbf{N}$$

where $\mathbf{N}$ is the outward unit normal vector.

The following example illustrates how Green's theorem can be used in connection with the normal derivative. Additional examples are found in the problem set.

Example 6 Green's formula for the integral of the Laplacian

Suppose f is a scalar function with continuous first and second partial derivatives in the simply connected region D. If the piecewise smooth positively oriented closed curve C bounds D, show that

$$\iint_D \nabla^2 f\, dx\, dy = \oint_C \frac{\partial f}{\partial n}\, ds$$

where $\nabla^2 f = f_{xx} + f_{yy}$ is the Laplacian of f and $\frac{\partial f}{\partial n} = \nabla f \cdot \mathbf{N}$ is the normal derivative of f.

Solution Let $u = -\dfrac{\partial f}{\partial y}$ and $v = \dfrac{\partial f}{\partial x}$. Then we have $\nabla^2 f = f_{xx} + f_{yy} = \dfrac{\partial v}{\partial x} - \dfrac{\partial u}{\partial y}$.

$$
\iint_D \nabla^2 f \, dx \, dy = \iint_D \left(\frac{\partial v}{\partial x} - \frac{\partial u}{\partial y} \right) dx \, dy
$$

$$
= \oint_C (u \, dx + v \, dy) \qquad \text{Green's theorem}
$$

$$
= \oint_C \left(u \frac{dx}{ds} + v \frac{dy}{ds} \right) ds
$$

$$
= \oint_C \left(-\frac{\partial f}{\partial y} \frac{dx}{ds} + \frac{\partial f}{\partial x} \frac{dy}{ds} \right) ds
$$

$$
= \oint_C \left(f_x \frac{dy}{ds} - f_y \frac{dx}{ds} \right) ds
$$

$$
= \oint_C \nabla f \cdot \left(\frac{dy}{ds} \mathbf{i} - \frac{dx}{ds} \mathbf{j} \right) ds
$$

$$
= \oint_C \nabla f \cdot \mathbf{N} \, ds \quad \text{Where } \mathbf{N} = \frac{dy}{ds} \mathbf{i} - \frac{dx}{ds} \mathbf{j}
$$
$$
\text{is the outward unit normal vector to } C.
$$

$$
= \oint_C \frac{\partial f}{\partial n} ds
$$

PROBLEM SET 13.4

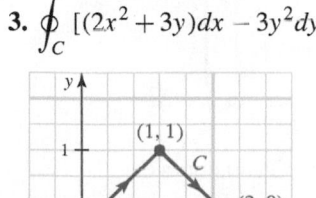
Level 1

In Problems 1-6, use Green's theorem to evaluate the given line integral around the indicated closed curve C. Then check your answer by parameterizing C.

1. $\oint_C (y^2 dx + x^2 dy)$ **2.** $\oint_C (y^3 dx - x^3 dy)$

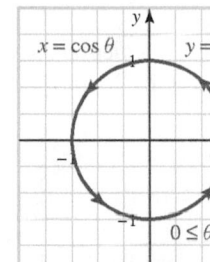

3. $\oint_C [(2x^2 + 3y)dx - 3y^2 dy]$ **4.** $\oint_C (y^2 dx + 3xy^2 dy)$

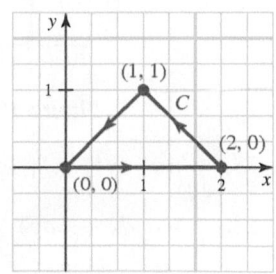

5. $\oint_C 4xy \, dx$ **6.** $\oint_C (4y \, dx - 3x \, dy)$

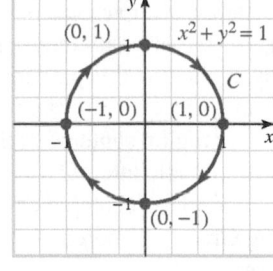 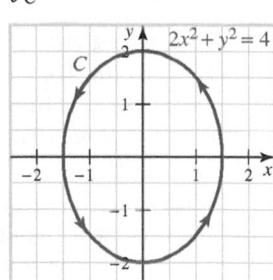

Use Green's theorem to evaluate

$$
\oint_C (2y \, dx - x \, dy)
$$

around the closed curve C in Problems 7-10.

7. **8.**

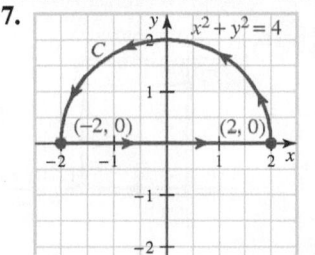 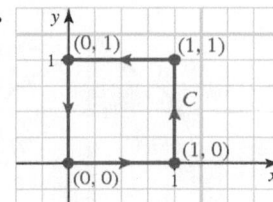

9. 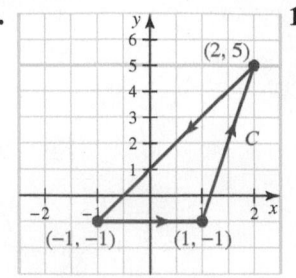   **10.**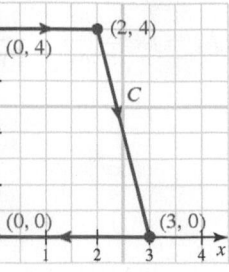

17. 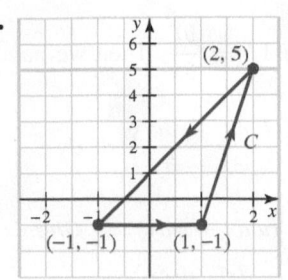   **18.**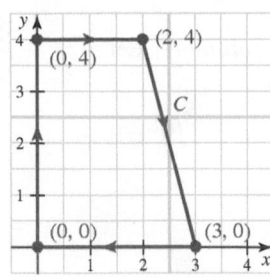

11. Use Green's theorem to evaluate

$$\oint_C (e^x \, dx - \sin x \, dy)$$

around the closed curve C in Figure 13.28.

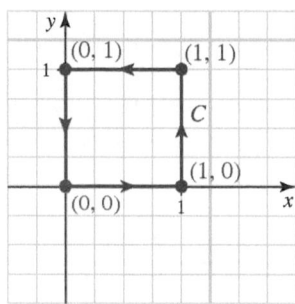

Figure 13.28   Curve C

12. Express the area of the rectangle with vertices $(-a, 0)$, $(a, 0)$, $(-a, b)$, (a, b) as a path integral. Verify the area formula, by evaluating this path integral directly. *Hint:* Use Theorem 13.7.

13. Express the area of an annular region centered at the origin, $0 < r < R$, by using a double integral.

14. Express the area of an annular region centered at the origin, $0 < r < R$ by using Green's Theorem for multiply-connected regions.

Use Green's theorem to evaluate

$$\oint_C (x \sin x \, dx - e^{y^2} \, dy)$$

around the closed curve C in Problems 15-18.

15. 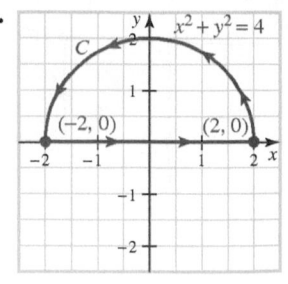   **16.**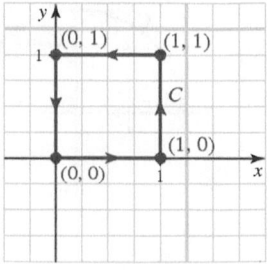

Use Green's theorem to evaluate

$$\oint_C [(x + y) \, dx - (3x - 2y) \, dy]$$

around the closed curve C in Problems 19-22.

19. 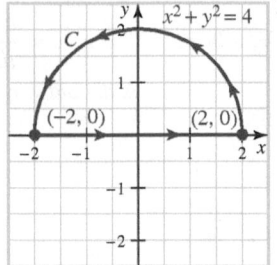   **20.**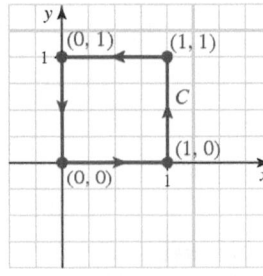

21. 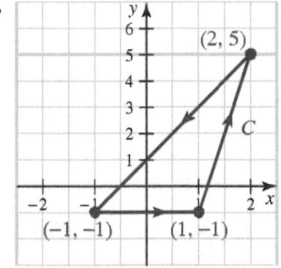   **22.**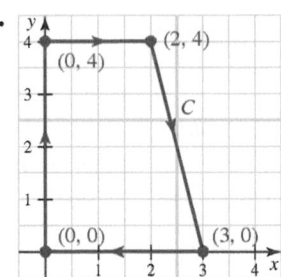

Use Green's theorem to evaluate the line integrals given in Problems 23-26 for the curve C given in Figure 13.29.

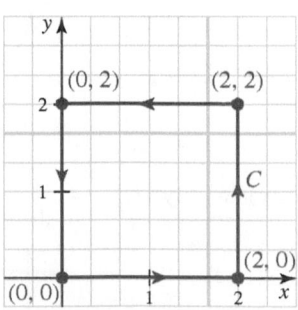

Figure 13.29   Given curve C

23. $\oint_C [(x - y^2) \, dx + 2xy \, dy]$

24. $\oint_C [(y^2 \, dx + x \, dy)$

25. $\oint_C [\sin x \cos y \, dx + \cos x \sin y \, dy]$

26. $\oint_C \left[2x \tan^{-1} y \, dx - \frac{x^2 y^2}{1+y^2} \, dy \right]$

Use Green's theorem to evaluate the line integrals given in Problems 27-30 for the curve C given in Figure 13.30.

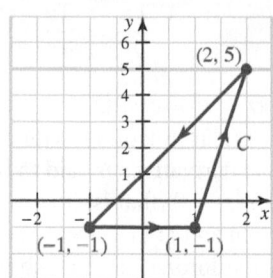

Figure 13.30 Given curve C

27. $\oint_C [2x \, dx + 2y \, dy]$

28. $\oint_C -y \, dx + x \, dy$

29. $\oint_C \sin y \, dx + x \cos y \, dy$

30. $\oint_C [2y \, dx + x \, dy]$

Level 2

31. Use Green's theorem to find the work done by the force field

$$\mathbf{F}(x, y) = (3y - 4x)\mathbf{i} + (4x - y)\mathbf{j}$$

when an object moves once counterclockwise around the ellipse $4x^2 + y^2 = 4$.

32. Find the work done when an object moves in the force field

$$\mathbf{F}(x, y) = y^2 \mathbf{i} + x^2 \mathbf{j}$$

counterclockwise along the circular path $x^2 + y^2 = 2$.

Use Theorem 13.7 to find the area enclosed by the regions described in Problems 33-36, and then check by using an appropriate formula.

33. circle $x^2 + y^2 = 4$

34. triangle with vertices $(0, 0), (1, 1)$, and $(0, 2)$

35. trapezoid with vertices $(0, 0), (4, 0), (1, 3)$, and $(0, 3)$

36. semicircle $y = \sqrt{4 - x^2}$

37. Evaluate the line integral

$$\oint_C (x^2 y \, dx - y^2 x \, dy)$$

where C is the boundary of the region between the x-axis and the semicircle $y = \sqrt{a^2 - x^2}$, traversed counterclockwise (including the x-axis).

38. Evaluate the line integral

$$\oint_C (3y \, dx - 2x \, dy)$$

where C is the cardioid $r = 1 + \sin \theta$, traversed counterclockwise.

39. Show that

$$\oint_C [(5 - xy - y^2) \, dx - (2xy - x^2) \, dy] = 3\overline{x}$$

where C is the square $0 \leq x \leq 1, 0 \leq y \leq 1$ traversed counterclockwise and $\overline{x}$ is the x-coordinate of the centroid of the square.

40. Find the work done by the force field $\mathbf{F}(x, y) = (x + 2y^2)\mathbf{j}$ as an object moves once counterclockwise about the circle $(x - 2)^2 + y^2 = 1$.

41. Let D be the region bounded by the Jordan curve C, and let A be the area of D. If $(\overline{x}, \overline{y})$ is the centroid of D, show that

$$\overline{x} = \frac{1}{2A} \oint_C x^2 \, dy \quad \text{and} \quad \overline{y} = \frac{1}{2A} \oint_C x^2 \, dy$$

where C is traversed counterclockwise.

42. Use Theorem 13.7 and the polar transformation formulas $x = r \cos \theta, r = r \sin \theta$ to obtain the area formula in polar coordinates, namely,

$$A = \frac{1}{2} \int_{\theta_1}^{\theta_2} r^2 \, d\theta = \frac{1}{2} \int_{\theta_1}^{\theta_2} [g(\theta)]^2 \, d\theta$$

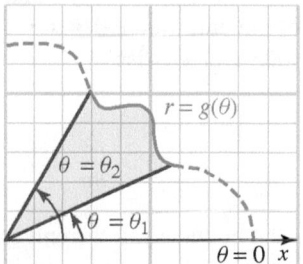

43. Evaluate

$$\oint_C \left[\left(\frac{-y}{x^2} + \frac{1}{x} \right) dx + \frac{1}{x} \, dy \right]$$

where C is any Jordan curve that does not touch or cross the y-axis, traversed counterclockwise.

44. Evaluate

$$\oint_C \frac{x \, dx + y \, dy}{x^2 + y^2}$$

where C is any Jordan curve whose interior does not contain the origin, traversed counterclockwise.

45. Evaluate the line integral

$$\oint_C \frac{x\,dx + y\,dy}{x^2 + y^2}$$

where C is any piecewise smooth Jordan curve enclosing the origin, traversed counterclockwise.

46. Evaluate

$$\oint_C \frac{-y\,dx + (x-1)\,dy}{(x-1)^2 + y^2}$$

where C is any Jordan curve whose interior does not contain the point $(1, 0)$, traversed counterclockwise.

47. Evaluate

$$\oint_C \frac{-(y+2)\,dx + (x-1)\,dy}{(x-1)^2 + (y+2)^2}$$

where C is any Jordan curve whose interior does not contain the point $(1, -2)$, traversed counterclockwise.

48. Evaluate

$$\oint_C \frac{\partial z}{\partial n}\,ds$$

where $z(x, y) = 2x^2 + 3y^2$, and C is the circular path $x^2 + y^2 = 16$, traversed counterclockwise.

49. Evaluate

$$\oint_C \frac{\partial f}{\partial n}\,ds$$

where $f(x, y) = x^2 y - 2xy + y^2$, and C is the boundary of the unit square $0 \le x \le 1, 0 \le y \le 1$, traversed counterclockwise.

50. Evaluate

$$\oint_C x\,\frac{\partial x}{\partial n}\,ds$$

where C is the boundary of the unit square $0 \le x \le 1$, $0 \le y \le 1$, traversed counterclockwise.

51. If C is a Jordan curve, show that

$$\oint_C [(x - 3y)\,dx + (2x - y^2)\,dy] = 5A$$

where A is the area of the region D enclosed by C.

52. Use line integration to find the area of the region bounded by the curve $C: x = \cos^3 t, y = \sin^3 t$ for $0 \le t \le 2\pi$.

Level 3

53. Prove the following theoretical application of Green's theorem: Let $\mathbf{F}(x, y) = u(x, y)\mathbf{i} + v(x, y)\mathbf{j}$ be continuously differentiable on the simply connected region D. Then $\mathbf{F}$ is conservative if and only if

$$\frac{\partial u}{\partial y} = \frac{\partial v}{\partial x}$$

throughout D.

54. Recall that a scalar function f with continuous first and second partial derivatives is said to be *harmonic* in a region D if $\nabla^2 f = 0$ (that is, if $f_{xx} + f_{yy} = 0$). If f is such a function and D is a simply connected region enclosed by the Jordan curve C, show that

$$\iint_D (f_x^2 + f_y^2)\,dx\,dy = \oint_C f\,\frac{\partial f}{\partial n}\,ds$$

55. *Historical Quest George Green (1793-1841) was the son of a baker, who worked in his father's mill and studied mathematics and physics in his spare time, using only books he obtained from the library. In 1828, he published a memoir titled "An Essay on the Application of Mathematical Analysis to the Theories of Electricity and Magnetism," which contains the result that now bears his name. Very few copies of the essay were printed and distributed, so few people knew of Green's results. In 1833, at the age of 40, he entered Cambridge University and graduated just four years before his death. His 1828 paper was discovered and publicized in 1845 by Sir William Thomson, (later known as Lord Kelvin, 1824-1907), and Green finally received proper credit for his work. In this Quest, use* Green's theorem to prove the following two important results, known as **Green's formulas.** In both cases, D is a simply connected region enclosed by the Jordan curve C.

a. Green's first formula

$$\iint_D [f\,\nabla^2 g + \nabla f \cdot \nabla g]\,dx\,dy = \oint_C f\,\frac{\partial g}{\partial n}\,ds$$

b. Once again start with the line integral and derive what is known as **Green's second formula.**

$$\iint_D [f\,\nabla^2 g - g\,\nabla^2 f]\,dx\,dy$$
$$= \oint_C \left(f\,\frac{\partial g}{\partial n} - g\,\frac{\partial f}{\partial n}\right)ds$$

56. *Historical Quest*

**Mikhail Ostrogradsky
(1801-1862)**

Karl Smith library

The result known as "Green's theorem" in the West is called "Ostrogradsky's theorem" in Russia, after Mikhail Ostrogradsky. Although Ostrogradsky published over 80 papers during a successful career as a mathematician, today he is known, even in his homeland, only for his version of Green's theorem, which appeared as part of a series of results presented to the Academy of Sciences in 1828. In this *Quest* you are asked to prove Green-Ostrogradsky's theorem in the plane for the non-standard region D shown in Figure 13.31.

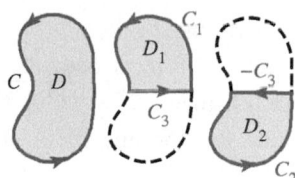

Figure 13.31 Green's theorem cut by a line L

Specifically, suppose that the line C_3 "cuts" the region D into two standard subregions D_1 and D_2. Apply Green's theorem to D_1 and D_2, then combine the results to show that the theorem applies to the non-standard region D. *Hint*: The key is what happens along the "cut line" C_3.

57. Extend Green's theorem to a "triply connected" region (two holes), such as the one shown in Figure 13.32.

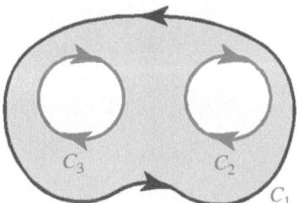

Figure 13.32 A triply connected region

58. Suppose $\mathbf{F} = M(x,y)\mathbf{i} + N(x,y)\mathbf{j}$ is continuously differentiable in a doubly connected region R and that

$$\frac{\partial N}{\partial x} = \frac{\partial M}{\partial y}$$

throughout R. How many distinct values of I are there for the integral

$$I = \oint_C [M(x,y)\,dx + N(x,y)\,dy]$$

where C is a piecewise smooth Jordan curve in R?

59. Answer the question in Problem 58 for the case where R is triply-connected (two holes). See Figure 13.32.

60. Let D be a simply connected region in the plane with a piecewise smooth closed boundary C. Complete the proof of Theorem 13.7 by showing that the area of D is given by

$$A = \oint_C x\,dy \quad \text{and} \quad A = -\oint_C y\,dx$$

13.5 SURFACE INTEGRALS

IN THIS SECTION: *Surface integration, flux integrals, integrals over parametrically defined surfaces*
A *surface integral* is a generalization of the line integral in which we integrate over regions in space bounded by a surface. The rest of this chapter deals with surface integrals.

Surface Integration

A surface is said to be **smooth** if there is a nonzero normal vector at each of its points, and it is **piecewise smooth** if it is composed of a finite number of smooth pieces.

We begin by defining the surface integral of a continuous scalar function $g(x,y,z)$ over a piecewise smooth surface S, as shown in Figure 13.33**a**.

Partition S into n subregions, the kth of which has area ΔS_k, and let $P_k^*(x_k^*, y_k^*, z_k^*)$ be a point chosen arbitrarily from the kth subregion, for $k = 1, 2, \cdots, n$ (see Figure 13.33**b**). Form the Riemann sum

$$\sum_{k=1}^{n} g(P_k^*)\Delta S_k$$

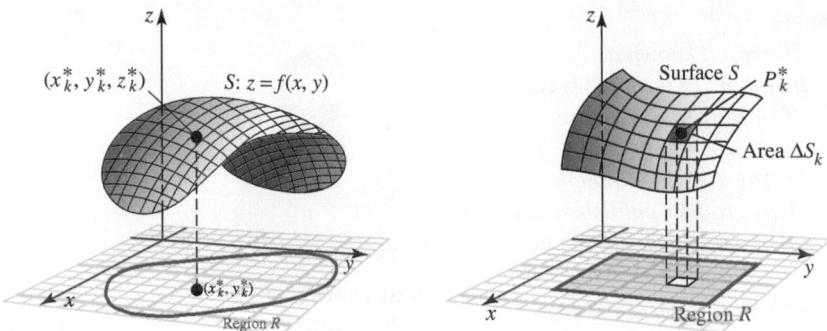

a. A piecewise smooth surface S **b.** Construction of a Riemann sum

Figure 13.33   Definition of a surface integral

and take the limit as the largest of the ΔS_k tends to 0, just as we did with surface area in Section 12.4. If this limit exists, it is called the *surface integral of g over S* and is denoted by

$$\iint_S g(x,y,z)\,dS$$

Recall from Section 12.4 that when the surface S projects onto the region R in the xy-plane and S has the representation $z = f(x,y)$, then $dS = \sqrt{f_x^2 + f_y^2 + 1}\,dA$, where dA is either $dx\,dy$ or $dy\,dx$. We can now state the formula for evaluating a surface integral.

SURFACE INTEGRAL   Let S be a surface defined by $z = f(x,y)$ and R be its projection on the xy-plane. If f, f_x, and f_y are continuous in R and g is a continuous function of three variables on S, then the **surface integral** of g over S is

$$\iint_S g(x,y,z)\,dS = \iint_R g(x,y,f(x,y))\sqrt{[f_x(x,y)]^2 + [f_y(x,y)]^2 + 1}\,dA$$

Example 1   Evaluating a surface integral

Evaluate the surface integral

$$\iint_S g\;dS$$

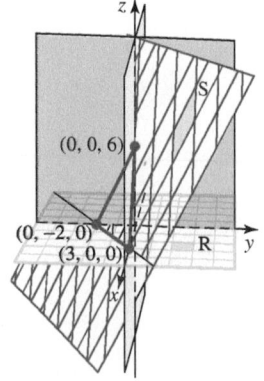

Figure 13.34 The portion of the plane that lies above the unit square R

for the surface S shown in Figure 13.34 where $g(x,y,z) = xz + 2x^2 - 3xy$ and S is that portion of the plane $2x - 3y + z = 6$ that lies over the unit square R: $2 \le x \le 3$, $2 \le y \le 3$.

Solution   First, note that the equation of the plane can be written as $z = f(x,y)$ where $f(x,y) = 6 - 2x + 3y$. We have $f_x(x,y) = -2$ and $f_y(x,y) = 3$, so that

$$dS = \sqrt{f_x^2 + f_y^2 + 1}\,dA = \sqrt{(-2)^2 + (3)^2 + 1}\,dA = \sqrt{14}\,dA$$

Consequently,

$$\iint_S g\;dS = \iint_S (xz + 2x^2 - 3xy)\sqrt{14}\,dA$$

$$= \iint_R [x(6 - 2x + 3y) + 2x^2 - 3xy]\sqrt{14}\,dy\,dx \qquad \textit{Since } z = 6 - 2x + 3y$$

$$= \sqrt{14} \iint\limits_{S} 6x \, dy \, dx$$

$$= 6\sqrt{14} \int_{2}^{3} \int_{2}^{3} x \, dy \, dx$$

$$= 6\sqrt{14} \int_{2}^{3} x \, dx$$

$$= 15\sqrt{14}$$

If the function g defined on S is simply $g(x, y, z) = 1$, then the surface integral gives the surface area of S.

SURFACE AREA FORMULA   If S is a piecewise smooth surface, its area is given by

$$A = \iint\limits_{S} dS$$

A useful application of surface integrals is to find the center of mass of a thin curved lamina whose shape is part of a given surface S, as shown in Figure 13.35.

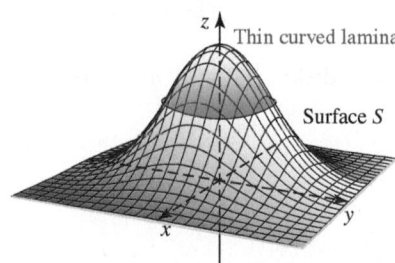

Figure 13.35   A thin lamina whose shape is the surface S

If $\rho(x, y, z)$ is the density (mass per unit area) at each point (x, y, z) on the lamina, then the *total mass, m,* of the lamina is given by the surface integral

$$\text{MASS OF A LAMINA} \qquad m = \iint\limits_{S} \rho(x, y, z) \, dS$$

and the center of mass of the surface is the point $C(\overline{x}, \overline{y}, \overline{z})$ where

$$\text{CENTER OF MASS} \quad \overline{x} = \frac{1}{m} \iint\limits_{S} x \rho \, dS, \quad \overline{y} = \frac{1}{m} \iint\limits_{S} y \rho \, dS, \quad \overline{z} = \frac{1}{m} \iint\limits_{S} z \rho \, dS$$

These formulas may be derived by essentially the same approach used in our previous work with centroids and moments of inertia of solid regions. (See, for example, Sections 6.5 and 12.6.)

Example 2   Mass of a curved lamina

Find the mass of a lamina of density $\rho(x, y, z) = z$ in the shape of the hemisphere $z = (a^2 - x^2 - y^2)^{1/2}$, as shown in Figure 13.36.

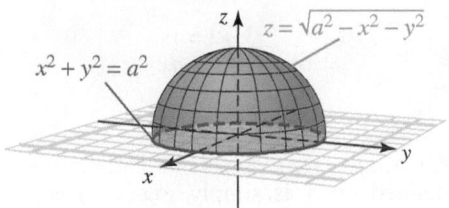

Figure 13.36   The hemisphere $z = \sqrt{a^2 - x^2 - y^2}$

Solution   We begin by calculating dS.

$$z_x = \frac{1}{2}(a^2 - x^2 - y^2)^{-1/2}(-2x)$$
$$= -x(a^2 - x^2 - y^2)^{-1/2}$$

$$z_y = \frac{1}{2}(a^2 - x^2 - y^2)^{-1/2}(-2y)$$
$$= -y(a^2 - x^2 - y^2)^{-1/2}$$

$$dS = \sqrt{z_x^2 + z_y^2 + 1}\, dA$$
$$= \sqrt{\frac{x^2}{a^2 - x^2 - y^2} + \frac{y^2}{a^2 - x^2 - y^2} + 1}\, dA$$
$$= \sqrt{\frac{a^2}{a^2 - x^2 - y^2}}\, dA$$
$$= a(a^2 - x^2 - y^2)^{-1/2}\, dA$$

The surface projects onto the disk $x^2 + y^2 \le a^2$ in the xy-plane, so the mass of the hemisphere S is given by

$$
\begin{aligned}
m &= \iint\limits_{S} \rho(x, y, z)\, dS \\
&= \iint\limits_{S} z\, dS \\
&= \iint\limits_{R} (a^2 - x^2 - y^2)^{1/2} a(a^2 - x^2 - y^2)^{-1/2} dA \\
&= a \iint\limits_{R} dA \\
&= \pi a^3 \qquad \textit{Since this integral represents the area of a circle of radius } a.
\end{aligned}
$$

Flux Integrals

An important application of surface integrals involves the flow of a fluid through a surface. To discuss such applications, we define a surface S to be **orientable** if S has a unit normal vector field **N** that varies continuously over S. Most common surfaces, such as spheres, cones, cylinders, ellipsoids, and paraboloids are orientable, but it is not difficult to construct a fairly simple surface that is not. For example, the **Möbius strip**, formed by twisting a long rectangular strip before joining the ends, is not orientable (see Figure 13.37**a**). If S is an orientable surface, there will be two possibilities for **N**, which can be thought of as **outward** (pointing toward the exterior of S) or **inward** (pointing toward the interior), as illustrated in Figure 13.37**b**.

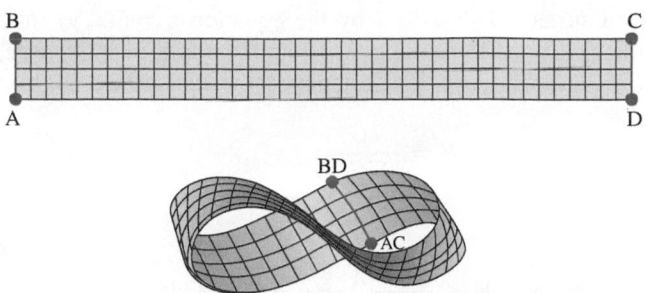

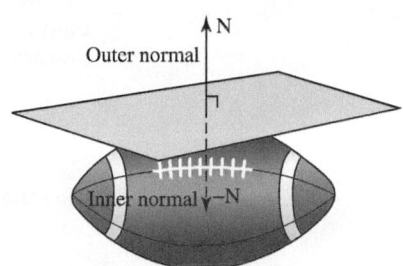

a. Möbius Strip; a one-sided (non-orientable) surface

b. Football; a two-sided (orientable) surface. This is also a closed surface.

Figure 13.37 Surfaces in space

Consider a fluid flowing steadily through a surface S with the continuous unit normal field $\mathbf{N}$. The *flux density* measures the quantity of fluid crossing the surface per unit area in unit time. In elementary physics, this quantity is shown to be proportional to the amount of field lines passing through a surface. Think of water flowing through a net as shown in Figure 13.38.

If ΔS is the area of a small patch of the surface S, then the amount of fluid crossing this patch in unit time may be estimated by the volume of the cylinder of height $\mathbf{F} \cdot \mathbf{N}$, and base area S (see Figure 13.39); that is,

$$\Delta V \approx (\mathbf{F} \cdot \mathbf{N}) \Delta S$$

Figure 13.38 Water flowing through a net

© Kendall Hunt Publishing Company

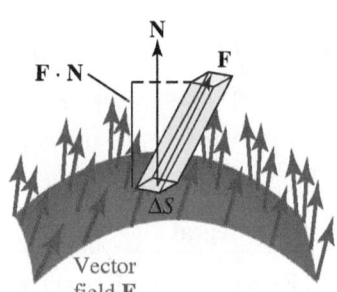

Vector field $\mathbf{F}$

Figure 13.39 The cylinder of fluid crossing the patch of area ΔS has height $\mathbf{F} \cdot \mathbf{N}$ and volume $(\mathbf{F} \cdot \mathbf{N}) \Delta S$

Thus, we can measure the total volume crossing the surface in unit time by computing the surface integral of $\mathbf{F} \cdot \mathbf{N}$, which is called a *flux integral*.

FLUX INTEGRAL   Let $\mathbf{F}$ be a vector field whose components have continuous partial derivatives on the surface S, which is oriented by the unit normal field $\mathbf{N}$. Then the **flux** of $\mathbf{F}$ across S is given by the surface integral

$$\text{Flux} = \iint\limits_{S} \mathbf{F} \cdot \mathbf{N} \, dS$$

Suppose we have a surface S described by the equation $z = f(x, y)$ that projects onto a region D in the xy-plane. Then, we can define $G(x, y, z) = z - f(x, y)$ and an upward unit normal (the one with positive $\mathbf{k}$-component) to the surface is given by

$$\mathbf{N} = \frac{\nabla G}{\|\nabla G\|}$$

Therefore,

$$\mathbf{F} \cdot \mathbf{N} \, dS = \mathbf{F} \cdot \left(\frac{\nabla G}{\|\nabla G\|} \right) \sqrt{f_x^2 + f_y^2 + 1} \, dA$$

and since $\|\nabla G\| = \sqrt{(-f_x)^2 + (-f_y)^2 + 1}$, the flux integral of $\mathbf{F}$ over S can be written as

$$\iint\limits_S \mathbf{F} \cdot \mathbf{N} \, dS = \iint\limits_R \mathbf{F}(x, y, f(x, y)) \cdot \nabla G \, dA$$

$$= \iint\limits_R \mathbf{F}(x, y, f(x, y)) \cdot \langle -f_x, -f_y, 1 \rangle \, dA$$

Similarly, a downward unit normal (negative $\mathbf{k}$-component) to the surface is given by

$$\mathbf{N} = \frac{-\nabla G}{\|\nabla G\|}$$

and the flux integral has the form

$$\iint\limits_S \mathbf{F} \cdot \mathbf{N} \, dS = \iint\limits_R \mathbf{F}(x, y, f(x, y)) \cdot \langle f_x, f_y, -1 \rangle \, dA$$

Note that it is always necessary to specify the orientation of $\mathbf{N}$ (upward or downward) when describing a flux integral.

Example 3   Evaluating a flux integral

Compute the flux integral $\iint\limits_S \mathbf{F} \cdot \mathbf{N} \, dS$, where $\mathbf{F} = xy\mathbf{i} + z\mathbf{j} + (x + y)\mathbf{k}$ and S is the triangular surface cut off from the plane $x + y + z = 1$ by the coordinate planes. Assume $\mathbf{N}$ is the upward unit normal.

Solution   Let $f(x, y) = z = 1 - x - y$. Then $f_x = -1, f_y = -1$, and the flux integral is

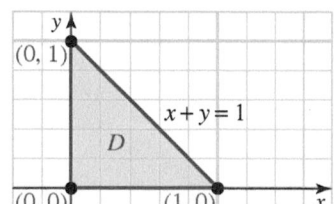

Figure 13.40 Projection D of $x + y + z = 1$ on the xy-plane

$$\iint\limits_S \mathbf{F} \cdot \mathbf{N} dS = \iint\limits_D \mathbf{F}(x, y, f(x, y)) \cdot \langle -f_x, -f_y, 1 \rangle \, dA$$

where D is the projection of S on the xy-plane. Setting $z = 0$, we find that D is the triangular region bounded by the lines $x + y = 1, x = 0$, and $y = 0$, as shown in Figure 13.40.
Thus, we have

$$\iint\limits_S \mathbf{F} \cdot \mathbf{N} \, dS = \iint\limits_D \langle xy, 1 - x - y, x + y \rangle \cdot \langle -(-1), -(-1), 1 \rangle \, dA$$

$$= \int_0^1 \int_0^{1-x} [xy + (1 - x - y) + x + y] \, dy \, dx$$

$$= \int_0^1 \int_0^{1-x} [xy + 1] \, dy \, dx$$

$$= \int_0^1 \left[\frac{1}{2}xy^2 + y \right]_0^{1-x} dx$$

$$= \int_0^1 \frac{1}{2}(x^3 - 2x^2 - x + 2)\, dx$$

$$= \frac{13}{24}$$

Example 4   Computing heat flow as a flux integral

Let R be the region that is bounded above by the paraboloid $z = 9 - x^2 - y^2$ and below by the xy-plane. The surface is shown in Figure 13.41.

   Experiments indicate that the velocity of heat flow is given by the vector field $\mathbf{H} = -K\nabla T$, where $T(x, y, z) = 2x + y - 3z^2$ is the temperature at each point $P(x, y, z)$ in the region and K is a constant (the *heat conductivity*, which is obtained experimentally for each different substance). Find the total heat flow $\int\int_R \mathbf{H} \cdot \mathbf{N}\, dS$ out of the region (that is, $\mathbf{N}$ is the outer unit normal, the one pointing away from the origin).

Solution   Note that $\nabla T = \langle 2, 1, -6z \rangle$ and that the surface S is composed of a top surface S_1 and a bottom surface S_2. The top surface is S_1: $z = 9 - x^2 - y^2$ and the bottom surface is the disk S_2: $x^2 + y^2 \le 9$, which is also the projection D of S_1 on the xy-plane.

   For S_1: Let $G = z + x^2 + y^2 - 9$; then, $\nabla G = \langle 2x, 2y, 1 \rangle$. The top surface S_1 has upward pointing normal, and the flux integral over S_1 is

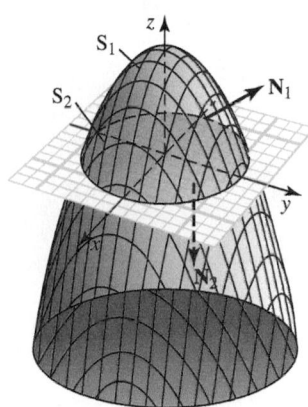

Figure 13.41   The surface S bounded by the region R

$$\iint\limits_{S_1} \mathbf{H} \cdot \mathbf{N}_1\, dS = \iint\limits_{S_1} -K\nabla T \cdot \nabla G\, dS$$

$$= \iint\limits_{D} -K\langle 2, 1, -6z \rangle \cdot \langle 2x, 2y, 1 \rangle\, dA$$

$$= \iint\limits_{D} -K[4x + 2y - 6(9 - x^2 - y^2)]\, dA \qquad \textit{Since } z = 9 - x^2 - y^2 \textit{ on } S$$

$$= -K \int_0^{2\pi} \int_0^3 [4r\cos\theta + 2r\sin\theta - 6(9 - r^2)]\, r\, dr\, d\theta$$

$$= -K \int_0^{2\pi} \left[36\cos\theta + 18\sin\theta - \frac{243}{2} \right] d\theta$$

$$= 243\pi K$$

For S_2: The outer normal $\mathbf{N}_2 = -\mathbf{k} = \langle 0, 0, -1 \rangle$ points downward and the flux integral is

$$\iint\limits_{S_2} \mathbf{H} \cdot \mathbf{N}_2\, dS = \iint\limits_{D} -K\langle 2, 1, -6z \rangle \cdot \langle 0, 0, -1 \rangle\, dA$$

$$= -K \iint\limits_{D} 6z\, dA$$

$$= -K \iint\limits_{D} 0\, dA \qquad \textit{Since } z = 0 \textit{ on } S_2$$

$$= 0$$

Thus, the total heat flow is

$$\iint\limits_{S} \mathbf{H} \cdot \mathbf{N} \, dS = \iint\limits_{S_1} \mathbf{H} \cdot \mathbf{N}_1 \, dS + \iint\limits_{S_2} \mathbf{H} \cdot \mathbf{N}_2 \, dS$$
$$= 243\pi K + 0$$
$$= 243\pi K$$

Integrals over Parametrically Defined Surfaces

In Section 12.4, we showed that if a surface S is defined parametrically by the vector function

$$\mathbf{R}(u, v) = x(u, v)\mathbf{i} + y(u, v)\mathbf{j} + z(u, v)\mathbf{k}$$

over a region D in the uv-plane, the surface area of S is given by the integral

$$\iint\limits_{D} \|\mathbf{R}_u \times \mathbf{R}_v\| \, du \, dv$$

Similarly, if f is continuous on D, the surface integral of f over D is given by

$$\iint\limits_{S} f(x, y, z) \, dS = \iint\limits_{D} f(\mathbf{R}) \|\mathbf{R}_u \times \mathbf{R}_v\| \, du \, dv$$

Example 5　Surface integral for a surface defined parametrically

Evaluate $\iint\limits_{S} (x + y + z) \, dS$ where S is the surface defined parametrically by

$$\mathbf{R}(u, v) = (2u + v)\mathbf{i} + (u - 2v)\mathbf{j} + (u + 3v)\mathbf{k}$$

for $0 \leq u \leq 1$, $0 \leq v \leq 2$.

Solution

$$\iint\limits_{S} (x + y + z) \, dS = \iint\limits_{D} f(\mathbf{R}) \|\mathbf{R}_u \times \mathbf{R}_v\| \, du \, dv$$

First, we need to find the component parts for the surface integral on the right. Using $\mathbf{R} = x\mathbf{i} + y\mathbf{j} + z\mathbf{k}$, we see that $x = 2u + v$, $y = u - 2v$, and $z = u + 3v$, and since $f(x, y, z) = x + y + z$, we have

$$f(\mathbf{R}) = f(2u + v, u - 2v, u + 3v)$$
$$= (2u + v) + (u - 2v) + (u + 3v)$$
$$= 4u + 2v$$

Next, we find that $\mathbf{R}_u = 2\mathbf{i} + \mathbf{j} + \mathbf{k}$ and $\mathbf{R}_v = \mathbf{i} - 2\mathbf{j} + 3\mathbf{k}$ so

$$\mathbf{R}_u \times \mathbf{R}_v = \begin{vmatrix} \mathbf{i} & \mathbf{j} & \mathbf{k} \\ 2 & 1 & 1 \\ 1 & -2 & 3 \end{vmatrix} = (3 + 2)\mathbf{i} - (6 - 1)\mathbf{j} + (-4 - 1)\mathbf{k} = 5\mathbf{i} - 5\mathbf{j} - 5\mathbf{k}$$

Thus, $\|\mathbf{R}_u \times \mathbf{R}_v\| = \sqrt{5^2 + (-5)^2 + (-5)^2} = 5\sqrt{3}$. We now substitute these values into the surface integral formula:

$$
\begin{aligned}
\iint\limits_{S} (x + y + z) = \iint\limits_{D} f(\mathbf{R}) \, \|\mathbf{R}_u \times \mathbf{R}_v\| \, du \, dv \\
= \int_0^2 \int_0^1 (4u + 2v)(5\sqrt{3}) \, du \, dv \\
= 5\sqrt{3} \int_0^2 \left[2u^2 + 2uv\right]_0^1 dv \\
= 5\sqrt{3} \int_0^2 (2 + 2v) \, dv \\
= 5\sqrt{3} \left[2v + v^2\right]_0^2 \\
= 5\sqrt{3}(8) \\
= 40\sqrt{3}
\end{aligned}
$$

For a flux integral over a surface S parameterized by $\mathbf{R}(u, v)$ that projects onto a region D in the uv-plane, we have

$$
\begin{aligned}
\iint\limits_{S} \mathbf{F} \cdot \mathbf{N} \, dS = \iint\limits_{D} \mathbf{F} \cdot \mathbf{N} \, \|\mathbf{R}_u \times \mathbf{R}_v\| \, dA \\
= \iint\limits_{D} \mathbf{F} \cdot \left(\frac{\mathbf{R}_u \times \mathbf{R}_v}{\|\mathbf{R}_u \times \mathbf{R}_v\|}\right) \|\mathbf{R}_u \times \mathbf{R}_v\| \, dA \\
= \iint\limits_{D} \mathbf{F} \cdot (\mathbf{R}_u \times \mathbf{R}_v) \, dA
\end{aligned}
$$

The use of this formula is demonstrated in the next example.

Example 6   Computing flux through a parameterized surface

Find the flux of the vector field $\mathbf{F} = z\mathbf{i} + x\mathbf{j} + (y + z)\mathbf{k}$ through the parameterized surface

$$
\mathbf{R}(u, v) = (uv)\mathbf{i} + (u - v)\mathbf{j} + (2u + v)\mathbf{k}
$$

over the triangular region D in the uv-plane that is bounded by $u = 0$, $v = 0$, and $u + v = 1$.

Solution   First, note from $\mathbf{R}(u, v)$ that $x = uv$, $y = u - v$, and $z = 2u + v$, so we have

$$
\begin{aligned}
\mathbf{F}(u, v) = (2u + v)\mathbf{i} + (uv)\mathbf{j} + [(u - v) + (2u + v)]\mathbf{k} \\
= (2u + v)\mathbf{i} + (uv)\mathbf{j} + (3u)\mathbf{k}
\end{aligned}
$$

Moreover, since $\mathbf{R}_u = v\mathbf{i} + \mathbf{j} + 2\mathbf{k}$ and $\mathbf{R}_v = u\mathbf{i} - \mathbf{j} + \mathbf{k}$, it follows that

$$
\begin{aligned}
\mathbf{R}_u \times \mathbf{R}_v = \begin{vmatrix} \mathbf{i} & \mathbf{j} & \mathbf{k} \\ v & 1 & 2 \\ u & -1 & 1 \end{vmatrix} \\
= 3\mathbf{i} - (v - 2u)\mathbf{j} + (-v - u)\mathbf{k}
\end{aligned}
$$

Thus, the flux is given by

$$\iint_S \mathbf{F} \cdot \mathbf{N} \, dS = \iint_D \mathbf{F} \cdot (\mathbf{R}_u \times \mathbf{R}_v) \, dA$$

$$= \int_0^1 \int_0^{1-u} \langle 2u + v, uv, 3u \rangle \cdot \langle 3, 2u - v, -(u + v) \rangle \, dv \, du$$

$$= \int_0^1 \int_0^{1-u} (2u^2 v - 3u^2 - uv^2 - 3uv + 6u + 3v) \, dv \, du$$

$$= \int_0^1 \frac{1}{6}(u - 1)(8u^3 - u^2 - 16u - 9) \, du$$

$$= \frac{137}{120}$$

PROBLEM SET 13.5

Level 1

Problems 1-6, evaluate $\int\int_S xy \, dS$.

1. $S: z = 2 - y, \ 0 \le x \le 2, \ 0 \le y \le 2$
2. $S: z = 4 - x - y, \ 0 \le x \le 4, \ 0 \le y \le 4$
3. $S: z = x + 3, \ 0 \le x \le 5, \ 0 \le y \le 5$
4. $S: z = 5 - x - y, \ 0 \le x \le 2, \ 0 \le y \le 2$
5. $S: z = 5, \ x^2 + y^2 \le 1$
6. $S: z = 10, \ \dfrac{x^2}{4} + \dfrac{y^2}{1} \le 1$

In Problems 7-10, evaluate $\int\int_S (x^2 + y^2) \, dS$.

7. $S: z = 4 - x - 2y, \ 0 \le x \le 4, \ 0 \le y \le 2$
8. $S: z = 4 - x, \ 0 \le x \le 2, \ 0 \le y \le 2$
9. $S: z = 4, \ x^2 + y^2 \le 1$
10. $S: z = xy, \ x^2 + y^2 \le 4, \ x \ge 0, \ y \ge 0$
11. Evaluate $\int\int_S 7 \, dS$, where the domain S is the disk $y^2 + z^2 \le 1$ in the plane $x = 4$.
12. Evaluate $\int\int_S 7 \, dS$, where S represents the planar domain $x^2 + z^2 \le 9, \ x \ge 0, \ z \ge 0$

Let S be the hemisphere $x^2 + y^2 + z^2 = 4$, with $z \ge 0$, in Problems 13-18. Evaluate each surface integral.

13. $\displaystyle\iint_S z \, dS$
14. $\displaystyle\iint_S z^2 \, dS$
15. $\displaystyle\iint_S (x - 2y) \, dS$
16. $\displaystyle\iint_S (5 - 2x) \, dS$
17. $\displaystyle\iint_S (x^2 + y^2)z \, dS$
18. $\displaystyle\iint_S (x^2 + y^2) \, dS$

In Problems 19-24, suppose S is the portion of the paraboloid $z = x^2 + y^2$ for which $z \le 4$. Evaluate the given surface integral.

19. $\displaystyle\iint_S z \, dS$
20. $\displaystyle\iint_S (4 - z) \, dS$
21. $\displaystyle\iint_S (4z + 1) \, dS$
22. $\displaystyle\iint_S 7 \, dS$
23. $\displaystyle\iint_S \sqrt{1 + 4z} \, dS$
24. $\displaystyle\iint_S \dfrac{dS}{\sqrt{1 + 4z}}$

25. Evaluate $\iint_S (x^2 + y^2) \, dS$, where S is the surface of the hemisphere defined in Figure 13.42.

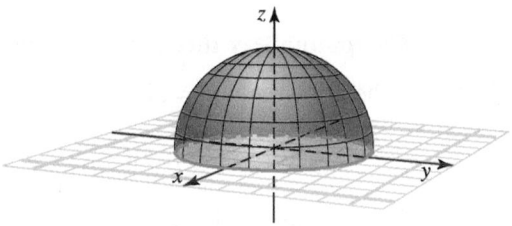

Figure 13.42   Surface $S: z = \sqrt{1 - x^2 - y^2}$

26. Evaluate $\iint_S dS$, where S is the surface of the hemisphere defined in Figure 13.42.
27. Evaluate $\iint_S dS$, where S is the portion of the plane defined in Figure 13.43.

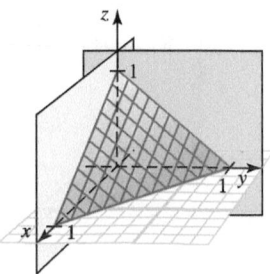

Figure 13.43   Surface $S: x + y + z = 1$ with $x \ge 0, y \ge 0,$ $z \ge 0$

28. Evaluate $\iint\limits_S 2x \, dS$, where S is the portion of the plane defined in Figure 13.43.

29. Evaluate $\iint\limits_S 2 \, dS$ where S is the portion of the cone defined in Figure 13.44.

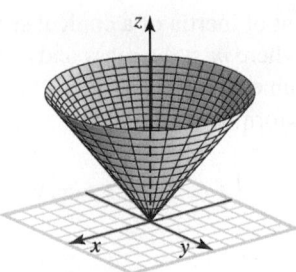

Figure 13.44 Surface S: $z^2 = x^2 + y^2$ with $0 \le z \le 4$

30. Evaluate $\iint\limits_S 7 \, dS$ where S is the portion of the surface $z = xy + 10$ that is contained in the cylinder $x^2 + y^2 = 1$.

31. Evaluate $\iint\limits_S (x^2 + y^2 + z^2) \, dS$, where S is defined in Figure 13.45.

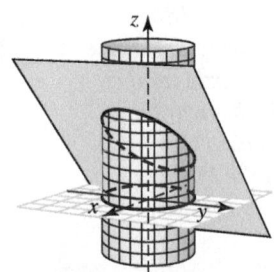

Figure 13.45 Surface S is the portion of the plane $z = x + 1$ that lies inside the cylinder $x^2 + y^2 = 1$

32. Prove that the area of the parameterized helicoid $r(u, v) = (u \cos v, u \sin v, v)$ defined over the rectangle $0 \le u \le 1$, $0 \le v \le \pi/2$ is greater than $\pi/2$.

Level 2

Evaluate $\iint\limits_S \mathbf{F} \cdot \mathbf{N} \, dS$ for the vector fields $\mathbf{F}$ and surfaces S given in Problems 33-40. Assume $\mathbf{N}$ is the outward directed normal field.

33. $\mathbf{F} = x\mathbf{i} + 2y\mathbf{j} - 3z\mathbf{k}$, and S is that part of the plane $15x - 12y + 3z = 6$ that lies above the unit square $0 \le x \le 1, 0 \le y \le 1$.

34. $\mathbf{F} = x\mathbf{i} + 2y\mathbf{j} + z\mathbf{k}$, and S is the triangular region bounded by the intersection of the plane $x + 2y + z = 1$ with the positive coordinate planes.

35. $\mathbf{F} = x\mathbf{i} + y\mathbf{j}$, and S is the hemisphere $z = \sqrt{1 - x^2 - y^2}$.

36. $\mathbf{F} = x\mathbf{i} + y\mathbf{j} + 2z\mathbf{k}$, and S is the surface of the cube bounded by the planes $x = 0$, $x = 1$, $y = 0$, $y = 1$, $z = 0$, and $z = 1$.

37. $\mathbf{F} = x^2\mathbf{i} + y^2\mathbf{j} + z^2\mathbf{k}$, and S is the portion of the plane $z = y + 1$ that lies inside the cylinder $x^2 + y^2 = 1$.

38. $\mathbf{F} = 2x\mathbf{i} - 3y\mathbf{j}$, and S is the part of the hemisphere given by $x^2 + y^2 + z^2 = 5$, for $z \ge 1$.

39. $\mathbf{F} = y\mathbf{i} + z^2\mathbf{j} - 2z\mathbf{k}$ and S is the part of the surface cut from the cylinder $z = 3 - x^2$ by $y = 0, y = 1$, and $z = 0$.

40. $\mathbf{F} = x^2\mathbf{i} + y^2\mathbf{j} + z^2\mathbf{k}$ and S is the part of the cone $z = \sqrt{x^2 + y^2}$ below the plane $z = 2$.

41. Evaluate

$$\iint\limits_S u \, v \, dS$$

where S is the surface defined by $\mathbf{R}(u, v) = \cos v\mathbf{i} + \sin v\mathbf{j} - u\mathbf{k}, 0 \le u \le 1$, $0 \le v \le \pi/2$.

42. Evaluate

$$\iint\limits_S (3x - y + 2z) \, dS$$

where S is the surface determined by $\mathbf{R}(u, v) = u\mathbf{i} + u\mathbf{j} - v\mathbf{k}, 0 \le u \le 1, 1 \le v \le 2$.

43. Evaluate

$$\iint\limits_S (x - y^2 + z) \, dS$$

where S is the surface defined by $\mathbf{R}(u, v) = u^2\mathbf{i} + v\mathbf{j} + u\mathbf{k}, 0 \le u \le 1, 0 \le v \le 1$.

44. Evaluate

$$\iint\limits_S 7 \, dS$$

where S is the surface defined by $\mathbf{R}(u, v) = u \cos v\mathbf{i} + u \sin v\mathbf{j} + v\mathbf{k}, 0 \le u \le 1$, $0 \le v \le \pi/2$.

45. Evaluate

$$\iint\limits_S (x^2 + y - z) \, dS$$

where S is the surface defined by $\mathbf{R}(u, v) = u\mathbf{i} - u^2\mathbf{j} + v\mathbf{k}, 0 \le u \le 2, 0 \le v \le 1$.

46. Evaluate

$$\iint\limits_S (x + y + z) \, dS$$

where S is the surface of the cube $0 \le x \le 1$, $0 \le y \le 1, 0 \le z \le 1$.

47. Evaluate

$$\iint_S \mathbf{F} \cdot \mathbf{N} \, dS$$

where $\mathbf{F} = x^2\mathbf{i} + z\mathbf{k}$ and S is the parametric surface $x = \sin u \cos v$, $y = \sin u \sin v$, $z = \cos u$ for $0 \le u \le \pi, 0 \le v \le 2\pi$.

48. Evaluate

$$\iint_S \mathbf{F} \cdot \mathbf{N} \, dS$$

where $\mathbf{F} = x\mathbf{i} + y\mathbf{j} + z^4\mathbf{k}$ and S is the parametric surface $x = u \cos v$, $y = u \sin v, z = u$ for $0 \le u \le 2$, $0 \le v \le 2\pi$.

In Problems 49-54 find the mass of the homogeneous lamina that has the shape of the given surface S.

49. S is the surface $z = 4 - x - 2y$, with $z \ge 0$, $x \ge 0$, $y \ge 0$, $\rho = x$.

50. S is the surface $z = 10 - 2x - y$, with $z \ge 0$, $x \ge 0$, $y \ge 0$, $\rho = y$.

51. S is the surface $z = x^2 + y^2$, with $z \le 1$; $\rho = z$. *Hint*: Use cylindrical coordinates.

52. S is the surface $z = 1 - x^2 - y^2$, with $z \ge 0$; $\rho = x^2 + y^2 + z^2$. *Hint*: Use cylindrical coordinates.

53. S is the surface $x^2 + y^2 + z^2 = 5$, with $z \ge 1$; $\rho = \theta^2$.

54. S is the triangular surface with vertices $(1, 0, 0)$, $(0, 1, 0)$, and $(0, 0, 1)$; $\rho = x + y$.

55. A fluid with constant density ρ flows with velocity $\mathbf{V} = xy\mathbf{i} + yz\mathbf{j} + xz\mathbf{k}$. Find the outward rate of fluid flow through the paraboloid $z = 9 - x^2 - y^2$, where z is positive.

56. The temperature at each point (x, y, z) in a region D is $T = 3x^2 + 3z^2$. If the heat conductivity is $K = 5.8$, find the rate of heat flow outward across the cylinder $x^2 + y^2 = 4$ for $0 \le y \le 3$.

Level 3

57. a. A lamina has the shape of the portion of the sphere $x^2 + y^2 + z^2 = a^2$ that lies within the cone $z = \sqrt{x^2 + y^2}$. Determine the mass of the lamina if $\rho(x, y, z) = x^2y^2z$.

b. Let S be the spherical shell centered at the origin with radius a, and let C be the right circular cone whose vertex is at the origin and whose axis of symmetry coincides with the z-axis. Suppose the vertex angle of the cone is ϕ_0, with $0 \le \phi_0 < \pi/2$. Determine the mass of that portion of the sphere that is enclosed in the intersection of S and C. Assume $\rho(x, y, z) = 1$.

58. Recall the formula

$$I_z = \iint_S (x^2 + y^2) \, \rho(x, y, z) \, dS$$

for the moment of inertia about the z-axis. Show that the moment of inertia of a conical shell about its axis is $\frac{1}{2}ma^2$, where m is the mass and a is the radius of the cone. Assume $\rho(x, y, z) = 1$.

59. Recall the formula

$$I_z = \iint_S (x^2 + y^2) \, \rho(x, y, z) \, dS$$

for the moment of inertia about the z-axis. Show that the moment of inertia of a spherical shell of uniform density about its diameter is $\frac{2}{3}ma^2$, where m is the mass and a is the radius. Assume $\rho(x, y, z) = 1$.

60. 𝔥istorical 𝔔uest

August Möbius studied under Karl Gauss (1777-1855; see 𝔥istorical 𝔔uest Problem 67 Chapter 5 Supplementary), as well as Gauss' own teacher Johann Pfaff (1765-1825). Möbius was a professor at the University of Leipzig, and is best known for his work in topology, especially for his conception of the Möbius strip (a two-dimensional surface with only one side).

Karl Smith library

**August Möbius
(1790-1868)**

© 2013 Yesaulov Vadym. Used under license from Shutterstock, Inc.

The following parametric surface is called a **Möbius strip**:

$$x = \cos v + u \cos \frac{v}{2} \cos v$$

$$y = \sin v + u \cos \frac{v}{2} \sin v$$

$$z = u \sin \frac{v}{2}$$

where $-\frac{1}{2} \le u \le \frac{1}{2}, 0 \le v \le 2\pi$. Construct a three-dimensional model of a Möbius strip and show that a Möbius strip is not orientable. If you have access to a CAS program, sketch the graph of this surface.

13.6 STOKES' THEOREM AND APPLICATIONS

IN THIS SECTION: *Stokes' theorem, theoretical applications of Stokes' theorem, physical interpretation of Stokes' theorem*

In Section 13.4, we observed that Green's theorem can be written

$$\oint_C \mathbf{F} \cdot d\mathbf{R} = \iint_A (\text{curl } \mathbf{F} \cdot \mathbf{k})\, dA$$

where A is the plane region bounded by the closed curve C. Stokes' theorem is a generalization of this result to surfaces in space and their boundaries.

Stokes' Theorem

Before stating Stokes' theorem, we need to explain what is meant by a *compatible orientation*. We say that the orientation of a closed path C on the surface S is **compatible with the orientation on S** if the positive direction on C is *counterclockwise* in relation to the outward normal vector $\mathbf{N}$ of the surface (see Figure 13.46). That is, if a person walks around C in a positive (counterclockwise) direction with her head pointing in the direction of $\mathbf{N}$, then the surface will always be on her left.

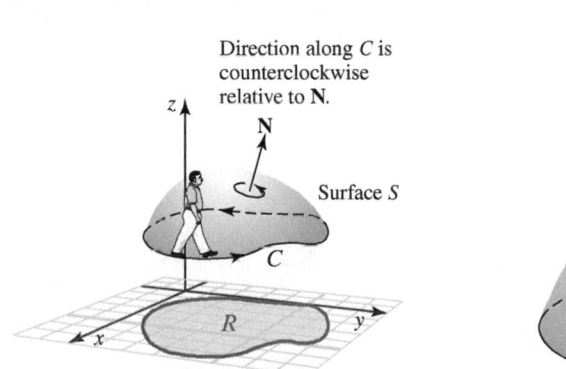
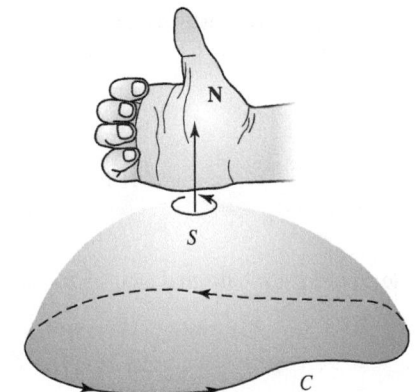

a. Positive direction (counterclockwise)   **b.** Fingers point in the positive direction

Figure 13.46   Compatible orientation

Theorem 13.8   Stokes' theorem

Let S be an oriented surface with unit normal vector field $\mathbf{N}$, and assume that S is bounded by a piecewise smooth Jordan curve C whose orientation is compatible with that of S. If $\mathbf{F}$ is a vector field that is continuously differentiable on S, then

$$\oint_C \mathbf{F} \cdot d\mathbf{R} = \iint_S (\text{curl } \mathbf{F} \cdot \mathbf{N})\, dS$$

Proof: A proof assuming $\mathbf{F}$, S, and C are "well behaved" is given in Appendix B.   ♦

Example 1   Using Stokes' theorem to evaluate a line integral

Evaluate $\oint_C \left(\frac{1}{2}y^2 dx + z\, dy + x\, dz\right)$ where C is the curve of intersection of the plane $x + z = 1$ and the ellipsoid $x^2 + 2y^2 + z^2 = 1$, oriented counterclockwise as viewed from above (see Figure 13.47).

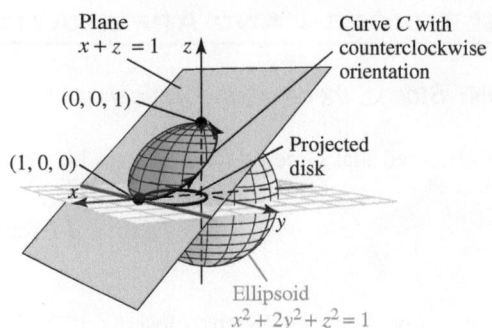

Figure 13.47 The curve C

Solution If we set $\mathbf{F} = \frac{1}{2}y^2\mathbf{i} + z\mathbf{j} + x\mathbf{k}$, the given line integral can be expressed as $\oint_C \mathbf{F} \cdot d\mathbf{R}$

According to Stokes' theorem, we have

$$\oint_C \mathbf{F} \cdot d\mathbf{R} = \iint_S (\text{curl } \mathbf{F} \cdot \mathbf{N})\, dS$$

There are many surfaces bounded by the curve C. We choose the surface S that is part of the plane $x + z = 1$. We find the required parts of the surface integral; namely, curl $\mathbf{F}$, $\mathbf{N}$, and dS.

$$\text{curl } \mathbf{F} = \begin{vmatrix} \mathbf{i} & \mathbf{j} & \mathbf{k} \\ \frac{\partial}{\partial x} & \frac{\partial}{\partial y} & \frac{\partial}{\partial z} \\ \frac{1}{2}y^2 & z & x \end{vmatrix} = -\mathbf{i} - \mathbf{j} - y\mathbf{k}$$

The upward unit normal vector to the plane $x + z = 1$ is $\mathbf{N} = \frac{1}{\sqrt{2}}(\mathbf{i} + \mathbf{k})$, so that

$$\text{curl } \mathbf{F} \cdot \mathbf{N} = \langle -1. -1, -y \rangle \cdot \left\langle \frac{1}{\sqrt{2}}, 0, \frac{1}{\sqrt{2}} \right\rangle$$

$$= -\frac{1}{\sqrt{2}} - \frac{y}{\sqrt{2}}$$

$$= \frac{-1}{\sqrt{2}}(1 + y)$$

and since $z = 1 - x$ on S, we have $z_x = -1, z_y = 0$, and

$$dS = \sqrt{z_x^2 + z_y^2 + 1}\, dA = \sqrt{(-1)^2 + (0)^2 + 1}\, dA = \sqrt{2}\, dA$$

Finally, to describe C we substitute $z = 1 - x$ into the equation for the ellipsoid:

$$x^2 + 2y^2 + z^2 = 1$$
$$x^2 + 2y^2 + (1 - x)^2 = 1$$
$$x^2 - x + y^2 = 0$$
$$\left(x - \frac{1}{2}\right)^2 + y^2 = \frac{1}{4}$$

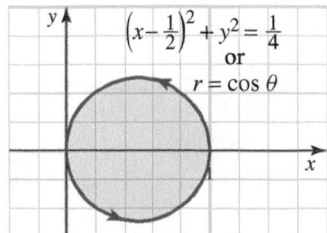

Figure 13.48 The projection D of the surface S on the xy-plane

Thus, the projection of C on the xy-plane is the disk D bounded by the circle $\left(x - \frac{1}{2}\right)^2 + y^2 = \frac{1}{4}$ shown in Figure 13.48. Note that this circle has the polar equation $r = \cos\theta$.

We now calculate the desired line integral using Stokes' theorem.

$$\oint_C \left(\frac{1}{2} y^2 \, dx + z \, dy + x \, dz \right) = \oint_C \mathbf{F} \cdot d\mathbf{R}$$

$$= \iint_S (\text{curl } \mathbf{F} \cdot \mathbf{N}) \, dS \qquad \textit{Stokes' theorem}$$

$$= \iint_D \frac{-1}{\sqrt{2}} (1 + y)(\sqrt{2} \, dA)$$

$$= -\int_0^\pi \int_0^{\cos\theta} (1 + r \sin\theta) \, r \, dr \, d\theta \qquad \textit{Change to polar coordinates}$$

$$= -\int_0^\pi \left[\frac{1}{2} \cos^2\theta + \frac{1}{3} \cos^3\theta \sin\theta \right] d\theta$$

$$= -\frac{\pi}{4}$$

Example 2   Verifying Stokes' theorem for a particular surface

Let S be the portion of the plane $x + y + z = 1$ that lies in the first octant, and let C be the boundary of S, traversed counterclockwise as viewed from above. Verify Stokes' theorem for the surface S and the vector field $\mathbf{F} = -\frac{3}{2} y^2 \mathbf{i} - 2xy\mathbf{j} + yz\mathbf{k}$

Solution   The surface S is the triangle with vertices $(1, 0, 0), (0, 1, 0), (0, 0, 1)$, whose boundary C is traversed in that order, as shown in Figure 13.49.

We will show that the line integral $\oint_C \mathbf{F} \cdot d\mathbf{R}$ and the surface integral $\iint_S (\text{curl } \mathbf{F} \cdot \mathbf{N}) dS$ have the same value.

I. *Evaluation of* $\oint_C \mathbf{F} \cdot d\mathbf{R}$

The three edges of the boundary curve C are expressed as

$$E_1: \ x + y = 1, z = 0$$
$$E_2: \ y + z = 1, x = 0$$
$$E_3: \ x + z = 1, y = 0$$

We traverse C in a counterclockwise direction (see Figure 13.49).
Edge E_1: Parameterize with $x = 1 - t, y = t, z = 0$, for $0 \le t \le 1$, so
$\mathbf{R}(t) = (1 - t)\mathbf{i} + t\mathbf{j}$ and $d\mathbf{R} = -dt\mathbf{i} + dt\mathbf{j}$. Finally, in terms of the parameter t, we have $\mathbf{F}(t) = -\frac{3}{2} t^2 \mathbf{i} - 2t(1 - t)\mathbf{j}$.

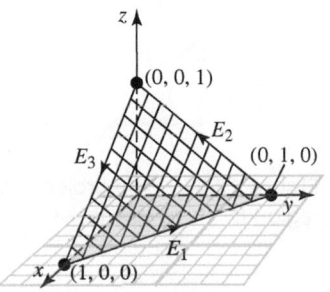

Figure 13.49 The curve C bounds the triangular surface S

$$\int_{E_1} \mathbf{F} \cdot d\mathbf{R} = \int_0^1 \left(\frac{3}{2} t^2 - 2t + 2t^2 \right) dt = \int_0^1 \left(\frac{7}{2} t^2 - 2t \right) dt = \left(\frac{7}{6} t^3 - t^2 \right) \Big|_0^1 = \frac{1}{6}$$

Edge E_2: Parameterize with $x = 0, y = 1 - s, z = s$, for $0 \le s \le 1$, so
$\mathbf{R}(s) = (1 - s)\mathbf{j} + s\mathbf{k}$ and $d\mathbf{R} = -ds\mathbf{j} + ds\mathbf{k}$. In terms of the parameter s, we have $\mathbf{F}(s) = -\frac{3}{2}(1 - s)^2 \mathbf{i} + (1 - s)s\mathbf{k}$.

$$\int_{E_2} \mathbf{F} \cdot d\mathbf{R} = \int_0^1 (1 - s)s \, ds = \left(\frac{s^2}{2} - \frac{s^3}{3} \right) \Big|_0^1 = \frac{1}{6}$$

Edge E_3: Parameterize with $x = r$, $y = 0$, $z = 1 - r$, for $0 \leq r \leq 1$, so $\mathbf{R}(r) = r\mathbf{i} + (1 - r)\mathbf{k}$ and $d\mathbf{R} = dr\mathbf{i} - dr\mathbf{k}$. In terms of the parameter r, we have $\mathbf{F}(r) = \mathbf{0}$.

$$\int_{E_3} \mathbf{F} \cdot d\mathbf{R} = 0$$

Combining these results, we find

$$\int_C \mathbf{F} \cdot d\mathbf{R} = \int_{E_1} \mathbf{F} \cdot d\mathbf{R} + \int_{E_2} \mathbf{F} \cdot d\mathbf{R} + \int_{E_3} \mathbf{F} \cdot d\mathbf{R} = \frac{1}{6} + \frac{1}{6} + 0 = \frac{1}{3}$$

II. *Evaluation of $\iint_S (\text{curl } \mathbf{F} \cdot \mathbf{N}) \, dS$*

$$\text{curl } \mathbf{F} = \begin{vmatrix} \mathbf{i} & \mathbf{j} & \mathbf{k} \\ \frac{\partial}{\partial x} & \frac{\partial}{\partial y} & \frac{\partial}{\partial z} \\ -\frac{3}{2}y^2 & -2xy & yz \end{vmatrix} = z\mathbf{i} + y\mathbf{k}$$

The triangular region S on the surface of the plane $x + y + z = 1$ has outward unit normal vector $\mathbf{N} = \frac{1}{\sqrt{3}}(\mathbf{i} + \mathbf{j} + \mathbf{k})$. Because $z = 1 - x - y$, we find

$$\text{curl } \mathbf{F} \cdot \mathbf{N} = (z\mathbf{i} + y\mathbf{k}) \cdot \frac{1}{\sqrt{3}}(\mathbf{i} + \mathbf{j} + \mathbf{k})$$
$$= \frac{1}{\sqrt{3}}(1 - x - y + y)$$
$$= \frac{1}{\sqrt{3}}(1 - x)$$

and since $z_x = -1$, $z_y = -1$, it follows that

$$dS = \sqrt{z_x^2 + z_y^2 + 1} \, dA = \sqrt{(-1)^2 + (-1)^2 + 1} dA = \sqrt{3} \, dA$$

Finally, the surface S projects onto the triangular region D in the xy-plane that is bounded by the lines $x = 0$, $y = 0$, and $x + y = 1$ (see Figure 13.50).

Thus, we have

$$\iint_S (\text{curl } \mathbf{F} \cdot \mathbf{N}) \, dS = \iint_D \frac{1}{\sqrt{3}}(1 - x)\sqrt{3} \, dA$$
$$= \int_0^1 \int_0^{1-x} (1 - x) \, dy \, dx$$
$$= \int_0^1 (1 - x)[(1 - x) - 0] \, dx$$
$$= -\frac{(1 - x)^3}{3} \Big|_0^1$$
$$= \frac{1}{3}$$

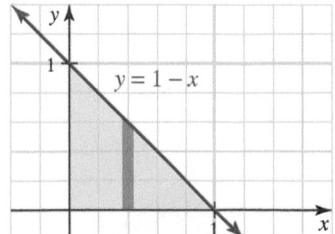

Figure 13.50 The projection region in the *xy*-plane

In conclusion, we see that for this example,

$$\oint_C \mathbf{F} \cdot d\mathbf{R} = \frac{1}{3} = \iint_S (\text{curl } \mathbf{F} \cdot \mathbf{N}) \, dS$$

as claimed by Stokes' theorem.

Sometimes it is possible to exchange a particularly difficult surface integration over one surface for a less difficult integration over another surface with the same boundary curve. Suppose two surfaces S_1 and S_2 are bounded by the same curve C and induce the same orientation on C. Stokes' theorem tells us that

$$\iint\limits_{S_1} (\text{curl } \mathbf{F} \cdot \mathbf{N}_1) \, dS_1 = \oint_C \mathbf{F} \cdot d\mathbf{R} = \iint\limits_{S_2} (\text{curl } \mathbf{F} \cdot \mathbf{N}_2) \, dS_2$$

for any function $\mathbf{F}$ whose components have continuous partial derivatives on both S_1 and S_2.

Example 3 Using Stokes' theorem to evaluate a surface integral

Evaluate $\iint\limits_{S} (\text{curl } \mathbf{F} \cdot \mathbf{N}) \, dS$, where $\mathbf{F} = x\mathbf{i} + y^2\mathbf{j} + ze^{xy}\mathbf{k}$ and S is that part of the surface $z = 1 - x^2 - 2y^2$ with $z \geq 0$.

Solution By setting $z = 0$ in the equation of the surface, we find that the boundary curve C for S is the ellipse $x^2 + 2y^2 = 1$ and the upward unit normal vector $\mathbf{N}$ on S induces a counterclockwise orientation on C, as shown in Figure 13.51.

Let S^* be the elliptical disk defined by $x^2 + 2y^2 \leq 1, z = 0$. We see that S and S^* have the same boundary and the same orientation. Since the upward unit normal to S^* is $\mathbf{k}$, we have

$$\iint\limits_{S} (\text{curl } \mathbf{F} \cdot \mathbf{N}) \, dS = \iint\limits_{S^*} (\text{curl } \mathbf{F} \cdot \mathbf{k}) \, dS^*$$

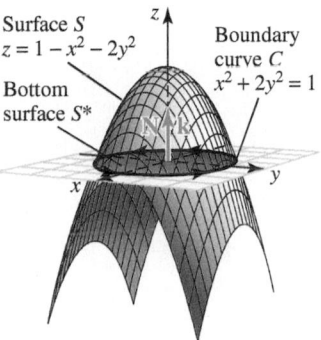

Surface S
$z = 1 - x^2 - 2y^2$

Bottom
surface S^*

Boundary
curve C
$x^2 + 2y^2 = 1$

Figure 13.51 Surface and boundary curve

and we use this second integral to calculate the required integral. We obtain

$$\text{curl } \mathbf{F} = \begin{vmatrix} \mathbf{i} & \mathbf{j} & \mathbf{k} \\ \frac{\partial}{\partial x} & \frac{\partial}{\partial y} & \frac{\partial}{\partial z} \\ x & y^2 & ze^{xy} \end{vmatrix} = zxe^{xy}\mathbf{i} - zye^{xy}\mathbf{j}$$

Since $z = 0$ on S^*, we have curl $\mathbf{F} \cdot \mathbf{k} = 0$ on S^* and

$$\iint\limits_{S} (\text{curl } \mathbf{F} \cdot \mathbf{N}) \, dS = \iint\limits_{S^*} (\text{curl } \mathbf{F} \cdot \mathbf{k}) \, dS^* = \iint\limits_{S^*} 0 \, dS^* = 0$$

Theoretical Applications of Stokes' Theorem

In physics and other applied areas, Stokes' theorem is often used as a device for establishing general properties. For instance, we now prove the curl criterion for a conservative vector field in $\mathbb{R}^3$ stated without proof in Section 13.3 (Theorem 13.4), and whose proof was promised at this time.

Proof of Theorem 13.4

If $\mathbf{F}$ is conservative, let f be a scalar potential function, so that $\nabla f = \mathbf{F}$. Then curl $\mathbf{F} = \nabla \times \mathbf{F} = \nabla \times (\nabla f) = \mathbf{0}$. (See Problem 52, Section 13.1 for a proof of this property.)

Conversely, if curl $\mathbf{F} = \mathbf{0}$, we will show that $\mathbf{F}$ is conservative by proving that it is independent of path. Let P_1 and P_2 be any two points in the simply connected region D and let C_1 and C_2 be two nonintersecting paths in D from P_1 to P_2. Let C be the Jordan curve from P_1 back to P_1 formed by C_1 followed by $-C_2$. Since D is simply connected (no holes), there is a piecewise smooth surface S whose boundary is C. Then, by Stokes' theorem

$$\oint_C \mathbf{F} \cdot d\mathbf{R} = \int_{C_1} \mathbf{F} \cdot d\mathbf{R} - \int_{C_2} \mathbf{F} \cdot d\mathbf{R}$$
$$= \iint_S \operatorname{curl} \mathbf{F} \cdot \mathbf{N}\, dS$$
$$= 0$$

Also, we see

$$\int_{C_1} \mathbf{F} \cdot d\mathbf{R} = \int_{C_2} \mathbf{F} \cdot d\mathbf{R}$$

It follows that the line integral $\oint_C \mathbf{F} \cdot d\mathbf{R}$ is independent of path, so $\mathbf{F}$ must be conservative.

$\square$

Maxwell's equations describing the behavior of electromagnetic phenomena were discussed briefly at the end of Section 13.1. The following example shows how Stokes' theorem can be used to establish one of Maxwell's equations, the *current density equation*.

Example 4   Maxwell's current density equation

In physics, it is shown that if I is the current crossing any surface S bounded by the closed curve C, then

$$\oint_C \mathbf{H} \cdot d\mathbf{R} = I \qquad \text{and} \qquad \iint_S \mathbf{J} \cdot \mathbf{N}\, dS = I$$

where $\mathbf{H}$ is the magnetic intensity, and $\mathbf{J}$ is the electric current density. Use this information to derive Maxwell's current density equation $\operatorname{curl} \mathbf{H} = \mathbf{J}$.

Solution   Equating the two equations of current, we obtain

$$\oint_C \mathbf{H} \cdot d\mathbf{R} = I = \iint_S \mathbf{J} \cdot \mathbf{N}\, dS$$

By Stokes' theorem, $\oint_C \mathbf{H} \cdot d\mathbf{R} = \iint_S (\operatorname{curl} \mathbf{H} \cdot \mathbf{N})\, dS$. Equating the two surface integrals that equal $\oint_C \mathbf{H} \cdot d\mathbf{R}$, we have

$$\iint_S \mathbf{J} \cdot \mathbf{N}\, dS = \iint_S (\operatorname{curl} \mathbf{H} \cdot \mathbf{N})\, dS$$

or, equivalently,

$$\iint_S (\mathbf{J} - \operatorname{curl} \mathbf{H}) \cdot \mathbf{N}\, dS = 0$$

Because this equation holds for *any* surface S bounded by C, it can be shown that

$$\mathbf{J} - \operatorname{curl} \mathbf{H} = \mathbf{0}$$
$$\mathbf{J} = \operatorname{curl} \mathbf{H}$$

Physical Interpretation of Stokes' Theorem

If $\mathbf{V}$ is the velocity field of a fluid flow, then $\operatorname{curl} \mathbf{V}$ measures the tendency of the fluid to rotate, or swirl (see Figure 13.52).

If the fluid flows across the surface S, the rotational tendency usually will vary from point to point on the surface, and the surface integral $\iint_S (\operatorname{curl} \mathbf{V} \cdot \mathbf{N})\, dS$ provides a measure of the *cumulative* rotational tendency over the entire surface.

Stokes' theorem tells us that this cumulative measure of rotational tendency equals the line integral $\oint_C \mathbf{V} \cdot d\mathbf{R}$. To interpret this line integral, recall that it can be written $\oint_C \mathbf{V} \cdot \mathbf{T}\, ds$ in terms of the arc length parameter s and the unit tangent $\mathbf{T}$ to the curve. Since the line integral sums the *tangential component* of the velocity field $\mathbf{V}$, it measures the rate of flow of fluid mass around C and for this reason is called the **circulation of F** around C. If curl $\mathbf{V} = \mathbf{0}$, the circulation is zero and $\mathbf{V}$ is said to be **irrotational**. To summarize:

Figure 13.52 The tendency of a fluid to swirl across the surface S is measured by curl $\mathbf{V} \cdot \mathbf{N}$

© Kendall Hunt Publishing Company

$$\underbrace{\iint_S (\operatorname{curl} \mathbf{V} \cdot \mathbf{N})\, dS}_{\substack{\text{The cumulative tendency} \\ \text{of a fluid to swirl across} \\ \text{the surface } S}} = \underbrace{\oint_C \mathbf{V} \cdot \mathbf{T}\, ds}_{\substack{\text{The circulation of} \\ \text{a fluid around the} \\ \text{boundary curve } C}}$$

PROBLEM SET 13.6

Level 1

1. **What does this say?** State Green's theorem and Stokes' theorem. Compare the hypotheses for the two theorems.
2. **What does this say?** Prove that Green's theorem represents indeed a particular case of Stokes' theorem. What is the bounded surface S in case of Green's theorem? What does the normal vector field $\mathbf{N}$ to this surface represent in this case?

Verify Stokes' theorem for the vector functions and surfaces given in Problems 3-7.

3. $\mathbf{F} = z\mathbf{i} + 2x\mathbf{j} + 3y\mathbf{k}$; S is the upper hemisphere $z = \sqrt{9 - x^2 - y^2}$.
4. $\mathbf{F} = (y + z)\mathbf{i} + x\mathbf{j} + (z - x)\mathbf{k}$; S is the triangular region of the plane $x + 2y + z = 3$ in the first octant.
5. $\mathbf{F} = (x + 2z)\mathbf{i} + (y - x)\mathbf{j} + (z - y)\mathbf{k}$; S is the triangular region with vertices $(3, 0, 0)$, $(0, \frac{3}{2}, 0)$, $(0, 0, 3)$.
6. $\mathbf{F} = 2xy\mathbf{i} + z^2\mathbf{k}$; S is the portion of the paraboloid $y = x^2 + z^2$, with $y \leq 4$.
7. $\mathbf{F} = 2y\mathbf{i} - 6z\mathbf{j} + 3x\mathbf{k}$; S is the portion of the paraboloid $z = 4 - x^2 - y^2$ above the xy-plane.
8. **What does this say?** We showed that Green's theorem relates a double integral to a path integral, which is a very significant result. Can you make an analogy with Stokes' theorem? What type of integrals does Stokes' theorem involve, and what is the connection between them?
9. **What does this say?** Let S be an oriented surface with unit normal vector field $\mathbf{N}$, and assume that S is bounded by a piecewise smooth Jordan curve C whose orientation is compatible with that of S. Let $\mathbf{F}$ be a vector field that is continuously differentiable on S. What is the **physical** meaning of the integral:

$$\oint_C \mathbf{F} \cdot d\mathbf{R}?$$

10. **What does this say?** Let S be an oriented surface with unit normal vector field $\mathbf{N}$, and assume that S is bounded by a piecewise smooth Jordan curve C whose orientation is compatible with that of S. If $\mathbf{F}$ is a vector field that is continuously differentiable on S, then what is the **physical** meaning of the integral

$$\iint_S (\operatorname{curl} \mathbf{F} \cdot \mathbf{N})\, dS\ ?$$

11. **What does this say?** Stokes' theorem is often referred to as the Kelvin-Stokes theorem in physics. Do physicists and mathematicians refer to the same theorem? Perform a library search that justifies the name of the Kelvin-Stokes theorem.
12. **What does this say?** Two of the four Maxwell equations involve curls of 3-D vector fields and their differential and integral forms are related by the Kelvin–Stokes theorem. Perform a library search that identifies these two Maxwell equations, their names, and the formulas that express them.

Use Stokes' theorem to evaluate the line integrals given in Problems 13-26.

13. $\oint_C (y\, dx + x\, dy + z\, dz)$, where C is the circle described in Figure 13.53.

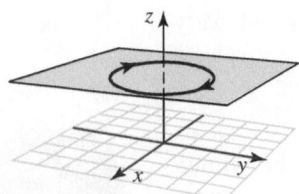

Figure 13.53 C is $x^2 + y^2 = 1$ in the plane $z = 1$ traversed clockwise when viewed from the origin

14. $\oint_C (x^3 y^2 \, dx + dy + z^2 \, dz)$, where C is the circle described in Figure 13.53.

15. $\oint_C (z \, dx + x \, dy + y \, dz)$, where C is the triangle described in Figure 13.54.

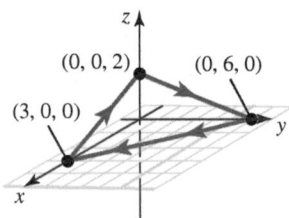

Figure 13.54 C is a triangle traversed in the order shown by the arrows

16. $\oint_C (y \, dx + z \, dy + x \, dz)$, where C is the triangle described in Figure 13.54.

17. $\oint_C [2xy^2 z \, dx + 2x^2 yz \, dy + (x^2 y^2 - 2z) \, dz]$, where C is the curve given by $x = \cos t$, $y = \sin t$, $z = \sin t$, $0 \le t \le 2\pi$, traversed in the direction of increasing t.

18. $\oint_C (y \, dx - 2x \, dy + z \, dz)$, where C is described in Figure 13.55.

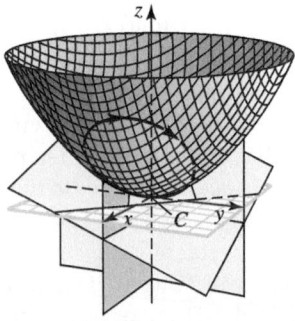

Figure 13.55 C is the intersection of the surface $z = x^2 + y^2$ and the plane $x + y + z = 1$ considered counterclockwise when viewed from the origin

19. $\oint_C (y \, dx + z \, dy + x \, dz)$, where C is the intersection of the plane $x + y = 2$ and the surface $x^2 + y^2 + z^2 = 2(x + y)$, traversed counterclockwise as viewed from the origin.

20. $\oint_C (y \, dx + z \, dy + y \, dz)$, where C is described in Figure 13.56.

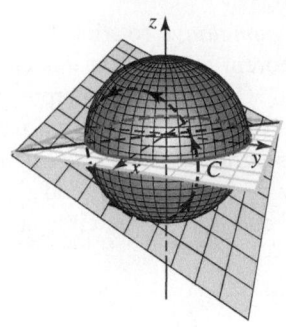

Figure 13.56 C is the intersection of the sphere $x^2 + y^2 + z^2 = 4$ and $x + y + z = 0$, traversed counterclockwise when viewed from above

21. $\oint_C (x \, dx + y \, dy + z \, dz)$, where C is the intersection of the plane $z = 4$ and the surface $x^2 + y^2 = z$, traversed counterclockwise as viewed from the origin.

22. $\oint_C [(z + \cos x) dx + (x + y^2) dy + (y + e^z) dz]$ where C is the intersection of the sphere $x^2 + y^2 + z^2 = 4$ and the cone $z = \sqrt{x^2 + y^2}$, traversed counterclockwise as viewed from above.

23. $\oint_C (3y \, dx + 2z \, dy - 5x \, dz)$, where C is the intersection of the xy-plane and the hemisphere $z = \sqrt{1 - x^2 - y^2}$, traversed counterclockwise as viewed from above.

24. $\oint_C (2z \, dx + 3x \, dy - 5x \, dz)$, where C is the intersection of the xy-plane and the hemisphere $z = \sqrt{1 - x^2 - y^2}$, traversed counterclockwise as viewed from above.

25. $\oint_C \mathbf{F} \cdot d\mathbf{R}$, where $\mathbf{F} = \langle y^2 + z^2, x^2 + y^2, x^2 + y^2 \rangle$ and C is the triangle $(1, 0, 0)$, $(0, 1, 0)$, $(0, 0, 1)$, traversed in that order.

26. $\oint \int_C \mathbf{F} \cdot d\mathbf{R}$, where $\mathbf{F} = \langle x - z, y - x, z - y \rangle$ and C is the boundary of the triangular region with vertices $(12, 0, 0)$, $(0, 3, 0)$, $(0, 0, 12)$ traversed counterclockwise as viewed from above.

Level 2

In Problems 27–34, evaluate the integral that appears on the right hand side of Stokes' theorem, namely

$$\iint_S (\text{curl } \mathbf{F} \cdot \mathbf{N}) \, dS$$

for the prescribed vector fields and surfaces. In each case, use the upward unit normal for S.

27. $\mathbf{F} = \langle 6x^2 e^{yz}, 2x^3 z e^{yz}, 2x^3 y e^{yz} \rangle$, and S is the part of the cone $z = \sqrt{x^2 + y^2}$ that lies above the inner loop of the limaçon $r = 1 + 2\cos\theta$.

28. $\mathbf{F} = x\mathbf{i} + y^2\mathbf{j} + xyz\,\mathbf{k}$, and S is the part of the paraboloid $z = 4 - x^2 - y^2$ with $z \ge 0$. Use the upward unit normal vector.

29. $\mathbf{F} = xy\mathbf{i} - z\mathbf{j}$, and S is the surface of the cube $0 \leq x \leq 1, 0 \leq y \leq 1, 0 \leq z \leq 1$ except for the face where $z = 0$.

30. $\mathbf{F} = y\mathbf{i} + z\mathbf{j} + x\mathbf{k}$, and S is the part of the plane $x + y + z = 1$ that lies in the first octant.

31. $\mathbf{F} = xy\mathbf{i} + x^2\mathbf{j} + z^2\mathbf{k}$, and S is the part of the plane $z = y$ that is inside the paraboloid $z = x^2 + y^2$.

32. $\mathbf{F} = xz\mathbf{i} + y^2\mathbf{j} + x^2\mathbf{k}$, and S is the part of the plane $x + y + z = 3$ inside the cylinder $9x^2 + y^2 = 9$.

33. $\mathbf{F} = 4y\mathbf{i} + z\mathbf{j} + 2y\mathbf{k}$, and S is the hemisphere $z = \sqrt{4 - x^2 - y^2}$.

34. $\mathbf{F} = \langle x\tan^{-1}e^{-x}, y\ln(1 + y^{3/2}), ze^{-1/z}\rangle$, and S is the part of the sphere $x^2 + y^2 + z^2 = 9$ that lies inside the cone $z = \sqrt{2x^2 + 2y^2}$.

In Problems 35-38, use Stokes' theorem to evaluate the line integral

$$\oint_C [(1 + y)z\,dx + (1 + z)x\,dy + (1 + x)y\,dz]$$

for the given closed path C.

35. C is the boundary of the standard unit circle in the xy-plane.

36. C is the elliptic path $x = 2\cos\theta$, $y = \sin\theta, z = 1$ for $0 \leq \theta \leq 2\pi$.

37. C is the boundary of the triangle with vertices $(1, 0, 0)$, $(0, 1, 0)$, $(0, 0, 1)$.

38. C is *any* closed path in the plane $2x - 3y + z = 1$.

39. **What does this say?** Explain the role of Stokes' theorem in thermodynamics. Use the library or other resources as appropriate.

40. **What does this say?** Explain why Stokes' theorem is of particular interest to a mechanical engineering major. Use the library or other resources as appropriate.

41. **What does this say?** Perform a library search to make yourself familiar to the notion of *differential form*, in particular: *0-form, 1-form*, and *2-form*. Identify the 1-forms and 2-forms that intervene in the statement of Green's theorem, as a particular example of Stokes' theorem. (*Note:* this problem is particularly recommended to Mathematics majors).

42. **What does this say?** Perform a reference search on differential forms and how the generalized Stokes' theorem can be expressed in terms of *n-forms*. (*Note:* this problem is particularly recommended to Mathematics majors).

In Problems 43-46, the vector field $\mathbf{V}$ represents the velocity of a fluid flow. In each case, find the circulation

$$\oint_C \mathbf{V} \cdot d\mathbf{R}$$

around the boundary C, assuming a counterclockwise orientation as viewed from above.

43. $\mathbf{V} = x\mathbf{i} + (z - x)\mathbf{j} + y\mathbf{k}$, and C is the intersection of the cylinder $x^2 + y^2 = y$ and the hemisphere $z = \sqrt{1 - x^2 - y^2}$.

44. $\mathbf{V} = y\mathbf{i} + (x^2 + y^2)\mathbf{j} + (x + y)\mathbf{k}$, and C is the triangle with vertices $(0, 0, 0)$, $(1, 0, 0)$, $(0, 1, 0)$.

45. $\mathbf{V} = (e^{x^2} + z)\mathbf{i} + (x + \sin y^3)\mathbf{j} + [y + \ln(\tan^{-1}z)]\mathbf{k}$, and C is the intersection of the sphere $x^2 + y^2 + z^2 = 1$ and the cone $z = \sqrt{x^2 + y^2}$.

46. $\mathbf{V} = y^2\mathbf{i} + \tan^{-1}z\,\mathbf{j} + (x^2 + 1)\mathbf{k}$, and C is the intersection of the plane $z = y$ and the cylinder $x^2 + y^2 = 2x$.

Use Stokes' theorem and technology to evaluate the line integrals given in Problems 47-49 correct to the nearest tenth.

47. $\oint_C [(x + y)^2dx - (x - y)^2dy + z^2\,dz]$ where C is defined in Figure 13.57.

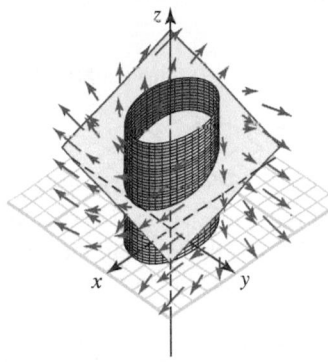

Figure 13.57 C is the intersection of the elliptic cylinder $4x^2 + 9y^2 = 36$ and the plane $x + y + 2z = 2$, traversed counterclockwise when viewed from the origin

48. $\oint_C [(x + y)^3dx - (x - y)^3dy + z^3\,dz]$ where C is defined in Figure 13.57.

49. $\oint_C (y^2\,dx + x^2\,dy + z^2\,dz)$ where C is defined in Figure 13.58.

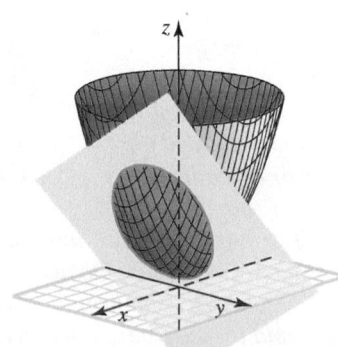

Figure 13.58 C is the intersection of the elliptic paraboloid $z = 2x^2 + 3y^2$ and the plane $z = 3y + 5x + 2$, traversed clockwise when viewed from the origin

50. **What does this say?** As you recall, Gauss' theorem can be applied for domains that have "holes", that is, those that are not simply connected. Do you think Stokes'

theorem can be generalized in an analogous way? Justify your answer by writing a one-page essay about your idea.

51. Let $\mathbf{F} = y^2\mathbf{i} + xy\mathbf{j} + xz\mathbf{k}$, and suppose S is the hemisphere $x^2 + y^2 + z^2 = 1$ with $z \geq 0$. Use Stokes' theorem to express

$$\iint\limits_{S} (\text{curl } \mathbf{F} \cdot \mathbf{N})\, dS$$

as a line integral, and then evaluate the surface integral by evaluating this line integral.

52. Let $\mathbf{F} = z\mathbf{i} + x\mathbf{j} + y\mathbf{k}$, and suppose S is a smooth surface in $\mathbb{R}^3$ whose boundary is given by $x = 2\cos\theta$, $y = 3\sin\theta$, $z = \sin\theta$, $0 \leq \theta \leq 2\pi$. Use Stokes' theorem to evaluate

$$\iint\limits_{S} (\text{curl } \mathbf{F} \cdot \mathbf{N})\, dS$$

53. **Think Tank Problem** Let C_1 and C_2 be positively oriented smooth closed curves in the plane $5x + 3y + 2z = 4$ that contain the same area, and let $\mathbf{F} = \langle 4z, -3x, 2y \rangle$. Either prove that

$$\oint_{C_1} \mathbf{F} \cdot d\mathbf{R} = \oint_{C_2} \mathbf{F} \cdot d\mathbf{R}$$

or find a counterexample.

54. *Historical Quest*

George Stokes was an English mathematical physicist who made important contributions to fluid mechanics, including the Navier-Stokes equations, which are important for modeling fluid flow. Most of his research was done before 1850, after which he held the Lucasian chair of mathematics at Cambridge for the better part of a half-century. William Thomson (Lord Kelvin) knew the result now known as Stokes' theorem in 1850 and sent it to Stokes as a challenge. Stokes proved the theorem, and then included it as an exam question in 1854. One of the students taking this particular examination was James Clerk Maxwell (Historical Quest Problems 57-60, Section 13.7), who derived the famous electromagnetic wave equations ten years later. For this *Quest*, write a history of the Lucasian chair, which was deeded in 1663 as a gift of Henry Lucas. The first and second appointees were Isaac Barrow (*Historical Quest* Supplementary Problem 92,

**George Stokes
(1819-1903)**

Chapter 1) and Isaac Newton (*Historical Quest* Supplementary Problem 87, Chapter 1); and the chair is currently held by Michael Green.

Level 3

55. Let S be the ellipsoid $\dfrac{x^2}{4} + \dfrac{y^2}{9} + z^2 = 1$, and let $\mathbf{F}$ be a vector field whose component functions have continuous partial derivatives on S. Use Stokes' theorem to show that

$$\iint\limits_{S} (\text{curl } \mathbf{F} \cdot \mathbf{N})\, dS = 0$$

Does it matter that S is an ellipsoid? State and prove a more general result based on what you have discovered in the first part of this problem.

56. **Faraday's law** of electromagnetism says that if $\mathbf{E}$ is the electric intensity vector in a system, then

$$\oint_C \mathbf{E} \cdot d\mathbf{R} = -\frac{\partial \phi}{\partial t}$$

around any closed curve C where t is time and ϕ is the total magnetic flux directed outward through any surface S bounded by C. Given that

$$\phi = \iint\limits_{S} \mathbf{B} \cdot \mathbf{N}\, dS$$

where $\mathbf{B}$ is the magnetic flux density, show that curl $\mathbf{E} = -\dfrac{\partial \mathbf{B}}{\partial t}$. *Hint*: It can be shown that

$$\frac{\partial}{\partial t} \iint\limits_{S} \mathbf{B} \cdot \mathbf{N}\, dS = \iint\limits_{S} \frac{\partial \mathbf{B}}{\partial t} \cdot \mathbf{N}\, dS$$

57. **Ampère's circuital law** In 1826, André-Marie Ampère formulated the following law: *The line integral of the magnetic field (assumed to be steady-state) around some closed loop is equal to a magnetic constant μ_0 times the algebraic sum of the currents which pass through the loop.* In symbols,

$$\oint_C \mathbf{B} \cdot d\mathbf{R} = \mu_0 \iint\limits_{S} \mathbf{J} \cdot \mathbf{N}\, dS$$

where C is the closed path in the plane traversed counterclockwise, $\mathbf{B}$ is the magnetic B-field, and $\mathbf{J}$ is the total current density through the surface S enclosed by the curve C, including both free and bound current. Derive Ampère's circuital law given that $\mu_0\mathbf{J} = \text{curl } \mathbf{B}$.

58. The current I flowing across a surface S bounded by the closed curve C is given by

$$I = \iint\limits_{S} \mathbf{J} \cdot \mathbf{N}\, dS$$

where $\mathbf{J}$ is the current density. Given that $\mu\mathbf{J} = \operatorname{curl} \mathbf{B}$, where $\mathbf{B}$ is magnetic flux density and μ is a constant, show that

$$\oint_{C} \mathbf{B} \cdot d\mathbf{R} = \mu I$$

Suppose f and g are functions of x, y, and z with continuous first- and second-order partial derivatives and C is a closed curve bounding the surface S. Use Stokes' theorem to verify the formulas given in Problems 59-60.

59. $\displaystyle\oint_{C} (f\nabla g) \cdot d\mathbf{R} = \iint\limits_{S} (\nabla f \times \nabla g) \cdot \mathbf{N}\, dS$

60. $\displaystyle\oint_{C} (f\nabla g + g\nabla f) \cdot d\mathbf{R} = 0$

13.7 DIVERGENCE THEOREM AND APPLICATIONS

IN THIS SECTION: *The divergence theorem, applications of the divergence theorem, physical interpretation of divergence*
In this section, we develop the divergence theorem, one of the great theorems of elementary calculus. We will also see the divergence theorem can be used to derive theoretical results and to study the properties of fluid dynamics.

The Divergence Theorem

We used Green's theorem to show that $\displaystyle\int_{C} \mathbf{F} \cdot \mathbf{N}\, ds = \iint\limits_{D} \operatorname{div} \mathbf{F}\, dA$, where D is a simply connected domain with the closed boundary curve C. The **divergence theorem** (also known as **Gauss' theorem**) is a generalization of this form of Green's theorem that relates an integral over a closed surface to a volume integral.

Theorem 13.9 The divergence theorem

Let S be a smooth, orientable surface that encloses a solid region R in $\mathbb{R}^3$. If $\mathbf{F}$ is a continuous vector field whose components have continuous partial derivatives in an open set containing R, then

$$\iint\limits_{S} \mathbf{F} \cdot \mathbf{N}\, dS = \iiint\limits_{R} \operatorname{div} \mathbf{F}\, dV$$

where $\mathbf{N}$ is the outward unit normal field for the surface S.

Proof: An important special case is proved in Appendix B. ◆

Example 1 Evaluating a surface integral using the divergence theorem

Evaluate $\iint\limits_{S} \mathbf{F} \cdot \mathbf{N}\, dS$, where $\mathbf{F} = x^2\mathbf{i} + xy\mathbf{j} + x^3y^3\mathbf{k}$ and S is the surface of the tetrahedron bounded by the plane $x + y + z = 1$ and the coordinate planes, with outward unit normal vector $\mathbf{N}$ (see Figure 13.59).

Solution We will use the divergence theorem. Note that

$$\operatorname{div} \mathbf{F} = \frac{\partial}{\partial x}(x^2) + \frac{\partial}{\partial y}(xy) + \frac{\partial}{\partial z}(x^3y^3) = 2x + x + 0 = 3x$$

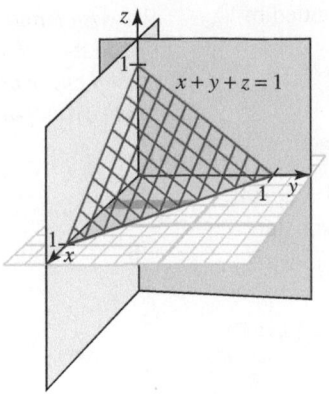
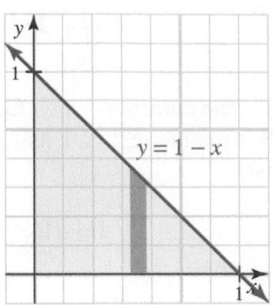

a. The surface S **b.** Projected region D in the xy-plane

Figure 13.59   A tetrahedron in $\mathbb{R}^3$

The tetrahedron is the set R: all (x, y, z) such that $0 \le z \le 1 - x - y$ whenever $0 \le y \le 1 - x$ for $0 \le x \le 1$. This projects onto the triangular region in the xy-plane described by D: all (x, y) such that $0 \le y \le 1 - x$ for $0 \le x \le 1$ (see Figure 13.59**b**.) Then, by applying the divergence theorem, we find that

$$\iint\limits_{S} \mathbf{F} \cdot \mathbf{N} \, dS = \iiint\limits_{R} \text{div } \mathbf{F} \, dV$$

$$= \int_0^1 \int_0^{1-x} \int_0^{1-x-y} 3x \, dz \, dy \, dx$$

$$= \int_0^1 \int_0^{1-x} 3x(1 - x - y) \, dy \, dx$$

$$= 3 \int_0^1 \left[x(1-x)y - \frac{1}{2}xy^2 \right]_0^{1-x} dx$$

$$= 3 \int_0^1 \left[x(1-x)^2 - \frac{1}{2}x(1-x)^2 \right] dx$$

$$= \frac{1}{8}$$

Example 2   Verifying the divergence theorem for a particular solid

Let $\mathbf{F} = 2x\mathbf{i} - 3y\mathbf{j} + 5z\mathbf{k}$, and let S be the hemisphere $z = \sqrt{9 - x^2 - y^2}$ together with the disk $x^2 + y^2 \le 9$ in the xy-plane. Verify the divergence theorem.

Solution   The solid is shown in Figure 13.60.

We will show that the surface integral $\iint\limits_{S} \mathbf{F} \cdot \mathbf{N} \, dS$ and the triple integral

$\iiint\limits_{R} \text{div } \mathbf{F} \, dV$ have the same value, where R is the solid bounded by S.

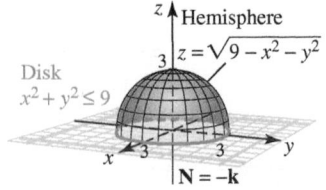

Figure 13.60   The surface S

I. *Evaluation of* $\iint\limits_{S} \mathbf{F} \cdot \mathbf{N} \, dS$

The surface S consists of two parts: S_1, the disk on the bottom of the hemisphere, and S_2, the hemisphere. We will consider these separately and then use

$$\iint\limits_{S} \mathbf{F} \cdot \mathbf{N} \, dS = \iint\limits_{S_1} \mathbf{F} \cdot \mathbf{N_1} \, dS + \iint\limits_{S_2} \mathbf{F} \cdot \mathbf{N_2} \, dS$$

The surface S_1: The disk $x^2 + y^2 \le 9$ with $z = 0$ has outward (downward) unit normal $\mathbf{N} = -\mathbf{k}$, so

$$\iint\limits_{S_1} \mathbf{F} \cdot \mathbf{N_1}\, dS_1 = \iint\limits_{S_1} \langle 2x, -3y, 5z \rangle \cdot \langle 0, 0, -1 \rangle\, dS$$

$$= \iint\limits_{S_1} (-5z)\, dS$$

$$= 0 \qquad \textit{Because } z = 0 \textit{ on } S_1$$

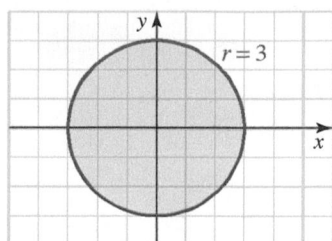

Figure 13.61 The hemisphere projects onto the disk $r \le 3$ in polar form

The surface S_2: Since $z = \sqrt{9 - x^2 - y^2}$, we have

$$z_x = \frac{-x}{\sqrt{9 - x^2 - y^2}} \qquad \text{and} \qquad z_y = \frac{-y}{\sqrt{9 - x^2 - y^2}}$$

and the projection of S_2 onto the xy-plane is the disk D: $x^2 + y^2 \le 9$ or $r \le 3$ in polar form. (See Figure 13.61.)

We find that

$$\iint\limits_{S_2} \mathbf{F} \cdot \mathbf{N_2}\, dS_2 = \iint\limits_{D} \langle 2x, -3y, 5z \rangle \cdot \left\langle -\left(\frac{-x}{\sqrt{9 - x^2 - y^2}} \right), -\left(\frac{-y}{\sqrt{9 - x^2 - y^2}} \right), 1 \right\rangle dA$$

$$= \iint\limits_{D} \left[\frac{2x^2 - 3y^2}{\sqrt{9 - x^2 - y^2}} + 5z \right] dA$$

$$= \iint\limits_{D} \left[\frac{2x^2 - 3y^2}{\sqrt{9 - x^2 - y^2}} + 5\sqrt{9 - x^2 - y^2} \right] dA \qquad \textit{Since } z = \sqrt{9 - x^2 - y^2} \textit{ on } S$$

$$= \int_0^{2\pi} \int_0^3 \left[\frac{2r^2 \cos^2 \theta - 3r^2 \sin^2 \theta}{\sqrt{9 - r^2}} + 5\sqrt{9 - r^2} \right] r\, dr\, d\theta \qquad \textit{Changing to polar coordinates}$$

$$= \int_0^{2\pi} [81 - 90 \sin^2 \theta]\, d\theta$$

$$= 72\pi$$

Adding the surface integrals for S_1 and S_2, we obtain

$$\iint\limits_{S} \mathbf{F} \cdot \mathbf{N}\, dS = \iint\limits_{S_1} \mathbf{F} \cdot \mathbf{N_1}\, dS + \iint\limits_{S_2} \mathbf{F} \cdot \mathbf{N_2}\, dS$$

$$= 0 + 72\pi$$

$$= 72\pi$$

II. *Evaluation of* $\displaystyle\iiint\limits_{R} \text{div}\, \mathbf{F}\, dV$

We begin with div $\mathbf{F}$.

$$\text{div}\, \mathbf{F} = \frac{\partial}{\partial x}(2x) + \frac{\partial}{\partial y}(-3y) + \frac{\partial}{\partial z}(5z) = 2 - 3 + 5 = 4$$

Therefore, $\displaystyle\iiint\limits_{R} \text{div}\, \mathbf{F}\, dV = \iiint\limits_{R} 4\, dV$, but $\displaystyle\iiint\limits_{R} dV$ is just the volume

of the hemisphere $z = \sqrt{9 - x^2 - y^2}$. A hemisphere of radius 3 has volume $\frac{1}{2} \left[\frac{4}{3} \pi (3)^3 \right] = 18\pi$, so

$$\iiint_R \operatorname{div} \mathbf{F} \, dV = \iiint_R 4 \, dV = 4V = 4(18\pi) = 72\pi$$

In conclusion, we see that for this example,

$$\iint_S \mathbf{F} \cdot \mathbf{N} \, dS = 72\pi = \iiint_R \operatorname{div} \mathbf{F} \, dV$$

as required by the divergence theorem.

STOP The divergence theorem applies only to closed surfaces. However, if we wish to evaluate $\iint_{S_1} \mathbf{F} \cdot \mathbf{N} \, dS$ where S_1 is *not* closed, we may be able to find a closed surface S that is the union of S_1 and some other surface S_2. Then, if the hypotheses of the divergence theorem are satisfied by $\mathbf{F}$ and S, we have

$$\iint_{S_1} \mathbf{F} \cdot \mathbf{N} \, dS + \iint_{S_2} \mathbf{F} \cdot \mathbf{N} \, dS = \iint_S \mathbf{F} \cdot \mathbf{N} \, dS = \iiint_R \operatorname{div} \mathbf{F} \, dV$$

where R is the solid region bounded by S. Thus, if we can compute $\iiint_R \operatorname{div} \mathbf{F} \, dV$ and $\iint_{S_2} \mathbf{F} \cdot \mathbf{N} \, dS$, we can compute $\iint_{S_1} \mathbf{F} \cdot \mathbf{N} \, dS$, by the equation

$$\iint_{S_1} \mathbf{F} \cdot \mathbf{N} \, dS = \iiint_R \operatorname{div} \mathbf{F} \, dV - \iint_{S_2} \mathbf{F} \cdot \mathbf{N} \, dS$$

This equation can also be used as a device for trading the evaluation of a difficult surface integral for that of an easier volume integral. Here is an example of this procedure.

Example 3   Evaluating a surface integral over an open surface

Evaluate $\iint_S \mathbf{F} \cdot \mathbf{N} \, dS$, where $\mathbf{F} = xy\mathbf{i} - z^2\mathbf{k}$ and S is the surface of the upper five faces of the unit cube $0 \leq x \leq 1$, $0 \leq y \leq 1$, $0 < z \leq 1$, as shown in Figure 13.62.

Solution   Note that the surface S is not closed, but we can close it by adding the missing face S_m, thus forming a closed surface S^* that satisfies the conditions of the divergence theorem. The strategy is to evaluate the surface integral on S^* and then subtract the surface integral of the added face S_m.

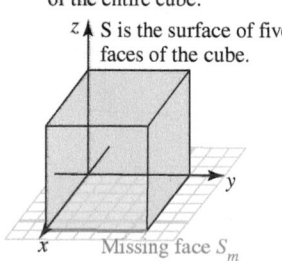

S^* is the closed surface of the entire cube.

S is the surface of five faces of the cube.

Missing face S_m

Figure 13.62   An open surface: a cube with a missing face

$$\iint_{S^*} \mathbf{F} \cdot \mathbf{N} \, dS = \iiint_{\text{CUBE}} \operatorname{div} \mathbf{F} \, dV$$

$$= \int_0^1 \int_0^1 \int_0^1 (y - 2z) \, dx \, dy \, dz$$

$$= \int_0^1 \int_0^1 (y - 2z) \, dy \, dz$$

$$= \int_0^1 \left(\frac{1}{2} - 2z \right) dz$$

$$= -\frac{1}{2}$$

Also, because the outward unit normal vector to the added face S_m is $\mathbf{N} = -\mathbf{k}$ and $z = 0$ on this face, it follows that

$$\iint_{S_m} \mathbf{F} \cdot \mathbf{N} \, dS = \iint_{S_m} (xy\mathbf{i} - z^2\mathbf{k}) \cdot (-\mathbf{k}) \, dS = \iint_{S_m} z^2 \, dS = 0$$

We obtain

$$\iint_S \mathbf{F} \cdot \mathbf{N} \, dS = \iiint_{\text{CUBE}} \text{div } \mathbf{F} \, dV - \iint_{S_m} \mathbf{F} \cdot \mathbf{N} \, dS = -\frac{1}{2} - 0 = -\frac{1}{2}$$

Applications of the Divergence Theorem

Like Stokes' theorem, the divergence theorem is often used for theoretical purposes, especially as a tool for deriving general properties in mathematical physics. Consider the vector field shown in Figure 13.63.

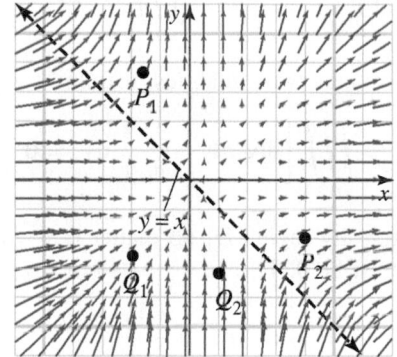

Figure 13.63 Interactive   A vector field F

Notice that the vectors that end near P_1 or P_2 are shorter than vectors that start next to P_1 or P_2, respectively. This means that the net flow is outward near these points, so div $\mathbf{F}(P_k) > 0$ and P_k is a *source* for any point where $y > -x$. On the other hand, we see that the incoming vectors near Q_1 or Q_2 are longer than the outgoing vectors. This means the net flow at these points is inward; that is div $\mathbf{F}(Q_k) < 0$ and Q_k is a *sink* for any point where $y < -x$. We can verify these observations if we also know that $\mathbf{F} = x^2\mathbf{i} + y^2\mathbf{j}$ to find div $\mathbf{F} = 2x + 2y$. Thus, div $\mathbf{F} > 0$ if $2x + 2y > 0$ or if $y > -x$. See Section 13.1.

The following example deals with an important property of fluid dynamics.

Example 4   Continuity equation of fluid dynamics

Suppose a fluid with density $\rho(x, y, z, t)$ flows in some region of space with velocity $\mathbf{F}(x, y, z, t)$ at the point (x, y, z) at time t. Assuming there are no sources or sinks (see Section 13.1), show that

$$\text{div } \rho \mathbf{F} = -\frac{\partial \rho}{\partial t}$$

Solution   Let D be a fixed domain enclosed by the surface S with outward normal $\mathbf{N}$. By physics, in the absence of sink/source terms, the rate of change of the total mass m inside D equals the total flux of mass flowing into the domain through the boundary S. The flux density represents the amount (quantity or volume) of fluid flowing out of D across S in unit time is $\iint_S \mathbf{F} \cdot \mathbf{N} \, dS$. On the other hand, the mass is given by the triple integral of the density over the volume D, and consequently, the change of mass with respect to time can be written as

$$\frac{dm}{dt} = \iiint_D \frac{\partial \rho}{\partial t} \, dV$$

whenever the fluid's density varies with time. The preceding step moves the derivative with respect to time inside the integral, which is possible because the volume does not change with time.

Then,

$$\iint\limits_S \rho\mathbf{F} \cdot \mathbf{N}\, dS = -\iiint\limits_D \frac{\partial \rho}{\partial t}\, dV$$

Note that the difference in sign is due to the fact that the flow is inward. On the other hand, by the divergence theorem, it follows that

$$\iint\limits_S \rho\mathbf{F} \cdot \mathbf{N}\, dS = \iiint\limits_D \text{div } \rho\mathbf{F}\, dV$$

Therefore, from the previous two equalities, it follows

$$\iiint\limits_D \left[\text{div } \rho\mathbf{F} + \frac{\partial \rho}{\partial t} \right] dV = 0$$

This equation must hold for any region D, no matter how small, which means that the integrand of the integral must be 0. That is,

$$\text{div } \rho\mathbf{F} = -\frac{\partial \rho}{\partial t}$$

It is known that the total heat contained in a body occupying a region D, with uniform density ρ and specific heat σ, is $\iiint\limits_D \sigma\rho T\, dV$, where T is the temperature. Thus, the amount of heat leaving D per unit of time is given by the derivative

$$-\frac{\partial}{\partial t}\left[\iiint\limits_D \sigma\rho T\, dV \right] = \iiint\limits_D -\sigma\rho\frac{\partial T}{\partial t}\, dV$$

In the following example, we use this result to obtain an important formula from mathematical physics.

Example 5   Derivation of the heat equation

Let $T(x,y,z,t)$ be the temperature at each point (x,y,z) in a solid body D at time t. Given that the velocity of heat flow in the body is $\mathbf{F} = -K\nabla T$ for a positive constant K (called the **thermal conductivity**), show that

$$\frac{\partial T}{\partial t} = \frac{K}{\sigma\rho}\nabla^2 T$$

where σ is the specific heat of the body and ρ is its density.

Solution   Let S be the closed surface that bounds D. Because $\mathbf{F}$ is the velocity of heat flow, the amount of heat leaving D per unit time is $\iint\limits_S \mathbf{F} \cdot \mathbf{N}\, dS$, and the divergence theorem applies:

$$\iint\limits_S \mathbf{F} \cdot \mathbf{N}\, dS = \iiint\limits_D \text{div }(-K\nabla T)\, dV$$

$$= \iiint\limits_D (-K\nabla \cdot \nabla T)\, dV$$

$$= \iiint\limits_D -K\nabla^2 T\, dV$$

Since this is the amount of heat leaving D per unit of time, it must equal the heat integral from physics derived just before this example. Thus,

$$\iiint\limits_{D} -K\nabla^2 T \, dV = \iiint\limits_{D} -\sigma\rho\frac{\partial T}{\partial t} \, dV$$

This equation holds not only for the body as a whole, but for every part of the body, no matter how small. Thus, we can shrink the body to a single point, and it then can be shown that when this occurs the integrands are equal; that is,

$$-K\nabla^2 T = -\sigma\rho\frac{\partial T}{\partial t}$$

$$\frac{\partial T}{\partial t} = \frac{K}{\sigma\rho}\nabla^2 T$$

Recall from Section 13.4 that the *normal derivative* $\partial g/\partial n$ of a scalar function g defined on the closed surface S is the directional derivative of g in the direction of the outward unit normal vector $\mathbf{N}$ to S; that is,

$$\frac{\partial g}{\partial n} = \nabla g \cdot \mathbf{N}$$

We will use this equation in the following example, which involves a generalization of a property we first obtained for $\mathbb{R}^2$ in Example 6 of Section 13.4.

Example 6   Derivation of Green's first identity

Show that if f and g are scalar functions such that $\mathbf{F} = f\nabla g$ is continuously differentiable in the solid domain D bounded by the closed surface S, then

$$\iiint\limits_{D} [f\nabla^2 g + \nabla f \cdot \nabla g] \, dV = \iint\limits_{S} f\,\frac{\partial g}{\partial n} \, dS$$

This is called **Green's first identity**.

Solution   We will apply the divergence theorem to the vector field $\mathbf{F}$ (note that $\mathbf{F}$ is continuously differentiable), but first we need to express div $\mathbf{F}$ in a more useful form.

$$\text{div}\,(f\nabla g) = \nabla \cdot (f\nabla g)$$

$$= \left[\frac{\partial}{\partial x}\mathbf{i} + \frac{\partial}{\partial y}\mathbf{j} + \frac{\partial}{\partial z}\mathbf{k}\right] \cdot \left[f\frac{\partial g}{\partial x}\mathbf{i} + f\frac{\partial g}{\partial y}\mathbf{j} + f\frac{\partial g}{\partial z}\mathbf{k}\right]$$

$$= \frac{\partial}{\partial x}\left[f\frac{\partial g}{\partial x}\right] + \frac{\partial}{\partial y}\left[f\frac{\partial g}{\partial y}\right] + \frac{\partial}{\partial z}\left[f\frac{\partial g}{\partial z}\right]$$

$$= \left[\frac{\partial f}{\partial x}\frac{\partial g}{\partial x} + f\frac{\partial^2 g}{\partial x^2}\right] + \left[\frac{\partial f}{\partial y}\frac{\partial g}{\partial y} + f\frac{\partial^2 g}{\partial y^2}\right] + \left[\frac{\partial f}{\partial z}\frac{\partial g}{\partial z} + f\frac{\partial^2 g}{\partial z^2}\right]$$

$$= \left[\frac{\partial f}{\partial x}\frac{\partial g}{\partial x} + \frac{\partial f}{\partial y}\frac{\partial g}{\partial y} + \frac{\partial f}{\partial z}\frac{\partial g}{\partial z}\right] + f\left[\frac{\partial^2 g}{\partial x^2} + \frac{\partial^2 g}{\partial y^2} + \frac{\partial^2 g}{\partial z^2}\right]$$

$$= (\nabla f) \cdot (\nabla g) + f\nabla^2 g$$

This calculation gives us the first step in the following computation.

$$
\iiint\limits_{D} [f\nabla^2 g + \nabla f \cdot \nabla g]\, dV = \iiint\limits_{D} \text{div}\,(f\nabla g)\, dV
$$

$$
= \iint\limits_{S} (f\nabla g) \cdot \mathbf{N}\, dS \qquad \textit{Divergence theorem}
$$

$$
= \iint\limits_{S} f(\nabla g \cdot \mathbf{N})\, dS
$$

$$
= \iint\limits_{S} f\, \frac{\partial g}{\partial n}\, dS \qquad \textit{Because } \nabla g \cdot \mathbf{N} = \frac{\partial g}{\partial n} \textit{ by definition}
$$

Physical Interpretation of Divergence

In Section 13.6, we used Stokes' theorem to give an interpretation of the curl as a measure of the tendency of a fluid to swirl (the circulation). Our last example gives an analogous interpretation of divergence. In particular, we show that the net rate of fluid mass flowing away (that is, "diverging") from point P_0 is given by div $\mathbf{F}_0$. This is the reason P_0 is a *source* if div $\mathbf{F}_0 > 0$ (mass flowing out from P_0) and a *sink* if div $\mathbf{F}_0 < 0$ (mass flowing back into P_0).

Example 7   Physical interpretation of divergence

Let $\mathbf{F} = \rho\mathbf{V}$ be the flux density associated with a fluid of density ρ flowing with velocity $\mathbf{V}$ and let P_0 be a point inside a solid region where the conditions of the divergence theorem are satisfied. Prove that

$$
\text{div } \mathbf{F}_0 = \lim_{r \to 0} \frac{1}{V(r)} \iint\limits_{S(r)} \mathbf{F} \cdot \mathbf{N}\, dS
$$

where div $\mathbf{F}_0$ denotes the value of div $\mathbf{F}$ at P_0, and $S(r)$ is a sphere centered at P_0 with volume $V(r) = \frac{4}{3}\pi r^3$.

Solution   Applying the divergence theorem to the solid sphere (ball) $B(r)$ with surface $S(r)$, we obtain

$$
\iint\limits_{S(r)} \mathbf{F} \cdot \mathbf{N}\, dS = \iiint\limits_{B(r)} \text{div } \mathbf{F}\, dV
$$

The mean value theorem (for triple integrals) tells us that

$$
\frac{1}{V(r)} \iiint\limits_{B(r)} \text{div } \mathbf{F}\, dV = \text{div } \mathbf{F}^*
$$

where div $\mathbf{F}^*$ denotes the value of div $\mathbf{F}$ at some point P^* in the ball $B(r)$. Combining these results, we find that

$$
\iint\limits_{S(r)} \mathbf{F} \cdot \mathbf{N}\, dS = \iiint\limits_{B(r)} \text{div } \mathbf{F}\, dV = V(r)\text{div } \mathbf{F}^*
$$

or

$$
\frac{1}{V(r)} \iint\limits_{S(r)} \mathbf{F} \cdot \mathbf{N}\, dS = \text{div } \mathbf{F}^*
$$

Since the point P^* is inside the ball $B(r)$ centered at P_0, it follows that $P^* \to P_0$ as $r \to 0$, so div $\mathbf{F}^* \to$ div $\mathbf{F}_0$ and we have

$$\lim_{r \to 0} \frac{1}{V(r)} \iint_{S(r)} \mathbf{F} \cdot \mathbf{N} \, dS = \lim_{r \to 0} \text{div } \mathbf{F}^* = \text{div } \mathbf{F}_0$$

as claimed.

PROBLEM SET 13.7

Level 1

1. **What does this say?** As you recall, Green's theorem establishes a relationship between a double integral and a path integral. What type of relationship does the divergence theorem establish?
2. **What does this say?** State the formula given by the divergence theorem and explain the physical meaning of its left and right hand sides.
3. **What does this say?** In vector calculus, the divergence theorem is also referred to as Gauss' theorem. Perform a library search to identify the contribution that Gauss had to the statement of this theorem. Explain.
4. **What does this say?** In vector calculus, the divergence theorem is also referred to as Ostrogradsky's theorem (sometimes, Gauss-Ostrogradsky's theorem or Green-Ostrogradsky theorem or Gauss-Green-Ostrogradsky theorem). Perform a library search to identify the contribution that Ostrogradsky had to the statement of this theorem. Explain.

Verify the divergence theorem for the vector function $\mathbf{F}$ *and solid D given in Problems 5-10. Assume* $\mathbf{N}$ *is the unit normal vector pointing away from the origin.*

5. $\mathbf{F} = xz\mathbf{i} + y^2\mathbf{j} + 2z\mathbf{k}$;
   D is the ball $x^2 + y^2 + z^2 \le 4$.
6. $\mathbf{F} = x\mathbf{i} - 2y\mathbf{j}$;
   D is the interior of the paraboloid
   $z = x^2 + y^2, 0 \le z < 9$.
7. $\mathbf{F} = x\mathbf{i} + y\mathbf{j} + z\mathbf{k}$
   D is the interior of the ellipsoid
   $\frac{x^2}{a^2} + \frac{y^2}{b^2} + \frac{z^2}{c^2} \le 1, a, b, c$ constants.
8. $\mathbf{F} = x\mathbf{i} + y\mathbf{j} + z\mathbf{k}$
   D is the ball $x^2 + y^2 + z^2 \le a^2, a$ constant.
9. $\mathbf{F} = 2y^2\mathbf{j}$; D is the tetrahedron bounded by the coordinate planes and the plane $x + 4y + z = 8$.
10. $\mathbf{F} = 3x\mathbf{i} + 5y\mathbf{j} + 6z\mathbf{k}$;
    D is the tetrahedron bounded by the coordinate planes and the plane $2x + y + z = 4$.

Classify the points R, S, T, and U shown in Problems 11-16 as sinks, sources or neither.

11. 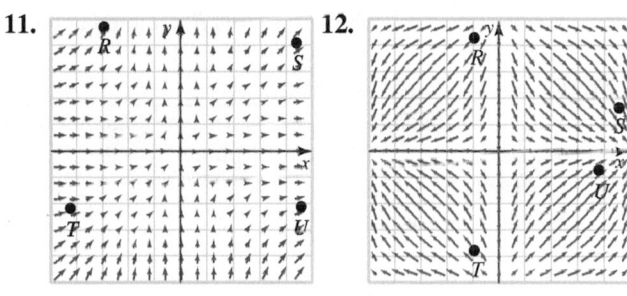 12.

13. 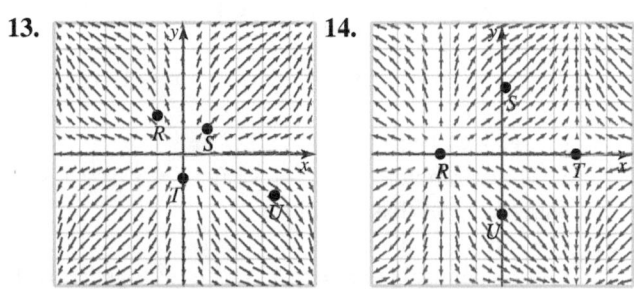 14.

15. 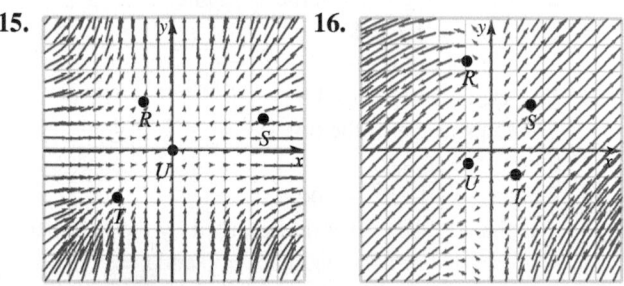 16.

In Problems 17-22 you are given the vector functions for Problems 11-16, respectively. Verify your answers for those problems. Hint: use the divergence of $\mathbf{F}$.

17. $\mathbf{F}(x, y) = x^2\mathbf{i} + y^2\mathbf{j}$
18. $\mathbf{F}(x, y) = (\sin x)\mathbf{i} - (\sin y)\mathbf{j}$
19. $\mathbf{F}(x, y) = (\sin x)\mathbf{i} + (\sin y)\mathbf{j}$
20. $\mathbf{F}(x, y) = (\cos x)\mathbf{i} + (\sin y)\mathbf{j}$
21. $\mathbf{F}(x, y) = 2x^2\mathbf{i} + 3y^2\mathbf{j}$
22. $\mathbf{F}(x, y) = xy\mathbf{i} + y^2\mathbf{j}$

Use the divergence theorem in Problems 23-40 to evaluate the surface integral $\iint_S \mathbf{F} \cdot \mathbf{N}\, dS$ for the given choice of $\mathbf{F}$ and closed boundary surface S. Assume $\mathbf{N}$ is the outward unit normal vector field.

23. $\mathbf{F} = x\mathbf{i} + y\mathbf{j} + z\mathbf{k}$;
    S is the cylinder $x^2 + y^2 = 1$, for $0 \le z \le 1$, with the two disk ends included.
24. $\mathbf{F} = xyz\mathbf{j}$;
    S is the cylinder $x^2 + y^2 = 9$, for $0 \le z \le 5$, with the two disk ends included.
25. $\mathbf{F} = x\mathbf{i} + y\mathbf{j} + z\mathbf{k}$;
    S is the cube $0 \le x \le 1, 0 \le y \le 1, 0 \le z \le 1$.
26. $\mathbf{F} = x^3\mathbf{i} + y^3\mathbf{j} + z^3\mathbf{k}$;
    S is the standard sphere of radius R centered at the origin.
27. $\mathbf{F} = (\cos yz)\mathbf{i} + e^{xz}\mathbf{j} + 3z^2\mathbf{k}$;
    S is the hemisphere $z = \sqrt{4 - x^2 - y^2}$ together with the disk $x^2 + y^2 \le 4$ in the xy-plane.
28. $\mathbf{F} = \text{curl}[e^{xz}\mathbf{i} - 4\mathbf{j} + (\sin xyz)\mathbf{k}]$;
    S is the ellipsoid $2x^2 + 3y^2 + 7z^2 = 1$.
29. $\mathbf{F} = (x^2 + y^2 - z^2)\mathbf{i} + x^2 y\mathbf{j} + 3z\mathbf{k}$;
    S is the surface of the unit cube $0 \le x \le 1, 0 \le y \le 1, 0 \le z \le 1$.
30. $\mathbf{F} = 2y\mathbf{i} - z\mathbf{j} + 3x\mathbf{k}$;
    S is the surface of the unit cube $0 \le x \le 1, 0 \le y \le 1, 0 \le z \le 1$.
31. $\mathbf{F} = x\mathbf{i} + y\mathbf{j} + z\mathbf{k}$;
    S is the paraboloid $z = x^2 + y^2$ for $0 \le z \le 9$, together with the upper disk that closes it.
32. $\mathbf{F} = \text{curl}(y\mathbf{i} + x\mathbf{j} - z\mathbf{k})$;
    S is the hemisphere $z = \sqrt{4 - x^2 - y^2}$ together with the disk $x^2 + y^2 \le 4$ in the xy-plane.
33. $\mathbf{F} = x^2\mathbf{i} + y^2\mathbf{j} + z^2\mathbf{k}$;
    S is the sphere $x^2 + y^2 + z^2 = 4$.
34. $\mathbf{F} = xyz\mathbf{i} + xyz\mathbf{j} + xyz\mathbf{k}$;
    S is the surface of the box $0 \le x \le 1, 0 \le y \le 2, 0 \le z \le 3$.
35. $\mathbf{F} = x\mathbf{i} + y\mathbf{j} + (z^2 - 1)\mathbf{k}$;
    S is the surface of a solid bounded by the cylinder $x^2 + y^2 = 4$ and the planes $z = 0$ and $z = 1$.
36. $\mathbf{F} = (x + 10y^2z^2)\mathbf{i} + (y + 10x^2z^2)\mathbf{j} + (z + 10x^2y^2)\mathbf{k}$;
    S is the hemispherical surface $z = \sqrt{1 - x^2 - y^2}$ together with the disk $x^2 + y^2 \le 1$ in the xy-plane.
37. $\mathbf{F} = xy^2\mathbf{i} + yz^2\mathbf{j} + x^2y\mathbf{k}$;
    S is the surface bounded above by the sphere $\rho = 2$ and below by the cone $\phi = \frac{\pi}{4}$ (in spherical coordinates). *Note*: S is the surface of an "ice cream cone."
38. $\mathbf{F} = xy^2\mathbf{i} + yz^2\mathbf{j} + x^2z\mathbf{k}$;
    S is the surface (in spherical coordinates) with top $\rho = 2, 0 \le \phi \le \frac{\pi}{4}, 0 \le \theta \le 2\pi$, and bottom $0 \le \rho \le 2, \phi = \frac{\pi}{4}, 0 \le \theta \le 2\pi$.
39. $\mathbf{F} = x^3\mathbf{i} + y^3\mathbf{j} + 3a^2z\mathbf{k}$ (constant $a > 0$);
    S is the surface bounded by the cylinder $x^2 + y^2 = a^2$ and the planes $z = 0$ and $z = 1$.

40. $\mathbf{F} = x\mathbf{i} + y\mathbf{j} + z\mathbf{k}$
    S is the surface of a solid bounded by the cylinder $x^2 + y^2 = b^2$ and the planes $z = 0$ and $z = 1$.

Level 2

41. Suppose that S is a closed surface that encloses a solid region D. Show that the volume of D is given by
$$V(D) = \frac{1}{3}\iint_S (x\mathbf{i} + y\mathbf{j} + z\mathbf{k}) \cdot \mathbf{N}\, dS$$
where $\mathbf{N}$ is an outward unit normal vector to S.
42. Use the formula in Problem 41 to find the volume of the hemisphere
$$z = \sqrt{R^2 - x^2 - y^2}$$
43. Use the divergence theorem to evaluate
$$\iint_S \|\mathbf{R}\|\ \mathbf{R} \cdot \mathbf{N}\, dS$$
where $\mathbf{R} = x\mathbf{i} + y\mathbf{j} + z\mathbf{k}$ and S is the sphere $x^2 + y^2 + z^2 = a^2$, with constant $a > 0$.
44. **Think Tank Problem** Let $\mathbf{F} = \langle f(y,z), g(x,z), h(x,y)\rangle$ and let S be the surface of a solid G. Either prove that the flux of $\mathbf{F}$ across S is zero or find a counterexample.
45. **Think Tank Problem** Let $f(x,y,z)$ be a differentiable nonzero scalar function. State an additional property of f that will guarantee
$$\iint_S f\,\nabla f \cdot \mathbf{N}\, dS = \iiint_G \|\nabla f\|^2\, dV$$
for any solid region G bounded by the closed oriented surface S.
46. The moment of inertia about the z-axis of a solid D of constant density $\rho = a$ is given by
$$I_z = \iiint_T a(x^2 + y^2)\, dV$$
Express this integral as a surface integral over the surface S that bounds D.

Level 3

47. Let u be a scalar function with continuous second partial derivatives in a region containing the solid region D, with closed boundary surface S. Show
$$\iint_S \frac{\partial u}{\partial n}\, dS = \iiint_D \nabla^2 u\, dV$$

48. In Problem 47, let $u = x + y + z$ and $v = \frac{1}{2}(x^2 + y^2 + z^2)$. Evaluate

$$\iint_S (u\nabla v) \cdot \mathbf{N}\, dS$$

where S is the boundary of the cube $0 \le x \le 1$, $0 \le y \le 1, 0 \le z \le 1$.

49. Let f and g be scalar functions such that $\mathbf{F} = f\nabla g$ is continuously differentiable in the region D, which is bounded by the closed surface S. Prove *Green's second identity* using the divergence theorem:

$$\iiint_D (f\nabla^2 g - g\nabla^2 f)\, dV$$

$$= \iint_S \left(f\frac{\partial g}{\partial n} - g\frac{\partial f}{\partial n} \right) dS$$

50. Show that if g is harmonic in the region D, then

$$\iint_S \frac{\partial g}{\partial n}\, dS = 0$$

where the closed surface S is the boundary of D. (Recall that g harmonic means $\nabla^2 g = 0$.)

51. Show that

$$\iint_S \mathbf{F} \cdot \mathbf{N}\, dS = 0$$

if S is a closed surface and $\mathbf{F} = \text{curl } \mathbf{U}$ throughout the interior of S for some vector field $\mathbf{U}$ with continuous second partial derivatives. A vector field $\mathbf{U}$ with this property is said to be a **vector potential** for $\mathbf{F}$.

52. In our derivation of the heat equation in this section, we assumed that the coefficient of thermal conductivity K is constant (no sinks or sources). If $K = K(x, y, z)$ is a variable, show that the heat equation becomes

$$K\nabla^2 T + \nabla K \cdot \nabla T = \sigma\rho \frac{\partial T}{\partial t}$$

An electric charge q located at the origin produces the electric field

$$\mathbf{E} = \frac{q\mathbf{R}}{4\pi\epsilon\, \|\mathbf{R}\|^3}$$

*where $\mathbf{R} = x\mathbf{i} + y\mathbf{j} + z\mathbf{k}$ and ϵ is a physical constant, called the **electric permittivity**. Use this constant in Problems 53-56.*

53. Show that

$$\iint_S \mathbf{E} \cdot \mathbf{N}\, dS = 0$$

if the closed surface S does not enclose the origin. This is **Gauss' law**.

54. Show that

$$\iint_S \mathbf{E} \cdot \mathbf{N}\, dS = \frac{q}{\epsilon}$$

in the case where the closed surface S encloses the origin. Note that the divergence theorem does not apply directly to this case.

55. Gauss' law can be expressed as

$$\iint_S \mathbf{D} \cdot \mathbf{N}\, dS = q$$

where $\mathbf{D} = \epsilon\mathbf{E}$ is the electric flux density, with electric intensity $\mathbf{E}$, permittivity ϵ, and q a constant. Show that $\text{div } \mathbf{D} = Q$, where Q is the *charge density*; that is,

$$\iiint_V Q\, dV = q$$

56. Use Gauss' law (see Problem 53) to find the charge contained in the solid hemisphere $x^2 + y^2 + z^2 \le a^2$, $z \ge 0$, if the electric field is $\mathbf{E}(x, y, z) = x\mathbf{i} + y\mathbf{j} + z\mathbf{k}$.

ℌistorical Ǭuest (*Problems* 57-60)

*James Clerk Maxwell was one of the greatest physicists of all time. Using the experimental discoveries of Michael Faraday as a basis, he was able to express the governing rules for electrical and magnetic fields in precise mathematical form. In 1871, he published his **Theory of***

Karl Smith library

James Clerk Maxwell (1831-1879)

Heat and Magnetism, *which formed the basis for modern electromagnetic theory and contributed to quantum theory and special relativity. Maxwell was influential in convincing other mathematicians and scientists to use vectors, and was interested in areas as diverse as the behavior of light and the statistical behavior of molecular motion. He was sometimes referred to as dp/dt because in thermodynamics, **dp/dt = JCM**, a unit of measurement named for him. For this Ǭuest you are to derive **Maxwell's equation for the electric***

intensity E:

$$(\nabla \cdot \nabla)\mathbf{E} = \mu\sigma\frac{\partial \mathbf{E}}{\partial t} + \mu\epsilon\frac{\partial^2 \mathbf{E}}{\partial t^2}$$

To derive this equation, you need to know that curl $\mathbf{E} = -\frac{\partial \mathbf{B}}{\partial t}$ and curl $\mathbf{H} = \sigma\mathbf{E} + \epsilon\frac{\partial \mathbf{E}}{\partial t}$ where $\mathbf{E}$ is electric intensity, $\mathbf{B}$ is magnetic flux density, $\mathbf{H}$ is magnetic intensity, and σ, ϵ, and μ are positive constants.

57. Use the fact that $\mathbf{B} = \mu\mathbf{H}$ to show that

$$\text{curl (curl } \mathbf{E}) = -\mu\frac{\partial}{\partial t}(\text{curl } \mathbf{H})$$

58. Next, show that for any vector field $\mathbf{F}\langle f, g, h \rangle$,

$$\text{curl (curl } \mathbf{F}) = \nabla(\text{div } \mathbf{F}) - \nabla \cdot \nabla\mathbf{F}$$

59. Use the formula in Problem 58 to show that

$$\nabla(\text{div } \mathbf{E}) - (\nabla \cdot \nabla)\mathbf{E} = -\mu\frac{\partial}{\partial t}\left(\sigma\mathbf{E} + \epsilon\frac{\partial \mathbf{E}}{\partial t}\right)$$

60. Complete the derivation of **Maxwell's electric intensity equation**, assuming that the charge density Q is 0 so that div $\mathbf{E} = 0$. (See Problem 55.)

CHAPTER 13 REVIEW

M*athematics is much like the Mississippi.*
      There are side-shoots and dead ends and minor tributaries; but the mainstream is
  there, and you can find it where the current—the mathematical power—is strongest.

Ian Stewart

From Here to Infinity, Oxford University Press, Oxford, 1996, p. 11.

Proficiency Examination

Concept Problems

1. What is a vector field?

2. What is the divergence of a vector field?

3. What is the curl of a vector field?

4. What is the del operator?

5. What is Laplace's equation?

6. What is the difference between the Riemann integral and a line integral? Discuss.

7. What is the formula for a line integral of a vector field?

8. How do we find work as a line integral?

9. What is the formula for a line integral in terms of arc length parameter?

10. State the fundamental theorem for line integrals.

11. Define a conservative vector field.

12. What is the scalar potential of a conservative vector field?

13. What is a Jordan curve?

14. State Green's theorem.

15. How can you use Green's theorem to find area as a line integral?

16. What is a normal derivative?

17. Define a surface integral.

18. What is the formula for a surface integral of a surface defined parametrically?

19. What is a flux integral?

20. State Stokes' theorem.

21. State the conservative vector field theorem.

22. State the divergence theorem.

Practice Problems

23. Show that $yz\mathbf{i} + xz\mathbf{j} + xy\mathbf{k}$ is conservative and find a scalar potential function.

24. Compute div $\mathbf{F}$ and curl $\mathbf{F}$ for $\mathbf{F} = x^2 y\mathbf{i} - e^{yz}\mathbf{j} + \frac{1}{2}x\mathbf{k}$.

25. Use Green's theorem to evaluate the line integral $\oint_C \mathbf{F} \cdot d\mathbf{R}$, where $\mathbf{F} = (2x + y)\mathbf{i} + 3y^2\mathbf{j}$ and C is the boundary of the triangle T with vertices $(-1, 2), (0, 0), (1, 2)$, traversed in the given order.

26. Use Stokes' theorem to evaluate the line integral $\oint_C \mathbf{F} \cdot d\mathbf{R}$, where $\mathbf{F} = 2y\mathbf{i} + z\mathbf{j} + y\mathbf{k}$ and C is the intersection of the plane $z = x + 2$ and the sphere $x^2 + y^2 + z^2 = 4z$, traversed counterclockwise as viewed from above.

27. Use the divergence theorem to evaluate the surface integral $\iint_S \mathbf{F} \cdot \mathbf{N} \, dS$, where $\mathbf{F} = x^2\mathbf{i} + (y + z)\mathbf{j} - 2z\mathbf{k}$, and S is the surface of the unit cube $0 \leq x \leq 1, 0 \leq y \leq 1, 0 \leq z \leq 1$.

28. Evaluate $\oint_C \dfrac{x\,dx + y\,dy}{(x^2 + y^2)^2}$, where C is the path shown in Figure 13.64, traversed counterclockwise.

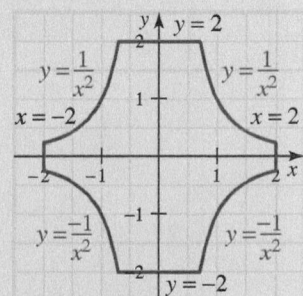

Figure 13.64 Curve C

29. An object with mass m travels counterclockwise (as viewed from above) in the circular orbit $x^2 + y^2 = 9$, $z = 2$, with angular speed ω. The mass is subject to a centrifugal force $\mathbf{F} = m\omega^2 \mathbf{R}$, where $\mathbf{R} = x\mathbf{i} + y\mathbf{j} + z\mathbf{k}$. Show that $\mathbf{F}$ is conservative, and find a scalar potential function for $\mathbf{F}$.

30. Find the work done as an object moves in the force field given in Problem 29 from point $(3, 0, 2)$ to $(-3, 0, 2)$ along the circle $x^2 + y^2 = 9$ in the plane $z = 2$.

Supplementary Problems*

In Problems 1-6, determine whether the given vector field is conservative, and if it is, find a scalar potential function.

1. $\mathbf{F} = 2\mathbf{i} - 3\mathbf{j}$

2. $\mathbf{F} = xy^{-2}\mathbf{i} + x^{-2}y\mathbf{j}$

3. $\mathbf{F} = y^{-3}\mathbf{i} + (-3xy^{-4} + \cos y)\mathbf{j}$

4. $\mathbf{F} = y^2\mathbf{i} + 2xy\mathbf{j}$

5. $\mathbf{F} = \left(\dfrac{1}{y} + \dfrac{y}{x^2}\right)\mathbf{i} - \left(\dfrac{x}{y^2} - \dfrac{1}{x}\right)\mathbf{j}$

6. $\mathbf{F} = [2x\tan^{-1}\left(\frac{y}{x}\right) - y]\mathbf{i} + [2y\tan^{-1}\left(\frac{y}{x}\right) + x]\mathbf{j}$

In Problems 7-12, find $\int_C \mathbf{F} \cdot d\mathbf{R}$, where C is the curve $\mathbf{R}(t) = t\mathbf{i} + t^2\mathbf{j}$, $1 \leq t \leq 2$. Note that these are the same as the vector fields in Problems 1-6.

7. $\mathbf{F} = 2\mathbf{i} - 3\mathbf{j}$

8. $\mathbf{F} = xy^{-2}\mathbf{i} + x^{-2}y\mathbf{j}$

9. $\mathbf{F} = y^{-3}\mathbf{i} + (-3xy^{-4} + \cos y)\mathbf{j}$

10. $\mathbf{F} = y^2\mathbf{i} + 2xy\mathbf{j}$

11. $\mathbf{F} = \left(\dfrac{1}{y} + \dfrac{y}{x^2}\right)\mathbf{i} - \left(\dfrac{x}{y^2} - \dfrac{1}{x}\right)\mathbf{j}$

12. $\mathbf{F} = [2x\tan^{-1}\left(\frac{y}{x}\right) - y]\mathbf{i} + [2y\tan^{-1}\left(\frac{y}{x}\right) + x]\mathbf{j}$

In Problems 13-22, find div $\mathbf{F}$ and curl $\mathbf{F}$.

13. $\mathbf{F} = x\mathbf{i} + y\mathbf{j} + z\mathbf{k}$

14. $\mathbf{F} = \left(\tan^{-1}\frac{y}{x}\right)\mathbf{i} - 3\mathbf{j} + z^2\mathbf{k}$

15. $\mathbf{F} = xy\mathbf{i} + yz\mathbf{j} + xz\mathbf{k}$

16. $\mathbf{F} = 2xz\mathbf{i} + 2yz^2\mathbf{j} - \mathbf{k}$

17. $\mathbf{F} = ax\mathbf{i} + by\mathbf{j} + c\mathbf{k}$ for constants a, b, c

18. $\mathbf{F} = (e^x \sin y)\mathbf{i} + (e^x \cos y)\mathbf{j} + \mathbf{k}$

19. $\mathbf{F} = ax\mathbf{i} + by\mathbf{j} + cz\mathbf{k}$ for constants a, b, c

20. $\mathbf{F} = \mathbf{i} + (xyz)\mathbf{j} + \mathbf{k}$

21. $\mathbf{F} = \frac{1}{r}(x\mathbf{i} + y\mathbf{j} + z\mathbf{k})$, where $r = \sqrt{x^2 + y^2 + z^2}$, $r \neq 0$

22. $\mathbf{F} = (xy\sin z)\mathbf{i} + (x^2\cos yz)\mathbf{j} + (z\sin xy)\mathbf{k}$

Evaluate

$$\int_C [(3x + 2y)\,dx - (2x + 3y)\,dy]$$

for each of the paths given in Problems 23-26.

23.
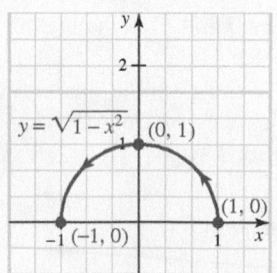

24.
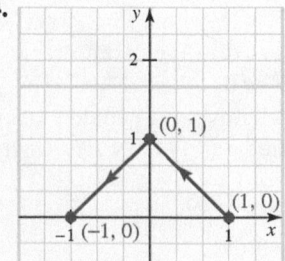

25.
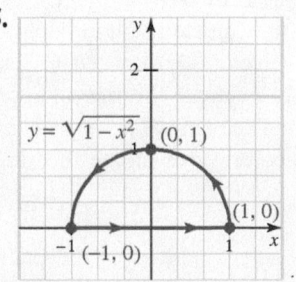

26.
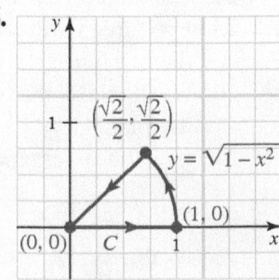

---

*The supplementary problems are presented in a somewhat random order, not necessarily in order of difficulty.

Evaluate the line integrals in Problems 27-30 by parameterization.

27. $\int_C [(\sin \pi y) \, dx + (\cos \pi x) \, dy]$, where C is the line segment from $(1,0)$ to $(\pi,0)$, followed by the line segment from $(\pi,0)$ to (π,π).

28. $\int_C [(2 \cos \pi y) \, dx + (3 \sin \pi x) \, dy]$, where C is the line segment from $(1,0)$ to $(\pi,0)$, followed by the line segment from $(\pi,0)$ to (π,π).

29. $\int_C (y \, dx + x \, dy + dz)$, where C is the arc of the helix $x = t, y = 3 \sin t, z = 2 \cos t$, for $0 \le t \le \frac{\pi}{2}$.

30. $\int_C (z \, dx - x \, dy + dz)$, where C is the arc of the helix $x = 3 \sin t, y = \cos t, z = t$, for $0 \le t \le \frac{\pi}{4}$.

In Problems 31-48, evaluate the line integral or the surface integral. In each surface integral, assume that $\mathbf{N}$ is the outward unit normal vector field.

31. $\int_C [yz \, dx + xz \, dy + (xy + 2) \, dz]$ where C is the curve $\mathbf{R}(t) = \tan^{-1} t \, \mathbf{i} + t^2 \mathbf{j} - 3t \mathbf{k}, 0 \le t \le 1$.

32. $\int_C [(x^2 + y) \, dx + xz \, dy - (y + z) \, dz]$ where C is the curve $\mathbf{R}(t) = t \mathbf{i} + t^2 \mathbf{j} + 2 \mathbf{k}, 0 \le t \le 1$.

33. $\iint_S (3x^2 + y - 2z) \, dS$ where S is the surface $\mathbf{R}(u, v) = u \mathbf{i} + (u + v) \mathbf{j} + u \mathbf{k}, 0 \le u \le 1, 0 \le v \le 1$.

34. $\iint_S \mathbf{F} \cdot \mathbf{N} \, dS$ where $\mathbf{F} = 3x \mathbf{i} + z^2 \mathbf{j} - 2y \mathbf{k}$ and S is the surface of the hemisphere $z = \sqrt{4 - x^2 - y^2}$.

35. $\int_C (x \, dx + x \, dy - y \, dz)$ where C is the curve $\mathbf{R}(t) = t \mathbf{i} + t^2 \mathbf{j} + t \mathbf{k}, 0 \le t \le 1$.

36. $\int_C (x^2 \, dx - 3y^2 \, dz)$ where C is the line segment from $(0, 1, 1)$ to $(1, 1, 2)$.

37. $\int_C (x^2 \, dx + y \, dy)$ where C is the curve $\mathbf{R}(t) = (t \sin t) \mathbf{i} + (1 - t \cos t) \mathbf{j}, 0 \le t \le 2\pi$.

38. $\int_C (xy \, dx - x^2 \, dy)$ where C is the square with vertices $(1, 0), (0, 1), (-1, 0), (0, -1)$ traversed counterclockwise.

39. $\int_C (y \, dx + x \, dy - 2 \, dz)$ where C is the curve of intersection of the cylinder $x^2 + y^2 = 2x$ and the plane $x = z$, traversed counterclockwise as viewed from above.

40. $\int_C [(y + z) \, dx + (x + z) \, dy + (x + y) \, dz]$ where C is the curve of intersection of the sphere $x^2 + (y - 3)^2 + z^2 = 9$ and the plane $x + 2y + z = 3$, traversed counterclockwise as viewed from above.

41. Evaluate $\int_C \mathbf{F} \cdot d\mathbf{R}$, where

$$\mathbf{F} = \left\langle \frac{x}{\sqrt{x^2 + y^2}}, \frac{-y}{\sqrt{x^2 + y^2}} \right\rangle$$

and C is the quarter circle path $x^2 + y^2 = a^2$, traversed from $(a, 0)$ to $(0, a)$.

42. Evaluate $\int_C \mathbf{F} \cdot d\mathbf{R}$, where $\mathbf{F} = (y - 2z) \mathbf{i} + x \mathbf{j} - 2xy \mathbf{k}$ and C is the path given by $\mathbf{R}(t) = t \mathbf{i} + t^2 \mathbf{j} - \mathbf{k}$ for $1 \le t \le 2$.

43. $\oint_C (-2y \, dx + 2x \, dy + dz)$ where C is the circle $x^2 + y^2 = 1$ in the plane $z = 3$, traversed counterclockwise as viewed from above.

44. $\iint_S (\text{curl } y \mathbf{i}) \cdot \mathbf{N} \, dS$ where S is the hemisphere $z = \sqrt{1 - x^2 - y^2}$.

45. $\iint_S (2x^3 \mathbf{i} + y^3 \mathbf{j} + z^3 \mathbf{k}) \cdot \mathbf{N} \, dS$ where S is the surface of the ellipsoid $2x^2 + y^2 + z^2 = 1$.

46. $\iint_S (y^2 \mathbf{i} + y^2 \mathbf{j} + yz \mathbf{k}) \cdot \mathbf{N} \, dS$ where S is the surface of the tetrahedron bounded by the plane $2x + 3y + z = 1$ and the coordinate planes, in the first octant.

47. $\iint_S \nabla \phi \cdot \mathbf{N} \, dS$, where $\phi(x, y, z) = 2x + 3y$ and S is the portion of the plane $ax + by + cz = 1$ $(a > 0, b > 0, c > 0)$ that lies in the first octant.

48. $\iint_S \nabla \phi \cdot \mathbf{N} \, dS$, where $\phi(x, y, z) = 3x + 2y$ and S is the portion of the plane $2x + y + z = 1$ that lies in the first octant.

In Problems 49-54 find $\iint_S \mathbf{F} \cdot \mathbf{N} \, dS$. Assume $\mathbf{N}$ is the outward unit normal vector field for S.

49. $\mathbf{F} = x \mathbf{i} + y \mathbf{j} + z \mathbf{k}$ and S is the surface of the unit cube $0 \le x \le 1, 0 \le y \le 1, 0 \le z \le 1$.

50. $\mathbf{F} = x^2 \mathbf{i} + y^2 \mathbf{j} + x^2 \mathbf{k}$ and S is the surface of the unit cube $0 \le x \le 1, 0 \le y \le 1, 0 \le z \le 1$.

51. $\mathbf{F} = 2yz \mathbf{i} + (\tan^{-1} xz) \mathbf{j} + e^{xy} \mathbf{k}$ and S is the surface of the sphere $x^2 + y^2 + z^2 = 1$.

52. $\mathbf{F} = x \mathbf{i} - 4 \mathbf{j} + 3 \mathbf{k}$ and S is the paraboloid $y = x^2 + z^2$ with $x^2 + z^2 < 9$. The disk $x^2 + z^2 = 9$ is omitted; that is, the paraboloid is open on the right.

53. $\mathbf{F} = xyz \mathbf{i} + xyz \mathbf{j} + xyz \mathbf{k}$ and S is the surface of the five faces of the unit cube $0 \le x \le 1, 0 \le y \le 1, 0 < z \le 1$, missing $z = 0$.

54. $\mathbf{F} = xy \mathbf{i} - 2z \mathbf{j}$ and S is the surface defined parametrically by $\mathbf{R}(u, v) = u \mathbf{i} + v \mathbf{j} + u \mathbf{k}$ for $0 \le u \le 1, 0 \le v \le 1$.

Find all real numbers c for which each vector field in Problems 55-58 is conservative.

55. $\mathbf{F}(x,y) = (\sqrt{x} + 3xy)\mathbf{i} + (cx^2 + 4y)\mathbf{j}$

56. $\mathbf{F}(x,y) = \left(\dfrac{cy}{x^3} + \dfrac{y}{x^2}\right)\mathbf{i} + \left(\dfrac{1}{x^2} - \dfrac{1}{x}\right)\mathbf{j}$

57. $\mathbf{F}(x,y,z) = e^{yz/x}\left[\left(\dfrac{cyz}{x^2}\right)\mathbf{i} + \left(\dfrac{z}{x}\right)\mathbf{j} + \left(\dfrac{y}{x}\right)\mathbf{k}\right]$

58. $\mathbf{F}(x,y,z) = (x + xyz)\mathbf{i} + (cx^2 + 4z)\mathbf{k}$

59. Let $\mathbf{F} = (y^2 + x^{-2}ye^{x/y})\mathbf{i} + (2xy + z - x^{-1}e^{x/y})\mathbf{j} + y\mathbf{k}$. Is $\mathbf{F}$ conservative?

60. Determine the most general function $u(x,y)$ for which the vector field $\mathbf{F} = u(x,y)\mathbf{i} + (2ye^x + y^2e^{3x})\mathbf{j}$ is conservative.

61. Show that if the vector field
$$\mathbf{F}(x,y,z) = M(x,y,z)\mathbf{i} + N(x,y,z)\mathbf{j} + P(x,y,z)\mathbf{k} \text{ is conservative, then}$$

$$\frac{\partial P}{\partial y} = \frac{\partial N}{\partial z} \qquad \frac{\partial M}{\partial z} = \frac{\partial P}{\partial x} \qquad \frac{\partial N}{\partial x} = \frac{\partial M}{\partial y}$$

62. A force field

$$\mathbf{F}(x,y) = (3x^2 + 6xy^2)\mathbf{i} + (6x^2y + 4y^2)\mathbf{j}$$

acts on an object moving in the plane.
Show that $\mathbf{F}$ is conservative, and find a scalar potential for $\mathbf{F}$. How much work is done as the object moves from $(1,0)$ to $(0,1)$ along any path connecting these points?

Figure 13.65 Force field

63. Let f and g be differentiable functions of one variable. Show that the vector field

$$\mathbf{F} = [f(x) + y]\mathbf{i} + [g(y) + x]\mathbf{j}$$

is conservative, and find the corresponding potential function.

64. If S is a closed surface in a region R and $\mathbf{F}$ is a twice continuously differentiable vector field on R, show that

$$\iint\limits_{S} (\text{curl } \mathbf{F} \cdot \mathbf{N})\, dS = 0$$

where $\mathbf{N}$ is the outward unit normal vector to S.

65. Find the work done when an object moves in the force field $\mathbf{F} = 2x\mathbf{i} - (x+z)\mathbf{j} + (y-x)\mathbf{k}$ along the path given by $\mathbf{R}(t) = t^2\mathbf{i} + (t^2 - t)\mathbf{j} + 3\mathbf{k}$, $0 \le t \le 1$.

66. Show that the force field $\mathbf{F} = yz^2\mathbf{i} + (xz^2 - 1)\mathbf{j} + (2xyz - 1)\mathbf{k}$ is conservative and determine the work done when an object moves in the force field from the origin to the point $(1,0,1)$.

67. Find a region $\mathbf{R}$ in the plane where the vector field $\mathbf{F} = \dfrac{1}{x+y}(\mathbf{i} + \mathbf{j})$ is conservative. Then evaluate $\int_C \mathbf{F} \cdot d\mathbf{R}$, where C is any path in R from the point $P_0(a,b)$ to $P_1(c,d)$.

68. If u is a scalar function and $\mathbf{F}$ is a continuously differentiable vector field, show that
$\text{curl}(u\mathbf{F}) = u\,\text{curl } \mathbf{F} + (\nabla u \times \mathbf{F})$.

69. Show that $\text{div}(\mathbf{F} \times \mathbf{G}) = \mathbf{G} \cdot \text{curl } \mathbf{F} - \mathbf{F} \cdot \text{curl } \mathbf{G}$, for any continuously differentiable vector fields $\mathbf{F}$ and $\mathbf{G}$.

70. A vector field $\mathbf{F}$ is *incompressible* in a region D if div $\mathbf{F} = 0$ throughout D. If $\mathbf{F}$ and $\mathbf{G}$ are both conservative vector fields in D, show that $\mathbf{F} \times \mathbf{G}$ is incompressible.

71. If $\mathbf{F} = \text{curl } \mathbf{G}$, show that $\mathbf{F}$ is incompressible. (See Problem 70.)

72. Suppose $\mathbf{F} = f(x,y,z)\mathbf{A}$, where $\mathbf{A}$ is a constant vector and f is a scalar function. Show that curl $\mathbf{F}$ is orthogonal to $\mathbf{A}$ and to ∇f.

73. If $\mathbf{A}$ is a constant vector and $\mathbf{F}$ is a continuously differentiable vector field, show that $\text{div}(\mathbf{A} \times \mathbf{F}) = -\mathbf{A} \cdot \text{curl } \mathbf{F}$.

74. Evaluate the line integral $\displaystyle\int_C \frac{x\,dx - y\,dy}{x^2 - y^2}$, where C is any path in the xy-plane that is interior to the region $x > 0$, $y < x$, $y > -x$ and connects the point $(5,4)$ to $(2,0)$.

75. Evaluate $\displaystyle\oint_C \left(\frac{-y}{x^2}\,dx + \frac{1}{x}\,dy\right)$, where C is the closed path $(x-2)^2 + y^2 = 1$, traversed once counterclockwise.

76. Find a region in the plane where the vector field

$$\mathbf{F} = \left(\frac{1+y^2}{x^3}\right)\mathbf{i} - \left(\frac{y+x^2y}{x^2}\right)\mathbf{j}$$

is conservative, and find a scalar potential for $\mathbf{F}$.

77. Think Tank Problem Use the results of Problem 76 to evaluate $\int_C \mathbf{F} \cdot d\mathbf{R}$, where $\mathbf{F}$ is defined in Problem 76 and C is a path from $(1, 1)$ to $(3, 4)$. Are there any limitations on the path C? Explain.

78. Think Tank Problem Let $u(x, y)$ and $v(x, y)$ be functions of two variables with continuous partial derivatives everywhere in the plane, and suppose that u and v satisfy the equation

$$\frac{\partial u}{\partial y} = \frac{\partial v}{\partial x}$$

for all (x, y). Are u and v necessarily harmonic? Either show that they are or find a counterexample.

79. Consider the line integral $\displaystyle\int_C \left(\frac{dx}{y} + \frac{dy}{x}\right)$, where C is the closed triangular path formed by the lines $y = 2x$, $x + 2y = 5$, and $x = 2$, traversed counterclockwise. First evaluate the line integral directly (by parameterizing C) and then by using Green's theorem.

80. A certain closed path C in the plane $2x + 2y + z = 1$ is known to project onto the unit circle $x^2 + y^2 = 1$ in the xy-plane. Let c be a constant, and let $\mathbf{R} = x\mathbf{i} + y\mathbf{j} + z\mathbf{k}$. Use Stokes' theorem to evaluate

$$\oint_C (c\mathbf{k} \times \mathbf{R}) \cdot d\mathbf{R}$$

81. Show that if the scalar function w is harmonic, then

$$\nabla \cdot (w\nabla w) = \|\nabla w\|^2$$

82. Let $w = x - y + 2z$, and let S be the surface of the sphere $x^2 + y^2 + z^2 = 9$. Use the results of Problem 81 to evaluate

$$\iint_S w \frac{\partial w}{\partial n} d\mathbf{S}$$

83. A particle moves along a curve C in space that is given parametrically by $\mathbf{R}(t) = x(t)\mathbf{i} + y(t)\mathbf{j} + z(t)\mathbf{k}$ for $a \leq t \leq b$. A force field $\mathbf{F}$ is applied to the particle in such a way that $\mathbf{F}$ is always perpendicular to the path C. How much work is performed by the force field $\mathbf{F}$ when the particle moves from the point where $t = a$ to the point where $t = b$?

84. If $\mathbf{D}$ is the electric displacement field, then div $\mathbf{D} = \phi$, where ϕ is the **charge density**. A region of space is said to be **charge-free** if $\phi = 0$ there. Describe the charge-free regions of the electric displacement field $\mathbf{D} = 2x^2\mathbf{i} + 3y^2\mathbf{j} - 2z^2\mathbf{k}$.

85. A satellite weighing 10,000 kg travels in a circular orbit 7,000 km from the center of the earth. How much work is done by gravity on the satellite during half a revolution?

86. If $\mathbf{F}$ is a conservative force field, the scalar function f such that $\mathbf{F} = -\nabla f$ is called the potential energy (See Problem 59, Section 13.3). Suppose an object with mass 10 g moves in the force field in such a way that its speed decreases from 3 cm/s to 2.5 cm/s. What is the corresponding change in the potential energy of the object?

87. Find the work done by the force field $\mathbf{F} = 2xyz\mathbf{i} + \left(x^2z - \frac{1}{z}\tan^{-1}\frac{y}{z}\right)\mathbf{j} + \left(x^2y + \frac{y}{z^2}\tan^{-1}\frac{y}{z}\right)\mathbf{k}$ in moving an object along the circular helix $\mathbf{R}(t) = (\sin \pi t)\mathbf{i} + (\cos \pi t)\mathbf{j} + (2t + 1)\mathbf{k}$ for $0 \leq t \leq \frac{1}{2}$.

88. Find the work done when an object moves against the force field $\mathbf{F} = 4y^2\mathbf{i} + (3x + y)\mathbf{j}$ from $(1, 0)$ to $(-1, 0)$ along the top half of the ellipse $x^2 + \dfrac{y^2}{k^2} = 1$. Which value of k minimizes the work?

89. Evaluate the line integral

$$\oint_C \frac{-y\,dx + x\,dy}{x^2 + y^2}$$

where C is the limaçon given in polar coordinates by $r = 3 + 2\cos\theta$, $0 \le \theta \le 2\pi$, traversed counterclockwise.

90. Suppose f and g are both harmonic in the region R with boundary surface S. Show that

$$\iint_S f\,\frac{\partial g}{\partial n}\,dS = \iint_S g\,\frac{\partial f}{\partial n}\,dS$$

91. Suppose f and g are both harmonic in the region R with boundary surface S. Show that

$$\iint_S f\,\frac{\partial f}{\partial n}\,dS = \iiint_D \|\nabla f\|^2\,dV$$

92. Evaluate the surface integral $\displaystyle\iint_S \frac{\partial f}{\partial n}\,dS$, where S is the surface of the unit sphere $x^2 + y^2 + z^2 = 1$ and f is a scalar field such that $\|\nabla f\|^2 = 3f$ and $\operatorname{div}(f\,\nabla f) = 7f$.

93. Evaluate the surface integral $\displaystyle\iint_S dS$, where S is the torus

$$\mathbf{R}(u, v) = [(a + b\cos v)\cos u]\mathbf{i} + [(a + b\cos v)\sin u]\mathbf{j} + (b\sin v)\mathbf{k} \text{ for } 0 < b < a \text{ and } 0 \le u \le 2\pi, 0 \le v \le 2\pi.$$

94. Evaluate $\iint_S \mathbf{F} \cdot \mathbf{N}\,dS$, where $\mathbf{F} = x\mathbf{i} + y\mathbf{j} + z\mathbf{k}$ and S is the closed cubic surface with a corner block removed, as shown in Figure 13.66.

95. Think Tank Problem If you remove the opposite corner in the previous problem, how will the answer change?

96. Show that a lamina that covers a standard region D in the plane with density $\rho = 1$ has moment of inertia

$$I = \frac{1}{3}\oint_C (-y^3 dx + x^3 dy)$$

with respect to the z-axis, where C is the boundary curve of D.

97. Use vector analysis to find the centroid of the conical surface $z = \sqrt{x^2 + y^2}$ between $z = 0$ and $z = 3$.

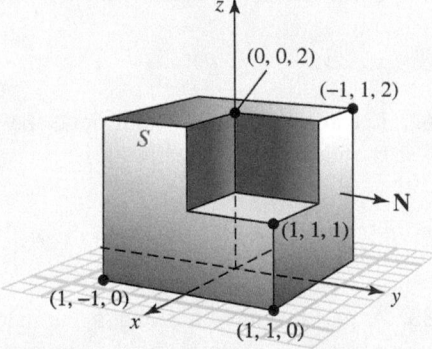

Figure 13.66   Cube with corner removed

98. Show that $\operatorname{curl}(\operatorname{curl}\mathbf{F}) = \nabla(\operatorname{div}\mathbf{F}) - \nabla^2\mathbf{F}$ if the components of $\mathbf{F}$ have continuous second-order partial derivatives.

99. Putnam Examination Problem A force acts on the element ds of a closed plane curve. The magnitude of this force is $r^{-1}ds$, where r is the radius of curvature at the point considered, and the direction of the force is perpendicular to the curve; it points to the convex side. Show that the system of such forces acting on all elements of the curve keeps it in equilibrium.

CHAPTER 13 GROUP RESEARCH PROJECT*

Working In small groups is typical of most work environments, and this book seeks to develop skills with group activities. We present a group project at the end of each chapter. These projects are to be done in groups of three or four students.

Reconstruction of Fossils

© Kendall Hunt Publishing Company

Reconstructing three dimensional surfaces (bone structure, for example) from point samples is a well-studied problem in computer graphics. For example, a bone fragment has been found and to construct a proper bone structure a process is needed to scan, fill holes, and remesh the fragments with existing models. One approach to this problem is to transform volume integrals into surface integrals and compute a discrete approximation using the oriented point samples.

This approach is carried out in three steps:

1. The point-normal pairs are splatted into a voxel grid.
2. The voxel grid is convolved with an integration filter.
3. The reconstructed surface is extracted as an iso-surface of the voxel grid.

This procedure involves the use of Stokes' theorem.

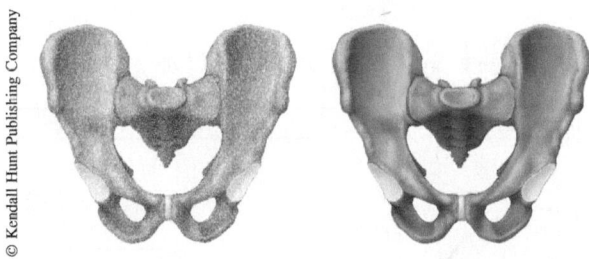

© Kendall Hunt Publishing Company

Extended paper for further study. Investigate the process of reconstructing bone fragments into complete bones as they existed when the specimen was living.

---

*Project is motivated by ''Reconstruction of Solid models from Oriented Point Sets,'' by Michael Kazhdan. © The Eurographics Association 2005.

Cumulative Review—Chapters 11-13 (multivariable calculus)

If you look at the table of contents, you will find cumulative reviews at the end of Chapters 5, 8, 10, and 13. There are two major divisions for a calculus course, single variable (Chapters 1-10) and multivariable (Chapters 11-13). These cumulative reviews were chosen to reflect the major divisions of a beginning calculus course: derivatives, integrals, vector calculus, and multivariable calculus. These summaries are provided here to help you see the big picture, and to focus your attention on the important ideas in each section of this book.

Table 13.1   Comparison of important integral theorems

Riemann Integral (Section 5.3)	*Line Integral* (Section 13.2)
$\int_a^b f(x)\,dx$ Subdivision on the x-axis 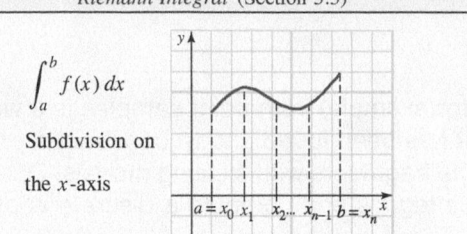	$\int_C f(x, y, z)\,ds$ Subdivisions on a curve C in space 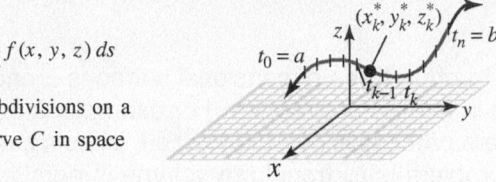
Fundamental Theorem of Calculus (Section 5.4)	*Fundamental Theorem for Line Integrals* (Section 13.3)
$\int_a^b f(x)\,dx = F(b) - F(a)$ if F is any antiderivative of f	$\int_C \mathbf{F} \cdot d\mathbf{R} = f(Q) - f(P)$ if $\mathbf{F}$ is conservative with scalar potential f; that is, $\nabla f = \mathbf{F}$

Green's Theorem (Section 13.4):
$\mathbf{F}(x, y) = M(x, y)\mathbf{i} + N(x, y)\mathbf{j}$ is continuously differentiable on the simply connected region D with positive oriented piecewise boundary curve C. Then,

$$\oint_C (M\,dx + N\,dy) = \iint_D \left(\frac{\partial N}{\partial x} - \frac{\partial M}{\partial y} \right) dA$$

Alternate forms include $\quad \oint_C \mathbf{F} \cdot \mathbf{T}\,ds = \iint_D (\text{curl } \mathbf{F} \cdot \mathbf{k})\,dA \quad$ and $\quad \oint_C \mathbf{F} \cdot \mathbf{N}\,ds = \iint_D \text{div } \mathbf{F}\,dA$

Double integral (Section 12.1)	*Surface integral* (Section 13.5)
$\iint_R f(x, y)\,dx\,dy$ Partition of R into mn cells in the xy-plane 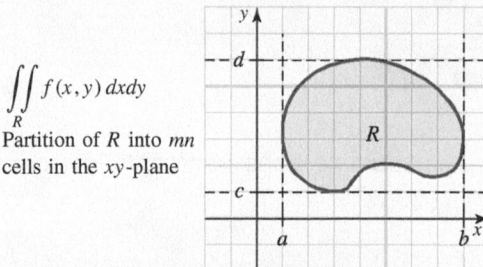	$\iint_S g(x, y, z)\,dS$ $= \iint_R g(x, y, z)\sqrt{f_x^2 + f_y^2 + 1}\,dA$ Partition of the surface S into n subregions. 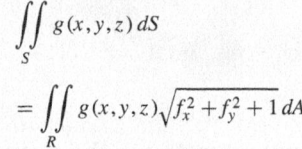 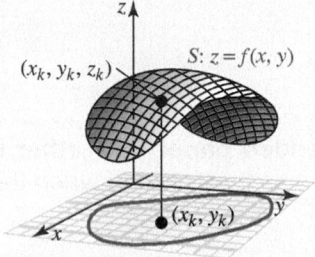
Stokes' Theorem (Section 13.6)	*The Divergence Theorem* (Section 13.7) Also known as Gauss' theorem.
$\oint_C \mathbf{F} \cdot d\mathbf{R} = \iint_R (\text{curl } \mathbf{F} \cdot \mathbf{N})\,dS$ where C is the positively oriented boundary curve of the surface S with outward unit normal field $\mathbf{N}$ 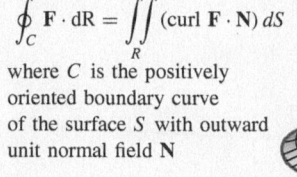 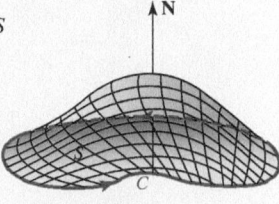	$\iint_S \mathbf{F} \cdot \mathbf{N}\,dS = \iiint_D \text{div } \mathbf{F}\,dV$ where D is the solid region with closed boundary surface S. 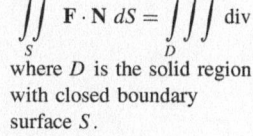 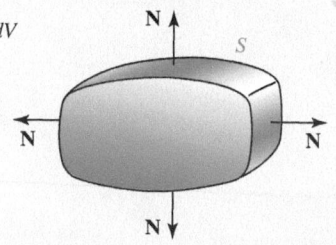

In the following statements, assume that all required conditions are satisfied. Main results are collected here (without hypotheses) so that you can compare and contrast various results and conclusions.

Scalar function: $f(x, y, z)$

Vector field: $\mathbf{F}(x, y, z) = u(x, y, z)\mathbf{i} + v(x, y, z)\mathbf{j} + w(x, y, z)\mathbf{k}$

Notation:

Del operator: $\nabla = \dfrac{\partial}{\partial x}\mathbf{i} + \dfrac{\partial}{\partial y}\mathbf{j} + \dfrac{\partial}{\partial z}\mathbf{k}$

Gradient: $\nabla f = \dfrac{\partial f}{\partial x}\mathbf{i} + \dfrac{\partial f}{\partial y}\mathbf{j} + \dfrac{\partial f}{\partial z}\mathbf{k} = f_x\mathbf{i} + f_y\mathbf{j} + f_z\mathbf{k}$

Laplacian: $\nabla^2 f = \dfrac{\partial^2 f}{\partial x^2} + \dfrac{\partial^2 f}{\partial y^2} + \dfrac{\partial^2 f}{\partial z^2} = f_{xx} + f_{yy} + f_{zz}$

Normal derivative: $\dfrac{\nabla f}{\partial n} = \nabla f \cdot \mathbf{N}$

Derivatives of a vector field: $\operatorname{div} \mathbf{F} = \dfrac{\partial u}{\partial x} + \dfrac{\partial v}{\partial y} + \dfrac{\partial w}{\partial z}$ *This is a scalar derivative.*

$$= \nabla \cdot \mathbf{F}$$

If P is a point (x_0, y_0, z_0), then P is a source if $\operatorname{div} \mathbf{F} > 0$

sink if $\operatorname{div} \mathbf{F} < 0$

$\mathbf{F}$ is incompressible if $\operatorname{div} \mathbf{F} = 0$

$$\operatorname{curl} \mathbf{F} = \begin{vmatrix} \mathbf{i} & \mathbf{j} & \mathbf{k} \\ \frac{\partial}{\partial x} & \frac{\partial}{\partial y} & \frac{\partial}{\partial z} \\ u & v & w \end{vmatrix} \qquad \textit{This is a vector derivative.}$$

$$= \nabla \times \mathbf{F}$$

$\mathbf{F}$ is irrotational if $\operatorname{curl} \mathbf{F} = \mathbf{0}$

Line Integral   If $f(x, y, z)$ is defined on the smooth curve C with parametric equations $x = x(t)$, $y = y(t)$, $z = z(t)$, then the line integral of f over C is given by

$$\int_C f(x, y, z)\, ds = \int_a^b f[x(t), y(t), z(t)]\sqrt{[x'(t)]^2 + [y'(t)]^2 + [z'(t)]^2}\, dt$$

Green's theorem expresses an important relationship between a line integral over a Jordan (simple closed) curve in the plane and a double integral over the region bounded by the curve. Let D be a simply connected region with a positively oriented piecewise smooth boundary C. Then if the vector field $\mathbf{F}(x, y) = M(x, y)\mathbf{i} + N(x, y)\mathbf{j}$ is continuously differentiable on D, we have

$$\oint_C (M\, dx + N\, dy) = \iint_D \left(\frac{\partial N}{\partial x} - \frac{\partial M}{\partial y} \right) dA$$

Conservative Vector Fields and Path Independence   Let $\mathbf{F}$ be a continuous vector field on the open connected set D. Then the following three conditions are either all true or all false:

a. $\mathbf{F}$ is conservative on D; that is, $\mathbf{F} = \nabla f$ for some function f defined on D.

b. $\oint_C \mathbf{F} \cdot d\mathbf{R} = 0$ for every piecewise smooth closed curve in D.

c. $\int_C \mathbf{F} \cdot d\mathbf{R}$ is independent of path within D

Also, if D is simply connected (no holes), then $\mathbf{F}$ is conservative if and only if $\operatorname{curl} \mathbf{F} = \mathbf{0}$ in D.

Fundamental Theorem for Line Integrals   Let C be a piecewise smooth curve that is parameterized by the vector function $\mathbf{R}(t)$ for $a \leq t \leq b$. If f is a differentiable function of two or three variables whose gradient $\mathbf{F} = \nabla f$ is continuous on C, then

$$\int_C \mathbf{F} \cdot d\mathbf{R} = f(Q) - f(P)$$

where $Q = \mathbf{R}(b)$ and $P = \mathbf{R}(a)$ are the endpoints of C.

EVALUATION OF LINE INTEGRALS: $\displaystyle\int_C \mathbf{F} \cdot d\mathbf{R}$

Step 1 Check to see whether $\mathbf{F}$ is conservative; if it is, then

$$\oint_C \mathbf{F} \cdot d\mathbf{R} = 0 \text{ if } C \text{ is closed}$$

$$\int_C \mathbf{F} \cdot d\mathbf{R} = f(Q) - f(P) \qquad \textit{Initial point P, terminal point Q}$$

Step 2 If $\mathbf{F}$ is not conservative, and C is a closed curve bounding a surface S, use Stokes' theorem (or Green's theorem in $\mathbb{R}^2$) to equate the given integral to a surface integral, namely,

$$\int_C \mathbf{F} \cdot d\mathbf{R} = \iint_S (\text{curl } \mathbf{F} \cdot \mathbf{N}) \, dS$$

Step 3 As a last resort, parameterize $\mathbf{F}$ and $\mathbf{R}$.

Let $\mathbf{R}(t) = x(t)\mathbf{i} + y(t)\mathbf{j} + z(t)\mathbf{k}$ for $a \leq t \leq b$.

$$\int_C f(x,y,z) \, dx = \int_a^b f[x(t),y(t),z(t)] \frac{dx}{dt} \, dt$$

$$\int_C f(x,y,z) \, ds = \int_a^b f[x(t),y(t),z(t)] \sqrt{[x'(t)]^2 + [y'(t)]^2 + [z'(t)]^2} \, dt$$

Surface Integral   Let S be a surface defined by $z = f(x,y)$ and R its projection on the xy-plane.

If f, f_x, and f_y are continuous in R and g is continuous on S, then the surface integral of g over S is

$$\iint_S g(x,y,z) \, dS = \iint_R g(x,y,f(x,y)) \sqrt{[f_x(x,y)]^2 + [f_y(x,y)]^2 + 1} \, dA$$

Stokes' theorem   Let S be an oriented surface with unit normal vector field $\mathbf{N}$, and assume that S is bounded by a piecewise smooth Jordan curve C whose orientation is compatible with that of S. If $\mathbf{F}$ is a vector field that is continuously differentiable on S, then

$$\oint_C \mathbf{F} \cdot d\mathbf{R} = \iint_S (\text{curl } \mathbf{F} \cdot \mathbf{N}) \, dS$$

The divergence theorem   Let S be a smooth, orientable surface that encloses a solid region D in $\mathbb{R}^3$. If $\mathbf{F}$ is a continuous vector field whose components have continuous partial derivatives in an open set containing D, then

$$\iint_S \mathbf{F} \cdot \mathbf{N} \, dS = \iiint_D \text{div } \mathbf{F} \, dV$$

where $\mathbf{N}$ is the outward unit normal field for the surface S.

EVALUATION OF FLUX INTEGRALS: $\iint\limits_{S} \mathbf{F} \cdot \mathbf{N} \, dS$

Step 1 If the surface S is a closed surface bounding the solid region D, use the divergence theorem to write the flux integral as a triple integral.

$$\iint\limits_{S} \mathbf{F} \cdot \mathbf{N} \, dS = \iiint\limits_{D} \text{div } \mathbf{F} \, dV$$

Step 2 If Step 1 does not apply, parameterize $\mathbf{F}$, $\mathbf{N}$ and dS. In the special case where S has the form $z = f(x, y)$, we have

$$\iint\limits_{S} \mathbf{F} \cdot \mathbf{N} \, dS = \iint\limits_{R} \mathbf{F}(x, y, f(x, y)) \cdot \langle -f_x, -f_y, 1 \rangle \, dA$$

for an upward normal and

$$\iint\limits_{S} \mathbf{F} \cdot \mathbf{N} \, dS = \iint\limits_{R} \mathbf{F}(x, y, f(x, y)) \cdot \langle f_x, f_y, -1 \rangle \, dA$$

for a downward normal.

Cumulative Review Problems—Chapters 11-13

1. ■ *What does this say?* Suppose you tell a fellow college student that you are about to finish a calculus course. The student has not had any mathematics beyond high school and asks you, "What is calculus?" Answer this question using your own words.

2. ■ *What does this say?* There are three great fundamental ideas in calculus: the limit, the derivative, and the integral. In your own words, explain each of these concepts.

3. ■ *What does this say?* What is a vector-valued function? Why are vectors and vector-valued functions important to your study of calculus?

4. ■ *What does this say?* Chapters 11-13 were concerned with functions of several variables. This is often referred to as **multivariable calculus**. In your own words, discuss what is meant by multivariable calculus.

5. ■ *What does this say?* Compare or contrast the notation $\int_a^b f(x) \, dx$ and $\int_C f(x, y, z) \, ds$.

6. ■ *What does this say?* Compare or contrast the fundamental theorem of calculus with the fundamental theorem for line integrals.

In Problems 7-10, assume that all required conditions are satisfied. Fill in the blanks to complete each formula.

7. **a.** Del operator: $\nabla = $ _____ **b.** Gradient: $\nabla f = $ _____ **c.** Laplacian: $\nabla^2 f = $ _____

8. **a.** Normal derivative: $\dfrac{\partial f}{\partial n} = $ _____ **b.** $\mathbf{F}$ is incompressible if

9. **a.** div $\mathbf{F}$ _____ **b.** curl $\mathbf{F} = $ _____

10. If P is a point (x_0, y_0, z_0), then P is a

   a. source if _____ **b.** sink if _____

Find $f_x, f_y,$ and f_{xy} for the functions whose equations are given in Problems 11-16.

11. $f(x, y) = 2x^2 + xy - 5y^3$ 12. $f(x, y) = x^2 e^{y/x}$ 13. $f(x, y) = \dfrac{x^2 - y^2}{x - y}$

14. $f(x, y) = y \sin^2 x + \cos xy$ 15. $f(x, y) = e^{x+y}$ 16. $f(x, y) = \dfrac{x^2 + y^2}{x - y}$

17. Find $f_{xx} - f_{xy} + f_{yy}$ where $f(x, y) = x^2 y^3 + x^3 y^2$.   18. Find $f_{xx} + f_{yy}$ (the Laplacian) where $f(x, y) = e^{xy}$.

Evaluate the integrals in Problems 19-26.

19. $\displaystyle\int_0^1 \int_x^{3x} e^{y-2x}\,dy\,dx$

20. $\displaystyle\int_0^2 \int_0^{\sqrt{x}} 2x^3\,dy\,dx$

21. $\displaystyle\int_0^1 \int_0^z \int_y^{y-z} (x+y+z)\,dx\,dy\,dz$

22. $\displaystyle\int_0^{15\pi} \int_0^{\pi} \int_0^{\sin\phi} \rho^3 \sin\phi\,d\rho\,d\theta\,d\phi$

23. $\displaystyle\iint_R e^{x+y}\,dA,\ 0\le x \le 1;\ 0 \le y \le 1$

24. $\displaystyle\iint_R ye^{xy}\,dA,\ R{:}\,0\le x \le 1;\ 0 \le y \le 2$

25. $\displaystyle\iint_R \sin(x+y)\,dA, 0\le x \le \frac{\pi}{2}; 0 \le y \le \frac{\pi}{4}$

26. $\displaystyle\iint_R x\sin xy\,dA,\ R{:}\,0\le x \le \pi;\ 0 \le y \le 1$

In Problems 27-28, evaluate the given integral by converting to polar coordinates.

27. $\displaystyle\int_0^2 \int_0^{\sqrt{4-y^2}} \frac{1}{\sqrt{9-x^2-y^2}}\,dx\,dy$

28. $\displaystyle\int_2^4 \int_0^{\sqrt{4y-y^2}} \frac{1}{\sqrt{x^2+y^2}}dx\,dy$

Sketch the region of integration in Problems 29-30, and then compute the integral in two ways:
   a. *with the given order of integration, and*   **b.** *with the order of integration reversed.*

29. $\displaystyle\int_0^1 \int_x^{2x} e^{y-x}\,dy\,dx$

30. $\displaystyle\int_0^4 \int_0^{\sqrt{x}} 3x^5\,dy\,dx$

Evaluate the line integrals in Problems 31-36.

31. $\displaystyle\int_C (y^2z\,dx + 2xyz\,dy + xy^2\,dz)$, where C is any path from $(0,0,0)$ to $(1,1,1)$

32. $\displaystyle\int_C (5xy\,dx + 10yz\,dy + z\,dz)$, where C is given by $x=t^2$, $y=t$, $z=2t^3$ for $0\le t \le 1$

33. $\displaystyle\oint_C \mathbf{F}\cdot d\mathbf{R}$, where $\mathbf{F}=yz\mathbf{i}-x\mathbf{k}$, and C is the boundary of the triangle $(1,1,1)$, $(1,0,1)$, $(0,0,1)$, traversed once counterclockwise as viewed from the origin

34. $\displaystyle\oint_C (x^3+y^3)\,ds$, where C is given by $\mathbf{R}(t)=(\cos^3 t)\mathbf{i}+(\sin^3 t)\mathbf{j}$, for $0\le t \le 2\pi$

35. $\displaystyle\int_C \left\{\left[\frac{x}{y}+\frac{3y}{x^2+y^2}\right]dx + e^{-xy}\,dy\right\}$ where C is the line segment from $(0,1)$ to $(1,1)$.

36. $\displaystyle\int_C \left\{\left[\tan^{-1}\frac{y}{x}-\frac{xy}{x^2+y^2}\right]dx + \left[\frac{x^2}{x^2+y^2}+e^{-y}(1-y)\right]dy\right\}$ where C is any smooth curve from $(1,1)$ to $(-1,2)$.

37. What is a conservative vector field?
38. Let $f(x,y)=y^2-x^2$, and let $P_0(1,1)$.
   a. Find the gradient of f at P_0.
   b. Find the directional derivative of f in the direction of P_0 toward the origin.
   c. Find the direction from P_0 in which the directional derivative has the largest value. What is the magnitude of the largest directional derivative at P_0?
39. Find the maximum and minimum values of $f(x,y)=x+2y$ subject to the constraint $x^2+y^2=1$.
40. Use Lagrange multipliers to find the dimensions of the box with largest volume if the total surface area is 256 in.²
41. What is a level curve of the function defined by $f(x,y)$?

42. Find the equations for the tangent plane and the normal line to $z = x^2 + y^2 + \sin xy$ at $P_0(0, 2, 4)$. The graph of this surface is shown in Figure 13.67.

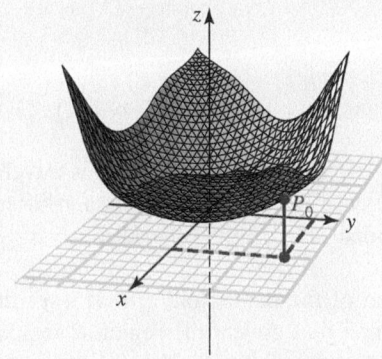

Figure 13.67 Graph of $z = x^2 + y^2 + \sin xy$

43. Classify the points A, B, and C shown in Figure 13.68 as sinks, sources or neither.

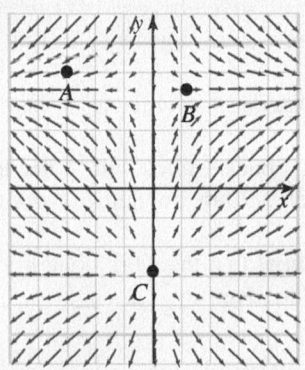

Figure 13.68 Force field

44. Sketch the level curves for the graph of $f(x, y) = xy$ in the first quadrant.

45. Find the parametric equations for the tangent line to the curve of intersection of the paraboloid $z = x^2 + y^2$ and the ellipsoid $2x^2 + 2y^2 + z^2 = 8$ at $P_0(-1, 1, 2)$.

46. Find the absolute extrema of $f(x, y) = x^2 - 4xy + y^3 + 4y$ on S is the square region $0 \le x \le 2$, $0 \le y \le 2$.

47. What are the dimensions of the closed rectangular box of fixed volume V_0 that has minimum surface area?

48. A manufacturer is planning to sell a new product at the price of \$350 per unit and estimates that if x thousand dollars is spent on development and y thousand dollars is spent on promotion, consumers will buy approximately

$$\frac{250y}{y+2} + \frac{100x}{x+5}$$

units of the product. If manufacturing costs for this product are \$150 per unit, how much should the manufacturer spend on development and how much on promotion to generate the largest possible profit, given the following circumstances?

a. Unlimited funds are available.

b. The manufacturer has only \$11,000 to spend on development and promotion of the new product.

49. The diameter of the base and the height of a closed right circular cylinder are measured, and the measurements are known to have errors of at most 0.5 cm. If the diameter and height are taken to be 4 cm and 8 cm, respectively, find bounds for the propagated error in

a. the volume V of the cylinder.

b. the surface area S of the cylinder.

50. A heat-seeking missile moves in a portion of space where the temperature (in degrees Celsius) at the point (x, y, z) is given by

$$T(x, y, z) = \frac{1}{10}(x^2 + y + z^3)$$

(x, y, and z are measured in kilometers).
   a. Find the rate at which the temperature is changing as (x, y, z) moves from the point $P_0(-2, 9, 1)$ toward $Q(1, -3, 5)$.
   b. If the heat-seeking missile is at P_0, in what direction will it travel to maximize the rate of heat increase?
   c. What is the maximal rate of increase (in degrees Celsius per kilometer)?
51. Find the volume of the solid bounded above by the surface $z = x^2 + y^2 + 1$, below by the xy-plane, and on the sides by the cylinder $x^2 + y^2 = 1$.
52. Find the surface area of that portion of the paraboloid $z = x^2 + y^2$ that lies below the plane $z = 16$.
53. Find the center of mass of the lamina that covers the region R inside the circle $x^2 + y^2 = 4$ in the first quadrant, given that the density at any point (x, y) is $\rho(x, y) = x + y$.
54. A force field $\mathbf{F}(x, y) = (x - 2y)\mathbf{i} + (y - 2x)\mathbf{j}$ acts on an object moving in the plane. Show that $\mathbf{F}$ is conservative, and find a scalar potential for $\mathbf{F}$. How much work is done as the object moves from $(1, 0)$ to $(0, 1)$ along any path connecting these points?
55. A particle of weight w is acted on only by the constant gravitational force $\mathbf{F} = -w\mathbf{k}$. How much work is done in moving the weight along the helical path given by $x = \cos t$, $y = \sin t$, $z = t$ for $0 \le t \le 2\pi$?
56. Find the moment of inertia of a rectangular homogeneous lamina with dimensions h and l about an axis L through its center of mass.
57. Find the volume of the solid D, the region that remains in the spherical solid $\rho \le 4$ after the solid cone $\phi \le \frac{\pi}{6}$ has been removed.
58. **Modeling Problem** A person whirls a bucket filled with water in a circle of radius 3 ft at the rate of 1 revolution per second. If the bucket and water weigh 30 lb, how much work is done by the force that keeps the bucket moving in a circular path?
59. Let $\mathbf{R} = x\mathbf{i} + y\mathbf{j} + z\mathbf{k}$ and $r = \|\mathbf{R}\|$. Show that the work done in moving an object from a distance r_1 to a distance r_2 in the central force field $\mathbf{F} = \mathbf{R}/r^3$ is given by

$$W = \frac{1}{r_1} - \frac{1}{r_2}$$

60. For constant a where $0 \le a \le R$, the plane $z = R - a$ cuts off a "cap" from the hemisphere

$$z = \sqrt{R^2 - x^2 - y^2}$$

Use a double integral in polar coordinates to find the volume of the cap. Use technology to solve for a in terms of R correct to the nearest hundredth.

CHAPTER 14

INTRODUCTION TO DIFFERENTIAL EQUATIONS

*A*mong all the mathematical disciplines the theory of differential equations is the most important ... It furnishes the explanation of all those elementary manifestations of nature which involve time ...

Sophus Lie
Leipziger Berichte, 47 (1895)

PREVIEW

We introduced and examined separable differential equations in Section 5.6 and first-order linear differential equations in Section 7.6. We examined applications such as orthogonal trajectories, flow of a fluid through an orifice, escape velocity of a projectile, carbon dating, learning curves, population models, dilution problems, and the flow of current in an *RL* circuit. In this chapter, we will extend our study of first-order differential equations by examining *homogeneous* and *exact* equations and then investigate *second-order linear differential equations*.

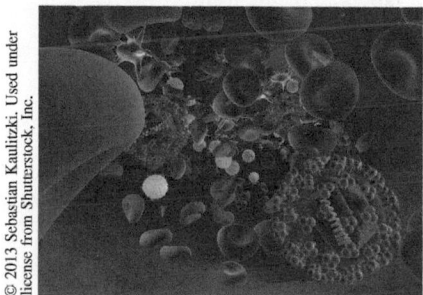

© 2013 Sebastian Kaulitzki. Used under license from Shutterstock, Inc.

The HIV virus invades a white blood cell. Differential equations are used to predict the spread of a disease through a population.

PERSPECTIVE

The study of differential equations is such an extensive topic that even a brief survey of its methods and applications usually occupies a full course. Our goal in this chapter is to preview such a course by introducing some useful techniques for solving differential equations and by examining a few important applications.

1111

14.1 FIRST-ORDER DIFFERENTIAL EQUATIONS

IN THIS SECTION: *Review of separable differential equations, homogeneous differential equations, review of first-order linear differential equations, exact differential equations, Euler's method*

We have discussed several different kinds of first-order differential equations so far in this text. In this section, we review our previous methods and introduce two new forms, *homogeneous* and *exact* differential equations.

Review of Separable Differential Equations

Recall that a differential equation is just an equation involving derivatives or differentials. In particular, an ***n*th-order differential equation** in the dependent variable y with respect to the independent variable x is an equation in which the highest derivative of y that appears is $\dfrac{d^n y}{dx^n}$. A **general solution** of a differential equation is an expression that completely characterizes all possible solutions of the equation, and a **particular solution** is obtained by assigning specific values to the constants that appear in the general solution. An **initial value problem** involves solving a given differential equation subject to one or more initial conditions, such as $y(x_0) = y_0$, $y'(x_0) = y_1, \cdots$.

In Section 5.6, we defined a **separable differential equation** as one that can be written in the form

$$\frac{dy}{dx} = \frac{g(x)}{f(y)}$$

and observed that such an equation can be solved by separating the variables and integrating each side; that is

$$\int f(y)\,dy = \int g(x)\,dx$$

Example 1   Separable differential equation

Find the general solution of the differential equation

$$\frac{dy}{dx} = e^{-y} \sin x$$

Solution   Separate the variables and integrate:

$$\frac{dy}{dx} = e^{-y} \sin x$$
$$e^y\,dy = \sin x\,dx$$
$$\int e^y\,dy = \int \sin x\,dx$$
$$e^y = -\cos x + C \qquad \textit{Combine constants of integration.}$$

This can also be written as $y = \ln|C - \cos x|$.

From the standpoint of applications, one of the most important separable differential equations is

$$\frac{dy}{dx} = ky$$

which occurs in the study of exponential growth and decay. You might review Table 7.2 on page 550, which gives the solution for uninhibited growth or decay, logistic (or inhibited) growth, and limited growth functions. In this section, we consider an application of separable differential equations from chemistry.

Example 2 Chemical conversion

Experiments in chemistry indicate that under certain conditions, two substances A and B will convert into a third substance C in such a way that the rate of conversion with respect to time is jointly proportional to the unconverted amounts of A and B.

For simplicity, assume that one unit of C is formed from the combination of one unit of A and one unit of B, and assume that initially there are α units of A, β units of B, and no units of C present. Set up and solve a differential equation for the amount $Q(t)$ of C present at time t, assuming $\alpha \neq \beta$.

Solution Since each unit of C is formed from one unit of A and one unit of B, it follows that at time t, $\alpha - Q(t)$ units of A and $\beta - Q(t)$ units of B remain unconverted. The specific rate condition can be expressed mathematically as

$$\frac{dQ}{dt} = k(\alpha - Q)(\beta - Q)$$

where k is a constant ($k > 0$ because $Q(t)$ is increasing).

To solve this equation, we separate the variables and integrate.

$$\int \frac{dQ}{(\alpha - Q)(\beta - Q)} = \int k \, dt$$

$$\int \frac{1}{\alpha - \beta} \left[\frac{-1}{\alpha - Q} + \frac{1}{\beta - Q} \right] dQ = \int k \, dt \qquad \textit{Partial fraction decomposition}$$

$$\frac{1}{\alpha - \beta} [\ln(\alpha - Q) - \ln(\beta - Q)] = kt + C_1$$

$$\ln \left(\frac{\alpha - Q}{\beta - Q} \right) = (\alpha - \beta)kt + C_2$$

$$\frac{\alpha - Q}{\beta - Q} = Me^{(\alpha - \beta)kt} \qquad \text{where } M = e^{C_2}$$

$$\alpha - Q = \beta M e^{(\alpha - \beta)kt} - QMe^{(\alpha - \beta)kt}$$

$$QMe^{(\alpha - \beta)kt} - Q = \beta M e^{(\alpha - \beta)kt} - \alpha$$

$$Q = \frac{\beta M e^{(\alpha - \beta)kt} - \alpha}{Me^{(\alpha - \beta)kt} - 1}$$

The initial condition tells us that $Q(0) = 0$, so that

$$0 = \frac{\beta M e^0 - \alpha}{Me^0 - 1}$$

$$0 = \beta M - \alpha$$

$$M = \frac{\alpha}{\beta}$$

Thus, $Q(t) = \dfrac{\beta \frac{\alpha}{\beta} e^{(\alpha - \beta)kt} - \alpha}{\frac{\alpha}{\beta} e^{(\alpha - \beta)kt} - 1} = \dfrac{\alpha \beta [e^{(\alpha - \beta)kt} - 1]}{\alpha e^{(\alpha - \beta)kt} - \beta}$

Homogeneous Differential Equations

Sometimes a first-order differential equation that is not separable can be put into separable form by a change of variables. A differential equation of the form

$$M(x, y) \, dx + N(x, y) \, dy = 0$$

is called a **homogeneous differential equation** if it can be written in the form

$$\frac{dy}{dx} = f \left(\frac{y}{x} \right)$$

In other words, dy/dx is isolated on one side of the equation and the other side can be expressed as a function of y/x. We can then solve the differential equation by substitution. To see how to solve such an equation, set $v = y/x$, so that

$$vx = y$$

$$\frac{d}{dx}(vx) = \frac{d}{dx}(y) \qquad \text{\textit{Take the derivative of both sides.}}$$

$$v + x\frac{dv}{dx} = \frac{dy}{dx} \qquad \text{\textit{Product rule}}$$

$$v + x\frac{dv}{dx} = f(v) \qquad \text{\textit{Substitute} } \frac{dy}{dx} = f\left(\frac{y}{x}\right) = f(v)$$

$$x\frac{dv}{dx} = f(v) - v$$

$$\frac{dv}{f(v) - v} = \frac{dx}{x}$$

The equation can now be solved by integrating both sides; remember to express your answer in terms of the original variables x and y (use $v = y/x$).

Example 3   Homogeneous differential equation

Find the general solution of the equation $2xy\,dx + (x^2 + y^2)\,dy = 0$.

Solution   First, show that the equation is homogeneous by writing it in the form $\frac{dy}{dx} = f\left(\frac{y}{x}\right)$:

$$2xy\,dx + (x^2 + y^2)dy = 0$$

$$\frac{dy}{dx} = \frac{-2xy}{x^2 + y^2}$$

$$= \frac{-2\left(\frac{y}{x}\right)}{1 + \left(\frac{y}{x}\right)^2}$$

$$= \frac{-2v}{1 + v^2} \qquad \text{\textit{Let} } v = \frac{y}{x}$$

We now use the procedure outlined just before this example, and we let $f(v) = \frac{-2v}{1 + v^2}$.

$$\frac{dv}{f(v) - v} = \frac{dx}{x}$$

$$\frac{dv}{\frac{-2v}{1+v^2} - v} = \frac{dx}{x} \qquad \text{\textit{Substitute} } f(v) = \frac{-2v}{1+v^2}$$

$$\frac{dv}{\frac{-2v-v-v^3}{1+v^2}} = \frac{dx}{x}$$

$$-\int \frac{(1+v^2)\,dv}{v^3 + 3v} = \int \frac{dx}{x} \qquad \text{\textit{Integrate both sides.}}$$

$$\int \left[\frac{\frac{1}{3}}{v} + \frac{\frac{2}{3}v}{v^2 + 3}\right] dv = -\int x^{-1}dx \qquad \text{\textit{Partial fraction decomposition}}$$

$$\frac{1}{3}\ln|v| + \frac{2}{3}\left[\frac{1}{2}\ln|v^2 + 3|\right] = -\ln|x| + C_1$$

$$\frac{1}{3}\ln|v(v^2 + 3)| + \ln|x| = C_1$$

$$\ln\left|\frac{y}{x}\left[\left(\frac{y}{x}\right)^2 + 3\right]\right| + \ln|x^3| = 3C_1 \qquad \text{\textit{Multiply both sides by 3 and substitute} } v = \frac{y}{x}.$$

$$\ln \left| \left[\left(\frac{y}{x} \right)^3 + 3 \left(\frac{y}{x} \right) \right] \cdot x^3 \right| = 3C_1 \qquad \textit{Simplify and use properties of logarithms.}$$

$$y^3 + 3x^2 y = e^C \qquad \textit{Let } C = 3C_1.$$

This is the general solution of the given differential equation.

Review of First-Order Linear Differential Equations

In Section 7.6, we considered differential equations of the form

$$\frac{dy}{dx} + p(x)y = q(x)$$

Such an equation is said to be **first-order linear,** and we showed that its general solution is given by

$$y = \frac{1}{I(x)} \left[\int I(x)q(x) \, dx + C \right]$$

where $I(x)$ is the *integrating factor*

Note that the coefficient of dy/dx is 1. If it is not, then divide by that nonzero coefficient.

$$I(x) = e^{\int p(x) \, dx}$$

Example 4   First-order differential equation

Find the solution of the differential equation

$$\frac{dy}{dx} + y \tan x = \sec x$$

that passes through the point $(\pi, 2)$.

Solution   Comparing the given first-order linear differential equation to the general first-order form, we see

$$p(x) = \tan x \qquad \text{and} \qquad q(x) = \sec x$$

The integrating factor is

$$I(x) = e^{\int \tan x \, dx} = e^{-\ln|\cos x|} = e^{\ln \left| (\cos x)^{-1} \right|} = (\cos x)^{-1} = \sec x$$

and the general solution is

$$y = \frac{1}{\sec x} \left[\int (\sec x)(\sec x) \, dx + C \right]$$

$$= \cos x \left[\tan x + C \right]$$

$$= \sin x + C \cos x$$

The initial condition gives

$$2 = \sin \pi + C \cos \pi$$

$$2 = -C$$

so

$$y = \sin x - 2 \cos x$$

First-order linear differential equations appear in a variety of applications. In Section 7.6, we showed how first-order linear equations may be used to model mixture (dilution) problems as well as problems involving the current in an *RL* circuit (one with only a resistor, an inductor, and an electromotive force). In this section, we consider the motion of a body that falls in a resisting medium.

Example 5   Motion of a body falling in a resisting medium

Consider an object with mass *m* that is initially at rest and is dropped from a great height (for example, from an airplane). Suppose the body falls in a straight line and the only forces acting on it are the downward force of the earth's gravitational attraction and a resisting upward force due to air resistance in the atmosphere. Assume that the resisting force is proportional to the velocity *v* of the falling body. Find equations for the velocity and position of the body's motion. Assume the distance $s(t)$ is measured down from the drop point.

Solution   The downward force is the weight *mg* of the body and the upward force is $-kv$, where *k* is a positive constant (the negative sign indicates that the force is directed upward). According to Newton's second law, the sum of the forces acting on a body at any time equals the product *ma*, where *a* is the acceleration of the body, that is

$$\underbrace{ma}_{\substack{\textit{Sum of forces} \\ \textit{on the body}}} = \underbrace{mg}_{\substack{\textit{Force due} \\ \textit{to gravity}}} - \underbrace{kv}_{\substack{\textit{Resisting} \\ \textit{force}}}$$

Now

$$m\frac{dv}{dt} = mg - kv \qquad \textit{Since } a = \frac{dv}{dt}$$

$$\frac{dv}{dt} = g - \frac{k}{m}v$$

$$\frac{dv}{dt} + \frac{k}{m}v = g$$

This is a first-order linear differential equation where $p(t) = \dfrac{k}{m}$ and $q(t) = g$. The integrating factor is

$$I(t) = e^{\int k/m\,dt} = e^{kt/m}$$

so that the solution is

$$v = \frac{1}{e^{kt/m}}\left[\int e^{kt/m}(g)\,dt + C\right] = e^{-kt/m}\left[\frac{ge^{kt/m}}{k/m} + C\right] = \frac{mg}{k} + Ce^{-kt/m}$$

Because $v = 0$ when $t = 0$ (the body is initially at rest), it follows that

$$0 = \frac{mg}{k} + Ce^0 = \frac{mg}{k} + C$$

Solving, we obtain $C = -\dfrac{mg}{k}$, so

$$v = \frac{mg}{k} + \left(-\frac{mg}{k}\right)e^{-kt/m}$$

Now, to find the position $s(t)$, we use the fact that $v(t) = \dfrac{ds}{dt}$:

$$\frac{ds}{dt} = \frac{mg}{k} - \frac{mg}{k}e^{-kt/m}$$

$$\int ds = \int \left[\frac{mg}{k} - \frac{mg}{k}e^{-kt/m}\right] dt$$

$$s(t) = \frac{mg}{k}t - \frac{mg}{k}\frac{e^{-kt/m}}{-\frac{k}{m}} + C$$

$$= \frac{mg}{k}t + \frac{m^2g}{k^2}e^{-kt/m} + C$$

Because $s(0) = 0$ (the position s is measured from the point where the object is dropped), we find that

$$0 = \frac{mg}{k}(0) + \frac{m^2g}{k^2}e^0 + C \qquad \text{so that} \qquad -\frac{m^2g}{k^2} = C$$

Thus, the position is

$$s(t) = \frac{mg}{k}t + \frac{m^2g}{k^2}\left(e^{-kt/m} - 1\right) \qquad\qquad \blacksquare$$

In the problem set, you are asked to show that no matter what the initial velocity may be, the velocity reached by the object in the long run (as $t \to \infty$) is mg/k.

Exact Differential Equations

Sometimes a first-order differential equation can be written in the general form

$$M(x,y)\,dx + N(x,y)\,dy = 0$$

where the left side is an exact differential, namely,

$$df = M(x,y)\,dx + N(x,y)\,dy$$

for some function f. In this case, the given differential equation is appropriately called **exact** and since $df = 0$, its general solution is given by $f(x,y) = C$.

But how can we tell whether a particular first-order equation is exact, and if it is, how can we find f? Since

$$df = \frac{\partial f}{\partial x}dx + \frac{\partial f}{\partial y}dy$$

for a total differential (see Section 11.4), we must have

$$df = \frac{\partial f}{\partial x}dx + \frac{\partial f}{\partial y}dy = M(x,y)\,dx + N(x,y)\,dy$$

In particular, suppose M and N are continuously differentiable in a simply connected region, with

$$\frac{\partial f}{\partial x} = M(x,y) \qquad \text{and} \qquad \frac{\partial f}{\partial y} = N(x,y)$$

This will be true if and only if f satisfies the *cross-derivative test*

$$\frac{\partial N}{\partial x} = \frac{\partial M}{\partial y}$$

and then the function $f(x, y)$ is found by partial integration, exactly as we found the potential function of a conservative vector field in Section 13.3. The procedure for identifying and then solving an exact differential equation is illustrated in Example 6.

Example 6   Exact differential equation

Find the general solution for $(2xy^3 + 3y)\, dx + (3x^2y^2 + 3x)\, dy = 0$.

Solution   Let $M(x, y) = 2xy^3 + 3y$ and $N(x, y) = 3x^2y^2 + 3x$, and apply the cross-derivative test

$$\frac{\partial M}{\partial y} = 6xy^2 + 3 \qquad \text{and} \qquad \frac{\partial N}{\partial x} = 6xy^2 + 3$$

Because $\dfrac{\partial M}{\partial y} = \dfrac{\partial N}{\partial x}$, the equation is exact. To obtain a general solution, we must find a function f such that

$$\frac{\partial f}{\partial x} = 2xy^3 + 3y \qquad \text{and} \qquad \frac{\partial f}{\partial y} = 3x^2y^2 + 3x$$

To find f, we integrate the first partial on the left with respect to x:

$$f(x, y) = \int (2xy^3 + 3y)\, dx = x^2y^3 + 3xy + u(y)$$

where u is a function of y. Taking the partial derivative of f with respect to y and comparing the result with $\dfrac{\partial f}{\partial y}$, we obtain

$$\frac{\partial f}{\partial y} = \frac{\partial}{\partial y}[x^2y^3 + 3xy + u(y)] = 3x^2y^2 + 3x + u'(y)$$

so that

$$3x^2y^2 + 3x = 3x^2y^2 + 3x + u'(y)$$

$$0 = u'(y)$$

This implies that u is a constant. Taking $u = 0$ (any other choice for the constant can be absorbed in the C below), we have $f = x^2y^3 + 3xy$, and the general solution to the exact differential equation is

$$x^2y^3 + 3xy = C$$

A summary of the strategies for identifying and solving various kinds of first-order differential equations is displayed in Table 14.1

Table 14.1   Summary of strategies for first-order differential equations

Form of Equation	Method	Solution
$\dfrac{dy}{dx} = \dfrac{g(x)}{f(y)}$	Separate the variables	$\displaystyle\int f(y)\, dy = \int g(x)\, dx$
$\dfrac{dy}{dx} = f\left(\dfrac{y}{x}\right)$	Homogeneous—use a change of variable $v = y/x$	$\displaystyle\int \frac{dv}{f(v) - v} = \int \frac{dx}{x}$
$\dfrac{dy}{dx} + p(x)y = q(x)$	Use the integrating factor $I(x) = e^{\int p(x)\, dx}$	$y = \dfrac{1}{I(x)}\left[\displaystyle\int I(x)q(x)\, dx + C\right]$
$M(x, y)\, dx + N(x, y)\, dy = 0$, where $\dfrac{\partial M}{\partial y} = \dfrac{\partial N}{\partial x}$	Exact—use partial integration to find f, where $\dfrac{\partial f}{\partial x} = M$ and $\dfrac{\partial f}{\partial y} = N$	$f(x, y) = C$

Euler's Method

In Section 5.6, we introduced direction fields as a means for obtaining a "picture" of various solutions to a differential equation, but sometimes we need more than a rough graph of a solution. We now consider approximating a solution by numerical means. **Euler's method** is a simple procedure for obtaining a table of approximate values for the solution of a given initial value problem*

$$\frac{dy}{dx} = f(x,y) \qquad y(x_0) = y_0$$

The key idea in Euler's method is to increment x_0 by a small quantity h and then to estimate $y_1 = y(x_1)$ for $x_1 = x_0 + h$ by assuming x and y change by so little over the interval $[x_0, x_1]$ that $f(x,y)$ can be replaced with $f(x_0, y_0)$ for this interval. Solving the approximating initial value problem we obtain,

$$y - y_0 = f(x_0, y_0)(x - x_0)$$

In other words, we are approximating the solution curve $y = y(x)$ near (x_0, y_0) by the tangent line to the curve at this point, as shown in Figure 14.1**a**.

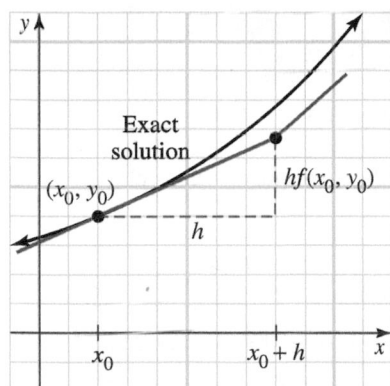

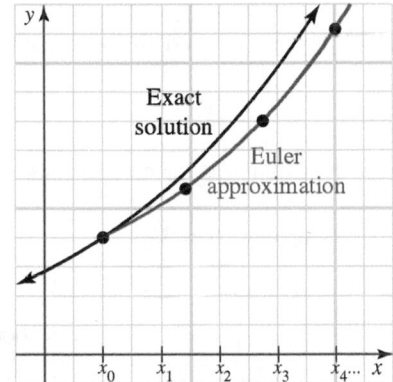

a. The first Euler approximation   **b.** A sequence of Euler approximations

Figure 14.1   Graphical representation of Euler's method

We then repeat this process with (x_1, y_1) assuming the role of (x_0, y_0) to obtain an approximation of the solution $y = y(x)$ over the interval $x_1 \le x \le x_2$ where $x_2 = x_1 + h$ and

$$y_2 = y_1 + hf(x_1, y_1)$$

Continuing in this fashion, we obtain a sequence of line segments that approximates the shape of the solution curve as shown in Figure 14.1**b**. Euler's method is illustrated in the following example.

Example 7   Euler's method

Use Euler's method with $h = 0.1$ to estimate the solution of the initial value problem

$$\frac{dy}{dx} = x + y^2 \qquad y(0) = 1$$

over the interval $0 \le x \le 0.5$.

---

*In our discussion of Euler's method, we assume that the given initial value problem has a unique solution. It can be shown that such an initial value problem always has a unique solution if f and $\partial f/\partial y$ are both continuous in a neighborhood of (x_0, y_0). The proof of this result is beyond the scope of this text but can be found in most elementary differential equations texts.

Solution   Before using Euler's method, we might first look at a graphical solution. The slope field is shown in Figure 14.2**a**, and the particular solution through the point $(0, 1)$ is shown in Figure 14.2**b**.

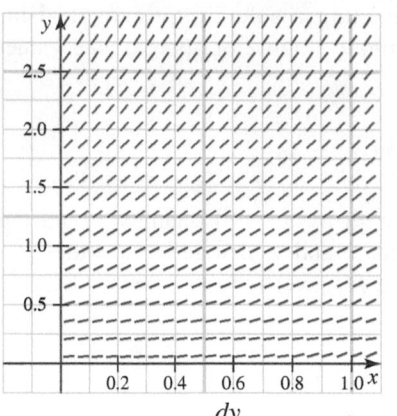 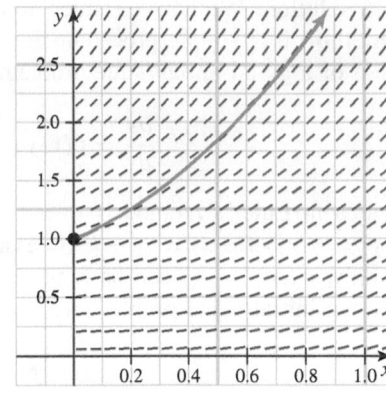

a. Slope field of $\dfrac{dy}{dx} = x + y^2$ **b.** Particular solution through $(0, 1)$

Figure 14.2 Interactive   Graphical solution using a direction field

To use Euler's method for this example, we note

$$f(x, \ y) = x + y^2, \ x_0 = 0, \ y_0 = 1, \text{ and } h = 0.1$$

We show the calculator (or computer) solution correct to four decimal places:

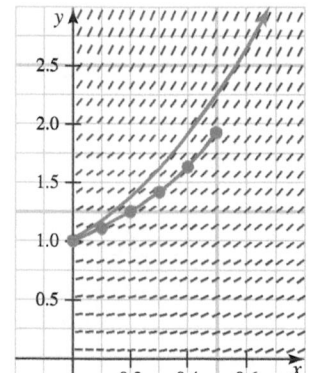

Figure 14.3 Interactive
Solution by Euler's method

$$y_0 = y(0) = 1$$

$$y_1 = y_0 + hf(x_0, y_0) = 1 + 0.1(0 + 1^2) = 1.1000$$

$$y_2 = y_1 + hf(x_1, y_1) = 1.1000 + 0.1(0.1 + 1.1000^2) = 1.2310$$

$$y_3 = y_2 + hf(x_2, y_2) = 1.2310 + 0.1(0.2 + 1.2310^2) \approx 1.4025$$

$$y_4 = y_3 + hf(x_3, y_3) = 1.4025 + 0.1(0.3 + 1.4025^2) \approx 1.6292$$

$$y_5 = y_4 + hf(x_4, y_4) = 1.6292 + 0.1(0.4 + 1.6292^2) \approx 1.9347$$

These points can be plotted to approximate the solution, as shown in Figure 14.3. Notice that we plotted these points by superimposing them on the direction field shown in Figure 14.2**b**.

Euler's method has educational value as the simplest numerical method for solving ordinary differential equations, and can be found in most computer-assisted programs. However, as you might guess by looking at Figure 14.3, as you move away from (x_0, y_0), the error may accumulate. The Euler method can be improved in a variety of ways, most notably by a collection of procedures known as the *Runge-Kutta* and *predictor-corrector* methods. These methods are studied in more advanced courses.

PROBLEM SET 14.1

Level 1

Find the general solution of the differential equations in Problems 1-6 by separating variables.

1. $xy \, dx = (x - 5)dy$ **2.** $\dfrac{dy}{dx} = y \tan x$

3. $(e^{2x} + 9)\dfrac{dy}{dx} = y$ **4.** $y\dfrac{dy}{dx} = e^{x-3y} \cos x$

5. $9 \, dx - x\sqrt{x^2 - 9} \, dy = 0$

6. $xy\dfrac{dy}{dx} = x^2 + y^2 + x^2y^2 + 1$

In Problems 7-12, solve the given first-order linear initial value problem. Compare your answer to the given graphical solution.

7. $\dfrac{dy}{dx} + 2xy = 4x$,

passing through $(0,0)$

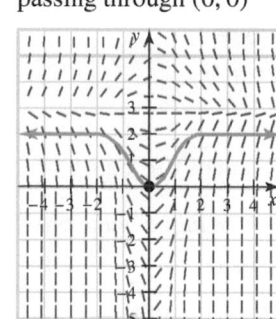

8. $\dfrac{dy}{dx} + \dfrac{y}{x} = \dfrac{\sin x}{x}$,

passing through $(-2, 0)$

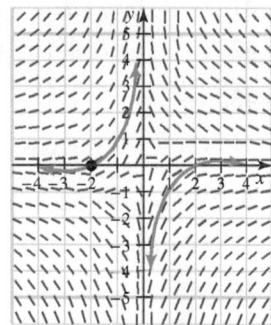

9. $\dfrac{dy}{dx} + y = \cos x$

passing through $(0,0)$

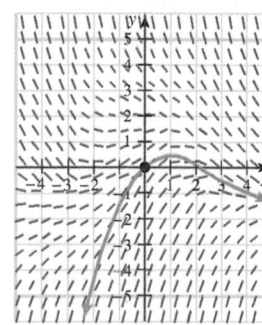

10. $\dfrac{dy}{dx} + y = \dfrac{e^x}{1 + e^{2x}}$,

passing through $(0, 2)$

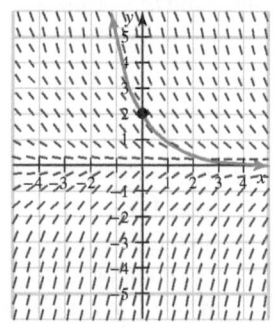

11. $x\dfrac{dy}{dx} - 2y = x^3$,

passing through $(2, -1)$

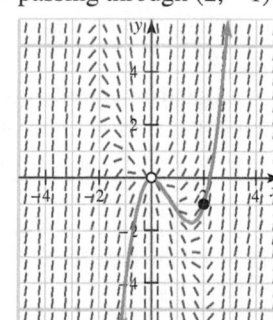

12. $\dfrac{dy}{dx} - 3xy = 5xe^{x^2}$,

passing through $(0, -3)$

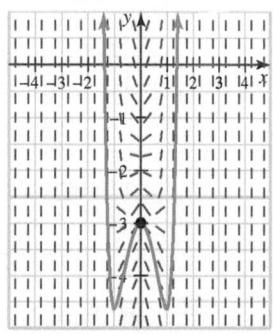

Show that each differentiable equation in Problems 13-18 is homogeneous and then find the general solution.

13. $(3x - y)dx + (x + 3y)dy = 0$.

14. $xy\,dx - (2x^2 + y^2)\,dy = 0$

15. $(3x - y)\,dx + (x - 3y)\,dy = 0$

16. $(x^2 + y^2)\,dx - 2xy\,dy = 0$

17. $(-6y^2 + 3xy + 2x^2)\,dx + x^2 dy = 0$

18. $x\,dy - (y + \sqrt{xy})\,dx = 0$

Show the differential equations in Problems 19-24 are exact and find the general solution.

19. $(3x^2y + \tan y)\,dx + (x^3 + x\sec^2 y)\,dy = 0$

20. $(3x^2 - 10xy)\,dx + (2y - 5x^2 + 4)\,dy = 0$

21. $\left[\dfrac{1}{1+x^2} + \dfrac{2x}{x^2 + y^2}\right]dx + \left[\dfrac{2y}{x^2 + y^2} - e^{-y}\right]dy = 0$

22. $(2xy^3 + 3y - 3x^2)\,dx + (3x^2y^2 + 3x)\,dy = 0$

23. $[2x\cos 2y - 3y(1 - 2x)]\,dx$
$\quad - [2x^2 \sin 2y + 3(2 + x - x^2)]\,dy = 0$

24. $[(x + xy - 3)(1 + y) - x^2\sqrt{y}]\,dx$
$\quad + \left[x^2(y + 1) - 3x - \dfrac{x^3}{6\sqrt{y}}\right]dy = 0$

Level 2

25. Consider the differential equation

$$\frac{dy}{dx} = x + y$$

  a. Find the particular solution that contains the point $(1, 2)$.

  b. Use the direction field for the given differential equation to sketch the solution through $(1, 2)$. Compare the result with part **a**.

  c. Use Euler's method to approximate a solution for $x_0 = 1$, $y_0 = 2$, and $h = 0.2$. Compare this result with part **b**.

26. Consider the differential equation

$$\frac{dy}{dx} = x^2 - y$$

  a. Find the particular solution that contains the point $(2, 1)$.

  b. Use the direction field for the given differential equation to sketch the solution through $(2, 1)$. Compare the result with part **a**.

  c. Use Euler's method to approximate a solution for $x_0 = 2$, $y_0 = 1$, and $h = 0.2$.

Estimate a solution for Problems 27-30 using Euler's method. For each of these problems, a direction field is given. Superimpose the segments from Euler's method on the given direction field.

27. $\dfrac{dy}{dx} = \dfrac{x+y}{y-x}$

passing through $(0, 1)$
for $0 \le x \le 0.5, h = 0.1$

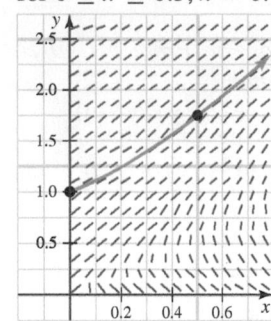

28. $\dfrac{dy}{dx} = 2x(x^2 - y)$,

passing through $(0, 4)$
$0 \le x \le 1, h = 0.2$

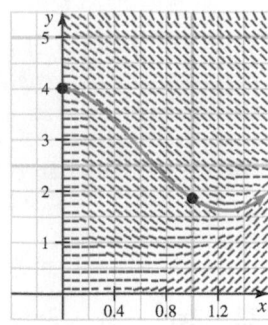

29. $\dfrac{dy}{dx} = \dfrac{5x - 3xy}{1 + x^2}$
passing through $(0, 0)$

$0 \le x \le 0.5, h = 0.1$

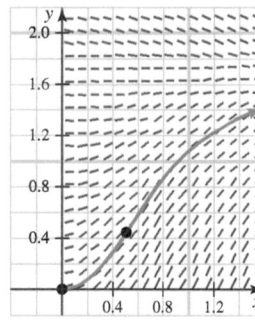

30. $\dfrac{dy}{dx} = \dfrac{y^2 + 2x}{3y^2 - 2xy}$,
passing through $(0, 1)$

$0 \le x \le 0.5, h = 0.1$

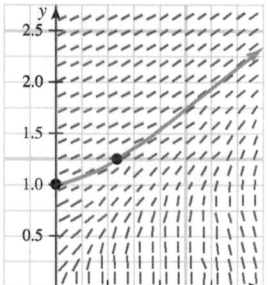

An integrating factor of the differential equation
$M \, dx + N \, dy = 0$ is a function $g(x, y)$ such that

$$g(x,y)M(x,y)\,dx + g(x,y)N(x,y)\,dy = 0$$

is exact. In Problems 31-32, find an integrating factor of
the specified type for the given differential equation, then
solve the equation.

31. $y \, dx + (y - x) \, dy = 0$; $g(x, y) = y^n$

32. $(x^2 + y^2) \, dx + (3xy) \, dy = 0$; $g(x, y) = x^n$

Identify each equation given in Problems 33-44 as separable, homogeneous, first-order linear, or exact, and then
solve. A change of variable may be needed to put the given
equation into one of these forms.

33. $(2xy^2 + 3x^2y - y^3)dx + (2x^2y + x^3 - 3xy^2) \, dy = 0$

34. $(1 + x) \, dy + \sqrt{1 - y^2} \, dx = 0$

35. $\left(\dfrac{2x}{y} - \dfrac{y^2}{x^2}\right) dx + \left(\dfrac{2y}{x} - \dfrac{x^2}{y^2} + 3\right) dy = 0$

36. $(x^2 - xy - x + y) \, dx - (xy - y^2)dy = 0$

37. $e^{y-x} \sin x \, dx - \csc x \, dy = 0$

38. $x^2 \dfrac{dy}{dx} + 2xy = \sin x$

39. $(3x^2 - y \sin xy) \, dx - (x \sin xy) \, dy = 0$

40. $\dfrac{dy}{dx} = \dfrac{y}{x} + x \cos \dfrac{y}{x}$ Hint. Let $u = \dfrac{y}{x}$

41. $(y - \sin^2 x) \, dx + (\sin x) \, dy = 0$

42. $(y^3 - y) \, dx + (x^2 + x) \, dy = 0$

43. $x \dfrac{dy}{dx} = y - \sqrt{x^2 + y^2}$

44. $(2x + \sin y - \cos y) \, dx + (x \cos y + x \sin y) \, dy = 0$

In Problems 45-52, solve the given initial value problem.

45. $\left(x \sin^2 \dfrac{y}{x} - y\right) dx + x \, dy = 0; x = \dfrac{4}{\pi}, y = 1$

46. $y(5x - y) \, dx - x(5x + 2y) \, dy = 0; x = 1, y = 1$

47. $x \dfrac{dy}{dx} - 3y = x^3; x = 1, y = 1$

48. $\dfrac{dy}{dx} = 1 + 3y \tan x; x = 0, y = 2$

49. $[\sin(x^2 + y) + 2x^2 \cos(x^2 + y)] \, dx$
$\qquad + [x \cos(x^2 + y)] \, dy = 0; x = 0, y = 0$

50. $y \dfrac{dy}{dx} = e^{x+2y} \sin x; x = 0, y = 0$

51. $ye^x \, dy = (y^2 + 2y + 2) \, dx; x = 0, y = -1$

52. $(2xy + y^2 + 2x) \, dx + (x^2 + 2xy - 1) \, dy = 0;$
$\qquad x = 1, y = 3$

53. In the chemical conversion analyzed in Example 2,
what happens to $Q(t)$ as $t \to \infty$ in the case where
$\alpha \ne \beta$? What if $\alpha = \beta$?

54. a. Find an equation for the velocity of the falling body
in Example 5 in the case where the body has initial
velocity $v_0 \ne 0$.

   b. Show that for any initial velocity (zero or nonzero)
the velocity reached by the object "in the long run"
(as $t \to \infty$) is always mg/k.

55. A population of foxes grows logistically until author-
ities decide to allow hunting at the constant rate of h
foxes per month. The population $P(t)$ is then modeled
by the differential equation

$$\dfrac{dP}{dt} = P(k - \ell P) - h$$

where $P(0) = P_0$.

   a. Solve this equation in terms of k, ℓ, h, and P_0 for the
case where $h < k^2/(4\ell)$.

   b. What happens to $P(t)$ as $t \to \infty$?

56. A chemical in a solution diffuses from a compartment
where the concentration is $C_0(t) = 7e^{-t}$ across a
membrane with diffusion coefficient $k = 1.75$.

The concentration $C(t)$ in the second compartment is modeled by

$$\frac{dC}{dt} = 1.75[C_0(t) - C(t)]$$

Solve this equation for $C(t)$, assuming that $C(0) = 0$.

57. The formula for the solution of a first-order linear equation requires the differential equation to be linear in x. Suppose, instead, that the equation is linear in y; that is, it can be written in the form

$$\frac{dx}{dy} + R(y)x = S(y)$$

   a. Find a formula for the general solution of an equation that is first-order linear in y.
   b. Use the formula obtained in part **a** to solve
$$y\,dx - 2x\,dy = y^4 e^{-y}\,dy$$

58. **Modeling Problem** A man is pulling a heavy sled along the ground by a rope of fixed length L. Assume the man begins walking at the origin of a coordinate plane and that the sled is initially at the point $(0, L)$. The man walks to the right (along the positive x-axis), dragging the sled behind him. When the man is at point M, the sled is at S as shown in Figure 14.4. Find a differential equation for the path of the sled and solve this differential equation.

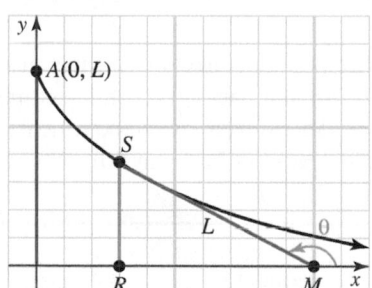

Figure 14.4 Sled problem

59. A **Riccati equation** is a differential equation of the form

$$\frac{dy}{dx} = P(x)y^2 + Q(x)y + R(x)$$

Suppose we know that $y = u(x)$ is a solution of a given Riccati equation.

   a. Change variables by setting

$$z = \frac{1}{y - u(x)}$$

Show that the given equation is transformed by this change of variables into the separable form

$$\frac{dz}{dx} + [2P(x)u(x) + Q(x)]z = -P(x)$$

   b. Solve the first-order linear equation in z obtained in part **a** and explain how to find the general solution of the given Riccati equation.
   c. Use the method outlined in parts **a** and **b** to solve the Riccati equation

$$\frac{dy}{dx} = \frac{1}{x^2}y^2 + \frac{2}{x}y - 2$$

     Hint: There is a solution of the general form $y = Ax$.

60. **Journal Problem** (*American Mathematical Monthly*)* Find all solutions of the Riccati equation (see Problem 59)

$$u' = u^2 + \frac{a}{x}u - b$$

$a, b \neq 0$, that are real rational functions of x.

14.2 SECOND-ORDER HOMOGENEOUS LINEAR DIFFERENTIAL EQUATIONS

IN THIS SECTION: *Linear independence, solutions of the equation $ay'' + by' + cy = 0$, higher-order homogeneous linear equations, undamped versus damped motion of a mass on a spring, reduction of order*

Linear Independence

A **linear differential equation** is one having the general form

$$a_n(x)y^{(n)} + a_{n-1}(x)y^{(n-1)} + \cdots + a_0(x)y = R(x)$$

---

and if $a_n(x) \neq 0$, it is said to be of **order** n. It is *homogeneous* if $R(x) = 0$ and *non-homogeneous* if $R(x) \neq 0$. In this section we focus attention on the homogeneous case, and we examine nonhomogeneous equations in the next section.

To characterize all solutions of the homogeneous equation

$$a_n(x)y^{(n)} + a_{n-1}(x)y^{(n-1)} + \cdots + a_0(x)y = 0$$

we require the following definition.

LINEAR DEPENDENCE AND INDEPENDENCE   The functions $y_1, y_2, \cdots,$ y_n are said to be **linearly independent** if the equation

$$C_1 y_1 + C_2 y_2 + \cdots + C_n y_n = 0$$

for constants $C_1, C_2, \cdots, C_n$ has only the trivial solution $C_1 = C_2 = \cdots = C_n = 0$ for all x in the interval I. Otherwise the y_k's are **linearly dependent**.

■ **W**hat this says: A set of functions is *linearly independent* if none of the y_i can be written as a linear equation of the remaining y_j's $(j \neq i)$. Recall that for two functions to be equal, they must be equal at every point x.

The functions $y_1 = \cos x$ and $y_2 = x$ are linearly independent because the only way we can have $C_1 \cos x + C_2 x = 0$ for all x is for C_1 and C_2 both to be 0. However, $y_1 = 1$, $y_2 = \sin^2 x$, and $y_3 = \cos 2x$ are linearly dependent, because

$$C_1(1) + C_2(\sin^2 x) + C_3(\cos 2x) = 0 \quad \text{for} \quad C_1 = 1, C_2 = -2, C_3 = -1$$

It can be quite difficult to determine whether a given collection of functions $y_1, y_2, \cdots, y_n$ is linearly independent using the definition. An alternate approach to settling the issue of linear independence involves the following determinant function, which is named after Josef Hoëné de Wronski (see the *Historical Quest* in Problem 50).

WRONSKIAN   The **Wronskian** $W(y_1, y_2, \cdots, y_n)$ of n functions $y_1, y_2, \cdots, y_n$ having $n - 1$ derivatives on an interval I is defined to be the determinant function

$$W(y_1, y_2, \cdots, y_n) = \begin{vmatrix} y_1 & y_2 & \cdots & y_n \\ y_1' & y_2' & \cdots & y_n' \\ \vdots & \vdots & & \vdots \\ y_1^{(n-1)} & y_2^{(n-1)} & \cdots & y_n^{(n-1)} \end{vmatrix}$$

Theorem 14.1   Determining linear independence with the Wronskian

Suppose the functions $a_n(x), a_{n-1}(x), \cdots, a_0(x)$ in the nth-order homogeneous linear differential equation

$$a_n(x)y^{(n)} + a_{n-1}(x)y^{(n-1)} + \cdots + a_0(x)y = 0$$

are all continuous on a closed interval $[c, d]$. Then solutions $y_1, y_2, \cdots, y_n$ of this differential equation are linearly independent if and only if the Wronskian is nonzero; that is,

$$W(y_1, y_2, \cdots, y_n) \neq 0$$

throughout the interval $[c, d]$.

Proof: The proof can be found in most differential equations texts. ◆

REMARK: It can be shown that the Wronskian $W(y_1, y_2, \cdots, y_n)$ of n solutions of an nth-order homogeneous linear differential equation is either identically zero or never zero.

Example 1   Showing linear independence

The functions $y_1 = e^{-x}, y_2 = xe^{-x}$, and $y_3 = e^{3x}$ are solutions of a certain homogeneous linear differential equation with constant coefficients. Show that these solutions are linearly independent.

Solution

$$
\begin{aligned}
W(e^{-x}, xe^{-x}, e^{3x}) &= \begin{vmatrix} e^{-x} & xe^{-x} & e^{3x} \\ -e^{-x} & (1-x)e^{-x} & 3e^{3x} \\ e^{-x} & (x-2)e^{-x} & 9e^{3x} \end{vmatrix} \\
&= e^{-x}[9e^{3x}(1-x)e^{-x} - 3e^{3x}(x-2)e^{-x}] \\
&\quad - xe^{-x}[-9e^{3x}e^{-x} - 3e^{-x}e^{3x}] \\
&\quad + e^{3x}[-e^{-x}(x-2)e^{-x} - e^{-x}(1-x)e^{-x}] \\
&= 16e^x
\end{aligned}
$$

Because $16e^x \neq 0$, the functions are linearly independent. ∎

The general solution of an nth-order homogeneous linear differential equation with constant coefficients can be characterized in terms of n linearly independent solutions. Here is the theorem that applies to the second-order case.

Theorem 14.2   Characterization theorem

Suppose y_1 and y_2 are linearly independent solutions of the differential equation $ay'' + by' + cy = 0$, that is, $W(y_1, y_2) \neq 0$. Then the general solution of the equation is $y = C_1 y_1 + C_2 y_2$ for arbitrary constants C_1 and C_2.

Proof: We will prove that, if y_1 and y_2 are linearly independent solutions, then $y = C_1 y_1 + C_2 y_2$ is also a solution. However, the proof that all solutions are of this form is beyond the scope of this book. Suppose y_1 and y_2 are solutions, so that

$$
\begin{aligned}
ay_1''(x) + by_1'(x) + cy_1(x) &= 0 \\
ay_2''(x) + by_2'(x) + cy_2(x) &= 0
\end{aligned}
$$

If $y = C_1 y_1 + C_2 y_2$, we have

$$
\begin{aligned}
ay'' + by' + cy &= a[C_1 y_1'' + C_2 y_2''] + b[C_1 y_1' + C_2 y_2'] + c[C_1 y_1 + C_2 y_2] \\
&= C_1[ay_1'' + by_1' + cy_1] + C_2[ay_2'' + by_2' + cy_2] \\
&= 0 + 0 \\
&= 0
\end{aligned}
$$

Thus, $y = C_1 y_1 + C_2 y_2$ is also a solution. ◆

Solutions of the Equation $ay'' + by' + cy = 0$

Thanks to the characterization theorem, we now know that once we have two linearly independent solutions y_1, y_2 of the equation $ay'' + by' + cy = 0$, we have them all because the general solution can be characterized as $y = C_1 y_1 + C_2 y_2$. Therefore, the whole issue of how to represent the solution of a second-order homogeneous linear equation with constant coefficients depends on finding two linearly independent solutions.

Recall that the general solution of the first-order equation $y' + ay = 0$ is $y = Ce^{-ax}$. Therefore, it is not unreasonable to expect the second-order equation $ay'' + by' + cy = 0$ to have one or more solutions of the form $y = e^{rx}$. If $y = e^{rx}$, then $y' = re^{rx}$ and $y'' = r^2 e^{rx}$, and by substituting these derivatives into the equation $ay'' + by' + cy = 0$, we obtain

$$ay'' + by' + cy = 0$$

$$a(r^2 e^{rx}) + b(re^{rx}) + ce^{rx} = 0$$

$$e^{rx}(ar^2 + br + c) = 0$$

$$ar^2 + br + c = 0 \qquad e^{rx} \neq 0$$

Thus, $y = e^{rx}$ is a solution of the given second-order differential equation if and only if $ar^2 + br + c = 0$. This equation is called the **characteristic equation** of $ay'' + by' + cy = 0$ and $ar^2 + br + c$ is the **characteristic polynomial**.

Example 2   Characteristic equation with distinct real roots

Find the general solution of the differential equation $y'' + 2y' - 3y = 0$.

Solution   Begin by solving the characteristic equation:

$$r^2 + 2r - 3 = 0$$
$$(r - 1)(r + 3) = 0$$
$$r = 1, -3$$

The particular solutions are $y_1 = e^x$ and $y_2 = e^{-3x}$. Next, determine whether these equations are linearly independent by looking at the Wronskian.

$$W(e^x, e^{-3x}) = \begin{vmatrix} e^x & e^{-3x} \\ e^x & -3e^{-3x} \end{vmatrix} = e^x(-3e^{-3x}) - e^x e^{-3x} = -4e^{-2x}$$

Because $W(e^x, e^{-3x}) = -4e^{-2x} \neq 0$, the functions are linearly independent, and the characterization theorem tells us that the general solution is

$$y = C_1 y_1 + C_2 y_2 = C_1 e^x + C_2 e^{-3x}$$

When a characteristic polynomial $r^2 + ar + b$ does not factor as easily as the one in Example 2, it may be necessary to find the roots r_1 and r_2 of the characteristic equation by applying the quadratic formula to obtain

$$r = \frac{-b \pm \sqrt{b^2 - 4ac}}{2a}$$

The *discriminant* for this equation, $b^2 - 4ac$, figures prominently in the following theorem.

Theorem 14.3   Solution of $ay'' + by' + cy = 0$

If the characteristic equation $ar^2 + br + c = 0$ of the homogeneous linear differential equation $ay'' + by' + cy = 0$ has roots

$$r_1 = \frac{-b + \sqrt{b^2 - 4ac}}{2a} \quad \text{and} \quad r_2 = \frac{-b - \sqrt{b^2 - 4ac}}{2a}$$

then the general solution of the differential equation can be expressed in exactly one of the three following forms, depending on the sign of the discriminant $b^2 - 4ac$:

$b^2 - 4ac > 0$: The roots r_1 and r_2 are real and distinct and the general solution is

$$y = C_1 e^{r_1 x} + C_2 e^{r_2 x}$$

for arbitrary constants C_1 and C_2.

$b^2 - 4ac = 0$: The roots r_1 and r_2 are real and equal, $r_1 = r_2 = -\dfrac{b}{2a}$. The general solution is

$$y = C_1 e^{r_1 x} + C_2 x e^{r_2 x} = (C_1 + C_2 x) e^{r_1 x}$$

$b^2 - 4ac < 0$: The roots r_1 and r_2 are complex conjugates, $r_1 = \alpha + \beta i$ and $r_2 = \alpha - \beta i$, where $\alpha = -\dfrac{b}{2a}, \beta = \dfrac{\sqrt{4ac - b^2}}{2a}$, and $i = \sqrt{-1}$. The general solution is

$$y = e^{\alpha x}(C_1 \cos \beta x + C_2 \sin \beta x)$$

Proof: We have just seen that the solutions of the differential equation of the form $y = e^{rx}$ correspond to the solutions of the characteristic equation $ar^2 + br + c = 0$. The quadratic formula characterizes the three cases according to the discriminant of this equation, namely, $b^2 - 4ac$. For each case, we must find two linearly independent solutions.

$b^2 - 4ac > 0$: Let $y_1 = e^{r_1 x}$, and $y_2 = e^{r_2 x}$. Then

$$W(e^{r_1 x}, e^{r_2 x}) = \begin{vmatrix} e^{r_1 x} & e^{r_2 x} \\ r_1 e^{r_1 x} & r_2 e^{r_2 x} \end{vmatrix} = r_2 e^{(r_1 + r_2)x} - r_1 e^{(r_1 + r_2)x} = (r_2 - r_1) e^{(r_1 + r_2)x}$$

Because $r_2 \neq r_1$ and $e^{(r_1 + r_2)x} > 0$, we see $W(e^{r_1 x}, e^{r_2 x}) \neq 0$, so the functions are linearly independent. The characterization theorem tells us that the general solution is

$$y = C_1 y_1 + C_2 y_2 = C_1 e^{r_1 x} + C_2 e^{r_2 x}$$

$b^2 - 4ac = 0$: In this case, the characteristic equation has one repeated root, $r = r_1 = r_2 = -\frac{b}{2a}$. Thus, $y_1 = e^{rx}$ is one solution, and it can be shown that $y_2 = x e^{rx}$ is a second, linearly independent solution (see Problem 45), so the general solution is

$$y = C_1 e^{rx} + C_2 x e^{rx}$$

$b^2 - 4ac < 0$: We must show that $y_1 = e^{\alpha x} \cos \beta x$ and $y_2 = e^{\alpha x} \sin \beta x$ are linearly independent and satisfy the given differential equation. You are asked to verify these facts in Problem 55. ◆

Example 3   Characteristic equation with repeated roots

Find the general solution of the differential equation $y'' + 4y' + 4y = 0$.

Solution   Solve the characteristic equation

$$r^2 + 4r + 4 = 0$$
$$(r + 2)^2 = 0$$
$$r = -2 \quad \textit{(multiplicity 2)}$$

The roots are $r_1 = r_2 = -2$. Thus, $y_1 = e^{-2x}$ and $y_2 = xe^{-2x}$, so the general solution of the differential equation is

$$y = C_1 e^{-2x} + C_2 xe^{-2x}$$

Example 4 Characteristic equation with complex roots

Find the general solution of the differential equation $2y'' + 3y' + 5y = 0$.

Solution Solve the characteristic equation:

$$2r^2 + 3r + 5 = 0$$
$$r = \frac{-3 \pm \sqrt{9 - 4(2)(5)}}{2(2)}$$
$$= \frac{-3 \pm \sqrt{31}i}{4}$$

The roots are

$$r_1 = -\frac{3}{4} + \frac{\sqrt{31}}{4}i, \quad r_2 = -\frac{3}{4} - \frac{\sqrt{31}}{4}i$$

Thus, the general solution is

$$y = e^{(-3/4)x}\left[C_1 \cos\frac{\sqrt{31}}{4}x + C_2 \sin\frac{\sqrt{31}}{4}x \right]$$

Example 5 Second-order initial value problem

Solve $4y'' + 12y' + 9y = 0$ subject to $y(0) = 3$ and $y'(0) = -2$.

Solution Solve

$$4r^2 + 12r + 9 = 0$$
$$(2r + 3)^2 = 0$$
$$r_1 = r_2 = -\frac{3}{2}$$

Thus, the general solution is

$$y = C_1 e^{(-3/2)x} + C_2 xe^{(-3/2)x}$$

Because $y(0) = 3$ we have

$$3 = C_1 e^0 + C_2(0)e^0$$
$$3 = C_1$$

Since $y'(0) = -2$, we find y':

$$y' = -\frac{3}{2}C_1 e^{(-3/2)x} - \frac{3}{2}C_2 xe^{(-3/2)x} + C_2 e^{(-3/2)x}$$
$$y'(0) = -\frac{3}{2}C_1 e^0 - \frac{3}{2}C_2 (0) e^0 + C_2 e^0$$
$$-2 = -\frac{3}{2}C_1 + C_2$$
$$-2 = -\frac{3}{2}(3) + C_2 \qquad \textit{Because } C_1 = 3$$
$$\frac{5}{2} = C_2$$

Thus, the particular solution is

$$y = 3e^{(-3/2)x} + \frac{5}{2}xe^{(-3/2)x}$$

Higher-Order Homogeneous Linear Differential Equations

Homogeneous linear differential equations of order 3 or more with constant coefficients can be handled in essentially the same way as the second-order equations we have analyzed. As in the second-order case, some of the roots of the characteristic equation may be real and distinct, some may be real and repeated, and some may occur in complex conjugate pairs. But now, roots of the characteristic equation may occur more than twice, and when this happens, the linearly independent solutions are obtained by multiplying by increasing powers of x. For example, if 2 is a root of multiplicity 4 in the characteristic equation, the corresponding linearly independent solutions are $e^{2x}, xe^{2x}, x^2e^{2x}$, and x^3e^{2x}. The procedure for obtaining the general solution of an nth-order linear homogeneous equation with constant coefficients is illustrated in the next two examples.

Example 6   Characteristic equation with repeated roots

Solve $y^{(4)} - 5y''' + 6y'' + 4y' - 8y = 0$.

Solution   Solve the characteristic equation:

$$r^4 - 5r^3 + 6r^2 + 4r - 8 = 0$$

Because this is 4th degree, we use synthetic division and the rational root theorem (or a calculator) to find the roots $-1, 2, 2$, and 2. The general solution is

$$y = C_1e^{-x} + C_2e^{2x} + C_3xe^{2x} + C_4x^2e^{2x}$$

Example 7   Characteristic equation with repeated roots (some not real)

Solve $y^{(7)} + 8y^{(5)} + 16y''' = 0$.

Solution   Solve the characteristic equation:

$$r^7 + 8r^5 + 16r^3 = 0$$
$$r^3(r^4 + 8r^2 + 16) = 0$$
$$r^3(r^2 + 4)^2 = 0$$
$$r = 0 \ (\textit{multiplicity 3}), r = \pm 2i \quad (\textit{multiplicity 2})$$

The roots (showing multiplicity) are $0, 0, 0, 2i, 2i, -2i, -2i$. The general solution is

$$y = C_1 + C_2x + C_3x^2 + C_4\cos 2x + C_5\sin 2x + C_6x\cos 2x + C_7x\sin 2x$$

> **STOP** *The general solution of an nth-order linear homogeneous differential equation involves n arbitrary constants $C_1, C_2, \cdots, C_n$. Always check to make sure your solution has the correct number of constants.*

Undamped Versus Damped Motion of a Mass on a Spring

To illustrate an application of a second-order homogeneous linear differential equation, we will consider the motion of an oscillating spring, which we illustrate in Figure 14.5.

Suppose an object suspended at the end of a spring is pulled and then released. A **harmonic oscillator** system is a system that, when displaced from its equilibrium position, experiences a restoring force proportional to the displacement. **Hooke's law** in physics says that a spring that is stretched or compressed x units from its natural length tends to restore itself to its natural length by a force whose magnitude F is proportional to x. Specifically, the restoring vector force is $F(x(t)) = -kx(t)$, where the constant

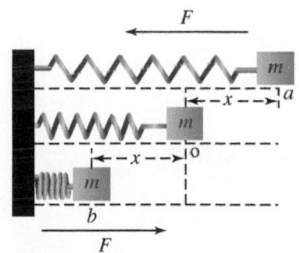

Figure 14.5 Interactive
Oscillating spring

of proportionality k is called the **spring constant** and depends on the stiffness of the spring. According to Newton's second law of motion, the force acting on the object is ma, where $a = x''(t)$ is the acceleration of the object. In the absence of nonconservative external forces, such as friction, the system's motion will be **undamped**. That is, the system will perform sinusoidal oscillations about the equilibrium point, with a constant amplitude and a constant frequency (which does not depend on the amplitude). This motion is governed by the second-order homogeneous equation

$$mx''(t) = -k_1 x(t) \qquad k_1 > 0$$
$$mx''(t) + k_1 x(t) = 0$$

Next, we consider another case: suppose the object is connected to a dashpot, or a device that imposes a damping force. A good example is a shock absorber in a car, which forces a spring to move through a fluid (see Figure 14.6).

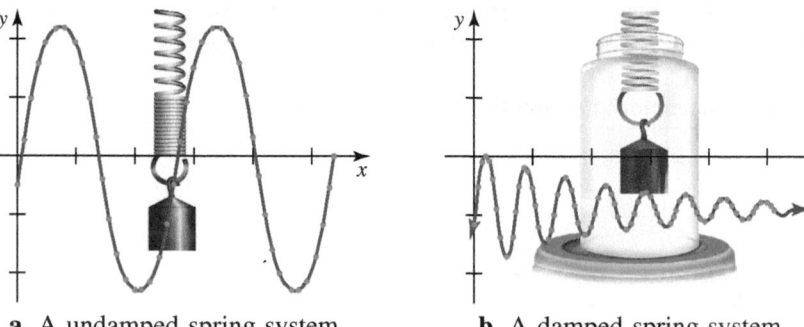

a. A undamped spring system   **b.** A damped spring system

Figure 14.6   Spring systems

Experiments indicate that the shock absorber introduces a damping force proportional to the velocity $v = x'(t)$. Thus, the total force in this case is $-k_1 x(t) - k_2 x'(t)$, and Newton's second law tells us

$$mx''(t) = -k_1 x(t) - k_2 x'(t) \qquad k_1 > 0, k_2 > 0$$
$$mx''(t) + k_2 x'(t) + k_1 x(t) = 0$$

The constant k_2 is called the *damping constant* to contrast it from the spring constant k_1. The characteristic equation is

$$mr^2 + k_2 r + k_1 = 0$$

$$r = \frac{-k_2 \pm \sqrt{k_2^2 - 4k_1 m}}{2m}$$

The three subcases that can occur correspond to different kinds of motion for the object on the spring.

overdamping: $k_2^2 - 4k_1 m > 0$
In this case, both roots are real and negative, and the solution is of the form

$$x(t) = C_1 e^{r_1 t} + C_2 e^{r_2 t}$$

where $r_1 = -\dfrac{k_2}{2m} + \dfrac{1}{2m}\sqrt{k_2^2 - 4k_1 m}$ and $r_2 = -\dfrac{k_2}{2m} - \dfrac{1}{2m}\sqrt{k_2^2 - 4k_1 m}$.
Note that the motion dies out eventually as $t \to \infty$:

$$\lim_{t \to \infty} x(t) = \lim_{t \to \infty} \left(C_1 e^{r_1 t} + C_2 e^{r_2 t} \right) = 0, \text{ because } r_1 < 0 \text{ and } r_2 < 0$$

Overdamping is illustrated in Figure 14.7**a**.

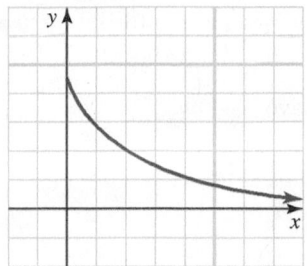

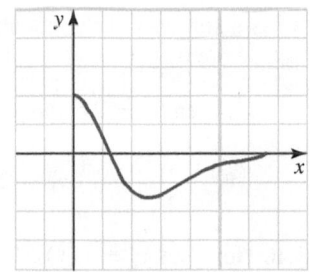

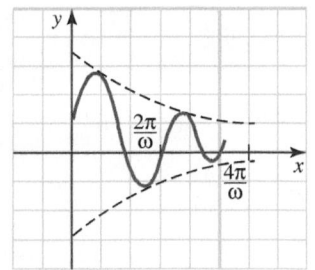

a. Overdamping $k_2^2 - 4k_1m > 0$ **b.** Critical damping $k_2^2 - 4k_1m = 0$ **c.** Underdamping $k_2^2 - 4k_1m < 0$

Figure 14.7   Damping motion

critical damping: $k_2 - 4k_1m = 0$
The solution has the form

$$x(t) = (C_1 + C_2t)e^{rt}$$

where $r_1 = r_2 = r = -\dfrac{k_2}{2m}$. In this case, the motion also eventually dies out (as $t \to \infty$), because $r < 0$. This solution is shown in Figure 14.7**b**.

underdamping: $k_2 - 4k_1m < 0$
In this case, the characteristic equation has complex roots, and the solution has the form

$$x(t) = e^{-k_2t/(2m)}\left[C_1 \cos\left(\frac{t}{2m}\sqrt{4k_1m - k_2^2}\right) + C_2 \sin\left(\frac{t}{2m}\sqrt{4k_1m - k_2^2}\right)\right]$$

This can be written as

$$x(t) = Ae^{\alpha t}\cos(\omega t - C)$$

where $A = \sqrt{C_1^2 + C_2^2}$; $\alpha = -\dfrac{k_2}{2m}$; $\omega = \dfrac{1}{2m}\sqrt{4k_1m - k_2^2}$; $C = \tan^{-1}\dfrac{C_2}{C_1}$. Because k_2 and m are both positive, α must be negative and we see that $x(t) \to 0$ as $t \to \infty$. Notice that as the motion dies out, it oscillates with frequency $2\pi/\omega$, as shown in Figure 14.7**c**.

Reduction of Order

Theorem 14.2 applies even when a, b, and c are functions of x instead of constants. In other words, the general solution of the equation

$$a(x)y'' + b(x)y' + c(x)y = 0$$

can be expressed as

$$y = C_1y_1 + C_2y_2$$

where $y_1(x)$ and $y_2(x)$ are any two linearly independent solutions. Sometimes, we can find one solution y_1 by observation and then obtain a second linearly independent solution y_2 by assuming that $y_2 = vy_1$, where $v(x)$ is a twice differentiable function of x. This procedure is called **reduction of order** because it involves solving the given second-order equation by solving two related first-order equations. The basic ideas of reduction of order are illustrated in the following example.

Example 8   Reduction of order

Show that $y = x^2$ is a solution of the equation $y'' - \dfrac{3}{x}y' + \dfrac{4}{x^2}y = 0$ for $x > 0$, then find the general solution.

Solution   If $y_1 = x^2$, then $y_1' = 2x$ and $y_1'' = 2$. We have

$$y_1'' - \frac{3}{x}y_1' + \frac{4}{x^2}y_1 = 2 - \frac{3}{x}(2x) + \frac{4}{x^2}(x^2) = 0$$

so y_1 is a solution. Next, let $y_2 = v(x)x^2$ and

$$y_2' = 2xv + x^2v' \qquad \text{and} \qquad y'' = 2v + 4xv' + x^2v''$$

Substitute these derivatives into the given equation:

$$(2v + 4xv' + x^2v'') - \frac{3}{x}(2xv + x^2v') + \frac{4}{x^2}(vx^2) = 0$$

$$x^2v'' + xv' = 0$$

$$\frac{v''}{v'} = \frac{-1}{x}$$

Integrate both sides of this equation:

$$\ln|v'| = -\ln|x|$$

$$v' = \frac{1}{|x|}$$

$$v = \ln|x| = \ln x$$

since $x > 0$.

Thus, $y_2 = vx^2 = x^2 \ln x$ is a second solution. To show that the solutions $y_1 = x^2$ and $y_2 = x^2 \ln x$ are linearly independent, we compute the Wronskian:

$$W(x^2, \ x^2 \ln x) = \begin{vmatrix} x^2 & x^2 \ln x \\ 2x & x + 2x \ln x \end{vmatrix} = x^3 \neq 0 \qquad \textit{Since } x > 0$$

It follows that the general solution of the given differential equation is

$$y = C_1 x^2 + C_2 x^2 \ln x$$

PROBLEM SET 14.2

Level 1

Find the general solution of the second-order homogeneous linear differential equations given in Problems 1-14.

1. $y'' + y' = 0$

2. $y'' + y' - 2y = 0$

3. $y'' + 6y' + 5y = 0$

4. $y'' + 4y = 0$

5. $y'' - y' - 6y = 0$

6. $y'' + 8y' + 16y = 0$

7. $2y'' - 5y' - 3y = 0$

8. $3y'' + 11y' - 4y = 0$

9. $y'' - y = 0$

10. $6y'' + 13y' + 6y = 0$

11. $y'' + 11y = 0$

12. $y'' - 4y' + 5y = 0$

13. $7y'' + 3y' + 5y = 0$

14. $2y'' + 5y' + 8y = 0$

Find the general solution of the given higher-order homogeneous linear differential equations in Problems 15-20.

15. $y''' + y'' = 0$

16. $y''' + 4y' = 0$

17. $y^{(4)} + y''' + 2y'' = 0$

18. $y^{(4)} + 10y'' + 9y = 0$

19. $y''' + 2y'' - 5y' - 6y = 0$

20. $y^{(4)} + 2y''' + 2y'' + 2y' + y = 0$

Find the particular solution that satisfies the differential equations in Problems 21-26 subject to the specified initial conditions.

21. $y'' - 10y' + 25y = 0$;
  $y(0) = 1$, $y'(0) = -1$

22. $y'' + 6y' + 9y = 0$;
  $y(0) = 4$, $y'(0) = -3$

23. $y'' - 12y' + 11y = 0$
  $y(0) = 3$, $y'(0) = 11$

24. $y'' + 4y' + 5y = 0$;
  $y(0) = -2$, $y'(0) = 1$

25. $y''' + 10y'' + 25y' = 0$;
  $y(0) = 3$, $y'(0) = 2$, $y''(0) = -1$

26. $y^{(4)} - y''' = 0$; $y(0) = 3$,
  $y'(0) = 0$, $y''(0) = 3$, $y'''(0) = 4$

In Problems 27-32, find the Wronskian, W, of the given set of functions and show that $W \neq 0$.

27. $\{e^{-2x}, e^{3x}\}$

28. $\{e^{2x}, e^{-x}\}$

29. $\{e^{-x}, xe^{-x}\}$

30. $\{2e^x, e^{-x}\}$

31. $\{e^{-x} \cos x, e^{-x} \sin x\}$

32. $\{xe^x \cos x, xe^x \sin x\}$

Level 2

In Problems 33-38, a second-order differential equation and one solution $y_1(x)$ are given. Use reduction of order to find a second solution $y_2(x)$. Show that y_1 and y_2 are linearly independent and find the general solution.

33. $y'' + 6y' + 9y = 0$; $y_1 = e^{-3x}$

34. $2y'' - y' - 6y = 0$; $y_1 = e^{2x}$

35. $xy'' + 4y' = 0$; $y_1 = 1$

36. $x^2 y'' + xy' - 4y = 0$; $y_1 = x^{-2}$

37. $x^2 y'' + 2xy' - 12y = 0$; $y_1 = x^3$

38. $(1 - x)^2 y'' - (1 - x)y' - y = 0$, $y_1 = 1 - x$

Suppose a 16-lb weight stretches a vertical spring 8 in. from its natural length. Find a formula for the position of the weight as a function of time for the situations described in Problems 39-44.

39. The weight is pulled down an additional 6 in. and is then released with an initial upward velocity of 8 ft/s.

40. The weight is pulled down 10 in. and is released with an initial upward velocity of 6 ft/s.

41. The weight is raised 8 in. above the equilibrium point and the compressed spring is then released.

42. The weight is raised 12 in. above the equilibrium point and the compressed spring is then released.

43. The weight is pulled 6 in. below the equilibrium point and is then released. The weight is connected to a dashpot that imposes a damping force of magnitude $0.4|v|$ at all times.

44. The weight is pulled 12 in. below the equilibrium point and is then released. The weight is connected to a dashpot that imposes a damping force of magnitude $0.08|v|$ at all times.

45. If $a^2 = b$, one solution of

$$y'' - 2ay' + by = 0$$

is $y_1 = e^{ax}$. Use reduction of order to show that $y_2 = xe^{ax}$ is a second solution, and then show that y_1 and y_2 are linearly independent.

46. Write the solution of the second-order equation

$$y'' - 2ay' + (a^2 - b)y = 0$$

$b > 0$, in terms of the functions $\cosh kx$ and $\sinh kx$, where $k = \sqrt{b}$.

47. Verify that $y_1 = x^{-3/2}$ is a solution to

$$4x^2 y'' + 12xy' + 3y = 0$$

$x > 0$.

48. Verify that $y_1 = e^{2x}$, $y_2 = xe^{2x}$, and $y_3 = x^2 e^{2x}$ are linearly independent solutions of the third-order differential equation

$$y''' - 6y'' + 12y' - 8y = 0$$

49. Modeling Problem A 100 lb object is projected vertically upward from the surface of the earth with initial velocity 150 ft/s.

  a. Modeling the object's motion with negligible air resistance, how long does it take for the object to return to earth?

  b. Change the model to assume air resistance equal to half the object's velocity. Before making any computation, does your intuition tell you the object takes less or more time to return to earth this time than in part **a**? Now set up and solve a differential equation to actually determine the round-trip time. Were you right?

50. *𝔥istorical 𝔔uest*

Karl Smith library

Joseph Hoëné (1778-1853) adopted the name Wronski when he was 32 years old, around the time he was married. Today, he is remembered for determinants now known as Wronskians, named by Thomas Muir (1844-1934) in 1882. Wronski's main work was in the philosophy of mathematics.

Joseph Hoëné de Wronski (1778-1853)

For years his mathematical work, which contained many errors, was dismissed as unimportant, but in recent years closer study of his work revealed that he had some significant mathematical insight. For this 𝔔uest you are asked to write a paper on one of the great philosophical issues in the history of mathematics. Here are a few quotations to get you started: Mathematics is discovered:

" . . . what is physical is subject to the laws of mathematics, and what is spiritual to the laws of God, and the laws of mathematics are but the expression of the thoughts of God." Thomas Hill, *The Uses of Mathesis; Bibliotheca Sacra*, p. 523.

"Our remote ancestors tried to interpret nature in terms of anthropomorphic concepts of their own creation and failed. The efforts of our nearer ancestors to interpret nature on engineering lines proved equally inadequate. Nature has refused to accommodate herself to either of these man-made molds. On the other hand, our efforts to interpret nature in terms of the concepts of pure mathematics have, so far, proved brilliantly successful . . . from the intrinsic evidence of His creation, the Great Architect of the Universe now begins to appear as a pure mathematician." James H. Jeans, *The Mysterious Universe*, p. 142. *Mathematics is invented*: "There is an old Armenian saying, 'He who lacks sense of the past is condemned to live in the narrow darkness of his own generation.' Mathematics without history is mathematics stripped of its greatness: for, like the other arts—and mathematics is one of the supreme arts of civilization—it derives its grandeur from the fact of being a human creation." G. F. Simmons, *Differential Equations with Applications and Historical Notes*, second edition, McGraw-Hill, Inc., 1991, p. xix.

Discuss whether the significant ideas in mathematics are *discovered or invented*.

Level 3

51. **Almost homogeneous equations**. Sometimes a differential equation is not quite homogeneous but becomes homogeneous with a linear change of variable. Specifically, consider a differential equation of the form

$$\frac{dy}{dx} = f\left(\frac{ax + by + c}{rx + sy + t}\right)$$

a. Suppose $ax \neq br$. Make the change of variable $x = X + A$ and $y = Y + B$ where A and B satisfy

$$\begin{cases} aA + bB + c = 0 \\ rA + sB + t = 0 \end{cases}$$

Show that with these choices for A and B the differential equation becomes homogeneous.

b. Apply the procedure outlined in part **a** to solve the differential equation

$$\frac{dy}{dx} = \left(\frac{-3x + y + 2}{x + 3y - 5}\right)$$

52. If there is no damping and there are no external forces, the motion of an object of mass m attached to a spring with spring constant k is governed by the differential equation $mx'' + kx = 0$. Show that the general solution of this equation is given by

$$x(t) = A \cos\left(\sqrt{\frac{k}{m}}\, t + C\right)$$

$k > 0$ where A and C are constants.
This is called **simple harmonic motion** with frequency $\frac{1}{2\pi}\sqrt{\frac{k}{m}}$.

53. Consider the motion of an object of mass m on a spring with spring constant k_1 and damping constant k_2 for the case where there is critical damping. Describe the motion of the object assuming that it begins at rest: $x'(0) = 0$ at $x(0) = x_0$. How is this different from the case where the object begins at $x(0) = 0$ with initial velocity $x'(0) = v_0$? How do initial velocity and initial displacement affect the motion?

54. If $x(0) = x'(0)$ in Problem 53, what is the maximum displacement? Show that the time at which the maximum displacement occurs is independent of the initial displacement $x(0)$. Assume $x(0) \neq 0$.

55. Suppose the characteristic equation of the differential equation

$$ay'' + by' + cy = 0$$

has complex conjugate roots, $r_1 = \alpha + \beta i$ and $r_2 = \alpha - \beta i$. Show that $y_1 = e^{\alpha x} \cos \beta x$ and $y_2 = e^{\alpha x} \sin \beta x$ are both solutions and are linearly independent. This is Theorem 14.3 for the case where $b^2 - 4ac < 0$.

Pendulum motion *Suppose a ball of mass m is suspended at the end of a rod of length L and is set in motion swinging back and forth like a pendulum, as shown in Figure 14.8.*

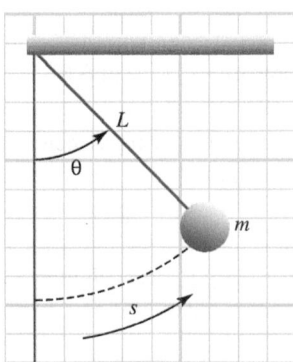

Figure 14.8 Interactive   Pendulum motion

In Problems 56-57, let θ be the angle between the rod and the vertical at time t, so that the displacement of the ball from the equilibrium position is $s = L\theta$ and the acceleration of the ball's motion is

$$\frac{d^2s}{dt^2} = L\frac{d^2\theta}{dt^2}$$

56. Use Newton's second law to show that

$$mL\frac{d^2\theta}{dt^2} + mg \sin \theta = 0$$

Assume that air resistance and the mass of the rod are negligible.

57. When the displacement is small (θ close to 0), $\sin \theta$ may be replaced by θ. In this case, solve the resulting differential equation

$$\frac{d^2\theta}{dt^2} + \frac{g}{L}\theta = 0$$

How is the motion of the pendulum like the simple harmonic motion discussed in Problem 52?

58. The motion of a pendulum subject to frictional damping proportional to its velocity is modeled by the differential equation

$$mL \frac{d^2\theta}{dt^2} + kL\frac{d\theta}{dt} + mg \sin \theta = 0$$

where L is the length of the pendulum, m is the mass of the bob at its end, and θ is the angle the pendulum arm makes with the vertical (see Figure 14.9). Assume θ is small, so $\sin \theta$ is approximately equal to θ.

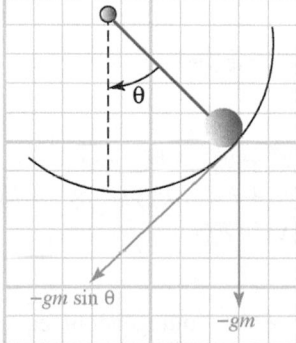

Figure 14.9 Damped Pendulum

a. Solve the resulting differential equation for the case where $k^2 \geq 4gm^2/L$. What happens to $\theta(t)$ as $t \to \infty$?

b. If $k^2 < 4gm^2/L$, show that

$$\theta(t) = Ae^{-kt/(2m)} \cos\left(\sqrt{\frac{B}{L}}t + C\right).$$

for constants A, B, and C. What happens to $\theta(t)$ as $t \to \infty$ in this case?

c. For the situation in part **b**, show that the time difference between successive vertical positions is approximately

$$T = 2\pi m\sqrt{\frac{L}{4gm^2 - k^2L}}$$

59. Path of a projectile with variable mass A rocket starts from rest and moves vertically upward, along a straight line. Assume the rocket and its fuel initially weigh w kilograms and that the fuel initially weighs w_f kilograms. Further assume that the fuel is consumed at a constant rate of r kilograms per second (relative to the rocket). Finally, assume that gravitational attraction is the only external force acting on the rocket.

a. If $m(t)$ is the mass of the rocket and fuel at time t and $s(t)$ is the height above the ground at time t, it can be shown that

$$m(t)s''(t) + m'(t)v_0 + m(t)g = 0$$

where v_0 is the velocity of the exhaust gas in relation to the rocket. Express $m(t)$ in terms of w, r, v_0, and g, then integrate this differential equation to obtain the velocity $s'(t)$. Note that $s'(0) = 0$ because the rocket starts from rest.

b. Integrate the velocity $s'(t)$ to obtain $s(t)$.

c. At what time is all the fuel consumed?

d. How high is the rocket at the instant the fuel is consumed?

60. Journal Problem (*Canadian Mathematical Bulletin*)* by Murray S. Klamkin. Solve the differential equation

$$x^4y'' - (x^3 + 2axy)y' + 4ay^2 = 0$$

---

*Problem 331 in the *Canadian Mathematical Bulletin*, Vol. 26, 1983, p. 126.

14.3 SECOND-ORDER NONHOMOGENEOUS LINEAR DIFFERENTIAL EQUATIONS

IN THIS SECTION: *Nonhomogeneous equations, method of undetermined coefficients, variation of parameters, an application to RLC circuits*

Nonhomogeneous Equations

Next we will see how to solve a nonhomogeneous second-order linear equation of the general form $ay'' + by' + cy = F(x)$, where $F(x) \neq 0$. The key to our results is the following theorem.

Theorem 14.4   Characterization of the general solution of $ay'' + by' + cy = F(x)$

Let y_p be a particular solution of the nonhomogeneous second-order linear equation $ay'' + by' + cy = F(x)$. Let y_h be the general solution of the related homogeneous equation $ay'' + by' + cy = 0$. Then the general solution of $ay'' + by' + cy = F(x)$ is given by the sum

$$y = y_h + y_p$$

■ **W**hat this says We can obtain the general solution of the nonhomogeneous equation $ay'' + by' + cy = F(x)$ by finding the general solution y_h of the related homogeneous equation $ay'' + by' + cy = 0$ and just one particular solution y_p of the given nonhomogeneous equation.

Proof: First, the sum $y = y_h + y_p$ is a solution of the nonhomogeneous equation $ay'' + by' + cy = F(x)$, because

$$
\begin{aligned}
ay'' + by' + cy &= a(y_h + y_p)'' + b(y_h + y_p)' + c(y_h + y_p) \\
&= ay_h'' + ay_p'' + by_h' + by_p' + cy_h + cy_p \\
&= (ay_h'' + by_h' + cy_h) + (ay_p'' + by_p' + cy_p) \\
&= 0 + F(x) \\
&= F(x)
\end{aligned}
$$

Conversely, if y is any solution of the nonhomogeneous equation, then $y - y_p$ is a solution of the related homogeneous equation because

$$
\begin{aligned}
a(y - y_p)'' + b(y - y_p)' + c(y - y_p) &= a(y'' - y_p'') + b(y' - y_p') + c(y - y_p) \\
&= (ay'' + by' + cy) - (ay_p'' + by_p' + cy_p) \\
&= F(x) - F(x) \\
&= 0
\end{aligned}
$$

Thus, $y - y_p = y_h$ (because it is a solution of the homogeneous equation). Therefore, because y was *any* solution of the nonhomogeneous equation, it follows that $y = y_h + y_p$ is the general solution of the nonhomogeneous equation. ◆

We can use the methods of the preceding section to find the general solution of the related homogeneous equation. We now develop two methods for finding particular solutions of the nonhomogeneous equation.

Method of Undetermined Coefficients

Sometimes it is possible to find a particular solution y_p of the nonhomogeneous equation $ay'' + by' + cy = F(x)$ by assuming a **trial solution** $\bar{y}_p$ of the same general form as $F(x)$. This procedure, called the **method of undetermined coefficients**, is illustrated in the following three examples, each of which has the related homogeneous equation $y'' + y' - 2y = 0$ with the general solution $y_h = C_1 e^x + C_2 e^{-2x}$.

Example 1   Method of undetermined coefficients

Find $\bar{y}_p$ and the general solution for $y'' + y' - 2y = 2x^2 - 4x$.

Solution   The right side $F(x) = 2x^2 - 4x$ is a quadratic polynomial. Because derivatives of a polynomial are polynomials of lower degree, it seems reasonable to consider a trial solution that is also a polynomial of degree 2. That is, we "guess" that this equation has a particular solution $\bar{y}_p$ of the general form $\bar{y}_p = A_1 x^2 + A_2 x + A_3$. To find the constants A_1, A_2, and A_3, calculate

$$\bar{y}_p' = 2A_1 x + A_2 \qquad \text{and} \qquad \bar{y}_p'' = 2A_1$$

Substitute the values for $\bar{y}_p, \bar{y}_p'$ and $\bar{y}_p''$ into the given equation:

$$y'' + y' - 2y = 2x^2 - 4x$$
$$2A_1 + 2A_1 x + A_2 - 2(A_1 x^2 + A_2 x + A_3) = 2x^2 - 4x$$
$$-2A_1 x^2 + (2A_1 - 2A_2)x + (2A_1 + A_2 - 2A_3) = 2x^2 - 4x$$

Since this must be true for *every* x, we see that this is true only when the coefficients of each power of x on each side of the equation match, so that

$$\begin{cases} -2A_1 = 2 & (x^2 \ terms) \\ 2A_1 - 2A_2 = -4 & (x \ terms) \\ 2A_1 + A_2 - 2A_3 = 0 & (constant \ terms) \end{cases}$$

Solve this system of equations simultaneously to find $A_1 = -1, A_2 = 1$, and $A_3 = -\frac{1}{2}$. Thus, a particular solution of the given nonhomogeneous equation is

$$y_p = A_1 x^2 + A_2 x + A_3 = -x^2 + x - \frac{1}{2}$$

and the general solution for the nonhomogeneous equation is

$$y = y_h + y_p = C_1 e^x + C_2 e^{-2x} - x^2 + x - \frac{1}{2}$$

Comment: *Notice that even though the constant term is zero in the polynomial function $F(x)$, we cannot assume that $y = A_1 x^2 + A_2 x$ is a suitable trial solution. In general, all terms of the same or lower degree that could possibly lead to the given right-side function $F(x)$ must be included in the trial solution.*

Example 2   Method of undetermined coefficients

Solve $y'' + y' - 2y = \sin x$.

Solution   Because the trial solution $\bar{y}_p$ is to be "like" the right-side function $F(x) = \sin x$, it seems that we should choose the trial solution to be $\bar{y}_p = A_1 \sin x$, but a sine function can have either a sine or a cosine in its derivatives, depending on how many derivatives are taken. Thus, to account for the $\sin x$, it is necessary to have *both* $\sin x$ and $\cos x$ in the trial solution, so we set

$$\bar{y}_p = A_1 \sin x + A_2 \cos x$$

Differentiating, we find

$$\overline{y}_p' = A_1 \cos x - A_2 \sin x \qquad \text{and} \qquad \overline{y}_p'' = -A_1 \sin x - A_2 \cos x$$

Substitute the values into the given equation:

$$y'' + y' - 2y = \sin x$$
$$(-A_1 \sin x - A_2 \cos x) + (A_1 \cos x - A_2 \sin x) - 2(A_1 \sin x + A_2 \cos x) = \sin x$$
$$(-3A_1 - A_2) \sin x + (A_1 - 3A_2) \cos x = \sin x$$

This gives the system

$$\begin{cases} -3A_1 - A_2 = 1 & (\sin x \text{ terms}) \\ A_1 - 3A_2 = 0 & (\cos x \text{ terms}) \end{cases}$$

with the solution $A_1 = -\frac{3}{10}$, $A_2 = -\frac{1}{10}$. Thus, the particular solution of the nonhomogeneous equation is $y_p = -\frac{3}{10} \sin x - \frac{1}{10} \cos x$, and the general solution is

$$y = y_h + y_p = C_1 e^x + C_2 e^{-2x} - \frac{3}{10} \sin x - \frac{1}{10} \cos x \qquad \blacksquare$$

Note: It can be shown that if y_1 is a solution of $ay'' + by' + cy = F(x)$ and y_2 is a solution of $ay'' + by' + cy = G(x)$, then $y_1 + y_2$ is a solution of $ay'' + by' + cy = F(x) + G(x)$. (See Problem 60.) This is called the **principle of superposition**. For instance, by combining the results of Examples 1 and 2, we see that a particular solution of the nonhomogeneous linear equation

$$y'' + y' - 2y = 2x^2 - 4x + \sin x$$

is

$$\overbrace{y_p = -x^2 + x - \frac{1}{2}}^{\text{Solution for } F = 2x^2 - 4x} \underbrace{- \frac{3}{10} \sin x - \frac{1}{10} \cos x}_{\text{Solution for } G = \sin x}$$

Example 3   Method of undetermined coefficients

Solve $y'' + y' - 2y = 4e^{-2x}$.

Solution   At first glance (looking at $F(x) = 4e^{-2x}$), it may seem that the trial solution should be $\overline{y}_p = Ae^{-2x}$. However, $y_1 = e^{-2x}$ is a solution of $y_1'' + y_1' - 2y_1 = 0$, so it cannot possibly also satisfy the given nonhomogeneous equation.

To deal with this situation, multiply the usual trial solution by x and consider the trial solution $\overline{y}_p = Axe^{-2x}$. Differentiating, we find

$$\overline{y}_p' = A(1 - 2x)e^{-2x} \qquad \text{and} \qquad \overline{y}_p'' = A(4x - 4)e^{-2x}$$

and by substituting into the given equation, we obtain

$$y'' + y' - 2y = 4e^{-2x}$$
$$A(4x - 4)e^{-2x} + A(1 - 2x)e^{-2x} - 2Axe^{-2x} = 4e^{-2x}$$
$$(4Ax - 4A + A - 2Ax - 2Ax)e^{-2x} = 4e^{-2x}$$
$$-3Ae^{-2x} = 4e^{-2x}$$
$$A = -\frac{4}{3}$$

Thus, $y_p = -\frac{4}{3}xe^{-2x}$, so that the general solution is

$$y = y_h + y_p = C_1e^x + C_2e^{-2x} - \frac{4}{3}xe^{-2x}$$

The procedure illustrated in Examples 1-3 can be applied to a differential equation $y'' + ay' + by = F(x)$ only when $F(x)$ has one of the following forms:

(i) $F(x) = P_n(x)$, a polynomial of degree n
(ii) $F(x) = P_n(x)e^{kx}$
(iii) $F(x) = e^{kx}[P_n(x)\cos\alpha x + Q_n(x)\sin\alpha x]$ where $Q_n(x)$ is another polynomial of degree n.

We can now describe the **method of undetermined coefficients**.

METHOD OF UNDETERMINED COEFFICIENTS To solve $ay'' + by' + cy = F(x)$ when $F(x)$ is one of the forms listed above, follow these steps:

Step 1 The solution is of the form $y = y_h + y_p$, where y_h is the general solution of the related homogeneous equation and y_p is a particular solution.

Step 2 Find y_h by solving the homogeneous equation

$$ay'' + by' + cy = 0$$

Step 3 Find y_p by picking an appropriate trial solution $\bar{y}_p$:

Form of $F(x)$	Corresponding trial expression $\bar{y}_p$
i. $P_n(x) = c_nx^n + \cdots + c_1x + c_0$	$A_nx^n + \cdots + A_1x + A_0$
ii. $P_n(x)e^{kx}$	$[A_nx^n + A_{n-1}x^{n-1} + \cdots + A_0]e^{kx}$
iii. $e^{kx}[P_n(x)\cos\alpha x$	$e^{kx}[(A_nx^n + \cdots + A_0)\cos\alpha x$
$\quad + Q_n(x)\sin\alpha x]$	$\quad + (B_nx^n + \cdots + B_0)\sin\alpha x]$

Step 4 If no term in the trial expression $\bar{y}_p$ appears in the general homogeneous solution y_h, the particular solution can be found by substituting $\bar{y}_p$ into the equation $ay'' + by' + cy = F(x)$ and solving for the undetermined coefficients.

Step 5 If any term in the trial expression $\bar{y}_p$ appears in y_h, multiply $\bar{y}_p$ by x^k, where k is the smallest integer such that no term in $x^k\bar{y}_p$ is a solution of $ay'' + by' + cy = 0$. Then proceed as in step 4, using $x^k\bar{y}_p$ as the trial solution.

Example 4 Finding trial solutions

Determine a suitable trial solution for undetermined coefficients in each of the given cases.

a. $y'' - 4y' + 4y = 3x^2 + 4e^{-2x}$ **b.** $y'' - 4y' + 4y = 5xe^{2x}$

c. $y'' + 2y' + 5y = 3e^{-x}\cos 2x$

Solution

a. The related homogeneous equation $y'' - 4y' + 4y = 0$ has the characteristic equation $r^2 - 4r + 4 = 0$, which has the root 2 of multiplicity two. Thus, the general homogeneous solution is

$$y_h = C_1e^{2x} + C_2xe^{2x}$$

The part of the trial solution for the nonhomogeneous equation that corresponds to $3x^2$ is $A_0 + A_1x + A_2x^2$ and the part that corresponds to $4e^{-2x}$ is Be^{-2x}. Since neither part includes terms in y_h, we apply the principle of superposition to conclude that

$$\overline{y}_p = A_0 + A_1x + A_2x^2 + Be^{-2x}$$

b. We know from part **a** that the general homogeneous solution is

$$y_h = C_1e^{2x} + C_2xe^{2x}$$

The expected trial solution for $5xe^{2x}$ would be $(A_0 + A_1x)e^{2x}$, but part of this expression is contained in y_h. If we multiply by x again, part of $(A_0 + A_1x)xe^{2x}$ is still contained in y_h, so we multiply by x again to obtain

$$\overline{y}_p = (A_0 + A_1x)x^2e^{2x}$$

c. The related homogeneous equation $y'' + 2y' + 5y = 0$ has the characteristic equation $r^2 + 2r + 5 = 0$. This has complex conjugate roots $r = -1 \pm 2i$, so the general homogeneous solution is

$$y_h = e^{-x}[C_1 \cos 2x + C_2 \sin 2x]$$

Ordinarily, the trial solution for the nonhomogeneous equation would be of the form $e^{-x}[A \cos 2x + B \sin 2x]$, but part of this expression is in y_h. Therefore, we multiply by x to obtain the trial solution

$$\overline{y}_p = xe^{-x}[A \cos 2x + B \sin 2x]$$

Variation of Parameters

The method of undetermined coefficients applies only when the coefficients a, b, and c are constant in the nonhomogeneous linear equation $ay'' + by' + cy = F(x)$ and the function $F(x)$ has the same general form as a solution of a second-order homogeneous linear differential equation with constant coefficients. Even though many important applications are modeled by differential equations of this type, there are other situations that require a more general procedure.

Our next goal is to examine a method of J. L. Lagrange (see *Historical Quest* Problem 60 in Section 11.8) called **variation of parameters**, which can be used to find a particular solution of any nonhomogeneous equation

$$y'' + P(x)y' + Q(x)y = F(x) \qquad \textit{Note the leading coefficient is 1.}$$

where P, Q, and F are continuous.

To use variation of parameters, we must be able to find two linearly independent solutions $y_1(x)$ and $y_2(x)$ of the related homogeneous equation,

$$y'' + P(x)y' + Q(x)y = 0$$

In practice, if $P(x)$ and $Q(x)$ are not both constants, these may be difficult to find, but once we have them, we assume there is a solution of the nonhomogeneous equation of the form

$$y_p = uy_1 + vy_2$$

Differentiating this expression y_p, we obtain

$$y_p' = u'y_1 + v'y_2 + uy_1' + vy_2'$$

To simplify, assume that

$$u'y_1 + v'y_2 = 0$$

Remember, we need to find only *one* particular solution y_p, and if imposing this side condition makes it easier to find such a y_p, so much the better! With the side condition, we have

$$y_p' = uy_1' + vy_2'$$

and by differentiating again, we obtain

$$y_p'' = uy_1'' + u'y_1' + vy_2'' + v'y_2'$$

Next, we substitute our expressions for y_p' and y_p'' into the given differential equation:

$$\begin{aligned} F(x) &= y_p'' + P(x)y_p' + Q(x)y_p \\ &= (uy_1'' + u'y_1' + vy_2'' + v'y_2') + P(x)(uy_1' + vy_2') + Q(x)(uy_1 + vy_2) \end{aligned}$$

This can be rewritten as

$$u[y_1'' + P(x)y_1' + Q(x)y_1] + v[y_2'' + P(x)y_2' + Q(x)y_2] + u'y_1' + v'y_2' = F(x)$$

Because y_1 and y_2 are solutions of $y'' + P(x)y' + Q(x)y = 0$, we have

$$u\underbrace{[y_1'' + P(x)y_1' + Q(x)y_1]}_{0} + v\underbrace{[y_2'' + P(x)y_2' + Q(x)y_2]}_{0} + u'y_1' + v'y_2' = F(x)$$

or

$$u'y_1' + v'y_2' = F(x)$$

Thus, the parameters u and v must satisfy the system of equations

$$\begin{cases} u'y_1 + v'y_2 = 0 \\ u'y_1' + v'y_2' = F(x) \end{cases}$$

Solve this system to obtain

$$u' = \frac{-y_2 F(x)}{y_1 y_2' - y_2 y_1'} \qquad \text{and} \qquad v' = \frac{y_1 F(x)}{y_1 y_2' - y_2 y_1'}$$

where in each case the denominator is not zero, because it is the Wronskian of the linearly independent solutions y_1, y_2 of the related homogeneous differential equation. Integrating, we find

$$u(x) = \int \frac{-y_2 F(x)}{y_1 y_2' - y_2 y_1'} dx \qquad \text{and} \qquad v(x) = \int \frac{y_1 F(x)}{y_1 y_2' - y_2 y_1'} dx$$

and by substituting into the expression

$$y_p = uy_1 + vy_2$$

we obtain a particular solution of the given differential equation. Here is a summary of the procedure we have described.

☠ *The coefficient of y''*
must be 1 for this method
to work. ☠

VARIATION OF PARAMETERS   To find the general solution of
$y'' + P(x)y' + Q(x)y = F(x)$:

Step 1 Find the general solution $y_h = C_1 y_1 + C_2 y_2$ to the homogeneous equation.
Step 2 Set $y_p = uy_1 + vy_2$ and substitute into the formulas:

$$u' = \frac{-y_2 F(x)}{y_1 y_2' - y_2 y_1'} \quad \text{and} \quad v' = \frac{y_1 F(x)}{y_1 y_2' - y_2 y_1'}$$

Step 3 Integrate u' and v' to find u and v.
Step 4 A particular solution is $y_p = uy_1 + vy_2$, and the general solution is
$y = y_h + y_p$.

These ideas are illustrated in the next example.

Example 5   Variation of parameters

Solve $y'' + 4y = \tan 2x$.

Solution   Notice that this problem cannot be solved by the method of undetermined coefficients, because the right-side function $F(x) = \tan 2x$ is not one of the forms for which the procedure applies.

To apply variation of parameters, begin by solving the related homogeneous equation $y'' + 4y = 0$. The characteristic equation is $r^2 + 4 = 0$ with roots $r = \pm 2i$. These complex roots have $\alpha = 0$ and $\beta = 2$ so that the general solution is

$$y = e^0[C_1 \cos 2x + C_2 \sin 2x] = C_1 \cos 2x + C_2 \sin 2x$$

This means $y_1(x) = \cos 2x$ and $y_2(x) = \sin 2x$. Set $y_p = uy_1 + vy_2$, where

$$u' = \frac{-y_2 F(x)}{y_1 y_2' - y_2 y_1'} = \frac{-\sin 2x \tan 2x}{2 \cos 2x \cos 2x + 2 \sin 2x \sin 2x} = -\frac{\sin^2 2x}{2 \cos 2x}$$

and

$$v' = \frac{y_1 F(x)}{y_1 y_2' - y_2 y_1'} = \frac{\cos 2x \tan 2x}{2 \cos 2x \cos 2x + 2 \sin 2x \sin 2x} = \frac{1}{2} \sin 2x$$

Integrating, we obtain

$$u(x) = \int -\frac{\sin^2 2x}{2 \cos 2x} \, dx \qquad\qquad v(x) = \int \frac{1}{2} \sin 2x \, dx$$
$$= -\frac{1}{2} \int \frac{1 - \cos^2 2x}{\cos 2x} \, dx \qquad\qquad = \frac{1}{2} \int \sin 2x \, dx$$
$$= -\frac{1}{2} \int (\sec 2x - \cos 2x) \, dx \qquad\qquad = -\frac{1}{4} \cos 2x + C_2$$
$$= -\frac{1}{2} \left[\frac{1}{2} \ln |\sec 2x + \tan 2x| - \frac{\sin 2x}{2} \right] + C_1$$

We now take $C_1 = C_2 = 0$ since $C_1 y_1$ and $C_2 y_2$ are solutions of the homogenerous equation. Thus, a particular solution is

$$y_p = uy_1 + vy_2$$
$$= \left[-\frac{1}{4} \ln |\sec 2x + \tan 2x| + \frac{1}{4} \sin 2x \right] \cos 2x + \left(-\frac{1}{4} \cos 2x \right) \sin 2x$$
$$= -\frac{1}{4} (\cos 2x) \ln |\sec 2x + \tan 2x|$$

Finally, the general solution is

$$y = y_h + y_p = C_1 \cos 2x + C_2 \sin 2x - \frac{1}{4}(\cos 2x) \ln|\sec 2x + \tan 2x|$$

An Application to *RLC* Circuits

An important application of second-order linear differential equations is in the analysis of electric circuits. Consider a circuit with constant resistance R, inductance L, and capacitance C. Such a circuit is called an **RLC circuit** and is illustrated in Figure 14.10.

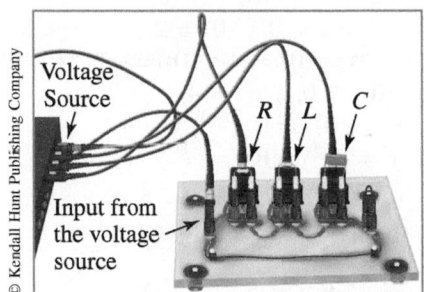
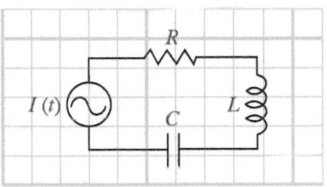

Figure 14.10   An *RLC* circuit

If $I(t)$ is the current in the circuit at time t and $Q(t)$ is the total charge on the capacitor, it is shown in physics that IR is the voltage drop across the resistance, Q/C is the voltage drop across the capacitor, and $L\,dI/dt$ is the voltage drop across the inductance. According to Kirchhoff's second law for circuits, the impressed voltage $E(t)$ in a circuit is the sum of the voltage drops, so that the current $I(t)$ in the circuit satisfies

$$L\frac{dI}{dt} + RI + \frac{Q}{C} = E(t)$$

By differentiating both sides of this equation and using the fact that $I = \dfrac{dQ}{dt}$, we can write

$$L\frac{d^2I}{dt^2} + R\frac{dI}{dt} + \frac{1}{C}I = \frac{dE}{dt}$$

For instance, suppose the voltage input is sinusoidal—that is, $E(t) = A\sin\omega t$. Then we have $dE/dt = A\omega\cos\omega t$, and the second-order linear differential equation used in the analysis of the circuit is

$$L\frac{d^2I}{dt^2} + R\frac{dI}{dt} + \frac{1}{C}I = A\omega\cos\omega t$$

To solve this equation, we proceed in the usual way, solving the related homogeneous equation and then finding a particular solution of the nonhomogeneous system. The solution to the related homogeneous equation is called the **transient current** because the current described by this solution usually does not last very long. The part of the nonhomogeneous solution that corresponds to transient current 0 is called the **steady-state current**. Several problems dealing with *RLC* circuits are outlined in the problem set.

PROBLEM SET 14.3

Level 1

In Problems 1-10, find a trial solution $\bar{y}_p$ for use in the method of undetermined coefficients.

1. $y'' - 6y' = e^{2x}$
2. $y'' - 6y' = e^{3x}$
3. $y'' + 6y' + 8y = 2 - e^{2x}$
4. $y'' + 6y' + 8y = 2 + e^{-3x}$
5. $y'' + 2y' + 2y = e^{-x}$
6. $y'' + 2y' + 2y = \cos x$
7. $y'' + 2y' + 2y = e^{-x} \sin x$
8. $2y'' - y' - 6y = x^2 e^{2x}$
9. $y'' + 4y' + 5y = e^{-2x}(x + \cos x)$
10. $y'' + 4y' + 5y = (e^{-x} \sin x)^2$

For the differential equation

$$y'' + 6y' + 9y = F(x)$$

find a trial solution $\bar{y}_p$ for undetermined coefficients for each choice of $F(x)$ given in Problems 11-18.

11. $3x^3 - 5x$ 12. $2x^2 e^{4x}$
13. $x^3 \cos x$ 14. $xe^{2x} \sin 5x$
15. $e^{2x} + \cos 3x$ 16. $2e^{3x} + 8xe^{-3x}$
17. $4x^3 - x^2 + 5 - 3e^{-x}$ 18. $(x^2 + 2x - 6)e^{-3x} \sin 3x$

Use the method of undetermined coefficients to find the general solution of the nonhomogeneous differential equations given in Problems 19-30.

19. $y'' + y' = -3x^2 + 7$
20. $y'' + 6y' + 5y = 2e^x - 3e^{-3x}$
21. $y'' + 8y' + 15y = 3e^{2x}$
22. $y'' - 5y' - 3y = 5e^{3x}$
23. $y'' + 2y' + 2y = \cos x$
24. $y'' - 6y' + 13y = e^{-3x} \sin 2x$
25. $7y'' + 6y' - y = e^{-x}(x + 1)$
26. $2y'' + 5y' = e^x \sin x$
27. $y'' - y' = x^3 - x + 5$
28. $y'' - y' = (x - 1)e^x$
29. $y'' + 2y' + y = (4 + x)e^{-x}$
30. $y'' - y' - 6y = e^{-2x} + \sin x$

Use variation of parameters to solve the differential equations given in Problems 31-40.

31. $y'' + y = \tan x$ 32. $y'' + 8y' + 16y = xe^{-2x}$
33. $y'' - y' - 6y = x^2 e^{2x}$ 34. $y'' - 3y' + 2y = \dfrac{e^x}{1 + e^x}$
35. $y'' + 4y = \sec 2x \tan 2x$ 36. $y'' + y = \sec^2 x$
37. $y'' + 2y' + y = e^{-x} \ln x$ 38. $y'' - 4y' + 4y = \dfrac{e^{2x}}{1 + x}$
39. $y'' - y' = \cos^2 x$ 40. $y'' - y' = e^{-2x} \cos e^{-x}$

Level 2

In each of Problems 41-48, use either undetermined coefficients or variation of parameters to find the particular solution of the differential equation that satisfies the specified initial conditions.

41. $y'' - y' = 2\cos^2 x$; $y(0) = y'(0) = 0$
42. $y'' - y' = e^{-2x} \cos e^{-x}$; $y(0) = y'(0) = 0$
43. $y'' + 9y = 4e^{3x}$; $y(0) = 0, y'(0) = 2$
44. $y'' - 6y = x^2 - 3x$; $y(0) = 3, y'(0) = -1$
45. $y'' + 9y = x$; $y(0) = 0, y'(0) = 4$
46. $y'' + y' = 2\sin x$; $y(0) = 0, y'(0) = -4$
47. $y'' - 4y' - 12y = 3e^{5x}$; $y(0) = \frac{18}{7}, y'(0) = -\frac{1}{7}$
48. $y'' + y = \cot x$; $y\left(\frac{\pi}{2}\right) = 0, y'\left(\frac{\pi}{2}\right) = 5$
49. Find the general solution of the differential equation

$$y'' + y' - 6y = F(x)$$

where F is the function whose graph is shown in Figure 14.11. *Hint*: Set up and solve two separate differential equations.

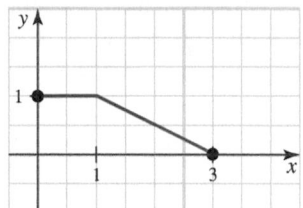

Figure 14.11   Graph of F

50. Find the general solution of the differential equation

$$y'' + 5y' + 6y = G(x)$$

where G is the function whose graph is shown in Figure 14.12. *Hint*: Set up and solve three separate differential equations.

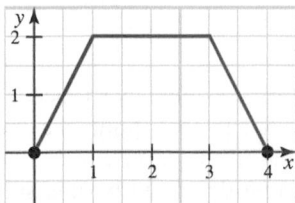

Figure 14.12   Graph of G

51. Find the steady-state current and the transient current in an *RLC* circuit with $L = 4$ henries, $R = 8$ ohms, $C = \frac{1}{8}$ farad, and $E(t) = 16 \sin t$ volts. You may assume that $I(0) = 0$ and $I'(0) = 0$.

52. Work Problem 51 for the case where $E(t) = 10te^{-t}$.

53. Find the steady-state current and the transient current in an *RLC* circuit with $L = 1$ henry, $R = 3$ ohms, $C = \frac{1}{2}$ farad, and $E(t) = 16t$ volts. You may assume that $I(0) = 0$ and $I'(0) = 0$.

54. Find the steady-state current and the transient current in an *RLC* circuit with $L = 1$ henry, $R = 10$ ohms, $C = \frac{1}{9}$ farad, and $E(t) = 9\cos t$ volts. You may assume that $I(0) = 0$ and $I'(0) = 0$.

55. Work Problem 54 for the case where $E(t) = 5t \sin t$.

56. Find the transient and the steady-state current in an *RLC* circuit with $L = 1$ henry, $R = 10$ ohms, $C = \frac{1}{9}$ farad, and $E(t) = 9 \sin t$ volts. You may assume that $I(0) = 0$ and $I'(0) = 0$.

Level 3

57. Theorem 14.4 is also valid when the coefficients a and b are continuous functions of x. Solve the differential equation

$$x^2 y'' - 3xy' + 4y = x \ln x$$

by completing the following steps:

a. The related homogeneous equation

$$x^2 y'' - 3xy' + 4y = 0$$

has one solution of the form $y_1 = x^n$. Find n, then use reduction of order to find a second, linearly independent solution y_2.

b. Use variation of parameters to find a particular solution y_p of the given nonhomogeneous equation. Then use Theorem 14.4 to write the general solution. (*Hint*: Divide the differential equation by x^2.)

58. Repeat Problem 57 for the equation

$$(1 + x^2)y'' + 2xy' - 2y = \tan^{-1} x$$

59. Consider the following equation:

$$y' = -\frac{x}{y}$$

What type of equation is it? Is this a separable differential equation? Is the equation homogeneous? What is the easiest method to solve this equation? Describe the trajectories for all solutions of this equation, as well as their orthogonal trajectories. See Section 5.6 for terminology.

60. The principle of superposition Let y_1 be a solution to the second-order-linear differential equation

$$y'' + P(x)y' + Q(x)y = F_1(x)$$

and let y_2 satisfy

$$y'' + P(x)y' + Q(x)y = F_2(x)$$

Show that $y_1 + y_2$ satisfies the differential equation

$$y'' + P(x)y' + Q(x)y = F_1(x) + F_2(x)$$

CHAPTER 14 REVIEW

*T*he die is cast; I have written my book; it will be read either in the present age or by posterity, it matters not which; it may well await a reader, since God has waited six thousand years for an interpreter of his words.

Johannes Kepler

From James R. Newman, *The World of Mathematics, Volume I*, (New York: Simon and Schuster, 1956), p. 220

Proficiency Examination

Concept Problems

1. What is a separable differential equation?
2. What is a homogeneous differential equation?
3. What is the form of a first-order linear differential equation?
4. What is an exact differential equation?
5. Describe Euler's method.
6. Define what it means for a set of functions to be linearly independent.
7. What is the Wronskian, and how is it used to test for linear independence?
8. **a.** What is the characteristic equation of $ay'' + by' + cy = 0$?
   b. What is the general solution of a second-order homogeneous equation?
9. Describe the form of the general solution of a second-order nonhomogeneous equation.
10. Describe the method of undetermined coefficients.
11. Describe the method of variation of parameters.

Practice Problems

Solve the differential equations in Problems 12-18.

12. $\dfrac{dy}{dx} = \sqrt{\dfrac{1-y^2}{1+x^2}}$

13. $\dfrac{x}{y^2}dx - \dfrac{x^2}{y^3}dy = 0$

14. $\dfrac{dy}{dx} = \dfrac{2x+y}{3x}$

15. $xy\,dy = (x^2 - y^2)\,dx$

16. $x^2 dy - (x^2 + y^2)dx = 0$

17. $y'' + 2y' + 2y = \sin x$

18. $(3x^2 e^{-y} + y^{-2} + 2xy^{-3})\,dx + (-x^3 e^{-y} - 2xy^{-3} - 3x^2 y^{-4})\,dy = 0$

19. A spring with spring constant $k = 30$ lb/ft hangs in a vertical position with its upper end fixed and an 8-lb object attached to its lower end. The object is pulled down 4 in. from the equilibrium position of the spring and is then released. Find the displacement $x(t)$ of the object, assuming that air resistance is $0.8v$, where $v(t)$ is the velocity of the object at time t.

20. An *RLC* circuit has inductance $L = 0.1$ henry, resistance $R = 25$ ohms, and capacitance $C = 200$ microfarads (i.e., 200×10^{-6} farad). If there is a variable voltage source of $E(t) = 50\cos 100t$ in the circuit, what is the current $I(t)$ at time t? Assume that when $t = 0$, there is no charge and no current flowing.

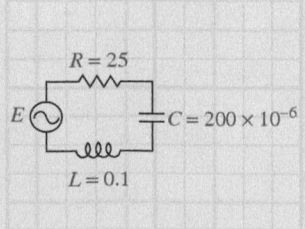

Supplementary Problems*

1. Consider the differential equation $\dfrac{dy}{dx} = 2x - 3y$
   a. Use the slope field to sketch the particular solution that passes through the point $(0, 1)$
   b. Use Euler's method to find a solution for the initial value problem in part **a** for $0 \le x \le 1$ and $h = 0.2$.

2. Consider the differential equation $\dfrac{dy}{dx} = x^2 y$
   a. Use the slope field to sketch the particular solution that passes through the point $(1, 2)$
   b. Use Euler's method to find a solution for the initial value problem in part **a** for $1 \le x \le 2$ and $h = 0.2$.

Find the general solution of the given differential equations in Problems 3-26.

3. $x\,dx + y\,dy = 0$

4. $y\,dx - x\,dy = 0$

5. $y' = \dfrac{-7x}{y}$

6. $y' = \dfrac{x^2}{y^2}$

7. $dy = (y \cos x)\,dx$

8. $y'' - y' = 0$

9. $\dfrac{dy}{dx} = \dfrac{-4x}{y^3}$

10. $x\,dy + (3y - xe^{x^2})\,dx = 0$

11. $y^2\,dy - \left(x^2 + \dfrac{y^3}{x}\right)dx = 0$

12. $\dfrac{x}{y}\,dy + \left(\dfrac{3y}{x} - 5\right)dx = 0$

13. $y\dfrac{dy}{dx} = e^{2x - y^2}$

14. $dy = (3y + e^x + \cos x)\,dx$

15. $dy - (5y + e^{5x}\sin x)\,dx = 0$

16. $x^2\,dy = (x^2 - y^2)\,dx$

17. $(1 - xe^y)\,dy = e^y\,dx$

18. $xy\,dx + (1 + x^2)\,dy = 0$

19. $x\,dy = (y + \sqrt{x^2 - y^2})\,dx$

20. $2xye^{x^2}\,dx - e^{x^2}\,dy = 0$

21. $(y - xy)\,dx + x^3\,dy = 0$

22. $\dfrac{dy}{dx} = 2y \cot 2x + 3\csc 2x$

23. $\left(4x^3y^3 + \dfrac{1}{x}\right)dx + \left(3x^4y^2 - \dfrac{1}{y}\right)dy = 0$

24. $(-y \sin xy + 2x)\,dx + (3y^2 - x \sin xy)\,dy = 0$

25. $\sin y\,dx + (e^x + e^{-x})\sin y\,dy = 0$

26. $\dfrac{dy}{dx} + 2y \cot x + \sin 2x = 0$

In Problems 27-32, find
a. *the general solution y_h of the related homogeneous equation*
b. *a particular solution y_p of the nonhomogeneous equation*

27. $y'' - 9y = 1 + x$

28. $y'' - 2y' + y = e^x$

29. $y'' - 5y' + 6y = x^2 e^x$

30. $y'' + 2y' + y = \sinh x$

31. $y'' - 3y' + 2y = x^3 e^x$

32. $y'' + y = \sec x$

33. Solve $\dfrac{d^2y}{dx^2} = e^{-3x}$ subject to the initial conditions $y(0) = y'(0) = 0$.

34. Use the substitution $p = \dfrac{dy}{dx}$ to solve the differential equation

$$\frac{d^2y}{dx^2} = \left(\frac{dy}{dx}\right)^3 + \frac{dy}{dx}$$

35. Solve the system of differential equations

$$\begin{cases} \dfrac{dx}{dt} = -2x \\[2mm] \dfrac{dy}{dt} = -3y + 2x \\[2mm] \dfrac{dz}{dt} = 3y \end{cases}$$

subject to the initial conditions $x(0) = 1, y(0) = 0, z(0) = 0$.

---

*The supplementary problems are presented in a somewhat random order, not necessarily in order of difficulty.

36. Find a curve $y = f(x)$ that passes through $(1,2)$ and has the property that the y-intercept of the tangent line at each point $P(x,y)$ is equal to y^2.

37. Find a curve that passes through the point $(1,2)$ and has the property that the length of the part of the tangent line between each point $P(x,y)$ on the curve and the y-axis is equal to the y-intercept of the tangent line at P.

38. Find a curve that passes through $(1,2)$ and has the property that the normal line at any point $P(x,y)$ on the curve and the line joining P to the origin form an isosceles triangle with the x-axis as its base.

Orthogonal trajectories *Recall from Section 5.6 that the orthogonal trajectories of a given family of curves are another family with the following property: every time a member of the second family intersects a member of the given family, the tangent lines to the two curves at the common point intersect at right angles. Find the orthogonal trajectories for the families of curves in Problems 39-42.*

39. $e^x + e^{-y} = C$ **40.** $3x^2 + 5y^2 = C$

41. $x^2 - y^2 = Cx$ **42.** $y = \dfrac{Cx}{x^2 + 1}$

43. Use reduction of order to find the general solution of the **Legendre equation**

$$(1 - x^2)y'' - 2xy' + 2y = 0$$

given that $y_1 = x$ is one solution.

44. Find functions $x = x(t)$ and $y = y(t)$ that satisfy the linear system

$$\frac{dx}{dt} = x + y \qquad \frac{dy}{dt} = x^2 - y^2$$

Sketch the solution curve $(x(t), y(t))$ that contains the point $(0,2)$. *Hint:* Note that $\dfrac{dy}{dx} = \dfrac{y'(t)}{x'(t)}$.

45. Solve the second-order differential equation $xy'' + 2y' = x$ by setting $p = y'$ to convert it to a first-order equation in p.

46. Find an equation for a curve for which the radius of curvature is proportional to the slope of the tangent line at each point $P(x,y)$.

47. Solve the differential equation

$$\frac{dx}{dy} - \frac{x}{y} = ye^y$$

48. A ship weighing 64,000 tons [mass = (64,000)(2,000)/32 slugs] starts from rest and is driven by the constant thrust of its propellers. Suppose the propellers supply 250,000 lb of thrust and the resistance of the water is $12{,}000v$ lb, where $v(t)$ is the velocity of the ship. Set up and solve a differential equation for $v(t)$.

49. A 10-ft uniform chain (see Figure 14.13) of mass m is hanging over a peg so that 3 ft are on one side and 7 ft are on the other. Set up and solve a differential equation to find how long it takes for the chain to slide off the peg. Neglect the friction.

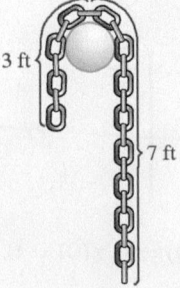

Figure 14.13 Chain on a peg

50. When an object weighing 10 lb is suspended from a spring, the spring is stretched 2 in. from its equilibrium position. The upper end of the spring is given a motion of $y = 2(\sin t + \cos t)$ ft. Find the displacement $x(t)$ of the object.

51. Suppose we wish to measure the deflection of a beam (see Figure 14.14).

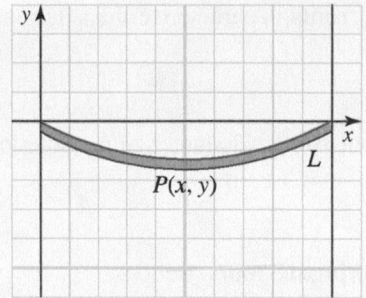

a. Deflection is the amount of sag   **b.** Modeling the deflection of a beam

Figure 14.14   Wood beam deflection

Assume the load on the beam is W pounds per foot and one end of the beam is at the origin, and the other end of the beam is at $(L, 0)$. Then the deflection y of the beam at point x is modeled by the differential equation

$$EI\frac{d^2y}{dx^2} = WLx - \frac{1}{2}Wx^2$$

for positive constants E and I.

a. Solve this equation to obtain the deflection $y(x)$. Note that $y(0) = y(L) = 0$.
b. What is the maximum deflection in the beam?

52. Consider the **Clairaut equation**

$$y = xy' + f(y')$$

a. Differentiate both sides of this equation to obtain

$$[x + f'(y')]y'' = 0$$

Because one of the two factors in the product on the left must be 0, we have two cases to consider. What is the solution if $y'' = 0$? This is the *general solution.*

b. What is the solution if

$$x + f'(y') = 0$$

(This solution is called the *singular solution.*) *Hint*: Use the parameterization $y' = t$.

c. Find the general and singular solutions for the Clairaut equation

$$y = xy' + \sqrt{4 + (y')^2}$$

53. A differential equation of the form

$$x^2y'' + Axy' + By = F(x)$$

with $x \neq 0$ is called an **Euler equation**. To find a general solution of the related homogeneous equation

$$x^2y'' + Axy' + By = 0$$

we assume that there are solutions of the form $y = x^m$.

a. Show that m must satisfy

$$m^2 + (A - 1)m + B = 0$$

This is the characteristic equation for the Euler equation.

b. Distinct real roots. Characterize the solutions of the related homogeneous equation in the case where

$$(A - 1)^2 > 4B$$

c. Repeated real roots. Characterize the solutions of the related homogeneous equation in the case where

$$(A - 1)^2 = 4B$$

d. Complex conjugate roots. Suppose

$$(A - 1)^2 < 4B$$

and that $\alpha \pm \beta i$ are the roots of the characteristic equation. Verify that $y_1 = x^\alpha \cos(\beta \ln |x|)$ is one solution. Use reduction of order to find a second solution y_2, then characterize the general solution.

54. Find the general solution of the Euler equation (see Problem 53)

$$x^2 y'' + 7xy' + 9y = \sqrt{x}$$

Hint: Use variation of parameters.

55. In certain biological studies, it is important to analyze *predator-prey* relationships. Suppose $x(t)$ is the prey population and $y(t)$ is the predator population at time t. Then these populations change at rates

$$\frac{dx}{dt} = a_{11}x - a_{12}y \qquad \frac{dy}{dt} = a_{21}x - a_{22}y$$

where a_{11} is the natural growth rate of the prey, a_{12} is the predation rate, a_{21} measures the food supply of the predators, and a_{22} is the death rate of the predators. Outline a procedure for solving this system.

56. Consider the almost homogeneous differential equation

$$\frac{dy}{dx} = f\left(\frac{ax + by + c}{rx + sy + t}\right)$$

with $as = br$ (recall Problem 51, Section 14.2).

a. Let $u = \dfrac{ax + by}{a}$. Show that $u = \dfrac{rx + sy}{r}$ and $\dfrac{du}{dx} = \dfrac{a}{b}\left(\dfrac{du}{dx} - 1\right)$.

b. Verify that by making the change of variable suggested in part **a**, you can rewrite the given differential equation in the separable form

$$\frac{du}{1 + \frac{b}{a}f\left(\frac{au + c}{ru + t}\right)} = a\,dx$$

c. Use the procedure outlined in parts **a** and **b** to solve the almost homogeneous differential equation

$$\frac{dy}{dx} = \frac{2x + y - 3}{4x + 2y + 5}$$

57. Putnam Examination Problem Find all solutions of the equation

$$yy'' - 2(y')^2 = 0$$

that pass through the point $x = 1, y = 1$.

58. **Putnam Examination Problem** A coast artillery gun can fire at any angle of elevation between 0° and 90° in a fixed vertical plane. If air resistance is neglected and the muzzle velocity is constant ($v(0) = v_0$), determine the set H of points in the plane and above the horizontal that can be hit.

59. **Putnam Examination Problem** Show that

$$x + \frac{2}{3}x^3 + \frac{2 \cdot 4}{3 \cdot 5}x^5 + \frac{2 \cdot 4 \cdot 6}{3 \cdot 5 \cdot 7}x^7 + \cdots = \frac{\sin^{-1} x}{\sqrt{1-x^2}}$$

60. **Book Report** "We often hear that mathematics consists mainly of 'proving theorems.' Is a writer's job mainly that of 'writing sentences'? A mathematician's work is mostly a tangle of guesswork, analogy, wishful thinking and frustration, and proof, far from being the core of discovery, is more often than not a way of making sure that our minds are not playing tricks. Few people, if any, had dared write this out loud before Davis and Hersh. Theorems are not to mathematics what successful courses are to a meal. The nutritional analogy is misleading. To master mathematics is to master an intangible view...." This quotation comes from the introduction to the book *The Mathematical Experience* by Philip J. Davis and Reuben Hersh (Boston: Houghton Mifflin, 1981). Read this book and prepare a book report.

CHAPTER 14 GROUP RESEARCH PROJECT*

Working in small groups is typical of most work environments, and this book seeks to develop skills with group activities. We present a group project at the end of each chapter. These projects are to be done in groups of three or four students.

Save the Perch Project

Happy Valley Pond is currently populated by yellow perch. A map is shown in Figure 14.15. Water flows into the pond from two springs and evaporates from the pond, as shown by the table in the margin.

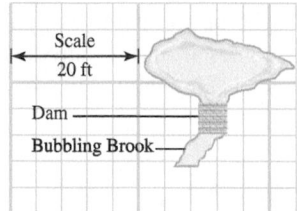

Spring	Dry Season	Rainy Season
A	50 gal/h	60 gal/h
B	60 gal/h	75 gal/h
Evaporation	110 gal/h	75 gal/h

Figure 14.15   Happy Valley Pond is fed by two springs A and B

Unfortunately, spring B has become contaminated with salt and is now 10% salt, which means that 10% of a gallon of water from spring B is salt. The yellow perch will start to die if the concentration of salt in the pond rises to 1%. Assume that the salt will not evaporate but will mix thoroughly with the water in the pond. There was no salt in the pond before the contamination of spring B. Your group has been called upon by the Happy Valley Bureau of Fisheries to try to save the perch.

Extended paper for further study. Your paper is not limited to the following questions, but should include the number of gallons of water in the pond when the water level is exactly even with the top of the spillover dam. The following table gives a series of measurements of the depth of the pond at the indicated points when the water level was exactly even with the top of the spillover dam.

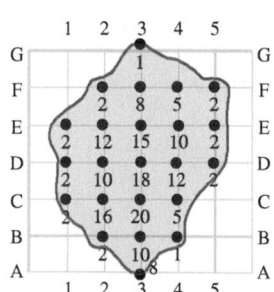

Depth of Happy Valley Pond (in feet)

	1	2	3	4	5
A			8		
B		2	10	1	
C	2	16	20	5	
D	2	10	18	12	2
E	2	12	15	10	2
F		2	8	5	2
G			1		

Figure 14.16   Happy Valley Pond is fed by two springs A and B

Let $t = 0$ hours correspond to the time when spring B became contaminated. Assume it is the dry season and that at time $t = 0$ the water level of the pond was exactly even with the top of the spillover dam. Write a differential equation for the amount of salt in the pond after t hours. Draw a graph of the amount of salt in the pond versus time for the next 3 months. How much salt will there be in the pond in the long run, and do the fish die? If so, when do they start to die? It is very difficult to find where the contamination of spring B originates, so the Happy Valley Bureau of Fisheries proposed to flush the pond by running 100 gal of pure water per hour through the pond. Your report should include an analysis of this plan and any modifications or improvements that could help save the perch.

---

*Adapted from a group project by Diane Schwartz of Ithaca College, New York.

APPENDICES

APPENDIX A: INTRODUCTION TO THE THEORY OF LIMITS

In Section 2.1 we defined the limit of a function as follows: The notation

$$\lim_{x \to c} f(x) = L$$

is read "the limit of $f(x)$ as x approaches c is L" and means that the function values $f(x)$ can be made arbitrarily close to L by choosing x sufficiently close to c but not equal to c.

This informal definition was valuable because it gave you an intuitive feeling for the limit of a function and allowed you to develop a working knowledge of this fundamental concept. For theoretical work, however, this definition will not suffice, because it gives no precise, quantifiable meaning to the terms "arbitrarily close to L" and "sufficiently close to c." The following definition derived from the work of Cauchy and Weierstraß, gives precision to the limit definition, and was also first stated in Section 2.1.

LIMIT OF A FUNCTION (formal definition) The limit statement

$$\lim_{x \to c} f(x) = L$$

means that for each number $\epsilon > 0$, there corresponds a number $\delta > 0$ with the property that

$$|f(x) - L| < \epsilon \quad \text{whenever} \quad 0 < |x - c| < \delta$$

Behind the formal language is a fairly straightforward idea. In particular, to establish a specific limit, say $\lim_{x \to c} f(x) = L$, a number $\epsilon > 0$ is chosen first to establish a desired degree of proximity to L, and then a number $\delta > 0$ is found that determines how close x must be to c to ensure that $f(x)$ is within ϵ units of L.

The situation is illustrated in Figure A.1, which shows a function that satisfies the conditions of the definition. Notice that whenever x is within δ units of c (but not equal to c), the point $(x, f(x))$ on the graph of f must lie in the rectangle (shaded region) formed by the intersection of the horizontal band of width 2ϵ centered at L and the vertical band of width 2δ centered at c. The smaller the ϵ-interval around the proposed limit L, generally the smaller the δ-interval will need to be in order for $f(x)$ to lie in the ϵ-interval. If such a δ can be found no matter how small ϵ is, then L must be the limit.

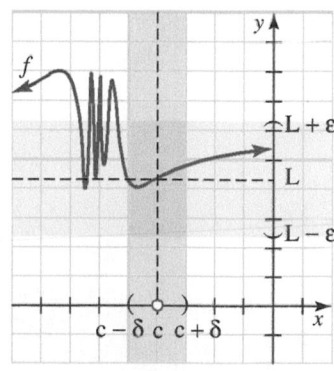

Figure A.1 The epsilon-delta definition of limit

The Believer/Doubter Format

The limit process can be thought of as a "contest" between a "believer" who claims that $\lim_{x \to c} f(x) = L$ and a "doubter" who disputes this claim. The contest begins with the doubter choosing a positive number ϵ and the believer countering with a positive number δ.

When will the believer win the argument and when will the doubter win? As you can see from Figure A.1, if the believer has the "correct limit" L, then no matter how the doubter chooses ϵ, the believer can find a δ so that the graph of $y = f(x)$ for

$c - \delta < x < c + \delta$ will lie inside the heavily shaded rectangle in Figure A.1 (with the possible exception of $x = c$); that is, inside the rectangular region

$$R : \text{width } [c - \delta, c + \delta]; \text{height } [L - \epsilon, L + \epsilon]$$

On the other hand, if the believer tries to defend an incorrect value ω (for "wrong") as the limit, then it will be possible for the doubter to choose an ϵ so that at least part of the curve $y = f(x)$ will lie outside the heavily shaded rectangle R, no matter what value of δ is chosen by the believer (see Figure A.2).

There is some ε > 0 so that no matter what δ is chosen, part of the graph will be outside this region.

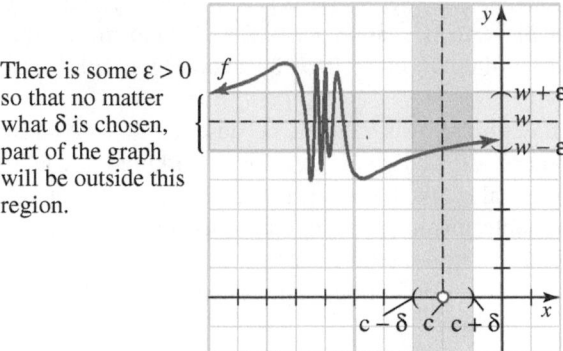

Figure A.2 False limit scenario

To avoid an endless chain of ϵ-δ challenges, the believer usually tries to settle the issue by producing a formula relating the choice of δ to the doubter's ϵ that will satisfy the requirement no matter what ϵ is chosen. The believer/doubter format is used in Example 1 to verify a given limit statement (believer "wins") and in Example 2 to show that a certain limit does not exist (doubter "wins").

Example 1 Verifying a limit claim (believer wins)

Show that $\lim\limits_{x \to 2}(2x + 1) = 5$.

Solution Let $f(x) = 2x + 1$. To verify the given limit statement, we begin by having the doubter choose a positive number ϵ. To help the believer, choose $\delta > 0$, and note the computation in the following box.

> $|(2x + 1) - 5| = |2x - 4| = 2|x - 2|$
> Thus, if $0 < |x - c| < \delta$ or $0 < |x - 2| < \delta$
> then $|(2x + 1) - 5| < 2\delta$

The believer *wants* $|f(x) - L| < \epsilon$, so we see that the believer should choose $2\delta = \epsilon$, or $\delta = \frac{\epsilon}{2}$.

With the information shown in this box, we can now make the following argument, which uses the formal definition of limit:

Let $\epsilon > 0$ be given. The believer chooses $\delta = \frac{\epsilon}{2} > 0$ so that

$$|f(x) - L| = |(2x + 1) - 5| < 2\delta < 2\left(\frac{\epsilon}{2}\right) = \epsilon$$

whenever $0 < |x - 2| < \delta$, and the limit statement is verified. The graphical relationship between ϵ and δ is shown in Figure A.3.

Notice that no matter what ϵ is chosen by the doubter, by choosing a number that is one-half of that ϵ, the believer will force the function to stay within the shaded portion of the graph, as shown in Figure A.3.

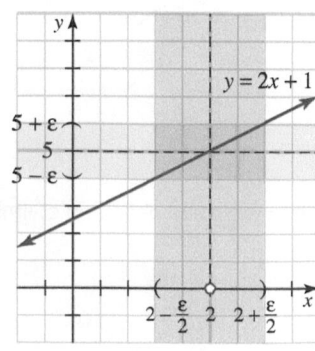

Figure A.3 Verifying $\lim\limits_{x \to 2}(2x + 1) = 5$: Given $\epsilon > 0$, choose $\delta = \frac{\epsilon}{2}$

Example 2 Disproving a limit claim (doubter wins)

Determine whether $\displaystyle\lim_{x\to 2}\frac{2x^2 - 3x - 2}{x - 2} = 6$.

Solution   Let $f(x) = \dfrac{2x^2 - 3x - 2}{x - 2} = \dfrac{(2x + 1)(x - 2)}{x - 2} = 2x + 1,\ x \neq 2$

Once again, the doubter will choose a positive number ϵ and the believer must respond with a δ. As before, the believer does some preliminary work with $f(x) - L$:

$$\left|\frac{2x^2 - 3x - 2}{x - 2} - 6\right| = \left|\frac{2x^2 - 3x - 2 - 6x + 12}{x - 2}\right|$$

$$= \left|\frac{2x^2 - 9x + 10}{x - 2}\right|$$

$$= |2x - 5|$$

The believer wants to write this expression in terms of $x - c = x - 2$. This example does not seem to "fall into place" as did Example 1. Let's analyze the situation shown in Figure A.4.

The doubter observes that if $\epsilon > 0$ is small enough, at least part of the line $y = 2x + 1$ for $0 < |x - 2| < \delta$ lies outside the shaded rectangle

$$R:\ \text{width } [2 - \delta, 2 + \delta];\ \text{height } [6 - \epsilon, 6 + \epsilon]$$

regardless of the value of $\delta > 0$. For instance, suppose $\epsilon < 1$. Then, if δ is any positive number, we have

$$\left|f\left(2 - \frac{\delta}{2}\right) - 6\right| = \left|\left[2\left(2 - \frac{\delta}{2}\right) + 1\right] - 6\right| = |-\delta - 1| > 1$$

so $|f(x) - 6| < \epsilon$ is not satisfied for all $0 < |x - 2| < \delta$ (in particular, not for $x = 2 - \frac{\delta}{2}$). Thus, $\displaystyle\lim_{x\to 2}\frac{2x^2 - 3x - 2}{x - 2} = 6$ is false.

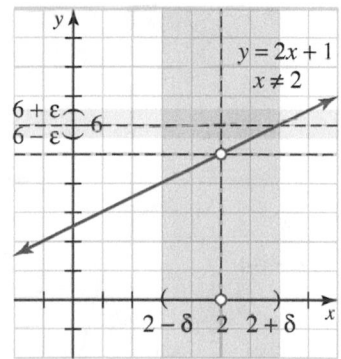

Figure A.4 Example of an incorrect limit statement

The believer/doubter format is a useful device for dramatizing the way certain choices are made in epsilon-delta arguments, but it is customary to be less "chatty" in formal mathematical proofs.

Example 3 An epsilon-delta proof of a limit of a rational function

Show that $\displaystyle\lim_{x\to 2}\frac{x^2 - 2x + 2}{x - 4} = -1$.

Solution   Let $f(x) = \dfrac{x^2 - 2x + 2}{x - 4}$. We have

$$|f(x) - L| = \left|\frac{x^2 - 2x + 2}{x - 4} - (-1)\right|$$

$$= \left|\frac{x^2 - x - 2}{x - 4}\right|$$

$$= |x - 2|\underbrace{\left|\frac{x + 1}{x - 4}\right|}$$

This must be less than the given ϵ
whenever x is near 2, but not equal to 2.

Certainly $|x - 2|$ is small if x is near 2, and the factor $\left|\dfrac{x + 1}{x - 4}\right|$ is not large (it is close to $\frac{3}{2}$). Note that if $|x - 2|$ is small it is reasonable to assume

$$|x - 2| < 1 \quad \text{so that} \quad 1 < x < 3$$

Let $g(x) = \dfrac{x + 1}{x - 4} = 1 + \dfrac{5}{x - 4}$; $g'(x) = -\dfrac{5}{(x - 4)^2} < 0$ on $(1, 3)$. Thus $g(x)$ is decreasing on $(1, 3)$ and

$$g(3) = \frac{4}{-1} < \frac{x + 1}{x - 4} < \frac{2}{-3} = g(1)$$

Hence, if $|x - 2| < 1$, then

$$-4 < \frac{x + 1}{x - 4} < 4 \qquad \textit{Because } -\tfrac{2}{3} < 4$$

and

$$\left|\frac{x + 1}{x - 4}\right| < 4$$

Now let $\epsilon > 0$ be given. If simultaneously

$$|x - 2| < \frac{\epsilon}{4} \quad \text{and} \quad \left|\frac{x + 1}{x - 4}\right| < 4$$

then

$$|f(x) - L| = |x - 2|\left|\frac{x + 1}{x - 4}\right| < \frac{\epsilon}{4}(4) = \epsilon$$

Thus, we have only to take δ to be the smaller of the two numbers 1 and $\frac{\epsilon}{4}$ in order to guarantee that

$$|f(x) - L| < \epsilon$$

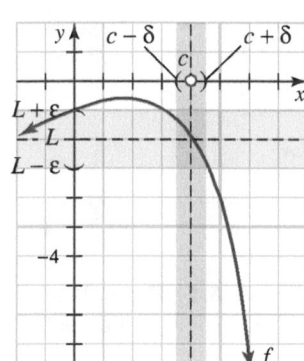

Figure A.5

$$\lim_{x \to 2} \frac{x^2 - 2x + 2}{x - 4} = -1$$

That is, given $\epsilon > 0$, choose δ to be the smaller of the numbers 1 and $\frac{\epsilon}{4}$. We write this as $\delta = \min\left(1, \frac{\epsilon}{4}\right)$. We can confirm this result by looking at the graph of $y = f(x)$ in Figure A.5.

Example 4   An epsilon-delta proof that a limit does not exist

Show that $\lim\limits_{x \to 0} \dfrac{1}{x}$ does not exist.

Solution   Let $f(x) = \dfrac{1}{x}$ and L be any number. Suppose that $\lim\limits_{x \to 0} f(x) = L$. Look at the graph of f, as shown in Figure A.6.

It would seem that no matter what value of ϵ is chosen, it would be impossible to find a corresponding δ. Consider the absolute value expression required by the definition of limit: If

$$|f(x) - L| < \epsilon, \text{ or for this example, } \left|\frac{1}{x} - L\right| < \epsilon$$

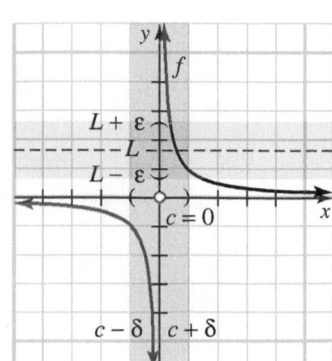

Figure A.6 $\lim\limits_{x \to 0} \dfrac{1}{x}$

$$-\epsilon < \frac{1}{x} - L < \epsilon \quad \textit{Property of absolute value, Table 1.1. p. 17.}$$

$$L - \epsilon < \frac{1}{x} < L + \epsilon$$

If $\epsilon = 1$ (not a particularly small ϵ), then

$$\left| \frac{1}{x} \right| < |L| + 1$$

$$|x| > \frac{1}{|L| + 1}$$

and thus x is not close to zero, which proves (since L was chosen arbitrarily) that $\lim\limits_{x \to 0} \frac{1}{x}$ does not exist. In other words, since $|x|$ can be chosen so that $0 < |x| < \frac{1}{|L| + 1}$, then $\frac{1}{|x|}$ will be very large, and it will be impossible to squeeze $\frac{1}{x}$ between $L - \epsilon$ and $L + \epsilon$ for any L.

Selected Theorems with Formal Proofs

Next, we will prove several theoretical results using the formal definition of the limit. The next two theorems are useful tools in the development of calculus. The first states that the points on a graph that are on or above the x-axis cannot possibly "tend toward" a point *below* the axis, as shown in Figure A.7.

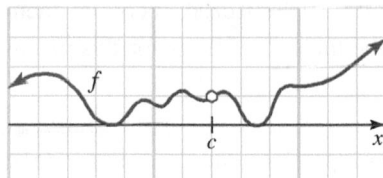

If $f(x) \geq 0$ for all x near c, then $\lim\limits_{x \to c} f(x) \geq 0$.

Figure A.7 Limit limitation theorem

Theorem A.1   Limit limitation theorem

Suppose $\lim\limits_{x \to c} f(x)$ exists and $f(x) \geq 0$ throughout an open interval containing the number c, except possibly at c itself. Then

$$\lim_{x \to c} f(x) \geq 0$$

Proof: Let $L = \lim\limits_{x \to c} f(x)$. To show that $L \geq 0$, assume the contrary; that is, assume $L < 0$. According to the definition of limit (with $\epsilon = -L$), there is a number $\delta > 0$ such that

$$|f(x) - L| < -L \text{ whenever } 0 < |x - c| < \delta$$

Thus,

$$f(x) - L < -L$$
$$f(x) < 0$$

whenever $0 < |x - c| < \delta$, which contradicts the hypothesis that $f(x) \geq 0$ throughout an open interval containing c (with the possible exception of $x = c$). The contradiction forces us to reject the assumption that $L < 0$, so $L \geq 0$, as required. ♦

☙ *It may seem reasonable to conjecture that if $f(x) > 0$ throughout an open interval containing c, then $\lim_{x \to c} f(x) > 0$. This is not necessarily true, and the most that can be said in this situation is that $\lim_{x \to c} f(x) \geq 0$, if the limit exists. For example, if*

$$f(x) = \begin{cases} x^2 & \text{for } x \neq 0 \\ 1 & \text{for } x = 0 \end{cases}$$

then $f(x) > 0$ for all x, but $\lim_{x \to 0} f(x) = 0$. ☙

Useful information about the limit of a given function f can often be obtained by examining other functions that bound f from above and below. For example, in Example 6, Section 2.1, we discovered

$$\lim_{x \to 0} \frac{\sin x}{x} = 1$$

by using a table. We justified the limit statement by using a geometric argument to show that

$$\cos x \leq \frac{\sin x}{x} \leq 1$$

for all x near 0 and then noting that since $\cos x$ and 1 both tend toward 1 as x approaches 0, the function $\frac{\sin x}{x}$ which is "squeezed" between them, must converge to 1 as well. Theorem A.2 provides the theoretical basis for this method of proof.

Theorem A.2   The squeeze rule

If $g(x) \leq f(x) \leq h(x)$ for all x in an open interval containing c (except possibly at c itself) and if

$$\lim_{x \to c} g(x) = \lim_{x \to c} h(x) = L$$

then

$$\lim_{x \to c} f(x) = L$$

(This is stated, without proof, in Section 2.2.)

Proof: Let $\epsilon > 0$ be given. Since $\lim_{x \to c} g(x) = L$ and $\lim_{x \to c} h(x) = L$, there are positive numbers δ_1 and δ_2 such that

$$|g(x) - L| < \epsilon \quad \text{and} \quad |h(x) - L| < \epsilon$$

whenever

$$0 < |x - c| < \delta_1 \quad \text{and} \quad 0 < |x - c| < \delta_2$$

respectively. Let δ be the smaller of the numbers δ_1 and δ_2. Then, if x is a number that satisfies $0 < |x - c| < \delta$, we have

$$-\epsilon < g(x) - L \leq f(x) - L \leq h(x) - L < \epsilon$$

and it follows that $|f(x) - L| < \epsilon$. Thus, $\lim_{x \to c} f(x) = L$, as claimed.   ◆

The geometric interpretation of the squeeze rule is shown in Figure A.8. Notice that since $g(x) \leq f(x) \leq h(x)$, the graph of f is "squeezed" between those of g and h in the neighborhood of c. Thus, if the bounding graphs converge to a common point P as x approaches c, then the graph of f must also converge to P as well.

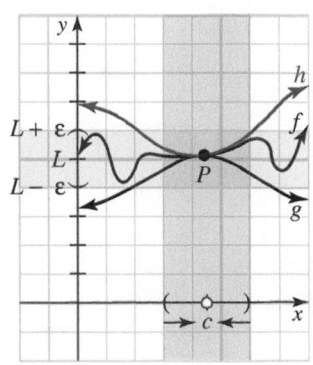

Figure A.8   The squeeze rule

APPENDIX B: SELECTED PROOFS

Chain Rule (Section 3.5)

Suppose f is a differentiable function of u and u is a differentiable function of x. Then $f[u(x)]$ is a differentiable function of x and

$$\frac{df}{dx} = \frac{df}{du}\frac{du}{dx}.$$

Proof: Define an auxiliary function g by

$$g(t) = \frac{f[u(x) + t] - f[u(x)]}{t} - \frac{df}{du}$$

if $t \neq 0$ and $g(t) = 0$ if $t = 0$. You can verify that g is continuous at $t = 0$. Notice that for $t = \Delta u$ and $\Delta u \neq 0$,

$$g(\Delta u) = \frac{f[u(x) + \Delta u] - f[u(x)]}{\Delta u} - \frac{df}{du}$$

$$g(\Delta u) + \frac{df}{du} = \frac{f[u(x) + \Delta u] - f[u(x)]}{\Delta u}$$

$$\left[g(\Delta u) + \frac{df}{du}\right]\Delta u = f[u(x) + \Delta u] - f[u(x)] \qquad \textit{Formula 1}$$

We now use the definition of the derivative for f.

$$\frac{df(u(x))}{dx} = \lim_{\Delta x \to 0} \frac{f[u(x + \Delta x)] - f[u(x)]}{\Delta x}$$

$$= \lim_{\Delta x \to 0} \frac{f[u(x) + \Delta u] - f[u(x)]}{\Delta x} \qquad \textit{Where } \Delta u = u(x + \Delta x) - u(x).$$

$$= \lim_{\Delta x \to 0} \frac{\left[g(\Delta u) + \frac{df}{du}\right]\Delta u}{\Delta x} \qquad \textit{Substituting previous (formula 1)}$$

$$= \lim_{\Delta x \to 0} \left[g(\Delta u) + \frac{df}{du}\right]\frac{\Delta u}{\Delta x} \qquad \textit{Rewriting previous formula}$$

$$= \lim_{\Delta x \to 0} \left[g(\Delta u) + \frac{df}{du}\right]\lim_{\Delta x \to 0}\frac{\Delta u}{\Delta x}$$

$$= \left[\lim_{\Delta x \to 0} g(\Delta u) + \lim_{\Delta x \to 0}\frac{df}{du}\right]\lim_{\Delta x \to 0}\frac{\Delta u}{\Delta x}$$

$$= \left[0 + \frac{df}{du}\right]\frac{du}{dx}$$

$$= \frac{df}{du}\frac{du}{dx} \qquad\qquad\qquad\qquad\qquad\qquad \blacklozenge$$

Cauchy's Generalized Mean Value Theorem (Section 4.2)

Let f and g be functions that are continuous on the closed interval $[a, b]$ and differentiable on the open interval (a, b). If $g(b) \neq g(a)$ and $g'(x) \neq 0$ on (a, b), then

$$\frac{f(b) - f(a)}{g(b) - g(a)} = \frac{f'(c)}{g'(c)}$$

for at least one number c between a and b. (*Note:* This is a generalization of the Mean Value Theorem stated and proved earlier.)

Proof: We begin by defining a special function, just as in the proof of the MVT as presented in Section 4.2. Specifically, let

$$F(x) = f(x) - f(a) - \frac{f(b) - f(a)}{g(b) - g(a)} \left[g(x) - g(a) \right]$$

for all x in the closed interval $[a, b]$. In the proof of the MVT in Section 4.2 we showed that F satisfies the hypotheses of Rolle's theorem, which means that $F'(c) = 0$ for at least one number c in (a, b). For this number c, we have

$$0 = F'(c) = f'(c) - \frac{f(b) - f(a)}{g(b) - g(a)} g'(c)$$

and the result follows from this equation.   ◆

L'Hôpital's Rule* (Section 4.5)

For any number a, let f and g be functions that are differentiable on an open interval (a, b), where $g'(x) \neq 0$. If

$$\lim_{x \to a^+} f(x) = 0, \quad \lim_{x \to a^+} g(x) = 0, \quad \text{and} \quad \lim_{x \to a^+} \frac{f'(x)}{g'(x)} \text{ exists, then}$$

$$\lim_{x \to a^+} \frac{f(x)}{g(x)} = \lim_{x \to a^+} \frac{f'(x)}{g'(x)}$$

Proof: First, define auxiliary functions F and G by

$$F(x) = f(x) \text{ for } a < x \leq b \text{ and } F(a) = 0$$
$$G(x) = g(x) \text{ for } a < x \leq b \text{ and } G(a) = 0$$

These definitions guarantee that $F(x) = f(x)$ and $G(x) = g(x)$ for $a < x \leq b$ and that $F(a) = G(a) = 0$. Thus, if w is any number between a and b, the functions F and G are continuous on the closed interval $[a, w]$ and differentiable on the open interval (a, w). According to the Cauchy generalized mean value theorem, which we can use since $G'(x) \neq 0$ guarantees that $G(w) \neq G(a)$, there exists a number t between a and w for which

$$\frac{F(w) - F(a)}{G(w) - G(a)} = \frac{F'(t)}{G'(t)}$$

$$\frac{F(w)}{G(w)} = \frac{F'(t)}{G'(t)} \qquad \textit{Because } F(a) = G(a) = 0$$

$$\frac{f(w)}{g(w)} = \frac{f'(t)}{g'(t)} \qquad \textit{Because } F(x) = f(x) \textit{ and } G(x) = g(x)$$

$$\lim_{w \to a^+} \frac{f(w)}{g(w)} = \lim_{w \to a^+} \frac{f'(t)}{g'(t)}$$

$$\lim_{w \to a^+} \frac{f(w)}{g(w)} = \lim_{t \to a^+} \frac{f'(t)}{g'(t)} \qquad \textit{Because } t \textit{ is "trapped" between } a \textit{ and } w$$

$$\lim_{x \to a^+} \frac{f(x)}{g(x)} = \lim_{x \to a^+} \frac{f'(x)}{g'(x)}$$

---

*This is a special case of l'Hôpital's rule. The other cases can be found in most advanced calculus textbooks.

Limit Comparison Test (Section 8.4)

Suppose $a_k > 0$ and $b_k > 0$ for all sufficiently large k and that

$$\lim_{k \to \infty} \frac{a_k}{b_k} = L$$

where L is finite and positive ($0 < L < \infty$). Then Σa_k and Σb_k either both converge or both diverge.

Proof: Assume that $\displaystyle\lim_{k \to \infty} \frac{a_k}{b_k} = L$, where $L > 0$. Using $\epsilon = \dfrac{L}{2}$ in the definition of the limit of a sequence, we see that there exists a number N so that

$$\left| \frac{a_k}{b_k} - L \right| < \frac{L}{2} \qquad \text{whenever } k > N$$

$$-\frac{L}{2} < \frac{a_k}{b_k} - L < \frac{L}{2}$$

$$\frac{L}{2} < \frac{a_k}{b_k} < \frac{3L}{2}$$

$$\frac{L}{2}b_k < a_k < \frac{3L}{2}b_k \quad b_k > 0$$

This is true for all $k > N$. Now we can complete the proof by using the direct comparison test. Suppose Σb_k converges. Then the series $\displaystyle\sum \frac{3L}{2}b_k$ also converges, and the inequality $a_k < \dfrac{3L}{2}b_k$ tells us that the series Σa_k must also converge since it is dominated by a convergent series. Similarly, if Σb_k diverges, the inequality

$$0 < \frac{L}{2}b_k < a_k$$

tells us that Σa_k dominates the divergent series $\sum \frac{L}{2}b_k$, and it follows that Σa_k also diverges. Thus Σa_k and Σb_k either both converge or both diverge. ◆

Taylor's Theorem (Section 8.8)

If f and all its derivatives exist in an open interval I containing c, then for each x in I

$$f(x) = f(c) + \frac{f'(c)}{1!}(x - c) + \frac{f''(c)}{2!}(x - c)^2 + \cdots + \frac{f^{(n)}(c)}{n!}(x - c)^n + R_n(x)$$

where the remainder function $R_n(x)$ is given by

$$R_n(x) = \frac{f^{(n+1)}(z_n)}{(n + 1)!}(x - c)^{n+1}$$

for some z_n that depends on x and lies between c and x.

Proof: We will prove Taylor's theorem by showing that if f and its first $n + 1$ derivatives are defined in an open interval I containing c, then for each fixed x in I

$$f(x) = f(c) + \frac{f'(c)}{1!}(x - c) + \frac{f''(c)}{2!}(x - c)^2 + \cdots + \frac{f^{(n)}(c)}{n!}(x - c)^n + \frac{f^{(n+1)}(z_n)}{(n + 1)!}(x - c)^{n+1}$$

where z_n is some number between x and c. In our proof, we will apply Cauchy's generalized mean value theorem to the auxiliary functions F and G defined for all t in I as follows:

$$F(t) = f(x) - f(t) - \frac{f'(t)}{1!}(x - t) - \frac{f''(t)}{2!}(x - t)^2 - \cdots - \frac{f^{(n)}(t)}{n!}(x - t)^n$$

$$G(t) = \frac{(x - t)^{n+1}}{(n+1)!}$$

Note that $F(x) = G(x) = 0$, and thus Cauchy's generalized mean value theorem tells us

$$\frac{F'(z_n)}{G'(z_n)} = \frac{F(x) - F(c)}{G(x) - G(c)} = \frac{F(c)}{G(c)}$$

for some number z_n between x and c. Rearranging the sides of this equation and finding the derivatives gives

$$\frac{F(c)}{G(c)} = \frac{F'(z_n)}{G'(z_n)}$$

$$= \frac{\dfrac{-f^{(n+1)}(z_n)}{n!}(x - z_n)^n}{\dfrac{-(x - z_n)^n}{n!}}$$

$$= f^{(n+1)}(z_n)$$

$$F(c) = f^{(n+1)}(z_n)G(c)$$

$$f(x) - f(c) - \frac{f'(c)}{1!}(x - c) - \frac{f''(c)}{2!}(x - c)^2 - \cdots - \frac{f^{(n)}(c)}{n!}(x - c)^n = f^{(n+1)}(z_n)\frac{(x - c)^{n+1}}{(n+1)!} \qquad \textit{By substitutions}$$

$$f(x) = f(c) + \frac{f'(c)}{1!}(x - c) + \frac{f''(c)}{2!}(x - c)^2$$

$$+ \cdots + \frac{f^{(n)}(c)}{n!}(x - c)^n + f^{(n+1)}(z_n)\frac{(x - c)^{n+1}}{(n+1)!}$$

$\blacklozenge$

Sufficient Condition for Differentiability (Section 11.4)

If f is a function of x and y and f, f_x, and f_y are continuous in a disk D centered at (x_0, y_0), then f is differentiable at (x_0, y_0).

Proof: If (x, y) is a point in D, we have

$$f(x, y) - f(x_0, y_0) = f(x, y) - f(x_0, y) + f(x_0, y) - f(x_0, y_0)$$

The function $f(x, y)$ with y fixed satisfies the conditions of the mean value theorem, so that

$$f(x, y) - f(x_0, y) = f_x(x_1, y)(x - x_0)$$

for some number x_1 between x and x_0, and similarly, there is a number y_1 between y and y_0 such that

$$f(x_0, y) - f(x_0, y_0) = f_y(x_0, y_1)(y - y_0)$$

Substituting these expressions we obtain:

$$
\begin{aligned}
f(x,y) - f(x_0, y_0) &= \left[f(x,y) - f(x_0, y) \right] + \left[f(x_0, y) - f(x_0, y_0) \right] \\
&= \left[f_x(x_1, y)(x - x_0) \right] + \left[f_y(x_0, y_1)(y - y_0) \right] \\
&= \left[f_x(x_1, y)(x - x_0) \right] + \underbrace{f_x(x_0, y_0)(x - x_0) - f_x(x_0, y_0)(x - x_0)}_{\text{This is zero.}} \\
&\quad + \left[f_y(x_0, y_1)(y - y_0) \right] + \underbrace{f_y(x_0, y_0)(y - y_0) - f_y(x_0, y_0)(y - y_0)}_{\text{This is also zero.}} \\
&= f_x(x_0, y_0)(x - x_0) + f_y(x_0, y_0)(y - y_0) \\
&\quad + \left[f_x(x_1, y) - f_x(x_0, y_0) \right](x - x_0) + \left[f_y(x_0, y_1) - f_y(x_0, y_0) \right](y - y_0)
\end{aligned}
$$

Let $\epsilon_1(x, y)$ and $\epsilon_2(x, y)$ be the functions

$$
\epsilon_1(x, y) = f_x(x_1, y) - f_x(x_0, y_0) \quad \text{and} \quad \epsilon_2(x, y) = f_y(x_0, y_1) - f_y(x_0, y_0)
$$

Then since x_1 is between x and x_0, and y_1 is between y and y_0, and the partial derivatives f_x and f_y are continuous at (x_0, y_0), we have

$$
\begin{aligned}
\lim_{(x,y) \to (x_0, y_0)} \epsilon_1(x, y) &= \lim_{(x,y) \to (x_0, y_0)} \left[f_x(x_1, y) - f_x(x_0, y_0) \right] = 0 \\
\lim_{(x,y) \to (x_0, y_0)} \epsilon_2(x, y) &= \lim_{(x,y) \to (x_0, y_0)} \left[f_y(x_0, y_1) - f_y(x_0, y_0) \right] = 0
\end{aligned}
$$

so that f is differentiable at (x_0, y_0), as required. ♦

Change of Variables Formula for Multiple Integration (Section 12.8)

Suppose f is a continuous function of a region D in $\mathbb{R}^2$ and let D be the image of the domain D^* under the change of variables $x = g(u, v)$, $y = h(u, v)$ where g and h are continuously differentiable on D^*. Then

$$
\iint_D f(x, y) \, dy \, dx = \iint_{D^*} f\left[g(u, v), h(u, v) \right] \underbrace{\left| \frac{\partial(x, y)}{\partial(u, v)} \right|}_{\text{absolute value of Jacobian}} dv \, du
$$

Proof: A proof of this theorem is found in advanced calculus texts, but we can provide a geometric argument that makes this formula plausible in the special case where $f(x, y) = 1$. In particular, we will show that in order to find the area of a region D in the xy-plane using the change of variables $x = X(u, v)$ and $y = Y(u, v)$, it is reasonable to use the formula for area, A:

$$
A = \iint_D dy \, dx = \iint_{D^*} \left| \frac{\partial(x, y)}{\partial(u, v)} \right| dv \, du
$$

where D^* is the region in the uv-plane that corresponds to D.

Suppose the given change of variables has an inverse $u = u(x, y), v = v(x, y)$ that transforms the region D in the xy-plane into a region D^* in the uv-plane. To find the area of D^* in the uv-coordinate system, it is natural to use a rectangular grid, with vertical lines $u = $ constant and horizontal lines $v = $ constant, as shown in Figure. B.9**a**. In the xy-plane, the equations $u = $ constant and $v = $ constant will be families of parallel curves, which provide a curvilinear grid for the region D (Figure B.9**b**).

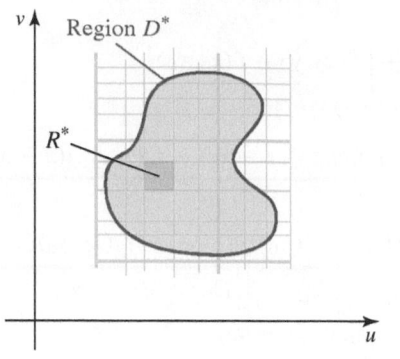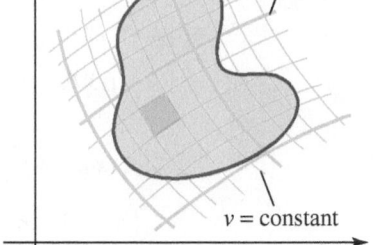

a. A rectangular grid in the uv-plane   **b.** Corresponding grid in the xy-plane is curvilinear

Figure B.9 Change of variables

Next, let R^* be a typical rectangular cell in the uv-grid that covers D^*, and let R be the corresponding set in the xy-plane (that is, R^* is the image of R under the given change of variables). Then, as shown in Figure B.10, if R^* has vertices $A^*(\overline{u}, \overline{v})$, $B^*(\overline{u} + \Delta u, \overline{v})$, $C^*(\overline{u}, \overline{v} + \Delta v)$, and $D^*(\overline{u} + \Delta u, \overline{v} + \Delta v)$, the set R will be the interior of a curvilinear rectangle with vertices

$$A\left[X(\overline{u}, \overline{v}), Y(\overline{u}, \overline{v})\right], B\left[X(\overline{u} + \Delta u, \overline{v}), Y(\overline{u} + \Delta u, \overline{v})\right]$$
$$C\left[X(\overline{u}, \overline{v} + \Delta v), Y(\overline{u}, \overline{v} + \Delta v)\right], D\left[X(\overline{u} + \Delta u, \overline{v} + \Delta v), Y(\overline{u} + \Delta u, \overline{v} + \Delta v)\right]$$

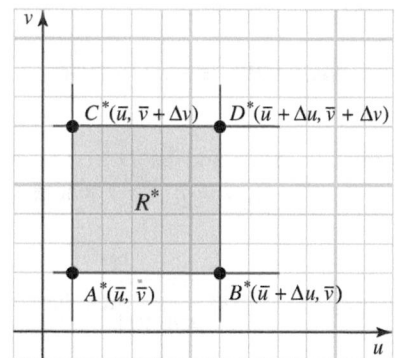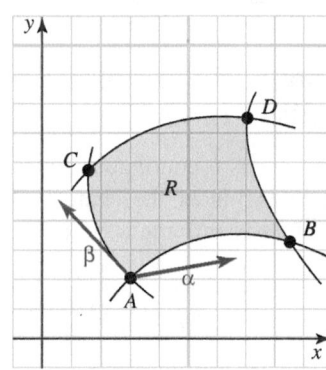

a. R^* is a typical rectangular cell in D^* **b.** The image of R^*

Figure B.10 Detail of Figure B.9

Note that the curved side of R joining A to B may be approximated by the secant vector

$$\mathbf{AB} = [X(\overline{u} + \Delta u, \overline{v}) - X(\overline{u}, \overline{v})]\,\mathbf{i} + [Y(\overline{u} + \Delta u, \overline{v}) - Y(\overline{u}, \overline{v})]\,\mathbf{j}$$

and by applying the mean value theorem, we find that

$$\mathbf{AB} = \left[\frac{\partial x}{\partial u}(a, \overline{v})\Delta u\right]\mathbf{i} + \left[\frac{\partial y}{\partial u}(b, \overline{v})\Delta u\right]\mathbf{j}$$

for some numbers a, b between $\overline{u}$ and $\overline{u} + \Delta u$. If Δu is very small, a and b are approximately the same as $\overline{u}$, and we can approximate $\mathbf{AB}$ by the vector

$$\alpha = \left[\frac{\partial x}{\partial u}\Delta u\right]\mathbf{i} + \left[\frac{\partial y}{\partial u}\Delta u\right]\mathbf{j}$$

where the partials are evaluated at the point $(\overline{u}, \overline{v})$. Similarly, the curved side of R joining A and C can be approximated by the vector

$$\boldsymbol{\beta} = \left[\frac{\partial x}{\partial v} \Delta v \right] \mathbf{i} + \left[\frac{\partial y}{\partial v} \Delta v \right] \mathbf{j}$$

The area of the curvilinear rectangle R is approximately the same as that of the parallelogram determined by $\boldsymbol{\alpha}$ and $\boldsymbol{\beta}$; that is,

$$
\begin{aligned}
\|\boldsymbol{\alpha} \times \boldsymbol{\beta}\| &= \begin{vmatrix} \mathbf{i} & \mathbf{j} & \mathbf{k} \\ \frac{\partial x}{\partial u} \Delta u & \frac{\partial y}{\partial u} \Delta u & 0 \\ \frac{\partial x}{\partial v} \Delta v & \frac{\partial y}{\partial v} \Delta v & 0 \end{vmatrix} \\
&= \left\| \begin{vmatrix} \mathbf{i} & \mathbf{j} & \mathbf{k} \\ \frac{\partial x}{\partial u} & \frac{\partial y}{\partial u} & 0 \\ \frac{\partial x}{\partial v} & \frac{\partial y}{\partial v} & 0 \end{vmatrix} \Delta v \Delta u \right\| \\
&= \left| \frac{\partial x}{\partial u} \frac{\partial y}{\partial v} - \frac{\partial y}{\partial u} \frac{\partial x}{\partial v} \right| \Delta v \Delta u \\
&= \left| \frac{\partial(x, y)}{\partial(u, v)} \right| \Delta v \Delta u
\end{aligned}
$$

By adding the contributions of all cells in the partition of D, we can approximate the area of D as follows:

APPROXIMATE AREA OF $D = \Sigma$ APPROXIMATE AREA OF CURVILINEAR RECTANGLES

$$= \sum \left| \frac{\partial(x, y)}{\partial(u, v)} \right| \Delta v \Delta u$$

Finally, using a limit to "smooth out" the approximation, we find

$$A = \iint_D dy\, dx = \lim \sum \left| \frac{\partial(x, y)}{\partial(u, v)} \right| \Delta v \Delta u = \iint_{D^*} \left| \frac{\partial(x, y)}{\partial(u, v)} \right| \Delta v \Delta u \qquad \blacklozenge$$

Stokes' Theorem (Section 13.6)

Let S be an oriented surface with unit normal vector N, and assume that S is bounded by a simple, closed, piecewise smooth curve C whose orientation is compatible with that of S. If F is a continuous vector field whose components have continuous partial derivatives on an open region containing S and C, then

$$\int_C \mathbf{F} \cdot d\mathbf{R} = \iint_S (\text{curl } \mathbf{F} \cdot \mathbf{N}) \, dS$$

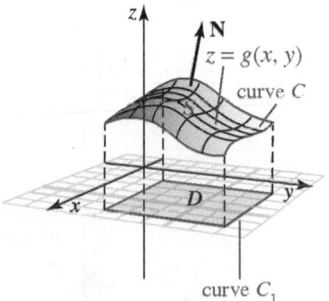

Figure B.11 Stokes' theorem

Proof: The general proof requires the methods of advanced calculus. However, a proof for the case where S is a graph and $\mathbf{F}$, S and C are "well behaved" can be given. Let S be given by $z = g(x, y)$ where (x, y) is in a region D of the xy-plane. Assume g has continuous second-order partial derivatives. Let C_1 be the projection of C in the xy-plane, as shown in Figure B.11. Also let

$$\mathbf{F}(x, y, z) = f_1(x, y, z)\mathbf{i} + f_2(x, y, z)\mathbf{j} + f_3(x, y, z)\mathbf{k}$$

where the partial derivatives of f_1, f_2, and f_3 are continuous.

We will evaluate each side of Stokes' theorem separately and show that the results for each are the same. If $x = x(t), y = y(t)$, and $z = z(t)$ for $a \le t \le b$ and $\mathbf{R}(t) = x(t)\mathbf{i} + y(t)\mathbf{j} + z(t)\mathbf{k}$, then

$$\int_C \mathbf{F} \cdot d\mathbf{R} = \int_a^b \left(f_1 \frac{dx}{dt} + f_2 \frac{dy}{dt} + f_3 \frac{dz}{dt} \right) dt$$

$$= \int_a^b \left[f_1 \frac{dx}{dt} + f_2 \frac{dy}{dt} + f_3 \left(\frac{\partial z}{\partial x} \frac{dx}{dt} + \frac{\partial z}{\partial y} \frac{dy}{dt} \right) \right] dt \qquad \text{Chain rule}$$

$$= \int_a^b \left[\left(f_1 + f_3 \frac{\partial z}{\partial x} \right) \frac{dx}{dt} + \left(f_2 + f_3 \frac{\partial z}{\partial y} \right) \frac{dy}{dt} \right] dt$$

$$= \int_{C_1} \left[\left(f_1 + f_3 \frac{\partial z}{\partial x} \right) dx + \left(f_2 + f_3 \frac{\partial z}{\partial y} \right) dy \right]$$

$$= \iint_D \left[\frac{\partial}{\partial x} \left(f_2 + f_3 \frac{\partial z}{\partial y} \right) - \frac{\partial}{\partial y} \left(f_1 + f_3 \frac{\partial z}{\partial x} \right) \right] dA \qquad \text{Green's theorem}$$

$$= \iint_D \left[\left(\frac{\partial f_2}{\partial x} + \frac{\partial f_2}{\partial z} \frac{\partial z}{\partial x} + \frac{\partial f_3}{\partial x} \frac{\partial z}{\partial y} + \frac{\partial f_3}{\partial z} \frac{\partial z}{\partial x} \frac{\partial z}{\partial y} + f_3 \frac{\partial^2 z}{\partial y \partial x} \right) \right.$$

$$\left. - \left(\frac{\partial f}{\partial y} + \frac{\partial f_1}{\partial z} \frac{\partial z}{\partial y} + \frac{\partial f_3}{\partial y} \frac{\partial z}{\partial x} + \frac{\partial f_3}{\partial z} \frac{\partial z}{\partial y} \frac{\partial z}{\partial x} + f_3 \frac{\partial^2 z}{\partial y \partial x} \right) \right] dA$$

$$= \iint_D \left[-\frac{\partial f_3}{\partial y} \frac{\partial z}{\partial x} + \frac{\partial f_2}{\partial z} \frac{\partial z}{\partial x} - \frac{\partial f_1}{\partial z} \frac{\partial z}{\partial y} + \frac{\partial f_3}{\partial x} \frac{\partial z}{\partial y} + \frac{\partial f_2}{\partial x} - \frac{\partial f_1}{\partial y} \right] dA$$

Next, we start over by evaluating the flux integral:

$$\iint_S \operatorname{curl} \mathbf{F} \cdot dS = \iint_D \left\langle \left(\frac{\partial f_3}{\partial y} - \frac{\partial f_2}{\partial z} \right), \left(\frac{\partial f_1}{\partial z} - \frac{\partial f_3}{\partial x} \right), \left(\frac{\partial f_2}{\partial x} - \frac{\partial f_1}{\partial y} \right) \right\rangle \cdot \left\langle \frac{-\partial z}{\partial x}, \frac{-\partial z}{\partial y}, 1 \right\rangle dA$$

$$= \iint_D \left[-\frac{\partial f_3}{\partial y} \frac{\partial z}{\partial x} + \frac{\partial f_2}{\partial z} \frac{\partial z}{\partial x} - \frac{\partial f_1}{\partial z} \frac{\partial z}{\partial y} + \frac{\partial f_3}{\partial x} \frac{\partial z}{\partial y} + \frac{\partial f_2}{\partial x} - \frac{\partial f_1}{\partial y} \right] dA$$

Since these results are the same, we have

$$\int_C \mathbf{F} \cdot d\mathbf{R} = \iint_S (\operatorname{curl} \mathbf{F} \cdot \mathbf{N}) \, dS \qquad \blacklozenge$$

Divergence Theorem (Section 13.7)

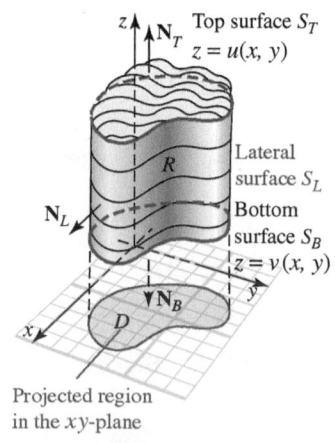

Figure B.12 A standard solid region in $\mathbb{R}^3$

Let R be a region in space bounded by a smooth, orientable closed surface S. If $\mathbf{F}$ is a continuous vector field whose components have continuous partial derivatives in R, then

$$\iint_S \mathbf{F} \cdot \mathbf{N} \, dS = \iiint_R \operatorname{div} \mathbf{F} \, dV$$

where $\mathbf{N}$ is an outward unit normal to the surface S. A standard solid region in $\mathbb{R}^3$ is shown in Figure B.12.

Proof: Let $\mathbf{F}(x,y,z) = f(x,y,z)\mathbf{i} + g(x,y,z)\mathbf{j} + h(x,y,z)\mathbf{k}$. If we state the divergence theorem using this notation for $\mathbf{F}$, we have

$$\iint_S f(\mathbf{i} \cdot \mathbf{N}) dS + \iint_S g(\mathbf{j} \cdot \mathbf{N}) dS + \iint_S h(\mathbf{k} \cdot \mathbf{N}) dS = \iiint_R \left(\frac{\partial f}{\partial x} + \frac{\partial g}{\partial y} + \frac{\partial h}{\partial z} \right) dV$$

This result can be verified by proving

$$\iint_S f(\mathbf{i} \cdot \mathbf{N}) dS = \iiint_R \frac{\partial f}{\partial x} dV \qquad \iint_S g(\mathbf{j} \cdot \mathbf{N}) dS = \iiint_R \frac{\partial g}{\partial y} dV \qquad \iint_S h(\mathbf{k} \cdot \mathbf{N}) dS = \iiint_R \frac{\partial h}{\partial z} dV$$

Since the proof of each of these is virtually identical, we will show the verification for the last of these three; the other two can be done in a similar fashion. We will evaluate this third integral by separately evaluating the left and right sides to show they are the same.

We will restrict our proof to a standard region; that is, a solid bounded above by a surface S_T with equation $z = u(x, y)$ and below by a surface S_B with equation $z = v(x, y)$. We assume that both S_T and S_B project onto the region D in the xy-plane. The lateral surface S_L of the region is the set of all (x, y, z) such that $v(x, y) \leq z \leq u(x, y)$ on the boundary of D, as shown in Figure B.12.

The upward flux across the top surface S_T is

$$\iint\limits_{S_T} h(\mathbf{k} \cdot \mathbf{N}_T) dS = \iint\limits_{D} h\langle 0, 0, 1\rangle \cdot \left\langle -\frac{\partial u}{\partial x}, -\frac{\partial u}{\partial y}, 1 \right\rangle dA$$

$$= \iint\limits_{D} h(x, y, z) dA$$

$$= \iint\limits_{D} h(x, y, u(x, y)) dA$$

Similarly, the downward flux across the bottom surface S_B is

$$\iint\limits_{S_B} h(\mathbf{k} \cdot \mathbf{N}_B) dS = \iint\limits_{D} h\langle 0, 0, 1\rangle \cdot \left\langle \frac{\partial v}{\partial x}, \frac{\partial v}{\partial y}, -1 \right\rangle dA$$

$$= -\iint\limits_{D} h(x, y, z) dA$$

$$= -\iint\limits_{D} h(x, y, v(x, y)) dA$$

Because the outward unit normal $\mathbf{N}_L$ is horizontal on the lateral surface S_L, it is perpendicular to $\mathbf{k}$, and

$$\iint\limits_{S_L} h(\mathbf{k} \cdot \mathbf{N}) dS = 0$$

To find the flux integral over the entire closed surface S, add the separate flux integrals over the top, the bottom, and the sides:

$$\iint\limits_{S} h(\mathbf{k} \cdot \mathbf{N}) dS = \iint\limits_{S_T} h(\mathbf{k} \cdot \mathbf{N}_T) dS + \iint\limits_{S_B} h(\mathbf{k} \cdot \mathbf{N}_B) dS + \iint\limits_{S_L} h(\mathbf{k} \cdot \mathbf{N}_L) dS$$

$$= \iint\limits_{D} h\left[x, y, u(x, y)\right] dA - \iint\limits_{D} h\left[x, y, v(x, y)\right] dA + 0$$

$$= \iint\limits_{D} h\left[x, y, u(x, y)\right] dA - \iint\limits_{D} h\left[x, y, v(x, y)\right] dA$$

Next, we start over by evaluating the triple integral $\iiint\limits_{R} \dfrac{\partial h}{\partial z} dV$. Notice that we can describe the solid S as the set of all (x, y, z) for (x, y) in D and $v(x, y) \leq z \leq u(x, y)$. Thus,

$$\iiint\limits_{R} \frac{\partial h}{\partial z}\, dV = \int\!\!\int\limits_{D} \left[\int_{v(x,y)}^{u(x,y)} \frac{\partial h}{\partial z}(x,y,z)\, dz \right] dA$$

$$= \iint\limits_{D} \{ h\left[x,y,u(x,y)\right] - h\left[x,y,v(x,y)\right] \}\, dA$$

$$= \iint\limits_{S} h(\mathbf{k} \cdot \mathbf{N})\, dS$$

By applying similar arguments to the integrals $\iiint\limits_{R} \dfrac{\partial f}{\partial x}\, dV$ and $\iiint\limits_{R} \dfrac{\partial g}{\partial y}\, dV$ and adding, we obtain

$$\iiint\limits_{R} \left(\frac{\partial f}{\partial x} + \frac{\partial g}{\partial y} + \frac{\partial h}{\partial z} \right) dV = \iint\limits_{S} f(\mathbf{i} \cdot \mathbf{N})\, dS + \iint\limits_{S} g(\mathbf{j} \cdot \mathbf{N})\, dS + \iint\limits_{S} h(\mathbf{k} \cdot \mathbf{N})\, dS$$

or

$$\iiint\limits_{R} \operatorname{div} \boldsymbol{F}\, dV = \iint\limits_{S} \mathbf{F} \cdot \mathbf{N}\, dS \qquad \blacklozenge$$

APPENDIX C: SIGNIFICANT DIGITS

Throughout this book, you will frequently find approximate (decimal) answers. Sometimes your answer may not exactly agree with the answer found in the back of the book. This does not necessarily mean that your answer is incorrect, particularly if your answer is very close to the given answer. To use your calculator intelligently and efficiently, you should become familiar with its functions and practice doing problems with the same calculator, whenever possible.

Significant Digits

Applications involving measurements can never be exact. It is particularly important that you pay attention to the accuracy of your measurements when you use a computer or calculator, because using available technology can give you a false sense of security about the accuracy in a particular problem. For example, if you measure a triangle and find the sides are approximately 1.2 and 3.4 and then find the ratio of $1.2/3.4 \approx 0.35294117$, it appears that the result is more accurate than the original measurements! Some discussion about accuracy and significant digits is necessary.

The digits known to be correct in a number obtained by a measurement are called **significant digits**. The digits $1, 2, 3, 4, 5, 6, 7, 8$ and 9 are always significant, whereas the digit 0 may or may not be significant.

- Zeros that come between two other digits are significant, as in 203 or 10.04.
- If the zero's only function is to place the decimal point, it is not significant, as in

$$\underbrace{0.00000}\ 23 \quad \text{or} \quad 23,\underbrace{000}$$
$$\textit{Placeholders} \qquad\qquad \textit{Placeholders}$$

If a zero does more than fix the decimal point, it is significant, as in

$$0.0023\ \underbrace{0} \qquad\qquad \text{or} \qquad 23,\underbrace{000}.01$$
$$\qquad\qquad \uparrow \qquad\qquad\qquad \textit{These are significant.}$$
$$\textit{This digit is significant.}$$

This second rule can, of course, result in certain ambiguities, such as 23,000 (measured to the *exact* unit). To avoid such confusion, we use scientific notation in this case:

$$2.3 \times 10^4 \text{ has two significant digits,}$$
$$2.3000 \times 10^4 \text{ has five significant digits.}$$

Numbers that come about by counting are considered to be exact and are correct to any number of significant digits.

When you compute an answer using a calculator, the answer may have 10 or more digits. In the technology notes we generally show the 10 or 12 digits that result from the numerical calculation but frequently the number in the answer section will have only 5 or 6 digits in the answer. It seems clear that if the first 3 or 4 nonzero digits of the answer coincide, you probably have the correct method of doing the problem.

However, you might ask why there are discrepancies, and how many digits you should use when you write down your final answer. Roughly speaking, the significant digits in a number are the digits that have meaning. To clarify the concept, we must for the moment assume that we know the exact answer. We then assume that we have been able to compute an approximation to this exact answer.

Usually we do this by some sort of iterative process, in which the answers are getting closer and closer to the exact answer. In such a process, we hope that the number of significant digits in our approximate answer is increasing at each trial. If our approximate answer is, say, 6 digits long (some of those digits might even be zero), and the difference between our answer and the exact answer is 4 units or less in the last place, then the first 5 digits are significant.

For example, if the exact answer is 3.14159 and our approximate answer is 3.14162, then our answer has 6 significant digits but is correct to 5 significant digits. Note that saying that our answer is correct to 5 significant digits does not guarantee that all of those 5 digits exactly match the first 5 digits of the exact answer. In fact, if an exact answer is 6.001 and our computed answer is 5.997, then our answer is correct to 3 significant digits and not one of them matches the digits in the exact answer. Also note that it may be necessary for an approximation to have more digits than are actually significant for it to have a certain number of significant digits. For example, if the exact answer is 6.003 and our approximation is 5.998, then it has 3 significant digits, but only if we consider the total 4-digit number and do not strip off the last nonsignificant digit.

Again, suppose you know that all digits are significant in the number 3.456; then you know that the exact number is less than 3.4565 and at least 3.4555. Some people may say that the number 3.456 is correct to 3 decimal places. This is the same as saying that it has 4 significant digits.

Why bother with significant digits? If you multiply (or divide) two numbers with the same number of significant digits, then the product will generally be at least twice as long, but will have roughly the same number of significant digits as the original factors. You can then dispense with the unneeded digits. In fact, to keep them would be misleading about the accuracy of the result.

Frequently we can make an educated guess of the number of significant digits in an answer. For example, if we compute an iterative approximation such as

$$2.3123, \qquad 2.3125, \qquad 2.3126, \qquad 2.31261, \qquad 2.31262, \qquad \cdots$$

we would generally conclude that the answer is 2.3126 to 5 significant digits. Of course, we may very well be wrong, and if we continued iterating, the answer might end up as 2.4.

Rounding and Rules of Computations Used in this Book

In hand and calculator computations, rounding a number is done to reduce the number of digits displayed and make the number easier to comprehend. Furthermore, if you suspect that the digit in the last place is not significant, then you might be tempted to round and remove this last digit. This can lead to error. For example, if the computed value is 0.64 and the true value is known to be between 0.61 and 0.67, then the computed value has only 1 significant digit. However, if we round it to 0.6 and the true value is really 0.66, then 0.6 is not correct to even 1 significant digit. In the interest of making the text easier to read, we have used the following rounding procedure:

ROUNDING PROCEDURE   To round off numbers,

Step 1 Increase the last retained digit by 1 if the remainder is greater than or equal to 5; or,

Step 2 Retain the last digit unchanged if the remainder is less than 5.

Elaborate rules for computation of approximate data can be developed when needed (in chemistry, for example), but in this text we will use three simple rules:

RULES FOR SIGNIFICANT DIGITS

Addition-subtraction: Add or subtract in the usual fashion, and then round off the result so that the last digit retained is in the column farthest to the right in which both given numbers have significant digits.

Multiplication-division: Multiply or divide in the usual fashion, and then round off the results to the smaller number of significant digits found in either of the given numbers.

Counting numbers: Numbers used to count or whole numbers used as exponents are considered to be correct to any number of significant digits.

ROUNDING RULE

We use the following rounding procedure in problems requiring rounding by involving several steps: *Round only once, at the end. That is, do not work with rounded results, because round-off errors can accumulate. If the problem does not ask for an approximate answer, but requires a calculator result, we will show the calculator result and leave it to the reader to write the appropriate number of digits.*

Calculator Experiments

You should be aware that you are much better than your calculator at performing certain computations. For example, almost all calculators will fail to give the correct answer to

$$10.0^{50} + 911.0 - 10.0^{50}$$

Calculators will return the value of 0, but you know at a glance that the answer is 911.0. We must reckon with this poor behavior on the part of calculators, which is called *loss of accuracy* due to *catastrophic cancellation*. In this case, it is easy to catch the error immediately, but what if the computation is so complicated (or hidden by other computations) that we do not see the error?

First, we want to point out that the order in which you perform computations can be very important. For example, most calculators will correctly conclude that

$$10.0^{50} - 10.0^{50} + 911.0 = 911$$

So, for what do we need to watch? If you subtract two numbers that are close to each other in magnitude, you *may* obtain an inaccurate result. When you have a sequence of multiplications and divisions in a string, try to arrange the factors so that the result of each intermediate calculation stays as close as possible to 1.0.

Second, since a calculator performs all computations with a finite number of digits, it is unable to do exact computations involving nonterminating decimals. This enables us to see how many digits the calculator actually uses when it computes a result. For example, the computation

$$(7.0/17.0)(17.0)$$

should give the result 7.0, but on many calculators it does not. The size of the answer gives an indication of how many digits "Accuracy" the calculator uses internally. That is, the calculator may display decimal numbers that have 10 digits, but use 12 digits internally. If the answer to the above computation is something similar to 1.0 EE − 12, then the calculator is using 12 digits internally.

Research in numerical computing demonstrated that analyzing a numerical technique requires a combined theoretical and computational approach. Theory is needed to guide the performance and interpretation of the numerical techniques, while computation is necessary to verify the robustness and stability of the numerical technique.

The scientific literature regarding use of technology in teaching mathematics demonstrates the need for teaching students the importance of using calculators effectively (e.g., see: Paige, R., Seshaiyer, P., Toda, M., *Student Misconceptions Caused by Misuse of Technology*, *International Journal for Technology in Mathematics Education*, vol. 14, no.4, (2007)). The paper reviews the concept of round-off error, and then presents some experimental results on the performance of a variety of computing aids for solving Calculus problems. The authors performed a statistical analysis of data collected from 215 students in freshman calculus classes at Texas Tech University and reported the findings of this analysis. Among other tasks, students were asked to evaluate the following limit using their TI calculators:

$$\lim_{k \to \infty} \left(1 + \tfrac{1}{k}\right)^k$$

To their surprise, many students found that above a certain specific value of k, their calculator answers were *no longer* approaching the true value e. Instead, they obtained the incorrect answer 1. Naturally, as any calculus student would, they got confused and came to various conclusions, some of which were: (a) they must be doing something wrong; (b) the teacher must have made a mistake; (c) their calculator was no good and they need to buy a better one.

When this problem was addressed to us by the students, we pointed out, of course, that the problem was because of inexact calculator arithmetic. You might want to try comparing the evaluation of $\left(1 + 10^{-2N}\right)^{10^{2N}}$ for $N = 1, 2, \cdots 10$ on your calculator with that of your classmates. On some calculators the display 1 will be reached for $N = 4$, but on others it may not be reached until $N = 8$.

Not implementing numerical techniques in a correct and sensible fashion can lead to major disasters. One such example is the Patriot Missile failure, in Dharan, Saudi Arabia, on February 25, 1991 (GAO/IMTEC-92-96, 1992). This incident resulted in 28 deaths and was ultimately attributable to poor handling of inexact computer round-off arithmetic. Another example is the explosion of the Ariane 5 rocket just 40 seconds after lift-off on its maiden voyage off French Guiana, on June 4, 1996 (Ariane 501, 1996; Gleick, 1996). The destroyed rocket and its cargo were valued at $500 million. It has been our experience that these kinds of problems also arise in a day-to-day mathematics classroom with just a desk calculator or a digital computer, although at not such a large scale as the Patriot Missile or Ariane 5 failures.

Trigonometric Evaluations

In many problems you will be asked to compute the values of trigonometric functions such as the sine, cosine, or tangent. In calculus, trigonometric arguments are usually assumed to be measured in radians. You must make sure the calculator is in radian mode. If it is in radian mode, then the sine of a small number will almost be equal to that number. For example,

$$\sin(0.00001) = 1\text{EE-}5 \quad \text{(which is } 0.00001\text{)}$$

If not, then you are not using radian mode. Make sure you know how to put your calculator in radian mode.

Graphing Blunders

When you are using the graphing features, you must always be careful to choose reasonable scales for the domain (horizontal scale) and range (vertical scale). If the scale is too large, you may not see important wiggles. If the scale is too small, you may not see important behavior elsewhere in the plane. Of course, knowing the techniques of graphing discussed in Chapter 4 will prevent you from making such blunders. Some calculators may have trouble with curves that jump suddenly at a point. An example of such a curve would be

$$y = \frac{e^x}{x}$$

which jumps at the origin. Try plotting this curve with your calculator using different horizontal and vertical scales, making sure that you understand how your calculator handles such graphs.

APPENDIX D: SHORT TABLE OF INTEGRALS

Elementary forms

BASIC RULES

1. Constant rule* $\displaystyle\int 0 \, du = 0 + C$ **2. Power rule** $\displaystyle\int u^n \, du = \frac{u^{n+1}}{n+1}; \quad n \neq -1$

$$\int u^n \, du = \ln|u|; \quad n = -1$$

3. Natural exponential rule $\displaystyle\int e^u \, du = e^u$ **4. Logarithmic rule** $\displaystyle\int \ln|u| \, du = u \ln|u| - u$

TRIGONOMETRIC RULES

5. $\displaystyle\int \sin u \, du = -\cos u$ **6.** $\displaystyle\int \cos u \, du = \sin u$

7. $\displaystyle\int \tan u \, du = -\ln|\cos u| = \ln|\sec u|$ **8.** $\displaystyle\int \cot u \, du = \ln|\sin u|$

9. $\displaystyle\int \sec u \, du = \ln|\sec u + \tan u|$ **10.** $\displaystyle\int \csc u \, du = -\ln|\csc u + \cot u|$

11. $\displaystyle\int \sec^2 u \, du = \tan u$ **12.** $\displaystyle\int \csc^2 u \, du = -\cot u$

13. $\displaystyle\int \sec u \tan u \, du = \sec u$ **14.** $\displaystyle\int \csc u \cot u \, du = -\csc u$

EXPONENTIAL RULE

15. $\displaystyle\int a^u \, du = \frac{a^u}{\ln a} \qquad a > 0, a \neq 1$

HYPERBOLIC RULES

16. $\displaystyle\int \cosh u \, du = \sinh u$ **17.** $\displaystyle\int \sinh u \, du = \cosh u$

18. $\displaystyle\int \tanh u \, du = \ln \cosh u$ **19.** $\displaystyle\int \coth u \, du = \ln|\sinh u|$

20. $\displaystyle\int \operatorname{sech} u \, du = \tan^{-1}(\sinh u) \text{ or } 2 \tan^{-1} e^u$ **21.** $\displaystyle\int \operatorname{csch} u \, du = \ln\left|\tanh \frac{u}{2}\right|$

INVERSE RULES

22. $\displaystyle\int \frac{du}{\sqrt{a^2 - u^2}} = \sin^{-1}\frac{u}{a}$ **23.** $\displaystyle\int \frac{du}{\sqrt{u^2 - a^2}} = \cosh^{-1}\frac{u}{a}$

24. $\displaystyle\int \frac{du}{a^2 + u^2} = \frac{1}{a}\tan^{-1}\frac{u}{a}$ **25.** $\displaystyle\int \frac{du}{a^2 - u^2} = \begin{cases} \frac{1}{a}\tanh^{-1}\frac{u}{a} & \text{if } \left|\frac{u}{a}\right| < 1 \\ \frac{1}{a}\coth^{-1}\frac{u}{a} & \text{if } \left|\frac{u}{a}\right| > 1 \end{cases}$

26. $\displaystyle\int \frac{du}{u\sqrt{u^2 - a^2}} = \frac{1}{a}\sec^{-1}\left|\frac{u}{a}\right|$ **27.** $\displaystyle\int \frac{du}{u\sqrt{a^2 - u^2}} = -\frac{1}{a}\operatorname{sech}^{-1}\left|\frac{u}{a}\right|$

$$= -\frac{1}{a}\ln\left|\frac{a + \sqrt{a^2 - u^2}}{u}\right|$$

---

*Notice that this formula shows the addition of a constant, C. When using an integral table or technology, the constant will usually not be shown, so it is important that you remember to insert the constant of integration each time you evaluate an integral, even when getting it from a table or from technology.

28. $\int \dfrac{du}{\sqrt{a^2 + u^2}} = \ln\left(u + \sqrt{a^2 + u^2}\right)$ **29.** $\int \dfrac{du}{u\sqrt{a^2 + u^2}} = -\dfrac{1}{a}\ln\left|\dfrac{\sqrt{a^2+u^2}+a}{u}\right|$

$\qquad\qquad = \sinh^{-1}\dfrac{u}{a}$ $\qquad\qquad\qquad = -\dfrac{1}{a}\operatorname{csch}^{-1}\left|\dfrac{u}{a}\right|$

Linear and quadratic forms

INTEGRALS INVOLVING $au + b$

30. $\int (au+b)^n\,du = \dfrac{(au+b)^{n+1}}{(n+1)a}$

31. $\int u(au+b)^n\,du = \dfrac{(au+b)^{n+2}}{(n+2)a^2} - \dfrac{b(au+b)^{n+1}}{(n+1)a^2}$

32. $\int u^2(au+b)^n\,du = \dfrac{(au+b)^{n+3}}{(n+3)a^3} - \dfrac{2b(au+b)^{n+2}}{(n+2)a^3} + \dfrac{b^2(au+b)^{n+1}}{(n+1)a^3}$

33. $^*\displaystyle\int u^m(au+b)^n\,du = \begin{cases} \dfrac{u^{m+1}(au+b)^n}{m+n+1} + \dfrac{nb}{m+n+1}\displaystyle\int u^m(au+b)^{n-1}\,du & \text{or} \\[2ex] \dfrac{u^m(au+b)^{n+1}}{(m+n+1)a} - \dfrac{mb}{(m+n+1)a}\displaystyle\int u^{m-1}(au+b)^n\,du & \text{or} \\[2ex] \dfrac{-u^{m+1}(au+b)^{n+1}}{(n+1)b} + \dfrac{m+n+2}{(n+1)b}\displaystyle\int u^m(au+b)^{n+1}\,du \end{cases}$

34. $\int \dfrac{du}{au+b} = \dfrac{1}{a}\ln|au+b|$

35. $\int \dfrac{u\,du}{au+b} = \dfrac{u}{a} - \dfrac{b}{a^2}\ln|au+b|$

36. $\int \dfrac{u^2\,du}{au+b} = \dfrac{(au+b)^2}{2a^3} - \dfrac{2b(au+b)}{a^3} + \dfrac{b^2}{a^3}\ln|au+b|$

37. $\int \dfrac{u^3\,du}{au+b} = \dfrac{(au+b)^3}{3a^4} - \dfrac{3b(au+b)^2}{2a^4} + \dfrac{3b^2(au+b)}{a^4} - \dfrac{b^3}{a^4}\ln|au+b|$

INTEGRALS INVOLVING $u^2 + a^2$

38. $\int \dfrac{du}{u^2+a^2} = \dfrac{1}{a}\tan^{-1}\dfrac{u}{a}$ **39.** $\int \dfrac{u\,du}{u^2+a^2} = \dfrac{1}{2}\ln(u^2+a^2)$

40. $\int \dfrac{u^2\,du}{u^2+a^2} = u - a\tan^{-1}\dfrac{u}{a}$ **41.** $\int \dfrac{u^3\,du}{u^2+u^2} = \dfrac{u^2}{2} - \dfrac{a^2}{2}\ln(u^2+a^2)$

42. $\int \dfrac{du}{u(u^2+a^2)} = \dfrac{1}{2a^2}\ln\left(\dfrac{u^2}{u^2+a^2}\right)$ **43.** $\int \dfrac{du}{u^2(u^2+a^2)} = -\dfrac{1}{a^2u} - \dfrac{1}{a^3}\tan^{-1}\dfrac{u}{a}$

44. $\int \dfrac{du}{u^3(u^2+a^2)} = -\dfrac{1}{2a^2u^2} - \dfrac{1}{2a^4}\ln\left(\dfrac{u^2}{u^2+a^2}\right)$

INTEGRALS INVOLVING $u^2 - a^2,\ u^2 > a^2$

45. $\int \dfrac{du}{u^2-a^2} = \dfrac{1}{2a}\ln\left|\dfrac{u-a}{u+a}\right|$ or $-\dfrac{1}{a}\coth^{-1}\dfrac{u}{a}$

46. $\int \dfrac{u\,du}{u^2-a^2} = \dfrac{1}{2}\ln|u^2-a^2|$

*When the integral is given as another integral, then do not add the constant until the form no longer involves an integration.

47. $\displaystyle\int \frac{u^2 du}{u^2 - a^2} = u + \frac{a}{2}\ln\left|\frac{u-a}{u+a}\right|$ **48.** $\displaystyle\int \frac{u^3\,du}{u^2 - a^2} = \frac{u^2}{2} + \frac{a^2}{2}\ln\left|u^2 - a^2\right|$

49. $\displaystyle\int \frac{du}{u(u^2 - a^2)} = \frac{1}{2a^2}\ln\left|\frac{u^2 - a^2}{u^2}\right|$ **50.** $\displaystyle\int \frac{du}{u^2(u^2 - a^2)} = \frac{1}{a^2 u} + \frac{1}{2a^3}\ln\left|\frac{u-a}{u+a}\right|$

51. $\displaystyle\int \frac{du}{u^3(u^2 - a^2)} = \frac{1}{2a^2 u^2} - \frac{1}{2a^4}\ln\left|\frac{u^2}{u^2 - a^2}\right|$

INTEGRALS INVOLVING $a^2 - u^2, u^2 < a^2$

52. $\displaystyle\int \frac{du}{a^2 - u^2} = \frac{1}{2a}\ln\left|\frac{a+u}{a-u}\right|$ or $\frac{1}{a}\tanh^{-1}\frac{u}{a}$

53. $\displaystyle\int \frac{u\,du}{a^2 - u^2} = -\frac{1}{2}\ln\left|a^2 - u^2\right|$

54. $\displaystyle\int \frac{u^2 du}{a^2 - u^2} = -u + \frac{a}{2}\ln\left|\frac{a+u}{a-u}\right|$

55. $\displaystyle\int \frac{u^3\,du}{a^2 - u^2} = -\frac{u^2}{2} - \frac{a^2}{2}\ln\left|a^2 - u^2\right|$

56. $\displaystyle\int \frac{du}{u(a^2 - u^2)} = \frac{1}{2a^2}\ln\left|\frac{u^2}{a^2 - u^2}\right|$

57. $\displaystyle\int \frac{du}{u^2(a^2 - u^2)} = -\frac{1}{a^2 u} + \frac{1}{2a^3}\ln\left|\frac{a+u}{a-u}\right|$

58. $\displaystyle\int \frac{du}{u^3(a^2 - u^2)} = -\frac{1}{2a^2 u^2} + \frac{1}{2a^4}\ln\left|\frac{u^2}{a^2 - u^2}\right|$

59. $\displaystyle\int \frac{du}{(a^2 - u^2)^2} = \frac{u}{2a^2(a^2 - u^2)} + \frac{1}{4a^3}\ln\left|\frac{a+u}{a-u}\right|$

60. $\displaystyle\int \frac{u\,du}{(a^2 - u^2)^2} = \frac{1}{2(a^2 - u^2)}$

61. $\displaystyle\int \frac{u^2\,du}{(a^2 - u^2)^2} = \frac{u}{2(a^2 - u^2)} - \frac{1}{4a}\ln\left|\frac{a+u}{a-u}\right|$

62. $\displaystyle\int \frac{u^3\,du}{(a^2 - u^2)^2} = \frac{a^2}{2(a^2 - u^2)} + \frac{1}{2}\ln\left|a^2 - u^2\right|$

63. $\displaystyle\int \frac{du}{u(a^2 - u^2)^2} = \frac{1}{2a^2(a^2 - u^2)} + \frac{1}{2a^4}\ln\left|\frac{u^2}{a^2 - u^2}\right|$

64. $\displaystyle\int \frac{du}{u^2(a^2 - u^2)^2} = -\frac{1}{a^4 u} + \frac{u}{2a^4(a^2 - u^2)} + \frac{3}{4a^5}\ln\left|\frac{a+u}{a-u}\right|$

65. $\displaystyle\int \frac{du}{u^3(a^2 - u^2)^2} = -\frac{1}{2a^4 u^2} + \frac{1}{2a^4(a^2 - u^2)} + \frac{1}{a^6}\ln\left|\frac{u^2}{a^2 - u^2}\right|$

INTEGRALS INVOLVING $au^2 + bu + c$

66. $\displaystyle\int \frac{du}{au^2 + bu + c} = \begin{cases} \dfrac{2}{\sqrt{4ac - b^2}}\tan^{-1}\dfrac{2au + b}{\sqrt{4ac - b^2}} \\[3mm] \dfrac{1}{\sqrt{b^2 - 4ac}}\ln\left|\dfrac{2au + b - \sqrt{b^2 - 4ac}}{2au + b + \sqrt{b^2 - 4ac}}\right| \end{cases}$

67. $\displaystyle\int \frac{u\,du}{au^2 + bu + c} = \frac{1}{2a}\ln\left|au^2 + bu + c\right| - \frac{b}{2a}\int \frac{du}{au^2 + bu + c}$

68. $\displaystyle\int \frac{u^2\,du}{au^2 + bu + c} = \frac{u}{a} - \frac{b}{2a^2}\ln\left|au^2 + bu + c\right| + \frac{b^2 - 2ac}{2a^2}\int \frac{du}{au^2 + bu + c}$

69. $\displaystyle\int \frac{u^m\,du}{au^2 + bu + c} = \frac{u^{m-1}}{(m-1)a} - \frac{c}{a}\int \frac{u^{m-2}\,du}{au^2 + bu + c} - \frac{b}{a}\int \frac{u^{m-1}du}{au^2 + bu + c}$

70. $\displaystyle\int \frac{du}{u(au^2 + bu + c)} = \frac{1}{2c} \ln\left|\frac{u^2}{au^2 + bu + c}\right| - \frac{b}{2c} \int \frac{du}{au^2 + bu + c}$

71. $\displaystyle\int \frac{du}{u^2(au^2 + bu + c)} = \frac{b}{2c^2} \ln\left|\frac{au^2 + bu + c}{u^2}\right| - \frac{1}{cu} + \frac{b^2 - 2ac}{2c^2} \int \frac{du}{au^2 + bu + c}$

72. $\displaystyle\int \frac{du}{u^n(au^2 + bu + c)} = -\frac{1}{(n-1)cu^{n-1}} - \frac{b}{c} \int \frac{du}{u^{n-1}(au^2 + bu + c)} - \frac{a}{c} \int \frac{du}{u^{n-2}(au^2 + bu + c)}$

73. $\displaystyle\int \frac{du}{(au^2 + bu + c)^2} = \frac{2au + b}{(4ac - b^2)(au^2 + bu + c)} + \frac{2a}{4ac - b^2} \int \frac{du}{au^2 + bu + c}$

74. $\displaystyle\int \frac{u\,du}{(au^2 + bu + c)^2} = -\frac{bu + 2c}{(4ac - b^2)(au^2 + bu + c)} - \frac{b}{4ac - b^2} \int \frac{du}{au^2 + bu + c}$

75. $\displaystyle\int \frac{u^2\,du}{(au^2 + bu + c)^2} = \frac{(b^2 - 2ac)u + bc}{a(4ac - b^2)(au^2 + bu + c)} + \frac{2c}{4ac - b^2} \int \frac{du}{au^2 + bu + c}$

76. $\displaystyle\int \frac{u^m\,du}{(au^2 + bu + c)^n} = \frac{-u^{m-1}}{(2n - m - 1)a(au^2 + bu + c)^{n-1}}$

$$- \frac{(n-m)b}{(2n - m - 1)a} \int \frac{u^{m-1}\,du}{(au^2 + bu + c)^n}$$

$$+ \frac{(m-1)c}{(2n - m - 1)a} \int \frac{u^{m-2}\,du}{(au^2 + bu + c)^n}$$

Radical forms

INTEGRALS INVOLVING $\sqrt{au + b}$

77. $\displaystyle\int \frac{du}{\sqrt{au + b}} = \frac{2\sqrt{au + b}}{a}$

78. $\displaystyle\int \frac{u\,du}{\sqrt{au + b}} = \frac{2(au - 2b)}{3a^2} \sqrt{au + b}$

79. $\displaystyle\int \frac{u^2\,du}{\sqrt{au + b}} = \frac{2(3a^2u^2 - 4abu + 8b^2)}{15a^3} \sqrt{au + b}$

80. $\displaystyle\int \frac{du}{u\sqrt{au + b}} = \begin{cases} \dfrac{1}{\sqrt{b}} \ln\left|\dfrac{\sqrt{au + b} - \sqrt{b}}{\sqrt{au + b} + \sqrt{b}}\right| \\[3mm] \dfrac{2}{\sqrt{-b}} \tan^{-1} \sqrt{\dfrac{au + b}{-b}} \end{cases}$

81. $\displaystyle\int \frac{du}{u^2\sqrt{au + b}} = -\frac{\sqrt{au + b}}{bu} - \frac{a}{2b} \int \frac{du}{u\sqrt{au + b}}$

82. $\displaystyle\int \sqrt{au + b}\,du = \frac{2\sqrt{(au + b)^3}}{3a}$

83. $\displaystyle\int u\sqrt{au + b}\,du = \frac{2(3au - 2b)}{15a^2} \sqrt{(au + b)^3}$

84. $\displaystyle\int u^2\sqrt{au + b}\,du = \frac{2(15a^2u^2 - 12abu + 8b^2)}{105a^3} \sqrt{(au + b)^3}$

INTEGRALS INVOLVING $\sqrt{u^2 + a^2}$

85. $\displaystyle\int \sqrt{u^2 + a^2}\,du = \frac{u\sqrt{u^2 + a^2}}{2} + \frac{a^2}{2} \ln\left(u + \sqrt{u^2 + a^2}\right)$

86. $\displaystyle\int u\sqrt{u^2 + a^2}\,du = \frac{(u^2 + a^2)^{3/2}}{3}$

87. $\displaystyle\int u^2\sqrt{u^2 + a^2}\,du = \frac{u(u^2 + a^2)^{3/2}}{4} - \frac{a^2u\sqrt{u^2 + a^2}}{8} - \frac{a^4}{8} \ln\left(u + \sqrt{u^2 + a^2}\right)$

88. $\displaystyle\int u^3\sqrt{u^2+a^2}\,du = \frac{(u^2+a^2)^{5/2}}{5} - \frac{a^2(u^2+a^2)^{3/2}}{3}$

89. $\displaystyle\int \frac{du}{\sqrt{u^2+a^2}} = \ln\left(u+\sqrt{u^2+a^2}\right)$ or $\sinh^{-1}\dfrac{u}{a}$

90. $\displaystyle\int \frac{u\,du}{\sqrt{u^2+a^2}} = \sqrt{u^2+a^2}$

91. $\displaystyle\int \frac{u^2\,du}{\sqrt{u^2+a^2}} = \frac{u\sqrt{u^2+a^2}}{2} - \frac{a^2}{2}\ln(u+\sqrt{u^2+a^2})$

92. $\displaystyle\int \frac{u^3\,du}{\sqrt{u^2+a^2}} = \frac{(u^2+a^2)^{3/2}}{3} - a^2\sqrt{u^2+a^2}$

93. $\displaystyle\int \frac{du}{u\sqrt{u^2+a^2}} = -\frac{1}{a}\ln\left|\frac{a+\sqrt{u^2+a^2}}{u}\right|$

94. $\displaystyle\int \frac{du}{u^2\sqrt{u^2+a^2}} = -\frac{\sqrt{u^2+a^2}}{a^2 u}$

95. $\displaystyle\int \frac{du}{u^3\sqrt{u^2+a^2}} = -\frac{\sqrt{u^2+a^2}}{2a^2u^2} + \frac{1}{2a^3}\ln\left|\frac{a+\sqrt{u^2+a^2}}{u}\right|$

96. $\displaystyle\int \frac{\sqrt{u^2+a^2}}{u}\,du = \sqrt{u^2+a^2} - a\ln\left|\frac{a+\sqrt{u^2+a^2}}{u}\right|$

97. $\displaystyle\int \frac{\sqrt{u^2+a^2}}{u^2}\,du = -\frac{\sqrt{u^2+a^2}}{u} + \ln\left(u+\sqrt{u^2+a^2}\right)$

INTEGRALS INVOLVING $\sqrt{u^2-a^2}$, $a > 0$

98. $\displaystyle\int \frac{du}{\sqrt{u^2-a^2}} = \ln\left|u+\sqrt{u^2-a^2}\right|$

99. $\displaystyle\int \frac{u\,du}{\sqrt{u^2-a^2}} = \sqrt{u^2-a^2}$

100. $\displaystyle\int \frac{u^2\,du}{\sqrt{u^2-a^2}} = \frac{u\sqrt{u^2-a^2}}{2} + \frac{a^2}{2}\ln\left|u+\sqrt{u^2-a^2}\right|$

101. $\displaystyle\int \frac{u^3\,du}{\sqrt{u^2-a^2}} = \frac{(u^2-a^2)^{3/2}}{3} + a^2\sqrt{u^2-a^2}$

102. $\displaystyle\int \frac{du}{u\sqrt{u^2-a^2}} = \frac{1}{a}\sec^{-1}\left|\frac{u}{a}\right|$

103. $\displaystyle\int \frac{du}{u^2\sqrt{u^2-a^2}} = \frac{\sqrt{u^2-a^2}}{a^2 u}$

104. $\displaystyle\int \frac{du}{u^3\sqrt{u^2-a^2}} = \frac{\sqrt{u^2-a^2}}{2a^2u^2} + \frac{1}{2a^3}\sec^{-1}\left|\frac{u}{a}\right|$

105. $\displaystyle\int \sqrt{u^2-a^2}\,du = \frac{u\sqrt{u^2-a^2}}{2} - \frac{a^2}{2}\ln\left|u+\sqrt{u^2-a^2}\right|$

106. $\displaystyle\int u\sqrt{u^2-a^2}\,du = \frac{(u^2-a^2)^{3/2}}{3}$

107. $\displaystyle\int u^2\sqrt{u^2-a^2}\,du = \frac{u(u^2-a^2)^{3/2}}{4} + \frac{a^2 u\sqrt{u^2-a^2}}{8} - \frac{a^4}{8}\ln\left|u+\sqrt{u^2-a^2}\right|$

108. $\displaystyle\int u^3\sqrt{u^2-a^2}\,du = \frac{(u^2-a^2)^{5/2}}{5} + \frac{a^2(u^2-a^2)^{3/2}}{3}$

109. $\displaystyle\int \frac{\sqrt{u^2-a^2}}{u}\,du = \sqrt{u^2-a^2} - a\sec^{-1}\left|\frac{u}{a}\right|$

INTEGRALS INVOLVING $\sqrt{a^2-u^2}$, $a > 0$

110. $\displaystyle\int \frac{du}{\sqrt{a^2-u^2}} = \sin^{-1}\frac{u}{a}$

111. $\displaystyle\int \frac{u\,du}{\sqrt{a^2-u^2}} = -\sqrt{a^2-u^2}$

112. $\displaystyle\int \frac{u^2\,du}{\sqrt{a^2-u^2}} = -\frac{u\sqrt{a^2-u^2}}{2} + \frac{a^2}{2}\sin^{-1}\frac{u}{a}$

113. $\displaystyle\int \frac{u^3\,du}{\sqrt{a^2-u^2}} = \frac{(a^2-u^2)^{3/2}}{3} - a^2\sqrt{a^2-u^2}$

114. $\displaystyle\int \frac{du}{u\sqrt{a^2-u^2}} = -\frac{1}{a}\ln\left|\frac{a+\sqrt{a^2-u^2}}{u}\right|$ or $-\frac{1}{a}\operatorname{sech}^{-1}\left|\frac{u}{a}\right|$

115. $\displaystyle\int \frac{du}{u^2\sqrt{a^2-u^2}} = -\frac{\sqrt{a^2-u^2}}{a^2u}$

116. $\displaystyle\int \frac{du}{u^3\sqrt{a^2-u^2}} = -\frac{\sqrt{a^2-u^2}}{2a^2u^2} - \frac{1}{2a^3}\ln\left|\frac{a+\sqrt{a^2-u^2}}{u}\right|$

117. $\displaystyle\int \sqrt{a^2-u^2}\,du = \frac{u\sqrt{a^2-u^2}}{2} + \frac{a^2}{2}\sin^{-1}\frac{u}{a}$

118. $\displaystyle\int u\sqrt{a^2-u^2}\,du = -\frac{(a^2-u^2)^{3/2}}{3}$

119. $\displaystyle\int u^2\sqrt{a^2-u^2}\,du = -\frac{u(a^2-u^2)^{3/2}}{4} + \frac{a^2u\sqrt{a^2-u^2}}{8} + \frac{a^4}{8}\sin^{-1}\frac{u}{a}$

120. $\displaystyle\int u^3\sqrt{a^2-u^2}\,du = \frac{(a^2-u^2)^{5/2}}{5} - \frac{a^2(a^2-u^2)^{3/2}}{3}$

121. $\displaystyle\int \frac{\sqrt{a^2-u^2}}{u}\,du = \sqrt{a^2-u^2} - a\ln\left|\frac{a+\sqrt{a^2-u^2}}{u}\right|$

Trigonometric forms

INTEGRALS INVOLVING $\cos au$

122. $\displaystyle\int \cos au\,du = \frac{\sin au}{a}$

123. $\displaystyle\int u\cos au\,du = \frac{\cos au}{a^2} + \frac{u\sin au}{a}$

124. $\displaystyle\int u^2\cos au\,du = \frac{2u}{a^2}\cos au + \left(\frac{u^2}{a} - \frac{2}{a^3}\right)\sin au$

125. $\displaystyle\int u^3\cos au\,du = \left(\frac{3u^2}{a^2} - \frac{6}{a^4}\right)\cos au + \left(\frac{u^3}{a} - \frac{6u}{a^3}\right)\sin au$

126. $\displaystyle\int u^n\cos au\,du = \frac{u^n\sin au}{a} - \frac{n}{a}\int u^{n-1}\sin au\,du$

127. $\displaystyle\int \cos^2 au\,du = \frac{u}{2} + \frac{\sin 2au}{4a}$

128. $\displaystyle\int u\cos^2 au\,du = \frac{u^2}{4} + \frac{u\sin 2au}{4a} + \frac{\cos 2au}{8a^2}$

129. $\displaystyle\int \cos^3 au\,du = \frac{\sin au}{a} - \frac{\sin^3 au}{3a}$

130. $\displaystyle\int \cos^4 au\,du = \frac{3u}{8} + \frac{\sin 2au}{4a} + \frac{\sin 4au}{32a}$

INTEGRALS INVOLVING sin *au*

131. $\displaystyle\int \sin au\, du = -\frac{\cos au}{a}$

132. $\displaystyle\int u \sin au\, du = \frac{\sin au}{a^2} - \frac{u \cos au}{a}$

133. $\displaystyle\int u^2 \sin au\, du = \frac{2u}{a^2} \sin au + \left(\frac{2}{a^3} - \frac{u^2}{a}\right) \cos au$

134. $\displaystyle\int u^3 \sin au\, du = \left(\frac{3u^2}{a^2} - \frac{6}{a^4}\right) \sin au + \left(\frac{6u}{a^3} - \frac{u^3}{a}\right) \cos au$

135. $\displaystyle\int u^n \sin au\, du = -\frac{u^n \cos au}{a} + \frac{n}{a} \int u^{n-1} \cos au\, du$

136. $\displaystyle\int \sin^2 au\, du = \frac{u}{2} - \frac{\sin 2au}{4a}$

137. $\displaystyle\int u \sin^2 au\, du = \frac{u^2}{4} - \frac{u \sin 2au}{4a} - \frac{\cos 2au}{8a^2}$

138. $\displaystyle\int \sin^3 au\, du = -\frac{\cos au}{a} + \frac{\cos^3 au}{3a}$

139. $\displaystyle\int \sin^4 au\, du = \frac{3u}{8} - \frac{\sin 2au}{4a} + \frac{\sin 4au}{32a}$

INTEGRALS INVOLVING sin *au* **and cos** *au*

140. $\displaystyle\int \sin au \cos au\, du = \frac{\sin^2 au}{2a}$

141. $\displaystyle\int \sin au \cos bu\, du = -\frac{\cos(a-b)u}{2(a-b)} - \frac{\cos(a+b)u}{2(a+b)}$

142. $\displaystyle\int \sin^n au \cos au\, du = \frac{\sin^{n+1} au}{(n+1)a}$

143. $\displaystyle\int \cos^n au \sin au\, du = -\frac{\cos^{n+1} au}{(n+1)a}$

144. $\displaystyle\int \sin^2 au \cos^2 au\, du = \frac{u}{8} - \frac{\sin 4au}{32a}$

145. $\displaystyle\int \frac{du}{\sin au \cos au} = \frac{1}{a} \ln |\tan au|$

146. $\displaystyle\int \frac{du}{\sin^2 au \cos au} = \frac{1}{a} \ln \left|\tan\left(\frac{\pi}{4} + \frac{au}{2}\right)\right| - \frac{1}{a \sin au}$

147. $\displaystyle\int \frac{du}{\sin au \cos^2 au} = \frac{1}{a} \ln \left|\tan \frac{au}{2}\right| + \frac{1}{a \cos au}$

INTEGRALS INVOLVING tan *au*

148. $\displaystyle\int \tan au\, du = -\frac{1}{a} \ln |\cos au| \ \text{ or } \ \frac{1}{a} \ln |\sec au|$

149. $\displaystyle\int \tan^2 au\, du = \frac{\tan au}{a} - u$

150. $\displaystyle\int \tan^3 au\, du = \frac{\tan^2 au}{2a} + \frac{1}{a} \ln |\cos au|$

151. $\displaystyle\int \tan^n au\, du = \frac{\tan^{n-1} au}{(n-1)a} - \int \tan^{n-2} au\, du$

152. $\displaystyle\int \tan^n au \sec^2 au\, du = \frac{\tan^{n+1} au}{(n+1)a}$

INTEGRALS INVOLVING cot *au*

153. $\displaystyle\int \cot au\, du = \frac{1}{a} \ln |\sin au|$

154. $\displaystyle\int \cot^2 au\, du = -\frac{\cot au}{a} - u$

155. $\displaystyle\int \cot^3 au\, du = -\frac{\cot^2 au}{2a} - \frac{1}{a} \ln |\sin au|$

156. $\displaystyle\int \cot^n au\, du = -\frac{\cot^{n-1} au}{(n-1)a} - \int \cot^{n-2} au\, du$

157. $\displaystyle\int \cot^n au\, \csc^2 au\, du = -\frac{\cot^{n+1} au}{(n+1)a}$

INTEGRALS INVOLVING sec *au*

158. $\displaystyle\int \sec au\, du = \frac{1}{a} \ln |\sec au + \tan au| = \frac{1}{a} \ln \left| \tan\left(\frac{au}{2} + \frac{\pi}{4}\right) \right|$

159. $\displaystyle\int \sec^2 au\, du = \frac{\tan au}{a}$

160. $\displaystyle\int \sec^3 au\, du = \frac{\sec au \tan au}{2a} + \frac{1}{2a} \ln |\sec au + \tan au|$

161. $\displaystyle\int \sec^n au\, du = \frac{\sec^{n-2} au \tan au}{a(n-1)} + \frac{n-2}{n-1} \int \sec^{n-2} au\, du$

162. $\displaystyle\int \sec^n au \tan au\, du = \frac{\sec^n au}{na}$

INTEGRALS INVOLVING csc *au*

163. $\displaystyle\int \csc au\, du = \frac{1}{a} \ln |\csc au - \cot au| = \frac{1}{a} \ln \left| \tan \frac{au}{2} \right|$

164. $\displaystyle\int \csc^2 au\, du = -\frac{\cot au}{a}$

165. $\displaystyle\int \csc^3 au\, du = -\frac{\csc au \cot au}{2a} + \frac{1}{2a} \ln \left| \tan \frac{au}{2} \right|$

166. $\displaystyle\int \csc^n au\, du = -\frac{\csc^{n-2} au \cot au}{a(n-1)} + \frac{n-2}{n-1} \int \csc^{n-2} au\, du$

167. $\displaystyle\int \csc^n au \cot au\, du = -\frac{\csc^n au}{na}$

Inverse trigonometric forms

INTEGRALS INVOLVING INVERSE TRIGONOMETRIC FUNCTIONS, *a* > 0

168. $\displaystyle\int \cos^{-1} \frac{u}{a}\, du = u \cos^{-1} \frac{u}{a} - \sqrt{a^2 - u^2}$

169. $\displaystyle\int u \cos^{-1} \frac{u}{a}\, du = \left(\frac{u^2}{2} - \frac{a^2}{4}\right) \cos^{-1} \frac{u}{a} - \frac{u\sqrt{a^2 - u^2}}{4}$

170. $\displaystyle\int u^2 \cos^{-1} \frac{u}{a}\, du = \frac{u^3}{3} \cos^{-1} \frac{u}{a} - \frac{(u^2 + 2a^2)\sqrt{a^2 - u^2}}{9}$

171. $\displaystyle\int \frac{\cos^{-1} \frac{u}{a}}{u}\, du = \frac{\pi}{2} \ln |u| - \int \frac{\sin^{-1} \frac{u}{a}}{u}\, du$

172. $\displaystyle\int \frac{\cos^{-1} \frac{u}{a}}{u^2}\, du = -\frac{\cos^{-1} \frac{u}{a}}{u} + \frac{1}{a} \ln \left| \frac{a + \sqrt{a^2 - u^2}}{u} \right|$

173. $\displaystyle\int \left(\cos^{-1}\frac{u}{a}\right)^2 du = u\left(\cos^{-1}\frac{u}{a}\right)^2 - 2u - 2\sqrt{a^2 - u^2}\,\cos^{-1}\frac{u}{a}$

174. $\displaystyle\int \sin^{-1}\frac{u}{a}\,du = u\sin^{-1}\frac{u}{a} + \sqrt{a^2 - u^2}$

175. $\displaystyle\int u\sin^{-1}\frac{u}{a}\,du = \left(\frac{u^2}{2} - \frac{a^2}{4}\right)\sin^{-1}\frac{u}{a} + \frac{u\sqrt{a^2 - u^2}}{4}$

176. $\displaystyle\int u^2 \sin^{-1}\frac{u}{a}\,du = \frac{u^3}{3}\sin^{-1}\frac{u}{a} + \frac{(u^2 + 2a^2)\sqrt{a^2 - u^2}}{9}$

177. $\displaystyle\int \frac{\sin^{-1}\frac{u}{a}}{u}\,du = \frac{u}{a} + \frac{\left(\frac{u}{a}\right)^3}{2\cdot 3\cdot 3} + \frac{1\cdot 3\left(\frac{u}{a}\right)^5}{2\cdot 4\cdot 5\cdot 5} + \frac{1\cdot 3\cdot 5\left(\frac{u}{a}\right)^7}{2\cdot 4\cdot 6\cdot 7\cdot 7} + \cdots$

178. $\displaystyle\int \frac{\sin^{-1}\frac{u}{a}}{u^2}\,du = -\frac{\sin^{-1}\frac{u}{a}}{u} - \frac{1}{a}\ln\left|\frac{a + \sqrt{a^2 - u^2}}{u}\right|$

179. $\displaystyle\int \left(\sin^{-1}\frac{u}{a}\right)^2 du = u\left(\sin^{-1}\frac{u}{a}\right)^2 - 2u + 2\sqrt{a^2 - u^2}\,\sin^{-1}\frac{u}{a}$

180. $\displaystyle\int \tan^{-1}\frac{u}{a}\,du = u\tan^{-1}\frac{u}{a} - \frac{a}{2}\ln(u^2 + a^2)$

181. $\displaystyle\int u\tan^{-1}\frac{u}{a}\,du = \frac{1}{2}\left(u^2 + a^2\right)\tan^{-1}\frac{u}{a} - \frac{au}{2}$

182. $\displaystyle\int u^2 \tan^{-1}\frac{u}{a}\,du = \frac{u^3}{3}\tan^{-1}\frac{u}{a} - \frac{au^2}{6} + \frac{a^3}{6}\ln(u^2 + a^2)$

Exponential and logarithmic forms

INTEGRALS INVOLVING e^{au}

183. $\displaystyle\int e^{au}\,du = \frac{e^{au}}{a}$

184. $\displaystyle\int ue^{au}\,du = \frac{e^{au}}{a}\left(u - \frac{1}{a}\right)$

185. $\displaystyle\int u^2 e^{au}\,du = \frac{e^{au}}{a}\left(u^2 - \frac{2u}{a} + \frac{2}{a^2}\right)$

186. $\displaystyle\int u^n e^{au}\,du = \frac{u^n e^{au}}{a} - \frac{n}{a}\int u^{n-1} e^{au}\,du$

187. $\displaystyle\int \frac{e^{au}}{u}\,du = \ln|u| + \frac{au}{1\cdot 1!} + \frac{(au)^2}{2\cdot 2!} + \frac{(au)^3}{3\cdot 3!} + \cdots$

188. $\displaystyle\int \frac{e^{au}}{u^n}\,du = \frac{-e^{au}}{(n-1)u^{n-1}} + \frac{a}{n-1}\int \frac{e^{au}}{u^{n-1}}\,du$

189. $\displaystyle\int \frac{du}{p + qe^{au}} = \frac{u}{p} - \frac{1}{ap}\ln\left|p + qe^{au}\right|$

190. $\displaystyle\int \frac{du}{(p + qe^{au})^2} = \frac{u}{p^2} + \frac{1}{ap(p + qe^{au})} - \frac{1}{ap^2}\ln\left|p + qe^{au}\right|$

191. $\displaystyle\int \frac{du}{pe^{au} + qe^{-au}} = \begin{cases} \frac{1}{a\sqrt{pq}}\tan^{-1}\left(\sqrt{\frac{p}{q}}\,e^{au}\right),\ p > 0,\, q > 0 \\[2mm] \frac{1}{2a\sqrt{-pq}}\ln\left|\frac{e^{au} - \sqrt{-\frac{q}{p}}}{e^{au} + \sqrt{-\frac{q}{p}}}\right|,\ p > 0,\, q < 0 \end{cases}$

192. $\displaystyle\int e^{au}\sin bu\,du = \frac{e^{au}(a\sin bu - b\cos bu)}{a^2 + b^2}$

193. $\displaystyle\int e^{au}\cos bu\,du = \frac{e^{au}(a\cos bu + b\sin bu)}{a^2 + b^2}$

194. $\displaystyle\int ue^{au}\sin bu\,du = \frac{ue^{au}(a\sin bu - b\cos bu)}{a^2+b^2} - \frac{e^{au}\left[(a^2-b^2)\sin bu - 2ab\cos bu\right]}{(a^2+b^2)^2}$

195. $\displaystyle\int ue^{au}\cos bu\,du = \frac{ue^{au}(a\cos bu + b\sin bu)}{a^2+b^2} - \frac{e^{au}\left[(a^2-b^2)\cos bu + 2ab\sin bu\right]}{(a^2+b^2)^2}$

INTEGRALS INVOLVING $\ln|u|$

196. $\displaystyle\int \ln|u|\,du = u\ln|u| - u$

197. $\displaystyle\int (\ln|u|)^2\,du = u\,(\ln|u|)^2 - 2u\ln|u| + 2u$

198. $\displaystyle\int (\ln|u|)^n\,du = u\,(\ln|u|)^n - n\int (\ln|u|)^{n-1}\,du$

199. $\displaystyle\int u\ln|u|\,du = \frac{u^2}{2}\left(\ln|u| - \frac{1}{2}\right)$

200. $\displaystyle\int u^m\ln|u|\,du = \frac{u^{m+1}}{m+1}\left(\ln|u| - \frac{1}{m+1}\right)$

Trigonometric Functions

The geometric definition of an angle as two rays with a common endpoint (called the **vertex**) is sometimes referred to as a **geometric angle**. In higher mathematics, an **angle** is defined as the amount of rotation of a ray from one position, called its **initial side** to another position, called its **terminal side**. If the rotation is a counterclockwise direction, it is called a **positive angle** and if the rotation is in a clockwise direction, it is called a **negative angle**. If a coordinate system is drawn so that the origin of the coordinate system is at the vertex of the angle, and the positive x-axis coincides with the initial side, then the angle is said to be in **standard position**. Figure E.13 shows a **unit circle** (circle with radius 1) and a standard position angle.

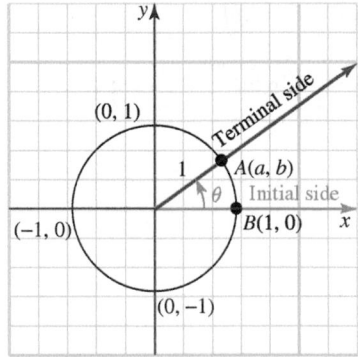

Figure E.13 Unit circle with angle θ

Let θ be any angle in standard position and let $P(x, y)$ be any point on the terminal side of the angle a distance of r from the origin ($r \neq 0$). Then

$$\cos\theta = \frac{x}{r} \qquad \sin\theta = \frac{y}{r} \qquad \tan\theta = \frac{y}{x}$$

$$\sec\theta = \frac{r}{x} \qquad \csc\theta = \frac{r}{y} \qquad \cot\theta = \frac{x}{y}$$

Radians and Degrees

$360° = 2\pi$ radians$= 1$ revolution

$$1 \text{ radian} = \left(\frac{180}{\pi}\right)^{\circ} \approx 57.2957\cdots \text{ degrees}$$

$$1 \text{ degree} = \left(\frac{\pi}{180}\right) \approx 0.0174532\cdots \text{ radian}$$

Angle measures, unless stated otherwise, are in radian measure. For example, an angle of 6 means that the angle has a measure of 6 **radians**, whereas if you want to denote an angle with measure six degrees, then the word **degrees** or degree symbol, as in 6°, must be stated.

In calculus, it is often necessary to determine when a given function is not defined. For example, in Problem 18, Problem Set 1.2 in the textbook, we need to know when

$$f(x) = (3\tan x + \sqrt{3})(3\tan x - \sqrt{3})$$

is not defined. From the definition given above, $f\left(\frac{\pi}{2}\right)$ is not defined since $\tan x$ is not defined for $x = \frac{\pi}{2}$ (or for any odd multiple of $\frac{\pi}{2}$).

Inverse Trigonometric Functions

Inverse Function	Domain	Range
$y = \arccos x$ or $y = \cos^{-1} x$	$-1 \leq x \leq 1$	$0 \leq y \leq \pi$
$y = \arcsin x$ or $y = \sin^{-1} x$	$-1 \leq x \leq 1$	$-\frac{\pi}{2} \leq y \leq \frac{\pi}{2}$
$y = \arctan x$ or $y = \tan^{-1} x$	all reals	$-\frac{\pi}{2} < y < \frac{\pi}{2}$
$y = \text{arccot}\, x$ or $y = \cot^{-1} x$	all reals	$0 < y < \pi$
$y = \text{arcsec}\, x$ or $y = \sec^{-1} x$	$|x| \geq 1$	$0 \leq y \leq \pi, y \neq \frac{\pi}{2}$
$y = \text{arccsc}\, x$ or $y = \csc^{-1} x$	$|x| \geq 1$	$-\frac{\pi}{2} \leq y \leq \frac{\pi}{2}, y \neq 0$

The principal values of the inverse trigonometric functions are those values defined by these inverse trigonometric functions. These are the values obtained when using a calculator.

Evaluating Trigonometric Functions

If an angle θ is not a quadrantal angle, then the **reference angle**, which is denoted by $\overline{\theta}$, is defined as the acute angle the terminal side of θ makes with the x-axis.

> **REDUCTION PRINCIPLE**   If t represents any of the six trigonometric functions, then $t(\theta) = \pm t(\overline{\theta})$ where the sign depends on the quadrant and $\overline{\theta}$ is the reference angle of θ.

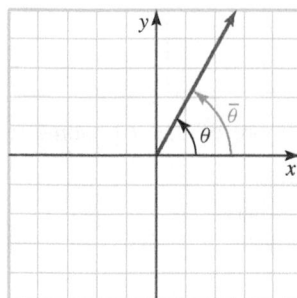

Quadrant I
All are positive in
the first quadrant

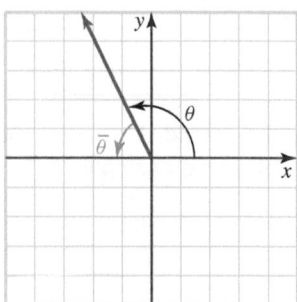

Quadrant II
Sine and cosecant
are positive

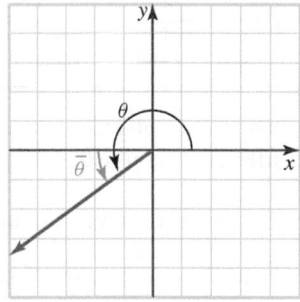

Quadrant III
Tangent and cotangent
are positive

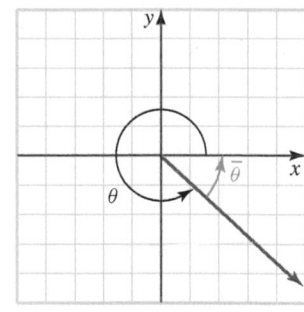

Quadrant IV
Cosine and secant
are positive

By calculator
Trigonometric Functions

Function	*Algebraic*	or	*Graphing*		Function	*Algebraic*	or	*Graphing*		
$\cos x$	$\boxed{x}\ \boxed{\cos}$		$\boxed{\cos}\ \boxed{x}\ \boxed{\text{ENTER}}$		$\sec x$	$\boxed{x}\ \boxed{\cos}\ \boxed{1/x}$		$\boxed{\cos}\ \boxed{x}\ \boxed{\text{ENTER}}\ \boxed{x^{-1}}$		
$\sin x$	$\boxed{x}\ \boxed{\sin}$		$\boxed{\sin}\ \boxed{x}\ \boxed{\text{ENTER}}$		$\csc x$	$\boxed{x}\ \boxed{\sin}\ \boxed{1/x}$		$\boxed{\sin}\ \boxed{x}\ \boxed{\text{ENTER}}\ \boxed{x^{-1}}$		
$\tan x$	$\boxed{x}\ \boxed{\tan}$		$\boxed{\tan}\ \boxed{x}\ \boxed{\text{ENTER}}$		$\cot x$	$\boxed{x}\ \boxed{\tan}\ \boxed{1/x}$		$\boxed{\tan}\ \boxed{x}\ \boxed{\text{ENTER}}\ \boxed{x^{-1}}$		

Inverse Trigonometric Functions

Function	Algebraic	or	Graphing
$\cos^{-1}x$	$\boxed{x}\ \boxed{\text{inv}}\ \boxed{\cos}$		$\boxed{\cos^{-1}}\ \boxed{x}\ \boxed{\text{ENTER}}$
$\sin^{-1}x$	$\boxed{x}\ \boxed{\text{inv}}\ \boxed{\sin}$		$\boxed{\sin^{-1}}\ \boxed{x}\ \boxed{\text{ENTER}}$
$\tan^{-1}x$	$\boxed{x}\ \boxed{\text{inv}}\ \boxed{\tan}$		$\boxed{\tan^{-1}}\ \boxed{x}\ \boxed{\text{ENTER}}$

Function	Algebraic	or	Graphing
$\sec^{-1}x$	$\boxed{x}\ \boxed{1/x}\ \boxed{\text{inv}}\ \boxed{\cos}$		$\boxed{\cos^{-1}}\ \boxed{x}\ \boxed{\text{ENTER}}\ \boxed{x^{-1}}$
$\csc^{-1}x$	$\boxed{x}\ \boxed{\text{inv}}\ \boxed{\sin}\ \boxed{1/x}$		$\boxed{\sin^{-1}}\ \boxed{x}\ \boxed{\text{ENTER}}\ \boxed{x^{-1}}$
$\cot^{-1}x$			

If $x > 0$: $\boxed{x}\ \boxed{1/x}\ \boxed{\text{inv}}\ \boxed{\tan}$ $\boxed{\tan^{-1}}\ \boxed{x}\ \boxed{\text{ENTER}}\ \boxed{x^{-1}}$

If $x < 0$: $\boxed{x}\ \boxed{1/x}\ \boxed{\text{inv}}\ \boxed{\tan}\ \boxed{+}\ \boxed{\pi}\ \boxed{=}$

or $\boxed{\tan^{-1}}\ \boxed{x}\ \boxed{\text{ENTER}}\ \boxed{x^{-1}}\ \boxed{+}\ \boxed{\pi}\ \boxed{=}$

By table

Angle θ	0	$\frac{\pi}{6}$	$\frac{\pi}{4}$	$\frac{\pi}{3}$	$\frac{\pi}{2}$	π	$\frac{3\pi}{2}$
$\cos\theta$	1	$\frac{\sqrt{3}}{2}$	$\frac{\sqrt{2}}{2}$	$\frac{1}{2}$	0	-1	0
$\sin\theta$	0	$\frac{1}{2}$	$\frac{\sqrt{2}}{2}$	$\frac{\sqrt{3}}{2}$	1	0	-1
$\tan\theta$	0	$\frac{\sqrt{3}}{3}$	1	$\sqrt{3}$	undefined	0	undefined
$\sec\theta$	1	$\frac{2}{\sqrt{3}}$	$\frac{2}{\sqrt{2}}$	2	undefined	-1	undefined
$\csc\theta$	undefined	2	$\frac{2}{\sqrt{2}}$	$\frac{2}{\sqrt{3}}$	1	undefined	-1
$\cot\theta$	undefined	$\sqrt{3}$	1	$\frac{1}{\sqrt{3}}$	0	undefined	0

This table of exact values can be extended by using one or more of the trigonometric identities as illustrated by the following example.

Example 1   Exact values

Find the exact value of $\sin\left(-\frac{11\pi}{12}\right)$.

Solution   First, use the reduction principle to write

$$\sin\left(-\frac{11\pi}{12}\right) = \underset{\uparrow}{-}\sin\frac{\pi}{12}$$

Negative because the angle is in Quadrant III

Then, notice that since $\frac{\pi}{12} = \frac{\pi}{4} - \frac{\pi}{6}$, we have

$$
\begin{aligned}
-\sin\frac{\pi}{12} &= -\sin\left(\frac{\pi}{4} - \frac{\pi}{6}\right) \\
&= -\left[\sin\frac{\pi}{4}\cos\frac{\pi}{6} - \cos\frac{\pi}{4}\sin\frac{\pi}{6}\right] && \textit{Identity 18 below} \\
&= -\left[\frac{\sqrt{2}}{2}\cdot\frac{\sqrt{3}}{2} - \frac{\sqrt{2}}{2}\cdot\frac{1}{2}\right] && \textit{Exact values} \\
&= -\left[\frac{\sqrt{6}-\sqrt{2}}{4}\right] \\
&= \frac{\sqrt{2}-\sqrt{6}}{4}
\end{aligned}
$$

Trigonometric Graphs

Trigonometric Functions

Cosine

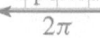

$y = \cos x$

Sine

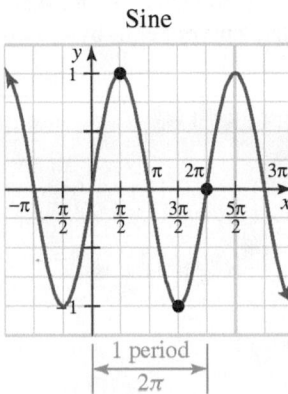

$y = \sin x$

Tangent

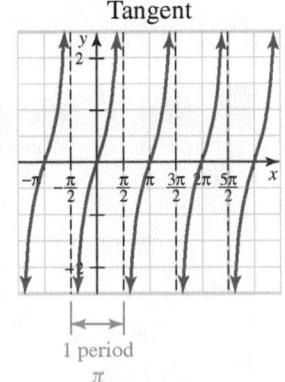

$y = \tan x$

Secant

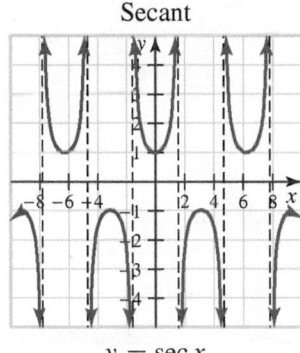

$y = \sec x$

Cosecant

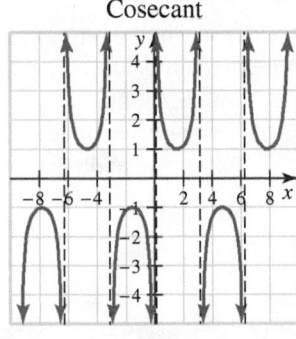

$y = \csc x$

Cotangent

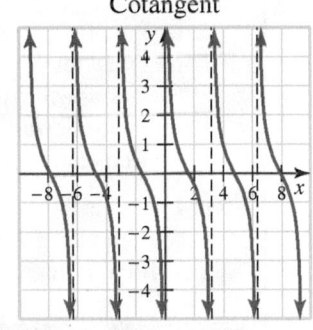

$y = \cot x$

Inverse Trigonometric Functions

Arccosine

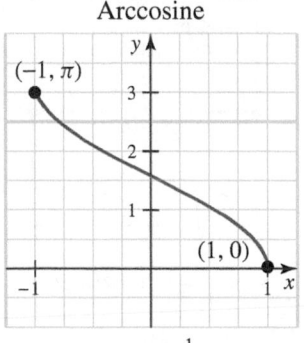
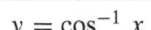

$y = \cos^{-1} x$

Arcsine

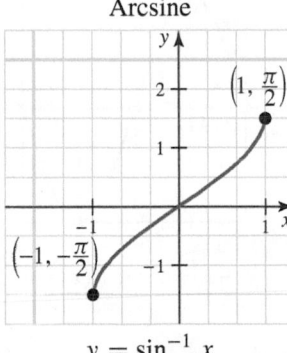

$y = \sin^{-1} x$

Arctangent

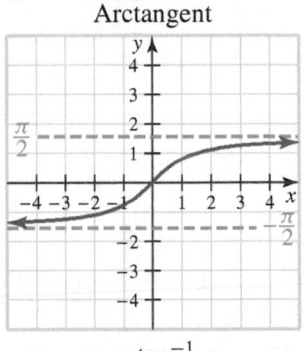

$y = \tan^{-1} x$

Arcsecant

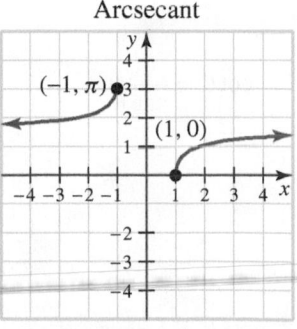

$y = \sec^{-1} x$

Arccosecant

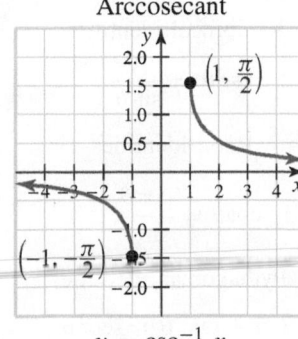

$y = \csc^{-1} x$

Arccotangent

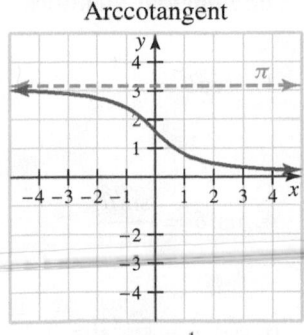

$y = \cot^{-1} x$

General Trigonometric Graphs

You are expected to be able to graph trigonometric functions. Problem Set 1.2 (Problems 45-48) and Problem Set 1.5 (Problems 41 and 42) should give you some practice. To graph a general trigonometric curve, begin by drawing a frame.

GENERAL TRIG CURVES One period for the cosine, sine, and tangent are shown in their frames.

Step 1 Algebraically, put the equation into one of the general forms shown under the graphs.

Step 2 Identify (by inspection) the following values: (h, k), a, and b.
   Calculate the period, p: $p = \dfrac{2\pi}{b}$ for cosine and sine

   $$p = \dfrac{\pi}{b} \text{ for tangent.}$$

Step 3 Draw the frame.

   a. Plot (h, k); this is the starting point.
   b. Draw **amplitude**, $|a|$; the *height of the frame* is $2|a|$: up a units from (h, k) and down a units from (h, k).*
   c. Draw **period**, p; *the length of the frame* is p.
      Remember, $p = \frac{2\pi}{b}$ for cosine and sine, and $p = \frac{\pi}{b}$ for tangent.

Step 4 Locate the critical values using the frame as a guide, and then sketch the appropriate graphs. **You do NOT need to know the coordinates of these critical values.**

Cosine Frame	Sine Frame	Tangent Frame
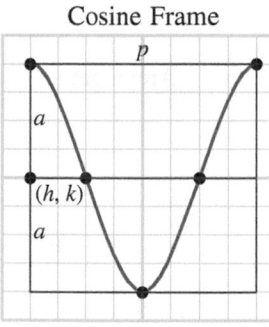	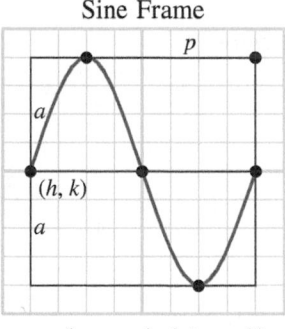	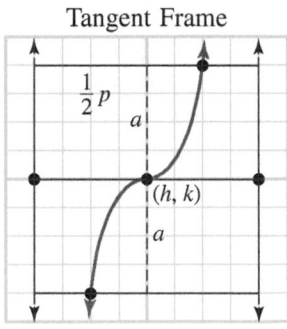
$y - k = a \cos b(x - h)$	$y - k = a \sin b(x - h)$	$y - k = a \tan b(x - h)$

Trigonometric Identities

Fundamental identities

Reciprocal identities

1. $\sec \theta = \dfrac{1}{\cos \theta}$ 2. $\csc \theta = \dfrac{1}{\sin \theta}$ 3. $\cot \theta = \dfrac{1}{\tan \theta}$

Ratio identities

4. $\tan \theta = \dfrac{\sin \theta}{\cos \theta}$ 5. $\cot \theta = \dfrac{\cos \theta}{\sin \theta}$

Pythagorean identities

6. $\cos^2 \theta + \sin^2 \theta = 1$ 7. $1 + \tan^2 \theta = \sec^2 \theta$ 8. $\cot^2 \theta + 1 = \csc^2 \theta$

---

*For now, assume $a > 0$; $a < 0$ is considered in Chapter 4.

Cofunction identities

9. $\cos\left(\frac{\pi}{2} - \theta\right) = \sin\theta$ 10. $\sin\left(\frac{\pi}{2} - \theta\right) = \cos\theta$ 11. $\tan\left(\frac{\pi}{2} - \theta\right) = \cot\theta$

Opposite-angle identities

12. $\cos(-\theta) = \cos\theta$ 13. $\sin(-\theta) = -\sin\theta$ 14. $\tan(-\theta) = -\tan\theta$

Addition laws

15. $\cos(\alpha + \beta) = \cos\alpha\cos\beta - \sin\alpha\sin\beta$
16. $\cos(\alpha - \beta) = \cos\alpha\cos\beta + \sin\alpha\sin\beta$
17. $\sin(\alpha + \beta) = \sin\alpha\cos\beta + \cos\alpha\sin\beta$
18. $\sin(\alpha - \beta) = \sin\alpha\cos\beta - \cos\alpha\sin\beta$
19. $\tan(\alpha + \beta) = \dfrac{\tan\alpha + \tan\beta}{1 - \tan\alpha\tan\beta}$ 20. $\tan(\alpha - \beta) = \dfrac{\tan\alpha - \tan\beta}{1 + \tan\alpha\tan\beta}$

Double-angle identities

21. $\cos 2\theta = \cos^2\theta - \sin^2\theta$ 22. $\sin 2\theta = 2\sin\theta\cos\theta$

$\qquad\qquad = 2\cos^2\theta - 1$

$\qquad\qquad = 1 - 2\sin^2\theta$ 23. $\tan 2\theta = \dfrac{2\tan\theta}{1 - \tan^2\theta}$

Half-angle identities

24. $\cos\frac{1}{2}\theta = \pm\sqrt{\dfrac{1 + \cos\theta}{2}}$ 25. $\sin\frac{1}{2}\theta = \pm\sqrt{\dfrac{1 - \cos\theta}{2}}$ 26. $\tan\frac{1}{2}\theta = \dfrac{1 - \cos\theta}{\sin\theta}$

The sign in identities 24 and 25 is determined by the

quadrant in which $\frac{1}{2}\theta$ lies.

$\qquad\qquad\qquad\qquad\qquad\qquad\qquad\qquad\qquad = \dfrac{\sin\theta}{1 + \cos\theta}$

Product-to-sum identities

27. $2\cos\alpha\cos\beta = \cos(\alpha - \beta) + \cos(\alpha + \beta)$
28. $2\sin\alpha\sin\beta = \cos(\alpha - \beta) - \cos(\alpha + \beta)$
29. $2\sin\alpha\cos\beta = \sin(\alpha + \beta) + \sin(\alpha - \beta)$
30. $2\cos\alpha\sin\beta = \sin(\alpha + \beta) - \sin(\alpha - \beta)$

Sum-to-product identities

31. $\cos\alpha + \cos\beta = 2\cos\left(\frac{\alpha+\beta}{2}\right)\cos\left(\frac{\alpha-\beta}{2}\right)$
32. $\cos\alpha - \cos\beta = -2\sin\left(\frac{\alpha+\beta}{2}\right)\sin\left(\frac{\alpha-\beta}{2}\right)$
33. $\sin\alpha + \sin\beta = 2\sin\left(\frac{\alpha+\beta}{2}\right)\cos\left(\frac{\alpha-\beta}{2}\right)$
34. $\sin\alpha - \sin\beta = 2\sin\left(\frac{\alpha-\beta}{2}\right)\cos\left(\frac{\alpha+\beta}{2}\right)$

Hyperbolic identities

35. $\operatorname{sech} x = \dfrac{1}{\cosh x}$ 36. $\operatorname{csch} x = \dfrac{1}{\sinh x}$ 37. $\coth x = \dfrac{1}{\tanh x}$

38. $\tanh x = \dfrac{\sinh x}{\cosh x}$ 39. $\coth x = \dfrac{\cosh x}{\sinh x}$

40. $\cosh^2 x - \sinh^2 x = 1$ 41. $1 - \tanh^2 x = \operatorname{sech}^2 x$ 42. $\coth^2 x - 1 = \operatorname{csch}^2 x$

43. $\cosh(-x) = \cosh x$ 44. $\sinh(-x) = -\sinh x$ 45. $\tanh(-x) = -\tanh x$

46. $\cosh(x \pm y) = \cosh x \cosh y \pm \sinh x \sinh y$

47. $\sinh(x \pm y) = \sinh x \cosh y \pm \cosh x \sinh y$

48. $\tanh(x \pm y) = \dfrac{\tanh x \pm \tanh y}{1 \pm \tanh x \tanh y}$ 49. $\cosh 2x = \cosh^2 x + \sinh^2 x$

$$= 2\cosh^2 x - 1$$

$$= 1 + 2\sinh^2 x$$

50. $\sinh 2x = 2\sinh x \cosh x$ 51. $\tanh 2x = \dfrac{2\tanh x}{1 + \tanh^2 x}$

52. $\cosh \dfrac{1}{2}x = \pm\sqrt{\dfrac{\cosh x + 1}{2}}$ 53. $\sinh \dfrac{1}{2}x = \pm\sqrt{\dfrac{\cosh x - 1}{2}}$

54. $\tanh \dfrac{1}{2}x = \dfrac{\cosh x - 1}{\sinh x}$ 55. $\sinh^{-1} x = \ln(x + \sqrt{x^2 + 1})$

$$= \dfrac{\sinh x}{\cosh x + 1}$$ 56. $\operatorname{csch}^{-1} x = \ln\left(\dfrac{1 + \sqrt{1 + x^2}}{x}\right)$ if $x > 0$

57. $\cosh^{-1} x = \ln\left(x + \sqrt{x^2 - 1}\right)$ if $x \geq 1$

58. $\operatorname{sech}^{-1} x = \ln\left(\dfrac{1 + \sqrt{1 - x^2}}{x}\right)$ if $0 < x \leq 1$

59. $\tanh^{-1} x = \dfrac{1}{2}\ln\left(\dfrac{1 + x}{1 - x}\right)$ if $-1 < x < 1$

60. $\coth^{-1} x = \dfrac{1}{2}\ln\left(\dfrac{x + 1}{x - 1}\right)$ if $|x| > 1$

APPENDIX F: DETERMINANTS

Suppose we decide to look for a *formula* solution to a system of equations. This formula method for solving a system of equations is known as *Cramer's rule*. Consider,

$$\begin{cases} a_{11}x_1 + a_{12}x_2 = b_1 \\ a_{21}x_1 + a_{22}x_2 = b_2 \end{cases}$$

Solve this system by using linear combinations.

$$\begin{array}{r} a_{22} \\ -a_{12} \end{array} \begin{cases} a_{11}x_1 + a_{12}x_2 = b_1 \\ a_{21}x_1 + a_{22}x_2 = b_2 \end{cases}$$

$$+ \begin{cases} a_{11}a_{22}x_1 + a_{12}a_{22}x_2 = b_1 a_{22} \\ -a_{12}a_{21}x_1 - a_{12}a_{22}x_2 = -b_2 a_{12} \end{cases}$$

$$a_{11}a_{22}x_1 - a_{12}a_{21}x_1 = b_1 a_{22} - b_2 a_{12}$$

If $a_{11}a_{22} - a_{12}a_{21} \neq 0$, then

$$x_1 = \frac{b_1 a_{22} - b_2 a_{12}}{a_{11}a_{22} - a_{12}a_{21}}$$

The difficulty with using this formula is that it is next to impossible to remember. To help with this matter, we introduce the concept of determinants.

Determinants

A matrix is a rectangular array of numbers, and associated with each square matrix is a real number called its **determinant**. The **order** of a determinant is the number of rows (and columns) of the matrix (only square matrices have determinants).

DETERMINANT OF ORDER 2 If A is the 2×2 matrix $\begin{bmatrix} a & b \\ c & d \end{bmatrix}$, then the **determinant** of A is defined to be the number $ad - bc$.

Some notations for this determinant are $\det A$, $\begin{vmatrix} a & b \\ c & d \end{vmatrix}$, and $|A|$. We will generally use $|A|$.

☠ *Do not confuse $|A|$ with $[A]$; the symbol $|A|$ is a determinant or a real number, and $[A]$ is an array or a matrix of real numbers.* ☠

Example 1   Evaluating determinants of order 2

Evaluate:

a. $\begin{vmatrix} 4 & -2 \\ -1 & 3 \end{vmatrix}$ **b.** $\begin{vmatrix} 2 & 2 \\ 2 & 2 \end{vmatrix}$ **c.** $\begin{vmatrix} 1 & 0 \\ 0 & 1 \end{vmatrix}$ **d.** $\begin{vmatrix} \cos\theta & -\sin\theta \\ \sin\theta & \cos\theta \end{vmatrix}$

e. $\begin{vmatrix} a_{11} & a_{12} \\ a_{21} & a_{22} \end{vmatrix}$ **f.** $\begin{vmatrix} b_1 & a_{12} \\ b_2 & a_{22} \end{vmatrix}$ **g.** $\begin{vmatrix} a_{11} & b_1 \\ a_{21} & b_2 \end{vmatrix}$

Solution

a. $\begin{vmatrix} 4 & -2 \\ -1 & 3 \end{vmatrix} = 4 \cdot 3 - (-2)(-1)$ **b.** $\begin{vmatrix} 2 & 2 \\ 2 & 2 \end{vmatrix} = 2 \cdot 2 - 2 \cdot 2$

$$= 12 - 2 \qquad\qquad\qquad\qquad\qquad = 0$$

$$= 10$$

c. $\begin{vmatrix} 1 & 0 \\ 0 & 1 \end{vmatrix} = 1 \cdot 1 - 0 \cdot 0$

$= 1$

d. $\begin{vmatrix} \cos\theta & -\sin\theta \\ \sin\theta & \cos\theta \end{vmatrix} = \cos^2\theta - (-\sin^2\theta)$

$= \cos^2\theta + \sin^2\theta$

$= 1$

e. $\begin{vmatrix} a_{11} & a_{12} \\ a_{21} & a_{22} \end{vmatrix} = a_{11}a_{22} - a_{12}a_{21}$

f. $\begin{vmatrix} b_1 & a_{12} \\ b_2 & a_{22} \end{vmatrix} = b_1 a_{22} - b_2 a_{12}$

g. $\begin{vmatrix} a_{11} & b_1 \\ a_{21} & b_2 \end{vmatrix} = b_2 a_{11} - b_1 a_{21}$

The solution to the general system

$$\begin{cases} a_{11}x_1 + a_{12}x_2 = b_1 \\ a_{21}x_1 + a_{22}x_2 = b_2 \end{cases}$$

can now be stated using determinant notation:

$$x_1 = \frac{\begin{vmatrix} b_1 & a_{12} \\ b_2 & a_{22} \end{vmatrix}}{\begin{vmatrix} a_{11} & a_{12} \\ a_{21} & a_{22} \end{vmatrix}} \qquad x_2 = \frac{\begin{vmatrix} a_{11} & b_1 \\ a_{21} & b_2 \end{vmatrix}}{\begin{vmatrix} a_{11} & a_{12} \\ a_{21} & a_{22} \end{vmatrix}}$$

Notice that the denominator of both variables is the same. This determinant is called the **determinant of the coefficients** for the system and is denoted by $|D|$. The numerators are also found by looking at $|D|$. For x_1, replace the first column of $|D|$ by the constant numbers b_1 and b_2, respectively. Denote this by $|D_1|$. Similarly, let $|D_2|$ be the determinant formed by replacing the coefficients of x_2 in $|D|$ by the constant numbers b_1 and b_2, respectively.

Properties of Determinants

In order to efficiently evaluate determinants of various dimensions, we need a preliminary observation. If we delete the first row and first column of the matrix,

$$A = \begin{bmatrix} a_{11} & a_{12} & a_{13} \\ a_{21} & a_{22} & a_{23} \\ a_{31} & a_{32} & a_{33} \end{bmatrix} \qquad a_{ij} \text{ is the entry in the } i\text{th row and the } j\text{th column}$$

We obtain the 2×2 matrix $\begin{bmatrix} a_{22} & a_{23} \\ a_{32} & a_{33} \end{bmatrix}$. The determinant of this 2×2 matrix is referred to as the **minor** associated with entry a_{11}, which is the element in the first row and first column.

Similarly,

The minor of a_{12} of $A = \begin{bmatrix} a_{11} & a_{12} & a_{13} \\ a_{21} & a_{22} & a_{23} \\ a_{31} & a_{32} & a_{33} \end{bmatrix}$ is $\begin{vmatrix} a_{21} & a_{23} \\ a_{31} & a_{33} \end{vmatrix}$.

The minor of a_{13} of $A = \begin{bmatrix} a_{11} & a_{12} & a_{13} \\ a_{21} & a_{22} & a_{23} \\ a_{31} & a_{32} & a_{33} \end{bmatrix}$ is $\begin{vmatrix} a_{21} & a_{22} \\ a_{31} & a_{32} \end{vmatrix}$.

Example 2   Finding minors

Consider $\begin{vmatrix} 2 & -4 & 0 \\ 1 & -3 & -1 \\ 6 & 5 & 3 \end{vmatrix}$ and note that the a_{11} entry is 2; the a_{21} entry is 1; and the a_{31} entry is 6. Find the minors of 2, 1, and 6.

Solution   The minor of 2 is $\begin{vmatrix} -3 & -1 \\ 5 & 3 \end{vmatrix} = (-3)(3) - (5)(-1) = -4$

The minor of 1 is $\begin{vmatrix} -4 & 0 \\ 5 & 3 \end{vmatrix} = (-4)(3) - (5)(0) = -12$

The minor of 6 is $\begin{vmatrix} -4 & 0 \\ -3 & -1 \end{vmatrix} = (-4)(-1) - (-3)(0) = 4$

The **cofactor** of an entry a_{ij} is $(-1)^{i+j}$ times the minor of the a_{ij} entry, This says that if the sum of the row and column numbers is even, the cofactor is the same as the minor. If the sum of the row and column numbers of an entry is odd, the cofactor of that entry is the opposite of its minor.

Example 3   Finding cofactors

Find the cofactor of the a_{45} entry (which is a 2) for the determinant:

$$\begin{vmatrix} 1 & 32 & -3 & 4 & 5 \\ -6 & 7 & 8 & 9 & 10 \\ 11 & -10 & 1 & -8 & -7 \\ -6 & -5 & -9 & 3 & 2 \\ 1 & 0 & 3 & 5 & 12 \end{vmatrix}$$

Solution   Locate the a_{45} entry mentally; delete the entries in the 4th row, 5th column:

$$\begin{vmatrix} 1 & 32 & -3 & 4 & 5 \\ -6 & 7 & 8 & 9 & 10 \\ 11 & -10 & 1 & -8 & -7 \\ -6 & -5 & -9 & 3 & 2 \\ 1 & 0 & 3 & 5 & 12 \end{vmatrix}$$

Since $4 + 5$ is odd, the sign is negative; therefore, the cofactor of 2 is:

$$-\begin{vmatrix} 1 & 32 & -3 & 4 \\ -6 & 7 & 8 & 9 \\ 11 & -10 & 1 & -8 \\ 1 & 0 & 3 & 5 \end{vmatrix}$$

We know how to evaluate a 2×2 determinant, but we have not yet evaluated higher-order determinants, such as the one shown in Example 3.

DETERMINANT OF ORDER n   A **determinant of order n** is a real number whose value is the sum of the products obtained by multiplying each element of a row (or column) by its cofactor.

Example 4   Evaluating determinants of order 3

Evaluate $\begin{vmatrix} 1 & 4 & -1 \\ -2 & 0 & 2 \\ 3 & 1 & 2 \end{vmatrix}$ by *expanding* it about the first row.

Solution $\begin{vmatrix} 1 & 4 & -1 \\ -2 & 0 & 2 \\ 3 & 1 & 2 \end{vmatrix} = 1\begin{vmatrix} 0 & 2 \\ 1 & 2 \end{vmatrix} - 4\begin{vmatrix} -2 & 2 \\ 3 & 2 \end{vmatrix} + (-1)\begin{vmatrix} -2 & 0 \\ 3 & 1 \end{vmatrix}$

$= 1(0 - 2) - 4(-4 - 6) + (-1)(-2 - 0)$

$= -2 + 40 + 2$

$= 40$

The value of the determinant in Example 4 is the same regardless of the row or column that is chosen for evaluation. Try at least one other row or column of Example 4 to show that you obtain the same value of 40.

The method of evaluating determinants by rows or columns is not very efficient for higher-order determinants. The following theorem considerably simplifies the work in evaluating determinants.

DETERMINANT REDUCTION   If $|A'|$ is a determinant obtained from a determinant $|A|$ by multiplying any row by a constant k and adding the result to any other row (entry by entry), then $|A'| = |A|$. The same result holds for columns.

Example 5   Evaluating a determinant using reduction

Expand $\begin{vmatrix} 1 & -2 & -5 \\ 2 & -1 & 0 \\ -4 & 5 & 6 \end{vmatrix}$ by using determinant reduction.

Solution   We wish to obtain some row or column with two zeros. Add twice the second column to the first column, replacing the original first column with the result.

$$\begin{vmatrix} 1 & -2 & -5 \\ 2 & -1 & 0 \\ -4 & 5 & 6 \end{vmatrix} = \begin{vmatrix} -3 & -2 & -5 \\ 0 & -1 & 0 \\ 6 & 5 & 6 \end{vmatrix}$$

Next, expand about the second row (do not even bother to write down the products that are zero):

$$\begin{vmatrix} -3 & -2 & -5 \\ 0 & -1 & 0 \\ 6 & 5 & 6 \end{vmatrix} = \overset{\substack{position\ sign \\ \downarrow \\ +}}{} \overset{\substack{entry \\ \downarrow \\ (-1)}}{} \begin{vmatrix} -3 & -5 \\ 6 & 6 \end{vmatrix}$$

Now that we have a 2×2 determinant, we finish off the multiplication:

$$+ (-1) \begin{vmatrix} -3 & -5 \\ 6 & 6 \end{vmatrix} = -(-18 + 30) = -12$$

Example 6   Evaluating a higher-order determinant

Evaluate $\begin{vmatrix} 2 & -3 & 2 & 5 & 0 \\ 4 & 2 & -1 & 4 & 0 \\ 5 & 1 & 0 & -2 & 0 \\ 6 & 2 & 3 & 6 & 0 \\ 3 & 4 & 6 & 1 & -2 \end{vmatrix}$ using determinant reduction.

Solution   First, notice that all the entries in the fifth column except one are zeros, so begin by expanding about the fifth column. The nonzero entry is located in position a_{55} and $5 + 5 = 10$, which is even, so the leading coefficient is $+1$:

$$\begin{vmatrix} 2 & -3 & 2 & 5 & 0 \\ 4 & 2 & -1 & 4 & 0 \\ 5 & 1 & 0 & -2 & 0 \\ 6 & 2 & 3 & 6 & 0 \\ 3 & 4 & 6 & 1 & -2 \end{vmatrix} = +(-2) \begin{vmatrix} 2 & -3 & 2 & 5 \\ 4 & 2 & -1 & 4 \\ 5 & 1 & 0 & -2 \\ 6 & 2 & 3 & 6 \end{vmatrix}$$

Next, obtain a row or column with all entries zero, except one. We work toward obtaining zeros in column 3. Multiply row 2 by 2 and add it to the first row. Multiply row 2 by 3 and add it to the fourth row.

$$= -2 \begin{vmatrix} 10 & 1 & 0 & 13 \\ 4 & 2 & -1 & 4 \\ 5 & 1 & 0 & -2 \\ 18 & 8 & 0 & 18 \end{vmatrix}$$ *Expand along the third column.*

$$= -2 \left[-(-1) \begin{vmatrix} 10 & 1 & 13 \\ 5 & 1 & -2 \\ 18 & 8 & 18 \end{vmatrix} \right]$$ *Now, we work toward obtaining zeros in the middle row.*

$$= (-2) \begin{vmatrix} 5 & 1 & 15 \\ 0 & 1 & 0 \\ -22 & 8 & 34 \end{vmatrix}$$ *In the 3 by 3 matrix add −5 times the second column to the first column and then add 2 times the second column to the third column.*

$$= -2(170 + 330)$$

$$= -1,000$$

Many mathematical software packages can compute determinants of any order. For determinants of high order this type of computational tool brings a huge advantage to real world problems.

APPENDIX G: ANSWERS TO SELECTED PROBLEMS

Chapter 1

1.1 What is Calculus?, page 11

1. limit, derivative, and integral   **3.** The assumptions are continually refined.

9. $\frac{1}{3}$ **11.** 1   **13.** 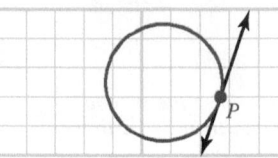   **15.**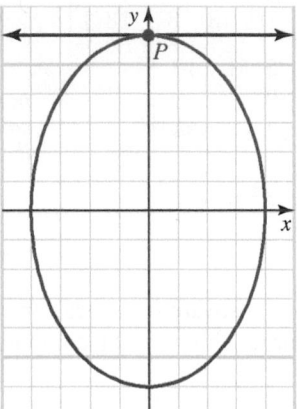

17. 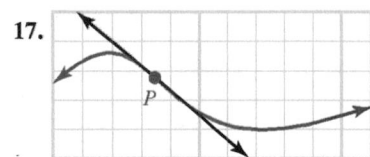   **19.** 16 inches   **21.** 62.5 pounds

23. 20 pounds   **25.** 13°C

27. 240   **29.** $3,685   **31.** $405,045

33. $25   **35.** 2   **37.** 1   **39.** 0   **41.** 3 square units   **43.** 6 or 7 square units   **45.** 6 square units   **47. a.** 0.40
b. 0.37   **c.** $\frac{1}{3}$ **49.** 21°C   **51. a.** 120   **b.** 11.88 years   **53.** 2,000 lb/in.^{2}

1.2 Preliminaries, page 25

1. a. $[-3, 4]$ **b.** $3 \le x \le 5$ **c.** $-2 \le x < 1$ **d.** $(2, 7]$

3. a. 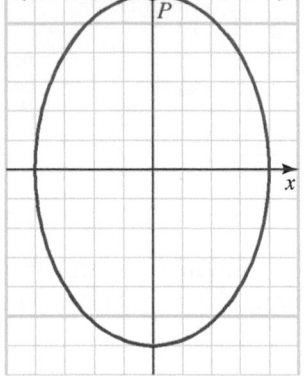   **b.**   **c.**

d.   **5. a.** 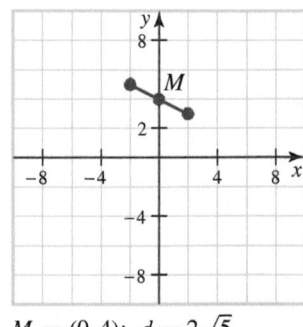   **b.**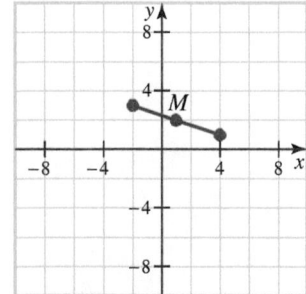

$M = (0, 4); \ d = 2\sqrt{5}$ $M = (1, 2); \ d = 2\sqrt{10}$

7. $x = 0, 1$ **9.** $x = \dfrac{b \pm \sqrt{b^2 + 12c}}{6}$ **11.** $w = -2, 5$ **13.** $\emptyset$ **15.** $x = \frac{7\pi}{6}, \frac{11\pi}{6}$ **17.** $x = \frac{3\pi}{4}, \frac{5\pi}{4}, \frac{\pi}{3}, \frac{5\pi}{3}$

19. $\left(-\infty, -\frac{5}{3}\right)$ **21.** $\left(-\frac{5}{3}, 0\right)$ **23.** $(-8, -3]$ **25.** $[-1, 3]$ **27.** $[7.999, 8.001]$ **29.** $(x + 1)^2 + (y - 2)^2 = 9$

31. $x^2 + (y - 1.5)^2 = 0.0625$ **33.** 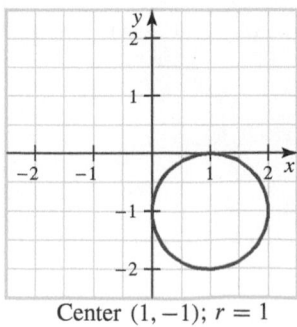   **35.**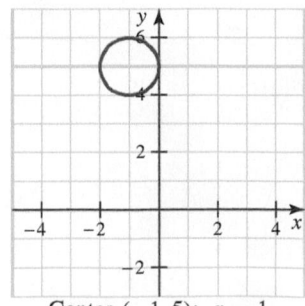

Center $(1, -1); \ r = 1$ Center $(-1, 5); \ r = 1$

37. $\dfrac{\sqrt{2}-\sqrt{6}}{4}\approx-0.2588$ **39.** $2-\sqrt{3}\approx 0.2679$

45. a.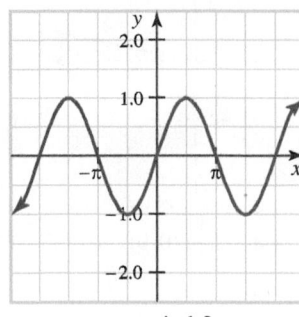
period 2π

b.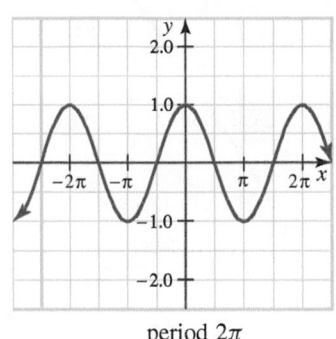
period 2π

c.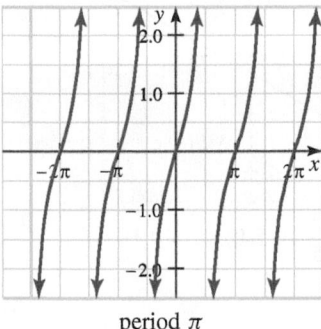
period π

47. a.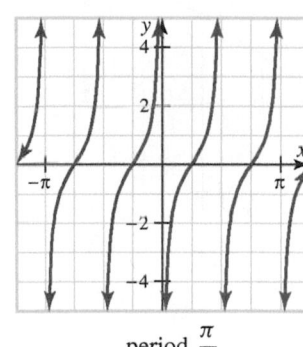
period $\dfrac{\pi}{2}$

b.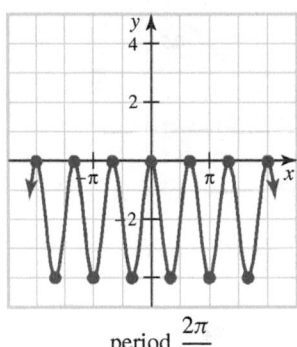
period $\dfrac{2\pi}{3}$

49.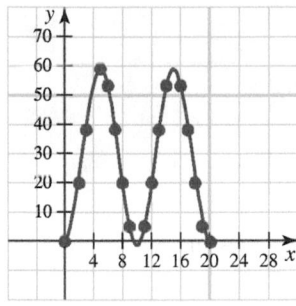
$A=29, B=30, C=\dfrac{\pi}{5}, D=5$

51. $x=\dfrac{\sqrt{2}\pm\sqrt{4\sqrt{2}-2}}{2},\quad \dfrac{-\sqrt{2}\pm\sqrt{2+4\sqrt{2}}\,i}{2}$

1.3 Lines in the Plane; Parametric Equations, page 33

3. $2x+y-5=0$ **5.** $2y-1=0$ **7.** $x+2=0$ **9.** $8x-7y-56=0$ **11.** $3x+y-5=0$
13. $3x+4y-1=0$ **15.** $4x+y+3=0$

17.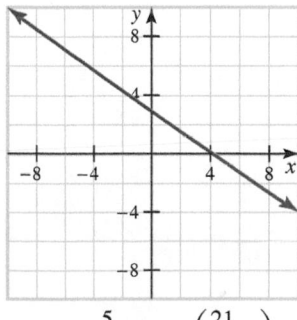
$m=-\dfrac{5}{7}; (0,3), \left(\dfrac{21}{5},0\right)$

19.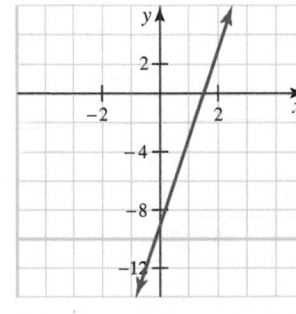
$m=6.001; (1.50025,0), (0,-9.003)$

21.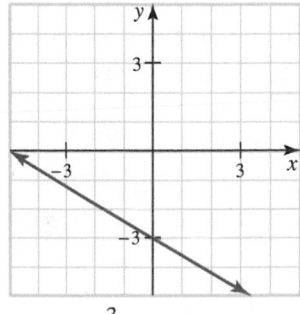
$m=-\dfrac{3}{5}; (-5,0), (0,-3)$

23.

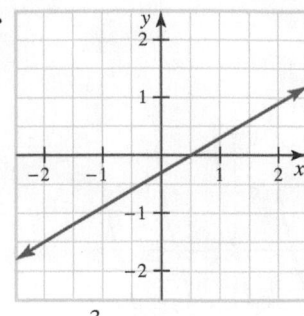

$m = \dfrac{3}{5}; (0.5, 0), (0, -0.3)$

25.

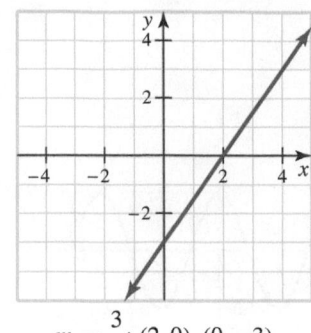

$m = \dfrac{3}{2}; (2, 0), (0, -3)$

27.

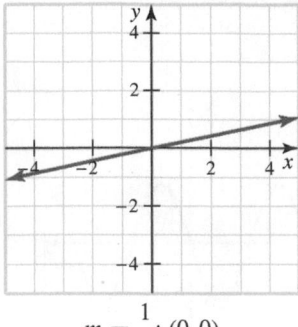

$m = \dfrac{1}{5}; (0, 0)$

29.

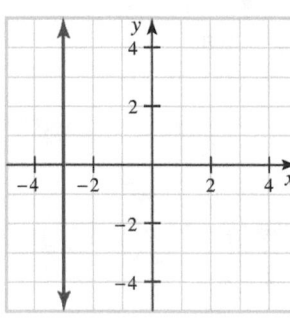

no slope; $(-3, 0)$

31.

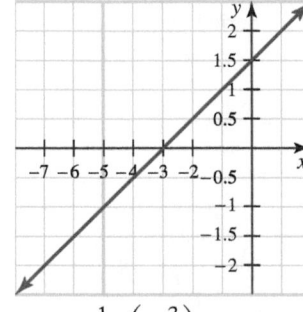

$m = \dfrac{1}{2}; \left(0, \dfrac{3}{2}\right), (-3, 0)$

33.

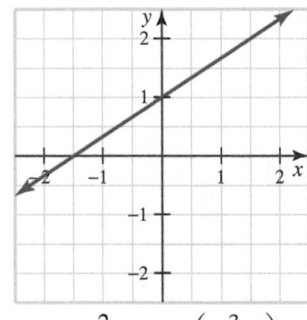

$m = \dfrac{2}{3}; (0, 1), \left(-\dfrac{3}{2}, 0\right)$

35.

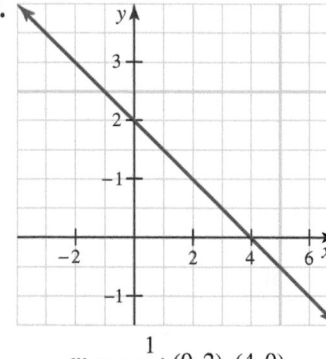

$m = -\dfrac{1}{2}; (0, 2), (4, 0)$

37.

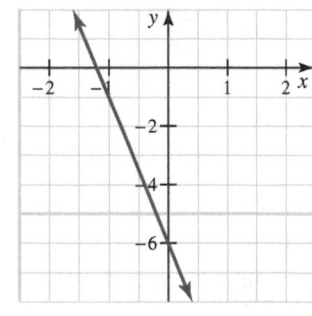

$m = -5; (0, -6), \left(-\dfrac{6}{5}, 0\right)$

39. $y = 0$ and $y = 6$

41. $D(6, 6)$, $E(2, 16)$, or $F(0, -10)$; three parallelograms can be found. **43. a.** $-38.2°F$ **b.** $-17.8°C$ **c.** $-40°$

45. a. \$40 **b.** 254 miles **47.**

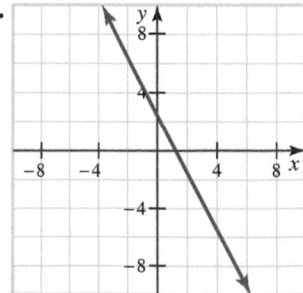

49. a. $t = \dfrac{(A - y)}{10}$ **b.** 3 hours

51. 289 vehicles **53.** (22.6, 26.5) or (26.5, 22.6) using a calculator **59.** $m = \dfrac{3}{4}$

1.4 Functions and Graphs, page 48

1. $D = (-\infty, \infty)$; $f(-2) = -1$; $f(1) = 5$; $f(0) = 3$ **3.** $D = (-\infty, 0) \cup (0, \infty)$; $f(-1) = -2$; $f(1) = 2$; $f(2) = \frac{5}{2}$
5. $D = \left(\frac{1}{2}, \infty\right)$; $f(1) = 1$; $f\left(\frac{1}{2}\right)$ is undefined; $f(13) = \frac{1}{125}$ **7.** $D = (-\infty, \infty)$; $f(-1) = \sin 3 \approx 0.1411$; $f\left(\frac{1}{2}\right) = 0$;

$f(1) = \sin(-1) \approx -0.8415$ **9.** $D = (-\infty, \infty)$; $f(3) = 4$; $f(1) = 2$; $f(0) = 4$ **11.** 9 **13.** $10x + 5h$ **15.** -1 **17.** $-\dfrac{1}{x(x+h)}$

19. a. not equal **b.** equal **21. a.** not equal **b.** equal **23. a.** even **b.** even **25. a.** even **b.** even **27.** $(f \circ g)(x) = 4x^2 + 1$;
$(g \circ f)(x) = 2x^2 + 2$ **29.** $(f \circ g)(x) = \sin(2x + 3)$; $(g \circ f)(x) = 2\sin x + 3$ **31.** $u(x) = 2x^2 - 1$; $g(u) = u^4$ **33.** $u(x) = 5x - 1$;
$g(u) = \sqrt{u}$ **35.** $u(x) = x^2$; $g(u) = \tan u$ **37.** $u(x) = \sin x$; $g(u) = \sqrt{u}$ **39.** $P(5, f(5))$; $Q(x_0, f(x_0))$ **41.** $-\frac{1}{3}, 2$
43. $\pm\sqrt{10}, \pm 2\sqrt{3}, \pm 2\sqrt{5}$ **45.** $\pm 4, \pm 5$ **47.** $0, \pm\sqrt{3}$ **49. a.** 64 units **b.** 44 units **51. a.** $S(0) \approx 25.344$ cm³/s
b. $S(0.6 \times 10^{-2}) \approx 19.008$ cm³/s **53.**

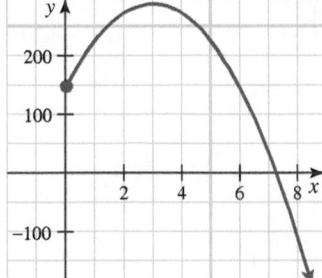

a. 144 ft **b.** $t = 3 + 3\sqrt{2} \approx 7.2$ seconds
c. 3 seconds, 288 ft

55. a. $D = (-\infty, 0) \cup (0, \infty)$ **b.** n is a positive integer **c.** 7 minutes **d.** $n = 12$ **e.** time approaches 3 minutes but never gets there
57.

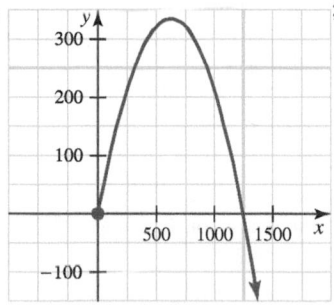

; maximum distance when $\alpha = 45°$ **59.** $\frac{5}{6}$

1.5 Inverse Functions; Inverse Trigonometric Functions, page 59

3. Yes **5.** No **7.** No **9.** No **11.** $\{(5, 4), (3, 6), (1, 7), (4, 2)\}$ **13.** $y = \frac{1}{2}x - \frac{3}{2}$ **15.** $y = \sqrt{x + 5}$ **17.** $y = (x - 5)^2$
19. $y = \dfrac{3x + 6}{2 - 3x}$ **21. a.** $\frac{\pi}{3}$ **b.** $\frac{2\pi}{3}$ **23. a.** $-\frac{\pi}{4}$ **b.** $\frac{5\pi}{6}$ **25. a.** $\frac{3\pi}{4}$ **b.** $-\frac{\pi}{4}$ **27. a.** $\frac{\pi}{3}$ **b.** $-\frac{\pi}{3}$ **29.** $\frac{\sqrt{3}}{2}$ **31.** 3 **33.** $-\frac{2\sqrt{6}}{5}$

37.

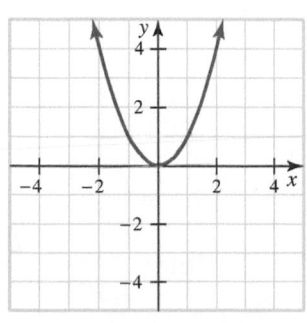

no inverse exists

39.

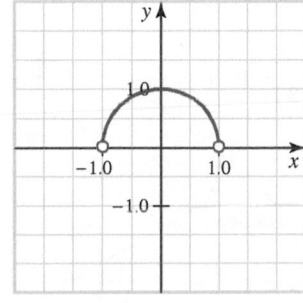

no inverse exists

41.

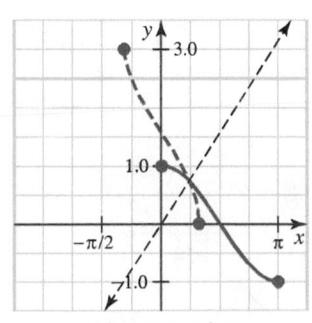

inverse exists

43. $2x\sqrt{1 - x^2}$

45. $\dfrac{\sqrt{1 - x^2}}{x}$ **47.** 1 **51. a.** 0.9805 **b.** 0.7284 **c.** 0.3964 **d.** 2.5657 **53.** $\theta = \tan^{-1}\frac{5}{x} - \tan^{-1}\frac{2}{x}$ **55. a.** 3.141592654; conjecture:
it is π **57.** $A = 0, B = -1$; $\csc^{-1}(-9.38) \approx -0.1068$

Chapter 1 Proficiency Examination, page 62

18. a. $6x + 8y - 37 = 0$ **b.** $3x + 10y - 41 = 0$ **c.** $3x - 28y - 12 = 0$ **d.** $2x + 5y - 24 = 0$ **e.** $4x - 3y + 8 = 0$

19.
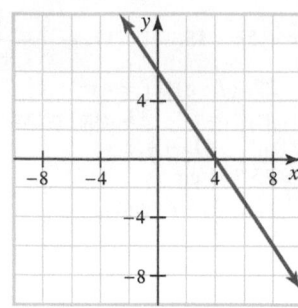

20.
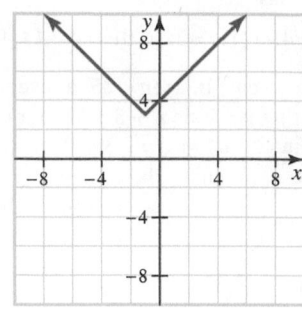

21.
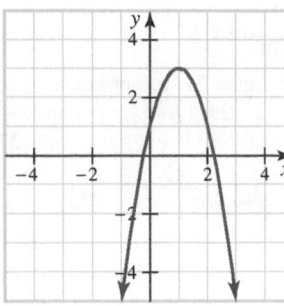

22.
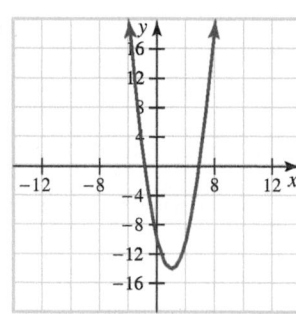

23.
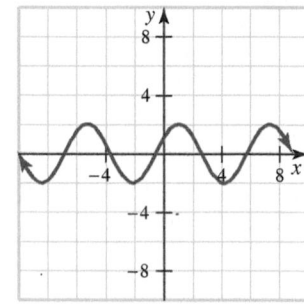

24.
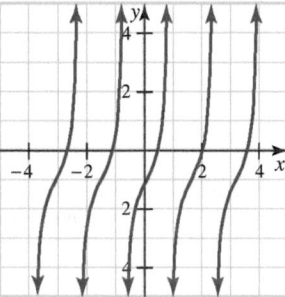

25.
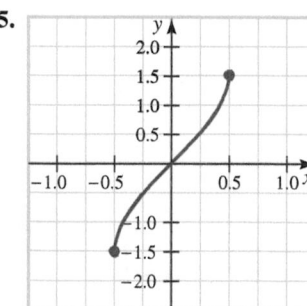

26.
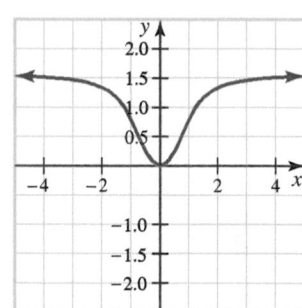

27. $-\frac{3}{2}, 1$

28. The two functions are not the same.

29. $(f \circ g)(x) = \sin\sqrt{1 - x^2}$; $(g \circ f)(x) = |\cos x|$

30. $V = \frac{2}{3}x(12 - x^2)$

Chapter 1 Supplementary Problems, page 63

1. $y = \frac{4}{5}x - 5$; $m = \frac{4}{5}$; $b = -5$; $a = \frac{25}{4}$ **3.** $\frac{x}{5} + \frac{y}{3} = 1$; $m = -\frac{3}{5}$; $b = 3$; $a = 5$

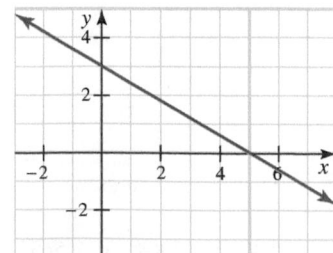

5. $\left(\frac{5}{8}, \infty\right)$ **7.** $(-\infty, -2] \cup [3, \infty)$ **9.** $P = 30$; $A = 30$ **11.** $P \approx 27.1$; $A = 40$ **13.** $5x - 3y + 3 = 0$ **15.** $(x - 5)^2 + (y - 4)^2 = 16$
17. $\{x \neq 0\}$; $2, -2, 2$; none are zeros   **19.** $(-\infty, \infty)$; $-1, -1, -1$; none are zeros   **21.** 5   **23.** $\dfrac{-1}{2x(x + h)}$

25. $3x^2, 9x^2$ **27.** u, u **29.** $x = \dfrac{-13 \pm \sqrt{409}}{12}$ **31.** $x = -2, -1, 1, 2$ **33.** does not exist   **35.** $y = -\sqrt{x-1}, x > 1$

37.

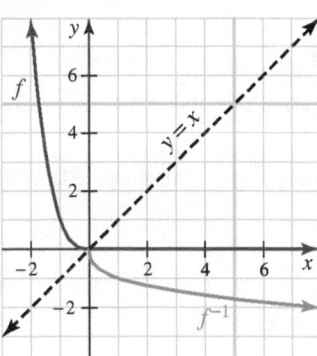

39.

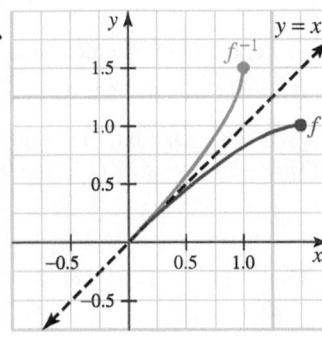

41. $\sqrt[3]{\frac{1}{2}(x+7)}$ **43.** $\sin^{-1} x^2$

45. $2x^2 - 1$ **47.** 0 **49.** $\frac{5\pi}{6}$ **51.** $A = -\frac{1}{4}$ and $B = \frac{3}{4}$ **53. a.** function; not one-to-one; D: $(-\infty, \infty)$; R: $[0, \infty)$
b. not a function   **c.** function; one-to-one; D: $(-\infty, \infty)$; R: $(-\infty, \infty)$ **d.** not a function   **e.** function; not one-to-one;
D: $(-\infty, -2) \cup (-2, 2) \cup (2, \infty)$; R: $(-\infty, 1) \cup (1, \infty)$ **55.** The altitudes meet at $\left(\frac{34}{23}, \frac{3}{23}\right)$. The medians meet at $\left(2, \frac{5}{3}\right)$.
57. 4,000; 5,200; in 3 years   **59. a.** $\frac{\sqrt{11}}{4}$ **b.** $\frac{3}{5}$ **c.** $\frac{63}{65}$ **63.** $\dfrac{x+1}{x-1}$; $(-\infty, 1) \cup (1, \infty)$ **65. a.** False; $\tan^{-1} 1 = \frac{\pi}{4}$,
but $\dfrac{\sin^{-1} 1}{\cos^{-1} 1}$ is not defined.   **b.** False; $\tan^{-1} 1 = \frac{\pi}{4}$, but $\dfrac{1}{\tan 1} = \cot 1 \approx 0.64209$.   **c.** true   **d.** true   **e.** true   **67. a.** about 3.3562 m
b. about $59.39°$ **69. a.** $(-\infty, 300) \cup (300, \infty)$ **b.** $0 \le x \le 100$ since x represents a percentage   **c.** 120 worker-hours   **d.** 300
e. 60%   **71. a.** $V = \frac{256}{3}\pi$; $S = 64\pi$ **b.** $V = 30$; $S = 62$ **c.** $V = 16\pi$; $S = 24\pi$ **d.** $V = 15\pi$; $S = 3\pi\sqrt{34}$
75. $c = -\frac{4}{5}$; $(-2, 0), (2, 0)$

77.

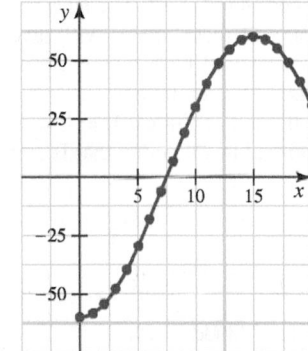

$A = 0, B = 60, C = \dfrac{\pi}{15}, D = 7.5$

79. $\frac{\pi}{4} - \tan^{-1}\frac{5}{12} \approx 0.3906$

81.

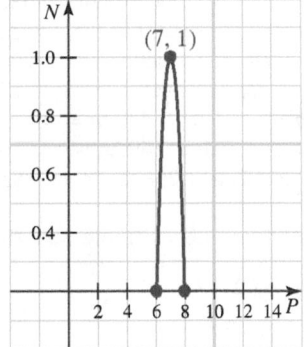

a. $6 < p < 8$ **b.** $N(7) = 1$ thousand dollars

83.

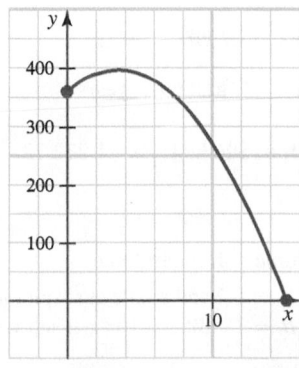

The glass should be taken in on the 11th
or 12th day

85. $C(x) = 4x^2 + \dfrac{1000}{x}$

97. $f(x) = x$; This is problem 5 in the morning session for 1960.

Chapter 2

2.1 The Limit of a Function, page 83

1. a. 0   **b.** 2   **c.** 6   **3. a.** 2   **b.** 7   **c.** 7.5   **5. a.** 6   **b.** 6   **c.** 6   **7.** 15   **9.** does not exist   **11.** $\frac{2}{3}$ **13.** $\lim\limits_{x \to 2} g(x) = 4$
15. $\lim\limits_{x \to 0^+} F(x) = 0$ **17.** $\lim\limits_{x \to 2\pi} t(x) = 1$ **19.** 0.00   **21.** 0.00   **23.** -0.50 **25.** does not exist   **27. a.** does not exist   **b.** -0.32
29. a. 0.24   **b.** 0.00   **31.** 3.14   **33.** 0.00   **35.** -1.00 **37.** 0.00   **39.** 1.00   **41.** 3.00   **43.** does not exist   **45.** does not exist
47. does not exist   **49. a.** $-32t + 40$ **b.** 40 ft/s   **c.** $v(3) = -56$ ft/s   **d.** $V(1.25) = 0$ at highest point of trajectory   **51.** 228

2.2 Algebraic Computation of Limits, page 94

1. -9 **3.** -8 **5.** $-\frac{1}{2}$ **7.** 2   **9.** $\frac{\sqrt{3}}{9}$ **11.** 4   **13.** -1 **15.** $\frac{1}{2}$ **17.** $\frac{1}{2}$ **19.** 0   **21.** 2   **23.** $\frac{5}{2}$ **25.** 0   **27.** 0   **29.** 0   **35.** -1 **37.** 0
39. the limit does not exist   **41.** the limit does not exist   **51.** the limit does not exist   **53.** 4   **55.** 8

2.3 Continuity, page 104

1. Temperature is continuous, so TEMPERATURE $= f$(TIME) would be a continuous function. The domain could be midnight to midnight say, $0 \le t \le 24$.   **3.** The domain of the function will be integers (end of each day), so there is no way to examine what happens since the elements of the domain are not discrete; i.e., the difference between two elements of the domain are not close to 0. Thus the function is not continuous.   **5.** The charges (range of the function) consist of rational numbers only (dollars and cents to the nearest cent), so the function CHARGE $= f$(MILEAGE) would be a step function (that is, not continuous). The domain would consist of the mileage from the beginning of the trip to its end.   **7.** There are no suspicious points and no points of discontinuity with a polynomial.   **9.** The denominator factors to $x(x-1)$, so suspicious points would be $x = 0, 1$. There will be a hole discontinuity at $x = 0$ and a pole discontinuity at $x = 1$.   **11.** $x = 0$ is a suspicious point and is a point of discontinuity.   **13.** $t = 0$ and $t = -1$ are suspicious points; when the fraction is simplified it becomes $\dfrac{1 - 2t}{t(t+1)}$, which is discontinuous at both points.   **15.** $x = 1$ is a suspicious point; there are no points of discontinuity.   **17.** The sine and cosine are continuous on the reals, but the tangent is discontinuous at $x = (n + \frac{1}{2})\pi$, n any integer. Each of these values will have a pole discontinuity.   **19.** The function has suspicious points at $x = n\pi$, n any integer. Each of these values will have a pole discontinuity.
21. 3   **23.** π **25.** no value works   **27. a.** continuous   **b.** discontinuous   **29.** discontinuous   **31.** continuous   **39.** $a = b = 2$
41. $a = 1; b = -18/5$ **43.** $a = b = 5$ **45.** $a = 0; b = 3$ **47.** $(f+g)(x) = 4x$ if $x \ne 0$ and $(f+g)(0) = 0 = 4 \cdot 0$, so $(f+g)(x) = 4x$ for all x **49.** Answers vary, $f(x) = \frac{1}{2}$ for x rational and $f(x) = 1 - x$ for x irrational is a function that is continuous only at $x = \frac{1}{2}$.   **57. a.** 1.25872   **b.** 0.785398

2.4 Exponential and Logarithmic Functions, page 119

1.

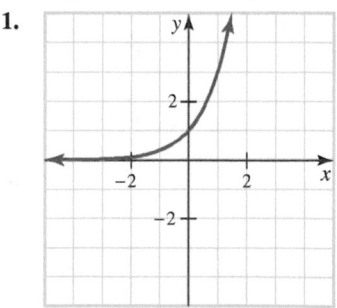

3.

5. 0   **7.** 2   **9.** 32   **11.** 1   **13.** 125892.54

15. ± 90.0177313 **17.** 0.322197023   **19.** 4

21. 2, -1 **23.** 3   **25.** 2, $-\frac{5}{3}$ **27.** $-\frac{3}{2}$ **29.** 1

31. a. 0   **b.** e^{-1} **33. a.** e **b.** 2   **35.** 27   **37.** 0.4

39. logarithmic   **41.** exponential

43. 3 m below surface   **45. a.** $0.69/r$ years

47. 6.9%   **49. a.** $1,000.2^{(-1/3)}$; $\frac{1}{60}$ **b.** 2 hr 39 min   **51. a.** 30.12%   **b.** 77.69%   **c.** 7.81%   **53. a.** $263.34   **b.** $1,232.72

Chapter 2 Proficiency Examination, page 122

19. $\frac{3}{2}$ **20.** $-\frac{1}{4}$ **21.** $\frac{1}{4}$ **22.** $\frac{9}{5}$ **23.** 0   **24.** -1 **25. a.** exponential   **b.** exponential   **c.** logarithmic   **d.** logarithmic
26. There are poles at both suspicious points $t = 0$ and -1.   **27.** There is a pole at suspicious point $x = -2$ and a hole at suspicious point $x = 1$.   **28. a.** 11 years, 3 quarters   **b.** 11 years, 6 months   **c.** 11 years, 166 days   **29.** $A = -1; B = 1$ **30.** $f(x) = x + \sin x - \frac{1}{\sqrt{x+3}}$ is continuous on $[0, \infty)$ with $f(0) < 0$ and $f(\pi) > 0$. Root location theorem gives result.

Chapter 2 Supplementary Problems, page 123

1. π **3.** 31   **5.** $216\sqrt{2}$ **7.** 1.50   **9.** 9,783.23   **11.** 2.81   **13.** $\frac{5}{2}$ **15.** ± 2 **17.** 16   **19.** $\sqrt{13}$ **21.** $x = 6, x = 3$ **23.** 5   **25.** $\frac{1}{9}$
27. 2   **29.** $\frac{3}{2}$ **31.** -1 **33.** $\frac{1}{2e}$ **35.** e^4 **37.** 5   **39.** $\frac{3}{2}$ **41.** $\sin 1$ **43.** 1   **45.** 3   **47.** 0   **49.** $\frac{1}{\sqrt{2x}}$ **51.** $-\frac{4}{x^2}$ **53.** $-\sin x$ **55.** $\frac{1}{x}$

57. $\sec^2 x$ **59.** $\frac{1}{x \ln 2}$ **61.** pole at $x = 8$ is not removable   **63.** continuous on $[-5, 5]$ **65.** continuous on $[-5, 5]$
67. a. continuous on $[0, 5]$ **b.** discontinuous at $x = -2$ **c.** continuous on $[-5, 5]$

69. a. 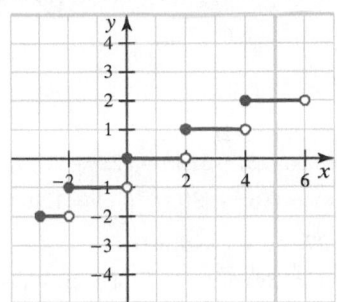   **b.** 1 **c.** The limit exists for any number that is not an even integer.

71. $A = 2, B = 4$ **73.** $-6; 4$ **75. a.**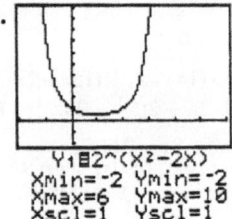

b. It cross the y-axis at $(0, 1)$. As $x \to \infty$, $y \to \infty$. As $x \to -\infty$, $y \to \infty$ **c.** The smallest value of $y = E$ is $E = 0.5$.
77. a. $k = 0.25 \ln 2$, approximately 29.7%   **b.** approximately 82.3%   **c.** approximately 7.96%   **85.** Answers vary.
95. a. The wind chill for 20 mi/h is $3.75°$ and for 50 mi/h is $-7°$.   **b.** $v \approx 25.2$ **c.** at $v = 4$, $T \approx 91.4$; at $v = 45$, $T \approx 868$
99. This is problem A1 in the morning session of 1956.

Chapter 3

3.1 An Introduction to the Derivative: Tangents, page 141

1. Some describe it as a five-step process:

   1. Find $f(x)$
   2. Find $f(x + \Delta x)$
   3. Find $f(x + \Delta x) - f(x)$
   4. Find $(f(x + \Delta x) - f(x))/\Delta x$
   5. Find $\lim_{\Delta x \to 0} (f(x + \Delta x) - f(x))/\Delta x$

3. Continuity does not imply differentiability but differentiability implies continuity.

5. 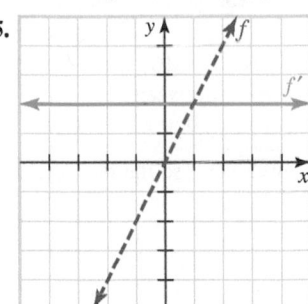   **7.** 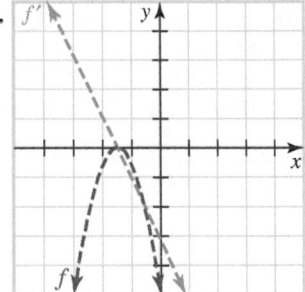   **9.**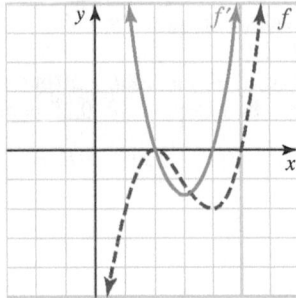

11. a. 0   **b.** 0   **13. a.** 2   **b.** 2   **15. a.** $-\Delta x$ **b.** 0   **17.** 0; differentiable for all x **19.** 3; differentiable for all x
21. $6t$; differentiable for all t **23.** $-1/(2x^2)$; differentiable for all $x \neq 0$ **25.** $2x - y + 3 = 0$ **27.** $3s - 4t + 1 = 0$
29. $x - y - 1 = 0$ **31.** $x + 3y - 15 = 0$ **33.** $125x - 5y - 249 = 0$ **35.** 2   **37.** 0   **39. a.** -3.9 **b.** -4

41. The derivative is 0 when $x = \frac{1}{2}$; the graph has a horizontal tangent at $\left(\frac{1}{2}, -\frac{1}{4}\right)$.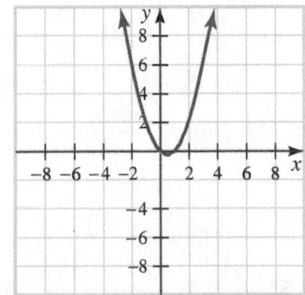

43. The limit of the difference quotient from the left is -1 and from the right is $+1$, so the limit does not exist.   **45.** yes   **47.** 4
49. $\frac{1}{2}$ **51.** $\frac{1}{4}$ **53. a.** $2x$; same graph, but shifted vertically   **b.** $2x$ **55.** For $x < 0$ the tangent lines have negative slope and for

---

$x > 0$ the tangent lines have positive slope. There is a corner at $x = 0$ and there is no tangent at $x = 0$, so there is no derivative at that point. **57. a.** $x_1 = x_0 + E$, $y_1 = y_0(1 + E/A)$ **b.** $E(3x_0^2 + \frac{3y_0^3}{A} - \frac{2x_0 y_0}{A} - 2y_0) + E^2(3x_0 + \frac{3y_0^3}{A^2} - \frac{2y_0}{A}) + E^3(1 + \frac{y_0^3}{A^3}) + 0 = 0$
c. $A = -\frac{3y^3 - 2xy}{3x^2 - 2y}$ **d.** See Figure 3.14 in textbook. $y - 0.9 = 0.7(x - 0.5)$ **59.** $(0, -Ac^2)$

3.2 Techniques of Differentiation, page 152

1. (11) 0 (12) 1 (13) 2 (14) $4x$ (15) $-2x$ (16) $-2x$ **3.** (25) 2 (26) $6x$ (27) $3s^2$ (28) $-2t$ (29) 1 **5. a.** $12x^3$
b. -1 **7. a.** $3x^2$ **b.** 1 **9.** $2t + 2t^{-3} - 20t^{-5}$ **11.** $-14x^{-3} + \frac{2}{3}x^{-1/3}$ **13.** $1 - x^{-2} + 14x^{-3}$ **15.** $-32x^3 - 12x^2 + 2$
17. $\frac{22}{(x+9)^2}$ **19.** $4x^3 + 12x^2 + 8x$ **21.** $f'(x) = 5x^4 - 15x^2 + 1$; $f''(x) = 20x^3 - 30x$; $f'''(x) = 60x^2 - 30$; $f^{(4)}(x) = 120x$
23. $f'(x) = 4x^{-3}$; $f''(x) = -12x^{-4}$; $f'''(x) = 48x^{-5}$; $f^{(4)}(x) = -240x^{-6}$ **25.** $18x - 14$ **27.** $7x + y + 9 = 0$
29. $6x + y - 6 = 0$ **31.** $x - 6y + 5 = 0$ **33.** $(1,0)$ and $(\frac{4}{3}, -\frac{1}{27})$ **35.** $(\frac{3}{2}, \frac{4}{27})$ **37.** $(9,6)$ **39. a.** $4x - 5$ **41.** $2x - y - 2 = 0$
43. a. $4x + y - 1 = 0$ **b.** $x = 0$ **45.** $(0,0)$ and $(4,64)$ **47.** The equation is not satisfied. **49.** The equation is not satisfied.
51. If n is the degree of the polynomial, $P^{(n+1)}(x) = 0$ for every value of x.

3.3 Derivatives of Trigonometric, Exponential, and Logarithmic Functions, page 160

1. $\cos x - \sin x$ **3.** $2t - \sin t$ **5.** $\sin 2t$ **7.** $-\sqrt{x}\sin x + \frac{1}{2}x^{-\frac{1}{2}}\cos x - x\csc^2 x + \cot x$ **9.** $-x^2\sin x + 2x\cos x$
11. $\frac{x\cos x - \sin x}{x^2}$ **13.** $e^t \csc t(1 - \cot t)$ **15.** $x + 2x\ln x$ **17.** $2e^x \sin x$ **19.** $e^{-x}(\cos x - \sin x)$ **21.** $\frac{\sec^2 x - 2x\sec^2 x + 2\tan x}{(1 - 2x)^2}$
23. $\frac{t\cos t + 2\cos t - 2 - \sin t}{(t+2)^2}$ **25.** $\frac{1}{\cos x - 1}$ **27.** $\frac{2\cos x - \sin x - 1}{(2 - \cos x)^2}$ **29.** $\frac{-2}{(\sin x - \cos x)^2}$ **31.** $-\sin x$ **33.** $-\sin\theta$
35. $2\sec^2\theta\tan\theta$ **37.** $\sec^3\theta + \sec\theta\tan^2\theta$ **39.** $-\sin x - \cos x$ **41.** $-2e^x \sin x$ **43.** $-\frac{1}{4}t^{-3/2}\ln t$ **45.** $4x - 2y - \pi + 2 = 0$
47. $\sqrt{3}x - 2y + \left(1 - \frac{\sqrt{3}\pi}{6}\right) = 0$ **49.** $2x - y = 0$ **51.** $x - y + 1 = 0$ **53.** $A = -\frac{1}{2}$, $B = \frac{1}{2}$ **55.** One example is $f(x) = x^2 \sin\frac{1}{x}$
for $x \neq 0$, $f(0) = 0$.

3.4 Rates of Change: Modeling Rectilinear Motion, page 170

3. 1 **5.** $\frac{1}{2}$ **7.** $\frac{141}{2}$ **9.** 1 **11.** 0 **13.** -6 **15. a.** $2t - 2$ **b.** 2 **c.** 2 **d.** always accelerating **17. a.** $3t^2 - 18t + 15$ **b.** $6t - 18$
c. 46 **d.** decelerating on $[0,3)$; accelerating on $(3,6]$ **19. a.** $-2t^{-2} - 2t^{-3}$ **b.** $4t^{-3} + 6t^{-4}$ **c.** $\frac{20}{9}$ **d.** accelerating on $[1,3]$
21. a. $-3\sin t$ **b.** $-3\cos t$ **c.** 12 **d.** decelerating on $[0, \frac{\pi}{2})$ and on $(\frac{3\pi}{2}, 2\pi]$; accelerating on $(\frac{\pi}{2}, \frac{3\pi}{2})$ **23.** quadratic model
25. exponential model **27.** quadratic model **29.** cubic model **31.** logarithmic model **33. a.** -6 **b.** The decline will be the same
each year. **35.** 136 units **37. a.** 9.945 m/min **b.** 9.960 m **39. a.** 64 ft/s **b.** 336 ft **c.** $-32t + 64$ ft/s **d.** -160 ft/s **41.** 30 ft
43. 144 ft **45.** $v_0 = 24$ ft/s; 126 ft **47. a.** $200t + 50\ln t + 450$ newspapers per year **b.** 1,530 newspapers per year **c.** 1,635
newspapers **49. a.** 0.2 ppm/yr **b.** 0.15 ppm **c.** 0.25 ppm **51.** 91 thousand/hr **53. a.** 20 persons per mo **b.** 0.39% per mo
55. $\frac{dP}{dT} = -\frac{4\pi\mu^2 N}{9k}T^{-2}$ **57. a.** $v(t) = -7\sin t$; $a(t) = -7\cos t$ **b.** 2π **c.** 14 **59.** $V'(x) = 3x^2 = 0.5\,S(x)$, so the rate of change
is half the surface area.

3.5 The Chain Rule, page 178

3. $6(3x - 2)$ **5.** $\frac{-8x}{(x^2-9)^3}$ **7.** $\left[\tan\left(3x + \frac{6}{x}\right) + \left(3x + \frac{6}{x}\right)\sec^2\left(3x + \frac{6}{x}\right)\right]\left[3 - \frac{6}{x^2}\right]$ **9. a.** $3u^2$ **b.** $2x$ **c.** $6x(x^2 + 1)^2$ **11. a.** $7u^6$
b. $-8 - 24x$ **c.** $-56(3x + 1)(5 - 8x - 12x^2)^6$ **13.** $25(5x - 2)^4$ **15.** $8(3x - 1)(3x^2 - 2x + 1)^3$ **17.** $4\cos(4\theta + 2)$
19. $(-2x + 3)e^{-x^2+3x}$ **21.** $e^{\sec x}\sec x\tan x$ **23.** $(2t + 1)\exp(t^2 + t + 5)$ **25.** $5x\cos 5x + \sin 5x$ **27.** $6(1 - 2x)^{-4}$
29. $\frac{1}{2}\left(\frac{x^2+3}{x^2-5}\right)^{-1/2}\left[\frac{-16x}{(x^2-5)^2}\right]$ **31.** $\frac{12x^3 + 5}{3x^4 + 5x}$ **33.** $\frac{\cos x - \sin x}{\sin x + \cos x}$ **35.** $\frac{2}{9}$ **37.** 1, 7 **39.** 0, $\frac{2}{3}$ **41. a.** 1 **b.** $\frac{3}{2}$ **c.** 1.5
43. a. $\frac{9}{40}$ **b.** $-\frac{3}{8}$ **c.** $\frac{3}{8}$ **45.** $-\frac{2}{3}$ **47.** decreasing by 6 lb/wk **49. a.** increasing by 0.035 lux/s **b.** 7.15 m **51.** $T'(2) \approx 5.23$;
it is getting hotter **53.** $I'(\theta) = \frac{2a\pi I_0}{\lambda\beta^3}\sin\beta(\beta\cos\beta - \sin\beta)\cos\frac{\theta}{\lambda}$ **55.** $6x + y + 15 = 0$ **59. a.** $\frac{d}{dx}f'[f(x)] = f''(f(x))f'(x)$ and
$\frac{d}{dx}f[f'(x)] = f'(f'(x))f''(x)$

3.6 Implicit Differentiation, page 190

1. $-\frac{x}{y}$ **3.** $-\frac{y}{x}$ **5.** $-\frac{y^2}{x^2}$ **7.** $\frac{2x - y\sin xy}{x\sin xy}$ **9.** $(2e^{2x} - x^{-1})y$ **11. a.** $-\frac{2x}{3y^2}$ **b.** $\frac{-2x}{3(12 - x^2)^{2/3}} = -\frac{2x}{3y^2}$ **13. a.** y^2
b. $\frac{1}{(x-5)^2} = y^2$ **15.** $\frac{1}{\sqrt{-x^2 - x}}$ **17.** $\frac{x}{(x^2 + 2)\sqrt{x^2 + 1}}$ **19.** $\frac{-1}{\sqrt{e^{-2x} - 1}}$ **21.** $\frac{-1}{x^2 + 1}$ **23.** $\frac{2y}{(1 - y^2)^{-1/2} + 1 - 2x}$

25. $2x - 3y + 13 = 0$ **27.** $(\pi + 1)x - y + \pi = 0$ **29.** $y = 0$ **31.** $y' = 0$ **33.** $y' = \frac{5}{4}$ **35.** $x - 1 = 0$ **37.** $y'' = -\dfrac{49}{100y^3}$

39. a. $y' = 2x$ **b.** $y' = 2^x \ln 2$ **c.** $y' = e^x$ **d.** $y' = ex^{e-1}$ **43.** $y\left[\dfrac{5x^9}{3(x^{10}+1)} + \dfrac{28x^6}{9(x^7-3)}\right]$ **45.** $y(1 + \ln x)$ **47. a.** $-\dfrac{a^2 v}{b^2 u}$

b. $-\dfrac{b^2 u}{a^2 v}$ **49.** $(4, 3)$ and $(-4, -3)$ **51.** $x + y - 2A = 0$ tangent line; $x - y = 0$ normal line **53.** $(2, 0)$, $\left(-\frac{1}{4}, \frac{\sqrt{3}}{4}\right)$, $\left(-\frac{1}{4}, -\frac{\sqrt{3}}{4}\right)$

3.7 Related Rates and Applications, page 198

1. -2 **3.** 2 **5.** $\frac{4}{5}$ **7.** $-\frac{1}{4}$ **9.** $\frac{1}{4}$ **11.** 15 **13.** 8 **15.** $\frac{4}{5}$ **17.** -20 **19.** $\sqrt{2}$ **21.** 0 **23.** -3 **25.** -3 ft/s **27.** increasing at rate of 48π in.^{2}/s **29.** 126 units/yr **31.** 54 people per year **35.** $\frac{3}{16\pi}$ in./s **37.** 2.5 ft/s **39.** 50 mi/h **41.** 1.2 m/min **43.** 30 ft/s **45.** $-\frac{5}{64\pi}$ in./min; -2.5 in.^{2}/min **47.** 0.71 ft/min **49. a.** 0.00139 atmospheres/s (approximately) **b.** after 66 seconds (approximately); 0.006125 atmospheres/s **51.** 8 ft/s **53.** $\frac{4}{15}$ rad/s **55. a.** $V(y) = 7.5y^2 + 20y$, where y is the depth of the water **b.** 3.5 in./min **57.** 1.28 ft/min **59. b.** 48π km/min

3.8 Linear Approximation and Differentials, page 211

1. $6x^2 dx$ **3.** $x^{-1/2} dx$ **5.** $(\cos x - x \sin x)\, dx$ **7.** $\dfrac{3x \sec^2 3x - \tan 3x}{2x^2}\, dx$ **9.** $\cot x\, dx$ **11.** $\frac{e^x}{x}(1 + x \ln x)\, dx$

13. $\dfrac{(x-3)(x^2 \sec x \tan x + 2x \sec x) - x^2 \sec x}{(x-3)^2}\, dx$ **15.** $\dfrac{x+13}{2(x+4)^{3/2}}\, dx$ **19.** 0.995; by calculator 0.9949874371

21. 217.69; by calculator 217.7155882 **23.** 0.06 or 6% **25.** 0.03 or 3% **27.** 0.05 parts per million **29.** reduced by $12{,}000$ units **31.** $\$6.00$; the actual increase is $\$6.15$. **33.** 0.00245 mg/cc^3 **35.** $\pm 2\%$ **37.** S increases by 2% and V increases by 3% **39.** 0.0525 ft **41.** -6.93 (or about 7) particles/unit area **43. a.** 472.7 **b.** 468.70 **45.** 1.2 units **47.** -1.4142 **49.** 0.5671 **51.** 1.27100 **53. b.** $x \approx 35.56684$ **57.** $\Delta x = -3$, $f(97) = 9.85$; by calculator 9.848857802; if $\Delta x = 16$, $f(97) \approx 9.89$ **59.** No, as any positive choice for x_0 quickly leads to a negative x_1 for $f(x) = x^{1/2}$, and negative numbers are not in the domain of $f(x)$.

Chapter 3 Proficiency Examination, page 215

17. $3x^2 + \frac{3}{2}x^{1/2} - 2\sin 2x$ **18.** $-x[\cos(3 - x^2)][\sin(3 - x^2)]^{-1/2}$ **19.** $\dfrac{-y}{x + 3y^2}$ **20.** $\frac{1}{2}xe^{-\sqrt{x}}(4 - \sqrt{x})$ **21.** $\dfrac{\ln 1.5}{x(\ln 3x)^2}$

22. $\dfrac{3}{\sqrt{1 - (3x+2)^2}}$ **23.** $\dfrac{2}{1 + 4x^2}$ **24.** $y\left[\dfrac{2x}{(x^2-1)\ln(x^2-1)} - \dfrac{1}{3x} - \dfrac{9}{3x-1}\right]$ **25.** 0 **26.** $2(2x - 3)(40x^2 - 48x + 9)$

27. $1 - 6x$ **28.** $14x - y - 6 = 0$ **29.** tangent line $y - \frac{1}{2} = \frac{\pi}{4}(x - 1)$; normal line $y - \frac{1}{2} = -\frac{4}{\pi}(x - 1)$ **30.** 2π ft^2/s

Chapter 3 Supplementary Problems, page 216

1. $4x^3 + 6x - 7$ **3.** $\dfrac{-4x}{(x^2-1)^{1/2}(x^2-5)^{3/2}}$ **5.** $\dfrac{4x - y}{x - 2}$ **7.** $\dfrac{(x^3+1)^4(46x^3+1)}{3x^{2/3}}$ **9.** $\dfrac{5\cos 5x}{2\sqrt{\sin 5x}}$ **11.** $(4x + 5)\exp(2x^2 + 5x - 3)$

13. $3^{2-x}(1 - x \ln 3)$ **15.** $\dfrac{y(1 + xye^{xy})}{x(1 - xye^{xy})}$ **17.** $e^{\sin x}(\cos x)$ **19.** $-\dfrac{2^y + y2^x \ln 2}{x2^y \ln 2 + 2^x}$ **21.** $e^{-x}\left(\dfrac{1}{2x\sqrt{\ln 2x}} - \sqrt{\ln 2x}\right)$

23. $-[\sin(\sin x)]\cos x$ **25.** $\frac{x}{4y}$ **27.** $e^{1-2x}(1 - 2x)$ **29.** $-\dfrac{\csc^2 \sqrt{x}\cot\sqrt{x}}{\sqrt{x}}$ **31.** $20x^3 - 60x^2 + 42x$ **33.** $-\dfrac{2(3y^3 + 4x^2)}{9y^5}$

35. $\dfrac{4x^2 \sin y - 2\cos^2 y}{\cos^3 y}$ **37.** $y + 3 = 0$ **39.** $2\pi x + 4y - \pi^2 = 0$ **41.** $x + y - 2 = 0$ **43.** $x - 2y - 1 = 0$ **45. a.** $-4x$

b. $y - 4 = 0$ **c.** $(\frac{2}{3}, \frac{28}{9})$ **47.** $2(1 - 2x^4)(\sin x^2) + 10x^2(\cos x^2)$ **49.** $-\dfrac{4x^3}{(x^4 - 2)^{4/3}(x^4 + 1)^{2/3}}$

51. $y' = \dfrac{x + 2y}{y - 2x}$; $y'' = \dfrac{5y^2 - 20xy - 5x^2}{(y - 2x)^3}$ **53.** Many examples exist. One such is $y = |x - 5|$. **55.** $(3, -9)$ and $(-3, -9)$

59. 7.5% per year in 2014 **61.** 0.31 ppm/yr **65.** -24 **67.** $\frac{32}{73}$ rad/s **71.** 1.87217123 **75.** 72.0625 **77.** -0.012 rad/min

79. The derivative does not exist at 0. **81.** -36 cm^2/hr **83.** 2.4×10^6 m/s^2 **85.** 1.8π mm^2/mm **87. a.** $C(x) = \dfrac{13800}{x} + 4x$

b. $-\$1.12$ approximately **89.** $\dfrac{12}{17{,}000}$ radians/s **95.** For the velocity, $\dfrac{dx}{dt} = \dfrac{12\pi x \sin\theta}{2\cos\theta - x}$; for acceleration,

$\dfrac{d^2 x}{dt^2} = \dfrac{144\pi^2 x}{(2\cos\theta - x)^3}\left[2\sin^2\theta\cos\theta + (2\cos\theta - x)(1 - \frac{1}{2}x\cos\theta)\right]$ **97.** This is Putnam Problem 1, morning session in 1946.

99. This is Putnam Problem 6, morning session in 1946.

Chapter 4

4.1 Extreme Values of a Continuous Function, page 235

1.

endpoints	critical numbers
$m = f(-3) = -34$ $M = f(3) = 26$	$x = 5$ is not in the domain.

3.

endpoints	critical numbers
$m = f(-1) = -4$ $f(3) = 0$	$M = f(0) = 0$ $m = f(2) = -4$

5.

endpoints	critical numbers
$M = f(-3) = 9$ $M = f(3) = 9$	$f(0) = 0$ $m = f(-2) = -16$ $m = f(2) = -16$

7.

endpoints	critical numbers
$m = f(-1) = -2$ $M = f(1) = 0$	$M = f(0) = 0$ $f\left(\frac{4}{5}\right) \approx -0.0819$

9.

endpoints	critical numbers
$m = h(0) = 0$ $h(2) = 2e^{-2}$	$M = h(1) = e^{-1}$

11.

endpoints	critical numbers
$M = f(-1) = 1$ $M = f(1) = 1$	$m = f(0) = 0$

13.

endpoints	critical numbers
$f(0) = 1$ $m = f(2) \approx 0.41067$	$f(0) = 1$ $M = f\left(\frac{\pi}{3}\right) = \frac{5}{4}$

15. Step 1: Find the value of the function at the endpoints of an interval. **Step 2:** Find the critical points; that is, points at which the derivative of the function is zero or undefined. **Step 3:** Find the value of the function at each critical point. **Step 4:** State the absolute extrema.

17. The maximum value is 1 and the minimum value is 0. **19.** The maximum value is 48 and the minimum value is -77. **21.** The maximum value is $\frac{353}{3}$ and the minimum value is $-\frac{863}{3}$. **23.** The function is not continuous and there is no maximum or minimum. **25.** The maximum value is $\frac{\sqrt{2}}{2}e^{-\pi/4}$ and the minimum value is $-\frac{\sqrt{2}}{2}e^{-5\pi/4}$. **27.** The maximum value is 9 and the minimum value is 5. **29.** There is no smallest value because x is negative and when $x \to 0^-$ the values get smaller without limit. **31.** The smallest value is -1. **33.** The smallest value is $\sqrt{5} - \ln\left(\frac{1+\sqrt{5}}{2}\right)$. **35.** The largest value is 0 and the smallest value is -5. **37.** The largest value is 20 and the smallest value is -20. **39.** The largest value is 0 and the smallest value is $-\sqrt[6]{108}$. **41.** The largest value is approximately 0.501 and the smallest value is 0. *Answers to* $43 - 48$ *may vary.*

43. a. $f(x) = \begin{cases} x^2 & \text{for } -1 < x < 1 \\ x+2 & \text{for } 1 \le x < 2 \end{cases}$ The minimum is 0 but there is no maximum.

b. $f(x) = \begin{cases} -x^2 & \text{for } -1 < x < 1 \\ -x+2 & \text{for } 1 \le x < 2 \end{cases}$ The maximum is 1 but there is no minimum.

c. $f(x) = \begin{cases} \sin x & \text{for } -\pi < x \le \frac{\pi}{2} \\ 0.5 & \text{for } \frac{\pi}{2} < x < 3 \end{cases}$ The maximum is 1 and the minimum is -1. **d.** $f(x) = \begin{cases} \sin x & \text{for } 0 < x < \frac{\pi}{2} \\ 0.5 & \text{for } \frac{\pi}{2} \le x < 3 \end{cases}$

There is no maximum and there is no minimum. **45. a.** $f(x) = \sin x$ on $(-\pi, 1)$. The minimum is -1 but there is no maximum. **b.** $f(x) = \sin x$ on $(0, 4)$. The maximum is 1 but there is no minimum. **c.** $f(x) = \sin x$ on $(0, 2\pi)$. The maximum is 1 and the minimum is -1. **d.** $f(x) = \sin x$ on $(-1, 1)$. There is no maximum and no minimum. **47.** On $[-1, 1]$, let $f(x) = \begin{cases} x^2 & x \ne 0 \\ 0 & x = 0 \end{cases}$

This function has no maximum. **49.** The maximum value for the velocity is -15. **51.** $x = 4$, $y = 4$ **53.** $x = \frac{20}{3}$, $y = 60$, $P = 1200^2$ **55.** $x = y = \frac{1}{4}P$ **57. a.** The greatest difference occurs at $x = \frac{1}{2}$. **b.** The greatest difference occurs at $x = 1/\sqrt{3}$. **c.** The greatest difference occurs at $x = \left(\frac{1}{n}\right)^{1/(n-1)}$.

4.2 The Mean Value Theorem, page 242

3. $c = 1$ **5.** $c = \sqrt{\frac{7}{3}}$ **7.** $c = \sqrt[3]{\frac{5}{4}}$ **9.** $c = \frac{9}{4}$ **11.** $c = \sqrt{3} - 1$ **13.** $c = \sin^{-1}\left(\frac{2}{\pi}\right)$ **15.** $c = \ln(e - 1)$ **17.** $c = \dfrac{3}{4\ln 2}$

19. $c = \dfrac{\sqrt{4 - \pi}}{\sqrt{\pi}}$ **21.** does not apply **23.** applies **25.** does not apply **27.** does not apply **29.** applies

31. $f(x) = 8x^3 - 6x + 10$ **33.** no; yes **35.** $f(x) - g(x) = 8$ **37.** $c = 2.5$ and $c = 6.25$; Rolle's theorem **41.** does not apply **43.** does not apply **47.** does not apply **55.** does not apply

4.3 Using Derivatives to Sketch the Graph of a Function, page 258

5. The blue curve is the function and the red curve is the derivative.

7. 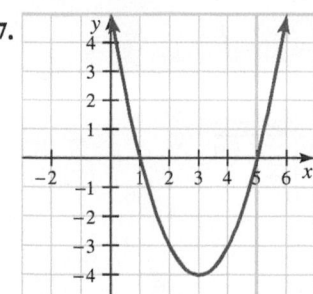   **9.**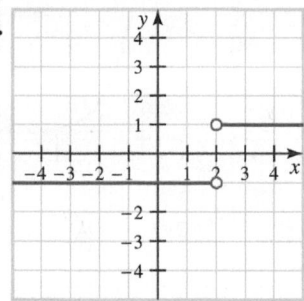

11. a. critical numbers: $x = \pm 3$

    b. increasing on $(-\infty, -3)$ and $(3, \infty)$, decreasing on $(-3, 3)$

    c. critical points: $(3, -16)$, relative minimum; $(-3, 20)$, relative maximum

    d. $x = 0$; concave down on $(-\infty, 0)$ and concave up on $(0, \infty)$

e.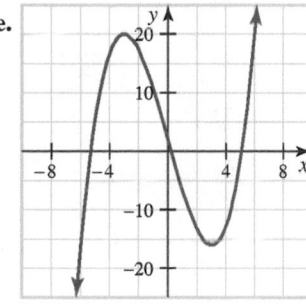

13. a. approximate critical numbers (by calculator or computer software)

    $x \approx -17.862$, $x \approx -7.789$, $x \approx 5.263$, $x \approx 16.388$

    b. increasing on $(-\infty, -17.862) \cup (-7.789, 5.263) \cup (16.388, \infty)$,

    decreasing on $(-17.862, -7.789) \cup (5.265, 16.388)$

    c. critical points: $(-17.862, 115,300)$, relative maximum; $(-7.789, -339,000)$, relative minimum; $(5.265, 188,100)$, relative maximum; $(16.388, -431,900)$, relative minimum

    d. approximate second order critical numbers $x \approx -13.862$, $x \approx -1.1965$, $x \approx 12.059$; concave down on $(-\infty, -13.862) \cup (-1.1965, 12.059)$ and concave up on $(-13.862, -1.1965) \cup (12.059, \infty)$

e.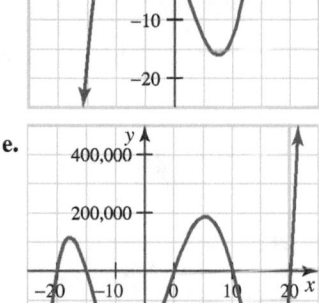

15. a. critical numbers $x = \pm 1$

    b. increasing on $(-\infty, -1) \cup (1, \infty)$, decreasing on $(-1, 0) \cup (0, 1)$

    c. critical points: $(-1, -2)$, relative maximum; $(1, 2)$, relative minimum

    d. no second order critical numbers as $x = 0$ not in domain; concave up for $x > 0$ and concave down for $x < 0$

e.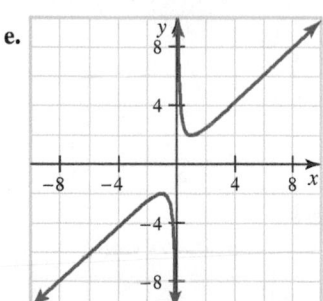

17. a. critical number: $x = e^{-1}$

    b. increasing on (e^{-1}, ∞), decreasing on $(0, e^{-1})$

    c. critical point $(e^{-1}, -e^{-1})$. relative minimum

    d. no second order critical numbers; concave up on $(0, \infty)$

e.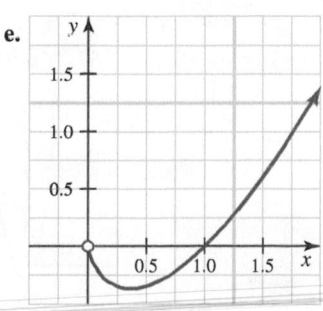

19.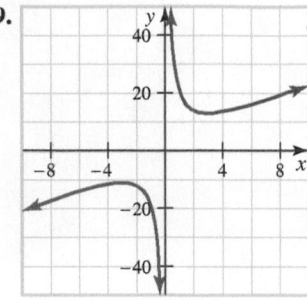

21.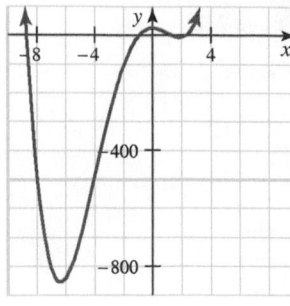

23.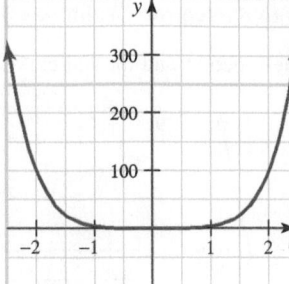

25.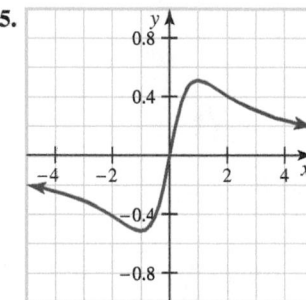

27.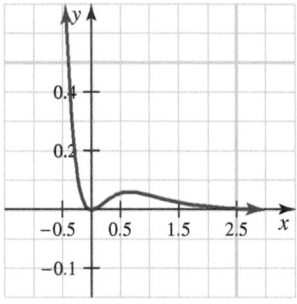

29.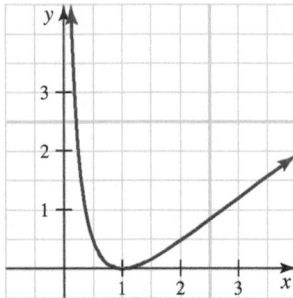

31.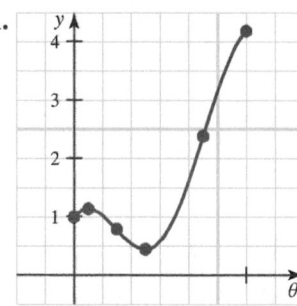

33.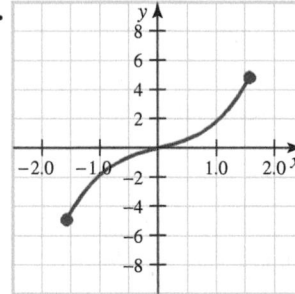

35. At $x = \frac{1}{2}$, relative maximum; at $x = 1$, relative minimum

37. At $x = 4$, relative minimum

39. relative minimum at 1, relative maximum at -9

41. relative minimum at $\frac{\pi}{2}$; relative maximum at $\frac{\pi}{6}$

Answers for Problems 42-45 may vary.

43.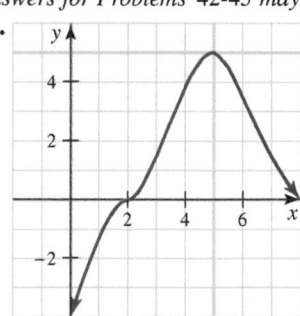

45.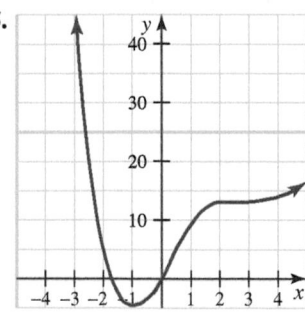

47.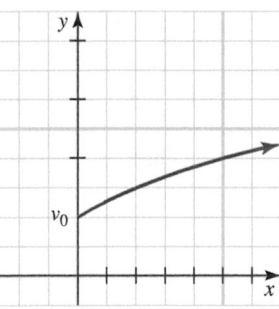

49. a. $-(2)^3 + 6(2)^2 + 13(2) = 42$; $-\frac{1}{3}(2)^3 + \frac{1}{2}(2)^2 + 25(2) = 49\frac{1}{3}$
b. $N(x) = -x^3 + 6x^2 + 13x - \frac{1}{3}(4 - x)^3 + \frac{1}{2}(4 - x)^2 + 25(4 - x)$ **c.** $N'(x) = -2x^2 + 5x = 0$ when $x = 2.5$, so the optimum time for the break is 10:30 A.M. **51.** Maximum deflection at $x = \dfrac{2\ell}{3}$ **55.** $y'' = 2A$

57. $f(x) = -3x^3 + 9x^2 - 1$

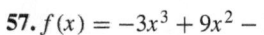

4.4 Curve Sketching with Asymptotes: Limits Involving Infinity, page 271

5. 0   **7.** 3   **9.** 9   **11.** 1   **13.** 0   **15.** $-\frac{1}{2}$ **17.** ∞ **19.** 1   **21.** $-\infty$ **23.** 0   **25.** asymptotes: $x = 7$, $y = -3$; graph rising on $(-\infty, 7) \cup (7, \infty)$; concave up on $(-\infty, 7)$; concave down on $(7, \infty)$; no critical points; no points of inflection;

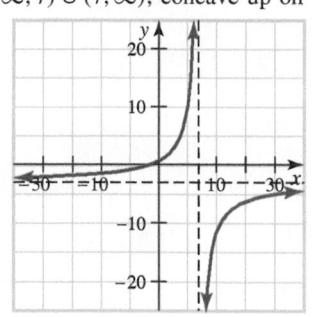

27.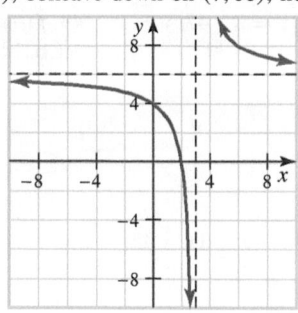

29.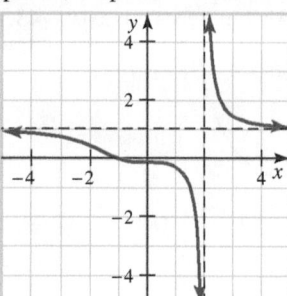

31.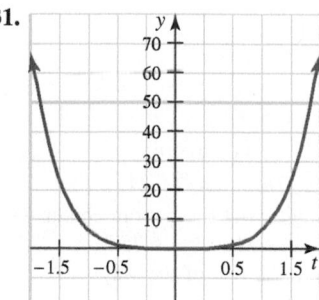

33.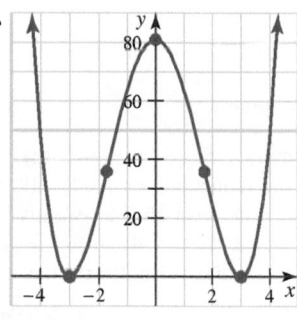

35.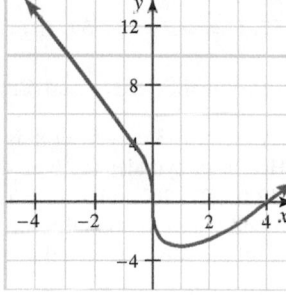

37.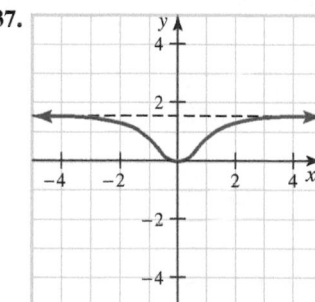

39.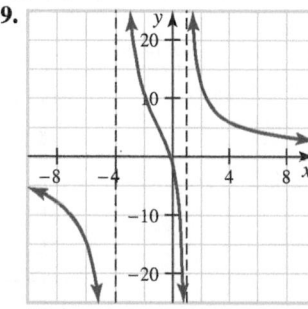

41.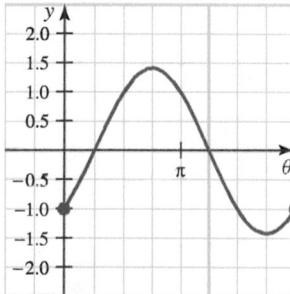

43.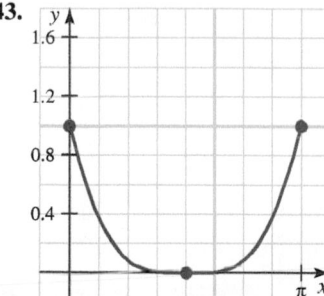

45.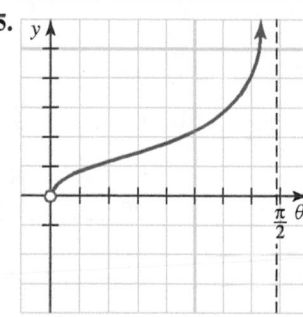

47.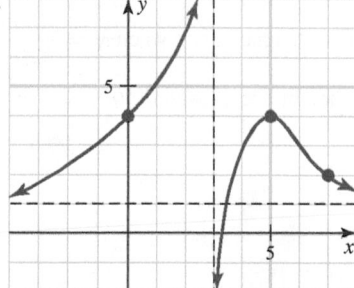

49. population will be largest after 24 minutes; population approaches 5,000 in the long run; inflection point at 49 minutes, after which the rate at which the population decreases per minute starts to decrease;

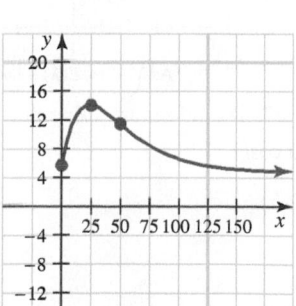

51. $a = \frac{9}{5}$, $b = \frac{3}{5}$ **53.** False   **55.** False

4.5 l'Hôpital's Rule, page 280

1. a. The limit is not an indeterminate form. The correct limit is $\frac{2}{\pi}$. **b.** The limit is not an indeterminate form. The correct limit is $\frac{2}{\pi}$. **3.** $\frac{3}{2}$ **5.** 10 **7.** ∞ **9.** $\frac{1}{2}$ **11.** $\frac{1}{2}$ **13.** $\frac{3}{5}$ **15.** $\frac{1}{2}$ **17.** $\frac{3}{2}$ **19.** 0 **21.** 0 **23.** 0 **25.** 0 **27.** e^{-6} **29.** 0 **31.** ∞ **33.** ∞
35. 1 **37.** e **39.** 0 **41.** limit does not exist **43.** $\frac{1}{120}$ **45.** $y = 0$ **47.** $y = e^2$ **49.** $f(a) = g(a) = 0$
51. a. 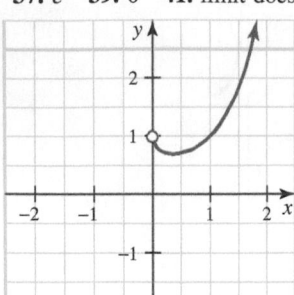 **b.** 1 **53. a.** 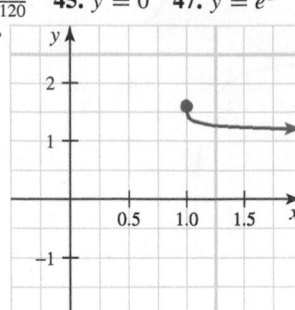 **b.** $e^{1/6}$

55. $a = -2$, $b = 7/3$ **57.** $C = -B = 6$ **59.** $E = -7$, $F = \frac{331}{6}$

4.6 Optimization in the Physical Sciences and Engineering, page 291

1. $y(-4) = 12$ is the maximum. **3.** $y(-5) = 8$ is the maximum. **5.** $y(-5) = 8$ is the maximum. **7.** 25 **9.** -9
15. $10\sqrt{2} \times 5\sqrt{2}$ **17.** Height of cylinder is $2h = \frac{40}{3}\sqrt{3}$ and the radius is $r = \frac{20}{3}\sqrt{6}$ **19.** $r = 20\sqrt{2}$, $2h = 40\sqrt{2}$ **21.** $L\sqrt{2}$
23. 500 ft **25.** 33 ft × 49 ft **27.** 200 miles after 1.5 hr. **31.** 1,117 feet after 2 seconds **33. a.** origin is at the bottom center with x-axis (w) pointing right and y-axis (h) pointing up. **b.** 18 ft **c.** 54 ft **d.** 16 ft **e.** 10 ft **35.** 11,664 in.^{3} = 6.75 ft^3
37. a. 72 minutes rowing all the way **b.** 84 minutes rowing to a point 4.5 miles from B and running the rest of the way
39. $60.00 **41.** approximately 403.1 ft^3 when $\theta = 1.153$ **43. b.** When $p = 12$, the largest value of f is 6. **45.** $x = \dfrac{Md}{\sqrt{4m^2 - M^2}}$

47. a. $T'(x) = -T\left[\dfrac{c}{(kx+c)x} + \ln p\right]$ **b.** $x = \dfrac{-c\ln p + \sqrt{c^2(\ln p)^2 - 4kc\ln p}}{2k\ln p}$ **c.**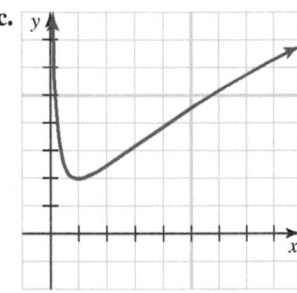

49. a. The maximum height occurs when $x = \dfrac{mv^2}{32(m^2+1)}$

b. The maximum height occurs when $y'(m) = 0$ or when $m = \dfrac{v^2}{32x_0}$. **51.**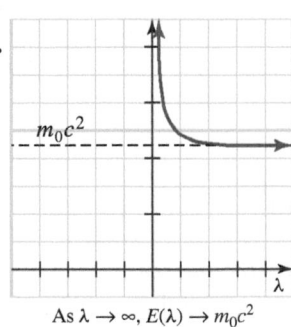

As $\lambda \to \infty$, $E(\lambda) \to m_0 c^2$

53. $r \approx 3.84$ cm and $h \approx 7.67$ cm. **55.** $15\sqrt{3}$ cm is a minimum. **57.** The minimum value occurs when $T \approx 4°$C.
59. a. 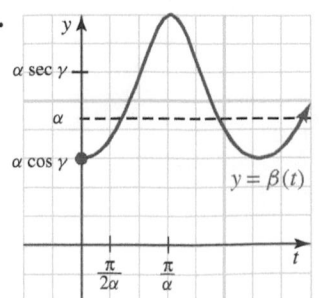 **b.** $\beta_{max} = \alpha\sec\gamma$, $\beta_{min} = \alpha\cos\gamma$

4.7 Optimization in Business, Economics, and the Life Sciences, page 307

1. $x = 26$ **3.** $x = 1875$ **5.** $x = 20$ **7.** $2x - 8 - 50x^{-2}$ **9.** -0.05 **11.** $2x - 50 - 200x^{-2}$ **13.** \$25

15. a. $R(x) = x\left(\dfrac{380 - x}{20}\right)$; $C(x) = 5x + \dfrac{x^2}{50}$; $P(x) = -0.07x^2 + 14x$ **b.** The maximum profit occurs when the price is \$14 per item. The maximum profit is \$700.   **17.** Profit is maximized when $x = \$60$.   **19. a.** 1.22 million people per year   **b.** The percentage rate is 2%.   **21.** 208 years from now.   **23.** 400 cases   **25. a.** $x = 8$ **c.**

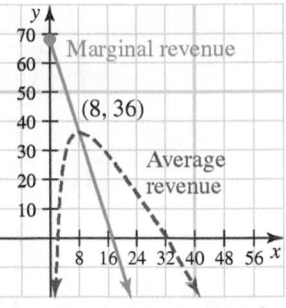

27. Since the optimum solution is over 100 years, you should will the book to your heirs so they can sell it in 117.19 years.   **29.** Sell the boards at a price of \$42.   **31.** Lower the fare \$250.   **33.** Plant 80 total trees.   **35.** 62 vines   **37. a.** The most profitable time to conclude the project is 10 days from now.   **b.** Assume R is continuous on [0,10] and that the glass is coming in at a constant rate throughout the time period.   **39. a.** $R(x) = xp(x) = \dfrac{bx - x^2}{a}$ on $[0, b]$; R is increasing on $\left(0, \dfrac{b}{2}\right)$ and decreasing on $\left(\dfrac{b}{2}, b\right)$ **b.**

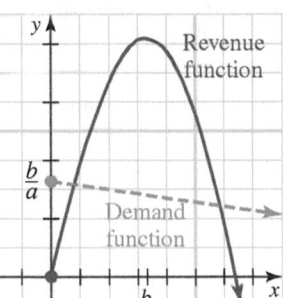

41. $v = \dfrac{kv_1}{k - 1}$

43. a. $x = r$ is a maximum.   **b.**

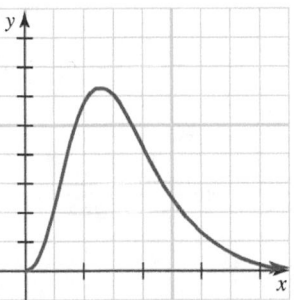

The largest survival percentage is 60.56% and the smallest survival percentage is 22%.

c.

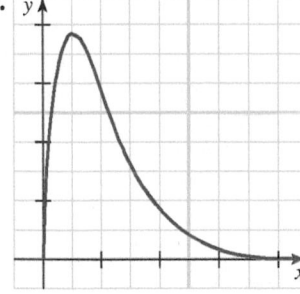

45. $v = 39$ **47. a.**

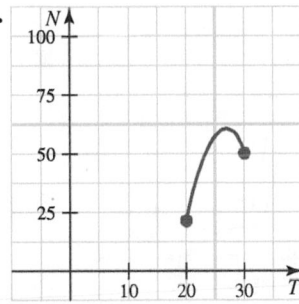

b.

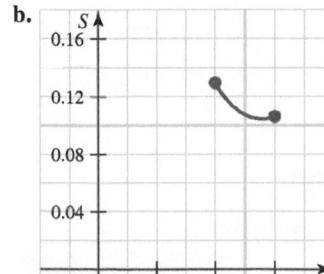

The largest hatching occurs when $T \approx 23.58$ and the smallest when $T = 30$

c.

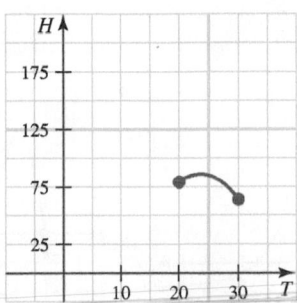

$S'(T) = -(-0.06T + 1.67)(-0.03T^2 + 1.67T - 13.67)^{-2}$ **49. a.** $C = Sx + \dfrac{pQ}{nx}$; $x = \sqrt{\dfrac{pQ}{nS}}$ **59.** $\theta \approx 0.9553$; this is about 55°.

Chapter 4 Proficiency Examination, page 313

18. 2   **19.** $-\frac{1}{2}$ **20.** 0   **21.** e^6

22.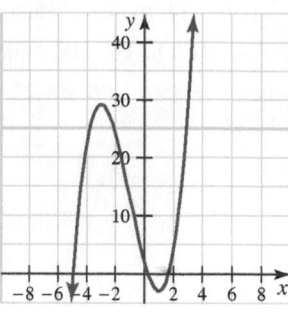

23.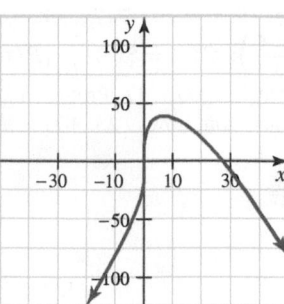

24.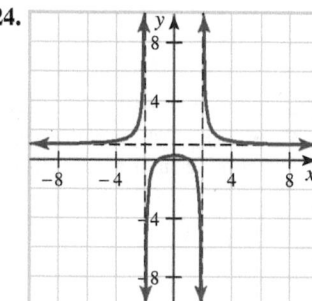

25.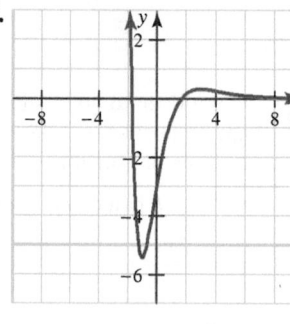

26.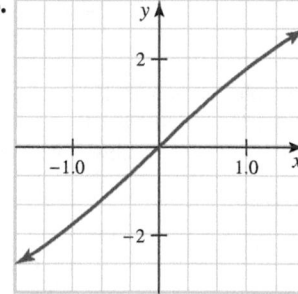

27.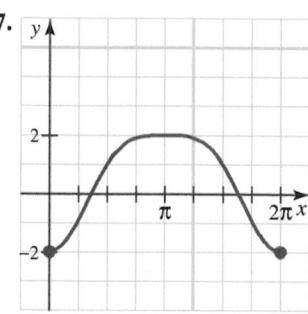

28. absolute maximum at $(0.4, 5.005)$; absolute minimum at $(1, 4)$

29. 19 in. × 19 in. × 10 in.   **30.** 7

Chapter 4 Supplementary Problems, page 314

1.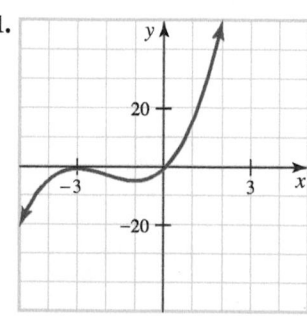

3.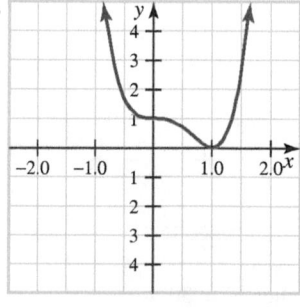

5.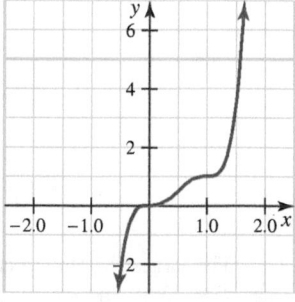

7.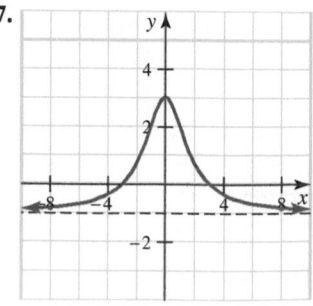

9.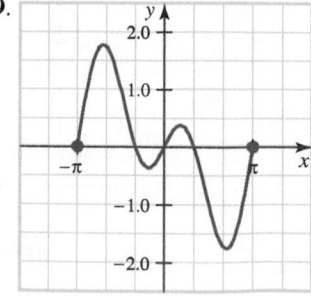

11.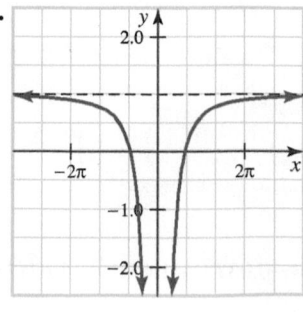

13. 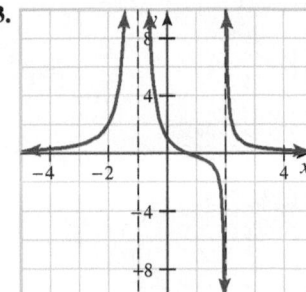   **15.** 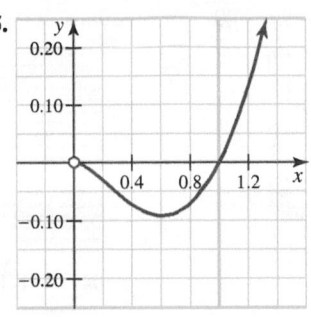   **17.**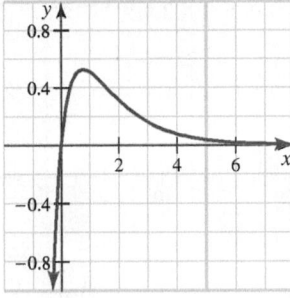

19. 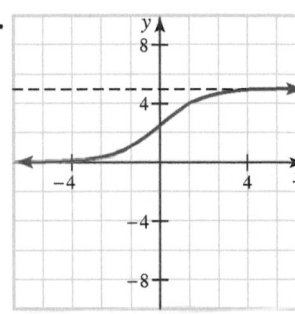   **21.** C **23.** B **25.** maximum $f(0) = 12$; minimum $f(2) = -4$

27. maximum $f(\sqrt{3}/2) \approx 0.68$; minimum $f(0) = 0$ **29.** 0   **31.** $-\frac{1}{2}$ **33.** 0   **35.** does not exist   **37.** 1   **39.** 9   **41.** 0   **43.** 2   **45.** 0
47. ∞ **49.** $\ln 5$ **51.** 0   **53.** does not exist (oscillates)   **55.** f is function and g is derivative

59. 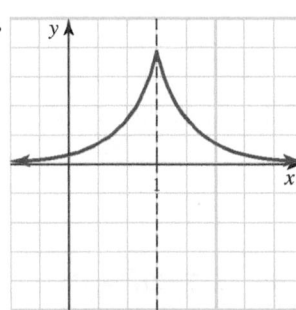   **61.** $A = \frac{1}{2}$, $B = 0$, $C = -\frac{3}{2}$, $D = 0$

The derivative does not exist at $x = 1$.

63. The maximum profit of \$108,900 is reached when 165 units are rented at \$740 each.   **65.** The maximum yield is 6,125 lb for 35 trees per acre.   **67.** 28,072 ft of pipe laid on the shore gives the minimum cost.   **69.** The dimensions are 20 ft by 30 ft.   **71.** The price is \$90 and the maximum profit is \$1,100.   **75.** $x_0 + y_0 = \sqrt{2}$ **79.** $x = \sqrt{ab}$ leads to a relative minimum; $x = -\sqrt{ab}$ leads to a relative maximum.   **83. a.** $f'(x) > 0$ on $(-\frac{3}{2}, -\frac{1}{3}) \cup (1, 2)$ **b.** $f'(x) < 0$ on $(-\frac{1}{3}, 1)$ **c.** $f''(x) > 0$ on $(\frac{1}{3}, 2)$ **d.** $f''(x) < 0$ on $(-\frac{3}{2}, \frac{1}{3})$ **e.** $f'(x) = 0$ at $x = -\frac{1}{3}, 1$ **f.** f' exists everywhere   **g.** $f''(x) = 0$ at $x = \frac{1}{3}$ **85. a.** $f'(x) > 0$ on $(-2, -0.876) \cup (0.876, 2)$ **b.** $f'(x) < 0$ on $(-0.876, 0) \cup (0, 0.876)$ **c.** $f''(x) > 0$ on $(0, 2)$ **d.** $f''(x) < 0$ on $(-2, 0)$ **e.** $f'(x) = 0$ at $x \approx \pm 0.876$ **f.** $f'(x)$ does not exist at $x = 0$ **g.** $f''(x) \neq 0$ **87. a.** ∞ **b.** ∞ **c.** $-\infty$ **d.** $-\infty$ **e.** ∞ **f.** $-\infty$ **89.** Distance is minimized to be approximately 1.7812 when $x \approx 0.460355$.   **91.** relative minima at $(0, 0)$ and approximately $(4.8, -107.9)$; relative maxima at $(1, 0.95)$ and approximately $(-0.8, 0.7)$.   **93.** $k = \lim_{\epsilon \to 0} \dfrac{a^\epsilon - 1}{\epsilon} = \lim_{\epsilon \to 0} \dfrac{a^\epsilon \ln a}{1} = \ln a$ **95.** This is Putnam Problem 1 of the morning session of 1941.   **97.** This is Putnam Problem 2 of the morning session of 1985.   **99.** This is Putnam Problem 1 of the morning session of 1961.

Chapter 5

5.1 Antidifferentiation, page 334

1. $2x + C$ **3.** $x^2 + 3x + C$ **5.** $t^4 + t^3 + C$ **7.** $\frac{1}{2} \ln |x| + C$ **9.** $2u^3 - 3 \sin u + C$ **11.** $\tan \theta + C$ **13.** $-2 \cos \theta + C$
15. $5 \sin^{-1} y + C$ **17.** $\frac{1}{3}x^3 + \frac{2}{5}x^{5/2} + C$ **19.** $\frac{2}{5}u^{5/2} - \frac{2}{3}u^{3/2} - \frac{1}{9}u^{-9} + C$ **21.** $-t^{-1} + \frac{1}{2}t^{-2} - \frac{1}{3}t^{-3} + C$
23. $\frac{4}{5}x^5 + \frac{20}{3}x^3 + 25x + C$ **25.** $-x^{-1} - \frac{3}{2}x^{-2} + \frac{1}{3}x^{-3} + C$ **27.** $x + \ln |x| + 2x^{-1} + C$ **29.** $x - \sin^{-1} x + C$

31. $F(x) = \frac{1}{3}x^3 + \frac{3}{2}x^2$

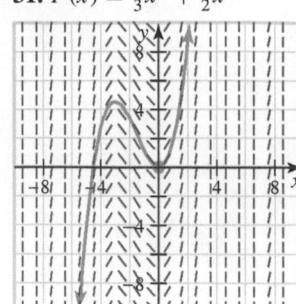

33. $F(x) = \frac{1}{2}x^2 + 4x^{3/2} + 9x - 40$

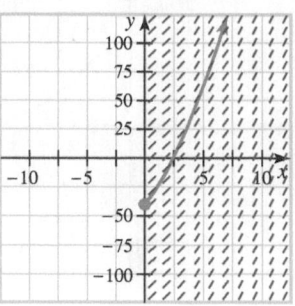

35. $F(x) = \ln|x| - x^{-1} - 1$

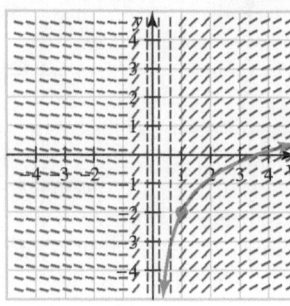

37. $F(x) = \frac{1}{2}x^2 + e^x + 1$

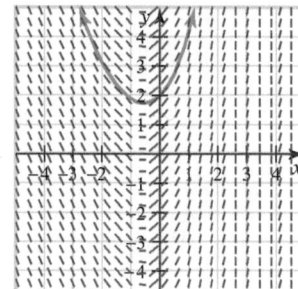

39. $F(x) = 2\sqrt{x} - 4x + 2$

a.

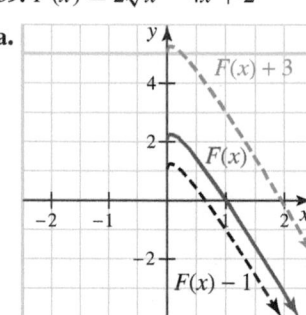

b. $C_0 = -\frac{9}{4}$

41. \$744   **43.** 10,128

45. $k = 20$ ft/s²   **47.** $a(t) \approx 4.3$ ft/s²

49. $k = -32$ ft/s²   **51.** \$203.19

53. $\frac{14}{3}$ **55.** $1 + \ln 2$ **57.** $\frac{\pi}{6}$

5.2 Area as the Limit of a Sum, page 342

1. 6   **3.** 21   **5.** 65   **7.** 225   **9.** 9,800   **11.** $\frac{1}{2}$ **13.** 1   **15.** 0   **17.** 3   **19. a.** 3.5   **b.** 3.25   **21. a.** 2.719   **b.** 2.588   **23.** 1.183
25. 0.415   **27.** 5.030   **29.** 20   **31.** 75   **33.** 3   **35.** true   **37.** true   **39.** false   **45.** 2 square units
47. 1 square unit   **49.** 3.4 square units   **51.** 1 square unit

5.3 Riemann Sums and the Definite Integral, page 354

1. 2.25   **3.** 10.75   **5.** −0.875   **7.** 1.183   **9.** 1.942   **11.** 1.75   **13.** 6.75   **15.** −0.125   **17.** 0.791   **19.** 1.512   **21.** 28.875
23. 1.896   **25.** 0.556   **27.** 22.125   **29.** 1.896   **31.** 0.714   **33.** $-\frac{1}{3}$ **35.** $\frac{3}{2}$ **37.** $-\frac{1}{2}$ **43.** 1;1   **45.** 10
47.

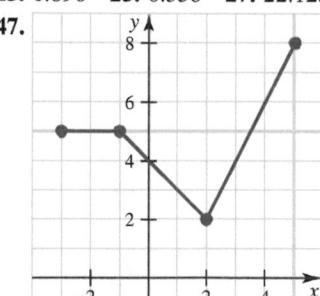

$\int_{-3}^{5} f(x)\,dx = \frac{71}{2}$ **53.** 2.3; $\|P\| = 1.1$ **57.** false

5.4 The Fundamental Theorems of Calculus, page 362

1. 140   **3.** $16 + 8a$ **5.** $\frac{15}{4}a$ **7.** $\frac{3}{8}c$ **9.** 18   **11.** $\frac{5}{8} + \pi^2$ **13.** $\dfrac{2\pi(2^{\pi+1} - 1)}{\pi + 1}$ **15.** $\frac{272}{15}$ **17.** 2   **19.** $\frac{a\pi}{2}$ **21.** $\frac{5}{2}$ **23.** 4   **25.** 5

27. $\frac{8}{3}$ **29.** 1   **31.** $e - \frac{3}{2}$ **33.** $3\ln 2 - \frac{1}{2}$ **35.** $(x-1)\sqrt{x+1}$ **37.** $\dfrac{\sin t}{t}$ **39.** $\dfrac{-1}{\sqrt{1+3x^2}}$ **49.** $c = \frac{a}{4}$ **51.** 6.45

53. a. relative minimum.   **b.** $g'''(1) = 0$ **c.** $x \approx 0.75$ **d.**

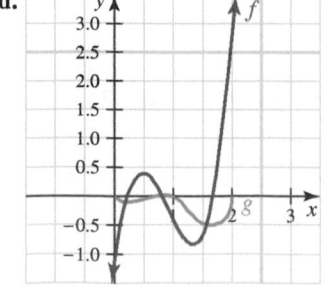

57. $8x^7 - 12x^3$

59. $F'(x) = f(v)\dfrac{dv}{dx} - f(u)\dfrac{du}{dx}$

5.5 Integration by Substitution, page 369

1. a. 32 **b.** $2\sqrt{3}-2$ **3. a.** 0 **b.** 0 **5. a.** $\frac{128}{5}$ **b.** $\frac{128}{5}$ **7. a.** $\frac{2}{9}\sqrt{2}\,x^{9/2}+C$ **b.** $\frac{1}{9}(2x^3-5)^{3/2}+C$ **9.** $\frac{1}{10}(2x+3)^5+C$
11. $\frac{1}{3}(x^3-\sin 3x)+C$ **13.** $\cos(4-x)+C$ **15.** $\frac{1}{6}(t^{3/2}+5)^4+C$ **17.** $-\frac{1}{2}\cos(3+x^2)+C$ **19.** $\frac{1}{4}\ln(2x^2+3)+C$
21. $\frac{1}{6}(2x^2+1)^{3/2}+C$ **23.** $\frac{2}{3}e^{x^{3/2}}+C$ **25.** $\frac{1}{3}(x^2+4)^{3/2}+C$ **27.** $\frac{1}{2}(\ln x)^2+C$ **29.** $2\ln(\sqrt{x}+7)+C$ **31.** $\ln(e^t+1)+C$
33. $\frac{2}{6}\ln 3$ **35.** 0 **37.** $e-e^{1/2}$ **39.** $\frac{1}{2}\ln 2$ **41.** $\ln(1+e)-\ln 2$ **43. a.** We take 1 Frdor as the variable, so the note from the students reads "Because of illness I cannot lecture between Easter and Michaelmas." **b.** The Dirichlet function is defined as a function so that $f(x)$ equals a determined constant c (usually 1) when the variable x takes a rational value, and another constant d (usually 0) when this variable is irrational. This famous function is one which is discontinuous everywhere. **45.** $\frac{16}{3}$ **47.** $\sqrt{10}-\sqrt{2}$
49. 0 **51.** 0 **53. a.** true **b.** true **c.** false **55.** $F(x)=-\frac{1}{3}\ln|1-3x^2|+5$ **57.** 2 ft³; $\frac{1}{4}$ in.
59. a. $L(t)=0.03\sqrt{36+16t-t^2}+3.82$

b.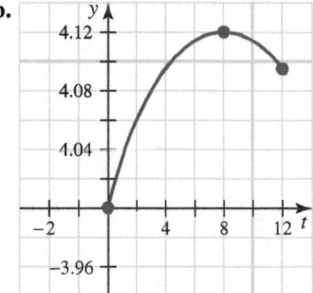

The highest level is 4.12 ppm.

c.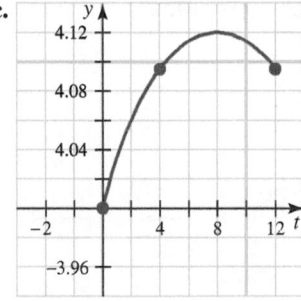

It is the same as 11 A.M.
$(t=4)$ at 7 P.M. $(t=12)$

5.6 Introduction to Differential Equations, page 382

9. $x^2+y^2=8$ **11.** $y=\tan(x-\frac{3\pi}{4})$ **13.** $x^{3/2}-y^{3/2}=7$ **15.**

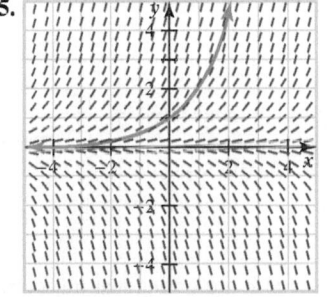

17. 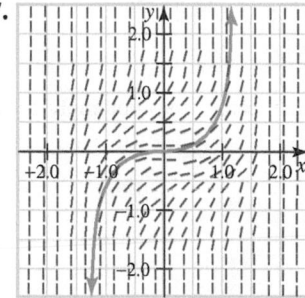 **19.** 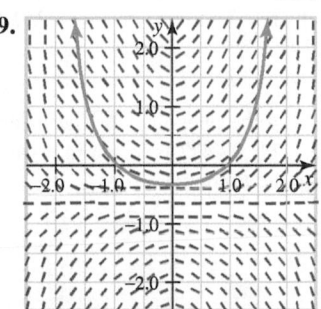 **21.** $y=Be^{(3/2)x^2}$ **23.** $2(1-x^2)^{3/2}+3y^2=C$

25. $\cos x+\sin y=C$ **27.** $-\frac{1}{3}(1-y^2)^{3/2}=\frac{1}{2}(\ln x)^2+C$ **29.** $xy=C$ **31.** $y=Cx$ **33.**

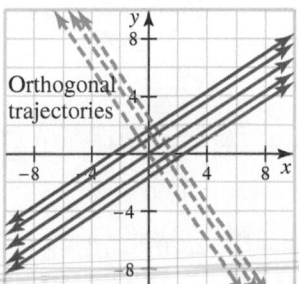

Orthogonal trajectories

35.
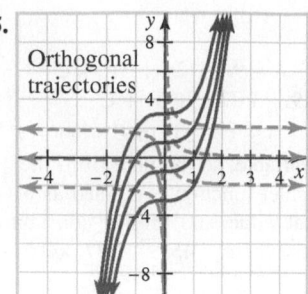

37.
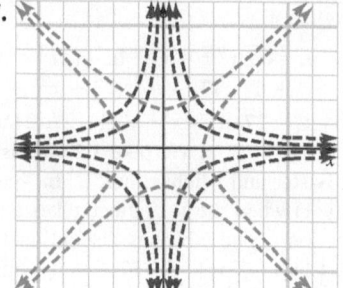

39.
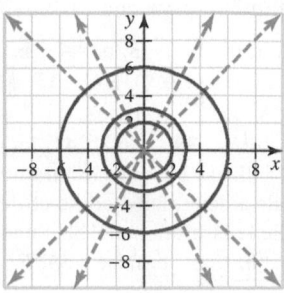

41. $\dfrac{dQ}{dt} = kQ$ **43.** $\dfrac{dT}{dt} = c(T - T_m)$ **45.** $\dfrac{dQ}{dt} = kQ(P - Q)$ **47.** $XY = C$

49. a.
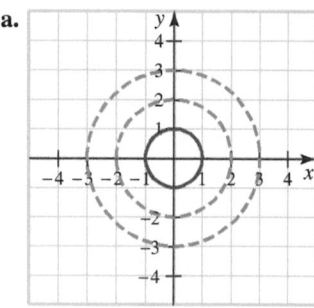

b.
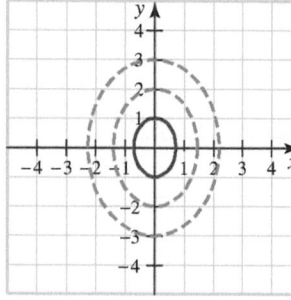

c.
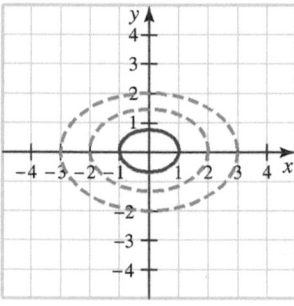

d.
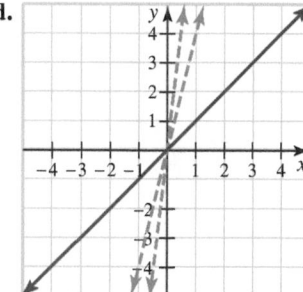

e.
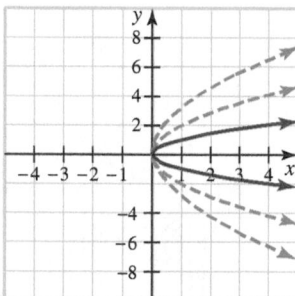

f.
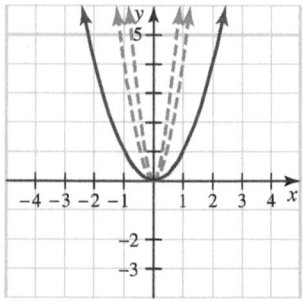

a and d are orthogonal trajectories; b and e are orthogonal trajectories; c and f are orthogonal trajectories
51. approximately 10,523 years   **53.** 5 min   **55. a.** 98.5 ft/s   **b.** $s \approx 3{,}956.067$ mi; $h \approx 352$ ft   **57.** 2 hr and 53 min

5.7 The Mean Value theorem for Integrals; Average Value, page 389

1. 1.55 is in the interval.   **3.** 1.055 is in the interval.   **5.** $\sqrt{5}$ is in the interval.   **7.** The mean value theorem does not apply because the function is discontinuous at 0.   **9.** 0.0807 is in the interval.   **11.** 0.4427 is in the interval.
13. The mean value theorem does not apply because the function is discontinuous at 0.
15. $A = 25$ **17.** $A = \dfrac{38}{3}$ **19.** $A \approx 1.839$

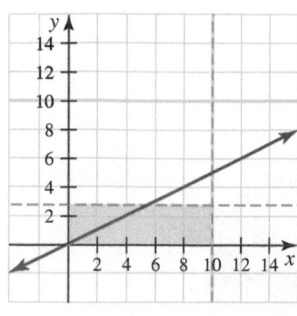

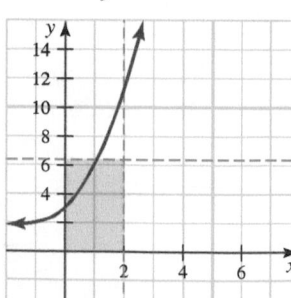

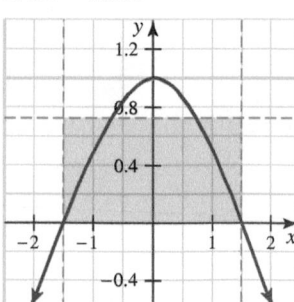

21. $\dfrac{3}{2}$ **23.** $-\dfrac{17}{4}$ **25.** 0   **27.** $\dfrac{1}{2} - \dfrac{3}{4}\ln\dfrac{5}{3}$ **29.** $\dfrac{1}{\pi}\left(4 - 2\sqrt{2}\right)$ **31.** $\dfrac{4}{3}$ **33.** $\dfrac{45}{28}$ **35.** -10 **37.** $\dfrac{1}{6}\left(27 - 7^{3/2}\right)$ **39.** $\dfrac{1}{3}\left(\sqrt{10} - 1\right)$
41. $\dfrac{3\pi}{4}$ **43.** \$16.10   **45.** \$318   **47.** 63°F   **49.** $-\dfrac{g}{2}(t_1 + t_0) + v_0$ **51.** 45.2° F   **53.** 18 months   **55.** 14 minutes   **57.** $\sec^2 x$
59. $A = 0$, $B = k$, $C = 0$, $g(\lambda) = kte^{-\lambda t}$

5.8 Numerical Integration: The Trapezoidal Rule and Simpson's Rule, page 398

3. a. 10 **b.** 12.5 **c.** 12.67 **5. a.** 11.81 **b.** 12.65 **c.** 12.67 **7. a.** 0.16 **b.** 0.129 **c.** 0.125 **9. a.** 0.135 **b.** 0.125 **c.** 0.125
11. Trapezoidal rule, 2.375; Simpson's rule, 2.33333; exact value, $\frac{7}{3}$ **13.** Trapezoidal rule, 5.146; Simpson's rule, 5.252; exact value,
$\frac{16}{3}$ **15. a.** 0.7828 **b.** 0.785 **17. a.** 2.037871 **b.** 2.048596 **19. a.** 0.584 **b.** 0.594 **21.** $A \approx 0.775$; the exact answer is between
$0.775 - 0.05$ and $0.775 + 0.05$ **23.** $A \approx 0.455$; the exact answer is between $0.455 - 0.0005$ and $0.455 + 0.0005$ **25.** $A \approx 3.25$; the
exact answer is between $3.25 - 0.01$ and $3.25 + 0.01$ **27.** $A \approx 0.44$; the exact answer is between $0.44 - 0.01$ and $0.44 + 0.01$
29. a. $n = 164$ **b.** $n = 18$ **31. a.** $n = 184$ **b.** $n = 22$ **33. a.** $n = 82$ **b.** $n = 8$ **35.** 3.1 **37.** $n = 578$ **39.** 100 yd^3 **41.** 50.38
mi **43.** 79.17 **45.** The order of convergence is n^2. **47.** The order of convergence is n^4. **49.** Simpson's rule yields 20.06095
approximately because the error is 0 since $f^{(4)}(x) = 0$ for cubics. **51.** Answers vary, but the calculated value should be
approximately 314 cm^2. **53.** Left endpoint, 2.0841; Trapezoid 2.5525; Newton-Cotes, 2.5975; Exact, 2.5958 from computer **55.** $\frac{45}{4}$
57. $\frac{1}{\sqrt{3}}$ **59.** Error is zero for third degree polynomial.

5.9 An Alternative Approach: The Logarithm as an Integral, page 406

1. $B(x) = y$ if and only if $\log_b y = x$. **3.** $e = E(1)$; $e - 1$ is the area under the graph of $E(x)$ from 0 to 1. **5.** e^{x+1} **7.** 5 **9.** $\sqrt{e}$
11. $E(x \ln 10)$ **13.** $E(x \ln 8)$ **15.** $E(\sqrt{2} \ln x)$ **27.** $\dfrac{\ln x + 2}{2\sqrt{x}}$ **29.** $3x^2 - 3^x \ln 3$ **31.** $\cot x$ **33.** $\dfrac{2x}{(x^2 - 4) \ln 2}$ **35.** $\cos x$ **37.** 2
39. $\frac{1}{2} \ln \frac{7}{3}$ **41.** $\frac{90}{\ln 10}$ **43.** $\ln |e^x + \sin x| + C$ **51.** $-12xe^{-3\sin^2(2x^2+1)} \sin(4x^2 + 2)$ **55.** Simpson's rule: 1.0958 and $\ln 3 \approx 1.0986$.
57. a. 18 **b.** 0.000113

Chapter 5 Proficiency Examination, page 409

17. $-\frac{3}{5}$ **18.** $x^5 \sqrt{\cos(2x + 1)}$ **19.** $\frac{1}{2} \tan^{-1}(2x) + C$ **20.** $-\frac{1}{2}e^{-x^2} + C$ **21.** $\frac{17}{3}$ **22.** $-\frac{35}{3}$ **23.** $\frac{1}{2}$ **24.** 0 **25.** 36 **26.** 14.99

27. The percent is approximately 1.33×10^{-184}, which exceeds the accuracy of most calculators and measuring devices. **28.** 0

29. 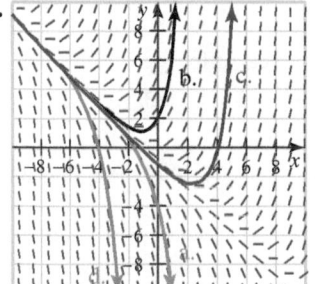 **e.** $y = -x - 1$ **30. a.** $n \geq 26$ **b.** $n \geq 4$

Chapter 5 Supplementary Problems, page 410

1. 25 **3.** 6 **5.** 1,710 **7.** $\frac{6 - 2\sqrt{2}}{3}$ **9.** $\sqrt{3} - 1$ **11.** $-\frac{3}{4}(1 - \sqrt[3]{9})$ **13.** $2x^{5/2} - \frac{4}{3}x^{3/2} + 2x^{1/2} + C$ **15.** $3 \sin^{-1} x + \sqrt{1 - x^2} + C$
17. $-\ln |\sin x + \cos x| + C$ **19.** $\frac{2}{7}x^{7/2} + \frac{1}{2}x^2 + \frac{2}{3}x^{3/2} + C$ **21.** $x + C$ **23.** $\frac{1}{3}x^3(x^2 + 1)^{3/2} + C$ **25.** $-\frac{1}{15}(1 - 5x^2)^{3/2} + C$
27. $\frac{2}{5}(1 + \ln 2x)^{5/2} - \frac{2}{3}(1 + \ln 2)(1 + \ln 2x)^{3/2} + C$ **29.** 0 **31.** 3 **33.** $\frac{7}{3} + \ln 2$ **35.** $\frac{1}{10,240} \sin^{10} 2x + C$

37. 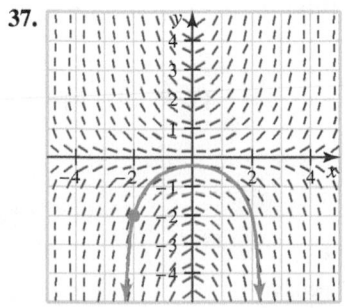 **39.** 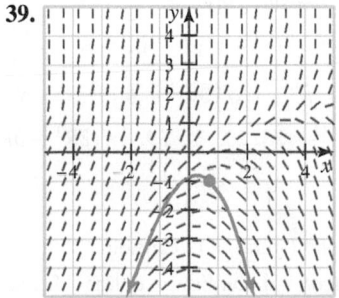 **41.**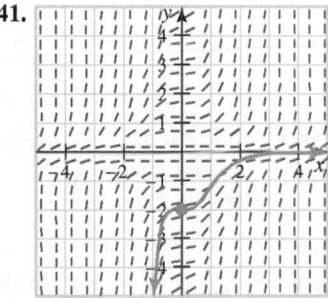

43. $\ln 4$ **45.** $\frac{1}{4}(e^8 - 1)$ **47.** $f(t) = -\frac{1}{16} \sin 4t + \frac{1}{4} \cos 2t + \frac{5}{4}t + \frac{5}{4} - \frac{5}{8}\pi$ **49.** $y = 1 - \frac{1}{x - C}$ **51.** $\tan y + \cot x = C$
53. $y = Ce^{(x + \frac{1}{3}x^3)}$ **55.** $y = \tan^{-1}(C - \ln |\cos x|)$ **57.** $y = 4 + \frac{1}{C - x}$ **59.** $\frac{4}{\pi}(\sqrt{2} - 1)$ **61.** Trapezoidal rule with $n = 6$, 1.9541;
exact, 2 **63.** 0.9089 **65.** 0.216 with $n = 6$ **67.** $G(100) = 29$, $G(1,000) = 177$, $G(10,000) = 1,245$ **69.** 1.0141 by trapezoidal
rule **71. b.** -8 m/s **73.** $R(x) = 1,575x - \frac{5}{3}x^3$; $R(5) = \$7,666.67$ **75.** 2.33 ft **77.** $y = \frac{1}{3}(x^2 + 5)^{3/2} + 1$ **79.** 4.45 ppm
81. 126 people **83.** \$1.32 per pound. **85.** about 2 min, 45 seconds **89. a.** 51.7% **b.** about $17\frac{1}{2}$ years **91.** 12.19 million,

35 years   **93.** $9\sqrt{2}$ ft/s   **95.** 7.182%; the percentage lost is always the same.   **97.** $x = \frac{3}{2}t^2 - 35\sin t + C_1$, $y = 14\sin t - \frac{1}{2}t^2 + C_2$
99. This is Putnam Problem 1 from the morning session in 1958.

Cumulative Review Problems—Chapters 1-5, page 418

5. $\frac{7}{5}$ **7.** $\frac{1}{2}$ **9.** 0   **11.** 1   **13.** 1   **15.** $6(x^2+1)^2(3x-4)(2x-1)^2$ **17.** $-\dfrac{2x+3y}{3x+2y}$ **19.** $3(\sin x + \cos x)^2(\cos x - \sin x)$

21. $\dfrac{e^{-x}(1 - x\ln 3x)}{x\ln 5}$ **23.** 5   **25.** $\sqrt{10} - 3$ **27.** $\ln(e^x + 2) + C$ **29.** 1.812   **31.** $4x - y - 3 = 0$ **33.** $27x - y + 26 = 0$

35. $\ln|y| = \tan^{-1}x + \frac{1}{2}\ln(x^2+1) + C$ **37.** $y = -\ln|\cos x + e^{-5} - 1|$ **39.** $36x(3x^2+1)$ **41.** $f(x) = -x^2 + 5x$

43.

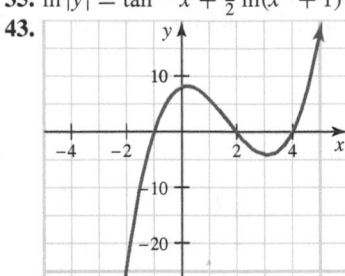

45. $f(6) = 8$ **47.** $\frac{0.04\pi}{3}$ **49. a.** $\frac{9\pi}{4} - 8$ **b.** 3   **c.** $4x + y - 20 - \frac{9\pi}{4} = 0$ **d.** $x = 0$ and $x = 7$

51. 17/164 rad/s   **53.** 75 mi/h   **55. a.** $y = -\frac{4}{\pi}x + 2$ **b.** $y = -2x + \pi$

c. $x\sin^{-1}(2/\pi)$; mean value theorem   **57. a.** 40 g   **b.** 36 days

59. The percentage left is so tiny, approximately 10^{-200}, that it exceeds the accuracy of most calculators and measuring devices.

Chapter 6

6.1 Area Between Two Curves, page 428

1. $\int_a^b [f(x) - g(x)]dx$ **3.** $\int_a^b [g(x) - f(x)]dx$ **5.** $\int_a^b [g(x) - f(x)]dx$ **7.** $\int_a^b [f(x) - g(x)]dx$

9. $\int_a^c [g(x) - f(x)]dx + \int_c^b [f(x) - g(x)]dx$ **11.** $\int_a^c [g(x) - f(x)]dx + \int_c^b [f(x) - g(x)]dx$ **13.** vertical strip; $\frac{125}{48}$

15. vertical strip; 2   **17.** horizontal strip; $\frac{125}{6}$ **19.** 1   **21.** 255   **23.** $\ln 2$ **25.** $\frac{1}{12}$ **27.** $\frac{8}{3}$ **29.** $\frac{324}{5}$ **31.** $\frac{9}{2}$ **33.** $\frac{253}{12}$ **35.** $\frac{5}{2}$

37. $\frac{323}{12}$ **39.** $e^2 - e^{-1}$ **41.** $\frac{131}{4}$ **43.** $2\ln 1.6 - \sin^{-1}0.6$ **45.** $k \approx 0.34$ **49.** \$108,000   **51.** \$33,913.06

53. 2,595; accumulated units   **55.** \$3,069; accumulated profits   **59.** The V values are approximately 72.5299, 196.539, 344.337, 502.655, 660.972, 808.77, 932.78, 1,005.31

6.2 Volume, page 443

The given volumes are all in cubic units.
1. 9   **3.** $\frac{1}{30}$ **5.** $\frac{32}{3}\sqrt{3}$ **7.** $\frac{\sqrt{3}}{2}$ **9.** $\frac{64\pi}{15}$ **11.** $\frac{\pi}{8}(1 - \frac{\pi}{4})$ **13.** $\frac{\pi}{2}$ **15.** π **17.** $\frac{\pi}{2}$ **19.** $\frac{\pi}{12}$ **21.** $\frac{\pi}{2}$ **23.** $\frac{\pi}{2}(1 - \ln 2)$

25. a. $\pi\int_0^4 (4-x)^2 dx$ **b.** $2\pi\int_0^4 x(4-x)dx$ **c.** $\pi\int_0^4 (24 - 10x + x^2)dx$ **d.** $2\pi\int_0^4 (x+2)(4-x)dx$

27. a. $2\pi\int_0^2 y\sqrt{4 - y^2}dy$ **b.** $\pi\int_0^2 (4 - y^2)dy$ **c.** $2\pi\int_0^2 (y+1)\sqrt{4 - y^2}dy$ **d.** $\pi\int_0^2 [4 - y^2 + 4\sqrt{4 - y^2}]dy$ **29. a.** $\pi\int_0^1 (e^{-x})^2 dx$

b. $2\pi\int_0^1 xe^{-x}dx$ **c.** $\pi\int_0^1 (e^{-2x} + 2e^{-x})dx$ **d.** $2\pi\int_0^1 (x+2)e^{-x}dx$ **31. a.** $2\pi\int_0^{\pi/2} y(1 - \sin y)dy$ **b.** $\pi\int_0^{\pi/2} [1^2 - \sin^2 y)]dy$

c. $2\pi\int_0^{\pi/2} (y+1)(1 - \sin y)dy$ **d.** $\pi\int_0^{\pi/2} [5 - 4\sin y - \sin^2 y]dy$ **33. a.** $V = \pi\int_0^1 [(\sqrt{x})^2 - (x^2)^2 dx]$

b. $V = \pi\int_0^1 [(\sqrt{y})^2 - (y^2)^2 dy]$ **35. a.** $V = \pi\int_0^2 [(x^2+1)^2 - 1^2]dx$ **b.** $V = \pi\int_1^5 \left[(2)^2 - (\sqrt{y-1})^2\right]dy$

37. a. $V = \pi\int_{1.138}^{3.566} [(\ln x)^2 - (0.1x^2)^2]dx$ **b.** $V = \pi\int_{0.130}^{1.271} [(\sqrt{10y})^2 - (e^y)^2]dy$ **39.** $\frac{2\pi}{35}$ **41.** $\frac{\pi}{3}$ **43.** 144   **45.** 36   **47.** $\frac{32\sqrt{3}}{3}$

49. 90,000,000 ft^3 **51. a.** $\frac{2\pi k^4}{3}$ **b.** $\frac{4\pi k^4}{3}$ **53.** $\frac{128\pi\sqrt{2}}{9}$ **55.** 20.13   **59.** $\frac{1}{3}\pi(2R^3 - 3R^2h + h^3)$

6.3 Polar Forms and Area, page 456

3. a. lemniscate   **b.** circle   **c.** rose (3 petals)   **d.** none (spiral)   **e.** cardioid   **f.** line   **g.** lemniscate
h. limaçon   **5. a.** rose (4 petals)   **b.** circle   **c.** limaçon   **d.** cardioid   **e.** line   **f.** line   **g.** rose (5 petals)
h. line   **7.**

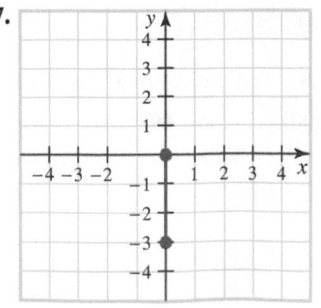

9.

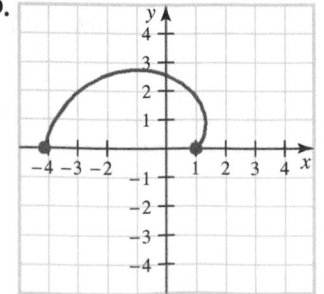

11.

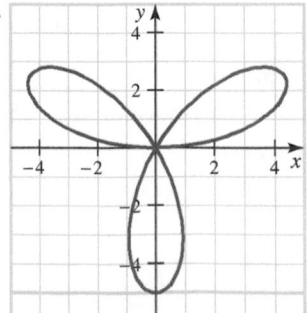

13.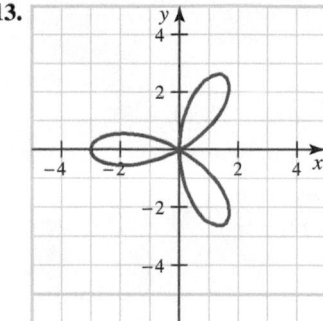

15.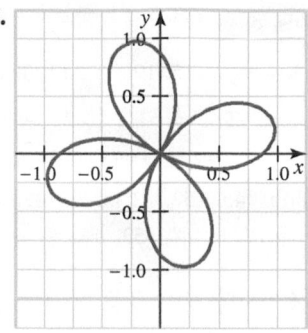

17.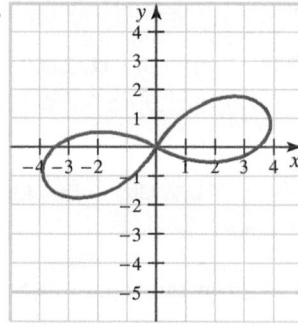

19.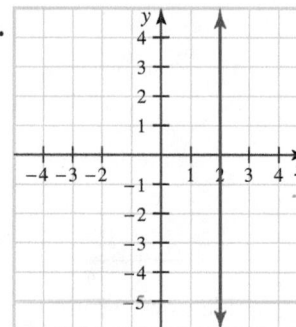

21.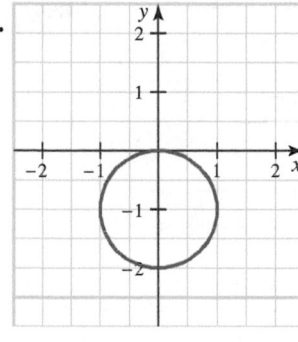

23. $P_1(2, \frac{\pi}{6})$, $P_2(2, \frac{5\pi}{6})$

25. $P_1(0,0)$,

$$P_n\left((3n+1)\pi, \frac{\pi}{3}\right) \text{ for even } n,$$

$$P_n\left((3n+1)\pi, \frac{4\pi}{3}\right) \text{ for odd } n$$

27. $P_1(0,0)$, $P_2(1, \frac{\pi}{4})$

29. There are no intersection points.

31. $\frac{\pi}{24} + \frac{\sqrt{3}}{16}$ **33.** $\frac{\sqrt{3}}{4}$ **35.** $\frac{1}{8}(\pi + 2)$ **37.** $\frac{16\pi^3}{5}$

39.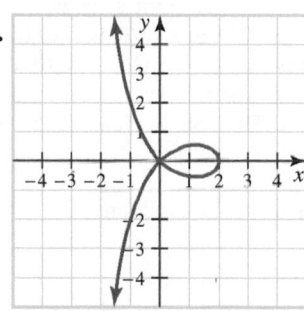

41.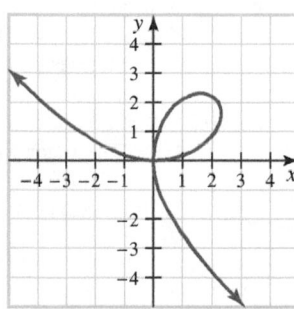

43. $\frac{\pi a^2}{4}$ **45.** $a^2(2 - \frac{\pi}{4})$ **47.** 4π

49. $4\pi + 12\sqrt{3}$ **51.** 0.0674

53. The maximum value of x is $9\sqrt{3}/4$

55.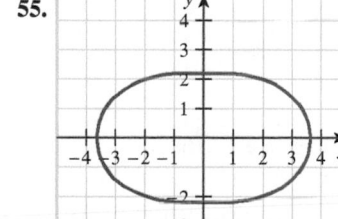

59. a. $-\cot\theta$ **b.** $\dfrac{1 - \cos\theta}{\sin\theta}$ **c.** $\frac{1}{3}$

6.4 Arc Length and Surface Area, page 466

1. $3\sqrt{10}$ **3.** $2\sqrt{5}$ **5.** $\frac{1}{3}(10\sqrt{5} - 2)$ **7.** 12   **9.** $\frac{331}{120}$ **11.** $\frac{3}{8} + \ln 2$ **13.** 1   **15.** $\frac{169}{24}$ **17.** $\frac{80}{3}$ **19.** $\frac{14}{3}$ **21.** $\frac{3}{2}$
23. $12\pi\sqrt{5}$ **25.** $\frac{49\pi}{3}$ **27.** $\frac{515\pi}{64}$ **29.** π **31.** $\frac{\sqrt{10}}{3}\left[e^{3\pi/2} - 1\right]$ **33.** $\frac{1}{3}(5^{3/2} - 8)$ **35.** 4   **37.** 25π **39.** $\frac{2\pi}{3}(3 - \sqrt{3})$
41. approximately 3.820   **43. a.** $\frac{1.505\pi}{18}$ **b.** 127.4   **45.** 29.80   **47.** 27.12   **49.** 6   **51.** approximately 14.0543
57. $C[b^{2n} - a^{2n}] - D[b^{2(1-n)} - a^{2(1-n)}]$ **59.** $f(x) = \pm\ln|\cos x|$

6.5 Physical Applications: Work, Liquid Force, and Centroids, page 480

5. 12,750 ft-lb   **7.** 6,500 ft-lb   **9.** $\frac{10}{3}$ ft-lb   **11.** 2,000 ft-lb   **13.** 4 ergs   **15.** 192 lb   **17.** 304 lb   **19.** 59.7 lb

21. $F = 64.5 \int_0^{1/24} 2(x + \frac{1}{24})\sqrt{(\frac{1}{24})^2 - x^2}dx$ **23.** $F = 57 \int_2^5 5xdx$ **25.** $F = 51.2 \int_{-2}^0 2(x + 3)\sqrt{4 - x^2}dx$ **27.** $(0, -\frac{18}{5})$

29. $\left(\frac{1}{\ln 2}, \frac{1}{4\ln 2}\right)$ **31.** $\left(\frac{7}{3}, \frac{1}{2}\ln 2\right)$ **33.** $V = \frac{200\pi}{3}$ **35.** $V = 8\pi(\frac{4}{3} + \pi)$ **37. a.** 4,284 ft-lb   **b.** 12,603 ft-lb

39. approximately 345,800 ft-lb   **41.** $7,920\pi$ ft-lb   **43.** 2,932.8 lb   **45.** 152,381 mi-lb   **47. a.** 7.5 ergs   **b.** 37.5 ergs

49. The total force on the bottom is about 85,169 lb.   **51.** $I_x = \frac{64}{15}, I_y = \frac{256}{7}$ **53.** $\rho = \frac{2}{3}\sqrt{6}$ **57.** $\pi L^2(2s + L)$

6.6 Applications to Business, Economics, and Life Sciences, page 492

5. a. 75   **b.** 432   **7. a.** 0.75   **b.** 0   **9.** $468.00

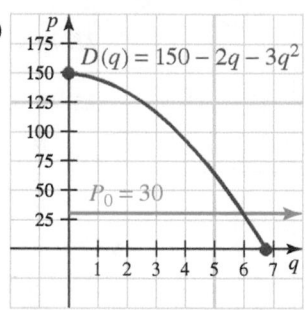

11. $12.47

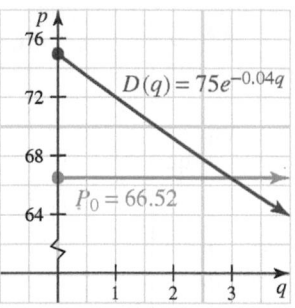

13. $6.25   **15.** $2.82   **17.** $42.67   **19.** $5.33   **21. a.** approximately 10 weeks   **b.** $14,857   **c.** In geometric terms, the net earnings are represented by the area between the curve $y = R'(t)$ and the horizontal line $y = 676$ from $t = 0$ to $t = 10$.

23. a. $FV = \$10,171.84$ **b.** $PV = \$8,001.46$ **25. a.** The machine will be profitable for 9 years.   **b.** $12,150; in geometric terms, the net earnings are represented by the area of the region between the curves $y = R'(x)$ and $y = C'(x)$ from $x = 0$ to $x = 9$.

27. a. The second plan is more profitable for the first 18 years.   **b.** $7,776; in geometric terms, the net extra profit generated by the second plan is the area of the region between the curves $y = P_2(x)$ and $y = P_1(x)$ from $x = 0$ to $x = 18$.   **29. a.** $x = 20$ is a maximum.

b. $400   **31. a.** The revenue function is $R(q) = qp(q) = \frac{1}{4}q(10 - q)^2$; the marginal revenue function is $R'(q) = \frac{1}{4}(10 - q)(10 - 3q)$.

b. $q = 2$ **c.** The consumer's surplus is $8.67.   **33.** The consumer's surplus is approximately $2.67.   **35.** The number of people entering the fair during the prescribed time period is 1,220 people.   **37.** $1,040,256   **39.** $4,081,077   **41. a.** The second plan will be more profitable for 12 years.   **b.** $100,800   **c.**

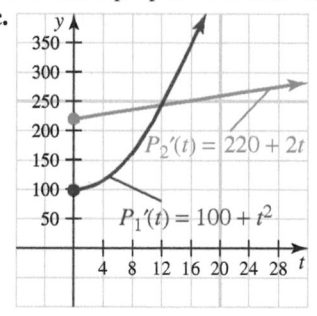

43. a. The second plan will be more profitable for about 24.5 years.

b. The excess profit is about $3.2 million.

c.

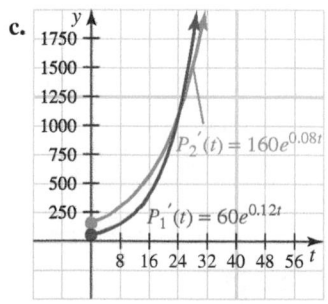

45. a. The machine will be profitable for 8 years.   **b.** The net earnings are $7,168.

c.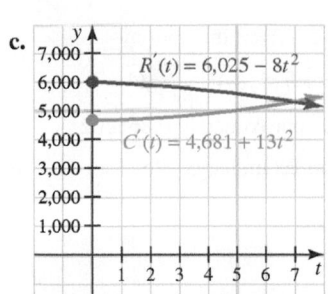

47. a. $\frac{184}{3}$ **b.**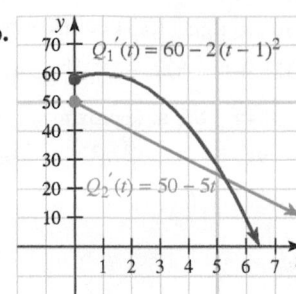

49. approximately 208,128 **51.** 0.04π cm^3/s

53. approximately 98 people **55.** approximately 515.48 billion barrels **57.** approximately 4,207

59. a. $f(t_k)e^{-rt_{k-1}}$ **b.** $\sum_{k=1}^{n} f(t_k)e^{-rt_{k-1}}\Delta_n t$, where $\Delta_n t = \frac{N}{n}$ **c.** $PV = \lim_{n\to\infty} \sum_{k=1}^{n} f(t_k)e^{-rt_{k-1}}\Delta_n \ t = \int_{0}^{N} f(t)e^{-rt}\,dt$

Chapter 6 Proficiency Examination, page 497

20. The definite integrals could represent the following:

    A. Disks revolved about the x-axis.
    B. Disks revolved about the y-axis.
    C. Slices taken perpendicular to the x-axis.
    D. Slices taken perpendicular to the y-axis
    E. Mass of a lamina with density π
    F. Washers taken along the x-axis
    G. Washers taken along the y-axis

a. All but E are formulas for volumes of solids. **b.** A, B, F, G **c.** F, G **d.** C, D **e.** A, F **f.** B, G **21.** $8\sqrt{2}$ **22.** $\frac{5}{12}$ **23.** $\frac{256}{3}$
24. a. $\frac{32}{3}$ **b.** $\frac{256\pi}{5}$ **c.** $\frac{128\pi}{3}$ **25.** $a^2(\frac{\pi}{2}-1)$ **26.** $\frac{1}{27}(13^{3/2}-8)$ **27.** $\frac{\pi}{6}(5^{3/2}-1)$ **28.** $m=\frac{5}{12}$, $M_y=\frac{13}{60}$,
$M_x = \frac{3}{140}$; the centroid is $(\frac{13}{25}, \frac{9}{175})$ **29.** $F = 128\int_{0}^{2}(5-x)\sqrt{1-(1+x)^2}\,dx$ **30.** \$7,377.37

Chapter 6 Supplementary Problems, page 498

1.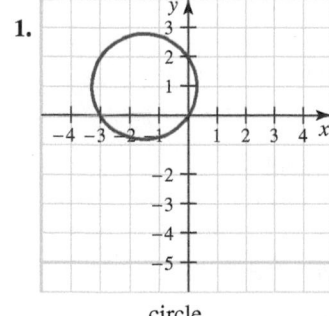

circle

3.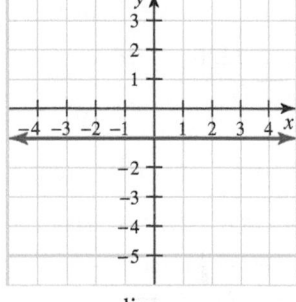

line

5.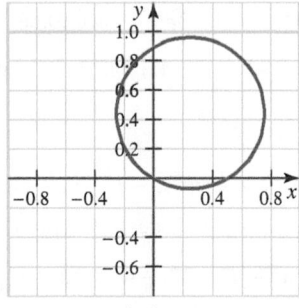

circle

7.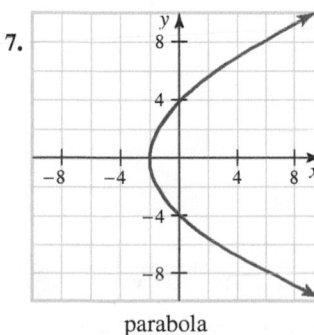

parabola

9.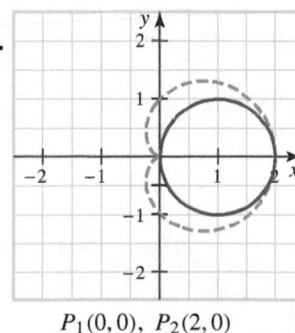

$P_1(0,0)$, $P_2(2,0)$

11.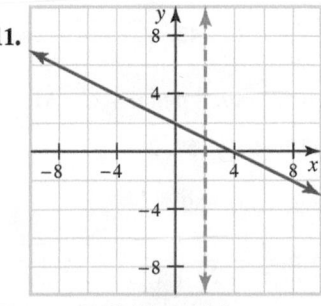

$P_1(2.24, 0.464)$

13. $\frac{9}{4}$ **15.** $\frac{4}{15}$ **17.** 8π **19.** $\frac{1}{2}$ **21.** $a^2(\frac{\sqrt{3}}{2}+\frac{\pi}{3})$ **23.** $\frac{81\pi}{4}$ **25.** $\frac{\pi}{2}(\sqrt{3}-\sqrt{2})$ **27.** 3π **29.** $\frac{3\pi}{10}$ **31.** 4π **33.** 8.6322 **35.** $\frac{\pi}{240}$
37. 225 in-lb **39.** $V_1 = \frac{16}{15}$; $V_2 = 2$; S_2 has greater volume. **41.** $S = \pi r\sqrt{r^2+h^2}$ **43.** 30,264 ft-lb **45.** $\frac{1,881\pi}{2}$ ft-lb
47. 14.04π ft-lb **49.** The spring should be stretched about 14 in. **51.** 5.625×10^{-5} joules **53.** $(\frac{7}{3}, \frac{1}{3})$ **55. a.** $\frac{4}{3}\pi ab^2$
b. $\frac{4}{3}\pi a^2 b$ **57.** $\frac{\pi b^2}{3a^2}(r^3 - 3a^2 r + 2a^3)$ **59.** \$7,191.64 **61.** 309,030 people **63. a.** 3 units **b.** \$31.50 **65.** 17,500 ft-lb **67.** 135 lb

69. $F(t) = \frac{t}{2} + \frac{2}{3}$ **71.** $6{,}176\pi$ ft-lb **73.** 11.12 cm from the right, 10.34 cm from the top, and 5.66 cm from the bottom

75. a. $\frac{35\pi}{3}$ ft-lb **b.** $\frac{70\pi}{9}$ ft-lb **77. a.** approximately 1,321 ft-lb **b.** $F = \int_D^{D+h} 62.4x[B\frac{x-D}{h}]dx$, where $h = \sqrt{A^2 - B^2/4}$

79. $S = 2\pi\sqrt{2}$ **81.** $(4{,}000)^2 P\left(\frac{1}{4{,}000} - \frac{1}{s}\right)$ mi-lb **83.** 354π **85. a.** approximately -0.644391 **b.** approximately 0.865202

87. approximately 1.728×10^{12} ft-lb **89.** $2\pi(3 - \sqrt{5})$ **91.** $\frac{\pi}{4}$ **93.** $\frac{\sqrt{3}}{8}(e^2 - 1)$ **95.** This is Putnam Examination Problem 1 of the morning session in 1939. **97.** This is Putnam Examination Problem 1 of the morning session in 1993.

Chapter 7

7.1 Review of Substitution and Integration by Table, page 516

1. $2(x^2 + 5x)^{1/2} + C$ **3.** $\ln|\ln x| + C$ **5.** $\frac{1}{4}\tan^{-1}\frac{x^2}{2} + C$ **7.** $-\frac{1}{5}(1 + \cot x)^5 + C$ **9.** $-\frac{1}{4}(x^4 - 2x^2 + 3)^{-1} + C$

11. $\ln(x^2 + x + 1) + C$ **13.** $\frac{\sqrt{x^2-a^2}}{a^2 x} + C$ **15.** $\frac{1}{2}x^2(\ln x - \frac{1}{2}) + C$ **17.** $a^{-1}e^{ax}(x - a^{-1}) + C$

19. $\frac{1}{2}x\sqrt{x^2 + 1} - \frac{1}{2}\ln(x + \sqrt{x^2 + 1}) + C$ **21.** $\frac{1}{4}\sqrt{4x^2 + 1} + C$ **23.** $\frac{-4\sin 5x - 5\cos 5x}{41e^{4x}} + C$ **25.** $b^{-1}\ln|1 + bx| + C$

27. $\frac{1}{20}(x + 1)^4(4x - 1) + C$ **29.** $\frac{1}{4}e^{4x}(x - \frac{1}{4}) + C$ **31.** $x - \frac{1}{2}\ln(1 + e^{2x}) + C$ **33.** $\sec\frac{x}{2}\tan\frac{x}{2} + \ln|\sec\frac{x}{2} + \tan\frac{x}{2}| + C$

35. $\frac{3x}{8} - \frac{\sin 2x}{4} + \frac{\sin 4x}{32} + C$ **37.** $\frac{x}{2}\sqrt{9 - x^2} + \frac{9}{2}\sin^{-1}\frac{x}{3} + C$ **41.** $-\frac{1}{5}\cos^5 x + \frac{1}{7}\cos^7 x + C$ **43.** $\frac{x}{8} - \frac{1}{32}\sin 4x + C$

45. $4\left[\frac{x^{1/2}}{2} - x^{1/4} + \ln\left(x^{1/4} + 1\right)\right] + C$ **47.** $-6(2 + \tan^3 t)^{-1} + C$ **49.** $\frac{9\pi}{2}(9 - \ln 10)$ **51.** $\frac{\pi}{35}(369\sqrt[3]{3} - 66\sqrt[3]{2})$ **53.** $\frac{3}{2}\ln^2 2$

55. $\frac{\pi}{32}(18\sqrt{5} - \ln(\sqrt{5} + 2))$

7.2 Integration by Parts, page 521

1. $-\frac{1}{4}e^{-2x}(2x + 1) + C$ **3.** $\frac{x^2}{4}(2\ln x - 1) + C$ **5.** $x\sin^{-1}x + \sqrt{1 - x^2} + C$ **7.** $\frac{4}{25}e^{-3x}\sin 4x - \frac{3}{25}e^{-3x}\cos 4x + C$

9. $\frac{x^3}{9}(3\ln x - 1) + C$ **11.** $\frac{x}{2}[\sin(\ln x) - \cos(\ln x)] + C$ **13.** $x\ln(x^2 + 1) - 2x + 2\tan^{-1}x + C$ **15.** $e^{-x}\left(\frac{1}{1-x} + 1\right) + C$

17. $\frac{32}{3}\ln 2 - \frac{28}{9}$ **19.** $e - 2$ **21.** $\frac{1}{4}(e^{2\pi} - 1)$ **25.** $\frac{1}{2}\cos^2 x(1 - 2\ln(\cos x)) + C$ **27.** $-(2 + \cos x)(\ln(2 + \cos x) + 1) + C$

29. $\frac{1}{2}(\ln|x^2 - 1| + x^2) + C$ **31.** $\frac{x}{2} + \frac{1}{4}\sin 2x + C$ **33.** $\frac{1}{4}x^2 + \frac{1}{4}x\sin 2x + \frac{1}{8}\cos 2x + C$ **35.** $\frac{x^{n+1}}{n+1}\left(\ln x - \frac{1}{n+1}\right) + C$

37. 177 units **39.** $13{,}212 **41.** $2\pi(1 - 3e^{-2})$ **43.** $(0.68, 1.27)$ **45.** $2y^{1/2} = \frac{2}{3}x^{3/2}(\ln x - \frac{2}{3}) + C$

47. $y = \exp(1 - \cos x - x\sin x)$ **49.** -3 **51.** 343,335ft $\approx$ 65 miles **53.** 0.07 **55.** $-\frac{1}{\sqrt{2}}\ln\left|\cos x + \sqrt{\cos^2 x - \frac{1}{2}}\right| + C$

7.3 Trigonometric Methods, page 530

5. $\sin x - \frac{1}{3}\sin^3 x + C$ **7.** $\frac{1}{3}\sin^3 x - \frac{1}{5}\sin^5 x + C$ **9.** $-\frac{2}{3}(\cos t)^{3/2} + C$ **11.** $-e^{\cos x} + C$ **13.** $\frac{x}{8} - \frac{\sin 4x}{32} + C$

15. $-\frac{1}{2}\ln|\cos 2\theta| + C$ **17.** $\frac{1}{6}\tan^6 x + \frac{1}{4}\tan^4 x + C$ **19.** $2\tan x - x + C$ **21.** $\frac{1}{2}\sec u\tan u - \frac{1}{2}\ln|\sec u + \tan u| + C$

23. $\frac{3}{4}(\tan x)^{4/3} + C$ **25.** $\frac{1}{4}\sin^2 x^2 + C$ **27.** $\frac{1}{4}\sec^3 t\tan t - \frac{5}{8}\sec t\tan t + \frac{3}{8}\ln|\sec t + \tan t| + C$ **29.** $-\frac{1}{3}\csc^3 x + C$

31. $-\csc x + C$ **33.** $\frac{9}{4}\sin^{-1}(\frac{2t}{3}) + \frac{1}{2}t\sqrt{9 - 4t^2} + C$ **35.** $\sqrt{4 + x^2} + \ln\left(\sqrt{4 + x^2} + x\right) + C$ **37.** $\ln|x + \sqrt{x^2 - 7}| + C$

39. $\sin^{-1}\frac{x}{\sqrt{5}} + C$ **41.** $-\frac{\sqrt{4 - x^2}}{4x} + C$ **43.** $\sqrt{x^2 - 4} - 2\sec^{-1}\frac{x}{2} + C$ **45.** $\frac{1}{3}\ln\left|\frac{x + 4}{\sqrt{9 - (x + 1)^2}}\right| + C$

47. $\ln\left|\sqrt{x^2 - 2x + 6} + x - 1\right| + C$ **49.** $\frac{1}{4}\tan^4 u + C$ **51.** $\frac{1}{2}$ **53.** 15.5031 **55.** $\frac{1}{4}\sin 2x - \frac{1}{16}\sin 8x + C$

57. $\frac{1}{8}\sin 4x - \frac{1}{8}\sin 2x - \frac{1}{40}\sin 10x + C$ **59.** 2

7.4 Method of Partial Fractions, page 539

1. $\frac{-1}{3x} + \frac{1}{3(x - 3)}$ **3.** $3 - \frac{1}{x}$ **5.** $\frac{4}{x} + \frac{-8}{2x + 1}$ **7.** $\frac{3}{x} + \frac{-1}{x^2} + \frac{1}{x + 1} + \frac{-2}{(x + 1)^2}$ **9.** $\frac{-4}{9x^2} + \frac{17}{27x} - \frac{13}{9(x + 3)^2} + \frac{10}{27(x + 3)}$

11. $\frac{-1}{4(x - 1)} + \frac{1}{4(x + 1)} + \frac{1}{2(x^2 + 1)}$ **13.** $\frac{1}{x} - \frac{1}{3(x + 1)} - \frac{1}{3(2x - 1)}$ **15.** $-9\ln|x| - x^{-1} + 6\ln|x + 1| + 5\ln|x - 1| + C$

17. $x - \frac{5}{3}\ln|x + 2| + \frac{2}{3}\ln|x - 1| + C$ **19.** $x + \frac{1}{2}\ln|x - 1| - \frac{1}{2}\ln|x + 1| - \tan^{-1}x + C$ **21.** $\ln|x + 1| + (x + 1)^{-1} + C$

23. $-\frac{1}{2}\ln|x| + \frac{1}{3}\ln|x + 1| + \frac{1}{6}\ln|x - 2| + C$ **25.** $\ln|x + 2| - \ln|x + 1| - \frac{2}{x + 2} + C$ **27.** $3\ln|x - 1| + 2\ln|x + 3| + C$

29. $\ln|x^3 - x^2 + 4x - 4| + C$ **33.** $\frac{1}{7}\ln|e^x - 3| - \frac{1}{7}\ln(2e^x + 1) + C$ **35.** $\frac{1}{1+\cos x} + C$ **37.** $\ln|\tan x + 4| + C$

39. $3x^{1/3} + 6x^{1/6} + 6\ln|x^{1/6} - 1| + C$ **41.** $-\frac{1}{5}\ln\left|\tan\frac{x}{2} - 3\right| + \frac{1}{5}\ln\left|3\tan\frac{x}{2} + 1\right| + C$ **43.** $-\ln|\sin x + \cos x| + C$

45. $-\ln(1 - \sin x) + C$ **47.** $\frac{1}{\tan\frac{x}{2} - 2} + C$ **49.** $\frac{1}{2}\ln|\ln x - 3| - \frac{1}{2}\ln|\ln x - 1| + C$ **51.** $\ln 2$ **53. a.** 1.0888 **b.** 7.6402

55. $\frac{1}{2}\ln|x^2 - 9| + C$

7.5 Summary of Integration Techniques, page 543

1. $\frac{1}{2x^2(x-1)^2} + C$ **3.** $\frac{1}{4}\ln|\sec 2x^2 + \tan 2x^2| + C$ **5.** $\ln|\sin e^x| + C$ **7.** $-\ln|\cos(\ln x)| + C$ **9.** $\frac{1}{2}\tan^{-1} e^{2t} + C$

11. $x + \frac{1}{2}\ln(x^2 + 9) - \frac{8}{3}\tan^{-1}\frac{x}{3} + C$ **13.** $x - 2\ln|1 - e^x| + C$ **15.** $x - \frac{1}{2}\ln(e^{2x} + 1) + C$ **17.** $\tan^{-1}(x + 1) + C$

19. $\frac{1}{2}e^{-x}(\sin x - \cos x) + C$ **21.** $-\cos x + \frac{1}{3}\cos^3 x + C$ **23.** $\frac{1}{5}\cos^5 x - \frac{1}{3}\cos^3 x + C$ **25.** $\frac{1}{16}x - \frac{1}{64}\sin 4x + \frac{1}{48}\sin^3 2x + C$

27. $\frac{1}{8}\tan^8 x + \frac{1}{6}\tan^6 x + C$ **29.** $\sqrt{1 - x^2} - \ln\left|\frac{1 + \sqrt{1 - x^2}}{x}\right| + C$ **31.** $\ln(\sin x + \sqrt{1 + \sin^2 x}) + C$

33. $-\frac{\cos^5 x}{5} + \frac{2\cos^3 x}{3} - \cos x + C$ **35.** π **37.** $\frac{1}{2}\ln(\sqrt{5} + 2) - \frac{1}{2}\ln(\sqrt{2} + 1) + \sqrt{5} - \frac{\sqrt{2}}{2}$ **39.** $\frac{1}{27}\left(\frac{5}{8}\sqrt{7} + 2\right)$

41. $\frac{1,792 - 64\sqrt{2}}{15}$ **43.** $\ln(\sqrt{1 + e^{2x}} + e^x) + C$ **45.** $3\ln|x| - \ln(x^2 + x + 1) + \frac{4}{\sqrt{3}}\tan^{-1}\left(\frac{2x + 1}{\sqrt{3}}\right) + C$

47. $3\ln|x + 1| - 2(x + 1)^{-1} + C$ **49.** $5\ln|x - 7| + 2(x + 2)^{-1} + C$ **53.** 2.9579 **55.** $\frac{\pi^2}{4}$

59. $\frac{x^3}{3}(\ln x)^3 - \left(\frac{x^3}{3}\right)(\ln x)^2 + 2\frac{x^3}{9}(\ln x) - 2\frac{x^3}{27} + C$

7.6 First-Order Differential Equations, page 553

1. $\frac{x^3}{3} + C$ **3.** $\frac{(5x + 1)^{3/2}}{90} + C$ **5.** $Ce^{0.02t}$ **7.** $\frac{x^2}{5} + Cx^{-3}$ **9.** $-5x^{-3} + Cx^{-2}$ **11.** $\frac{1}{3}x^{-2}e^{x^3} + Cx^{-2}$

13. $\left(\frac{1}{2}x + \frac{1}{2x}\right)\tan^{-1} x - \frac{1}{2} + \frac{C}{x}$ **15.** $-\cos x \ln|\cos x| + C\cos x$ **17.** $y = (x^3 + 125)^{1/3}$ **19.** $e^{2(\sqrt{x} - 1)}$

21. $y = x^2 - 1$ for $x > -1$ **23.** $y = \frac{1}{x^2 + 1}\left[2x\sin x + (1 - x^2)\cos x\right]$ **25.** $y = 2x^3 - 6x^2$ **27.** $2X^2 + Y^2 = C$ **29.** $Y = CX$

31. approximately \$17.0723 trillion **33.** approximately 2,310,000 **35. a.** $30 - 20e^{-t/15}$ **b.** 4 min, 19 sec

37. In the "long run" (as $t \to \infty$) the concentration will be $\lim_{t\to\infty}\frac{\alpha}{\beta}(1 - e^{-\beta t}) = \frac{\alpha}{\beta}$; the "half-way point" is reached when $t = \frac{\ln 2}{\beta}$

39. 62.4 million **41.** $\frac{k}{r}(1 - e^{-rt})$ **43.** approximately one minute **45. a.** $y(x) = -\frac{k}{24}(x^4 - 2Lx^3 + L^3 x)$ **b.** The maximum deflection of approximately $-0.0130kL^4$ occurs at $x = L/2$. **c.** The maximum deflection of approximately $-0.0054kL^4$ occurs at $x \approx 0.4215L$. It is less than that in the previous case. **47. a.** $S_1(t) = 200(1 - e^{-t/100})$ **b.** $S_2(t) = 200 - 2te^{-t/100} - 200e^{-t/100}$

c. The maximum excess is $S(100) \approx 73.58$ lb. **49.** $P(t) = P_\infty\exp\left[-\left(\ln\frac{P_\infty}{P_0}\right)e^{-kt}\right]$ **51.** $x = \frac{1}{2}(y + 1)e^y + \frac{Ce^y}{y + 1}$

53. a. $\frac{3}{2}(1 - e^{-2t})$ **b.** $e^{-2t}(1 - \cos t)$ **55.** approximately 9.872 days **57.** $y = Bx^{1/2}$

59. a. $v(t) = \frac{-mg}{k} + \left(\frac{mg}{k} + v_0\right)e^{-kt/m}$; $s(t) = \frac{-mg}{k}t + \frac{m}{k}\left(\frac{mg}{k} + v_0\right)(1 - e^{-kt/m})$

b. The object reaches its maximum height at $t_{max} = \frac{m}{k}\ln\left(1 + \frac{kv_0}{mg}\right)$. The maximum height is $s_{max} = \frac{mv_0}{k} - \frac{m^2g}{k^2}\ln\left(1 + \frac{kv_0}{mg}\right)$ **c.** 83 ft; approximately 5.5 seconds **d.** It will take longer.

7.7 Improper Integrals, page 566

3. $\frac{1}{2}$ **5.** diverges **7.** 100 **9.** diverges **11.** $\frac{1}{10}$ **13.** $\frac{1}{9}$ **15.** diverges **17.** $\frac{2}{e}$ **19.** $5e^{10}$ **21.** diverges **23.** diverges

25. diverges **27.** 0 **29.** $\frac{5}{4}$ **31.** 2 **33.** $\frac{3}{2}\left[(1 - e^{-1})^{2/3} - (1 - e)^{2/3}\right]$ **35.** -1 **37.** 1 **39.** 2 **41.** diverges **43.** $\frac{1}{4}$

45. 100,000 millirads **47.** $(p - 1)^{-1}(\ln 2)^{1-p}$ if $p > 1$ and diverges if $p \leq 1$ **49.** $(1 - p)^{-1}(-\ln 2)^{1-p}$ if $p > 1$ and

diverges if $p \leq 1$ **51.** l'Hôpital's rule misapplied **57. b.** $\frac{s^2 - 4}{(s^2 + 4)^2}$ **59.** $F''(s)$; $(-1)^n F^{(n)}(s)$

7.8 Hyperbolic and Inverse Hyperbolic Functions, page 574

1. 3.6269 **3.** -0.7616 **5.** 0.0000 **7.** 1.1995 **9.** 0.7500 **11.** 2.2924 **13.** $3\cosh 3x$ **15.** $-4x\sinh(1-2x^2)$

17. $-x^{-2}\cosh x^{-1}$ **19.** $\dfrac{3x^2}{\sqrt{1+x^6}}$ **21.** $|\sec x|$ **23.** $\sec x$ **25.** $\dfrac{x-\sqrt{1+x^2}\,\sinh^{-1}x}{x^2\sqrt{1+x^2}}$ **27.** $\cosh^{-1}x$ **29.** $-\frac{1}{2}\sinh(1-x^2)+C$

31. $-\cosh x^{-1}+C$ **33.** $\frac{1}{3}\cosh^{-1}\frac{3t}{4}+C$ **35.** $\sinh^{-1}(\sin x)+C$ **37.** $\frac{1}{3}\tanh^{-1}x^3+C$ **39.** $\coth^{-1}3-\coth^{-1}2$
41. $\cosh^{-1}e^2-\cosh^{-1}e$ **43.** $\frac{1}{2}\tanh 1$ **47.** The curve is rising for all x. It is concave up for $x<0$ and concave down for $x>0$.
51. b. $\cosh 2x+\sinh 2x$ **53.** $(e-e^{-1})a$ **55.** $\frac{\pi}{2}(e^2-e^{-2}+4)$

Chapter 7 Proficiency Examination, page 576

12. a. $\frac{1}{2}\ln 3$ **b.** $\frac{4}{3}$ **c.** $\frac{1}{2}\ln 3$ **13.** $2\sqrt{x^2+1}+3\sinh^{-1}x+C$ **14.** $-\frac{x}{2}\cos 2x+\frac{1}{4}\sin 2x+C$ **15.** $-\frac{1}{2}\cosh(1-2x)+C$

16. $\sin^{-1}\frac{x}{2}+C$ **17.** $\frac{1}{4}\ln(x^2+1)+\frac{1}{2}\tan^{-1}x+\frac{1}{2}\ln|x-1|+C$ **18.** $\frac{x^2}{2}+\frac{1}{2}\ln|x^2-1|+C$ **19.** $6\ln 2-\frac{9}{4}$ **20.** $\frac{1}{9}\ln\frac{5}{8}+\frac{1}{6}$

21. $\frac{1}{2}\ln\frac{2}{3}$ **22.** $\dfrac{2\sqrt{2}-1}{3}$ **23.** $\frac{1}{4}$ **24.** 2 **25.** diverges **26.** $\frac{1}{2}$ **27.** $\dfrac{1}{\sqrt{\tanh^{-1}2x}}\left[\dfrac{1}{1-4x^2}\right]$ **28.** $2\pi(3-\sqrt{5})$

29. $y=\dfrac{x+1}{e^x}(\ln|x+1|+1)$ **30.** 485.2 lb

Chapter 7 Supplementary Problems, page 577

1. $(1-x^2)^{-1}$ **3.** e^{-2x} **5.** $x\cosh x+\sinh x+(e^x-e^{-x})\cosh(e^x+e^{-x})$ **7.** $-\frac{x}{2}\sqrt{4-x^2}+2\sin^{-1}\frac{x}{2}+C$

9. $-\ln|x|-x^{-1}+\ln|x-2|+C$ **11.** $4x^{1/4}-4\ln(1+x^{1/4})+C$ **13.** $\frac{x^3}{3}\tan^{-1}x-\frac{x^2}{6}+\frac{1}{6}\ln(1+x^2)+C$

15. $\frac{1}{2}e^x\sqrt{4-e^{2x}}+2\sin^{-1}\left(\frac{e^x}{2}\right)+C$ **17.** $-\frac{1}{5}(1+x^{-2})^{5/2}+\frac{1}{3}(1+x^{-2})^{3/2}+C$ **19.** $-2\sqrt{1-\sin x}+C$

21. $\frac{x}{2}[\sin(\ln x)-\cos(\ln x)]+C$ **23.** $\ln(2+\cosh x)+C$ **25.** $\frac{1}{6}(2x^3\cot^{-1}x+x^2-\ln(1+x^2))+C$ **27.** $\frac{1}{3}\ln|x^3+6x+1|+C$

29. $2\cos\sqrt{x+2}+2\sqrt{x+2}\sin\sqrt{x+2}+C$ **31.** $\frac{1}{4}\ln(x^4+4x^2+3)+C$ **33.** $\sqrt{5-x^2}-\sqrt{5}\ln\left|\dfrac{\sqrt{5}+\sqrt{5-x^2}}{x}\right|+C$

35. $\frac{1}{3}\sqrt{x^2+4}(x^2-8)+C$ **37.** $\dfrac{\sec^3 x}{3}+C$ **39.** $\frac{3}{2}x^{2/3}-2x^{1/2}+3x^{1/3}-6x^{1/6}+6\ln(x^{1/6}+1)-C$

41. $-\ln|1+\cos x-\sin x|+C$ **43.** $-2e^{x/2}-3e^{x/3}-6e^{x/6}-6\ln|e^{x/6}-1|+C$ **45.** $\frac{1}{3}\tan^3 x+\tan x+C$
47. $-\frac{1}{2}\ln|x+1|+2\ln|x+2|-\frac{3}{2}\ln|x+3|+C$ **49.** $y=\dfrac{1}{x+1}[x+C]$ **51.** $y=ke^{2\sqrt{x}}$

53. $\ln|\sec y+\tan y|=\frac{1}{2}\tan^2 x+\ln|\cos x|+C$ **55.** 1 **57.** 2 **59.** $\frac{3}{4}(3^{2/3}-1)$ **61.** -1 **63.** $T=T_0+Be^{kt},B$ constant

67. $\left(\frac{1}{3}\pi-\ln\frac{2}{\sqrt{3}},1\right)$ **69.** 2.85 **71.** $\frac{8}{15}(1+\sqrt{2})$ **73.** $\pi\tan^{-1}16$ **75.** $\pi(e^2-\frac{1}{16}e^{-2}+\frac{1}{16})$ **77.** $A_1=A_2=\ln|a|$

79. $3|a|+\dfrac{b^2}{8|a|}\ln 2$ **85.** $\frac{1}{2};\frac{3}{4}\ln 2$ **87.** $\frac{3\pi}{16}$ **89.**

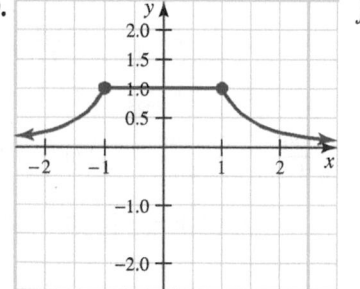

$\int_{-\infty}^{\infty}f(x)\,dx=4$ **91.** $\frac{3}{40}$

93. a. $v(t)=\left(\frac{W-B}{k}\right)\left(1-e^{-(kg/W)t}\right)$ **b.** $s(t)=\dfrac{W-B}{k}\left[t+\dfrac{W}{kg}e^{-(kg/W)t}-\dfrac{W}{kg}\right]$ **c.** 269 m **95.** approximately 5.6 years

97. This is Putnam Examination Problem 1 in the morning session of 1968.
99. This is Putnam Examination Problem B5 in the afternoon session of 1985.

Chapter 8

8.1 Sequences and Their Limits, page 594

3. $0, 2, 0, 2, 0$ **5.** $1, \frac{1}{2}, \frac{1}{3}, \frac{1}{4}, \frac{1}{5}$ **7.** $\frac{4}{3}, \frac{7}{4}, \frac{10}{5}, \frac{13}{6}, \frac{16}{7}$ **9.** $256, 16, 4, 2, \sqrt{2}$ **11.** $1, 3, 13, 183, 33673$ **13.** 5 **15.** -7 **17.** 0 **19.** $\frac{1}{2}$ **21.** 4
23. 0 **25.** 1 **27.** 1 **29.** 1 **31.** $\frac{1}{2}$ **33.** 1 **35.** 1 **47. a.** converges to 0 **b.** converges to 1 **c.** If $n \to \infty$, $m \to 0$
49. $\frac{\pi}{4}$ **51.** 99 **53.** 1,000

8.2 Introduction to Infinite Series: Geometric Series, page 603

3. 5 **5.** 3 **7.** diverges **9.** $\frac{5}{27}$ **11.** $\frac{1}{e^{(1/5)} - 1}$ **13.** $-\frac{2}{45}$ **15.** $\frac{1}{3}$ **17.** $\frac{16}{63}$ **19.** $2(2 + \sqrt{2})$ **21.** $\frac{1}{2}(4 + 3\sqrt{2})$ **23.** converges to 1
25. converges to 1 **27.** diverges **29.** converges to 1 **31.** $\frac{1}{99}$ **33.** $\frac{52}{37}$ **35. a.** $A = 1$, $B = 1$ **b.** $\frac{1}{2}$ **37.** 0 **39.** 9.14 **41.** $\frac{583}{120}$
43. $132,335.83 **45.** 1,500 **47.** 3 ft **49.** $10,000 **51.** approximately 30.8 units **53.** 33 **55. a.** 4 miles **59.** 3

8.3 The Integral Test; p-series, page 612

3. $p = 3$; converges **5.** $p = \frac{1}{3}$; diverges **7.** converges **9.** diverges **11.** converges **13.** diverges **15.** diverges **17.** diverges
19. converges **21.** diverges **23.** converges **25.** diverges **27.** diverges **29.** diverges **31.** converges **33.** converges
35. diverges **37.** diverges **39.** diverges **41.** diverges **43.** converges **45.** converges **47.** diverges **49.** diverges
51. converges if $p > 1$ and diverges if $p \leq 1$ **53.** converges if $p > 1$ and diverges if $p \leq 1$
55. converges if $p > 1$ and diverges if $p \leq 1$ **57.** false

8.4 Comparison Tests, page 618

1. converges if $|r| < 1$ and diverges if $|r| \geq 1$ **3.** geometric; converges **5.** geometric; diverges **7.** p-series; diverges
9. p-series; converges **11.** geometric; diverges **13.** converges **15.** diverges **17.** diverges **19.** converges **21.** converges
23. converges **25.** diverges **27.** converges **29.** converges **31.** converges **33.** diverges **35.** diverges **37.** converges
39. converges **41.** diverges **43.** converges **45.** converges **47.** converges **49.** diverges

8.5 The Ratio Test and the Root Test, page 625

3. converges **5.** diverges **7.** converges **9.** converges **11.** diverges **13.** converges **15.** converges **17.** converges **19.** diverges
21. converges **23.** converges **25.** converges **27.** diverges **29.** converges **31.** diverges **33.** converges **35.** converges
37. diverges **39.** converges **41.** converges **43.** converges **45.** converges **47.** converges for $0 \leq x < 1$
49. converges for $0 \leq x \leq 0.5$ **51.** converges for $x \geq 0$ **53.** converges for $0 \leq x < a^{-1}$ **55.** The integral also converges.

8.6 Alternating Series; Absolute and Conditional Convergence, page 636

3. converges conditionally **5.** diverges **7.** converges absolutely **9.** converges absolutely **11.** converges conditionally
13. diverges **15.** converges absolutely **17.** converges absolutely **19.** converges conditionally **21.** converges conditionally
23. converges conditionally **25.** converges absolutely **27.** diverges **29.** diverges **31. a.** $\frac{51}{64}$ **b.** $S_7 \approx 0.800$ **33. a.** $-\frac{115}{144}$
b. $S_{45} \approx -0.823$ (using software) **35. a.** $\frac{1,549}{1,728}$ **b.** $S_{13} \approx 0.902$ **37.** $[-1, 1)$ **39.** $(-\infty, \infty)$ **41.** $(-\infty, \infty)$ **43.** 0.1667
45. 0.4809 **47.** It converges absolutely for $p > 1$ and conditionally for $p \leq 1$ **49.** 0 **53.** Follow the outline found in the preceding
problem. The complete solution can be found on page 89 of Volume 83 (1983) of *School Science and Mathematics* published by Wiley.

8.7 Power Series, page 647

1. $(-1, 1)$ **3.** $(-1, 1)$ **5.** $(\frac{8}{3}, \frac{10}{3})$ **7.** $(-\frac{13}{3}, -\frac{5}{3})$ **9.** $x = 1$ **11.** $(-1, 3)$ **13.** $[1, \frac{5}{3})$ **15.** $(-7, 7)$ **17.** $x = 0$ **19.** $[-1, 1]$
21. $(-\infty, \infty)$ **23.** $x = 0$ **25.** $(-\infty, \infty)$ **27.** $[-1, 1]$ **29.** $R = 1$ **31.** $R = e$ **33.** $R = 4$ **35.** $R = \frac{1}{|a|}$ **37.** $\sum_{k=1}^{\infty} \frac{kx^{k-1}}{2^k}$

39. $\sum_{k=1}^{\infty} k(k+2)x^{k-1}$ **41.** $\sum_{k=1}^{\infty} kx^{k-1}$ **43.** $\sum_{k=1}^{\infty} \frac{x^k}{k2^{k-1}}$ **45.** $\sum_{k=0}^{\infty} \frac{(k+2)x^{k+1}}{k+1}$ **47.** $\sum_{k=0}^{\infty} \frac{x^{k+1}}{(k+1)!}$

49. $\sum_{k=2}^{\infty} \frac{1}{2}k(k-1)x^{k-2}$; $(-1, 1)$ **51.** $R = \infty$ **59.** $|x| > 1$

8.8 Taylor and Maclaurin Series, page 661

3. $e^{2x} = \sum_{k=0}^{\infty} \frac{(2x)^k}{k!}$ **5.** $e^{x^2} = \sum_{k=0}^{\infty} \frac{x^{2k}}{k!}$ **7.** $\sin x^2 = \sum_{k=0}^{\infty} \frac{(-1)^k x^{4k+2}}{(2k+1)!}$ **9.** $\sin ax = \sum_{k=0}^{\infty} \frac{(-1)^k (ax)^{2k+1}}{(2k+1)!}$

11. $\cos 2x^2 = \sum\limits_{k=0}^{\infty} \dfrac{(-1)^k (2x^2)^{2k}}{(2k)!}$ **13.** $x^2 \cos x = \sum\limits_{k=0}^{\infty} \dfrac{(-1)^k x^{2k+2}}{(2k)!}$ **15.** $x^2 + 2x + 1$ is its own Maclaurin series.

17. $xe^x = \sum\limits_{k=0}^{\infty} \dfrac{x^{k+1}}{k!}$ **19.** $e^x + \sin x = 1 + 2x + \dfrac{x^2}{2!} + \dfrac{x^4}{4!} + \dfrac{2x^5}{5!} + \dfrac{x^6}{6!} + \dfrac{x^8}{8!} + \dfrac{2x^9}{9!} + \cdots$ **21.** $\dfrac{1}{1+4x} = \sum\limits_{k=0}^{\infty} (-4)^k x^k$

23. $\dfrac{1}{a+x} = \sum\limits_{k=0}^{\infty} \dfrac{(-1)^k x^k}{a^{k+1}}$ **25.** $\ln(3+x) = \ln 3 + \sum\limits_{k=0}^{\infty} \dfrac{(-1)^k x^{k+1}}{(k+1)3^{k+1}}$ **27.** $e^x \approx e + e(x-1) + \frac{1}{2!}e(x-1)^2 + \frac{1}{3!}e(x-1)^3$

29. $\cos x \approx \cos\dfrac{\pi}{3} - (x - \dfrac{\pi}{3})\sin\dfrac{\pi}{3} - \dfrac{(x-\frac{\pi}{3})^2}{2!}\cos\dfrac{\pi}{3} + \dfrac{(x-\frac{\pi}{3})^3}{3!}\sin\dfrac{\pi}{3}$ **31.** $\tan x \approx 0 + 1 \cdot x + \dfrac{0x^2}{2!} + \dfrac{2x^3}{3!}$

33. $\dfrac{1}{2-x} \approx -\dfrac{1}{3} + \dfrac{1}{9}(x-5) - \dfrac{1}{27}(x-5)^2 + \dfrac{1}{81}(x-5)^3$ **35.** $\dfrac{3}{2x-1} \approx 1 - \dfrac{2}{3}(x-2) + \dfrac{4}{9}(x-2)^2 - \dfrac{8}{27}(x-2)^3$

37. $\sqrt{1+x} = 1 + \frac{1}{2}x - \frac{1}{8}x^2 + \frac{1}{16}x^3 - \frac{5}{128}x^4 + \cdots$ Interval of convergence is $[-1,1]$.

39. $(1+x)^{2/3} = 1 + \frac{2}{3}x - \frac{1}{9}x^2 + \frac{4}{81}x^3 - \cdots$ Interval of convergence is $[-1,1]$.

41. $\dfrac{x}{\sqrt{1-x^2}} = x + \dfrac{1}{2}x^3 + \dfrac{3}{8}x^5 + \dfrac{5}{16}x^7 + \cdots$ Interval of convergence is $(-1,1)$.   **43.** $\sinh x = \sum\limits_{k=0}^{\infty} \dfrac{x^{2k+1}}{(2k+1)!}$

45. $\sqrt[3]{e} \approx 1.396; n = 4$ **47.** $\dfrac{1}{x^2 - 3x + 2} = \sum\limits_{k=0}^{\infty}\left[1 - \dfrac{1}{2^{k+1}}\right]x^k$ **49.** $\sin x \cos x = \dfrac{1}{2}\sin 2x = \sum\limits_{k=0}^{\infty} \dfrac{(-1)^k 2^{2k} x^{2k+1}}{(2k+1)!}$

51. $\ln\left[\dfrac{1+2x}{1-3x+2x^2}\right] = \sum\limits_{k=0}^{\infty}[(-1)^k 2^{k+1} + 2^{k+1} + 1]\dfrac{x^{k+1}}{k+1}$ **53. a.** 2   **b.** 1

Chapter 8 Proficiency Examination, page 664

21. e **22. a.** 0   **b.** converges   **23.** diverges   **24.** diverges   **25.** diverges   **26.** diverges   **27.** converges absolutely   **28.** $(-1,1)$
29. $\sin 2x = \sum\limits_{k=0}^{\infty} \dfrac{(-1)^k (2x)^{2k+1}}{(2k+1)!}$ **30.** $\dfrac{1}{x-3} = -\sum\limits_{k=0}^{\infty} \left(\dfrac{2}{5}\right)^{k+1}\left(x - \dfrac{1}{2}\right)^k$

Chapter 8 Supplementary Problems, page 665

1. diverges   **3.** converges to e^{-2} **5.** converges to 1   **7.** converges to $\frac{5}{3}$ **9.** converges to 0   **11.** converges to 0   **13.** converges to e^4
15. converges to 0   **17.** $\frac{1}{3}$ **19.** $\dfrac{e}{3-e}$ **21.** $\frac{1}{2}$ **23.** $\dfrac{15-e}{9(6-e)}$ **25.** $\frac{1}{4}$ **27.** diverges   **29.** diverges   **31.** converges   **33.** converges
35. converges   **37.** converges   **39.** diverges   **41.** diverges   **43.** converges   **45.** diverges   **47.** converges absolutely   **49.** converges
absolutely   **51.** diverges   **53.** converges absolutely   **55.** converges absolutely   **57.** $(0,2)$ **59.** $(-1,1)$ **61.** $(-\infty,\infty)$
63. $[-3,-1)$ **65.** $x = 0$ **67.** $(-\infty,\infty)$ **69.** $x^3 \sin x = \sum\limits_{k=0}^{\infty}(-1)^k \dfrac{x^{2k+4}}{(2k+1)!}$ **71.** $x^2 + \tan^{-1} x = x + x^2 + \sum\limits_{k=1}^{\infty} \dfrac{(-1)^k x^{2k+1}}{2k+1}$

73. $\dfrac{11x-1}{2+x-3x^2} = \sum\limits_{k=0}^{\infty}\left[2 + \dfrac{5(-1)^{k+1}}{2}\left(\dfrac{3}{2}\right)^k\right]x^k$ **75.** 0.000022   **77.** $\tan x = x + \dfrac{x^3}{3} + \dfrac{2x^5}{15} + \cdots;$

$\ln|\cos x| = -\dfrac{1}{2}x^2 - \dfrac{1}{12}x^4 - \dfrac{1}{45}x^6 - \cdots$ **79. a.** $a_k = (-1)^k (4k+1)\left[\dfrac{1 \cdot 3 \cdot 5 \cdots (2k-1)}{2 \cdot 4 \cdot 6 \cdots (2k)}\right]^3$ **83. c.** $\sqrt{2}-1$ **d.** $\frac{3}{4}$

87. $\int \sqrt{x^3+1}\,dx = x + \frac{1}{8}x^4 - \frac{1}{56}x^7 + \frac{1}{160}x^{10} + \cdots$ **91.** 0.19867   **93.** $R_3(x) < 10^{-12}$ **95.** $|x| < 0.187$
97. This is Putnam Problem A2 in the morning session of 1984.
99. This is Putnam Problem A2 in the morning session of 1982.

Cumulative Review Problems—Chapters 6-8, page 671

7. $\frac{1}{2}x\ln x - \frac{1}{2}x + C$ **9.** $\frac{1}{2}(\tan^{-1} x)^2 + C$ **11.** $2\sin^{-1}\left(\frac{x}{2}\right) + \frac{1}{2}x\sqrt{4-x^2} + C$ **13.** $3\ln|\sec t + \tan t| + 2\ln|\sec t| + C$
15. $\dfrac{2}{\sqrt{3}}\tan^{-1}\left(\dfrac{2x+1}{\sqrt{3}}\right) + C$ **17.** $x\cos^{-1}(-x) + \sqrt{1-x^2} + C$ **19.** $\dfrac{1}{e^x + e^{-x}} + C$ **21.** $\frac{1}{4}$ **23.** 2   **25. a.** $\dfrac{3}{x-1} + \dfrac{2}{(x-1)^2} - \dfrac{5}{2x+3}$
b. $3\ln|x-1| - \dfrac{2}{x-1} - \dfrac{5}{2}\ln|2x+3| + C$ **c.** diverges   **27.** converges   **29.** diverges   **31.** converges   **33.** -6 **35.** $[-1,1]$
37. $y = x\tan^{-1} x + Cx$ **39.** $\sin x^2 = \sum\limits_{k=0}^{\infty} \dfrac{(-1)^k x^{4k+2}}{(2k+1)!}$ **41.** $\dfrac{3x-2}{5+2x} = \dfrac{3}{2} - \dfrac{19}{10}\sum\limits_{k=0}^{\infty}\left(-\dfrac{2x}{5}\right)^k$ **43.** 16,937 ft-lb   **45. a.** $\pi\ln\frac{9}{4}$
b. $\frac{64\pi}{3}$ **c.** $\pi\left[\ln\frac{9}{4} + 4\right]$ **47.** 4   **51.** 2   **55.** $\frac{1}{2} + \frac{11\sqrt{3}}{160}$ **57.** $\frac{162\pi}{5}$ **59.** approximately $(0.71, 0.12)$

Chapter 9

9.1 Vectors in $\mathbb{R}^2$, page 684

1.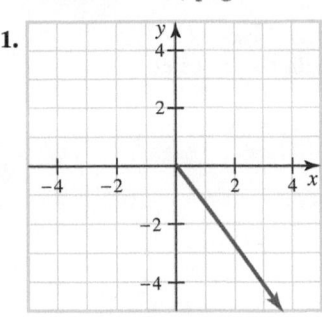

3.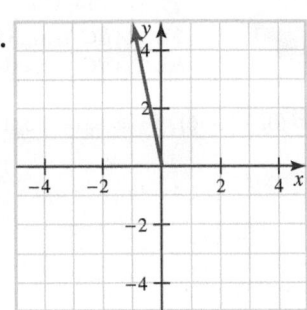

5. 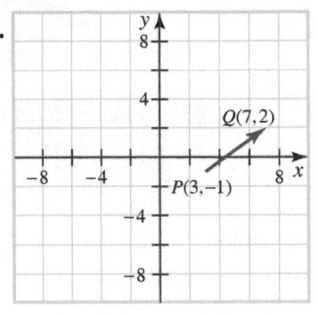 $\langle 4, 3\rangle$; $\|\mathbf{PQ}\| = 5$

7. 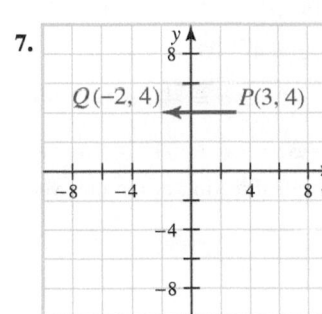 $\langle -5, 0\rangle$; $\|\mathbf{PQ}\| = 5$ **9.** $\mathbf{PQ} = 2\mathbf{i}$; $\|\mathbf{PQ}\| = 2$ **11.** $\mathbf{PQ} = 4\mathbf{i} + 2\mathbf{j}$; $\|\mathbf{PQ}\| = 2\sqrt{5}$

13. $\frac{1}{\sqrt{2}}(\mathbf{i} + \mathbf{j})$ **15.** $\frac{3}{5}\mathbf{i} - \frac{4}{5}\mathbf{j}$ **17.** $s = 6$; $t = 24$ **19.** $s = -1$; $t = -5$ **21.** $17\mathbf{i} - 18\mathbf{j}$

23. $35\mathbf{i} - 35\mathbf{j}$ **25.** $(3, 2)$ **27.** $(2, 4), (-4, -2)$ **29.** $\frac{\sqrt{3}}{2}\mathbf{i} + \frac{1}{2}\mathbf{j}$ **31.** $\frac{4}{\sqrt{17}}\mathbf{i} - \frac{1}{\sqrt{17}}\mathbf{j}$

33. $\frac{5}{\sqrt{26}}\mathbf{i} + \frac{1}{\sqrt{26}}\mathbf{j}$ **35.** $(3, 10)$ **37. a.** $6\mathbf{i} + 3\mathbf{j}$ **b.** $10\mathbf{i} + 5\mathbf{j}$

39. $\|\mathbf{v}\| = \sqrt{\cos^2\theta + \sin^2\theta} = 1$ **41.** Not necessarily equal   **43. a.** This is the set of all points on the circle with center (x_0, y_0) and radius 1.   **b.** This is the set of all points on or interior to the circle with center (x_0, y_0) and radius r.   **45.** $a = -2t$, $b = t$, and $c = t$, for any t **47.** $-6\mathbf{i} + 3\mathbf{j}$ **49.** 5.22 mi/h; N53.6°W

9.2 Coordinates and Vectors in $\mathbb{R}^3$, page 694

1. a. $\langle 2, 2, 4\rangle$ **b.** $\langle 6, -8, -2\rangle$ **c.** $\langle 10, -\frac{15}{2}, \frac{5}{2}\rangle$ **d.** $\langle 2, 9, 11\rangle$ **3. a.** $\langle 1, -3, 8\rangle$ **b.** $\langle 1, -1, 2\rangle$ **c.** $\langle \frac{5}{2}, -5, \frac{25}{2}\rangle$ **d.** $\langle 2, -7, 19\rangle$

5. $d = \sqrt{149}$ 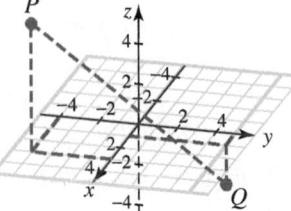   **7.** $d = \sqrt{382}$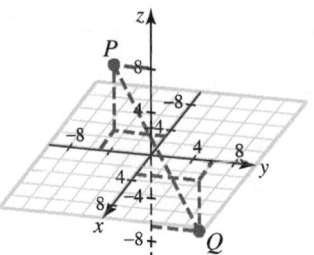

9. $x^2 + y^2 + z^2 = 1$ **11.** $x^2 + (y - 4)^2 + (z + 5)^2 = 9$ **13.** $C(0, 1, -1)$; $r = 2$ **15.** $C(3, -1, 1)$; $r = 1$

17. 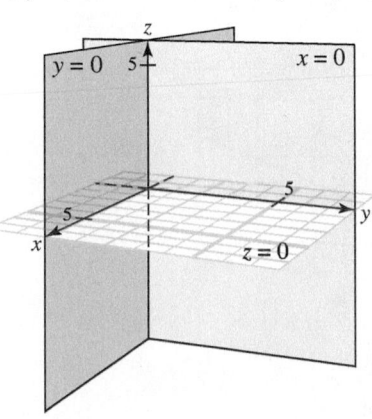   **19.**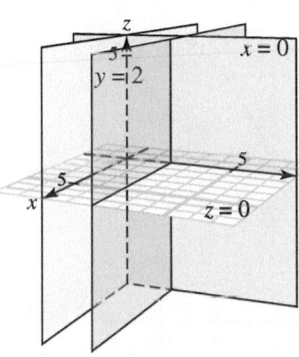

21. $\mathbf{PQ} = -2\mathbf{i} + 2\mathbf{j} + \mathbf{k}$; $\|\mathbf{PQ}\| = 3$

23. $\mathbf{PQ} = -4\mathbf{i} - 4\mathbf{j} - 4\mathbf{k}$; $\|\mathbf{PQ}\| = 4\sqrt{3}$

25. $-7\mathbf{i} - 16\mathbf{k}$

27. $13\mathbf{i} - 4\mathbf{j} + 19\mathbf{k}$

29. $\frac{1}{\sqrt{14}}\langle 3, -2, 1\rangle$

31. $\frac{1}{5\sqrt{2}}\langle -5, 3, 4\rangle$

33.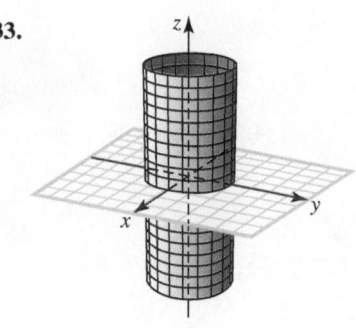

35.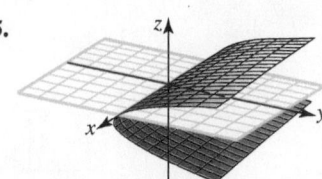

37. $(2x + 1)^2 + (2y - 5)^2 + 4z^2 = 46$ **39.** $\sqrt{3}$

41. 66   **43.** $6\mathbf{i} + 12\mathbf{j} - 3\mathbf{k}$ **45.** $\sqrt{46}(3\mathbf{i} + 2\mathbf{j} + \mathbf{k})$

47. no   **49.** yes

51. $\triangle ABC$ is right but not isosceles.

53. $\triangle ABC$ is neither right nor isosceles.

55. collinear   **57.** $\mathbf{F}_1 = \frac{25}{3}\langle 0, -2, -6\rangle$, $\mathbf{F}_2 = \frac{25}{3}\langle -\sqrt{3}, 1, -6\rangle$, $\mathbf{F}_3 = \frac{25}{3}\langle \sqrt{3}, 1, -6\rangle$

59. $P\left(\frac{1}{3}, \frac{7}{3}, \frac{13}{3}\right)$

9.3 The Dot Product, page 703

3. -16 **5.** 5   **7.** orthogonal   **9.** orthogonal   **11.** 11   **13.** 0   **15.** $-4\mathbf{i} - 16\mathbf{j} + 7\mathbf{k}$ **17.** $\sqrt{321}$ **19.** $\sqrt{3}$ **21.** 14   **23.** $71°$

25. $114°$ **27.** $1; \mathbf{k}$ **29.** $0; \mathbf{0}$ **31.** $\pm\langle \frac{1}{\sqrt{2}}, 0, \frac{1}{\sqrt{2}}\rangle$ **33.** $\mathbf{u} = \frac{-1}{\sqrt{17}}(2\mathbf{i} + 3\mathbf{j} - 2\mathbf{k})$ **35.** $x = -1, y = -1, z = 4$ **37.** $a = \frac{3}{2}$

39. $\cos\alpha = \frac{2}{\sqrt{38}}$; $\alpha \approx 1.24$ or $71°$; $\cos\beta = \frac{-3}{\sqrt{38}}$; $\beta \approx 2.08$ or $119°$; $\cos\gamma = \frac{-5}{\sqrt{38}}$; $\gamma \approx 2.52$ or $144°$ **41.** $\cos\alpha = \frac{1}{\sqrt{2}}$; $\alpha \approx 0.79$ or $45°$; $\cos\beta = \frac{-4}{5\sqrt{2}}$; $\beta \approx 2.17$ or $124°$; $\cos\gamma = \frac{3}{5\sqrt{2}}$; $\gamma \approx 1.13$ or $65°$ **43. a.** 26   **b.** $\frac{26}{35}$ **c.** $\frac{49}{26}$ **d.** $\frac{-26}{25}$

45. $\cos\theta = -\frac{1}{6}$; $-\frac{1}{6}(2\mathbf{i} + \mathbf{j} - \mathbf{k})$ **47.** 10   **49.** The force has magnitude $\|\mathbf{F}\| \approx 49.53$ lb and points in the direction of the unit vector $\langle 0.979, -0.025, 0.202\rangle$ **51. a.** 500 ft-lb   **b.** $500\sqrt{2}$ ft-lb

9.4 The Cross Product, page 714

1. $\mathbf{k}$ **3.** $-2\mathbf{i} + 4\mathbf{j} + 3\mathbf{k}$ **5.** $-2\mathbf{i} + 25\mathbf{j} + 14\mathbf{k}$ **7.** $-2\mathbf{i} + 16\mathbf{j} + 11\mathbf{k}$ **9.** $-14\mathbf{i} - 4\mathbf{j} - \mathbf{k}$ **11.** $\frac{\sqrt{3}}{2}$ **13.** $\frac{\sqrt{3}}{2}$

15. $\pm\left(\frac{1}{\sqrt{14}}\mathbf{i} + \frac{3}{\sqrt{14}}\mathbf{j} - \frac{2}{\sqrt{14}}\mathbf{k}\right)$ **17.** $\pm\left(\frac{-16}{\sqrt{386}}\mathbf{i} + \frac{7}{\sqrt{386}}\mathbf{j} + \frac{9}{\sqrt{386}}\mathbf{k}\right)$ **19.** $\sqrt{26}$ **21.** $2\sqrt{59}$ **23.** $\frac{1}{2}\sqrt{3}$ **25.** $\frac{3}{2}\sqrt{3}$

27. a. does not exist   **b.** scalar   **29. a.** scalar   **b.** vector   **31.** 5   **33.** 8   **37.** $s = 2$ **39.** Cross product is not an associative operation.   **41.** $\mathbf{w} = \mathbf{0}$ **43.** $\mathbf{T} = -40\sqrt{3}\mathbf{i}$ **45.** $40°$

9.5 Lines in $\mathbb{R}^3$, page 724

3. $x = 1 + 3t, y = -1 - 2t, z = -2 + 5t$;   **5.** $x = 1 + t, y = -1 + 2t, z = 2 + t$;   **7.** $x = 1 + t, y = -3 - 3t, z = 6 - 5t$;

$\dfrac{x-1}{3} = \dfrac{y+1}{-2} = \dfrac{z+2}{5}$ $\dfrac{x-1}{1} = \dfrac{y+1}{2} = \dfrac{z-2}{1}$ $\dfrac{x-1}{1} = \dfrac{y+3}{-3} = \dfrac{z-6}{-5}$

9. $x = 11t, y = 4 - 3t, z = -3 + 5t$;   **11.** $x = 3 + t, y = -1, z = 0$ **13.** $(0, -6, -3), (8, 0, -1), (12, 3, 0)$

$\dfrac{x}{11} = \dfrac{y-4}{-3} = \dfrac{z+3}{5}$ **15.** $(0, 4, 9), (8, 0, -3), (6, 1, 0)$ **17.** parallel   **19.** parallel   **21.** $(1, 2, 3)$

23. $y = x - 1, 0 \leq x \leq 3$

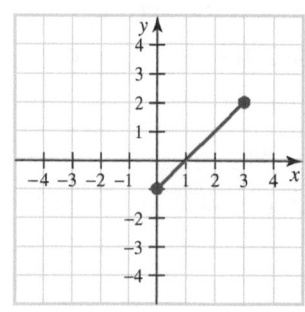

25. $y = \dfrac{4}{3}x - \dfrac{1}{225}x^2, 0 \leq x \leq 180$

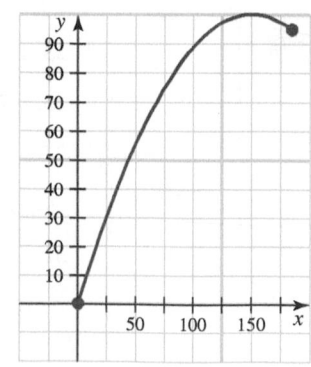

27. $y = x^{2/3}, x \geq 0$

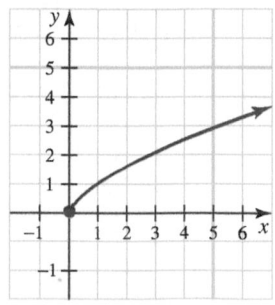

29. $x^2 + y^2 = 9,$

$-3 \le x \le 3$

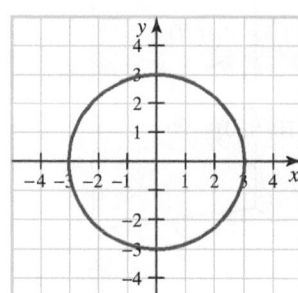

31. $(x-1)^2 + (y+2)^2 = 1,$

$0 \le x \le 2$

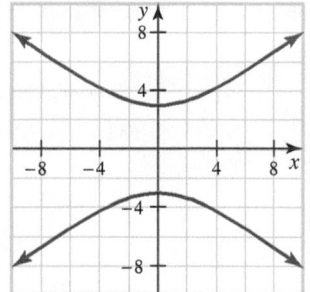

33. $\dfrac{y^2}{9} - \dfrac{x^2}{16} = 1,$

$-\infty < x < \infty$

35. $y = \ln x, x > 0$

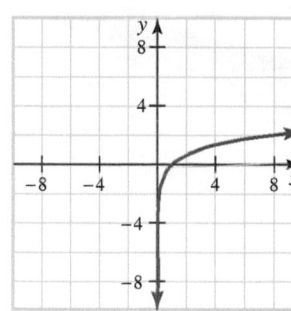

37. $\pm \frac{1}{\sqrt{21}}(4\mathbf{i} + 2\mathbf{j} + \mathbf{k})$ **39.** $x = 3\cos t, y = 3\sin t, 0 \le t < 2\pi$

41. $x = 3\cos t, y = 2\sin t, 0 \le t < 2\pi$

43. $x = 4\sec t, y = 3\tan t, 0 \le t < 2\pi, t \ne \dfrac{\pi}{2}, t \ne \dfrac{3\pi}{2}$

45. parabolic arc **47.** $y = b\left(1 - \dfrac{x^2}{16a^2}\right)$ **51.** $a = 0, b = 5$

53. intersect at $(-1, 3, 2)$; $\theta \approx 27°$ **55.** $x = -1 - 23t, y = 3 + 2t, z = 1 - 19t$

57. The lines are perpendicular and intersect at $P(x_0, y_0, z_0)$.

9.6 Planes in $\mathbb{R}^3$, page 733

3. $x - 2y + 4z + 7 = 0$ **5.** $5x - 3y + 4z - 4 = 0$ **7.**

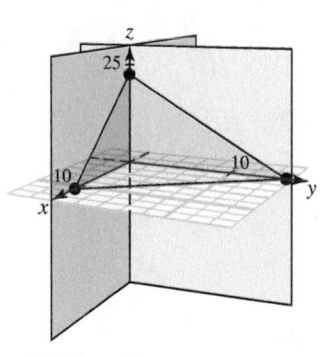

9.

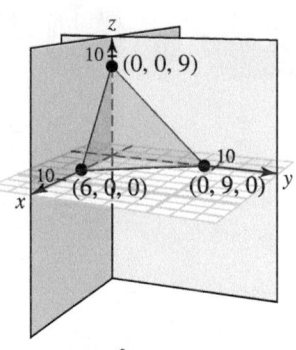

11. $x - z + 1 = 0$ **13.** $x + 2y - 3z - 6 = 0$ **15.** $x = 0$ **17.** $\pm \frac{1}{\sqrt{38}}(5\mathbf{i} - 3\mathbf{j} + 2\mathbf{k})$ **19.** $\frac{10}{\sqrt{38}}$ **21.** $\dfrac{2a^2 + 1}{|a|\sqrt{14}}$ **23.** $\frac{29}{\sqrt{1,265}}$

25. 2 **27.** $\frac{47}{5}$ **29.** $\frac{31}{13}$ **31.** $\frac{4\sqrt{3}}{\sqrt{14}}$ **33.** $\frac{65}{\sqrt{122}}$ **35.** $\frac{2}{\sqrt{6}}$ **37. b.** $\frac{1}{\sqrt{6}}$ **39.** $(x+1)^2 + (y-2)^2 + (z-4)^2 = \frac{1}{38}(2x - 5y + 3z - 7)^2$

41. a. $\cos \varphi = \dfrac{aA + bB + cC}{\sqrt{a^2 + b^2 + c^2}\sqrt{A^2 + B^2 + C^2}}$ **b.** $\theta \approx 38°$ **43.** $3\mathbf{i} - 2\mathbf{j} - 11\mathbf{k}$ **45.** $\cos \alpha = -\frac{4}{\sqrt{26}}, \alpha \approx 2.47$ or $142°$;

$\cos \beta = \frac{3}{\sqrt{26}}, \beta \approx 0.94$ or $54°$; $\cos \gamma = \frac{1}{\sqrt{26}}, \gamma \approx 1.37$ or $79°$ **47.** $\dfrac{x-1}{2} = \dfrac{y+5}{-3} = \dfrac{z-3}{1}$ **49.** $\pm \frac{1}{\sqrt{2}}(\mathbf{i} - \mathbf{k})$

51. $2x - y + 3z + 3 = 0$ **53.** $P(9, -5, 12)$ **55.** $\dfrac{x-2}{1} = \dfrac{y-3}{1} = \dfrac{z-1}{1}$

9.7 Quadric Surfaces, page 740

3. B   **5.** A   **7.** K   **9.** G   **11.** I   **13.** C   **15.**

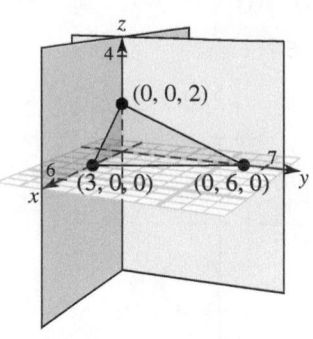

17.

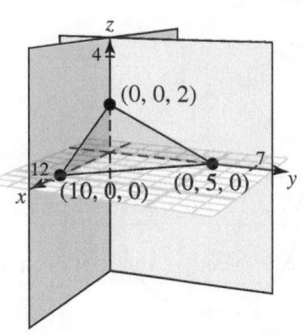

19.

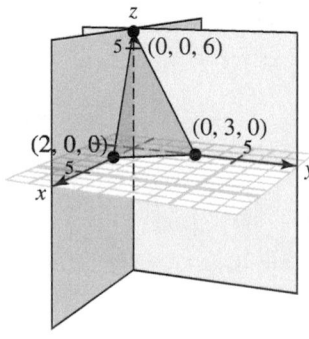

21.

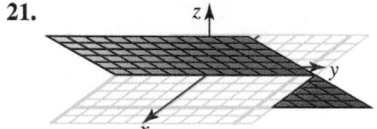

23.

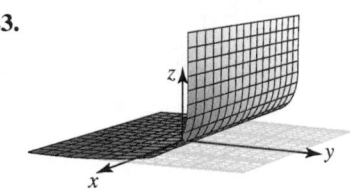

25.

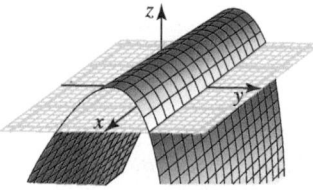

27.

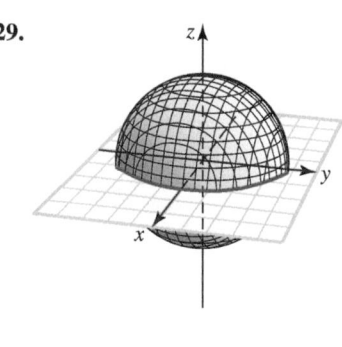

29.

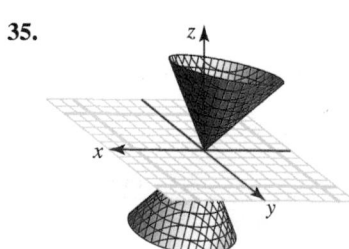

31.

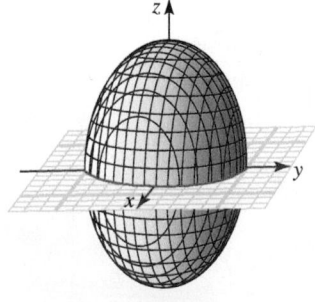

33.

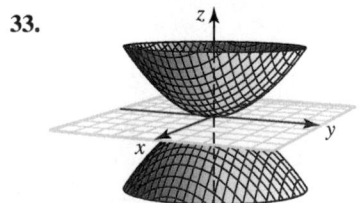

35.

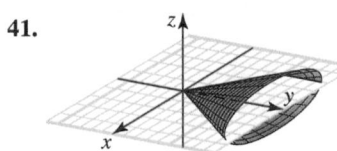

37.

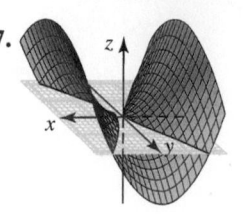

39.

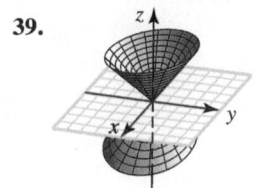

41.

43.

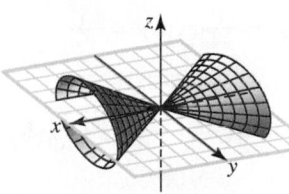

45. ellipsoid centered at $(1, -2, 3)$ with semi-axes $2, 1$, and 3.

47. ellipsoid centered at $(\frac{9}{14}, -2, -\frac{7}{6})$ with semi-axes approximately $1.31, 3.46$, and 2.00.

49. elliptic paraboloid with vertex $(\frac{1}{2}, \frac{9}{2}, -\frac{1}{2})$ and the y-axis as the axis of symmetry.

51. $\frac{8}{45}x^2 + \frac{7}{20}y^2 = 1$; this is an ellipse in the xy-plane.

55. $x^2 + 2y^2 = 1$; this is the equation of an ellipse.   **57. a.** The equatorial radius is $a = 6{,}378.2$ and the polar radius is $b = 6{,}356.5$.
b. The volume is approximately 1.08319×10^{12} km^3.   **59.** $20x^2 + 36y^2 + 36z^2 = 45$ is an ellipsoid.

Chapter 9 Proficiency Examination, page 742

16. a. $13\mathbf{i} - 12\mathbf{j} + 2\mathbf{k}$ **b.** 1 **c.** 12 **d.** $2\mathbf{i} + 3\mathbf{j} + 5\mathbf{k}$ **e.** $\frac{12}{13}(3\mathbf{i} - 2\mathbf{j})$ **f.** $\frac{6\sqrt{14}}{7}$ **17. a.** 40 **b.** not possible   **c.** $25\mathbf{i} - 10\mathbf{j} - 15\mathbf{k}$

d. not possible   **18.** $\frac{x}{1} = \frac{y+2}{-6} = \frac{z-1}{4}$ **19.** $2x + 3z - 11 = 0$ **20.** $x = -\frac{13}{2} - 10t, y = 5 + 6t, z = 2t$

21. $17x + 19y + 13z - 25 = 0$ **22.** $\cos\alpha = \frac{-2}{\sqrt{14}}, \alpha \approx 2.13$ or $122°$; $\cos\beta = \frac{3}{\sqrt{14}}, \beta \approx 0.64$ or $37°$; $\cos\gamma = \frac{1}{\sqrt{14}}, \gamma \approx 1.30$ or $74°$

23. a. skew   **b.** intersect at $(2, 2, 3)$ **24.** 6 **25.** $\sqrt{2}$ **26. a.** $\frac{2\sqrt{30}}{15}$ **b.** $\frac{5\sqrt{6}}{\sqrt{35}}$ **c.** $\frac{6}{\sqrt{26}}$ **27.** 168.4 mi/h

28. a.

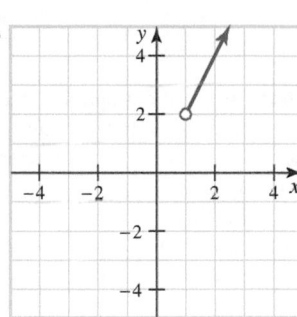

b. $\frac{y}{x} = \frac{2^{t+1}}{2^t} = 2$, or $y = 2x, x \geq 1$ **29. a. i.**

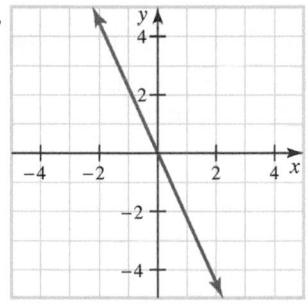

ii. $\frac{x}{4} = t$, so $x = 4t, \frac{y}{-9} = t$, so $y = -9t$ **b. i.**

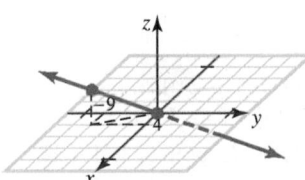

ii. $x = 4t, y = -9t, z = t$

30. a. elliptic cone   **b.** hyperbolic paraboloid   **c.** hyperboloid of two sheets   **d.** hyperboloid of one sheet   **e.** plane   **f.** line
g. sphere   **h.** elliptic (circular) paraboloid   **i.** hyperboloid of two sheets

Chapter 9 Supplementary Problems, page 744

1. $8\sqrt{2} + \sqrt{14}$ **3.** $A \approx 125°, B \approx 26°, C \approx 30°$ **5.** $x = 4 + 2t, y = -3 - t, z = 2 + t$; $\frac{x-4}{2} = \frac{y+3}{-1} = \frac{z-2}{1}$; answers vary for the
two points   **7.** $x = 1 + 2t, y = 4, z = t$; $\frac{x-1}{2} = \frac{z}{1}$ and $y = 4$; answers vary for the two points   **9.** $x = 3 + 8t, y = 4 - 7t$,
$z = -1 + 3t$; $\frac{x-3}{8} = \frac{y-4}{-7} = \frac{z+1}{3}$ **11.** $z = 0$ **13.** $3x + 4y - z + 13 = 0$ **15.** $5x - 2y + 3z - 32 = 0$ **17.**
$3x + 2y - 11z + 19 = 0$ **19.** $\sqrt{21}$ **21.** $14\mathbf{i} + 7\mathbf{j} - 13\mathbf{k}$ **23.** $\sqrt{94}$ **25.** $\frac{5}{\sqrt{21}}$

27.

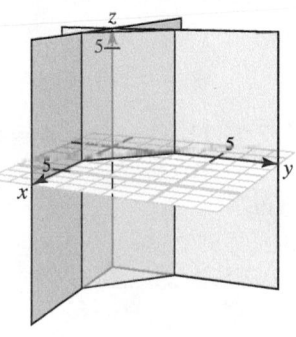

29.

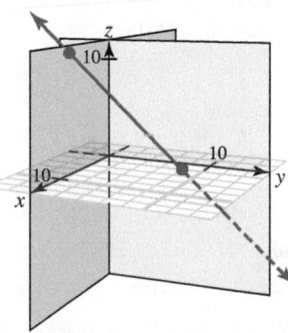

31.

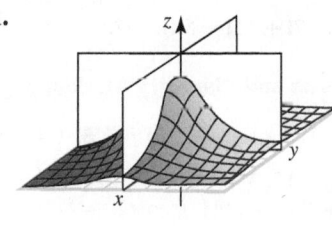

33.

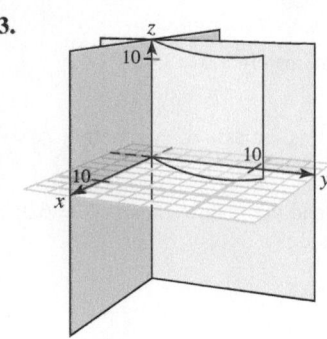

35. $\frac{1}{\sqrt{26}}$ **37.** $\frac{2}{\sqrt{3}}$ **39.** 3; $6\mathbf{i}-10\mathbf{j}-10\mathbf{k}$ **43.** $\cos\alpha = \frac{-3}{\sqrt{46}}$; $\cos\beta = \frac{6}{\sqrt{46}}$; $\cos\gamma = \frac{1}{\sqrt{46}}$

45. $\theta = 74°$ **47.** The center is $(0, -\frac{3}{2}, \frac{1}{2})$ and the radius is $\frac{3}{2}$. **49.** $P(3, -2, 1)$

51. $37X + 20Y - 21Z = 61$ and $67X - 20Y + 99Z = 121$ **53.** $\pm\left(\frac{3}{\sqrt{19}}\mathbf{i} + \frac{1}{\sqrt{19}}\mathbf{j} + \frac{3}{\sqrt{19}}\mathbf{k}\right)$

55. approximately 2,600 ft-lb **57.** 13

59. $\frac{1}{2}(\|\mathbf{u}\times\mathbf{v}\| + \|\mathbf{u}\times\mathbf{w}\| + \|\mathbf{v}\times\mathbf{w}\| + \|(\mathbf{u}-\mathbf{v})\times(\mathbf{w}-\mathbf{v})\|)$

63. $\cos\theta\sin\phi - \sin\theta\cos\phi = \sin(\phi-\theta)$ **65.** $a_1 + 2b_1 + c_1 = 0$ **67.** $\frac{9}{\sqrt{2}}$ **69.** $\frac{1}{2}\sqrt{26}$

71. $2x - 3y - 7z = 0$ **73.** $\frac{x+1}{1} = \frac{y-2}{1}$ and $z = 0$ **75.** $\mathbf{N} = \frac{\pm 1}{\sqrt{37}}(-6\mathbf{i}+\mathbf{j})$

77. $\mathbf{r} = \mathbf{v} + \mathbf{w}t$ is the set of position vectors in the line $x = a + At$, $y = b + Bt$, and $z = c + Ct$. **79.** The region inside a sphere with center (a, b, c) and radius r **85.** $M(\frac{1}{3}, \frac{16}{3}, 1)$ **91.** yes **97.** This is Putnam Problem 4 of the morning session of 1939. **99.** This is Putnam Problem 2 of the morning session of 1983.

Chapter 10

10.1 Introduction to Vector Functions, page 758

3. $t \neq 0$ **5.** $t \neq (n + \frac{1}{2})\pi, n$ an integer **7.** $t \neq \frac{n\pi}{2}, n$ an integer **9.** $t > 0$ **11.** a parabolic cylinder; in $\mathbb{R}^2$ the graph is a parabola in the xy-plane **13.** cylinder; in $\mathbb{R}^2$ the graph is a circle in the xy-plane **15.** plane; in $\mathbb{R}^2$ the graph is a line in the xz-plane parallel to and four units below the x-axis **17.** a circular helix; in $\mathbb{R}^2$ the graph is a circle in the xy-plane **19.** the curve is the intersection of the parabolic cylinder $y = (1 - x)^2$ with the plane $x + z = 1$; in $\mathbb{R}^2$ the graph is a parabola in the xy-plane
21. $(7t - 3)\mathbf{i} - 10\mathbf{j} + (2t^2 - \frac{3}{t})\mathbf{k}$ **23.** $3t - 2t^2$ **25.** $(1 - t)\sin t$ **27.** $-t^2 e^t \mathbf{i} + (t^2 \sin t)\mathbf{j} + (2te^t + 5\sin t)\mathbf{k}$
29. $(4te^t - t^2 + t + 10\sin t)\mathbf{i} + (2e^t t^2 + 1)\mathbf{k}$ **31.** $t^2 e^t - t^3 e^t - 2e^t - \frac{5}{t}\sin t$ **33.** $x^2 + y^2 + z^2 = 8$ **35.** $\mathbf{F}(t) = t\mathbf{i} + t^2\mathbf{j} + 2\mathbf{k}$
37. $\mathbf{F}(t) = 2t\mathbf{i} + (1 - t)\mathbf{j} + (\sin t)\mathbf{k}$ **39.** $\mathbf{F}(t) = t^2\mathbf{i} + t\mathbf{j} + \sqrt{9 - t^2 - t^4}\mathbf{k}$ **41.** $2\mathbf{i} - 3\mathbf{j} + e\mathbf{k}$ **43.** 0 **45.** $-\mathbf{i} + e^{-1}\mathbf{j}$ **47.** $\frac{3}{2}\mathbf{i} + \mathbf{j}$
49. continuous for all t **51.** continuous for all $t \neq 0, t \neq -1$ **53.** continuous for all $t \neq 0$ **55.** $x - y + z = 1$

10.2 Differentiation and Integration of Vector Functions, page 769

3. $\mathbf{F}'(t) = \mathbf{i} + 2t\mathbf{j} + (1 + 3t^2)\mathbf{k}$ **5.** $\mathbf{F}'(s) = (1 + \ln s)\mathbf{i} + 5s^{-1}\mathbf{j} - e^s(\ln s + s^{-1})\mathbf{k}$ **7.** $\mathbf{F}'(t) = 2t\mathbf{i} - t^{-2}\mathbf{j} + 2e^{2t}\mathbf{k}$;
$\mathbf{F}''(t) = 2\mathbf{i} + 2t^{-3}\mathbf{j} + 4e^{2t}\mathbf{k}$ **9.** $\mathbf{F}'(s) = (\cos s)\mathbf{i} - (\sin s)\mathbf{j} + 2s\mathbf{k}$; $\mathbf{F}''(s) = (-\sin s)\mathbf{i} - (\cos s)\mathbf{j} + 2\mathbf{k}$ **11.** $f'(x) = -9x^2 - 2x$
13. $g'(x) = \frac{4x}{\sqrt{1 + 4x^2}}$ **15.** $\mathbf{V}(1) = \mathbf{i} + 2\mathbf{j} + 2\mathbf{k}$; $\mathbf{A}(1) = 2\mathbf{j}$; speed is 3; direction of motion is $\frac{1}{3}\mathbf{i} + \frac{2}{3}\mathbf{j} + \frac{2}{3}\mathbf{k}$
17. $\mathbf{V}(\frac{\pi}{4}) = -\frac{\sqrt{2}}{2}\mathbf{i} + \frac{\sqrt{2}}{2}\mathbf{j} + 3\mathbf{k}$; $\mathbf{A}(\frac{\pi}{4}) = -\frac{\sqrt{2}}{2}\mathbf{i} - \frac{\sqrt{2}}{2}\mathbf{j}$; speed is $\sqrt{10}$; direction of motion is $-\frac{1}{2\sqrt{5}}\mathbf{i} + \frac{1}{2\sqrt{5}}\mathbf{j} + \frac{3}{\sqrt{10}}\mathbf{k}$
19. $\mathbf{V}(\ln 2) = 2\mathbf{i} - \frac{1}{2}\mathbf{j} + 8\mathbf{k}$; $\mathbf{A}(\ln 2) = 2\mathbf{i} + \frac{1}{2}\mathbf{j} + 16\mathbf{k}$; speed is $\frac{\sqrt{273}}{2}$; direction of motion is $\frac{1}{\sqrt{273}}(4\mathbf{i} - \mathbf{j} + 16\mathbf{k})$ **21.** $\mathbf{F}'(0) = 2\mathbf{j}$;
$\mathbf{F}'(1) = 2\mathbf{i} + 2\mathbf{j} + 5\mathbf{k}$; $\mathbf{F}'(-1) = -2\mathbf{i} + 2\mathbf{j} + \mathbf{k}$ **23.** $\mathbf{F}'(0) = \mathbf{0}$; $\mathbf{F}'(\frac{\pi}{2}) = \pi\mathbf{i} - \mathbf{j} - \frac{\pi^2}{4}\mathbf{k}$ **25.** $x = -1 - 3t, y = 1 + 2t, z = -1 - t$
27. $\frac{t^2}{2}\mathbf{i} - \frac{e^{3t}}{3}\mathbf{j} + 3t\mathbf{k} + \mathbf{C}$ **29.** $(t\ln t - t)\mathbf{i} - \frac{1}{2}t^2\mathbf{j} + 3t\mathbf{k} + \mathbf{C}$ **31.** $\frac{t^2}{2}\left(\ln t - \frac{1}{2}\right)\mathbf{i} - \cos(1 - t)\mathbf{j} + \frac{t^2}{2}\mathbf{k} + \mathbf{C}$

33. $\mathbf{R}(t) = (\frac{1}{3}t^3 + 1)\mathbf{i} + (-\frac{1}{2}e^{2t} + \frac{9}{2})\mathbf{j} + (\frac{2}{3}t^{3/2} - 1)\mathbf{k}$ **35.** $\mathbf{R}(t) = \left(\frac{4}{3}t^{3/2} + 1\right)\mathbf{i} + (\sin t + 1)\mathbf{j}$

37. $\mathbf{V}(t) = \langle 2 + \sin t, 0, 3 + t\cos t - \sin t\rangle$; $\mathbf{R}(t) = \langle 2 + 2t - \cos t, -2, -1 + 3t + 2\cos t + t\sin t\rangle$ **39.** $\mathbf{R}(t) = e^t\mathbf{i} + (\frac{1}{3}t^3 - 1)\mathbf{j}$
41. $-180t^2$ **43.** $3t^2(-7\mathbf{i} + 11\mathbf{j} + 5\mathbf{k})$ **47.** $a = \frac{\pi}{4}$ **51.** not smooth **53.** false

10.3 Modeling Ballistics and Planetary Motion, page 779

1. 4.6 sec; 481 ft **3.** 129.9 sec; 73,175 m **5.** 1.9 sec; 41 m **7.** 1.5 sec; 148 ft **9.** 21.3 sec; 6,257 ft **11.** $\mathbf{V} = \sqrt{5}\mathbf{u}_r$; $\mathbf{A} = \mathbf{0}$

13. $\mathbf{V} = \mathbf{u}_\theta$; $\mathbf{A} = -\mathbf{u}_r$ **15.** $\mathbf{V} = (2\cos 2t)\mathbf{u}_r + (2\sin 2t)\mathbf{u}_\theta$; $\mathbf{A} = (-8\sin 2t)\mathbf{u}_r + (8\cos 2t)\mathbf{u}_\theta$
17. $\mathbf{V} = -10\sin(2t + 1)\mathbf{u}_r + 10[1 + \cos(2t + 1)]\mathbf{u}_\theta$; $\mathbf{A} = [-40\cos(2t + 1) - 20]\mathbf{u}_r - 40\sin(2t + 1)\mathbf{u}_\theta$
19. $\mathbf{V} = (-a\omega\sin\omega t)\mathbf{i} + (a\omega\cos\omega t)\mathbf{j}$ **21.** $|a\omega|$ **25.** 140 m/s **27.** 21.7° **29.** The maximum height is 85 ft; the ball will land at a distance of 568 ft; the distance to the fence is 560 ft **31.** 52.3 ft **33.** $x(t) = 0$, $y(t) = -\frac{1}{2}gt^2 + v_0 t + s_0$ **35.** 5.5 sec

37. 233.24 ft/s $\approx$ 159 mi/h **39.** 18.13 ft **41.** $\mathbf{V}(t) = -3\sin t\mathbf{u}_r + (4 + 3\cos t)(3t^2)\mathbf{u}_\theta$;
$\mathbf{A}(t) = [-3\cos t - (4 + 3\cos t)9t^4]\mathbf{u}_r + [(4 + 3\cos t)6t - 18t^2\sin t]\mathbf{u}_\theta$ **43.** 21.8° **45.** 471.77 ft/s; 37.2° **51.** They are the same since they have the same semimajor axis. **53.** 142.41 m/s **55.** $\alpha = \frac{\pi}{4} - \frac{\beta}{2}$

10.4 Unit Tangent and Principal Unit Normal Vectors; Curvature, page 794

3. $\mathbf{T}(t) = \dfrac{2}{\sqrt{4+9t^2}}\mathbf{i} + \dfrac{3t}{\sqrt{4+9t^2}}\mathbf{j}$; $\mathbf{N}(t) = \dfrac{-3t}{\sqrt{4+9t^2}}\mathbf{i} + \dfrac{2}{\sqrt{4+9t^2}}\mathbf{j}$ **5.** $\mathbf{T}(t) = \frac{1}{\sqrt{2}}[(\cos t - \sin t)\mathbf{i} + (\cos t + \sin t)\mathbf{j}]$;

$\mathbf{N}(t) = -\frac{1}{\sqrt{2}}[(\sin t + \cos t)\mathbf{i} + (\sin t - \cos t)\mathbf{j}]$ **7.** $\mathbf{T}(t) = \frac{1}{\sqrt{2}}(-\sin t\,\mathbf{i} + \cos t\,\mathbf{j} + \mathbf{k})$; $\mathbf{N}(t) = -\cos t\,\mathbf{i} - \sin t\,\mathbf{j}$

9. $\mathbf{T}(t) = \dfrac{1}{\sqrt{1+4t^4}}(\mathbf{i} + 2t^2\mathbf{k})$; $\mathbf{N}(t) = \dfrac{1}{\sqrt{1+4t^4}}(-2t^2\mathbf{i} + \mathbf{k})$ **11.** $4\sqrt{5}$ **13.** $\frac{1}{27}(13^{3/2} - 4^{3/2})$ **15.** $\sqrt{41}\pi$ **17.** $\kappa = 0$ **19.** $\kappa = 2$

21. $\kappa = 2^{-3/2}$ **23.** $\mathbf{R}(s) = \left\langle \frac{\sqrt{2}-s}{\sqrt{2}}, \frac{s}{\sqrt{2}} \right\rangle$ **25.** $\mathbf{R}(s) = \left\langle 2 - \frac{3s}{\sqrt{26}}, 1 + \frac{s}{\sqrt{26}}, -\frac{4s}{\sqrt{26}} \right\rangle$ **29.** $\mathbf{T}(\frac{\pi}{3}) = \frac{1}{\sqrt{10}}(\mathbf{i}-3\mathbf{j})$ **31. a.** $\mathbf{T}(\pi) = \frac{1}{\sqrt{2}}(-\mathbf{i}+\mathbf{k})$

b. $\frac{1}{2}$ **c.** $\sqrt{2}\pi$ **33.** $(0,3)$ and $(0,-3)$ **35.** $\rho(0) = \frac{1}{6}$; $\rho(1) = \rho(-1) = \frac{1}{24}$

37. $\kappa = \dfrac{1}{2^{3/2}(1 - \cos t)^{1/2}}$

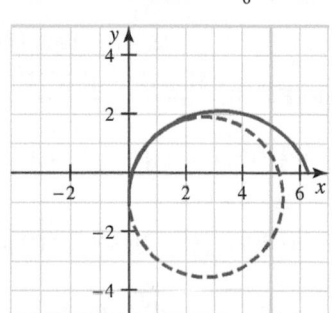

39. $\kappa = \dfrac{2\sqrt{9t^4 + 9t^2 + 1}}{(1 + 4t^2 + 9t^4)^{3/2}}$ **41.** $\kappa = \dfrac{2}{(1 + 4x^2)^{3/2}}$

43. $\kappa = \dfrac{6x^5}{(x^6 + 4)^{3/2}}$ **45.** $\kappa = \dfrac{1}{\sqrt{2}e^{\theta}}$ **47.** $\kappa = 1$ **49.** $\kappa = \dfrac{3}{2^{3/2}\sqrt{r}}$

53. c. $\lim\limits_{s\to\infty}\kappa(s) = \infty$, so the spiral keeps winding tighter and tighter. **59.** The total length of the ramp is about 7 units.

10.5 Tangential and Normal Components of Acceleration, page 801

1. $A_T = \dfrac{4t}{\sqrt{1+4t^2}}$; $A_N = \dfrac{2}{\sqrt{1+4t^2}}$ **3.** $A_T = \dfrac{t}{\sqrt{1+t^2}}$; $A_N = \dfrac{t^2+2}{\sqrt{1+t^2}}$ **5.** $A_T = \dfrac{4t}{\sqrt{2+4t^2}}$; $A_N = \dfrac{2}{\sqrt{1+2t^2}}$

7. $A_T = 0$; $A_N = 2$ **9.** $A_T = -\dfrac{\sin t \cos t}{\sqrt{1+\cos^2 t}}$; $A_N = \sqrt{\dfrac{2}{1+\cos^2 t}}$ **11.** $A_T = -\dfrac{13}{\sqrt{10}}$; $A_N = \dfrac{11}{\sqrt{10}}$

13. $A_T = -\dfrac{2}{\sqrt{26}}$; $A_N = \dfrac{16}{\sqrt{26}}$ **15.** $A_T = -\dfrac{19}{\sqrt{14}}$; $A_N = \sqrt{\dfrac{59}{14}}$ **17.** $A_T = -\dfrac{11}{\sqrt{14}}$; $A_N = \sqrt{\dfrac{19}{14}}$ **19.** $\frac{5}{\sqrt{8}}$ **21.** $\frac{5}{2\sqrt{6}}$ **23.** 0

25. $\frac{1}{2}$ **27.** $A_T = \dfrac{4t}{\sqrt{5+4t^2}}$; $A_N = \dfrac{2\sqrt{5}}{\sqrt{5+4t^2}}$ **29.** $A_T = 0$; $A_N = 2$ **31.** $A_T = \sqrt{3}e^t$; $A_N = \sqrt{2}e^t$

33. The maximum speed is 8 at $t = \frac{n\pi}{2}$, n any integer; the minimum speed is 6 at $t = \frac{(2n+1)\pi}{4}$, n any integer.

35. $\mathbf{V}(t) = (-2\sin 2t)\mathbf{i} + (2\cos 2t)\mathbf{j}$; $\mathbf{A}(t) = (-4\cos 2t)\mathbf{i} - (4\sin 2t)\mathbf{j}$; $A_T = 0$; $A_N = 4$

37. $A_T\left(\frac{\sqrt{399}}{8}\right) = \frac{2\sqrt{399}}{5}$; $A_N\left(\frac{\sqrt{399}}{8}\right) = 0.4$ **39.** $\omega = 0.52$ rev/s **41.** 1,815 ft **43.** Since the mass does not influence the largest speed, the result would not change. **45.** Since the mass does not influence the largest speed, the result would not change. **47.** $28.4°$ **49.** $\frac{\pi^2\omega^2}{1{,}920}W$ **51.** 42.43 units/s **55.** $A_T = \dfrac{\mathbf{V}\cdot\mathbf{A}}{\|\mathbf{V}\|} = 0$ and $A_N = \dfrac{\|\mathbf{V}\times\mathbf{A}\|}{\|\mathbf{V}\|} = g$

57. b. about 36,000 km **c.** 3.3367 km/s, or about 12,000 km/h

Chapter 10 Proficiency Examination, page 805

24.

$s = 2\pi\sqrt{10}$ **25.** $\mathbf{F}' = \dfrac{1}{(1+t)^2}\mathbf{i} + \dfrac{t\cos t - \sin t}{t^2}\mathbf{j} + (-\sin t)\mathbf{k}$;

$\mathbf{F}'' = -\dfrac{2}{(1+t)^3}\mathbf{i} + \dfrac{-t^2\sin t - 2t\cos t + 2\sin t}{t^3}\mathbf{j} - (\cos t)\mathbf{k}$

26. $(3 - 6\ln 2)\mathbf{i} + \frac{45}{4}\mathbf{j} + \left(6\ln 2 - \frac{9}{4}\right)\mathbf{k}$ **27.** $(e^t - t)\mathbf{i} - \left(\frac{t^4}{12} + 2\right)\mathbf{j} + \left(\frac{3t^2}{2} + 3t\right)\mathbf{k}$ **28.** $\mathbf{V} = \mathbf{i} + 2\mathbf{j} + e^t(t+1)\mathbf{k}$; speed is

$\sqrt{5 + e^{2t}(t+1)^2}$; $\mathbf{A} = e^t(t+2)\mathbf{k}$ **29.** $\mathbf{T} = \dfrac{2t\mathbf{i} + 3\mathbf{j} - 3\mathbf{k}}{\sqrt{4t^2 + 18}}$; $\mathbf{N} = \dfrac{3\mathbf{i} - t\mathbf{j} + t\mathbf{k}}{\sqrt{2t^2 + 9}}$; $A_T = \dfrac{2\sqrt{2}t}{\sqrt{2t^2 + 9}}$; $A_N = \dfrac{6}{\sqrt{2t^2 + 9}}$; $\kappa = \dfrac{3}{(2t^2 + 9)^{3/2}}$

30. a. The maximum height is 9.77 ft. **b.** $\frac{25}{16}$ sec; range is 67.7 ft

Chapter 10 Supplementary Problems, page 806

1. $2\mathbf{k}$ **3.** -5 **5.** $\mathbf{i} - \mathbf{j}$ **7.** $\mathbf{F}'(t) = (1+t)e^t\mathbf{i} + 2t\mathbf{j}$; $\mathbf{F}''(t) = (2+t)e^t\mathbf{i} + 2\mathbf{j}$ **9.** $\mathbf{F}'(t) = \langle 2t, \frac{3}{2}t^{1/2}, -3t^{-4} \rangle$; $\mathbf{F}''(t) = \langle 2, \frac{3}{4}t^{-1/2}, 12t^{-5} \rangle$
11. $\mathbf{F}'(t) = \langle -2t^{-2}, -2, (1-t)e^{-t} \rangle$; $\mathbf{F}''(t) = \langle 4t^{-3}, 0, (t-2)e^{-t} \rangle$ **13.** $\mathbf{F}'(t) = (2t + e^t)\mathbf{i} + (1-t)e^{-t}\mathbf{j} + e^{t+1}\mathbf{k}$;
$\mathbf{F}''(t) = (2 + e^t)\mathbf{i} + (t-2)e^{-t}\mathbf{j} + e^{t+1}\mathbf{k}$

15.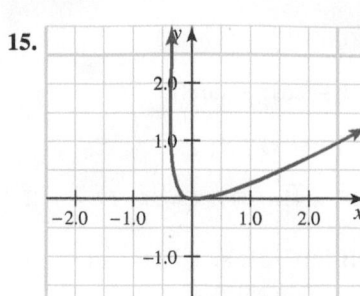

17.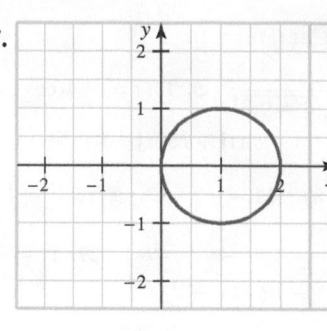

19. The graph is a circular helix of radius 2, traversed clockwise. The curve begins $(t = 0)$ at $(0, 2, 0)$ and rises 10π units with each revolution.

21. The graph is the line $z = -5x$ in the plane $y = 3$.

23. The graph is the intersection of the cylinder $y = x^2 + 1$ and the plane $y = z + 1$. **25.** The graph is the intersection of the plane $z = -2x$ and the paraboloid $y = x^2$. **27. a.** no **b.** yes **c.** yes **29.** $2\mathbf{F} \cdot \mathbf{F}'$ **31.** $\dfrac{\|\mathbf{F}\|^2 \mathbf{F}' - (\mathbf{F} \cdot \mathbf{F}')\mathbf{F}}{\|\mathbf{F}\|^3}$ **33.** 0

35. $(e - e^{-1})\mathbf{i} + 6\mathbf{k}$ **37.** $(te^t - e^t)\mathbf{i} + \dfrac{\cos 2t}{2}\mathbf{j} + \dfrac{t^3}{3}\mathbf{k} + \mathbf{C}$ **39.** $(te^t - e^t)\mathbf{i} + \frac{1}{4}t^2(2\ln t - 1)\mathbf{j} + \frac{3}{2}t^2\mathbf{k} + \mathbf{C}$

41. $\mathbf{V}(t) = \mathbf{i} - \mathbf{j}$; $\frac{ds}{dt} = \sqrt{2}$; $\mathbf{A}(t) = \mathbf{0}$ **43.** $\mathbf{V}(t) = (t\cos t + \sin t)\mathbf{i} + (1 - t)e^{-t}\mathbf{j} + \mathbf{k}$;
$\frac{ds}{dt} = \sqrt{t^2\cos^2 t + t\sin 2t + \sin^2 t + (1 - t)^2 e^{-2t} + 1}$; $\mathbf{A}(t) = (2\cos t - t\sin t)\mathbf{i} + e^{-t}(t - 2)\mathbf{j}$ **45.** $\mathbf{T}(t) = \dfrac{\mathbf{i} - 2t\mathbf{j}}{\sqrt{4t^2 + 1}}$;

$\mathbf{N}(t) = \dfrac{-2t\mathbf{i} - \mathbf{j}}{\sqrt{4t^2 + 1}}$ **47.** $\mathbf{T}(t) = \frac{1}{5}(-4\sin t\mathbf{i} - 3\mathbf{j} + 4\cos t\mathbf{k})$; $\mathbf{N}(t) = (-\cos t)\mathbf{i} + (-\sin t)\mathbf{k}$ **49.** $A_T = \dfrac{4t + e^{2t}}{\sqrt{4 + 4t^2 + e^{2t}}}$;

$A_N = 2\sqrt{\dfrac{t^2 e^{2t} - 2te^{2t} + 2e^{2t} + 4}{4 + 4t^2 + e^{2t}}}$; $\kappa = \dfrac{2\sqrt{t^2 e^{2t} - 2te^{2t} + 2e^{2t} + 4}}{(4 + 4t^2 + e^{2t})^{3/2}}$ **51.** $A_T = 0$; $A_N = 4$; $\kappa = \frac{1}{8}$ **53.** $\kappa = \frac{1}{2}$ **55.** $\kappa = \frac{e}{\sqrt{2}}$

57. $\kappa = \frac{1}{6}$ **59.** $x\mathbf{i} - x^{-1}\mathbf{j} + e^{-x}\mathbf{k}$ **61.** $(\sin x)\mathbf{i} - (\cos 2x)\mathbf{j} + e^{-x}\mathbf{k}$ **63.** $\mathbf{F}(t) = (\frac{1}{2}t^2 + 1)\mathbf{i} + (\frac{1}{2}t^2 + 2)\mathbf{j} - (\frac{1}{2}t^2 + 3)\mathbf{k}$

65. $\mathbf{F}(t) = \left[\frac{1}{2}(3 - \cos 2t)\right]\mathbf{i} + \left[\frac{e^t}{2}(\sin t + \cos t) - \frac{1}{2}\right]\mathbf{j} - [3\ln|t + 1| + 3]\mathbf{k}$ **67.** $t \neq 1$

69. $\mathbf{F}(t) = (2t - \sin t + 1)\mathbf{i} + (\frac{3}{2}t^2 + \cos t - 1)\mathbf{j} + \left(\dfrac{t^4}{24\pi} + \dfrac{t^2}{2}\right)\mathbf{k}$ **71. a.** $x = 1 + s$, $y = 1 + s$, $z = 1 + 3s$

b. $(-2, -2, -8)$ **73.** $\mathbf{R}(t) = \langle 2t^4 + 1, 2t^2 + 2\rangle$ **75.** $\mathbf{T}(s) = \dfrac{1}{\sqrt{5}}\left[\left(-\sin\dfrac{s}{a}\right)\mathbf{i} + \left(\cos\dfrac{s}{a}\right)\mathbf{j} + 2\mathbf{k}\right]$; $\mathbf{N}(s) = -\left(\cos\dfrac{s}{a}\right)\mathbf{i} - \left(\sin\dfrac{s}{a}\right)\mathbf{j}$

77. $4,537.5$ lb **79. a.** $\mathbf{V}(t) = e^{-t}[(-\cos t - \sin t)\mathbf{i} + (\cos t - \sin t)\mathbf{j} - \mathbf{k}]$; speed is $e^{-t}\sqrt{3}$; $\mathbf{A}(t) = e^{-t}[2\sin t\mathbf{i} - 2\cos t\mathbf{j} + \mathbf{k}]$;
b. $\kappa = \frac{\sqrt{2}}{3}e^t$ **83.** $\mathbf{V}(t) = -a\sin at\mathbf{u}_r - ae^{-at}(1 + \cos at)\mathbf{u}_\theta$;
$\mathbf{A}(t) = -a^2[\cos at + e^{-2at}(1 + \cos at)]\mathbf{u}_r + a^2 e^{-at}[1 + \cos at + 2\sin at]\mathbf{u}_\theta$

85. 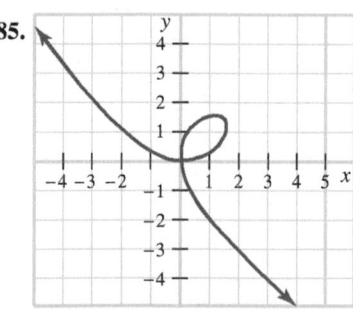 $x = \frac{2}{3} - \frac{5}{9}t$, $y = \frac{4}{3} - \frac{4}{9}t$

87. $20.888\ \mu\text{m}$ **89. a.** $\mathbf{V}(0) = 10\mathbf{u}_r$ **b.** $\mathbf{V}(0.25) = 10\mathbf{u}_r + 5\pi\mathbf{u}_\theta$ **95. a.** continuous **b.** continuous **c.** discontinuous for some choices of h and $\mathbf{F}$ **d.** continuous **97.** This is Putnam Problem 6ii of the morning session of 1939. **99.** This is Putnam Problem 6 of the afternoon session of 1946.

Cumulative Review Problems, page 816

5. -2 **7.** e^{-6} **9.** -1 **11.** $3\sin^2 x\cos x + 2\sec^2 x$ **13.** $\dfrac{1}{\sqrt{1 - x^2}} - \dfrac{2}{x^2 + 1}$ **15.** $\dfrac{y^3 + 2xe^{-y}}{x^2 e^{-y} - 3xy^2}$ **17.** $\frac{1}{3}\sin^3 x - \frac{1}{5}\sin^5 x + C$

19. $-\cot x + \csc x + C$ **21.** $\sin^{-1}(x - 1) + C$ **23. a.** yes **b.** yes **c.** no **25.** $2\sqrt{2} - 1 - \cos 1 - \sin 1$ **27.** $\frac{4\pi}{3}\ln 4$
29. a. $2x + 7y - z - 15 = 0$ **b.** $2x + 5y - 2z + 17 = 0$ **31.** converges **33.** diverges **35.** converges absolutely **37. a.** converges
to $5e^{-1}$ **b.** converges to $\frac{\pi}{2}$ **39.** $\mathbf{F}'(t) = 2\mathbf{i} - 3e^{-3t}\mathbf{j} + 4t^3\mathbf{k}$; $\mathbf{F}''(t) = 9e^{-3t}\mathbf{j} + 12t^2\mathbf{k}$ **41.** $\mathbf{T} = \dfrac{1}{\sqrt{13}}[(2\cos 2t)\mathbf{i} - (2\sin 2t)\mathbf{j} + 3\mathbf{k}]$;

$\mathbf{N} = -(\sin 2t)\mathbf{i} - (\cos 2t)\mathbf{j}$ **43. a.** $\displaystyle\sum_{k=0}^{\infty} \frac{(-1)^k x^{2k+2}}{k!}$ **b.** 0.189 **45.** $y = \dfrac{x^2}{2} - \dfrac{x}{2} + \dfrac{1}{4} + \dfrac{7}{4}e^{-2x}$ **47.** $x = -2$ **49.** $1{,}562.5$ ft-lb
51. tangent line $27x - y = 0$; normal line $x + 27y - 730 = 0$ **53.** $(2x + 1)(x^4 + 2x^3 + 2x^2 + x)$ **55.** $8\pi\,\text{cm}^3/\text{s}$
57. a. $v(t) = 400 - 395e^{-0.08t}$ **b.** 400 ft/s

Chapter 11

11.1 Functions of Several Variables, page 826

3. a. 0 **b.** 0 **c.** 0 **d.** 2 **e.** 48 **f.** $2t^3$ **g.** $t^4 + t^5$ **h.** $t - t^2$ **5.** The domain is $x - y \geq 0$, and the range is $f(x, y) \geq 0$.   **7.** The domain is $uv \geq 0$, and the range is $f(u, v) \geq 0$.   **9.** The domain is $y - x > 0$, and the range is $\mathbb{R}$.   **11.** The domain is $\mathbb{R}^2$, and the range is $f(x, y) \geq 0$.   **13.** The domain is $x^2 - y^2 > 0$, and the range is $f(x, y) > 0$.

15. $C = 0$: $2x - 3y = 0$
$\quad\ C = 1$: $2x - 3y = 1$
$\quad\ C = 2$: $2x - 3y = 2$
$\quad\ C = 3$: $2x - 3y = 3$

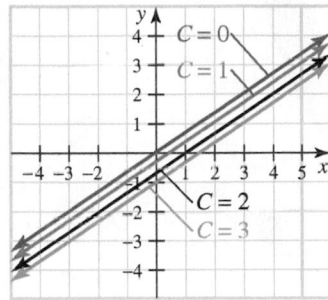

17. $C = 0$: $x^3 - y = 0$
$\quad\ C = 1$: $x^3 - y = 1$
$\quad\ C = 2$: $x^3 - y = 2$
$\quad\ C = 3$: $x^3 - y = 3$

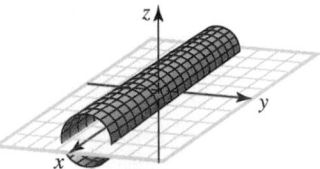

19. $C = 1$: $x^2 + \dfrac{y^2}{4} = 1$
$\quad\ C = 2$: $x^2 + \dfrac{y^2}{4} = 2$
$\quad\ C = 3$: $x^2 + \dfrac{y^2}{4} = 3$
$\quad\ C = 4$: $x^2 + \dfrac{y^2}{4} = 4$

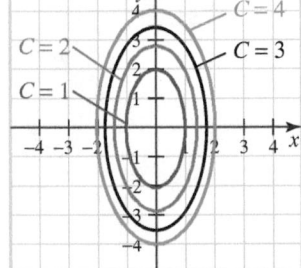

21. This is a cylinder, $y^2 + z^2 = 1$, which has the x-axis as its axis.

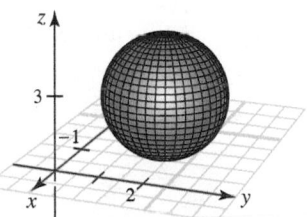

23. This is a plane, $x + y - z = 1$; its trace in the xy-plane is the line $x + y = 1$; its trace in the xz-plane is $x - z = 1$; and its trace in the yz-plane is $y - z = 1$.

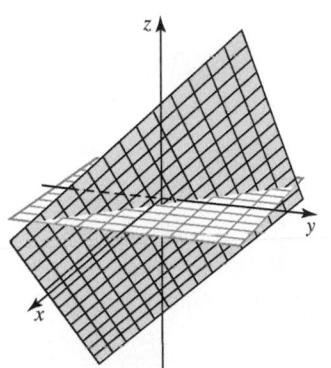

25. This is a sphere, $(x + 1)^2 + (y - 2)^2 + (z - 3)^2 = 4$; its cross sections in the planes $x = -1$, $y = 2$, and $z = 3$ are circles.

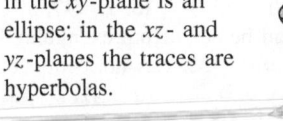

27. This is an ellipsoid; traces are ellipses in all three coordinate planes.

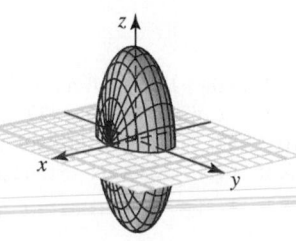

29. This is a hyperboloid of one sheet; the trace in the xy-plane is an ellipse; in the xz- and yz-planes the traces are hyperbolas.

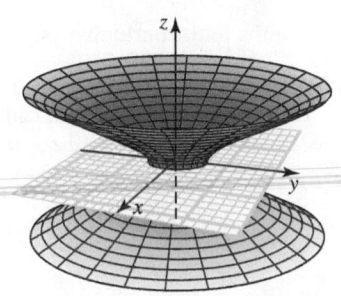

31. This is an elliptic
paraboloid; its traces in
the xz- and yz-planes
are parabolas; the
trace in the xy-plane is
a point (the origin).

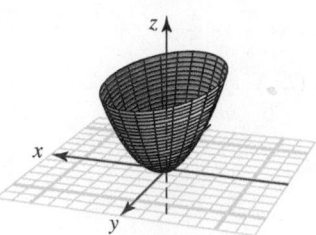

33. This is an elliptic cone;
the traces in the xz- and
yz-planes are pairs of
lines; the trace in the
xy-plane is the origin if
$z = 0$ and an ellipse
if $z \neq 0$.

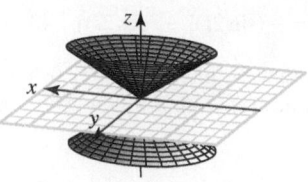

35. D   **37.** B   **39.** A

41.

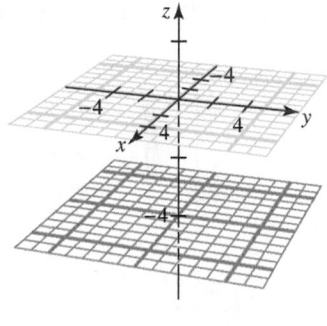

43.

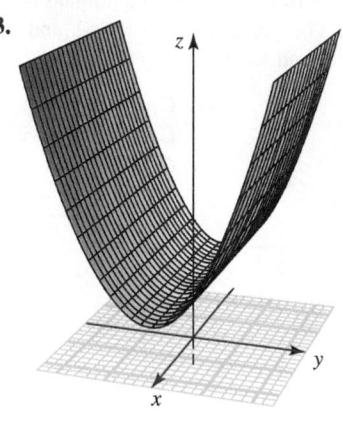

45.

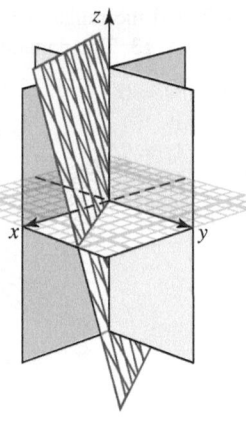

47.

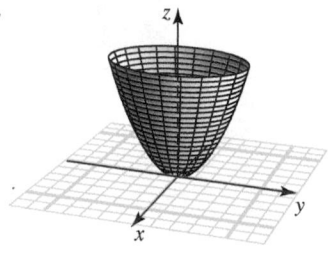

49.

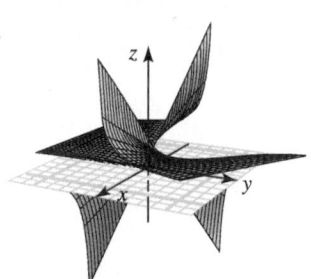

51.

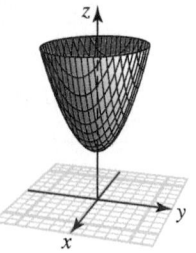

53. Equipotential curves are ellipses.

$E = 1$: $x^2 + 2y^2 = 46$

$E = 2$: $x^2 + 2y^2 = \dfrac{37}{4}$

$E = 3$: $x^2 + 2y^2 = \dfrac{22}{9}$

55. $L = \dfrac{d_o d_i}{d_i + d_o}$, $d_o > 0$, $d_i > 0$; the level curves are hyperbolas.   **57.** $y = \dfrac{100}{x}$ **59.** $R = (60 - \frac{x}{5} + \frac{y}{20})x + (50 - \frac{x}{10} + \frac{y}{20})y$

11.2 Limits and Continuity, page 836

3. 5   **5.** -2 **7.** 1   **9.** 1   **11.** 0   **13.** 1   **15.** 1   **17.** 50   **19.** 4   **21.** 10   **23.** The limit does not exist.   **25.** 1   **27.** -1 **29.** 1
31. The limit does not exist as can be seen from examining behavior on line $y = mx$.   **33.** The limit does not exist as can be seen
by examining behavior on line $y = 0$.   **35.** f is not continuous at $(0, 0)$.   **37. a.** $\frac{11}{10}$ **b.** The limit does not exist as can be seen by
examining behavior on $x = 0$ and $y = 0$.   **39.** 0   **41.** 2   **43.** Yes   **45.** 0   **47.** e **49.** 0   **51.** The limit does not exist.   **53.** Choose
$\delta = \dfrac{\epsilon}{3}$.   **55.** Choose $\delta = \dfrac{\epsilon}{\sqrt{2}}$.

11.3 Partial Derivatives, page 846

3. $f_x = 3x^2 + 2xy + y^2$; $f_y = x^2 + 2xy + 3y^2$; $f_{xx} = 6x + 2y$; $f_{yx} = 2x + 2y$ **5.** $f_x = \frac{1}{y}$; $f_y = -\frac{x}{y^2}$; $f_{xx} = 0$; $f_{yx} = -\frac{1}{y^2}$

7. $f_x = \frac{2}{2x + 3y}$; $f_y = \frac{3}{2x + 3y}$; $f_{xx} = \frac{-4}{(2x + 3y)^2}$; $f_{yx} = \frac{-6}{(2x + 3y)^2}$ **9. a.** $f_x = 2x(\cos x^2)(\cos y)$; $f_y = -(\sin x^2)(\sin y)$

b. $f_x = \cos(x^2 \cos y)(2x \cos y)$; $f_y = \cos(x^2 \cos y)(-x^2 \sin y)$ **11.** $f_x = \frac{3x}{(3x^2 + y^4)^{1/2}}$; $f_y = \frac{2y^3}{(3x^2 + y^4)^{1/2}}$

13. $f_x = xe^{x+y}(x + 2)\cos y$; $f_y = x^2 e^{x+y}(\cos y - \sin y)$ **15.** $f_x = \frac{y}{\sqrt{1 - x^2 y^2}}$; $f_y = \frac{x}{\sqrt{1 - x^2 y^2}}$

17. $f_x = y^2 + yz$; $f_y = 2xy + z^3 + xz$; $f_z = 3yz^2 + xy$ **19.** $f_x = \frac{1}{z}$; $f_y = \frac{2y}{z}$; $f_z = -\frac{x + y^2}{z^2}$ **21.** $f_x = \frac{1}{x + y^2 + z^3}$;

$f_y = \frac{2y}{x + y^2 + z^3}$; $f_z = \frac{3z^2}{x + y^2 + z^3}$ **23.** $z_x = -\frac{2x}{9z}$; $z_y = \frac{y}{2z}$ **25.** $z_x = \frac{6xy}{2z - y^3}$; $z_y = \frac{3x^2 + 3y^2 z}{2z - y^3}$

27. $z_x = -\frac{1}{2x^{3/2}\cos xz} - \frac{z}{x}$; $z_y = -\frac{2y}{x \cos xz}$ **29. a.** -4 **b.** 4 **31. a.** $-\frac{\pi^2}{4}$ **b.** $-\frac{\pi^2}{4}$

33. $f_x = -(x^2 + 2x + 1)$; $f_y = y^2 + 2y + 1$ **35.** $f_x = 6xy$; $f_{xx} = 6y$; $f_y = 3x^2 - 3y^2$; $f_{yy} = -6y$; thus $f_{xx} + f_{yy} = 0$.

37. $f_x = f_{xx} = f$; $f_{yy} = -f$ since $(\sin y)_{yy} = -\sin y$; thus $f_{xx} + f_{yy} = 0$.   **41.** $2(\sin z - x \cos z)$ **43. a.** yes

b. both negative   **c.** paint and paint brushes   **45. a.** $C_m = -0.67\sigma(T - t)m^{-1.67}$ **b.** $C_T = \sigma m^{-0.67}$

c. $C_t = -\sigma m^{-0.67}$ **47. a.** $\frac{\partial Q}{\partial K} = 80K^{-1/3}L^{2/5}$; $\frac{\partial Q}{\partial L} = 48K^{2/3}L^{-3/5}$ **b.** They are both negative.   **49. a.** 9   **b.** 14

51. a. no   **b.** no   **c.** yes   **59. a.** $\frac{\partial A}{\partial a} = \frac{1}{2}b \sin \gamma$; $\frac{\partial A}{\partial b} = \frac{1}{2}a \sin \gamma$; $\frac{\partial A}{\partial \gamma} = \frac{1}{2}ab \cos \gamma$ **b.** $\frac{\partial a}{\partial \gamma} = -\frac{2A \cos \gamma}{b \sin^2 \gamma}$

11.4 Tangent Planes, Approximations, and Differentiability, page 857

3. $3x + y - \sqrt{10}z = 0$ **5.** $2x + 4y - z - 4 = 0$ **7.** $x - y + 4z - \pi = 0$ **9.** $df = 10xy^3 dx + 15x^2 y^2 dy$

11. $df = y(\cos xy)dx + x(\cos xy)dy$ **13.** $df = -\frac{y}{x^2}dx + \frac{1}{x}dy$ **15.** $df = ye^x dx + e^x dy$

17. $df = 9x^2 dx - 8y^3 dy + 5dz$ **19.** $df = 2z^2 \cos(2x - 3y)dx - 3z^2 \cos(2x - 3y)dy + 2z \sin(2x - 3y)dz$

21. $f, f_x,$ and f_y are continuous, so the function is differentiable.   **23.** $f, f_x,$ and f_y are continuous, so the function is differentiable.

25. $f, f_x,$ and f_y are continuous, so the function is differentiable.   **27.** $f_x = 1 + 2y + \lambda y$; $f_y = 2x + \lambda x$; $f_\lambda = xy - 10$

29. $f_x = 2x - 3\lambda$; $f_y = 2y - 2\lambda$; $f_\lambda = 6 - 3x - 2y$ **31.** 37.04   **33.** 0   **35.** 2.691   **37.** $z = 9$ **41.** $\frac{1}{2}(1 - x - y)$

43. \$1.14   **45. a.** $R = \left(\frac{4,000 - p}{500}\right)p + \left(\frac{3,000 - q}{450}\right)q$ **b.** The revenue is increased by approximately \$180.

47. R increases by approximately 11%   **49.** 0.0360 cal   **51.** The manufacturer should decrease the level of unskilled labor by about 2.4 hours.   **53.** The correct period is approximately 0.0014 seconds more than the computed period.   **55.** The maximum possible error in the measurement of S is $\frac{12}{490}$lb.   **57.** The function is not continuous at $(0, 0)$.   **59.** A increases by about 1%.

11.5 Chain Rules, page 866

5. $8e^{2t}(1 + e^{6t})$ **7.** $\frac{-3 \sin^3 3t + 6 \sin 3t}{\cos^2 3t}$ **9.** $\frac{\partial F}{\partial u} = 2u \sin^2 v + 2u - 4v$; $\frac{\partial F}{\partial v} = 2u^2 \sin v \cos v - 4u + 8v$

11. $\frac{\partial F}{\partial u} = v^2 + v$; $\frac{\partial F}{\partial v} = 2uv + u$ **13.** $\frac{\partial w}{\partial s} = \frac{\partial w}{\partial x}\frac{\partial x}{\partial s} + \frac{\partial w}{\partial y}\frac{\partial y}{\partial s} + \frac{\partial w}{\partial z}\frac{\partial z}{\partial s}$; $\frac{\partial w}{\partial t} = \frac{\partial w}{\partial x}\frac{\partial x}{\partial t} + \frac{\partial w}{\partial y}\frac{\partial y}{\partial t} + \frac{\partial w}{\partial z}\frac{\partial z}{\partial t}$

15. $\frac{\partial w}{\partial s} = \frac{\partial w}{\partial x}\frac{\partial x}{\partial s} + \frac{\partial w}{\partial y}\frac{\partial y}{\partial s} + \frac{\partial w}{\partial z}\frac{\partial z}{\partial s}$; $\frac{\partial w}{\partial t} = \frac{\partial w}{\partial x}\frac{\partial x}{\partial t} + \frac{\partial w}{\partial y}\frac{\partial y}{\partial t} + \frac{\partial w}{\partial z}\frac{\partial z}{\partial t}$; $\frac{\partial w}{\partial u} = \frac{\partial w}{\partial x}\frac{\partial x}{\partial u} + \frac{\partial w}{\partial y}\frac{\partial y}{\partial u} + \frac{\partial w}{\partial z}\frac{\partial z}{\partial u}$

17. $\frac{dw}{dt} = \cos(xyz)[-3yz - e^{1-t}xz + 4xy]$ **19.** $\frac{dw}{dt} = (e^{x^3 + yz})\left[-\frac{6x^2}{t^2} + 2ty + \frac{2z}{2t - 3}\right]$

21. $\frac{\partial w}{\partial r} = \frac{2s + t \cos(rt)}{2 - z}$; $\frac{\partial w}{\partial t} = \frac{(2 - z)[r \cos(rt)] + 2st(x + y)}{(2 - z)^2}$ **23.** $\frac{dy}{dx} = \frac{3x(x^2 - y)^{1/2} + 2xy}{\frac{3}{2}(x^2 - y)^{1/2} - x^2}$

25. $\frac{dy}{dx} = \frac{(1 - \cos y)(1 + x^2) - y}{(1 + x^2)(-x \sin y + \tan^{-1} x)}$ **27.** $\frac{dy}{dx} = \frac{y}{x}$ **29. a.** $\frac{z^2}{2}$ **b.** $\frac{4}{x^3 y}$ **c.** $\frac{4}{xy^3}$ **31. a.** $\frac{2z^3}{x^2 y^2}$ **b.** $\frac{2z^2(x + z)}{x^4}$

c. $\frac{2z^2(y + z)}{y^4}$ **33. a.** $\frac{\sec^2 y \cos x}{4z^3}$ **b.** $\frac{2z^2 \sin x - \cos^2 x}{4z^3}$ **c.** $\frac{4z^2 \sec^2 y \tan y - \sec^4 y}{4z^3}$

35. $\frac{\partial^2 z}{\partial u^2} = a^2 z_{xx}$; $\frac{\partial^2 z}{\partial v^2} = b^2 z_{yy}$ **37.** volume and surface area both decreasing

39. a. $\dfrac{\partial C}{\partial a} = \dfrac{[1 - t(b-a)]e^{-at} - e^{-bt}}{(b-a)^2}$; $\dfrac{\partial C}{\partial b} = \dfrac{[(b-a)t+1]e^{-bt} - e^{-at}}{(b-a)^2}$; $\dfrac{\partial C}{\partial t} = \dfrac{1}{b-a}[be^{-bt} - ae^{-at}]$

b. $\dfrac{dC}{dt} = \left[\dfrac{(1 - bt + \ln b)(\frac{1}{b}) - e^{-bt}}{(bt - \ln b)^2}\right](-\ln b) + \dfrac{t}{bt - \ln b}\left[\dfrac{-\ln b}{bt} + be^{-bt}\right]$

41. The monthly demand for bicycles will be increasing at the rate of about 54 bicycles per month (4 months from now).

43. The joint resistance is decreasing at the approximate rate of 0.3471 ohms/second. **51.** $\dfrac{d^2 z}{d\theta^2} = y^2 f_{xx} + x^2 f_{yy} - 2xy f_{xy} - x f_x - y f_y$

57. $\dfrac{\partial z}{\partial x} = -\dfrac{F_x}{F_z} = -\dfrac{2x + 2yz}{2xy + e^z}$; $\dfrac{\partial z}{\partial y} = -\dfrac{F_y}{F_z} = -\dfrac{2xz + 3y^2}{2xy + e^z}$ **59. a.** The degree is $n = 3$.

11.6 Directional Derivatives and the Gradient, page 881

1. $(2x - 2y)\mathbf{i} - 2x\mathbf{j}$ **3.** $\left(-\dfrac{y}{x^2} + \dfrac{1}{y}\right)\mathbf{i} + \left(\dfrac{1}{x} - \dfrac{x}{y^2}\right)\mathbf{j}$ **5.** $e^{3-y}(\mathbf{i} - x\mathbf{j})$ **7.** $\cos(x + 2y)(\mathbf{i} + 2\mathbf{j})$

9. $e^{y+3z}(\mathbf{i} + x\mathbf{j} + 3x\mathbf{k})$ **11.** $\dfrac{1}{\sqrt{2}}$ **13.** $\dfrac{5\sqrt{2}}{8}$ **15.** 0 **17.** $\mathbf{N_u} = \pm\dfrac{1}{\sqrt{3}}(\mathbf{i} - \mathbf{j} + \mathbf{k})$;
the tangent plane is $x - y + z - 3 = 0$ **19.** $\mathbf{N_u} = \pm\dfrac{1}{\sqrt{3}}(-\mathbf{i} - \mathbf{j} + \mathbf{k})$; the tangent plane is $x + y - z - \dfrac{\pi}{2} = 0$

21. $\mathbf{N_u} = \pm\dfrac{1}{\sqrt{3}}(\mathbf{i} - \mathbf{j} + \mathbf{k})$; the tangent plane is $x - y + z = 0$ **23.** $\mathbf{N_u} = \pm\dfrac{1}{\sqrt{46}}(3\mathbf{i} + 6\mathbf{j} + \mathbf{k})$;
the tangent plane is $3x + 6y + z - 3 = 0$ **25.** $3\mathbf{i} + 2\mathbf{j}$; $\sqrt{13}$ **27.** $27(\mathbf{i} + \mathbf{j})$; $27\sqrt{2}$ **29.** $\dfrac{1}{5}(\mathbf{i} + 2\mathbf{j})$; $\dfrac{1}{\sqrt{5}}$

31. $2(5\mathbf{i} + 2\mathbf{j} + 5\mathbf{k})$; $6\sqrt{6}$ **33.** $\pm\dfrac{a\mathbf{i} + b\mathbf{j}}{\sqrt{a^2 + b^2}}$ **35.** $\pm\dfrac{b^2 x_0 \mathbf{i} + a^2 y_0 \mathbf{j}}{\sqrt{b^4 x_0^2 + a^4 y_0^2}}$ **37.** $\sqrt{3} + 1$ **39.** $-\dfrac{6e}{\sqrt{17}}$ **41.** $\dfrac{11\sqrt{3}}{15}$

43. $2\sqrt{3}$ in the direction of $\mathbf{u} = \dfrac{1}{\sqrt{3}}(\mathbf{i} + \mathbf{j} + \mathbf{k})$ **45.** in the direction corresponding to $\dfrac{3}{2}\mathbf{i} - \dfrac{5}{2}\mathbf{j}$

47. approximately 8.06 **49.** $5\sqrt{10}(\mathbf{i} - 3\mathbf{j})$ **55.** $D_\mathbf{u}f = f_x\cos\theta + f_y\sin\theta$; for $f(x,y) = xy^2 e^{x-2y}$, $D_\mathbf{u}f(1,3) = 6e^{-7}$

11.7 Extrema of Functions of Two Variables, page 893

3. $(1,0)$, relative maximum; $(-1,0)$, relative minimum **5.** A minimum occurs at $(2,3)$.

7. A maximum occurs at $(0,0)$. **9.** $(0,0)$, relative minimum **11.** $(\sqrt{3},0)$, relative maximum; $(-\sqrt{3},0)$, saddle point

13. $(0,0)$, relative minimum; $(1,0)$, saddle point; $(-1,0)$, saddle point; $(0,1)$, relative maximum; $(0,-1)$,
relative maximum **15.** $(2^{-1/3}, 2^{-1/3})$, relative minimum **17.** $(0,3)$, saddle point; $(0,9)$, relative minimum; $(-2,9)$,
saddle point; $(-2,3)$, relative maximum **19.** $(\frac{3}{2}, 2)$, saddle point **21.** $(6,2)$, relative minimum; $(8.985, -2.447)$,
saddle point **23.** 0 is minimum, $\frac{3}{2}$ is maximum **25.** -14 is minimum, 0 is maximum

27. -5 is minimum, 7 is maximum **29.** $y = 1.62x + 0.68$ **31.** $y = -0.02x + 5.54$ **33.** $(0, \pm 2, 0)$

35. $x = y = \sqrt[3]{2V}, z = \sqrt[3]{0.25V}$ **37.** The product is maximized when all three numbers are 18. **39.** The temperature is greatest
$(13°C)$ at $(3,2)$ and is least $(-\frac{1}{7}°C)$ at $(\frac{2}{7}, \frac{8}{7})$. **41.** The revenue is maximized when 10,500 units of A and 6,500 units of B are
produced. **43.** The owner should charge \$2.70 for California water and \$2.50 for New York water. **45.** 200 machines should be
supplied to the domestic market and 300 to the foreign market. **47.** approximately 2.12

49. a.

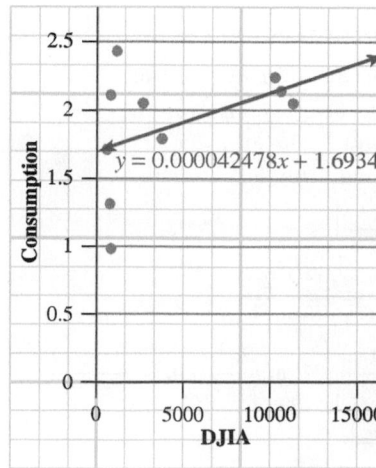

b. $y = 0.000041971x + 1.6978$

51. a. $k = 30.951, m = 0.5718$

55. The minimum time of travel occurs when
Dick waits 0.4 miles from the line AS and Mary
waits 1.6 miles from the finish line.

57. a. minimum **b.** saddle point

11.8 Lagrange Multipliers, page 904

1. $\frac{25}{16}$ **3.** 288 **5.** $\frac{81}{2}$ **7.** 2 **9.** 5 **11.** -4 **13.** 1.8478 **15.** $\frac{5}{2}\ln\frac{4}{3}$ **17.** $\frac{8}{7}$ **19.** $\frac{25}{7}$

21. $10\sqrt{3}$ is the constrained maximum and $-10\sqrt{3}$ is the constrained minimum. **23.** $\dfrac{|D|}{\sqrt{A^2+B^2+C^2}}$

25. The nearest point is $\left(\frac{1}{3},\frac{1}{6},\frac{1}{6}\right)$ and the distance is $\frac{1}{\sqrt{6}}$. **27.** $x=z=3,\ y=6$; the largest product is 324. **29.** $\frac{200}{3}$ **31.** The maximum area is obtained when the rectangle is a square with each side 80 yd. **33.** The radius $x=1$ in. and the height $y=4$ in. **35.** \$2,000 to development and \$6,000 to promotion gives the maximum sales of approximately 1,039 units. **37.** $x\approx 13.866$ ft and $y\approx 8.017$ ft **39.** $x=y=z=\sqrt[3]{C}$ **41.** The farmer should apply 4.24 acre-ft of water and 1.27 lb of fertilizer. **45.** $\frac{56}{3}$ **47.** $\frac{121}{24}$ **49. a.** \$3,000 for development and \$5,000 for promotion **b.** The estimated increase in profit is \$30.61. **c.** \$4,000 should be spent on development and \$6,000 should be spent on promotion. **d.** $\lambda=0$ **51.** $\frac{2}{3}$ **53.** $\dfrac{8abc}{3\sqrt{3}}$ **57. b.** If we drop the condition that $\alpha+\beta=1$ then $x=\dfrac{k\alpha}{p(\alpha+\beta)}$, $y=\dfrac{k\beta}{q(\alpha+\beta)}$; if k is increased by one unit, the maximum output is estimated to increase by

$$\frac{dQ}{dk}=c\left(\frac{k}{\alpha+\beta}\right)^{\alpha+\beta-1}\left(\frac{\alpha}{p}\right)^{\alpha}\left(\frac{\beta}{q}\right)^{\beta}$$

Chapter 11 Proficiency Examination, page 908

21. $f_{xy}=f_{yx}=\dfrac{1}{(1-x^2y^2)^{3/2}}$ **22.** $\dfrac{dw}{dt}=6\pi^2$ **23. a.** $\nabla f=\mathbf{i}+3\mathbf{k}$ **b.** $D_{\mathbf{u}}f=-\frac{2}{\sqrt{5}}$ **c.** $\dfrac{\mathbf{i}+3\mathbf{k}}{\sqrt{10}}$; $\|\nabla f\|=\sqrt{10}$

24. $\lim\limits_{(x,y)\to(0,0)}f(x,y)$ along the line $y=x$ is $\frac{1}{2}$, which is not $f(0,0)$. **25.** $f_x=-\frac{1}{x}$, $f_y=\frac{1}{y}$, $f_{xx}=\frac{1}{x^2}$, $f_{xy}=0$, $f_{yy}=-\frac{1}{y^2}$

27. $D_{\mathbf{u}}f=-32\left(1+\sqrt{3}\right)$ **28.** $(0,0)$, saddle point; $(9,3)$, relative maximum; $(-9,-3)$, relative maximum

29. maximum of 12, minimum of 3 **30.** maximum of $\frac{49}{4}$, minimum of $-\frac{9}{4}$

Chapter 11 Supplementary Problems, page 909

1. the closed disk of radius 4 centered at the origin. **3.** $-1\le x\le 1, -1\le y\le 1$ **5.** $f_x=1, f_y=-1$

7. $f_x=2xy-\frac{y}{x^2}\cos\frac{y}{x}$; $f_y=x^2+\frac{1}{x}\cos\frac{y}{x}$ **9.** $f_x=6x^2y+3y^2-\frac{y}{x^2}$; $f_y=2x^3+6xy+\frac{1}{x}$

11. For $c=2$, $x^2-y=2$ is a parabola opening up, with vertex at $(0,-2)$. For $c=-2$, $x^2-y=-2$ is a parabola opening up, with vertex at $(0,2)$. **13.** For $c=0$, we get the origin and half-line $y=0,\ x<0$. For $c=1$, $\sqrt{x^2+y^2}=1$ is a semi-circle (to the right of the y-axis). For $c=-1$, $|y|=1$ is a pair of half-lines, 1 unit above or below the x-axis, to the left of the y-axis. **15.** For both values of c, the surface is an ellipsoid. **17.** 0 **19.** Different values are obtained along different curves.

21. $\dfrac{dz}{dt}=ye^t(-t^{-2}+t^{-1})+(x+2y)\sec^2 t$

23. $\dfrac{\partial z}{\partial u}=\left(\tan\frac{x}{y}+\frac{x}{y}\sec^2\frac{x}{y}\right)v+\left(-\frac{x^2}{y^2}\sec^2\frac{x}{y}\right)v^{-1}$; $\dfrac{\partial z}{\partial v}=\left(\tan\frac{x}{y}+\frac{x}{y}\sec^2\frac{x}{y}\right)u+\left(-\frac{x^2}{y^2}\sec^2\frac{x}{y}\right)(-uv^{-2})$

25. $\dfrac{\partial z}{\partial x}=\frac{z}{3z+1}$; $\dfrac{\partial z}{\partial y}=\frac{2z}{3z+1}$ **27.** $\dfrac{\partial z}{\partial x}=-e^{x-z}$; $\dfrac{\partial z}{\partial y}=-e^{y-z}$ **29.** $f_{xx}=\sin x\cos(\cos x)$; $f_{yx}=0$

31. $f_{xx}=\dfrac{xy^3}{(1-x^2y^2)^{3/2}}$; $f_{yx}=(1-x^2y^2)^{-3/2}$ **33.** $f_{xx}=2e^{x^2+y^2}(2x^2+1)$; $f_{yx}=4xye^{x^2+y^2}$

35. normal line: $\frac{x-1}{2}=\frac{y-1}{-2}=\frac{z}{-1}$; tangent plane: $z=2x-2y$ **37.** normal line: $\frac{x-1}{16}=\frac{y-1}{-7}$ and $z=1$; tangent plane: $16(x-1)-7(y-1)=0$ **39.** relative minimum at $(3,-1)$ **41.** $(1,0)$, saddle point; $(0,1)$, saddle point; $\left(\frac{2}{3},\frac{2}{3}\right)$, relative maximum; $(1,1)$, saddle point **43.** $(0,9)$, relative minimum; $(0,3)$, saddle point; $(-2,9)$, saddle point; $(-2,3)$, relative maximum **45.** The largest value is 2 and the smallest value is -2.

47. The largest value is 10 and the smallest value is $-9/4$. **49.** $\dfrac{dz}{dt}=(2x-3y^2)(2)-6xy(2t)$

51. $\dfrac{\partial z}{\partial x}=e^{u^2-v^2}(8xu^2+4x-12uvx)$; $\dfrac{\partial z}{\partial y}=e^{u^2-v^2}(12yu^2+6y+8uvy)$ **53.** $\dfrac{dy}{dx}=-1$ at $(1,1)$

55. normal line: $\dfrac{x}{2}=\dfrac{y-1}{2}=\dfrac{z-3}{-1}$; tangent plane: $2x+2y-z+1=0$ **57.** $g(x,y,z)=x^2y+y^2z+z^2x$

59. $\dfrac{dw}{dt}=\left[\dfrac{2x}{1+x^2+y^2}\right]\left(\dfrac{2t}{1+t^2}\right)+\left[\dfrac{2y}{1+x^2+y^2}-\dfrac{2}{1+y^2}\right]e^t$ **61.** $D_{\mathbf{u}}f=\frac{1}{\sqrt{5}}(-16\ln 2-12)$ **63. a.** $D_{\mathbf{u}}f=-\frac{18}{\sqrt{6}}$

b. $\frac{1}{\sqrt{86}}(-6\mathbf{i}+\mathbf{j}+7\mathbf{k})$ **65.** 324 **67.** $\sqrt{3}$ **69. a.** $t_m=\dfrac{-1+\sqrt{1+4k\gamma r^2}}{4k\gamma}$ **b.** $M(r)=\sqrt{\dfrac{k\gamma}{\pi}}\dfrac{e^{-z/2}}{\sqrt{z-1}}$

c. The darkest part of the eyespot pattern occurs near the center.

71. a.

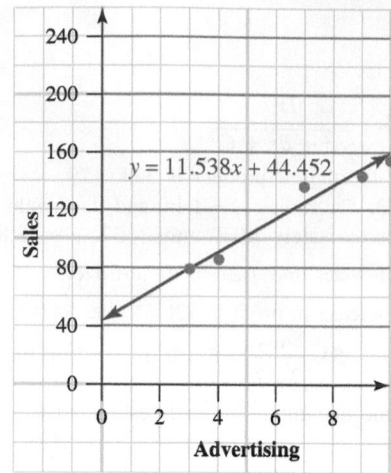

$y = 11.538x + 44.452$ **b.** approximately $102,142

73. $-\frac{2}{3}$ **75.** $\pm\frac{1}{3}(2\mathbf{i} + 2\mathbf{j} - \mathbf{k})$ **77.** $-\dfrac{\sin^2(x+z)\cos(x+y) + \sin^2(x+y)\cos(x+z)}{\sin^3(x+z)}$ **79.** $f(2,3) = 18$ is a relative minimum.

81. The minimum $\dfrac{1}{a^2 + b^2 + c^2}$ is attained for $x = \dfrac{a}{a^2 + b^2 + c^2}$; $y = \dfrac{b}{a^2 + b^2 + c^2}$; $z = \dfrac{c}{a^2 + b^2 + c^2}$.

87. $\theta = 2$ and $r = \sqrt{A_0}$ **89.** approximately -0.64π cm^3 **93.** $z_{xx} = z_{xy} = z_{yy} = 0$

97. This is Putnam Problem 5 in the afternoon session of 1946.

Chapter 12

12.1 Double Integration over Rectangular Regions, page 928

3. $\frac{13}{3}$ **5.** $\frac{3}{2}$ **7.** $(e^2 - 1)\ln 2 + 2$ **9.** $12(e - 1)$ **11.** $\frac{15}{2}\ln 3 - 10\ln 2 + \frac{1}{2}$ **13.** 32   **15.** 24   **17.** 24   **19.** $\frac{7}{6}$ **21.** $\frac{4}{3\sqrt{2}}\tan^{-1}\left(\frac{1}{\sqrt{2}}\right)$

23. -1 **25.** $4\ln 2$ **27.** 1   **29.** 8   **31.** $3\ln 2$ **33.** $2(e - 1)\ln 2 - \frac{3}{4}(e - 3)$ **35.** $\frac{32}{9}$ **37.** $\sqrt{3} - \frac{4}{3}\sqrt{2} + \frac{1}{3}$ **39.** 3   **41.** $\frac{\pi^2}{4}$

43. $\iint\limits_{R} 20{,}000\,dA$ **45.** $M = \iint\limits_{R} \rho(x, y)\,dA$ **47.** $\frac{1}{9}(e^6 - 1)$ **49.** $1/4\,(13\ln 13 - 15\ln 5 - 8\ln 2)$ **51.** $f(x, y) \geq 2$ and area of base is 1

53. 1.44   **59.** The first integral is $\frac{1}{2}$ and the second integral is $-\frac{1}{2}$.

12.2 Double Integration over Nonrectangular Regions, page 938

3. $\frac{32}{3}$ **5.** $3\pi - 2\sqrt{3}$ **7.** $\frac{5}{12}$ **9.** $\frac{1}{6}\sin^3 2$ **11.** $\frac{\pi}{3} - \frac{\sqrt{3}}{2}$ **13.** $\frac{1}{2}e - 1$ **15.** $\sqrt{2} - 1$ **17.** $\frac{1}{2}$ **19.** $\frac{5}{12}$ **21.** $\frac{3}{2}\ln 5$ **23.** 2   **25.** $\frac{32}{3}$

27. $\frac{2}{3}$ **29.** $\frac{1}{3}(8\pi + 2\sqrt{3})$ **31.**

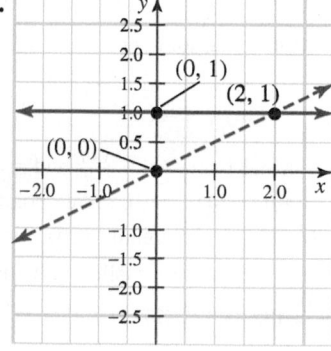

$\int_0^2 \int_{x/2}^1 f(x, y)\,dy\,dx$

33.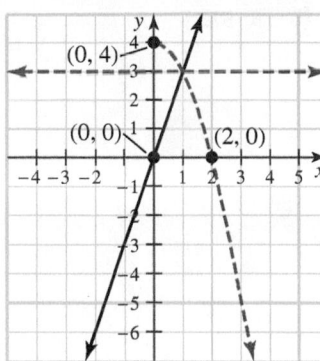
Intersection: $(1, 3)$; $\int_0^1 \int_0^{3x} f(x, y)\, dy\, dx + \int_1^2 \int_0^{4-x^2} f(x, y)\, dy\, dx$

35.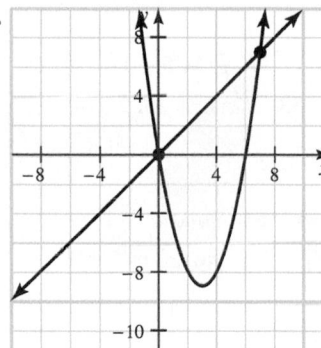
Intersections: $(0, 0)$ and $(7, 7)$;

$$\int_{-9}^{0} \int_{3-\sqrt{9+y}}^{3+\sqrt{9+y}} f(x, y)\, dx\, dy + \int_{0}^{7} \int_{y}^{3+\sqrt{9+y}} f(x, y)\, dx\, dy$$

37. $V = \int_0^{7/3} \int_0^{(7-3x)/2} (7 - 3x - 2y)\, dy\, dx$ **39.** $V = 8\int_0^{\sqrt{3}} \int_0^{\sqrt{3-x^2}} \sqrt{7 - x^2 - y^2}\, dy\, dx$

41. $V = 8\int_0^{a} \int_0^{(b/a)\sqrt{a^2-x^2}} c\sqrt{1 - \dfrac{x^2}{a^2} - \dfrac{y^2}{b^2}}\, dy\, dx$

43. $V = 8\left(\dfrac{\pi\sqrt{2}}{3} - \int_0^1 \int_0^1 \sqrt{2 - x^2 - y^2}\, dy\, dx\right)$ **45.** πab **47.** 1 **49.** $\int_1^8 \int_{y^{1/3}}^{y} f(x, y)\, dx\, dy$ **51.** $\frac{5}{24}$ **53.** 22 **55.** 18π

57. $\frac{128\pi}{3}$ **59.** The value of the integral is between 0.55 and 4.08.

12.3 Double Integrals in Polar Coordinates, page 948

1. $\dfrac{\pi}{4}\left(\dfrac{1}{e} - \dfrac{1}{e^9}\right)$

3. $\dfrac{32\pi}{3}$

5. 0

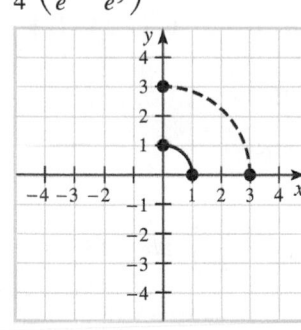   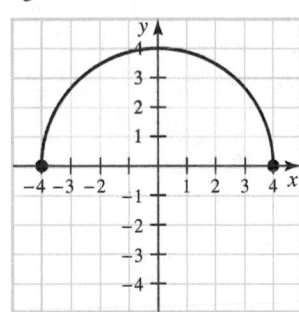   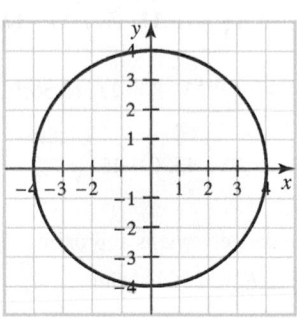

7. 2

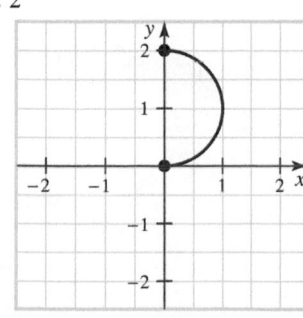

9. 16π **11.** 6π **13.** 4π **15.** $\frac{\pi}{2}$ **17.** $\frac{\pi}{3} + \frac{\sqrt{3}}{2}$ **19.** $\frac{5\pi}{4} - 2$ **21.** π

23. $\pi - \frac{3\sqrt{3}}{2}$ **25.** $\frac{a^4\pi}{4}$ **27.** $\pi\ln 2$ **29.** $2\pi a(1 - \ln 2)$ **31.** 0 **33.** $\pi(e^9 - 1)$

35. $\pi(\frac{3}{2}\ln 3 + \ln 2 - 1)$ **37.** 9 **39.** $\frac{\pi}{4}(e^4 - 1)$ **41.** 4 **43.** $\frac{\pi-2}{2}$ **45.** 8π

47. $\frac{8}{3}$

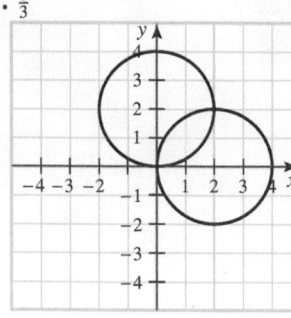

49. $\frac{a^2}{8}(\pi - 2)$ **51.** $\frac{256\pi}{3}$ **53.** $\frac{16}{9}(3\pi - 4)$ **55.** $\frac{5\pi}{32}$ **57.** $\frac{\pi}{4} - \frac{1}{2}$

12.4 Surface Area, page 958

5. $4\sqrt{21}$ **7.** $\sqrt{6}$ **9.** $\frac{\pi}{6}(5\sqrt{5} - 1)$ **11.** $\frac{7}{4}$ **13.** $2\sqrt{17} + \frac{1}{2}\ln(4 + \sqrt{17})$ **15.** $8\sqrt{65} + \ln(8 + \sqrt{65})$ **17.** $\frac{\pi}{6}(5\sqrt{5} - 1)$ **19.** 2π
21. $16\pi(2 - \sqrt{2})$ **23.** $15\sqrt{2} + \frac{5}{2}\ln(2\sqrt{2} + 3)$ **25.** $\frac{\pi}{6}(17\sqrt{17} - 1)$ **27.** $24,000\sqrt{26}$ **31.** 36π **33.** $\sqrt{2}\pi h^2$ **35.** $2(2 - \sqrt{3})$
37. 72 **39.** $S = \dfrac{D^2}{2ABC}\sqrt{A^2 + B^2 + C^2}$ **41.** $\frac{a^2}{6}\sqrt{14}$ **43.** $\frac{3\sqrt{3}\pi a^2}{4}$ **45.** $S = \int_0^1 \int_0^y \sqrt{\csc^2 y + z^{-2}}\, dz\, dy$
47. $S = \int_0^{2\pi} \int_0^{\sqrt{\pi/2}} \sqrt{4r^2 \sin^2 r^2 + 1}\, r\, dr\, d\theta$ **49.** $S = \int_1^5 \int_{5/x}^{6-x} \sqrt{(2x + 5y)^2 + (2y + 5x)^2 + 1}\, dy\, dx$ **51.** $4|u|\sqrt{u^2 + 1}$
53. $2|v|\sqrt{9u^4 + 1}$ **55.** $2\pi\sqrt{6} - \frac{8\pi}{3}$ **57. a.** $(\cos v)\mathbf{i} - (\sin v)\mathbf{j} - u\mathbf{k}$ **b.** $\frac{b}{2}[\ln(a + \sqrt{1 + a^2}) + a\sqrt{1 + a^2}]$ **59.** $4\pi R^2$

12.5 Triple Integrals, page 970

3. 45 **5.** 18 **7.** 60 **9.** 45 **11.** $\frac{68}{9}$ **13.** $-\frac{15}{\pi}$ **15.** $\frac{1}{8}(5 - e^2)$ **17.** $6\pi - 18\ln 2$ **19.** 8 **21.** $\frac{1}{720}$ **23.** 0 **25.** $\frac{1}{2}e^2 - e$ **27.** $\frac{1}{6}$
29. $\frac{4\pi}{3}$ **31.** 4π **33.** $3\sqrt{6}\pi$ **35.** $\frac{16}{3}$ **37.** $\int_0^1 \int_0^{1-y} \int_0^{1-x-y} f(x, y, z)\, dz\, dx\, dy$ **39.** $\int_0^1 \int_0^{1-z} \int_0^{1-y-z} f(x, y, z)\, dx\, dy\, dz$
41. $\int_0^2 \int_0^{\sqrt{4-z^2}} \int_0^{\sqrt{4-y^2-z^2}} f(x, y, z)\, dx\, dy\, dz$ **43.** $\int_0^1 \int_{\sqrt[3]{x}}^1 \int_0^{1-y} f(x, y, z)\, dz\, dy\, dx$ **45.** 32π **47.** 32π **49.** $\frac{3\pi}{2}$ **51.** $\frac{4}{3}\pi R^3$
53. $\frac{4\pi abc}{3}$ **57.** $\frac{1 - \cos \pi^3}{6}$ **59.** $\frac{14}{9}$

12.6 Mass, Moments, and Probability Density Functions, page 982

5. $(\frac{3}{2}, 2)$ **7.** $(1, \frac{8}{5})$ **9.** $(\frac{1}{4}, \frac{4}{5})$ **11.** $(1, 1, 1)$ **13.** $(0, \frac{24}{5\pi})$ **15.** $(7, \frac{5}{3})$ **17.** $\left(\dfrac{4\ln 2 - 2}{(\ln 2)^2}, \dfrac{\ln 2}{3}\right)$ **19. a.** $(0, \frac{3a}{2\pi})$ **b.** $\left(-\frac{8a}{3\pi^2}, \frac{4a}{3\pi}\right)$
21. $\left(\dfrac{3e^4 + 1}{4(e^2 + 1)}, \dfrac{e^2 - 1}{e^2 + 1}\right)$ **23.** $\frac{8}{15}$ **25.** $(0.56, 0.56)$ **27.** $(\frac{3a}{4}, \frac{3b}{8})$ **29.** $\frac{a^4 \pi}{64}$ **33.** $\left(\frac{3a}{8}, \frac{3a}{8}, \frac{3a}{8}\right)$ **35.** $1 - \frac{5}{2e}$ **37.** 75%
39. $(\frac{11}{12}, \frac{67}{18})$ **41.** $3(e - 2)$ **43.** 0 **45.** $\sqrt{\frac{a^2 + b^2}{6}}$ **47.** 2.4107 **49. a.** $t_m = \dfrac{x_0^2}{2k}$; $C_m(x_0) = \sqrt{\dfrac{2}{\pi}} C_0 \dfrac{e^{-1/2}}{x_0}$

b. $x_m \leq 1.9358$; the danger zone is approximately 1.9 miles **c.** $AV = \dfrac{1}{A} \int_0^{1.9358} \int_0^{x^2/(2k)} \dfrac{C_0}{\sqrt{k\pi t}} \exp\left(-\dfrac{x^2}{4kt}\right) dt\, dx$ **53.** $5,400\,\pi\delta$
57. Using integration, $F = 4.39741\, Gm$

12.7 Cylindrical and Spherical Coordinates, page 995

3. a. $(4, \frac{\pi}{2}, \sqrt{3})$ **b.** $\left(\sqrt{19}, \frac{\pi}{2}, \cos^{-1}\left(\frac{\sqrt{3}}{\sqrt{19}}\right)\right)$ **5. a.** $(\sqrt{5}, \tan^{-1} 2, 3)$ **b.** $\left(\sqrt{14}, \tan^{-1} 2, \cos^{-1}\left(\frac{3}{\sqrt{14}}\right)\right)$ **7. a.** $\left(-\frac{3}{2}, \frac{3\sqrt{3}}{2}, -3\right)$
b. $\left(3\sqrt{2}, \frac{2\pi}{3}, \frac{3\pi}{4}\right)$ **9. a.** $(\sqrt{2}, \sqrt{2}, \pi)$ **b.** $\left(\sqrt{\pi^2 + 4}, \frac{\pi}{4}, \cos^{-1}\left(\frac{\pi}{\sqrt{\pi^2 + 4}}\right)\right)$ **11. a.** $(\frac{3}{2}, \frac{\sqrt{3}}{2}, -1)$ **b.** $(\sqrt{3}, \frac{\pi}{6}, -1)$
13. a. $(\sin 3 \cos 2, \sin 3 \sin 2, \cos 3)$ **b.** $(\sin 3, 2, \cos 3)$
15. $z = r^2 \cos 2\theta$ **17.** $r = \dfrac{6z}{\sqrt{4 - 13\cos^2 \theta}}$ or $9r^2 \cos^2 \theta - 4r^2 \sin^2 \theta + 36z^2 = 0$

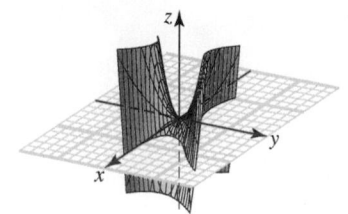

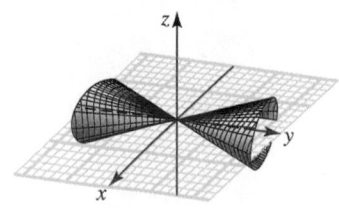

19. $\phi = \dfrac{\pi}{4}$

21. $\rho = \dfrac{4\cot\phi\,\csc\phi}{3 - 2\cos^2\theta}$ or $4\rho\cos\phi = \rho^2\sin^2\phi\,(\cos^2\theta + 3\sin^2\theta)$

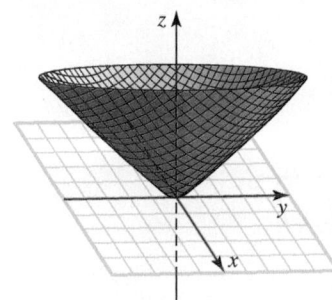

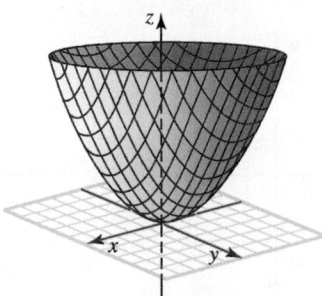

23. $z = 2xy$

25. $z = x^2 - y^2$

27. $xz = 1$

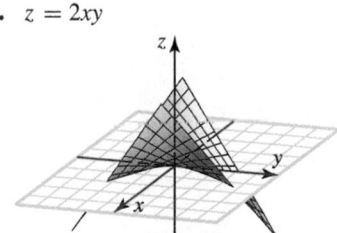

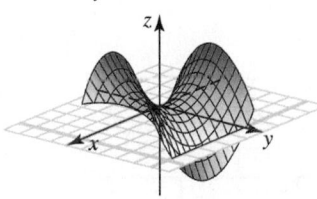

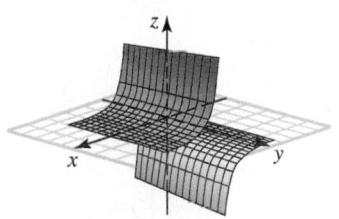

29. $\frac{16}{3}$ **31.** 2π **33.** 8π **35.** $\frac{7}{60}$ **37.** $\dfrac{hR^4\pi}{2}$ **39.** $\dfrac{a^6}{3}$ **41.** $(0, 0, \frac{27}{4})$ **43. a.** $m = \int_0^{2\pi} \int_0^3 \int_0^{\sqrt{9-r^2}} (r^2\sin\theta\cos\theta + z)\,r\,dz\,dr\,d\theta$

b. $\bar{x} = \frac{1}{m}\int_0^{2\pi} \int_0^3 \int_0^{\sqrt{9-r^2}} r\cos\theta(r^2\sin\theta\cos\theta + z)\,r\,dz\,dr\,d\theta$ **c.** $I_z = \int_0^{2\pi} \int_0^3 \int_0^{\sqrt{9-r^2}} r^2(r^2\sin\theta\cos\theta + z)\,r\,dz\,dr\,d\theta$

45. $\frac{128\pi}{15}$ **47.** The average value of $\theta = \pi$, The average value of $\phi = \frac{\pi}{2}$.   **49.** $\frac{16\pi\sqrt{2}}{5}$ **51.** 6π **53.** $\frac{7\pi}{2}$ **55.** $\frac{5\pi}{2}$ **59.** $\frac{8\pi}{5}$

12.8 Jacobians: Change of Variables, page 1005

3. 2   **5.** $2u$ **7.** $-2e^{2u}$ **9.** $-e^{2u}$ **11.** -9 **13.** ue^{uv} **15.** $\frac{1}{11}$ **17.** -1 **19.** $-u$ **21.** $\frac{1}{8xy}$

23. $A(0,5) \rightarrow (5,-5)$ $B(6,5) \rightarrow (11,1)$
$C(6,0) \rightarrow (6,6)$ $O(0,0) \rightarrow (0,0)$

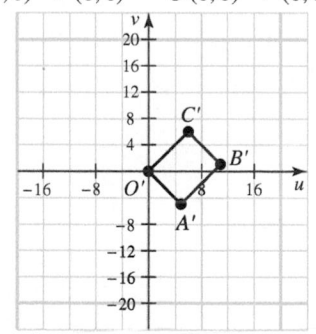

This transformation represents a rotation combined with a dilation.

25. $A(5,0) \rightarrow (5,0)$ $B(7,4) \rightarrow (7,-4)$
$C(2,4) \rightarrow (2,-4)$ $O(0,0) \rightarrow (0,0)$
Note that the new figure represents the same parallelogram reflected with respect to the x-axis.

27. $dx\,dy = u\,du\,dv$ **29.** $3\ln 2$ **31.** 0   **33.** 0   **35.** $\frac{5}{6}$

37. $\frac{625}{6}$ **39.** $\frac{25}{2}\tan^{-1}5 - \frac{5}{4}\ln 26$ **41.** 1   **43.** $\frac{2}{3}(e-1)$

45. $\frac{ab\pi}{4}(1 - e^{-1})$ **47.** $\frac{\sqrt{5}\pi}{10}(1 - e^{-20})$ **49.** $(1.14, 1.57)$

51. $\frac{4}{3}\pi abc$ **53.** $\frac{1}{4}(49\ln 7 - \frac{75}{2}\ln 5 - 27\ln 3 + 6)$ **55.** $\frac{\pi}{\sqrt{6}}$

59. $C = 4\int_0^a \dfrac{1}{a\sqrt{a^2 - x^2}}\sqrt{a^4 + (b^2 - a^2)x^2}\,dx$

Chapter 12 Proficiency Examination, page 1008

22. $e^{-\sqrt{3}/2} + \frac{\sqrt{3}}{2} - 1$ **23.** 0   **24.** 36   **25.** $\frac{\pi}{4}\sin 1$ **26.** $\frac{abc}{6}$ **27.** $(1 - e^{-1/2})^2$ **28.** $\frac{128\pi}{15}$ **29.** $2\pi\sqrt{2}$ **30.** $\frac{1}{3}(e^2 + \frac{8}{e} - 3)$

Chapter 12 Supplementary Problems, page 1010

1. $\frac{2}{3}$ **3.** $\frac{27}{2}$ **5.** $\frac{3}{2}$

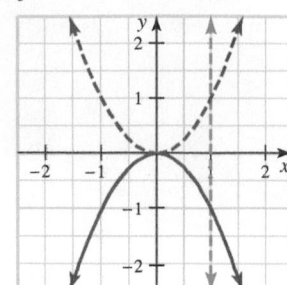
     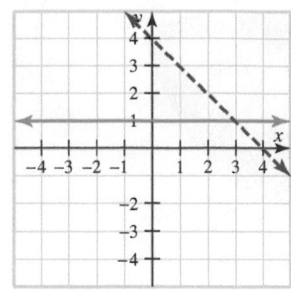
     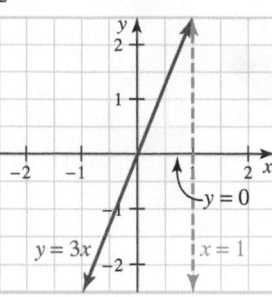

7. $\frac{\pi}{4}\ln 2$ **9.** $\dfrac{8\sqrt{2}-4}{15}$ **11.** $\frac{3}{2}(e^3-1)$

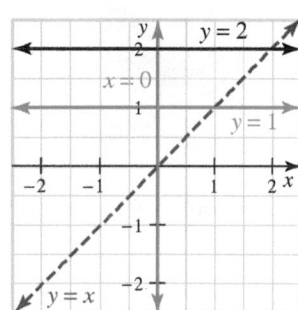
     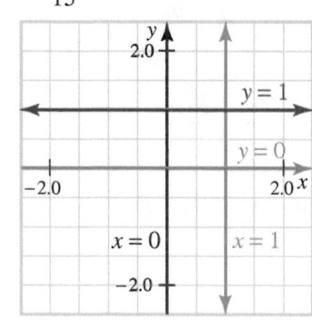
     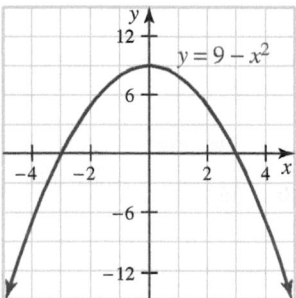

13. $\dfrac{1-\cos 1}{12}$ **15.** $\displaystyle\int_0^2 \int_{x^2}^{2x} f(x,y)\,dy\,dx$

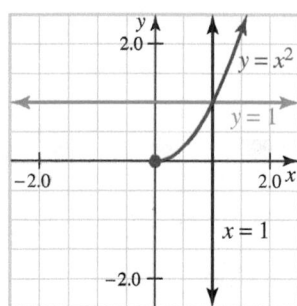
     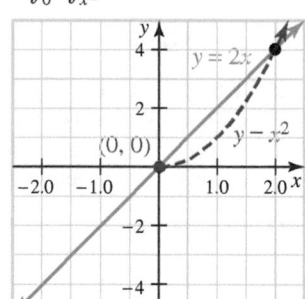

17. $\displaystyle\int_0^4 \int_{-\sqrt{y}}^{\sqrt{y}} f(x,y)\,dx\,dy + \int_4^9 \int_{-\sqrt{y}}^{6-y} f(x,y)\,dx\,dy$ **19.** $\frac{1}{3}(e-1)$ **21.** 10 **23.** $\frac{2}{3}\pi^4(2-\sqrt{2})$

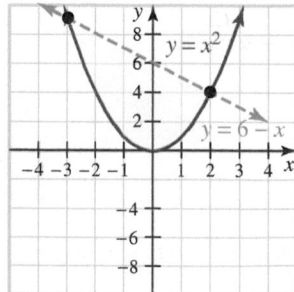

25. $\frac{31}{30}$ **27.** 0 **29.** 12π **31.** $\frac{2^{2n+2}\pi}{n+1}$ **33.** $\frac{2}{3}+2\ln 2$ **35.** $\frac{32}{3}$ **37.** $\frac{81\pi}{8}$ **39.** $\frac{1}{5}$ **41.** 11 **43.** $\frac{2\pi}{15}$ **45.** $2\pi(\ln 2+2\sqrt{2}-2)$

47. $m=\displaystyle\int_{\pi/12}^{5\pi/12} \int_{\sqrt{2}a}^{2a\sqrt{\sin 2\theta}} \rho r\,dr\,d\theta$ **49.** a^2 **51.** $\int_0^1 \int_x^{4-3x} f(x,y)\,dy\,dx$ **53.** $\sqrt{3}-\frac{\pi}{3}$ **55.** $\pi+3\sqrt{3}$ **57.** $\frac{2}{3}$ **59.** $\frac{\pi}{2}$

61. $\frac{\pi}{6}(37\sqrt{37}-1)$ **63.** $\frac{9\pi}{4}$ **65.** $\frac{4(3\pi-4)}{9}$ **69.** $\dfrac{2\pi a^5}{15}(\cos^3\phi_0-3\cos\phi_0+2)$ **71.** $\frac{2}{5}$ **73.** $\frac{3}{4}H$ **75.** $32xyz$ **77.** $\frac{3\pi-7}{243}$

79. b. $\frac{\partial(u,v)}{\partial(x,y)} = -2$; $\iint\limits_{D}(u+v)\,du\,dv = \frac{7}{3}$ **81.** $(\frac{1}{3A}, \frac{1}{3B}, \frac{1}{3C})$ **83. a.** $(\frac{16}{35}, \frac{1}{2})$ **b.** $\frac{11}{240}$ **85.** $\frac{32a^3}{9}$ **87.** $\frac{x^2}{a^2} + \frac{y^2}{b^2} + \frac{z^2}{c^2} = R^2$

89. $\frac{4a^3\pi}{35}$ **91.** $(0,0,0)$ **93.** $(\frac{1-\ln 2}{4-\pi}, \frac{1-\ln 2}{4-\pi}, \frac{1-\ln 2}{4-\pi})$ **95.** $(0, 0, \frac{3\pi-96}{8\pi+96})$ **97.** This is Putnam Problem 5 in the morning session of 1942.
99. This is Putnam Problem 5 in the morning session of 1958.

Chapter 13

13.1 Properties of a Vector Field: Divergence and Curl, page 1026

3. div $\mathbf{F} = 3x + 3z^2$; curl $\mathbf{F} = y\mathbf{k}$ **5.** div $\mathbf{F} = 2$; curl $\mathbf{F} = \mathbf{0}$ **7.** At $(2,-1,3)$ div $\mathbf{F} = 0$; curl $\mathbf{F} = \mathbf{0}$ **9.** At $(1,2,3)$ div $\mathbf{F} = 7$;
curl $\mathbf{F} = \mathbf{j} - 3\mathbf{k}$ **11.** At $(3,2,0)$ div $\mathbf{F} = 2 - 2e^{-6}$; curl $\mathbf{F} = -3\mathbf{i} + 3e^{-6}\mathbf{k}$ **13.** div $\mathbf{F} = \cos x - \sin y$; curl $\mathbf{F} = \mathbf{0}$
15. div $\mathbf{F} = 0$; curl $\mathbf{F} = \mathbf{0}$ **17.** div $\mathbf{F} = \frac{1}{\sqrt{x^2+y^2}}$; curl $\mathbf{F} = \mathbf{0}$ **19.** div $\mathbf{F} = 2x - 2z$; curl $\mathbf{F} = \mathbf{0}$ **21.** div $\mathbf{F} = 0$; curl $\mathbf{F} = -\mathbf{i} - \mathbf{j} - \mathbf{k}$

23. div $\mathbf{F} = xe^{xy} - \frac{x}{x^2+z^2}$; curl $\mathbf{F} = \left(\frac{1}{z} - \frac{z}{x^2+z^2}\right)\mathbf{j} + ye^{xy}\mathbf{k}$ **25.** div $\mathbf{F} = \frac{2}{\sqrt{x^2+y^2+z^2}}$; curl $\mathbf{F} = \mathbf{0}$ **27.** harmonic

29. harmonic **31.** $6xyz^2 + 2xy^3$ **33.** $6x\mathbf{j} - 3y\mathbf{k}$ **35.** $2z + 3x$ **39. a.** $(bz - cy)\mathbf{i} + (cx - az)\mathbf{j} + (ay - bx)\mathbf{k}$ **b.** div $\mathbf{V} = 0$;
curl $\mathbf{V} = 2\boldsymbol{\omega}$ **41.** II only **59.** div $(\nabla fg) = f\,\mathrm{div}(\nabla g) + 2\nabla f \cdot \nabla g + g\,\mathrm{div}(\nabla f)$

13.2 Line Integrals, page 1036

3. $-\frac{4}{3}$ **5.** 1 **7.** $\frac{26}{3}$ **9.** $2\left(\sqrt{2}-1\right)$ **11.** 6 **13.** $\frac{\sqrt{10}}{9}$ **15. a.** $-\frac{1}{3}$ **b.** $-\frac{1}{3}$ **17.** $\frac{104}{3}$ **19.** $-\frac{5}{12}$ **21.** 4π **23.** 12 **25.** 12
27. a. 5π **b.** $\frac{\pi}{2}(a^2+1)$ **29. a.** 2π **b.** 2π **31.** $\frac{\pi}{2}$ **33.** $\frac{116}{297}$ **35.** $-\frac{1}{3}$ **37.** 0 **39.** $\frac{1}{\sqrt{2}}$ **41.** 2π **43.** $-2a^2$ **45.** $\frac{\pi}{2} - \frac{4}{3}$

47. $\frac{1}{35}$ **49.** $\frac{25}{6}$ **51.** $\left(\frac{3\sqrt{2}\pi}{16}, 0, 0\right)$ **53.** $(\frac{12}{5}, \frac{44}{25}, \frac{14}{15})$ **55.** 11,000 ft-lb **57.** 0 **59.** No; answers vary.

13.3 The Fundamental Theorem and Path Independence, page 1047

3. not conservative **5.** x^2y^3 **7.** not conservative **9.** 0 **11.** 0 **13.** $\frac{\pi}{8}$ **15.** $-\frac{4}{3}$ **17.** 0 **19.** 32 **21.** $x^2 - xy + \frac{1}{3}y^3$; $\frac{1}{3}$ **23.** x^2y; 1
25. $\frac{x}{y+1}$; $\frac{1}{2}$ **27.** ze^{xy} **29.** $\frac{1}{4}(x^2 + y^2 + z^2)^2$ **31.** $\frac{1}{2}x^2y^2 + xyz + y^3z$ **33.** $\cos 1 + \sin 1 + 1$ **35.** $\frac{\pi}{4}$ **37.** $\frac{13}{4}$ **39.** 8 **41.** $\frac{3\pi}{2} - 2$

43. 8 **45.** 32 **47.** Cx^{-5} **49. a.** $\frac{kmM}{r}$ **b.** $\frac{kmM}{\sqrt{a_2^2 + b_2^2 + c_2^2}} - \frac{kmM}{\sqrt{a_1^2 + b_1^2 + c_1^2}}$ **51. a.** π; π **53. a.** $a(e^{-1} - 3)$;

the path does not matter.

13.4 Green's Theorem, page 1059

1. 0 **3.** 3 **5.** 0 **7.** -6π **9.** -18 **11.** $-\sin 1$ **13.** $\pi(R^2 - r^2)$ **15.** 0 **17.** 0 **19.** -8π **21.** -24 **23.** 16 **25.** 0 **27.** 0
29. 0 **31.** 2π **33.** 4π **35.** $\frac{15}{2}$ **37.** $-\frac{\pi a^4}{4}$ **43.** 0 **45.** 0 **47.** 0 **49.** 3

57. $\oint_C (M\,dx + N\,dy) = \iint\limits_D \left(\frac{\partial N}{\partial x} - \frac{\partial M}{\partial y}\right)dA = \oint_{C_1}(M\,dx + N\,dy) + \oint_{C_2}(M\,dx + N\,dy) + \oint_{C_3}(M\,dx + N\,dy)$

13.5 Surface Integrals, page 1072

1. $4\sqrt{2}$ **3.** $\frac{625\sqrt{2}}{4}$ **5.** 0 **7.** $\frac{160\sqrt{6}}{3}$ **9.** $\frac{\pi}{2}$ **11.** 7π **13.** 8π **15.** 0 **17.** 16π **19.** $\frac{\pi}{60}(391\sqrt{17} + 1)$ **21.** $\frac{\pi}{10}((17)^{5/2} - 1)$
23. 36π **25.** $\frac{4\pi}{3}$ **27.** $\frac{\sqrt{3}}{2}$ **29.** $32\pi\sqrt{2}$ **31.** $\frac{7\pi\sqrt{2}}{4}$ **33.** -6 **35.** $\frac{4\pi}{3}$ **37.** π **39.** $-8\sqrt{3}$ **41.** $\frac{\pi^2}{16}$ **43.** $-\frac{19\ln(\sqrt{5}+2)}{192} + \frac{17\sqrt{5}}{32} - \frac{1}{12}$
45. $-\frac{1}{8}\left(4\sqrt{17} + \ln(4 + \sqrt{17})\right)$ **47.** $\frac{4\pi}{3}$ **49.** $\frac{16\sqrt{6}}{3}$ **51.** $\frac{\pi}{60}(25\sqrt{5} + 1)$ **53.** $\frac{8}{3}\pi^3(5 - \sqrt{5})$ **55.** $\frac{243\pi\rho}{2}$ **57. a.** $\frac{\pi a^7}{192}$
b. $2\pi a^2(1 - \cos\phi_0)$

13.6 Stokes' Theorem and Applications, page 1081

3. 18π **5.** $\frac{9}{2}$ **7.** -8π **13.** 0 **15.** -18 **17.** 0 **19.** $2\sqrt{2}\pi$ **21.** 0 **23.** -3π **25.** $\frac{1}{3}$ **27.** 0 **29.** $-\frac{1}{2}$ **31.** 0 **33.** -16π **35.** π
37. $\frac{3}{2}$ **43.** $-\frac{\pi}{4}$ **45.** $\frac{\pi}{2}$ **47.** 0 **49.** $\frac{47\pi}{16}\sqrt{\frac{3}{2}}$ **51.** 0

13.7 Divergence Theorem and Applications, page 1093

5. $\frac{64\pi}{3}$ **7.** $4\pi abc$ **9.** $\frac{128}{3}$ **11.** R, source; S, source; T, sink; U, source **13.** R, source; S, source; T, source; U, source

15. R, source; S, source; T, sink; U, neither **17.** R, source; S, source; T, sink; U, source **19.** R, source; S, source; T, source; U, source

21. R, source; S, source; T, sink; U, neither **23.** 3π **25.** 3 **27.** 24π **29.** $\frac{13}{3}$ **31.** $\frac{243\pi}{2}$ **33.** 0 **35.** 12π **37.** $\frac{8(16-7\sqrt{2})\pi}{15}$

39. $\frac{9\pi a^4}{2}$ **43.** $4\pi a^4$ **45.** div $\nabla f = 0$

Chapter 13 Proficiency Examination, page 1097

23. $f = xyz$ **24.** div $\mathbf{F} = 2xy - ze^{yz}$; curl $\mathbf{F} = ye^{yz}\mathbf{i} - \frac{1}{2}\mathbf{j} - x^2\mathbf{k}$ **25.** -2 **26.** $-4\pi\sqrt{2}$ **27.** 0 **28.** 0 **29.** $\phi = \frac{m\omega^2}{2}(x^2 + y^2 + z^2)$

30. 0

Chapter 13 Supplementary Problems, page 1098

1. $\mathbf{F}$ is conservative; $f = 2x - 3y$ **3.** $\mathbf{F}$ is conservative; $f = xy^{-3} + \sin y$ **5.** $\mathbf{F}$ is not conservative **7.** -7 **9.** $\sin 4 - \sin 1 - \frac{31}{32}$
11. $\frac{5}{2}$ **13.** div $\mathbf{F} = 3$; curl $\mathbf{F} = 0$ **15.** div $\mathbf{F} = x + y + z$; curl $\mathbf{F} = -y\mathbf{i} - z\mathbf{j} - x\mathbf{k}$ **17.** div $\mathbf{F} = a + b$; curl $\mathbf{F} = 0$
19. div $\mathbf{F} = a + b + c$; curl $\mathbf{F} = 0$ **21.** div $\mathbf{F} = \frac{2}{r}$; curl $\mathbf{F} = 0$ **23.** -2π **25.** -2π **27.** $\pi \cos \pi^2$ **29.** $\frac{3\pi}{2} - 2$ **31.** $-\frac{3\pi}{4} - 6$
33. $\sqrt{2}$ **35.** $\frac{5}{6}$ **37.** $2\pi(\pi - 1)$ **39.** 0 **41.** $-a$ **43.** 4π **45.** $\frac{6\sqrt{2}\pi}{5}$ **47.** $\frac{2a+3b}{2abc}$ **49.** 3 **51.** 0 **53.** $\frac{3}{4}$ **55.** $\frac{3}{2}$ **57.** -1 **59.** yes,
as curl $\mathbf{F} = 0$ **63.** $xy + \int f(x)dx + \int g(y)dy$ **65.** $\frac{5}{6}$ **67.** $\mathbf{F}$ is conservative in any region of the plane where $x + y \neq 0$; $\ln|\frac{c+d}{a+b}|$
75. 0 **77.** $-\frac{67}{9}$; C is any curve that does not intersect the y-axis **79.** $\frac{5}{4} + \ln\frac{2}{9}$ **83.** 0 **85.** 0 **87.** $\frac{\pi}{4} - \frac{1}{2}\ln 2$ **89.** 2π **93.** $4\pi^2 ab$
95. 21 (same answer as for the previous problem.) **97.** $(0,0,2)$ **99.** This is Putnam Problem 6i of the morning session of 1948.

Cumulative Review Problems—Chapters 11-13, page 1107

11. $f_x = 4x + y$; $f_y = x - 15y^2$; $f_{xy} = 1$ **13.** $f_x = 1$; $f_y = 1$; $f_{xy} = 0$ **15.** $f_x = f_y = f_{xy} = f = e^{x+y}$ **17.** $2x^3 + 2y^3$

19. $e + e^{-1} - 2$ **21.** $-\frac{3}{8}$ **23.** $(e - 1)^2$ **25.** 1 **27.** $(3 - \sqrt{5})\frac{\pi}{2}$ **29.**

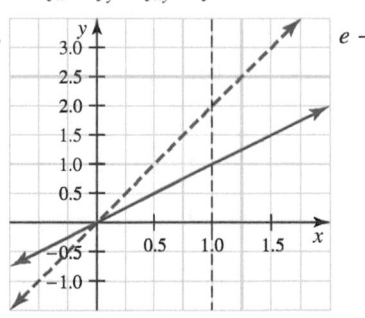

$e - 2$ **31.** 1 **33.** $\frac{1}{2}$

35. Answers vary with choice of path; $\frac{1}{2} + \frac{3\pi}{4}$ **39.** maximum $\sqrt{5}$, minimum $-\sqrt{5}$ **43.** A, sink; B, source; C, source
45. $x = -1 + t$, $y = 1 + t$, $z = 2$ **47.** $x = y = z = \sqrt[3]{V_0}$ **49. a.** 10π cm³ **b.** 8π cm² **51.** $\frac{3\pi}{2}$ **53.** $\left(\frac{3}{16}(\pi + 2), \frac{3}{16}(\pi + 2)\right)$
55. $-2\pi\omega$ **57.** $\frac{64}{3}(\sqrt{3} + 2)\pi$

Chapter 14

14.1 First-Order Differential Equations, page 1120

1. $\ln|y| = x + \ln|x - 5|^5 + C$ **3.** $\ln|y| = -\frac{1}{18}[\ln(e^{2x} + 9) - 2x] + C$ **5.** $y = 3\sec^{-1}|\frac{x}{3}| + C$ **7.** $y = 2\left(1 - e^{-x^2}\right)$
9. $y = \frac{1}{2}(\cos x + \sin x) - \frac{1}{2}e^{-x}$ **11.** $y = x^3 - \frac{9}{4}x^2$ **13.** $-\ln\sqrt{x^2 + y^2} - \frac{1}{3}\tan^{-1}\frac{y}{x} = C$ **15.** $(y + x)^2(y - x) = C$
17. $\frac{y - x}{3y + x} = Cx^8$ **19.** $x^3y + x\tan y = C$ **21.** $\tan^{-1}x + \ln(x^2 + y^2) + e^{-y} = C$ **23.** $x^2\cos 2y - 3xy + 3x^2y - 6y = C$
25. a. $y = -x - 1 + 4e^{x-1}$ **b.**

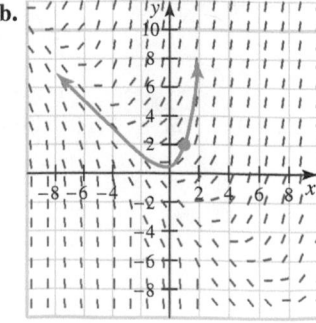

c.

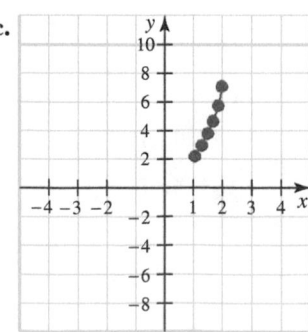

27.
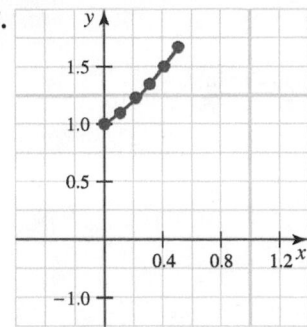

29.
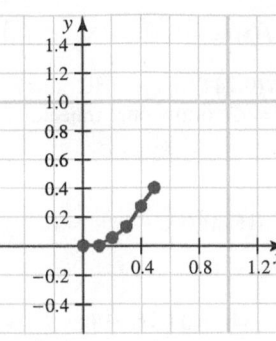

31. $\frac{x}{y} + \ln|y| = C$ **33.** $x^2y^2 + x^3y - xy^3 = C$

35. $\frac{x^2}{y} + \frac{y^2}{x} + 3y = C$

37. $e^{-y} = \frac{1}{2}e^{-x} + \frac{1}{10}e^{-x}(2\sin 2x - \cos 2x) + C$

39. $x^3 + \cos xy = C$

41. $y = (\csc x + \cot x)(x - \sin x + C)$

43. $\sqrt{x^2 + y^2} + y = B$ **45.** $\cot \frac{y}{x} = \ln\left(\frac{\pi x}{4}\right) + 1$

47. $y = x^3 \ln x + x^3$ **49.** $x \sin(x^2 + y) = 0$ **51.** $\frac{1}{2}\ln(y^2 + 2y + 2) - \tan^{-1}(y+1) + e^{-x} = 1$ **53.** If $\alpha > \beta$,

$\lim\limits_{t\to\infty} Q(t) = \beta$; if $\alpha \le \beta$, $\lim\limits_{t\to\infty} Q(t) = \alpha$ **55. a.** $P(t) = \dfrac{r_1(P_0 - r_2) - r_2(P_0 - r_1)e^{-Dt}}{(P_0 - r_2) - (P_0 - r_1)e^{-Dt}}$, where $D = \sqrt{k^2 - 4h\ell}$,

$r_1 = (k + D)/2\ell$, and $r_2 = (k - D)/2\ell$ **b.** $\lim\limits_{t\to\infty} P(t) = r_1$ **57. a.** $x = \dfrac{1}{I(y)}\left[\displaystyle\int I(y)S(y)dy + C\right]$, where

$I(y) = \exp\left(\int R(y)dy\right)$ **b.** $x = -y^2[(y+1)e^{-y} + C]$ **59. b.** $y = u(x) + \dfrac{1}{z(x)}$ **c.** $y = \dfrac{3Cx + 2x^4}{3C - x^3}$

14.2 Second-Order Homogeneous Linear Differential Equations, page 1132

1. $y = C_1 + C_2 e^{-x}$ **3.** $y = C_1 e^{-5x} + C_2 e^{-x}$ **5.** $y = C_1 e^{3x} + C_2 e^{-2x}$ **7.** $y = C_1 e^{(-1/2)x} + C_2 e^{3x}$

9. $y = C_1 e^x + C_2 e^{-x}$ **11.** $y = C_1 \cos\sqrt{11}x + C_2 \sin\sqrt{11}x$ **13.** $y = e^{(-3/14)x}\left[C_1 \cos\left(\frac{\sqrt{131}}{14}x\right) + C_2 \sin\left(\frac{\sqrt{131}}{14}x\right)\right]$

15. $y = C_1 + C_2 x + C_3 e^{-x}$ **17.** $y = C_1 + C_2 x + e^{-(1/2)x}\left[C_3 \cos\left(\frac{\sqrt{7}}{2}x\right) + C_4 \sin\left(\frac{\sqrt{7}}{2}x\right)\right]$

19. $y = C_1 e^{2x} + C_2 e^{-3x} + C_3 e^{-x}$ **21.** $y = e^{5x}(1 - 6x)$ **23.** $y = \frac{11}{5}e^x + \frac{4}{5}e^{11x}$ **25.** $y = \frac{94}{25} - \frac{19}{25}e^{-5x} - \frac{9}{5}xe^{-5x}$

27. $5e^x$ **29.** e^{-2x} **31.** e^{-2x} **33.** $y = e^{-3x}(C_1 + C_2 x)$ **35.** $y = C_1 + C_2 x^{-3}$ **37.** $y = C_1 x^3 + C_2 x^{-4}$

39. $y = \frac{1}{2}\cos(4\sqrt{3}t) - \frac{2}{\sqrt{3}}\sin(4\sqrt{3}t)$ **41.** $y = -\frac{2}{3}\cos(4\sqrt{3}t)$ **43.** $y \approx e^{-0.4t}[0.5\cos 6.9t + 0.03\sin 6.9t]$

49. a. $\frac{75}{8}$ seconds   **b.** approximately 7.80 seconds   **51. b.** $\ln\sqrt{(10x-11)^2 + (10y-13)^2} + \frac{1}{3}\tan^{-1}\left(\frac{10y-13}{10x-11}\right) = C$

53. maximum displacement occurs when $t = 0$; maximum displacement occurs when $t = \dfrac{2m}{k_2}$

57. The equation is identical to the one in Problem 52 with $\frac{k}{m} = \frac{g}{L}$ **59. a.** $m(t) = \dfrac{w - rt}{g}$; $s'(t) = -v_0 \ln\left(\dfrac{w - rt}{w}\right) - gt$

b. $s(t) = \dfrac{v_0(w - rt)}{r}\ln\left(\dfrac{w - rt}{w}\right) - \dfrac{1}{2}gt^2 + v_0 t$ **c.** $t = \dfrac{w_f}{r}$

d. $s\left(\dfrac{w_f}{r}\right) = \dfrac{v_0(w - w_f)}{r}\ln\left(\dfrac{w - w_f}{w}\right) - \dfrac{1}{2}\dfrac{gw_f^2}{r^2} + \dfrac{v_0 w_f}{r}$

14.3 Second-Order Nonhomogeneous Linear Differential Equations, page 1144

1. $\overline{y}_p = Ae^{2x}$ **3.** $\overline{y}_p = A + Be^{2x}$ **5.** $\overline{y}_p = Ae^{-x}$ **7.** $\overline{y}_p = xe^{-x}(A\cos x + B\sin x)$

9. $\overline{y}_p = (A + Bx)e^{-2x} + xe^{-2x}(C\cos x + D\sin x)$ **11.** $\overline{y}_p = A_3 x^3 + A_2 x^2 + A_1 x + A_0$

13. $\overline{y}_p = (A_3 x^3 + A_2 x^2 + A_1 x + A_0)\cos x + (B_3 x^3 + B_2 x^2 + B_1 x + B_0)\sin x$ **15.** $\overline{y}_p = A_0 e^{2x} + B_0 \cos 3x + C_0 \sin 3x$

17. $\overline{y}_p = A_3 x^3 + A_2 x^2 + A_1 x + A_0 + B_0 e^{-x}$ **19.** $y = C_1 + C_2 e^{-x} - x^3 + 3x^2 + x$ **21.** $y = C_1 e^{-3x} + C_2 e^{-5x} + \frac{3}{35}e^{2x}$

23. $y = e^{-x}(C_1 \cos x + C_2 \sin x) + \frac{1}{5}\cos x + \frac{2}{5}\sin x$ **25.** $y = C_1 e^{x/7} + C_2 e^{-x} - e^{-x}\left(\frac{1}{16}x^2 + \frac{15}{64}x\right)$

27. $y = C_1 + C_2 e^x - \left(\frac{1}{4}x^4 + x^3 + \frac{5}{2}x^2 + 10x\right)$ **29.** $y = e^{-x}\left(C_1 + C_2 x + 2x^2 + \frac{1}{6}x^3\right)$

31. $y = C_1 \cos x + C_2 \sin x - \cos x \ln|\sec x + \tan x|$ **33.** $y = C_1 e^{3x} + C_2 e^{-2x} - \frac{1}{32}(8x^2 + 12x + 13)e^{2x}$

35. $y = C_1 \cos 2x + C_2 \sin 2x + \frac{1}{2}x\cos 2x - \frac{1}{4}\sin 2x \ln|\cos 2x|$ **37.** $y = e^{-x}(C_1 + C_2 x) + \frac{1}{4}x^2(2\ln x - 3)e^{-x}$

39. $y = C_1 + C_2 e^x - \frac{1}{20}\sin 2x - \frac{1}{2}x - \frac{1}{5}\cos^2 x$ **41.** $y = -\frac{4}{5} + \frac{6}{5}e^x - \frac{1}{10}\sin 2x - x - \frac{2}{5}\cos^2 x$

43. $y = -\frac{2}{9}\cos 3x + \frac{4}{9}\sin 3x + \frac{2}{9}e^{3x}$ **45.** $y = \frac{35}{27}\sin 3x + \frac{1}{9}x$ **47.** $y = 2e^{-2x} + e^{6x} - \frac{3}{7}e^{5x}$

49. $y = C_1 e^{2x} + C_2 e^{-3x} + G(x)$ where $G(x) = \begin{cases} -\frac{1}{6} & \text{for } 0 \le x < 1 \\ \frac{x}{12} - \frac{17}{72} & \text{for } 1 \le x \le 3 \end{cases}$

51. $I(t) = e^{-t} \left[-\frac{4}{5} \cos t - \frac{12}{5} \sin t \right] + \frac{4}{5} \cos t + \frac{8}{5} \sin t$ **53.** $I(t) = -16e^{-t} + 8e^{-2t} + 8$

55. $I(t) \approx 0.312e^{-t} - 0.015e^{-9t} - 0.297 \cos t - 0.067 \sin t + t(0.244 \cos t + 0.305 \sin t)$ **57. a.** $y_2 = x^2 \ln x$
b. $y = C_1 x^2 + C_2 x^2 \ln x + x(\ln x + 2)$ **59.** solutions $x^2 + y^2 = C$, orthogonal trajectories $Y = CX$

Chapter 14 Proficiency Examination, page 1146

12. $\sin^{-1} y = \sinh^{-1} x + C$ **13.** $y = Bx$ **14.** $y = x + C\sqrt[3]{x}$ **15.** $x^2(x^2 - 2y^2) = C$

16. $\frac{2}{\sqrt{3}} \tan^{-1} \left[\frac{2}{\sqrt{3}} \left(\frac{y}{x} - \frac{1}{2} \right) \right] = \ln|x| + C$ **17.** $y = e^{-x}(C_1 \cos x + C_2 \sin x) - \frac{2}{5} \cos x + \frac{1}{5} \sin x$

18. $x^3 e^{-y} + xy^{-2} + x^2 y^{-3} = C$ **19.** $x(t) \approx e^{-1.60t}[0.33 \cos 10.84t + 0.05 \sin 10.84t]$

20. $I(t) \approx e^{-125t}[-0.562t \cos 185.4t + 2.79 \sin 185.4t] + 0.562 \cos 100t - 0.899 \sin 100t$

Chapter 14 Supplementary Problems, page 1147

1. a. 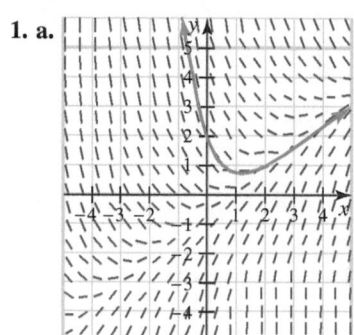   **b.**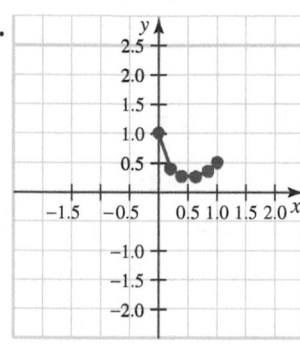

3. $x^2 + y^2 = C, C \geq 0$

5. $7x^2 + y^2 = C, y \neq 0, C > 0$

7. $y = ke^{\sin x}, k$ any real constant

9. $\frac{1}{4}y^4 = -2x^2 + C$ **11.** $\frac{y^3}{3x^3} = \ln|x| + C$

13. $e^{y^2} = e^{2x} + C$ **15.** $y = -e^{5x} \cos x + Ce^{5x}$

17. $x = e^{-y}(y + C)$ **19.** $\frac{y}{x} = \sin(\ln|x| + C)$

21. $y = Be^{(1-2x)/(2x^2)}$ **23.** $x^4 y^3 + \ln|x| - \ln|y| = C$ **25.** $y = -\tan^{-1} e^x + C$ **27. a.** $y_h = C_1 e^{3x} + C_2 e^{-3x}$
b. $y_p = -\frac{1}{9}(x + 1)$ **29. a.** $y_h = C_1 e^{2x} + C_2 e^{3x}$ **b.** $y_p = \left(\frac{1}{2}x^2 + \frac{3}{2}x + \frac{7}{4} \right) e^x$ **31. a.** $y_h = C_1 e^x + C_2 e^{2x}$
b. $y_p = -\left(\frac{1}{4}x^4 + x^3 + 3x^2 + 6x \right) e^x$ **33.** $y = \frac{1}{9} \left(e^{-3x} + 3x - 1 \right)$

35. $x(t) = e^{-2t}, y(t) = 2e^{-2t} - 2e^{-3t}, z(t) = -3e^{-2t} + 2e^{-3t} + 1$ **37.** $y^2 = 5x - x^2$ **39.** $e^{X+Y} = 1 + Ke^X$

41. $Y(Y^2 + 3X^2) = K$ **43.** $y = C_1 x + C_2 \left[\frac{x}{2} \ln \left| \frac{x-1}{x+1} \right| + 1 \right]$ **45.** $y = \frac{1}{6}x^2 - \frac{C_1}{x} + C_2$ **47.** $x = y[e^y + C]$

49. approximately 0.62 sec   **51. a.** $y(t) = \frac{Wx}{24EI}(4Lx^2 - x^3 - 3L^3)$ **b.** $y_{\max} \approx \frac{WL^4}{EI}(-0.045)$

53. b. $y = C_1 x^{m_1} + C_2 x^{m_2}$ **c.** $y = x^{m_0}(C_1 + C_2 \ln|x|)$ **d.** $y = x^\alpha [C_1 \cos(\beta \ln|x|) + C_2 \sin(\beta \ln|x|)]$

57. This is Problem 2 of the afternoon session of the 1938 Putnam Examination.

59. This is Problem 6ii of the morning session of the 1948 Putnam Examination.

INDEX